ALGAE ABSTRACTS
A Guide to the Literature

Volume 1
To 1969

Volume 1: To 1969
Volume 2: 1970–1972

ALGAE ABSTRACTS

A Guide to the Literature

Volume 1
To 1969

SPRINGER SCIENCE+BUSINESS MEDIA, LLC

.

Library of Congress Catalog Card Number 73-84004

ISBN 978-1-4757-0410-5   ISBN 978-1-4757-0408-2 (eBook)
DOI 10.1007/978-1-4757-0408-2

Softcover reprint of the hardcover 1st edition

INTRODUCTION

Algae Abstracts is the first in a series of bibliographies on water resources and pollution published by IFI/Plenum Data Corporation in cooperation with the Water Resources Scientific Information Center (WRSIC). It is produced wholly from the information base comprising material abstracted and indexed for *Selected Water Resources Abstracts.*

The bibliography is divided into volumes according to the publication dates of the source documents. Volume 1 contains 569 abstracts covering publication dates up to and including 1969; Volume 2 contains 730 abstracts covering the years 1970 to 1972. The material included in this bibliography represents computer selections based on the presence of a form of the word "alga" somewhere in the referenced citation. Substantively, the material typifies WRSIC's "centers of competence" approach to information support of the Office of Water Resources Research (OWRR) of the Department of the Interior. Most of the references in this bibliography are the work of the center of competence on eutrophication at the University of Wisconsin.

The indexes refer to the WRSIC accession number, which follows each abstract. The *Significant Descriptor Index* is made up of a fraction of the total descriptors and identifiers by which each paper has been indexed. It represents weighted terms that best describe the information content; this status is indicated by the asterisks which precede them. The *General Index* includes all the remaining descriptors and identifiers by which each paper in this bibliography has been indexed. Through permutation, each word in a multiple-word descriptor or identifier is made to file in its normal alphabetical order, giving access to each abstract by all conceivable routes. To use the indexes, scan the middle rank for a few keywords describing your subject matter interest, then note the WRSIC accession numbers at the right margin. These numbers locate the full record in the bibliography, which is arranged in ascending accession number order.

Each volume, including a complete *Author Index,* is a self-contained entity, offering ready access to the pertinent literature for the time period it spans.

CONTENTS

ABSTRACTS

Publication dates up to and including 1969

The abstracts are presented in order of increasing accession number.
This number appears as the last line of each entry.

INTERTIDAL COMMUNITIES AS MONITORS OF POLLUTION,

WASHINGTON UNIV. DEPT. OF CIVIL ENGINEERING.

R. T. OGLESBY, AND DAVID JARMISON.

ASCE PROC, J SANIT ENG DIV, VOL 94, NO SA3, PAP 6008, PP 541-550, JUN 1968. 5
FIG, 14 REF, 1 APPEND.

    DESCRIPTORS:
     *POLLUTANTS, *ESTUARIES, *TIDAL WATERS, ECOSYSTEMS, ESTUARINE
     ENVIRONMENT, MARINE ALGAE, SALINITY, TEMPERATURE, MONITORING, AERIAL
     PHOTOGRAPHY, INFRARED RADIATION, DYE RELEASES, BEACHES, SANDS, BIOTA,
     WASTE WATER(POLLUTION), CURRENTS(WATER).

    IDENTIFIERS:
     *INTERTIDAL COMMUNITIES, *WASTE DISCHARGES, FLORA, FAUNA, *BIOLOGICAL
     COMMUNITIES, INTERTIDAL ALGAE, PUGET SOUND, FLOURESCEIN DYE.

    ABSTRACT:
     ECOLOGICAL SURVEYS OF INTERTIDAL ZONES ARE PROPOSED TO BE THE MOST
     ECONOMICAL WAY TO ASSESS THE BIOLOGIC IMPACT OF MANY WASTES DISCHARGED
     INTO ESTUARIES AND COASTAL MARINE ENVIRONMENTS. SPECIES DIVERSITY AND
     COMMUNITY STRUCTURE ARE THE MEASURES SELECTED; THESE ARE DETERMINED
     THROUGH A CAREFULLY DESIGNED AND STATISTICALLY VALID PROGRAM OF
     SAMPLING. AERIAL PHOTOGRAPHY IS USED IN DEVELOPING THE SAMPLING PROGRAM
     WITH INFRARED EKTACHROME FILM PROVIDING DATA ON DISTRIBUTION AND, TO
     SOME EXTENT, THE NATURE OF INTERTIDAL FLORA. DYE TRACERS IN THE
     EFFLUENT ARE IMAGED, ON TIME-SEQUENCE, COLOR AERIAL-PHOTOGRAPHS, AND
     PROVIDE INFORMATION ON PATTERNS OF WASTE DISPERSAL. THE METHODOLOGY
     PROPOSED IS APPLICABLE ONLY WHERE EITHER SPATIAL OR TEMPORAL CONTROLS
     CAN BE ESTABLISHED.

    FIELD 05C

    ACCESSION NO. W68-00010

REMOVAL OF ALGAL NUTRIENTS FROM RAW WASTEWATER WITH LIME,

METCALF AND EDDY, BOSTON, MASS.

J. C. BUZZELL, AND C. N. SAWYER.

WATER POLL CONTROL FED JOUR, VOL 39, NO 10, R16-R24 OCT 1967.

    DESCRIPTORS:
     *ACTIVATED SLUDGE, ALGAE, EUTROPHICATION, LIME, *NITROGEN, HYDROGEN ION
     CONCENTRATION, *PHOSPHORUS, SEWAGE TREATMENT, *OXIDATION LAGOONS.

    IDENTIFIERS:
     *NUTRIENTS, PONDS, PRIMARY TREATMENT, SECONDARY TREATMENT,
     STABILIZATION PONDS, SEWAGE LAGOONS.

    ABSTRACT:
     A LABORATORY EVALUATION OF THE APPLICATION OF LIME TO RAW DOMESTIC
     WASTEWATER AS A MEANS OF ENHANCING THE EFFECTIVENESS OF PRIMARY
     TREATMENT, SPECIFICALLY TO REDUCE PHOSPHORUS LEVELS. LIME TREATMENT
     EFFECTED AN 80- TO 90-PERCENT REMOVAL OF TOTAL PHOSPHORUS, WITH GREATER
     THAN 97-PERCENT REMOVAL OF SOLUBLE, INORGANIC FORMS. LIME TREATMENT
     REMOVED 50 TO 70 PERCENT OF THE BOD, APPROXIMATELY 25 PERCENT OF THE
     TOTAL NITROGEN, AND DESTROYED 99.9 PERCENT OF THE COLIFORM BACTERIA.
     LIME DOSAGE WAS CONTROLLED BY PH. THE VOLUME OF SLUDGE PRODUCED WAS
     APPROXIMATELY 1 PERCENT OF THE VOLUME OF WASTEWATER TREATED. THE LIME
     TREATMENT EFFLUENTS HAD NITROGEN: PHOSPHORUS RATIOS OF 25:1 OR GREATER.
     PRELIMINARY STUDIES INDICATED THAT STABILIZATION PONDS SHOULD SERVE
     WELL IN POLISHING THE LIME-TREATED WASTEWATER, REMOVING 75 TO 80
     PERCENT OF THE REMAINING PHOSPHORUS.

    FIELD 05D

    ACCESSION NO. W68-00012

LAKE EUTROPHICATION PROBLEM AND PROGRESS,

SOUTH DAKOTA STATE UNIV., BROOKINGS.

A. C. FOX.

MIN 152ND MTG, MO BAS INTER-AGENCY COMM, APPEND F, PP F-1/F-6, MAR 1968. 4
  REF.

DESCRIPTORS:
  *EUTROPHICATION, SEQUENCE, LAKES, CHEMICALS, FISHING, SPORT FISHING,
  RECREATION, SILT, NUTRIENTS, AQUATIC POPULATIONS, OLIGOTROPHY, *FISH
  POPULATIONS, PLANT POPULATIONS, ROUGH FISH, NITROGEN, PHOSPHORUS,
  DIATOMS, *ALGAE, CHLOROPHYTA, CYANOPHYTA, LAND MANAGEMENT, WATER
  POLLUTION SOURCES, DREDGING, *POLLUTION ABATEMENT, DIKES, SOIL
  CONSERVATION.

IDENTIFIERS:
  DESMINS, FLAGELLATES.

ABSTRACT:
  THE SUBSTANTIAL INFLOW OF SILT AND NUTRIENTS CAUSES A LAKE TO AGE
  RAPIDLY AND TO MOVE FROM OLIGOTROPHY TO EUTROPHY AND SO AFFECT THE
  GAME-FISH POPULATION. THE ADDITION OF NITROGEN AND PHOSPHORUS CAUSES
  LAKES TO CHANGE FROM ONE ONLY SPARSELY POPULATED, MOSTLY BY DESMINS, TO
  ONE IN WHICH DIATOMS REPLACE THE DESMINS, THESE IN TURN REPLACED BY
  FLAGELLATES AND GREEN ALGAE, AND FINALLY BY BLUE-GREEN ALGAE,
  ALARMINGLY IN TREMENDOUS 'BLOOMS'. THE ZOOLOGICAL SEQUENCE IS FROM THE
  GENERALLY OLIGOTROPHI TROUT LAKES, TO THE LAKES POPULATED BY BASS,
  PIKE, PERCH, AND PANFISH, TO THE EUTROPHIC POPULATION OF ROUGH FISH.
  MOST OF THE CAUSATIVE NUTRIENTS ARE THE RESULT OF AGRICULTURAL
  POLLUTION RESULTING FROM POOR LAND PRACTICES AND LIVESTOCK FEEDLOT
  OPERATIONS. ONCE THERE IS EUTROPHY, REMEDIES ARE EXPENSIVE, SUCH AS
  EFFORTS TO DILUTE THE LAKE BY DREDGING OR DIKING. THE CONTROL OF INFLOW
  OF HARMFUL NUTRIENTS AND SILT CAN BE DONE INEXPENSIVELY, AS BY
  INITIATING SOIL CONSERVATION PRACTICES AND BY TREATING DOMESTIC AND
  INDUSTRIAL SEWAGE. CONTROL BY HARVESTING ALGAE IS NOT WITHIN REACH AS
  YET BUT FISH CAN BE HARVESTED AND MARKETED ECONOMICALLY TO REDUCE THE
  TOTAL AMOUNT OF NUTRIENTS IN THE WATER. CONTROL BEST STARTS WITH LAKES
  IN EARLY EUTROPHY.

FIELD 05C

ACCESSION NO. W68-00172

THE USE OF ALGAE IN REMOVING NUTRIENTS FROM DOMESTIC SEWAGE,

WASHINGTON UNIV., SEATTLE.

R. H. BOGAN.

TRANS 1960 SEM ON ALGAE AND METRO WASTES, ROBT A TAFT SANITARY ENGG CENTER,
  CINCINNATI, OHIO, TECH REPT W61-3, PP 140-147, 1961. 8 P.

DESCRIPTORS:
  *EUTROPHICATION, *ALGAE, NUTRIENTS, DOMESTIC WASTES, *SEWAGE TREATMENT,
  SEWAGE, BIOLOGICAL TREATMENT, SOLID WASTES, MICROBIOLOGY, ACTIVATED
  SLUDGE, RATES, CHLORELLA, HYDROGEN ION CONCENTRATION, PLANT TISSUES,
  SCENEDESMUS, LIMNOLOGY, WATER PROPERTIES, WATER TEMPERATURE,
  ALKALINITY, *PHOTOSYNTHESIS, LIGHT PENETRATION, PHOSPHORUS, LAGOONS.

IDENTIFIERS:
  STIGLEOCLONIUS, INSOLUBLE WASTES, CULTURE MEDIA.

ABSTRACT:
  THE SOLUBLE ORGANIC PHOSPHORUS IN SEWAGE MAY BE CONVERTED BY BIOLOGICAL
  OR CHEMICAL MEANS INTO RECOVERABLE INSOLUBLE MATTER. THE LATTER MEANS
  HAS RECEIVED MORE ATTENTION TO DATE. THE RATE OF REMOVAL BY BIOLOGICAL
  MEANS IS A FUNCTION OF CELL-TISSUE SYNTHESIS WHICH VARIES MARKEDLY WITH
  THE TYPE OF ORGANISM. THE MIXED MICROBIAL CULTURES PROVIDED BY THE
  ACTIVATED SLUDGE PROCESS APPEAR TO BE THE MOST RAPID MECHANISM.
  CHLORELLA, SCENEDESMUS, AND STIGLEOCLONIUS WERE GROWN IN RAW AND
  TREATED SEWAGE FOR THE EXPERIMENTS, AND THE AUTHOR STUDIED THEIR ROLE
  IN ADJUSTING PH. TEMPERATURE, CELL-TISSUE CONCENTRATION, COMPOSITION OF
  CULTURE MEDIA, AND ALKALINITY INFLUENCE THE RATE OF PHOTOSYNTHETIC PH
  ADJUSTMENT. WITH ADEQUATE LIGHT, THE MINIMUM VALUE BEING 100-200 F C,
  RAPID EXTRACTION OF PHOSPHORUS BY PHOTOSYNTHETIC BIOLOGICAL ACTIVITY IS
  POSSIBLE. A HIGH-RATE PROCESS WAS DEVELOPED IN THE LABORATORY BY WHICH
  SOLUBLE PHOSPHORUS REDUCTIONS OF 90 PERCENT OR MORE WERE ACHIEVED IN
  CONTACT TIMES OF AS LITTLE AS 6-12 HOURS. TRANSFERRING THE PILOT-PLANT
  STUDY TO LAGOONAL WATERS 3-4 FT DEEP CREATED LIGHTING
  PROBLEMS--SUNLIGHT COULD NOT PENETRATE DEEPLY ENOUGH THROUGH THE ALGAL
  GROWTH TO CAUSE AN ADEQUATE PHOTOSYNTHETIC RESPONSE.

FIELD 05D

2

ACCESSION NO. W68-00248

INDICES OF GREAT LAKES EUTROPHICATION,

A. M. BEETON.

PUBL GREAT LAKES RES DIV, (IN PRESS), VOL 14, 1966.

DESCRIPTORS:
*EUTROPHICATION, NUTRIENTS, NITROGEN, PHOSPHORUS, PLANKTON, DISSOLVED
OXYGEN, HYPOLIMNION, GREAT LAKES, LAKE ERIE, STATISTICS, GROWTH RATES,
ALGAE, FRESH WATER FISH.

ABSTRACT:
INDICES OF EUTROPHICATION WERE CITED: (1) INCREASES IN NITROGEN AND
PHOSPHORUS; (2) CHANGES IN SPECIES COMPOSITION AND AN INCREASE IN THE
ABUNDANCE OF PLANKTON; (3) DECREASES IN THE DISSOLVED OXYGEN CONTENT OF
BOTTOM WATERS; (4) CHANGES IN THE FISH POPULATION; (5) THE REPLACEMENT
OF BOSMINA COREGONI BY B. LONGIROSTRIS; AND (6) EXTENSIVE GROWTHS OF
CLADOPHORA. OTHER CHANGES SUCH AS INCREASES IN TDS AND MAJOR IONS ARE
REGARDED AS REPRESENTATIVE OF ENVIRONMENTAL CHANGES AND NOT NECESSARILY
INDICES OF EUTROPHICATION. THERE ARE FEW OFFSHORE DATA ON NUTRIENTS
(NITROGEN AND PHOSPHORUS) FROM THE GREAT LAKES OTHER THAN LAKE ERIE,
AND EVEN THESE DATA ARE QUESTIONABLE. CHANGES IN THE RATES OF GROWTH OF
FISH SHOULD BE VIEWED WITH CAUTION WHEN RELATING THEM TO EUTROPHICATION
INASMUCH AS MANY ENVIRONMENTAL VARIABLES MAY BE OF INFLUENCE AS WELL AS
AN INCREASE IN NUTRIENTS.

FIELD 05C

ACCESSION NO. W68-00253

PLANKTONIC ALGAE AS INDICATORS OF LAKE TYPES, WITH SPECIAL REFERENCE TO THE
DESMIDIACEAE,

MINNESOTA UNIV., MINNEAPOLIS.

A. J. BROOK.

LIMNOL OCEANOGR, VOL 10, PP 403-411, 1965. 9 P.

DESCRIPTORS:
*EUTROPHICATION, AQUATIC ALGAE, *LAKES, *CLASSIFICATION, DESMIDS,
OLIGOTROPHIC, *PHYTOPLANKTON, BRITISH ISLES, BIOINDICATORS, *ESTIMATING
EQUATIONS, PERIOD OF GROWTH.

ABSTRACT:
AFTER REVIEWING THE WORK OF NYGARD (1949) ON THE 'QUOTIENT SYSTEM' OF
LAKE CLASSIFICATION, THE AUTHOR DESCRIBES THE USE OF DESMIDS AS
INDICATORS OF LAKE TYPES. NYGARD STATES THAT QUOTIENTS CAN ONLY BE
DETERMINED FROM SAMPLES COLLECTED DURING THE TIME OF GREATEST ALGAL
DEVELOPMENT, FROM MID-MAY THROUGH SEPT. A BIG PROBLEM APPEARS TO BE THE
CORRECT IDENTIFICATION OF ALL OF THE PLANKTON SPECIES. NYGARD'S
'COMPOUND QUOTIENT' (1949 AND 1953) IS: MYXOPHYCEAE + CHLOROCOCCALEA +
CENTRALES + EUGLENINEAS/DESMIDIACEAS. SLIGHT ERRORS IN DETERMINING THE
NUMBER OF DESMIDS WILL ALTER GREATLY THE DIMENSIONLESS NUMBERS DERIVED
FROM THE QUOTIENTS AND THUS WILL BIAS THE INVESTIGATOR AS TO THE
TROPHIC STATUS OF THE LAKES. EUTROPHIC WATERS SOMETIMES CONTAIN AS MANY
PHYTOPLANKTON SPECIES AS DO OLIGOTROPHIC WATERS. IN THE AUTHOR'S
OPINION, THE COMPOSITION OF THE PHYTOPLANKTON IS NOT NECESSARILY
ALTERED BY AN INCREASE IN NUTRIENTS; FURTHER, THAT A BLOOM MAY NOT FORM
IF THE INCREASE IN NUTRIENTS IS A VERY SLOW ONE. A STUDY OF 300 LAKES
IN THE BRITISH ISLES INDICATES THAT DESMIDS ARE MOST COMMON (59
PERCENT) IN OLIGOTROPHIC WATERS, BUT ARE NOT LIMITED TO THEM; THEY
OCCUR, ALSO (24 PERCENT) IN WATERS CLASSED AS EUTROPHIC AND HAVING
QUOTIENTS OF 2.0 OR MORE.

FIELD 05C

ACCESSION NO. W68-00255

A SYNERGISTIC APPROACH TO PHOSPHORUS REMOVAL,

DORR-OLIVER INC, STANFORD, CONNECTICUT.

ROBERT J. SHERWOOD.

J OF CIV ENG, AM SOC CIV ENG, PP 32-35, MAY 1968. 4 P, 3 FIG, 1 TAB, OF REF.

DESCRIPTORS:
*EUTROPHICATION, SEWAGE, *PHOSPHORUS, PHOSPHATES, *SEWAGE TREATMENT,
DOMESTIC WASTES, *ALGAE, WATER PROPERTIES, BIOCHEMICAL OXYGEN DEMAND,
*BACTERIA, ACTIVATED SLUDGE, IRON COMPOUNDS, LIME, OXYGEN, TERTIARY
TREATMENT, COST COMPARISONS.

ABSTRACT:
THE PHOSPHORUS IN SEWAGE HAS BEEN HELD RESPONSIBLE FOR THE EXCESSIVE
ALGAL GROWTHS IN LAKES INTO WHICH DOMESTIC SEWAGE IS DISCHARGED, BUT
THE EXTENT TO WHICH IT MUST BE ELIMINATED IS NOT KNOWN. THE DESIGN
FACTORS IN A PHOSPHATE-REMOVAL PLANT ARE: THE RATIO OF SOLUBLE TO TOTAL
PHOSPHATE; WATER ALKALINITY AND HARDNESS; BOD REMOVAL REQUIRED;
PHOSPHORUS REMOVAL REQUIRED; AND THE RATIO OF SOLUBLE BOD TO TOTAL BOD.
THE BIOCHEMICAL REMOVAL OF PHOSPHORUS FROM WASTES BY THE ACTIVATED
SLUDGE METHOD IS EFFECTIVE IN CERTAIN RATIOS OF INFLUENT BOD TO
PHOSPHORUS. THE REMOVAL RATE IS HIGH WHEN THE SLUDGE AGE IS 0.5-1 DAY,
BUT IS GREATLY REDUCED WHEN THE AGE IS 15-25 DAYS. ALGAE CAN HELP IN
SEWAGE STABILIZATION IN THIS CYCLE: ALGAE PRODUCE OXYGEN WHICH COMBINES
WITH ORGANIC MATTER; THIS, BY BACTERIAL ACTION RELEASES CARBON DIOXIDE,
AMMONIA, PHOSPHATE, AND WATER, WHICH ARE THEN USED BY THE ALGAE TO
PRODUCE MORE OXYGEN. TERTIARY TREATMENT BY THE ADDITION OF CHEMICALS IS
EFFECTIVE, BUT ITS EXPENSE INCREASES GREATLY AFTER THE INITIAL
PHOSPHORUS REMOVAL BY THE FIRST APPLIED DOSES. LIME IS A LESS EXPENSIVE
ADDITIVE THAN ALUM OR AN IRON SALT AND IS ALSO EFFECTIVE IN PRIMARY
PHOSPHATE COAGULATION.

FIELD 05C

ACCESSION NO. W68-00256

INDUCED EUTROPHICATION - A GROWING WATER RESOURCE PROBLEM,

FWPCA, US DEPT. OF THE INTERIOR, CINCINNATI.

A. F. BARTSCH.

TRANS 1960 SEM ON ALGAE AND METROPOLITAN WASTES, ROBT A TAFT SANITARY ENG
CENTER, TECH REPT W61-3, PP 6-9, 1961. 4 P.

DESCRIPTORS:
*EUTROPHICATION, SEWAGE, FISH POPULATIONS, PLANKTON, *CYANOPHYTA,
WASHINGTON, DC, *ESTUARIES, HUMAN POPULATION, SEWAGE EFFLUENTS, AQUATIC
POPULATIONS, NUISANCE ALGAE, DUCKS(WILD), OYSTERS, CHANNELS, POLLUTION
ABATEMENT, HUMAN POPULATION.

ABSTRACT:
A FAMILIAR THREAD OF SIMILARITY AND REPETITION CAN BE NOTED IN EACH
CASE OF EUTROPHICATION. SOMETIMES IT INCLUDES ALL, BUT MORE COMMONLY
ONLY SEVERAL, OF THE FOLLOWING STEPS: (1) INTRODUCTION OF RAW OR
TREATED, SEWAGE; (2) REPLACEMENT OF PRIZED DEEP-WATER TROUT OR
WHITEFISH BY LESS DESIRABLE KINDS; (3) INCREASE IN PLANKTON OR
FREE-FLOATING PLANT AND ANIMAL LIFE; AND (4) EXPLOSIVE SEASONAL
APPEARANCE OF THE CYANOPHYTE, OSCILLATORIA RUBESCENS. THE POTOMAC RIVER
BELOW WASHINGTON, DC, IS ANOTHER OF THE SUSCEPTIBLE BODIES OF WATER.
HERE IT IS A TIDAL REACH, WHERE THE WATER WASHES BACK AND FORTH, AND IS
AS MUCH LIKE A LAKE AS A RIVER. THE IMMEDIATE TRIBUTARY LAND AREA IS
OCCUPIED BY A CONCENTRATED SEWERED POPULATION OF 1.7 MILLION PEOPLE.
LIMITED INTERMITTENT STUDIES DURING THE LAST FEW YEARS REVEAL THAT
ALGAL POPULATIONS ARE MUCH GREATER NOW THAN IN THE PAST. BLOOMS OF
CYANOPHYTES ARE COMMON AND MAY BECOME OBJECTIONABLE IN SOME OF THE BAYS
AND COVES. LONG ISLAND DUCKLINGS ALSO CONTRIBUTE TO THE ENRICHMENT. THE
AREA NORMALLY RAISES FINE OYSTERS BUT THE OYSTER BEDS BECOME POLLUTED
UNLESS CHANNELS ARE DREDGED TO ALLOW THE FLUSHING OF THE BAY BY THE
TIDES.

FIELD 05C

ACCESSION NO. W68-00461

SCIENTIFIC STUDIES AND CHEMICAL TREATMENT OF THE MADISON LAKES,

CITY OF MADISON, MADISON, WISC.

BERNHARD DOMOGALLA.

A SYMPOSIUM ON HYDROBIOLOGY, UNIV OF WISCONSIN PRESS, MADISON, PP 303-309,
   1941. 7 P, 3 FIG, 4 TAB, 8 REF.

   DESCRIPTORS:
      *EUTRIPHICATION, *LAKES, WISCONSIN, COPPER SULPHATE, NUISANCE ALGAE,
      *ALGAL CONTROL, WEED CONTROL, *INVESTIGATIONS, ZOO-PLANKTON, SPRAYING,
      BIOCONTROL, INFECTION, NUTRIENTS, ALGAE, NITROGEN, PHOSPHORUS,
      BICARBONATES, *AQUATIC POPULATIONS.

   IDENTIFIERS:
      SAPROLEGNIA.

   ABSTRACT:
      STUDIES OF CHEMICAL TREATMENTS APPLIED TO THE MADISON LAKES INDICATED:
      (1) SPRAYING WITH COPPER SULPHATE AND OTHER CHEMICAL TREATMENTS WAS
      FOUND TO BE THE MOST EFFECTIVE AND ECONOMICAL METHOD OF CONTROLLING
      GROWTHS OF OBNOXIOUS ALGAE AND WEEDS. (2) STUDIES SHOWED THAT THE
      CHEMICAL TREATMENTS AS THEY WERE MADE DID NOT SERIOUSLY INJURE THE
      ZOOPLANKTON. (3) THE CHEMICAL TREATMENTS FOR CONTROL OF ALGAE ALSO
      CONTROLLED SAPROLEGNIA. (4) CHEMICAL SPRAYING ALSO BROUGHT 'SWIMMER'S
      ITCH' AND FUNGUS INFECTIONS UNDER CONTROL. (5) STUDIES OVER AN 18-YR
      PERIOD INDICATE THAT NITROGEN COMPOUNDS, PHOSPHORUS, AND BICARBONATES
      ARE AMONG THE FACTORS THAT TEND TO PROMOTE THE EXCESSIVE GROWTH OF
      ALGAE AND LARGE AQUATIC PLANTS. CHEMICAL TREATMENT WITH COPPER SULPHATE
      REDUCED THE NUMBERS OF ALGAE IN THE MADISON LAKES BUT DID NOT ELIMINATE
      THEM. THESE ALGAE IDENTIFIED IN LAKE MONONA IN 1940 WERE, IN DESCENDING
      ORDER OF ABUNDANCE: ANABAENA, MICROCYSTIS, PEDIASTRUM, LYNGBA,
      STAURASTRUM, FRAGILARIA, HYDRODICTYON, AND PANDORINA.

   FIELD 05C

   ACCESSION NO. W68-00468

FERTILIZATION AND ALGAE IN LAKE SEBASTICOOK, ME.

FWPCA, DEPT. OF THE INTERIOR, CINCINNATI.

REPT OF THE TECH ADVISORY AND INVESTIGATIONS ACTIVITIES OF THE TECH SERVICES
   PROGRAM, FED WTR POLL CONTR ADM, ROBT A TAFT ENGG CENTER, CINCINNATI, OHIO,
   122 P, 1966.

   DESCRIPTORS:
      *EUTROPHICATION, MAINE, NUTRIENTS, AQUATIC POPULATIONS, WATER QUALITY,
      LAKE MORPHOMETRY, DOMESTIC WASTES, PLANT GROWTH SUBSTANCES, LIMNOLOGY,
      PLANT GROWTH REGULATORS, CYANOPHYTA, ENVIRONMENTAL EFFECTS,
      CURRENTS(WATER), INDUSTRIAL WASTE, CORES, DIATOMS, ODOR-PRODUCING
      ALGAE, CHLOROPHYLL, CHEMTROL, COPPER SULFATE, LAKES, NITROGEN.

   IDENTIFIERS:
      MICROCYSTIS, ANABAENA.

   ABSTRACT:
      IN 1965, A STUDY WAS MADE OF LAKE SEBASTICOOK, ME, BY THE TECH ADVISORY
      AND INVESTIGATIONS ACTIVITY AND THE ME WTR IMPR COMM. THE AREA OF THE
      LAKE IS 1,736 HA, ITS MAX DEPTH IS 17.7 M, AND ITS MEAN DEPTH IS 6 M.
      SAMPLING WAS DONE IN FEB, MAY, JUL, AND OCT IN THE EFFORT TO DETERMINE
      THE MAJOR SOURCES OF NUTRIENTS, TO ASSESS THEIR SIGNIFICANCE, AND TO
      RECOMMEND THEIR CONTROL. DATA INDICATE THAT ABOUT 98% OF THE NUTRIENTS
      COME FROM SOURCES OTHER THAN LAND DRAINAGE, THE BULK (75%) FROM
      DOMESTIC AND INDUSTRIAL (INCLUDING WOOLEN MILLS AND A POTATO-PROCESSING
      PLANT) WASTES. EXTENSIVE BLOOMS OF THE BLUE-GREEN ALGAE, MICROCYSTIS
      AERUGINOSA AND ANABAENA SPP, PRODUCE OBJECTIONABLE MATS AND ODORS. THE
      AVERAGE DETENTION TIME IN THE LAKE IS 3.5 YR. PHOSPHORUS IS JUDGED TO
      BE THE MOST SIGNIFICANT NUTRIENT AND IS FOLLOWED IN IMPORTANCE BY
      NITROGEN. THE PHOSPHORUS INPUT EXCEEDS ITS OUTPUT. A LIMITED ANALYSIS
      OF A 48 CM CORE SHOWED AN INCREASE IN DIATOMS IN THE TOP FEW CM.
      CHLOROPHYLL CONTENT FLUCTUATES WIDELY. STATEMENTS BY RESIDENTS AND
      OTHERS INDICATE THAT MOST OF THE CHANGES IN THE LAKE OCCURRED WITHIN
      THE PAST 10-20 YR. CHEMICAL CONTROL BY THE USE OF COPPER SULFATE IS NOT
      RECOMMENDED BECAUSE IT WOULD BE ONLY TEMPORARY AND TOO EXPENSIVE, UP TO
      $40,000/YR.

   FIELD 05C

   ACCESSION NO. W68-00470

CHANGES OF EUTROPHICATION AND BIO-PRODUCTION MEASURED BY THE BIOMASSTITRE OF
TEST ALGAE (GERMAN),

BUNDESGESUNDHEITSAMT, INSTITUT FUR WASSER-, BODEN- UND LUFTHYGIENE, BAD
GODESBERG, W. GERMANY.

G. BRINGMANN, AND R. KUHN.

GESUNDHEITS-INGENIEUR, HEFT 2.79, PP 50-54, 1958. 5 P, 10 FIG, 1 TAB.

DESCRIPTORS:
*EUTROPHICATION, TEST ANALYSIS, *LABORATORY TESTS, TEST PROCEDURES,
NURTRIENTS, RIVERS, LAKES, *GROWTH RATES, *WATER TEMPERATURE, SEASONAL,
TEST PROCEDURES, SCENEDESMUS, *ALGAE.

IDENTIFIERS:
BERLIN(WEST), GERMANY.

ABSTRACT:
THE ARTICLE DESCRIBES A CONTINUATION OF THE WORK OF BRINGMANN AND KUHN
IN WHICH A BIO-ASSAY TECHNIQUE WAS USED TO ESTIMATE THE INFLUENCES OF
ENRICHMENT ON THE STREAMS AND OTHER BODIES OF WATER IN THE BERLIN
(WEST) WATER NET. BY COMPARING WATER TEMPERATURES AND THE BIOPRODUCTION
OF THE TEST ALGAE, THE AUTHORS NOTED THAT THE HIGHEST PRODUCTION
OCCURRED DURING THE WINTER WHEN WATER TEMPERATURES WERE LOW, AND VICE
VERSA. AS BASED UPON MEAGER DATA--ABOUT 9 SAMPLINGS A YEAR FOR
TEMPERATURE AND 7 SAMPLINGS A YEAR TO DETERMINE
BIO-PRODUCTION--TEMPERATURE AND PRODUCTION WERE USUALLY FOUND TO BE OUT
OF PHASE. VALUES WERE COMPARED FOR THE INTERVAL FROM MAY 1956 TO APRIL
1957. SCENEDESMUS WAS USED AGAIN AS THE TEST ORGANISM.

FIELD 05C

ACCESSION NO. W68-00472

THE ALGAE-TITRE AS A MEASURE OF EUTROPHICATION OF WATER AND MUD (GERMAN),

INSTITUT FUR WASSER-BODEN-UND LUFTHYGIENE, BERLIN-DAHLEM, GERMANY.

G. BRINGMANN, AND R. KUHN.

GESUNDEITS-INGENIEUR, VOL 77, PP 374-381, 1956. 8 P.

DESCRIPTORS:
*EUTROPHICATION, ALGAE, GERMANY, VOLUMETRIC ANALYSIS, LIMNOLOGY, GROWTH
RATES, *LABORATORY TESTS, *SCENEDESMUS, DIATOMACEOUS EARTH, MUD, LAKES,
CULTIVATION, CHANNEL MORPHOLOGY, SEWAGE INPUT, BIOLOGICAL PROPERTIES,
*WATER QUALITY, TURBIDITY, PLANT POPULATIONS.

IDENTIFIERS:
BERLIN, *ALGAE-TITRE.

ABSTRACT:
ALGAE OF THE GENUS SCENEDESMUS WERE INJECTED INTO MUD SAMPLES AND INTO
WATER SAMPLES FILTERED FREE OF ALGAE. THESE WERE THEN CULTIVATED UNDER
CONSTANT CONDITIONS FOR UNIFORM PERIODS, AFTER WHICH THE TURBIDITY, A
FUNCTION OF THE NUMBER OF ORGANISMS, WAS DETERMINED. THE NEPHELOMETRIC
VALUE IS GENERALLY EQUIVALENT TO MG/1 OF INFUSORIAL SOIL. THE
EQUIVALENT VALUE OF INFUSORIAL SOIL SERVED AS A MEASURE OF TURBIDITY
AND IS TERMED 'ALGAE-TITRE.' THE RELATION OF THE ALGAE-TITRE TO THE
DISCHARGE CHANNEL AND TO THE SEWAGE INPUT WAS DEMONSTRATED. EXPERIMENTS
SHOWED THAT ALGAE-TITRE MAY SERVE AS A BIOLOGICAL MEASURE OF THE
EUTROPHICATION OF WATER AND MUD. DATA WERE DERIVED FROM SAMPLES TAKEN
AT 41 PLACES IN THE BERLIN WATER NET DURING 14 DAYS FROM JULY 1955 TO
JAN 1956. THE AVERAGE VALUES OF ALGAE-TITRE WERE THEN COMPARED ON
SEVERAL BASES. THE AUTHORS HAVE USED THE METHOD IN ROUTINE TESTS OF
WATER QUALITY.

FIELD 05C

ACCESSION NO. W68-00475

EUTROPHICATION PROBLEMS AND PROGRESS,

PACIFIC NORTHWEST WATER LABORATORY, FWPCA, CORVALLIS, OREGON.

A. F. BARTSCH.

MIN 152ND MTG, MO BAS INTER-AGENCY COMM, BISMARCK, N DAK, APPEND G, PP
  G-1/G-7, MAR 1968.

DESCRIPTORS:
  *EUTROPHICATION, LAKES, *NUTRIENTS, UNITED STATES, PHOSPHORUS,
  NITROGEN, SEWAGE, INDUSTRIAL WASTES, DOMESTIC WASTES, SEWAGE TREATMENT,
  HARVESTING OF ALGAE, BOTTOM SEDIMENTS, WEED CONTROL, SOLAR RADIATION,
  *WATER POLLUTION CONTROL, ALGICIDES, BIOCONTROL, *WATERSHED MANAGEMENT,
  *ENVIRONMENTAL EFFECTS, CHEMTROL, ALGAE.

IDENTIFIERS:
  CRATER LAKE, ORE, GROWTH RATES, WATER PROPERTIES.

ABSTRACT:
  THERE ARE ALL DEG OF EUTROPHICATION IN THE 100,000 SMALL LAKES OF THE
  USA. CITIZENS AND THEIR AGENCIES ARE INCREASINGLY AWARE OF THE THREAT.
  THE DENSE ALGAL GROWTH OF ADVANCED EUTROPHICATION IS THE RESULT OF
  ADDITION OF CERTAIN NUTRIENTS TO WATER: ONLY .4 OZ OF PHOSPHORUS OR 1
  LB OF NITROGEN MIXED WITH AN ACRE FT OF WATER WILL SUPPORT AN
  OBJECTIONABLE GROWTH. MAN CONTRIBUTES GREAT QUANTITIES OF NUTRIENTS IN
  SEWAGE, INDUSTRIAL WASTES, AND THE FERTILIZERS WASHED FROM FARMLANDS. 4
  NUTRIENT-CONTROL MEASURES ARE: (1) MECHANICAL REMOVAL OF PLANTS FROM
  THE WATER; (2) CONTROLLING PLANT GROWTH BY THE ADDITION OF CERTAIN
  CHEMICALS; (3) BIOLOGICAL CONTROL BY INTRODUCING PLANT-KILLING FUNGI
  AND VIRI; AND (4) ECOLOGICAL CONTROL BY: (A) REDUCING THE INPUT OF
  PHOSPHORUS AND NITROGEN (SEWAGE DIVERSION, PERHAPS); (B) TREATMENT OF
  SEWAGE TO REMOVE MOST OF THE PHOSPHORUS BEFORE DISCHARGE TO A LAKE; (C)
  DILUTION OF THE NUTRIENTS ALREADY PRESENT BY FLUSHING OF LAKE (MOSES
  LAKE, WASH); AND (D) REMOVING NUTRIENTS BY HARVESTING ALGAE (NOT NOW
  FEASIBLE) AND WATERWEEDS (NOW POSSIBLE). ESTIMATES ARE THAT IF
  WATERWEED HARVEST CAN BE 3X THE NUTRIENT INPUT, EUTROPHICATION CAN BE
  CONTROLLED. BOTTOM-SEDIMENT/NUTRIENT INTERCHANGE, AND THE EFFECT OF
  REDUCING INSOLATION ARE BEING STUDIED.

FIELD 05C

ACCESSION NO. W68-00478

ALGAE BLOOMS IN LAKE ZOAR, CONN,

GENERAL DYNAMICS CORP., GROTON, CONN.; WATER RESOURCES COMM., HARTFORD, CONN.

R. J. BENOIT, AND J. J. CURRY.

TRANS 1960 SEM ON ALGAE AND METRO WASTES, ROBT A TAFT SANITARY ENGG CENTER,
  CINCINNATI, OHIO, TECH REPT W61-3, PP 18-22, 1961. 5 P.

DESCRIPTORS:
  *EUTROPHICATION, *POLLUTION ABATEMENT, DRAINAGE SYSTEMS, FLOW RATES,
  CONN STREAM FLOW, ALGAE, AQUATIC ALGAE, CURRENTS(WATER), LIMNOLOGY,
  LAKES, *PILOT PLANTS, *NUISANCE ALGAE, ROOTED AQUATIC PLANTS,
  *RESERVOIRS, HYDROELECTRIC PLANTS, WEED CONTROL, WATER PROPERTIES,
  PHOSPHORUS, CHEMICAL ANALYSIS.

IDENTIFIERS:
  HYDRODICTYON, ELODEA, CERATOPHYLLUM, MICROCYSTIS, ANABAENA.

ABSTRACT:
  LAKE ZOAR, CONN, WAS IMPOUNDED IN 1919 BEHIND THE STEVENSON DAM ACROSS
  THE HOUSATONIC RIVER. THE DRAINAGE AREA ABOVE THAT POINT IS ABOUT 1,545
  SQ MI. DURING HIGH-FLOW STAGE (2,500 CFS) THE LAKE WATERS ARE REPLACED
  EVERY 9 DAYS; DURING LOW-FLOW PERIODS (AUG AND SEPT) THE REPLACEMENT
  TIME IS 40 DAYS. BY 1947 THE INCREASE IN ALGAL GROWTH HAD BEEN SO GREAT
  AS TO CREATE A SERIOUS NUISANCE. ROOTED AQUATIC PLANTS WERE FIRST
  DESIGNATED AS THE NUISANCE-CAUSING ORGANISMS BUT ALGAL BLOOMS WERE
  FOUND TO BE EQUALLY IMPORTANT. HYDRODICTYON OCCURS IN MID-JUL IN GREAT
  QUANTITIES IN THE SHALLOW, QUIET WATERS AND IS INTIMATELY ASSOCIATED
  WITH SUCH AQUATIC WEEDS AS ELODEA AND CERATOPHYLLUM. CURRENTS MOVE
  FLOATING MATS OF WEEDS FROM ONE PLACE TO ANOTHER. IN THE EARLY SUMMER,
  LARGE QUANTITIES OF MICROCYSTIS AND ANABAENA APPEAR IN THE UPPERMOST
  4-FT ZONE OF LAKE WATER. PHOSPHORUS CONCENTRATIONS MEASURED IN THE
  WATERS OF THE LAKE AND SOURCE STREAMS RANGE FROM 12-41 PPB AND AVERAGE
  ABOUT 25 PPB. THE POLLUTION-CONTROL AGENCIES CONSTRUCTED A PILOT PLANT
  ON THE PLAN SUGGESTED BY LEA, ROHLICH, AND KATZ (1954). IT WAS PARTLY
  DESTROYED BY A FLOOD AND, DESPITE APPARENTLY SUCCESSFUL OPERATION, WAS
  NOT REBUILT BECAUSE AN INVESTIGATION BY AN INDEPENDENT AGENCY FOUND AN
  UNEXPLAINED INCREASE IN PHOSPHORUS CONTENT.

FIELD 05C

ACCESSION NO. W68-00479

NUTRIENTS AND ALGAE IN LAKE SEBASTICOOK, MAINE,

FWPCA, CINCINNATI, OHIO.

K. M. MACKENTHUN, L. E. KEUP, AND R. K. STEWART.

JOUR WATER POLL CONTROL FED, 40, 2 PART 2, R72-R81, FEB 1968. 10 P, 6 FIG, 3 TAB, 9 REF.

DESCRIPTORS:
   ALGAE, EUTROPHICATION, LAKES, MICROORGANISMS, NITROGEN, NUTRIENTS, PHOSPHORUS, PLANKTON, PLANTS(ORGANISMS).

IDENTIFIERS:
   LAKE SEBASTICOOK, ME, MAINE, PERIODIC VARIATIONS, SEASONAL VARIATIONS, SOURCES.

ABSTRACT:
   FOUR SEASONAL FIELD STUDIES WERE MADE ON LAKE SEBASTICOOK, MAINE, AND ITS SURROUNDING AREA, TO DETERMINE THE CAUSE OF NUISANCE ALGAL GROWTHS, IDENTIFY MAJOR NUTRIENTS SOURCES, ASSESS THEIR SIGNIFICANCE, AND RECOMMEND CONTROL MEASURES. IN FEBRUARY TOTAL NITROGEN VALUES WERE 3.3 MG/1 IN THE LAKE'S SURFACE WATERS AND 6.2 MG/1 IN THE PROFUNDAL WATERS. VALUES WERE LOWER DURING OTHER SEASONS. THERE WAS NO REDUCTION IN NITROGEN PASSING THROUGH THE LAKE. THE LAKE RECEIVED 8,000 LB (3,630 KG) OF TOTAL PHOSPHORUS ANNUALLY, DISCHARGED 4,150 LB (1,880 KG), AND RETAINED 48 PERCENT. PHYTOPLANKTON WERE AS HIGH AS 2,260 LB/SURFACE ACRE (2,540 KG/HA) (WET WEIGHT) DURING THE MAXIMUM GROWING PERIOD. AN 80 PERCENT REDUCTION IN CONTROLLABLE INFLOWING PHOSPHORUS WOULD CORRECT THE NUISANCE CONDITIONS.

FIELD 05C

ACCESSION NO. W68-00481

PLANKTON ALGAE AS INDICATORS OF THE SANITARY CONDITION OF A STREAM,

FWPCA, CINCINNATI, OHIO.

F. J. BRINLEY.

BIOLOGICAL STUDIES, OHIO RIVER POLLUTION SURVEY II, SEWAGE WKS. J, VOL 14, PP 152-159, 1942. 8 P.

DESCRIPTORS:
   *EUTROPHICATION, PLANKTON, *ALGAE, CHEMICAL PROPERTIES, BACTERIA, ORGANIC POLLUTANTS, *STREAMS, EUGLENA, NUTRIENTS, PLANT POPULATIONS, HISTORY, WATER PURIFICATION, *WATER UTILIZATION, DECOMPOSING ORGANIC MATTER, *STREAM POLLUTION.

IDENTIFIERS:
   *DOMESTIC WATER USE.

ABSTRACT:
   THE AUTHOR COMPARES THE POPULATION OF CERTAIN SPECIES OF PLANKTONIC ALGAE WITH THE CHEMICAL AND BACTERIOLOGICAL INDICES OF THE DEG OF ORGANIC POLLUTION IN A FLOWING STREAM. THE COMPARISON SHOWS THAT LARGE NUMBERS OF CHRYSOCOCCUS AND CRYPTOMONAS INDICATE THAT THE DECOMPOSITION OF THE ORGANIC MATTER IN THE STREAM HAS BEEN COMPLETED BY NATURAL PROCESSES AND THE STREAM MAY BE CONSIDERED CLEAN BUT NOT THAT IT CAN BE PUT TO DOMESTIC USE WITHOUT FURTHER TREATMENT. LARGE NUMBERS OF EUGLENA, TRACHELOMONAS AND PHACOTUS INDICATE THAT THE WATER UPSTREAM IS HEAVILY POLLUTED, AND THAT BACTERIA HAVE CHANGED THE ORGANIC MATTER TO AVAILABLE PLANT FOOD TO PRODUCE A RICH MEDIUM. THE PLANKTON POPULATION SERVES AS A BETTER INDICATOR OF THE HISTORY OF A STREAM THAN IT DOES OF THE SANITARY CONDITION OF THE STREAM AT ANY SINGLE PLACE. MOST WATERS, EVEN HEAVILY POLLUTED ONES, WILL CONTAIN A FEW OF THESE FORMS: THEREFORE, THEIR PRESENCE DOES NOT NECESSARILY INDICATE PURIFIED WATER NOR DOES THEIR ABSENCE INDICATE A POLLUTED CONDITION.

FIELD 05B

ACCESSION NO. W68-00483

FERTILIZATION OF LAKES--GOOD OR BAD,

MICH DEPT OF CONSERV.

ROBERT C. BALL.

MICH CONSERV, VOL 17, NO 9, PP 7-14, SEPT 1948. 8P, 1 FIG.

DESCRIPTORS:
*EUTROPHICATION, *LAKES, NUTRIENTS, OXYGEN REQUIREMENTS, ICED LAKES,
*FERTILIZATION, WATER FLEAS, LARVAE, GROWTH RATES, *FRESH WATER FISH,
MICHIGAN, TROUT, WINTERKILLING, FISHKILL, NUISANCE ALGAE, ALGAE,
SEASONAL, LIMNOLOGY, LIGHT PENETRATION, OXYGEN, SPORT FISHING, ON-SITE
INVESTIGATIONS.

IDENTIFIERS:
ARTIFICIAL FERTILIZATION, COMMERCIAL FERTILIZERS.

ABSTRACT:
STUDIES OF 4 LAKES IN THE NORTHERN PART OF THE SOUTHERN PENINSULA OF
MICHIGAN INDICATE THAT INORGANIC FERTILIZERS APPLIED TO THE WATERS MAY
BENEFIT THE FISH POPULATIONS IN SOME LAKES, BUT EXERT A HARMFUL EFFECT
IN OTHERS. 2 PAIRS OF LAKES WERE STUDIED, 1 OF EACH PAIR RECEIVED THE
INORGANIC FERTILIZER AND THE OTHER DID NOT. THE EXPERIMENTS HAD BEEN
CARRIED ON FOR 3 YEARS BY 1948. THE ADDITION OF COMMERCIAL FERTILIZER
TO SMALL PONDS INCREASES THE NUMBER OF ORGANISMS UPON WHICH YOUNG FISH
FEED. THE NUTRIENTS IN THE FERTILIZER PROVIDE FOOD FOR ALGAE, THE FOOD
FOR WATER FLEAS AND INSECT LARVAE; THESE ORGANISMS IN TURN, ARE A FOOD
SOURCE FOR BOTH YOUND AND ADULT GAME FISH. THE ADDITION OF FERTILIZER
SEEMS TO INCREASE THE TOTAL NUMBER OF SMALL GAME FISH BUT DOES NOT
RESULT IN GROWTH TO SIZES OF INTEREST TO GAME FISHERMEN. ARTIFICIAL
FERTILIZER INCREASES THE GROWTH OF ALGAE. IT ALSO FAVORS THE GROWTH OF
LESS DESIRABLE SPECIES OF GAME FISH AT THE EXPENSE OF TROUT. THE HEAVY
ALGAL GROWTH OF OVER-FERTILIZED LAKES, TENDS TO DEVELOP A SERIOUS
OXYGEN DEFICIENCY WHEN ICED OVER IN WINTER; EXCLUSION OF AIR AND MUCH
OF THE SUNLIGHT CAUSES THE ALGAE TO DIE AND, IN THEIR DECAY, TO USE UP
THE OXYGEN NEEDED BY FISH; SEVERE WINTERKILL OF GAME FISH RESULTS.
UNTIL MORE IS KNOWN, THE USE OF FERTILIZERS IN SMALL LAKES SHOULD BE
AVOIDED.

FIELD 05C

ACCESSION NO. W68-00487

ELEVEN YEARS OF CHEMICAL TREATMENT OF THE MADISONLAKES--ITS EFFECT ON FISH AND
FISH FOODS,

CITY OF MADISON AND WIS STATE LAB OF HYGIENE.

BERNHARD DOMOGALLA.

TRANS AMER FISH SOC, VOL 65, PP 115-121, 1935. 7 P, 1 FIG, 3 TAB.

DESCRIPTORS:
*EUTROPHICATION, *LAKES, WISCONSIN, ALGAE, *NUISANCE ALGAE, SPRAYING,
SEASONAL, *ALGAL CONTROL, FISHKILL, ROOTED AQUATIC PLANTS, GLACIAL
LAKES, PLANKTON, EPIDEMICS.

IDENTIFIERS:
SECCHI DISC, VISIBILITY TESTS, CONTROL METHODS.

ABSTRACT:
LAKE MONONA IS THE SECOND OF FOUR GLACIAL LAKES IN THE VALLEY OF THE
YAHARA RIVER NEAR MADISON, WIS. COMPLAINTS OF OBNOXIOUS ODORS AND ALGAL
GROWTHS LED IN 1925 TO STUDIES OF POSSIBLE REMEDIAL MEASURES. IT WAS
FOUND THAT PHOSPHORUS, NITROGEN, AND BICARBONATES WERE THE NUTRIENTS
THAT CAUSED ALGAL OVERGROWTHS. TO CONTROL THESE GROWTHS, COPPER
SULPHATE WAS FIRST APPLIED BY DRAGGING A BAG FILLED WITH IT THROUGH THE
WATER. MUCH MORE EFFECTIVE WAS THE SYSTEMATIC SPRAYING OF THE LAKE WITH
THE SAME CHEMICAL. THE SPRAYING REDUCED THE GROWTH OF CLADOPHORA AND
ULOTHRIX IN MAY; ANABAENA IN JUN; MICROCYSTIS, HYDRODICTYON, AND
APHANIZOMENON IN JUL AND AUG; AND APHANIZOMENON, CLADOPHORA, AND
ULOTHRIX IN SEPT. ROOTED WEEDS WERE CUT FROM UNDER 600 AC OF THE 5.4 SQ
MI OF THE LAKE. WEEDS IN THE DEEPER WATERS WERE NOT AFFECTED BY THE
CHEMICAL SPRAYING. VISIBILITY TESTS SHOWED DEPTHS OF 4.3 M IN LAKE
MONONA AS COMPARED WITH 0.6 AND 2.4 M IN OTHER LAKES IN THE GROUP. AN
EPIZOOTIC AMONG FISHES INFECTED WITH SAPROLEGNIA WAS PARTICULARLY
SEVERE IN 1930; HOWEVER, THE SPRAYING OF LAKE MONONA GREATLY REDUCED
THE FISHKILL THERE.

FIELD 05C

ACCESSION NO. W68-00488

NUTRIENT BUDGET: RATIONAL ANALYSIS OF EUTROPHICATION IN A CONNECTICUT LAKE,

DEPT OF SOILS CLIMATOL, THE CONN AGR EXP STATION, NEW HAVEN, CONN.

C. R. FRINK.

ENVIRONMENTAL SCI AND TECH, VOL 1, NO 5, PP 425-428, MAY 1967. 4 P.

DESCRIPTORS:
*EUTROPHICATION, *LAKES, CONNECTICUT, WATER POLLUTION, *NUTRIENT
REQUIREMENTS, *BOTTOM SEDIMENTS, ON-SITE DATA COLLECTIONS, AQUATIC
PLANTS, WATERSHED, ALGAE, AQUATIC VEGETATION, NITROGEN, PHOSPHOROUS,
AVAILABLE NUTRIENTS, SEDIMENT CORES.

IDENTIFIERS:
*SEDIMENTS, *NUTRIENT BUDGET, CYCLING NUTRIENTS.

ABSTRACT:
THE SIGNIFICANCE OF NUTRIENT CONTAMINATION OF WATER BY MAN CAN BE
EVALUATED ONLY BY COMPARISON WITH ALL OTHER SOURCES. IN AN EXEMPLARY
EUTROPHIC LAKE IN NORTHWESTERN CONNECTICUT, THE NUTRIENT INPUT FROM A
LARGELY FORESTED WATERSHED WITH NO OVERT SOURCE OF POLLUTION WAS
ADEQUATE TO SUPPORT THE OBSERVED ABUNDANT VEGETATIVE GROWTH. IN
ADDITION, A VAST ACCUMULATION OF NUTRIENTS WAS FOUND IN THE LAKE BOTTOM
SEDIMENTS; THE UPPER CENTIMETER OF SEDIMENT CONTAINS AT LEAST 10 TIMES
THE ESTIMATED ANNUAL INPUT OF NITROGEN AND PHOSPHORUS. MOREOVER, THIS
RESERVOIR OF NUTRIENTS IN THE SEDIMENTS SHOULD BE CAPABLE OF SUPPORTING
PLANT GROWTH FOR SOME TIME EVEN IF ALL NUTRIENTS COULD BE EXCLUDED FROM
THE LAKE. APPARENTLY, THE ABUNDANT WEEDS AND ALGAE IN THIS LAKE ARE THE
RESULT OF NATURAL EUTROPHICATION WHICH MAN WILL BE HARD PRESSED TO
ALTER.

FIELD 05C

ACCESSION NO. W68-00511

EUTROPHICATION OF LAKES BY DOMESTIC DRAINAGE,

A. D. HASLER.

ECOL, VOL 28, PP 383-395, 1947. 13 P, 1 TAB, 2 FIG.

DESCRIPTORS:
*EUTROPHICATION, ALGAE, AQUATIC PLANTS, BIBLIOGRAPHIES, CYANOPHYTA,
FRESH WATER FISH, LAKES, *LAKE ZURICH, *MUNICIPAL WASTES, NUTRIENTS,
NUISANCE ALGAE, OLIGOTROPHY, PHYTOPLANKTON, *REVIEWS, *WATER POLLUTION
EFFECTS.

ABSTRACT:
THE REPORTS OF PURPOSEFUL FERTILIZATION OF LAKES ARE VERY MEAGER, BUT
INADVERTENT EUTROPHICATION HAS BEEN RECORDED FOR MANY LAKES,
PRINCIPALLY THOSE IN EUROPE. A CLASSIC EXAMPLE IS LAKE ZURICH IN WHICH
2 BASINS ARE SEPARATED BY A NARROW PASSAGE. THE LARGER UNTERSEE (141 M
DEEP) WAS ONCE DECIDEDLY OLOGOTROPHIC BUT NOW TRENDS STRONGLY TOWARD
EUTROPHY. THE SHALLOWER OBERSEE (50 M DEEP) SHOWS VIRTUALLY NO EFFECTS
OF EUTROPHY. THE UNTERSEE IS RIMMED BY NUMEROUS COMMUNITIES (COMBINED
POPULATION OF 110,000) (ZURICH CONTRIBUTES NO SEWAGE) BUT THE OBERSEE
LACKS SUCH COMMUNITIES. EVIDENCES OF EUTROPHICATION OF THE UNTERSEE
DATE BACK TO LAST CENTURY AND ARE THE INCREASE OF NUISANCE ALGAE; THE
DISAPPEARANCE OF DESIRABLE FISH AND REPLACEMENT BY COARSE FISH; AND
INCREASES IN THE CHLORIDE ION. THE AUTHOR REVIEWS THE FERTILIZATION
PROCESS IN NUMEROUS LAKES IN EUROPE AND THE UNITED STATES. HE CONCLUDES
THAT FERTILIZATION ALMOST CERTAINLY HASTENS THE EXTINCTION OF A LAKE
AND THUS LIMITS ITS BENEFITS TO MAN TO A RELATIVELY SHORT PERIOD.
(BYRNE-WISC)

FIELD 05C, 02H

ACCESSION NO. W68-00680

BIOLOGICAL REMOVAL OF NUTRIENTS FROM TREATED SEWAGE: LABORATORY EXPERIMENTS,

WISCONSIN UNIV.

G. P. FITZGERALD, AND G. A. ROHLICH.

VERH INTER VER LIMNOL, VOL 15, PP 597-608, 1964. 12 P, 3 FIG, 2 TAB, 26 REF.

DESCRIPTORS:
*EUTROPHICATION, LABORATORY TESTS, BIOLOGICAL TREATMENT,
BIBLIOGRAPHIES, CHLORELLA, *SEWAGE, EFFLUENTS, CARBON DIOXIDE, *ALGAE,
*HYDROGEN ION CONCENTRATION, PHOSPHORUS, NITROGEN, GROWTH RATES,
STABILIZATION PONDS, SEWAGE DISPOSAL.

ABSTRACT:
THE AUTHORS DEMONSTRATED BY LABORATORY TESTS THAT IN A RELATIVELY SHORT
PERIOD OF TIME (13 DAYS) THE NUTRITIONAL VALUE OF TREATED SEWAGES, BOTH
SECONDARY EFFLUENT AND STABILIZATION POND WATERS, WAS MORE CLOSELY
RELATED TO THE AMOUNTS OF AMMONIA NITROGEN THAN TO PHOSPHATE
PHOSPHORUS. THEIR STUDIES SHOWED THAT THE TYPE AND QUANTITY OF ALGAL
GROWTH ATTAINED IN A STABILIZATION POND WILL BE CONTROLLED BY THE
ENVIORNMENT. 1% INOCULATIONS FROM LAKE WATERS WITH VARIED ALGAL
POPULATIONS INTO STABILIZATION POND WATER HAD NO APPARENT EFFECT ON
LATER GROWTH OF THE ALGAE. THE EXPERIMENTS INDICATED THAT THE RATE OF
NUTRIENT UTILIZATION AND GROWTH OF CHLORELLA IN PRIMARY OR SECONDARY
EFFLUENTS APPEARS TO BE INDEPENDENT OF THE ORGANIC MATTER PRESENT.
BECAUSE THE CARBON DIOXIDE PRESENT EFFECTS PH OF THE SEWAGE, THE CARBON
DIOXIDE USED BY ALGAE RESULTS IN HIGHER PH VALUES AND IN THE
CONSIDERABLE LOSS OF SOLUBLE PHOSPHATE PHOSPHORUS IN THE MEDIUM. THE
PRECIPITATED PHOSPHORUS CAN BE REDISSOLVED WHEN THE PH OF THE MEDIUM
DECREASES. (BYRNE-WISC)

FIELD 05D

ACCESSION NO. W68-00855

THE INFLUENCE OF THE MINERAL COMPOSITION OF THE MEDIUM ON THE GROWTH OF
PLANKTONIC ALGAE. PART I. METHODS AND CULTURE MEDIA,

LONDON UNIV., QUEEN MARY COLL., ENGLAND.

S. P. CHU.

J ECOL, VOL 30, NO 2, PP 284-325, AUG 1942. 42 P, 8 FIG, 22 TAB, 43 REF.

DESCRIPTORS:
*EUTROPHICATION, PHYTOPLANKTON, BIBLIOGRAPHIES, *ALGAE, *CHEMICAL
PROPERTIES, NUTRIENTS, LAKES, TESTS, NITROGEN, PHOSPHORUS, CALCIUM,
MAGNESIUM, SODIUM, SILICA, DIATOMS, RESISTANCE, *PLANT GROWTH
REGULATORS, FRESH WATER LAKES, LABORATORY TESTS.

ABSTRACT:
THE AUTHOR REPORTS THAT ALL OF THE ALGAE USED IN A TEST SERIES, WITH
BUT ONE EXCEPTION, GROW EQUALLY WELL WHEN THE MEDIA WERE SUPPLIED WITH
NITROGEN SOURCES FROM NITRATE SALTS OR AMONIUM SALTS, SO LONG AS THE
NITROGEN CONCENTRATION WAS WITHIN THE OPTIMUM RANGE; IN LOWER NITROGEN
CONCENTRATIONS, GROWTH WAS GENERALLY BETTER WHEN THE SOURCE WAS A
NITRATE SALT. THE EXPERIMENT WAS ONE IN WHICH 14 PLANKTONIC ALGAE WERE
MAINTAINED IN CULTURE MEDIA SIMULATING THE CHEMICAL COMPOSITIONS OF
NATURAL FRESH WATERS. ALGAE WERE FOUND TO RESPOND WITH CONSIDERABLE
DIFFERENCE TO CONCENTRATIONS OF CALCIUM, MAGNESIUM, POTASSIUM, SODIUM,
AND SILICA. THE CALCIUM REQUIREMENT WAS OFTEN LOWER IN MEDIA WITH
HIGHER MAGNESIUM CONCENTRATIONS, WHILE EXCESSIVE POTASSIUM INCREASES
TOLERANCES FOR HIGHER CONCENTRATIONS OF CALCIUM AND MAGNESIUM.
APPRECIABLE QUANTITIES OF SILICA WERE FOUND TO BE NECESSARY ONLY FOR
THE GROWTH OF DIATOMS; SOME ALGAE, IN FACT, WERE INHIBITED BY THE
PRESENCE OF DISSOLVED SILICA. THERE WAS A WIDER RANGE IN PHOSPHORUS
CONCENTRATION WHEN NITROGEN WAS DERIVED FROM A NITRATE SOURCE RATHER
THAN FROM AN AMMONIUM SALT. (BYRNE-WISC)

FIELD 05C

ACCESSION NO. W68-00856

RADIOACTIVE PHOSPHORUS AND EXCHANGE OF LAKE NUTRIENTS,

F. R. HAYES, AND C. C. COFFIN.

ENDEAVOUR, VOL 10, PP 78-81, 1951. 4 P, 3 FIG, 8 REF.

DESCRIPTORS:
*EUTROPHICATION, *LAKES, ACID LAKES, BIBLIOGRAPHIES, *PHOSPHORUS
RADIOISOTOPE, ON-SITE DATA COLLECTIONS, FISHES, PLANKTON, ALGAE, FROGS,
MOSSES, HYPOLIMNION, THERMOCLINE, *ION EXCHANGE, WATER CIRCULATION,
NUTRIENT ABSORPTION, *CHEMICAL PROPERTIES, EPILMNION, BOTTOM SEDIMENTS.

ABSTRACT:
THE AUTHOR'S CONCLUDE FROM FIELD EXPERIMENTS THAT THE DISTRIBUTION OF
IONS IN WATER TAKES PLACE FAR MORE RAPIDLY (FROM 10-100 TIMES) THAN
PREVIOUSLY SUPPOSED, AND THAT THE REMOVAL OF MATERIALS FROM THE WATERS
DOES NOT TAKE PLACE SEPARATELY, BUT PERHAPS IS AN EXCHANGE WITH
SUBSTANCES IN CONTACT WITH THE WATER. PHOSPHORUS 32 WAS ADDED FIRST TO
THE EPILIMNION OF A HIGHLY STRATIFIED ACID-BOG LAKE LESS THAN 1 AC IN
AREA AND 22 FT IN DEPTH. THE UPTAKE BY ZOOPLANKTON, SPONGES, SPHAGNUM,
AND ALGAE WAS IMMEDIATE; FISHES, FUNDULUS, AND NOTEMIGONUS SHOWED NO
MEASURABLE UPTAKE FOR SEVERAL DAYS, SUGGESTING THAT THEIR INTAKE WAS BY
FEEDING ON PLANKTON; NO EVIDENCE WAS FOUND THAT THE ADDITIVE REACHED
THE HYPOLIMNION. CERTAIN PLANTS CAN CONCENTRATE PHOSPHORUS UP TO 40,000
TIMES ITS CONCENTRATION IN THE SURROUNDING WATER. IN THE SECOND
EXPERIMENT, THE TRACER WAS ADDED TO THE HYPOLIMNION OF THE SAME LAKE.
THE ISOTOPE MOVED NO MORE THAN 5 MPH THROUGH THE COLD BOTTOM WATER AND
THE MOVEMENT WAS LARGELY LATERAL. AN 8-AC PRIMITIVE LAKE, WITH LITTLE
VEGETATION AND NO SPHAGNUM, WAS THE SITE FOR THE THIRD EXPERIMENT. ITS
WATERS WERE COMPLETELY MIXED FROM TOP TO BOTTOM. THE TRACER WAS ADDED
AT THE SURFACE, AND THE SURFACE AND BOTTOM WATERS, BOTTOM SEDIMENTS,
AND PLANTS QUICKLY REACHED EQUILBRIUM. (BYRNE-WISC)

FIELD 05C

ACCESSION NO. W68-00857

ON THE KINETICS OF PHOSPHORUS EXCHANGE IN LAKES,

DALHOUSIE UNIV., NOVA SCOTIA, CANADA.

F. R. HAYES, J. A. MCCARTER, M. L. CAMERON, AND D. A. LIVINGSTONE.

J ECOL, VOL 40, PP 202-216, 1952. 15 P, 7 FIG, 2 TAB.

DESCRIPTORS:
*EUTROPHICATION, OLIGOTROPHIC, *LAKES, *PHOSPHORUS RADIOISOTOPE,
*TURNOVERS, NUTRIENTS, PHOSPHORUS, INFLUENT SEWAGE, EQUILIBRIUM,
MATHEMATICAL MODELS, ON-SITE DATA COLLECTIONS, BOTTOM SEDIMENTS,
BIBLIOGRAPHIES, ABSORPTION, FISHES, ALGAE.

ABSTRACT:
ANY SINGLE OR PERIODIC APPLICATION OF A FERTILIZER IS NOT LIKELY TO
HAVE ANY MORE THAN A BRIEF EFFECT ON THE WATERS TO WHICH IT WAS
APPLIED, HOWEVER GREAT ITS QUANTITY. THIS CONCLUSION WAS REACHED FROM
STUDIES OF THE DATA COLLECTED AFTER PHOSPHORUS 32 WAS ADDED TO THE
WATER OF A SMALL LAKE NEAR HALIFAX, NOVA SCOTIA. THE TRACER WAS RAPIDLY
LOST BY UPTAKE BY AQUATIC ORGANISMS AND BOTTOM SEDIMENTS (MUD). THE
ADDED PHOSPHORUS INCREASED THE NORMAL WATER CONTENT BY ONLY 0.25%. THE
ACTIVE EXCHANGE OF THE PHOSPHORUS RADIOISOTOPE AND THE PHOSPHORUS IN
THE LAKE ORGANISMS AND BOTTOM SEDIMENTS SOON BECAME A SINGLE
EQUILIBRIUM SYSTEM. THE TURNOVER TIME FOR PHOSPHORUS IN THE WATER WAS
5.4 DAYS, AND THAT FOR THE PHOSPHORUS IN ORGANISMS AND MUD WAS 39 DAYS.
ONLY THE SUPERFICIAL LAYER 1 MM THICK WAS INVOLVED IN THE INTERCHANGE.
EUTROPHICATION CAN BE INDUCED BY THE CONSTANT ADDITION OF SEWAGE TO A
LAKE; THIS KEEPS THE NUTRIENTS CONTINUOUSLY ABOVE THE EQUILIBRIUM LEVEL
FOR ORGANISMS AND MUDS. STOPPAGE OF SEWAGE DISCHARGE WOULD LOWER THE
PHOSPHORUS LEVEL AND THE LAKE WOULD BECOME OLIGOTROPHIC. (BYRNE-WISC)

FIELD 05C

ACCESSION NO. W68-00858

ANTIBIOTIC ASPECTS OF COPPER TREATMENT OF LAKES,

WISCONSIN UNIV., DEPARTMENT OF ZOOLOGY.

ARTHUR D. HASLER.

TRANS WISCONSIN ACAD OF SCI, ARTS AND LETTERS, VOL 39, PP 97-103, 1949. 7 P, 18 REF.

DESCRIPTORS:
*EUTROPHICATION, *LAKES, *COPPER SULFATE, *ALGAL CONTROL, *TOXINS, ALGAE, RESISTANCE, FISH, AMPHIBIANS, INVERTEBRATES, AQUATIC POPULATIONS, BIBLIOGRAPHIES, BOTTOM SEDIMENTS, WATERSHED MANAGEMENT, HARVESTING OF ALGAE, WATER ZONING.

IDENTIFIERS:
PHYSIOLOGICAL EFFECTS.

ABSTRACT:
LAKES, IN A COMPARATIVELY SHORT TIME, MOVE TOWARD A EUTROPHIC STAGE IN WHICH MASSIVE ALGAL DEVELOPMENTS MAY BE ANTICIPATED. SUCH ALGAL GROWTHS INHIBIT MANY OF THE RECREATIONAL AND OTHER HUMAN USES MADE OF THE LAKES AND OF THEIR WATERS. THE SPRAY APPLICATION OF COPPER SULFATE SOLUTION HAS BEEN FOUND EFFECTIVE IN INHIBITING THE DEVELOPMENT OF ALGAL BLOOMS. HOWEVER, THE ADDITION OF COPPER THROUGH SPRAYS MAY INTRODUCE A TOXIC CONDITION. THE RESULTS OF SHORT EXPOSURES OF FISH TO COPPER SALT TOXICITY DO NOT REFLECT THE IMPACT OF TOXIC EFFECTS OVER LONGER PERIODS. ALGAE, FISH, INVERTEBRATES, AND AMPHIBIANS ARE ALL SUBJECT TO INHIBITION OR DESTRUCTION BY THE ABSORPTION OF COPPER SALTS. THE AUTHOR SUGGESTS THAT STUDIES BE UNDERTAKEN TO SUBSTITUTE OTHER CONTROL SUBSTANCES FOR A PERMANENT POISON THAT HAS BEEN EMPLOYED TO COMBAT A TEMPORARY NUISANCE. HARVESTING OF CROPS OF ALGAE AND LARGE AQUATIC PLANTS MIGHT BE A CONTROL MECHANISM, AS MIGHT ZONING OF LAKES FOR CERTAIN SPECIFIC USES, AND VARIOUS KINDS OF WATERSHED CONTROLS. (BYRNE-WISC)

FIELD 05C

ACCESSION NO. W68-00859

LIMNOLOGICAL STUDIES IN CONNECTICUT. 7. A CRITICAL EXAMINATION OF THE SUPPOSED RELATIONSHIP BETWEEN PHYTOPLANKTON PERIODICITY AND CHEMICAL CHANGES IN LAKE WATERS,

YALE UNIV., OSBORN ZOOL LAB.

G. EVELYN HUTCHINSON.

ECOL, VOL 25, NO 1, PP 3-26, JAN 1944. 24 P, 6 FIG, 37 REF.

DESCRIPTORS:
*EUTROPHICATION, *LAKES, PHYTOPLANKTON, ALGAE, SEASONAL, CYANOPHYTA, *INORGANIC COMPOUNDS, PHOSPHORUS, BIBLIOGRAPHIES, NITRATES, NUTRIENTS, PHOSPHATES, SILICATES, IRON OXIDES, *PLANT POPULATIONS, REVIEWS, LIGHT INTENSITY, WATER TEMPERATURE, DIATOMS, CHLOROPHYTA.

IDENTIFIERS:
DESMIDS.

ABSTRACT:
AS THE RESULT OF HIS STUDIES OF LINSLEY POND, A SMALL LAKE IN CONN, THE AUTHOR CONCLUDES THAT THERE IS IN GENERAL, NO CLEAR CUT CORRELATION BETWEEN THE CHEMICAL CONDITION OF THE WATER AND THE QUALITATIVE COMPOSITION OF THE PHYTOPLANKTON. THE PHYSIOLOGICAL CONDITION OF A POPULATION AND ITS RELATION TO POPULATIONS OF OTHER SPECIES IS LIKELY TO EXPLAIN MANY OF THE APPARENT INCONSISTENCIES BETWEEN THE SEASONS AND DIFFERENT LAKES. THE SPRINGTIME APPEARANCE OF DYNOBRYON DIVERGENS IN LINSLEY POND IS TO BE CORRELATED WITH A RISE IN THE RATIO OF NITRATE TO PHOSPHATE, BUT IS INDEPENDENT OF VARIATIONS IN SOLUBLE SILICATE. INVARIABLY THE DYNOBRYON POPULATION SEEMS TO INCREASE AFTER THE MAJOR COMPONENTS OF THE SPRING MAXIMUM HAVE DECLINED. THE POND GIVES NO EVIDENCE THAT THE APPEARANCE OF ASTERIONELLA FORMOSA DEPENDS ON THE LEVEL OF INORGANIC NUTRITION. THE BLUE-GREEN ALGAL BLOOMS OF LATE SUMMER ARISE WHEN THE INORGANIC NUTRIENTS ARE PRACTICALLY EXHAUSTED. IN SMALL EUTROPHIC LAKES, ALL OF THE INORGANIC NUTRIENTS, EXCEPT PHOSPHORUS AND COMBINED NITROGEN, ARE NORMALLY PRESENT IN GREAT EXCESS. AS FOR IRON, ONLY PARTICULATE FERRIC IRON IS LIKELY TO BE IMPORTANT. (BYRNE-WISC)

FIELD 05C

ACCESSION NO. W68-00860

ECOLOGICAL STUDY OF THE EFFECTS OF STRIP MINING ON THE MICROBIOLOGY OF STREAMS,

KENTUCKY UNIV., LEXINGTON.

HARRY D. NASH, AND RALPH H. WEAVER.

RESEARCH REPORT 18, KENTUCKY WATER RESOURCES INST., LEXINGTON, 1968. 35 P, 19 TAB.

DESCRIPTORS:
STRIP MINES, DRAINAGE, FERROBACILLUS, WATER TEMPERATURE, HYDROGEN ION CONCENTRATION, FUNGI, YEASTS, PYRITE, ALGAE, CYANOPHYTA, DOMINANT ORGANISM, NUTRIENTS, RAINFALL-RUNOFF RELATIONSHIPS, GEOGRAPHICAL REGIONS, KENTUCKY, ACID STREAMS, MICROFLORA.

IDENTIFIERS:
TAXONOMIC GROUPS, MOLECULAR FILTER, CHEMOSYNTHETIC, SAPROPHYTIC, PLATE COUNT, FILAMENTOUS, UNICELLULAR, NATURAL RECOVERY.

ABSTRACT:
THE MICROFLORA OF CANE BRANCH OF BEAVER CREEK IN MCCREARY COUNTY, KENTUCKY, WHICH DRAINS AN AREA THAT WAS STRIP-MINED BETWEEN 1955 AND 1959, WAS STUDIED AND COMPARED WITH THAT OF HELTON BRANCH WHICH DRAINS A COMPARABLE AREA WHERE THERE HAS BEEN NO MINING. DIFFERENCES INCLUDE: THE ESTABLISHMENT OF FERROBACILLUS FERROOXIDANS, FOR WHICH PROCEDURES WERE DEVELOPED FOR DIRECT COLONY ISOLATION FROM THE STREAM; FEWER SAPROPHYTIC BACTERIA; MORE NUMEROUS AND MORE DIVERSIFIED FILAMENTOUS AND UNICELLULAR FUNGI; AND CHARACTERISTIC DIFFERENCES IN ALGAL FLORA. REPRESENTATIVES OF 42 GENERA OF FILAMENTOUS FUNGI WERE IDENTIFIED. OF THESE, 21 WERE ISOLATED ONLY FROM CANE BRANCH. REPRESENTATIVES OF FIVE GENERA OF UNICELLULAR FUNGI WERE FOUND. ONE, RHODOTORULA, WAS FOUND CONSISTENTLY IN CANE BRANCH BUT NEVER IN HELTON BRANCH. FROM 1966 TO 1968, BUMILLERIA APPEARS TO HAVE ESTABLISHED ITSELF AS THE DOMINATE ALGA IN CANE BRANCH AT SOME DISTANCE DOWNSTREAM FROM THE STRIP-MINE DRAINAGE AREA. SEASONAL DIFFERENCES IN THE MICROFLORA APPEAR TO BE RELATIVELY INSIGNIFICANT, EXCEPT FOR THE ALGAE. (AUTHOR)

FIELD 05C

ACCESSION NO. W68-00891

FACTORS AFFECTING ACCELERATED EUTROPHICATION OF FLORIDA LAKES,

FLORIDA UNIV., GAINESVILLE, FLORIDA WATER RESOURCES RESEARCH CENTER.

HUGH D. PUTNAM.

OFFICE OF WATER RESOURCES RESEARCH, PROJECT COMPLETION REPORT A-002-FLA, AUGUST 1968. 123 PP, 36 FIG, 28 TAB, 37 REF.

DESCRIPTORS:
*EUTROPHICATION, ALGAL BLOOMS, WATER BLOOM, AQUATIC ALGAE, AGING(BIOLOGICAL), *AQUATIC PRODUCTIVITY, LIMNOLOGY, *NUTRIENTS, BIO MASS, OLIGOTROPHY, *PRIMARY PRODUCTIVITY, TROPIC LEVEL, WATER QUALITY, WASTE ASSIMILATIVE CAPACITY, STANDING CROP, ANNUAL TURNOVER, FISH POPULATIONS.

ABSTRACT:
EUTROPHICATION HAS BEEN INVESTIGATED IN AN EXPERIMENTAL LAKE UNDER STRESS WITH A PREDETERMINED NUTRIENT LOADING RATE (500 MG N AND 42.7 MG P/M 3 YR). OBJECTIVES ARE TO DETERMINE VALID PARAMETERS IN ASSESSING TROPHIC LEVELS IN SHALLOW, SUBTROPICAL LAKES AND TO DEVELOP NEW CRITERIA IF NECESSARY; TO CONSTRUCT A MATHEMATICAL MODEL OF EUTROPHICATION USING SYSTEMS ANALYSES; AND TO IDENTIFY CONTROLLING FACTORS IN EUTROPHICATION. THUS FAR THE EXPERIMENTAL LAKE HAS PHOSPHORUS AND AMMONIA NITROGEN IN EXCESS OF LEVELS IN AN ADJACENT CONTROL LAKE, BUT NO CLEAR CUT BIOLOGICAL DIFFERENCES BETWEEN THE TWO LAKES HAVE BEEN NOTED. PRIMARY PRODUCTION AND PLANKTON CHLOROPHYLL HAVE REMAINED LOW (1 - 190 MGC PER M 3 DAY - 1, 1 - 18 MG CHL A/M 3). BIOASSAYS FOR LIMITING NUTRIENTS REVEAL THAT PHOSPHORUS COMMONLY LIMITS PRODUCTION IN THE EXPERIMENTAL LAKE WHILE MANY NUTRIENTS (PHOSPHORUS, TRACE METALS, VITAMINS, SILICA) HAVE CYCLIC BEHAVIORS IN THE CONTROL LAKE. THE MICROBIOTA OF THE LAKE IS VARIED, BUT A MEASURABLE CROP OF DINOBRYON, SYNURA, PERIDINIUM UMBONATUM AND STENTOR AMETHYSTINUM PERSISTS AT TIMES. TERTIARY TROPHIC LEVELS ARE BEING INVESTIGATED USING THE PLANKTIVOROUS FISH LABIDESTHES SICCULUS. CHANGES IN ZOOPLANKTON BIOMASS AND THE LABIDESTHES POPULATION ARE BEING MONITORED. (AUTHOR)

FIELD 02H, 05C

ACCESSION NO. W68-00912

EUTROPHICATION IS BEGINNING IN LAKE MICHIGAN,

FEDERAL WATER POLLUTION CONTROL ADMIN.

A. F. BARTSCH.

WATER AND WASTE ENG, VOL 5, NO 9, PP 84-87, SEPT 1968. 4 P, 3 FIG, 1 TAB, 3
REF.

DESCRIPTORS:
*EUTROPHICATION, *LAKE MICHIGAN, ALGAE, NUISANCE ALGAE, PLANKTON,
BENTHIC FAUNA, SLUDGE WORMS, BLOODWORMS, WATER POLLUTION.

IDENTIFIERS:
ACCELERATED EUTROPHICATION.

ABSTRACT:
A STUDY OF THE PHYSICAL, CHEMICAL, AND BIOLOGICAL STATUS OF LAKE
MICHIGAN REVEALS SOME EARLY SYMPTOMS OF ACCELERATED EUTROPHICATION. IN
THE OFFSHORE AREAS, TWO CHANGES IN SPECIES COMPOSITION OF ZOOPLANKTON
WERE SEEN. BOSMIA LONGIROSTRIS REPLACED B. COREGONI, AND DIAPTOMUS
OREGONENIS HAS APPEARED. BOTTOM-DWELLING ANIMAL POPULATIONS HAVE
INCREASED MARKEDLY. TOTAL DISSOLVED SOLIDS INCREASED 30 MG PER L IN THE
LAST NINETY YEARS, SULFATE 13 MG PER L, AND CHLORIDE 6 MG PER L.
BETWEEN 1955 AND 1966, OXYGEN CONTENT DECREASED SLIGHTLY. IN THE
INSHORE AREAS, CHANGES ARE DRASTIC. ATTACHED AND FLOATING ALGAE,
INCLUDING LARGE GROWTHS OF CLADOPHORA, ARE FREQUENTLY AT NUISANCE
LEVELS. AT GREEN BAY, PHOSPHATE CONCENTRATION VARIES FROM 0.07-0.6 MG
PER L, AND IN MILWAUKEE HARBOR IT AVERAGES 0.44 MG PER L AND IS
SOMETIMES AS HIGH 1.4 MG PER L. A SHIFT OF BOTTOM POPULATIONS TO DENSE
CONCENTRATIONS OF POLLUTION-TOLERANT SPECIES SUCH AS BLOODWORMS AND
SLUDGE WORMS HAS OCCURRED. LAKE MICHIGAN'S RETENTION OF NITROGEN IS 81%
AND OF PHOSPHATE, 95%, SO THE PROBLEM WILL NOT BE SOLVED QUICKLY BY
ELIMINATING INPUT. IMMEDIATE CONTROL OF NUTRIENT INPUT IS, HOWEVER,
NECESSARY TO PREVENT FURTHER DETERIORATION. (KNAPP-USGS)

FIELD 05C

ACCESSION NO. W68-01244

THE EFFECTS OF STRIP MINING ON THE MICROBIOLOGY OF A STREAM FREE FROM DOMESTIC
POLLUTION,

KENTUCKY UNIVERSITY, DEPARTMENT OF MICROBIOLOGY, LEXINGTON.

RALPH H. WEAVER, AND HARRY D. NASH.

REPRINTS OF PAP 2D SYMP ON COAL MINE DRAINAGE RES, MELLON INST, PP 80-97, MAY
1968. 16 P, 5 TAB, 17 REF.

DESCRIPTORS:
*ACID MINE WATER, *AQUIATIC MICROORGANISMS, *ALGAE, *AQUATIC BACTERIA,
*AQUATIC FUNGI, YEASTS, HYDROGEN ION CONCENTRATION, IRON, SULFATE,
TEMPERATURE, KENTUCKY.

IDENTIFIERS:
ACID WASTE MICROFLORA, MCCREARY COUNTY, KENTUCKY.

ABSTRACT:
THE MICROFLORA OF CANE BRANCH, MCCREARY CO., KENTUCKY, WHICH DRAINS A
STRIP-MINED AREA, AND HELTON BRANCH, WHICH DRAINS A SIMILAR BUT
NON-MINED AREA, WERE STUDIED TO DETERMINE THE BIOLOGICAL EFFECTS OF
ACID MINE DRAINAGE. NUMBERS AND TYPES OF BACTERIA, FUNGI, YEASTS, AND
ALGAE, FROM BOTH SURFACE AND BOTTOM SAMPLES, AS WELL AS TEMPERATURE AND
PH WERE STUDIED TO LEARN THE CHANGES OF ECOLOGY WITH DILUTION AND
RECOVERY. THESE DATA ARE TABULATED BY SEASON AND STREAM. IRON OXIDIZING
BACTERIA APPEAR TO BE INDIGENOUS IN ACID MINE WATER AND DIRECTLY
ASSOCIATED WITH ACID FORMATION. FERROBACILLUS FERROXIDANS WAS FOUND IN
CANE BRANCH IN ALL SEASONS, BUT ONLY IN SUMMER IN HELTON BRANCH. THE
LOWERING OF PH TO 3.0-4.1 FROM 6.3-6.7, WITH AN INCREASE IN SULFATES
AND ALMOST TOTAL ELIMINATION OF BICARBONATE ALKALINITY, RESULTED IN THE
ALTERATION OF MICROFLORA IN CANE BRANCH, PRIMARILY THE ESTABLISHMENT OF
F. FERROXIDANS. THE ROLE OF FUNGI IN STREAM CHEMISTRY IS AS YET
UNKNOWN. CANE BRANCH ALGAL GROWTH AND DIVERSITY OF TYPES INCREASED WITH
DISTANCE FROM THE SURFACE OF POLLUTION WHERE BUMILLERIA, WHICH IS FOUND
ONLY IN STREAMS WITH ACID MINE WASTE, WAS THE ONLY GENUS PRESENT.
(KNAPP-USGS)

FIELD 05C

ACCESSION NO. W69-00096

THE HIDDEN FLORA OF A LAKE,

HEBREW UNIVERSITY OF JERUSALEM.

MENACHEM RAHAT, AND INKA DOR.

HYDROBIOL, VOL 31, FASC 2, PP 186-192, 1968. 8 P, 1 TAB, 27 REF.

DESCRIPTORS:
*LAKES, PLANTS, *ALGAE, *SALINITY, IRRIGATION EFFECTS, DRAINAGE
EFFECTS, *SAMPLING, ALGAE TOXINS, SALINE WATER, EQUILIBRIUM, BIOLOGICAL
COMMUNITIES, *AQUATIC ALGAE, HALOPHYTES.

ABSTRACT:
A SURVEY WAS CONDUCTED TO DETERMINE HIDDEN ALGAL FLORA OF LAKE KINNERET
(SEA OF GALILEE) IN THE JORDAN RIFT VALLEY, ISRAEL, WHICH WOULD
DOMINATE IN A RISE OF SALINITY DUE TO DISTURBANCE OF PRESENT
EQUILIBRIUM FROM DRAINAGE FOR IRRIGATION. THE CHRYSOMONAD PRYMNESIUM
PARVUM WHICH CAUSED TOXIC BLOOMS WAS OF PARTICULAR INTEREST. WATER
SAMPLES WERE SELECTIVELY ENRICHED FOR HALOPHILIC ALGAE WITH AN
INORGANIC SALT MEDIUM. A RADICAL CHANGE WAS OBSERVED. ALGAE RARE OR
UNKNOWN TO THE LAKE WERE PRESENT. REPRESENTATION OF THE ALGAL GROUPS
WAS AS FOLLOWS: 7 SPECIES BELONGED TO CHLOROPHYCEAE, 8 SPECIES TO
DRATOMEAE AND 15 TO CYANOPHYCEAE. PRYMNESIUM PARVUM WAS PRESENT IN 6 OF
THE 27 SAMPLES. THUS IT WAS INDICATED THAT A SUFFICIENT RISE OF THE
LAKES SALINITY WOULD RESULT IN A PREDOMINANCE OF HALOPHILIC ALGAE AND
THE PRESENCE OF PRYMNESIUM PARVUM FORMS A POTENTIAL DANGER OF TOXIC
BLOOMS. (AFFLECK-ARIZ)

FIELD 05C, 02H

ACCESSION NO. W69-00360

AN EARTH-SMELLING COMPOUND ASSOCIATED WITH BLUE-GREEN ALGAE AND ACTINOMYCETES,

CALIFORNIA UNIV., BERKELEY.

LLOYD L. MEDSKER, DAVID JENKINS, AND JEROME F. THOMAS.

ENVIRON SCI TECHNOL, VOL 2, NO 6, PP 461-64, JAN 1968. 4 P, 4 FIG, 14 REF.

DESCRIPTORS:
*WATER ANALYSIS, *ODORS, *ODOR PRODUCING ALGAE, EUTROPHICATION,
ALCOHOLS, ACTINOMYCETES, CYANOPHYTA, SPECTROSCOPY, GAS CHROMATOGRAPHY,
ANALYTICAL TECHNIQUES, WATER POLLUTION SOURCES.

IDENTIFIERS:
PROTON MAGNETIC RESONANCE SPECTROMETRY, OSCILLATORIA TENUIS, RAMMAN
SPECTROSCOPY, MASS SPECTROMETRY, SYMPLOCA MUSCORUM.

ABSTRACT:
A COMPOUND, WITH EARTHY-MUSTY ODOR SIMILAR TO THAT OFTEN CAUSING
PROBLEMS IN FOOD AND WATER SUPPLIES, WAS ISOLATED BY PREPARATIVE GAS
CHROMATOGRAPHY FROM A CULTURE OF ACTINOMYCETES (NO 18) AND CULTURES OF
2 SPECIES OF CYANOPHYTES, SYMPLOCA MUSCORUM AND OSCILLATORIA TENUIS.
ACTINOMYCETES WERE GROWN IN MEDIUM (M SUB 1 B SUB 2); CYANOPHYTES, IN
HUGHES MEDIUM 11 (16 LITERS AT 26-28 DEG C AND 200-300 FT-CANDLES OF
LIGHT). THE FORMER YIELDED CA 200 MICROG/1 AFTER 8-12 DAYS AT ROOM
TEMP; SYMPLOCA, 1 MG/VESSEL AFTER 4-6 WK. CYANOPHYTE CULTURES WERE
SHOWN TO BE FREE OF CONTAMINATING ACTINOMYCETES. MASS SPECTRUM OF THE
COMPOUND SHOWED IT TO BE IDENTICAL WITH THAT OF GEOSMIN (C SUB 12 H SUB
22 O). VARIOUS ANALYTICAL TECHNIQUES (DEGRADATION, BROMINATION,
HYDROGENATION, ANALYTICAL GAS CHROMATOGRAPHY, HIGH RESOLUTION MASS
SPECTROMETRY, INFRARED SPECTROMETRY AND PROTON MAGNETIC RESONANCE
SPECTROMETRY) SHOWED THAT GEOSMIN IS A DIMETHYL SUBSTITUTED, SATURATED,
TERTIARY ALCOHOL WITH 2 RINGS IN WHICH THE HYDROXYL GROUP IS VERY
STERICALLY HINDRED. STRUCTURE OF THE RING SYSTEM IS UNKNOWN, BUT RAMMAN
SPECTROSCOPY INDICATES THAT IT MAY BE SIMILAR TO ALPHA-PINENE. GEOSMIN
IS A COLORLESS, VISCOUS LIQUID, WITH CAMPHORISH ODOR (WHEN
CONCENTRATED) WHICH BECOMES EARTHY UPON DILUTION. A THRESHOLD ODOR OF
0.05 MICROG/1 IS REPORTED. (EICHHORN-WISC)

FIELD 05A, 05C

ACCESSION NO. W69-00387

PLANKTONIC PHOTOSYNTHESIS AND THE ENVIRONMENT OF CALCIUM CARBONATE DEPOSITION IN LAKES,

MINNESOTA UNIV., MINNEAPOLIS. LIMNOLOGICAL RESEARCH CENTER.

ROBERT O. MEGARD.

INTERIM REP 2, UNIV MINNESOTA, JAN 1968. 47 P, 7 FIG, 6 TAB, 33 REF. OWRR PROJ A-008-MINN.

DESCRIPTORS:
*PHOTOSYNTHESIS, *CALCIUM CARBONATE, AQUATIC ENVIRONMENT, AQUATIC PRODUCTIVITY, EUTROPHICATION, *LAKES, LIMNOLOGY, *PLANKTON, ALGAE, BIBLIOGRAPHIES, CHEMICAL PRECIPITATION, IONS, BIOCHEMICAL OXYGEN DEMAND, MINNESOTA, WATER CHEMISTRY, BIODEGRADATION.

ABSTRACT:
CALCIUM AND ALKALINITY CONCENTRATIONS INCREASE IN A PRODUCTIVE LAKE DURING WINTER AND THE WATER AT ALL DEPTHS BECOMES UNDERSATURATED WITH CALCIUM CARBONATE. SOON AFTER ICE MELT, THE ENTIRE WATER COLUMN BECOMES OVERSATURATED WITH CALCIUM CARBONATE, AND CALCIUM AND ALKALINITY CONCENTRATIONS DECREASE DURING SPRING, SUMMER, AND AUTUMN. WHEN THE LAKE BECOMES THERMALLY STRATIFIED, CARBONATE OVERSATURATION INCREASES AT DEPTHS IN WHICH PHOTOSYNTHESIS OCCURS, BUT THE WATER AT GREATER DEPTHS BECOMES UNDERSATURATED. CARBONATE SATURATION APPEARS CONTROLLED BY THE BALANCE BETWEEN CARBON DIOXIDE ASSIMILATION BY PLANKTONIC PHOTOSYNTHESIS AND CARBON DIOXIDE RELEASE BY RESPIRATION. SEASONAL CHANGES OF CALCIUM CONCENTRATIONS ARE BALANCED BY CHEMICALLY EQUIVALENT ALKALINITY CHANGES, BUT MAGNESIUM CONCENTRATIONS DO NOT FLUCTUATE. PHOTOSYNTHESIS OCCURS IN BOTH THE EPILIMNION AND THE METALIMNION OF A DEEPER, LESS PRODUCTIVE LAKE. THESE ARE THE ONLY DEPTHS OVERSATURATED WITH CALCIUM CARBONATE DURING OPEN-WATER SEASON.

FIELD 02H, 05C

ACCESSION NO. W69-00446

LIMNOLOGY, PRIMARY PRODUCTIVITY, AND CARBONATE SEDIMENTATION OF MINNESOTA LAKES,

MINNESOTA UNIV., MINNEAPOLIS.

ROBERT O. MEGARD.

LIMNOLOGICAL RES CENTER, INTERIM REP 1, JAN 1967. 69 P, 25 FIG, 11 TAB, 7 REF. OWRR PROJECT A-008-MINN.

DESCRIPTORS:
SEDIMENTATION RATES, *EUTROPHICATION, *LAKES, *PRIMARY PRODUCTIVITY, *PLANKTON, ALGAE, *PHOTOSYNTHESIS, PRODUCTIVITY, WATER QUALITY, *LIMNOLOGY, MINNESOTA, WATER ANALYSIS, CALCIUM CARBONATE, HYPOLIMNION.

ABSTRACT:
THE REPORT DESCRIBES INVESTIGATIONS OF THE REGIONAL LIMNOLOGY OF MINN. THE WORK HAS INVOLVED 3 PHASES: (1) A MOBILE LABORATORY WAS DESIGNED AND OUTFITTED; (2) STANDARD FIELD AND LABORATORY METHODS WERE MODIFIED TO MAKE THEM MORE COMPATIBLE WITH THE MOBILITY OF THE LABORATORY; AND (3) A STUDY WAS BEGUN OF THE PRIMARY PRODUCTIVITY AND GENERAL LIMNOLOGY OF 14 LAKES IN VARIOUS PARTS OF THE STATE. THE LABORATORY PERMITS WATER SAMPLES TO BE ANALYZED AT LAKESIDE BEFORE NUTRIENT CONCENTRATIONS ARE ALTERED BY BIOLOGICAL ACTIVITY AND PROVIDES FACILITIES FOR MEASURING THE GROWTH RATES OF ALGAE AND DETERMINING THE FERTILITY OF LAKE WATER. THE OXYGEN LIGHT-DARK-BOTTLE METHOD IS BEING USED TO COMPARE ALGAL PRODUCTIVITY. THE RESPONSE OF THE CARBONATE SATUROMETER IN FRESH WATER WAS TESTED, AND THE INSTRUMENT WAS USED TO DETERMINE THE DEGREE OF CALCIUM CARBONATE SATURATION AT DIFFERENT TIMES OF THE YEAR AND AT VARIOUS DEPTHS IN THE LAKES. THE SEASONAL CHEMICAL CHANGES IN THE LAKES ARE DESCRIBED AND THE PRODUCTIVITY DATA ASSEMBLED, BUT THERE ARE NOT YET ENOUGH DATA FOR A DETAILED COMPARISON OF THE LAKES.

FIELD 02H, 05C

ACCESSION NO. W69-00632

SURFACE-WATER CHEMISTRY OF SOME MINNESOTA LAKES, WITH PRELIMINARY NOTES ON
   DIATOMS,

MINNESOTA UNIV., MINNEAPOLIS.

ROBERT C. BRIGHT.

LIMNOLOGICAL RES CENTER, BELL MUSEUM NATUR HISTORY, UNIV MINN INTERIM REP 3,
   APR 1968. 51 FIG, 44 REF, APPEND, DISC. OWRR PROJECT B-001-MINN.

   DESCRIPTORS:
      BIBLIOGRAPHIES, *MINNESOTA, *LAKES, *DIATOMS, *ALGAE, ECOLOGY, SOUTH
      DAKOTA, FIELD INVESTIGATIONS, WATER POLLUTION, WATER SAMPLING, *WATER
      ANALYSIS, *CHEMICAL ANALYSIS, PH, SAMPLING, ON-SITE DATA COLLECTIONS,
      LIMNOLOGY.

   ABSTRACT:
      A COMPREHENSIVE FIELD STUDY WAS MADE OF DIATOM ECOLOGY IN LAKES OF MINN
      AND ADJACENT PARTS OF SOUTH DAKOTA. THE PROGRAM CONSISTED OF 4 PHASES:
      (1) IDENTIFICATION AND DETERMINATION OF THE GEOGRAPHIC DISTRIBUTION AND
      RELATIVE ABUNDANCE OF DIATOMS IN THE REGION, (2) CORRELATION OF DIATOM
      DISTRIBUTION AND RELATIVE ABUNDANCE WITH REGIONAL PATTERNS OF
      SURFACE-WATER CHEMISTRY, (3) CORRELATION OF INFORMATION FROM THE FIRST
      2 PHASES WITH THE COMPOSITION OF THE DIATOM FLORA IN SURFACE SEDIMENTS
      FROM EACH LAKE, AND (4) APPLICATION OF CONCLUSIONS FROM THE FIRST THREE
      PARTS TO FOSSIL DIATOM FLORAS AND PALEOECOLOGY OF LAKES. MINNESOTA AND
      ADJACENT SOUTH DAKOTA COMPRISE THE BEST AREA IN NORTH AMERICA FOR SUCH
      A BROAD STUDY BECAUSE OF THE WIDE VARIETY OF LAKE TYPES AND THEIR WIDE
      RANGE OF SALINITY. THE ENSUING DISCUSSIONS DEAL MAINLY WITH FALL
      SURFACE-WATER CHEMISTRY, POSSIBLE ORIGINS OF ITS REGIONAL PATTERN, ITS
      CORRELATION WITH SIMILAR DATA FROM OTHER AREAS, AND ITS CONTROLLING
      FACTORS.

   FIELD 02H, 05C

   ACCESSION NO. W69-00659

THE ECOLOGICAL SIGNIFICANCE OF CELLULOLYTIC BACTERIA IN QUABBIN RESERVOIR,

CLARK UNIV., WORCESTER, MASS.

JOHN T. REYNOLDS.

PROJ COMPLETION REP, JAN 1968. 11 P. OWRR PROJECT A-007-MASS.

   DESCRIPTORS:
      *ECOLOGY, BACTERIA, ALGAE, PHYTOPLANKTON, *LAKES, *CELLULOSE,
      DISTRIBUTION PATTERNS, MEMBRANE PROCESSES, MASSACHUSETTS.

   ABSTRACT:
      ECOLOGICAL SIGNIFICANCE OF CELLULOLYTIC BACTERIA IN A LARGE RESERVOIR
      IN CENTRAL MASSACHUSETTS WAS STUDIED. BACTERIA WERE ENUMERATED USING
      MEMBRANE FILTER TECHNIQUES: CELLULOLYTIC FORMS WERE GROWN OUT ON A
      MODIFICATION OF SANBORN'S CHINA BLUE-CELLULOSE DEXTRIN AGAR, AND
      HETEROTROPHS ON A MINERAL SALT-MUD EXTRACT AGAR AND ON MPH AGAR.
      SUSPENDED CELLULOSE AND HEMICELLULOSE WERE CONCENTRATED BY FILTRATION
      AND ESTIMATED BY SPECTROPHOTOMETRIC DETERMINATIONS AFTER TREATMENT WITH
      PHENOL AND SULFURIC ACID. ALGAE WERE ENUMERATED AND TYPED USING
      STANDARD PROCEDURES, AND ESTIMATES OF THE AMOUNTS OF CELLULOSE PRODUCED
      BY THE PHYTOPLANKTON OBSERVED WERE MADE AFTER DETERMINATIONS OF THE
      AMOUNTS OF CELLULOSE PRESENT IN KNOWN CONCENTRATIONS OF ALGAE
      REPRESENTATIVE OF THE TYPES FOUND IN THE RESERVOIR. TOTAL NUMBERS OF
      BACTERIA VARIED BETWEEN LESS THAN 1 PER ML OF SAMPLE TO A HIGH OF 170
      PER ML OF SAMPLE. PRESUMPTIVE CELLULOLYTIC BACTERIA ACCOUNTED FOR FROM
      LESS THAN 0.5 TO 52% OF THE TOTAL BACTERIAL POPULATIONS MEASURED IN
      SEVERAL DIFFERENT AREAS BEFORE, DURING AND AFTER AN ALGAL BLOOM.
      NUMBERS OF PRESUMPTIVE CELLULOLYTIC BACTERIA APPEAR TO HAVE NO
      CONSISTANT RELATION WITH ESTIMATES OF AMOUNTS OF CELLULOSE. ESTIMATES
      OF CELLULOSE CONTENT OF THE WATER VARIED BETWEEN 360 TO 2750 MICROGRAMS
      PER LITER OF SAMPLE.

   FIELD 02H, 05C

   ACCESSION NO. W69-00976

THE EFFECT OF SELECTED HERBICIDES UPON THE GROWTH OF PLANKTONIC FRESH WATER
ALGAE AND THEIR PERSISTENCE IN SURFACE WATERS,

LOUISIANA POLYTECHNIC INSTITUTE, RUSTON, BOTANY AND BACTERIOLOGY DEPT.

OTTO WASMER, JR.

LOUISIANA WATER RESOUR RES INST, LOUISIANA STATE UNIV, BATON ROUGE, MAY 1967.
19 P, 11 FIG, 5 REF. OWRR PROJECT A-003-LA.

DESCRIPTORS:
*2-4 D, *HERBICIDES, *PHYTOPLANKTON, *PHYTOTOXICITY, ALGAL POISONING,
SCENEDESMUS, CHLORELLA, *EUGLENA, ALGAL CONTROL, BACTERICIDES, GROWTH
RATES.

ABSTRACT:
THE PURPOSES OF THE INVESTIGATION WERE TO DETERMINE EFFECT OF SELECTED
HERBICIDES UPON THE RATE OF GROWTH OF CERTAIN FRESH WATER PLANKTONIC
ALGAE; TO DETERMINE EFFECT OF SELECTED HERBICIDES UPON THE PHYSIOLOGY
OF THESE ALGAE; AND TO DETERMINE MORPHOLOGICAL CHANGES IN ALGAE EXPOSED
TO HERBICIDE PRESENCE. DUE TO LIMITATIONS IMPOSED BY TIME, CERTAIN
RESTRICTIONS IN THE INVESTIGATIONAL PROCEDURES WERE NECESSARY. A SIGNAL
SPECIES OF FRESH WATER ALGAE WAS UTILIZED IN THE COMPLETE
INVESTIGATION, ALTHOUGH 2 OTHER SPECIES WERE EXAMINED IN THE
MORPHOLOGICAL STUDIES. A SINGLE HERBICIDE WAS USED IN ALL PHASES OF THE
STUDY. FINALLY, THE PHYSIOLOGICAL STUDY WAS CONFINED TO THE EFFECT OF
HERBICIDE PRESENCE ON RATE OF RESPIRATION IN ALGAL CELLS.

FIELD 05G

ACCESSION NO. W69-00994

POLLUTION OF WATERWAYS IN THE ELLIOT LAKE AREA, ONTARIO,

WESTERN ONTARIO UNIV. LONDON, ONTARIO, CANADA.

RICHARD P. W. MCCUTCHEON.

ONTARIO GEOGR, NO 2, PP 25-33, 1968. 9 P, 1 FIG, 5 REF.

DESCRIPTORS:
*WATER POLLUTION, *MINE WASTES, *RADIOACTIVE WASTES, *MUNICIPAL WASTES,
AQUATIC ALGAE, URANIUM RADIOISOTOPES, RADIUM RADIOISOTOPES.

IDENTIFIERS:
ELLIOT LAKE, ONTARIO.

ABSTRACT:
RAPID DEVELOPMENT OF URANIUM MINING IN THE ELLIOT LAKE AREA, ONTARIO,
CAUSED WIDESPREAD RADIOLOGICAL AND SEWAGE POLLUTION OF THE SERPENT
RIVER SYSTEM. THE WIDELY USED ACID LEACHING PROCESS OF URANIUM ORE...
TREATMENT PRODUCES LARGE AMOUNTS OF WASTE, BOTH SOLID AND LIQUID. FOR
EACH TON OF ORE, NEARLY 1 TON OF TAILINGS AND AT LEAST 2 TONS AND AS
MUCH AS 5 TONS OF WATER, ACIDS, AND NEUTRALIZERS ARE PRODUCED. NOT ALL
THE URANIUM, AND NONE OF THE RADIUM-226, IS EXTRACTED FROM THE ORE, SO
THE TAILINGS PILES CONTRIBUTE SOME RADIOACTIVITY TO WATER PASSING
THROUGH THEM. SOME LAKES AND STREAMS CONTAIN MORE THAN THE SAFE AMOUNT
OF RADIUM. LEAKS IN THE REFINING SYSTEMS ALSO CONTRIBUTE POLLUTION. ALL
SEWAGE IS TREATED, BUT LAKE AND STREAM POLLUTION BY INCREASED OXYGEN
DEMAND AND INCREASE OF NUTRIENTS IS COMMON AND HAS MADE ONE FORMER
SWIMMING AREA OBNOXIOUS. NO FISH CAN LIVE IN ANGEL AND HORNE LAKES, AND
ALGAE ARE PRESENT IN EXCESSIVE AMOUNTS. (KNAPP-USGS)

FIELD 05G

ACCESSION NO. W69-01165

EFFECTS OF WASTES ON RECEIVING WATER--SOME COMPUTATIONAL AND SCIENTIFIC PROBLEMS.

PP 49-56 OF ENTRY NO. W69-01269 IN 05G.

DESCRIPTORS:
BIOCHEMICAL OXYGEN DEMAND, ALGAL TOXINS, ORGANIC WASTES, POLLUTANT IDENTIFICATION, POLLUTANTS, QUALITY CONTROL, WATER POLLUTION, *WATER POLLUTION SOURCES, SEPARATION TECHNIQUES, *ANALYTICAL TECHNIQUES, CLASSIFICATION, *MODEL STUDIES, WASTE IDENTIFICATION, BIOLOGICAL PROPERTIES, PHYSICAL PROPERTIES, WASTE WATER(POLLUTION), WATER QUALITY, WATER POLLUTION EFFECTS, IMPAIRED WATER QUALITY, WASTES.

ABSTRACT:
COMPUTATIONAL AND SCIENTIFIC PROBLEMS INVOLVED IN DETERMINING THE EFFECT OF WASTES ON RECEIVING WATER ARE DISCUSSED. THE NEED FOR AN EFFICIENT AND FLEXIBLE MEANS OF CALCULATION OF POLLUTANTS IS NOTED AND PROBLEMS INVOLVED IN DEVISING SUCH A MEANS ARE DISCUSSED. TWO BASIC PROBLEMS ARE PRESENTED. THE FIRST IS THE CONCEPTUAL PROBLEM OF DEVISING A MODEL WHICH COMPREHENDS RELEVANT RELATIONSHIPS AND IS AMENDABLE TO RAPID CALCULATION. THE SECOND IS THE DETERMINATION OF THE SPECIFIC NUMERICAL EFFECT OF A CHANGE IN ONE VARIABLE ON THE OTHERS. IT IS CONCLUDED THAT ACCURATE KNOWLEDGE OF PHYSICAL AND BIOLOGICAL RELATIONSHIPS IN WATER BODIES IS AN ESSENTIAL PREREQUISITE TO THE DESIGN AND OPERATION OF OPTIMUM WASTE-DISPOSAL SYSTEMS. RESEARCH GEARED TO EXPLORING A MORE DEFINITELY EMPIRICAL APPROACH TO WATER PHENOMENA IS ENCOURAGED.

FIELD 05C

ACCESSION NO. W69-01273

MAJOR SOURCES OF NUTRIENTS FOR ALGAL GROWTH IN WESTERN LAKE ERIE,

FEDERAL WATER POLLUTION CONTROL ADMIN., GROSSE ILE, MICH.

GEORGE L. HARLOW.

PUBLICATION NUMBER 15, GREAT LAKES RES DIV, UNIV MICH, PP 389-394, 1966. 6 P, 2 FIG, 5 TAB, 9 REF.

DESCRIPTORS:
*EUTROPHICATION, *LAKE ERIE, *NUTRIENTS, ALGAE, WATER POLLUTION SOURCES, NITROGEN COMPOUNDS, PHOSPHATES, GREAT LAKES, MICHIGAN.

IDENTIFIERS:
LAKE ST CLAIR, ROUGE RIVER, HURON RIVER, RAISIN RIVER, DETROIT RIVER, MAUMEE RIVER.

ABSTRACT:
DUE TO AN EXCESS OF NUTRIENTS, WESTERN LAKE ERIE HAS DEVELOPED PROLIFIC BLOOMS OF ALGAE CONTRIBUTING TO AND INDICATING THE ACCELERATED ENRICHMENT OF THIS VALUABLE NATURAL RESOURCE. NUTRIENT CONCENTRATIONS AND QUANTITIES INCREASE AS THE WATERS FROM LAKE ST CLAIR PASS BY THE METROPOLITAN COMPLEX OF DETROIT TO WESTERN LAKE ERIE. THE SOURCES OF WASTE WHICH CONTRIBUTE TO THESE INCREASES ARE PRESENTED, AS WELL AS THE RELATIVE QUANTITIES FROM EACH WASTE SOURCE. MUNICIPAL WASTES CONTRIBUTE 89% OF TOTAL PHOSPHATES AND 86% OF TOTAL NITROGEN RECEIVED INTO LAKE ERIE FROM SOUTHEAST MICHIGAN, WHEREAS THE CONTRIBUTION OF PHOSPHATE FROM LAND DRAINAGE IS A MINOR FACTOR. NITROGEN RUNOFF FROM LANDS CONTRIBUTES MORE SIGNIFICANTLY TO NUTRIENT LOADING THAN DOES PHOSPHATE FROM THE SAME SOURCE. IN LAKE ERIE, CONCENTRATIONS OF NUTRIENT COMPOUNDS CLOSER TO THE MOUTH OF THE DETROIT RIVER ARE HIGHER THAN AT STATIONS FURTHER OFFSHORE, EXCEPT IN THE CASE OF TWO OFFSHORE STATIONS THOUGHT TO BE INFLUENCED BY THE MAUMEE RIVER. RESULTS OF NUTRIENT DETERMINATIONS FROM ALL STATIONS REPORTED FOR LAKE ERIE SHOW CONCENTRATIONS GREATER THAN THOSE THOUGHT TO BE CRITICAL FOR TRIGGERING BLOOMS OF ALGAE.

FIELD 05B

ACCESSION NO. W69-01445

EUTROPHICATION AND WATER POLLUTION,

FEDERAL WATER POLLUTION CONTROL ADMINISTRATION, WASHINGTON.

EDWARD J. MARTIN, AND LEON W. WEINBERGER.

PUBLICATION NUMBER 15, GREAT LAKES RES DIV, UNIV MICH, PP 451-469, 1966. 19
P, 4 FIG, 8 TAB, 45 REF.

DESCRIPTORS:
*EUTROPHICATION, *NUTRIENTS, POLLUTANTS, POLLUTION ABATEMENT, NUISANCE
ALGAE, PHOSPHATES, NITRATES, RIVERS, SEWAGE TREATMENT, WATER POLLUTION
SOURCES, PRIMARY TREATMENT, SECONDARY TREATMENT, TERTIARY TREATMENT,
ELECTRODIALYSIS, ION EXCHANGE, ACTIVATED SLUDGE, DENITRIFICATION.

ABSTRACT:
AUTHORS DEFINE THE MECHANISM OF EUTROPHICATION AND EVALUATE PROBABLE
SOURCES AND NATURE OF THE POLLUTIONAL LOADS RESPONSIBLE FOR ACCELERATED
EUTROPHICATION. THIS INVOLVES CONSIDERATIONS OF TYPES AND QUANTITIES OF
NUTRIENT MATERIAL, ESPECIALLY NITROGEN AND PHOSPHORUS, IN TERMS OF
CONTROL AND THE FREQUENCY AND INTENSITY OF ALGAL BLOOMS. ON RIVERS,
BETWEEN 1961 AND 1963, 49% OF 130 STATIONS REPORTED ALGAL COUNTS WHICH
COULD BE DEFINED AS BLOOMS; 55% OF 345 STATIONS REPORTED VALUES OF
PHOSPHATE EQUALING OR EXCEEDING 0.2 MG/1. WASTES RESULTING FROM
MUNICIPAL, AGRICULTURAL, AND INDUSTRIAL USES OF WATER CONTRIBUTE
SIGNIFICANT AMOUNTS OF NUTRIENTS THAT MAY PROMOTE EUTROPHICATION IN
WATERWAYS. ABOUT 235 TO 2,350 TONS OF PHOSPHATE ARE DISCHARGED DAILY TO
RECEIVING WATERS FROM MUNICIPAL WASTE SOURCES IN THE US ALONE. MAXIMUM
EFFICIENCES OF SEVEN TREATMENT PROCESSES FOR REMOVING NITROGEN AND
PHOSPHORUS ARE REVIEWED. 50% OF TOTAL PHOSPHORUS AND 80% OF TOTAL
NITROGEN ARE REMOVED BY USE OF AN ACTIVATED-SLUDGE PROCESS, WHILE 99+%
OF THE PHOSPHORUS AND NITROGEN ARE REMOVED WITH A TREATMENT SEQUENCE
USING ION EXCHANGE. THESE DATA INDICATE THAT TREATMENT CAPABILITIES ARE
AVAILABLE TO CONTROL THE DISCHARGE OF NUTRIENT COMPOUNDS FROM POINT
SOURCES OF WASTES AND TO TREAT THEM TO ALMOST ANY DESIRED DEGREE.

FIELD 05F

ACCESSION NO. W69-01453

THE BIOTIC RELATIONSHIPS WITHIN WATER BLOOMS,

WISCONSIN UNIV., MADISON. HYDRAULIC AND SANITARY LAB.

GEORGE P. FITZGERALD.

D F JACKSON, ALGAE AND MAN, PLENUM PRESS, NEW YORK, PP 300-306, 1964. 7 P, 41
REF.

DESCRIPTORS:
*EUTROPHICATION, SUCCESSION, WATER QUALITY, ALGAL CONTROL, ALGICIDES,
DOMINANT ORGANISMS, GRAZING, WATER POLLUTION EFFECTS, DIATOMS,
CHLOROPHYTA, CYANOPHYTA, NUISANCE ALGAE, WISCONSIN, NUTRIENTS,
INDICATOR SPECIES, EUGLENA, CHLAMYDOMONAS.

IDENTIFIERS:
*BLOOMS, *BIOTIC FACTORS, SPECIES DIVERSITY, LAKE WAUBESA, LYNGBYA,
MICROCYSTIS, APHANIZOMENON, ANABAENA, MELOSIRA, CERATIUM, OSCILLATORIA,
EUDORINA.

ABSTRACT:
ALGAL SUCCESSIONS, IN WHICH NEARLY UNISPECIFIC POPULATIONS OF ALGAE
ALTERNATE WITH MORE HIGHLY DIVERSIFIED ASSEMBLAGES OF SPECIES, ARE
CHARACTERISTIC OF MORE NATURAL AQUATIC ENVIRONMENTS AS WELL OF
IMPOUNDMENTS, SUCH AS SEWAGE STABILIZATION PONDS, WITH NO LACK OF THE
USUALLY CONSIDERED NUTRIENTS. THE USUAL SEQUENCE OF SPECIES, ESPECIALLY
IN EUTROPHIC ENVIRONMENTS, CONSISTS OF A VERNAL BLOOM OF DIATOMS, AN
EARLY SUMMER GROWTH OF GREEN ALGAE, AND A LATE SUMMER BLOOM OF
CYANOPHYTES (BLUE-GREEN ALGAE). AFTER TREATMENT OF A WATER BLOOM WITH
ALGICIDES TO ALLEVIATE NUISANCE CONDITIONS, THE RAPID REPLACEMENT OF
DOMINANT ALGAL SPECIES BY OTHERS EMPHASIZES THAT FACTORS WHICH
INFLUENCE DOMINANCE MAY BE VERY SUBTLE. HIGH POLLUTIONAL LOADS FROM
SEWAGE SOURCES NOT ONLY INCREASE ALGAL GROWTH, BUT MAY ALSO INFLUENCE
SPECIES COMPOSITION AND ALGAL SUCCESSION. EVIDENCE SUGGESTS THAT
SECONDARY EFFLUENTS MAY ACT TO SUPPRESS SUCCESSION AND REINFORCE A
REDUCTION IN SPECIES DIVERSITY LONGER THAN WOULD NORMALLY BE EXPECTED.
SINCE NUTRIENT STATUS OF AN AQUATIC ENVIRONMENT DOES NOT ALWAYS EXPLAIN
THE COMPOSITION OF A BLOOM, ONE MUST OFTEN CONSIDER THE EFFECTS OF
BIOTIC FACTORS SUCH AS GRAZING BY THE FAUNA, ACTIVITIES OF BACTERIA AND
FUNGI, AND THE EXCRETION OF BIOLOGICALLY ACTIVE COMPOUNDS BY THE ALGAE
THEMSELVES. (WIS)

FIELD 05C

ACCESSION NO. W69-01977

EFFECTS OF PESTICIDES ON FRESHWATER ORGANISMS,

CLEMSON UNIV., CLEMSON. DEPT. OF ENTOMOLOGY AND ZOOLOGY.

JOHN K. REED, LAMAR E. PRIESTER, AND RUDOLPH PRINS.

WATER RESOURCES INSTITUTE, TECHNICAL COMPLETION REPORT MAY 1, 1965 - JUNE 30, 1968. 31 P, 4 TAB. OWRR PROJECT A-001-SC.

DESCRIPTORS:
ALGAE, PROTOZOA, *AQUATIC INSECTS, *AQUATIC DRIFT, FISH FOOD ORGANISMS, FISH PHYSIOLOGY, FISH TAXONOMY, DDT, PHOSPHOTHIOATE PESTICIDES, HERBICIDES, *PESTICIDE RESIDUES, DIATOMS, CRAYFISH, POLLUTANT IDENTIFICATION, *WATER POLLUTION EFFECTS, GAS CHROMATOGRAPHY.

IDENTIFIERS:
SOUTH CAROLINA, LYOPHILIZATION, *PESTICIDE METABOLISM, *PESTICIDE ACCUMULATION.

ABSTRACT:
LISTS OF ARTHROPODS, REPTILES, AMPHIBIANS AND FISH FOUND IN NORTHWESTERN SOUTH CAROLINA STREAMS BY THE AUTHORS ARE GIVEN. THE ECOLOGY OF THESE FORMS WAS AFFECTED BY THE PRESENCE OF ORGANIC PESTICIDE POLLUTION IN CERTAIN STREAMS. THE POLLUTANTS FOUND WERE DDT, DDE, BHC, TOXAPHENE, ALDRIN AND DIELDRIN. THE HIGHEST LEVEL OF PESTICIDE WAS IN THE ORGANISMS, THEN BOTTOM MUD WITH THE LEAST IN WATER. LABORATORY STUDIES SHOWED SELECTED SPECIES OF ALGAE AND PROTOZOA CONCENTRATED DDT AND PARATHION FROM WATER BUT BREAKDOWN PRODUCTS WERE NOT DETECTABLE. TRIFLURALIN WAS BOTH CONCENTRATED FROM WATER AND METABOLIZED BY GOLDFISH. DRIFTING FOOD ORGANISMS OF FISH IN THE FIELD WERE FOUND TO BE INFLUENCED BY THE SEASON OF THE YEAR, WEATHER, AND POLLUTION.

FIELD 05C

ACCESSION NO. W69-02782

PRIMARY PRODUCTIVITY STUDIES IN ONONDAGA LAKE, NEW YORK,

SYRACUSE UNIV., N. Y. DEPT OF CIVIL ENGINEERING.

DANIEL F. JACKSON.

FOR PUBLICATION IN 'XVII VERHANDLUNGEN INTERNATIONALE VEREINIGUNG FUR THEORETISCHE UND ANGEWANDTE LIMNOLOGIE', 1968. 19 P. NY DEPT. OF HEALTH C-19638.

DESCRIPTORS:
*EUTROPHICATION, *WATER POLLUTION SOURCES, *WATER POLLUTION EFFECTS, *PRIMARY PRODUCTIVITY, ALGAE, BIOMASS, CALCIUM COMPOUNDS, CHLORIDES, SALINE LAKES, HEAVY METALS, INDUSTRIAL WASTES, LAKES, PHOTOSYNTHESIS, PHYTOPLANKTON, CISCO, CARP, DIATOMS, CHLOROPHYTA, CHLAMYDOMONAS, CHLORELLA.

IDENTIFIERS:
*ONONDAGA LAKE, ONONDAGA COUNTY, SYRACUSE, WARBURG RESPIROMETER, CLADOPHORA, ENTEROMORPHA, LEPOCINCLIS, CHLOROGONIUM, SCENEDESMUS, CYCLOTELLA, STEPHANODSICUS.

ABSTRACT:
ONONDAGA LAKE, WHOSE SOUTH END IS WITHIN THE CITY LIMITS OF SYRACUSE, NY, IS SMALL, BRACKISH, AND GROSSLY POLLUTED. IT RECEIVES DAILY 200,000,000 LITERS OF PARTIALLY TREATED SEWAGE, 172 METRIC TONS OF CALCIUM CHLORIDE, 32 METRIC TONS OF CALCIUM SULFATE, ONE METRIC TON OF GREASE AND OIL, AND 22 KILOGRAMS OF HEAVY METALS. TO ESTABLISH A BASE LINE FOR DETERMINING EXTENT OF EUTROPHICATION, AUTHOR ESTABLISHED TWO STATIONS: I AT NORTH END OF LAKE, WHICH IS SUBJECT TO INORGANIC POLLUTION; AND V, A SOUTHERN STATION INFLUENCED BY OUTFALL FROM A PRIMARY TREATMENT SEWAGE PLANT. BETWEEN MAY AND OCTOBER 1967, MONTHLY DETERMINATIONS WERE MADE OF ALGAL COUNTS, ASH-FREE DRY WEIGHTS, AND RESPIRATORY AND PHOTOSYNTHETIC RATES (BY WARBURG MANOMETRY). PHYTOPLANKTON WAS DOMINATED BY SPECIES OF CHLOROPHYTES AND DIATOMS; VERY CYANOPHYTES WERE COUNTED. PATTERNS OF DOMINANCE DID NOT DIFFER BETWEEN STATIONS. AT 1% LEVEL OF CONFIDENCE, THERE WERE SIGNIFICANT DIFFERENCES BETWEEN STATIONS. AS COMPARED WITH I, VALUES AT V WERE GREATER IN THE FOLLOWING (SAMPLE NUMBERS IN PARENTHESES): (N=3) NITRATE-NITROGEN, AMMONIUM-NITROGEN, PHOSPHATES; (N=6) ALGAL NUMBERS, ASH-FREE DRY WEIGHTS; (N=98) PHOTOSYNTHETIC RATE, RESPIRATORY RATE. DETAILED LIMNOLOGICAL STUDIES ARE NEEDED BEFORE THE EXPENDITURE OF LARGE SUMS FOR POLLUTION ABATEMENT PROJECTS.

FIELD 05C, 02H

ACCESSION NO. W69-02959

ADSORPTION OF PHOSPHORUS BY LAKE SEDIMENT,

CONNECTICUT AGRICULTURAL EXPERIMENTAL STATION, NEW HAVEN.

ROBERT D. HARTER.

SOIL SCI SOC OF AMER PROC, VOL 32, NO 4, PP 514-518, JULY-AUG 1968. 5 P, 5
FIG, 1 TAB, 12 REF.

DESCRIPTORS:
*ADSORPTION, *PHOSPHORUS COMPOUNDS, *LAKES, *SEDIMENTS,
*EUTROPHICATION, ALGAE, CLAYS, SILICATES.

IDENTIFIERS:
EXTRACTABLE PHOSPHORUS.

ABSTRACT:
TO STUDY THE MECHANISM OF SORPTION OF PHOSPHOROUS BY EUTROPHIC LAKE
SEDIMENTS, BETWEEN O AND 2.2 MG P WAS ADDED TO O.1-G SEDIMENT SAMPLES
AT A 1:50 SEDIMENT-SOLUTION RATIO. AFTER EQUILIBRIUM HAD BEEN
ESTABLISHED, P REMAINING IN SOLUTION WAS DETERMINED, AND ADSORBED P WAS
EXTRACTED WITH O.5N AMMONIUM FLUORIDE AND O.1N NAOH. WHEREAS ALL P
ADSORBED INTO THE NAOH-EXTRACTABLE FRACTION APPEARED TO OCCUR AS AN
IRON PHOSPHATE, AMMONIUM FLUORIDE APPARENTLY EXTRACTED P BONDED BY TWO
DIFFERENT MECHANISMS. WHEN LESS THAN ABOUT O.1 MG P WAS ADDED, AMMONIUM
FLUORIDE EXTRACTED A TIGHTLY BONDED FORM OF P, PROBABLY OCCURRING AS AN
ALUMINUM PHOSPHATE. WHEN MORE THAN O.1 MG P WAS ADDED, ADDITIONAL P IN
A MORE LOOSELY BONDED FORM WAS ABSORBED INTO THE AMMONIUM
FLUORIDE-EXTRACTABLE FRACTION. THE LOOSELY BONDED P APPEARED TO BE
INDEPENDENT OF AL CONTENT OF THE SEDIMENT, AND COULD BE REMOVED BY
SUCCESSIVE WATER EXTRACTION. THE CAPABILITY OF THE SEDIMENT TO ADSORB
CONSIDERABLE LOOSELY BONDED P MEANS THAT LARGE INFLUXES OF P INTO THE
LAKE MAY BE HELD TEMPORARILY AND SUBSEQUENTLY RELEASED TO GROWING
PLANTS AND ALGAE. (KNAPP-USGS)

FIELD 02H, 05C

ACCESSION NO. W69-03075

NUTRIENT SOURCES FOR ALGAE AND AQUATIC WEEDS,

WISCONSIN UNIV., MADISON. ENGINEERING EXPERIMENT STATION.

GEORGE P. FITZGERALD.

IN: DETECTION OF LIMITING OR SURPLUS NUTRIENTS IN ALGAE, 9 DECEMBER 1968. 21
P. WP - 297.

DESCRIPTORS:
*ALGAE, *CYCLING NUTRIENTS, *ENZYMES, *ESSENTIAL NUTRIENTS, *NITROGEN
CYCLE, ANALYTICAL TECHNIQUES, AQUATIC ALGAE, AQUATIC MICROBIOLOGY,
AQUATIC PRODUCTIVITY, AQUATIC WEEDS, BALANCE OF NATURE, BIOASSAY,
CHLOROPHYTA, CYANOPHYTA, ENVIRONMENTAL EFFECTS, EUTROPHICATION,
LIMNOLOGY, NITROGEN COMPOUNDS, NITROGEN FIXATION, NUTRIENT
REQUIREMENTS, PHOSPHORUS COMPOUNDS, PHYSIOLOGICAL ECOLOGY,
PHYTOPLANKTON, POLLUTANT IDENTIFICATION, RAIN, RAINFALL-RUNOFF
RELATIONSHIPS, RAIN WATER, ROOTED AQUATIC PLANTS, SEWERS, WATER
POLLUTION, WATER POLLUTION CONTROL, WATER POLLUTION EFFECTS, WATER
POLLUTION SOURCES.

IDENTIFIERS:
LAKE MENDOTA(WIS).

ABSTRACT:
SIMPLE BIOASSAYS FOR LIMITING OR SURPLUS NITROGEN AND PHOSPHORUS
CONDITIONS IN ALGAE OR AQUATIC WEEDS HAVE BEEN USED TO FOLLOW THE
NUTRITION OF ALGAE AND AQUATIC WEEDS DURING SUMMER OF 1968. OF
PARTICULAR INTEREST HAVE BEEN THE SOURCES OF NITROGEN AND PHOSPHORUS
AVAILABLE TO THESE PLANTS. IT HAS BEEN SHOWN THAT DURING SUMMER RAIN
COULD BE THE MAJOR SOURCE OF AVAILABLE NITROGEN TO CLADOPHORA SP IN
LAKE MENDOTA, WIS. INCREASES IN PHOSPHORUS ASSOCIATED WITH RAINFALLS
WERE DETECTED BUT WERE NOT AS DRAMATIC AS INCREASES IN NITROGEN. IN
ADDITION, STUDIES HAVE SHOWN THAT MIXED BLOOMS OF PLANKTONIC ALGAE DO
NOT SHARE THEIR NUTRIENTS WITH OTHER ALGAE EVEN WHEN ONE SPECIES MAY
HAVE A SURPLUS AND ANOTHER BE NUTRIENT-LIMITED. COMPARISONS OF SURFACE
AND SUBSURFACE PHYTOPLANKTON HAVE INDICATED THAT AT CERTAIN TIMES
SURFACE PLANKTON COULD BE NUTRIENT-LIMITED WHILE THE SAME SPECIES FROM
SUBSURFACE SOURCES HAD ADEQUATE OR SURPLUS NUTRIENTS. FALL OVERTURN AS
A NUTRIENT SOURCE HAS BEEN USED TO DEMONSTRATE HOW SIMILAR CHANGES CAN
OCCUR IN THE NUTRITION OF DIFFERENT TYPES OF ALGAE. THE RESULTS OF
THESE STUDIES ARE USED TO POINT OUT SOME OF THE IMPORTANT FACTORS
INVOLVED IN FIELD STUDIES OF THE NUTRITION OF AQUATIC PLANTS.

FIELD 05B

ACCESSION NO. W69-03185

FIELD AND LABORATORY EVALUATIONS OF BIOASSAYS FOR NITROGEN AND PHOSPHORUS WITH ALGAE AND AQUATIC WEEDS,

WISCONSIN UNIV., MADISON. ENGINEERING EXPERIMENTAL STATION.

GEORGE P. FITZGERALD.

IN: DETECTION OF LIMITING OR SURPLUS NUTRIENTS IN ALGAE, 9 DECEMBER 1968. 12 P. WP - 297.

DESCRIPTORS:
*CHLOROPHYTA, *CYCLING NUTRIENTS, *ENZYMES, *ESSENTIAL NUTRIENTS, *NITROGEN CYCLE, *NUTRIENT REQUIREMENTS, ALGAE, ANALYTICAL TECHNIQUES, AQUATIC ALGAE, AQUATIC MICROBIOLOGY, AQUATIC PRODUCTIVITY, AQUATIC WEEDS, BIOASSAY, CYANOPHYTA, ENVIRONMENTAL EFFECTS, EUTROPHICATION, LIMNOLOGY, NITROGEN COMPOUNDS, NITROGEN-FIXATION, PHOSPHORUS COMPOUNDS, PHYSIOLOGICAL ECOLOGY, PHYTOPLANKTON, POLLUTANT IDENTIFICATION, RAIN, RAINFALL-RUNOFF RELATIONSHIPS, RAIN WATER, ROOTED AQUATIC PLANTS, WATER POLLUTION, WATER POLLUTION CONTROL, WATER POLLUTION EFFECTS, WATER POLLUTION SOURCES, SEWERS.

ABSTRACT:
THE RATE OF AMMONIA ABSORPTION IN THE DARK AND THE AMOUNT OF ORTHOPHOSPHATE EXTRACTED BY BOILING WATER TREATMENTS HAVE BEEN USED TO FOLLOW THE TRANSIENT NATURE OF THE NITROGEN AND PHOSPHORUS NUTRITION OF ALGAE AND AQUATIC WEEDS. THE MEASUREMENT OF THE ENZYMATIC ACTIVITY OF ALGAL ALKALINE PHOSPHATASE ALSO USED TO FOLLOW PHOSPHORUS NUTRITION HAS BEEN SHOWN TO BE INHIBITED BY THE PRESENCE OF 5 MG/L OR MORE OF PHOSPHATE-PHOSPHORUS, BUT THE ACTUAL ENZYMATIC ACTIVITY OF THE ALGAE IS NOT INHIBITED BY PHOSPHATE-PHOSPHORUS AND IS REDUCED ONLY BY GROWTH (OR DILUTION) UNDER CONDITIONS OF SURPLUS AVAILABLE PHOSPHORUS. IT IS SUGGESTED THAT ONLY TERMINAL PORTIONS OF AQUATIC WEEDS BE USED FOR NUTRITIONAL BIOASSAYS BECAUSE OF NUTRITIONAL DIFFERENCES BETWEEN YOUNG AND OLD PORTIONS OF THE SAME PLANT. AVAILABLE NITROGEN COMPOUNDS RELATED TO RAINFALL HAVE BEEN SHOWN TO INFLUENCE THE NITROGEN NUTRITIONAL STATUS OF SPIROGYRA AND CLADOPHORA. IN APPLICATIONS OF BIOASSAYS THE IMPORTANCE OF TESTING EACH SPECIES OF PLANTS SEPARATELY IS SHOWN BY CONTRASTING RESULTS OBTAINED WITH NITROGEN-FIXING (PHOSPHORUS-LIMITED) AND NON-FIXING (NITROGEN-LIMITED) BLUE-GREEN ALGAE FROM THE SAME ENVIRONMENT.

FIELD 05A

ACCESSION NO. W69-03186

SOME FACTORS IN THE COMPETITION OR ANTAGONISM BETWEEN BACTERIA, ALGAE, AND
    AQUATIC WEEDS,

    WISCONSIN UNIV., MADISON. ENGINEERING EXPERIMENT STATION.

    GEORGE P. FITZGERALD.

    IN: DETECTION OF LIMITING OR SURPLUS NUTRIENTS IN ALGAE, 9 DECEMBER 1968. 19
    P. WP-297.

        DESCRIPTORS:
        *ALGAE, *ALGAL TOXINS, *AQUARIA, *AQUATIC WEEDS, *INHIBITION,
        *PHYSIOLOGICAL ECOLOGY, *PHYTOTOXICITY, ALGAL CONTROL, ALGAL POISONING,
        ANALYTICAL TECHNIQUES, AQUATIC ALGAE, AQUATIC BACTERIA, AQUATIC
        MICROBIOLOGY, BALANCE OF NATURE, BIOASSAY, CHLOROPHYTA, CYANOPHYTA,
        DIATOMS, CYCLING NUTRIENTS, ENVIRONMENTAL EFFECTS, ENZYMES, ESSENTIAL
        NUTRIENTS, NITROGEN COMPOUNDS, NUISANCE ALGAE, NUTRIENT REQUIREMENTS,
        PHOSPHORUS COMPOUNDS, PHYTOPLANKTON, PLANKTON, POLLUTANT
        IDENTIFICATION, ROOTED AQUATIC PLANTS, SEWERS, SEWAGE, SEWAGE
        EFFLUENTS, SEWAGE TREATMENT, WATER POLLUTION, WATER POLLUTION EFFECTS,
        WATER POLLUTION SOURCES.

        ABSTRACT:
        FIELD OBSERVATIONS OF CHANGES IN THE POPULATIONS OF AQUATIC WEEDS AND
        PHYTOPLANKTON HAVE CONFIRMED THAT AQUATIC WEEDS HAVE ANTAGONISTIC
        ACTIVITY AGAINST PHYTOPLANKTON. NUTRITIONAL STUDIES IN THE LABORATORY
        INDICATE THAT CULTURES OF THE AQUATIC WEEDS MYRIOPHYLLUM SP,
        CERATOPHYLLUM SP, AND DUCKWEED (LEMNA MINOR L); LIQUID CULTURES OF
        BARLEY (HORDEUM VULGARE L, DICKSON VARIETY); AND CULTURES OF THE
        FILAMENTOUS GREEN ALGAE, CLADOPHORA SP AND PITHOPHORA OEDOGONIUM (MONT)
        WITHROCK WILL REMAIN RELATIVELY FREE OF EPIPHYTES OR COMPETING
        PHYTOPLANKTON IF THE CULTURES ARE NITROGEN LIMITED. FIELD OBSERVATIONS
        OF CLADOPHORA SP HAVE CONFIRMED THAT THE GROWTH OF EPIPHYTES ON THE
        CLADOPHORA IS RELATED TO CONDITIONS OF SURPLUS AVAILABLE NITROGEN
        COMPOUNDS. IT IS PROPOSED THAT THIS ANTAGONISTIC ACTIVITY MAY BE DUE TO
        A 'NITROGEN SINK' EFFECT IN WHICH THE AQUATIC WEEDS OR FILAMENTOUS
        GREEN ALGAE PREVENT THE GROWTH OF CONTAMINATING ALGAE BY COMPETITION
        FOR THE LIMITED NITROGEN COMPOUNDS AVAILABLE. HOWEVER, THE PRESENCE OF
        BACTERIA-SIZED ORGANISMS WHICH HAVE SELECTIVE TOXICITY TO CERTAIN ALGAE
        INDICATES THAT PERHAPS MULTIPLE FACTORS EXIST. AUTHOR DISCUSSES THE
        ECOLOGICAL IMPLICATIONS, SUCH AS NITROGEN-LIMITED GROWTH, OF
        ASSOCIATIONS OF CERTAIN ALGAE WITH BACTERIA HAVING SELECTIVE TOXICITIES
        TO OTHER ALGAE UNDER CERTAIN ENVIRONMENTAL CONDITIONS.

    FIELD 05C

    ACCESSION NO. W69-03188

STUDIES OF THE ANALYSIS OF PHOSPHATES IN ALGAL CULTURES,

    WISCONSIN UNIV., MADISON. WATER CHEMISTRY LAB.

    G. FRED LEE, NICHOLAS L. CLESCERI, AND GEORGE P. FITZGERALD.

    INT J AIR WATER POLLUTION, VOL 9, PP 715-722, 1965. 8 P, 1 FIG, 5 TAB, 8 REF.

        DESCRIPTORS:
        *ALGAE, *ANALYTICAL TECHNIQUES, *CYCLING NUTRIENTS, *EUTROPHICATION,
        *PHOSPHATES, *PHYSIOLOGICAL ECOLOGY, AQUATIC ALGAE, AQUATIC
        MICROBIOLOGY, CHLOROPHYTA, CYANOPHYTA, ESSENTIAL NUTRIENTS,
        ENVIRONMENTAL EFFECTS, NITROGEN COMPOUNDS, NUTRIENT REQUIREMENTS,
        PHOSPHORUS COMPOUNDS, PHOSPHORUS, PHYTOPLANKTON.

        ABSTRACT:
        TOTAL PHOSPHATE ANALYSIS PROCEDURES WERE EVALUATED. IT WAS FOUND THAT
        THE STANDARD METHODS (1960) PROCEDURE OF ACID HYDROLYSIS IN 1 ML
        SULFURIC ACID (30%) WAS SUITABLE FOR PYROPHOSPHATES AND
        TRIPOLYPHOSPHATES BUT NOT FOR TOTAL PHOSPHATE ANALYSIS OF ALGAE OR
        ALGAL EXTRACTS. ASHING (600 DEG C, 1 HR), PERCHLORIC ACID DIGESTION
        (SULFURIC ACID PLUS NITRIC ACID PLUS PERCHLORIC ACID), AND SULFURIC
        PLUS NITRIC ACID DIGESTION GAVE SATISFACTORY RESULTS. SUPERNATANTS FROM
        THE GREEN ALGA CHLORELLA PYRENOIDOSA (WIS 2005) DID NOT AFFECT
        ORTHOPHOSPHATE ANALYSES, BUT A CONSISTANT ERROR (LOSS) WAS FOUND WHEN
        THE SUPERNATANT OF THE BLUE-GREEN ALGA MICROCYSTIS AERUGINOSA (WIS
        1036) WAS TESTED. (EICHHORN-WIS)

    FIELD 05A

    ACCESSION NO. W69-03358

THE EFFECT OF ALGAE ON BOD MEASUREMENTS,

WISCONSIN UNIV., MADISON. HYDRAULIC LAB., AND WISCONSIN UNIV, MADISON. SANITARY LAB.

GEORGE P. GITZGERALD.

J WATER POLLUTION CONTROL FED, VOL 36, NO 12, PP 1524-1542, DEC 1964. 19 P, 7 FIG, 4 TAB, 18 REF. ENGINEERING EXPERIMENT STATION REPRINT NO. 728.

DESCRIPTORS:
*BIOCHEMICAL OXYGEN DEMAND, *CHLORELLA, *PHOTOSYNTHETIC OXYGEN, ALGAE, ANALYTICAL TECHNIQUES, AQUATIC ALGAE, AQUATIC BACTERIA, AQUATIC MICROBIOLOGY, BIOASSAY, CHLOROPHYTA, CYANOPHYTA, ENVIRONMENTAL EFFECTS, ESSENTIAL NUTRIENTS, FREEZE DRYING, FREEZE-THAW TESTS, HARVESTING OF ALGAE, NUISANCE ALGAE, NUTRIENT REQUIREMENTS, PHYSIOLOGICAL ECOLOGY, PHYTOPLANKTON, SEWAGE BACTERIA, SEWAGE TREATMENT, WATER POLLUTION, WATER POLLUTION CONTROL, WATER POLLUTION EFFECTS, WATER POLLUTION TREATMENT.

ABSTRACT:
THE 5-DAY, 20 DEG C BOD OF THE GREEN ALGA CHLORELLA PYRENOIDOSA (WIS 2005) WAS COMPARED WHEN THE ALGAE HAD BEEN GROWN IN DIFFERENT CULTURE MEDIA. THE RELATIVE GROWTH AND RELATIONSHIPS BETWEEN GROWTH MEASUREMENTS ARE PRESENTED. THE ACTIVITY OF CHLORELLA IN BOD TESTS WAS INDEPENDENT OF THE VOLUME OF SETTLED SEWAGE SEED. ALGAE FROM PURE CULTURES HAD THE SAME BOD AS ALGAE FROM CULTURES SEEDED WITH SEWAGE BACTERIA; THE BOD OF CULTURE SUPERNATANTS WAS DECREASED TO ONE-FIFTH BY BACTERIA ADDED TO THE CULTURES. DEAD ALGAE HAD 4 TIMES THE BOD OF LIVE ALGAE. OXYGEN PRODUCED BY ALGAE AFTER 2-HR INCUBATION IN LIGHT WAS USED TO FOLLOW THE PHOTOSYNTHETIC CAPACITY OF ALGAE DURING INCUBATION TESTS IN THE DARK FOR AS LONG AS 16 DAYS. CHLORELLA COULD STILL PRODUCE OXYGEN AFTER 16 DAYS IN THE DARK AND FOR AT LEAST 7 DAYS AFTER THE BOD BOTTLES BECAME ANAEROBIC. THE BOD OF CHLORELLA VARIED BETWEEN 0.09 AND 0.19 MG OXYGEN USED PER MILLIGRAM (DRY WEIGHT) OF SUSPENDED SOLIDS PER LITER FOR 5 DAYS AT 20 DEG C. THE RESULTS FOR ALGAE FROM DIFFERENT CULTURE MEDIA AND DIFFERENT SPECIES ARE PRESENTED. (EICHHORN-WISC)

FIELD 05A

ACCESSION NO. W69-03362

DETECTION OF LIMITING OR SURPLUS NITROGEN IN ALGAE AND AQUATIC WEEDS,

WISCONSIN UNIV., MADISON. WATER CHEMISTRY LAB.

GEORGE P. FITZGERALD.

J PHYCOL, VOL 4, NO 2, PP 121-126, 1968. 6 P, 4 FIG, 1 TAB, 13 REF.

DESCRIPTORS:
*ALGAE, *ANALYTICAL TECHNIQUES, *AQUATIC WEEDS, *BIOASSAY, *ESSENTIAL NUTRIENTS, *NITROGEN COMPOUNDS, AQUATIC ALGAE, AQUATIC MICROBIOLOGY, CHLAMYDOMONAS, CHLOROPHYTA, CYANOPHYTA, DIATOMS, CYCLING NUTRIENTS, ENVIRONMENTAL EFFECTS, AMMONIA, AMMONIUM COMPOUNDS, NITROGEN FIXATION, NUTRIENT REQUIREMENTS, NUTRIENTS, PHYSIOLOGICAL ECOLOGY, PHYTOPLANKTON, ROOTED AQUATIC PLANTS, BIOINDICATORS.

ABSTRACT:
THE RATE OF AMMONIUM-NITROGEN ABSORPTION BY ALGAE AND AQUATIC WEEDS IN THE DARK HAS BEEN SHOWN TO BE 4 TO 5 TIMES GREATER FOR PLANTS WHICH ARE NITROGEN-LIMITED THAN FOR PLANTS WITH SUFFICIENT AVAILABLE NITROGEN. EIGHT SPECIES OF GREEN ALGAE, 2 BLUE-GREEN ALGAE, 2 DIATOMS, AND 3 AQUATIC WEEDS WERE USED TO DEMONSTRATE THE USEFULNESS OF THE TEST IN DETERMINING IF AVAILABLE NITROGEN WAS IN SURPLUS OR LIMITED SUPPLY IN A PARTICULAR ENVIRONMENT. THE TEST WAS SHOWN NOT TO DIFFERENTIATE BETWEEN BLUE-GREEN ALGAE CAPABLE OF FIXING NITROGEN (4 SPECIES) FROM MEDIA WITH NITRATE-NITROGEN OR WITHOUT COMBINED NITROGEN. THE FACTORS INFLUENCING THE RESULTS OF AMMONIUM-NITROGEN ABSORPTION TESTS HAVE BEEN INVESTIGATED. IN ORDER TO DIFFERENTIATE BETWEEN PLANTS WITH SUFFICIENT AVAILABLE NITROGEN AND THOSE WHICH ARE NITROGEN-LIMITED, THE RATE OF AMMONIUM-NITROGEN ABSORPTION (0.1 MG NITROGEN) OVER 1-HR INCUBATION IN THE DARK BY 10-20 MG OF ALGAE OR AQUATIC WEED TISSUES IS MEASURED. THE RELATIVELY SIMPLE ANALYSIS FOR AMMONIUM-NITROGEN IN THE SAMPLES MAKES IT VERY EASY TO FOLLOW THE CHANGING NITROGEN NUTRITION OF PLANTS IN CULTURES WITH A LIMITED NITROGEN SUPPLY OR IN THE PRESENCE OF POSSIBLE NITROGEN SOURCES. (EICHHORN-WISC)

FIELD 05C

ACCESSION NO. W69-03364

COPPER IN LAKE MUDS FROM LAKES OF THE MADISON AREA,

WISCONSIN UNIV., MADISON. STATE LAB. OF HYGIENE.

M. STARR NICHOLS, THERESA HENKEL, AND DOROTHY MCNALL.

TRANS WIS ACAD SCI ARTS LETT, VOL 38, PP 333-350, 1946. 18 P, 6 FIG, 9 TAB, 4 REF.

DESCRIPTORS:
*LAKES, *MUD, *COPPER, COPPER SULFATE, ALGAL CONTROL, ALGICIDES, TRACE ELEMENTS, CORES, WISCONSIN, BOTTOM SEDIMENT, BENTHIC FAUNA.

IDENTIFIERS:
*MADISON(WIS), NUISANCE ODORS, LAKE MENDOTA(WIS), LAKE MONONA(WIS), LAKE WINGRA(WIS), LAKE KEGONSA(WIS), LAKE WAUBESA(WIS), PRECIPITATION.

ABSTRACT:
THE AMOUNT OF COPPER SULFATE APPLIED ANNUALLY FROM 1925-1944 TO MADISON, WIS, LAKES IS GIVEN. MUCH OF THE COPPER SULFATE ADDED TO LAKE WATERS OF NOTABLE ALKALINITY WAS PRECIPITATED AS A BASIC COPPER COMPOUND. FROM LAKE MENDOTA, LAKE MONONA, LAKE WINGRA, LAKE WAUBESA AND LAKE KEGONSA THE BOTTOM MUDS AT VARIOUS WATER DEPTHS WERE ANALYZED FOR TOTAL AND SOLUBLE COPPER. CORE SAMPLES OF LAKE MUDS FROM THE VARIOUS LAKES WERE ALSO TAKEN. LAKE MONONA HAS RECEIVED COPPER SULFATE TREATMENT FOR ALGAL CONTROL FOR A MUCH LONGER PERIOD THAN THE OTHER LAKES, AND THIS IS REFLECTED IN THE HIGHER CONCENTRATIONS OF COPPER FOUND IN THE LAYERS OF MUD OF LAKE MONONA. THE GREATEST CONCENTRATION OF COPPER IS FOUND IN THE DEEPER PART OF THE LAKE. THE COPPER CONTENT OF THE LAKE MONONA CORE PENETRATED TO A DEPTH OF 8 FEET. ACTION BY BURROWING FAUNA MAY HAVE CAUSED MIXING OF THE UPPER LAYERS TO THIS DEPTH. IT APPEARS THAT BY FAR THE GREATEST AMOUNT OF COPPER APPLIED REMAINS AS A DEPOSIT IN THE MUD OF THE LAKE. (EICHHORN-WISC)

FIELD 05A

ACCESSION NO. W69-03366

STREAM POLLUTION BY ALGAL NUTRIENTS,

HARVARD UNIV., CAMBRIDGE, MASS. DIV. OF ENGINEERING AND APPLIED PHYSICS.

WERNER STUMM, AND JAMES J. MORGAN.

SANITARY ENGINEERING REPRINT NO. 45. TRANS 12TH ANN CONF SANITARY ENG, UNIV KANSAS, PP 16-26, 1962. 11 P, 3 FIG, 2 TAB, 19 REF.

DESCRIPTORS:
*STREAMS, *WATER POLLUTION EFFECTS, *CYCLING NUTRIENTS, *EUTROPHICATION, ALGAE, AQUATIC ALGAE, AQUATIC BACTERIA, AQUATIC MICROBIOLOGY, BALANCE OF NATURE, WISCONSIN, BIOASSAY, BIOCHEMICAL OXYGEN DEMAND, ENVIRONMENTAL EFFECTS, ENVIRONMENTAL SANITATION, ESSENTIAL NUTRIENTS, HARVESTING OF ALGAE, INHIBITION, LIGHT PENETRATION, NITROGEN COMPOUNDS, PHOSPHORUS COMPOUNDS, NUTRIENT REQUIREMENTS, PHOTOSYNTHETIC OXYGEN, WATER POLLUTION CONTROL, WATER POLLUTION SOURCES, WATER POLLUTION TREATMENT, POLLUTION IDENTIFICATION, SEWAGE, SEWAGE DISPOSAL, SEWAGE EFFLUENTS, SEWAGE LAGOONS.

IDENTIFIERS:
MADISON(WIS).

ABSTRACT:
SINCE ORGANISMS ARE CONSUMED AS WELL AS PRODUCED, SIZE OF STANDING BIOMASS MAY BEAR LITTLE RELATION TO ORGANISM ACTIVITY RATE. AEROBIC BIOLOGICAL SEWAGE TREATMENT MINERALIZES OXIDIZABLE ORGANIC SUBSTANCES BUT ELIMINATES ONLY 20-50% OF NITROGEN AND PHOSPHORUS COMPOUNDS. IN SEWAGE TREATMENT, CARBON BECOMES LIMITING BEFORE NITROGEN AND PHOSPHORUS ARE INCORPORATED INTO SLUDGE. INCREASE OF PHOSPHORUS FROM DETERGENTS MAY REACH POINT WHERE PHOSPHORUS COULD NO LONGER BE CONSIDERED A LIMITING FACTOR. SEWAGE STABILIZATION PONDS REPRESENT MIXTURE OF BACTERIAL DECOMPOSITION AND ALGAL GROWTH, BUT, TO REMOVE NITROGEN AND PHOSPHORUS, ALGAE MUST BE SEPARATED FROM POND EFFLUENT. DECOMPOSITION OF PHYTOPLANKTON WHICH SETTLED TO BOTTOM OF LAKE HELPS INCREASE NUTRIENT CONTENT OF BOTTOM WATERS DURING STAGNATION PERIODS. NOT ALL NUTRIENTS ARE REGENERATED. MADISON, WIS, LAKES RETAIN 30-60% OF NITROGEN RECEIVED. CONTRASTED WITH STABILIZATION PONDS, LAKE STAGNATION TENDS TO SEPARATE AUTOTROPHS (ALGAE) FROM HETEROTROPHS (BACTERIA AND ANIMALS). RELATION BETWEEN FERTILIZATION AND ALGAE IS INDICATED BY ASSUMPTION THAT EVERY ION OF PHOSPHORUS ADDED TO WATER, IF COMPLETELY UTILIZED FOR PHOTOSYNTHETIC PRODUCTION, CAUSES USE OF 16 NITROGEN ATOMS AND 106 ATOMS OF CARBON ALGAL PROTOPLASM. COMPLETE OXIDATION OF ORGANIC MATTER CONTAINING ONE ATOM OF PHOSPHORUS REQUIRES 150 MOLECULES OF OXYGEN. NUTRIENT REMOVAL METHODS ARE REVIEWED. (EICHHORN-WISC)

FIELD 05C

ACCESSION NO. W69-03369

EFFECT OF WATER SAMPLE PRESERVATION METHODS ON THE RELEASE OF PHOSPHORUS FROM ALGAE,

WISCONSIN UNIV., MADISON. WATER CHEMISTRY LAB.

G. P. FITZGERALD, AND S. L. FAUST.

ENGINEERING EXPERIMENT STATION REPRINT NO. 995. LIMNOL OCEANOGR, VOL 12, NO 2, PP 332-334, APRIL 1967. 3 P, 2 TAB, 9 REF.

DESCRIPTORS:
*ANALYTICAL TECHNIQUES, *CYCLING NUTRIENTS, *EUTROPHICATION, *PHOSPHORUS COMPOUNDS, *WATER POLLUTION SOURCES, ALGAE, ALGAL CONTROL, ALGAL POISONING, ALGICIDES, AQUATIC ALGAE, BIOASSAY, CHLOROPHYTA, CYANOPHYTA, DIATOMS, DISINFECTION, ENVIRONMENTAL EFFECTS, ESSENTIAL NUTRIENTS, FREEZE DRYING, FREEZE-THAW TESTS, INHIBITION, NUTRIENT REQUIREMENTS, PHOSPHORUS, PHOSPHATES, PHYSIOLOGICAL ECOLOGY, PHYTOPLANKTON.

ABSTRACT:
RECOMMENDED METHODS FOR PRESERVING WATER SAMPLES BEFORE ANALYSIS WERE USED TO DEMONSTRATE THAT IF ALGAE ARE NOT REMOVED FROM THE SAMPLES BEFORE PRESERVATION, PHOSPHATE WILL BE RELEASED BY TREATED ALGAE. THE AMOUNT OF PHOSPHATE-PHOSPHORUS (IN MILLIGRAMS PER 100 MG ALGAE) RELEASED BY DIFFERENT ALGAL SPECIES VARIED FROM 0.01-0.017 FOR OVERNIGHT REFRIGERATION AT 3-5 DEG C TO 0.38-0.78 AFTER 60-MIN EXTRACTION IN BOILING WATER BATH. FREEZING (-15 DEG C) OVERNIGHT, CHLOROFORM SATURATION, AND AN ORGANIC MERCURY ALGICIDE ALSO CAUSED THE RELEASE OF PHOSPHATE-PHOSPHORUS. IT IS CONCLUDED THAT THE RELEASE OF PHOSPHORUS FROM ALGAE COULD BE SIGNIFICANT UNLESS THE ALGAE ARE REMOVED BEFORE THE SAMPLES ARE PRESERVED FOR STORAGE. (EICHHORN-WISC)

FIELD 05B

ACCESSION NO. W69-03370

THE EFFECT OF DILUTION MEDIA ON THE BOD OF ALGAE,

WISCONSIN UNIV., MADISON. HYDRAULIC LAB; AND WISCONSIN UNIV., MADISON. SANITARY LAB.

MARY M. ALLEN, GEORGE P. FITZGERALD, AND GERARD A. ROHLICH.

ENGINEERING EXPERIMENT STATION REPRINT NO. 697. J WATER POLLUTION CONTROL FED, VOL 36, NO 8, PP 1049-1056, AUG 1964. 8 P, 3 TAB, 11 REF.

DESCRIPTORS:
*BIOCHEMICAL OXYGEN DEMAND, *PHOTOSYNTHETIC OXYGEN, *CHLORELLA, ALGAE, ANALYTICAL TECHNIQUES, AQUATIC ALGAE, AQUATIC BACTERIA, AQUATIC MICROBIOLOGY, BIOASSAY, CHLOROPHYTA, CYANOPHYTA, DIATOMS, ENVIRONMENTAL EFFECTS, ESSENTIAL NUTRIENTS, HARVESTING OF ALGAE, NUISANCE ALGAE, NUTRIENT REQUIREMENTS, PHYSIOLOGICAL ECOLOGY, PHYTOPLANKTON, SEWAGE, SEWAGE BACTERIA, SEWAGE TREATMENT, WATER POLLUTION, WATER POLLUTION CONTROL, WATER POLLUTION EFFECTS, WATER POLLUTION TREATMENT.

ABSTRACT:
THE 5-DAY, 20 DEG C BOD OF CHLORELLA PYRENOIDOSA (WIS 2005) AVERAGED 0.11 MG OF OXYGEN USED PER MILLIGRAM OF ALGAE WHEN TESTED IN BOD DILUTION WATER BUT WAS 0.25 MG OF OXYGEN USED PER MILLIGRAM OF ALGAE WHEN TESTED IN FRESH LAKE WATER. THE PHOTOSYNTHETIC CAPACITY (MILLIGRAMS OF OXYGEN PRODUCED IN 2 HOURS IN LIGHT PER MILLIGRAM OF ALGAE) OF ALGAE INCUBATED 5 DAYS IN THE DARK IN LAKE WATER WAS NEARLY THE SAME AS ALGAE INCUBATED IN DILUTION WATER. THE ADDITION OF MAJOR OR MINOR NUTRIENTS AND ADDED BUFFER CAPACITY TO STANDARD DILUTION WATER DID NOT INCREASE THE ALGAL BOD. THE ORGANISMS PRESENT IN LAKE WATERS HAD LITTLE EFFECT ON ALGAL BOD. WHEN BOD WAS MEASURED IN ALGAL CULTURE MEDIUM, THE SAME RATE AS IN DILUTION WATER WAS OBTAINED. DEIONIZED LAKE WATER MADE UP TO STANDARD DILUTION WATER GAVE SAME BOD OF ALGAE AS IN DILUTION WATER PREPARED WITH DISTILLED WATER. IT IS CONCLUDED THAT SOME HEAT-STABLE FACTOR THAT CANNOT PASS THROUGH ION EXCHANGE COLUMN AFFECTED THE ALGAL BOD OR THAT THE NATURAL RATIOS OF IONS IN LAKE WATERS ARE ESSENTIAL FOR MAXIMUM ACTIVITY OF ALGAE IN BOD. (EICHHORN-WISC)

FIELD 05A

ACCESSION NO. W69-03371

EXTRACTIVE AND ENZYMATIC ANALYSES FOR LIMITING OR SURPLUS PHOSPHORUS IN ALGAE,

WISCONSIN UNIV., MADISON. WATER CHEMISTRY LAB; AND WISCONSIN UNIV., MADISON. MCARDLE MEMORIAL LAB.

GEORGE P. FITZGERALD, AND THOMAS C. NELSON.

ENGINEERING EXPERIMENT STATION REPRINT NO. 909. J PHYCOL, VOL 2, NO 1, PP 32-37, 1966. 6 P, 1 FIG, 2 TAB, 33 REF.

DESCRIPTORS:
*ANALYTICAL TECHNIQUES, *BIOASSAY, *ENZYMES, *NUTRIENT REQUIREMENTS, *PHOSPHORUS COMPOUNDS, ALGAE, AQUATIC ALGAE, AQUATIC MICROBIOLOGY, AQUATIC PRODUCTIVITY, AQUATIC WEEDS, CHLOROPHYTA, CYANOPHYTA, DIATOMS, CYCLING NUTRIENTS, ENVIRONMENTAL EFFECTS, ESSENTIAL NUTRIENTS, NUTRIENTS, PHOSPHORUS, PHOSPHATES, PHYSIOLOGICAL ECOLOGY, PHYTOPLANKTON, ROOTED AQUATIC PLANTS, EUTROPHICATION.

ABSTRACT:
AN EXTRACTIVE PROCEDURE FOR DETECTION OF SURPLUS-STORED PHOSPHORUS (LUXURY CONSUMPTION) IN ALGAE AND AN ENZYMATIC ANALYSIS FOR CONDITIONS OF PHOSPHORUS-LIMITED GROWTH IN ALGAE HAVE BEEN EVALUATED. A SIMPLE 60-MIN BOILING WATER EXTRACTION OF ALGAE KNOWN TO CONTAIN SURPLUS PHOSPHORUS SEPARATES ESSENTIAL PHOSPHORUS COMPOUNDS AND SURPLUS-STORED PHOSPHORUS COMPOUNDS. SURPLUS PHOSPHORUS COMPOUNDS CAN BE MEASURED IN THE EXTRACT AS ORTHOPHOSPHATE. EXTRACTS OF ALGAE LIMITED IN THEIR GROWTH BY THE AMOUNT OF AVAILABLE PHOSPHORUS CONTAIN LITTLE OR NO ORTHOPHOSPHATE. LIMITATION OF ALGAL GROWTH BY PHOSPHORUS SUPPLY INDUCES THE ENZYME ALKALINE PHOSPHATASE. THE ACTIVITY OF THIS ENZYME CAN BE MEASURED AT PH 9 USING PHOSPHORUS-NITROPHENYLPHOSPHATE AS SUBSTRATE. ALGAE WHICH WERE PHOSPHORUS-LIMITED AND CONTAINED NO EXTRACTABLE ORTHOPHOSPHATE HAVE AS MUCH AS 25 TIMES MORE ALKALINE PHOSPHATASE ACTIVITY THAN ALGAE WITH SURPLUS AVAILABLE PHOSPHORUS. COMPARATIVE RESULTS ARE PRESENTED OF EXTRACT AND ENZYME ANALYSES OF 9 SPECIES OF ALGAE FROM LABORATORY CULTURES AND ALGAE FROM 3 LAKES BEFORE AND AFTER INCUBATION IN LIMITING OR SURPLUS PHOSPHORUS IN THE LABORATORY. CONCENTRATIONS OF MORE THAN 0.08 MG PHOSPHATE-PHOSPHORUS PER 100 MG OF ALGAE EXTRACTED INDICATE ALGAE CONTAINED SURPLUS PHOSPHORUS. (EICHHORN-WISC)

FIELD 05A

ACCESSION NO. W69-03373

STRIPPING EFFLUENTS OF NUTRIENTS BY BIOLOGICAL MEANS,

WISCONSIN UNIV., MADISON. SANITARY LAB.

GEORGE P. FITZGERALD.

TRANS OF THE SEMINAR ON ALGAE AND METROPOLITAN WASTES, ROBERT A TAFT SAN ENGR CTR, CINCINNATI, 1961. PP 136-139, 5 FIG.

DESCRIPTORS:
*NUTRIENT REQUIREMENTS, *EUTROPHICATION, *WATER POLLUTION CONTROL, *WATER POLLUTION TREATMENT, ALGAE, ALGAL CONTROL, ANALYTICAL TECHNIQUES, AQUATIC ALGAE, AQUATIC MICROBIOLOGY, BALANCE OF NATURE, BIOASSAY, BIOCHEMICAL OXYGEN DEMAND, CHLOROPHYTA, CYCLING NUTRIENTS, ENVIRONMENTAL EFFECTS, ESSENTIAL NUTRIENTS, ENVIRONMENTAL ENGINEERING, HARVESTING ALGAE, NITROGEN COMPOUNDS, PHOSPHORUS COMPOUNDS, PHOTOSYNTHETIC OXYGEN, PHYSIOLOGICAL ECOLOGY, PHYTOPLANKTON, SEWAGE, SEWAGE BACTERIA, SEWAGE DISPOSAL, SEWAGE EFFLUENTS, SEWAGE LAGOONS, SEWAGE TREATMENT, WATER POLLUTION, WATER POLLUTION SOURCES.

ABSTRACT:
LABORATORY EXPERIMENTS WITH THE GREEN ALGA CHLORELLA PYRENOIDOSA (WIS 2005) HAVE SHOWN THIS ALGA WOULD FIRST ABSORB THE AMMONIUM-NITROGEN, THEN NITRITE-NITROGEN AND NITRATE-NITROGEN FROM SECONDARY SEWAGE EFFLUENTS. LESS THAN 0.5 MG/LITER OF ANY OF THE NITROGEN OR PHOSPHORUS OF THE EFFLUENT REMAINED IN SOLUTION AFTER 17 DAYS OF CULTURE. GROWTH OF CHLORELLA PARALLELED NITROGEN AND PHOSPHORUS REMOVAL IN PRIMARY AND SECONDARY EFFLUENT. GROWTH IN SECONDARY EFFLUENT COULD BE STIMULATED BY THE ADDITION OF CARBON DIOXIDE TO AIR. THE OPERATION OF A 1/2-ACRE STABILIZATION POND FOR NUTRIENT REMOVAL BY ALGAE IN THE POND INDICATED THAT A SUCCESSION OF ALGAL SPECIES TOOK PLACE DESPITE A CONTINUOUS SUPPLY OF NUTRIENTS. THE AVERAGE NITROGEN REMOVAL THROUGHOUT THE YEAR IS ABOUT 30%, WITH SUMMER REMOVALS REACHING ABOUT 70%. THERE WERE ONLY 33 DAYS DURING 1956 AND 76 DAYS DURING 1957 WHEN NITROGEN REMOVAL EXCEEDED 50%. PHOSPHORUS REMOVAL COINCIDED WITH PERIODS OF HIGH PH. DURING EARLY WINTER, POND EFFLUENT CONTAINED HIGHER LEVELS OF PHOSPHORUS THAN INFLUENT DUE TO DISSOLUTION OF PRECIPITATED PHOSPHORUS. (EICHHORN-WISC)

FIELD 05F

ACCESSION NO. W69-03374

RELATIVE ABSORPTION OF STRONTIUM AND CALCIUM BY CERTAIN ALGAE,

ARIZONA UNIV., TUCSON. DEPT. OF AGRICULTURAL CHEMISTRY AND SOILS.

WALLACE H. FULLER, AND JAMES E. HARDCASTLE.

REPRINT FROM SOIL SCI SOC AMER PROC, VOL 31, NO 6, PP 772-774, NOV-DEC 1967. 3 P, 2 FIG, 2 TAB.

DESCRIPTORS:
*SOIL ALGAE, *AQUATIC ALGAE, ALGAE, ARID CLIMATES, SEMIARID CLIMATES, *ABSORPTION, *CALCIUM, SOIL CONTAMINATION, *STRONTIUM RADIOSOTOPES, CYTOLOGICAL STUDIES.

IDENTIFIERS:
NUTRIENT MEDIA.

ABSTRACT:
SOIL ALGAE APPEAR IN ABUNDANCE IN ARID AND SEMIARID LANDS WHERE THEY FORM SURFACE CRUSTS OF VARYING DENSITY AND CLING TO STONE SURFACES AND LIME OUTCROPS. THE UPTAKE OF RADIOSTRONTIUM BY 7 FRESH-WATER AND DESERT ALGAE WITH RESPECT TO THEIR FUNCTION IN CONCENTRATING RADIOACTIVE CONTAMINATION PRODUCTS OF URANIUM FISSION WAS EVALUATED. IN ADDITION, THE INTERRELATIONSHIPS BETWEEN THE UPTAKE BY ALGAE OF STRONTIUM AND CALCIUM WERE STUDIED. THE OBJECTIVE WAS TWOFOLD: (1) TO EVALUATE THE CAPACITY OF CERTAIN ARID LAND ALGAE TO DISTINGUISH BETWEEN CALCIUM AND STRONTIUM IN THEIR METABOLISM, AND (2) TO EVALUATE THE EXTENT TO WHICH CALCIUM ENTERS INTO A COMPETITIVE POSITION IN THE ABSORPTION AND CONCENTRATION OF STRONTIUM BY ALGAE CELLS. UPTAKE ALWAYS FAVORED CALCIUM OVER STRONTIUM WHEN THE TWO ELEMENTS WERE PRESENT IN NUTRIENT MEDIA TOGETHER THEREBY MODIFYING THE HAZARD OF STRONTIUM ENTERING THE FOOD CHAIN. (AFFLECK-ARIZ)

FIELD 05B

ACCESSION NO. W69-03491

ENVIRONMENTAL REQUIREMENTS OF BLUE-GREEN ALGAE.

FEDERAL WATER POLLUTION CONTROL ADMINISTRATION, CORVALLIS, ORE.; PACIFIC NORTHWEST WATER LAB, AND WASHINGTON UNIV., SEATTLE.

PROCEEDINGS OF A SYMPOSIUM CORVALLIS ORE., SEPT 23-24, 1966. OCT, 1967, 111 P.

DESCRIPTORS:
*CYANOPHYTA, *EUTROPHICATION, *PLANKTON, ALGAE, *NUTRIENT REQUIREMENTS.

IDENTIFIERS:
*BLUE-GREEN ALGAE, *ALGA BLOOMS, ALGAE PHYSIOLOGY.

ABSTRACT:
THE PROBLEM OF ACCELERATED EUTROPHICATION HAS MANY FACETS, AND SOME ARE BEING EXAMINED THROUGH RESEARCH TO FIND THE KEYS FOR SOLUTION. THIS SYMPOSIUM WAS HELD TO FULFILL THE NEED TO UNDERSTAND BETTER THE ENVIRONMENTAL REQUIREMENTS OF BLUE-GREEN ALGAE. IT CONTAINS EIGHT PAPERS BY SCIENTISTS WHO ARE CONCERNED WITH THE ECOLOGICAL CONTROL OF ALGAE BLOOMS. CURRENT KNOWLEDGE ON THE SUBJECT HAS BEEN EVALUATED AND GAPS IDENTIFIED FOR FUTURE INVESTIGATION. THE PAPERS ARE FOLLOWED BY A GENERAL DISCUSSION AND A SPECIFIC DISCUSSION OF NUTRIENT MEASUREMENT AND NUISANCE CONTROL. (SEE ALSO W69-03513 THRU W69-03518)

FIELD 05C

ACCESSION NO. W69-03512

WHY STUDY BLUE-GREEN ALGAE,

WASHINGTON UNIV., SEATTLE. DEPT. OF ZOOLOGY.

W. T. EDMONSON.

SYMPOSIUM ON ENVIRONMENTAL REQUIREMENTS OF BLUE-GREEN ALGAE. FEDERAL WATER
POLLUTION CONTROL ADMIN, PACIFIC NORTHWEST WATER LAB, CORVALLIS, ORE, PP
1-6, OCT 1967. 6 P, 11 REF.

DESCRIPTORS:
*CYANOPHYTA, *EUTROPHICATION, *PHYSIOLOGICAL ECOLOGY, *WATER POLLUTION
EFFECTS, ALGAE, ALGAL CONTROL, HARVESTING, AQUATIC ALGAE, AQUATIC
ANIMALS, AQUATIC MICROBIOLOGY, AQUATIC PRODUCTIVITY, BALANCE OF NATURE,
CYCLING NUTRIENTS, ENVIRONMENTAL EFFECTS, ENVIRONMENTAL SANITATION,
ESSENTIAL NUTRIENTS, INHIBITION, LIGHT INTENSITY, LIMNOLOGY, NITROGEN
COMPOUNDS, NUISANCE ALGAE, NUTRIENT REQUIREMENTS, PHOSPHORUS COMPOUNDS,
PHOTOSYNTHETIC OXYGEN, PHYTOPLANKTON, PLANKTON, POLLUTANT
IDENTIFICATION, SEWAGE EFFLUENTS, SEWAGE DISPOSAL, WATER POLLUTION
CONTROL, WATER POLLUTION SOURCES.

IDENTIFIERS:
LAKE WASHINGTON.

ABSTRACT:
IN 1950, L WASHINGTON HAD ONLY 1,500,000 CUBIC MICRONS PER MILLILITER
OF PHYTOPLANKTON, 15% OF WHICH WERE BLUE-GREEN ALGAE. IN 1962, IT HAD
10,500,000 CUBIC MICRONS PER MILLILITER OF WHICH 95% WERE BLUE-GREEN
ALGAE, AND THE APPEARANCE OF THE LAKE STIMULATED THE CONSTRUCTION OF AN
EXPENSIVE SEWAGE DIVERSION PROGRAM. THE BLUE-GREENS ATTRACT ATTENTION
BECAUSE OF THE PUBLIC NUISANCE THEY CAUSE DUE TO THEIR ABILITY TO FLOAT
AND TO CONCENTRATE DOWNWIND; THEY HAVE SPECIAL NUTRITIONAL PROPERTIES,
ARE CONSPICUOUS IN EXTREME KINDS OF HABITATS, AND ARE IMPORTANT IN
FORMING AND STABILIZING SOIL. THE COLONIAL SPECIES OF BLUE-GREENS ARE
LESS READILY EATEN BY SOME ZOOPLANKTON, WHICH MAY BE A FACTOR IN THEIR
SURVIVAL. THE RELATIONSHIP BETWEEN HIGH NUTRIENT WATERS AND PRESENCE OF
BLUE-GREEN ALGAE IS DISCUSSED, AND IT IS POINTED OUT THAT THE
BLUE-GREEN, OSCILLATORIA AGARDHII, REMAINED IN L WASHINGTON, WHEN
NITROGEN AND PHOSPHORUS LEVELS WERE VERY LOW. THE KEY TO DOMINANCE OF
ALGAL SPECIES MAY BE IN THE PROPORTION OF NUTRIENTS MORE THAN IN
QUANTITY. MORE INFORMATION IS NEEDED CONCERNING ANTIBIOTIC SUBSTANCES
PRODUCED BY ALGAE. (FOR MAIN ENTRY SEE W69-03512)

FIELD 05C

ACCESSION NO. W69-03513

PROBLEMS IN THE LABORATORY CULTURE OF PLANKTONIC BLUE-GREEN ALGAE,

MANCHESTER COLLEGE, NORTH MANCHESTER, IND.

WILLIAM R. EBERLY.

SYMPOSIUM ON ENVIRONMENTAL REQUIREMENTS OF BLUE-GREEN ALGAE, FEDERAL WATER
  POLLUTION CONTROL ADMIN, PACIFIC NORTHWEST WATER LAB, CORVALLIS, ORE, PP
  7-34, OCT 1967. 28 P, 15 FIG, 1 TAB, 16 REF.

DESCRIPTORS:
  *CYANOPHYTA, *PHYSIOLOGICAL ECOLOGY, *PHYTOPLANKTON, *CULTURES, ALGAE,
  AQUATIC PRODUCTIVITY, CYCLING NUTRIENTS, ENVIRONMENTAL EFFECTS,
  ESSENTIAL NUTRIENTS, EUTROPHICATION, INHIBITION, LIGHT INTENSITY, LIGHT
  PENETRATION, LIMNOLOGY, NITROGEN COMPOUNDS, NUTRIENT REQUIREMENTS,
  PHOTOSYNTHETIC, OXYGEN, PLANKTON, WATER POLLUTION EFFECTS.

IDENTIFIERS:
  ALGAL TAXONOMY, ALGAE IDENTIFICATION.

ABSTRACT:
  SURFACE BLOOMS ARE USUALLY DOMINATED BY EITHER CHROOCOCCALES
  (MICROCYSTIS) OR FILAMENTOUS FORMS (ANABAENA OR APHANIZOMENON), WHEREAS
  DEEP-WATER BLOOMS ARE NEARLY ALWAYS OSCILLATORIA. THE AUTHOR IS
  CONCERNED WITH THE OSCILLATORIA ASSOCIATED WITH DEEP-WATER OXYGEN
  MAXIMA (10-12 DEG C, LIGHT 1% OF SURFACE INTENSITY). GRAPHS ARE
  PRESENTED SHOWING THE RATES OF GROWTH (OD) FOR A NUMBER OF OSCILLATORIA
  ISOLATES AND THE EFFECTS OF INITIAL INOCULATION LEVEL, PH, SOIL
  EXTRACT, TEMPERATURE, AND LIGHT INTENSITY IN ASM AND ASM-1 MEDIA. THE
  TAXONOMIC IMPLICATIONS OF THESE FINDINGS ARE DISCUSSED. ALSO CONSIDERED
  IS THE PROBLEM OF HOW ALGAE IN NATURE MAINTAIN THEIR POSITION, WHEREAS
  IN LABORATORY CULTURES CONSIDERABLE TURBULENCE IS REQUIRED TO KEEP THEM
  SUSPENDED. OTHER PROBLEMS REVIEWED ARE THE RELATION BETWEEN BACTERIA
  AND ALGAE, THE ALMOST UNIALGAL BLOOMS OCCURRING IN NATURE, AND THE
  RELATIONS BETWEEN LIGHT AND TEMPERATURE. AUTHOR POINTS OUT THAT WITH AN
  INCREASE OF NUTRIENTS (EUTROPHICATION) LIGHT AND TEMPERATURE OPTIMAL
  MIGHT BE RAISED WITH THE RESULT THAT THE MAXIMUM CONCENTRATION OF ALGAE
  WOULD MOVE CLOSER TO THE SURFACE. SEASONAL DOWNWARD MOVEMENTS OF A
  BLOOM OF OSCILLATORIA DURING THE SUMMER DEPLETION OF NUTRIENTS IS
  DESCRIBED. (FOR MAIN ENTRY SEE W69-03512)

FIELD 05C

ACCESSION NO. W69-03514

ASPECTS OF THE NITROGEN NUTRITION OF SOME NATURALLY OCCURRING POPULATIONS OF BLUE-GREEN ALGAE,

ALASKA UNIV., COLLEGE. INST. OF MARINE SCIENCE.

VERA A. (DUGDALE) BILLAUD.

SYMPOSIUM ON ENVIRONMENTAL REQUIREMENTS OF BLUE-GREEN ALGAE, FEDERAL WATER POLLUTION CONTROL ADMIN, PACIFIC NORTHWEST WATER LAB, CORVALLIS, ORE. PP 35-53, OCT 1967. 19 P, 5 FIG, 6 TAB, 19 REF.

DESCRIPTORS:
*CYANOPHYTA, *NITROGEN COMPOUNDS, *NUTRIENT REQUIREMENTS, *PHYSIOLOGICAL ECOLOGY, ALGAE, ANALYTICAL TECHNIQUES, AQUATIC ALGAE, AQUATIC MICROBIOLOGY, AQUATIC WEEDS, BALANCE OF NATURE, BIOASSAY, CHLOROPHYTA, DIATOMS, CYCLING NUTRIENTS, ENVIRONMENTAL EFFECTS, ESSENTIAL NUTRIENTS, EUTROPHICATION, INHIBITION, LIGHT INTENSITY, LIGHT PENETRATION, LIMNOLOGY, NITROGEN CYCLE, NITROGEN FIXATION, NUTRIENTS, PHOTOSYNTHETIC OXYGEN, PHYTOPLANKTON, PLANKTON, ROOTED AQUATIC PLANTS, ALASKA.

IDENTIFIERS:
SMITH LAKE(ALASKA).

ABSTRACT:
IT IS WELL ESTABLISHED THAT NITROGEN-FIXING BLUE-GREEN ALGAE INCREASE THE NITROGEN CONTENT OF WATERS, AND IT IS ASSUMED THAT THIS NITROGEN IS AVAILABLE UPON DEGRADATION, TO OTHER ALGAL SPECIES. EXPERIMENTS IN SMITH LAKE, CENTRAL ALASKA, INDICATE THAT AN ANABAENA POPULATION DEVELOPS WITHIN A FEW DAYS AFTER ICE BREAKUP. DURING THE SUMMER, A STEADY, LOWER LEVEL OF APHANIZOMENON GROWTH OCCURS. THE INITIAL NITROGEN USED BY ANABAENA IS AMMONIUM-NITROGEN; THEN NITROGEN FIXATION BECOMES IMPORTANT, AT ONE POINT ACCOUNTING FOR HALF THE NITROGEN ASSIMILATED. THE FIRST ANABAENA FILAMENTS GROWING ON AMMONIUM-NITROGEN HAVE NO HETEROCYSTS; THEY APPEAR AT THE TIME NITROGEN FIXATION OCCURS. THE POPULATION OF APHANIZOMENON IS VERY STABLE AND APPEARS TO BE CONTROLLED BY GRAZING BY DENSE POPULATION OF ZOOPLANKTON. NITROGEN FIXATION IN SIGNIFICANT RATES ONLY OCCURS DURING THE ANABAENA BLOOM. IN ANOTHER LAKE, APHANIZOMENON OCCURS IN ANOXIC, DEEPER WATERS. THE PRESENCE OF COMBINED NITROGEN SOURCES DOES NOT APPEAR TO PREVENT NITROGEN-FIXING BLOOMS FROM DEVELOPING. (FOR MAIN ENTRY SEE W69-03512)

FIELD 05C

ACCESSION NO. W69-03515

ENVIRONMENTAL REQUIREMENTS OF THERMOPHILIC BLUE-GREEN ALGAE,

OREGON UNIV., EUGENE, DEPT. OF BIOLOGY.

RICHARD W. CASTENHOLZ.

SYMPOSIUM ENVIRONMENTAL REQUIREMENTS OF BLUE-GREEN ALGAE, FEDERAL WATER POLLUTION CONTROL ADMIN, PACIFIC NORTHWEST WATER LAB, CORVALLIS, ORE. PP 55-79, OCT 1967. 25 P, 4 FIG, 1 TAB, 53 REF.

DESCRIPTORS:
*CYANOPHYTA, *PHYSIOLOGICAL ECOLOGY, *SALINE WATER, *THERMAL WATER, ALGAE, AQUATIC ALGAE, AQUATIC MICROBIOLOGY, BALANCE OF NATURE, BIOASSAY, CHLOROPHYTA, DIATOMS, ENVIRONMENTAL EFFECTS, INHIBITION, LIGHT, LIGHT INTENSITY, LIGHT PENETRATION, LIMNOLOGY, PHOTOSYNTHETIC OXYGEN, TEMPERATURE, THERMAL POLLUTION, THERMAL SPRINGS, THERMAL STRESS.

ABSTRACT:
THERMOPHILIC ALGAE ARE DESIGNATED AS THOSE WHOSE OPTIMUM TEMPERATURE IS ABOVE 45 DEG C. THE GREEN MATS FORMED IN HOT SPRINGS ABOVE 45 DEG C MAY PERSIST BECAUSE OF THE ABSENCE OF GRAZERS. THERMOPHILIC ALGAE MUST BE ABLE TO WITHSTAND THE VERY HIGH LIGHT INTENSITIES OF THE SHALLOW SPRING WATERS. THERE IS A GREAT SIMILARITY IN THE FLORA OF HOT SPRINGS THROUGHOUT THE WORLD. HOT SPRINGS PROVIDE VERY CONSTANT TEMPERATURE HABITATS, BUT SOME BLUE-GREEN THERMOPHILES CAN WITHSTAND TEMPERATURE SHOCKS AS GREAT AS ROOM TEMPERATURE TO 70 DEG C WITHOUT EFFECT. THE HIGHEST TEMPERATURE FOR CONSISTENT SURVIVAL APPEARS TO BE 74-75 DEG C, BUT THIS LIMIT MAY DEPEND UPON WATER CHEMISTRY. DETAILS OF OBSERVATIONS ON TEMPERATURES AS RELATED TO HABITATS ARE PRESENTED AND REVIEWED. THE CORRELATIONS OF GROWTH AND LIGHT INTENSITIES FOUND IN FIELD AND LABORATORY STUDIES ARE DISCUSSED. (FOR MAIN ENTRY SEE W69-03512)

FIELD 05C

ACCESSION NO. W69-03516

GROWTH REQUIREMENTS OF BLUE-GREEN ALGAE AS DEDUCED FROM THEIR NATURAL
  DISTRIBUTION,

HUMBOLDT STATE COLL., ARCATA, CALIF.

WILLIAM C. VINYARD.

SYMPOSIUM ON ENVIRONMENTAL REQUIREMENTS OF BLUE-GREEN ALGAE, FEDERAL WATER
  POLLUTION CONTROL ADMIN, PACIFIC WATER LAB, CORVALLIS, ORE, PP 81-85, OCT
  1968. 5 P, 2 REF.

DESCRIPTORS:
  *EUTROPHICATION, *CYANOPHYTA, *PHYSIOLOGICAL ECOLOGY, ALGAE, ANALYTICAL
  TECHNIQUES, AQUATIC ALGAE, AQUATIC MICROBIOLOGY, BALANCE OF NATURE,
  CYCLING NUTRIENTS, ENVIRONMENTAL EFFECTS, ESSENTIAL NUTRIENTS,
  CHLOROPHYTA, DIATOMS, INHIBITION, LIGHT, LIGHT INTENSITY, LIGHT
  PENETRATION, LIMNOLOGY, NUTRIENT REQUIREMENTS, NUTRIENTS,
  PHOTOSYNTHETIC OXYGEN, PHYTOPLANKTON, SALINE WATER, ECOLOGICAL
  DISTRIBUTION, CALIFORNIA.

ABSTRACT:
  ORGANIC POLLUTION ALMOST ALWAYS INSURES AN OVERABUNDANCE OF BLUE-GREEN
  ALGAE. ALSO, EXCESSIVE GROWTH OF CYANOPHYTES OCCURS DURING PERIODS WHEN
  HIGH TEMPERATURES AND LIGHT INTENSITIES PREVAIL. BECAUSE NOT ALL
  SPECIES OF CYANOPHYTES HAVE NUISANCE SIGNIFICANCE, THEIR TAXONOMY
  ASSUMES SOME IMPORTANCE. AUTHOR INDICATES THAT KNOWLEDGE RELATING TO
  TAXONOMY IS DEFICIENT AND THAT MODERN STUDIES RELATING SYSTEMATICS TO
  PHYSIOLOGICAL AND OTHER CULTURAL REQUIREMENTS ARE NEEDED. INFORMATION
  PERTAINING TO THE GEOGRAPHICAL DISTRIBUTION OF THESE ALGAE IS ALSO
  REQUIRED. SPECIES OF BLUE-GREENS PREDOMINATE IN SUMP PONDS OF NON-BRINE
  OIL FIELDS CONTAINING CRUDE OILS AND TARS, SOME OCCURRING AT THE BOTTOM
  UNDER THE HEAVY SCUM OF BLACK TAR, CONDITIONS WHICH PRESENT THE ALGAE
  WITH LITTLE OR NO LIGHT AND MINIMAL POSSIBILITY OF GASEOUS DIFFUSION
  INTO OR OUT OF THE MEDIUM. IN FRESHWATER LAGOON, A RELATIVELY COOL
  COASTAL ENVIRONMENT NEAR ORICK, CALIF, A NEARLY CONTINUOUS BLOOM,
  INCLUDING MICROCYSTIS, COELOSPHAERIUM AND ANABAENA, OCCURRED FOR SOME
  FIVE YEARS. (FOR MAIN ENTRY SEE W69-03512)

FIELD 05C

ACCESSION NO. W69-03517

RECENT ADVANCES IN THE PHYSIOLOGY OF BLUE-GREEN ALGAE,

SCRIPPS INSTITUTION OF OCEANOGRAPHY, LA JOLLA, CALIF.

OSMUND HOLM-HANSON.

SYMPOSIUM ON ENVIRONMENTAL REQUIREMENTS OF BLUE-GREEN ALGAE, FEDERAL WATER
  POLLUTION CONTROL ADMIN, NORTHWEST WATER LAB, CORVALLIS, ORE, PP 87-96, OCT
  1967. 10 P.

DESCRIPTORS:
  *AQUATIC MICROBIOLOGY, *PHYSIOLOGICAL ECOLOGY, *CYANOPHYTA, ALGAE,
  ANALYTICAL TECHNIQUES, AQUATIC ALGAE, BIOASSAY, CYCLING NUTRIENTS,
  ENVIRONMENTAL EFFECTS, ESSENTIAL NUTRIENTS, EUTROPHICATION, FREEZE
  DRYING, FREEZE-THAW TESTS, FREEZING, LIGHT INTENSITY, LIGHT
  PENETRATION, LIMNOLOGY, NITROGEN COMPOUNDS, NITROGEN FIXATION,
  NUTRIENTS, NUTRIENT REQUIREMENTS, PHOTOSYNTHETIC OXYGEN, PHYTOPLANKTON,
  CHLOROPHYTA.

ABSTRACT:
  SOME OF THE SUBJECTS DISCUSSED ARE THE MORPHOLOGY OF BLUE-GREEN ALGAE
  COMPARED TO RELATED ALGAE OR BACTERIA, THE UNDEFINED MANNER IN WHICH
  BLUE-GREEN ALGAE MOVE, AND THEIR MICROELEMENT NUTRITION. PRESENTED ARE
  DETAILS OF WORK ON THE RESISTANCE OF BLUE-GREEN ALGAE TO FREEZING AND
  FREEZE-DRYING, ESPECIALLY ALGAE ISOLATED IN ANARCTICA. ADDITIONS OF
  HORSE SERUM, GLYCEROL, MILK POWDER, ETC, MAY HELP THE SURVIVAL OF GREEN
  ALGAE; BUT THEY DO NOT HELP BLUE-GREENS. THE QUESTION IS RAISED AS TO
  WHAT FORM OF CARBON THE BLUE-GREENS CAN UTILIZE; AT PH 11 MOST CARBON
  IS PRESENT AS CARBONATE IONS. RESULTS OF EXPERIMENTS ON THE ORGANIC
  NUTRITION OF NOSTOC ARE PRESENTED WHICH INDICATE THAT THIS ALGA CAN
  UTILIZE GLUCOSE IN THE LIGHT BUT LITTLE OR NONE IN THE DARK. (FOR MAIN
  ENTRY SEE W69-03512)

FIELD 05C

ACCESSION NO. W69-03518

'PREDICTING DIURNAL VARIATIONS IN DISSOLVED OXYGEN CAUSED BY ALGAE IN ESTUARINE
WATERS, PART I',

STANFORD UNIV., STANFORD, CALIFORNIA.

RICHARD C. BAIN, JR.

PROCEEDINGS OF THE NATIONAL SYMPOSIUM ON ESTUARINE POLLUTION, AUGUST 23-25,
1967, PP 250-279.

DESCRIPTORS:
*PHYTOPLANKTON, *DISSOLVED OXYGEN, *ALGAE, *DIURNAL DISTRIBUTION,
*CALIFORNIA, ESTIMATING EQUATIONS, OXYGENATION, STANDING CROP,
MICROENVIRONMENT, AQUATIC LIFE, ENVIRONMENT, MICROORGANISMS, WATER
CHEMISTRY, PLANTS, SOUTHWEST U. S., REGIONS, PACIFIC COAST REGIONS,
GEOGRAPHICAL REGIONS, PHOTOSYNTHESIS, CHEMICAL REACTIONS, ZOOPLANKTON,
ANIMALS, AQUATIC ANIMALS, PHOTOSYNTHETIC OXYGEN, GASES, OXYGEN, OXYGEN
DEMANDS, EUTROPHICATION.

ABSTRACT:
EUTROPHIC ENVIRONMENTS ARE OFTEN DOMINATED BY PLANKTONIC ALGAL
POPULATIONS (PHYTOPLANKTON) WHICH CAN CAUSE DIURNAL VARIATIONS IN
DISSOLVED OXYGEN CONCENTRATIONS THROUGH RESPIRATORY ACTIVITY AND
PHOTOSYNTHESIS. PHOTOSYNTHETIC OXYGENATION AND RESPIRATORY
DEOXYGENATION RATES OF ESTUARINE PHYTOPLANKTON WERE MEASURED AT VARIOUS
STANDING CROP (CHLOROPHYLL) LEVELS. OXYGEN PRODUCTION AND CONSUMPTION
RATES FOR ACTIVELY GROWING PHYTOPLANKTON POPULATIONS WERE RELATED TO
STANDING CROP AT 20 C AND NONLIMITING LIGHT. VARIATIONS IN ALGAL
PHOTOSYNTHETIC PRODUCTION RATE AS RELATED TO LIGHT ADAPTION, AGE OF
CELLS, NUTRITION, TEMPERATURE, AND ALGAL TYPE ARE DISCUSSED.
LIGHT-PRODUCTION RELATIONSHIPS (BASED ON OCEANOGRAPHIC LITERATURE) WERE
USED TO ESTIMATE TOTAL PRODUCTION OF A WELL-MIXED SYSTEM.
STREETER-PHELPS EQUATIONS WERE MODIFIED TO INCLUDE PHYTOPLANKTON
PRODUCTION AND RESPIRATION RATES IN FORMULATIONS DESIGNED TO PREDICT
DISSOLVED OXYGEN CONCENTRATIONS OVER A 24 HOUR PERIOD. AN EXAMPLE IS
GIVEN, AND THE RESULTING DISSOLVED OXYGEN PREDICTION IS COMPARED WITH
FIELD MEASUREMENTS FROM A TIDAL REACH OF THE SAN JOAQUIN RIVER,
CALIFORNIA.

FIELD 05C

ACCESSION NO. W69-03611

SEATTLE'S EFFORTS IN RESTORATION OF BAYS AND ESTUARIES,

GARY W. ISAAC, AND CURTIS P. LEISER.

32ND NORTH AMERICAN WILDLIFE AND NATURAL RESOURCES CONFERENCE, SAN FRANCISCO,
CALIFORNIA, TRANSACTIONS, MARCH 13-15, 1967, PP 127-137.

DESCRIPTORS:
*WASHINGTON, *SEWAGE TREATMENT, *WATER POLLUTION CONTROL, PACIFIC
NORTHWEST U. S., CIVIL ENGINEERING, WASTE TREATMENT, ENGINEERING,
ALGAE, PLANTS, CONTROL, POLLUTION ABATEMENT, ABATEMENT, PACIFIC COAST
REGION, REGIONS, GEOGRAPHICAL REGIONS, BAYS, BODIES OF WATER.

ABSTRACT:
CITIZENS IN THE SEATTLE AREA, WHICH IS COMPRISED OF 14 CITIES,
INITIATED THE LEGISLATION AND VOTED INTO BEING A METROPOLITAN MUNICIPAL
CORPORATION, SPECIFICALLY TO DEAL WITH THE POLLUTION PROBLEM, BUT ABLE
ALSO TO EXPAND TO ALLEVIATE OTHER REGIONAL DIFFICULTIES AS THEY WOULD
ARISE. AN OPERATING TREATMENT AGENCY AND FOUR ENGINEERING FIRMS
COMPOSED A MASTER PLAN WHICH IS DETAILED IN THIS PAPER. THE COST OF THE
PLAN WAS ESTIMATED AT $125,000,000. THE POLLUTION CONTROL PROGRAM WILL
CONTINUE IN THREE MAJOR PARTS: (1) EMISSION CONTROLS APPLIED TO
INDUSTRIAL WASTE DISCHARGES AT THE SOURCE, (2) TREATMENT PLANT CONTROL
AND MONITORING BY PERSONNEL ATTACHED PERMANENTLY TO PLANT OPERATIONS,
(3) A SEPARATE FORCE TO MONITOR THE RECEIVING WATERS FOR BIOLOGICAL
NUTRIENT AND PHYSICAL PARAMETERS.

FIELD 05G, 05D

ACCESSION NO. W69-03683

'NUTRIENT ASSIMILATION IN A VIRGINIA TIDAL SYSTEM',

STANFORD UNIV., CALIF.

MORRIS L. BREHMER.

PROCEEDINGS OF THE NATIONAL SYMPOSIUM ON ESTUARINE POLLUTION, AUGUST 23-25, 1967, PP 218-237.

DESCRIPTORS:
*WATER PROPERTIES, *NUTRIENTS, *VIRGINIA, NITROGEN COMPOUNDS, PHOSPHORUS, NUISANCE ALGAE, SALTS, TIDES, SOUTHEAST U. S., REGIONS, GEOGRAPHICAL REGIONS, COASTAL PLAINS, ATLANTIC COASTAL PLAIN, INORGANIC COMPOUNDS, METALS, PLANKTON, AQUATIC LIFE, ZOOPLANKTON, AQUATIC ANIMALS, ANIMALS, APPALACHIAN MOUNTAIN REGION.

ABSTRACT:
THIS STUDY WAS CONDUCTED TO ASSESS NUTRIENT ASSIMILATION AND PHYTOPLANKTON RESPONSE IN THE TIDAL JAMES RIVER ESTUARY AND IN THE NANSEMOND ESTUARY, A TRIBUTARY TO THE JAMES. THE ASSIMILATION OF NITROGEN AND PHOSPHORUS WAS MEASURED FOR A PERIOD OF ONE YEAR IN EACH OF THE ABOVE TWO RIVER SYSTEMS. IT WAS DETERMINED THAT THE NUTRIENT ASSIMILATION CAPACITY OF ESTUARINE WATERS VARIES SEASONALLY, BEING GREATEST IN THE WINTER, EVEN THOUGH FRESH WATER DISCHARGE LEVELS MAY REMAIN NEARLY CONSTANT. DATE INDICATE THAT WATER CONTAINING DISSOLVED SOLIDS OF MARINE ORIGIN MAY BE ABLE TO ASSIMILATE HIGHER NUTRIENT LEVELS THAN FRESH WATER WITHOUT PRODUCING AQUATIC NUISANCE CONDITIONS.

FIELD 05C

ACCESSION NO. W69-03695

THE EFFECT OF INDUCED TURBULENCE ON THE GROWTH OF ALGAE,

GEORGIA INST. OF TECH., ATLANTA. SCHOOL OF ENGINEERING AND GEORGIA INST. OF TECH., ATLANTA. WATER RESOURCES CENTER.

LAWRENCE W. OLINGER.

GEORGIA INST TECH, WRC REP 0468, DEC 1968. 81 P, 20 FIG, 2 TAB, 47 REF, 4 APPEND. 5T1-WP-62-03 AND 04, FWPCA.

DESCRIPTORS:
*ALGAE, *PHOTOSYNTHESIS, *GROWTH RATES, *TURBULENCE, CHLORRELA, SCENEDESMUS, KINETICS, ECOSYSTEMS.

IDENTIFIERS:
ALGAL GROWTH KINETICS, MONOD GROWTH EQUATION, INCIDENT LIGHT UTILIZATION.

ABSTRACT:
A LABORATORY STUDY WAS MADE OF THE RELATIONSHIP BETWEEN TURBULENCE AND UTILIZATION OF INCIDENT LIGHT BY ALGAL CULTURES. IT WAS FOUND THAT THE MAXIMUM GROWTH RATE CONSTANT OF THE MONOD GROWTH EQUATION INCREASED WITH INCREASING TURBULENCE AND THAT THE RATE AT WHICH CONSTANT GROWTH OCCURRED WAS A FUNCTION OF TURBULENCE. WITH CONSTANT ILLUMINATION OF 600 FT - CANDLES AND VARIABLE TURBULENCE, THE GROWTH RATE CONSTANT VARIED FROM 1.65 TO 8.5. (KNAPP-USGS)

FIELD 05C

ACCESSION NO. W69-03730

MICROBIOLOGY OF OCEANS AND ESTUARIES,

MIAMI UNIV., FLA. INST. OF MARINE SCIENCE.

E. J. F. WOOD.

AMSTERDAM, ELSEVIER PUBL CO OCEANOGR SER, 1967. 319 P, 29 FIG, 8 PLATE, 19 TAB, 521 REF.

DESCRIPTORS:
*MICROBIOLOGY, *AQUATIC MICROBIOLOGY, *ESTUARIES, *MARINE MICROORGANISMS, MARINE ALGAE, MARINE ANIMALS, MARINE BACTERIA, MARINE FUNGI.

IDENTIFIERS:
ESTUARINE MICROBIOLOGY.

ABSTRACT:
A TEXTBOOK ON MARINE AND ESTUARINE MICROBIOLOGY IS PRIMARILY INTENDED FOR UNDERGRADUATE STUDENTS IN MARINE BIOLOGY AND OCEANOGRAPHY AND MAY ALSO BE OF USE TO RESEARCH WORKERS IN MARINE MICROBIOLOGY. THE PROBLEMS USUALLY ENCOUNTERED IN STUDY, INCLUDING SAMPLING TECHNIQUES, TAXONOMY, AND QUANTIFICATION, ARE DISCUSSED THOROUGHLY. ENVIRONMENTAL RELATIONSHIPS AND MODIFICATIONS BY ORGANISMS OF THEIR ENVIRONMENTS ARE EXPLAINED. TAXONOMY IS DISCUSSED WITH REFERENCES TO THE MOST SIGNIFICANT TAXONOMIC WORKS AND WORKERS IN EACH GROUP. NUTRITION AND REPRODUCTION ARE CONSIDERED IN DETAIL. ECONOMIC ASPECTS SUCH AS FOULING, THE RELATION OF BORERS TO FUNGI, MICROBIAL CORROSION, FERMENTATION, ROTTING OF CORDAGE AND STRUCTURES, FISH SPOILAGE, SHELLFISH POISONING, AND GEOBIOLOGICAL ASPECTS AND ORE FORMATION ARE DISCUSSED. (KNAPP-USGS)

FIELD 02L, 05C

ACCESSION NO. W69-03752

WATER POLLUTION IN THE GREAT LAKES BASIN,

FEDERAL WATER POLLUTION CONTROL ADMINISTRATION, GREAT LAKES REGION, CHICAGO, ILL.

H. W. POSTON.

LIMNOS, VOL 1, NO 1, PP 6-11, SPRING 1968. 6 P, 6 PHOTO.

DESCRIPTORS:
*GREAT LAKES, *WATER POLLUTION, *WATER POLLUTION CONTROL, WATER POLLUTION EFFECTS, EUTROPHICATION, FISHKILL, ALGAE, OXYGEN SAG, WATER QUALITY, CHLORIDES, PHOSPHORUS, SULFATES, HARDNESS(WATER).

IDENTIFIERS:
GREAT LAKES BASIN, BACTERIAL POLLUTION.

ABSTRACT:
THE CAUSES OF WATER POLLUTION OF THE GREAT LAKES BASIN, PRESENTLY AVAILABLE REMEDIES, AND RECOMMENDATIONS FOR ACTION IN POLLUTION CONTROL ARE DISCUSSED. THE LAKES, PARTICULARLY LAKE ERIE, ARE BECOMING EUTROPHIC BECAUSE OF WASTE DISCHARGES. TERTIARY TREATMENT WITH PHOSPHORUS REMOVAL WILL ALLEVIATE THE PROBLEM AND ALSO REDUCE BIOCHEMICAL OXYGEN DEMAND. CHLORIDES, SULFATES, AND HARDNESS ARE INCREASING. BACTERIAL POLLUTION IS A SERIOUS PROBLEM IN TRIBUTARIES, AND IN LOCAL ZONES IN THE GREAT LAKES NEAR POPULATION CENTERS WHERE RECREATIONAL DEMAND IS ALSO HIGHEST. ELIMINATION OF COMBINED SEWERS AS WELL AS BETTER SEWAGE TREATMENT WILL HELP. CHEMICAL CONTAMINATION CAUSED BY INDUSTRIAL WASTE DISPOSAL IS WIDESPREAD. OXYGEN LEVELS ARE LOW IN STREAMS, SMALL LAKES, AND THE CENTRAL BASIN OF LAKE ERIE. AN UNUSUAL FORM OF POLLUTION IS THE OVER-POPULATION OF ALEWIVES IN LAKE MICHIGAN. PERIODIC FISHKILLS LITTER THE SHORE. ONE OF THE MAIN DIFFICULTIES IN SOLVING GREAT LAKES PROBLEMS IS DIVERSITY AND NEED FOR COORDINATION OF GOVERNMENT AGENCIES IN THE BASIN. CREATION OF THE WATER QUALITY STANDARDS AND THE GREAT LAKES BASIN COMMISSION SHOULD HELP. (KNAPP-USGS)

FIELD 05B, 05C

ACCESSION NO. W69-03948

NUTRIENT AND POLLUTANT RESPONSE OF ESTUARINE BIOTAS,

STANFORD UNIV., CALIF.

JAMES B. LACKEY.

IN PROCEEDINGS OF THE NATIONAL SYMPOSIUM ON ESTUARINE POLLUTION, AUGUST
23-25, 1967, PP 188-217.

DESCRIPTORS:
*NUTRIENTS, *MARINE MICROORGANISMS, *POLLUTANTS, RECREATION, WATER
SPORTS, ENVIRONMENT, AQUATIC ENVIRONMENT, ESTUARINE ENVIRONMENT, WATER
QUALITY, SEWAGE BACTERIA, ENVIRONMENTAL EFFECTS, AQUATIC LIFE, AQUATIC
MICROORGANISMS, MICROORGANISMS, SESTON, INVERTEBRATES, SHELLFISH, WASTE
DISPOSAL, SEWAGE DISPOSAL, WATER CHEMISTRY, ECOLOGY, WATER POLLUTION,
WATER POLLUTION CONTROL, CONTROL, ALGAE, PLANTS, COMMERCIAL FISHING,
FISHING, INDUSTRIES, COMMERCIAL SHELLFISHING, ANIMALS, AQUATIC ANIMALS.

ABSTRACT:
THE BEHAVIOR OF MICROBIOTA--ALGAE AND PROTOZOA--AS AFFECTED BY
COMMERCIAL, RECREATIONAL, AND METROPOLITAN USES OF ESTUARINE WATERS ARE
OUTLINED. SOME OF THE FINDINGS PRESENTED AT THE SYMPOSIUM ARE AS
FOLLOWS: (1) ESTUARIES AND BAYS ARE OF GREAT ECONOMIC AND RECREATIONAL
VALUE, FOR NURSERIES, FISHING GROUNDS, SHELLFISHERIES, CRUSTACEAN
PRODUCTION, BOATING, AND BATHING. (2) HUMAN ACTIVITY OFTEN MODIFIES
THEIR PREFERRED ECOLOGY BY ADDING NUTRIENTS OR POLLUTANTS. (3) THE
SUSPENDED (PLANKTON) MICROSCOPIC PLANTS AND ANIMALS, AND THOSE OF THE
INTERFACE (BENTHOS) ARE MOST EASILY STUDIED QUALITATIVELY AND
QUANTITATIVELY. (4) SOME POLLUTANTS, SUCH AS SILT, SHARPLY REDUCE THE
MICROSCOPIC POPULATIONS. (5) THE ROLE OF THE ENGINEER WITH REGARD TO
ESTUARINE STUDIES IS THAT OF SEEING THAT ADEQUATE BIOLOGICAL STUDIES
ARE MADE, SO THAT HE IN TURN CAN DESIGN AND CONSTRUCT THE NECESSARY
TREATMENT PLANTS AND OUTFALLS WHICH WILL PREVENT OVERENRICHMENT OR
POLLUTION.

FIELD 05C, 02L

ACCESSION NO. W69-04276

EXPERIMENTS ON THE UTILIZATION OF NITROGEN IN FRESH WATER,

FRESHWATER BIOLOGICAL ASSOCIATION, AMBLESIDE (ENGLAND).

WINIFRED PENNINGTON.

J ECOL, VOL 30, PP 326-340, 1942. 1 FIG, 8 TAB, 18 REF, 2 APPEND, DISC.

DESCRIPTORS:
*NITROGEN COMPOUNDS, *NITROGEN UTILIZATION, *FRESH WATER, ALGAE,
NITRATE AMMONIUM, NITRITE, OXIDATION-REDUCTION POTENTIAL, CYCLING
NUTRIENTS, EUTROPHICATION.

ABSTRACT:
THE EXPERIMENTS DESCRIBED WERE CARRIED OUT ON MIXED CULTURES OF ALGAE
AND BACTERIA IN A RICH CULTURE SOLUTION CONTAINING NITRATE OR AMMONIUM.
NO ATTEMPT WAS MADE TO DIFFERENTIATE BETWEEN THE ROLES OF ALGAE AND
BACTERIA IN NITROGEN UTILIZATION. AMMONIUM AND NITRATE WERE BOTH
UTILIZED, BUT AMMONIUM APPEARED TO BE USED MORE RAPIDLY. NITRITE WAS
PRODUCED FROM BOTH AMMONIUM AND NITRATE, AND THERE WAS EVIDENCE THAT
AMMONIUM WAS PRODUCED FROM NITRATE AND NITRATE FROM AMMONIUM. IN SOME
EXPERIMENTS THERE WAS A MARKED LOSS OF NITROGEN FROM THE CULTURES,
APPARENTLY DUE TO LIBERATION OF NITROGEN. (KONRAD-WISC)

FIELD 05C, 02K, 02H

ACCESSION NO. W69-04521

DISTRIBUTION, ENVIRONMENTAL REQUIREMENTS AND SIGNIFICANCE OF CLADOPHORA IN THE
GREAT LAKES,

ONTARIO WATER RESOURCES COMMISSION, TORONTO, BIOLOGY BRANCH.

JOHN H. NEIL, AND GLENN E. OWEN.

PUBLICATION NUMBER 11, GREAT LAKES RES DIV, MICHIGAN UNIV, PP 113-121, 1964.
4 FIG, 2 TAB, 2 REF.

DESCRIPTORS:
*ALGAE, *ALGAL CONTROL, *ESSENTIAL NUTRIENTS, *EUTROPHICATION,
*CLADOPHORA, *NUISANCE ALGAE, *SEWAGE, ALGICIDES, AQUATIC ALGAE,
AQUATIC PRODUCTIVITY, BIOASSAY, CYCLING NUTRIENTS, ENVIRONMENTAL
EFFECTS, HARVESTING ALGAE, INHIBITION, LIGHT PENETRATION, LIMNOLOGY,
NITROGEN COMPOUNDS, NUTRIENT REQUIREMENTS, NUTRIENTS, PHOSPHORUS
COMPOUNDS, PHYSIOLOGICAL ECOLOGY, POLLUTION IDENTIFICATION, SEWAGE
DISPOSAL, WATER POLLUTION, WATER POLLUTION CONTROL, WATER POLLUTION
EFFECTS, WATER POLLUTION SOURCES.

ABSTRACT:
EXCESSIVE GROWTHS OF CLADOPHORA SP ALONG CERTAIN SECTIONS OF THE GREAT
LAKES SHORELINE CREATE SERIOUS NUISANCE CONDITIONS WHICH AFFECT THE USE
OF WATER FOR RECREATIONAL, INDUSTRIAL, AND MUNICIPAL PURPOSES.
INFORMATION ON THE ECOLOGY OF THIS ALGAE WAS COLLECTED AS PART OF A
STUDY DIRECTED TOWARD THE DEVELOPMENT OF CONTROL MEASURES. THE PRESENCE
OF CLADOPHORA SP IS DEPENDENT ON A SUITABLE SUBSTRATE FOR ATTACHMENT,
WATER MOVEMENT, ADEQUATE LIGHT, AND NUTRIENTS IN EXCESS OF THOSE
NORMALLY AVAILABLE IN THE WATERS OF THE UPPER GREAT LAKES. LAKES
ONTARIO AND ERIE HAVE SUFFICIENT INHERENT FERTILITY TO SUPPORT MARGINAL
GROWTHS, BUT, WHERE LOCAL NUTRIENT SOURCES ARE AVAILABLE, PRODUCTION
INCREASES. APPLICATION OF PHOSPHORUS TO A LOCATION PROVIDING SUITABLE
PHYSICAL CONDITIONS BUT DEVOID OF CLADOPHORA SP RESULTED IN THE
ESTABLISHMENT OF A SIZEABLE AREA OF GROWTH. THE RESULTS OF ATTEMPTS AT
CONTROL ARE ALSO DISCUSSED. (FITZGERALD-WIS)

FIELD 02H, 05C

ACCESSION NO. W69-04798

UNUSUAL PHOSPHORUS SOURCE FOR PLANKTON ALGAE,

TEXAS A AND M UNIV., COLLEGE STATION. DEPT. OF BIOLOGY.

WALTER ABBOTT.

ECOLOGY, VOL 38, NO 1, P 152, 1957.

DESCRIPTORS:
*PHOSPHORUS COMPOUNDS, *PHYTOPLANKTON, *WATER POLLUTION SOURCES,
*CYCLING NUTRIENTS, ALGAE, ANALYTICAL TECHNIQUES, CLAY LOAM, CLAYS,
COLLOIDS, EUTROPHICATION, IMPOUNDMENTS, LAKES, LOAMS, RESERVOIRS,
TEXAS.

IDENTIFIERS:
HARRIS COUNTY, TEX, POLYPHOSPHATES, ORGANIC PHOSPHATES.

ABSTRACT:
ALTHOUGH LAKE HOUSTON, A NEWLY IMPOUNDING LAKE IN EASTERN HARRIS
COUNTY, TEXAS, SUPPORTS PHYTOPLANKTON POPULATIONS FLUCTUATING
TWENTY-FOLD, PHOSPHATE-PHOSPHORUS IN CENTRIFUGATE FROM SAMPLES OF WATER
FROM TEN STATIONS WAS NOT DEMONSTRABLE BY A TECHNIQUE SENSITIVE TO 1
GAMMA/LITER. IN A SERIES OF 18 SAMPLINGS, WATER CONTAINING COLLOIDAL
MATTER, AFTER OXIDATION WITH PERCHLORIC ACID, CONTAINED A MEAN OF 85
GAMMA/LITER (RANGE, 32-285), HIGHEST VALUES OCCURRING IMMEDIATELY AFTER
HEAVY RAINS. FILTRATE FROM PASSAGE THROUGH SEITZ NO. 14 FILTER,
SIMILARLY TREATED, SHOWED ONLY BARELY DETECTABLE TRACES OF
PHOSPHATE-PHOSPHORUS. AUTHOR OFFERS THESE DATA AS EVIDENCE THAT
PLANKTONIC ALGAE MAY DERIVE NUTRITIVE PHOSPHORUS FROM COMPLEX
POLYPHOSPHATES OR IN THE SUSPENDED COLLOIDAL CLAY, ORIGINATING FROM
SOILS OF SURROUNDING WATERSHEDS, WITHOUT BENEFIT OF INTERMEDIATE STAGE
OF DISSOLVED PHOSPHATE. (EICHHORN-WIS)

FIELD 05B

ACCESSION NO. W69-04800

ALGAL GROWTH AQUEOUS FACTORS OTHER THAN NITROGEN AND PHOSPHORUS,

ROBERT A. TAFT SANITARY ENGINEERING CENTER, CINCINNATI, OHIO.

KENNETH M. MACKENTHUN, AND WILLIAM MARCUS INGRAM.

PUBLICATION NO WP-24, FEDERAL WATER POLLUTION CONTROL ADMIN, DEPT OF INTERIOR, WASHINGTON, DC, 1966.

DESCRIPTORS:
*ALGAE, *BIBLIOGRAPHIES, *EUTROPHICATION, *PLANT GROWTH, ANALYTICAL TECHNIQUES, ANTIBIOTICS, BIOASSAY, NORON, CALCIUM, CHLORINE, COBALT, COPPER, ECOLOGY, ENVIRONMENTAL EFFECTS, GRAZING, IRON, MANGANESE, MOLYBDENUM, PLANT GROWTH REGULATORS, PREDATION, PRODUCTIVITY, TRACE ELEMENTS, SODIUM, VIRUSES, VITAMINS, WATER CHEMISTRY, WATER POLLUTION CONTROL, WATER POLLUTION SOURCES, WATER QUALITY.

IDENTIFIERS:
AUTOINHIBITORS, EXTRACELLULAR PRODUCTS, HORMONES, AUXINS, METABOLITES, INHIBITION, SILICON, VANADIUM, ZINC, GALLIUM.

ABSTRACT:
ALTHOUGH RESEARCHERS HAVE EMPHASIZED THE ROLE IN EUTROPHICATION OF THE MAJOR PLANT NUTRIENTS, NITROGEN AND PHOSPHORUS, THERE IS LITTLE DOUBT THAT OTHER CHEMICAL AND BIOTIC INFLUENCES ON PRODUCTION OF MACROPHYTES AND PHYTOPHANKTON ARE IMPORTANT, THOUGH POORLY UNDERSTOOD. THIS SELECTED BIBLIOGRAPHY OF 399 ENTRIES, COMPILED FOR ENGINEERS AND SCIENTISTS CONFRONTED WITH APPLIED CONTROL INVESTIGATIONS RELATED TO POLLUTANTS, FOCUSES ATTENTION ON SUCH FACTORS AND THEIR EFFECTS ON ALGAE. THE REPORT IS ORGANIZED INTO SECTIONS, AS FOLLOWS, WITH NUMBERS OF CITATIONS INDICATED: GENERAL REFERENCES, 15; ANTIBIOTICS, 31; AUTOINHIBITORS, 11; ENVIRONMENTAL FACTORS, 25; EXTRACELLULAR PRODUCTS, 31; HORMONES AND AUXINS, 13; INHIBITING METABOLITES, 24; PREDATION AND GRAZING, 24; GENERAL REFERENCES ON TRACE METALS AND COMPOUNDS, 29; BORON, 14; CALCIUM, 2; CHLORINE, 5; COBALT, 7; COPPER, 5; GALLIUM, 1; IRON, 2; MANGANESE, 5; MOLYBDENUM, 12; SILICON, 7; SODIUM, 8; VANADIUM, 3; ZINC, 17; MIXED ELEMENTS, 15; VIRUSES AND PARASITES, 11; VITAMIN ASSAY METHODS, 22; VITAMIN OCCURRENCE, 22; ORGANISMIC REQUIREMENTS FOR VITAMINS, 27; SYNTHESES OF VITAMINS, 11. CITATIONS ARE DATED FROM 1903 TO 1966, OF WHICH 69 CITATIONS ARE FROM 1960-1966. (EICHHORN-WIS)

FIELD 05C, 10

ACCESSION NO. W69-04801

MINERAL REQUIREMENTS FOR THE GROWTH OF ANABAENA SPIROIDES IN VITRO,

OREGON STATE UNIV., CORVALLIS. DEPT. OF BOTANY.

SHERRY L. VOLK, AND HARRY K. PHINNEY.

CAN J BOT, VOL 46, PP 619-630, MAY 1968. 4 FIG, 4 TAB, 67 REF.

DESCRIPTORS:
*ALGAE, *ANALYTICAL TECHNIQUES, *BIOASSAY, *ESSENTIAL NUTRIENTS, *NUTRIENT REQUIREMENTS, *PHYTOPLANKTON, AQUATIC ALGAE, AQUATIC MICROBIOLOGY, AQUATIC PRODUCTIVITY, CYANOPHYTA, ENVIRONMENTAL EFFECTS, EUTROPHICATION, LIMNOLOGY, NITROGEN COMPOUNDS, NITROGEN FIXATION, NUISANCE ALGAE, NUTRIENTS, PHOSPHORUS COMPOUNDS, PHYSIOLOGICAL ECOLOGY, PLANKTON, OREGON.

IDENTIFIERS:
KLAMATH LAKE.

ABSTRACT:
ANABAENA SPIROIDES KLEBAHN ISOLATED FROM UPPER KLAMATH LAKE, OREGON, WAS CULTURED IN GERLOFF'S MODIFICATION OF CHU'S NO 10 MEDIUM. VARIATIONS OF THE CONCENTRATIONS OF NUTRIENTS WERE STUDIED TO DETERMINE MINIMAL REQUIREMENTS FOR MAXIMUM GROWTH. MAXIMUM GROWTH IN THE NUTRITION TESTS VARIED BETWEEN 30 AND 80 MG/LITTER DRY WEIGHT. IRON PROVIDED IN THE ALKALINE-STABLE CHELATE ETHYLENEDIAMINE DI-ORTHO-HYDROXYPHENYLACETATE WAS REQUIRED IN LOWER CONCENTRATION (0.3 PPM) THAN IN GERLOFF'S MEDIUM (IRON CITRATE, 1.2 PPM). THE MINIMUM CONCENTRATION OF NUTRIENTS PROVIDING FOR MAXIMUM GROWTH: MAGNESIUM, 0.125 PPM; PHOSPHORUS, 0.52 PPM; CALCIUM, 10 PPM; SULFUR, 3 PPM; POTASSIUM, 2.5 PPM. NITROGEN WAS FIXED BY THIS ORGANISM. UREA INHIBITED GROWTH (AT 16 PPM NITROGEN) WHILE GLYCINE MIGHT HAVE STIMULATED GROWTH. CARBONATE DID NOT AFFECT GROWTH. NO REQUIREMENT OF MINOR ELEMENTS WAS FOUND FOR THE AMOUNT OF GROWTH ATTAINED. (FITZGERALD-WIS)

FIELD 05C

ACCESSION NO. W69-04802

DEVELOPMENT AND STATUS OF POND FERTILIZATION IN CENTRAL EUROPE.

WISCONSIN UNIV., MADISON. DEPT. OF ZOOLOGY.

JOHN C. NEESS.

TRANS AMER FISH SOC, VOL 76, PP 335-358, 1949. 24 P, 1 FIG, 73 REF.

DESCRIPTORS:
*CYCLING NUTRIENTS, *ENVIRONMENTAL EFFECTS, *ESSENTIAL NUTRIENTS,
*FISH, *FISHERIES, ALGAE, AQUATIC ALGAE, AQUATIC MICROBIOLOGY, AQUATIC
WEEDS, BALANCE OF NATURE, EUTROPHICATION, FISH FARMING, FISH ROOD
ORGANISMS, FISH GENETICS, FISH HANDLING FACILITIES, FISH MANAGEMENT,
LIMNOLOGY, NITROGEN COMPOUNDS, NITROGEN CYCLE, NUTRIENT REQUIREMENTS,
LIME, PHOSPHORUS COMPOUNDS, PHYSIOLOGICAL ECOLOGY, PLANKTON, SEWAGE,
SEWAGE LAGOONS.

IDENTIFIERS:
EUROPE.

ABSTRACT:
A REVIEW OF RECENT EUROPEAN LITERATURE ON POND FERTILIZATION IS OFFERED
AS A FOUNDATION FOR CERTAIN PHASES OF AMERICAN POND-CULTURE TECHNIQUE.
THE FERTILIZING SUBSTANCES, LIME, POTASSIUM, PHOSPHORUS, NITROGEN, AND
CARBON, ARE DISCUSSED INDIVIDUALLY IN THE LIGHT OF THEIR USEFULNESS IN
INCREASING FISH YIELDS AND WITH REGARD TO THE IMPORTANT MECHANISMS
THROUGH WHICH THEY ARE ABLE TO EFFECT THESE INCREASES. EUROPEAN
TECHNIQUES IN THE EMPLOYMENT OF FERTILIZER ARE DESCRIBED WHERE THESE
ARE DIFFERENT FROM CORRESPONDING AMERICAN PROCEDURES. IN ADDITION, AN
ATTEMPT HAS BEEN MADE TO INTEGRATE DIVERGENT ASPECTS OF THE SUBJECT
THROUGH A CONSIDERATION OF CERTAIN FUNDAMENTALS OF TROPHIC METABOLISM
IN PONDS, THE FUNCTION OF POND SOILS AS CONSERVATORS OF NUTRIENTS, AND
MICROBIOLOGICAL PROCESSES RELATING TO THE THEORY OF FERTILIZATION. A
SYNOPSIS OF THE HISTORY OF POND FERTILIZATION IN CENTRAL EUROPE AND A
GENERAL CRITICISM OF RESEARCH METHODS ARE INCLUDED. (FITZGERALD-WIS)

FIELD 05G

ACCESSION NO. W69-04804

NITROGEN AND PHOSPHORUS IN WATER; AN ANNOTATED BIBLIOGRAPHY OF THEIR BIOLOGICAL
EFFECTS.

PUBLIC HEALTH SERVICE, WASHINGTON, D. C.

PUBLIC HEALTH SERVICE PUBLICATION NO 1305, WASHINGTON, D. C., 1965. 111 PP.

DESCRIPTORS:
*BIBLIOGRAPHIES, *EUTROPHICATION, *NITROGEN COMPOUNDS, *NUTRIENTS,
*PHOSPHORUS COMPOUNDS, *WATER POLLUTION EFFECTS, ALGAE, ANIMALS,
AQUATIC PLANTS, BIOMASS, FERTILIZATION, LAKES, LIMNOLOGY, MIDGES,
NITRATES, NITROGEN, NITROGEN CYCLE, NITROGEN FIXATION, PHOSPHATES,
PHOSPHORUS, PRODUCTIVITY, STANDING CROP, STREAMS, WATER CHEMISTRY,
WATER POLLUTION CONTROL, WATER POLLUTION SOURCES, WATER QUALITY.

IDENTIFIERS:
DUCKS, NITRATE POISONING.

ABSTRACT:
THIS SELECTED AND ANNOTATED BIBLIOGRAPHY OF 171 ENTRIES WAS INTENDED BY
ITS AUTHOR FOR ENGINEERS AND SCIENTISTS WHO ARE FACED WITH PREDICTING
LIMNOLOGICAL CHANGES RESULTING FROM PLANT NUTRIENT LOADINGS TO STANDING
BODIES OF WATER, EMPHASIZING CRITICAL CONCENTRATION VALUES FOR ALGAL
DEVELOPMENT AND EFFECTS OF FERTILICATION UPON AQUATIC LIFE. IT IS
INDEXED FOR MAJOR CATEGORIES WITH NUMBERS OF CITATIONS INDICATED AS
FOLLOWS: ALGAE, 86; ANIMALS, 4; CHEMISTRY, 12; EUTROPHICATION, 5;
ARTIFICIAL FERTILIZATION, 32; FERTILIZATION BY DUCKS, 4; LAKES, 18;
MIDGES, 1; NITRATE POISONING, 8; NITROGEN FIXATION BY ALGAE, 7;
NITROGEN AND PHOSPHORUS, 194; ORGANISM STANDING CROPS, 11; HIGHER
AQUATIC PLANTS, 6; PRODUCTION, 12. DATES CITED RANGE FROM 1918 TO 1965
WITH 70 REFERENCES FROM 1960-1965. PUBLICATION INCLUDES A GENERAL
INTRODUCTION, SUBJECT INDEX, AUTHOR INDEX, AND LISTS OF DEFINITIONS AND
EQUIVALENTS AT BOTH BEGINNING AND END OF THE BIBLIOGRAPHY.
(EICHHORN-WIS)

FIELD 05C, 10

ACCESSION NO. W69-04805

OBSERVATIONS ON THE RESPONSE OF SOME BENTHONIC ORGANISM TO POWER STATION
COOLING,

S. MARKOWSKI.

JOURNAL OF ANIMAL ECOLOGY, VOL 29, NO 2, 1960, PP 349-357.

DESCRIPTORS:
*POWERPLANTS, *THERMAL POLLUTION, *TEMPERATURE, *COOLING WATER,
CHEMICAL PROPERTIES, ECOLOGY, SALINITY, BENTHOS, AQUATIC LIFE, ALGAE,
PLANTS, WATER PROPERTIES, WATER TEMPERATURE, OUTLETS, HYDROELECTRIC
PLANTS, ELECTRIC POWERPLANTS, ENGINEERING STRUCTURES, INDUSTRIAL
PLANTS, STRUCTURES, AFTERBAYS, WATER TYPES, INTAKES.

ABSTRACT:
THE AUTHOR REPORTS OBSERVATIONS MADE ON THE SETTLING OF BENTHONIC
ORGANISMS ON EXPERIMENTAL SLABS PLACED IN THE INTAKE AND IN THE OUTFALL
WATER OF THE POWER STATION LOCATED AT THE CAVENDISH DOCK. DATA INDICATE
THAT OUTFALL AREA IS MORE CONDUCIVE TO GROWTH AS THE BENTHONIC FORMS
APPEAR EARLIER THAN IN THE INTAKE. NO ALGAE GROWTH WAS FOUND IN THE
INTAKE, BUT THERE WAS VERY PROLIFIC GROWTH IN THE OUTFALL. SPECIFIC
COMPOSITION OF BENTHONIC INVERTEBRATES WERE SIMILAR IN BOTH INTAKE AND
OUTFALL SLABS. IN ADDITION, DENSER ANIMAL POPULATIONS WERE FOUND IN THE
INTAKE THAN IN THE OUTFALL. GENERAL CHARACTERISTICS OF THE ENVIRONMENTS
AND THE FACTORS INTRODUCED BY THE POWER STATION INTO THE AQUATIC MEDIUM
ARE DISCUSSED.

FIELD 05C

ACCESSION NO. W69-05023

SEASONAL VARIATION IN CONTENT OF NITROGENOUS COMPOUNDS AND PHOSPHATE IN THE
WATER OF TAKASUKA POND, SAITAMA, JAPAN,

TOKYO IMPERIAL UNIV. (JAPAN). GEOGRAPHICAL INST.

S. YOSHIMURA.

ARCHIV FUR HYDROBIOLOGIE, VOL 24, PP 155-176, 1932. 6 FIG, 13 TAB, 72 REF.

DESCRIPTORS:
*NITROGEN COMPOUNDS, *PHOSPHORUS COMPOUNDS, *CYCLING NUTRIENTS, WATER
CHEMISTRY, WATER ANALYSIS, PHOSPHORUS, PONDS, PHYTOPLANKTON,
EUTROPHICATION, LAKES, LIMNOLOGY, AQUATIC ENVIRONMENT, AQUATIC
PRODUCTIVITY, HYPOLIMNION, ALGAL CONTROL, SEASONAL, ANNUAL TURNOVER.

IDENTIFIERS:
TAKASUKA POND(JAPAN), LAKE MENDOTA(WIS), PHYTOPLANKTON CROPS, TROPHIC,
TROPHOLYTIC, NITROGEN ANNUAL BUDGET, PHOSPHORUS ANNUAL BUDGET.

ABSTRACT:
SMALL, EUTROPHIC TAKASUKA POND, JAPAN, HAS BEEN ANALYZED BOTH
VERTICALLY AND SEASONALLY (1928 TO 1931) FOR NITRATE, FREE AMMONIA,
ALBUMINOID NITROGEN AND SOLUBLE PHOSPHATE, AND THE DATA CORRELATED WITH
PHYTOPLANKTON CYCLES. ORGANIC NITROGEN IS INVERSELY PROPORTIONAL TO
INORGANIC NITROGEN. INORGANIC NITROGEN AND SOLUBLE PHOSPHATE ARE
GENERATED AT THE END OF STAGNATION CAUSED BY DECOMPOSITION OF
WARM-WATER PHYTOPLANKTON AND SOME AQUATIC PLANTS. SEASONAL CHANGE IN
SOLUBLE PHOSPHATE IN INFLOWING WATERS AFFECTS THE PHYTOPLANKTON
PRODUCTIVITY. DATA CONCERNING THE HYPOLIMNION IN THE STAGNANT PERIOD
(APRIL TO END OF OCTOBER) ARE PRESENTED. CALCULATIONS ARE OFFERED FOR
ANNUAL BUDGETS FOR NITROGEN AND PHOSPHORUS IN THE TROPHIC ZONE,
TROPHOLYTIC ZONE, AND POND AS A WHOLE. COMPARISONS ARE MADE WITH LAKES
AND PONDS OF THE WORLD, INCLUDING LAKE MENDOTA, WISCONSIN. THE LIMITING
FACTOR FOR PHYTOPLANKTON CROP IN TAKASUKA IS SAID TO BE 'PHOSPHATE AT
TIMES AND AMMONIA IN OTHERS.' (MCCOY-WISCONSIN)

FIELD 02H, 05C

ACCESSION NO. W69-05142

DEMONSTRATION OF A TOXIN FROM APHANIZOMENON FLOS-AQUAE (L.) RALFS,

NEW HAMPSHIRE UNIV., DURHAM; NATIONAL MARINE WATER QUALITY LAB., KINGSTON, R. I.

PHILIP J. SAWYER, JOHN H. GENTILE, AND JOHN J. SASNER, JR.

CANADIAN JOURNAL OF MICROBIOLOGY, VOL 14, NO 11, PP 1199-1204, 1968. 7 P, 1 TAB, 4 FIG, 23 REF. OWRR PROJECT A-013-NH.

DESCRIPTORS:
CYANAPHYTA, ALGAL CONTROL, *ALGAL TOXINS MEMBRANE, NORTHEAST U.S.

IDENTIFIERS:
AQUATIC ALGAE, WATER POLLUTION, *TOXICITY, BIOLOGICAL MEMBRANES, HUMID AREA.

ABSTRACT:
A POTENT TOXIN WAS EXTRACTED FROM A NATURAL POPULATION OF THE BLUE-GREEN ALGA, APHANIZOMENON FLOS-AQUAE. THE TOXIN IS THERMO- AND ACID-STABLE; ALKALINE LABILE; SOLUBLE IN WATER AND ETHANOL; INSOLUBLE IN ACETONE, ETHER, AND CHLOROFORM; AND READILY DIALYZABLE. PRELIMINARY STUDIES SHOW THAT THE ALGA CONTAINS A TOXIN WHICH, WHEN RELEASED FROM LYSED CELLS, OPERATES AT THE MEMBRANE LEVEL, DESTROYING EXCITABILITY WITHOUT ALTERATION OF THE TRANSMEMBRANE RESTING POTENTIAL.

FIELD 05C

ACCESSION NO. W69-05306

NITROGEN CYCLE IN SURFACE AND SUBSURFACE WATERS,

WISCONSIN UNIV., MADISON.

S. WITZEL, E. MCCOY, O. J. ATTOE, L. B. POLKOWSKI, AND K. T. CRABTREE.

WATER RESOURCES CENTER, UNIV., OF WISCONSIN, TECHNICAL COMPLETION REPORT, DECEMBER 1968. 65 P, 15 TAB, 12 FIG, 27 REF. OWRR PROJECT B-004-WIS.

DESCRIPTORS:
*DOMESTIC ANIMALS, *WASTES, FERTILIZERS, *ESSENTIAL NUTRIENTS, *NITRIFICATION, CROPS, TOXICITY, SOIL POROSITY, IRRIGATION, GROUND WATER, RUNOFF, FROZEN GROUND, SOIL EROSION, WATER POLLUTION, *DENITRIFICATION, PHOSPHORUS COMPOUNDS, ALGAE, AQUATIC PLANTS.

IDENTIFIERS:
*NITROGEN CYCLE, *FARM WASTE, GROUND WATER, *NITRATES, EUTROPHICATION, *WATER POLLUTION SOURCES, AGRICULTURAL WATERSHEDS.

ABSTRACT:
AUTOTROPHIC AND HETEROTROPHIC NITRIFICATION HAVE BEEN STUDIED WITH 191 SAMPLES OF SHALLOW WATER FROM STREAMS, FARM PONDS AND ADJACENT SOILS. OF 47 CHOSEN FOR REPEATED TESTS, 45 PRODUCED NO SUB 2-N RANGING 5-154 MICROGRAM/ML, AV. 48 MICROGRAM/ML. ONLY 2 PRODUCED NO SUB 3-N AT 33 AND 46 MICROGRAM/ML. OF 167 STOCK CULTURES OF SOIL FUNGI, THE MAIN PRODUCERS WERE IN THE ASPERGILLUS FLAVUS-ORYZAE AND A. WENTII GROUPS (75% YIELDED 65-100 MICROGRAM/ML OF NO SUB 3-N) AND THE PENICILLIUM GENUS (21 OF 24 SPECIES YIELDED 7-19 MICROGRAM/ML OF NO SUB 3-N FROM NO SUB 2-N, NOT FROM ORGANIC N). NITRIFIERS OF NO SUB 2, NO SUB 3 TYPE WERE FOUND IN 5 OTHER GENERA. RESIDUAL NO SUB 3 FOLLOWING CROP MATURITY MIGRATES DOWNWARD FROM 12 INCHES TO AQUIFER DEPTH AT 20 FEET OVER WINTER ON WAUPUN AND PLAINFIELD SOILS, RESPECTIVELY. ONE WISCONSIN COMMUNITY HAD 86 WELLS WITH 34.5% UNSAFE CONTAINING HIGH NO SUB 3 AND ANOTHER HAD 550 WELLS WITH 1/3 UNSAFE. SURFACE WATERS RECEIVED LESS THAN 5.7 LB. N AND 2.53 LB. P PER A. IN FLOOD FLOWS FROM A 1346 A. WATERSHED; 3.62 LB. N AND 1.14 LB. P PER A. FROM 3 FARM AREAS TOTALING 246 A. LANCASTER PLOTS RECEIVING 15 TONS DAIRY COW WASTES PER A. LOST 19.8% MORE N AND 11.3% MORE P WHEN APPLIED ON FROZEN GROUND IN A YEAR OF HIGH WINTER RUNOFF.

FIELD 05B, 05G

ACCESSION NO. W69-05323

THE EFFECTS OF ALGAE ON WATER QUALITY,

SYRACUSE UNIV., N. Y. DEPT. OF CIVIL ENGINEERING.

DANIEL F. JACKSON.

WATER QUALITY RES SYMP, LST PROC, NEW YORK STATE DEPT HEALTH, FEB 20, 1964.
23 P, 5 FIG, 2 TAB, 53 REF.

DESCRIPTORS:
*ALGAE, *WATER QUALITY, *EUTROPHICATION, COLOR, TASTE, ODOR, TOXINS,
DISSOLVED OXYGEN, LAKES, RIVERS, LIGHT PENETRATION, FLUORESCENCE,
PHOTOSYNTHESIS, BLUE-GREEN ALGAE, PHYTOPLANKTON, OHIO RIVER,
PENNSYLVANIA, WATER SUPPLY.

IDENTIFIERS:
PHOTOSYNTHETIC RATE, PHYTOPLANKTON DYNAMICS, EXTRACELLULAR PRODUCTS,
ALGAL CELLS.

ABSTRACT:
HISTORICAL OCCURRENCES OF COLORED WATER CAUSED BY ALGAL GROWTH ARE
CITED. INVESTIGATIONS (1797) REVEALED THAT VOLVOCALES CAUSED RED WATER,
WITH FIRST RECORDED STUDY ON PHYTOPLANKTONIC DEVELOPMENT (OSCILLATORIA)
CONDUCTED BY DECANDOLLE (1825). IN PYMATUNING RESERVOIR, PENNSYLVANIA,
CYANOPHYTES BLOOM ANNUALLY, RESTRICTING LIGHT PENETRATION TO LESS THAN
46 CM DURING JULY AND AUGUST, RETARDING GROWTH OF ROOTED AQUATIC
PLANTS, AND EXPOSING BOTTOM SOIL TO WAVE-ACTION EROSION AND CONSEQUENT
SUSPENSION OF NUTRIENT-RICH MATERIALS. TABULAR DATA ON ODORS AND TASTES
ARE DISCUSSED. AUTHOR DETERMINED PERCENTAGE OF VIABLE CELLS TO VALIDATE
COMMON ASSUMPTION THAT DEGREE OF ODOR IS ASSOCIATED WITH QUANTITY OF
ALGAL CELLS. AVERAGE VIABLE FRACTION FOR 1,700 MEASUREMENTS ON OHIO
RIVER SAMPLES (1960-61) WAS 59%; MEASUREMENTS WERE BASED ON CHLOROPHYLL
FLUORESCENCE. DEAD ALGAL CELLS, WHICH CREATE ODORS AND TASTE, SHOULD BE
DETERMINED IN PLANKTON ALGAE STUDIES. DISCUSSION OF ALGAL CONTRIBUTION
TO DISSOLVED OXYGEN INDICATES THAT EACH SPECIES HAS ITS OWN
MINIMUM/MAXIMUM PHOTOSYNTHETIC RATE INFLUENCED BY ENVIRONMENT, AND
TOTAL OXYGEN PRODUCTION FOLLOWS A PHYLOGENETIC HIERARCHY. LITERATURE
REVIEW AND EXPERIMENTS SHOW THAT ALGAE CONTRIBUTE SIGNIFICANT AMOUNTS
OF DISSOLVED OXYGEN. AUTHOR'S REVIEW OF ALGAL EXCRETION AND SECRETION
INCLUDES DISCUSSION OF TOXINS. HE DISCUSSES THEORETICAL EQUATION TO
EVALUATE PHYTOPLANKTON DYNAMICS REPORTED BY RILEY AND VON ARX (1949).
(KERRIGAN-WIS)

FIELD 05C

ACCESSION NO. W69-05697

USE OF POTASSIUM PERMANGANATE FOR CONTROL OF PROBLEM ALGAE,

WISCONSIN UNIV., MADISON. HYDRAULIC AND SANITARY LAB.

GEORGE P. FITZGERALD.

J AMER WATER WORKS ASSOC, VOL 58, NO 5, PP 609-614, 1966. 1 FIG, 1 TAB, 20 REF.

DESCRIPTORS:
*NUISANCE ALGAE, *ALGICIDES, *COPPER SULFATE, *WATER QUALITY CONTROL, ALGAL CONTROL, SLIME, TASTE-PRODUCING ALGAE, ODOR-PRODUCING ALGAE, DIATOMS, CYANOPHYTA, CHLOROPHYTA, IRON, MANGANESE, EUTROPHICATION, WATER POLLUTION CONTROL.

IDENTIFIERS:
*POTASSIUM PERMANGANATE, ALGICIDAL TESTS, ALGISTATIC TESTS, SLIMEPRODUCING ALGAE, PHOTOMICROGRAPHS, MICROCYSTIS AERUGINOSA, ANABAENA CIRCINALIS, GLOEOTRICHIA ECHINULATE, OSCILLATORIA RUBESCENS, OSCILLATORIA CHALYBIA, HYDRODICTYON RETICULATUM, DICTYOSPHAERIUM PULCHELLUM, GOMPHONEMA, DINOBRYON.

ABSTRACT:
AUTHOR TREATED FOLLOWING SPECIES OF ALGAE WITH VARIOUS CONCENTRATIONS OF POTASSIUM PERMANGANATE AND COPPER SULFATE TO DETERMINE CONCENTRATIONS REQUIRED TO KILL CELLS AFTER 12 AND 72 HOURS' TREATMENT AND TO PREVENT GROWTH: (BLOOMFORMING CYANOPHYTES) MICROCYSTIS AERUGINOSA, ANABAENA CIRCINALIS, GLOEOTRICHIA ECHINULATA, OSCILLATORIA RUBESCENS; (FILTER-CLOGGING CYANOPHYTE) OSCILLATORIA CHALYBIA; (FILTER-CLOGGING CHLOROPHYTES) HYDRODICTYON RETICULATUM, DICTYOSPHAERIUM PULCHELLUM; (DIATOM) PROBABLY GOMPHONEMA. TOXICITY OF COPPER SULFATE WAS HIGHLY VARIABLE; CYANOPHYTES EXCEPT OSCILLATORIA WERE KILLED AT CONCENTRATIONS OF 0.1 PPM; O RUBESCENS WAS KILLED AFTER 4 HOURS AT 0.4 PPM; GROWTH OF O CHALYBIA AND HYDRODICTYON WAS INHIBITED. BY 0.1-0.2 PPM BUT ALGICIDE REQUIRED 0.6-0.8 PPM FOR 72 HOURS; INHIBITION OF DICTYOSPHAERIUM AND THE DIATOM REQUIRED 0.4 PPM, WHILE ALGICIDE REQUIRED MORE THAN 8 PPM AND 72 HOURS. WITH POTASSIUM PERMANGANATE, ALL SPECIES EXCEPT DICTYOSPHAERIUM WERE KILLED BY 1-5 PPM. AUTHOR SPECULATES THAT VARIATIONS IN EFFECTIVENESS OF COPPER MAY RESULT FROM VARIATIONS IN METABOLIC RATES, METABOLIC PROCESSES, OR RATES OF ABSORPTION AMONG THE SEVERAL SPECIES. HE SUGGESTS THAT POTASSIUM PERMANGANATE TREATMENT OF RAW-WATER RESERVOIRS MAY BE MULTIPLY EFFECTIVE FOR ALGICIDE FOR ELIMINATION OF TASTES AND ODORS RESULTING FROM DECOMPOSED ALGAE AND OTHER SOURCES AND FOR REMOVAL OF IRON AND MANGANESE. (EICHHORN-WIS)

FIELD 05F

ACCESSION NO. W69-05704

LIST OF REFERENCES ON CONTROL OF AQUATIC PLANTS, INCLUDING ALGAE,

CHIPMAN CHEMICAL CO., INC., BOUND BROOK, N. J. RESEARCH DIV.

MARGARET GREENWALD.

(INTERNAL PUBLICATION), CHIPMAN CHEMICAL CO, INC, BOUND BROOK, NJ, JAN 1956. 22 P, 716 REF.

DESCRIPTORS:
*BIBLIOGRAPHIES, *ALGAL CONTROL, *AQUATIC WEED CONTROL, *EUTROPHICATION, AQUATIC PLANTS, HERBICIDES, PESTICIDES, HARVESTING, WATER QUALITY CONTROL, WATER POLLUTION CONTROL, WATER POLLUTION EFFECTS.

IDENTIFIERS:
BIOCIDES.

ABSTRACT:
ALTHOUGH THIS BIBLIOGRAPHY COMPRISES SOME 716 ENTRIES, IN ITS PREFACE THE COMPILER STATES THAT IT IS NEITHER COMPLETE NOR SELECTED. HOWEVER, EMPHASIS WAS PLACED ON RECENT WORKS. 263 OF THE DATED WORKS WERE PUBLISHED BETWEEN 1950 AND 1966. EARLIEST CITATION IS DATED 1896, AND THE LIST IS ALMOST EXCLUSIVELY FROM THE LITERATURE IN ENGLISH. IN MOST CASES, BIOLOGICAL AND ECOLOGICAL STUDIES WERE OMITTED, BUT SOME HANDBOOKS, GENERAL TEXTS, AND A FEW MISCELLANEOUS REFERENCES ARE INCLUDED WHERE THEY HAVE BEEN USEFUL IN PLANT IDENTIFICATION OR THE UNDERSTANDING OF CONDITIONS ASSOCIATED WITH PLANT CONTROL OR GIVE INFORMATION ON PROBLEMS RELATED TO PLANT CONTROL (E.G., TOXICITY OF HERBICIDES TO ANIMAL ORGANISMS). (EICHHORN-WIS)

FIELD 05G, 04A

ACCESSION NO. W69-05705

SUPPLEMENT TO THE LIST OF REFERENCES ON CONTROL OF AQUATIC PLANTS, INCLUDING ALGAE,

CHIPMAN CHEMICAL CO., INC., BOUND BROOK, N. J. RESEARCH DIV.

MARGARET GREENWALD.

(INTERNAL PUBLICATION), CHIPMAN CHEMICAL CO, INC, BOUND BROOK, N. J., JAN 1957. 12 P, 258 REF.

DESCRIPTORS:
*BIBLIOGRAPHIES, *ALGAL CONTROL, *AQUATIC WEED CONTROL, *EUTROPHICATION, AQUATIC PLANTS, HERBICIDES, PESTICIDES, HARVESTING, WATER QUALITY CONTROL, WATER POLLUTION CONTROL, WATER POLLUTION EFFECTS, FUNGI, AQUATIC FUNGI.

IDENTIFIERS:
BIOCIDES.

ABSTRACT:
THIS SUPPLEMENT RECORDS CORRECTIONS TO COMPILER'S ORIGINAL BIBLIOGRAPHY, AND ADDS 258 ENTRIES TO THE ORIGINAL 716. ORIGINAL COVERAGE OF AQUATIC PLANTS AND ALGAE IS EXTENDED TO INCLUDE A FEW REFERENCES TO AQUATIC FUNGI. NEW CITATIONS ARE FOR LITERATURE PUBLISHED BETWEEN 1922 AND 1957, 134 HAVING BEEN PUBLISHED DURING THE PERIOD 1950-1957. (EICHHORN-WIS)

FIELD 05G, 04A

ACCESSION NO. W69-05706

NEARSHORE PHYTOPLANKTON POPULATIONS IN THE GRAND HAVEN, MICHIGAN VICINITY DURING THERMAL BAR CONDITIONS,

MICHIGAN UNIV., ANN ARBOR. GREAT LAKES RESEARCH DIV.

E. F. STOERMER.

PROC. 11TH CONF. GREAT LAKES RES., VOL 11, PP 137-150, APR 1968. 14 P, 13 FIG, 1 TAB, 17 REF.

DESCRIPTORS:
*PHYTOPLANKTON, *LAKE MICHIGAN, *WATER CIRCULATION, NUISANCE ALGAE, THERMAL POLLUTION, DIFFUSION, DIATOMS, CHLOROPHYTA, CHRYSOPHYTA, DINOFLAGELLATES, CYANOPHYTA.

IDENTIFIERS:
*THERMAL BAR.

ABSTRACT:
IN LATE APRIL 1967, THE THERMAL BAR PHENOMENON HAD PROFOUND EFFECTS ON BOTH THE ABUNDANCE AND SPECIES COMPOSITION OF PHYTOPLANKTON IN NEARSHORE LAKE MICHIGAN WATERS IN THE VICINITY OF GRAND HAVEN, MICHIGAN. LANDWARD OF THE THERMAL BAR POPULATIONS AVERAGED BETWEEN 1500 AND 2000 CELLS/ML COMPARED WITH AVERAGE VALUES OF ABOUT 400 CELLS/ML IN WATER ENTERING THE LAKE FROM THE GRAND RIVER AND ABOUT 350 CELLS/ML IN WATER OUTSIDE THE THERMAL BAR. THE HIGHEST POPULATIONS OBSERVED, OCCURRED IN THE INTERFACE BETWEEN INSHORE AND OFFSHORE WATERS. THE RELATIVE ABUNDANCE OF MOST SPECIES OF ALGAE WAS STRIKINGLY DISSIMILAR IN THE DIFFERENT WATER MASSES. FOUR GENERAL TYPES OF SPECIES DISTRIBUTION WERE NOTED: (1) SPECIES WHICH WERE ABUNDANT IN THE HARBOR FLORA, RARE IN THE INSHORE WATERS, AND VERY RARE OR LACKING IN THE OFFSHORE WATERS; (2) SPECIES WHICH WERE ABUNDANT IN THE INSHORE WATERS, RARE IN THE HARBOR FLORA, AND RARE OR LACKING IN THE OFFSHORE WATERS; (3) SPECIES WHICH WERE ABUNDANT IN THE OFFSHORE WATERS, RARE IN THE INSHORE AREA AND LACKING IN THE HARBOR FLORA; (4) SPECIES WHICH WERE MOST ABUNDANT IN THE INTERFACE WATER WITH SECONDARY ABUNDANCE PEAKS IN EITHER OF THE OTHER HABITATS. IT APPEARS THAT THE PRIMARY EFFECT OF THERMAL ADDITIONS TO LAKE MICHIGAN IS THROUGH CONSTRAINT ON CIRCULATION WHICH TRAPS NUTRIENTS NEAR SHORE.

FIELD 02H, 05C

ACCESSION NO. W69-05763

DEATH OF A LAKE,

QUEST, VOL 6, NO 4, PP 6-15, DEC 1968. 10 P, 9 FIG.

DESCRIPTORS:
*LAKES, *EUTROPHICATION, FERTILIZATION, FISH, *LIMNOLOGY, FISHKILL,
NUTRIENTS, *ALGAE, *ALGAL CONTROL, ALGAL POISONING, ALGICIDES, AQUATIC
ALGAE, AQUATIC ENVIRONMENT, AQUATIC PLANTS, AQUATIC POPULATIONS,
CONSERVATION, WATER QUALITY, WATER POLLUTION, WATER RESOURCES.

IDENTIFIERS:
ALGAL NUTRIENTS.

THE NATURAL PROCESS OF LAKE EUTROPHICATION IS BEING ACCELERATED BY THE
ACTIVITIES OF MAN. SEWAGE DISCHARGES, WASTE EFFLUENTS, AND AGRICULTURAL
FERTILIZERS DRAINING INTO A LAKE PRODUCE AN OVERABUNDANCE OF NUTRIENTS
IN THE LAKE. THE ABUNDANCE OF NUTRIENTS GREATLY STIMULATES THE GROWTH
OF ALGAE AND OTHER WATER PLANTS, BUT SETS OFF A DESTRUCTIVE CHAIN
REACTION. AT FIRST, THE ABUNDANT FOOD CHAIN WILL BENEFIT THE FISH, BUT
IN SEVERE CASES, THE ALGAE MAY GROW IN SUCH NUMBERS THAT WHEN THEY DIE,
THEY DEPLETE THE OXYGEN SUPPLY AND RELEASE TOXIC SUBSTANCES. THE RESULT
MAY BE A LOSS OF A SPARKLING FRESH LAKE, SLIME INFESTATION, BAD FLAVOR
AND ODOR, DISCOLORATION OF WATER, STAGNATION, AND MASSIVE FISHKILL.
CLEAN-WATER AND POLLUTED-WATER ALGAE ARE DESCRIBED AND METHODS FOR
ALGAE CONTROL ARE COVERED. LIMNOLOGY STUDIES AT WILLIAMS LAKE AND ROCK
LAKE, WASH, ARE DISCUSSED. MANY BODIES OF WATER APPEAR DESTINED TO DIE
LONG BEFORE THEIR NATURAL TIME, PRODUCING AN ECONOMIC LOSS AMOUNTING TO
MILLIONS OF DOLLARS. THE LOSS IN AESTHETIC VALUES OF LAKES CANNOT BE
MEASURED. (USBR)

FIELD 05C, 02H

ACCESSION NO. W69-05844

AVAILABILITY OF IRON AND MANGANESE IN SOUTHERN WISCONSIN LAKES FOR THE GROWTH
OF MICROCYSTIS AERUGINOSA,

WISCONSIN UNIV., MADISON. DEPT. OF BOTANY.

GERALD C. GERLOFF, AND FOLKE SKOOG.

ECOLOGY, VOL 38, NO 4, PP 551-556, OCT 1957. 7 TAB, 10 REF.

DESCRIPTORS:
*EUTROPHICATION, *IRON, *LAKES, *MANGANESE, *WISCONSIN, ALGAE, CALCIUM,
CHELATION, CYANOPHYTA, NUISANCE ALGAE, NUTRIENT REQUIREMENTS,
NUTRIENTS, TOXICITY, WATER CHEMISTRY, WATER POLLUTION EFFECTS.

IDENTIFIERS:
*NUTRIENT AVAILABILITY, *ALGAL GROWTH, *MICROCYSTIS AERUGINOSA,
ANTAGONISMS, ALGAL NUTRITION, ALGAL PHYSIOLOGY, LAKE MENDOTA(WIS), LAKE
MONONA(WIS), LAKE WAUBESA(WIS), SPAULDING POND(WIS), LAKE DELAVAN(WIS),
PEWAUKEE LAKE(WIS), NORTH LAKE(WIS), LAKE NAGAWICKA(WIS).

CELLULAR CONTENTS WERE COMPARED OF IRON AND MANGANESE IN MICROCYSTIS
AERUGINOSA, COLLECTED FROM BLOOMS IN SEVEN SOUTHERN WISCONSIN LAKES,
WITH CRITICAL LEVELS OF THESE ELEMENTS FOR MAXIMUM GROWTH AS DETERMINED
IN THE LABORATORY. CRITICAL LEVELS WERE APPROXIMATELY 100 PPM IRON AND
4 PPM MANGANESE. RANGE OF ALGAL IRON CONTENT FOR LAKES SAMPLED WAS
164-543 PPM; MANGANESE, 13-232 PPM. FOR 36 SAMPLES FROM EUTROPHIC LAKE
MENDOTA COLLECTED AT SURFACE AND AT 10 AND 0.5 METERS ABOVE BOTTOM AT
12 DATES DURING ALL SEASONS FOR ONE YEAR, IRON CONCENTRATIONS IN WATER
HAD RANGE 0.000-0.083 PPM; MANGANESE, 0.035-0.673 PPM. CELLULAR
CONTENTS OF IRON AND MANGANESE WERE SUFFICIENTLY IN EXCESS OF SUCH
CRITICAL VALUES THAT IT SEEMS UNLIKELY THAT EITHER IS IMPORTANT IN
DEVELOPMENT OF MICROCYSTIS BLOOMS IN LAKES. ANTAGONISTIC RELATIONSHIPS
BETWEEN MANGANESE AND IRON AND BETWEEN CALCIUM AND MANGANESE WERE
DEMONSTRATED AND CONTRIBUTED TO DIFFICULTIES IN DETERMINING LEVELS OF
IRON AND MANGANESE REQUIRED FOR NUTRITION. SUCH ANTAGONISMS WERE
CONSIDERED AS INSUFFICIENTLY SERIOUS TO SIGNIFICANTLY AFFECT ACCURACY
OF THEIR STUDIES AND CONCLUSIONS DERIVED THEREFROM. MANGANESE APPEARS
TO ATTAIN CONCENTRATIONS IN LAKE WATER SUFFICIENTLY HIGH TO BE TOXIC TO
ALGAE WERE IT NOT FOR ANTAGONISM OF CALCIUM, ALSO PRESENT IN RELATIVELY
HIGH CONCENTRATIONS. (EICHHORN-WISC)

FIELD 05C, 02H

ACCESSION NO. W69-05867

CELL CONTENTS OF NITROGEN AND PHOSPHORUS AS A MEASURE OF THEIR AVAILABILITY FOR
  GROWTH OF MICROCYSTIS AERUGINOSA,

WISCONSIN UNIV., MADISON. DEPT OF BOTANY.

GERALD C. GERLOFF, AND FOLKE SKOOG.

ECOLOGY, VOL 35, NO 3, PP 348-353, JULY 1954. 5 FIG, 7 TAB, 8 REF.

DESCRIPTORS:
  *CHEMICAL ANALYSIS, *NITROGEN, *PHOSPHORUS, ALGAE, CYANOPHYTA,
  NITRATES, NITROGEN COMPOUNDS, NUISANCE ALGAE, NUTRIENT REQUIREMENTS,
  NUTRIENTS, PHOSPHORUS COMPOUNDS, WATER CHEMISTRY, WATER POLLUTION
  EFFECTS, WATER QUALITY, WISCONSIN, LAKES, EUTROPHICATION.

IDENTIFIERS:
  *CHEMICAL COMPOSITION, *MICROCYSTIS AERUGINOSA, *NUTRIENT AVAILABILITY,
  ALGAL GROWTH, ALGAL NUTRITION, ALGAL PHYSIOLOGY, KJELDAHL PROCEDURE,
  LUXURY CONSUMPTION, NUTRIENT REMOVAL, SPAULDING'S POND(WIS), LAKE
  WAUBESA(WIS), GREEN LAKE(WIS).

ABSTRACT:
  QUANTITATIVE RELATIONSHIPS WERE STUDIED BETWEEN TOTAL NITROGEN AND
  PHOSPHORUS CONTENTS OF MICROCYSTIS AERUGINOSA GROWN IN NUTRIENT
  SOLUTIONS CONTAINING THESE ELEMENTS IN QUANTITIES VARYING FROM HIGHLY
  DEFICIENT TO EXCESSIVE. CELLULAR CONTENT OF NITROGEN OR PHOSPHORUS
  INCREASES WITH EXTERNAL SUPPLY OVER A WIDE RANGE, FOR A MAJOR PORTION
  OF WHICH THE AMOUNT OF GROWTH REMAINED PRACTICALLY CONSTANT. THUS,
  ABOVE FAIRLY CONSTANT CRITICAL LEVELS OF NITROGEN AND PHOSPHORUS
  REQUIRED FOR MAXIMUM YIELD, LARGE CELLULAR EXCESSES ACCUMULATED,
  REPRESENTING LUXURY CONSUMPTION. ALTHOUGH NOT INFLUENCING GROWTH, THESE
  EXCESSES REFLECT ABUNDANCE OF EXTERNAL SUPPLIES. NITROGEN AND
  PHOSPHORUS CONTENTS OF MICROCYSTIS COLLECTED FROM BLOOMS IN THREE
  WISCONSIN LAKES, WHEN COMPARED WITH CULTURES GROWN IN LABORATORY,
  SHOWED THAT CELLULAR CONTENTS OF THESE ELEMENTS IN COLLECTED SAMPLES
  WERE IN RANGE OF LUXURY CONSUMPTION AND SUFFICIENTLY ABOVE CRITICAL
  LEVELS TO SUGGEST THAT NEITHER PHOSPHORUS NOR NITROGEN LIMITED ALGAL
  GROWTH IN THESE LAKES. NITROGEN OCCURRED IN RELATIVELY SMALLER EXCESS
  THAN PHOSPHORUS. CELLULAR EXCESSES WERE SUFFICIENT TO PERMIT
  CONSIDERABLE GROWTH WHEN ALGAE WERE TRANSFERRED TO MEDIA DEFICIENT IN
  THESE NUTRIENTS. EFFICACY OF PHOSPHORUS REMOVAL FROM CULTURAL EFFLUENTS
  MAY BE DOUBTFUL UNLESS LEVELS ARE ALSO MUCH REDUCED, BOTH IN OTHER
  INFLOWING WATER AND IN 'SINKS' WITHIN LAKES. (EICHHORN-WISC)

FIELD 05C, 02H

ACCESSION NO. W69-05868

USE POLLUTION TO BENEFIT MANKIND,

VITRO/HANFORD ENGINEERING SERVICES, RICHLAND, WASH.

J. LEON POTTER.

OCEAN IND, VOL 4, NO 5, PP 94-97, MAY 1969. 4 P, 1 FIG, 2 TAB, 4 REF.

DESCRIPTORS:
  *WASTE TREATMENT, *AQUICULTURE, ALGAE, YEASTS, BACTERIA, FISH,
  PROTEINS, OIL WASTES, OILY WATER, SULFITE LIQUORS, ULTIMATE DISPOSAL,
  WATER REUSE, FISH FARMING.

IDENTIFIERS:
  *OIL SPILLS, PROTEIN AQUICULTURE.

ABSTRACT:
  ORGANIC WASTE, HEATED WATER, SEWAGE NUTRIENTS, OIL WASTES, AND GARBAGE
  MAY BE USED AS NUTRIENTS FOR GROWING ALGAE, BACTERIA, AND YEASTS TO
  PRODUCE PROTEINS. WASTE LOW-GRADE PETROLEUM IS USED IN A PLANT AT
  LAUERA, FRANCE. A POUND OF BACTERIA, FEEDING ON CRUDE SO WORTHLESS IT
  IS USUALLY BURNED, CAN PRODUCE 10 LBS OF PROTEIN A DAY. WASTE EFFLUENTS
  RICH IN CARBOHYDRATES, SUGARS AND INORGANIC NUTRIENTS MAY BE COLLECTED
  AND SEPARATED INTO SOLIDS AND LIQUIDS. THE SOLIDS ARE GOOD FERTILIZERS
  AND ANIMAL FEED ADDITIVES, WHILE THE SOLUBLE NUTRIENTS MAY BE USED BY
  TORULA YEAST FOR PROTEIN PRODUCTION. HEATED WATER DISCHARGES MAY BE
  USED TO INCREASE THE PRODUCTION OF FISH. THE COST OF PROTEIN PRODUCED
  FROM WASTES IS ABOUT 3 CENTS PER LB, COMPARED WITH FARM-GROWN PROTEIN
  PRICES OF ABOUT 10 CENTS. ANALYSES OF TORULA YEAST AND ALGAE ARE
  TABULATED. (KNAPP-USGS)

FIELD 05D, 05E

ACCESSION NO. W69-05891

RIVER WATER QUALITY FOR ARTIFICIAL RECHARGE,

NEBRASKA UNIV., LINCOLN. DEPT. OF CIVIL ENGINEERING.

LEON S. DIRECTO, AND MERLIN E. LINDAHL.

J AMER WATER WORKS ASS, VOL 61, NO 4, PP 175-180, APR 1969. 6 P, 8 FIG, 1 TAB, 8 REF. RES SUPPORTED BY NEBR UNIV. WATER RESOURCES CENTER.

DESCRIPTORS:
*WATER QUALITY, *ARTIFICIAL RECHARGE, *DATA COLLECTIONS, PIT RECHARGE, WATER SUPPLY, MUNICIPAL WATER, SEDIMENT LOAD, CONDUCTIVITY, CALCIUM, MAGNESIUM, ALKALINITY, SULFATES, CHLORIDES, SODIUM, IRON, DISSOLVED SOILDS, PHOSPHATES, NITROGEN, ALGAE, COLIFORMS.

IDENTIFIERS:
PLATTE RIVER, LINCOLN(NEBR).

ABSTRACT:
THE PLATTE RIVER WAS SAMPLED IN 1966-67 AT 3 LOCATIONS NEAR ASHLAND, NEBRASKA TO INVESTIGATE SEASONAL VARIATIONS IN ITS PHYSICAL, CHEMICAL, AND BIOLOGICAL QUALITIES IN A STUDY OF THE FEASIBILITY OF USING THE RIVER WATER FOR RECHARGE OF THE AQUIFER SUPPLYING THE WELLS FOR THE CITY OF LINCOLN. ALL THE SAMPLES WERE TAKEN USING THE EQUAL TRANSIT RATE METHOD OF DEPTH INTEGRATION. STREAMFLOW DATA WERE OBTAINED FROM USGS GAGING STATION RECORDS. THE MEAN SPECIFIC CONDUCTANCE OF 439 MICROMHOS/CM WAS EXCEEDED 60% OF THE TIME. THE MEAN SOLIDS CONCENTRATION WAS 1090 MG/1. THIS IS EXPECTED TO CAUSE CLOGGING PROBLEMS IN USE OF THE WATER FOR RECHARGE UNLESS SOLIDS ARE REMOVED. THE SODIUM ABSORPTION RATIO AVERAGED 1.05 MG/1 AND IRON CONCENTRATION 0.41 MG/1, NEITHER OF WHICH SHOULD CAUSE ANY PROBLEM. PHOSPHATE CONCENTRATION AVERAGED 1.07 MG/1 WITH AN ALGAE COUNT OF 910, WHICH COULD RESULT IN HEAVY ALGAL GROWTH ON SAND RECHARGE BEDS. COLIFORM DENSITY WAS 6,610/100 ML AT ASHLAND. ALL THE COLLECTED DATA, INCLUDING TEMPERATURE, PH, ALKALINITY, AND HARDNESS ARE SHOWN GRAPHICALLY. RIVER WATER IS SIMILAR IN CHEMICAL QUALITY TO THE GROUNDWATER IN THE WELL FIELD, AND SHOULD CAUSE NO COMPATIBILITY PROBLEMS. THE HIGH SOLIDS, ALGAE, AND COLIFORM CONCENTRATIONS MAY CAUSE CLOGGING. (KNAPP-USGS)

FIELD 05B, 04B

ACCESSION NO. W69-05894

LIMNOLOGICAL INVESTIGATIONS OF TEXAS IMPOUNDMENTS FOR WATER QUALITY MANAGEMENT PURPOSES - THE USE OF ALGAL CULTURES TO ASSESS THE EFFECTS OF NUTRIENT ENRICHMENT ON THE HIGHLAND LAKES OF THE COLORADO RIVER, TEXAS,

TEXAS UNIV., AUSTIN.

BERT A. FLOYD, E. GUS FRUH, AND ERNST M. DAVIS.

TEXAS CENTER FOR RESEARCH IN WATER RESOURCES, TECHNICAL PROJECT REPORT NO 2, JANUARY, 1969. 45 P, 3 TAB, 4 FIG, 23 REF. OWRR PROJECT B-020-TEX.

DESCRIPTORS:
HIGHLAND LAKES, COLORADO RIVER, TEXAS, *NUTRIENT ENRICHMENT TESTS, *NITROGEN, *PHOSPHORUS, *IRON, *LIMNOLOGY, *ALGAE.

ABSTRACT:
THIS STUDY WAS AN INVESTIGATION OF THE EFFECTS OF INORGANIC NITROGEN, PHOSPHORUS, AND IRON ON ALGAL GROWTH IN THE HIGHLAND LAKES OF THE COLORADO RIVER, TEXAS. RIVER WATER FROM THE UPPER AND LOWER REACHES OF THE RESERVOIR SYSTEM WAS USED AS MEDIA FOR UNIALGAL CULTURES IN LABORATORY CONTROLLED GROWTH EXPERIMENTS. DIFFERENCES IN GROWTH WERE MEASURED BETWEEN A CONTROL AND SAMPLES ENRICHED WITH THE THREE NUTRITIVE SUBSTANCES. THESE MEASUREMENTS SHOWED INCREASED ALGAL GROWTH WITH THE ADDITION OF NITROGEN AND PHOSPHORUS, BUT NO INCREASE WITH EITHER OF THE THREE ELEMENTS SEPARATELY. ALSO, DIFFERENCES IN GROWTH AND NUTRIENT REQUIREMENTS WERE FOUND BETWEEN GREEN AND BLUE-GREEN SPECIES. FINALLY, AN INCREASE IN THE FERTILITY OF THE WATER AS IT FLOWS THROUGH THE RESERVOIR SYSTEM WAS NOTED.

FIELD 05C

ACCESSION NO. W69-06004

MEASURING ESTUARINE POLLUTION,

FEDERAL WATER POLLUTION CONTROL ADMINISTRATION, WASHINGTON, D. C. ESTUARINE
AND OCEANOGRAPHIC PROGRAMS BRANCH.

T. A. WASTLER.

OCEANOLOGY INT, PP 43-45, MAY-JUNE 1969. 3 P, 3 PHOTO.

DESCRIPTORS:
*ESTUARIES, *ESTUARINE ENVIRONMENT, *INSTRUMENTATION, *POLLUTION
IDENTIFICATION, *POLLUTION ABATEMENT, TIDAL WATERS, MIXING, SEA WATER,
FRESH WATER, SHORES, WATER CIRCULATION, FLOODS, DENSITY DURRENTS,
MEASUREMENTS, SEWAGE EFFLUENTS, TEMPERATURE, OXYGEN, SALINITY, ALGAE,
MECHANICAL EQUIPMENT.

IDENTIFIERS:
*ESTUARINE POLLUTION.

ABSTRACT:
THIS ARTICLE DESCRIBES BRIEFLY THE BASIC ELEMENTS OF ESTUARY POLLUTIONS
AND THE COMMON TECHNIQUES USED FOR THE DETECTION AND MAPPING OF THESE
POLLUTANTS. IN GENERAL THE STUDY SHOWS THAT ACTUAL TECHNIQUES OF DATA
COLLECTION IN THE ESTUARINE ZONE MUST BE DESIGNED TO COPE WITH THE
UNIQUE DEMANDS OF THE ESTUARINE ENVIRONMENT. THE INCREASING
SOPHISTICATION OF INSTRUMENTATION AND DATA ANALYSIS PROCEDURES,
COMBINED WITH THE NATIONAL CONCERN ABOUT ESTUARINE POLLUTION, OFFERS
HOPE, ACCORDING TO THE AUTHOR, THAT A VIABLE NATIONAL BASIC DATA
COLLECTION SYSTEM MAY BE DEVELOPED SOON TO SURVEY THE COASTAL ZONES.
(GABRIEL-USGS)

FIELD 05A

ACCESSION NO. W69-06203

RAIN AS A SOURCE OF VITAMIN (B-12),

WASHINGTON UNIV., ST LOUIS, MO. DEPT. OF BOTANY.

BRUCE C. PARKER.

NATURE, VOL 219, NO 5154, PP 617-618, AUG 10, 1968. 1 TAB, 15 REF.

DESCRIPTORS:
*RAIN, *VITAMINS, ALGAE, BIOASSAY, BIOINDICATORS, CHLAMYDOMONAS,
CARBON, ECOSYSTEMS, EUGLENA, EUTROPHICATION, ORGANIC MATTER,
PHYTOPLANKTON, PRIMARY PRODUCTIVITY, SOIL CHEMISTRY, WATER CHEMISTRY,
WATER POLLUTION EFFECTS, WATER POLLUTION SOURCES.

IDENTIFIERS:
*VITAMIN B-12, ALGAL GROWTH, CYANOCOBALAMIN, EUGLENA GRACILIS Z,
MICROBIOLOGICAL ASSAY, SPECIES COMPOSITION.

ABSTRACT:
FORMS AND CONCENTRATIONS OF VITAMIN B-12 IN NATURAL AQUATIC ECOSYSTEMS
ARE IMPORTANT IN THAT THEY MAY LIMIT PRIMARY PRODUCTIVITY AND INFLUENCE
THE SPECIES COMPOSITION OF PHYTOPLANKTONIC POPULATIONS. AUTHOR REPORTS
STUDIES STEMMING FROM THE OBSERVATION THAT AFTER SPRING RAINS,
PHYTOPLANKTONIC CHLAMYDOMONAS BECAME DOMINANT ORGANISMS IN SMALL
EXPERIMENTAL PONDS CONCOMMITANTLY WITH INCREASES IN THEIR VITAMIN
CONTENT. RAIN, COLLECTED IN STAINLESS STEEL CONTAINERS, WAS FILTERED
THROUGH MILLIPORE FILTERS (PORES, 2 MICRONS), AND ASSAYED FOR VITAMIN
B-12 MICROBIOLOGICALLY UTILIZING EUGLENA GRACILIS, Z STRAIN. IN NINE
COLLECTIONS TAKEN DURING 13 MONTHS, CONCENTRATIONS RANGED FROM
NEGLIGIBLE TO 20.0 PICOGRAMS/MILLILITER. FRAGMENTARY EVIDENCE SUPPORTS
THE HYPOTHESIS THAT VITAMIN ORIGINATES FROM SOIL PARTICLES BORNE INTO
ATMOSPHERE. AUTHOR POINTS OUT THAT A HYPOTHETICAL LAKE CONTAINING
EUGLENA, AND RICH IN ALL NUTRIENTS EXCEPT B-12, WOULD SUPPORT GROWTH OF
A MILLION CELLS/SQUARE CENTIMETER OF LAKE SURFACE AFTER A RAIN OF 1
CENTIMETER CONTAINING 20 PICOGRAMS OF VITAMIN/MILLILITER. (EICHHORN-WIS)

FIELD 05B

ACCESSION NO. W69-06273

HARVESTING UNDERWATER WEEDS,

WISCONSIN UNIV., MADISON. DEPT. OF MECHANICAL ENGINEERING.

DONALD F. LIVERMORE.

WATER WORKS ENG, PP 118-120, 151, FEB 1954. 4 FIG, 4 REF.

DESCRIPTORS:
*AQUATIC WEED CONTROL, *EUTROPHICATION, *HARVESTING, AQUATIC PLANTS,
AQUATIC WEEDS, ECONOMIC FEASIBILITY, LAKES, NUISANCE ALGAE, WATER
POLLUTION CONTROL, NUTRIENTS, WATER POLLUTION EFFECTS, WATER POLLUTION
SOURCES, WISCONSIN.

IDENTIFIERS:
LAKE MENDOTA(WIS), LAKE MONONA(WIS), LAKE KEGONSA(WIS), LAKE
WAUBESA(WIS), MADISON(WIS), UNIVERSITY OF WISCONSIN, YAHARA RIVER(WIS),
NUTRIENT REMOVAL.

ABSTRACT:
ONE APPROACH TO EUTROPHICATION CONTROL IS REPEATED HARVESTING OF A
NUTRIENT-LADEN CROP, ITS REMOVAL FROM THE WATER, AND ITS UTILIZATION,
TRANSFORMATION, OR TRANSFER FROM THE THREATENED DRAINAGE BASIN. AUTHOR
PRESENTS DESIGN FOR A MECHANICAL HARVESTER, DEVELOPED AT UNIVERSITY OF
WISCONSIN, WHICH WILL CUT AND COLLECT UNDERWATER WEEDS. HE DESCRIBES
OPERATION OF PROTOTYPE MODEL WHICH WAS FUNCTIONAL TO DEPTH OF FIVE
FEET, OPERATED AT SPEEDS OF 1-2 MILES/HOUR, AND COULD REMOVE OVER FOUR
TONS OF DRAINED WEEDS/HOUR. COST OF OPERATING MACHINERY OF THIS TYPE
WOULD NECESSARILY BE QUITE HIGH. AUTHOR ESTIMATES, FOR ERA WHEN THIS
PAPER WAS WRITTEN, THAT INITIAL INVESTMENT FOR ONE HARVESTING MACHINE,
THREE 'SHUTTLE' BARGES FOR TRANSPORTATION OF HARVESTED PLANTS, AND
DOCKS AND UNLOADING FACILITIES WOULD BE IN RANGE OF $20,000 AND THAT AN
ANNUAL OPERATING BUDGET OF $10,000-$15,000 WOULD BE REQUIRED TO COVER
FIXED AND OPERATING CHARGES. MECHANICAL HARVESTING OF MACROPHYTES WOULD
NOT ONLY REMOVE PLANT NUTRIENTS BUT WOULD ALSO ELIMINATE SOME
UNDESIRABLE END EFFECTS OF EUTROPHICATION BY CLEARING BOATING AND
SWIMMING AREAS AND BY ELIMINATING SCUM AREAS WHICH DEVELOP FROM
ENTRAPMENT BY EMERGENT PLANTS OF ALGAE AND OTHER FLOATING DEBRIS.
AUTHOR SUGGESTS THAT HARVESTED LACUSTRINE PLANTS MAY HAVE POTENTIAL
COMMERCIAL VALUE. (EICHHORN-WIS)

FIELD 05G, 04A

ACCESSION NO. W69-06276

COBALT AS AN ESSENTIAL ELEMENT FOR BLUE-GREEN ALGAE,

WISCONSIN UNIV., MADISON. DEPT. OF BOTANY.

OSMUND HOLM-HANSEN, GERALD C. GERLOFF, AND FOLKE SKOOG.

PHYSIOLOGIA PLANTARUM, VOL 7, PP 665-675, 1954. 3 FIG, 5 TAB, 8 REF.

DESCRIPTORS:
*COBALT, *CYANOPHYTA, *ESSENTIAL NUTRIENTS, *TRACE ELEMENTS, ALGAE,
EUTROPHICATION, MOLYBDENUM, NITROGEN FIXATION, NUISANCE ALGAE,
NUTRIENTS, VITAMINS, WATER POLLUTION EFFECTS, WATER POLLUTION SOURCES,
VITAMIN B.

IDENTIFIERS:
*CYANOCOBALAMIN, ALGAL GROWTH, ALGAL NUTRITION, ALGAL PHYSIOLOGY,
CALOTHRIX PARIETINA, COCCOCHLORIS PENIOCYSTIS, DIPLOCYSTIS AERUGINOSA,
NOSTOC MUSCORUM, VANADIUM, VITAMIN B-12.

ABSTRACT:
AUTHORS, EXPERIMENTING WITH NOSTOC MUSCORUM, DEMONSTRATED THAT COBALT
IS AN ESSENTIAL NUTRIENT ELEMENT FOR THIS CYANOPHYTE AS WELL AS FOR
THREE ADDITIONAL SPECIES: NITROGEN-FIXING CALOTHRIX PARIETINA; AND THE
NON-FIXERS, COCCOCHLORIS PENIOCYSTIS AND DIPLOCYSTIS AERUGINOSA. IN
NOSTOC, QUANTITATIVE REQUIREMENT IS QUITE LOW, INCREASES IN YIELD AND
NITROGEN CONTENT OBTAINING FROM COBALT CONCENTRATIONS FROM 0.002 TO
0.40 MICROGRAMS/LITER, THE LATTER LEVEL ORDINARILY REQUIRED FOR OPTIMUM
GROWTH. THESE DATA AND OTHER EVIDENCE INDICATE THAT THE FOLLOWING
ESSENTIALITY SPECIFICATIONS HAVE BEEN MET: ELEMENT HAS POSITIVE EFFECT
ON TOTAL GROWTH; IT EXERTS A POSITIVE PHYSIOLOGICAL EFFECT ON ALGAE; IT
IS NOT REPLACEABLE BY ANOTHER ELEMENT; DEFICIENCY IS REVERSIBLE IN
INCIPIENT STAGES; AND RESPONSE CAN BE DEMONSTRATED IN REPRESENTATIVE
NUMBER OF SPECIES. THERE IS EVIDENCE THAT COBALT REQUIREMENT IS
INDEPENDENT OF NITROGEN SOURCE ALTHOUGH IT IS MUCH EASIER TO
DEMONSTRATE COBALT RESPONSE IN THE TWO NITROGEN FIXERS THAN IN THE TWO
NON-FIXERS, A CONDITION WHICH QUITE LIKELY RESULTS FROM LOWER
REQUIREMENT IN THE LATTER. PRELIMINARY EXPERIMENTS SHOW THAT ADDITION
OF AS LITTLE AS 0.075 MICROGRAM OF VITAMIN B-12 PER LITER OF CULTURE
ELIMINATES REQUIREMENT FOR ADDED COBALT AND RESULTS IN OPTIMUM YIELD
REFERABLE TO ADDED COBALT. (EICHHORN-WIS)

FIELD 05A

ACCESSION NO. W69-06277

WASTES, WATER, AND WISHFUL THINKING: THE BATTLE OF LAKE ERIE,

CASE WESTERN RESERVE UNIV., CLEVELAND.

ARNOLD W. REITZE, JR.

CASE W RES L REV, VOL 20, NO 1, PP 5-86, NOV 1968. 82 P, 448 REF.

DESCRIPTORS:
*OHIO, *LAKE ERIE, *WATER POLLUTION CONTROL, *POLLUTION ABATEMENT, POLLUTANTS, WASTES, THERMAL POLLUTION, TOXINS, PESTICIDE RESIDUES, PHOSPHATES, ALGAE, SEWAGE, WATER POLLUTION, WATER LAW, WATER POLLUTION EFFECTS, WATER POLLUTION SOURCES, WATER POLLUTION TREATMENT, FINANCING, GRANTS, FEDERAL GOVERNMENT, STATE GOVERNMENTS, WATER QUALITY STANDARDS.

ABSTRACT:
LAKE ERIE IS SERIOUSLY POLLUTED, AND ITS CONTINUED USE AS A PUBLIC WATER SUPPLY IS IN JEOPARDY. THE POLLUTANTS INCLUDE SEWAGE, PHOSPHATES, ORGANIC CHEMICALS, AND PETROLELUM PRODUCTS. THE FEDERAL GOVERNMENT HAS PASSED POLLUTION STATUTES WHICH DATE BACK TO THE 19TH CENTURY, BUT THE FIRST SERIOUS ATTEMPT TO CLEAN UP LAKE ERIE BEGAN IN 1965, UNDER THE FEDERAL WATER POLLUTION CONTROL ACT. UNDER THIS ACT, THE STATES ARE REQUIRED TO SET WATER QUALITY STANDARDS FOR THEIR NAVIGABLE WATERWAYS. EACH JURISDICTION MUST MAINTAIN THE EXISTING QUALITY OF THE WATER AND MAY NOT ALLOW TREATABLE WASTES TO BE DISCHARGED INTO THE WATER IN AN UNTREATED STATE. OHIO HAS ESTABLISHED POLLUTION STANDARDS, BUT THESE STANDARDS ARE CRITICIZED BECAUSE THEY ASSUME THE WATER DOES NOT HAVE TO BE PURE AND BECAUSE THEY ARE TOO VAGUE AND INCONSISTENT. THE POLLUTION ABATEMENT OF LAKE ERIE REQUIRES THE EFFORTS OF FEDERAL AND STATE ADMINISTRATIVE AGENCIES AND OF THE COURTS. THE ATTEMPT AT POLLUTION ABATEMENT OVER THE LAST THREE YEARS HAS ENDED IN FAILURE. OHIO EXPENDS ONLY $239,000 YEARLY ON ITS RIVER SANITATION PROGRAM; APPROXIMATELY TEN BILLION DOLLARS ARE NEEDED TO CLEAN LAKE ERIE. (HOFFMAN-FLA).

FIELD 05G, 05B, 06E

ACCESSION NO. W69-06305

THE AGING OF LAKES,

WISCONSIN UNIV., MADISON. WATER CHEMISTRY LAB.

G. FRED LEE, AND E. GUS FRUH.

IND WATER ENG, VOL 3, NO 2, PP 26-30, FEB 1966. 2 FIG, 12 REF.

DESCRIPTORS:
*EUTROPHICATION, *LAKES, ALGAE, AQUATIC WEEDS, CYANOPHYTA, FISHKILL, FERTILIZATION, LIMITING FACTORS, NITROGEN COMPOUNDS, NUTRIENTS, ODOR, OLIGOTROPHY, OXYGEN, OXYGEN DEMAND, PHOSPHORUS COMPOUNDS, PLANT GROWTH, PRIMARY PRODUCTIVITY, RIVERS, TASTE, THERMAL STRATIFICATION, WASTE TREATMENT, WATER QUALITY, WATER POLLUTION CONTROL, WATER POLLUTION EFFECTS, WATER POLLUTION SOURCES.

IDENTIFIERS:
*CULTURAL EUTROPHICATION, RECREATIONAL USE, WASTE EFFLUENTS.

ABSTRACT:
A LAKE'S FERTILIZATION IS DETERMINED TO A LARGE EXTENT BY THE RATE OF NUTRIENT INFLUX. THIS SITUATION IS ANALOGOUS TO SOILS; MORE FERTILIZER, MORE CROPS. WHEN LAKES HAVE EXCESSIVE ALGAE AND/OR ROOTED AQUATIC VEGETATION ALONG THE SHORELINE, IT IS A SIGN THAT THIS FERTILIZATION PROCESS IS IN MOTION. FREQUENTLY, THE INFLUENCE OF MAN IN THE LAKE'S WATERSHED RESULTS IN INCREASED RATES OF NUTRIENT INFLUX WHICH BRING ABOUT AN ACCELERATED RATE OF STANDING CROP PRODUCTION IN THE LAKE. THIS RESULTS IN A DETERIORATION OF THE WATER QUALITY. DOMESTIC AND INDUSTRIAL WATER TREATMENT COSTS ARE INCREASED BECAUSE OF ADDED COST TO REMOVE RESULTANT TASTE AND ODORS AND INCREASED NUMBERS OF ORGANICS FROM THE EUTROPHIC WATER. SUCH RECREATIONAL PURSUITS AS BOATING, FISHING, AND SWIMMING CAN ALSO BE SERIOUSLY IMPAIRED. ODOR ARISING FROM DECAYING MATS OF ALGAE CAN ALSO SERIOUSLY DECREASE THE ECONOMIC VALUE OF SHORELINE PROPERTY. THE LARGE PRODUCTIVITY OF EUTROPHIC LAKES CAN LEAD TO THE ABSENCE OF DISSOLVED OXYGEN IN CERTAIN PARTS OF THE HYPOLIMNION, ELIMINATING SUCH COLD-WATER SPORT FISHES AS TROUT AND CISCO. THUS, THE CULTURAL EUTROPHICATION OF NATURAL WATERS HAS BECOME ONE OF THE MAJOR WATER MANAGEMENT PROBLEMS OF TODAY. (LEE-WIS)

FIELD 02H, 05C

ACCESSION NO. W69-06535

SIMULATION OF AN AQUATIC ECOSYSTEM,

WASHINGTON STATE UNIV., PULLMAN. DEPT. OF ZOOLOGY; AND WASHINGTON STATE
    UNIV., PULLMAN. DEPT. OF INFORMATION SCIENCE.

RICHARD A. PARKER.

BIOMETRICS, VOL 24, NO 4, PP 803-822, DEC 1968. 9 FIG, 3 TAB, 18 REF.

DESCRIPTORS:
    *SYSTEMS ANALYSIS, *ECOSYSTEMS, *SIMULATION ANALYSIS, WATER POLLUTION
    EFFECTS, LAKES, PHOSPHORUS, TROPHIC LEVEL, PRODUCTIVITY, MODEL STUDIES,
    SALMON, AQUATIC ENVIRONMENT, PHOSPHATES, ALGAE, DAPHNIA, GROWTH RATES,
    PREDATION.

IDENTIFIERS:
    KOOTENAY LAKE, BRITISH COLUMBIA, CANADA, AQUATIC ECOSYSTEM MODEL,
    SIMULATION MODEL, PHOSPHORUS EFFECT, CLADECERA.

ABSTRACT:
A DETERMINISTIC MODEL SIMULATING AQUATIC ECOSYSTEM OF KOOTENAY LAKE,
BRITISH COLUMBIA, WAS DEVELOPED CONSIDERING ONLY CONDITIONS IN UPPER 10
METERS OF SOUTH ARM, WHICH RECEIVES HIGH PHOSPHATE LOAD FROM TRIBUTARY
RIVER. SOLAR RADIATION WAS EXPRESSED AS WEEKLY PHOTOPERIOD CHANGES;
SEASONAL TEMPERATURE CURVE DERIVED FROM MEANS OF MEASUREMENTS AT THREE
DEPTHS. FOUR DIFFERENTIAL EQUATIONS SIMULATING CHANGE RATES IN
PHOSPHORUS CONCENTRATION, ALGAE ABUNDANCE, CLADOCERAN DENSITY, AND
KOKANEE SALMON BIOMASS WERE SIMULTANEOUSLY APPLIED. SEVERAL APPROPRIATE
CONSTANTS WERE USED. CHANGE IN PHOSPHATE CONCENTRATION WAS ASSUMED TO
BE LINEAR FUNCTION OF PHOSPHATE INPUT TO SOUTH ARM, OUTPUT FROM ARM,
AND CHANGE RATES IN ALGAL AND CLADOCERAN POPULATIONS. CHANGES IN
BIOLOGICAL COMPONENTS WERE DERIVED ESSENTIALLY FROM DIFFERENCE BETWEEN
GROWTH AND DEATH RATES, WHICH WERE ASSUMED TO BE LINEARLY DEPENDENT ON
COMBINATION OF FACTORS SUCH AS PHOTOPERIOD, PHOSPHATE LEVEL,
TEMPERATURE, AND PREDATION. MODEL SIMULATED ALGAL AND CLADOCERAN CYCLES
REASONABLY WELL BUT NOT PHOSPHATE CYCLE. 80% PHOSPHATE LEVEL REDUCTION
DELAYED ALGAL PEAK 9 WEEKS AND DECREASED OTHER COMPONENTS MORE. 30%
ALGAL GROWTH RATE REDUCTION DELAYED ALGAL AND CLADOCERAN PEAKS BY 1-2
WEEKS AND DECREASED KOKANEE ABUNDANCE. 30% ALGAL DEATH RATE DECREASE
PRODUCED NO SIGNIFICANT CHANGES. INCREASING CLADOCERAN'S ALGAL
PREDATORY RATES HARDLY DISPLACED VARIOUS PEAKS BUT DEPRESSED ALL
POPULATIONS. (TERAGUCHI-WIS)

FIELD 02H, 05C

ACCESSION NO. W69-06536

APPLICATION OF AN ELECTRONIC PARTICLE COUNTER IN ANALYZING NATURAL POPULATIONS OF PHYTOPLANKTON,

CORNELL UNIV., ITHACA, N. Y. DEPT. OF BOTANY.

HUGH F. MULLIGAN, AND JOHN M. KINGSBURY.

LIMNOL OCEANOGR, VOL 13, NO 3, PP 499-506, JULY 1968. 6 FIG, 3 TAB, 20 REF.

DESCRIPTORS:
*PHYTOPLANKTON, *PONDS, *EUTROPHICATION, ALGAE, BIOMASS, CHLOROPHYLL, NEW YORK.

IDENTIFIERS:
*COULTER COUNTER, *POPULATION DYNAMICS, CELLULAR SIZE, CHLORELLA VULGARIS, CORNELL UNIVERSITY, ELECTRONIC PARTICLE COUNTING, EUDORINA ELEGANS, EUGLENA GRACILIS, FOREST CENTRIFUGE, GLENODINIUM PULVISCULUS, HAEMOCYTOMETER COUNTING, ITHACA(NY), GLENODINIUM QUADRIDENS, OOCYSTIS PUSILLA, OOCYSTIS SOLITARIA, PANDORINA MORUM, PEDIASTRUM BORYANUM, PLEODORINA CALIFORNICA, SCENEDESMUS BIJUGA, SCHROEDERIA JUDAYI, SEDGWICK-RAFTER METHOD, SPHAEROCYSTIS SCHROETERI.

ABSTRACT:
UTILIZING COULTER COUNTER WITH ORIFICE OF 100-MICRON DIAMETER, AUTHORS ANALYZED, FROM 10 SMALL MORPHOMETRICALLY IDENTICAL PONDS, PHYTOPLANKTONIC POPULATIONS BY 10 SIZE INTERVALS (MEAN VOLUME, IN MICRONS, BY INTERVAL: 9.1, 23.8, 49.5, 89.3, 191, 396, 715, 1530, 3170, 5720). THEY COMPARED DETERMINATIONS OF BIOMASS BY COULTER COUNTING (B) IN MILLIGRAMS/LITER, WITH CHLOROPHYLL-A (C) IN MILLIGRAMS/CUBIC METER AND DRY WEIGHT (D) IN MILLIGRAMS/LITER. FOR EXAMPLE, IN A SINGLE POND FOR FIVE DATES, RATIO B:C HAD VALUES 0.215, 0.270, 0.94, 0.302, 0.66; AND B:D, 1.77, 1.76, 2.32, 1.82, 1.81. WHILE COUNTER CANNOT DISTINGUISH LIVING FROM DEAD PARTICLES, NOR SPECIES FROM SPECIES, ITS USE IN SIZING AND COUNTING PLANKTONIC POPULATIONS YIELDS VALUES OF BIOMASS WITH TIME WHICH CONSTITUTE HIGHLY RESOLVED AND CHARACTERISTIC PROFILE OF POND DYNAMICS. PEAK BIOMASSES CAN OFTEN BE CORRELATED WITH A DOMINANT PHYTOPLANKTER, OBSERVABLE BY DIRECT MICROSCOPIC EXAMINATION. BIOMASS DETERMINED IN THIS WAY IS MORE PRECISE THAN DETERMINATIONS OF CHLOROPHYLL OR DRY WEIGHTS, AND METHOD IS MORE EFFICIENT THAN ESTIMATIONS UTILIZING HAEMOCYTOMETER. AUTHORS EMPHASIZE THAT METHOD IS PARTICULARLY EFFECTIVE WHEN USED IN CONJUNCTION WITH STANDARD METHODS. (EICHHORN-WIS)

FIELD 07B, 05C

ACCESSION NO. W69-06540

CULTURAL EUTROPHICATION IS REVERSIBLE,

WISCONSIN UNIV., MADISON. LAB. OF LIMNOLOGY.

ARTHUR D. HASLER.

BIOSCIENCE, VOL 19, NO 5, PP 425-431, MAY 1969. 3 FIG, 1 TAB, 7 REF.

DESCRIPTORS:
*EUTROPHICATION, *FERTILIZATION, *LIMNOLOGY, *PHYTOPLANKTON, FISH, GREAT LAKES, LEGISLATION, NITROGEN, NUISANCE ALGAE, PHOSPHORUS, WASHINGTON, WISCONSIN, CALIFORNIA.

IDENTIFIERS:
*CULTURAL EUTROPHICATION, LAKE MONONA(WIS), LAKE TAHOE(CALIF), LAKE WASHINGTON(WASH), ZURICHSEE(SWITZERLAND).

ABSTRACT:
VARIOUS HUMAN ACTIVITIES CONTRIBUTE TO EXCESSIVE ENRICHMENT (CULTURAL EUTROPHICATION) OF WATERS. AMONG THE SYMPTOMS OF CULTURAL EUTROPHICATION ARE NUISANCE BLOOMS OF ALGAE, INCREASED NUTRIENT LEVELS, DEPLETION OF HYPOLIMNETIC OXYGEN, INCREASED TURBIDITY, AND CHANGES IN THE SPECIES-COMPOSITIONS OF PHYTOPLANKTON, INVERTEBRATES, AND FISHES. A CASE HISTORY IS PRESENTED (ZURICHSEE) DOCUMENTING CHANGES IN LIMNOLOGICAL CONDITIONS, AS WELL AS ADDITIONAL RELEVANT DATA FROM NORTH AMERICAN LAKES. AS METHODS OF SOLVING THE PROBLEMS OF CULTURAL EUTROPHICATION, AUTHOR SUGGESTS THAT MORE EXTENSIVE DISSEMINATION OF INFORMATION, INVOLVEMENT OF ALL CITIZENS AND CITIZEN GROUPS IN WORKSHOPS AND SEMINARS, AND ENLIGHTENED LEGISLATION. THE WISCONSIN RESOURCES LAW (1965) IS GIVEN AS AN EXAMPLE OF LEGISLATION AND THE PHILOSOPHY OF ZONING IS DISCUSSED. THREE EXAMPLES ARE GIVEN (LAKES MONONA, TAHOE, WASHINGTON) WHERE SUCCESS IS CURBING CULTURAL EUTROPHICATION HAS BEEN ACHIEVED. METHODS OF CONTROL AND HARVESTING AQUATIC PLANTS AND FISH ARE DISCUSSED; ALSO CHEMICAL CONTROL AND SEWAGE UTILIZATION. TABLE GIVES ESTIMATES OF NITROGEN AND PHOSPHORUS INPUT TO WISCONSIN SURFACE WATERS. SEE ALSO W68-00680. (VOIGTLANDER-WIS)

FIELD 05G, 02H

ACCESSION NO. W69-06858

THE REMOVAL OF DISSOLVED PHOSPHATE FROM LAKE WATERS BY BOTTOM DEPOSITS,

A. V. HOLDEN.

VERH INTERNAT VEREIN LIMNOL, VOL 14, PP 247-251, JULY 1961. 4 FIG, 2 REF.

DESCRIPTORS:
*PHOSPHATE, *LAKES, *CYCLING NUTRIENTS, *SEDIMENT-WATER INFERFACE,
MUD-WATER INTERFACE, PHOSPHORUS, ADSORPTION, EUTROPHICATION, ALGAE,
MACROAQUATIC PLANTS, FERTILIZATION.

ABSTRACT:
PHOSPHATE FERTILIZER WAS ADDED TO SCOTTISH LOCHS AND PHOSPHATE LOSS
FROM THE WATER MEASURED. AT LOW LEVELS OF PHOSPHATE ADDED (45
MICROGRAMS/LITER) THE EXPOTENTIAL EXPRESSION DC/DT = KC DESCRIBED
PHOSPHATE DECLINE; AT HIGHER LEVELS, THE EXPRESSION WAS ALTERED TO
ACCOUNT FOR PHOSPHATE SATURATION OF THE SEDIMENT. REMOVAL RATES WERE
EXPRESSED AS KD VALUES, MICROGRAMS PHOSPHATE/CENTIMETER SQUARED, PER
MILLIGRAMS PHOSPHATE/LITER PER DAY. A KD VALUE OF 6.5 FOR LOCH
KINARDOCHY AS COMPARED WITH 0.6 FOR CORES OVERLAIN BY WATER (30 TO 40
CM) AND LACKING ALGAE OR MACROPHYTES, SUGGESTED THAT ATTACHED FLORA
LARGELY ACCOUNTED FOR PHOSPHATE UPTAKE IN THE LOCH. IN LABORATORY
EXPERIMENTS INVOLVING HIGH PHOSPHATE UPTAKE (700 MICROGRAMS/CM
SQUARED), PHOSPHATE INCREASE UP TO 200 MILLIGRAM PHOSPHATE PERCENT
OCCURRED IN THE SEDIMENT AEROBIC ZONE (UPPER 20 MILLIMETERS); SOME
PHOSPHATE PENETRATED TO 80 MILLIMETERS, INTO ANAEROBIC ZONE. ABOUT 6%
OF PHOSPHATE RETAINED IN THE AEROBIC ZONE WAS DECINORMAL HYDROCHLORIC
ACID EXTRACTABLE. DEVELOPMENT OF ANEROBIC ZONES DECREASED
ACID-EXTRACTABLE PHOSPHORUS, ALTHOUGH AUTHOR DOES NOT STATE WHETHER
REOXIDATION OCCURRED BEFORE EXTRACTION. PHOSPHORUS GRADIENTS FROM
PRODUCTIVE, UNFERTILIZED LOCHS, WERE SIMILAR TO FERTILIZED LOCHS
INCUBATED 48 MONTHS. (ARMSTRONG-WIS)

FIELD 02H, 05C

ACCESSION NO. W69-06859

PROVISIONAL ALGAL ASSAY PROCEDURE,

JOINT INDUSTRY-GOVERNMENT TASK FORCE ON EUTROPHICATION, NEW YORK.

G. P. FITZGERALD, JOSEPH SHAPIRO, C. R. GOLDMAN, O. R. ARMSTRONG, AND JACK
MYERS.

PUBLISHED BY JOINT INDUSTRY/GOVERNMENT TASK FORCE ON EUTROPHICATION, P O BOX
3011, GRAND CENTRAL STATION, NEW YORK, 62 PP, FEB 1969. 7 FIG, 5 TAB, 26
REF.

DESCRIPTORS:
*ALGAE, *BIOASSAY, NUTRIENTS, EUTROPHICATION, WATER POLLUTION EFFECTS,
WATER POLLUTION SOURCES, BIOINDICATORS, NITROGEN FIXATION.

IDENTIFIERS:
*ALGAL GROWTH, NUTRIENT AVAILABILITY, ALGAL GROWTH POTENTIAL,
SELENASTRUM CAPRICORNUTUM, ANABAENA FLOS-AQUAE, MICROCYSTIS AERUGINOSA,
ALGAL CULTURES, CHEMOSTATS, CARBON-14.

ABSTRACT:
THE PROCEDURE DESCRIBED WAS DEVELOPED BY A TEAM OF INTERNATIONAL
EXPERTS UNDER THE SPONSORSHIP OF THE JOINT INDUSTRY/GOVERNMENT TASK
FORCE ON EUTROPHICATION AND REPRESENTS THE FIRST STEP TOWARD DEVELOPING
RELIABLE AND REPRODUCIBLE TESTS TO DETERMINE THE CAPACITY OF VARIOUS
AQUATIC ENVIRONMENTS TO GROW ALGAE. IT IS EMPHASIZED THAT THE
PROCEDURES PRESENTED ARE PROVISIONAL AND ARE NOT STANDARDIZED TESTS.
THREE PROCEDURES ARE DESCRIBED: A BOTTLE TEST, A CONTINUOUS-FLOW
CHEMOSTAT TEST, AND AN IN SITU TEST. DETAILED DESCRIPTIONS OF EQUIPMENT
REQUIRED, LABORATORY PROCEDURES, ANALYTICAL TECHNIQUES, AND SAMPLE
PREPARATION ARE GIVEN. INDICATOR ORGANISMS SELECTED FOR USE IN THE
TESTS ARE SELENASTRUM CAPRICORNUTUM, A GREEN ALGA; ANABAENA FLOS-AQUAE,
A BLUE-GREEN ALGA CAPABLE OF FIXING NITROGEN; AND MICROCYSTIS
AERUGINOSA, A BLUE-GREEN ALGA NOT CAPABLE OF FIXING NITROGEN.
(UTTORMARK-WIS)

FIELD 05A, 05C

ACCESSION NO. W69-06864

ALGAL CULTURE: FROM LABORATORY TO PILOT PLANT.

CARNEGIE INSTITUTION OF WASHINGTON PUBLICATION 600, 357 PP, WASHINGTON, DC,
MARCH 1961. 103 FIG, 59 TAB, 297 REF.

DESCRIPTORS:
*ALGAE, *RESEARCH AND DEVELOPMENT, BIBLIOGRAPHIES, CHLORELLA,
CHLOROPHYTA, DIATOMS, LIGHT, LIPIDS, NUTRIENT REQUIREMENTS, ORGANIC
MATTER, PHOTOSYNTHESIS, TEMPERATURE, TROPICAL REGIONS.

IDENTIFIERS:
*ALGAL CULTURE, *LABORATORY STUDIES, *MASS CULTURE, *PILOT-PLANT
STUDIES, ALGAL GROWTH, ALGAL NUTRITION, ALGAL PHYSIOLOGY, AMERICAN
RESEARCH AND DEVELOPMENT CORP(MASS), ARTHUR D LITTLE INC(MASS),
CARNEGIE DEPT OF PLANT BIOLOGY, CARNEGIE INSTITUTION OF WASHINGTON,
CHEMICAL COMPOSITION(ENGLAND), FOOD SOURCES, GREENHOUSE STUDIES, GROWTH
KINETICS, INDUSTRIAL RAW MATERIALS(ISRAEL), LIGHT-ENERGY CONVERSION,
MINERAL NUTRITION, RESEARCH CORPORATION OF NY, STANFORD RESEARCH
INSTITUTE(CALIF), STEROLS.

ABSTRACT:
RESEARCH SPONSORED BY THE INSTITUTION IS SUMMARIZED, BOTH IN-HOUSE AND
CONTRACTUAL, ON MASS CULTURE OF ALGAE IN ITS BASIC ASPECTS AND IN THE
DEVELOPMENTAL AND ENGINEERING PHASES OF HARVESTING AND UTILIZATION AS
FOOD. THE SCOPE IS INDICATED BY THE SUBJECT-MATTER OF CHAPTERS BY
VARIOUS AUTHORS: CURRENT STATUS OF LARGE-SCALE CULTURE OF ALGAE; NEED
FOR NEW FOOD-SOURCES; BIOLOGY OF ALGAE SUMMARIZED; ALGAL GROWTH
CHARACTERISTICS AS APPLIED TO MASS CULTURE; EFFICIENCY OF LIGHT-ENERGY
CONVERSION IN CHLORELLA; PHOTOSYNTHESIS BY CHLORELLA IN FLASHING LIGHT;
EFFECT OF DIURNALLY INTERMITTENT LIGHT ON CHLORELLA; INORGANIC
NUTRITION OF ALGAE; LABORATORY EXPERIMENTS ON CHLORELLA CULTURE AT
INSTITUTION'S LABORATORY; PRODUCTION OF ORGANIC MATTER BY CHLOROPHYTES
AND DIATOMS; NON-STERILE GREENHOUSE CULTURE; ALGAL ACCUMULATION OF
LIPIDS; BRITISH EXPERIMENTS; TROPICAL CULTIVATION OF ALGAL COMPLEXES;
ISRAELI EXPERIMENTS; CHLORELLA GROWTH KINETICS; PRE-PILOT-PLANT AND
PILOT-PLANT STUDIES IN MASS CULTURE; CHEMICAL COMPOSITION OF ALGAE;
NUTRITIONAL VALUES; ALGAE AS INDUSTRIAL RAW MATERIALS; STEROL CONTENT
OF ALGAE. A SELECTED GENERAL BIBLIOGRAPHY AND LIST OF LITERATURE CITED
IS INCLUDED. (EICHHORN-WIS)

FIELD 05C

ACCESSION NO. W69-06865

MADISON METROPOLITAN SEWERAGE DIST V COMM ON WATER POLLUTION (DISCHARGE OF
SEWAGE INTO LAKE WATER).

260 WIS 229, 50 NW 2D 424-439 (1951).

DESCRIPTORS:
*WISCONSIN, *SEWAGE EFFLUENTS, *WATER POLLUTION CONTROL, *LOCAL
GOVERNMENTS, STATE GOVERNMENTS, BIOCHEMICAL OXYGEN DEMAND,
ODOR-PRODUCING ALGAE, POLLUTION ABATEMENT, SEWAGE TREATMENT, SEWAGE
DISPOSAL, WASTE DISPOSAL, WATER POLLUTION, LAKES, STREAMS, SEWAGE,
REMEDIES, JUDICIAL DECISIONS, LEGAL ASPECTS.

ABSTRACT:
PETITIONER SEWERAGE DISTRICT, A PUBLIC CORPORATION, DISCHARGED THE
EFFLUENT FROM ITS SEWAGE TREATMENT PLANTS INTO SURROUNDING LAKES. THE
INORGANIC CHEMICALS CONTAINED IN THE EFFLUENT CONTRIBUTED SUBSTANTIALLY
TO THE ABNORMAL GROWTH OF A TYPE OF ALGAE WHICH CREATED OBJECTIONABLE
CONDITIONS IN THOSE LAKES. RESPONDENT, A PUBLIC AGENCY ACTING UNDER
POWERS GRANTED BY THE LEGISLATURE ORDERED PETITIONER TO MODIFY AND
IMPROVE ITS SEWAGE DISPOSAL SYSTEM SO AS TO ELIMINATE THE ALGAE
NUISANCE. THE CIRCUIT COURT AFFIRMED THE LOWER COURT'S ORDER AND
PETITIONER APPEALED. THE SUPREME COURT HELD THAT THE RIGHT OF A
MUNICIPALITY TO EMPTY ITS SEWAGE INTO A STREAM OR RIVER WAS MERELY A
LEGISLATIVE LICENSE REVOCABLE WHENEVER PUBLIC HEALTH AND SAFETY
REQUIRED. THE RESPONDENT DID NOT HAVE TO STATE EXPRESSLY THAT APPELLANT
WAS CREATING A NUISANCE WHEN ITS FINDINGS SUBSTANTIALLY DEMONSTRATED
THAT A NUISANCE EXISTED AS A RESULT OF APPELLANT'S OPERATIONS.
(GABRIELSON-FLA)

FIELD 06E, 05B

ACCESSION NO. W69-06909

UPTAKE AND ASSIMILATION OF NITROGEN IN MICROECOLOGICAL SYSTEMS,

GEOLOGICAL SURVEY, MENLO PARK, CALIF.

G. G. EHRLICH, AND K. V. SLACK.

ASTM STP NO. 448, SYMP ON MICROORGANIC MATTER IN WATER, AMER SOC TESTING
MATER 71ST ANNU MEETING, JUNE 23-28, 1968, SAN FRANCISCO, PP 11-23, FEB
1969. 13 P, 5 FIG, 2 TAB, 29 REF.

DESCRIPTORS:
*NITROGEN COMPOUNDS, *NUTRIENTS, *ALGAE, *MICROORGANISMS,
BIODEGRADATION, NITRIFICATION, DENITRIFICATION, ORGANIC COMPOUNDS,
AMMONIA, NITRITES, NITRATES, YEASTS, BACTERIA, AQUATIC BACTERIA, WATER
POLLUTION EFFECTS.

IDENTIFIERS:
NITROGEN ASSIMILATION.

ABSTRACT:
A STUDY OF BIOLOGICAL NITROGEN REMOVAL UTILIZED 2 RECIRCULATING
LABORATORY STREAMS. THE SOLE NITROGEN SOURCE IN 1 STREAM WAS CALCIUM
NITRATE AND IN THE OTHER WAS YEAST EXTRACT. NITRITE AND NITRATE
INCREASED SLIGHTLY. BACTERIA AND COMBINED NITROGEN DECREASED RAPIDLY
AFTER THE INITIAL RISE. IN THE STREAM WITH THE INORGANIC NITROGEN
MEDIUM, ABOUT HALF OF THE NITRATE DISAPPEARED IN 2 WEEKS, AND NITRATE
WAS UNDETECTABLE AFTER THE 3RD WEEK. BACTERIAL COUNTS WERE RELATIVELY
LOW AND CONSTANT. PERIPHYTON INCREASED IN BOTH STREAMS AS NUTRIENTS
DECREASED, REACHING QUASI-STEADY STATES AFTER ABOUT 4 WEEKS. THE RATE
OF BIOMASS INCREASE WAS GREATER IN THE INORGANIC MEDIUM. NITRATE WAS
DIRECTLY ASSIMILATED BY ALGAE. ORGANIC NITROGEN WAS CONVERTED TO
AMMONIA BY PROTEOLYTIC BACTERIA. AMMONIA FROM ORGANIC COMPOUNDS WAS
PARTLY ASSIMILATED BY ALGAE AND PARTLY NITRIFIED BY BACTERIA. NITRATE
OF BACTERIAL ORIGIN WAS ASSIMILATED BY ALGAE. NITRIFICATION APPARENTLY
WAS NOT OF MAJOR IMPORTANCE IN CONVERTING ORGANIC NITROGEN TO ALGAL
BIOMASS. LABORATORY STREAMS SIMULATE NATURAL STREAMS IN DIURNAL CYCLES
OF PH AND DISSOLVED OXYGEN, IN PERIPHYTON GROWTH RATE AND BIOMASS, AND
IN ASSIMILATION OF NUTRIENTS. USE OF MODEL SYSTEMS TO STUDY NATURAL
PROCESSES UNDER CONTROLLED CONDITIONS APPEARS JUSTIFIED. (KNAPP-USGS)

FIELD 05C, 05B

ACCESSION NO. W69-06970

EUTROPHICATION -- CAUSES AND EFFECTS,

MICHIGAN UNIV., ANN ARBOR. DEPT. OF SANITARY AND WATER RESOURCES ENGINEERING.

J. A. BORCHARDT.

J AMER WATER WORKS ASS, VOL 61, NO 6, PP 272-275, JUNE 1969. 4 P, 2 FIG, 1
TAB, 1 PHOTO, 4 REF.

DESCRIPTORS:
*EUTROPHICATION, *WATER POLLUTION SOURCES, *LAKES, NUTRIENTS,
FERTILIZERS, PHOSPHATES, NITRATES, ALGAE, GREAT LAKES.

IDENTIFIERS:
EUTROPHICATION PREVENTION, ECOLOGIC IMBALANCES.

ABSTRACT:
THE CAUSES AND EFFECTS OF ACCELERATED EUTROPHICATION OF LAKES AND
STREAMS ARE OUTLINED. THE IMPORTANT NATURAL FACTORS INVOLVED ARE
GEOLOGY, LAKE BASIN GEOMETRY AND SIZE, TYPE AND SIZE OF WATERSHED, AND
LATITUDE. THE MOST IMPORTANT SINGLE VARIABLE IS MAN'S TREATMENT OF THE
LAKE. ACCELERATED EUTROPHICATION IS A RESULT OF ACCELERATED ADDITION OF
PHOSPHORUS AND NITROGEN NUTRIENTS, NEARLY ALWAYS CAUSED BY SOME FORM OF
HUMAN ACTIVITY. FERTILIZERS, MUNICIPAL WASTES, INDUSTRIAL WASTES, AND
RECREATIONAL WASTES ARE THE MOST COMMON NUTRIENT SOURCES. THE MOST
VISIBLE AND OBNOXIOUS RESULT OF EUTROPHICATION IS ALGAE GROWTH.
CORRECTION OF EUTROPHIC CONDITIONS IS ALMOST IMPOSSIBLE. PREVENTION IS
NECESSARY TO AVOID THE PROBLEM IN THE FUTURE. (KNAPP-USGS)

FIELD 05C, 02H

ACCESSION NO. W69-07084

AN ECOLOGICAL STUDY OF THE ALGAE OF THE RIVER MOOSI, HYDERABAD (INDIA) WITH
SPECIAL REFERENCE TO WATER POLLUTION - 1. PHYSICO-CHEMICAL COMPLEXES,

OSMANIA UNIV., HYDERABAD, (INDIA). DEPT. OF BOTANY.

V. VENKATESWARLU.

HYDROBIOLOGIA, VOL 33, NO 1, PP 117-143, MAR 1969. 26 P, 18 FIG, 2 TAB, 55
REF.

DESCRIPTORS:
*ECOLOGY, *ALGAE, *RIVERS, *WATER POLLUTION, *WATER POLLUTION SOURCES,
IRRIGATION, AQUATIC ALGAE, WATER ANALYSIS, CARBONATES, NITRATES,
CHLORIDES, AMMONIUM COMPOUNDS, OXYGEN, SILICATES, ORGANIC MATTER, AIR
TEMPERATURE, AEROBIC BACTERIA, WATER TEMPERATURE, IRON.

IDENTIFIERS:
*INDIA, MOOSI RIVER ECOLOGICAL STUDY.

ABSTRACT:
THIS INVESTIGATION IS CONCERNED WITH THE ECOLOGY OF THE RIVER MOOSI,
HYDERABAD (INDIA) USING THE DATA EXTENDED OVER A PERIOD OF TWO YEARS
WITH THE COLLECTION OF MONTHLY SAMPLES. IT DEALS MAINLY WITH THE
PHYSICO-CHEMICAL COMPLEXES PRESENT IN THE UNPOLLUTED AND POLLUTED
WATERS OF THE RIVER. THE INTERRELATIONSHIPS AND FLUCTUATIONS EXISTING
BETWEEN THE FOLLOWING FACTORS HAVE BEEN INVESTIGATED: (1) CHLORIDE AND
WATER CURRENT; (2) NITRATE, DISSOLVED OXYGEN, AND FREE AMMONIA; (3)
NITRATE AND ALBUMINOID AMMONIA; (4) FREE AMMONIA AND ALBUMINOID
AMMONIA; (5) DISSOLVED OXYGEN, OXIDIZABLE ORGANIC MATTER, AND WATER
TEMPERATURE; AND (6) TOTAL IRON AND PH. (GABRIEL-USGS)

FIELD 05B, 02I

ACCESSION NO. W69-07096

THE ROLE OF ALGAE IN THE BIOLOGICAL TREATMENT OF WATER (FRENCH),

IRCHA (FRANCE). MICROBIOLOGY SERVICE.

R. CABRIDENC, AND H. LEPAILLEUR.

TERRES AND EAUX, REV INT DE L'HYDRAUL, VOL 22, NO 58, PP 12-18, JAN-MAR 1969.
7 P, 7 FIG, 1 TAB, 25 REF.

DESCRIPTORS:
*ALGAE, *BIOLOGICAL TREATMENT, *WATER TREATMENT, *POLLUTANT
IDENTIFICATION, *POLLUTANT ABATEMENT, EFFLUENTS, STREAMFLOW,
DESALINATION, TERTIARY TREATMENT, LAGOONS, OXYGENATION, ABSORPTION,
ECOLOGY, WATER PURIFICATION, CARBON, BICARBONATES, PHOTOSYNTHESIS,
NUTRIENTS, MICROBIOLOGY.

IDENTIFIERS:
WATER PURIFICATION BY ALGAE.

ABSTRACT:
THE ROLE OF ALGAE, AS A DESALINATION AND PURIFICATION AGENT, WAS
INVESTIGATED ON THE BASIS OF EARLIER PUBLICATIONS AND RECENT
EXPERIMENTS CONDUCTED BY THE AUTHORS. THE STUDY SHOWS THAT OXYGEN
LIBERATED IN THE COURSE OF PHOTOSYNTHESIS LEADS TO THE OXYGENATION OF
AQUATIC MEDIA AND CONSIDERABLE REDUCTION OF PATHOGENIC GERMS. THE STUDY
ALSO SHOWS THAT THERE IS COMPETITION BETWEEN ALGAE AND OTHER
MICROORGANISMS; HOWEVER, IN THE PRESENT STATE OF KNOWLEDGE, THE PROBLEM
OF INTERACTION DUE TO THE PRODUCTION BY ALGAE OF INHIBITING OR
ACTIVATING SUBSTANCES IS STILL NOT CLEARLY UNDERSTOOD. THE DEGREE OF
PURIFICATION OF WATER BY THE LAGOONING PROCESS, IN MANY CASES IS
EQUIVALENT TO, OR GREATER THAN THAT OBTAINED BY OTHER BIOLOGICAL
PROCESSES. (GABRIEL-USGS)

FIELD 05C

ACCESSION NO. W69-07389

MICROBIAL TRANSFORMATIONS OF MINERALS,

RENSSELAER POLYTECHNIC INST., TROY, N. Y. DEPT. OF BIOLOGY.

HENRY L. EHRLICH.

PROC RUDOLFS RES CONF, RUTGERS UNIV, NEW BRUNSWICK, NJ. PRINCIPLES AND
APPLICATIONS IN AQUATIC MICROBIOLOGY, HEUKELEKIAN, H AND DONDERO, NORMAN C
(EDS), JOHN WILEY AND SONS, NEW YORK, PP 43-60, 1964. 5 FIG, 43 REF, DISC.

DESCRIPTORS:
  *MICROBIOLOGY, *MINERALOGY, *MICROORGANISMS, *WATER POLLUTION, *WATER
  PURIFICATION, *BACTERIA, *ENZYMES, CORROSION, OXIDATION,
  REDUCTION(CHEMICAL), HYDROLYSIS, CHELATION, FUNGI, ALGAE, PROTOZOA,
  LICHENS, PLANTS, DIATOMS, CORAL, MOLLUSKS, MINE ACIDS, COAL MINES,
  ACIDITY, PYRITE, SULFUR COMPOUNDS, MOLYBDENUM, ROTIFERS, ECOSYSTEMS,
  PHOTOSYNTHESIS, ENVIRONMENT, AQUEOUS SOLUTIONS, TERRESTRIAL HABITATS,
  IRON COMPOUNDS, MANGANESE, TROPHIC LEVEL.

IDENTIFIERS:
  *MINERAL TRANSFORMATION, METAZOA, RADIOLARIA, FORAMINIFERA,
  THIOBACILLUS-FERROBACILLUS, THIOBACILLUS DENITRIFICANS, W VIRGINIA,
  PENNSYLVANIA, MARCASITE, CHALCOCITE, COVELLITE, SPHALERITE, MILLERITE,
  ORPIMENT, TETRAHEDRITE, BORNITE, CHALCOPYRITE, CHEMOSYNTHESIS,
  CONCRETIONS, ARTHROBACTER, PYROPHOSPHATE, ORTHOPHOSPHATE, DISMUTATION,
  TANNINS.

ABSTRACT:
  BIOLOGICAL PROCESSES OF SOME MICROORGANISMS, ESPECIALLY BACTERIA, CAUSE
  TRANSFORMATION OF MINERAL MATTER. INVOLVED ARE DIRECT ENZYMIC
  INTERACTION--OXIDATIONS AND REDUCTIONS, HYDROLYSIS, CHELATE
  DESTRUCTION--AND, INDIRECT NONENZYMIC INTERACTION--INSOLUBLE MATTER
  CORROSION BY METABOLICALLY PRODUCED ACID, INORGANIC IONS PRECIPITATION
  BY OTHER, METABOLICALLY-PRODUCED INORGANIC IONS, ADSORPTION ONTO CELL
  SURFACES, AND METAL CHELATE FORMATION. THE BACTERIAL CHEMOSYNTHETIC
  AUTOTROPHS USE MINERAL OXIDATION AS A CHEMICAL ENERGY SOURCE, REDUCING
  POWER FOR THE SYNTHESIS OF ORGANIC MATTER FROM CARBON DIOXIDE AND
  WATER. OXIDATION OF METAL SULFIDES AND OF FE (II) TO FE (III) TAKES
  PLACE. BACTERIAL PHOTOSYNTHETIC AUTOTROPHS USE MINERAL OXIDATION MERELY
  AS A SOURCE OF REDUCING POWER IN CARBON DIOXIDE ASSIMILATION, DEPENDING
  ON RADIATION FOR ENERGY. THE ENZYME, INORGANIC PYROPHOSPHATASE, PRESENT
  IN BACTERIA, IS PROBABLY HYDROLYZED TO ORTHOPHOSPHATE. PYROPHOSPHATE
  CAN COMPLEX MN (III) AND HYDROLYSIS COULD LEAD TO THE ION
  PRECIPITATION. ENZYMIC BREAKDOWN OF THE CHELATING AGENT BY BACTERIA MAY
  RESULT IN PRECIPITATING THE METAL IONS. THE NONENZYMIC INTERACTIONS
  DEPEND ON END PRODUCTS OF METABOLISM, ACIDS WHICH ACT ON ACID-SOLUBLE
  MINERALS. SOME PRECIPITATIONS MAY FORM AMORPHOUS OR CRYSTALLINE
  AGGLOMERATES. IN ADSORPTION, IRON AND MANGANESE OXIDES POSSIBLY FORM
  INTO ENCRUSTATIONS. IN CHELATION A METABOLICALLY PRODUCED ORGANIC
  COMPOUND MAY CO MPLEX ONE OR ANOTHER METAL ION. (JONES-WISC)

FIELD 05C

ACCESSION NO. W69-07428

PROBLEMS OF BIOLOGICAL OCEANOGRAPHY.

AKADEMIYA NAUK URSR. INST. OF BIOLOGY OF SOUTHERN SEAS.

AVAILABLE FROM CLEARINGHOUSE AS AEC-TR-6940 AT $3.00 IN PAPER COPY AND $0.65
IN MICROFICHE. TRANSL VOPR BIOOKEANOGRAFFI MATERIALY II MEZHDUNARODNOGA
OKEANOGR KONGR 30 MAY-9 JUNE 1966. NAUKOVA DUMKA, KIEV, 1967. US ATOMIC
ENERGY COMM, TRANSLATION SERIES AEC-TR-6940. 94 PP, 13 FIG, 18 TAB, 99 REF.

DESCRIPTORS:
*CONFERENCES, *OCEANOGRAPHY, *TRANSLATIONS, ALGAE, CRUSTACEANS,
BENTHOS, BRINE SHRIMP, ECOSYSTEMS, FISH, PHYTOPLANKTON, PLANKTON,
RADIOACTIVITY, RADIOACTIVITY EFFECTS, RADIOECOLOGY, SORPTION, WATER
POLLUTION EFFECTS, WATER POLLUTION SOURCES.

IDENTIFIERS:
*BIOLOGICAL OCEANOGRAPHY, *TRANSACTIONS, BLACK SEA, CONTACT ZONES,
ARTEMIA, BALANUS, DESORPTION, EMBRYOLOGY, ENERGY TRANSFORMATIONS,
HETEROTROPHY, ORGANIC POLLUTION, MARINE ORGANISMS, MARINE ASSOCIATIONS,
MOSCOW, PETROLEUM POLLUTION, PHOTOSYNTHETIC PIGMENTS, PRODUCTION
BIOLOGY, RUSSIA, SECOND INTERNATIONAL OCEANOGRAPHIC CONGRESS,
HYPONEUSTON.

ABSTRACT:
TROPICAL COVERAGE OF 16 INDIVIDUAL CONTRIBUTIONS ARE AS FOLLOWS:
PLANKTONIC ECOSYSTEMS OF THE BLACK SEA; PRODUCTION RATE OF
HETEROTROPHS; TRANSFORMATION OF SUBSTANCES AND ENERGY BY AMPHIBIONTIC
CRUSTACEA IN BLACK SEA; ENERGY TRANSFORMATION BY ARTEMIA, SEASONAL
DYNAMICS OF ALGAL PHOTOSYNTHETIC PIGMENTS IN BLACK SEA; BENTHOS
DEVELOPMENT IN MEDITERRANEAN BASIN; ECOLOGICO-GEOGRAPHIC DEVELOPMENT OF
CONTACT ZONES IN SOUTHERN SEAS; INFLUENCE OF PETROLEUM PRODUCTS ON
PHYTOPLANKTON OF SEAS; INFLUENCE OF ORGANIC POLLUTION ON DEVELOPMENT OF
BALANUS; MARINE RADIOECOLOGY AND OCEANOGRAPHY; RADIOECOLOGICAL
SIGNIFICANCE OF HYPONEUSTON; EFFECTS OF RADIOACTIVITY ON EMBRYOLOGY OF
FISHES; KINETICS OF ACCUMULATION AND EXCHANGE OF RADIOISOTOPES BY
MARINE ALGAE; ARTIFICIAL RADIOACTIVITY OF MARINE ORGANISMS;
DISTRIBUTION OF RADIOACTIVITY IN MARINE ORGANISMS AND ASSOCIATIONS; AND
SORPTION AND DESORPTION OF RADIONUCLIDES IN BOTTOMS OF SHALLOW SEAS.
(EICHHORN-WISC)

FIELD 05C

ACCESSION NO. W69-07440

GROWTH CHARACTERISTICS OF ALGAE IN RELATION TO THE PROBLEMS OF MASS CULTURE,

CARNEGIE INSTITUTION OF WASHINGTON, STANFORD, CALIF. DEPT. OF PLANT BIOLOGY.

JACK MYERS.

EXTRACTED FROM ALGAL CULTURE: FROM LABORATORY TO PILOT PLANT, JOHN S BURLEW,
EDITOR. CARNEGIE INSTITUTION OF WASHINGTON PUBLICATION 600, WASHINGTON, DC.
4 FIG, 2 TAB.

DESCRIPTORS:
*GROWTH RATES, *ALGAE, PHOTOSYNTHESIS, EUTROPHICATION, LIGHT,
TEMPERATURE, CARBON DIOXIDE, NUTRIENTS, BIOLOGICAL PROPERTIES, WATER
POLLUTION EFFECTS.

IDENTIFIERS:
*MASS CULTURE, *ALGAL GROWTH, CHLORELLA PYRENOIDOSA(EMERSON STRAIN),
AUTOINHIBITORS.

ABSTRACT:
BIOLOGICAL SIGNIFICANCE OF THE CONDITIONS AFFECTING ALGAL GROWTH IS
DISCUSSED WITH SPECIAL REFERENCE TO CHLORELLA PYRENOIDOSA (EMERSON
STRAIN). CONSIDERATION IS GIVEN TO THE EFFECT OF CONTROLLED AND
MEASURABLE CONDITIONS, AND THESE RESULTS ARE EXTRAPOLATED TO DESCRIBE
THE CHARACTERISTICS OF HIGH-DENSITY CULTURES. THE EFFECT OF LIGHT AS A
LIMITING FACTOR IS DISCUSSED ALSO. IN PARTICULAR, GROWTH RATES ARE
DESCRIBED AS FUNCTIONS OF (1) TEMPERATURE, (2) LIGHT INTENSITY, (3)
CONCENTRATION OF CARBON DIOXIDE, (4) COMPONENTS OF THE NUTRIENT MEDIUM,
(5) AUTOINHIBITORS, AND (6) CHARACTERISTICS OF THE ORGANISM. A SPECIFIC
GROWTH RATE OF 1.96 PER DAY IS CONSIDERED TO BE ABOUT THE MAXIMUM RATE
FOR C PYRENOIDOSA AND IT IS CONCLUDED THAT THIS VALUE IS NOT
ESTABLISHED BY LIMITATIONS OF PHOTOSYNTHESIS. (UTTORMARK-WISC)

FIELD 05C

ACCESSION NO. W69-07442

PREDICTING DISSOLVED OXYGEN VARIATIONS CAUSED BY ALGAE,

FEDERAL WATER POLLUTION CONTROL ADMINISTRATION, ALAMEDA, CALIF. CENTRAL
PACIFIC RIVER BASINS PROJECT.

RICHARD C. BAIN, JR.

ASCE PROC, J SANIT ENG DIV, VOL 94, NO SA5, PAP 6155, PP 867-881, OCT 1968.
15 P, 8 FIG, 2 TAB, 26 REF, APPEND.

DESCRIPTORS:
*DISSOLVED OXYGEN, *ALGAE, *RESPIRATION, *ESTUARIES, *EUTROPHICATION,
*CALIFORNIA, WATER POLLUTION, REAERATION, NUTRIENTS, OXYGEN SAG,
AERATION, BIOCHEMICAL OXYGEN DEMAND, PHOTOSYNTHETIC OXYGEN.

IDENTIFIERS:
STREETER-PHELPS EQUATION, DISSOLVED OXYGEN VARIATIONS, SAN JOAQUIN
RIVER, POTOMAC ESTUARY, SAN FRANCISCO BAY.

ABSTRACT:
EUTROPHIC ENVIRONMENTS ARE OFTEN DOMINATED BY PLANKTONIC ALGAL
POPULATIONS (PHYTOPLANKTON) WHICH CAUSE DIURNAL VARIATIONS IN DO
CONCENTRATIONS THROUGH RESPIRATORY ACTIVITY AND PHOTOSYNTHESIS.
PHOTOSYNTHETIC OXYGENATION AND RESPIRATORY DEOXYGENATION RATES OF
ESTUARINE PHYTOPLANKTON WERE MEASURED AT VARIOUS STANDING CROP
(CHLOROPHYLL) LEVELS. OXYGEN PRODUCTION AND CONSUMPTION RATES FOR
ACTIVELY GROWING PHYTOPLANKTON POPULATIONS WERE RELATED TO STANDING
CROP AT 20 DEG C AND NON-LIMITING LIGHT. VARIATIONS IN ALGAL
PHOTOSYNTHETIC PRODUCTION RATE, AS RELATED TO LIGHT ADAPTION, AGE OF
CELLS, NUTRITION, TEMPERATURE AND ALGAL TYPE, ARE CONSIDERED. THE
STREETER-PHELPS EQUATION WAS MODIFIED TO INCLUDE PHYTOPLANKTON
PRODUCTION AND RESPIRATION RATES IN FORMULATIONS DESIGNED TO PREDICT DO
CONCENTRATIONS OVER A 24-HR PERIOD. AN EXAMPLE IS GIVEN, AND THE
RESULTING PREDICTIONS ARE COMPARED WITH FIELD MEASUREMENTS FROM A TIDAL
REACH OF THE SAN JOAQUIN RIVER, CALIFORNIA. (KNAPP-USGS)

FIELD 05C

ACCESSION NO. W69-07520

THE GALLOPING GHOST OF EUTROPHY,

NEW HAMPSHIRE WATER SUPPLY AND POLLUTION CONTROL COMMISSION.

TERRENCE P. FROST.

APPALACHIA, VOL XXXVII, NO 1, PP 25-36, JUNE 1968. 1 FIG, 4 PLATES.

DESCRIPTORS:
*EUTROPHICATION, *LAKES, GLACIATION, URBANIZATION, DOMESTIC WASTES,
NUTRIENTS, PHOSPHORUS, ALGAE, ALGICIDES, HARVESTING OF ALGAE,
ENVIRONMENT, CYANOPHYTA.

IDENTIFIERS:
*AESTHETIC DETERIORATION, *ADVANCED WASTE TREATMENT, *NUTRIENT REMOVAL,
AGRICULTURAL DRAINAGE, APPALACHIAN MOUNTAIN CLUB HUTS, TEGERNSEE,
GERMANY, EXPANDING TECHNOLOGY, OVERPOPULATION.

ABSTRACT:
THE INCREASED INFLUENCE OF MAN ON THE ENVIRONMENT HAS CAUSED A CHANGE
IN THE DEFINITION OF EUTROPHICATION FROM A NATURAL AGING PROCESS OF
LAKES DUE TO TOPOGRAPHY, DEGREE OF FERTILITY, SEDIMENT LOADS FROM
MINERAL AND ORGANIC SOLIDS, SUBSURFACE GEOLOGY, AND WEATHERING, TO AN
ACCELERATED AGING DUE TO MAN'S INFLUENCE. INCREASING URBANIZATION ALONG
WITH OUR CHANGING LIFE STYLE, IS SUGGESTED AS THE CAUSE FOR 'GALLOPING
EUTROPHY.' THE EFFECT IS THE PROLIFERATION OF ALGAL SPECIES, HARMLESS
AND DIVERSIFIED IN SMALL NUMBERS; NOXIOUS AND CONFINED TO PROFUSION OF
A FEW HARDY BLUE-GREEN SPECIES IN OVERPOPULATION. THE RESULT IS
AESTHETIC DETERIORATION, FISH KILLS, SEPTIC ODORS, DISCOLORATION OF
OBJECTS IN WATER, ORGANIC DEBRIS BUILDUPS ON BOTTOMS, SCUMS AND MATS.
SUGGESTED METHODS OF CONTROL INCLUDE CHEMICAL TREATMENT WITH ALGICIDES
OR HARVESTING ORGANISMS, BUT THESE ELIMINATE SYMPTOMS WITHOUT REACHING
THE CAUSES. REAL NEED IS NUTRIENT REMOVAL FROM SEWAGE EFFLUENT BY A
VARIETY OF TECHNIQUES, COAGULATION (LIME OR ALUM), REVERSE OSMOSIS, ION
EXCHANGE, ELECTRODIALYSIS, DISTILLATION, HYDROPHONICS, AND IRRIGATION.
MENTION IS ALSO MADE OF METHODS AND IMPROVEMENTS IN WASTE DISPOSAL AT
APPALACHIAN MOUNTAIN CLUB HUTS AND THE TEGERNSEE RECREATION AREA IN
GERMANY. AUTHOR'S CONCLUSION: THE HEART OF THE PROBLEM LIES IN
UNHARNESSED TECHNOLOGY AND PEOPLE POLLUTION. (KETELLE-WISC)

FIELD 05C

ACCESSION NO. W69-07818

SEAWEED EXTRACTS AS FERTILISERS,

PORTSMOUTH COLL. OF TECH. (ENGLAND). SCHOOL OF PHARMACY.

G. BLUNDE, S. B. CHALLEN, AND D. L. WOODS.

J SCI FOOD AGRIC, VOL 19, PP 289-293, JUNE 1968, 1 FIG. 8 TAB, 26 REF.

DESCRIPTORS:
*FERTILIZERS, *CYCLING NUTRIENTS, EUTROPHICATION, ALGAE, PLANT GROWTH
SUBSTANCES, PLANT GROWTH, CARBOHYDRATES.

IDENTIFIERS:
*SEAWEED EXTRACTS, ASCOPHYLLUM NODOSUM, ALGINIC ACID, FUCUS
VESICULOSUS, GROWTH FACTORS, LAMINARIA SACCHARINA, MANNITOL, MUSTARD
GROWTH TEST, ORGANIC GROWTH FACTORS, POLYSACCHARIDES.

ABSTRACT:
IN THE MUSTARD-GROWTH TEST, THE GROWTH-PROMOTING EFFECTS OF AQUEOUS
EXTRACTS OF LAMINARIA SACCHARINA, FUCUS VESICULOSUS AND ASCOPHYLLUM
NODOSUM WERE LARGELY DUE TO THE METAL CATIONS PRESENT, BUT THE EFFECTS
WERE MODIFIED BY ORGANIC SUBSTANCES IN THE EXTRACTS. AMINO ACIDS AND
MANNITOL IN SEAWEED EXTRACTS HAD LITTLE EFFECT ON PLANT GROWTH, AND
COMPOUNDS EXTRACTED WITH ORGANIC SOLVENTS WERE ONLY PARTIALLY
RESPONSIBLE FOR THE MODIFICATION OF THE GROWTH-PROMOTING EFFECT. THE
POLYSACCHARIDE ALGINIC ACID AND ITS SALTS SEEMED TO BE THE MAIN ORGANIC
COMPOUNDS RESPONSIBLE FOR REDUCING THE EFFECTS OF METALS AND MAY HAVE
COMPETED WITH THE PLANTS BY ION EXCHANGE FOR THE METALS IN THE EXTRACT.
(KONRAD-WISC)

FIELD 05C

ACCESSION NO. W69-07826

ALGAE AND MAN.

SYRACUSE UNIV., N.Y. DEPT. OF CIVIL ENGINEERING.

FROM SYMP ON ALGAE AND MAN, NATO ADVANCED STUDY INST, LOUISVILLE, KY, 1962.
DANIEL F. JACKSON, (ED). PLENUM PRESS, NEW YORK, 1964. 90 FIG, 48 TAB, 1000
REF.

DESCRIPTORS:
*EUTROPHICATION, *ALGAE, BIOASSAY, BIOLOGICAL COMMUNITIES, CULTURES,
DIATOMS, ECOLOGY, NUISANCE ALGAE, PHAEOPHYTA, PHOTOSYNTHESIS,
PHYTOPLANKTON, PRIMARY PRODUCTIVITY, SYSTEMATICS, WATER POLLUTION
EFFECTS, WATER POLLUTION SOURCES, WATER SUPPLY.

IDENTIFIERS:
BENTHIC ALGAE, BLOOMS, CLOSED SYSTEMS, CYTOLOGY, ENVIRONMENTAL
CONDITIONS, EXTRACELLULAR PRODUCTS, GAS EXCHANGE, GREEN PLANTS, MASS
CULTURE, MEDICINE, METABOLIC PATTERNS, MICROALGAE, MICRONUTRIENTS,
PHYCOLOGY, SPACE RESEARCH, TOXIC ALGAE.

ABSTRACT:
THIS COMPENDIUM IS BASED ON A SERIES OF LECTURES PRESENTED AT THE NATO
ADVANCED STUDY INSTITUTE, THE FIRST TO BE HELD IN THE UNITED STATES,
WHICH MET AT THE UNIVERSITY OF LOUISVILLE, KENTUCKY, JULY 22-AUGUST 11,
1962. THE SERIES BEARS ON THE ROLE OF ALGAE IN EUTROPHICATION OF
NATURAL WATERS AND ON THEIR POSSIBLE UTILIZATION FOR CLOSED
LIFE-SUPPORT SYSTEMS IN THE SPACE AGE. THE SCOPE OF THE WORK IS
INDICATED BY NAMES OF AUTHORS AND TOPICS OF THEIR CONTRIBUTIONS: G. W.
PRESCOTT, SYSTEMATICS; C VAN DEN HOEK, TAXONOMY; TYGE CHRISTENSEN,
GROSS CLASSIFICATION; MARGARET ROBERTS, CYTOLOGY OF PHAEOPHYTA; G. E.
FOGG, ENVIRONMENTAL CONDITIONS AND ALGAL METABOLIC PATTERNS; CLYDE
EYSTER, MICRONUTRIENT REQUIREMENTS; A. G. WURTZ, PROBLEMS OF CULTURE;
F. E. ROUND, ECOLOGY OF BENTHIC ALGAE; RUTH PATRICK, DIATOM
COMMUNITIES; JAMES B. LACKEY, ECOLOGY OF PLANKTONIC ALGAE; JACOB
VERDUIN, PRINCIPLES OF PRIMARY PRODUCTIVITY; C. MERVIN PALMER, ALGAE IN
WATER SUPPLIES; OTTO SKULBERG, EUTROPHICATION AND BIOASSAY; GEORGE P.
FITZGERALD, BIOTIC RELATIONS IN BLOOMS; PAUL R. GORHAM, TOXIC ALGAE;
MARCEL LEFEVRE, EXTRACELLULAR PRODUCTS; DAVID AND MORTON SCHWIMMER,
ALGAE AND MEDICINE; RICHARD J. BENOIT, MASS CULTURE FOR GAS EXCHANGE;
AND F. EVENS, FUTURE OF PHYCOLOGY. WSELECTED INDIVIDUAL CONTRIBUTIONS
TO THE VOLUME WILL BE ABSTRACTED INDIVIDUALLY. (SEE ALSO W69-01977).
(EICHHORN-WISC)

FIELD 05C

ACCESSION NO. W69-07832

ALGAL PROBLEMS RELATED TO THE EUTROPHICATION OF EUROPEAN WATER SUPPLIES, AND A
BIO-ASSAY METHOD TO ASSESS FERTILIZING INFLUENCES OF POLLUTION ON INLAND
WATERS,

NORSK INSTITUTT FOR VANNFORSKNING, BLINDERN.

OLAV M. SKULBERG.

ALGAE AND MAN, JACKSON, DANIEL F, ED, PLENUM PRESS, N Y, PP 262-299, 1964. 16
FIG, 9 TAB, 76 REF.

DESCRIPTORS:
    *ALGAE, *BIOASSAY, *EUTROPHICATION, *FERTILIZATION, *INLAND WATERWAYS,
    *WATER POLLUTION EFFECTS, *WATER SUPPLY, BIOINDICATORS, CALCIUM,
    CHLORIDES, CONDUCTIVITY, CULTURES, HYDROGEN ION CONCENTRATION, IRON,
    LAKES, MAGNESIUM, NUTRIENTS, NUISANCE ALGAE, OLIGOTROPHY, ODOR,
    PHYTOPLANKTON, SULFATES, SEDIMENTATION, SEDIMENTS, TASTE, WATER
    CHEMISTRY, WATER QUALITY.

IDENTIFIERS:
    *EUROPE, ALGAL GROWTH, ANKISTRODESMUS, BACILLARIOPHYCEAE,
    CHLOROPHYCEAE, CHRYSOPHYCEAE, CARTERIA, COMPARATIVE STUDIES, CRUCIGENIA
    RECTANGULARIS, DICERAS CHODATI, DINOPHYCEAE, FILTER BLOCKING,
    HETERETROPHS, ITALY, LAKE BORREVANNET NORWAY, LAKE LUGANO ITALY, LAKE
    MARIDALSVANNET NORWAY, LAKE ZURICH, SWITZERLAND, NITELV RIVER NORWAY,
    NORWAY, OSCILLATORIA RUBESCENS, OSLO NORWAY, PERMANGANATE VALUES,
    SCHIZOPHYCEAE, SELENASTRUM CAPRICORNUTUM, TABELLARIA FLOCCULOSA.

ABSTRACT:
    THE AUTHOR LISTS AMONG THE PROBLEMS CAUSED BY EXCESSIVE ALGAL GROWTH,
    FILTER-BLOCKING, IMPARTATION OF TASTE AND ODOR TO WATER SUPPLIES,
    DISCOLORATION OF WATER, PROVISION OF NUTRIENTS FOR UNDESIRABLE
    HETEROTROPHS IN PIPES, AND FORMATION OF SEDIMENTS IN LAKES. TO
    ILLUSTRATE DIFFERENCES IN TROPHIC STATUS AMONG LAKES, HE COMPARES TWO
    LAKES IN NORWAY: LAKE MARIDALSVANNET (3.9 SQUARE KILOMETERS; MAXIMUM
    DEPTH, 45 METERS) WHICH PROVIDES OSLO WITH DRINKING WATER, IS
    OLIGOTROPHIC, AND SHOWS LITTLE ANNUAL VARIATION IN ITS LOW
    CONCENTRATIONS OF DISSOLVED NUTRIENTS. LAKE BORREVANNET (1.8 SQUARE KM;
    MAX DEPTH, 16 METERS) IS EUTROPHIC AND SHOWS IMPORTANT ANNUAL CHANGES
    WITH RESPECT TO CHEMISTRY AND WATER QUALITY. AUTHOR STATES THAT LAKE
    LUGANO, NORTHERN ITALY, HAS EUTROPHIED DUE TO POLLUTION ENRICHMENT,
    CITING CLASSIC EXAMPLE OF LAKE ZURICH AS EVIDENCE THAT LACUSTRINE
    TROPHIC CHANGES ARE BEST RECORDED IN THE SEDIMENTS. UTILIZING
    SELENASTRUM (OR ANKISTRODESMUS) CAPRICORNUTUM, AN ORGANISM REQUIRING
    ONLY A FEW DAYS TO ATTAIN PEAK GROWTH, AUTHOR DEVISED A BIOASSAY TO
    ASSESS EFFECTS AND CONCENTRATIONS OF NUTRIENTS IN NATURAL WATERS. SUCH
    AN APPROACH WOULD APPEAR TO HAVE ITS GREATEST VALUE IN COMPARATIVE
    LIMNOLOGICAL STUDIES. (SEE ALSO W69-07832). (EICHHORN-WISC)

FIELD 05C

ACCESSION NO. W69-07833

TRANSFORMATIONS IN INFILTRATION PONDS AND IN THE SOIL LAYERS IMMEDIATELY
  UNDERNEATH,

RESEARCH INSTITUTE FOR PUBLIC HEALTH ENGINEERING TNO, THE HAGUE
  (NETHERLANDS). WATER, SOIL, AND AIR DIV.

J. K. BAARS.

PROC RUDOLFS RES CONF, RUTGERS UNIV, NEW BRUNSWICK, NJ. PRINCIPLES AND
APPLICATIONS IN AQUATIC MICROBIOLOGY, HEUKELEKIAN, H AND DONDERO, NORMAN C
(EDS), JOHN WILEY AND SONS, INC, NEW YORK, PP 344-365, 1964. 8 FIG, 2 TAB,
9 REF, DISC.

DESCRIPTORS:
  *INFILTRATION, *PONDS, *SOIL, GROUNDWATER BASINS, WATER SUPPLY, MUD,
  BENTHOS, HYDROGEN, BACTERIA, AEROBIC BACTERIA, METABOLISM, CANALS,
  SANDS, CHLORINATION, MICROORGANISMS, CHEMICAL ANALYSIS, TEMPERATURE,
  NITRATES, PLANTS, AMMONIA, HARDNESS(WATER), ACIDITY, OXYGEN, ALGAE,
  SELF-PURIFICATION, BACTERIOPHAGE, RESERVOIRS, SPORES, NITRITES, IRON,
  ORGANIC MATTER, SUBSOIL, VELOCITY, OXIDATION, DENITRIFICATION,
  SULFATES, REDUCTION(CHEMICAL), FERMENTATION, METHANE, HYDROGENATION,
  CHEMICAL OXYGEN DEMAND, BIOCHEMICAL OXYGEN DEMAND, POTASSIUM COMPOUNDS,
  CARBON DIOXIDE, PROTOZOA, SEWAGE, SLUDGE, E COLI.

IDENTIFIERS:
  *TRANSFORMATIONS, AMSTERDAM, HAARLEM, LEYDEN, THE HAGUE,
  MINERALIZATION, CLAY LENSES, RHINE RIVER, AMSTERDAM RHINE, POLDERS,
  COMPOSITION, PSEUDOMONAS, MYCOBACTERIUM, BACILLUS SUBTILIS, BACILLUS
  MYCOIDES, BACILLUS MESENTERICUS, DIATOMEA, CLADOPHORA, FLAGELLATA,
  CILIATES, RANUNCULUS CIRCINATUS, POTAMOGETON PUSILLUS.

ABSTRACT:
WATER FROM STORAGE PONDS, FILTERED THROUGH BOTTOM MUD AND A BODY OF
SAND, WAS MEASURED FOR CHLORIDE, FORMS OF NITROGEN, OXYGEN, ORGANIC
MATTER, ALGAE, AND BACTERIAL CONTENT. TO STUDY BENTHOS, WATER WAS
ANALYZED FROM WELLS CONSTRUCTED AT DISTANCES FROM A POND IN THE
DIRECTION OF FLOW. NOT ONLY DISSOLVED ORGANIC MATTER, AS DETERMINED BY
THE POTASSIUM PERMANGANATE METHOD, IS MINERALIZED, BUT MUCH MORE
OXIDANT (FREE OXYGEN PLUS NITRATE OXYGEN) IS USED, APPARENTLY IN THE
BENTHOS. THE HIGH OXIDANT CONSUMPTION INDICATES THAT THE REDUCTION OF
ORGANIC MATTER IS INTENSE. ORGANIC SUBSTANCES, DETECTABLE BY TASTE, ARE
NOT COMPLETELY ELIMINATED. WHEN THERE IS A CONSTANT SUPPLY OF NEW
OXIDANT, DEAD BIOLOGICAL MATERIALS MAY BE MINERALIZED AEROBICALLY. IF
THIS SUPPLY IS TERMINATED, ANAEROBIC TRANSFORMATIONS MAY DOMINATE.
PROBABLY BACTERIA ARE STRONGLY ADSORBED TO THE SAND GRAINS IN
CONCENTRATIONS RESULTING IN INTENSE STRUGGLE FOR LIFE AND DECREASE
RAPIDLY WITH INCREASING DISTANCE FROM THE POND. SAMPLES FROM THE BOTTOM
OF THE PONDS DOWNWARD SHOWED PRONOUNCED DECREASE IN BACTERIA AT
0.5-METER DEPTH. A SMALL AMOUNT OF SILT IN THE WATER MAY CAUSE CLOGGING
OF THE SOIL. THE INTERMITTENT INFILITRATION SYSTEM AFFORDS THESAME
PERMEABILITY OF THE SAND EACH YEAR WITH THE CONSEQUENT USE OF AVAILABLE
PURIFYING ACTORS. (JONES-WIS)

FIELD 05D

ACCESSION NO. W69-07838

EFFECTS OF TEMPERATURE ON THE SORPTION OF RADIONUCLIDES BY A BLUE-GREEN ALGA,

DU PONT DE NEMOURS (E. I.) AND CO., AIKEN, S. C. SAVANNAH RIVER LAB.

R. S. HARVEY.

AVAILABLE FROM CLEARINGHOUSE AS CONF 670503 AT $3.00 IN PAPER COPY AND $0.65
IN MICROFICHE. SYMPOSIUM ON RADIOECOLOGY, PROC 2ND NATL SYMP, ANN ARBOR,
MICH, PP 266-269, MAY 15-17, 1967. 7 FIG, 2 TAB, 5 REF.

DESCRIPTORS:
  *ALGAE, *RADIOISOTOPES, *SORPTION, *TEMPERATURE, SOUTH CAROLINA,
  CESIUM, STRONTIUM RADIOISOTOPES, ZINC RADIOISOTOPES, IRON, MANGANESE,
  BENTHIC FLORA, STREAMS, CULTURES, COBALT RADIOISOTOPES.

IDENTIFIERS:
  SAVANNAH RIVER LABORATORY, PLECTONEMA BORYANUM, REACTOR.

ABSTRACT:
  THE EFFECTS OF TEMPERATURE DIFFERENTIAL (15 DEGREES CELSIUS) ON
  SORPTION OF CESIUM-137, ZINC-65, IRON-59, COBALT-57, AND MANGANESE-54,
  BY THE FILAMENTOUS CYANOPHYTE, PLECTONEMA BORYANUM, ARE REPORTED. THE
  ALGA WAS COLLECTED FROM REACTOR EFFLUENT STREAMS. UNIALGAL CULTURES
  WERE DEVELOPED. CULTURES WERE GROWN CONCURRENTLY AT FOUR WATER
  TEMPERATURES AND SAMPLES WEIGHED AND RADIO-ASSAYED AFTER EXPOSURES OF
  INCREASING TIMES. THIS SPECIES GREW BEST BETWEEN 30 C AND 40 C. GROWTH
  WAS NOT AFFECTED BY LOW CONCENTRATIONS OF RADIONUCLIDES IN THE MEDIUM.
  RADIONUCLIDE CONCENTRATIONS AT THE VARIOUS WATER TEMPERATURES WERE
  COMPARED. FOR A GIVEN WATER TEMPERATURE, SORPTION LEVELS DIFFERED FOR
  THE RADIONUCLIDES STUDIED BECAUSE OF VARIANCES IN SPECIFIC ACTIVITY AND
  BIOLOGICAL DEMAND FOR VARIOUS ELEMENTS. THE ESSENTIAL ELEMENTS OF
  MANGANESE-54, XENON-65, COBALT-57, AND IRON-59, WERE SORBED TO HIGHER
  LEVELS THAN WERE THE NONESSENTIAL ELEMENTS, STRONTIUM-85 AND
  CESIUM-137, PERHAPS DUE IN PART TO THEIR PHYSICAL STATE; ONLY
  CESIUM-137 AND STRONTIUM-85 WERE PRESENT, MAINLY IN IONIC FORM. SINCE
  SORPTION LEVELS FOR THE VARIOUS RADIONUCLIDES WERE RAISED OR LOWERED BY
  FACTORS LESS THAN 2.5 BY THE TEMPERATURE DIFFERENTIAL, THE CONCLUSION
  WAS DRAWN THAT NONLETHAL VARIATIONS IN WATER TEMPERATURE HAVE NO MAJOR
  INFLUENCE ON THE SORPTION BY P BORYANUM OF THE RADIONUCLIDES TESTED.

FIELD 05C

ACCESSION NO. W69-07845

ACCUMULATION OF RADIUM-226 IN TWO AQUATIC ECOSYSTEMS,

UTAH STATE UNIV., LOGAN. DEPT. OF WILDLIFE RESOURCES.

SUSAN S. MARTIN, WILLIAM T. HELM, AND WILLIAM F. SIGLER.

AVAILABLE FROM CLEARINGHOUSE AS CONF 670503 AT $3.00 IN PAPER COPY AND $0.65
IN MICROFICHE. SYMPOSIUM ON RADIOECOLOGY, PROC 2ND NATL SYMP, ANN ARBOR,
MICH, PP 307-318, MAY 15-17, 1967. 10 FIG, 14 REF, DISC. CONF-670503.

DESCRIPTORS:
*AQUATIC ENVIRONMENTS, *ECOSYSTEMS, *RADIUM RADIOISOTOPES, *URANIUM
RADIOISOTOPES, RIVERS, MILLS, COLORADO, WASTES, BIOLOGICAL COMMUNITIES,
SEDIMENTS, BIOTA, FISH, UTAH, ALGAE, INSECTS, INVERTEBRATES, BACKGROUND
RADIATION, FLOCCULATION, POLLUTANTS, SAMPLING.

IDENTIFIERS:
ANIMAS RIVER, DURANGO, SAN MIGUEL RIVER, DOLORES RIVER, URAVAN, COTTUS
SPP, RHINICHTHYS OSCULUS, SUCKERS, ICTALURUS PUNCTATUS, TRICHOPTERA,
ODONATA, EPHEMEROPTERA, PLECOPTERA, DIPTERA, COLEOPTERA, HEMIPTERA,
MOLLUSCA, RAFFINATE, THORIUM, TAILINGS.

ABSTRACT:
RADIUM-226, IN URANIUM PROCESSING WASTES, SERIOUSLY AFFECTS STREAM
BIOTA. ITS ACCUMULATION IN BIOTIC COMPONENTS WAS FOLLOWED FOR THREE
YEARS. THE ANIMAS RIVER, COLORADO, AFTER DIVERSION OF THE MOST TOXIC OF
A MILL'S EFFLUENTS, SHOWED EXCELLENT RECOVERY BIOLOGICALLY. WATER,
SEDIMENT, ALGAE, INVERTEBRATES, AND FISH, WERE SAMPLED ABOVE AND BELOW
MILL SITES. COMPARISON OF BACKGROUND LEVELS AND RADIUM ACCUMULATION
SHOWED DIFFERENCES IN UPTAKE BY VARIOUS FISH SPECIES. A CONCENTRATION
FACTOR (THE RATIO OF THE CONCENTRATION OF A PARTICULAR RADIONUCLIDE IN
THE ORGANISM TO ITS CONCENTRATION IN THE AQUEOUS MEDIUM) SHOWED THE
MEAN RADIUM-226 CONTENT DECREASING WITH DOWNSTREAM DISTANCE. DURING THE
LAST TWO YEARS, RADIUM CONTENT OF SEDIMENTS AVERAGED ONLY ABOUT THREE
TIMES BACKGROUND LEVELS, EARLIER SAMPLES CONTAINED UP TO TWENTY TIMES,
WHILE STILL EARLIER WORK INDICATED UP TO SEVERAL HUNDRED TIMES THE
BACKGROUND LEVELS OF RADIUM-226. THE SAN MIGUEL-DOLORES RIVER SYSTEM
WAS IN POOR BIOLOGICAL CONDITION DUE PRIMARILY TO CHEMICAL POLLUTION
FROM A URANIUM MILL. BIOTA SAMPLES CONTAINED INCREASING AMOUNTS OF
RADIUM-226 AS DISTANCE BELOW THE MILL INCREASED, ALTHOUGH RADIUM-226
CONTENT OF THE WATER AND SEDIMENTS DECREASED WITH DISTANCE. APPARENTLY
ORGANISMS COULD NOT SURVIVE LONG ENOUGH IN THE UPPER POLLUTION ZONE TO
ACCUMULATE THE AMOUNTS OF RADIUM-226 WHICH MIGHT OTHERWISE BE FOUND.
(SEE ALSO VOL. 2, NO. 18, FIELD 5C, W69-07441) (JONES-WISC)

FIELD 05C

ACCESSION NO. W69-07846

EFFECT OF HANFORD REACTOR SHUTDOWN ON COLUMBIA RIVER BIOTA,

BATTELLE-NORTHWEST, RICHLAND, WASH. PACIFIC NORTHWEST LAB.

D. G. WATSON, C. E. CUSHING, C. C. COUTANT, AND W. L. TEMPLETON.

AVAILABLE FROM CLEARINGHOUSE AS CONF 670503 AT $3.00 IN PAPER COPY AND $0.65
IN MICROFICHE. SYMP ON RADIOECOLOGY, PROC 2ND NATIONAL SYMP, MAY 15-17,
1967, ANN ARBOR, MICH, NELSON, DANIEL J AND EVANS, FRANCIS C (EDS). US
ATOMIC ENERGY COMM, DOC CONF 670503, PP 291-299. 8 FIG, 23 REF, DISC.

DESCRIPTORS:
*BIOTA, *COLUMBIA RIVER, CHROMIUM, PHOSPHORUS RADIOISOTOPES, TROPHIC
LEVEL, ZINC RADIOISOTOPES, IRON, SEDIMENTS, FISH, RADIOECOLOGY,
EFFLUENTS, ECOSYSTEMS, PLANKTON, ALGAE, PERIPHYTON, INVERTEBRATES,
ADSORPTION, PHYTOPLANKTON, SUCKERS, SHINERS, CADDISFLIES, TEMPERATURE,
FLOW, DIATOMS, MANGANESE.

IDENTIFIERS:
*HANFORD(WASH), *REACTOR, RADIONUCLIDES, ASSIMILATION, BA-LA,
HALF-LIFE, PTYCHOCHEILUS OREGONENSIS, ACROCHEILUS ALUTACEUS, ULOTHRIX,
CLADOPHORA, PROSOPIUM WILLIAMSONI.

ABSTRACT:
CLOSURE OF THE HANFORD REACTORS FOR AN EXTENDED PERIOD CAUSED RAPID AND
EXTENSIVE DECLINE IN CONCENTRATION OF RADIONUCLIDES. CHROMIUM-51
(CR-51) AND PHOSPHORUS-32 (P-32) DECREASED TWO TO THREE ORDERS OF
MAGNITUDE IN LOWER TROPHIC LEVELS. THE RAPID CHANGE OF P-32 WAS
PROBABLY DUE TO ITS RELATIVE SHORT PHYSICAL HALF-LIFE AND ITS RAPID
TURNOVER IN AQUATIC ORGANISMS. THIS WAS CHIEFLY A BIOLOGICAL PROCESS,
NOT MERELY SURFACE ADSORPTION. ALTHOUGH CR-51 WAS ABUNDANT, ITS LOW
CONCENTRATION IN HIGHER LEVELS INDICATES ITS MINOR BIOLOGICAL
IMPORTANCE. ITS RELATIVELY HIGH CONCENTRATIONS IN PLANKTON, ALGAE, AND
INVERTEBRATES, WAS PROBABLY DUE TO ADSORPTION RATHER THAN ASSIMILATION.
DECLINES IN CONCENTRATIONS OF ZINC-65 (ZN-65), MANGANESE-54, AND
IRON-59, WERE MUCH LESS. THEIR LESSER DECLINE IN BIOTA AS OPPOSED TO
RIVER WATER SUGGESTED THAT THESE NUCLIDES EITHER ARE TURNED OVER AT A
VERY SLOW RATE IN THE ORGANISMS, OR THAT THEY WERE AVAILABLE FROM PARTS
OF THE ECOSYSTEM OTHER THAN WATER. IN FISH P-32 WAS LOST RAPIDLY, ZN-65
SLOWLY. THE SLOWER RATE OF CHANGE OF RADIONUCLIDES BY FISH INDICATED
DIFFERENCE IN UPTAKE ROUTES. NEAR EQUILIBRIUM CONCENTRATIONS OF
RADIONUCLIDES IN MOST RIVER ORGANISMS WERE APPROACHED WITHIN TWO OR
THREE WEEKS AFTER RESUMPTION OF REACTOR OPERATION.
(JONES-WISC)

FIELD 05C, 05B

ACCESSION NO. W69-07853

THE ROLE OF TUBIFICID WORMS IN THE TRANSFER OF RADIOACTIVE PHOSPHOROUS IN AN AQUATIC ECOSYSTEM,

WESTERN MICHIGAN UNIV., KALAMAZOO.

BERT K. WHITTEN, AND CLARENCE J. GOODNIGHT.

AVAILABLE FROM CLEARINGHOUSE AS CONF 670503 AT $3.00 IN PAPER COPY AND $0.65 IN MICROFICHE. SYMP ON RADIOECOLOGY, PROC 2ND NATIONAL SYMP, MAY 15-17, 1967, ANN ARBOR, MICH, NELSON, DANIEL J AND EVANS, FRANCIS C (EDS). US ATOMIC ENERGY COMM, DOC CONF 670503, PP 270-277. 2 FIG, 2 TAB, 20 REF.

DESCRIPTORS:
*TUBIFICIDS, *ECOSYSTEMS, *PHOSPHORUS RADIOISOTOPES, *TRANSFER, WATER, BACTERIA, SEDIMENTS, MINNOWS, DETRITUS, CYCLING NUTRIENTS, PLANKTON, ALGAE, INVERTEBRATES, TURBIDITY, WORMS, SUNFISHES.

IDENTIFIERS:
RADIOAUTOGRAPHS, HALF-LIFE, CONCENTRATION FACTOR, LIMNODRILUS SPP, AEOLOSOMA HEMPRICHI, ESCHERICHIA COLI, MYRIOPHYLLUM, LEPOMIS MACROCHIRUS, PIMEPHALES NOTATUS.

ABSTRACT:
ACCUMULATION OF RADIOPHOSPHORUS BY TUBIFICID WORMS (COMMON HABITAT BOTTOM SEDIMENTS OF STREAMS AND LAKES) FROM WATER, BACTERIA, AND SEDIMENT, WAS STUDIED. TUBIFICID WORMS TOOK UP MORE PHOSPHORUS FROM WATER AND BACTERIA THAN FROM SEDIMENT. THESE WORMS WERE ABLE TO ACCUMULATE PHOSPHORUS FROM BOTH SOLUBLE AND ORGANIC PARTICULATE SOURCES. RADIOAUTOGRAPHS DEMONSTRATED THAT MUCH OF THE ACTIVITY WAS IN THE TISSUES OF THESE WORMS AND NOT SIMPLY ADSORBED. NO SIGNIFICANT ACCUMULATION FROM RADIOACTIVE ORTHOPHOSPHATE, WHICH WAS ADSORBED ONTO STERILE SEDIMENTS, WAS OBSERVED. BLUEGILLS AND BLUNTNOSE MINNOWS FED TUBIFICID WORMS LABELLED WITH RADIOPHOSPHORUS ACCUMULATED RADIOPHOSPHORUS IN THEIR TISSUES. THEORETICAL CALCULATIONS BASED ON THESE FEEDING STUDIES DEMONSTRATED THAT AN EQUILIBRIUM WAS APPROACHED AFTER TWO WEEKS IN THE TISSUES OF THESE FISH AND WAS MAINTAINED THROUGH THE DURATION OF THE EXPERIMENT. TUBIFICID WORMS, BECAUSE OF THEIR ABUNDANCE IN SOME AQUATIC ECOSYSTEMS, MAY HAVE AN IMPORTANT ROLE IN TRANSFER OF RADIOACTIVE PHOSPHORUS FROM WATER AND DETRITUS TO OTHER COMPONENTS OF THE AQUATIC ECOSYSTEM, SUCH AS FISH. THESE WORMS APPEAR TO FUNCTION IN CONJUNCTION WITH BACTERIA IN THE RECYCLING OF RADIOPHOSPHORUS FROM DETRITUS AND SEDIMENTS. THIS EXPERIMENTATION IS INTENDED TO SUGGEST POSSIBLE RELATIONSHIPS WHICH MAY OCCUR IN THE ENVIRONMENT.
(JONES-WISC)

FIELD 05C

ACCESSION NO. W69-07861

RADIONUCLIDE CYCLING BY PERIPHYTON: AN APPARATUS FOR CONTINUOUS IN SITU
MEASUREMENTS AND INITIAL DATA ON ZINC-65 CYCLING,

BATTELLE-NORTHWEST, RICHLAND, WASH. PACIFIC NORTHWEST LAB.

C. E. CUSHING, AND N. S. PORTER.

AVAILABLE FROM CLEARINGHOUSE AS CONF 670503 AT $3.00 IN PAPER COPY AND $0.65
IN MICROFICHE. SYMP ON RADIOECOLOGY, PROC 2ND NATIONAL SYMP, MAY 15-17,
1967, ANN ARBOR, MICH, NELSON, DANIEL J AND EVANS, FRANCIS C (EDS). US
ATOMIC ENERGY COMM, DOC CONF-670503, PP 285-290. 3 FIG, 1 TAB, 12 REF.

DESCRIPTORS:
  *PERIPHYTON, *CYCLING NUTRIENTS, *ZINC RADIOISOTOPES, STREAMFLOW,
  SAMPLING, ALGAE, ECOSYSTEMS, ENVIRONMENT, RETENTION, PHOTOPERIODISM,
  LIGHT INTENSITY, VELOCITY, TEMPERATURE, CHEMICAL ANALYSIS,
  PHYSIOLOGICAL ECOLOGY, OXYGEN, CARBON DIOXIDE, CARBON RADIOISOTOPES,
  DETRITUS, COLUMBIA RIVER, ELECTRONICS, SCALING.

IDENTIFIERS:
  *RADIONUCLIDE, ULOTHRIX, UPTAKE, HALF-LIFE, DETECTOR.

ABSTRACT:
  THIS PART OF THE COLUMBIA RIVER PROGRAM CONCERNS THE ROLE OF THE
  PERIPHYTON COMMUNITY IN CYCLING RADIONUCLIDES. DIFFICULTIES IN STUDYING
  A LARGE RIVER WITH EXTREMELY UNSTABLE HYDROGRAPHY HAVE LED TO
  DEVELOPMENT OF A SYSTEM HAVING ADVANTAGES OF LABORATORY CONTROLLED
  EXPERIMENTS WHILE MAINTAINING SOME SEMBLANCE OF NATURAL CONDITIONS. IT
  PERMITS THE CONTINUOUS MEASUREMENT OF UPTAKE AND CYCLING OF
  RADIONUCLIDES BETWEEN STREAM PERIPHYTON AND A CONTROLLED AQUEOUS
  ENVIRONMENT. THE SYSTEM WAS DESIGNED TO AVOID INHERENT DIFFICULTIES IN
  AQUARIA STUDIES, THAT IS, THE NECESSITY OF DESTROYING OR SUB-SAMPLING
  THE COMMUNITY FOR SEQUENTIAL ANALYSES, THE LACK OF A CONTINUOUS FLOW OF
  WATER OVER ALGAE TO SIMULATE STREAM CONDITIONS, AND INABILITY TO
  MAINTAIN A FIXED AMBIENT RADIONUCLIDE CONCENTRATION BECAUSE OF
  IMMEDIATE UPTAKE AND RECYCLING BY THE ORGANISMS. APPARATUS PERMITS
  EVALUATION OF THE EFFECT OF INDIVIDUAL PHYSICAL AND CHEMICAL
  ENVIRONMENTAL FACTORS IN EITHER AN OPEN ONE-PASS SYSTEM OR IN A CLOSED
  RECIRCULATING SYSTEM. PRELIMINARY RESULTS OF SIX UPTAKE AND TWO
  RETENTION EXPERIMENTS USING ZINC-65 IN THE CLOSED SYSTEM ARE PRESENTED.
  DESPITE COMMUNITY TYPE, THAT IS MATURE OR 'YOUNG', THE TIME OF
  APPROXIMATE EQUILIBRIUM IS AROUND 20 HOURS AND THE ACTIVITY ACCUMULATED
  ON A WEIGHT BASIS IS ABOUT 0.021 NANNOCURIES/MILLIGRAM DRY WEIGHT.
  (JONES-WISC)

FIELD 05C, 07B

ACCESSION NO. W69-07862

REMOVAL OF NITROGENOUS COMPOUNDS FROM WASTEWATERS,

CARNEGIE-MELLON UNIV., PITTSBURGH, PA.

WILLIAM R. SAMPLES.

WATER REUSE, AMER INST CHEM ENG PROGR SYMP, SER NO 78, VOL 63, PP 223-229, 1967. 7 P, 1 FIG, 35 REF.

DESCRIPTORS:
*WASTE WATER TREATMENT, *WATER REUSE, *NITROGEN, NITROGEN COMPOUNDS, OXIDATION, BIODEGRADATION, NITRIFICATION, DENITRIFICATION, NUTRIENTS, TERTIARY TREATMENT, ALGAE, AMMONIA.

IDENTIFIERS:
NITROGEN REMOVAL.

ABSTRACT:
NORMAL TREATMENT OF OUR WATERBORNE WASTES HAS NOT IN THE PAST BEEN CONCERNED WITH REMOVAL OF NITROGENOUS MATERIALS. IN RECENT YEARS, HOWEVER, WASTE DISCHARGES CONTAINING COMPOUNDS OF NITROGEN HAVE BEEN INCRIMINATED IN THE FERTILIZATION OF STREAMS AND LAKES, CAUSING POSSIBLE HEALTH HAZARDS, AND IN GENERAL, CAUSING A DECREASE IN WATER QUALITY FOR MANY USES. A REVIEW OF MANY INVESTIGATIONS INTO MECHANISMS FOR REMOVAL OF NITROGENOUS MATERIALS IN WASTE EFFLUENTS IS GIVEN. THE MECHANISMS REVIEWED INCLUDE BOTH BIOLOGICAL AND CHEMICAL METHODS. NITROGENOUS MATERIALS MAY BE REMOVED FROM WASTE EFFLUENTS BY ANY ONE OF SEVERAL METHODS. THE MOST APPLICABLE AT THE PRESENT TIME APPEAR TO BE DENITRIFICATION, INCORPORATION BY ALGAE, AND AMMONIA STRIPPING. THE SELECTION OF THE PROPER METHOD FOR NITROGEN REMOVAL IN ANY PARTICULAR CASE WILL DEPEND ON THE LOCAL CIRCUMSTANCES, INCLUDING OTHER TREATMENTS REQUIRED, DEGREE OF NITROGEN REMOVAL DESIRED, CHEMICAL QUALITY OF THE WATER, FURTHER USES FOR THE WATER, AND MANY OTHERS. THE REMOVAL OF NITROGEN FROM EFFLUENTS WILL BE EXPENSIVE BUT WILL BY JUSTIFIABLE IN AN INCREASING NUMBER OF INSTANCES. (KNAPP-USGS)

FIELD 05D

ACCESSION NO. W69-08053

A DESIGN PROCEDURE FOR BIOLOGICAL NITRIFICATION AND DENITRIFICATION,

TEXAS UNIV., AUSTIN.

W. WESLEY ECKENFELDER, JR.

WATER REUSE, AMER INST CHEM ENG PROGR SYMP, SER NO 78, VOL 63, PP 230-234, 1967. 5 P, 6 FIG, 1 TAB, 6 REF, APPEND.

DESCRIPTORS:
*DENITRIFICATION, *BIODEGRADATION, *WATER REUSE, *RECLAIMED WATER, NITROGEN COMPOUNDS, NITROGEN, AMMONIA, NUTRIENTS, BIOCHEMICAL OXYGEN DEMAND, TERTIARY TREATMENT, ALGAE, BACTERIA, ACTIVATED SLUDGE.

IDENTIFIERS:
NITROGEN REMOVAL.

ABSTRACT:
THE REMOVAL OF NITROGEN FROM SEWAGE AND INDUSTRIAL WASTEWATERS IS ASSUMING INCREASING IMPORTANCE AS THE WASTE LOADINGS FROM URBAN AND INDUSTRIALIZED AREAS INCREASE. UNOXIDIZED NITROGEN EXERTS ON OXYGEN DEMAND ON THE RECEIVING WATERS AND OXIDIZED NITROGEN SERVES AS A NUTRIENT SOURCE FOR ALGAL GROWTH. IT IS POSSIBLE TO REMOVE NITROGEN BY BIOLOGICAL MEANS BY FIRST OXIDIZING AMMONIA TO NITRATES FOLLOWED BY DENITRIFICATION IN WHICH MICROORGANISMS REDUCE THE NITRATE TO NITROGEN GAS. STUDIES WHICH HAVE BEEN CONDUCTED ON THIS PROCESS TO DATE IN LABORATORY, PILOT-PLANT, AND, TO A LIMITED EXTENT, PLANT-SCALE INVESTIGATIONS ARE REVIEWED. THE THEORY OF THE PROCESS OPERATION IS DEVELOPED FROM BASIC MICROBIOLOGICAL CONSIDERATIONS. A DESIGN EXAMPLE FOR A SEWAGE TREATMENT PLANT IS PRESENTED TO ILLUSTRATE THE DEVELOPMENT OF THE PROCESS. (KNAPP-USGS)

FIELD 05D

ACCESSION NO. W69-08054

INFLUENCE OF ENVIRONMENTAL FACTORS ON THE CONCENTRATIONS OF ZN-65 BY AN
EXPERIMENTAL COMMUNITY,

BUREAU OF COMMERCIAL FISHERIES, BEAUFORT, N. C. RADIOBIOLOGICAL LAB.

T. DUKE, J. WILLIS, T. PRICE, AND K. FISCHLER.

AVAILABLE FROM CLEARINGHOUSE AS CONF 670503 AT $3.00 IN PAPER COPY AND $0.65
IN MICROFICHE. SYMP ON RADIOECOLOGY, PROC 2ND NATIONAL SYMP, MAY 15-17,
1967, ANN ARBOR, MICH, NELSON, DANIEL J AND EVANS, FRANCIS C (EDS). US
ATOMIC ENERGY COMM, DOC CONF 670503, PP 355-362. 3 FIG, 1 TAB, 18 REF.

DESCRIPTORS:
    *ZINC RADIOISOTOPES, *ENVIRONMENTAL EFFECTS, SALINITY, TEMPERATURE, SEA
WATER, OYSTERS, CLAMS, CRABS, NORTH CAROLINA, SEDIMENTS, RADIOACTIVITY,
MARINE ANIMALS, ESTUARINE ENVIRONMENT, FALLOUT, NUCLEAR EXPLOSIONS,
BIOTA, SPECTROMETERS, BACTERIA, ALGAE, SANDS, HYDROGEN ION
CONCENTRATION.

IDENTIFIERS:
    *CONCENTRATIONS, AEQUIPECTEN IRRADIANS, PIVERS ISLAND(N C), DETECTOR,
BEAUFORT(N C), CRASSOSTREA VIRGINICA, MERCENARIA MERCENARIA, PANOPEUS
HERBSTII.

ABSTRACT:
    CAPACITY OF ORGANISMS TO CONCENTRATE ELEMENTS FROM SEAWATER IS AFFECTED
BY CHANGES IN ENVIRONMENTAL FACTORS THAT ALTER THE PHYSIOLOGICAL
CONDITION OF THE ORGANISMS OR THE PHYSICAL-CHEMICAL PROPERTIES OF
ELEMENTS. THIS KNOWLEDGE IS NEEDED TO EVALUATE POTENTIAL HAZARDS FROM
RADIOACTIVE POLLUTION. ZINC-65 (ZN-65) ENTERS THE BIOCHEMICAL CYCLE OF
THE ESTUARINE ENVIRONMENT IN FALLOUT FROM NUCLEAR WEAPONS AND IN
EFFLUENT FROM NUCLEAR REACTORS, OCCURRING IN MEASURABLE AMOUNTS IN
ESTUARINE BIOTA, SEDIMENTS, AND WATER. A POLYFACTORIAL APPROACH IS
REQUIRED TO STUDY THE CONCENTRATION OF RADIOISOTOPES. DIFFERENCES IN
SALINITY, TEMPERATURE, PH, AND ZINC IN SEAWATER SIGNIFICANTLY AFFECTED
THE CONCENTRATION OF ZN-65 OF A COMMUNITY OF OYSTERS, CLAMS, MUD CRABS,
AND SCALLOPS, AND THEIR SEDIMENT SUBSTRATE. A FACTORIAL ANALYSIS OF
VARIANCE SHOWED NO INTERACTIONS AMONG THE ENVIRONMENTAL FACTORS. HIGH
SALINITY AND ZINC CONCENTRATION SUPPRESSED THE CONCENTRATION OF ZN-65
IN ANIMALS AND SEDIMENT, WHEREAS HIGH TEMPERATURE AND PH HAD THE
OPPOSITE EFFECT. TEMPORAL VARIATION IN CONCENTRATION FACTORS FOR
CONTROL ANIMALS AND SEDIMENT WAS SIGNIFICANT DURING THE EXPERIMENT.
SUCH VARIATION MUST BE CONSIDERED WHEN REPORTING CONCENTRATION FACTORS
FOR RADIONUCLIDES IN MARINE ORGANISMS. DATA FROM THIS EXPERIMENT
REPRESENT A FIRST STEP TOWARD PREDICTION OF THE FATE OF A RADIOISOTOPE
RELEASED INTO AN ESTUARY.
(JONES-WISC)

FIELD 05C

ACCESSION NO. W69-08267

DISTRIBUTION OF RADIONUCLIDES IN THE ENVIRONMENT OF ENIWETOK AND BIKINI ATOLS,
AUGUST 1964,

WASHINGTON UNIV., SEATTLE. COLL. OF FISHERIES.

A. D. WELANDER.

SYMP ON RADIOECOLOGY, PROC 2ND NATL SYMP, MAY 15-17, 1967, ANN ARBOR, MICH,
NELSON, DANIEL J AND EVANS, FRANCIS C (EDS). US ATOMIC ENERGY COMM, DOC
CONF 670503, PP 346-354. 8 TAB, 7 REF, DISC. AVAILABLE FROM CLEARINGHOUSE
AS CONF 670503 AT $3.00 IN PAPER COPY AND $0.65 IN MICROFICHE.

DESCRIPTORS:
*RADIOISOTOPES, *DISTRIBUTION, ANIMALS, PLANTS, BOTTOM SEDIMENTS,
SOILS, NUCLEAR EXPLOSIONS, COBALT RADIOISOTOPES, CESIUM, STRONTIUM
RADIOISOTOPES, MANGANESE, ALGAE, IRON, BIRDS, RADIOACTIVITY,
INVERTEBRATES, PLANKTON, FISH, SEA WATER, GAMMA RAYS, RADIOCHEMICAL
ANALYSIS, TRITIUM, CARBON RADIOISOTOPES, BIOTA, CLAMS, FALLOUT, PACIFIC
OCEAN, SPECTROMETERS.

IDENTIFIERS:
*ENIWETOK ATOLL, *BIKINI ATOLL, RUTHENIUM-106, ANTIMONY-125,
GROUNDWATER, BISMUTH-207, VERTEBRATES, PLUTONIUM-239, RATS,
ZIRCONIUM-95-NIOBIUM, EUROPIUM-155, SILVER-110M, CERIUM, GUETTARDA,
IPOMOEA, PISONIA, PANDANUS, MUSCLE, BONE, LEAVES, SHELL, RHODIUM-102.

ABSTRACT:
RADIONUCLIDE ANALYSES WERE MADE OF MORE THAN 2000 SAMPLES OF ANIMALS,
PLANTS, WATER, BOTTOM SEDIMENTS, AND SOILS, COLLECTED AT ENIWETOK AND
BIKINI ATOLLS, SITE OF 59 NUCLEAR TESTS BETWEEN 1946 AND 1958.
COBALT-60 WAS FOUND IN ALL SAMPLES BUT WAS DOMINANT IN THE MARINE
ENVIRONMENT. CESIUM-137 (CS-137) AND STRONTIUM-90, HOWEVER,
PREDOMINATED IN TERRESTRIAL ENVIRONMENTS. ALL SAMPLES CONTAINED TRACES
OF MANGANESE-54, ALTHOUGH THERE WERE SLIGHTLY LARGER AMOUNTS OF THIS
RADIONUCLIDE IN LAND PLANTS AND LAND INVERTEBRATES. RUTHENIUM-106 AND
ANTIMONY-125 WERE DETECTED IN SIGNIFICANT AMOUNTS IN GROUNDWATER AND
SOIL, WITH TRACES IN ANIMALS AND PLANTS. TRACE AMOUNTS OF BISMUTH-207
WERE ALSO FOUND IN MOST SAMPLES, AND CERIUM-144 USUALLY IN ALGAE, SOILS
AND LAND PLANTS. THERE WERE COMPARATIVELY LARGE AMOUNTS OF IRON-55 IN A
NUMBER OF SAMPLES, ESPECIALLY IN VERTEBRATES. PLUTONIUM-239 WAS FOUND
IN SOIL AND IN SKINS OF RATS AND BIRDS, IN THE FEW SAMPLES ANALYZED.
SOILS CONTAINED HIGHEST LEVELS OF RADIOACTIVITY, FOLLOWED BY
INVERTEBRATES, GROUNDWATER, SHOREBIRDS, PLANTS, RATS, PLANKTON, ALGAE,
FISH, BOTTOM SEDIMENTS, SEAWATER, AND SEABIRDS. IN LAND SAMPLES, CS-137
PROVED TO BE BEST MEASURE OF DISTRIBUTION OF RADIOACTIVITY IN
BIOENVIRONMENTAL SAMPLES. HIGHEST RADIOACTIVITY WAS FOUND IN SAMPLES
COLLECTED NEAR TEST SITES.
(JONES-WIS)

FIELD 05C

ACCESSION NO. W69-08269

MEASUREMENTS OF BACKGROUND RADIATION IN AQUATIC HABITATS IN ALASKA,

DARTMOUTH COLL., HANOVER, N. H. DEPT. OF BIOLOGICAL SCIENCES; AND ARMY
  TERRESTRIAL SCIENCES CENTER, HANOVER, N. H.

G. E. LIKENS, AND P. L. JOHNSON.

SYMP ON RADIOECOLOGY, PROC 2ND NATL SYMP, MAY 15-17, 1967, ANN ARBOR, MICH,
  NELSON, DANIEL J AND EVANS, FRANCIS C (EDS). US ATOMIC ENERGY COMM, DOC
  CONF 670503, PP 319-328. 3 FIG, 2 TAB, 21 REF, DISC. AVAILABLE FROM
  CLEARINGHOUSE AS CONF 670503 AT $3.00 IN PAPER COPY AND $0.65 IN MICROFICHE.

DESCRIPTORS:
  *BACKGROUND RADIATION, *AQUATIC HABITATS, *ALASKA, LAKES, RIVERS, HOT
  SPRINGS, TERRESTRIAL HABITATS, DEPTH, AIR, SEDIMENTS, GEOLOGIC
  FORMATIONS, ARCTIC, TUNDRA, SCALING, GRAVELS, MUD, SANDS, SHALES,
  BOTTOM SEDIMENTS, CATTAILS, RADIOACTIVITY, INTERFACES, ORGANIC MATTER,
  RADIOISOTOPES, GAMMA RAYS, URANIUM RADIOISOTOPES, RADIUM RADIOISOTOPES,
  POTASSIUM RADIOISOTOPES, SPECTROMETERS, EROSION, FALLOUT, LITTORAL,
  ALGAE, BIOTA, GENETICS, PLANKTON.

IDENTIFIERS:
  DETECTOR, ALASKA, HUDEUC LAKE, PULLIN LAKE, GRAPHITE LAKE, OLD JOHN
  LAKE, PINGO LAKE, TWELVE MILE LAKE, BRANT LAKE, IMIKPUK LAKE, NORTH
  MEADOW LAKE, GYTTJA, DUFF, OOZE, COBBLES, EQUISETUM, THORIUM-232, YUKON
  RIVER, NEBESNA RIVER, MOOSE CREEK, ARCTIC CIRCLE HOT SPRINGS, RADON,
  CENTRAL CITY(COLO), MUTATION, BETA RAYS, COSMIC RADIATION, STEESE
  HIGHWAY, POINT BARROW.

ABSTRACT:
  HIGH HUMAN WHOLE-BODY COUNTS OF ESKIMO POPULATION IN NORTHERN ALASKA
  INTRODUCED THROUGH FALLOUT PARTICLES FROM NUCLEAR WEAPONS TESTING
  PROMPTED STUDY OF BACKGROUND RADIATION IN VARIOUS ALASKAN LAKES.
  MEASUREMENTS OF IONIZING RADIATION, MADE DURING 1965 IN 9 LAKES, 3
  RIVERS, 1 HOT SPRINGS AREA, AND ADJACENT TERRESTRIAL SUBSTRATES IN
  ALASKA, INDICATED THAT BACKGROUND RADIATION IN THESE HABITATS VARIED
  GREATLY. THE LAKES WERE CHARACTERIZED BY A SMALL AMOUNT OF IONIZING
  RADIATION AT MID-DEPTHS AND INCREASING QUANTITIES NEAR THE AIR AND
  SEDIMENT BOUNDARIES. MOREOVER, RADIOACTIVITY OF SEDIMENTS NEAR SHORE
  WAS 3.2 TIMES GREATER AS THAT NEAR THE CENTER. THIS PATTERN OF
  BACKGROUND RADIATION HAS BEEN RELATIVELY CONSISTENT IN NATURAL LAKES OF
  DIVERSE GEOGRAPHIC AREAS; HENCE, A MODEL FOR BACKGROUND RADIATION IN
  FRESHWATER LAKES IS PROPOSED. VARIATIONS IN THIS MODEL WERE A FUNCTION
  OF LOCAL DIFFERENCES IN GEOLOGIC SUBSTRATES AND INPUT OF ALLOCHTHONOUS
  MATERIALS. RADIOACTIVITY FROM SEDIMENTS IN ARCTIC TUNDRA PONDS WAS DUE
  TO NATURALLY OCCURRING RADIONUCLIDES WITH PROBABLE EFFECTS ON BIOTA.
  AUTHORS SPECULATE THAT ORGANISMS LIVING IN OR NEAR SUBLITTORAL
  SEDIMENTS OF LAKES IN RECENT TIMES MIGHT UNDERGO MORE FREQUENT GENETIC
  CHANGES THAN PLANKTON OR ORGANISMS IN DEEPWATER SEDIMENTS. (JONES-WISC)

FIELD 05C

ACCESSION NO. W69-08272

THE PHOSPHORUS AND ZINC CYCLES AND PRODUCTIVITY OF A SALT MARSH,

GEORGIA UNIV., ATHENS. DEPT. OF ZOOLOGY.

L. R. POMEROY, R. E. JOHANNES, E. P. ODUM, AND B. ROFFMAN.

SYMP ON RADIOECOLOGY, PROC 2ND NATL SYMP, MAY 15-17, 1967, ANN ARBOR, MICH, NELSON, DANIEL J AND EVANS, FRANCIS C (EDS). US ATOMIC ENERGY COMM, DOC CONF 670503, PP 412-419. 3 FIG, 1 TAB, 33 REF.

DESCRIPTORS:
   *ZINC RADIOISOTOPES, *PHOSPHORUS RADIOISOTOPES, *SALT MARSHES, *CYCLING
   NUTRIENTS, GEORGIA, SEDIMENTS, BACTERIA, DETRITUS, ESTUARIES, PLANKTON,
   ECOSYSTEMS, SINKS, TRACERS, CRABS, SHRIMP, MULLETS, FISH, ALGAE.

IDENTIFIERS:
   DUPLIN RIVER(GA), SPARTINA ALTEMIFLORA, ORGANISMS, TURSIOPS TRUNCATUS,
   BREVOORTIA TYRRANUS, CONCENTRATION PROCESSES.

ABSTRACT:
   BY SYNTHESIZING RESULTS OF EXPERIMENTS USING PHOSPHORUS-32 (P-32) AND
   ZINC-65 (ZN-65) WITH EARLIER WORK IN GEORGIA SALT MARSHES, THE CYCLES
   OF THE ELEMENTS P AND ZN ARE DESCRIBED QUANTITATIVELY. MARSH GRASS AND
   SEDIMENTS DOMINATE BOTH CYCLES. THE UPPERMOST METER OF SEDIMENTS
   CONTAINS ENOUGH P TO SUPPORT SPARTINA PRODUCTION FOR 500 YEARS AND
   ENOUGH ZN FOR 5000 YEARS. SPARTINA PRODUCTION REMOVES P AND ZN FROM
   SUBSURFACE (REDUCED) SEDIMENTS AND INTRODUCES THEM INTO THE WATER (VIA
   BACTERIAL UTILIZATION AND SUBSEQUENT UTILIZATION BY DETRITUS FEEDERS)
   AT A RATE THAT REPLACES TOTAL WATER P IN A MONTH AND TOTAL WATER ZN IN
   A YEAR. SIGNIFICANT PART OF P IS EXPORTED FROM THE MARSH IN ORGANISMS
   AND DETRITUS. THE P AND ZN IN ESTUARINE WATER ARE IN EQUILIBRIUM WITH
   PLANKTON, BACTERIA, AND SURFACE (OXIDIZED) SEDIMENTS. EQUILIBRIUM
   STRONGLY FAVORS THE SEDIMENTS. SUBSURFACE-SEDIMENT COMPARTMENT OF P AND
   ZN IS PROBABLY REPLACED BY CONVERSION OF OXIDIZED SEDIMENTS TO REDUCED
   SEDIMENTS THROUGH CREEK MEANDERING. OXIDIZED SEDIMENTS ARE IN
   EQUILIBRIUM WITH WATER RECEIVING INPUTS OF P AND ZN FROM LAND AND SEA.
   TRANSFER OF P AND ZN FROM DEEP SEDIMENTS TO WATER BY SPARTINA EXPLAINS
   THE HIGH CONCENTRATION OF THESE ELEMENTS IN THE WATER.
   (JONES-WIS)

FIELD 05C

ACCESSION NO. W69-08274

A COMPARISON BETWEEN THE UPTAKE OF RADIOACTIVE AND STABLE ZINC BY A MARINE
UNICELLULAR ALGA,

COMITATO NAZIONALE PER L'ENERGIA NUCLEARE, LA SPEZIA (ITALY).

M. BERNHARD, AND A. ZATTERA.

SYMP ON RADIOECOLOGY, PROC 2ND NATL SYMP, MAY 15-17, 1967, ANN ARBOR, MICH, NELSON, DANIEL J AND EVANS, FRANCIS C (EDS). US ATOMIC ENERGY COMM, DOC CONF 670503, PP 389-398. 5 FIG, 7 REF, DISC. AVAILABLE FROM CLEARINGHOUSE AS CONF 670503 AT $3.00 IN PAPER COPY AND $0.65 IN MICROFICHE.

DESCRIPTORS:
   *ZINC RADIOISOTOPES, *MARINE ALGAE, *RESINS, SEA WATER, PHYSICOCHEMICAL
   PROPERTIES, TRACERS, DISTRIBUTION, CHELATION, DETERGENTS,
   RADIOACTIVITY, AMINO ACIDS.

IDENTIFIERS:
   *RADIONUCLIDE UPTAKE, *PHAEODACTYLUM TRICORNUTUM, EDTA, LIGURIAN SEA,
   EUROPEAN MEDITERRANEAN.

ABSTRACT:
   ORGANIC SUBSTANCES OCCURRING IN SEAWATER OR INTRODUCED AS CONTAMINANTS,
   E.G., DETERGENTS, MAY CHELATE ZINC AND INFLUENCE ITS AVAILABILITY TO
   ALGAE. UPTAKE OF ZINC-65 (ZN-65) AND STABLE ZINC BY A MARINE
   UNICELLULAR ALGA (PHAEODACTYLUM TRICORNUTUM) AND BY A CHELATING RESIN
   (CHELEX-100) WAS STUDIED IN BATCH CULTURES DURING A CERTAIN TIME
   INTERVAL. IN ALL EXPERIMENTS, THE DISTRIBUTION PATTERN OF ZN-65 AND
   STABLE ZINC BETWEEN PARTICULATE MATTER (ALGAE OR CHELATING RESIN) AND
   NATURAL AND ARTIFICIAL SEAWATER MEDIUM DIFFERED. WHEN ZN-65 WAS ADDED
   AS IONIC ZINC, PROPORTIONATELY MUCH MORE RADIOACTIVE ZINC WAS TAKEN UP
   THAN STABLE ZINC. HOWEVER, WHEN ZN-65 IS ADDED AS ZN-EDTA-COMPLEX TO A
   BATCH CONTAINING SEAWATER AND CHELEX, STABLE ZINC WAS INITIALLY TAKEN
   UP AT A HIGHER RATE THAN WAS ZN-65. THESE PRELIMINARY RESULTS DO NOT
   EXPLAIN THE DIFFERENCE IN DISTRIBUTION OF RADIOACTIVE AND STABLE ZINC,
   BUT THE DATA SHOW THAT UPTAKE RATE OF THE STABLE ISOTOPE DOES NOT
   NECESSARILY FOLLOW UPTAKE RATE OF THE RADIOACTIVE ISOTOPE IF THE TWO
   ARE PRESENT IN DIFFERENT PHYSICAL-CHEMICAL STATES. THIS POSSIBILITY
   MUST BE TAKEN INTO ACCOUNT IN TRACER STUDIES AND IN PREDICTIONS OF THE
   DISTRIBUTION OF RADIOISOTOPES IN ENVIRONMENT BASED ON SPECIFIC ISOTOPE
   CONTENT APPROACH. (JONES-WIS)

FIELD 05C

ACCESSION NO. W69-08275

ALGAE: AMOUNTS OF DNA AND ORGANIC CARBON IN SINGLE CELLS,

CALIFORNIA UNIV., SAN DIEGO, LA JOLLA. INST. OF MARINE RESOURCES.

OSMUND HOLM-HANSEN.

SCIENCE, VOL 163, PP 87-88, JANUARY 1969. 1 FIG, 19 REF.

DESCRIPTORS:
*ALGAE, *CARBON, *CARBON CYCLE, *CYTOLOGICAL STUDIES, CULTURES,
FLUOROMETRY, BIOMASS, OCEANOGRAPHY.

IDENTIFIERS:
*DNA, *ORGANIC CARBON, *CELL SIZE, EUKARYOTES, MONOCHRYSIS LUTHERI,
NAVICULA PELLICULOSA, GONYAULAX POLYEDRA, 3,5-DIAMINOBENZOIC ACID
DIHYDROCHLORIDE, INFRARED GAS ANALYSIS, COULTER COUNTER, EUGLENA
GRACILIS, CACHONINA NIEI, SKELETONEMA COSTATUM, DUNALIELLA TERTIOLECTA,
AMPHIDINIUM CARTERI, SYRACOSPHAERA ELONGATA, THALASSIOSIRA FLUVIATILIS,
DITYLUM BRIGHTWELLII, BIOLOGICAL OCEANOGRAPHY, ECOLOGIGAL TECHNIQUES.

ABSTRACT:
UTILIZING INFRARED GAS ANALYSIS FOR CARBON AND FLUOROMETRY FOR
DESOXYRIBONUCLEIC ACID (DNA), AUTHOR ANALYZED THE FOLLOWING VARIOUSLY
SIZED SPECIES OF UNICELLULAR ALGAE: MONOCHRYSIS LUTHERI, NAVICULA
PELLICULOSA, GONYAULAX POLYEDRA, CACHONINA NIEI, SKELETONEMA COSTATUM,
DUNALIELLA TERTIOLECTA, AMPHIDINIUM CARTERI, SYRACOSPHAERA ELONGATA,
THALASSIOSIRA FLUVIATILIS, DITYLUM BRIGHTWELLII. CARBON CONTENT VARIED
FROM 10 TO 6000 PICOGRAMS PER CELL, AND TOTAL CELLULAR DNA IS DIRECTLY
PROPORTIONAL TO CELL SIZE. CONTENT OF DNA IS EQUAL TO APPROXIMATELY
1-3% OF CELLULAR ORGANIC CARBON. DATA AND RELATIONSHIPS DEVELOPED IN
THIS STUDY SUGGEST THAT DNA-DETERMINATIONS ARE OF POTENTIAL VALUE IN
DETERMINING LIVING BIOMASS IN ECOLOGICAL STUDIES. (EICHHORN-WIS)

FIELD 02K, 05C

ACCESSION NO. W69-08278

LAKE TERMINOLOGY: WATER BLOOM,

MICHIGAN STATE UNIV., EAST LANSING.

J. O. VEATCH, AND C. R. HUMPHRYS.

BULL MICH AGRIC COLL EXP STATION, EAST LANSING, P 241, 1964. 1 FIG.

DESCRIPTORS:
*EUTROPHICATION, ALGAE, DUCKS, COLOR, FISHKILL, LAKES, ODOR, TOXICITY,
WATER POLLUTION EFFECTS, WATER QUALITY, CATTLE.

IDENTIFIERS:
*DEFINITIONS, *WATER BLOOM, TOXIC ALGAE, RECREATIONAL USE.

ABSTRACT:
THIS LEXICON OF LAKE TERMINOLOGY DEFINES A WATER BLOOM AS: 'A PROLIFIC
GROWTH OF PLANKTON. A BLOOM OF ALGAE MAY BE SO DENSE THAT IT IMPARTS A
GREENISH, YELLOWISH, OR BROWNISH COLOR TO THE WATER. THE GROWTH MAY BE
SO CONCENTRATED IN SOME PARTS OF A LAKE THAT IT INTERFERES WITH
SWIMMING AND BOATING. THE ALGAE NOT ONLY IMPARTS A DISAGREEABLE ODOR,
BUT IT MAY BE A CAUSE OF FISH MORTALITY, AND SOME SPECIES MAY BE
POISONOUS TO CATTLE AND DUCKS AND A MENACE TO DRINKING WATER SUPPLIES.'
THIS ENTRY INCLUDES AN AERIAL PHOTOGRAPH OF AN ALGAL BLOOM CONCENTRATED
IN BAY OF A LAKE. (EICHHORN-WIS)

FIELD 05C

ACCESSION NO. W69-08279

PRACTICAL ALGAE CONTROL METHODS FOR NEW HAMPSHIRE WATER SUPPLIES,

NEW HAMPSHIRE WATER SUPPLY AND POLLUTION CONTROL COMMISSION, CONCORD.

TERRENCE P. FROST.

J NEW HAMPSHIRE WATER WORKS ASSOC, 7 P, APRIL 1960. 6 REF.

DESCRIPTORS:
*ALGAE CONTROL, *WATER SUPPLY, *NEW HAMPSHIRE, ALGAE, ODOR, TASTE,
LIGHT TEMPERATURE, NUTRIENTS, COPPER SULFATE, STRATIFICATION.

IDENTIFIERS:
*CORRECTIVE TREATMENT, *PREVENTATIVE TREATMENT, ALGAE NUISANCE, ALGAE
BLOOM, ALGAE GROWTH, WATER DENSITY, CONVECTION CURRENTS, ALGAE
TREATMENT COSTS, DINOBRYON, CONCORD(N H), PEANACOOK LAKE(N H).

ABSTRACT:
AUTHOR BRIEFLY REVIEWS CHARACTERISTICS OF ALGAE AND ASSOCIATED PROBLEMS
IN NEW HAMPSHIRE WATER SUPPLIES, OUTLINING PREVENTATIVE MEASURES
TOGETHER WITH EASIEST AND MOST ECONOMICAL CONTROLS USED TO DATE.
SYMPTOMS ARE UNDESIRABLE TASTE AND ODOR IN WATER SUPPLIES, BLOOMS IN
RESERVOIRS, FOULED PIPES, AND CLOGGED FILTERS. OCCURRENCE OF ALGAE IS
INFLUENCED BY LIGHT, TEMPERATURE, CONVECTION CURRENTS, DENSITY
STRATIFICATION (SPRING AND FALL TURNOVERS CIRCULATE BOTTOM MATERIALS
AND ALGAE, OFTEN INITIATING ALGAL BLOOM), AND NUTRIENTS WHICH PROMOTE
ALGAL GROWTH. IDEALLY, EACH WATERWORKS SHOULD PERIODICALLY EXAMINE
SAMPLES MICROSCOPICALLY TO MONITOR ALGAE SPECIES AND APPLY PREVENTATIVE
TREATMENT BEFORE NUISANCE LEVELS ARE REACHED. ALGAL CONTROL IN MOST OF
NEW HAMPSHIRE'S 76 PUBLIC SURFACE WATER SUPPLIES IS PRACTICABLE AND
INEXPENSIVE. PENACOOK LAKE, SUPPLYING WATER FOR CONCORD, HAS A PROBLEM
CAUSED PRIMARILY BY THE SPECIES DINOBRYON; THIS LAKE COULD BE TREATED
WITH COPPER SULFATE TO A DEPTH OF TEN FEET (3,380 ACRE FEET) FOR $340.
MUNICIPALITIES CAN OBTAIN HELP FROM NEW HAMPSHIRE'S WATER POLLUTION
COMMISSION OR SANITARY ENGINEERING DIVISION FOR IDENTIFICATION OF
ORGANISMS AND RECOMMENDED DOSAGES OF COPPER SULFATE. USE OF
CHLORINATION AND COPPER SULFATE SEEM THE BEST AVAILABLE MEANS AND ONLY
ACCEPTABLE CHEMICALS FOR TREATING THEIR WATER SUPPLIES. (KETELLE-WIS)

FIELD 05C, 05F

ACCESSION NO. W69-08282

THE 'WORKING' OF THE MADISON LAKES,

WILLIAM TRELEASE.

WISCONSIN ACAD SCI, ARTS AND LETTERS, VOL 7, PP 121-129, 1889. 1 FIG, 1
PLATE, 60 REF.

DESCRIPTORS:
*EUTROPHICATION, CYANOPHYTA, ODOR, SCUM, FISHING, WISCONSIN,
BIBLIOGRAPHIES, LAKES, NUISANCE ALGAE, BACTERIA, WATER POLLUTION
EFFECTS.

IDENTIFIERS:
*MADISON(WIS), *'WORKING' OF LAKES, *WATER-BLOOM, 'BREAKING' OF LAKES,
CHROOCOCCACEAE, NOSTOCHINEAE, CLATHROCYSTIS AERUGINOSA, COELOSPHAERIUM
KUTZINGIANUM, ANABAENA FLOS AQUAE, ANABAENA MENDOTAE, ANABAENA HASSALI,
LYNGBYA NOLLEI, GLOCOTRICHIA PISUM, NOSTOC VERRUCOSUM, LAKE
MENDOTA(WIS), LAKE MONONA(WIS), LAKE WAUBESA(WIS), ANABAENA CIRCINALIS.

ABSTRACT:
THE MADISON (WISCONSIN) LAKES WERE OBSERVED FOR SIGNS OF EUTROPHICATION
FROM 1882-1887. SYMPTOMS OBSERVED INCLUDED GREENISH-YELLOW SCUM OR FINE
GRANULES SUSPENDED IN WATER, ODORS ASSOCIATED WITH ALGAL DECOMPOSITION,
STAINING OF SHORELINES, AND DECREASED FISH BITING. THE ALGAE GENERALLY
APPEARED IN MID-SUMMER AND PERSISTED FOR ONLY A FEW DAYS BEFORE BEING
BROKEN UP AND DISPERSED BY BREEZES. SUCH PHENOMENA HAVE BEEN REFERRED
TO AS THE 'WORKING' OR 'BREAKING' OF LAKES AND ARE SYNONYMOUS WITH THE
EUROPEAN TERM 'WATER-BLOOM'. THE SYMPTONS, CAUSED BY THE OCCURRENCE OF
MEMBERS OF THE ALGAE GROUP CYANOPHYCEAE, ARE FAVORED IN STAGNANT WATER
BODIES AND USUALLY OCCUR AFTER PROLONGED HOT SPELLS. THE TWO PRINCIPAL
SUBDIVISIONS OF THE GROUP ARE CHROOCOCCACEAE AND NOSTOCHINEAE. OF THE
FORMER, CLATHROCYSTIS AERUGINOSA AND COELOSPHAERIUM KUTZINGIANUM WERE
THE PRINCIPAL PROBLEM SPECIES IN THE MADISON LAKES, AND FROM THE
LATTER, ANABAENA FLOS AQUAE AND ANABAENA CIRCINALIS WERE MOST COMMON.
ANABAENA MENDOTAE, WHICH IS MORE TOLERANT OF COLD, PREDOMINATES LATE IN
THE SEASON. PAPER INCLUDES A LIST OF SIXTY REFERENCES ON WATER-BLOOMS.
ONLY PRINCIPAL PAPERS TREATING RECURRENCE OF BACTERIA IN QUANTITY ARE
INCLUDED, MOST OF THEM HAVE LITTLE BOTANICAL VALUE BUT SERVE AS A
NUCLEUS FOR OTHER REFERENCES AVAILABLE IN 1889. (KETELLE-WIS)

FIELD 02H, 05C

ACCESSION NO. W69-08283

OLEFINS OF HIGH MOLECULAR WEIGHT IN TWO MICROSCOPIC ALGAE,

HOUSTON UNIV., TEX. DEPT. OF CHEMISTRY; HOUSTON UNIV., TEX. DEPT. OF
BIOLOGY; AND HOUSTON UNIV., TEX. DEPT. OF BIOPHYSICAL SCIENCES.

E. GELPI, J. ORO, H. J. SCHNEIDER, AND E. O. BENNETT.

SCIENCE, VOL 161, NO 3842, PP 700-701, AUGUST 1968. 2 FIG, 1 TAB, 23 REF.

DESCRIPTORS:
*ALGAE, CYANOPHYTA, CHRYSOPHYTA, CULTURES, DIAGENESIS, SEDIMENTS, GAS
CHROMATOGRAPHY, LIPIDS, OIL, OIL SHALE.

IDENTIFIERS:
*BOTRYOCOCCUS BRAUNII, *ANACYSTIS MONTANA, *HYDROCARBONS, *OLEFINS,
GREEN RIVER FORMATION, CHLOROPHYLL DERIVATIVES, RECENT SEDIMENTS,
PRECAMBRIAN SEDIMENTS, CHLORELLA PYRENOIDOSA, FUCALES, ALKANES,
ALKENES, ALIPHATIC HYDROCARBONS, PETROLEUM CRUDES, MASS SPECTROMETRY,
N-HEPTADECANE, MONOENES, DIENES, TRIENES.

ABSTRACT:
UTILIZING GAS CHROMATOGRAPHY AND MASS SPECTROMETRY, AUTHORS STUDIED THE
HYDROCARBON COMPOSITION OF TWO SPECIES OF ALGAE: BOTRYOCOCCUS BRAUNII,
A GOLDEN-BROWN (CHRYSOPHYTE), AND ANACYSTIS MONTANA, A BLUE-GREEN
(CYANOPHYTE). HYDROCARBONS OF THE CHRYSOPHYTE ARE ALKENES WITH ONE,
TWO, OR THREE DOUBLE BONDS; AND ODD NUMBERS OF CARBON ATOMS (IN RANGE,
17-33). DIOLEFINS WITH 27, 29 OR 31 CARBON ATOMS PREDOMINATE, THAT OF
29 CARBONS BEING MAJOR COMPONENT. DISTRIBUTION IN THE CYANOPHYTE
COMPRISES PRIMARILY MONOENES WITH CARBON ATOMS IN RANGE, 19-29, THE
MAJOR COMPONENT BEING A MONOENE OF 27 CARBONS. HEPTADECANE, ONLY
PARAFFIN FOUND, CONSTITUTED 11.5% OF HYDROCARBONS FOUND IN ANACYSTIS.
CONSIDERATION OF INFORMATION DEVELOPED IN THIS STUDY, TOGETHER WITH
DATA REPORTED ELSEWHERE IN THE LITERATURE, LED AUTHORS TO THE FOLLOWING
CONCLUSIONS: HYDROCARBONS OF HIGH MOLECULAR WEIGHT ARE FOUND IN
CONTEMPORARY COUNTERPARTS OF ANCIENT MICROSCOPIC ORGANISMS; SUCH
HYDROCARBONS ARE PRESENT IN HIGH CELLULAR CONCENTRATIONS (0.1-0.3% OF
DRY CELL WEIGHT); THEY POSSESS HIGH DEGREE OF UNSATURATION; THEY SHOW
MARKED ODD OVER EVEN PREDOMINANCE; AND THEY SHOW ACTUAL AND POTENTIAL
SIMILARITIES WITH PATTERNS OF ANCIENT AND MODERN SEDIMENTS AND OIL
SHALES. (EICHHORN-WIS)

FIELD 05A

ACCESSION NO. W69-08284

EUTROPHICATION,

J. FOEHRENBACH.

JOURNAL WPCF, VOL 41, NO 6, PP 1029-1036, JUNE 1969. 53 REF.

DESCRIPTORS:
*EUTROPHICATION, *REVIEWS, DOCUMENTATION, BIBLIOGRAPHIES, PHOSPHORUS
COMPOUNDS, NITROGEN COMPOUNDS, BIOINDICATORS, PRODUCTIVITY, WASTE
TREATMENT, WATER POLLUTION EFFECTS, WATER POLLUTION SOURCES, CYCLING
NUTRIENTS, COST COMPARISONS, ECONOMICS, ALGAE, ALGAL CONTROL,
ALGICIDES, SEDIMENTS, DISSOLVED OXYGEN, OXYGEN DEMAND, CYANOPHYTA,
DIATOMS, SAMPLING, LAKE MICHIGAN.

IDENTIFIERS:
CULTURAL EUTROPHICATION, SPECIES DIVERSITY, CHEMICAL PRECIPITATION.

ABSTRACT:
AUTHOR REVIEWS 53 STUDIES CONTRIBUTED TO THE EUTROPHICATION LITERATURE
DURING THE PERIOD 1966-1968. SUBJECT COVERAGE, WITH NUMBER OF
LITERATURE CITATIONS PER SUBJECT IN PARENTHESES, ARE: USE OF FERTILIZED
WATER TO INCREASE BIOLOGICAL PRODUCTION (8); IDENTIFICATION OF
NUTRIENTS (6); OXYGEN KINETICS (3); ALGAL CONTROL (2); SOURCES OF
NITROGEN AND PHOSPHORUS (4); ROLE OF SEDIMENTS (2); RESULTS OF HUMAN
CULTURAL ACTIVITIES (2); EFFECTS OF POLLUTIONAL LOAD (2); CYANOPHYTE
BLOOMS AND DECREASE IN ALGAL SPECIES DIVERSITY (2); EUTROPHICATION
INDICATORS IN LAKE MICHIGAN (2); SAMPLING AND DIATOM-INDICATORS (2);
NUTRIENTS, THEIR CRITICAL LEVELS AND CYCLING (4); ADVANCED WASTE
TREATMENT TECHNIQUES WITH COST ESTIMATES (7); CHEMICAL TREATMENT FOR
NUTRIENT REMOVAL WITH COST ESTIMATES (7). (EICHHORN-WIS)

FIELD 05C

ACCESSION NO. W69-08518

SORPTION OF RADIOACTIVE NUCLIDES BY SARGASSUM FLUITANS AND S NATANS,

TEXAS A AND M UNIV., COLLEGE STATION; AND KANSAS UNIV., LAWRENCE.

JOHN E. SIMEK, J. A. DAVIS, C. E. DAY III, AND ERNEST E. ANGINO.

AVAILABLE FROM CLEARINGHOUSE FOR SCI AND TECH INFORMATION AS CONF-670503.
SYMP ON RADIOECOLOGY, PROC 2ND NATL SYMP, MAY 15-17, 1967, ANN ARBOR, MICH.
NELSON, DANIEL J AND EVANS, FRANCIS C (EDS). US ATOMIC ENERGY COMM, DOC
CONF 670503, P 505-508. 2 FIG, 5 REF. DISC.

DESCRIPTORS:
*RADIOISOTOPES, *ALGAE, *SORPTION, ABSORPTION, GULF OF MEXICO, ATLANTIC
OCEAN, RADIOACTIVITY, GAMMA RAYS, POTASSIUM RADIOISOTOPES, MANGANESE,
CESIUM, RADIUM RADIOISOTOPES, URANIUM RADIOISOTOPES.

IDENTIFIERS:
*SARGASSUM FLUITANS, *SARGASSUM NATANS, SEAWEED,
ZIRCONIUM-95-NIOBIUM-95, PRE-BOMB MATERIAL, RUTHENIUM-106-RHODIUM-106,
RUTHENIUM-103-RHODIUM-103, CERIUM-144-PRASEODYMIUM-144.

ABSTRACT:
SARGASSUM NATANS AND S FLUITANS, FREE-FLOATING MARINE ALGAE, CAN
EFFECTIVELY CONCENTRATE CERTAIN RADIONUCLIDES. IN THE EVALUATION OF
RADIONUCLIDES FOUND ASSOCIATED WITH THESE TWO SPECIES, AN ANSWER WAS
SOUGHT AS TO THE MANNER OF THIS ASSOCIATION. FLOATING SAMPLES WERE
COLLECTED FROM THE GULF OF MEXICO AND FROM NORTH ATLANTIC AREAS. AN
ANALYSIS OF SARGASSUM MATERIAL COLLECTED PRIOR TO 1940 (PRE-BOMB)
PROVIDED A POSITIVE MEANS OF COMPARING THE ACTIVITY ADDED TO THE
SPECTRUM BY NUCLEAR ACTIVITY. TESTS UTILIZING ACID, BASE, AND NEUTRAL
WASHES, WERE DEVISED TO DETERMINE WHETHER ADSORPTION OR ABSORPTION WAS
THE MAJOR NUCLIDE-PLANT ASSOCIATION. WASHINGTON WITH UNCONTAMINATED,
DOUBLY DISTILLED WATER, RESULTED IN REMOVAL OF ONLY SMALL AMOUNTS OF
POTASSIUM-40 (K-40), PROBABLY DUE SIMPLY TO A WASHING OUT OF SEA SALTS.
SEPARATE, SHORT TERM (UP TO 1 HOUR) WASHED OF SAMPLES WITH SLIGHTLY
ACIDIC AND BASIC SOLUTION FAILED TO INDICATE ANY ACTIVITY OTHER THAN
THAT OF K-40. HOWEVER, ACID AND BASE WASHES FOR 24 HOURS PRODUCED
DEFINITE EVIDENCE OF ACTIVITY; OVER THIS PERIOD, BREAKDOWN OF THE PLANT
MATERIAL BECAME OBVIOUS. RESULTS SUGGEST THAT RADIONUCLIDES IN QUESTION
ARE ABSORBED, OR METABOLICALLY FIXED, RATHER THAN ADSORBED TO SARGASSUM
NATANS AND S FLUITANS.
(JONES-WIS)

FIELD 05C

ACCESSION NO. W69-08524

THE NATURE OF THE DISTRIBUTION OF TRACE ELEMENTS IN LONGNOSE ANCHOVY (ANCHOA LAMPROTAENIA HILDEBRAND), ATLANTIC THREAD HERRING (OPISTHONEMA OGLINUM LASUEUR), AND ALGA (UDOTEA FLABELLUM LAMOUROUX),

PUERTO RICO NUCLEAR CENTER, MAYAGUEZ.

ROBERT Y. TING, AND V. ROMAN DE VEGA.

AVAILABLE FROM CLEARINGHOUSE FOR SCI AND TECH INFORMATION AS CONF-670503.
SYMP ON RADIOECOLOGY, PROC 2ND NATL SYMP, MAY 15-17, 1967, ANN ARBOR, MICH.
NELSON, DANIEL J AND EVANS, FRANCIS C (EDS). US ATOMIC ENERGY COMM, DOC
CONF 670503, P 527-534. 5 FIG, 15 REF, DISC.

DESCRIPTORS:
    *ALGAE, *HERRINGS, COPPER, IRON, COBALT, MANGANESE, ATLANTIC OCEAN,
    PUERTO RICO, SPECTROSCOPY, DISTRIBUTION, CALCIUM, TROPHIC LEVELS, FOOD
    WEBS.

IDENTIFIERS:
    *TRACE ELEMENTS, *ANCHOA LAMPROTAENIA, *OPISTHONEMA OGLIMUN, UDOTEA
    FLABELLUM, ZINC, NICKEL, PUERTO RICO, LA PARGUERA, JOYUDA.

ABSTRACT:
    QUALITATIVE AND QUANTITATIVE INFORMATION ON STABLE ELEMENTS CONTAINED
    IN ORGANISMS AND THEIR ENVIRONMENT IS ESSENTIAL IN PREDICTING
    DISTRIBUTION PATTERNS OF ARTIFICIALLY INTRODUCED RADIOISOTOPES OF THE
    SAME ELEMENTS. LITTLE IS KNOWN OF THE DISTRIBUTION AND VARIABILITY OF
    VARIOUS ELEMENTS AND THEIR RADIOISOTOPES WITHIN SPECIES OF MARINE
    ORGANISMS SAMPLED FROM THE SAME AREA AT A GIVEN TIME. TO STUDY THEIR
    TRANSFER THROUGH FOOD WEBS AND TO EVALUATE CONCENTRATION FACTORS AT
    EACH LEVEL, TRACE AMOUNTS OF THE ELEMENTS COPPER, IRON, AND ZINC, WERE
    DETERMINED IN LONGNOSE ANCHOVY (ANCHOA LAMPROTAENIA HILDEBRAND) AND IN
    ALGA (UDOTEA FLABELLUM LAMOUROUX) FROM LA PARGUERA AND JOYUDA, PUERTO
    RICO. LEVELS OF COBALT, COPPER, IRON, MANGANESE, NICKEL, AND ZINC IN
    ATLANTIC THREAD HERRING (OPISTHONEMA OGLINUM LASUEUR) FROM LA PARGUERA,
    WERE ALSO DETERMINED. ANALYSES WERE ACCOMPLISHED WITH ATOMIC ABSORPTION
    SPECTROSCOPY. CONCENTRATIONS OF THE ELEMENTS IN INDIVIDUALS OF THE SAME
    SPECIES SAMPLED WITHIN THE SAME LOCALITY AT THE SAME TIME DID NOT
    APPROXIMATE NORMAL DISTRIBUTION; HOWEVER, DISTRIBUTIONS OF THE
    LOG-TRANSFORMED CONCENTRATIONS CLOSELY APPROXIMATED A NORMAL CURVE IN
    MOST CASES. WITH EXPANDED NUCLEAR TECHNOLOGY, RADIOISOTOPES IN THE
    ENVIRONMENT MAY APPROACH LEVELS FOR WHICH KNOWLEDGE OF THE DISTRIBUTION
    PATTERNS OF THESE MATERIALS WILL BE OF PRIMARY IMPORTANCE.
    (JONES-WIS)

FIELD 05C

ACCESSION NO. W69-08525

PHOSPHORUS TURNOVER BY CORAL REEF ANIMALS,

GEORGIA UNIV., ATHENS. DEPT. OF ZOOLOGY; AND NORTH CAROLINA UNIV., CHAPEL
   HILL. DEPT. OF ENVIRONMENTAL SCIENCES AND ENGINEERING.

L. R. POMEROY, AND E. J. KUENZLER.

AVAILABLE FROM CLEARINGHOUSE FOR SCI AND TECH INFORMATION AS CONF-670503.
   SYMP ON RADIOECOLOGY, PROC 2ND NATL SYMP, MAY 15-17, 1967, ANN ARBOR, MICH,
   NELSON, DANIEL J AND EVANS, FRANCIS C (EDS). US ATOMIC ENERGY COMM, DOC
   CONF 670503, PP 474-482. 1 FIG, 3 TAB, 34 REF, DISC.

   DESCRIPTORS:
      *PHOSPHORUS, *CORAL, *REEFS, *MARINE ANIMALS, FISH, ECOSYSTEMS, CYCLING
      NUTRIENTS, TEMPERATURE, LIGHT QUALITY, CHLOROPHYLL, ZOOPLANKTON, ALGAE,
      POLYCHAETA, CRUSTACEA, MOLLUSCA, ECHINODERMATA, PACIFIC OCEAN.

   IDENTIFIERS:
      ENIWETOK ATOLL, TURNOVER TIME, TRIDACNA CROCEA, WOODS HOLE
      OCEANOGRAPHIC INSTITUTION, ACROPORA HYACINTHUS, POCILLOPORA EYDOUXI,
      PORITES LOBATA, LEPTASTREA PURPUREA, FUNGIA FUNGITES, ZOOXANTHELLAE,
      ACANTHURUS TRIOSTEGUS, CONUS EBRAEUS.

   ABSTRACT:
      CORAL REEFS ARE PARTICULARLY INTERESTING ECOSYSTEMS IN WHICH TO STUDY
      NUTRIENT CYCLING, BECAUSE OF THEIR GREAT DIVERSITY OF LIFE AND THEIR
      HIGH RATES OF ORGANIC PRODUCTION IN A NUTRIENT-DEPLETED ENVIRONMENT.
      PHOSPHORUS (P) CONTENT AND ELIMINATION RATES WERE MEASURED FOR DOMINANT
      REEF ANIMALS AT ENIWETOK ATOLL, AND FLUX OF P THROUGH THE ORGANISMS WAS
      EXPRESSED AS TURNOVER TIME. HERBIVOROUS FISHES APPEAR TO RECEIVE IN
      THEIR DIET JUST ENOUGH P FOR GROWTH AND REPRODUCTION; THIS IS REFLECTED
      IN A SOMEWHAT LOWER EXCRETION RATE AND LONGER TURNOVER TIME THAN WOULD
      BE PREDICTED FOR ORGANISMS OF THEIR SIZE, PARTICULARLY THE JUVENILES.
      CARNIVOROUS ANIMALS AND DEPOSIT FEEDERS RECEIVE EXCESS P IN THEIR DIET,
      AND THEIR TURNOVER TIME DOES NOT DIFFER FROM ECOSYSTEMS WITH A MORE
      ABUNDANT P SUPPLY. OF 5 SPECIES OF REEF CORALS EXAMINED, 4 SHOWED A
      VERY LONG TURNOVER TIME AND LITTLE FLUX OF P TO THE ENVIRONMENT.
      TRIDACNA CROCEA SHOWED A TYPICAL TURNOVER TIME FOR A MOLLUSC OF ITS
      SIZE, BUT SEEMED TO LOSE MOST P VIA INCORPORATION IN LIVING
      ZOOXANTHELLAE, WHICH ARE SUBSEQUENTLY LOST. THIS SAMPLING OF THE FLUX
      OF P IS NECESSARILY INCOMPLETE ALTHOUGH ORGANISMS KNOWN TO BE DOMINANT
      HAVE BEEN SELECTED FOR STUDY.
      (JONES-WIS)

   FIELD 05C

   ACCESSION NO. W69-08526

NATURAL AND POLLUTION SOURCES OF IODINE, BROMINE, AND CHLORINE IN THE GREAT
   LAKES,

MICHIGAN UNIV., ANN ARBOR. DEPT. OF METEOROLOGY AND OCEANOGRAPHY.

MARY A. TIFFANY, JOHN W. WINCHESTER, AND RONALD H. LOUCKS.

J WATER POLLUT CONTR FEDERATION, VOL 41, NO 7, P 1319-1329, JULY 1969. 11 P,
   10 FIG, 5 TAB, 13 REF. CONTRACT NO AT(11-1)-1705AEC).

   DESCRIPTORS:
      *WATER QUALITY, *GREAT LAKES, *TRACE ELEMENTS, *WATER POLLUTION
      EFFECTS, CHLORIDES, HALOGENS, ALGAE, WATER POLLUTION SOURCES, WATER
      CHEMISTRY, WATER ANALYSIS.

   IDENTIFIERS:
      *BROMINE, *IODINE.

   ABSTRACT:
      THIS STUDY INVOLVES THE DETERMINATION OF TRACE ELEMENTS I, BR, AND CL
      IN THE GREAT LAKES, USING NEUTRON ACTIVATION ANALYSIS OF 90 WATER
      SAMPLES FROM LAKE SUPERIOR AND ITS TRIBUTARY STREAMS, LAKE MICHIGAN,
      SOUTHERN LAKE HURON, LAKE ST. CLAIR, WESTERN LAKE ERIE, AND NORTHERN
      LAKE ONTARIO. POSSIBLE POLLUTION BY BROMINE THROUGH AN ATMOSPHERIC
      ROUTE IS OF INTEREST BECAUSE OF THE COMBUSTION OF LEADED GASOLINE.
      IODINE DEFICIENCY IN LAKE WATER MAY BE RELATED TO THYROID DISORDERS
      AMONG MARINE FISH WHICH HAVE BECOME ADAPTED TO FRESH WATER, AND ALGAE
      MAY OFFER COMPETITION FOR THE AVAILABLE IODINE. CHLORINE IS A
      NOTICEABLE CONTAMINANT EXCEPT IN LAKE SUPERIOR. THE LAKE SUPERIOR
      STREAMS APPEAR TO REPRESENT A GOOD AVERAGE OF THE ATMOSPHERIC INPUTS OF
      I, BR, AND CL. (KNAPP-USGS)

   FIELD 05A, 02K

   ACCESSION NO. W69-08562

BIOLOGICAL CONCENTRATION OF PESTICIDES BY ALGAE,

NORTH TEXAS STATE UNIV., DENTON. DEPT. OF BIOLOGY.

B. DWAIN VANCE, AND WAYMON DRUMMOND.

J AMER WATER WORKS ASS, VOL 61, NO 7, P 360-362, JULY 1969. 3 P, 2 TAB, 11
REF. GRANT NO CC00269(PHS).

DESCRIPTORS:
*PESTICIDE RESIDUES, *ALGAE, BIOASSAY, BIOINDICATORS, FISHKILL,
HAZARDS, PESTICIDE KINETICS, PESTICIDE TOXICITY, POLLUTANTS, WATER
POLLUTION EFFECTS, ALDRIN, DIELDRIN, ENDRIN, DDT.

IDENTIFIERS:
PESTICIDE BIOACCUMULATION.

ABSTRACT:
ALGAE, THE BASE CONSTITUENTS OF THE AQUATIC FOOD CHAIN, CONCENTRATE
PESTICIDES AND ARE MUCH MORE RESISTANT TO TOXIC EFFECTS THAN THE HIGHER
ORGANISMS THAT EAT THEM. ALGAL CULTURES WERE GROWN IN THE PRESENCE OF
DDT, ALDRIN, DIELDRIN, AND ENDRIN, IN CONCENTRATIONS OF 5-20 MICROGRAM
PER ML. THE ALGAE CONCENTRATED THE PESTICIDES 100 TIMES OR MORE IN ALL
THE TESTS. THE DATA ARE TABULATED. PRACTICALLY NO DEGRADATION OF THE
PESTICIDES OCCURRED. (KNAPP-USGS)

FIELD 05C, 05B

ACCESSION NO. W69-08565

CHEMICAL FERTILIZERS ON LAKE WATERS (IN GERMAN),

VON HEINZ AMBUHL.

SUMMARIES IN ENGLISH AND FRENCH. GAS- UND WASSERFACH, VOL 107, NO 14, P
357-363, APR 1966. 6 FIG, 2 TAB, 18 REF.

DESCRIPTORS:
*EUTROPHICATION, *NUTRIENTS, *LAKES, PONDS, FERTILIZERS, ALGAE,
NITROGEN, PHOSPHORUS, WATER POLLUTION EFFECTS, WATER POLLUTION SOURCES,
WATER POLLUTION CONTROL, OXYGEN BALANCE, SEWAGE, RUNOFF.

IDENTIFIERS:
LAKE LUCERNE, FERTILIZED FIELDS, SUBTERRANEAN DRAINAGE, MECHANICAL AND
BIOLOGICAL PURIFICATION, SWITZERLAND.

ABSTRACT:
THE POLLUTION OF LAKES AND PONDS IS IN A LARGE PART DUE TO INORGANIC
SALTS PROMOTING ALGAL GROWTH AND SUBSEQUENT PROCESSES OF DECOMPOSITION.
AS REVEALED BY STUDIES OF LAKE LUCERNE, A FEW PARTS PER MILLION OF
FERTILIZER SALTS, PARTICULARLY PHOSPHATES, ARE SUFFICIENT TO STIMULATE
THE GROWTH OF ALGAE AND UPSET THE OXYGEN BALANCE. A LARGE FRACTION OF
EUTROPHYING SUBSTANCES IS BROUGHT TO LAKES FROM FERTILIZED FIELDS VIA
RUN-OFF AND SUBTERRANEAN DRAINAGE. THE AMOUNT OF PHOSPHORUS FROM THESE
SOURCES IS EQUAL TO THAT SUPPLIED BY THE NORMAL DISCHARGES OF SEWAGE;
THE AMOUNT OF NITROGEN DELIVERED FROM FIELDS EXCEEDS THAT CONTRIBUTED
BY SEWAGE. ON THE AVERAGE, SWISS LAKES RECEIVE 0.3 TO 0.5 KG OF
PHOSPHORUS AND ABOUT 45 KG OF NITROGEN PER HECTARE, PER YEAR; THE
LATTER FIGURE IS DERIVED FROM A VERY WIDE AMPLITUDE. THIS INFLOW OF
NUTRIENTS IS SUFFICIENT TO EUTROPHY LAKES, BUT THE ONLY CORRECTIVE
MEASURE AVAILABLE AT THIS TIME IS PURIFICATION OF THE SEWAGE BY
MECHANICAL AND BIOLOGICAL METHODS ELIMINATING PLANT NUTRIENTS.
(WILDE-WIS)

FIELD 05C

ACCESSION NO. W69-08668

OBSERVATIONS AND EXPERIENCES IN THE CONTROL OF ALGAE,

TERRENCE P. FROST.

J NEW HAMPSHIRE WATER WORKS ASSOC, 10 P, DEC 1963. 1 TAB, 12 REF.

DESCRIPTORS:
*NUISANCE ALGAE, *ALGAL CONTROL, *NEW HAMPSHIRE, *COPPER SULFATE,
SEWAGE EFFLUENTS, SMELTS, FISHKILL, CRUSTACEANS, BOTTOM SEDIMENTS.

IDENTIFIERS:
*NEW HAMPSHIRE WATER POLLUTION COMMISSION, ALGAL BLOOMS, NUTRIENT LOAD,
SYNTHETIC DETERGENTS, FISH MANAGERS, FISH YIELDS, UNDERWATER SHOCK
WAVE, ANABAENA, PHYGON XL(DICHLONE), NEW HAMPSHIRE, LAKE WINNISQUAM,
LONG POND, CONCORD, BOSTON LOT RESERVOIR, LEBANON, WADLEIGH STATE PARK,
LAKE SHATUTAKEE, MASCOMA LAKE, GLEN LAKE.

ABSTRACT:
AUTHOR OFFERS VARIOUS OBSERVATIONS AND EXPERIENCES RELATED TO ALGAL
CONTROL OPERATIONS IN NUTRIENT RICH WATERS BY THE NEW HAMPSHIRE WATER
POLLUTION COMMISSION. SINCE 1961, ABOUT 6,000 ACRES/YEAR IN LAKE
WINNISQUAM AND OTHERS HAVE BEEN TREATED. OVERENRICHMENT OF THE LAKES IS
DUE TO EFFLUENTS FROM MUNICIPAL AND INDUSTRIAL TREATMENT PLANTS NOT
EQUIPPED TO STRIP NUTRIENTS BECAUSE OF INCREASED OPERATIONAL COSTS. IN
CONTROLLING ALGAE WITH COPPER SULFATE, THE COMMISSION HAS ENCOUNTERED
SOME OPPOSITION FROM FISH MANAGERS WARY OF ITS EFFECTS ON FISH
POPULATIONS. TESTS RUN ON ADULT SMELT AND EGGS SHOWED LEVELS APPLIED IN
ALGAL CONTROL WERE NOT LETHAL TO FISH AND MAY IN FACT INCREASE FISH
YIELDS IN SOME LAKES. A SIDE EFFECT OF COPPER SULFATING IS THE
INCREASED CRUSTACEAN POPULATION WHICH IS SUPPRESSED BY LARGE ALGAL
COMMUNITIES. SO FAR IN NEW HAMPSHIRE, SAMPLING OF BOTTOM MUDS HAS SHOWN
NO ACCUMULATIONS OF COPPER EXCEEDING THE LEVELS IN UNTREATED LAKES.
AUTHOR MENTIONS MANY SUCCESSFUL PROJECTS AND ONLY ONE CATASTROPHE, A
MASSIVE FISHKILL IN BOSTON LOT RESERVOIR, RESULTING FROM A
MISCALCULATION OF VOLUME AND DOUBLING DOSE OF COPPER SULFATE. THE
COMMISSION PLANS TO EXPERIMENT WITH COPPER SULFATING IN THE FUTURE FOR
PREVENTIVE CONTROL. (KETELLE-WIS)

FIELD 05G

ACCESSION NO. W69-08674

RELATION OF PHOSPHATES TO EUTROPHICATION,

NORTH CAROLINA UNIV., CHAPEL HILL. DEPT. OF ENVIRONMENTAL SCIENCES AND
ENGINEERING.

CHARLES M. WEISS.

J AMER WATER WORKS ASS, VOL 61, NO 8, P 387-391, AUG 1969. 5 P, 2 FIG, 4 TAB,
22 REF.

DESCRIPTORS:
*EUTROPHICATION, *LAKES, *PHOSPHATES, NUTRIENTS, ALGAE, WATER POLLUTION
EFFECTS, WATER POLLUTION SOURCES, AQUATIC PRODUCTIVITY, LIMNOLOGY,
PLANKTON, PRIMARY PRODUCTIVITY, TROPHIC LEVEL.

IDENTIFIERS:
LAKE TAHOE.

ABSTRACT:
THE RELATIONSHIP OF PHOSPHATES IN WATER TO PRIMARY PRODUCTION IS
EXAMINED, AND THE PHOSPHORUS CYCLE IN NATURAL LAKE BIOLOGICAL SYSTEMS
IS OUTLINED. ALTHOUGH HIGHER THAN NORMAL MACRONUTRIENT LEVELS WILL
GENERALLY PRODUCE AN INCREASE IN LEVELS OF PRODUCTIVITY OR SHIFT OF
TROPHIC STATUS, THE SPECIFIC INTERRELATIONSHIPS OF EACH NUTRIENT FACTOR
ARE EXCEEDINGLY COMPLEX AND REMAIN UNCLEAR. THE BIOLOGICAL STORAGE OF
PHOSPHORUS BY ALGAE IN EXCESS OF IMMEDIATE GROWTH REQUIREMENT, ITS
SECRETION AS SOLUBLE ORGANIC PHOSPHORUS, AND ITS REUSE WOULD APPEAR TO
BE A SIGNIFICANT FACTOR IN SATISFYING NUTRIENT LEVELS, PARTICULARLY FOR
THOSE SPECIES WITH LOW PHOSPHORUS REQUIREMENTS. INCREASES IN TROPHIC
STATUS ARE GENERALLY ASSOCIATED WITH INVASION OF BLUE-GREEN ALGAE,
WHICH HAVE CAPABILITY FOR NITROGEN FIXATION WITH RESULTING ENHANCEMENT
OF NITROGEN LEVELS AND ACCELERATION OF THE RATE OF EUTROPHICATION. THE
RATE OF TROPHIC RESPONSE OF EACH LAKE OR RESERVOIR TO CHANGES IN
MACRONUTRIENT LEVELS INVOLVES NOT ONLY NUTRIENT CHANGE, BUT ALSO THE
TOTAL CHEMICAL AND PHYSICAL CHARACTERISTICS OF THE PARTICULAR BASIN.
THE CHEMICAL NATURE OF THE PARTICULAR PHOSPHORUS SPECIES, ITS
ASSOCIATION WITH COLLOIDAL OR PARTICULATE MATERIALS AND THE SPECIFIC
ASSOCIATED PHYTOPLANKTON ALL CONTRIBUTE TO THE RATE OF UTILIZATION OF
THE PHOSPHATE POOL AND THUS THE RATE OF EUTROPHICATION. (KNAPP-USGS)

FIELD 05C, 05B

ACCESSION NO. W69-09135

TRANSLOCATION OF PHOSPHORUS IN A TROUT STREAM ECOSYSTEM,

MICHIGAN STATE UNIV., EAST LANSING. DEPT. OF FISHERIES AND WILDLIFE; AND
  INSTITUTE FOR FISHERIES RESEARCH, ANN ARBOR, MICH.

ROBERT C. BALL, AND FRANK F. HOOPER.

PROC 1ST NAT SYMP ON RADIOECOLOGY, COLORADO STATE UNIV, FORT COLLINS, SEPT
  10-15, 1961. RADIOECOLOGY, 1963, REINHOLD PUB CORP, NY AND AMER INST OF
  BIOLOGICAL SCIENCES, WASH, DC, SCHULTZ, VINCENT AND KLEMENT JR, ALFRED W
  (EDS), P 217-228. 4 FIG, 7 TAB, 16 REF.

  DESCRIPTORS:
    *TRANSLOCATION, *PHOSPHORUS RADIOISOTOPES, *TROUT, *STREAMS,
    *ECOSYSTEMS, RADIOACTIVITY, BIOTA, CYCLES, MICHIGAN, INVERTEBRATES,
    FISH, TEMPERATURE, TRACERS, SAMPLING, PERIPHYTON, AQUATIC PLANTS,
    BACKGROUND RADIATION, GEOCHEMISTRY, SPRING WATERS, PHOSPHORUS,
    ORGANOPHOSPHORUS COMPOUNDS, BACTERIA, DIATOMS, DETRITUS, CYCLING
    NUTRIENTS, LIGHT INTENSITY, FERTILIZATION, GROWTH RATES, EQUILIBRIUM,
    CHARA, ALGAE, AMMOCETES, LAMPREYS, OMNIVORES.

  IDENTIFIERS:
    MICHIGAN, CHEBOYGAN COUNTY, STURGEON RIVER, FLUORESCEIN, MACROPHYTES,
    GROUND SOLIDS, UPTAKE, CONSUMER ORGANISMS, SIMULIUM, PHYSA, SALMO
    TRUTTA, ENTOSPHENUS LAMOTTENII, GAMMARUS, CORDULEGASTER.

  ABSTRACT:
    TO EXPLORE MOVEMENT OF PHOSPHORUS IN A COLD-WATER STREAM AND
    RELATIONSHIPS BETWEEN PHOSPHORUS AND STREAM PRODUCTIVITY, SPIKES OF
    APPROXIMATELY 23 MILLICURIES OF PHOSPHORUS-32 (P-32) WERE ADDED TO THE
    WATER DURING THREE SUMMERS. THE LOSS PATTERN VARIED FROM YEAR TO YEAR,
    WITH MUCH UPTAKE APPARENTLY ATTRIBUTED TO PERIPHYTON AND THREE AQUATIC
    MACROPHYTES. P-32 RATE OF LOSS FROM PLANTS TO WATER DECREASED WITH TIME
    AND CONCENTRATION APPROACHED EQUILIBRIUM LEVEL IN 15 TO 20 DAYS AFTER
    SPIKE WAS RELEASED. DECREASE CHANGE RATES SUGGESTED THAT PLANTS WERE
    REMOVING RE-CYCLED PHOSPHORUS FROM THE STREAM WATER. HIGHEST UPTAKE BY
    PERIPHYTON OCCURRED IN 1958 AND 1960; LOWEST IN 1959. THIS PATTERN WAS
    REVERSED BY AQUATIC MACROPHYTES. CONCENTRATION CURVES FOR CONSUMER
    ORGANISMS REFLECTED DIFFERENCES IN METABOLIC TURNOVER RATES AND FOOD
    RELATIONSHIPS. IN SIMULIUM, CONCENTRATIONS QUICKLY ROSE TO A HIGH
    LEVEL, SEEMINGLY DURING PASSAGE OF THE SPIKE. OTHER SMALL FILTER
    FEEDERS AND PERIPHYTON SCRAPERS ALSO ACCUMULATED P-32 RAPIDLY, THEN
    LOST IT QUICKLY. ACCUMULATING P-32 WAS SLOWER FOR LARGE OMNIVOROUS
    STREAM INSECTS AND SMALLER PREDACIOUS FORMS. DIFFERENCES IN CIRCULATION
    THROUGH THE ECOSYSTEM, AS NOTED DURING THE 3-YEAR TEST PERIOD, MAY BE
    DUE TO DISTRIBUTION OF P-32 BETWEEN SOLUBLE AND PARTICULATE PHASES.
    (JONES-WIS)

  FIELD 05C

  ACCESSION NO. W69-09334

THE PHOSPHORUS PROBLEM,

DEPARTMENT OF THE INTERIOR, WASHINGTON, D.C.

KENNETH M. MACKENTHUN.

J AMER WATER WORKS ASS, VOL 60, NO 9, P 1047-1054, SEPT 1968. 2 TAB, 32 REF.

DESCRIPTORS:
*PHOSPHORUS, *NUTRIENTS, *EUTROPHICATION, *TERTIARY TREATMENT, NITROGEN, AQUATIC PLANTS, BOTTOM SEDIMENTS, DREDGING, HARVESTING OF ALGAE, FISH HARVEST, VITAMINS.

IDENTIFIERS:
*WASTE WATER(MUNICIPAL AND INDUSTRIAL), *PHOSPHORUS SOURCES, *NUISANCE PLANT GROWTHS, *NUTRIENT CONTROL, *PHOSPHORUS CONCENTRATIONS, MICRONUTRIENTS, AQUATIC ECOSYSTEM, MIDGE LARVAE, LAKE STRATIFICATION, SURFACE AREA(LAKES), PHAEDACTYLUM, WASTE WATER DIVERSION, INSTREAM TREATMENT, BOTTOM SEALING, LAKE DILUTION.

ABSTRACT:
REDUCTION OF WASTEWATER PHOSPHORUS INFLOWS TO RECEIVING WATERS IS NECESSARY TO CHECK ACCELERATING CULTURAL EUTROPHICATION. MINIMIZING PHOSPHORUS INFLOWS TO SURFACE WATERS IS RECOMMENDED BECAUSE IT IS THE MOST EASILY REDUCED OF ALL CONSTITUENTS IN WASTE WATERS ESSENTIAL TO AQUATIC PLANT GROWTH. THE AUTHOR SUGGESTS THAT PREVENTION OF BIOLOGICAL NUISANCES REQUIRES TOTAL PHOSPHORUS CONCENTRATIONS NOT EXCEEDING 100 MICROGRAMS/LITER AT ANY POINT IN FLOWING WATER; AND 50 MICROGRAMS/LITER WHERE WATER ENTERS A STANDING BODY (LAKE OR RESERVOIR). THE MOST EFFECTIVE MEANS OF CONTROLLING PHOSPHORUS LEVELS IN LAKES NOW ARE TREATMENT OF NUTRIENT POINT SOURCES, WASTE WATER DIVERSIONS AROUND LAKES, AND DILUTION WITHIN LAKES WHERE FEASIBLE. ONCE NUTRIENTS ARE COMBINED WITHIN THE ECOSYSTEM OF THE RECEIVING WATER, THEIR REMOVAL IS TEDIOUS AND EXPENSIVE. RESULTS OF HARVESTING OF AQUATIC CROPS, DREDGING, AND OTHER MEANS TO REMOVE NUTRIENTS AFTER THEY HAVE REACHED RECEIVING WATERS SHOULD BE COMPARED WITH INFLOWING NUTRIENT QUANTITIES TO EVALUATE EFFECTIVENESS. THE PRESERVATION OF SURFACE WATERS IN A USABLE STATE WILL DEPEND IN THE LONG RUN ON MAXIMUM NUTRIENT REMOVAL AT THE WASTE SOURCES. (KETELLE-WIS)

FIELD 05C

ACCESSION NO. W69-09340

WATER-QUALITY MANAGEMENT AND LAKE EUTROPHICATION: THE LAKE WASHINGTON CASE,

WASHINGTON UNIV., SEATTLE. DEPT. OF ZOOLOGY.

W. THOMAS EDMONDSON.

WATER RESOURCES MANAGEMENT AND PUBLIC POLICY, WASHINGTON UNIV PRESS, SEATTLE,
P 139-178, 1968. 2 FIG, 1 TAB, 104 REF.

DESCRIPTORS:
*LAKES, *EUTROPHICATION, *SEWAGE EFFLUENTS, *NUTRIENTS, *NITROGEN,
PHOSPHORUS, SULFATES, VITAMINS, ALGAE, GROWTH RATES, COPPER SULFATE,
BOTTOM SEDIMENTS, LAKE MICHIGAN, WASHINGTON, DISSOLVED OXYGEN.

IDENTIFIERS:
*EFFLUENT DIVERSION, *WATER QUALITY MANAGEMENT, *LAKE WASHINGTON,
*SEWAGE FERTILIZATION EFFECTS, *BIOLOGICAL PRODUCTIVITY, AQUATIC
NUISANCES, TRANSPARENCY, AGRICULTURAL RUNOFF, NUTRIENT BUDGETS,
COMPARATIVE STUDIES, NUTRIENT REMOVAL, LAKE RECOVERY, WISCONSIN, LAKE
MONONA, LAKE MENDOTA, LAKE WAUBESA, LAKE KEGONSA, LAKE ZURICH,
OSCILLATORIA RUBESCENS, CLADOPHORA, DINOBRYON DIVERGENS, SWITZERLAND,
LAKE LUCERNE, LAKE CONSTANCE, WASHINGTON, PUGET SOUND, BARE
LAKE(ALASKA), LAKE LYNGBY SO(DENMARK), LAKE TAHOE(CALIF), LAKE
MAGGIORE(ITALY), SPECIES DOMINANCE.

ABSTRACT:
LAKE WASHINGTON'S CASE PARALLELS OTHER LAKES WORLDWIDE WHERE CULTURAL
EUTROPHICATION HAS LED TO INCREASED BIOLOGICAL PRODUCTIVITY, LIMITING
WATER USES. FIRST DEFINITE INDICATION OF ENRICHMENT APPEARED IN 1955
WITH ABUNDANCE OF OSCILLATORIA RUBESCENS, FOLLOWED BY REDUCTIONS IN
TRANSPARENCY AND DISSOLVED OXYGEN LEVELS IN WATER. PRESENT SOLUTION TO
THE PROBLEM HAS BEEN ALMOST TOTAL DIVERSION OF EFFLUENTS FROM THE LAKE,
COMPLETED IN 1967 AT A COST OF $85 MILLION (INCLUDING NEW TREATMENT
FACILITIES). EVIDENCE OF THE ROLE OF SEWAGE IN FERTILIZATION AND LAKE
DETERIORATION COMES FROM MANY SOURCES: DIRECT OBSERVATION; URBANIZATION
FOLLOWED BY SYMPTOMS OF INCREASED LAKE PRODUCTIVITY; NUTRIENT BUDGET
STUDIES, INDICATING RELATIVE NUTRIENT PERCENTAGES FROM DIFFERENT
SOURCES; COMPARISON OF SIMILAR LAKES IN DIFFERENT ENVIRONMENTS; CHANGES
RESULTING FROM SEWAGE EFFLUENT DIVERSIONS AROUND LAKES; BIOASSAY
TECHNIQUES OF LAKE WATER SAMPLES; AND PHYSIOLOGICAL STUDIES OF ALGAE
GROWN IN NUTRIENT SOLUTIONS. THE RELATIVE IMPORTANCE OF VARIOUS
NUTRIENTS AND MAXIMUM ALLOWABLE CONCENTRATIONS IN LAKES ARE DEPENDENT
ON SEVERAL INTERRELATED FACTORS INCLUDING DIVERSE EFFECTS OF DIFFERENT
COMBINATIONS OF NUTRIENTS, AND VARIETY IN REACTIONS BY SPECIES TO
NUTRIENTS. IMPROVED METHODS OF NUTRIENT REMOVAL FROM SEWAGE WILL
UNDOUBTEDLY DEVELOP, BUT AT PRESENT, EFFLUENT DIVERSION IS THE ONLY
SURE METHOD OF CONTROLLING ENRICHMENT OF LAKES. (KETELLE-WIS)

FIELD 05G, 02H

ACCESSION NO. W69-09349

THE ROLE OF ALGAE IN DEGRADING DETERGENT SURFACE ACTIVE AGENTS,

TEXAS UNIV., AUSTIN. DEPT. OF CIVIL ENGINEERING.

ERNST M. DAVIS, AND E. F. GLOYNA.

J WATER POLLUT CONTR FEDERATION, VOL 41, NO 8, PART 1, P 1494-1504, AUG 1969. 11 P, 4 FIG, 5 TAB, 21 REF. GRANT NO NSF-GU-1963.

DESCRIPTORS:
*BIODEGRADATION, *ALGAE, *AQUATIC BACTERIA, DETERGENTS, SURFACTANTS, ALKYLBENZENE SULFONATES, LINEAR ALKYLATE SULFONATES, WATER POLLUTION EFFECTS, STABILIZATION, LAGOONS, TOXICITY, WASTE WATER TREATMENT, WASTE WATER DISPOSAL.

IDENTIFIERS:
*ABS, LAS.

ABSTRACT:
THE RATES OF DEGRADATION OF 3 TYPICAL NONIONIC AND 2 ANIONIC SURFACTANTS BY GREEN AND BLUE-GREEN ALGAE WERE EVALUATED. ADDITIONAL DATA WERE GATHERED FOR ORGANISMS OBTAINED FROM RAW WASTEWATER, ACTIVATED SLUDGE, AND WASTE STABILIZATION POND WATER. RESULTS SHOW THAT ALGAE CONTRIBUTE TO A SMALL EXTENT IN THE DEGRADATION OF COMMON HOUSEHOLD DETERGENTS. MOST DEGRADATION IS ATTRIBUTED TO BACTERIA OR OTHER ORGANISMS. ALGAE DID NOT DEGRADE ALKYL BENZENE SULFONATE WHILE BACTERIA DID (0.45 MG/1/DAY). THE OTHER ANIONIC SURFACTANT, LINEAR ALKYL SULFONATE, WAS DEGRADED IN SMALL AMOUNTS BY SOME OF THE DIFFERENT ALGAL FORMS (0.97 MG/1/DAY BY BACTERIA). NONIONIC DETERGENTS WERE MORE EASILY DEGRADED. (KNAPP-USGS)

FIELD 05D

ACCESSION NO. W69-09454

EFFECT OF LIGHT INTENSITY ON PHOTOSYNTHESIS BY THERMAL ALGAE ADAPTED TO NATURAL AND REDUCED SUNLIGHT,

INDIANA UNIV., INDIANAPOLIS. DEPT. OF MICROBIOLOGY.

THOMAS D. BROCK, AND M. LOUISE BROCK.

LIMNOL AND OCEANOGR, VOL 14, NO 3, P 334-341, MAY 1969. 8 P, 7 TAB, 19 REF.

DESCRIPTORS:
*PHOTOSYNTHESIS, *ALGAE, *THERMAL SPRINGS, *LIGHT INTENSITY, SPRING WATER, WATER CHEMISTRY, WATER TEMPERATURE, CHEMICAL ANALYSIS, ANALYTICAL TECHNIQUES, CLOUD COVER, SPECTROSCOPY, ULTRAVIOLET RADIATION, TRANSMISSIVITY, CHLOROPHYLL.

ABSTRACT:
THE EFFECT OF LIGHT INTENSITY ON THERMAL ALGAE OF HOT SPRINGS OF YELLOWSTONE NATIONAL PARK WAS INVESTIGATED BY MEASURING AUTORADIOGRAPHY OF PHOTOSYNTHETIC ACTIVITY OF CELLS AT DIFFERENT LEVELS IN THE ALGAE MAT. THE RATE OF PHOTOSYNTHESIS DECREASES PROGRESSIVELY WITH DECREASING LIGHT, ALTHOUGH THE MOST EFFICIENT USE IS AT 7-14% OF FULL SUNLIGHT. THE CHLOROPHYLL CONTENT OF THE ALGAE AT THE SURFACE OF THE MAT INCREASES RAPIDLY; HOWEVER, THE CHLOROPHYLL CONTENT OF THE MAT INCREASES SLOWLY OR NOT AT ALL. ALTHOUGH INDIVIDUAL ALGAE CELLS CAN ADAPT RAPIDLY TO CHANGES IN LIGHT, THE ENTIRE POPULATION, BECAUSE OF ITS EXISTENCE IN COMPACT MATS, ADAPTS SLOWLY. AT THE LATITUDE OF YELLOWSTONE THERE IS SUFFICIENT LIGHT THROUGHOUT THE WHOLE YEAR TO STIMULATE THE ALGAE GROWTH EVEN AT CRITICAL TEMPERATURES. (GABRIEL-USGS)

FIELD 02K, 05A

ACCESSION NO. W69-09676

EUTROPHICATION FACTORS IN NORTH CENTRAL FLORIDA LAKES,

FLORIDA UNIV., GAINESVILLE. ENGINEERING AND INDUSTRIAL EXPERIMENT STATION.

P. L. BREZONIK, W. H. MORGAN, E. E. SHANNON, AND H. D. PUTNAM.

ENGINEERING PROGRESS IN FLORIDA. BULL SERIES NO 134 AND WATER RESOURCES RESEARCH CENTER PUB NO 5, GAINESVILLE, FLORIDA, AUG 1969. 101 P, 50 FIG, 43 TAB, 124 REF, 4 APPEND. GRANT DON 16, 010(FWPCA). OWRR PROJECT A-002-FLA.

DESCRIPTORS:
WATER QUALITY, *EUTROPHICATION, ALGAL BLOOMS, NUTRIENTS, LIMNOLOGY, DYSTROPHY, OLIGOTROPHY, *PRIMARY PRODUCTIVITY, *TROPHIC LEVEL, WATER STORAGE, EVAPORATION, WATER CIRCULATION, SEDIMENTS.

IDENTIFIERS:
SEEPAGE LOSSES, DIURNAL VARIATIONS.

ABSTRACT:
A SMALL LAKE IN AN UNINHABITED DRAINAGE BASIN IN NORTH FLORIDA IS BEING USED AS A MODEL FOR THE EUTROPHICATION RESEARCH DESCRIBED IN THIS REPORT. AN ADJACENT LAKE SERVES AS THE CONTROL. BACKGROUND DATA ON THE CHEMISTRY AND BIOLOGY OF THE LAKES WERE OBTAINED IN ORDER TO BE CERTAIN OF THEIR SIMILARITY AND TROPHIC STATUS. WATER AND NUTRIENT BUDGETS WERE DETERMINED AND NUTRIENT ENRICHMENT (TREATED SEWAGE EFFLUENT) HAS BEEN CONTINUOUS SINCE MARCH 1967. BOTH ROUTINE MONITORING AND SPECIAL (10 DAY) STUDIES WERE CARRIED OUT IN BOTH LAKES COMPRISING THE EXPERIMENTAL SYSTEM. IN 1968 THE RESEARCH PROGRAM WAS EXTENDED TO INCLUDE OTHER LAKES OF VARIOUS TYPES AND EXHIBITING VARYING TROPHIC STAGES IN NORTH CENTRAL FLORIDA. THE EUTROPHIC LAKES IN THE OKLAHOMA CHAIN ARE INCLUDED. CHANGES IN PARAMETERS HAVE BEEN AND ARE CONTINUING TO BE DETERMINED AND RELATED TO THE NUTRIENT FLUX. SYSTEMS ANALYSIS TECHNIQUES HAVE BEEN APPLIED. APPENDICES INCLUDE METHODS OF CHEMICAL ANALYSES; BIOLOGICAL PROCEDURES, A LISTING OF PROTOZOA, MICROSCOPIC ALGAE AND SULFUR BACTERIA RECORDED FROM THE EXPERIMENTAL AND CONTROL LAKES ON FIVE DATES, AND THE MULTIPLE REGRESSION ANALYSES PERFORMED ON TEN DAY STUDY DATA AND ROUTINE MONTHLY DATA.

FIELD 05C, 02H

ACCESSION NO. W69-09723

NATURAL RADIOACTIVITY IN THE FOOD WEB OF THE BANDED SCULPIN COTTUS CAROLINAE
   (GILL),

LOUISVILLE UNIV., KY. DEPT. OF BIOLOGY.

W. L. MINCKLEY, J. E. CRADDOCK, AND L. A. KRUMHOLZ.

PROC 1ST NAT SYMP ON RADIOECOLOGY, COLO STATE UNIV, FORT COLLINS, SEPT 10-15,
   1961. RADIOECOLOGY, 1963, REINHOLD PUB CORP, NY AND AMER INST OF BIOLOGICAL
   SCIENCES, WASH DC, SCHULTZ, VINCENT AND KLEMENT JR, ALFRED W (EDS)
   P229-236. 8 FIG, 4 TAB, 13 REF.

DESCRIPTORS:
   *SCULPINS, *RADIOACTIVITY, *FOOD WEBS, KENTUCKY, LIFE HISTORY STUDIES,
   GAMMA RAYS, SPECTROMETERS, CYANOPHYTA, URANIUM RADIOISOTOPES, STRONTIUM
   RADIOISOTOPES, CESIUM, FALLOUT, SILTS, DETRITUS, BOTTOM SEDIMENTS,
   CHUTES, EDDIES, HABITATS, CRAYFISH, GROWTH STAGES, DIATOMS, ALGAE,
   METABOLISM, ISOPODS, AMPHIPODA, FISH, GROWTH RATES.

IDENTIFIERS:
   *COTTUS CAROLINAE, MEAD COUNTY(KY), THORIUM-232, BISMUTH-214,
   ACTINIUM-228, SWIFT RIFFLES, ZIRCONIUM-95, NIOBIUM-95, RUTHENIUM-106,
   RHODIUM-106, CERIUM-144, BETA RADIOACTIVITY, GONIOBASIS, DECAPODA,
   PLECOPTERA, EPHEMEROPTERA, TIPULIDAE, CHIRONOMIDAE, HALF-LIFE.

ABSTRACT:
   CONSIDERATION OF LIFE HISTORY AND RADIOECOLOGY OF BANDED SCULPIN
   (COTTUS CAROLINAE) REVEALS THAT MOST RADIOACTIVITY RESULTS FROM NATURAL
   DECAY; SOME, FROM FALLOUT. BIOTA OF THE ECOSYSTEM ACCUMULATE MEASURABLE
   AMOUNTS. IN THE FOOD WEB OF COTTUS CAROLINAE, RADIOACTIVITY
   CONCENTRATES AT THE PRODUCER LEVEL, WITH DECREASED AMOUNTS AT HIGHER
   TROPHIC LEVELS INVOLVING ANIMALS. ANIMALS, FED MOSTLY ON DIATOMS AND
   BLUE-GREEN ALGAE, GENERALLY CONTAINED THE MOST RADIOACTIVITY.
   VERTEBRATES, AND LARGER INVERTEBRATES CONTAINED RADIOACTIVITY WITH
   LEVELS VARYING ACCORDING TO SIZE OF INDIVIDUAL ANIMAL. SMALLER ANIMALS
   CONTAINED MORE RADIOACTIVE MATERIALS PER UNIT WEIGHT THAN LARGER ONES.
   ACTUAL METABOLIC CHANGES SEEM TO BE AS IMPORTANT IN DETERMINING AMOUNT
   OF RETENTION AND ACCUMULATION OF RADIOMATERIALS AS ARE AMOUNTS OF THE
   RADIOMATERIALS INGESTED. SIMILARLY, CUMULATIVE AMOUNTS OF RADIOACTIVITY
   TAKEN IN AS FOOD, COMPARED WITH EVER-INCREASING BODY BURDEN, ALLOW
   CALCULATION OF PERCENTAGE OF INGESTED RADIOACTIVITY RETAINED IN THE
   BODY. PERCENTAGES ARE RELATIVELY HIGH DURING EARLY LIFE OF SCULPINS,
   DECREASING JUST BEFORE MATURITY, AND INCREASING SLOWLY THROUGHOUT
   REMAINDER OF LIFE. LACK OF MAJOR VARIATIONS IN THIS LEVEL OF RETENTION
   INDICATE A CERTAIN METABOLIC NEED, OR SOMEWHAT CONSTANT METABOLIC
   ACCUMULATION, WHICH COMPRISES APPROXIMATELY 2% OF TOTAL RADIOMATERIALS
   INGESTED. (JONES-WIS)

FIELD 05C

ACCESSION NO. W69-09742

MODE OF FILAMENTOUS GROWTH OF LEUCOTHRIX MUCOR IN PURE CULTURE AND IN NATURE,
AS STUDIED BY TRITIATED THYMIDINE AUTORADIOGRAPHY,

INDIANA UNIV., INDIANAPOLIS. DEPT. OF MICROBIOLOGY.

THOMAS D. BROCK.

JOURNAL OF BACTERIOLOGY, VOL 93, NO 3, P 985-990, MARCH 1967. 3 FIG, 5 TAB,
11 REF.

DESCRIPTORS:
*BACTERIA, *CULTURES, *TRITIUM, EPIPHYTOLOGY, MARINE ALGAE, ECOLOGY,
NUTRIENTS, SEAWATER, ORGANIC COMPOUNDS, HABITATS, STATISTICAL METHODS.

IDENTIFIERS:
*LEUCOTHRIX MUCOR, *GROWTH, *FILAMENTS, *AUTORADIOGRAPHY, ANTITHAMNION
SARNIENSE, TRITIATED GLUCOSE, GONIDIA FORMATION, LONG ISLAND SOUND(NY),
ICELAND, TRITIATED THYMIDINE, NARRAGANSETT BAY(RI), CAPE
REYKJANES(ICELAND), SUDURNES(ICELAND), NON-PARAMETRIC TESTS, ONE-SAMPLE
RUNS TEST, CLUSTER ANALYSIS, REYKJAVIK(ICELAND).

ABSTRACT:
GROWTH MODE OF LEUCOTHRIX MUCOR FILAMENTS WAS MEASURED BY
AUTORADIOGRAPHY WITH TRITIATED THYMIDINE. STUDIES WERE PERFORMED ON L
MUCOR IN PURE CULTURES IN FREE SUSPENSION, AS AN EPIPHYTE OF PURE
CULTURE OF THE RED ALGA ANTITHAMNION SARNIENSE, AND AS AN EPIPHYTE OF
RED ALGAE IN THE SEA. STATISTICAL ANALYSES OF GROWING CELLS
DISTRIBUTION WAS PERFORMED BY USE OF THE NONPARAMETRIC ONE-SAMPLE RUNS
TEST AND A CLUSTER ANALYSIS ADAPTED FROM QUADRAT ANALYSES OF PLANT
ECOLOGISTS. NO EVIDENCE OF PREFERENTIAL GROWTH AT BASE OR TIP OF L
MUCOR FILAMENTS WAS OBTAINED IN ANY OF THESE STUDIES. HOWEVER, IN
NATURE, BUT NOT IN THE LABORATORY, THERE WERE REGIONS OF L MUCOR
FILAMENTS WHICH WERE NONGROWING OR DORMANT. SUCH NONGROWING REGIONS
COULD INCORPORATE TRITIATED GLUCOSE. IN LABORATORY CULTURE OF
LEUCOTHRIX ON ANTITHAMNION, LEUCOTHRIX CANNOT GROW ALONE IN THE MEDIUM
(ASP-6) IN WHICH THE ALGA GROWS, BUT ONLY WHEN ATTACHED TO ALGA. THE
ALGAE SEEMINGLY RELEASED NUTRIENTS INTO THE WATER WHERE COMPLETE MIXING
OCCURS. ALL REGIONS OF BACTERIAL FILAMENTS SHOULD HAVE EQUAL ACCESS TO
THE NUTRIENT SUPPLY AND THERE SHOULD BE NO PREFERENTIAL GROWTH AT THE
BASE. IN NATURE, EPIPHYTIC BACTERIA PROBABLY DERIVE THEIR NUTRIENTS
FROM SEAWATER. (JONES-WIS)

FIELD 05C

ACCESSION NO. W69-09755

DISTRIBUTION OF PESTICIDES IN SURFACE WATERS,

VIRGINIA POLYTECHNIC INST., BLACKSBURG, VA.

PAUL H. KING, H. H. YEH, PIERRE S. WARREN, AND CLIFFORD W. RANDALL.

J AMER WATER WORKS ASS, VOL 61, NO 9, P 483-486, SEPT 1969. 4 P, 7 FIG, 2
TAB, 7 REF.

DESCRIPTORS:
*PESTICIDE KINETICS, *PESTICIDE REMOVAL, *ADSORPTION, SOILS, COALS,
DISTRIBUTION PATTERNS, ACTIVATED CARBON, CLAYS, ALGAE, WATER
PURIFICATION, WATER QUALITY, WATER CHEMISTRY.

IDENTIFIERS:
PESTICIDE SORPTION.

ABSTRACT:
SEVERAL PAPERS ON PESTICIDE DISTRIBUTION IN WATER, PESTICIDE SORPTION
IN SOIL, AND PESTICIDE REMOVAL TECHNOLOGY ARE REVIEWED AND SUMMARIZED,
AND EXPERIMENTAL DATA ARE PRESENTED ON PESTICIDE SORPTION BY SOILS,
ALGAE, AND ACTIVATED CARBON. THE PESTICIDES STUDIED ARE LINDANE AND
PARATHION. EXPERIMENTAL RESULTS ARE SHOWN BY SORPTION ISOTHERMS. IN
GENERAL, HIGH-CLAY SOILS ADSORBED ABOUT TWICE AS MUCH PESTICIDE AS
SANDY SOILS. ALGAE SORBED ABOUT 10 TIMES AS MUCH AS SOILS. COAL
ADSORBED ABOUT 2 1/2 TIMES AS MUCH AS SOIL, AND ACTIVATED CARBON
ADSORBED ABOUT 4 ORDERS OF MAGNITUDE MORE THAN SOIL. (KNAPP-USGS)

FIELD 05B, 05G

ACCESSION NO. W69-09884

RADIOECOLOGICAL SURVEILLANCE OF THE WATERWAYS AROUND A NUCLEAR FUELS
REPROCESSING PLANT,

NEW YORK STATE DEPT. OF HEALTH, ALBANY. DIV. OF LABORATORIES AND RESEARCH;
STATE UNIV. OF NEW YORK, ALBANY. DEPT. OF BIOLOGICAL SCIENCES; AND STATE
UNIV. COLL., FREDONIA, N.Y.

N. I. SAX, PAUL C. LEMON, ALLEN H. BENTON, AND JACK J. GABAY.

RADIO HEALTH DATA AND REP, VOL 10, NO 7, P 289-296, JULY 1969. 8 P, 3 FIG, 6
TAB, 26 REF. USPHS SERVICE GRANT, NAT CENTER FOR RADIOLOGICAL HEALTH.

DESCRIPTORS:
*MONITORING, *RADIOISOTOPES, *STREAMS, BIOINDICATORS, *NEW YORK,
NUCLEAR WASTES, WATER POLLUTION SOURCES, PATH OF POLLUTANTS, AQUATIC
LIFE, ALGAE, SILTS, ECOLOGY, AQUATIC ENVIRONMENT.

IDENTIFIERS:
CATTARAUGUS CREEK(NY).

ABSTRACT:
A 3-YR STUDY OF THE AQUATIC ECOSYSTEM AROUND A NUCLEAR FUELS
REPROCESSING PLANT, LOCATED IN WESTERN NEW YORK STATE, WAS CONDUCTED TO
FIND AND EVALUATE NATURAL INDICATORS OF ENVIRONMENTAL CONTAMINATION.
THE STUDY COVERED PREOPERATIONAL AND EARLY POST-OPERATIONAL PHASES. THE
ECOLOGICAL VECTORS SELECTED FOR STUDY WERE ALGAE, SILT, AND FISH, ALL
FROM CATTARAUGUS CREEK AND ITS TRIBUTARIES. SAMPLES WERE COLLECTED AT
SEVERAL POINTS UPSTREAM AND DOWNSTREAM FROM THE PLANT EFFLUENT AND
QUANTITATIVELY ANALYZED BY GAMMA-RAY SPECTROMETRY. ALL VECTORS ANALYZED
INDICATED PROCESS OF UPTAKE AND CONCENTRATION OF RUTHENIUM-RHODIUM-106,
CESIUM-137, CESIUM-134, AND/OR ZIRCONIUM-NIOBIUM-95, AND SOMETIMES
COBALT-60. IN ADDITION, THE CONCEPT OF USING NATURAL INDICATORS NOT
NATIVE TO THE STREAMS UNDER STUDY WAS TESTED BY TRANSLOCATING FRESH
WATER CLAMS FROM CHAUTAUQUA LAKE, 60 MI SOUTHWEST OF THE SITE, AND
PLACING THEM IN THE STREAMS AROUND THE PLANT. THE CLAMS NOT ONLY
THRIVED IN THEIR NEW ENVIRONMENT BUT UPON ANALYSIS SHOWED DEFINITE
INTERACTION BY CONCENTRATION OF THE ABOVE RADIOISOTOPES, WITH THE
SHELLS SHOWING APPROXIMATELY TWICE THE RADIOACTIVITY OF THE SOFT PARTS
OF THE CLAM. THIS CONCEPT MIGHT PROVE TO BE A SENSITIVE INDICATOR OF
ENVIRONMENTAL CONTAMINATION. (KNAPP-USGS)

FIELD 05C, 05A, 05B

ACCESSION NO. W69-10080

THE MEASUREMENT OF CHLOROPHYLL, PRIMARY PRODUCTIVITY, PHOTOPHOSPHORYLATION, AND
MACROMOLECULES IN BENTHIC ALGAL MATS,

INDIANA UNIV., BLOOMINGTON. DEPT. OF MICROBIOLOGY.

THOMAS D. BROCK, AND M. LOUISE BROCK.

LIMNOLOGY AND OCEANOGRAPHY, VOL 12, NO 4, P 600-605, OCT 1967. 1 FIG, 2 TAB,
13 REF.

DESCRIPTORS:
*BENTHIC FLORA, *CHLOROPHYLL, *PRIMARY PRODUCTIVITY, *MEASUREMENT,
PROTEINS, RADIOACTIVITY, CARBON RADIOISOTOPES, PHOSPHORUS
RADIOISOTOPES, ALGAE, BIOCHEMISTRY, ECOLOGY, SAMPLING, THERMAL SPRINGS,
EFFLUENTS, CYANOPHYTA.

IDENTIFIERS:
*PHOTOPHOSPHORYLATION, *MACROMOLECULES, NUCLEIC ACID, RIBONUCLEIC ACID
SYNTHESIS.

ABSTRACT:
THIS IS PART OF A STUDY OF THE BIOCHEMICAL ECOLOGY OF THERMAL
ENVIRONMENTS. ALONG THE GRADIENTS OF HOT-SPRING EFFLUENTS, ALGAE
DEVELOP ON THE BOTTOMS OF SINTER-LINED CHANNELS. THICKNESS AND
CONSISTENCY OF THE MAT IS DETERMINED BY TEMPERATURE, FLOW RATE, AND
TYPE OF FLOW--WHETHER TURBULENT OR LAMINAR. IN CHANNELS WITH LAMINAR
FLOW, THE ALGAL MATS ARE RELATIVELY UNIFORM OVER SEVERAL SQUARE
DECIMETERS. METHODS ARE DESCRIBED FOR MEASURING PRIMARY PRODUCTIVITY,
CHLOROPHYLL, PROTEIN, AND NUCLEIC ACID IN ALGAL CORES. THE CORES ARE
HOMOGENIZED AFTER INCUBATION WITH A RADIOACTIVE ISOTOPE ALLOWING
REPRESENTATIVE SUBSAMPLING. ISOTOPE SELF-ABSORPTION IS NOT A PROBLEM,
BECAUSE ONLY A SMALL FRACTION OF THE CORE IS ACTUALLY FILTERED. THE
METHOD ELIMINATES SAMPLING VARIABILITY SINCE CHLOROPHYLL IS ESTIMATED
IN THE SAME SAMPLES USED TO DETERMINE CARBON-14 (C-14) UPTAKE. ACID OR
ALKALI EXTRACTION WAS USED TO INDICATE THE BIOCHEMICAL FRACTIONS INTO
WHICH THE C-14 WAS INCORPORATED. TOTAL PHOSPHATE CONTENT OF THE SPRING
WATER WAS DETERMINED CHEMICALLY FOR CALCULATION OF SPECIFIC
RADIOACTIVITY OF PHOSPHATE-32 (P-32) ION. RATES OF PHOTOPHOSPHORYLATION
AND RIBONUCLEIC ACID SYNTHESIS WERE MEASURED USING P-32. ALL WORK WAS
DONE WITH BLUE-GREEN ALGAE, AND METHOD SHOULD NOT BE APPLIED TO OTHER
ALGAE WITHOUT PRELIMINARY TESTING. (JONES-WIS)

FIELD 05C, 07B

ACCESSION NO. W69-10151

QUANTITATIVE RELATIONS OF THE FEEDING AND GROWTH OF DAPHNIA PULEX OBTUSA (KURZ)
  SCOURFIELD,

DEPARTMENT OF WATER TECHNOLOGY, PRAGUE (CZECHOSLOVAKIA).

N. M. KRYUTCHKOVA, AND V. SLADECEK.

HYDROBIOLOGIA, VOL 33, NO 1, P 47-64, 1969. 9 FIG, 4 TAB, 51 REF.

DESCRIPTORS:
  *FEEDING RATES, *GROWTH RATES, *DAPHNIA, PRODUCTIVITY, BIOINDICATORS,
  SECONDARY PRODUCTIVITY, AQUATIC PRODUCTIVITY, WATER POLLUTION EFFECTS,
  YEASTS, CRUSTACEA, ALGAE.

IDENTIFIERS:
  *DAPHNIA PULEX OBTUSA, SCENEDESMUS QUADRICAUDA, ECOLOGICAL
  EFFICIENCIES, TROPHIC ECOLOGY, CLADOCERA.

ABSTRACT:
  AUTHORS FED INDIVIDUAL NEWBORN CLADOCERANS, DAPHNIA PULEX OBTUSA, THREE
  CONCENTRATIONS OF YEAST (5, 10, 20 MILLIGRAMS/LITER), OR ALGAE,
  SCENEDESMUS QUADRICAUDA, 5, 10 15 MILLIGRAMS/LITER) IN TEN REPLICATED
  EXPERIMENTS PER TREATMENT. DURATION OF LIFE, GROWTH, AND FECUNDITY WERE
  RECORDED FOR EACH DAPHNID. WITH INCREASING CONCENTRATIONS OF YEAST,
  MEAN TOTAL EGG PRODUCTION INCREASED FROM 101 TO 192; WITH ALGAE, FROM
  154 TO 270. MEAN BODY-LENGTH OF DAPHNID FED YEAST ATTAINED 2.4
  MILLIMETERS; ALGAE FED, 2.9 MILLIMETERS. OF TOTAL PRODUCTION (BODY
  WEIGHT + EGG WEIGHT), 66-75% WAS UTILIZED FOR REPRODUCTION IN ENTIRE
  PERIOD TO DEATH, BEGINNING 5-6 DAYS AFTER INCEPTION OF SEXUAL MATURITY.
  RELATIVE AVERAGE DAILY DAPHNID GROWTH (ADG) WAS GREATER IN EARLY LIFE,
  DECREASING WITH AGE; INCREASED WITH INCREASING CONCENTRATIONS OF FOOD,
  AND WAS GREATER WITH ALGAL FOOD THAN WITH YEAST. AUTHORS CLAIM THAT
  VALUES FOR ADG (RANGE: 0.23-0.46 FOR ALL TREATMENTS) ARE HIGH AND
  INDICATE THAT THIS DAPHNIA IS ADAPTED FOR LIFE IN POLLUTED WATER. FOR
  INCREASING ALGAL CONCENTRATIONS, AVERAGE DAILY RATIO INCREASED (39.1%,
  58.8%, 75.9% DAPHNID DRY WEIGHT); EFFICIENCY OF FOOD-UTILIZATION FOR
  GROWTH DECREASED (56.1%, 47.9%, 35.7%). FILTRATION RATE FOR LIFE SPAN
  DECREASED WITH INCREASING ALGAL CONCENTRATION FROM 842.3 TO 400
  MILLILITERS/MILLIGRAM DRY DAPHNID WEIGHT. (EICHHORN-WIS)

FIELD 05C

ACCESSION NO. W69-10152

FIELD AND EXPERIMENTAL WINTER LIMNOLOGY OF THREE COLORADO MOUNTAIN LAKES,

COLORADO UNIV., BOULDER. DEPT. OF BIOLOGY.

ROBERT W. PENNAK.

ECOLOGY, VOL 49, NO 3, P 505-520, LATE SPRING 1968. 8 FIG, 15 TAB, 36 REF.

DESCRIPTORS:
*LIMNOLOGY, *COLORADO, *LAKES, *MOUNTAINS, *WINTER, CYANOPHYTA, PLANKTON, COPEPODS, EUTROPHICATION, OLIGOTROPHY, TEMPERATURE, PHOTOSYNTHESIS, RAINBOW TROUT, DIATOMS, ANAEROBIC CONDITIONS, ALGAE, ROTIFERS, SESTON, RESPIRATION, ALTITUDE, SAMPLING, ZOOPLANKTON, PHYTOPLANKTON, ICE, BACTERIA, TRIPTON, BOTTOM SEDIMENTS, SOLAR RADIATION, OXYGEN, CARBON DIOXIDE, NANNOPLANKTON.

IDENTIFIERS:
*FIELD STUDIES, *EXPERIMENTAL LIMNOLOGY, CLADOCERAN, MONTANE ZONE, ALPINE ZONE, BOULDER(COLO), TEA LAKE(COLO), BLACK LAKE(COLO), PASS LAKE(COLO), MICRO-ALGAE, FLAGELLATES, CILIATES, COPEPODIDS, NAUPLII, TERRAMYCIN, TETRACYCLINE, STREPTOMYCIN, MILLIPORE FILTRATION, MESOTROPHY.

ABSTRACT:
PHYSICAL, CHEMICAL, AND PLANKTONIC CONDITIONS WERE STUDIED DURING TWO WINTERS IN COLORADO MOUNTAIN LAKES--BLACK LAKE (MESOTROPHIC), PASS LAKE (HIGHLY OLIGOTROPHIC), TEA LAKE (SHALLOW, EUTROPHIC, AND PONDLIKE). WINTER TEMPERATURES OF LOWER WATERS OF BLACK AND PASS LAKES ARE ABOVE 4 DEG C AND AS HIGH AS 5.4 DEG C. SUBSTRATE AND BOTTOM WATERS REACH EQUILIBRIUM IN LATE FEBRUARY OR MARCH. THICK SNOW COVERED BLACK AND PASS LAKES AND NO PHOTOSYNTHESIS OCCURRED FOR 5-7 MONTHS, BUT TEA LAKE HAD LITTLE SNOW AND INTERMITTENT PHOTOSYNTHESIS DURING ALL WINTER. BLACK AND PASS LAKES ARE SUMMER-OLIGOTROPHIC AND WINTER-EUTROPHIC. THEY BECAME SO HIGHLY ANAEROBIC BY MARCH AND APRIL THAT TROUT POPULATIONS DIED. ALL THREE LAKES HAD NEGLIGIBLE WINTER POPULATIONS OF DIATOMS AND GREEN AND BLUE-GREEN ALGAE, BUT POPULATIONS OF MICRO-ALGAE ATTAINED WINTER MAXIMA OF 1-14.5 MILLION CELLS PER LITER, WITH NO CONSISTENT SEASONAL PATTERN. WINTER COPEPOD AND CLADOCERAN POPULATIONS WERE NEGLIGIBLE. ROTIFER POPULATIONS WERE USUALLY DENSE, ESPECIALLY DURING DECEMBER AND JANUARY, BEFORE ONSET OF ANAEROBIOSIS. SESTON VARIED MORE WIDELY DURING MONTHS OF OPEN WATER. WINTER PLANKTON HAD AVERAGE RESPIRATORY RATE IN SITU OF ABOUT TWICE THAT IN CORRESPONDING SAMPLES KEPT IN DARK REFRIGERATOR AT 3 DEG C. (JONES-WIS)

FIELD 02H, 05C

ACCESSION NO. W69-10154

COPPER SULPHATE AIR SPRAY CURES LAKE ALGAE PROBLEM,

SEWER AND WATER ENGINEER, SUDBURY (ONTARIO).

D. D. BROUSE.

WATER AND POLLUTION CONTROL, VOL 104, NO 4, P 25-27, APR 1966. 4 FIG.

DESCRIPTORS:
*COPPER SULFATE, *ALGAL CONTROL, WATER SUPPLY, WATER QUALITY, WATER POLLUTION CONTROL, NUISANCE ALGAE, TASTE, ODOR.

IDENTIFIERS:
*AIR SPRAY, SUDBURY(ONTARIO), LAKE RAMSEY(ONTARIO), ALGAE COUNTS, THRESHOLD ODOR NUMBERS, CROP-DUSTING AIRCRAFT, WATER SAMPLING, ONTARIO WATER RESOURCES COMMISSION, PEST CONTROL ORGANIZATION OF TORONTO, FILAMENTOUS ALGAE, NAVICULA, APHANIZOMENON, ONTARIO(CANADA).

ABSTRACT:
SINCE 1960, SUDBURY, ONTARIO, BORDERING LAKE RAMSEY, HAS TWICE EXPERIENCED SEVERE TASTE AND ODOR PROBLEMS IN ITS MUNICIPAL WATER SUPPLY. LAKE RAMSEY HAS A SURFACE AREA OF 2,023 ACRES, DEPTHS TO 65 FEET, AND LITTLE OUTFLOW. IN OCTOBER 1965, THE ALGAE COUNT PEAKED AT 33,400 AREAL STANDARD UNITS, APHANIZOMENON PREDOMINATING, ACCOMPANIED BY A THRESHOLD ODOR NUMBER OF 200. APHANIZOMENON'S ABILITY TO WITHSTAND WINTER TEMPERATURES NECESSITATED IMMEDIATE ACTION. ON NOVEMBER 6-8, CROPDUSTING AIRCRAFT SPRAYED 20 TONS OF COPPER SULFATE OVER THE LAKE SURFACE RESULTING IN COPPER CONTENT OF 0.20 PARTS PER MILLION (PPM). THE IMMEDIATE AREA OF MUNICIPAL WATER INTAKE WAS AVOIDED. BY NOVEMBER 14TH, THE ALGAE COUNT HAD BEEN REDUCED TO 2,500 AND THE ODOR NUMBER TO 24. ON NOVENBER 22ND, THE ALGAE COUNT HAD REBOUNDED TO 5,000 AND ODOR NUMBER TO 140, WHILE THE COPPER CONCENTRATION REMAINED 0.20 PPM. IN LATE NOVEMBER, A SECOND TREATMENT BROUGHT THE MAXIMUM AVERAGE COPPER CONCENTRATION TO 0.36 PPM. ONE WEEK LATER, AFTER FREEZING OVER, THE LAKE'S ALGAE COUNT WAS BACK DOWN TO 2,000 AND THE ODOR NUMBER TO 17. BY THE END OF DECEMBER, THE WATER WAS SATISFACTORY FOR MUNICIPAL USE WITH ONLY SLIGHT NOTICEABLE ODOR (ALGAE COUNT: 1,300, ODOR NUMBER: 4). (KETELLE-WIS)

FIELD 05G

ACCESSION NO. W69-10155

ALGAE CONTROL WITH COPPER SULFATE,

PORTLAND WATER DISTRICT, MAINE.

W. D. MONIE.

WATER AND SEWAGE WORKS, VOL 103, P 392-397, SEPT 1956. 7 FIG, 8 REF.

DESCRIPTORS:
*ALGAL CONTROL, *COPPER SULFATE, WATER SUPPLY, RESERVOIRS, WATER QUALITY, NEW JERSEY, TEMPERATURE, ORGANIC MATTER, HARDNESS, CARBON DIOXIDE, FISH.

IDENTIFIERS:
*MICROSCOPIC EXAMINATION, *CRITICAL TEMPERATURES, *COPPER DOSAGE, *COPPER APPLICATION METHODS, ALGAE IDENTIFICATION, ALGAE COUNTS, ALGAL GROWTH, SEDGWICK-RAFTER METHOD, TREATMENT TIMING, ALGAE CONTROL REQUIREMENTS, THRESHOLD ODOR TESTS, ASTERIONELLA, APHANIZOMENON, SYNURA, ANABAENA, MELOSIRA, MALLOMONAS, ULOTHRIX, DAVID MONIE TEST, SUMMIT(NJ), CANOE BROOK RESERVOIR(NJ).

ABSTRACT:
ALGAL CONTROL SHOULD BE PRACTICED AS A PREVENTATIVE MEASURE AND NOT A CURE. THE IMPORTANT FACTORS OF CONTROL WITH COPPER SULFATE ARE: (1) ABILITY TO DETERMINE WHEN TREATMENT IS NECESSARY, (2) CORRECT DOSAGE OF COPPER SULFATE, (3) APPLICATION AIMED FOR UNIFORM DISTRIBUTION. THE FIRST STEP IS ALGAE IDENTIFICATION AND COUNTING THROUGH MICROSCOPIC ANALYSIS, COMMONLY USING THE SEDGWICK-RAFTER METHOD. WATER TEMPERATURE DATA ARE USEFUL IN REVEALING CRITICAL TEMPERATURE RANGES IN WHICH GROWTH OF VARIOUS SPECIES IS STIMULATED. MONITORING TEMPERATURE ALLOWS ANTICIPATION OF PERIOD OF RAPID ALGAL GROWTH AND SPECIES COMPOSITION. TIMING OF TREATMENT AND DOSAGE ARE IMPORTANT CONSIDERATIONS, PARTIALLY DEPENDENT ON CHEMICAL CHARACTERISTICS OF WATER, TEMPERATURE, AND TYPE OF ORGANISMS. NO FIXED DOSAGE CAN BE PRESCRIBED, BUT SEVEN FACTORS MAY BE USED AS GUIDELINES: ALGAL SPECIES, AMOUNT OF ORGANIC MATTER, WATER HARDNESS, CARBONIC ACID CONTENT, TEMPERATURE, FISH SPECIES, AND TOTAL VOLUME TO BE TREATED. A SIMPLE, PRACTICAL LABORATORY TEST (DAVID MONIE TEST), DEVELOPED BY THE AUTHOR FOR DETERMINING CORRECT COPPER SULFATE DOSAGE WITHOUT REGARD TO WATER TEMPERATURE AND COMPOSITION AND--WITHIN LIMITS--THE TYPE AND AMOUNT OF ORGANISMS PRESENT, IS DESCRIBED. FINALLY, BOTTOM TOPOGRAPHY IS CRITICAL IN DETERMINING DOSAGES FOR ACHIEVING EQUITABLE DISTRIBUTION. (KETELLE-WIS)

FIELD 05G

ACCESSION NO. W69-10157

EXTRACELLULAR PRODUCTION IN RELATION TO GROWTH OF FOUR PLANKTONIC ALGAE AND OF
  PHYTOPLANKTON POPULATIONS FROM LAKE ONTARIO,

SCARBOROUGH COLL., TORONTO (ONTARIO).

C. NALEWAJKO, AND L. MARIN.

CANADIAN JOURNAL OF BOTANY, VOL, 47, P 405-413, 1969. 8 FIG, 3 TAB, 9 REF.

DESCRIPTORS:
  *PLANKTON, *ALGAE, *PHYTOPLANKTON, *LAKE ONTARIO, *GROWTH RATES,
  ENVIRONMENTAL EFFECTS, PHOTOSYNTHESIS, DIATOMS, CHLORELLA, CARBON
  RADIOISOTOPES, LIGHT INTENSITY, CULTURES, RADIOACTIVITY, CARBON
  DIOXIDE, WATER POLLUTION EFFECTS, WATER POLLUTION SOURCES.

IDENTIFIERS:
  *EXTRACELLULAR PRODUCTION, *CARBON FIXATION, *EXCRETION, STEPHANODISCUS
  TENUIS, ASTERIONELLA FORMOSA, MELOSIRA BINDERANA, GENERATION TIME,
  CHLORELLA PYRENOIDOSA, DIATOMA ELONGATUM, MELOSIRA ISLANDICA, NITZSCHIA
  ACICULARIS, NITZSCHIA DISSIPATA, SYNEDRA ACUS, SINKING RATES, GRAZING
  PRESSURE, POPULATION DENSITY.

ABSTRACT:
  AUTHORS PRESENT DATA ON RELATIONSHIP BETWEEN PHOTOSYNTHESIS AND
  EXTRACELLULAR PRODUCTION AND RELATIVE GROWTH RATE IN FOUR PLANKTONIC
  ALGAE. THREE ARE DIATOMS ABUNDANT IN LAKE ONTARIO; THE FOURTH SPECIES,
  CHLORELLA PYRENOIDOSA CHICK, WAS INCLUDED FOR COMPARISON. ALL CULTURES
  WERE GROWN UNDER LOW LIGHT AND STIRRED CONTINUOUSLY. AT VARIOUS STAGES
  OF GROWTH THE CARBON-14 METHOD WAS USED TO MEASURE PHOTOSYNTHESIS AND
  EXCRETION ON EXPOSURE TO HIGH LIGHT INTENSITY. ADDITIONALLY,
  PHOTOSYNTHESIS AND EXCRETION WERE MEASURED DURING SPRING (1966)
  INCREASE OF DIATOMS IN LAKE ONTARIO. GROWTH RATES OF PLANKTONIC ALGAL
  POPULATIONS CANNOT BE MEASURED SINCE FACTORS SUCH AS SINKING AND
  GRAZING MAY MASK THE TRUE RATE. HOWEVER, SUCH DATA WERE THOUGHT TO BE
  USEFUL FOR COMPARISON WITH LABORATORY EXPERIMENTS, PARTICULARLY SINCE
  TWO OF THE SPECIES STUDIED IN CULTURE WERE OF QUANTITATIVE IMPORTANCE
  IN THE BLOOM. IN THE FOUR SPECIES, BOTH CARBON FIXATION AND EXCRETION
  (ON BASIS OF ASH-FREE DRY WEIGHT) INCREASE WITH RELATIVE GROWTH RATE OF
  CULTURES MEASURED IN LOGARITHMIC (TO BASE 10) UNITS. IN NATURAL
  POPULATIONS, PERCENTAGE EXCRETION VALUES ARE POSTIVELY CORRELATED WITH
  RELATIVE GROWTH RATES, HOWEVER, ENVIRONMENTAL FACTORS MAY BE OF
  RELATIVELY GREATER IMPORTANCE AND DETERMINE THE EXTENT OF EXCRETION.
  (JONES-WISCONSIN)

FIELD 05C, 02H

ACCESSION NO. W69-10158

ALGAL RESPIRATION IN A EUTROPHIC ENVIRONMENT,

PENNSYLVANIA STATE UNIV., UNIVERSITY PARK. DEPT. OF CIVIL ENGINEERING.

ARCHIE J. MCDONNELL, AND R. RUPERT KOUNTZ.

JOURNAL OF WATER POLLUTION CONTROL FEDERATION, VOL 38, NO 5, P 841-847, MAY 1966. 5 FIG, 3 TAB, 9 REF.

DESCRIPTORS:
*ALGAE, *RESPIRATION, *EUTROPHICATION, *DISSOLVED OXYGEN, PENNSYLVANIA, BIOCHEMICAL OXYGEN DEMAND, TEMPERATURE, FLOW, SAMPLING, WASTE WATER DISPOSAL, DIURNAL, EFFLUENTS, NITROGEN, PHOSPHATES, ALKALINITY, FISH, STREAMS, WATER POLLUTION EFFECTS, WATER POLLUTION SOURCES, NUTRIENTS.

IDENTIFIERS:
SPRING CREEK(PA), STATE COLLEGE(PA), FLORA.

ABSTRACT:
PRESENT-DAY COMMUNITY DEVELOPMENT, ESPECIALLY IN SMALL WATERSHEDS, SUCH AS SPRING CREEK, PENNSYLVANIA, HAS COMPOUNDED PROBLEMS OF STREAM POLLUTION CRITERIA. IN MANY STREAM SITUATIONS, DEPLETION OF DISSOLVED OXYGEN RESOURCES IS A RESULT NOT ONLY OF BIOCHEMICAL OXYGEN DEMAND (BOD) BUT ALSO OF THE EFFECT OF COMMUNITY RESPIRATION. EXACTLY HOW MUCH EFFECT EACH OF THESE FACTORS HAS ON THE DISSOLVED OXYGEN LEVEL OF THE STREAM IS DIFFICULT TO EVALUATE. WITH STATISTICAL METHODS OF REGRESSION AND VARIANCE ANALYSIS, IT WAS POSSIBLE TO DEFINE THE BOD PARAMETER AS ONE OF SECONDARY ENGINEERING SIGNIFICANCE AND TO EVALUATE QUANTITATIVELY THE NOCTURNAL OXYGEN DEMANDS OF A HETEROGENEOUS STREAM FLORA. COMMUNITY PLANT RESPIRATION VARIES DIRECTLY WITH STREAM TEMPERATURE AND INVERSELY WITH STREAM FLOW. CHEMICAL ANALYSIS OF CONTRIBUTING SOURCES INDICATED NUTRIENT CONCENTRATIONS SUFFICIENT TO INDUCE AND MAINTAIN A EUTROPHIC ENVIRONMENT. ON THIS BASIS, WHERE POPULATION-TO-RECEIVING WATER RESOURCE RATIO IS HIGH, ANALYSIS FOR SIGNIFICANCE OF PARAMETERS OTHER THAN BOD SHOULD BE MADE AND THE FINDINGS MANIFESTED IN THE TREATMENT PROCESS. THE SOLUTION APPEARS TO LIE IN THE REMOVAL OF NITROGEN AND PHOSPHORUS FROM THE WASTEWATER TREATMENT PLANT EFFLUENT BEFORE IT ENTERS THE CREEK WATERS, WHICH PRACTICALLY IMPLIES TOTAL EXCLUSION. (JONES-WIS)

FIELD 05B

ACCESSION NO. W69-10159

TEMPERATURE OPTIMA FOR ALGAL DEVELOPMENT IN YELLOWSTONE AND ICELAND HOT SPRINGS,

INDIANA UNIV., BLOOMINGTON. DEPT. OF BACTERIOLOGY.

T. D. BROCK, AND M. LOUISE BROCK.

NATURE, VOL 209, NO 5024, P 733-734, FEB 1966. 2 FIG, 7 REF.

DESCRIPTORS:
*ALGAE, *TEMPERATURE, *HOT SPRINGS, HABITATS, ALKALINE WATER, ACIDIC WATER, TRAVERTINE, CALCIUM CARBONATE, BACTERIA, BIOMASS, CHLOROPHYLL, PROTEINS, LIGHT, NUTRIENTS, HYDROSTATIC PRESSURE, CARBON DIOXIDE, CYANOPHYTA, PHYSIOLOGICAL ECOLOGY, THERMAL WATER.

IDENTIFIERS:
*YELLOWSTONE NATIONAL PARK, *ICELAND HOT SPRINGS, SINTER, NUCLEIC ACID, SISJOTHANDI(ICELAND), ENZYME-SUBSTRATE COMPLEXES, RIBONUCLEIC ACID, ORCINOL METHOD.

ABSTRACT:
IN ANALYZING ENVIRONMENTAL FACTORS AFFECTING GROWTH, STUDY OF HABITATS WHERE ONLY A SINGLE FACTOR VARIES IS DESIRABLE. IN HOT SPRINGS TEMPERATURE IS THE ONLY VARIABLE AND ITS RELATION TO BIOLOGICAL DEVELOPMENT CAN BE MEASURED DIRECTLY. IN THE YELLOWSTONE SPRINGS, ALKALINE, ACID, AND TRAVERTINE-DEPOSITING, HIGH IN CALCIUM CARBONATE, EXHIBIT THE GREATEST BIOLOGICAL DEVELOPMENT. OVER SILICEOUS SINTER CONES, WHERE CONSTANT AND UNIFORM THERMAL GRADIENTS MAY EXIST, BLUE-GREEN ALGAE PROLIFERATE. POOLS WERE CHOSEN WHERE INITIAL CHEMICAL COMPOSITION OF WATER FLOWING OVER ALL ORGANISMS ALONG THE THERMAL GRADIENT IS IDENTICAL. THE EXTENT OF ALGAL DEVELOPMENT IS INFLUENCED BY AMOUNT AND RATE OF WATER FLOW AND, SINCE SPRINGS DIFFERED IN THIS FACTOR, THEY ALSO DIFFERED IN ALGAL CONTENT, WHILE EXHIBITING SIMILAR THERMAL GRADIENTS. SAMPLES WERE TAKEN WHENEVER POSSIBLE FROM AREAS HAVING IDENTICAL AMOUNTS AND RATES OF WATER FLOW. CHLOROPHYLL CONTENT WAS ESTIMATED AND RIBONUCLEIC ACID AND PROTEIN ASSAYS MADE. THESE RESULTS WERE EXPRESSED SHOWING RELATIONSHIP BETWEEN TEMPERATURE OF SAMPLE AND BIOCHEMICAL CONTENT PER UNIT AREA. SPECIES NUMBER AND COMPOSITION CHANGED ALONG THE GRADIENT. THE OPTIMUM TEMPERATURE FOR ALGAL DEVELOPMENT IN THESE ALKALINE SPRINGS IS 51-56 DEGREES C. TEMPERATURES WERE LOWER IN ICELAND SPRINGS AND HYDROGEN ION CONCENTRATION HIGHER. (JONES-WIS)

FIELD 05C

ACCESSION NO. W69-10160

THE HABITAT OF LEUCOTHRIX MUCOR, A WIDESPREAD MARINE MICROORGANISM,

INDIANA UNIV., BLOOMINGTON. DEPT. OF BACTERIOLOGY.

THOMAS D. BROCK.

LIMNOLOGY AND OCEANOGRAPHY, VOL 11, NO 2, P 303-307, APRIL 1966. 2 FIG, 7 REF.

DESCRIPTORS:
*MARINE MICROORGANISMS, *HABITATS, EPIPHYTOLOGY, MARINE ALGAE, ECOLOGY, CULTURES, WASHINGTON, WAVES(WATER), CONNECTICUT, RHODE ISLAND, TEMPERATURE, NUTRIENT REQUIREMENTS, PHYSIOCOCHEMICAL PROPERTIES, PHYSIOLOGICAL ECOLOGY, RHODOPHYTA, CHLOROPHYTA.

IDENTIFIERS:
*LEUCOTHRIX MUCOR, MORPHOLOGY, FRIDAY HARBOR(WASH), ANTITHAMNION SARNIENSE, RHODOCHORTON, BANGIA FUSCO-PURPUREA, SPHACELARIA, GONIDIA, CALLOPHYLLIS HAENOPHYLLA, TIDAL CURRENT, PUGET SOUND(WASH), LONG ISLAND SOUND, NARRAGANSETT BAY, CAPE REYKJANES, FAXAFLOI FJORD, ICELAND, DNA BASE COMPOSITION.

ABSTRACT:
LEUCOTHRIX MUCOR IS A LARGE, WIDESPREAD MARINE MICROORGANISM WITH CHARACTERISTIC ORPHOLOGICAL FEATURES RECOGNIZABLE IN NATURAL COLLECTIONS. IT GROWS AS AN EPIPHYTE ON MARINE ALGAE, OCCURRING MOST EXTENSIVELY ON RHODOPHYTES (RED ALGAE) AND ON FILAMENTOUS GREEN ALGAE. AUTHOR DESCRIBES METHOD OF ISOLATION AND SUCCESS IN EXTENSIVE EPIPHYTIC GROWN ON ALGAE IN THE LABORATORY, CONCLUDING THAT THE ALGA NOT ONLY PROVIDES A SUBSTRATUM FOR ATTACHMENT OF LEUCOTHRIX, BUT ALSO NUTRIENTS FOR ITS GROWTH. THE ALGAL CULTURES, WHICH HAVE BEEN MAINTAINED THROUGH SUCCESSIVE TRANSFERS MADE OVER SEVERAL MONTHS, ARE APPARENTLY NOT HARMED IN ANY WAY BY ATTACHMENT AND GROWTH OF L MUCOR. IN NATURE, A WIDE VARIETY OF FILAMENTOUS AND LEAFY RHODOPHYTES ARE COLONIZED WITH L MUCOR. WHERE WATER IS STILL OR SLOW MOVING, L MUCOR IS RARE, BUT IT OCCURS IN EXTREMELY HIGH DENSITIES ON RHODOPHYTES GROWING IN RAPIDLY MOVING WATER. ALL ISOLATES IN PURE CULTURE HAVE BEEN REMARKABLY SIMILAR IN PHYSIOLOGICAL AND MORPHOGENETIC BEHAVIOR, HAVING SIMILAR TEMPERATURE OPTIMA AND NUTRITIONAL REQUIREMENTS. SIX STRAINS HAVE DEOXYRIBONUCLEIC ACID DEMONSTRATING IDENTICAL BASE COMPOSITION. THUS, THE SPECIES, AS DEFINED MORPHOLOGICALLY, COMPRISES A HOMOGENEOUS GROUP OF STRAINS PHYSIOLOGICALLY AND BIOCHEMICALLY. (JONES-WIS)

FIELD 05C, 02L

ACCESSION NO. W69-10161

THE APPLICATION OF MICRO-AUTORADIOGRAPHIC TECHNIQUES TO ECOLOGICAL STUDIES,

INDIANA UNIV., BLOOMINGTON. DEPT. OF MICROBIOLOGY.

M. LOUISE BROCK, AND THOMAS D. BROCK.

INTERNATIONAL ASSOCIATION OF THEORETICAL AND APPLIED LIMNOLOGY,
   COMMUNICATIONS, NO 15, P 1-29, 1968. 6 FIG, 2 TAB, 38 REF.

DESCRIPTORS:
  *ECOLOGY, *RADIOACTIVITY TECHNIQUES, *RADIOISOTOPES, LIMNOLOGY,
  OCEANOGRAPHY, TRACERS, BACTERIA, GROWTH RATES, TRITIUM, PHOTOGRAPHY,
  CARBON RADIOISOTOPES, PHOSPHORUS RADIOISOTOPES, ALGAE, EMULSIONS,
  BACKGROUND RADIATION, IODINE RADIOISOTOPES, NUTRIENT REQUIREMENTS.

IDENTIFIERS:
  *MICRO-AUTORADIOGRAPHY, PHYSIOLOGY, BIOCHEMICAL EXTRACTION, MILLIPORE
  FILTERS, PHASE MICROSCOPY, STAINING, STRIPPING FILM, SYNECHOCOCCUS,
  GRAIN COUNT(AUTORADIOGRAPHY), LEUCOTHRIX MUCOR, ANTITHAMNION SARNIENSE,
  QUANTITATION, GROWTH MODES.

ABSTRACT:
  AUTHORS DESCRIBE AUTORADIOGRAPHY IN DETAIL, NOT ONLY AS AN ADJUNCT TO
  OTHER RADIOACTIVITY EXPERIMENTS, BUT AS A PRINCIPAL TECHNIQUE IN
  LIMNOLOGY, OCEANOGRAPHY, AND RELATED ENVIRONMENTALLY ORIENTED
  DISCIPLINES. TECHNIQUE CAN BE USED AS A SOLE METHOD TO STUDY NUTRITION
  AND PHYSIOLOGY IN THE LABORATORY AS WELL AS IN THE NATURAL ENVIRONMENT.
  SIMPLE BIOCHEMICAL EXTRACTION METHODS ARE DESCRIBED. AUTORADIOGRAPHIC
  RESULTS OBTAINED FROM EXPERIMENTS BASED ON SUCH METHODS ILLUSTRATE THE
  ANALYSIS OF AUTORADIOGRAMS AND DEMONSTRATE THE VALIDITY OF
  AUTORADIOGRAPHIC TECHNIQUES. DETAILED INFORMATION DELINEATES THE USE OF
  AUTORADIOGRAPHY IN EVALUATING MODE AND RATE OF BACTERIA GROWTH.
  DETAILED METHODS FOR PREPARATION OF SPECIMENS, HANDLING OF LIQUID
  AUTORADIOGRAPHIC EMULSION, AND DEVELOPMENT OF AUTORADIOGRAMS ARE GIVEN
  WITH SPECIAL EMPHASIS ON LIMNOLOGICAL APPLICATIONS. COMPARISONS ARE
  MADE WITH CONVENTIONAL METHODS FOR ASSAY OF RADIOACTIVE ISOTOPES.
  TRITIUM WAS THE MOST SUITABLE RADIOACTIVE ISOTOPE FOR THEIR
  EXPERIMENTS. LIQUID EMULSION (NTB-2) AND STRIPPING FILM ARE COMPARED.
  QUALITATIVE EVALUATION CAN BE ACCOMPLISHED WITH TWO-EXPOSURE
  PHOTOGRAPHS BY FOCUSING ON GRAINS TO PHOTOGRAPH THEM AT HALF THE TOTAL
  REQUIRED EXPOSURE, THEN REFOCUSING ON CELLS TO PHOTOGRAPH THEM DURING
  OTHER HALF OF TOTAL REQUIRED EXPOSURE. RELATIVE QUANTITATION AND
  ABSOLUTE QUANTITATION ARE DISCUSSED. (JONES-WISC)

FIELD 07B, 05C

ACCESSION NO. W69-10163

ALGAE AND PHOSPHORUS IN LAKE MINNETONKA,

MINNESOTA UNIV., MINNEAPOLIS. LIMNOLOGICAL RESEARCH CENTER.

ROBERT O. MEGARD.

LIMNOLOGICAL RESEARCH CENTER, MINNESOTA UNIV, MINNEAPOLIS, INTERIM REPORT NO
    4, 27 P, DECEMBER 1968. 9 FIG, 3 REF. OWRR PROJECT A-016-MINN.

DESCRIPTORS:
    *ALGAE, *PHOSPHORUS, *MINNESOTA, *EUTROPHICATION, POLLUTION ABATEMENT,
    GROWTH RATES, CHLOROPHYLL, PHOTOSYNTHESIS, CHEMICAL ANALYSIS, BACTERIA,
    NUTRIENTS, TEMPERATURE, EPILIMNION, NITROGEN COMPOUNDS, SEWAGE
    EFFLUENTS, LIGHT INTENSITY, LAKES, WATER POLLUTION SOURCES, WATER
    POLLUTION EFFECTS, WATER POLLUTION CONTROL, NITRATES, NITROGEN FIXATION.

IDENTIFIERS:
    *LAKE MINNETONKA(MINN), LAKE MANAGEMENT PROGRAM, ALGAL DENSITY,
    APHANIZOMENON, MELOSIRA, LYNGBYA.

ABSTRACT:
    DATA INDICATE THAT PHOSPHORUS IS THE CRITICAL NUTRIENT FOR ALGAE IN
    LAKE MINNETONKA, MINNESOTA, DURING SUMMER. THE RELATIONSHIP BETWEEN
    ALGAE AND NITROGEN WAS NOT STUDIED IN DETAIL BUT NITRATE IS APPARENTLY
    NOT CRITICAL AT ANY TIME. THIS IS PRESUMABLY DUE TO SYNTHESIS OF
    NITRATE FROM DISSOLVED GASEOUS NITROGEN BY DOMINANT ALGAS IN SUMMER.
    PHOSPHORUS PROBABLY WAS NOT A LIMITING NUTRIENT IN SPRING, AT WHICH
    TIME ANOTHER SUBSTANCE, POSSIBLY NITROGEN, WAS LIMITING. DETAILED
    ANALYSES OF NUTRIENT BUDGETS AND WATER BALANCES WILL BE REQUIRED TO
    DETERMINE WHAT MUST BE DONE TO RESTORE LAKE MINNETONKA, AND REQUISITE
    TIME AND COST OF PROGRAMS. INITIAL OBJECTIVE OF AN INTERIM PROGRAM
    SHOULD BE TO REDUCE INFLUX OF PHOSPHORUS FROM THE WATERSHED. MOST OF
    PHOSPHORUS COMES FROM SEWAGE EFFLUENTS, CONTROLLABLE EITHER BY ADVANCED
    TREATMENT OR BY DIVERSION. DETERIORATION OF THE LAKE WILL BE RETARDED
    EVEN IF ONLY MODERATE QUANTITIES OF NUTRIENTS ARE PREVENTED FROM
    ENTERING. ALL MUNICIPALITIES MUST COOPERATE. LAKE MINNETONKA WILL
    IMPROVE ONLY WHEN ANNUAL NUTRIENT LOSSES EXCEED INFLUX. FURTHER STUDY
    CAN RESOLVE WHETHER ADVANCED SEWAGE TREATMENT WITH ABATEMENT OF OTHER
    SOURCES OF NUTRIENTS, OR SEWAGE DIVERSION, WILL BE REQUIRED. (JONES-WIS)

FIELD 05C, 02H

ACCESSION NO. W69-10167

ARTIFICIAL EUTROPHICATION OF LAKE WASHINGTON,.

WASHINGTON UNIV., SEATTLE. DEPT. OF ZOOLOGY.

W. T. EDMONDSON, G. C. ANDERSON, AND DONALD R. PETERSON.

LIMNOLOGY AND OCEANOGRAPHY, VOL 1, P 47-53, 1956. 3 FIG, 1 TAB, 27 REF.

DESCRIPTORS:
*WASHINGTON, *LAKES, *EUTROPHICATION, SEWAGE, BIOTA, PHYTOPLANKTON,
DIATOMS, DINOFLAGELLATES, ALGAE, HYPOLIMNION, OXYGEN, PRODUCTIVITY,
PHOSPHATES, CHLOROPHYLL, EPILIMNION, EFFLUENT, TEMPERATURE,
PHOTOSYNTHESIS, CARBON DIOXIDE, DEPTH, THERMOCLINE, ZOOPLANKTON,
TRIPTON, BACTERIA, SEDIMENTATION, COPEPODS, WATER POLLUTION SOURCES,
WATER POLLUTION EFFECTS.

IDENTIFIERS:
*LAKE WASHINGTON(WASH), OSCILLATORIA RUBESCENS, SWITZERLAND,
PERIDINIUM, OSCILLATORIA AGARDHI, PHORMIDIUM, APHANIZOMENON FLOS-AQUAE,
SECCHI DISC TRANSPARENCY, ANABAENA LEMMERMANNI, APHANOCAPSA, BOSMINA
LONGISPINA, ZURICHSEE(SWITZERLAND), LINSLEY POND(CONN), BOSMINA
COREGONI LONGISPINA.

ABSTRACT:
LAKE WASHINGTON HAS BEEN RECEIVING INCREASING AMOUNTS OF TREATED
SEWAGE, AND APPEARS TO BE RESPONDING BY CHANGES IN TYPE AND QUANTITY OF
BIOTA. IN 1933 AND 1950, DOMINANT PHYTOPLANKTON ORGANISMS WERE ANABAENA
AND VARIOUS DIATOMS AND DINOFLAGELLATES, BUT IN 1955, APPARENTLY FOR
THE FIRST TIME, LARGE POPULATIONS OF BLUE-GREEN ALGA, OSCILLATORIA
RUBESCENS, OCCURRED, A SPECIES, WHICH CONSTITUTES NUISANCE BLOOMS IN
MANY LAKES. AN INTERESTING ECOLOGICAL PROBLEM EXISTS IN CONNECTION WITH
THE TWO SPECIES OF OSCILLATORIA IN LAKE WASHINGTON, O AGARDHI AND O
RUBESCENS. THE REPLACEMENT OF ONE BY ANOTHER MAY IMPLY A DISTINCT, BUT
PERHAPS SUBTLE DIFFERENCE IN ECOLOGICAL REQUIREMENTS. A GREAT INCREASE
IN THE HYPOLIMNETIC OXYGEN DEFICIT IS CONSIDERED AS EVIDENCE OF
INCREASED PRODUCTIVITY; THE DEFICIT WAS 1.18 MICROGRAMS PER SQUARE
CENTIMETER PER MONTH IN 1933, 2.00 IN 1950, AND 3.13 IN 1955. DECREASE
IN OXYGEN IS FAIRLY CLOSELY RELATED TO INCREASE IN HYPOLIMNETIC
PHOSPHATE CONCENTRATION BETWEEN MEASUREMENTS, MUCH LESS CLOSELY RELATED
TO THE CHLOROPHYLL CONCENTRATION IN THE EPILIMNION. FURTHER STUDY OF
LAKE WASHINGTON AS ITS EUTROPHICATION PROCEEDS OR, IF EFFLUENTS ARE
DIVERTED, OF THE EXTENT TO WHICH THE LAKE REGAINS ITS FORMER, MORE
OLIGOTROPHIC, CONDITION WILL BE VALUABLE. (JONES-WIS)

FIELD 05C, 02H

ACCESSION NO. W69-10169

EUTROPHICATION OF LAKES AND RIVERS: ITS ORIGIN AND PREVENTION (IN GERMAN),

KANTONALES LABORATORIUM, ZURICH (SWITZERLAND).

E. A. THOMAS.

VIERTELJAHRSSCHRIFT DER NATURFORSCHENDEN GESELLSCHAFT, ZURICH, VOL 107, NO 3,
P 127-140, 1962. 24 REF. IN GERMAN.

DESCRIPTORS:
*LAKES, *RIVERS, *EUTROPHICATION, RUNOFF, FERTILIZERS, NUTRIENTS,
NITRATES, PHOSPHATES, PHYTOPLANKTON, IRON COMPOUNDS, ALGAE, PLANTS,
WATER POLLUTION EFFECTS, WATER POLLUTION SOURCES, WATER POLLUTION
CONTROL, LIMNOLOGY, DIVERSION, STREAMS.

, IDENTIFIERS:
ALUMINUM SULFATE, IRON CHLORIDE, IRON SULFATE, CLADOPHORA BLANKETS,
HYDRODICTYON, MACROPHYTES.

ABSTRACT:
RUNOFF, EVEN WHEN DEPRIVED OF ITS MECHANICAL AND BIOGENIC CONSTITUENTS
BY PURIFICATION, PROVIDES RICH FERTILIZER FOR ALGAE AND HIGHER PLANTS
OF STREAMS AND LAKES. NITRATES AND PHOSPHATES ARE PARTICULARLY
IMPORTANT STIMULANTS WHICH INDUCE VARIOUS HARMFUL CONSEQUENCES. REMOVAL
OF PHOSPHATES FROM THE RUNOFF BY THE USE OF SUITABLE CHEMICAL REACTIONS
CONSTITUTES A DIRECT CONTROL MEASURE INHIBITING EUTROPHICATION. THE
POSSIBILITY OF DIVERSION IF DISSOLVED PHOSPHATES FOR PRODUCTION OF FARM
CROPS IS SUGGESTED. THE METHOD OF EUTROPHICATION CONTROL MUST BE
DETERMINED BY THE LIMNOLOGICAL CHARACTERISTICS OF WATER IN QUESTION.
(WILDE-WISC)

FIELD 05C

ACCESSION NO. W69-10170

PSEUDOMONAS AERUGINOSA FOR THE EVALUATION OF SWIMMING POOL CHLORINATION AND ALGICIDES,

WISCONSIN UNIV., MADISON. WATER CHEMISTRY LAB.

G. P. FITZGERALD, AND M. E. DERVARTANIAN.

APPLIED MICROBIOLOGY, VOL 17, NO 3, P 415-421, MARCH 1969. 1 FIG, 4 TAB, 19 REF.

DESCRIPTORS:
*PSEUDOMONAS, *SWIMMING POOLS, *BACTERIA, *BACTERICIDES, *CHLORINATION, *AQUATIC BACTERIA, *AQUATIC MICROBIOLOGY, ALGICIDES, BIOASSAY, DISINFECTION, ENVIRONMENTAL EFFECTS, ENVIRONMENTAL SANITATION, INHIBITION, CHLORINE, NITROGEN COMPOUNDS, SEWAGE, SEWAGE BACTERIA, SEWAGE TREATMENT, WATER POLLUTION, WATER POLLUTION EFFECTS, ALGAL CONTROL, ANALYTICAL TECHNIQUES, AQUATIC ALGAE.

IDENTIFIERS:
PSEUDOMONAS AERUGINOSA.

ABSTRACT:
CONCENTRATIONS OF AMMONIA AND THE CHLORINE STABILIZER, CYANURIC ACID, WHICH COULD BE EXPECTED IN SWIMMING POOLS, DECREASED THE RATE OF KILL BY CHLORINE (Cl) OF THE POTENTIAL PATHOGEN, PSEUDOMONAS AERUGINOSA. THE EFFECT OF CYANURIC ACID INCREASED AS THE CONCENTRATION OF Cl DECREASED, A FACT OF SIGNIFICANCE FROM A PUBLIC HEALTH VIEW. QUATERNARY AMMONIUM ALGICIDES HAD LITTLE EFFECT ON THE KILL RATE OF Cl, BUT AN ORGANIC MERCURY ALGICIDE HAD A SYNERGISTIC EFFECT WITH Cl WHEN Cl-ACTIVITY WAS STRESSED BY THE ADDITION OF AMMONIA OR THE USE OF 100 TIMES THE NORMAL CONCENTRATION OF BACTERIA. THE EFFECT OF NATURAL WATERS, RAIN, BEACHES, AND SWIMMING POOLS ON THE KILL RATE OF 0.5 MILLIGRAMS Cl/LITER INDICATED THAT A TREATMENT TIME OF ONE HOUR OR MORE WAS REQUIRED TO KILL 99.9% OF PSEUDOMONAS CELLS AT CONCENTRATION OF A MILLION CELLS PER LITER. THE SYNERGISM OF Cl AND THE ORGANIC MERCURY ALGICIDE WAS ALSO DEMONSTRATED WITH THESE WATERS AND WITH SEWAGE TREATMENT PLANT EFFLUENTS. AUTHOR DISCUSSES THE NECESSITY OF DEVELOPING AND USING LABORATORY TESTS WHICH SIMULATE CONDITIONS IN SWIMMING POOLS WITH HEAVY LOADS OF SWIMMERS, AS OPPOSED TO TESTS IN CHLORINE DEMAND-FREE CONDITIONS. (FITZGERALD-WIS)

FIELD 05F

ACCESSION NO. W69-10171

STUDIES ON MORPHOGENESIS IN A BLUE-GREEN ALGA. I. EFFECT OF INORGANIC NITROGEN SOURCES ON DEVELOPMENTAL MORPHOLOGY OF ANABAENA DOLIOLUM,

RAJASTHAN UNIV., JAIPUR (INDIA). DEPT. OF BOTANY.

H. N. SINGH, AND B. S. SRIVASTAVA.

CANADIAN JOURNAL OF MICROBIOLOGY, VOL 14, P 1341-1346, 1968. 5 FIG, 2 TAB, 12 REF.

DESCRIPTORS:
*PHYSIOLOGICAL ECOLOGY, *ENVIRONMENTAL EFFECTS, *MORPHOLOGY, *ALGAE, AQUATIC MICROBIOLOGY, CYANOPHYTA, ESSENTIAL NUTRIENTS, NITROGEN COMPOUNDS, AMMONIA, NITRATES, NITRITES, NUTRIENT REQUIREMENTS, NUTRIENTS, PHYTOPLANKTON, INHIBITION.

IDENTIFIERS:
*ANABAENA DOLIOLUM.

ABSTRACT:
SPORE GERMINATION, HETEROCYST PRODUCTION, HORMOGONE FORMATION, AND SPORULATION ARE THE MORPHOGENETIC STAGES IN DEVELOPMENT OF ANABAENA DOLIOLUM. IN BASAL MEDIUM, SPORULATION IS SIMULTANEOUS WHILE HETEROCYST FORMATION IS SEQUENTIAL. NITRATE-, NITRITE-, AND AMMONIUM-NITROGEN INHIBIT SPORULATION AND HETEROCYST FORMATION, DEGREE OF INHIBITION DEPENDING ON CONCENTRATION AND SOURCE OF INORGANIC NITROGEN. NITRATE AND NITRITE INDUCE LYSIS WHICH IS CONCENTRATION DEPENDENT AND CIRCUMSCRIBED BY TIME, I E, UP TO A CERTAIN STAGE THE LYTIC EVENTS ARE PREVENTABLE BY TRANSFER TO BASAL MEDIUM, BUT BEYOND THIS STAGE THEY BECOME INEVITABLE AND COMPLETE LYSIS OCCURS EVEN IN ABSENCE OF NITRATE OR NITRITE. SEQUENTIAL DIFFERENTIATION AND SPATIAL RELATIONS OF HETEROCYSTS IN A GROWING FILAMENT ARE CHARACTERISTIC OF EACH NITROGEN SOURCE AND INDICATIVE OF THE POLARITY INVOLVED IN GRADIENT OF HETEROCYST FORMATION. (FITZGERALD-WIS)

FIELD 05C

ACCESSION NO. W69-10177

EXCESSIVE WATER FERTILIZATION,

WISCONSIN DEPT. OF RESOURCE DEVELOPMENT, MADISON. WATER RESOURCES DIV.

RICHARD B. COREY, ARTHUR D. HASLER, G. FRED LEE, F. H. SCHRAUFNAGEL, AND
   THOMAS L. WIRTH.

REPORT TO WATER SUBCOMMITTEE, NATURAL RESOURCES COMMITTEE OF STATE AGENCIES,
   MADISON, WISCONSIN, 50 P, JANUARY 1967. 7 TAB, 131 REF.

   DESCRIPTORS:
      *WATER, *FERTILIZATION, *EUTROPHICATION, LAKES, STREAMS, ALGAE, FISH,
      WISCONSIN, SEWAGE, TROUT, NITROGEN, PHOSPHORUS, DETERGENTS, RUNOFF,
      PERCOLATION, ROADS, ROOFS, PRECIPITATION(ATMOSPHERIC), WETLANDS,
      SEEPAGE, INDUSTRIAL WASTES, TERTIARY TREATMENT, CHEMICAL PRECIPITATION,
      HARVESTING, ZONING, DIVERSION, DREDGING, IMPOUNDMENTS, WEEDS, SPORTS,
      PROPERTY VALUES, TASTE, ODOR, BASE FLOW, LAKE MENDOTA(WIS), BOTTOM
      SEDIMENTS, MUNICIPAL WASTES, WATER POLLUTION EFFECTS, WATER POLLUTION
      SOURCES, WATER POLLUTION CONTROL.

   IDENTIFIERS:
      MANURE, MADISON(WIS), FLUSHING WATERWAYS, MILWAUKEE(WIS), CHICAGO(ILL),
      ARTIFICIAL CIRCULATION, NUISANCES.

   ABSTRACT:
      DETERIORATION OF WISCONSIN WATERS WILL PROBABLY INCREASE. EXCESSIVE
      FERTILIZATION WILL BECOME THE IMPORTANT PUBLIC PROBLEM IN WATER
      RESOURCES. DOMESTIC SEWAGE CONSTITUTES A MAJOR SOURCE OF NITROGEN AND
      PHOSPHORUS, APPARENTLY THE MAJOR CAUSATIVE ELEMENTS OF EUTROPHICATION.
      ACCELERATED USE OF SYNTHETIC DETERGENTS ACCENTUATES A PROBLEM TO WHICH
      MANY OTHER FACTORS CONTRIBUTE: RUNOFF AND UNDERGROUND PERCOLATION FROM
      RURAL LANDS; MANURE APPLIED ON FROZEN SOIL; RUNOFF FROM ROOFS AND
      ROADS; 'WASHING' OF THE ATMOSPHERE BY PRECIPITATION; AND RELEASE OF
      NITROGEN AND PHOSPHORUS RESULTING FROM DEVELOPMENT OF WETLANDS FOR
      AGRICULTURAL AND URBAN USE. IMPROVED REMOVAL OF NUTRIENTS FROM SEWAGE
      APPEARS POSSIBLE. ADDING TERTIARY TREATMENT FOR EFFLUENTS; PERFECTING
      USE OF CHEMICAL PRECIPITANTS; BIOCHEMICALLY REMOVING NUTRIENTS BY
      GROWING AND HARVESTING ALGAE IN EFFLUENT LAGOONS; REMOVING NUTRIENTS BY
      MODIFICATION OF ACTIVATED SLUDGE PROCESSES; LIQUIDIZING AND STORING
      WINTER MANURE UNTIL SPRING--ALL MAY PROVE EFFECTIVE. ENTIRE DRAINAGE
      BASINS, SHORES AND FRONTAGES, MAY REQUIRE REZONING. DIVERSION OF
      EFFLUENTS, FLUSHING WATERWAYS WITH CLEAR WATER, CHEMICALLY CONTROLLING
      ALGAE AND AQUATIC PLANTS, DREDGING SHALLOW PORTIONS OF LAKES,
      ARTIFICIALLY CIRCULATING ENTIRE LAKES, DRAWING NUTRIENT-RICH SURPLUS
      WATER FROM LAKE BOTTOMS, IMPOUNDING STREAMS AND CONTROLLING SUBSEQUENT
      RELEASES--ALL SHOW PROMISE. (JONES-WIS)

   FIELD 05C

   ACCESSION NO. W69-10178

THE IMPORTANCE OF EXTRACELLULAR PRODUCTS OF ALGAE IN FRESHWATER,

UNIVERSITY COLL., LONDON (ENGLAND). DEPT. OF BOTANY.

G. E. FOGG, AND D. F. WESTLAKE.

VERH INT VEREIN THEOR ANGEW LIMNOL, VOL 12, P 219-232, 1953. 3 FIG, 2 TAB, 20
  REF.

DESCRIPTORS:
   *ALGAE, *FRESH WATER, *PEPTIDES, COPPER, IRON, IONS, PHOSPHATES,
   ORGANIC COMPOUNDS, TOXICITY, CHLOROPHYTA, LAKES, RESERVOIRS, NITROGEN,
   GROWTH, AMINO ACIDS, CHLAMYDOMONAS, CHLORELLA, WISCONSIN, LAKE
   MICHIGAN, PHYTOPLANKTON, DIATOMS, ECOLOGY, SOIL, CULTURES, CYANOPHYTA,
   WATER POLLUTION SOURCES, WATER POLLUTION EFFECTS.

IDENTIFIERS:
   *EXTRACELLULAR PRODUCTS, ANABAENA CYLINDRICA, ZINC, XANTHOPHYCEAE,
   BACILLARIOPHYCEAE, MYXOPHYCEAE, COMPLEX FORMATION, GROWTH-PROMOTING
   PROPERTIES, CHLAMYDOMONAS MOEWUSII, CHLORELLA PYRENOIDOSA, TRIBONEMA
   AEQUALE, NAVICULA PELLICULOSA, LAKE MENDOTA(WIS), ENGLAND,
   LONDON(ENGLAND), ESTHWAITE(ENGLAND), LOWESWATER(ENGLAND), BARNES SOUTH
   RESERVOIR(ENGLAND), WINDERMERE NORTH BASIN(ENGLAND), LOUGHRIGG
   TARN(ENGLAND), BASSENTHWAITE(ENGLAND), TARN TARN(ENGLAND), SOLWAY
   FIRTH(ENGLAND), LITTLE HAWESWATER(ENGLAND), STAINES RESERVOIR(ENGLAND),
   GLOEOTRICHIA NATANS, OSCILLATORIA, BOTRYOCOCCUS, VOLVOX, UROGLENA,
   APHANIZOMEMON, MICROCYSTIS, COELOSPHAERIUM.

ABSTRACT:
   EXPERIMENTS WITH PARTIALLY PURIFIED PREPARATIONS OF THE EXTRACELLULAR
   POLYPEPTIDE, PRODUCED DURING NORMAL GROWTH OF THE CYANOPHYTE
   (BLUE-GREEN ALGA), ANABAENA CYLINDRICA, SHOW THAT THIS MATERIAL FORMS
   COMPLEXES WITH VARIOUS IONS INCLUDING THOSE OF COPPER, ZINC, FERRIC
   IRON, PHOSPHATE AND CERTAIN ORGANIC SUBSTANCES. AUTHORS SUGGEST THAT
   SUCH COMPLEX FORMATION MAY HAVE BIOLOGICALLY IMPORTANT EFFECTS, AND, AS
   AN EXAMPLE, THEY SUGGEST THAT COMPLEX FORMATION BETWEEN EXTRACELLULAR
   POLYPEPTIDE AND CUPRIC ION CONSIDERABLY REDUCES TOXICITY OF THE LATTER
   TOWARD ANABAENA CYLINDRICA. THIS MAY EXPLAIN THE ERRATIC RESULTS OFTEN
   OBTAINED WITH COPPER SULPHATE USED AS AN ALGICIDE. ALGAE REPRESENTING
   OTHER CLASSES (CHLOROPHYCEAE, XANTHOPHYCEAE, BACILLARIOPHYCEAE) ALSO
   PRODUCE EXTRACELLULAR PEPTIDE. THAT RELATIVELY SUBSTANTIAL AMOUNTS OF
   PEPTIDE-NITROGEN OCCUR DISSOLVED IN LAKE WATERS HAS BEEN CONFIRMED FOR
   A NUMBER OF ENGLISH LAKES AND RESERVOIRS. AUTHORS SUGGEST THAT THIS
   PEPTIDE MAY ORIGINATE PARTIALLY FROM LIVING ALGAE AND THAT BY FORMING
   COMPLEXES WITH OTHER DISSOLVED SUBSTANCES, IT MAY EXERT IMPORTANT
   EFFECTS ON GROWTH OF AQUATIC ORGANISMS. SPECIFIC SUBSTANCES MAY NOT BE
   NECESSARY, AND IN LAKES, COMPLEX-FORMING SUBSTANCES DERIVED FROM
   ORGANIC DECAY OR LIVING ALGAE, MAY EFFECTIVELY PRODUCE THE REQUISITE
   ENVIRONMENT FOR CYANOPHYTES.(JONES-WIS)

FIELD 05C, 02H

ACCESSION NO. W69-10180

CHANGES IN THE OXYGEN DEFICIT OF LAKE WASHINGTON,

WASHINGTON UNIV., SEATTLE. DEPT. OF ZOOLOGY.

W. T. EDMONDSON.

VERH INT VEREIN THEOR ANGEW LIMNOL, VOL 16, P 153-158, DEC 1966. 3 FIG, 1
    TAB, 8 REF.

    DESCRIPTORS:
        *WASHINGTON, *OXYGEN, *FLUCTUATION, SEWAGE TREATMENT, SOLAR RADIATION,
    CHLOROPHYLL, PRODUCTIVITY, PHOTOSYNTHESIS, ALGAE, HYPOLIMNION,
    EPILIMNION, PHYTOPLANKTON, SESTON, CRUSTACEANS, DIATOMS, CHLOROPHYTA,
    DEPOSITION(SEDIMENTS), WATERSHEDS(BASINS), WATER POLLUTION EFFECTS,
    CYANOPHYTA, EURTROPHICATION.

    IDENTIFIERS:
        *LAKE WASHINGTON(WASH), *OXYGEN DEFICIT, SECCHI DISC TRANSPARENCY,
    MYXOPHYCEA, OSCILLATORIA, ANABAENA, DIAPTOMUS ASHLANDI, EPISCHURA
    NEVADENSIS, DIAPHANOSOMA LEUCHTENBER GIANUM, CYCLOPS BICUSPIDATUS,
    REPRODUCTIVE RATE.

    ABSTRACT:
        RECORDS OF RECENT CHANGES IN OXYGEN REGIME IN LAKE WASHINGTON SHOW
    THAT, THOUGH INCREASING AMOUNTS OF EFFLUENT FROM SEWAGE TREATMENT
    PLANTS WERE RECEIVED AND OXYGEN DEFICIT RATE, IN GENERAL WAS HIGHER IN
    1956 TO 1964 THAN IN 1933 AND 1950, THE INCREASE FLUCTUATED AND, IN
    SOME YEARS, RATE WAS SIMILAR TO THAT OF 1950. CHANGES IN DEFICIT DO NOT
    APPEAR TO BE ATTRIBUTABLE TO CORRESPONDING CHANGES IN PRODUCTIVITY. NO
    CLEAR RELATION APPEARS BETWEEN DEFICIT AND CHLOROPHYLL OR SOLAR
    RADIATION INPUT. THUS THE LAKE HAS CONTINUED TO SHOW SIGNS OF
    ENRICHMENT BY INCREASED PRIMARY PRODUCTIVITY AND ALGAL ABUNDANCE, WHILE
    OXYGEN DEFICIT AND ITS RATE OF DEVELOPMENT DURING THE SUMMER HAVE NOT
    CHANGED PROPORTIONATELY. A PARTIAL EXPLANATION MAY LIE IN QUALITATIVE
    CHANGES IN CHARACTER OF PHYTOPLANKTON. A RELATIVELY SMALL FRACTION OF
    THE 1950 POPULATION CONSISTED OF COLONIAL MYXOPHYCEAE, WHEREAS IN
    1961-1964, THEY DOMINATED THE POPULATION. THE RATE OF DEPOSITION OF
    MATERIAL TO THE BOTTOM, WHOSE SUBSTANCES SUPPORT THE OXYGEN DEFICIT,
    APPEARS NOT TO HAVE INCREASED CORRESPONDINGLY TO PRODUCTIVITY OR CROP.
    THERE IS NO INDICATION THAT CHANGES IN OXYGEN DEFICIT ARE ATTRIBUTABLE
    TO CHANGES IN WATERSHED OF LAKE. (JONES-WIS)

    FIELD 05C, 02H

    ACCESSION NO. W69-10182

SEA PLANTS FOR FOOD,

SOUTHAMPTON UNIV. (ENGLAND).

J. E. G. RAYMONT.

THE NEW SCIENTIST, P 10-11, JULY 18, 1957. 1 FIG.

    DESCRIPTORS:
        *MARINE PLANTS, *ALGAE, CHLORINATION, SODIUM COMPOUNDS, PHOTOSYNTHESIS,
    CARBON DIOXIDE, NITRATES, PHOSPHATES.

    IDENTIFIERS:
        *PHAEODACTYLUM, *GAS EFFLUENT, FOOD PRODUCTION, SOUTHAMPTON UNIVERSITY.

    ABSTRACT:
        A MARINE ALGAE CROP OF PHAEODACTYLUM HAS BEEN EXPERIMENTALLY GROWN BUT
    NOT SUBJECTED TO THOROUGH CHEMICAL ANALYSIS WHICH WOULD ESTIMATE ITS
    FOOD VALUE, OR PROTEIN PRODUCED PER NUTRIENT INTAKE. THE DRY WEIGHT OF
    THE HARVESTED CROP WAS 1/3 GRAM PER LITER CULTURE WHICH WAS A FAVORABLE
    BEGINNING RESULT. HEAT STERILIZATION WAS THE PREFERRED PROCEDURE IN THE
    FILTERING AND STERILIZING STEPS. A COMMERCIAL AMMONIUM SULFATE WAS USED
    AS FERTILIZER. ALGAE GROWTH DEPENDED ON A STRONG ENOUGH LIGHT SOURCE
    AND WAS THEREFORE GROWN IN SHALLOW AREAS AND IN THE WINTER BY
    ARTIFICIAL LIGHTING. CARBON DIOXIDE FOR THE PHOTOSYNTHESIS WAS SUPPLIED
    AS SUCCESSFULLY BY EFFLUENT WASTE GASES OF PARTIALLY PURIFIED WASTE
    WATER AS BY CYLINDERS. THE EFFLUENT CARBON DIOXIDE IS HOPED TO DECREASE
    THE NEED FOR NITRATE AND PHOSPHATE. (SHERMAN-VANDERBILT)

    FIELD 03C, 05D

    ACCESSION NO. W70-00161

LAKE MICHIGAN POLLUTION AND CHICAGO'S SUPPLY,

DEPARTMENT OF WATER AND SEWERS, CHICAGO, ILL.

HYMAN H. GERSTEIN.

J AMERICAN WATER WORKS ASSOCIATION, VOL 57, P 841-857, JULY 1965. 7 FIG, 7 TAB.

DESCRIPTORS:
*LAKE MICHIGAN, *WATER SUPPLY, INTAKES, NUTRIENTS, QUALITY CONTROL, TREATMENT FACILITIES, FILTRATION, SEWAGE, COLIFORMS, EUTROPHICATION, ODOR, AMMONIUM COMPOUNDS, ACTIVATED CARBON, CHLORINE, PHENOLS, OIL WASTES, HYDROGEN ION CONCENTRATION, FLUORIDES, ALKYL BENZENE SULFONATES, WIND VELOCITY, TASTE, COAGULATION, PLANKTON, MONITORING, TEMPERATURE, DETERGENTS, PHOSPHORUS COMPOUNDS, DIATOMS, BIOCHEMICAL OXYGEN DEMAND, DISSOLVED OXYGEN, COLOR, DISSOLVED SOLIDS, BIODEGRADATION, ALGAE, IRON, SULFATES.

IDENTIFIERS:
*POLLUTION, *CHICAGO(ILL), COOK COUNTY(ILL), CALUMET RIVER(ILL), POOLS, WIND DIRECTION, TABELLARIA, FRAGILARIA, ASTERIONELLA, FILTER CLOGGING, STEPHANODISCUS HANTZSCHII, STEPHANODISCUS BINDERANUS, CLADOCERA, WATER MASSES, U S PUBLIC HEALTH SERVICE, GREAT LAKES-ILLINOIS RIVER BASINS PROJECT, INDIANA HARBOR SHIP CANAL.

ABSTRACT:
ALTHOUGH NO SEWAGE AND INDUSTRIAL WASTES ARE DISCHARGED ALONG 30 MILES OF CHICAGO'S LAKE FRONT, GROSS POLLUTION EXISTS IN THE SOUTHERN END OF LAKE MICHIGAN. DATA OVER 40 YEARS REVEAL QUALITY AT THE INTAKES, AND, FROM 1950 TO 1964, ILLUSTRATE POLLUTIONAL TRENDS. A GENERAL INCREASE OF EUTROPHICATION INDICATORS AND INCREASE IN NUMBER AND INTENSITY OF PERIODS WHEN POLLUTANTS SERIOUSLY AFFECTED THE INTAKE AT THE SOUTH PLANT APPEAR IN THE LAST TEN YEARS. SLUGS OF WASTES MAY HAVE ABNORMAL ODORS, LIKE THOSE OF OIL REFINERY WASTES DILUTED WITH LAKE WATER. SINCE 1948, SAMPLES HAVE BEEN COLLECTED ONE DAY EACH WEEK AT ESTABLISHED POINTS. POLLUTIONAL POOLS, INFLUENCED BY WIND-INDUCED LAKE CURRENTS, WERE OBSERVABLE IN A SERIES OF TASTE AND ODOR INCIDENTS. THE PRINCIPAL AGENT IN TREATMENT FOR REMOVAL OF TASTE AND ODOR IS ACTIVATED CARBON. IN TREATING HIGHLY POLLUTED WATERS, APPLICATION OF CHLORINE SUFFICIENT TO DECOMPOSE AMMONIA-NITROGEN MAY NOT ADEQUATELY DESTROY THE BACTERIA, WHEREUPON BACTERIOLOGICALLY UNSAFE WATER MAY RESULT. INCREASING PLANKTON GROWTH NECESSITATES MORE FREQUENT WASHING OF FILTERS; INCREASED GROWTH OF FILAMENTOUS ALGAE TOGETHER WITH APPEARANCE OF NEW SPECIES OF DIATOMS HAVE CAUSED ADDITIONAL PROBLEMS WITH FILTERS. AUTHOR PRESENTS, AS PRACTICAL AND DESIRABLE, A SERIES OF CRITERIA FOR QUALITY OF CHICAGO'S WATER SUPPLY. (JONES-WISCONSIN)

FIELD 05B, 05C, 05G

ACCESSION NO. W70-00263

DEATH OF A LAKE,

DAVID C. FLAHERTY.

QUEST, VOL 6, NO 4, P 7-15, 1968. 9 FIG.

DESCRIPTORS:
*LAKES, *ALGAE, *EUTROPHICATION, OLIGOTROPHY, NUTRIENTS, WATER
TEMPERATURE, DISSOLVED OXYGEN, FISHKILLS, LIMNOLOGY, ALGAL TOXINS,
ALGAL CONTROL, TRACE ELEMENTS, WASHINGTON, NUCLEAR POWERPLANTS, WATER
POLLUTION EFFECTS, PHOSPHORUS, NITROGEN.

IDENTIFIERS:
*CULTURAL EUTROPHY, *NUTRIENT BALANCE, *LAKE AGING, NUTRIENT SOURCES,
WATER SAMPLING, ALGAL GROWTH, OXYGEN DEPLETION, NUCLEAR ACTIVATION
TECHNIQUES, ROCK LAKE(WASH), WILLIAMS LAKE(WASH), CHANNELED
SCABLANDS(WASH), WASHINGTON STATE UNIVERSITY.

ABSTRACT:
ALGAE ARE THE PRIMARY PRODUCERS WHICH INITIATE THE FOOD CHAIN IN LAKES
LEADING UP TO FISH AND MAN. IN NORMAL LAKES, A ROUGH BALANCE EXISTS
BETWEEN INCOMING NUTRIENTS, ALGAL NUMBERS, AND THEIR PREDATORS.
HOWEVER, THIS BALANCE IS UPSET BY MAN'S INFLUENCE, HASTENING CHANGE
FROM OLIGOTROPHIC TO EUTROPHIC CONDITIONS. DETRIMENTAL EFFECTS OF LARGE
NUMBERS OF UNDESIRABLE ALGAL FORMS, INCLUDE OXYGEN DEPLETION RESULTING
IN FISHKILLS; FLAVOR, ODOR, AND COLOR IN WATER SUPPLIES; INDUCED
CORROSION AND SLIME IN INDUSTRIAL WATER SUPPLIES; AND ENDANGERING OF
OTHER AQUATIC AND TERRESTRIAL LIFE FORMS FROM TOXIC SUBSTANCES. STUDIES
CONDUCTED BY DR. WILLIAM FUNK, WASHINGTON STATE UNIVERSITY CIVIL
ENGINEERING PROFESSOR, ON ROCK AND WILLIAMS LAKES IN THE SCABLANDS
BASALTS OF EASTERN WASHINGTON ARE DESCRIBED. THE BASALTS ARE RELATIVELY
LOW IN NITROGEN AND PHOSPHORUS, AND DECOMPOSE SLOWLY, WITH CONSEQUENT
LOW NATURAL NUTRIENT INCOME. HOWEVER, THESE LAKES ARE ENDANGERED BY
HUMAN ACTIVITIES. PRESENT STUDIES MEASURE DISSOLVED OXYGEN,
TEMPERATURE, LIGHT PENETRATION, ALGAL SPECIES, AND CONCENTRATIONS OF
VARIOUS ELEMENTS INCLUDING PHOSPHORUS AND NITROGEN. NUCLEAR ACTIVATION
TECHNIQUES ARE USED TO DETECT TRACE ELEMENTS WHICH DR. FUNK SUSPECTS
MAY HAVE A CRITICAL ROLE IN ALGAL GROWTH. (KETELLE-WISCONSIN)

FIELD 05C, 02H

ACCESSION NO. W70-00264

PRIMARY PRODUCTION IN LABORATORY STREAMS,

OREGON STATE UNIV., CORVALLIS. DEPT. OF FISH AND GAME MANAGEMENT; OREGON
STATE UNIV., CORVALLIS. DEPT. OF BOTANY; AND OREGON STATE UNIV., CORVALLIS.
PACIFIC COOPERATIVE WATER POLLUTION AND FISHERIES RESEARCH LABS.

C. DAVID MCINTIRE, ROBERT L. GARRISON, HARRY K. PHINNEY, AND CHARLES E.
WARREN.

LIMNOLOGY AND OCEANOGRAPHY, VOL 9, NO 1, P 92-102, JAN 1964. 5 FIG, 3 TAB, 20
REF.

DESCRIPTORS:
*PRIMARY PRODUCTIVITY, *STREAMS, *LABORATORY TESTS, RESPIRATION,
ENVIRONMENT, PHOTOSYNTHESIS, OXYGEN, TROPHIC LEVEL, LIGHT INTENSITY,
ECOLOGY, LOTIC ENVIRONMENT, BIOMASS, FLOW RATES, DEPTH, DISSOLVED
OXYGEN, ALGAE, VELOCITY, WATER QUALITY, DISSOLVED SOLIDS, HYDROGEN ION
CONCENTRATION, TEMPERATURE, NITROGEN, NUTRIENTS, ORGANIC MATTER,
PIGMENTS, CHLOROPHYLL, DIFFUSION, CYANOPHYTA.

IDENTIFIERS:
COMMUNITY, SPECIFIC CONDUCTANCE, SPECIES, MELOSIRA VARIANS, SYNEDRA
ULNA, OEDOGONIUM, PHORMIDIUM RETZII, SURFACE DIFFUSION RATE.

ABSTRACT:
SIX LABORATORY STREAMS WERE USED TO STUDY PRIMARY PRODUCTION AND
COMMUNITY RESPIRATION IN SIMPLE COMMUNITIES UNDER DIFFERENT
ENVIRONMENTAL CONDITIONS. A PHOTOSYNTHESIS-RESPIRATION CHAMBER WAS
DEVELOPED WHICH CAN BE USED TO RELATE PRIMARY PRODUCTION TO VARIOUS
ENVIRONMENTAL CONDITIONS AND TO VERIFY DIRECT MEASUREMENT OF PRIMARY
PRODUCTION IN LABORATORY OR NATURAL STREAMS. GROSS PRIMARY PRODUCTION
IN LABORATORY STREAMS UNDER 6000 LUX OF ILLUMINATION RANGED FROM 2.4 TO
4.7 GRAMS OXYGEN/SQUARE METER/DAY. COMMUNITY RESPIRATION RANGED FROM
1.6 TO 4.2 GRAMS OXYGEN/SQUARE METER/DAY. RATIOS OF PHOTOSYNTHESIS TO
RESPIRATION USUALLY VARIED BETWEEN 1.0 AND 2.0. ILLUMINATION INTENSITY
WAS APPROXIMATELY LINEARLY RELATED TO PRIMARY PRODUCTION IN RANGE,
0-4000 LUX, MAXIMUM PHOTOSYNTHETIC RATES OBTAINING AT ABOUT 21,000 LUX.
COMMUNITIES DEVELOPING ON THE SUBSTRATE WERE SEEDED NATURALLY BY
SPECIES ENTERING STREAMS THROUGH THE WATER SUPPLY. COMPOSITION OF PLANT
COMMUNITIES REMAINED SURPRISINGLY CONSTANT, USUALLY VARYING ONLY IN
RELATIVE ABUNDANCE OF DIFFERENT SPECIES. THE DIATOMS, MELOSIRA VARIANS
AND SYNEDRA ULNA CONSISTENTLY DOMINATED; OEDOGONIUM SPP WERE SOMETIMES
VERY ABUNDANT DURING SUMMER. GENERA OF BLUE-GREEN ALGAE WERE
UBIQUITOUS, FREQUENTLY GROWING ATTACHED TO SIDES OF TROUGHS, NEAR THE
AIR-WATER INTERFACE. (JONES-WISCONSIN)

FIELD 05C

ACCESSION NO. W70-00265

CHARACTERISTICS OF HYPERTROPHIC LAKES AND CANALS IN CITIES,

STATE ZOOLOGICAL INST. AND MUSEUM, HAMBURG (WEST GERMANY). HYDROBIOLOGICAL
STATION.

HUBERT CASPERS.

INTERNATIONAL ASSOCIATION OF THEORETICAL AND APPLIED LIMNOLOGY, PROCEEDINGS,
VOL 15, P 631-638, FEB 1964. 2 FIG, 2 TAB, 13 REF.

DESCRIPTORS:
*LAKES, *EUTROPHICATION, *CANALS, *CITIES, BIOLOGICAL COMMUNITIES,
SEWAGE, DETRITUS, TUBIFICIDS, PLANKTON, DIATOMS, CYANOPHYTA,
PHYTOPLANKTON, ROTIFERS, CHLOROPHYTA, COPEPODS, CYCLES, CHLAMYDOMONAS,
ALGAE, ZOOPLANKTON, FISH, CHRYSOPHYTA, SESTON, SELF-PURIFICATION.

IDENTIFIERS:
*HYPERTROPHIC WATERS, HAMBURG(GERMANY), CHIRONOMIDS, CLADOCERA,
APHANIZOMENON FLOS-AQUAE, STEPHANODISCUS HANTZSCHII, DIDINIUM,
PARAMECIUM, PHACUS, GLENODINIUM, CRYPTOMONAS, CILIATA, ASTERIONELLA
FORMOSA, RECOLONIZATION, ALSTER LAKE(GERMANY), GERMANY.

ABSTRACT:
ALSTER LAKE IN HAMBURG, GERMANY, IS AN EXAMPLE OF A TYPICAL URBAN LAKE
AND DEMONSTRATES THE SPECIAL CONDITIONS IN CITY WATERS, WHICH ARE
SPECIALIZED HABITATS AND WHERE HYDROBIOLOGICAL STATUS CAN BE MORE
READILY ANALYZED THAN COMPLEX SITES OF 'HEALTHIER' WATERS. URBAN WATERS
MAY CONSTITUTE A RESTRICTED FACIES OF EUTROPIC WATER WHERE DESTRUCTION
OF ORIGINAL COMMUNITIES IS TYPICAL AND NO NEW EQUILIBRIUM DEVELOPS.
RATHER, THE RELATIONSHIP BETWEEN BUILDUP AND BREAKDOWN OF NUTRITIVE
MATERIAL, BETWEEN PRODUCERS AND CONSUMERS, IS ALWAYS IN A STATE OF
TEMPORARY EQUILIBRIUM, AND IS CONSTANTLY IN DANGER OF BECOMING
UNBALANCED. THE ORIGINAL COMMUNITY LOSES ITS MOST IMPORTANT ZONE OF
REGENERATION, THE SUBMERGED PLANTS ALONG BANKS; BOTTOM FAUNA IS
BURDENED WITH ANAEROBIC PROCESSES RESULTING FROM INFLOW OF DETRITUS,
LEAVING ONLY A FEW SPECIES WHOSE MASS DEVELOPMENT IS INHIBITED. THE
SURVIVING PLANKTON TENDS TO INSTABILITY AND EXTREME VARIATIONS FROM
YEAR TO YEAR. NEVERTHELESS, THERE ARE CERTAIN DEFINITE AND REGULAR
EVENTS WHICH MAKE URBANIZED WATERS A UNIQUE EXPERIMENTAL SITUATION,
SINCE THE RESTRICTED FACIES OF EUTROPHIC WATER AFFORDS US LIMNOLOGICAL
INSIGHT INTO FORCED PRODUCTION PROCESSES AND THEIR ECOLOGICAL EFFECTS.
AUTHOR BELIEVES THAT CONDITIONS IN URBAN WATERS IMPROVE WHEN PLANTS
RECOLONIZE THE BANKS. (JONES-WISCONSIN)

FIELD 05C, 02H

ACCESSION NO. W70-00268

THE CONTROL OF WATER WEEDS.

AGRICULTURAL RESEARCH COUNCIL, KIDLINGTON (ENGLAND). WEED RESEARCH
  ORGANIZATION.

E. C. S. LITTLE.

WEED RESEARCH, VOL 8, NO 2, P 79-105, 1968. 363 REF.

DESCRIPTORS:
  *AQUATIC WEEDS, *AQUATIC PLANTS, *AQUATIC WEED CONTROL, WATER
  CONSERVATION, WATER CONTROL, FERTILIZERS, SEWAGE EFFLUENT, NUTRIENTS,
  LAKES, CHANNELS, REVIEWS, BIBLIOGRAPHIES, HERBICIDES, FISH, FRESH
  WATER, ALGAE, PONDS, WATER HYACINTH, CHEMICAL CONTROLS, PARAQUAT,
  DIQUAT, SODIUM ARSENITE, COPPER SULPHATE, MONURON, AMMONIA, DALAPON,
  2-4-5-T UREAS, SEEDS, PROTEINS, MICROORGANISMS, FERMENTATION,
  SURFACTANTS, FLOATING PLANTS, FERNS, ALLIGATORWEED, FORMULATION,
  EMULSIFIERS, SOIL TEXTURE, SILAGE, CHEMICALS, DITCHES, PERSISTENCE,
  IRRIGATION WATER, HARVESTING, MECHANICAL CONTROL, SPRAYING, WATER
  LEVELS, MAMMALS, BIRDS, SNAILS, INSECTS, FUNGI, WATER QUALITY, SOIL
  STERILANTS, EUTROPHICATION, RIVERS, BIOCONTROL, TOXICITY, DRAWDOWN,
  2-4-D, AMINOTRIAZOLE.

IDENTIFIERS:
  TRIAZINE, DICHLOBENIL.

ABSTRACT:
  WATER WEEDS ARE POSING INCREASING PROBLEMS IN MANY COUNTRIES WHICH
  DEPEND ON WATER CONTROL FOR DEVELOPMENT OF AGRICULTURAL, POWER, AND
  TRANSPORT RESOURCES. THE UNITED STATES, BESIDES HAVING ITS SHARE OF
  DIFFICULTIES FROM WATER WEEDS, IS ALSO CONCERNED WITH AQUATIC WEED
  IMPAIRMENT OF INCREASINGLY POPULAR RECREATIONAL ASPECTS OF WATER. HEAVY
  WATER-WEED INFESTATION IS EXPECTED WHEN FERTILE LAND IS SUBMERGED TO
  FORM LAKES, OR WHEN LAKES AND CHANNELS BECOME SILTED. THIS PROBLEM IS
  ACCENTUATED IN DEVELOPED COUNTRIES BY EXTRA PLANT NUTRIENTS REACHING
  WATER SUPPLIES FROM FERTILIZER AND SEWAGE EFFLUENT. AUTHOR PRESENTS A
  COMPREHENSIVE REVIEW OF THE WORLD'S LITERATURE ON AQUATIC WEED CONTROL
  SINCE 1960, TO PROVIDE A GUIDE TO RESEARCH WITH PRIMARY ATTENTION TO
  THOSE PLANTS CAUSING PROBLEMS IN WARM ENVIRONMENTS. THE LITERATURE
  CITATIONS ARE GROUPED AS FOLLOWS: REVIEWS, IDENTIFICATION, GENERAL
  RECOMMENDATIONS, IMPORTANT WATER WEEDS, CHEMICALS USED IN AQUATIC WEED
  CONTROL, CONTROL TECHNIQUES, BIOLOGICAL CONTROL, UTILIZATION OF WATER
  WEEDS, TOXICOLOGY OF HERBICIDES TO FISH, HERBICIDES RESIDUES, AND WATER
  AND ITS EFFECT. AUTHOR INDICATES THE NEED FOR MORE RESEARCH IN
  BIOLOGICAL CONTROL AND UTILIZATION OF WATER WEEDS WHICH MIGHT BE USEFUL
  IN REGIONS WITHOUT THE FINANCIAL RESOURCES TO DEAL WITH THE PROBLEM.
  (SIMSIMAN-WISCONSIN)

FIELD 05G, 04A

ACCESSION NO. W70-00269

CHANGES IN LAKE WASHINGTON FOLLOWING AN INCREASE IN THE NUTRIENT INCOME,

WASHINGTON UNIV., SEATTLE. DEPT. OF ZOOLOGY.

W. T. EDMONDSON.

INTERNATIONAL ASSOCIATION OF THEORETICAL AND APPLIED LIMNOLOGY, PROCEEDINGS,
   VOL 14, P 167-175, JULY 1961. 5 FIG, 1 TAB, 12 REF.

DESCRIPTORS:
   *NUTRIENTS, *PRODUCTIVITY, NUISANCE ALGAE, SEWAGE EFFLUENTS,
   PHOSPHORUS, EUTROPHICATION, WASHINGTON, LAKES, WATER POLLUTION EFFECTS.

IDENTIFIERS:
   *LAKE WASHINGTON(WASH), *LAKE CHANGES, *NUTRIENT CONCENTRATION,
   *NUTRIENT BUDGET, POTENTIAL NUTRIENT CONCENTRATION, AREAL
   INCOME(NUTRIENTS), LAKE VOLUME, MEAN DEPTH, DILUTION, OSCILLATORIA
   RUBESCENS, MADISON LAKES(WIS), CEDAR RIVER(WASH), ZURICHSEE(SWITZ),
   TURLERSEE(SWITZ), GREIFENSEE(SWITZ), PAFFIKERSEE(SWITZ),
   AEGERISEE(SWITZ), FURESO(SWITZ).

ABSTRACT:
   A DENSE BLOOM OF OSCILLATORIA RUBESCENS IN 1955 PROMPTED A STUDY OF
   LAKE WASHINGTON TO DETERMINE EFFECTS OF A GROWING HUMAN POPULATION ON
   LAKE EUTROPHICATION. ONE PHASE WAS AN ENGINEERING SURVEY MADE TO
   ESTABLISH THE MAGNITUDE OF SEWAGE EFFLUENT AND OTHER SOURCES OF
   NUTRIENTS. INCREASES IN NUTRIENTS FROM 1950-1955 WERE SMALL COMPARED
   WITH LIMNOLOGICAL CHANGES. IN ORDER TO COMPARE THE NUTRIENT STATUS OF
   VARIOUS LAKES, A SIMPLE NUMERICAL BASIS OF COMPARISON OF NUTRIENT RATE
   SUPPLY FROM OUTSIDE SOURCES IS DESIRABLE. SYSTEMS WHICH INCORPORATE A
   CONSIDERATION OF THE POTENTIAL CONCENTRATION OF NUTRIENTS IN LAKES ARE
   PREFERABLE TO THOSE RELATING ONLY INCOME TO AREA. POTENTIAL
   CONCENTRATION CAN BE RELATED TO THE AREAL INCOME AND THE MEAN DEPTH OF
   THE LAKE AS FOLLOWS: AREAL INCOME/MEAN DEPTH=POTENTIAL CONCENTRATION
   (GRAMS/SQUARE METER/METER=GRAMS/CUBIC METER). THE PRESENT WORK DEALS
   WITH ONLY A PORTION OF THE PROBLEM--COMPARISON OF INCOME OF PHOSPHORUS
   AMONG LAKES. THE AUTHOR PRESENTS COMPUTATIONS FOR SEVERAL LAKES IN
   WISCONSIN AND EUROPE IN ADDITION TO LAKE WASHINGTON, ALSO SHOWS GRAPHS
   OF AREAL INCOME PLOTTED AGAINST POTENTIAL CONCENTRATION. RESULTS OF
   SUCH CALCULATIONS INDICATE THAT POTENTIAL CONCENTRATION IS MORE CLOSELY
   RELATED TO ALGAL STANDING CROP THAN IS THE AREAL INCOME.
   (KETELLE-WISCONSIN)

FIELD 05C, 06G,

ACCESSION NO. W70-00270

ISOLATION AND CULTURE OF TOXIC STRAINS OF ANABAENA FLOS-AQUAE (LYNGB.) DE BREB.,

NATIONAL RESEARCH COUNCIL OF CANADA, OTTAWA (ONTARIO).

P. R. GORHAM, J. MCLACHLAN, U. T. HAMMER, AND W. K. KIM.

INTERNATIONAL ASSOCIATION OF THEORETICAL AND APPLIED LIMNOLOGY, PROCEEDINGS, VOL 15, P 796-804, FEBRUARY 1964. 1 FIG, 3 TAB, 22 REF.

DESCRIPTORS:
 *ALGAE, *TOXICITY, *CULTURES, *ISOLATION, LIMNOLOGY, PUBLIC HEALTH, PHYSICOCHEMICAL PROPERTIES, GENETICS, BACTERIA, PEPTIDES, ENVIRONMENTAL EFFECTS, HETEROGENEITY, IRON, MANGANESE, WATERFOWL, CYANOPHYTA.

IDENTIFIERS:
 *ANABAENA FLOS-AQUAE, *STRAINS, PHYCOLOGISTS, VETERINARIANS, MICROCYSTIS AERUGINOSA, ENDOTOXIN, FAST-DEATH FACTOR, DOSAGE, SPECIES SUSCEPTIBILITY, ANABAENA LEMMERMANNII, OTTAWA(ONTARIO), BURTON LAKE(SASKATCHEWAN), HUMBOLDT(SASKATCHEWAN), MICE, MORPHOLOGICAL CHARACTERS, TRIS.

ABSTRACT:
 TOXIC WATERBLOOMS OF PLANKTONIC BLUE-GREEN ALGAE HAVE PUZZLED PHYCOLOGISTS, LIMNOLOGISTS, VETERINARIANS, AND PUBLIC HEALTH OFFICIALS FOR MORE THAN 80 YEARS. THE AUTHORS OF THIS PAPER SUCCEEDED IN ISOLATING TOXIC AND NON-TOXIC STRAINS OF ANABAENA FLOS-AQUAE FROM A TOXIC BLOOM IN BURTON LAKE, SASKATCHEWAN. OF FOURTEEN STRAINS ISOLATED IN 1960 AND 1961, EIGHT WERE TOXIC AND SIX NON-TOXIC. THEY SHOWED, FOR THE FIRST TIME (ALTHOUGH LONG SUSPECTED), THAT THIS ALGA CAN BE POISONOUS. THE FAST-DEATH FACTOR PRODUCED IS OFTEN SECRETED INTO THE SURROUNDING WATER. THIS AND OTHER PROPERTIES OF ANABAENA VERY FAST-DEATH FACTOR INDICATE THAT IT IS DIFFERENT FROM MICROCYSTIS FAST-DEATH FACTOR. ONE SAMPLE COLLECTED IN 1960 CONTAINED A VERY FAST-DEATH FACTOR, KILLING MICE IN ONE TO TWO MINUTES AFTER INTRAPERITONEAL INJECTION WITH A MINIMAL LETHAL DOSE. DEATH WAS PRECEDED BY SYMPTOMS OF PARALYSIS, TREMORS, AND MILD CONVULSIONS. THE SALT CONCENTRATIONS GIVING BEST GROWTH OF ONE STRAIN OF ANABAENA FLOS-AQUAE IN UNIALGAL CULTURE BEAR LITTLE RESEMBLANCE TO THOSE OF THE LAKE WATER FROM WHICH IT WAS ISOLATED. INTERESTING CHANGES IN GROWTH HABIT (GRADUAL LOSS OF COLONIAL HABIT AND FILAMENT COILING) WERE OBSERVED IN CULTURING ANABAENA FLOS-AQUAE, ILLUSTRATING HOW VARIABLE SOME MORPHOLOGICAL CHARACTERS CAN BE. (JONES-WISCONSIN)

FIELD 05C

ACCESSION NO. W70-00273

CONTRIBUTION TO THE ECOLOGY OF WATER-BLOOM-FORMING BLUE-GREEN ALGAE
 -APHANIZOMENON FLOS AQUAE AND MICROCYSTIS AERUGINOSA,

CESKOSLOVENSKA AKADEMIE VED, PRAGUE. HYDROBIOLOGICAL STATION.

J. HRBACEK.

INTERNATIONAL ASSOCIATION OF THEORETICAL AND APPLIED LIMNOLOGY, PROCEEDINGS,
 VOL 15, P 837-846, FEB 1964. 8 FIG, 2 TAB, 6 REF.

DESCRIPTORS:
 *ALGAE, *CYANOPHYTA, *ECOLOGY, *EUTROPHICATION, PONDS, RESERVOIRS,
 BACKWATER, FERTILIZATION, NITROGEN COMPOUNDS, PHOSPHORUS COMPOUND,
 PERCHES, FISH, DIATOMS, CARP, PLANKTON, METABOLISM, BIOLOGICAL
 COMMUNITIES, DAPHNIA, NANNOPLANKTON, NUTRIENTS, DEPTH, DOMESTIC
 ANIMALS, SAMPLING, NUISANCE ALGAE.

IDENTIFIERS:
 *APHANIZOMENON FLOS AQUAE, *MICROCYSTIS AERUGINOSA, BOHEMIA, ELBE
 RIVER, CELAKOVICE(CZECHOSLOVAKIA), BLATNA(CZECHOSLOVAKIA), CYPRINID
 FISHES, CLOROCOCCALES, DAPHNIA PULICARIS, DAPHNIA CUCULLATA, DAPHNIA
 HYALINA, DAPHNIA MAGNA, CZECHOSLOVAKIA.

ABSTRACT:
 KNOWLEDGE OF INDIVIDUAL FACTORS WHICH EITHER PROMOTE OR INHIBIT
 WATER-BLOOM-FORMING CYANOPHYTES (BLUE-GREEN ALGAE), APHANIZOMENON FLOS
 AQUAE, AND MICROCYSTIS AERUGINOSA IS NOT EXTENSIVE. THIS REPORT IS
 BASED ON A COMPARATIVE STUDY OF WATER-BODIES IN BOHEMIA. TEN BACKWATERS
 IN THE INUNDATION AREA OF THE ELBE RIVER, SIX FISH PONDS, AND THREE
 WATER RESERVOIRS WERE SAMPLED AT MONTHLY INTERVALS FROM APRIL TO
 SEPTEMBER. IN ORDER TO EXAMINE THE RELATIVE IMPORTANCE OF NITROGEN AND
 PHOSPHORUS COMPOUNDS FROM EACH OF THE THREE CATEGORIES OF WATER-BODIES,
 THE FOLLOWING WERE SELECTED: ONE WITH WATER BLOOM CONSISTING OF   .
 APHANIZOMENON WITH OCCASIONALLY SOME MICROCYSTIS; ONE WITHOUT WATER
 BLOOM, AND, INSOFAR AS POSSIBLE, ONE WITH MICROCYSTIS WATER BLOOM. IN
 PONDS, BACKWATERS, AND RESERVOIRS WITH AND WITHOUT WATER BLOOMS, NO
 IMPORTANT DIFFERENCES IN CONTENT OF PHOSPHORUS AND NITROGEN COMPOUNDS
 WERE FOUND. IN ONE BLOOM-CONTAINING RESERVOIR, WATER SHOWED
 SIGNIFICANTLY SMALLER CONCENTRATIONS OF PHOSPHORUS COMPOUNDS THAN IN
 PONDS. AN ASSOCIATION EXISTS BETWEEN CYANOPHYTES AND DAPHNIA. DEPTH HAD
 INCREASING EFFECT ON DEVELOPMENT OF WATER BLOOM OF THESE TWO
 CYANOPHYTES. IN PONDS, INCREASED FISH STOCK ACCOMPANIED DECREASED
 APHANIZOMENON BLOOMS, WHILE IN SOME WATER-BODIES, LARGER FISHSTOCK WAS
 ASSOCIATED WITH DEVELOPMENT OF MICROCYSTIS AERUGINOSA BLOOM.
 (JONES-WISCONSIN)

FIELD 05C, 02H

ACCESSION NO. W70-00274

THE USE OF INVERTEBRATES IN A WATER QUALITY INVESTIGATION: A BIOLOGICAL STUDY OF THE MIAMI RIVER, OHIO,

STANFORD UNIV., CALIF. DEPT. OF CIVIL ENGINEERING.

P. JONATHAN YOUNG.

PROC 3RD ANNU AMER WATER RESOURCES CONF, NOV 8-10, 1967, SAN FRANCISCO, CALIF, P 304-313, 1967. 10 P, 2 FIG, 2 TAB, 7 REF.

DESCRIPTORS:
*WATER POLLUTION EFFECTS, *BIOINDICATORS, *WATER QUALITY, *DISSOLVED OXYGEN, *OHIO, WATER TEMPERATURE, SLUDGE WORMS, BLOODWORMS, AQUATIC BACTERIA, ALGAE, MAYFLIES, CADDISFLIES.

IDENTIFIERS:
GREAT MIAMI RIVER(OHIO).

ABSTRACT:
A BIOLOGICAL STUDY WAS MADE OF THE GREAT MIAMI RIVER, OHIO TO DETERMINE THE FACTORS CAUSING VARIATIONS IN AQUATIC LIFE. SAMPLING SITES WERE CHOSEN TO VARY AS LITTLE AS POSSIBLE IN ALL CHARACTERISTICS EXCEPT WATER QUALITY. THE VARIETY OF LIFE WAS MOST RELATED TO DISSOLVED OXYGEN, AND WATER TEMPERATURE WAS THE NEXT MOST IMPORTANT FACTOR. THE WATER POLLUTION CONDITIONS WERE DIVIDED INTO 4 ZONES. VERY POLLUTED: HERE, BACTERIA AND PROTOZOANS ARE ABUNDANT. ONLY A FEW SPECIES OF ANIMALS EXIST (SLUDGE WORMS, BLOOD WORMS, AND MAGGOTS) AND THESE LIVE ON DECAYING MATTER OR BACTERIA. GREEN PLANTS ARE USUALLY ABSENT. SEWAGE FUNGUS MAY OCCUR. POLLUTED: BACTERIA ARE STILL ABUNDANT AND OXYGEN IS LOW. GREEN ALGAE BEGIN TO APPEAR AND BLOOM, AND SOME OF THE MORE TOLERANT INVERTEBRATES AND ROOTED PLANTS ARE FOUND. SEMI-HEALTHY: ALGAE BLOOMS HAVE USUALLY DISAPPEARED. MOST OF THE POLLUTED-FAVORING ORGANISMS ARE DECREASING IN ABUNDANCE WHILE LESS TOLERANT SPECIES BEGIN TO APPEAR. DISSOLVED OXYGEN MAY AVERAGE 5 PPM OR MORE. HEALTHY: THE ANIMAL AND PLANT LIFE IS MORE OR LESS NORMAL. DECOMPOSITION CONTINUES AT A MUCH REDUCED RATE. DISSOLVED OXYGEN IS HIGH AND APPROACHES SATURATION VALUES. THREE INSECTS WERE CHOSEN TO SEPARATE THE CLASSIFICATION ZONES. THE LARVA OF THE DAMSELFLY ISHNURA WAS CHOSEN TO SEPARATE 'POLUTED' FROM 'VERY POLLUTED.' THE LARVA OF THE MAYFLY BAETIS WAS CHOSEN TO SEPARATE 'SEMI-HEALTHY' FROM 'POLLUTED'; AND THE CADDIS FLY LARVA HYDROPSYCHE WAS CHOSEN TO SIGNIFY THE BEGINNING OF THE 'HEALTHY' ZONE. ALONG THE GREAT MIAMI RIVER, THE PROGRESSION OF THESE INVERTEGRATES WAS FOUND TO BE VERY CONSISTENT. (KNAPP-USGS)

FIELD 05C

ACCESSION NO. W70-00475

BACTERIA, CARBON DIOXIDE, AND ALGAL BLOOMS,

WYANDOTTE CHEMICAL CORP., MICH.

L. E. KUENTZEL.

J WATER POLLUT CONTR FEDERATION, VOL 41, NO 10, P 1737-1747, OCT 1969. 11 P, 51 REF.

DESCRIPTORS:
*EUTROPHICATION, *ALGAE, *NUTRIENTS, *CARBON DIOXIDE, *REVIEWS, BIBLIOGRAPHIES, BACTERIA, DECOMPOSING ORGANIC MATTER, WASTE WATER(POLLUTION).

IDENTIFIERS:
ALGAL BLOOMS.

ABSTRACT:
A RECENT LITERATURE SURVEY ON THE SUBJECT OF LAKE EUTROPHICATION, AND ALGAL BLOOMS AND NUTRIENTS AS CAUSATIVE FACTORS, REVEALS THAT THERE IS A DECIDED LACK OF PUBLISHED INFORMATION ON THE ROLE OF THE MAJOR NUTRIENT, CARBON DIOXIDE. GREAT EMPHASIS IS PLACED ON PHOSPHORUS AND NITROGEN AS CAUSATIVE AGENTS IN THE PRODUCTION OF UNWANTED ALGAL GROWTHS, YET A SATISFACTORY ACCOUNTING FOR THE SOURCE OF THE TREMENDOUS AMOUNTS OF CARBON DIOXIDE REQUIRED FOR MASSIVE ALGAL BLOOMS HAS NOT BEEN FOUND IN THE LITERATURE. THE PURPOSES OF THIS PAPER ARE (A) TO PROVIDE SOME LOGISTICS ON CARBON DIOXIDE REQUIREMENTS FOR SUCH BLOOMS, (B) TO ESTABLISH THE FACT THAT SUCH QUANTITIES OF CARBON DIOXIDE ARE NOT AVAILABLE FROM THE ATMOSPHERE AND DISSOLVED SALTS, AND (C) TO POINT OUT THE ROLE PLAYED BY BACTERIA AND DECOMPOSABLE ORGANIC MATTER. (KNAPP-USGS)

FIELD 05C, 10

ACCESSION NO. W70-00664

ECOLOGICAL FACTORS AND THE DISTRIBUTION OF CLADOPHORA GLOMERATA IN THE GREAT LAKES,

WISCONSIN UNIV., MILWAUKEE. DEPT. OF BOTANY.

RICHARD P. HERBST.

THE AMER MIDLAND NATUR, VOL 82, NO 1, P 90-98, JULY 1969. 9 P, 3 FIG, 1 TAB, 15 REF.

DESCRIPTORS:
*ECOLOGY, *EUTROPHICATION, *GREAT LAKES, *ALGAE, NUTRIENTS, PHOSPHORUS, POPULATION, POLLUTANTS, MAPPING, HYDROGEN SULFIDE, CITIES, LAKE HURON, LAKE ERIE, LAKE ONTARIO, LAKE MICHIGAN, WATER TEMPERATURE, PHOSPHATES, TURBIDITY.

IDENTIFIERS:
CLADOPHORA GLOMERATA.

ABSTRACT:
NUTRIENT ENRICHMENT IN THE GREAT LAKES HAS PROVIDED FERTILE AREAS FOR GROWTH OF ALGAL NUISANCES. ONE OF THESE SPECIES, CLADOPHORA GLOMERATA, HAS BECOME A MAJOR PROBLEM FOR MANY CITIES BORDERING THE GREAT LAKES. ECOLOGICAL FACTORS CONCERNING ITS GROWTH IN MILWAUKEE'S HARBOR WERE STUDIED, AND ITS DISTRIBUTION IN THE GREAT LAKES DETERMINED. PHOSPHORUS LEVELS APPEAR TO BE CLOSELY LINKED WITH CLADOPHORA INCREASES. (GABRIEL-USGS)

FIELD 05C, 02H

ACCESSION NO. W70-00667

EFFECT OF PHOTOSYNTHESIS ON OXYGEN SATURATION,

TECHNICAL SCHOOL, NIS (YUGOSLAVIA).

LAZAR R. IGNJATOVIC.

J WATER POLLUT CONTR FEDERATION, VOL 40, NO 5, PART 2, P 151-161, MAY 1968. 11 P, 3 FIG, 2 TAB, 9 REF.

DESCRIPTORS:
*PHOTOSYNTHESIS, *OXYGEN, *SATURATION, TEMPERATURE, PRESSURE, MIXING, EVAPORATION, SUPERSATURATION, RESERVOIRS, ALGAE, BODIES OF WATER, DISSOLVED OXYGEN, NORTH CAROLINA.

IDENTIFIERS:
*OXYGEN SUPERSATURATION, VERTICAL MIXING, ROANOKE RAPIDS RESERVOIR(NC).

ABSTRACT:
TEMPERATURE CHANGES, PRESSURE CHANGES, VERTICAL MIXING OF STRATIFIED WATER, AND EVAPORATION ARE DISCUSSED AND EVALUATED IN ORDER TO CLARIFY THE PHENOMENON OF OXYGEN SUPERSATURATION IN NATURAL BODIES OF WATER. DATA FROM ROANOKE RAPIDS RESERVOIR IN NORTH CAROLINA WERE USED. SUPERSATURATION WITH DISSOLVED OXYGEN MAY BE AN HIGH AS 25% AT A 5 FT DEPTH AND AT AN AVERAGE TEMPERATURE OF 25 DEG C. THESE CHANGES CANNOT ACCOUNT FOR OBSERVED SUPERSATURATION. SOME SIGNIFICANCE TO THE SUPERSATURATION IS ATTACHED TO THE EFFECTS OF SUNSHINE AND ALGAL GROWTH. PHOTOSYNTHESIS IS THE ONLY FORCE CAPABLE OF PRODUCING THE SUPERSATURATION OF SUCH A MAGNITUDE. (CARSTEA-USGS)

FIELD 05C, 05B

ACCESSION NO. W70-00683

THE UPTAKE, ACCUMULATION AND EXCHANGE OF STRONTIUM-90 BY OPEN SEA PHYTOPLANKTON,

MIAMI UNIV., FLA. INST. OF MARINE SCIENCE.

E. F. CORCORAN, AND J. F. KIMBALL, JR.

PROC 1ST NAT SYMP ON RADIOECOLOGY, COLORADO STATE UNIV, FT COLLINS, SEPT 10-15, 1961. RADIOECOLOGY, REINHOLD PUB CORP, N Y AND AMER INSTITUTE OF BIOLOGICAL SCIENCES, WASHINGTON, D C, SCHULTZ, VINCENT AND KLEMENT ALFRED W JR (EDS), P 187-191, 1963. 5 FIG, 4 TAB, 7 REF.

DESCRIPTORS:
*STRONTIUM RADIOISOTOPES, *PHYTOPLANKTON, RADIOACTIVITY, WASTE DISPOSAL, NUCLEAR REACTORS, POWERPLANTS, SHIPS, SUBMARINES, EFFLUENTS, ION EXCHANGE, RESINS, LEAKAGE, CESIUM, NUCLEAR EXPLOSIONS, ABSORPTION, ADSORPTION, ATLANTIC OCEAN, DIATOMS, NERITIC ALGAE, IONS, WATER POLLUTION EFFECTS, WATER POLLUTION SOURCES, RADIOECOLOGY.

IDENTIFIERS:
*RADIONUCLIDE UPTAKE, *RADIONUCLIDE ACCUMULATION, *RADIONUCLIDE EXCHANGE, *OPEN SEA, WEAPONS TESTING, NUCLEAR SHIP SAVANNAH, ZIRCONIUM-95, RUBIDIUM-106, NIOBIUM-95, GYMNODINIUM SIMPLEX, FLAGELLATES, ELEUTHERA ISLAND, NITZCHIA SERIATA, NAVICULA, FLORIDA CURRENT, YTTRIUM-90, KATODINIUM ROTUNDATA.

ABSTRACT:
AUTHORS USED FIVE CULTURES OF OCEANIC PHYTOPLANKTON TO DETERMINE THEIR ABILITY TO CONCENTRATE STRONTIUM (SR) FROM SEA WATER, WITH THE OBJECTIVE OF ASCERTAINING WHETHER RESULTS OF STUDIES WITH NERITIC AND FRESHWATER FORMS CAN APPLY TO NERITIC ORGANISMS. THE EXPERIMENTS WERE SHORT-TERMED, AND WERE CONDUCTED DURING ACTIVE GROWTH PHASES TO ELIMINATE POSSIBILITY OF NUTRIENT DEFICIENCIES AND OVER-POPULATION, TO MINIMIZE BACTERIAL INTERFERENCE, AND TO EMULATE NATURAL CONDITIONS AS CLOSELY AS POSSIBLE. ALL CELLS GROWN IN MEDIA CONTAINING STRONTIUM-90-YTTRIUM-90 BECAME RADIOACTIVE. UPTAKE OF SR WAS PROPORTIONAL TO THE CONCENTRATION OF SR IN THE MEDIUM, CELLS BEING ABLE TO CONCENTRATE RADIOACTIVE SR FROM 18.6 TO 381.7 TIMES ITS AMBIENT CONCENTRATION. CELL GROWTH, IN MEDIA CONTAINING UP TO TEN TIMES NORMAL CONCENTRATION OF SR, WAS NOT INHIBITED. MEDIA CONTAINING HIGH CONCENTRATIONS OF SR HAD LITTLE EFFECT ON ABILITY OF CELLS TO ACCUMULATE SR-90. ADSORPTION WAS FOUND TO PLAY AN IMPORTANT ROLE IN ACCUMULATION OF SR-90. WHILE THESE RESULTS ARE PRELIMINARY, IT SHOULD BE NOTED THAT PRIMARY PRODUCERS IN MARINE ENVIRONMENTS CAN ACCUMULATE SR-90 FROM SEA WATER. CONTROLLED LABORATORY EXPERIMENTS HAVE THE ADVANTAGE OF ALLOWING COMPARISONS BETWEEN VARIOUS PHYTOPLANKTERS AND RADIONUCLIDES. (JONES-WISCONSIN)

FIELD 05C

ACCESSION NO. W70-00707

THE ROLE OF PHYTOPLANKTON IN THE CYCLING OF RADIONUCLIDES IN THE MARINE
  ENVIRONMENT,

BUREAU OF COMMERCIAL FISHERIES, BEAUFORT, N.C.

T. R. RICE.

PROC 1ST NAT SYMP ON RADIOECOLOGY, COLORADO STATE UNIV, FORT COLLINS, SEPT
  10-15, 1961. RADIOECOLOGY, REINHOLD PUB CORP, N Y AND AMER INSTITUTE OF
  BIOLOGICAL SCIENCES, WASHINGTON, D C, SCHULTZ, VINCENT AND KLEMENT, ALFRED
  W JR (EDS). P 179-185, 1963. 5 FIG, 8 TAB, 23 REF.

  DESCRIPTORS:
    *PHYTOPLANKTON, *CYCLING, *RADIOISOTOPES, SEA WATER, FOOD CHAINS,
    PHOSPHORUS RADIOISOTOPES, COBALT RADIOISOTOPES, URANIUM RADIOISOTOPES,
    NUCLEAR EXPLOSIONS, STRONTIUM RADIOISOTOPES, CESIUM, PHYSICOCHEMICAL
    PROPERTIES, ADSORPTION, COLLOIDS, ZINC RADIOISOTOPES, CALCIUM,
    POTASSIUM, CLAMS, SIZE, ALGAE, DIATOMS, TRACE ELEMENTS, METABOLISM,
    IONS, CHLORELLA, WATER POLLUTION EFFECTS, WATER POLLUTION SOURCES,
    NICHES, RADIOECOLOGY.

  IDENTIFIERS:
    *MARINE ENVIRONMENT, CERIUM-144, NITZSCHIA CLOSERIUM, ARTEMIA,
    CARTERIA, NORTH TEMPERATE ZONE, PLUTONIUM, HALF-LIFE(RADIONUCLIDE),
    FILTER-FEEDING ANIMALS, RUTHENIUM-106, PLATYMONAS, THALASSIOSIRA,
    AMPHIDINIUM KLEBSI, PORPHYRIDIUM CRUENTUM, RADIONUCLIDE UPTAKE,
    NANNOCHLORIS, THORACOMONAS, YTTRIUM, 'BIOLOGICAL DILUTION',
    PHYSIOLOGICAL CONDITION, MERCENARIA MERCENARIA.

  ABSTRACT:
    THE IMPORTANCE OF PHYTOPLANKTON IN CYCLING RADIONUCLIDES IN OCEANS
    DEPENDS UPON NUMBER AND CONCENTRATION OF RADIONUCLIDES ACCUMULATED,
    LENGTH OF TIME THEY ARE RETAINED BY THE PHYTOPLANKTON, SIZE OF
    PHYTOPLANKTON CELLS, NUMBERS OF CELLS PRESENT IN THE WATER, AND
    EFFICIENCY OF DIGESTION OF PHYTOPLANKTERS BY FILTER-FEEDING ANIMALS.
    THE AVAILABILITY OF A RADIONUCLIDE IN SEA WATER DEPENDS LARGELY UPON
    ITS PHYSICAL STATE (COLLOIDAL OR PARTICULATE) AS WELL AS ITS IONIC
    STATE. ACCUMULATION IS DEPENDENT ON AMOUNTS OF THE ELEMENT PRESENT IN
    SEA WATER AND THE QUANTITIES OF OTHER METABOLICALLY SIMILAR ELEMENTS.
    BINARY FISSION OF RADIOACTIVE PHYTOPLANKTERS IN NONACTIVE WATER RESULTS
    IN 50% REDUCTION OF ACTIVITY PER CELLS WITH EACH DIVISION, TERMED AS
    'BIOLOGICAL DILUTION.' DIFFERENT SPECIES OF PHYTOPLANKTON, WHICH
    ACCUMULATE A RADIONUCLIDE TO THE SAME LEVEL AND ARE REMOVED FROM THE
    WATER WITH THE SAME EFFICIENCY BY A FILTER-FEEDING ANIMAL, MAY NOT
    RESULT IN THE ANIMAL'S BECOMING EQUALLY RADIOACTIVE, FOR ALL SPECIES OF
    PHYTOPLANKTON ARE NOT DIGESTED WITH EQUAL EFFICIENCY. BRINE SHRIMP
    (ARTEMIA) CONCENTRATE RADIOCOBALT TO HIGHER LEVELS FED UPON RADIOACTIVE
    CARTERIA CELLS, THAN WHEN RADIOACTIVITY IS DERIVED FROM THE WATER.
    (JONES-WISCONSIN)

  FIELD 05C

  ACCESSION NO. W70-00708

115

OBSERVATIONS ON DEEPWATER PLANTS IN LAKE TAHOE, CALIFORNIA AND NEVADA,

NEVADA FISH AND GAME DEPT., RENO; AND CALIFORNIA STATE DEPT. OF FISH AND
  GAME, SACRAMENTO. INLAND FISHERIES BRANCH.

TED C. FRANTZ, AND ALMO J. CORDONE.

ECOLOGY, VOL 48, NO 5, P 709-714, 1967. 3 FIG, 3 TAB, 17 REF.

DESCRIPTORS:
   *DEEP WATER, *LAKES, *PLANTS, *CALIFORNIA, *NEVADA, DEPTH, ALGAE,
   MOSSES, CHARA, FISH, LIMNOLOGY, LAKE TROUT, CRAYFISH, STONEFLIES,
   SNAILS, ECOLOGY, EUTROPHICATION, DOMESTIC WASTES, SHALLOW WATER,
   CHLOROPHYTA, CHRYSOPHYTA, CYANOPHYTA, SAMPLING, DENSITY, LIGHT, MUD,
   SILTS, SANDS, GRAVELS, SEDIMENTS, WATER POLLUTION SOURCES.

IDENTIFIERS:
   *LAKE TAHOE(CALIF-NEV), HYDROPHYTES, LIVERWORTS, MORPHOMETRY,
   PACIFASTACUS LENIUSCULUS, ANACHARIS CANADENSIS, MYRIOPHYLLUM,
   POTAMOGETON CRISPUS, MICROSPORA WITTROCKII, SPIROGYRA, XANTHOPHYCEAE,
   MYXOPHYCEAE, SALVELINUS NAMAYCUSH, SUBSTRATE TYPE.

ABSTRACT:
   AQUATIC STUDIES OF LAKE TAHOE HAVE DISCLOSED A REMARKABLE PLANT
   ZONATION. WITH SURFACE AREA NOT UNUSUALLY LARGE, IT HAS MAXIMUM DEPTH
   OF 1645 FEET, MEAN DEPTH OF 1027 FEET. DEEPWATER MACROSCOPIC
   HYDROPHYTES IN LAKE TAHOE, FOUND TO DEPTHS OF 500 FEET, CONSISTED OF
   ALGAE, MOSSES, AND LIVERWORTS. MOST WERE CONCENTRATED AT DEPTHS OF
   200-350 FEET. ONLY CHARA OCCASIONALLY INVADES AREAS SHALLOW AS 20 FEET;
   OTHER DEEPWATER HYDROPHYTES ARE RESTRICTED TO DEPTHS BELOW 50 FEET. THE
   LIST OF IDENTIFIED SPECIES, PROBABLY INCOMPLETE, INCLUDES SIX ALGAE,
   TEN MOSSES, AND TWO LIVERWORTS. ALTHOUGH LIGHT TRANSMISSION PROBABLY
   CONTROLS MAXIMUM DEPTH OF PLANT OCCURRENCE, FACTORS LIMITING INSHORE
   DISTRIBUTION ARE LESS CLEAR. SUBSTRATE TYPE AT LAKE'S SOUTH END SEEMS
   TO CONTROL SHOREWARD EXTENSION OF PLANTS. THESE DEEPWATER PLANT BEDS
   ARE NOT ONLY UNUSUAL SCIENTIFIC INTEREST BUT APPEAR IMPORTANT IN THE
   LIFE HISTORY OF TAHOE'S MAJOR GAME FISH, THE LAKE TROUT, WHOSE MAXIMUM
   CONCENTRATION COINCIDES WITH THESE PLANT ZONES; PLANTS APPARENTLY
   PROVIDE SHELTER FOR CRAYFISH, NONGAME FISHES, AND SMALL LAKE TROUT.
   LONGTERM RESEARCH ON ECOLOGY OF THESE DEEPWATER HYDROPHYTE COMMUNITIES
   IS NEEDED. LAKE TAHOE IS THREATENED WITH EUTROPHICATION FROM DOMESTIC
   WASTES. (JONES-WISCONSIN)

FIELD 02H, 05C

ACCESSION NO. W70-00711

SEASONAL SELECTION OF ESTUARINE BACTERIA BY WATER TEMPERATURE,

RHODE ISLAND UNIV., KINGSTON. NARRAGANSETT MARINE LAB.

JOHN MCN SIEBURTH.

JOURNAL EXPERIMENTAL MARINE BIOLOGY AND ECOLOGY, VOL 1, P 98-121, 1967. 11
   FIG, 4 TAB, 39 REF.

DESCRIPTORS:
   *BACTERIA, *ESTUARINE ENVIRONMENT, *WATER TEMPERATURE, *SEASONAL, RHODE
   ISLAND, SAMPLING, ECOLOGY, AMMONIA, ENZYMES, CLASSIFICATION, GROWTH
   RATES, SEAWATER, INCUBATION, PSEUDOMONAS, SUBSURFACE-WATERS, BOTTOM
   SEDIMENTS, SURFACE WATERS, DETRITUS, LIGHT INTENSITY, TURBIDITY, ALGAE,
   DIATOMS, CYCLING NUTRIENTS, PHYSIOLOGICAL ECOLOGY, WATER POLLUTION
   EFFECTS, PHYTOPLANKTON.

IDENTIFIERS:
   NARRAGANSETT BAY(RHODE ISLAND), TAXONOMIC TYPES, HETERETROPHY,
   BACTERIAL MEDIUM, ARTHROBACTER, FLORA, AMBIENT TEMPERATURE, THERMAL
   SENSITIVITY, GENERA, 'FLAVOBACTERIUM-CYTOPHAGA', VIBRIOS,
   ACHROMOBACTERS, GRAM-POSITIVE BACTERIA, SARCINE-LIKE, BACILLUS-LIKE,
   MICROCOCCUS-LIKE, PSYCHROPHILES, PLEMORPHIC, MYCELOID, CORYNEFORM RODS,
   COCCI, ARTHROSPORES, GRAM-NEGATIVE BACTERIA, COLONY DEVELOPMENT.

ABSTRACT:
   HETERETROPHIC PLANKTONIC BACTERIA WERE OBSERVED IN TEMPERATURE
   NARRAGANSETT BAY, RHODE ISLAND, TO ESTABLISH WHETHER SEASONAL WATER
   TEMPERATURE CHANGES ARE SUFFICIENT TO SELECT THERMAL AND TAXONOMIC
   TYPES. POPULATION CHANGES WERE EXAMINED IN SEMI-MONTHLY SAMPLES AND
   GENERA DETERMINED FOR 2500+ ISOLATES; GROWTH-TEMPERATURE SPECTRA OF 600
   REPRESENTATIVE ISOLATES AT 20 TEMPERATURES, 2C (C= DEGREES CELSIUS)
   APART, WERE EXAMINED. GROWTH RANGE OF NATURAL BACTERIAL POPULATIONS·
   VARIED WITH WATER TEMPERATURE. ALL THERMAL TYPES WERE ISOLATED AND
   POPULATION TRENDS OBSERVED. MESOPHILIC ISOLATES (GROWTH ABOVE 10C;
   OPTIMUM HAVE ABOVE 30C) WERE ONLY OBTAINED ON PLATES AT 36C WHEN WATER
   TEMPERATURE EXCEEDED 10C. SEVERAL MESOPHILES WERE COLD-SENSITIVE AND
   LABILE BELOW 18C. OBLIGATE PSYCHROPHILES (GROWTH BELOW 16-20C; OPTIMUM
   9-10C) WERE ONLY OBTAINED ON OC PLATES WHERE WATER TEMPERATURES WERE
   BELOW 10C. TWO ADDITIONAL THERMAL TYPES WERE OBTAINED ON BOTH O AND 18C
   PLATES. ORGANISMS WHICH GREW AT OC (OPTIMUM 10-30C) WERE CONSIDERED
   FACULTATIVE PSYCHROPHILES; THOSE WHICH GREW AT OC (NARROWER OPTIMUM,
   18-20C), PSYCHROTOLERANT. BESIDES THESE MAJOR THERMAL TYPES, A FEW
   ISOLATES OF SEVERAL TAXONOMIC GROUPS EXHIBITED TENDENCY FOR MULTIPLE
   TEMPERATURE OPTIMA. SEASONAL SELECTION OF THERMAL TYPES OF WATER
   TEMPERATURE OCCURRED IN ALL TAXONOMIC GROUPINGS. (JONES-WISCONSIN)

FIELD 05C

ACCESSION NO. W70-00713

THE ISOLATION, PURIFICATION, AND NUTRIENT SOLUTION REQUIREMENTS OF BLUE-GREEN ALGAE,

WISCONSIN UNIV., MADISON. DEPT. OF BOTANY.

GERALD C. GERLOFF, GEORGE P. FITZGERALD, AND FOLKE SKOOG.

IN SYMPOSIUM ON THE CULTURING OF ALGAE, CHARLES F. KETTERING FOUNDATION, DAYTON, OHIO, P 27-44, 1950. 6 TAB.

DESCRIPTORS:
*ISOLATION, *CYANOPHYTES, *NUTRIENT REQUIREMENTS, CULTURES, LAKES, ALGAE, GROWTH RATES, BACTERIA, ULTRAVIOLET RADIATION, ACTINOMYCETES, HYDROGEN ION CONCENTRATION, NITRATES, POTASSIUM, CALCIUM, MAGNESIUM, IRON, ORGANIC MATTER, NITROGEN FIXATION, VOLUME, STREAMS, WATER POLLUTION EFFECTS, EUTROPHICATION, NUISANCE ALGAE.

IDENTIFIERS:
*PURIFICATION, CULTURE MEDIA, MICROCYSTIS AERUGINOSA, APHANIZOMENON FLOS-AQUAE, GLOEOTRICHIA ECHINULATA, COCCOCHLORIS PENIOCYSTIS, MADISON(WIS), CHICAGO NATURAL HISTORY MUSEUM, NOSTOC MUSCORUN, NUTRIENT CONCENTRATION, WISCONSIN, SULFUR, PHOSPHORUS.

ABSTRACT:
PROCEDURES AND PROGRESS ARE DESCRIBED IN ISOLATING AND CULTURING REPRESENTATIVE CYANOPHYTES (BLUE-GREEN ALGAE) AND MAKING THEM BACTERIA-FREE. EACH SPECIES ISOLATED WAS DESIGNATED AS A PERMANENT SPECIMEN IN CHICAGO NATURAL HISTORY MUSEUM COLLECTION. COCCOCHLORIS PENIOCYSTIS REQUIRES AN UNUSUALLY HIGH PH FOR MAXIMUM GROWTH. SYSTEMATIC CONCENTRATION VARIATION OF ONE ELEMENT AT A TIME OF THE BASIC CULTURE SOLUTION SHOWED THAT HIGHER NITRATE CONCENTRATIONS WERE SOLELY RESPONSIBLE FOR INCREASED GROWTH. RELATIONSHIP OF PHOSPHORUS, SULFUR, POTASSIUM, MAGNESIUM, CALCIUM, AND IRON WERE TESTED. THE DEMONSTRATION INDICATES THAT 24 SPECIES OF CYANOPHYTES CAN BE CULTURED CONTINUOUSLY IN INORGANIC SOLUTIONS OF KNOWN COMPOSITION, EVIDENCES STRONGLY THAT, AS A GROUP, THEY DO NOT REQUIRE ORGANIC SUBSTANCES OR UNKNOWN FACTORS FOR NORMAL GROWTH. IT APPEARS THAT NITROGEN SUPPLY IS AN IMPORTANT FACTOR IN THEIR GROWTH UNDER NATURAL CONDITIONS. COCCOCHLORIS PENIOCYSTIS APPARENTLY DOES NOT FIX ATMOSPHERIC NITROGEN. THE LABORATORY DATA PROBABLY CANNOT BE APPLIED DIRECTLY TO FIELD CONDITIONS SINCE CONCENTRATIONS OF ESSENTIAL ELEMENTS AS WELL AS VOLUMES OF SOLUTION FROM WHICH ALGAE EXTRACT THEM, ARE CERTAINLY IMPORTANT. NEVERTHELESS, LABORATORY DATA SHOULD HELP CONSIDERABLY IN DETERMINING REQUIREMENTS FOR OPTIMUM GROWTH OF CYANOPHYTES IN LAKES AND STREAMS. (JONES-WISCONSIN)

FIELD 05C

ACCESSION NO. W70-00719

MODIFIED FILTER MEDIA FROM REMOVAL OF WATER POLLUTANTS,

C. D. AGARWAL, AND A. V. S. PRABHAKARA RAO.

WATER RESEARCH, VOL 2, NO 1, P 43-45, JAN 1968. 6 REF.

DESCRIPTORS:
*WATER POLLUTION TREATMENT, *FILTRATION, *WATER POLLUTION CONTROL, *FILTERS, SEWAGE EFFLUENTS, ELECTROLYTES, ZETA POTENTIAL, BACTERIA, PROTOZOA, ALGAE, SOLID WASTES, TERTIARY TREATMENT.

IDENTIFIERS:
POLYELECTROLYTE-COATED FILTERS, BAUXITE.

ABSTRACT:
A SIMPLE FILTER WAS DEVELOPED FOR EFFICIENT REMOVAL OF MICRON-SIZED PARTICLES. THESE PARTICLES INCLUDE SOME BACTERIA, PROTOZOA, ALGAE, CYSTS, AND EVEN SOME WORMS. SEWAGE PARTICLES CARRY SMALL ELECTROKINETIC CHARGES (SURFACE ZETA POTENTIALS), WHICH ARE GENERALLY NEGATIVE. THE EFFICIENCY OF FILTRATION CAN BE GREATLY IMPROVED IF THE FILTER MEDIA ARE COATED WITH CATIONIC POLYELECTROLYTES TO RETAIN THE NEGATIVELY CHARGED PARTICLES. THE EXPERIMENTAL RESULTS OBTAINED BY USING BAUXITE COATED FILTERS WERE COMPARABLE WITH THOSE USING POLYELECTROLYTES. THE ECONOMICS OF THE DESIGNED CRITERIA FOR THIS FILTRATION PROGRESS ARE EVALUATED. (CARSTEA-USGS)

FIELD 05D, 05G

ACCESSION NO. W70-01027

STUDIES ON NATURAL FACTORS AFFECTING PHOSPHATE ABSORPTION AND ITS UTILIZATION
  BY ALGAE,

JOSEPH SHAPIRO.

WATER RESEARCH, VOL 2, NO 1, P 21-23, JAN 1968. 2 FIG, 3 REF.

DESCRIPTORS:
  *EUTROPHICATION, *NUTRIENTS, *WATER POLLUTION SOURCES, *PHOSPHATES,
  ALGAE, WATER POLLUTION TREATMENT, WATER QUALITY CONTROL, SEWAGE
  EFFLUENTS, ABSORPTION, LAKES.

IDENTIFIERS:
  *PHOSPHATE ABSORPTION.

ABSTRACT:
  THE ACCUMULATION OF NUTRIENT ELEMENTS IN WATER, PARTICULARLY NITROGEN
  AND PHOSPHORUS, FAVORS THE PROCESS OF EUTROPHICATION. PREVIOUS
  EXPERIMENTS HAVE SHOWN THAT DIFFERENT FACTORS ARE PRESENT IN NATURAL
  WATERS WHICH INHIBIT OR FACILITATE THE USE OF PHOSPHATE BY ALGAE. THE
  'PHOSPHATE SPARING FACTOR' WAS STUDIED BY USING LAKE WATER AND
  MICROCYSTIS AERUGINOSA AS THE TEST ORGANISM. ION EXCHANGE RESINS WERE
  USED TO DETERMINE AN IONIC FACTOR THAT FACILITATED THE USE OF PHOSPHATE
  BY MICROCYSTIS. ADDITION OF SALTS INCREASED THE RATES OF PHOSPHATE
  UPTAKE BY THE ALGAE IN DILUTED LAKES; HOWEVER, NITRATE SEEMED TO BE THE
  MOST IMPORTANT FACTOR. MORE EXPERIMENTAL WORK USING DIFFERENT ALGAE AND
  CHEMICAL COMPONENTS ARE IN PROGRESS. (CARSTEA-USGS)

FIELD 05C, 05B

ACCESSION NO. W70-01031

THE INFLUENCE OF ALGAL ANTIBIOSIS ON THE ECOLOGY OF MARINE MICROORGANISMS,

RHODE ISLAND UNIV., KINGSTON. NARRAGANSETT MARINE LAB.

JOHN MCNEILL SIEBURTH.

ADVANCES IN MICROBIOLOGY OF THE SEA, ACADEMIC PRESS, NEW YORK, VOL 1, P
  63-94, 1968. 8 FIG, 122 REF.

DESCRIPTORS:
  *ALGAE, *ECOLOGY, *MARINE MICROORGANISMS, PHYTOPLANKTON, BACTERICIDES,
  ALGAL TOXINS, LARVAE, PHENOLS, SHELLFISH, OYSTERS, INDUSTRIES,
  CHELATION, BACTERIA, STREPTOCOCCUS, SECONDARY PRODUCTIVITY, SULFIDES,
  SEA WATER, PHAWOPHYTA, CRUSTACEANS, HERBIVORES, CHLOROPHYLL, ENZUMES,
  PEPTIDES, BIOCHEMISTRY, HYDROGEN ION CONCENTRATION, OXIDATION, DIATOMS,
  EPIPHYTOLOGY, METABOLISM, PRIMARY PRODUCTIVITY, FLUORESCENCE, ISOTROPY,
  TRITIUM, SYMBIOSIS, TEMPERATURE, COPEPODS, PROTOZOA, ZOOPLANKTON, SOLAR
  RADIATION, PSEUDOMONAS, PARASITISM, PHYSIOLOGICAL ECOLOGY, WATER
  POLLUTION SOURCES, WATER POLLUTION EFFECTS.

IDENTIFIERS:
  *ANTIBIOSIS, AUTOLYSIS, EXCRETION, DEPURATION, PHYCOCOLLOIDS, TANNINS,
  FLAGELLATES, PHARMACEUTICALS, EXTRACELLULAR PRODUCTS, CHEMICAL
  RESISTANCE, EUPHAUSIA, PHAEOCYSTIS, SARGASSUM, FRONDS, HYDROCARBONS,
  BACTERIAL ATTACHMENT, ECOTOCARPUS, POLYSIPHONIA, ORGANIC AGGREGATES,
  SEASONAL EFFECTS.

ABSTRACT:
  AUTHOR REVIEWS THE ROLE IN MARINE ECOLOGY OF ANTIBIOSIS RESULTING FROM
  EXCRETION OF SUBSTANCES SECONDARILY EXCRETED BY ALGAE, A SUBJECT
  CONTRIBUTING TO THE MUCH LARGER PICTURE OF MICROBIAL INTERACTIONS.
  PHOTOSYNTHETIC PLANTS ARE PRIMARILY IMPORTANT TO HETEROTROPHIC
  MICROBIOTA AS PROVIDERS OF PRIMARY NUTRITIONAL SUBSTANCES. CONVERSELY,
  HETEROTROPHIC BACTERIA MAY MARKEDLY INFLUENCE THE NUTRITION AND SUCCESS
  OF MARINE PLANTS, PRESUMABLY RESULTING FROM MODIFICATION OF EXCRETED
  AND AUTOLYTIC PRODUCTS BY DEPOLYMERIZATION AND MINERALIZATION INTO
  USABLE NUTRIENTS, PRODUCTION OF GROWTH AND MORPHOGENETIC FACTORS FOR
  ALGAE, AND BY DETOXIFICATION OF INHIBITORY SUBSTANCES. WHEN POSSIBLE,
  SUCH PROCESSES SHOULD BE STUDIED IN CONCERT, RATHER THAN SEPARATELY.
  REALIZATION THAT ALGAL PRODUCTS MAY BENEFIT, MODIFY, AND EVEN SUPPRESS
  EARLY LIFE STAGES OF MARINE ANIMALS REEMPHASIZES THE IMPORTANCE OF
  INTERDISCIPLINARY ATTACKS ON ECOLOGY. FIELD STUDIES PROMISE TO
  DEMONSTRATE FACTORS WHICH INFLUENCE SUPPRESSION, SELECTION, DOMINANCE,
  AND SUCCESSION OF MARINE MICROORGANISMS. ANY SUCH STUDIES MUST INCLUDE
  CONSIDERATION OF EFFECTS OF EXTRACELLULAR SECRETIONS. AUTHOR SUMMARIZES
  HIS KNOWLEDGE AND UNDERSTANDING OF THE NATURE AND INFLUENCE OF
  INHIBITORY SECONDARY PLANT MATERIALS IN MARINE ENVIRONMENTS TO
  ENCOURAGE CONTINUED INVESTIGATION OF ALGAL ANTIBIOSIS. (JONES-WISCONSIN)

FIELD 05C

ACCESSION NO. W70-01068

STUDIES ON ALGAL SUBSTANCES IN THE SEA. III. THE PRODUCTION OF EXTRACELLULAR ORGANIC MATTER BY LITTORAL MARINE ALGAE,

RHODE ISLAND UNIV., KINGSTON. NARRAGANSETT MARINE LAB.

JOHN MCN SIEBURTH.

J EXPERIMENTAL MARINE BIOLOGY AND ECOLOGY, VOL 3, P 290-309, 1969. 1 FIG, 18 TAB, 26 REF.

DESCRIPTORS:
*OCEANS, *MARINE ALGAE, *LITTORAL, *ORGANIC MATTER, EXUDATION, TIDES, PHOTOSYNTHESIS, SOLAR RADIATION, SALINITY, RAINFALL, TEMPERATURE, COLORIMETRY, CARBOHYDRATES, PHENOLS, NITROGEN, RESPIRATION, CARBON, PRIMARY PRODUCTIVITY, OXYGEN, METABOLISM, WIND VELOCITY, HUMIDITY, BACTERIA, HYDROGEN ION CONCENTRATION, SULFATES, RESPIRATION, FOOD CHAINS.

IDENTIFIERS:
*ALGAL SUBSTANCES, *EXTRACELLULAR, FUCUS VESICULOSUS, CHONDRUS CRISPUS, ASCOPHYLLUM NODOSUM, LAMINARIA DIGITATA, LAMINARIA AGARDHII, DESICCATION, IMMERSION, NARRAGANSETT MARINE LABORATORY(RHODE ISLAND), DILUTION RATE, FRUITING BODIES, ULVA LACTUCA VAR LATISSIMA, POLYSIPHONIA HARVEYI, EPIPHYTIC BACTERIA, SUBLITTORAL, YELLOW MATERIAL, DUMONTIA INCRASSATA, EMERSION.

ABSTRACT:
SUNLIT AND DARKENED FUCUS VESICULOSUS ACCUMULATE NO DISSOLVED ORGANIC MATTER IN CLOSED SYSTEMS, IN WHICH PROBLEM OF BACTERIAL GROWTH WAS NOT ENTIRELY OVERCOME. OPEN SYSTEMS, WITH SEAWATER DILUTIONS, CONSISTENTLY ACCUMULATED APPRECIABLE QUANTITIES OF EXTRACELLULAR ORGANIC CARBON. EXUDATION IN FUCUS, COUPLED DIRECTLY WITH PHOTOSYNTHESIS, INCREASED WITH SOLAR RADIATION. TWO LAMINARIA PRODUCED EXUDATION IN THE DARK. REDUCING SALINITY OF SEAWATER APPARENTLY DIMINISHED ORGANIC MATTER EXUDATION RATE BY FUCUS. ORGANIC CARBON IS LOST UPON RE-IMMERSION OF DISSICATED FUCUS IN SEAWATER. ALGAL CLUMPS ON EXPOSED ROCKS LOST SOME 39 MILLIGRAMS CARBON/100 GRAMS. LIGHT NATURAL RAINFALL MORE EFFECTIVELY EXTRACTS ORGANIC CARBON THAN ARTIFICIAL RAINFALL. NATURE OF EXUDATES, DETERMINED BY COLOROMETRIC ASSAY, REVEALED THAT CARBOHYDRATES ARE MOST ABUNDANT, FOLLOWED BY EQUAL AMOUNTS OF NITROGENUOUS AND POLYPHENOLIC MATERIAL. DURING SPRING CONDITIONS, FUCUS EXUDES APPROXIMATELY 30% OF TOTAL OR 40% OF NET CARBON FIXED DAILY. SINCE BIOMASS OF CARBON IN FUCUS BEDS CAN EXCEED 1000 GRAMS/SQUARE METER AND ALGAE MINIMALLY FIX 6.5 GRAMS/SQUARE METER/DAY, FUCUS EXUDATION OF ORGANIC MATTER EQUAL 5-7 GRAMS CARBON/SQUARE METER/DAY. (JONES-WISCONSIN)

FIELD 05B

ACCESSION NO. W70-01073

STUDIES ON ALGAL SUBSTANCES IN THE SEA. I. GELBSTOFF (HUMIC MATERIAL) IN
TERRESTRIAL AND MARINE WATERS,

NORWEGIAN INST. OF SEAWEED RESEARCH, TRONDHEIM.

JOHN MCN SIEBURTH, AND ARNE JENSEN.

J EXPERIMENTAL MARINE BIOLOGY AND ECOLOGY, VOL 2, P 174-189, 1968. 8 FIG, 1
TAB, 24 REF.

DESCRIPTORS:
*OCEANS, *MARINE ALGAE, *SEAWATER, *FRESH WATER, *HUMIC ACIDS, *COLOR,
RIVERS, BOGS, PHENOLS, PIGMENTS, ORGANIC MATTER, BIOMASS, BENTHIC
FLORA, PHAEOPHYTA, POLAR REGIONS, SPECTROPHOTOMETRY, CHROMATOGRAPHY,
HYDROGEN ION CONCENTRATION, FLUORESCENCE, CARBOHYDRATES, TEMPERATURE,
SALINITY, ULTRAVIOLET RADIATION, SULFATES, CELLULOSE, INDUSTRIES, WATER
POLLUTION SOURCES.

IDENTIFIERS:
*GELBSTOFF, *ALGAL SUBSTANCES, SUBPOLAR WATERS, TYNSET(NORWAY), NID
RIVER(NORWAY), GAULA RIVER(NORWAY), FLAKK(NORWAY), RANHEIM(NORWAY),
TRONDHEIMSFJORD(NORWAY), SMOLA(NORWAY), HITRA(NORWAY), FROYA(NORWAY),
FUCUS VESCICULOSUS, ASCOPHYLLUM NODOSUM, ECTOCARPUS CONFERVOIDES, NYLON
COLUMN TECHNIQUE, PARTICULATE MATTER, BLUE FLUORESCING MATERIAL, YELLOW
MATERIAL, FRACTIONATION.

ABSTRACT:
DISSOLVED YELLOW COLOURING MATTER OF TERRESTRIAL AND MARINE WATERS WAS
CONCENTRATED IN NYLON COLUMNS. A CONCENTRATION FACTOR OF 10,000 WAS
OBTAINED WITH A RECOVERY OF APPROXIMATELY 70%. THE METHOD ALLOWED
ISOLATION OF A REFERENCE 'GELBSTOFF' OF MARINE ORIGIN USED TO ESTIMATE
GELBSTOFF CONCENTRATIONS IN BOG (17 MILLIGRAMS/LITER (MG/L)), RIVER
(ABOUT 1 MG/L), AND SEA WATER (0.003-0.8 MG/L). SEASONAL AND
GEOGRAPHICAL VARIATION OF GELBSTOFF CONCENTRATION IN SEA WATER WAS
OBSERVED. METHODS BASED ON SPECTRA AND DIFFERENTIAL SPECTRA WERE
INADEQUATE FOR CHARACTERIZATION OF DIFFERENT TYPES OF YELLOW MATERIAL
FROM TERRESTRIAL AND SEA WATER; THEY ARE APPARENTLY DERIVED FROM
POLYPHENOLS. FAIR SEPARATION OF GELBSTOFF WAS EFFECTED FROM DIFFERENT
SOURCES BY A METHOD OF TWO-DIMENSIONAL PAPER CHROMATOGRAPHY, WHICH THEY
DEVELOPED. MARINE GELBSTOFF COULD BE DISTINGUISHED FROM BOTH RIVER AND
BOG WATER PIGMENTS. A CONSIDERABLE FRACTION OF TERRESTRIAL GELBSTOFF
PRECIPITATED RAPIDLY IN CONTACT WITH SEA WATER. CONTENT AND
CHARACTERISTICS OF MARINE GELBSTOFF ARE VARIABLE, AND IS UNDETECTABLE
DURING JANUARY, FEBRUARY, AND MARCH. PART OF THE YELLOW COLOR OF SEA
WATER IS DUE TO PARTICULATE ORGANIC MATTER, WHICH IS LARGELY AMORPHOUS
AND SHOWS POLYPHENOL STAINING. PRECIPITATING GELBSTOFF MAY BE A MAJOR
SOURCE OF THESE ORGANIC AGGREGATES. (JONES-WISCONSIN)

FIELD 05B

ACCESSION NO. W70-01074

RECOVERY OF A SALT MARSH IN PEMBROKESHIRE, SOUTH-WEST WALES, FROM POLLUTION BY
CRUDE OIL,

FIELD STUDIES COUNCIL, PEMBROKE (ENGLAND). OIL POLLUTION RESEARCH UNIT.

E., B. COWELL, AND J. M. BAKER.

BIOLOGICAL CONSERVATION, VOL 1, NO 4, P 291-296, JULY 1969. 2 FIG, 2 TAB, 8
REF.

DESCRIPTORS:
*ALGAL POISONING, *MARSH PLANTS, *OILY WATER, AQUATIC ALGAE, MARINE
ALGAE, OIL, MARSH, POLLUTANTS.

IDENTIFIERS:
*SALT MARSH, *OIL POLLUTION, *RECOVERY FROM POLLUTION, *KUWAIT CRUDE,
BENTLASS SALT MARSH.

ABSTRACT:
AN ACCOUNT OF THE BIOLOGICAL EFFECTS OF A SPILL OF KUWAIT CRUDE OIL ON
THE PLANTS AND ALGAE OF THE BENTLASS SALT MARSH NEAR PEMBROKE, S.W.
WALES. SOME SPECIES OF THE MARSH PLANTS WERE MORE AFFECTED BY THE OIL
THAN OTHERS. SPILL OCCURRED IN JANUARY 1967 BUT BY JUNE 1968, MOST OF
THE SPECIES AFFECTED WERE SHOWING SOME SIGNS OF RECOVERY WITH THE
EXCEPTION OF TRIGLOCHIN MARITIMA. NO ANIMAL OBSERVATIONS.
(KATZ-WASHINGTON)

FIELD 05C

ACCESSION NO. W70-01231

A COMPOSITE RATING OF ALGAE TOLERATING ORGANIC POLLUTION,

FEDERAL WATER POLLUTION CONTROL ADMINISTRATION, CINCINNATI, OHIO.

C. M. PALMER.

JOURNAL OF PHYCOLOGY, VOL 5, NO 1, P 78-82, 1969. 11 TAB, 29 REF.

DESCRIPTORS:
*ALGAE, *AQUATIC ALGAE, *AQUATIC ENVIRONMENT, POLLUTANTS, BIOLOGICAL
COMMUNITIES, WATER POLLUTION EFFECTS, DIATOMS.

IDENTIFIERS:
*POLLUTION-TOLERANT ALGAE, *BLUE GREEN ALGAE, *EUGLENA, PIGMENTED
FLAGELLATES.

ABSTRACT:
FROM INFORMATION ON POLLUTION-TOLERANT ALGAE COMPILED FROM REPORTS FROM
165 AUTHORS, THE GENERA AND SPECIES MOST OFTEN REFERRED TO AS
SIGNIFICANT FALL INTO A RELATIVELY STABLE SERIES. DIATOMS, PIGMENTED
FLAGELLATES, GREEN, AND BLUE-GREEN ALGAE ARE ALL WELL REPRESENTED AMONG
THE POLLUTION-TOLERANT GENERA AND SPECIES. THE TOP 8 GENERA ARE
EUGLENA, OSCILLATORIA, CHLAMYDOMONAS, SCENEDESMUS, CHLORELLA,
NITZSHCIA, NAVICULA, AND STIGEOCLONIUM, AND THE TOP 5 SPECIES, EUGLENA
VIRIDIS, NITZSCHIA PALAE, OSCILLATORIA LIMOSA, SCENEDESMUS QUADRICAUDA,
AND OSCILLATORIA TENUIS. IN SOME GENERA, E.G., EUGLENA, A SINGLE
SPECIES IS FAR MORE SIGNIFICANT THAN ALL OTHERS AS A POLLUTION-TOLERANT
FORM. IN OTHER GENERA, E.G., OSCILLATORIA, ONLY A SLIGHT DIFFERENCE
DISTINGUISHES THE POLLUTION TOLERANCE OF 2 OR MORE SPECIES. ALGAL GENUS
AND SPECIES POLLUTION INDICES ARE PRESENTED FOR USE IN RATING WATER
SAMPLES WITH HIGH ORGANIC POLLUTION. (KATZ-WASHINGTON)

FIELD 05C

ACCESSION NO. W70-01233

STUDY OF THE TOXICITY OF CHEMICAL PRODUCTS WITH RESPECT TO THE MARINE FOOD
CHAIN (FRENCH),

CENTRE D'ETUDES ET DE RECHERCHES DE BIOLOGIE ET D'OCEANOGRAPHIE MEDICALE,
NICE (FRANCE).

M. AUBERT, R. CHARRA, AND G. MALARA.

REVUE INTERNATIONALE D'OCEANOGRAPHIE MEDICALE, VOL XIII-XIV, P 45-72, 1969.
14 FIG, 7 TAB, 7 REF.

DESCRIPTORS:
*DETERGENTS, *INSECTICIDES, *TOXICITY, *WATER POLLUTION, FOOD CHAINS,
AQUATIC ALGAE, BRINE SHRIMP, FISH, MUSSELS.

IDENTIFIERS:
*HYDROCARBON, *TOXICITY THRESHOLD.

ABSTRACT:
THE THRESHOLD TOXICITIES OF 5 CHEMICAL POLLUTANTS (3 DETERGENTS, 1
INSECTICIDE, 1 HYDROCARBON) WERE DETERMINED FOR ORGANISMS FROM 3 LEVELS
OF THE FOOD CHAIN (PHYTOPLANKTON, ZOOPLANKTON, FISH). THE RESULTS SHOW
THAT THE TOXICITY OF THE CHEMICALS VARIES FROM ONE ORGANISM TO ANOTHER
AND FROM ONE LIFE STAGE TO ANOTHER IN THE SAME ORGANISM.
(TYNER-WASHINGTON)

FIELD 05C

ACCESSION NO. W70-01466

STUDY OF THE BIODEGRADABILITY OF TOXIC CHEMICAL PRODUCTS WITH RESPECT TO THE
   MARINE FOOD CHAIN (FRENCH),

   CENTRE D'ETUDES ET DE RECHERCHES DE BIOLOGIE ET D'OCEANOGRAPHIE MEDICALE,
      NICE (FRANCE).

   M. AUBERT, AND J. P. GAMBAROTTA.

   REVUE INTERNATIONALE D'OCEANOGRAPHIE MEDICALE, VOL XIII-XIV, P 73-105, 1969.
      19 FIG, 3 TAB, 8 REF.

      DESCRIPTORS:
         *BIODEGRADATION, *DETERGENTS, *INSECTICIDES, *MARINE BACTERIA,
         *TOXICITY, *WATER POLLUTION, AQUATIC ALGAE, BRINE SHRIMP, FISH, FOOD
         CHAINS, MUSSELS.

      IDENTIFIERS:
         *HYDROCARBON.

      ABSTRACT:

         TEST SOLUTIONS OF 5 CHEMICAL POLLUTANTS (3 DETERGENTS, 1 INSECTICIDE, 1
         HYDROCARBON) AT CONCENTRATIONS TWICE THE PREVIOUSLY DETERMINED
         THRESHOLD DOSES FOR THE ORGANISMS IN QUESTION WERE SUBJECTED TO
         DEGRADATION BY MARINE MICROORGANISMS FOR 3, 6, OR 9 DAYS PRIOR TO THE
         INTRODUCTION OF THE TEST ORGANISMS. THESE ORGANISMS WERE DRAWN FROM
         DIFFERENT LEVELS OF THE FOOD CHAIN; PRIMARY PRODUCERS, 2 SPECIES OF
         PHYTOPLANKTON; HERBIVORES, ZOOPLANKTON AND MUSSELS; CARNIVORES, FISH.
         IN SOME CASES BIOLOGICAL DEGRADATION OF THE CHEMICAL POLLUTANT CAUSED A
         REDUCTION IN TOXICITY. (TYNER-WASHINGTON)

      FIELD 05C

      ACCESSION NO. W70-01467

EFFECTS OF 'TORREY CANYON' TYPE POLLUTION ON THE UNICELLULAR MARINE ALGA
   PHAEODACTYLUM TRICORNUTUM (FRENCH),

   LABORATOIRE DE PHYSIOLOGIE GENERALE ET COMPAREE DU MUSEUM, PARIS (FRANCE);
      AND LABORATOIRE DE PHYSIOLOGIE DES ETRES MARINS DE L'INSTITUT
      OCEANOGRAPHIQUE, PARIS (FRANCE).

   J. C. LACAZE.

   REVUE INTERNATIONALE D'OCEANOGRAPHIE MEDICALE, VOL XIII-XIV, P 157-179, 1969.
      3 FIG, 2 TAB, 36 REF.

      DESCRIPTORS:
         *DETERGENTS, *DIATOMS, *OIL, *OIL WASTES, *SURFACTANTS, *WATER
         POLLUTION, CLEANING, DISASTERS, OILY WATER, MARINE ALGAE, SHIPS.

      IDENTIFIERS:
         *CRUDE OIL, *GAMOSOL, *OIL SPILLS, *TORREY CANYON, AROMATIC SOLVENTS,
         PHAEODACTYLUM TRICORNUTUM.

      ABSTRACT:

         THE AUTHOR STUDIED THE EFFECTS OF BOTH CRUDE OIL (OF THE SORT CARRIED
         BY THE 'TORREY CANYON') AND GAMOSOL (A DETERGENT USED TO DISPERSE OIL)
         ON PHAEODACTYLUM TRICORNUTUM, A MARINE DIATOM. A 10% REDUCTION IN
         GROWTH WAS OBSERVED IN CULTURES PREPARED WITH 1% PETROLEUM EXTRACTS.
         AFTER AERATION THE REDUCTION IN GROWTH WAS 5%, SOME TOXIC VOLATILE
         COMPOUNDS HAVING BEEN ELIMINATED AND THE OXYGEN CONCENTRATION
         INCREASED. PETROLEUM EXTRACTS WHICH HAD BEEN PREVIOUSLY RAISED TO A
         TEMPERATURE ABOVE 100 DEG C EXHIBITED AN INCREASED TOXICITY (35%
         REDUCTION IN GROWTH). GAMOSOL (14% NONIONIC SURFACTANTS AND 86%
         AROMATIC SOLVENTS) IS MORE TOXIC (50% REDUCTION IN GROWTH AFTER
         EXPOSURE TO 5.6 PPM) THAN THE SURFACE ACTIVE FRACTION (50% REDUCTION IN
         GROWTH AFTER EXPOSURE TO 8 PPM). HOWEVER, THIS TOXICITY VARIES BECAUSE
         OF THE HIGH VOLATILITY OF THE SOLVENT IN DILUTE SOLUTION. A 30 MINUTE
         EXPOSURE TO 40 PPM GAMOSOL KILLED THE CELLS. (TYNER-WASHINGTON)

      FIELD 05C, 05B

      ACCESSION NO. W70-01470

PRACTICAL ASPECTS OF THE DESIGN OF WASTE STABILIZATION PONDS,

W. D. BARLOW.

PROC 9TH STH MUNIC INDUSTR WASTE CONF, P 65-70, 1960.

DESCRIPTORS:
*LAGOONS, *DOMESTIC WASTES, WASTE WATER TREATMENT, INDUSTRIAL WASTES,
NEUTRIENTS, BACTERIA, ALGAE, TREATMENT FACILITIES, TEST PROCEDURES,
METALS.

IDENTIFIERS:
*FINISHING WASTES.

ABSTRACT:
LAGOONS CAN BE DESIGNED TO PROVIDE COMPLETE TREATMENT FOR ANY WASTE
WATER, DOMESTIC AND/OR INDUSTRIAL, PROVIDED IT CONTAINS THE NECESSARY
NUTRIENTS AND IS FREE FROM SUBSTANCES WHICH ARE TOXIC TO BACTERIA OR
ALGAE. THE AUTHOR DISCUSSES THE BASIC PROCESSES INVOLVED; FACTORS
AFFECTING THE DESIGN AND OPERATION OF THE LAGOON (PARTICULARLY UNDER
CONDITIONS IN NORTH AND SOUTH CAROLINA); PERFORMANCE OF THE LAGOON,
INCLUDING METHODS OF MEASUREMENT; AND PRECAUTIONS TO BE TAKEN TO AVOID
HEALTH HAZARDS, POLLUTION AND OTHER NUISANCE PROBLEMS. IN DISCUSSION,
REFERENCE WAS MADE TO LAGOONS USED FOR TREATING MIXED WASTE WATERS AND
20-30 PER CENT OF DOMESTIC SEWAGE, AND ESPECIALLY TO THE PROBLEMS
ENCOUNTERED WHICH APPEAR TO BE CAUSED BY TOXIC METALS IN THE WASTE
WATER. (LIVENGOOD-NORTH CAROLINA STATE UNIV)

FIELD 05F

ACCESSION NO. W70-01519

NUTRIENT LIMITATION OF SUMMER PHYTOPLANKTON GROWTH IN CAYUGA LAKE,

CORNELL UNIV., ITHACA, N.Y.

D. H. HAMILTON, JR.

LIMNOLOGY AND OCEANOGRAPHY, VOL 14, NO 4, P 579-590, JULY 1969. 3 FIG, 9 TAB,
32 REF.

DESCRIPTORS:
*PHYTOPLANKTON, *LIMITING FACTORS, *NUTRIENTS, *PRIMARY PRODUCTIVITY,
PHOSPHATES, SILICATES, CITRATES, BIOASSAY, NEW YORK, LAKES, INHIBITION,
ALGAE, DIATOMS, EUTROPHICATION, CARBON RADIOISOTOPES, CHLOROPHYLL,
HYDROGEN ION CONCENTRATION, MAGNESIUM, CALCIUM.

IDENTIFIERS:
*ENRICHMENT EXPERIMENTS, *NUTRIENT LIMITATION, CAYUGA LAKE(NY),
ASTERIONELLA FORMOSA, SYNEDRA, MELOSIRA, DINOBRYON, FRAGILARIA,
RADIOCARBON METHOD, FINGER LAKES(NY), RODHE'S SOLUTION VIII, NUTRIENT
LIMITATION, CARBON-14.

ABSTRACT:
AUTHOR DESCRIBES EXPERIMENTS PERFORMED TO DETERMINE THE EFFECT OF
DEFINED CULTURE MEDIUM COMPONENTS ON NATURAL PHYTOPLANKTON POPULATIONS.
SAMPLES COLLECTED FROM CAYUGA LAKE (ITHACA, NEW YORK) AT 2-3 DAY
INTERVALS FROM 20 JULY THROUGH 18 AUGUST WERE PLACED IN AN ENRICHMENT
MEDIUM (RODHE'S SOLUTION VIII). ASSIMILATION WAS MEASURED WITH
RADIOCARBON (C-14). MAJOR ALGAL CULTURE COMPONENTS WERE ASTERIONELLA
FORMOSA AND UNIDENTIFIED SPECIES OF SYNEDRA, MELOSIRA, DINOBRYON, AND
FRAGILARIA. RESULTS INDICATED AN ENHANCEMENT EFFECT WITH CALCIUM
NITRATE, DUE APPARENTLY TO THE CALCIUM ELEMENT. MAGNESIUM SULFATE AND
FERRIC CITRATE-CITRIC ACID WERE IRREGULARLY INHIBITORY. PHOSPHATE
FAILED TO STIMULATE CULTURES AND OCASIONALLY PROVED INHIBITING.
NUTRIENT COMBINATIONS YIELDED VARIABLE RESULTS; AUTHOR STATES THIS MAY
BE DUE TO PHENOMENA PECULIAR TO THE SAMPLE BOTTLES OR TO CHANGING
COMPOSITION AND ALTERNATIVELY, DUE TO PHYSIOLOGY OF THE PHYTOPLANKTON
POPULATIONS. TABULAR DATA INCLUDE RESULTS AND SUMMARY OF ENRICHMENT
EXPERIMENTS. (VOIGHTLANDER-WISCONSIN)

FIELD 02H, 05C

ACCESSION NO. W70-01579

EXTENT OF DAMAGE TO COASTAL HABITATS DUE TO THE TORREY CANYON INCIDENT,

FURZEBROOK RESEARCH STATION, WAREHAM (ENGLAND).

D. S. RANWELL.

BIOLOGICAL EFFECTS OF OIL POLLUTION ON LITTORAL COMMUNITIES. SUPPLEMENT TO
   VOLUME 2 OF FIELD STUDIES. LONDON, FIELD STUDIES COUNCIL, JULY, 1968. P
   39-47.

   DESCRIPTORS:
      *EMULSIFIERS, *MORTALITY, *MARINE ALGAE, *MARSH PLANTS, *MARINE
      ANIMALS, *OILY WATER, LICHENS, TIDAL EFFECTS, DISPERSION, OIL-WATER
      INTERFACES, PERSISTENCE, WATER POLLUTION TREATMENT, WATER POLLUTION
      EFFECTS.

   IDENTIFIERS:
      *TORREY CANYON, SUSCEPTIBILITY, REGENERATION, CLIFF VEGETATION, SAND
      DUNES.

   ABSTRACT:
      THE FULL RANGE OF OIL-CONTAMINATED HABITATS RESULTING FROM THE TORREY
      CANYON SPILL WERE INVESTIGATED FROM THE AIR AND GROUND. DECONTAMINATION
      MEASURES AND TIDAL ACTION EFFECTIVELY CLEARED MOST OIL FROM THE CORNISH
      COAST, ALTHOUGH SOME PERSISTED THROUGHOUT 1967. OVER 1 1/2 MILLION
      GALLONS OF EMULSIFIERS WERE USED. THEY RAPIDLY CLEARED THE WATER
      SURFACE BUT THEY DISTRIBUTED CONTAMINATION BEYOND THE ORIGINAL
      CONTAMINATION AREA, BOTH IN DEPTH IN SAND AND WATER. TOXIC ELEMENTS IN
      THE EMULSIFIERS DESTROYED SOME MARINE FLORA AND FAUNA. THE GREATEST
      DAMAGE OCCURRED ON THE ROCKY SHORELINES WHERE MUCH OIL WAS STRANDED.
      INTER-TIDAL ALGAE AND LICHENS SUFFERED THE MOST EXTENSIVELY AMONG
      PLANTS BECAUSE OF A CUT-OFF OF LIGHT AND AIR. THE MOST TOXIC FRACTIONS
      OF CRUDE OIL ARE VOLATILE AND DISAPPEAR RAPIDLY. MANUAL REMOVAL OF OIL
      WAS MUCH PREFERRED OVER CHEMICAL REMOVAL METHODS, AS IT DID NOT ADD TO
      POLLUTION. AT LEAST 100 SPECIES OF ALGAE, LICHENS, AND FLOWERING PLANTS
      WERE FILLED BY OIL OR EMULSIFIER POLLUTION.
      (SJOLSETH-WASHINGTON)

   FIELD 05C

   ACCESSION NO.   W70-01777

EXPERIMENTS WITH SOME DETERGENTS AND CERTAIN INTERTIDAL ALGAE,

UNIVERSITY COLL. OF WALES, ABERYSTWYTH. DEPT. OF BOTANY.

A. D. BONEY.

BIOLOGICAL EFFECTS OF OIL POLLUTION ON LITTORAL COMMUNITIES. SUPPLEMENT TO
   VOLUME 2 OF FIELD STUDIES. LONDON, FIELD STUDIES COUNCIL, JULY, 1968. P
   55-72.

   DESCRIPTORS:
      *DETERGENTS, *MARINE ALGAE, WATER POLLUTION EFFECTS, LETHAL LIMIT,
      GROWTH RATES, SPORES.

   IDENTIFIERS:
      *TORREY CANYON, *DETERGENT CONCENTRATIONS, SURVIVAL, SUSCEPTIBILITY,
      PHYSIOLOGICAL OBSERVATIONS, CHRONIC EXPOSURE, ACUTE EXPOSURE.

   ABSTRACT:
      THE DIRECT EFFECTS OF DETERGENT APPLICATION FOLLOWING THE TORREY CANYON
      DISASTER ON SEVEN TEST SPECIES OF MARINE ALGAE WERE STUDIED. SIX
      DETERGENTS WERE USED EACH AT SEVERAL DIFFERENT CONCENTRATIONS. THE
      RECEPTACLE TISSUES OF ASCOPHYLLUM NODOSUM WERE DAMAGED. CELL DAMAGE WAS
      EXTENSIVE IN CLADOPHORA RUPESTRIS. BRYOPSIS HYPNOIDES WERE EASILY
      KILLED. THE CEPT PHASES OF PRASINOCLADUS MARINUS WERE DESTROYED BY
      DETERGENTS. GROWTH AND SPORE PRODUCTION OF ACROHAETIUM INFESTANS WERE
      RETARDED. NO IMMEDIATE CELL DAMAGE WAS NOTICED AMONG POLYSIPHONIA
      LANOSA AND PORPHYRA UMBILICALIS. BOTH ACUTE AND CHRONIC EXPOSURES WERE
      UTILIZED. (SJOLSETH-WASHINGTON)

   FIELD 05C

   ACCESSION NO. W70-01779

BIOLOGICAL CONSEQUENCES OF OIL POLLUTION AND SHORE CLEANSING,

UNIVERSITY COLL. OF SWANSEA (WALES). DEPT. OF ZOOLOGY.

A. NELSON-SMITH.

BIOLOGICAL EFFECTS OF OIL POLLUTION ON LITTORAL COMMUNITIES. SUPPLEMENT TO
VOLUME 2 OF FIELD STUDIES. LONDON, FIELD STUDIES COUNCIL, JULY, 1968. P
73-80.

DESCRIPTORS:
*OILY WATER, *MARINE ALGAE, *EMULSIFIERS, WATER POLLUTION EFFECTS,
MORTALITY, GASTROPODS, MARINE PLANTS, ON-SITE DATA COLLECTIONS.

IDENTIFIERS:
*TORREY CANYON, REPOPULATION, SUSCEPTIBILITY.

ABSTRACT:
THE EFFECTS OF OIL SPILLS IN DIFFERENT AREAS WERE INVESTIGATED TO
PROVIDE INTERESTING COMPARISONS. OIL SPILLED INTO A SMALL COVE YIELDED
EXTENSIVE SHORE POLLUTION AND MORTALITIES. NO EMULSIFIERS WERE USED,
AND IN LESS THAN A YEAR THE ANIMAL POPULATIONS WERE AGAIN NORMAL.
GASTROPOD MOLLUSCS APPEARED TO BE MOST SERIOUSLY AFFECTED. MARINE ALGAE
POPULATIONS WERE ALSO REDUCED. MISUSE OF EMULSIFIERS EMBEDDED THE OIL
MORE FIRMLY AND RESULTED IN INCREASED MORTALITIES. EMULSIFIERS ENABLE
THE OIL TO WET THE SURFACES OF SHORE ORGANISMS AND TO PENETRATE THEIR
SYSTEMS AND ALSO TO SPREAD THE OIL TO PREVIOUSLY UNREACHED AREAS. TIME
OF YEAR OF OIL SPILL AND MOBILITY OF ORGANISMS REGULATE THE RATE AT
WHICH THE MEMBERS OF THE SHORE COMMUNITY ARE ABLE TO RECOVER FROM OIL
POLLUTION. (SJOLSETH-WASHINGTON)

FIELD 05C

ACCESSION NO. W70-01780

PHYSIOLOGY OF GRANULAR LEUKOCYTES IN FISH BLOOD,

L. I. SMIRNOVA.

TRANSLATION FROM VOPROSY IKHTIOLOGII. PROBLEMS OF ICHTHYOLOGY, VOL 8, NO 5,
1968. P 748-755.

DESCRIPTORS:
*FISH PHYSIOLOGY, *ALGAE, *THERMAL POLLUTION, *TOXICITY.

IDENTIFIERS:
*FISH HEMATOLOGY, *GRANULAR LEUCOCYTES, *MUSCULAR EXERTION,
*2-METHYL-5-VINYLPYRIDINE, *GLUE GREEN ALGAE, BREAM, WHITE BREAM,
LEUKOCYTIC COMPOSITION, ALTERNATING CURRENT.

ABSTRACT:
A CONNECTION WAS ESTABLISHED BETWEEN THE LEUCOCYTIC COMPOSITION OF THE
BLOOD IN FED AND STARVING BREAM. AN INCREASE IN THE NUMBER OF
GRANULOCYTES CIRCULATING IN THE BLOOD WAS OBSERVED IN SOME FAVORABLE
AND SOME STRESS CONDITIONS, SUCH AS INTENSIVE FEEDING, TEMPERATURE
INCREASE, INCREASED ACTIVITY, ELECTRICAL STIMULATION AND EXPOSURE TO
2-METHYL-5-VINYLPYRIDINE, BLUE GREEN ALGAE. WITH THE CHANGES IN THE
WHITE BLOOD CELL COMPOSITION, THERE WERE PARALLEL REDUCTIONS IN THE
RESISTANCE OF RED CELLS AND THEIR SUBSEQUENT DESTRUCTION.
(KATZ-WASHINGTON)

FIELD 05C

ACCESSION NO. W70-01788

AN INVESTIGATION BY RAPID CARBON-14 BIOASSAY OF FACTORS AFFECTING THE CULTURAL
EUTROPHICATION OF LAKE TAHOE, CALIFORNIA-NEVADA,

CALIFORNIA UNIV., DAVIS. DEPT OF ZOOLOGY; AND ENGINEERING-SCIENCE INC.,
OAKLAND, CALIF.

CHARLES R. GOLDMAN, AND RALF C. CARTER.

JOURNAL WATER POLLUTION CONTROL FEDERATION, VOL 37, NO 7, P 1044-1059, 1965.
11 FIG, 5 TAB, 27 REF.

DESCRIPTORS:
*LAKES, *CARBON RADIOISOTOPES, *BIOASSAY, *EUTROPHICATION, *CALIFORNIA,
*NEVADA, LIMNOLOGY, PRIMARY PRODUCTIVITY, CHLOROPHYLL, SAMPLING,
PHYTOPLANKTON, RADIOACTIVITY, PHOTOSYNTHESIS, CHEMICAL ANALYSIS, TRACE
ELEMENTS, TEMPERATURE, IRON, NUTRIENTS, LIGHT PENETRATION, SEWAGE
EFFLUENTS, OXYGEN, SOLAR RADIATION, PERIPHYTON, BACTERIA, SPECTROSCOPY,
COPPER, NITROGEN, PHOSPHORUS, ALGAE, WATER POLLUTION EFFECTS, WATER
CHEMISTRY.

IDENTIFIERS:
*LAKE TAHOE(CALIF-NEV), TRANSPARENCY, LIGHT MEASUREMENTS, EUPHOTIC
ZONE, PLANT CAROTENOIDS, CARBON FIXATION, PELAGIC REGION, MONOMICTIC,
DEPTH EFFECTS, IONIC CONCENTRATION, SEASONAL EFFECTS.

ABSTRACT:
ATTEMPTS WERE MADE TO DETERMINE SIGNIFICANT VARIABILITY IN THE PRESENT
FERTILITY OF LAKE TAHOE, TO ESTIMATE MORE PRECISELY THE LAKE'S
PRODUCTIVITY BY A MORE COMPREHENSIVE SAMPLING PROGRAM THAN PREVIOUSLY
POSSIBLE, AND, WITH CARBON-14 BIOASSAY METHODS, TO DETERMINE
EUTROPHICATION POTENTIAL OF INFLOW, BOTH NATURAL AND THAT ENRICHED WITH
SEWAGE EFFLUENT. THE EXTREME SENSITIVITY OF THE CARBON-14 METHOD
PROVIDES A MEANS OF PREDICTING LEVELS OF NUTRIENT ADDITION WHICH MIGHT
CAUSE OBJECTIONAL EUTROPHICATION. ASSUMPTION THAT THE PHYSIOLOGICAL
CONDITION OF PLANKTON IS VARIABLE IN DIFFERENT PARTS OF THE LAKE IS
SUPPORTED BY RESULTS OF STUDY. AREAL ESTIMATES OF PRODUCTIVITY WERE
DETERMINED BY INTEGRATING PHOTOSYNTHETIC CURVES RESULTING FROM
CONSIDERATION OF LIGHT EXTINCTION, VARIATION IN TEMPERATURE, AND THE
DISTRIBUTION OF PLANKTON WITH DEPTH. A SECONDARY SEWAGE EFFLUENT, WHEN
DILUTED TO 1/3,700 OF ORIGINAL STRENGTH, STIMULATES BIOACTIVITY IN LAKE
TAHOE SUFFICIENTLY TO PRODUCE A 1% DECREASE IN LIGHT TRANSMITTANCE
WITHIN 12 DAYS. MEASUREMENTS OF PRIMARY PRODUCTIVITY AROUND THE LAKE'S
MARGIN SHOW THAT, DESPITE EXTENSIVE MIXING, LOCALIZED SIGNS OF
EUTROPHICATION ARE EVIDENT. THE EXTREME SENSITIVITY OF LAKE TAHOE
WATERS TO ANY CHANGE IN NUTRIENT REGIME CLEARLY EVIDENCES THE POTENTIAL
OF EVEN TREATED EFFLUENT FOR SPEEDING EUTROPHICATION. (JONES-WISCONSIN)

FIELD 05C, 02H

ACCESSION NO. W70-01933

CHANGES IN THE BIOLOGY OF THE LOWER GREAT LAKES,

OHIO STATE UNIV., COLUMBUS. NATURAL RESOURCES INST.

CHARLES A. DAMBACH.

BULLETIN OF THE BUFFALO SOCIETY OF NATURAL SCIENCES, VOL 25, NO 1, P 1-17, 1969. 19 REF.

DESCRIPTORS:
*BIOLOGY, *GREAT LAKES, *LAKE ERIE, *LAKE MICHIGAN, *LAKE ONTARIO, LAKE HURON, AESTHETICS, ECOLOGY, ECONOMICS, OHIO, COMMERCIAL FISHING, PUBLIC HEALTH, WALLEYE, VEGETATION, WILD RICE, SILTS, PLANKTON, MAYFLIES, FAUNA, FISH, OLIGOCHAETES, MIDGES, SNAILS, PHOSPHORUS, NUTRIENTS, ALGAE, DIATOMS, CHLOROPHYTA, DISSOLVED OXYGEN, PIKE, CARP, DRUM(FRESHWATER), CISCO, LAKE TROUT, EUTROPHICATION, STRIPED BASS, CYANOPHYTA, WATER POLLUTION EFFECTS, ELECTRIC POWERPLANTS.

IDENTIFIERS:
HEXAGENIA, CHIRONOMIDAE, PROCLADIUS, CHIRONOMUS PROMOSUS, TRICHOPTERA, LEECHES, FINGERNAIL CLAMS, WHITE FISH, ALEWIFE, GIZZARD SHAD, SEA LAMPHREY, COHO SALMON.

ABSTRACT:
DRAMATIC BIOLOGICAL CHANGES HAVE APPEARED IN BOTTOM FAUNA AND AMONG CERTAIN FISHES OF THE LOWER GREAT LAKES. OF SPECIAL SIGNIFICANCE IS ABUNDANT INCREASE, SINCE 1959, OF THE MIDGE PROCLADIUS, A SUPPOSEDLY MORE POLLUTION-TOLERANT FORM, WHILE CHIRONOMUS PROMOSUS HAS DECREASED, SUGGESTING THAT POLLUTION ZONES HAVE EXTENDED FURTHER INTO THE LAKES. THE MAYFLY IS NOW RARE. BENTHIC FAUNA IS NOW DOMINATED BY OLIGOCHAETES AND MIDGES, WITH SOME FINGERNAIL CLAMS, SNAILS, AND LEECHES ON THE INCREASE. CHEMICAL CONDITIONS PROBABLY PROVIDE A MORE RELIABLE INDEX TO CHANGES THAN PLANKTON DATA, BUT ARE DIFFICULT TO RELATE. SPECIES COMPOSITION, ONCE DOMINATED BY DIATOMS, ARE NOW DOMINATED BY BLUE-GREEN ALGAE. DECLINE OF CERTAIN HIGH QUALITY FISHES, NOTABLY THE BLUE PIKE AND WALLEYE PIKE, IS LARGELY RESPONSIBLE FOR THE ACCELERATED PUBLIC INTEREST IN CORRECTIVE MEASURES. RELATIVE SIGNIFICANCE OF ENVIRONMENT VERSUS OVERFISHING IS DEBATABLE. INCREASE IN EUTROPHICATION RATE OF THE GREAT LAKES, ESPECIALLY LAKE ERIE, IS SIGNIFICANT. HUMAN TECHNOLOGY CAN SO MODIFY THE ENVIRONMENT THAT BIOLOGICAL POPULATIONS ARE SIGNIFICANTLY AFFECTED. BIOLOGISTS WITH REQUISITE KNOWLEDGE, SHOULD DEVELOP BETTER GUIDELINES FOR WEIGHING COSTS OF EACH INCREMENT OF DEGRADATION AND EACH INCREMENT OF IMPROVEMENT. (JONES-WISCONSIN)

FIELD 05C, 02H

ACCESSION NO. W70-01943

SULFIDE PRODUCTION IN WASTE STABILIZATION PONDS,

TEXAS UNIV., AUSTIN. ENVIRONMENTAL HEALTH ENGINEERING RESEARCH LAB.

EARNEST F. GLOYNA, AND ERNESTO ESPINO.

J SANIT ENG DIV, PROC AMER SOC CIVIL ENG, VOL 95, NO SA3, PROC PAPER 6637, P 607-628, JUNE 1969. 19 FIG, 4 TAB, 17 REF.

DESCRIPTORS:
*SULFIDES, *OXIDATION LAGOONS, *DESIGN, *SULFUR BACTERIA, STABILIZATION, WASTE WATER, ALGAE, DIFFUSION, SULFUR.

IDENTIFIERS:
*SULFATE ION, *DESULFOVIBRIO, FACULTATIVE, BOD, DETENTION TIME.

ABSTRACT:
THE PRODUCTION AND RELEASE OF HYDROGEN SULFIDE FROM WASTE STABILIZATION PONDS AND REQUIRED DESIGN MODIFICATIONS, RESULTING FROM INCREASED SULFATE OR HYDROGEN SULFIDE CONTENT ARE DISCUSSED. A 6-FT DEEP FACULTATIVE MODEL POND WAS OPERATED UNDER SEVEN DIFFERENT OPERATING CONDITIONS. THE OBJECTIVE OF THE EXPERIMENTS WAS TO STUDY THE RELATIONSHIPS BETWEEN SULFIDE CONCENTRATION. (LEDBETTER-TEXAS)

FIELD 05D

ACCESSION NO. W70-01971

REMOVAL OF NITROGEN AND PHOSPHORUS FROM WASTE WATER,

STANFORD UNIV., CALIF.

ROLF ELIASSEN, AND GEORGE TCHOBANOGLOUS.

ENVIRON SCI TECHNOL, VOL 3, NO 6, P 536-541, JUNE 1969. 2 TAB, 15 REF.

DESCRIPTORS:
*NUTRIENTS, *NITROGEN, *PHOSPHORUS, EFFICIENCIES, COST COMPARISONS, TREATMENT, BIOLOGICAL TREATMENT, ALGAL CONTROL, AMMONIA, ION EXCHANGE, REVERSE OSMOSIS, DISTILLATION, ELECTRODIALYSIS, SORPTION, CHEMICAL PRECIPITATION, ULTIMATE DISPOSAL.

IDENTIFIERS:
*NUTRIENT REMOVAL, *EUTROPHICATION, ANAEROBIC DENITRIFICATION, LAND APPLICATION, ELECTROCHEMICAL TREATMENT, AMMONIA STRIPPING, ALGAE HARVESTING.

ABSTRACT:
THE PRINCIPAL SOURCES OF NITROGEN AND PHOSPHORUS IN WASTEWATERS ARE TABULATED WITH THE AMOUNTS FROM EACH SOURCE; DOMESTIC AND AGRICULTURAL OPERATIONS ARE ON A PAR WITH INDUSTRIAL OUTPUT FOR NITROGEN DISCHARGES, BUT PHOSPHORUS RELEASES BY INDUSTRY ARE NOT AVAILABLE. NITROGEN REMOVAL MAY BE DONE BY THE FOLLOWING PROCESSES WITH REMOVAL EFFICIENCIES AND COST IN DOLLARS PER MILLION GALLONS IN PARENTHESES: AMMONIA STRIPPING (80-98, 9-25), ANAEROBIC DENITRIFICATION (60-95, 25-30), ALGAE HARVESTING (50-90, 20-35) AND CONVENTIONAL BIOLOGICAL TREATMENT (30-50, 30-100). A SIMILAR LIST FOR PHOSPHORUS REMOVAL FOLLOWS: MODIFIED ACTIVATED SLUDGE (60-80, 30-100), CHEMICAL PRECIPITATION (88-95, 10-70), CHEMICAL PRECIPITATION WITH FILTRATION (95-98, 70-90), AND SORPTION (90-98, 40-70). IN ADDITION, THE CONVENTIONAL BIOLOGICAL TREATMENT REMOVES 10-30 PERCENT OF THE PHOSPHORUS. A LIST OF THE TREATMENT SCHEMES FOR REMOVAL OF BOTH NITROGEN AND PHOSPHORUS IS THE FOLLOWING: ION EXCHANGE (80-92, 86-98, 170-300), ELECTROCHEMICAL TREATMENT (80-85, 4-8 FOR POWER), ELECTRODIALYSIS (30-50, 100-250), REVERSE OSMOSIS (65-95, 250-400), DISTILLATION (90-98, 400-1000), AND LAND APPLICATION (60-90 PHOSPHORUS, 75-150). ULTIMATE DISPOSAL METHODS MAY BE CLASSIFIED INTO 4 GROUPS: DUMPING; SUBSURFACE INJECTION; CONVERSION AND DUMPING; AND CONVERSION, PRODUCT RECOVERY, AND DUMPING. (LEDBETTER-TEXAS)

FIELD 05D

ACCESSION NO. W70-01981

EFFECTS OF INSECTICIDES ON ALGAE,

ONTARIO WATER RESOURCES COMMISSION, TORONTO.

A. E. CHRISTIE.

WATER AND SEWAGE WORKS, VOL 116, NO 5, P 172-176, MAY 1969. 9 TAB, 31 REF.

DESCRIPTORS:
*ALGAE, *TOXICITY, *INSECTICIDES, METABOLISM, RADIOACTIVITY TECHNIQUES, CHLORELLA, OXIDATION LAGOONS.

IDENTIFIERS:
*ALGAE TOXICITY, CARBON-14, DDT, MALATHION, SEVIN.

ABSTRACT:
A STUDY WAS CONDUCTED OF THE TOXICITY AND METABOLISM OF DDT, SEVIN, MALATHION TO CULTURES OF CHLORELLA PYRENOIDOSA AND ALGAE FROM A WASTE STABILIZATION POND. THE INSECTICIDES IN THE APPROPRIATE SOLVENTS WERE ADDED TO CULTURE FLASKS OF ALGAL MEDIUM AND THE NUMBER OF ALGAL CELLS COUNTED AFTER VARIOUS NUMBER OF DAYS INCUBATION. CARBON-14 TAGGED INSECTICIDES WERE USED TO MEASURE UPTAKE AND METABOLISM OF THE INSECTICIDES BY ALGAE. DDT EXHIBITED NO TOXIC PROPERTIES UP TO 100 MG/L AND WAS ONLY SLIGHTLY DEGRADED. SEVIN IS TOXIC AT CONCENTRATIONS ABOVE 0.1 MG/L AND IS NOT ALTERED APPRECIABLY IN ACIDIC MEDIA. MALATHION ALTERS THE MAKEUP OF MIXED CULTURE BUT IS NOT PERSISTENTLY INHIBITORY. (LEDBETTER-TEXAS)

FIELD 05D

ACCESSION NO. W70-02198

QUANTITATIVE CATION REQUIREMENTS OF SEVERAL GREEN AND BLUE-GREEN ALGAE,

WISCONSIN UNIV., MADISON. INST. OF PLANT DEVELOPMENT, AND WISCONSIN UNIV., MADISON. DEPT. OF BOTANY.

G. C. GERLOFF, AND K. A. FISHBECK.

JOURNAL OF PHYCOLOGY, VOL 5, NO 2, P 109-114, 1969. 3 FIG, 3 TAB, 23 REF.

DESCRIPTORS:
*ALGAE, *CHLOROPHYTA, *CYANOPHYTA, *NUTRIENT REQUIREMENTS, ANALYTICAL TECHNIQUES, WATER POLLUTION EFFECTS, PHYTOPLANKTON, PLANKTON, AQUATIC MICROBIOLOGY, CYCLING NUTRIENTS, ESSENTIAL NUTRIENTS, CALCIUM, MAGNESIUM, POTASSIUM, HARVESTING OF ALGAE, LIMNOLOGY, NUTRIENTS, PHYSIOLOGICAL ECOLOGY, CHLORELLA, SCENEDESMUS.

IDENTIFIERS:
*CATIONS, DRAPARNALDIA, DRAPARNALDIA PLUMOSA, STIGEOCLONIUM, MICROCYSTIS, STIGEOCLONIUM TENUE, MICROCYSTIS AERUGINOSA, NOSTOC, NOSTOC MUSCORUM, ABSORPTION, SPECTROMETRY.

ABSTRACT:
AUTHORS EVALUATED REQUIREMENTS OF SIX SPECIES OF GREEN AND BLUE-GREEN ALGAE (CHLORELLA PYRENOIDOSA, SCENEDESMUS QUADRICAUDA, DRAPARNALDIA PLUMOSA, STIGEOCLONIUM TENUE, MICROCYSTIS AERUGINOSA, NOSTOC MUSCORUM) FOR CALCIUM, MAGNESIUM, AND POTASSIUM IN TERMS OF CRITICAL CELL CONCENTRATIONS (CCC) OF THE THREE CATIONS. 'CRITICAL CELL CONCENTRATION' IS DEFINED AS THE MINIMAL CELL CONTENT OF AN ELEMENT WHICH PERMITS MAXIMAL OR NEAR-MAXIMAL GROWTH OF AN ORGANISM. AS COMPARED WITH REQUIREMENTS OF ANGIOSPERM CROP PLANTS, CALCIUM NEEDS OF ALL SIX SPECIES WERE EXTREMELY LOW (CCC 0.06% OR LESS ON OVEN-DRY BASIS); REQUIREMENTS FOR MAGNESIUM WERE EQUAL TO OR SLIGHTLY LESS THAN IN AGIOSPERMS (0.15-0.30%) WITH EXCEPTION OF SCENEDESMUS (0.05%); AND REQUIREMENTS FOR POTASSIUM VARIED GREATLY, FROM CCC LESS THAN AVERAGE VALUES ESTABLISHED FOR ANGIOSPERMS (0.25-0.50%) TO VALUES EQUAL TO OR GREATER THAN MEAN VALUES FOR HIGHER PLANTS (0.80-2.40%). THE RESULTS SUGGEST DIFFERENCES IN PHYSIOLOGY AND FUNCTION OF CATIONS IN THE ALGAE STUDIED AND IN ANGIOSPERMS WHICH SEEM WORTHY OF FURTHER INVESTIGATION. (FITZGERALD-WIS)

FIELD 05C

ACCESSION NO. W70-02245

TRISODIUM NITRILOTRIACETATE AND ALGAE,

ONTARIO WATER RESOURCES COMMISSION, TORONTO. DIV. OF RESEARCH.

A. E. CHRISTIE.

ONTARIO WATER RESOURCES COMMISSION, DIV OF RESEARCH PAPER NO 2023, JUNE 1969. 11 P, 4 TAB, 6 REF.

DESCRIPTORS:
*ALGAE, ALGAL CONTROL, ALGICIDES, AQUATIC ALGAE, AQUATIC MICROBIOLOGY, BIOASSAY, CHLOROPHYTA, CHLORELLA, CYCLING NUTRIENTS, ESSENTIAL NUTRIENTS, INHIBITION, NITROGEN COMPOUNDS, NUTRIENT REQUIREMENTS, NUTRIENTS, POLLUTION IDENTIFICATION, SEWAGE, SEWAGE EFFLUENTS, SEWAGE TREATMENT, WATER POLLUTION CONTROL, WATER POLLUTION EFFECTS, WATER POLLUTION TREATMENT, DETERGENTS, TOXICITY, ACTIVATED SLUDGE, EUTROPHICATION, BIODEGRADATION, AMINO ACIDS.

IDENTIFIERS:
*TRISODIUM NITRILOTRIACETATE, NTA, CHLORELLA PYRENOIDOSA, CARBOXYLIC ACIDS, AMINO CARBOXYLIC ACIDS.

ABSTRACT:
AUTHOR REPORTS RESULTS OF PRELIMINARY TESTS TO DETERMINE THE EFFECT OF THE POTENTIAL DETERGENT BUILDER, TRISODIUM NITRILOTRIACETATE, ON GROWTH OF THE GREEN ALGA, CHLORELLA PYRENOIDOSA (IND 395). INCREASES IN NITROGEN (N) FROM 0.14 TO 14 MILLIGRAMS N/LITER RESULTED IN THREEFOLD INCREASES IN CELL COUNTS FOR N EITHER FROM TRISODIUM NITRILOTRIACETATE OR FROM SODIUM NITRATE. TRISODIUM NITRILOTRIACETATE IN CONCENTRATION OF 275 MILLIGAMS/LITER WAS NOT TOXIC TO CULTURES OF CHLORELLA AT AN INITIAL CONCENTRATION OF 1 MILLION CELLS/MILLILITER. THE FILTRATE OF AN ACTIVATED SLUDGE, CAPABLE OF DEGRADING TRISODIUM NITRILOTRIACETATE, WOULD SUPPORT THE GROWTH OF CHLORELLA. (FITZGERALD-WIS)

FIELD 05C

130

ACCESSION NO. W70-02248

DEMONSTRATION OF THE ANTAGONISTIC ACTION OF LARGE AQUATIC PLANTS ON ALGAE AND
ROTIFERS,

WISCONSIN UNIV., MADISON. DEPT. OF ZOOLOGY.

ARTHUR D. HASLER, AND ELIZABETH JONES.

ECOLOGY, VOL 30, NO 3, P 359-364, JULY 1949. 1 FIG, 4 TAB, 6 REF.

DESCRIPTORS:
*AQUATIC PLANTS, *ALGAE, *ROTIFERS, PLANKTON, BASS, PHYTOPLANKTON,
PONDS, ZOOPLANKTON, MINNOWS, PEDALFERS, SCENEDESMUS, CHLAMYDOMONAS,
TEMPERATURE, CARBON DIOXIDE, HYDROGEN ION CONCENTRATION, BIOCARBONATES,
DISSOLVED OXYGEN, DETRITUS, BACTERIA, WATER POLLUTION EFFECTS,
LIMNOLOGY, CYANOPHYTA.

IDENTIFIERS:
*ANTAGONISTIC ACTION, *AQUATIC MACROPHYTES, ANACHARIS CANADENSIS,
MACROPHYTES, POTAMOGETON FOLIOSUS, EXPERIMENTAL LIMNOLOGY, BRACHIONUS,
SYNCHAETA PECTINATA, POLYARTHRA VULGARIS, KARATELLA COHCLEARIS,
ASPLANCHA, LECANE, BOSMINA, NAVICULA, FRAGILARIA, CYCLOTELLA,
FLAGELLATES, ANKISTRODESMUS, GOMPHENEMA, SPIROGYRA.

ABSTRACT:
PLANTS AND ANIMALS VARY GREATLY IN DENSITY AND SPECIES COMPOSITION AND
ARE CONTROLLED BY EXTREMELY COMPLEX FACTORS. MACROPHYTES MAY HAVE
CERTAIN COMPETITIVE ADVANTAGES OVER ALGAE WHEN THEIR POPULATIONS ATTAIN
SPECIFIC DENSITIES. IN EXPERIMENTS REPORTED HERE, AUTHORS AIMED AT
EVIDENCE CONFIRMING OR CONTRADICTING PREVIOUS CASUAL OBSERVATIONS
BEARING ON SUCH ANTAGONISMS. 'SILO PONDS' ERECTED WITHIN A LARGE
HATCHERY POND, WERE DENSELY PLANTED WITH ANACHARIS CANADENSIS
ASSOCIATED SLIGHTLY WITH POTAMOGETON FOLIOSUS. DURING SUMMER,
PHYTOPLANKTERS AND ZOOPLANKTERS WERE COUNTED WEEKLY. ZOOPLANKTON OF ALL
PONDS WERE SUBJECT TO SIMILAR DEGREES OF PREDATION SINCE EACH CONTAINED
50 FINGERLING LARGEMOUTH BASS. FATHEAD MINNOWS WERE MAINTAINED IN EACH
POND AS FOOD FOR THE BASS. DENSE GROWTHS OF AQUATIC MACROPHYTES IN THE
'SILO PONDS' HAD STATISTICALLY SIGNIFICANT INHIBITING EFFECT UPON
PHYTOPLANKTON AND ROTIFERS, WHEREAS PLANKTONIC CRUSTACEA WERE NOT
AFFECTED. GROWTH OF BASS VARIED SO GREATLY THAT IT PRECLUDED
DETERMINATION OF EFFECT OF AQUATIC PLANTS ON FISH GROWTH. LISTS OF
DOMINANT PLANKTONIC GENERA IN PLANTED AND CONTROL (PLANT-FREE) PONDS
ARE GIVEN. DATA ON TEMPERATURE, FREE CARBON DIOXIDE, PH, BICARBONATE
ION, AND DISSOLVED OXYGEN OF EXPERIMENTAL AND CONTROL PONDS ARE
REPORTED. (JONES-WIS)

FIELD 05C

ACCESSION NO. W70-02249

THE OVERALL PICTURE OF EUTROPHICATION,

TEXAS UNIV., AUSTIN. ENVIRONMENTAL HEALTH ENGINEERING RESEARCH LAB.

E. GUS FRUH.

JOURNAL WATER POLLUTION CONTROL FEDERATION, VOL 39, NO 9, P 1449-1463, 1967.
1 FIG, 7 TAB, 66 REF.

DESCRIPTORS:
*EUTROPHICATION, LAKES, ORGANIC WASTES, OXYGEN, STANDING CROP, ALGAE,
ZOOPLANKTON, FISH, BACTERIA, BENTHIC FAUNA, PHOTOSYNTHESIS, BIOASSAYS,
AQUATIC WEEDS, THERMAL STRATIFICATION, EPILIMNION, HYPOLIMNION, CYCLING
NUTRIENTS, TASTE, ODOR, COLOR, TURBIDITY, HYDROGEN SULFIDE, IRON,
MAGNESIUM, COOLING, PROPERTY VALUES, RECREATION DEMAND, IRRIGATION,
TRACE ELEMENTS, NITROGEN FIXATION, ATMOSPHERE, GROUNDWATER,
WATERSHEDS(BASINS), RUNOFF, PHOSPHORUS, NITROGEN, AGRICULTURAL
WATERSHEDS, FORESTS, WASTE WATER(POLLUTION), LAND MANAGEMENT, COPPER
SULFATE, SODIUM ARSENITE, HARVESTING, WATER POLLUTION, VITAMINS.

IDENTIFIERS:
SHORELINE VEGETATION, STORMWATER, DRAINAGE.

ABSTRACT:
NITROGEN AND PHOSPHORUS, IMPORTANT IN PLANT GROWTH, ARE SUBJECT TO
CONTROL AND MODIFICATION. THE IMMEDIATE QUESTION WHICH CONFRONTS WATER
QUALITY INVESTIGATORS IS WHETHER NITROGEN AND PHOSPHORUS ARE REALLY THE
LIMITING FACTORS IN PARTICULAR AQUATIC ENVIRONMENTS WHICH THEY ARE
INVESTIGATING. SUCH DETERMINATIONS CAN BE ACCOMPLISHED BY VARIOUS
BIOASSAY TECHNIQUES. IN LABORATORY CULTURES OF ALGAE, TRACE ELEMENTS,
PREVIOUSLY OVERLOOKED, ARE ESSENTIAL. LIKEWISE, DIFFERENT SOILS HAVE
BEEN ADDED TO CULTURE MEDIA TO SUPPLY VARIOUS NECESSARY ORGANICS AND
GROWTH FACTORS. TODAY, KNOWN VITAMINS AND ORGANIC COMPOUNDS ARE USED.
WATER QUALITY INVESTIGATORS SHOULD DETERMINE WHETHER SUCH
MICRONUTRIENTS ARE IN SUFFICIENT SUPPLY IN THE HYDROSPHERE UNDER
VARIOUS ECOLOGICAL CONDITIONS. KNOWLEDGE OF SUCH FACTS PERMITS A
LOGICAL APPROACH FOR IDENTIFICATION OF PRIMARY NUTRIENT SOURCES. SOME
NUTRIENT SOURCES, INCLUDING RAINFALL, GROUNDWATER, AND NORMAL WATERSHED
DRAINAGE, CANNOT BE CONTROLLED. IF WASTEWATER EFFLUENTS CAUSE
EUTROPHICATION, DIVERSION OR TERTIARY TREATMENT CAN BE APPLIED. IF
AGRICULTURAL AND/OR URBAN DRAINAGE ARE PRIMARY NUTRIENT SOURCES, LITTLE
CAN, FOR ALL PRACTICAL PURPOSES, BE ACCOMPLISHED. POSSIBLE ECONOMIC
APPROACHES UNDER INVESTIGATION IN LAKES ARE SELECTIVE WITHDRAWAL OF
HYPOLIMNETIC WATERS, DILUTION, AERATION OF THE HYPOLIMNION, AND MIXING
OF HYPOLIMNETIC WATERS WITH THE EPILIMNION. (JONES-WIS)

FIELD 05C

ACCESSION NO. W70-02251

ALGAL GROWTH AND SEWAGE EFFLUENT IN THE POTOMAC ESTUARY,

JOHNS HOPKINS UNIV., BALTIMORE, MD. DEPT. OF SANITARY ENGINEERING AND WATER RESOURCES.

JOSEPH SHAPIRO, AND ROBERTO RIBEIRO.

JOURNAL WATER POLLUTION CONTROL FEDERATION, VOL 37, NO 7, P 1034-1042, JULY 1965. 6 FIG, 2 TAB, 7 REF.

DESCRIPTORS:
*ALGAE, *SEWAGE EFFLUENTS, *ESTUARIES, FRESH WATER, LAKES, SILTS, PHOSPHATES, NITROGEN, WASTE WATER(POLLUTION), SCENEDESMUS, CYANOPHYTA, CHLOROPHYTA, CHLOROPHYLL, BACTERIA, TRACE ELEMENTS, VITAMINS, AMINO ACIDS, CALCIUM, AMMONIA, NITROGEN FIXATION, IRON COMPOUNDS, LIGHT PENETRATION, WATER POLLUTION EFFECTS, EUTROPHICATION.

IDENTIFIERS:
*ALGAL GROWTH, *POTOMAC RIVER, PLECTONEMA, NOSTOC PALUDOSUM, BLEACHING, WASHINGTON(DC).

ABSTRACT:
ADDITION OF EFFLUENTS FROM SECONDARY WASTEWATER TREATMENT PLANTS TO POTOMAC RIVER GREATLY INCREASES GROWTH OF BOTH GREEN (CHLOROPHYTES) AND BLUE-GREEN ALGAE (CYANOPHYTES) IN PROPORTION TO THE QUANTITY OF EFFLUENT ADDED. AS LITTLE AS 5% EFFLUENT IS EFFECTIVE, AND 40% IS NOT SUPRAOPTIMAL. SINCE CYANOPHYTES ARE ABLE TO PROVIDE THEIR OWN NITROGEN SUPPLY THROUGH FIXATION OF MOLECULAR NITROGEN, PHOSPHATE IS THE SOLE NUTRIENT RESPONSIBLE FOR LIMITING THEIR GROWTH. CHLOROPHYTES REQUIRE BOTH PHOSPHATE AND AMMONIUM-NITROGEN FOR GROWTH. REMOVAL OF AMMONIA FROM EFFLUENTS WILL CONTROL CHLOROPHYTES ONLY, BUT PHOSPHATE REMOVAL WILL LIMIT CYANOPHYTES AS WELL. PARTIAL REMOVAL OF PHOSPHORUS FROM EFFLUENTS WILL CONTROL BOTH ALGAL GROUPS TO A DEGREE COMMENSURATE WITH EXTENT OF REMOVAL. NORMAL PHOSPHORUS CONCENTRATIONS OF THE POTOMAC RIVER ARE SUCH THAT INCREASES IN THESE CONCENTRATIONS WILL STIMULATE FURTHER ALGAL GROWTH. THE HYPOTHESIS THAT TRACE SUBSTANCES ARE NOT LIMITING WAS VERIFIED. SILT, AFFECTING LIGHT PENETRATION AND ADSORPTION OF PHOSPHATES OR AMMONIA ONTO SILT PARTICLES MAY BOTH LEAD TO REDUCED ALGAL GROWTHS. SOLUTION OF THE SILT PROBLEM, THROUGH BUILDING OF DAMS, MUST BE TIED TO PROGRAMS FOR NUTRIENT CONTROL. (JONES-WIS)

FIELD 02L, 05C

ACCESSION NO. W70-02255

COOLING WATER CHEMICAL TREATMENTS,

BETZ LABS., INC., TREVOSE, PA. COOLING WATER SERVICES.

JOHN M. DONOHUE.

IND WATER ENG (IWE) 1ST ANNU AIR/WATER ENG BUYERS GUIDE, P 70, 1969.

DESCRIPTORS:
*WATER TREATMENT, *COOLING WATER, *CORROSION, *CHEMICAL REACTIONS, *WATER REUSE, DISSOLVED OXYGEN, PITTING(CORROSION), RUSTING, METALS, INHIBITORS, SILTS, ORGANIC MATTER, IRON OXIDES, BACTERIA, ALGAE, FUNGI, SULFUR COMPOUNDS, CARBONATES.

IDENTIFIERS:
*METAL CORROSION, COOLING WATER TREATMENT, ON-STREAM DESLUDGING.

ABSTRACT:
CORROSION OF METALS IS PRIMARILY CAUSED BY OXYGEN THAT IS DISSOLVED IN WATER AND THEN RELEASED AS THE COOLING WATER INCREASES IN TEMPERATURE. THE MOST COMMON TYPES OF METAL ATTACK ARE PITTING OF THE METAL SURFACE, GENERAL WASTING AWAY CAUSED BY LOW WATER PH, AND GOUGING OF METAL SURFACES. CHROMATE-BASED TREATMENTS ARE THE MOST EFFECTIVE AND WIDELY USED OF THE INHIBITORS FOR CONTROLLING WATERSIDE ATTACK. COOLING WATER ANTIFOULANTS ARE USED TO PREVENT DEPOSITION OF SILT, IRON OXIDE, ORGANIC MATTER, AND GENERAL COOLING-WATER DEBRIS. THE TECHNIQUES OF ON-STREAM DESLUDGING OF COOLING SYSTEMS WITH SYNTHETIC POLYMERS HAVE BEEN SUCCESSFULLY USED IN COOLING OPERATIONS. BACTERIA, FUNGI, AND ALGAE CAN INTERFERE WITH WATER FLOW, DECREASE HEAT TRANSFER, AND ACCELERATE CORROSION. BIOCIDES DEVELOPED AND USED IN COOLING SYSTEMS INCLUDE ORGANIC SULFUR COMPOUNDS, ORGANO-METALLICS, CARBONATES, DIAMINES, COPPER, CHLORINE, HYPOCHLORITES, ORGANIC CHLORINE DONORS AND BROMINE COMPOUNDS. (CARSTEA-USGS)

FIELD 05D, 08G

ACCESSION NO. W70-02294

FURTHER PLANKTONIC ALGAE OF AUCKLAND SEWAGE TREATMENT PONDS AND OTHER WATERS,

BOLIVAR LABS., SALISBURY (AUSTRALIA).

A. HAUGHEY.

NEW ZEAL J MARINE AND FRESHWATER RES, VOL 3, NO 2, P 245-261, JUNE 1969. 17
P, 6 FIG, 13 REF.

DESCRIPTORS:
*ALGAE, *PLANKTON, *SEWAGE, *SEWAGE TREATMENT, *EUTROPHICATION, PONDS,
POLLUTANT IDENTIFICATION, BIOLOGY, WATER POLLUTION, WATER POLLUTION
SOURCES, ALGAL CONTROL, BIOINDICATORS.

IDENTIFIERS:
*NEW ZEALAND, SEWAGE TREATMENT POND ALGAE.

ABSTRACT:
SEVENTEEN SPECIES OF ALGAE, NEWLY RECORDED BETWEEN 1965 AND 1968 FROM
SEWAGE TREATMENT PONDS AROUND AUCKLAND CITY, ARE DESCRIBED TOGETHER
WITH ANOTHER 11 SPECIES FOUND IN EUTROPHIC WATERS OF THE AREA. THESE
LATTER NATURAL WATERS ARE OFTEN OF A TRANSIENT NATURE AND USUALLY
HEAVILY POLLUTED IN SOME WAY. (GABRIEL-USGS)

FIELD 05C

ACCESSION NO. W70-02304

EFFECT OF HALOGENS ON ALGAE - I. CHLORELLA SOROKINIANA,

TECHNION - ISRAEL INST. OF TECH., HAIFA. SANITARY ENGINEERING LAB.

YEHUDA KOTT, AND J. EDLIS.

SPONSORED BY MEKOROTH WATER CO. WATER RESEARCH, VOL 3, P 251-256, 1969. 4
FIG, 1 TAB, 19 REF.

DESCRIPTORS:
*HALOGENS, *ALGICIDES, CHLORINE, ALGAL CONTROL, CHLORELLA,
DISINFECTION, MORTALITY.

IDENTIFIERS:
*KILL RATES, *BROMINE, IODINE, PHOTO REACTIVITY, ALGAL METABOLISM.

ABSTRACT:
BROMINE WITHIN THE CONCENTRATION RANGE OF 0.2 TO 1.0 MG/L CAUSED A
DECREASE IN CHLORELLA NUMBER AFTER 2 - 5 HR CONTACT, AFTER WHICH TIME
REGROWTH OCCURRED. A SECOND DOSE OF BROMINE AT THE TIME OF REGROWTH
BROUGHT ABOUT TOTAL ALGAL KILL. USING CHLORINE AT THE SAME
CONCENTRATION HAD AN ALGISTATIC EFFECT, WHILE A SECOND INTRODUCTION OF
CHLORINE HAD NO ADDITIONAL EFFECT AT ALL. USING A MIXTURE OF CHLORINE
AND BROMINE SOLUTIONS CAUSED RAPID ALGAL DEATH. RESULTS OF HALOGENATION
IN THE DARK DEMONSTRATED THE SUPERIORITY OF CHLORINE UNDER THESE
CONDITIONS. IT IS ASSUMED THAT PHOTOREACTIVITY OF BROMINE MIGHT BE THE
REASON FOR ITS BETTER ALGICIDAL EFFECTS IN THE LIGHT.
(SJOLSETH-WASHINGTON)

FIELD 05C

ACCESSION NO. W70-02363

EFFECT OF HALOGENS ON ALGAE - II. CLADOPHORA SP.,

TECHNION - ISRAEL INST. OF TECH., HAIFA. SANITARY ENGINEERING LAB.

NACHUM BETZER, AND YEHUDA KOTT.

SPONSORED BY MEKOROTH WATER CO. WATER RESEARCH, VOL 3, P 257-264, 1969. 7 FIG, 5 TAB, 8 REF.

DESCRIPTORS:
*HALOGENS, ALGICIDES, CYTOLOGICAL STUDIES, CHLORINE, COPPER SULFATE, ALGAL CONTROL, MORTALITY, CHLOROPHYLL, ALGAE, BIOASSAY, VIABILITY.

IDENTIFIERS:
*CLADOPHORA, BROMINE, FILAMENTOUS ALGAE, KILL RATES, PHOTO REACTIVITY, CELL COLOR, CELL DAMAGE.

ABSTRACT:
HALOGENATION OF NATURAL TUFTS GAVE NO FRUITFUL RESULTS. USING ISOLATED FILAMENTS OF CLADOPHORA SP. SHOWED THAT HALOGENS CAUSE CHANGES IN CELL COLOR FROM GREEN TO YELLOW OR EVEN TO COLORLESS, VERY OFTEN DAMAGE TO CELL WALL. LEACHING OF CELL CONTENT ALSO OCCURRED, OR CONTRACTION OF CYTOPLASM WAS OBSERVED. USING DARK FIELD TECHNIQUE IT WAS OBSERVED THAT THE NORMAL RED FLUORESCENCE SEEN IN HEALTHY CELLS DISAPPEARED. VERY HIGH CONCENTRATIONS (50 MG/L AND OVER) CAUSED SEVERE DEFORMATION OF CELLS. IN THE ABSENCE OF A RECOMMENDED PROCEDURE FOR MEASURING VIABILITY OF FILAMENTOUS ALGAE, ONE HUNDRED CELLS WERE COUNTED IN EACH FILAMENT AND A BIOASSAY WAS PERFORMED AT THE END OF EACH EXPERIMENT. ADDITION OF 5 MG/L OF HALOGENS HAS CAUSED SOME EFFECT ON THE ALGAE, WHILE ADDITION OF COPPER SULPHATE AT THIS CONCENTRATION SHOWED NO EFFECT AT ALL. WITH HIGHER CONCENTRATIONS OF HALOGENS OR COPPER SULPHATE, CELL DAMAGE WAS CAUSED. TEN MG/L CHLORINE FOR 2 HR, 10 MG/L BROMINE FOR 10 HR OR 10 MG/L COPPER SULPHATE FOR 4 DAYS OF CONTACT CAUSED COMPLETE ALGAL KILL. (SJOLSETH-WASHINGTON)

FIELD 05C

ACCESSION NO. W70-02364

ALGICIDAL EFFECT OF BROMINE AND CHLORINE ON CHLORELLA PYRENOIDOSA,

TECHNION - ISRAEL INST. OF TECH., HAIFA. SANITARY ENGINEERING LAB.

YEHUDA KOTT, GALILA HERSHKOVITZ, A. SHEMTOB, AND J. B. SLESS.

APPLIED MICROBIOLOGY, 14 (1), P 8-11, 1966. 4 TAB, 3 FIG, 16 REF.

DESCRIPTORS:
*BIOASSAY, *HALOGENS, *GROWTH RATES, *CHLORELLA, CHLORINE, ALGICIDES, ALGAL CONTROL, DISINFECTION, MODE OF ACTION, MORTALITY, INHIBITION, ALGAE.

IDENTIFIERS:
BROMINE, BROMAMINE, CHLORAMINE.

ABSTRACT:
CHLORELLA PYRENOIDOSA WAS FOUND TO GROW RAPIDLY IN TAP WATER. PEAK GROWTH WAS REACHED AFTER 2 TO 3 DAYS. CHLORINE AND BROMINE, ADDED TO SUCH WATER IN CONCENTRATIONS OF .18 TO .42 PPM, WERE SHOWN TO BE EFFECTIVE INHIBITORS OF ALGAL GROWTH. BROMINE AND BROMAMINE WERE PRIMARILY ALGICIDAL, WHEREAS CHLORINE AND CHLORAMINES WERE MAINLY ALGISTATIC. IT IS ASSUMED THAT THE MECHANISMS OF ACTION OF THESE HALOGENS ON CHLORELLA ARE NOT THE SAME. THE NEED TO INVESTIGATE THE MODE OF ACTION OF THESE HALOGENS IS STRESSED. (SJOLSETH-WASHINGTON)

FIELD 05C

ACCESSION NO. W70-02370

CHLORINATION AS A MEANS OF CONTROLLING EARTHY ODORS IN RESERVOIRS,

TECHNION - ISRAEL INST. OF TECH., HAIFA. SANITARY ENGINEERING LAB.

M. REBHUN, M. A. FOX, AND J. B. SLESS.

THE JERUSALEM CONFERENCE FOR WATER QUALITY RESEARCH, P 1-8, 1969. 5 FIG, 10
REF.

DESCRIPTORS:
*ODOR-PRODUCING ALGAE, *TASTE-PRODUCING ALGAE, ALGAE, CHLORINATION,
WATER QUALITY CONTROL.

IDENTIFIERS:
*OSCILLATORIA, NATIONAL WATER CARRIER(ISRAEL), THRESHOLD ODOR NUMBERS.

ABSTRACT:
SEASONAL APPEARANCE OF BLUE-GREEN ALGAE (OSCILLATORIA SPP) PRODUCES
EARTHY ODORS AND TASTES IN THE WATER OF THE TSALMON AND BETH NATUFA
RESERVOIRS OF THE NATIONAL WATER CARRIER. FIELD TRIALS, IN WHICH WATER
FROM THESE RESERVOIRS WAS CHLORINATED, SHOWED THAT CHLORINATION WAS NOT
COMPLETELY EFFECTIVE AS A MEANS OF CONTROLLING THE EARTHY ODOR,
ALTHOUGH A REDUCTION OF THE ODOR LEVEL WAS ALWAYS OBSERVED, WHEN THE
EARTHY ODOR WAS THE SOLE ODOR PRESENT IN THE WATER. A MODEL SYSTEM
EMPLOYING A SPECIES OF OSCILLATORIA, WAS USED TO PROVIDE SAMPLES OF
WATER HAVING DIFFERENT LEVELS OF THRESHOLD ODOR, AND THE EFFECT OF
CHLORINATION INVESTIGATED. IT WAS FOUND THAT CHLORINATION REDUCED THE
INITIAL ODOR TO AN ASYMPTOTIC VALUE WHICH PROVED TO BE DIRECTLY
PROPORTIONAL TO THE INITIAL ODOR LEVEL. THEY CONCLUDED THAT COMPLETE
CONTROL OF THE EARTHY ODOR PRODUCED BY OSCILLATORIA IS NOT FEASIBLE BY
USE OF CHLORINE ALONE, BUT THAT ODOR LEVELS MAY BE REDUCED BY
CHLORINATION TO AN EXTENT WHERE OTHER METHODS (SUCH AS DILUTION) ARE
MORE READILY APPLICABLE. (SJOLSETH-WASHINGTON)

FIELD 05G, 05C

ACCESSION NO. W70-02373

EXTRACELLULAR PRODUCTS OF PHYTOPLANKTON PHOTOSYNTHESIS,

WESTFIELD COLL., LONDON (ENGLAND). DEPT OF BOTANY; AND UNIVERSITY COLL.,
LONDON. DEPT. OF BOTANY.

G. E. FOGG, CZESLAWA NALEWAJKO, AND W. D. WATT.

PROCEEDINGS ROYAL SOCIETY (B), VOL 162, P 517-534, 1965. 6 FIG, 4 TAB, 31 REF.

DESCRIPTORS:
*PHYTOPLANKTON, *PHOTOSYNTHESIS, BICARBONATES, RADIOACTIVITY, CARBON
RADIOISOTOPES, LAKES, ORGANIC MATTER, SEA WATER, PRIMARY PRODUCTIVITY,
LIGHT INTENSITY, TROPHIC LEVEL, SAMPLING TEMPERATURE, TRACERS,
CHLORELLA, HYDROGEN ION CONCENTRATION, ALKALINITY, DIATOMS,
TEMPERATURE, WEATHER, EPILIMNION, ALGAE, DENSITY, SEASONAL, MARINE
PLANTS, WATER POLLUTION SOURCES.

IDENTIFIERS:
*EXTRACELLULAR PRODUCTS, PHOTIC ZONE, GLYCOLLIC ACID, CHEMOTROPHY,
PARTICULATE MATTER, BLELHAM TARN(ENGLAND), CHLORELLA PYRENOIDOSA,
FIXATION, TRINGFORD(ENGLAND), ANABAENA CYLINDRICA, NORTH SEA, FIRTH OF
CLYDE(SCOTLAND), SCOTLAND, ENGLAND, LAKE WINDERMERE(ENGLAND), CARBON-14.

ABSTRACT:
FOLLOWING EXPOSURE IN SITU (3-24 HOURS) OF LAKE OR SEA WATER SAMPLES,
WITH ADDED RADIOACTIVE BICARBONATE, RADIOCARBON WAS FOUND IN
PHYTOPLANKTERS AND IN DISSOLVED ORGANIC MATTER. AMOUNTS IN THE WATER
VARIED BETWEEN 7 AND 50% OF TOTAL CARBON FIXED IN PHOTIC ZONE OF THE
WATER COLUMN. THIS PRODUCTION OF LABELLED EXTRACELLULAR ORGANIC MATTER
OCCURRED UNDER VARIOUS CONDITIONS AND WITH DIVERSE PHYTOPLANKTERS.
ESTIMATION OF PRIMARY PRODUCTIVITY FROM FIXATION OF CARBON-14 IN
PARTICULATES ONLY ARE PROBABLY ERRONEOUSLY LOW. LABELLED ORGANIC
SUBSTANCES IN WATER ARE APPARENTLY LIBERATED BY INTACT
PHOTOSYNTHESIZING CELLS RATHER THAN FROM CELLS BROKEN DURING
FILTRATION. GLYCOLLIC ACID IS POSSIBLY A PRINCIPAL SUBSTANCE. OVER A
WIDE RANGE OF LIGHT INTENSITIES, LIBERATION OF EXTRACELLULAR PRODUCTS
BY A GIVEN PHYTOPLANKTON POPULATION WAS PROPORTIONAL TO THE CARBON
FIXED IN CELLS, TENDING TO BE RELATIVELY GREATER AT LOW LIGHT
INTENSITIES AND AT LIGHT INTENSITIES HIGH ENOUGH TO INHIBIT
PHOTOSYNTHESIS, WHEN AS MUCH AS 95% OF TOTAL ORGANIC RADIOCARBON MIGHT
BE EXTRACELLULAR. POPULATION DENSITY, PERIOD OF EXPOSURE TO LABELLED
BICARBONATE, AND SPECIES DIFFERENCES INFLUENCE SUCH EXCRETION.
EXTENSIVE EXTRACELLULAR LIBERATION, BY PHYTOPLANKTON, OF PHOTOSYNTHETIC
PRODUCTS HAS IMPORTANT IMPLICATIONS FOR STUDIES OF AQUATIC TROPHIC
RELATIONSHIPS. (JONES-WISCONSIN)

FIELD 05B

ACCESSION NO. W70-02504

INTERRELATIONS OF DISSOLVED ORGANIC MATTER AND PHYTOPLANKTON,

MICHIGAN UNIV., ANN ARBOR. DEPT OF ZOOLOGY.

GEORGE W. SAUNDERS.

THE BOTANICAL REVIEW, VOL 23, NO 6, P 389-409, 1957. 4 TAB, 131 REF.

DESCRIPTORS:
*ORGANIC MATTER, *PHYTOPLANKTON, LIMNOLOGY, ZOOPLANKTON, ALGAE,
COLLOIDS, TRACE ELEMENTS, DISSOLVED SOLIDS, NUTRIENTS, VITAMINS,
TOXINS, INHIBITORS, SUCCESSION, CHLORELLA, CHELATION, OCHROMONAS,
BACTERIA, FUNGI, AQUATIC PLANTS, ASSAY, DISTRIBUTION, DENSITY,
CHROMATOGRAPHY, RADIOACTIVITY, TRACERS, WATER POLLUTION EFFECTS,
PHYSIOLOGICAL ECOLOGY, ALGICIDES.

IDENTIFIERS:
*DISSOLVED ORGANIC MATTER, ORGANIC DEBRIS, ORGANIC GROWTH FACTORS,
VITAMIN B1, VITAMIN B6, VITAMIN B12, BIOTIN, HISTIDINE, URACIL, REDUCED
SULFUR COMPOUNDS, ANTIBIOTIC EFFECTS, STIMULANTS, GONYAULAX CATENELLA,
GONYAULAX TAMERENSIS, GYMNODINIUM BREVIS, MYCROSYSTIS, ALGASTATIC
SUBSTANCES, ALGADYNAMIC SUBSTANCES, PHORMIDINE, PANDORININE,
SCENEDESMINE, MICROCRUSTACEANS, CYTOTOXIC AGENTS, AUXOTROPHY,
AUTOTROPHY, HETERETROPHY, NUTRITION, PHYSIOLOGY.

ABSTRACT:
LIMNOLOGISTS ATTEMPT TO EXPLAIN DISTRIBUTION AND INTERRELATIONSHIPS OF
FRESHWATER ORGANISMS FROM THE MORE CONVENTIONAL PHYSICAL AND CHEMICAL
POINTS OF VIEW. THE BIOCHEMIST AND PHYSIOLOGIST ARE BEGINNING TO
UNDERSTAND BASIC METABOLIC PATHWAYS OF SYNTHESIS AND DEGRADATION.
MICRO-QUANTITIES OF BOTH ORGANIC AND INORGANIC MATTER ARE SIGNIFICANT
TO GROWTH, BEHAVIOR, AND SUCCESS OF LIVING ORGANISMS. WITH THE NEW
INFORMATION FROM BIOCHEMISTRY AND NEW TOOLS AND TECHNIQUES, SUCH AS
PAPER CHROMATOGRAPHY, CHELATE COMPLEXES, RADIOACTIVE TRACERS, AND
MICROBIOLOGICAL ASSAY, LIMNOLOGISTS MAY COMPLETE THE UNIFIED PICTURE.
THE ATTACK MAY BE INITIATED BY USE OF PURE CULTURES IN THE LABORATORY
TO DETERMINE THE BASIC REQUIREMENTS OF THE MORE SIGNIFICANT ALGAL FORMS
IN TERMS OF THEIR ABILITY TO UTILIZE VARIOUS ORGANIC SUBSTRATES,
REQUIREMENT FOR GROWTH FACTORS, REACTION TOWARD ANTIBIOTICS, AND THEIR
REQUIREMENT FOR TRACE METALS. LIMNOLOGISTS IN THE FIELD MAY DETERMINE
YEARLY CYCLES OF SPECIFIC ORGANIC NUTRIENTS, VITAMINS, AND TRACE
METALS. MANY QUESTIONS MAY BE ANSWERED BY CORRELATING LABORATORY AND
FIELD STUDIES WITH ANALYSES OF POPULATIONS AND THEIR SUCCESSION IN
NATURE. THESE TOOLS SHOULD BE CONSIDERED AND USED TO USHER IN A NEW ERA
IN LIMNOLOGY AND THE INVESTIGATION OF INTERRELATIONS OF DISSOLVED
ORGANIC MATTER, PHYTOPLANKTON, AND ZOOPLANKTON. (JONES-WISCONSIN)

FIELD 05C, 02H

ACCESSION NO. W70-02510

DESIGN PRINCIPLES OF WASTE STABILIZATION PONDS,

NOVA SCOTIA TECHNICAL COLL., HALIFAX.

D. THIRUMURTHI.

J SANIT ENG DIV, PROC AMER SOC CIVIL ENG, VOL 95, NO SA2, PROC PAPER 6515, P
311-330, APR 1969. 8 FIG, 6 TAB, 22 REF.

DESCRIPTORS:
*ALGAE, *DISPERSIONS, *ENVIRONMENTAL ENGINEERING, *SANITARY
ENGINEERING, *SEWAGE TREATMENT, *TRACERS, *WATER POLLUTION, DESIGN,
DESIGN CRITERIA, OXIDATION LAGOONS PHOTOSYNTHESIS.

IDENTIFIERS:
*MODEL TESTS, DYES.

ABSTRACT:
WASTE STABILIZATION PONDS WILL BE MORE EFFECTIVELY EMPLOYED AS
POLLUTION CONTROL SYSTEMS IF SUITABLE DESIGN FORMULA AND CORRESPONDING
DESIGN PARAMETERS ARE MADE AVAILABLE TO DESIGN ENGINEERS. TO ACCOMPLISH
THIS OBJECTIVE, AN ANALOGY HAS BEEN DRAWN BETWEEN A FIRST-ORDER
CHEMICAL REACTOR AND A WASTE STABILIZATION POND. THIS ANALOGY HAS
RESULTED IN THE ADAPTATION OF A CHEMICAL ENGINEERING REACTOR DESIGN
EQUATION FOR DESIGNING STABILIZATION PONDS. TWO DESIGN PARAMETERS,
FIRST-ORDER BOD REMOVAL COEFFICIENT AND DISPERSION INDEX, WERE
CONSIDERED AS THE BASIS OF THE NEW DESIGN APPROACH. SYNTHETIC SEWAGE
WAS USED TO DETERMINE THE BOD REMOVAL KINETICS. A CONTROL POND, WITH
TAP WATER AS THE MEDIUM AND SODIUM CHLORIDE AS A TRACER, WAS USED TO
CALCULATE THE DISPERSION INDEX. THE INTERPRETATION OF THE RESULTS
DEMONSTRATE THE SUCCESSFUL APPLICABILITY OF THE PROPOSED DESIGN
FORMULA. AN EXAMPLE DESIGN PROBLEM HAS BEEN WORKED OUT.
(LEDBETTER-TEXAS)

FIELD 05D

ACCESSION NO. W70-02609

LIMITATION OF ALGAL GROWTH IN SOME CENTRAL AFRICAN WATERS,

MALAWI UNIV., LIMBE.

BRIAN MOSS.

LIMNOLOGY AND OCEANOGRAPHY, VOL 14, NO 4, P 591-601, JULY 1969. 11 P, 3 FIG,
4 TAB, 20 REF.

DESCRIPTORS:
*ALGAE, *PRODUCTIVITY, *EUTROPHICATION, *WATER CHEMISTRY, POLLUTANTS,
ALGAL CONTROL, NITRATES, PHOSPHATES, SULFATES, SEASONAL, LAKES,
SAMPLING, PONDS, RIVERS, RESERVOIRS, AIR TEMPERATURE, RAIN, GEOLOGY,
MAPPING, NUTRIENTS.

IDENTIFIERS:
*AFRICA, MALAWI.

ABSTRACT:
LEVELS OF INORGANIC NUTRIENTS (NITRATE, PHOSPHATE, SULFATE) AVAILABLE
FOR ALGAL GROWTH IN NINE BODIES OF WATER IN MALAWI WERE INVESTIGATED,
USING ENRICHMENT BIOASSAYS WITH NATURALLY PRESENT ALGAE AS TEST
ORGANISMS. NITRATE OR NITRATE PLUS PHOSPHATE WERE POTENTIALLY LIMITING
TO ALGAL GROWTH. THERE WAS NO EVIDENCE OF SULFATE LIMITATION. THE
RELEVANCE OF ENRICHMENT BIOASSAYS TO WHOLE WATER BODY SITUATIONS IS
CONSIDERED. A MODEL, ILLUSTRATING INVALID USE OF ASSAYS IN THE
INTERPRETATION OF SEASONAL SUCCESSION IN ALGAL COMMUNITIES, IS
DISCUSSED. (GABRIEL-USGS)

FIELD 05C, 05A

ACCESSION NO. W70-02646

BUOYANCY AND SINKING CHARACTERISTICS OF FRESHWATER PHYTOPLANKTON,

RHODE ISLAND UNIV., KINGSTON.

THEODORE J. SMAYDA.

TECHNICAL COMPLETION REPORT, RHODE ISLAND WATER RESOURCES CENTER, 1969. 7 P.
OWRR PROJECT B-006-RI.

DESCRIPTORS:
*FLOTATION, *SINKING, *PHYTOPLANKTON, *DIATOMS, *BLUE-GREEN ALGA,
COPPER SULFATE.

ABSTRACT:
THE SINKING RATES OF SELECTED FRESHWATER PHYTOPLANKTONS WERE
DETERMINED. THE SPECIES WERE THE DIATOMS ASTERIONALLA FORMOSA,
TABELLARIA FLOCCULOSA, AND THE BLUE-GREEN ALGA APHANIZOMANON
FLOS-AQUAE. THE SINKING RATES OF SOME BRACKISH-COASTAL WATER DIATOMS
WHICH HAVE MORPHOLOGICAL HOMOLOGUES IN FRESHWATER WERE ALSO
INVESTIGATED TO EXAMINE THE INFLUENCE OF MORPHOLOGY ON SINKING RATE.
SINKING RATES OF THE FRESHWATER FORMS USED WERE FOUND TO APPROXIMATE
THOSE (APPROXIMATELY 0.2 TO 1 METERS DAY-1) FOR MARINE PHYTOPLANKTERS.
INCREASED CULTURE AGE AND COLONY (CHAIN) SIZE FAVORED INCREASED SINKING
RATES. SINKING COULD BE HEIGHTENED BY THE ADDITION OF CUSO4, FORMALIN
AND KI. KILLING BY HEAT ALSO HEIGHTENED SINKING. THE ADDITION OF 0.1%
BUTANOL AND ISOPROPANOL ALSO HEIGHTENED SINKING OF THE DIATOMS USED,
BUT ONLY THE BUTANOL TREATED BLUE-GREEN ALGA CELLS RESPONDED THUSLY.
THESE ALGAE REMAINED VIABLE DESPITE THE ALCOHOL TREATMENT. THEIR
HEIGHTENED SINKING RATE (AND, CONVERSELY, THEREFORE ONE OF THEIR
SUSPENSION AIDES) APPEARS TO RELATED IN TOTO (OR PRIMARILY) TO A CHANGE
IN TOTAL CELL DENSITY DUE TO OSMOTIC DISRUPTIONS AND/OR DISTURBANCE OF
THE CELL MEMBRANE ELECTROKINETICS WHICH CAN INFLUENCE THE HYDROFUGE OR
HYDROPHILIC PROPERTIES OF THE CELL. THE RESULTS WITH CUCO4 ADDITION
SUGGEST THAT IN ADDITION TO CONTROL OF NOXIOUS WATER BLOOMS THROUGH
GROWTH CESSATION BY THIS AGENT, SINKING OF CERTAIN PHYTOPLANKTERS MAY
ALSO BE PROMOTED LEADING TO SETTLING OUT OF THE TROPHOGENIC ZONE INTO
THE APHOTIC LAYERS WHERE REMINERALIZATION OCCURS.

FIELD 05G

ACCESSION NO. W70-02754

PHYTOPLANKTON FLORA OF NEWFOUND AND WINNISQUAM LAKES, NEW HAMPSHIRE,

NEW HAMPSHIRE UNIV., DURHAM. DEPT. OF BOTANY.

G. K. GRUENDLING, AND A. C. MATHIESON.

RHODORA, VOL 71 (787): P 444-447, 1969. 42 FIG, 117 REF. OWRR PROJECT
A-010-NH.

DESCRIPTORS:
*ALGAE, *TROPHIC LEVELS, AQUATIC ENVIRONMENT, WATER QUALITY, NUTRIENTS,
NORTHEAST U S, ECOLOGICAL DISTRIBUTION, BIOLOGICAL INDICATORS, NEW
HAMPSHIRE.

IDENTIFIERS:
*NEWFOUND LAKE, *WINNISQUAM LAKE, WATER AND PLANTS, WATER POLLUTION
EFFECTS, ECOLOGY, WATER QUALITY CHARACTERIZATION.

ABSTRACT:
A TOTAL OF 185 TAXA OF FRESH WATER ALGAE (PRIMARILY PHYTOPLANKTON) WERE
IDENTIFIED AT WINNISQUAM (PARTIALLY EUTROPHIC) AND NEWFOUND
(OLIGOTROPHIC) LAKES. OF THE 142 TAXA RECORDED AT WINNISQUAM LAKE 85
WERE FOUND EXCLUSIVELY AT THIS LOCATION. THE GREEN ALGAE CONTRIBUTED
THE GREATEST NUMBER OF SPECIES, BUT THE BULK OF THE STANDING CROP WAS
COMPOSED OF BLUE-GREEN ALGAE. A TOTAL OF 100 TAXA WERE RECORDED AT
NEWFOUND LAKE, AND 43 OF THESE WERE FOUND EXCLUSIVELY AT NEWFOUND LAKE.
THE LARGEST NUMBER OF SPECIES WERE GREEN ALGAE. DURING THE SUMMER THE
PHYTOPLANKTON FLORA AT WINNISQUAM LAKE WAS PRIMARILY COMPOSED OF
MEMBERS OF THE CHLOROPHYCEAE AND CYANOPHYCEAE. MEMBERS OF THE
BACILLARIOPHYCEAE, CHRYSOPHYCEAE AND CHLOROPHYCEAE (MAINLY DESMIDS)
CONTRIBUTED THE LARGEST NUMBER OF SPECIES DURING THE SPRING AND FALL.
DIATOMS WERE THE MAJOR COMPONENT OF THE WINTER FLORA. THE SUMMER AND
FALL FLORA AT NEWFOUND LAKE WAS PRIMARILY COMPOSED OF MEMBERS OF THE
CYANOPHYCEAE, CHLOROPHYCEAE (ESPECIALLY DESMIDS) AND CHRYSOPHYCEAE. THE
DIATOMS AND GOLDEN-BROWN ALGAE CONTRIBUTED THE LARGEST NUMBER OF
SPECIES DURING THE WINTER AND SPRING.

FIELD 05C, 02H

ACCESSION NO. W70-02764

ALGAE IN RELATION TO MINE WATER,

WEST VIRGINIA UNIV., MORGANTOWN. WATER RESEARCH INST.

H. D. BENNETT.

PB-188 906. WEST VIRGINIA WATER RESEARCH INSTITUTE, JOURNAL REPORT 1,
(REPRINT FROM CASTANEA VOL 34, NO 3, P 306-328, SEPT 1969). OWRR A-007-WVA.

DESCRIPTORS:
*ALGAE, *ACID MINE WATER, *HABITATS, *ENVIRONMENTAL EFFECTS, *LIMITING
FACTORS, BENTHIC FLORA, BENTHOS, MINE WATER, WATER POLLUTION, WATER
POLLUTION EFFECTS, ACID STREAMS, MINE DRAINAGE, CHLOROPHYTA,
CHRYSOPHYTA, EUGLENOPHYTA, CYANOPHYTA, DIATOMS, AQUATIC HABITAT,
POPULATION.

IDENTIFIERS:
*CYTOLOGICAL EFFECTS, ALGAL POPULATION.

ABSTRACT:
AN ANNUAL CYCLE OF BIMONTHLY SAMPLING WAS DONE TO OBTAIN INFORMATION ON
SPECIES OF ALGAE TOLERANT TO ACID MINE WATER AND ON CHEMICAL FACTORS
RELATED TO ECOLOGY OF ALGAE. NITRATE, PHOSPHATE, AND CALCIUM APPEARED
NOT TO BE LIMITING OR MODIFYING. CORRELATION WAS LACKING BETWEEN
FACTORS EXCEPT ACIDITY, IRON, AND PH. A RANGE FOR THESE CHARACTERIZED
EACH HABITAT, AND THEY APPEARED CONTROLLING ON ALGAL POPULATION IN
MORE-ACID STREAMS. THE TOTAL NUMBER OF GENERA AND SPECIES COMPARED
FAVORABLY WITH NUMBERS REPORTED FROM 'UNPOLLUTED' WATERS. NEARLY HALF
THE TOTAL SPECIES FOUND IN MORE-ACID STREAMS EQUALLY WERE IN LESS-ACID
CREEKS, RIVERS, AND PONDS. ALGAE CHARACTERISTIC OF MINE WATER WERE
THOSE COMMON TO THE OTHER HABITATS. REDUCTION IN NUMBERS OF SPECIES AT
HIGHER ACIDITY AND LOWER PH WAS PRIMARILY IN SPECIES LESS COMMON TO A
RANGE OF HABITATS. THE RANGE OF ACIDITY AND PH AT A SITE IS
CHARACTERIZED BY A RANGE IN SPECIES NUMBER AND ABUNDANCE A CHANGE IN
ABUNDANCE BEING DEPENDENT ON THE SPECIES. SOME SUCH AS EUGLENA
MUTABILIS SCH., EUNOTIA TENELLA (GRUN) CLEVE, AND PINNULARIA BRAUNII
(GRUN) CLEVE ARE MOST ABUNDANT IN MINE-POLLUTED WATER. (DODSON-WEST
VIRGINIA)

FIELD 05C

ACCESSION NO. W70-02770

PHYTOPLANKTON POPULATIONS IN RELATION TO TROPHIC LEVELS OF LAKES IN NEW
HAMPSHIRE, U.S.A.

NEW HAMPSHIRE UNIV., DURHAM. DEPT. OF BOTANY.

G. K. GRUENDLING, AND A. C. MATHIESON.

AVAILABLE FROM THE CLEARINGHOUSE AS PB-188 912, $3.00 IN PAPER COPY, $0.65 IN
MICROFICHE. COMPLETION REPORT, UNIVERSITY OF NEW HAMPSHIRE WATER RESOURCES
RESEARCH CENTER, RESEARCH REPORT NO 1, JUNE 1969. 81 P, 15 FIG, 9 TAB. OWRR
PROJECT A-010-NH.

DESCRIPTORS:
  *ALGAE, AQUATIC ENVIRONMENT, *AQUATIC ALGAE, WATER QUALITY, *TROPHIC
  LEVELS, NUTRIENTS, NORTHEAST U.S., ECOLOGICAL DISTRIBUTION, FREQUENCY
  DISTRIBUTION, POPULATION, BIOLOGICAL INDICATORS, NEW HAMPSHIRE.

IDENTIFIERS:
  PLANT GROWTH, WATER AND PLANTS, WATER POLLUTION EFFECTS, ECOLOGY,
  EFFECTS OF POLLUTION, WATER QUALITY CONTROL, WINNISQUAM LAKE, NEWFOUND
  LAKE.

ABSTRACT:
  THE RELATIONSHIP BETWEEN COMPOSITION, PERIODICITY AND ABUNDANCE OF
  PHYTOPLANKTON SPECIES AND THE TROPHIC STATUS OF TWO NEW HAMPSHIRE LAKES
  (WINNISQUAM AND NEWFOUND) WERE INVESTIGATED. NEWFOUND LAKE IS
  OLIGOTROPHIC, WHILE WINNISQUAM IS UNDERGOING RAPID EUTROPHICATION. THE
  DIFFERENCES IN THE TOTAL STANDING CROP AT THE TWO LAKES WERE RELATED
  WITH THEIR DIFFERENCES IN NUTRIENT CONCENTRATIONS. ORTHOPHOSPHATE
  APPEARED TO BE THE PRIMARY LIMITING FACTOR OF ALGAL GROWTH AT NEWFOUND
  LAKE, WHILE LIGHT, TEMPERATURE AND POSSIBLE EXTRACELLULAR PRODUCTS WERE
  THE PRIMARY LIMITING FACTORS AT WINNISQUAM LAKE. A DISCUSSION OF THE
  FACTORS INFLUENCING THE PERIODICITY, ABUNDANCE AND DISTRIBUTION OF
  PHYTOPLANKTON SPECIES AT THE TWO LAKES IS PRESENTED, AND THE RESULTS
  ARE COMPARED WITH PREVIOUS FINDINGS. THE SPECIES COMPOSITION AT THE
  LAKES WAS DEPENDENT UPON THEIR TROPHIC STATUS. DOMINANT SPECIES WERE
  CONSIDERED TO BE THE BEST INDICATORS OF TROPHIC LEVELS. A SUMMARY OF
  POSSIBLE RARE INDICATOR ORGANISMS IS ALSO INCLUDED. THE PHYTOPLANKTON
  QUOTIENTS PROPOSED BY VARIOUS AUTHORS WERE APPLIED TO THE LAKES. MOST
  OF THE INDICES CORRECTLY DESIGNATED THE EUTROPHIC LAKE, BUT NOT THE
  OLIGOTROPHIC ONE.

FIELD 05C, 02H

ACCESSION NO. W70-02772

CALIFORNIA UNIV., BERKELEY. SANITARY ENGINEERING RESEARCH LAB. AND PACIFIC
NORTHWEST WATER LAB. CORVALLIS, OREG.

PROCEEDINGS OF THE EUTROPHICATION-BIOSTIMULATION ASSESSMENT WORKSHOP, JUNE
19-21, 1969, CALIFORNIA UNIV, BERKELEY, SANITARY ENGINEERING RESEARCH LAB
AND NATIONAL EUTROPHICATION RESEARCH PROGRAM, CORVALLIS, OREGON, PACIFIC
NORTHWEST WATER LAB P I-IV + 281 P. 93 FIG, 43 TAB, 341 REF.

DESCRIPTORS:
EUTROPHICATION, *CONFERENCES, WATER POLLUTION EFFECTS, ALGAL CONTROL,
WATER POLLUTION, TOXICITY, LAKES, DESIGN CRITERIA, REMOTE SENSING,
MINNESOTA, SEDIMENT-WATER INTERFACES, CALIFORNIA, BIOASSAY,
SCENEDESMUS, PHYSIOLOGICAL ECOLOGY, KINETICS, FARM WASTES, BENTHOS,
NITRATES, FLUORESCENCE, NITROGEN FIXATION, ENVIRONMENTAL EFFECTS,
NUTRIENT REQUIREMENTS, COMMUNITIES(BIOLOGICAL), CARBON RADIOISOTOPES,
PRODUCTIVITY, RIVERS, TRACE ELEMENTS.

IDENTIFIERS:
*BIOSTIMULATION, ALGAL GROWTH, WATER POLLUTION ASSESSMENT, SANITARY
ENGINEERING RESEARCH LABORATORY(CALIF), NATIONAL EUTROPHICATION
RESEARCH PROGRAM, CONTINUOUS CULTURE, CHEMOSTATS, BATCH CULTURES,
ACETYLENE REDUCTION TECHNIQUE, HETEROTROPHY, FLAGELLATES, LAKE
CLASSIFICATION, SHAGAWA LAKE(MINN).

ABSTRACT:
EUTROPHICATION MAY BE THE MOST SERIOUS AND PRESSING WATER QUALITY
PROBLEM WE FACE, AND NEW SCRUTINY OF RESEARCH APPROACHES IS REQUIRED.
THE DELIBERATIONS ARE PRESENTED OF A GROUP OF WORKERS IN EUTROPHICATION
HELD AT BERKELEY, CALIFORNIA, DURING 19-21 JUNE 1969. THE SESSIONS WERE
SPONSORED BY THE SANITARY ENGINEERING RESEARCH LABORATORY, UNIVERSITY
OF CALIFORNIA, AND THE NATIONAL EUTROPHICATION RESEARCH PROGRAM, US
DEPARTMENT OF THE INTERIOR. SCOPE OF THE REPORT IS INDICATED BY
SUBJECTS OF INDIVIDUAL CONTRIBUTIONS: NEED FOR ASSAYS; CHEMOSTAT
ASSAYS; BATCH ASSAYS FOR DETERMINING ALGAL GROWTH POTENTIAL; ALGAL
GROWTH ASSESSMENT BY FLUORESCENCE TECHNIQUES; KINETIC ASSESSMENT OF
ALGAL GROWTH; FACTORS IN UTILIZATION BY SCENEDESMUS OF NITRATE FROM
AGRICULTURAL WASTES; ENVIRONMENTAL AND NUTRITIONAL REQUIREMENTS FOR
ALGAE; ASSESSMENT OF NITROGEN FIXATION BY ACETYLENE REDUCTION;
HETEROTROPHY IN ALGAL FLAGELLATES; PHYSIOLOGICAL ECOLOGY; PHYSIOLOGICAL
ECOLOGY OF BENTHIC ALGAL COMMUNITIES; MEASUREMENT OF ALGAL GROWTH WITH
RADIOCARBON; LACUSTRINE CLASSIFICATION; PRODUCTIVITY OF RIVERS;
MEASUREMENT OF TRACE ELEMENTS IN THE HYDROSPHERE; DESIGN APPLICATIONS
OF BIOSTIMULATION AND TOXICITY CRITERIA; REMOTE SENSING IN WATER
RESOURCES MANAGEMENT; NUTRIENT EXCHANGE BETWEEN SEDIMENT AND WATER; AND
SHAGAWA LAKE (MINNESOTA) EUTROPHICATION RESEARCH PROJECT. AN INDEX IS
PROVIDED. (EICHHORN-WISCONSIN)

FIELD 05A, 05C

ACCESSION NO. W70-02775

ALGAL GROWTH ASSESSMENTS BY FLUORESCENCE TECHNIQUES,

FEDERAL WATER POLLUTION CONTROL ADMINISTRATION, ALAMEDA, CALIF.

RICHARD C. BAIN, JR.

PROC OF THE EUTROPHICATION-BIOSTIMULATION ASSESSMENT WORKSHOP, JUNE 19-21, 1969, CALIFORNIA UNIV, BERKELEY, SANITARY ENGINEERING RESEARCH LAB AND NAT EUTROPHICATION RESEARCH PROGRAM, CORVALLIS, ORE, PAC NORTHWEST WATER LAB, P 39-55. 9 FIG, 20 REF, APPEND.

DESCRIPTORS:
*ALGAL GROWTH, *BIOASSAYS, *FLUOROMETRY, CALIFORNIA, NUTRIENTS, EUTROPHICATION, CHLOROPHYLL, FLUORESCENCE, LAKES, RIVERS, NITROGEN, BIOMASS, SEWAGE, FARM WASTES, CARBON CYCLE, PHOSPHORUS, PHOSPHATE, NITRATES, INDUSTRIAL WASTES, WATER POLLUTION EFFECTS, DENITRIFICATION, SPECTROPHOTOMETRY.

IDENTIFIERS:
SAN JOAQUIN RIVER(CALIF), MERCED RIVER(CALIF), TUOLUMNE RIVER(CALIF), STANISLAUS RIVER(CALIF), WATER POLLUTION ASSESSMENT, AMERICAN RIVER(CALIF), CLEAR LAKE(CALIF), EEL RIVER(CALIF), JOINT INDUSTRY/GOVERNMENT TASK FORCE ON EUTROPHICATION, PROVISIONAL ALGAL ASSAY PROCEDURE, BATCH CULTURES, CHEMOSTATS, CHLOROPHYLL A, PHOTOBIOLOGY, ALGAL GROWTH POTENTIAL, SAN FRANCISCO BAY.

ABSTRACT:
AUTHOR DESCRIBES PROCEDURES FOR ASSESSING ALGAL GROWTH POTENTIAL, IN JOINT INDUSTRY/GOVERNMENT TASK FORCE'S PROVISIONAL ALGAL ASSAY PROCEDURE, BY ESTIMATING ALGAL CHLOROPHYLL FLUOROMETRICALLY OR SPECTROPHOTOMETRICALLY. WATER SMAPLES ARE FILTERED THROUGH GLASS FIBER FILTERS, PRESERVED WITH MAGNESIUM CARBONATE, AND FILTERS KEPT FROZEN IN DARK REFRIGERATOR. CHLOROPHYLL CONCENTRATION, EXTRACTED INTO SOLUTION BY GRINDING THE FILTER IN ACETONE, IS ESTIMATED BY SPECTROPHOTOMETRY (AT WAVELENGTHS 630, 645 AND 665 NANOMETERS) OR FLUOROMETRY. CORRECTION FOR PHAEOPHYTIN IS ACCOMPLISHED BY A SECOND READING AFTER TREATMENT OF SAMPLE WITH HYDROCHLORIC ACID. ILLUSTRZTIVE DATA FOR VARIOUS CALIFORNIA WATERS RECEIVING SEWAGE, AGRICULTURAL AND INDUSTRIAL WASTES ARE DELINEATED IN THE REPORT. IN BATCH ASSAYS, PEAK FLUORESCENCE DUE TO CHLOROPHYLL, CHARACTERISTICALLY APPEARS WITHINTHE FIRST WEEK, NOT NECESSARILY PARALLELLING PEAK BIOMASSES. DATA ANALYSIS FROM SUCH ASSAYS TOGETHER WITH THOSE FOR NUTRIENT ANALYSIS PERMITS INTERPRETIVE APPLICATION OF THE PROCEDURE. THUS FOR 44 DATA POINTS FROM VARIOUS STREAMS OF SAN JOAQUIN VALLEY, IN WHICH NITROGEN IS THOUGHT TO BE LIMITING, REGRESSION ANALYSIS OF RELATIONSHIP BETWEEN TOTAL N (MILLIGRAMS/LITER) AND CHLOROPHYLL (MICROGRAMS/LITER) YIELDS RATIO, TOTAL N/CHLOROPHYLL, WITH 95% CONFIDENCE INTERVAL OF 4.6-10.8, IN GOOD AGREEMENT WITH LITERATURE VALUES FOR ALGAL CELLS WITH NITROGEN LIMITINC. (EICHHORN-WISCONSIN)

FIELD 05A

ACCESSION NO. W70-02777

CONTINUOUS-FLOW (CHEMOSTAT) ASSAYS,

CALIFORNIA UNIV., BERKELEY. SANITARY ENGINEERING RESEARCH LAB.

DONALD B. PORCELLA.

PROCEEDINGS OF THE EUTROPHICATION-BIOSTIMULATION ASSESSMENT WORKSHOP, JUNE 19-21, 1969, CALIFORNIA UNIV, BERKELEY, SANITARY ENGINEERING RESEARCH LAB AND NATIONAL EUTROPHICATION RESEARCH PROGRAM, CORVALLIS, ORE, PACIFIC NORTHWEST WATER LAB, P 7-22. 6 FIG, 1 TAB, 30 REF.

DESCRIPTORS:
*BIOASSAYS, NUTRIENTS, PHOSPHORUS, KINETIC THEORY, ENVIRONMENTAL FACTORS, BIOMASS, GROWTH RATES, ALGAE, OLIGOTROPHY, LIMITING FACTORS, CHLORELLA, ANALYTICAL TECHNIQUES, CYANOPHYTA, DIATOMS, EUTROPHICATION, WATER POLLUTION EFFECTS, SYSTEMS ANALYSIS, SUCCESSION, ECOSYSTEMS, CYCLING NUTRIENTS.

IDENTIFIERS:
*CONTINUOUS CULTURE, *CHEMOSTATS, TURBIDOSTATS, NUTRIENT LIMITED GROWTH, LUXURY CONSUMPTION, PROVISIONAL ALGAL ASSAY, PROCEDURES, SELENASTRUM CAPRICORNUTUM, SELENASTRUM GRACILE, GROWTH LIMITATION, SELENASTRUM, ALGAL GROWTH, ANABAENA, BIOSTIMULATION, WATER POLLUTION ASSESSMENT.

ABSTRACT:
BECAUSE THEIR DYNAMIC PROPERTIES MORE NEARLY APPROXIMATE ECOLOGICAL CONDITIONS THAN DO BATCH CULTURES, CONTINUOUS CULTURES OPERATING IN CHEMOSTAT MODE (CONSTANT FLOW-RATE) HAVE BEEN UTILIZED IN A VARIETY OF STUDIES OF ALGAL GROWTH, INCLUDING ASSESSMENT OF BIOSTIMULATORY RESPONSES. REQUIREMENTS FOR STEADY STATE CONTINUOUS FLOW SYSTEMS AND CHARACTERISTICS OF IDEAL CHEMOSTATS WITH ILLUSTRATIONS OF IDEALIZED CURVES FROM AN EXPERIMENT CONCERNING RELATIVE CHANGE IN ALGAL CELL MATERIAL AND LIMITING PHOSPHORUS CONCENTRATIONS ARE INCLUDED. IN TRANSIENT PHASES OF CHEMOSTAT OPERATION, PHOSPHORUS CONCENTRATIONS ARE LESS THAN THOSE CALCULATED THEORETICALLY, PROBABLY DUE TO 'LUXURY' CONSUMPTION BY ALGAE. CONSIDERATIONS OF MASS BALANCES AND KINETICS OF ALGAL GROWTH, USEFUL IN DERIVATION OF CONTINUOUS CULTURES THEORY, NEED BE APPLIED TO NATURAL ECOSYSTEMS TO ELICIT USEFUL INFORMATION ON PRODUCTIVITY, NUTRIENT CYCLING, AND SUCCESSIONAL RELATIONSHIPS. BASED ON CHEMOSTAT EXPERIMENTS DESCRIBED, UTILIZING BOTH SELENASTRUM GRACILE AND NATURAL ALGAL POPULATIONS AS 'SEED', ARE AREAS REQUIRING FURTHER STUDY, INCLUDING THE FOLLOWING PROBLEMS: MULTIPLE LIMITING FACTORS; USE OF NATURAL 'SEED' POPULATIONS; EXCESSIVE WASHOUT AND EXTINCTION OF CULTURES IN BIOASSAY OF OLIGOTROPHIC WATERS; DEVELOPMENT OF SYNTHETIC MEDIA; AND PROFILES OF BIOMASS ANALYSIS AND NUTRIENT CONCENTRATION DURING TRANSIENT GROWTH IN ORDER TO ESTABLISH APPROPRIATE KINETIC PARAMETERS. (SEE W70-02775). (EICHHORN-WISCONSIN)

FIELD 05A

ACCESSION NO. W70-02779

STRUCTURAL CHARACTERISTICS OF BENTHIC ALGAL COMMUNITIES IN LABORATORY STREAMS,

OREGON STATE UNIV., CORVALLIS. DEPT. OF BOTANY.

C. DAVID MCINTIRE.

ECOLOGY, VOL 49, NO 3, P 520-537, 1968. 5 FIG, 6 TAB, 31 REF.

DESCRIPTORS:
*BENTHIC FLORA, *ALGAE, *BIOLOGICAL COMMUNITIES, *LABORATORIES,
*STREAMS, LIGHT INTENSITY, CURRENTS(WATER), VELOCITY, DIATOMS,
SEASONAL, BIOMASS, PIGMENTS, PRODUCTIVITY, PERIPHYTON, CHLOROPHYTA,
CHRYSOPHYTA, CYANOPHYTA, SAMPLING, PHYTOPLANKTON, ORGANIC MATTER,
CHLOROPHYLL, DYNAMICS, TEMPERATURE, SILTS, STANDING CROP, ECOLOGY,
PHOTOSYNTHESIS, RESPIRATION, CARBON DIOXIDE, NUTRIENTS, DISSOLVED
OXYGEN, LOTIC ENVIRONMENT.

IDENTIFIERS:
SPECIES DIVERSITY, AUTECOLOGY, MORPHOLOGY, EXPORT LOSSES,
CHEMOSYNTHESIS, ABUNDANCE.

ABSTRACT:
EFFECTS OF LIGHT INTENSITY AND CURRENT VELOCITY ON SPECIES COMPOSITION
AND ECOLOGICAL PROPERTIES OF BENTHIC ALGAL COMMUNITIES WERE
INVESTIGATED IN LABORATORY STREAMS. OF 15 DIATOM TAXA STUDIED, ONLY
MELOSIRA VARIANS, MERIDION CIRCULARE, AND NAVICULA RADIOSA WERE MORE
ABUNDANT IN STREAMS RECEIVING 700-FOOT-CANDLE ILLUMINATION THAN IN
THOSE RECEIVING 150-FOOT-CANDLE. ACHNANTHES EXIGUA, A MINUTISSIMA,
MERIDION CIRCULARE, RHOICOSPHENIA CURVATA, AND NAVICULA RADIOSA WERE
INDIFFERENT TO CURRENT VELOCITY, WHICH HAD POSITIVE EFFECT ON ABUNDANCE
OF NITZSCHIA LINEARIS, ACHNANTHES LANCEOLATA, NAVICULA CRYPTOCEPHALA, N
MINIMA, N SEMINULUM, SYNEDRA ULNA, GOMPHONEMA PARVULUM, G ANGUSTATUM,
COCCONEIS PLACENTULA, AND THE LANCEOLATE NITZSCHIA. MELOSIRA VARIANS
EXHIBITED NEGATIVE RESPONSE TO CURRENT VELOCITY; SIX TAXA, OTHER THAN
DIATOMS, ANABAENA VARIABILIS AND TRIBONEMA MINOR WERE MORE ABUNDANT IN
STREAMS WITH HIGHER LIGHT INTENSITY, BUT ONLY ONE SPECIES, PHORMIDIUM
RETZII, WAS MORE ABUNDANT IN STREAMS WITH CURRENT THAN IN STANDING
WATER. AT A PARTICULAR SEASON, LIGHT INTENSITY, AND CURRENT
VELOCITY--LABORATORY STREAM CONDITIONS--ALLOW ESTABLISHMENT OF ALGAL
COMMUNITIES WITH COMPARATIVELY UNIQUE SPECIES COMPOSITION AND
CHARACTERISTIC BIOMASS, PIGMENT CONCENTRATION, AND INCREMENT OF EXPORT.
TO UNDERSTAND FACTORS INFLUENCING PRODUCTIVITY IN THESE COMMUNITIES,
SOME KNOWLEDGE IS NECESSARY OF AUTECOLOGIES OF COMMUNITY CONSTITUENTS,
AND MECHANISMS REGULATING SPECIES COMPOSITION AND FLORAL DIVERSITY.
(JONES-WISCONSIN)

FIELD 05C

ACCESSION NO. W70-02780

CLADOPHORA GLOMERATA AND CONCOMITANT ALGAE IN THE RIVER SKAWA. DISTRIBUTION AND
CONDITIONS OF APPEARANCE,

WYZSZA SZKOLA ROLNICZA, OLSZTYN-KORTOWA (POLAND). KATEDRA BOTANIKI.

HENRYK CHUDYBA.

ACTA HYDROBIOLOGICA, VOL 10, NO 1-2, P 39-84, 1968. 27 FIG, 12 TAB, 82 REF.

DESCRIPTORS:
*CLADOPHORA, ALGAE, AQUATIC ALGAE, AQUATIC MICROBIOLOGY, AQUATIC
PRODUCTIVITY, CHLOROPHYTA, GONOPHYTA, DIATOMS, ENVIRONMENTAL EFFECTS,
LIMNOLOGY, ECOLOGY, ECOLOGICAL DISTRIBUTION.

IDENTIFIERS:
*SKAWA RIVER(POLAND), MONTANE RIVER COURSE, CARPATHIAN SUBMONTANE
REGION, GRADIENTS(RIVER), THALLI, SCHIZOMYCETES, CYANOPHYTA, FUNGI,
BACILLARIOPHYCEAE, CHAROPHYTA, RHODOPHYTA.

ABSTRACT:
RESULTS OF 137 SAMPLES COLLECTED DURING SEPTEMBER 1-27, 1963 FROM THE
RIVER SKAWA, POLAND, INDICATED THAT TWO GROUPS CAN BE DISTINGUISHED BY
THALLI: C GLOMERATA RHEOBENTHICUM FROM PLACES WITH CURRENT; C GLOMERATA
LIMNOBENTHICUM FROM PLACES WITH NO CURRENT. THE THALLI APPEARED YELLOW
OR BROWN AS A RESULT OF ATTACHED DIATOMS. THE RAPIDITY OF CURRENT DID
NOT PLAY ANY SIGNIFICANT ROLE IN FORMATION OF THE COMMUNITY OF
ACCOMPANYING ALGAE. EPIPHYTES WERE MAINLY DIATOMS, FOLLOWED BY GREEN
AND BLUE-GREEN ALGAS. PENNATAE DIATOMS WERE MOST COMMON; NAVICULA AND
NITZSCHIA WERE THE MOST COMMON GENUS. GROUPS OF ALGAE ACCOMPANYING C
GLOMERATA AT ALL SAMPLING PLACES FORM ONE TYPE OF COMMUNITY AND A
SINGLE PHYTOCENOSE WITH A SIMILAR FLORISTIC COMPOSITION AND STRUCTURE.
(FITZGERALD-WISCONSIN)

FIELD 05C

ACCESSION NO. W70-02784

INTERPRETATION OF RADIONUCLIDE UPTAKE FROM AQUATIC ENVIRONMENTS,

OAK RIDGE NATIONAL LAB., TENN. RADIATION ECOLOGY SECTION.

DANIEL J. NELSON.

NUCLEAR SAFETY, VOL 5, NO 2, P 196-199, 1964. 12 REF.

DESCRIPTORS:
*RADIOISOTOPES, *AQUATIC ENVIRONMENT, CHLORELLA, TRACERS, PHOTOSYNTHESIS, CARBON RADIOISOTOPES, FISH, CLAMS, STRONTIUM RADIOISOTOPES, TENNESSEE RIVER, TROPHIC LEVEL, ALGAE, CADDISFLIES, FOOD CHAINS, STABLE ISOTOPES, RADIOACTIVE WASTE DISPOSAL, OCEANS, FOODS, ABSORPTION, HAZARDS, CESIUM, HYDROGEN, DEUTERIUM, TRITIUM.

IDENTIFIERS:
*UPTAKE, SPECIFIC ACTIVITY, CHLORELLA PYRENOIDOSA, BIOLOGICAL HALF-LIFE, CLINCH RIVER(TENNESSEE), PHYSICAL HALF-LIFE, HYDROPSYCHE COCKERELLI, BODY BURDENS, POMOXIS ANNULARIS, GASTROINTESTINAL TRACT.

ABSTRACT:
SPECIFIC ACTIVITIES IN AQUATIC ORGANISMS HELP EXPLAIN DIFFERENCES IN RADIONUCLIDE UPTAKE. DATA EVOLVED BY ROUTINE RADIOLOGICAL MONITORING ESTABLISHES SAFETY STANDARDS FOR RADIONUCLIDE RELEASES TO FRESH-WATER HABITATS. EQUILIBRIUM LEVELS OF RADIONUCLIDES BETWEEN ORGANISMS AND ENVIRONMENT ARE EXPECTED ONLY FOR ELEMENTS WITH SHORT BIOLOGICAL HALF-LIFE IN TISSUE. METHOD OF SPECIFIC ACTIVITIES IS BASED ON ASSUMPTION THAT IF THE DISTRIBUTION OF A STABLE ELEMENT IS KNOWN BETWEEN AN ORGANISM AND ITS ENVIRONMENT, IT MAY BE USED TO PREDICT RADIONUCLIDE CONCENTRATIONS IN WATER ORGANISMS RECEIVING CONSTANT RADIONUCLIDE RELEASES. THE RELATION OF STRONTIUM AND STRONTIUM-90 IN FISH MAY BE USED TO CALCULATE EXPECTED BODY BURDENS. SPECIFIC ACTIVITY RELATION MIGHT BE A CRITERION FOR LIMITING RADIOACTIVE WASTE DISPOSAL IN OCEANS. WHILE SPECIFIC ACTIVITY OF A CHEMICAL ELEMENT IN ENVIRONMENT OF HUMAN FOOD ORGANISMS IS BELOW THE PERMISSIBLE SPECIFIC ACTIVITY FOR THE ELEMENT IN A HUMAN BODY, MAN CANNOT OBTAIN MORE THAN HIS PERMISSIBLE BODY BURDEN FROM THE SEAS. DIRECT APPLICATION OF THE SPECIFIC ACTIVITY CONCEPT APPLIES ONLY TO THOSE RADIONUCLIDES WHICH CONSTITUTE AN ABSORPTION HAZARD. WITH RADIONUCLIDES FOR WHICH ABSORPTION IS NOT CRITICAL, SPECIFIC ACTIVITIES MAY BE RELATED TO IRRADIATION OF THE GASTROINTESTINAL TRACT BY ITS CONTENTS.
(JONES-WISCONSIN)

FIELD 05A

ACCESSION NO. W70-02786

FERTILIZATION OF LAKES BY AGRICULTURAL AND URBAN DRAINAGE,

MASSACHUSETTS INST. OF TECH., CAMBRIDGE. DEPT. OF SANITARY CHEMISTRY.

CLAIR N. SAWYER.

NEW ENGLAND WATER WORKS ASSOCIATION, VOL 61, NO 2, P 109-127, JUNE 1947. 5 FIG, 6 TAB, 6 REF.

DESCRIPTORS:
  *SEWAGE DISPOSAL, *AGRICULTURAL WATERSHEDS, *RUNOFF, *FERTILIZATION, LAKES, SEWAGE TREATMENT, SEWAGE BACTERIA, NITROGEN, PHOSPHORUS, ALGAE, COPPER SULFATE, EFFLUENTS, MICROCYSTIS AERUGINOSA.

IDENTIFIERS:
  *URBAN DRAINAGE, *AGRICULTURAL DRAINAGE, BIOLOGICAL RESPONSE, FERTILIZATION SURVEY, MADISON LAKES(WIS), LAKE MENDOTA(WIS), LAKE WAUBESA(WIS), LAKE KEGONSA(WIS).

ABSTRACT:
  THE REPORTED INVESTIGATIONS DISCLOSE THAT AGRICULTURAL DRAINAGE CONTRIBUTES ANNUALLY ABOUT 4,500 POUNDS OF NITROGEN AND 255 POUNDS OF PHOSPHORUS PER SQUARE MILE OF THE WATERSHED. BIOLOGICALLY TREATED SEWAGE SUPPLIES ANNUALLY APPROXIMATELY 6.0 OF NITROGEN AND 1.2 POUNDS OF PHOSPHORUS PER CAPITA, THUS EQUATING THE ENRICHMENT OF LAKE WATER FROM 1 SQUARE MILE OF AGRICULTURAL DRAINAGE TO 750 PERSONS FOR NITROGEN AND TO 212 PERSONS FOR PHOSPHORUS. IN 1946, LAKES MONONA, WAUBESA, AND KEGONSA, LOCATED ON THE OUTSKIRTS OF THE CITY OF MADISON, RECEIVED APPROXIMATELY 1,300 TONS OF NITROGEN AND 215 TONS OF PHOSPHORUS IN ORGANIC AND INORGANIC FORMS. THE LAKES RETAINED FROM 30 TO 60% OF NITROGEN RECEIVED. THE PHORPHORUS CONCENTRATION IN LAKES WAUBESA AND KEGONSA ATTAINED THE HIGH LEVEL ABOVE 0.25 PARTS PER MILLION (PPM) AS COMPARED WITH 0.01-0.02 PPM CONCENTRATION FOUND IN LESS POLLUTED LAKES OF SOUTHERN WISCONSIN. IMPORTANCE OF THE HYPOLIMNION, THE NITROGEN:PHOSPHORUS RATIO, AND A 'ROUND-THE-CALENDAR' RECORDS IS STRESSED. MICROCYSTIS AERUGINOSA WAS A PARTICULARLY OBNOXIOUS BY-PRODUCT OF EUTROPHICATION, EITHER BECAUSE OF ITS INHERENT REPRODUCTIVE POTENTIAL OR RESISTANCE TO COPPER SULFATE. OTHER DATA OF THIS REPORT HAVE CONSIDERABLE IMMEDIATE AS WELL AS HISTORICAL INTEREST. (WILDE-WISCONSIN)

FIELD 05B

ACCESSION NO. W70-02787

EFFECTS OF ACID MINE WASTES ON PHYTOPLANKTON IN NORTHERN ONTARIO LAKES,

ONTARIO WATER RESOURCES COMMISSION, TORONTO. DIV. OF RESEARCH.

M. G. JOHNSON, M. F. P. MICHALSKI, AND A. E. CHRISTIE.

ONTARIO WATER RESOURCES COMMISSION, DIVISION OF RESEARCH PUBLICATION NO 30, (1968). 44 P, 6 FIG, 5 TAB, 15 REF, APPEND A, B, C.

DESCRIPTORS:
*ALGAE, *ANALYTICAL TECHNIQUES, *AQUATIC ALGAE, AQUATIC MICROBIOLOGY, AQUATIC MICROORGANISMS, AQUATIC PLANTS, AQUATIC PRODUCTIVITY, BALANCE OF NATURE, BIOASSAY, CHLOROPHYTA, CYANOPHYTA, DIATOMS, ENVIRONMENTAL EFFECTS, ESSENTIAL NUTRIENTS, EUTROPHICATION, BICARBONATES, CARBON, CARBONATES, CARBON CYCLE.

IDENTIFIERS:
*ACID MINE WATER, *NUTRIENT REQUIREMENTS, *POLLUTANTS, INHIBITION, LIMNOLOGY, ACIDITY, ACID SOILS, ACIDIC WATER, ACIDS, PHOTOSYNTHESIS, PHOTOSYNTHETIC OXYGEN, PHYSIOLOGICAL ECOLOGY, PHYTOPLANKTON, PLANKTON, POLLUTION ABATEMENT, POLLUTANT IDENTIFICATION, WATER POLLUTION CONTROL, WATER POLLUTION EFFECTS, WATER POLLUTION SOURCES.

ABSTRACT:
STUDIES WERE CARRIED OUT DURING 1965 THROUGH 1967 ON THREE NORTHERN ONTARIO LAKES; TWO (QUIRKE AND PECORS LAKES) WERE CONTAMINATED BY QUANTITIES OF FREE MINERAL ACIDITY FROM URANIUM MILLING WASTES AND ONE (DUNLOP LAKE) WAS UNAFFECTED. DIFFERENCES IN CHEMICAL COMPOSITION IN THE AFFECTED LAKES, INCLUDING LOW PH VALUES AND INCREASED CONCENTRATIONS OF SULFATE ION, NITRATE ION, AND CALCIUM ION WERE DIRECTLY RELATED TO PROCESSES FOR THE EXTRACTION OF URANIUM AND SUBSEQUENT TREATMENT OF WASTES. NITRATES WERE IN GREATER SUPPLY IN THE CONTAMINATED LAKES THAN IN DUNLOP LAKE, WHILE PHOSPHORUS AND SILICA OCCURRED IN SIMILAR CONCENTRATIONS IN ALL THREE LAKES. LOWER PHYTOPLANKTON POPULATIONS AND INDICES OF DIVERSITY WERE FOUND IN QUIRKE AND PECORS LAKES THAN IN DUNLOP. AVERAGE PRIMARY PRODUCTIVITIES IN THE THREE LAKES WERE 126, 71 AND 34 MILLIGRAMS CARBON/SQUARE METER PER DAY, RESPECTIVELY. IN SITU AREAL AND VOLUMETRIC MEASUREMENTS IN LABORATORY AND FIELD BIOASSAYS CONFIRMED THE IMPORTANCE OF INORGANIC CARBON IN LIMITING PRIMARY PRODUCTIVITY. AUTHORS CONCLUDED THAT INORGANIC CARBON LIMITS PRIMARY PRODUCTIVITY IN THE LAKES CONTAMINATED BY ACID MINE WASTES. HOWEVER, THEY DID NOT SEPARATE THE EFFECTS OF REDUCED PH ALONE FROM THE EFFECTS OF REDUCED INORGANIC CARBON IN ARRIVING AT THAT CONCLUSION. (FITZGERALD-WISCONSIN)

FIELD 05C, 05B

ACCESSION NO. W70-02792

NUTRIENT-PHYTOPLANKTON RELATIONSHIPS IN EIGHT SOUTHERN ONTARIO LAKES,

ONTARIO WATER RESOURCES COMMISSION, TORONTO. DIV. OF RESEARCH.

A. E. CHRISTIE.

ONTARIO WATER RESOURCES COMMISSION, DIV OF RESEARCH PUBLICATION NO 32, OCT 1968. 37 P, 10 FIG, 7 TAB, 23 REF.

DESCRIPTORS:
*ALGAE, *PHYTOPLANKTON, *NUTRIENTS, *PHOSPHORUS, *NITROGEN,
*ALKALINITY, LAKES, PRODUCTIVITY, PHYTOPLANKTON.

IDENTIFIERS:
*ONTARIO, *CANADA, KUSHOG LAKE(ONTARIO), TWELVE MILE LAKE(ONTARIO),
GULL LAKE(ONTARIO), BALSAM LAKE(ONTARIO), STURGEON LAKE(ONTARIO),
BUCKHORN LAKE(ONTARIO), CLEAR LAKE(ONTARIO), RICE LAKE(ONTARIO).

ABSTRACT:
A CHAIN OF EIGHT SOUTHERN ONTARIO LAKES WAS SAMPLED FROM APRIL TO
SEPTEMBER 1967, APPROXIMATELY EVERY TWO WEEKS, TO DEPTHS COMMENSURATE
WITH THE PHOTOTROPHIC ZONE. EACH LAKE WAS SAMPLED AT TWO SITES.
ANALYSIS OF ALL SAMPLES INCLUDED SUSPENDED AND DISSOLVED SOLIDS; TOTAL
ALKALINITY; PH; NITRATE, AMMONIA, ORGANIC NITROGEN; ORTHO AND ACID
HYDROLIZABLE (STANDARD METHOD'S 'TOTAL') PHOSPHORUS; ALGAL GENERA
(REPORTED AS AREAL STANDARD UNITS); AND CHLOROPHYLL A. A FEW SAMPLES
WERE ANALYZED FOR CALCIUM, MAGNESIUM, IRON, SODIUM, POTASSIUM, SULFATE,
AND CHLORINE. ONE LABORATORY GROWTH STUDY ON A SINGLE SAMPLE WAS
REPORTED WHERE GLUCOSE, SODIUM BICARBONATE, AND A COMBINATION OF THE
TWO WAS ADDED TO FLASKS CONTAINING AN 'OLIGOTROPHIC' LAKE WATER SAMPLE.
BASED ON THIS SINGLE LABORATORY EXPERIMENT AND A POSITIVE CORRELATION
BETWEEN ALKALINITY AND ALGAL STANDING CROPS IN THE OTHER LAKE WATER
SAMPLES, IT WAS CONCLUDED THAT THE UPPER RELATIVELY OLIGOTROPHIC LAKES
CONTAINED ADEQUATE NITROGEN AND PHOSPHORUS, BUT COULD SUPPORT GREATER
ALGAL CROPS IF ADDITIONAL INORGANIC OR BIODEGRADABLE ORGANIC CARBON
WERE SUPPLIED. IN THE LOWER LAKES THE CARBON WAS CONCLUDED TO BE IN
AMPLE SUPPLY. IT WAS CONCLUDED FURTHER THAT ENTRY OF NITROGEN AND
PHOSPHORUS INTO ALL EIGHT LAKES SHOULD BE CONTROLLED IN THE FUTURE.
(GERHOLD-WISCONSIN)

FIELD 05C

ACCESSION NO. W70-02795

OBJECTIONABLE ALGAE WITH REFERENCE TO THE KILLING OF FISH AND OTHER ANIMALS,

MICHIGAN STATE UNIV., EAST LANSING. DEPT. OF BOTANY AND PLANT PATHOLOGY.

G. W. PRESCOTT.

HYDROBIOLOGIA, VOL 1, P 1-13, 1948. 6 TAB, 14 REF.

DESCRIPTORS:
*NUISANCE ALGAE, *EUTROPHICATION, CYANOPHYTA, LAKES, RESERVOIRS, SCUM,
NITROGEN, PHOSPHORUS, CARBONATES, TEMPERATURE, SHALLOW WATER, HYDROGEN
SULFIDE, COPPER SULFATE, CHLOROPHYTA, CRUSTACEANS, SOIL EROSION, SEWAGE
EFFLUENTS, FARM WASTES. POPULATION, INDUSTRIES, METHANE, ALKALINITY,
HYDROGEN ION CONCENTRATION, TOXINS, OXYGEN, DECOMPOSING ORGANIC MATTER,
WATER POLLUTION EFFECTS, PHOTOSYNTHESIS, IOWA, CHLORINE.

IDENTIFIERS:
HYDROXYLAMINE, ANABAENA, MICROCYSTIS, GLOEOTRICHIA ECHINULATA,
DINOBRYON, SYNURA UVELLA, FRAGILARIA, TABELLARIA, MELOSIRA, SYNEDRA,
COELOSPHAERIUM, SPIROGYRA, CLADOPHORA, DIAPTOMUS, CYCLOPS, GOMPHONEMA,
PHORMIDIUM, APHANIZOMENON FLOS-AQUAE, LYNGBYA, STORM LAKE(IOWA), CENTER
LAKE(IOWA), DIAMOND LAKE(IOWA).

ABSTRACT:
MOST TROUBLESOME ALGAE IN LAKES AND RESERVOIRS ARE SPECIES OF
CYANOPHYTES (BLUE-GREEN ALGAE), WHICH MULTIPLY RAPIDLY AND TEND TO
FLOAT HIGH IN THE WATER, FORMING SURFACE SCUMS. UPON DECOMPOSITION,
MALODOROUS GASES AND DISAGREEABLE TASTES ARE PRODUCED. CRITICAL
CONDITIONS ARISING FROM A SUPERABUNDANT GROWTH OF ALGAE OCCUR IN LAKES
WHICH ARE SHALLOW, WARM, AND RICH IN NITROGEN, PHOSPHORUS, AND
HALF-BOUND CARBONATES. DENSE 'BLOOMS' ARE PARTICULARLY ANNOYING AND
OFTEN CAUSE ECONOMIC LOSS BY RUINING RECREATIONAL SITES, SPOILING
DRINKING WATER, CLOGGING SAND FILTERS, KILLING FISH (DIRECTLY OR
INDIRECTLY), AND KILLING ANIMALS WHICH DRINK THE HEAVILY INFESTED
WATERS. EXPERIMENTS INDICATE THAT FISH MAY BE POISONED BY PRODUCTS OF
PROTEIN DECOMPOSITION, INCLUDING HYDROXYLAMINE AND HYDROGEN SULFIDE,
WHICH ARE RELEASED WHEN DENSE MASSES OF CYANOPHYTES DECAY. FACTORS
WHICH SHOULD BE TAKEN INTO CONSIDERATION FOR EFFECTIVE TREATMENT OF
LAKES WITH COPPER SULPHATE ARE LISTED. ADVANTAGES OF APPLYING COPPER
SULPHATE BY SPRAY METHOD ARE POINTED OUT. A LABORATORY EXPERIMENT TO
TEST THE ALGICIDAL ACTION OF COPPER IS DESCRIBED. DATA INDICATE THAT
FISH CAN WITHSTAND MUCH GREATER CONCENTRATIONS OF COPPER SULPHATE THAN
THOSE NEEDED TO CONTROL ALGAE. (JONES-WISCONSIN)

FIELD 05C

ACCESSION NO. W70-02803

MINERAL NUTRITION OF PHYTOPLANKTON,

WOODS HOLE OCEANOGRAPHIC INSTITUTION, MASS.

BOSTWICK H. KETCHUM.

ANNUAL REVIEW OF PLANT PHYSIOLOGY, VOL 5, P 55-74, 1954. 2 TAB, 126 REF.

DESCRIPTORS:
*PHYTOPLANKTON, *NUTRIENT REQUIREMENTS, ECOLOGY, CHLORELLA, ALGAE, ORGANIC MATTER, CULTURES, CARBON, METABOLISM, NITROGEN, PHOSPHORUS, SULFUR, CALCIUM, MAGNESIUM, POTASSIUM, SODIUM, IRON, MANGANESE, TRACE ELEMENTS, CHELATION, PHOTOSYNTHESIS, CARBON DIOXIDE, BICARBONATES, CARBONATES, TEMPERATURE, HYDROGEN ION CONCENTRATION, ALKALINITY, LIGHT INTENSITY, CYANOPHYTA, IONS, AMMONIA, NITRATES, NITRITES, DINOFLAGELLATES, CHLOROPHYTA, HUMIC ACIDS.

IDENTIFIERS:
HETEROTROPHIC, CHEMOTROPHIC, AUTOTROPHIC, SILICON, PEDIASTRUM BORYANUM, STAURASTRUM PARADOXUM, BOTRYOCCUS BRAUNII, FRAGILARIA CROTONENSIS, NITZSCHIA, ASTERIONELLA GRACILLIMA, GROWTH FACTORS.

ABSTRACT:
PHYTOPLANKTON NUTRITION WAS STUDIED FOR UNDERSTANDING ECOLOGICAL CONDITIONS FOR GROWTH. USE OF ALGAL ORGANIC MATERIAL AS A POSSIBLE FOOD SOURCE HAS STIMULATED EXTENSIVE STUDIES OF MASS CULTURES. FOR COMPARISONS OF RESULTS, NUTRIENT REQUIREMENTS ARE DEFINED AS: ABSOLUTE, NORMAL, MINIMUM, AND OPTIMUM CONCENTRATIONS. OPTIMUM CONCENTRATION WILL PERMIT MAXIMUM GROWTH RATE, REPRODUCTION, OR PHOTOSYNTHESIS OF ALGAL POPULATIONS. THIS IS BASED UPON THE CONCEPT THAT AN INADEQUATE CONCENTRATION MAY RESULT IN DECREASED ASSIMILATION RATE AND AN EXCESSIVE CONCENTRATION MAY BE INHIBITORY OR TOXIC. PURE, OR AT LEAST, UNIALGAL CULTURES, ARE NEEDED FOR THIS STUDY. MOST NUTRITIONAL INVESTIGATIONS HAVE BEEN MADE ON ALGAE SPECIES EASY TO CULTURE WHICH PERHAPS DIFFER IN NUTRITION AND PHYSIOLOGY FROM THOSE DIFFICULT TO CULTURE. MAJOR ABSOLUTE REQUIREMENTS INCLUDE CARBON, NITROGEN, PHOSPHORUS, SULPHUR, POTASSIUM, AND MAGNESIUM; IRON AND MANGANESE ARE REQUIRED IN SMALLER AMOUNTS, SODIUM IS UNESSENTIAL AND CALCIUM REQUIREMENT IS CONTRADICTORY. SILICON IS A MAJOR REQUIREMENT FOR DIATOM GROWTH BUT UNNECESSARY FOR OTHER FORMS. VARIOUS OTHER ELEMENTS--ZINC, BORON, COBALT, MOLYBDENUM, AND COPPER MAY BE NECESSARY AS TRACE ELEMENTS. NORMAL REQUIREMENTS OF VARIOUS ALGAE ARE GIVEN FOR CARBON, NITROGEN, PHOSPHORUS, UNDER CERTAIN CONDITIONS. (JONES-WISCONSIN)

FIELD 05C

ACCESSION NO. W70-02804

MICRONUTRIENT REQUIREMENTS FOR GREEN PLANTS, ESPECIALLY ALGAE,

CHARLES F. KETTERING RESEARCH LAB., YELLOW SPRINGS, OHIO.

CLYDE EYSTER.

ALGAE AND MAN, JACKSON, DANIEL F, EDITOR, PLENUM PRESS, NY, P 86-119, 1964.
11 FIG, 2 TAB, 174 REF.

DESCRIPTORS:
    *NUTRIENT REQUIREMENTS, *PLANT GROWTH, *ALGAE, ELEMENTS(CHEMICAL),
    CHLOROPHYTA, CYANOPHYTA, CHLORELLA, MANGANESE, PHOTOSYNTHESIS, OXYGEN,
    CARBON DIOXIDE, SCENEDESMUS, CHLOROPHYLL, IRON, CHELATION, ENZYMES,
    CHLORINE, LIGHT INTENSITY, NITRATE, EUGLENA, CALCIUM, AMMONIUM SALTS,
    NITROGEN FIXATION, BORON, AZOTOBACTER, MOLYBDENUM, COBALT, COPPER,
    SODIUM, MAGNESIUM, POTASSIUM, SULFUR, PHOSPHORUS, HYDROGEN.

IDENTIFIERS:
    MACRONUTRIENTS, MICRONUTRIENTS, CHLORELLA PYRENOIDOSA, AUTOTROPHIC
    GROWTH, HETEROTROPHIC GROWTH, ANKISTRODESMUS, NOSTOC MUSCORUM,
    SCENEDESMUS OBLIQUUES, SCENEDESMUS QUADRICAUDA, PORPHYRIDIUM CRUENTUM,
    LEMNA MINOR, PHOTOPHOSPHORYLATION, VANADIUM, ZINC, EUGLENA GRACILIS,
    ANABAENA CYLINDRICA, BACILLUS SUBTILIS, ASPERGILLUS ORYZAE NIGER,
    SILICON, PHANEROGAMS.

ABSTRACT:
    FIVE ELEMENTS, MANGANESE, IRON, CHLORINE, ZINC, AND VANADIUM ARE
    GENERALLY REQUIRED BY ALGAE AND GREEN PLANTS FOR PHOTOSYNTHESIS.
    MANGANESE AND CHLORINE HAVE A ROLE IN OXYGEN EVOLUTION. VANADIUM
    ACCELERATES PHOTOSYNTHESIS UNDER HIGH LIGHT INTENSITY BUT THIS EFFECT
    HAS BEEN REPORTED ONLY FOR SCENEDESMUS OBLIQUUS AND CHLORELLA. IRON IS
    THE METAL PART OF AT LEAST TWO PLANT CYTOCHROMES WHICH OCCUR IN THE
    CHLOROPLASTS OF GREEN CELLS AND PROBABLY FACILITATES TRANSFER OF
    ELECTRONS DURING PHOTOSYNTHESIS. ROLE OF ZINC IN PHOTOSYNTHESIS IS
    UNCERTAIN; IT MAY AID IN HYDROGEN TRANSFER SINCE SINC IS KNOWN TO BE
    THE METAL CONSTITUENT OF VARIOUS DEHYDROGENASES. NITROGEN FIXATION IN
    ALGAE HAVING THE CAPACITY TO FIX GASEOUS NITROGEN, REQUIRES THE
    FOLLOWING ELEMENTS: BORON, CALCIUM, IRON, MOLYBDENUM, AND PERHAPS
    COBALT; THESE ELEMENTS MAY BE REQUIRED FOR ONE OR MORE OTHER FUNCTIONS.
    CRITICAL CONCENTRATION OF AN ELEMENT FOR ITS TWO OR MORE FUNCTIONS IS
    USUALLY MUCH HIGHER FOR NITROGEN FIXATION. ALGAE WHICH ARE FORCED TO
    FIX NITROGEN HAVE HIGH BORON AND CALCIUM REQUIREMENTS. PLACE OF COBALT
    UTILIZATION IS UNCERTAIN. SOME EVIDENCE INDICATES THAT SMALL
    CONCENTRATIONS OF MANGANESE, CALCIUM, BORON, COBALT, COPPER, AND
    SILICON ARE REQUIRED FOR OTHER METABOLIC FUNCTIONS. (SEE
    W69-07832). (JONES-WISCONSIN)

FIELD 05C

ACCESSION NO. W70-02964

ENVIRONMENTAL CONDITIONS AND THE PATTERN OF METABOLISM IN ALGAE,

WESTFIELD COLL., LONDON (ENGLAND). DEPT. OF BOTANY.

G. E. FOGG.

ALGAE AND MAN, DANIEL F. JACKSON, EDITOR, PLENUM PRESS, NY, P 77-85, 1964. 22
    REF.

DESCRIPTORS:
    *ALGAE, *METABOLISM, *ENVIRONMENTS, VARIABILITY, CHLORELLA, CARBON,
    TEMPERATURE, LIGHT INTENSITY, NUTRIENT REQUIREMENTS, NITROGEN, NITROGEN
    FIXATION, PHOTOCHEMICAL, CARBON DIOXIDE, SYNTHESIS, ENZYMES,
    CHLOROPHYTA, RHODOPHYTA, SCENEDESMUS, CHLAMYDOMONAS, ECOLOGY,
    PHOSPHORUS, SULPHUR, HYDROGEN ION CONCENTRATION, TIME.

IDENTIFIERS:
    BOTRYOCOCCUS BRAUNII, METABOLIC PATTERN, ANABAENA CYLINDRICA,
    GLYCOLYSIS, TRICARBOXYLIC ACID CYCLE, AUTORADIOGRAPHY, MONODUS
    SUBTERRANEUS, NAVICULA PELLICULOSA, XANTHOPHYCEAE, BACILLARIOPHYCEAE,
    NITZSCHIA PALEA, MYXOPHYCEAE, CHLAMYDOMONAS MOEWUSSI, FAT ACCUMULATION.

ABSTRACT:
A SINGLE STRAIN OF ALGAL SPECIE MAY SHOW REMARKABLE VARIATIONS IN
INTENSITY AND PATTERN OF ITS METABOLIC ACTIVITIES, DEPENDING ON
EXPOSURE CONDITIONS. EFFECTS OF LIGHT INTENSITY, TEMPERATURE, AND
HYDROGEN ION CONCENTRATION ON RATES OF INDIVIDUAL METABOLIC PROCESSES
AND ON THEIR FINAL GROWTH IN CELL NUMBERS ARE KNOWN, YET THE
DIFFERENTIAL EFFECTS THAT THESE CONDITIONS MAY HAVE ON DIFFERENT
PROCESSES ARE LARGELY UNKNOWN. TIME FACTOR IS SIGNIFICANT; TEMPERATURE
CHANGES MAY PRODUCE MARKED ALTERATIONS IN THE BALANCE OF MAJOR ANABOLIC
ALGAL PROCESSES. AVAILABILITY OF METABOLITES MAY AFFECT METABOLIC
PATTERN, APART FROM CHANGES ON OVERALL METABOLIC RATE. IN ALGAL
POPULATIONS GROWING EXPONENTIALLY, SYNTHESIS OF PROTEINS AND OTHER
PROTOPLASMIC CONSTITUENTS PREDOMINATE AND DIRECTLY UTILIZE
INTERMEDIATES OF THE PHOTOSYNTHETIC CARBON CYCLE. A SIMILAR EFFECT IS
PERHAPS INVOLVED IN EXCRETION OF GLYCOLIC ACID FROM PHOTOSYNTHESIZING
CELLS. ALGAE IS ABLE TO PRODUCE ADAPTIVE ENZYMES AND METABOLIC PATTERN
IS DETERMINED PRIMARILY BY RELATIVE ACTIVITIES OF THE VARIOUS ENZYME
SYSTEMS. IT IS UNCLEAR WHETHER FAT ACCUMULATION IN NITROGEN-DEFICIENT
CULTURES RESULTS FROM HALTING OF THE DEVELOPMENTAL CYCLE AT A POINT AT
WHICH FAT SYNTHESIS PREDOMINATES. FACTORS AFFECTING FAT ACCUMULATION
ARE DISCUSSED. (SEE W69-07832). (JONES-WISCONSIN)

FIELD 05C

ACCESSION NO. W70-02965

INSECTICIDES AND ALGAE: TOXICITY AND DEGRADATION,

ONTARIO WATER RESOURCES COMMISSION, TORONTO. DIV. OF RESEARCH.

A. E. CHRISTIE.

ONTARIO WATER RESOURCES COMMISSION, DIV OF RESEARCH PUBLICATION NO 27, 1968.
26 P, 9 TAB, 31 REF.

DESCRIPTORS:
*ALGAE, *ALGAL CONTROL, *ALGICIDES, ANALYTICAL TECHNIQUES, AQUATIC
ALGAE, AQUATIC ENVIRONMENT, AQUATIC MICROBIOLOGY, BALANCE OF NATURE,
BIOASSAY, CHLOROPHYTA, ENVIRONMENTAL EFFECTS, INHIBITION, LIMNOLOGY,
PESTICIDES, PESTICIDE REMOVAL, PESTICIDE RESIDUES, PHYSIOLOGICAL
ECOLOGY, PHYTOPLANKTON, PHYTOTOXICITY, PLANKTON, POLLUTANT
IDENTIFICATION, SEWAGE, SEWAGE DISPOSAL, SEWAGE LAGOONS, SEWAGE
TREATMENT, STABILIZATION PONDS, WATER POLLUTION CONTROL, WATER
POLLUTION EFFECTS, DDT.

IDENTIFIERS:
CHLORELLA PYRENOIDOSA, PESTICIDE DEGRADATION, SEVIN, MALATHION,
ACETONE, ETHANOL.

ABSTRACT:
RELATIONSHIPS BETWEEN THREE INSECTICIDES (DDT, SEVIN AND MALATHION) AND
ALGAE WITH RESPECT TO TOXICITY AND DEGRADATION WERE INVESTIGATED USING
MATERIALS OBTAINED FROM A NEARBY WASTE STABILIZATION POND AND AXENIC
CULTURES OF CHLORELLA PYRENOIDOSA AT PH 6.0 AND 9.0. DDT EXHIBITED NO
TOXIC PROPERTIES TO EITHER SYSTEM AT CONCENTRATIONS UP TO 100
MILLIGRAM/LITER AND WAS BROKEN DOWN ONLY TO A LIMITED EXTENT BY ALGAE.
MALATHION EXHIBITED NO PERSISTENT INHIBITORY EFFECT TO ALGAE AND
APPEARED TO BE METABOLIZED FAIRLY READILY IN COMPARISON TO THE OTHER
TWO INSECTICIDES STUDIED. SEVIN AT 100 MILLIGRAM/LITER WAS MORE TOXIC
TO THE MIXED POND ALGAE THAN TO CHLORELLA. THE TOXICITY OF SEVIN TO
CHLORELLA WAS MORE PRONOUNCED AT PH 6 THAN AT PH 9 (30 AND 15%
INHIBITION, RESPECTIVELY). THE EFFECT OF SEVIN COULD BE DETECTED AT 0.1
MILLIGRAM/LITER, BUT INCREASING CONCENTRATS HAD LITTLE INCREASED
EFFECTIVENESS. (FITZGERALD-WISCONSIN)

FIELD 05G

ACCESSION NO. W70-02968

THE ORIGIN AND QUANTITIES OF PLANT NUTRIENTS IN LAKE MENDOTA,

WISCONSIN UNIV., MADISON.

GERARD A. ROHLICH, AND WILLIAM L. LEA.

LAKE INVESTIGATIONS COMMITTEE REPORT, WISCONSIN UNIV, MADISON, 1949. 8 P, 6
TAB.

DESCRIPTORS:
*EUTROPHICATION, *AGRICULTURAL WASTES, *ALGAE, *LAKES, *TRIBUTARIES,
SAMPLING, FLOW, ANALYSIS, FERTILIZATION, PLANT GROWTH SUBSTANCES,
SEWAGE, SULFATE, MANGANESE, IRON, CARBON DIOXIDE.

IDENTIFIERS:
*INLAND, HORIZONTAL SAMPLING, VERTICAL SAMPLING, LAKE MENDOTA(WIS),
LAKE MONONA(WIS), LAKE WAUBESA(WIS), LAKE KEGONSA(WIS), WEED BEDS,
PHOSPHORUS, NITROGEN BALANCE.

ABSTRACT:
TO OBTAIN A COMPREHENSIVE PICTURE OF EUTROPHICATION DYNAMICS OF LAKE
MENDOTA, WISCONSIN, THREE SERIES OF STATIONS--INLAND, HORIZONTAL, AND
VERTICAL-- WERE ESTABLISHED ON THE LAKE AND ITS TRIBUTARIES. ANALYSES
OF SAMPLES COLLECTED AT INLAND STATIONS ENABLED DETERMINATION OF
NUTRIENTS RETAINED IN THE LAKE. SAMPLING STATIONS OF THE HORIZONTAL
SERIES FORMED TRANSECTS FROM UPPER PARTS OF STREAMS TO THEIR MOUTHS AND
INTO THE LAKE. THESE STATIONS PROVIDED INFORMATION ON THE MODIFICATION
OF ORIGINAL NUTRIENT CONCENTRATION BY WEED BEDS AND ALGAE. VERTICAL
SERIES STATIONS WERE LOCATED IN TWO 65-FOOT-DEEP AND ONE 80-FOOT-DEEP
HOLES. SAMPLING OF THREE WATER HORIZONS--SURFACE, THERMOCLINE AT 10
METERS, AND ABOUT 0.5 METERS ABOVE THE BOTTOM--AIMED TO REVEAL
VARIATION IN NUTRIENT CONCENTRATION AND AQUATIC BIOMASS WITH DEPTH.
LAKE'S INFLOW-OUTFLOW BALANCE CONSTITUTED 130 PLUS OR MINUS 1.2 CUBIC
FEET/SECOND. ANALYSES OF CHEMICAL FACTORS INCLUDED PH, ALKALINITY,
DISSOLVED OXYGEN, BIOCHEMICAL OXYGEN DEMAND, SOLUBLE PHOSPHORUS, TOTAL
PHOSPHORUS, FREE AMMONIA NITROGEN, TOTAL ORGANIC NITROGEN, NITRITE,
NITRATE, SULFATES, SILICATES, MANGANESE, AND IRON. DURING THE YEAR
17,460 POUNDS SOLUBLE PHOSPHORUS ENTERED THE LAKE, THE AMOUNT
CORRESPONDING TO AN APPLICATION OF 1.8 POUNDS/ACRE OF LAKE SURFACE.
ENRICHMENT IN ORGANIC NITROGEN, LARGELY IN NITRATE FORM, AMOUNTED TO
259,720 POUNDS, EQUIVALENT TO FERTILIZATION OF 26.7 POUNDS/ACRE.
(WILDE-WISCONSIN)

FIELD 05C

ACCESSION NO. W70-02969

IN VITRO RESPONSES OF SOFT WATER ALGAE TO FERTILIZATION,

ONTARIO WATER RESOURCES COMMISSION, TORONTO. DIV. OF RESEARCH.

A. E. CHRISTIE.

DIVISION OF RESEARCH PAPER NO 2024, AUG 1969. 16 P, 2 FIG, 4 TAB, 9 REF.

DESCRIPTORS:
*BIOASSAY, PHYTOPLANKTON, LAKES, EUTROPHICATION, SAMPLING, SURFACES WATERS, CARBON, FERTILITY, ALGAE, NITROGEN, PHOSPHORUS, WATER POLLUTION EFFECTS.

IDENTIFIERS:
*TWELVE MILE LAKE(CANADA), *ONTARIO, *CANADA, NUTRIENT LEVELS, SOFT WATER.

ABSTRACT:
A 'PRELIMINARY' SINGLE SAMPLE FROM OLIGOTROPHIC TWELVE MILE LAKE, ONTARIO, WAS BIOASSAYED BY SPIKING WITH SODIUM BICARBONATE, POTASSIUM NITRATE, AND POTASSIUM DIHYDROGEN PHOSPHATE, OR COMBINATIONS OF THESE COMPOUNDS. BOTTLES WERE CARRIED 21 DAYS AT 20 C UNDER A DIURNAL CYCLE OF 14 HOURS LIGHT AND 10 HOURS DARK UNDER THE EXTREMELY HIGH LIGHT INTENSITY OF 4000-FOOT-CANDLES. THE SAMPLE'S INDIGINOUS POPULATION SERVED AS THE ASSAY RESPONSE ORGANISMS, AND WERE MEASURED BY AREAL STANDARD UNITS. THE ONLY RESPONSE BY THE ALGAE IN THE SAMPLE WAS TO A COMBINATION OF NITROGEN AND PHOSPHORUS. (GERHOLD-WISCONSIN)

FIELD 05C

ACCESSION NO. W70-02973

STUDIES ON CHEMICALS WITH SELECTIVE TOXICITY TO BLUE-GREEN ALGAE,

WISCONSIN UNIV., MADISON. WATER CHEMISTRY LAB.

GEORGE P. FITZGERALD, GERALD C. GERLOFF, AND FOLKE SKOOG.

SEWAGE AND INDUSTRIAL WASTES, VOL 24, NO 7, P 888-896, 1952. 1 FIG, 6 TAB, 7 REF.

DESCRIPTORS:
*TOXICITY, *CYANOPHYTA, *CHEMCONTROL, EUTROPHICATION, BASS, WATER POLLUTION CONTROL, MINNOWS, NUISANCE ALGAE, SUNFISHES, ALGAE, DAPHNIA, LAKES, CHLOROPHYTA, STREAMS, DIATOMS, RESERVOIRS, CULTURES, SEWAGE EFFLUENTS, WISCONSIN, ALGICIDES, HYDROLYSIS, COPPER SULFATE, FISHES, HERBICIDES, MIDGES, PESTICIDES.

IDENTIFIERS:
*SELECTIVE TOXICITY, CHEMICAL STRUCTURE, 2,3-DICHLORONAPHTHOQUINONE, MICROCYSTIS AERUGINOSA, APHANIZOMENON FLOS-AQUAE, MINIMUM LETHAL DOSES, CHU'S MEDIA, QUINONE DERIVATIVES, ALKYL SUBSTITUTED IMIDAZOLINES, BIOLOGICAL ACTIVITY, QUARTERNARY, AMMONIUM COMPOUNDS, MINIMUM LETHAL EXPOSURE, PHENANTHROQUINONE, MADISON(WIS), ELODEA, REDOX COMPOUNDS, CERATOPHYLLUM, MYROPHYLLUM, RHIZOCLONIUM, PHOTOBIOLOGY.

ABSTRACT:
PROCEDURES USED AND RESULTS OF TOXICITY TESTS OF 300 ORGANIC CHEMICALS TO MICROCYSTIS AERUGINOSA, A BLOOM-PRODUCING CYANOPHYTE (BLUE-GREEN ALGA) ARE DESCRIBED. MANY COMPOUNDS WERE TOXIC IN CONCENTRATIONS IN THE RANGE 1-10 PARTS/MILLON (PPM); SEVERAL QUINONE-LIKE COMPOUNDS ARE TOXIC BELOW 1 PPM, AND ONE, 2,3-DICHLORONAPTHOQUINONE, IS LETHAL AT CONCENTRATIONS OF 0.002 PPM. APPARENTLY SELECTIVELY TOXIC TO NUISANCE CYANOPHYTES, IT IS NONTOXIC TO MOST GREEN ALGAE AND HIGHER PLANTS, AND HAS NO SHORTTERM DELETERIOUS EFFECTS ON SOME FISHES AND INVERTEBRATES EVEN WHEN PRESENT IN EXCESS OF SATURATION CONCENTRATIONS. FURTHER STUDY OF THIS COMPOUND TO RULE OUT INJURIOUS EFFECTS TO DESIRABLE SPECIES AFTER LONG EXPOSURES ARE RECOMMENDED. AGE OF CULTURES OF MICROCYSTIS INFLUENCE ITS RESPONSE TO THE COMPOUND; CENTRIFUGED CELLS FROM BOTH 6-DAY AND 19-DAY CULTURES WERE MARKEDLY LESS SENSITIVE WHEN RESUSPENDED IN OLD MEDIUM. THIS EFFECT WAS PARTIALLY COUNTERACTED AFTER FORTIFICATION OF OLD MEDIUM WITH NUTRIENTS. ESTIMATES OF MINIMUM LETHAL EXPOSURES INDICATE THAT COMPOUND IS EFFECTIVELY TOXIC AFTER 8-10 MINUTES TREATMENT OF CELLS. LIMITED FIELD TESTS IN BARRELS, TANKS, AND LAGOONS, CONTAINING DENSE GROWTHS OF NUISANCE ALGAE, PRINCIPALLY APHANIZOMENON FLOS-AQUAE, IN LAKE WATER OR SEWAGE EFFLUENT, INDICATE THAT COMPOUND IS EFFECTIVE IN LOW CONCENTRATIONS (0.01-0.1 PPM). (EICHHORN-WISCONSIN)

FIELD 05G

ACCESSION NO. W70-02982

EVALUATION OF WATER QUALITY TRENDS, SHENANDOAH RIVER, VIRGINIA, FRONT ROYAL TO
  BERRYVILLE,

FEDERAL WATER POLLUTION CONTROL ADMINISTRATION, CHARLOTTESVILLE, VA. MIDDLE
  ATLANTIC REGION.

JOHN W. BAUMEISTER.

FEDERAL WATER POLLUTION CONTROL ADMINISTRATION, MIDDLE ATLANTIC REGION
  TECHNICAL REPORT 2, DEC 1968. 14 P, 5 FIG, 5 TAB, 20 REF.

  DESCRIPTORS:
    *WATER QUALITY, *WATER POLLUTION, *RIVERS, WATER CHEMISTRY, WATER
    POLLUTION SOURCES, PHOSPHORUS, NITROGEN COMPOUNDS, EUTROPHICATION,
    PRODUCTIVITY, ALGAE, AQUATIC LIFE.

  IDENTIFIERS:
    SHENANDOAH RIVER.

  ABSTRACT:
    PHYSICAL AND CHEMICAL DATA COLLECTED AT THE BERRYVILLE SURVEILLANCE
    STATION WERE EVALUATED TO DETERMINE POLLUTION AND EUTROPHICATION TRENDS
    IN THE MAIN STEM OF THE SHENANDOAH DOWNSTREAM FROM FRONT ROYAL.
    CHEMICAL QUALITY WAS WITHIN THE RAW WATER CRITERIA ESTABLISHED FOR
    PUBLIC WATER SUPPLIES; HOWEVER, TOTAL·PHOSHORUS AND NITROGEN CONTENT OF
    RIVER WATER WAS SUFFICIENTLY HIGH TO INDICATE A FERTILE STREAM CAPABLE
    OF PRODUCING ABUNDANT AQUATIC ORGANISMS. THIRTEEN YEARS OF BOTTOM
    SAMPLES COLLECTED OVER A 30 YR PERIOD IN THIS REACH SHOW A GRADUAL
    INCREASE IN THE PRODUCTIVITY OF AQUATIC INVERTEBRATES. FOR THE MOST
    PART, THESE SAMPLES CONTAINED CLEAN WATER AND FACULTATIVE FORMS WHICH
    INDICATE THE PRESENCE OF RELATIVELY CLEAN WATER. PHYTOPLANKTON NUMBERS
    AND POPULATION COMPOSITION DURING THE PERIOD 1962-1965 INDICATE A
    CHANGE FROM A DIATOM DOMINATED POPULATION TO ONE DOMINATED BY GREEN
    ALGAE. NUISANCE ALGAE WERE NOT PRESENT IN QUANTITIES SUFFICIENT TO
    CAUSE WATER TREATMENT PROBLEMS. (KNAPP-USGS)

  FIELD 05A

  ACCESSION NO. W70-03068

ALGAL GROWTH AND PRIMARY PRODUCTIVITY IN A THERMAL STREAM,

WASHINGTON UNIV., SEATTLE. DEPT. OF ZOOLOGY.

JOHN G. STOCKNER.

JOURNAL FISHERIES RESEARCH BOARD OF CANADA, VOL 25, NO 10, P 2037-2058, 1968.
13 FIG, 7 TAB, 25 REF.

DESCRIPTORS:
*PRIMARY PRODUCTIVITY, *ALGAE, *THERMAL WATER, PERIPHYTON, BIOMASS,
PHYSIOLOGICAL ECOLOGY, CYANOPHYTA, DIFFUSION, RESPIRATION, PHOSPHATES,
ECOSYSTEMS, ORGANIC MATTER, WATER POLLUTION, THERMAL POLLUTION,
STREAMS, NUTRIENTS, LIGHT, WASHINGTON, WATER CHEMISTRY, OXYGEN,
HYDROGEN ION CONCENTRATION, PHOTOSYNTHESIS, ALKALINITY, NITRATES,
CONDUCTIVITY, CHLORIDES, CARBONATES, SULFATES, BICARBONATES, POTASSIUM,
IRON, SODIUM, CALCIUM, MAGNESIUM, PHOSPHORUS.

IDENTIFIERS:
*ALGAL GROWTH, *THERMOBIOLOGY, *DIURNAL OXYGEN CURVES, CALORIMETRY,
SOLARIMETRY, YELLOWSTONE NATIONAL PARK, MT LASSEN NATIONAL PARK, MT
RAINIER NATIONAL PARK, OHANAPECOSH HOT SPRINGS(WASH), SCHIZOTHRIX,
PHORMIDIUM, WINKLER TECHNIQUE, OXYGEN ELECTRODES, ATOMIC ABSORPTION
SPECTROPHOTOMETRY, TOTAL DISSOLVED SOLIDS, HEDRIODISCUS, CALOPARYPHUS.

ABSTRACT:
TROPHIC RELATIONSHIPS IN FLOWING WATER ARE NOT AS WELL UNDERSTOOD AS
THOSE IN MORE STATIC ENVIRONMENTS. BECAUSE OF THEIR RELATIVELY CONSTANT
CONDITIONS, STREAMS DRAINING THERMAL SPRINGS ARE IDEAL SITES FOR SUCH
STUDIES. ESTIMATES OF PRIMARY PRODUCTION AND RESPIRATION OF BLUE-GREEN
ALGAE IN STREAM DRAINING ONE OF THE OHANAPECOSH HOT SPRINGS,
WASHINGTON, WERE OBTAINED THROUGHOUT 1966 FROM DAILY OXYGEN CURVES.
RATES OF GROSS PRIMARY PRODUCTIVITY WERE SLIGHTLY HIGHER THAN PUBLISHED
VALUES FROM OTHER AQUATIC ECOSYSTEMS, BUT WERE LESS THAN RECENT
ESTIMATES, DERIVED FROM MEASUREMENT OF RADIOCARBON UPTAKE FROM OTHER
THERMAL WATERS. CALCULATED RATIOS, PRODUCTION/RESPIRATION, WERE IN
RANGE 1.1-5.0. AS GROWING SEASON PROGRESSED, PHOTOSYNTHETIC EFFECTS ON
IONIC COMPOSITION OF EFFLUENT WATER, AS WELL AS MORPHOLOGICAL CHANGES
IN ALGAL MAT, WERE OBSERVABLE. CHANGES IN ALGAL BIOMASS DURING
1964-1965, TOGETHER WITH GROWTH RATES IN DENUDED AREAS BY TWO METHODS
OF HARVEST WERE STUDIED. ACCELERATED GROWTH OCCURRED IN MAY AND JUNE,
ASSOCIATED WITH MAXIMAL VALUES FOR PRIMARY PRODUCTION, WHEREAS GROWTH
WAS MINIMAL IN WINTER. SIGNIFICANT DIFFERENCES IN GROWTH BETWEEN YEARS
WERE ATTRIBUTED TO DIFFERENCES IN HARVESTING METHODS, AND TO PRODUCTION
OF AUTO-INHIBITORY TOXINS BY SENESCENT CELLS IN MATURE MAT. ANNUAL
ENERGY BUDGET FOR MAT IS DERIVED FROM 1966 ESTIMATES.
(EICHHORN-WISCONSIN)

FIELD 05C

ACCESSION NO. W70-03309

CONTROL OF BLUE-GREEN ALGAE BLOOMS WITH 2,3-DICHLORONAPHTHOQUINONE,

WISCONSIN UNIV., MADISON. DEPT. OF BOTANY.

GEORGE P. FITZGERALD, AND FOLKE SKOOG.

SEWAGE AND INDUSTRIAL WASTES, VOL 26, NO 9, P 1136-1140, SEPT 1954. 1 FIG, 1 REF.

DESCRIPTORS:
*ALGICIDES, *CYANOPHYTA, *EUTROPHICATION, WISCONSIN, WATER QUALITY CONTROL, LAKES, WATER POLLUTION SOURCES, CHLOROPHYTA, NUISANCE ALGAE, FISH, ZOOPLANKTON, SNAILS, APPLICATION METHODS.

IDENTIFIERS:
*2,3-DICHLORONAPHTHOQUINONE, MICROCYSTIS, APHANIZOMENON, ANABAENA, GLOEOTRICHIA, JANESVILLE(WIS), SPAULDING'S POND(WIS), HYDRODICTYON, NAJAS, ANACHARIS, LAKE WAUBESA(WIS), MACROPHYTES, ALGAL GROWTH.

ABSTRACT:
METHODS FOR APPLICATION OF THE SELECTIVE ALGICIDE, 2,3-DICHLORONAPHTHOQUINONE (2,3-CNQ), AND RESULTS OF A LARGE-SCALE TEST OF ITS EFFECTIVENESS IN CONTROLLING EXCESSIVE GROWTHS OF CYANOPHYTES (BLUE-GREEN ALGAE) IN SPAULDING'S POND, WISCONSIN, ARE REPORTED. THIS HIGHLY EUTROPHIC LAKE OF 27 ACRES NORMALLY SUPPORTS A CONTINUOUS BLOOM OF ALGAE, PRINCIPALLY MICROCYSTIS AND ALPHANIZOMENON, DURING SUMMER. SPRAY APPLICATIONS OF 2,3-CNQ SUSPENSIONS, YIELDING CONCENTRATIONS OF 30-55 PARTS PER BILLION IN THE LAKE, CLUMPED AND KILLED HEAVY GROWTHS OF CYANOPHYTES. SUCH CHEMICAL TREATMENT HAD NO OBSERVABLE HARMFUL EFFECTS ON CHLOROPHYTES (GREEN ALGAE), HIGHER AQUATIC PLANTS, FISH, OR ZOOPLANKTON. SOME INSTANCES OF RAPIDLY RECURRING CYANOPHYTE BLOOMS REQUIRED REPEATED APPLICATIONS. LASTING EFFECTS OF TREATMENT WITH 2,3-CNQ MAY BE INDIRECTLY INCREASED AS A RESULT OF VIGOROUS GROWTH OF CHLOROPHYTES AND MACROPHYTES WHICH FOLLOWS SUPPRESSION OF CYANOPHYTES. (EICHHORN-WISCONSIN)

FIELD 05G, 05C

ACCESSION NO. W70-03310

MICROBIOLOGY OF SEWAGE LAGOONS-EFFECTS OF INDUSTRIAL WASTES ON LAGOON ECOLOGY,

NORTH DAKOTA UNIV., GRAND FORKS. SCHOOL OF MEDICINE.

JOHN W. VENNES.

AVAILABLE FROM THE CLEARINGHOUSE AS PB-189 159, $3.00 IN PAPER COPY $0.65 IN MICROFICHE. COMPLETION REPORT, AUG 1969. 82 P, 3 FIG, 41 TAB, 35 REF. OWRR PROJECT NO A-016-NDAK.

DESCRIPTORS:
*MICROBIOLOGY, *SEWAGE LAGOONS, *INDUSTRIAL WASTES, SEWAGE TREATMENT, NORTH DAKOTA, COLIFORMS, OXIDATION LAGOONS, BIOCHEMICAL OXYGEN DEMAND, PHOTOSYNTHESIS, HYDROGEN SULFIDE, LAKES, ALGAE, ENTERIC BACTERIA, METHANE BACTERIA, CHEMICAL ANALYSIS, WATER POLLUTION, CHLORELLA, SCENEDESMUS.

IDENTIFIERS:
*MICROBIAL ECOLOGY, PURPLE SULFUR BACTERIA, POPULATION EQUIVALENTS(WASTES), RED RIVER VALLEY(NORTH DAKOTA), SULFATE REDUCTION, POTATO PROCESSING WASTES, CHEESE MANUFACTURING WASTES, STARCH MANUFACTURING WASTES, WATER QUALITY STANDARDS, GRAFTON(ND), LAKOTA(ND), HARVEY(ND), CHROMATIUM VINOSUM, THIOCAPSA FLORIDANA, ENTEROCOCCI, PARK RIVER(ND), SHEYENNE RIVER(ND), THIOPEDIA.

ABSTRACT:
STATE AND FEDERAL LEGISLATION, REGULATING EFFLUENT DISCHARGE TO NATURAL WATERS, IS SUCH THAT FOOD PROCESSING INDUSTRIES IN NORTH DAKOTA MUST DISCHARGE WASTES INTO MUNICIPAL LAGOONS RESULTING IN FREQUENT OVERLOADS OF THESE SYSTEMS. LAGOON AT GRAFTON, DESIGNED FOR 5-DAY BIOCHEMICAL OXYGEN DEMAND (BOD-5) OF 30 POUNDS/ACRE PER DAY, RECEIVES 2'40, LARGELY FROM TWO POTATO PROCESSING PLANTS. INCREASED LOADINGS FAVOR (AS DOMINANT MICROORGANISMS IN THESE LAGOONS) THE PURPLE SULFUR BACTERIA (PSB), WHICH PHOTOSYNTHETICALLY UTILIZE HYDROGEN SULFIDE AND VARIOUS VOLATILE ORGANIC ACIDS FOR GROWTH. MOST REDUCTION IN BOD-5 AT GRAFTON IS ATTRIBUTABLE TO THREE GENERA OF PSB: CHROMATIUM, THIOPEDIA, THIOCAPSA. INCREASED LOADING DOES NOT SIGNIFICANTLY MULTIPLY NUMBERS OF ENTERIC ORGANISMS HERE. A NATURAL LAKE, USED FOR SEWAGE STABILIZATION IN LAKOTA, RECEIVES WHEY; IT RECOVERS TO AN ALGAL PHASE DURING SUMMER, AND SHOWS NO INCREASE IN NUMBERS OF ENTERIC ORGANISMS. AT HARVEY, AN AERATED CONTINUOUS DISCHARGE LAGOON, RECEIVING DOMESTIC WASTES AND INTERMITTENT CREAMERY WASTES, REDUCES BOD-5 TO NEAR STREAM LEVELS, BUT CANNOT REDUCE COLIFORMS TO ACCEPTABLE NUMBERS. IF THE LATTER TREATMENT IS TO SUCCEED IN NORTHERN CLIMATES, CONSIDERATION MUST BE GIVEN TO LONGER DETENTION TIMES, GREATER DELIVERY OF OXYGEN TO CELLS, CHLORINATION, OR REEVALUATION OF STREAM STANDARDS. (EICHHORN-WISCONSIN)

FIELD 05C

ACCESSION NO. W70-03312

A LABORATORY METHOD FOR THE STUDY OF MARINE BENTHIC DIATOMS,

OREGON STATE UNIV., CORVALLIS. DEPT. OF BOTANY.

C. DAVID MCINTIRE, AND BARRY L. WULFF.

LIMNOLOGY AND OCEANOGRAPHY, VOL 14, NO 5, P 667-678, 1969. 6 FIG, 4 TAB, 20 REF.

DESCRIPTORS:
*MARINE ALGAE, *DIATOMS, *BENTHIC FLORA, LIGHT INTENSITY, BIOMASS, DISTRIBUTION, ECOSYSTEMS, POPULATION, OREGON, INTERTIDAL AREAS, PHYSIOLOGICAL ECOLOGY, LITTORAL, ESTUARIES, PHOTOSYNTHESIS, RESPIRATION, PIGMENTS, CHLOROPHYLL, TEMPERATURE, OXYGEN, SALINITY.

IDENTIFIERS:
DESICCATION, YAQUINA BAY(ORE), CAROTENOIDS, SPECIES, DIVERSITY INDEX, ACHNANTHES, ACTINOPTYCHUS, AMPHIPLEURA RUTILANS, AMPHORA, AULACODISCUS, BACILLARIA, BIDDULPHIA, CHAETOCEROS, COCCONEIS, COSCINODISCUS, DIMEROGRAMMA, DIPLONEIS, EUNOTOGRAMMA, FRAGILARIA, GOMPHONEMA, GYROSIGMA, MELOSIRA, NAVICULA, NITZSCHIA, SKELETONEMA, PLAGIOGRAMMA, PLEUROSIGMA, RHAPHONEIS, RHOICOSPHENIA, SURIRELLA, SYNEDRA.

ABSTRACT:
EFFECTS OF LIGHT INTENSITY AND EXPOSURE TO DESICCATION ON VERTICAL DISTRIBUTION AND GROWTH OF MARINE BENTHIC DIATOM POPULATIONS WERE INVESTIGATED IN A LABORATORY MODEL ECOSYSTEM AND RESPIROMETER CHAMBER. DIATOM FLORA DEVELOPED WAS SIMILAR TO THAT FROM FIELD STATIONS IN LOWER YAQUINA BAY, OREGON. VERTICAL DISTRIBUTION OF MANY SPECIES WAS CLOSELY RELATED TO LIGHT INTENSITY AND PERIOD OF EXPOSURE TO DESICCATION. BIOMASS ACCUMULATED RAPIDLY ON SUBSTRATES SUBJECTED TO HIGH LIGHT INTENSITIES WITHOUT EXPOSURE TO DESICCATION. COMMUNITIES ACCLIMATED TO VARIOUS LIGHT INTENSITIES AND PERIOD OF DESICCATION RESPONDED DIFFERENTLY TO CHANGES IN LIGHT INTENSITY IN RESPIROMETER CHAMBER. EXPERIMENTAL RESULTS SHOW THAT LABORATORY APPARATUS DESCRIBED CAN BE USEFUL IN STUDY OF SIMPLIFIED INTERTIDAL COMMUNITIES. SYSTEM PROVIDES SOME LABORATORY CONTROL OVER SALINITY, LIGHT INTENSITY, TIDAL CYCLE, AND PROBABLY TEMPERATURE. MOVEMENTS AND GRAZING ACTIVITIES OF SMALL MARINE INVERTEBRATES COULD BE STUDIED. IT IS EMPHASIZED THAT USE OF SUCH LABORATORY ECOSYSTEMS IS SUBJECT TO CERTAIN LIMITATIONS RESULTING FROM SIMPLIFICATIONS OF NATURE. LABORATORY ECOSYSTEMS ARE BEST USED TO GAIN INFORMATION TO SUPPLEMENT AND HELP UNDERSTAND CONCURRENT OBSERVATIONS IN THE FIELD. (JONES-WISCONSIN)

FIELD 05C

ACCESSION NO. W70-03325

EFFECT OF SURFACE/VOLUME RELATIONSHIP, CO-SUB-2 ADDITION, AERATION, AND MIXING
ON NITRATE UTILIZATION BY SCENEDESMUS CULTURES IN SUBSURFACE AGRICULTURAL
WASTE WATERS,

CALIFORNIA STATE DEPT. OF WATER RESOURCES, FRESNO; AND FEDERAL WATER
  POLLUTION CONTROL ADMINISTRATION, FRESNO, CALIF.

RANDALL BROWN, AND JAMES ARTHUR.

PROCEEDINGS OF THE EUTROPHICATION-BIOSTIMULATION ASSESSMENT WORKSHOP, JUNE
  19-21, 1969, CALIFORNIA UNIV, BERKELEY, SANITARY ENGINEERING RESEARCH LAB
  AND NATIONAL EUTROPHICATION RESEARCH PROGRAM, CORVALLIS, ORE, PAC NORTHWEST
  WATER LAB, P 80-97. 12 FIG, 1 TAB, 1 REF.

  DESCRIPTORS:
    *SURFACE, *VOLUME, *CARBON DIOXIDE, *AERATION, *NITRATES, *CULTURES,
    *SCENEDESMUS, *SUBSURFACE WATERS, *AGRICULTURE, *WASTES, ALGAE, ASSAY,
    DISSOLVED SOLIDS, CHEMICAL ANALYSIS, PHOSPHATES, IRON COMPOUNDS,
    HYDROGEN ION CONCENTRATION, ALKALINITY, CALIFORNIA, LIGHT INTENSITY,
    TEMPERATURE, IONS, SPECTROPHOTOMETRY, SUSPENDED LOAD, FLUORESCENCE,
    SUMPS.

  IDENTIFIERS:
    STRIPPING, FIREBAUGH CENTER(CALIF), SCENEDESMUS QUADRICAUDA, CELL
    COUNTS, SPECIES DETERMINATION, ABSORBANCE, COENOBIA, SWIRLING, VOLATILE
    SOLIDS.

  ABSTRACT:
    A METHOD OF REMOVING NITRATE-NITROGEN FROM SUBSURFACE AGRICULTURAL
    DRAINAGE WATER--ALGAE STRIPPING--ENCOURAGES ALGAL GROWTH, NITRATE IS
    CONVERTED TO CELLULAR MATERIALS, AND ALGAE SEPARATED FROM LIQUID PHASE
    BY CENTRIFUGATION, FLOCCULATION, ETC. THIS WATER, HIGH IN DISSOLVED
    SALTS, IS NOT A COMPLETE MEDIA FOR SCENEDESMUS (THE ALGA USED) GROWTH.
    LABORATORY STUDIES ARE CONDUCTED TO DETERMINE NUTRIENT ADDITIONS
    NECESSARY FOR OPTIMUM GROWTH (AND NITROGEN REMOVAL). ONE EXPERIMENT
    DETERMINED EFFECT OF FLASK SIZE, SURFACE/VOLUME RATIO, AND TYPE OF
    MIXING (INCLUDING AERATION AND CARBON DIOXIDE ADDITION). SURFACE/VOLUME
    RATIO SHOWED MAXIMUM GROWTH OR NITROGEN REMOVAL OCCURRED IN
    250-MILLILITER FLASKS CONTAINING 125 MILLILITERS OF CULTURE. COMPRESSED
    AIR ALONE EFFECTED A SIMILAR GROWTH RATE AS COMPRESSED AIR ENRICHED
    WITH CARBON DIOXIDE, NEITHER ADDITIVE AFFECTING ULTIMATE YIELD. IN THE
    AGRICULTURAL DRAINAGE WATER USED, CARBON WAS PROBABLY A LIMITING
    ELEMENT AND RATE OF EXCHANGE ACROSS THE SURFACE OF THE CULTURE WAS NOT
    SUFFICIENTLY RAPID WITH SWIRLING ONLY. THE 7.3 - 7.6 PH RANGE
    DIFFERENCE AFFECTED NUTRIENT AVAILABILITY. IN DETERMINING OPTIMUM
    AMOUNT OF IRON AND PHOSPHORUS FOR NITROGEN REMOVAL, TYPE OF MIXING WAS
    SIGNIFICANT. MEASUREMENT OF THE NITRATE PARAMETER BY THE SPECIFIC ION
    ELECTRODE METHOD CAN PRODUCE A RAPID AND RELIABLE ESTIMATE OF ALGAL
    ACTIVITY. (SEE W70-02775).
    (JONES-WISCONSIN)

  FIELD 05D

  ACCESSION NO. W70-03334

ESTABLISHMENT OF TOWN SANITARY DISTRICTS.

  WIS STAT ANN SECS 60.305-60.315 (1957), AS AMENDED, (SUPP 1969) 60.316 (SUPP
    1969).

  DESCRIPTORS:
    *WISCONSIN, *ADMINISTRATIVE AGENCIES, *WATER WORKS, *WASTE DISPOSAL,
    DOMESTIC WASTES, SEWERS, SEWAGE TREATMENT, SEWAGE DISPOSAL, STORM
    DRAINS, SURFACE DRAINAGE, ALGAL CONTROL, AQUATIC LIFE, WATER
    CONSERVATION, PUBLIC HEALTH, WATER POLLUTION, CONTRACTS, REGULATION,
    RIGHT-OF-WAY, ESTIMATED COSTS, LEGAL ASPECTS, LEGISLATION, SEWAGE
    DISTRICTS.

  ABSTRACT:
    THE TOWN SANITARY DISTRICT COMMISSIONS ARE REQUIRED TO PROJECT, PLAN,
    CONSTRUCT, AND MAINTAIN SYSTEMS OF WATER WORKS AND GARBAGE OR REFUSE
    DISPOSAL, INCLUDING SANITARY SEWERS, SURFACE SEWERS, OR STORM WATER
    SEWERS. THEY ARE AUTHORIZED TO PROVIDE FOR SEWAGE COLLECTION AND THE
    CHEMICAL TREATMENT OF WATER FOR CONTROL OF NUISANCE PRODUCING AQUATIC
    GROWTHS. THEY ARE TO UNDERTAKE IMPROVEMENTS FOR THE PROMOTION OF PUBLIC
    HEALTH. THEY MAY REQUIRE THE INSTALLATION OF PRIVATE SEWAGE SYSTEMS.
    CHARGES FOR SERVICES AND METHODS OF FINANCING CONSTRUCTION ARE
    DISCUSSED. THE DEPARTMENT OF RESOURCE DEVELOPMENT MAY ESTABLISH TOWN
    SANITARY DISTRICTS IN AREAS WHERE IT FINDS EXISTING FACILITIES
    INADEQUATE AND WHERE A MENACE TO HEALTH OR POLLUTION OF SURFACE WATERS
    EXISTS. THIS DETERMINATION BY THE DEPARTMENT IS SUBJECT TO JUDICIAL
    REVIEW. (DUSS-FLORIDA)

  FIELD 05E

ACCESSION NO. W70-03344

BIOLOGICAL N2 FIXATION IN LAKES,

WISCONSIN UNIV., MADISON. WATER RESOURCES CENTER.

ROBERT H. BURRIS.

AVAILABLE FROM THE CLEARINGHOUSE AS PB-189 163, $3.00 IN PAPER COPY, $0.65 IN
MICROFICHE. WISCONSIN WATER RESOURCES CENTER, TECHNICAL REPORT OWRR
B-020-WIS, 1969. 6 P. OWRR PROJECT B-020-WIS.

DESCRIPTORS:
*NITROGEN-FIXATION, *NUTRIENTS, *LAKES, *ALGAE, *NITROGEN, WISCONSIN.

IDENTIFIERS:
*ANALYTICAL METHODS, *BIOCHEMISTRY METHODS, SEASONS, ACETYLENE
REDUCTION TEST.

ABSTRACT:
ACETYLENE REDUCTION BY NITROGEN-FIXING ORGANISMS WAS USED TO ESTIMATE
POTENTIAL NITROGEN FIXATION BY BLUE-GREEN ALGAE IN LAKES. THE METHOD
GIVES A USEFUL INDEX OF NITROGEN FIXATION WITH REGARD TO THE ROLE OF
NITROGEN-FIXING AGENTS IN THE EUTROPHICATION PROCESS. LAKE MENDOTA
(EUTROPHIC), MADISON, CRYSTAL AND TROUT LAKES (OLIGOTROPHIC), VILAS
COUNTY, WISCONSIN, AND LITTLE ARBOR VITAE LAKE (EUTROPHIC) VILAS
COUNTY, WISCONSIN, WERE SAMPLED AT POINTS 100 FEET FROM SHORE AND AT
MID-LAKE DURING THE SUMMER OF 1968. ANALYSES OF SAMPLES INDICATED THAT
ACETYLENE REDUCTION VARIES DRASTICALLY AMONG LAKES IN A GIVEN AREA
(LITTLE ARBOR VITAE LAKE FIXED 125-380 NANOMOLES C SUB 2 H SUB 4 N/30
MINUTES TO LESS THAN ONE NANOMOLES FOR TROUT LAKE ON JULY 28, 1968),
AND VARIES MARKEDLY WITH TIME IN A SPECIFIC LAKE (FOR LITTLE ARBOR
VITAE LAKE, WHEN SAMPLES WERE COLLECTED ON SEPTEMBER 10, 1969, THE
ACETYLENE REDUCTION PER LITER OF WATER OR PER MG N HAD DECREASED 10 TO
20 FOLD FROM THOSE OF JULY 28, 1968). ACETYLENE REDUCTION IN LAKES WAS
STUDIED WITH RESPECT TO CHANGES IN SEASON, PRESENCE OF
NATURAL-OCCURRING ALGAE (GLOEOTTICHIA COLONIES CHIEFLY), VARIATION IN
DEPTHS, AND DIFFERENCE IN LIGHT. SAMPLES TAKEN FROM GREEN BAY,
WISCONSIN WERE ALSO ANALYSED. (KERRIGAN-WISCONSIN)

FIELD 05C, 02H

ACCESSION NO. W70-03429

CHEMICAL AND DETRITAL FEATURES OF PALOUSE RIVER, IDAHO, RUNOFF FLOWAGE,

EASTERN NEW MEXICO UNIV., PORTALES. DEPT. OF BIOLOGICAL SCIENCES.

PHILIP A. BUSCEMI.

OIKOS, VOL 20, NO 1, P 119-127, 1969. 3 FIG, 3 TAB, 30 REF.

DESCRIPTORS:
*RUNOFF, *GROUNDWATER, *SESTON, *SEDIMENTS, ORGANIC MATTER,
CURRENTS(WATER), WATER POLLUTION SOURCES, SURFACE RUNOFF, RAINFALL,
SNOWMELT, SURFACE WATERS, STREAMS, HYDROGEN ION CONCENTRATION, IRON,
MAGNESIUM, NITRATES, PHOSPHATES, ALGAE, DETRITUS, PULP WASTES,
LIVESTOCK, IDAHO.

IDENTIFIERS:
TREE BARK, CHARCOAL, ALGAL GROWTH, STREAM CONCOURSE, PALOUSE
RIVER(IDAHO).

ABSTRACT:
WATER AND SEDIMENT SAMPLES WERE COLLECTED AT FOUR STATIONS ESTABLISHED
ON PALOUSE RIVER, IDAHO. DETERMINATIONS INCLUDED PH, NITRATE,
PHOSPHATE, IRON, MANGANESE, HARDNESS, SUSPENDED DETRITUS (DRIFTING
SESTON), AND ORGANIC CONTENT OF SEDIMENTARY DEPOSITS. THE RESULTS SHOW
INFLUENCE OF RAINFALL, MELT WATER, STREAM INFLOW, AND MILL POND ON
SESTON CONCENTRATION. VARIATION IN CHEMICAL COMPOSITION OF WATER
INDICATED THE EFFECT OF EITHER MELT WATER RUNOFF OR SUBTERRANEAN
DISCHARGE. THE INCREASED POLLUTION OF WATER AND SEDIMENT ENRICHMENT IN
ORGANIC MATTER WERE CORRELATED WITH LOCATIONS OF LUMBER CAMPS AND
CATTLE GRAZING. (WILDE-WISCONSIN)

FIELD 05B, 05C

ACCESSION NO. W70-03501

THE MINERAL NUTRITION OF MICROCYSTIS AERUGINOSA,

WISCONSIN UNIV., MADISON. DEPT. OF BOTANY.

GERALD C. GERLOFF, GEORGE P. FITZGERALD, AND FOLKE SKOOG.

AMERICAN JOURNAL OF BOTANY, VOL 39, P 26-32, 1952. 7 TAB, 8 REF.

DESCRIPTORS:
*ALGAE, *NUTRIENTS, *NUTRIENT REQUIREMENTS, GROWTH RATES, LABORATORY
TESTS, CULTURES, SODIUM COMPOUNDS, NITROGEN.

IDENTIFIERS:
*MICROCYSTIS AERUGINOSA(KUTZ), CULTURE MEDIA, CHU NO 10 SOLUTION,
MINIMUM NUTRIENTS.

ABSTRACT:
A MICROCYSTIS AERUGINOSA (KUTZ) CULTURE WAS MADE AXENIC BY ULTRAVIOLET
LIGHT AND GROWN IN BASAL MEDIUM OF CHU NO 10 SOLUTION, MODIFIED SO THAT
CATIONS WERE FURNISHED AS CHLORIDE SALTS, AND ANION AS THE SODIUM
SALTS. MINIMUM ELEMENTAL REQUIREMENTS FOR OPTIMUM GROWTH YIELD WAS
DETERMINED BY USE OF THIS MEDIUM. SMALL AMOUNTS OF SODIUM WERE REPORTED
TO BE ESSENTIAL FOR THE GROWTH OF MICROCYSTIS BUT A BENEFICIAL EFFECT
OF EITHER SODIUM CARBONATE OR SODIUM SILICATE WAS FOUND TO BE DUE
PRIMARILY TO PH CONTROL. MICROCYSTIS EXHIBITED AN OPTIMUM GROWTH AT PH
OF APPROXIMATELY 10; AT THIS PH, SODIUM NITRATE, SODIUM NITRITE OR
AMMONIUM CHLORIDE WERE EQUALLY EFFECTIVE NITROGEN SOURCES. WHEN MINIMUM
NUTRIENT CONCENTRATIONS WERE DETERMINED EACH IN TURN, A NEW MEDIUM
COMPOSED OF THESE CONCENTRATIONS WAS FOUND TO SUPPORT MAXIMUM YIELDS AS
COMPARED TO THE SLIGHTLY MODIFIED CHU NO 10 MEDIUM. COMPOSITION OF THIS
MEDIUM, IN MILLIGRAMS/LITER, WAS: NITROGEN 13.6; PHOSPHORUS 0.18;
SULFUR 0.83; POTASSIUM 1.13; MAGNESIUM 0.13; CALCIUM 0.25; IRON 0.06;
AND THE 'USUAL' AMOUNTS OF SODIUM CARBONATE AND SODIUM SILICATE.
DISADVANTAGES OF THIS MEDIUM FOR ROUTINE USE WERE THAT GROWTH LAG, PALE
CELLS, AND REQUIREMENTS FOR A 'LARGER' INOCULUM WAS EXHIBITED.
(GERHOLD-WISCONSIN)

FIELD 05C

ACCESSION NO. W70-03507

THE SHAGAWA LAKE, MINNESOTA, EUTROPHICATION RESEARCH PROJECT,

FEDERAL WATER POLLUTION CONTROL ADMINISTRATION, CORVALLIS, OREG. PACIFIC
NORTHWEST WATER LAB.

ROBERT M. BRICE, AND CHARLES F. POWERS.

PROCEEDINGS OF THE EUTROPHICATION-BIOSTIMULATION ASSESSMENT WORKSHOP, JUNE
19-21, 1969, CALIFORNIA UNIV, BERKELEY, SANITARY ENGINEERING RESEARCH LAB
AND NATIONAL EUTROPHICATION RESEARCH PROGRAM, CORVALLIS, ORE, PAC NORTHWEST
WATER LAB, P 258-269. 5 FIG, 4 TAB.

DESCRIPTORS:
*MINNESOTA, *EUTROPHICATION, SEWAGE EFFLUENTS, TERTIARY TREATMENT,
ALGAE, WATER QUALITY, CONIFEROUS FORESTS, DECIDUOUS FORESTS,
PHOSPHORUS, NITROGEN, CYANOPHYTA, CHLOROPHYTA, DIATOMS, WEIGHT, PILOT
PLANTS, IRON, CHLORELLA, SODIUM, POTASSIUM, SILICATES, DISSOLVED
SOLIDS, CALCIUM, PRODUCTIVITY, STANDING CROP, CHLOROPHYLL, PHOSPHATES,
ADDITIVES.

IDENTIFIERS:
*SHAGAWA LAKE(MINN), ELY(MINN), BLOOMS, CHLORELLA PYRENOIDOSA,
MICROCYSTIS AERUGINOSA, BURNTSIDE RIVER(MINN).

ABSTRACT:
THIS PROJECT SOUGHT TO DETERMINE WHETHER SHAGAWA LAKE WOULD RECOVER
FROM ITS PRESENTLY ADVANCED STATE OF EUTROPHICATION IF SECONDARY SEWAGE
EFFLUENT PRESENTLY ENTERING THE LAKE WERE CHANGED TO NUTRIENT-STRIPPED
TERTIARY. ALTHOUGH DATA ARE INCOMPLETE AND CONCLUSIONS PRELIMINARY,
STUDIES SHOWED LAKE WATER CONTAINING 5% SECONDARY SEWAGE EFFLUENT HAD
APPRECIABLY HIGHER ALGAL PRODUCTIVITY THAN SHAGAWA LAKE WATER ONLY OR
WATER CONTAINING 5% TERTIARY SEWAGE EFFLUENT, THE LATTER ADDITION
MAKING NO APPRECIABLE DIFFERENCE IN THE LAKE WATER. BURNTSIDE RIVER
WATER (80% SURFACE INFLOW) CONTAINING 2% SECONDARY SEWAGE EFFLUENT HAD
APPRECIABLY HIGHER ALGAL PRODUCTIVITY THAN CORRESPONDING MIXTURES
CONTAINING TERTIARY EFFLUENT. FURTHER TESTING OF SHAGAWA LAKE WATER AND
BURNTSIDE RIVER IS CONTINUING, USING OTHER CONCENTRATIONS OF ADDITIVES
TO EXPLORE MORE COMPLETELY THE BIOSTIMULATION ACHIEVED THROUGH SEWAGE
FERTILIZATION. ADDITIONALLY, EXPERIMENTS WILL BE CARRIED OUT IN WHICH
VARIOUS CONCENTRATIONS OF NITROGEN AND PHOSPHORUS ARE ADDED TO
BURNTSIDE RIVER WATER, TO DETERMINE WHETHER EITHER OR BOTH ELEMENTS ACT
TO LIMIT ALGAL PRODUCTION. ANTICIPATED RESULTS OF THESE EXPERIMENTS
WILL PROVIDE INFORMATION NECESSARY TO DECISIONS ON FUTURE MANAGEMENT
POLICIES FOR SHAGAWA LAKE. (SEE W70-02775).
(JONES-WISCONSIN)

FIELD 05C, 02H

ACCESSION NO. W70-03512

EFFECTS OF FIVE HERBICIDES ON THREE GREEN ALGAE,

NORTH TEXAS STATE UNIV., DENTON. DEPT. OF BIOLOGY.

B. DWAIN VANCE, AND DAVID L. SMITH.

TEXAS JOURNAL OF SCIENCE, VOL XX, P 329-337, APR 1969. 1 TAB, 41 REF.

DESCRIPTORS:
*HERBICIDES, *GROWTH RATES, *ALGAE, 2-4-D, AMINOTRIAZOLE, 2-4-5-T,
SCENEDESMUS, CHLAMYDOMONAS, CHLORELLA, MORTALITY, BIOASSAY, INHIBITION,
MODE OF ACTION, PHOTOSYNTHESIS.

IDENTIFIERS:
SIMAZINE 80 W, DACTHAL.

ABSTRACT:
SCREENING TESTS WERE MADE TO DETERMINE IF THE HERBICIDES SIMAZINE 80W,
DACTHAL, AMITROL-T, 2,4-D, AND 2,4,5-T WERE TOXIC TO BACTERIA-FREE
UNIALGAL CULTURES OF SCENEDESMUS QUADRICAUDA, CHLAMYDOMONAS EUGAMETOS,
AND CHLORELLA PYRENOIDOSA TX 7-11-05. DACTHAL, SIMAZINE, 2,4-D AND
2,4,5-T SHOWED NO TOXIC EFFECTS ON ANY OF THE ALGAE AT CONCENTRATIONS
UP TO 200 UG/ML. SIMAZINE INCREASED THE GROWTH OF C. EUGAMETOS IN
CONCENTRATIONS UP TO 200 UG/ML. AMITROL-T WAS INHIBITORY TO ALL THE
ALGAE AT HIGH CONCENTRATIONS BUT SHOWED NO EFFECT AT CONCENTRATIONS
BELOW 150 UG/ML. (SJOLSETH-WASHINGTON)

FIELD 05C

ACCESSION NO.   W70-03519

BIOLOGICAL CONCENTRATION OF PESTICIDES BY ALGAE,

NORTH TEXAS STATE UNIV., DENTON. DEPT. OF BIOLOGY.

B. DWAIN VANCE, AND WAYMON DRUMMOND.

JOURNAL AMERICAN WATER WORKS ASSOCIATION, VOL 61, NO 7, P 360-362, JULY 1969.
2 TAB, 11 REF. US PUBLIC HEALTH SERV RESEARCH GRANT CC 00269.

DESCRIPTORS:
*ALGAE, *PESTICIDE TOXICITY, *PESTICIDE REMOVAL, *FOOD WEBS, DDT,
CHLORINATED HYDROCARBON PESTICIDES, ALDRIN, ENDRIN, DIELDRIN,
SCENEDESMUS, GAS CHROMATOGRAPHY, RESISTANCE, BIOASSAY, GROWTH RATES,
PESTICIDE RESIDUES, DETERIORATION.

IDENTIFIERS:
*BIOLOGICAL CONCENTRATION, ANABAENA CYLINDRICA, MICROCYSTIS AERUGINOSA,
OEDOGNIUM.

ABSTRACT:
UNIALGAL CULTURES OF MICROCYSTIS AERUGINOSA, ANABAENA CYLINDRICA,
SCENEDESMUS QUADRICAUDA, AND OEDOGONIUM SP. WERE GROWN IN THE PRESENCE
OF ALDRIN, DIELDRIN, ENDRIN AND DDT IN CONCENTRATIONS OF 5, 10 AND 20
UG/ML. THESE ALGAE ARE EFFECTIVE IN ACCUMULATING PESTICIDES FROM THE
MEDIUM IN WHICH THEY GROW. IN NO CASE WAS CONCENTRATION LESS THAN 100
FOLD. ALGAE WERE SHOWN TO ACCUMULATE DETECTABLE QUANTITIES WITHIN 30
MINUTES AFTER EXPOSURE. THE ALGAE WERE SHOWN TO BE HIGHLY RESISTANT TO
THE PESTICIDE, EXCEPT THAT DIELDRIN, ALDRIN, AND ENDRIN ARE ALGICIDAL
TO M. AERUGINOSA AT CONCENTRATION OF LESS THAN 5 UG/ML. THE AUTHORS
FEEL THAT THE ABILITY OF ALGAE TO CONCENTRATE PESTICIDES WITHOUT
THEMSELVES SUCCUMBING MAY BE A POTENTIAL DANGER TO ORGANISMS HIGHER UP
THE FOOD CHAIN. (SJOLSETH-WASHINGTON)

FIELD 05C

ACCESSION NO.   W70-03520

A METHOD FOR PREDICTING THE EFFECTS OF LIGHT INTENSITY ON ALGAL GROWTH AND
PHOSPHOROUS ASSIMILATION,

ABBOTT LABS., NORTH CHICAGO, ILL. DEPT. OF SANITARY ENGINEERING; AND MICHIGAN
UNIV., ANN ARBOR.

H. S. AZAD, AND J. A. BORCHARDT.

JOURNAL WATER POLLUTION CONTROL FEDERATION, VOL 41, NO 11, PART 2, P
R392-R404, NOV 1969. 10 FIG, 14 REF.

DESCRIPTORS:
*ALGAE, *CHLORELLA, *SCENEDESMUS, *LIGHT, PHOSPHORUS, PHOTOSYNTHESIS,
LIGHT INTENSITY, STABILIZATION, PONDS, PENETRATION, ABSORPTION.

IDENTIFIERS:
*BEER-LAMBERT LAW, *FIXED-DENSITY STUDIES, *FIXED-LIGHT TESTS,
*EFFECTIVE-AVERAGE-LIGHT-INTENSITY, PHOTO-ELECTRIC TURBIDOSTAT, LIGHT
EXTINCTION COEFFICIENT, ALGAL CELL DENSITY.

ABSTRACT:
RESEARCH IS DEVOTED TO OBTAINING A BETTER UNDERSTANDING OF THE
INTERACTION OF LIGHT INTENSITY ON THE RATE OF ALGAL GROWTH, AND
PHOSPHOROUS UPTAKE BY ALGAE IN AN AQUEOUS ENVIRONMENT. GREEN ALGAE
CHLORELLA AND SCENEDESMUS WERE GROWN UNDER A CAREFULLY CONTROLLED
ENVIRONMENT. THE LIGHT INTENSITY WAS VARIED BY TWO METHODS: (1)
MAINTENANCE OF A CONSTANT BIOMASS DENSITY BY USING A
PHOTO-ELECTRIC-TURBIDOSTAT APPARATUS PROVIDING FOR VARIABLE LIGHT
INTENSITY CONTROL AT THE SOURCES; (2) MAINTENANCE OF THE INCIDENT
LUMINOUS ENERGY AT A CONSTANT VALUE PROVIDING FOR VARIABLE ALGAE CELL
DENSITY. THE BEER-LAMBERT EXPRESSION (I SUB O/I = E TO THE POWER CAD)
FAILED TO PREDICT THE LIGHT PENETRATION DATA. HOWEVER, AN EMPIRICAL
RELATIONSHIP IN WHICH THE FAMILIAR COEFFICIENT OF EXTINCTION WAS
REPLACED BY AN EXPONENTIAL VARIABLE DID PROVIDE THE NECESSARY
AGREEMENT. ALGAL GROWTH RATE VS. THE CULTURE-FACE-LIGHT INTENSITY
YIELDS A CURVE THAT IMPLIES THE EXISTENCE OF A LIGHT SATURATION POINT.
WHEN GROWTH RATE VS EFFECTIVE AVERAGE LIGHT INTENSITY WAS PLOTTED, A
MORE LOGICAL EFFECT OF LIGHT WAS PRODUCED. (D'AREZZO-TEXAS)

FIELD 05C, 05D

ACCESSION NO. W70-03923

ANIONIC AND NONIONIC SURFACTANT SORPTION AND DEGRADATION BY ALGAE CULTURES,

TEXAS UNIV., AUSTIN. DEPT. ENVIRONMENTAL HEALTH ENGINEERING.

ERNST M. DAVIS, AND EARNEST F. GLOYNA.

JOURNAL OF THE AMERICAN OIL CHEMISTS' SOCIETY, VOL 46, NO 11, P 604-608, NOV
1969. 2 FIG, 3 TAB, 9 REF.

DESCRIPTORS:
*SURFACTANTS, *SORPTION, *ALGAE, CHLOROPHYTA, CYANOPHYTA, DETERGENTS,
SPECTROSCOPY.

IDENTIFIERS:
*LINEAR ALKYL SULFONATES, *AXENIC CULTURES, *ALKYL POLYETHOXYLATE,
*ALKYL PHENOL POLYETHOXYLATE, ORGANIC EXTRACTIONS, INFRARED
SPECTROSCOPY, BLUE-GREEN ALGAE, GREEN ALGAE.

ABSTRACT:
INVESTIGATION WAS MADE TO DETERMINE EFFECTS OF AXENIC CULTURES OF 5
SPECIES OF CYANOPHYTA ALGAE AND 3 SPECIES OF CHLOROPHYTA ALGAE, WHICH
ARE COMMON TO WASTE STABILIZATION PONDS, ON IONIC AND ANIONIC
SURFACTANTS CONTAINED IN THE BULK OF HOUSEHOLD DETERGENT PRODUCTS.
TESTS MADE TO DETERMINE THE SORPTION CHARACTERISTICS OF THE SURFACTANTS
BY THE ALGAE SHOWED RELEASES AND DEGRADATION OF UP TO 99% OF SOME OF
THE COMPONENT PARTS OF THE SURFACTANT MOLECULE. COMPARATIVE TESTS USING
A HETEROGENEOUS MICROCOSM ASSOCIATED WITH LABORATORY ACCLIMATED
STABILIZATION POND WATER SHOWED CAPABILITY OF GREATER REDUCTION OF ALL
SURFACTANT COMPONENTS THAN THE INDIVIDUAL ALGAE SPECIES. THE TESTS
CONFIRMED THE FEASIBILITY OF USING INFRARED SPECTROPHOTOMETRIC ANALYSIS
TECHNIQUES AND THE ADEQUACY OF ORGANIC SOLVENTS FOR EXTRACTION OF
SURFACTANTS PRIOR TO COMPLEXATION IN CONCENTRATIONS OF AND BELOW 50
MG/L. (D'AREZZO-TEXAS)

FIELD 05D

164

ACCESSION NO. W70-03928

DISTRIBUTION, CHARACTERIZATION, AND NUTRITION OF MARINE MICROORGANISMS FROM THE
ALGAE POLYSIPHONIA LANOSA AND ASCOPHYLLUM NODOSUM,

MCGILL UNIV., MONTREAL (QUEBEC). DEPT. OF MICROBIOLOGY AND IMMUNOLOGY; AND
NEW BRUNSWICK UNIV., FREDERICTON. DEPT. OF BIOLOGY.

E. C. S. CHAN, AND ELIZABETH A. MCMANUS.

CANADIAN JOURNAL OF MICROBIOLOGY, VOL 15, NO 5, P 409-420, 1969. 5 FIG, 4
TAB, 38 REF.

DESCRIPTORS:
*MARINE MICROORGANISMS, *ALGAE, *NUTRIENT REQUIREMENTS, *DISTRIBUTION,
BACTERIA, SAMPLING, SEA WATER, SEASONAL, PSEUDOMONAS, TEMPERATURE,
AMINO ACIDS, INTERTIDAL AREAS, CARBOHYDRATES, YEASTS, ENZYMES,
CULTURES, HYDROGEN SULFIDE, ELECTRON MICROSCOPY, BIOCHEMISTRY,
VITAMINS, EPIPHYTOLOGY.

IDENTIFIERS:
*POLYSIPHONIA LANOSA, *ASCOPHYLLUM NODOSUM, VIBRIO, FLAVOBACTER,
ESCHERICHIA, SARCINA, STAPHYLOCOCCUS, ACHROMOBACTER(ALKALIGENES),
RHODOTORULA, GROWTH FACTORS, COLONIAL MORPHOLOGY, CELLULAR MORPHOLOGY,
FLAGELLATION, IODINE TEST, GRAM REACTION, CYTOPHAGA, PROTEUS, SERRATIA,
BAY OF FUNDY(CANADA), RHYZOIDS.

ABSTRACT:
BACTERIAL POPULATIONS OF TWO ASSOCIATED LITTORAL MARINE ALGAE,
POLYSIPHONIA LANOSA AND ASCOPHYLLUM NODOSUM, AND THEIR ENVIRONMENTAL
SEAWATER WERE STUDIED QUANTITATIVELY. SAMPLINGS SHOWED NUMBERS OF
BACTERIA ON P LANOSA AND IN THE SEAWATER REMAINED RELATIVELY CONSTANT
WHILE THOSE ON A NODOSUM DECREASED IN MID-SUMMER, AFTER AN APPARENT
SPRING MAXIMUM, AND THEN GRADUALLY INCREASED. PURE CULTURES OF 25
ISOLATES WERE STUDIED AND IDENTIFIED TO GENUS LEVEL. THERE WAS A
PREPONDERANCE OF THE GENERA VIBRIO AND FLAVOBACTER (8 EACH). THREE OF
THE ESCHERICHIA GROUP, 2 OF PSEUDOMONAS, ONE EACH OF GENERA SARCINA,
STAPHYLOCOCCUS, AND ACHROMOBACTER (OR ALKALIGENES), AND A PINK YEAST
(RHODOTORULA) MADE UP THE OTHER ISOLATES. OPTIMUM TEMPERATURE FOR
GROWTH OF MOST ISOLATES WAS AROUND 21C. FIVE ISOLATES FAILED TO GROW AT
30C; 20 DID NOT GROW AT 37C; OF THE 5 GROWING AT 37C, DEVELOPMENT WAS
FEEBLE IN EVERY CASE BUT ONE. STUDIES ON GROSS NUTRITIONAL REQUIREMENTS
OF THE MARINE ORGANISMS SHOWED THAT 22 OF 25 ISOLATES REQUIRED A
SUPPLEMENT OF AMINO ACIDS IN GROWTH MEDIUM. TWO WERE ABLE TO GROW ON
BASAL MEDIUM; ONE ISOLATE GREW ONLY IN A MEDIUM SUPPLEMENTED WITH
GROWTH FACTORS IN ADDITION TO AMINO ACIDS (JONES-WISCONSIN)

FIELD 05C

ACCESSION NO. W70-03952

AVAILABILITY OF MUD PHOSPHATES FOR THE GROWTH OF ALGAE,

H. L. GOLTERMAN, C. C. BAKELS, AND J. J. JAKOBS-MOGELIN.

VERHANDLUNGEN DER INTERNATIONALEN VEREINIGUNG FUR THEORETISCHE UND ANGEWANDTE
LIMNOLOGIE, VOL 17, P 467-479, 1969. 10 FIG, 3 TAB, 10 REF, DISCUSSION.

DESCRIPTORS:
*EUTROPHICATION, *MUD(LAKE), *PHOSPHORUS, *AQUATIC ALGAE, *BIOASSAY,
SCENEDESMUS, PHOSPHATES, WATER POLLUTION SOURCES.

IDENTIFIERS:
*DUTCH LAKES, *ALLOCHTHONOUS PHOSPHATES, ULTRA-VIOLET LIGHT
STERILIZATION, ASTERIONELLA, SCENEDESMUS OBLIQUUS, GROWTH CONSTATS,
RIVER RHINE.

ABSTRACT:
AVILABILITY OF LAKE MUDS PHOSPHORUS TO ALGAE WAS DETERMINED BY
EMPLOYING SCENEDESMUS OBLIQUUS AS TEST ORGANISM. CULTURE SOLUTIONS
CONTAINED POTASSIUM PHOSPHATE IN DIFFERENT CONCENTRATIONS, CALCIUM
PHOSPHATES, IRON PHOSPHATES, AND LAKE MUDS AS THE ONLY SOURCE OF
PHOSPHORUS. THE GROWTH CONSTANTS WERE OBTAINED FROM CULTURES WITH
INORGANIC PHOSPHATES. BASED ON THE NUMBER OF CELLS OF SCENEDESMUS, A
CONSIDERABLE FRACTION OF MUD PHOSPHORUS WAS AVAILABLE TO ALGAE.
DEPENDING ON THE ORIGIN OF SEDIMENT, FROM 7 TO 30.5% OF PHOSPHORUS WAS
USED FOR CELL GROWTH. STERILIZATION WITH ULTRA-VIOLET LIGHT HAD NO
SIGNIFICANT INFLUENCE ON THE GROWTH OF ORGANISMS. RESULTS SUGGESTED
THAT SCENEDESMUS IS UTILIZING LARGELY ALLOCHTHONOUS PHOSPHATES, LEAVING
SUFFICIENT SUPPLY OF INORGANIC PHOSPHORUS FOR ASTERIONELLA ALGAE.
(WILDE-WISCONSIN)

FIELD 05C, 02H

ACCESSION NO. W70-03955

RELATION BETWEEN FILTERING RATE, TEMPERATURE, AND BODY SIZE IN FOUR SPECIES OF
DAPHNIA,

YALE UNIV., NEW HAVEN, CONN. OSBORN MEMORIAL LABS.

CAROLYN W. BURNS.

LIMNOLOGY AND OCEANOGRAPHY, VOL 14, NO 5, P 693-700, 1969. 2 FIG, 2 TAB, 18
REF, DISCUSSION.

DESCRIPTORS:
*GRAZING, *TEMPERATURE, *DAPHNIA, PHYSIOLOGICAL ECOLOGY, LAKES,
ENVIRONMENTAL EFFECTS, WATER POLLUTION EFFECTS, EPILIMNION,
CONNECTICUT, EUGLENA, CHLAMYDOMONAS, PHOSPHORUS RADIOISOTOPES,
REGRESSION ANALYSIS, ALGAE, PONDS, PREDATION.

IDENTIFIERS:
*FILTERING RATE, *BODY SIZE, *COMPARATIVE STUDIES, DAPHNIA SCHODLERI,
DAPHNIA PULEX, DAPHNIA GALEATA MENDOTAE, DAPHNIA MAGNA, SELENASTRUM,
RHODOTORULA, TEMPERATURE COEFFICIENTS, NORTH AMERICA, CENTRAL AMERICA,
RATE EQUATIONS, ANALYSIS OF VARIANCE, FOOD CONCENTRATION, MEXICO,
PREDICTIVE EQUATIONS, ZOOGEOGRAPHY, THERMOBIOLOGY.

ABSTRACT:
FILTERING RATES (F) OF CLADOCERANS CAN BE MEASURED BY DETERMINING THEIR
UPTAKE OF RADIOPHOSPHORUS FROM LABELLED FOOD. F OF FOUR DAPHNID
SPECIES, FED ON LABELLED YEAST, WERE MEASURED AT TEMPERATURES OF 15C,
20C, AND 25C. MAXIMUM F INCREASED WITH RISING TEMPERATURE AND BODY SIZE
IN ALL SPECIES, BUT A GENERAL EQUATION RELATING F TO BODY LENGTH (L)
COULD NOT BE DERIVED. BODY WEIGHT (W) CAN BE DEDUCED FROM L WITH
EQUATIONS OF TYPE, LOG W = LOG A + B LOG L, WHERE A AND B ARE
EMPIRICALLY DERIVED COEFFICIENTS. DATA ARE TABULATED FOR A AND B OF ALL
SPECIES, CONSIDERED COLLECTIVELY AND INDIVIDUALLY. F, EXPRESSED IN
MILLILITERS WATER FILTERED PER HOUR/MILLIGRAM DAPHNID DRY WEIGHT,
DIFFERED AMONG SPECIES. F FOR ADULT DAPHNIA SCHODLERI (DS) AND DAPHNIA
PULEX (DP) WERE SIMILAR, AND AT 20C, WERE SLIGHTLY HIGHER THAN AT 15C
OR 25C. F FOR DAPHNIA MAGNA (DM) AND DAPHNIA GALEATA MENDOTAE (DG)
INCREASED WITH INCREASING TEMPERATURE. TEMPERATURE COEFFICIENTS (Q-10)
FOR F, CHANGING IN RANGE 15-25C WERE, FOR FOUR SPECIES: DG-2.71,
DM-2.38, DP-0.94, DS-0.90. THERMAL DETERMINANTS OF MAXIMUM F MAY
DETERMINE SUCCESS OF DAPHNIDS OCCUPYING WARM EPILIMNETIC LACUSTRINE
WATERS IN SUMMER. (EICHHORN-WISCONSIN)

FIELD 05C

ACCESSION NO. W70-03957

CONCEPTS OF EUTROPHICATION AND TROPHIC BIOLOGY,

OESTERREICHISCHE AKADEMIE DER WISSENSCHAFTEN, VIENNA. LUNZ BIOLOGICAL STATION.

FRANZ RUTTNER.

FUNDAMENTALS OF LIMNOLOGY, UNIVERSITY OF TORONTO PRESS, P 130-131, 159-165, 1963. 2 FIG.

DESCRIPTORS:
*LAKES, *EUTROPHICATION, *BIOMASS, *PRODUCTIVITY, CYANOPHYTA, NUISANCE ALGAE, LIMNOLOGY, CHLOROPHYTA, PLANKTON, PHYTOPLANKTON, WATER POLLUTION EFFECTS, WISCONSIN ZOOPLANKTON, TRIPTON, NANNOPLANKTON, NUTRIENTS, OLIGOTROPHY, TEMPERATURE, LIGHT INTENSITY, DAPHNIA, CONNECTICUT, ROTIFERS.

IDENTIFIERS:
*CONCEPTS, *TROPHIC BIOLOGY, ANABAENA FLOS AQUAE, APHANIZOMENON FLOS AQUAE, GLOEOTRICHIA ECHINULATA, BOTRYOCOCCUS, PLANKTON DISTRIBUTION, WISCONSIN LAKE SURVEY, NITROGEN CONTENT, LAKE ERKEN(SWEDEN), LAKE MENDOTA(WIS), NET PLANKTON, CHLOROPHYLL CONTENT, CENTRIFUGE METHOD, CALCULATED VOLUME METHOD, CYCLOTELLA COMENSIS, RHODOMONAS LACUSTRIS, ENOTOMOSTRACA, DIAPTOMIDS, BOSMINA, CARINTHIA, SUNDA ISLANDS.

ABSTRACT:
THIS STANDARD TEXTBOOK OF LIMNOLOGY INCLUDES A DISCUSSION OF CONCEPTS PERTAINING TO EUTROPHICATION AND TROPHIC ECOLOGY OF LAKES. MECHANICAL FACTORS INFLUENCING PLANKTON DISTRIBUTION INCLUDE DENSITY OF WATER AND PLANKTON, DIFFERENTIAL WATER VISCOSITY, AND TURBULENT EDDY DIFFUSION CURRENTS. THE FOLLOWING DESCRIBES FORMATION OF NUISANCE BLOOMS, 'IN CALM WEATHER THE BLUE-GREEN ALGAE OF THE PLANKTON RISE TO THE SURFACE IN CONSEQUENCE OF THEIR 'GAS VACUOLES' AND THERE AGGREGATE INTO THICK MASSES, OFTEN FORMING A 'WATER BLOOM' RECOGNIZABLE FROM AFAR BY ITS BLUE-GREEN COLORATION. ANABAENA FLOS AQUAE AND APHANIZOMENON FLOS AQUAE REVEAL BY THEIR NAMES THAT THEY BELONG TO THE GROUP OF ALGAE THAT CAN PRODUCE THIS STRIKING PHENOMENON.' BASED UPON A COMPOSITE DEFINITION BY OTHERS, PRODUCTION CAN BE CONSIDERED AS THE TOTAL AMOUNT OF ORGANIC MATTER, REPRESENTING BALANCE BETWEEN ASSIMILATION AND DISSIMILATION, WHICH FORMS IN A DEFINED PERIOD FROM RAW MATERIALS SUPPLIED. STANDING CROP IS DEFINED AS THE INSTANTANEOUS LIVING POPULATION. FERTILITY IS THE AVERAGE BIOMASS OF A UNIT VOLUME OF WATER WITHIN THE TROPHOGENIC ZONE, AND IS GOVERNED BY LIMITING NUTRIENTS, THEIR CONVERSION TO ORGANIC MATTER OCCURRING IN PROPORTION TO TEMPERATURE AND ILLUMINATION. DISCUSSION INCLUDES A CRITIQUE OF TECHNIQUES FOR DETERMINATION OF PLANKTONIC BIOMASS. (EICHHORN-WISCONSIN)

FIELD 02H, 05C

ACCESSION NO. W70-03959

RX FOR AILING LAKES--A LOW PHOSPHATE DIET,

INTERNATIONAL JOINT COMMISSION-UNITED STATES AND CANADA.

ENVIRONMENTAL SCIENCE AND TECHNOLOGY, VOL 3, NO 12, P 1243-1245, 1969. 2 FIG.

DESCRIPTORS:
 *PHOSPHATES, *LAKES, *DETERGENTS, *TERTIARY TREATMENT, *GREAT LAKES,
 CONTROL, COSTS, EUTROPHICATION, NITRATES, LAKE ERIE, LAKE ONTARIO,
 POLLUTION ABATEMENT, OLIGOTROPHY, DEPTH, PHYTOPLANKTON, ZOOPLANKTON,
 PHYSICOCHEMICAL PROPERTIES, DOMESTIC WASTES, SEWAGE, INDUSTRIAL WASTES,
 AGRICULTURE, ST LAWRENCE RIVER, ALGAE, COLIFORMS, DISSOLVED OXYGEN,
 DISSOLVED SOLIDS, TEMPERATURE, COLOR, TASTE, HYDROGEN ION
 CONCENTRATION, IRON, RADIOACTIVITY.

IDENTIFIERS:
 CANADA, MESOTROPHY, LAKE NORRVIKEN, LAKE MENDOTA, LAKE FURES, LAKE
 SEBASTICOOK, LAKE WASHINGTON, LAKE MALAREN, LAKE ANNECY, LAKE VANERN,
 LAKE CONSTANCE, PFAFFIKERSEE, TURLERSEE, BALDEGGERSEE, GREIFENSEE,
 ZURICHSEE, MOSES LAKE, HALLWILLERSEE.

ABSTRACT:
 STUDY WAS INITIATED IN 1964 WHEN THE INTERNATIONAL JOINT COMMISSION OF
 THE U S AND CANADA ESTABLISHED ADVISORY GROUPS ON STATUS OF POLLUTION
 IN LAKES ERIE AND ONTARIO AND SEGMENTS OF THE ST LAWRENCE RIVER. REPORT
 RECOMMENDS TECHNICAL AND LEGISLATIVE MACHINERY FOR CONTROL MEASURES.
 DETERGENTS' PHOSPHATE CONTENT SHOULD BE REDUCED IMMEDIATELY TO MINIMUM
 PRACTICAL LEVELS, WITH COMPLETE REPLACEMENT OF PHOSPHORUS WITH LESS
 INNOCUOUS SUBSTANCES NO LATER THAN 1972. 80% REMOVAL OF PHOSPHATES FROM
 ALL EFFLUENTS SHOULD BE EFFECTED BY 1972 IN THE LAKE ERIE BASIN AND BY
 1975 IN LAKE ONTARIO. TREATMENT OF WASTE EFFLUENTS FOR PHOSPHATE
 REMOVAL MUST BE IN ADDITION TO, NOT A SUBSTITUTE FOR DETERGENT
 REFORMULATION. PHOSPHORUS AND NITROGEN ARE RECOGNIZED AS THE MAJOR
 NUTRIENTS RESPONSIBLE FOR EUTROPHICATION; IT IS APPARENT THAT PHOSPHATE
 IS THE CONTROLLING FACTOR IN ENRICHMENT OF LOWER GREAT LAKES. EFFICIENT
 AND RELATIVELY INEXPENSIVE METHODS ARE AVAILABLE FOR 80-95% REMOVAL OF
 PHOSPHORUS DURING SEWAGE TREATMENT, WHEREAS COMPARABLE ELIMINATION OF
 NITROGEN COMPOUNDS IS NOT YET FEASIBLE. COSTS FOR PHOSPHATE REMOVAL AT
 TREATMENT PLANTS WOULD BE REDUCED BY ONE-HALF TO TWO-THIRDS WITH
 REPLACEMENT OF PHOSPHATE DETERGENT BUILDERS. (JONES-WISCONSIN)

FIELD 02H, 05C

ACCESSION NO. W70-03964

REPORT OF AN ALGAL BLOOM IN VIET-NAM,

HUNGARIAN UNIV. OF AGRICULTURAL SCIENCES, GODOLLO (HUNGARY). INST. OF BOTANY AND PLANT PHYSIOLOGY.

T. HORTOBAGYI.

ACTA BIOLOGICA ACADEMIAE SCIENTIARUM HUNGARICAE, VOL 20, NO 1, P 23-34, 1969. 5 FIG, 23 REF.

DESCRIPTORS:
*ALGAE, TEMPERATURE, CYANOPHYTA, CHLOROPHYTA, SCENEDESMUS, BENTHOS, CONVECTION, WAVES(WATER), FISH, SEWAGE, EUGLENOPHYTA, DIATOMS, BIOLOGICAL COMMUNITIES, BACKWATER, PONDS, ALASKA.

IDENTIFIERS:
*BLOOM, *VIETNAM, LITTLE LAKE(HANOI), RED RIVER(VIETNAM), CHLOROCOCCUS, SPINES, SCHIZOMYCOPHYTA, CAULOBACTERIALES, CHLOROCOCCALES, CONJUGATOPHYCAEA, PLANCTOMYCES, BACILLARIOPHYTA, GLOECAPSA, MERISMOPEDIA, MICROCYSTIS, ANKISTRODESMUS, PEDIASTRUM, TETRAEDRON, TETRASTRUM, TISZA(HUNGARY), ZIMONA(HUNGARY), ALGAL TAXONOMY, SPIRULINA, CHODATELLA, MIRACANTHA, BUZSAK(HUNGARY), JAMUNA(INDIA), HORMOGONALES, APHANOCAPSA, COELOSPHAERIUM, CHLORONOSTOC, PELODICTYON, SYNECHOCYSTIS, ANABAENA, LYNGBYA, OSCILLATORIA, ROMERIA, ASTASIA, SCHROEDERIA, COELASTRUM, DICTYOSPHAERIUM, GLOEOACTINUM, GOLENKINIA, NEPHROCHLAMYS, NEPHROCYTIUM, OOCYSTIS, SIDEROCELIS, STAURASTRUM.

ABSTRACT:
AN ALGAL BLOOM OF VIVID GREEN COLOR DEVELOPED IN LITTLE LAKE, HANOI (WATER TEMPERATURE 13-26C). THE LAKE, AN ISOLATED BACKWATER OF THE RED RIVER, CONTAINS CONTAMINATED WATER. THE BLOOM WAS RICH IN SPECIES AND NEW TAXONS WITH CYANOPHYTA AND CHLOROPHYTA FROM BOTH SURFACE AND BOTTOM DOMINATING. SCENEDESMUS VARIABILITY WAS PROMINENT. A SAMPLE INDICATED 115 SPECIES WITHOUT BACILLARIOPHYCEAE. THE NUMBER OF THE SCENEDESMUS SPECIES WAS LARGER THAN IN THE BLOOM. CYANOPHYTA DOMINATED WITH 82% IN THE ALGAL BLOOM AND 92% IN THE BOTTOM SAMPLE; CHLOROPHYTA WITH 12% AND 6.8%, RESPECTIVELY. NO DEFINITE LINE OF DEMARCATION EXISTS BETWEEN THE TWO BIOTOPES BECAUSE THE ALGAL BLOOM DEVELOPED ABOVE THE BOTTOM AND BOTTOM PLANTS ARE EASILY RAISED NEAR THE SURFACE BY WAVES, CONVECTIONAL STREAMING AND FISH MOVEMENT. THE ALGAE OF THE TWO BIOTOPES ARE DIFFERENT, THE PHYTOBIOCOENOSES BEING DISSIMILAR. NUMBER OF COMMON TAXONS IS 51% OF THE TOTAL; UNDESCRIBED TAXONS IS 33%. PARALLEL MORPHOSES ARE WELL REPRESENTED BY SOME SCENEDESMUS SPECIES. AMONG SPINY FORMS OF THE CHLOROCOCCALES, THOSE OF THIN SPINES ARE LESS FREQUENT. PROTUBERANCES BECOME MORE DEVELOPED IN HANOI AND TOWARD THE TROPICS. (JONES-WISCONSIN)

FIELD 05C

ACCESSION NO. W70-03969

ALGAE, MAN, AND THE ENVIRONMENT.

PROCEEDINGS OF AN INTERNATIONAL SYMPOSIUM HELD AT SYRACUSE UNIVERSITY, JUNE
18-30, 1967. DANIEL F JACKSON, EDITOR. NEW YORK, SYRACUSE UNIV PRESS, 1968.
554 P.

DESCRIPTORS:
*ALGAE, *ENVIRONMENTAL EFFECTS, *HUMAN POPULATION, *EUTROPHICATION,
*CONFERENCES, NUISANCE ALGAE, PUBLICATIONS, WATER POLLUTION EFFECTS,
NEW YORK, CHLOROPHYTA, NUTRIENTS, ECOSYSTEMS, PHYTOPLANKTON,
PHOTOSYNTHESIS, CYANOPHYTA, CYTOLOGICAL STUDIES, ELECTRON MICROSCOPY,
WASTE TREATMENT, LAKES, WATER QUALITY, VIRUSES, ALGICIDES, FISH,
STREAMS, PRODUCTIVITY.

IDENTIFIERS:
*PHYCOLOGY, NY STATE SCIENCE AND TECHNOLOGY FOUND, SYRACUSE(NY),
SYRACUSE UNIVERSITY, FUNDAMENTAL STUDIES, APPLIED STUDIES, ALGAL
PHYSIOLOGY, ALGAL BIOCHEMISTRY, ALGAL TAXONOMY, ALGAL PHYLOGENY,
CLADOPHORA, ALGAL GENETICS, RUSSIA, BAVARIA, MORICHES BAY(NY), ONONDAGA
LAKE(NY).

ABSTRACT:
PHYCOLOGICAL STUDIES ARE IMPORTANT NOT ONLY BECAUSE OF NUISANCE EFFECTS
OF ALGAE FOR MAN, BUT ALSO FOR THEIR POTENTIAL BENEFITS. TO COMPILE
INFORMATION ON BASIC AND APPLIED PHYCOLOGY, A SYMPOSIUM--WHICH THIS
VOLUME RECORDS--WAS SPONSORED BY NEW YORK STATE SCIENCE AND TECHNOLOGY
FOUNDATION AND SYRACUSE UNIVERSITY, NEW YORK (18-30 JUNE 1967). VOLUME
IS ORGANIZED INTO THREE SECTIONS: FUNDAMENTAL PHYCOLOGY, APPLIED
PHYCOLOGY, AND ALGAL STUDIES IN NEW YORK STATE. THE FOLLOWING TOPICAL
LIST OF 27 PAPERS BY 32 CONTRIBUTORS INDICATES ITS SCOPE: HISTORICAL
PHYCOLOGY (1927-1967); TAXONOMY AND PHYLOGENY IN ALGAL BIOCHEMISTRY AND
PHYSIOLOGY; ALGAL REQUIREMENTS FOR MICRONUTRIENTS; PHOSPHATE METABOLISM
OF GREEN ALGAE; NITROGEN INPUT TO AQUATIC ECOSYSTEMS; LIGHT AND
TEMPERATURE EFFECTS ON ALGAE; PHYTOPLANKTONIC PHOTOSYNTHESIS;
PHOTOSYNTHESIS OF CLADOPHORA; BIOLOGY OF FILAMENTOUS CONJUGATING ALGAE;
APPROACH TO MODERN ALGAL TAXONOMY; TRENDS IN ALGAL GENETICS; CYTOLOGY
OF CYANOPHYTES (BLUE-GREEN ALGAE); ULTRASTRUCTURE OF CYANOPHYTES;
RUSSIAN EXPERIMENTS IN ALGAL CULTURE; PHYCOLOGICAL MEDICINE; RUSSIAN
SOIL ALGOLOGY; HARVESTING WASTE-GROWN MICROALGAE; WASTEWATER TREATMENT
AND ALGAL GROWTH; REGENERATION OF BAVARIAN LAKES; NUTRIENT REMOVAL,
VIRAL DISEASES IN CYANOPHYTES; ALGICIDAL EFFECTS; ALGAE-FISH
RELATIONSHIPS; STREAM ASSIMILATION (NEW YORK); PRODUCTIVITY, MORICHES
BAY (NEW YORK); ALGAL ENVIRONMENT, ONONDAGA LAKE (NEW YORK); REVIEW OF
ALGAL LITERATURE (NEW YORK STATE). AN INDEX IS PROVIDED. (SEE
W70-03974). (EICHHORN-WISCONSIN)

FIELD 05C

ACCESSION NO. W70-03973

ONONDAGA LAKE, NEW YORK--AN UNUSUAL ALGAL ENVIRONMENT,

SYRACUSE UNIV., N. Y. DEPT. OF CIVIL ENGINEERING.

DANIEL F. JACKSON.

ALGAE, MAN, AND THE ENVIRONMENT, SYRACUSE UNIV PRESS, NEW YORK, P 515-524,
1968.

DESCRIPTORS:
*ALGAE, *AQUATIC ENVIRONMENTS, *EUTROPHICATION, *MEROMIXIS, *WATER
POLLUTION EFFECTS, BIOINDICATORS, NUISANCE ALGAE, NEW YORK, SEWAGE
EFFLUENTS, SEWAGE TREATMENT, INDUSTRIAL WASTES, WATER CHEMISTRY,
DIATOMS, CHLOROPHYTA, EUGLENOPHYTA, CYANOPHYTA, CHRYSOPHYTA, BIOASSAYS,
SALINE WATER, CHLORELLA, CHLAMYDOMONAS, EUGLENA.

IDENTIFIERS:
*ONONDAGA LAKE(NY), EUNOTIA, AMPHORA, GOMPHONEMA, PINNULARIA,
NITZSCHIA, NAVICULA, SYRACUSE(NY), ENTEROMORPHA INTESTINALIS,
CLADOPHORA, CYCLOTELLA, HALOCHLOROCOCCUM, SYNECHOCOCCUS, CARTERIA,
GONIUM, ANKISTRODESMUS, MICROSPORA, PEDIASTRUM, SCENEDESMUS,
LEPOCINCLIS, ANABAENA, ANACYSTIS, GLOEOTRICHIA, GLOEOCAPSA, STAURONEIS.

ABSTRACT:
FED BY SALINE SPRINGS, LAKE ONONDAGA HAS HIGHEST KNOWN CHLORIDE CONTENT
(TYPICALLY 1460 MILLIGRAMS/LITER) OF NEW YORK LAKES AND IS PROBABLY
MEROMICTIC. ITS MORPHOMETRIC, GEOGRAPHIC AND HISTORICAL FEATURES ARE
DESCRIBED, AND WATER QUALITY DATA ARE TABULATED TO INCLUDE COMPARISONS
WITH OTHER LAKES. ALTHOUGH SUBJECT TO EXTENSIVE EUTROPHICATION SINCE
1863, IT FAILS TO BLOOM WITH PLANKTONIC CYANOPHYTES, BUT ANNUAL BLOOMS
OF CHLOROPHYTES AND EUGLENOPHYTES HAVE OCCURRED DURING SUMMERS SINCE
1962. LAKE RECEIVES INDUSTRIAL POLLUTANTS. TWO CHARACTERISTIC AREAS ARE
OBSERVABLE, BASED ON DISTRIBUTION OF ATTACHED ALGAE, CLADOPHORA AND
ENTEREMORPHA; VARIATIONS IN THE IONIC RATIO, SODIUM/CALCIUM, ARE
THOUGHT TO CONTROL THEIR DISTRIBUTION. LAKE SUPPORTS ABUNDANT GROWTH OF
DIATOMS, DISTRIBUTED AMONG 28 GENERA. IN A STUDY TO EVALUATE EFFECTS ON
THE LAKE OF REPLACING PRIMARY TREATMENT PLANTS WITH IMPROVED SECONDARY
TREATMENT, GROWTH OF TEN STRAINS OF CHLOROPHYTES, TWO EUGLENOPHYTES,
EIGHT DIATOMS, AND FIVE CYANOPHYTES CULTURED IN FOUR CONCENTRATIONS OF
ENRICHED LAKE WATER WAS COMPARED WITH THEIR GROWTH IN NUTRIENT MEDIA.
BIOASSAY DATA ARE TABULATED; SUMMARIZED, THEY INDICATE THAT LAKE'S
WATER WILL NOT SUPPORT GROWTH OF PLANKTONIC CYANOPHYTES, AND THAT SOME
COMPONENT STIMULATES GROWTH OF DIATOM, CYCLOTELLA, AND THE
CHLOROPHYTES: CHLAMYDOMONAS, HALOCHLOROCOCCUS, AND CHLORELLA (MILFORD
STRAIN). (SEE ALSO W70-03973). (EICHHORN-WISCONSIN)

FIELD 05C, 02H

ACCESSION NO. W70-03974

EUTROPHICATION: CAUSES, CONSEQUENCES, CORRECTIVES.

NATIONAL ACADEMY OF SCIENCES, WASHINGTON, D.C.

PROCEEDINGS OF SYMPOSIUM HELD JUNE 11-15, 1967, WISCONSIN UNIV, MADISON.
GERARD A ROHLICH, CHAIRMAN, PLANNING COMMITTEE FOR THE INTERNATIONAL
SYMPOSIUM ON EUTROPHICATION. NAS STANDARD BOOK NO 309-01700-9. PRINTING AND
PUBLISHING OFFICE, NATIONAL ACADEMY OF SCIENCES, WASHINGTON, DC, 1969. 661
P.

DESCRIPTORS:
*EUTROPHICATION, *WATER POLLUTION SOURCES, *WATER POLLUTION EFFECTS,
*WATER POLLUTION CONTROL, *CONFERENCES, STREAMS, ESTUARIES, LAKES,
GREAT LAKES, PHYSIOLOGICAL ECOLOGY, WATER CHEMISTRY, GEOGRAPHICAL
REGIONS, CYCLING NUTRIENTS, ZOOPLANKTON, ALGAE, AQUATIC PLANTS,
PALEOLIMNOLOGY, MATHEMATICAL MODELS, BACTERIA, PHYTOPLANKTON,
GEOCHEMISTRY, BENTHOS, FISH, ECOSYSTEMS.

IDENTIFIERS:
NATIONAL ACADEMY OF SCIENCES(USA), UNIVERSITY OF WISCONSIN, ASIA,
NATIONAL RESEARCH COUNCIL(USA), US ATOMIC ENERGY COMMISSION, NATIONAL
SCIENCE FOUNDATION, OFFICE OF NAVAL RESEARCH, US DEPARTMENT OF
INTERIOR, MADISON(WIS), EUROPE, WATER POLLUTION ASSESSMENT, URBAN
DRAINAGE, AGRICULTURAL DRAINAGE, NORTH AMERICA, FOREST DRAINAGE.

ABSTRACT:
IN 1965, A PLANNING COMMITTEE APPOINTED BY NATIONAL ACADEMY OF
SCIENCES-NATIONAL RESEARCH COUNCIL (NAS-NRC) RECOMMENDED THAT '--AN
INTERNATIONAL SYMPOSIUM ON EUTROPHICATION BE HELD IN ORDER THAT THE
PRESENT WORLDWIDE STATE OF KNOWLEDGE AND UNDERSTANDING OF THIS
PHENOMENON CAN BE DISCUSSED IN OPEN FORUM AND RECOMMENDATIONS DEVELOPED
FOR THE EFFECTIVE MANAGEMENT OF PROBLEMS AND FOR THE COURSE OF FUTURE
RESEARCH.' THAT SYMPOSIUM MET AT UNIVERSITY OF WISCONSIN, MADISON,
11-15 JUNE 1967, SPONSORED BY NAS-NRC, ATOMIC ENERGY COMMISSION,
DEPARTMENT OF INTERIOR, NATIONAL SCIENCE FOUNDATION, AND OFFICE OF
NAVAL RESEARCH. APPROXIMATELY 600 PERSONS FROM 12 COUNTRIES ATTENDED.
THIS REPORT OF PROCEEDINGS INCLUDES AN INTRODUCTORY ADDRESS,
'EUTROPHICATION, PAST AND PRESENT', BY YALE UNIVERSITY PROFESSOR G. E.
HUTCHINSON, AND FOUR SECTIONS OF TECHNICAL PRESENTATIONS AS FOLLOWS
(NUMBERS OF CONTRIBUTIONS TO EACH INDICATED IN PARENTHESES):
GEOGRAPHICAL CONCEPTS (9); DETECTION AND MEASUREMENT (7); PREVENTION
AND CORRECTION (8); SCIENTIFIC CONTRIBUTIONS FROM EUTROPHICATION
RESEARCH (8). TWO PAPERS, PRESENTED AT CONFERENCE, ARE NOT PUBLISHED
HERE. THE REPORT ALSO INCLUDES AN INTRODUCTION, SUMMARY, AND SERIES OF
SPECIFIC RECOMMENDATIONS FORMULATED BY THE ORGANIZING COMMITTEE: 12
RELATING TO EDUCATION AND INFORMATION AND 8 TO RESEARCH. THE COMMITTEE
ALSO RECOMMENDED THAT NAS-NRC FORM A PERMANENT COMMITTEE TO IMPLEMENT
ITS RECOMMENDATIONS. (EICHHORN-WISCONSIN)

FIELD 05C

ACCESSION NO. W70-03975

EUTROPHICATION: CAUSES, CONSEQUENCES, CORRECTIVES.

NATIONAL ACADEMY OF SCIENCES, WASHINGTON, D.C.

PROCEEDINGS OF SYMPOSIUM HELD JUNE 11-15, 1967, WISCONSIN UNIV, MADISON.
GERARD A ROHLICH, CHAIRMAN, PLANNING COMMITTEE FOR THE INTERNATIONAL
SYMPOSIUM ON EUTROPHICATION. NAS STANDARD BOOK NO 309-01700-9. PRINTING AND
PUBLISHING OFFICE, NATIONAL ACADEMY OF SCIENCES, WASHINGTON, DC, 1969. 661
P.

DESCRIPTORS:
*EUTROPHICATION, *WATER POLLUTION SOURCES, *WATER POLLUTION EFFECTS,
*WATER POLLUTION CONTROL, *CONFERENCES, STREAMS, ESTUARIES, LAKES,
GREAT LAKES, PHYSIOLOGICAL ECOLOGY, WATER CHEMISTRY, GEOGRAPHICAL
REGIONS, CYCLING NUTRIENTS, ZOOPLANKTON, ALGAE, AQUATIC PLANTS,
PALEOLIMNOLOGY, MATHEMATICAL MODELS, BACTERIA, PHYTOPLANKTON,
GEOCHEMISTRY, BENTHOS, FISH, ECOSYSTEMS.

IDENTIFIERS:
NATIONAL ACADEMY OF SCIENCES(USA), UNIVERSITY OF WISCONSIN, ASIA,
NATIONAL RESEARCH COUNCIL(USA), US ATOMIC ENERGY COMMISSION, NATIONAL
SCIENCE FOUNDATION, OFFICE OF NAVAL RESEARCH, US DEPARTMENT OF
INTERIOR, MADISON(WIS), EUROPE, WATER POLLUTION ASSESSMENT, URBAN
DRAINAGE, AGRICULTURAL DRAINAGE, NORTH AMERICA, FOREST DRAINAGE.

ABSTRACT:
IN 1965, A PLANNING COMMITTEE APPOINTED BY NATIONAL ACADEMY OF
SCIENCES-NATIONAL RESEARCH COUNCIL (NAS-NRC) RECOMMENDED THAT '--AN
INTERNATIONAL SYMPOSIUM ON EUTROPHICATION BE HELD IN ORDER THAT THE
PRESENT WORLDWIDE STATE OF KNOWLEDGE AND UNDERSTANDING OF THIS
PHENOMENON CAN BE DISCUSSED IN OPEN FORUM AND RECOMMENDATIONS DEVELOPED
FOR THE EFFECTIVE MANAGEMENT OF PROBLEMS AND FOR THE COURSE OF FUTURE
RESEARCH.' THAT SYMPOSIUM MET AT UNIVERSITY OF WISCONSIN, MADISON,
11-15 JUNE 1967, SPONSORED BY NAS-NRC, ATOMIC ENERGY COMMISSION,
DEPARTMENT OF INTERIOR, NATIONAL SCIENCE FOUNDATION, AND OFFICE OF
NAVAL RESEARCH. APPROXIMATELY 600 PERSONS FROM 12 COUNTRIES ATTENDED.
THIS REPORT OF PROCEEDINGS INCLUDES AN INTRODUCTORY ADDRESS,
'EUTROPHICATION, PAST AND PRESENT', BY YALE UNIVERSITY PROFESSOR G. E.
HUTCHINSON, AND FOUR SECTIONS OF TECHNICAL PRESENTATIONS AS FOLLOWS
(NUMBERS OF CONTRIBUTIONS TO EACH INDICATED IN PARENTHESES):
GEOGRAPHICAL CONCEPTS (9); DETECTION AND MEASUREMENT (7); PREVENTION
AND CORRECTION (8); SCIENTIFIC CONTRIBUTIONS FROM EUTROPHICATION
RESEARCH (8). TWO PAPERS, PRESENTED AT CONFERENCE, ARE NOT PUBLISHED
HERE. THE REPORT ALSO INCLUDES AN INTRODUCTION, SUMMARY, AND SERIES OF
SPECIFIC RECOMMENDATIONS FORMULATED BY THE ORGANIZING COMMITTEE: 12
RELATING TO EDUCATION AND INFORMATION AND 8 TO RESEARCH. THE COMMITTEE
ALSO RECOMMENDED THAT NAS-NRC FORM A PERMANENT COMMITTEE TO IMPLEMENT
ITS RECOMMENDATIONS. (EICHHORN-WISCOSNIN)

FIELD 05C

ACCESSION NO. W70-03975

PHYSIOLOGICAL ECOLOGY,

OREGON STATE UNIV., CORVALLIS. DEPT. OF BOTANY.

HARRY K. PHINNEY.

PROCEEDINGS OF THE EUTROPHICATION-BIOSTIMULATION ASSESSMENT WORKSHOP, JUNE
  19-21, 1969, CALIFORNIA UNIV, BERKELEY, SANITARY ENGINEERING RESEARCH LAB
  AND NATIONAL EUTROPHICATION RESEARCH PROGRAM, CORVALLIS, ORE, PAC NORTHWEST
  WATER LAB, P 141-145. 1 FIG, 5 REF.

DESCRIPTORS:
  *PHYSIOLOGICAL ECOLOGY, *AQUATIC ENVIRONMENTS, RIVERS, ENVIRONMENTAL
  EFFECTS, PERIPHYTON, WATER POLLUTION EFFECTS, OREGON, BIOINDICATORS,
  LIGHT INTENSITY, EUTROPHICATION, GRAZING, ALGAE, WATER CHEMISTRY,
  BIOMASS, BENTHOS, SALINITY, SNAILS, SULFATES, NITRATES, PHOSPHATES,
  POTASSIUM, IRON, BICARBONATE, ZINC, COBALT, CALCIUM, MAGNESIUM,
  CURRENTS(WATER).

IDENTIFIERS:
  BERRY CREEK(ORE), OREGON STATE UNIVERSITY, EXPERIMENTAL STREAMS,
  SEASONAL EFFECTS, THERMAL EFFECTS, APHANIZOMENON FLOS AQUAE, METOLIUS
  RIVER(ORE), COMMUNITY STRUCTURE, OXYTREMA, ORGANIC ENRICHMENT,
  INTERACTIONS.

ABSTRACT:
  PHYSIOLOGICAL ECOLOGY MAY BE DEFINED AS THE STUDY OF PHYSIOLOGICAL
  RESPONSES OF ORGANISMS TO VARIATIONS IN ECOLOGICAL CONDITIONS
  EXPERIENCED IN NATURE OR TO THOSE WHICH CAN BE MANIPULATED IN THE
  LABORATORY. BOTH LABORATORY AND FIELD APPROACHES TO ECOLOGY SUFFER FROM
  BASIC PHILOSOPHICAL AND METHODOLOGICAL DEFECTS. PHYSIOLOGICAL
  ECOLOGISTS ATTEMPT TO RETAIN A BALANCED PERSPECTIVE, AND MAINTAIN AN
  AWARENESS OF THESE PITFALLS IN INTERPRETING DATA AND CONSTRUCTING
  GENERALIZATIONS. SPECIFIC STUDIES ARE CITED WHICH ILLUSTRATE THESE
  PRINCIPLES FOR A VARIETY OF AQUATIC ENVIRONMENTS (RIVERS, LABORATORY
  STREAMS, AND ESTUARIES) AND FOR A VARIETY OF INFLUENCING FACTORS
  (CHEMICAL, PHYSICAL, AND BIOTIC). BECAUSE OF COMPLEXITIES IN
  ELUCIDATING CAUSAL FACTORS INFLUENCING SPECIES ABUNDANCE, GREAT CAUTION
  SHOULD BE EXERTED IN IDENTIFYING ALGAL SPECIES AS INDICATORS OF
  POLLUTION. ACCUMULATION OF INFORMATION RELATIVE TO PHYSIOLOGICAL
  ECOLOGY IS SLOW, LABORIOUS, AND FRAUGHT WITH DIFFICULTIES HAMPERING
  CORRECT INTERPRETATION. IT IS UNLIKELY THAT 'CRASH' PROGRAMS DESIGNED
  TO CHARACTERIZE THE PHYSIOLOGICAL ECOLOGY OF CULTURAL EUTROPHICATION
  CAN PROVIDE SOLUTIONS TO OUR PRESENT CRISIS IN TIME TO PREVENT THE
  FORESEEABLE DISASTER. (SEE W70-02775).
  (EICHHORN-WISCONSIN)

FIELD 05C

ACCESSION NO. W70-03978

C-14 UPTAKE AS A SENSITIVE MEASURE OF THE GROWTH OF ALGAL CULTURES,

CALIFORNIA UNIV., DAVIS. DEPT. OF ZOOLOGY; AND CALIFORNIA UNIV., DAVIS. INST. OF ECOLOGY.

CHARLES R. GOLDMAN, MILTON G. TUNZI, AND RICHARD ARMSTRONG.

PROCEEDINGS OF THE EUTROPHICATION-BIOSTIMULATION ASSESSMENT WORKSHOP, JUNE 19-21, 1969, CALIFORNIA UNIV, BERKELEY, SANITARY ENGINEERING RESEARCH LAB AND NATIONAL EUTROPHICATION RESEARCH PROGRAM, CORVALLIS, ORE, PAC NORTHWEST WATER LAB, P 158-170. 6 FIG, 1 TAB, 11 REF.

DESCRIPTORS:
*CARBON RADIOISOTOPES, *ALGAE, *CULTURES, PHYTOPLANKTON, CARBON, RADIOACTIVITY, CHLORELLA, SAMPLING, EFFLUENTS, TRIBUTARIES, ALKALINITY, HYDROGEN ION CONCENTRATION, NUTRIENTS, PHOSPHORUS, NITROGEN, IRON, GROWTH RATES, CHLOROPHYTA, CYANOPHYTA, CHRYSOPHYTA, PHOTOSYNTHESIS.

IDENTIFIERS:
*GROWTH, *UPTAKE, FIXATION, CELL COUNT, OPTICAL DENSITY, PARTICULATE ORGANIC CARBON, LAKE TAHOE(CALIF), CHLORELLA PYRENOIDOSA, SELENASTRUM GRACILE, CARBON ASSIMILATION, BACILLARIOPHYCEAE, DINOPHYCEAE.

ABSTRACT:
RESULTS OF COMPARISONS TO DETERMINE RELIABILITY OF CARBON-14 EXPERIMENTS WITH OTHER METHODS EMPLOYED FOR MEASURING GROWTH IN CULTURES ARE SUMMARIZED. THESE INCLUDE MAKING CELL COUNTS, MEASURING THE INCREASE IN LIGHT EXTINCTION AS THE CELLS MULTIPLY, AND DETERMINING THE BIOMASS OF CELLS PRODUCED. RESULTS OF A CHLORELLA EXPERIMENT PROVIDE A VALUABLE EVALUATION OF THESE METHODS. THE CELL COUNT WAS QUITE SUCCESSFUL; THE OPTICAL DENSITY METHOD WAS FASTEST AND EASIEST BUT WAS MAINLY APPLICABLE TO MEASURING CONCENTRATION OF DENSE CULTURES. THE CARBON-14 METHOD CAN EASILY BE USED FOR MEASURING ALGAL CELL GROWTH IN BOTH NUTRIENT-RICH AND NUTRIENT-POOR WATERS. CARBON-14 CAN DETECT GROWTH IN WATER WITH NATURAL PHYTOPLANKTON WHEN CELL COUNT AND OPTICAL DENSITY METHODS DO NOT GIVE EVIDENCE OF ANY CHANGE IN ALGAL MASS; CARBON-14 DATA GIVE A CARBON FIXATION RATE, WHEREAS CELL COUNT AND OPTICAL DENSITY MEASUREMENTS ARE STATIC MEASURES OF MASS. IN ADDITION, CARBON-14 EXPERIMENTS YIELD VALUES PROPORTIONAL TO THE MASS OF CELLS IN A CULTURE. THE COMBINED EFFECT OF NITROGEN AND PHOSPHORUS WAS TO STIMULATE PHOTOSYNTHESIS, THE GREATEST STIMULATION RESULTING FROM THE MAXIMAL ADDITION OF THE TWO NUTRIENTS. (SEE W70-02775).
(JONES-WISCONSIN)

FIELD 05C

ACCESSION NO. W70-03983

CARBON SOURCES IN ALGAL POPULATIONS AND ALGAL COMMUNITY STRUCTURE,

OKLAHOMA WATER RESOURCES RESEARCH INST., STILLWATER.

TROY C. DORRIS.

AVAILABLE FROM THE CLEARINGHOUSE AS PB-189 523, $3.00 IN PAPER COPY, $0.65 IN
MICROFICHE. RESEARCH PROJECT TECHNICAL COMPLETION REPORT, OKLAHOMA WRSIC
INSTITUTE (1969). 10 P, 1 FIG, 2 TAB. OWRR PROJECT NO B-005-OKLA.

DESCRIPTORS:
*CARBON, *ALGAE, *POPULATION, *BIOLOGICAL COMMUNITIES, EFFLUENTS,
EUTROPHICATION, PHYTOPLANKTON, DECOMPOSING ORGANIC MATTER, OKLAHOMA,
RESERVOIRS, SAMPLING, INDUSTRIAL WASTES, MUNICIPAL WASTES, SEWAGE,
ADSORPTION, ACTIVATED CARBON, FILTERS, SEASONAL, STRATIFICATION,
BIOCHEMICAL OXYGEN DEMAND, CARBON RADIOISOTOPES, DETERGENTS,
ANTIFREEZE, OXIDATION, PHOTOSYNTHESIS, OIL WASTES, GARBAGE DUMPS,
PIGMENTS, DAMS, PHYSICOCHEMICAL PROPERTIES, CHLOROPHYLL.

IDENTIFIERS:
ARKANSAS RIVER(OKLA), TULSA(OKLA), POLLUTION, FOSSIL, CIMARRON
RIVER(OKLA), BIXBY(OKLA), KEYSTONE RESERVOIR(OKLA), CAROTENOIDS.

ABSTRACT:
A PORTION OF A STUDY TO DETERMINE SOURCE AND DISTRIBUTION OF ORGANIC
COMPOUNDS IN THE KEYSTONE RESERVOIR, OKLAHOMA, AND THE DEVELOPMENT OF
PHYTOPLANKTON POPULATIONS IN RESPONSE TO EUTROPHICATION PROCESSES
RESULTING FROM DECOMPOSITION OF ORGANIC COMPOUNDS IN EFFLUENTS ENTERING
THE CIMARRON ARM OF THE RESERVOIR IS DESCRIBED. ORGANIC COMPOUNDS FROM
INDUSTRIAL AND MUNICIPAL SEWAGE EFFLUENTS WERE SAMPLED BY ADSORPTION ON
ACTIVATED CARBON FILTERS AT SIX STATIONS TO DEVELOP INFORMATION ON
SEASONAL CHANGES IN QUANTITY AND NATURE OF ORGANIC COMPOUNDS, EFFECT OF
STRATIFICATION ON MOVEMENT AND CHANGES OF COMPOUNDS, AND POLLUTION OF
THE ARKANSAS RIVER BY EFFLUENTS FROM TULSA. THE EFFORT TO IDENTIFY
PARTICULAR COMPOUNDS, TRACE THEIR PROGRESS THROUGH THE RESERVOIR AND
DETERMINE EXTENT OF THEIR DEGRADATION BY PASSAGE THROUGH THE RESERVOIR
IS EXPECTED TO BE COMPLETED WITH FURTHER STUDY. EXTRACTION OF ADSORBED
COMPOUNDS FROM CARBON FILTERS HAS BEEN ACCOMPLISHED BUT DATA WERE
INCOMPLETE. PHYTOPLANKTON STUDIES ON THE CIMARRON ARM OF THE RESERVOIR
WERE DESIGNED TO STUDY EFFECTS OF EUTROPHICATION OF THE WATER BY
DOMESTIC AND INDUSTRIAL EFFLUENTS. THE FULL SPECTRUM OF PHOTOSYNTHETIC
AND OTHER PIGMENTS WAS MEASURED AS EACH SAMPLE WAS TAKEN. EFFORTS WERE
MADE IN SPECIES IDENTIFICATION. (JONES-WISCONSIN)

FIELD 05B

ACCESSION NO. W70-04001

PROGRESS AND DEVELOPMENTS IN TREATMENT OF INTEGRATED TEXTILE MILL WASTES,

COLORADO STATE UNIV., FORT COLLINS. NATURAL RESOURCES CENTER.

T. A. ALSPAUGH.

PROCEEDINGS 12TH SOUTHERN MUNICIPAL INDUSTRIAL WASTE CONFERENCE, NORTH CAROLINA STATE UNIV, RALEIGH, NC, P 46-62, 1964.

DESCRIPTORS:
*ACTIVATED SLUDGE, *BIOCHEMICAL OXYGEN DEMAND, *COLOR, WASTE WATER TREATMENT, TREATMENT FACILITIES, AERATION, LAGOONS, DOMESTIC WASTES, SEDIMENTATION, CHLORINATION, SLUDGE, DETERGENTS, SPRAYING, ALGAE, CENTRIFUGATION, FILTRATION, ANEROBIC DIGESTION, INCINERATION, IRON COMPOUNDS, SULFATES.

IDENTIFIERS:
*ANTI-FOAM AGENTS, *TEXTILE WASTES, FIBER WASTES, DYEING WASTES.

ABSTRACT:
A FULL-SCALE PLANT WAS CONSTRUCTED TO TREAT 21 MIL. GAL. PER WEEK BY THE 24-HOUR HIGH-SOLIDS ACTIVATED-SLUDGE PROCESS IN WHICH THE LONGER AERATION PERIOD AND THE HIGHER SOLIDS CONCENTRATE IN THE AERATION TANK BOTH ASSIST IN REMOVING THE HIGHLY-SOLUBLE BIOCHEMICAL OXYGEN DEMAND OF TEXTILE WASTE WATERS, BUFFERING SHOCK LOADS, AND PROVIDING EFFICIENT TREATMENT AS REGARDS REMOVAL OF COLOR. THE TEXTILE WASTE WATERS PASS THROUGH TWO STORAGE LAGOONS IN SERIES (TO PROVIDE EQUALIZATION AND SOME REMOVAL OF LINT AND DYE PASTE) AND ARE MIXED WITH 10% OF SCREENED AND DEGRITTED DOMESTIC SEWAGE BEFORE ENTRY INTO THE AERATION TANK. AFTER FINAL SEDIMENTATION THE EFFLUENT IS CHLORINATED AND REAERATED BEFORE DISCHARGE TO A CREEK. OPERATING DATA ARE TABULATED SHOWING AN AVERAGE BIOCHEMICAL OXYGEN DEMAND REMOVAL OF 89-90% AND COLOR REMOVAL OF 50-75%. THE PROBLEM OF FOAMING IS ACCENTUATED BY THE HIGH CONCENTRATE OF SYNTHETIC DETERGENTS IN TEXTILE WASTE WATERS; THE MOST SUCCESSFUL TREATMENT METHODS ARE SPRAYING WITH CRUDE WASTE WATER, ALONE OR WITH THE ADDITION OF ANTI-FOAMING AGENT ON TO THE ROTOR BLADES, AND POSSIBLY THE REMOVAL AND DESTRUCTION OF FOAM BY A VACUUM PUMP. THE POORLY-SETTLING SLUDGE WAS FOUND TO CONTAIN HEAVY GROWTHS OF ALGAE. THE PROBLEM COULD BE OVERCOME BY ALLOWING THE SLUDGE TO BECOME SEPTIC BEFORE RETURN TO THE AERATION TANK, OR BY INTRODUCING A SHOCK LOAD OF ORGANIC ALGAE. HOWEVER, THESE ALGAE HAVE GOOD CHARACTERISTICS FOR REMOVAL OF BIOCHEMICAL OXYGEN DEMAND AND COLOR. EXCESS SLUDGE FROM THE TREATMENT OF TEXTILE WASTE WATERS IS DIFFICULT TO SETTLE AND CONCENTRATE. RATES OF VACUUM FILTRATION AND CENTRIFUGATION ARE LOW COMPARED WITH SEWAGE SLUDGE. AT PRESENT THE SLUDGE IS DISCHARGED TO THE MUNICIPAL SEWER. OTHER PROMISING METHODS OF DISPOSAL APPEAR TO BE VACUUM FILTRATION AND CENTRIFUGATION. (LIVENGOOD-NORTH CAROLINA STATE UNIV)

FIELD 05D

ACCESSION NO. W70-04060

APPLICATION OF RADIOISOTOPE TECHNIQUES TO A CRITICAL WATER RESOURCES PROBLEM
AREA - NAMELY NUTRITIONAL POLLUTION,

OKLAHOMA UNIV., NORMAN. DEPT. OF CIVIL ENGINEERING AND SANITARY SCIENCE; AND
ARKANSAS UNIV., FAYETTEVILLE. DEPT. OF CIVIL ENGINEERING.

GEORGE W. REID, ROBERT A. GEARHEART, JAMES M. ROBERTSON, AND ROBERT M.
SWEAZY.

AVAILABLE FROM THE CLEARINGHOUSE AS PB-189 793, $3.00 IN PAPER COPY, $0.65 IN
MICROFICHE. OKLAHOMA STATE UNIVERSITY WATER RESOURCES RESEARCH INSTITUTE
PROJECT COMPLETION REPORT, 1969. 14 P, 1 FIG, 1 TAB, 1 APPEND. OWRR PROJ NO
A-011-OKLA.

DESCRIPTORS:
*EUTROPHICATION, *NUTRIENTS, *PHOSPHORUS COMPOUNDS, ALGAE, PHOSPHATES,
NITRATES, SEWAGE EFFLUENTS, TERTIARY TREATMENT, WATER POLLUTION
SOURCES, WATER POLLUTION EFFECTS, LABORATORY TESTS, MODEL STUDIES,
RADIOACTIVITY TECHNIQUES, TAGGING.

IDENTIFIERS:
NUTRIENT REQUIREMENTS(ALGAE).

ABSTRACT:
MINIMAL CONCENTRATIONS OF NITROGEN AND PHOSPHORUS NECESSARY FOR THE
GROWTH OF ALGAE WERE STUDIED. A RADIOACTIVE PHOSPHORUS METHOD WAS USED
AS ONE OF THE PARAMETERS FOR BIOLOGICAL MASS PRODUCED AND PHOSPHORUS
UPTAKE RATE IN ALGA CULTURING UNIT. ALL OTHER NECESSARY INORGANIC AND
ORGANIC NUTRIENTS, INCLUDING VITAMIN B-12 AND OTHER GROWTH METABOLITES,
WERE SUPPLIED IN EXCESS TO ALLOW MAXIMUM ALGAE REACTION TO THE NITROGEN
AND PHOSPHORUS. THE RELATIONSHIP BETWEEN NITROGEN, PHOSPHORUS AND ALGAL
GROWTH IS EXPRESSED BY A MULTIPLE REGRESSION ANALYSIS PROCEDURE. A
BACTERIAL BLOOM OCCURS WITHIN THE SECOND OR THIRD DAY IN CONTRAST TO
THE ALGAE GROWTH WHICH USUALLY REQUIRES 6 TO 8 DAYS. THE SOLUBLE
PHOSPHATE IS TAKEN UP 80-90% IN THE FIRST DAY. AT A PHOSPHATE
CONCENTRATION OF 0.10 MG/L, THE ALGAL BLOOM CONDITION THRESHOLD IS
REACHED. REGARDLESS OF THE NITROGEN CONCENTRATION, THIS NARROW RANGE OF
SENSITIVITY TO THE PHOSPHORUS CONCENTRATION DEMONSTRATES THE PROBLEM IN
ATTEMPTING TO CONTROL NUTRITIONAL POLLUTION. (KNAPP-USGS)

FIELD 05A, 05C

ACCESSION NO. W70-04074

INVESTIGATION OF SOME METHODS FOR INCREASING THE DIGESTIBILITY IN VITRO OF
MICROALGAE,

ROYAL INST. OF TECH., STOCKHOLM (SWEDEN). DIV. OF APPLIED BIOCHEMISTRY.

GUDMUND HEDENSKOG, LENNART ENEBO, JITKA VENDLOVA, AND BOHUMIR PROKES.

BIOTECHNOLOGY AND BIOENGINEERING, VOL 11, NO 1, P 37-51, 1969. 6 FIG, 4 TAB,
18 REF.

DESCRIPTORS:
*ALGAE, *MICROORGANISMS, SCENEDESMUS, PROTEINS, WALLS, STRUCTURE, AMINO
ACIDS, MECHANICAL EQUIPMENT, ENZYMES, CHEMICAL DEGRADATION, VITAMINS,
CELLULOSE, CHLORELLA, PEPTIDES, BACTERIA, SNAILS, WEIGHT, VACUUM
DRYING, BOILING.

IDENTIFIERS:
*DIGESTIBILITY, *IN VITRO, IN VIVO, BALL-MILL, MEICELASE, HYDROGEN
PEROXIDE, RATS, TRYPSIN, SCENEDESMUS QUADRICAUDA, SCENEDESMUS OBLIQUUS,
HELIX POMATIA, CHLORELLA PYRENOIDOSA, NITROGEN UTILIZATION, ROLLER
DRYING, SPRAY-DRIED, GLASS BEADS, TRICHODERMA, DISINTEGRATION.

ABSTRACT:
TO INCREASE AVAILABILITY OF CELL BOUND PROTEIN IN SCENEDESMUS ALGAE,
MECHANICAL, ENZYMATIC, AND CHEMICAL METHODS OF DEGRADING THE CELL WALL
STRUCTURE WERE INVESTIGATED. MECHANICAL TREATMENT INVOLVED THE USE OF A
BALL-MILL; ALGAE SUSPENSION TOGETHER WITH GLASS BEADS WAS MILLED IN A
WATER-COOLED CHAMBER EQUIPPED WITH ROTATING DISKS. THE CHEMICAL TEST
INCLUDED A CELLULOLYTIC ENZYME (MEICELASE) EMPLOYING HYDROGEN PEROXIDE.
USING BATCHES IN THE BALL-MILL EXPERIMENTS A COMPLETE DISINTEGRATION
WAS ACHIEVED IN A DISINTEGRATOR. TRIALS WERE ALSO PERFORMED WITH A
CONTINUOUS DISINTEGRATOR AND DEPENDENCE OF DISINTEGRATION ON BEAD SIZE
AND FLOW RATE STUDIED. DISINTEGRATION DETERMINED BY MICROSCOPIC CELL
COUNT WAS COMPARED TO INCREASE OF PEPSIN DIGESTIBILITY. THE MEICELASE
TREATMENT CAUSED A SLIGHT INCREASE OF PEPSIN DIGESTIBILITY, MEASURED
AFTER 3 HOUR PEPSIN INCUBATION. NO INCREASE OF PEPSIN DIGESTIBILITY WAS
DETECTED WITH HYDROGEN PEROXIDE TREATMENT. AFTER THE BALL-MILL
DISINTEGRATION, 95% OF CONTAMINATING BACTERIA WERE KILLED AND YIELDS OF
EXTRACTABLE PROTEINS WERE HIGHER. CAPACITY OF AVAILABLE CONTINUOUS
BALL-MILLS IS SUCH THAT THEY COULD BE USED ON A PILOT-PLANT SCALE AND
ENERGY COST OF DISINTEGRATION WOULD BE OF THE SAME MAGNITUDE AS THAT OF
SEPARATION. (JONES-WISCONSIN)

FIELD 05G

ACCESSION NO. W70-04184

WHAT'S KILLING OUR LAKES,

MINNESOTA UNIV., MINNEAPOLIS. DEPT. SOIL SCIENCE.

LOWELL D. HANSON, AND WILLIAM E. FENSTER.

CROPS AND SOILS, VOL 22, P 13-15, 1969.

DESCRIPTORS:
*LAKES, STREAMS, FISH, LAND, ECOSYSTEMS, EUTROPHICATION, OLIGOTROPHY, SEWAGE, ALGAE, SCUM, WEEDS, NUTRIENTS, SEPTIC TANKS, EFFLUENTS, SWAMPS, FOREST SOILS, MINNESOTA, WISCONSIN, MICHIGAN, PHOSPHORUS, AGRICULTURAL WATERSHEDS, ORGANIC MATTER, SEDIMENTS, FERTILIZERS, WATER POLLUTION, SNOW, RUNOFF, DRAINAGE WATER, GRASSLANDS, SOIL EROSION, GROUNDWATER, POTHOLES.

IDENTIFIERS:
*KILLING, ANIMAL MANURES, LAKE MINNETONKA(MINN), SOIL MINERALS, PLANT RESIDUES, MORRIS(MINN), SOIL PERCOLATION.

ABSTRACT:
PHOSPHORUS SOURCES FROM AGRICULTURAL WATERSHEDS ARE SOIL MINERALS AND ORGANIC MATTER; FERTILIZERS, MANURES, AND PLANT RESIDUES; SEWAGE AND SEDIMENT. IF SOIL REACTS WITH PHOSPHORUS-CONTAINING MATERIALS, SEEPING WATER WILL BE STRIPPED OF PHOSPHORUS. SNOW-MELT CARRIES HIGHER AMOUNTS OF PHOSPHORUS THAN RUNOFF WATERS DURING OTHER SEASONS. SEPTIC TANK WASTE WATER DOES NOT ALWAYS EVAPORATE AS EXPECTED AND SEEPS INTO GROUNDWATER AND LAKES. BOTTOM SEDIMENTS SERVE AS A 'PHOSPHORUS SINK' REDUCING THE EXPLOSIVE BIOLOGICAL EFFECTS OF PHOSPHORUS IN WATER. SOLUTION OF ENVIRONMENTAL PROBLEMS MAY INVOLVE EXTENSIVE LAND-FORMING OPERATIONS TO FORCE SOIL FILTRATION OF RUNOFF WATER. IMMEDIATE IMPROVEMENTS CAN BE EFFECTED BY NOT SPREADING MANURE ON SLOPING FROZEN LAND, DIVERTING FEEDLOT AND GRAIN STORAGE RUNOFF INTO SEEPAGE AREAS, AND STORING MANURE IN AREAS WHERE RUNOFF IS NOT A PROBLEM; LARGE FEED LOTS SHOULD BE LOCATED ON LEVEL, FINE TEXTURED SOILS RATHER THAN SHORELINE SLOPES OF A STREAM OR LAKE. LOCATION DECISIONS IN AGRICULTURE WILL INCREASINGLY BE GUIDED BY POLLUTION CONTROL AGENCY REGULATIONS AND ZONING RESTRICTIONS. LAND USE PLANNING HAS LONG RANGE CONSEQUENCES. COOPERATION OF SOIL SCIENTISTS, AGRICULTURAL ENGINEERS, AND FARMERS' INNOVATIVE TALENTS WILL RESULT IN QUICKER SOLUTIONS. (JONES-WISCONSIN)

FIELD 02H, 05C

ACCESSION NO. W70-04193

THE UPTAKE OF ORGANIC SOLUTES IN LAKE WATER,

UPPSALA UNIV. (SWEDEN). INST. OF LIMNOLOGY.

RICHARD T. WRIGHT, AND JOHN E. HOBBIE.

LIMNOLOGY AND OCEANOGRAPHY, VOL 10, NO 1, P 22-28, JAN 1965. 3 FIG, 3 TAB, 15 REF.

DESCRIPTORS:
*LAKES, *ORGANIC MATTER, *PHYSIOLOGICAL ECOLOGY, RADIOACTIVITY TECHNIQUES, ANALYTICAL TECHNIQUES, WATER POLLUTION EFFECTS, PHYTOPLANKTON, DIFFUSION, ALGAE, BACTERIA, EUTROPHICATION, GLUCOSE, CARBON RADIOISOTOPES, DINOFLAGELLATES, BIOASSAY, BIOINDICATORS.

IDENTIFIERS:
*HETEROTROPHY, *BIOLOGICAL UPTAKE, *MICROBIAL ECOLOGY, MAXIMUM UPTAKE VELOCITY, ACTIVE TRANSPORT, BACTERIAL POPULATIONS, ALGAL POPULATIONS, HETEROTROPHIC POTENTIAL, RELATIVE HETEROTROPHIC POTENTIAL, SWEDEN, ENZYME KINETICS, LINEWEAVER-BURK EQUATION, MICHAELIS KINETICS, FLAGELLATES, LAKE ERKEN(SWEDEN), ACETATE, LAKE NORRVIKEN(SWEDEN), ACETATE BACTERIA, GLUCOSE BACTERIA, CHLAMYDOMONADS.

ABSTRACT:
METHODS EXIST FOR ESTIMATING, WITH RADIOCARBON (C-14) LABELLED ORGANIC COMPOUNDS, THEIR ASSIMILATION BY PLANKTONIC MICROBIOTA. BASED ON LINEWEAVER-BURK MODIFICATION OF MICHAELIS ENZYME KINETICS, A MODIFICATION OF THOSE METHODS IS DESCRIBED WHICH GIVES BETTER ESTIMATES OF HETEROTROPHIC POTENTIAL, PARTICULARLY IN CASES WHERE UPTAKE IS MEDIATED VIA PROCESSES OF ACTIVE TRANSPORT. THE TREATMENT YIELDS A LINEAR RELATIONSHIP SUCH THAT A PLOT OF THE EXPRESSION, (C)(U)(T)/C, VERSUS A HAS A SLOPE WHOSE INVERSE ESTIMATES MAXIMUM UPTAKE VELOCITY (WHERE: C- COUNTS/MINUTE FROM 1 MICROCURIE C-14 IN COUNTING DEVICE; U- MICROCURIES ADDED TO SAMPLE; T- INCUBATION TIME (HOURS); C- COUNTS/MINUTE FROM FILTERED ORGANISMS; A- MILLIGRAMS/LITER OF ADDED SUBSTRATE). WHERE UPTAKE RESULTS FROM DIFFUSION INTO CELL WITH INSTANTANEOUS METABOLISM OF SUBSTRATE, A DIFFUSION CONSTANT CAN BE DEFINED FROM KNOWLEDGE OF UPTAKE VELOCITIES AT TWO DIFFERENT SUBSTRATE CONCENTRATIONS. METHOD WAS TESTED IN TWO SWEDISH LAKES SUBJECT TO EUTROPHICATION. OVER A WIDE RANGE OF SUBSTRATES, ESTIMATES APPARENTLY SHOW EXISTENCE OF TWO UPTAKE MECHANISMS: A TRANSPORT SYSTEM, SATURATED AT LOW SUBSTRATE CONCENTRATIONS, AND A DIFFUSION PROCESS WHOSE VELOCITY INCREASES STEADILY WITH INCREASING SUBSTRATE. EVIDENCE SUGGESTS THAT BACTERIAL POPULATIONS ARE RESPONSIBLE FOR 'TRANSPORT-TYPE' PHENOMENA, AND ALGAL POPULATIONS, FOR UPTAKE BY DIFFUSION. (EICHHORN-WISCONSIN)

FIELD 02H, 05C

ACCESSION NO. W70-04194

LIBERATION OF ORGANIC ACIDS BY GREEN UNICELLULAR ALGAE,

MOSCOW STATE UNIV. (USSR). DEPT. OF MICROBIOLOGY.

I. V. MAKSIMOVA, AND M. N. PIMENOVA.

TRANSLATED FROM MIKROBIOLOGIYA, VOL 37, NO L, P 77-86, JAN-FEB 1969. MICROBIOLOGY, VOL 38, NO 1, P 64-70, 1969. 7 FIG, 4 TAB, 17 REF.

DESCRIPTORS:
*ALGAE, *CHLORELLA, *ORGANIC ACIDS, BIOMASS, ORGANIC COMPOUNDS, CARBOHYDRATES.

IDENTIFIERS:
*GREEN ALGAE, CHLORELLA VULGARIS, CHLORELLA PYRENOIDOSA, VOLATILE ACIDS, NON-VOLATILE ACIDS, KETOACIDS, TAMIYA'S MINERAL MEDIUM, GLYOXYLIC ACID.

ABSTRACT:
AN INVESTIGATION OF THE QUALITATIVE COMPOSITION OF ORGANIC ACIDS LIBERATED TO THE MINERAL MEDIUM BY DIFFERENT STRAINS OF CHLORELLA VULGARIS AND CHLORELLA PYRENOIDOSA. FORMIC, ACETIC, LACTIC, PYRUVIC, A-KETOGLUTARIC, ACETOACETIC IN SOME CASES COMPRISED 25% OF THE EXTRACELLULAR COMPOUND. THE FILTRATES OF THE ALGAE ALWAYS CONTAINED GLYOXIC ACID. THE TOTAL QUANTITY OF ACIDS AND THEIR REPRESENTATION VARIED WITH THE PHASE OF GROWTH AND SIZE OF THE BIOMASS. ANALYSIS OF CELLS INDICATED THAT THE ACIDS WERE NOT DERIVED FROM EXTRACELLULAR COMPOUNDS. SECRETION OF NONVOLATILE ACIDS, GRADUAL ACCUMULATION OF VOLATILE ACIDS, AND THE SHARP INCREASE IN KETOACIDS AT HIGH DENSITY OF THE CULTURE SUGGESTED THAT THE PROCESSES ARE CONTROLLED BY DIFFERENT MACHANISMS. (WILDE-WISCONSIN)

FIELD 05C

180

ACCESSION NO. W70-04195

GRAVITY FILTRATION OF ALGAL SUSPENSIONS,

ONTARIO WATER RESOURCES COMMISSION, TORONTO. DIV. OF RESEARCH.

R. H. G. ANDREWS.

ONTARIO WATER RESOURCES COMMISSION, DIVISION OF RESEARCH PUBLICATION NO 21,
AUG 1968. 41 P, 22 FIG, 6 TAB.

DESCRIPTORS:
*WATER, *FILTERS, *ALGAE, NUISANCE ALGAE, INTAKES, SANDS, FLOW RATES,
TURBIDITY, CHLORELLA, SCENEDESMUS, EUGLENA.

IDENTIFIERS:
*WATER SUPPLIES, GREEN ALGAE, HEAD LOSSES, OOCISTUS, FILTRATION MEDIA,
ANTHRACITE, GRAVITY FILTERS.

ABSTRACT:
THIS STUDY WAS CONCERNED WITH EFFECTIVENESS OF DIFFERENT FILTER MEDIA
PREVENTING THE CLOGGING OF WATER SUPPLIES BY ALGAE: CHLORELLA,
SCENEDESMUS, OOCISTUS, AND EUGLENA. THE DUAL-MEDIA AND ANTHRACITE
FILTERS PROVED COMPARATIVELY MOST EFFECTIVE, BUT THEIR MAXIMUM
RETENTION WAS ONLY 50% OF EUGLENA AND 18% OF OTHER ALGAE. THERE WAS NO
CONSISTENT EFFECT OF THE RATE OF FLOW ON ALGAE REMOVED. THE USE OF
NEPHELOMETRIC TURBIDIMETER FAILED TO PROVIDE A SIGNIFICANT CORRELATION
BETWEEN ALGAE COUNTS AND RECORDED TURBIDITIES. THE CONCLUSION IS THAT
GRAVITY FILTERS ALONE SHOULD NOT BE CONSIDERED FOR REMOVAL OF ALGAE
FROM RAW WATER SUPPLIES. (WILDE-WISCONSIN)

FIELD 05F

ACCESSION NO. W70-04199

SPECIAL ASPECTS OF NITROGEN FIXATION BY BLUE-GREEN ALGAE,

BRISTOL UNIV. (ENGLAND). DEPT. OF BOTANY; AND WESTFIELD COLL., LONDON
(ENGLAND). DEPT. OF BOTANY.

ROSALIE M. COX, AND P. FAY.

PROCEEDINGS OF THE ROYAL SOCIETY (B), VOL 172, NO 1029, P 357-366, 1969. 2
FIG, 3 TAB, 30 REF.

DESCRIPTORS:
*NITROGEN FIXATION, *ALGAE, LIGHT, CARBON DIOXIDE, INHIBITION, NITROGEN
FIXING BACTERIA, RADIOACTIVITY TECHNIQUE, PHOTOSYNTHESIS.

IDENTIFIERS:
*ANABAENA CYLINDRICA, ATP, CMU, NITROGEN-15, PYRUVATE, ACETYLENE
REDUCTION, REDUCTANT.

ABSTRACT:
MORE THAN 90% INHIBITION OF CARBON DIOXIDE FIXATION BY
P-CHLOROPHENYL-1, 1-DIMETHYLUREA (CMU) FAILED TO INFLUENCE NITROGEN
FIXATION BY NITROGEN-STARVED CELLS OF ANABAENA CYLINDRICA, BUT RETARDED
THE PROCESS ABOUT 50% IN NORMAL CELLS OF THE ORGANISM. THIS SUGGESTED
THAT NITROGEN-FIXATION IN ANABAENA IS INDEPENDENT OF REDUCING POTENTIAL
GENERATED BY THE PHOTO-ELECTRON TRANSPORT, BUT IS STIMULATED BY
PHOTOSYNTHETICALLY PRODUCED CARBON SKELETONS. INCREASED LIGHT AUGMENTED
NITROGEN-FIXATION IN THE PRESENCE AND IN ABSENCE OF CMU THUS INDICATING
THE ESSENTIAL PART OF ATP IN THE PROCESS. THE ESTABLISHED 3:1 RATIO OF
PYRUVATE DECARBOXYLATION TO NITROGEN-FIXATION IS IN AGREEMENT WITH THE
HYPOTHESIS THAT PYRUVATE DONATES HYDROGEN FOR NITROGEN REDUCTION, AND
THAT IN BLUE-GREEN ALGAE THE AVAILABILITY OF THE REDUCTANT IS NOT
INFLUENCED BY PHOTOSYNTHESIS. (WILDE-WISCONSIN)

FIELD 05C

ACCESSION NO. W70-04249

A STUDY OF THE PROFUNDAL BOTTOM FAUNA OF LAKE WASHINGTON,

WASHINGTON UNIV., SEATTLE. DEPT. OF ZOOLOGY.

RUDOLPH N. THUT.

ECOLOGICAL MONOGRAPHS, VOL 39, NO 1, P 79-110, 1969. 27 FIG, 5 TAB, 34 REF.

DESCRIPTORS:
*BENTHIC FAUNA, NITROGEN, PHOSPHORUS, PHYTOPLANKTON, DIATOMS,
ZOOPLANKTON, DINOFLAGELLATES, ALGAE, DISSOLVED OXYGEN, HYPOLIMNION,
EPILIMNION, LAKE ERIE, INSECTS, DIPTERA, CRUSTACEA, MOLLUSCS,
OLIGOCHAETES, GASTROPODS.

IDENTIFIERS:
*LAKE WASHINGTON(WASH), MACROFAUNA, GYTTJA, OSCILLATORIA RUBESCENS,
ZURICHSEE, FILAMENTOUS ALGAE, FILAMENTOUS BLUE-GREEN ALGAE, OCULAR
MICROMETER, SUCROSE-FLOTATION TECHNIQUE, METALIMNION, BATHYTHERMOGRAPH,
ARTHROPODA, CHIRONOMIDAE, AQUATIC OLIGOCHAETES.

ABSTRACT:
A REVIEW OF THE MACROFAUNA IN THE PROFUNDAL ZONE OF LAKE WASHINGTON
BETWEEN SEPTEMBER 1963 AND SEPTEMBER 1964 AND DISCUSSION. ALTHOUGH
SEWAGE DIVERSION BEGAN IN 1963, PHOSPHATE AND OXYGEN VALUES INDICATED
THE LAKE WAS STILL IN THE EUTROPHIC PHASE. TEN STATIONS WERE CHOSEN AT
5-METER DEPTH INTERVALS FROM 10 TO 55 METERS AND SAMPLED APPROXIMATELY
MONTHLY WITH AN EKMAN DREDGE. 24 SPECIES WERE RECOGNIZED FROM THE
PROFUNDAL ZONE, EACH PRESENTED SEPARATELY, WITH THE EXCEPTION OF
OLIGOCHAETA SPECIES, ALONG WITH THEIR DEPTH AND POPULATION DYNAMICS
THROUGHOUT THE YEAR. THE CHIRONOMIDAE WERE MOST NUMEROUS OF BOTTOM
FAUNA CONSTITUENTS (ABOUT 45% OF THE TOTAL). 13 SPECIES WERE FOUND
DURING THE STUDY, INCLUDING PREDATORS, DEPOSIT-FEEDERS, AND
FILTER-FEEDERS. LARVAE WERE MOST COMMON AT THE SHALLOW-WATER STATIONS
AND BECAME PROGRESSIVELY DIMINISHED WITH INCREASE IN DEPTH. THE
OLIGOCHAETA COMPRISED ABOUT 1/2 OF THE TOTAL NUMBER AND 1/3 OF THE
TOTAL BIOMASS OF THE PROFUNDAL FAUNA. FOUR SPECIES WERE IDENTIFIED. THE
OLIGOCHAETA WERE FOUND IN THE GREATEST NUMBERS AND BIOMASS AT THE
GREATEST DEPTH SAMPLED; THEIR ABUNDANCE PROGRESSIVELY DECLINED WITH
DECREASE IN DEPTH. AMPHIPODA AND SPHAERIDAE WERE PRESENT BUT IN SMALLER
NUMBERS. (HASKINS-WISCONSIN)

FIELD 02H, 05C

ACCESSION NO. W70-04253

WATER POLLUTION BY NUTRIENTS-SOURCES, EFFECTS AND CONTROL, PAPERS PRESENTED AT
1966 ANNUAL MEETING OF MINNESOTA CHAPTER SOIL CONSERVATION SOCIETY OF
AMERICA.

MINNESOTA UNIV., MINNEAPOLIS. WATER RESOURCES RESEARCH CENTER.

AVAILABLE FROM THE CLEARINGHOUSE AS PB-189 794, $3.00 IN PAPER COPY, $0.65 IN
MICROFICHE. WRRC BULLETIN 13, MINNESOTA WATER RESOURCES RESEARCH CENTER,
JUNE 1969. 79 P. OWRR PROJECT A-999-MINN.

DESCRIPTORS:
*WATER POLLUTION SOURCES, *WATER POLLUTION EFFECTS, *EUTROPHICATION,
*NUTRIENTS, ALGAE, FISH POPULATION, *WATER POLLUTION CONTROL, FARM
WASTES, MUNICIPAL WASTES, SEPTIC TANKS, RECREATION WASTES, WATER
QUALITY.

IDENTIFIERS:
*NUTRIENT SOURCES.

ABSTRACT:
THE BULLETIN INCLUDES THE PAPERS PRESENTED AT A CONFERENCE ON 'NUTRIENT
POLLUTION - SOURCES, EFFECTS AND CONTROL' HELD IN MINNEAPOLIS,
MINNESOTA ON JANUARY 8, L969. THE CONFERENCE WAS PLANNED AS THE ANNUAL
MEETING OF THE MINNESOTA CHAPTER, SOIL CONSERVATION SOCIETY OF AMERICA.
THE PAPERS, ALL INDIVIDUALLY ABSTRACTED, INCLUDE THE FOLLOWING TITLES:
NUTRIENTS AND OTHER FORMS OF POLLUTION, DIAGNOSING POLLUTION IN LAKE
MINNETONKA, EFFECT OF EUTROPHICATION ON FISH AND RELATED ORGANISMS,
HEALTH ASPECTS, SURFACE AND GROUNDWATERS, ANIMAL WASTE DISPOSAL
PROBLEMS AND TRENDS IN MINNESOTA, MANAGING LIVESTOCK WASTES TO CONTROL
POLLUTION, RUNOFF AND SEDIMENT AS NUTRIENT SOURCES, CONTROLLING
NUTRIENTS AND ORGANIC TOXICANTS IN RUNOFF, TREATMENT OF MUNICIPAL
WASTES, SEPTIC TANK EFFLUENTS, WATER POLLUTION IN RECREATIONAL AREAS -
SOURCES AND CONTROL, AND SETTING WATER QUALITY STANDARDS AND REGULATING
NUTRIENT SOURCES, IMPLEMENTING POLLUTION CONTROL.

FIELD 05B, 05C, 05G

ACCESSION NO. W70-04266

DIAGNOSING POLLUTION IN LAKE MINNETONKA,

MINNESOTA UNIV., MINNEAPOLIS. LIMNOLOGICAL RESEARCH CENTER.

ROBERT O. MEGARD.

WATER POLLUTION BY NUTRIENTS--SOURCES, EFFECTS AND CONTROL, WATER RESOURCES
RESEARCH CENTER, UNIVERSITY OF MINNESOTA, MINNEAPOLIS, WRRC BULLETIN 13, P
9-14, JUNE 1969. 1 FIG, 1 TAB, 2 REF.

DESCRIPTORS:
*PHOSPHORUS, *NITROGEN, *ALGAE, SEWAGE PLANTS, CHLOROPHYLL,
PHOTOSYNTHESIS, TRIBUTARIES, WATERSHED MANAGEMENT, SEDIMENTS.

IDENTIFIERS:
LAKE MINNETONKA(MINN), NUTRIENT TRAPS, NUTRIENT INFLUX, NUTRIENT
SOURCES, PHOSPHORUS BUDGET, APHANIZOMINON FLOS AQUAE.

ABSTRACT:
INCREASED MONETARY SPENDING ON WASTE DISPOSAL AND WATERSHED MANAGEMENT
IS SUGGESTED AS A RECOURSE TO DETERIORATING QUALITY OF LAKES. AN
EXAMPLE, IS WIDENING THE SCOPE OF SEWAGE TREATMENT TO INCLUDE
OBJECTIVES OF MAINTAINING RECREATIONAL AND AESTHETIC VALUES OF LAKES AS
WELL AS MINIMIZING HEALTH HAZARDS. MEASURING ABUNDANCE AND DAILY GROWTH
OF ALGAE BY CHLOROPHYLL ANALYSIS AND PHOTOSYNTHESIS EXPERIMENTS
RESPECTIVELY, IS A METHOD OF COMPARING LAKES AND ASSESSING
EFFECTIVENESS OF NUTRIENT ABATEMENT PROGRAMS. DATA WAS GATHERED FROM
LOWER LAKE OF LAKE MINNETONKA ADJACENT TO MINNEAPOLIS, MINNESOTA, AND A
DIRECT CORRELATION BETWEEN PHOSPHORUS CONCENTRATION WAS DISCOVERED
DURING JULY AND AUGUST. EACH POUND OF PHOSPHORUS WAS CONSIDERED
RESPONSIBLE FOR 500 LBS OF DRY ALGAE DURING THE TWO MONTHS. DATA ON
INFLUX OF PHOSPHORUS FROM TRIBUTARIES AND CONCENTRATION IN THE LAKE
INDICATED THAT SEDIMENTS WERE THE MAJOR NUTRITION TRAP. THERE WAS DOUBT
WHETHER PHOSPHORUS RELEASED BY SEDIMENTS BETWEEN MAY AND AUGUST WAS
AVAILABLE TO ALGAE. DATA INDICATED THAT REDUCING PHOSPHORUS INFLUX BY
ONE-HALF WOULD SUBSTANTIALLY CONTROL LOWER LAKE'S EUTROPHICATION BASED
ON THE ASSUMPTION THAT A LARGE PORTION OF PHOSPHORUS IS REMOVED BY
BIOGEOCHEMICAL PROCESS. (SEE W70-04266). (BANNERMAN-WISCONSIN)

FIELD 05C, 02H

ACCESSION NO.   W70-04268

STUDY OF BACTERIA ASSOCIATED WITH MARINE ALGAE IN CULTURE. I. PRELIMINARY
DETERMINATION OF SPECIES (FRENCH),

STATION MARINE D'ENDOUME, MARSEILLE (FRANCE).

B. R. BERLAND, M. G. BIANCHI, AND S. Y. MAESTRINI.

MARINE BIOLOGY, VOL 2, NO 4, P 350-355, 1969. 3 TAB, 12 REF.

DESCRIPTORS:
*BACTERIA, *MARINE ALGAE, *CULTURES, PSEUDOMONAS, METABOLISM, KINETICS,
NITROGEN COMPOUNDS.

IDENTIFIERS:
SPECIES DETERMINATION, TAXONOMY, FLAVOBACTERIUM, ACHROMOBACTER, GULF OF
MARSEILLES, ASTERIONELLA JAPONICA, CHAETOCEROS LAUNDERI, LAUDERIA
BOREALIS, LEPTOCYLINDRUS DANICUS, PHAEODACTYLUM TRICORNUTUM,
STICHOCHRYSIS IMMOBILIS, VIBRIO, MORPHOLOGY, CULTURAL CHARACTERISTICS,
TERMARY COMPOUNDS, MACROMOLECULES, ANTIBIOTICS, AGARBACTERIUM,
XANTHOMONAS, MICROCOCCUS, STAPHYLOCOCCUS.

ABSTRACT:
BACTERIA STRAINS HAVE BEEN ISOLATED FROM MARINE ALGAE CULTURES AND
ASSIGNED TENTATIVE GENERA AND SPECIES. OBSERVATIONS AND TESTS WERE
CONDUCTED USING BERGEY'S MANUAL AND SPECIFIC PAPERS ON MARINE BACTERIA
AS TAXONOMIC KEYS. THE GENERA PSEUDOMONAS, FLAVOBACTERIUM, AND
ACHROMOBACTER APPEAR MOST IMPORTANT, BASED ON THE NUMBER OF SPECIES.
THIS DOES NOT IMPLY THAT THEY ACTUALLY DOMINATE IN NUMBER OF GERMS AND
IN METABOLIC IMPORTANCE. A LATER STUDY WILL PERMIT DETERMINATION OF THE
PRECISE NATURE AND DIVISION OF SPECIES WHICH ACCOMPANY AN ALGA IN
CULTURE AND SUBSEQUENTLY THE SAME ALGA IN ITS NATURAL ENVIRONMENT. IT
MAY BE NECESSARY TO USE SOME CULTURE MEDIUM MORE DEFINITIVE AND LESS
PLETHORIC, THE MEDIUM CURRENTLY USED BEING TOO RICH IN ORGANIC MATTER
AND NOT CORRESPONDING TO CONDITIONS OF THE NATURAL MEDIUM OF THE
ORGANISMS; THE LITERATURE SHOWS THAT THEIR NUTRITIONAL NEEDS ARE VERY
SIMPLE. THE STUDY IS UNDERTAKEN TO DETERMINE PRECISELY THEIR METABOLIC
NEEDS AND HOW THESE ARE SATISFIED IN A CULTURE OF ALGAE.
(JONES-WISCONSIN)

FIELD 02H, 05C

ACCESSION NO.   W70-04280

PHYTOPLANKTON AND RELATED WATER-QUALITY CONDITIONS IN AN ENRICHED ESTUARY,

GEOLOGICAL SURVEY, TACOMA, WASH.

EUGENE B. WELCH.

JOURNAL WATER POLLUTION CONTROL FEDERATION, VOL 40, NO 10, P 1711-1727, 1968.
10 FIG, 27 REF.

DESCRIPTORS:
*ANALYTICAL TECHNIQUES, *AQUATIC ALGAE, *AQUATIC MICROBIOLOGY, *AQUATIC
PRODUCTIVITY, BIOCHEMICAL OXYGEN DEMAND, CYCLING NUTRIENTS, DIATOMS,
ENVIRONMENTAL EFFECTS, ESSENTIAL NUTRIENTS, EUTROPHICATION, LIGHT
INTENSITY, LIGHT PENETRATION, LIMNOLOGY, NITROGEN COMPOUNDS, NUISANCE
ALGAE, NUTRIENT REQUIREMENTS, NUTRIENTS, PHOSPHORUS COMPOUNDS,
PHOTOSYNTHETIC OXYGEN, PHYSIOLOGICAL ECOLOGY, PHYTOPLANKTON, ESTUARIES,
POLLUTANT IDENTIFICATION, SEWAGE, SEWAGE DISPOSAL, SEWAGE EFFLUENTS,
WATER POLLUTION CONTROL, WATER POLLUTION EFFECTS, WATER POLLUTION
SOURCES.

IDENTIFIERS:
*SEATTLE(WASH), RENTON TREATMENT PLANT, DUWAMISH ESTUARY(WASH).

ABSTRACT:
AMMONIA, SOLUBLE PHOSPHATE, AND TOTAL PHOSPHATE CONCENTRATIONS WERE
OBSERVED TO INCREASE IN THE DUWAMISH ESTUARY, WASHINGTON, FOLLOWING
INITIAL DISCHARGE OF EFFLUENT FROM THE RENTON TREATMENT PLANT AT
SEATTLE. A PHYTOPLANKTON BLOOM DOMINATED BY MARINE SPECIES OCCURRED IN
THE LOWER ESTUARY IN AUGUST 1965, ABOUT 1.5 MONTHS FOLLOWING THE
NUTRIENT INCREASE. A BLOOM DID NOT OCCUR IN 1964, PRIOR TO EFFLUENT
DISCHARGE FROM THE TREATMENT PLANT, BUT SOME EVIDENCE SHOWS THAT ALGAL
ACTIVITY WAS GREAT IN AUGUST 1963. THE NUTRIENT INCREASE IN 1965
PROBABLY WAS NOT THE SOLE FACTOR CAUSING THE BLOOM DURING THAT SUMMER
BECAUSE: (A) THE PRE-EFFLUENT NUTRIENT CONCENTRATIONS IN 1963 AND 1964
WERE RELATIVELY HIGH; (B) A BLOOM ALSO OCCURRED AT THE FURTHEST
DOWNSTREAM STATION WHERE NO INCREASE IN NUTRIENTS WAS APPARENT IN 1965
OVER THAT OF THE PRECEDING TWO YEARS; (C) A BLOOM PROBABLY OCCURRED IN
1963, BEFORE THE ADDITION OF NUTRIENTS FROM THE PLANT; AND (D) THE
BLOOM MAXIMUM DID NOT OCCUR IN 1965 UNTIL ABOUT 1.5 MONTHS AFTER THE
NUTRIENT INCREASE, WHEN DISCHARGE AND TIDAL EXCHANGE CONDITIONS WERE
MINIMUM. BLOOM TIMING SEEMED RELATED MOST CLOSELY TO HYDROGRAPHIC
CONDITIONS. (FITZGERALD-WISCONSIN)

FIELD 02L, 05C

ACCESSION NO. W70-04283

COMPETITION BETWEEN PLANKTONIC BACTERIA AND ALGAE FOR ORGANIC SOLUTES,

UPPSALA UNIV. (SWEDEN). INST. OF LIMNOLOGY.

JOHN E. HOBBIE, AND RICHARD T. WRIGHT.

MEMORIE DELL'INSTITUTO ITALIANO DI IDROBIOLOGIA, SUPPL 18, P 175-185, 1965. 4 FIG, 1 TAB, 5 REF.

DESCRIPTORS:
*COMPETITION, *PHYTOPLANKTON, *AQUATIC BACTERIA, *ALGAE, ORGANIC MATTER, LAKES, DIFFUSION, CARBON RADIOISOTOPES, CHLAMYDOMONAS, BIOASSAYS, WATER POLLUTION EFFECTS, ANALYTICAL TECHNIQUES.

IDENTIFIERS:
*ORGANIC SOLUTES, *HETEROTROPHY, MICROBIAL ECOLOGY, SWEDEN, LAKE ERKEN(SWEDEN), ACTIVE TRANSPORT, CARBON-14, ENZYME KINETICS, GLUCOSE, MECHAELIS-MENTEN EQUATION, ACETATE, GLYCOLLIC ACID.

ABSTRACT:
SOPHISTICATED INVESTIGATIVE TECHNIQUES, IN ADDITION TO THE TRADITIONAL METHODS OF DIRECT COUNTS AND ENRICHMENT CULTURES, ARE REQUIRED FOR BETTER UNDERSTANDING OF ROLES OF MICROORGANISMS IN THE DYNAMICS OF LACUSTRINE METABOLISM. A PROBLEM OF CURRENT INTEREST IS THE IDENTIFICATION OF ORGANIC SOLUTES UTILIZED BY HETEROTROPHIC BACTERIA AND ALGAE, AND ESTIMATION OF THEIR CONCENTRATIONS AND UPTAKE RATES. DATA ON ASSIMILATION OF RADIOCARBON-LABELED ORGANIC SUBSTRATES CAN BE ANALYZED ON BASES OF THEORIES OF DIFFUSION AND ENZYME KINETICS. SUCH METHODS ARE ILLUSTRATED WITH DATA FROM STUDIES ON MODEL CULTURES (CHLAMYDOMONAS AND PLANKTONIC BACTERIA) AND SAMPLES FROM LAKE ERKEN (SWEDEN). MEASUREMENTS OF VELOCITY UPTAKE OF RADIOISOTOPE OVER A RANGE OF SUBSTRATE CONCENTRATIONS REVEAL TWO MECHANISMS: AN ACTIVE TRANSPORT SYSTEM, ATTRIBUTABLE TO BACTERIA, AND ANOTHER SYSTEM IN ALGAE WHICH · FOLLOWS DIFFUSION KINETICS. THE LATTER IS EFFECTIVE AT SUBSTRATE CONCENTRATIONS ABOVE 500 MICROGRAMS/LITER. IN NATURAL WATERS, BACTERIA EFFECTIVELY REDUCE ACETATE AND GLUCOSE CONCENTRATIONS BELOW 20 MICROGRAMS/LITER, APPARENTLY EFFECTING DRASTIC LIMITATION OF ALGAL HETEROTROPHY. IN NATURE, OTHER COMPOUNDS ARE AVAILABLE TO ALGAE AND HIGHER TEMPERATURE MAY INCREASE UPTAKE RATES, SO THAT ALGAL HETEROTROPHY IS NOT NECESSARILY PRECLUDED. (EICHHORN-WISCONSIN)

FIELD 02H, 05C

ACCESSION NO. W70-04284

A NEW METHOD FOR THE STUDY OF BACTERIA IN LAKES: DESCRIPTION AND RESULTS,

NORTH CAROLINA STATE UNIV., RALEIGH. DEPT. OF ZOOLOGY; AND GORDON COLL., WENHAM, MASS.

JOHN E. HOBBIE, AND RICHARD T. WRIGHT.

MITTEILUNGEN INTERNATIONALE VEREINIGUNG FUR THEORETISCHE UND ANGEWANDTE LIMNOLOGIE, VOL 14, P 64-71, 1968. 5 FIG, 10 REF.

DESCRIPTORS:
*LAKES, *BACTERIA, *ANALYTICAL TECHNIQUES, WATER POLLUTION EFFECTS, ALGAE, ECOSYSTEMS, PHYSIOLOGICAL ECOLOGY, AQUATIC BACTERIA, CARBON RADIOISOTOPES, ENZYMES, KINETICS, PHOTOSYNTHESIS, ORGANIC MATTER, ORGANIC COMPOUNDS, METABOLISM, BIOINDICATORS, BIOASSAYS.

IDENTIFIERS:
*HETEROTROPHY, AUTOTROPHY, CARBON-14, RADIOCARBON UPTAKE TECHNIQUES, GLUCOSE, MICHAELIS-MENTEN EQUATION, SWEDEN, LAKE ERKEN(SWEDEN), LINEWEAVER-BURK EQUATION, MAXIMUM UPTAKE VELOCITY, MICHAELIS CONSTANT, LAKE NORRVIKEN(SWEDEN), LAKE EKOLN(SWEDEN), LAKE LOTSJON(SWEDEN), PERMEASES, CHAMYDOMONADS, LAPPLAND(SWEDEN), ACETATE, TURNOVER TIMES.

ABSTRACT:
ALTHOUGH BACTERIA ARE IMPORTANT IN LACUSTRINE METABOLISM, DETAILS OF THEIR POPULATION DYNAMICS--INCLUDING GENERATION TIMES AND TURNOVER RATES--ARE POORLY UNDERSTOOD. RADIOCARBON-LABELED ORGANIC COMPOUNDS HAVE BEEN USED PREVIOUSLY TO DETERMINE HETEROTROPHIC POTENTIAL OF NATURAL WATERS, BUT CONCENTRATIONS OF SUBSTRATE ADDED HAVE BEEN SUFFICIENTLY LARGE AS TO MAKE INTERPRETATION OF RESULTANT DATA AMBIGUOUS. BASED UPON THEORETICAL TREATMENT ADAPTED FROM LINEWEAVER-BURK MODIFICATION OF MICHAELIS-MENTEN ENZYME KINETICS, AND ON ADDITIONS OF VERY LOW CONCENTRATIONS OF LABELED ORGANIC SUBSTRATES, A NEW METHOD FOR STUDY OF BACTERIAL ASSIMILATION IS DESCRIBED. THAT DERIVATION YIELDS RELATIONSHIP, $T/F = (K + S)/V) + A/V$. HERE, T IS INCUBATION TIME (HOURS); F, PERCENTAGE UPTAKE OF ISOTOPE BY ORGANISMS; K, CONSTANT (SIMILAR TO MICHAELIS CONSTANT); S, SUBSTRATE CONCENTRATION IN NATURE; A, ADDED SUBSTRATE CONCENTRATION; V, MAXIMUM VELOCITY. PLOT OF T/F VERSUS A USUALLY RESULTS IN LINEAR RELATIONSHIP FROM WHICH CAN BE ESTIMATED MAXIMUM UPTAKE VELOCITY (V), TURNOVER TIME, AND APPROXIMATE NATURALLY OCCURRING SUBSTRATE CONCENTRATION. METHOD IS ILLUSTRATED WITH DATA FROM A SERIES OF SWEDISH LAKES. DURING A SINGLE YEAR, V MAY VARY BY 2 ORDERS OF MAGNITUDE WITHIN ONE LAKE; BY 4, BETWEEN LAKES. V IS ALSO A VERY SENSITIVE INDICATOR OF POLLUTION. (EICHHORN-WISCONSIN)

FIELD 02H, 05C

ACCESSION NO. W70-04287

ANNUAL AUTOTROPHIC PRODUCTION OF PHYTOPLANKTON IN LAKE BAIKAL,

K. K. VOTINTSEV, AND G. I. POPOVSKAYA.

TRANSLATED FROM RUSSIAN. ORIGINAL P 205-208. DOKLADY AKADEMII NAUK SSSR, VOL 176, NO 1, P 634-637, 1967. 3 FIG, 1 TAB, 13 REF.

DESCRIPTORS:
*PHYTOPLANKTON, *LAKES, ORGANIC MATTER, SEASONAL, ALGAE, PRIMARY PRODUCTIVITY, HARVESTING OF ALGAE, BIOMASS.

IDENTIFIERS:
SEASONAL DISTRIBUTION, LAKE BAIKAL, PHOTOSYNTHETIC LAYER, MELOSIRA BAICALENSIS, GYMNODINIUM BAICALENSE, CYCLOTELLA MINUTA.

ABSTRACT:
THIS PIONEERING SURVEY OF ORGANIC MATTER PRODUCTION IN LAKE BAIKAL DISCLOSED A WIDE PLACE-AND-TIME VARIATION IN THE BIOMASS OF PHYTOPLANKTON, REPRESENTED LARGELY BY MELOSIRA BAICALENSIS, CYCLOTELLA MINUTA, AND GYMNODINIUM BAICALENSE. ACCORDING TO PRELIMINARY ESTIMATES, THE LAKE HAS AN ANNUAL PRODUCTIVITY OF 157 MILLION TONS (WET WEIGHT) OF HARVESTABLE ALGAE, AND 136 MILLION TONS OF NON-HARVESTABLE ALGAE. THE STUDY OF PRIMARY PRODUCTIVITY BY THE 'FLASK' METHOD, USING OXYGEN, WAS CONFINED TO THE UPPER 25 METERS OF THE PHOTOSYNTHETIC LAYER. THE MEAN ANNUAL BIOMASS PER SURFACE SQUARE METER OF THIS LAYER APPROACHED 7 GRAMS. THE RATIO OF THE AVERAGE ANNUAL PRIMARY PRODUCTION TO THE MEAN ANNUAL INDEX OF PHYTOPLANKTON BIOMASS FOR THE ENTIRE LAKE WAS EQUAL TO 464. PREVIOUS STUDIES HAVE SHOWN THAT VALUES FOR GROSS PHOTOSYNTHESIS ARE CLOSE TO THOSE PERMITTING EFFECTIVE PRODUCTIVITY OF ALGAE. INFORMATION ON FOOD REQUIREMENTS OF MAIN REPRESENTATIVES OF ZOOPLANKTON OF LAKE BAIKAL THUS FAR IS NOT AVAILABLE. (WILDE-WISCONSIN)

FIELD 02H, 05C

ACCESSION NO. W70-04290

DEVELOPMENT OF A METHOD FOR THE TOTAL COUNT OF MARINE BACTERIA ON ALGAE,

MCGILL UNIV., MONTREAL (QUEBEC). DEPT. OF MICROBIOLOGY AND IMMUNOLOGY; AND
NEW BRUNSWICK UNIV., FREDERICTON. DEPT. OF BIOLOGY.

E. C. S. CHAN, AND ELIZABETH A. MCMANUS.

CANADIAN JOURNAL OF MICROBIOLOGY, VOL 13, P 295-301, 1967. 2 FIG, 3 TAB, 10
REF.

DESCRIPTORS:
*MARINE BACTERIA, *ALGAE, *ANALYSIS, *ANALYTICAL TECHNIQUES,
TEMPERATURE, PIPTIDES, INTERTIDAL AREAS, SOIL BACTERIA, LABORATORY
TESTS.

IDENTIFIERS:
*ALGAE HOMOGENIZATION, *BACTERIA COUNT, WARING BLENDOR MORTALITY, CELL
WALL, HEXOSAMINE, ASCOPHYLLUM NODOSUM, POLYSIPHONIA LANOSA, POINT
LEPREAU(CANADA), NEW BRUNSWICK(CANADA), SERVALL OMNI-MIXER HOMOGENIZER.

ABSTRACT:
IN AN INITIAL ATTEMPT TO DEVELOP A TECHNIQUE FOR ENUMERATION OF
BACTERIA ON LITTORAL ALGAE, IT WAS OBSERVED THAT HOMOGENIZATION OF
ALGAL SAMPLES IN THE WARING BLENDOR RESULTED IN MORTALITY OF MARINE
BACTERIA SAMPLES, APPARENTLY DUE TO SINGULAR SUSCEPTIBILITY TO
MANIPULATIONS. FURTHER WORK ON THIS LETHAL EFFECT OF BLENDING SHOWED
TWO DETRIMENTAL FACTORS, HEAT AND MECHANICAL INJURY OF THE CELLS, WERE
IMPLICATED: DUE TO THE INCREASE IN TEMPERATURE DURING HOMOGENIZATION
(TEMPERATURES SLIGHTLY BEYOND 35C RESULT IN RAPID BACTERIAL KILLING)
AND THE HIGH SPEED ROTATION OF THE KNIFE-BLADE ASSEMBLY OF THE WARING
BLENDOR. THESE FINDINGS RESULTED IN A SATISFACTORY TECHNIQUE FOR
OBTAINING HOMOGENATES OF MARINE MATERIALS TO OBTAIN TOTAL BACTERIA
COUNT BY USING A SERVALL OMNI-MIXER HOMOGENIZER AT SPEEDS OF ABOUT 5000
REVOLUTIONS PER MINUTE AND TEMPERATURE OF HOMOGENIZATION NOT EXCEEDING
30C. WITH THIS METHOD, IT WAS POSSIBLE TO MAKE AN EXTENSIVE STUDY OF
ECOLOGY OF MARINE BACTERIA ON POLYSIPHONIA LANOSA AND ASCOPHYLLUM
NODOSUM, THE FORMER ALGA GROWING EPIPHYTICALLY ON ASCOPHYLLUM.
(JONES-WISCONSIN)

FIELD 05A

ACCESSION NO. W70-04365

YEASTS FROM THE NORTH SEA.

MIAMI UNIV., FLA. INST. OF MARINE SCIENCES; MIAMI UNIV., FLA. DEPT. OF
  MICROBIOLOGY; AND BIOLOGISCHE ANSTALT HELGOLAND (WEST GERMANY).

S. P. MEYERS, D. G. AHEARN, W. GUNKEL, AND F. J. ROTH, JR.

MARINE BIOLOGY, VOL 1, NO 2, P 118-123, 1967. 6 FIG, 1 TAB, 19 REF.

DESCRIPTORS:
  *YEASTS, POPULATION, MARINE PLANTS, DINOFLAGELLATES, OCEANS,
  BIOINDICATORS, FUNGI, ESTUARIES, BACTERIA, SAMPLING, SEASONAL,
  HYDROGRAPHY, FLORIDA, THERMOCLINE, CURRENTS(WATER), PHYTOPLANKTON,
  ZOOPLANKTON, HYDROGEN ION CONCENTRATION, ALGAE.

IDENTIFIERS:
  *NORTH SEA, MYCOTA, DEBARYOMYCES HANSENII, RHODOTORULA RUBRA, CANDIDA
  DIDDENSII, CANDIDA ZEYLANOIDES, CANDIDA KRUSEI, CANDIDA OBTUSA, CANDIDA
  TROPICALIS, NOCTILUCA MILIARIS, HELGOLAND(GERMANY), AUREOBASIDIUM
  PULLULANS, HANSENIASPORA UVARUM, RADULESPORES, ZYMOLOGICAL,
  ELBE(GERMANY), SPOROBOLOMYCES ROSEUS, AUTOLYSIS, ASCOSPOROGENOUS.

ABSTRACT:
MYCOLOGICAL EXAMINATION OF ESTUARINE AND OPEN OCEAN ENVIRONMENTS HAVE
REVEALED DIVERSE POPULATIONS OF YEAST OF VARIOUS TAXA AND PHYSIOLOGICAL
GROUPS. EVIDENCE INDICATES THAT THESE MICROORGANISMS ARE NORMAL
COMPONENTS OF OCEAN BIOTA, AND THAT CERTAIN SPECIES MAY SERVE AS
BIO-INDICATORS FOR SPECIFIC WATER MASSES AND CONDITIONS. DATA ARE
NEEDED ON THE MYCOTA ASSOCIATED WITH ALGAL 'BLOOMS' AND THEIR EFFECT ON
GROWTH STIMULATION AND REPRODUCTION OF FUNGI. YEASTS WERE ISOLATED FROM
TWELVE ESTABLISHED SITES IN THE NORTH SEA FROM 1964 TO 1966. A
PERCENTAGE FREQUENCY OF 99% WITH POPULATIONS VARYING FROM LESS THAN 10
TO MORE THAN 3000 VIABLE CELLS/LITER WAS OBSERVED. THIS MYCOTA WAS
CHARACTERIZED BY CONSIDERABLE SPATIAL AND TEMPORAL FLUCTUATIONS, WITH
THE DOMINANT YEAST PRESENT THE ASCOSPOROGENOUS SPECIES, DEBARYOMYCES
HANSENII. THIS TAXON, AS WELL AS OTHER COMMON NORTH SEA YEASTS,
RHODOTORULA RUBRA AND CANDIDA DIDDENSII, HAVE BEEN REPORTED FREQUENTLY
FROM OTHER MARINE LOCALES. NOTEWORTHY CONCENTRATIONS OF YEASTS,
ESPECIALLY D HANSENII, WERE OBSERVED DURING SUMMER, OFTEN IN
ASSOCIATION WITH VARIOUS STAGES OF DEVELOPMENT OF THE DINOFLAGELLATE,
NOCTILUCA MILIARIS. POPULATION DYNAMICS OF THE NORTH SEA YEASTS ARE
DISCUSSED IN RELATION TO SIMILAR STUDIES OF OTHER MARINE ENVIRONMENTS.
(JONES-WISCONSIN)

FIELD 05C

ACCESSION NO. W70-04368

THE PHYSIOLOGY OF AN ALGAL NUISANCE,

WESTFIELD COLL., LONDON (ENGLAND). DEPT. OF BOTANY.

G. E. FOGG.

PROCEEDINGS ROYAL SOCIETY B, VOL 173, P 175-189, 1969. 4 FIG, 84 REF.

DESCRIPTORS:
*NUISANCE ALGAE, CYANOPHYTA, VITAMINS, PEPTIDES, ORGANIC MATTER, TRACE ELEMENTS, LIGHT INTENSITY, CHLAMYDOMONAS, CHLORELLA, ENZYMES, CARBON DIOXIDE, GROWTH RATES, PLANKTON, OXYGEN, NITROGEN FIXATION, NITRATES, ELECTRON MICROSCOPY, AMMONIA, SPECTROMETERS, DIFFUSION, TURBULENCE, TOXINS, BACTERIA, COPPER SULFATE, CIRCULATION.

IDENTIFIERS:
*PHYSIOLOGY, SPIROGYRA, MYXOPHYCEAE, MICROCYSTIS, COELOSPHAERIUM, OSCILLATORIA, ANABAENA, APHANIZOMENON, GLOEOTRICHIA, POLYPEPTIDES, PHOTOASSIMILATION, BERKELSE LAKE(NETHERLANDS), HETEROCYSTS, GAS-VACUOLES, LLANGORS LAKE(WALES), HALOBACTERIUM, RESPIROMETER, PROTOPLASM, WATER-BLOOMS, LAKE GEORGE(UGANDA), RHINE-MEEUSE DELTA(NETHERLANDS), HOHLSPINDELN, MYXOPHYCEAN BLOOMS.

ABSTRACT:
PHYSIOLOGY OF MICROCYSTIS, COELOSPHAERIUM, OSCILLATORIA, ANABAENA, APHANIZOMENON, AND GLOEOTRICHIA ARE DISCUSSED RELATIVE TO DEVELOPMENT OF BLOOMS. SINCE THE ALGAE FORM GELATINOUS MASSES, LOCAL EXHAUSTION OF AMMONIA MAY FORCE CELLS TO UTILIZE THE NITROGEN MOLECULE, AGGRAVATING EUTROPHICATION. REMARKABLE FEATURES OF THE PLANKTONIC BLUE-GREEN ALGAE · ARE CELLS CONTAINING GAS-VACUOLES. THESE GAS-FILLED CAVITIES IN PROTOPLASM APPEAR AS INDEFINITE, IRREGULAR, REFRACTIVE BODIES. THE RELATIVELY HIGH BUOYANT DENSITY OF THEIR MEMBRANES SHOWS THEM AS PROBABLY AT VARIANCE TO OTHER CELLULAR MEMBRANES. THEY ARE SPECIFIC STRUCTURES WITH DEFINITE SELECTIVE VALUE. PERHAPS GAS-VACUOLE FORMATION IN DIM LIGHT CAUSES A CELL OR COLONY TO RISE; INHIBITION OF GAS-VACUOLE FORMATION, AND CONTINUED GROWTH AND CELL DIVISION NEARER THE SURFACE SUBSEQUENTLY CAUSING IT TO BECOME HEAVIER AND SINKING. THIS UP AND DOWN MOVEMENT WOULD CONTRIBUTE TO MAXIMUM ABSORPTION OF MINERAL NUTRIENTS. NEAR THE SURFACE LIGHT INHIBITION OF PHOTOSYNTHESIS MAY RESULT IN INABILITY TO 'DILUTE OUT' THE GAS-VACUOLES BY GROWTH, THUS THE ALGA REMAINS FLOATING AND BECOMES MORIBUND. COPPER SULPHATE USAGE FOR CONTROL OF MYXOPHYCEAN WATER-BLOOMS IS QUESTIONABLE. INHIBITING FORMATION OF THESE GAS-VACUOLES OR THEIR DESTRUCTION BY ULTRASONIC RADIATION, OR OTHER MEANS, MIGHT EFFECTIVELY CONTROL THE NUISANCE. ARTIFICIAL CIRCULATION MAY BREAK DOWN CHEMICAL STRATIFICATION IN THE WATER. (JONES-WISCONSIN)

FIELD 05C

ACCESSION NO. W70-04369

PERIPHYTON GROWTH ON ARTIFICIAL SUBSTRATES IN A RADIOACTIVELY CONTAMINATED LAKE,

OAK RIDGE NATIONAL LAB., TENN. RADIATION ECOLOGY SECTION.

ERNEST C. NEAL, BERNARD C. PATTEN, AND CHARLES E. DEPOE.

ECOLOGY, VOL 48, NO 6, P 918-924, 1967. 1 FIG, 1 TAB, 17 REF.

DESCRIPTORS:
*PERIPHYTON, *RADIOACTIVITY, *LAKES, PRODUCTIVITY, RADIOISOTOPES,
TEMPERATURE, DEPTH, BACTERIA, SLIME, ALGAE, CYANOPHYTA, DIATOMS,
CHLOROPHYTA, LIGHT INTENSITY, SUCCESSION, ZINC RADIOISOTOPES, CESIUM,
COBALT RADIOISOTOPES, SEASONAL, RADIOCHEMICAL ANALYSIS, MONITORING,
PIGMENTS, PHYTOPLANKTON, WASTE WATER(POLLUTION), PHOTOSYNTHESIS,
CHLAMYDOMONAS, CHLORELLA, EUGLENA, SCENEDESMUS, PROTOZOA, NEMATODES,
SURFACES.

IDENTIFIERS:
*GROWTH DYNAMICS, *BACTERIAL COLONIZATION, *ARTIFICIAL SUBSTRATES,
*RADIONUCLIDE ACCUMULATION, RUTHENIUM-106, WHITE OAK LAKE(TENN),
ACTIVITY-DENSITY, VERTICAL, MICROCYSTIS, OSCILLATORIA, NAVICULA,
CYMBELLA, LYNGBYA, FRAGILARIA, OEDOGONIUM, SPIROGYRA, CHAETOPHORA,
STIGEOCLONIUM, ANKISTRODESMUS, APHANOCAPSA, CHROOCOCCUS, CLOSTERIUM,
GLEOCAPSA, SPIRULINA, ULOTHRIX, CHIRONOMIDS, MELTON HILL DAM(TENN),
COBALAMIN.

ABSTRACT:
PERIPHYTON COLONIZATION, BIOMASS DEVELOPMENT AND RADIONUCLIDE
ACCUMULATION WERE STUDIED BY SUSPENDING POLYETHYLENE TAPE VERTICALLY
FOR PERIODS TO 9 WEEKS IN A RADIOACTIVELY CONTAMINATED LAKE. BIOMASS
GROWTH WAS NEARLY COMPLETE ON UPPER TAPE SECTIONS AFTER 2 WEEKS, BUT
FULL DEVELOPMENT ON LOWER SECTIONS TOOK LONGER; MAXIMUM BIOMASS
OCCURRED AT 25.4 TO 50.8 CENTIMETERS DEPTH. BLUE-GREENS WERE INITIAL
ALGAL DOMINANTS, SUCCEEDED BY DIATOMS AND FILAMENTOUS GREENS. SPECIES
SUCCESSION CONTINUED BEYOND BIOMASS EQUILIBRIUM, WITH BLUE-GREENS
IMPORTANT IN DEEPER ZONES AND GREENS DEVELOPING BEST IN UPPER, LIGHTED
REGIONS. RADIOISOTOPE CONCENTRATIONS EQUILIBRATED RAPIDLY IN BOTH
ARTIFICIAL SUBSTRATES AND PERIPHYTON BIOMASS. ZINC-65 WAS CONCENTRATED
HIGHLY BY THE POLYETHYLENE TAPE, CESIUM-137 MODERATELY, AND COBALT-60
AND RUTHENIUM-106 NEGLIGIBLY, IF AT ALL. AMBIENT CONCENTRATIONS OF
ZINC-65 IN WATER AND PERIPHYTON WERE BELOW DETECTION. THE OTHER
ISOTOPES CONTRIBUTED ABOUT EQUALLY TO BIOMASS RADIOACTIVITY. BIOMASS
ACTIVITY-DENSITIES OF RUTHENIUM-106 AND CESIUM-137 INCREASED SLIGHTLY
WITH DEPTH; THIS PATTERN WAS PRONOUNCED FOR COBALT-60. FACTORS
CONSIDERED TO ACCOUNT FOR OBSERVED VERTICAL DISTRIBUTIONS INCLUDE:
DIFFERENT RADIOISOTOPE CONCENTRATIONS IN SURFACE AND DEEP WATER; A
HYPERBOLIC RATHER THAN LINEAR RELATIONSHIP BETWEEN ACTIVITY-DENSITY AND
BIOMASS; AND CONCENTRATION OF COBALT-60 BY BLUE-GREEN ALGAE AS AN
ESSENTIAL ELEMENT, OR BY MICROBIOTA FOR USE IN CODALMIN SYNTHESIS.
(JONES-WISCONSIN)

FIELD 05C

ACCESSION NO. W70-04371

BIODEGRADABILITY OF POTENTIAL ORGANIC SUBSTITUTES FOR PHOSPHATES,

ROBERT A. TAFT SANITARY ENGINEERING CENTER, CINCINNATI, OHIO. CINCINNATI
WATER RESEARCH LAB.

ROBERT L. BUNCH, AND M. B. ETTINGER.

PROCEEDINGS 22ND INDUSTRIAL WASTE CONFERENCE, LAFAYETTE, INDIANA, P 393-396,
1967. 1 FIG, 14 REF.

DESCRIPTORS:
*WATER QUALITY, *EUTROPHICATION, *DETERGENTS, *PHOSPHATES, CHELATION,
WASTE TREATMENT, AQUATIC ALGAE, AQUATIC PLANTS, TESTING, BIODEGRADATION.

IDENTIFIERS:
*COMPLEXONES, *ALGAL BLOOMS, NTA, EDTA, HEIDA, LAKE TAHOE(CALIF).

ABSTRACT:
THE UNDESIRABLE EUTROPHICATION OF WATERS MAY BE CONSIDERABLY REDUCED BY
REPLACING PHOSPHATES IN INDUSTRIAL PROCESSES AND DETERGENTS WITH
BIODEGRADABLE CHELATING AGENTS, SUCH AS NITRILOACETIC ACID (NTA) AND
HYDROXYETHYLIMINODIACETIC ACID (HEIDA). THE USE OF DETERGENTS
CONTAINING NON-PHOSPHATE BUILDERS IS SUGGESTED, ESPECIALLY FOR EXTREME
SITUATIONS, SUCH AS LAKE TAHOE. THE USE OF NON-BIODEGRADABLE AGENTS
WHICH COMPLEX HEAVY METALS WILL DEPEND UPON THEIR EFFECT ON WASTE
TREATMENTS, SOFTENING WATER, DRINKING WATER, AND CALCIUM METABOLISM OF
MAN. (WILDE-WISCONSIN)

FIELD 05G

ACCESSION NO. W70-04373

SOME PROBLEMS REMAINING IN ALGAE CULTURING,

STATION D'HYDROBIOLOGIE APPLIQUEE, LE PARACLET PAR BOVES (FRANCE).

A. G. WURTZ.

ALGAE AND MAN, JACKSON, DANIEL F (EDITOR), PLENUM PRESS, NEW YORK, P 120-137, 1964.

DESCRIPTORS:
*ANALYTICAL TECHNIQUES, *ALGAE, *GROWTH RATES, *BIOINDICATORS, *WATER POLLUTION, *WASTE TREATMENT, FOOD CHAINS, CHLORELLA, CHLOROPHYLL, PHYSIOLOGICAL ECOLOGY, CYANOPHYTA, EUTROPHICATION, NUISANCE ALGAE, NITROGEN, PHOSPHATES, HYDROGEN ION CONCENTRATION, CYCLING NUTRIENTS, MUD-WATER INTERFACES, EUGLENOPHYTA, TOXICITY, BIOASSAYS, DIATOMS, EUGLENA, SPHAEROTILUS.

IDENTIFIERS:
*ALGAL CULTURES, *AXENIC CULTURES, PEDIASTRUM, SELF-PURIFICATION, ALGAL ENUMERATION, SCENEDESMUS, GROWTH EQUATIONS, ANKISTRODESMUS, HETEROTROPHY, CONTINUOUS CULTURES, CHLAMYDOMONAS, BATCH CULTURES, OPTICAL DENSITY, SOIL-WATER TECHNIQUES, PHYCOLOGY, MICROCYSTIS, GLOEOTRICHIA, APHANIZOMENON, MICRONUTRIENTS, DINOPHYCEAE, CRYPTOPHYCEAE, CHROMATIUM, OSCILLATORIA, ANTIBIOTIC TREATMENT, CHU'S MEDIUM, ULTRAVIOLET STERILIZATION, CONSTRUCTIVE SYNECOLOGY, TABELLARIA.

ABSTRACT:
NUMEROUS PROBLEMS EXIST AMONG THE INTERFACES BETWEEN PURE-CULTURE STUDY OF ALGAE AND PHYCOLOGICAL ECOLOGY, ESPECIALLY AS REGARDS THE QUALITY OF MAN'S ENVIRONMENT. WHILE ATTAINMENT OF AXENIC CULTURES IS, OF ITSELF, A VALID GOAL, INCREASING USE OF SUCH CULTURES, COUPLED WITH ECOLOGICAL INSIGHTS, MAY YIELD USEFUL INFORMATION PERTAINING TO MANY PERPLEXING PRACTICAL PROBLEMS. EQUATIONS DESCRIBING RELATIONSHIPS AMONG GROWTH, YIELD, AND BIOMASS IN BATCH AND CONTINUOUS CULTURES ARE REVIEWED, AS ARE METHODS FOR EVALUATING ALGAL GROWTH. THERE REMAINS A BASIC NEED FOR STANDARDIZED TECHNIQUES FOR ASSESSING GROWTH IN ALGAL CULTURES. SOME ASPECTS OF TECHNIQUES FOR AXENIC CULTURING ARE REVIEWED. DISCREPANCIES BETWEEN INFORMATION ON MINERAL NUTRITION OF BLUE-GREEN ALGAE, AS DEVELOPED FROM STUDIES OF AXENIC CULTURES, AND FIELD OBSERVATIONS OF CONDITIONS LEADING TO NUISANCE BLOOMS NEED FURTHER STUDY AND CLARIFICATION. THE SOIL-WATER TECHNIQUE, WHICH HAS BEEN INSTRUMENTAL IN SUCCESSFUL CULTURING OF SOME 500 STRAINS OF VARIOUS ALGAE, RAISES FURTHER ECOLOGICAL QUESTIONS, INCLUDING THOSE OF HETEROTROPHY AND CONDITIONS AT THE SOIL-WATER INTERFACE. OTHER SIMILAR QUESTIONS CONSIDERED IN SOME DETAIL IN THIS REPORT INCLUDE INFLUENCE OF INNOCULUM SIZE ON POPULATION GROWTH; ECOLOGICAL IMPLICATIONS OF ALGAL CULTURE; SELF-PURIFICATION OF WASTES BY ALGAE; ALGAL INDICATORS OF WATER QUALITY; AND ROLE OF ALGAE IN FOOD CHAINS. (SEE W69-07832).
(EICHHORN-WISCONSIN)

FIELD 05C

ACCESSION NO. W70-04381

ANALYTICAL CHEMISTRY OF PLANT NUTRIENTS,

WISCONSIN UNIV., MADISON. WATER CHEMISTRY LAB.

G. FRED LEE.

EUTROPHICATION: CAUSES, CONSEQUENCES, CORRECTIVES. PRINTING AND PUBLISHING
OFFICE, NATIONAL ACADEMY OF SCIENCES, WASHINGTON, D C, 1969. P 646-658. 3
FIG, 3 TAB, 22 REF.

DESCRIPTORS:
*ANALYTICAL TECHNIQUES, *AQUATIC WEEDS, *WATER ANALYSIS, *SAMPLING,
PLANT PHYSIOLOGY, EUTROPHICATION, DATA COLLECTION, CHEMICAL ANALYSIS,
RELIABILITY, WATER POLLUTION, WATER POLLUTION EFFECTS, NITROGEN,
PHOSPHORUS, ALGAE, NUTRIENTS, WATER PROPERTIES.

IDENTIFIERS:
*NUTRIENT ELEMENTS, PHOSPHORUS RELEASE, SAMPLE PRESERVATION, WATER
DYNAMICS, WATER CONSTITUENTS, ANALYTICAL CHEMISTRY, WASTE WATER
EFFECTS, LAKE MENDOTA(WIS), BLACK EARTH CREEK(WIS).

ABSTRACT:
VARIOUS STEPS MUST BE TAKEN TO ACHIEVE RATIONAL MANAGEMENT OF WATER
BASINS INVADED OR THREATENED BY NOXIOUS AQUATIC WEEDS. THE FIRST PHASE
OF AN EUTROPHICATION STUDY INVOLVES A DECISION ON THE METHOD, LOCATION,
AND FREQUENCY OF SAMPLING, AND THE MANNER OF SAMPLE PRESERVATION. NEXT,
IS THE SELECTION OF METHODS OF ANALYSES APPROPRIATE TO THE REQUIRED
ACCURACY OF RESULTS. FINALLY, A CRITICAL EVALUATION OF RESULTS OBTAINED
AS TO THEIR GENERAL RELIABILITY AND STATISTICAL SIGNIFICANCE. DETAILED
INFORMATION ON SOURCES, FORMS, AND METHODS OF ANALYSIS APPLICABLE TO
NITROGEN AND PHOSPHORUS ARE GIVEN. ADDITION OF MERCURIC CHLORIDE AND
FREEZING TO -20C, RATHER THAN THE USE OF CHLOROFORM AND HYDROGEN
SULFATE, IS RECOMMENDED FOR THE PRESERVATION OF NITROGEN FORMS AND
PHOSPHORUS COMPOUNDS IN STORED WATER SAMPLES. A WARNING IS GIVEN ABOUT
POSSIBILITY PHOSPHATE LOSS DUE TO SORPTION BY GLASS OR PLASTIC
CONTAINERS. (SEE W70-03975). (WILDE-WISCONSIN)

FIELD 05A, 07B

ACCESSION NO. W70-04382

GEOCHEMISTRY OF EUTROPHICATION,

GENERAL DYNAMICS CORP., GROTON, CONN. ELECTRIC BOAT DIV.

RICHARD J. BENOIT.

EUTROPHICATION: CAUSES, CONSEQUENCES, CORRECTIVES, PRINTING AND PUBLISHING
OFFICE, NATIONAL ACADEMY OF SCIENCES, WASHINGTON, D C, 1969. P 614-630. 3
FIG, 6 TAB, 30 REF.

DESCRIPTORS:
*GEOCHEMISTRY, *EUTROPHICATION, *WATER STRUCTURE, ATMOSPHERE, SEA
WATER, FRESH WATER, SORPTION, SODIUM, ALGAE, PHOSPHORUS, TRACE
ELEMENTS, PLANKTON, LAKES, RIVERS, GREAT LAKES.

IDENTIFIERS:
*EQUILIBRIUM MODEL, LITHOSPHERE, SORPTION REACTIONS, SOLUBILITY
EQUILIBRIUM THEORY.

ABSTRACT:
THE TENDENCY OF THE LAND-BASED HYDROSPHERE TO ATTAIN SIMILAR
COMPOSITION UNDER THE COMBINED INFLUENCES OF THE BUFFERING ACTION OF
ION-EXCHANGE REACTIONS AND INTEGRATION OF TRIBUTARIES IS STRESSED. THE
EQUILIBRIUM MODEL, BASED ON SOLUBILITY EQUILIBRIA OF COMMON
CONSTITUENTS OF THE LITHOSPHERE AND CARBON DIOXIDE CONCENTRATION EQUAL
TO 35,000 ATMOSPHERES, GIVES A REASONABLY SATISFACTORY AGREEMENT WITH
THE ACTUAL COMPOSITION OF WATER OF THE GREAT LAKES. WHILE SORPTION
REACTIONS MAY CAUSE SOME DEVIATIONS FROM THE EQUILIBRIUM MODELS, THE
MAJOR DISCREPANCIES IN THE CHEMICAL COMPOSITION OF WATERS ARE
ATTRIBUTED TO MAN'S ACTIVITY. THIS IS PARTICULARLY TRUE OF THE CONTENTS
OF SODIUM AND PHOSPHORUS, THE TWO ELEMENTS THAT MAY LIMIT OR UNDULY
PROMOTE THE ALGAL GROWTH. TRACE ELEMENTS MAY ALSO ALTER THE TROPHIC
LEVEL OF A LAKE TO CREATE A NUISANCE OR EVEN LEAD TO A DEGRADATION OF A
RESOURCE. (SEE W70-03975). (WILDE-WISCONSIN)

FIELD 05C

ACCESSION NO. W70-04385

THE GREAT AND DIRTY LAKES,

GLADWIN HILL.

IN: CONTROLLING POLLUTION: THE ECONOMICS OF A CLEANER AMERICA, ED. MARSHALL
I. GOLDMAN, ENGLEWOOD CLIFFS, N.J.: PRENTICE-HALL, INC., 1967, P. 43-48,
AND SATURDAY REIVEW, OCT. 23, 1965.

DESCRIPTORS:
*ALGAE, *INDUSTRIAL WASTES, *MUNICIPAL WASTES, *GREAT LAKES, *WATER
POLLUTION, SEWAGE, AESTHETICS, AEROBIC CONDITIONS, ANAEROBIC
CONDITIONS, WATER SUPPLY.

IDENTIFIERS:
*DEVOLUTION, *INTERSTATE POLLUTION, *METROPOLITAN AREAS, BLONDIN.

ABSTRACT:
THE POLLUTION OF THE GREAT LAKES IS DISCUSSED. FOCUSING ON NIAGRA
FALLS, THE SOURCES OF POLLUTION FROM DULUTH TO BUFFALO ARE TRACED.
SPECIAL ATTENTION IS GIVEN TO THE CHICAGO AND DETROIT METROPOLITAN
AREAS. THE GREAT LAKES ARE TREATED AS ONE CONTINUOUS SYSTEM AND
SUGGESTS INTERSTATE POLLUTION ABATEMENT AS THE ONLY POSSIBLE RECOURSE,
REJECTING INTERNATIONAL EFFORTS AS CUMBERSOME AND INTRASTATE EFFORTS AS
INEFFECTUAL. WHILE RECOGNIZING THE MUNICIPAL WASTE AS A MAJOR POLLUTION
SOURCE, THE AUTHOR CONCENTRATES ON INDUSTRIAL WASTES FROM THE OIL,
STEEL, PAPER, SOAP, CHEMICAL AND AUTOMOTIVE INDUSTRIES. HE CITES THE
CHICAGO AND DETROIT METROPOLITAN AREAS AS THE MAIN DISCHARGERS OF
AMMONIA NITROGEN, PHENOLS, CYANIDE, OIL, PHOSPHATES, CHLORIDES,
SUSPENDED AND SETTLEABLE SOLIDS AND NITROGEN COMPOUND WASTES INTO THE
LAKES. CHICAGO'S EFFORTS TO MAINTAIN A USABLE WATER SUPPLY AND
DETROIT'S OUTDATED SEWAGE TREATMENT FACILITIES ARE ALSO REVIEWED. THE
DEATH OF LAKE ERIE IS TREATED ALONG WITH A DISCUSSION OF THE DEVOLUTION
OF AQUATIC LIFE WHICH ACCOMPANIED IT. THE EFFORTS OF PRIVATE
INDIVIDUALS AND CONSERVATION GROUPS AND THOSE OF VARIOUS STATE AND
FEDERAL GOVERNMENTS ARE TREATED. (RICHMOND-CHICAGO)

FIELD 05G, 05B

ACCESSION NO. W70-04430

THE FUTURE OF THE LAKE,

WASHINGTON UNIV., SEATTLE. DEPT. OF ZOOLOGY.

W. T. EDMONDSON.

MUNICIPALITY OF METROPOLITAN SEATTLE, METRO QUARTERLY, FALL 1965. 4 P, 1 FIG.

DESCRIPTORS:
*ALGAE, *WATER POLLUTION TREATMENT, LAKES, SEWAGE, DRAINAGE DISTRICTS,
DEBT, COSTS, CITIES.

IDENTIFIERS:
*LAKE WASHINGTON(WASH), SEATTLE(WASH), REVENUE BONDS, METRO
ACT(SEATTLE).

ABSTRACT:
ORIGINALLY SEATTLE, WASHINGTON, RELEASED NON-TREATED SEWAGE INTO PUGET
SOUND AND TREATED SEWAGE INTO LAKE WASHINGTON. SWIMMING IN THE SOUND
WAS DESTROYED, AND LAKE WASHINGTON BECAME EUTROPHIC. THE METRO ACT
ENACTED IN 1957 PROVIDED FOR THE ESTABLISHMENT OF A METROPOLITAN
MUNICIPAL CORPORATION WITH THE POWER TO PERFORM SEWAGE DISPOSAL, WATER,
PARKS, TRANSPORTATION, COMPREHENSIVE PLANNING AND/OR GARBAGE DISPOSAL.
IN 1958 THE SEWAGE DISPOSAL PROVISO WAS APPROVED BY 60% VOTE. JULY 1966
SHOULD SEE THE END OF SEWAGE DISCHARGE INTO LAKE WASHINGTON. $120
MILLION IN REVENUE BONDS HAVE BEEN SOLD TO DATE, TO BE REPAID BY A $2
MONTHLY CHARGE PER RESIDENTIAL CONNECTION AND A $2 MONTHLY CHARGE PER
900 CUBIC FEET OF WATER USED BY INDUSTRIAL AND COMMERCIAL ENTERPRISES.
PROMPT IMPROVEMENT IN THE LAKE IS EXPECTED WITHIN A FEW YEARS AFTER
FINAL DIVERSION IN 1966 AS THE WATERSHED IS RELATIVELY POOR IN
DISSOLVED MINERALS INCLUDING NUTRIENTS NEEDED BY ALGAE, AND AS THERE
ARE NO MASSIVE DEPOSITS OF ACCUMULATED SOFT SEDIMENTS ON THE LAKE
BOTTOM RICH IN NUTRIENTS AVAILABLE TO ALGAE, THE PROGNOSIS IS GOOD FOR
THE RECOVERY OF LAKE WASHINGTON. (POWERS-WISCONSIN)

FIELD 05C, 02H

ACCESSION NO. W70-04455

OUTWITTING THE PATIENT ASSASSIN: THE HUMAN USE OF LAKE POLLUTION,

NORTHWESTERN UNIV., EVANSTON, ILL. TECHNOLOGICAL INST.

HAROLD B. GOTAAS.

BULLETIN OF THE ATOMIC SCIENTISTS, P 8-10, MAY 1969.

DESCRIPTORS:
    *NUTRIENTS, *BENEFIT-COST ANALYSIS, *LAKE ERIE, PHOSPHORUS, FISH
    STOCKING, COMMERCIAL FISH, SALMON, ALGAE, FISH.

IDENTIFIERS:
    ANTIPOLLUTION PROGRAMS, BIOLOGICAL BALANCE, OVERFISHING, COHO SALMON,
    ALEWIFE, NUTRIENT REMOVAL.

ABSTRACT:
    SOME ASPECTS OF THE OCTOBER 1968 DEPARTMENT OF INTERIOR REPORT ON LAKE
    ERIE ARE CHALLENGED. THE EMPHASIS ON NUTRIENT REMOVAL WHICH WOULD
    RESULT IN IMMEDIATE EXPENDITURE OF $1.1 BILLION TO CONTROL MUNICIPAL
    POLLUTION AND $285 MILLION FOR CURBING INDUSTRIAL CONTAMINATION IS
    QUESTIONED IN LIGHT OF THE EXPECTED BENEFITS. SEVERAL EFFICACIOUS
    ALTERNATIVES FOR REVIVING THE GREAT LAKES ARE SUGGESTED: (1) FURTHER
    NUTRIENT RELATIONSHIPS AND COST VERSUS BENEFITS STUDIES SHOULD BE
    UNDERTAKEN BEFORE MONEY IS SPENT TO BUILD EXPENSIVE TREATMENT
    FACILITIES; (2) HARVEST ALGAE AS A POTENTIAL SOURCE OF FOOD THUS
    PREVENTING IT FROM CONTRIBUTING TO THE ORGANIC WASTE LOAD; (3) PREVENT
    THE INTRODUCTION OF ALL TOXIC MATERIALS ENABLING DESIRABLE BIOLOGICAL
    BALANCES TO BE ESTABLISHED; (4) SEED THE LAKES WITH DESIRABLE FISH TO
    ESTABLISH A FOOD CHAIN WHICH WOULD PERMIT THE GROWTH AND REMOVAL OF
    NUTRIENTS AS WELL AS FISH FOR FOOD NEEDS AND SPORT; AND (5) ADOPT WATER
    QUALITY STANDARDS THAT ARE REALISTIC AS TO THEIR COSTS, BENEFITS, AND
    TIME PRIORITIES. (HASKINS-WISCONSIN)

FIELD 05C, 02H

ACCESSION NO. W70-04465

ALGAE FROM WESTERN LAKE ERIE,

OHIO STATE UNIV., COLUMBUS. DEPT. OF BOTANY AND PLANT PATHOLOGY.

CLARENCE E. TAFT, AND W. JACK KISHLER.

OHIO JOURNAL OF SCIENCE, VOL 68, NO 2, P 80-83, 1968. 9 FIG, 7 REF.

DESCRIPTORS:
    *ALGAE, *LAKE ERIE, CHLOROPHYTA, CYANOPHYTA, UNITED STATES, HABITATS.

IDENTIFIERS:
    *NEW SPECIES, *WESTERN LAKE ERIE(OHIO), GONGROSIRA STAGNALIS,
    NEPHROCYTIUM OBESUM W AND G S WEST, RADIOCOCCUS NIMBATUS, URONEMA
    ELONGATUM HODGETTS, CHROOCOCCUS PRESCOTTII, CALOTHRIX FUSCA,
    MICROCOLEUS LACUSTRIS (RAB) FARLOW, CLADOPHORA, TAXONOMIC DESCRIPTIONS,
    GONGROSIRA LACUSTRIS BRAND, GONGROSIRA DEBARYANA RAB, OOCYSTIS,
    OSCILLATORIA, ULOTHRIX, EUCAPSIS ALPINA CLEMENTS AND SCHANTZ, SCYTONEMA
    ALATUM (CARM) BORZI, SCYTONEMA MYOCHROUS.

ABSTRACT:
    FOUR SPECIES OF ALGAE IN THE CHLOROPHYTA (GONGROSIRA STAGNALIS,
    NEPHROCYTIUM OBESUM, RADIOCOCCUS NIMBATUS, AND URONEMA ELONGATUM) AND
    FIVE IN THE CYANOPHYTA (CHROOCOCCUS PRESCOTTI, CALOTHRIX FUSCA,
    MICROCOLEUS LACUSTRIS, SCYTONEMA ALATUM, AND SCYTONEMA MYOCHROUS) ARE
    NEWLY REPORTED FOR WESTERN LAKE ERIE. GONGROSIRA STAGNALIS (G S WEST)
    SCHMIDLE, COLLECTED FROM THE BASAL FRAGMENTS OF OLD CLADOPHORA, APPEARS
    TO BE A NEW RECORD FOR THE UNITED STATES. NEPHROCYTIUM OBESUM W AND G S
    WEST, WHICH IS REPORTED AS OFTEN HAVING A SHALLOWLY SCROBICULATE WALL,
    IS UNIQUE AND MERITS FURTHER INTENSIVE STUDY. THESE NEW ALGAE
    OCCASIONALLY APPEAR IN TEACHING AND RESEARCH COLLECTIONS AT STONE
    LABORATORY, PUT-IN-BAY, OHIO. THE HABITAT OF G STAGNALIS ON
    CALCIUM-ENCRUSTED CLADOPHORA FILAMENTS IS UNIQUE. (JONES-WISCONSIN)

FIELD 05C

ACCESSION NO. W70-04468

A SIMPLE PIPETTE CONTROL FOR ISOLATING PLANKTONIC ALGAE,

RHODE ISLAND UNIV., KINGSTON. GRADUATE SCHOOL OF OCEANOGRAPHY.

BRENDA J. BOLEYN.

CANADIAN JOURNAL OF MICROBIOLOGY, VOL 13, P 1129-1130, 1967. 1 FIG, 4 REF.

DESCRIPTORS:
*ALGAE, *ANALYTICAL TECHNIQUES, *PLANKTON, *LABORATORY TESTS,
*ISOLATION, *CONTROL, METHODOLOGY, CULTURES, TESTING, CAPILLARY
CONDUCTIVITY, BACTERIA, DIATOMS, TEST PROCEDURES.

IDENTIFIERS:
*PIPETTE, UNIALGAL, NITZSCHIA, ALGAE ISOLATION, CAPILLARITY,
CAPILLARITY CONTROL DEVICE.

ABSTRACT:
A DEVICE FOR SIMPLIFYING THE ISOLATION OF PLANKTON OR OTHER SMALL
PARTICLES HAS BEEN DEVELOPED. IT OPERATES BY CONTROLLING THE
CAPILLARITY OF THE FINE-TIPPED PIPETTE ROUTINELY USED FOR PHYTOPLANKTON
ISOLATION. THE INWARD AND OUTWARD FLOW OF FLUID MAY BE CONTROLLED
SMOOTHLY AND COMPLETELY. THERE ARE A NUMBER OF ADVANTAGES IN THIS
DEVICE OVER THE UNMODIFIED ISOLATING PIPETTE. A STEADY HAND IS EASILY
MAINTAINED SINCE THE HAND GUIDING THE PIPETTE NEED NOT CONTROL
CAPILLARITY ALSO. THE OPERATOR IS ABLE TO NUDGE AND STEER RELUCTANT OR
ADHERING FORMS INTO THE PIPETTE, ESPECIALLY WHERE LONG SLENDER CELLS,
SETOSE FORMS, OR CHAIN-FORMERS MIGHT EASILY CLUSTER AND OCCLUDE THE
TIP. A PIPETTE WITH AN ORIFICE OF SMALL OR LARGE SIZE MAY BE USED,
MAKING IT HELPFUL TO ISOLATE LARGE OR PECULIARLY SHAPED CELLS. THE
EXPLOSIVE NATURE OF PIPETTE DELIVERY IS LESSENED AND BUBBLING IS
PREVENTED. MOTILE ORGANISMS MAY BE FOLLOWED AND CAUGHT READILY. THE
ENTIRE PROCEDURE IS HASTENED, DIMINISHING DANGER OF THERMAL SHOCK,
ESPECIALLY IMPORTANT FOR WINTER FORMS. ALSO THE RISK OF PRODUCING
APPRECIABLE SALINITY CHANGES IS MINIMIZED. (JONES-WISCONSIN)

FIELD 05A

ACCESSION NO. W70-04469

THE EFFECT OF IMPOUNDMENTS ON NATURAL WATERS (IN GERMAN),

AKADEMIYA NAUK SSSR, LENINGRAD (USSR).

W. I. SHADIN.

VERH. INTERNAT. VEREIN, LIMNOL., VOL. IV, P 792-805, JULY 1961.

DESCRIPTORS:
*LAKES, *RESERVOIRS, ZOOPLANKTON, PHYTOPLANKTON, TEMPERATURE,
*EPILIMNION, *HYPOLIMNION, ALGAE, FUNGI, BACTERIA, FISH, LIMNOLOGY,
HYDROLOGIC ASPECTS, AQUATIC ENVIRONMENT, WATER PROPERTIES, BIOLOGICAL
PROPERTIES, ECOLOGY, STRATIFICATION.

IDENTIFIERS:
WATER QUALITY PROPERTIES.

ABSTRACT:
THE BUILDING OF IMPOUNDMENT QUALITATIVELY CHANGES THE BIOLOGICAL AND
ECOLOGICAL SYSTEM OF WATER. A SURVEY OF RESPONSE OF WATER QUALITY TO
THE BUILDING OF RESERVOIRS AND REVIEW OF WATER QUALITY PROPERTIES IN
LAKES AND RESERVOIRS IN DIFFERENT GEOLOGIC AND CLIMATIC CONDITIONS IS
PRESENTED IN THE PAPER. THE WATER QUALITY AND BIOLOGICAL SYSTEM OF
WATER IN IMPOUNDMENT IS DEPENDENT ON MANY FACTORS, SUCH AS HYDROLOGIC
REGIME, CLIMATIC AND REGIONAL CONDITIONS, PURPOSE OF RESERVOIR, ETC.
THE TEMPERATURE DISTRIBUTION INFLUENCES THE DETENTION (THROUGH-FLOW)
TIME AND FORMATION OF DENSITY LAYERS WITH DIFFERENT WATER QUALITY
PARAMETERS. THE CHEMICAL REGIME IS MAINLY DEPENDENT ON SLUDGE
FORMATION. THE AMOUNT OF BACTERIA IN WATER HAS TWO MAXIMUMS DURING A
YEAR. THE FORMATION OF ZOOPLANKTON AND PHYTOPLANKTON IS FURTHER
DISCUSSED. THE PAPER WAS PRESENTED AS A LEADING LECTURE AT XIV-TH
INTERNATIONAL LIMNOLOGICAL CONFERENCE IN 1961. (NOVOTNY-VANDERBILT)

FIELD 02H, 05C

ACCESSION NO. W70-04475

IMPOUNDMENT DESTRATIFICATION FOR RAW WATER QUALITY CONTROL USING EITHER
  MECHANICAL OR DIFFUSED AIR PUMPING,

FEDERAL WATER POLLUTION CONTROL ADMINISTRATION, CINCINNATI, OHIO.

J. M. SYMONS, W. H. IRWIN, E. L. ROBINSON, AND G. G. ROBECK.

J. AMERICAN WATER WORKS ASSOC., VOL 59, NO 10, P 1268-1291, OCTOBER 1967. 20
  FIG., 5 TAB., 19 REF.

  DESCRIPTORS:
    *THERMAL STRATIFICATION, *AERATION, *IMPOUNDMENTS, *RESERVOIRS,
    *MIXING, *WATER QUALITY CONTROL, WATER QUALITY, WATER POLLUTION,
    NITROGEN, PHOSPHORUS, AIR, PHYTOPLANKTON, ALGAE, EFFICIENCIES,
    DISSOLVED OXYGEN, OXYGEN, LAKES.

  ABSTRACT:
    TWO METHODS ARE COMPARED FOR ARTIFICIALLY DESTRATIFYING LAKES AND
    RESERVOIRS BOTH OF WHICH BROKE UP THE THERMAL STRATIFICATIONS PATTERN
    IN THE LAKES SUCCESSFULLY AND IMPROVED WATER QUALITIES. OTHER
    CONCLUSIONS REACHED WERE THAT MULTIPLE MIXING STARTED IN THE SPRING
    WILL MAINTAIN GOOD QUALITY WATER IN AN IMPOUNDMENT, AND IS SUPERIOR TO
    A SINGLE MIXING IN MIDSUMMER OR EARLY FALL. SOME RISE IN NITROGEN AND
    PHOSPHORUS CONCENTRATIONS OCCURRED IN SURFACE WATER DURING
    DESTRATIFICATION, BUT THE SURFACE POPULATION OF PHYTOPLANKTON AND THE
    PHYTOPLANKTON STANDING CROP DECLINED DURING MIXING. ALGAL POPULATIONS
    INCREASED AGAIN WHEN MIXING STOPPED. DATA TAKEN IN THIS STUDY AND FROM
    5 OTHER STUDIES IN THE LITERATURE SHOWED A DESTRATIFICATION EFFICIENCY
    RANGING FROM 0.1 TO 1.2%, AND AVERAGING 0.5%; AND AN ENERGY INPUT/UNIT
    OF IMPOUNDMENT WATER VOLUME RANGING FROM 0.7 TO 4.9 KW-HR/ACRE-FT. AND
    AVERAGING 2 KW-HR/ACRE-FT. (NOVOTNY-VANDERBILT)

  FIELD 05F, 02H

  ACCESSION NO. W70-04484

TREATMENT OF ALGAE AND WEEDS IN LAKES AT MADISON,

  HEALTH DEPT., MADISON, WIS. LAB.

  BERNARD P. DOMOGALLA.

  ENGINEERING NEWS-RECORD, VOL 97, NO 24, P 950-954, 1926. 8 FIG, 3 TAB.

    DESCRIPTORS:
      *COPPER SULFATE, *EUTROPHICATION, *ALGAE, *AQUATIC WEEDS, LAKES,
      ALGICIDES, AQUATIC WEED CONTROL, MECHANICAL CONTROL, WATER POLLUTION
      CONTROL, CHEMCONTROL, ARSENIC COMPOUNDS, NITROGEN, PHOSPHORUS.

    IDENTIFIERS:
      *MADISON LAKES(WIS), *AQUATIC WEED CUTTING, *WEED EXTRACTING STEEL
      CABLES, ALGAL SUCCESSION, ANABAENA, CLADOPHORA, PEDIASTRUM,
      HYDRODICTYON, POTAMOGETON, RANUNCULUS.

    ABSTRACT:
      FIELD AND LABORATORY INVESTIGATIONS OF HIGHLY EUTROPHIC LAKES IN THE
      VICINITY OF MADISON YIELDED SEVERAL CONCLUSIONS. THE COPPER SULFATE
      DRAGGING METHOD PROVED TO BE MOST EFFECTIVE FOR SUPPRESSION OF
      BOTTOM-ANCHORED ALGAE ON WIND EXPOSED AREAS. THE SURFACE ALGAE AND
      THOSE INHABITING SHALLOW WATERS WERE BEST CONTROLLED BY SPRAYING WITH
      COPPER SULFATE. WEEDS IN OPEN WATER WERE ELIMINATED BY WEED-CUTTING
      MACHINES. STEMS AND ROOTS OF WEEDS IN SHALLOW WATERS AND NEAR SHORE
      LINES WERE EXTRACTED BY STEEL CABLES WITH CLIPS AND SWIVELS. SEVERAL
      SPECIES OF WATER WEEDS WERE ERADICATED BY SODIUM ARSENITE. THE STUDY
      DISCLOSED A SEASONAL SUCCESSION OF AQUATIC PLANTS AND A PRONOUNCED
      RELATIONSHIP BETWEEN ERADICATION TREATMENTS AND BIOLOGICAL FIXATION OF
      AVAILABLE FORMS OF NITROGEN AND PHOSPHORUS. (WILDE-WISCONSIN)

    FIELD 02H, 05C

    ACCESSION NO. W70-04494

ALGAL NUTRITION AND EUTROPHICATION,

HASKINS LABS., NEW YORK.

LUIGI PROVASOLI.

EUTROPHICATION: CAUSES, CONSEQUENCES, CORRECTIVES, PRINTING AND PUBLISHING
OFFICE, NATIONAL ACADEMY OF SCIENCES, WASHINGTON, D C, 1969, P 574-593. 2
FIG, 3 TAB, 72 REF.

DESCRIPTORS:
*EUTROPHICATION, *ALGAE, *NUTRIENTS, NITROGEN, PHOSPHORUS, SODIUM,
POTASSIUM, TRACE METALS, VITAMINS, WATER POLLUTION, PHYSIOLOGICAL
ECOLOGY, DIATOMS, OLIGOTROPHY, CYANOPHYTA, CHLOROPHYTA, GREAT LAKES,
BIOINDICATORS, FISH, WASHINGTON, LAKES, INDIANA, OCHROMONAS,
PHYTOPLANKTON, SEAWEEDS.

IDENTIFIERS:
*ALGAL NUTRITION, DITYLUM, CHAETOCEROS, SKELETONEMA, THALASSIOSIRA,
COCCOLITHUS, SARGASSO SEA, SYNURA, GYMNODINIUM, CHEMICAL SYMBIOSIS,
MICROBIAL ECOLOGY, DINOBRYON, VOLVOX, PANDORINA, SPIROGYRA,
SYNECHOCOCCUS, CHROOCOCCUS, OSCILLATORIA, NOSTOC, ANABAENA, ANACYSTIS,
MICROCYSTIS, BOSMINA, ASTERIONELLA, MELOSIRA, SYNEDRA, FRAGILARIA,
CERATIUM, SWEDEN, APHANIZOMENON, RHODOMONAS, CRYPTOMONADS,
CHRYSOMONADS, EUGLENIDS, DINOFLAGELLATES.

ABSTRACT:
PREVALENT EMPHASIS ON NITROGEN AND PHOSPHORUS NOTWITHSTANDING, LITTLE
DOUBT EXISTS THAT OTHER NUTRIENTS PLAY ROLES IN EUTROPHICATION OF
NATURAL WATERS. ALSO, ACCENT IS ON ASPECTS OF BIOLOGICAL OVERPRODUCTION
IN EUTROPHICATION, WHEREAS MORE VISIBLE MANIFESTATIONS ARE CHANGES IN
COMMUNITY STRUCTURE, PRODUCING LESS DESIRABLE SPECIES VIA ALTERATIONS
IN FOOD CHAINS. FROM THESE VIEWPOINTS, UNFORTUNATELY LITTLE IS KNOWN
REGARDING ECOLOGICALLY IMPORTANT SPECIES. SELECTED NUTRITIONAL ASPECTS
OF ALGAE ARE REVIEWED AS REGARDS NITROGEN, PHOSPHORUS, SODIUM,
POTASSIUM, TRACE METALS, VITAMINS, AND CHEMICAL SYMBIOSIS. IN SOME
SITUATIONS, GROWTH OF CHLOROPHYTES (GREEN ALGAE) ARE FAVORED BY HIGH
CONCENTRATIONS OF NITROGEN, PHOSPHORUS, AND POTASSIUM. WHERE
CHLOROPHYTES ARE REPLACED, NITROGEN-FIXING BLUE-GREEN ALGAE MAY TAKE
OVER BECAUSE NITROGEN IS TOO SCARCE TO MEET REQUIREMENTS OF
CHLOROPHYTES. IT IS KNOWN THAT BOTH SODIUM AND POTASSIUM INCREASE
DURING EUTROPHICATION. SUCH CHANGES MAY BE FUNCTIONAL AS WELL AS
INDICATIVE OF EUTROPHICATION. ALTHOUGH TRACE ELEMENTS ARE OF
ACKNOWLEDGED SIGNIFICANCE IN ALGAL NUTRITION, FEW RELIABLE ECOLOGICAL
DATA ARE AVAILABLE, AND ADDITIONAL RESEARCH IS REQUIRED. VITAMINS ARE
LIKELY TO BE RICH IN EUTROPHIC ENVIRONMENTS, A MATTER OF IMPORTANCE
SINCE 197 OF 204 ALGAL SPECIES STUDIED REQUIRE VITAMINS, AND SINCE A
PATTERN OF VITAMIN REQUIREMENTS IS BEGINNING TO EMERGE FOR VARIOUS
TAXONOMIC GROUPS. (SEE W70-03975).
(EICHHORN-WISCONSIN)

FIELD 05C

ACCESSION NO. W70-04503

ALGAE AND METROPOLITAN WASTES.

PUBLIC HEALTH SERVICE, CINCINNATI, OHIO. DIV. OF WATER SUPPLY AND POLLUTION
CONTROL; AND ROBERT A. TAFT SANITARY ENGINEERING CENTER, CINCINNATI, OHIO.

TRANSACTIONS OF SEMINAR ON ALGAE AND METROPOLITAN WASTES HELD AT CINCINNATI,
OHIO, APRIL 27-29, 1960, SPONSORED BY DIV OF WATER SUPPLY AND POLLUTION
CONTROL AND THE ROBERT A TAFT SANITARY ENGINEERING CENTER. BARTSCH, ALFRED
F, CHAIRMAN. ROBERT A TAFT SANITARY ENGINEERING CENTER, SEC TR W61-3, 1961.
162 P.

DESCRIPTORS:
*CONFERENCES, *EUTROPHICATION, *ALGAE, PLANNING, RIVERS, LAKES,
LIMNOLOGY, WISCONSIN, URBANIZATION, CONNECTICUT, OREGON, WASHINGTON,
PRIMARY PRODUCTIVITY, ECOSYSTEMS, ILLINOIS, PHOSPHORUS, DRAINAGE,
AGRICULTURAL WATERSHEDS, NITROGEN FIXATION, CYANOPHYTA, SEWAGE
EFFLUENTS, MINNESOTA, WATER POLLUTION EFFECTS, ALGICIDES, WATER QUALITY.

IDENTIFIERS:
*URBAN WASTES, MADISON(WIS), LAKE ZOAR(CONN), KLAMATH LAKE(ORE), LAKE
WASHINGTON(WASH), ALGAL NUTRITION, ALGAL PHYSIOLOGY, HETEROTROPHY,
MICRONUTRIENTS, BADFISH RIVER(WIS), DETROIT LAKES(MINN), NUTRIENT
REMOVAL, MASS COMMUNICATION.

ABSTRACT:
IN APRIL 1960, A SEMINAR MET IN CINCINNATI, OHIO, TO CONSIDER THE
PROBLEM, ALGAE AND METROPOLITAN WASTES. THE MEETING, OF WHICH THIS
VOLUME CONSTITUTES THE RECORDED TRANSACTIONS, WAS SPONSORED BY DIVISION
OF WATER SUPPLY AND POLLUTION CONTROL AND ROBERT A TAFT SANITARY
ENGINEERING CENTER, BOTH AGENCIES OF THE US PUBLIC HEALTH SERVICE.
DISCUSSIONS AT THE SEMINAR WERE LIMITED TO PREVENTION AND CONTROL OF
OBJECTIONABLE ALGAL BLOOMS ARISING FROM ENRICHMENT OF THE HYDROSPHERE
BY URBAN AND OTHER WASTES. THE CONFERENCE WAS ORGANIZED INTO PANEL
DISCUSSIONS AS FOLLOWS: STATEMENT OF THE PROBLEM, GROWTH
CHARACTERISTICS OF ALGAE, SOURCES OF NUTRIENTS, METHODS OF PREVENTION
OR CONTROL. TOPICAL COVERAGE OF SECTIONS, WITH NUMBER OF CONTRIBUTIONS
TO EACH RECORDED IN PARENTHESES, WILL INDICATE SCOPE AND COVERAGE OF
MEETING: INTRODUCTION (1); SPECIFIC PROBLEMS IN LAKES (4), AND IN
RIVERS (1); ALGAL NUTRITION (3); MEASUREMENT OF PRODUCTIVITY (2); LAND
DRAINAGE (2); WASTES (1); LIMNOLOGICAL RELATIONSHIPS (3); NUTRIENT
LIMITATION (6); ALGICIDE USE (1). THE VOLUME INCLUDES A SUMMATION OF
RESEARCH NEEDS AND TEXT OF A BANQUET ADDRESS BY A JOURNALIST ON THEME
OF COMMUNICATION BETWEEN SCIENTIST AND LAYMAN. (SEE ALSO W70-04507).
(EICHHORN-WISCONSIN)

FIELD 05C

ACCESSION NO. W70-04506

SPECIFIC PROBLEMS IN RIVERS: ALGAE IN RIVERS OF THE UNITED STATES,

PUBLIC HEALTH SERVICE, CINCINNATI, OHIO. DIV. OF WATER SUPPLY AND POLLUTION
CONTROL.

C. MERVIN PALMER.

TRANSACTIONS OF THE SEMINAR ON ALGAE AND METROPOLITAN WASTES, HELD AT
CINCINNATI, OHIO, APRIL 27-29, 1960, SPONSORED BY DIVISION OF WATER SUPPLY
AND POLLUTION CONTROL AND THE ROBERT A TAFT SANITARY ENGINEERING CENTER.
ALGAE AND METROPOLITAN WASTES, U S DEPARTMENT OF HEALTH, EDUCATION, AND
WELFARE, DIVISION OF WATER SUPPLY AND POLLUTION CONTROL; AND ROBERT A TAFT
SANITARY ENGINEERING CENTER, CINCINNATI, OHIO, TECHNICAL REPORT, SEC TR
W61-3, P 34-38, 1961. 3 FIG, 4 REF.

DESCRIPTORS:
*RIVERS, *ALGAE, *SAMPLING, *MISSISSIPPI RIVER, *OHIO RIVER, NUISANCE
ALGAE, WATER QUALITY, SEWAGE EFFLUENTS, DIATOMS, MISSOURI RIVER,
COLUMBIA RIVER, COLORADO RIVER, TENNESSEE RIVER, HUDSON RIVER, DELAWARE
RIVER, WATER POLLUTION EFFECTS.

IDENTIFIERS:
*RIVER STUDIES, *NATIONAL WATER QUALITY NETWORK, ALGAE COUNTS, PLANKTON
ANALYSES, AGRICULTURAL RUNOFF, ARKANSAS RIVER, MERRIMACK RIVER, RED
RIVER, SAVANNAH RIVER, SNAKE RIVER, RIO GRANDE RIVER, POTOMAC RIVER.

ABSTRACT:
INVESTIGATION OF ALGAE IN RIVERS OF THE UNITED STATES HAVE BEEN LACKING
DUE TO THE GENERAL IMPRESSION THAT THEY ARE RELATIVELY SCARCE. HOWEVER,
MORE DATA IS NEEDED TO DETERMINE THEIR SIGNIFICANCE AS POLLUTION
INDICATORS, STREAM PURIFIERS, NUISANCE GROWTHS, AND FOOD FOR FISH.
BASIC INFORMATION SHOULD BE COLLECTED FOR EXTENDED TIME PERIODS DURING
EACH MONTH OF THE YEAR FOR AS MANY RIVERS IN REPRESENTATIVE LOCATIONS
AS POSSIBLE. SUCH DATA SHOULD YIELD NECESSARY BACKGROUND INFORMATION
FOR STUDIES AND AID IN DETERMINING EFFECTS OF PARTICULAR FACTORS ON
RIVER BIOTA. SINCE 1957, THE NATIONAL WATER QUALITY NETWORK PROGRAM OF
THE PUBLIC HEALTH SERVICE HAS CARRIED ON SUCH SAMPLING PROGRAM FOR 16
RIVERS AND THE GREAT LAKES. FOR THE FIRST TWO YEARS THE AVERAGE COUNT
PER MONTHLY SAMPLES WAS 3,625 ALGAE/MILLILITER, LOWEST COUNTS OCCURRING
IN NOVEMBER, DECEMBER, AND MARCH, HIGHEST IN APRIL, SEPTEMBER, AND
OCTOBER. HIGHEST PLANKTON COUNTS OCCURRED IN RIVERS ENRICHED BY
AGRICULTURAL RUNOFF AND SEWAGE EFFLUENTS, PARTICULARLY WHERE TOXIC
INDUSTRIAL WASTES WERE LACKING. THE SAME FEW GENERA OF ALGAE ARE FOUND
TO DOMINATE RIVERS ACROSS THE COUNTRY. TWO COMPARISON STUDIES AT THE
ROBERT A TAFT SANITARY ENGINEERING CENTER DEAL WITH GREEN ALGAE OF THE
MISSISSIPPI AND OHIO RIVERS AND NUMBER OF GENERA OF DIATOMS IN SEVERAL
RIVER SYSTEMS. (SEE W70-04506). (KETELLE-WISCONSIN)

FIELD 05G

ACCESSION NO. W70-04507

ALGAE AS INDICATORS OF POLLUTION,

ACADEMY OF NATURAL SCIENCES OF PHILADELPHIA, PA. DEPT. OF LIMNOLOGY.

RUTH PATRICK.

THIRD SEMINAR, 1962. BIOLOGICAL PROBLEMS IN WATER POLLUTION, ROBERT A TAFT
SANITARY ENGINEERING CENTER, CINCINNATI, OHIO, PUBLIC HEALTH SERVICE
PUBLICATION NO 999-WP-25, P 225-231, 1965. 4 FIG, 1 TAB, 33 REF.

DESCRIPTORS:
*ALGAE, *BIOINDICATORS, *WATER POLLUTION, DIATOMS, MARYLAND,
PENNSYLVANIA, RIVERS, STREAMS, EUGLENA, AQUATIC POPULATIONS,
EUTROPHICATION, CYANOPHYTA, SEWAGE EFFLUENTS, CHLOROPHYTA, RHODOPHYTA,
PHYSIOLOGICAL ECOLOGY, INDUSTRIAL WASTES, HYDROGEN ION CONCENTRATION,
IRON COMPOUNDS, HYDROGEN SULFIDE, LEAD, COPPER, AMMONIA, CONDUCTIVITY,
CHROMIUM, PHENOLS, NITROGEN, CALCIUM, DISSOLVED OXYGEN.

IDENTIFIERS:
COMMUNITY STRUCTURE, RIDLEY CREEK(PA), NOBS CREEK(MD), BACK RIVER(MD),
OSCILLATORIA, CATHERWOOD DIATOMETER, SAVANNAH RIVER, POPULATION
DYNAMICS, ALGAL COMMUNITIES, MICROBIAL ECOLOGY, WATER QUALITY CRITERIA,
CYANIDE, NAPHTHENIC ACID.

ABSTRACT:
EXTENSIVE LITERATURE HAS DEVELOPED REGARDING ALGAL INDICATORS OF
POLLUTION. POLLUTION CONCEPT REFERS TO VARIOUS PHYSICAL AND CHEMICAL
CHANGES IN THE HYDROSPHERE PRODUCED BY HUMAN ACTIVITIES. EARLIEST
RECOGNIZABLE FORM OF POLLUTION RESULTED FROM ORGANIC CONTAMINATION, AND
SPECIES WERE CLASSIFIED ACCORDING TO DEGREES OF ORGANIC POLLUTION THEY
TOLERATED. LITTLE ATTENTION WAS FOCUSED ON POPULATION DYNAMICS OF
VARIOUS SPECIES OR ON HOLISTIC ASPECTS OF ALGAL COMMUNITIES. BECAUSE OF
INCREASING DIVERSITY AND COMPLEXITY OF POLLUTIONAL LOADS AND BECAUSE
ENVIRONMENT STRONGLY AFFECTS THEIR OCCURRENCE AND POPULATION SIZE,
ASSESSING POLLUTION BY PRESENCE OR ABSENCE OF A FEW SPECIES IS INCLINED
TO BE MISLEADING. THUS, WELL ESTABLISHED POPULATIONS OF SEVERAL ALGAL
SPECIES MAY INDICATE PRESENCE OF PARTICULAR FORMS OF POLLUTION, BUT
THEIR ELIMINATION OR DECREASE MAY RESULT FROM CAUSES VERY DIFFERENT
THAN CHANGES IN POLLUTIONAL LOAD. BECAUSE ALGAL COMMUNITIES INTEGRATE
EFFECTS OF INFLUENCING FACTORS, MORE RECENT STUDY OF POLLUTION HAS BEEN
DIRECTED TO CONSIDER STRUCTURE OF THE ENTIRE ALGAL COMMUNITY AS
INDICATIVE OF POLLUTION. EXAMPLES ARE CITED OF SEVERAL SUCH APPROACHES,
CONSIDERING VARIOUS FORMS OF POLLUTION AT A NUMBER OF DIFFERENT
LOCALITIES. DISCUSSION FOLLOWING THE PAPER DEVELOPED INFORMATION THAT
EUGLENA AND OSCILLATORIA ARE OFTEN CITED AS GENERA MOST TOLERANT TO
SEWAGE AND RELATED CONDITIONS. (EICHHORN-WISCONSIN)

FIELD 05C

ACCESSION NO. W70-04510

LAKE EUTROPHICATION -- A NATURAL PROCESS,

GEOLOGICAL SURVEY, ALBANY, N.Y.

PHILLIP E. GREESON.

WATER RESOURCES BULLETIN, VOL 5, NO 4, P 16-30, DECEMBER 1969. 15 P, 6 FIG, 2
TAB, 44 REF.

DESCRIPTORS:
*EUTROPHICATION, *LAKES, NUTRIENTS, WATER POLLUTION EFFECTS,
PRODUCTIVITY, SEDIMENTS, URBANIZATION, WATER POLLUTION TREATMENT,
ALGAE, WATER MANAGEMENT(APPLIED), LIMNOLOGY.

IDENTIFIERS:
EUTROPHICATION TREATMENT.

ABSTRACT:
LAKE EUTROPHICATION IS AN ECONOMIC, RECREATIONAL, AND AESTHETIC PROBLEM
THAT AFFECTS EVERY LAKE OF THE WORLD. EUTROPHICATION IS THE NATURAL
PROCESS OF LAKE AGING, AND PROGRESSES IRRESPECTIVE OF MAN'S ACTIVITIES.
POLLUTION, HOWEVER, CAN HASTEN THE NATURAL RATE OF AGING AND SHORTEN
THE LIFE EXPECTANCY OF A BODY OF WATER. THE EUTROPHICATION OF A LAKE
CONSISTS OF THE GRADUAL PROGRESSION FROM ONE LIFE STAGE TO ANOTHER
BASED ON THE DEGREE OF NOURISHMENT OR PRODUCTIVITY. THE EXTINCTION OF A
LAKE IS ATTRIBUTED TO ENRICHMENT BY NUTRITIVE MATERIALS, BIOLOGICAL
PRODUCTIVITY, DECAY, AND SEDIMENTATION. PRESENTLY USED METHODS FOR
RETARDING EUTROPHICATION ARE THE ABATEMENT OF CULTURAL ENRICHMENT,
TREATMENT OF EUTROPHIC SYMPTOMS, AND CONTROL OF FUNDAMENTAL CAUSES.
(KNAPP-USGS)

FIELD 05C, 02H

ACCESSION NO. W70-04721

URBAN EFFECTS ON QUALITY OF STREAMFLOW,

TEXAS UNIV., AUSTIN. ENVIRONMENTAL HEALTH ENGINEERING RESEARCH LAB.

E. GUS FRUH.

IN: EFFECTS OF WATERSHED CHANGES ON STREAMFLOW, WATER RESOURCES SYMPOSIUM NO
2, AUSTIN, TEXAS, OCTOBER 1968, P 255-282, UNIVERSITY OF TEXAS PRESS,
AUSTIN AND LONDON, 1969. 28 P, 22 FIG, 5 TAB, 15 REF. NSF GRANT GU-1963,
AND TEXAS WATER QUALITY BD CONTRACT NO. 68-69-281.

DESCRIPTORS:
*RESERVOIRS, *URBANIZATION, *WATER QUALITY, *STRATIFICATION, *TEXAS,
WATER POLLUTION SOURCES, WATER POLLUTION EFFECTS, DISSOLVED OXYGEN,
AQUATIC BACTERIA, ALGAE.

IDENTIFIERS:
AUSTIN(TEX), COLORADO RIVER(TEX).

ABSTRACT:
THE EFFECTS OF IMPOUNDMENTS AND URBANIZATION ON THE WATER QUALITY OF
THE COLORADO RIVER OF TEXAS WERE STUDIED IN THE RESERVOIRS NEAR AUSTIN,
TEXAS. LAKE TRAVIS, UPSTREAM FROM AUSTIN, IS LARGE AND DEEP AND HAS NO
SIGNIFICANT INPUT OF POLLUTION. LAKE AUSTIN, THE NEXT RESERVOIR
DOWNSTREAM, IS MUCH SMALLER AND SHALLOWER, AND RECEIVES SOME
RECREATIONAL AND URBAN RUNOFF POLLUTION. TOWN LAKE, IN AUSTIN, IS SMALL
AND NARROW WITH PREDOMINANTLY RIVER CHARACTERISTICS, AND RECEIVES SOME
URBAN RUNOFF POLLUTION. OXYGEN CONCENTRATION-WATER DEPTH DATA FOR THE 3
RESERVOIRS ARE TABULATED. IN LAKE TRAVIS, EVEN IN THE WINTER, THE
TEMPERATURE VARIED WITH DEPTH. DURING THE SUMMER, THE OXYGEN FIRST
BECAME DEPLETED AT THE THERMOCLINE REGION. THROUGHOUT THE FALL, VARIOUS
DEPTHS OF LAKE TRAVIS BECAME REAERATED, BUT OXYGEN-DEPLETED WATERS WERE
STILL PASSING THROUGH THE PENSTOCKS IN NOVEMBER. IN WINTER, OXYGEN WAS
PRESENT AT ALL DEPTHS. IN LAKE AUSTIN, THE SUMMER OXYGEN CONCENTRATION
IN THE EPILIMNION VARIED AROUND SATURATION. DISSOLVED OXYGEN DECREASED
STEADILY IN THE HYPOLIMNION DURING THE SUMMER AND WAS CONSISTENTLY
LOWER ABOVE THE SEDIMENTS. AFTER AUTUMN TURNOVER, OXYGEN REMAINED
UNIFORM FROM TOP TO BOTTOM. HIGH NUMBERS OF TOTAL AND COLIFORM BACTERIA
WERE FOUND IN LAKE AUSTIN DURING THE SPRING FOLLOWING PERIODS OF
INTENSIVE RAINFALL. THE URBAN STREAM, BARTON CREEK, HAD SIGNIFICANTLY
HIGHER CONCENTRATIONS OF SOLUTES, NUTRIENTS, AND BACTERIA THAN TOWN
LAKE ABOVE THE STREAM'S ENTRANCE, PARTICULARLY DURING THE SPRING
RAINFALL PERIOD. ALL OF AUSTIN'S URBAN STREAMS ENTER TOWN LAKE AND LAKE
AUSTIN, WITH THE STREAMS FROM THE MORE HIGHLY DEVELOPED AREAS ENTERING
TOWN LAKE. (KNAPP-USGS)

FIELD 04C, 05B

ACCESSION NO. W70-04727

BIOLOGICAL RESPONSES TO NUTRIENTS-EUTROPHICATION: PROBLEMS IN FRESHWATER,

TEXAS UNIV., AUSTIN.

E. GUS FRUH.

ADVANCES IN WATER QUALITY IMPROVEMENT, (EDITORS: GLOYNA, E. F. AND
ECKENFELDER, W. W., JR.), AUSTIN, TEXAS, UNIV OF TEXAS PRESS, 1968. P
49-64, 7 TAB, 82 REF.

DESCRIPTORS:
*EUTROPHICATION, *NITROGEN COMPOUNDS, *PHOSPHORUS COMPOUNDS,
*NUTRIENTS, *BIOASSAY, *RUNOFF, *LAKES, *IMPOUNDMENTS, TERTIARY
TREATMENT, ALGAE, DOMESTIC WASTES.

IDENTIFIERS:
*MICRONUTRIENTS, *NUTRIENT INTERACTIONS, *RADIOCARBON BIOASSAY,
*NUTRIENT TYPES, MICROQUANTITIES, MACROQUANTITIES, ALGAL BLOOMS,
LIEBIG'S LAW OF THE MINIMUM, VOISIN'S LAW OF THE MAXIMUM.

ABSTRACT:
THE PURPOSE OF THE PRESENTATION IS TO PROVIDE A FRAMEWORK OF REFERENCE
ON THE EUTROPHICATION PROBLEMS OF FRESHWATER. IMPORTANT INSIGHTS ON
EUTROPHICATION HAVE COME FROM THE FIELDS OF BOTANY, LIMNOLOGY, AND
SANITARY ENGINEERING. DISCUSSED ARE THE TYPES AND IMPACTS OF CRITICAL
NUTRIENTS. SPECIAL ATTENTION IS FOCUSED ON NITROGEN, AND PHOSPHORUS,
INCLUDING A DETAILED DISCUSSION OF SPECIFIC SOURCES, NAMELY RAINFALL,
GROUNDWATER, FOREST RUNOFF, AGRICULTURAL RUNOFF, URBAN DRAINAGE AND
WASTEWATER EFFLUENTS. ANALYSIS IS MADE OF THE RELATIONSHIP OF NUISANCE
CONDITIONS TO NITROGEN AND PHOSPHORUS. EXAMINATION IS MADE OF THE ROLE
OF ORGANIC AND INORGANIC MICRONUTRIENTS; THE NATURE OF NUTRIENT
INTERACTIONS WITH VARIOUS PHYSICAL, CHEMICAL, AND BIOLOGICAL VARIABLES.
VARIOUS APPROACHES TO CONTROL EUTROPHICATION ARE REPORTED. CONCLUSIONS
ARE REACHED THAT OTHER THAN TERTIARY TREATMENT, DIVERSION OF THE
NUTRIENT-RICH SOURCES APPEARS THE MOST FEASIBLE EUTROPHICATION CONTROL.
HOWEVER, LEACHING OR BACTERIAL DECOMPOSITION OF NUTRIENTS FROM THE
SEDIMENTS SEEMS THE LIMITING FACTOR IN LAKE RECOVERY RATES. MANY
PROCESSES IN AQUATIC ENVIRONMENTS ARE ATTRIBUTED TO BACTERIA, HOWEVER,
CURRENT ANALYTICAL TECHNIQUES ARE INADEQUATE FOR THE QUANTITATIVE
EVALUATION OF BACTERIAL SPECIES AND NUMBER.
(D'AREZZO-TEXAS)

FIELD 05D

ACCESSION NO. W70-04765

BASIS FOR WASTE STABILIZATION POND DESIGNS,

TEXAS UNIV., AUSTIN.

EARNEST F. GLOYNA.

ADVANCES IN WATER QUALITY IMPROVEMENT, (EDITORS: GLOYNA, E. F., AND
  ECKENFELDER, W. W., JR.), AUSTIN, TEXAS, UNIV OF TEXAS PRESS, 1968. P
  397-408, 1 FIG, 1 TAB, 19 REF.

  DESCRIPTORS:
    *WASTE TREATMENT, *STABILIZATION, *PONDS, *DESIGN CRITERIA, FACILITIES,
    EQUATIONS, AEROBIC CONDITIONS, ANAEROBIC CONDITIONS, ALGAE, ALGAL
    CONTROL, ECOLOGICAL DISTRIBUTION, ACTIVATED SLUDGE, MAINTENANCE COSTS,
    CONSTRUCTION COSTS.

  IDENTIFIERS:
    *WASTE STABILIZATION PONDS, *ALGAL PHYSIOLOGY, *DESIGN THEORY AND
    CALCULATIONS, PHOTOSYNTHESIS, ILLUMINATION, TEMPERATURE, NUTRIENTS,
    FACULTATIVE POND DESIGNS, BOD, SLUDGE.

  ABSTRACT:
    MANY PROBLEMS ARISING FROM USE OF WASTE STABILIZATION PONDS ARE
    TRACEABLE TO INADEQUATE ENGINEERING, POOR MAINTENANCE AND LACK OF
    OPERATIONAL SUPERVISION. HOWEVER, THERE IS A MORE FUNDAMENTAL NEED:
    DESIGN AND ENGINEERING FOR SYSTEMS USING STABILIZATION PONDS MUST BE
    BASED ON AN UNDERSTANDING OF THE PRINCIPLES OF ALGAL PHYSIOLOGY.
    GENERALLY, ALGAE GROWING IN A STREAM OR A WASTE STABILIZATION POND ARE
    IN A HIGHLY COMPETITIVE ENVIRONMENT. ONE OR MORE OF THE FACTORS
    NECESSARY FOR PHOTOSYNTHESIS ARE INTERRELATED, AND UNICELLULAR ALGAE,
    IN PARTICULAR WILL REACT RAPIDLY TO ENVIRONMENTAL CHANGES.
    ILLUMINATION, TEMPERATURE, AND NUTRIENTS--THE FACTORS WHICH AFFECT
    GROWTH RATE ARE EXAMINED CRITICALLY. THE PHENOMENA OF ALGAL PHYSIOLOGY,
    TOGETHER WITH ACTUAL TREATMENT OBJECTIVES ARE SYNTHESIZED IN GOOD
    DESIGN. IN THE FINAL ANALYSIS, THE SPECIFIC DESIGN OF A WASTE
    STABILIZATION POND DEPENDS HEAVILY UPON THE SPECIFICALLY-IMPOSED
    TREATMENT OBJECTIVES AND REQUIREMENTS. A POND SYSTEM MAY BE DESIGNED TO
    RECEIVE UNTREATED SEWAGE OR INDUSTRIAL WASTES, PRIMARY OR SECONDARY
    TREATMENT EFFLUENTS OR EXCESS ACTIVATED SLUDGES AND SETTLEABLE SOLIDS.
    CERTAIN CRITICAL FACTORS IN DESIGN DEVELOPMENT ARE DISCUSSED, BASED ON
    OPERATIONAL EXPERIENCES AND RESEARCH WHICH PROVIDE VITAL INSIGHTS INTO
    ENVIRONMENTAL FACTORS GOVERNING WASTE STABILIZATION PROCESSES.
    (D'AREZZO-TEXAS)

  FIELD 05D

  ACCESSION NO. W70-04786

ADVANCES IN ANAEROBIC POND SYSTEMS DESIGN,

CALIFORNIA UNIV., BERKELEY.

W. J. OSWALD.

ADVANCES IN WATER QUALITY IMPROVEMENT, (EDITORS: GLOYNA, E. F., AND
ECKENFELDER, W. W., JR.), AUSTIN, TEXAS, UNIV OF TEXAS PRESS, 1968. P
409-426, 6 FIG, 2 TAB, 11 REF.

DESCRIPTORS:
*WASTE TREATMENT, *STABILIZATION, *PONDS, *ENVIRONMENTAL EFFECTS,
*BIOLOGICAL TREATMENT, *DESIGN CRITERIA, DISSOLVED OXYGEN, TEMPERATURE,
BIOCHEMICAL OXYGEN DEMAND(BOD), EVAPORATION, FERMENTATION, METHANE,
ALGAL CONTROL, BACTERIA, COLIFORMS.

IDENTIFIERS:
*ANAEROBIC WASTE PONDS, *DEPTH EFFECTS, *HYDRAULIC LOAD EFFECTS,
*ORGANIC LOAD EFFECTS, *DETENTION TIME EFFECTS, *FACULATIVE PONDS,
*AEROBIC PONDS, *ANAEROBIC PONDS, DESIGN FOR ENVIRONMENTAL CONTROL,
RECIRCULATION, BOD REMOVAL, GAS PRODUCTION.

ABSTRACT:
THE BIOLOGICAL ACTIONS AND INTERACTIONS IN WASTE STABILIZATION PONDS
HAVE BEEN COMPLEX ENOUGH TO DETER SCIENTIFIC INVESTIGATION AND THUS TO
FOSTER THE WIDESPREAD USE OF EMPIRICAL DESIGN FORMULATIONS WHICH ARE
BASED ON LIMITED DATA. RECOGNITION OF THE FACT THAT SEDIMENTATION AND
FOUR BASIC BIOLOGICAL REACTIONS INTERACT IN PONDS, MAKES POSSIBLE A
RATIONAL DESIGN OF PONDS. PRECISE FORMULAS HAVE NOT YET BEEN DEVELOPED
GOVERNING ALL VARIABLES. HOWEVER, THE RELATIONSHIPS BETWEEN TEMPERATURE
AND METHANE FERMENTATION, DEPTH AND TEMPERATURE, AND BETWEEN DEPTH AND
OXYGENATION ARE WELL-ESTABLISHED. METHANE FERMENTATION IS THE S.INE QUA
NON FOR AEROBIC AND FACULTATIVE PONDS; THEREFORE, DESIGNS MAXIMIZING
TEMPERATURE UNIFORMITY AND MINIMIZING OXYGEN INTRUSION ARE MOST
SUCCESSFUL. IN SERIES PONDS, WHICH RECYCLE OXYGENATED WATER, THE
PREVENTION OF ODOR IN THE INITIAL ANAEROBIC POND IS DONE EASILY BY
OVERLAYING THE PRIMARY ANAEROBIC POND WITH OXYGENATED WATER. FOR
DESTRUCTION OF COLIFORM BACTERIA, BOTH TIME AND TEMPERATURE ARE VITAL
FACTORS. AT NORMAL POND TEMPERATURES, OBTAINING TIME INCREMENTS BY
DEPTH INCREASES IS EASIER THAN GETTING TEMPERATURE INCREMENTS BY DEPTH
DECREASES. WASTES HAVING HIGH CARBON/NITROGEN RATIOS MUST BE OXIDIZED
AND AERATED BEFORE FERMENTATION. (D'AREZZO-TEXAS)

FIELD 05D

ACCESSION NO. W70-04787

WASTE STABILIZATION POND PRACTICES IN THE UNITED STATES,

FEDERAL WATER POLLUTION CONTROL ADMINISTRATION, DALLAS, TEX.

JEROME H. SVORE.

ADVANCES IN WATER QUALITY IMPROVEMENT, (EDITORS: GLOYNA, E. F., AND
ECKENFELDER, W. W., JR.), AUSTIN, TEXAS, UNIV OF TEXAS PRESS, 1968. P
427-434, 2 FIG, 2 TAB, 8 REF, 1 PLATE.

DESCRIPTORS:
*STABILIZATION, *PONDS, *DESIGN, *EFFICIENCIES, *CHLORINATION, *ODOR,
*WASTE TREATMENT, ALGAE, DISINFECTION, BIOCHEMICAL OXYGEN DEMAND(BOD),
AERATION, COLIFORMS, ALGAL CONTROL, DISSOLVED OXYGEN.

IDENTIFIERS:
*AERATED LAGOONS, *MIDWEST(US), *SOUTHWEST(US), *DESIGN CRITERIA, POND
EFFLUENT, COLORADO, WEST(US), STABILIZATION POND LOADINGS, ODOR
PERSISTENCE.

ABSTRACT:
THE DESIGN, OPERATIONS, AND GRADUAL ACCEPTANCE OF WASTE STABILIZATION
PONDS IN THE SOUTHWEST DURING THE PAST 25 YEARS IS TRACED. RECENT
SURVEYS IN COLORADO HAVE SUBSTANTIATED GENERAL EXPERIENCE IN THE
MIDWEST, CLASSIFYING THE PREDOMINANT DEFECTS IN DESIGN, CONSTRUCTION,
AND OPERATION INTO THREE CLASSES: (A) LIQUID DEPTH NOT MAINTAINED, (B)
WATER SURFACE INADEQUATE, AND (C) PRIMARY SOLIDS NOT EVENLY
DISTRIBUTED. A COMPARISON IS MADE OF CURRENT METHODS USED THROUGHOUT
THE SOUTHERN HALF OF THE UNITED STATES REGARDING THE EXCESSIVE ALGAE
PROBLEM. THE DETERMINATION OF LAGOON TREATMENT EFFICIENCY MUST BE MADE
FROM THE VIEWPOINT OF REGIONAL WATER PRIORITIES AND OBJECTIVES. THE
REDUCTION IN REQUIRED CHLORINE DISINFECTION USAGE RESULTING FROM
EFFICIENT POND OPERATION IS DISCUSSED. SOME REASONS FOR THE TREND
TOWARD GREATER USE OF AERATED LAGOONS ARE PRESENTED. WHILE THERE ARE
STILL ISOLATED AREAS WHERE OPINION EXISTS THAT WASTE STABILIZATION
PONDS ARE DANGEROUS, UNDESIRABLE AND, A HEALTH HAZARD, THE MAJOR TREND
IS TOWARD RECOGNITION THAT IF PONDS ARE DESIGNED AND OPERATED PROPERLY
(IN SERIES), THEY WILL PROVIDE MORE TREATMENT PER UNIT COST THAN ANY
OTHER SINGLE PROCESS. (D'AREZZO-TEXAS)

FIELD 05D

ACCESSION NO. W70-04788

EXTRA-DEEP PONDS,

TECHNION - ISRAEL INST. OF TECH., HAIFA; AND WATER PLANNING FOR ISRAEL LTD.,
TEL AVIV.

ALBERTO M. WACHS, AND ANDRE BEREND.

ADVANCES IN WATER QUALITY IMPROVEMENT, (EDITORS GLOYNA, E. F., AND
ECKENFELDER, W. W., JR.), AUSTIN, TEXAS, UNIV OF TEXAS PRESS, 1968. P
450-456, 5 FIG, 1 TAB, 9 REF.

DESCRIPTORS:
*PONDS, *STABILIZATION, *THERMAL STRATIFICATION, *SEASONAL, *SULFIDES,
TEMPERATURE, CLIMATIC DATA, STRATIFICATION, DISSOLVED OXYGEN, ALGAE,
BIOCHEMICAL OXYGEN DEMAND(BOD), MIXING, WINDS.

IDENTIFIERS:
*THERMAL GRADIENT, *BOD REMOVAL, *ISRAEL, *ALGAL CONCENTRATION,
'EXTRA-DEEP PONDS', ASHKELON(ISRAEL), TEMPERATURE PROFILES, VOLATILE
SUSPENDED SOLIDS, MULTI-CELL SERIAL SYSTEM, PH, FACULTATIVE POND,
ANAEROBIC POND.

ABSTRACT:
AN EXPERIMENTAL UNIT WAS BUILT BY TAHAL, LTD, NEAR THE WASTEWATER
TREATMENT FACILITY, ASHKELON, ISRAEL, TO OBTAIN PILOT PLANT PERFORMANCE
DATA ON 'EXTRA-DEEP' WASTE STABILIZATION PONDS, (I.E., DEPTH GREATER
THAN 8 FEET). THE EXPERIENCE GAINED SHOWS THAT WHEN LOADINGS ACCEPTABLE
TO THE ISRAEL CLIMATE WERE USED DURING THE WARM SEASON, A PRONOUNCED
THERMAL GRADIENT CAUSED THE UPPER PART OF THE POND TO BEHAVE AS A
'SEPARATE' FACULTATIVE POND 'FLOATING' ON AN UNDERLYING ANAEROBIC POND.
THIS STRATIFICATION WAS ENHANCED BY THE FACT THAT BOTH INLET AND OUTLET
WERE LOCATED AT THE SURFACE. EFFLUENT WAS RICH IN ALGAE. DURING THE
WARM SEASON, ALGAL CONCENTRATIONS WERE MUCH LOWER AT OR BELOW DEPTHS OF
1 METER; THEREFORE, PLACING OF OUTLET AT THAT DEPTH WOULD DRAW OFF
BETTER EFFLUENT INSOFAR AS ALGAL DENSITY WAS CONCERNED, BUT IT WAS
DEFICIENT IN DISSOLVED OXYGEN AT TIMES. THE THERMAL GRADIENT
DISAPPEARED DURING THE COLD SEASON. MIXING WAS DUE MAINLY TO WIND
ACTION AND INFLUENCED BY POND GEOMETRY. DISSOLVED OXYGEN WAS NOT ALWAYS
PRESENT IN THE UPPER POND LAYERS BUT IT DID CONTAIN CONSIDERABLE ALGAE.
DURING FALL, A POND TURNOVER OCCURRED RESULTING IN MIXING OF LOWER AND
UPPER LAYERS. THIS CAUSED A SURFACE INCREASE OF SULFIDES CONCENTRATION,
A DROP IN PH, BUT NO ODORS. DURING BOTH COLD AND WARM SEASONS, BOD
REMOVAL WAS ABOUT 80%. (D'AREZZO-TEXAS)

FIELD 05D

ACCESSION NO. W70-04790

THE ADAPTATION OF PLANKTON ALGAE. III. WITH SPECIAL CONSIDERATION OF THE
   IMPORTANCE IN NATURE,

ROYAL DANISH SCHOOL OF PHARMACY, COPENHAGEN. DEPT. OF BOTANY.

E. STEEMANN NIELSEN, AND ERIK G. JORGENSEN.

PHYSIOLOGIA PLANTARUM, VOL 21, NO 3, P 647-654, 1968. 2 TAB, 20 REF.

   DESCRIPTORS:
      *PLANKTON, *ALGAE, LIGHT INTENSITY, TEMPERATURE, PRIMARY PRODUCTIVITY,
      SURFACES, ARCTIC, ENZYMES, CHLOROPHYLL, PHOTOSYNTHESIS, CHLORELLA,
      GRAZING, ZOOPLANKTON, DIATOMS, RESPIRATION, PIGMENTS, OCEANS, ORGANIC
      MATTER, DIURNAL, LATITUDINAL STUDIES.

   IDENTIFIERS:
      PHOTOOXIDATION, CHLORELLA VULGARIS, CHLORELLA VULGARIS(COLUMBIA),
      CHLORELLA PYRENOIDOSA, CHLORELLA SACCHAROPHILA, CHLORELLA ELLIPSOIDEA,
      CHLORELLA LUTEOVIRIDIS, SKELETONEMA COSTATUM, PHYSIOLOGICAL ADJUSTMENT,
      ASTERIONELLA FORMOSA, CYCLOTELLA MENEGHINIANA, PHOTIC ZONE,
      HELSINGOR(DENMARK), SALINO-CLINE.

   ABSTRACT:
      ADAPTATION (PHYSIOLOGICAL ADJUSTMANT) OF PLANKTON ALGAE, IN NATURAL
      STATES, TO DIFFERENT LIGHT INTENSITIES AND TEMPERATURES IS DESCRIBED.
      ALTHOUGH DIFFERENT SPECIES VARY IN EXTENT OF ADAPTATION, THEY ALL ADAPT
      TO SOME DEGREE, AND ADAPTATION IS SIGNIFICANT FOR PLANKTON POPULATIONS
      AT ALL DEPTHS. SHADE ADAPTATION OF ALGAE IN THE LOWER PART OF THE
      PHOTIC ZONE IS OFTEN OF MINOR IMPORTANCE TO THE INTEGRAL PRIMARY
      PRODUCTION PER UNIT OF SURFACE, EXCEPT ON OCCASIONS WHEN THE BULK OF
      THE ALGAE IS FOUND IN THIS LOWER PART. IT IS SHOWN THAT THE I-SUB-K (A
      LIGHT INTENSITY DEFINING ONSET OF LIGHT-SATURATED PHOTOSYNTHESIS) OF
      SURFACE PLANKTON DURING SUMMER IN THE ARCTIC IS HIGH DESPITE LOW
      TEMPERATURES DUE TO INCREASE OF THE ENZYME QUANTITIES PER CELL. DAILY
      FLUCTUATIONS IN I-SUB-K ARE PARTIALLY DUE TO PHOTOOXIDATION AND
      PARTIALLY BECAUSE THE PERIODS FOR THE PRODUCTION OF CHLOROPHYLL AND
      PHOTOSYNTHETIC ENZYMES ARE MUTUALLY DISPLACED. DESPITE DAILY
      FLUCTUATIONS IN I-SUB-K, THE DIFFERENCE BETWEEN 'SUN' AND 'SHADE'
      PHYTOPLANKTON IS DISTINCT. THE DIURNAL VARIATION IN PHOTOSYNTHETIC
      ACTIVITY OF PHYTOPLANKTON MAY TAKE PLACE ESPECIALLY AT LOW LATITUDES,
      AND NEAR THE SURFACE; AND TO SOME EXTENT AT HIGHER LATITUDES, AND IN
      LOWER PHOTIC ZONE. (JONES-WISCONSIN)

   FIELD 02H, 05C

   ACCESSION NO. W70-04809

206

EUTROPHICATION TESTING PROGRAM FOR MINNESOTA'S 'SKY BLUE WATERS',

SCHOELL AND MADSON, INC., HOPKINS, MINN.

JAMES R. ORR.

WATER AND SEWAGE WORKS, VOL 115, NO 3, P 97-101, 1968. 2 FIG, 2 TAB, 10 REF.

DESCRIPTORS:
*EUTROPHICATION, *MINNESOTA, RECREATION, SEWAGE DISPOSAL, WATER
POLLUTION, POLLUTION ABATEMENT, COLIFORMS, WATER QUALITY, PHOSPHORUS,
NITROGEN, FLOW, CHEMICAL ANALYSIS, NUTRIENTS, EFFLUENTS, ALGAE, WEEDS,
CARBON DIOXIDE, VITAMINS, NITRATES, INORGANIC COMPOUNDS, SURFACE
RUNOFF, UNDERFLOW, PRECIPITATION(ATMOSPHERIC), DRAINAGE, FERTILIZERS,
FARM WASTES, SOIL EROSION, FROZEN GROUND.

IDENTIFIERS:
LAKE MINNETONKA(MINN), MINNEAPOLIS-ST PAUL(MINN).

ABSTRACT:
THE GREATEST SOURCE OF NUTRIENTS TO LAKE MINNETONKA, ONE OF MINNESOTA'S
BEST KNOWN LAKES, IS SEWAGE TREATMENT PLANT EFFLUENT. A MAJOR PART OF
THE PROBLEM WOULD BE SOLVED BY REMOVING NUTRIENTS FROM EFFLUENTS OR
SUPPRESSING SEWAGE EFFLUENT DISCHARGE. LOSS OF LARGE QUANTITIES OF
WASTEWATER OUTSIDE THE WATERSHED COULD AGGRAVATE THE PROBLEM OF
MAINTAINING LAKE LEVEL. ANOTHER APPROACH MAY BE CONTROL OF EXCESSIVE
WEED AND ALGAE PRODUCTION BY CHEMICAL TREATMENT OR MECHANICAL
HARVESTING, HOWEVER, CONTINUED USE OF ALGICIDES MAY EVENTUALLY HAVE A
TOXIC EFFECT ON FISH AND OTHER AQUATIC LIFE; MECHANICAL HARVESTING IS
IMPRACTICAL. ADVANCED WASTE TREATMENT OF SEWAGE DISCHARGED TO LAKE,
WITH EMPHASIS ON REMOVAL OF PHOSPHORUS SHOULD BE CONSIDERED.
MISCELLANEOUS NUTRIENT SOURCES, OTHER THAN SEWAGE PLANT EFFLUENT,
REPRESENT APPROXIMATELY 2/3 TOTAL NITROGEN AND APPROXIMATELY 1/5 TOTAL
PHOSPHORUS INFLOW. WATERSHED DWELLERS CAN MINIMIZE THESE SOURCES BY
PREVENTING DECAYING VEGETATION BEING WASHED INTO THE LAKE BY RAINFALL,
DISCONTINUING COMMERCIAL AND MANURE FERTILIZER USAGE, AND PRACTICING
SOIL EROSION CONTROL. MUNICIPALITIES CAN ASSIST BY KEEPING PUBLIC LANDS
CLEAN, LOCATING WASTE DUMPS TO AVERT DRAINAGE INTO THE LAKE,
ELIMINATING SOIL ABSORPTION SEWAGE DISPOSAL FACILITIES ADJACENT TO
LAKE, AND BY PUBLICIZING THE PROBLEM. (JONES-WISCONSIN)

FIELD 02H, 05C

ACCESSION NO. W70-04810

DISTINCTION BETWEEN BACTERIAL AND ALGAL UTILIZATION OF SOLUBLE SUBSTANCES IN THE SEA,

MARINE LAB., ABERDEEN (SCOTLAND); AND INDIANA UNIV., INDIANAPOLIS. DEPT. OF MICROBIOLOGY.

A. L. S. MUNRO, AND T. D. BROCK.

JOURNAL GENERAL MICROBIOLOGY, VOL 51, P 35-42, 1968. 3 FIG, 1 TAB, 14 REF, 3 PLATES.

DESCRIPTORS:
*AQUATIC BACTERIA, *ALGAE, *ANALYTICAL TECHNIQUES, *MARINE MICROORGANISMS, PHYSIOLOGICAL ECOLOGY, ORGANIC MATTER, WATER POLLUTION EFFECTS, DIATOMS, CARBON RADIOISOTOPES, KINETICS, TRITIUM, CHLOROPHYLL.

IDENTIFIERS:
*HETEROTROPHY, *ORGANIC SOLUTES, *MARINE ENVIRONMENTS, MICROBIAL ECOLOGY, FLUORESCENCE MICROSCOPY, ACETATE, CARBON-14, AUTORADIOGRAPHY, MICHAELIS-MENTEN EQUATION, LOCH EWE(SCOTLAND), SCOTLAND, PHAEOPIGMENTS, ACRIDINE ORANGE, LINEWEAVER-BURKE EQUATION.

ABSTRACT:
STAINING WITH ACRIDINE ORANGE AND DIRECT OBSERVATION WITH FLUORESCENCE MICROSCOPY DEMONSTRATES NUMEROUS VIABLE BACTERIA AND DIATOMS ATTACHED TO SAND GRAINS OF A LITTORAL BEACH TO DEPTHS EXCEEDING 10 CENTIMETERS. KINETIC ANALYSIS OF UPTAKE OF RADIOCARBON-LABELLED ACETATE OVER RANGE, 10-5000 MICROGRAMS/LITER, DEMONSTRATES TWO UPTAKE MECHANISMS: ONE, SATURABLE AT LOWER CONCENTRATIONS (MICHAELIS-MENTEN KINETICS), AND ANOTHER, DESCRIBABLE AS DIFFUSION, THE LATTER HAVING BEEN ATTRIBUTED BY OTHERS TO ALGAL UPTAKE. AUTORADIOGRAPHY, HOWEVER, REVEALS THAT BACTERIA ALONE ARE RESPONSIBLE FOR UPTAKE OF TRITIATED ACETATE. IN THOSE SITUATIONS WHERE MICHAELIS-MENTEN KINETICS APPLIED, MAXIMUM ATTAINABLE UPTAKE RATES (V) WERE MUCH GREATER THAN ANY PREVIOUSLY REPORTED FOR WATER, SUCH VALUES FOR V BEING ATTRIBUTABLE TO LARGER MICROBIAL POPULATIONS IN SAND. SINCE REASONS FOR EXISTENCE OF TWO TYPES OF KINETICS ARE OBSCURE, IT IS DANGEROUS TO MAKE CONCLUSIONS REGARDING MICROBIAL HETEROTROPHY BASED ON KINETIC ANALYSIS ALONE. BASED ON THESE STUDIES, IT SEEMS CLEAR THAT ALGAL HETEROTROPHY IS NEGLIGIBLE IN SEA WATERS, AND THAT AUTORADIOGRAPHY IS AN EFFECTIVE TOOL IN STUDIES OF ENERGETICS AND TROPHIC RELATIONSHIPS IN MICROBIAL ECOLOGY. (EICHHORN-WISCONSIN)

FIELD 05C

ACCESSION NO. W70-04811

EUTROPHICATION AND DETERGENTS,

SOUTHAM BUSINESS PUBLICATIONS LTD., DON MILLS (ONTARIO). READER SERVICE DEPT; AND TORONTO UNIV. (ONTARIO).

P. H. JONES, AND TOM DAVEY.

WATER AND POLLUTION CONTROL, VOL 106, NO 9, P 22-25, 46, SEPTEMBER 1968.

DESCRIPTORS:
*EUTROPHICATION, PHOSPHORUS, *DETERGENTS, DOMESTIC WASTES, POLLUTANTS, WATER POLLUTION SOURCES, WATER QUALITY, ALGAE, SEWAGE EFFLUENTS.

IDENTIFIERS:
POLYPHOSPHATES, NUTRIENT BALANCE, FOAMING DETERGENTS, PUBLIC APPREHENSION.

ABSTRACT:
DETERGENTS ARE ATTACKED AS A MAJOR NUTRIENT SOURCE IN EUTROPHICATION. SOLUBLE POLYPHOSPHATES SERVING AS FILLER IN DETERGENTS HELP UPSET NUTRIENT BALANCES IN LAKES AND STIMULATE ALGAL GROWTH. OTHER SIGNIFICANT SOURCES OF BIOLOGICALLY ACTIVE PHOSPHORUS ARE ANIMAL AND HUMAN WASTE. BECAUSE MOST DETERGENTS CONTAIN 50% POLYPHOSPHATES, SEWAGE EFFLUENT PHOSPHATE HAS DOUBLED OR TRIPLED. SUBSTITUTE FILLERS FOR POLYPHOSPHATES, WHICH CAN ACT AS CHELATING AGENTS AND CONTROL PH, ARE NOW AVAILABLE. REPLACING PHOSPHORUS IN DETERGENTS WOULD HAVE AN INSTANTANEOUS BENEFICIAL EFFECT AT A COST OF AN EXTRA 15-20 CENTS A BOX. PUBLIC PRESSURE COULD FORCE THE DESIRED CHANGE. EXCESS PHOSPHORUS ACCELERATES A LAKE'S NATURAL AGING PROCESS. (BANNERMAN-WISCONSIN)

FIELD 05B

ACCESSION NO. W70-05084

THE INFLUENCE OF NITROGEN ON HETEROCYST PRODUCTION IN BLUE-GREEN ALGAE,

BUREAU OF COMMERCIAL FISHERIES, ANN ARBOR, MICH. BIOLOGICAL LAB.

ROANN E. OGAWA, AND JOHN F. CARR.

LIMNOLOGY AND OCEANOGRAPHY, VOL 14, NO 3, P 342-351, 1969. 7 FIG, 4 TAB, 26 REF.

DESCRIPTORS:
  *ALGAE, *CYANOPHYTA, *NITROGEN FIXATION, NUTRIENTS, LAKE ERIE.

IDENTIFIERS:
  *HETEROCYSTS, ANABAENA VARIABILIS, MICROCYSTIS AERUGINOSA,
  APHANIZOMENON FLOS-AQUAE, OSCILLATORIA, ANABAENA CYLINDRICA B629,
  ANABAENA INAEQUALIS 381, ANABAENA FLOS AQUAE, TOLYPOTHRIX DISTORTA,
  GLOEOTRICHIA ECHINULATA LB 1303.

ABSTRACT:
  INDIRECT EVIDENCE IS PRESENTED SUGGESTING INVOLVEMENT OF HETEROCYSTS IN
  UTILIZATION OF ATMOSPHERIC NITROGEN AS SOLE NITROGEN SOURCE FOR ALGAE.
  SEVEN HETEROCYSTOUS (KNOWN ATMOSPHERIC NITROGEN-FIXING) BLUE-GREEN
  ALGAE WERE GROWN IN A MODIFIED CHU NO 10 MEDIUM DEVOID OF COMBINED
  NITROGEN; TWO NON-HETEROCYSTOUS (NON-ATMOSPHERIC-FIXING) BLUE-GREENS
  DID NOT GROW IN THIS MEDIUM. HETEROCYSTS WERE PRODUCED SHORTLY AFTER
  INOCULA DEVOID OF HETEROCYSTS WERE PLACED IN NITORGEN-FREE MEDIUM.
  PRODUCTION WAS GREATEST WHEN ATMOSPHERIC NITORGEN SERVED AS SOLE
  NITROGEN SOURCE, AND LEAST WHEN AMMONIA-NITROGEN SERVED AS SOLE
  NITROGEN-SOURCE. NITRATE-NITROGEN PRODUCED AN INTERMEDIATE NUMBER OF
  HETEROCYSTS. WHEN MEDIUM NITRATE-NITROGEN CONTENT WAS VARIED AS SOLE
  NITROGEN SOURCE, NUMBERS OF HETEROCYSTS PRODUCED WERE INVERSELY
  PROPORTIONAL TO THE NITROGEN CONCENTRATION. THEY DID NOT DEVELOP IN THE
  ABSENCE OF PHOSPHORUS, BUT NO EFFORT WAS MADE TO DETERMINE THE CRITICAL
  PHOSPHORUS CONCENTRATION FOR THEIR PRODUCTION. PRODUCTION OF
  HETEROCYSTS IN CIRCUMSTANCES OF LOW COMBINED NITROGEN MAY PROVIDE AN
  ECOLOGICAL ADVANTAGE. THE RELATIVE NUMBERS OF HETEROCYSTS IN
  FIELD-COLLECTED SAMPLES INDICATE THE RELATIVE AMOUNT OF AVAILABLE
  NITROGEN AND A CONTINUING SUPPLY OF AVAILABLE PHOSPHORUS.
  (GERHOLD-WISCONSIN)

FIELD 05C

ACCESSION NO. W70-05091

PHYTOPLANKTON AS AN AGENT OF THE SELF-PURIFICATION OF CONTAMINATED WATERS,

FOREIGN TECHNOLOGY DIV., WRIGHT-PATTERSON AFB, OHIO.

G. G. VINBERG, AND T. N. SIVKE.

AVAILABLE FROM THE CLEARINGHOUSE AS TT-67-62992, AND AD-659 310, $3.00 IN
PAPER COPY, $0.65 IN MICROFICHE. TRANSLATION BY FOREIGN TECHNOLOGY
DIVISION, WRIGHT-PATTERSON AIR FORCE BASE, FTD-MT-66-13, AUGUST 18, 1967,
37 P. TRANS. FROM TRUDY VSESOYUZNOGO GIDROBIOLOGICHESKOGO OBSHCHESTVA, VOL
7, P 3-23, 1956.

DESCRIPTORS:
  *PHYTOPLANKTON, *SELF-PURIFICATION, *WASTE WATER(POLLUTION), *WATER
  PURIFICATION, ALGAE, SEWAGE, ANAEROBIC, OXYGEN, NITRIFICATION,
  COLIFORMS, PONDS, CLIMATES, PHOTOSYNTHESIS, PRODUCTIVITY, SYMBIOSIS,
  RESERVOIRS, LAKES, REAERATION, SESTON, CHLOROPHYLL, CYCLES, UNITED
  STATES, CHLORELLA, BIOCHEMICAL OXYGEN DEMAND, OXIDATION LAGOONS,
  CALIFORNIA, TEXAS, DEPTH, RETENTION, NORTH DAKOTA, ICE, SNOW, CARP,
  SALMON, IRRIGATION, FILTRATION, ORGANIC MATTER, AMMONIA, HYDROGEN ION
  CONCENTRATION, ANTIBIOTICS, SYMBIOSIS, EUGLENA, LIGHT INTENSITY,
  NUTRIENTS, TEMPERATURE.

IDENTIFIERS:
  PISCICULTURAL PONDS, SVISLOCH' RIVER(USSR), MINSK(USSR), ENGLAND,
  OSCILLATORIA, AUSTRALIA, SANTA ROSA(CALIF), LYUBLIN(USSR),
  LYUBERETS(USSR), MUNICH(GERMANY), CHLORELLA PYRENOIDOSA, BICHOROMATE,
  DRY WEIGHT, B COLI COMMUNAE, EUGLENE GRACILIS, PROTOCOCCALES.

ABSTRACT:
  PLANKTONIC ALGAE GROWN ON A LARGE SCALE ON UNDILUTED CITY SEWAGE
  SHARPLY ACCELERATES SELF-PURIFICATION PROCESSES; IT IS EXPRESSED IN
  FASTER INITIAL LOWERING OF THE BIOCHEMICAL OXYGEN DEMAND, EARLY
  TERMINATION OF THE ANAEROBIC PHASE, APPEARANCE OF FREE OXYGEN, AND IN
  THE ACCELERATED ONSET OF NITRIFICATION. ACCUMULATION OF A LARGE
  QUANTITY OF ORGANIC SUBSTANCES SYNTHESIZED BY THE ALGAE IN THE
  COMPOSITION OF THE BODIES OF THE LIVING CELLS IS NOT REFLECTED IN THE
  MAGNITUDE OF THE BIOCHEMICAL OXYGEN DEMAND. THE DEATH RATE OF COLIFORM
  BACTERIA INCREASED SHARPLY WITH THE LARGE-SCALE DEVELOPMENT OF GREEN
  ALGAE IN SELF-CLEANING SEWAGE. EUGLENA AND CHLORELLA ARE COMPARED.
  PONDS, FILLED WITH UNDILUTED SEWAGE, ARE THE SIMPLEST METHOD TO USE
  GREEN ORGANISMS AS AGENTS OF SELF-PURIFICATION AND IS AN ESPECIALLY
  EFFECTIVE METHOD OF WATER RECLAMATION, PARTICULARLY APPLICABLE TO WARM,
  DRY, CLIMATIC REGIONS. FURTHER STUDY OF FAVORABLE CONDITIONS FOR
  DEVELOPMENT OF PHOTOSYNTHESIZING PLANKTON ORGANISMS WILL ALLEVIATE
  PROBLEMS IN LARGE-SCALE CULTIVATION OF ALGAE AS A METHOD OF UTILIZING
  SEWAGE AND PARTICULARLY SHOULD CLARIFY THE APPLICABILITY OF USING ALGAE
  GROWN ON SEWAGE TO INCREASE THE PRODUCTIVITY OF PISCICULTURAL PONDS.
  (JONES-WISCONSIN)

FIELD 05F

ACCESSION NO. W70-05092

MEASUREMENT AND DETECTION OF EUTROPHICATION,

CENTRAL COASTAL REGIONAL WATER QUALITY CONTROL BOARD, SAN LUIS OBISPO, CALIF.

THOMAS E. BAILEY.

JOURNAL OF THE SANITARY ENGINEERING DIVISION, PROCEEDINGS OF THE AMERICAN
  SOCIETY OF CIVIL ENGINEERS, VOL 93, NO SA6, P 121-132, 1967. 6 FIG.

DESCRIPTORS:
  *CORRELATION ANALYSIS, *FLUOROMETRY, *FLUORESCENCE, *CHLOROPHYLL,
  *EUTROPHICATION, PRIMARY PRODUCTIVITY, PHYTOPLANKTON, PHOTOSYNTHESIS,
  BIOMASS, CULTURES, PLANT PHYSIOLOGY, ALGAE, DIURNAL DISTRIBUTION,
  OXYGEN, DISSOLVED OXYGEN, MEASUREMENT, ESTUARIES, SANITARY ENGINEERING,
  WATER TREATMENT, WASTE TREATMENT.

IDENTIFIERS:
  SACRAMENTO-SAN JOAQUIN DELTA, CALIFORNIA WATER PLAN, DELTA WATER
  FACILITIES(CALIF), OXYGEN DYNAMICS, SACRAMENTO-SAN JOAQUIN ESTUARY,
  ENVIRONMENTAL RELATIONSHIPS, FLUOROMETRIC CHLOROPHYLL TECHNIQUE.

ABSTRACT:
  PRIMARY PRODUCTIVITY, AN IMPORTANT INDICATOR OF EUTROPHICATION, WAS
  MEASURED INDIRECTLY AS CHLOROPHYLL CONCENTRATION OF PHYTOPLANKTON
  STANDING CROP. PIGMENT CONCENTRATIONS WERE DETERMINED
  SPECTROPHOTOMETRICALLY AS ABSORBANCE WITH A 90% ACETONE SOLVENT.
  FLUORESCENT DYE TRACER TECHNIQUES WERE MODIFIED TO MEASURE CHLOROPHYLL
  FLUOROMETRICALLY. LIGHT SCATTERING, QUENCHING, AND VARIATION IN
  PHYTOPLANKTON WERE CONSIDERED TO INFLUENCE FLUORESCENCE MEASURED BY THE
  TURNER MODEL 111 FLUOROMETER, ESTABLISHING CHLOROPHYLL AS A GOOD
  INDICATOR OF EUTROPHICATION BASED ON RELATIONSHIP WITH PHYTOPLANKTON
  STANDING CROP AND PHOTOSYNTHETIC ACTIVITY. AN EMPIRICAL EQUATION WAS
  DEFINED FOR ESTIMATING PHOTOSYNTHESIS FROM CHLOROPHYLL DATA. DEPENDENCE
  OF DISSOLVED OXYGEN SUPERSATURATION ON PLANT METABOLISM RELATED DIURNAL
  OXYGEN CONCENTRATION TO CHLOROPHYLL. FLUOROMETRIC TECHNIQUE OF
  MEASURING CHLOROPHYLL IS A FAST AND FLEXIBLE METHOD FOR MEASURING
  PRIMARY ENVIRONMENTAL RELATIONSHIP AND BIOLOGICAL MONITORING IN SUCH
  INSTALLATIONS AS SEWAGE TREATMENT PLANTS. ONCE PRIMARY ENVIRONMENTAL
  CHLOROPHYLL RELATIONSHIPS ARE DEFINED, THE FLUOROMETRIC TECHNIQUE CAN
  BE USED ALMOST WITHOUT LIMITATION. (BANNERMAN-WISCONSIN)

FIELD 02L, 05C

ACCESSION NO. W70-05094

GLUCOSE AND ACETATE IN FRESHWATER: CONCENTRATIONS AND TURNOVER RATES,

NORTH CAROLINA STATE UNIV., RALEIGH. DEPT. OF ZOOLOGY.

JOHN E. HOBBIE.

PROCEEDINGS OF INTERNATIONAL BIOLOGICAL PROGRAM SYMPOSIUM HELD IN AMSTERDAM AND NIEUWERSLUIS, OCTOBER 10-16, 1966. CHEMICAL ENVIRONMENT IN THE AQUATIC HABITAT, GOLTERMAN, H L AND CLYMO, R S, EDITORS, N V NOORD-HOLLANDSCHE UITGEVERS MAATSCHAPPIJ, AMSTERDAM, 1967. P 245-251, 5 FIG, 9 REF.

DESCRIPTORS:
*CHEMICAL ANALYSIS, ORGANIC MATTER, WATER POLLUTION EFFECTS, ANALYTICAL TECHNIQUES, KINETICS, AQUATIC BACTERIA, ALGAE, PHYSIOLOGICAL ECOLOGY, CARBON RADIOISOTOPES, LAKES, PHYTOPLANKTON, BIOASSAYS.

IDENTIFIERS:
*GLUCOSE, *ACETATE, *FRESHWATER ENVIRONMENTS, *TURNOVER RATES, *HETEROTROPHY, ORGANIC SOLUTES, SWEDEN, LAKE ERKEN(SWEDEN), ANNUAL CYCLES, MICROBIAL ECOLOGY, RADIOCARBON UPTAKE TECHNIQUES, MICHAELIS-MENTEN EQUATION, RELATIVE HETEROTROPHIC POTENTIAL, LINEWEAVER-BURK EQUATION, UPTAKE VELOCITY, MAXIMUM UPTAKE VELOCITY, SUBSTRATE CONCENTRATIONS.

ABSTRACT:
ORGANIC SOLUTES IN THE HYDROSPHERE ARE EXTREMELY DIVERSE, AND THEIR DYNAMIC PROPERTIES HAVE GREATER ECOLOGICAL SIGNIFICANCE THAN DO THEIR INSTANTANEOUS CONCENTRATIONS. USEFUL INFORMATION REGARDING HETEROTROPHIC POTENTIAL OF LACUSTRINE BACTERIA CAN BE DERIVED FROM ANALYSIS, ACCORDING TO ENZYME-TYPE KINETICS ORIGINALLY PROPOSED BY MICHAELIS AND MENTEN, OF THEIR UPTAKE OF RADIOCARBON-LABELED SUBSTRATES. SUCH TREATMENT YIELDS ESTIMATES OF UPTAKE VELOCITY; MAXIMUM UPTAKE VELOCITY (V); TURNOVER TIME (T); AND RELATIVE MEASURE (K + S) OF CONCENTRATIONS OF NATURALLY-OCCURRING SUBSTRATES CONCENTRATION (S), WHERE K IS ANALOGOUS TO MICHAELIS-MENTEN CONSTANT. THESE KINETIC PARAMETERS, FOLLOWED OVER ONE ANNUAL CYCLE, ARE DESCRIBED FOR GLUCOSE AND ACETATE IN LAKE ERKEN, SWEDEN. V, VARYING 40-FOLD, FOLLOWED ALGAL CYCLES CLOSELY, PEAKING IN JUNE AND SEPTEMBER. THAT CONSTANT REMOVAL AND SUPPLY OF BOTH COMPOUNDS ARE IN BALANCE, AND THAT THEY DO NOT ACCUMULATE IS INDICATED BY SEASONALLY CONSTANT VALUE FOR (K + S), NATURALLY-OCCURRING CONCENTRATIONS (IN MICROGRAMS/LITER) REMAINING AT APPROXIMATELY 6 FOR GLUCOSE, AND 10 FOR ACETATE. T VARIED FROM 10 HOURS (SUMMER) TO APPROXIMATELY 1000 HOURS (WINTER). WHEN T = 10, SIZEABLE FRACTION OF NATURALLY-OCCURRING DISSOLVED ORGANIC CARBON IN LAKE IS CYCLED, ILLUSTRATING ECOLOGICAL VALUE OF KNOWING T AS COMPARED WITH KNOWING S. (EICHHORN-WISCONSIN)

FIELD 05C

ACCESSION NO. W70-05109

INVESTIGATION OF THE ODOR NUISANCE OCCURRING IN THE MADISON LAKES PARTICULARLY LAKES MONONA, WAUBESA AND KEGONSA FROM JULY 1942 TO JULY 1943,

PUBLIC HEALTH SERVICE, MADISON, WIS.; AND WISCONSIN UNIV., MADISON. DEPT. OF CIVIL ENGINEERING.

C. N. SAWYER, J. B. LACKEY, AND A. T. LENZ.

REPORT TO GOVERNOR'S COMMITTEE, MADISON, WISCONSIN, 1943. 79 P. 15 FIG, 4 TAB, 22 REF, APPENDIX.

DESCRIPTORS:
*ODOR-PRODUCING ALGAE, *NUISANCE ALGAE, *WISCONSIN, *LAKES, PRODUCTIVITY, NITROGEN, PHOSPHORUS, DRAINAGE, MARSHES, FARMS, CYANOPHYTA, PLANKTON, ZOOPLANKTON, EUTROPHICATION, STORM DRAINS, SEWAGE EFFLUENTS, STREAMFLOW, MEASUREMENT, WATER POLLUTION, FERTILIZATION, INDUSTRIAL WASTES, GARBAGE DUMPS, OIL WASTES, INLETS(WATERWAYS), ORGANIC MATTER, POTASSIUM, SULFATES, CRUSTACEANS, DIATOMS, COPPER SULFATE, DINOFLAGELLATES, EUGLENOPHYTA, CHLOROPHYTA, CHLAMYDOMONAS, FUNGI.

IDENTIFIERS:
*MADISON(WIS), LAKE MONONA(WIS), LAKE WAUBESA(WIS), LAKE KEGONSA(WIS), BLOOMING, LAKE MENDOTA(WIS), LAKE WINGRA(WIS), ANABAENA SPIROIDES, MICROCYSTIS AERUGINOSA, CRYPTOMONADIDA, MASTIGOPHORA, CRYPTOPHYCEAE, CHRYSOPHYCEAE, VOLVOCALES, SARCODINA, CILIATES, CYCLOTELLA, APHANIZOMENON FLOS AQUAE, MELOSIRA, ASTERIONELLA FORMOSA, FRAGILARIA, STEPHANODISCUS, PANDORINA.

ABSTRACT:
MADISON LAKES, ESPECIALLY THE SHALLOW DOWNSTREAM LAKES BELOW THE CITY, HAVE BEEN TROUBLED DURING WARM SEASONS WITH OFFENSIVE ODORS AND UNSIGHLTY SHORELINES. THE MAJOR CAUSE OF THE OFFENSIVE CONDITIONS ARE A DIRECT RESULT OF ACCUMULATIONS OF ALGAL GROWTHS IN ADVANCED STAGES OF DECOMPOSITION. THESE GROWTH STIMULATIONS ARE RELATED TO FERTILIZATION OF THE WATER WITH MINERALS, SUCH AS NITROGEN, PHOSPHORUS, POTASSIUM, AND POSSIBLY OTHER ENRICHING ELEMENTS AND COMPOUNDS. DRAINAGE FROM RICH AGRICULTURAL AREAS AND URBAN COMMUNITIES INTO SHALLOW LAKES ENHANCES ALGAL PROBLEMS. NITROGEN IS SHOWN TO BE ONE OF THE MAJOR COMPONENTIAL ELEMENTS, WITH PHOSPHORUS MINOR. NUISANCES IN INLAND LAKES ARE LARGELY MEMBERS OF THE PLANT KINGDOM. DOMESTIC SEWAGE, ALTHOUGH HIGHLY PURIFIED BY MODERN TREATMENT METHODS, IS RICH IN PHOSPHORUS, NITROGEN, POTASSIUM, AND POSSIBLY OTHER ELEMENTS. THE MOST OBNOXIOUS BLOOMS ARE THOSE OF BLUE-GREEN ALGAE, PARTICULARLY ANABAENA AND MICROCYSTIS. EXCESSIVE AMOUNTS OF PLANKTON ARE RELATED TO AGRICULTURAL DRAINAGE. LAKES MONONA, WAUBESA, AND KEGONSA ARE ADDITIONALLY ENRICHED BY STORM SEWERS, OTHER THAN RAIN WATER, AND BY SEWAGE TREATMENT PLANT EFFLUENT. (SEE W70-05113). (JONES-WISCONSIN)

FIELD 02H, 05C

ACCESSION NO. W70-05112

INVESTIGATION OF THE ODOR NUISANCE OCCURRING IN THE MADISON LAKES PARTICULARLY
LAKES MONONA, WAUBESA AND KEGONSA FROM JULY 1943 TO JULY 1944,

PUBLIC HEALTH SERVICE, MADISON, WIS.; AND WISCONSIN UNIV., MADISON. DEPT. OF
CIVIL ENGINEERING.

C. N. SAWYER, J. B. LACKEY, AND A. T. LENZ.

REPORT TO GOVERNOR'S COMMITTEE, MADISON, WISCONSIN, 1945. 92 P. 23 FIG, 29
TAB, APPENDIX.

DESCRIPTORS:
*ALGAE, *LAKES, *EUTROPHICATION, *HYDROLOGY, CHEMICAL PROPERTIES,
WISCONSIN, SEWAGE EFFLUENTS, NITROGEN, PHOSPHORUS.

IDENTIFIERS:
*TRUAX FIELD(MADISON) EFFLUENTS, *MADISON LAKES, YAHARA RIVER(WIS),
STORKWEATHER CREEK(WIS), NINE SPRINGS CREEK(WIS), DOOR CREEK(WIS),
DUTCH MILL DRAINAGE DITCH(WIS), LAKE MONONA(WIS), LAKE WAUBESA(WIS),
LAKE KEGONSA(WIS).

ABSTRACT:
THE THREE PARTS OF THIS VOLUMINOUS REPORT PRESENT RESULTS OF DETAILED
HYDROLOGICAL, CHEMICAL, AND BIOLOGICAL INVESTIGATIONS OF THE THREE
MADISON LAKES AND THEIR TRIBUTARIES. THE STUDY CONFIRMED THAT THE
OBJECTIONAL GROWTH OF ALGAE AND LAKE BLOOM ARE PRIMARILY INDUCED BY
INORGANIC FORMS OF NITROGEN AND PHOSPHORUS OF SEWAGE EFFLUENTS.
NUUISANCE BLOOMS ARE ATTRIBUTED CHIEFLY TO BLUE-GREEN ALGAE. THE
RESULTS OF LABORATORY EXPERIMENTS INDICATED THAT REMOVAL OF SOLUBLE
NITROGEN AND PHOSPHORUS FROM SEWAGE WOULD MATERIALLY DECREASE THE ALGAL
DENSITY. THE CRITICAL CONTENTS OF NITROGEN AND PHOSPHORUS IN THE LAKE
WATERS WERE ESTABLISHED AS 0.30 AND 0.015 PPM, RESPECTIVELY. A
SUGGESTION IS MADE TO DUPLICATE LABORATORY TRIALS IN ARTIFICIAL PONDS
TREATED WITH SEWAGE AND SEWAGE FREED FROM ITS NITROGEN AND PHOSPHORUS.
(SEE W70-05112). (WILDE-WISCONSIN)

FIELD 02H, 05C

ACCESSION NO. W70-05113

RAPID POLAROGRAPHIC DETERMINATION OF CHLORINE, BROMINE AND IODINE IN ALGAE,

TRIESTE UNIV. (ITALY). INST. OF THE SCIENCE OF PROPERTIES AND QUALITIES OF
COMMERCIAL PRODUCTS.

C. CALZOLARI, L. GABRIELLI, AND G. PERTOLDI MARLETTA.

ANALYST, VOL 94, NO 1122, P 774-779, SEPTEMBER 1969. 6 P, 1 FIG, 4 TAB, 15
REF.

DESCRIPTORS:
*ALGAE, *CHEMICAL ANALYSIS, *CHLORINE, *ANALYTICAL TECHNIQUES,
*POLAROGRAPHIC ANALYSIS, PHOTOMETRY, HALOGENS.

IDENTIFIERS:
BROMINE, IODINE.

ABSTRACT:
A RAPID METHOD IS DESCRIBED FOR THE POLAROGRAPHIC DETERMINATION OF
CHLORINE, BROMINE AND IODINE IN ALGAE AFTER A SINGLE COMBUSTION
PROCEDURE. FOR THE DETERMINATION OF CHLORINE AND IODINE, THE ANODIC
WAVES OF THEIR IONS WERE RECORDED. FOR BROMINE, ON THE OTHER HAND, THE
BROMIDE IONS WERE OXIDIZED QUANTITATIVELY TO BROMATE IONS, AND THE
CATHODIC WAVE FOR THESE IONS WAS RECORDED. THE METHOD HAS BEEN TESTED
ON SEVERAL SPECIES OF ALGAE FROM THE GULF OF TRIESTE. (KNAPP-USGS)

FIELD 05A

ACCESSION NO. W70-05178

THE EFFECT OF LIGHT ON NITRATE AND NITRITE ASSIMILATION BY CHLORELLA AND
  ANKISTRODESMUS,

UNIVERSITY COLL., LONDON (ENGLAND). DEPT. OF BOTANY.

I. MORRIS, AND J. AHMED.

PHYSIOLOGIA PLANTARUM, VOL 22, NO 6, P 1166-1174, 1969. 5 TAB, 20 REF.

DESCRIPTORS:
  *ALGAE, *LIGHT, *NITRATES, *NITRITES, AEROBIC CONDITIONS, ANAEROBIC
  CONDITIONS, CHLORELLA.

IDENTIFIERS:
  CARBON DIOXIDE, CHLORELLA PYRENOIDOSA, ANKISTRODESMUS BRAUNII,
  DUNALIELLA TERTIOLECTA, DCMU INHIBITOR.

ABSTRACT:
  ASSIMILATION OF NITRATE AND NITRITE NITROGEN BY CHLORELLA PYRENOIDOSA
  AND ANKISTRODESMUS BRAUNII WAS OBSERVED UNDER ILLUMINATION IN AEROBIC
  AND ANAEROBIC MEDIA IN THE ABSENCE OF CARBON DIOXIDE. THE PRESENCE OF
  CARBON DIOXIDE FAILED TO INFLUENCE THE ASSIMILATION OF CHLORELLA, BUT
  STIMULATED THAT OF ANKISTRODESMUS. THE RATIOS OF OXYGEN:NITRATE AND
  OXYGEN:NITRITE VARIED IN DIFFERENT TRIALS AND WERE HIGHER THAN
  CALCULATED VALUES FOR OXYGEN OF 2.0 AND 1.5 EXPECTED IN REDUCTION OF
  NITRATE AND NITRITE, RESPECTIVELY. OXYGEN EVOLUTION WAS COMPLETELY
  ARRESTED BY 3-(3'-4'-DICHLOROPHENYL)-1-1 DIMETHYL UREA (DCMU) AT A
  CONCENTRATION OF 0.000004M. HOWEVER, NITRITE ASSIMILATION BY BOTH ALGAE
  AND NITRATE ASSIMILATION BY ANKISTRODESMUS WERE LESS DEPRESSED BY THE
  INHIBITOR. (WILDE-WISCONSIN)

FIELD 02K, 05C

ACCESSION NO. W70-05261

SUMMARY OF LITERATURE ON AQUATIC WEED CONTROL,

TORONTO UNIV. (ONTARIO).

J. MURRAY SPEIRS.

CANADIAN FISH CULTURIST, VOL 3, NO 4, P 20-32, 1948. 55 REF.

DESCRIPTORS:
  *AQUATIC WEED CONTROL, *BIBLIOGRAPHIES, CATTAILS, BULRUSH, DREDGING,
  COPPER SULPHATE, PONDWEEDS, SODIUM ARSENITE, FISH, ANIMALS, ALGAE,
  CUTTING MANAGEMENT, POTABLE WATER, HERBICIDES, CHARA, ALKALINITY,
  BULLHEADS, EELS, KILLIFISHES, TROUT, SUCKERS, BASS, WATER HYACINTH,
  FROGS, CRAYFISH, MOSQUITOES, BACTERIA, IRRIGATION, WILLOW TREES,
  FERTILIZERS, BURNING.

IDENTIFIERS:
  *LITERATURE, ELEOCHARIS, 2,4-D, AMMATE, CHLORAMINE, DE-K-PRUF-21,
  BENOCLORS, NIGROSINE DYE, SODIUM CHLORATE, SANTOBRITE, CLADOPHORA,
  POTAMOGETON SPP, LAKE MENDOTA(WIS), ANACHARIS, NAIAS, SCIRPUS,
  NYMPHAEA, CRASSIPES EICHORNIA, TYPHA, BRASENIA, DOW CONTACT,
  SPARGANIUM, SALIX, GROWTH REGULATORS, MYRIOPHYLLUM, WATER BLOOMS,
  CHAINING.

ABSTRACT:
  SEVERAL METHODS ARE AVAILABLE TO CONTROL PARTICULAR AQUATICS. TO
  CONTROL EMERGENT AQUATICS SUCH AS CATTAILS, BULRUSHES AND SPIKERUSHES,
  2,4-D OR AMMATE ARE EFFECTIVE FOR PERMANENT CONTROL; DREDGING AND USE
  OF A FLAME THROWER ARE ADEQUATE TEMPORARY CONTROLS. TO CONTROL ALGAE,
  COPPER SULPHATE APPEARS TO BE EFFICIENT; CHLORAMINE MAY BE EQUALLY
  USEFUL BUT REQUIRES ADDITIONAL TESTING; FOR ALGAE COATING CONCRETE OR
  WOODEN WALLS OF FISH TANKS OR RESERVOIRS, DE-K-PRUF-21 IS SUGGESTED.
  THE MOST EFFECTIVE CONTROL FOR PONDWEEDS APPEARS TO BE SODIUM ARSENITE
  BUT CAUTION MUST BE EXERCISED WHERE THE WATER MIGHT BE A DRINKING
  SUPPLY FOR MAMMALS (INCLUDING MAN). BENOCLORS ARE EFFECTIVE CONTROLS
  WHERE FISH AND OTHER AQUATIC ANIMALS ARE UNIMPORTANT; NIGROSINE DYE MAY
  BE USED TO SHADE OUT THE PONDWEEDS WHERE THE BLACK COLOR GIVEN THE
  WATER IS ACCEPTABLE; IN SOME WATERS, PONDWEEDS HAVE BEEN CONTROLLED BY
  ADDING ENOUGH FERTILIZER TO PRODUCE A WATER BLOOM SHADING OUT THE
  PONDWEEDS; THIS METHOD MAY HAVE MERIT WHERE DENSE ALGAL GROWTHS ARE
  PREFERABLE TO PONDWEED GROWTHS. WATER LILIES ARE PARTICULARLY RESISTANT
  TO CHEMICAL CONTROL METHODS, AND PERSISTENT CUTTING APPEARS THE ONLY
  SATISFACTORY METHOD. (JONES-WISCONSIN)

FIELD 05C, 04A

ACCESSION NO. W70-05263

THE ECOLOGY OF A RESERVOIR,

PAUL A. ERICKSON, AND JOHN T. REYNOLDS.

NATURAL HISTORY, P 48-53, NOVEMBER 1969. 4 FIG.

DESCRIPTORS:
*WATER QUALITY CONTROL, *ECOLOGY, *RESERVOIR, MASSACHUSETTS, POTABLE
WATER, WATER POLLUTION, BACTERIA, ALGAE, NUTRIENTS, INDICATORS, ODOR,
TASTE, FILTERS, WATER TREATMENT, DISSOLVED OXYGEN, EUTROPHICATION,
FISH, LAND, VALUE, LEAVES, FERTILIZERS, PESTICIDES, CELLULOSE, CYCLES,
HARVESTING, DREDGING, MONITORING, MICROBIOLOGY, FOOD CHAINS.

IDENTIFIERS:
QUABBIN RESERVOIR(MASS), BOSTON(MASS).

ABSTRACT:
TECHNIQUES DESIGNED TO SLOW THE EUTROPHICATION PROCESS ARE: REDUCTION
OF NUTRIENTS BY CONTROLING SEWAGE WASTE DISPOSAL, AND CHEMICAL
FERTILIZERS POSSIBLY LEACHING INTO THE WATERSHED, AND BY BACTERIAL
DECOMPOSITION OF BOTTOM SEDIMENTS; BY MECHANICAL HARVESTING OF ALGAL
POPULATIONS, ROOTED AND BLOSSOMING PLANTS; DREDGING; REMOVAL OF
DISSOLVED NUTRIENTS BY PHYSICAL AND CHEMICAL AGENTS; REMOVAL OF ALGAL
POPULATIONS BY CHEMICAL AGENTS AND REDUCTION BY INCREASING FISH
POPULATIONS. THE FINANCIAL AND PHYSICAL PROBLEMS ASSOCIATED WITH THEIR
ACTUAL APPLICATION ARE IMMENSE, AS A RESERVOIR IS THE FOCUS OF
SOCIO-POLITICAL FACTORS WHICH MUST BE WEIGHED IN ANY PROCEDURE. ANOTHER
POSSIBILITY FOR SLOWING DOWN EUTROPHICATION LIES IN THE UNIQUE STUDIES,
CLASSIFIED AS 'MICROBIAL INTERVENTION.' IF THE RESERVOIR ECOSYSTEM
COULD BE MANAGED TO FAVOR BACTERIA GROUPS THAT ALREADY COMPETE WITH
ALGAE FOR DECOMPOSED ORGANIC NUTRIENTS, THE ALGAL FOOD CHAIN WOULD BE
BROKEN AND ALGAL POPULATIONS OF UNDESIRABLE PROPORTIONS COULD NOT
OCCUR, THUS MAXIMIZING THE RESERVOIR'S SELF-CLEANSING CAPABILITY.
MICROBIAL INTERVENTION WOULD TEND TO SOOTHE THE POLITICAL AND SOCIAL
PULSE, FOR IT IS A 'NATURAL PROCESS.' PRECISE INFORMATION DESCRIBING
INTERACTIONS OF ALGAL AND BACTERIAL POPULATIONS MUST BE UNDERSTOOD
BEFORE THEY CAN BE MANIPULATED FOR HUMAN BENEFITS. (JONES-WISCONSIN)

FIELD 05C, 05G

ACCESSION NO. W70-05264

ALGAL FLOCCULATION WITH SYNTHETIC ORGANIC POLYELECTROLYTES,

NOTRE DAME UNIV., IND. DEPT. OF CIVIL ENGINEERING.

MARK W. TENNEY, WAYNE F. ECHELBERGER, JR., RONALD G. SCHUESSLER, AND JOSEPH
L. PAVONI.

APPLIED MICROBIOLOGY, VOL 18, NO 6, P 965-971, 1969. 8 FIG, 11 REF.

DESCRIPTORS:
*ALGAE, *FLOCCULATION, *WASTE WATER TREATMENT, ELECTROLYTES, CHEMICAL
REACTIONS, ELECTROCHEMISTRY, RESISTIVITY, CHLORELLA, SCENEDESMUS.

IDENTIFIERS:
*POLYELECTROLYTES, *SYNTHETIC ORGANIC POLYELECTROLYTES, CATIONIC
ELECTROLYTES, ALGAL SURFACE CHARGE, ELECTROSTATIC REPULSION, DOW C-31,
DOW A-21, NALCO N-670.

ABSTRACT:
CATONIC, ANIONIC, AND NONIONIC SYNTHETIC POLYELECTROLYTES WERE USED IN
ATTEMPTS TO REMOVE ALGAE FROM WASTE WATERS. UNDER CONDITIONS OF THE
STUDY, FLOCCULATION OF ALGAE WAS CAUSED ONLY BY CATIONIC ELECTROLYTES.
THE MECHANISM OF FLOCCULATION IS INTERPRETED AS BRIDGING PHENOMENA
BETWEEN ALGAL CELLS AND THE MATRIX OF LINEARY EXTENDED POLYMER CHAINS.
THE FLOCCULATION REQUIRES A REDUCTION IN THE ALGAL SURFACE CHARGE TO A
LEVEL AT WHICH THE EXTENDED POLYMERS CAN OVERCOME ELECTROSTATIC
REPULSION. THE RATE OF APPLICATION OF CATIONIC FLOCCULANTS DEPENDS UPON
PH OF THE MEDIUM, ALGAL CONCENTRATION, AND ALGAL GROWTH PHASE.
(WILDE-WISCONSIN)

FIELD 05D

ACCESSION NO. W70-05267

ON THE BIOLOGICAL SIGNIFICANCE OF PHOSPHATE ANALYSIS; COMPARISON OF STANDARD
AND NEW METHODS WITH A BIOASSAY,

MINNESOTA UNIV., MINNEAPOLIS. LIMNOLOGICAL RESEARCH CENTER.

WILLIAM CHAMBERLAIN, AND JOSEPH SHAPIRO.

LIMNOLOGY AND OCEANOGRAPHY, VOL 14, NO 6, P 921-927, 1969. 2 FIG, 2 TAB, 13
REF.

DESCRIPTORS:
*WATER CHEMISTRY, *WATER POLLUTION EFFECTS, *PHOSPHATES, *ANALYTICAL
TECHNIQUES, *BIOASSAYS, ORGANOPHOSPHORUS COMPOUNDS, SOLVENT
EXTRACTIONS, PHOSPHORUS COMPOUNDS, LAKES, SEWAGE EFFLUENTS, ARSENIC
COMPOUNDS, SPECTROPHOTOMETRY, ALGAE, MINNESOTA, SODIUM ARSENITE.

IDENTIFIERS:
ARSENATE INTERFERENCE, DISSOLVED INORGANIC PHOSPHATES, SOLUBLE REACTIVE
PHOSPHORUS, PHOSPHATE(BIOLOGICALLY AVAILABLE), MICROCYSTIS AERUGINOSA,
LAKE COMO(MINN), LAKE VADNAIS(MINN), MILLE LAC(MINN), DEMING
LAKE(MINN), ORCHARD LAKE(MINN), LAKE OWASSO(MINN), LAKE
JOSEPHINE(MINN), GRAY'S BAY(MINN), SMITH'S BAY(MINN), PHOSPHATE
DETERMINATION METHODS.

ABSTRACT:
THE POSSIBILITY EXISTS THAT STANDARD ANALYTICAL METHODS MIGHT
OVERESTIMATE DISSOLVED INORGANIC PHOSPHATE, THUS MODIFIED TECHNIQUES
HAVE BEEN INTRODUCED TO OPERATE AN OPERATIONALLY DEFINED FRACTION OF
PHOSPHORUS COMPOUND, 'SOLUBLE REACTIVE PHOSPHORUS.' AVAILABILITY OF
VARIOUS ANALYTICAL PROCEDURES RAISES THE QUESTION OF WHICH ONE MEASURES
THE PHOSPHORUS FRACTION WHICH IS IMPORTANT AS A PLANT NUTRIENT. TWO
SEWAGE EFFLUENTS AND WATER FROM 11 MINNESOTA LAKES WERE ANALYZED BY
FOUR PROCEDURES: HARVEY METHOD, 6-SECOND METHOD, SOLVENT EXTRACTION,
STEPHENS METHOD. ANALYTICAL RESULTS WERE COMPARED WITH MAGNITUDE OF
PHOSPHATE UPTAKE DURING A BIOASSAY EMPLOYING PHOSPHORUS-STARVED CELLS
OF BLUE-GREEN ALGA, MICROCYSTIS AERUGINOSA. WATERS COULD BE ARRANGED
INTO TWO GROUPS BASED UPON MAGNITUDE OF DIFFERENCE BETWEEN RESULTS OF
HARVEY METHOD AND EXTRACTION PROCEDURE. WATERS OF GROUP II, WHERE SUCH
DIFFERENCES WERE LARGE, ALL WERE HIGH IN ARSENATE (16-56 MICROGRAMS
ARSENATE-ARSENIC/LITER), ARSENATE INTERFERENCE FROM SUCH CONCENTRATIONS
BEING SUFFICIENTLY LARGE TO ACCOUNT FOR DISCREPANCIES IN PHOSPHATE
DETERMINATIONS. EXTRACTION PROCEDURE IS LESS SENSITIVE TO ARSENATE
INTERFERENCE THAN ARE OTHER METHODS. IN ANALYZING LAKE WATER FOR
PHOSPHATES, IT IS ADVISABLE TO ASSUME THAT ARSENIC IS PRESENT.
PHOSPHATE ANALYSES RESULTING FROM EXTRACTION PROCEDURE WERE IN GOOD
AGREEMENT WITH BIOLOGICALLY AVAILABLE PHOSPHATES AS DETERMINED BY
BIOASSAY. (EICHHORN-WISCONSIN)

FIELD 05C, 05A

ACCESSION NO. W70-05269

THERMOPHILIC BLUE-GREEN ALGAE AND THE THERMAL ENVIRONMENT,

OREGON UNIV., EUGENE. DEPT. OF BIOLOGY.

RICHARD W. CASTENHOLZ.

BACTERIOLOGICAL REVIEWS, VOL 33, NO 4, P 476-504, 1969. 4 FIG, 4 TAB, 187 REF.

DESCRIPTORS:
*CYANOPHYTA, *ALGAE, *THERMAL SPRINGS, *OREGON, THERMOPHILIC BACTERIA, THERMAO POLLUTION, NUTRIENTS, TEMPERATURE, PHOTOSYNTHESIS, LIGHT INTENSITY.

IDENTIFIERS:
*YELLOWSTONE PARK, *ICELAND, MASTIGOCLADUS LAMINOSUS, PHORMIDIUM LAMINOSUM, OSCILLATORIA TENUIS, OSCILLATORIA TEREBRIFORMIS, MATS, SYNECHOCOCCUS LIVIDUS, BLUE-GREEN ALGAE, FILAMENTS.

ABSTRACT:
THE BLUE-GREEN ALGAE IN GENERAL HAVE HIGHER TEMPERATURE OPTIMA THAN OTHER ALGAE AND SOME OF THEM, THE THERMOPHILES, AND INHABITANTS OF HOT SPRINGS AT TEMPERATURES AS HIGH AS 73-74C. IT IS NO SURPRISE, THEREFORE, THAT THERMAL POLLUTION MAY FAVOR THE GROWTH OF THE BLUE-GREEN ALGAE. WATERS FROM HOT SPRINGS GENERALLY POSSESS MUCH GREATER SOLUTE (NUTRIENT) CONCENTRATIONS THAN SURFACE WATERS AND IT IS PROBABLE THAT MANY THERMAL STREAMS CAUSE EUTROPHICATION IN THE BODIES OF WATER INTO WHICH THEY DRAIN. THE PAPER COVERS THE FOLLOWING TOPICS: DISTRIBUTION OF THERMAL WATERS; DISTRIBUTION OF SPECIES (UPPER AND LOWER TEMPERATURE LIMITS, CLASSIFICATION AND GEOGRAPHICAL DISTRIBUTION, PROBLEMS OF SURVIVAL AND TRANSPORT); STUDIES OF NATURAL POPULATIONS (MAT FORMATION AND STABILITY, MOVEMENTS OF FILAMENTS AND MATS, MEASUREMENTS OF PHOTOSYNTHESIS AND GROWTH); CULTIVATION OF THERMOPHILIC CYANOPHYTES (MEDIUMS AND NUTRITION, ISOLATION AND MAINTENANCE, RATES OF GROWTH, RATES OF PHOTOSYNTHESIS, AND RESPIRATION RATES IN CULTURE); RESPONSES TO TEMPERATURE AND LIGHT INTENSITY (OPTIMAL TEMPERATURE AND LIGHT INTENSITY, EFFECTS OF LIGHT AND TEMPERATURE ON PIGMENTATION, GROWTH AND SURVIVAL AT SUBOPTIMAL TEMPERATURES). (GERHOLD-WISCONSIN)

FIELD 05B

ACCESSION NO. W70-05270

DDT REDUCES PHOTOSYNTHESIS BY MARINE PHYTOPLANKTON,

STATE UNIV. OF NEW YORK, STONY BROOK. DEPT. OF ELECTRICAL ENGINEERING.

CHARLES F. WURSTER, JR.

SCIENCE, VOL 159, P 1474-1475, MARCH 1968. 2 FIG, 21 REF.

DESCRIPTORS:
*WATER POLLUTION EFFECTS, *PHOTOSYNTHESIS, *PHYTOPLANKTON, *DDT, *MARINE ALGAE, TOXICITY, DIATOMS, ESTUARIES, PHYSIOLOGICAL ECOLOGY, MASSACHUSETTS, CHLOROPHYTA, DINOFLAGELLATES, ALGAE, CALIFORNIA, FLORIDA, PESTICIDE RESIDUES, ECOSYSTEMS, FOOD CHAINS, BIOLOGICAL COMMUNITIES, BIOASSAYS, EUTROPHICATION.

IDENTIFIERS:
SKELETONEMA COSTATUM, COCCOLITHUS HUXLEYI, PYRAMIMONAS, PERIDINIUM TROCHOIDEUM, VINEYARD SOUND(MASS), WOODS HOLE(MASS). RADIOCARBON UPTAKE TECHNIQUES, MICROBIAL ECOLOGY, COCCOLITHOPHORES, SEAWATER MEDIUM F.

ABSTRACT:
WIDELY DISTRIBUTED IN THE HYDROSPHERE, THE INSECTICIDE DDT HAS POTENTIALLY GREAT AND DIVERSE ECOLOGICAL EFFECTS. LITTLE IS KNOWN ABOUT ITS INFLUENCE ON PHYTOPLANKTON. OVER RANGE OF CONCENTRATIONS 1-500 PPB, ITS EFFECT ON PHOTOSYNTHESIS, UNDER CONTROLLED CONDITIONS OF LIGHT AND TEMPERATURE, WAS DETERMINED FOR A NATURAL POPULATION OF PHYTOPLANKTON AND FOR FOUR SPECIES OF LABORATORY CULTURED MARINE PHYTOPLANKTERS: SKELETONEMA COSTATUM (DIATOM), COCCOLITHUS HUXLEYI (COCCOLITHOPHORE), PYRAMIMONAS (GREEN ALGA), PERIDINIUM TROCHOIDEUM (DINOFLAGELLATE). WHEN PHOTOSYNTHESIS (MEASURED AS PERCENTAGE OF RADIOCARBON UPTAKE BY UNTREATED CONTROLS) IS PLOTTED AGAINST INCREASING DOSAGE OF DDT, CHARACTERISTIC SIGMOID CURVES RESULT, WITH PHOTOSYNTHESIS DECREASING APPROXIMATELY LINEARLY IN DOSAGE RANGE 1-100 PPB DDT. SIGMOID DOSE-RESPONSE CURVES APPARENTLY INDICATE ABSENCE OF THRESHOLD LEVELS BELOW WHICH NO EFFECTS OCCUR. TESTING RELATIONSHIP IN SKELETONEMA BETWEEN CELL CONCENTRATIONS AND TOXICITY REVEALED THAT PHOTOSYNTHESIS DIMINISHED TO 50% OF CONTROLS WHEN FINAL CELLULAR CONCENTRATIONS WERE REDUCED 100-FOLD. FOR CHEMICAL REASONS, EFFECTIVE DOSES OF DDT IN THESE EXPERIMENTS MAY BE LOWER THAN QUANTITIES ADDED INITIALLY. MOREOVER, BECAUSE DDT BECOMES AVAILABLE NATURALLY AT STEADY RATES, GIVEN CONCENTRATION IN NATURE WOULD HAVE GREATER ECOLOGICAL SIGNIFICANCE THAN SAME INITIAL EXPERIMENTAL CONCENTRATION. SELECTIVE TOXIC STRESS BY DDT MAY CREATE FLORAL IMBALANCE, THEREBY AGGRAVATING LESS DESIRABLE MANIFESTATION OF EUTROPHICATION. (EICHHORN-WISCONSIN)

FIELD 05C

ACCESSION NO. W70-05272

REVIEW ON TOXIGENIC ALGAE,

HEBREW UNIV., JERUSALEM (ISRAEL).

MOSHE SHILO.

VERHANDLUNGEN DER INTERNATIONALEN VEREINIGUNG FUR THEORESTISCHE UND
ANGEWANDTE LIMNOLOGIE, VOL 15, P 782-795, 1964. 90 REF, DISCUSSION.

DESCRIPTORS:
*TOXICITY, *ALGAE, *REVIEWS, HABITATS, CLAMS, MUSSELS, SHELLFISH,
DINOFLAGELLATES, CYANOPHYTA, CHRYSOPHYTA, CALIFORNIA, UNITED STATES,
BRACKISH WATER, ESTUARIES, BACTERIA, MOLLUSKS, VITAMIN B, RED TIDE,
SALINITY, FLORIDA, TEXAS, PHOSPHORUS, MIGRATION, DIURNAL, ALGICIDES,
CLOSTRIDIUM, FISH, AMPHIBIANS, ANABAENA, CYTOLOGICAL STUDIES,
PHYSICOCHEMICAL PROPERTIES.

IDENTIFIERS:
GONYAULAX, GONYAULAX VENIFICUM, PYRODINIUM, GYMNODINIUM BREVE, CANADA,
EUROPE, SOUTH AFRICA, ISRAEL, MICROCYSTIS, PRYMNESIUM PARVUM,
SYNERGISTIC EFFECTS, AXENIC CULTURES, MONOALGAL, PORTUGAL, TAPES,
JAPAN, AMPHIDINIUM, BLOOMS, MYTILUS, TRICHODESMIUM ERYTHREUM,
SAXIDOMUS, CLOSTRIDIUM BOTULINUS, VENERUPIS, OSTREA, ARTHROPODS.

ABSTRACT:
TOXIGENIC ALGAE ARE FOUND IN MARINE, BRACKISH AND FRESH WATER
WORLDWIDE. HIGH CONCENTRATIONS OF TOXIN-FORMING ORGANISMS, STORED IN
CLAMS, MUSSELS, AND SHELLFISH CAUSE FOOD POISONING. DIFFERENT TOXONOMIC
GROUPS CONTAIN TOXIN-FORMING ALGAE--SOME DINOFLAGELLATES, CERTAIN
BLUE-GREEN ALGAE, AND ONE SPECIES OF CHRYSOPHYCEAE. RELATIONSHIP
BETWEEN ALGAL APPEARANCE AND TOXIC PHENOMENA DID NOT PROVE
UNEQUIVOCALLY THAT TOXIN-FORMATION IS CAUSED BY ALGAE THEMSELVES. MANY
FORMS IN TOXIC TIDES AND BLOOMS ARE NOW AVAILABLE IN AXENIC CULTURE.
MASS DEVELOPMENT OF TOXIC ALGAE BEING SPORADIC, INVESTIGATORS SEEK FOR
CRITICAL COMBINATION OF PHYSICOCHEMICAL FACTORS TRIGGERING THEM. BLOOMS
HAVE BEEN ASSOCIATED WITH UPWELLING OF DEEP, COOL, NUTRIENT-RICH WATER,
HEAVY RAINFALL WITH RIVER DISCHARGE, SEWAGE POLLUTION, PHOSPHORUS
CONCENTRATIONS, GROWTH-PROMOTING SUBSTANCES FORMED BY MICROORGANISMS,
SALINITY, ACTIVE OR PASSIVE CONCENTRATIONS OF THE CELLS AFTER GROWTH.
PROPERTIES AND BEHAVIOR OF ALGAL TOXINS ARE DESCRIBED. LACK OF
CORRELATION BETWEEN CONCENTRATION OF POTENTIALLY TOXIC ALGAE AND LEVEL
OF TOXINS IN WATER IS EMPHASIZED. INTRA-AND EXTRA-CELLULAR
CONCENTRATIONS OF TOXIC PRINCIPLES IN AXENIC PRYMNESIUM CULTURES VARIED
WITH STAGES OF GROWTH. OPTIMAL GROWTH CONDITIONS WERE NOT NECESSARILY
OPTIMAL FOR TOXIN BIOSYNTHESIS. SHIFTING CONCENTRATIONS OF CO-FACTORS,
SUCH AS STREPTOMYCIN, SPERMINE, AND OTHERS AS YET UNKNOWN MAY ALTER
TOXIC EFFECT WITHOUT CHANGES IN TOXIN LEVEL. (JONES-WISCONSIN)

FIELD 05B

ACCESSION NO. W70-05372

DISTRIBUTION AND ABUNDANCE OF PHYTOPLANKTON AND ROTIFERS IN A MAIN STEM
MISSOURI RIVER RESERVOIR,

BUREAU OF SPORT FISHERIES AND WILDLIFE, YANKTON, S. DAK. NORTH CENTRAL
RESERVOIR INVESTIGATIONS.

PATRICK L. HUDSON, AND BRUCE C. COWELL.

PROCEEDINGS OF SOUTH DAKOTA ACADEMY OF SCIENCE, VOL 45, P 84-106, 1966. 7
FIG, 5 TAB, 16 REF.

DESCRIPTORS:
*RESERVOIRS, *PHYTOPLANKTON, *ROTIFERS, *DISTRIBUTION, MISSOURI RIVER,
SOUTH DAKOTA, NEBRASKA, ALGAE, NANNOPLANKTON, DIATOMS.

IDENTIFIERS:
*LEWIS AND CLARK LAKE (S D), ABUNDANCE, BLUE-GREEN ALGAE, GREEN ALGAE,
FRAGILARIA, ASTERIONELLA, MELOSIRA, MICROCYSTIS, KERATELLA, POLYARTHRA,
ASPLANCHNA, BRANCHIONUS.

ABSTRACT:
WEEKLY SAMPLES WERE TAKEN AT TEN STATIONS IN LEWIS AND CLARK LAKE
RESERVOIR, BORDERING SOUTH DAKOTA-NEBRASKA, FROM MAY TO OCTOBER;
ADDITIONAL SAMPLES WERE OBTAINED FROM AN AUTOMATIC PLANKTON SAMPLER AT
THE DISCHARGE. MEAN NUMBERS OF NET PHYTOPLANKTON INCREASED FROM
HEADWATERS TO THE DAM; VOLUMES OF NET PHYTOPLANKTON PER LITER OF WATER
VARIED IRREGULARLY THROUGHOUT THE RESERVOIR. DATA FROM THE RESERVOIR
DISCHARGE INDICATE NANNOPLANKTON COMPRISED 84% OF PHYTOPLANKTON VOLUME.
DISCHARGE OF PHYTOPLANKTON TOTALLED 9,058 METRIC TONS FOR ONE YEAR WITH
PEAK DISCHARGES OCCURRING IN MARCH, MAY AND AUGUST. PHYTOPLANKTON
DISCHARGE DOES NOT APPEAR TO BE DIRECTLY RELATED TO VOLUME OF WATER
DISCHARGED. DIATOMS WERE DOMINANT PHYTOPLANKTER IN ALL MONTHS EXCEPT
AUGUST (GREEN ALGAE) AND OCTOBER (BLUE-GREEN ALGAE). ABUNDANCE OF
ROTIFERS INCREASED FROM HEADWATERS DOWNSTREAM. DOMINANT ROTIFERS IN
TERMS OF MONTHS PRESENT AND MEAN NUMBERS PER LITER WERE POLYARTHRA,
ASPLANCHNA AND KERATELLA; TEN ADDITIONAL GENERA WERE NOTED. SEASONAL
ABUNDANCE FIGURES INDICATE A SINGLE PEAK IN LATE SPRING--EARLY SUMMER.
HIGHEST TOTAL MEAN VALUES OCCURRED IN APRIL AND JUNE; LOWEST TOTALS
WERE IN NOVEMBER AND JANUARY. GRAPHICAL AND TABULAR DATA INCLUDE
MONTHLY AVERAGES OF PHYTOPLANKTON AND ROTIFER DENSITIES.
(VOIGTLANDER-WISCONSIN)

FIELD 02H, 05C

ACCESSION NO. W70-05375

AIRBORNE FLUOROMETER APPLICABLE TO MARINE AND ESTUARINE STUDIES,

GEOLOGICAL SURVEY, WASHINGTON, D.C.; AND PERKIN-ELMER CORP., NORWALK, CONN.

GEORGE E. STOERTZ, WILLIAM R. HEMPHILL, AND DAVID A. MARKLE.

MARINE TECHNOLOGY SOCIETY JOURNAL, VOL 3, NO 6, P 11-26, 1969. 24 FIG, 16 REF.

DESCRIPTORS:
*INSTRUMENTATION, *SAMPLING, *MEASUREMENT, *FLUOROMETRY, DYE RELEASES,
CURRENTS(WATER), REMOTE SENSING, SEA WATER, CHLOROPHYLL, OIL, ALGAE,
PACIFIC OCEAN.

IDENTIFIERS:
*FRAUNHOFER LINE DISCRIMINATOR, RHODAMINE WT DYE, AIRBORNE TESTS, OIL
SPILLS, GEOLOGICAL SS POLARIS, CHLOROPHYLL FLUORESCENCE, FABRY-PEROT
FILTERS, SOLAR FRAUNHOFER LINES, ATTENUATION MEASUREMENT APPARATUS,
HYDROGEN F LINE.

ABSTRACT:
THIS STUDY WAS CONCERNED WITH PERFORMANCE AND POSSIBLE IMPROVEMENTS OF
THE FRAUNHOFER LINE DISCRIMINATOR (FLD). THIS FLUOROMETER IS THE FIRST
TO PROVIDE DIRECT MEASUREMENTS OF FLUORESCENCE IN BROAD DAYLIGHT AND
FROM A REMOTE PLATFORM, THUS GREATLY FACILITATING STUDIES OF CURRENT
FLOW AND RATE OF DISPERSION OF RHODAMINE AND OTHER TAGGING DYES. THE
SMALLEST DETECTABLE INCREMENT OF RHODAMINE WT IN 0.5 METER DEPTH WAS 1
PPB. TEMPERATURE SIGNIFICANTLY INFLUENCED THE FLD RESULTS. SHIPBOARD
AND HELICOPTER TESTS OF THE FLD AND PLOTTING OF FLIGHT PATHS BY
TRACKING RADAR WERE SUCCESSFUL. THE DYE WAS DETECTABLE IN AIRBORNE
TESTS AT ALTITUDES EXCEEDING 5000 FEET. A NEW DESIGN HAVING ONLY ONE
PHOTOMULTIPLIER SHOULD ELIMINATE THE NEED FOR MONITORING STANDARD
TARGETS. ADDITIONAL IMPROVEMENTS MAY ENABLE DETECTION OF CHLOROPHYLL
FLUORESCENCE, OIL SPILLS OR LEAKAGES, ALGAL BLOOMS AND INDICATION OF
POLLUTION, AND FACILITATE DISPERSION STUDIES. (WILDE-WISCONSIN)

FIELD 07B, 05A

ACCESSION NO. W70-05377

THE ADAPTATION OF PLANKTON ALGAE. IV. LIGHT ADAPTATION IN DIFFERENT ALGAL
  SPECIES,

ROYAL DANISH SCHOOL OF PHARMACY, COPENHAGEN. DEPT. OF BOTANY.

ERIK G. JORGENSEN.

PHYSIOLOGIA PLANTARUM, VOL 22, NO 6, P 1307-1315, 1969. 4 FIG, 2 TAB, 15 REF.

DESCRIPTORS:
  *PLANKTON, *ALGAE, *LIGHT INTENSITY, CHLORELLA, CHLOROPHYTA, DIATOMS,
  PIGMENTS, CHLOROPHYLL, PHOTOSYNTHESIS, GROWTH RATE, ORGANIC MATTER,
  SCENEDESMUS, CHLAMYDOMONAS, FRESHWATER, MARINE ALGAE, ENZYMES.

IDENTIFIERS:
  *ADAPTATION, *SPECIES, CYCLOTELLA MENEGHINIANA, LIGHT-SATURATED RATE,
  CELL VOLUME, NITZSCHIA CLOSTERIUM, NITZSCHIA PALEA, CHLOROPLASTS,
  SCENEDESMUS OBLIQUUS, SCENEDESMUS QUADRICAUDA, CHLORELLA PYRENOIDOSA,
  CHLORELLA VULGARIS, MONODUS SUBTERRANEUS, ANKISTRODESMUS FALCATUS,
  CHLAMYDOMONAS MOEWUSII, SKELETONEMA COSTATUM, SYNECHOCOCCUS ELONGATUS.

ABSTRACT:
TWO TYPES OF PLANKTON ALGAE ADAPT TO LIGHT INTENSITIES
DIFFERENTLY--CHLORELLA AND CYCLOTELLA. THE CHLORELLA TYPE IS MOSTLY
FOUND AMONG GREEN ALGAE AND ADAPTS TO A NEW LIGHT INTENSITY CHIEFLY BY
CHANGING THE PIGMENT CONTENT. THE CELLS ADAPTED TO A HIGH LIGHT
INTENSITY HAVE A LOWER CHLOROPHYLL A CONTENT PER CELL THAN CELLS
ADAPTED TO LOW LIGHT INTENSITY, THE LATTER WITH RATHER LOW LIGHT
SATURATION. THE LIGHT-SATURATED RATE OF PHOTOSYNTHESIS IS USUALLY LOWER
FOR CELLS ADAPTED TO A HIGH LIGHT INTENSITY THAN FOR CELLS ADAPTED TO A
LOW LIGHT INTENSITY. ACTUAL PHOTOSYNTHESIS (PHOTOSYNTHESIS AT THE LIGHT
INTENSITY WHERE THE CELLS ARE GROWN) IS NOT MUCH HIGHER AT HIGH LIGHT
INTENSITY THAN AT LOW. THE CYCLOTELLA, PREVALENT AMONG DIATOMS, ADAPTS
ONLY BY CHANGING THE LIGHT-SATURATION RATE. CHLOROPHYLL CONTENT IS THE
SAME IN CELLS GROWN AT LOW AND HIGH INTENSITIES. LIGHT SATURATION FOR
CELLS GROWN AT A LOW LIGHT INTENSITY IS RATHER HIGH. THE
LIGHT-SATURATED RATE IS MUCH HIGHER IN THE CASE EXAMINED AT HIGH
INTENSITY THAN AT LOW. ACTUAL PHOTOSYNTHESIS IS CONSIDERABLY HIGHER FOR
CELLS GROWN AT HIGH LIGHT INTENSITIES OPPOSED TO LOW. TRANSITIONAL
TYPES OCCUR. (JONES-WISCONSIN)

FIELD 02K, 05C

ACCESSION NO. W70-05381

ALGAE AND PHOTOSYNTHESIS IN SHAGAWA LAKE, MINNESOTA,

MINNESOTA UNIV., MINNEAPOLIS. LIMNOLOGICAL RESEARCH CENTER.

ROBERT O. MEGARD.

UNIVERSITY OF MINNESOTA, MINNEAPOLIS, LIMNOLOGICAL RESEARCH CENTER, INTERIM
REPORT NO 5, APRIL 1969. 5 FIG, 10 TAB, 8 REF.

DESCRIPTORS:
*ALGAE, *PHOTOSYNTHESIS, *MINNESOTA, PHYTOPLANKTON, CYANOPHYTA,
DIATOMS, STRATIFICATION, OXYGEN, HYDROGEN ION CONCENTRATION,
TEMPERATURE, DEPTH, DENSITY, CHLOROPHYLL, ICE, SEASONAL, DIURNAL, LIGHT
INTENSITY, CARBON, SEWAGE TREATMENT, TERTIARY TREATMENT, EFFLUENTS,
LIMNOLOGY, WATER CHEMISTRY, RESPIRATION, BATHYMETRY, SALINITY,
PHOSPHORUS, PRODUCTIVITY, CHLOROPHYTA, CHRYSOPHYTA, SCENEDESMUS,
STANDING CROP.

IDENTIFIERS:
*SHAGAWA LAKE(MINN), APHANIZOMENON FLOS AQUAE, STEPHANODISCUS ASTREA,
LAKE MINNETONKA(MINN), MORPHOMETRY, ALGAE BLOOMS, CLEARWATER
LAKE(MINN), TROUT LAKE(MINN), KIMBALL LAKE(MINN), CRYPTOPHYTA,
PYROPHYTA, CHROOMONAS ACUTA, SCHROEDERIA SETIGERA, MELOSIRA AMBIGUA,
CERATIUM HIRUDINELLA, ANKISTRODESMUS, GLENODINIUM PULVISCULUS,
SCENEDESMUS QUADRICAUDA.

ABSTRACT:
SHAGAWA LAKE (MINNESOTA) PHYTOPLANKTON WAS DOMINATED BY APHANIZOMENON
FLOS AQUAE IN SUMMER AND STEPHANODISCUS ASTREA VAR MINUTULA IN WINTER,
MAINLY CONCENTRATED IN DEPTHS ABOVE 4 METERS. CHLOROPHYLL CONTENT OF
APHANIZOMENON FILAMENTS IN DIFFERENT PARTS AT THE LAKE SURFACE WERE
SIMILAR ON ANY DATE BUT DECREASED 40% IN TWO WEEKS, A POSSIBLE RESPONSE
TO RECENT PHOTIC HISTORY OR POPULATION AGING. CHLOROPHYLL
CONCENTRATIONS AND ALGAL DENSITIES BENEATH MARCH ICE WERE GREATER THAN
IN SOME PARTS OF THE LAKE DURING SUMMER. WHEN ALGAE FROM DIFFERENT
LOCALITIES WERE INCUBATED AT 0.5 METER DEPTH (LIGHT INTENSITIES OPTIMAL
FOR PHOTOSYNTHESIS ON SUNNY DAYS), AVERAGE DAILY RATE OF PHOTOSYNTHETIC
CARBON ASSIMILATION WAS 42 MICROGRAMS CARBON/MICROGRAM CHLOROPHYLL,
IDENTICAL TO THE RATE AT THE SAME TEMPERATURE IN LAKE MINNETONKA. THE
DENSE ALGAL POPULATIONS IN SURFACE WATERS ABSORBED SO MUCH LIGHT THAT
THERE WAS LITTLE PHOTOSYNTHESIS BELOW 3 METERS. ALGAL PHOTOSYNTHETIC
CAPACITY BENEATH ALMOST ONE METER CLEAR ICE NEARLY EQUALLED THAT OF
CLOUDY SUMMER DAYS. EFFECTS OF ALGAL PHOTOSYNTHETIC CAPACITY WERE
MEASURED ON VARYING AMOUNTS OF EFFLUENT FROM SECONDARY AND TERTIARY
WASTE TREATMENT PLANTS. PHOTOSYNTHESIS WAS STIMULATED SLIGHTLY BY
TERTIARY EFFLUENT, BUT HALTED BY 20% MUNICIPAL WATER. (SEE ALSO
W69-00659, W69-10167). (JONES-WISCONSIN)

FIELD 02H, 05C

ACCESSION NO. W70-05387

222

ALGAE IN WATER SUPPLIES,

ROBERT A. TAFT SANITARY ENGINEERING CENTER, CINCINNATI, OHIO.

C. MERVIN PALMER.

U S PUBLIC HEALTH SERVICE PUBLICATION NO 657, 1959. 88 P. 55 FIG, 13 TAB, 170
   REF, 6 PLATES, APPENDIX.

   DESCRIPTORS:
      *ALGAE, *WATER SUPPLY, *CONTROL, TASTE, ODOR, FILTERS, PLANKTON, WATER
      POLLUTION, SURFACE WATERS, RESERVOIRS, CONTROL, BIBLIOGRAPHIES,
      ANALYTICAL TECHNIQUES, DIATOMS, SAMPLING, PHYSICOCHEMICAL PROPERTIES,
      CENTRIFUGATION, MICROSCOPY, NETS, NANNOPLANKTON, PIGMENTS, CHLOROPHYTA,
      CYANOPHYTA, CHRYSOPHYTA, RHODOPHYTA, EUGLENOPHYTA, SLIME, COLOR,
      CORROSION, DISTRIBUTION SYSTEMS, COAGULATION, WATER SOFTENING,
      TOXICITY, MARINE ALGAE, FRESH WATER, PARASITISM, WATER TREATMENT,
      INDICATORS, WASTE TREATMENT, COPPER SULFATE, ALGICIDES, PHOTOSYNTHESIS.

   IDENTIFIERS:
      *IDENTIFICATION, *SIGNIFICANCE, ALGAL KEY, CLUMP COUNT, DESMIDS,
      FLAGELLATES, MESOPLANKTON, MICROPLANKTON, PSEUDOPLANKTON,
      CRYPTOPHYCEAE, VOLVOCALES, INDUSTRIAL USES, CLEAN WATER, GENUS,
      SPECIES, TOUCH SENSATION.

   ABSTRACT:
      CONTINUED SURVEILLANCE OF ALGAL POPULATIONS UNDER A PLANNED AND UNIFORM
      APPROACH IS NEEDED TO PUT INFORMATION IN AN UNDERSTANDABLE AND USEFUL
      FORM. THIS MANUAL IS AN ATTEMPT TO APPRAISE NUISANCE ORGANISM PROBLEMS
      AND FURNISH INFORMATION FOR REMEDYING SOME DIFFICULTIES. IT DEALS WITH
      THE ECOLOGY AND LIFE HISTORY OF ALGAE AND GIVES INFORMATION ON FILTER
      CLOGGING, MAT-FORMING ALGAE, ATTACHED FORMS, ALGICIDES, AND ALGAL
      CONTROL. LARGE QUANTITIES OF ALGAL MATERIAL CAUSE SERIOUS DIFFICULTIES
      IN WATER TREATMENT PLANTS. CERTAIN DIATOMS ALMOST INVARIABLY REDUCE THE
      LENGTH OF FILTER RUNS. SYNURA, IS A TASTE AND ODOR PRODUCER. LOW
      CONCENTRATIONS OF MOST ALGAE ARE OFTEN AN ASSET IN RAW WATERS. A
      SIMPLIFIED KEY OF 289 SPECIES WITH DRAWINGS OF MOST IMPORTANT IS GIVEN.
      THE ALGAE ARE CONSIDERED AND DISPLAYED ACCORDING TO THEIR SIGNIFICANCE
      TO SANITARY SCIENTISTS AND TECHNICIANS RATHER THAN WITH REGARD TO THEIR
      EVOLUTIONARY RELATIONSHIP. MOST ALGAE OF IMPORTANCE IN WATER SUPPLIES
      MAY BE CHARACTERIZED IN FOUR GENERAL GROUPS: BLUE-GREEN ALGAE, GREEN
      ALGAE, DIATOMS, AND PIGMENTED FLAGELLATES. A FEW MISCELLANEOUS FORMS DO
      NOT FIT INTO THESE FOUR GROUPS. POLLUTED WATER ALGAE, CLEAN WATER
      ALGAE, PLANKTON, AND OTHER SURFACE WATER ALGAE ARE INCLUDED.
      (JONES-WISCONSIN)

FIELD 05F

ACCESSION NO.   W70-05389

ON THE THEORY OF ADDING NUTRIENTS TO LAKES WITH THE OBJECT OF INCREASING TROUT PRODUCTION,

DALHOUSIE UNIV., HALIFAX (NOVA SCOTIA). ZOOLOGICAL LAB.

F. R. HAYES.

CANADIAN FISH CULTURIST, VOL 10, P 32-37, 1951. 13 REF.

DESCRIPTORS:
*NUTRIENTS, *LAKES, *TROUT, *FISH MANAGEMENT, FERTILIZATION, ALGAE, PLANKTON, NITROGEN, PHOSPHORUS, POTASSIUM, SELECTIVITY, POISONS, OXYGEN, TRACERS, PHOSPHORUS RADIOISOTOPES, IONS, HUMIC ACIDS, CARP, ICE, IRON COMPOUNDS, HYDROGEN ION CONCENTRATION, ACIDITY, MUD, HYDROGEN SULFIDE.

IDENTIFIERS:
PRINCE EDWARD ISLAND(CANADA), NOVA SCOTIA(CANADA), PENGUINS, LAKE ZURICH(SWITZERLAND), WHITEFISH, PUNCHBOWL LAKE(CANADA).

ABSTRACT:
WHEN NUTRIENT SALTS ARE ADDED TO LAKES FOR FERTILIZATION, 90% MAY DISAPPEAR IN AN EXCHANGE REACTION, WITHIN A FEW WEEKS. THE REMAINING 10% IS LIKELY TO REMAIN IN OR AVAILABLE TO THE WATER CONTRIBUTING TO ALGAL GROWTH, LEADING TO ANIMAL PLANKTON GROWTH, AND THENCE TO FISH GROWTH. NONE OF THE EXTRA MATERIAL IS PROBABLY SEEN DURING THE SECOND SUMMER. WHILE INFORMATION IS LESS COMPLETE REGARDING POTASSIUM AND NITROGEN, THAN PHOSPHORUS, THE VIEW IS THAT ABOUT THE SAME PERCENTAGE OF EACH WOULD SURVIVE IN WATER AFTER EQUILIBRATION. AS A CONSERVATION MEASURE, FERTILIZATION OF LAKES DOES NOT APPEAR TO BE AN ECONOMICALLY FEASIBLE PROCEDURE, ALTHOUGH IT CONSTITUTES AN INTERESTING EXPERIMENT FOR A PRIVATELY CONTROLLED LAKE. IF THERE ARE NO COARSE FISH TO REPLACE TROUT UNDER HIGH FERTILITY CONDITIONS, LARGE TROUT YIELDS MAY BE EXPECTED. THIS SEEMS THE SITUATION IN PRINCE EDWARD ISLAND. ALTERNATIVELY COARSE FISH MIGHT BE SELECTIVELY POISONED IN CONJUNCTION WITH A FERTILIZATION PROGRAM. SELECTIVE POISONING, BY REMOVING COARSE FISH FROM SHALLOWS WHILE LEAVING TROUT IN THE DEPTHS, HAS BEEN FOUND PROMISING. IT MAY BE THAT TROUT SURVIVE WHERE THEY DO, BECAUSE THEY CAN TOLERATE MORE ADVERSE CONDITIONS THAN COMPETING FISH. (JONES-WISCONSIN)

FIELD 02H, 05C

ACCESSION NO. W70-05399

LIMNOLOGICAL RELATIONS OF INDIAN INLAND WATERS WITH SPECIAL REFERENCE TO
WATERBLOOMS,

UNIVERSITY COLL., LONDON (ENGLAND). DEPT. OF BOTANY.

R. N. SINGH.

VERHANDLUNGEN DER INTERNATIONALEN VEREINIGUNG FUR THEORETISCHE UND ANGEWANDTE
LIMNOLOGIE, VOL 12, P 831-836, 1953. 16 REF.

DESCRIPTORS:
*LIMNOLOGY, *WATER, *RIVERS, *LAKES, *PONDS, LIGHT, TEMPERATURE,
FERTILIZATION, FISH, NUISANCE ALGAE, ODOR, TASTE, DEPTH, CIRCULATION,
DIURNAL, WINDS, CURRENTS(WATER), WATER POLLUTION, DRAINAGE, ORGANIC
MATTER, SEWAGE, PHYTOPLANKTON, PRODUCTIVITY, CHLOROPHYTA, HYDROGEN ION
CONCENTRATION, NITRATES, PHOSPHATES, CYANOPHYTA, OXIDATION-REDUCTION
POTENTIAL, BOTTOM SEDIMENTS, EUGLENOPHYTA, TOXICITY.

IDENTIFIERS:
*INDIA, *BLOOMS, RIVER GANGES(INDIA), RAJGHAT(INDIA), BOMBAY(INDIA),
BENARES(INDIA), MACROPHYTES, CLADOPHORA, OEDOGONIA, SPIROGYRA,
MICROCYSTIS AERUGINOSA, ANABAENOPSIS, RACIBORSKII, RAPHIDIOPSIS,
ANABAENA, WOLLEA, OSCILLATORIA, SPIRULINA, VOLVOCALES, GLENODINIUM,
TRACHELOMONAS.

ABSTRACT:
INDIAN WATERS ARE INFESTED WITH WATER BLOOMS. EXPERIMENTS WITH BLOOM
TYPES AND THEIR EFFECT ON ANIMALS PROVED THAT INDIAN WATER BLOOMS ARE
NEVER TOXIC ALTHOUGH THE SAME ALGAL SPECIES ARE POISONOUS IN OTHER
COUNTRIES. BLUE-GREEN ALGAE ENCOURAGE GROWTH OF FISH WHICH FEED UPON
THEM DIRECTLY OR INDIRECTLY, PROVIDING AN EXCELLENT SOURCE OF
FERTILIZATION OF A POND. THE BLOOM CAN ALSO BE UTILIZED AS GREEN
MANURE; DRIED BOTTOM MUDS ARE USED AS A MANURE SOURCE. THE ALGAE BLOOMS
ARE INITAILLY SAPROPHYTIC--SIGNIFICANT IN VIEW OF THEIR SUDDEN
APPEARANCE AFTER A LONG ABSENCE. IN THE GANGES, THE ALGAL FLORA IS
TYPICALLY CHLOROPHYCEAN, BUT A MARKED CHANGE OCCURS WHEN THE RIVER
RECEIVES SEWAGE OR EFFLUENT, THEN A MYXOPHYCEAN BLOOM IS FORMED.
SAMPLES OF ALGAE BLOOM DRIED FOR FERTILIZATION OF HATCHERY PONDS HAVE
BEEN PREPARED WHICH CAN BE INOCULATED AND GROWN SUCCESSFULLY. A
BIOLOGICAL METHOD OF CONTROL BY INTRODUCING FISH AND CRUSTACEANS FOR
FOOD UTILIZATION MIGHT BE ENVISAGED. WHEN BLOOM DISINTEGRATION HAS
REACHED A MAXIMUM AND TEMPERATURE IS HIGH, A POND MAY BE COVERED WITH
DEAD FISHES INVOLVING ECONOMIC LOSS. (JONES-WISCONSIN)

FIELD 02H, 05C

ACCESSION NO. W70-05404

THE EUTROPHICATION OF LAKE KLOPEINER (IN GERMAN),

BIOLOGISCHE STATION LUNZ (AUSTRIA).

I. FINDENEGG.

OESTERREICHISCHE WASSERWIRTSCHAFT, VOL 17, NO 7-8, P 175-181, 1965. 4 FIG, 3 TAB.

DESCRIPTORS:
*EUTROPHICATION, TEMPERATURE STRATIFICATION, DEPTH, PLANKTON, LIGHT PENETRATION, SEASONAL, PRIMARY PRODUCTIVITY, CALCIUM, NITRATES, AMMONIA, PHOSPHATES, HYDROGEN ION CONCENTRATION, PHYTOPLANKTON, ODOR, WINDS, EPILIMNION, HYPOLIMNION, LIGHT INTENSITY, CONDUCTIVITY, ALGAE, SEWAGE, DIATOMS, CYANOPHYTA, CARBON RADIOISOTOPES, POTASSIUM, OXYGEN.

IDENTIFIERS:
*LAKE KLOPEINER(AUSTRIA), LAKE WORTHER(AUSTRIA), VERTICAL, CHALK, METALIMNION, OSCILLATORIA RUBESCENS, DINOBRYON, UROGLENA AMERICANA, CYCLOTELLA, CERATIUM HIRUNDINELLA, PERIDINIUM WILLEI, CHLOROCOCCALES, ANKISTRODESMUS, OOCYSTIS, CRYOMONAS EROSA, COSMARIUM, CHROCOCCUS LIMNETICUS, LYNGBYA LIMNETICA, ANABAENA FLOS AQUAE, TABELLARIA, MELOSIRA, FRAGILARIA CROTONENSIS, APHANIZOMENON.

ABSTRACT:
LAKE KLOPEINER, AUSTRIA, DESPITE HEAVY SEWAGE INFLUX, HAS NOT PROGRESSED FAR IN EUTROPHICATION, ALTHOUGH CONSIDERABLE AMOUNTS OF NUTRIENTS HAVE ACCUMULATED IN THE DEPTHS AS A RESULT OF MEROMIXIS. THIS MEROMICTIC BEHAVIOR IS THE KEY TO UNDERSTANDING THE STRATIFICATION OF OXYGEN AND ITS CHANGES SINCE 1931. THE ZONE FREE OF OXYGEN EXPANDED UPWARD DURING THE COURSE OF THE DECADES AND LAY BETWEEN 20 AND 35 METERS IN 1935 AND IS NOW ELEVATED TO A LEVEL BETWEEN 10-25 METERS. THIS MEROMICTIC BEHAVIOR IN WINTER AND THE STRICT SEPARATION OF BOTH EPI- AND HYPOLIMNION IN SUMMER GIVE A SPECIAL CHARACTER TO THE COURSE OF EUTROPHICATION. THE LAKE HAS AN UNUSUAL THERMIC POSITION: SELDOM DISTURBED BY WINDS AND THEN ONLY FOR SHORT PERIODS, IT HAS NO STRONG CURRENTS; LOCAL BREEZES DISTRIBUTE THE SURFACE WATER, WARMED BY SUNSHINE, ONLY WITHIN A THIN LAYER OF EPILIMNION. AN INCREASE IN AMOUNT OF ALGAE IS DUE TO THE IMMIGRATION OF A SPECIES OF PLANKTON, THE BLOOD ALGA (OSCILLATORIA RUBESCENS), NOT OBSERVED BEFORE. THIS SPECIES IS REGARDED AS AN INDICATOR OF INCREASING EUTROPHICATION ACCELERATED BY DOMESTIC SEWAGE. THERE IS ALSO A DECREASE OF TRANSPARENCY.
(JONES-WISCONSIN)

FIELD 02H, 05C

ACCESSION NO. W70-05405

ENVIRONMENTAL REQUIREMENTS OF FRESH-WATER PLANKTON ALGAE,

UPPSALA UNIV. (SWEDEN). INST. FOR PHYSIOLOGICAL BOTANY.

WILHELM RODHE.

SYMBOLAE BOTANICAE UPSALIENSES, VOL 10, NO 1, P 1-149, 1948. 33 FIG, 30 TAB, 96 REF.

DESCRIPTORS:
*ENVIRONMENT, *FRESH WATER, *PLANKTON, *ALGAE, SCENEDESMUS, LIGHT, CULTURES, IRON COMPOUNDS, DIATOMS, CYANOPHYTA, PHOSPHATES, CHLOROPHYLL, MAGNESIUM, POTASSIUM, TEMPERATURE, EPILIMNION, LIGHT INTENSITY, SEASONAL, PHYSIOLOGICAL ECOLOGY, ECOTYPES, CHLOROPHYTA, PHOTOMETRY, CHLORELLA, NUTRIENTS, TRACE ELEMENTS, CHLAMYDOMONAS, CHRYSOPHYTA, NITRATES, PLANT GROWTH, BICARBONATES, DEPTH.

IDENTIFIERS:
*REQUIREMENTS, SCENEDESMUS, ABSORPTION, CELL CONCENTRATION, UROGLENA, ANKISTRODESMUS, CHLOROCOCCALES, VOLVOCALES, CONJUGATAE, HETEROKONTAE, GLOEOTRICHIA, ASTERIONELLA, DINOBRYON, MELOSIRA, ANEBODA(SWEDEN), UPPSALA(SWEDEN), LAKE ERKEN(SWEDEN), EXTINCTION VALUES, EUDORINA, CULTURE MEDIUM, COELASTRUM, PEDIASTRUM.

ABSTRACT:
THIS WORK ELUCIDATES THE INFLUENCE OF SOME ENVIRONMENTAL FACTORS ON FRESHWATER PLANKTON ALGAE. THE REQUIREMENTS OF SOME ALGAE, ESPECIALLY SCENEDESMUS QUADRICAUDA, WERE STUDIED IN CULTURES UNDER CONTROLLED CONDITIONS AND COMPARED WITH LAKE CONDITIONS. SUBSTANCE PRODUCTION, CELL MULTIPLICATION, AND CHLOROPHYLL FORMATION ARE RATHER INDEPENDENT ASPECTS OF THE DEVELOPMENT OF A CULTURE. PARALLEL DETERMINATIONS SHOULD BE CARRIED OUT FOR THE DIFFERENT PARAMETERS. ALPHA A-DIPYRIDYL AND O-PHENANTHROLINE ARE SENSITIVE AND RELIABLE REAGENTS FOR DETERMINATION OF IRON IN WATER. IN LIGHT, ONE PART CRITRIC ACID TO ONE PART FERRIC CITRATE HAS A STABILIZING EFFECT SUFFICIENT FOR ALGAL CULTURES. A CULTURE SOLUTION WAS COMPOSED SUITABLE FOR UNLIMITED CULTIVATION OF ABOUT 40 SPECIES AND FORMS BELONGING TO CHLOROCOCCALES, VOLVOCALES, CONJUGATAE, HETEROKONTAE, AND DIATOMS. THE PHOSPHATE DEPENDENCE OF GROWTH CAN DIFFER FOR VARIOUS PLANKTON ALGAE. EXPERIMENTAL RESULTS WITH DINOBRYON AND UROGLENA SUGGEST THAT PHOSPHORUS, EVEN IN ORDINARY LAKES, MAY BECOME A MAXIMUM FACTOR FOR SOME ALGAE; ECOLOGICAL OBSERVATIONS CONFIRM THIS SUGGESTION. THE COMPETITION OF ALGAE FOR AVAILABLE NITROGEN IONS IS A FACTOR IN NATURAL CONDITIONS WITH DIFFERENT ALGAL DENSITIES. IN AVERAGE LAKES MAGNESIUM AND POTASSIUM NEVER LIMIT THE DEVELOPMENT OF SCENEDESMUS QUADRICAUDA. MELOSIRA ISLANDICA SSP HELVETICA PREFERS LOW TEMPERATURES; ANKISTRODESMUS FALCATUS HIGHER TEMPERATURES. (JONES-WISCONSIN)

FIELD 02H, 05C

ACCESSION NO. W70-05409

PHOSPHATE FERTILIZATION OF PONDS,

ALABAMA AGRICULTURAL EXPERIMENT STATION, AUBURN.

H. S. SWINGLE, B. C. GOOCH, AND H. R. RABANAL.

PROCEEDINGS OF THE 17TH ANNUAL CONFERENCE, SOUTHEASTERN ASSOCIATION OF GAME AND FISH COMMISSIONERS, HELD AT HOT SPRINGS, ARKANSAS, SEPTEMBER 29-OCTOBER 2, P 213-218, 1963. 1 TAB, 14 REF.

DESCRIPTORS:
*PONDS, *FERTILIZATION, *PHOSPHATE, FISH, PLANKTON, AQUATIC WEED CONTROL, INSECT CONTROL, MOSQUITOS, ALGAL CONTROL, BASS, CARP, CATFISHES, NITROGEN, POTASSIUM.

IDENTIFIERS:
PHOSPHATE FERTILIZERS, BLUEGILLS, MICROCYSTIS SCUMS, PITHIOPHORA ALGAE.

ABSTRACT:
REINFORCEMENT OF PHOSPHATE FERTILIZATION OF FISH PONDS WITH NITROGEN AND POTASSIUM PRODUCED NO SIGNIFICANT INCREASE IN THE BIOMASS OF PLANKTON OR FISH YIELD. IN PONDS PREVIOUSLY TREATED WITH 8-8-2 FERTILIZER, MICROCYSTIS SCUMS CONTINUED TO BE A PROBLEM EVEN WHEN FERTILIZATION WAS LIMITED TO PHOSPHATES. IN SOME INSTANCES, APPLICATION OF PHOSPHATE FERTILIZERS ALONE DELAYED RESPONSE FOR 30 TO 60 DAYS AND PERMITTED THE GROWTH OF PITHOPHORA ALGAE. (WILDE-WISCONSIN)

FIELD 02H, 05C

ACCESSION NO. W70-05417

EFFECTS OF ACID MINE WASTES ON PHYTOPLANKTON IN NORTHERN ONTARIO LAKES,

ONTARIO WATER RESOURCES COMMISSION, TORONTO.

M. G. JOHNSON, M. F. P. MICHALSKI, AND A. E. CHRISTIE.

ONTARIO WATER RESOURCES COMMISSION, DIVISION OF RESEARCH - PUBLICATION NO 30, 1969. 41 P, 5 TAB, 6 FIG, 15 REF, 3 APPEND.

DESCRIPTORS:
*ACID MINE WATER, ACIDIC WATER, HYDROGEN ION CONCENTRATION, WATER CHEMISTRY, *PHYTOPLANKTON, CARBON DIOXIDE, NUTRIENT REQUIREMENTS, POPULATIONS, ENVIRONMENTAL EFFECTS, *PRIMARY PRODUCTIVITY, WATER POLLUTION EFFECTS, BIOASSAYS, ALGAE, COMPARATIVE PRODUCTIVITY, AQUATIC PRODUCTIVITY, DISTRIBUTION, SPECIES DIVERSITY.

IDENTIFIERS:
ONTARIO LAKES, QUIRKE LAKE, PECORS LAKE, DUNLOP LAKE, *INORGANIC CARBON.

ABSTRACT:
STUDIES WERE CARRIED OUT ON THREE NORTHERN ONTARIO LAKES; TWO (QUIRKE AND PECORS LAKES) WERE CONTAMINATED BY FREE MINERAL ACIDITY AND ONE (DUNLOP LAKE) WAS UNAFFECTED. LOWER PHYTOPLANKTON POPULATIONS AND INDICES OF DIVERSITY WERE FOUND IN QUIRKE AND PECORS LAKES THAN IN DUNLOP LAKE. MANY SPECIES OF BACILLARIOPHYCEAE, CHRYSOPHYCEAE AND MYXOPHYCEAE DEVELOPED IN THE REFERENCE LAKE BUT WERE ABSENT OR OCCURRED IN EXTREMELY LOW NUMBERS IN THE CONTAMINATED LAKES. AVERAGE PRIMARY PRODUCTIVITIES IN DUNLOP, QUIRKE AND PECORS LAKES WERE 126, 71 AND 34 MG C M-(2) DAY-(1), RESPECTIVELY. IN SITU AREAL AND VOLUMETRIC MEASUREMENTS IN LABORATORY AND FIELD BIOASSAYS CONFIRMED THE' IMPORTANCE OF INORGANIC CARBON IN LIMITING PRIMARY PRODUCTIVITY. IT IS CONCLUDED THAT INORGANIC CARBON LIMITS PRIMARY PRODUCTIVITY IN THE LAKES CONTAMINATED BY ACID MINE WASTES, WITH THE REDUCED PH AND INORGANIC CARBON DECREASING BOTH THE NUMBER OF SPECIES AND TOTAL NUMBER OF PHYTOPLANKTON. (SJOLSETH AND KATZ-WASHINGTON)

FIELD 05C

ACCESSION NO. W70-05424

THE POTENTIAL ROLE OF FISHERY MANAGEMENT IN THE REDUCTION OF CHAOBORID MIDGE POPULATIONS AND WATER QUALITY ENHANCEMENT,

S. F. COOK, JR.

CALIFORNIA VECTOR VIEWS, VOL 15, NO 7, P 63-70, JULY 1968. 2 FIG, 11 REF.

DESCRIPTORS:
FISH MANAGEMENT, STREAM IMPROVEMENT, *WATER QUALITY CONTROL, *MIDGES, FISH FOOD ORGANISMS, NUTRIENTS, ENERGY TRANSFER, *FOOD CHAINS, FOOD WEBS, FORAGE FISH, PLANKTON, ALGAE, *ALGAL CONTROL, PREDATION, COMPETITION, *INSECT CONTROL.

IDENTIFIERS:
*FISHERY MANAGEMENT, *CHAOBORID MIDGES, CLEAR LAKE(CALIF), NUTRIENT ENERGY PATHWAYS, TROPHIC RELATIONSHIPS.

ABSTRACT:
INFORMATION HAS BEEN ACQUIRED INDICATING THAT AT LEAST PARTIAL BIOLOGICAL CONTROL OF CHAOBORID MIDGES AND PLANKTONIC ALGAE MAY BE POSSIBLE WITH THE USE OF FISH. TWO BASIC CONCEPTS PERTAINING TO THIS RELATIONSHIP HAVE BEEN EXPLORED; CONTROL THROUGH DIRECT INFLUENCE OF PREDATION AND CONTROL THROUGH COMPETITION FOR NUTRIENT ENERGY. THE EMPHASIS IS PLACED ON THE LATTER CONCEPT, AND MEANS FOR ITS IMPLEMENTATION ARE DISCUSSED. THE WORK HAS CENTERED PRIMARILY UPON THE TROPHIC RELATIONSHIPS BETWEEN THE MIDGES AND PLANKTON AND TWO SPECIES OF ATHERINID SILVERSIDES AND THE THREADFIN SHAD. (SJOLSETH-WASHINGTON)

FIELD 05G, 05C

ACCESSION NO. W70-05428

REMOVAL OF SELECTED CONTAMINANTS FROM WATER BY SORPTION OF COAL,

VIRGINIA POLYTECHNIC INST., BLACKSBURG. WATER RESOURCES RESEARCH CENTER.

PAUL H. KING, FRANCIS R. MCNEICE, AND PIERRE S. WARREN.

AVAILABLE FROM THE CLEARINGHOUSE AS PB-190 802, $3.00 IN PAPER COPY, $0.65 IN MICROFICHE. VIRGINIA POLYTECHNIC INSTITUTE WATER RESOURCES RESEARCH CENTER BULLETIN 32, NOVEMBER 1969. 75 P, 34 FIG, 3 TAB, 37 REF. OWRR PROJ NO A-015-VA.

DESCRIPTORS:
*TERTIARY TREATMENT, *PHOSPHORUS COMPOUNDS, *NUTRIENTS, *PESTICIDES, *WATER POLLUTION CONTROL, EUTROPHICATION, ALGAE, FILTRATION, SORPTION, FLOCCULATION, COAL, ACTIVATED CARBON.

IDENTIFIERS:
COAL CONTACT PROCESSES, PESTICIDE REMOVAL.

ABSTRACT:
REMOVAL OR LIMITATION OF SOLUBLE PHOSPHORUS IN WASTEWATER EFFLUENTS IS A KEY FACTOR IN ALLEVIATING THE AESTHETIC AND ECONOMIC PROBLEMS ASSOCIATED WITH EXCESSIVE ALGAL GROWTH. VERY DILUTE CONCENTRATIONS OF LONG-LIVED PESTICIDES CAN IN TIME BE LETHAL TO ANIMALS EXPOSED TO THEM. IF DANGEROUS OR POTENTIALLY DANGEROUS CONCENTRATIONS OF PESTICIDES EXIST IN WATER UTILIZED FOR HUMAN CONSUMPTION, METHODS MUST BE DEVELOPED TO REMOVE THEM EASILY AND ECONOMICALLY. ACTIVATED CARBON IS VERY EFFECTIVE IN ADSORBING ORGANIC COMPOUNDS FROM DILUTE SOLUTION. THE HIGH COST OF THIS ADSORBENT, HOWEVER, OFTEN PROHIBITS ITS USE IN LARGE-SCALE TREATMENT PROCESSES. COAL IS RELATIVELY LOW IN COST AND THUS IS AN ATTRACTIVE FILTER MEDIA IF THE REMOVAL OF DISSOLVED MATERIAL BY SORPTION CAN BE ACCOMPLISHED CONCURRENTLY WITH THE ELIMINATION OF SUSPENDED SOLIDS BY FILTRATION. ALTHOUGH COAL IS A POORER SORBENT THAN IS ACTIVATED CARBON, IT IS CONSIDERABLY MORE ECONOMICAL AND CAN BE INCINERATED AFTER EXPOSURE TO A WASTEWATER STREAM. THE EXTENT OF UPTAKE OF SELECTED PHOSPHORUS CONTAINING COMPOUNDS AND SYNTHETIC ORGANIC PESTICIDES BY COAL WERE EVALUATED IN BOTH BATCH CONTINUOUS FLOW OPERATIONS. THE KINETICS OF THE UPTAKE PROCESS AS WELL AS THE CAPACITY OF THE COAL FOR THE SORBATE AT EQUILIBRIUM ARE REPORTED. (KNAPP-USGS)

FIELD 05D

ACCESSION NO. W70-05470

EUTROPHICATION TRENDS IN A CHAIN OF ARTIFICIAL LAKES IN MADRAS (INDIA),

HYDROLOGICAL RESEARCH STATION, KILPAUK (INDIA).

A. SREENIVASAN.

ENVIRONMENTAL HEALTH, VOL 11, P 392-401, 1969. 5 TAB, 22 REF.

DESCRIPTORS:
*RESERVOIRS, *EUTROPHICATION, NITROGEN, OXYGEN, OXIDATION, CARBON DIOXIDE, ALKALINITY, ALGAE.

IDENTIFIERS:
INDIA, METHYL ORANGE, ALGAL BLOOMS.

ABSTRACT:
EUTROPHICATION OF FOUR CONSECUTIVE ARTIFICIAL LAKES--OOTY, SANDYNULLA, PYKARA AND GLENMORGAN FOREBAY, INDIA,--IS DISCUSSED. MORPHOMETRIC DATA, GENERAL CONDITIONS OF THE LAKES FOR THE YEARS 1965-1967, THE HYDROLOGICAL CONDITIONS OF GLENMORGAN RESERVOIR, AND THE QUALITY OF INFLUENT WATER INTO OOTY LAKE AND ITS OUTFLOW, ARE FURNISHED. DUE TO DAMAGE OF A WASTE DISPOSAL PIPE, SEWAGE HAS DISCHARGED INTO OOTY MAKING IT HYPER-EUTROPHIC. DURING 1962-1963, OXYGEN DEPLETION BELOW 2.0 WAS NOTED. DURING 1965-1966 STEEP GRADIENTS IN FREE CARBON DIOXIDE AND METHYL ORANGE ALKALINITY WERE NOTED AND THE ORGANIC NITROGEN CONTENT REACHED 7.6-18.0 MILLIGRAMS/LITER. OOTY LAKE PLANKTON IS DENSE YEAR ROUND WITH MICROCYSTIS AERUGINOSA OCCURRING AS A PERMANENT BLOOM. CARBON DIOXIDE ACCUMULATES IN THE BOTTOM OF SANDYNULLA RESERVOIR, BUT NOT AS SEVERELY AS IN OOTY. METHYL ORANGE ALKALINITY GRADIENT WAS NOTED FROM SURFACE TO BOTTOM DURING CERTAIN PERIODS, MORE OR LESS REFLECTING THE INFLUENCE OF OOTY LAKE. OXYGEN DEPLETION IN THE BOTTOM WAS NOTED AND ORGANIC NITROGEN CONTENT WAS 2.1-15.0 PPM. PYKARA LAKE SHOWED NO CHANGE. GLENMORGAN FOREBAY SHOWS THAT IT IS NOT INFLUENCED BY THE WATERS FROM THE OTHER LAKES. THE REMOVAL OF BOTTOM SEDIMENTS FROM OOTY IS SUGGESTED TO ALLEVIATE THE PROBLEM. (HASKINS-WISCONSIN)

FIELD 02H, 05C

ACCESSION NO. W70-05550

FACTORS INVOLVED IN DISPOSAL OF SEWAGE EFFLUENTS TO LAKES,

MASSACHUSETTS INST. OF TECH., CAMBRIDGE. DEPT. OF SANITARY CHEMISTRY.

CLAIR N. SAWYER.

SEWAGE AND INDUSTRIAL WASTES, VOL 26, NO 3, P 317-328, 1954. 6 FIG, 3 TAB, 10 REF, DISCUSSION.

DESCRIPTORS:
*LAKES, *SEWAGE EFFLUENTS, *SEWAGE DISPOSAL, ALGAE, AGRICULTURAL WATERSHEDS, NUTRIENTS, FERTILIZATION, PHYSICAL PROPERTIES, CLIMATE, NITROGEN, PHOSPHORUS.

IDENTIFIERS:
ALGAL BLOOMS, ALGAE NUTRITION, MICROCYSTIS AERUGINOSA.

ABSTRACT:
THE SIGNIFICANCE OF FERTILIZATION AS THE CAUSE OF ALGAL BLOOMING AND THE VARIOUS PHYSICAL, CHEMICAL, AND BIOCHEMICAL FACTORS THAT DETERMINE THE SAFE LOADINGS WHICH MAY BE APPLIED ARE DISCUSSED. FUNDAMENTAL CONSIDERATIONS ARE SOURCES OF FERTILIZING MATTER REACHING LAKES INCLUDING GROUNDWATER, SURFACE RUNOFF, AND SEWAGE EFFLUENTS OR INDUSTRIAL WASTES; PERMISSIBLE NUTRIENT OR FERTILIZING LEVELS; AND FACTORS CONSIDERED OF IMPORTANCE TO LAKE BEHAVIOR. CLIMATE; PHYSICAL FACTORS OF THE LAKE INCLUDING AREA, MORPHOLOGY, DEPTH, TEMPERATURE, AND COLOR; BIOLOGICAL FACTORS; AND CHEMICAL FACTORS INCLUDING ESSENTIAL ELEMENTS, CARBON DIOXIDE, PH, AND FEEDBACK FROM BOTTOM DEPOSITS ARE PRESENTED. THE CONTOUR OF THE LAKE DETERMINES TO A GREAT EXTENT WHEN AND HOW ALGAL GROWTH DEVELOPS. FOR LAKES THAT DO NOT STRATIFY, ALL NUTRIENTS ARE AVAILABLE; IN STRATIFIED LAKES, ELEMENTS BELOW THE THERMOCLINE ARE SEALED OFF FROM THE ACTIVE BIOLOGICAL ZONE. ECOLOGICAL RELATIONSHIPS ARE AN IMPORTANT FACTOR IN DETERMINING LAKE BEHAVIOR. NITROGEN IS NEEDED IN THE GREATEST AMOUNT AS A FERTILIZING ELEMENT. DISCUSSIONS OF THE PAPER BY PROFESSOR HAROLD B GOTAAS OF UNIVERSITY OF CALIFORNIA, BERKELEY, AND PROFESSOR JAMES B LACKEY OF THE UNIVERSITY OF FLORIDA, GAINESVILLE, ARE INCLUDED. (HASKINS-WISCONSIN)

FIELD 05C, 02H

ACCESSION NO. W70-05565

THE CHEMICAL CONTROL OF AQUATIC NUISANCES,

WISCONSIN COMMITTEE ON WATER POLLUTION, MADISON.

KENNETH M. MACKENTHUN.

COMMITTEE ON WATER POLLUTION, MADISON, WISCONSIN, JANUARY 1958, 64 P, 22 FIG, 3 TAB.

DESCRIPTORS:
*AQUATIC WEED CONTROL, *HERBICIDES, *COPPER SULFATE, *SODIUM ARSENITE, NUISANCE ALGAE, SNAILS, LEGISLATION, WISCONSIN, CHEMCONTROL, PONDWEEDS, ROOTED AQUATIC PLANTS, RATES OF APPLICATION, APPLICATION TECHNIQUES, SWIMMING, WATERFOWL, BIRDS, COSTS, COST-SHARING.

IDENTIFIERS:
SCHISTOSOME DERMATITIS, COPPER CARBONATE, CERCARIAE, WATER MILFOIL COONTAIL, WILD CELERY, BUR REED, DUCKWEED, WHITE WATER LILY, SWIMMERS' ITCH.

ABSTRACT:
A DETAILED DESCRIPTION OF METHODS OF CONTROL OF AQUATIC ROOTED VEGETATION, ALGAE, AND CERCARIAE, CAUSING SWIMMERS' ITCH, IS GIVEN. THE BIOMASS OF AQUATIC WEEDS WAS REDUCED BY SODIUM ARSENITE TREATMENTS IN SOME 80 LAKES WITH AN APPROXIMATE APPLICATION EFFICIENCY OF 150 GALLONS OF HERBICIDE PER HOUR. ONE OR MORE ALGAE-ERADICATING TREATMENTS WERE GIVEN TO ABOUT 40 LAKES, AN APPLICATION OF 250 POUNDS OF COMMERCIAL COPPER SULFATE CONSUMING ONE HOUR. CONTROL OF CERCARIAE WAS SUCCESSFULLY ACHIEVED IN A LIMITED NUMBER OF LAKES WITH A MIXTURE OF COPPER SULFATE AND COPPER CARBONATE. THE EFFICIENCY OF CONTROL MEASURES WAS MATERIALLY AUGMENTED BY THE USE OF SPECIALLY DESIGNED EQUIPMENT, INCLUDING STEEL BARGES, PORTABLE ALUMINUM ALLOY BARGES, SPRAY BOATS AND ASSEMBLIES, LOADING BOOMS, AND GRAVITY FLOW INJECTORS. (WILDE-WISCONSIN)

FIELD 05C

ACCESSION NO. W70-05568

WASTE WATER RECLAMATION PROJECT FOR ANTELOPE VALLEY AREA,

LOS ANGELES DEPT. OF COUNTY ENGINEER, CALIF.

JOHN A. LAMBIE.

AVAILABLE FROM THE CLEARINGHOUSE AS PB-191 067, $3.00 IN PAPER COPY, $0.65 IN
    MICROFICHE. LOS ANGELES COUNTY DEPARTMENT OF ENGINEER REPORT, AUGUST 1968.
    225 P, 14 FIG, 32 TAB, 71 REF. FWPCA PROJECT 17080---8/68.

    DESCRIPTORS:
      *WATER REUSE, *RECLAIMED WATER, *CALIFORNIA, WATER QUALITY, AESTHETICS,
      RECREATION, FISHING, ALGAE, NUTRIENTS, TURBIDITY, TERTIARY TREATMENT,
      SAMPLING, MONITORING, WATER DEMAND, WATER SUPPLY, ARID LANDS, LIMNOLOGY.

    IDENTIFIERS:
      LOS ANGELES COUNTY(CALIF).

    ABSTRACT:
      AN ECONOMICALLY FEASIBLE WASTE WATER RENOVATION PROCESS WAS DEVELOPED
      IN LOS ANGELES COUNTY. THE TERTIARY TREATED PRODUCT WATER IS
      PATHOGENICALLY SAFE, ESTHETICALLY PLEASING AND SUITABLE FOR FISH LIFE.
      PRIMARY USE OF THE WATER WILL BE FOR AN AQUATIC RECREATION PARK, BUT
      OTHER ANTICIPATED USES INCLUDE SOIL RECLAMATION, IRRIGATION, AND
      INDUSTRIAL APPLICATIONS. THESE REUSES OF WASTE WATER WILL RESULT IN THE
      CONSERVATION OF THE PRESENTLY DIMINISHING FRESH WATER SUPPLY IN THIS
      ARID AREA. ALL RESEARCH AND TESTING WAS DONE WITH THE KNOWLEDGE THAT
      ULTIMATELY THE PUBLIC MUST ACCEPT THE CONCEPT OF REUSE OF RENOVATED
      WASTE WATER FOR THE PROJECT TO BE A TOTAL SUCCESS. ESTHETICS OF THE
      PRODUCT WATER, INCLUDING CLARITY, COLOR AND ODOR WERE IMPORTANT ASPECTS
      OF THE TREATMENT PROCESS. TEST DATA DEMONSTRATE THAT BACTERIOLOGICAL
      AND VIRAL REQUIREMENTS CAN BE MET. FISH HAVE SUCCESSFULLY SURVIVED AND
      PROPAGATED IN THE TEST PONDS. ALGAL GROWTH AND NUTRIENT LEVELS OF WATER
      IN THE PILOT PONDS ARE CONSIDERED WITHIN ACCEPTABLE LIMITS.

    FIELD 05D, 05F

    ACCESSION NO. W70-05645

QUANTITATIVE MICRO-DETERMINATION OF LIPID CARBON IN MICROORGANISMS,

    CALIFORNIA UNIV., SAN DIEGO, LA JOLLA. INST. OF MARINE RESOURCES; AND SCRIPPS
      INSTITUTION OF OCEANOGRAPHY, LA JOLLA, CALIF. DIV. OF MARINE BIOLOGY.

    O. HOLM-HANSEN, J. COOMBS, B. E. VOLCANI, AND P. M. WILLIAMS.

    ANALYTICAL BIOCHEMISTRY, VOL 19, NO 3, P 561-568, 1967. 1 FIG, 1 TAB, 17 REF.

    DESCRIPTORS:
      *ANALYTICAL TECHNIQUES, *MICROORGANISMS, *LIPIDS, LIGHT, ALGAE,
      TEMPERATURE, CARBON DIOXIDE, OXIDATION, CHEMICALS, CARBON, GRAVIMETRY.

    IDENTIFIERS:
      CHLOROFORM-METHANE EXTRACTS, LIPID CARBON, CHLOROPLASTS, INFRARED
      ANALYSIS.

    ABSTRACT:
      THE DESCRIBED PROCEDURE FOR THE QUANTITATIVE DETERMINATION OF LIPID IN
      MICROORGANISMS EMPLOYS COMBUSTION OF THE SOLVENT EXTRACTS TO CARBON
      DIOXIDE AND MEASUREMENT OF THE LIBERATED CARBON DIOXIDE BY INFRARED
      ANALYSIS. THE METHOD PERMITS ANALYSES OF SAMPLES CONTAINING AS LITTLE
      AS 20 MICROGRAMS OF LIPID, AND THE RESULTS ARE COMPARABLE TO THOSE
      OBTAINED BY GRAVIMETRIC AND COLORIMETRIC METHODS. (WILDE-WISCONSIN)

    FIELD 07B, 05A

    ACCESSION NO. W70-05651

BIOASSAY OF SEAWATER. IV. THE DETERMINATION OF DISSOLVED BIOTIN IN SEAWATER
USING C-14 UPTAKE BY CELLS OF AMPHIDINIUM CARTERI,

CALIFORNIA UNIV., SAN DIEGO, LA JOLLA. INST. OF MARINE RESOURCES.

A. F. CARLUCCI, AND S. B. SILBERNAGEL.

CANADIAN JOURNAL OF MICROBIOLOGY, VOL 3, P 979-986, 1967. 2 FIG, 3 TAB, 11
REF.

DESCRIPTORS:
*BIOASSAY, *SEA WATER, *VITAMINS, VITAMIN B, ALGAE, CULTURES, CHEMICAL
ANALYSIS, ANALYTICAL TECHNIQUES, DINOFLAGELLATES, MARINE BACTERIA,
WATER CHEMISTRY, CARBON RADIOISOTOPES, PACIFIC OCEAN, CARBON DIOXIDE,
LIGHT, TEMPERATURE.

IDENTIFIERS:
*BIOTIN, *AMPHIDINIUM CARTERI, *RADIOCARBON UPTAKE TECHNIQUES, WATER
POLLUTION IDENTIFICATION, ORGANIC SOLUTES, VITAMIN B-1, VITAMIN B-12,
THIAMINE, SERRATIA MARINORUBRA, GYRODINIUM COHNII, MICROBIAL ECOLOGY,
COULTER COUNTER, GROWTH INHIBITION, CYANOCOBALAMIN, MUTANT BACTERIA,
CYCLOTELLA NANA, GEIGER COUNTER.

ABSTRACT:
THREE VITAMINS OF POTENTIAL IMPORTANCE IN MARINE MICROBIAL ECOLOGY ARE
BIOTIN, CYANOCOBALAMIN (B-12), AND THIAMINE (B-1). MUTANT MARINE
BACTERIA AND HETEROTROPHIC DINOFLAGELLATES HAVE BEEN EMPLOYED IN
BIOASSAYS TO ESTIMATE BIOTIN PREVIOUSLY, BUT SUCH ASSAYS HAVE BEEN
RELATIVELY INSENSITIVE. BIOTIN DISSOLVED IN SEAWATER CAN BE DETERMINED
BY RAPID, SENSITIVE, AND PRECISE TECHNIQUE DESCRIBED. UPTAKE OF
RADIOCARBON-LABELED CARBON DIOXIDE CELLS OF THE MARINE DINOFLAGELLATE,
AMPHIDINIUM CARTERI, DURING 2 HOURS FOLLOWING 94 HOURS OF PREINCUBATION
UNDER CAREFULLY CONTROLLED CONDITIONS IS PROPORTIONAL TO BIOTIN
CONCENTRATION. CELLS CAN BE ENUMERATED IN HEMACYTOMETER OR WITH
ELECTRONIC PARTICLE COUNTER TO EVALUATE GROWTH RESPONSES OF
DINOFLAGELLATE POPULATIONS TO VITAMIN CONCENTRATIONS, BUT INCUBATION
MUST BE EXTENDED TO 144 OR 168 HOURS TO OBTAIN PROPORTIONAL
DOSE-RESPONSE CURVES. BIOTIN CAN BE ASSAYED IN RANGE OF CONCENTRATIONS,
0.2-6.0 NANOGRAMS/LITER, A SENSITIVITY CONSIDERABLY GREATER THAN THOSE
FOR MICROBIOLOGICAL ASSAYS PREVIOUSLY DESCRIBED. SIX REPLICATED
DETERMINATIONS AT 1 AND 3 NANOGRAMS/LITER YIELDED STANDARD DEVIATIONS
OF 0.09 AND 0.28 RESPECTIVLY. FROM THE NORTH PACIFIC OCEAN, 45 SAMPLES
CONTAINED BIOTIN IN CONCENTRATION RANGE, 0.65-3.0 NANOGRAMS/LITER.
AVAILABILITY OF TECHNIQUES FOR ASSAY OF ECOLOGICALLY IMPORTANT
CONCENTRATIONS OF VITAMINS MAKES POSSIBLE A STUDY OF THEIR ROLE IN
MARINE ENVIRONMENTS. (EICHHORN-WISCONSIN)

FIELD 05A

ACCESSION NO. W70-05652

DEVELOPMENT OF A SYMBIOTIC ALGAL-BACTERIAL SYSTEM FOR NUTRIENT REMOVAL FROM
WASTEWATER,

OHIO STATE UNIV., COLUMBUS. DEPT. OF CIVIL ENGINEERING.

F. J. HUMENIK, AND G. P. HANNA, JR.

24TH ANNUAL PURDUE INDUSTRIAL WASTE CONFERENCE, MAY 6-8, PURDUE UNIVERSITY,
LAFAYETTE, INDIANA, 1969. 16 P, 7 FIG, 1 TAB, 15 REF.

DESCRIPTORS:
*WASTE WATER TREATMENT, *NUTRIENTS, ALGAE, BACTERIA, CULTURES, CHEMICAL
OXYGEN DEMAND, NITROGEN, PHOSPHATES, BIOMASS, SYMBIOSIS, OXYDATION
LAGOONS.

IDENTIFIERS:
*CHLORELLA PYRENOIDOSA, SYNTHETIC SEWAGE FEED, SYMBIOTIC RELATIONSHIPS.

ABSTRACT:
AN ALGAL-BACTERIAL CULTURE, DEVELOPED BY INOCULATION OF WASTEWATER WITH
CHLORELLA PYRENOIDOSA, PROVED TO BE EFFICIENT IN REMOVING CERTAIN
NUTRIENTS FROM THE EFFLUENTS. UNDER OPTIMUM CONDITIONS, THE
ALGAL-BACTERIAL FLOC SETTLED VERY RAPIDLY, YIELDING A CLEAR SUPERNAUT.
THE AVERAGE REMOVAL OF PHOSPHATES, HOWEVER, WAS LESS THAN 3%. MAXIMUM
REMOVAL OF CHEMICAL OXYGEN DEMAND AND ORGANIC NITROGEN WAS OBTAINED
DURING UNAERATED OPERATION WITH HARVESTING. THE BIOMASS OXIDATION
REQUIRED 1.27 MILLIGRAMS OF OXYGEN/MILLIGRAM OF SOLIDS. SUPPLEMENTAL
AERATION FAILED TO IMPROVE EITHER REMOVAL OR CONSERVATION OF NUTRIENTS.
THE TERM 'SYMBIOTIC CULTURE' WAS INTRODUCED ON THE BASIS OF A
STEADY-STATE EQUILIBRIUM IN THE OXYGEN CONCENTRATION, MAINTAINED BY
PHOTOSYNTHETIC OXYGENATION AND TOTAL RESPIRATION. (WILDE-WISCONSIN)

FIELD 05D

ACCESSION NO. W70-05655

CULTURAL EUTROPHICATION WITH SPECIAL REFERENCE TO LAKE WASHINGTON,

WASHINGTON UNIV., SEATTLE. DEPT. OF ZOOLOGY.

W. T. EDMONDSON.

MITTEILUNGEN INTERNATIONALE VEREINIGUNG FUR THEORETISCHE UND ANGEWANDTE
LIMNOLOGIE, VOL 17, P 19-32, 1969. 5 FIG, 37 REF, DISCUSSION.

DESCRIPTORS:
*LAKES, *EUTROPHICATION, *WASHINGTON, URBANIZATION, WATER POLLUTION,
SEDIMENTS, PALEOLIMNOLOGY, NUTRIENTS, PRODUCTIVITY, SEDIMENTATION
RATES, STRATIGRAPHY, BIOINDICATORS, ALGAE, OLIGOTROPHY, DIVERSION,
CORES, SEWAGE EFFLUENTS, DIATOMS, PHOSPHORUS, ORGANIC MATTER, NITROGEN,
CHLOROPHYLL.

IDENTIFIERS:
*LAKE WASHINGTON(WASH), *CULTURAL INFLUENCES, MICROCYSTIS, DREDGE
SAMPLES, ANABAENA, SEASONAL EFFECTS, DECOMPOSITION RATES, CHRONOLOGY,
CLADOCERANS, CHYDORID CLADOCERANS, SPECIES COMPOSITION, MICROFOSSILS,
COMPETITIVE EXCLUSION PRINCIPLE, OSCILLATORIA RUBESCENS, PISTON CORER,
FRAGILARIA, MELOSIRA, ASTERIONELLA, ARAPHIDINEAE, CENTRIC DIATOMS,
BOSMINA, DIAPHANOSOMA, OSCILLAXANTHIN.

ABSTRACT:
EUTROPHICATION OF LAKE WASHINGTON BY URBAN SEWAGE AND ITS SUBSEQUENT
RECOVERY AFTER INSTALLATION OF TREATMENT FACILITIES AND EFFLUENT
DIVERSION ARE WELL DOCUMENTED. PALEOLIMNOLOGICAL EVIDENCE BEARING ON
THIS CASE HISTORY IS DESCRIBED. OF MANY DIATOMS OCCURRING IN CORES,
FRAGILARIA CROTONENSIS DOMINATES RECENTLY; MELOSIRA ITALICA, IN THE
DEEPER (OLDER) SEDIMENTS; AND ASTERIONELLA FORMOSA OCCURS AT ALL
DEPTHS. NO SPECIES, ABUNDANT AT ANY DEPTH, IS MISSING FROM SIGNIFICANT
CORE SECTIONS. RELATIVE COMPOSITION OF CENTRIC VERSUS ARAPHIDINATE
GROUPS OF DIATOMS CORRELATE WITH KNOWN LAKE'S EUTROPHICATION HISTORY,
LARGE DECREASES IN ARAPHIDINEAE COINCIDING WITH DIVERSION OF RAW
SEWAGE. USING SPECIES GROUPS MAY OVERCOME INHERENT DISADVANTAGES POISED
BY CONSIDERING SINGLE SPECIES AS BIOINDICATORS. ANIMAL FOSSILS HAVE
BEEN GIVEN LITTLE STUDY, ALTHOUGH BOSMINA LONGIROSTRIS--COMMON
CLADOCERAN IN NATURALLY OR CULTURALLY EUTROPHIC LAKES--IS KNOWN TO BE
ABUNDANT THROUGHOUT SEDIMENTS, WITH INCREASING ABUNDANCE SEEMINGLY
PARALLELING PERIOD OF ENRICHMENT. BASED UPON REASONABLY RELIABLE MARKER
LAYER, SEDIMENTATION SINCE 1916 HAS PROCEEDED AT RATE BETWEEN 2.3 AND
3.1 MILLIMETERS/YEAR. SURFICIAL SEDIMENTS TEND TO BE DISTINCTLY
ENRICHED IN PHOSPHORUS, AFTER INTERPOSED MINIMUM IN 1930-1940, WHEN
SEWAGE WAS DIVERTED. IN GENERAL, EXTENSIVE LIMNOLOGICAL DATA ARE SCANTY
FOR LAKES WITH HISTORY OF EUTROPHICATION, MAKING ELUCIDATION OF
PALEOLIMNOLOGICAL CORRELATIVES DIFFICULT. (EICHHORN-WISCONSIN)

FIELD 05C, 02H

ACCESSION NO. W70-05663

EUTROPHICATION OF LAKES AND STREAMS, CAUSE AND PREVENTION (IN GERMAN),

KANTONALES LABORATORIUM, ZURICH (SWITZERLAND).

E. A. THOMAS.

MONATSBULLETIN DES SCHWEIZERISCHEN VEREINS VON GAS- UND WASSERFACHMANNERN, NO 6/7, P 3-12, 1963. 7 FIG, 24 REF.

DESCRIPTORS:
*EUTROPHICATION, *LAKES, *RIVERS, *ALGAE, WATER POLLUTION CONTROL, WATER POLLUTION EFFECTS, WATER POLLUTION SOURCES, SEWAGE, PHOSPHORUS, CHEMCONTROL, INDUSTRIAL WASTES, SEWAGE TREATMENT.

IDENTIFIERS:
RING TUBE, CANALIZATION, ALUMINUM SULFATE, FERRICHLORIDE, IRON SULFATE, GALVANIZING WASTES, LAKE ZURICH(SWITZERLAND), SWITZERLAND, KILCHBERG(SWITZERLAND), ZOLLIKON(SWITZERLAND), LAKE TEGERN(BAVARIA), LAKE SCHLIER(BAVARIA), LAKE ZELL(AUSTRIA), LAKE HALL WILER(SWITZERLAND).

ABSTRACT:
WATER POLLUTION WAS RECOGNIZED IN SOME COUNTRIES A CENTURY AGO WHEN ZURICH KANTON, SWITZERLAND, ENACTED A LAW IN 1881. WHILE MECHANICAL AND BIOLOGICAL TREATMENT OF SEWAGE ARE THE MOST GENERALLY USED METHODS OF WATER PURIFICATION AND EUTROPHICATION REVERSAL, THE RESULTANT EFFLUENTS STILL PROMOTE ALGAE AND HIGHER WATER PLANTS. LAKES SURROUNDED BY COMMUNITIES BENEFIT BY SEWAGE CANALIZATION CIRCLING THE LAKE, A PROCESS USED FOR 50 YEARS FOR A PORTION OF ZURICH AND ADJACENT COMMUNITIES OF KILCHBERG AND ZOLLIKON, TOTALING A 50,000 POPULATION. LARGE LAKES AND THOSE WITH IRREGULAR SHORELINES RECEIVING SEWAGE AS WELL AS NUTRIENT-RICH TRIBUTARY STREAMS REQUIRE PHOSPHATE REMOVAL FROM SEWAGE BY ALUMINUM SULFATE, BY FERRICHLORIDE, OR BY A COMBINATION OF IRON CHLORIDE AND IRON SULFATE. TESTS TO PRECIPITATE PHOSPHATE IN SEWAGE PROVED PROMISING, ESPECIALLY BY A PRECIPITANT PRODUCED BY SWISS SODA PLANT, ZURZACH, FROM PICKLE WASTES, A PRODUCT OF GALVANIZATION (89.9% PHOSPHATE ELIMINATED) AT 1/3 TO 1/4 COST OF PURE IRON CHLORIDE. THE PRODUCT ALSO SERVED TO ELIMINATE INDUSTRIAL WASTES AND TO PRODUCE A MARKETABLE COMMODITY. CHEMICAL CONTROL OF NUISANCES, CHANNELING THE HYPOLIMNION FROM A LAKE TO REMOVE NUTRIENTS, AND OXYGENATION EFFECTS ARE DISCUSSED. (AUEN-WISCONSIN)

FIELD 05C

ACCESSION NO. W70-05668

WASTE TREATMENT FOR THE CONTROL OF HETEROTROPHIC AND AUTOTROPHIC ACTIVITY IN
   RECEIVING WATERS,

HARVARD UNIV., CAMBRIDGE, MASS. DEPT. OF WATER RESOURCES.

WERNER STUMM, AND MARK W. TENNEY.

PROCEEDINGS 12TH SOUTHERN MUNICIPAL INDUSTRIAL WASTE CONFERENCE, RALEIGH,
   NORTH CAROLINA, P 97-111, 1963. 5 FIG, 1 TAB, 15 REF.

   DESCRIPTORS:
      *WASTE TREATMENT, *WATERS, *ORGANISMS, STREAMS, WATER POLLUTION,
      ANIMALS, BACTERIA, FUNGI, ALGAE, FLOATING PLANTS, ROOTED AQUATIC
      PLANTS, SELF-PURIFICATION, BIOCHEMICAL OXYGEN DEMAND, TOXICITY,
      OXYGEN-REDUCTION POTENTIAL, SOLAR RADIATION, ENERGY, CARBON DIOXIDE,
      PHOTOSYNTHESIS, GROWTH RATE, AEROBIC CONDITIONS, ANAEROBIC CONDITIONS,
      BIOLOGICAL TREATMENT, FERTILIZATION, RIVERS, SEWAGE, PHOSPHORUS,
      NITROGEN, CARBON, ALUMINUM.

   IDENTIFIERS:
      MATHEMATICAL STUDIES, AUTOTROPHIC, HETEROTROPHIC.

   ABSTRACT:
      THE FEASIBILITY OF AEROBIC BIOLOGICAL WASTE TREATMENT WITH
      COMPLEMENTARY CHEMICAL TREATMENT FOR FLOCCULATION OF DISPERSED
      MICROORGANISMS AND REMOVAL OF PHOSPHATES HAS BEEN EXPERIMENTALLY
      CORROBORATED. RESULTS WERE OBTAINED ON THE EFFICIENCY OF ORGANIC CARBON
      REMOVED BY BIOLOGICAL AND SUBSEQUENT CHEMICAL TREATMENT WITH ALUM.
      CHEMICAL TREATMENT ALONE (WITHOUT PRIOR BIOLOGICAL SUBSTRATE
      UTILIZATION) DOES NOT REMOVE ORGANIC CARBON TO AN APPRECIABLE EXTENT.
      BACTERIA, IN THE FLOCCULATION PROCESS, ACT AS HYDROPHILIC COLLOIDS. THE·
      STOICHIOMETRIC RELATION BETWEEN COAGULANT DOSE AND NUMBER OF
      MICROORGANISMS (OR ORGANIC CARBON OF BIOLOGICAL SOLIDS) IS INDICATED.
      THE REQUIRED TOTAL COAGULANT DEMAND IS A FUNCTION OF BOTH THE
      CONCENTRATION OF DISPERSED MICROORGANISMS AND THE CONCENTRATION OF
      PHOSPHORUS PRESENT. RESULTS ILLUSTRATE THAT CHEMICAL COAGULATIVE
      METHODS CAN SERVE AS A VALUABLE COMPLEMENT TO THE BIOLOGICAL SUBSTRATE
      UTILIZATION PROCESSES. THE COMPLEMENTARY CHEMICAL AND BIOLOGICAL
      TREATMENT IS MORE ECONOMICAL THAN THE CONVENTIONAL BIOLOGICAL
      TREATMENT; IT IS CERTAINLY MORE ECONOMICAL THAN A CONVENTIONAL
      BI9LOGICAL TREATMENT FOLLOWED BY A THIRD STAGE CHEMICAL TREATMENT. IN
      THOSE CASES WHERE CHEMICAL TREATMENT IS MORE BENEFICIAL THAN BIOLOGICAL
      TREATMENT, THE PROPOSED TREATMENT COMBINATION APPEARS TO BE, EVEN FROM
      AN ECONOMICAL POINT OF VIEW, THE MOST EXPEDIENT WASTE PURIFICATION
      ARRANGEMENT. (JONES-WISCONSIN)

   FIELD 05D

   ACCESSION NO. W70-05750

USE OF ALGAL CULTURE METHODS IN EUTROPHICATION RESEARCH (IN NORWEGIAN),

NORSK INSTITUTT FOR VANNFORSKNING, BLINDERN.

OLAV SKULBERG.

ENGLISH SUMMARY. NORDISKE JORDBRUKSFORSKERES FORENING, VOL 47, P 211-215, 1965. 1 FIG, 2 REF.

DESCRIPTORS:
*EUTROPHICATION, ALGAE, CULTURES, CHLOROPHYTA, CYANOPHYTA, BRACKISH WATER, OLIGOTROPHY, TEMPERATURE, LIGHT INTENSITY, CHLOROPHYLL, NUTRIENTS, CONDUCTIVITY, AGRICULTURAL WATERSHEDS, SEWAGE, WATER POLLUTION, CHEMICAL ANALYSIS, PHYSICAL PROPERTIES.

IDENTIFIERS:
BACILLARIOPHYCEAE, SELENASTRUM CAPRICORNUTUM PRINTZ, OSCILLATORIA VAUCHER, SKELETONEMA COSTATUM(GREV) CL, CHLORELLA OVALIS BUTCHER, NJOER LAKE(NORWAY), HOBOL RIVER(NORWAY), VAN LAKE(NORWAY).

ABSTRACT:
UNIALGAL CULTURES ARE USED IN A PROCEDURE TO MEASURE THE FERTILITY OF WATER RECEIVING POLLUTION WITH PLANT NUTRIENTS. THE TECHNIQUE DEVELOPED IS CONSIDERED A SUPPLEMENT TO CHEMICAL AND PHYSICAL METHODS FOR THE HANDLING OF EUTROPHICATION PROBLEMS. SPECIES FROM THE CLASSES OF CHLOROPHYCEAE, CYANOPHYCEAE, AND BACILLARIOPHYCEAE WERE USED AS TEST ALGAE. OF THESE, SELENASTRUM CAPRICORNUTUM, WHICH OCCURS WIDELY IN OLIGOTROPHIC AND EUTROPHIC FRESH WATER LOCALITIES UNDER NATURAL CONDITIONS, WAS USED FOR EXPERIMENTATION MOST EXTENSIVELY. OTHER SUITABLE ALGAE HAVE BEEN OSCILLATORIA VAUCHER SP, SKELETONEMA COSTATUM (GREV) CL, AND CHLORELLA OVALIS BUTCHER. THE METHOD CAN BE USED TO DETERMINE THE EFFICIENCY OF THE WATER TREATMENT PLANTS IN REDUCING PLANT NUTRIENTS IN THE EFFLUENT, AND CAN MEASURE THE TROPHIC LEVEL OF WATER MASS. EXPERIMENTAL METHODS ARE NECESSARY TO GIVE A DETAILED UNDERSTANDING OF THE BIOLOGICAL PHENOMENA WHICH FOLLOW EUTROPHICATION. WITH CAREFUL AND CRITICAL USE OF ALGAL CULTURE EXPERIMENTS, IT WOULD BE POSSIBLE TO DEFINE NUMERICAL RELATIONS BETWEEN THE CONDITIONS OF THE WATER, DEGREE OF POLLUTION AND THE BIOLOGICAL SITUATION.
(JONES-WISCONSIN)

FIELD 05C, 07B

ACCESSION NO. W70-05752

FAYETTEVILLE GREEN LAKE, NEW YORK. V. STUDIES OF PRIMARY PRODUCTION AND
  ZOOPLANKTON IN A MEROMICTIC MARL LAKE,

CORNELL UNIV., ITHACA, N.Y. DIV. OF BIOLOGICAL SCIENCES.

D. A. CULVER, AND G. J. BRUNSKILL.

LIMNOLOGY AND OCEANOGRAPHY, VOL 14, NO 6, P 862-873, 1969. 7 FIG, 4 TAB, 26
  REF.

DESCRIPTORS:
  *LAKES, *NEW YORK, *MEROMIXIS, *PRIMARY PRODUCTIVITY, *ZOOPLANKTON,
  SULFUR BACTERIA, DIEL MIGRATION, CARBON DIOXIDE, HYDROGEN ION
  CONCENTRATION, PHOSPHATES, NITRATES, DAPHNIA, SILICA, THERMAL
  STRATIFICATION, SULFIDES, AMMONIUM COMPOUNDS, OXIDATION REDUCTION
  POTENTIAL, DISSOLVED OXYGEN, CHEMICAL ANALYSIS, ORGANIC MATTER, ALGAE,
  IRON, HYDROGEN SULFIDE, CYANOPHYTA, CHLOROPHYTA, LIGHT PENETRATION.

IDENTIFIERS:
  *FAYETTEVILLE GREEN LAKE(NY), *MARL LAKES, CHLOROBIUM, FILINIA,
  MONIMOLIMNION, CHEMOCLINE, MIXOLIMNION, RADIOCARBON UPTAKE TECHNIQUES,
  BACTERIAL PLATE, BACTERIAL PIGMENTS, DIAPTOMUS, EPISCHURA,
  ORTHOCYCLOPS, KERATELLA, POLYARTHRA, CONOCHILUS, PYCNOCLINE,
  PHOTOLITHOTROPHIC BACTERIA, DESULFOVIBRIO, MESOTHERMY, MELOSIRA,
  SYNEDRA, RHODOMONAS, CYCLOTELLA, CHROMATIUM.

ABSTRACT:
  MEROMIXIS OF FAYETTEVILLE GREEN LAKE, NEW YORK, INFLUENCES ITS TROPHIC
  STRUCTURE IN THE SENSE THAT BACTERIAL PRIMARY PRODUCTION RESEMBLES
  TOTAL PRIMARY PRODUCTION IN EUTROPHIC LAKES, WHILE ALGAL PRODUCTION IS
  MORE OLIGOTROPHIC. SEVEN DETERMINATIONS OF RADIOCARBON UPTAKE IN LIGHT
  AND DARK BOTTLES WERE MADE BETWEEN OCTOBER 1966 AND THE FOLLOWING JULY.
  ANNUAL PRIMARY PRODUCTION WAS ESTIMATED AT 290 GRAMS CARBON/SQUARE
  METER, OF WHICH APPROXIMATELY 83% WAS BACTERIAL, PRIMARILY DUE TO WELL
  DEVELOPED PLATE OF PHOTOSYNTHETIC SULFUR OXIDIZERS IN THE CHEMOCLINE
  (18-20 METERS). ALGAL PRODUCTIVITY MAY BE LIMITED BY UNFAVORABLE RATIO,
  MONOVALENT CATIONS:DIVALENT CATIONS, WHICH HAS VALUE OF 0.06 IN THIS
  LAKE, WHEREAS HIGHER RATIOS, IN RANGE 2-30, ARE CONSIDERED OPTIMAL.
  ZOOPLANKTONIC BIOMASS WAS DOMINATED BY COPEPODS. DENSE CONCENTRATIONS
  OF ZOOPLANKTERS OCCUR NEAR THE BACTERIAL PLATE, WHERE THEY PROBABLY
  FEED ON BACTERIA OR ON MIXOLIMNETICALLY PRODUCED SESTON, WHICH TENDS TO
  ACCUMULATE IN THE CHEMOCLINE. QUANTITATIVE SAMPLING OF ZOOPLANKTON WITH
  CLARKE-BUMPUS SAMPLER YIELDED EVIDENCE FOR DIEL MIGRATIONS OF MANY
  SPECIES. (EICHHORN-WISCONSIN)

FIELD 02H, 05C

ACCESSION NO. W70-05760

THE EUTROPHICATION OF LAKES AND THE RADICAL CHANGE OF THEIR MATERIAL CONTENT
(IN GERMAN),

WALDEMAR OHLE.

LIMNOLOGISKA FOREINGEN I FINLAND, LIMNOLOGISYMPOSION, HELSINKI, 10-23, P
1-16, 1964. 10 REF.

DESCRIPTORS:
*EUTROPHICATION, *LAKES, NUTRIENTS, FERTILIZATION, BACTERIA, ALGAE,
PLANKTON, TUBIFICIDS, CALCIUM, SILICATES, DIATOMS, WINDS, NITROGEN,
PHOSPHATES, METHANE, MUD, OXYGEN, ANAEROBIC CONDITIONS, HYDROGEN
SULFIDE, IRON, FUNGI, CLAYS, BRACKISH WATER, FISHERIES, LIMNOLOGY.

IDENTIFIERS:
CHIRONOMIDS, LAKE CONSTANCE, WHITEFISH, WURTTEMBERG(GERMANY),
PLON(GERMANY), LAKE SCHLIER(GERMANY), LAKE TEGERN(GERMANY), GERMANY,
FINLAND, SEWAGE RING TUBES.

ABSTRACT:
MOST NORTH GERMAN LAKES HAVE BEEN SUBJECT TO EUTROPHICATION SINCE EARLY
1930'S DUE TO DOMESTIC SEWAGE INFLOWS AND ALLUVIAL SOIL PARTICLES. AT
THE END OF SUMMER STAGNATION, HYPOLIMNETIC OXYGEN IS REPLACED BY
HYDROGEN SULFIDE IN SOLUTION. THE SEDIMENTS, ACCUMULATING LAKE
NUTRIENTS, ARE REGARDED AS RESULTANT OF AUTOCHTHON DYNAMICS OF
METABOLISM AND INFLUENCES OF REGIONAL SURROUNDINGS. 3% TO 5% OF THE
MINERAL FERTILIZER WASHED FROM SLOPING FIELDS ARE EXTREMELY HIGH
AMOUNTS FOR LAKES WITH NO DENSE PLANT GROWTH, ESPECIALLY BECAUSE OF
PHOSPHATE ENRICHMENT. HIGHER AMOUNTS OF PHOSPHORUS ARE ADDED BY SEWAGE,
SUPPLEMENTED BY DETERGENTS. PHOSPHATE IS A DOMINANT PRODUCTION FACTOR
WITH NITROGEN COMPOUNDS OF SECONDARY IMPORTANCE. WHEN LARGE AMOUNTS OF
ORGANIC SUBSTANCES HAVE ACCUMULATED SO THAT DISSOLVED OXYGEN IN DEEP
WATER IS CONSUMED BY BACTERIAL ACTIVITY, LARGER AMOUNTS OF NUTRIENTS
ARE RELEASED FROM MUD INTO HYPOLIMNION. THE MORE BACTERIAL METHANE
OXIDATION DEVELOPS, THE HIGHER UTILIZATION OF OXYGEN, THUS MORE
NUTRIENTS ARE RECIRCULATED FROM THE MUD INTO THE WATER UNDER ANAEROBIC
CONDITIONS. THE INCREASING PRIMARY PRODUCTION IS INTENSIFIED
EXPONENTIALLY. ALL SEWAGE SHOULD BE DIVERTED FROM LAKES BY VARIOUS
MEANS AND CHEMICAL TREATMENT COORDINATED WITH BIOLOGICAL PURIFICATION.
(JONES-WISCONSIN)

FIELD 05C, 02H

ACCESSION NO. W70-05761

INCREASING WASTEWATER FLOW VELOCITY BY USING CHEMICAL ADDITIVES,

WESTERN CO. OF NORTH AMERICA. RICHARDSON, TEX. AND SOUTHERN METHODIST UNIV.,
DALLAS, TEX.

J. L. OVERFIELD, H. R. CRAWFORD, J. K. BAXTER, L. J. HARRINGTON, AND I. W.
SANTRY, JR.

JOURNAL OF THE WATER POLLUTION CONTROL FEDERATION, VOL 41, NO 9, SEPT 1969. P
1570-1585, 17 FIG, 1 TAB.

DESCRIPTORS:
*ADDITIVES, *FLOW RATE, *WASTE WATER, *ALGAE, *SLUDGE, STORMS,
TURBULENCE, PILOT PLANTS, BACTERIA, SEDIMENTATION, TOXICITY,
COST-BENEFIT ANALYSIS.

IDENTIFIERS:
*POLYMERS, *BIOLOGICAL LIFE, *ALGAE BLOOMS, *FISH TOXICITY TEST, SLUDGE
DRYING, SOLIDS-SETTLING, ECONOMIC ADVANTAGE, TEMPERATURE-EFFECT.

ABSTRACT:
PRACTICAL METHODS HAVE BEEN DEVELOPED TO INCREASE WASTE WATER FLOW
VELOCITY BY USING CHEMICAL ADDITIVES WITHOUT INCREASING THE ENERGY
NEEDED TO MOVE THE LIQUID. EXISTING SEWER LINE FLOW CAN BE INCREASED BY
USING POLYMER ADDITIVES THAT WOULD INCREASE THE SOLID REMOVAL
EFFECTIVENESS AT TREATMENT PLANTS. A 100 FT LONG 6 IN SEWER LINE TEST
FACILITY ALONG WITH 25,000-GAL CONTROLLED-TEMPERATURE RESERVOIR IS USED
TO DETERMINE THE INCREASED WATER FLOW WITH ADDITION OF CHEMICAL
ADDITIVES. THERE IS A RAPID INCREASE UP TO 24 TIMES THE NORMAL FLOW.
FLOW COULD BE INCREASED BY A FACTOR OF 2 OR MORE BY USING
CONCENTRATIONS OF ADDITIVES BETWEEN 45 AND 200 ML/L. THE COST/BENEFIT
ANALYSIS FOR A GIVEN EXAMPLE SHOWS THAT THE COST OF RECTIFYING OVERFLOW
FROM A CONSTRICTION BY USING AN ADDITIVE INJECTION SYSTEM, IS LESS THAN
HALF THE PRORATED COST OF CONSTRUCTING A PARALLEL PIPE SYSTEM TO
RELIEVE THE CONDITION. HOWEVER, IF THE OVERFLOWS OCCUR TEN TIMES A
YEAR, THE ADDITIVE INJECTION SYSTEM BECOMES MORE EXPENSIVE. BIOCHEMICAL
STUDIES CONDUCTED INDICATE THAT ADDITIVES DO NOT AFFECT ADVERSELY THE
WASTE WATER BACTERIA, FISH, OR PROMOTE ALGAE GROWTH. FURTHER, THE
ADDITIVES IMPROVE WASTE WATER SETTLING AND SLUDGE DRYING RATES.
ADDITIVE INJECTION SYSTEM, THROUGH PREFERABLE ECONOMICALLY IN MANY
CASES, CANNOT BE APPLIED AS A LONG TERM SOLUTION. (SHANKAR-TEXAS)

FIELD 05D

ACCESSION NO. W70-05819

AN INVESTIGATION INTO UPWARD-FLOW FILTRATION.·

WEST HERTFORDSHIRE MAIN DRAINAGE AUTHORITY, RICKMANSWORTH (ENGLAND).

RICHARD WOOD, WILLIAM S. SMITH, AND J. K. MURRAY.

WATER POLLUTION CONTROL, VOL 67, NO 4, 1968. P 421-428, 3 FIG, 7 TAB, 2 REF.

DESCRIPTORS:
*SEWAGE TREATMENT, *FILTRATION, *BIOCHEMICAL OXYGEN DEMAND, *ACTIVATED
SLUDGE, BIOLOGICAL TREATMENT, PILOT PLANTS, HUMUS, REAERATION,
EFFLUENTS, FLOW RATES, FILTERS, ALGAE.

IDENTIFIERS:
*UPWARD FLOW, PUTRESCIBILITY TEST, BODY IMMEDIUM FILTER, BACK-WASHING.

ABSTRACT:
SEWAGE TREATMENT AUTHORITIES ARE INCREASINGLY BEING REQUIRED TO PRODUCE
HIGHER QUALITY EFFLUENTS FROM INCREASING VOLUMES OF STRONGER SEWAGES.
ENLARGED ACTIVATED SLUDGE PLANT IS DESIGNED TO PRODUCE FULLY NITRIFIED
EFFLUENT OF LOW SOLIDS CONTENT, TO WHICH COULD BE ADDED ADDITIONAL
ACTIVATED SLUDGE SOLIDS. THIS MAKES IT POSSIBLE TO CARRY OUT AN
INVESTIGATION OF UPWARD-FLOW SAND FILTRATION. A BODY IMMEDIUM FILTER IS
USED TO DETERMINE THE EFFECTIVENESS IN FILTERING A HIGH-QUALITY
EFFLUENT CONTAINING A WIDE RANGE OF ACTIVATED SLUDGE CONCENTRATIONS.
SATISFACTORY REMOVALS OF COARSER ACTIVATED SLUDGE ARE OBTAINED. BY
DOSING THE FILTER WITH ARTIFICIALLY HIGH PROPORTIONS OF WELL-OXIDIZED
ACTIVATED SLUDGE, HIGH PERCENTAGE REMOVALS ARE OBTAINED. ALTHOUGH THE
UNIT IS MANUALLY CONTROLLED, THIS TYPE OF FILTER LENDS ITSELF TO
AUTOMATED OPERATION AND PROVIDED THAT THE FEED LIQUOR IS OF HIGH
QUALITY AND ONLY CONTAMINATED WITH SOLIDS, VERY HIGH QUALITY EFFLUENTS
CAN BE PRODUCED. (SHANKAR-TEXAS)

FIELD 05D

ACCESSION NO. W70-05821

THE ECOLOGY OF PERIPHYTON IN WESTERN LAKE SUPERIOR: PART I - TAXONOMY AND
DISTRIBUTION,

MINNESOTA UNIV., MINNEAPOLIS. WATER RESOURCES RESEARCH CENTER.

JACKSON L. FOX, THERON O. ODLAUG, AND THEODORE A. OLSON.

AVAILABLE FROM THE CLEARINGHOUSE AS PB-191 214, $3.00 IN PAPER COPY, $0.65 IN
MICROFICHE. MINNESOTA UNIVERSITY WATER RESOURCES RESEARCH CENTER BULLETIN
14, AUGUST 1969. 127 P, 50 FIG, 37 PLATE, 24 TAB, 79 REF, APPEND. OWRR
PROJECT NO A-011-MINN.

DESCRIPTORS:
*LAKE SUPERIOR, *PERIPHYTON, *LIMNOLOGY, *WATER QUALITY, BIOINDICATORS,
AQUATIC LIFE, AQUATIC ALGAE, WATER POLLUTION EFFECTS.

IDENTIFIERS:
PERIPHYTON INVENTORY(LAKE SUPERIOR), WATER QUALITY INDICATORS.

ABSTRACT:
THE PLANT PORTION OF THE EPILITHIC PERIPHYTON OF THE WESTERN ARM OF
LAKE SUPERIOR WAS FOUND TO CONSIST SOLELY OF REPRESENTATIVES FROM THREE
PHYLA OF ALGAE, THE CHRYSOPHYTA, THE CHLOROPHYTA, AND THE CYANOPHYTA.
MEMBERS OF THE PHYLUM CHRYSOPHYTA WERE THE MOST ABUNDANT ORGANISMS.
DIATOMS COMPRISED OVER 90 PER CENT OF THE TOTAL NUMBER OF ORGANISMS.
THE PREDOMINANT GENERA WERE FOUND TO BE SYNEDRA, ACHNANTHES, NAVICULA,
CYMBELLA, AND GOMPHONEMA. THE MEAN TOTAL COUNTS OF ORGANISMS IN THE
NATURALLY OCCURRING PERIPHYTON OF STONY POINT BAY, THE PRIMARY SAMPLING
AREA, RANGED FROM 497,000 PER SQUARE CENTIMETER OF ROCK SURFACE IN 1966
TO 1,470,000 PER SQUARE CENTIMETER IN 1967. THE BIOMASS OF THE
NATURALLY OCCURRING STONY POINT BAY PERIPHYTON, IN TERMS OF DRY WEIGHT,
WAS 153 GRAMS PER SQUARE METER IN 1965. THE ORGANISMS FOUND IN THE
PERIPHYTON OF THE WESTERN ARM OF LAKE SUPERIOR ARE INDICATIVE OF CLEAN
WATER. THE EXTENSIVE SHALLOW WATER AREA OF LAKE SUPERIOR SUPPORTS LARGE
QUANTITIES OF ATTACHED ALGAE, WHICH, AS PRIMARY PRODUCERS, FORM THE
FIRST LINK IN THE FOOD CHAIN. (KNAPP-USGS)

FIELD 02H, 05C

ACCESSION NO. W70-05958

GROWTH OF CHLORELLA PYRENOIDOSA 71105 IN ACTIVATED SLUDGE WASTE EFFLUENT,

GENERAL DYNAMICS CORP., GROTON, CONN. ELECTRIC BOAT DIV.

DONALD E. LEONE.

JOURNAL WATER POLLUTION CONTROL FEDERATION, VOL 41, NO 1, P 51-55, JAN 1969.
5 TAB, 9 REF.

DESCRIPTORS:
*ALGAE, *ACTIVATED SLUDGE, *EFFLUENTS, WASTE WATER RECLAMATION, WASTE
WATER TREATMENT, *NUTRIENTS.

IDENTIFIERS:
MASS CULTURE OF ALGAE, PACKED CELL VOLUME.

ABSTRACT:
PROCESSED WASTE EFFLUENT WAS STUDIED, USING CHLORELLA PYRENOIDOSA
71105, AS A METHOD FOR ALGAE GROWTH AND POSSIBLY WASTE WATER
RECLAMATION. TREATED WASTE EFFLUENT, CENTRIFUGED AND FILTERED TWICE
THROUGH CHARCOAL FOR REMOVAL OF COLOR, WAS SUPPLEMENTED WITH IRON
AND/OR UREA AND COMPARED WITH DL-61 AND EB STANDARD ALGAE MEDIA. WHILE
PRELIMINARY DATA REVEALED DEFICIENT MAGNESIUM, PHOSPHORUS, AND SULFATES
AND SUFFICIENT CALCIUM, NITROGEN, AND IRON FOR NORMAL ALGAL NUTRIENTS;
SUPPLEMENTING WASTE EFFLUENT WITH IRON WAS BENEFICIAL. DATA REVEALED
THAT EFFLUENT ALONE AND EFFLUENT SUPPLEMENTED WITH UREA SHOWED A
SUBSTANTIAL LOWERING OF DOUBLING TIME AND REDUCED PCV (PACKED CELL
VOLUME). THE WASTE EFFLUENT CONTAINING ADDITIONAL IRON AND UREA SHOWED
HIGHER CHLOROPHYLL CONTENT, ALTHOUGH FEWER CELLS WERE PRODUCED, WHEN
COMPARED WITH EFFLUENT SUPPLEMENTED WITH IRON AND DILUTED STANDARD
MEDIA. RESULTS INDICATED THAT PROCESSED WASTE EFFLUENT SUPPLEMENTED
WITH IRON (0.56 MG/L TOTAL) WAS A GOOD MEDIUM FOR CHLORELLA PYRENOIDOSA
71105 AND PRODUCED AS MUCH PCV AS THE DILUTED STANDARD MEDIA.
(MORGAN-TEXAS)

FIELD 05D

ACCESSION NO. W70-06053

THE BIOLOGICAL EFFECT OF COPPER SULPHATE TREATMENT ON LAKE ECOLOGY,

WISCONSIN COMMITTEE ON WATER POLLUTION, MADISON.

KENNETH M. MACKENTHUN, AND HAROLD L. COOLEY.

WISCONSIN ACADEMY OF SCIENCES, ARTS AND LETTERS, VOL 41, P 177-187, 1952. 7
TAB, 18 REF.

DESCRIPTORS:
*BIOLOGY, *EFFECTS, *COPPER SULFATE, *TREATMENT, *LAKES, *ECOLOGY,
BOTTOM SEDIMENTS, MUD, BENTHOS, WISCONSIN, SAMPLING, CONTOURS, DEPTH,
OLIGOCHAETES, TOXICITY, FISH, BLOODWORMS, ALGAL CONTROL.

IDENTIFIERS:
LAKE MENDOTA(WIS), LAKE MONONA(WIS), LAKE NAGAWICKA(WIS), LAKE
PEWAUKEE(WIS), CHIRONOMUS, PROCLADIUS, CORETHRA PUNCTIPENNIS, PISIDIUM
IDAHOENSE, TENDIPES PLUMOSUS, BEAVER DAM LAKE(WIS), DELAVAN LAKE(WIS).

ABSTRACT:
THESE STUDIES WERE DESIGNED TO DETERMINE THE EFFECT OF ACCUMULATED
COPPER IN BOTTOM MUDS, APPLIED AS COPPER SULFATE FOR ALGAL CONTROL, ON
BOTTOM DWELLING ORGANISMS WHICH ARE AVAILABLE AS FISH FOOD. ALTHOUGH
THE TOXIC LIMIT OF COPPER SULFATE, PRECIPITATED AND ACCUMULATED IN
BOTTOM MUDS, UPON CERTAIN TYPES OF BOTTOM-DWELLING ORGANISMS COULD NOT
BE ACCURATELY DETERMINED IN THE TIME ALLOTTED, LABORATORY TESTS
INDICATE THAT IT IS NEAR 9000 PARTS/MILLION COPPER ON A DRY-WEIGHT
BASIS. RESULTS INDICATE THAT COPPER ACCUMULATION IN BOTTOM MUDS, DUE TO
USAGE OF COPPER SULFATE TO CONTROL ALGAE IN HARD WATER LAKES, IS
CONSIDERABLY LOWER IN CONCENTRATION THAN THE AMOUNTS EXPERIMENTALLY
DETERMINED TO HAVE A DELETERIOUS EFFECT ON THE PROFUNDAL
BOTTOM-DWELLING ORGANISMS STUDIED. THESE STUDIES INDICATED THAT
DIFFERENCES OCCURRING IN THE POPULATION DENSITY OF BOTTOM ORGANISMS IN
THE FOUR WISCONSIN LAKES STUDIED ARE DUE TO ECOLOGICAL VARIABLES WITHIN
THESE SEPARATE BODIES OF WATER. (JONES-WISCONSIN)

FIELD 02H, 05C

ACCESSION NO. W70-06217

LIMNOLOGICAL ASPECTS OF RECREATIONAL LAKES,

ROBERT A. TAFT SANITARY ENGINEERING CENTER, CINCINNATI, OHIO. DIV. OF WATER
   SUPPLY AND POLLUTION CONTROL.

KENNETH M. MACKENTHUN, WILLIAM M. INGRAM, AND RALPH PORGES.

AVAILABLE FROM THE SUPERINTENDENT OF DOCUMENTS, U.S. GOVERNMENT PRINTING
   OFFICE, WASHINGTON, DC, 20402, PRICE $4.25.U S DEP OF HEALTH, EDUCATION,
   AND WELFARE, PUBLIC HEALTH SERVICE PUBLICATION NO 1167, 1964. 176 P, 11
   FIG, 9 TAB, 52 PLATES, 448 REF.

   DESCRIPTORS:
      *LIMNOLOGY, *LAKES, *RECREATION, RESERVOIRS, PONDS, TEMPERATURE, LIGHT
      PENETRATION, DISSOLVED OXYGEN, BICARBONATES, HYDROGEN ION
      CONCENTRATION, NUISANCE ALGAE, AQUATIC PLANTS, BOTTOM SEDIMENTS,
      ANIMALS, FISH, INFLOW, OUTLETS, NUTRIENTS, PRODUCTIVITY,
      PHOTOSYNTHESIS, EUTROPHICATION, TOXICITY, MIDGES, MOSQUITOES, SAMPLING,
      CONTROL, INDUSTRIAL WASTES, TENNESSEE, SHORE-LINE COVER, SNAILS,
      WISCONSIN, HUMAN DISEASE, GAME BIRDS, TECHNOLOGY, POPULATION.

   IDENTIFIERS:
      LEECHES, SCHISTOSOME CERCARIAE, TVA RESERVOIRS, ANOPHELES
      QUADRIMACULATUS, MADISON(WIS), PSOROPHORA CILIATA, ARTIFICIAL KEY.

   ABSTRACT:
      THIS BOOK CONSIDERS THE MANY PROBLEMS ASSOCIATED WITH THE RECREATIONAL
      USE OF LAKES, RESERVOIRS, AND PONDS, AND IS DIRECTED TOWARD
      INTERPRETATION OF PROBLEMS AND MANAGEMENT OF THE RECREATIONAL LAKE
      PHENOMENA. CHAPTERS CONSIDER THE GENERAL PROBLEM OF AQUATIC NUISANCES;
      REVIEW THE ECOLOGY OF LAKES, RESERVOIRS AND PONDS AND PRESENT
      INFORMATION ON BIOTIC PRODUCTION, LEADING TO AN UNDERSTANDING OF THE
      SCOPE AND MAGNITUDE OF BASIC NUISANCE PROBLEMS; DISCUSS NUTRIENTS FROM
      SEWAGE AND RUNOFF AND THEIR IMPACT ON BIOLOGICAL GROWTHS; REVIEW PLANT
      AND ANIMAL PESTS AFFECTING RECREATIONAL WATER; PRESENT SIMPLE KEYS WITH
      ILLUSTRATIONS FOR IDENTIFICATION OF SOME COMMON AQUATIC PLANTS; DISCUSS
      THE MECHANICS OF ESTABLISHING A SAMPLING PROGRAM FOR A LAKE, RESERVOIR,
      OR POND TO DETERMINE THE PRESENT STATE OF BIOLOGICAL GROWTHS AND
      POSSIBLY PREDICT FUTURE TRENDS; AND PRESENT INFORMATION FOR CONTROL OR
      ALLEVIATION OF EXCESSIVE PRODUCTION OF BIOLOGICAL NUISANCES. KNOWLEDGE
      OF THE MUTUAL RELATIONS BETWEEN ORGANISMS AND THEIR ENVIRONMENT IS
      NECESSARY FOR INSTIGATION OF AN EFFECTIVE CONTROL PROGRAM. A GLOSSARY
      IS INCLUDED. (JONES-WISCONSIN)

   FIELD 02H, 05C

   ACCESSION NO. W70-06225

SOME QUANTITATIVE ASPECTS OF ALGAL GROWTH IN LAKE MENDOTA,

WISCONSIN UNIV., MADISON. DEPT. OF ZOOLOGY.

DONALD E. WOHLSCHLAG, AND ARTHUR D. HASLER.

ECOLOGY, VOL 32, NO 4, P 581-593, 1951. 1 FIG, 5 TAB, 27 REF.

DESCRIPTORS:
    *ALGAE, *PLANT GROWTH, *WISCONSIN, ECOLOGY, CHEMICAL ANALYSIS,
    PHYTOPLANKTON, CYANOPHYTA, SEDIMENTS, PLANT POPULATIONS, WEATHER,
    WATER, MOVEMENT, CULTURES, BAYS, STREAMS, WEEDS, LITTORAL, WINDS.

IDENTIFIERS:
    *LAKE MENDOTA(WIS), PELAGIC AREAS, HORIZONTAL, BLOOMS, GELATINOUS
    MATTER.

ABSTRACT:
    DIVERSE AREAS OF LAKE MENDOTA SHOW DIFFERENCES IN QUANTITIES OF ALGAE,
    BOTH SEASONALLY AND INTRASEASONALLY. VARIOUS BAY AREAS AND STREAM
    MOUTHS, WEEDY LITTORAL, AND PELAGIC STATIONS WITHIN BAYS HAVE
    CHARACTERISTIC ALGAL NUMBERS. CERTAIN STATIONS RETAIN CHARACTERISTICS
    FROM ONE SEASON TO THE NEXT, OTHERS VARY ON A SEASONAL BASIS. WHERE
    STATION INDIVIDUALITY IS INDICATED, IT IS PRESENT OVER AND ABOVE
    TEMPORAL VARIATION. ALGAL CULTURES DERIVED FROM SEDIMENTS EXHIBITED
    CHARACTERISTIC DIFFERENCES FROM BAY TO BAY AND FROM STATION TO STATION
    WITHIN BAYS. GENERALLY, THE SAME STATION INDIVIDUALITIES PRESENT IN
    CULTURES WERE PRESENT IN CORRESPONDING COLLECTIONS FROM OVERLYING
    WATERS. COARSE, SANDY SEDIMENTS FROM THE SAME LOCALE PRODUCED MORE
    ALGAE WHEN OBTAINED PRIOR TO ICE-FREE SPRING CIRCULATION. THE POSSIBLE
    SIGNIFICANCE OF BLUE-GREEN ALGAE TO SUBSEQUENT BLOOMS IS DISCUSSED.
    ALGAL CONCENTRATIONS APPEAR GREATER AT A LITTORAL AREA AFTER 48-HOUR
    ONSHORE WIND THAN AT A SIMILAR UNEXPOSED AREA. CHLOROPHYLL, CHEMICAL,
    AND MICROSCOPIC DATA SUGGEST THAT SUDDEN APPEARANCE OF DIPLOCYSTIS
    AEROGINOSA IS ACCOMPANIED BY INCREASE IN GELATINOUS MATTER
    PROPORTIONATELY GREATER THAN INCREASE IN ACTIVITY REPRESENTED BY
    CHLOROPHYLL INCREASE. WIND PRODUCED SEDIMENT AGITATION IS A POSSIBLE
    AGENT IN PRODUCTION OF BLUE-GREEN BLOOMS RELATIVE TO SEASONAL
    SUCCESSION OF ALGAL BLOOMS. (JONES-WISCONSIN)

FIELD 02H, 05C

ACCESSION NO. W70-06229

KINETICS OF OXYGEN UPTAKE BY DEAD ALGAE,

HOWARD UNIV., WASHINGTON, D.C. DEPT. OF CIVIL ENGINEERING.

MAN M. VARMA, AND FRANCIS DIGIAND.

JOURNAL OF THE WATER POLLUTION CONTROL FEDERATION, VOL 40, NO 4, P 613-626,
APRIL 1968. 11 FIG, 6 TAB, 16 REF.

DESCRIPTORS:
    *AGE, *ALGAE, *BIODEGRADATION, *OXYGEN DEMAND, LABORATORY TESTS,
    LAGOONS, OXIDATION, PHOTOSYNTHESES, PLANTS, TEMPERATURE, *WASTE WATER
    TREATMENT.

IDENTIFIERS:
    *UPTAKE, CELLS, CONCENTRATION, COMPOSITION, MATERIAL BALANCES,
    ORGANISMS.

ABSTRACT:
    THE RATE OF OXYGEN UPTAKE BY DEAD ALGAL CELLS IS INDEPENDENT OF THE
    ALGAL CELL CONCENTRATION. THE RATE OF UPTAKE BY YOUNG CELLS IS GREATER
    THAN THAT BY OLD CELLS. RATES INCREASE WITH TEMPERATURE UP TO ABOUT 35
    DEG C, THEN DECREASE SHARPLY. INCREASED BACTERIAL POPULATION WILL
    INCREASE THE BIODEGRADATION OF THE DEAD ALGAE. UPTAKE RATES OBSERVED
    RANGED FROM 0.008 MICRO L O SUB 2/MIN/MG ALGAE FOR OLD CELLS AT 20 DEG
    C TO 0.294 MICRO L O SUB 2/MIN/MG ALGAE FOR YOUNG CELLS AT 35 DEG C.
    MATERIAL BALANCES SHOWED THAT THE TOTAL ACTUAL OXYGEN UPTAKE WAS LESS
    THAN THE TOTAL OXYGEN REQUIRED FOR COMPLETE DEGRADATION OF THE
    SUBSTRATE BECAUSE OXIDATION WAS INCOMPLETE. (AGUIRRE-TEXAS)

FIELD 05D

ACCESSION NO. W70-06600

OVERLOADED OXIDATION PONDS--TWO CASE STUDIES,

KANSAS UNIV., LAWRENCE. DEPT. OF CIVIL ENGINEERING.

ROSS E. MCKINNEY.

JOURNAL OF THE WATER POLLUTION CONTROL FEDERATION, VOL 40, NO 1, P 49-56, JAN 1968. 9 TAB, 7 REF.

DESCRIPTORS:
*AERATION, *ALGAE, *LAGOONS, *OXIDATION LAGOONS, BIOCHEMICAL OXYGEN DEMAND, CHEMICAL OXYGEN DEMAND, DISSOLVED OXYGEN, ODOR, ORGANIC LOADING, *WASTE WATER TREATMENT.

IDENTIFIERS:
*AERATION TUBES, *DIFFUSERS, *RAYTOWN(MO), LEE'S SUMMIT(MO), OXYGEN TRANSFER, PH, SUSPENDED SOLIDS.

ABSTRACT:
THE USE OF A DIFFUSED AERATION SYSTEM AS A MEANS OF INCREASING THE CAPACITY AND IMPROVING THE PERFORMANCE OF OVERLOADED OXIDATION PONDS WAS EVALUATED IN A ONE YEAR DUAL STUDY OF THE PONDS AT LEE'S SUMMIT AND RAYTOWN, MO. THE SUPPLEMENTAL AERATION SYSTEM CONSISTED OF POLYETHYLENE TUBE DIFFUSERS, A COMMON HEADER, AND AN AIR COMPRESSOR. DATA ACCUMULATED DURING THE TEST PERIOD INCLUDE BOD, COD, SUSPENDED SOLIDS, PH, AND DISSOLVED OXYGEN. BIOCHEMICAL OXYGEN DEMAND (BOD) AND COD REMOVAL AVERAGED 80% AND 54% AT LEE'S SUMMIT, RESPECTIVELY. THE POND COVERED 25 ACRES, WITH AN AVERAGE DEPTH OF 5 FEET AND AN AVERAGE DETENTION PERIOD OF 30 DAYS. THE BOD5 LOAD TO THE POND WAS ESTIMATED TO BE 2550 LBS/DAY. THE OXIDATION POND AT RAYTOWN CONSISTED OF A FIRST-STAGE AERATED LAGOON WITH TWO 15-HP AERATORS FOLLOWED BY A SECOND-STAGE LAGOON, WHERE THE PLASTIC AERATION TUBING WAS INSTALLED, ESTIMATED TO RECEIVE AN ORGANIC LOAD OF 1000 LBS BOD5/DAY. ORGANIC REMOVAL EFFICIENCIES WERE SIMILAR TO THE LEE'S SUMMIT SYSTEM AS A STATISTICAL BASIS. THE PLASTIC TUBING AERATION UNITS CAN PRODUCE AN EFFLUENT UNDER 30 MG/L BOD5 AT A LOADING OF 1000 PEOPLE/D/AC. BUT, THE STUDY INDICATES THAT THE AERATION TUBING SHOULD BE USED WITH CAUTION. IT CAN ASSIST OVERLOADED OXIDATION PONDS BY IMPROVING MIXING, BUT IS NOT A CURE-ALL. CLOGGING OF THE AERATION TUBING WAS A PROBLEM RELIEVED BY TREATMENT WITH HYDROCHLORIC ACID, AND THE BASIC MECHANISM OF TREATMENT WAS INCREASED ALGAL METABOLISM THROUGH BETTER MIXING RATHER THAN DIRECT OXYGEN TRANSFER. (AGUIRRE-TEXAS)

FIELD 05D, 05G

ACCESSION NO. W70-06601

BIOLOGICAL EXTRACTION OF NUTRIENTS,

MICHIGAN UNIV., ANN ARBOR. DEPT. OF CIVIL ENGINEERING.

JACK A. BORCHARDT, AND HARDAM S. AZAD.

JOURNAL OF THE WATER POLLUTION CONTROL FEDERATION, VOL 40, NO 10, P 1739-1754, OCT 1968. 13 FIG, 22 REF.

DESCRIPTORS:
*ALGAE, *PHOSPHORUS, *PHOSPHATES, ENERGY, DI-URNAL DISTRIBUTION, TEMPERATURE, METABOLISM, *WASTE WATER TREATMENT, LIGHT, MICROORGANISMS, PLANTS.

IDENTIFIERS:
*BIOLOGICAL EXTRACTION, GROWTH, SHOCK LOADING.

ABSTRACT:
STUDIES OF PHOSPHATE UPTAKE FROM SYNTHETIC SECONDARY WASTE WATER BY ALGAE IN CONTINUOUS-CULTURE LABORATORY UNITS SHOWED THREE REGIMES OF PHOSPHATE UPTAKE: (1) GROWTH DEPENDENT ON PHOSPHATE CONCENTRATION BELOW 0-1.5 MG/L PHOSPHATE (0-3% OF ALGAL DRY WEIGHT), (2) STORAGE OF PHOSPHATE AT 1.5 TO 4.5 MG/L (3-9%), AND (3) SATURATION AT PHOSPHATE LEVELS GREATER THAN 4.5 MG/L (9%), AT AN ORGANISM DENSITY OF 50 MG/L. AT 15 MG/L THE GROWTH RATE ALSO DECLINED WHEN PHOSPHATE CONCENTRATION WAS LOWERED BEYOND 3% OF THE ALGAL DRY WEIGHT. AS TEMPERATURES BECOME LOWER, HIGHER PHOSPHATE CONCENTRATIONS WERE NEEDED TO GROW THE SAME MASS OF PROTOPLASM. ABOUT TWO-THIRDS OF THE PHOSPHATE UPTAKE ACHIEVED UNDER CONSTANT LIGHT CONDITIONS IS ACHIEVED UNDER A DI-URNAL TYPE LIGHT CYCLE. AFTER SHOCK LOADING PHOSPHATE STARVED CULTURES CAN EXHIBIT PHOSPHATE UPTAKES OF UP TO 20% BY WEIGHT. (HANCUFF-TEXAS)

FIELD 05D, 05C

ACCESSION NO. W70-06604

IMPROVEMENTS IN TREATMENT DESIGN FOR ENHANCING WASTE WATER QUALITY,

WASHINGTON UNIV., SEATTLE. DEPT. OF CIVIL ENGINEERING.

DALE A. CARLSON, AND H. KIRK WILLARD.

WASHINGTON UNIVERSITY WATER RESEARCH CENTER FINAL PROJECT REPORT FOR FY 1968,
    JULY 11, 1968. 17P, 13 FIG, 20 REF, APPEND. OWRR PROJECT NO A-016-WASH.

DESCRIPTORS:
    *TERTIARY TREATMENT, *AEROBIC TREATMENT, OXYGEN DEMAND, BIOCHEMICAL
    OXYGEN DEMAND, SUSPENDED LOAD, SOLIDS CONTACT PROCESS, SEWAGE
    TREATMENT, TRICKLING FILTERS, WASTE WATER TREATMENT, PHOSPHATES,
    NITRATES, ALGAE.

IDENTIFIERS:
    SOLIDS REMOVAL.

ABSTRACT:
    A NEW TWO-PHASE BIOLOGICAL TERTIARY WASTE TREATMENT SYSTEM INCLUDES AN
    AEROBIC BACTERIAL GROWTH UNIT AND A PARTICULATE REMOVAL UNIT. IN THE
    REMOVAL UNIT, THE EFFECTS OF SUSPENDED SOLIDS LOADING ON A COMMUNITY OF
    ATTACHED PARTICULATE FEEDING ORGANISMS WERE DETERMINED. A BIOMASS YIELD
    OF 16% OF THE INFLUENT SUSPENDED SOLIDS FOR THE PARTICULATE FEEDERS
    PRODUCED A LOW EFFLUENT SOLIDS CONCENTRATION OF ONLY 0.069 MG SUSPENDED
    SOLIDS PER MG B. O. D. REMOVED. THIS DENSE MATERIAL WAS FOUND TO SETTLE
    RAPIDLY. TOTAL DISSOLVED ORGANIC CARBON WAS OVER 90% DESTROYED. SIMILAR
    REMOVAL VALUES OF 90 TO 94% WERE MEASURED FOR B. O. D. THE DECREASES OF
    BOTH ORGANIC CARBON AND B. O. D. WERE HIGHER AT LOWER LOADING RATES.
    MOST NITROGEN COMPOUNDS WERE OXIDIZED TO NITRATES. AS EXPECTED, THE
    PERCENT OF NITRIFICATION DECREASED WITH INCREASED LOADING. PHOSPHATES
    WERE NOT REMOVED EXCEPT FOR THE AMOUNT INCORPORATED INTO WASTED CELL
    BIOMASS RANGING IN MAGNITUDE FROM 22 TO 36%. (KNAPP-USGS)

FIELD 05D

ACCESSION NO. W70-06792

FUNCTION OF SOLIDS IN ANAEROBIC LAGOON TREATMENT OF WASTE WATER,

MELBOURNE WATER SCIENCE INST. (AUSTRALIA).

C. D. PARKER, AND G. P. SKERRY.

JOURNAL OF THE WATER POLLUTION CONTROL FEDERATION, VOL 40, NO 2, P 192-204,
    FEBRUARY, 1968. 13 TAB, 4 REF.

DESCRIPTORS:
    *ANAEROBIC CONDITIONS, *EFFLICIENCIES, *LAGOONS, *SLUDGE, ALGAE,
    BIOCHEMICAL OXYGEN DEMAND, MIXING, SEWAGE TREATMENT, WASTE WATER
    TREATMENT.

IDENTIFIERS:
    *ANAEROBIC TREATMENT, *AUSTRALIA, LABORATORY-SCALE STUDIES, PLANT-SCALE
    STUDIES, REMOVAL.

ABSTRACT:
    FIELD STUDIES OF ANAEROBIC LAGOONS IN AUSTRALIA HAVE SHOWN THAT SOLIDS
    FARTHER FROM THE INLET OF A LAGOON ARE MORE ACTIVE IN PURIFICATION THAN
    THOSE DEPOSITED CLOSER TO THE INLET, ALTHOUGH IN MANY CASES THEY SHOW
    LOWER GAS YIELDS. LAGOON SLUDGE PURIFICATION INDEX (LBS BOD/ACRE/DAY/LB
    VS) VALUES RANGED FROM 0.0053 DURING THE WINTER AT A TYPICAL POND INLET
    TO 0.00902 DURING THE SUMMER AT THE POND OUTLET. THE ORGANIC LOAD TO
    THE POND DURING THIS PERIOD WAS ABOUT 482 LBS BOD/DAY/ACRE, AND THE
    SLUDGE GAS YIELD VARIED FROM 1.2 TO 7.7 ML/DAY/GVS. THE PRESENCE OF
    ALGAE IN POND SUPERNATANT APPARENTLY DOES NOT REDUCE THE SLUDGE'S BOD
    REMOVAL CAPACITY. WHERE METHANE FERMENTATION IS INHIBITED, BOD CAN
    STILL BE REMOVED BY SULFATE REDUCTION. LABORATORY STUDIES SHOWED THAT A
    SUBSTANTIAL INCREASE (ALMOST 30%) IN BOD REMOVAL CAN BE ACHIEVED BY
    MIXING SLUDGE WITH THE SUPERNATANT, AND THAT ALGAE PROLIFERATE READILY
    IN MEDIA WITH HIGH ORGANIC CONTENTS. (AGUIRRE-TEXAS)

FIELD 05D

ACCESSION NO. W70-06899

USE OF UNICELLULAR ALGAE FOR PURIFYING WASTE WATERS FROM MAN-MADE FIBER PLANTS
(IN RUSSIAN),

G. M. P. MORDVINTSEVA, V. V. GRABOVSKAYA, AND V. K. MARINICH.

KHIMICHESKIE VOLOKNA, VOL. 3, P 61-62, 1966.

DESCRIPTORS:
*ALGAE, IRON COMPOUNDS, HYDROGEN SULFIDE, BIOCHEMICAL OXYGEN DEMAND,
NUTRIENTS, WASTE WATER TREATMENT, *FIBER(PLANT).

IDENTIFIERS:
*VISCOSE RAYON PLANT WASTES, *MAN-MADE FIBER PLANTS, NYLON-6, *VISCOSE
RAYON, ZINC COMPOUNDS, CARBON DISULFIDE.

ABSTRACT:
LABORATORY AND PILOT-PLANT EXPERIMENTS ARE REPORTED ON THE USE OF
UNICELLULAR GREEN ALGAE (CHLORELLA AND ANKISTRODESMUS) TO TREAT WASTE
WATERS FROM THE MANUFACTURE OF NYLON-6 AND VISCOSE RAYON. GROWTH OF THE
ALGAE COULD BE STIMULATED BY ADDITION OF NUTRIENTS SUCH AS UREA,
MAGNESIUM SULPHATE, AND POTASSIUM PHOSPHATE. IT WAS FOUND THAT
TREATMENT, WHICH WAS COMPLETE WITHIN 6 DAYS UNDER LABORATORY CONDITIONS
AND 3 DAYS IN THE OPEN AIR, EFFECTED COMPLETE REMOVAL OF ZINC, IRON,
HYDROGEN SULPHIDE, CARBON DISULPHIDE, AND SULPHIDES, REDUCED THE BOD BY
90 PERCENT, AND INCREASED THE DISSOLVED-OXYGEN CONCENTRATION BY 100-400
PER CENT, BUT HAD NO EFFECT ON CALCIUM AND SULPHATES. (LIVENGOOD -
WORK-NORTH CAROLINA STATE UNIV)

FIELD 05D

ACCESSION NO. W70-06925

AN INTRODUCTION TO THE LIMNOLOGY OF HARTBEESPOORT DAM WITH SPECIAL REFERENCE TO
THE EFFECT OF INDUSTRIAL AND DOMESTIC POLLUTION,

NATIONAL INST. FOR WATER RESEARCH, PRETORIA (SOUTH AFRICA).

B. R. ALLANSON, AND J. M. T. M. GIESKES.

HYDROBIOLOGIA, VOL 18, P 77-94, 1961. 11 REF.

DESCRIPTORS:
*RESERVOIRS, *LIMNOLOGY, *INDUSTRIAL WASTES, *WATER POLLUTION EFFECTS,
*EUTROPHICATION, DISSOLVED OXYGEN, HYDROGEN ION CONCENTRATION,
EPILIMNION, PHOTOSYNTHESIS, ALGAE, BOTTOM SEDIMENTS, THERMOCLINE,
HYPOLIMNION, ANIMALS, DENSITY, IRON, PHOSPHORUS, SILICA, SAMPLING,
CHEMICAL ANALYSIS, PHYSICAL PROPERTIES, CARP, TILAPIA, TEMPERATURE,
NUTRIENTS, PHYTOPLANKTON, CYANOPHYTA, NITROGEN, OLIGOCHAETES, DIPTERA.

IDENTIFIERS:
*HARTBEESPOORT DAM(SOUTH AFRICA), TRANSPARENCY, MARGINAL VEGETATION
FAUNA, SANDY BOTTOMS FAUNA, MARGINAL VEGETATION, MICROCYSTIS, ANABAENA,
EUCYCLOPS, NAIS, CRICOTOPUS, AUSTROCLOEON, ENALLAGMA, PSEUDAGRION,
BRACHYTHEMIS, CHYDORUS, PLEUROXUS, TANYPUS, TANYTARSUS, LIMNODRILUS,
EUCYCLOPS, ALONA, CHAOBORUS.

ABSTRACT:
THE CORRELATED DATA OF THIS STUDY OF WATER MASS IN SOUTH AFRICA
CONTRIBUTE TO KNOWLEDGE OF POLLUTION EFFECTS IN A RESERVOIR.
HARTBEESPOORT DAM IS EUTROPHIC WITH A LOW TRANSPARENCY DEVELOPED DURING
THE PAST 25 YEARS. MONOMICTIC WITH COMPLETE CIRCULATION DURING WINTER
MONTHS, LOW DISSOLVED OXYGEN CONCENTRATIONS THROUGHOUT THE RESERVOIR
OFTEN RESULT. NUTRIENTS ARE CONTRIBUTED BY THE JUKSKEI-CROCODILE RIVER
SYSTEM AND TO A LESSER DEGREE FROM THE MAGALIES RIVER. RISES IN THE
EPILIMNETIC PH WERE MAINLY CAUSED BY PHOTOSYNTHETIC ACTIVITY OF ALGAE
IN THE PRESENCE OF MODERATELY BASIC WATER. BOTTOM SEDIMENTS IN THE
DEEPER PORTION OF THE RESERVOIR UNDERWENT RAPID ANAEROBIASIS AFTER THE
ESTABLISHMENT OF A THERMOCLINE AND OXYCLINE IN SUMMER. FAUNA DENSITY IN
THE BOTTOM SEDIMENTS DECREASES RAPIDLY WITH THE ESTABLISHMENT OF THE
HYPOLIMNION. SEDIMENTS ARE RICH IN IRON AND DURING SUMMER THERE WAS AN
INCREASE IN THE FERROUS COMPONENT. THE SOLUBLE AND ACID SOLUBLE
PHOSPHORUS INCREASES IN THE HYPOLIMNION AFTER A WINTER MINIMUM. SILICA
CONTENT IS COMPARABLE WITH PRODUCTIVE WATER MASSES IN OTHER PARTS OF
THE WORLD. FAUNAL ASSOCIATIONS ARE TYPICAL OF OTHER MODERATELY BASIC
AND HIGHLY PRODUCTIVE WATER IN THE SAME GEOGRAPHICAL REGION OF SOUTH
AFRICA. (JONES-WISCONSIN)

FIELD 05C

ACCESSION NO. W70-06975

A COMPOSITE RATING OF ALGAE TOLERATING ORGANIC POLLUTION,

ROBERT A. TAFT WATER RESEARCH CENTER, CINCINNATI, OHIO. ADVANCED WASTE
  TREATMENT RESEARCH LAB.

C. MERVIN PALMER.

JOURNAL OF PHYCOLOGY, VOL 5, NO 1, P 78-82, 1969. 5 P, 11 TAB, 29 REF.

DESCRIPTORS:
  *ALGAE, *BIOINDICATORS, *WATER POLLUTION EFFECTS, *REVIEWS, *WATER
  QUALITY, WATER POLLUTION, ORGANIC MATTER, BIBLIOGRAPHIES.

IDENTIFIERS:
  POLLUTION-TOLERANT ALGAE.

ABSTRACT:
  A REVIEW OF POLLUTION-TOLERANT ALGAE USING REPORTS FROM 165 AUTHORS
  SHOWS THAT THE GENERA AND SPECIES MOST OFTEN REFERRED TO AS SIGNIFICANT
  FALL INTO A RELATIVELY STABLE SERIES. DIATOMS, PIGMENTED FLAGELLATES,
  GREEN, AND BLUE-GREEN ALGAE ARE ALL WELL REPRESENTED AMONG THE
  POLLUTION-TOLERANT GENERA AND SPECIES. EUGLENA, OSCILLATORIA,
  CHLAMYDOMONAS, SCENEDESMUS, CHLORELLA, NITZSCHIA, NAVICULA, AND
  STIGEOCLONIUM ARE THE MOST TOLERANT GENERA, AND EUGLENA VIRIDIS,
  NITZSCHIA PALEA, OSCILLATORIA LIMOSA, SCENEDESMUS QUADRICAUDA, AND
  OSCILLATORIA TENUIS ARE THE MOST TOLERANT SPECIES. IN SOME GENERA, FOR
  EXAMPLE EUGLENA, A SINGLE TOLERANT SPECIES IS FAR MORE SIGNIFICANT THAT
  ALL OTHERS. IN OTHER GENERA, AS IN OSCILLATORIA, ONLY A SLIGHT
  DIFFERENCE DISTINGUISHES THE POLLUTION TOLERANCE OF TWO OR MORE
  SPECIES. ALGAL POLLUTION INDICES ARE PRESENTED FOR USE IN RATING WATER
  SAMPLES WITH HIGH ORGANIC POLLUTION. (KNAPP-USGS)

FIELD 05C

ACCESSION NO. W70-07027

NITRIFICATION AND DENITRIFICATION IN A MODEL WASTE STABILIZATION POND,

TEXAS UNIV., AUSTIN. CENTER FOR RESEARCH IN WATER RESOURCES.

JORGE AGUIRRE, AND EARNEST F. GLOYNA.

ENVIRONMENTAL HEALTH ENGINEERING RESEARCH LABORATORY, REPORT EHE-05-6701,
  CRWR-19, JUNE 1967. 82 P, 22 FIG, 5 TAB, 38 REF. FWQA CONTRACT WP-00688-38.

DESCRIPTORS:
  *DENITRIFICATION, EFFICIENCIES, *NITRIFICATION, NITROGEN COMPOUNDS,
  NITROGEN CYCLE, ALGAE, BIOLOGICAL TREATMENT, BOTTOM SEDIMENTS,
  NUTRIENTS, *WASTE WATER TREATMENT, *OXIDATION LAGOONS, MILK, *MODEL
  STUDIES.

IDENTIFIERS:
  *FACULTATIVE LAGOONS, LABORATORY MODEL, COD REMOVAL, MILK SUBSTRATE,
  NITROGEN REMOVAL, *MILK WASTES.

ABSTRACT:
  A LABORATORY MODEL OF A FACULTATIVE WASTE STABILIZATION POND WAS USED
  TO INVESTIGATE NITRIFICATION AND DENITRIFICATION OF MILK WASTES AS A
  FUNCTION OF DEPTH. THE MODEL WAS OPERATED ON A SEMI-CONTINUOUS FLOW
  BASIS USING A DRY MILK PRODUCT AS FEED. THE ORGANIC AND HYDRAULIC
  LOADING WERE KEPT CONSTANT AT 50 LB BOD5/ACRE/DAY AND 3 INCHES/DAY (5
  GAL/D), RESPECTIVELY. GENERAL POND CHARACTERISTICS WERE EVALUATED BY
  MEASURING D.O., PH, ORP AND TEMPERATURE PROFILES; COD AND THE DIFFERENT
  FORMS OF NITROGEN WERE MONITORED AT FREQUENT INTERVALS DURING THE TEST
  PERIOD TO DETERMINE DEPTH PROFILES OF THESE PARAMETERS AS WELL COD
  REMOVAL EFFICIENCIES AND NITROGEN TRANSPORATIONS. A RAPID REMOVAL OF
  ORGANICS OCCURRED IN OR NEAR THE BOTTOM SEDIMENT LAYER WITH A
  CORRESPONDINGLY HIGH DEGREE OF AMMONIFICATION. ABOVE THE SLUDGE LAYER
  THE CONTENTS REMAINED ESSENTIALLY UNCHANGED EXCEPT FOR MINOR VARIATIONS
  DUE TO ALGAL DENSITY FLUCTUATIONS. DETECTABLE CONCENTRATIONS OF
  NITRATES WERE NOT FOUND DURING THE TEST PERIOD, BUT A SUBSTANTIAL
  AMOUNT OF NITROGEN WAS LOST FROM THE LIQUID FRACTION. COD AND NITROGEN
  REMOVALS FROM THE WASTE WATER, AS MEASURED BY THE OVERFLOW, WERE 85%
  AND 44% RESPECTIVELY. (AGUIRRE-TEXAS)

FIELD 05D

ACCESSION NO. W70-07031

SULFIDE PRODUCTION IN WASTE STABILIZATION PONDS,

TEXAS UNIV., AUSTIN. CENTER FOR RESEARCH IN WATER RESOURCES.

ERNESTO ESPINO DE LA O, AND EARNEST F. GLOYNA.

ENVIRONMENTAL HEALTH ENGINEERING RESEARCH LABORATORY, REPORT EHE-04-6802, CRWR NO. 26, MAY 1967. 156 P, 50 FIG, 33 TAB, 75 REF. FWQA CONTRACT WP-00688-03.

DESCRIPTORS:
*HYDROGEN SULFIDE, *ORGANIC LOADING, *SULFATES, *SULFIDES, ALGAE, GASES, ODOR, OXIDATION-REDUCTION POTENTIAL, SULFUR BACTERIA, *WASTE WATER TREATMENT, *OXIDATION LAGOONS, *MODEL STUDIES.

IDENTIFIERS:
DETENTION TIME, LABORATORY MODELS, BOD REMOVAL RATES, DESIGN APPLICATIONS, FACULTATIVE PONDS, INFRARED SPECTROSCOPY, SULFUR COMPOUNDS, WASTE STABILIZATION PONDS.

ABSTRACT:
LABORATORY MODELS, UNDER CONTROLLED ENVIRONMENTAL CONDITIONS, WERE USED TO INVESTIGATE THE PRODUCTION AND RELEASE OF HYDROGEN SULFIDE FROM WASTE STABILIZATION PONDS. A 6 FOOT DEEP MODEL FACULTATIVE POND WAS STUDIED UNDER SEVEN DIFFERENT OPERATING CONDITIONS. THE EXPERIMENTS EVALUATED THE RELATIONSHIP BETWEEN SULFIDE CONCENTRATION IN THE POND AND FOUR OPERATING CHARACTERISTICS: DETENTION, CONCENTRATION OF SULFATE ION IN THE INFLUENT, BOD SURFACE LOAD AND SULFATE ION CONCENTRATION IN TERMS OF SURFACE LOAD. AT CONSTANT DETENTION TIME AND BOD SURFACE LOAD, A LINEAR RELATIONSHIP WAS FOUND TO APPLY BETWEEN SULFIDE CONCENTRATION IN THE POND AND SULFATE CONCENTRATION IN THE INFLUENT. INCREASES IN SULFIDE CONCENTRATIONS IN THE POND FROM ZERO TO 1 AND 2 MG/L WERE ACCOMPANIED WITH DROPS IN OXIDATION-REDUCTION POTENTIALS OF 80 AND 120MV, RESPECTIVELY. FURTHER INCREASES IN SULFIDE CONCENTRATION AFFECTED THE ORP IN A LESSER WAY. SULFIDE CONCENTRATIONS IN THE POND OF ABOUT 7 MG/L WERE ACCOMPANIED BY AN ALMOST COMPLETE DISAPPEARANCE OF ALGAE, BUT THE BOD REMOVAL RATES WERE NOT APPRECIABLY AFFECTED. INFRARED SPECTROSCOPY WAS USED TO IDENTIFY METHANE, HYDROGEN SULFIDE AND CARBON DIOXIDE IN THE GASEOUS EMISSIONS FROM THE POND. (AGUIRRE-TEXAS)

FIELD 05D

ACCESSION NO. W70-07034

TEMPERATURE AND MANGANESE AS DETERMINING FACTORS IN THE PRESENCE OF DIATOM OR BLUE-GREEN ALGAL FLORAS IN STREAMS,

ACADEMY OF NATURAL SCIENCES OF PHILADELPHIA, PA. DEPT. OF LIMNOLOGY.

RUTH PATRICK, BOWMAN CRUM, AND JOHN COLES.

PROCEEDINGS OF THE NATIONAL ACADEMY OF SCIENCES, VOL 64, NO 2, P 472-478, 1969. 2 TAB, 10 REF.

DESCRIPTORS:
*STREAMS, *ALGAE, *DIATOMS, TEMPERATURE, MANGANESE, CULTURES, RECIRCULATED WATER, NITRATES, PHOSPHATES, SODIUM, WATER POLLUTION EFFECTS, LABORATORY TESTS.

IDENTIFIERS:
GREEN ALGAE, BLUE-GREEN ALGAE, TEMPERATURE EXPERIMENTS, RECYCLED WATER.

ABSTRACT:
TRIALS CONDUCTED IN CULTURE BOXES INDICATED THAT AT AN AVERAGE TEMPERATURE OF 34-38C BLUE-GREEN ALGAE TEND TO REPLACE DIATOMS. BOTH BLUE-GREEN AND GREEN ALGAE, COMMONLY FOUND IN POLLUTED WATERS, WERE FAVORED BY MEDIA WITH MANGANESE CONCENTRATION OF ONLY A FEW PARTS PER BILLION. AN AVERAGE CONCENTRATION OF MANGANESE OF 0.02-0.043 MILLIGRAM/LITER IN NATURAL STREAMS AND 0.04-0.28 MILLIGRAM/LITER IN RECYCLED WATER EXPERIMENT FAVORED DOMINATION OF DIATOM FLORA. INCREASE IN NITRATE AND PHOSPHATE CONTENT OF WATER FAILED TO INFLUENCE THE FLORA OF FREE-FLOWING STREAMS, BUT PROMOTED THE GROWTH OF BLUE-GREEN AND GREEN ALGAE IN RECYCLED WATER OF A LOW MANGANESE CONTENT. (WILDE-WISCONSIN)

FIELD 05C

ACCESSION NO. W70-07257

EUTROPHICATION IN NORTH AMERICA,

WASHINGTON UNIV., SEATTLE.

W. T. EDMONDSON.

EUTROPHICATION: CAUSES, CONSEQUENCES, CORRECTIVES, P 124-149. PRINTING AND
PUBLISHING OFFICE, NATIONAL ACADEMY OF SCIENCES, WASHINGTON, D C, 1969. 6
FIG, 4 TAB, 70 REF.

DESCRIPTORS:
*EUTROPHICATION, *NUISANCE ALGAE, *SEWAGE, WATER POLLUTION SOURCES,
WATER POLLUTION EFFECTS, PHOSPHORUS, NITROGEN, CHLOROPHYLL,
PRODUCTIVITY, REGIONAL ANALYSIS, RESERVOIRS.

IDENTIFIERS:
NUTRIENT BUDGETS, LAKE SEBASTICOOK(ME), LAKE WINNISQUAM(N H), MADISON
LAKES(WIS), LAKE WASHINGTON(WASH), NUISANCE CONDITIONS.

ABSTRACT:
EXAMPLES OF ARTIFICIAL ENRICHMENT ARE GIVEN. WELL-DOCUMENTED HISTORIES
OF THE EFFECTS OF ARTIFICIAL ENRICHMENT ARE NOT COMMON. CASE HISTORIES
ARE AVAILABLE FOR FOUR LAKES OR LAKE SYSTEMS: LAKE SEBASTICOOK, MAINE;
LAKE WINNISQUAM, NEW HAMPSHIRE; THE MADISON, WISCONSIN, LAKES; AND LAKE
WASHINGTON, WASHINGTON. IN EACH CASE, ARTIFICIAL ENRICHMENT OCCURRED BY
SEWAGE EFFLUENT INPUT; SEBASTICOOK ALSO RECEIVED LARGE INPUTS OF
INDUSTRIAL WASTES HIGH IN NUTRIENTS AND ORGANIC MATTER. ARTIFICIAL
ENRICHMENT IN ALL CASES HAS RESULTED IN LOWERED TRANSPARENCY,
MYXOPHYCEAN BLOOMS AND NUISANCE CONDITIONS OF SURFACE SCUM AND ODOR.
DIVERSION OF SEWAGE EFFLUENTS HAS LED TO IMPROVED CONDITIONS IN LAKE
WASHINGTON AND THE MADISON LAKES MONONA AND WAUBESA, ALTHOUGH LAKE
MENDOTA APPEARS TO BE DETERIORATING. ONE OF THE MAJOR FEATURES OF THE
INVESTIGATIONS OF ARTIFICIAL ENRICHMENT IN NORTH AMERICAN LAKES IS THE
DEVELOPMENT OF NUTRIENT BUDGETS. THIS APPROACH NEEDS FURTHER
DEVELOPMENT AND REFINEMENT, AND WITH CONTINUED DEVELOPMENT OF BIOASSAY
TECHNIQUES CAN BE EXPECTED TO AID FUTURE INVESTIGATIONS. (SEE ALSO
W70-03975). (VOIGTLANDER-WISCONSIN)

FIELD 05C

ACCESSION NO. W70-07261

EFFECTS OF INCREASED TEMPERATURES ON AQUATIC ORGANISMS,

ACADEMY OF NATURAL SCIENCES OF PHILADELPHIA, PA. DEPT. OF LIMNOLOGY.

JOHN CAIRNS, JR.

PROCEEDINGS OF THE 10TH INDUSTRIAL WASTE CONFERENCE, PURDUE UNIVERSITY,
ENGINEERING BULLETIN, VOL. 40, NO. 1, 1956. P 346, 4 FIG, 3 REF.

DESCRIPTORS:
*MICROORGANISMS, *ALGAE, *DOMINANT ORGANISMS, *ACCLIMATIZATION,
DIATOMS, THERMAL POLLUTION, TEMPERATURE, CHEMICAL PROPERTIES, ECOLOGY,
ECOLOGICAL DISTRIBUTION, THERMAL EFFECTS.

IDENTIFIERS:
GREEN ALGAE, BLUE-GREEN ALGAE, HEAT DEATH.

ABSTRACT:
THE POPULATION OF MICROORGANISMS OF A RIVER IS COMPOSED OF THE SPECIES
THAT ARE BEST FITTED TO THE LOCAL ECOLOGICAL CONDITIONS. BECAUSE THERE
IS A FIERCE COMPETITION FOR SPACE, A SPECIES MAY BE ELIMINATED BY AN
UNFAVORABLE, RATHER THAN LETHAL, FACTOR, SUCH AS INCREASED TEMPERATURE.
A MIXED POPULATION OF ALGAE WAS TAKEN FROM A NORMAL HEALTHY STREAM AND
GRADUALLY EXPOSED TO AN INCREASED TEMPERATURE. AT 20 DEG C, A LARGE
NUMBER OF SPECIES OF DIATOMS WERE PRESENT AND DOMINATED THE CULTURES.
AT 30 DEG C, ONLY TWO SPECIES OF DIATOMS WERE NUMEROUS. ABOVE 30 DEG C,
A FEW INDIVIDUAL DIATOMS SURVIVED, BUT GREEN ALGAE PREDOMINATED AT 30
DEG C TO 35 DEG C. IN THE RANGE OF 35 DEG C TO 40 DEG C, BLUE-GREEN
ALGAE WERE DOMINANT. THERE ARE STRONG INDICATIONS THAT THE COMPOSITION
OF THE WATER IS AN IMPORTANT FACTOR IN HEAT DEATH OF AQUATIC ORGANISMS,
SINCE THE CHEMICAL BALANCE IS USUALLY QUITE DIFFERENT IN HEATED WATERS
FROM THAT OF A TYPICAL STREAM. MICROORGANISMS CAN BE ACCLIMATIZED TO
VERY HIGH TEMPERATURES IF THE CHANGES ARE SMALL AND GRADUAL.
(SPEAKMAN-VANDERBILT)

FIELD 05C

ACCESSION NO. W70-07313

RESEARCH TO SAVE AMERICA'S LAKES,

FEDERAL WATER POLLUTION CONTROL ADMINISTRATION, CORVALLIS, OREG. PACIFIC NORTHWEST WATER LAB.

THOMAS E. MALONEY, AND ALFRED F. BARTSCH.

IN: WATER - 1969, CHEMICAL ENGINEERING PROGRESS SYMPOSIUM SERIES 97, VOL 65, PUBLISHED BY AMERICAN INSTITUTE OF CHEMICAL ENGINEERS, NEW YORK, P 278-280, 1969. 3 P.

DESCRIPTORS:
*EUTROPHICATION, *LAKES, *WATER POLLUTION CONTROL, *REVIEWS, ALGAE, WEEDS, ECOLOGY, ENVIRONMENT, CHEMCONTROL, BIOCONTROL, INTEGRATED CONTROL MEASURES, PHYSICAL CONTROL, WATER QUALITY.

IDENTIFIERS:
*EUTROPHICATION CONTROL.

ABSTRACT:
THE MATURATION PROCESS OF A LAKE IS CALLED EUTROPHICATION AND IS GREATLY ACCELERATED BY MAN'S ACTIVITIES WHICH INCREASE THE NUTRIENT INPUT TO THE LAKE. THESE ACTIVITIES INCLUDE AGRICULTURAL DEVELOPMENT, URBANIZATION, AND THE DISCHARGE OF SEWAGE, INDUSTRIAL WASTE, AND TREATMENT PLANT EFFLUENTS TO THE LAKES. WITH ADVANCING OF EUTROPHICATION, SYMPTOMS APPEAR INCLUDING: AN INCREASE IN PRODUCTIVITY OF ALGAE, WATERWEEDS, AND OTHER LIVING MATTER; THE NITROGEN AND PHOSPHORUS CONCENTRATIONS INCREASE; DISSOLVED OXYGEN IS DEPLETED IN THE LOWER DEPTHS; THE MORE DESIRABLE FISH SPECIES BECOME EXTINCT, AND ROUGH FISH SPECIES APPEAR; AND ROOTED VEGETATION BEGINS TO GROW IN FROM THE LAKE MARGIN AND EVENTUALLY FILLS THE LAKE. (KNAPP-USGS)

FIELD 05C, 02H

ACCESSION NO. W70-07393

REMOVAL OF NITROGEN AND PHOSPHORUS,

STANFORD UNIV., CALIF. DEPT. OF CIVIL ENGINEERING.

ROLF ELIASSEN, AND GEORGE TCHOBANOGLOUS.

PROCEEDINGS OF THE INDUSTRIAL WASTE CONFERENCE, 23RD, P 35-48, 1968. 5 TAB, 16 REF.

DESCRIPTORS:
*COST COMPARISONS, *EFFICIENCIES, *NITROGEN, *PHOSPHORUS, ALGAE, BIOLOGICAL TREATMENT, WASTE WATER TREATMENT.

IDENTIFIERS:
*REMOVAL PROCESSES, CHEMICAL TREATMENT, OVERFERTILIZATION, PHYSICAL TREATMENT, UNIT OPERATIONS, UNIT PROCESSES.

ABSTRACT:
THE NEED TO REMOVE NITROGEN AND PHOSPHORUS FROM WASTE WATERS STEMS FROM THE FACT THAT THEY ARE CAPABLE OF STIMULATING AND PROMOTING THE UNDESIRABLE GROWTH OF ALGAE AND AQUATIC PLANTS. THE IMPORTANT FORMS OF NITROGEN ARE AMMONIA, NITRITE AND NITRATE, WHILE PHOSPHORUS MAY OCCUR AS SOLUBLE INORGANIC PHOSPHATES. NITROGEN AND PHOSPHORUS REMOVAL METHODS MAY BE CLASSIFIED AS BIOLOGICAL, CHEMICAL AND PHYSICAL. BIOLOGICAL METHODS INCLUDE ACTIVATED SLUDGE, TRICKLING FILTERS AND ANAEROBIC TREATMENT. CHEMICAL AND PHYSICAL PROCESSES INCLUDE ION EXCHANGE, ELECTRODIALYSIS, REVERSE OSMOSIS AND DISTILLATION. TYPICAL REMOVAL EFFICIENCIES RANGE BETWEEN 40 TO 98 PERCENT AND 30 TO 98 PERCENT FOR NITROGEN AND PHOSPHORUS COMPOUNDS, RESPECTIVELY. IN PLANNING AND DESIGNING FACILITIES TO BE USED FOR THE REMOVAL OF NITROGEN AND PHOSPHORUS, KEY FACTORS WHICH MUST BE CONSIDERED ARE: (1) WHETHER COMPOUNDS OF ONE OR BOTH ARE TO BE REMOVED, (2) THE USE TO BE MADE OF THE TREATED WASTE WATER, (3) THE AVAILABLE MEANS FOR DISPOSING OF THE ULTIMATE CONTAMINANTS, AND (4) THE ECONOMIC FEASIBILITY OF THE SELECTED PROCESS OR PROCESSES. THE ESTIMATED REMOVAL COSTS PER MILLION GALLONS RANGE FROM $4 TO $1000 FOR BOTH NITROGEN AND PHOSPHORUS DEPENDING ON THE PROCESS USED AND DEGREE OF TREATMENT REQUIRED. (AGUIRRE-TEXAS)

FIELD 05D

ACCESSION NO. W70-07469

THE MANAGEMENT AND DISPOSAL OF DAIRY MANURE,

WASHINGTON STATE UNIV., PULLMAN. COLL. OF ENGINEERING.

DONALD E. PROCTOR.

PROCEEDINGS OF THE INDUSTRIAL WASTE CONFERENCE, 23RD, 1968. P 554-566, 8 FIG.

DESCRIPTORS:
  *ANIMAL WASTES, MANAGEMENT, *FARM MANAGEMENT, *DAIRY INDUSTRY, *ALGAE,
  *CATTLE, SLURRIES, SPRAYING, ACTIVATED SLUDGE, FARM WASTES.

IDENTIFIERS:
  *MANURE, ANAEROBIC LAGOON, AERATED LAGOON.

ABSTRACT:
  DAIRY MANURE CAN BE EITHER AN ASSET OR A LIABILITY DEPENDING ON THE
  FARMER'S MANAGEMENT POLICIES. INCREASED DEMAND FOR LIVESTOCK-DERIVED
  PRODUCTS, SPECIALIZATION OF FARM OPERATIONS, CONFINEMENT REARING,
  CHEAPER CHEMICAL FERTILIZERS, URBAN SPRAWL AND FARM AREA ENCROACHMENT,
  AND HIGHER AESTHETIC STANDARDS ARE ALL FACETS OF THE CHANGING PROBLEM
  OF MANURE DISPOSAL. OF THESE CHANGES, CONFINEMENT REARING IS MOST
  SIGNIFICANT. NOT ONLY IS THE MANURE CONCENTRATED INTO A SMALLER AREA,
  BUT RAINFALL RUNOFF CAN TREBLE THE WASTE VOLUME TO BE HANDLED. TWO
  WASHINGTON STATE DAIRY FARMS RECEIVED FEDERAL DEMONSTRATION PROJECT
  GRANTS, THE KNOTT DAIRY FARM OF THE WASHINGTON STATE UNIVERSITY
  RECEIVING A GRANT FROM THE FWPCA, AND THE MONROE HONOR FARM OF THE
  STATE OF WASHINGTON INSTITUTIONAL FARM INDUSTRIES RECEIVING A GRANT
  FROM THE PUBLIC HEALTH SERVICE. THE TWO-YEAR PROJECT AT THE KNOTT DAIRY
  FARM PROPOSED TO (1) DEMONSTRATE THE CAPABILITIES OF AN ANAEROBIC
  LAGOON FOR FIRST STAGE TREATMENT OF DAIRY MANURE, (2) DEMONSTRATE THE
  COMPARATIVE CAPABILITIES AND ECONOMICS OF ACTIVATED SLUDGE AND
  NATURALLY AERATED LAGOONS FOR SECOND STAGE TREATMENT AND (3) DETERMINE
  WHETHER IT IS POSSIBLE AND PRACTICAL TO REDUCE THE NITROGEN AND
  PHOSPHORUS CONTENT OF THE TREATED EFFLUENT BY ALGAE PROPAGATION AND
  HARVESTING FOR USE AS CATTLE FEED. THE MONROE HONOR FARM PROJECT
  PROPOSED TO (1) DEMONSTRATE THE CAPABILITIES OF AN ANAEROBIC LAGOON FOR
  FIRST STAGE TREATMENT OF DAIRY MANURE. (MAKELA-TEXAS)

FIELD 05D

ACCESSION NO. W70-07491

STABILIZATION POND TREATMENT OF SLAUGHTER-HOUSE WASTES,

JAPAN PUBLIC NUISANCE LAB., TOKYO.

ISAMU HORASAWA.

PROCEEDINGS OF THE INDUSTRIAL WASTE CONFERENCE, 23RD, 1968. P 1178-1185, 3
  FIG, 6 TAB, 9 REF.

DESCRIPTORS:
  *HOGS, *LAGOONS, ALGAE, ORGANIC LOADING, WASTE WATER TREATMENT, WATER
  UTILIZATION, INDUSTRIAL WASTES, OXIDATION LAGOONS.

IDENTIFIERS:
  *JAPAN, *SLAUGHTERHOUSE WASTES, EXPERIMENTAL STUDIES, LAGOON DESIGN,
  SCALE-UP FACTORS, WASTE COMPOSITION.

ABSTRACT:
  WASTES DISCHARGED FROM ABOUT 900 SLAUGHTERHOUSES IN JAPAN, WHICH KILL
  APPROXIMATELY 6,000,000 ANIMALS PER YEAR, REQUIRE A HIGH DEGREE OF
  TREATMENT TO PREVENT WATER POLLUTION AND AESTHETIC NUISANCES OF BLOOD
  COLOR AND OFFENSIVE ODOR. THIS EXPERIMENTAL STUDY WAS DIRECTED TOWARD
  THE USE OF THE STABILIZATION POND PROCESS BY SMALL SLAUGHTERHOUSES. THE
  AVERAGE WATER CONSUMPTION BY SLAUGHTERHOUSES LOCATED THROUGHOUT JAPAN
  IS ABOUT 848 LITERS PER HOG UNIT AND THE AVERAGE BOD LOADING PER HOG
  UNIT IS 1.255 KG. FOUR EXPERIMENTAL STABILIZATION POND UNITS WERE RUN
  BOTH IN THE LABORATORY AND EXPOSED TO OUTSIDE WINTER CONDITIONS. THE
  BOD LOADING ON THE UNITS RANGED FROM 47 TO 239 MG/L.THESE STUDIES
  SHOWED THAT, WITHIN THE LIMITS OF LABORATORY SCALE TESTING, BOD
  LOADINGS UP TO 100 MG/L (100 G/M2) CAN BE TOLERATED WITHOUT DISTINCT
  AESTHETIC NUISANCE. THE DOMINANT ALGAE IN THE PONDS WAS CHLORELLA. THE
  RESPONSE TO VARIATION IN WEATHER CONDITIONS APPEARED TO BE DEPENDENT TO
  SOME EXTENT UPON THE BOD LOADING OF THE UNIT. THE MORE HEAVILY LOADED
  UNITS SHOWED AN INDIFFERENT RESPONSE TO WEATHER CHANGE. BIOCHEMICAL
  OXYGEN DEMAND (BOD) LOADING FOR STABILIZATION PONDS IS REPORTED TO
  RANGE FROM LESS THAN 10 MG/L UP TO 150 MG/L DEPENDING ON THE DGREEE OF
  TREATMENT REQUIRED. (AGUIRRE-TEXAS)

FIELD 05D

ACCESSION NO. W70-07508

THE RESPONSE OF FRESH-WATER PROTOZOAN COMMUNITIES TO HEATED WASTE WATERS,

VIRGINIA POLYTECHNIC INST., BLACKSBURG. DEPT. OF BIOLOGY.

JOHN CAIRNS, JR.

CHESAPEAKE SCIENCE, VOL 10, NO 3 AND 4, P 177-185, 1969. 12 REF, 8 FIG.

DESCRIPTORS:
*WATER POLLUTION EFFECTS, WATER POLLUTION SOURCES, *THERMAL POLLUTION,
THERMAL POWERPLANTS, *THERMAL STRESS, THERMAL WATER, *PROTOZOA, ALGAE,
DIATOMS, ANIMAL BEHAVIOR, *THERMODYNAMIC BEHAVIOR.

IDENTIFIERS:
*TEMPERATURE SHOCK, *PROTOZOAN COMMUNITIES, POTOMAC RIVER, SAVANNAH
RIVER, SHOCK EFFECTS.

ABSTRACT:
THE RESPONSE OF FRESH-WATER PROTOZOAN COMMUNITIES EXPOSED TO BOTH
SEVERE ACUTE TEMPERATURE SHOCKS AS WELL AS SMALL GRADUAL LONG-TERM
INCREASES ARE DISCUSSED. THE FORMER EXPERIMENTS WERE CARRIED OUT IN
PLASTIC TROUGHS WITH A CONSTANT FLOW OF UNFILTERED LAKE WATER. SEVERE
ACUTE SHOCKS (SOME TO NEARLY 50 C) RESULTED IN A MARKED REDUCTION IN
NUMBER OF SPECIES PRESENT. HOWEVER, RECOVERY WAS QUITE RAPID (A MATTER
OF A FEW DAYS) ONCE THE TEMPERATURE STRESS CEASED. OBSERVATIONS OF THE
EFFECTS OF SMALL GRADUAL LONG-TERM INCREASES WERE MADE ON THE PROTOZOAN
COMMUNITIES OF THE SAVANNAH AND POTOMAC RIVERS EACH OF WHICH RECEIVED
HEATED WASTE WATER DISCHARGES. EACH OF THESE STUDIES COVERED A PERIOD
IN EXCESS OF NINE YEARS AND OBSERVATIONS ARE STILL BEING MADE. THERE
WAS NO EVIDENCE THAT INDICATED THE PROTOZOAN COMMUNITIES OF THESE
RIVERS HAD BEEN DEGRADED BY THE SMALL GRADUAL TEMPERATURE INCREASES
RESULTING FROM THE DISCHARGE OF HEATED WASTE WATERS. HOWEVER, THERE IS
EVIDENCE THAT COMPETITIVE EXCLUSION OF ALGAL SPECIES BY OTHER MORE
TOLERANT ALGAL SPECIES MAY CAUSE QUALITATIVE SHIFTS IN THE COMMUNITY
STRUCTURE WHICH MAY BE UNDESIRABLE. (SJOLSETH-WASHINGTON)

FIELD 05C

ACCESSION NO. W70-07847

EFFECTS OF NITROGEN LIMITATION ON THE GROWTH AND COMPOSITION OF UNICELLULAR
ALGAE IN CONTINUOUS CULTURE,

SCHOOL OF AEROSPACE MEDICINE, BROOKS AFB, TEX.

B. RICHARDSON, D. M. ORCUTT, H. A. SCHWERTNER, CARA L. MARTINEZ, AND HAZEL E.
WICKLINE.

APPLIED MICROBIOLOGY, VOL 18, NO 2, P 245-250, AUGUST 1969, 3 FIG, 4 TAB, 19
REF.

DESCRIPTORS:
*GROWTH, *ALGAE, LIPIDS, NUTRIENTS, CHLORELLA, CULTURES.

IDENTIFIERS:
*FAT ACCUMULATION, *LABORATORY STUDIES, *NITROGEN LIMITATION, BATCH
CULTURES, CHLORELLA SOROKINIANA, COMPOSITION OF ALGAE, CONTINUOUS
CULTURES, FATTY ACIDS, OOCYSTIS POLYMORPHA, UNICELLULAR ALGAE.

ABSTRACT:
FAT ACCUMULATION TAKES PLACE IN MANY ALGAE AS A RESPONSE TO EXHAUSTION
OF THE NITROGEN SUPPLY AND MAY PROVIDE A MEANS OF ENCHANCING THE
POTENTIAL FOOD VALUE OF ALGAE. CHEMOSTATIC CONTINUOUS CULTURES OF
CHLORELLA SOROKINIANA AND OOCYSTIS POLYMORPHA WERE SUBJECTED TO
SUCCESSIVE REDUCTIONS IN INFLUENT NITROGEN. AS CELLULAR NITROGEN
CONTENT DECREASED FROM ABOUT 10 TO 4%, OXYGEN EVOLUTION, CARBON DIOXIDE
UPTAKE, CHLOROPHYLL CONTENT, AND TISSUE PRODUCTION WERE DRASTICALLY
REDUCED, BUT TOTAL LIPID CONTENT WAS ESSENTIALLY UNCHANGED. IN
BATCH-CULTURED CELLS, NITROGEN COULD BE REDUCED TO 3% OF DRY WEIGHT,
CAUSING AN INCREASE IN TOTAL FATTY ACIDS AND PRONOUNCED CHANGES IN THE
COMPOSITION OF THE FATTY ACID FRACTION. THESE RESULTS SUGGEST THAT
CELLULAR NITROGEN MUST FALL TO ABOUT 3% OF DRY WEIGHT BEFORE
APPRECIABLE INCREASES IN LIPID SYNTHESIS CAN OCCUR. ALL NITROGEN IS
THEN APPARENTLY COMPLETELY BOUND IN ESSENTIAL CELL CONSTITUENTS, AND
CARBON SUBSEQUENTLY FIXED IS CONVERTED INTO LIPID PRODUCTS. NITROGEN
LIMITATION MAY BE USEFUL IN INCREASING THE FOOD QUALITY OF
BATCH-CULTURED CELLS, BUT THE TECHNIQUE HAS LITTLE VALUE FOR CONTINUOUS
CULTURE SYSTEMS PER SE. (AGUIRRE-TEXAS)

FIELD 05D, 05C

ACCESSION NO. W70-07957

ALGAE HANDLED EFFICIENTLY BY AUGUSTA WATER DISTRICT,

AUGUSTA WATER DISTRICT, MAINE.

SIDNEY S. ANTHONY.

WATER AND SEWAGE WORKS, VOL 116, P 185-189, 1969. 1 FIG, 6 PHOTOS.

DESCRIPTORS:
*ALGAE, *MAINE, *WATER SUPPLY, COPPER SULFATE, RESERVOIRS, WATER
POLLUTION, MICROORGANISMS, HISTORY, TASTE, ODOR, DEEP WELLS, WATER
TREATMENT, MECHANICAL EQUIPMENT, PUMPING PLANTS, MAINE.

IDENTIFIERS:
*AUGUSTA(MAINE), CARLETON POND(MAINE), SYNURA, ASTERIONELLA, TYPHOID
FEVER, KENNEBEC RIVER(MAINE), COBBOSSEECONTEE LAKE(MAINE), BARGE.

ABSTRACT:
THE HISTORY OF THE AUGUSTA, MAINE, WATER SUPPLY AFTER 1870 IS
DESCRIBED. AT THAT TIME THE KENNEBEC RIVER WAS USED AS THE SOURCE OF
WATER; IN THE WINTER OF 1902-1903 WIDESPREAD TYPHOID FEVER OUTBREAKS,
TRACEABLE TO THE WATER SUPPLY, CAUSED THE INAUGURATION OF A WATER
DISTRICT. CARLTON POND, THE PRINCIPAL RESERVOIR, WAS CHOSEN BY THE
AUGUSTA WATER DISTRICT AND PLACED IN OPERATION IN 1906. THE POND,
HOWEVER, IS SUBJECT TO PERIODIC INFESTATIONS OF ALGAE, PRIMARILY
SYNURA, WITH ASTERIONELLA ALSO PRESENT. TO ALLEVIATE THE TASTE AND ODOR
PROBLEM CAUSED BY ALGAE, CONTROL MEASURES ARE TAKEN. A SPECIAL BARGE
AND PUMPS WERE DEVELOPED FOR APPLICATION OF COPPER SULFATE (THE ONLY
CHEMICAL USED) IN DILUTION OF 0.35 MILLIGRAMS/LITER. THE TREATMENT IS
COMPLETED IN ONE DAY, IN THE FALL, WHEN WATER TURBULENCE IS MINIMAL,
AND APPLICATION IS MADE TO THE ENTIRE LAKE AND SHORELINE. ONE TREATMENT
IS USUALLY ADEQUATE TO ELIMINATE THE ALGAE. (JONES-WISCONSIN)

FIELD 05F

ACCESSION NO. W70-08096

EFFECTS OF INSECTICIDES ON ALGAE,

ONTARIO WATER RESOURCES COMMISSION, TORONTO. DIV. OF RESEARCH.

A. E. CHRISTIE.

WATER AND SEWAGE WORKS, VOL 116, P 172-176, 1969. 6 TAB, 31 REF.

DESCRIPTORS:
*ALGAE, *INSECTICIDES, TOXICITY, DEGRADATION(DECOMPOSITION), DDT,
ALKALINITY, STABILIZATION, WASTE STORAGE, PONDS, HYDROLYSIS,
METABOLISM, CHEMICAL REACTIONS, CHLORELLA, TRACERS, RADIOACTIVITY,
CARBON RADIOISOTOPES, HYDROGEN ION CONCENTRATION, SCENEDESMUS, EUGLENA,
BACTERIA, YEASTS, ACTINOMYCETES, PONDING, WASTE TREATMENT.

IDENTIFIERS:
SEVIN, MALATHION, CHLORELLA PYRENOIDOSA, SCHRODERIA, THIMET,
TRICHODERMA.

ABSTRACT:
DDT, SEVIN, AND MALATHION WERE STUDIED TO DETERMINE THEIR EFFECT ON
ALGAL POPULATIONS ASSOCIATED WITH WASTE STABILIZATION PONDS AND THEIR
POSSIBLE DEGRADATION IN AN ALGAL ENVIRONMENT. THE PESTICIDES VARY IN
DEGREE OF TOXICITY TO ALGAE AND IN EXTENT TO WHICH THEY ARE DEGRADED IN
PRESENCE OF ALGAE. DDT EXHIBITED NO TOXIC PROPERTIES UP TO
CONCENTRATIONS OF 100 MILLIGRAMS/LITER AND RECEIVED ONLY SLIGHT
DEGRADATION. SEVIN IS TOXIC AT CONCENTRATIONS ABOVE 0.1
MILLIGRAMS/LITER AND IS NOT ALTERED APPRECIABLY IN ACIDIC MEDIA.
MALATHION APPEARS TO RECEIVE EXTENSIVE CONVERSION AND, ALTHOUGH CAPABLE
OF ALTERING COMPOSITION OF A MIXED ALGAL COMMUNITY, DID NOT DISPLAY
PERSISTENT INHIBITORY EFFECT. EXTENDING THIS TO THE ALKALINE WASTE
STABILIZATION POND AND ITS ALGAL COMMUNITY, ENTRY OF DDT, UP TO 100
MILLIGRAMS/LITEE, IS NOT LIKELY TO BE TOXIC NOR WILL THE MATERIAL BE
DEGRADED BY ALGAE. A SIMILAR QUANTITY OF SEVIN WOULD SERIOUSLY REDUCE
EFFICIENCY OF THE POND WITH CONVERSION MORE LIKELY TO OCCUR BY ALKALINE
HYDROLYSIS THAN ALGAL METABOLIC PROCESSES. AN EQUIVALENT SLUG OF
MALATHION, ALTHOUGH TEMPORARILY INTERRUPTING OPERATION OF THE POND,
COULD BE BROKEN DOWN RAPIDLY BY CHEMICAL AND METABOLIC REACTIONS.
(JONES-WISCONSIN)

FIELD 05C

252   ACCESSION NO. W70-08097

ALGAL REMOVAL IN UNIT PROCESSES,

ENVIRONMENTAL CONTROL ADMINISTRATION, CINCINNATI, OHIO; AND IOWA UNIV, IOWA
CITY. DEPT. OF ENVIRONMENTAL HEALTH; AND IOWA UNIV., IOWA CITY. DEPT. OF
CIVIL ENGINEERING.

RONALD R. SPEEDY, NEIL B. FISHER, AND DONALD B. MCDONALD.

JOURNAL AMERICAN WATER WORKS ASSOCIATION, VOL 61, P 289-292, 1969. 1 FIG, 5
TAB, 6 REF.

DESCRIPTORS:
*ALGAL CONTROL, *POTABLE WATER, CYANOPHYTA, SULFUR BACTERIA, COLOR,
IRON BACTERIA, TASTE, FILTERS, IOWA, DIATOMS, EUGLENA, CHLAMYDOMONAS,
SAMPLING, SCENEDESMUS, CHLOROPHYTA, CHLORELLA, CHLORINE, LIME,
COAGULATION.

IDENTIFIERS:
*UNIT PROCESSES, GASTROENTERITIS, SYNURA, ANABAENA, IOWA RIVER,
CYCLOTELLA, FRAGILARIA, SYNEDRA, NAVICULA, APHANIZOMENON, OSCILLATORIA,
MELOSIRA, ACTINASTRUM, COELASTRUM, PEDIASTRUM, WESTELLA,
STEPHANODISCUS, ANKISTRODESMUS, CHLOROMONAD, ALUM.

ABSTRACT:
A DETERMINATION OF THE EFFICIENCY OF ALGAL REMOVAL IN VARIOUS UNIT
OPERATIONS OF POTABLE WATER TREATMENT PROCESS WAS MADE. SAMPLES WERE
COLLECTED AT VARIOUS LOCATIONS THROUGHOUT THE TREATMENT PLANT AND
EXAMINED TO DETERMINE KINDS AND NUMBERS OF ALGAL FORMS PRESENT. THE
RESULTS INDICATE A WIDE DISPARITY IN THE EFFICIENCY OF ALGAL REMOVAL
AMONG THE SEVERAL UNIT PROCESSES STUDIED. NUMEROUS ORGANISMS PASS THE
FILTERS INTO FINISHED TAP WATER; COUNTS AS HIGH AS 44
ORGANISMS/MILLILITER WERE OBTAINED. CYCLOTELLA AND CHLORELLA APPEAR
WITH THE GREATEST FREQUENCY IN THE PLANT EFFLUENT. THE USE OF ALUM AS A
COAGULANT SEEMS MODERATELY EFFECTIVE FOR ALGAL REMOVAL, BUT LIME
APPEARS TO BE MUCH MORE EFFICIENT DUE TO FORMATION OF HEAVY FLOC WHICH
ENTRAPS THE CELLS UPON SETTLING. BOTH TYPES OF FILTERS EFFECT A HIGH
DEGREE OF ALGAL REMOVAL AT THE START OF A FILTER RUN; THE SAND
CONTINUES TO SHOW FAIRLY EFFICIENT REMOVAL UNTIL THE HEAD LOSS GOES
ABOVE SIX FEET. THE ANTHRAFILT FILTER SHOWS A HIGHER NUMBER OF ALGAE
PASSING THROUGH THE FILTER. THE IMPORTANCE OF MAINTAINING A CHLORINE
RESIDUAL IS SHOWN BY THE NUMBER OF ALGAE FOUND IN THE FINISHED TAP
WATER. (JONES-WISCONSIN)

FIELD 05F

ACCESSION NO. W70-08107

THE INHIBITORY EFFECTS OF COPPER ON MARINE PHYTOPLANKTON,

DOW CHEMICAL CO., FREEPORT, TEX, TEXAS DIV.

E. F. MANDELLI.

MARINE SCIENCE, VOL 14, P 47-57, 1969. 6 FIG, 1 TAB, 21 REF.

DESCRIPTORS:
*MARINE WATER, *COPPER, *ALGAE, SEA WATER, TEMPERATURE, SALINITY,
BIOASSAY, ANALYTICAL TECHNIQUES, LETHAL LIMITS, DINOFLAGELLATES,
DIATOMS.

IDENTIFIERS:
*COPPER CHLORIDE, UPTAKE, FREUNDLICH ISOTHERM, BLUE-GREEN ALGAE, GREEN
ALGAE.

ABSTRACT:
THE GROWTH OF SEVERAL SPECIES OF DINOFLAGELLATES AND DIATOMS IN BATCH
CULTURES UNDER CONTINUOUS ILLUMINATION WAS INHIBITED AT 20C BY
CONCENTRATION OF COPPER BETWEEN 0.055 AND 0.265 MICROGRAM/MILLILITER.
COCCOCHLORIS ELABANS FAILED TO GROW AT 0.03 MICROGRAM/MILLILITER AND
40C, WHEREAS DUNALIELLA TERTIOLECTA SURVIVED 0.6 MICROGRAM/MILLILITER
AT 30C. SKELETONEMA COSTATUM IN LIGHT-DARK CULTURES WITH MAINTENANCE OF
CONSTANT NUMBER OF CELLS AND COPPER IONS WAS INHIBITED AT 0.05
MICROGRAM/MILLILITER. A POSITIVE CORRELATION WAS FOUND BETWEEN THE LOG
OF COPPER UPTAKE:BIOMASS OF ALGAE RATIO AND THE TEMPERATURE, AND A
NEGATIVE CORRELATION BETWEEN THE SAME RATIO AND SALINITY.
(WILDE-WISCONSIN)

FIELD 05C

ACCESSION NO. W70-08111

GLENODININE, AN ICHTHYOTOXIC SUBSTANCE PRODUCED BY A DINOFLAGELLATE, PERIDINIUM POLONICUM (IN JAPANESE),

TOKYO UNIV. (JAPAN). LAB. OF MARINE BIOCHEMISTRY.

YOSHIRO HASHIMOTO, TOMOTOSHI OKAICHI, LE DUNG DANG, AND TAMAO NOGUCHI.

BULLETIN OF THE JAPANESE SOCIETY OF SCIENTIFIC FISHERIES, VOL 34, NO. 6, 1968. P. 528-534, 7 FIGS, 1 TAB, 8 REFS.

DESCRIPTORS:
WATER POLLUTION SOURCES, WATER POLLUTION EFFECTS, DINOFLAGELLATES, FISHKILL, FISH TOXINS, BIOASSAY, AQUATIC ALGAE, *ALGAL TOXINS, KILLIFISHES, *TOXICITY, ISOLATION.

IDENTIFIERS:
LAKE SAGAMI, JAPAN, *GLENODININE, *PERIDINIUM POLONICUM, PLANKTON BLOOMS.

ABSTRACT:
IN SEPTEMBER 1962, EXTENSIVE WATER BLOOMS OF A DINOFLAGELLATE IDENTIFIED AS PERIDINIUM POLONICUM, ACCOMPANIED BY MASS MORTALITY OF FISH, OCCURRED IN LAKE SAGAMI NEAR TOKYO. AN ATTEMPT WAS MADE TO ISOLATE THE SUSPECTED ICHTHYOTOXIC SUBSTANCE FROM THE PLANKTON COLLECTED FROM THE LAKE. THE SUBSTANCE WAS CONFIRMED TO BE AN ALKALOID AND TO CONTAIN A SULFHYDRYL RADICAL. THE AUTHORS DESIGNATED IT AS GLENODININE. BIOASSAYS SHOWED GLENODININE TO BE EXTREMELY TOXIC TO BOTH FRESHWATER AND MARINE FISHES. (SJOLSETH-WASHINGTON)

FIELD 05C

ACCESSION NO. W70-08372

BIOLOGICAL PARAMETERS FOR THE OPERATION AND CONTROL OF OXIDATION PONDS-II,

DEPARTMENT OF INDUSTRIES, CAPE TOWN (SOUTH AFRICA). DIV. OF SEA FISHERIES.

U. S. TSCHORTNER.

WATER RESEARCH, VOL 2, P 327-346, 1968. 9 FIG, 7 TAB, 24 REF.

DESCRIPTORS:
*ALGAE, *BIOLOGICAL PROPERTIES, CONTROL, *OPERATION AND MAINTENANCE, *OXIDATION LAGOONS, ABSORPTION, BIODEGRADATION, CHEMICAL OXYGEN DEMAND, CHLOROPHYLL, LAGOONS, PONDS, OXIDATION, WASTE WATER TREATMENT.

ABSTRACT:
OXIDATION PONDS ARE AN ECONOMICAL MEANS OF TREATMENT FOR DOMESTIC SEWAGE AND INDUSTRIAL WASTE EFFLUENTS, AND ARE WIDELY USED IN THE LESS POPULATED AREAS OF WARMER CLIMATES. THEIR ACTION RESEMBLES THE NATURAL PURIFICATION PROCESS OF POLLUTED RIVER. THEY REQUIRE LESS ATTENTION AND SKILLED STAFF THAN MECHANICAL SYSTEMS. THE SUCCESSFUL PERFORMANCE OF STABILIZATION, HOWEVER, DEPENDS ON THE BIOLOGICAL BALANCE OF THE SYSTEM. THE PRESENT STUDY DESCRIBES THE DEVELOPMENT OF PARAMETERS FOR THE DETERMINATION OF THE OXYGEN-PRODUCING COMMUNITY (PARAMETERS A AND AP) AND THE EFFICIENCY OF THE POND ON THE BREAKING DOWN OF THE LOAD (PARAMETER E). PARAMETER A REVEALS THE RELATIVE FRACTION OF CHLOROPHYLL-BEARING ORGANISMS OF THE TOTAL LOAD AND ALLOWS A DIRECT MEASUREMENT OF AUTOTROPHIC POND PERFORMANCE. PARAMETER AP REVEALS THE RATE OF PHOTOSYNTHESIS AND DOES NOT DEPEND ON VARIATIONS OF CHLOROPHYLL-A CONTENT OF CELLS. PARAMETER E IS BASED ON THE METABOLISM OF ALGAE AND CRUSTACEOUS; IT ESTIMATES THE CONTRIBUTION OF BOTH TO THE MICROPOPULATION AND INDICATES THE EFFECTIVENESS OF ALGAE IN REMOVING POLLUTION. COMPARISON OF THE COD CONTRIBUTED BY DISSOLVED AND SUSPENDED SOLIDS, WITH THE CALCULATED VALUES FOR ALGAE AND NON-ALGAL MATTER OF THE SUSPENDED FRACTION, REVEALED THAT IN A WELL-BALANCED SYSTEM MORE THAN HALF OF THE TOTAL COD IS CONTRIBUTED BY ALGAL CELLS, THE FIGURE IN THE OVERLOADED PONDS BEING ONLY 20%. (AGUIRRE-TEXAS)

FIELD 05D

ACCESSION NO. W70-08392

EFFECT OF ORGANIC COMPOUNDS ON PHOTOSYNTHETIC OXYGENATION--I. CHLOROPHYLL
  DESTRUCTION AND SUPPRESSION OF PHOTOSYNTHETIC OXYGEN PRODUCTION,

TEXAS UNIV., AUSTIN. ENVIRONMENTAL HEALTH ENGINEERING RESEARCH LAB.

JU-CHANG HUANG, AND EARNEST F. GLOYNA.

WATER RESEARCH, VOL 2, P 347-366, 1968. 28 FIG, 3 TAB, 19 REF.

DESCRIPTORS:
  *ALGAE, *CHLOROPHYLL, WATER POLLUTION EFFECTS, *ORGANIC COMPOUNDS,
  *OXYGENATION, *PHOTOSYNTHESIS, CHLORELLA, PESTICIDES, PESTICIDE
  TOXICITY, PHOTOSYNTHETIC OXYGEN, PRODUCTIVITY, *WASTE WATER TREATMENT.

IDENTIFIERS:
  *CHLORELLA PYRENOIDOSA.

ABSTRACT:
  THIS PAPER DESCRIBES THE INHIBITION OF CHLOROPHYLL SYNTHESIS AND THE
  DESTRUCTION OF CHLOROPHYLL IN CHLORELLA PYRENOIDOSA DUE TO THE PRESENCE
  OF SELECTED ORGANIC COMPOUNDS. EMPHASIS IS DIRECTED TOWARD AN
  EXPLANATION OF LABORATORY TECHNIQUES AND AN EVALUATION OF CHANGES IN
  PHOTOSYNTHETIC OXYGEN PRODUCTION DUE TO THE ADDITION OF SELECTED
  COMPOUNDS. CHANGES IN EXTRACTABLE CHLOROPHYLL CONCENTRATIONS ARE
  PRESENTED IN TERMS OF EXPOSURE TIME AND CONCENTRATION OF CHEMICAL
  ADDITIVES. SELECTED ALGAL CULTURES WERE GROWN FOR A TEST PERIOD OF 72
  HOURS IN TEST TUBES UNDER CONSTANT TEMPERATURE CONDITIONS, WITH
  CONTINUOUS ILLUMINATION, AND UNDER CONTROLLED NUTRIENT SUPPLIES. UNDER
  SUCH CONDITIONS THE CHLOROPHYLL CONTENT DECREASES MARKEDLY IN RELATION
  TO THE CONCENTRATION OF THE ORGANIC ADDITIVE. FOR THE COMPOUNDS
  EVALUATED, THE CHLOROPHYLL REDUCTION SEEMS TO BE A FUNCTION OF THE
  SUBSTITUTED GROUPS AND THEIR RELATIVE POSITIONS. ALSO, A DIRECT
  MANOMETRIC TECHNIQUE IS DESCRIBED FOR MEASUREMENT OF PHOTOSYNTHETIC GAS
  EXCHANGE. RESULTS SHOW THAT OXYGEN PRODUCTION IS SUPPRESSED BY
  CHLORINATED AND NITRATED PHENOLS. (AGUIRRE-TEXAS)

FIELD 05D

ACCESSION NO. W70-08416

ADVANCES IN WATER POLLUTION RESEARCH.

PROCEEDINGS 4TH INTERNATIONAL CONFERENCE ON WATER POLLUTION RESEARCH, HELD IN
  PRAGUE, CZECHOSLOVAKIA, APRIL 21-25, 1969: LONDON, PERGAMON PRESS, LTD,
  1969. 936 P.

DESCRIPTORS:
  *WATER RESOURCES, *WATER POLLUTION CONTROL, *RESEARCH EQUIPMENT,
  *ANALYTICAL TECHNIQUES, WATER QUALITY, DISSOLVED OXYGEN, BIOCHEMICAL
  OXYGEN DEMAND, SUSPENDED LOAD, WASTES, SEWAGE, ALGAE, MICROORGANISMS,
  PHOSPHATES, HYDROGENATION, AERATION, FLOCCULATION, FILTRATION, MODEL
  STUDIES, COMPUTERS, NUTRIENTS, EUTROPHICATION.

IDENTIFIERS:
  *WATER POLLUTION RESEARCH.

ABSTRACT:
  THIS VOLUME CONTAINS THE 53 ORIGINAL PAPERS PRESENTED AT THE FOURTH
  INTERNATIONAL CONFERENCE ON WATER POLLUTION RESEARCH HELD IN PRAGUE IN
  APRIL, 1969. AMONG THE TOPICS DISCUSSED ARE THE SELF-PURIFICATION OF
  POLLUTED RIVERS THAT HAVE BEEN ARTIFICIALLY CANALIZED, THE FACTORS THAT
  INFLUENCE POLLUTION UNDER THESE CONDITIONS, AND METHODS FOR PREDICTING
  THE EXTENT OF POLLUTION AND THE DISTRIBUTION OF DISSOLVED OXYGEN IN
  RIVERS. COMPUTERS MAY BE USED TO ANALYZE VARIOUS ALTERNATIVE MEANS OF
  OBTAINING RIVER WATER OF REQUIRED QUALITY. THE MOST IMPORTANT
  CONTRIBUTION TOWARDS POLLUTION CONTROL IS MADE BY PURIFYING SEWAGE AND
  OTHER WASTE WATER. PURIFIED SEWAGE IS AN IMPORTANT SOURCE OF THE
  MINERAL NUTRIENT REQUIREMENTS OF MICRORGANISMS ESPECIALLY ALGAE, AND
  POLLUTION CONTROL MUST INCLUDE NUTRIENT REMOVAL FROM EFFLUENTS.
  (WOODARD-USGS)

FIELD 05G, 05C, 05D

ACCESSION NO. W70-08627

MODIFIED FILTER MEDIA FOR REMOVAL OF WATER POLLUTANTS,

INDIAN INST. OF TECH., KANPUR. DEPT. OF CIVIL ENGINEERING.

G. D. AGRAWAL, AND A. V. S. PRABHAKARA RAO.

IN: ADVANCES IN WATER POLLUTION RESEARCH, PROCEEDINGS 4TH INTERNATIONAL
CONFERENCE ON WATER POLLUTION RESEARCH, HELD IN PRAGUE, CZECHOSLOVAKIA,
APRIL 21-25, 1969: LONDON, PERGAMON PRESS, LTD, P 299-307, 1969. 9 P, 5
FIG, 10 REF.

DESCRIPTORS:
*WATER POLLUTION CONTROL, *FILTRATION, *PARTICLE SIZE, *CLAYS, *SILTS,
SANDS, ALGAE, PLANKTON, MICROORGANISMS, TURBIDITY, ANALYTICAL
TECHNIQUES, LABORATORY TESTS, FOREIGN RESEARCH.

IDENTIFIERS:
*FILTER MEDIA, ALUMINA-COATED SAND, FERRIC OXIDE-COATED SAND, CRUSHED
BAUXITE, ACID-WASHED BAUXITE, KANPUR(INDIA).

ABSTRACT:
EXPERIMENTS WERE MADE TO DEVELOP A SIMPLE FILTER, WITH MODIFIED MEDIA,
FOR EFFICIENT REMOVALS OF MICRO SIZED PARTICLES, WHETHER USED FOR
DISPOSAL TO EXCLUDE PARTICULATE POLLUTANTS FROM EFFLUENTS OR FOR WATER
SUPPLY TO REMOVE PARTICULATE CONTAMINANTS. TWO PYREX GLASS TUBES WERE
USED TO CONTAIN THE FILTER COLUMNS. THE FILTER BED FOR ALL THE RUNS WAS
A 7.6 CM LAYER OF SHINGLE AND COARSE SAND SUPPORTING AN 18 CM LAYER OF
THE MEDIA UNDER STUDY. FILTER MEDIA CONSISTED OF UNTREATED SAND, SAND
COATED WITH ALUMINA, SAND COATED WITH FERRIC OXIDE, CRUSHED LOW-GRADE
BAUXITE AND BAUXITE WASHED WITH ACID. WATER WITH LESS THAN 10 MG/L.
TURBIDITY COULD BE PRODUCED BY FILTRATION OF RAW STREAM WATER (INFLUENT
TURBIDITY 46 MG/L.) OR RAW POND WATER (INFLUENT TURBIDITY 32 MG/L.) AND
REDUCED THE TOTAL BACTERIAL NUMBERS BY ABOUT 50%. (WOODWARD-USGS)

FIELD 05D

ACCESSION NO.   W70-08628

FACTORS INFLUENCING PHOSPHATE USE BY ALGAE,

MINNESOTA UNIV., MINNEAPOLIS. LIMNOLOGICAL RESEARCH CENTER.

JOSEPH SHAPIRO, WILLIAM CHAMBERLAIN, AND JUDITH BARRETT.

PARTIALLY SUPPORTED BY FWQA RESEARCH GRANT. IN: ADVANCES IN WATER POLLUTION
RESEARCH, PROCEEDING 4TH INTERNATIONAL CONFERENCE ON WATER POLLUTION
RESEARCH, HELD IN PRAGUE, CZECHOSLOVAKIA, APRIL 21-25, 1969: LONDON,
PERGAMON PRESS, LTD, P 149-160, 1969. 12 P, 10 FIG, 3 TAB, 13 REF.

DESCRIPTORS:
*EUTROPHICATION, *LAKES, *ALGAE, *GROWTH RATES, *PHOSPHATES, LABORATORY
TESTS, MINNESOTA, NUTRIENTS, WATER POLLUTION EFFECTS.

IDENTIFIERS:
*MICROCYSTIS AERUGINOSA, *LAKE WATERS.

ABSTRACT:
MICROCYSTIS AERUGINOSA FROM LAKES IN MINNESOTA, WAS USED TO SHOW THE
RELATIONSHIPS OF PHOSPHORUS TO ALGAE GROWTH IN NATURAL WATER SYSTEMS.
THE RAPID RATE OF ABSORPTION OF PHOSPHATE BY ALGAE SUSPENDED IN LAKE
WATER IS LARGELY DUE TO THE INORGANIC SALTS OF THE WATER AND NOT TO AN
ORGANIC FACTOR. (WOODARD-USGS)

FIELD 05C, 02H

ACCESSION NO.   W70-08637

METABOLISM OF LINDANE BY UNICELLULAR ALGAE,

STATE UNIVERSITY COLL., BUFFALO, N.Y. GREAT LAKES LAB.

ROBERT A. SWEENEY.

PROCEEDINGS OF THE 12TH CONFERENCE ON GREAT LAKES RESEARCH, P 98-102, 1969. 3
FIG, 18 REF. USPHS GRANT IT-WP-3901-A.

DESCRIPTORS:
PESTICIDE RESIDUES, ALGAE, GREAT LAKES, METABOLISM, *CHLORELLA,
*CHLAMYDOMONAS LABORATORY TESTS, GAS CHROMATOGRAPHY, DEGRADATION, WATER
POLLUTION EFFECTS, AQUATIC ENVIRONMENT, CHLORINATED HYDROCARBON
PESTICIDES, PLANT PHYSIOLOGY.

IDENTIFIERS:
*LINDANE, *ALGAE METABOLISM, DETOXIFICATION, LINDANE METABOLISM.

ABSTRACT:
LINDANE IS AN INSECTICIDE THAT HAS BEEN DETECTED IN THE GREAT LAKES
BASIN. THE METABOLISM OF THIS PESTICIDE BY CHLORELLA VULGARIS AND
CHLAMYDOMONAS REINHARDTII IN AXENIC CULTURE WAS ASCERTAINED. EACH
SPECIES WAS CULTURED IN A SOLUTION INITIALLY CONTAINING 4.5, 2.3, OR
0.0 PPM LINDANE. USING CHROMATOGRAPHIC TECHNIQUES, A MARKED DECREASE IN
THE LINDANE CONCENTRATION WITH TIME WAS NOTED. A SUBSTANCE WAS OBSERVED
IN EXTRACTIONS OF THE SOLUTIONS CONTAINING ALGAE AND LINDANE WHICH WAS
NOT PRESENT IN EXTRACTS FROM EITHER ALGAE-FREE CULTURES CONTAINING
LINDANE OR FROM ALGAE IN A LINDANE-FREE MEDIUM. THIS SUBSTANCE WAS
IDENTIFIED AS 1, 3, 4, 5, 6 --PENTACHLORO-CYCLOHEX-L-ENE, A NON-TOXIC
LINDANE METABOLITE. THIS INDICATED THAT CHLORELLA AND CHLAMYDOMONAS CAN
METABOLIZE LINDANE. IT ALSO MAY EXPLAIN IN PART THE RELATIVELY LOW
CONCENTRATION OF THIS PESTICIDE IN THE GREAT LAKES.
(SJOLSETH-WASHINGTON)

FIELD 05C

ACCESSION NO. W70-08652

RELATIONSHIPS BETWEEN BACTERIA AND PHYTOPLANKTON,

BYELORUSSIAN STATE UNIV., MINSK (USSR). DEPT. OF INVERTEBRATE ZOOLOGY.

YU S POTAENKO, AND T M MIKHEEVA.

TRANS. FROM MIKROBIOLOGIYA, VOL 38, NO 4, P 722-727. JULY-AUG 1969.
MICROBIOLOGY, VOL 38, NO 4, P 603-607, 1969. 3 FIG, 37 REF.

DESCRIPTORS:
*BACTERIA, *PHYTOPLANKTON, *LAKES, ALGAE, MICROORGANISMS, BIOMASS,
DIATOMS, CYANOPHYTA, IRON, ALGICIDES, ZOOPLANKTON, FISH.

IDENTIFIERS:
BYELORUSSIA(USSR), POLYCHLOROPINENE.

ABSTRACT:
SEVERAL FACTORS MAY BE OPERATIONAL SIMULTANEOUSLY IN THE
CHARACTERISTICS OF THE RELATIONSHIP BETWEEN BACTERIA AND PHYTOPLANKTON.
STUDY OF BACTERIOPLANKTON AND PHYTOPLANKTON OF EIGHT BYELORUSSIAN LAKES
IN DIFFERENT YEARS SHOWED THERE WERE EXTREMELY DIVERSE RELATIONSHIPS
BETWEEN ALGAE AND OTHER MICROORGANISMA: GROWTH OF ALGAE AND BACTERIA
PROCEEDED IN PARALLEL, MAXIMUM ALGAL DEVELOPMENT CORRESPONDED TO
MINIMUM BACTERIAL POPULATION AND VICE VERSA, OR THE GROWTH OF THESE
GROUPS OF ORGANISMS WAS INDEPENDENT. NO DEFINITE RELATIONSHIP WAS
OBSERVED BETWEEN BACTERIAL GROWTH AND THE PRESENCE OF A GIVEN TOXONOMIC
ALGAL GROUP. THE USUAL CYCLES FOR INDIVIDUAL SPECIES AND ALGAL GROUPS
WERE REESTABLISHED WHEN THE DISRUPTED EQUILIBRIUM AMONG DIFFERENT
ELEMENTS OF THE FOOD CHAINS WAS RESTORED A YEAR AFTER THE FISH KILL.
FROM A STUDY OF VARIATION IN BACTERIOPLANKTON GROWN IN THE NAROCHANSKII
LAKES AND FROM DATA IN THE LITERATURE OF THE SEASONAL DYNAMICS OF
BACTERIAL POPULATION IN A NUMBER OF OTHER LAKES, IT CAN BE CONCLUDED
THAT THE BACTERIOPLANKTON DISTRIBUTION AND POPULATION CHARACTERISTIC OF
OTHER EUTROPHIC LAKES IN THE SOVIET UNION IS RESTORED WITHIN 1.5 YEARS
AFTER TREATMENT WITH PCP COMPOUND. (JONES-WISCONSIN)

FIELD 02H, 05C

ACCESSION NO. W70-08654

BIODEGRADABLE DETERGENTS - SOFT DETERGENTS,

HART CHEMICAL LTD., GUELPH (ONTARIO).

H. ZIMMERMAN.

CANADIAN TEXTILE JOURNAL, VOL. 82, P 43-45, JANUARY 22, 1965. 3 FIG.

DESCRIPTORS:
*ALKYLBENZENE SULFONATES, *BIODEGRADATION, DETERGENTS, WATER POLLUTION, WASTE WATERS, BIOCHEMICAL OXYGEN DEMAND, FISH, ALGAE, FOAMING, SEWAGE TREATMENT.

IDENTIFIERS:
*TEXTILE MILL WASTES, ALFOL ALCOHOLS.

ABSTRACT:
THE ALKYL ARYL SULPHONATES ARE NON-BIODEGRADABLE, WHEREAS SOAPS, THE FATTY ALCOHOL SULPHATES AND OTHER FAT BASED DETERGENTS, ARE READILY BIODEGRADABLE. THESE BIODEGRADABLE SYNTHETIC DETERGENTS ARE AVAILABLE, BUT THEIR USE DOES NOT SOLVE THE OVER-ALL PROBLEMS OF WATER POLLUTION. NEARLY ALL WASTE WATERS FROM TEXTILE OPERATIONS HAVE A HIGH BOD WHICH AFFECTS FISH LIFE. THE DEGRADATION OF AN ALKYL ARYL SULPHONATE IN A SEWAGE TREATMENT PROCESS IS ONLY ABOUT 50 PERCENT, AND IN RIVER WATER IS ONLY ABOUT 75 PERCENT IN 30 DAYS. THE CHEMISTRY OF ALKYL BENZENE SULPHONATES(ABS) AND THE ALFOL ALCOHOL PROCESS IS DISCUSSED. THE MAJOR NUISANCES FROM ABS RESIDUES ARE FOAMING AND ALGAL GROWTHS. THERE IS LITTLE EVIDENCE THAT ABS RESIDUES ARE TOXIC TO HUMANS OR FISH, BUT THE FOAMING ADVERSELY AFFECTS SEWAGE TREATMENT. (LONON-NORTH CAROLINASTATE)

FIELD 05B

ACCESSION NO. W70-08740

HEAT AS A POLLUTANT,

GENERAL ELECTRIC CO., RICHLAND, WASH.

ROBERT T. JASKE.

OREGON STATE UNIV. WATER RESOURCES. RESEARCH INSTITUTE WATER QUALITY CONTROL SEMINAR PROCEEDINGS, SESSION 7, P 61-82, NOVEMBER 1964. 10 FIG.

DESCRIPTORS:
*THERMAL POLLUTION, *TEMPERATURE, HYDROGEN ION CONCENTRATION, NUTRIENTS, ALGAE, FISH, STRATIFICATION, FLOCCULATION, TURBINES, ECONOMIC EFFICIENCY, HEAT BUDGET, SOCIAL ASPECTS.

ABSTRACT:
THE EFFECTS OF INCREASED TEMPERATURE ON PH, NUTRIENTS, ALGAE, TASTE, AND FISH ARE DISCUSSED. THE ADVANTAGES AND DISADVANTAGES OF STRATIFICATION, FLOCCULATION, AND OTHER VARIATIONS IN THE FLUID MECHANICS OF STREAMS ARE CONSIDERED. A SPECIFIC EXAMPLE OF A TYPICAL HEAT-PRODUCING INDUSTRIAL PROCESS OF A TURBINE GENERATOR PLANT IS EXAMINED. THE EFFICIENCY AND THE OVERALL PLANT ECONOMY IS DETERMINED FOR VARYING TEMPERATURES OF THE EFFLUENT. PROBLEMS ARISING FROM STREAM SIZE AND THE ERECTION OF DAMS AND RESERVOIRS ARE MENTIONED. THREE PRIMARY CONCERNS OF FURTHER RESEARCH INCLUDE: (1) IMPROVED ATTENTION TO HEAT BUDGETS AND MATHEMATICAL TREATMENT OF AIR-WATER TRANSPORT PHENOMENA; (2) CLOSER IDENTIFICATION OF THE SPECIFIC CHANGES ASSOCIATED WITH THE TEMPERATURE PARAMETER; AND (3) IMPROVED STANDARDIZATION AND METHODS OF ECONOMIC ANALYSIS FOR IDENTIFICATION OF THE SIGNIFICANT ECONOMIC FACTORS INVOLVED. (OSBORNE-VANDERBILT)

FIELD 05C

ACCESSION NO. W70-08834

NUTRIENT REMOVAL FROM SEWAGE EFFLUENTS BY ALGAL ACTIVITY,

NATIONAL INST. FOR WATER RESEARCH, PRETORIA (SOUTH AFRICA).

J. HEMENS, AND G. J. STANDER.

ADVANCES IN WATER POLLUTION RESEARCH, PROCEEDINGS, FOURTH INTERNATIONAL
CONFERENCE ON WATER POLLUTION RESEARCH, PRAGUE, APR 21-25, 1969, LONDON,
PERGAMON PRESS, 1969, P 701-716.

DESCRIPTORS:
*ALGAE, *HYDROGEN ION CONCENTRATION, *NUTRIENTS, *TERTIARY TREATMENT,
BIOLOGICAL TREATMENT, NITROGEN, PHOSPHORUS, SEWAGE EFFLUENTS.

IDENTIFIERS:
*ALGAL ACTIVITY, *NUTRIENT REMOVAL.

ABSTRACT:
RESULTS OBTAINED IN REMOVING NITROGEN AND PHOSPHORUS FROM SEWAGE
EFFLUENT BASED ON A MODIFICATION OF THE METHODS REPORTED BY OTHER
WORKERS IS DESCRIBED. THE RESULTS SUGGEST THAT INSTEAD OF ATTEMPTING TO
REMOVE NUTRIENTS FROM SEWAGE EFFLUENT BY ASSIMILATION INTO NEW ALGAL
CELLS IT MAY BE MORE PRACTICAL, UNDER SUITABLE CLIMATIC CONDITIONS, TO
EXPLOIT THE EFFECTS OF PH INCREASE DUE TO ALGAL PHOTOSYNTHESIS AND TO
AVOID THE PROBLEMS OF LIGHT UTILIZATION BY DENSE ALGAL CULTURES. THE
BASIC REQUIREMENTS FOR THIS SYSTEM WOULD BE A LEVEL OF PHOTOSYNTHESIS
SUFFICIENT TO PRODUCE THE PH INCREASE REQUIRED TO PRECIPITATE PHOSPHATE
AND TO PROMOTE LOSS OF AMMONIA TO THE ATMOSPHERE. THE DECREASE IN
AMMONIA AND PHOSPHATE REMOVAL OBSERVED DURING THE NIGHT MIGHT BE
OVERCOME BY THE PROVISION OF SUFFICIENT FLUORESCENT LIGHT ABOVE THE
EXPERIMENTAL DISCS TO LESSEN THE PH DECREASE. ALTERNATIVELY, PH CONTROL
BY THE ADDITION OF ALKAKI MIGHT BE POSSIBLE, PARTICULARLY SINCE IT WAS
OBSERVED THAT THE REDUCTION IN ALKALINITY DURING DAYLIGHT REDUCED THE
AMOUNT OF LIME REQUIRED TO REACH A PH OF 10.3 BY AS MUCH AS 75% WHEN
COMPARED TO THAT NECESSARY TO ACHIEVE THE SAME INCREASE IN THE
UNTREATED EFFLUENT. (AGUIRRE-TEXAS)

FIELD 05D

ACCESSION NO. W70-08980

WATER QUALITY IN RELATION TO PRODUCTIVITY OF LAKE ASHTABULA RESERVOIR IN
SOUTHEASTERN NORTH DAKOTA,

NORTH DAKOTA WATER RESOURCES RESEARCH INST., FARGO.

JOHN J. PETERKA.

AVAILABLE FROM NTIS AS PB-193 683, $3.00 IN PAPER COPY, $0.65 IN MICROFICHE.
RESEARCH PROJECT TECHNICAL COMPLETION REPORT, WI-221-002-69, MARCH 1969. 23
P, 4 TAB, 7 FIG, 18 REF. OWRR PROJECT A-004-NDAK(1).

DESCRIPTORS:
RESERVOIR, *PRIMARY PRODUCTIVITY, *EUTROPHICATION, WATER QUALITY,
*NUTRIENTS, LIMNOLOGY, WATER TEMPERATURE, *ALGAE BLOOMS, CYANOPHYTA,
NORTH DAKOTA.

IDENTIFIERS:
LAKE ASHTABULA.

ABSTRACT:
THIS STUDY ATTEMPTS TO RELATE THE CHEMICAL AND PHYSICAL CHARACTERISTICS
OF THE LAKE WATER TO STANDING CROPS AND PRODUCTION RATES OF
PHYTOPLANKTON. THE AVERAGE GROSS PHOTOSYNTHESIS RATE WAS 41 PERCENT OF
THE OPTIMAL RATE. PHYTOPLANKTON STANDING CROPS AVERAGED $ UG
CHLOROPHYLL/LITER IN THE EUPHOTIC ZONE. OPTIMAL PHOTOSYNTHESIS PER UNIT
CHLOROPHYLL AVERAGED 13.5 MG O2/MG CHLOROPHYLL/HOUR. GROSS PRODUCTION
PER UNIT CHLOROPHYLL DECREASED WITH INCREASE IN PHYTOPLANKTON STANDING
CROP. ANNUAL GROSS PHOTOSYNTHESIS WAS 1.4 GC/M2/DAY. THREE EMPIRICAL
EQUATIONS WERE USED TO PREDICT PHOTOSYNTHESIS IN THE EUPHOTIC ZONE.
GROSS PHOTOSYNTHESIS WAS RELATED TO STANDING CROP AND WATER
TEMPERATURE, BUT NOT TO CO2 OR MACRO-NUTRIENTS. HIGH NUTRIENT LEVELS
AND CONTINUAL CIRCULATION OF THE WATER DURING ICE-FREE PERIODS MADE THE
LAKE EUTROPHIC. (PETERKA-NORTH DAKOTA STATE)

FIELD 5C, 2H

ACCESSION NO. W70-09093

EFFECTIVE PHOSPHORUS REMOVAL BY THE ADDITION OF ALUM TO THE ACTIVATED SLUDGE
  PROCESS,

  METROPOLITAN SANITARY DISTRICT OF GREATER CHICAGO, ILL.

  DAVID R. ZENZ, AND JOSEPH R. PIVNICKA.

  PAPER PRESENTED AT THE 24TH ANNUAL PURDUE INDUSTRIAL WASTES CONFERENCE, MAY
    6-8, 1969. 59 P, 34 FIG, 4 TAB, 21 REF.

      DESCRIPTORS:
        *PHOSPHORUS, *PHOSPHATES, *PHOSPHORUS COMPOUNDS, *CHEMICAL
        PRECIPITATION, *CHEMICAL DEGRADATION, *BIOLOGICAL TREATMENT,
        *BIODEGRADATION, *BIOCHEMISTRY, *BIOCHEMICAL OXYGEN DEMAND, *WASTE
        WATER TREATMENT, *WATER POLLUTION CONTROL, *EUTROPHICATION, *ALGAL
        CONTROL, *TERTIARY TREATMENT, SUSPENDED LOAD, PILOT PLANTS, ACTIVATED
        SLUDGE, TREATMENT FACILITIES.

      IDENTIFIERS:
        *PHOSPHORUS REMOVAL, *ALUM, *ALUMINUM SULFATE, *ACTIVATED SLUDGE
        PROCESS, HANOVER TREATMENT PLANT.

      ABSTRACT:
        THE OBJECTIVE OF THIS EXPERIMENTAL RESEARCH WAS TO EVALUATE, UNDER
        ACTUAL WASTE WATER TREATMENT PLANT OPERATING CONDITIONS, A COMBINED
        CHEMICAL-BIOLOGICAL PROCESS CAPABLE OF PRODUCING AN EFFLUENT LOW IN
        PHOSPHORUS, BOD AND SUSPENDED SOLIDS. TESTS WERE CONDUCTED AT THE
        HANOVER PARK WATER RECLAMATION PLANT IN COOK COUNTY, ILLINOIS. THE
        PLANT CONSISTS OF A 2 MGD CONVENTIONAL ACTIVATED SLUDGE FACILITY AND A
        TERTIARY FACILITY CAPABLE OF TREATING THE ENTIRE FLOW FROM THE
        SECONDARY FACILITY. THE SECONDARY PLANT WAS SEPARATED INTO TWO DISTINCT
        SYSTEMS FOR THE TESTS. ALUM WAS ADDED TO THE INFLUENT OF THE AERATION
        TANK OF ONE SYSTEM BY A FEED PUMP. EFFLUENT SAMPLES, TAKEN
        AUTOMATICALLY AS 24-HOUR COMPOSITES, WERE ANALYZED FOR BOD, SS, VS, PER
        CENT TOTAL SOLIDS, PER CENT VOLATILE SOLIDS, ALKALINITY, S.V.I AND
        NH3-N. AN AUTOANALYZER WAS USED FOR PHOSPHORUS DETERMINATIONS. IT WAS
        FOUND THAT THE PROCESS CAN PRODUCE EFFLUENT PHOSPHORUS CONCENTRATIONS
        OF 2.30 MG/L (TOTAL P) AND 0.83 MG/L (SOLUBLE P) AT A AL:TOTAL P RATIO
        (MOLAR BASIS) OF 1.54 AND A AL:SOLUBLE P RATIO OF 1.85. THIS IS 83 AND
        93 PER CENT REMOVAL, RESPECTIVELY. BOUND PHOSPHORUS WAS NOT RELEASED
        DURING ANAEROBIC DIGESTION. ADDITIONS OF LARGE AMOUNTS OF ALUM DID NOT
        ADVERSELY AFFECT BOD REMOVAL EFFICIENCY, BUT REMOVED HIGHER MICROBIAL
        LIFE FORMS IN THE MIXED LIQUOR. THE SUSPENDED SOLIDS INCREASED AT AN
        ALUM CONCENTRATION OF 252 MG/L. WASTE SLUDGE INCREASED, AND S.V.I
        DECREASED. ALUM DID NOT UPSET THE ANAEROBIC BIOLOGICAL TREATMENT
        PROCESS. (POERTNER)

      FIELD 05D

      ACCESSION NO. W70-09186

  BAFFLED BIOLOGICAL BASIS FOR TREATING POULTRY PLANT WASTES,

    SYRACUSE UNIV., N.Y.

    NELSON L. NEMEROW.

    JOURNAL OF THE WATER POLLUTION CONTROL FEDERATION, VOL 41, NO 9, P 1602-1612,
      SEPTEMBER 1969. 8 FIG, 9 TAB.

      DESCRIPTORS:
        *BIOLOGICAL TREATMENT, *OXIDATION, LAGOONS, *BAFFLES, *WASTE WATER
        TREATMENT, *POULTRY, INDUSTRIAL WASTES, LAGOONS, PONDS, PHOTOSYNTHESIS,
        COSTS, EFFICIENCY, BIOCHEMICAL OXYGEN DEMAND, ALGAE, SEDIMENTATION,
        SETTLING BASINS.

      IDENTIFIERS:
        *CHICKENS, FOOD WASTES, POULTRY WASTES.

      ABSTRACT:
        A POULTRY PLANT IN MILLSBOROUGH, DELAWARE, PROCESSED 10,000 CHICKENS
        PER HOUR WITH A WASTE WATER OF 40,000 GPH, AND AN EFFLUENT OF 2,500 LB
        BOD5/DAY AT AN AVERAGE BOD OF 630 MG/L. BECAUSE THE AREA IS
        COMMERCIALLY AND RECREATIONALLY OF GREAT VALUE A PROGRAM WAS INITIATED
        TO REDUCE THE WASTE WATER CONCENTRATION AT A MAXIMUM COST OF $100,000.
        ADEQUATE SCREENING FOLLOWED BY A TWO-STAGE OXIDATION POND PLANT
        UTILIZING OVER AND UNDER CONTACT BAFFLES IN THE FIRST STAGE FOLLOWED BY
        CHLORINATION PROVIDED A 85 TO 95% BOD REDUCTIONS. THE FIRST STAGE
        CONSISTS OF A BAFFLED HIGH-RATE DEEP POND. THE SECOND STATE IS A
        SHALLOW SYNTHETIC BASIN. LOADINGS OF OVER 200 LB/DAY/ACRE RESULTED IN
        HIGH EFFICIENCY AND COLIFORM COUNTS OF LESS THAN 10/1000 ML. THE FINAL
        COST WAS $90,000. (HANCUFF-TEXAS)

      FIELD 05D

      ACCESSION NO. W70-09320

ANIONIC AND NONIONIC SURFACTANT SORPTION AND DEGRADATION BY ALGAE CULTURES,

TEXAS UNIV., AUSTIN. ENVIRONMENTAL HEALTH ENGINEERING RESEARCH LAB.

ERNST M. DAVIS, AND EARNEST F. GLOYNA.

JOURNAL OF THE AMERICAN OIL CHEMISTS' SOCIETY, VOL 46, NO 11, P 604-608, 1969. 9 REF, 2 FIG, 3 TAB. NSF GRANT NSF-GU-1963.

DESCRIPTORS:
*LINEAR ALKYLATE SULFONATES, *SURFACTANTS, WATER POLLUTION SOURCES, *DETERGENTS, *DEGRADATION(DECOMPOSITION), CYANOPHYTA, CHLOROPHYTA, *ALGAE, SORPTION, WASTE WATER TREATMENT, CULTURES, CHLORELLA, WATER POLLUTION EFFECTS.

IDENTIFIERS:
*ALKYL POLY ETHOXYLATE, *ALKYL PHENOL POLYETHOXYLATE, ANIONIC SURFACTANTS, NONIONIC SURFACTANTS, INFRARED SPECTRA.

ABSTRACT:
DEGRADATION OF THREE SURFACTANTS HAS BEEN DETERMINED BY ORGANIC EXTRACTION PROCEDURES AND INFRARED SPECTROSCOPY. AXENIC CULTURES OF FIVE SPECIES OF BLUE-GREEN ALGAE AND THREE SPECIES OF GREEN ALGAE WHICH ARE COMMON TO WASTE STABILIZATION PONDS WERE TEST ORGANISMS. ANALYTICAL DATA ARE SHOWN COMPARING THE EFFECTS PRODUCED BY THE ALGAE CULTURES AND A HETEROGENEOUS MICROCOSM. LINEAR ALKYL SULFONATE WAS THE ANIONIC SURFACTANT COMPOUND TESTED. AN ALKYL POLYETHOXYLATE AND AN ALKYL PHENOL POLYETHOXYLATE WERE THE NONIONIC TEST SURFACTANTS. SORPTION OF THE COMPOUNDS BY THE ALGAE USUALLY WAS FOLLOWED BY RELEASE AND DEGRADATION OF UP TO 99% OF SOME OF THE COMPONENT PARTS OF THE SURFACTANT MOLECULE. (SJOLSETH-WASHINGTON)

FIELD 05C

ACCESSION NO. W70-09438

POTOMAC RIVER WATER QUALITY NETWORK, COMPILATION OF DATA-WATER YEAR 1969.

INTERSTATE COMMISSION ON THE POTOMAC RIVER BASIN, WASHINGTON, D.C.

INTERSTATE COMMISSION ON THE POTOMAC RIVER BASIN DATA COMPILATION REPORT, 1969. 92 P, 2 FIG, 3 MAP.

DESCRIPTORS:
*DATA COLLECTIONS, *WATER QUALITY, BIOCHEMICAL OXYGEN DEMAND, SUSPENDED LOAD, TURBIDITY, DATA STORAGE AND RETRIEVAL, COLIFORMS, DISSOLVED OXYGEN, ALGAE, NUTRIENTS, SOLUTES, CHLORIDES, CALCIUM, CARBONATES, PHOSPHATES.

IDENTIFIERS:
POTOMAC RIVER BASIN.

ABSTRACT:
WATER QUALITY DATA FROM THE POTOMAC RIVER BASIN ARE SUMMARIZED. THE WATER STATISTICS ARE OBTAINED FROM SAMPLES TAKEN AT 80 STREAM SAMPLING POINTS IN THE BASIN. THE SAMPLES WERE TAKEN AND ANALYZED BY AGENCIES, INDUSTRIES AND MUNICIPALITIES. AN EXPLANATION OF PARAMETERS IS INCLUDED. THE DATA OBTAINED THIS WATER YEAR WERE STORED IN THE FEDERAL WATER POLLUTION CONTROL ADMINISTRATION COMPUTER- STORET. (KNAPP-USGS)

FIELD 05A, 07A

ACCESSION NO. W70-09557

THE PERSISTENCE OF PESTICIDES IN IMPOUNDED WATERS,

KENTUCKY WATER RESOURCES INST., LEXINGTON.

ROBERT A. LAUDERDALE.

AVAILABLE FROM NTIS AS PB-194 056, $3.00 IN PAPER COPY, $0.65 IN MICROFICHE.
KENTUCKY UNIVERSITY WATER RESOURCES INSTITUTE RESEARCH REPORT NO 17, 1969.
39 P, 8 FIG, 5 TAB, 8 REF. OWRR PROJECT NO A-002-KY(4).

DESCRIPTORS:
*WATER POLLUTION EFFECTS, *PESTICIDES, RESIDUES, *PESTICIDE KINETICS,
*PERSISTENCE, EVAPORATION, BIODEGRADATION, ADSORPTION, ALGAE, CLAYS,
ALDRIN, DIELDRIN, DDT, AERATION, SILTS.

IDENTIFIERS:
CHLORDANE.

ABSTRACT:
THE PERSISTENCE OF THE INSECTICIDES ALDRIN, DIELDRIN, CHLORDANE, AND
DDT IN WATER WERE MEASURED. A SMALL AREA WAS SPRAYED WITH THESE
COMPOUNDS, AND THE INSECTICIDES WERE COLLECTED IN THE RUNOFF WATER FROM
THE AREA. LABORATORY EXPERIMENTS WERE PERFORMED TO DETERMINE THE
EFFECTIVENESS OF AERATION, ADSORPTION ON SILT, AND ADSORPTION ON ALGAE
IN REMOVING THE PESTICIDES FROM WATER. ALL OF THE PESTICIDES WERE FOUND
IN SAMPLES OF WATER COLLECTED FOR THE FULL PERIOD OF THE TESTS. THE
AMOUNTS WHICH WERE FOUND IN A SMALL POND INTO WHICH THE SURFACE WATER
DRAINED WAS CONSISTENTLY LESS THAN THAT FOUND IN THE SURFACE RUNOFF
WATER ITSELF. THE REDUCED CONCENTRATIONS IN THE POND WATER WERE
APPARENTLY THE RESULT OF A COMBINATION OF SEDIMENTATION, ADSORPTION ON
ALGAE OR SILT, OR VOLATILIZATION OF THE INSECTICIDE FROM THE WATER INTO
THE ATMOSPHERE. IN GENERAL, THE EFFICIENCY OF REMOVAL OF INSECTICIDE
FROM WATER BY ANY ONE OF THESE PROCESSES APPEARS TO BE LOW. HOWEVER,
BECAUSE OF THE LARGE RATIO OF ABSORBANT TO INSECTICIDE, THE OVERALL
REMOVAL IS SUBSTANTIAL. (KNAPP-USGS)

FIELD 05B, 05G

ACCESSION NO. W70-09768

PRELIMINARY OBSERVATIONS OF THE HYDROMECHANICS, NUTRIENT CYCLES AND
EUTROPHICATION STATUS OF LAKE KINNERET (LAKE TIBERIAS),

LIMNOLOGICAL LAB., TABIGHA (ISRAEL).

C. SERRUYA, S. SERRUYA, AND T. BERMAN.

VERHANDLUNGEN DER INTERNATIONALEN VEREINIGUNG FUR THEORETISCHE UND ANGEWANDTE
LIMNOLOGIE, VOL 17, P 342-351, 1969. 7 FIG, 1 TAB, 8 REF.

DESCRIPTORS:
*CYCLING NUTRIENTS, *EUTROPHICATION, *HYDRODYNAMICS, LAKES, INTERNAL
WATER, WAVES(WATER), DIURNAL, EPILIMNION, WINDS, THERMOCLINE,
TEMPERATURE, CURRENTS(WATER), HYPOLIMNION, ALGAE, NITROGEN, PHOSPHORUS,
OXYGEN, CARBON, ENZYMES, SAMPLING, BOTTOM SEDIMENTS, SODIUM, POTASSIUM,
IRON, MANGANESE, CALCIUM, PIGMENTS.

IDENTIFIERS:
*LAKE KINNERET(ISRAEL).

ABSTRACT:
EXISTENCE OF INTERNAL WAVES IN LAKE KINNERET (ISRAEL) WAS CONFIRMED. IN
EARLY MAY 1968, WITH WELL DEVELOPED THERMOCLINE, CHARACTERISTIC
INTERNAL WAVES WERE OBSERVED, AVERAGING 11 HOURS, OCCURRING REGULARLY
AT SEMI-DIURNAL INTERVALS, CORRELATED SOMEWHAT WITH PERIODIC WEST WIND.
IN THE HYPOLIMNION, LARGE QUANTITIES OF AMMONIUM-NITROGEN AND
ORTHOPHOSPHATE CONCENTRATIONS INCREASINGLY ACCUMULATED FROM MAY TO
DECEMBER. RATIO OF TOTAL AVAILABLE NITROGEN TO ORTHOPHOSPHATES IS HIGH
IN FEBRUARY AND DECREASES RAPIDLY TO MID-SUMMER. ASSIMILATION RATE AND
REGENERATION IN EPILIMNION IS ABOUT FOUR TIMES GREATER IN SPRING THAN
SUMMER. UNDER PREVAILING LAKE CONDITIONS, DURING THE DECOMPOSITION OF
ORGANIC MATTER IN HYPOLIMNION, FOR EVERY PHOSPHORUS ATOM LIBERATED AS
PHOSPHATE, 580 ATOMS OF OXYGEN ARE USED FOR OXIDATION OF CORRESPONDING
TOTAL QUANTITY OF ORGANIC MATTER. STUDY OF THE EPOLIMNION STRONGLY
SUGGESTS THAT THE HIGH OXYGEN:PHOSPHORUS RATIO REFLECTS THE HIGH
ORGANIC CONTENT OF LAKE ALGAE AND PARTICULARLY THEIR HIGH
NITROGEN:PHOSPHORUS RATIO. HIGH LEVEL OF ORGANIC MATTER INDICATES THAT
SEDIMENTS ARE PROVIDING QUANTITIES OF NUTRIENTS, PARTICULARLY
AMMONIUM-NITROGEN. BIWEEKLY SAMPLES OF ALKALINE MONOPHOSPHOESTERASES
ARE MEASURED. IN BOTTOM DEPOSITS, CONTENT OF ORGANIC CARBON AND
KJELDAHL NITROGEN IS RATHER HIGH. THERE ARE INDICATIONS OF RECENT
ENHANCED EUTROPHICATION, SUPPORTED BY RESULTS OF PIGMENT EXTRACTION
FROM SEDIMENTS. (JONES-WISCONSIN)

FIELD 02H, 05C

ACCESSION NO. W70-09889

A SIMPLE, INEXPENSIVE ALGAL CHEMOSTAT,

NORTH CAROLINA STATE UNIV., RALEIGH. DEPT. OF ZOOLOGY.

EDWARD J. CARPENTER.

LIMNOLOGY AND OCEANOGRAPHY, VOL 13, NO 4, P 720-721, 1968. 1 FIG, 2 REF.

DESCRIPTORS:
*ANALYTICAL TECHNIQUES, *ALGAE, PLANKTON, DIATOMS, MARINE ALGAE, FRESH
WATER, FLOW RATES, PUMPS, ELECTROLYSIS, CULTURES.

IDENTIFIERS:
*CHEMOSTAT, PHYSIOLOGY, FLAGELLATES.

ABSTRACT:
A SIMPLE CHEMOSTAT HAS BEEN DEVELOPED WHICH DOES NOT USE EXPENSIVE
MECHANICAL ACTION PUMPS OR ELABORATE GRAVITY FLOW SYSTEMS. IN ADDITION,
THE ENTIRE APPARATUS CAN BE EASILY STERILIZED AS A UNIT IF THE RUBBER
STOPPERS IN THE CHEMOSTAT ARE LEFT AJAR DURING AUTOCLAVING. AN
ELECTROLYSIS PUMP GENERATES A GAS MIXTURE WHICH DISPLACES THE CULTURE
MEDIUM IN THE MEDIUM RESERVOIR INTO THE ALGAE CULTURE VESSEL. THE FLOW
RATE OF THE CULTURE MEDIUM IS REGULATED BY A VARIABLE RESISTOR IN THE
CIRCUIT AND INDIRECTLY MONITORED WITH A MILLIAMMETER. MIXING AND
AERATION OF ALGAE AND MEDIUM IS ACCOMPLISHED BY PUMPING STERILE,
COTTON-FILTERED AIR THROUGH THE CULTURE VESSEL. AN OVERFLOW VESSEL
EFFECTIVELY PREVENTS BACK CONTAMINATION OF THE CULTURE VESSEL BY
BACTERIA AS A SUSPENSION OF THE ORGANISMS UNDER CULTURE CONTINUALLY
FLOWS OUT. THE VOLUME OF MEDIUM WITHIN THE CULTURE VESSEL IS REGULATED
THROUGH SIPHON ACTION BY ADJUSTING THE HEIGHT OF THE OVERFLOW TUBE.
THIS CHEMOSTAT HAS BEEN USED SUCCESSFULLY IN SEVERAL NUTRIENT LIMITED,
AXENIC EXPERIMENTS ON TWO PLANKTONIC MARINE DIATOMS AND ONE FRESHWATER
FLAGELLATE. (JONES-WISCONSIN)

FIELD 07B, 05C

ACCESSION NO. W70-09904

ON USING INDUSTRIAL AND DOMESTIC WASTES IN AQUACULTURE,

MARYLAND UNIV., SOLOMONS. NATURAL RESOURCES INST.

J. A. MIHURSKY.

AGRICULTURAL ENGINEERING, VOL 50, NO 11, P 667-689, 1969. 1 FIG, 3 TAB, 18
REF.

DESCRIPTORS:
*WASTE DISPOSAL, *AQUACULTURE, *ULTIMATE DISPOSAL, DOMESTIC WASTES,
INDUSTRIAL WASTES, SEWAGE WASTES, BYPRODUCTS, FERTILIZERS, ALGAE,
ZOOPLANKTON, FISH, SHELLFISH, STEAM, ELECTRIC POWER, HEATED WATER,
PROTEINS, AQUATIC MICROORGANISMS, TEMPERATURE, ACTIVATED SLUDGE,
FISHERIES, CARP, CLAMS, PONDS, RUNNING WATERS, FRESH WATER, BRACKISH
WATER, ECONOMICS, CYCLING NUTRIENTS, FOODS.

IDENTIFIERS:
PETROLEUM WASTES, MERCENARIA MERCENARIA, AQUACULTURE, RECYCLING WASTES.

ABSTRACT:
THREE WASTE MATERIALS HAVE THE GREATEST POTENTIAL FOR RECYCLING THROUGH
AQUATIC SYSTEMS INTO USEFUL NUTRIENTS--PETROLEUM WASTES, SEWAGE WASTES,
AND WASTE HEAT. PROTEIN DEVELOPED AND SOLD IN FRANCE IS PRODUCED BY
MICROORGANISMS GROWING ON WASTE LOW-GRADE PETROLEUM. LARGE-SCALE USE OF
MICROBIOLOGICALLY DERIVED PROTEINS SEEMS POSSIBLE. SEWAGE CONVERTED TO
INORGANIC MATERIALS MAY PROVIDE FERTILIZER FOR ALGAL PRODUCTION. WASTE
HEAT MAY ASSIST IN PROVIDING OPTIMAL ENVIRONMENTAL CONDITIONS FOR
BIOLOGICAL GROWTH AT ALL TROPHIC LEVELS. IN VARIOUS WORLD SITUATIONS
SEWAGE OR SEWAGE PRODUCTS ARE USED TO STIMULATE FISHERY PRODUCTION.
ALGAE GROWN IN MASS AIDED BY WATER FROM TREATED SEWAGE AND CONCENTRATED
BY EVAPODRYING TECHNIQUE PRODUCES FOOD FOR CHICKENS. WASTE HEAT FROM
STEAM ELECTRIC STATIONS IS USED TO RAISE CARP IN LATITUDES FARTHER
NORTH THAN USUAL AND TO EXTEND THE GROWING SEASON IN SOME LOCALITIES. A
WASTE FLUE GAS WAS USED TO SUPPLY CARBON DIOXIDE GAS TO MASS-CULTURE AN
ALGAL FOOD SUPPLY. WILD OR NATURAL WATER IS PREFERABLE TO POND CULTURE.
RUNNING WATER SYSTEMS PERMIT DENSE STOCKING, WASTE REMOVAL AND
CONTINUED DISSOLVED OXYGEN RENEWAL. COSTS OF PRODUCING FISH PROTEIN
CONCENTRATE THROUGH AQUACULTURE PLUS WASTE UTILIZATION WILL BE
SUBSIDIZED BY SOCIETY'S NEED TO RECYCLE ITS WASTE MATERIALS INTO USEFUL
FOODSTUFFS. (JONES-WISCONSIN)

FIELD 05E

ACCESSION NO. W70-09905

RECOVERY OF A SALT MARSH IN PEMBROKESHIRE, SOUTH-WEST WALES, FROM POLLUTION BY
    CRUDE OIL,

    K. J. M. BAKER, AND E. B. COWELL.

    BIOLOGICAL CONSERVATION, VOL. 1, NO. 4, JULY 1969, P 291-295. 2 FIG, 2 TAB, 8
    REF.

        DESCRIPTORS:
            RESISTANCE, WATER POLLUTION EFFECTS, *SALT MARSHES, OILY WATER, *OIL,
            *MARINE ALGAE, WATER POLLUTION SOURCES, ON-SITE DATA COLLECTIONS,
            *ON-SITE INVESTIGATIONS, FREQUENCY, *FREQUENCY ANALYSIS.

        IDENTIFIERS:
            *RECOVERY, OIL SPILL, *OIL POLLUTION, *FILAMENTOUS GREEN ALGAE,
            HALIMONE PORTULACOIDES, SUAEDA MARITIMA, ASTER TRIPOLIUM, PEMBROKE,
            SOUTH-WEST WALES.

        ABSTRACT:
            BENTLASS SALT MARSH, NEAR PEMBROKE, SOUTH-WEST WALES, WAS POLLUTED BY
            KUWAIT CRUDE OIL IN JANUARY 1967. FREQUENCY DATA TAKEN IN JUNE 1966,
            JUNE 1967, AND JUNE 1968, SHOW THAT SOME SPECIES WERE MORE AFFECTED
            THAN OTHERS BY THE OIL. IN JUNE 1967, SUAEDA MARITIMA, SALICORNIA SPP.,
            HALIMIONE PORTULACOIDES, AND FILAMENTOUS GREEN ALGAE, SHOWED THE
            GREATEST REDUCTION OF FREQUENCY, WHILE ASTER TRIPOLIUM, COCHLEARIA
            SPP., TRIGLOCHIN MARITIMA, PUCCINELLIA MARITIMA, JUNCUS GERARDI,
            LIMONIUM HUMILE, AND SPARTINA TOWNSENDII SHOWED SOME REDUCTION. AT THE
            TIME FREQUENCIES OF FESTUCA RUBRA, PANTAGO MARITIMA, ARMERIA MARITIMA,
            ARTEMISIA MARITIMA, GLAUX MARITIMA, AND SPERGULARIA SPP., WERE NOT
            SIGNIFICANTLY LOWER THAN THE PRE-SPILL LEVELS. BY JUNE 1968 MOST OF THE
            SPECIES WHICH WERE AFFECTED IN 1967 SHOWED SOME RECOVERY, THE EXCEPTION
            BEING TRIGLOCHIN MARITIMA, THE ILL-EFFECT ON WHICH HAD INCREASED. THE
            SPECIES WITH THE GREATEST COVERAGE, FESTUCA RUBRA IN THE UPPER MARSH,
            PUCCINELLIA MARITIMA IN THE MID-MARSH, AND SPARTINA TOWNSENDII IN THE
            LOWER MARSH, HAD ALL RECOVERED MORE OR LESS COMPLETELY BY JUNE 1968.
            (SJOLSETH-WASHINGTON)

    FIELD 05C

    ACCESSION NO.   W70-09976

CONTROL OF AQUATIC PLANT NUISANCES (ESPECIALLY ALGAE) WITH SOME SUBSTITUTED
    PHENYLUREAS,

    IVONNE J. LE COSQUINO DE BUSSY.

    VERHANDLUNGEN DER INTERNATIONALEN VEREINIGUNG FUR THEORETISCHE UND ANGEWANDTE
    LIMNOLOGIE, VOL 17, P 539-545, 1969. 1 FIG, 4 TAB, 9 REF.

        DESCRIPTORS:
            *AQUATIC PLANTS, *NUISANCE ALGAE, *ALGICIDES, *PHENOLS, *WATER QUALITY
            CONTROL, CRUSTACEANS, ROTIFERS, SWIMMING POOLS, CHLAMYDOMONAS, DAPHNIA,
            DIATOMS, CHLORELLA, SCENEDESMUS, OXYGEN, COLIFORMS, SEDIMENTS, MUD,
            SANDS, SLUDGE, RECREATION.

        IDENTIFIERS:
            *PHENYLUREAS, DIURON, FLAGELLATES, MONURON, TRANSPARENCY, CYCLOPS,
            SCENEDESMUS OBLIQUUS, DACTYLOCOCCUS, ESCHERICHIA COLI, OOCYSTIS, LEMNA,
            SPIRODELA, ELODEA.

        ABSTRACT:
            THE REQUIREMENT IN THE NETHERLANDS FOR A GOOD MEANS OF CONTROLLING
            ALGAE NUISANCE IN OPEN-AIR FLOW-THROUGH SWIMMING POOLS OR IN
            RECREATIONAL PONDS, FOSTERED RESEARCH WITH SUBSTITUTED PHENYLUREAS.
            THESE HERBICIDES ARE INHIBITORS OF PHOTOSYNTHESIS. HIGHER SUBMERGED
            WATER PLANTS AND FILAMENTOUS ALGAE CAN BE CONTROLLED VERY WELL BY THE
            APPLICATION OF DIURON IN EARLY APRIL IN DOSAGES VARYING FROM 0.2 TO 0.4
            PPM. IN ORDER TO CONTROL MICROALGAE IN SWIMMING POOLS AND TO OBTAIN A
            TRANSPARENCY OF 100-250 CENTIMETERS A HIGHER DOSE IS NECESSARY, THE
            GROWTH OF CRUSTACEA, WHICH EAT SMALL ALGAE BEING DESIRABLE AT THE SAME
            TIME. THE ADDITION OF DIURON SHOULD BE EFFECTED IN CLEAR WATER
            PREFERABLY WITH AN INITIAL DOSE OF ONE PPM. IN SMALLER SWIMMING POOLS
            THE CONCENTRATION OF DIURON MUST BE ABOVE 0.6 PPM IN ORDER TO OBTAIN A
            TRANSPARENCY OF 120-240 CENTIMETERS. IN LARGE SWIMMING POOLS (6000
            CUBIC METERS) A TRANSPARENCY FROM 170-270 CENTIMETERS COULD BE
            MAINTAINED EVEN DURING THE TIME WHEN THE DIURON IN THE WATER WAS NO
            LONGER DETECTABLE. (JONES-WISCONSIN)

    FIELD 05G

    ACCESSION NO.   W70-10161

ALGAL RECORDS FOR THREE INDIANA SEWAGE STABILIZATION PONDS,

ROBERT A. TAFT WATER RESEARCH CENTER, CINCINNATI, OHIO. ADVANCED WASTE
    TREATMENT RESEARCH LAB.

C. MERVIN PALMER.

PROCEEDINGS OF THE INDIANA ACADEMY OF SCIENCE, VOL 78, P 139-145, 1968. 3
    TAB, 8 REF.

DESCRIPTORS:
    *SEWAGE LAGOONS, *ALGAE, *SEWAGE TREATMENT, *INDIANA, CHLOROPHYTA,
    CYANOPHYTA, DIATOMS, SEASONAL, EUGLENA, CHLAMYDOMONAS, CHLORELLA,
    SCENEDESMUS, SAMPLING, OXYDATION LAGOONS.

IDENTIFIERS:
    BLUE-GREEN ALGAE, NITZSCHIA, BIOLOGICAL STUDIES, ALGAL IDENTIFICATION.

ABSTRACT:
    THREE SEWAGE STABILIZATION PONDS IN SOUTHEASTERN INDIANA WERE AMONG
    SEVERAL THROUGHOUT THE UNITED STATES SELECTED FOR BIOLOGICAL STUDIES,
    WITH PARTICULAR EMPHASIS ON THE ALGAL FLORA. ALGAL IDENTIFICATIONS HAVE
    BEEN RECORDED FROM 376 SAMPLES COLLECTED FROM THE PONDS DURING A PERIOD
    FROM MAY 1962 TO AUGUST 1968. ALTHOUGH CERTAIN GENERA WERE FOUND
    FREQUENTLY IN ALL THREE PONDS, EACH POND HAD A DISTINCTIVE ALGAL FLORA.
    GREEN ALGAE WERE INVARIABLY THE MOST ABUNDANT; HOWEVER, FLAGELLATES
    WERE ALSO PROMINENT. OF A TOTAL OF 64 GENERA OF THE MOST SIGNIFICANT
    AND ABUNDANT ALGAE THERE WERE 29 GREEN ALGAE, 19 FLAGELLATES, 10
    BLUE-GREEN ALGAE, AND 6 DIATOMS. SOME GENERA WERE LIMITED TO THE SUMMER
    SEASON, WHILE OTHERS WERE MOST PROMINENT IN SPRING AND FALL OR IN
    WINTER. THE POLLUTION-TOLERANT ALGAE, EUGLENA AND NITZSCHIA, WERE
    ABUNDANT AND PERSISTENT IN ALL THREE PONDS. THE BIOCHEMICAL OXYGEN
    DEMAND AT THE INTAKES WAS ABOUT 500 PPM AND FOR THE EFFLUENTS LESS THAN
    50 PPM. THE FIRST VIRUS INFECTING A BLUE-GREEN ALGA WAS ISOLATED FROM
    ONE OF THE PONDS IN RIPLEY COUNTY. (JONES-WISCONSIN)

FIELD 05D

ACCESSION NO. W70-10173

MANAGEMENT OF AQUATIC VASCULAR PLANTS AND ALGAE,

CORNELL UNIV., ITHACA, N.Y. DEPT. OF BOTANY.

HUGH F. MULLIGAN.

EUTROPHICATION: CAUSES, CONSEQUENCES, CORRECTIVES, P 464-482. PRINTING AND
   PUBLISHING OFFICE, NATIONAL ACADEMY OF SCIENCES, WASHINGTON, D.C., 1969. 2
   TAB, 110 REF.

   DESCRIPTORS:
      *AQUATIC WEED CONTROL, *AQUATIC PLANTS, *HERBICIDES, *VASCULAR TISSUES,
      *ALGAE, ECOSYSTEMS, CHARA, BENTHIC FLORA, FLOATING PLANTS,
      PHYTOPLANKTON, AQUATIC INSECTS, FISH, GAME BIRDS, ANIMALS, FOOD CHAINS,
      OXYGEN, ROTIFERS, WEEDS, WATER HYACINTH, ALLIGATORWEED, CHEMCONTROL,
      MECHANICAL CONTROL, BIOCONTROL, COPPER SULFATE, SODIUM ARSENITE,
      DIQUAT, TOXICITY, SNAILS, MAMMALS, TILAPIA, CARP, FERTILIZATION,
      NORTHEAST US, NEW YORK, TENNESSEE VALLEY, 2-4-D.

   IDENTIFIERS:
      CERATOPHYLLUM, MYRIOPHYLLUM, EICHHORNIA, POTAMOGETON, TRAPA, PISTIA,
      ALTERNANTHERA, HETERANTHERA, NAJAS, SILVEX, ENDOTHAL, POTASSIUM
      PERMANGANATE, FENAC, SILVER NITRATE, SIMAZIN, NUPHAR, AGASICLES,
      MANATEE, MARISA.

   ABSTRACT:
      CHEMICAL TREATMENTS OFFER, AT BEST, TEMPORARY RELIEF FROM OVERABUNDANCE
      OF AQUATIC VEGETATION. EVEN WHEN EXCELLENT RESULTS ARE OBTAINED,
      TREATMENTS MUST BE REPEATED ANNUALLY. PRESENTLY, AQUATIC HERBICIDES ARE
      NONSELECTIVE, USUALLY KILLING ALL PLANTS BY CONTACT. IT WOULD BE MORE
      EFFICACIOUS TO USE AQUATIC HERBICIDES POSSESSING SELECTIVE TOXICITY
      AGAINST SPECIFIC AQUATIC WEEDS, THUS PERMITTING THE REESTABLISHMENT OF
      DESIRABLE PLANTS. THE COST OF DEVELOPING AQUATIC HERBICIDES WOULD BE
      GREATER AS APPOSED TO AGRICULTURAL HERBICIDES BECAUSE OF THE
      SPECIALIZED TESTING REQUIREMENTS, AND GENERALLY, THE POTENTIAL MARKET
      WOULD NOT BE AS GREAT. MECHANICAL CONTROL IS VERY PROMISING: IT
      INVOLVES HAND-PULLING OR CUTTING AND HARVESTING PLANTS FROM RESTRICTED
      AREAS. BIOLOGICAL CONTROL HAS MET WITH ONLY LIMITED SUCCESS, BUT IS AN
      AVENUE OF PROMISING POTENTIAL. THIS APPROACH WOULD INCLUDE INTRODUCING
      ORGANISMS THAT ARE INIMICAL TO THE TARGET ORGANISMS AND MANIPULATING
      THE EXISTING AQUATIC ENVIRONMENT. IN A CONCEPT, THIS TYPE OF CONTROL
      USES BIOLOGY AGAINST BIOLOGY. AMONG THE ORGANISMS UTILIZED IN CONTROL
      OF AQUATIC PLANTS ARE FISH, SNAILS, MAMMALS, AND BEETLES. BIOLOGICAL
      CONTROL IS THE ONLY METHOD THAT COULD PROVIDE A PERMANENT SOLUTION TO
      THE PROBLEM OF EXCESSIVE AQUATIC VEGETATION. (SEE ALSO W70-03975)
      (JONES-WISCONSIN)

   FIELD 02I, 05C

   ACCESSION NO. W70-10175

TREATMENT OF WASTE FROM POLYESTER MANUFACTURING. OPERATIONS,

FIBER INDUSTRIES, INC., CHARLOTTE, N.C.; AND DAVIS AND FLOYD ENGINEERS, INC.,
GREENWOOD, S.C.

C. E. STEINMETZ, AND W. J. DAY.

CHAMICAL ENGINEERING PROGRESS, SYMPOSIUM SERIES VOL 65, NO 97, 1969. P
188-190.

DESCRIPTORS:
*WATER REUSE, *RECLAIMED WATER, ORGANIC COMPOUNDS, CHEMICAL WASTES, ION
EXCHANGE, BIOCHEMICAL OXYGEN DEMAND, CHEMICAL OXYGEN DEMAND, COLOR,
ACTIVATED SLUDGE, ALGAE, WASTE WATER TREATMENT.

IDENTIFIERS:
*POLYESTER FIBERS, MAN-MADE FIBERS, METHYL ALCOHOL, ETHYLENE GLYCOL,
POLYMERS, DIMETHYL TEREPHTHALATE.

ABSTRACT:
THIS IS A PROGRESS REPORT TO MAY 1969, ON A WATER REUSE PROJECT,
JOINTLY UNDERTAKEN BY THE SHELBY PLANT OF FIBER INDUSTRIES, AND THE
FWPCA. THIS PLANT APPLIES A LUBRICANT, OR SO-CALLED 'FIBER FINISH', TO
THE POLYESTER FIBER MANUFACTURED THERE. IT IS OF A PROPRIETARY NATURE
BUT FOR PURPOSES OF WASTE TREATMENT SCHEMES MAY BE CONSIDERED A LONG
CHAIN FATTY ACID OR DERIVATIVE THEREOF. THIS MATERIAL, BLOW DOWN WATER
FROM THE POWER PLANT, AND SMALL AMOUNTS OF CHEMICAL WASTES, INCLUDING
METHYL ALCOHOL, ETHYLENE GLYCOL, DIMETHYL TEREPHTHALATE, LOW MOLECULAR
WEIGHT POLYMER AND TITANIUM DIOXIDE, ION EXCHANGE RECHARGE CHEMICALS,
HOUSEKEEPING AND LABORATORY WASTES ARE THE ESSENTIAL WASTE PRODUCTS. IN
A PILOT PLANT STUDY THE WASTE WATER PREVIOUSLY TREATED IN AN EXTENDED
AERATION ACTIVATED SLUDGE SYSTEM IS FURTHER TREATED. AS RECEIVED IT HAS
A COLOR OF 50 UNITS, C.O.D. OF 280-300 PPM AND B.O.D. OF 8-24 PPM.
ALGAE ARE REMOVED BY SCREENING AND A CARBON UNIT REDUCES THE COLOR TO
10 UNITS, C.O.D. TO 125 PPM AND B.O.D. TO 4 PPM. ONE HALF OF THIS WATER
IS DISCHARGED AND THE REST RETURNED TO THE SYSTEM AFTER COOLING AND
REMOVAL OF CHROMIUM AND ZINC SALTS. (WORK-NORTH CAROLINA STATE)

FIELD 05D

ACCESSION NO. W70-10433

A REVIEW OF THE LITERATURE OF 1966 ON WASTE WATER AND WATER POLLUTION CONTROL,

C. M. WEISS.

JOURNAL WATER POLLUTION CONTROL FEDERATION, VOL 39, NO 7, P 1049-1154, JULY
1967. 917 REF.

DESCRIPTORS:
*WASTE WATER TREATMENT, *BIOLOGICAL TREATMENT, *WATER POLLUTION
CONTROL, FISH, TOXICITY, FRESH WATER, PULP WASTES, PESTICIDE RESIDUES,
HERBICIDES, ESTUARIES, WATER QUALITY, DETERGENTS, ALGAL CONTROL,
INDUSTRIAL WASTES, THERMAL POLLUTION, METHODOLOGY, INSTRUMENTATION, SEA
WATER.

IDENTIFIERS:
METAL TOXICITY.

ABSTRACT:
A REVIEW OF PAPERS PUBLISHED DURING 1966 IS PRESENTED. MATERIAL COVERED
INCLUDES: BIOLOGICAL STUDIES EVALUATING WATER POLLUTION, CAUSES OF FISH
MORTALITY, TOXIC NATURE OF SUBSTANCES IN FRESH WATER, WATER QUALITY,
DETERGENTS AS RELATED TO ALGAL GROWTH AND FISH DAMAGE, DOMESTIC AND
INDUSTRIAL WASTE TREATMENT, METAL TOXICITY ASSAY, EFFECT OF WASTES FROM
THE PULPING AND FOREST INDUSTRIES, THE POTENTIAL HAZARDS FROM PESTICIDE
AND HERBICIDE RESIDUES, EFFECT OF THERMAL DISCHARGES, PHYSIOLOGICAL AND
PATHOLOGICAL CONDITIONS IN FISH CAUSED BY TOXIC COMPOUNDS, METHODOLOGY
AND INSTRUMENTATION, AND ESTUARINE AND MARINE POLLUTION.
(LIVENGOOD-NORTH CAROLINA STATE)

FIELD 05D, 10

ACCESSION NO. W70-10437

BEVERLY HILLS WATER DEPARTMENT SOLVES ALGAE PROBLEM WITH KMNO-SUB-4,

BEVERLY HILLS MUNICIPAL WATER DEPT., CALIF.

MARY JO WITT.

WATER AND SEWAGE WORKS, VOL 115, P 400-403, 1968. 2 FIG, 6 PHOTOS.

DESCRIPTORS:
*POTABLE WATER, *WATER QUALITY CONTROL, *ALGAE, COPPER SULFATE,
CHLORINE, WELLS, CALIFORNIA.

IDENTIFIERS:
*POTASSIUM PERMANGANATE, BEVERLY HILLS(CALIF).

ABSTRACT:
BEVERLY HILLS (CALIFORNIA) OPERATES TWO LIME SOFTENING AND FILTRATION
PLANTS FOR THEIR WELL WATER. PLANTS NO 1 AND NO 2 TREAT AN AVERAGE OF
5.5 MGD AND 2.5 MGD, RESPECTIVELY. IN MARCH 1963, A 900,000 GAL
SEDIMENTATION BASIN IN PLANT NO 1 WAS REPLACED BY A 250,000 GAL BASIN.
COPPER SULFATE AND CHLORINE HAD CONTROLLED ALGAE AND BACTERIAL GROWTHS
IN THE OLD BASIN BUT CONVERTED ALGAE AND BACTERIA TO A GREENISH-BLACK
TAR-LIKE SCUM IN THE NEW BASIN. IN JULY 1964, A 1.8 MILLIGRAM/LITER
DOSE OF POTASSIUM PERMANGANATE WAS ADDED TO THE BASIN ALONG WITH
CHLORINE AND COPPER SULFATE RESULTING IN NOTICEABLE IMPROVEMENT WITHIN
A DAY OR SO; CONTINUED TREATMENT ELIMINATED THE GROWTH. FOR MAXIMUM
CONTACT, POTASSIUM PERMANGANATE WAS FED INTO THE PUMP DISCHARGE LINE AT
THE FURTHEST WELL. SINCE STARTING POTASSIUM PERMANGANATE TREATMENT,
COPPER SULFATE APPLICATION WAS ELIMINATED; SHUTDOWNS FOR CLEAN-OUT ARE
NECESSARY ONLY EVERY 9-12 MONTHS, AND THERE IS PRACTICALLY NO ALGAE IN
THE SETTLING BASIN. DURING THE SUMMER AND FALL LARGE PIECES OF ALGAE
WOULD BREAK LOOSE FROM THE CLARIFIER AND BASIN WALLS IN PLANT NO 2. IN
OCTOBER 1965, A 1.2-1.5 MILLIGRAM/LITER CONTINUING DOSE OF POTASSIUM
PERMANGANATE, FED TO THE WELL FARTHEST FROM THE PLANT, FREED THE GROWTH
FROM THE BASIN WALLS. (HASKINGS-WISCONSIN)

FIELD 05F

ACCESSION NO. W71-00110

SEASONAL DYNAMICS OF THE REDUCING SUGAR CONTENT OF WATER IN THE KIEV RESERVOIR
AND INFLOWING RIVERS,

AKADEMIYA NAUK URSR, KIEV. INSTYTUT HIDROBIOLOGII.

G. A. YENAKI.

TRANSLATION OF GIDROBIOLOGICHESKIY ZHURNAL, VOL 5, NO 3, 1969.
HYDROBIOLOGICAL JOURNAL, VOL 5, NO 3, P 47-50, 1969. 4 P, 2 FIG, 1 TAB, 13
REF.

DESCRIPTORS:
*RESERVOIRS, *EUTROPHICATION, *CARBOHYDRATES, *NUTRIENTS,
*BIODEGRADATION, ALGAE, BACTERIA, AQUATIC MICROORGANISMS, SAMPLING,
WATER ANALYSIS, WATER QUALITY, POLLUTANT IDENTIFICATION.

IDENTIFIERS:
KIEV RESERVOIR(USSR).

ABSTRACT:
CARBOHYDRATE-LIKE COMPOUNDS IN NATURAL WATERS PLAY AN IMPORTANT ROLE IN
PROCESSES WHICH OCCUR WITHIN RESERVOIRS AND IN PARTICULAR IN PROCESSES
OF FORMATION AND TRANSFORMATION OF ORGANIC SUBSTANCES IN NATURAL
WATERS. MONTHLY DETERMINATIONS WERE MADE IN 1965 AND 1966 OF THE
REDUCING SUGAR CONTENT OF WATER IN THE KIEV RESERVOIR AND THE RIVERS
FLOWING INTO IT--THE DNIEPER, PRIPYAT', AND TETEREV. DETERMINATION OF
SUGARS WAS CONDUCTED ON FILTERED WATER BY A SPECTROPHOTOMETRICAL
METHOD. THE HIGH CONTENT OF REDUCING SUGARS IN WATER OF THE KIEV
RESERVOIR IS ONE OF THE MAIN FACTORS RESPONSIBLE FOR THE INTENSIVE
DEVELOPMENT OF MICROORGANISMS, THE INTENSIFICATION OF BIOCHEMICAL
PROCESSES, AND THE TRANSFORMATION OF BIOGENIC ELEMENTS. THIS CONDITION
INVOLVES AN INCREASE IN THE NUMBER OF PLANKTONIC ORGANISMS AND THE
OCCURRENCE OF 'BLOOMING' OF THE WATER. THE CONTENT OF REDUCING SUGARS
IN WATER OF THE KIEV RESERVOIR IN 1965 AND 1966 VARIED FROM 0.26-1.33
MG/LITER, AND IN THE RIVERS FLOWING INTO THE RESERVOIR THIS VALUE WAS
EQUIVALENT TO 0.17-1.63 MG/LITER. THE GLUCOSE CARBON WAS EQUIVALENT TO
0.7-6.7% AND 0.9-4.1% OF THE TOTAL ORGANIC CARBON, RESPECTIVELY. THE
MAXIMUM QUANTITY OF SUGARS WAS FOUND IN SPRING AND SUMMER, AND THE
MINIMUM QUANTITY WAS FOUND IN WINTER. THE INCREASE IN CONTENT OF
REDUCING SUGARS IN THE WATER OF THE KIEV RESERVOIR WAS ONE OF THE
FACTORS WHICH CAUSED 'BLOOMING' OF THE WATER IN THE SPRING-AUTUMN
PERIOD OF 1966. (KNAPP-USGS)

FIELD 05B

ACCESSION NO. W71-00221

BOTTOM FAUNA AND EUTROPHICATION,

COPENHAGEN UNIV., HILLEROD (DENMARK). FRESHWATER-BIOLOGICAL LAB.

PETUR M. JONASSON.

EUTROPHICATION: CAUSES, CONSEQUENCES, CORRECTIVES, P 274-305. PRINTING AND
  PUBLISHING OFFICE, NATIONAL ACADEMY OF SCIENCES, WASHINGTON, DC, 1969. 10
  FIG, 3 TAB, 47 REF.

DESCRIPTORS:
  *BENTHIC FAUNA, *EUTROPHICATION, *NUTRIENTS, LAKES, ECOLOGY, PRIMARY
  PRODUCTIVITY, ALGAE, PLANKTON, SUBMERGED PLANTS, PHYSICOCHEMICAL
  PROPERTIES, PHYTOPLANKTON, DISTRIBUTION, DEPTH, OXYGEN, ALKALINITY,
  HYDROGEN ION CONCENTRATION, STANDING CROP, LIGHT, BIOMASS, RESPIRATION,
  LIFE CYCLES, OLIGOCHAETES, MOLLUSKS, STRATIFICATION, DIATOMS,
  CHLOROPHYTA, CYANOPHYTA, DAPHNIA, SNAILS.

IDENTIFIERS:
  LAKE ERSOM(DENMARK).

ABSTRACT:
  BOTTOM FAUNA FITS INTO AN ECOLOGICAL PATTERN SET BY PRIMARY PRODUCTION
  OF ALGAE, SUBMERGED MACROPHYTES, AND PHYSICAL AND CHEMICAL FACTORS. THE
  PHYSICOCHEMICAL FACTORS AFFECT SIZE AND SEASONAL TREND OF PRIMARY
  PRODUCTION. IN TURN, PRIMARY PRODUCTION DETERMINES RANGE OF VARIOUS
  PHYSICAL AND CHEMICAL FACTORS AT THE BOTTOM. PHYTOPLANKTON PRODUCTION,
  VERTICAL DISTRIBUTION AND ABUNDANCE OF MACROPHYTES ARE IMPORTANT
  DETERMINANTS OF OXYGEN REGIME, ALKALINITY, PH, AND OTHER FACTORS
  ESPECIALLY IMPORTANT IN RELATION TO VERTICAL AND TEMPORAL VARIATION.
  BOTTOM FAUNA IS CORRESPONDINGLY AFFECTED IN QUALITY, QUANTITY, AND
  DEPTH DISTRIBUTION. ABOVE THE THERMOCLINE IS A RICH FAUNA WITH HIGH
  OXYGEN DEMANDS; BELOW THE THERMOCLINE ARE A FEW SPECIALISTS TOLERATING
  LOW OXYGEN TENSION. THESE QUALITATIVE CONTRASTS ARE PARALLELED BY
  QUANTITATIVE CONTRASTS. INCREASING NUTRIENTS TO EPILIMNION RESULTS IN
  INCREASED STANDING CROP AND PHYTOPLANKTON PRODUCTION. MACROPHYTES CLIMB
  TO COMPENSATE FOR DECREASED ILLUMINATION. LAKE CHANGES INTO ONE DEVOID
  OF SUBMERGED VEGETATION, DOMINATED BY NYMPHOIDS SHOWING ONLY THE LOWER
  PEAK OF ANIMAL BIOMASS. INCREASING NUTRIENTS TO HYPOLIMNION CAUSES
  RAPIDLY DEVELOPING OXYGEN DEFICIT, LOWERING RESPIRATORY ACTIVITY,
  REDUCING GROWTH PERIODS, AND OTHER ADVERSE EFFECTS ON BOTTOM FAUNA.
  DURING FINAL STAGE OF EUTROPHICATION, SESSILE MUD-DWELLERS MAY DROP TO
  ALMOST ZERO OR BECOME RESTRICTED TO OLIGOCHAETES. (JONES-WISCONSIN)

FIELD 05C

ACCESSION NO. W71-00665

PHYSIOLOGICAL-ECOLOGICAL STUDIES OF BENTHIC ALGAE IN LABORATORY STREAMS,

OREGON STATE UNIV., CORVALLIS. DEPT. OF BOTANY.

C. DAVID MCINTIRE.

JOURNAL WATER POLLUTION CONTROL FEDERATION, VOL 40, NO 11, PART 1, P
1940-1952, 1968. 3 FIG, 2 TAB, 18 REF.

DESCRIPTORS:
*PHYSIOLOGICAL ECOLOGY, *BENTHIC FLORA, *ALGAE, *HYDRAULIC MODELS,
LABORATORY TESTS, PERIPHYTON, STREAMS, PHOTOSYNTHESIS, RESPIRATION,
MICROORGANISMS, LIGHT INTENSITY, CARBON DIOXIDE, DISSOLVED OXYGEN,
TEMPERATURE, VELOCITY, CYANOPHYTA, BIOMASS, INSECTS, DIATOMS,
CHLOROPHYTA, CHRYSOPHYTA, GRAZING, SNAILS, AGE, TURBIDITY, DIPTERA,
MAYFLIES, DETRITUS.

IDENTIFIERS:
*RESPIROMETER CHAMBER.

ABSTRACT:
PHYSIOLOGICAL ECOLOGY OF BENTHIC ALGAL COMMUNITIES WERE STUDIED IN
LABORATORY STREAMS. RELATIVE ABUNDANCE OF ALGAL SPECIES DEPENDED ON
SEASONAL FLUCTUATIONS IN CHEMICAL AND PHYSICAL PROPERTIES OF THE CREEK
WATER IN THE EXPERIMENTAL STREAMS AND COMBINATIONS OF LIGHT INTENSITY
AND CURRENT VELOCITY. PERIPHYTON ACCUMULATED ON RUBBLE AND GRAVEL
GROWING SURFACES MOST RAPIDLY AT RELATIVELY HIGH LIGHT INTENSITIES AND
CURRENT VELOCITIES, BUT WAS RETARDED AT HIGH DENSITIES OF GRAZING
SNAILS. EXPORT RATE OF COMMUNITY MATERIAL DEPENDED ON FACTORS
INFLUENCING BIOMASS ACCUMULATION, COMMUNITY AGE, CURRENT VELOCITY AND
SILT LOAD. WITH COMMUNITIES DEVELOPED AT A LIGHT INTENSITY OF 550-FOOT
CANDLE, INCREASE IN CONCENTRATION OF MOLECULAR CARBON DIOXIDE MARKEDLY
INCREASED PHOTOSYNTHETIC RATE AT ILLUMINATION INTENSITIES NEAR AND
ABOVE 1000-FOOT CANDLE. AT LIGHT INTENSITIES BELOW 500-FT C, A
TEMPERATURE INCREASE OF 10C HAD NO SIGNIFICANT EFFECT ON PHOTOSYNTHETIC
RATE, BUT AT INTENSITIES ABOVE 1000-FT C, THE TEMPERATURE COEFFICIENT
VARIED BETWEEN 1.3 AND 1.6. COMMUNITY RESPIRATION INCREASED
EXPONENTIALLY WITH TEMPERATURE INCREASES IN AIR-SATURATED WATER.
RESPIRATORY RATES DECREASED RAPIDLY AS DISSOLVED OXYGEN WAS REDUCED
BELOW AIR-SATURATION. EXPERIMENTS WITH LABORATORY COMMUNITIES SUGGESTED
THAT, IN CERTAIN ECOLOGICAL INVESTIGATIONS, THE BENTHIC PERIPHYTON
COMMUNITY MAY BE VIEWED AS A SINGLE FUNCTIONAL UNIT OR QUASI-ORGANISM.
(JONES-WISCONSIN)

FIELD 07B, 05C

ACCESSION NO. W71-00668

PRIMARY PRODUCTIVITY, CHEMO-ORGANOTHROPHY, AND NUTRITIONAL INTERACTIONS OF
EPIPHYTIC ALGAE AND BACTERIA ON MACROPHYTES IN THE LITTORAL OF A LAKE,

MICHIGAN STATE UNIV., EAST LANSING. DEPT. OF BOTANY AND PLANT PATHOLOGY.

HAROLD LEROY ALLEN.

AVAILABLE FROM NTIS AS COO-1599-25(PT. 2), $3.00 IN PAPER COPY, $0.95 IN
MICROFICHE. 1969. 186 P. PH D THESIS.

DESCRIPTORS:
*LAKES, *PRODUCTIVITY, MICHIGAN, *EPIPHYTOLOGY, NUTRIENTS, PLANT
PATHOLOGY, *ALGAE.

IDENTIFIERS:
LAWRENCE LAKE(MICHIGAN), *MACROPHYTES.

ABSTRACT:
ASSESSMENT OF EPIPHYTIC ALGAL AND BACTERIAL IN SITU COMMUNITY
METABOLISM, AND PHYSIOLOGICAL-NUTRITIONAL INTERRELATIONSHIPS OF
MACROPHYTE-EPIPHYTE SYSTEMS, WERE INVESTIGATED IN THE LITTORAL ZONE OF
A SMALL TEMPERATE LAKE FROM APRIL 1968 THROUGH MAY 1969. ANNUAL PRIMARY
PRODUCTIVITY, CHEMO-ORGANOTROPHIC UTILIZATION OF DISSOLVED ORGANIC
COMPOUNDS, AND FIELD AND LABORATORY STUDIES OF MACROPHYTE-EPIPHYTE
INTERACTIONS WERE MONITORED BY CARBON-14 TECHNIQUES. QUALITATIVE AND
QUANTITATIVE PHOTOSYNTHETIC PIGMENT COMPOSITION, AND A BRIEF TAXONOMIC
EXAMINATION OF THE SESSILE COMPLEX, ACCOMPANIED MEASUREMENT OF FIELD
PARAMETERS. PRODUCTIVITY MEASUREMENTS OF EPIPHYTIC ALGAE ON ARTIFICIAL
SUBSTRATA COLONIZED IN EMERGENT (SCIRPUS ACUTUS MUHL.) AND SUBMERGENT
(NAJAS FLEXILIS L. AND CHARA SPP.) MACROPHYTIC VEGETATION SITES WERE
COMPARED OVER AN ANNUAL PERIOD WITH PIGMENT (CHLOROPHYLL A AND TOTAL
PLANT CAROTENOIDS) ESTIMATES OF BIOMASS.

FIELD 02H, 05C

ACCESSION NO. W71-00832

TREATMENT OF DAIRY MANURE BY LAGOONING,

WASHINGTON STATE UNIV., PULLMAN.

SURINDER K. BHAGAT, AND DONALD E. PROCTOR.

JOURNAL WATER POLLUTION CONTROL FEDERATION, VOL 41, NO 5, 1969, P 785-795. 9
FIG, 7 TAB, 6 REF.

DESCRIPTORS:
*FARM WASTES, *FARM LAGOONS, *BIODEGRADATION, ALGAE, BIOCHEMICAL OXYGEN
DEMAND, ANAEROBIC DIGESTION, AEROBIC TREATMENT, CHEMICAL OXYGEN DEMAND,
STORAGE CAPACITY, EFFLUENTS, CONSTRUCTION.

IDENTIFIERS:
*DAIRY MANURE, TOTAL SOLIDS, VOLITILE SOLIDS, NON-DEGRADABLE SOLID.

ABSTRACT:
BECAUSE OF HIGH SOLIDS CONTENT OF DAIRY MANURE WASTE, ANAEROBIC LAGOONS
CAN BE USED SATISFACTORILY AS A PRIMARY WASTE TREATMENT. AVERAGE
REMOVALS OF BOD, COD, TS, AND VS ABOVE 86 PERCENT CAN BE ACCOMPLISHED
WITH AN APPLIED LOADING OF 70 LB VS/DAY/1000 CU. FT. (1120 G/DAY/CU.
M.). AN ANAEROBIC LAGOON CAN ACT AS A SEDIMENTATION, FLOTATION, AND
ANAEROBIC DIGESTION PROCESS UNIT WHILE SIMULTANEOUSLY PROVIDING
LONG-TERM STORAGE FOR NON-DEGRADABLE SOLID RESIDUE. THE EFFLUENT FROM
THE ANAEROBIC LAGOON RETAINS MOST OF THE NUTRIENTS PRESENT IN THE RAW
MANURE WASTE AND THUS HAS FERTILIZER VALUE. THE EFFLUENT HAS ORGANIC
MATTER WHICH CAN BE OXIDIZED. THE EFFLUENT CAN BE APPLIED TO A FIELD OR
SUBJECTED TO FURTHER TREATMENT. THE SECONDARY TREATMENT CAN BE AN
AERATED LAGOON, OXIDATION DITCH, OR AN OXIDATION POND. THE RESULTS OF
THE BATCH TYPE AEROBIC TREATMENT INDICATED THAT AN EFFLUENT BOD AT 20
MG/L CAN BE ACHIEVED BY A 24-HR. AERATION PERIOD. (CHRISTENBURY-IOWA
STATE)

FIELD 05D

ACCESSION NO. W71-00936

NATURAL FILTERS FOR AGRICULTURAL WASTES,

SOIL CONSERVATION SERVICE, WASHINGTON, D.C.

W. E. BULLARD, JR.

SOIL CONSERVATION, VOL 34, NO 4, NOVEMBER 1968, P 75-77. 2 FIG.

DESCRIPTORS:
*FARM WASTES, *SPRINKLER IRRIGATION, *ORGANIC WASTES, *WASTE WATER
DISPOSAL, AIR POLLUTION, WATER POLLUTION, ODOR, EFFLUENTS, SEWAGE
EFFLUENTS, DILUTION, DECOMPOSING ORGANIC MATTER, INSECTS, MITES,
BACTERIS, FUNGI, ALGAE, NITRITES, NITRATES, DETERGENTS, PHOSPHATES.

IDENTIFIERS:
*BIOLOGIC 'DISPOSERS', MICROSCOPIC ORGANISMS, PAPERMILL WASTE EFFLUENT,
CHEESE FACTORY WASTE WATERS.

ABSTRACT:
FARMERS AND PROCESSORS OF FARM PRODUCTS ARE FINDING THAT THE WASTE
PRODUCTS OF THEIR OPERATIONS GENERALLY CAN BE RETURNED TO THE LAND WITH
LESS HAZARD TO THE ENVIRONMENT THAN WHEN DISCHARGED INTO STREAMS. BY
COMPLETING THE NATURAL CYCLE OF GROWTH, DEATH, AND DECAY ON THE LAND
WHERE CROPS ARE PRODUCED THEY MAKE USE OF A LEGION OF 'DISPOSER'
ORGANISMS IN THE SOIL CAPABLE OF DECOMPOSING ORGANIC WASTES ON SITE.
WHEN SPRAYED ON GRASS OR CROPS, THE EFFLUENT SERVES THE DUAL PURPOSE OF
IRRIGATING AND FERTILIZING THE FIELD, THUS, AGRICULTURE HAS THE
POTENTIAL MEANS OF DISPOSING OF ITS OWN WASTES AND PREVENTING OR
REDUCING ENVIRONMENTAL POLLUTION. NUMEROUS EXAMPLES ARE GIVEN OF
RETURNING SEWAGE TREATMENT EFFLUENTS, PULP AND PAPERMILL EFFLUENTS, AND
CHEESE FACTORY WASTE WATERS, AMONG OTHERS, TO THE LAND BY SPRINKLER
IRRIGATION. THE PRINCIPLE BEHIND THESE SUCCESSFUL OPERATIONS IS THAT OF
GETTING MATERIAL PRODUCED FROM THE LAND BACK ONTO THE LAND WHERE THEY
CAN BE USED AGAIN IN PRODUCTION. (WHITE-IOWA STATE)

FIELD 05D

ACCESSION NO. W71-00940

REQUIREMENTS FOR MICROBIAL REDUCTION OF FARM ANIMAL WASTES,

SOUTH DAKOTA STATE UNIV., BROOKINGS. DEPT. OF BACTERIOLOGY.

EDWARD C. BERRY.

PROCEEDINGS NATIONAL SYMPOSIUM ON ANIMAL WASTE MANAGEMENT, ASAE PUBLICATION NO SP-0366, MICHIGAN STATE UNIVERSITY, MAY 1966, P 56-58. 2 TAB, 1 FIG, 13 REF.

DESCRIPTORS:
*FARM WASTES, *LAGOONS, *BIODEGRADATION, *SEWAGE BACTERIA, ODOR, ANAEROBIC CONDITIONS, AEROBIC CONDITIONS, E. COLI, BACTERIA, FUNGI, ACTINOMYCETES, PROTOZOA, ALGAE.

IDENTIFIERS:
*CHEMICAL ENVIRONMENT, *PHYSICAL ENVIRONMENT, PHAGE, FERMENTOR, FACULTATIVE, MICROAEROPHILIC, INOCULATION.

ABSTRACT:
THE CARDINAL PRINCIPLE ON WHICH ALL SANITATION REDUCTION WORK IS BASED IS TO PROVIDE AN ENVIRONMENT IN WHICH THE MICRO-ORGANISMS CAN BRING ABOUT CONVERSION OF UNDESIRABLE MATERIAL TO A NON-OFFENSIVE AND STABLE STATE IN THE SHORTEST POSSIBLE TIME. TO BRING THIS ABOUT IT IS NECESSARY TO CONSIDER (A) THE WASTES WE WANT REDUCED AND THEIR END PRODUCTS AND (B) THE ORGANISMS THAT WE WANT TO PERFORM THIS CHORE FOR US. THE MICRO-ORGANISMS INVOLVED IN MANURE REDUCTION ARE TO BE FOUND IN THE FOLLOWING GROUPS: (A) BACTERIA (AEROBIC, ANAEROBIC, MICROAEROPHILIC, FACULTATIVE OR OBLIGATE), (B) FUNGI, (C) ACTINOMYCETE, (D) PROTOZOA, (E) ALGAE, AND (F) PHAGE. EACH ORGANISM FINDS ITS OPTIMUM ENVIRONMENT UNDER FAIRLY RESTRICTED ENVIRONMENTAL CONDITIONS. THIS PAPER GIVES EXAMPLES OF THE MOST COMMON MICRO-ORGANISMS AND THE EFFECT THE ENVIRONMENT HAS ON THEIR ACTIVITY.
(CHRISTENBURY-IOWA STATE)

FIELD 05D

ACCESSION NO. W71-02009

ECOLOGICAL CHANGES OF APPLIED SIGNIFICANCE INDUCED BY THE DISCHARGE OF HEATED WATERS,

ASTON UNIV., BIRMINGHAM (ENGLAND). DEPT. OF BIOLOGICAL SCIENCES.

H. A. HAWKES.

IN: ENGINEERING ASPECTS OF THERMAL POLLUTION, CHAPTER 2, P 15-57, 1969. 4 TAB, 3 FIG, 84 REF.

DESCRIPTORS:
*TEMPERATURE EFFECTS, *THERMAL POLLUTION, *ECOLOGY, ALGAE, BACTERIA, COLIFORMS, FISH, FUNGI, HEAT, WATER POLLUTION, STREAMS, RIVERS, LAKES, PHYSIOLOGICAL ECOLOGY, LIMITING FACTORS, ECOSYSTEMS, MICROORGANISMS, AQUATIC MICROORGANISMS, ENZYMES.

ABSTRACT:
THE GENERAL REVIEW OF ECOLOGICAL ASPECTS OF THERMAL POLLUTION IS DISCUSSED. TEMPERATURE OF WATER IS AN ECOLOGICAL FACTOR WHICH INFLUENCES ALL FORMS OF BIOLOGICAL LIFE IN WATER. THE MINIMUM, OPTIMUM, AND HARMFUL TEMPERATURE RANGES FOR ALL KINDS OF BIOTA IN WATERS IS DISCUSSED. THE TEMPERATURE INFLUENCES THE MICROBIOLOGICAL ORGANISMS BY ITS EFFECTS ON ENZYMES, WHICH ARE INACTIVATED AT HIGH TEMPERATURES. THE MOST OBVIOUS EFFECT OF THE RAISED WATER TEMPERATURE ON FISH IS THE DIRECT AUTECOLOGICAL ONE, AFFECTING THE DISTRIBUTION OF EACH SPECIES ACCORDING TO THEIR TEMPERATURE TOLERANCE RANGES. THE OPTIMUM TEMPERATURE RANGES FOR FISH ARE PRESENTED, AND INDIRECT EFFECTS OF RAISED TEMPERATURE ON FISH ARE DISCUSSED (DECREASE OF OXYGEN LEVEL, INCREASE OF TOXICITY EFFECTS, ETC.). THE PAPER ALSO DISCUSSES THE TEMPERATURE EFFECTS ON MACRO-INVERTEBRATES, ALGAE, BACTERIA, AND FUNGI.
(NOVOTNY-VANDERBILT)

FIELD 05C

ACCESSION NO. W71-02479

BIOLOGICAL ASPECTS OF THERMAL POLLUTION.

VANDERBILT UNIV., NASHVILLE, TENN. DEPT. OF ENVIRONMENTAL AND WATER RESOURCES
ENGINEERING.

PROCEEDINGS OF THE NATIONAL SYMPOSIUM ON THERMAL POLLUTION, SPONSORED BY THE
FEDERAL WATER POLLUTION CONTROL ADMINISTRATION AND VANDERBILT UNIVERSITY,
PORTLAND, OREGON, JUNE 3-5, 1968, PRICE: $7.95. PETER A. KRENKEL AND FRANK
L. PARKER, EDITORS, VANDERBILT UNIVERSITY PRESS, NASHVILLE, 1969. 351 P.

DESCRIPTORS:
*THERMAL POLLUTION, *THERMAL POWERPLANTS, *ECOLOGY, *BIOTA,
*TEMPERATURE, *AQUATIC ENVIRONMENT, *AQUATIC LIFE, *AQUATIC
POPULATIONS, ALGAE, BENTHOS, ZOOPLANKTON, FRESHWATER FISH, ESTUARINE
ENVIRONMENT, IMPOUNDED WATERS, FRESHWATER ALGAE, ANADROMOUS FISHES.

ABSTRACT:
THE PROCEEDINGS OF THE NATIONAL SYMPOSIUM ON THERMAL POLLUTION -
BIOLOGICAL CONSIDERATIONS - ARE INCLUDED IN THIS BOOK. A BRIEF
DISCUSSION OF THE ENGINEERING PROBLEMS OF THERMAL POLLUTION INCLUDES
STRATIFICATION OF IMPOUNDED WATERS, THE CONTRIBUTION OF HEAT FROM POWER
PLANTS, COOLING WATER REQUIREMENTS, NATURAL HEAT DISSIPATION
MECHANISMS, AND ALTERNATIVE METHODS FOR COOLING WATER. REPORTS AND
DISCUSSIONS CONCERNING THE EFFECTS OF THERMAL POLLUTION ON FRESHWATER
ALGAE, MARINE ZOOPLANKTON, BRITISH FRESHWATER FISHES, ANADROMOUS
FISHES, AND FRESHWATER BENTHOS ARE PRESENTED. THE DEVELOPMENT OF
THERMAL REQUIREMENTS OF FRESHWATER FISHES, THEORETICAL CONSIDERATIONS
OF THERMAL POLLUTION ON MARINE FISHES AT VARIOUS LIFE STAGES, AND THE
POTENTIAL EFFECT OF THERMAL ALTERATION ON MARINE AND ESTUARINE BENTHOS
ARE ALSO DISCUSSED. MOST PRESENTATIONS INCLUDE REVIEWS OF PAST AND
CURRENT LITERATURE AND RECOMMENDATIONS, AND THE FINAL PRESENTATION
DEALS EXCLUSIVELY WITH RESEARCH NEEDS FOR THERMAL POLLUTION. (SEE ALSO
W71-02494 AND W71-02496) (SPEAKMAN-VANDERBILT)

FIELD 05C, 05B, 05D

ACCESSION NO. W71-02491

ASPECTS OF THE POTENTIAL EFFECT OF THERMAL ALTERATION ON MARINE AND ESTUARINE
BENTHOS,

OREGON STATE UNIV., NEWPORT. YAQUINA BILOGICAL LAB.

JOEL W. HEDGPETH, AND JEFFERSON J. GONOR.

IN: BIOLOGICAL ASPECTS OF THERMAL POLLUTION, CHAPTER 4, P 80-118, 1969. 7
FIG, 1 TAB, 91 REF.

DESCRIPTORS:
*BENTHOS, *TEMPERATURE, *THERMAL POLLUTION, LABORATORY ANIMALS,
ACCLIMATIZATION, MARINE ANIMALS, AQUATIC LIFE, AQUATIC ANIMALS, BENTHIC
FAUNA, AQUATIC PLANTS, MARINE ALGAE, BENTHIC FLORA, ESTUARIES,
*ESTUARINE ENVIRONMENT, BAYS, *FLUCTUATION, RESISTANCE, LIFE CYCLES,
GROWTH STAGES.

IDENTIFIERS:
TEMPERATURE RANGES, TEMPERATURE REQUIREMENTS.

ABSTRACT:
LABORATORY EXPERIMENTS ON THERMAL TOLERANCES, DEATH POINTS, AND THE
LIFE, WITHOUT REFERENCE TO THE NATURAL CONDITIONS, INCLUDING PREVIOUS
TEMPERATURE EXPERIENCE AND STATE OF TIDE OR SEASON AT WHICH
EXPERIMENTAL MATERIAL WAS GATHERED, HAVE QUESTIONABLE UTILITY IN
REFERENCE TO WHAT THE ORGANISMS MAY ACTUALLY DO IN NATURE. INTENSIVE
FIELD STUDIES WITH IN SITU MEASUREMENTS OF THE ENVIRONMENT AND THE
ORGANISMS, COMBINED WITH CONTINUOUS MONITORING, ESPECIALLY OF THERMAL
GRADIENTS, ARE NEEDED. FIELD EVIDENCE INDICATES THAT SOME INTERTIDAL
HERBIVORES ARE WELL ADJUSTED TO THE TEMPERATURE RANGES FROM SEAWATER TO
RATHER HIGH AIR TEMPERATURES FOR VARYING PERIODS OF TIME, AND THAT
INDEED SUCH A TEMPERATURE RANGE MAY BE AN ECOLOGICAL REQUIREMENT,
RATHER THAN AN ENVIRONMENT TO BE ENDURED. HOW EXTENSIVE THIS MAY BE,
AND TO WHAT DEGREE IT MAY APPLY ALSO TO SUBTIDAL ORGANISMS, REMAINS TO
BE DETERMINED. THERE IS EVIDENCE INDICATING THAT SOME MARINE ORGANISMS
DO NOT FLOURISH IN A STABLE TEMPERATURE REGIME BUT REQUIRE THE
VARIATION AROUND THE STATISTICAL MEAN. HOWEVER, SUCH TEMPERATURE
REQUIREMENTS ARE NOT YET ESTABLISHED FOR MANY ORGANISMS, SINCE MOST
EXPERIMENTS INVOLVING LABORATORY CULTURE ARE OF COMPARATIVELY SHORT
DURATION. IN CONTRAST TO THE OBSERVED ABILITY OF MANY MARINE ORGANISMS
TO WITHSTAND WIDE RANGES OF ENVIRONMENTAL TEMPERATURES AT SOME STAGES
OF THEIR LIFE CYCLES, THERE IS THE GROWING BODY OF EVIDENCE THAT
COMPARATIVELY SMALL FLUCTUATIONS IN OCEANIC TEMPERATURES MAY INFLUENCE
THE DISTRIBUTION AND ABUNDANCE OF MANY SPECIES, ESPECIALLY THOSE THAT
OCCUR IN LARGE POPULATIONS. (SEE ALSO W71-02491) (SPEAKMAN-VANDERBILT)

FIELD 05C

ACCESSION NO. W71-02494

SOME EFFECTS OF TEMPERATURE ON FRESHWATER ALGAE,

ACADEMY OF NATURAL SCIENCES OF PHILADELPHIA, PA. DEPT. OF LIMNOLOGY.

RUTH PATRICK.

IN: BIOLOGICAL ASPECTS OF THERMAL POLLUTION, CHAPTER 7, P 161-185, 1969. 53
  REF.

DESCRIPTORS:
  *ALGAE, *TEMPERATURE, DIATOMS, PHOTOSYNTHESIS.

IDENTIFIERS:
  *BLUE-GREEN ALGAE, *GREEN ALGAE, *RED ALGAE, *TEMPERATURE TOLERANCE,
  OPTIMUM GROWTH TEMPERATURE, CONDENSER PAGGAGE.

ABSTRACT:
  MOST OF THE COMMON FRESHWATER ALGAE BELONG TO THE GREEN ALGAE, DIATOMS,
  RED ALGAE, AND BLUE-GREEN ALGAE GROUPS. IN EACH OF THESE MAJOR GROUPS
  THERE ARE MANY SPECIES. THESE SPECIES COVER A WIDE RANGE OF
  TEMPERATURES FROM THOSE THAT PREFER COOL-WATER CONDITIONS TO THOSE THAT
  PREFER VERY WARM WATER CONDITIONS. EACH SPECIES HAS A RANGE OF
  TEMPERATURE THAT IT CAN TOLERATE AND A RANGE IN WHICH ITS GROWTH IS
  OPTIMUM. IN GENERAL, THE BLUE-GREEN ALGAE HAVE MORE SPECIES THAT PREFER
  TEMPERATURES FROM 35C UPWARD, WHEREAS THE GREEN ALGAE HAVE A RELATIVELY
  LARGE NUMBER OF SPECIES THAT GROW BEST IN TEMPERATURES RANGING UP TO
  35C. MOST DIATOM SPECIES PREFER TEMPERATURES BELOW 30C. THE EFFECT OF
  ARTIFICIALLY INCREASING THE TEMPERATURE REGIME OF A SPECIES TENDS TO
  INCREASE GROWTH AND PHOTOSYNTHESIS SO LONG AS LIGHT IS SUFFICIENT FOR
  THESE FUNCTIONS AND THE LIMITS OF TEMPERATURE TOLERANCE ARE NOT
  REACHED. AS ONE APPROACHES THE LIMITS OF TEMPERATURE TOLERANCE FOR A
  SPECIES, CELL DIVISION IS REPRESSED, AS IS PHOTOSYNTHESIS, AND THE
  FORMATION OF REPRODUCTIVE CELLS MAY BE REPRESSED. THE PATTERN OF GROWTH
  MAY BE GREATLY ALTERED. THE TEMPERATURES AT WHICH SPECIES MAY
  SUCCESSFULLY GROW IN THE LABORATORY MAY BE HIGHER THAN THOSE WHICH
  OCCUR IN NATURE BECAUSE OF LACK OF COMPETITION WITH OTHER SPECIES UNDER
  NATURAL CONDITIONS AND THE EFFECT OF PREDATOR PRESSURE. STUDIES MADE
  CONCERNING THE EFFECT ON ALGAE OF PASSING THEM THROUGH A CONDENSER
  INDICATE THAT IF THE TEMPERATURE DOES NOT EXCEED 34-34.5C, LITTLE, IF
  ANY, HARM IS DONE. (SEE ALSO W71-02491) (SPEAKMAN-VANDERBILT)

FIELD 05C

ACCESSION NO. W71-02496

AUTOECOLOGICAL AND SAPROBIOLOGICAL INVESTIGATIONS OF FRESH WATER CILIATES (IN
  GERMAN),

BONN UNIV. (GERMANY). ZOOLOGICAL INST.

HARTMUT BICK.

ENGLISH SUMMARY. HYDROBIOLOGIA, VOL 31, NO 1, P 17-36, 1968. 13 REF.

DESCRIPTORS:
  *INDICATORS, *PROTOZOA, *ENVIRONMENT, *ORGANIC MATTER, WATER QUALITY,
  TEMPERATURE, HYDROGEN ION CONCENTRATION, OXYGEN, CARBON DIOXIDE,
  AMMONIA, NITRITES, HYDROGEN SULFIDE, SALINITY, LAKES, NUTRIENTS,
  BACTERIA, ALGAE, BRACKISH WATER, ALKALINE WATER, SODIUM.

IDENTIFIERS:
  *CILIATES, FLAGELLATES, SAPROBIEN SYSTEM, CILIATE AUTOECOLOGY.

ABSTRACT:
  ECOLOGICAL VALENCIES OF 31 SP OF FRESH WATER CILIATES ARE PRESENTED IN
  TABULATED FORM. ENVIRONMENTAL FACTORS INCLUDE TEMPERTURE, PH, OXYGEN,
  CARBON DIOXIDE, FREE AND BOUND AMMONIA, NITRITES, HYDROGEN SULFIDE, AND
  THE TOTAL CONCENTRATION OF SALTS IN BRACKISH WATERS AND INLAND ALKALINE
  LAKES ENRICHED IN SODIUM. THIS RECORD IS SUPPLEMENTED BY A NUMBER OF
  BACTERIA, ACCOMPANYING DIFFERENT CILIATES. AS FAR AS POSSIBLE, THE
  OPTIMUM RANGES OF SITE FACTORS ARE GIVEN FOR DIFFERENT SPECIES. THE
  AUTOECOLOGICAL RESULTS ARE RELATED TO THE INDICATOR VALUE OF SPECIES
  FOR THE SAPROBIEN SYSTEM OF MONITORING WATER QUALITY. (WILDE-WISCONSIN)

FIELD 05G, 02H

ACCESSION NO. W71-03031

RAPIDLY LABELED POLYPHOSPHATES AND METAPHOSPHATES IN THE BLUE-GREEN ALGA
   ANACYSTIS NIDULANS (IN GERMAN),

TECHNISCHE HOCHSCHULE, HANOVER (WEST GERMANY). INST. FOR BOTANY.

R. NIEMEYER, AND G. RICHTER.

ENGLISH SUMMARY. ARCHIV FUR MIKROBIOLOGIE, VOL 69, P 54-59, 1969. 3 FIG, 10
   REF.

   DESCRIPTORS:
      *ANALYTICAL TECHNIQUES, *PHOSPHATES, *ALGAE, CHROMOTOGRAPHY,
      RADIOCHEMICAL ANALYSIS, TRACERS.

   IDENTIFIERS:
      *ANACYSTIS NIDULANS, OLIGOPHOSPHATES, METAPHOSPHATES, PULSE-LABELING,
      POLYPHOSPHATES, FRACTIONATION.

   ABSTRACT:
      THE CELLS OF THE BLUE-GREEN ALGA ANACYSTIS NIDULANS WERE PULSE-LABELED
      WITH PHOSPHORUS-32-ORTHOPHOSPHATE AND GROWN FOR 20 HOURS IN
      PHOSPHORUS-FREE MEDIUM. MOST OF THE PHOSPHORUS-32 WAS INCORPORATED INTO
      INORGANIC PHOSPHATES WHICH WERE ISOLATED BY JOINING EXTRACTION WITH THE
      NUCLEIC ACID AND FRACTIONATION ON METHYLATED SERUM ALBUMIN AND
      KIESELGUR. THE COLUMN CHROMTOGRAPHY OF THE ISOLATE REVEALED 8 PHOSPHATE
      RESIDUES INCLUDING OLIGOPHOSPHATES AND CIRCULAR METAPHOSPHATES OF THUS
      FAR UNKNOWN ORIGIN AND FUNCTION. THE TRIMETAPHOSPHATE INCORPORATED
      CONSIDERABLY MORE RADIOACTIVITY THAN OTHER LOW-MOLECULAR INORGANIC
      PHOSPHATES. (WILDE-WISCONSIN)

   FIELD 07B, 05C

   ACCESSION NO. W71-03033

ELECTROKINETIC PHENOMENA IN THE FILTRATION OF ALGAL SUSPENSIONS,

CORNELL, HOWLAND, HAYES AND MERRYFIELD, SEATTLE, WASH.; AND MICHIGAN UNIV.,
   ANN ARBOR. DEPT. OF CIVIL ENGINEERING.

GERALD W. FOESS, AND JACK A. BORCHARDT.

JOURNAL AMERICAN WATER WORKS ASSOCIATION, VOL 61, P 333-338, 1969. 10 FIG, 1
   TAB, 18 REF.

   DESCRIPTORS:
      *ALGAE, *SANDS, *FILTRATION, FILTERS, ELECTROPHORESIS, CHLORELLA,
      SCENEDESMUS, ADSORPTION, SEWAGE TREATMENT, PARTICLE SIZE.

   IDENTIFIERS:
      *ALGAL SUSPENSIONS, ELECTROKINETIC EFFECTS, ANN ARBOR(MICH), SAND BEDS.

   ABSTRACT:
      THIS INVESTIGATION REVEALED THAT THE ATTACHMENT MECHANISM FOR DISCRETE
      ALGAE IN SAND FILTRATION INVOLVES A SURFACE AFFINITY BETWEEN THE ALGAL
      CELLS AND SAND GRAINS, AN INTERACTION WHICH CAN BE CHEMICALLY
      CONTROLLED. A LOWERING OF THE SUSPENSION PH VALUE INCREASES ADSORPTION
      OF THE PARTICLES AND THEREBY SIGNIFICANTLY AUGMENTS THE REMOVAL OF
      SUSPENDED MATTER. THE REMOVAL IS ALSO INCREASED DUE TO A REDUCTION OF
      THE REPULSIVE ENERGY OF SAND BY ITS COATING WITH POSITIVELY CHARGED
      SUBSTANCES. THESE MODIFICATIONS OF SAND AND SUSPENDED PARTICLE
      PROPERTIES PROMISE TO INCREASE FILTER REMOVAL EFFICIENCIES BEYOND THOSE
      POSSIBLE WITH NORMAL DOSES OF COAGULANTS AND WITH MINIMAL HEAD LOSS
      BUILDUP. (WILDE-WISCONSIN)

   FIELD 05D

   ACCESSION NO. W71-03035

CULTURAL EUTROPHICATION: THE NATURE OF ·THE PROBLEM AND SOME RECOMMENDATIONS FOR
ITS STUDY AND CONTROL,

FLORIDA UNIV., GAINESVILLE. DEPT. OF ENVIRONMENTAL ENGINEERING.

PATRICK L. BREZONIK.

DEPARTMENT OF ENVIRONMENTAL ENGINEERING, UNIVERSITY OF FLORIDA, GAINESVILLE,
UNDATED. 26 P, 5 TAB, 51 REF. OWRR PROJECT B-004-FLA(2).

DESCRIPTORS:
*LAKES, *EUTROPHICATION, *WATER POLLUTION CONTROL, *WATER POLLUTION
SOURCES, *WATER POLLUTION EFFECTS, WATER UTILIZATION, ANALYTICAL
TECHNIQUES, NUTRIENTS, NITROGEN, PHOSPHORUS, AGRICULTURE, WATER
MANAGEMENT(APPLIED), GEOCHEMISTRY, SOIL TYPES, HYDROLOGY, CLIMATES,
TROPHIC LEVEL, NITROGEN FIXATION, FOREST, WINDS, LIGHT, TEMPERATURE,
HYDROGEN ION CONCENTRATION, DISSOLVED OXYGEN, CHLOROPHYLL, ALGAE,
CONDUCTIVITY, SEDIMENTS, BENTHIC FAUNA, RECREATION, BIOASSAY, FUNGI,
VIRUSES.

IDENTIFIERS:
*CULTURAL EUTROPHICATION, TROPHIC INDICATORS, NUTRIENT BUDGET.

ABSTRACT:
THE CAUSAL CHANGES IN WATER QUALITY FROM PRE-INDUSTRIAL MAN TO PRESENT
HIGH-DENSITY POPULATIONS ARE DESCRIBED. OF DESIRABLE TECHNIQUES,
MATHEMATICAL MODELS AND BIOASSAYS MAKE POSSIBLE COMPARISONS OF THE
NUTRITIONAL STATUS OF THE DIFFERENT SPECIES OF ALGAE MICROSCOPICALLY,
BY CYTOCHEMICAL STAINING OF ENZYMES. THE FIELD OF PALEOLIMNOLOGY POINTS
OUT THE DIFFERENTIATION OF NATURAL AND CULTURAL EFFECTS ON LAKES. STUDY
OF SEDIMENT CORES GIVES A BASIS FOR COMPARISON IN EUTROPHICATION
DEVELOPMENT. BLOOM-FORMING BLUE-GREEN ALGAE CHARACTERISTICS POINT TO
NEED OF KNOWLEDGE OF ALGAL PHYSIOLOGY. THE BEST SOLUTION TO THE PROBLEM
OF CULTURAL EUTROPHICATION IS ELIMINATION OF NUTRIENTS FROM SOURCE
WATERS. IN RESTORING LAKES, METHODS OF PREVENTING NUTRIENT RELEASE FROM
SEDIMENTS, AND RESEARCH ON EFFECTIVENESS OF PROVIDING GAME FISH
SPAWNING AND FEEDING AREAS IS NEEDED. REFINING THE TOOLS FOR LAKE
MANAGEMENT, INCLUDING LEGAL DEVELOPMENTS IN SHORELAND CORRIDOR ZONING
ARE PREREQUISITES. GREATER ACCURACY IS NEEDED IN PREDICTING RESPONSE OF
A LAKE TO A GIVEN NUTRIENT INPUT OR TO CHANGES IN LAND UTILIZATION TO
MAKE POSSIBLE LAKE BASIN ZONING AND DEVELOPMENT OF LAND USE PATTERNS ON
A RATIONAL BASIS. (JONES-WISCONSIN)

FIELD 05C

ACCESSION NO. W71-03048

UTILIZATION OF ALKOXYSILANES FOR REDUCTION OF THE ABUNDANCE OF PHYTOPLANKTON
DURING 'BLOOMS' OF WATER BASINS (IN RUSSIAN),

MOSCOW STATE UNIV. (USSR).

N. S. STROGANOV, V. G. KHOBOTIEV, L. V. KOLOSOVA, AND M. A. KADINA.

DOKLADY AKADEMII NAUK SSSR, VOL 181, NO 5, P 1257-1259, 1968. 2 FIG, 2 TAB,
13 REF.

DESCRIPTORS:
*HERBICIDES, *ALGAE, TOXICITY, SCENEDESMUS, CHLORELLA, DAPHNIA,
DETERIORATION, WATER TREATMENT, DEGRADATION(DECOMPOSITION).

IDENTIFIERS:
*ALGAL BLOOMS, *ORGANO-SILICATES, *ALKOXYSILANES, SCENEDESMUS
QUADRICAUDA, CHLORELLA VULGARIS, DAPHNIA MAGNA, LIMNAEA STAGNALIS,
OUSPENSKII'S MEDIUM.

ABSTRACT:
A SEARCH FOR HERBICIDES SUPPRESSING ALGAL BLOOMS, BUT HARMLESS TO
ZOOPLANKTERS AND OTHER HYDROBIONTS LEAD TO INVESTIGATION OF
ORGANO-SILICATES: VENYLTRIETHOXYSILANE, TETRAETHOXYSILANE, AND
TRIFLUOROPROPENYL-(METHYL)-DIETHOXYSILANE. THE TEST ORGANISIMS INCLUDED
SCENEDESMUS QUADRICAUDE, CHLORELLA VULGARIS, DAPHNIA MAGNA, AND LIMNAEA
STAGNALIS. THE COUNT OF SURVIVED SPECIMENS SHOWED THAT THE
CONCENTRATION OF 0.01 MG/LITER OF ALL INVESTIGATED COMPOUNDS ELIMINATED
93% TO 96% OF ALGAL CELLS WITHOUT DAMAGING EITHER DAPHNIA OR MOLLUSKS.
AMONG THE ADVANTAGES OF THE ALKOXYSILANES IS THEIR RELATIVELY RAPID
DISINTEGRATION IN WATER. (WILDE-WISCONSIN)

FIELD 05F, 05C

ACCESSION NO. W71-03056

THE EFFECTS OF DIELDRIN ON DIATOMS,

KANSAS UNIV., LAWRENCE. DEPT. OF ZOOLOGY.

JOHN CAIRNS, JR.

MOSQUITO NEWS, VOL 28, P 172-179, 1968. 1 TAB, 1 FIG, 6 REF. USPHS GRANT EF-00266.

DESCRIPTORS:
*DIELDRIN, *DIATOMS, LABORATORY TESTS, INSECTICIDES, WATER POLLUTION EFFECTS, ALGAE.

IDENTIFIERS:
*EFFECTS, NAVICULA SEMINULUM VAR.HUSTEDTII.

ABSTRACT:
THE EFFECTS OF DIELDRIN ON NAVICULA SEMINULUM WAS DESCRIBED. AT A DIELDRIN CONCENTRATION OF 12.8 PPM THE NUMBER OF DIATOMS WAS ONLY HALF THAT IN THE CONTROLS AT THE END OF THE 5-DAY GROWTH PERIOD. THERE WAS NO INCREASE IN THE NUMBER OF CELLS AT 32.0 PPM, AND A CONCENTRATION OF 1.8 PPM WAS ONLY 10.6 PERCENT DIFFERENT FROM THE CONTROLS. (WAHTOLA-WASHINGTON)

FIELD 05C

ACCESSION NO. W71-03183

THE ISOLATION OF PROTOZOA AND ALGAE FROM THEIR BACTERIAL CONTAMINANTS,

MASSACHUSETTS UNIV., WALTHAM. DEPT. OF ENVIRONMENTAL SCIENCES.

ROBERT A. COLER, AND HAIM B. GUNNER.

PHYTON, VOL 26, NO 2, P 191-194, NOVEMBER 1969. 3 FIG, 11 REF. OWRR PROJECT A-020-MASS(3).

DESCRIPTORS:
*ISOLATION, *BACTERIA, *PROTOZOA, *ALGAE, LETHAL LIMIT, ELECTROPHORESIS, MICROORGANISMS, ELECTRIC CURRENTS, *WASTE TREATMENT, WASTE WATER TREATMENT.

ABSTRACT:
A TECHNIQUE IS OUTLINED FOR THE ISOLATION OF CILIATED PROTOZOA AND GREEN AND BLUE-GREEN ALGAE FROM BACTERIA. THE PROCEDURE IS DESIGNED TO EXPLOIT BOTH THE ELECTROPHORETIC MOBILITY OF SELECTED CILIATED PROTOZOA AND FLAGELLATED ALGAE AND THE APPARENTLY GREATER VULNERABILITY OF BACTERIA OVER PROTOZOA AND ALGAE TO THE LETHAL AFFECTS OF AN ELECTRIC FIELD.

FIELD 05F, 05D

ACCESSION NO. W71-03775

MOVEMENT OF ALGAL- AND FUNGAL-BOUND RADIOSTRONTUM AS CHELATE COMPLEXES IN A
  CALCAREOUS SOIL,

ARIZONA UNIV., TUCSON.

WALLACE H. FULLER, AND MICHAEL F. L'ANNUNZIATA.

ATOMIC ENERGY COMMISSION CONTRACT AT (11-1)-947. SOIL SCIENCE, VOL 107, NO 3,
  MARCH 1969, P 223-230. 3 TAB, 5 FIG, 7 REF.

DESCRIPTORS:
  *CHELATION, *STRONTIUM RADIOISOTOPES, *ALGAE, *FUNGI, *CALCAREOUS SOIL,
  SOIL WATER, SOIL WATER MOVEMENT, LEACHING, ARID LANDS, DEPTH,
  CYANOPHYTA, CHLOROPHYTA, SIEROZEMS, SOIL MICROBIOLOGY, DISTRIBUTION
  PATTERNS, SOIL PROFILES.

IDENTIFIERS:
  *RADIOSTRONTIUM, *ALGAL-BOUND RADIOSTRONTIUM, *FUNGAL-BOUND
  RADIOSTRONTIUM, *MICROBIAL ACTIVITY, *RADIOSTRONTIUM COMPLEXES, DTPA,
  SOIL WATER PROFILE.

ABSTRACT:
  THIS REPORT EVALUATES THE EFFECT OF MICROBIAL ACTIVITY ON MOVEMENT OF
  ALGAL- AND FUNGAL-BOUND RAIOSTRONTIUM, THE EFFECT OF DTPA ON SR SUPER
  89 MOBILITY ADSORBED BY ALGAL AND FUNGAL CELLS IN A CALCAREOUS SOIL,
  AND EXTRA-CELLULAR ALGAL EXCRETIONS AS SR SUPER 89 COMPLEXING AGENTS.
  LEACHING WITH A WATER-FORMALDEHYDE AND WATER SOLUTION CAUSED LITTLE SR
  SUPER 89-BOUND MOVEMENT BY ASPERGILLUS NIGER (FUNGI), RHISOPUS ARRHIZUS
  (FUNGI), OSCILLATORIA BIJUGA (BLUE-GREEN ALGA) AND CHLOROCOCCUM SP. (A
  GREEN ALGA) IN MOHAVE SANDY LOAM. LEACHING WITH 0.067 M DTPA SOLUTION
  DISPLACED FROM 0.2 TO 1.6 PERCENT OF THE ALGAL- AND FUNGAL-BOUND SR
  SUPER 89 BELOW A 20-CM SOIL DEPTH. THE SR SUPER 89 PROFILE DISTRIBUTION
  REMAINING IN THE LEACHED SOIL CONTRASTED WITH THE DTPA-TREATED AND THE
  WATER-TREATED SOIL COLUMNS. WHEREVER DTPA WAS ADDED SR SUPER 89 MOVED
  DOWNWARD, WHEREAS IN THE WATER-TREATED CONTROL SR SUPER 89 REMAINED IN
  THE UPPER 2.5 CM. NEGATIVELY CHARGED SR SUPER 89 COMPLEXES FORMED FROM
  EXCRETIONS OF THE DESERT SOIL ALGA ANACYSTIS MENEGH ARE SHOWN. TABLES
  INDICATE SR SUPER 89 MOVEMENT WITH RESPECT TO ALGAL AND FUNGAL TISSUE,
  BIOLOGICALLY BOUND SR SUPER 89 FROM DTPA TREATMENT AND TOTAL
  EXTRACTABLE SR SUPER 89. FIGURES ARE INCLUDED OF EXCHANGEABLE AND FIXED
  SR SUPER 89 WITH SOIL DEPTH FOR VARIOUS BIOLOGICAL SOIL TREATMENTS, AND
  EVIDENCE OF SR SUPER 89 IN A CELL-FREE ANACYSTIS MEDIUM AS NEGATIVE
  COMPLEXES. (POPKIN-ARIZONA)

FIELD 05B, 02G

ACCESSION NO. W71-04067

BIODEGRADABILITY OF POTENTIAL ORGANIC SUBSTITUTES FOR PHOSPHATES,

ROBERT A. TAFT SANITARY ENGINEERING CENTER, CINCINNATI, OHIO. CINCINNATI
WATER RESEARCH LAB.

ROBERT L. BUNCH, AND M. B. ETTINGER.

INDUSTRIAL WASTE CONFERENCE, 22ND, MAY 2, 3, 4, 1967, PURDUE UNIVERSITY, VOL
52, NO 3, P 393-396, JULY 1968. 1 FIG, 14 REF.

DESCRIPTORS:
*ALGAE, *BIODEGRADATION, *PHOSPHATES, *EUTROPHICATION, NUTRIENTS,
NITROGEN, WASTE WATER TREATMENT, PHOSPHOROUS, RUNOFF, DOMESTIC WASTES,
DETERGENTS, INDUSTRIAL WASTES, *CHELATION.

IDENTIFIERS:
URBAN RUNOFF.

ABSTRACT:
EUTROPHICATION WILL BE AN INCREASINGLY IMPORTANT PROBLEM IN THE FUTURE
AS MORE WATERS ARE IMPOUNDED AND REUSED. CONSIDERATION SHOULD BE GIVEN
NOT ONLY TO IMPROVING METHODS OF NUTRIENT REMOVAL FROM SEWAGE
INFLUENTS, BUT ALSO TO FINDING ACCEPTABLE SUBSTITUTES FOR PHOSPHATES IN
INDUSTRIAL PROCESSES AND DETERGENT FORMULATIONS. PRIVATE INDUSTRY CAN
MAKE A CONTRIBUTION BY DEVELOPING ECONOMICALLY PRACTICAL MATERIALS
WHICH WILL BE CONVERTED INTO INNOCUOUS PRODUCTS AT A RATE SUFFICIENTLY
RAPID TO PREVENT SIGNIFICANT ACCUMULATION OF POLLUTANTS. IT APPEARS
THAT BIODEGRADABLE CHELATING AGENTS MAY BE CONSIDERED AS A PARTIAL
SUBSTITUTE FOR PHOSPHATES IN CERTAIN TYPES OF DETERGENT FORMULATIONS.
THE RELATIONSHIP BETWEEN THE STRUCTURE OF CHELATING AGENTS AND THE LOSS
OF CHELATING ABILITY DUE TO BIOLOGICAL ACTION REMAINS TO BE ANSWERED.
RESULTS ON COMPOUNDS NOW UNDER STUDY SHOULD GIVE A BETTER UNDERSTANDING
OF THE TYPE OF CHELATING COMPOUNDS THAT WOULD MINIMIZE ENVIRONMENTAL
CONTAMINATION. SO FAR, ONLY COMPOUNDS CONTAINING ONLY ORE NITROGEN
MOLECULE HAVE SHOW BIODEGRADABILITY. IT DOES NOT SEEM DESIRABLE TO USE
NON-BIODEGRADABLE CHELATING AGENTS THAT STRONGLY COMPLEX HEAVY METALS
UNLESS IT HAS BEEN DEMONSTRATED THAT THEY DO NOT INTERFERE WITH WASTE
TREATMENT, COAGULATION, WATER SOFTENING, CALCIUM METABOLISM OF MAN, OR
CONVEY METALS INTO THE DRINKING WATER. (ELLIS-TEXAS)

FIELD 05D, 05C

ACCESSION NO. W71-04072

PERIPHYTON BIOMASS-CHLOROPHYLL RATIO AS AN INDEX OF WATER QUALITY,

FEDERAL WATER POLLUTION CONTROL ADMINISTRATION, CINCINNATI, OHIO. ANALYTICAL
QUALITY CONTROL LAB.

CORNELIUS I. WEBER, AND BEN H. MCFARLAND.

AVAILABLE FROM: EPA WQO, ANALYTICAL QUALITY CONTROL LABORATORY, 1014
BROADWAY, CINCINNATI, OHIO 45202. MARCH 1969. 19 P, 8 TAB, 20 REF.

DESCRIPTORS:
*WATER QUALITY, *BIOINDICATORS, *CHLOROPHYLL, *OHIO RIVER, *ANALYTICAL
TECHNIQUES, BIOMASS, WATER POLLUTION, ALGAE, PERIPHYTON, PHOSPHORUS,
HYDROGEN ION CONCENTRATION, ALKALINITY, HARDNESS(WATER), CHLORIDES,
ORGANIC MATTER, CARBON, WATER POLLUTION EFFECTS, BACTERIA, PERIPHYTON,
OHIO.

IDENTIFIERS:
*AUTOTROPHIC INDEX, CHLOROPHYLL A, SHAYLER RUN, ORGANIC CARBON,
CINCINNATI(OHIO).

ABSTRACT:
A REVIEW WAS MADE OF THE LITERATURE REGARDING THE RELATIONSHIP OF
BIOMASS AND CHLOROPHYLL AS A FIRST STEP IN DETERMINING THE POTENTIAL
USE OF THE BIOMASS-CHLOROPHYLL RATIO AS AN INDEX OF WATER QUALITY.
LABORATORY STUDIES WERE THEN BEGUN IN 1967 ON SAMPLES TAKEN AT TWO
STATIONS ABOVE AND BELOW THE MAJOR WASTE OUTFALL IN THE OHIO RIVER AT
CINCINNATI. IN 1968, SAMPLES TAKEN BY THE FWPCA NEWTOWN FISH TOXICOLOGY
LABORATORY FROM A SMALL CREEK (SHAYLER RUN) WERE INCLUDED IN THE STUDY.
SINCE CHLOROPHYLL A IS THE ONLY FORM OF CHLOROPHYLL FOUND IN ALL ALGAE,
THE BIOMASS-CHLOROPHYLL WHICH IS TERMED THE AUTOTROPHIC INDEX SHOULD BE
BASED EXCLUSIVELY ON THIS PIGMENT. THE AUTOTROPHIC INDICES FOR THE
SAMPLES TAKEN AT THE TWO OHIO RIVER STATIONS WERE SIGNIFICANTLY
DIFFERENT. THE DATA OBTAINED AT THE UPSTREAM STATION WERE SIMILAR TO
THOSE FROM SHAYLER RUN, AND INDICATED A PREDOMINANTLY AUTOTROPHIC
PERIPHYTON. IN CONTRAST TO THE UPSTREAM STATION, THE HIGH INDICES AT
THE STATION BELOW THE WASTE OUTFALL SUGGESTED A 'CONSUMER-TYPE'
COMMUNITY. THE AUTOTROPHIC INDEX SHOWS CONSIDERABLE PROMISE AS A TOOL
IN WATER QUALITY STUDIES. DATA ON WATER CHEMISTRY, AUTOTROPHIC INDICES,
COMPOSITION OF PERIPHYTON, AND TEST METHODS ARE INCLUDED.
(LITTLE-BATTELLE)

FIELD 05A, 05C

ACCESSION NO. W71-04206

ALGAL NUTRIENT RESPONSES IN AGRICULTURAL WASTE WATER,

FEDERAL WATER QUALITY ADMINISTRATION, FRESNO, CALIF.; AND CALIFORNIA STATE
DEPT. OF WATER RESOURCES, FRESNO.

JAMES F. ARTHUR, RANDALL L. BROWN, BRUCE A. BUTTERFIELD, AND JOEL C. GOLDMAN.

IN: COLLECTED PAPERS REGARDING NITRATES IN AGRICULTURAL WASTE WATERS, FEDERAL
WATER QUALITY ADMINISTRATION WATER POLLUTION CONTROL RESEARCH SERIES 13030
ELY, 12/69, P 123-141, DECEMBER 1969. 19 P, 8 FIG, 1 TAB, 12 REF. FWQA
PROJECT 13030 ELY.

DESCRIPTORS:
*ALGAE, *WASTE WATER TREATMENT, *WATER POLLUTION CONTROL,
*DENITRIFICATION, *BIODEGRADATION, *DRAINAGE WATER, CALIFORNIA,
DRAINAGE PROGRAMS, ANAEROBIC BACTERIA, ANEROBIC CONDITIONS, LAGOONS.

IDENTIFIERS:
*CENTRAL VALLEY(CALIFORNIA), *BACTERIAL DENITRIFICATION.

ABSTRACT:
ALGAL ASSIMILATION OF NUTRIENTS INTO CELLULAR MATERIAL WITH SUBSEQUENT
REMOVAL FROM THE GROWTH MEDIUM IS A FEASIBLE PROCESS TO REMOVE NITRATES
FROM IRRIGATION WASTE WATER. THE EFFICIENCY OF THE PROPOSED SYSTEM IS
GREATLY ENHANCED IF AS MANY VARIABLES AS POSSIBLE ARE OPTIMIZED,
LEAVING ONLY NITROGEN THE LIMITING NUTRIENT. ORTHOPHOSPHATE ADDITIONS
OF 2.0-3.0 MG/LITER P ARE REQUIRED THE YEAR ROUND TO REMOVE 20.0
MG/LITER NITRATE-NITROGEN FROM THE GROWTH MEDIUM. IRON AND CARBON ALSO
HAVE BEEN FOUND TO BE LIMITING ALGAL GROWTH AND NITROGEN ASSIMILATION
DURING PART OF THE YEAR. (KNAPP-USGS)

FIELD 05D, 05B

ACCESSION NO. W71-04554

THE EFFECTS OF NITROGEN REMOVAL ON THE ALGAL GROWTH POTENTIAL OF SAN JOAQUIN
VALLEY AGRICULTURAL TILE DRAINAGE EFFLUENTS,

CALIFORNIA STATE DEPT. OF WATER RESOURCES, FRESNO; AND FEDERAL WATER QUALITY
ADMINISTRATION, ALAMEDA, CALIF.

RANDALL L. BROWN, RICHARD C. BAIN, JR., AND MILTON G. TUNZI.

IN: COLLECTED PAPERS REGARDING NITRATES IN AGRICULTURAL WASTE WATERS, FEDERAL
WATER QUALITY ADMINISTRATION WATER POLLUTION CONTROL RESEARCH SERIES 13030
ELY, 12/69, P 143-155, DECEMBER 1969. 13 P, 4 FIG, 4 TAB, 3 REF. FWQA
PROJECT 13030 ELY.

DESCRIPTORS:
*ALGAE, *WASTE WATER TREATMENT, *WATER POLLUTION CONTROL,
*DENITRIFICATION, *BIODEGRADATION, *DRAINAGE WATER, CALIFORNIA,
DRAINAGE PROGRAMS, ANAEROBIC BACTERIA, ANAEROBIC CONDITIONS, LAGOONS.

IDENTIFIERS:
*CENTRAL VALLEY(CALIFORNIA), *BACTERIAL DENITRIFICATION.

ABSTRACT:
LABORATORY CULTURE EXPERIMENTS WERE MADE TO DETERMINE THE EFFECTIVENESS
OF THE TWO BIOLOGICAL PROCESSES UNDER INVESTIGATION, ALGAL STRIPPING
AND BACTERIAL DENITRIFICATION, FOR REMOVING THE ALGAL GROWTH POTENTIAL
OF THE TILE DRAINAGE WATER WHEN ADDED TO POTENTIAL RECEIVING WATERS IN
THE SACRAMENTO-SAN JOAQUIN DELTA. ALGAL GROWTH POTENTIAL TESTS IN TWO
DIFFERENT LABORATORIES INDICATE THAT NITRATE-RICH AGRICULTURAL
DRAINAGE, WHEN MIXED WITH SAN JOAQUIN RIVER DELTA WATER, STIMULATES
ALGAL GROWTH. SELECTIVE REMOVAL OF NITRATE-NITROGEN BY ANAEROBIC
DENITRIFICATION OR REMOVAL OF NUTRIENTS BY ALGAL CELLS GROWN IN SHALLOW
PONDS YIELDED COMPARABLE BIOASSAY RESULTS. EUTROPHICATION DUE TO
AGRICULTURAL WASTE WATERS CAN BE CONTROLLED BY TREATMENT.
(KNAPP-USGS)

FIELD 05D, 05B

ACCESSION NO. W71-04555

HARVESTING OF ALGAE GROWN IN AGRICULTURAL WASTE WATERS,

CALIFORNIA STATE DEPT. OF WATER RESOURCES, FRESNO; AND BUREAU OF RECLAMATION,
FRESNO, CALIF.

BRUCE A. BUTTERFIELD, AND JAMES R. JONES.

IN: COLLECTED PAPERS REGARDING NITRATES IN AGRICULTURAL WASTE WATERS, FEDERAL
WATER QUALITY ADMINISTRATION WATER POLLUTION CONTROL RESEARCH SERIES 13030
ELY, 12/69, P 157-163, DECEMBER 1969. 7 P, 1 TAB. FWQA PROJECT 13030 ELY.

DESCRIPTORS:
*ALGAE, *WASTE WATER TREATMENT, *WATER POLLUTION CONTROL,
*DENITRIFICATION, *BIODEGRADATION, *DRAINAGE WATER, CALIFORNIA,
DRAINAGE PROGRAMS, ANAEROBIC BACTERIA, ANAEROBIC CONDITIONS, LAGOONS.

IDENTIFIERS:
*ALGAE HARVESTING, *CENTRAL VALLEY(CALIF.).

ABSTRACT:
THROUGH LABORATORY AND FIELD TESTING OF ALGAE HARVESTING METHODS FOR
DENITRIFICATION, IT WAS SHOWN THAT EFFECTIVE CONCENTRATION CAN BE
ACCOMPLISHED USING THE FLOCCULATION-SEDIMENTATION PROCESS TO REMOVE
90-95 PERCENT OF THE SUSPENDED SOLIDS FROM ALGAE-LADEN AGRICULTURAL
WASTE WATER. DEWATERING AND DRYING CAN BE ACCOMPLISHED BUT THE
EFFICIENCIES OF THE UNITS TESTED WERE LOW. THE NEED TO RECIRCULATE THE
WATER WOULD INCREASE THE OVERALL COST; HOWEVER, IT IS BELIEVED BETTER
RESULTS WILL BE ACHIEVED IN LARGER CAPACITY UNITS.
(KNAPP-USGS)

FIELD 05D, 05B

ACCESSION NO. W71-04556

COMBINED NUTRIENT REMOVAL AND TRANSPORT SYSTEM FOR TILE DRAINAGE FROM THE SAN
JOAQUIN VALLEY,

CALIFORNIA STATE DEPT. OF WATER RESOURCES, FRESNO; AND FEDERAL WATER QUALITY
ADMINISTRATION, FRESNO, CALIF.; AND CALIFORNIA UNIV., BERKELEY. DEPT. OF
SANITARY ENGINEERING AND PUBLIC HEALTH.

JOEL C. GOLDMAN, JAMES F. ARTHUR, WILLIAM J. OSWALD, AND LOUIS A. BECK.

IN: COLLECTED PAPERS REGARDING NITRATES IN AGRICULTURAL WASTE WATERS, FEDERAL
WATER QUALITY ADMINISTRATION WATER POLLUTION CONTROL RESEARCH SERIES 13030
ELY, 12/69, P 165-186, DECEMBER 1969. 22 P, 4 FIG, 1 TAB, APPEND. FWQA
PROJECT 13030 ELY.

DESCRIPTORS:
*ALGAE, *WASTE WATER TREATMENT, *WATER POLLUTION CONTROL,
*DENITRIFICATION, *BIODEGRADATION, *DRAINAGE WATER, CALIFORNIA,
DRAINAGE PROGRAMS, ANAEROBIC BACTERIA, ANAEROBIC CONDITIONS, LAGOONS.

IDENTIFIERS:
*ALGAE HARVESTING, *CENTRAL VALLEY(CALIF.).

ABSTRACT:
CURRENT PLANS CALL FOR TREATMENT OF AGRICULTURAL WASTE WATER FOR
NUTRIENT (NITROGEN) REMOVAL FROM THE PROPOSED SAN LUIS AND MASTER
DRAINS PRIOR TO DISCHARGE INTO THE BAY-DELTA AREA. OF THE SEVERAL
TREATMENT PROCESSES BEING INVESTIGATED, THE ALGAE STRIPPING PROCESS IS
ESTIMATED TO REQUIRE BETWEEN 6,000 AND 12,000 ACRES OF LAND TO
ACCOMPLISH THIS TASK. BECAUSE EVERY BODY OF WATER IS A POTENTIAL ALGAL
GROWTH SYSTEM, THERE ARE SEVERAL ALTERNATIVES WHICH WILL GREATLY REDUCE
THE TOTAL COST OF TREATMENT AND PERHAPS IMPROVE THE OVERALL EFFICIENCY
OF NUTRIENT REMOVAL. IN-LINE TREATMENT USING THE DRAIN AND THE DRAINAGE
RESERVOIRS FOR ALGAE GROWTH MAY BE AN ECONOMICAL AND PRACTICAL METHOD
FOR NUTRIENT REMOVAL. BECAUSE KESTERSON RESERVOIR IS AN INTEGRAL PART
OF THE PROPOSED DRAINAGE SYSTEM AND CONTAINS THE REQUIRED AREA NEEDED
FOR TREATMENT BY ALGAE STRIPPING, IT SEEMS LOGICAL TO USE IT AS A
DUAL-PURPOSE TREATMENT AND STORAGE RESERVOIR. KESTERSON RESERVOIR, IF
MODIFIED AS SUGGESTED, HAS THE POTENTIAL TO PROVIDE NITROGEN REMOVALS
IN EXCESS OF 90%. (KNAPP-USGS)

FIELD 05D, 05G, 05B

ACCESSION NO. W71-04557

SOME MICROBIAL-CHEMICAL INTERACTIONS AS SYSTEMS PARAMETERS IN LAKE ERIE,

OHIO STATE UNIV., COLUMBUS. MICROBIAL AND CELLULAR BIOLOGY.

P. R. DUGAN, J. I. FREA, AND R. M. PFISTER.

IN: SYSTEMS ANALYSIS FOR GREAT LAKES WATER RESOURCES, P 21-28, OCTOBER 1969.
8 P, 2 FIG, 16 REF. OWRR PROJECT A-999-OHIO(3).

DESCRIPTORS:
*SYSTEMS ANALYSIS, DATA COLLECTIONS, *ALGAE CONTROL, *LAKE ERIE,
BIOCHEMICAL OXYGEN DEMAND, BACTERIA, ECOLOGY, WATER POLLUTION EFFECTS,
LAKES, *CYANOPHYTA, *EUTROPHICATION, WATER QUALITY.

ABSTRACT:
SOME MICROBIAL-CHEMICAL INTERACTIONS AS SYSTEMS PARAMETERS IN LAKE ERIE
WERE PRESENTED. ONE OF THE MAJOR PROBLEMS CITED WAS THE INCREASED
GROWTH RATE OF BLUE GREEN ALGAE AND OTHER MICROORGANISMS IN THE LAKE
WHICH HAVE LED TO OBJECTIONS BECAUSE OF DECREASED RECREATIONAL VALUE,
MORTALITY OF FISH AND DOMESTIC ANIMALS, CLOGGING OF WATER SUPPLY INTAKE
FILTERS, AND DEPLETION OF OXYGEN IN THE WATER. FOUR GENERAL PARAMETERS
WERE CONSIDERED IN RELATIONSHIP TO ACCELERATED GROWTH OF BLUE GREEN
ALGAE: (1) AMOUNT OF LIGHT (ENERGY); (2) NITROGEN SUPPLY; (3) CO_2 OR
CO_3; AND (4) MINERALS. SEVERAL OBSERVATIONS WERE MADE FROM DATA
COLLECTED IN THE WESTERN BASIN OF LAKE ERIE DURING THE SPRING AND
SUMMER OF 1969. USING BOD AS AN EXAMPLE OF RECYCLING, ROLE OF BACTERIA
IN MAKING NUTRIENTS AVAILABLE FOR ALGAE GROWTH WAS SHOWN SIGNIFICANT;
AND ALGAE, ONCE ABOVE A CRITICAL CONCENTRATION SIGNIFICANTLY ENRICH
THEIR OWN ENVIRONMENTS WITH ORGANICS, WHICH INDICATED A SPIRALING
INCREASE IN RATE OF EUTROPHICATION. SUGGESTIONS FOR DECREASING ALGAE
AND BACTERIA POPULATION IN THE LAKE WERE: PREVENTING ORGANIC AND
MINERAL NUTRIENTS FROM ENTERING THE WATER COLUMN AND REMOVING SEDIMENTS
OF RELATIVELY HIGH ORGANIC CONTENT PHYSICALLY OR ALLOWING THEM TO
DECREASE NATURALLY. (KRISS-CORNELL)

FIELD 06A, 05C, 02H

ACCESSION NO. W71-04758

POULTRY PROCESSING WASTE TREATMENT AT MILLSBORO, DELAWARE,

SYRACUSE UNIV., N.Y. DEPT. OF CIVIL ENGINEERING.

NELSON L. NEMEROW.

PROCEEDINGS, INDUSTRIAL WASTE CONFERENCE, 22ND, MAY 2, 3, 4, 1967, PURDUE
   UNIVERSITY, VOL LII, NO 3, JULY 1968, P 526-536. 4 FIG, 4 TAB, 1 REF.

   DESCRIPTORS:
      *POULTRY, *OXIDATION LAGOONS, *ORGANIC LOADING, ALGAE, AMMONIA,
      SEDIMENTATION, FLOTATION, BIODEGRADATION, CHLORINATION, COLIFORM,
      BACTERIA, BIOCHEMICAL OXYGEN DEMAND, DELAWARE, *WASTE WATER TREATMENT.

   IDENTIFIERS:
      MILLSBORO(DELAWARE), SWAN CREEK, INDIAN RIVER.

   ABSTRACT:
      9,000 TO 10,000 CHICKENS WITH AN AVERAGE WEIGHT OF 3.75 LBS ARE
      PROCESSED DURING AN 8-11 HOUR WORKING DAY AT MILLSBORO, DELAWARE. THE
      WASTES FROM THIS OPERATION ARE TREATED IN A TWO STAGE OXIDATION POND
      WITH EFFLUENT CHLORINATION. SAMPLES WERE COLLECTED IN APRIL, JUNE, AND
      DECEMBER OF 1966, AND MARCH OF 1967. THE BOD LOADING WAS FOUND TO BE
      1390 LBS/DAY, WITH A LOADING ON THE FIRST POND OF 935 LB BOD/ACRE. THIS
      POND EFFECTED A 72.5 PERCENT REDUCTION IN BOD. THE SECOND POND HAD A
      LOADING OF 79 LB BOD/ACRE, WITH AN ADDITIONAL BOD REDUCTION OF 9.5%
      WHEN ALGAE ARE NOT REMOVED FROM THE EFFLUENT, AND 70% WHEN ALGAE ARE
      REMOVED. THE OVERALL REMOVAL, BASED ON THE 4 SAMPLES PER YEAR, WAS 92%
      FOR THE FILTERED EFFLUENT, AND 75% WHEN THE ALGAE ARE NOT FILTERED OUT.
      (LOWRY-TEXAS)

   FIELD 05D

   ACCESSION NO. W71-05013

THE EFFECT OF VARYING AMOUNTS AND RATIOS OF NITROGEN AND PHOSPHATES ON ALGAE
   BLOOMS,

ACADEMY OF NATURAL SCIENCES OF PHILADELPHIA, PA. DEPT. OF LIMNOLOGY.

RUTH PATRICK.

PROCEEDINGS, INDUSTRIAL WASTE CONFERENCE, 21ST, MAY 3, 4, 5, 1966, PURDUE
   UNIVERSITY, VOL L, NO 2, MARCH 1966, P 41-51. 6 TAB, 6 REF.

   DESCRIPTORS:
      *NITROGEN, *PHOSPHATES, *ALGAE, *EUTROPHICATION, DIATOMS, MICROBIOLOGY,
      NITRATES, AMMONIA, WATER POLLUTION EFFECTS, WASTE WATER TREATMENT,
      PENNSYLVANIA.

   IDENTIFIERS:
      *GLUCOSE, WATERLOO MILLS(PA).

   ABSTRACT:
      A SERIES OF EXPERIMENTS WAS UNDERTAKEN AT THE WATERLOO MILLS FIELD
      RESEARCH STATION TO TRY TO DETERMINE THE EFFECT OF NITROGEN AS NITRATES
      OR AMMONIUM AND PHOSPHATES IN VARYING AMOUNTS AND RATIOS ON NATURAL
      COMMUNITIES OF DIATOMS. SOME EXPERIMENTS WERE CARRIED OUT TO DETERMINE
      THE EFFECTS OF ADDING SMALL AMOUNTS OF GLUCOSE TO THESE ENRICHED
      WATERS. CERTAIN SPECIES INCREASED IN ABUNDANCE WITH INCREASES IN
      NITROGEN AS NITRATES AND/OR A AMMONIUM AND WITH INCREASES IN PHOSPHATES
      AND SOME OF THEM WITH THE ADDITION OF SMALL AMOUNTS OF GLUCOSE.
      HOWEVER, THE HEAVY DIATOM GROWTHS WHICH OFTEN OCCUR IN STREAMS ENRICHED
      WITH EFFLUENTS FROM SEWAGE TREATMENT PLANTS DID NOT OCCUR WHICH
      INDICATES OTHER FACTORS ARE IMPORTANT TO THE GROWTH OF DIATOMS AND
      HENCE, ALGAL BLOOMS. (ELLIS-TEXAS)

   FIELD 05C, 05D

   ACCESSION NO. W71-05026

APPLICATION OF INDUSTRIAL PHOTOSYNTHESIS PROCESS TO WASTE WATER RENOVATION
SYSTEMS,

NORTH AMERICAN AVIATION, INC., EL SEGUNDO, CALIF. S AND ID LIFE SCIENCES.

R. H. T. MATTONI, H. NUGENT MYRICK, AND E. C. KELLER, JR.

PROCEEDINGS, INDUSTRIAL WASTE CONFERENCE, 20TH, MAY 4, 5, 6, 1965.
ENGINEERING BULLETIN OF PURDUE UNIVERSITY, VOL XLIX, NO 4, P 684-705. 17
FIG, 1 TAB.

DESCRIPTORS:
*ALGAE, *BIOMASS, WATER PURIFICATION, *PROTOZOA, *FUNGI, *BACTERIA,
HYDROGEN ION CONCENTRATION, TEMPERATURE, ALKALINITY, BICARBONATES,
BACTERICIDE, KINETICS, HARDNESS, COLIFORMS, CYTOLOGICAL STUDIES,
EVAPORATION, MICROORGANISMS, CALIFORNIA, *PHOTOSYNTHESIS, WASTE WATER
TREATMENT.

IDENTIFIERS:
INDUSTRIAL PHOTOSYNTHESIS, MEAN DOUBLING TIME, PACKED CELL VOLUMES,
BACTERIAL PLATE COUNTS, LANCASTER(CALIF).

ABSTRACT:
INDUSTRIAL PHOTOSYNTHESIS REFERS TO THE CONTROLLED GROWTH OF COMPLEX
POPULATIONS OF MICROORGANISMS, MAINLY UNICELLULAR ALGAE, ON COMPLEX
ORGANIC SUBSTRATES SUCH AS PRIMARY AND SECONDARY WASTE WATERS. THE GOAL
OF THIS PROCESS IS TO PRODUCE SALABLE BIOMASS IN ADDITION TO INCREASING
THE WATER QUALITY TO LEVELS ACCEPTABLE FOR IRRIGATIONAL OR RECREATIONAL
USES. THREE FULL-SCALE TANK REACTORS WERE USED TO PERFORM EXPERIMENTS
ON SELECTION PROCESSES, PH DEPENDENCE, AND TEMPERATURE DEPENDENCE. THE
NATURAL SELECTION OF A SIMPLE ARRAY OF SPECIES FROM A COMPLEX SEED WAS
DEMONSTRATED. PH, HOWEVER, WAS SHOWN TO HAVE LITTLE EFFECT ON THE
RELATIVE SPECIES FREQUENCY PROFILES. IT WAS ALSO DETERMINED THAT WHEN
REACTORS ARE OPERATED IN AN IDENTICAL FASHION, THEY DO PRODUCE
IDENTICAL POPULATIONS OF MICROORGANISMS. THE EXPERIMENT WAS SITUATED IN
AN AREA KNOWN FOR HIGH TEMPERATURE AND CLOUD FREE DAYS. DURING THE
EXPERIMENT, HOWEVER, THE REGION WAS EXPERIENCING ITS ANNUAL TEMPERATURE
MINIMUMS, WITH ICE FORMING ON THE TOPS OF THE REACTORS. IN SPITE OF
THIS, GROWTH RATES OBTAINED WERE COMPARABLE TO GROWTH RATES DURING THE
SPRING, SUMMER, AND AUTUMN, DEMONSTRATING THESE SYSTEMS TO BE
APPLICABLE EVEN WHERE FREEZING TEMPERATURES ARE ENCOUNTERED DURING PART
OF THE DIURNAL CYCLE. FINALLY, THE FINAL REDUCTIONS OF BACTERIA WERE
SHOWN TO BE A FUNCTION OF ALGAE GROWTH. THIS WAS ATTRIBUTED TO
BACTERIOCIDAL ENZYMES EXCRETED BY THE ALGAE. (LOWRY-TEXAS)

FIELD 05D

ACCESSION NO. W71-05267

A TECHNICAL ASSESSMENT OF CURRENT WATER QUALITY CONDITIONS AND FACTORS
AFFECTING WATER QUALITY IN THE UPPER POTOMAC ESTUARY,

ENVIRONMENTAL PROTECTION AGENCY, ANNAPOLIS, MD. CHESAPEAKE TECHNICAL SUPPORT
LAB.

NORBERT A. JAWORSKI, DONALD W. LEAR, JR., AND JOHAN A. AALTO.

WATER QUALITY STATUS REPORT. CTSL TECHNICAL REPORT NO 5, MAR, MARCH 1969. 55
P, 17 FIG, 6 TAB, 21 REF.

DESCRIPTORS:
*BIOCHEMICAL OXYGEN DEMAND, *NUTRIENTS, *DISSOLVED OXYGEN, *ALGAE,
COLIFORMS, EUTROPHICATION(PHYTOPLANKTON), ESTUARIES, WATER POLLUTION
SOURCES, WATER POLLUTION EFFECTS, NITROGEN, COLIFORMS, WATER QUALITY.

IDENTIFIERS:
*POTOMAC RIVER ESTUARY, NUTRIENT-PHYTOPLANKTON RELATIONSHIPS.

ABSTRACT:
WATER QUALITY CONDITIONS IN THE UPPER POTOMAC ESTUARY INCLUDING BOD,
DISSOLVED OXYGEN, PHOSPHORUS, NITROGEN AND COLIFORM DENSITIES, ARE
PRESENTED AND DISCUSSED. THE SOURCES AND RELATIONSHIPS OF NUTRIENTS TO
ALGAL STANDING CROP ARE STUDIED. MAJOR SOURCES OF CARBONACEOUS AND
NITROGENOUS OXYGEN DEMAND AS WELL AS THE DEMAND FROM THE ALGAL STANDING
CROP ON DISSOLVED OXYGEN ARE EVALUATED. (AALTO-CHESAPEAKE TECHNICAL
SUPPORT LABORATORY)

FIELD 05B, 05C, 02L

ACCESSION NO. W71-05407

WATER QUALITY AND WASTE WATER LOADINGS UPPER PO.TOMAC ESTUARY DURING 1969,

ENVIRONMENTAL PROTECTION AGENCY, ANNAPOLIS, MD. CHESAPEAKE TECHNICAL SUPPORT
  LAB.

NORBERT A. JAWORSKI.

WATER QUALITY STATUS REPORT. CTSL TECHNICAL REPORT NO 27, MAR, NOV 1969. 62
  P, 14 FIG, 9 TAB, 4 REF, APPEND.

  DESCRIPTORS:
    *BIOCHEMICAL OXYGEN DEMAND, *PHOSPHORUS, *NITROGEN, *COLIFORMS,
    *DISSOLVED OXYGEN, ESTUARIES, ALGAE, WATER POLLUTION SOURCES, WATER
    POLLUTION EFFECTS, WATER QUALITY.

  IDENTIFIERS:
    *POTOMAC RIVER ESTUARY.

  ABSTRACT:
    FOR THE FIRST EIGHT MONTHS OF 1969, ABOUT 55% OF THE BOD ENTERING THE
    POTOMAC ESTUARY FROM ALL MAJOR SOURCES WAS FROM WASTE WATER DISCHARGES
    IN THE WASHINGTON AREA. ALSO THESE DISCHARGES CONTRIBUTED 86% OF THE
    TOTAL PHOSPHORUS AND 66% OF THE TOTAL NITROGEN. TOTAL CARBON ENTERING
    THE ESTUARY FROM ALL SOURCES WAS MAINLY IN THE FORM OF INORGANIC
    CARBON. DISSOLVED OXYGEN LEVELS WERE MEASURED. AS A RESULT OF INCREASED
    CHLORINATION OF THE WASTE WATER TREATMENT FACILITY EFFLUENTS, COLIFORM
    DENSITIES WERE SIGNIFICANTLY LOWER. TO FACILITATE THE DETERMINATION OF
    WASTE WATER DISCHARGE LOADINGS, THE ESTUARY WAS ZONED INTO FIFTEEN MILE
    REACHES. MAXIMUM WASTE DISCHARGE LOADINGS FOR BOD (ORGANIC CARBON),
    NITROGEN AND PHOSPHORUS HAVE BEEN DETERMINED FOR ZONE I IN THE
    WASHINGTON AREA. PRELIMINARY ESTIMATES OF LOADINGS HAVE BEEN
    ESTABLISHED FOR ZONE II DOWNSTREAM FROM THE WASHINGTON AREA.
    (AALTO-CHESAPEAKE TECHNICAL SUPPORT LABORATORY)

  FIELD 05B, 05C, 02L

  ACCESSION NO. W71-05409

PARAMETERS AFFECTING THE GROWTH OF ULVA LATISSIMA IN A POLLUTED ESTUARY,

NORTHEASTERN UNIV., BOSTON, MASS.

THOMAS D. WAITE, AND CONSTANTINE GREGORY.

NORTH EASTERN REGIONAL ANTIPOLLUTION CONFERENCE, ANNUAL, PROCEEDINGS,
  KINGSTON, RHODE ISLAND, JULY 22-25, 1969, P 12-15.

  DESCRIPTORS:
    *ESTUARIES, *HARBORS, *SEWAGE, WATER POLLUTION EFFECTS, *ALGAE.

  IDENTIFIERS:
    *ULVA LATISSIMA, *AMMONIA NITROGEN.

  ABSTRACT:
    ULVA LATISSIMA, A MARINE FLORA THAT HAS PRODUCED SEVERE POLLUTION
    PROBLEMS DUE TO ITS UNRESTRAINED GROWTH IN ESTUARIES AND HARBORS AS A
    RESULT OF DOMESTIC SEWAGE FERTILIZING THE MARINE ENVIRONMENT WAS
    STUDIED. THE CORRELATION BETWEEN AMMONIA NITROGEN CONCENTRATION IN
    SEAWATER AND RATE OF GROWTH OF ULVA LATISSIMA WAS INVESTIGATED. THE
    GREATEST GROWTH RATE IS REALIZED AT 0.7 MG/L AMMONIA NITROGEN
    CONCENTRATION. REDUCTION OF AMMONIA CONCENTRATION TO LESS THAN 0.5 MG/L
    IS NECESSARY TO REGULATE EXCESSIVE GROWTH OF ULVA LATISSIMA.
    (ENSIGN-PAI)

  FIELD 05C

  ACCESSION NO. W71-05500

SOME BIOGEOCHEMICAL CONSIDERATIONS ON THE RADIOACTIVE CONTAMINATION OF MARINE
    BIOTA AND ENVIRONMENTS,

INTERNATIONAL LABORATORY OF MARINE RADIOACTIVITY, MONTE CARLO (MONACO).

RINNOSUKE FUKAI.

SOME BIOGEOCHEMICAL CONSIDERATIONS ON THE RADIOACTIVE CONTAMINATION OF MARINE
    BIOTA AND ENVIRONMENTS, PAPER, PERGAMON PRESS, NEW YORK, 1969, P 391-394. 1
    FIG, 1 TAB, 7 REF.

DESCRIPTORS:
    *RADIOACTIVITY, *TRACE ELEMENTS, *AQUATIC LIFE, WATER POLLUTION
    EFFECTS, MARINE ANIMALS, ALGAE, MOLLUSKS, FISH, RADIOACTIVE WASTES.

IDENTIFIERS:
    *BIOGEOCHEMICAL DATA, *ABUNDANCE RATIOS.

ABSTRACT:
    RADIOACTIVE CONTAMINATION OF MARINE ENVIRONMENTS AND BIOTA IS A LINK IN
    THE BIOGEOCHEMICAL PROCESS IN THE SEA. THE IMPORTANCE OF THE
    ACCUMULATION OF BIOGEOCHEMICAL DATA ON THE DISTRIBUTIONS OF TRACE
    ELEMENTS IN MARINE BIOTA IN RELATION TO THE PREDICTION OF RADIOACTIVE
    CONTAMINATION IS DISCUSSED. THE REASON FOR OBTAINING THE 'STANDARD
    ABUNDANCE' OF TRACE ELEMENTS FOR CERTAIN GROUPS OF MARINE ORGANISMS IS
    EXPLAINED AND THE IMPORTANCE OF THE INTERPRETATION OF THE ANALYTICAL
    RESULTS AND COMPUTATIONS OF THE ABUNDANCE RATIO IS MENTIONED.
    (ENSIGN-PAI)

FIELD 05A, 05C

ACCESSION NO. W71-05531

UTILIZATION OF ALKOXYSILANES FOR REDUCTION OF THE ABUNDANCE OF PHYTOPLANKTON
    DURING 'BLOOMS' OF WATER BASINS (IN RUSSIAN),

MOSCOW STATE UNIV. (USSR).

N. S. STROGANOV, V. G. KHOBOTIEV, L. V. KOLOSOVA, AND M. A. KADINA.

DOKLADY ADADEMII NAUK SSSR, VOL 181, NO 5, P 1257-1259, 1968. 2 FIG, 2 TAB,
    13 REF.

DESCRIPTORS:
    *HERBICIDES, *ALGAE, TOXICITY, SCENEDESMUS, CHLORELLA, DAPHNIA,
    DETERIORATION, WATER TREATMENT, DEGRADATION(DECOMPOSITION).

IDENTIFIERS:
    *ALGAL BLOOMS, *ORGANO-SILICATES, *ALKOXYSILANES, SCENEDESMUS
    QUADRICAUDA, CHLORELLA VULGARIS, DAPHNIA MAGNA, LIMNAEA STAGNALIS,
    OUSPENSKII'S MEDIUM.

ABSTRACT:
    A SEARCH FOR HERBICIDES SUPPRESSING ALGAL BLOOMS, BUT HARMLESS TO
    ZOOPLANKTERS AND OTHER HYDROBIONTS LEAD TO INVESTIGATION OF
    ORGANO-SILICATES: VENYLTRIETHOXYSILANE, TETRAETHOXYSILANE, AND
    TRIFLUOROPROPENYL-(METHYL)-DIETHOXYSILANE. THE TEST ORGANISM INCLUDED
    SCENEDESMUS QUADRICAUDE, CHLORELLA VULGARIS, DAPHNIA MAGNA, AND LIMNAEA
    STAGNALIS. THE COUNT OF SURVIVED SPECIMENS SHOWED THAT THE
    CONCENTRATION OF 0.01 MG/LITER OF ALL INVESTIGATED COMPOUNDS ELIMINATED
    93% TO 96% OF ALGAL CELLS WITHOUT DAMAGING EITHER DAPHNIA OR MOLLUSKS.
    AMONG THE ADVANTAGES OF THE ALKOXYSILANES IS THEIR RELATIVELY RAPID
    DISINTEGRATION IN WATER. (WILDE-WISCONSIN)

FIELD 05F, 05C

ACCESSION NO. W71-05719

CHEMICAL AND BIOLOGICAL REACTIONS FROM LAGOONS USED FOR CATTLE,

WISCONSIN UNIV., MADISON.

S. A. WITZEL, ELIZABETH MCCOY, AND RICHARD LEHNER.

ASAE PAPER NO 64-417. TRANSACTIONS OF THE AMERICAN SOCIETY OF AGRICULTURAL
ENGINEERS, VOL 8, P 449-451, 1965. 1 FIG, 16 REF.

DESCRIPTORS:
*OXIDATION LAGOONS, *BIOCHEMICAL OXYGEN DEMAND, *ANAEROBICS, ALGAE,
BACTERIA, WATER POLLUTION SOURCES, ODOR, FARM WASTE, SLUDGE, SOLID
WASTES, DECOMPOSTING ORGANIC MATTER, LAGOONS, CATTLE.

IDENTIFIERS:
ORGANIC NITROGEN, BACTERIOLOGICAL STUDY, AEROBICITY.

ABSTRACT:
AN EXPERIMENTAL LAGOON TO RECEIVE THE WASTES AS LIQUID MANURE FROM SIX
BULLS WAS CONSTRUCTED. IN AN EXPERIMENTAL BARN THE MANURE FROM THE
BULLS WAS WASHED DAILY INTO TWO GUTTERS 24 IN. WIDE AT THE TOP. THE
GUTTERS HAD A CAPACITY OF 2000 GALLONS. THE GUTTERS WERE FLUSHED EVERY
SEVEN DAYS INTO A LAGOON. THE CIRCULAR LAGOON HAD A 60 FT DIAMETER AT
THE TOP, A 40 FT DIAMETER AT THE BOTTOM AND WAS 5 FT DEEP. LIQUID
MANURE SAMPLES WERE TAKEN FROM THE BARN GUTTER AND FROM THE LAGOON
MANURE LIQUID AND SLUDGE. BOD TESTS WERE CONDUCTED ON ALL SAMPLES AND
THE PERCENT BOD REDUCTION FROM GUTTER TO LAGOON WAS CALCULATED. OTHER
TESTS WERE MADE TO DETERMINE PERCENT TOTAL SOLIDS REMOVAL, ORGANIC
NITROGEN CONTENT, PH VALUES, AND TEMPERATURE. EXTENSIVE TESTS WERE MADE
AND REPORTED ON BACTERIA CONTENT, BOTH QUANTITATIVE AND QUALITATIVE.
FIVE CONCLUSIONS OF THE STUDY ARE MENTIONED. (PARKER-IOWA STATE)

FIELD 05D

ACCESSION NO. W71-05742

REMOVAL OF CONTAMINANTS FROM LAKE ONTARIO BY NATURAL PROCESSES,

CANADIAN OCEANOGRAPHIC DATA CENTRE, OTTAWA (ONTARIO).

H. E. SWEERS.

IN: PROCEEDINGS TWELFTH CONFERENCE ON GREAT LAKES RESEARCH, MAY 5-7, 1969,
UNIVERSITY OF MICHIGAN, ANN ARBOR: INTERNATIONAL ASSOCIATION FOR GREAT
LAKES RESEARCH, P 734-741, 1969. 8 P, 4 FIG, 7 REF.

DESCRIPTORS:
*SELF-PURIFICATION, *WATER POLLUTION CONTROL, *LAKE ONTARIO, *WATER
CIRCULATION, PATH OF POLLUTANTS, NUTRIENTS, ALGAE, STRATIFIED FLOW,
MIXING, FLOW, MODEL STUDIES, MATHEMATICAL MODELS, GREAT LAKES, THERMAL
STRATIFICATION.

IDENTIFIERS:
REMOVAL TIME(POLLUTANTS).

ABSTRACT:
A MODEL IS DEVELOPED TO CALCULATE THE REMOVAL TIME OF A CONSERVATIVE
CONTAMINANT FROM A LAKE, TAKING SUMMER STRATIFICATION INTO ACCOUNT. THE
BASIC ASSUMPTIONS OF THE MODEL WERE COMPARED WITH CONDITIONS ACTUALLY
OCCURRING IN LAKE ONTARIO, AND IT IS SHOWN THAT STRATIFICATION HAS
LITTLE EFFECT ON THE CALCULATED REMOVAL TIME. THE RESULTS ARE
EXTRAPOLATED TO THE BEHAVIOR OF NON-CONSERVATIVE PARAMETERS. A SHARP
REDUCTION IN THE RATE OF INPUT OF NUTRIENTS COULD RESULT IN A MARKED
DECREASE IN ALGAL GROWTH WITHIN A YEAR AFTER SUCH MEASURES BECOME
EFFECTIVE. (KNAPP-USGS)

FIELD 05G, 05B, 02H

ACCESSION NO. W71-05878

INVESTIGATIONS OF DAILY VARIATIONS IN CHEMICAL BACTERIOLOGICAL AND BIOLOGICAL PARAMETERS AT TWO LAKE ONTARIO LOCATIONS NEAR TORONTO, PART I--CHEMISTRY,

ONTARIO WATER RESOURCES COMMISSION, REXDALE.

T. G. BRYDGES.

IN: PROCEEDINGS TWELFTH CONFERENCE ON GREAT LAKES RESEARCH, MAY 5-7, 1969, UNIVERSITY OF MICHIGAN, ANN ARBOR: INTERNATIONAL ASSOCIATION FOR GREAT LAKES RESEARCH, P 750-759, 1969. 10 P, 4 FIG, 1 TAB, 5 REF.

DESCRIPTORS:
*LAKE ONTARIO, *WATER QUALITY, *MONITORING, BACTERIA, WATER CHEMISTRY, PHOSPHORUS, NITROGEN, AMMONIA, ALGAE, NUTRIENTS, SEWAGE, WASTE WATER(POLLUTION), WATER POLLUTION SOURCES, MUNICIPAL WASTES, GREAT LAKES, *POLLUTANT IDENTIFICATION.

IDENTIFIERS:
*TORONTO(ONTARIO).

ABSTRACT:
AN INVESTIGATION OF DAILY VARIATIONS IN CHEMICAL, BACTERIOLOGICAL AND ALGAL PARAMETERS WAS CARRIED OUT USING SAMPLES COLLECTED FIVE DAYS A WEEK FROM TORONTO HARBOUR AND FROM THE INTAKE OF THE R. C. HARRIS FILTRATION PLANT. THE INTAKE EXTENDS 2500 M FROM SHORE AND IS 12 M DEEP IN 23 M OF WATER. THE STUDY BEGAN IN JULY 1968. WEEKLY DEPTH AND SEDIMENT SAMPLES WERE COLLECTED TO SUPPLEMENT THE DAILY DATA. FOURTEEN CHEMICAL PARAMETERS WERE MONITORED. TOTAL PHOSPHORUS CONCENTRATION WAS THE BEST INDICATOR OF THE PRESENCE OF URBAN DRAINAGE. AMMONIA AND ORGANIC NITROGEN CONCENTRATIONS IN THE HARBOR VARY INVERSELY; CONSEQUENTLY KJELDAHL NITROGEN IS A BETTER MEASURE OF WATER QUALITY THAN EITHER OF THEM. SURFACE NITRATE CONCENTRATIONS VARY WITH WIND SPEED AND BIOLOGICAL ACTIVITY AND DO NOT ALWAYS REFLECT THE EFFECTS OF RUNOFF. CHLOROPHYLL AND SOLUBLE PHOSPHORUS CONCENTRATIONS IN THE HARBOR VARY INVERSELY. PHOSPHORUS WAS APPARENTLY NOT LIMITING ALGAL GROWTH, BUT THERE WERE INSUFFICIENT DATA TO TEST THE RELATIONSHIP FOR THE OPEN LAKE. DAILY DATA WERE REQUIRED TO DEFINE SOME INTER-PARAMETER RELATIONSHIPS WHICH WERE NOT DEFINED BY DATA COLLECTED AT WEEKLY INTERVALS. (KNAPP-USGS)

FIELD 05A, 05B, 02H

ACCESSION NO. W71-05879

75 YEARS OF IMPROVEMENT IN WATER SUPPLY QUALITY,

AMERICAN WATER WORKS ASSOCIATION, NEW YORK.

ABEL WOLMAN.

IN: WATER, HEALTH AND SOCIETY, BLOOMINGTON, INDIANA UNIVERSITY PRESS, P 93-103, 1969.

DESCRIPTORS:
*WATER SUPPLY, *WATER QUALITY, *WATER TREATMENT, FILTRATION, CHLORINATION, COAGULATION, WATERSHEDS, SEWERAGE ALGAE, POTABLE WATER.

IDENTIFIERS:
*AMERICAN WATER WORKS ASSOCIATION, CHEMICAL INDUSTRY, ODOR, TASTE, CONSUMER.

ABSTRACT:
THIS ARTICLE DISCUSSES THE VARIOUS EFFORTS OF THE AMERICAL WATER WORKS ASSOCIATION, IN BOTH TECHNOLOGICAL AND SCIENTIFIC FIELDS, TO PROVIDE WATER OF MAXIMUM USEFULNESS AND SATISFACTION TO THE CONSUMER OVER THE LAST 75 YEARS. ADVANCES IN WATER SUPPLY SUCH AS EFFECTIVE COAGULATION, SEDIMENTATION, FILTRATION, DISINFECTION AND THE USE OF CHLORINATION ARE SURVEYED AS MILESTONES IN WATER SUPPLY QUALITY. SPECIFIC WATER QUALITY ANALYSIS DIFFICULTIES ARE ASSESSED FROM THE POINTS OF VIEW OF BOTH THE WATER DIAGNOSTICIAN AND THE CONSUMER. PARTICULAR ATTENTION IS GIVEN TO THE IDENTIFICATION OF SOURCES AND CAUSES OF TASTE AND ODOR, AND THE RELATIONSHIP OF THESE TO ALGAE AND TO THE BURGEONING CHEMICAL INDUSTRY. THE USE OF SUCH REMEDIES AS POWDERED ACTIVATED CARBON, OZONE, CHLORINATION, AND CHLORINE DIOXIDE IS DISCUSSED. ADVANCES IN THE WATER SUPPLY QUALITY ARE ATTRIBUTED TO THE DEVELOPMENT OF STANDARD METHODS OF ANALYSIS AND TO THE JUDICIOUS USE OF COMMERCIAL EQUIPMENT. A RAPID TREND IS HYPOTHESIZED TOWARD THE LITERAL TAILORING OF WATER FOR THE DOMESTIC, INDUSTRIAL, OR FARMING CONSUMER AS REQUIREMENTS OF WATER SUPPLY QUALITY BECOME MORE NUMEROUS AND COMPLEX. (MURPHY-RUTGERS)

FIELD 06E, 05G

ACCESSION NO. W71-05943

INVESTIGATIONS OF THE TECHNIQUES OF PURIFICATION OF ALGAL CULTURES: ELABORATION
OF METHODS APPLICABLE TO DESMIDS, FILAMENTOUS CYANOPHYTES, AND DIATOMS, (IN
FRENCH),

CENTRE NATIONAL DE LA RECHERCHE SCIENTIFIQUE, GIF-SUR-YVETTE (FRANCE). CENTRE
DE RECHERCHES HYDROBIOLOGIQUES.

M. TASSIGNY, G. LAPORTE, AND R. POURRIOT.

ENGLISH SUMMARY. ANNALES DE L'ISTITUT PASTEUR, VOL 117, NO 1, P 64-75, 1969.
2 TAB, 22 REF.

    DESCRIPTORS:
     *LABORATORY TESTS, *ALGAE, CULTURES, DIATOMS, CYANOPHYTA, BIOASSAY,
     TEST PROCEDURES.

    IDENTIFIERS:
     *PURE ALGAL CULTURES.

    ABSTRACT:
     SEVERAL METHODS OF OBTAINING PURE CULTURES OF ALGAE WERE SUBJECTED TO
     CRITICAL REVIEW: DILUTION AND USE OF PIPETTES, DILUTION FOLLOWING AN
     ULTRASONIC TREATMENT, USE OF ANTIBIOTICS, AND GERMINATION OF STERILIZED
     SPORES. EACH OF THESE METHODS OFFERED A POSSIBILITY OF OBTAINING PURE
     CULTURES OF A LIMITED NUMBER OF SPECIES. THE SIMPLE TECHNIQUE OF
     DILUTION AND USE OF PIPETTES PROVIDED THE GREATEST CHANCE OF SUCCESS. A
     PROPER CHOICE OF THE METHOD AND CERTAIN MODIFICATIONS PERMITTED
     ISOLATION OF MICROSTERIAE, CLOSTERIUM, DRAPARNALDIA, NITZSCHIA, AND
     SEVERAL FILAMENTOUS CYANOPHYCEAE. (WILDE-WISCONSIN)

    FIELD 07B, 05A

    ACCESSION NO. W71-06002

PRACTICAL ASPECTS OF THE DESIGN OF WASTE STABILIZATION PONDS,

WOOTON (L.E.) AND CO., RALEIGH, N.C.

W. D. BARLOW.

NINTH SOUTHERN MUNICIPAL AND INDUSTRIAL WASTE CONFERENCE PROCEEDINGS (NORTH
CAROLINA STATE UNIVERSITY, RALEIGH, NORTH CAROLINA, APRIL 7-8 1960) P
65-74, TECHNICAL PRESS, RALEIGH, NORTH CAROLINA, 1960.

    DESCRIPTORS:
     *LAGOONS, *DOMESTIC WASTES, WASTE WATER TREATMENT, INDUSTRIAL WASTES,
     NUTRIENTS, BACTERIA, ALGAE, TREATMENT FACILITIES, TEST PROCEDURES,
     METALS.

    IDENTIFIERS:
     *FINISHING WASTES.

    ABSTRACT:
     LAGOONS CAN BE DESIGNED TO PROVIDE COMPLETE TREATMENT FOR ANY WASTE
     WATER, DOMESTIC AND/OR INDUSTRIAL, PROVIDED IT CONTAINS THE NECESSARY
     NUTRIENTS AND IS FREE FROM SUBSTANCES WHICH ARE TOXIC TO BACTERIA OR
     ALGAE. THE AUTHOR DISCUSSES THE BASIC PROCESSES INVOLVED; FACTORS
     AFFECTING THE DESIGN AND OPERATION OF THE LAGOON (PARTICULARLY UNDER
     CONDITIONS IN NORTH AND SOUTH CAROLINA); PERFORMANCE OF THE LAGOON,
     INCLUDING METHODS OF MEASUREMENT; AND PRECAUTIONS TO BE TAKEN TO AVOID
     HEALTH HAZARDS, POLLUTION AND OTHER NUISANCE PROBLEMS. IN DISCUSSION,
     REFERENCE WAS MADE TO LAGOONS USED FOR TREATING MIXED WASTE WATERS AND
     20-30 PERCENT OF DOMESTIC SEWAGE, AND ESPECIALLY TO THE PROBLEMS
     ENCOUNTERED WHICH APPEAR TO BE CAUSED BY TOXIC METALS IN THE WASTE
     WATER. (LIVENGOOD — NORTH CAROLINA STATE UNIVERSITY)

    FIELD 05D

    ACCESSION NO. W71-06737

NATION-WIDE RESEARCH ON ANIMAL WASTE DISPOSAL,

FEDERAL WATER POLLUTION CONTROL ADMINISTRATION, CHICAGO, ILL. LAKE MICHIGAN
  BASIN OFFICE.

JACOB O. DUMELLE.

IN: PROCEEDINGS OF FARM ANIMAL WASTE AND BY-PRODUCT MANAGEMENT CONFERENCE,
  UNIVERSITY EXTENSION, UNIVERSITY OF WISCONSIN, P 80-81, NOVEMBER 6-7, 1969.

DESCRIPTORS:
    *FARM WASTES, *RESEARCH AND DEVELOPMENT, GRANTS, ALGAE, AIR POLLUTION,
    WATER POLLUTION, SOIL CONTAMINATION, NUTRIENTS.

IDENTIFIERS:
    *FWPCA, ACTIVATED ALGAE, FEEDLOTS, OXIDATION DITCH.

ABSTRACT:
    THE ARTICLE GIVES BRIEF DESCRIPTIONS OF RESEARCH PROJECTS WHICH THE
    FEDERAL WATER POLLUTION CONTROL ADMINISTRATION IS HELPING TO FUND. ONE
    SUCH PROJECT UNDER WAY IN CALIFORNIA IS TRYING TO DETERMINE THE
    PRACTICABILITY OF PRODUCING AND HARVESTING ALGAE TO REMOVE NUTRIENTS
    FROM AGRICULTURAL DRAINAGE WATERS. OTHER PROJECTS INVOLVE CATTLE
    FEEDLOT RUNOFF, AND DAIRY WASTE WATERS. BESIDES RESEARCH ON TREATMENT
    METHODS, SOME PROJECTS ARE TRYING TO FIND OUT HOW MUCH NUTRIENT RUNS
    OFF, AND HOW MUCH GETS INTO WATER. (WHITE-IOWA STATE)

FIELD 05G, 05D

ACCESSION NO. W71-06821

SOME PROBLEMS IN THE TESTING OF MATERIALS WITH ALGAE,

CENTRAL LAB. TNO, DELFT (NETHERLANDS).

H. J. HUECK, AND D. M. M. ADEMA.

SUMMARIES IN GERMAN, ENGLISH, FRENCH AND SPANISH. MATERIAL UND ORGANISMEN,
  VOL 2, NO 2, P 141-152, 1967. 1 FIG, 4 TAB, 15 REF.

DESCRIPTORS:
    *ALGICIDES, *PAINTS, *TEST PROCEDURES, ANTIFOULING MATERIALS, TESTING,
    BIODEGRADATION, ALGAE, MARINE ALGAE, GROWTH RATES, ANALYTICAL
    TECHNIQUES, LABORATORY EQUIPMENT, ASBESTOS CEMENT, CHLORELLA, CARBON
    DIOXIDE.

IDENTIFIERS:
    *TEST DEVELOPMENT, *TEST PROBLEMS, HORMIDIUM SP.

ABSTRACT:
    A FEW CONSIDERATIONS ARE GIVEN ABOUT THE TESTING OF ALGICIDES
    INCORPORATED IN PAINTS AND STONY MATERIALS. TWO EXPERIMENTAL DESIGNS
    ARE DESCRIBED: ONE FOR TESTING ALGICIDAL PAINTS WITH AQUATIC ALGAE, AND
    ONE FOR TESTING ALGICIDAL ASBESTOS-CEMENT WITH AERIAL ALGAE. FOR THE
    FIRST PURPOSE A STATIC TEST IS UNDESIRABLE BECAUSE OF THE BUILDING UP
    OF TOXIC CONCENTRATIONS IN THE MEDIUM WHEREAS THE ACTUAL TOXICITY AT
    THE SURFACE OF THE MATERIAL SHOULD BE MEASURED. A DEVICE IS PROPOSED
    THAT USES A CONTINUOUSLY INNOCULATED MEDIUM FLOWING SLOWLY THROUGH A
    TEST AQUARIUM. FOR THE TESTING OF ASBESTOS-CEMENT AND OTHER ALKALINE
    STONY MATERIALS WITH AERIAL ALGAE IT IS FOUND THAT THE AVAILABILITY OF
    CO2 NEAR THE SURFACE MAY BE A LIMITING FACTOR. ANOTHER DEVICE IS
    DESCRIBED TO TEST SAMPLES IN A FLOW OF AIR ENRICHED WITH 5% CO2.
    (SJOLSETH-WASHINGTON)

FIELD 05C

290

ACCESSION NO. W71-07339

EXISTING AND POTENTIAL PROBLEMS OF EXCESSIVE EUTROPHICATION IN THE TENNESSEE
VALLEY,

TENNESSEE VALLEY AUTHORITY, CHATTANOOGA. WATER QUALITY BRANCH.

EUGENE B. WELCH.

PROCEEDINGS OF THE 7TH ANNUAL SANITARY AND WATER RESOURCES ENGINEERING
CONFERENCE, VANDERBILT UNIVERSITY, NASHVILLE, TENNESSEE, P 45-73, 1968. 7
FIG, 8 REF.

DESCRIPTORS:
*EUTROPHICATION, *TENNESSEE VALLEY AUTHORITY PROJECT, *RESERVOIRS,
TEMPERATURE, DISSOLVED OXYGEN, DENSITY, ROOTED AQUATIC PLANTS, ALGAE,
CHLOROPHYLL, HYPOLIMNION, PHOSPHORUS, NITROGEN, PHYTOPLANKTON,
PRODUCTIVITY, TURBIDITY, SURFACES, DEPTH, DISTRIBUTION, TIME,
EPILIMNION, THERMOCLINE, LIGHT PENETRATION, STANDING CROP, INDUSTRIAL
WASTES, RAINFALL, ALABAMA.

IDENTIFIERS:
CHEROKEE RESERVOIR(TENN), HOLSTON RIVER(TENN), POTOMOGETON PECTINATUS,
POTOMOGETON CRISPUS, HETERAMTHERA, PICKWICK RESERVOIR(TENN), NAJAS,
PHOTIC ZONE, DOUGLAS RESERVOIR(TENN), NORRIS RESERVOIR(TENN), FRENCH
BROAD RIVER(TENN), CHICKAMAUGA RESERVOIR(TENN), FLORENCE(ALA).

ABSTRACT:
PATTERN OF HYPOLIMNETIC OXYGEN DEPLETION IN THREE TVA SYSTEM RESERVOIRS
APPEARS RELATED TO PHYTOPLANKTON PRODUCTIVITY, WHICH IS AFFECTED BY
SEVERAL ENVIRONMENTAL FACTORS. IN RESERVOIRS STUDIED, NUTRIENT
(NITROGEN AND PHOSPHORUS) CONTENT POTENTIAL LIMITS PRODUCTION ONLY
DURING INTENSE SUMMER STRATIFICATION. THERMAL STRATIFICATION AFFECTING
NUTRIENT DISTRIBUTION AND LIGHT AVAILABILITY, TURBIDITY AFFECTING LIGHT
PENETRATION, AND WATER DETENTION TIME MAY AT VARIOUS PERIODS AND PLACES
CONTROL PRODUCTION REGARDLESS OF NUTRIENT CONTENT. FIELD AND LABORATORY
EVIDENCE SHOWS THAT GROWTH OF TWO NUISANCE MACROPHYTES IN TENNESSEE
VALLEY IS PROBABLY NOT CONTROLLED BY NITROGEN AND PHOSPHORUS CONTENT OF
WATER. NUTRIENT REQUIREMENTS ARE APPARENTLY LESS THAN AMOUNTS MEASURED
AND AVAILABLE. ORGANIC SEDIMENT IS RELATIVELY MORE IMPORTANT AS
NUTRITION SOURCES FOR MAXIMUM GROWTH THAN SURROUNDING WATER. AMOUNT OF
AVAILABLE LIGHT AT LEAST PARTLY EXPLAINS ANNUAL AND SEASONAL
VARIABILITY IN NUISANCE MACROPHYTE GROWTH. LIGHT REACHING STREAM OR
LAKE BOTTOM AND AVAILABLE FOR PLANT GROWTH DURING GROWING SEASON FROM
YEAR TO YEAR IS AFFECTED BY INTERACTION OF INCIDENT LIGHT, WATER DEPTH,
AND TURBIDITY OR RELATIVE PENETRATION. THESE, OR RELATED FACTORS, WERE
MOST FAVORABLE DURING THE TWO YEARS WHEN NUISANCE MACROPHYTE PROBLEMS
WERE GREATEST IN PICKWICK RESERVOIR, AND WERE RELATED TO PLANT GROWTH
IN HOLSTON RIVER DURING 1967. (JONES-WISCONSIN)

FIELD 05C, 02H

ACCESSION NO. W71-07698

GLENODININE, AN ICHTHYOTOXIC SUBSTANCE PRODUCED BY A DINOFLAGELLATE, PERIDINIUM
POLONICUM,

TOKYO UNIV. (JAPAN). LAB. OF MARINE BIOCHEMISTRY.

YOSHIRO HASHIMOTO, TOMOTOSHI OKAICHI, LE DUNG DANG, AND TAMAO NOGUCHI.

BULLETIN OF THE JAPANESE SOCIETY OF SCIENTIFIC FISHERIES, VOL 34, NO 6, P
528-534, JUNE 1968. 7 FIG, 1 TAB, 8 REF.

DESCRIPTORS:
*FISHKILL, *TOXINS, *ALGAL POISONING, *ALGAL TOXINS, *PHYTOTOXICITY,
*FISH DISEASES, *FISH TOXINS, TOXICITY, HYDROGEN ION CONCENTRATION,
DISSOLVED OXYGEN, ALGAE, PHYTOPLANKTON, ALKALINE WATER, ALKALINITY,
FISH FARMING, FISH FOOD ORGANISMS, FISH PHYSIOLOGY, RED TIDE, ORGANIC
COMPOUNDS, DINOFLAGELLATES.

IDENTIFIERS:
*PERIDINIUM SP., *GLENODININE, ALKALOID TOXINS, IBOGAALKALOID, ORGANIC
TOXINS.

ABSTRACT:
MASS MORTALITIES ASSOCIATED WITH PHYTOPLANKTON BLOOMS SOMETIMES OCCUR
IN JAPANESE FISH CULTURE OPERATIONS. THE AUTHORS OPINE THAT THIS MAY BE
THE FIRST REPORT IDENTIFYING THE TOXIC SPECIES AND ITS ASSOCIATED
TOXIN. THE OFFENDER, GLENODININE, WAS ISOLATED FROM THE DINOFLAGELLATE,
PERIDINIUM POLONICUM. THE TOXIN IS AN ALKALOID SHOWING SOME RESEMBLANCE
TO IBOGAALKALOID, AND HAS THE SULFHYDRYL FUNCTION IN THE MOLECULE.
FISHKILLS OCCURRED IN THE AFTERNOON OR EVENING WHEN DISSOLVED OXYGEN
WAS HIGH AND PH LEVELS OF 8.7-9.2 WERE PRESENT. ALTHOUGH MORTALITIES
SOMETIMES OCCURRED AFTER BLOOMS RATHER THAN CONCURRENTLY, LETHAL LEVELS
OF GLENODININE WERE FOUND IN THE WATER DURING THE MORTALITIES. THE
SUBSTANCE IS APPARENTLY TOXIC ONLY IN ALKALINE MEDIUM, CREATED BY HIGH
PHYTOPLANKTONIC PHOTOSYNTHETIC ACTIVITY. (LEGORE-WASHINGTON)

FIELD 05C, 05A

ACCESSION NO. W71-07731

MANURE LAGOONS......DESIGN CRITERIA AND MANAGEMENT,

MARYLAND UNIV., COLLEGE PARK. DEPT. OF AGRICULTURAL ENGINEERING.

HARRY J. EBY.

ASAE PAPER NO 61-935. AGRICULTURAL ENGINEERING JOURNAL, VOL 43, P 698-701,
714-715, DEC 1962. 6 FIG, 1 TAB, 19 REF.

DESCRIPTORS:
*FARM LAGOONS, *DESIGN CRITERIA, WATER TEMPERATURE, SEWAGE TREATMENT,
AEROBIC BACTERIA, AQUATIC PLANTS, ANAEROBIC BACTERIA, ALGAE,
BIOCHEMICAL OXYGEN DEMAND, OXIDATION LAGOONS, SLUDGE, PHOTOSYNTHETIC
OXYGEN, FARM WASTES, WASTE WATER TREATMENT.

IDENTIFIERS:
*SITE SELECTION, LOADING.

ABSTRACT:
CRITERIA TO BE CONSIDERED WHEN DESIGNING A LAGOON FOR TREATMENT OF
WASTES PRODUCED BY ANIMALS IN CONFINEMENT IS DISCUSSED. IT MENTIONS
SITUATIONS WHERE LAGOONS WOULD NOT BE FEASIBLE. SEVEN CRITERIA FOR SITE
SELECTION ARE GIVEN. THE PHYSICAL, CHEMICAL AND BIOLOGICAL FACTORS
DISCUSSED INCLUDE TEMPERATURE, LIGHT, SPECIFIC GRAVITY, MIXING,
NUTRITIONAL EFFECTS, PH EFFECTS, TOXIC EFFECTS, AND INTERRELATIONSHIP
OF BIOLOGICAL SPECIES. ALSO MENTIONED IS THE ALGAL-BACTERIAL
RELATIONSHIP. DESIGN FACTORS FOR SIZE AND VOLUME ARE GIVEN. THE ARTICLE
CONCLUDES WITH MANAGEMENT PROBLEMS ENCOUNTERED SUCH AS FLOATING DEBRIS,
OVERLOADING, INTERMITTENT LOADING, AQUATIC WEEDS AND SLUDGE BUILD-UP.
(PARKER-IOWA STATE)

FIELD 05D

ACCESSION NO. W71-08221

TREATMENT OF POTATO PROCESSING WASTES AT SALADA FOODS LTD., ALLISTON,

ONTARIO WATER RESOURCES COMMISSION, TORONTO.

T. D. ARMSTRONG, AND B. I. BOYKO.

PROCEEDINGS, ONTARIO INDUSTRIAL WASTE CONFERENCE, 16TH, JUNE 1969, NIAGARA
FALLS, ONTARIO, P 188-208.

DESCRIPTORS:
*POTATOES, *INDUSTRIAL WASTES, SCREENS, SEDIMENTATION, ORGANIC LOADING,
ACTIVATED SLUDGE, AERATION, DISSOLVED OXYGEN, SLUDGE, OXIDATION
LAGOONS, ALGAE, ADMINISTRATION, MAINTENANCE, *WASTE WATER TREATMENT.

IDENTIFIERS:
*POTATO PROCESSING WASTES, AEROBIC LAGOONS, ANAEROBIC LAGOONS.

ABSTRACT:
THE SALADA PLANT AT ALLISTON, ONTARIO, WAS CONSTRUCTED IN 1959. AT THIS
TIME, 5000 LBS/HR OF POTATOES WERE PROCESSED INTO POTATO FLAKES DURING
A 24 HOUR DAY. SINCE THAT TIME NEW PRODUCTS HAVE BEEN ADDED, AND WATER
CONSUMPTION, INITIALLY 200,000 GPD NOW RANGES FROM 630,000 TO 750,000
GPD DURING THE AUGUST TO MAY PROCESSING SEASON. THE COMPANY AND THE
TOWN WERE TO CONSTRUCT SECONDARY TREATMENT FACILITIES JOINTLY. THE
INITIAL INSTALLATION WAS DESIGNED FOR A WASTE FLOW OF 150,000 GPD AND
INCLUDED SCREENING, SEDIMENTATION, AN ANAEROBIC LAGOON, AND TWO AEROBIC
LAGOONS. THE LAGOON FILLED FOR 5 1/2 MONTHS, AND EFFLUENT ON THE FIRST
DAY OF OVERFLOW WAS 376,000 GALLONS WITH CORRESPONDING 6010 LBS BOD.
THE PLANT WAS OVERLOADED BOTH HYDRAULICALLY AND ORGANICALLY, THE
ANAEROBIC LAGOON WAS CHANGED TO AN AEROBIC LAGOON, TWO CONCRETE
CLARIFIERS WERE ADDED, EACH WITH 2.75 HOURS DETENTION TIME, AND BUBBLE
GUN AERATION DEVICES WERE INSTALLED. SINCE THE PLANT WAS STILL
OVER-LOADED, A COMPREHENSIVE SURVEY WAS ORDERED, AND FROM THE RESULTS
OF THIS STUDY, A 6.2 HOUR DETENTION TIME ACTIVATED SLUDGE BASIN WAS
BUILT. TOTAL COST OF THE TREATMENT PLANT HAD RISEN TO $350,000.
ALTHOUGH PROBLEMS WERE STILL ENCOUNTERED, THIS TREATMENT FACILITY
PROVIDED ADEQUATE TREATMENT. HAD THE NECESSARY LONG RANGE PLANNING BEEN
DONE EARLIER, A SERIOUS POLLUTIONAL PROBLEM MAY HAVE BEEN AVOIDED.
(LOWRY-TEXAS)

FIELD 05D

ACCESSION NO. W71-08970

STUDIES ON ZINC IN FRESHWATER ECOSYSTEMS,

OBERLIN COLL., OHIO.

EDWARD J. KORMONDY.

AVAILABLE FROM THE NATIONAL TECHNICAL INFORMATION SERVICE AS COO-1499-3,
$3.00 IN PAPER COPY, $0.95 IN MICROFICHE. REPORT COO-1499-3, JAN 1969. 32
P. CONTRACT AT(11-1)-1499.

DESCRIPTORS:
*ZINC RADIOISOTOPES, *AQUATIC INSECTS, *AQUATIC PLANTS, AQUATIC
ENVIRONMENT, ECOSYSTEMS, AQUATIC HABITATS, DRAGONFLIES, ALGAE,
SEDIMENTS, SOIL-WATER-PLANT RELATIONSHIPS, NUCLEAR WASTES, ABSORPTION,
CHELATION, METABOLISM, AQUARIA, WATER POLLUTION EFFECTS, ADSORPTION.

IDENTIFIERS:
*ZINC.

ABSTRACT:
ZINC-65 IS SIGNIFICANTLY CONCENTRATED IN THE AQUATIC ENVIRONMENT FROM
NUCLEAR WASTES. THE PRESENT EXPERIMENTS WITH MICROCOSMS (DRAGONFLY
LARVAE, ALGAE, WATER, AND SOIL) SHOW A DEPENDENCE OF UPTAKE ON
TEMPERATURE, VOLUME OF THE AQUARIUM AND LIGHT; WHICH SUGGESTS THAT
METABOLIC ACTIVITY PLAYS A ROLE. IN PARTICULAR, UPTAKE BY AQUATIC
PLANTS WAS AFFECTED BY PHOTOSYNTHESIS. ONLY ABOUT 1% AS MUCH UPTAKE BY
PLANTS OCCURRED IN THE PRESENCE OF THE CHELATING AGENT EDTA.
PRELIMINARY STUDIES SHOWED VARIATIONS IN CHELATION BY SOIL BETWEEN
VARIOUS SOIL TYPES. THE BINDING OF ZINC BY SOIL WAS REDUCED BY THE
PRESENCE OF PLANTS. (BOPP-NSIC)

FIELD 05C

ACCESSION NO. W71-09005

HEALTH PHYSICS DIVISION ANNUAL PROGRESS REPORT FOR PERIOD ENDING JULY 31, 1969:
SYSTEMS ECOLOGY, WATERSHED AQUATIC HABITAT INTERACTIONS.

OAK RIDGE NATIONAL LAB., TENN.

AVAILABLE FROM THE NATIONAL TECHNICAL INFORMATION SERVICE AS ORNL-4446, $3.00
IN PAPER COPY, $0.95 IN MICROFICHE. REPORT ORNL-4446, (P 137-162), JULY
1969.

DESCRIPTORS:
*ECOLOGY, *MATHEMATICAL MODELS, *WATERSHED MANAGEMENT, GEOCHEMISTRY,
LAKES, CONIFEROUS FORESTS, ALGAE, INSECTS, PHOSPHORUS RADIOISOTOPES,
TECHNOLOGY, NATURAL RESOURCES, ECOLOGICAL DISTRIBUTION, STOCHASTIC
PROCESSES, ECOSYSTEMS, AQUATIC HABITATS, WATER POLLUTION EFFECTS.

ABSTRACT:
THE PURPOSE OF THE ECOLOGICAL SYSTEMS ANALYSIS PROGRAM IS TO DEVELOP A
METHODOLOGY FOR MAKING USEFUL PREDICTIONS ABOUT THE OUTCOME OF
ENVIRONMENTAL INTERACTIONS. PROGRESS IS REPORTED ON THE FOLLOWING
PROJECTS: PERTIENCE OF RECENT ECOLOGICAL LITERATURE TO THE MODELING OF
ECOSYSTEMS, BIOGEOCHEMICAL ECOLOGY RESEARCH COLLECTION, PARAMETER
IDENTIFICATION IN SYSTEMS ECOLOGY, NUMERICAL METHODS IN SYSTEMS
ECOLOGY, A VERSATILE COMPARTMENT SYSTEM SIMULATION PROGRAM, A
STOCHASTIC MODEL OF FEEDING IN A FOREST CENTIPEDE, A MODEL FOR CEDAR
BOG LAKE (MINNESOTA), ANALYSIS OF THE CHANGE-IN-RATIO MODEL FOR
ESTIMATING POPULATION ABUNDANCE, AND SYSTEMS ANALYSIS OF BALSAM
FORESTS. THE WALKER BRANCH WATERSHED PROJECT WAS ORGANIZED TO ASSESS
THE POTENTIAL ECOLOGICAL IMPACT OF NEW TECHNOLOGY BEING APPLIED TO THE
FIELD OF NATURAL RESOURCE MANAGEMENT. A COMPARTMENT MODEL IS
ILLUSTRATED FOR EVALUATION OF TRANSFER RATES AND CHARACTERISTICS OF THE
TRANSFER FUNCTIONS AMONG THE PRINCIPAL COMPONENTS OF THE WATERSHED
ECOSYSTEM. A MATERIAL BALANCE STUDY USING P-32 AND PERIPHYTIC ALGAE IN
A CLOSED SYSTEM IN THE LABORATORY WAS CONDUCTED TO OBTAIN DATA ON
UPTAKE AND TURNOVER OF PHOSPHORUS BY STREAM DIATOMS.
(BOPP-NSIC)

FIELD 05C

ACCESSION NO. W71-09012

RADIOLOGICAL PHYSICS DIVISION. ANNUAL REPORT, JULY 1968-JUNE 1969.

ARGONNE NATIONAL LAB., ILL.

AVAILABLE FROM THE NATIONAL TECHNICAL INFORMATION SERVICE AS ANL-7615, $3.00
IN PAPER COPY, $0.95 IN MICROFICHE. REPORT ANL-7615, 1969. 233 P.

DESCRIPTORS:
*WATER POLLUTION EFFECTS, *FALLOUT, *ECOSYSTEMS, GREAT LAKES, TRACE
ELEMENTS, RADIUM RADIOISOTOPES, THORIUM RADIOISOTOPES, URANIUM
RADIOISOTOPES, MARINE ALGAE, WATER POLLUTION SOURCES.

IDENTIFIERS:
*CESIUM RADIOISOTOPES.

ABSTRACT:
INCLUDES 46 ARTICLES OF WHICH THREE CONCERN RADIOISOTOPES IN AQUATIC
ECOSYSTEMS: CONCENTRATIONS OF TRACE ELEMENTS IN GREAT LAKES FISH;
CONCENTRATION OF RADIUM, THORIUM, AND URANIUM BY TROPICAL ALGAE; AND
BEHAVIOR OF FALLOUT CESIUM-137 IN AQUATIC AND TERRESTRIAL SYSTEMS.
(BOPP-NSIC)

FIELD 05G, 05C

ACCESSION NO. W71-09013

THE USE OF AERO-HYDRAULIC GUNS IN THE BIOLOGICAL TREATMENT OF ORGANIC WASTES,

PROCTOR AND REDFERN, TORONTO (ONTARIO).

C. S. DUTTON, AND C. P. FISHER.

PROCEEDINGS, INDUSTRIAL WASTE CONFERENCE, 21, VOL L, NO 2, MARCH 1966. P
403-423, 7 FIG.

DESCRIPTORS:
*OXIDATION LAGOONS, *ORGANIC WASTES, *OXYGENATION, AEROBIC TREATMENT,
ANAEROBIC CONDITIONS, CANNERIES, ACTIVATED SLUDGE, BIOCHEMICAL OXYGEN
DEMAND, OXIDATION-REDUCTION POTENTIAL, AQUATIC ALGAE, *BIOLOGICAL
TREATMENT, *WASTE WATER TREATMENT.

IDENTIFIERS:
AERO-HYDRAULIC GUN, ALLISTON(ONTARIO).

ABSTRACT:
THE AERATED LAGOON TREATMENT OF ORGANIC WASTES IS AN ECONOMICAL AND
INCREASINGLY POPULAR PROCESS. HOWEVER, THE DEGREE OF TREATMENT HAS BEEN
LIMITED BY THE NATURE OF DEVICES AVAILABLE FOR ACCOMPLISHING
OXYGENATION. RECENT EXPERIMENTS UTILIZED AERO-HYDRAULIC GUNS IN SIMPLE
AERATED LAGOONS FOR TREATMENT OF ORGANIC WASTES. THE GUNS CONSISTED OF
A POLYETHYLENE STACK PIPE, BUBBLE GENERATOR, AND AN AIR SUPPLY THAT
PERIODICALLY RELEASED NUMEROUS SMALL BUBBLES FROM THE LAGOON BOTTOM.
THE RESULTING FLOW PATTERN OF LIQUID FROM THE LAGOON BOTTOM INVOLVED A
STRONG RADIAL CURRENT AT THE SURFACE. OXYGENATION CAPACITY OF THE GUNS
WERE DERIVED IN A SERIES OF CURVES THAT PLOTTED LBS O2/HR AGAINST FREE
AIR CONSUMPTION. CORRECTION FACTORS PERMITTED APPLICATION OF THE CURVES
TO ACTUAL DESIGNS. A GUN CAN ALSO FUNCTION AS A POSITIVE SLUDGE RETURN
PUMP BY SEALING UP ONE OF THE PORTS AND CONNECTING A SUCTION PIPE. AT A
POTATO WASTE TREATMENT FACILITY IN ALLISTON, ONTARIO, SOME 790,000 GPD
OF POTATO WASTE WAS DISCHARGED DAILY ALONG WITH A LOAD OF 5,000 LBS BOD
AND A LIKE AMOUNT OF SUSPENDED SOLIDS. RECOMMENDATIONS FOR THIS
FACILITY WERE: (1) TWO CONVENTIONAL SETTLING BASINS, EACH 24 FT SQUARE
EQUIPPED WITH A SLUDGE SCRAPER, (2) A SIMPLE AERATED LAGOON EQUIPPED
WITH 127 GUNS, AND (3) TWO ALGAE LAGOONS. DR. C. P. FISHER STUDIED THE
SYSTEM AND FOUND BOD REMOVAL ACHIEVED BY THE AERATED LAGOON WAS
APPROXIMATELY 90% WITH LITTLE OR NO MEASURABLE RESIDUAL DISSOLVED
OXYGEN. THERE WAS NO DOUBT THAT THE THOROUGH MIXING PRODUCED BY THE
GUNS WAS RESPONSIBLE TO A GREATER OR LESSER DEGREE FOR THE HIGH
REMOVALS. OTHER STUDIES WERE CONDUCTED ON POULTRY AND FRUIT CANNING
WASTES. (BURDETTE-TEXAS)

FIELD 05D

ACCESSION NO. W71-09546

FOOD CANNERY WASTE TREATMENT BY LAGOONS AND DITCHES AT SHEPPARTON, VICTORIA, AUSTRALIA,

MELBOURNE WATER SCIENCE INST. (AUSTRALIA).

C. D. PARKER.

PROCEEDINGS, INDUSTRIAL WASTE CONFERENCE, 21, VOL L, NO 2, MARCH 1966, P 284-302, 5 FIG, 13 TAB, 6 REF.

DESCRIPTORS:
*CANNERIES, *INDUSTRIAL WASTES, PEACHES, APRICOTS, PEARS, CITRUS FRUITS, TOMATOES, *AEROBIC TREATMENT, *ANAEROBIC DIGESTION, BIOCHEMICAL OXYGEN DEMAND, ACTIVATED SLUDGE, COST COMPARISONS, SEDIMENTS, ALGAE, HYDROGEN ION CONCENTRATION, ELECTRIC POWER, SEWAGE DISPOSAL, *WASTE WATER TREATMENT, *OXIDATION LAGOONS.

IDENTIFIERS:
SHEPPARTION(AUSTRALIA), OXIDATION DITCHES, SOUP, FOOD WASTES.

ABSTRACT:
CANNERY WASTE IS CHARACTERIZED BY HIGH BOD AND HIGH ACIDITY. THE USUAL METHOD OF TREATMENT IS SPRAY OR FLOOD IRRIGATION OR AEROBIC TYPE LAGOONS WITH HEAVY DOSAGE OF SODIUM NITRATE FOR ODOR CONTROL. IN SHEPPARTON, VICTORIA, EXPANSIONS IN THE FACILITIES OF SHEPPARTON PRESERVING COMPANY, THE LARGEST CANNERY IN THE SOUTHERN HEMISPHERE, AND THE BUILDING OF A SOUP CANNERY IN THE AREA CAUSED PROBLEMS IN THE TREATMENT OF THE RESULTING SEWAGE. THERE WAS A PRESSING NEED FOR AN EFFICIENT MEANS OF PURIFYING, WITHOUT CAUSING NUISANCE, THE LARGE SEASONAL FLOWS OF HIGHLY POLLUTED FOOD WASTES FROM TWO CANNERIES ON AREAS CLOSELY ADJACENT TO A PROSPEROUS AND RAPIDLY DEVELOPING URBAN CENTER. THE SHORT SEASONAL NATURE OF CANNERY OPERATIONS AND CONSEQUENT HIGH CAPITAL COST MADE THE PROBLEM EVEN MORE DIFFICULT. IT WAS DECIDED TO EXPERIMENT WITH ANAEROBIC AND AEROBIC TYPE LAGOONS AND OXIDATION DITCHES. THE WASTE FROM EACH FIRM IS MIXED WITH SEWAGE PLANT EFFLUENT AND TREATED IN THE NEW PROCESSING UNIT WITH VERY GOOD RESULTS WHICH ARE DESCRIBED IN GREAT DETAIL. THE TOTAL COST OF THE NEW UNIT, EXCLUDING THE COST OF INFLOW AND OUTFLOW PIPELINES AND LABOR AND MAINTAINENCE, WAS $155,000 TO THE PRESERVING CO., AND $143,000 (AMERICAN DOLLARS) TO THE SOUP CANNERY PLUS ANNUAL COSTS OF $5,364 AND $8,192 RESPECTIVELY, EXCLUDING PIPELINES, PUMPING STATIONS, AND MAINTAINENCE LABOR. THE NET RESULT WAS ALMOST COMPLETE PURIFICATION AT COSTS THE INDUSTRIES COULD ACCEPT. (LOWRY-TEXAS)

FIELD 05D

ACCESSION NO. W71-09548

REPORT TO THE CONTRA COSTA COUNTY WATER AGENCY ON A STUDY OF THE EFFECTS OF THE PROPOSED FEDERAL SAN LUIS INTERCEPTOR DRAIN AND THE STATE SAN JOAQUIN VALLEY MASTER DRAIN.

METCALF AND EDDY, INC., PALO ALTO, CALIF.

OCT 30, 1964. 93 P, 4 FIG, 32 TAB, 37 REF.

DESCRIPTORS:
*AGRICULTURAL CHEMICALS, *DELTAS, *DRAINAGE EFFECTS, *WATER QUALITY, *FARM WASTES, *INDUSTRIAL WASTES, MUNICIPAL WASTES, NUTRIENTS, PESTICIDES, RECREATION, ALGAE, OXYGEN DEMAND, WATER POLLUTION, MUNICIPAL WATER, WATER REQUIREMENTS, WATER CONVEYANCE, WATER COSTS, ARTIFICIAL WATER COURSES, CALIFORNIA.

IDENTIFIERS:
*CONTRA COSTA COUNTY, WATER DETERIORATION, DRAIN, SAN JOAQUIN VALLEY MASTER DRAIN, SAN FRANCISCO BAY-DELTA.

ABSTRACT:
THE PROBABLE EFFECTS ON THE OFFSHORE WATERS OF CONTRA COSTA COUNTY, CALIFORNIA BY THE PROPOSED DRAIN WHICH WILL CONVEY AGRICULTURAL SUBSURFACE WASTE WATERS FROM THE SAN JOAQUIN VALLEY AND WASTES FROM THE SAN LUIS DIVISION OF THE CENTRAL VALLEY PROJECT ARE DISCUSSED. THE POINT OF DISCHARGE WOULD BE THE SAN JOAQUIN RIVER IN THE VICINITY OF THE ANTIOCH BRIDGE. DISCHARGE IS EXPECTED TO BEGIN IN 1969 AT A RATE OF 75 CFS AND INCREASE TO 900 CFS BY 1995. SERIOUS DETERIORATION OF THE QUALITY OF OFFSHORE WATERS CAN BE EXPECTED BY 1975. ACQUATIC LIFE AND RECREATIONAL USE OF THE OFFSHORE WATERS COULD BE SERIOUSLY AFFECTED BY THE DISCHARGE OF NUTRIENTS AND PESTICIDES FROM THE DRAIN. NUTRIENTS ARE EXPECTED TO HAVE A GREATER EFFECT THAN PESTICIDES. THE NUTRIENTS IN THE DRAIN WILL PRODUCE GROWTHS OF ALGAE IN THE DRAIN WHICH BY THEIR DEATH AND DECAY WILL EXERT AN OXYGEN DEMAND IN THE RIVER AND BAY VARYING FROM 7500 LBS PER DAY IN 1975 TO 27,000 LBS PER DAY IN 1995. INDUSTRY AND THE CITY OF ANTIOCH WILL BE ABLE TO USE RIVER WATER A LESSER PROPORTION OF THE TIME TO SATISFY HIGH QUALITY REQUIREMENTS. THE DISCHARGE OF THE DRAIN WILL HAVE NO SIGNIFICANT EFFECT ON WATER FOR AGRICULTURAL USE. (POERTNER)

FIELD 06G, 05B

ACCESSION NO. W71-09577

SPECIFIC COMPOSITION OF PHYTOPLANKTON IN A LAKE WARMED UP BY WASTEWATER FROM A
THERMAL POWER STATION AND IN LAKES WITH NORMAL TEMPERATURES,

INSTYTUT RYBACTWA SRODLADOWEGO, OLSZTYN-KORTOWO (POLAND).

J. POLTORACKA.

TRANSLATION FROM POLISH AEC-TR-7194. ACTA SOCIETATIS BOTANICORUM POLONIAE,
POLISH, VOL. 37, NO. 2, 1968, P. 297-325, 5 FIG., 1 TAB., 5 REF.

DESCRIPTORS:
*THERMAL POWERPLANTS, *LAKES, TEMPERATURE, PHYTOPLANKTON, ALGAE, HEATED
WATER, POPULATION, WATER POLLUTION EFFECTS.

IDENTIFIERS:
SPECIES, TAXONS.

ABSTRACT:
A PRELIMINARY EVALUATION OF THE SPECIFIC COMPOSITION OF PHYTOPLANKTON
WAS CARRIED OUT IN LAKES LICHEN, MIKORZYN AND SLESIN, CHARACTERIZED BY
DIFFERENT TEMPERATURES CAUSED BY THE ACTIVITY OF THE POWERPLANT AT
KONIN. THE HIGHEST WATER TEMPERATURES WERE DISPLAYED BY LAKE LICHEN
(7.4-27.5C), AND THE LOWEST TEMPERATURES PREVAILED IN LAKE SLESIN
(0.8-20.7C). DURING THE ENTIRE PERIOD OF INVESTIGATION IN ALL THREE
LAKES, 305 SPECIES WERE FOUND AND 414 TAXONS WERE IDENTIFIED.
PREDOMINANT AMONG THEM WERE BACILLARIOPHYCEAE (44 PERCENT),
CHLOROPHYCEAE (34 PERCENT) AND CYNONPHYCEAE (11 PERCENT). THE HIGHEST
NUMBER OF PHYTOPLANKTON COMPONENTS (285) WERE FOUND IN LAKE LICHEN, AND
THEIR LOWEST NUMBER (197) WAS ASCERTAINED IN LAKE SLESIN, WHICH WAS
ASSOCIATED WITH GREAT DIFFERENCES IN THE SHARE OF CHLOROPHYCEAE WHOSE
HIGHEST NUMBER WAS FOUND IN THE WARMEST, LAKE LICHEN, WHEREAS THEIR
LOWEST NUMBER WAS DISCOVERED IN LAKE SLESIN, THE COLDEST LAKE. THE
NUMBER OF SPECIES OF ALGAE IN LAKE LICHEN WAS RATHER STABLE·THROUGHOUT
THE YEAR BUT FLUCTUATED SEASONALLY IN OTHER LAKES. A MOST MARKED
CORRELATION BETWEEN THE NUMBER OF THE TAXONS OF ALGAE AND SURFACE WATER
TEMPERATURE WAS OBSERVED IN LAKE LICHEN. WITH RESPECT TO THE SIZE OF
INDIVIDUAL POPULATION, 14 PLANKTONIC SPECIES WERE DOMINANT AT VARIOUS
PERIODS IN DIFFERENT LAKES. (UPADHYAYA-VANDERBILT)

FIELD 05C, 02H

ACCESSION NO. W71-09767

UPTAKE OF RADIOACTIVE NUCLIDES BY AQUATIC ORGANISMS: THE APPLICATION OF THE
EXPONENTIAL MODEL,

TOKYO UNIV. (JAPAN). DEPT. OF FISHERIES.

Y. HIYAMA, AND M. SHIMIZU.

ENVIRONMENTAL CONTAMINATION BY RADIOACTIVE MATERIALS, VIENNA, INTERNATIONAL
ATOMIC ENERGY (1969), P 463-476.

DESCRIPTORS:
*MARINE ANIMALS, *RADIOISOTOPES, ABSORPTION, MARINE ALGAE, MARINE FISH,
INVERTEBRATES, CRUSTACEANS, MOLLUSKS, STRONTIUM RADIOISOTOPES, ZINC
RADIOISOTOPES, IODINE RADIOISOTOPES, COBALT RADIOISOTOPES, WATER
POLLUTION EFFECTS.

IDENTIFIERS:
CESIUM RADIOISOTOPES, CADMIUM RADIOISOTOPES, CERIUM RADIOISOTOPES,
ECHINODERMS.

ABSTRACT:
A COMPARTMENTAL MODEL WAS USED TO CORRELATE EXPERIMENTAL DATA ON
RADIONUCLIDES (CS, SR, ZN, CD, CE, I, AND CO) UPTAKE BY MARINE ALGAE,
FISH AND INVERTEBRATES. FOR SOME SYSTEMS THE CONCENTRATION FACTORS
VARIED WITH TIME AS MUCH AS 100-FOLD. (BOPP-NSIC)

FIELD 05C

ACCESSION NO. W71-09850

INFLUENCE OF THE PHYSIOCOCHEMICAL FORM OF RUTHENIUM ON CONTAMINATION OF MARINE
ORGANISMS, (IN FRENCH),

DEPARTEMENT DE LA PROTECTION SANITAIRE, CHERBOURG (FRANCE). SECTION DE
RADIO-ECOLOGIE.

PIERRE GUEGUENIAT, PIERRE BOVARD, AND JACQUES ANCELLIN.

COMPTES RENDUS, SERIES D, VOL 268, FEB 10, 1969, P 976-979.

DESCRIPTORS:
*MARINE ANIMALS, *MARINE PLANTS, RADIOISOTOPES, PHYSICOCHEMICAL
PROPERTIES, ANION ADSORPTION, COLLOIDS, AQUARIA, ABSORPTION, WATER
POLLUTION EFFECTS, CATION EXCHANGE, CATION ADSORPTION, HYDROLYSIS,
MARINE ALGAE, INVERTEBRATES, NUCLEAR WASTES, WATER POLLUTION SOURCES.

IDENTIFIERS:
*RUTHENIUM RADIOISOTOPES.

ABSTRACT:
THE DERIVATIVES OF RUTHENIUM NITROSYLS FROM EFFLUENTS OF FUEL
REPROCESSING PLANTS WERE FOUND IN VARIOUS PHYSICOCHEMICAL FORMS IN SEA
WATER - AS PRECIPITATES, IN SOLUTION, OR IN COLLOIDAL FORM. THE FORM IN
WHICH THE RUTHENIUM WAS PRESENT DEPENDED UPON THE AMOUNTS OF OTHER
SUSPENDED MATERIAL IN THE SEA WATER. EXPERIMENTS IN AQUARIA SHOWED THAT
RUTHENIUM PRESENT IN PRECIPITATES CAUSED MORE CONTAMINATION OF MARINE
ALGAE AND INVERTEBRATES AS COMPARED WITH SOLUBLE (CATIONIC) RUTHENIUM.
(BOPP-NSIC)

FIELD 05B, 05C

ACCESSION NO. W71-09863

WATER QUALITY INFLUENCES ON OUTDOOR RECREATION IN THE LAKE ONTARIO BASIN,

BUREAU OF OUTDOOR RECREATION, ANN ARBOR, MICH.

ROBERT J. HENLEY.

IN: PROCEEDINGS, TENTH CONFERENCE ON GREAT LAKES RESEARCH, ANN ARBOR,
BRAUN-BRUMFIELD, INC, 1967, P 427-440.

DESCRIPTORS:
*WATER QUALITY, *RECREATION, *SWIMMING, *WATER POLLUTION EFFECTS,
BACTERIA, ALGAE, FISH, SEWERS, RUNOFF, WASTES, WATER SUPPLY, EFFLUENTS,
STREAMS, LAKES, RIVERS, RESERVOIRS, BOATING, EUTHROPHICATION.

IDENTIFIERS:
*LAKE ONTARIO, WATER TURBIDITY, INDUSTRIAL PLANTS, WATER SPORTS.

ABSTRACT:
THE RELATIONSHIPS BETWEEN WATER QUALITY AND THE USES OF RECREATIONAL
WATER RESOURCES IN THE LAKE ONTARIO BASIN ARE SURVEYED. THERE IS A
DISCUSSION OF WATER QUALITY INDICES (HIGH BACTERIA COUNTS, ALGAE
MASSES, FISH MORTALITY AND WATER TURPIDITY). RECREATION VALUE OF WATER
IS ANALYZED IN LIGHT OF THE INFLUENCE OF POOR WATER QUALITY ON
SWIMMING, WITH PARTICULAR ATTENTION TO SITUATIONS IN WHICH SWIMMING HAS
BEEN BANNED BY PUBLIC OFFICIALS. THE VALUE OF POLLUTION CONTROL TO SUCH
WATER RECREATION IS STRESSED. PREDICTIONS ARE MADE FOR THE FUTURE
DEMAND OF THE PUBLIC FOR SWIMMING AND OTHER WATER-RELATED RECREATION
ACTIVITIES AND INDICATIONS POINT TO AN INCREASE IN THIS DEMAND.
(MURPHY-RUTGERS)

FIELD 05C

ACCESSION NO. W71-09880

DISSOLVED ORGANIC MATTER AND PHYTOPLANKTONIC PRODUCTIVITY IN MARL LAKES,

MICHIGAN STATE UNIV., HICKORY CORNERS. W. K. KELLOGG BIOLOGICAL STATION.

ROBERT G. WETZEL.

MITTEILUNGEN INTERNATIONALE VEREINIGUNG FUR THEORETISCHE UND ANGEWANDTE
LIMNOLOGIE, VOL 14, P 261-270, 1968. 1 FIG, 1 TAB, 40 REF. NSF GRANT
GB-1452.

DESCRIPTORS:
*PHYTOPLANKTON, *ALGAE, *PRIMARY PRODUCTIVITY, *ORGANIC MATTER,
NUTRIENTS, LAKES, WATER PROPERTIES, BACTERIA, BIOASSAY, METABOLISM,
ABSORPTION, COMPETITION, OLIGOTROPHY, VITAMINS, HEAVY METALS, TOXICITY,
DIATOMS, MARL.

IDENTIFIERS:
*MARL LAKES, DISSOLVED ORGANIC MATTER, ARCTIC ALGAE, AUTOTROPHIC
METABOLISM, HETEROTROPHY, CROOKED LAKE(IND), LITTLE CROOKED LAKE(IND).

ABSTRACT:
THE HIGH EFFICIENCY OF BACTERIA IN UTILIZATION OF DISSOLVED ORGANIC
MATTER DEPRIVES ALGAE OF ENERGY MATERIAL AND DEPRESSES THEIR GROWTH. AN
IMPORTANT PART IN THE COMPETITION OF THE TWO GROUPS OF ORGANISMS IS
PLAYED BY CHELATION INFLUENCING THE AVAILABILITY OF NUTRIENTS, THEIR
RATIO, AND POTENCY OF TOXICANTS. THESE RELATIONSHIPS WERE ELUCIDATED BY
BIOASSAYS CONDUCTED SIMULTANEOUSLY IN TWO CONNECTED HARD WATER LAKES
CONTAINING DIFFERENT CONCENTRATION OF DISSOLVED ORGANIC MATTER.
(WILDE-WISCONSIN)

FIELD 05C, 02H

ACCESSION NO. W71-10079

EFFECTS OF ZOOPLANKTON ON PHOTOSYNTHESIS BY ALGAE IN LAKES,

RICHMOND UNIV., VA. DEPT. OF BIOLOGY.

JOHN W. BISHOP.

VIRGINIA POLYTECHNIC INSTITUTE, BLACKSBURG, VIRGINIA, WATER RESOURCES
RESEARCH CENTER, BULLETIN NO 31, P 89-93, 1969. OWRR PROJECT A-002-VA(1).

DESCRIPTORS:
*ZOOPLANKTON, *PHOTOSYNTHESIS, *ALGAE, LAKES, OXYGEN, EUTROPHICATION,
ALGAL CONTROL, BIOCONTROL, TURBULENCE, COPEPODS, ROTIFERS, CHLOROPHYLL,
PIGMENTS, IMPOUNDMENTS, RESERVOIRS, ANALYTICAL TECHNIQUES, ON-SITE
TESTS.

IDENTIFIERS:
ASTERIONELLA SPIROIDES, CLADOCERANS, WESTHAMPTON LAKE(VA).

ABSTRACT:
POSSIBILITY OF BIOLOGICAL CONTROL OF ALGAE IN NUTRIENT ENRICHED WATERS
PROMPTED STUDY OF ZOOPLANKTON AS POTENTIAL REGULATORS OF ALGAE.
EXPERIMENTS WERE CONDUCTED IN AQUARIA WHERE ASSEMBLAGES OF ALGAE WERE
CONFINED (ONE IN PAIRED AQUARIA IN THE LAKE WITH AND ONE WITHOUT
MIGRATORY ZOOPLANKTON). EFFECTS OF THE AQUARIA ON THREE PHYSICAL
FACTORS OF LAKE WATER, LIGHT INTENSITY, TEMPERATURE, AND TURBULENCE
WERE COMPARED. DIFFERENCES BETWEEN QUANTITIES AND RATES OF ALGAL
PHOTOSYNTHESIS IN THE AQUARIA WERE USED AS ESTIMATES OF ZOOPLANKTON
EFFECT ON ALGAE. THE CHLOROPHYLL METHOD FOR MEASUREMENT OF ALGAL
CONCENTRATION CORRECTED FOR PHEO-PIGMENTS. RATE OF RADIOACTIVE CARBON
UPTAKE AS MEASURE OF PHOTOSYNTHESIS WAS STUDIED, USING A LIQUID
SCINTILLATION COUNTER AND BY THE OXYGEN EVOLUTION METHOD WHICH WAS
SUFFICIENTLY ACCURATE AND EASIER FOR ROUTINE EXPERIMENTS. PHEO-PIGMENTS
CONSISTED OF NEARLY 52% OF THE ALGAL PIGMENTS ANALYZED AND INCREASED
FROM THE SURFACE TOWARD LAKE BOTTOM; ROTIFERS WERE ABOUT 60% OF
ZOOPLANKTON DURING MID=SUMMER; CLADOCERANS WERE NOT MORE THAN &%.
EXPERIMENTS TO ASCERTAIN EFFECTS OF AQUARIA AND ZOOPLANKTON ON
PHOTOSYNTHESIS ARE TO BE CONTINUED.
(JONES-WISCONSIN)

FIELD 05C

ACCESSION NO. W71-10098

EFFECT OF AGRICULTURE ON WATER QUALITY,

FEDERAL WATER POLLUTION CONTROL ADMINISTRATION, EVANSVILLE, IND. LOWER OHIO
  BASIN OFFICE.

T. R. SMITH.

IN: 2ND COMPENDIUM OF ANIMAL WASTE MANAGEMENT, JUNE 1969, PAPER NO. 11, 11 P,
  9 REF.

DESCRIPTORS:
  WATER QUALITY CONTROL, *NUTRIENTS, *PESTICIDES, FARM WASTES, SILT,
  EROSION, SEDIMENTATION, RUNOFF, FERTILIZERS, NITROGEN, PHOSPHORUS,
  ALGAE, LIVESTOCK, WATER POLLUTION EFFECTS.

IDENTIFIERS:
  FEEDLOTS, WABASH RIVER BASIN.

ABSTRACT:
  THE MAIN SOURCES OF AGRICULTURAL ASSOCIATED WATER POLLUTION IN HUMID
  REGIONS ARE: (1) SILT FROM SOIL EROSION; (2) FERTILIZERS, MAINLY
  PHOSPHORUS AND NITROGEN COMPOUNDS (3) PESTICIDES; AND (4) ORGANIC
  WASTES FROM FEEDLOTS. IN ARID REGIONS, IRRIGATION RETURN FLOWS ARE A
  PROBLEM. THIS PAPER DISCUSSES THE EFFECT OF EACH OF THESE SOURCES CAN
  HAVE ON WATER QUALITY AND SUGGESTS SOME PREVENTATIVE MEASURES. IT IS
  NECESSARY THAT AGRICULTURALISTS PLAN TO CONTROL POLLUTIONAL EFFECTS ON
  THEIR ACTIVITY. (CHRISTENBURY-ISU)

FIELD 05G, 05C

ACCESSION NO. W71-10376

DEMONSTRATE FEASIBILITY OF THE USE OF ULTRASONIC FILTRATION IN TREATING THE
OVERFLOWS FROM COMBINED AND/OR STORM SEWERS.

ACOUSTICA ASSOCIATES, INC., LOS ANGELES, CALIF.

AVAILABLE FROM THE NATIONAL TECHNICAL INFORMATION SERVICE AS PB-201 745,
  $3.00 IN PAPER COPY, $0.95 IN MICROFICHE. FINAL REPORT SEPTEMBER 22, 1969,
  86 P, 10 FIG, 1 TAB, 3 APPEND. FWPCA PROGRAM 11020--09/67, CONTRACT NO.
  14-12-23.

DESCRIPTORS:
  *ULTRASONICS, *SUSPENDED LOAD, *FILTRATION, ALGAE, COLLOIDS,
  FLOCCULATION, MIXING, ECONOMIC FEASIBILITY, TECHNICAL FEASIBILITY, COST
  ANALYSIS, SLUDGE, INCINERATION, STORM RUN-OFF, SEDIMENTATION,
  BIOCHEMICAL OXYGEN DEMAND, WASTE WATER TREATMENT, OVERFLOW, SEWERS,
  STORM DRAINS, SEWAGE EFFLUENTS, *SEWAGE TREATMENT, STORM RUNOFF.

IDENTIFIERS:
  BACKWASHING, *SUSPENDED SOLIDS, *COMBINED SEWERS, STORMWATER OVERFLOWS.

ABSTRACT:
  THE FEASIBILITY, BOTH ECONOMIC AND TECHNICAL, OF USING ULTRASONIC
  FILTRATION AS AN AID TO REMOVING SUSPENDED IMPURITIES WAS INVESTIGATED.
  INPUTS TO THE TEST APPARATUS INCLUDED DRINKING WATER, SIMULATED SEWAGE,
  PRIMARY AND SECONDARY SEWAGE EFFLUENTS, ALGAE POND EFFLUENT, COARSELY
  SCREENED, RAW SEWAGE DILUTED 2:1, 5:1, AND 10:1 WITH SIMULATED STORM
  DRAINAGE, AND DELIBERATELY CONCENTRATED RAW SEWAGE. DIRT, SAND, AND
  LEAVES WERE ADDED IN SOME TEST RUNS WITH NO APPARENT EFFECT ON
  PERFORMANCE. 20 AND 50 MICRON FILTER ELEMENTS, OPERATED AT
  APPROXIMATELY 10 GPM/FT2 WITH HEAD LOSSES FROM 1 TO 4 PSI, REDUCED THE
  BOD AND SUSPENDED SOLIDS OF RAW SEWAGE DILUTED IN VARYING DEGREES WITH
  WATER BY AN AVERAGE OF 40 AND 70% RESPECTIVELY. SYSTEM PERFORMANCE WAS
  INTERMEDIATE BETWEEN THAT OF PRIMARY AND SECONDARY TREATMENT. OTHER
  BENEFITS INCLUDED: (1) APPLICATION OF ULTRASONIC ENERGY INCREASED THE
  TOTAL QUANTITY OF EFFLUENT FILTERED BETWEEN BACKWASH CYCLES FROM 4 TO
  18 TIMES THAT TREATED WITH SIMILAR EQUIPMENT WITHOUT ULTRASONICS
  APPLIED; AND (2) APPLICATION OF ULTRASONIC ENERGY RESTORED THE FILTER
  ELEMENTS TO 'LIKE-NEW' CONDITION WITHOUT NECESSITATING FREQUENT FILTER
  REPLACEMENT. COST FIGURES FOR AN ULTRASONIC FILTRATION SYSTEM CAPABLE
  OF HANDLING 1.4 MGD, EXHIBITING PERFORMANCE MIDWAY BETWEEN PRIMARY AND
  SECONDARY TREATMENT, AND OCCUPYING ONLY SEVERAL HUNDRED SQUARE FEET,
  RANGED BETWEEN 2.8 AND 3.5 CENTS/1000 GALLONS. BY COMPARISON, COMBINED
  PRIMARY-SECONDARY TREATMENT COST FIGURES WERE REPORTED AS 10 CENTS/1000
  GALLONS EXCLUDING THE COST OF THE CONSIDERABLE AMOUNT OF LAND INVOLVED.
  (LOWRY-TEXAS)

FIELD 05D

ACCESSION NO. W71-10654

BIOLOGICAL PROBLEMS IN WATER POLLUTION,

ROBERT A TAFT SANITARY ENGINEERING CENTER, CINCINNATI, OHIO.

CLARENCE M. TARZWELL.

AVAILABLE FROM THE NATIONAL TECHNICAL INFORMATION SERVICE AS PB-196 627,
$3.00 IN PAPER COPY, $0.95 IN MICROFICHE. TRANSACTIONS, SEMINAR ON
BIOLOGICAL PROBLEMS IN WATER POLLUTION, TAFT ENGINEERING CENTER, APR 23-25,
1956. 272 P.

DESCRIPTORS:
  *WATER POLLUTION, *AQUATIC BIOLOGY, *BIOINDICATORS, *BIOASSAY,
  TOXICITY, AQUATIC MAMMALS, ALGAE, FISH, DIATOMS, FUNGI, INSECTS,
  BACTERIA, PROTOZOA, SEWAGE, LAKES, NUTRIENTS.

IDENTIFIERS:
  *INDICATOR SPECIES, MICROORGANISMS, FRESH WATER BIOLOGY, SUSPENDED
  SEDIMENTS, MARINE BIOLOGY, STREAM POLLUTION, *WATER POLLUTION
  DETECTION.

ABSTRACT:
  BIOLOGISTS ENGAGED IN POLLUTION INVESTIGATIONS AND RESEARCH OFTEN WORK
  ALONE AND ARE SOMEWHAT ISOLATED. FOR SOME TIME, THEREFORE, THERE HAD
  BEEN RECOGNIZED A NEED FOR A CONFERENCE OF THOSE ENGAGED IN THE STUDY
  OF BIOLOGICAL PROBLEMS IN WATER POLLUTION CONTROL, TO ACQUAINT THEM
  WITH CURRENT DEVELOPMENTS AND NEW METHODS OF APPROACH, AND TO ENABLE
  THEM TO BECOME ACQUAINTED WITH OTHER WORKERS IN THE FIELD. THE FIRST
  SUCH GATHERING WAS HELD AS A SEMINAR AT THE ROBERT A. TAFT SANITARY
  ENGINEERING CENTER, APRIL 23 - 27, 1956. THE SEMINAR CONSISTED OF PANEL
  DISCUSSIONS AND WAS PLANNED SO THAT MOST OF THE TIME WAS DEVOTED TO
  COMMENTARY FROM THE FLOOR WITH ONLY SHORT PRESENTATIONS BY PANEL
  MEMBERS. SUBJECTS DISCUSSED WERE (1) USE AND VALUE OF BIOASSAYS; (2)
  USE AND VALUE OF BIOLOGICAL INDICATORS OF POLLUTION; (3) CURRENT
  INVESTIGATIONS IN WATER POLLUTION BIOLOGY; (4) WATER QUALITY CRITERIA
  FOR AQUATIC LIFE; AND (5) TRAINING OF SANITARY AQUATIC BIOLOGISTS.

FIELD 05C

ACCESSION NO. W71-10786

THE DEPENDENCE OF PRIMARY PRODUCTION ON THE COMPOSITION OF PHYTOPLANKTON
(ZAVISIMOST PERVICHNOI PRODUKTSH OT SOSTAVA FITOPLANKTONA),

NAVAL OCEANOGRAPHIC OFFICE WASHINGTON, D.C.

I. L. PYRINA.

AVAILABLE FROM THE NATIONAL TECHNICAL INFORMATION SERVICE AS AD-718 176,
$3.00 IN PAPER COPY, $0.95 IN MICROFICHE. NAVAL OCEANOGRAPHIC TRANSLATION
201, 1963. TRANSLATION: PERVICHNAYA PRODUKTSIYA MORAY I VNUTVENNIKH, P
308-313, 1961.

DESCRIPTORS:
  *PLANKTON, *PHOTOSYNTHESIS, PHYTOPLANKTON, ALGAE, TEMPERATURE, SOLAR
  RADIATION.

IDENTIFIERS:
  MARINE BIOLOGY, *PRIMARY BIOLOGICAL PRODUCTIVITY.

ABSTRACT:
  THE DEPENDENCE OF PRIMARY PRODUCTION ON THE COMPOSITION OF
  PHYTOPLANKTON IS BASED ON SPECIAL OBSERVATIONS CARRIED OUT IN THE
  SUMMER OF 1958 IN THE IVAN'KOVSKOYE, RYBINSKOYE AND KUYBYSHEVSKOYA
  WATER BASINS, AS WELL AS IN CERTAIN SECTORS OF THE VOLGA RIVER. IN
  ADDITION TO THE PRIMARY PRODUCTION, THE SPECIES' COMPOSITION AND THE
  BIOMASS OF PHYTOPLANKTON WAS DETERMINED, THE ENERGY OF SOLAR LIGHT
  STRIKING THE WATER SURFACE, THE TEMPERATURE AND TRANSPARENCY OF WATER.
  THE INSTRUMENTS AND METHODS USED ARE BRIEFLY DESCRIBED, ILLUSTRATING
  GRAPHICALLY THE VARIATIONS OF PHOTOSYNTHESIS IN VARIOUS ALGAL GROUPS
  AND THE UTILIZATION OF LIGHT ENERGY BY VARIOUS PHYTOPLANKTON SPECIES IN
  THE ABOVE-MENTIONED WATER BASINS.

FIELD 05C, 02I

ACCESSION NO. W71-10788

PHOTOSYNTHETIC RECLAMATION OF AGRICULTURAL SOLID AND LIQUID WASTES--SECOND PROGRESS REPORT,

CALIFORNIA UNIV., BERKELEY. SANITARY ENGINEERING RESEARCH LAB.

GORDON L. DUGAN, CLARENCE G. GOLUEKE, WILLIAM J. OSWALD, AND CHARLES E. RIXFORD.

CALIFORNIA UNIVERSITY, SANITARY ENGINEERING RESEARCH LABORATORY, REPORT NO. 70-1, 165 P, 24 FIG, 55 TAB, 51 REF. US PUBLIC HEALTH SERVICE 5RO1 UI 00566-03.

DESCRIPTORS:
*FARM WASTES, *POULTRY, WASTE WATER TREATMENT, SEDIMENTATION, OXIDATION LAGOONS, ANAEROBIC DIGESTION, ALGAE, SLUDGE, METHANE, ORGANIC LOADING, HYDROGEN ION CONCENTRATION, TEMPERATURE, CENTRIFUGATION, COAGULATION, DEWATERING, *COST ANALYSIS.

IDENTIFIERS:
VOLATILE SOLIDS, GRIT.

ABSTRACT:

A 36-WEEK STUDY WAS INITIATED TO PROVIDE INFORMATION ON THE ECONOMICS OF TREATING ANIMAL WASTES. CHICKENS WERE CHOSEN, MAINLY FOR CONVENIENCE SAKE, AS THE WASTE PRODUCERS TO BE STUDIED, AND 113-TWENTY WEEK OLD LEGHORN PULLETS WERE PLACED IN CAGES. THE TREATMENT SYSTEM USED CONSISTED OF INCLINED TROUGHS UNDER THE CAGES WHICH WERE FLUSHED WITH WATER FROM A FLUSHING BUCKET, GRIT REMOVAL, SEDIMENTATION, AN OXIDATION LAGOON FOR THE SEDIMENTATION TANK SUPERNATANT, AND AN ANAEROBIC DIGESTER FOR THE SOLIDS. THE SYSTEMS APPROACH WAS USED, AND SYSTEM BALANCES WERE PERFORMED FOR TOTAL SOLIDS, VOLATILE SOLIDS, TOTAL UNOXIDIZED NITROGEN AND ENERGY FOR THE CHICKENS, SEDIMENTATION TANK, DIGESTER, AND ALGAE. ALL BALANCES WERE PERFORMED FROM WEEK 5 THROUGH WEEK 36 EXCEPT FOR THE DIGESTER, WHERE OPERATION WAS TERMINATED AT WEEK 24 AND THE SOLIDS DEWATERED, DRIED, AND STUDIED FOR POSSIBLE FURTHER REUSE. AN ECONOMIC ANALYSIS OF AN INTEGRATED SYSTEM OF 100,000 LAYING HENS REVEALED A COST OF APPROXIMATELY 2 CENTS/DOZEN EGGS FOR A SYSTEM BASED ON THE ONE TESTED. HOWEVER, CONSIDERATION OF AN EXTREMELY CONSERVATIVE ALGAE HARVESTING RATE OF 12 TONS/ACRE/YEAR AT A PRICE OF 5 CENTS/LB (DRY WEIGHT) DROPPED THE OVERALL WASTE HANDLING OUTLAY TO APPROXIMATELY 1 CENT/DOZEN. ON THIS BASIS, ADDITIONAL STUDIES WERE RECOMMENDED TO AID IN THE IMPLEMENTATION OF TREATMENT FACILITIES FOR ANIMAL WASTES AS SOON AS POSSIBLE. (LOWRY-TEXAS)

FIELD 05D

ACCESSION NO. W71-11375

SULFIDE PRODUCTION IN WASTE STABILIZATION PONDS,

TEXAS UNIV., AUSTIN. CENTER FOR RESEARCH IN WATER RESOURCES.

ERNESTO ESPINO DE LA O, AND E. F. GLOYNA.

TECHNICAL REPORT NO. EHE-04-6802, CRWR-26, MAY 1967, 156 P, 50 FIG, 33 TAB, 75 REF. FWPCA CONTRACT WP-00688-03.

DESCRIPTORS:
*OXIDATION PONDS, *HYDROGEN SULFIDE, *SULFIDES, ALGAE, ORGANIC LOADING, BIOCHEMICAL OXYGEN DEMAND, OXIDATION-REDUCTION POTENTIAL, DISSOLVED OXYGEN, ODORS, CORROSION, OXIDATION, PHOTOSYNTHESIS, ANALYTICAL TECHNIQUES, SPECTROSCOPY, COLORIMETRY, SAMPLING, METHANE, CARBON DIOXIDE, DIFFUSION, *WASTE WATER TREATMENT.

ABSTRACT:
EXPERIMENTS WERE CONDUCTED ON A 6 FT DEEP MODEL FACULTATIVE POND TO DETERMINE THE RELATIONSHIP BETWEEN SULFIDE CONCENTRATION IN THE POND AND FOUR OPERATING PARAMETERS: (1) DETENTION TIME, (2) CONCENTRATION OF SULFATE ION IN THE INFLUENT, (3) BOD SURFACE LOAD, AND (4) SULFATE ION CONCENTRATION IN TERMS OF SURFACE LOAD. ALL EXPERIMENTS WERE CONDUCTED UNDER CONTROLLED ENVIRONMENTAL CONDITIONS, ENCOMPASSING 7 DIFFERENT OPERATIONAL CONDITIONS. AT CONSTANT DETENTION TIME AND CONSTANT BOD SURFACE LOADING, A LINEAR RELATIONSHIP EXISTED BETWEEN SULFIDE CONCENTRATION IN THE POND AND SULFATE CONCENTRATION IN THE INFLUENT. INCREASES IN POND SULFIDE CONCENTRATION FROM NEAR ZERO TO 1 AND 2 MG/L, RESPECTIVELY, WERE ACCOMPANIED BY DROPS IN THE OXIDATION-REDUCTION POTENTIAL OF 80 AND 120 MICRON. FURTHER INCREASES IN SULFIDE CONCENTRATION AFFECTED THE ORP TO A MUCH LESSER EXTENT. THIS PHENOMENON ESTABLISHED ORP MEASUREMENT AS AN INDICATOR OF POTENTIAL ODOR TROUBLES AND/OR POND SYSTEM UPSET. ALGAE DISAPPEARED ALMOST ENTIRELY WHEN SULFIDE CONCENTRATIONS IN THE POND ROSE ABOVE 7.0 MG/L BUT BOD REMOVAL RATES WERE NOT DETECTABLY DIMINSIHED. INFRARED SPECTROSCOPIC ANALYSIS OF GASEOUS EMISSIONS FROM THE POND IDENTIFIED THREE OF THE COMPONENT GASES AS METHATE, HYDROGEN SULFIDE, AND CARBON DIOXIDE. INTERFERENCE FROM MOISTURE IN THE AIR PROHIBITED FURTHER IDENTIFICATION. (LOWRY-TEXAS)

FIELD 05D

ACCESSION NO. W71-11377

RADIOACTIVITY TRANSPORT IN WATER--EFFECTS OF ORGANIC POLLUTION ON RADIONUCLIDE TRANSPORT,

TEXAS UNIV., AUSTIN. CENTER FOR RESEARCH IN WATER RESOURCES.

KAZUHIRO FUTAGAWA, AND ERNEST F. GLOYNA.

TECHNICAL REPORT NO. 16, EHE-04-6803, CRWR-27, ORO-16, MAY 1968, 61 P, 14 FIG, 4 TAB, 31 REF. U.S. ATOMIC ENERGY COMMISSION CONTRACT AT-(11-1)-490.

DESCRIPTORS:
*RADIONUCLIDE, *DISSOLVED OXYGEN, *RIVERS, OXIDATION, PHOTOSYNTHESIS, RESPIRATION, SCOUR, SATURATION, MIXING, DIFFUSION, SEDIMENTATION, ABSORPTION, BIOCHEMICAL OXYGEN DEMAND, AERATION, TEMPERATURE, FLOW, ALGAE, BENTHOS, AEROBIC CONDITIONS, ANAEROBIC CONDITIONS, WATER POLLUTION SOURCES, PATH OF POLLUTANT.

ABSTRACT:
RADIONUCLIDE TRANSPORT IN WATER CAN BE AFFECTED BY THE PRESENCE OF ORGANIC POLLUTION IN TWO WAYS. ONE OF THESE WAYS IS RELATED TO THE LACK OF DISSOLVED OXYGEN CAUSED BY BIOLOGICAL DECOMPOSITION OF ORGANIC POLLUTANTS. THE OTHER MECHANISM IS DETERMINED BY THE EFFECTS OF THE INTERACTIONS OF THE RADIONUCLIDES WITH THE ORGANIC MATERIAL ITSELF. AS A PRELIMINARY STEP TOWARDS A MORE COMPLETE EXPLANATION OF THE DISSOLVED OXYGEN DEFICIT EFFECT, THE ECOLOGICAL ENVIRONMENT OF A RIVER SYSTEM WAS SIMULATED IN A SPECIALLY ADAPTED RESEARCH FLUME. DISTRIBUTION OF DISSOLVED OXYGEN AND RE-AERATION RATES WERE THEN MATHEMATICALLY MODELED USING DATA TAKEN FROM THE MODEL RIVER. USING A WATER DEPTH OF 1.0 FT, REAERATION RATE CONSTANTS WERE DETERMINED AT A FLOW OF 1.7 IN/MIN AND 28C. THE VALUES FOR THE RE-AERATION RATE CONSTANTS WERE 0.18 AND 0.23 RESPECTIVELY. WITH A 200 MG/L CONCENTRATION OF DEXTRINE IN THE INFLUENT, THE BIOLOGICAL OXIDATION RATE CONSTANT IN THE MODEL RIVER WAS DETERMINED TO BE 0.006, WHICH WAS QUITE LOW IN COMPARISON WITH VALUES DETERMINED ON FULL SCALE RIVERS. IT WAS ALSO DETERMINED THE VARIABILITY OF THE ENVIRONMENTAL FACTORS WHICH DETERMINE PHOTOSYNTHETIC OXYGEN PRODUCTION PRECLUDED THE USE OF ALGAL PRODUCED OXYGEN AS A RELIABLE OXYGEN SOURCE. HOWEVER, SINCE THE VARIOUS PHOTOSYNTHETIC-RESPIRATIONSHIPS CAN BE MEASURED IN A MODEL WITH SUITABLE ACCURACY, THE FUTURE, MORE INTENSIVE, RADIONUCLIDE TRANSPORT STUDIES CAN PROCEED. (LOWRY-TEXAS)

FIELD 05B

ACCESSION NO. W71-11381

FACTORS AFFECTING THE USE OF PURIFIED SEWAGE EFFLUENTS FOR COOLING PURPOSES,

JOHANNESBURG MUNICIPALITY (SOUTH AFRICA).

D. W. OSBORN.

WATER POLLUTION CONTROL, VOL 69, NO. 4, P 456-464, 5 FIG, 1 TAB, 1 REF.

DESCRIPTORS:
*COOLING WATER, *ELECTRIC POWER PLANTS, *WATER REUSE, COOLING TOWERS,
DAMS, ALGAE, HYDROGEN ION CONCENTRATION, TEMPERATURE, ALKALINITY,
NITRIFICATION, *CORROSION, CARBON DIOXIDE, SOLUBILITY, *PHOSPHATES,
CHEMICAL PRECIPITATION, FILTRATION, AMMONIA, WATER PURIFICATION, WASTE
WATER TREATMENT.

IDENTIFIERS:
*NITRIC ACID, *LANGELIER INDEX.

ABSTRACT:
ONE OF THE MORE IMPORTANT PROBLEMS INVOLVED IN USING PURIFIED SEWAGE
EFFLUENT AS COOLING WATER FOR POWER STATIONS HAS BEEN THE OXIDATION OF
AMMONIA TO NITRIC ACID. SERIOUS CORROSION DAMAGE TO CONCRETE AND METAL
FIXTURES HAS BEEN THE RESULT IN SEVERAL INSTANCES. IN ADDITION TO
CORROSION, PHOSPHATE PRECIPITATION FROM LOW ALKALINITY, LOW PH WATERS
HAS ALSO CAUSED MANY PROBLEMS IN LINES AND PUMPS. THEREFORE,
CONSTRUCTION OF THE NEW ORLANDO POWER STATION INCLUDED A LARGE DAM. THE
65 ACRE AREA, 180 X 10 6 GAL. VOLUME FORMED BY THE DAM SERVED AS A
MATURATION POND IN WHICH THE AMMONIA PRESENT IN THE WATER IS SLOWLY
DISTILLED OFF AND NOT CONVERTED TO NITRIC ACID. THE POND IS MAINTAINED
AT 24C THE YEAR ROUND, SO THAT AMMONIA AND BOD REMOVALS CAN CONTINUE
DURING BOTH SUMMER AND WINTER CONDITIONS. DURING THIS DETENTION PERIOD,
THE ALGAE ALSO RAISE THE PH LEVELS OF THE WATER TO THE POINT WHERE
PHOSPHATE IS PRECIPITATED OUT BEFORE IT CAN FOUL THE PIPES AND VALVES
OF THE COOLING TOWERS. THE OPERATIONAL EXPERIENCE NEEDED TO CORRECT
THESE PROBLEMS WAS GAINED OVER A 30 YEAR PERIOD. SUCCESSFUL UTILIZATION
OF WASTE WATER EFFLUENTS TO AUGMENT SCARCE WATER SUPPLIES HAS LONG BEEN
A WAY OF LIFE IN JOHANNESBURG, AND THE EXPERIENCE GAINED THERE HAS
CONTRIBUTED MUCH TO THE NEW TECHNOLOGY OF WATER RE-USE. (LOWRY-TEXAS)

FIELD 05D

ACCESSION NO. W71-11393

THE TOXICITY OF COPPER COMPLEXES TOWARDS SCENEDESMUS QUADRICAUDA BREB.,

MOSCOW STATE UNIV. (USSR). DEPT. OF HYDROBIOLOGY.

V. G. KHOBOT'EV, V. I. KAPKOV, AND E. G. RUKHADZE.

MICROBIOLOGY (USSR), VOL 38, NO 5, P 729-731, SEPTEMBER-OCTOBER 1969. 2 TAB,
8 REF.

DESCRIPTORS:
*COPPER, *TOXICITY, *ALGAE, *WATER POLLUTION EFFECTS, CHELATION,
SPECTROPHOTOMETRY, NUTRIENTS, HEAVY METALS.

IDENTIFIERS:
PROTOCOCCUS, SCENEDESMUS QUADRICAUDA.

ABSTRACT:
THE TOXIC EFFECTS OF COPPER ON PROTOCOCCAL ALGAE WERE STUDIED USING TWO
TYPES OF COPPER COMPOUNDS: CHELATES, THE MAJORITY OF WHICH ARE SOLUBLE
IN WATER (COPPER BIS-O-HYDROXYQUINOLINE, COPPER BIS-O-VANILLAL, AND
COPPER BIS-O-VANILLALMETHYLIMINE), AND A COMPLEX OF COPPER WITH
PYRIDINE (COPPER DICHLORIDE PYRIDINATE). CONCENTRATIONS OF 0.64-32 MG
COPPER/LITER OF MEDIUM FREE FROM K SUB 2 CO SUB 3 AND KH SUB 2 PO SUB 4
WERE USED. THE RESULTS SHOWED THE COPPER COMPLEXES TO HAVE A TOXIC
ACTION ON THE ALGAE. THE GREATEST TOXIC ACTION WAS SHOWN BY COPPER
DICHLORIDE PYRIDINATE, WHICH AT CONCENTRATIONS OF 0.4 MG COPPER/LITER
REDUCED THE NUMBER OF ALGA CELLS BY FOUR TIMES AFTER THE SEVENTH DAY AS
COMPARED WITH THE CONTROL. COPPER CHELATES ALSO SHOWED TOXIC ACTION BUT
AT SOMEWHAT LARGER CONCENTRATION. THE LEAST TOXIC WAS COPPER
BIS-O-HYDROXYQUINOLINE, WHICH AT ALL CONCENTRATIONS TESTED ONLY
INHIBITED GROWTH. (MORTLAND-BATTELLE)

FIELD 05C

ACCESSION NO. W71-11561

A COMBINATION OF RING-OVEN AND CIRCULAR CHROMATOGRAPHY OF TRACE METAL ANALYSIS,

DALHOUSIE UNIV., HALIFAX (NOVA SCOTIA). DEPT. OF CHEMISTRY.

R. W. FREI, AND C. A. STOCKTON.

MIKROCHIMICA ACTA, P 1196-1203, 1969. 1 FIG, 2 TAB, 20 REF.

DESCRIPTORS:
*CHROMATOGRAPHY, *COBALT, *COPPER, *IRON, TRACE ELEMENTS, ALGAE, SEA WATER, FRESH WATER.

IDENTIFIERS:
*NICKEL, DETECTION LIMITS, CHEMICAL INTERFERENCES.

ABSTRACT:
A CHROMATOGRAPHIC METHOD FOR METAL CHELATES OF PYRIDINE-2-ALDEHYDE-2QUINOLYLHYDRAZONE(PAQH) HAS BEEN ADAPTED TO THE RING-OVEN TECHNIQUE. ADVANTAGE HAS BEEN TAKEN OF THE LARGE DIFFERENCE IN RF VALUES OF THE COBALT, COPPER, NICKEL AND IRON(III) CHELATES ON ALUMINA, THE ACID STABILITY OF THE COBALT PAQH COMPLEX, SELECTIVE FORMATION OF CHLORIDE COMPLEXES OF THESE METALS AND THE USE OF SPECIFIC SPRAY REAGENTS FOR COPPER AND IRON, SUCH AS RUBEANIC ACID AND KSCN TO DEVELOP A SIMPLE AND RAPID FIELD METHOD FOR TRACE METAL ANALYSIS. QUANTITATIVE ESTIMATION OF THESE FOUR METALS WITH AN ACCURACY RANGING FROM 5.0 TO 12 PERCENT HAS BEEN CARRIED OUT IN LESS THAN 30 MINUTES. DETECTION LIMITS WERE FOUND TO BE ABOUT 0.01 MICROGRAM FOR NICKEL, 0.008 MICROGRAM FOR COPPER AND IRON AND 0.005 MICROGRAM FOR COBALT. OTHER IONS DID NOT INTERFERE IN UP TO 10-FOLD EXCESS. THE SUITABILITY OF THIS METHOD FOR FIELD WORK WAS TESTED BY MONITORING TRACE METALS IN FRESH AND SEA WATER SUPPLIES AND IN ALGAE. (LITTLE-BATTELLE)

FIELD 05A

ACCESSION NO. W71-11687

BIOLOGICAL HANDBOOK FOR ENGINEERS.

NATIONAL AERONAUTICS AND SPACE ADMINISTRATION, HUNTSVILLE, ALA. GEORGE C. MARSHALL SPACE FLIGHT CENTER.

NASA CONTRACTOR REPORT 61237, JUNE 1968. 225 P, 64 FIG, 74 TAB, 146 REF.

DESCRIPTORS:
*MICROBIOLOGY, *CLASSIFICATION, *CULTURES, *BIOASSAYS, PROTOZOA, BACTERIA, FUNGI, VIRUSES, ALGAE, LICHENS, ACTIMOMYCETES, CYTOLOGICAL STUDIES, FILTRATION, ULTRASONICS, DISINFECTION, SAMPLING, CENTRIFUGATION, SEDIMENTATION, CHEMICAL PRECIPITATION, LIFE CYCLES, GENETICS, E. COLI, ISOLATION.

IDENTIFIERS:
RICKETTSIAE, STERILIZATION, BACILLUS SUBTILIS, BACILLUS STEAROTHERMOPHILUS, PROTEUS VULGARIS, PSEUDOMONAS AERUGINOSA, STAPHYLOCOCCUS AUREUS, LACTOBACILLUS CASEI, BACILLUS THURINGIENSIS, AGARS, STAPHYLOCOCCUS EPIDERMIDIS, MICROCOCCUS, STREPTOCOCCUS, BREVIBACTERIUM, CORYNEBACTERIUM, ELECTROSTATIC PRECIPITATION, THERMAL PRECIPITATION.

ABSTRACT:
USEFUL MICROBIOLOGICAL BACKGROUND INFORMATION HAS BEEN COMPILED IN A HANDBOOK FOR ENGINEERS AND SCIENTISTS WORKING ON BIO-RELATED PROJECTS. IT IS INTENDED AS AN AID IN: (1) EVALUATING THE EFFECTS OF ENGINEERING PROCEDURES ON MICROBIAL LIFE; (2) DETERMINING THE EFFECTS OF DECONTAMINATION OF STERILIZATION PROCEDURES ON THE PERFORMANCE OF AN OVERALL SYSTEM; AND (3) UNDERSTANDING THE LANGUAGE OF MICROBIOLOGISTS. THE INFORMATION PRESENTED COVERS SUCH SUBJECTS AS: (1) THE VARIOUS CATEGORIES OF MICROORGANISMS, THEIR PHYSICAL CHARACTERISTICS, THEIR RELATIVE ABUNDANCES IN VARIOUS LOCATIONS, THEIR LIFE CYCLES, AND THEIR GENETICS; (2) THE METHODS AND APPLICATIONS OF DECONTAMINATION AND STERILIZATION BY WET OR DRY HEAT, CHEMICAL TREATMENT, RADIATION EXPOSURE, AND PRESSURE TREATMENT; (3) THE RELATIVE SUCCESS OF VARIOUS METHODS OF CONTAMINATION CONTROL, SUCH AS FILTRATION, CLEAN ROOMS, LAMINAR AIR FLOWS, AND SHIELDING; AND (4) THE THEORY AND TECHNIQUES OF BIOASSAY--THE DETECTING, IDENTIFYING, AND COUNTING OF MICROBES IN SPECIFIC LOCATIONS. THE FINAL SECTION COMPRISES A GLOSSARY OF COMMON MICROBIOLOGICAL TERMS. THE MANY REFERENCES CITED THROUGHOUT THE WORK GUIDE THE INTERESTED READER TO ADDITIONAL INFORMATION. (MORTLAND-BATTELLE)

FIELD 05A

ACCESSION NO. W71-11823

THERMAL EFFECTS AND NUCLEAR POWER STATIONS IN THE U.S.A.,

ARGONNE NATIONAL LAB., ILL. CENTER FOR ENVIRONMENTAL STUDIES; AND BATTELLE
MEMORIAL INST., RICHLAND, WASH. PACIFIC NORTHWEST LABS.

D. MILLER, J. V. TOKAR, AND R. E. NAKATANI.

AVAILABLE FROM THE NATIONAL TECHNICAL INFORMATION SERVICE AS CONF-700810-30,
$3.00 IN PAPER COPY, $0.95 IN MICROFICHE. REPORT NO IAEA-SM-146/30 (NO
DATE). 14 P, 1 TAB, 21 REF.

DESCRIPTORS:
*THERMAL POLLUTION, *NUCLEAR POWERPLANTS, *WATER QUALITY CONTROL,
*BIODEGRADATION, WATER UTILIZATION, FISH, ALGAE, MODEL STUDIES,
BIOASSAY, AQUATIC LIFE, COOLING WATER, SITES.

ABSTRACT:
OF 94 NUCLEAR POWER PLANTS IN THE U.S., 60.6 PERCENT ARE COOLED BY
ONCE-THROUGH SYSTEMS IN WHICH THE COOLING WATER IS RETURNED DIRECTLY TO
A NATURAL WATER SOURCE. THESE THERMAL DISCHARGES UNDOUBTEDLY MODIFY THE
BIOLOGICAL COMMUNITIES THAT ARE EXPOSED. ALTHOUGH GENERAL EFFECTS ARE
KNOWN, THERE ARE INSUFFICIENT DATA AVAILABLE AT PRESENT TO PERMIT
EFFECTIVE EVALUATION OF PLANT SITES AND DISCHARGE DETAILS ON THE BIOTIC
COMMUNITIES. IT IS THEREFORE NECESSARY THAT STUDIES BE UNDERTAKEN
RELATIVE TO POWER PLANT SITING AND OPERATION TO ENSURE COMPLIANCE WITH
WATER QUALITY STANDARDS, TO DEFINE MIXING ZONES, TO PREDICT TEMPERATURE
DISTRIBUTIONS IN RECEIVING WATER, TO ASSESS BIOLOGICAL PERTURBATIONS,
TO DESIGN INTAKE AND OUTFALL STRUCTURES TO MINIMIZE BIOLOGICAL DAMAGE,
TO CONTROL NUISANCE GROWTHS OF PLANTS AND ALGAE, AND TO DETERMINE
SUBLETHAL EFFECTS OF TEMPERATURE ON AQUATIC LIFE. (LITTLE-BATTELLE)

FIELD 05C, 08A

ACCESSION NO. W71-11881

A SURVEY OF THE BOTTOM FAUNA IN LAKES OF THE OKANAGAN VALLEY, BRITISH COLUMBIA,

FISHERIES RESEARCH BOARD OF CANADA, WINNIPEG (MANITOBA). FRESHWATER INST.

OLE A. SEATHER.

FISHERIES RESEARCH BOARD OF CANADA (MS) TECHNICAL REPORT NO 196 (UNDATED). 31
P, 3 FIG, 7 TAB, 15 REF.

DESCRIPTORS:
*LAKES, *BENTHIC FAUNA, *TROPHIC LEVELS, *SURVEYS, BIOLOGICAL
COMMUNITIES, BIOINDICATORS, LAKE MORPHOLOGY, ECOLOGICAL DISTRIBUTION,
MIDGES, WATER POLLUTION EFFECTS, OLIGOCHAETES, DIPTERA, ALGAE, INSECTS.

IDENTIFIERS:
TROPHIC COMMUNITIES, OKANAGAN LAKE(CANADA), SKAHA LAKE(CANADA), OSOYOOS
LAKE(CANADA), DEFORMED CHIRONOMIDS.

ABSTRACT:
BOTTOM FAUNA WERE STUDIED TO DETERMINE POLLUTION EFFECTS IN THREE LARGE
LAKES IN OKANAGAN BASIN, BRITISH COLUMBIA. A COMPARISON WITH EARLIER
DATA INDICATES LAKE OKANAGAN IS NOT ONLY POLLUTED IN CERTAIN AREAS, BUT
BOTTOM FAUNA OF WHOLE LAKE HAS CHANGED. THERE HAS BEEN SIGNIFICANT
INCREASE IN TOTAL NUMBERS OF BOTTOM ORGANISMS FOR STATIONS SHALLOWER
THAN 50 METERS AND SIGNIFICANT INCREASE IN TOTAL NUMBERS OF
CHIRONOMIDS, OLIGOCHAETES, PISIDIUM AND PERHAPS OTHERS. MOST STRIKING
CHANGE IS IN PERCENTAGE DISTRIBUTION BETWEEN GROUPS. OLIGOCHAETES HAVE
INCREASED FROM 15% OF MACROBENTHOS TO 50% TO 60% OF FAUNA, SUGGESTIVE,
TOGETHER WITH DEFORMED CHIRONOMIDS OCCURRENCE, OF INSECTICIDE
POLLUTION. THE LAKE AS A WHOLE IS STILL OLIGOTROPHIC ALTHOUGH NOT
ULTRAOLIGOTROPHIC. THE FEW SKAHA LAKE STATIONS WERE INSUFFICIENT FOR
DETAILED CHARACTERIZATION; IT HAS PROBABLY BEEN MODERATELY EUTROPHIC A
LONG TIME AND A DRAMATIC CHANGE IN BOTTOM FAUNA MAY HAVE OCCURRED QUITE
RECENTLY. OSOYOOS LAKE'S NORTHERN AND CENTRAL BASINS ARE, ACCORDING TO
COMPOSITION OF BOTTOM FAUNA, MODERATELY AND STRONGLY EUTROPHIC,
RESPECTIVELY. IN OKANAGAN AND SKAHA LAKES CHIRONOMIDS WITH DEFORMED
MOUTH-PARTS AND THICKENED BODY WALLS WERE FOUND. DEFORMED CHIRONOMIDS
MAY BE USEFUL IN DETECTING SOME FORMS OF POLLUTION. (JONES-WISCONSIN)

FIELD 05C

ACCESSION NO. W71-12077

EFFECTS OF TOXIC ORGANICS ON PHOTOSYNTHETIC REOXYGENATION,

TEXAS UNIV., AUSTIN. CENTER FOR RESEARCH IN WATER RESOURCES.

JU-CHANG HUANG, AND EARNEST F. GLOYNA.

TECHNICAL REPORT NO. EHE-07-6701, CRWR-20, AUGUST 1967, 163 P, 64 FIG, 24 TAB, 78 REF. EPA GRANT WP-00688-03.

DESCRIPTORS:
*REAERATION, *CHLOROPHYLL, *OXYGENATION, *TOXICITY, ORGANIC COMPOUNDS, *PHENOLS, PESTICIDES, OXIDATION LAGOONS, ALGAE, PHOTOSYNTHESIS, DISSOLVED OXYGEN, LABORATORY TESTS, TEMPERATURE, WASTE WATER TREATMENT.

ABSTRACT:
THE RELATIVE TOXICITY OF 33 PHENOLIC-TYPE COMPOUNDS AND 8 PESTICIDES WAS INVESTIGATED TO DETERMINE THE EXTENT TO WHICH CHLOROPHYLL INHIBITION BY CERTAIN INDUSTRIAL WASTES DECREASES STREAM DISSOLVED OXYGEN CONTENTS. PHENOLIC-TYPE GROUPS INCLUDED THOSE WITH MONO-AND POLY-SUBSTITUTED GROUPS SUCH AS -BR, -CL, -CH3, -NO2, -NH2, AND -OH. PESTICIDES STUDIED INCLUDED DDT, LINDANE, 2,4-D, SODIUM SALT OF 2, 4-D, 2, 45-T, DOW SODIUM TCA, DOW ESTERON 99, AND DOW FORMULA 40. ORGANIC COMPOUND TOXICITY WAS EVALUATED IN TWO WAYS. FIRST, TEST TUBE ALGAL CULTURES WERE GROWN AND CHEMOSTATED AT 25C UNDER CONTINUOUS ILLUMINATION FOR 72 HOURS. KNOP'S SOLUTION, INCLUDING THE HUTNER-EDTA MICRO ELEMENT SYSTEM, WAS USED AS CULTURE MEDIUM, AND A 5% CARBON DIOXIDE IN AIR GAS WAS BLOWN THROUGH TO SUPPLY THE INORGANIC CARBON SOURCE AND THE MIXING. SECONDLY, WARBURG MANOMETRIC TECHNIQUES WERE USED TO MEASURE GAS EXCHANGE OCCURING WHEN CARBONATE-BICARBONATE BUFFER IS USED AS A SUSPENDING FLUID. RESULTS SHOWED THE DECREASE IN CHLOROPHYLL CONTENT TO BE GENERALLY LOGARITHMIC TO TOXIC ORGANIC CONCENTRATION, WITH SUBSEQUENT SUPPRESSED OXYGEN PRODUCTION. FULL SCALE TESTS SUBSTANTIATED THESE CONCLUSIONS, THEREFORE A TOXICITY COMPENSATION FACTOR MUST BE INTRODUCED INTO DESIGN EQUATIONS FOR WASTE STABILIZATION PONDS RECEIVING TOXIC ORGANICS. (LOWRY-TEXAS)

FIELD 05C, 05D

ACCESSION NO. W71-12183

METHODS FOR DETERMINING THE PRIMARY PRODUCTIVITY OF EPIPELIC AND EPIPSAMMIC ALGAL ASSOCIATIONS,

BRISTOL UNIV. (ENGLAND). DEPT. OF BOTANY.

M. HICKMAN.

LIMNOLOGY AND OCEANOGRAPHY, VOL 14, NO 6, P 936-941, 1969. 7 TAB, 13 REF.

DESCRIPTORS:
PRIMARY PRODUCTIVITY, *BIOLOGICAL COMMUNITIES, *ALGAE, *METHODOLOGY, CARBON RADIOISOTOPES, DIATOMS, MEASUREMENT, SEDIMENTS, SAMPLING, DEPTH, SEA WATER, FRES WATER, LITTORAL, DISSOLVED OXYGEN, DIURNAL, MIGRATION PATTERNS.

IDENTIFIERS:
*EPIPSAMMIC ALGAE, *EPIPELIC ALGAE, ALGAL ASSOCIATIONS, SHEAR WATER(ENGLAND), AMPHORA OVALIS, OPEPHORA MARTYI, ABBOT'S POND(SOMERSET, ENGLAND), SAND GRAINS, TIDAL CYCLES.

ABSTRACT:
QUANTITATIVE TECHNIQUES OF DEFINED ACCURACY WERE DEVELOPED FOR ESTIMATING PRIMARY PRODUCTIVITY OF BOTH EPIPELIC AND EPIPSAMMIC ALGAL ASSOCIATIONS OF BOTH FRESH AND MARINE SHALLOW AND DEEP-WATER SITES. FOR THE EPIPSAMMIC ALGAL ASSOCIATION SAMPLES WERE TAKEN FROM A RICH DIATOMAL FLORA ATTACHED TO SAND GRAIN SURFACES IN THE LITTORAL ZONE OF A SMALL LAKE. THE ALGAE WERE SEPARATED FROM THE EPIPELIC ALGAE AND DETRITAL MATERIAL AND A C-14 SOURCE ADDED. THE ALGAE WERE REMOVED FROM THE SAND GRAINS BEFORE ASSAY FOR C-14 BY SONIFICATION. THE EPIPELIC ALGAL ASSOCIATION WAS STUDIED BY TRAPPING THE ALGAE WITH A DOUBLE LAYER OF LENS TISSUE AND C-14 METHOD FOR ASSAY. HARVESTING TIME IS CRITICAL, SINCE EPIPELIC POPULATIONS SHOW MARKED DIURNAL MIGRATION RHYTHMS. AFTER INCUBATION, ALGAE WERE REMOVED FROM LENS TISSUE. SINCE 87.5% OF THE EPIPELIC POPULATION WAS TRAPPED IN THE TISSUE AND 85.8% OF THE TRAPPED POPULATION REMOVED FROM IT, PRODUCTIVITY ESTIMATES WERE ATTRIBUTABLE TO 73.3% OF TOTAL POPULATION. THE PRIMARY PRODUCTIVITY OF BOTH EPIPSAMMIC AND EPIPELIC ALGAE CAN ALSO BE STUDIED WITH OXYGEN TECHNIQUES. THE ROUTINE METHODS FOR BOTH ASSOCIATIONS, THE EPIPELIC AND EPIPSAMMIC, WERE STATISTICALLY DEFINED. (JONES-WISCONSIN)

FIELD 05C, 05A

ACCESSION NO. W71-12870

REMOVAL OF PHOSPHOROUS FROM A SEWAGE LAGOON BY HARVESTING ALGA,

MISSISSIPPI STATE UNIV., STATE COLLEGE, DEPT. OF CIVIL ENGINEERING.

J. E. LITTLE.

MASTER'S THESIS, AUGUST 1969. 41 P, 11 FIG, 2 TAB, 30 REF.

DESCRIPTORS:
*EUTROPHICATION, *PHOSPHOROUS, ALGAE, *HARVESTING OF ALGAE, EFFLUENT, QUALITY CONTROL, OXIDATION LAGOON, NITROGEN, NUTRIENTS, FILTRATION, MICROORGANISMS, ANALYTICAL TECHNIQUES, *WASTE WATER TREATMENT, TREATMENT FACILITIES, MISSISSIPPI.

ABSTRACT:
EUTROPHICATION, DEFINED AS THE PROCESS IN WHICH SURFACE BODIES OF WATER ARE ENRICHED BY THE EXCESSIVE ADDITION OF NUTRIENTS, AFFECTS AS MUCH AS 56% OF TOTAL MUNICIPAL SURFACE WATER SUPPLIES IN THE UNITED STATES. EXCESSIVE ALGAL GROWTH IN EUTROPHIED WATER CAUSED DIFFICULTIES IN TREATING SEWAGE AND REDUCED THE RECREATION VALUE OF LAKES AND STREAMS. BECAUSE EUTROPHICATION IS A PRODUCT OF NUTRIENT ENRICHMENT, IT WAS DECIDED THAT CONTROL MEASURES SHOULD BE BASED ON EITHER REMOVAL, REDUCTION, OR PREVENTION OF THE ADDITION OF NUTRIENTS. MAKING USE OF THE SEWAGE LAGOON RESEARCH FACILITY AT MISSISSIPPI STATE UNIVERSITY, AN ATTEMPT WAS MADE TO SEE WHETHER THE PHOSPHOROUS CONTENT OF AN OXIDATION POND EFFLUENT COULD BE BROUGHT TO ACCEPTABLE LEVELS THROUGH HARVESTING THE ALGAE CONTAINED IN THE POND EFFLUENT. A SIX MONTH STUDY REVEALED THAT: (1) HARVESTING ALGAE REDUCED THE PHOSPHATE CONCENTRATION TO 91% MAXIMUM REMOVAL, (2) HARVESTING ALGAE WILL NOT REMOVE ENOUGH PHOSPHOROUS FROM OXIDATION POND EFFLUENT TO CONTROL ALGAL GROWTHS IN RECEIVING STREAMS AND LAKES UNLESS A LARGE QUANTITY OF ALMOST PHOSPHOROUS-FREE WATER IS AVAILABLE FOR DILUTION. (3) A FILTER WITH A PORE SIZE OF 45 MICRON OR LESS APPEARS TO REMOVE ALL INSOLUBLE PHOSPHATE FROM OXIDATION POND EFFLUENT, BUT ADEQUATE FLOW IS DIFFICULT TO ATTAIN, (4) IT DOES NOT APPEAR THAT THE METHOD STUDIED IS PRACTICAL BRCAUSE A SIGNIFICANT DEGREE OF PHOSPHOROUS REMOVAL IS ACHIEVED ONLY WHEN BOD REMOVAL EFFICIENCY IS LOW, THEREBY DEFEATING THE ORIGINAL PURPOSE OF OXIDATION PONDS. (ADKINS-TEXAS)

FIELD 05D, 05C

ACCESSION NO. W71-13313

THE MECHANISM OF PHOSPHOROUS REMOVAL IN ACTIVATED SLUDGE,

MAINE UNIV., ORONO. DEPT. OF CIVIL ENGINEERING.

RALPH E. OULTON.

MASTER'S THESIS, UNIVERSITY OF MAINE, JANUARY 1969. 87 PAGES, 31 FIG, 20 REF.

DESCRIPTORS:
*EUTROPHICATION, *ALGAE, NUTRIENTS, NITROGEN, PHOSPHOROUS, NITROGEN FIXING BACTERIA, *ACTIVATED SLUDGE, STORAGE, TEMPERATURE, MICROORGANISMS, NITROGEN COMPOUNDS, *PHOSPHORUS COMPOUNDS, ADSORPTION, REMOVAL, DEGRADATION, CHEMICAL OXYGEN DEMAND, *WASTE WATER TREATMENT.

IDENTIFIERS:
MIXED LIQUOR, SUSPENDED SOLIDS, *PHOSPHOROUS UPTAKE RATE.

ABSTRACT:
COMPOUNDS OF NITROGEN AND PHOSPHOROUS ARE THE MAJOR NUTRIENTS IN THE EUTROPHICATION PROCESS. IN ORDER TO CONTROL EUTROPHICATION, THE SUPPLY OF AT LEAST ONE OF THESE NUTRIENTS MUST BE CONTROLLED. NITROGEN-FIXING BACTERIA OFFER A VIRTUALLY UNLIMITED SUPPLY OF NITROGEN TO ALGAE; THEREFORE, PHOSPHORUS IS THE NUTRIENT WHICH CAN BEST BE LIMITED. TESTS WERE MADE TO DETERMINE WHETHER THE UPTAKE OF SOLUBLE PHOSPHOROUS BY ACTIVATED SLUDGE MIXED LIQUOR SUSPENDED SOLIDS IS RELATED TO THE VARIATION IN STORED ENERGY PRODUCTS WITHIN THE CELLS COMPRISING THE MLSS. RESULTS OF THE ANALYTICAL TESTS PERFORMED REVEALED SEVERAL FACTS, (1) AN ACCIDENTAL INCREASE IN TEMPERATURE FROM 25 DEG C TO 55 DEG C CAUSED AN INCREASE IN THE SOLUBLE PHOSPHOROUS UPTAKE RATE FROM 40 MG/L TO 80 MG/L, (2) REMOVAL OF SOLUBLE PHOSPHORUS FROM SOLUTION IS RELATED TO THE CONTINUED METABOLISM OF THE ACTIVATED SLUDGE CELLS EVEN AFTER ALL AVAILABLE EXTERNAL ENERGY SOURCES ARE EXHAUSTED, (3) THE FIRST TOTAL SOLUBLE PHOSPHORUS CONCENTRATION HAD NO EFFECT ON THE PHOSPHORUS UPTAKE RATE IN THE SYSTEM STUDIED, (4) THE PHOSPHORUS REMOVAL AND USE OF STORAGE PRODUCTS AFTER THE EXHAUSTION OF EXTERNAL COD WAS INDEPENDENT OF THE FIRST MLSS CONCENTRATION AND THE TEMPERATURE VARIATIONS STUDIED. (ATKINS-TEXAS)

FIELD 05D, 05C

ACCESSION NO. W71-13335

THE RATE AND EXTENT OF ALGAL DECOMPOSITION IN ANAEROBIC WATERS,

KENTUCKY UNIV., LEXINGTON.

EDWARD G. FOREE, AND PERRY L. MCCARTY.

PROCEEDINGS, INDUSTRIAL WASTE CONFERENCE, 24TH, MAY 6-8, 1969, P 13-36, 13
FIG, 8 TAB, 9 REF.

DESCRIPTORS:
*ALGAE, *LABORATORY TESTS, *ANAEROBIC CONDITIONS, LIGHT INTENSITY,
EUTROPHICATION, NUTRIENTS, *FERMENTATION, TOXICITY, CHEMICAL OXYGEN
DEMAND, HYDROGEN ION CONCENTRATION, TEMPERATURE, PROTEINS,
CARBOHYDRATES, LIPIDS, KINETICS, MATHEMATICAL MODELS, NEUTRALIZATION,
*DEGRADATION(DECOMPOSITION), METHANE, WATER POLLUTION EFFECTS.

IDENTIFIERS:
*METHANE FERMENTATION.

ABSTRACT:
ALGAE CULTURES DECOMPOSING UNDER DARK, ANAEROBIC CONSTANT TEMPERATURE
CONDITIONS WERE STUDIED IN THE LABORATORY TO EVALUATE THE EFFECTS OF
CELL COMPOSITION, PH, TEMPERATURE, AND TO DETECT ANY CHANGES IN METHANE
FERMENTATION AND SULGATE REDUCTION. IT WAS DETERMINED THAT
STABILIZATION OF ALGAE OCCURRED BY SULFATE REDUCTION AS LONG AS SULFATE
REMAINED IN SOLUTION. ONCE THE SULFATE SUPPLY WAS EXHAUSTED,
STABILIZATION CONTINUED THROUGH METHANE FERMENTATION. DARK, ANAEROBIC
DECAY OF ALGAE WAS DETERMINED AS: M= (MO - FMO) E-KT+ F MO, WHERE F=
REFRACTORY ORGANIC PORTION HAVING AN AVERAGE VALUE OF 0.40 + 0.15 FROM
15 TO 25C, ALGAL DECOMPOSITION RATE IS TEMPERATURE DEPENDENT: K/KO=ECK
(T-TO), WITH AN EXPECTED CK OF 0.055/C. PROTEIN, CARBOHYDRATE AND LIPID
CONTENTS OF ALGAE CAN BE RELIABLY CALCULATED AS A FUNCTION OF NITROGEN
CONTENT AND COD TO VOLATILE SOLIDS RATIO OF THE CELLS. (LOWRY-TEXAS)

FIELD 05C

ACCESSION NO. W71-13413

EXPERIENCE IN THE TREATMENT AND RE-USE OF INDUSTRIAL WASTE WATERS,

JOHNS HOPKINS UNIV., BALTIMORE, MD. DEPT. OF GEOGRAPHY AND ENVIRONMENTAL
ENGINEERING.

C. E. RENN.

PROCEEDINGS, INDUSTRIAL WASTE CONFERENCE, 24TH, MAY 6, 7, AND 8, 1969. P
962-968.

DESCRIPTORS:
*WATER REUSE, *WASTE WATER TREATMENT, *INDUSTRIAL WASTES, OXIDATION
PONDS, ALGAE, BENTHIC FAUNA, WATERSHEDS, WATER SHORTAGE FILTERS,
SNAILS, ODOR, COOLING TOWERS, FILTERS, EFFLUENT SAMPLING.

IDENTIFIERS:
HYDROGEN SULFIDE, IMHOFF TANKS, PRECIPITATES.

ABSTRACT:
THE WORLDS LARGEST ELECTRIC MACHINE TOOL PLANT REUSES IT'S WATER ON A
REGULAR BASIS. INITIALLY, THE PLANT HAD AN ADEQUATE WATER SUPPLY FROM 3
TO 5 WELLS PUMPING AT A RATE OF 100 TO 150 GPM. PLANT EXPANSION AND NEW
PROCESSES FORCED THE DEVELOPMENT OF A WATER REUSE SCHEME. WASTE WATER
FROM THE PLANT IS TREATED IN 2 IMHOFF TANKS AND A HIGH RATE TRICKLING
FILTER AND THEN STORED IN A LARGE POND. WATER IS DRAWN FROM THE POND,
FILTERED AND FED BACK TO THOSE PARTS OF THE PLANT THAT CAN USE IT OR BE
ADAPTED TO USE IT. THERE WERE NUMEROUS PROBLEMS ASSOCIATED WITH
PERFECTION THIS SYSTEM, THE POND GOES ANAEROBIC IN THE SUMMER, HYDROGEN
SULFIDE PRODUCED IN THE POND ESCAPED INTO THE PLANT, SNAILS COLLECTED
IN DEAD ENDS, ALGAL FILAMENTS CLOGGED HEAT EXCHANGERS, COOLING WATER
TUNNELS WERE STOPPED UP AND ODORS WERE EMITTED FROM THE EVAPORATIVE
COOLERS. APPROACHES TO THE SOLUTION OF THESE PROBLEMS, SUCCESSFUL AND
UNSUCCESSFUL, ARE PRESENTED AND THE RESULTS DISCUSSED. A UNIQUE PROFILE
SAMPLER WAS DEVELOPED AND USED TO DETERMINE THE WATER QUALITY. THE
CHANGES THAT OCCUR IN THE POND ARE DISCUSSED AND AN INDICATION OF THE
PROCESSES BRINGING ABOUT THESE CHANGES ARE PRESENTED. PREDICTABLY, THE
POND IS GRADUALLY FILLING WITH PRECIPITATES. IN THE PROCESS, THE
COMPANY HAS MADE A LOT OF USE OF A LITTLE WATER AND IS DISCHARGING A
BETTER EFFLUENT TO THE RECEIVING STREAM. (GOESSLING-TEXAS)

FIELD 05D

ACCESSION NO. W72-00027

WHAT IS POLLUTION,

M. I. GOLDMAN, AND R. SHOOP.

IN: CONTROLLING POLLUTION: THE ECONOMICS OF A CLEANER AMERICA, P 59-70,
   PRENTICE HALL INC., ENGLEWOOD CLIFFS, NEW JERSEY, 1967. 12 P.

   DESCRIPTORS:
      *WATER POLLUTION SOURCES, *WATER POLLUTION EFFECTS, *CLASSIFICATION,
      *WATER POLLUTION CONTROL, WATERCOURSES(LEGAL), ORGANIC WASTES, PULP
      WASTES, INDUSTRIAL WASTES, TOXINS, SOLID WASTES, RADIOACTIVE WASTES,
      HEATED WATER, THERMAL POLLUTION, BACTERIA, EFFLUENTS, OXYGEN DEMAND,
      AEROBIC TREATMENT, PHOTOSYNTHESIS, ALGAE, SEWAGE, MUNICIPAL WASTES,
      SEWAGE TREATMENT, TREATMENT FACILITIES, LEGAL ASPECTS.

   ABSTRACT:
      A BRIEF SURVEY IS MADE OF THE VARIOUS KINDS OF AIR AND WATER POLLUTION,
      THE MEANS OF MEASURING THESE POLLUTANTS, AND THE METHODS EMPLOYED TO
      CONTROL POLLUTION. POLLUTION IS A RELATIVE CONCEPT, AS POLLUTION OCCURS
      ONLY WHEN IMPURITIES RISE ABOVE A CERTAIN SPECIFIED LEVEL. WATER
      POLLUTION CAN BE CLASSIFIED IN FIVE ASPECTS: (1) PUTRESCIBLE MATERIALS
      ARE ORDINARY DOMESTIC AND INDUSTRIAL ORGANIC WASTES; (2) HEATED
      EFFLUENTS THAT CAUSE THERMAL POLLUTION; (3) TOXINS WHICH ARE NOT EASILY
      BROKEN DOWN BY BIOLOGICAL MEANS AND WHICH ARE NOT READILY NEUTRALIZED
      BY WATERCOURSES; (4) INERT WASTES THAT ENTER AS SOLIDS INTO WATER BUT
      ARE NOT INVOLVED IN CHEMICAL REACTIONS; AND (5) RADIOACTIVE SUBSTANCES.
      THE ARTICLE DESCRIBES THE PROCESS OF OXYGEN DEPLETION IN WATERS AND THE
      NATURAL AND ARTIFICIAL MEANS TO RESTORE AN OXYGEN BALANCE. PHYSICAL AND
      BIOLOGICAL MEANS TO MEASURE WATER POLLUTION ARE EXAMINED, ALONG WITH
      THE METHODS EMPLOYED TO TREAT THESE VARIOUS POLLUTANTS. THE PROCESS AND
      EFFECTIVENESS OF PRIMARY, SECONDARY, AND TERTIARY SEWAGE TREATMENT ARE
      DISCUSSED. (REES-FLORIDA)

   FIELD 05G, 05B

   ACCESSION NO. W72-00462

AQUATIC PLANT COMMUNITIES IN LAKE BUTTE DES MORTS: PHASE 2. EFFECTS OF HIGHER
   AQUATIC PLANTS, MARSH WATER AND MARSH SEDIMENTS ON PHYTOPLANKTON,

WISCONSIN STATE UNIV., OSHKOSH. DEPT. OF BIOLOGY.

WILLIAM E. SLOEY.

WISCONSIN DEPT OF NATURAL RESOURCES, MADISON, PROJECT COMPLETION REPORT,
   SUMMER 1969. 20 P, 3 FIG, 6 TAB, 4 REF.

   DESCRIPTORS:
      *AQUATIC PLANTS, *INHIBITORS, *MARSHES, *LAKES, *PHYTOPLANKTON,
      BIOASSAY, LABORATORY TESTS, PRODUCTIVITY, ABSORPTION, SEDIMENTS,
      ORGANIC MATTER, ALGAE, DIATOMS, CYANOPHYTA, CHLOROPHYTA, BULRUSH, PLANT
      GROWTH SUBSTANCES, CHLORELLA, CHLAMYDOMONAS, SCENEDESMUS, WISCONSIN.

   IDENTIFIERS:
      *SCIRPUS FLUVIATILIS, *LAKE BUTTE DES MORTS(WISC), GROWTH STIMULATION,
      SURFACE LITTER.

   ABSTRACT:
      TO VERIFY DIFFERENCES BETWEEN MARSH AND OPEN LAKE WATER IN LAKE BUTTE
      DES MORTS (WISCONSIN) AND TO DETERMINE IF SOME SUBSTANCE OR SUBSTANCES
      ARE PRODUCED IN MARSHES WHICH INHIBIT OR LIMIT GROWTH OF PLANKTONIC
      ALGAE, THE EFFECTS OF MARSH WATER, BOTTOM MUD, AND AQUATIC PLANT
      EXTRACTS UPON THE PHOTOSYNTHETIC RATE OF NATURAL PHYTOPLANKTON BY USE
      OF THE C-14 PRIMARY PRODUCTION TECHNIQUE WERE STUDIED. PARTIALLY
      DECAYED LITTER AND SURFACE ORGANIC PARTICLES INHIBITED PHYTOPLANKTON
      C-14 PRODUCTIVITY WHILE SUB-SURFACE LITTER AND DECAYED ORGANIC MATTER
      STIMULATED PHYTOPLANKTON. MILLIPORE FILTERED MARSH WATER SOMETIMES
      STIMULATED AND SOMETIMES INHIBITED C-14 UPTAKE. A FEW OF THE 16 NATIVE
      MARSH PLANTS TESTED INHIBITED C-14 UPTAKE WHEN MACERATED EXTRACTS OF
      THE PLANTS WERE EMPLOYED; THE MOST PRONOUNCED INHIBITOR WAS SCIRPUS
      FLUVIATILIS. THE PRESENT STANDS OF S FLUVIATILIS IN LAKE BUTTE DES
      MORTS SHOULD BE PROTECTED AND ENCOURAGED TO PROLIFERATE. SOME OTHER
      PLANT EXTRACTS GREATLY STIMULATED C-14 UPTAKE OF PHYTOPLANKTON, PERHAPS
      DUE TO THEIR PRODUCTION OF STIMULATORY SUBSTANCES OR TO EXTRACTION OF
      MINERALS AND/OR PLANT GROWTH SUBSTANCES FROM THE TISSUES DURING
      MACERATION. (JONES-WISCONSIN)

   FIELD 05C

   ACCESSION NO. W72-00845

MASS CULTURING OF ALGAE IN CONTROLLED LIGHT-DARK DETENTION SYSTEMS: I. EFFECTS OF CULTURE DENSITY,

NORWICH UNIV., NORTHFIELD, VT. DEPT. OF CIVIL ENGINEERING.

G. R. PYPER, W. J. WEBER, JR., AND J. A. BORCHARDT.

PROCEEDINGS, INDUSTRIAL WASTE CONFERENCE, 21ST, MAY 3, 4, AND 5, 1966, P 1003-1020, 8 FIG, 2 TAB, 20 REF. USPRS RES. GRANT WP-00155.

DESCRIPTORS:
*ALGAE, NITROGEN, PHOSPHORUS, *EUTROPHICATION, PHOTOSYNTHESIS, LAGOONS, LABORATORY TESTS, LIGHT INTENSITY, PHOTOPERIODISM, LIGHT PENETRATION, WATER POLLUTION EFFECTS, WATER POLLUTION CONTROL, *NUTRIENTS.

IDENTIFIERS:
*NUTRIENT REMOVAL, CONTINUOUS FLOW SYSTEM.

ABSTRACT:
THE POLLUTION-ACCELERATED AGING OF SURFACE WATERS HAS BECOME A MATTER OF CONCERN OVER THE PAST SEVERAL YEARS. CONTINUING INCREASES IN THE VOLUMES OF WASTEWATER INDICATE THAT MORE THOROUGH PURIFICATION OF THESE DISCHARGES MAY BE THE ONLY FEASIBLE MEASURE FOR THE PROTECTION OF OUR NATURAL WATERS. AQUATIC PLANTS CAN BE UTILIZED IN LAGOONS TO EXTRACT PHOSPHORUS AND NITROGEN FROM WASTEWATERS PRIOR TO DISCHARGE TO THE RECEIVING WATER. LABORATORY STUDIES HAVE BEEN MADE ON THE ALGAL GROWTH RATES AND YIELDS AS A FUNCTION OF DARK TO LIGHT RATIOS AND AS A FUNCTION OF ALGAL CONCENTRATION. THE GENERAL CONCLUSIONS ARE: (1) MASS CULTURES OF ALGAE CAN BE GROWN IN A CONTINUOUS SYSTEM CONSISTING OF A DARK RESERVOIR FROM WHICH THE CULTURE CAN BE CIRCULATED THROUGH A LIGHT CELL; (2) THERE IS A MAXIMUM LIGHT UTILIZATION RATE WHICH GIVES A MAXIMUM YIELD WITH RESPECT TO DARK TO LIGHT DETENTION TIMES; (3) GROWTH RATE VARIES LINEARLY WITH DARK TO LIGHT RATIOS OVER A BROAD RANGE OF VALUES; (4) NUTRIENT REMOVAL CHARACTERISTICS ARE NOT ADVERSELY AFFECTED BY LIGHT AND DARK CULTURING SYSTEMS; (5) NUTRIENT REMOVAL APPEARS TO BE A FUNCTION OF CULTURE DENSITY; (6) CURRENT WORK IS ONLY A BEGINNING AND MUST BE EXTENDED; AND (7) A MAJOR FACTOR TO BE STUDIED IS THE EFFECT OF FLOW VELOCITY THROUGH THE LIGHT CELL. (GOESSLING-TEXAS)

FIELD 05C, 05G

ACCESSION NO. W72-00917

ENVIRONMENTAL MONITORING ASSOCIATED WITH DISCHARGES OF RADIOACTIVE WASTE DURING 1969 FROM UKAEA ESTABLISHMENTS,

UNITED KINGDOM ATOMIC ENERGY AUTHORITY, HERWELL (ENGLAND). AUTHORITY HEALTH AND SAFETY BRANCH.

E. T. WRAY.

AVAILABLE FROM THE NATIONAL TECHNICAL INFORMATION SERVICE AS AHSB(RP)R-105. PRICE - $3.00, MICROFICHE $0.95. ATOMIC HEALTH AND SAFETY BRANCH REPORT NO. AHSB(RP)R-105.

DESCRIPTORS:
*MONITORING, *DISCHARGE MEASUREMENT, *EFFLUENTS, *RADIOISOTOPES, *WASTE DISPOSAL, *WASTE STORAGE, *WATER POLLUTION, WATER POLLUTION SOURCES, WATER POLLUTION CONTROL, RADIOACTIVE WASTE, WATER QUALITY ACT, PATH OF POLLUTANT, FOOD CHAINS, SAMPLING, SURVEYS, MARINE ALGAE, AQUATIC LIFE, MILK, ATLANTIC OCEAN, LANDFILLS, DISPOSAL, REGULATION, WATER POLICY, GOVERNMENTS, GASES, NUCLEAR WASTES.

IDENTIFIERS:
CONCENTRATION, ENVIRONMENTAL MONITORING.

ABSTRACT:
THIS REPORT SUMMARISES THE RESULTS OF THE MONITORING PROGRAMS CARRIED OUT IN CONNECTION WITH RADIOACTIVE WASTE DISCHARGES FROM THE PRINCIPAL ESTABLISHMENTS OF THE UKAEA IN 1969. THE INFORMATION GIVEN INCLUDES THE ESSENTIAL FREATURES OF THE AUTHORIZATIONS FOR DISCHARGES WHICH WERE MADE THE RESULTS OF ENVIRONMENTAL MONITORING. DERIVED WORKING LIMITS ARE QUOTED TO PROVIDE A BASIS FOR COMPARISON OF THE ENVIRONMENTAL MONITORING RESULTS. MOST ESTABLISHMENTS CARRIED OUT SAMPLING AND ANALYSIS ADDITIONAL TO THE MINIMUM REQUIRED BY STATUTE, AND SOME REFERENCE TO THIS HAS BEEN MADE IN THE REPORT. ONE OF THE PRINCIPAL METHODS OF ESTIMATING THE EFFECT OF GASEOUS DISCHARGES TO THE ENVIRONMENT IS BY THE SAMPLING AND ANALYSIS OF MILK. THE ACTIVITY FOUND IN THE VICINITY OF AUTHORITY ESTABLISHMENTS IS ALMOST ALWAYS DUE TO FALLOUT AND NOT TO CONTAMINATION OF THE ENVIRONMENT BY THE ESTABLISHMENT ITSELF. METHODS OF TREATMENT OF RADIOACTIVE WASTE IN THE UKAEA HAVE BEEN UNDER INVESTIGATION WITH THE OBJECT OF IMPROVING THE SYSTEM OF HANDLING AND DISCHARGE OF THE WASTE. THE APPENDIX TO THE REPORT CONTAINS A SUMMARY OF THE GENERAL PRINCIPLES GOVERNING THE NATURE OF THE POSSIBLE HAZARDS WHICH MIGHT ARISE AS A CONSEQUENCE OF THESE DISCHARGES. (HOUSER-NSIC)

FIELD 05B

ACCESSION NO. W72-00941

EVALUATION OF EFFECT OF IMPOUNDMENT ON WATER QUALITY IN CHENEY RESERVOIR,

COLORADO STATE UNIV., FORT COLLINS.

J. C. WARD, AND S. KARAKI.

RESEARCH REPORT NO 25, SEPT 1969. 67 P, 38 FIG, 19 TAB, 23 REF.

DESCRIPTORS:
EFFECTS, *IMPOUNDED WATERS, *WATER QUALITY, *DISSOLVED SOLIDS, DISSOLVED OXYGEN, EVAPORATION CONTROL, WATER BUDGET, WATER CHEMISTRY, WATER TEMPERATURE, HEAT BUDGET, TURBIDITY, *RESERVOIR EVAPORATION, SALINITY, ALGAE, STRATIFICATION, SALT BALANCE, BIBLIOGRAPHIES.

IDENTIFIERS:
USBR RESEARCH CONTRACTS, CHENEY DAM, KANS.

ABSTRACT:
A STUDY WAS CONDUCTED TO DETERMINE THE EFFECT OF IMPOUNDMENT ON THE QUALITY OF WATER IN CHENEY RESERVOIR NEAR WICHITA, KANS. THE RESULTS SHOWED THAT AN INCREASE IN THE DISSOLVED SOLIDS CONCENTRATION WAS DIRECTLY RELATED TO EVAPORATION. ON AN ANNUAL BASIS, 42% OF THE TOTAL INFLOW WAS EVAPORATED FROM THE RESERVOIR. EVAPORATION CONTROL WAS SUGGESTED FOR CONTROLLING DISSOLVED SOLIDS CONCENTRATION; HOWEVER, THIS WOULD CAUSE A POSSIBLY UNDESIRABLE INCREASE IN RESERVOIR TEMPERATURE OF 12 TO 19 DEG F. THE BIOLOGICAL ACTIVITY WITHIN THE RESERVOIR DID NOT SEEM TO AFFECT THE WATER QUALITY MATERIALLY, ALTHOUGH SOME ODORS WERE DETECTED. THE ODOR APPEARS TO HAVE STABILIZED AT A THRESHOLD NUMBER OF ABOUT 5. THE EFFECT OF THE INTERACTION BETWEEN MICRO-ORGANISMS AND NUTRIENTS WAS CHARACTERIZED IN THE ANALYSIS OF PHOSPHATES, NITRATES, AND SILICA CONCENTRATIONS IN THE RESERVOIR. THE DISSOLVED OXYGEN SATURATION DECREASED FROM 100% AT THE WATER SURFACE TO 82% AT 25 FT DEPTH. NO RESERVOIR STRATIFICATION WAS DETECTED DURING THE PERIOD OF DATA COLLECTION. (USBA)

FIELD 05F

ACCESSION NO. W72-01773

BIOLOGY OF WATER POLLUTION: A COLLECTION OF SELECTED PAPERS ON STREAM
POLLUTION, WASTE WATER, AND WATER TREATMENT.

FEDERAL WATER POLLUTION CONTROL ADMINISTRATION, WASHINGTON, D.C.

FEDERAL WATER POLLUTION CONTROL ADMINISTRATION PUBLICATION CWA-3, 1967. 290
P, 107 FIG, 58 TAB, 2 PHOTO, 530 REF.

DESCRIPTORS:
*WATER POLLUTION, WASTE DISPOSAL, *SEWAGE TREATMENT, *WASTE WATER
TREATMENT, *WASTE WATER DISPOSAL, *INDUSTRIAL WASTES, *WATER POLLUTION
CONTROL, FISH CONSERVATION, POLLUTION ABATEMENT, BIBLIOGRAPHIES,
ECOLOGY, BIOLOGY, FISH, PLANKTON, WATER QUALITY, ALGAE, AQUATIC LIFE,
STREAM POLLUTION, BACTERIA, IMPURITIES, WASTE TREATMENT.

ABSTRACT:
THIS BOOK OF SELECTED PUBLICATIONS ON BIOLOGY OF WATER POLLUTION, WATER
TREATMENT AND SEWAGE AND INDUSTRIAL WASTE TREATMENT CONTAINS SOME OF
THE MANY EXCELLENT AND BASIC PERTINENT BIOLOGICAL PAPERS THAT HAVE BEEN
COMMONLY INACCESSIBLE TO THE CONTEMPORARY INVESTIGATOR. THESE PAPERS
OFTEN QUOTED AND ARE A PORTION OF THE FOUNDATION UPON WHICH MODERN
AQUATIC ECOLOGICAL SCIENTIFIC THOUGHT AND DECISIONS ARE OFTEN BASED IN
SUMMATING WATER POLLUTION CONTROL INVESTIGATIONS. THIS COMPILED
COLLECTION WILL BE OF ASSISTANCE IN 3 PHASES OF WATER POLLUTION
ABATEMENT: (1) IT WILL PROVIDE A TECHNICAL SERVICE TO THE AQUATIC
ECOLOGIST THROUGH ASSEMBLAGE OF INFORMATIVE LITERATURE; (2) IT WILL
ILLUSTRATE MANY OF THE CONCEPTS UPON WHICH REGULATIONS HAVE BEEN
FORMULATED FOR THE PROTECTION OF AQUATIC LIFE; AND (3) IT WILL AID IN
THE TRAINING OF NEW ENVIRONMENTAL SCIENTISTS TO MEET TODAY'S AND
TOMORROW'S PERSONNEL NEEDS IN THE CONSERVATION OF OUR NATION'S NATURAL
RESOURCES. (SEE ALSO W72-01788 THRU W72-01819)

FIELD 05C, 05G

ACCESSION NO. W72-01786

SEWAGE, ALGAE AND FISH,

PUBLIC HEALTH SERVICE, CINCINNATI, OHIO.

FLOYD J. BRINLEY.

IN: BIOLOGY OF WATER POLLUTION, P 10-13. 1 FIG, 6 REF. COMPILED BY L. E.
KEUP, W. M. INGRAM, AND K. M. MACKENTHUN, FEDERAL WATER POLLUTION CONTROL
ADMINISTRATION, WASHINGTON, D C, 1967.

DESCRIPTORS:
*SEWAGE, *STREAMS, *WATER POLLUTION EFFECTS, *FISH, ALGAE, TROPHIC
LEVEL, PRODUCTIVITY, FOOD CHAINS, TOXICITY, WATER POLLUTION CONTROL,
DECOMPOSING ORGANIC MATTER, CARBON DIOXIDE.

IDENTIFIERS:
*OHIO RIVER BASIN.

ABSTRACT:
DISCHARGE OF RAW SEWAGE INTO STREAMS IMPARTS TOXICITY TO THE AREA
SURROUNDING THE OUTLET. AT A DISTANCE DOWNSTREAM, HOWEVER, SEWAGE
UNDERGOES BACTERIOLOGICAL DECOMPOSITION, LOSES ITS TOXICITY, AND ACTS
AS A FERTILIZER, STIMULATING THE GROWTH OF PLANKTON AND CONSEQUENT
INCREASE OF THE FISH POPULATION. SECONDARY SEWAGE TREATMENT ELIMINATES
TOXIC INGREDIENTS AND THE ENTIRE STREAM IS BENEFITED BY THE INFLOW OF
AVAILABLE NUTRIENTS. (SEE ALSO W72-01786) (WILDE-WISCONSIN)

FIELD 05C

ACCESSION NO. W72-01788

A HEAVY MORTALITY OF FISHES RESULTING FROM THE DECOMPOSITION OF ALGAE IN THE
YAHARA RIVER, WISCONSIN,

WISCONSIN DEPT. OF CONSERVATION, MADISON; AND WISCONSIN BOARD OF HEALTH,
MADISON.

KENNETH M. MACKENTHUN, AND ALFRED F. BARTSCH.

IN: BIOLOGY OF WATER POLLUTION, P 75-78. 3 FIG, 1 TAB, 2 REF. COMPILED BY L E
KEUP, W M INGRAM, AND K M MACKENTHUN, FEDERAL WATER CONTROL ADMINISTRATION,
WASHINGTON, D C, 1967.

DESCRIPTORS:
*FISHKILL, *CYANOPHYTA, *DECOMPOSING ORGANIC MATTER, TOXICITY,
DISSOLVED OXYGEN, ALGAE, WISCONSIN.

IDENTIFIERS:
*YAHARA RIVER(WIS), APHANIZOMENON FLOS AQUAE, OXYGEN DEPLETION.

ABSTRACT:
EARLY IN OCTOBER, 1946, A HEAVY MORTALITY OF ALL FISH SPECIES OCCURRED
IN THE YAHARA RIVER, WISCONSIN. BEFORE EXPIRING, THE FISH, INCLUDING
CARP, NORTHERN PIKE, WALLEYE PIKE, CRAPPIES, SUCKERS, AND EEL,
CONGREGATED NEAR THE SHORE BREATHING AT THE SURFACE. THE MORTALITY WAS
CORRELATED WITH THE CIRTICALLY LOW CONCENTRATION OF DISSOLVED
OXYGEN--UNDER 1.0 PPM; THIS DEFICIENCY RESULTED FROM THE DECOMPOSITION
OF AN ALGAL MASS, LARGELY APHANIZOMENON FLOS AQUAE. AS REVEALED BY
LABORATORY TRIALS, TOXIC SUBSTANCES RELEASED BY DECOMPOSING ALGAL
TISSUES WERE A CONTRIBUTING FACTOR. (SEE ALSO W72-01786)
(WILDE-WISCONSIN)

FIELD 05C

ACCESSION NO. W72-01797

SUGGESTED CLASSIFICATION OF ALGAE AND PROTOZOA IN SANITARY SCIENCE,

ROBERT A. TAFT SANITARY ENGINEERING CENTER, CINCINNATI, OHIO. INTERFERENCE
ORGANISMS STUDIES AND WATER SUPPLY AND WATER POLLUTION CONTROL RESEARCH.

C. MERVIN, PALMER, AND WILLIAM M. INGRAM.

IN: BIOLOGY OF WATER POLLUTION, P 79-83. 1 FIG, 1 TAB, 24 REF. COMPILED BY L.
E. KEUP, W. M. INGRAM, AND K. M. MACKENTHUN, FEDERAL WATER POLLUTION
CONTROL ADMINISTRATION, WASHINGTON, D.C., 1967.

DESCRIPTORS:
*ALGAE, *PROTOZOA, *SYSTEMATICS, *SANITARY ENGINEERING, MICROORGANISMS,
NUISANCE ALGAE, SEWAGE TREATMENT, SELF-PURIFICATION, DISSOLVED OXYGEN,
PHOTOSYNTHESIS.

IDENTIFIERS:
*PIGMENTED FLAGELLATES, *NON-PIGMENTED FLAGELLATES.

ABSTRACT:
TO MEET THE TAXONOMIC REQUIREMENTS OF SANITARY SCIENTISTS REGARDING THE
DEMARKATION LINE BETWEEN ALGAE AND PROTOZOA, IT IS RECOMMENDED TO
SEPARATE THE FLAGELLATES ON THE BASIS OF THEIR ABILITY TO PRODUCE
OXYGEN. ACCORDINGLY, THE FLAGELLATES WITH PHOTOSYNTHETIC PIGMENTS WOULD
BE ALGAE, AND WITHOUT THE PIGMENTS--PROTOZOA. A LIST OF FLAGELLATES OF
ALGAL AND PROTOZOAN TYPES IS INCLUDED. (SEE ALSO W72-01786)
(WILDE-WISCONSIN)

FIELD 05C, 02H

ACCESSION NO. W72-01798

AQUATIC ORGANISMS AS AN AID IN SOLVING WASTE DISPOSAL PROBLEMS,

ACADEMY OF NATURAL SCIENCES, PHILADELPHIA, PA. DEPT. OF LIMNOLOGY.

RUTH PATRICK.

IN: BIOLOGY OF WATER POLLUTION, P 108-110. 6 REF, DISCUSSION. COMPILED BY L E
   KEUP, W M INGRAM, AND K M MACKENTHUN, FEDERAL WATER POLLUTION CONTROL
   ADMINISTRATION, WASHINGTON, D C, 1967.

   DESCRIPTORS:
      *BIOASSAY, *AQUATIC LIFE, *WASTE DISPOSAL, *WATER POLLUTION EFFECTS,
      FOOD CHAINS, BACTERIA, DEGRADATION(DECOMPOSITION), ALGAE, OXYGENATION,
      SELF-PURIFICATION, LABORATORY TESTS, INVERTEBRATES, FISH, WATER
      ANALYSIS, INSECTS, SNAILS, TOXICITY, RIVERS, HISTOGRAMS.

   IDENTIFIERS:
      *RIVER SURVEYS.

   ABSTRACT:
      SEVERAL LABORATORY TESTS OR BIOASSAYS, SUPPLEMENTING DETERMINATIONS OF
      OXYGEN DEMANDS, ARE DESCRIBED AS AIDS IN SOLVING WASTE DISPOSAL
      PROBLEMS. THE SUGGESTED TEST ORGANISMS INCLUDE FISH, INSECTS, SNAILS,
      AND ALGAE, PARTICULARLY NITZSCHIA LINEARIS. THESE TESTS ARE INTENDED
      PRIMARILY TO DETECT THE TOXICITY OF POLLUTANTS. THE SECOND APPROACH,
      THE BIOLOGICAL SURVEY OF RIVERS, COMPRISES IDENTIFICATION OF ALL
      AQUATIC ORGANISMS, TOTAL BACTERIAL AND COLIFORM COUNTS, AND THE
      DETERMINATION OF BOD AND OTHER CHEMICAL CHARACTERISTICS OF WATER. (SEE
      ALSO W72-01786) (WILDE-WISCONSIN)

   FIELD 05C

   ACCESSION NO. W72-01801

CHEMICAL COMPOSITION OF ALGAE AND ITS RELATIONSHIP TO TASTE AND ODOR,

MICHIGAN UNIV., ANN ARBOR. SCHOOL OF PUBLIC HEALTH.

GERARD A. ROHLICH, AND WILLIAM B. SARLES.

IN: BIOLOGY OF WATER POLLUTION, P 232-235. COMPILED BY W. M. INGRAM, L. E.
   KEUP, AND K. M. MACKENTHUN, FEDERAL WATER POLLUTION CONTROL ADMINISTRATION,
   WASHINGTON, D. C., 1967. 4 TAB, 12 REF.

   DESCRIPTORS:
      *WATER POLLUTION SOURCES, *ALGAE, *TASTE, *ODOR, *AMINO ACIDS, POTABLE
      WATER, ACTINOMYCETES, PLANT GROWTH, DECOMPOSING ORGANIC MATTER,
      CHEMICAL ANALYSIS, PROTEINS, BACTERIA.

   IDENTIFIERS:
      ODORIFEROUS PRODUCTS, ORGANISMS' ODORS.

   ABSTRACT:
      THE OBJECTIONABLE TASTE AND ODOR OF DRINKING WATER IS LARGELY DUE TO
      DECOMPOSITION OF NUCLEO-PROTEINS, PHOSPHO-PROTEINS, LIPIDS, AND OTHER
      TISSUE CONSTITUENTS OF AQUATIC BIOTA, PARTICULARLY ALGAE IN ASSOCIATION
      WITH BACTERIA AND ACTINOMYCETES. THE ODOR OF DRINKING WATER, RANGING
      FROM VILE STENCH AND FISHY OR COD-LIVER OIL TO EARTHY, SWEET GRASSY AND
      AROMATIC, IS IMPARTED BY DIATOMECEAE, CYANOPHYCEAE, CHLOROPHYCEAE,
      ACTINOMYCES, AND PROTOZOA. THE ODOR MAY ALSO ORIGINATE FROM BACTERIAL
      ACTION LIBERATING AMMONIA FROM AMINO-ACIDS AND PRODUCING VARIOUS
      ALCOHOLS. ODOR MAY BE INTENSIFIED BY DECOMPOSITION OF ALGAE, ESPECIALLY
      FOLLOWING THEIR DESTRUCTION BY CHLORINATION, COPPER SULFATE, AND
      SIMILAR ERADICATING TREATMENTS. (SEE ALSO W72-01786) (WILDE-WISCONSIN)

   FIELD 05A, 05C

ACCESSION NO. W72-01812

AQUATIC BIOLOGY AND THE WATER WORKS ENGINEER,

JAMES B. LACKEY.

IN: BIOLOGY OF WATER POLLUTION, P 236-239. COMPILED BY W. M. INGRAM, L. E. KEUP, AND K. M. MACKENTHUN, FEDERAL WATER POLLUTION CONTROL ADMINISTRATION, WASHINGTON, D. C., 1967. 9 FIG.

DESCRIPTORS:
*SANITARY ENGINEERING, *WATER SUPPLY, *BIOLOGICAL PROPERTIES, *WATER POLLUTION SOURCES, TASTE, ODOR, ALGAE, RESERVOIRS, WATERSHEDS(BASINS), INSECTICIDES, HARDNESS(WATER), DIATOMS, FOULING, PLANKTON, SEWAGE, DOMESTIC WASTE, INDUSTRIAL WASTES, FARM WASTES.

IDENTIFIERS:
*BIOLOGICAL PROBLEMS, SCIOTO RIVER(OHIO), LAKE MENDOTA(WIS), LAKE MONONA(WIS), LAKE WAUBESA(WIS), LAKE KEGONSA(WIS).

ABSTRACT:
AN ENGINEER PROPOSING TO USE RIVER OR LAKE WATER FOR THE NEEDS OF A CITY OR AN INDUSTRY SHOULD CONSIDER THE PERTINENT PHYSICAL, CHEMICAL, AND BIOLOGICAL PROPERTIES OF THE WATER. THE USE OF WATER FOR EITHER HUMAN OR INDUSTRIAL CONSUMPTION MAY BE PRECLUDED BY A NUMBER OF UNFAVORABLE CHARACTERISTICS, SUCH AS HIGH CONCENTRATION OF NITRATES, PHOSPHATES, SEWAGE, AND SUBSEQUENT ALGAL BLOOMS, ABUNDANT PLANKTON WITH ITS HIGH BOD AND FILTER CLOGGING POTENTIAL, THE PRESENCE OF PESTICIDES, TOXIC INDUSTRIAL WASTES, OR PATHOGENIC ORGANISMS. (SEE ALSO W72-01786) (WILDE-WISCONSIN)

FIELD 05C

ACCESSION NO. W72-01813

PRE-TREATMENT BASIN FOR ALGAE REMOVAL,

MENASHA ELECTRIC AND WATER UTILITIES, WIS.

ANDREW J. MARX.

IN: BIOLOGY OF WATER POLLUTION, P 239-244. COMPILED BY W. M. INGRAM, L. E. KEUP, AND K. M. MACKENTHUN, FEDERAL WATER POLLUTION CONTROL ADMINISTRATION, WASHINGTON, D. C., 1967. 1 FIG, 2 PHOTOS.

DESCRIPTORS:
*WATER PURIFICATION, *ALGAL CONTROL, *WATER SUPPLY, *BASINS, WATER POLLUTION TREATMENT, COPPER SULFATE, ODOR, COLIFORMS, ECONOMICS.

IDENTIFIERS:
LAKE WINNEBAGO(WIS), MENASHA(WIS).

ABSTRACT:
FOLLOWING TRIALS WITH A SMALL PILOT PLANT, THE ELECTRIC AND WATER UTILITY COMPANY OF MENASHA, WISCONSIN CONSTRUCTED A 600 X 700 FT PRETREATMENT BASIN AT THE MOUTH OF FOX RIVER. THIS BASIN PERMITTED ALGAE CONTROL BY A SOLUTION OF COPPER SULFATE, PREPARED BY DISSOLVING 500 LBS OF COPPER SULFATE AND 1 QT OF SULFURIC ACID IN 1,000 GALLONS OF WATER IN A WOODEN TANK. THE TANK IS CONNECTED TO A SMALL LIQUID FEEDER WITH A METERING DEVICE TO DELIVER 2 PPM OF COPPER SULFATE. THIS WATER PRETREATMENT ERADICATED ALGAE AND WEEDS, DECREASED THE TURBIDITY, REDUCED ODOR AND COLIFORM ORGANISMS, REDUCED THE EXPENSE ON CHEMICALS, ELIMINATED THE ANCHOR ICE, AND GREATLY SIMPLIFIED THE MANAGEMENT. IN 1951 THE AGE OF THE INSTALLATION WAS 5 YEARS. (SEE ALSO W72-01786) (WILDE-WISCONSIN)

FIELD 05F, 05C

ACCESSION NO. W72-01814

PROTOZOA AND ACTIVATED SLUDGE,

MASSACHUSETTS INST. OF TECH., CAMBRIDGE. DEPT. OF SANITARY ENGINEERING.

ROSS E. MCKINNEY, AND ANDREW GRAM.

IN: BIOLOGY OF WATER POLLUTION, P 252-262. COMPILED BY W. M. INGRAM, L. E.
KEUP, AND K. M. MACKENTHUN, FEDERAL WATER POLLUTION CONTROL ADMINISTRATION,
WASHINGTON, D. C., 1967. 7 FIG, 1 TAB, 18 REF.

DESCRIPTORS:
*SEWAGE TREATMENT, *PROTOZOA, *ACTIVATED SLUDGE, *BACTERIA, METABOLISM,
GROWTH RATES, MICROORGANISMS, ALGAE, FUNGI, LABORATORY TESTS,
BIOCHEMICAL OXYGEN DEMAND, SEWAGE EFFLUENTS, SANITARY ENGINEERING.

IDENTIFIERS:
SARCODINA, MASTIGOPHORA, HOLOZOIC MASTIGOPHORA, HOLOPHYTIC
MASTIGOPHORA, CILIATA.

ABSTRACT:
TO DELINEATE THE ROLE OF PROTOZOA IN SLUDGE ACTIVATION, EXPERIMENTS
WERE CONDUCTED UNDER THE CONCEPT OF 'MINIMUM CONDITIONS.' THE LATTER
WERE ESTABLISHED BY A COMBINATION OF A BALANCED SOLUBLE SUBSTRATE,
BACTERIA FROM ACTIVATED SLUDGE, AND A PURE CULTURE OF PROTOZOA. THE
FLOC FORMED WAS THE RESULT OF METABOLIC ACTIVITY OF MICROORGANISMS IN
THE MIXED LIQUOR. THE FLAGELLATE AND CILIATE PROTOZOA INCLUDED
CHILOMONAS PARAMECIUM, EUGLENA GRACILIS, TETRAHYMENA FILII, AND
GLAUCOMA SCINTILLANS. THE TRIALS DEMONSTRATED THE GENERAL PRINCIPLES OF
COMPETITION OF PROTOZOA WITH BACTERIA AND THE IMPORTANCE OF PROTOZOA IN
SLUDGE ACTIVATION. THE SUCCESSION OF PROTOZOA--SARCODINA, THE
HOLOPHYTIC AND HOLOZOIC MASTIGOPHORA, AND THE FREE-SWIMMING AND STALKED
CILIATA--WAS ESTABLISHED TO SERVE AS A GUIDE TO BETTER OPERATIONS. THE
BACTERIAL FLOCCULATION WAS CONFIRMED IN THE CONTROL UNIT. (SEE ALSO
W72-01786) (WILDE-WISCONSIN)

FIELD 05D, 05C

ACCESSION NO. W72-01817

BIOLOGICAL FACTORS IN TREATMENT OF RAW SEWAGE IN ARTIFICIAL PONDS,

ROBERT A. TAFT SANITARY ENGINEERING CENTER, CINCINNATI, OHIO; AND SOUTH
DAKOTA STATE COLL., BROOKINGS. DEPT. OF ZOOLOGY.

A. F. BARTSCH, AND M. O. ALLUM.

IN: BIOLOGY OF WATER POLLUTION, P 262-269. COMPILED BY L. E. KEUP, W. M.
INGRAM, AND K. M. MACKENTHUN, FEDERAL WATER POLLUTION CONTROL
ADMINISTRATION, WASHINGTON, D. C., 1967. 6 FIG, 7 TAB, 18 REF.

DESCRIPTORS:
*SEWAGE TREATMENT, *OXIDATION LAGOONS, *ALGAE, *BACTERIA, *DISSOLVED
OXYGEN, SEWAGE LAGOONS, PLANT PHYSIOLOGY, ANIMAL PHYSIOLOGY, NORTH
DAKOTA, SOUTH DAKOTA, PHOTOSYNTHESIS, LIGHT PENETRATION, ON-SITE TESTS,
BIOCHEMICAL OXYGEN DEMAND, DIURNAL, CHLOROPHYLL.

IDENTIFIERS:
*PHOTOSYNTHETIC OXYGEN, LEMMON(SD), KADOKA(SD).

ABSTRACT:
FREEDOM FROM NUISANCE CONDITIONS AND THE ACCEPTABILITY OF RAW SEWAGE
PONDS TO MUNICIPALITIES AS TREATMENT FACILITIES ARE INTIMATELY
DEPENDENT UPON THE DISSOLVED OXYGEN SUPPLY IN SEWAGE. LIGHT AND DARK
BOTTLE ANALYSES AND THE ESTIMATED OXYGEN PRODUCTION BASED ON VERDUIN'S
FACTOR PERMIT ESTABLISHMENT OF SUITABLE LOADING AND DEPTH OF SEWAGE
LAGOONS. THE USE OF POND DESIGN EQUATIONS, HOWEVER, REQUIRES CAUTION
BECAUSE PROFILE VARIATIONS, WEATHER CONDITIONS, DIURNAL CHANGES, AND
OTHER FACTORS MAY IMPOSE CONSIDERABLE DEVIATION IN DESIGN. (SEE ALSO
W72-01786) (WILDE-WISCONSIN)

FIELD 05D, 05C

ACCESSION NO. W72-01818

TRICKLING FILTER ECOLOGY,

ROBERT A. TAFT SANITARY ENGINEERING CENTER, CINCINNATI, OHIO.

WILLIAM B. COOKE.

IN: BIOLOGY OF WATER POLLUTION, P 269-287. COMPILED BY L. E. KEUP, W. M.
    INGRAM, AND K. M. MACKENTHUN, FEDERAL WATER POLLUTION CONTROL
    ADMINISTRATION, WASHINGTON, D. C., 1967. 7 FIG, 3 TAB, 45 REF.

    DESCRIPTORS:
      *BIOLOGICAL COMMUNITIES, *SEWAGE TREATMENT, *TRICKLING FILTERS, *BIOTA,
      SEWAGE BACTERIA, SYSTEMATICS, FUNGI, ALGAE, DIATOMS, PROTOZOA,
      PLANKTON, WORMS, NEMATODES, ROTIFERS, LARVAE, SNAILS, MITES, MOLDS,
      GROWTH RATES, HABITATS.

    IDENTIFIERS:
      *ZOOGLEAL SLIME, DAYTON(OHIO), VANDALIA(OHIO), RHIZOPODS, SPIDERS, MILK
      WASTE TREATMENT, COLONY COUNTS, ARTHROPODS.

    ABSTRACT:
      THE POPULATIONS OF SEVERAL TRICKLING FILTERS IN THE USA AND ENGLAND ARE
      REVIEWED. THE COMPILED LIST OF THE COMMON MEMBERS OF THIS MAN-MADE
      ECOSYSTEM, EXCEEDING 200 SPECIES, INCLUDES HERBIVORES, CARNIVORES, AND
      SAPROBES, SERVING AS PRIMARY OR SECONDARY DECOMPOSERS. THIS POPULATION
      EFFECTIVELY REMOVES SOLIDS, COLLOIDS, AND DISSOLVED MATTER, THUS
      REDUCING BOD OF THE SEWAGE. NO CORRELATION BETWEEN THE SURFACE GROWTH
      AND TEMPERATURE OR PRECIPITATION WAS OBSERVED. THE FUNCTIONING OF THE
      POPULATION DEPENDS LARGELY ON PREFORMED ORGANIC MATTER. AT DEATH THE
      ORGANISMS ARE ATTACKED BY SCAVENGERS AND THUS REENTER THE METABOLIC
      SYSTEM. (SEE ALSO W72-01786) (WILDE-WISCONSIN)

    FIELD 05D, 05C

    ACCESSION NO. W72-01819

CHEMISTRY OF NITROGEN AND PHOSPHORUS IN WATER.

AMERICAN WATER WORKS ASSOCIATION, NEW YORK. WATER QUALITY DIV. COMMITTEE ON
    NUTRIENTS IN WATER.

    DESCRIPTORS:
      WASTE WATER TREATMENT, WATER TREATMENT, *ALGAE, *EUTROPHICATION,
      *NITROGEN, NITROGEN CYCLE, *PHOSPHORUS, PHOSPHORUS COMPOUNDS, WATER
      SUPPLY, WATER MANAGEMENT.

    ABSTRACT:
      OF THE MAJOR ELEMENTS ESSENTIAL TO ALGAL GROWTH, NITROGEN AND
      PHOSPHORUS ARE THE ONES MOST LIKELY TO BE FOUND IN LIMITED QUANTITIES
      IN NATURAL WATERS. SINCE THEY THEREFORE REPRESENT PROMISING WEAK LINKS
      IN ALGAL LIFE CYCLES, THEIR CHEMICAL STATES AND BEHAVIOR IN WATER ARE
      EXAMINED TO SEE HOW WATER TREATMENT MIGHT BENEFIT. LARGE SUPPLIES OF
      NITROGEN AND PHOSPHORUS ARE PRESENT IN MANY BODIES OF WATER EITHER IN
      THE SEDIMENTS, THE ATMOSPHERE ABOVE, OR IN THE FORM OF DISSOLVED GAS.
      THESE FORMS MAY BE AVAILABLE FOR THE GROWTH OF ALGAE AND OTHER AQUATIC
      PLANTS, BUT THE RATES AT WHICH THEY BECOME AVAILABLE IS SLOW. THESE
      RATES ARE IMPORTANT, HOWEVER, AS THEY TEND TO CONTROL THE AMOUNT OF
      VEGETABLE GROWTH WHICH CAN BE SUPPORTED. SOLUBLE NITROGEN AND
      PHOSPHORUS CONTAINED IN WASTEWATER EFFLUENTS ARE IN A READILY AVAILABLE
      FORM. IF DISCHARGED TO NATURAL WATERS, THEY CAN STIMULATE GROWTH FAR IN
      EXCESS OF THAT WHICH WOULD OCCUR NATURALLY. THUS, IN ASSESSING THE
      EXTENT OF NUTRIENT-RELATED PROBLEMS AND THEIR CONTROL, THE WATER
      MANAGER MUST EVALUATE THE SIGNIFICANCE OF THE READILY AVAILABLE FORMS
      AND ALSO THE FORMS WHICH MAY BE RELEASED SLOWLY FROM THE SUSPENDED
      PARTICLES AND THE SEDIMENTS. TO DO THIS, THE BODY OF WATER MUST BE
      CONSIDERED A CHEMICAL REACTOR WHICH IS CONTROLLED BY THE KINETICS OF A
      NUMBER OF PROCESSES. (GOESSLING-TEXAS)

    FIELD 05C, 05D, 02K

    ACCESSION NO. W72-01867

EUTROPHICATION: SMALL FLORIDA LAKES AS MODELS TO STUDY THE PROCESS,

FLORIDA UNIV., GAINSVILLE. DEPT. OF ENVIRONMENTAL ENGINEERING.

PATRICK L. BREZONIK, AND HUGH D. PUTNAM.

PROCEEDINGS OF THE SEVENTEENTH SOUTHERN WATER RESOURCES AND POLLUTION CONTROL
CONFERENCE, APRIL 16-18, 1968. P 315-333, 3 FIG, 8 TAB, 9 REF. GRANT NO.
16010DON.

DESCRIPTORS:
*EUTROPHICATION, *LIMNOLOGY, *ESSENTIAL NUTRIENTS, *PRIMARY
PRODUCTIVITY, AQUATIC ALGAE, TROPHIC LEVEL, WATER QUALITY, FLORIDA.

IDENTIFIERS:
*ANDERSON-CUE LAKE, MELROSE(FLORIDA).

ABSTRACT:
EUTROPHICATION IS A PROCESS OF LAKE AGING WHICH CAN BE ACCELERATED BY
NUTRIENT ENRICHMENT. THIS ENRICHMENT MANIFESTS ITSELF IN A VARIETY OF
LARGELY DELETERIOUS EFFECTS ON THE LACUSTRINE WATER QUALITY AND BIOTA.
RESPONSE OF LAKES TO INCREASED NUTRIENT LOADS IS HIGHLY INDIVIDUALISTIC
AND DEPENDENT ON NUMEROUS PHYSICAL, CHEMICAL AND BIOLOGICAL FACTORS
WHICH DEFINE THE LAKE'S ORIGINAL TROPHIC STRUCTURE. THE ASSIMILATIVE
CAPACITY OF A LAKE, I.E. THE NUTRIENT LOAD IT CAN ACCEPT WITHOUT
DEVIATING FROM CERTAIN IMPOSED WATER QUALITY STANDARDS AND CONDITIONS,
MUST BE KNOWN FOR INTELLIGENT WATER QUALITY MANAGEMENT. A SMALL
ISOLATED FLORIDA LAKE HAS RECEIVED A CONTROLLED AMOUNT OF NUTRIENT
LOADING. ANDERSON-CUE IS AN OLIGOTROPHIC LAKE OF 19.1 ACRES. A
PRELIMINARY INVESTIGATION INDICATED A SMALL NUTRIENT SUPPLY IN THE
LAKE, WHICH REFLECTED THE IMPOVERISHED SANDY SOILS, SPARSE LAKE BIOTA,
AND LOW PRIMARY PRODUCTIVITY. SEWAGE EFFLUENT ENRICHED WITH NITROGEN
AND PHOSPHOROUS COMPOUNDS HAS BEEN ADDED TO THE EXPERIMENTAL LAKE SINCE
MARCH 1967. THE LAKE'S RESPONSE TO THE INCREASED NUTRIENT FLUX IS BEING
DETERMINED BY MONITORING A VARIETY OF CHEMICAL AND BIOLOGICAL
PARAMETERS, INCLUDING PHYTOPLANKTON SPECIES COMPOSITION, PRIMARY
PRODUCTION, CHLOROPHYLL, NITROGEN AND PHOSPHOROUS FORMS, DISSOLVED
OXYGEN AND TRACE METALS. BIOASSAY METHODS HAVE INDICATED THAT
PHOSPHOROUS IS STILL THE LIMITING NUTRIENT. IN THE FIRST YEAR OF STUDY,
EFFECTS ON WATER QUALITY AND TROPHIC STATUS HAVE BEEN SMALL. (EPA
ABSTRACT)

FIELD 05B, 05C

ACCESSION NO. W72-01990

A PARTIAL CHECKLIST OF FLORIDA FRESH-WATER ALGAE AND PROTOZOA WITH REFERENCE TO
MCCLOUD AND CUE LAKES,

FLORIDA UNIV., GAINESVILLE. DEPT. OF ENVIRONMENTAL ENGINEERING.

JAMES B. LACKEY, AND ELSIE W. LACKEY.

AVAILABLE FROM THE NATIONAL TECHNICAL INFORMATION SERVICE AS PB-179 071,
$3.00 IN PAPER COPY, $0.95 IN MICROFICHE. WATER RESOURCES RESEARCH CENTER,
PUBLICATION NO. 3, BULLETIN SERIES NO. 131, FLORIDA ENGINEERING AND
INDUSTRIAL EXPERIMENT STATION, VOL 21, NO 11, NOV 1967, P 1-28. 16 TAB, 24
REF. OWRR B-004-FLA(7). FWPCA GRANT 16010 DON.

DESCRIPTORS:
*AQUATIC ALGAE, *PROTOZOA, *FLORIDA, *EUTROPHICATION, AQUATIC
MICROORGANISMS, TROPHIC LEVEL.

IDENTIFIERS:
CUE LAKE(FLORIDA), MCCLOUD LAKE(FLORIDA).

ABSTRACT:
FLORIDA IS A LAND WHERE THERE ARE MANY KINDS OF FRESH WATER AND MANY
KINDS OF CLIMATE. SINCE THERE IS USUALLY A DISTINCT RAINY SEASON,
BLOOMS TEND TO BE PREVALENT DURING OR JUST AFTER THIS SEASON. IN
EUTROPHIC WATERS, BLOOMS MAY OCCUR ANYTIME. ALL THESE MATTERS MAKE
FLORIDA WATERS AN EXTREMELY VALUABLE RESEARCH AREA FOR THE
MICROBIOLOGIST. THE MOST INTENSIVE WORK HAS BEEN DONE ON TWO LAKES NEAR
MELROSE, FLORIDA: CUE LAKE AND MCCLOUD LAKE. CUE LAKE BECAME SUBJECT TO
ROUTINE FERTILIZATION SINCE 1967. IN ADDITION TO THESE TWO LAKES, 33
OTHER LOCATIONS WERE SAMPLED. THE REPORT CONTAINS A LIST OF OCCURRING
SPECIES. A WIDE SPECTRUM OF ALGAE AND PROTOZOA WERE IDENTIFIED
EMPHASIZING THAT MOST OF THESE ORGANISMS ARE COSMOPOLITAN. MANY OF THEM
DO NOT CROSS BROAD ECOLOGICAL BOUNDARIES BUT NARROW BOUNDARIES AND EACH
SPECIES HAS A MAXIMUM SET OF CONDITIONS UNDER WHICH IT ATTAINS MAXIMUM
NUMBERS. IT IS USEFUL TO RECOGNIZE THE SPECIES WE SEE BECAUSE BY
CORRELATING NUMBERS OF INDIVIDUALS WITH KNOWN CONDITIONS WE MAY REACH
CONCLUSIONS REGARDING TAXONOMIC AND PHYSIOLOGICAL RELATIONSHIPS.

FIELD 05A, 05C

ACCESSION NO. W72-01993

PRELIMINARY RESEARCH IN THE LABORATORY ON EXPERIMENTAL BRACKISH ECOSYSTEMS,

J. C. LACAZE, C. HALLOPEAU, AND M. VOIGT.

BULL MUS NAT HIST NATUR (PARIS). 41(5): 1278-1289. ILLUS. 1969.

DESCRIPTORS:
ALGAE, BRACKISH, ECOSYSTEMS, FAUNA, FLORA, LABORATORY, MINERALS, POLLUTION.

ABSTRACT:
FOUR AQUATIC ECOSYSTEMS WERE STUDIED STARTING FROM ALGAE, SEDIMENT, AND WATER TAKEN FROM A BRACKISH MARSH OF THE ARCACHON REGION. THE IMPORTANT THING IS TO OBTAIN SYSTEMS POSSESSING CHARACTERISTIC STRUCTURE, BEHAVIOR AND REPLICABILITY. THE EXPERIMENT LASTED A YEAR (JULY 1966 - AUG. 1967) DURING WHICH TIME FAUNA, FLORA, NITRATES, PHOSPHATES, AND ALKALINITY WERE ANALYZED. SPECIAL IMPORTANCE WAS GIVEN TO THE MICRO-FLORA OF DIATOMS DEVELOPING ON THE IMMERSED SUBSTRATA. THE EXPERIMENTAL ECOSYSTEMS WERE STUDIED AS BIOLOGICAL MATERIAL FOR POLLUTION STUDIES.--COPYRIGHT 1971, BIOLOGICAL ABSTRACTS, INC.

FIELD 05A

ACCESSION NO. W72-02203

UTILIZATION OF ALGAE BY DAPHINIA AS INFLUENCED BY CELL SENESCENCE AND UV IRRADIATION,

MARYLAND UNIV., COLLEGE PARK. DEPT. OF ZOOLOGY.

R. G. STROSS, J. C. JONES, F. M. UNGER, AND J. M. VAIL.

PROCEEDINGS, INDUSTRIAL WASTE CONFERENCE, 20TH, MAY 4, 5, AND 6, 1965. P 706-714, 4 FIG, 5 TAB, 17 REF.

DESCRIPTORS:
WATER QUALITY, *ALGAL CONTROL, EUTROPHICATION, DAPHNIA, CHLAMYDOMONAS, LABORATORY TESTS, PARTICLE SIZE, POPULATION, GROWTH RATES, WATER POLLUTION CONTROL.

IDENTIFIERS:
ALGAL SENESCENCE, GROWTH INHIBITIONS.

ABSTRACT:
SUSPENSIONS OF ALGAE, BACTERIA AND DETRITUS ARE INGESTED BY MANY GROUPS OF ANIMALS. DAPHINIA ARE EQUIPPED WITH THORACIC LIMBS WHICH COLLECT PARTICLES FROM DILUTE SUSPENSIONS. THE NUMBER OF PARTICLES INGESTED DEPENDS ON SIZE, DENSITY AND OTHER FACTORS SUCH AS NUTRITIONAL DEFICIENCIES OR INDEGESTIBILITY. A STUDY HAS BEEN COMPLETED IN WHICH THE UTILIZATION OF CHLAMYDOMONAS REINHARDI (DANGEARD I.U. STRAINS NO. 89 AND 90) BY DAPHNIA PULEX DE GEER WAS MEASURED WHICH ALLOWED AN EVALUATION OF THE EFFECT OF SENESCENT AND UV IRRADIATED ALGAE ON THE INTRINSIC RATES OF INCREASE OF DAPHNIA POPULATIONS. THE RESULTS OF THESE TESTS WERE: (1) THE INTAKE OF ALGAL CELLS BY DAPHNIA DECREASED WITH THE AGE OF THE ALGAL CULTURE; (2) INHIBITION WAS PRODUCED BY THE CELLS AND THE PARTICLE-FREE MEDIUM OF THE CULTURE; (3) THE FEEDING OF SENESCENT CELLS REDUCED THE INTRINSIC RATE OF POPULATION INCREASE TO NEARLY ONE THIRD THAT OF CONTROL POPULATIONS; (4) BIRTH RATE OF DAPHNIA WAS MORE STRONGLY INFLUENCED THAN DEATH RATE; (5) IRRADIATION OF LOG-PHASE CELLS WITH FAR UV STIMULATES THE EFFECT OF CELL SENESCENCE ON GROWTH OF THE DAPHNIA; (6) LIPID PEROXIDES IN THE CELLS INCREASED WITH THE DOSE OF UV BUT MOST OF THE GROWTH INHIBITION RESULTED FROM THE MINIMUM DOSE; AND, (7) A HYPOTHETICAL POPULATION WAS CONSTRUCTED TO SHOW THE INFLUENCE OF ALGAL SENESCENCE ON DENSITY AND PERMISSABLE WASHOUT RATE OF DAPHNIA POPULATION IN EQUILIBRIUM WITH ITS FOOD SUPPLY. (GOESSLING-TEXAS)

FIELD 05G, 05C

ACCESSION NO. W72-02417

THE EFFECT OF LAS VEGAS WASH EFFLUENT UPON THE WATER QUALITY IN LAKE MEAD,

BUREAU OF RECLAMATION, DENVER, COLO. ENGINEERING AND RESEARCH CENTER.

D. A. HOFFMAN, P. R. TRAMUTT, AND F. C. HELLER.

AVAILABLE FROM THE NATIONAL TECHNICAL INFORMATION SERVICE AS PB-200 137.
$3.00 IN PAPER COPY, $0.95 IN MICROFICHE. REC-ERC-71-11, JAN 71, 32P.

DESCRIPTORS:
*NEVADA, *EUTROPHICATION, *LAKES, *WATER POLLUTION, DEGRADATION, SEWAGE
TREATMENT, NITROGEN COMPOUNDS, PHOSPHORUS COMPOUNDS, RESERVOIRS, ALGAE,
GROUND WATER, NITRATES, PHOSPHATES, LAS VEGAS BAY.

IDENTIFIERS:
DISSOLVED ORGANIC MATTER, LAS VEGAS, NEVADA, *LAKE MEAD, LAS VEGAS BAY.

ABSTRACT:
EFFLUENTS FROM THE CLARK COUNTY AND LAS VEGAS, NEV SEWAGE TREATMENT
PLANTS ARE THE PRINCIPAL SOURCES OF WATER FLOWING INTO LAS VEGAS WASH.
WATER IN THE WASH FLOWS INTO LAS VEGAS BAY, AN ARM OF THE BOULDER BASIN
REACH OF LAKE MEAD. A STUDY WAS CONDUCTED TO DETERMINE THE QUALITY OF
WATER IN LAS VEGAS WASH AND THE EFFECT OF THAT WATER ON THE WATER
QUALITY OF THE LAKE. LARGE AMOUNTS OF TOTAL DISSOLVED SOLIDS AND ALGAE
NUTRIENTS ENTER THE LAKE FROM THE WASH. CHLOROPHYLL A VALUES,
INDICATORS OF ALGAE BLOOMS, WERE 20-25 TIMES HIGHER IN LAS VEGAS BAY
THAN IN THE CONTROL STATIONS ELSEWHERE IN LAKE MEAD. AN UNACCOUNTABLE
INCREASE IN CATIONS AND ANIONS OCCURS IN THE WASH. THE SOURCE OF THE
ADDITION IS NOT KNOWN BUT MAY BE CAUSED BY UNMEASURED GROUND-WATER
INFLOW. METHODS OF COLLECTIONS AND ANALYSIS FOR NITROGEN AND PHOSPHORUS
ARE DESCRIBED.

FIELD 02H, 05C

ACCESSION NO. W72-02817

STUDIES ON ALGAE FROM EUTROPHIC AND OLIGOTROPHIC WATERS AND THEIR USE IN THE
BIO-ASSAY OF WATER QUALITY,

MICHIGAN STATE UNIV., EAST LANSING. DEPT. OF BOTANY AND PLANT PATHOLOGY.

BRIAN MOSS.

AVAILABLE FROM THE NATIONAL TECHNICAL INFORMATION SERVICE AS PB-205 805,
$3.00 IN PAPER COPY, $0.95 IN MICROFICHE. PROJECT COMPLETION REPORT 11 P.
OWRR A-034-MICH(1).

DESCRIPTORS:
ALGAE, *NUTRIENT REQUIREMENTS, BIOASSAY, EUTROPHICATION, OLIGOTROPHY,
LAKES, MICHIGAN, BICARBONATES, IRON, PHOSPHORUS, NITROGEN, WATER
QUALITY.

ABSTRACT:
THE NUTRITION OF SEVERAL SPECIES OF ALGAE FOUND IN EUTROPHIC AND
OLIGOTROPHIC LAKES AND THE USEFULNESS OF THESE ALGAE IN BIO-ASSAYS FOR
DETECTING LIMITING NUTRIENTS IN NATURAL WATERS WERE INVESTIGATED. THE
ROLES OF BICARBONATE, PH, MONOVALENT AND DIVALENT CATIONS, PHOSPHORUS,
NITROGEN AND IRON ARE DISCUSSED. (BAHR-MICHIGAN)

FIELD 05A, 05C

ACCESSION NO. W72-03183

PETROCHEMICAL EFFLUENTS TREATMENT PRACTICES, DETAILED REPORT,

ENGINEERING-SCIENCE, INC., TEXAS, AUSTIN.

EARNEST F. GLOYNA, AND DAVIS L. FORD.

AVAILABLE FROM NTIS AS PB-205 824, $3.00 IN PAPER COPY, $0.95 IN MICROFICHE.
WATER POLLUTION CONTROL RESEARCH SERIES, FEB 70, 249 P, 31 FIG, 46 TAB, 301
REF, 2 APPEND. FWPCA PROGRAM NO 12020 02/70, CONTRACT NO 14-12-461.

DESCRIPTORS:
*OIL, *OIL WASTES, *CHEMICAL WASTES, INDUSTRIAL WASTES, WATER ANALYSIS,
ANALYTICAL TECHNIQUES, SAMPLING, ORGANIC MATTER, TOXICITY, PHENOLS,
ODOR, COLOR, TURBIDITY, BIOCHEMICAL OXYGEN DEMAND, HYDROGEN ION
CONCENTRATION, ORGANIC LOADING, *WATER REUSE, RESEARCH AND DEVELOPMENT,
WASTE IDENTIFICATION, ALGAE, *WASTE WATER TREATMENT.

ABSTRACT:
500 NEW PETROLEUM PRODUCTS ARE INTRODUCED ON THE MARKET EACH YEAR,
MAKING THE PETROLEUM INDUSTRY AND ITS ALLIED PETROCHEMICL INDUSTRIES
SOME OF THE NATION'S FASTEST GROWING INDUSTRIES. THE PETROCHEMICAL
INDUSTRY IS PROJECTED AS INCREASING BY 9% PER YEAR THROUGH 1975. MOST
OF THE POLLUTIONAL LOAD FROM PETROCHEMICAL INDUSTRIES EMANATES FROM
PROCESS WASTE STREAMS. THE PROCESSES FOR REDUCING ITS POTENTIAL HARM TO
THE ENVIRONMENT ARE WELL DEVELOPED AND UNDERSTOOD. HOWEVER, CONSTANT
RESEARCH AND DEVELOPMENT EFFORTS ARE CONTINUALLY INTRODUCING NEW
POLLUTANTS INTO THE ENVIRONMENT. THEREFORE, DEVELOPMENT OF NEW PRODUCTS
SHOULD NOW INCLUDE DEVELOPMENT OF TREATMENT SCHEMES BY WHICH THE WASTES
FROM THE NEW PRODUCTS MAY BE ALLEVIATED. CARE MUST BE EXERCISED EVEN IN
THE SAMPLING PROCEDURES, SINCE MANY PETROLEUM-BASED CHEMICALS INTERFERE
WITH THE ACCEPTED ANALYTICAL TEST PROCEDURES. FINALLY, FEASIBILITY
STUDIES ARE NEEDED TO DETERMINE THE CAPABILITY OF THE PETROCHEMICAL
INDUSTRIES TO RE-USE MORE OF THEIR WASTEWATERS IN THE PLANT, AND
PHYSICAL, CHEMICAL, AND BIOLOGICAL TREATMENT ARE DISCUSSED. (SEE ALSO
W70-07511) (LOWRY-TEXAS)

FIELD 05D

ACCESSION NO. W72-03299

TASTE AND ODOR PROBLEMS IN NEW RESERVOIRS IN WOODED AREAS,

SEATTLE DEPT. OF WATER, WASH.

E. J. ALLEN.

JOURNAL OF THE AMERICAN WATER WORKS ASSOCIATION, VOL 52, NO 8, P 1027-1032,
AUGUST 1960. 1 FIG, 8 REF.

DESCRIPTORS:
*WATER TREATMENT, *TASTE, *ODOR, *RESERVOIRS, *DECIDUOUS TREES,
*DECOMPOSING ORGANIC MATTER, *ALGAE, *PHENOLS.

ABSTRACT:
THE NEARLY UNIVERSAL APPEARANCE OF TASTE AND ODORS IN RESERVOIRS DUE TO
MANY DIFFERENT CAUSES, ARE DISCUSSED WHICH INCLUDE DECAYING VEGETATION
ORIGINALLY PRESENT ON THE SITE, DECIDUOUS TREES ADJACENT TO THE
RESERVOIRS WITH DECAY OF FALLEN LEAVES, ALGAE FED BY THE LEAF
INFUSIONS, PHENOLS PRODUCED BY DECAYING LEAVES, (BEAVERS) BUILDING DAMS
AND DRAGGING VEGETABLE MATERIALS INTO THE RESERVOIR, WEED DECAY, WATER
COLLECTING IN ROAD DITCHES PROVIDING ENVIRONMENT FOR ALGAE GROWTHS.
TASTE AND ODOR PROBLEMS IN NEW RESERVOIRS IN WOODED AREAS CAN BE
AVOIDED BY RECOGNITION OF THE ENVIRONMENTAL CONDITIONS CONDUCIVE TO
THEIR PRODUCTION AND BY TAKING THE PROPER MEASURES TO ELIMINATE SUCH
CONDITIONS. (BEAN-AWWA)

FIELD 05F

ACCESSION NO. W72-03475

ALGAE AND OTHER INTERFERENCE ORGANISMS IN WATER OF THE SOUTH CENTRAL UNITED STATES,

ROBERT A. TAFT SANITARY ENGINEERING CENTER, CINCINNATI, OHIO.

C. M. PALMER.

JOURNAL OF THE AMERICAN WATER WORKS ASSOCIATION, VOL. 3, NO. 7, P 897-914, JULY 1960. 2 TAB, 91 REF.

DESCRIPTORS:
*WATER TREATMENT, *ALGAE, *ALGAL CONTROL, *TASTE, *ODOR, *NATURAL STREAMS, *DISTRIBUTION SYSTEMS, CHLORINATION, COPPER SULFATE, ACTIVATED CARBON, WATER SUPPLY.

IDENTIFIERS:
SOUTH - CENTRAL STATES.

ABSTRACT:
COVERING THE 6 SOUTH-CENTRAL STATES OF ALABAMA, ARKANSAS, LOUISIANA, MISSISSIPPI, OKLAHOMA, AND TEXAS, THE DIFFERENCES IN DIFFERENT WATERSHEDS ARE DISCUSSED ALONG WITH THE TYPES OF ORGANISMS MOST PREVALENT, THE DIFFICULTIES CAUSED IN TREATMENT PLANTS, IN IMPOUNDMENTS, AND BY INFESTATIONS IN DISTRIBUTION SYSTEMS. TASTES AND ODORS CAUSED ARE DISCUSSED, ALSO TOXIC ALGAE. CONTROL BY CHEMICALS SUCH AS COPPER SULFATE, ALGICIDES, CHLORINE AND CHLORINE-AMMONIA ARE DISCUSSED. THE COMMON USE OF ACTIVATED CARBON IS ELABORATED, ALSO MECHANICAL CONTROLS FOR REDUCING TASTE AND ODORS IN A FILTER PLANT. NINETY - ONE REFERENCES ARE NOTED. (BEAN-AWWA)

FIELD 05F

ACCESSION NO. W72-03562

USE OF ACTIVATED CARBON TO PREVENT WATER SUPPLY CONTAMINATION,

WEST VIRGINIA PULP AND PAPER COMPANY, COVINGTON, VA. CARBON TECHNICAL SERVICE LAB.

A. Y. HYNDSHAW.

WATER AND WASTES ENGINEERING, VOL 6, NO 2, FEBRUARY 1969, P 42-44.

DESCRIPTORS:
*ACTIVATED CARBON, *ADSORPTION, *WATER TREATMENT, *WASTE WATER TREATMENT, ALGAE, FILTRATION, NEW HAMPSHIRE, NEW YORK, VIRGINIA, ORGANIC COMPOUNDS, LAGOON, FILTERS, PESTICIDES, SLUDGE, DIGESTION, INDIANA.

IDENTIFIERS:
*CARBON FILTERS, LEBANON, BATAVIA, CLARKSBURG, BUFFALO, DUNREITH, OIL SPILLS.

ABSTRACT:
MOST ORGANICS THAT MIGHT BE POTENTIAL CONTAMINANTS OF POTABLE WATER SUPPLIES CAN BE EFFECTIVELY REMOVED OR INACTIVATED BY APPLICATION OF ACTIVATED CARBON. IN 1929, AFTER PHENOL-CONTAMINATED WATER CAUSED SPOILAGE OF SOME MEAT AT A CHICAGO MEAT PACKING COMPANY, IT WAS FOUND THAT POWDERED CARBON COULD BE USED TO PREVENT THIS TYPE OF CONTAMINATION. SINCE THAT TIME, USE OF ACTIVATED CARBON HAS EXPANDED GREATLY. TODAY MOST WATER TREATMENT PLANTS USE ACTIVATED CARBON TO REMOVE TASTE AND ODOR-CAUSING SUBSTANCES THAT ARE NOT REMOVED BY SUCH PROCESSES AS COAGULATION, FLOCCULATION, AND FILTRATION. APPLICATIONS OF CARBON TO IMPOUNDED SUPPLIES WILL ALSO CONTROL ALGAE GROWTHS. POWDERED CARBON HAS BEEN USED VERY EFFECTIVELY TO CORRECT CONDITIONS RESULTING FROM OIL SPILLS. THE VERY GREAT SURFACE AREAS ATTRACT AND HOLD LARGE QUANTITIES OF OIL. AFTER A FIRE IN A WAREHOUSE RELEASED LARGE AMOUNTS OF ETHYLENE GLYCOL INTO A RIVER IN CLARKSBURG, VIRGINIA, ACTIVATED CARBON WAS FED AT A RATE OF 10 MG/L UNTIL THE CONTAMINATION PASSED. THIS AMOUNT WAS SUFFICIENT TO ELIMINATE TASTES AND ODORS. IN BUFFALO, AN INDUSTRY USED 55 GALLON BARRELS WITH CARBON AS MAKESHIFT FILTERS TO ELIMINATE TRICRESYL PHOSPHATE FROM WATER BEFORE DISCHARGING IT INTO A STREAM. CARBON TREATMENT HAS ALSO BEEN USED TO ADSORB DDT AND SIMILAR PESTICIDES. ADDITION OF POWDERED CARBON AT THE GRIT CHAMBER IN WATER TREATMENT PLANTS HAS BEEN SHOWN TO REDUCE ATMOSPHERIC ODORS, ACCELERATE SETTLING, IMPROVE DIGESTION, AND REDUCE FOAM FORMATION. (BIGGS-TEXAS)

FIELD 05G, 05F

ACCESSION NO. W72-03641

SAND FILTRATION OF ALGAL SUSPENSIONS,

MICHIGAN UNIV., ANN ARBOR. DEPT. OF SANITARY ENGINEERING.

J. A. BORCHARDT, AND C. R. O'MELIA.

JOURNAL OF THE AMERICAN WATER WORKS ASSOCIATION, VOL 53, NO 12, P 1493-1502, DECEMBER 1961. 6 FIG, 3 TAB, 10 REF.

DESCRIPTORS:
*WATER TREATMENT, *ALGAE, *FILTRATION, HEAD LOSS, FILTERS, *SCENEDESMUS, FLOCCULATION, WATER POLLUTION TREATMENT.

ABSTRACT:
FILTRATION OF BOTH ALGAL AND FLOCCULENT PARTICLES THROUGH SAND BEDS WAS TESTED. EIGHT-FOOT PLEXIGLASS FILTERS WITH INSIDE DIAMETER OF 3 3/8 IN WERE UTILIZED WITH OTTAWA SAND DIVIDED IN THREE FRACTIONS, GRADED BY MICROSCOPIC MEASUREMENTS. ALL RUNS WERE MADE AFTER TAPPING THE FILTERS TO PRODUCE POROSITY OF 40%. ALGAE OF GENERA SCENEDESMUS, ANKISTRODESMUS AND ANABAENA WERE CULTURED WITH INORGANIC SALTS, AERATED AND TEMPERATURE CONTROLLED. HEAD LOSSES THROUGHOUT THE BEDS WERE DETERMINED BY 7 PIEZOMETER TUBES, AND ALGAE SAMPLES WERE COUNTED WITH AN ELECTRONIC PARTICLE COUNTER. MINIMUM REMOVAL EFFICIENCIES OF 33, 22 AND 10% WERE PRODUCED BY THE THREE SAND SIZES 0.316, 0.397, AND 0.524 MM RESPECTIVELY. REMOVAL OF THE STRINGY ANABAENA CONSISTENTLY WAS HIGHER THAN OF THE SMALLER MORE COMPACT ANKISTRODESMUS. ALGAL REMOVAL IN EVERY CASE DECREASED WITH TIME TO A CONSTANT MINIMUM VALUE. SIGNIFICANT NUMBERS OF ALGAE WERE FOUND IN EVERY SAMPLE THROUGHOUT THE RESEARCH. HEAD LOSS INCREASE WAS LINEAR WITH TIME THROUGHOUT THE GREATER PART OF EACH RUN. FUNCTIONALLY, THE BEHAVIOR OF THE FILTER BED IS THE SAME IN THE REMOVAL OF EITHER FLOCCULANT OR NON-FLOCCULANT PARTICLES. SOME SURFACE STRAINING TAKES PLACE FOLLOWED BY CONTINUOUS UNIFORM REMOVAL WITH DEPTH. THE PRESENCE OF FLOCCULENT MATERIAL ASSISTS IN THE ENTRAPMENT OF ALGAE CELLS, THEREFORE MULTIPLE ADDITIONS OF CHEMICALS AND CONTROLLED FLOCCULATION MAY BE MORE IMPORTANT THAN SEDIMENTATION CAPACITY. ADDITION OF CHEMICALS TO THE FILTER INFLUENT MIGHT BE BEST. (BEAN-AWWA)

FIELD 05F

ACCESSION NO. W72-03703

ALGAE AND OTHER INTERFERENCE ORGANISMS IN WATER SUPPLIES OF CALIFORNIA,

ROBERT A. TAFT SANITARY ENGINEERING CENTER, CINCINNATI, OHIO.

C. M. PALMER.

JOURNAL OF THE AMERICAN WATER WORKS ASSOCIATION, VOL. 53, NO. 10, P.1297-1312, OCTOBER 1961. 100 REF.

DESCRIPTORS:
*WATER TREATMENT, *ALGAE, *TASTE, *ODOR, *ALGICIDES, ACTIVATED CARBON, FILTRATION, *TOXICITY, *DISTRIBUTION SYSTEMS, WATER POLLUTION, WATER QUALITY.

ABSTRACT:
70% OF ALL PRECIPITATION IN CALIFORNIA OCCURS DURING THREE WINTER MONTHS. RAINFALL IN NORTHERN CALIFORNIA EXCEEDS 100 IN. PER YEAR; IN THE EXTREME SOUTH, 3 IN. OR LESS PER YEAR. WATER OF WINTER FLOWS MUST BE STORED AND SOME FROM WET YEARS MUST BE HELD FOR USE IN DRY YEARS, PARTICULARLY AS DRY PERIODS SOMETIMES EXTEND OVER A FEW YEARS. WATER IN VARIOUS AREAS VARIES GREATLY IN PHYSICO-CHEMICAL PROPERTIES, IN GENERAL BEING SOFT IN THE NORTH AND HARD TOWARD THE SOUTH. NEARLY 40% OF WATER UTILIZED IS DRAWN FROM WELLS AND SPRINGS. THE MANY TYPES OF ALGAE FOUND IN CALIFORNIA WATERS ARE LISTED. RESERVOIRS PROBLEMS, AQUATIC ANIMALS, PHYSICO-CHEMICAL CHANGES, TASTE AND ODOR CAUSES, TREATMENT PLANT PROBLEMS, INFESTATIONS IN DISTRIBUTION SYSTEMS, ALGICIDES FOR RESERVOIRS, ACTIVATED CARBON, SELECTIVE DRAFT, RESERVOIR AND CANAL COVERS, CONTROL OF WEEDS, WATERSHED CONTROL, CONTROL BY FILTRATION, CONTROL OF ORGANISMS IN DISTRIBUTION SYSTEMS AND TOXIC ALGAE ARE DISCUSSED. (BEAN - AWWA)

FIELD 05A, 05F

ACCESSION NO. W72-03746

REMOVAL OF ALGAE BY MICROSTRAINERS,

ONTARIO WATER RESOURCES COMMISSION, TORONTO (ONTARIO).

A. E. BERRY.

JOURNAL OF THE AMERICAN WATER WORKS ASSOCIATION, VOL 53, NO 12, P 1503-1508, DECEMBER 1961. 2 REF.

DESCRIPTORS:
*WATER TREATMENT, *ALGAE, FILTRATION, *SLIME, *ULTRAVIOLET RADIATION, *TREATMENT FACILITIES, FILTERS.

IDENTIFIERS:
*MICROSTRAINERS, HYPOCHLORITES.

ABSTRACT:
MICROSTRAINERS MAY SERVE AS THE SOLE FILTRATION PROCESS, AS TREATMENT AHEAD OF BOTH SLOW AND RAPID SAND FILTERS, OR AS FILTERS OF SEWAGE AND INDUSTRIAL EFFLUENTS AFTER TREATMENT. THEY ARE USED ON DRINKING WATER AND INDUSTRIAL WATER. SELECTION OF MICROSTRAINER MUST BE RELATED TO EFFICIENCIES IN REMOVAL OF ALGAE AND ECONOMICS IN COMPARISON WITH OTHER METHODS OF TREATMENT. SELECTION OF APPROPRIATE UNITS AND DESIGN OF THE INSTALLATION MUST BE RELATED TO THE FILTERABILITY INDEX OF THE WATER FOR SELECTION OF FABRIC, AND MAXIMUM CAPACITY IS DEPENDENT ON MAXIMUM SUBMERGENCE. IN THE UNITED STATES 12 INSTALLATIONS ARE REPORTED IN USE. IN CANADA 15 INSTALLATIONS ARE IN USE OR WILL BE SHORTLY. OF THESE, 13 ARE SERVING AS THE ONLY TREATMENT APART FROM DISINFECTION AND TWO ARE USED AHEAD OF RAPID SAND FILTERS. SOME DIFFICULTIES HAVE ARISEN. SOME UNITS HAD INADEQUATE MEANS OF CONTROLLING THE HEAD LOSS THROUGH THE FABRIC, WHICH WAS DAMAGED. WASHWATER PRESSURES WERE NOT EFFECTIVELY CONTROLLED, SLIMING AND BINDING OF THE FABRICS REQUIRED TREATMENT WITH SODIUM HYPOCHLORITE OR OTHER CHEMICALS AND IT HAS BEEN FOUND DESIRABLE TO EQUIP ALL UNITS WITH ULTRA-VIOLET LIGHTS AND TO OPERATE THEM CONTINUOUSLY. BLINDING OF THE FABRIC HAS BEEN ENCOUNTERED FROM SOME CAUSE OTHER THAN SLIMES. THE CAUSE REMAINS TO BE DETERMINED. (BEAN-AWWA)

FIELD 05F

ACCESSION NO. W72-03824

EXPERIENCES WITH MUNICIPAL DIATOMITE FILTERS EXPERIENCES IN NEW YORK,

NEW YORK STATE DEPT. OF HEALTH, LATHAN.

G. W. MOORE.

JOURNAL OF THE AMERICAN WATER WORKS ASSOCIATION, VOL. 54, NO. 12, P 1500-1504, DECEMBER 1962. 3 TAB.

DESCRIPTORS:
*WATER TREATMENT, *TURBIDITY, *IRON, *MANGANESE, *COLOR, *ALGAE, *COLLOIDS, FLOW RATES, *FILTERS, *NEW YORK, TREATMENT FACILITIES.

IDENTIFIERS:
*DIATOMITE FILTERS, FILTER AIDS.

ABSTRACT:
EXPERIENCES AT THE VARIOUS DIATOMITE FILTER INSTALLATIONS IN N.Y. ARE REVIEWED, AND THE N.Y. STATE HEALTH DEPARTMENT'S POSITION REGARDING DIATOMITE FILTRATION IS SUMMARIZED. IN GENERAL DIATOMITE FILTERS SHOULD BE CONSIDERED AS FINISHING FILTERS WHERE THE WATER IS NOT HEAVILY POLLUTED, THE AVERAGE TURBIDITY IS 10 OR LESS, AND MAXIMUM TURBIDITY DOES NOT EXCEED 30 FOR ANY APPRECIABLE TIME. THEY SHOULD BE APPROVED ONLY IF PRETREATMENT IS INCLUDED WHERE THERE IS A PROBLEM OF IRON REMOVAL, FIXED COLOR, HEAVY ALGAE LAODING, OR TURBIDITY IS IN FINE SUSPENSION OR COLLOIDAL FORM. AT LEAST TWO UNITS SHOULD BE PROVIDED. PERMISSIBLE RATE OF FILTRATION SHOULD DEPEND UPON THE CHARACTER OF THE RAW WATER AND THE DEGREE OR TYPE OF SUPERVISION AND LABORATORY CONTROL. THE NORMAL FILTER RATE SHOULD BE 1 GPM/SQ FT. EITHER HORIZONTAL OR VERTICAL ELEMENTS ARE ACCEPTABLE, ALSO PRESSURE OR VACUUM TYPES. THE DIATOMITE USED SHOULD BE ACCEPTABLE TO THE REVIEWING AUTHORITY. (BEAN-AWWA)

FIELD 05F

ACCESSION NO. W72-04145

USE OF POTASSIUM PERMANGANATE IN WATER TREATMENT,

CEDAR RAPIDS WATER WORKS, IOWA.

A. K. CHERRY.

JOURNAL OF THE AMERICAN WATER WORKS ASSOCIATION, VOL. 54, NO. 4, P 417-424,
   APRIL 1962. 1 FIG, 1 TAB.

DESCRIPTORS:
   *WATER TREATMENT, *ALGAE, *ACTINOMYCETES, *ODOR, ACTIVATED CARBON,
   AMMONIA, CHLORINE.

IDENTIFIERS:
   *POTASSIUM PERMANGANATE.

ABSTRACT:
   THRESHOLD ODORS OF RAW WATER, NORMALLY UNDER 200, ROSE TO AN AVERAGE OF
   896 IN JULY WITH A MAXIMUM ONE DAY OF 4,000. ALGAE COUNT AVERAGED
   48,600 PER ML. AVERAGE PERMANGANATE DOSAGE WAS 10.1 PPM, AND CARBON
   APPLICATION 51 PPM. FINISHED WATER ODORS WERE AS HIGH AS 14-18. THE
   MOST SUCCESSFUL TREATMENT WAS DISCONTINUANCE OF SOFTENING AND THE
   APPLICATION OF PERMANGANATE AT THE INTAKE, PERMITTING IT TO PASS
   THROUGH THE PRIMARY MIXERS AND CLARIFIERS WITHOUT ADDING OTHER
   CHEMICALS. ALUM WAS APPLIED IN THE SECONDARY MIXERS, WITH LIME FOR PH
   ADJUSTMENT. CARBON WAS APPLIED AS THE WATER PASSED TO THE FINAL
   CLARIFIERS. AMMONIA AND CHLORINE APPLICATION WAS CHANGED TO THE CLEAR
   WELL AFTER FILTRATION. BY THIS TREATMENT, ODOR WAS REDUCED FROM
   400-2,000 TO 3-5 IN FINISHED WATER. THE ABNORMALLY LARGE NUMBER OF
   KNOWN ODOR PRODUCING ALGAE WAS ASSUMED TO BE THE PRINCIPAL CAUSE OF THE
   UNDESIRABLE TASTES AND ODORS, HOWEVER, PART OF THE ODORS COULD BE
   TRACED TO ACTINOMYCETES. (BEAN-AWWA)

FIELD 05F

ACCESSION NO. W72-04155

ODOR CONTROL WITH CARBON AND PERMANGANATE AT DES MOINES,

DES MOINES WATER WORKS, IOWA.

JOURNAL OF THE AMERICAN WATER WORKS ASSOCIATION, VOL 60, NO 10, P 1195-1198,
   OCTOBER 1968. 2 TAB.

DESCRIPTORS:
   *WATER TREATMENT, ODOR, *ALGAE, *DECOMPOSING ORGANIC MATTER, *POTASSIUM
   COMPOUNDS, *ACTIVATED CARBON, ACTINOMYCETES, IOWA.

IDENTIFIERS:
   *POTASSIUM PERMANGANATE, *DES MOINES(IOWA).

ABSTRACT:
   AT PERIODS DURING THE YEAR THE THRESHOLD ODOR OF THE WATER SUPPLY AT
   DES MOINES ROSE TO LEVELS CAUSING WIDESPREAD COMPLAINTS OF CONSUMERS.
   THESE ODORS ARE TYPICALLY EARTHY OR MUSTY IN CHARACTER. INDUSTRIAL
   WASTE OR DOMESTIC SEWAGE CONTAMINATION ARE NOT PROBLEMS IN THE RACCOON
   RIVER AN OCCASIONAL SOURCE OF RAW WATER FOR DES MOINES. THE ODORS
   RESULT FROM GROWTH OF ALGAE OR WASHING OF EXTRACTS FROM DECAYED
   VEGETATION. ACTINOMYCETES HAVE BEEN ISOLATED WITH THE SAME ODOR
   CHARACTERISTICS AS THE RIVER WATER. POTASSIUM PERMANGANATE IS A
   VALUABLE ADJUNCT TO THE NORMAL FEED OF ACTIVATED CARBON DURING PERIODS
   OF RIVER INTAKE; WHEN THE THRESHOLD ODOR OF THE RACCOON RIVER EXCEEDS
   25, POTASSIUM PERMANGANATE IS FED. AT THRESHOLD ODOR LEVELS BELOW 25,
   EITHER MATERIAL WILL PRODUCE SATISFACTORY WATER, AND THE CHOICE IS
   PRIMARILY ONE OF CONVENIENCE. (BEAN-AWWA)

FIELD 05F

ACCESSION NO. W72-04242

ION-EXCHANGE FOR RECLAMATION OF REUSABLE SUPPLIES,

STANFORD UNIV., CALIF. DEPT. OF CIVIL ENGINEERING.

R. ELIASSEN, B. M. WYCKOFF, AND C. D. TONKIN.

JOURNAL OF THE AMERICAN WATER WORKS ASSOCIATION, VOL 57, NO 9, P 1113-1122,
SEPTEMBER 1965. 6 FIG, 3 TAB, 4 REF.

DESCRIPTORS:
*WATER TREATMENT, *WATER REUSE, *RECLAMATION, *GROUNDWATER, RECHARGE,
*ION EXCHANGE, *ORGANIC MATTER, *FILTRATION, DIATOMACEOUS EARTH, ALGAE,
TREATMENT FACILITIES.

IDENTIFIERS:
FILTER AIDS, DIATOMITE FILTERS.

ABSTRACT:
PILOT PLANT STUDIES ON SEWAGE AT PALO ALTO, CALIF., INDICATE THE MOST
ECONOMIC TREATMENT TO RECOVER REUSABLE WATER (POTABILITY NOT MENTIONED)
IS THE ION-EXCHANGE PROCESS. INORGANIC MATERIALS, SUCH AS NITROGEN AND
PHOSPHORUS COMPOUNDS, AND ORGANIC MATERIALS MEASURED BY THE CHEMICAL
OXYGEN DEMAND, ALKYL BENZENE SULFURATE AND COLOR TESTS, CAN BE REMOVED
IN ONE OPERATION, PROVIDING FILTRATION PRECEDES THE ION-EXCHANGE
PROCESS. WATER SUITABLE FOR MANY PURPOSES, INCLUDING SURFACE STORAGE,
CAN BE PRODUCED AT A COST COMPETITIVE WITH EXISTING SUPPLIES IN HIGH
COST AREAS. DIATOMACEOUS-EARTH FILTRATION, FOLLOWED BY ANION-EXCHANGE,
CAN BE ACCOMPLISHED AT A COST OF ABOUT 21 CENTS PER 1,000 GAL. ALTHOUGH
WATER RECLAIMED WITH ION-EXCHANGE WILL STILL CONTAIN PHOSPHOROUS AND
NITROGEN COMPOUNDS, THE PROBLEM OF ALGAE GROWTH IN RESERVOIRS WILL BE
MINIMIZED. A LESSER DEGREE OF NUTRIENT REMOVAL, AND CONSEQUENT
LOWER-COST TREATMENT, WOULD SUFFICE FOR RECHARGE OF GROUND WATER.
(BEAN-AWWA)

FIELD 05F

ACCESSION NO. W72-04318

ORGANIC MATTER AND FINISHED WATER QUALITY,

EAST BAY MUNICIPAL UTILITY DISTRICT, OAKLAND, CALIF.

J. J. CONNORS, AND R. B. BAKER.

JOURNAL OF THE AMERICAN WATER WORKS ASSOCIATION, VOL 61, NO 3, P 107-113,
MARCH 1969. 6 FIG, 2 TAB, 10 REF.

DESCRIPTORS:
*WATER TREATMENT, WATER QUALITY, *ORGANIC MATTER, ALGAE, COPPER
SULFATE, *CHEMICAL OXYGEN DEMAND.

IDENTIFIERS:
TOTAL ORGANIC CARBON.

ABSTRACT:
WATER QUALITY DETERIORATION CAN OCCUR BECAUSE OF ORGANIC MATTER AFTER
WATER HAS LEFT THE TREATMENT PLANT. SIGNIFICANT DIFFERENCES IN THE
AMOUNT OF ORGANIC MATTER AND ALGAL BIOMASS CAN OCCUR BUT NOT BE
INDICATED BY CHEMICAL ANALYSIS. COPPER SULFATE MAY NOT CONTROL TOTAL
ORGANIC MATTER. IN SOME RESERVOIRS THE TOTAL ORGANIC LOAD AS MEASURED
BY THE CHEMICAL OXYGEN DEMAND (COD) IS ESTABLISHED BY THE CHARACTER OF
THE RUNOFF. ORGANIC REMOVALS BY FILTRATION ARE SLIGHT COMPARED WITH
THOSE BY COAGULATION AND SEDIMENTATION. THE TOC APPARATUS FOR MEASURING
'TOTAL ORGANIC CARBON' WILL BE USEFUL FOR CONTINUED STUDIES. THE COD
TEST, WHATEVER IT MEASURES, CAN BE MADE TO YIELD GOOD PRECISION WITH
POTABLE WATERS. (BEAN-AWWA)

FIELD 05F

ACCESSION NO. W72-04495

BIOLOGICAL CONCENTRATION OF PESTICIDES BY ALGAE,

NORTH TEXAS STATE UNIV., DENTON. DEPARTMENT OF BIOLOGY.

B. D. VANCE, AND W. DRUMMOND.

JOURNAL OF THE AMERICAN WATER WORKS ASSOCIATION, VOL 61, NO 7, P 360-362,
JULY 1969. 2 TAB, 11 REF.

DESCRIPTORS:
*WATER QUALITY, *ALGAE, *PESTICIDES, *FISH, *FISH KILLS, *ROTIFERS,
*ECOLOGY, *ECOSYSTEMS.

IDENTIFIERS:
*POTENTIATION, *ORGANIC DEGRADATION.

ABSTRACT:
ALGAE ARE EXTREMELY EFFICIENT POTENTIATORS OF PESTICIDE RESIDUES.
CONCENTRATION OF MANY PESTICIDES BY CERTAIN ALGAE OCCURS QUITE RAPIDLY,
AND THERE IS APPARENTLY VERY LITTLE, IF ANY, DEGRADATION BY THE ALGAE.
THEREFORE THESE HIGHLY CONCENTRATED RESIDUES CAN BE BIOLOGICALLY
TRANSFERRED TO HIGHER MEMBERS OF THE FOOD CHAIN SUCH AS ROTIFERS AND
SMALL FISH. MOST HIGHER MEMBERS DO NOT APPEAR TO BE NEARLY AS RESISTANT
AS THE ALGAE TO THESE CHLORINATED COMPOUNDS. CONSEQUENTLY, FISH KILLS
AND SIMILAR CATASTROPHIES MAY BE BROUGHT ABOUT IN WATER CONTAINING AS
LITTLE AS 1 PPB OF A PESTICIDE. SUCH POTENTIATION AND BIOLOGICAL
TRANSFER HAS BEEN AMPLY DEMONSTRATED, BEGINNING WITH FISH AND
CONTINUING UP THE FOOD CHAIN TO MAN. IT IS IMPORTANT TO KNOW THE
EFFECTS OF SUCH CHEMICALS UPON ECOSYSTEMS AND HOW SUCH CHLORINATED
HYDROCARBON MOLECULES MAY INFLUENCE THE BASIC ECOLOGY OF OUR
ENVIRONMENT. IT WAS THOUGHT AT ONE TIME THAT INSECTICIDES DID NOT CAUSE
DEATH IN PLANTS, EVEN IN GROSS OVER-DOSAGES. HOWEVER, TESTS SHOW THAT
DIELDRIN, ALDRIN, AND ENDRIN ARE ALGICIDAL TO M. AERUGINOSA AT A
CONCENTRATION OF LESS THAN 5 NG/ML. (BEAN-AWWA)

FIELD 05B, 05C

ACCESSION NO. W72-04506

EUTROPHICATION OF SMALL RESERVOIRS IN THE GREAT PLAINS,

NEBRASKA UNIV., LINCOLN.

GARY L. HERGENRADER, AND MARK J. HAMMER.

AVAILABLE FROM THE NATIONAL TECHNICAL INFORMATION SERVICE AS PB-206 953,
$3.00 IN PAPER COPY, $0.95 IN MICROFICHE. MIMEO (UNDATED). 26 P, 4 FIG, 4
TAB, 6 REF. OWRR A-014-NEB(5).

DESCRIPTORS:
*EUTROPHICATION, *RESERVOIRS, *GREAT PLAINS, ALGAE, NEBRASKA,
TURBIDITY, STRATIFICATION, TEMPERATURE, CHLOROPHYLL, CYANOPHYTA,
AQUATIC PLANTS, PHOSPHORUS, DENSITY, WATER POLLUTION SOURCES, RUNOFF,
LIGHT PENETRATION, SAMPLING, PRIMARY PRODUCTIVITY, PHYTOPLANKTON,
EUGLENA, CHLOROPHYTA, NITRATES, AMMONIA, CARBON, RECREATION, WATER
POLLUTION CONTROL, ALKALINITY, DISSOLVED SOLIDS.

IDENTIFIERS:
*SALT VALLEY RESERVOIRS(NEB), *CLEAR WATER RESERVOIRS, *TURBID
RESERVOIRS, CYCLOTELLA, TRACHELOMONAS, MELOSIRA, STEPHANODISCUS,
ANABAENA, APHANIZOMENON, MICROCYSTIS, POTAMOGETON, POLYGONUM,
SAGITTARIA, NAJAS.

ABSTRACT:
LIMNOLOGICAL STUDIES OF FIVE RESERVOIRS IN THE SALT VALLEY WATERSHED
DISTRICT (EASTERN NEBRASKA) WERE INITIATED TO DETERMINE EXISTING
TROPHIC CONDITIONS, ESTIMATE EUTROPHICATION RATES, IF POSSIBLE, BY
MEASURING CHANGES IN SEVERAL PARAMETERS, IDENTIFY SOURCES OF NUTRIENT
INPUTS, AND EVALUATE PREVENTATIVE AND REMEDIAL MEASURES. DURING JUNE,
JULY, AND AUGUST, EACH OF THE STUDY RESERVOIRS WAS SAMPLED AT WEEKLY
INTERVALS; ONE LAKE WAS SAMPLED THROUGHOUT THE YEAR, BUT LESS
FREQUENTLY DURING ICE-COVER. WATER SAMPLES WERE ANALYZED FOR DISSOLVED
OXYGEN, TEMPERATURE, UNDERWATER LIGHT INTENSITY, ALKALINITY, PH,
HARDNESS, DISSOLVED AND SUSPENDED SOLIDS, IRON, COD, PHOSPHATES,
AMMONIA, NITRATE AND ORGANIC NITROGEN, CHLORIDE, SULFATE, AND
TURBIDITY; ALGAE IDENTIFIED, AND PRIMARY PRODUCTION DETERMINED. RUNOFF
WATERS IMPOUNDED IN THE RESERVOIRS CONTAIN SUFFICIENT NUTRIENT SALTS TO
SUPPORT ABUNDANT GROWTHS OF AQUATIC PLANTS. RESERVOIRS, LIGHT-LIMITED
BY SOIL TURBIDITY, SUPPORT NEITHER ABUNDANT GROWTHS OF AQUATIC PLANTS
NOR DENSE BLUE-GREEN ALGAL BLOOMS; CLEAR WATER RESERVOIRS ARE VERY
EUTROPHIC. EUTROPHICATION RATE IS VERY RAPID AND APPARENTLY DIRECTLY
RELATED TO AGE. CONTROL OF PHOTOSYNTHESIS THROUGH INHIBITION OF
SUNLIGHT PENETRATION BY THE ADDITION OF VARIOUS SUBSTANCES INTO THE
RESERVOIRS DIRECTLY OR TO THE WATER SURFACE SHOULD BE INVESTIGATED.
(JONES-WISCONSIN)

FIELD 05C, 02H

ACCESSION NO. W72-04761

BIOLOGICAL AND BACTERIOLOGICAL EVALUATION OF PILOT PLANT ARTIFICIAL RECHARGE
  EXPERIMENTS,

PRAGUE DEPT. OF WATER TECHNOLOGY (CZECHOSLOVAKIA).

A. MORAVCOVA, L. MASINOVA, AND V. BERNATOVA.

WATER RESEARCH, VOL 2, NO 4, P 265-276, 1968. 8 FIG, 5 TAB, 13 REF.

DESCRIPTORS:
  *ARTIFICIAL RECHARGE, COPPER, *ALGICIDES, ALGAE, *EVALUATION,
  FILTRATION.

IDENTIFIERS:
  SILVER.

ABSTRACT:
  THE EFFECT OF THE ALGICIDE CA-350 (A COMBINATION OF COPPER AND SILVER)
  WAS INVESTIGATED ON FOUR SAND FILTERS, MODELS OF AN ARTIFICIAL RECHARGE
  BASIN. USING 300 MICROGRAM/1 (PPB) CU IN A PREVENTIVE TREATMENT THE
  DEVELOPMENT OF FILAMENTOUS ALGAL FORMS WAS FULLY CONTROLLED WITH THE
  CONSERVATION OF A NORMAL BACTERIAL MICROFLORA. THE PROCESSES OF
  MINERALIZATION OF ORGANIC MATTER CONTINUED UNDER AEROBIC CONDITIONS.
  COPPER WAS GRADUALLY ELUTED FROM THE SAND LAYERS OF THE MODEL FILTERS,
  NEVER REACHING ITS MAXIMUM ALLOWABLE CONCENTRATION IN THE OUTFLOWING
  WATER. (SKOGERBOE-COLORADO STATE)

FIELD 04B, 05F

ACCESSION NO. W72-05143

ECOLOGY OF CLADOPHORA FRACTA AND CLADOPHORA GLOMERATA,

SYRACUSE UNIV., N.Y.

DANIEL F. JACKSON, AND SHUNN-DAR LIN.

FEDERAL WATER POLLUTION CONTROL ADMINISTRATION, NO 16010--05/68, FINAL
  REPORT, MAY 1968. 133 P, 10 FIG, 19 TAB, 93 REF, APPEND. FWPCA-WP 00782.

DESCRIPTORS:
  *ALGAE, *EUTROPHICATION, *PHOTOSYNTHESIS, *RESPIRATION, GREAT LAKES,
  NUISANCE ALGAE, WATER QUALITY, NUTRIENTS, PHOSPHORUS, NITROGEN, CALCIUM
  CHLORIDE, TEMPERATURE, METABOLISM, EPIPHYTOLOGY, ON-SITE TESTS,
  LABORATORY TESTS, LOTIC ENVIRONMENTS, LENTIC ENVIRONMENTS.

IDENTIFIERS:
  *CLADOPHORA FRACTA, CLADOPHORA GLOMERATA, BUTTERNUT CREEK(NY).

ABSTRACT:
  DETERMINATIONS OF RESPIRATION AND PHOTOSYNTHETIC RATES OF CLADOPHORA
  FRACTA AND CLADOPHORA GLOMERATA UNDER NATURAL AND LABORATORY
  CONDITIONS, AND NITRATE AND PHOSPHATE REQUIREMENTS UNDER LABORATORY
  CONDITIONS WERE MADE. IN THE LABORATORY, CLADOPHORA FRACTA RESPIRATION
  AND PHOTOSYNTHETIC RATES WERE GREATER WHEN OSCILLATED AT 120 STROKES
  PER MINUTE THAN WITH NO OSCILLATION. THE HIGHEST CLADOPHORA FRACTA
  METABOLIC RATES WERE OBSERVED IN THE MEDIUM LACKING CALCIUM CHLORIDE.
  MARKEDLY LOW RATES WERE OBTAINED IN MEDIA LACKING NITRATE AND
  PHOSPHATE. WITH SMALL INCREASES (0.00 TO 0.01 MG/L) OF EITHER NITRATE
  OR PHOSPHATE CONCENTRATION, CLADOPHORA GLOMERATE PHOTOSYNTHETIC RATE
  INCREASED SIGNIFICANTLY. WITH NO PHOSPHATE, CLADOPHORA GLOMERATE
  RESPIRATION RATE WAS ERRATIC. PHOTOSYNTHETIC RATE WAS NOT AFFECTED BY
  NITRATE CONCENTRATION VARYING FROM 0.01 TO 100.00 MG/L. IT IS CONCLUDED
  THAT PHOSPHORUS IS A LIMITING FACTOR OF CLADOPHORA GROWTH. CLADOPHORA
  METABOLIC RATES WERE NOT INFLUENCED BY PHOSPHATE CONCENTRATION LEVELS
  0.01 TO 5.00G/L IN NITRATE-FREE MEDIA. GENERALLY, CLADOPHORA GLOMERATE
  RESPIRATION AND PHOTOSYNTHETIC RATES INCREASED AS THE COMBINATION OF
  NITRATE AND PHOSPHATE DOSES INCREASED. CLADOPHORA GLOMERATE RESPIRATION
  AND PHOTOSYNTHETIC RATES UNDER NATURAL CONDITIONS WERE ABOUT THREE AND
  SIX TIMES, RESPECTIVELY, GREATER THAN UNDER LABORATORY CONDITIONS.
  (JONES-WISCONSIN)

FIELD 05C

ACCESSION NO. W72-05453

POWERS OF LAKE AUTHORITIES.

PUBLIC ACT NO. 416, CONNECTICUT LEGISLATIVE SERVICE, P 430 (1969). 1 P.

DESCRIPTORS:
*CONNECTICUT, *LAKES, *LOCAL GOVERNMENTS, *WATER QUALITY CONTROL,
ALGAE, AQUATIC WEEDS, ADMINISTRATION, SUPERVISORY CONTROL(POWER),
ADMINISTRATIVE AGENCIES.

ABSTRACT:
BY AMENDMENT TO CONNECTICUT STATUES, LEGISLATURES OF LOCAL GOVERNMENTS
ARE AUTHORIZED TO GRANT LAKE AUTHORITIES POWER TO CONTROL AND ABATE
ALGAE AND AQUATIC WEEDS, IN COOPERATION WITH THE STATE WATER RESOURCES
COMMISSION, AND STUDY WATER MANAGEMENT FOR RECOMMENDATIONS TO LOCAL
GOVERNMENTS. LAKE AUTHORITIES SHALL HAVE NO JURISDICTION IN MATTERS
SUBJECT TO REGULATION BY THE CONNECTICUT BOARD OF FISHERIES AND GAME.
(HART-FLORIDA)

FIELD 06E, 05G

ACCESSION NO. W72-05774

FORMAL DISCUSSION OF, 'THE RECLAMATION OF SEWAGE EFFLUENT FOR DOMESTIC USE,

METROPOLITAN WATER BOARD, LONDON (ENGLAND).

N. P. BURMAN.

PROCEEDINGS, INTERNATIONAL ASSOCIATION ON WATER POLLUTION RESEARCH
CONFERENCE, 3RD, MUNICH, GERMANY, 1966, P 19-23, 2 FIG, 1 TAB, 2 REF.

DESCRIPTORS:
*WATER PURIFICATION, *FLOW AUGMENTATION, *LAGOONS, ACTIVATED SLUDGE,
ALGAE, NUTRIENTS, VIRUSES, BACTERIA, E. COLI, PUBLIC HEALTH, PATHOGENIC
BACTERIA, TEMPERATURE, NITRATES, AMMONIA, *WATER REUSE, WASTE WATER
TREATMENT.

ABSTRACT:
SECONDARY EFFLUENT FROM A CONVENTIONAL ACTIVATED SLUDGE PLANT WAS
PASSED THROUGH A SERIES OF 3 LAGOONS. EFFLUENT FROM THE LAGOONS FLOWED
INTO THE RIVER LEE AT RYE MEADS. CHEMICAL, BACTERIOLOGICAL, AND
BIOLOGICAL QUALITY WERE OVER A PERIOD OF 2.5 YEARS. BACTERIAL
REDUCTIONS DEMONSTRATED A WIDE VARIABILITY WITH TEMPERATURE AND TYPE OF
ORGANISMS. IN GENERAL, E. COLI, TOTAL COLIFORM, AND FECAL STREPTOCOCCI
WERE THE MOST COMPLETELY REMOVED (IN EXCESS OF 90% ALWAYS). THE DEGREE
OF BACTERIAL IMPROVEMENT WAS NOT LINEAR, CHANGING WITH TEMPERATURE.
ENTEROVIRUSES WERE FOUND IN HALF THE SAMPLES OF THE LAGOON INFLUENT AND
ONLY TWICE IN THE LAGOON EFFLUENT, FULLY JUSTIFYING THE PLACE OF
MATURATION PONDS IN WATER RECLAMATION. QUESTIONS WERE ALSO RAISED ABOUT
THE FLOTATION PROCESS, AND TEMPERATURE OF THE ALGAE NUTRIENT STRIPPING
PROCESS, AS WELL AS ABS REMOVALS. (LOWRY-TEXAS)

FIELD 05D

ACCESSION NO. W72-06017

THE EFFECT OF ALGAL CONCENTRATION, LUMINOUS INTENSITY, TEMPERATURE, AND DIURNAL
CYCLE OR PERIODICITY UPON GROWTH OF MIXED ALGAL CULTURES FROM WASTE
STABILIZATION LAGOONS AS DETERMINED ON THE WARBURG APPARATUS,

MISSOURI UNIV., COLUMBIA. DEPT. OF CHEMICAL ENGINEERING.

R. H. LUEBBERS, AND D. N. PARIKH.

PROCEEDINGS, INDUSTRIAL WASTE CONFERENCE, 21ST, MAY 3-5, 1966, PART I, P
348-367. 13 FIG, 21 REF.

DESCRIPTORS:
*WASTE WATER TREATMENT, *OXIDATION LAGOONS, *ALGAE, LIGHT INTENSITY,
CARBON DIOXIDE, GROWTH RATES, TEMPERATURE, PILOT PLANTS, HYDROGEN ION
CONCENTRATION, ORGANIC LOADING, CENTRIFUGATION, RESPIRATION,
BIOCHEMICAL OXYGEN DEMAND, NITROGEN FIXATION.

IDENTIFIERS:
*OXYGEN PRODUCTION.

ABSTRACT:
THE NORMAL PERFORMANCE AND EFFECT OF VARIOUS OPERATING VARIABLES ON
WASTE STABILIZATION LAGOONS WERE INVESTIGATED IN A 2 STAGE EXPERIMENTAL
STUDY. THE RESULTS OF THE FIRST STAGE PILOT PLANT STUDY, PARTICULARLY
THE EFFECT OF SUPPLEMENTARY LIGHT ON ALGAL GROWTH PROMPTED THE USE OF
THE WARBURG APPARATUS. USING SUBSTRATE FROM THE PILOT PLANT LAGOONS,
TESTS RUNS WERE MADE BOTH WITH AND WITHOUT A SOURCE OF ACTINIC LIGHT,
AND WITH KOH, A 'CARBON DIOXIDE BUFFER', OR NO MATERIAL IN THE CENTER
WELL OF THE WARBURG FLASKS. THE STANDARD 5 DAY BOD OCCURRED IN 50 HOURS
AND THE NITROGEN PLATEAU AT 350 HOURS. O2 PRODUCTION IS ALMOST A LINEAR
FUNCTION OF ALGAL AND OTHER BIOTA CONCENTRATION. O2 PRODUCTION
INCREASED AS A LOGARITHMIC FUNCTION OF LIGHT CONCENTRATION UP TO 720
FT-CANDLES, REMAINED CONSTANT UNTIL 4500 FT-CANDLES, AND THEN DECREASED
WITH THE ALGAE DYING WITHIN 24 HOURS. O2 PRODUCTION VARIED WITH
TEMPERATURE, VERY LOW AT 10 DEG C, INCREASING RATE AT AN INCREASING
RATE TO ABOVE 20 DEG C, AND CONTINUES TO INCREASE AT A DECREASING RATE.
MAXIMUM RATE OF O2 PRODUCTION IS ABOVE 35 DEG C. O2 CONSUMPTION WITHOUT
LIGHT INCREASED LINEARLY WITH TEMPERATURE. BOTH DIURNAL CYCLE AND
PERIODICITY EFFECTED O2 PRODUCTION WITH A MAXIMUM RATE OF PRODUCTION
OBTAINED WITH A 55 TO 60% PERIODICITY. (MORGAN-TEXAS)

FIELD 05D, 05C

ACCESSION NO. W72-06022

FIELD STUDIES ON SEDIMENT-WATER ALGAL NUTRIENT INTERCHANGE PROCESSES AND WATER
QUALITY OF UPPER KLAMATH AND AGENCY LAKES, JULY 1967-MARCH 1969,

PACIFIC NORTHWEST WATER LAB., CORVALLIS, OREG.

A. R. GAHLER.

AVAILABLE FROM THE NATIONAL TECHNICAL INFORMATION SERVICE AS PB-207 643,
$3.00 IN PAPER COPY, $0.95 IN MICROFICHE. WORKING PAPER NO. 66, OCTOBER
1969, 33 P, 1 FIG, 9 TAB, 6 REF, APPEND. FWPCA PROGRAM 16010---10/69.

DESCRIPTORS:
*EUTROPHICATION, LAKE, *LAKE SEDIMENTS, *NUTRIENTS, NITROGEN COMPOUNDS,
PHOSPHOROUS COMPOUNDS, HYDROGEN ION CONCENTRATION, CONDUCTIVITY,
TEMPERATURE, DISSOLVED OXYGEN, PHYTOPLANKTON, WATER ANALYSIS, ALGAL
CONTROL, ALGAE, WATER QUALITY, SEDIMENT-WATER INTERFACES, OREGON,
SESSILE ALGAE, CYANOPHYTA.

IDENTIFIERS:
*UPPER KLAMATH LAKE(ORE.), AGENCY LAKE(ORE.), *OSCILLATORA.

ABSTRACT:
STUDIES OF ALGAL NUTRIENT INTERCHANGE BETWEEN SEDIMENT AND WATER UNDER
ENVIRONMENTAL CONDITIONS WERE CARRIED OUT IN UPPER KLAMATH LAKE,
OREGON, FROM JULY 1967 TO MARCH 1969. EXPERIMENTAL 'POOLS' OF LAKE
WATER IN CONTACT WITH THE SEDIMENT AND EXPERIMENTAL POOLS OF WATER NOT
EXPOSED TO THE SEDIMENT WERE COMPARED WITH THE OPEN LAKE FROM NOVEMBER
1967 TO JUNE 1968. WATER QUALITY MEASUREMENTS IN AGENCY AND UPPER
KLAMATH LAKES WERE MADE TO DETERMINE WHETHER INTERCHANGE PROCESSES
COULD BE OBSERVED DIRECTLY IN THE WATER, TO ESTABLISH CONDITIONS FOR
LABORATORY INTERCHANGE TESTS, AND TO COMPARE LAKE CONDITIONS WITH THE
EXPERIMENTAL POOLS. DATA WERE OBTAINED FROM JULY 1967 TO MARCH 1969 ON
PH, CONDUCTIVITY, TEMPERATURE, DISSOLVED OXYGEN, CHEMICAL COMPOSITION,
AND PHYTOPLANKTON. INTERCHANGE DEFINITELY OCCURRED WHEN OSCILLATORA
FLOATED TO THE LAKE SURFACE WITH ATTACHED SEDIMENT WHICH CONTAINED
SOLUBLE NITROGEN AND PHOSPHORUS COMPOUNDS. A PLASTIC-BOTTOMED POOL OF
WATER NOT EXPOSED TO SEDIMENT EXHIBITED HIGHER OXYGEN CONTENT UNDER THE
ICE AND HAD LESS PHYTOPLANKTON GROWTH IN SPRING THAN THE POOLS EXPOSED
TO SEDIMENTS. THE EFFECTS OF GAS EVOLUTION, WIND, CURRENTS, FISH,
BOATING, BENTHOS, DIFFUSION, ETC., ON THE SHALLOW LAKES WAS NOT
QUANTITATIVELY DETERMINED, BUT IT SEEMS QUITE PROBABLE THAT ANYTHING
THAT STIRS THE SEDIMENT CAUSES INTERCHANGE OF NUTRIENTS. (ALSO SEE
W72-06052)

FIELD 05C, 02H

ACCESSION NO. W72-06051

INTERIM REPORT, UPPER KLAMATH LAKE STUDIES OREGON,

FEDERAL WATER POLLUTION CONTROL ADMINISTRATION, CORVALLIS, OREG. PACIFIC
NORTHWEST WATER LAB.

W. E. MILLER, AND J. C. TASH.

AVAILABLE FROM THE NATIONAL TECHNICAL INFORMATION SERVICE AS PB-207 640,
$3.00 IN PAPER COPY, $0.95 IN MICROFICHE. WATER POLLUTION CONTROL RESEARCH
SERIES, PUBLICATION WP-20-8, SEPTEMBER 1967, 37 P, 3 REF, 14 TAB, 30 REF.
FWPCA PROGRAM 16010---09/67.

DESCRIPTORS:
*EUTROPHICATION, LAKES, LAKE SEDIMENTS, ALGAL CONTROL, *NUTRIENTS,
ALGAE, BENTHIC FAUNA, CHEMICAL ANALYSIS, SESSILE ALGAE, *PRIMARY
PRODUCTIVITY, OREGON, CYANOPHYTA.

IDENTIFIERS:
*UPPER KLAMATH LAKE(ORE.), *APHANIZOMENON FLOS-AQUAE, *CARBON-14
MEASUREMENTS.

ABSTRACT:
THE WATER QUALITY IN UPPER KLAMATH LAKE WATERSHED DURING THE PERIOD
MARCH 1965 TO APRIL 1966 IS DEFINED. THE SOURCES OF ALGAL NUTRIENTS AND
OTHER CONSTITUENTS IN WATER FLOWING INTO AND OUT OF UPPER KLAMATH LAKE
ARE IDENTIFIED AND THE QUANTITY OF THESE CONSTITUENTS IS COMPARED IN
PRISTINE STREAMS, CANALS, RIVERS, AGRICULTURAL DRAINAGE, AND SPRINGS.
CHEMICAL ANALYSES WERE MADE OF ALGAE, LAKE SEDIMENT AND BOTTOM FAUNA
SAMPLES. ALGAL SPECIES WERE IDENTIFIED AND MEASURED; DURING THE PERIOD
OF THE STUDY APHANIZOMENON FLOS-AQUAE REPRESENTED ABOUT 90-99 PERCENT
OF THE TOTAL ALGAL CROP DURING THE SUMMER. LIMITING NUTRIENT STUDIES
USING THE ISOTOPE CARBON-14 TO MEASURE VARIATIONS IN PRIMARY
PRODUCTIVITY WERE CONDUCTED IN SITU AND IN THE LABORATORY; DATA FROM
THESE STUDIES WERE DIFFICULT TO INTERPRET. (SEE ALSO W72-06051)

FIELD 05C, 02H

ACCESSION NO. W72-06052

MECHANISMS OF ANAEROBIC WASTE TREATMENT,

CALIFORNIA UNIV., BERKELEY.

GORDON L. DUGAN, AND WILLIAM J. OSWALD.

IN: POTATO WASTE TREATMENT, PROCEEDINGS OF A SYMPOSIUM SPONSORED BY FWPCA AND
IDAHO UNIVERSITY, MARCH 1968. P 5-17, 5 FIG, 1 TAB.

DESCRIPTORS:
*OXIDATION LAGOONS, *ANAEROBIC DIGESTION, *MUNICIPAL WASTES, METHANE
BACTERIA, ACID BACTERIA, SULFUR BACTERIA, ALGAE, MIXING, HYDROGEN ION
CONCENTRATION, TEMPERATURE, TOXICITY, DISSOLVED OXYGEN, LIGHT
INTENSITY, SLUDGE, DEPTH, ORGANIC LOADING, DESIGN CRITERIA, ODORS,
EFFICIENCIES, *WASTE WATER TREATMENT.

IDENTIFIERS:
*DETENTION TIME.

ABSTRACT:
ANAEROBIC TREATMENT OF VARIOUS MUNICIPAL AND INDUSTRIAL WASTES IS
PRACTICED MAINLY IN ANAEROBIC DIGESTERS, FACULTATIVE LAGOONS, AND
ANAEROBIC LAGOONS. DECOMPOSITION TAKES PLACE IN TWO PHASES, DURING
ANAEROBIC DIGESTION. THE FIRST, OR 'ACID' PHASE INVOLVES THE BREAKDOWN
OF SETTLED SLUDGE TO ORGANIC VOLATILE ACIDS AND ENERGY, AND THE SECOND
OR 'GAS' PHASE CONSISTS OF THE CONVERSION OF THE ORGANIC VOLATILE ACIDS
TO CO_2, CH_4, H_2, N_2, AND MINOR AMOUNTS OF OTHER GASES. DIGESTION
TEMPERATURES USUALLY LIE SOMEWHERE BETWEEN 15 AND 35 DEG C, WITH GAS
COLLECTION AND RE-USE AS A HEAT SOURCE BEING PRACTICED IN CONVENTIONAL
ANAEROBIC DIGESTION. DIGESTER LOADING IS EXPRESSED AS LBS VOLATILE
SOLIDS/FT3/DAY AND RANGE FROM 0.08 TO 0.1. FACULTATIVE PONDS ARE
INITIALLY LOADED AT FROM 20 TO 50 LBS BOD5/ACRE/DAY, BUT ONCE
ESTABLISHED THEY MAY BE LOADED AS HIGH AS 1000 LBS BOD5/ACRE/DAY.
DETENTION TIMES OF 3, 30, 90 AND 8-10 DAYS ARE COMMON FOR AEROBIC
PONDS, CONVENTIONAL DIGESTION, FACULTATIVE PONDS, AND ANAEROBIC PONDS
RESPECTIVELY, WITH EFFICIENCIES FOR SAME BEING 80% FOR AEROBIC PONDS,
85 TO 95% FOR FACULTATIVE PONDS, AND 70% FOR ANAEROBIC PONDS. LACK OF
CONTROL OF ODOR PROBLEMS HAS HINDERED APPLICATION OF ANAEROBIC PONDS,
AND RESEARCH WORK TO DEVELOP SUCH CONTROLS IS NEEDED.
(LOWRY-TEXAS)

FIELD 05D

ACCESSION NO. W72-06854

SAND FILTRATION OF PARTICULATE MATTER,

WYOMING UNIV., LARAMIE. DEPT. OF CIVIL ENGINEERING.

E. DAVIS, AND J. A. BORCHARDT.

JOURNAL SANITARY ENGINEERING DIVISION, AMERICAN SOCIETY OF CIVIL ENGINEERS,
VOL. 92, NO. SA5, P 47-60, PAPER 4940, OCTOBER 1966. 8 FIG, 2 TAB. USPHS
GRANT WP-00329.

DESCRIPTORS:
*WATER TREATMENT, *FILTRATION, SANITARY ENGINEERING, *FILTERS, *DESIGN,
COAGULATION, ALGAE, ACTIVATED CARBON, *FLOCCULATION, *HEAD LOSS.

IDENTIFIERS:
*FILTER MEDIA.

ABSTRACT:
THE PENETRATION OF PARTICULATE MATTER THROUGH SAND FILTERS CAN BE
ELIMINATED OR GREATLY REDUCED BY THE APPLICATION OF A COAGULATING
CHEMICAL SUPPLIED AS A SOLUTION DIRECTLY TO THE FILTER INFLUENT. AT
HIGH DENSITIES OF ORGANISMS, THE INABILITY TO REMOVE ALL CELLS OF
SELENASTOUM AND THE DIMINISHING RATE OF REMOVAL ASSOCIATED WITH
INCREASED COAGULANT DOSAGES INDICATES THE NECESSITY OF PRETREATMENT.
THE MODIFICATION OF THE PATTERN OF COAGULANT DEPOSIT, THE HEAD LOSS
INCREASE PATTERN AND THE INABILITY TO REMOVE ALL ALGAL CELLS AT HIGH
DENSITIES SEEM TO INDICATE THE IMPORTANCE OF THE CHEMICAL NATURE OF THE
FLOC IN ANY FILTRATION MECHANISM. THE LOCATION OF POWDERED ACTIVATED
CARBON DEPOSITED UNDER PLAIN FILTRATION WAS SHOWN TO BE LOGARITHMIC
WITH DEPTH. THE LOCATION OF DEPOSITED COAGULANT AND ACTIVATED CARBON
DECLINED LOGARITHMICALLY WITH DEPTH IN THE UPPER PART OF THE UNIFORM
SIZE FILTER MEDIA. (BEAN-AWWARF)

FIELD 05F

ACCESSION NO. W72-07334

ALGAE CONTROL,

WISCONSIN COMMITTEE ON WATER POLLUTION, MADISON.

K. M. MACKENTHUN.

PUBLIC WORKS, VOL 91, NO 9, P 114-116 AND 158, SEPTEMBER 1960. 2 FIG, 8 REF.

DESCRIPTORS:
WATER TREATMENT, *ALGAE, *COPPER SULFATE, *ALKALINITY, WATER QUALITY
CONTROL, *CONTROL, *ALGAL CONTROL, TOXICITY.

ABSTRACT:
DAMAGE CAUSED BY ALGAL GROWTHS IS DESCRIBED. THE DIFFERENCE IN
TREATMENT OF WATERS WITH LOW ALKALINITY AND THOSE WITH OVER 40 MG/L
METHYL ORANGE ALKALINITY IS DISCUSSED. THE APPLICATION OF COPPER
SULFATE, EQUIPMENT REQUIRED, CONTROL OF THE APPLICATION, AND THE
TOXICITY OF THE APPLICATION TO AQUATIC LIFE ARE DESCRIBED. MORE THAN
HALF THE STATES REPORT COMPLETE SUPERVISION OF FIELD APPLICATION OF
CHEMICALS, AND IN ADDITION ONE-FIFTH REPORT THAT THEY SPOT CHECK FIELD
APPLICATIONS. (BEAN-AWWARF)

FIELD 04A, 05F, 05G, 05C

ACCESSION NO. W72-07442

FIXATION OF ATMOSPHERIC NITROGEN BY NONLEGUMES IN WET MOUNTAIN MEADOWS,

AGRICULTURAL RESEARCH SERVICE, FORT COLLINS, COLO. SOIL AND WATER
CONSERVATION RESEARCH DIV.

L. K. PORTER, AND A. R. GRABLE.

AGRONOMY JOURNAL, VOL 61, NO 4, P 521-523, 1969, 1 FIG, 5 TAB, 8 REF.
JULY-AUGUST.

DESCRIPTORS:
*NITROGEN FIXATION, IRRIGATION, ORGANIC MATTER, *GRASSLANDS, MOUNTAINS,
*SOIL MICROORGANISMS, SOIL ALGAE, TURF, SOIL MICROBIOLOGY, WATER
POLLUTION SOURCES.

IDENTIFIERS:
*MOUNTAIN MEADOW, NITROGEN-15, SOD MATS, NON-LEGUMES, *NITROGEN FIXING
ORGANISMS.

ABSTRACT:
MOUNTAIN SOILS SUBJECTED TO EXCESSIVE IRRIGATION AND HIGH WATER TABLES
ACCUMULATE ORGANIC MATTER IN SOD MATS. MEADOW SOILS OFTEN CONTAIN TWO
OR THREE TIMES MORE N THAN EQUAL AREAS OF ADJACENT DRY SOILS. SYMBIOTIC
NITROGEN FIXATION BY LEGUMES IS ONE POSSIBLE SOURCE OF N FOR SOD MAT
FORMATION. FIXATION OF N2 BY FREE-LIVING ORGANISMS HAS ALSO BEEN
POSTULATED AND WAS VERIFIED BY THE LABORATORY STUDIES REPORTED HERE.
NITROGEN FIXATION BY SOD MATS CONTAINING NO LEGUMES WAS DETERMINED AT
18C IN ATMOSPHERES CONTAINING N2 15. BOTH PHOTOSYNTHETIC AND
NONPHOTOSYNTHETIC ORGANISMS APPEARED TO FIX N2. IN 10 DAYS, MATS IN THE
DARK FIXED FROM 0.76 TO 1.90 KG N2 HA-1 AND ILLUMINATED MATS FIXED FROM
3.72 TO 6.86 KG N HA-1. (SKOGERBOE-COLORADO STATE)

FIELD 05B, 03F

ACCESSION NO. W72-08110

SEWAGE NUTRIENT REMOVAL BY A SHALLOW ALGAL STREAM,

NATIONAL INST. FOR WATER RESEARCH, PRETORIA (SOUTH AFRICA).

J. HEMENS, AND M. H. MASON.

WATER RESEARCH, VOL. 2, NO. 4, P 277-287, 1968. 5 FIG, 1 TAB, 28 REF.

DESCRIPTORS:
*ALGAE, *NITROGEN, *PHOSPHORUS, *SEWAGE EFFLUENT, *SEWAGE TREATMENT,
*NUTRIENT REMOVAL, *WATER REUSE, SEASONAL, WATER TREATMENT.

IDENTIFIERS:
*SEWAGE NUTRIENT REMOVAL.

ABSTRACT:
THE USE OF ALGAE PONDS FOR REMOVING NITROGEN AND PHOSPHORUS BY
ASSIMILATION FROM SECONDARY SEWAGE EFFLUENT SUFFERS FROM THE
DISADVANTAGE OF INEFFICIENT LIGHT UTILIZATION BY THE DENSE ALGAL
CULTURES REQUIRED AND FROM THE PRACTICAL PROBLEM OF ECONOMICALLY
REMOVING THE ALGAE FROM SUSPENSION. AN EXPERIMENT IS DESCRIBED IN WHICH
THE PH INCREASE RESULTING FROM ALGAL PHOTOSYNTHESIS IN A SHALLOW STREAM
WAS USED TO REMOVE PHOSPHATE BY PRECIPITATION AND NITROGEN BY
ASSIMILATION AND LOSS TO THE ATMOSPHERE AS AMMONIA GAS. ASSOCIATION OF
PRECIPITATED PHOSPHATE WITH THE ALGAL CELLS PRODUCED A GRANULAR ALGAL
SEDIMENT EASILY REMOVED BY GRAVITY SETTLEMENT. REMOVAL OF NITROGEN AND
PHOSPHATE EXCEEDED 90 PER CENT IN THE WARMER SEASONS BUT AT WINTER
TEMPERATURES THE EFFICIENCY WAS LESS, DUE TO DECREASE IN THE ELEVATION
OF PH VALUE. THE METHOD APPEARS TO HAVE POSSIBLE APPLICATION FOR RURAL
COMMUNITIES IN EQUABLE CLIMATES WHERE WATER RE-USE IS DESIRABLE.
(SKOGERBOE-COLORADO STATE)

FIELD 05D

ACCESSION NO. W72-08133

DIATOMITE FILTRATION OF GOOD QUALITY SURFACE SUPPLIES,

JOHNS-MANVILLE PRODUCTS CORP., MANVILLE, N.J.

D. W. DAVIS, T. S. BROWN, S. SYROTYNSKI, AND E. HENDERSON.

CHEMICAL ENGINEERING PROGRESS SYMPOSIUM SERIES NO. 97, VOL 65, 1969, P
133-139, 3 FIG, 2 TAB, 10 REF.

DESCRIPTORS:
*DIATOMACEOUS EARTH, *FILTRATION, *SEPARATION TECHNIQUES, TURBIDITY,
SUSPENDED SOLIDS, ALGAE, PILOT PLANTS, HEAD LOSS, PERMEABILITY,
TEMPERATURE, OPERATION AND MAINTENANCE, DESIGN CRITERIA, *WATER
TREATMENT, *NEW YORK, WATER PURIFICATION.

IDENTIFIERS:
*MASSENA(NY).

ABSTRACT:
THE DESIGN OF DIATOMACEOUS EARTH FILTERS FOR WATER PURIFICATION STILL
REQUIRES PILOT PLANT DATA ON THE ACTUAL FILTRATION CHARACTERISTICS OF
EACH WATER TO BE FILTERED. IN ADDITION, THE TRADITIONAL APPROACH TO
DIATOMITE FILTRATION USES BODY FEED LEVELS EXPRESSED AS A RATIO OF
MILLIGRAMS/LITER OF BODY FEED TO THE MEASURED TURBIDITY. HOWEVER, FULL
SCALE OPERATIONAL DATA OBTAINED AT THE MASSENA, NEW YORK WATER
TREATMENT PLANT HAVE DEMONSTRATED THAT THE BODY FEED TO TURBIDITY
REQUIRED FOR ECONOMIC OPERATION IS MUCH HIGHER THAN THE RANGE OF 2-5
COMMONLY FOUND TO BE DESIRABLE AT OTHER LOCATIONS. SINCE THESE PROBLEMS
WERE ANTICIPATED, MEASUREMENTS OF THE WATER QUALITY BY DIRECT
MICROSCOPIC COUNTS WAS ALSO EMPLOYED, AND SUBSEQUENT INVESTIGATIONS
REVEALED THAT THE TOTAL COUNT VARIATIONS EXERTED A GREATER INFLUENCE ON
THE FILTER PERFORMANCE THAN THE TURBIDITY (OPTICALLY MEASURED). FURTHER
ITEMS FOR CONSIDERATION WHEN PLANNING AN EXPERIMENTAL PROGRAM TO OBTAIN
DESIGN DATA INCLUDE: (1) REVIEW OF HISTORICAL DATA; (2) USE OF A WIDE
ENOUGH VARIATION OF BODY FEED LEVELS AND FILTER AIDS TO PRECLUDE
EXCESSIVE EXTRAPOLATION; AND (3) TESTS SHOULD BE RUN AT TIMES OF
POOREST FEED WATER QUALITY. (LOWRY-TEXAS)

FIELD 05F

ACCESSION NO. W72-08686

ALGAL REMOVAL WITH A DIATOMACEOUS EARTH REVERSIBLE CAKE FILTER,

GRANGER FILTER CO., WAKEFIELD, MASS.

J. G. BROWN.

CHEMICAL ENGINEERING PROGRESS SYMPOSIUM SERIES NO. 97, VOL 65, 1969, P 128-132, 1 TAB, 7 REF.

DESCRIPTORS:
*ALGAE, *FILTRATION, *DIATOMACEOUS EARTH, SEPARATION TECHNIQUES, PRESSURE, FLOW RATES, SLUDGE, SLURRIES, AUTOMATIC CONTROL, PIPING SYSTEMS(MECHANICAL), CLEANING, COST ANALYSIS, WATER PURIFICATION, *WATER TREATMENT.

IDENTIFIERS:
*ALGAE HARVESTING, *DIATOMITE FILTRATION.

ABSTRACT:
LAGOON WATERS CONTAINING VARYING CONCENTRATIONS OF GREEN ALGAE WERE FILTERED THROUGH A STANDARD GRANGER C-33 REVERSIBLE FILTER WITH DIATOMITE PRECOAT AT HIGH VELOCITY DURING A 10 WEEK PILOT PLANT TEST. THE FILTER CLOTH COMPRISED 22 SQUARE FEET AND WAS MADE OF MONOFILAMENT POLYPROPYLENE. THE PILOT PLANT INCLUDED AN AUTOMATIC CYCLE CONTROLLER WITH AIR OPERATED VALVES, AND A DIATOMITE SLURRY HANDLING SYSTEM INCLUDING A SLUDGE PUMP, LIFT PUMP, AND AN AGITATED DIATOMITE SETTLING AND DECANTING TANK. THE SUCCESS OF THE SELF-PLUGGING BACKWASH OF THE REVERSIBLE FILTRATION PROCESS WAS DEMONSTRATED, AS WAS THE EFFECTIVENESS OF USING HIGH INTERNAL FILTER VELOCITIES, LOW PRESSURE DROPS ACROSS THE CAKE, AND SMOOTH MONOFILAMENT POLYPROPYLENE CLOTH. FILTRATE AVERAGED 10 PPM OF ALGAE, WHILE THE RECYCLED CONCENTRATED STREAM CONTAINED 500 PPM OF ALGAE. TOTAL COSTS FOR REMOVAL OF 0.54 LB ALGAE/1000 GALLONS USING 2.2 LB OF DIATOMITE/LB OF ALGAE WERE ESTIMATED AT 8.6 TO 13.6 CENTS/1000 GALLONS PROCESSED. AFTER 10 WEEKS OF NEARLY CONTINUOUS OPERATION, THE FILTER DID NOT REQUIRE CLEANING. THE EXPERIENCE IN OPERATION ACQUIRED FROM THE PILOT PLANT TESTS WILL HOPEFULLY MAKE THE ECONOMICS OF THE PROCESS EVEN MORE ATTRACTIVE. (LOWRY-TEXAS)

FIELD 05F

ACCESSION NO.   W72-08688

PRE-TREATMENT LICKS ALGAE PROBLEM,

OSHKOSH DEPT. OF PUBLIC WORKS, WIS.

J. STRAUSS.

WATER AND SEWAGE WORKS, VOL 110, NO 7, P 267-268, JULY 1963. 2 FIG.

DESCRIPTORS:
*WATER TREATMENT, SANITARY ENGINEERING, *FILTRATION, DESIGN, *ALGAE, COPPER SULFATE, SEDIMENTATION, ACTIVATED CARBON, CHLORINE, COSTS, TREATMENT FACILITIES, CLEANING, *ALGAL CONTROL, WISCONSIN.

IDENTIFIERS:
ALUM, *OSHKOSH(WIS), BACKWASHING.

ABSTRACT:
ON AN 8.8 MGD FILTER PLANT AT OSHKOSH, 9.8 PERCENT WASHWATER WAS SOMETIMES REQUIRED, BECAUSE ALGAE CLOGGED THE FILTERS. CHLORINE, ALUM, AND ACTIVATED CARBON APPLIED INSIDE THE PLANT REDUCED UNPLEASANT TASTES AND ODORS BUT DID NOT ELIMINATE SUFFICIENT QUANTITIES OF ALGAE. A 30 ACRE PRESETTLING BASIN WAS CONSTRUCTED IN A BAY NEAR THE PLANT, WITH BAFFLE WALLS TO INSURE MAXIMUM RETENTION. WATER ENTERS THE BASIN THROUGH A 36 IN. STEEL LINE TO 2,000 FT. OUT IN THE LAKE. COPPER SULFATE SOLUTION IS PUMPED INTO THE INTAKE LINE AS WATER ENTERS THE BASIN. AFTER THE PRE-SETTLING CHLORINE, ACTIVATED CARBON AND ALUM ARE APPLIED IN THE MIXING TANK. TREATMENT IN CONVENTIONAL SETTLING BASINS AND FILTRATION FOLLOW. DURING THE FIRST FIVE MONTHS OF OPERATION THE AVERAGE NUMBER OF FILTERS WASHED DAILY WAS 3.8 RATHER THAN THE PREVIOUS 18.6. BACKWASHING WATER WAS REDUCED TO 2.6 PERCENT AND COST OF CHEMICALS WAS REDUCED TO 70 PERCENT OF THE PREVIOUS COST. (BEAN-AWWARF)

FIELD 05F

ACCESSION NO.   W72-08859

MEASUREMENT AND CONTROL OF PARTICULATE MATTER IN FILTER EFFLUENTS,

FLINT WATER UTILITY, MICH.

R. M. HARWOOD.

JOURNAL OF THE AMERICAN WATER WORKS ASSOCIATION, VOL. 55, NO. 4, P 487-489, APRIL 1963. 1 REF.

DESCRIPTORS:
  *WATER TREATMENT, *FILTRATION, *ALGAE, *WATER QUALITY, *ACTIVATED CARBON, *ELECTROLYTES, FILTERS, *ANALYTICAL TECHNIQUES.

IDENTIFIERS:
  *MILLIPORE FILTERS.

ABSTRACT:
  THE SEARCH FOR IMPROVED QUALITY PUTS MORE AND MORE EMPHASIS ON MEASURMENT AND CONTROL. TYPE AA MILIPORE FILTERS, PORE SIZE 0.8 MICRONS WITH 9.6-SQ CM FILTER AREA AND SUCTION EQUAL TO 12 IN. MERCURY COLUMN ARE USED, TO MEASURE MATERIALS IN THE EFFLUENT OF FILTERS. THE MOST RAPID METHOD FOR DETECTION AND ANALYSIS IS THE SIMPLE PATCH TEST. THE FILTER CAKE IS MOUNTED AND COMPARED WITH PREVIOUS SAMPLES COLLECTED FROM THE SAME SYSTEM. THE PARTICULATE DISCOLORATION OF THE FILTER IS A ROUGH INDICATION OF THE CLEANLINESS OF THE SAMPLE. FOR ALGAE COUNT THE FILTER IS WASHED AND SUBJECTED TO MICROSCOPIC COUNTING, REPORTED IN ORGANISMS PER GALLON. THE DRY FILTER MAY BE PLACED ON THE MICROSCOPE STAGE FOR DIRECT COUNTING OF PARTICLES BY REFLECTED LIGHT. CARBON AND FLOC PARTICLES CAN BE VISUALLY IDENTIFIED. TIME OF FLOW THROUGH THE MEMBRANE FILTER WILL INDICATE THE FILTERABILITY. MOST OF THE ALGAE FORMS IN THE RAW WATER MAY APPEAR ON THE MEMBRANE FILTER, ALSO FINES OF THE ACTIVATED CARBON. POLYELECTROLYTES MAY BE USEFUL IN REDUCING PENETRATION IN THE PLANT FILTERS. (BEAN-AWWARF)

FIELD 05F, 05D

ACCESSION NO. W72-09693

RESEARCH ON ALGAL ODOR,

ROBERT A. TAFT SANITARY ENGINEERING CENTER, CINCINNATI, OHIO.

T. E. MALONEY.

JOURNAL OF THE AMERICAN WATER WORK ASSOCIATION, VOL. 55, NO. 4, P 481-486, APRIL 1963. 1 FIG, 2 TAB, 3 REF.

DESCRIPTORS:
  *WATER TREATMENT, *WATER QUALITY, *ALGAE, *ODOR, HYDROGEN ION CONCENTRATION.

ABSTRACT:
  PRELIMINARY STUDIES IN RESEARCH ON THE ROLE OF ALGAE IN CAUSING ODORS IN WATER SUPPLIES ARE PRESENTED. INVESTIGATIONS CARRIED OUT WITH CHLOROCOCCUM, A UNICELLULAR GREEN ALGA, IN PURE CULTURE INDICATE THAT THE ODORIFEROUS MATERIAL IS, FOR THE MOST PART, RETAINED WITHIN THE CELLS UNTIL THEIR DEATH OR DISINTEGRATION. THE ODOR PRODUCED BY THIS ALGA INCREASED NOTABLY UNDER EXPERIMENTAL CONDITIONS IN THE ACIDIC PH RANGE. FRACTIONIZATION OF DISINTEGRATED ALGAL CELLS REVEALED THAT A NUMBER OF CHEMICAL COMPOUNDS ARE INVOLVED IN THE ODORIFEROUS MATERIAL. (BEAN-AWWARF)

FIELD 05F, 05C

ACCESSION NO. W72-09694

EFFECTS OF DEPTH OF WASTE STABILIZATION POND PERFORMANCE,

VIRGINIA POLYTECHNIC INST., BLACKSBURG.

S. K. ANDERSON.

MASTER'S THESIS, 1969, 75 P, 17 FIG, 4 TAB, 37 REF.

DESCRIPTORS:
*OXIDATION LAGOONS, *WASTE WATER TREATMENT, ALGAE, PHOTOSYNTHESIS,
DEPTH, THERMOCLINE, CLIMATIC ZONES, ORGANIC LOADING, DISSOLVED OXYGEN,
HYDROGEN ION CONCENTRATION, BIOCHEMICAL OXYGEN DEMAND, VIRGINIA.

IDENTIFIERS:
*SUSPENDED SOLIDS, CHASE CITY(VA).

ABSTRACT:
AN 8 ACRE POND WITH A 3 FT DEPTH AND A 4 ACRE POND WITH A 6 FT DEPTH
WERE BOTH LOADED AT 26 LB BOD/ACRE/DAY. BOTH PONDS HAD BEEN OPERATING
FOR TWO YEARS ON DOMESTIC WASTES. FOLLOWING A 45 DAY ACCLIMATION
PERIOD, DISSOLVED OXYGEN, TEMPERATURE, AND PH MEASUREMENTS WERE TAKEN
DURING 24 HOUR TESTING PERIODS SPACED AT 2 WEEK INTERVALS FOR A 12 WEEK
PERIOD. BOD AND SUSPENDED SOLIDS CONCENTRATIONS WERE ALSO MEASURED
DURING EACH TESTING PERIOD. ALL POND LEVELS CONTAINED DISSOLVED OXYGEN
AT ALL TIMES, WITH THE DEEPER LEVELS SOMETIMES APPROACHING THE
ANAEROBIC STATE. MEANWHILE, ALGAL PHOTOSYNTHESIS PRODUCED LEVELS OF 20
MG/1 DISSOLVED OXYGEN IN THE UPPER LEVELS OF THE POND, WHILE RAISING TO
PH ABOVE 10. BECAUSE OF THE FREQUENT D.O. INTRUSIONS TO THE BOTTOM
LEVEL, METHANE FERMENTATION WAS NEVER ESTABLISHED. THE 6 FT POND DEPTH
IN THE DEEPER POND WAS SUFFICIENT TO INSURE THE ESTABLISHMENT OF A
DEFINITE THERMOCLINE UNDER EXISTING CLIMATIC CONDITIONS IN VIRGINIA.
BOTH THE BOD AND THE SUSPENDED SOLIDS IN THE EFFLUENT PRESENT AT
CONCENTRATIONS OF 30-40 PPM AND 350 PPM RESPECTIVELY, WERE ATTRIBUTED
TO ALGAE CARRIED OVER IN THE EFFLUENT. IN ACCORDANCE WITH OTHER RECENT
STUDIES AND WITH THEORY, IT WAS DEMONSTRATED THAT THE DEEPER POND
CONSISTENTLY PRODUCED A SUPERIOR EFFLUENT IN TERMS OF BOTH BOD AND SS.
(LOWRY-TEXAS)

FIELD 05D

ACCESSION NO. W72-10573

ALGAE OF THE LITORAL OF WEST COAST OF SAKHALIN,

T. F. SHCHAPOVA, AND V. B. VOZZHINSKAYA.

AVAILABLE FROM THE NATIONAL TECHNICAL INFORMATION SERVICE AS AD-724 266,
$3.00 IN PAPER COPY, $0.95 IN MICROFICHE. NAVAL OCEANOGRAPHIC OFFICE
TRANSLATION 339. TRANS. OF VODOROSLI LITORAL: ZAPADNOGO POBERZH'YA
SAKHALINA, TRUDY INSTITUTA OKEANOLOGII, VOL. 34, P 123-146, 1960. 5 FIG, 2
TAB, 33 REF.

DESCRIPTORS:
*ALGAE, *LITTORAL, *AQUATIC PLANTS, ISLANDS, ECOLOGICAL DISTRIBUTION,
BIOLOGICAL COMMUNITIES, SYSTEMATICS, MARINE PLANTS.

IDENTIFIERS:
*SAKHALIN ISLAND(USSR).

ABSTRACT:
THE WESTERN COAST OF SAKHALIN ISLAND WAS STUDIED ON AN EXPEDITION OF
THE INSTITUTE OF OCEANOLOGY USSR IN 1954. THE COASTAL BELT INCLUDING
THE LITTORAL AND SUBLITTORAL FROM KHOE IN THE NORTH TO IVANOVKA IN THE
SOUTH, ABOUT 600 KM, IS DESCRIBED POINTING OUT TYPE OF BOTTOM,
AMPLITUDE OF TIDES, EXPOSURE TO WAVES AND EFFECTS OF WARM OR COLD
CURRENTS. ABOUT 500 QUALITATIVE AND QUANTITATIVE SAMPLES WERE OBTAINED
AND A HERBARIUM OF 700 PLANTS WAS PREPARED. FROM 1860 TO 1926 VARIOUS
EXPEDITIONS AND INVESTIGATIONS SAMPLED ALGAE AT SEVERAL POINTS AND THE
BOTANICAL HISTORY IS GIVEN. AFTER THE RETURN OF SOUTH SAKHALIN TO THE
SOVIET UNION FROM JAPAN COMPREHENSIVE STUDIES OF ALGAE IN THE AREA WERE
UNDERTAKEN, CONCENTRATING ON THE ECOLOGY AND PRACTICAL APPLICATION OF
MACROPHYTES. A CONSIDERABLE PART OF THESE STUDIES HAS REMAINED
UNPUBLISHED. IN THIS WORK ALGAE ARE DEFINED TO THE SPECIES, ALSO THEIR
QUANTITY AND DISTRIBUTION ARE EVALUATED AND DESCRIBED. A COMPREHENSIVE
LIST OF ALGAE IN THE APPENDIX SUMS UP THE LOCATION OF SAMPLES, THEIR
SIZE, FORM, QUANTITY, AND DISTRIBUTION. (JONES-WISCONSIN)

FIELD 05C, 05B

ACCESSION NO. W72-10623

TOXICITY OF PETROCHEMICALS IN THE AQUATIC ENVIRONMENT,

TEXAS UNIV., AUSTIN. DEPT. OF CIVIL ENGINEERING.

J. F. MALINA, JR.

WATER AND SEWAGE WORKS, VOL 111, NO 10, OCTOBER 1964, P 456-460. 1 FIG, 2 TAB, 20 REF.

DESCRIPTORS:
*POLLUTANT IDENTIFICATION, *TOXICITY, *BIOASSAY, FISH, AQUATIC ENVIRONMENT, ORGANIC PESTICIDES, HYDROGEN ION CONCENTRATION, TEMPERATURE, CHLORINATED HYDROCARBON PESTICIDES, ALGAE, LABORATORY TESTS, WATER POLLUTION EFFECTS.

IDENTIFIERS:
*PETROCHEMICAL COMPOUNDS, MEAN TOLERANCE LIMIT, BLUEGILLS, CHLORELLA PYRENOIDOSA, BENZENE COMPOUNDS.

ABSTRACT:
THE ORGANIC CHEMICAL CONCENTRATION CARRIED IN MANY WATER COURSES HAS BEEN INCREASING AS A RESULT OF THE GROWING INDUSTRIAL, AGRICULTURAL, AND DOMESTIC APPLICATIONS OF PETROCHEMICAL PRODUCTS. THE TOXIC EFFECTS OF VARIOUS PETROCHEMICALS ON AQUATIC ORGANISMS WERE SUMMARIZED. THE RESULTS OF BIOASSAY TESTS INDICATING THE ACUTE TOXICITY OF A SPECIFIC COMPOUND ON ACCLIMATIZED TEST ANIMALS, NAMELY FISH, WERE ARRANGED IN TABULAR FORM. THE TOXICITY DATA WERE REPORTED AS THE 96 HOUR MEDIAN TOLERANCE LIMIT. IT WAS NOTED THAT IN CONDUCTING A TOXICITY BIOASSAY IT IS NECESSARY TO USE THE SAME SPECIES AND SIZE OF TEST ANIMAL AND THAT THE WATER MUST BE OF UNIFORM QUALITY AND THE PH AND TEMPERATURE MUST BE RELATIVELY CONSTANT. THE CHLORINATED AND PHOSPHORUS CONTAINING ORGANIC PESTICIDES WERE MORE TOXIC TO FISH AND INVERTEBRATES THAN ANY OF THE COMPOUNDS TESTED. ALKYL ARYL SULFONATE, LACTONITRIDE, SODIUM BUTYL MECAPTIDE, SODIUM CYANIDE AND TETRAETHYL LEAD WERE TOXIC TO BLUEGILLS IN CONCENTRATIONS LESS THAN 10 MG/L. ALSO, THE TOXICITY OF VARIOUS ORGANIC CHEMICALS TO A SPECIES OF GREEN ALGAE CHLORELLA PYRENOIDOSA WHICH HAD BEEN EVALUATED IN LABORATORY STUDIES WERE REPORTED. COMPOUNDS OF THE BENZENE SERIES EXHIBITED THE HIGHEST TOXICITY WITH NITROBENZENE BEING TOXIC AT CONCENTRATIONS OF 0.01 MG/L OR LESS. (GALWARDI-TEXAS)

FIELD 05C, 05B

ACCESSION NO. W72-11105

INFLUENCE OF SITE CHARACTERISTICS ON QUALITY OF IMPOUNDED WATER,

WASHINGTON UNIV., SEATTLE. DEPT. OF CIVIL ENGINEERING.

R. O. SYLVESTER, AND R. W. SEABLOOM.

JOURNAL OF THE AMERICAN WATER WORKS ASSOCIATION, VOL 57, NO 12, P 1528-1546, DEC. 1965. 11 FIG, 4 TAB, 24 REF.

DESCRIPTORS:
*RESERVOIR SITES, WATER RESOURCES DEVELOPMENT, WATER QUALITY, SOIL INVESTIGATIONS, *IMPOUNDED WATERS, IMPOUNDMENTS, *DISSOLVED OXYGEN, BIOCHEMICAL OXYGEN DEMAND, ALGAL CONTROL, *COLOR, *ALKALINITY, *NITRATES, NITROGEN, *CONDUCTIVITY, *LEACHING, IRON, TEMPERATURE, SOIL ANALYSIS, TASTE-PRODUCING ALGAE, HYDROGEN ION CONCENTRATION, *ORGANIC SOILS, MINERALOGY.

IDENTIFIERS:
WOOD, ALGAL COUNT, GREEN RIVER, WASHINGTON, HOWARD A. HANSON RESERVOIR, TACOMA(WASH).

ABSTRACT:
SOILS AT IMPOUNDMENT SITES SHOULD BE TESTED FOR THEIR EFFECT ON OVERLYING WATER. ORGANIC SOILS CREATE THE MOST SEVERE PROBLEMS BUT LEACHING THEM OR COVERING THEM WITH AS LITTLE AS 2 INCHES OF MINERAL SOIL MAY BE EFFECTIVE. TESTS WERE PERFORMED ON HOWARD A. HANSON RESERVOIR, NEAR TACOMA, WASHINGTON, TO ASCERTAIN PHYSICAL, CHEMICAL AND BIOLOGIC EFFECTS. BASED ON THESE STUDIES, RECOMMENDED WATER QUALITY TESTS OVER SOILS OF IMPOUNDMENTS SHOULD INCLUDE DO, COLOR, NITRATE, AMMONIA, ALGAL COUNTS AND PH MEASUREMENTS OF AT LEAST ONE MONTH DURATION. LEACHING AND EXCHANGE STUDIES WILL REVEAL IF SITE SHOULD BE ABANDONED OR IF COVER SOIL IS FEASIBLE. SHORELINES SHOULD BE INVESTIGATED FOR EFFECT ON TURBIDITY. COMPLETE WATER QUALITY ANALYSES UNDER VARIOUS FLOW CONDITIONS ARE NEEDED TO EVALUATE FUTURE RESERVOIR QUALITY AS RELATED TO THE UNDERLYING SOIL. (FLACK-AWWARF)

FIELD 05G, 05F, 02G

ACCESSION NO. W72-11906

MODIFICATION OF SOME PHYSIOLOGICAL METHODS FOR ESTIMATION OF WATER QUALITY AND
   THEIR RELATIONSHIP TO DEGREES OF SAPROBITY (MODIFIKACE VYBRANYCH
   FYZIOLOGICKYCH METOD STANOVENI KVALITY VODY A JEJICH VZTAH K STUPNUM SAPROB
   ITY),

VYSOKA SKOLA CHEMICO-TECHNOLOGICKA, PRAGUE (CZECHOSLOVAKIA). DEPT. OF WATER
   TECHNOLOGY.

D. MATULOVA.

VODNI HOSPODARSTVI, VOL 9, P 393-395, 1967. 1 TAB, 18 REF. ENGLISH SUMMARY.

   DESCRIPTORS:
      *WATER QUALITY, *BIOINDICATORS, *WATER ANALYSIS, *BIOASSAY,
      *COMPARATIVE PRODUCTIVITY, AQUATIC MICROBIOLOGY, EUTROPHICATION,
      MUNICIPAL WASTES, FRESHWATER ALGAE, WATER POLLUTION EFFECTS, DOMESTIC
      WASTES, INDUSTRIAL WASTES, ENVIRONMENTAL EFFECTS, ANALYTICAL
      TECHNIQUES, PHYSIOLOGICAL ECOLOGY, PLANT PHYSIOLOGY.

   IDENTIFIERS:
      *ALGAL CULTURE, *BIOMASS TITER, *BIOLOGICAL INDEX, *SAPROBITY, PRAGUE,
      CZECHOSLOVAKIA, LABORATORY PROCEDURES, STATISTICAL PROCEDURES, BMT.

   ABSTRACT:
      STUDIES WERE MADE WITH SEWAGE AND INDUSTRIAL WASTES COLLECTED FROM
      SEVERAL AREAS NEAR PRAGUE, ESPECIALLY WHERE THE BOTIC EMPTIES INTO THE
      RIVER VLTAVA, TO DETERMINE ITS EFFECT ON ALGAE. A TECHNIQUE WAS
      DEVELOPED TO EVALUATE ITS TOXICITY TO ALGAE. ON THE BASIS OF RESULTS
      OBTAINED USING THE BIOMASS TITER (BMT) AS A CRITERION, A BIOLOGICAL
      INDEX OF WATER QUALITY WAS OBTAINED. THE INDEX ENABLED THE CALCULATION
      OF DIFFERENT DEGREES OF SAPROBITY. THE VALUES OBTAINED BY THE
      LABORATORY EXPERIMENTS WERE COMPARED WITH FINDINGS IN THE FIELD. A GOOD
      CORRELATION OF BETA SAPROBIC TO POLYSAPROBIC WATERS WAS FOUND.
      (KATZ-WASHINGTON)

   FIELD 05C

   ACCESSION NO. W72-12240

THE BIO-FOULING MENACE IN COOLING SYSTEMS,

A. MENNIE, AND E. TEHLE.

EFFLUENT AND WATER TREATMENT JOURNAL, VOL 9, NO. 9, P 493-499, SEPTEMBER
   1969, 4 FIG.

   DESCRIPTORS:
      *COOLING WATER, *BIOMASS, *SLIME, ALGAE, FUNGI, BACTERIA, BACTERICIDES,
      CHLORINE, COSTS, *WASTE WATER TREATMENT.

   IDENTIFIERS:
      BIOCIDES, PRODUCTION LOSS, QUATERNARY AMMONIUM COMPOUNDS, PHENOLIC
      COMPOUNDS, ORGANO-METALLIC COMPOUNDS, AMINES, ENGLAND.

   ABSTRACT:
      THE DEVELOPMENT OF BIOLOGICAL GROWTHS IN INDUSTRIAL COOLING SYSTEMS HAS
      PRODUCED DIFFICULT AND SERIOUS PROBLEMS, AND STEPS MUST BE TAKEN TO
      CONTROL OR ELIMINATE BIOFOULING. EXCESSIVE GROWTH AND DEVELOPMENT OF
      ALGAE, FUNGI AND BACTERIA HAVE RESULTED IN BIOLOGICAL FOULING IN
      COOLING WATER SYSTEMS. TWO CASE-HISTORIES ILLUSTRATING HOW BIOLOLOGICAL
      FOULING CAN ACCUMULATE TO SHUT DOWN ESSENTIAL PRODUCTION PROCESSES WERE
      PRESENTED. ONE CONCERNING A RECIRCULATING COOLING SYSTEM IN A
      PHARMACEUTICAL PLANT WHERE THE COST OF EACH SHUTDOWN WAS ABOUT 1,000
      POUNDS PER HOUR, AND THE ANNUAL LOSS OF PRODUCTION ABOUT 12 HOURS.
      TREATMENT INCLUDED THE USE OF A SYNTHETIC ANTIFOULANT WHICH BROKE UP
      THE GROWTH DEBRIS, DISPERSING IT INTO A HARMLESS AND EASILY REMOVED
      CLOUD OF FINE PARTICLES. THE SECOND CASE INVOLVED A RECIRCULATING
      COOLING SYSTEM OF A CHEMICAL BY-PRODUCTS COMPLEX OF A STEELWORKS COKE
      PRODUCING PLANT. THE COST OF A COMPLETE CLEAN-UP TO RESTORE FULL
      OPERATING EFFICIENCIES WAS ASSESSED AT 3,000 POUNDS, OF WHICH HALF WAS
      LOSS OF PRODUCTION. TYPES OF BIOCIDES USED INCLUDE QUARTERNARY AMMONIUM
      COMPOUNDS, PHENOLIC COMPOUNDS, ORGANO-METALLIC COMPOUNDS AND AMINES.
      ONE COMMON METHOD OF INHIBITING ALGAL AND FUNGAL GROWTH IN LARGE
      COOLING SYSTEMS MAKES EFFECTIVE USE OF CHLORINE AS A OXIDIZING BIOCIDE.
      HOWEVER, THE USE OF CHLORINE AND RELATED CONTROL EQUIPMENT WAS SHOWN TO
      BE EXPENSIVE WHEREAS NON-OXIDIZING BIOCIDES CAN BE INTRODUCED TO THE
      LARGEST OF SYSTEMS AT MUCH LOWER COST BY DIRECT ADDITION TO THE COOLING
      TOWER SUMP. (GALWARDI-TEXAS)

   FIELD 05D, 05F

   ACCESSION NO. W72-12987

ADVANCES IN WATER TREATMENT,

OMAHA METROPOLITAN UTILITIES DISTRICT, NEBR.

J. F. ERDEI.

JOURNAL AMERICAN WATER WORKS ASSOCIATION, VOL 55, NO 7, P 845-856, JULY 1963.
  1 TAB, 16 REF.

DESCRIPTORS:
  *WATER TREATMENT, SANITARY ENGINEERING, *PATHOGENIC BACTERIA, ALGAE,
  ORGANIC MATTER, RADIOACTIVITY, PESTICIDES, TASTE, ODOR, COAGULATION,
  ELECTROLYTES, POTABLE WATER.

ABSTRACT:
  REVIEWS WERE CONDUCTED OF THE STATUS OF INCREASING WATER POLLUTION
  PROBLEMS. USE OF CHEMICALS, NEW SCIENTIFIC TOOLS, ANALYTICAL
  TECHNIQUES, TREATMENT PROCESSES, POLYELECTROLYTE USAGE, COAGULATION,
  JAR TEST, PATHOGENIC ORGANISMS, ALGAE, TRACE ELEMENTS, PESTICIDES,
  RADIOSOTOPES, AND TASTES AND ODORS, ARE DISCUSSED. MORE STUDY IS NEEDED
  ON METHODS OF TREATING DRINKING WATER, A GOOD SHARE OF WHICH SHOULD
  INVOLVE ORGANIC CONTAMINANTS. THERE IS NO ASSURANCE THAT SUCH HARMFUL
  SUBSTANCES IN SOLUTION WILL SIGNIFY THEIR PRESENCE BY PRODUCING TASTE
  OR ODOR, AND MANY INTERESTING OPPORTUNITIES FOR RESEARCH EXIST IN THESE
  AREAS. (BEAN-AWWARF)

FIELD 05F

ACCESSION NO. W72-13601

CHEMICAL ASPECTS OF ACTINOMYCETE METABOLITES AS CONTRIBUTORS OF TASTE AND ODOR,

IOWA UNIV., IOWA CITY. STATE HYGIENIC LAB.

R. L. MORRIS, J. D. DOUGHERTY, AND G. W. RONALD.

JOURNAL OF THE AMERICAN WATER WORKS ASSOCIATION, VOL 55, NO 10, P 1380-1390,
  OCTOBER 1963. 7 FIG, 1 TAB, 8 REF.

DESCRIPTORS:
  *WATER TREATMENT, *ACTINOMYCETES, *ALGAE, *TASTE, *ODOR, WATER QUALITY
  CONTROL, POLLUTANT IDENTIFICATION, IOWA.

IDENTIFIERS:
  *METABOLITES, *CEDAR RIVER(IOWA).

ABSTRACT:
  THE EXTREMELY SEVERE MUSTY TASTES AND ODORS THAT OCCURRED IN THE CEDAR
  RIVER WERE DUE TO CERTAIN METABOLITES OF THE ACTINOMYCETES,
  PARTICULARLY A 'NEUTRAL FRACTION'. NO CORRELATION EXISTED BETWEEN THE
  INTENSITY OF TASTE PROBLEMS AND THRESHOLD ODOR VALUES, AS THE BURNING,
  MUSTY TASTE OF CRITICAL FRACTIONS OF ACTIMYCETE CULTURES WERE FAR MORE
  PERSISTENT THAN THE ASSOCIATED ODOR PHENOMENA. METABOLITES WERE
  EXTRACTED, PURIFIED AND DETERMINATION MADE AS TO WHICH PRODUCED THE
  TYPICAL TASTES AND ODORS. THESE WERE PURIFIED BY ELUTRIATION WITH
  VARIOUS SOLVENTS. WHEN THE REMAINING ETHER SOLUTION WAS DRIED WITH
  SODIUM SULFATE A PALE YELLOW OIL CALLED THE 'NEUTRAL FRACTION' WAS
  RECLAIMED; IT HAD A DECIDEDLY MUSTY TASTE AND ODOR. VARIOUS 'NEUTRAL
  FRACTIONS' IN SPECTROGRAMS, SHOWED CORRELATIONS OF BOUND HYDROXYL AND
  CARBONYL GROUPS AT WAVELENGTHS 2.9-2.95 AND 5.8-5.85 MICRONS
  RESPECTIVELY. (BEAN-AWWARF)

FIELD 05F

ACCESSION NO. W72-13605

EFFECTS OF RIVER PHYSICAL AND CHEMICAL CHARACTERISTICS ON AQUATIC LIFE,

ACADEMY OF NATURAL SCIENCES OF PHILADELPHIA, PA.

R. PATRICK.

JOURNAL OF THE AMERICAN WATER WORKS ASSOCIATION, VOL 54, NO 5, P 544-550, MAY
1962. 2 REF.

DESCRIPTORS:
*AQUATIC LIFE, *AQUATIC POPULATIONS, *RIVERS, ALGAE, PROTOZOA, AQUATIC
INSECTS, FISH REPRODUCTION, INDUSTRIAL WASTES, ECOLOGY, DISSOLVED
OXYGEN, HARDNESS, TIDAL WATERS, ORGANIC MATTER, CONDUCTIVITY,
TEMPERATURE, BACTERIA, NUTRIENTS, ESTUARINE ENVIRONMENT, PONDS,
BREEDINGS, SALINE WATER-FRESHWATER INTERFACES, COLLOIDS.

IDENTIFIERS:
*RIVER REGIONS, *PHYSICAL CHARACTERISTICS, *CHEMICAL CHARACTERISTICS,
ORGANISMS, SUSPENDED SOLIDS, SALTWATER TONGUE, BENTHIC FORMS, EPIPHYTIC
FORMS, COLLOIDAL FRACTION, MULTIPLE-USES, SAVANNAH RIVER, GULF COAST.

ABSTRACT:
THE PHYSICAL AND CHEMICAL CHARACTERISTICS OF A RIVER HAVE IMPORTANT
EFFECTS ON ITS AQUATIC LIFE AND, IN TURN, AQUATIC LIFE CAN INDICATE
POLLUTION AFFECTS ON THE RIVER. HEAD WATER REGIONS HAVE RELATIVELY LOW
CAPACITY FOR ORGANIC ASSIMILATION, SERVE AS EXCELLENT BREEDING ZONES,
AND MUST BE MAINTAINED IN THEIR NATURAL STATE. THE MAIN TRUNK REGION
CAN ACCOMMODATE WASTES IF THEY ARE DISPERSED THROUGHOUT THE STREAM.
THIS SECTION IS RICH IN HABITATS AND POPULATIONS MAY BE LARGE. THE
TIDAL SECTION IS COMPLEX, EXHIBITING RETENTION, LIMITED MIXING,
SALTWATER INTERFACE EFFECTS, ALL OF WHICH AFFECT THE AQUATIC LIFE.
CHEMICAL AND PHYSICAL ASPECTS INCLUDE DISSOLVED OXYGEN, HARDNESS,
ORGANIC MATTER, TOTAL CONDUCTIVITY, TEMPERATURE, SUSPENDED SOLIDS,
WHICH AFFECT THE VARIETY AND NUMBERS OF SPECIES OF AQUATIC LIFE. IN
PLANNING THE USES OF A RIVER THE EFFECTS OF WASTES ON AQUATIC LIFE MUST
BE CONSIDERED. (FLACK-AWWARF)

FIELD 05C, 02L, 05B

ACCESSION NO. W73-00748

RELATIONSHIP OF IMPOUNDMENT TO WATER QUALITY,

GEOLOGICAL SURVEY, WASHINGTON, D.C. WATER QUALITY BRANCH.

S. K. LOVE.

JOURNAL OF THE AMERICAN WATER WORKS ASSOCIATION, VOL 53, NO 5, P 559-568, MAY
1961. 5 FIG, 1 TAB, 13 REF.

DESCRIPTORS:
*RESERVOIR OPERATIONS, DISSOLVED OXYGEN, DISSOLVED SOLIDS,
SEDIMENTATION RATES, MANGANESE, LOW-FLOW AUGMENTATION, TENNESSEE RIVER,
ALGAE, IRON, ALKALINITY, HARDNESS, RADIOACTIVITY, TEMPERATURE,
SALINITY, REAERATION, FISH, TASTE, ODOR, OXYGEN SAG, POTOMAC RIVER.

IDENTIFIERS:
*RESERVOIR EFFECTS, LAKE MEAD, DILUTION, ALGAL GROWTHS, OXYGEN
DEPLETION.

ABSTRACT:
EFFECTS OF IMPOUNDMENT ON WATER INCLUDE REDUCTION OF POLLUTANTS,
LEVELING OF VARIATIONS IN DISSOLVED SOLIDS, HARDNESS, PH AND
ALKALINITY, TEMPERATURE REDUCTION, SEDIMENT ENTRAPMENT, LOW FLOW
AUGMENTATION AND DILUTION, INCREASED ALGAE GROWTHS, REDUCED DISSOLVED
OXYGEN, INCREASE IN CARBON DIOXIDE, IRON, MANGANESE, AND ALKALINITY,
AND INCREASED DISSOLVED SOLIDS DUE TO EVAPORATION OR DISSOLVING OF
ROCKS. CASES OF BENEFICIAL AND DETRIMENTAL EFFECTS INCLUDE LAKE MEAD,
THE TENNESSEE VALLEY, SWITZERLAND LAKES, CALIFORNIA AND TEXAS. THE
POTOMAC RIVER IS ANALYZED IN SOME DETAIL. YET, MOST STUDIES OF
IMPOUNDMENT EFFECTS ARE MORE QUANTITATIVE THAN QUALITATIVE. THERE IS
NEED FOR BETTER STUDIES OF STREAM IMPOUNDMENTS SO THAT PROBLEMS WITH
PROPOSED IMPOUNDMENTS CAN BE RESOLVED IN ADVANCE. (FLACK-AWWARF)

FIELD 05C, 02H

ACCESSION NO. W73-00750

PREVENTION OF WATER SOURCE CONTAMINATION,

HENNINGSON, DURHAM AND RICHARDSON, INC., OMAHA, NEBR.

P. BOLTON.

JOURNAL OF THE AMERICAN WATER WORKS ASSOCIATION, VOL. 53, NO 10, P 1243-1250, OCTOBER 1961, 1 TAB.

DESCRIPTORS:
*WATER POLLUTION SOURCES, *WATER POLLUTION EFFECTS, *WATER SOURCES, *GROUNDWATER, DDT, MISSISSIPPI RIVER BASIN, URBANIZATION, ALGAL POISONING, GASOLINE, FLOODS, AQUIFERS, WATER LAW, NITRATES, TURBIDITY, FLOODWATER, REGULATION, WELL REGULATIONS.

IDENTIFIERS:
*MASTER PLANNING, *SOURCE CONTAMINATION, CYANIDE, NEW PRODUCT CONTROL, PLUMBING CODES, WELL DRILLING CODES.

ABSTRACT:
SAFE WATER SUPPLIES REQUIRE CONSIDERATIONS OF LOCATION, CONSTRUCTION AND OPERATION. INCREASED COMPLEXITY OF INDUSTRIALIZED CIVILIZATION HAS INTRODUCED NEW SOURCES OF WATER SUPPLY CONTAMINANTS. SOURCES OF CONTAMINATION OF SURFACE AND GROUND WATER ARE LISTED. CAUSES OF GROUNDWATER POLLUTION ARE OFTEN THE PROMISCUOUS STORAGE OF WASTES ON THE GROUND SURFACE NEAR A WELL FIELD OR ABOVE A WATER BEARING FORMATION. SUDDEN INTRODUCTION OF CONTAMINANTS IS A TYPICAL CAUSE OF SURFACE WATER POLLUTION. A NATIONAL MASTER PLAN FOR PREVENTING CONTAMINATION OF WATER RESOURCES IS PROPOSED. (FLACK-AWWARF)

FIELD 05B, 05G

ACCESSION NO. W73-00758

THE NEED FOR NUTRIENT CONTROL,

METCALF AND EDDY, INC., BOSTON, MASS.

C. N. SAWYER.

JOURNAL WATER POLLUTION CONTROL FEDERATION, VOL 40, NO 3, P 363-370, MARCH 1968. 5 FIG, 1 TAB, 11 REF.

DESCRIPTORS:
*WATER QUALITY, *NUTRIENTS, *EUTROPHICATION, NITROGEN, PHOSPHORUS, *NUTRIENT REMOVAL, SOIL EROSION, ALGAE, *WATER POLLUTION CONTROL, WATER QUALITY CONTROL.

ABSTRACT:
NUTRIENT CONTROL IS NEEDED TO ARREST EUTROPHICATION OF LAKES AND STREAMS. CULTURAL DRAINAGE IS A MAJOR CONTRIBUTOR OF NUTRIENTS AND A LOGICAL POINT OF ATTACK ON THE PROBLEM. SOIL EROSION CONTROL AND PREVENTION OF LEACHING FROM ANIMAL MANURES ARE HELPFUL AGRICULTURAL PRACTICES. PHOSPHORUS REMOVAL OFFERS THE MOST PROMISE FOR EUTROPHICATION CONTROL, SINCE SOME BLUE-GREEN ALGAE CAN FIX NITROGEN AND OTHER NUTRIENTS ARE PRESENT IN NATURAL WATERS OR EXIST IN QUANTITIES TOO SMALL FOR PRACTICE REMOVAL. CHEMICAL ANALYSIS OF THE LAKE WATER FREQUENTLY DURING THE YEAR CAN DETERMINE WHICH NUTRIENTS ARE CRITICAL, I.E., WHICH NUTRIENTS DECLINE IN CONCENTRATION AS PRIMARY PRODUCTIVITY INCREASES. (BEAN-AWWARF)

FIELD 05G, 05C

ACCESSION NO. W73-01122

ECOLOGY OF A EUTROPHIC ENVIRONMENT: MATHEMATICAL ANALYSIS OF DATA, (ECOLOGIE
D'UN MILIEU EUTROPHIQUE: TRAITEMENT MATHEMATIQUE DES DONNEES).

CENTRE UNIVERSITAIRE DE LUMINY, MARSEILLE (FRANCE). LABORATOIRE
D'HYDROBIOLOGIE MARINE.

DESCRIPTORS:
*PHYTOPLANKTON, *STATISTICAL METHODS, *ZOOPLANKTON, *BIOMASS,
*SUCESSION, *ISOLATION, DIATOMS, ECOLOGY, SALINITY, ORGANIC MATTER,
RIVERS, ESTUARINE ENVIRONMENT, WATER TEMPERATURE, *EUTROPHICATION,
AQUATIC LIFE, SESTON, SCENEDESMUS, MATHEMATICAL STUDIES, DATA
COLLECTIONS, MARINE ANIMALS, PRIMARY PRODUCTIVITY, CHLOROPHYLL,
PIGMENTS, MAGNESIUM COMPOUNDS, DISTRIBUTION PATTERNS, PERSISTENCE,
WATER PROPERTIES, PHYSICAL PROPERTIES, AQUATIC PRODUCTIVITY, MARINE
ALGAE, SALT TOLERANCE, CORRELATION ANALYSIS, CHRYSOPHYTA, CHLOROPHYTA,
FISH EGGS, LARVAE, CRUSTACEANS, COPEPODS, PHOSPHATES, NITRATES.

IDENTIFIERS:
*CHLOROPHYLL A, DATA INTERPRETATION, *RHONE RIVER, NAVICULA, MELOSIRA,
THALASSIOSIRA, COSCINDISCUS, BACTERIOSTRUM, LEPTOCYLINDRUS, CORYCAEUS,
OIKOPLEURA, FRITILLARIA, OBELIA, CLUSTER ANALYSIS, MULTIVARIATE
ANALYSIS, PRINCIPAL-COMPONENT ANALYSIS, PART CORRELATION, DECAPODS,
SPECIES DIVERSITY INDEX.

ABSTRACT:
NUMEROUS DATA ON PHYSICAL, CHEMICAL AND BIOLOGICAL PARAMETERS IN THE
DILUTION LAYER OF THE RHONE MOUTH HAVE BEEN STUDIED BY MULTIVARIATE
TECHNIQUES: PRINCIPAL-COMPONENT ANALYSIS, PART CORRELATION. A NEW
TECHNIQUE OF CLUSTER ANALYSIS IS ALSO PROPOSED. BY THESE MEANS, A VERY
EURYHALINE GROUP OF ZOOPLANKTON SPECIES HAS BEEN ISOLATED AND THE
EXTREMELY LOW SENSIBILITY OF PHYTOPLANKTON TOWARDS, SALINITY HAS BEEN
SHOWN. HOWEVER, TEMPERATURE SEEMS TO BE THE MOST IMPORTANT ECOLOGICAL
FACTOR. INSTABILITY AND EUTROPHY OF THIS AREA DO NOT APPEAR TO DISTURB
THE PHYTOPLANKTON CYCLE, WHICH OCCURS WITH ITS USUAL SUCCESSIONS.
CHLOROPHYLL A AND ORGANIC MATTER DO NOT SEEM OF VALUE FOR ESTIMATION OF
THE BIOMASS IN THE AREA STUDIED. (LONG-BATTELLE)

FIELD 05C, 05A

ACCESSION NO. W73-01446

THE INTERRELATIONSHIPS AMONG PLANKTON, ATTACHED ALGAE AND THE PHOSPHORUS CYCLE
IN ARTIFICIAL OPEN SYSTEMS,

TORONTO UNIV. (ONTARIO).

J. L. CONFER.

AVAILABLE FROM UNIV. MICROFILMS, INC., ANN ARBOR, MICH., 48106. PH. D.
THESIS, 1969.

DESCRIPTORS:
*PLANKTON, *ALGAE, *SESSILE ALGAE, *CYCLING NUTRIENTS, ECOSYSTEMS,
PHOSPHORUS, SCENEDESMUS, CHLOROPHYTA, BACTERIA, PHOSPHORUS
RADIOISOTOPES, INFLOW, CYANOPHYTA, LITTORAL, NUTRIENTS, MODEL STUDIES,
BIOASSAY, RADIOACTIVITY TECHNIQUES, ABSORPTION.

IDENTIFIERS:
*PHOSPHORUS CYCLE, GLOEOTRICHA, P-32.

ABSTRACT:
THE DISTRIBUTION AND RATES OF CIRCULATION OF PHOSPHORUS WERE STUDIED IN
200 LITER MICROECOSYSTEMS. THE SYSTEMS WERE MAINTAINED AS OPEN SYSTEMS
WITH A CONSTANT INFLUX OF WATER CONTAINING PHOSPHORUS. THE EFFECTS OF
DIFFERENT INFLOWING PHOSPHORUS CONCENTRATIONS WERE DETERMINED.
FILAMENTOUS ALGAL GROWTH ATTACHED TO THE SIDES OF THE TANKS HAD A
DOMINANT INFLUENCE ON THE PHOSPHORUS CONCENTRATION IN THE OPEN WATER.
THERE WAS A NEGATIVE CORRELATION BETWEEN THE AMOUNT OF ALGAE ATTACHED
TO THE SIDES AND PHOSPHORUS CONCENTRATION IN THE OPEN WATER. THE
ATTACHED GROWTH REMOVED PARTICLES AS SMALL AS BACTERIA AND AS LARGE AS
SCENEDESMUS FROM THE OPEN WATER. THE ESTIMATED RATE OF MOVEMENT FOR
VARIOUS PARTICLES RANGED FROM 14 TO 385 PERCENT PER DAY. WHEN THE
INFLOWING WATER WAS ENRICHED WITH PHOSPHORUS, THE SPECIES COMPOSITION
OF THE LITTORAL ALGAE WAS CHANGED. BLUE-GREEN ALGAE, PREDOMINANTLY
GLOEOTRICHA, REPLACED THE GREEN ALGAL CHARACTERISTIC OF UNFERTILIZED
PONDS. PHOSPHORUS CONTENT OF THE OPEN WATER WAS NOT INCREASED BY
ENRICHMENT BECAUSE THE BLUE-GREEN ALGAE ACCUMULATED A LARGER FRACTION
OF THE INFLOWING PHOSPHORUS THAN DID THE GREEN ALGAE. RESULTS OF
EXPERIMENTS IN WHICH RADIOACTIVE PHOSPHORUS WAS ADDED TO THE OPEN WATER
WERE SIMILAR TO THOSE OBTAINED PREVIOUSLY IN LAKES. THE KINETICS OF
P-32 LOSS FROM THE OPEN WATER ARE THEORETICALLY COMPATIBLE EITHER WITH
AN EQUILIBRIUM DISTRIBUTION BETWEEN THE OPEN WATER AND THE ATTACHED
ALGAE OR OTHER LITTORAL SOLIDS, OR WITH A STEADY STATE SYSTEM IN WHICH
THERE IS A NET MOVEMENT FROM THE OPEN WATER TO THE ATTACHED ALGAE OR
LITTORAL SOLIDS. (HOLOMAN-BATTELLE)

FIELD 05C, 05B

ACCESSION NO. W73-01454

KINETICS OF ALGAL BIOMASS PRODUCTION SYSTEMS WITH RESPECT TO INTENSITY AND
NITROGEN CONCENTRATION,

CALIFORNIA UNIV., BERKELEY.

G. SHELEF.

AVAILABLE FROM UNIVERSITY MICROFILMS, INC., ANN ARBOR, MICHIGAN 48106, PH D.
DISSERTATION, 1968, 265 P.

DESCRIPTORS:
*NITROGEN, *IRRADIATION, *ALGAE, WASTE TREATMENT, EUTHROPHICATION,
MODEL STUDIES, BIOMASS, *LIGHT INTENSITY.

IDENTIFIERS:
KINETIC MODELS, CHLORELLA PYRENOIDOSA, *ALGAE BIOMASS.

ABSTRACT:
THREE KINETIC MODELS TO DESCRIBE THE RELATIONSHIP BETWEEN THE
IRRADIANCE AND ALGAL GROWTH AND PRODUCTION WERE DEVELOPED AND
EVALUATED. THESE THREE WERE BASED UPON: (A) ENERGY BALANCE AND/OR A
'ZERO ORDER-FIRST ORDER' KINETIC RELATIONSHIP; (B) THE RECTANGULAR
HYPERBOLA FUNCTION; AND, (C) THE EXPONENTIAL FUNCTION. THE MODELS
INCORPORATED THE SPECTRAL DISTRIBUTION OF THE LIGHT SOURCE AS IT
AFFECTED THE ESSENTIAL PARAMETERS - LIGHT CONVERSION EFFICIENCY AND
LIGHT TRANSMISSION IN AN OPTICALLY DENSE ALGAL SUSPENSION. OF THE
THREE, THE MODEL BASED ON THE EXPONENTIAL RELATIONSHIP PROVED MOST
COMPATIBLE WITH THE EXPERIMENTAL RESULTS UNDER THE CONDITIONS OF A
CHEMOSTATIC ALGATRON REACTION. THE MODIFIED EXPONENTIAL FUNCTION BEST
DESCRIBED THE RELATIONSHIP BETWEEN NITRATE-NITROGEN CONCENTRATION AND
ALGAL SPECIFIC GROWTH RATE UNDER CONDITIONS OF LIGHT SATURATION IN A
CONTINUOUS CULTURE OF CHLORELLA PYRENOIDOSA. THE USE OF THE MEF WAS
APPLICABLE IN OPTICALLY DENSE ALGAL CULTURES IN DESCRIBING THE
RELATIONSHIP BETWEEN NITRATE-NITROGEN CONCENTRATION AND THE AVERAGE NET
ALGAL SPECIFIC GROWTH RATE OR NET ALGAL PRODUCTION. (ANDERSON-TEXAS)

FIELD 05C, 05B

ACCESSION NO. W73-02218

KENOSHA INCREASES PLANT CAPACITY WITH MICROSTRAINERS,

KENOSHA WATER DEPT., WIS.

O. F. NELSON.

WATER WORKS AND WASTES ENGINEERING, VOL 2, NO 7, P 43-46, JULY 1965, 3 FIG, 1
TAB.

DESCRIPTORS:
*WATER TREATMENT, *ALGAE, *FILTERS, SANDS, LAKE MICHIGAN, WATER QUALITY
CONTROL, FILTRATION, WATER PURIFICATION, MEMBRANES, SCREENS, TREATMENT
FACILITIES, *WISCONSIN.

IDENTIFIERS:
*SCREENING, *KENOSHA(WIS), *STRAINING, *MICROSTRAINERS.

ABSTRACT:
THE INSTALLATION OF MICROSTRAINERS AHEAD OF ALL OTHER TREATMENT AT THE
KENOSHA, WISCONSIN WATER TREATMENT PLANT HAS RESULTED IN EXTENDED
FILTER RUNS, 46 TO 97 PERCENT REMOVAL OF ALGAE, 25 PERCENT INCREASE IN
PLANT CAPACITY AND ELIMINATION OF TASTE AND ODOR PROBLEMS. THE PROCESS
COSTS ARE APPROXIMATELY $1.50 PER MILLION GALLONS OF WATER TREATED WHEN
THE PLANT IS PROCESSING 12 MILLION GALLONS PER DAY. (FLACK-AWWARF)

FIELD 05F

ACCESSION NO. W73-02426

SIGNIFICANT DESCRIPTOR INDEX

The starred descriptors indexed here have been selected as best describing the content of each paper. If additional references to these or other terms are desired, consult the *General Index*. The numbers at the right are the accession numbers appearing at the end of the abstracts.

SEWAGE, WATER POLLUTION EFFECTS,
OGY, NUTRIENTS, PLANT PATHOLOGY,
L, TEST PROCEDURES, SCENEDESMUS,
*PHOSPHORUS, *IRON, *LIMNOLOGY,
BERLIN,
AQUATIC WEEDS, *BIOASSAY, *ESSE/
CYCLING NUTRIENTS, *EUTROPHICAT/
WATER, *AQUATIC MICROORGANISMS,
*AQUATIC WEEDS, *INHIBITION, *P/
*LIMNOLOGY, FISHKILL, NUTRIENTS,
L NUTRIENTS, *EUTROPHICATION, */
BIOASSAY, *ESSENTIAL NUTRIENTS,
CHNIQUES, TEM/ *MARINE BACTERIA,
UTROPHICATION, *MEROMIXIS, *WAT/
MARINE MICRO/ *AQUATIC BACTERIA,
PLANKTON, *LABORATORY TESTS, *I/
COPPER SULFATE, *EUTROPHICATION,
ENTS, WASTE WATER RECLAMATION, /
AQUATIC ALGAE, AQUATIC MICROBIO/
S, ANALYTICAL TECHNIQUES, AQUAT/
TS, *ENVIRONMENTAL EFFECTS, *LI/
ENVIRONMENT, POLLUTANTS, BIOLOG/
ENTS, SURFACTA/ *BIODEGRADATION,
ODOR, *NATURA/ *WATER TREATMENT,
TIVATED CARBO/ *WATER TREATMENT,
*LABORATORIES, / *BENTHIC FLORA,
DEMAND, LABORATORY TESTS,/ *AGE,
ONTROL, *OPERATION AND MAINTENA/
LLUTION EFFECTS, *REVIEWS, *WAT/
ION, *PROTOZOA, *FUNGI, *BACTER/
TES, *EUTROPHICATION, NUTRIENTS/
E TREATMENT, *OXIDATION LAGOONS,
LLUTION, DIATOMS, MARYLAND, PEN/
ICATION, *PLANT GROWTH, ANALYTI/
ROPHICATION, WATER POLLUTION EF/
N, *FERTILIZATION, *INLAND WATE/
ATER TREATMENT, *POLLUTANT IDEN/
CYTOLOGICAL STUDIES, CULTURES, /
RINE, *ANALYTICAL TECHNIQUES, */
XATION, NUTRIENTS, LAKE ERIE.:
*LIGHT, PHOSPHORUS, PHOTOSYNTH/
S, BIOMASS, ORGANIC COMPOUNDS, /
CARBON,/ *CARBON RADIOISOTOPES,
MES, *ESSENTIAL NUTRIENTS, *NIT/
, PHYTOPLANKTON, BIBLIOGRAPHIES,
ARM MANAGEMENT, *DAIRY INDUSTRY,
TION EFFECTS, *ORGANIC COMPOUND/
, *NUTRIENT REQUIREMENTS, ANALY/
UTROPHICATION, PONDS, RESERVOIR/
DITY, *IRON, *MANGANESE, *COLOR,
ITY, WATER QUA/ WATER TREATMENT,
ER, *PO/ *WATER TREATMENT, ODOR,
TAL ENGINEERING, *SANITARY ENGI/
LIMATIZATION, / *MICROORGANISMS,
NGANESE, CULTURES, RE/ *STREAMS,
HYTOPLANKTON, *DISSOLVED OXYGEN,
HUMAN POPULATION, *EUTROPHICATI/
GANISMS, PHYTOPLANKTON, BACTERI/
PHY, NUTRIENTS, WATER T/ *LAKES,
ATER TREATMENT, WATER TREATMENT,
MICROB/ *NITROGEN, *PHOSPHATES,
NTHESIS, *RESPIRATION, GREAT LA/
ILTERS, *SCEN/ *WATER TREATMENT,
S EARTH, SEPARATION TECHNIQUES,/
HIGAN, WATER / *WATER TREATMENT,
ATION, *STRONTIUM RADIOISOTOPES,
COPPER, IRON, IONS, PHOSPHATES,/
ER TREATMENT, ELECTROLYTES, CHE/
TORS, */ *ANALYTICAL TECHNIQUES,
S, LAB/ *EUTROPHICATION, *LAKES,
ON, *NUTRIENTS, *TERTIARY TREAT/
LOGICAL ECOLOGY, *BENTHIC FLORA,
COBALT, MANGANESE, ATLANTIC OCE/
WAGE, EFFLUENTS, CARBON DIOXIDE,
CIPAL WASTES, *GREAT LAKES, *WA/
DEGRADATION(DECOMPOSITION), DDT/
ONS, BIOCHEMICAL OXY/ *AERATION,
OBIC CONDITIONS, LIGHT INTENSIT/
PHICATION, *AGRICULTURAL WASTES,
LA, CHLOROPHYTA, DIA/ *PLANKTON,
*HYDROLOGY, CHEMICAL PROPERTIES/
TES, AEROBIC CONDITIONS, ANAERO/
CYANOPHYTA, UNITED STATES, HABI/
S, ISLANDS, ECOLOGICAL DISTRIBU/
MUS, PROTEINS, WALLS, STRUCTURE/
*NITROGEN COMPOUNDS, *NUTRIENTS,
S, VARIABILITY, CHLORELLA, CARB/
TIVITY, *BIOLOGICAL COMMUNITIES,
PPER SULFATE, RESERVOIRS, WATER/
DE, *REVIEWS, / *EUTROPHICATION,

*ALGAE.: *ESTUARIES, *HARBORS, *   W71-05500
*ALGAE.: /Y, MICHIGAN, *EPIPHYTOL   W71-00832
*ALGAE.: /ER TEMPERATURE, SEASONA   W68-00472
*ALGAE.: /HMENT TESTS, *NITROGEN,   W69-06004
*ALGAE-TITRE.:   W68-00475
*ALGAE, *ANALYTICAL TECHNIQUES, *   W69-03364
*ALGAE, *ANALYTICAL TECHNIQUES, *   W69-03358
*ALGAE, *AQUATIC BACTERIA, *AQUAT   W69-00096
*ALGAE, *ALGAL TOXINS, *AQUARIA,   W69-03188
*ALGAE, *ALGAL CONTROL, ALGAL POI   W69-05844
*ALGAE, *ALGAL CONTROL, *ESSENTIA   W69-04798
*ALGAE, *ANALYTICAL TECHNIQUES, *   W69-04802
*ALGAE, *ANALYSIS, *ANALYTICAL TE   W70-04365
*ALGAE, *AQUATIC ENVIRONMENTS, *E   W70-03974
*ALGAE, *ANALYTICAL TECHNIQUES, *   W70-04811
*ALGAE, *ANALYTICAL TECHNIQUES, *   W70-04469
*ALGAE, *AQUATIC WEEDS, LAKES, AL   W70-04494
*ALGAE, *ACTIVATED SLUDGE, *EFFLU   W70-06053
*ALGAE, *ANALYTICAL TECHNIQUES, *   W70-02792
*ALGAE, *ALGAL CONTROL, *ALGICIDE   W70-02968
*ALGAE, *ACID MINE WATER, *HABITA   W70-02770
*ALGAE, *AQUATIC ALGAE, *AQUATIC   W70-01233
*ALGAE, *AQUATIC BACTERIA, DETERG   W69-09454
*ALGAE, *ALGAL CONTROL, *TASTE, *   W72-03562
*ALGAE, *ACTINOMYCETES, *ODOR, AC   W72-04155
*ALGAE, *BIOLOGICAL COMMUNITIES,   W70-02780
*ALGAE, *BIODEGRADATION, *OXYGEN   W70-06600
*ALGAE, *BIOLOGICAL PROPERTIES, C   W70-08392
*ALGAE, *BIOINDICATORS, *WATER PO   W70-07027
*ALGAE, *BIOMASS, WATER PURIFICAT   W71-05267
*ALGAE, *BIODEGRADATION, *PHOSPHA   W71-04072
*ALGAE, *BACTERIA, *DISSOLVED OXY   W72-01818
*ALGAE, *BIOINDICATORS, *WATER PO   W70-04510
*ALGAE, *BIBLIOGRAPHIES, *EUTROPH   W69-04801
*ALGAE, *BIOASSAY, NUTRIENTS, EUT   W69-06864
*ALGAE, *BIOASSAY, *EUTROPHICATIO   W69-07833
*ALGAE, *BIOLOGICAL TREATMENT, *W   W69-07389
*ALGAE, *CARBON, *CARBON CYCLE, *   W69-08278
*ALGAE, *CHEMICAL ANALYSIS, *CHLO   W70-05178
*ALGAE, *CYANOPHYTA, *NITROGEN FI   W70-05091
*ALGAE, *CHLORELLA, *SCENEDESMUS,   W70-03923
*ALGAE, *CHLORELLA, *ORGANIC ACID   W70-04195
*ALGAE, *CULTURES, PHYTOPLANKTON,   W70-03983
*ALGAE, *CYCLING NUTRIENTS, *ENZY   W69-03185
*ALGAE, *CHEMICAL PROPERTIES, NUT   W68-00856
*ALGAE, *CATTLE, SLURRIES, SPRAYI   W70-07491
*ALGAE, *CHLOROPHYLL, WATER POLLU   W70-08416
*ALGAE, *CHLOROPHYTA, *CYANOPHYTA   W70-02245
*ALGAE, *CYANOPHYTA, *ECOLOGY, *E   W70-00274
*ALGAE, *COLLOIDS, FLOW RATES, *F   W72-04145
*ALGAE, *COPPER SULFATE, *ALKALIN   W72-07442
*ALGAE, *DECOMPOSING ORGANIC MATT   W72-04242
*ALGAE, *DISPERSIONS, *ENVIRONMEN   W70-02609
*ALGAE, *DOMINANT ORGANISMS, *ACC   W70-07313
*ALGAE, *DIATOMS, TEMPERATURE, MA   W70-07257
*ALGAE, *DIURNAL DISTRIBUTION, *C   W69-03611
*ALGAE, *ENVIRONMENTAL EFFECTS, *   W70-03973
*ALGAE, *ECOLOGY, *MARINE MICROOR   W70-01068
*ALGAE, *EUTROPHICATION, OLIGOTRO   W70-00264
*ALGAE, *EUTROPHICATION, *NITROGE   W72-01867
*ALGAE, *EUTROPHICATION, DIATOMS,   W71-05026
*ALGAE, *EUTROPHICATION, *PHOTOSY   W72-05453
*ALGAE, *FILTRATION, HEAD LOSS, F   W72-03703
*ALGAE, *FILTRATION, *DIATOMACEOU   W72-08688
*ALGAE, *FILTERS, SANDS, LAKE MIC   W73-02426
*ALGAE, *FUNGI, *CALCAREOUS SOIL,   W71-04067
*ALGAE, *FRESH WATER, *PEPTIDES,   W69-10180
*ALGAE, *FLOCCULATION, *WASTE WAT   W70-05267
*ALGAE, *GROWTH RATES, *BIOINDICA   W70-04381
*ALGAE, *GROWTH RATES, *PHOSPHATE   W70-08637
*ALGAE, *HYDROGEN ION CONCENTRATI   W70-08980
*ALGAE, *HYDRAULIC MODELS, LABORA   W71-00668
*ALGAE, *HERRINGS, COPPER, IRON,   W69-08525
*ALGAE, *HYDROGEN ION CONCENTRATI   W68-00855
*ALGAE, *INDUSTRIAL WASTES, *MUNI   W70-04430
*ALGAE, *INSECTICIDES, TOXICITY,   W70-08097
*ALGAE, *LAGOONS, *OXIDATION LAGO   W70-06601
*ALGAE, *LABORATORY TESTS, *ANAER   W71-13413
*ALGAE, *LAKES, *TRIBUTARIES, SAM   W70-02969
*ALGAE, *LIGHT INTENSITY, CHLOREL   W70-05381
*ALGAE, *LAKES, *EUTROPHICATION,   W70-05113
*ALGAE, *LIGHT, *NITRATES, *NITRI   W70-05261
*ALGAE, *LAKE ERIE, CHLOROPHYTA,   W70-04468
*ALGAE, *LITTORAL, *AQUATIC PLANT   W72-10623
*ALGAE, *MICROORGANISMS, SCENEDES   W70-04184
*ALGAE, *MICROORGANISMS, BIODEGRA   W69-06970
*ALGAE, *METABOLISM, *ENVIRONMENT   W70-02965
*ALGAE, *METHODOLOGY, CARBON RADI   W71-12870
*ALGAE, *MAINE, *WATER SUPPLY, CO   W70-08096
*ALGAE, *NUTRIENTS, *CARBON DIOXI   W70-00664

346

```
DISTRIB/ *MARINE MICROORGANISMS,        *ALGAE, *NUTRIENT REQUIREMENTS, *    W70-03952
EN-FIXATION, *NUTRIENTS, *LAKES,        *ALGAE, *NITROGEN, WISCONSIN.: /G     W70-03429
UIREMENTS, GROWTH RATES, LABORA/        *ALGAE, *NUTRIENTS, *NUTRIENT REQ     W70-03507
SPHORUS, SODIU/ *EUTROPHICATION,        *ALGAE, *NUTRIENTS, NITROGEN, PHO     W70-04503
SEWAGE EFFLUENT, *SEWAGE TREATM/        *ALGAE, *NITROGEN, *PHOSPHORUS, *     W72-08133
WATER TREATMENT, *WATER QUALITY,        *ALGAE, *ODOR, HYDROGEN ION CONCE     W72-09694
ES, *DECOMPOSING ORGANIC MATTER,        *ALGAE, *PHENOLS.: /DECIDUOUS TRE     W72-03475
KILLS, *ROTIFE/ *WATER QUALITY,         *ALGAE, *PESTICIDES, *FISH, *FISH     W72-04506
TA, PHYTOPLANKTON, CYANOPHYTA, /        *ALGAE, *PHOTOSYNTHESIS, *MINNESO     W70-05387
TICIDE REMOVAL, *FOOD WEBS, DDT/        *ALGAE, *PESTICIDE TOXICITY, *PES     W70-03520
COMMUNITIES, EFFLUENTS/ *CARBON,        *ALGAE, *POPULATION, *BIOLOGICAL      W70-04001
RATES, *TURBULENCE, CHLORRELA, /        *ALGAE, *PHOTOSYNTHESIS, *GROWTH      W69-03730
*EUTROPHICATION, POLLUTION ABAT/        *ALGAE, *PHOSPHORUS, *MINNESOTA,      W69-10167
ARIO, *GROWTH RATES,/ *PLANKTON,        *ALGAE, *PHYTOPLANKTON, *LAKE ONT     W69-10158
, ECOLOGY, CHEMICAL ANALYSIS, P/        *ALGAE, *PLANT GROWTH, *WISCONSIN     W70-06229
ENERGY, DI-URNAL DISTRIBUTION,/         *ALGAE, *PHOSPHORUS, *PHOSPHATES,     W70-06604
*SANITARY ENGINEERING, MICROORG/        *ALGAE, *PROTOZOA, *SYSTEMATICS,      W72-01798
RGANIC MATTER, / *PHYTOPLANKTON,        *ALGAE, *PRIMARY PRODUCTIVITY, *O     W71-10079
S, *PHOSPHORUS, *NITROGEN, *ALK/        *ALGAE, *PHYTOPLANKTON, *NUTRIENT     W70-02795
ATION, *WATER CHEMISTRY, POLLUT/        *ALGAE, *PRODUCTIVITY, *EUTROPHIC     W70-02646
GE TREATMENT, *EUTROPHICATION, /        *ALGAE, *PLANKTON, *SEWAGE, *SEWA     W70-02304
, PHYTOPLANKTO/ *AQUATIC PLANTS,        *ALGAE, *ROTIFERS, PLANKTON, BASS     W70-02249
TION, *DISSOLVED OXYGEN, PENNSY/        *ALGAE, *RESPIRATION, *EUTROPHICA     W69-10159
, BIBLIOGRAPHIES, CHLORELLA, CH/        *ALGAE, *RESEARCH AND DEVELOPMENT     W69-06865
*EUTROPHICA/ *DISSOLVED OXYGEN,         *ALGAE, *RESPIRATION, *ESTUARIES,     W69-07520
, *WATER POLLUTION SO/ *ECOLOGY,        *ALGAE, *RIVERS, *WATER POLLUTION     W69-07096
, *TEMPERATURE, SOUTH CAROLINA,/        *ALGAE, *RADIOISOTOPES, *SORPTION     W69-07845
, MUSSELS, SHELLFISH/ *TOXICITY,        *ALGAE, *REVIEWS, HABITATS, CLAMS     W70-05372
IVER, *OHIO RIVER, NUI/ *RIVERS,        *ALGAE, *SAMPLING, *MISSISSIPPI R     W70-04507
ECTS, DRAINAGE / *LAKES, PLANTS,        *ALGAE, *SALINITY, IRRIGATION EFF     W69-00360
LF OF MEXICO, A/ *RADIOISOTOPES,        *ALGAE, *SORPTION, ABSORPTION, GU     W69-08524
RIES, FRESH WATER, LAKES, SILTS/        *ALGAE, *SEWAGE EFFLUENTS, *ESTUA     W70-02255
TIVES, *FLOW RATE, *WASTE WATER,        *ALGAE, *SLUDGE, STORMS, TURBULEN     W70-05819
ERS, ELECTROPHORESIS, CHLORELLA/        *ALGAE, *SANDS, *FILTRATION, FILT     W71-03035
NA, CHLOROPHYT/ *SEWAGE LAGOONS,        *ALGAE, *SEWAGE TREATMENT, *INDIA     W70-10173
NUTRIENTS, ECOSYSTEM/ *PLANKTON,        *ALGAE, *SESSILE ALGAE, *CYCLING      W73-01454
, ACTIVATED C/ *WATER TREATMENT,        *ALGAE, *TASTE, *ODOR, *ALGICIDES     W72-03746
WATER TREATMENT, *ACTINOMYCETES,        *ALGAE, *TASTE, *ODOR, WATER QUAL     W72-13605
OTOSYNTHESIS.:                          *ALGAE, *TEMPERATURE, DIATOMS, PH     W71-02496
METABOLISM, RADIOACTIVITY TECH/         *ALGAE, *TOXICITY, *INSECTICIDES,     W70-02198
CITY.:          *FISH PHYSIOLOGY,       *ALGAE, *THERMAL POLLUTION, *TOXI     W70-01788
ENVIRONMENT, WATER QUALITY, NUT/        *ALGAE, *TROPHIC LEVELS, AQUATIC      W70-02764
S, HABITATS, ALKALINE WATER, AC/        *ALGAE, *TEMPERATURE, *HOT SPRING     W69-10160
INTENSITY, SPR/ *PHOTOSYNTHESIS,        *ALGAE, *THERMAL SPRINGS, *LIGHT      W69-09676
OLATION, LIMNOLOGY, PUBLIC HEAL/        *ALGAE, *TOXICITY, *CULTURES, *IS     W70-00273
DS, P/ *WATER POLLUTION SOURCES,        *ALGAE, *TASTE, *ODOR, *AMINO ACI     W72-01812
, THERMOPHILIC BAC/ *CYANOPHYTA,        *ALGAE, *THERMAL SPRINGS, *OREGON     W70-05270
N, BIOMA/ *PRIMARY PRODUCTIVITY,        *ALGAE, *THERMAL WATER, PERIPHYTO     W70-03309
TASTE, ODOR, FILTERS, PLANKTON,/        *ALGAE, *WATER SUPPLY, *CONTROL,      W70-05389
T, LAKES, SEWAGE, DRAINAGE DIST/        *ALGAE, *WATER POLLUTION TREATMEN     W70-04455
CATION, COLOR, TASTE, ODOR, TOX/        *ALGAE, *WATER QUALITY, *EUTROPHI     W69-05697
CHELATION,/ *COPPER, *TOXICITY,         *ALGAE, *WATER POLLUTION EFFECTS,     W71-11561
WATER POLLUTION CONTROL, *DENIT/        *ALGAE, *WASTE WATER TREATMENT, *     W71-04555
WATER POLLUTION CONTROL, *DENIT/        *ALGAE, *WASTE WATER TREATMENT, *     W71-04556
WATER POLLUTION CONTROL, *DENIT/        *ALGAE, *WASTE WATER TREATMENT, *     W71-04554
WATER POLLUTION CONTROL, *DENIT/        *ALGAE, *WASTE WATER TREATMENT, *     W71-04557
 *WATER TREATMENT, *FILTRATION,         *ALGAE, *WATER QUALITY, *ACTIVATE     W72-09693
AQUATIC ALGAE, AQUATIC MICROBI/         *ALGAE, ALGAL CONTROL, ALGICIDES,     W70-02248
ATIC ALGAE, WATER QUALITY, *TRO/        *ALGAE, AQUATIC ENVIRONMENT, *AQU     W70-02772
IRONMENTAL EFFECTS, *MORPHOLOGY,        *ALGAE, AQUATIC MICROBIOLOGY, CYA     W69-10177
FISHKILL, / *PESTICIDE RESIDUES,        *ALGAE, BIOASSAY, BIOINDICATORS,      W69-08565
UNITIES, CULTU/ *EUTROPHICATION,        *ALGAE, BIOASSAY, BIOLOGICAL COMM     W69-07832
DETERG/ *SURFACTANTS, *SORPTION,        *ALGAE, CHLOROPHYTA, CYANOPHYTA,      W70-03928
ERIA/ *EUTROPHICATION, PLANKTON,        *ALGAE, CHEMICAL PROPERTIES, BACT     W68-00483
, NITROGEN, PHOSPHORUS, DIATOMS,        *ALGAE, CHLOROPHYTA, CYANOPHYTA,      W68-00172
OUNDS, PHOTOSYN/ *MARINE PLANTS,        *ALGAE, CHLORINATION, SODIUM COMP     W70-00161
LYTICAL TECHNIQUES, *PHOSPHATES,        *ALGAE, CHROMOTOGRAPHY, RADIOCHEM     W71-03033
E WATER, *WATER QUALITY CONTROL,        *ALGAE, COPPER SULFATE, CHLORINE,     W71-00110
, *NUTRIENTS, *DISSOLVED OXYGEN,        *ALGAE, COLIFORMS, EUTROPHICATION     W71-05407
NGINEERING, *FILTRATION, DESIGN,        *ALGAE, COPPER SULFATE, SEDIMENTA     W72-08859
HYTA, BIOASS/ *LABORATORY TESTS,        *ALGAE, CULTURES, DIATOMS, CYANOP     W71-06002
CULTURES, DIAGENESIS, SEDIMENTS/        *ALGAE, CYANOPHYTA, CHRYSOPHYTA,      W69-08284
 *HERBICIDES, *VASCULAR TISSUES,        *ALGAE, ECOSYSTEMS, CHARA, BENTHI     W70-10175
S, *MINNESOTA, *LAKES, *DIATOMS,        *ALGAE, ECOLOGY, SOUTH DAKOTA, FI     W69-00659
ENT REQUIREMENTS, *PLANT GROWTH,        *ALGAE, ELEMENTS(CHEMICAL), CHLOR     W70-02964
AVIOLET RADIA/ *WATER TREATMENT,        *ALGAE, FILTRATION, *SLIME, *ULTR     W72-03824
SULFIDE, BIOCHEMICAL OXYGEN DEM/        *ALGAE, IRON COMPOUNDS, HYDROGEN      W70-06925
 *ZOOPLANKTON, *PHOTOSYNTHESIS,         *ALGAE, LAKES, OXYGEN, EUTROPHICA     W71-10098
ISOLATION, *BACTERIA, *PROTOZOA,        *ALGAE, LETHAL LIMIT, ELECTROPHOR     W71-03775
LLA, CULTURES.:          *GROWTH,       *ALGAE, LIPIDS, NUTRIENTS, CHLORE     W70-07957
R TREATMENT, *OXIDATION LAGOONS,        *ALGAE, LIGHT INTENSITY, CARBON D     W72-06022
HIBITION, N/ *NITROGEN FIXATION,        *ALGAE, LIGHT, CARBON DIOXIDE, IN     W70-04249
URE, PRIMARY PRODUCT/ *PLANKTON,        *ALGAE, LIGHT INTENSITY, TEMPERAT     W70-04809
TROPHICATION, PHOTOSYNTHESIS, L/        *ALGAE, NITROGEN, PHOSPHORUS, *EU     W72-00917
PHOROUS, NITRO/ *EUTROPHICATION,        *ALGAE, NUTRIENTS, NITROGEN, PHOS     W71-13335
, *EUTROPHICATION, *GREAT LAKES,        *ALGAE, NUTRIENTS, PHOSPHORUS, PO     W70-00667
SANDS, FLOW R/ *WATER, *FILTERS,        *ALGAE, NUISANCE ALGAE, INTAKES,      W70-04199
S, *SEWAGE TRE/ *EUTROPHICATION,        *ALGAE, NUTRIENTS, DOMESTIC WASTE     W68-00248
HYTOPLANKTON, *AQUATIC BACTERIA,        *ALGAE, ORGANIC MATTER, LAKES, DI     W70-04284
ATION, LIGHT, TE/ *GROWTH RATES,        *ALGAE, PHOTOSYNTHESIS, EUTROPHIC     W69-07442
```

```
*CONFERENCES, *EUTROPHICATION,          *ALGAE, PLANNING, RIVERS, LAKES,      W70-04506
ALGAE,/ *ANALYTICAL TECHNIQUES,         *ALGAE, PLANKTON, DIATOMS, MARINE     W70-09904
ONMENT, *FRESH WATER, *PLANKTON,        *ALGAE, SCENEDESMUS, LIGHT, CULTU      W70-05409
L, PHOT/ *PHOSPHORUS, *NITROGEN,        *ALGAE, SEWAGE PLANTS, CHLOROPHYL      W70-04268
ALINITY/ *MARINE WATER, *COPPER,        *ALGAE, SEA WATER, TEMPERATURE, S      W70-08111
ITION) CYANOPHYTA, CHLOROPHYTA,         *ALGAE, SORPTION, WASTE WATER TRE      W70-09438
CHLOROPHYTA, SCENEDESMUS, BENTH/        *ALGAE, TEMPERATURE, CYANOPHYTA,       W70-03969
LORELLA, DAPHNIA, / *HERBICIDES,        *ALGAE, TOXICITY, SCENEDESMUS, CH      W71-03056
LORELLA, DAPHNIA, / *HERBICIDES,        *ALGAE, TOXICITY, SCENEDESMUS, CH      W71-05719
EUTROPHICATION, *LAKES, *RIVERS,        *ALGAE, WATER POLLUTION CONTROL,       W70-05668
WAGE TREATMENT, DOMESTIC WASTES,        *ALGAE, WATER PROPERTIES, BIOCHEM      W68-00256
ICATIO/ *NITROGEN, *IRRADIATION,        *ALGAE, WASTE TREATMENT, EUTHROPH      W73-02218
-5-/ *HERBICIDES, *GROWTH RATES,        *ALGAE, 2-4-D, AMINOTRIAZOLE, 2-4      W70-03519
L.:                                     *ALGAL ACTIVITY, *NUTRIENT REMOVA      W70-08980
*ALKOXYSILANES, SCENEDESMUS QU/         *ALGAL BLOOMS, *ORGANO-SILICATES,      W71-03056
*ALKOXYSILANES, SCENEDESMUS QU/         *ALGAL BLOOMS, *ORGANO-SILICATES,      W71-05719
LAKE TAHOE(CALIF)/ *COMPLEXONES,        *ALGAL BLOOMS, NTA, EDTA, HEIDA,       W70-04373
COPPER SULPHATE, NUISANCE ALGAE,        *ALGAL CONTROL, WEED CONTROL, *IN      W68-00468
SANCE ALGAE, SPRAYING, SEASONAL,        *ALGAL CONTROL, FISHKILL, ROOTED       W68-00488
CATION, *LAKES, *COPPER SULFATE,        *ALGAL CONTROL, *TOXINS, ALGAE, R      W68-00859
S, FORAGE FISH, PLANKTON, ALGAE,        *ALGAL CONTROL, PREDATION, COMPET      W70-05428
GRADIENT, *BOD REMOVAL, *ISRAEL,        *ALGAL CONCENTRATION, 'EXTRA-DEEP      W70-04790
TROL, *EUTROPH/ *BIBLIOGRAPHIES,        *ALGAL CONTROL, *AQUATIC WEED CON      W69-05706
TROL, *EUTROPH/ *BIBLIOGRAPHIES,        *ALGAL CONTROL, *AQUATIC WEED CON      W69-05705
NTS, *EUTROPHICATION, */ *ALGAE,        *ALGAL CONTROL, *ESSENTIAL NUTRIE      W69-04798
GY, FISHKILL, NUTRIENTS, *ALGAE,        *ALGAL CONTROL, ALGAL POISONING,       W69-05844
APHNIA, CHLAMYDO/ WATER QUALITY,        *ALGAL CONTROL, EUTROPHICATION, D      W72-02417
ASINS, WAT/ *WATER PURIFICATION,        *ALGAL CONTROL, *WATER SUPPLY, *B      W72-01814
LUTION CONTROL, *EUTROPHICATION,        *ALGAL CONTROL, *TERTIARY TREATME      W70-09186
YANOPHYTA, SULFUR BACTERIA, COL/        *ALGAL CONTROL, *POTABLE WATER, C      W70-08107
COPPER SULFATE/ *NUISANCE ALGAE,        *ALGAL CONTROL, *NEW HAMPSHIRE, *      W69-08674
WATER SUPPLY, RESERVOIRS, WATER/        *ALGAL CONTROL, *COPPER SULFATE,       W69-10157
ER QUALITY, WA/ *COPPER SULFATE,        *ALGAL CONTROL, WATER SUPPLY, WAT      W69-10155
TICAL TECHNIQUES, AQUAT/ *ALGAE,        *ALGAL CONTROL, *ALGICIDES, ANALY      W70-02968
ATURA/ *WATER TREATMENT, *ALGAE,        *ALGAL CONTROL, *TASTE, *ODOR, *N      W72-03562
WATER QUALITY CONTROL, *CONTROL,        *ALGAL CONTROL, TOXICITY.: /ITY,       W72-07442
TREATMENT FACILITIES, CLEANING,         *ALGAL CONTROL, WISCONSIN.: /STS,      W72-08859
BIOLOGICAL INDEX, *SAPROBITY, P/        *ALGAL CULTURE, *BIOMASS TITER, *      W72-12240
ES, *MASS CULTURE, *PILOT-PLANT/        *ALGAL CULTURE, *LABORATORY STUDI      W69-06865
, PEDIASTRUM, SELF-PURIFICATION/        *ALGAL CULTURES, *AXENIC CULTURES      W70-04381
IURNAL OXYGEN CURVES, CALORIMET/        *ALGAL GROWTH, *THERMOBIOLOGY, *D      W70-03309
OSA(EMERSON STRA/ *MASS CULTURE,        *ALGAL GROWTH, CHLORELLA PYRENOID      W69-07442
ITY, ALGAL GROWTH POTENTIAL, SE/        *ALGAL GROWTH, NUTRIENT AVAILABIL      W69-06864
INOSA, / *NUTRIENT AVAILABILITY,        *ALGAL GROWTH, *MICROCYSTIS AERUG      W69-05867
OMETRY, CALIFORNIA, NUTRIENTS, /        *ALGAL GROWTH, *BIOASSAYS, *FLUOR      W70-02777
ECTONEMA, NOSTOC PALUDOSUM, BLE/        *ALGAL GROWTH, *POTOMAC RIVER, PL      W70-02255
CEROS, SKELETONEMA, THALASSIOSI/        *ALGAL NUTRITION, DITYLUM, CHAETO      W70-04503
AN/ *WASTE STABILIZATION PONDS,         *ALGAL PHYSIOLOGY, *DESIGN THEORY      W70-04786
*OILY WATER, AQUATIC ALGAE, MAR/        *ALGAL POISONING, *MARSH PLANTS,       W70-01231
*PHYTOTOXIC/ *FISHKILL, *TOXINS,        *ALGAL POISONING, *ALGAL TOXINS,       W71-07731
, FUCUS VESICULOSUS, CHONDRUS C/        *ALGAL SUBSTANCES, *EXTRACELLULAR      W70-01073
S, TYNSET(NORWAY), / *GELBSTOFF,        *ALGAL SUBSTANCES, SUBPOLAR WATER      W70-01074
C EFFECTS, ANN ARBOR(MICH), SAN/        *ALGAL SUSPENSIONS, ELECTROKINETI      W71-03035
TOXINS, BIOASSAY, AQUATIC ALGAE,        *ALGAL TOXINS, KILLIFISHES, *TOXI      W70-08372
KILL, *TOXINS, *ALGAL POISONING,        *ALGAL TOXINS, *PHYTOTOXICITY, *F      W71-07731
U.S/ CYANAPHYTA, ALGAL CONTROL,         *ALGAL TOXINS MEMBRANE, NORTHEAST      W69-05306
WEEDS, *INHIBITION, *P/ *ALGAE,         *ALGAL TOXINS, *AQUARIA, *AQUATIC      W69-03188
GAL-BOUND RADI/ *RADIOSTRONTIUM,        *ALGAL-BOUND RADIOSTRONTIUM, *FUN      W71-04067
URES, ANTIFOULING MATERIALS, TE/        *ALGICIDES, *PAINTS, *TEST PROCED      W71-07339
AQUATIC PLANTS, *NUISANCE ALGAE,        *ALGICIDES, *PHENOLS, *WATER QUAL      W70-10161
OL, CHLORELLA, DISIN/ *HALOGENS,        *ALGICIDES, CHLORINE, ALGAL CONTR      W70-02363
, AQUAT/ *ALGAE, *ALGAL CONTROL,        *ALGICIDES, ANALYTICAL TECHNIQUES     W70-02968
ER QUALITY CON/ *NUISANCE ALGAE,        *ALGICIDES, *COPPER SULFATE, *WAT      W69-05704
ICATION, WISCONSIN, WATER QUALI/        *ALGICIDES, *CYANOPHYTA, *EUTROPH      W70-03310
I/ *ARTIFICIAL RECHARGE, COPPER,        *ALGICIDES, ALGAE, *EVALUATION, F      W72-05143
REATMENT, *ALGAE, *TASTE, *ODOR,        *ALGICIDES, ACTIVATED CARBON, FIL      W72-03746
ATMENT, *ALGAE, *COPPER SULFATE,        *ALKALINITY, WATER QUALITY CONTRO      W72-07442
N DEMAND, ALGAL CONTROL, *COLOR,        *ALKALINITY, *NITRATES, NITROGEN,      W72-11906
TRIENTS, *PHOSPHORUS, *NITROGEN,        *ALKALINITY, LAKES, PRODUCTIVITY,      W70-02795
ALGAL BLOOMS, *ORGANO-SILICATES,        *ALKOXYSILANES, SCENEDESMUS QUADR      W71-05719
ALGAL BLOOMS, *ORGANO-SILICATES,        *ALKOXYSILANES, SCENEDESMUS QUADR      W71-03056
ONIC SU/ *ALKYL POLY ETHOXYLATE,        *ALKYL PHENOL POLYETHOXYLATE, ANI      W70-09438
CULTURES, *ALKYL POLYETHOXYLATE,        *ALKYL PHENOL POLYETHOXYLATE, ORG      W70-03928
YL SULFONATES, *AXENIC CULTURES,        *ALKYL POLYETHOXYLATE, *ALKYL PHE      W70-03928
ENOL POLYETHOXYLATE, ANIONIC SU/        *ALKYL POLY ETHOXYLATE, *ALKYL PH      W70-09438
RADATION, DETERGENTS, WATER POL/        *ALKYLBENZENE SULFONATES, *BIODEG      W70-08740
VIOLET LIGHT STER/ *DUTCH LAKES,        *ALLOCHTHONOUS PHOSPHATES, ULTRA-      W70-03955
TED SLUDGE/ *PHOSPHORUS REMOVAL,        *ALUM, *ALUMINUM SULFATE, *ACTIVA      W70-09186
DGE/ *PHOSPHORUS REMOVAL, *ALUM,        *ALUMINUM SULFATE, *ACTIVATED SLU      W70-09186
, CHEMICAL INDUSTRY, ODOR, TAST/        *AMERICAN WATER WORKS ASSOCIATION      W71-05943
SOURCES, *ALGAE, *TASTE, *ODOR,         *AMINO ACIDS, POTABLE WATER, ACTI      W72-01812
*ULVA LATISSIMA,         *AMMONIA NITROGEN.:                    W71-05500
N UPTAKE TECHNIQUES, W/ *BIOTIN,        *AMPHIDINIUM CARTERI, *RADIOCARBO      W70-05652
ITROGEN-15, PYRUVATE, ACETYLENE/        *ANABAENA CYLINDRICA, ATP, CMU, N      W70-04249
HYCOLOGISTS, VETERINARIANS, MIC/        *ANABAENA FLOS-AQUAE, *STRAINS, P      W70-00273
*ANABAENA DOLIOLUM.:                  W69-10177
, *OLEFI/ *BOTRYOCOCCUS BRAUNII,        *ANACYSTIS MONTANA, *HYDROCARBONS      W69-08284
TES, METAPHOSPHATES, PULSE-LABE/        *ANACYSTIS NIDULANS, OLIGOPHOSPHA      W71-03033
LABORATORY-SCALE STUDIES, PLAN/         *ANAEROBIC TREATMENT, *AUSTRALIA,      W70-06899
CIES, *LAGOONS, *SLUDGE, ALGAE,/        *ANAEROBIC CONDITIONS, *EFFLICIEN      W70-06899
```

ONS, *BIOCHEMICAL OXYGEN DEMAND,
S, TOMATOES, *AEROBIC TREATMENT,
NSIT/ *ALGAE, *LABORATORY TESTS,
FECTS, *HYDRAULIC LOAD EFFECTS,/
ACULATIVE PONDS, *AEROBIC PONDS,
WASTES, MET/ *OXIDATION LAGOONS,
, TEM/ *MARINE BACTERIA, *ALGAE,
ALGAE, AQUATIC MICROBIOLOGY, */
GROWTH RATES, *BIOINDICATORS, */
INE BACTERIA, *ALGAE, *ANALYSIS,
LUTION EFFEC/ *LAKES, *BACTERIA,
WEEDS, *WATER ANALYSIS, *SAMPLI/
Y METHODS, SEASONS, ACETYLENE R/
, *LABORATORY TESTS, *I/ *ALGAE,
ANISMS, *LIPIDS, LIGHT, ALGAE, /
ICRO/ *AQUATIC BACTERIA, *ALGAE,
, *CHEMICAL ANALYSIS, *CHLORINE,
POLLUTION EFFECTS, *PHOSPHATES,
, *ESSENTIAL NUTRIENTS,/ *ALGAE,
NUTRIENTS, *EUTROPHICAT/ *ALGAE,
SOURCES, SEPARATION TECHNIQUES,
, *ENZYMES, *NUTRIENT REQUIREME/
NUTRIENTS, *EUTROPHICATION, *PH/
WEEDS, *BIOASSAY, *ESSE/ *ALGAE,
CARBON, *ELECTROLYTES, FILTERS,
TORS, *CHLOROPHYLL, *OHIO RIVER,
ES, *ALGAE, CHROMOTOGRAPHY, RAD/
ON CONTROL, *RESEARCH EQUIPMENT,
LANKTON, DIATOMS, MARINE ALGAE,/
ALGAE, AQUATIC MICROBIO/ *ALGAE,
A OGLIMUN, UDO/ *TRACE ELEMENTS,
DA).:
MANAGEMENT, *DAIRY INDUSTRY, */
CROPHYTES, ANACHARIS CANADENSIS/
S, FIBER WASTES, DYEING WASTES./
, DEPURATION, PHYCOCOLLOIDS, TA/
CYSTIS AERUGINOSA, BOHEMIA, ELB/
N-14/ *UPPER KLAMATH LAKE(ORE.),
DOMESTIC WAST/ *WASTE DISPOSAL,
TION, *P/ *ALGAE, *ALGAL TOXINS,
LIBRIUM, BIOLOGICAL COMMUNITIES,
TES, SEMIARID CLIM/ *SOIL ALGAE,
LOGY, */ *ANALYTICAL TECHNIQUES,
CATION, *MUD(LAKE), *PHOSPHORUS,
ENT, POLLUTANTS, BIOLOG/ *ALGAE,
*ALGAE, *ANALYTICAL TECHNIQUES,
RO/ *ALGAE, AQUATIC ENVIRONMENT,
DA, *EUTROPHICATION, AQUATIC MI/
*BIOASSAY, T/ *WATER POLLUTION
A, *BACTERICIDES, *CHLORINATION,
RFACTA/ *BIODEGRADATION, *ALGAE/
C/ *COMPETITION, *PHYTOPLANKTON,
AQUIATIC MICROORGANISMS, *ALGAE,
TICAL TECHNIQUES, *MARINE MICRO/
GAE, PROTOZOA, *AQUATIC INSECTS,
ATION, *MEROMIXIS, *WAT/ *ALGAE,
VIRONME/ *PHYSIOLOGICAL ECOLOGY,
S, *RADIUM RADIOISOTOPES, *URAN/
TRACERS, PHOTOS/ *RADIOISOTOPES,
BIOLOG/ *ALGAE, *AQUATIC ALGAE,
*ECOLOGY, *BIOTA, *TEMPERATURE,
ISMS, *ALGAE, *AQUATIC BACTERIA,
, RIVERS/ *BACKGROUND RADIATION,
FISH FOOD ORG/ ALGAE, PROTOZOA,
, AQUATIC / *ZINC RADIOISOTOPES,
WATER POLLUTION EFFE/ *BIOASSAY,
*RADIOACTIVITY, *TRACE ELEMENTS,
MPERATURE, *AQUATIC ENVIRONMENT,
NS, *RIVERS, ALGAE, PROTOZOA, A/
ANADENSIS/ *ANTAGONISTIC ACTION,
CHLORINATION, *AQUATIC BACTERIA,
ICAL ECOLOGY, *CYANOPHYTA, ALGA/
, *MARINE MICROO/ *MICROBIOLOGY,
ICAL TECHNIQUES, *AQUATIC ALGAE,
IC ALGAE, *AQUATIC MICROBIOLOGY,
QUATIC ALGAE, AGING(BROLOGICAL),
ROGEN, PHOSPHORUS, BICARBONATES,
NTROL, WATER CO/ *AQUATIC WEEDS,
S, PLANKTON, BASS, PHYTOPLANKTO/
SCULAR T/ *AQUATIC WEED CONTROL,
ATIC ENVIRONMENT, *AQUATIC LIFE,
*ALGICIDES, *PHENOLS, *WATER Q/
RSHES, *LAKES, *PHYTOPLANKTON, /
RADIOISOTOPES, *AQUATIC INSECTS,
GAE, PROTOZOA, A/ *AQUATIC LIFE,
CAL DISTRIBU/ *ALGAE, *LITTORAL,
LANTS, *HERBICIDES, *VASCULAR T/
*AQUATIC WEEDS, *AQUATIC PLANTS,
*AQUATIC WEED CONTROL, WATER CO/
*ALGAE, *ANALYTICAL TECHNIQUES,

*ANAEROBICS, ALGAE, BACTERIA, WAT   W71-05742
*ANAEROBIC DIGESTION, BIOCHEMICAL   W71-09548
*ANAEROBIC CONDITIONS, LIGHT INTE   W71-13413
*ANAEROBIC WASTE PONDS, *DEPTH EF   W70-04787
*ANAEROBIC PONDS, DESIGN FOR ENVI   W70-04787
*ANAEROBIC DIGESTION, *MUNICIPAL   W72-06854
*ANALYSIS, *ANALYTICAL TECHNIQUES   W70-04365
*ANALYTICAL TECHNIQUES, *AQUATIC   W70-04283
*ANALYTICAL TECHNIQUES, *ALGAE, *   W70-04381
*ANALYTICAL TECHNIQUES, TEMPERATU   W70-04365
*ANALYTICAL TECHNIQUES, WATER POL   W70-04287
*ANALYTICAL TECHNIQUES, *AQUATIC   W70-04382
*ANALYTICAL METHODS, *BIOCHEMISTR   W70-03429
*ANALYTICAL TECHNIQUES, *PLANKTON   W70-04469
*ANALYTICAL TECHNIQUES, *MICROORG   W70-05651
*ANALYTICAL TECHNIQUES, *MARINE M   W70-04811
*ANALYTICAL TECHNIQUES, *POLAROGR   W70-05178
*ANALYTICAL TECHNIQUES, *BIOASSAY   W70-05269
*ANALYTICAL TECHNIQUES, *BIOASSAY   W69-04802
*ANALYTICAL TECHNIQUES, *CYCLING   W69-03358
*ANALYTICAL TECHNIQUES, CLASSIFIC   W69-01273
*ANALYTICAL TECHNIQUES, *BIOASSAY   W69-03373
*ANALYTICAL TECHNIQUES, *CYCLING   W69-03370
*ANALYTICAL TECHNIQUES, *AQUATIC   W69-03364
*ANALYTICAL TECHNIQUES.: /TIVATED   W72-09693
*ANALYTICAL TECHNIQUES, BIOMASS,   W71-04206
*ANALYTICAL TECHNIQUES, *PHOSPHAT   W71-03033
*ANALYTICAL TECHNIQUES, WATER QUA   W70-08627
*ANALYTICAL TECHNIQUES, *ALGAE, P   W70-09904
*ANALYTICAL TECHNIQUES, *AQUATIC   W70-02792
*ANCHOA LAMPROTAENIA, *OPISTHONEM   W69-08525
*ANDERSON-CUE LAKE, MELROSE(FLORI   W72-01990
*ANIMAL WASTES, MANAGEMENT, *FARM   W70-07491
*ANTAGONISTIC ACTION, *AQUATIC MA   W70-02249
*ANTI-FOAM AGENTS, *TEXTILE WASTE   W70-04060
*ANTIBIOSIS, AUTOLYSIS, EXCRETION   W70-01068
*APHANIZOMENON FLOS AQUAE, *MICRO   W70-00274
*APHANIZOMENON FLOS-AQUAE, *CARBO   W72-06052
*AQUACULTURE, *ULTIMATE DISPOSAL,   W70-09905
*AQUARIA, *AQUATIC WEEDS, *INHIBI   W69-03188
*AQUATIC ALGAE, HALOPHYTES.: /QUI   W69-00360
*AQUATIC ALGAE, ALGAE, ARID CLIMA   W69-03491
*AQUATIC ALGAE, *AQUATIC MICROBIO   W70-04283
*AQUATIC ALGAE, *BIOASSAY, SCENED   W70-03955
*AQUATIC ALGAE, *AQUATIC ENVIRONM   W70-01233
*AQUATIC ALGAE, AQUATIC MICROBIOL   W70-02792
*AQUATIC ALGAE, WATER QUALITY, *T   W70-02772
*AQUATIC ALGAE, *PROTOZOA, *FLORI   W72-01993
*AQUATIC BIOLOGY, *BIOINDICATORS,   W71-10786
*AQUATIC BACTERIA, *AQUATIC MICRO   W69-10171
*AQUATIC BACTERIA, DETERGENTS, SU   W69-09454
*AQUATIC BACTERIA, *ALGAE, ORGANI   W70-04284
*AQUATIC BACTERIA, *AQUATIC FUNGI   W69-00096
*AQUATIC BACTERIA, *ALGAE, *ANALY   W70-04811
*AQUATIC DRIFT, FISH FOOD ORGANIS   W69-02782
*AQUATIC ENVIRONMENTS, *EUTROPHIC   W70-03974
*AQUATIC ENVIRONMENTS, RIVERS, EN   W70-03978
*AQUATIC ENVIRONMENTS, *ECOSYSTEM   W69-07846
*AQUATIC ENVIRONMENT, CHLORELLA,   W70-02786
*AQUATIC ENVIRONMENT, POLLUTANTS,   W70-01233
*AQUATIC ENVIRONMENT, *AQUATIC LI   W71-02491
*AQUATIC FUNGI, YEASTS, HYDROGEN   W69-00096
*AQUATIC HABITATS, *ALASKA, LAKES   W69-08272
*AQUATIC INSECTS, *AQUATIC DRIFT,   W69-02782
*AQUATIC INSECTS, *AQUATIC PLANTS   W71-09005
*AQUATIC LIFE, *WASTE DISPOSAL, *   W72-01801
*AQUATIC LIFE, WATER POLLUTION EF   W71-05531
*AQUATIC LIFE, *AQUATIC POPULATIO   W71-02491
*AQUATIC LIFE, *AQUATIC POPULATIO   W73-00748
*AQUATIC MACROPHYTES, ANACHARIS C   W70-02249
*AQUATIC MICROBIOLOGY, ALGICIDES,   W69-10171
*AQUATIC MICROBIOLOGY, *PHYSIOLOG   W69-03518
*AQUATIC MICROBIOLOGY, *ESTUARIES   W69-03752
*AQUATIC MICROBIOLOGY, *AQUATIC P   W70-04283
*AQUATIC PRODUCTIVITY, BIOCHEMICA   W70-04283
*AQUATIC PRODUCTIVITY, LIMNOLOGY,   W68-00912
*AQUATIC POPULATIONS.: /LGAE, NIT   W68-00468
*AQUATIC PLANTS, *AQUATIC WEED CO   W70-00269
*AQUATIC PLANTS, *ALGAE, *ROTIFER   W70-02249
*AQUATIC PLANTS, *HERBICIDES, *VA   W70-10175
*AQUATIC POPULATIONS, ALGAE, BENT   W71-02491
*AQUATIC PLANTS, *NUISANCE ALGAE,   W70-10161
*AQUATIC PLANTS, *INHIBITORS, *MA   W72-00845
*AQUATIC PLANTS, AQUATIC ENVIRONM   W71-09005
*AQUATIC POPULATIONS, *RIVERS, AL   W73-00748
*AQUATIC PLANTS, ISLANDS, ECOLOGI   W72-10623
*AQUATIC WEED CONTROL, *AQUATIC P   W70-10175
*AQUATIC WEED CONTROL, WATER CONS   W70-00269
*AQUATIC WEEDS, *AQUATIC PLANTS,   W70-00269
*AQUATIC WEEDS, *BIOASSAY, *ESSEN   W69-03364

AL/ *FARM WASTES, *FARM LAGOONS,
, ODOR,/ *FARM WASTES, *LAGOONS,
ION, *CARBOHYDRATES, *NUTRIENTS,
UTION CONTROL, *DENITRIFICATION,
TROPHICATION, NUTRIENTS/ *ALGAE,
UTION CONTROL, *DENITRIFICATION,
UTION CONTROL, *DENITRIFICATION,
UTION CONTROL, *DENITRIFICATION,
RPLANTS, *WATER QUALITY CONTROL,
SECTICIDES, *MARINE BACTERIA, */
BACTERIA, DETERGENTS, SURFACTA/
RATIOS.:
IO RIVER, *ANAL/ *WATER QUALITY,
TER POLLUTION, *AQUATIC BIOLOGY,
DISSO/ *WATER POLLUTION EFFECTS,
EFFECTS, *REVIEWS, *WAT/ *ALGAE,
CHNIQUES, *ALGAE, *GROWTH RATES,
DIATOMS, MARYLAND, PEN/ *ALGAE,
*BIOASSAY, *COM/ *WATER QUALITY,
C ORGANISMS, PAPERMILL WASTE EF/
LUTION / *WASTE WATER TREATMENT,
HOCK LOADING.:
*FISH TOXICITY TEST,/ *POLYMERS,
*OPERATION AND MAINTENA/ *ALGAE,
, LAGOONS, *BAFFLES, *WASTE WAT/
PITATION, *CHEMICAL DEGRADATION,
, *SEWAGE FERTILIZATION EFFECTS,
*METHODOL/ PRIMARY PRODUCTIVITY,
R(OHIO), LAKE MENDOTA(WIS), LAK/
TREATMENT, *TRICKLING FILTERS, /
TARY ENGINEERING, *WATER SUPPLY,
UCTION POTENTIAL, AQUATIC ALGAE,
ORIES, / *BENTHIC FLORA, *ALGAE,
*ALGAL CULTURE, *BIOMASS TITER,
*PONDS, *ENVIRONMENTAL EFFECTS,
NA CYLINDRICA, MICROCYSTIS AERU/
S/ *CARBON, *ALGAE, *POPULATION,
OLOGY, MAXIMUM U/ *HETEROTROPHY,
ATMENT, *POLLUTANT IDEN/ *ALGAE,
CTIONS, BLACK SEA, CONTACT ZONE/
*WASTE DISCHARGES, FLORA, FAUNA,
E, *LAKE MICHIGAN, *LAKE ONTARI/
TE, *TREATMENT, *LAKES, *ECOLOG/
, *SAPROBITY, P/ *ALGAL CULTURE,
ACTERIA, BACTER/ *COOLING WATER,
ATISTICAL METHODS, *ZOOPLANKTON,
OTOZOA, *FUNGI, *BACTER/ *ALGAE,
TA, NU/ *LAKES, *EUTROPHICATION,
TER POLLUTION ASSESSMENT, SANIT/
, PHOSPHORUS RADIOISOTOPES, TRO/
*THERMAL POWERPLANTS, *ECOLOGY,
E TREATMENT, *TRICKLING FILTERS,
Y, LAKE WAUBESA, LYNGB/ *BLOOMS,
ADIOCARBON UPTAKE TECHNIQUES, W/
OI), RED RIVER(VIETNAM), CHLORO/
DIVERSITY, LAKE WAUBESA, LYNGB/
GHAT(INDIA), BOMBAY(IND/ *INDIA,
ENTE/ *POLLUTION-TOLERANT ALGAE,
NKING, *PHYTOPLANKTON, *DIATOMS,
*RED ALGAE, *TEMPERATURE TOLERA/
ALGAE PHYSIOLOGY.:
CENTRATION, / *THERMAL GRADIENT,
DAPHNIA SCHOD/ *FILTERING RATE,
MONTANA, *HYDROCARBONS, *OLEFI/
LLUTION, *NUTRIENT REQUIREMENTS,

Y, ALGAL METABOLIS/ *KILL RATES,
M RADIOISOTOPES, *ALGAE, *FUNGI,
ONMENT, AQUATI/ *PHOTOSYNTHESIS,
SEMIARID CLIMATES, *ABSORPTION,
, *ALGAE, *DIURNAL DISTRIBUTION,
*WATER REUSE, *RECLAIMED WATER,
ON, *ESTUARIES, *EUTROPHICATION,
PES, *BIOASSAY, *EUTROPHICATION,
E/ *DEEP WATER, *LAKES, *PLANTS,
ELVE MILE LAKE(ONTARI/ *ONTARIO,
LVE MILE LAKE(CANADA), *ONTARIO,
UNITIE/ *LAKES, *EUTROPHICATION,
EACHES, APRICOTS, PEARS, CITRUS/
E/ *RESERVOIRS, *EUTROPHICATION,
ES, CULTURES, / *ALGAE, *CARBON,
ATES, *CULTU/ *SURFACE, *VOLUME,
ROPHICATION, *ALGAE, *NUTRIENTS,
PHAN/ *EXTRACELLULAR PRODUCTION,
, CLARKSBURG, BUFFALO, DUNREITH/
*EUTROPHICATION, *CALI/ *LAKES,
ULTURES, PHYTOPLANKTON, CARBON,/
RE.), *APHANIZOMENON FLOS-AQUAE,
OLOGICAL COMMUNITIES, EFFLUENTS/
CAL STUDIES, CULTURES, / *ALGAE,

*BIODEGRADATION, ALGAE, BIOCHEMIC   W71-00936
*BIODEGRADATION, *SEWAGE BACTERIA   W71-02009
*BIODEGRADATION, ALGAE, BACTERIA,   W71-00221
*BIODEGRADATION, *DRAINAGE WATER,   W71-04557
*BIODEGRADATION, *PHOSPHATES, *EU   W71-04072
*BIODEGRADATION, *DRAINAGE WATER,   W71-04554
*BIODEGRADATION, *DRAINAGE WATER,   W71-04556
*BIODEGRADATION, *DRAINAGE WATER,   W71-04555
*BIODEGRADATION, WATER UTILIZATIO   W71-11881
*BIODEGRADATION, *DETERGENTS, *IN   W70-01467
*BIODEGRADATION, *ALGAE, *AQUATIC   W69-09454
*BIOGEOCHEMICAL DATA, *ABUNDANCE   W71-05531
*BIOINDICATORS, *CHLOROPHYLL, *OH   W71-04206
*BIOINDICATORS, *BIOASSAY, TOXICI   W71-10786
*BIOINDICATORS, *WATER QUALITY, *   W70-00475
*BIOINDICATORS, *WATER POLLUTION   W70-07027
*BIOINDICATORS, *WATER POLLUTION,   W70-04381
*BIOINDICATORS, *WATER POLLUTION,   W70-04510
*BIOINDICATORS, *WATER ANALYSIS,   W72-12240
*BIOLOGIC 'DISPOSERS', MICROSCOPI   W71-00940
*BIOLOGICAL TREATMENT, *WATER POL   W70-10437
*BIOLOGICAL EXTRACTION, GROWTH, S   W70-06604
*BIOLOGICAL LIFE, *ALGAE BLOOMS,   W70-05819
*BIOLOGICAL PROPERTIES, CONTROL,   W70-08392
*BIOLOGICAL TREATMENT, *OXIDATION   W70-09320
*BIOLOGICAL TREATMENT, *BIODEGRAD   W70-09186
*BIOLOGICAL PRODUCTIVITY, AQUATIC   W69-09349
*BIOLOGICAL COMMUNITIES, *ALGAE,   W71-12870
*BIOLOGICAL PROBLEMS, SCIOTO RIVE   W72-01813
*BIOLOGICAL COMMUNITIES, *SEWAGE   W72-01819
*BIOLOGICAL PROPERTIES, *WATER PO   W72-01813
*BIOLOGICAL TREATMENT, *WASTE WAT   W71-09546
*BIOLOGICAL COMMUNITIES, *LABORAT   W70-02780
*BIOLOGICAL INDEX, *SAPROBITY, PR   W72-12240
*BIOLOGICAL TREATMENT, *DESIGN CR   W70-04787
*BIOLOGICAL CONCENTRATION, ANABAE   W70-03520
*BIOLOGICAL COMMUNITIES, EFFLUENT   W70-04001
*BIOLOGICAL UPTAKE, *MICROBIAL EC   W70-04194
*BIOLOGICAL TREATMENT, *WATER TRE   W69-07389
*BIOLOGICAL OCEANOGRAPHY, *TRANSA   W69-07440
*BIOLOGICAL COMMUNITIES, INTERTID   W68-00010
*BIOLOGY, *GREAT LAKES, *LAKE ERI   W70-01943
*BIOLOGY, *EFFECTS, *COPPER SULFA   W70-06217
*BIOMASS TITER, *BIOLOGICAL INDEX   W72-12240
*BIOMASS, *SLIME, ALGAE, FUNGI, B   W72-12987
*BIOMASS, *SUCESSION, *ISOLATION,   W73-01446
*BIOMASS, WATER PURIFICATION, *PR   W71-05267
*BIOMASS, *PRODUCTIVITY, CYANOPHY   W70-03959
*BIOSTIMULATION, ALGAL GROWTH, WA   W70-02775
*BIOTA, *COLUMBIA RIVER, CHROMIUM   W69-07853
*BIOTA, *TEMPERATURE, *AQUATIC EN   W71-02491
*BIOTA, SEWAGE BACTERIA, SYSTEMAT   W72-01819
*BIOTIC FACTORS, SPECIES DIVERSIT   W69-01977
*BIOTIN, *AMPHIDINIUM CARTERI, *R   W70-05652
*BLOOM, *VIETNAM, LITTLE LAKE(HAN   W70-03969
*BLOOMS, *BIOTIC FACTORS, SPECIES   W69-01977
*BLOOMS, RIVER GANGES(INDIA), RAJ   W70-05404
*BLUE GREEN ALGAE, *EUGLENA, PIGM   W70-01233
*BLUE-GREEN ALGA, COPPER SULFATE.   W70-02754
*BLUE-GREEN ALGAE, *GREEN ALGAE,   W71-02496
*BLUE-GREEN ALGAE, *ALGA BLOOMS,   W69-03512
*BOD REMOVAL, *ISRAEL, *ALGAL CON   W70-04790
*BODY SIZE, *COMPARATIVE STUDIES,   W70-03957
*BOTRYOCOCCUS BRAUNII, *ANACYSTIS   W69-08284
*BOTTOM SEDIMENTS, ON-SITE DATA C   W68-00511
*BROMINE, *IODINE.:   W69-08562
*BROMINE, IODINE, PHOTO REACTIVIT   W70-02363
*CALCAREOUS SOIL, SOIL WATER, SOI   W71-04067
*CALCIUM CARBONATE, AQUATIC ENVIR   W69-00446
*CALCIUM, SOIL CONTAMINATION, *ST   W69-03491
*CALIFORNIA, ESTIMATING EQUATIONS   W69-03611
*CALIFORNIA, WATER QUALITY, AESTH   W70-05645
*CALIFORNIA, WATER POLLUTION, REA   W69-07520
*CALIFORNIA, *NEVADA, LIMNOLOGY,   W70-01933
*CALIFORNIA, *NEVADA, DEPTH, ALGA   W70-00711
*CANADA, KUSHOG LAKE(ONTARIO), TW   W70-02795
*CANADA, NUTRIENT LEVELS, SOFT WA   W70-02973
*CANALS, *CITIES, BIOLOGICAL COMM   W70-00268
*CANNERIES, *INDUSTRIAL WASTES, P   W71-09548
*CARBOHYDRATES, *NUTRIENTS, *BIOD   W71-00221
*CARBON CYCLE, *CYTOLOGICAL STUDI   W69-08278
*CARBON DIOXIDE, *AERATION, *NITR   W70-03334
*CARBON DIOXIDE, *REVIEWS, BIBLIO   W70-00664
*CARBON FIXATION, *EXCRETION, STE   W69-10158
*CARBON FILTERS, LEBANON, BATAVIA   W72-03641
*CARBON RADIOISOTOPES, *BIOASSAY,   W70-01933
*CARBON RADIOISOTOPES, *ALGAE, *C   W70-03983
*CARBON-14 MEASUREMENTS.: /LAKE(O   W72-06052
*CARBON, *ALGAE, *POPULATION, *BI   W70-04001
*CARBON, *CARBON CYCLE, *CYTOLOGI   W69-08278

351

DIA PLUMOSA, STIGEOCLONIUM, MIC/
GEMENT, *DAIRY INDUSTRY, *ALGAE,
                    *METABOLITES,
IS LUTHE/ *DNA, *ORGANIC CARBON,
A, ALGAE, PHYTOPLANKTON, *LAKES,
TERIAL DENITRIFICATION.:
TERIAL DENITRIFICATION.:
                    *ALGAE HARVESTING,
                    *ALGAE HARVESTING,

IF), NUTRI/ *FISHERY MANAGEMENT,
PES, *ALGAE, *FUNGI, *CALCAREOU/
, DETERGENTS, INDUSTRIAL WASTES,
S, WATE/ *TOXICITY, *CYANOPHYTA,
ENT, *COOLING WATER, *CORROSION,
ENVIRONMENT, PHAGE, FERMENTOR, /
OSPHATES, *PHOSPHORUS COMPOUNDS,
POUNDS, *CHEMICAL PRECIPITATION,
R, WATER POLLUTION EFFECTS, ANA/
NALYTICAL TECHNIQUES, */ *ALGAE,
WATER SAMPLING, *WATER ANALYSIS,
LANKTON, BIBLIOGRAPHIES, *ALGAE,
IRCULATION, NUTRIENT ABSORPTION,
HOSPHORUS, ALGAE, CYANOPHYTA, N/
IS AERUGINOSA, *NUTRIENT AVAILA/
IONS, *PHYSICAL CHARACTERISTICS,
ES, WATER AN/ *OIL, *OIL WASTES,
C MATTER, ALGAE, COPPER SULFATE,
ES.:
NT LIMITED/ *CONTINUOUS CULTURE,
CALUMET RIVER(ILL),/ *POLLUTION,
ASTES.:
T LAKES, METABOLISM, *CHLORELLA,
ALGAE, GREAT LAKES, METABOLISM,

ASSAY, *HALOGENS, *GROWTH RATES,
N, / *BIOCHEMICAL OXYGEN DEMAND,
DEMAND, *PHOTOSYNTHETIC OXYGEN,
SEWAGE FEED, SYMBIOTIC RELATIO/
SS, ORGANIC COMPOUNDS, / *ALGAE,
PHOSPHORUS, PHOTOSYNTH/ *ALGAE,
*PONDS, *DESIGN, *EFFICIENCIES,
POOLS, *BACTERIA, *BACTERICIDES,
, */ *ALGAE, *CHEMICAL ANALYSIS,
IS, *FLUOROMETRY, *FLUORESCENCE,
*ENZYMES, *ESSENTIAL NUTRIENTS,
TY, *MEASUREMEN/ *BENTHIC FLORA,
ENT REQUIREMENTS, ANALY/ *ALGAE,
ECTS, *ORGANIC COMPOUND/ *ALGAE,
*WATER QUALITY, *BIOINDICATORS,
CITY, ORGANIC COMP/ *REAERATION,
ON, *RHONE RIVER, NAVICULA, MEL/
, *IRON, TRACE ELEMENTS, ALGAE,/
SYSTEM, CILIATE AUTOECOLOGY.:
LAKES, *EUTROPHICATION, *CANALS,
ALGAE, KILL RATES, PHOTO REACT/
, AQUATIC MICROBIOLOGY, AQUATIC/
OMERATA, BUTTERNUT CREEK(NY).:
TIAL NUTRIENTS, *EUTROPHICATION,
HICATION, AQUATIC ALGAE, *LAKES,
SSAYS, PROTOZOA,/ *MICROBIOLOGY,
URCES, *WATER POLLUTION EFFECTS,
OL, *FILTRATION, *PARTICLE SIZE,
R/ *SALT VALLEY RESERVOIRS(NEB),
EMENTS, ALGAE,/ *CHROMATOGRAPHY,
NUTRIENTS, *TRACE ELEMENTS, ALG/
DEMAND, *PHOSPHORUS, *NITROGEN,
RON, *MANGANESE, *COLOR, *ALGAE,
, *TURBIDITY, *IRON, *MANGANESE,
AL OXYGEN DEMAND, ALGAL CONTROL,
TER, *FRESH WATER, *HUMIC ACIDS,
DGE, *BIOCHEMICAL OXYGEN DEMAND,
INTER, CYANOPHYTA, / *LIMNOLOGY,
ORUS RADIOISOTOPES, TRO/ *BIOTA,
BACKWASHING, *SUSPENDED SOLIDS,
OD/ *FILTERING RATE, *BODY SIZE,
ORS, *WATER ANALYSIS, *BIOASSAY,
UATIC BACTERIA, *ALGAE, ORGANIC/
EDTA, HEIDA, LAKE TAHOE(CALIF)/
DIANS, PIVERS ISLAND(N C), DETE/
AENA FLOS AQUAE, APHANIZOMENON /
ALKALINITY, *NITRATES, NITROGEN,
MAN POPULATION, *EUTROPHICATION,
FECTS, *WATER POLLUTION CONTROL,
FECTS, *WATER POLLUTION CONTROL,
NSLATIONS, ALGAE, CRUSTACEANS, /
LGAE, PLANNING, RIVERS, LAKES, /
ECTS, ALGAL CON/ EUTROPHICATION,
RNMENTS, *WATER QUALITY CONTROL/
TURBIDOSTATS, NUTRIENT LIMITED/

*CATIONS, DRAPARNALDIA, DRAPARNAL   W70-02245
*CATTLE, SLURRIES, SPRAYING, ACTI   W70-07491
*CEDAR RIVER(IOWA).:   W72-13605
*CELL SIZE, EUKARYOTES, MONOCHRYS   W69-08278
*CELLULOSE, DISTRIBUTION PATTERNS   W69-00976
*CENTRAL VALLEY(CALIFORNIA), *BAC   W71-04555
*CENTRAL VALLEY(CALIFORNIA), *BAC   W71-04554
*CENTRAL VALLEY(CALIF.).:   W71-04556
*CENTRAL VALLEY(CALIF.).:   W71-04557
*CESIUM RADIOISOTOPES.:   W71-09013
*CHAOBORID MIDGES, CLEAR LAKE(CAL   W70-05428
*CHELATION, *STRONTIUM RADIOISOTO   W71-04067
*CHELATION.: /FF, DOMESTIC WASTES   W71-04072
*CHEMCONTROL, EUTROPHICATION, BAS   W70-02982
*CHEMICAL REACTIONS, *WATER REUSE   W70-02294
*CHEMICAL ENVIRONMENT, *PHYSICAL   W71-02009
*CHEMICAL PRECIPITATION, *CHEMICA   W70-09186
*CHEMICAL DEGRADATION, *BIOLOGICA   W70-09186
*CHEMICAL ANALYSIS, ORGANIC MATTE   W70-05109
*CHEMICAL ANALYSIS, *CHLORINE, *A   W70-05178
*CHEMICAL ANALYSIS, PH, SAMPLING,   W69-00659
*CHEMICAL PROPERTIES, NUTRIENTS,   W68-00856
*CHEMICAL PROPERTIES, EPILMNION,   W68-00857
*CHEMICAL ANALYSIS, *NITROGEN, *P   W69-05868
*CHEMICAL COMPOSITION, *MICROCYST   W69-05868
*CHEMICAL CHARACTERISTICS, ORGANI   W73-00748
*CHEMICAL WASTES, INDUSTRIAL WAST   W72-03299
*CHEMICAL OXYGEN DEMAND.: /ORGANI   W72-04495
*CHEMOSTAT, PHYSIOLOGY, FLAGELLAT   W70-09904
*CHEMOSTATS, TURBIDOSTATS, NUTRIE   W70-02779
*CHICAGO(ILL), COOK COUNTY(ILL),   W70-00263
*CHICKENS, FOOD WASTES, POULTRY W   W70-09320
*CHLAMYDOMONAS LABORATORY TESTS,   W70-08652
*CHLORELLA, *CHLAMYDOMONAS LABORA   W70-08652
*CHLORELLA PYRENOIDOSA.:   W70-08416
*CHLORELLA, CHLORINE, ALGICIDES,   W70-02370
*CHLORELLA, *PHOTOSYNTHETIC OXYGE   W69-03362
*CHLORELLA, ALGAE, ANALYTICAL TEC   W69-03371
*CHLORELLA PYRENOIDOSA, SYNTHETIC   W70-05655
*CHLORELLA, *ORGANIC ACIDS, BIOMA   W70-04195
*CHLORELLA, *SCENEDESMUS, *LIGHT,   W70-03923
*CHLORINATION, *ODOR, *WASTE TREA   W70-04788
*CHLORINATION, *AQUATIC BACTERIA,   W69-10171
*CHLORINE, *ANALYTICAL TECHNIQUES   W70-05178
*CHLOROPHYLL, *EUTROPHICATION, PR   W70-05094
*CHLOROPHYTA, *CYCLING NUTRIENTS,   W69-03186
*CHLOROPHYLL, *PRIMARY PRODUCTIVI   W69-10151
*CHLOROPHYTA, *CYANOPHYTA, *NUTRI   W70-02245
*CHLOROPHYLL, WATER POLLUTION EFF   W70-08416
*CHLOROPHYLL, *OHIO RIVER, *ANALY   W71-04206
*CHLOROPHYLL, *OXYGENATION, *TOXI   W71-12183
*CHLOROPHYLL A, DATA INTERPRETATI   W73-01446
*CHROMATOGRAPHY, *COBALT, *COPPER   W71-11687
*CILIATES, FLAGELLATES, SAPROBIEN   W71-03031
*CITIES, BIOLOGICAL COMMUNITIES,   W70-00268
*CLADOPHORA, BROMINE, FILAMENTOUS   W70-02364
*CLADOPHORA, ALGAE, AQUATIC ALGAE   W70-02784
*CLADOPHORA FRACTA, CLADOPHORA GL   W72-05453
*CLADOPHORA, *NUISANCE ALGAE, *SE   W69-04798
*CLASSIFICATION, DESMIDS, OLIGOTR   W68-00255
*CLASSIFICATION, *CULTURES, *BIOA   W71-11823
*CLASSIFICATION, *WATER POLLUTION   W72-00462
*CLAYS, *SILTS, SANDS, ALGAE, PLA   W70-08628
*CLEAR WATER RESERVOIRS, *TURBID   W72-04761
*COBALT, *COPPER, *IRON, TRACE EL   W71-11687
*COBALT, *CYANOPHYTA, *ESSENTIAL   W69-06277
*COLIFORMS, *DISSOLVED OXYGEN, ES   W71-05409
*COLLOIDS, FLOW RATES, *FILTERS,   W72-04145
*COLOR, *ALGAE, *COLLOIDS, FLOW R   W72-04145
*COLOR, *ALKALINITY, *NITRATES, N   W72-11906
*COLOR, RIVERS, BOGS, PHENOLS, PI   W70-01074
*COLOR, WASTE WATER TREATMENT, TR   W70-04060
*COLORADO, *LAKES, *MOUNTAINS, *W   W69-10154
*COLUMBIA RIVER, CHROMIUM, PHOSPH   W69-07853
*COMBINED SEWERS, STORMWATER OVER   W71-10654
*COMPARATIVE STUDIES, DAPHNIA SCH   W70-03957
*COMPARATIVE PRODUCTIVITY, AQUATI   W72-12240
*COMPETITION, *PHYTOPLANKTON, *AQ   W70-04284
*COMPLEXONES, *ALGAL BLOOMS, NTA,   W70-04373
*CONCENTRATIONS, AEQUIPECTEN IRRA   W69-08267
*CONCEPTS, *TROPHIC BIOLOGY, ANAB   W70-03959
*CONDUCTIVITY, *LEACHING, IRON, T   W72-11906
*CONFERENCES, NUISANCE ALGAE, PUB   W70-03973
*CONFERENCES, STREAMS, ESTUARIES,   W70-03975
*CONFERENCES, STREAMS, ESTUARIES,   W70-03975
*CONFERENCES, *OCEANOGRAPHY, *TRA   W69-07440
*CONFERENCES, *EUTROPHICATION, *A   W70-04506
*CONFERENCES, WATER POLLUTION EFF   W70-02775
*CONNECTICUT, *LAKES, *LOCAL GOVE   W72-05774
*CONTINUOUS CULTURE, *CHEMOSTATS,   W70-02779

IORATION, DRAIN, SAN JOAQUIN VA/
KALINITY, WATER QUALITY CONTROL,
, *LABORATORY TESTS, *ISOLATION,
LANKTON,/ *ALGAE, *WATER SUPPLY,
THERMAL POLLUTION, *TEMPERATURE,
ALGAE, FUNGI, BACTERIA, BACTER/
LANTS, *WATER REUSE, COOLING TO/
ICAL REACTION/ *WATER TREATMENT,
AL TEMPERATURES, *COPPER DOSAGE,
ICH ISOTHERM, BLUE-GREEN ALGAE,/
INATION, *CRITICAL TEMPERATURES,
SERVOIRS, WATER/ *ALGAL CONTROL,
WATER SUPPLY, WATER QUALITY, WA/
*ALGAL CONTROL, *NEW HAMPSHIRE,
ES, *ECOLOG/ *BIOLOGY, *EFFECTS,
ER QUA/ WATER TREATMENT, *ALGAE,
ON/ *NUISANCE ALGAE, *ALGICIDES,
*ALGAE, *AQUATIC WEEDS, LAKES,/
UATIC WEED CONTROL, *HERBICIDES,
*TOXIN/ *EUTROPHICATION, *LAKES,
RATURE, SALINITY/ *MARINE WATER,
LGAE,/ *CHROMATOGRAPHY, *COBALT,
R POLLUTION EFFECTS, CHELATION,/
NTROL, ALGICIDES,/ *LAKES, *MUD,
FISH, ECOSYSTEMS, / *PHOSPHORUS,
IVE TREATMENT, ALGAE NUISANCE, /
RY, *FLUORESCENCE, *CHLOROPHYLL/
TURE, ALKALINITY, NITRIFICATION,
WATER TREATMENT, *COOLING WATER,
GATION, COAGULATION, DEWATERING,
*NITROGEN, *PHOSPHORUS, ALGAE,/
), THORIUM-232, BISMUTH-214, AC/
AMICS, CELLULAR SIZE, CHLORELLA/
OSAGE/ *MICROSCOPIC EXAMINATION,
, *TORREY CANYON, AROMATIC SOLV/
ANCE, *LAKE AGING, NUTRIENT SOU/
, DREDG/ *LAKE WASHINGTON(WASH),
INDICATORS, NUTRIENT BUDGET.:
NONA(WIS), LAKE TAHOE(CALIF), L/
IONAL USE, WASTE EFFLUENTS.:
LOGICAL ECOLOGY, *PHYTOPLANKTON,
*CARBON RADIOISOTOPES, *ALGAE,
M, KI/ *BACTERIA, *MARINE ALGAE,
N DIOXIDE, *AERATION, *NITRATES,
PUBLIC HEAL/ *ALGAE, *TOXICITY,
, MARINE ALGAE, ECOL/ *BACTERIA,
*MICROBIOLOGY, *CLASSIFICATION,
GAL NUTRITION, ALGAL PHYSIOLOGY/
, *TRACE ELEMENTS, ALG/ *COBALT,
CONSIN, WATER QUALI/ *ALGICIDES,
GY, *SALINE WATER, *THERMAL WAT/
*NUTRIENT REQUIREMENTS, *PHYSI/
BIOLOGY, *PHYSIOLOGICAL ECOLOGY,
GY, *PHYTOPLANKTON, *CULTURES, /
ANKTON, ALGAE, *NUTRIENT REQUIR/
YSIOLOGICAL ECOLOGY, *WATER POL/
GY, ALGAE, ANA/ *EUTROPHICATION,
AGE, FISH POPULATIONS, PLANKTON,
NUTRIENTS, LAKE ERIE.:   *ALGAE,
INGS, *OREGON, THERMOPHILIC BAC/
WATER POLLUTION EFFECTS, LAKES,
MATTER, TOXICITY, D/ *FISHKILL,
ATION, PONDS, RESERVOIR/ *ALGAE,
NTS, CULTURES, LAKE/ *ISOLATION,
TS, ANALY/ *ALGAE, *CHLOROPHYTA,
HICATION, BASS, WATE/ *TOXICITY,
ON, *HYDRODYNAMICS, LAKES, INTE/
REAMS, *WATER POLLUTION EFFECTS,
ON, *PH/ *ANALYTICAL TECHNIQUES,
SENTIAL NUTRIENTS/ *CHLOROPHYTA,
*ALGAE, *ANALYTICAL TECHNIQUES,
SENTIAL NUTRIENTS, *NIT/ *ALGAE,
ER INFERFAC/ *PHOSPHATE, *LAKES,
L EFFECTS, *ESSENTIAL NUTRIENTS/
NKTON, *WATER POLLUTION SOURCES,
OMPOUNDS, *PHOSPHORUS COMPOUNDS,
N, ALGAE, PLANT G/ *FERTILIZERS,
OTOPES, STREAMFLOW/ *PERIPHYTON,
US RADIOISOTOPES, *SALT MARSHES,
LANKTON, *ALGAE, *SESSILE ALGAE,
ER, FOOD CHAINS/ *PHYTOPLANKTON,
ATION.:
*ALGAE, *CARBON, *CARBON CYCLE,
S, MANAGEMENT, *FARM MANAGEMENT,
TILE SOLIDS, NON-DEGRADABLE SOL/
S QUADRICAUDA, ECOLOGICAL EFFIC/
*FEEDING RATES, *GROWTH RATES,
LAKES, / *GRAZING, *TEMPERATURE,
R QUALITY, *ARTIFICIAL RECHARGE,
, BIOCHEMICAL OXYGEN DEMAND, SU/

*CONTRA COSTA COUNTY, WATER DETER   W71-09577
*CONTROL, *ALGAL CONTROL, TOXICIT   W72-07442
*CONTROL, METHODOLOGY, CULTURES,   W70-04469
*CONTROL, TASTE, ODOR, FILTERS, P   W70-05389
*COOLING WATER, CHEMICAL PROPERTI   W69-05023
*COOLING WATER, *BIOMASS, *SLIME,   W72-12987
*COOLING WATER, *ELECTRIC POWER P   W71-11393
*COOLING WATER, *CORROSION, *CHEM   W70-02294
*COPPER APPLICATION METHODS, ALGA   W69-10157
*COPPER CHLORIDE, UPTAKE, FREUNDL   W70-08111
*COPPER DOSAGE, *COPPER APPLICATI   W69-10157
*COPPER SULFATE, WATER SUPPLY, RE   W69-10157
*COPPER SULFATE, *ALGAL CONTROL,   W69-10155
*COPPER SULFATE, SEWAGE EFFLUENTS   W69-08674
*COPPER SULFATE, *TREATMENT, *LAK   W70-06217
*COPPER SULFATE, ALKALINITY, WAT   W72-07442
*COPPER SULFATE, *WATER QUALITY C   W69-05704
*COPPER SULFATE, *EUTROPHICATION,   W70-04494
*COPPER SULFATE, *SODIUM ARSENITE   W70-05568
*COPPER SULFATE, *ALGAL CONTROL,   W68-00859
*COPPER, *ALGAE, SEA WATER, TEMPE   W70-08111
*COPPER, *IRON, TRACE ELEMENTS, A   W71-11687
*COPPER, *TOXICITY, *ALGAE, *WATE   W71-11561
*COPPER, COPPER SULFATE, ALGAL CO   W69-03366
*CORAL, *REEFS, *MARINE ANIMALS,   W69-08526
*CORRECTIVE TREATMENT, *PREVENTAT   W69-08282
*CORRELATION ANALYSIS, *FLUOROMET   W70-05094
*CORROSION, CARBON DIOXIDE, SOLUB   W71-11393
*CORROSION, *CHEMICAL REACTIONS,   W70-02294
*COST ANALYSIS.: /ATURE, CENTRIFU   W71-11375
*COST COMPARISONS, *EFFICIENCIES,   W70-07469
*COTTUS CAROLINAE, MEAD COUNTY(KY   W69-09742
*COULTER COUNTER, *POPULATION DYN   W69-06540
*CRITICAL TEMPERATURES, *COPPER D   W69-10157
*CRUDE OIL, *GAMOSOL, *OIL SPILLS   W70-01470
*CULTURAL EUTROPHY, *NUTRIENT BAL   W70-00264
*CULTURAL INFLUENCES, MICROCYSTIS   W70-05663
*CULTURAL EUTROPHICATION, TROPHIC   W71-03048
*CULTURAL EUTROPHICATION, LAKE MO   W69-06858
*CULTURAL EUTROPHICATION, RECREAT   W69-06535
*CULTURES, ALGAE, AQUATIC PRODUCT   W69-03514
*CULTURES, PHYTOPLANKTON, CARBON,   W70-03983
*CULTURES, PSEUDOMONAS, METABOLIS   W70-04280
*CULTURES, *SCENEDESMUS, *SUBSURF   W70-03334
*CULTURES, *ISOLATION, LIMNOLOGY,   W70-00273
*CULTURES, *TRITIUM, EPIPHYTOLOGY   W69-09755
*CULTURES, *BIOASSAYS, PROTOZOA,   W71-11823
*CYANOCOBALAMIN, ALGAL GROWTH, AL   W69-06277
*CYANOPHYTA, *ESSENTIAL NUTRIENTS   W69-06277
*CYANOPHYTA, *EUTROPHICATION, WIS   W70-03310
*CYANOPHYTA, *PHYSIOLOGICAL ECOLO   W69-03516
*CYANOPHYTA, *NITROGEN COMPOUNDS,   W69-03515
*CYANOPHYTA, ALGAE, ANALYTICAL TE   W69-03518
*CYANOPHYTA, *PHYSIOLOGICAL ECOLO   W69-03514
*CYANOPHYTA, *EUTROPHICATION, *PL   W69-03512
*CYANOPHYTA, *EUTROPHICATION, *PH   W69-03513
*CYANOPHYTA, *PHYSIOLOGICAL ECOLO   W69-03517
*CYANOPHYTA, WASHINGTON, DC, *EST   W68-00461
*CYANOPHYTA, *NITROGEN FIXATION,   W70-05091
*CYANOPHYTA, *ALGAE, *THERMAL SPR   W70-05270
*CYANOPHYTA, *EUTROPHICATION, WAT   W71-04758
*CYANOPHYTA, *DECOMPOSING ORGANIC   W72-01797
*CYANOPHYTA, *ECOLOGY, *EUTROPHIC   W70-00274
*CYANOPHYTES, *NUTRIENT REQUIREME   W70-00719
*CYANOPHYTA, *NUTRIENT REQUIREMEN   W70-02245
*CYANOPHYTA, *CHEMCONTROL, EUTROP   W70-02982
*CYCLING NUTRIENTS, *EUTROPHICATI   W70-09889
*CYCLING NUTRIENTS, *EUTROPHICATI   W69-03369
*CYCLING NUTRIENTS, *EUTROPHICATI   W69-03370
*CYCLING NUTRIENTS, *ENZYMES, *ES   W69-03186
*CYCLING NUTRIENTS, *EUTROPHICATI   W69-03358
*CYCLING NUTRIENTS, *ENZYMES, *ES   W69-03185
*CYCLING NUTRIENTS, *SEDIMENT-WAT   W69-06859
*CYCLING NUTRIENTS, *ENVIRONMENTA   W69-04804
*CYCLING NUTRIENTS, ALGAE, ANALYT   W69-04800
*CYCLING NUTRIENTS, WATER CHEMIST   W69-05142
*CYCLING NUTRIENTS, EUTROPHICATIO   W69-07826
*CYCLING NUTRIENTS, *ZINC RADIOIS   W69-07862
*CYCLING NUTRIENTS, GEORGIA, SEDI   W69-08274
*CYCLING NUTRIENTS, ECOSYSTEMS, P   W73-01454
*CYCLING, *RADIOISOTOPES, SEA WAT   W70-00708
*CYTOLOGICAL EFFECTS, ALGAL POPUL   W70-02770
*CYTOLOGICAL STUDIES, CULTURES, F   W69-08278
*DAIRY INDUSTRY, *ALGAE, *CATTLE,   W70-07491
*DAIRY MANURE, TOTAL SOLIDS, VOLI   W71-00936
*DAPHNIA PULEX OBTUSA, SCENEDESMU   W69-10152
*DAPHNIA, PRODUCTIVITY, BIOINDICA   W69-10152
*DAPHNIA, PHYSIOLOGICAL ECOLOGY,   W70-03957
*DATA COLLECTIONS, PIT RECHARGE,   W69-05894
*DATA COLLECTIONS, *WATER QUALITY   W70-09557

```
*PHOTOSYNTHESIS, *PHYTOPLANKTON,        *DDT, *MARINE ALGAE, TOXICITY, DI    W70-05272
ENT, *TASTE, *ODOR, *RESERVOIRS,        *DECIDUOUS TREES, *DECOMPOSING OR    W72-03475
 *WATER TREATMENT, ODOR, *ALGAE,        *DECOMPOSING ORGANIC MATTER, *POT    W72-04242
 , *RESERVOIRS, *DECIDUOUS TREES,       *DECOMPOSING ORGANIC MATTER, *ALG    W72-03475
CITY, D/ *FISHKILL, *CYANOPHYTA,        *DECOMPOSING ORGANIC MATTER, TOXI    W72-01797
LIFORNIA, *NEVADA, DEPTH, ALGAE/        *DEEP WATER, *LAKES, *PLANTS, *CA    W70-00711
ALGAE, RECREATIONAL USE.:              *DEFINITIONS, *WATER BLOOM, TOXIC    W69-08279
EMATICAL MODELS, NEUTRALIZATION,        *DEGRADATION(DECOMPOSITION), METH    W71-13413
POLLUTION SOURCES, *DETERGENTS,         *DEGRADATION(DECOMPOSITION), CYAN    W70-09438
R QUAL/ *AGRICULTURAL CHEMICALS,        *DELTAS, *DRAINAGE EFFECTS, *WATE    W71-09577
TMENT, *WATER POLLUTION CONTROL,        *DENITRIFICATION, *BIODEGRADATION    W71-04557
TMENT, *WATER POLLUTION CONTROL,        *DENITRIFICATION, *BIODEGRADATION    W71-04554
TMENT, *WATER POLLUTION CONTROL,        *DENITRIFICATION, *BIODEGRADATION    W71-04555
TMENT, *WATER POLLUTION CONTROL,        *DENITRIFICATION, *BIODEGRADATION    W71-04556
NITRIFICATION, NITROGEN COMPOUN/        *DENITRIFICATION, EFFICIENCIES, *    W70-07031
, *WATER REUSE, *RECLAIMED WATE/        *DENITRIFICATION, *BIODEGRADATION    W69-08054
, SOIL EROSION, WATER POLLUTION,        *DENITRIFICATION, PHOSPHORUS COMP    W69-05323
FFECTS,/ *ANAEROBIC WASTE PONDS,        *DEPTH EFFECTS, *HYDRAULIC LOAD E    W70-04787
      *POTASSIUM PERMANGANATE,          *DES MOINES(IOWA).:                  W72-04242
EFFECTS, *BIOLOGICAL TREATMENT,         *DESIGN CRITERIA, DISSOLVED OXYGE    W70-04787
EATMENT, *STABILIZATION, *PONDS,        *DESIGN CRITERIA, FACILITIES, EQU    W70-04786
S, *MIDWEST(US), *SOUTHWEST(US),        *DESIGN CRITERIA, POND EFFLUENT,     W70-04788
RE, SEWAGE TREAT/ *FARM LAGOONS,        *DESIGN CRITERIA, WATER TEMPERATU    W71-08221
ZATION PONDS, *ALGAL PHYSIOLOGY,        *DESIGN THEORY AND CALCULATIONS,     W70-04786
TION, */ *STABILIZATION, *PONDS,        *DESIGN, *EFFICIENCIES, *CHLORINA    W70-04788
 *SULFIDES, *OXIDATION LAGOONS,         *DESIGN, *SULFUR BACTERIA, STABIL    W70-01971
SANITARY ENGINEERING, *FILTERS,         *DESIGN, COAGULATION, ALGAE, ACTI    W72-07334
DETENTION TIME.: *SULFATE ION,          *DESULFOVIBRIO, FACULTATIVE, BOD,    W70-01971
                                        *DETENTION TIME.:                   W72-06854
EFFECTS, *ORGANIC LOAD EFFECTS,         *DETENTION TIME EFFECTS, *FACULAT    W70-04787
LU/ *EUTROPHICATION, PHOSPHORUS,        *DETERGENTS, DOMESTIC WASTES, POL    W70-05084
 *GREAT LA/ *PHOSPHATES, *LAKES,        *DETERGENTS, *TERTIARY TREATMENT,    W70-03964
*WATER QUALITY, *EUTROPHICATION,        *DETERGENTS, *PHOSPHATES, CHELATI    W70-04373
POLLUTION EFFECTS, LETHAL LIMI/         *DETERGENTS, *MARINE ALGAE, WATER    W70-01779
AL, SUSCEPTIBIL/ *TORREY CANYON,        *DETERGENT CONCENTRATIONS, SURVIV    W70-01779
WASTES, *SURFACTANTS, *WATER P/         *DETERGENTS, *DIATOMS, *OIL, *OIL    W70-01470
CITY, *WATER POLLUTION, FOOD CH/        *DETERGENTS, *INSECTICIDES, *TOXI    W70-01466
NE BACTERIA, */ *BIODEGRADATION,        *DETERGENTS, *INSECTICIDES, *MARI    W70-01467
CTANTS, WATER POLLUTION SOURCES,        *DETERGENTS, *DEGRADATION(DECOMPO    W70-09438
N, *METROPOLITAN AREAS, BLONDIN/        *DEVOLUTION, *INTERSTATE POLLUTIO    W70-04430
 *SEPARATION TECHNIQUES, TURBID/        *DIATOMACEOUS EARTH, *FILTRATION,    W72-08686
ECHNIQUES,/ *ALGAE, *FILTRATION,        *DIATOMACEOUS EARTH, SEPARATION T    W72-08688
           *ALGAE HARVESTING,           *DIATOMITE FILTRATION.:              W72-08688
                                        *DIATOMITE FILTERS, FILTER AIDS.:    W72-04145
NTENSITY, BIOMAS/ *MARINE ALGAE,        *DIATOMS, *BENTHIC FLORA, LIGHT I    W70-03325
LIOGRAPHIES, *MINNESOTA, *LAKES,        *DIATOMS, *ALGAE, ECOLOGY, SOUTH     W69-00659
 CULTURES, RE/ *STREAMS, *ALGAE,        *DIATOMS, TEMPERATURE, MANGANESE,    W70-07257
FACTANTS, *WATER P/ *DETERGENTS,        *DIATOMS, *OIL, *OIL WASTES, *SUR    W70-01470
ATION, *SINKING, *PHYTOPLANKTON,        *DIATOMS, *BLUE-GREEN ALGA, COPPE    W70-02754
TICIDES, WATER POLLU/ *DIELDRIN,        *DIATOMS, LABORATORY TESTS, INSEC    W71-03183
ESTS, INSECTICIDES, WATER POLLU/        *DIELDRIN, *DIATOMS, LABORATORY T    W71-03183
UMMIT(MO), OXY/ *AERATION TUBES,        *DIFFUSERS, *RAYTOWN(MO), LEE'S S    W70-06601
D, BALL-MILL, MEICELASE, HYDROG/        *DIGESTIBILITY, *IN VITRO, IN VIV    W70-04184
S, *RADIOISOTOPES,/ *MONITORING,        *DISCHARGE MEASUREMENT, *EFFLUENT    W72-00941
NEERING, *SANITARY ENGI/ *ALGAE,        *DISPERSIONS, *ENVIRONMENTAL ENGI    W70-02609
C DEBRIS, ORGANIC GROWTH FACTOR/        *DISSOLVED ORGANIC MATTER, ORGANI    W70-02510
MPOUNDED WATERS, *WATER QUALITY,        *DISSOLVED SOLIDS, DISSOLVED OXYG    W72-01773
TION LAGOONS, *ALGAE, *BACTERIA,        *DISSOLVED OXYGEN, SEWAGE LAGOONS    W72-01818
MICAL OXYGEN DEMAND, *NUTRIENTS,        *DISSOLVED OXYGEN, *ALGAE, COLIFO    W71-05407
OSPHORUS, *NITROGEN, *COLIFORMS,        *DISSOLVED OXYGEN, ESTUARIES, ALG    W71-05409
TION, PHOTOSYNTH/ *RADIONUCLIDE,        *DISSOLVED OXYGEN, *RIVERS, OXIDA    W71-11381
 *BIOINDICATORS, *WATER QUALITY,        *DISSOLVED OXYGEN, *OHIO, WATER T    W70-00475
 , *RESPIRATION, *EUTROPHICATION,       *DISSOLVED OXYGEN, PENNSYLVANIA,     W69-10159
AL DISTRIBUTION/ *PHYTOPLANKTON,        *DISSOLVED OXYGEN, *ALGAE, *DIURN    W69-03611
RATION, *ESTUARIES, *EUTROPHICA/        *DISSOLVED OXYGEN, *ALGAE, *RESPI    W69-07520
*IMPOUNDED WATERS, IMPOUNDMENTS,        *DISSOLVED OXYGEN, BIOCHEMICAL OX    W72-11906
 *TASTE, *ODOR, *NATURAL STREAMS,       *DISTRIBUTION SYSTEMS, CHLORINATI    W72-03562
D CARBON, FILTRATION, *TOXICITY,        *DISTRIBUTION SYSTEMS, WATER POLL    W72-03746
OTTOM SEDIMENTS/ *RADIOISOTOPES,        *DISTRIBUTION, ANIMALS, PLANTS, B    W69-08269
 *ALGAE, *NUTRIENT REQUIREMENTS,        *DISTRIBUTION, BACTERIA, SAMPLING    W70-03952
OIRS, *PHYTOPLANKTON, *ROTIFERS,        *DISTRIBUTION, MISSOURI RIVER, SO    W70-05375
KTON, *DISSOLVED OXYGEN, *ALGAE,        *DIURNAL DISTRIBUTION, *CALIFORNI    W69-03611
 *ALGAL GROWTH, *THERMOBIOLOGY,         *DIURNAL OXYGEN CURVES, CALORIMET    W70-03309
 , EUKARYOTES, MONOCHRYSIS LUTHE/       *DNA, *ORGANIC CARBON, *CELL SIZE    W69-08278
LIZERS, *ESSENTIAL NUTRIENTS, */        *DOMESTIC ANIMALS, *WASTES, FERTI    W69-05323
                                        *DOMESTIC WATER USE.:                W68-00483
ATMENT, INDUSTRIAL WA/ *LAGOONS,        *DOMESTIC WASTES, WASTE WATER TRE    W71-06737
ATMENT, INDUSTRIAL WA/ *LAGOONS,        *DOMESTIC WASTES, WASTE WATER TRE    W70-01519
TION, / *MICROORGANISMS, *ALGAE,        *DOMINANT ORGANISMS, *ACCLIMATIZA    W70-07313
AGRICULTURAL CHEMICALS, *DELTAS,        *DRAINAGE EFFECTS, *WATER QUALITY    W71-09577
ENITRIFICATION, *BIODEGRADATION,        *DRAINAGE WATER, CALIFORNIA, DRAI    W71-04557
ENITRIFICATION, *BIODEGRADATION,        *DRAINAGE WATER, CALIFORNIA, DRAI    W71-04554
ENITRIFICATION, *BIODEGRADATION,        *DRAINAGE WATER, CALIFORNIA, DRAI    W71-04556
ENITRIFICATION, *BIODEGRADATION,        *DRAINAGE WATER, CALIFORNIA, DRAI    W71-04555
PHATES, ULTRA-VIOLET LIGHT STER/        *DUTCH LAKES, *ALLOCHTHONOUS PHOS    W70-03955
LANKTON, *LAKES, *CELLULOSE, DI/        *ECOLOGY, BACTERIA, ALGAE, PHYTOP    W69-00976
POLLUTION, *WATER POLLUTION SO/         *ECOLOGY, *ALGAE, *RIVERS, *WATER    W69-09061
TS, POT/ *WATER QUALITY CONTROL,        *ECOLOGY, *RESERVOIR, MASSACHUSET    W70-05264
WATERSHED MANAGEMENT, GEOCHEMIS/        *ECOLOGY, *MATHEMATICAL MODELS, *    W71-09012
PER SULFATE, *TREATMENT, *LAKES,        *ECOLOGY, BOTTOM SEDIMENTS, MUD,     W70-06217
```

POLLUTION, *THERMAL POWERPLANTS,   *ECOLOGY, *BIOTA, *TEMPERATURE, *   W71-02491
URE EFFECTS, *THERMAL POLLUTION,   *ECOLOGY, ALGAE, BACTERIA, COLIFO   W71-02479
PHYTOPLANKTON, BACTERI/ *ALGAE,   *ECOLOGY, *MARINE MICROORGANISMS,   W70-01068
LAKES, *ALGAE, NUTRIENTS, PHOS/   *ECOLOGY, *EUTROPHICATION, *GREAT   W70-00667
RESERVOIR/ *ALGAE, *CYANOPHYTA,   *ECOLOGY, *EUTROPHICATION, PONDS,   W70-00274
ES, *RADIOISOTOPES, LIMNOLOGY, /   *ECOLOGY, *RADIOACTIVITY TECHNIQU   W69-10163
, *FISH, *FISH KILLS, *ROTIFERS,   *ECOLOGY, *ECOSYSTEMS.: /STICIDES   W72-04506
FISH KILLS, *ROTIFERS, *ECOLOGY,   *ECOSYSTEMS.: /STICIDES, *FISH, *   W72-04506
RADIOISOTOPES, *TROUT, *STREAMS,   *ECOSYSTEMS, RADIOACTIVITY, BIOTA   W69-09334
TER POLLUTION EFFECTS, *FALLOUT,   *ECOSYSTEMS, GREAT LAKES, TRACE E   W71-09013
S, *URAN/ *AQUATIC ENVIRONMENTS,   *ECOSYSTEMS, *RADIUM RADIOISOTOPE   W69-07846
TOPES, *TRANSFER, / *TUBIFICIDS,   *ECOSYSTEMS, *PHOSPHORUS RADIOISO   W69-07861
, WATER POLL/ *SYSTEMS ANALYSIS,   *ECOSYSTEMS, *SIMULATION ANALYSIS   W69-06536
ITY STUDIES, *FIXED-LIGHT TESTS,   *EFFECTIVE-AVERAGE-LIGHT-INTENSIT   W70-03923
HUSTEDTII.:   *EFFECTS, NAVICULA SEMINULUM VAR.   W71-03183
MENT, *LAKES, *ECOLOG/ *BIOLOGY,   *EFFECTS, *COPPER SULFATE, *TREAT   W70-06217
ORUS, ALGAE,/ *COST COMPARISONS,   *EFFICIENCIES, *NITROGEN, *PHOSPH   W70-07469
*STABILIZATION, *PONDS, *DESIGN,   *EFFICIENCIES, *CHLORINATION, *OD   W70-04788
, ALGAE,/ *ANAEROBIC CONDITIONS,   *EFFLICIENCIES, *LAGOONS, *SLUDGE   W70-06899
TY MANAGEMENT, *LAKE WASHINGTON/   *EFFLUENT DIVERSION, *WATER QUALI   W69-09349
ON, / *ALGAE, *ACTIVATED SLUDGE,   *EFFLUENTS, WASTE WATER RECLAMATI   W70-06053
ITORING, *DISCHARGE MEASUREMENT,   *EFFLUENTS, *RADIOISOTOPES, *WAST   W72-00941
USE, COOLING TO/ *COOLING WATER,   *ELECTRIC POWER PLANTS, *WATER RE   W71-11393
ATER QUALITY, *ACTIVATED CARBON,   *ELECTROLYTES, FILTERS, *ANALYTIC   W72-09693
ALGAE, *MARSH PLANTS, *MARINE /   *EMULSIFIERS, *MORTALITY, *MARINE   W70-01777
ECT/ *OILY WATER, *MARINE ALGAE,   *EMULSIFIERS, WATER POLLUTION EFF   W70-01780
UTHENIUM-106, ANTIMONY-125, GRO/   *ENIWETOK ATOLL, *BIKINI ATOLL, R   W69-08269
T LIMITATION, CAYUGA LAKE(NY), /   *ENRICHMENT EXPERIMENTS, *NUTRIEN   W70-01579
AE, *ACID MINE WATER, *HABITATS,   *ENVIRONMENTAL EFFECTS, *LIMITING   W70-02770
TARY ENGI/ *ALGAE, *DISPERSIONS,   *ENVIRONMENTAL ENGINEERING, *SANI   W70-02609
ELLA, CARB/ *ALGAE, *METABOLISM,   *ENVIRONMENTS, VARIABILITY, CHLOR   W70-02965
TER QUA/ *INDICATORS, *PROTOZOA,   *ENVIRONMENT, *ORGANIC MATTER, WA   W71-03031
GY, *AL/ *PHYSIOLOGICAL ECOLOGY,   *ENVIRONMENTAL EFFECTS, *MORPHOLO   W69-10177
TEMPERATU/ *ZINC RADIOISOTOPES,   *ENVIRONMENTAL EFFECTS, SALINITY,   W69-08267
L NUTRIENTS/ *CYCLING NUTRIENTS,   *ENVIRONMENTAL EFFECTS, *ESSENTIA   W69-04804
EATMENT, *STABILIZATION, *PONDS,   *ENVIRONMENTAL EFFECTS, *BIOLOGIC   W70-04787
KTON, *ALGAE, SCENEDESMUS, LIGH/   *ENVIRONMENT, *FRESH WATER, *PLAN   W70-05409
PULATION, *EUTROPHICATI/ *ALGAE,   *ENVIRONMENTAL EFFECTS, *HUMAN PO   W70-03973
OCONTROL, *WATERSHED MANAGEMENT,   *ENVIRONMENTAL EFFECTS, CHEMTROL,   W68-00478
NIT/ *ALGAE, *CYCLING NUTRIENTS,   *ENZYMES, *ESSENTIAL NUTRIENTS, *   W69-03185
CHLOROPHYTA, *CYCLING NUTRIENTS,   *ENZYMES, *ESSENTIAL NUTRIENTS, *   W69-03186
NALYTICAL TECHNIQUES, *BIOASSAY,   *ENZYMES, *NUTRIENT REQUIREMENTS,   W69-03373
*WATER PURIFICATION, *BACTERIA,   *ENZYMES, CORROSION, OXIDATION, R   W69-07428
TON, PHYTOPLANKTON, TEMPERATURE,   *EPILIMNION, *HYPOLIMNION, ALGAE,   W70-04475
NS, SHEAR WA/ *EPIPSAMMIC ALGAE,   *EPIPELIC ALGAE, ALGAL ASSOCIATIO   W71-12870
*LAKES, *PRODUCTIVITY, MICHIGAN,   *EPIPHYTOLOGY, NUTRIENTS, PLANT P   W71-00832
E, ALGAL ASSOCIATIONS, SHEAR WA/   *EPIPSAMMIC ALGAE, *EPIPELIC ALGA   W71-12870
SORPTION REACTIONS, SOLUBILITY /   *EQUILIBRIUM MODEL, LITHOSPHERE,   W70-04385
NALYTICAL TECHNIQUES, *BIOASSAY,   *ESSENTIAL NUTRIENTS, *NUTRIENT R   W69-04802
TRIENTS, *ENVIRONMENTAL EFFECTS,   *ESSENTIAL NUTRIENTS, *FISH, *FIS   W69-04804
C ANIMALS, *WASTES, FERTILIZERS,   *ESSENTIAL NUTRIENTS, *NITRIFICAT   W69-05323
TION, */ *ALGAE, *ALGAL CONTROL,   *ESSENTIAL NUTRIENTS, *EUTROPHICA   W69-04798
ENTS, ALG/ *COBALT, *CYANOPHYTA,   *ESSENTIAL NUTRIENTS, *TRACE ELEM   W69-06277
QUES, *AQUATIC WEEDS, *BIOASSAY,   *ESSENTIAL NUTRIENTS, *NITROGEN C   W69-03364
E, *CYCLING NUTRIENTS, *ENZYMES,   *ESSENTIAL NUTRIENTS, *NITROGEN C   W69-03185
A, *CYCLING NUTRIENTS, *ENZYMES,   *ESSENTIAL NUTRIENTS, *NITROGEN C   W69-03186
OD/ *EUTROPHICATION, *LIMNOLOGY,   *ESSENTIAL NUTRIENTS, *PRIMARY PR   W72-01990
N, BRITISH ISLES, BIOINDICATORS,   *ESTIMATING EQUATIONS, PERIOD OF   W68-00255
ON, *CYANOPHYTA, WASHINGTON, DC,   *ESTUARIES, HUMAN POPULATION, SEW   W68-00461
TEMS, ESTUARINE EN/ *POLLUTANTS,   *ESTUARIES, *TIDAL WATERS, ECOSYS   W68-00010
T, *INSTRUMENTATION, *POLLUTION/   *ESTUARIES, *ESTUARINE ENVIRONMEN   W69-06203
OBIOLOGY, *AQUATIC MICROBIOLOGY,   *ESTUARIES, *MARINE MICROORGANISM   W69-03752
ED OXYGEN, *ALGAE, *RESPIRATION,   *ESTUARIES, *EUTROPHICATION, *CAL   W69-07520
TER POLLUTION EFFECTS, *ALGAE.:   *ESTUARIES, *HARBORS, *SEWAGE, WA   W71-05500
ILTS/ *ALGAE, *SEWAGE EFFLUENTS,   *ESTUARIES, FRESH WATER, LAKES, S   W70-02255
MPERATURE, *SEASONAL/ *BACTERIA,   *ESTUARINE ENVIRONMENT, *WATER TE   W70-00713
ALGAE, BENTHIC FLORA, ESTUARIES,   *ESTUARINE ENVIRONMENT, BAYS, *FL   W71-02494
                                      *ESTUARINE POLLUTION.:   W69-06203
NTATION, *POLLUTION/ *ESTUARIES,   *ESTUARINE ENVIRONMENT, *INSTRUME   W69-06203
ISONING, SCENEDESMUS, CHLORELLA,   *EUGLENA, ALGAL CONTROL, BACTERIC   W69-00994
LERANT ALGAE, *BLUE GREEN ALGAE,   *EUGLENA, PIGMENTED FLAGELLATES.:   W70-01233
SMUS, BACILLARIOPHYCEAE, CHLORO/   *EUROPE, ALGAL GROWTH, ANKISTRODE   W69-07833
N, COPPER SULPHATE, NUISANCE AL/   *EUTRIPHICATION, *LAKES, WISCONSI   W68-00468
ULATIONS, PLANKTON, *CYANOPHYTA/   *EUTROPHICATION, SEWAGE, FISH POP   W68-00461
, AQUATIC POPULATIONS, WATER QU/   *EUTROPHICATION, MAINE, NUTRIENTS   W68-00470
VOLUMETRIC ANALYSIS, LIMNOLOGY,/   *EUTROPHICATION, ALGAE, GERMANY,   W68-00475
LABORATORY TESTS, TEST PROCEDUR/   *EUTROPHICATION, TEST ANALYSIS, *   W68-00472
RUS, PHOSPHATES, *SEWAGE TREATM/   *EUTROPHICATION, SEWAGE, *PHOSPHO   W68-00256
LAKES, *CLASSIFICATION, DESMIDS/   *EUTROPHICATION, AQUATIC ALGAE, *   W68-00255
GEN, PHOSPHORUS, PLANKTON, DISS/   *EUTROPHICATION, NUTRIENTS, NITRO   W68-00253
S, DOMESTIC WASTES, *SEWAGE TRE/   *EUTROPHICATION, *ALGAE, NUTRIENT   W68-00248
CHEMICALS, FISHING, SPORT FISH/   *EUTROPHICATION, SEQUENCE, LAKES,   W68-00172
PRODUCTIV/ SEDIMENTATION RATES,   *EUTROPHICATION, *LAKES, *PRIMARY   W69-00632
UTANTS, POLLUTION ABATEMENT, NU/   *EUTROPHICATION, *NUTRIENTS, POLL   W69-01453
SOURCES, *WATER POLLUTION EFFE/   *EUTROPHICATION, *WATER POLLUTION   W69-02959
RIENTS, ALGAE, WATER POLLUTION /   *EUTROPHICATION, *LAKE ERIE, *NUT   W69-01445
TECHNIQUES, *CYCLING NUTRIENTS,   *EUTROPHICATION, *PHOSPHATES, *PH   W69-03358
S COMPOUNDS, *LAKES, *SEDIMENTS,   *EUTROPHICATION, ALGAE, CLAYS, SI   W69-03075
R QUALITY, ALGAL CONTROL, ALGIC/   *EUTROPHICATION, SUCCESSION, WATE   W69-01977
CONTRO/ *NUTRIENT REQUIREMENTS,   *EUTROPHICATION, *WATER POLLUTION   W69-03374

355

ION EFFECTS, *CYCLING NUTRIENTS,
TECHNIQUES, *CYCLING NUTRIENTS,
YSIOLOGICAL ECOLOGY, ALGAE, ANA/
, *NUTRIENT REQUIR/ *CYANOPHYTA,
COLOGY, *WATER POL/ *CYANOPHYTA,
ALGAE, NUISANCE ALGAE, PLANKTON/
IBLIOGRAPHIES, *ALGAE, *CHEMICA/
NKTON, ALGAE, SEASONAL, CYANOPH/
, BIOLOGICAL TREATMENT, BIBLIOG/
CUT, WATER POLLUTION, *NUTRIENT/
TER BLOOM, AQUATIC ALGAE, AGING/
LANTS, BIBLIOGRAPHIES, CYANOPHY/
S, UNITED STATES, PHOSPHORUS, N/
, CHEMICAL PROPERTIES, BACTERIA/
SULFATE, *ALGAL CONTROL, *TOXIN/
ES, BIBLIOGRAPHIES, *PHOSPHORUS/
N, ALGAE, *NUISANCE ALGAE, SPRA/
S, OXYGEN REQUIREMENTS, ICED LA/
AKES, *PHOSPHORUS RADIOISOTOPE,/
MENT, DRAINAGE SYSTEMS, FLOW RA/
SOURCES, *LAKES, NUTRIENTS, FE/
ALGAE, *RESPIRATION, *ESTUARIES,
*INLAND WATE/ *ALGAE, *BIOASSAY,
, BIOLOGICAL COMMUNITIES, CULTU/
ON, URBANIZATION, DOMESTIC WAST/
QUATIC WEEDS, CYANOPHYTA, FISHK/
CHLOROP/ *PHYTOPLANKTON, *PONDS,
ISH, *LIMNOLOGY, FISHKI/ *LAKES,
MANGANESE, *WISCONSIN, ALGAE, C/
*LIMNOLOGY, *PHYTOPLANKTON, FIS/
ATIC PLA/ *AQUATIC WEED CONTROL,
OR, TOX/ *ALGAE, *WATER QUALITY,
NALYTI/ *ALGAE, *BIBLIOGRAPHIES,
L CONTROL, *ESSENTIAL NUTRIENTS,
CONTROL, *AQUATIC WEED CONTROL,
CONTROL, *AQUATIC WEED CONTROL,
NDS, *NUTRIENT/ *BIBLIOGRAPHIES,
LOR, FISHKILL, LAKES, ODOR, TOX/
, ATMOSPHERE, SE/ *GEOCHEMISTRY,
OSPHATES, CHELA/ *WATER QUALITY,
SOURCES, *WATER POLLUTION EFFE/
CTIVITY, CYANOPHYTA, NU/ *LAKES,
*ALGAE, *AQUATIC ENVIRONMENTS,
NTAL EFFECTS, *HUMAN POPULATION,
SPHORUS, *AQUATIC ALGAE, *BIOAS/
SOURCES, *WATER POLLUTION EFFE/
, TERTIARY TREATMEN/ *MINNESOTA,
QUALI/ *ALGICIDES, *CYANOPHYTA,
URCES, *WATER POLLUTION EFFECTS,
SPHORUS COMPOUNDS, ALGAE, PHOSP/
S, WATER POLLUTION EFFECTS, PRO/
NDS, *PHOSPHORUS COMPOUNDS, *NU/
TS, NITROGEN, PHOSPHORUS, SODIU/
WEEDS, LAKES,/ *COPPER SULFATE,
, RIVERS, LAKES, / *CONFERENCES,
ANIZATION, WATER POLLUT/ *LAKES,
, OXIDATION, CARBO/ *RESERVOIRS,
TIFICATION, DEPTH, PLANKTON, LI/
ICAL PROPERTIES/ *ALGAE, *LAKES,
RY, *FLUORESCENCE, *CHLOROPHYLL,
ERGENTS, DOMESTIC WASTES, POLLU/
EATION, SEWAGE DISPOSAL, WATER /
ER POLLUTION SOURCES, *PHOSPHAT/
TON, *SEWAGE, *SEWAGE TREATMENT,
CARBON RADIOISOTOPES, *BIOASSAY,
ASTES, OXYGEN, STANDING CROP, A/
IFICATION, L/ *NUTRIENT REMOVAL,
S, RESERVOIRS,/ *NUISANCE ALGAE,
STES, *ALGAE, *LAKES, *TRIBUTAR/
, POLLUT/ *ALGAE, *PRODUCTIVITY,
S, ECOLOGY, PRI/ *BENTHIC FAUNA,
CONTROL, *WATER POLLUT/ *LAKES,
*NUTRIENTS, *BIODE/ *RESERVOIRS,
*ALGAE, WATER POLLUTION CONTRO/
S, FERTILIZATION, BACTERIA, ALG/
CHLOROPHYTA, CYANOPHYTA, BRACK/
ASTES, *WATER POLLUTION EFFECTS,
OLLUTION CONTROL, *REVIEWS, ALG/

*SEWAGE, WATER POLLUTION SOURC/
*GROWTH RATES, *PHOSPHATES, LAB/
LAKES, INTE/ *CYCLING NUTRIENTS,
TMENT, *WATER POLLUTION CONTROL,
ESERVOIR, *PRIMARY PRODUCTIVITY,
*NITROGEN, *PHOSPHATES, *ALGAE,
ION EFFECTS, LAKES, CYANOPHYTA,
E, *BIODEGRADATION, *PHOSPHATES,
ENTIAL NUTRIENTS, *PRIMARY PROD/
ATIC ALGAE, *PROTOZOA, *FLORIDA,
ATMENT, WATER TREATMENT, *ALGAE,

*EUTROPHICATION, ALGAE, AQUATIC A   W69-03369
*EUTROPHICATION, *PHOSPHORUS COMP   W69-03370
*EUTROPHICATION, *CYANOPHYTA, *PH   W69-03517
*EUTROPHICATION, *PLANKTON, ALGAE   W69-03512
*EUTROPHICATION, *PHYSIOLOGICAL E   W69-03513
*EUTROPHICATION, *LAKE MICHIGAN,   W68-01244
*EUTROPHICATION, PHYTOPLANKTON, B   W68-00856
*EUTROPHICATION, *LAKES, PHYTOPLA   W68-00860
*EUTROPHICATION, LABORATORY TESTS   W68-00855
*EUTROPHICATION, *LAKES, CONNECTI   W68-00511
*EUTROPHICATION, ALGAL BLOOMS, WA   W68-00912
*EUTROPHICATION, ALGAE, AQUATIC P   W68-00680
*EUTROPHICATION, LAKES, *NUTRIENT   W68-00478
*EUTROPHICATION, PLANKTON, *ALGAE   W68-00483
*EUTROPHICATION, *LAKES, *COPPER   W68-00859
*EUTROPHICATION, *LAKES, ACID LAK   W68-00857
*EUTROPHICATION, *LAKES, WISCONSI   W68-00488
*EUTROPHICATION, *LAKES, NUTRIENT   W68-00487
*EUTROPHICATION, OLIGOTROPHIC, *L   W68-00858
*EUTROPHICATION, *POLLUTION ABATE   W68-00479
*EUTROPHICATION, *WATER POLLUTION   W69-07084
*EUTROPHICATION, *CALIFORNIA, WAT   W69-07520
*EUTROPHICATION, *FERTILIZATION,   W69-07833
*EUTROPHICATION, *ALGAE, BIOASSAY   W69-07832
*EUTROPHICATION, *LAKES, GLACIATI   W69-07818
*EUTROPHICATION, *LAKES, ALGAE, A   W69-06535
*EUTROPHICATION, ALGAE, BIOMASS,   W69-06540
*EUTROPHICATION, FERTILIZATION, F   W69-05844
*EUTROPHICATION, *IRON, *LAKES, *   W69-05867
*EUTROPHICATION, *FERTILIZATION,   W69-06858
*EUTROPHICATION, *HARVESTING, AQU   W69-06276
*EUTROPHICATION, COLOR, TASTE, OD   W69-05697
*EUTROPHICATION, *PLANT GROWTH, A   W69-04801
*EUTROPHICATION, *CLADOPHORA, *NU   W69-04798
*EUTROPHICATION, AQUATIC PLANTS,   W69-05706
*EUTROPHICATION, AQUATIC PLANTS,   W69-05705
*EUTROPHICATION, *NITROGEN COMPOU   W69-04805
*EUTROPHICATION, ALGAE, DUCKS, CO   W69-08279
*EUTROPHICATION, *WATER STRUCTURE   W70-04385
*EUTROPHICATION, *DETERGENTS, *PH   W70-04373
*EUTROPHICATION, *WATER POLLUTION   W70-03975
*EUTROPHICATION, *BIOMASS, *PRODU   W70-03959
*EUTROPHICATION, *MEROMIXIS, *WAT   W70-03974
*EUTROPHICATION, *CONFERENCES, NU   W70-03973
*EUTROPHICATION, *MUD(LAKE), *PHO   W70-03955
*EUTROPHICATION, *WATER POLLUTION   W70-03975
*EUTROPHICATION, SEWAGE EFFLUENTS   W70-03512
*EUTROPHICATION, WISCONSIN, WATER   W70-03310
*EUTROPHICATION, *NUTRIENTS, ALGA   W70-04266
*EUTROPHICATION, *NUTRIENTS, *PHO   W70-04074
*EUTROPHICATION, *LAKES, NUTRIENT   W70-04721
*EUTROPHICATION, *NITROGEN COMPOU   W70-04765
*EUTROPHICATION, *ALGAE, *NUTRIEN   W70-04503
*EUTROPHICATION, *ALGAE, *AQUATIC   W70-04494
*EUTROPHICATION, *ALGAE, PLANNING   W70-04506
*EUTROPHICATION, *WASHINGTON, URB   W70-05663
*EUTROPHICATION, NITROGEN, OXYGEN   W70-05550
*EUTROPHICATION, TEMPERATURE STRA   W70-05405
*EUTROPHICATION, *HYDROLOGY, CHEM   W70-05113
*EUTROPHICATION, PRIMARY PRODUCTI   W70-05094
*EUTROPHICATION, PHOSPHORUS, *DET   W70-05084
*EUTROPHICATION, *MINNESOTA, RECR   W70-04810
*EUTROPHICATION, *NUTRIENTS, *WAT   W70-01031
*EUTROPHICATION, PONDS, POLLUTANT   W70-02304
*EUTROPHICATION, *CALIFORNIA, *NE   W70-01933
*EUTROPHICATION, LAKES, ORGANIC W   W70-02251
*EUTROPHICATION, ANAEROBIC DENITR   W70-01981
*EUTROPHICATION, CYANOPHYTA, LAKE   W70-02803
*EUTROPHICATION, *AGRICULTURAL WA   W70-02969
*EUTROPHICATION, *WATER CHEMISTRY   W70-02646
*EUTROPHICATION, *NUTRIENTS, LAKE   W71-00665
*EUTROPHICATION, *WATER POLLUTION   W71-03048
*EUTROPHICATION, *CARBOHYDRATES,   W71-00221
*EUTROPHICATION, *LAKES, *RIVERS,   W70-05668
*EUTROPHICATION, *LAKES, NUTRIENT   W70-05761
*EUTROPHICATION, ALGAE, CULTURES,   W70-05752
*EUTROPHICATION, DISSOLVED OXYGEN   W70-06975
*EUTROPHICATION, *LAKES, *WATER P   W70-07393
*EUTROPHICATION CONTROL.:   W70-07393
*EUTROPHICATION, *NUISANCE ALGAE,   W70-07261
*EUTROPHICATION, *LAKES, *ALGAE,   W70-08637
*EUTROPHICATION, *HYDRODYNAMICS,   W70-09889
*EUTROPHICATION, *ALGAL CONTROL,   W70-09186
*EUTROPHICATION, WATER QUALITY, *   W70-09093
*EUTROPHICATION, DIATOMS, MICROBI   W71-05026
*EUTROPHICATION, WATER QUALITY.: /   W71-04758
*EUTROPHICATION, NUTRIENTS, NITRO   W71-04072
*EUTROPHICATION, *LIMNOLOGY, *ESS   W72-01990
*EUTROPHICATION, AQUATIC MICROORG   W72-01993
*EUTROPHICATION, *NITROGEN, NITRO   W72-01867

356

OLLUTION, DEGRADATION,/ *NEVADA,
L/ *ALGAE, NITROGEN, PHOSPHORUS,
GAE, *HARVESTING OF ALGAE, EFFL/
S, NITROGEN, PHOSPHOROUS, NITRO/
Y AUTHORITY PROJECT, *RESERVOIR/
*ALGAE, *PHOSPHORUS, *MINNESOTA,
HYTOPLANKT/ *WASHINGTON, *LAKES,
RIENTS, WATER T/ *LAKES, *ALGAE,
ALGAE, / *WATER, *FERTILIZATION,
ERS, NUTRIENTS/ *LAKES, *RIVERS,
ENT, N/ *PHOSPHORUS, *NUTRIENTS,
TES, NUTRIENTS, ALGAE, WATER PO/
NTATION, BIBLIOGRAPHIES, PHOSPH/
S, *NUTRIENTS, *NITROGE/ *LAKES,
, SCUM, FISHING, WISCONSIN, BIB/
ES, PONDS, FERTILIZERS, ALGAE, /
, BIOLOGICAL COMMUNITIE/ *LAKES,
TS, *CARBON DIOXIDE, *REVIEWS, /
 *ALGAE, *CYANOPHYTA, *ECOLOGY,
LGAE, NUTRIENTS, PHOS/ *ECOLOGY,
N, PENNSY/ *ALGAE, *RESPIRATION,
TRIENTS, LIMNOLO/ WATER QUALITY,
ENVIRONMENT, WATER TEMPERATURE,
ORU/ *WATER QUALITY, *NUTRIENTS,
MENTS, ALGAL CONTROL, *NUTRIENT/
MENTS, *NUTRIENTS, NITROGEN COM/
EAT PLAINS, ALGAE, NEBRASKA, TU/
*RESPIRATION, GREAT LA/ *ALGAE,
ARGE, COPPER, *ALGICIDES, ALGAE,
AR PRODUCTION, *CARBON FIXATION,
AN, MONTANE ZON/ *FIELD STUDIES,
N FIXATION, *EXCRETION, STEPHAN/
CYLINDRICA, ZINC, XANTHOPHYCEA/
ONE, GLYCOLLIC ACID, CHEMOTROPH/
, CHONDRUS C/ *ALGAL SUBSTANCES,
FFECTS, *DETENTION TIME EFFECTS,
MODEL, COD REMOVAL, MILK SUBSTR/
S, TR/ *WATER POLLUTION EFFECTS,
WATER TEMPERATURE, SEWAGE TREAT/
LGAE, BIOCHEMICAL/ *FARM WASTES,
, */ *ANIMAL WASTES, MANAGEMENT,
DEGRADATION, ALGAE, BIOCHEMICAL/
ON, *ORGANIC WASTES, *WASTE WAT/
DATION, *SEWAGE BACTERIA, ODOR,/
RAINAGE EFFECTS, *WATER QUALITY,
OPMENT, GRANTS, ALGAE, AIR POLL/
ER TREATMENT, SEDIMENTATION, OX/
TES, EUTROPHIC/ *NITROGEN CYCLE,
UDIES, *NITROGEN LIMITATION, BA/
RL LAKES, CHLOROBIUM, FILINIA, /
APHNIA, PRODUCTIVITY, BIOINDICA/
SITY, EUTROPHICATION, NUTRIENTS,
LAKES, STREAMS, ALGAE, / *WATER,
GRICULTURAL WATERSHEDS, *RUNOFF,
OPLANKTON, FIS/ *EUTROPHICATION,
GAE, *BIOASSAY, *EUTROPHICATION,
PLANKTON, AQUATIC WEED/ *PONDS,
OXYGEN REQUIREMENTS, ICED LAKES,
 EUTROPHICATION, ALGAE, PLANT G/
UTRIENTS, WASTE WATER TREATMENT,
NOLOGY, CLADOCERAN, MONTANE ZON/
VERY, OIL SPILL, *OIL POLLUTION,
ITH/ *LEUCOTHRIX MUCOR, *GROWTH,
D, FERRIC OXIDE-COATED SAND, CR/

PARATIVE STUDIES, DAPHNIA SCHOD/
INTAKES, SANDS, FLOW R/ *WATER,
ILTRATION, SANITARY ENGINEERING,
, *ALGAE, *COLLOIDS, FLOW RATES,
ATER / *WATER TREATMENT, *ALGAE,
ATION, *WATER POLLUTION CONTROL,
TRO/ *WATER POLLUTION TREATMENT,
YS, */ *WATER POLLUTION CONTROL,
DEMAND, *ACT/ *SEWAGE TREATMENT,
ESIS, CHLORELLA/ *ALGAE, *SANDS,
 *ULTRASONICS, *SUSPENDED LOAD,
*ION EXCHANGE, *ORGANIC MATTER,
, *FILTERS, */ *WATER TREATMENT,
*SCEN/ *WATER TREATMENT, *ALGAE,
TY, *ACTIVATE/ *WATER TREATMENT,
TREATMENT, SANITARY ENGINEERING,
 SEPARATION TECHNIQUES,/ *ALGAE,
ES, TURBID/ *DIATOMACEOUS EARTH,

, *ALGAL TOXINS, *PHYTOTOXICITY,
CYTES, *MUSCULAR EXERTION, *2-M/
ITY, *ALGAE, *PESTICIDES, *FISH,
ALG/ *NUTRIENTS, *LAKES, *TROUT,
L POLLUTION, *TOXICITY.:

*EUTROPHICATION, *LAKES, *WATER P   W72-02817
*EUTROPHICATION, PHOTOSYNTHESIS,   W72-00917
*EUTROPHICATION, *PHOSPHOROUS, AL   W71-13313
*EUTROPHICATION, *ALGAE, NUTRIENT   W71-13335
*EUTROPHICATION, *TENNESSEE VALLE   W71-07698
*EUTROPHICATION, POLLUTION ABATEM   W69-10167
*EUTROPHICATION, SEWAGE, BIOTA, P   W69-10169
*EUTROPHICATION, OLIGOTROPHY, NUT   W70-00264
*EUTROPHICATION, LAKES, STREAMS,   W69-10178
*EUTROPHICATION, RUNOFF, FERTILIZ   W69-10170
*EUTROPHICATION, *TERTIARY TREATM   W69-09340
*EUTROPHICATION, *LAKES, *PHOSPHA   W69-09135
*EUTROPHICATION, *REVIEWS, DOCUME   W69-08518
*EUTROPHICATION, *SEWAGE EFFLUENT   W69-09349
*EUTROPHICATION, CYANOPHYTA, ODOR   W69-08283
*EUTROPHICATION, *NUTRIENTS, *LAK   W69-08668
*EUTROPHICATION, *CANALS, *CITIES   W70-00268
*EUTROPHICATION, *ALGAE, *NUTRIEN   W70-00664
*EUTROPHICATION, PONDS, RESERVOIR   W70-00274
*EUTROPHICATION, *GREAT LAKES, *A   W70-00667
*EUTROPHICATION, *DISSOLVED OXYGE   W69-10159
*EUTROPHICATION, ALGAL BLOOMS, NU   W69-09723
*EUTROPHICATION, AQUATIC LIFE, SE   W73-01446
*EUTROPHICATION, NITROGEN, PHOSPH   W73-01122
*EUTROPHICATION, LAKES, LAKE SEDI   W72-06052
*EUTROPHICATION, LAKE, *LAKE SEDI   W72-06051
*EUTROPHICATION, *RESERVOIRS, *GR   W72-04761
*EUTROPHICATION, *PHOTOSYNTHESIS,   W72-05453
*EVALUATION, FILTRATION.: /L RECH   W72-05143
*EXCRETION, STEPHANODISCUS TENUIS   W69-10158
*EXPERIMENTAL LIMNOLOGY, CLADOCER   W69-10154
*EXTRACELLULAR PRODUCTION, *CARBO   W69-10158
*EXTRACELLULAR PRODUCTS, ANABAENA   W69-10180
*EXTRACELLULAR PRODUCTS, PHOTIC Z   W70-02504
*EXTRACELLULAR, FUCUS VESICULOSUS   W70-01073
*FACULATIVE PONDS, *AEROBIC PONDS   W70-04787
*FACULTATIVE LAGOONS, LABORATORY   W70-07031
*FALLOUT, *ECOSYSTEMS, GREAT LAKE   W71-09013
*FARM LAGOONS, *DESIGN CRITERIA,   W71-08221
*FARM LAGOONS, *BIODEGRADATION, A   W71-00936
*FARM MANAGEMENT, *DAIRY INDUSTRY   W70-07491
*FARM WASTES, *FARM LAGOONS, *BIO   W71-00936
*FARM WASTES, *SPRINKLER IRRIGATI   W71-00940
*FARM WASTES, *LAGOONS, *BIODEGRA   W71-02009
*FARM WASTES, *INDUSTRIAL WASTES,   W71-09577
*FARM WASTES, *RESEARCH AND DEVEL   W71-06821
*FARM WASTES, *POULTRY, WASTE WAT   W71-11375
*FARM WASTE, GROUND WATER, *NITRA   W69-05323
*FAT ACCUMULATION, *LABORATORY ST   W70-07957
*FAYETTEVILLE GREEN LAKE(NY), *MA   W70-05760
*FEEDING RATES, *GROWTH RATES, *D   W69-10152
*FERMENTATION, TOXICITY, CHEMICAL   W71-13413
*FERTILIZATION, *EUTROPHICATION,   W69-10178
*FERTILIZATION, LAKES, SEWAGE TRE   W70-02787
*FERTILIZATION, *LIMNOLOGY, *PHYT   W69-06858
*FERTILIZATION, *INLAND WATERWAYS   W69-07833
*FERTILIZATION, *PHOSPHATE, FISH,   W70-05417
*FERTILIZATION, WATER FLEAS, LARV   W68-00487
*FERTILIZERS, *CYCLING NUTRIENTS,   W69-07826
*FIBER(PLANT).: /OXYGEN DEMAND, N   W70-06925
*FIELD STUDIES, *EXPERIMENTAL LIM   W69-10154
*FILAMENTOUS GREEN ALGAE, HALIMON   W70-09976
*FILAMENTS, *AUTORADIOGRAPHY, ANT   W69-09755
*FILTER MEDIA, ALUMINA-COATED SAN   W70-08628
*FILTER MEDIA.:   W72-07334
*FILTERING RATE, *BODY SIZE, *COM   W70-03957
*FILTERS, *ALGAE, NUISANCE ALGAE,   W70-04199
*FILTERS, *DESIGN, COAGULATION, A   W72-07334
*FILTERS, *NEW YORK, TREATMENT FA   W72-04145
*FILTERS, SANDS, LAKE MICHIGAN, W   W73-02426
*FILTERS, SEWAGE EFFLUENTS, ELECT   W70-01027
*FILTRATION, *WATER POLLUTION CON   W70-01027
*FILTRATION, *PARTICLE SIZE, *CLA   W70-08628
*FILTRATION, *BIOCHEMICAL OXYGEN   W70-05821
*FILTRATION, FILTERS, ELECTROPHOR   W71-03035
*FILTRATION, ALGAE, COLLOIDS, FLO   W71-10654
*FILTRATION, DIATOMACEOUS EARTH,   W72-04318
*FILTRATION, SANITARY ENGINEERING   W72-07334
*FILTRATION, HEAD LOSS, FILTERS,   W72-03703
*FILTRATION, *ALGAE, *WATER QUALI   W72-09693
*FILTRATION, DESIGN, *ALGAE, COPP   W72-08859
*FILTRATION, *DIATOMACEOUS EARTH,   W72-08688
*FILTRATION, *SEPARATION TECHNIQU   W72-08686
*FINISHING WASTES.:   W71-06737
*FINISHING WASTES.:   W70-01519
*FISH DISEASES, *FISH TOXINS, TOX   W71-07731
*FISH HEMATOLOGY, *GRANULAR LEUCO   W70-01788
*FISH KILLS, *ROTIFERS, *ECOLOGY,   W72-04506
*FISH MANAGEMENT, FERTILIZATION,   W70-05399
*FISH PHYSIOLOGY, *ALGAE, *THERMA   W70-01788

QUATIC POPULATIONS, OLIGOTROPHY,
*PHYTOTOXICITY, *FISH DISEASES,
*BIOLOGICAL LIFE, *ALGAE BLOOMS,
L EFFECTS, *ESSENTIAL NUTRIENTS,
ER QUALITY, *ALGAE, *PESTICIDES,
REAMS, *WATER POLLUTION EFFECTS,
TS, *ESSENTIAL NUTRIENTS, *FISH,
IDGES, CLEAR LAKE(CALIF), NUTRI/
ING ORGANIC MATTER, TOXICITY, D/
ING, *ALGAL TOXINS, *PHYTOTOXIC/
GHT TESTS, */ *BEER-LAMBERT LAW,
ERT LAW, *FIXED-DENSITY STUDIES,
MENT, ELECTROLYTES, CHE/ *ALGAE,
LATION, ALGAE, ACTIVATED CARBON,
C MI/ *AQUATIC ALGAE, *PROTOZOA,
TON, *DIATOMS, *BLUE-GREEN ALGA/
IVATED SLU/ *WATER PURIFICATION,
*SLUDGE, STORMS, T/ *ADDITIVES,
S, *ESTUARINE ENVIRONMENT, BAYS,
OLAR RADI/ *WASHINGTON, *OXYGEN,
RELATION ANALYSIS, *FLUOROMETRY,
OROPHYLL/ *CORRELATION ANALYSIS,
TATION, *SAMPLING, *MEASUREMENT,
TS, / *ALGAL GROWTH, *BIOASSAYS,
SMS, NUTRIENTS, ENERGY TRANSFER,
DE TOXICITY, *PESTICIDE REMOVAL,
Y ST/ *SCULPINS, *RADIOACTIVITY,
HODAMINE WT DYE, AIRBORNE TESTS/
-SITE INVESTIGATIONS, FREQUENCY,
IRON, IONS, PHOSPHATES,/ *ALGAE,
CEANS, *MARINE ALGAE, *SEAWATER,
SCENEDESMUS, LIGH/ *ENVIRONMENT,
OMPOUNDS, *NITROGEN UTILIZATION,
TER FLEAS, LARVAE, GROWTH RATES,
ER RATES, */ *GLUCOSE, *ACETATE,
UM, *ALGAL-BOUND RADIOSTRONTIUM,
, WATER PURIFICATION, *PROTOZOA,
STRONTIUM RADIOISOTOPES, *ALGAE,
, OXIDATION DITCH.:
NYON, AROMATIC SOLV/ *CRUDE OIL,
OUTHAMPTON UNIV/ *PHAEODACTYLUM,
BPOLAR WATERS, TYNSET(NORWAY), /
WATER STRUCTURE, ATMOSPHERE, SE/
OGAALKALOID, O/ *PERIDINIUM SP.,
UM, PLANKTO/ LAKE SAGAMI, JAPAN,

NVIRONMENTS, *TURNOVER RATES, */
TION, *2-METHYL-5-VINYLPYRIDINE,
XERTION, *2-M/ *FISH HEMATOLOGY,
ION, IRRIGATION, ORGANIC MATTER,
PHYSIOLOGICAL ECOLOGY, LAKES, /
DETERGENTS, *TERTIARY TREATMENT,
TRIAL WASTES, *MUNICIPAL WASTES,
WATER POLLUTION CONTROL, WATER /
ICHIGAN, *LAKE ONTARI/ *BIOLOGY,
PHOS/ *ECOLOGY, *EUTROPHICATION,
ATER POLLUTION / *WATER QUALITY,
U/ *EUTROPHICATION, *RESERVOIRS,
ATURE TOLERA/ *BLUE-GREEN ALGAE,
CHLORELLA PYRENOIDOSA, VOLATIL/
, ORGANIC MATTER, CURR/ *RUNOFF,
ENT, *WATER REUSE, *RECLAMATION,
LLUTION EFFECTS, *WATER SOURCES,
NIZATION, *ARTIFICIAL SUBSTRATE/
*ANALYTICAL TECHNIQUES, *ALGAE,
NOTRIAZOLE, 2-4-5-/ *HERBICIDES,
RELA, / *ALGAE, *PHOTOSYNTHESIS,
ESIS, EUTROPHICATION, LIGHT, TE/
URES, NUTRIENTS, RIVERS, LAKES,
*EUTROPHICATION, *LAKES, *ALGAE,
, *PHYTOPLANKTON, *LAKE ONTARIO,
VITY, BIOINDICA/ *FEEDING RATES,
NE, ALGIC/ *BIOASSAY, *HALOGENS,
APHY, ANTITH/ *LEUCOTHRIX MUCOR,
S, CHLORELLA, CULTURES.:
COUNT, OPTICAL DENSITY, PARTICU/
LGAE, E/ *MARINE MICROORGANISMS,
, *LI/ *ALGAE, *ACID MINE WATER,
STUDIES, CHLORINE, COPPER SULF/
ALGAL CONTROL, CHLORELLA, DISIN/
LLA, CHLORINE, ALGIC/ *BIOASSAY,
CLIDES, ASSIMILATION, BA-LA, HA/
N EFFECTS, *ALGAE.: *ESTUARIES,
TRANSPARENCY, MARGINAL VEGETAT/
OPHICATION, *PHOSPHOROUS, ALGAE,
C WEED CONTROL, *EUTROPHICATION,
ACTIVATED CARBON, *FLOCCULATION,
E, 2-4-D, AMINOTRIAZOLE, 2-4-5-/
TOTOXICITY, ALGAL POISO/ *2-4 D,
DIUM ARS/ *AQUATIC WEED CONTROL,

*FISH POPULATIONS, PLANT POPULATI   W68-00172
*FISH TOXINS, TOXICITY, HYDROGEN   W71-07731
*FISH TOXICITY TEST, SLUDGE DRYIN   W70-05819
*FISH, *FISHERIES, ALGAE, AQUATIC   W69-04804
*FISH, *FISH KILLS, *ROTIFERS, *E   W72-04506
*FISH, ALGAE, TROPHIC LEVEL, PROD   W72-01788
*FISHERIES, ALGAE, AQUATIC ALGAE,   W69-04804
*FISHERY MANAGEMENT, *CHAOBORID M   W70-05428
*FISHKILL, *CYANOPHYTA, *DECOMPOS   W72-01797
*FISHKILL, *TOXINS, *ALGAL POISON   W71-07731
*FIXED-DENSITY STUDIES, *FIXED-LI   W70-03923
*FIXED-LIGHT TESTS, *EFFECTIVE-AV   W70-03923
*FLOCCULATION, *WASTE WATER TREAT   W70-05267
*FLOCCULATION, *HEAD LOSS.: /OAGU   W72-07334
*FLORIDA, *EUTROPHICATION, AQUATI   W72-01993
*FLOTATION, *SINKING, *PHYTOPLANK   W70-02754
*FLOW AUGMENTATION, *LAGOONS, ACT   W72-06017
*FLOW RATE, *WASTE WATER, *ALGAE,   W70-05819
*FLUCTUATION, RESISTANCE, LIFE CY   W71-02494
*FLUCTUATION, SEWAGE TREATMENT, S   W69-10182
*FLUORESCENCE, *CHLOROPHYLL, *EUT   W70-05094
*FLUOROMETRY, *FLUORESCENCE, *CHL   W70-05094
*FLUOROMETRY, DYE RELEASES, CURRE   W70-05377
*FLUOROMETRY, CALIFORNIA, NUTRIEN   W70-02777
*FOOD CHAINS, FOOD WEBS, FORAGE F   W70-05428
*FOOD WEBS, DDT, CHLORINATED HYDR   W70-03520
*FOOD WEBS, KENTUCKY, LIFE HISTOR   W69-09742
*FRAUNHOFER LINE DISCRIMINATOR, R   W70-05377
*FREQUENCY ANALYSIS.: /TIONS, *ON   W70-09976
*FRESH WATER, *PEPTIDES, COPPER,.   W69-10180
*FRESH WATER, *HUMIC ACIDS, *COLO   W70-01074
*FRESH WATER, *PLANKTON, *ALGAE,   W70-05409
*FRESH WATER, ALGAE, NITRATE AMMO   W69-04521
*FRESH WATER FISH, MICHIGAN, TROU   W68-00487
*FRESHWATER ENVIRONMENTS, *TURNOV   W70-05109
*FUNGAL-BOUND RADIOSTRONTIUM, *MI   W71-04067
*FUNGI, *BACTERIA, HYDROGEN ION C   W71-05267
*FUNGI, *CALCAREOUS SOIL, SOIL WA   W71-04067
*FWPCA, ACTIVATED ALGAE, FEEDLOTS   W71-06821
*GAMOSOL, *OIL SPILLS, *TORREY CA   W70-01470
*GAS EFFLUENT, FOOD PRODUCTION, S   W70-00161
*GELBSTOFF, *ALGAL SUBSTANCES, SU   W70-01074
*GEOCHEMISTRY, *EUTROPHICATION, *   W70-04385
*GLENODININE, ALKALOID TOXINS, AL   W71-07731
*GLENODININE, *PERIDINIUM POLONIC   W70-08372
*GLUCOSE, WATERLOO MILLS(PA).:   W71-05026
*GLUCOSE, *ACETATE, *FRESHWATER E   W70-05109
*GLUE GREEN ALGAE, BREAM, WHITE B   W70-01788
*GRANULAR LEUCOCYTES, *MUSCULAR E   W70-01788
*GRASSLANDS, MOUNTAINS, *SOIL MIC   W72-08110
*GRAZING, *TEMPERATURE, *DAPHNIA,   W70-03957
*GREAT LAKES, CONTROL, COSTS, EUT   W70-03964
*GREAT LAKES, *WATER POLLUTION, S   W70-04430
*GREAT LAKES, *WATER POLLUTION, *   W69-03948
*GREAT LAKES, *LAKE ERIE, *LAKE M   W70-01943
*GREAT LAKES, *ALGAE, NUTRIENTS,   W70-00667
*GREAT LAKES, *TRACE ELEMENTS, *W   W69-08562
*GREAT PLAINS, ALGAE, NEBRASKA, T   W72-04761
*GREEN ALGAE, *RED ALGAE, *TEMPER   W71-02496
*GREEN ALGAE, CHLORELLA VULGARIS,   W70-04195
*GROUNDWATER, *SESTON, *SEDIMENTS   W70-03501
*GROUNDWATER, RECHARGE, *ION EXCH   W72-04318
*GROUNDWATER, DDT, MISSISSIPPI RI   W73-00758
*GROWTH DYNAMICS, *BACTERIAL COLO   W70-04371
*GROWTH RATES, *BIOINDICATORS, *W   W70-04381
*GROWTH RATES, *ALGAE, 2-4-D, AMI   W70-03519
*GROWTH RATES, *TURBULENCE, CHLOR   W69-03730
*GROWTH RATES, *ALGAE, PHOTOSYNTH   W69-07442
*GROWTH RATES, *WATER TEMPERATURE   W68-00472
*GROWTH RATES, *PHOSPHATES, LABOR   W70-08637
*GROWTH RATES, ENVIRONMENTAL EFFE   W69-10158
*GROWTH RATES, *DAPHNIA, PRODUCTI   W69-10152
*GROWTH RATES, *CHLORELLA, CHLORI   W70-02370
*GROWTH, *FILAMENTS, *AUTORADIOGR   W69-09755
*GROWTH, *ALGAE, LIPIDS, NUTRIENT   W70-07957
*GROWTH, *UPTAKE, FIXATION, CELL   W70-03983
*HABITATS, EPIPHYTOLOGY, MARINE A   W69-10161
*HABITATS, *ENVIRONMENTAL EFFECTS   W70-02770
*HALOGENS, ALGICIDES, CYTOLOGICAL   W70-02364
*HALOGENS, *ALGICIDES, CHLORINE,   W70-02363
*HALOGENS, *GROWTH RATES, *CHLORE   W70-02370
*HANFORD(WASH), *REACTOR, RADIONU   W69-07853
*HARBORS, *SEWAGE, WATER POLLUTIO   W71-05500
*HARTBEESPOORT DAM(SOUTH AFRICA),   W70-06975
*HARVESTING OF ALGAE, EFFLUENT, Q   W71-13313
*HARVESTING, AQUATIC PLANTS, AQUA   W69-06276
*HEAD LOSS.: /OAGULATION, ALGAE,   W72-07334
*HERBICIDES, *GROWTH RATES, *ALGA   W70-03519
*HERBICIDES, *PHYTOPLANKTON, *PHY   W69-00994
*HERBICIDES, *COPPER SULFATE, *SO   W70-05568

ENEDESMUS, CHLORELLA, DAPHNIA, /
C WEED CONTROL, *AQUATIC PLANTS,
ENEDESMUS, CHLORELLA, DAPHNIA, /
MANGANESE, ATLANTIC OCE/ *ALGAE,
, MICROCYSTIS AERUGINOSA, APHAN/
*MARINE ENVIRONMENTS, MICROBIAL/
R ENVIRONMENTS, *TURNOVER RATES,
, *MICROBIAL ECOLOGY, MAXIMUM U/
SWEDEN, LAKE/ *ORGANIC SOLUTES,
-14, RADIOCARBON UPTAKE TECHNIQ/
OADING, WASTE WATER TREATMENT, /
WATER, AC/ *ALGAE, *TEMPERATURE,
*ALGAE, *ENVIRONMENTAL EFFECTS,
ALGAE, *SEAWATER, *FRESH WATER,
ECOLOGY, *BENTHIC FLORA, *ALGAE,
BIC WASTE PONDS, *DEPTH EFFECTS,
.:

CUS BRAUNII, *ANACYSTIS MONTANA,
LING NUTRIENTS, *EUTROPHICATION,
RIENTS, *TERTIARY TREAT/ *ALGAE,
NG, *SULFATES, *SULFIDES, ALGAE/
AE, ORGANIC L/ *OXIDATION PONDS,
FLUENTS, CARBON DIOXIDE, *ALGAE,
*ALGAE, *LAKES, *EUTROPHICATION,
MANY), CHIRONOMIDS, CLADOCERA, /
NKTON, TEMPERATURE, *EPILIMNION,
LEI/ *YELLOWSTONE NATIONAL PARK,
, PHORMIDIUM/ *YELLOWSTONE PARK,
LGAL KEY, CLUMP COUNT, DESMIDS,/
, *DISSOLVED SOLIDS, D/ EFFECTS,
ER QUALITY, SOIL INVESTIGATIONS,
ERMAL STRATIFICATION, *AERATION,
NTS, *BIOASSAY, *RUNOFF, *LAKES,
ICELASE, HYDROG/ *DIGESTIBILITY,
IA), RAJGHAT(INDIA), BOMBAY(IND/
UDY.:
OONS, *ALGAE, *SEWAGE TREATMENT,
ENT, *ORGANIC MATTER, WATER QUA/
S, FRESH WATER BIOLOGY, SUSPEND/
R REUSE, *WASTE WATER TREATMENT,
REATMENT, *WASTE WATER DISPOSAL,
S, *WATER QUALITY, *FARM WASTES,
MENTATION, ORGANIC L/ *POTATOES,
COTS, PEARS, CITRUS/ *CANNERIES,
ION EF/ *RESERVOIRS, *LIMNOLOGY,
STES, *GREAT LAKES, *WA/ *ALGAE,
*MICROBIOLOGY, *SEWAGE LAGOONS,
UNDWATER BASINS, WATER SUPPLY, /
OXINS, *AQUARIA, *AQUATIC WEEDS,
HYTOPLANKTON, / *AQUATIC PLANTS,
*EUTROPHICATION, *FERTILIZATION,
TICAL SAMPLING, LAKE MENDOTA(WI/
ON, ALGAE, SEASONAL, CYANOPHYTA,
LAKE, PECORS LAKE, DUNLOP LAKE,
CONTROL, PREDATION, COMPETITION,
*/ *BIODEGRADATION, *DETERGENTS,
POLLUTION, FOOD CH/ *DETERGENTS,
CTIVITY TECH/ *ALGAE, *TOXICITY,
ION(DECOMPOSITION), DDT/ *ALGAE,
SUREMENT, *FLUOROMETRY, DYE REL/
TUARIES, *ESTUARINE ENVIRONMENT,
TAN AREAS, BLONDIN/ *DEVOLUTION,
ISCHARGES, FLORA, FAUNA, *BIOLO/
E, *ALGAL CONTROL, WEED CONTROL,
                         *BROMINE,
OSSES, HYPOLIMNION, THERMOCLINE,
AMATION, *GROUNDWATER, RECHARGE,
NSIN, ALGAE, C/ *EUTROPHICATION,
T TESTS, *NITROGEN, *PHOSPHORUS,
,/ *WATER TREATMENT, *TURBIDITY,
HROMATOGRAPHY, *COBALT, *COPPER,
MENT, EUTHROPHICATIO/ *NITROGEN,
OPLANKTON, *BIOMASS, *SUCESSION,
*ALGAE, LETHAL LIMIT, ELECTROP/
L/ *ALGAE, *TOXICITY, *CULTURES,
NT REQUIREMENTS, CULTURES, LAKE/
S, *PLANKTON, *LABORATORY TESTS,
*THERMAL GRADIENT, *BOD REMOVAL,
XPERIMENTAL STUDIES, LAGOON DES/
STRAINERS.:   *SCREENING,
OTO REACTIVITY, ALGAL METABOLIS/
NNETONKA(MINN), SOIL MINERALS, /
UTION, *RECOVERY FROM POLLUTION,
*ALGAE, *BIOLOGICAL COMMUNITIES,
MITATION, BA/ *FAT ACCUMULATION,
*PRIMARY PRODUCTIVITY, *STREAMS,
ES, DIATOMS, CYANOPHYTA, BIOASS/
DITIONS, LIGHT INTENSIT/ *ALGAE,
NALYTICAL TECHNIQUES, *PLANKTON,

*HERBICIDES, *ALGAE, TOXICITY, SC   W71-05719
*HERBICIDES, *VASCULAR TISSUES, *   W70-10175
*HERBICIDES, *ALGAE, TOXICITY, SC   W71-03056
*HERRINGS, COPPER, IRON, COBALT,   W69-08525
*HETEROCYSTS, ANABAENA VARIABILIS   W70-05091
*HETEROTROPHY, *ORGANIC SOLUTES,   W70-04811
*HETEROTROPHY, ORGANIC SOLUTES, S   W70-05109
*HETEROTROPHY, *BIOLOGICAL UPTAKE   W70-04194
*HETEROTROPHY, MICROBIAL ECOLOGY,   W70-04284
*HETEROTROPHY, AUTOTROPHY, CARBON   W70-04287
*HOGS, *LAGOONS, ALGAE, ORGANIC L   W70-07508
*HOT SPRINGS, HABITATS, ALKALINE   W69-10160
*HUMAN POPULATION, *EUTROPHICATIO   W70-03973
*HUMIC ACIDS, *COLOR, RIVERS, BOG   W70-01074
*HYDRAULIC MODELS, LABORATORY TES   W71-00668
*HYDRAULIC LOAD EFFECTS, *ORGANIC   W70-04787
*HYDROCARBON, *TOXICITY THRESHOLD   W70-01466
*HYDROCARBON.:   W70-01467
*HYDROCARBONS, *OLEFINS, GREEN RI   W69-08284
*HYDRODYNAMICS, LAKES, INTERNAL W   W70-09889
*HYDROGEN ION CONCENTRATION, *NUT   W70-08980
*HYDROGEN SULFIDE, *ORGANIC LOADI   W70-07034
*HYDROGEN SULFIDE, *SULFIDES, ALG   W71-11377
*HYDROGEN ION CONCENTRATION, PHOS   W68-00855
*HYDROLOGY, CHEMICAL PROPERTIES,   W70-05113
*HYPERTROPHIC WATERS, HAMBURG(GER   W70-00268
*HYPOLIMNION, ALGAE, FUNGI, BACTE   W70-04475
*ICELAND HOT SPRINGS, SINTER, NUC   W69-10160
*ICELAND, MASTIGOCLADUS LAMINOSUS   W70-05270
*IDENTIFICATION, *SIGNIFICANCE, A   W70-05389
*IMPOUNDED WATERS, *WATER QUALITY   W72-01773
*IMPOUNDED WATERS, IMPOUNDMENTS,   W72-11906
*IMPOUNDMENTS, *RESERVOIRS, *MIXI   W70-04484
*IMPOUNDMENTS, TERTIARY TREATMENT   W70-04765
*IN VITRO, IN VIVO, BALL-MILL, ME   W70-04184
*INDIA, *BLOOMS, RIVER GANGES(IND   W70-05404
*INDIA, MOOSI RIVER ECOLOGICAL ST   W69-07096
*INDIANA, CHLOROPHYTA, CYANOPHYTA   W70-10173
*INDICATORS, *PROTOZOA, *ENVIRONM   W71-03031
*INDICATOR SPECIES, MICROORGANISM   W71-10786
*INDUSTRIAL WASTES, OXIDATION PON   W72-00027
*INDUSTRIAL WASTES, *WATER POLLUT   W72-01786
*INDUSTRIAL WASTES, MUNICIPAL WAS   W71-09577
*INDUSTRIAL WASTES, SCREENS, SEDI   W71-08970
*INDUSTRIAL WASTES, PEACHES, APRI   W71-09548
*INDUSTRIAL WASTES, *WATER POLLUT   W70-06975
*INDUSTRIAL WASTES, *MUNICIPAL WA   W70-04430
*INDUSTRIAL WASTES, SEWAGE TREATM   W70-03312
*INFILTRATION, *PONDS, *SOIL, GRO   W69-07838
*INHIBITION, *PHYSIOLOGICAL ECOLO   W69-03188
*INHIBITORS, *MARSHES, *LAKES, *P   W72-00845
*INLAND WATERWAYS, *WATER POLLUTI   W69-07833
*INLAND, HORIZONTAL SAMPLING, VER   W70-02969
*INORGANIC COMPOUNDS, PHOSPHORUS,   W68-00860
*INORGANIC CARBON.: /AKES, QUIRKE   W70-05424
*INSECT CONTROL.: /ALGAE, *ALGAL   W70-05428
*INSECTICIDES, *MARINE BACTERIA,   W70-01467
*INSECTICIDES, *TOXICITY, *WATER   W70-01466
*INSECTICIDES, METABOLISM, RADIOA   W70-02198
*INSECTICIDES, TOXICITY, DEGRADAT   W70-08097
*INSTRUMENTATION, *SAMPLING, *MEA   W70-05377
*INSTRUMENTATION, *POLLUTION IDEN   W69-06203
*INTERSTATE POLLUTION, *METROPOLI   W70-04430
*INTERTIDAL COMMUNITIES, *WASTE D   W68-00010
*INVESTIGATIONS, ZOO-PLANKTON, SP   W68-00468
*IODINE.:   W69-08562
*ION EXCHANGE, WATER CIRCULATION,   W68-00857
*ION EXCHANGE, *ORGANIC MATTER, *   W72-04318
*IRON, *LAKES, *MANGANESE, *WISCO   W69-05867
*IRON, *LIMNOLOGY, *ALGAE.: /HMEN   W69-06004
*IRON, *MANGANESE, *COLOR, *ALGAE   W72-04145
*IRON, TRACE ELEMENTS, ALGAE, SEA   W71-11687
*IRRADIATION, *ALGAE, WASTE TREAT   W73-02218
*ISOLATION, DIATOMS, ECOLOGY, SAL   W73-01446
*ISOLATION, *BACTERIA, *PROTOZOA,   W71-03775
*ISOLATION, LIMNOLOGY, PUBLIC HEA   W70-00273
*ISOLATION, *CYANOPHYTES, *NUTRIE   W70-00719
*ISOLATION, *CONTROL, METHODOLOGY   W70-04469
*ISRAEL, *ALGAL CONCENTRATION, 'E   W70-04790
*JAPAN, *SLAUGHTERHOUSE WASTES, E   W70-07508
*KENOSHA(WIS), *STRAINING, *MICRO   W73-02426
*KILL RATES, *BROMINE, IODINE, PH   W70-02363
*KILLING, ANIMAL MANURES, LAKE MI   W70-04193
*KUWAIT CRUDE, BENTLASS SALT MARS   W70-01231
*LABORATORIES, *STREAMS, LIGHT IN   W70-02780
*LABORATORY STUDIES, *NITROGEN LI   W70-07957
*LABORATORY TESTS, RESPIRATION, E   W70-00265
*LABORATORY TESTS, *ALGAE, CULTUR   W71-06002
*LABORATORY TESTS, *ANAEROBIC CON   W71-13413
*LABORATORY TESTS, *ISOLATION, *C   W70-04469

E, *PILOT-PLANT/ *ALGAL CULTURE,
*EUTROPHICATION, TEST ANALYSIS,
ALYSIS, LIMNOLOGY, GROWTH RATES,
WATER TREATMENT, INDUSTRIAL WA/
WASTE WATER TREATMENT, / *HOGS,
E BACTERIA, ODOR,/ *FARM WASTES,
OBIC CONDITIONS, *EFFLICIENCIES,
CHEMICAL OXY/ *AERATION, *ALGAE,
WATER TREATMENT, INDUSTRIAL WA/
URIFICATION, *FLOW AUGMENTATION,
RAL EUTROPHY, *NUTRIENT BALANCE,
TH STIMUL/ *SCIRPUS FLUVIATILIS,
ATION, / *LAKE WASHINGTON(WASH),
ATA COLLECTIONS, *ALGAE CONTROL,
ONTARI/ *BIOLOGY, *GREAT LAKES,
TER POLLUTION / *EUTROPHICATION,
ROL, *POLLUTION ABATEMEN/ *OHIO,
TA, UNITED STATES, HABI/ *ALGAE,
TRIENTS, *BENEFIT-COST ANALYSIS,

RTHER(AUSTRIA), VERTICAL, CHALK/
GANIC MATTER, LAS VEGAS, NEVADA,
, HORIZONTAL, BLOOMS, GELATINOU/
OLOGY, *GREAT LAKES, *LAKE ERIE,
TAKES, NUTRIENTS, QUALITY CONTR/
N, NUISANCE ALG/ *PHYTOPLANKTON,
LGAE, PLANKTON/ *EUTROPHICATION,
GEMENT PROGRAM, ALGAL DENSITY, /
LANKTON, *ALGAE, *PHYTOPLANKTON,
KES, *LAKE ERIE, *LAKE MICHIGAN,
ATION, *WATER POLLUTION CONTROL,
ONITORING, BACTERIA, WATER CHEM/
NDUSTRIAL PLANTS, WATER SPORTS./
OGEN COM/ *EUTROPHICATION, LAKE,
NOLOGY, *WATER QUALITY, BIOINDI/
NCY, LIGHT MEASUREMENTS, EUPHOT/
ES, LIVERWORTS, MORPHOMETRY, PA/
NGES, *NUTRIENT CONCENTRATION, /
EFICIT, SECCHI DISC TRANSPARENC/
RIA RUBESCENS, SWITZERLAND, PER/
SION, *WATER QUALITY MANAGEMENT,
INFLUENCES, MICROCYSTIS, DREDG/
ASH), REVENUE BONDS, METRO ACT(/
A, GYTTJA, OSCILLATORIA RUBESCE/
        *MICROCYSTIS AERUGINOSA,
OPHYTA, FRESH WATER FISH, LAKES,
*NITROGEN-FIXATION, *NUTRIENTS,
HOSPHATES, LAB/ *EUTROPHICATION,
OLIGOTROPHY, NUTRIENTS, WATER T/
LEVELS, *SURVEYS, BIOLOGICAL CO/
CHNIQUES, WATER POLLUTION EFFEC/
ONTROL, *TOXIN/ *EUTROPHICATION,
BACTERIA, ALGAE, PHYTOPLANKTON,
*EUTROPHICATION, AQUATIC ALGAE,
MENT-WATER INFERFAC/ *PHOSPHATE,
IOASSAY, *EUTROPHICATION, *CALI/
, S/ BIBLIOGRAPHIES, *MINNESOTA,
EATMENT, *GREAT LA/ *PHOSPHATES,
, *PRODUCTIVITY, CYANOPHYTA, NU/
ATION, FISH, *LIMNOLOGY, FISHKI/
TON, URBANIZATION, WATER POLLUT/
GY, CHEMICAL PROPERTIES/ *ALGAE,
BIOTA, PHYTOPLANKT/ *WASHINGTON,
EFFLUENTS, *NUTRIENTS, *NITROGE/
*CITIES, BIOLOGICAL COMMUNITIE/
TS, *COPPER SULFATE, *TREATMENT,
OLLUTION CONTROL, *WATER POLLUT/
*NUTRIENTS, *BIOASSAY, *RUNOFF,
R QUALITY CONTROL/ *CONNECTICUT,
LGAE, C/ *EUTROPHICATION, *IRON,
FATE, ALGAL CONTROL, ALGICIDES,/
OPHYTA, / *LIMNOLOGY, *COLORADO,
RIMARY PRODUCTIVITY, *ZOOPLANKT/
OGICAL ECOLOGY, RADIOACTIVITY T/
ENTATION RATES, *EUTROPHICATION,
*EUTROPHICATION, OLIGOTROPHIC,
E,/ *LIMNOLOGY, *WATER, *RIVERS,
*EPIPHYTOLOGY, NUTRIENTS, PLANT/
VADA, DEPTH, ALGAE, *DEEP WATER,
LGAE, WATER PO/ *EUTROPHICATION,
C PLANTS, *INHIBITORS, *MARSHES,
RUNOFF, FERTILIZERS, NUTRIENTS/
LLUTION CONTRO/ *EUTROPHICATION,
PONDS, TEMPERATURE,/ *LIMNOLOGY,
PHYTOPLANKTON, TEMPERATURE, *E/
E DISPOSAL, ALGAE, AGRICULTURAL/
SORPTION, *PHOSPHORUS COMPOUNDS,
FERTILIZATION, ALG/ *NUTRIENTS,
N, *AGRICULTURAL WASTES, *ALGAE,
*REVIEWS, ALG/ *EUTROPHICATION,

*LABORATORY STUDIES, *MASS CULTUR   W69-06865
*LABORATORY TESTS, TEST PROCEDURE   W68-00472
*LABORATORY TESTS, *SCENEDESMUS,   W68-00475
*LAGOONS, *DOMESTIC WASTES, WASTE   W71-06737
*LAGOONS, ALGAE, ORGANIC LOADING,   W70-07508
*LAGOONS, *BIODEGRADATION, *SEWAG   W71-02009
*LAGOONS, *SLUDGE, ALGAE, BIOCHEM   W70-06899
*LAGOONS, *OXIDATION LAGOONS, BIO   W70-06601
*LAGOONS, *DOMESTIC WASTES, WASTE   W70-01519
*LAGOONS, ACTIVATED SLUDGE, ALGAE   W72-06017
*LAKE AGING, NUTRIENT SOURCES, WA   W70-00264
*LAKE BUTTE DES MORTS(WISC), GROW   W72-00845
*LAKE CHANGES, *NUTRIENT CONCENTR   W70-00270
*LAKE ERIE, BIOCHEMICAL OXYGEN DE   W71-04758
*LAKE ERIE, *LAKE MICHIGAN, *LAKE   W70-01943
*LAKE ERIE, *NUTRIENTS, ALGAE, WA   W69-01445
*LAKE ERIE, *WATER POLLUTION CONT   W69-06305
*LAKE ERIE, CHLOROPHYTA, CYANOPHY   W70-04468
*LAKE ERIE, PHOSPHORUS, FISH STOC   W70-04465
*LAKE KINNERET(ISRAEL).:   W70-09889
*LAKE KLOPEINER(AUSTRIA), LAKE WO   W70-05405
*LAKE MEAD, LAS VEGAS BAY.: /D OR   W72-02817
*LAKE MENDOTA(WIS), PELAGIC AREAS   W70-06229
*LAKE MICHIGAN, *LAKE ONTARIO, LA   W70-01943
*LAKE MICHIGAN, *WATER SUPPLY, IN   W70-00263
*LAKE MICHIGAN, *WATER CIRCULATIO   W69-05763
*LAKE MICHIGAN, ALGAE, NUISANCE A   W68-01244
*LAKE MINNETONKA(MINN), LAKE MANA   W69-10167
*LAKE ONTARIO, *GROWTH RATES, ENV   W69-10158
*LAKE ONTARIO, LAKE HURON, AESTHE   W70-01943
*LAKE ONTARIO, *WATER CIRCULATION   W71-05878
*LAKE ONTARIO, *WATER QUALITY, *M   W71-05879
*LAKE ONTARIO, WATER TURBIDITY, I   W71-09880
*LAKE SEDIMENTS, *NUTRIENTS, NITR   W72-06051
*LAKE SUPERIOR, *PERIPHYTON, *LIM   W70-05958
*LAKE TAHOE(CALIF-NEV), TRANSPARE   W70-01933
*LAKE TAHOE(CALIF-NEV), HYDROPHYT   W70-00711
*LAKE WASHINGTON(WASH), *LAKE CHA   W70-00270
*LAKE WASHINGTON(WASH), *OXYGEN D   W69-10182
*LAKE WASHINGTON(WASH), OSCILLATO   W69-10169
*LAKE WASHINGTON, *SEWAGE FERTILI   W69-09349
*LAKE WASHINGTON(WASH), *CULTURAL   W70-05663
*LAKE WASHINGTON(WASH), SEATTLE(W   W70-04455
*LAKE WASHINGTON(WASH), MACROFAUN   W70-04253
*LAKE WATERS.:   W70-08637
*LAKE ZURICH, *MUNICIPAL WASTES,   W68-00680
*LAKES, *ALGAE, *NITROGEN, WISCON   W70-03429
*LAKES, *ALGAE, *GROWTH RATES, *P   W70-08637
*LAKES, *ALGAE, *EUTROPHICATION,   W70-00264
*LAKES, *BENTHIC FAUNA, *TROPHIC   W71-12077
*LAKES, *BACTERIA, *ANALYTICAL TE   W70-04287
*LAKES, *COPPER SULFATE, *ALGAL C   W68-00859
*LAKES, *CELLULOSE, DISTRIBUTION   W69-00976
*LAKES, *CLASSIFICATION, DESMIDS,   W68-00255
*LAKES, *CYCLING NUTRIENTS, *SEDI   W69-06859
*LAKES, *CARBON RADIOISOTOPES, *B   W70-01933
*LAKES, *DIATOMS, *ALGAE, ECOLOGY   W69-00659
*LAKES, *DETERGENTS, *TERTIARY TR   W70-03964
*LAKES, *EUTROPHICATION, *BIOMASS   W70-03959
*LAKES, *EUTROPHICATION, FERTILIZ   W69-05844
*LAKES, *EUTROPHICATION, *WASHING   W70-05663
*LAKES, *EUTROPHICATION, *HYDROLO   W70-05113
*LAKES, *EUTROPHICATION, SEWAGE,   W69-10169
*LAKES, *EUTROPHICATION, *SEWAGE   W69-09349
*LAKES, *EUTROPHICATION, *CANALS,   W70-00268
*LAKES, *ECOLOGY, BOTTOM SEDIMENT   W70-06217
*LAKES, *EUTROPHICATION, *WATER P   W71-03048
*LAKES, *IMPOUNDMENTS, TERTIARY T   W70-04765
*LAKES, *LOCAL GOVERNMENTS, *WATE   W72-05774
*LAKES, *MANGANESE, *WISCONSIN, A   W69-05867
*LAKES, *MUD, *COPPER, COPPER SUL   W69-03366
*LAKES, *MOUNTAINS, *WINTER, CYAN   W69-10154
*LAKES, *NEW YORK, *MEROMIXIS, *P   W70-05760
*LAKES, *ORGANIC MATTER, *PHYSIOL   W70-04194
*LAKES, *PRIMARY PRODUCTIVITY, *P   W69-00632
*LAKES, *PHOSPHORUS RADIOISOTOPE,   W68-00858
*LAKES, *PONDS, LIGHT, TEMPERATUR   W70-05404
*LAKES, *PRODUCTIVITY, MICHIGAN,   W71-00832
*LAKES, *PLANTS, *CALIFORNIA, *NE   W70-00711
*LAKES, *PHOSPHATES, NUTRIENTS, A   W69-09135
*LAKES, *PHYTOPLANKTON, BIOASSAY,   W72-00845
*LAKES, *RIVERS, *EUTROPHICATION,   W69-10170
*LAKES, *RIVERS, ALGAE, WATER PO   W70-05668
*LAKES, *RECREATION, RESERVOIRS,   W70-06225
*LAKES, *RESERVOIRS, ZOOPLANKTON,   W70-04475
*LAKES, *SEWAGE EFFLUENTS, *SEWAG   W70-05565
*LAKES, *SEDIMENTS, *EUTROPHICATI   W69-03075
*LAKES, *TROUT, *FISH MANAGEMENT,   W70-05399
*LAKES, *TRIBUTARIES, SAMPLING, F   W70-02969
*LAKES, *WATER POLLUTION CONTROL,   W70-07393

TION,/ *NEVADA, *EUTROPHICATION,
S, *PHOSPHORUS/ *EUTROPHICATION,
NOPHYTA, FISHK/ *EUTROPHICATION,
OMAS/ *BACTERIA, *PHYTOPLANKTON,
ION, *NUTRIENT/ *EUTROPHICATION,
DOMESTIC WAST/ *EUTROPHICATION,
IC PRODUCTIVITY, EUTROPHICATION,
MENTS, ICED LA/ *EUTROPHICATION,
ATION, *WATER POLLUTION SOURCES,
N EFFECTS, PRO/ *EUTROPHICATION,
BACTERIA, ALG/ *EUTROPHICATION,
ALGAE, PRIMARY/ *PHYTOPLANKTON,
SONAL, CYANOPH/ *EUTROPHICATION,
, IRRIGATION EFFECTS, DRAINAGE /
, / *EUTROPHICATION, *NUTRIENTS,
ES/ *PERIPHYTON, *RADIOACTIVITY,
AE, *NUISANCE ALGAE, *WISCONSIN,
YSTEMS, EUTROPHICATION, OLIGOTR/
N, ALGAE,/ *THERMAL POWERPLANTS,
CE ALGAE, SPRA/ *EUTROPHICATION,
E, NUISANCE AL/ *EUTRIPHICATION,
        *NITRIC ACID,
TRATES, NITROGEN, *CONDUCTIVITY,
IDAY HARBOR(WASH), ANTITHAMNION/
MENTS, *AUTORADIOGRAPHY, ANTITH/
DANCE, BLUE-GREEN ALGAE, GREEN /
ROPHYTA, DIA/ *PLANKTON, *ALGAE,
HESIS, *ALGAE, *THERMAL SPRINGS,
ICATION, MODEL STUDIES, BIOMASS,
OBIC CONDITIONS, ANAERO/ *ALGAE,
ALGAE, *CHLORELLA, *SCENEDESMUS,
RIMARY PRODUCTI/ *PHYTOPLANKTON,
ABITATS, *ENVIRONMENTAL EFFECTS,
OUNTAINS, *WINTER, CYANOPHYTA, /
 *PRIMARY PROD/ *EUTROPHICATION,
DI/ *LAKE SUPERIOR, *PERIPHYTON,
RESERVOIRS, PONDS, TEMPERATURE,/
WATER POLLUTION EF/ *RESERVOIRS,
ES, *PONDS, LIGHT, TEMPERATURE,/
IS, PRODUCTIVITY, WATER QUALITY,
*EUTROPHICATION, *FERTILIZATION,
, *NITROGEN, *PHOSPHORUS, *IRON,
OPHICATION, FERTILIZATION, FISH,
XIFICATION, LINDANE METABOLISM./
FACTANTS, WATER POLLUTION SOURC/
 CULTURES, *ALKYL POLYETHOXYLAT/
CAL TECHNIQUES, *MICROORGANISMS,
MMATE, CHLORAMINE, DE-K-PRUF-21/
TION, T/ *OCEANS, *MARINE ALGAE,
DS, ECOLOGICAL DISTRIBU/ *ALGAE,
Y CONTROL/ *CONNECTICUT, *LAKES,
UENTS, *WATER POLLUTION CONTROL,
BONUCLEI/ *PHOTOPHOSPHORYLATION,
        LAWRENCE LAKE(MICHIGAN),
*TRUAX FIELD(MADISON) EFFLUENTS,
D CUTTING, *WEED EXTRACTING STE/
LAKE WAUBESA(WIS), LAKE KEGONSA/
KE MENDOTA(WIS), LAKE MONONA(WI/
S, *WATER-BLOOM, 'BREAKING' OF /
FATE, RESERVOIRS, WATER/ *ALGAE,
*V/ *VISCOSE RAYON PLANT WASTES,
 *EUTROPHICATION, *IRON, *LAKES,
ER TREATMENT, *TURBIDITY, *IRON,
D LAGOON.:
*SALT MARSHES, OILY WATER, *OIL,
RINE / *EMULSIFIERS, *MORTALITY,
FECTS, LETHAL LIMI/ *DETERGENTS,
C MATTER, EXUDATION, T/ *OCEANS,
WATER, *HUMIC ACIDS, */ *OCEANS,
R POLLUTION EFFECT/ *OILY WATER,
, PHYSICOC/ *ZINC RADIOISOTOPES,
SYNTHESIS, *PHYTOPLANKTON, *DDT,
 FLORA, LIGHT INTENSITY, BIOMAS/
ONAS, METABOLISM, KI/ *BACTERIA,
Y, *MARINE ALGAE, *MARSH PLANTS,
, / *PHOSPHORUS, *CORAL, *REEFS,
ABSORPTION, MARINE ALGAE, MARIN/
RADIOISOTOPES, PHYSICOCHEMICAL /
ION, *DETERGENTS, *INSECTICIDES,
IS, *ANALYTICAL TECHNIQUES, TEM/
*HETEROTROPHY, *ORGANIC SOLUTES,
NITZSCHIA CLOSERIUM, ARTEMIA, C/
, EPIPHYTOLOGY, MARINE ALGAE, E/
KTON, BACTERI/ *ALGAE, *ECOLOGY,
 *ALGAE, *ANALYTICAL TECHNIQUES,
NUTRIENT REQUIREMENTS, *DISTRIB/
TS, RECREATION, WAT/ *NUTRIENTS,
QUATIC MICROBIOLOGY, *ESTUARIES,
ION, SODIUM COMPOUNDS, PHOTOSYN/
YSICOCHEMICAL / *MARINE ANIMALS,

*LAKES, *WATER POLLUTION, DEGRADA   W72-02817
*LAKES, ACID LAKES, BIBLIOGRAPHIE   W68-00857
*LAKES, ALGAE, AQUATIC WEEDS, CYA   W69-06535
*LAKES, ALGAE, MICROORGANISMS, BI   W70-08654
*LAKES, CONNECTICUT, WATER POLLUT   W68-00511
*LAKES, GLACIATION, URBANIZATION,   W69-07818
*LAKES, LIMNOLOGY, *PLANKTON, ALG   W69-00446
*LAKES, NUTRIENTS, OXYGEN REQUIRE   W68-00487
*LAKES, NUTRIENTS, FERTILIZERS, P   W69-07084
*LAKES, NUTRIENTS, WATER POLLUTIO   W70-04721
*LAKES, NUTRIENTS, FERTILIZATION,   W70-05761
*LAKES, ORGANIC MATTER, SEASONAL,   W70-04290
*LAKES, PHYTOPLANKTON, ALGAE, SEA   W68-00860
*LAKES, PLANTS, *ALGAE, *SALINITY   W69-00360
*LAKES, PONDS, FERTILIZERS, ALGAE   W69-08668
*LAKES, PRODUCTIVITY, RADIOISOTOP   W70-04371
*LAKES, PRODUCTIVITY, NITROGEN, P   W70-05112
*LAKES, STREAMS, FISH, LAND, ECOS   W70-04193
*LAKES, TEMPERATURE, PHYTOPLANKTO   W71-09767
*LAKES, WISCONSIN, ALGAE, *NUISAN   W68-00488
*LAKES, WISCONSIN, COPPER SULPHAT   W68-00468
*LANGELIER INDEX.:   W71-11393
*LEACHING, IRON, TEMPERATURE, SOI   W72-11906
*LEUCOTHRIX MUCOR, MORPHOLOGY, FR   W69-10161
*LEUCOTHRIX MUCOR, *GROWTH, *FILA   W69-09755
*LEWIS AND CLARK LAKE (S D), ABUN   W70-05375
*LIGHT INTENSITY, CHLORELLA, CHLO   W70-05381
*LIGHT INTENSITY, SPRING WATER, W   W69-09676
*LIGHT INTENSITY.: /ENT, EUTROPH   W73-02218
*LIGHT, *NITRATES, *NITRITES, AER   W70-05261
*LIGHT, PHOSPHORUS, PHOTOSYNTHESI   W70-03923
*LIMITING FACTORS, *NUTRIENTS, *P   W70-01579
*LIMITING FACTORS, BENTHIC FLORA,   W70-02770
*LIMNOLOGY, *COLORADO, *LAKES, *M   W69-10154
*LIMNOLOGY, *ESSENTIAL NUTRIENTS,   W72-01990
*LIMNOLOGY, *WATER QUALITY, BIOIN   W70-05958
*LIMNOLOGY, *LAKES, *RECREATION,   W70-06225
*LIMNOLOGY, *INDUSTRIAL WASTES, *   W70-06975
*LIMNOLOGY, *WATER, *RIVERS, *LAK   W70-05404
*LIMNOLOGY, MINNESOTA, WATER ANAL   W69-00632
*LIMNOLOGY, *PHYTOPLANKTON, FISH,   W69-06858
*LIMNOLOGY, *ALGAE.: /HMENT TESTS   W69-06004
*LIMNOLOGY, FISHKILL, NUTRIENTS,   W69-05844
*LINDANE, *ALGAE METABOLISM, DETO   W70-08652
*LINEAR ALKYLATE SULFONATES, *SUR   W70-09438
*LINEAR ALKYL SULFONATES, *AXENIC   W70-03928
*LIPIDS, LIGHT, ALGAE, TEMPERATUR   W70-05651
*LITERATURE, ELEOCHARIS, 2,4-D, A   W70-05263
*LITTORAL, *ORGANIC MATTER, EXUDA   W70-01073
*LITTORAL, *AQUATIC PLANTS, ISLAN   W72-10623
*LOCAL GOVERNMENTS, *WATER QUALIT   W72-05774
*LOCAL GOVERNM, STATE GOVERNM   W69-06909
*MACROMOLECULES, NUCLEIC ACID, RI   W69-10151
*MACROPHYTES.:   W71-00832
*MADISON LAKES, YAHARA RIVER(WIS)   W70-05113
*MADISON LAKES(WIS), *AQUATIC WEE   W70-04494
*MADISON(WIS), LAKE MONONA(WIS),   W70-05112
*MADISON(WIS), NUISANCE ODORS, LA   W69-03366
*MADISON(WIS), *'WORKING' OF LAKE   W69-08283
*MAINE, *WATER SUPPLY, COPPER SUL   W70-08096
*MAN-MADE FIBER PLANTS, NYLON-6,   W70-06925
*MANGANESE, *WISCONSIN, ALGAE, CA   W69-05867
*MANGANESE, *COLOR, *ALGAE, *COLL   W72-04145
*MANURE, ANAEROBIC LAGOON, AERATE   W70-07491
*MARINE ALGAE, WATER POLLUTION SO   W70-09976
*MARINE ALGAE, *MARSH PLANTS, *MA   W70-01777
*MARINE ALGAE, WATER POLLUTION EF   W70-01779
*MARINE ALGAE, *LITTORAL, *ORGANI   W70-01073
*MARINE ALGAE, *SEAWATER, *FRESH   W70-01074
*MARINE ALGAE, *EMULSIFIERS, WATE   W70-01780
*MARINE ALGAE, *RESINS, SEA WATER   W69-08275
*MARINE ALGAE, TOXICITY, DIATOMS,   W70-05272
*MARINE ALGAE, *DIATOMS, *BENTHIC   W70-03325
*MARINE ALGAE, *CULTURES, PSEUDOM   W70-04280
*MARINE ANIMALS, *OILY WATER, LIC   W70-01777
*MARINE ANIMALS, FISH, ECOSYSTEMS   W69-08526
*MARINE ANIMALS, *RADIOISOTOPES,   W71-09850
*MARINE ANIMALS, *MARINE PLANTS,   W71-09863
*MARINE BACTERIA, *TOXICITY, *WAT   W70-01467
*MARINE BACTERIA, *ALGAE, *ANALYS   W70-04365
*MARINE ENVIRONMENTS, MICROBIAL E   W70-04811
*MARINE ENVIRONMENT, CERIUM-144,   W70-00708
*MARINE MICROORGANISMS, *HABITATS   W69-10161
*MARINE MICROORGANISMS, PHYTOPLAN   W70-01068
*MARINE MICROORGANISMS, PHYSIOLOG   W70-04811
*MARINE MICROORGANISMS, *ALGAE, *   W70-03952
*MARINE MICROORGANISMS, *POLLUTAN   W69-04276
*MARINE MICROORGANISMS, MARINE AL   W69-03752
*MARINE PLANTS, *ALGAE, CHLORINAT   W70-00161
*MARINE PLANTS, RADIOISOTOPES, PH   W71-09863

361

EA WATER, TEMPERATURE, SALINITY/
/ *FAYETTEVILLE GREEN LAKE(NY),
TTER, ARCTIC ALGAE, AUTOTROPHIC/
IC ALGAE, MAR/ *ALGAL POISONING,
IERS, *MORTALITY, *MARINE ALGAE,
/ *AQUATIC PLANTS, *INHIBITORS,
ORELLA PYRENOIDOSA(EMERSON STRA/
AL CULTURE, *LABORATORY STUDIES,

NATION, CYANIDE, NEW PRODUCT CO/
MANAGEMENT, GEOCHEMIS/ *ECOLOGY,
OROPHYLL, *PRIMARY PRODUCTIVITY
EL/ *INSTRUMENTATION, *SAMPLING,
C ENVIRONMENTS, *EUTROPHICATION,
, *ZOOPLANKT/ *LAKES, *NEW YORK,
BILITY, CHLORELLA, CARB/ *ALGAE,
:
REATMENT, ON-STREAM DESLUDGING./

*BIOLOGICAL COMMUNITIES, *ALGAE,
VOLUTION, *INTERSTATE POLLUTION,
Y, BIOCHEMICAL EXTRACTION, MILL/
M, *FUNGAL-BOUND RADIOSTRONTIUM,
ETEROTROPHY, *BIOLOGICAL UPTAKE,
BACTERIA, POPULATION EQUIVALEN/
INDUSTRIAL WASTES, SEWAGE TREAT/
OORGANISMS, *WATER POLLUTION, */
OGY, *ESTUARIES, *MARINE MICROO/
CULTURES, *BIOASSAYS, PROTOZOA,/
ELB/ *APHANIZOMENON FLOS AQUAE,
TERS.:
T AVAILA/ *CHEMICAL COMPOSITION,
ENT AVAILABILITY, *ALGAL GROWTH,
LTURE MEDIA, CHU NO 10 SOLUTION/
CTIONS, *RADIOCARBON BIOASSAY, /
ALGAE, / *ANALYTICAL TECHNIQUES,
TEINS, WALLS, STRUCTURE/ *ALGAE,
, */ *MICROBIOLOGY, *MINERALOGY,
N COMPOUNDS, *NUTRIENTS, *ALGAE,
T ORGANISMS, *ACCLIMATIZATION, /
AL TEMPERATURES, *COPPER DOSAGE/

NING, *KENOSHA(WIS), *STRAINING,
OVEMENT, *WATER QUALITY CONTROL,
SIGN CRITERIA/ AERATED LAGOONS,
ILK SUBSTRATE, NITROGEN REMOVAL,

, *MUNICIPAL / *WATER POLLUTION,
RADIOLARIA, FORAMINIFERA, THIO/
TER POLLUTION, */ *MICROBIOLOGY,
GAE, ECOLOGY, S/ BIBLIOGRAPHIES,
HYTA, / *ALGAE, *PHOTOSYNTHESIS,
SPOSAL, WATER / *EUTROPHICATION,
GE EFFLUENTS, TERTIARY TREATMEN/
UTION ABAT/ *ALGAE, *PHOSPHORUS,
NUI/ *RIVERS, *ALGAE, *SAMPLING,
ION, *IMPOUNDMENTS, *RESERVOIRS,
ICAL TECHNIQUES, CLASSIFICATION,
TMENT, *OXIDATION LAGOONS, MILK,
R TREATMENT, *OXIDATION LAGOONS,

EAMS, BIOINDICATORS, *NEW YORK,/
*LAKE ONTARIO, *WATER QUALITY
NT, *EFFLUENTS, *RADIOISOTOPES,/
ECOLOGY, *ENVIRONMENTAL EFFECTS,
PLANTS, *MARINE / *EMULSIFIERS,
D MATS, NON-LEGUMES, *NITROGEN /
*LIMNOLOGY, *COLORADO, *LAKES,
ALGAE, *BIOAS/ *EUTROPHICATION,
GAL CONTROL, ALGICIDES,/ *LAKES,
INE WASTES, *RADIOACTIVE WASTES,
WATER FISH, LAKES, *LAKE ZURICH,
*WA/ *ALGAE, *INDUSTRIAL WASTES,
N LAGOONS, *ANAEROBIC DIGESTION,
EMATOLOGY, *GRANULAR LEUCOCYTES,
ALGAE COUNTS, P/ *RIVER STUDIES,
, *ALGAL CONTROL, *TASTE, *ODOR,
*WATER POLLUTION, DEGRADATION,/
R, *LAKES, *PLANTS, *CALIFORNIA,
Y, *EUTROPHICATION, *CALIFORNIA,
MMISSION, ALGAL BLOOMS, NUTRIEN/
*NUISANCE ALGAE, *ALGAL CONTROL,
*ALGAE CONTROL, *WATER SUPPLY,
OHIO), GONGROSIRA STAGNALIS, NE/
OTOPES, *STREAMS, BIOINDICATORS,
RODUCTIVITY, *ZOOPLANKT/ *LAKES,
*COLLOIDS, FLOW RATES, *FILTERS,
SIGN CRITERIA, *WATER TREATMENT,
ND ALGAE.:
WATER AND PLANTS, WATER POLLUT/

*MARINE WATER, *COPPER, *ALGAE, S   W70-08111
*MARL LAKES, CHLOROBIUM, FILINIA,   W70-05760
*MARL LAKES, DISSOLVED ORGANIC MA   W71-10079
*MARSH PLANTS, *OILY WATER, AQUAT   W70-01231
*MARSH PLANTS, *MARINE ANIMALS, *   W70-01777
*MARSHES, *LAKES, *PHYTOPLANKTON,   W72-00845
*MASS CULTURE, *ALGAL GROWTH, CHL   W69-07442
*MASS CULTURE, *PILOT-PLANT STUDI   W69-06865
*MASSENA(NY).:   W72-08686
*MASTER PLANNING, *SOURCE CONTAMI   W73-00758
*MATHEMATICAL MODELS, *WATERSHED   W71-09012
*MEASUREMENT, PROTEINS, RADIOACTI   W69-10151
*MEASUREMENT, *FLUOROMETRY, DYE R   W70-05377
*MEROMIXIS, *WATER POLLUTION EFFE   W70-03974
*MEROMIXIS, *PRIMARY PRODUCTIVITY   W70-05760
*METABOLISM, *ENVIRONMENTS, VARIA   W70-02965
*METABOLITES, *CEDAR RIVER(IOWA).   W72-13605
*METAL CORROSION, COOLING WATER T   W70-02294
*METHANE FERMENTATION.:   W71-13413
*METHODOLOGY, CARBON RADIOISOTOPE   W71-12870
*METROPOLITAN AREAS, BLONDIN.: /E   W70-04430
*MICRO-AUTORADIOGRAPHY, PHYSIOLOG   W69-10163
*MICROBIAL ACTIVITY, *RADIOSTRONT   W71-04067
*MICROBIAL ECOLOGY, MAXIMUM UPTAK   W70-04194
*MICROBIAL ECOLOGY, PURPLE SULFUR   W70-03312
*MICROBIOLOGY, *SEWAGE LAGOONS, *   W70-03312
*MICROBIOLOGY, *MINERALOGY, *MICR   W69-07428
*MICROBIOLOGY, *AQUATIC MICROBIOL   W69-03752
*MICROBIOLOGY, *CLASSIFICATION, *   W71-11823
*MICROCYSTIS AERUGINOSA, BOHEMIA,   W70-00274
*MICROCYSTIS AERUGINOSA, *LAKE WA   W70-08637
*MICROCYSTIS AERUGINOSA, *NUTRIEN   W69-05868
*MICROCYSTIS AERUGINOSA, ANTAGONI   W69-05867
*MICROCYSTIS AERUGINOSA(KUTZ), CU   W70-03507
*MICRONUTRIENTS, *NUTRIENT INTERA   W70-04765
*MICROORGANISMS, *LIPIDS, LIGHT,   W70-05651
*MICROORGANISMS, SCENEDESMUS, PRO   W70-04184
*MICROORGANISMS, *WATER POLLUTION   W69-07428
*MICROORGANISMS, BIODEGRADATION,   W69-06970
*MICROORGANISMS, *ALGAE, *DOMINAN   W70-07313
*MICROSCOPIC EXAMINATION, *CRITIC   W69-10157
*MICROSTRAINERS, HYPOCHLORITES.:   W72-03824
*MICROSTRAINERS.:   *SCREE   W73-02426
*MIDGES, FISH FOOD ORGANISMS, NUT   W70-05428
*MIDWEST(US), *SOUTHWEST(US), *DE   W70-04788
*MILK WASTES.: /L, COD REMOVAL, M   W70-07031
*MILLIPORE FILTERS.:   W72-09693
*MINE WASTES, *RADIOACTIVE WASTES   W69-01165
*MINERAL TRANSFORMATION, METAZOA,   W69-07428
*MINERALOGY, *MICROORGANISMS, *WA   W69-07428
*MINNESOTA, *LAKES, *DIATOMS, *AL   W69-00659
*MINNESOTA, PHYTOPLANKTON, CYANOP   W70-05387
*MINNESOTA, RECREATION, SEWAGE DI   W70-04810
*MINNESOTA, *EUTROPHICATION, SEWA   W70-03512
*MINNESOTA, *EUTROPHICATION, POLL   W69-10167
*MISSISSIPPI RIVER, *OHIO RIVER,   W70-04507
*MIXING, *WATER QUALITY CONTROL,   W70-04484
*MODEL STUDIES, WASTE IDENTIFICAT   W69-01273
*MODEL STUDIES.: /ASTE WATER TREA   W70-07031
*MODEL STUDIES.: /IA, *WASTE WATE   W70-07034
*MODEL TESTS, DYES.:   W70-02609
*MONITORING, *RADIOISOTOPES, *STR   W69-10080
*MONITORING, BACTERIA, WATER CHEM   W71-05879
*MONITORING, *DISCHARGE MEASUREME   W72-00941
*MORPHOLOGY, *ALGAE, AQUATIC MICR   W69-10177
*MORTALITY, *MARINE ALGAE, *MARSH   W70-01777
*MOUNTAIN MEADOW, NITROGEN-15, SO   W72-08110
*MOUNTAINS, *WINTER, CYANOPHYTA,   W69-10154
*MUD(LAKE), *PHOSPHORUS, *AQUATIC   W70-03955
*MUD, *COPPER, COPPER SULFATE, AL   W69-03366
*MUNICIPAL WASTES, AQUATIC ALGAE,   W69-01165
*MUNICIPAL WASTES, NUTRIENTS, NUI   W68-00680
*MUNICIPAL WASTES, *GREAT LAKES,   W70-04430
*MUNICIPAL WASTES, METHANE BACTER   W72-06854
*MUSCULAR EXERTION, *2-METHYL-5-V   W70-01788
*NATIONAL WATER QUALITY NETWORK,   W70-04507
*NATURAL STREAMS, *DISTRIBUTION S   W72-03562
*NEVADA, *EUTROPHICATION, *LAKES,   W72-02817
*NEVADA, DEPTH, ALGAE, MOSSES, CH   W70-00711
*NEVADA, LIMNOLOGY, PRIMARY PRODU   W70-01933
*NEW HAMPSHIRE WATER POLLUTION CO   W69-08674
*NEW HAMPSHIRE, *COPPER SULFATE,   W69-08674
*NEW HAMPSHIRE, ALGAE, ODOR, TAST   W69-08282
*NEW SPECIES, *WESTERN LAKE ERIE(W70-04468
*NEW YORK, NUCLEAR WASTES, WATER   W69-10080
*NEW YORK, *MEROMIXIS, *PRIMARY P   W70-05760
*NEW YORK, TREATMENT FACILITIES.:   W72-04145
*NEW YORK, WATER PURIFICATION.: /   W72-08686
*NEW ZEALAND, SEWAGE TREATMENT PO   W70-02304
*NEWFOUND LAKE, *WINNISQUAM LAKE,   W70-02764

AL INTERFERENCES.:
AL CONTROL, *COLOR, *ALKALINITY,
DITIONS, ANAERO/ *ALGAE, *LIGHT,
YCLE, *FARM WASTE, GROUND WATER,
UME, *CARBON DIOXIDE, *AERATION,

*DENITRIFICATION, EFFICIENCIES,
RTILIZERS, *ESSENTIAL NUTRIENTS,
AERO/ *ALGAE, *LIGHT, *NITRATES,
KE ERIE.:   *ALGAE, *CYANOPHYTA,
COMPOUNDS, *NU/ *EUTROPHICATION,
BIBLIOGRAPHIES, *EUTROPHICATION,
ER, ALGAE,/ *NITROGEN COMPOUNDS,
COMPOUNDS, *CYCLING NUTRIENTS, /
UND WATER, *NITRATES, EUTROPHIC/
ILIZATION, *FRESH WATER, ALGAE,/
*ALGAE, *MICROORGANISMS, BIODEG/
, CARBON DIOXIDE, INHIBITION, N/
*BIOASSAY, *ESSENTIAL NUTRIENTS,
*ENZYMES, *ESSENTIAL NUTRIENTS,
*ENZYMES, *ESSENTIAL NUTRIENTS,
QUIREMENTS, *PHYSI/ *CYANOPHYTA,
CUMULATION, *LABORATORY STUDIES,
ROGEN-15, SOD MATS, NON-LEGUMES,
RGANIC MATTER, *GRASSLANDS, MOU/
LAKES, *ALGAE, *NITROGEN, WISCO/
ION, *NUTRIENTS, *LAKES, *ALGAE,
CHLOROPHYLL, PHOT/ *PHOSPHORUS,
GE, ALGAE, EUTROPHICATION, LIME,
E WATER TREATMENT, *WATER REUSE,
ANOPHYTA, N/ *CHEMICAL ANALYSIS,
XAS, *NUTRIENT ENRICHMENT TESTS,
FFLUENT, *SEWAGE TREATM/ *ALGAE,
WASTE TREATMENT, EUTHROPHICATIO/
COST COMPARISONS, *EFFICIENCIES,
ATMENT, *ALGAE, *EUTROPHICATION,
EUTROPHICATION, DIATOMS, MICROB/
ICAL OXYGEN DEMAND, *PHOSPHORUS,
IES, COST COMPARISO/ *NUTRIENTS,
ANKTON, *NUTRIENTS, *PHOSPHORUS,
, *SEWAGE EFFLUENTS, *NUTRIENTS,
        *PIGMENTED FLAGELLATES,
HANSENII, RHODOTORULA RUBRA, CA/
ITY CONTROL/ *THERMAL POLLUTION,
*NEW HAMPSHIRE, *COPPER SULFATE/
NDUSTRIAL), *PHOSPHORUS SOURCES,
CYANOPHYTA, LAKES, RESERVOIRS,/
OLLUTION SOURC/ *EUTROPHICATION,
NOLS, *WATER Q/ *AQUATIC PLANTS,
MINS, PEPTIDES, ORGANIC MATTER,/
S, *EUTROPHICATION, *CLADOPHORA,
PER SULFATE, *WATER QUALITY CON/
ATION, *LAKES, WISCONSIN, ALGAE,
LIMNOLOGY, LAKES, *PILOT PLANTS,
ES, PROD/ *ODOR-PRODUCING ALGAE,
RACTIONS, *RADIOCARBON BIOASSAY,
BON BIOASSAY, / *MICRONUTRIENTS,
S, CONNECTICUT, WATER POLLUTION,
TS.:   *SEDIMENTS,
TECHNIQUES, *BIOASSAY, *ENZYMES,
CYANOPHYTA, *NITROGEN COMPOUNDS,
CATION, *WATER POLLUTION CONTRO/
UTROPHICATION, *PLANKTON, ALGAE,
TIAL NUTRIENTS, *NITROGEN CYCLE,
*BIOASSAY, *ESSENTIAL NUTRIENTS,
OWTH, *MICROCYSTIS AERUGINOSA, /
ND LAKES, COLORADO RIVER, TEXAS,
SITION, *MICROCYSTIS AERUGINOSA,
TION, *ADVANCED WASTE TREATMENT,
LEASE, SAMPLE PRESERVATION, WAT/

TES, LABORA/ *ALGAE, *NUTRIENTS,
*MARINE MICROORGANISMS, *ALGAE,
        *ALGAL ACTIVITY,
CHLORELLA, ALGA/ *PHYTOPLANKTON,
TS, INHIBITIO/ *ACID MINE WATER,
OWTH, *ALGAE, ELEMENTS(CHEMICAL/
N, ANAEROBIC DENITRIFICATION, L/
LGAE, *CHLOROPHYTA, *CYANOPHYTA,
(NY), / *ENRICHMENT EXPERIMENTS,
LAKE/ *ISOLATION, *CYANOPHYTES,
OURCES, *NUISANCE PLANT GROWTHS,
WASHINGTON(WASH), *LAKE CHANGES,
HANGES, *NUTRIENT CONCENTRATION,
UTRIENT SOU/ *CULTURAL EUTROPHY,
W SYSTEM.:
AGE EFFLUENT, *SEWAGE TREATMENT,
PHICATION, NITROGEN, PHOSPHORUS,
EUTROPHICATION, OLIGOTR/ ALGAE,
HICATION, LAKE, *LAKE SEDIMENTS,

, LAKE SEDIMENTS, ALGAL CONTROL,
OGEN, PHOSPHORU/ *WATER QUALITY,
FFECTS, WATER POLLUTION CONTROL,
LGA/ *BIOCHEMICAL OXYGEN DEMAND,
TES, SIL/ WATER QUALITY CONTROL,
CE ALGAE, SEWAGE EFFLUENTS, PHO/
IEWS, / *EUTROPHICATION, *ALGAE,
TIARY TREATMENT, N/ *PHOSPHORUS,
TROPHICATION, *SEWAGE EFFLUENTS,
IZERS, ALGAE, / *EUTROPHICATION,
CES, *PHOSPHAT/ *EUTROPHICATION,
HYTOPLANKTON, *LIMITING FACTORS,
S, EFFICIENCIES, COST COMPARISO/
N, *ALK/ *ALGAE, *PHYTOPLANKTON,
*EUTROPHICATION, WATER QUALITY,
AE, *HYDROGEN ION CONCENTRATION,
LAMATION, WASTE WATER TREATMENT,
*BENTHIC FAUNA, *EUTROPHICATION,
*EUTROPHICATION, *CARBOHYDRATES,
S, GROWTH RATES, LABORA/ *ALGAE,
OGEN, WISCO/ *NITROGEN-FIXATION,
LUTION EFFECTS, *EUTROPHICATION,
, ALGAE, PHOSP/ *EUTROPHICATION,
S, *LAKE ERIE, PHOSPHORUS, FISH/
MS, BIODEG/ *NITROGEN COMPOUNDS,
S, *POLLUTANTS, RECREATION, WAT/
OPHICATION, *NITROGEN COMPOUNDS,
N / *EUTROPHICATION, *LAKE ERIE,
ABATEMENT, NU/ *EUTROPHICATION,
OMPOUNDS, PH/ *WATER PROPERTIES,
ORUS, N/ *EUTROPHICATION, LAKES,
AQUATIC PRODUCTIVITY, LIMNOLOGY,
ENT, SECONDARY TREATMENT, STABI/
SODIU/ *EUTROPHICATION, *ALGAE,
OMPOUNDS, *PHOSPHORUS COMPOUNDS,
MANAGEMENT, FERTILIZATION, ALG/
URES, C/ *WASTE WATER TREATMENT,
REATMENT, *PHOSPHORUS COMPOUNDS,
AE, CRUSTACEANS, / *CONFERENCES,
, *ORGANIC MATTER, EXUDATION, T/
, *FRESH WATER, *HUMIC ACIDS, */
TION, / *WATER ANALYSIS, *ODORS,
ALGAE, *WISCONSIN, *LAKES, PROD/
DUCING ALGAE, ALGAE, CHLORINATI/
WATER TREATMENT, *ALGAE, *TASTE,
LLUTION SOURCES, *ALGAE, *TASTE,
*ALGAE, *ALGAL CONTROL, *TASTE,
EES, / *WATER TREATMENT, *TASTE,
N, *EFFICIENCIES, *CHLORINATION,
EATMENT, *ALGAE, *ACTINOMYCETES,
EATMENT, *WATER QUALITY, *ALGAE,
*ACTINOMYCETES, *ALGAE, *TASTE,
TROPHICATION, / *WATER ANALYSIS,
, *SAMPLING, *MISSISSIPPI RIVER,

Y, *BIOINDICATORS, *CHLOROPHYLL,
ON CONTROL, *POLLUTION ABATEMEN/
ATER QUALITY, *DISSOLVED OXYGEN,
LLUTION, *KUWAIT C/ *SALT MARSH,
N ALGAE, / *RECOVERY, OIL SPILL,
ATIC SOLV/ *CRUDE OIL, *GAMOSOL,
:
P/ *DETERGENTS, *DIATOMS, *OIL,
DUSTRIAL WASTES, WATER AN/ *OIL,
ECTS, *SALT MARSHES, OILY WATER,
*WATER P/ *DETERGENTS, *DIATOMS,
ES, INDUSTRIAL WASTES, WATER AN/
*MARSH PLANTS, *MARINE ANIMALS,
*ALGAL POISONING, *MARSH PLANTS,
SIFIERS, WATER POLLUTION EFFECT/
NACYSTIS MONTANA, *HYDROCARBONS,
URCES, ON-SITE DATA COLLECTIONS,
SYRACUSE, WARBURG RESPIROMETER,/
ORA, GOMPHONEMA, PINNULARIA, NI/
S, S/ *TWELVE MILE LAKE(CANADA),
TARIO), TWELVE MILE LAKE(ONTARI/
ULATION, *RADIONUCLIDE EXCHANGE,
*BIOLOGICAL PROPERTIES, CONTROL,
ELEMENTS, *ANCHOA LAMPROTAENIA,
PHYTA, *ALGAE, *THERMAL SPRINGS,
GAE, HYDROGEN ION CONCENTRATION,
COMPOUNDS, / *ALGAE, *CHLORELLA,
RYOTES, MONOCHRYSIS LUTHE/ *DNA,
OPHYLL, WATER POLLUTION EFFECTS,
*POTENTIATION,
FIDES, ALGAE/ *HYDROGEN SULFIDE,
/ *POULTRY, *OXIDATION LAGOONS,
FFECTS, *HYDRAULIC LOAD EFFECTS,
COLOGY, RADIOACTIVITY T/ *LAKES,
, *ALGAE, *PRIMARY PRODUCTIVITY,

*NUTRIENTS, ALGAE, BENTHIC FAUNA,   W72-06052
*NUTRIENTS, *EUTROPHICATION, NITR   W73-01122
*NUTRIENTS.: /, WATER POLLUTION E   W72-00917
*NUTRIENTS, *DISSOLVED OXYGEN, *A   W71-05407
*NUTRIENTS, *PESTICIDES, FARM WAS   W71-10376
*NUTRIENTS, *PRODUCTIVITY, NUISAN   W70-00270
*NUTRIENTS, *CARBON DIOXIDE, *REV   W70-00664
*NUTRIENTS, *EUTROPHICATION, *TER   W69-09340
*NUTRIENTS, *NITROGEN, PHOSPHORUS   W69-09349
*NUTRIENTS, *LAKES, PONDS, FERTIL   W69-08668
*NUTRIENTS, *WATER POLLUTION SOUR   W70-01031
*NUTRIENTS, *PRIMARY PRODUCTIVITY   W70-01579
*NUTRIENTS, *NITROGEN, *PHOSPHORU   W70-01981
*NUTRIENTS, *PHOSPHORUS, *NITROGE   W70-02795
*NUTRIENTS, LIMNOLOGY, WATER TEMP   W70-09093
*NUTRIENTS, *TERTIARY TREATMENT,   W70-08980
*NUTRIENTS.: /TS, WASTE WATER REC   W70-06053
*NUTRIENTS, LAKES, ECOLOGY, PRIMA   W71-00665
*NUTRIENTS, *BIODEGRADATION, ALGA   W71-00221
*NUTRIENTS, *NUTRIENT REQUIREMENT   W70-03507
*NUTRIENTS, *LAKES, *ALGAE, *NITR   W70-03429
*NUTRIENTS, ALGAE, FISH POPULATIO   W70-04266
*NUTRIENTS, *PHOSPHORUS COMPOUNDS   W70-04074
*NUTRIENTS, *BENEFIT-COST ANALYSI   W70-04465
*NUTRIENTS, *ALGAE, *MICROORGANIS   W69-06970
*NUTRIENTS, *MARINE MICROORGANISM   W69-04276
*NUTRIENTS, *PHOSPHORUS COMPOUNDS   W69-04805
*NUTRIENTS, ALGAE, WATER POLLUTIO   W69-01445
*NUTRIENTS, POLLUTANTS, POLLUTION   W69-01453
*NUTRIENTS, *VIRGINIA, NITROGEN C   W69-03695
*NUTRIENTS, UNITED STATES, PHOSPH   W68-00478
*NUTRIENTS, BIO MASS, OLIGOTROPHY   W68-00912
*NUTRIENTS, PONDS, PRIMARY TREATM   W68-00012
*NUTRIENTS, NITROGEN, PHOSPHORUS,   W70-04503
*NUTRIENTS, *BIOASSAY, *RUNOFF, *   W70-04765
*NUTRIENTS, *LAKES, *TROUT, *FISH   W70-05399
*NUTRIENTS, ALGAE, BACTERIA, CULT   W70-05655
*NUTRIENTS, *PESTICIDES, *WATER P   W70-05470
*OCEANOGRAPHY, *TRANSLATIONS, ALG   W69-07440
*OCEANS, *MARINE ALGAE, *LITTORAL   W70-01073
*OCEANS, *MARINE ALGAE, *SEAWATER   W70-01074
*ODOR PRODUCING ALGAE, EUTROPHICA   W69-00387
*ODOR-PRODUCING ALGAE, *NUISANCE   W70-05112
*ODOR-PRODUCING ALGAE, *TASTE-PRO   W70-02373
*ODOR, *ALGICIDES, ACTIVATED CARB   W72-03746
*ODOR, *AMINO ACIDS, POTABLE WATE   W72-01812
*ODOR, *NATURAL STREAMS, *DISTRIB   W72-03562
*ODOR, *RESERVOIRS, *DECIDUOUS TR   W72-03475
*ODOR, *WASTE TREATMENT, ALGAE, D   W70-04788
*ODOR, ACTIVATED CARBON, AMMONIA,   W72-04155
*ODOR, HYDROGEN ION CONCENTRATION   W72-09694
*ODOR, WATER QUALITY CONTROL, POL   W72-13605
*ODORS, *ODOR PRODUCING ALGAE, EU   W69-00387
*OHIO RIVER, NUISANCE ALGAE, WATE   W70-04507
*OHIO RIVER BASIN.:   W72-01788
*OHIO RIVER, *ANALYTICAL TECHNIQU   W71-04206
*OHIO, *LAKE ERIE, *WATER POLLUTI   W69-06305
*OHIO, WATER TEMPERATURE, SLUDGE   W70-00475
*OIL POLLUTION, *RECOVERY FROM PO   W70-01231
*OIL POLLUTION, *FILAMENTOUS GREE   W70-09976
*OIL SPILLS, *TORREY CANYON, AROM   W70-01470
*OIL SPILLS, PROTEIN AQUICULTURE.   W69-05891
*OIL WASTES, *SURFACTANTS, *WATER   W70-01470
*OIL WASTES, *CHEMICAL WASTES, IN   W72-03299
*OIL, *MARINE ALGAE, WATER POLLUT   W70-09976
*OIL, *OIL WASTES, *SURFACTANTS,   W70-01470
*OIL, *OIL WASTES, *CHEMICAL WAST   W72-03299
*OILY WATER, LICHENS, TIDAL EFFEC   W70-01777
*OILY WATER, AQUATIC ALGAE, MARIN   W70-01231
*OILY WATER, *MARINE ALGAE, *EMUL   W70-01780
*OLEFINS, GREEN RIVER FORMATION,   W69-08284
*ON-SITE INVESTIGATIONS, FREQUENC   W70-09976
*ONONDAGA LAKE, ONONDAGA COUNTY,   W69-02959
*ONONDAGA LAKE(NY), EUNOTIA, AMPH   W70-03974
*ONTARIO, *CANADA, NUTRIENT LEVEL   W70-02973
*ONTARIO, *CANADA, KUSHOG LAKE(ON   W70-02795
*OPEN SEA, WEAPONS TESTING, NUCLE   W70-00707
*OPERATION AND MAINTENANCE, *OXID   W70-08392
*OPISTHONEMA OGLIMUN, UDOTEA FLAB   W69-08525
*OREGON, THERMOPHILIC BACTERIA, T   W70-05270
*ORGANI: /SIS, TASTE-PRODUCING AL   W72-11906
*ORGANIC ACIDS, BIOMASS, ORGANIC   W70-04195
*ORGANIC CARBON, *CELL SIZE, EUKA   W69-08278
*ORGANIC COMPOUNDS, *OXYGENATION,   W70-08416
*ORGANIC DEGRADATION.:   W72-04506
*ORGANIC LOADING, *SULFATES, *SUL   W70-07034
*ORGANIC LOADING, ALGAE, AMMONIA,   W71-05013
*ORGANIC LOAD EFFECTS, *DETENTION   W70-04787
*ORGANIC MATTER, *PHYSIOLOGICAL E   W70-04194
*ORGANIC MATTER, NUTRIENTS, LAKES   W71-10079

CATORS, *PROTOZOA, *ENVIRONMENT,
CEANS, *MARINE ALGAE, *LITTORAL,
LIMNOLOGY, ZOOPLANKTON, ALGAE, /
*WATER TREATMENT, WATER QUALITY,
DWATER, RECHARGE, *ION EXCHANGE,
MICROBIAL ECOLOGY, SWEDEN, LAKE/
MENTS, MICROBIAL/ *HETEROTROPHY,
M WASTES, *SPRINKLER IRRIGATION,
ROBIC TREAT/ *OXIDATION LAGOONS,
ION,/ *WASTE TREATMENT, *WATERS,
, SCENEDESMUS QU/ *ALGAL BLOOMS,
, SCENEDESMUS QU/ *ALGAL BLOOMS,
RIER(ISRAEL), THRESHOLD ODOR NU/
H LAKE(ORE.), AGENCY LAKE(ORE.),
                                   ALUM
TREATMENT, ALGAE, PHOTOSYNTHESI/
GESTION, *MUNICIPAL WASTES, MET/
INTENS/ *WASTE WATER TREATMENT,
FUR BACTERIA, STABIL/ *SULFIDES,
ING, ALGAE, AMMONIA, / *POULTRY,
OXYGEN DEMAND, *ANAEROBICS, ALG/
ISPOSAL, *WASTE WATER TREATMENT,
ES, *OXYGENATION, AEROBIC TREAT/
ERIA, *DISSO/ *SEWAGE TREATMENT,
DE, *SULFIDES, ALGAE, ORGANIC L/
XY/ *AERATION, *ALGAE, *LAGOONS,
ROL, *OPERATION AND MAINTENANCE,
TRIENTS, *WASTE WATER TREATMENT,
ACTERIA, *WASTE WATER TREATMENT,
ASTE WAT/ *BIOLOGICAL TREATMENT,
, *PHOSPHORUS, SEWAGE TREATMENT,
   *AGE, *ALGAE, *BIODEGRADATION,
SPARENC/ *LAKE WASHINGTON(WASH),

MIXING, ROANOKE RAPIDS RESERVO/
, PRESSURE, MI/ *PHOTOSYNTHESIS,
ATMENT, SOLAR RADI/ *WASHINGTON,
ION EFFECTS, *ORGANIC COMPOUNDS,
COMP/ *REAERATION, *CHLOROPHYLL,
DATION LAGOONS, *ORGANIC WASTES,
ULING MATERIALS, TE/ *ALGICIDES,
 POLLUTION CONTROL, *FILTRATION,
TREATMENT, SANITARY ENGINEERING,
OSPHATES,/ *ALGAE, *FRESH WATER,
AKE SAGAMI, JAPAN, *GLENODININE,
KALOID TOXINS, IBOGAALKALOID, O/
UALITY, BIOINDI/ *LAKE SUPERIOR,
ES, PRODUCTIVITY, RADIOISOTOPES/
*ZINC RADIOISOTOPES, STREAMFLOW/
, RESIDUES, *PESTICIDE KINETICS,
KINE/ *WATER POLLUTION EFFECTS,
EFFECTS, *PESTICIDES, RESIDUES,
TER QUALITY CONTROL, *NUTRIENTS,
SSAY, BIOINDICATORS, FISHKILL, /
SOILS, COA/ *PESTICIDE KINETICS,
EMOVAL, *ADSORPTION, SOILS, COA/
EMOVAL, *FOOD WEBS, DDT/ *ALGAE,
DT/ *ALGAE, *PESTICIDE TOXICITY,
LIZATION, *PESTICIDE METABOLISM,
SOUTH CAROLINA, LYOPHILIZATION,
OTHIOATE PESTICIDES, HERBICIDES,
HOSPHORUS COMPOUNDS, *NUTRIENTS,
*ROTIFE/ *WATER QUALITY, *ALGAE,
LERANCE LIMIT, BLUEGILLS, CHLOR/
LIGURIAN/ *RADIONUCLIDE UPTAKE,
OD PRODUCTION, SOUTHAMPTON UNIV/
OMPOSING ORGANIC MATTER, *ALGAE,
N, *TOXICITY, ORGANIC COMPOUNDS,
TS, *NUISANCE ALGAE, *ALGICIDES,
, MONURON, TRANSPARENCY, CYCLOP/
S, *CHEMICAL PRECI/ *PHOSPHORUS,
, *LAKES, *ALGAE, *GROWTH RATES,
TRIBUTION,/ *ALGAE, *PHOSPHORUS,
HY, RAD/ *ANALYTICAL TECHNIQUES,
ION, CARBON DIOXIDE, SOLUBILITY,
RIENTS/ *ALGAE, *BIODEGRADATION,
ION, DIATOMS, MICROB/ *NITROGEN,
TER PO/ *EUTROPHICATION, *LAKES,

IENTS, *WATER POLLUTION SOURCES,
IENTS, *SEDIMENT-WATER INFERFAC/
IC WEED/ *PONDS, *FERTILIZATION,
ISTRY, *WATER POLLUTION EFFECTS,
LING NUTRIENTS, *EUTROPHICATION,
Y, *EUTROPHICATION, *DETERGENTS,
*TERTIARY TREATMENT, *GREAT LA/
MIXED LIQUOR, SUSPENDED SOLIDS,
OF ALGAE, EFFL/ *EUTROPHICATION,
ROORGANISMS, NITROGEN COMPOUNDS,
S, / *BIOCHEMICAL OXYGEN DEMAND,

*ORGANIC MATTER, WATER QUALITY, T   W71-03031
*ORGANIC MATTER, EXUDATION, TIDES   W70-01073
*ORGANIC MATTER, *PHYTOPLANKTON,   W70-02510
*ORGANIC MATTER, ALGAE, COPPER SU   W72-04495
*ORGANIC MATTER, *FILTRATION, DIA   W72-04318
*ORGANIC SOLUTES, *HETEROTROPHY,   W70-04284
*ORGANIC SOLUTES, *MARINE ENVIRON   W70-04811
*ORGANIC WASTES, *WASTE WATER DIS   W71-00940
*ORGANIC WASTES, *OXYGENATION, AE   W71-09546
*ORGANISMS, STREAMS, WATER POLLUT   W70-05750
*ORGANO-SILICATES, *ALKOXYSILANES   W71-03056
*ORGANO-SILICATES, *ALKOXYSILANES   W71-05719
*OSCILLATORIA, NATIONAL WATER CAR   W70-02373
*OSCILLATORA.:   *UPPER KLAMAT   W72-06051
*OSHKOSH(WIS), BACKWASHING.:   W72-08859
*OXIDATION LAGOONS, *WASTE WATER   W72-10573
*OXIDATION LAGOONS, *ANAEROBIC DI   W72-06854
*OXIDATION LAGOONS, *ALGAE, LIGHT   W72-06022
*OXIDATION LAGOONS, *DESIGN, *SUL   W70-01971
*OXIDATION LAGOONS, *ORGANIC LOAD   W71-05013
*OXIDATION LAGOONS, *BIOCHEMICAL   W71-05742
*OXIDATION LAGOONS.: /R, SEWAGE D   W71-09548
*OXIDATION LAGOONS, *ORGANIC WAST   W71-09546
*OXIDATION LAGOONS, *ALGAE, *BACT   W72-01818
*OXIDATION PONDS, *HYDROGEN SULFI   W71-11377
*OXIDATION LAGOONS, BIOCHEMICAL O   W70-06601
*OXIDATION LAGOONS, ABSORPTION, B   W70-08392
*OXIDATION LAGOONS, MILK, *MODEL   W70-07031
*OXIDATION LAGOONS, *MODEL STUDIE   W70-07034
*OXIDATION, LAGOONS, *BAFFLES, *W   W70-09320
*OXIDATION LAGOONS.: /NCENTRATION   W68-00012
*OXYGEN DEMAND, LABORATORY TESTS,   W70-06600
*OXYGEN DEFICIT, SECCHI DISC TRAN   W69-10182
*OXYGEN PRODUCTION.:   W72-06022
*OXYGEN SUPERSATURATION, VERTICAL   W70-00683
*OXYGEN, *SATURATION, TEMPERATURE   W70-00683
*OXYGEN, *FLUCTUATION, SEWAGE TRE   W69-10182
*OXYGENATION, *PHOTOSYNTHESIS, CH   W70-08416
*OXYGENATION, *TOXICITY, ORGANIC   W71-12183
*OXYGENATION, AEROBIC TREATMENT,   W71-09546
*PAINTS, *TEST PROCEDURES, ANTIFO   W71-07339
*PARTICLE SIZE, *CLAYS, *SILTS, S   W70-08628
*PATHOGENIC BACTERIA, ALGAE, ORGA   W72-13601
*PEPTIDES, COPPER, IRON, IONS, PH   W69-10180
*PERIDINIUM POLONICUM, PLANKTON B   W70-08372
*PERIDINIUM SP., *GLENODININE, AL   W71-07731
*PERIPHYTON, *LIMNOLOGY, *WATER Q   W70-05958
*PERIPHYTON, *RADIOACTIVITY, *LAK   W70-04371
*PERIPHYTON, *CYCLING NUTRIENTS,   W69-07862
*PERSISTENCE, EVAPORATION, BIODEG   W70-09768
*PESTICIDES, RESIDUES, *PESTICIDE   W70-09768
*PESTICIDE KINETICS, *PERSISTENCE   W70-09768
*PESTICIDES, FARM WASTES, SILT, E   W71-10376
*PESTICIDE RESIDUES, *ALGAE, BIOA   W69-08565
*PESTICIDE REMOVAL, *ADSORPTION,   W69-09884
*PESTICIDE KINETICS, *PESTICIDE R   W69-09884
*PESTICIDE TOXICITY, *PESTICIDE R   W70-03520
*PESTICIDE REMOVAL, *FOOD WEBS, D   W70-03520
*PESTICIDE ACCUMULATION.: /LYOPHI   W69-02782
*PESTICIDE METABOLISM, *PESTICIDE   W69-02782
*PESTICIDE RESIDUES, DIATOMS, CRA   W69-02782
*PESTICIDES, *WATER POLLUTION CON   W70-05470
*PESTICIDES, *FISH, *FISH KILLS,   W72-04506
*PETROCHEMICAL COMPOUNDS, MEAN TO   W72-11105
*PHAEODACTYLUM TRICORNUTUM, EDTA,   W69-08275
*PHAEODACTYLUM, *GAS EFFLUENT, FO   W70-00161
*PHENOLS.: /DECIDUOUS TREES, *DEC   W72-03475
*PHENOLS, PESTICIDES, OXIDATION L   W71-12183
*PHENOLS, *WATER QUALITY CONTROL,   W70-10161
*PHENYLUREAS, DIURON, FLAGELLATES   W70-10161
*PHOSPHATES, *PHOSPHORUS COMPOUND   W70-09186
*PHOSPHATES, LABORATORY TESTS, MI   W70-08637
*PHOSPHATES, ENERGY, DI-URNAL DIS   W70-06604
*PHOSPHATES, *ALGAE, CHROMOTOGRAP   W71-03033
*PHOSPHATES, CHEMICAL PRECIPITATI   W71-11393
*PHOSPHATES, *EUTROPHICATION, NUT   W71-04072
*PHOSPHATES, *ALGAE, *EUTROPHICAT   W71-05026
*PHOSPHATES, NUTRIENTS, ALGAE, WA   W69-09135
*PHOSPHATE ABSORPTION.:   W70-01031
*PHOSPHATES, ALGAE, WATER POLLUTI   W70-01031
*PHOSPHATE, *LAKES, *CYCLING NUTR   W69-06859
*PHOSPHATE, FISH, PLANKTON, AQUAT   W70-05417
*PHOSPHATES, *ANALYTICAL TECHNIQU   W70-05269
*PHOSPHATES, *PHYSIOLOGICAL ECOLO   W69-03358
*PHOSPHATES, CHELATION, WASTE TRE   W70-04373
*PHOSPHATES, *LAKES, *DETERGENTS,   W70-03964
*PHOSPHOROUS UPTAKE RATE.:   W71-13335
*PHOSPHOROUS, ALGAE, *HARVESTING   W71-13313
*PHOSPHORUS COMPOUNDS, ADSORPTION   W71-13335
*PHOSPHORUS, *NITROGEN, *COLIFORM   W71-05409

TION, *NITROGEN, NITROGEN CYCLE,   *PHOSPHORUS, PHOSPHORUS COMPOUNDS W72-01867
OMPARISO/ *NUTRIENTS, *NITROGEN,   *PHOSPHORUS, EFFICIENCIES, COST C W70-01981
GAE, *PHYTOPLANKTON, *NUTRIENTS,   *PHOSPHORUS, *NITROGEN, *ALKALINI W70-02795
LANT GROWTHS, *NUTRIENT CONTROL,   *PHOSPHORUS CONCENTRATIONS, MICRO W69-09340
INE ANIMALS, FISH, ECOSYSTEMS, /   *PHOSPHORUS, *CORAL, *REEFS, *MAR W69-08526
WATER(MUNICIPAL AND INDUSTRIAL),   *PHOSPHORUS SOURCES, *NUISANCE PL W69-09340
, *STREAMS, *EC/ *TRANSLOCATION,   *PHOSPHORUS RADIOISOTOPES, *TROUT W69-09334
ICATION, *TERTIARY TREATMENT, N/   *PHOSPHORUS, *NUTRIENTS, *EUTROPH W69-09340
ICATION, POLLUTION ABAT/ *ALGAE,   *PHOSPHORUS, *MINNESOTA, *EUTROPH W69-10167
DI-URNAL DISTRIBUTION,/ *ALGAE,   *PHOSPHORUS, *PHOSPHATES, ENERGY, W70-06604
INUM SULFATE, *ACTIVATED SLUDGE/   *PHOSPHORUS REMOVAL, *ALUM, *ALUM W70-09186
ORUS COMPOUNDS, *CHEMICAL PRECI/   *PHOSPHORUS, *PHOSPHATES, *PHOSPH W70-09186
PRECI/ *PHOSPHORUS, *PHOSPHATES,   *PHOSPHORUS COMPOUNDS, *CHEMICAL W70-09186
ISONS, *EFFICIENCIES, *NITROGEN,   *PHOSPHORUS, ALGAE, BIOLOGICAL TR W70-07469
AS/ *EUTROPHICATION, *MUD(LAKE),   *PHOSPHORUS, *AQUATIC ALGAE, *BIO W70-03955
SP/ *EUTROPHICATION, *NUTRIENTS,   *PHOSPHORUS COMPOUNDS, ALGAE, PHO W70-04074
EWAGE PLANTS, CHLOROPHYLL, PHOT/   *PHOSPHORUS, *NITROGEN, *ALGAE, S W70-04268
EDIMENTS, *EUTROPH/ *ADSORPTION,   *PHOSPHORUS COMPOUNDS, *LAKES, *S W69-03075
GEN, HYDROGEN ION CONCENTRATION,   *PHOSPHORUS, SEWAGE TREATMENT, *O W68-00012
TREATM/ *EUTROPHICATION, SEWAGE,   *PHOSPHORUS, PHOSPHATES, *SEWAGE W68-00256
KES, ACID LAKES, BIBLIOGRAPHIES,   *PHOSPHORUS RADIOISOTOPE, ON-SITE W68-00857
PHICATION, OLIGOTROPHIC, *LAKES,   *PHOSPHORUS RADIOISOTOPE, *TURNOV W68-00858
ENZYMES, *NUTRIENT REQUIREMENTS,   *PHOSPHORUS COMPOUNDS, ALGAE, AQU W69-03373
LING NUTRIENTS, *EUTROPHICATION,   *PHOSPHORUS COMPOUNDS, *WATER POL W69-03370
, *PESTICI/ *TERTIARY TREATMENT,   *PHOSPHORUS COMPOUNDS, *NUTRIENTS W70-05470
OPHICATION, *NITROGEN COMPOUNDS,   *PHOSPHORUS COMPOUNDS, *NUTRIENTS W70-04765
*CHEMICAL ANALYSIS, *NITROGEN,   *PHOSPHORUS, ALGAE, CYANOPHYTA, N W69-05868
ENT ENRICHMENT TESTS, *NITROGEN,   *PHOSPHORUS, *IRON, *LIMNOLOGY, * W69-06004
FER, / *TUBIFICIDS, *ECOSYSTEMS,   *PHOSPHORUS RADIOISOTOPES, *TRANS W69-07861
MARSHES, */ *ZINC RADIOISOTOPES,   *PHOSPHORUS RADIOISOTOPES, *SALT W69-08274
UTRIENTS, / *NITROGEN COMPOUNDS,   *PHOSPHORUS COMPOUNDS, *CYCLING N W69-05142
*NITROGEN COMPOUNDS, *NUTRIENTS,   *PHOSPHORUS COMPOUNDS, *WATER POL W69-04805
KTON, *WATER POLLUTION SOURCES,/   *PHOSPHORUS COMPOUNDS, *PHYTOPLAN W69-04800
EWAGE TREATM/ *ALGAE, *NITROGEN,   *PHOSPHORUS, *SEWAGE EFFLUENT, *S W72-08133
-32.:   *PHOSPHORUS CYCLE, GLOEOTRICHA, P W73-01454
CULES, NUCLEIC ACID, RIBONUCLEI/   *PHOTOPHOSPHORYLATION, *MACROMOLE W69-10151
SPRINGS, *LIGHT INTENSITY, SPR/   *PHOTOSYNTHESIS, *ALGAE, *THERMAL W69-09676
TION, TEMPERATURE, PRESSURE, MI/   *PHOTOSYNTHESIS, *OXYGEN, *SATURA W70-00683
ORGANIC COMPOUNDS, *OXYGENATION,   *PHOTOSYNTHESIS, CHLORELLA, PESTI W70-08416
DIOACTIVITY, CA/ *PHYTOPLANKTON,   *PHOTOSYNTHESIS, BICARBONATES, RA W70-02504
), KADOKA(SD).:   *PHOTOSYNTHETIC OXYGEN, LEMMON(SD W72-01818
ION, MICROORGANISMS, CALIFORNIA,   *PHOTOSYNTHESIS, WASTE WATER TREA W71-05267
LGAE, TEMPERATURE, S/ *PLANKTON,   *PHOTOSYNTHESIS, PHYTOPLANKTON, A W71-10788
XYGEN, EUTROPHICA/ *ZOOPLANKTON,   *PHOTOSYNTHESIS, *ALGAE, LAKES, O W71-10098
EAT LA/ *ALGAE, *EUTROPHICATION,   *PHOTOSYNTHESIS, *RESPIRATION, GR W72-05453
*DDT,/ *WATER POLLUTION EFFECTS,   *PHOTOSYNTHESIS, *PHYTOPLANKTON, W70-05272
OPLANKTON, CYANOPHYTA, / *ALGAE,   *PHOTOSYNTHESIS, *MINNESOTA, PHYT W70-05387
TURBULENCE, CHLORRELA, / *ALGAE,   *PHOTOSYNTHESIS, *GROWTH RATES, * W69-03730
A, / *BIOCHEMICAL OXYGEN DEMAND,   *PHOTOSYNTHETIC OXYGEN, *CHLORELL W69-03371
, WATER TEMPERATURE, ALKALINITY,   *PHOTOSYNTHESIS, LIGHT PENETRATIO W68-00248
MICAL OXYGEN DEMAND, *CHLORELLA,   *PHOTOSYNTHETIC OXYGEN, ALGAE, AN W69-03362
PRODUCTIVITY, *PLANKTON, ALGAE,   *PHOTOSYNTHESIS, PRODUCTIVITY, W W69-00632
TE, AQUATIC ENVIRONMENT, AQUATI/   *PHOTOSYNTHESIS, *CALCIUM CARBONA W69-00446
TECHNOLOGY FOUND, SYRACUSE(NY),/   *PHYCOLOGY, NY STATE SCIENCE AND W70-03973
CAL CHARACTERIS/ *RIVER REGIONS,   *PHYSICAL CHARACTERISTICS, *CHEMI W73-00748
MENTOR, / *CHEMICAL ENVIRONMENT,   *PHYSICAL ENVIRONMENT, PHAGE, FER W71-02009
FLORA, *ALGAE, *HYDRAULIC MODEL/   *PHYSIOLOGICAL ECOLOGY, *BENTHIC W71-00668
ENTAL EFFECTS, *MORPHOLOGY, *AL/   *PHYSIOLOGICAL ECOLOGY, *ENVIRONM W69-10177
ENVIRONMENTS, RIVERS, ENVIRONME/   *PHYSIOLOGICAL ECOLOGY, *AQUATIC W70-03978
VITY T/ *LAKES, *ORGANIC MATTER,   *PHYSIOLOGICAL ECOLOGY, RADIOACTI W70-04194
AE, MICROCYSTIS, COELOSPHAERIUM/   *PHYSIOLOGY, SPIROGYRA, MYXOPHYCE W70-04369
IA, *AQUATIC WEEDS, *INHIBITION,   *PHYSIOLOGICAL ECOLOGY, *PHYTOTOX W69-03188
S, *EUTROPHICATION, *PHOSPHATES,   *PHYSIOLOGICAL ECOLOGY, AQUATIC A W69-03358
L/ *CYANOPHYTA, *EUTROPHICATION,   *PHYSIOLOGICAL ECOLOGY, *WATER PO W69-03513
MPOUNDS, *NUTRIENT REQUIREMENTS,   *PHYSIOLOGICAL ECOLOGY, ALGAE, AN W69-03515
TA, ALGA/ *AQUATIC MICROBIOLOGY,   *PHYSIOLOGICAL ECOLOGY, *CYANOPHY W69-03518
ATER, *THERMAL WAT/ *CYANOPHYTA,   *PHYSIOLOGICAL ECOLOGY, *SALINE W W69-03516
NKTON, *CULTURES, / *CYANOPHYTA,   *PHYSIOLOGICAL ECOLOGY, *PHYTOPLA W69-03514
A/ *EUTROPHICATION, *CYANOPHYTA,   *PHYSIOLOGICAL ECOLOGY, ALGAE, AN W69-03517
NOPHYTA, *PHYSIOLOGICAL ECOLOGY,   *PHYTOPLANKTON, *CULTURES, ALGAE, W69-03514
, *ALGAE, *DIURNAL DISTRIBUTION,   *PHYTOPLANKTON, *DISSOLVED OXYGEN W69-03611
LGAL POISO/ *2-4 D, *HERBICIDES,   *PHYTOPLANKTON, *PHYTOTOXICITY, A W69-00994
FICATION, DESMIDS, OLIGOTROPHIC,   *PHYTOPLANKTON, BRITISH ISLES, BI W68-00255
ATTER, SEASONAL, ALGAE, PRIMARY/   *PHYTOPLANKTON, *LAKES, ORGANIC M W70-04290
, *ALGAE, ORGANIC/ *COMPETITION,   *PHYTOPLANKTON, *AQUATIC BACTERIA W70-04284
LUTION EFFECTS, *PHOTOSYNTHESIS,   *PHYTOPLANKTON, *DDT, *MARINE ALG W70-05272
IBUTION, MISSOURI / *RESERVOIRS,   *PHYTOPLANKTON, *ROTIFERS, *DISTR W70-05375
CONCENTRATION, WATER CHEMISTRY,   *PHYTOPLANKTON, CARBON DIOXIDE, N W70-05424
N, *WASTE WATER(POLLUTION), *WA/   *PHYTOPLANKTON, *SELF-PURIFICATIO W70-05092
WATER CIRCULATION, NUISANCE ALG/   *PHYTOPLANKTON, *LAKE MICHIGAN, * W69-05763
TRIENTS, *NUTRIENT REQUIREMENTS,   *PHYTOPLANKTON, AQUATIC ALGAE, AQ W69-04802
SOURCES,/ *PHOSPHORUS COMPOUNDS,   *PHYTOPLANKTON, *WATER POLLUTION W69-04800
ION, *FERTILIZATION, *LIMNOLOGY,   *PHYTOPLANKTON, FISH, GREAT LAKES W69-06858
CATION, ALGAE, BIOMASS, CHLOROP/   *PHYTOPLANKTON, *PONDS, *EUTROPHI W69-06540
SOTOPES, SEA WATER, FOOD CHAINS/   *PHYTOPLANKTON, *CYCLING, *RADIOI W70-00708
STE D/ *STRONTIUM RADIOISOTOPES,   *PHYTOPLANKTON, RADIOACTIVITY, WA W70-00707
ROWTH RATES,/ *ALGAE, *ALGAE,   *PHYTOPLANKTON, *LAKE ONTARIO, *G W69-10158
CROORGANISMS, BIOMAS/ *BACTERIA,   *PHYTOPLANKTON, *LAKES, ALGAE, MI W70-08654
PRODUCTIVITY, *ORGANIC MATTER, /   *PHYTOPLANKTON, *ALGAE, *PRIMARY W71-10079
, *INHIBITORS, *MARSHES, *LAKES,   *PHYTOPLANKTON, BIOASSAY, LABORAT W72-00845

NKTON, ALGAE, / *ORGANIC MATTER,
BICARBONATES, RADIOACTIVITY, CA/
REEN ALGA/ *FLOTATION, *SINKING,
PHORUS, *NITROGEN, *ALK/ *ALGAE,
MENTS, ECOLOGY, CHLORELLA, ALGA/
, *NUTRIENTS, *PRIMARY PRODUCTI/
ODS, *ZOOPLANKTON, *BIOMASS, *S/
*ALGAL POISONING, *ALGAL TOXINS,
D, *HERBICIDES, *PHYTOPLANKTON,
IBITION, *PHYSIOLOGICAL ECOLOGY,
ENTED FLAGELLATES.:
RRENTS(WATER), LIMNOLOGY, LAKES,
BORATORY STUDIES, *MASS CULTURE,
GAE ISOLATION, CAPILLARITY, CAP/
*ALGAE, *ANALYTICAL TECHNIQUES,
Y, TEMPERATURE, PRIMARY PRODUCT/
IGH/ *ENVIRONMENT, *FRESH WATER,
TY, CHLORELLA, CHLOROPHYTA, DIA/
TROPHICATION, *LAKES, LIMNOLOGY,
, *LAKES, *PRIMARY PRODUCTIVITY,
R/ *CYANOPHYTA, *EUTROPHICATION,
PLANKTON, ALGAE, TEMPERATURE, S/
MENT, *EUTROPHICATION, / *ALGAE,
, *LAKE ONTARIO, *GROWTH RATES,/
, *CYCLING NUTRIENTS, ECOSYSTEM/
HEMICAL/ *NUTRIENT REQUIREMENTS,
Y, CHEMICAL ANALYSIS, P/ *ALGAE,
UM, SILICA, DIATOMS, RESISTANCE,
BIBLIOGRAPHIES, *EUTROPHICATION,
SPHATES, SILICATES, IRON OXIDES,
PTH, ALGAE/ *DEEP WATER, *LAKES,
HLORINE, *ANALYTICAL TECHNIQUES,
ATERS, ECOSYSTEMS, ESTUARINE EN/
TRIENTS, *MARINE MICROORGANISMS,
CAL TREATMENT, *WATER TREATMENT,
MENT, *POLLUTANT IDENTIFICATION,
E WATER, *NUTRIENT REQUIREMENTS,
, MUNICIPAL WASTES, GREAT LAKES,
ITY, *BIOASSAY, FISH, AQUATIC E/
GREEN ALGAE, *EUGLENA, PIGMENTE/
OUNTY(ILL), CALUMET RIVER(ILL),/
TION, *POLLUTION IDENTIFICATION,
 ERIE, *WATER POLLUTION CONTROL,
E ENVIRONMENT, *INSTRUMENTATION,
TER POLLUTION SOURCES, DREDGING,
STEMS, FLOW RA/ *EUTROPHICATION,
ANIC POLYELECTROLYTES, CATIONIC/
S, METHYL ALCOHOL, ETHYLENE GLY/
AE BLOOMS, *FISH TOXICITY TEST,/
M NODOSUM, VIBRIO, FLAVOBACTER,/
CHLORINATION, */ *STABILIZATION,
WASTE TREATMENT, *STABILIZATION,
WASTE TREATMENT, *STABILIZATION,
IOMASS, CHLOROP/ *PHYTOPLANKTON,
E, FISH, PLANKTON, AQUATIC WEED/
STRATIFICATION, *SEASONAL, *SUL/
, WATER SUPPLY, / *INFILTRATION,
NOLOGY, *WATER, *RIVERS, *LAKES,
ZE, CHLORELLA/ *COULTER COUNTER,
IES, EFFLUENTS/ *CARBON, *ALGAE,
NTROL, *ALGAE, COPPER SULFATE, /
R BACTERIA, COL/ *ALGAL CONTROL,
HILLS(CALIF).:
L TESTS, ALGISTATIC TESTS, SLIM/
AE, *DECOMPOSING ORGANIC MATTER,
NES(IOWA).:

C LAGOONS, ANAEROBIC LAGOONS.:
REENS, SEDIMENTATION, ORGANIC L/
ON.:
PHYTOPLANKTON RELATIONSHIPS.:

C PALUDOSUM, BLE/ *ALGAL GROWTH,
GANIC LOADING, ALGAE, AMMONIA, /
SEDIMENTATION, OX/ *FARM WASTES,
BAFFLES, *WASTE WATER TREATMENT,
*TEMPERATURE, *COOLING WATER, /
ISANCE, / *CORRECTIVE TREATMENT,
:   MARINE BIOLOGY,
ATTER, / *PHYTOPLANKTON, *ALGAE,
LIMNOLOGY, *ESSENTIAL NUTRIENTS,
ATION, WATER QUALITY/ RESERVOIR,
*LAKES, *NEW YORK, *MEROMIXIS,
, *LIMITING FACTORS, *NUTRIENTS,
*LABORATORY TESTS, RESPIRATION,/
N/ *BENTHIC FLORA, *CHLOROPHYLL,
MNOLOGY, DYSTROPHY, OLIGOTROPHY,
HERMAL WATER, PERIPHYTON, BIOMA/
ULATIONS, ENVIRONMENTAL EFFECTS,
UTRIENTS, BIO MASS, OLIGOTROPHY,

*PHYTOPLANKTON, LIMNOLOGY, ZOOPLA   W70-02510
*PHYTOPLANKTON, *PHOTOSYNTHESIS,   W70-02504
*PHYTOPLANKTON, *DIATOMS, *BLUE-G   W70-02754
*PHYTOPLANKTON, *NUTRIENTS, *PHOS   W70-02795
*PHYTOPLANKTON, *NUTRIENT REQUIRE   W70-02804
*PHYTOPLANKTON, *LIMITING FACTORS   W70-01579
*PHYTOPLANKTON, *STATISTICAL METH   W73-01446
*PHYTOTOXICITY, *FISH DISEASES, *   W71-07731
*PHYTOTOXICITY, ALGAL POISONING,   W69-00994
*PHYTOTOXICITY, ALGAL CONTROL, AL   W69-03188
*PIGMENTED FLAGELLATES, *NON-PIGM   W72-01798
*PILOT PLANTS, *NUISANCE ALGAE, R   W68-00479
*PILOT-PLANT STUDIES, ALGAL GROWT   W69-06865
*PIPETTE, UNIALGAL, NITZSCHIA, AL   W70-04469
*PLANKTON, *LABORATORY TESTS, *IS   W70-04469
*PLANKTON, *ALGAE, LIGHT INTENSIT   W70-04809
*PLANKTON, *ALGAE, SCENEDESMUS, L   W70-05409
*PLANKTON, *ALGAE, *LIGHT INTENSI   W70-05381
*PLANKTON, ALGAE, BIBLIOGRAPHIES,   W69-00446
*PLANKTON, ALGAE, PHOTOSYNTHESIS   W69-00632
*PLANKTON, ALGAE, *NUTRIENT REQUI   W69-03512
*PLANKTON, *PHOTOSYNTHESIS, PHYTO   W71-10788
*PLANKTON, *SEWAGE, *SEWAGE TREAT   W70-02304
*PLANKTON, *ALGAE, *PHYTOPLANKTON   W69-10158
*PLANKTON, *ALGAE, *SESSILE ALGAE   W73-01454
*PLANT GROWTH, *ALGAE, ELEMENTS(C   W70-02964
*PLANT GROWTH, *WISCONSIN, ECOLOG   W70-06229
*PLANT GROWTH REGULATORS, FRESH W   W68-00856
*PLANT GROWTH, ANALYTICAL TECHNIQ   W69-04801
*PLANT POPULATIONS, REVIEWS, LIGH   W68-00860
*PLANTS, *CALIFORNIA, *NEVADA, DE   W70-00711
*POLAROGRAPHIC ANALYSIS, PHOTOMET   W70-05178
*POLLUTANTS, *ESTUARIES, *TIDAL W   W68-00010
*POLLUTANTS, RECREATION, WATER SP   W69-04276
*POLLUTANT IDENTIFICATION, *POLLU   W69-07389
*POLLUTANT ABATEMENT, EFFLUENTS,   W69-07389
*POLLUTANTS, INHIBITION, LIMNOLOG   W70-02792
*POLLUTANT IDENTIFICATION.: /RCES   W71-05879
*POLLUTANT IDENTIFICATION, *TOXIC   W72-11105
*POLLUTION-TOLERANT ALGAE, *BLUE   W70-01233
*POLLUTION, *CHICAGO(ILL), COOK C   W70-00263
*POLLUTION ABATEMENT, TIDAL WATER   W69-06203
*POLLUTION ABATEMENT, POLLUTANTS,   W69-06305
*POLLUTION IDENTIFICATION, *POLLU   W69-06203
*POLLUTION ABATEMENT, DIKES, SOIL   W68-00172
*POLLUTION ABATEMENT, DRAINAGE SY   W68-00479
*POLYELECTROLYTES, *SYNTHETIC ORG   W70-05267
*POLYESTER FIBERS, MAN-MADE FIBER   W70-10433
*POLYMERS, *BIOLOGICAL LIFE, *ALG   W70-05819
*POLYSIPHONIA LANOSA, *ASCOPHYLLU   W70-03952
*PONDS, *DESIGN, *EFFICIENCIES, *   W70-04788
*PONDS, *DESIGN CRITERIA, FACILIT   W70-04786
*PONDS, *ENVIRONMENTAL EFFECTS, *   W70-04787
*PONDS, *EUTROPHICATION, ALGAE, B   W69-06540
*PONDS, *FERTILIZATION, *PHOSPHAT   W70-05417
*PONDS, *STABILIZATION, *THERMAL   W70-04790
*PONDS, *SOIL, GROUNDWATER BASINS   W69-07838
*PONDS, LIGHT, TEMPERATURE, FERTI   W70-05404
*POPULATION DYNAMICS, CELLULAR SI   W69-06540
*POPULATION, *BIOLOGICAL COMMUNIT   W70-04001
*POTABLE WATER, *WATER QUALITY CO   W71-00110
*POTABLE WATER, CYANOPHYTA, SULFU   W70-08107
*POTASSIUM PERMANGANATE, BEVERLY   W71-00110
*POTASSIUM PERMANGANATE, ALGICIDA   W69-05704
*POTASSIUM COMPOUNDS, *ACTIVATED   W72-04242
*POTASSIUM PERMANGANATE, *DES MOI   W72-04242
*POTASSIUM PERMANGANATE.:   W72-04155
*POTATO PROCESSING WASTES, AEROBI   W71-08970
*POTATOES, *INDUSTRIAL WASTES, SC   W71-08970
*POTENTIATION, *ORGANIC DEGRADATI   W72-04506
*POTOMAC RIVER ESTUARY, NUTRIENT-   W71-05407
*POTOMAC RIVER ESTUARY.:   W71-05409
*POTOMAC RIVER, PLECTONEMA, NOSTO   W70-02255
*POULTRY, *OXIDATION LAGOONS, *OR   W71-05013
*POULTRY, WASTE WATER TREATMENT,   W71-11375
*POULTRY, INDUSTRIAL WASTES, LAGO   W70-09320
*POWERPLANTS, *THERMAL POLLUTION,   W69-05023
*PREVENTATIVE TREATMENT, ALGAE NU   W69-08282
*PRIMARY BIOLOGICAL PRODUCTIVITY.   W71-10788
*PRIMARY PRODUCTIVITY, *ORGANIC M   W71-10079
*PRIMARY PRODUCTIVITY, AQUATIC AL   W72-01990
*PRIMARY PRODUCTIVITY, *EUTROPHIC   W70-09093
*PRIMARY PRODUCTIVITY, *ZOOPLANKT   W70-05760
*PRIMARY PRODUCTIVITY, PHOSPHATES   W70-01579
*PRIMARY PRODUCTIVITY, *STREAMS,   W70-00265
*PRIMARY PRODUCTIVITY, *MEASUREME   W69-10151
*PRIMARY PRODUCTIVITY, *TROPHIC L   W69-09723
*PRIMARY PRODUCTIVITY, *ALGAE, *T   W70-03309
*PRIMARY PRODUCTIVITY, WATER POLL   W70-05424
*PRIMARY PRODUCTIVITY, TROPIC LEV   W68-00912

367

URCES, *WATER POLLUTION EFFECTS,
 RATES, *EUTROPHICATION, *LAKES,
HEMICAL ANALYSIS, SESSILE ALGAE,
AKES, *EUTROPHICATION, *BIOMASS,
WAGE EFFLUENTS, PHO/ *NUTRIENTS,
WATER CHEMISTRY, POLLUT/ *ALGAE,
OLOGY, NUTRIENTS, PLANT/ *LAKES,
 MATTER, WATER QUA/ *INDICATORS,
ELECTROP/ *ISOLATION, *BACTERIA,
 *THERMAL STRESS, THERMAL WATER,
CTERIA, META/ *SEWAGE TREATMENT,
ION, AQUATIC MI/ *AQUATIC ALGAE,
Y ENGINEERING, MICROORG/ *ALGAE,
E, *BIOMASS, WATER PURIFICATION,
IVER, SAVAN/ *TEMPERATURE SHOCK,
ACTERIA, *BACTERICIDES, *CHLORI/

ROCYSTIS AERUGINOSA, APHANIZOME/
*AQUATIC LIFE, WATER POLLUTION /
ISOTOPES, LIMNOLOGY, / *ECOLOGY,
CKY, LIFE HISTORY ST/ *SCULPINS,
ITY, RADIOISOTOPES/ *PERIPHYTON,
*WATER POLLUTION, *MINE WASTES,
 *BIOTIN, *AMPHIDINIUM CARTERI,
TRIENTS, *NUTRIENT INTERACTIONS,
IMALS, PLANTS, BOTTOM SEDIMENTS/
RATURE, SOUTH CAROLINA,/ *ALGAE,
CATORS, *NEW YORK,/ *MONITORING,
LOGY, *RADIOACTIVITY TECHNIQUES,
HAINS/ *PHYTOPLANKTON, *CYCLING,
, ABSORPTION, GULF OF MEXICO, A/
SCHARGE MEASUREMENT, *EFFLUENTS,
E ALGAE, MARIN/ *MARINE ANIMALS,
ENT, CHLORELLA, TRACERS, PHOTOS/
*RIVERS, OXIDATION, PHOTOSYNTH/
AKE, *RADIONUCLIDE ACCUMULATION,
ONUCLIDE / *RADIONUCLIDE UPTAKE,
DE ACCUMULATION, *RADIONUCLIDE /
HALF-LIFE, DETECTOR.:
LUM TRICORNUTUM, EDTA, LIGURIAN/
IZATION, *ARTIFICIAL SUBSTRATES,
OSTRONTIUM, *MICROBIAL ACTIVITY,
IOSTRONTIUM, *FUNGAL-BOUND RADI/
UATIC ENVIRONMENTS, *ECOSYSTEMS,
, BIOINDICATORS, CHLAMYDOMONAS,/
XY/ *AERATION TUBES, *DIFFUSERS,
TION, BA-LA, HA/ *HANFORD(WASH),
NATION, *TOXICITY, ORGANIC COMP/
DS, CHEMICAL WAST/ *WATER REUSE,
, *BIODEGRADATION, *WATER REUSE,
TER QUALITY, AEST/ *WATER REUSE,
*WATER TREATMENT, *WATER REUSE,
C/ *SALT MARSH, *OIL POLLUTION,
ION, *FILAMENTOUS GREEN ALGAE, /
EMPERATURE,/ *LIMNOLOGY, *LAKES,
LLUTION EFFECTS/ *WATER QUALITY,
*BLUE-GREEN ALGAE, *GREEN ALGAE,
OSYSTEMS, / *PHOSPHORUS, *CORAL,
TMENT, OVERFERTILIZATION, PHYSI/
PTION, CELL CONCENTRATION, UROG/
GRAPHIES, CHLORELLA, CH/ *ALGAE,
URCES, *WATER POLLUTION CONTROL,
, ALGAE, AIR POLL/ *FARM WASTES,
NESSEE VALLEY AUTHORITY PROJECT,
ERATURE, HEAT BUDGET, TURBIDITY,
IAL WASTES, *WATER POLLUTION EF/
RBOHYDRATES, *NUTRIENTS, *BIODE/
ROGEN, OXYGEN, OXIDATION, CARBO/
ATION, *AERATION, *IMPOUNDMENTS,
ANKTON, TEMPERATURE, *E/ *LAKES,
R QUALITY, *STRATIFICATION, *TE/
IFERS, *DISTRIBUTION, MISSOURI /
WATER QUALITY CONTROL, *ECOLOGY,
CE ALGAE, ROOTED AQUATIC PLANTS,
, NEBRASKA, TU/ *EUTROPHICATION,
*WATER TREATMENT, *TASTE, *ODOR,
DEVELOPMENT, WATER QUALITY, SO/
LUTION, ALGAL GROWTHS, OXYGEN D/
OXYGEN, DISSOLVED SOLIDS, SEDIM/
NC RADIOISOTOPES, *MARINE ALGAE,
HICA/ *DISSOLVED OXYGEN, *ALGAE,
EUTROPHICATION, *PHOTOSYNTHESIS,
ISSOLVED OXYGEN, PENNSY/ *ALGAE,

LAKES, *WATER POLLUTION CONTROL,
ATORS, *WATER POLLUTION EFFECTS,
APHIES, PHOSPH/ *EUTROPHICATION,
AE, *NUTRIENTS, *CARBON DIOXIDE,
GAE, OLIGOTROPHY, PHYTOPLANKTON,
S, SHELLFISH/ *TOXICITY, *ALGAE,

*PRIMARY PRODUCTIVITY, ALGAE, BIO   W69-02959
*PRIMARY PRODUCTIVITY, *PLANKTON,   W69-00632
*PRIMARY PRODUCTIVITY, OREGON, CY   W72-06052
*PRODUCTIVITY, CYANOPHYTA, NUISAN   W70-03959
*PRODUCTIVITY, NUISANCE ALGAE, SE   W70-00270
*PRODUCTIVITY, *EUTROPHICATION, *   W70-02646
*PRODUCTIVITY, MICHIGAN, *EPIPHYT   W71-00832
*PROTOZOA, *ENVIRONMENT, *ORGANIC   W71-03031
*PROTOZOA, *ALGAE, LETHAL LIMIT,   W71-03775
*PROTOZOA, ALGAE, DIATOMS, ANIMAL   W70-07847
*PROTOZOA, *ACTIVATED SLUDGE, *BA   W72-01817
*PROTOZOA, *FLORIDA, *EUTROPHICAT   W72-01993
*PROTOZOA, *SYSTEMATICS, *SANITAR   W72-01798
*PROTOZOA, *FUNGI, *BACTERIA, HYD   W71-05267
*PROTOZOAN COMMUNITIES, POTOMAC R   W70-07847
*PSEUDOMONAS, *SWIMMING POOLS, *B   W69-10171
*PURE ALGAL CULTURES.:   W71-06002
*PURIFICATION, CULTURE MEDIA, MIC   W70-00719
*RADIOACTIVITY, *TRACE ELEMENTS,   W71-05531
*RADIOACTIVITY TECHNIQUES, *RADIO   W69-10163
*RADIOACTIVITY, *FOOD WEBS, KENTU   W69-09742
*RADIOACTIVITY, *LAKES, PRODUCTIV   W70-04371
*RADIOACTIVE WASTES, *MUNICIPAL W   W69-01165
*RADIOCARBON UPTAKE TECHNIQUES, W   W70-05652
*RADIOCARBON BIOASSAY, *NUTRIENT   W70-04765
*RADIOISOTOPES, *DISTRIBUTION, AN   W69-08269
*RADIOISOTOPES, *SORPTION, *TEMPE   W69-07845
*RADIOISOTOPES, *STREAMS, BIOINDI   W69-10080
*RADIOISOTOPES, LIMNOLOGY, OCEANO   W69-10163
*RADIOISOTOPES, SEA WATER, FOOD C   W70-00708
*RADIOISOTOPES, *ALGAE, *SORPTION   W69-08524
*RADIOISOTOPES, *WASTE DISPOSAL,   W72-00941
*RADIOISOTOPES, ABSORPTION, MARIN   W71-0985Q
*RADIOISOTOPES, *AQUATIC ENVIRONM   W70-02786
*RADIONUCLIDE, *DISSOLVED OXYGEN,   W71-11381
*RADIONUCLIDE EXCHANGE, *OPEN SEA   W70-00707
*RADIONUCLIDE ACCUMULATION, *RADI   W70-00707
*RADIONUCLIDE UPTAKE, *RADIONUCLI   W70-00707
*RADIONUCLIDE, ULOTHRIX, UPTAKE,   W69-07862
*RADIONUCLIDE UPTAKE, *PHAEODACTY   W69-08275
*RADIONUCLIDE ACCUMULATION, RUTHE   W70-04371
*RADIOSTRONTIUM COMPLEXES, DTPA,   W71-04067
*RADIOSTRONTIUM, *ALGAL-BOUND RAD   W71-04067
*RADIUM RADIOISOTOPES, *URANIUM R   W69-07846
*RAIN, *VITAMINS, ALGAE, BIOASSAY   W69-06273
*RAYTOWN(MO), LEE'S SUMMIT(MO), O   W70-06601
*REACTOR, RADIONUCLIDES, ASSIMILA   W69-07853
*REAERATION, *CHLOROPHYLL, *OXYGE   W71-12183
*RECLAIMED WATER, ORGANIC COMPOUN   W70-10433
*RECLAIMED WATER, NITROGEN COMPOU   W69-08054
*RECLAIMED WATER, *CALIFORNIA, WA   W70-05645
*RECLAMATION, *GROUNDWATER, RECHA   W72-04318
*RECOVERY FROM POLLUTION, *KUWAIT   W70-01231
*RECOVERY, OIL SPILL, *OIL POLLUT   W70-09976
*RECREATION, RESERVOIRS, PONDS, T   W70-06225
*RECREATION, *SWIMMING, *WATER PO   W71-09880
*RED ALGAE, *TEMPERATURE TOLERANC   W71-02496
*REEFS, *MARINE ANIMALS, FISH, EC   W69-08526
*REMOVAL PROCESSES, CHEMICAL TREA   W70-07469
*REQUIREMENTS, SCENEDESMUS, ABSOR   W70-05409
*RESEARCH AND DEVELOPMENT, BIBLIO   W69-06865
*RESEARCH EQUIPMENT, *ANALYTICAL   W70-08627
*RESEARCH AND DEVELOPMENT, GRANTS   W71-06821
*RESERVOIRS, TEMPERATURE, DISSOLV   W71-07698
*RESERVOIR EVAPORATION, SALINITY,   W72-01773
*RESERVOIRS, *LIMNOLOGY, *INDUSTR   W70-06975
*RESERVOIRS, *EUTROPHICATION, *CA   W71-00221
*RESERVOIRS, *EUTROPHICATION, NIT   W70-05550
*RESERVOIRS, *MIXING, *WATER QUAL   W70-04484
*RESERVOIRS, ZOOPLANKTON, PHYTOPL   W70-04475
*RESERVOIRS, *URBANIZATION, *WATE   W70-04727
*RESERVOIRS, *PHYTOPLANKTON, *ROT   W70-05375
*RESERVOIR, MASSACHUSETTS, POTABL   W70-05264
*RESERVOIRS, HYDROELECTRIC PLANTS   W68-00479
*RESERVOIRS, *GREAT PLAINS, ALGAE   W72-04761
*RESERVOIRS, *DECIDUOUS TREES, *D   W72-03475
*RESERVOIR SITES, WATER RESOURCES   W72-11906
*RESERVOIR EFFECTS, LAKE MEAD, DI   W73-00750
*RESERVOIR OPERATIONS, DISSOLVED   W73-00750
*RESINS, SEA WATER, PHYSICOCHEMIC   W69-08275
*RESPIRATION, *ESTUARIES, *EUTROP   W69-07520
*RESPIRATION, GREAT LAKES, NUISAN   W72-05453
*RESPIRATION, *EUTROPHICATION, *D   W69-10159
*RESPIROMETER CHAMBER.:   W71-00668
*REVIEWS, ALGAE, WEEDS, ECOLOGY,   W70-07393
*REVIEWS, *WATER QUALITY, WATER P   W70-07027
*REVIEWS, DOCUMENTATION, BIBLIOGR   W69-08518
*REVIEWS, BIBLIOGRAPHIES, BACTERI   W70-00664
*REVIEWS, *WATER POLLUTION EFFECT   W68-00680
*REVIEWS, HABITATS, CLAMS, MUSSEL   W70-05372

368

OROPHYLL A, DATA INTERPRETATION,
ERISTICS, *CHEMICAL CHARACTERIS/
UALITY NETWORK, ALGAE COUNTS, P/

FERTILIZERS, NUTRIENTS/ *LAKES,
ISSIPPI RIVER, *OHIO RIVER, NUI/
EMPERATURE,/ *LIMNOLOGY, *WATER,
CONTRO/ *EUTROPHICATION, *LAKES,
POLLUTION SO/ *ECOLOGY, *ALGAE,
ATIC LIFE, *AQUATIC POPULATIONS,
RADIONUCLIDE, *DISSOLVED OXYGEN,
WATER QUALITY, *WATER POLLUTION
I / *RESERVOIRS, *PHYTOPLANKTON,
LANKTO/ *AQUATIC PLANTS, *ALGAE,
*PESTICIDES, *FISH, *FISH KILLS,
POSAL, *AGRICULTURAL WATERSHEDS,
OMPOUNDS, *NUTRIENTS, *BIOASSAY,
SEDIMENTS, ORGANIC MATTER, CURR/

NOPHYTA, *PHYSIOLOGICAL ECOLOGY,
AINAGE / *LAKES, PLANTS, *ALGAE,
OPES, *PHOSPHORUS RADIOISOTOPES,
OVERY FROM POLLUTION, *KUWAIT C/
STANCE, WATER POLLUTION EFFECTS,
EAR WATER RESERVOIRS, *TURBID R/
ATION EFFECTS, DRAINAGE EFFECTS,
*AQUATIC WEEDS, *WATER ANALYSIS,
HIO RIVER, NUI /*RIVERS, *ALGAE,
ETRY, DYE REL/ *INSTRUMENTATION,
CTROPHORESIS, CHLORELLA/ *ALGAE,
ONS, *ENVIRONMENTAL ENGINEERING,
*ALGAE, *PROTOZOA, *SYSTEMATICS,
PLY, *BIOLOGICAL PROPERTIES, *W/
IOMASS TITER, *BIOLOGICAL INDEX,
ATANS, SEAWEED, ZIRCONIUM-95-NI/
NIUM-95-NI/ *SARGASSUM FLUITANS,
E, MI/ *PHOTOSYNTHESIS, *OXYGEN,
*FILTRATION, HEAD LOSS, FILTERS,
*AERATION, *NITRATES, *CULTURES,
PHOTOSYNTH/ *ALGAE, *CHLORELLA,
GROWTH RATES, *LABORATORY TESTS,
DES MORTS(WISC), GROWTH STIMUL/
NING, *MICROSTRAINERS.:
WEBS, KENTUCKY, LIFE HISTORY ST/
ALGAE, CULTURES, CH/ *BIOASSAY,
ZATION, *THERMAL STRATIFICATION,
ENVIRONMENT, *WATER TEMPERATURE,
PLANT, DUWAMISH ESTUARY(WASH).:
CIDS, */ *OCEANS, *MARINE ALGAE,
DOSUM, ALGINIC ACID, FUCUS VESI/
ATE, *LAKES, *CYCLING NUTRIENTS,
*RUNOFF, *GROUNDWATER, *SESTON,
LING NUTRIENTS.:
, *PHOSPHORUS COMPOUNDS, *LAKES,
UCTURE, 2,3-DICHLORONAPHTHOQUIN/
ION CONTROL, *LAKE ONTARIO, *WA/
POLLUTION), *WA/ *PHYTOPLANKTON,
DIATOMACEOUS EARTH, *FILTRATION,
S, ECOSYSTEM/ *PLANKTON, *ALGAE,
ER, CURR/ *RUNOFF, *GROUNDWATER,
STES, *LAGOONS, *BIODEGRADATION,
ATERSHEDS, *RUNOFF, *FERTILIZAT/
URAL/ *LAKES, *SEWAGE EFFLUENTS,
AL, ALGAE, AGRICULTURAL/ *LAKES,
ON CONTROL, *LOCAL / *WISCONSIN,
ESH SALTS, LAKES, SILTS/ *ALGAE,
ITROGE/ *LAKES, *EUTROPHICATION,
*ALGAE, *NITROGEN, *PHOSPHORUS,
TY MANAGEMENT, *LAKE WASHINGTON,
TREATMENT, *INDIANA, CHLOROPHYT/
ES, SEWAGE TREAT/ *MICROBIOLOGY,

, *PHOSPHORUS, *SEWAGE EFFLUENT,
GAE, NUTRIENTS, DOMESTIC WASTES,
SEWAGE, *PHOSPHORUS, PHOSPHATES,
ON CONTROL, PACIFI/ *WASHINGTON,
ROPHYT/ *SEWAGE LAGOONS, *ALGAE,
BIOCHEMICAL OXYGEN DEMAND, *ACT/
N, / *ALGAE, *PLANKTON, *SEWAGE,
INEERING, *SANITARY ENGINEERING,
WATER POLLUTION, WASTE DISPOSAL,
TIVATED SLUDGE, *BACTERIA, META/
OONS, *ALGAE, *BACTERIA, *DISSO/
TERS, / *BIOLOGICAL COMMUNITIES,
STORM DRAINS, SEWAGE EFFLUENTS,
ON EFFECTS, *FISH, ALGAE, TROPH/
OPHICATION, / *ALGAE, *PLANKTON,
N, *CLADOPHORA, *NUISANCE ALGAE,
MENT, BIBLIOGRAPHIES, CHLORELLA,

*RHONE RIVER, NAVICULA, MELOSIRA,   W73-01446
*RIVER REGIONS, *PHYSICAL CHARACT   W73-00748
*RIVER STUDIES, *NATIONAL WATER Q   W70-04507
*RIVER SURVEYS.:   W72-01801
*RIVERS, *EUTROPHICATION, RUNOFF,   W69-10170
*RIVERS, *ALGAE, *SAMPLING, *MISS   W70-04507
*RIVERS, *LAKES, *PONDS, LIGHT, T   W70-05404
*RIVERS, *ALGAE, WATER POLLUTION   W70-05668
*RIVERS, *WATER POLLUTION, *WATER   W69-07096
*RIVERS, ALGAE, PROTOZOA, AQUATIC   W73-00748
*RIVERS, OXIDATION, PHOTOSYNTHESI   W71-11381
*RIVERS, WATER CHEMISTRY, WATER P   W70-03068
*ROTIFERS, *DISTRIBUTION, MISSOUR   W70-05375
*ROTIFERS, PLANKTON, BASS, PHYTOP   W70-02249
*ROTIFERS, *ECOLOGY, *ECOSYSTEMS.   W72-04506
*RUNOFF, *FERTILIZATION, LAKES, S   W70-02787
*RUNOFF, *LAKES, *IMPOUNDMENTS, T   W70-04765
*RUNOFF, *GROUNDWATER, *SESTON, *   W70-03501
*RUTHENIUM RADIOISOTOPES.:   W71-09863
*SAKHALIN ISLAND(USSR).:   W72-10623
*SALINE WATER, *THERMAL WATER, AL   W69-03516
*SALINITY, IRRIGATION EFFECTS, DR   W69-00360
*SALT MARSHES, *CYCLING NUTRIENTS   W69-08274
*SALT MARSH, *OIL POLLUTION, *REC   W70-01231
*SALT MARSHES, OILY WATER, *OIL,   W70-09976
*SALT VALLEY RESERVOIRS(NEB), *CL   W72-04761
*SAMPLING, ALGAE TOXINS, SALINE W   W69-00360
*SAMPLING, PLANT PHYSIOLOGY, EUTR   W70-04382
*SAMPLING, *MISSISSIPPI RIVER, *O   W70-04507
*SAMPLING, *MEASUREMENT, *FLUOROM   W70-05377
*SANDS, *FILTRATION, FILTERS, ELE   W71-03035
*SANITARY ENGINEERING, *SEWAGE TR   W70-02609
*SANITARY ENGINEERING, MICROORGAN   W72-01798
*SANITARY ENGINEERING, *WATER SUP   W72-01813
*SAPROBITY, PRAGUE, CZECHOSLOVAKI   W72-12240
*SARGASSUM FLUITANS, *SARGASSUM N   W69-08524
*SARGASSUM NATANS, SEAWEED, ZIRCO   W69-08524
*SATURATION, TEMPERATURE, PRESSUR   W70-00683
*SCENEDESMUS, FLOCCULATION, WATER   W72-03703
*SCENEDESMUS, *SUBSURFACE WATERS,   W70-03334
*SCENEDESMUS, *LIGHT, PHOSPHORUS,   W70-03923
*SCENEDESMUS, DIATOMACEOUS EARTH,   W68-00475
*SCIRPUS FLUVIATILIS, *LAKE BUTTE   W72-00845
*SCREENING, *KENOSHA(WIS), *STRAI   W73-02426
*SCULPINS, *RADIOACTIVITY, *FOOD   W69-09742
*SEA WATER, *VITAMINS, VITAMIN B,   W70-05652
*SEASONAL, *SULFIDES, TEMPERATURE   W70-04790
*SEASONAL, RHODE ISLAND, SAMPLING   W70-00713
*SEATTLE(WASH), RENTON TREATMENT   W70-04283
*SEAWATER, *FRESH WATER, *HUMIC A   W70-01074
*SEAWEED EXTRACTS, ASCOPHYLLUM NO   W69-07826
*SEDIMENT-WATER INFERFACE, MUD-WA   W69-06859
*SEDIMENTS, ORGANIC MATTER, CURRE   W70-03501
*SEDIMENTS, *NUTRIENT BUDGET, CYC   W68-00511
*SEDIMENTS, *EUTROPHICATION, ALGA   W69-03075
*SELECTIVE TOXICITY, CHEMICAL STR   W70-02982
*SELF-PURIFICATION, *WATER POLLUT   W71-05878
*SELF-PURIFICATION, *WASTE WATER(W70-05092
*SEPARATION TECHNIQUES, TURBIDITY   W72-08686
*SESSILE ALGAE, *CYCLING NUTRIENT   W73-01454
*SESTON, *SEDIMENTS, ORGANIC MATT   W70-03501
*SEWAGE BACTERIA, ODOR, ANAEROBIC   W71-02009
*SEWAGE DISPOSAL, *AGRICULTURAL W   W70-02787
*SEWAGE DISPOSAL, ALGAE, AGRICULT   W70-05565
*SEWAGE EFFLUENTS, *SEWAGE DISPOS   W70-05565
*SEWAGE EFFLUENTS, *WATER POLLUTI   W69-06909
*SEWAGE EFFLUENTS, *ESTUARIES, FR   W70-02255
*SEWAGE EFFLUENTS, *NUTRIENTS, *N   W69-09349
*SEWAGE EFFLUENT, *SEWAGE TREATME   W72-08133
*SEWAGE FERTILIZATION EFFECTS, *B   W69-09349
*SEWAGE LAGOONS, *ALGAE, *SEWAGE   W70-10173
*SEWAGE LAGOONS, *INDUSTRIAL WAST   W70-03312
*SEWAGE NUTRIENT REMOVAL.:   W72-08133
*SEWAGE TREATMENT, *NUTRIENT REMO   W72-08133
*SEWAGE TREATMENT, SEWAGE, BIOLOG   W68-00248
*SEWAGE TREATMENT, DOMESTIC WASTE   W68-00256
*SEWAGE TREATMENT, *WATER POLLUTI   W69-03683
*SEWAGE TREATMENT, *INDIANA, CHLO   W70-10173
*SEWAGE TREATMENT, *FILTRATION, *   W70-05821
*SEWAGE TREATMENT, *EUTROPHICATIO   W70-02304
*SEWAGE TREATMENT, *TRACERS, *WAT   W70-02609
*SEWAGE TREATMENT, *WASTE WATER T   W72-01786
*SEWAGE TREATMENT, *PROTOZOA, *AC   W72-01817
*SEWAGE TREATMENT, *OXIDATION LAG   W72-01818
*SEWAGE TREATMENT, *TRICKLING FIL   W72-01819
*SEWAGE TREATMENT, STORM RUNOFF.:   W71-10654
*SEWAGE, *STREAMS, *WATER POLLUTI   W72-01788
*SEWAGE, *SEWAGE TREATMENT, *EUTR   W70-02304
*SEWAGE, ALGICIDES, AQUATIC ALGAE   W69-04798
*SEWAGE, EFFLUENTS, CARBON DIOXID   W68-00855

369

*ALGAE.: *ESTUARIES, *HARBORS,
EUTROPHICATION, *NUISANCE ALGAE,
LOOMS, CHLORELLA PYRENOIDOSA, M/
N FLOS AQUAE, STEPHANODISCUS AS/
OUNT, DESMIDS,/ *IDENTIFICATION,
TRATION, *PARTICLE SIZE, *CLAYS,
*SYSTEMS ANALYSIS, *ECOSYSTEMS,
S, *BLUE-GREEN ALGA/ *FLOTATION,

ER COURSE, CARPATHIAN SUBMONTAN/
TAL STUDIES, LAGOON DES/ *JAPAN,
R TREATMENT, *ALGAE, FILTRATION,
ACTER/ *COOLING WATER, *BIOMASS,
TIONS, *EFFLICIENCIES, *LAGOONS,
FLOW RATE, *WASTE WATER, *ALGAE,
L, *HERBICIDES, *COPPER SULFATE,
E, ARID CLIMATES, SEMIARID CLIM/
MATTER, *GRASSLANDS, MOUNTAINS,
SUPPLY, / *INFILTRATION, *PONDS,
ROLINA, / *ALGAE, *RADIOISOTOPES,
YANOPHYTA, DETERG/ *SURFACTANTS,
XICO, A/ *RADIOISOTOPES, *ALGAE,
EW PRODUCT CO/ *MASTER PLANNING,
*AERATED LAGOONS, *MIDWEST(US),
, LIGHT-SATURATED / *ADAPTATION,
ASTES, *WASTE WAT/ *FARM WASTES,
RITERIA, FACI/ *WASTE TREATMENT,
CATION, *SEASONAL, *SUL/ *PONDS,
ENTAL EFFECTS/ *WASTE TREATMENT,
*EFFICIENCIES, *CHLORINATION, */
N, *BIOMASS, *S/ *PHYTOPLANKTON,
*SCREENING, *KENOSHA(WIS)/
IANS, MIC/ *ANABAENA FLOS-AQUAE,
, *URBANIZATION, *WATER QUALITY,
ION, DECOMPOSING ORGANIC MATTER,
S, BACTERIA, ORGANIC POLLUTANTS,
S, *CYCLING NUTRIENTS, *EUTROPH/
HOSPHORUS RADIOISOTOPES, *TROUT,
K,/ *MONITORING, *RADIOISOTOPES,
IRATION,/ *PRIMARY PRODUCTIVITY,
RATURE, MANGANESE, CULTURES, RE/
ICAL COMMUNITIES, *LABORATORIES,
S, *FISH, ALGAE, TROPH/ *SEWAGE,
*FUNGI, *CALCAREOU/ *CHELATION,
LANKTON, RADIOACTIVITY, WASTE D/
N, *CALCIUM, SOIL CONTAMINATION,
TRATES, *CULTURES, *SCENEDESMUS,
METHODS, *ZOOPLANKTON, *BIOMASS,
ULTATIVE, BOD, DETENTION TIME.:
ROGEN SULFIDE, *ORGANIC LOADING,
DE, *ORGANIC LOADING, *SULFATES,
ESIGN, *SULFUR BACTERIA, STABIL/
DATION PONDS, *HYDROGEN SULFIDE,
ERMAL STRATIFICATION, *SEASONAL,
ES, *OXIDATION LAGOONS, *DESIGN,
E, *AERATION, *NITRATES, *CULTU/
CHLOROPHYTA, CYANOPHYTA, DETERG/
TS, *DIATOMS, *OIL, *OIL WASTES,
RC/ *LINEAR ALKYLATE SULFONATES,
*BENTHIC FAUNA, *TROPHIC LEVELS,
RS, STORMWATER OVE/ BACKWASHING,
AE, COLLOIDS, FLO/ *ULTRASONICS,
.:
ERICIDES, *CHLORI/ *PSEUDOMONAS,
TS/ *WATER QUALITY, *RECREATION,
ES, CATIONIC/ *POLYELECTROLYTES,
NG, MICROORG/ *ALGAE, *PROTOZOA,
NS, *ALGAE CONTROL, *LAKE ERIE,/
SIMULATION ANALYSIS, WATER POLL/
LORINATI/ *ODOR-PRODUCING ALGAE,
WATER POLLUTION SOURCES, *ALGAE,
EATMENT, *ACTINOMYCETES, *ALGAE,
EATMENT, *ALGAE, *ALGAL CONTROL,
TED C/ *WATER TREATMENT, *ALGAE,
DUOUS TREES, / *WATER TREATMENT,
ATS, ALKALINE WATER, AC/ *ALGAE,
TRATION, NU/ *THERMAL POLLUTION,
MMUNITIES, POTOMAC RIVER, SAVAN/
ESIS.: *ALGAE,
ALGAE, *GREEN ALGAE, *RED ALGAE,
L POWERPLANTS, *ECOLOGY, *BIOTA,
LLUTION, *ECOLOGY, ALGAE, BACTE/
LABORATORY ANIMALS, / *BENTHOS,
LGAE, *RADIOISOTOPES, *SORPTION,
POWERPLANTS, *THERMAL POLLUTION,
ICAL ECOLOGY, LAKES, / *GRAZING,
CT, *RESERVOIR/ *EUTROPHICATION,
N ION CONCENTRATION, *NUTRIENTS,
*EUTROPHICATION, *ALGAL CONTROL,
ATMENT, OXYGEN DEMAND, BIOCHEMI/

*SEWAGE, WATER POLLUTION EFFECTS,   W71-05500
*SEWAGE, WATER POLLUTION SOURCES,   W70-07261
*SHAGAWA LAKE(MINN), ELY(MINN), B   W70-03512
*SHAGAWA LAKE(MINN), APHANIZOMENO   W70-05387
*SIGNIFICANCE, ALGAL KEY, CLUMP C   W70-05389
*SILTS, SANDS, ALGAE, PLANKTON, M   W70-08628
*SIMULATION ANALYSIS, WATER POLLU   W69-06536
*SINKING, *PHYTOPLANKTON, *DIATOM   W70-02754
*SITE SELECTION, LOADING.:   W71-08221
*SKAWA RIVER(POLAND), MONTANE RIV   W70-02784
*SLAUGHTERHOUSE WASTES, EXPERIMEN   W70-07508
*SLIME, *ULTRAVIOLET RADIATION, *   W72-03824
*SLIME, ALGAE, FUNGI, BACTERIA, B   W72-12987
*SLUDGE, ALGAE, BIOCHEMICAL OXYGE   W70-06899
*SLUDGE, STORMS, TURBULENCE, PILO   W70-05819
*SODIUM ARSENITE, NUISANCE ALGAE,   W70-05568
*SOIL ALGAE, *AQUATIC ALGAE, ALGA   W69-03491
*SOIL MICROORGANISMS, SOIL ALGAE,   W72-08110
*SOIL, GROUNDWATER BASINS, WATER   W69-07838
*SORPTION, *TEMPERATURE, SOUTH CA   W69-07845
*SORPTION, *ALGAE, CHLOROPHYTA, C   W70-03928
*SORPTION, ABSORPTION, GULF OF ME   W69-08524
*SOURCE CONTAMINATION, CYANIDE, N   W73-00758
*SOUTHWEST(US), *DESIGN CRITERIA,   W70-04788
*SPECIES, CYCLOTELLA MENEGHINIANA   W70-05381
*SPRINKLER IRRIGATION, *ORGANIC W   W71-00940
*STABILIZATION, *PONDS, *DESIGN C   W70-04786
*STABILIZATION, *THERMAL STRATIFI   W70-04790
*STABILIZATION, *PONDS, *ENVIRONM   W70-04787
*STABILIZATION, *PONDS, *DESIGN,·   W70-04788
*STATISTICAL METHODS, *ZOOPLANKTO   W73-01446
*STRAINING, *MICROSTRAINERS.:   W73-02426
*STRAINS, PHYCOLOGISTS, VETERINAR   W70-00273
*STRATIFICATION, *TEXAS, WATER PO   W70-04727
*STREAM POLLUTION.: /TER UTILIZAT   W68-00483
*STREAMS, EUGLENA, NUTRIENTS, PLA   W68-00483
*STREAMS, *WATER POLLUTION EFFECT   W69-03369
*STREAMS, *ECOSYSTEMS, RADIOACTIV   W69-09334
*STREAMS, BIOINDICATORS, *NEW YOR   W69-10080
*STREAMS, *LABORATORY TESTS, RESP   W70-00265
*STREAMS, *ALGAE, *DIATOMS, TEMPE   W70-07257
*STREAMS, LIGHT INTENSITY, CURREN   W70-02780
*STREAMS, *WATER POLLUTION EFFECT   W72-01788
*STRONTIUM RADIOISOTOPES, *ALGAE,   W71-04067
*STRONTIUM RADIOISOTOPES, *PHYTOP   W70-00707
*STRONTIUM RADIOSOTOPES, CYTOLOGI   W69-03491
*SUBSURFACE WATERS, *AGRICULTURE,   W70-03334
*SUCESSION, *ISOLATION, DIATOMS,   W73-01446
*SULFATE ION, *DESULFOVIBRIO, FAC   W70-01971
*SULFATES, *SULFIDES, ALGAE, GASE   W70-07034
*SULFIDES, ALGAE, GASES, ODOR, OX   W70-07034
*SULFIDES, *OXIDATION LAGOONS, *D   W70-01971
*SULFIDES, ALGAE, ORGANIC LOADING   W71-11377
*SULFIDES, TEMPERATURE, CLIMATIC   W70-04790
*SULFUR BACTERIA, STABILIZATION,   W70-01971
*SURFACE, *VOLUME, *CARBON DIOXID   W70-03334
*SURFACTANTS, *SORPTION, *ALGAE,   W70-03928
*SURFACTANTS, *WATER POLLUTION, C   W70-01470
*SURFACTANTS, WATER POLLUTION SOU   W70-09438
*SURVEYS, BIOLOGICAL COMMUNITIES,   W71-12077
*SUSPENDED SOLIDS, *COMBINED SEWE   W71-10654
*SUSPENDED LOAD, *FILTRATION, ALG   W71-10654
*SUSPENDED SOLIDS, CHASE CITY(VA)   W72-10573
*SWIMMING POOLS, *BACTERIA, *BACT   W69-10171
*SWIMMING, *WATER POLLUTION EFFEC   W71-09880
*SYNTHETIC ORGANIC POLYELECTROLYT   W70-05267
*SYSTEMATICS, *SANITARY ENGINEERI   W72-01798
*SYSTEMS ANALYSIS, DATA COLLECTIO   W71-04758
*SYSTEMS ANALYSIS, *ECOSYSTEMS, *   W69-06536
*TASTE-PRODUCING ALGAE, ALGAE, CH   W70-02373
*TASTE, *ODOR, *AMINO ACIDS, POTA   W72-01812
*TASTE, *ODOR, WATER QUALITY CONT   W72-13605
*TASTE, *ODOR, *NATURAL STREAMS,   W72-03562
*TASTE, *ODOR, *ALGICIDES, ACTIVA   W72-03746
*TASTE, *ODOR, *RESERVOIRS, *DECI   W72-03475
*TEMPERATURE, *HOT SPRINGS, HABIT   W69-10160
*TEMPERATURE, HYDROGEN ION CONCEN   W70-08834
*TEMPERATURE SHOCK, *PROTOZOAN CO   W70-07847
*TEMPERATURE, DIATOMS, PHOTOSYNTH   W71-02496
*TEMPERATURE TOLERANCE, OPTIMUM G   W71-02496
*TEMPERATURE, *AQUATIC ENVIRONMEN   W71-02491
*TEMPERATURE EFFECTS, *THERMAL PO   W71-02479
*TEMPERATURE, *THERMAL POLLUTION,   W71-02494
*TEMPERATURE, SOUTH CAROLINA, CES   W69-07845
*TEMPERATURE, *COOLING WATER, CHE   W69-05023
*TEMPERATURE, *DAPHNIA, PHYSIOLOG   W70-03957
*TENNESSEE VALLEY AUTHORITY PROJE   W71-07698
*TERTIARY TREATMENT, BIOLOGICAL T   W70-08980
*TERTIARY TREATMENT, SUSPENDED LO   W70-09186
*TERTIARY TREATMENT, *AEROBIC TRE   W70-06792

US, *NUTRIENTS, *EUTROPHICATION,
PHOSPHATES, *LAKES, *DETERGENTS,
COMPOUNDS, *NUTRIENTS, *PESTICI/
, HORMIDIUM SP.:
                *TEST DEVELOPMENT,
ERIALS, TE/ *ALGICIDES, *PAINTS,
*WATER QUALITY, *STRATIFICATION,
OLS.:
EING WASTES./ *ANTI-FOAM AGENTS,

*ISRAEL, *ALGAL CONCENTRATION, /
*COOLING WATER, / *POWERPLANTS,
HYDROGEN ION CONCENTRATION, NU/
BIOTA, *TEM/ *THERMAL POLLUTION,
RPLANTS, *ECOLOGY, *BIOTA, *TEM/
IMALS, / *BENTHOS, *TEMPERATURE,
AE, BACTE/ *TEMPERATURE EFFECTS,
FFECTS, WATER POLLUTION SOURCES,
PERATURE, PHYTOPLANKTON, ALGAE,/
RPLANTS, *WATER QUALITY CONTROL/
      *FISH PHYSIOLOGY, *ALGAE,
POLLUTION, THERMAL POWERPLANTS,
Y, SPR/ *PHOTOSYNTHESIS, *ALGAE,
N, *IMPOUNDMENTS, *RESERVOIRS, /
L, *SUL/ *PONDS, *STABILIZATION,
PHILIC BAC/ *CYANOPHYTA, *ALGAE,
   *PRIMARY PRODUCTIVITY, *ALGAE,
OLOGICAL ECOLOGY, *SALINE WATER,
URVES, CALORIMET/ *ALGAL GROWTH,
ALGAE, DIATOMS, ANIMAL BEHAVIOR,
INE EN/ *POLLUTANTS, *ESTUARIES,

CEPTIBILITY.:
RUDE OIL, *GAMOSOL, *OIL SPILLS,
TRATIONS, SURVIVAL, SUSCEPTIBIL/
EGENERATION, CLIFF VEGETATION, /
                *HYDROCARBON,
OGY, *ALGAE, *THERMAL POLLUTION,
ISM, RADIOACTIVITY TECH/ *ALGAE,
*INSECTICIDES, *MARINE BACTERIA,
CH/ *DETERGENTS, *INSECTICIDES,
ROL, EUTROPHICATION, BASS, WATE/
ION, *CHLOROPHYLL, *OXYGENATION,
ON EFFECTS, CHELATION,/ *COPPER,
GAE, *ALGAL TOXINS, KILLIFISHES,
LIMNOLOGY, PUBLIC HEAL/ *ALGAE,
TATS, CLAMS, MUSSELS, SHELLFISH/
AQUATIC ALGAE, WATER POLLUTION,
S, ACTIVATED CARBON, FILTRATION,
IC E/ *POLLUTANT IDENTIFICATION,
TOXINS, *PHYTOTOXIC/ *FISHKILL,
*COPPER SULFATE, *ALGAL CONTROL,
YANOPHYTA, *ESSENTIAL NUTRIENTS,
ATER POLLUTION / *RADIOACTIVITY,
ENIA, *OPISTHONEMA OGLIMUN, UDO/
/ *WATER QUALITY, *GREAT LAKES,
ENGINEERING, *SEWAGE TREATMENT,
ZONE/ *BIOLOGICAL OCEANOGRAPHY,
TEMS, *PHOSPHORUS RADIOISOTOPES,
LEM, LEYDEN, THE HAGUE, MINERAL/
, / *CONFERENCES, *OCEANOGRAPHY,
ISOTOPES, *TROUT, *STREAMS, *EC/
LOGY, *EFFECTS, *COPPER SULFATE,
*SLIME, *ULTRAVIOLET RADIATION,
CULTURAL WASTES, *ALGAE, *LAKES,
COMMUNITIES, *SEWAGE TREATMENT,
, CHLORELLA PYRENOIDOSA, CARBOX/
GAE, ECOL/ *BACTERIA, *CULTURES,
QUAE, APHANIZOMENON / *CONCEPTS,
GOTROPHY, *PRIMARY PRODUCTIVITY,
, *AQUATIC ALGAE, WATER QUALITY,
ENT, WATER QUALITY, NUT/ *ALGAE,
ICAL CO/ *LAKES, *BENTHIC FAUNA,
ZATION, ALG/ *NUTRIENTS, *LAKES,
TION, *PHOSPHORUS RADIOISOTOPES,
*MADISON LAKES, YAHARA RIVER(WI/
ORUS RADIOISOTOPES, *TRANSFER, /
S(NEB), *CLEAR WATER RESERVOIRS,
OLOR, *ALGAE,/ *WATER TREATMENT,
*PHOTOSYNTHESIS, *GROWTH RATES,
ETATE, *FRESHWATER ENVIRONMENTS,
LAKES, *PHOSPHORUS RADIOISOTOPE,
IO, *CANADA, NUTRIENT LEVELS, S/
*WASTE DISPOSAL, *AQUACULTURE,
ILTRATION, ALGAE, COLLOIDS, FLO/
ENT, *ALGAE, FILTRATION, *SLIME,
N.:
SYNURA, ANABAENA, IOWA RIVER, /
IZOMENON FLOS-AQUAE, *CARBON-14/
LAKE(ORE.), *OSCILLATORA.:

*TERTIARY TREATMENT, NITROGEN, AQ   W69-09340
*TERTIARY TREATMENT, *GREAT LAKES   W70-03964
*TERTIARY TREATMENT, *PHOSPHORUS   W70-05470
*TEST DEVELOPMENT, *TEST PROBLEMS   W71-07339
*TEST PROBLEMS, HORMIDIUM SP.:   W71-07339
*TEST PROCEDURES, ANTIFOULING MAT   W71-07339
*TEXAS, WATER POLLUTION SOURCES,   W70-04727
*TEXTILE MILL WASTES, ALFOL ALCOH   W70-08740
*TEXTILE WASTES, FIBER WASTES, DY   W70-04060
*THERMAL BAR.:   W69-05763
*THERMAL GRADIENT, *BOD REMOVAL,   W70-04790
*THERMAL POLLUTION, *TEMPERATURE,   W69-05023
*THERMAL POLLUTION, *TEMPERATURE,   W70-08834
*THERMAL POWERPLANTS, *ECOLOGY, *   W71-02491
*THERMAL POLLUTION, *THERMAL POWE   W71-02491
*THERMAL POLLUTION, LABORATORY AN   W71-02494
*THERMAL POLLUTION, *ECOLOGY, ALG   W71-02479
*THERMAL POLLUTION, THERMAL POWER   W70-07847
*THERMAL POWERPLANTS, *LAKES, TEM   W71-09767
*THERMAL POLLUTION, *NUCLEAR POWE   W71-11881
*THERMAL POLLUTION, *TOXICITY.:   W70-01788
*THERMAL STRESS, THERMAL WATER, *   W70-07847
*THERMAL SPRINGS, *LIGHT INTENSIT   W69-09676
*THERMAL STRATIFICATION, *AERATIO   W70-04484
*THERMAL STRATIFICATION, *SEASONA   W70-04790
*THERMAL SPRINGS, *OREGON, THERMO   W70-05270
*THERMAL WATER, PERIPHYTON, BIOMA   W70-03309
*THERMAL WATER, ALGAE, AQUATIC AL   W69-03516
*THERMOBIOLOGY, *DIURNAL OXYGEN C   W70-03309
*THERMODYNAMIC BEHAVIOR.: /OZOA,   W70-07847
*TIDAL WATERS, ECOSYSTEMS, ESTUAR   W68-00010
*TORONTO(ONTARIO).:   W71-05879
*TORREY CANYON, REPOPULATION, SUS   W70-01780
*TORREY CANYON, AROMATIC SOLVENTS   W70-01470
*TORREY CANYON, *DETERGENT CONCEN   W70-01779
*TORREY CANYON, SUSCEPTIBILITY, R   W70-01777
*TOXICITY THRESHOLD.:   W70-01466
*TOXICITY.:   *FISH PHYSIOL   W70-01788
*TOXICITY, *INSECTICIDES, METABOL   W70-02198
*TOXICITY, *WATER POLLUTION, AQUA   W70-01467
*TOXICITY, *WATER POLLUTION, FOOD   W70-01466
*TOXICITY, *CYANOPHYTA, *CHEMCONT   W70-02982
*TOXICITY, ORGANIC COMPOUNDS, *PH   W71-12183
*TOXICITY, *ALGAE, *WATER POLLUTI   W71-11561
*TOXICITY, ISOLATION.: /QUATIC AL   W70-08372
*TOXICITY, *CULTURES, *ISOLATION,   W70-00273
*TOXICITY, *ALGAE, *REVIEWS, HABI   W70-05372
*TOXICITY, BIOLOGICAL MEMBRANES,   W69-05306
*TOXICITY, *DISTRIBUTION SYSTEMS,   W72-03746
*TOXICITY, *BIOASSAY, FISH, AQUAT   W72-11105
*TOXINS, *ALGAL POISONING, *ALGAL   W71-07731
*TOXINS, ALGAE, RESISTANCE, FISH,   W68-00859
*TRACE ELEMENTS, ALGAE, EUTROPHIC   W69-06277
*TRACE ELEMENTS, *AQUATIC LIFE, W   W71-05531
*TRACE ELEMENTS, *ANCHOA LAMPROTA   W69-08525
*TRACE ELEMENTS, *WATER POLLUTION   W69-08562
*TRACERS, *WATER POLLUTION, DESIG   W70-02609
*TRANSACTIONS, BLACK SEA, CONTACT   W69-07440
*TRANSFER, WATER, BACTERIA, SEDIM   W69-07861
*TRANSFORMATIONS, AMSTERDAM, HAAR   W69-07838
*TRANSLATIONS, ALGAE, CRUSTACEANS   W69-07440
*TRANSLOCATION, *PHOSPHORUS RADIO   W69-09334
*TREATMENT, *LAKES, *ECOLOGY, BOT   W70-06217
*TREATMENT FACILITIES, FILTERS.: / W72-03824
*TRIBUTARIES, SAMPLING, FLOW, ANA   W70-02969
*TRICKLING FILTERS, *BIOTA, SEWAG   W72-01819
*TRISODIUM NITRILOTRIACETATE, NTA   W70-02248
*TRITIUM, EPIPHYTOLOGY, MARINE AL   W69-09755
*TROPHIC BIOLOGY, ANABAENA FLOS A   W70-03959
*TROPHIC LEVEL, WATER STORAGE, EV   W69-09723
*TROPHIC LEVELS, NUTRIENTS, NORTH   W70-02772
*TROPHIC LEVELS, AQUATIC ENVIRONM   W70-02764
*TROPHIC LEVELS, *SURVEYS, BIOLOG   W71-12077
*TROUT, *FISH MANAGEMENT, FERTILI   W70-05399
*TROUT, *STREAMS, *ECOSYSTEMS, RA   W69-09334
*TRUAX FIELD(MADISON) EFFLUENTS,   W70-05113
*TUBIFICIDS, *ECOSYSTEMS, *PHOSPH   W69-07861
*TURBID RESERVOIRS, CYCLOTELLA, T   W72-04761
*TURBIDITY, *IRON, *MANGANESE, *C   W72-04145
*TURBULENCE, CHLORRELA, SCENEDESM   W69-03730
*TURNOVER RATES, *HETEROTROPHY, O   W70-05109
*TURNOVERS, NUTRIENTS, PHOSPHORUS W68-00858
*TWELVE MILE LAKE(CANADA), *ONTAR   W70-02973
*ULTIMATE DISPOSAL, DOMESTIC WAST   W70-09905
*ULTRASONICS, *SUSPENDED LOAD, *F   W71-10654
*ULTRAVIOLET RADIATION, *TREATMEN   W72-03824
*ULVA LATISSIMA, *AMMONIA NITROGE   W71-05500
*UNIT PROCESSES, GASTROENTERITIS,   W70-08107
*UPPER KLAMATH LAKE(ORE.), *APHAN   W72-06052
*UPPER KLAMATH LAKE(ORE.), AGENCY   W72-06051

MPOSITION, MATERIAL BALANCES, O/
TICAL DENSITY, PARTICU/ *GROWTH,
ELLA PYRENOIDOSA, BIOLOGICAL HA/
, BODY IMMEDIUM FILTER, BACK-WA/
OSYSTEMS, *RADIUM RADIOISOTOPES,
AINAGE, BIOLOGICAL RESPONSE, FE/
ZOAR(CONN), KLAMATH LAKE(ORE),/
TRATIFICATION, *TE/ *RESERVOIRS,
L, *AQUATIC PLANTS, *HERBICIDES,
RIVER(VIETNAM), CHLORO/ *BLOOM,
*WATER PROPERTIES, *NUTRIENTS,
*MAN-MADE FIBER PLANTS, NYLON-6,
-MADE FIBER PLANTS, NYLON-6, *V/
OCOBALAMIN, EUGLENA GRACILIS Z/
DICATORS, CHLAMYDOMONAS,/ *RAIN,
URES, CH/ *BIOASSAY, *SEA WATER,
ON, *NITRATES, *CULTU/ *SURFACE,
POLLUT/ *LAKES, *EUTROPHICATION,
WATER POLLUTION CONTROL, PACIFI/
N, SEWAGE TREATMENT, SOLAR RADI/
ION, SEWAGE, BIOTA, PHYTOPLANKT/
LTIMATE DISPOSAL, DOMESTIC WAST/
ENT, *EFFLUENTS, *RADIOISOTOPES,
EFFE/ *BIOASSAY, *AQUATIC LIFE,
*BIOLO/ *INTERTIDAL COMMUNITIES,
STRATIVE AGENCIES, *WATER WORKS,
L PHYSIOLOGY, *DESIGN THEORY AN/
*RADIOISOTOPES, *WASTE DISPOSAL,
ISMS, STREAMS, WATER POLLUTION,/
CROORGANISMS, ELECTRIC CURRENTS,
ICIENCIES, *CHLORINATION, *ODOR,
*PONDS, *DESIGN CRITERIA, FACI/
*PONDS, *ENVIRONMENTAL EFFECTS/
BIOINDICATORS, *WATER POLLUTION,
LGAE, YEASTS, BACTERIA, FISH, P/
USE, *NITROGEN, NITROGEN COMPOU/
S, ALGAE, BACTERIA, CULTURES, C/
YTOPLANKTON, *SELF-PURIFICATION,
TES, CHE/ *ALGAE, *FLOCCULATION,
AL TREATMENT, *WATER POLLUTION /
LER IRRIGATION, *ORGANIC WASTES,
OSYNTHESES, PLANTS, TEMPERATURE,
ORMS, T/ *ADDITIVES, *FLOW RATE,
BUTION, TEMPERATURE, METABOLISM,
D OXYGEN, ODOR, ORGANIC LOADING,
, *OXIDATION, LAGOONS, *BAFFLES,
TRY, *BIOCHEMICAL OXYGEN DEMAND,
TION POTENTIAL, SULFUR BACTERIA,
NT, BOTTOM SEDIMENTS, NUTRIENTS,
OSYNTHETIC OXYGEN, PRODUCTIVITY,
EATMENT, *WASTE WATER TREATMENT,
STE DISPOSAL, *SEWAGE TREATMENT,
LLUTION CONTROL, *DENIT/ *ALGAE,
HEMICAL OXYGEN DEMAND, DELAWARE,
LLUTION CONTROL, *DENIT/ *ALGAE,
LLUTION CONTROL, *DENIT/ *ALGAE,
IC ALGAE, *BIOLOGICAL TREATMENT,
ELECTRIC POWER, SEWAGE DISPOSAL,
AE, ADMINISTRATION, MAINTENANCE,
AL WASTES, OXIDAT/ *WATER REUSE,
RGANISMS, ANALYTICAL TECHNIQUES,
ADATION, CHEMICAL OXYGEN DEMAND,
HANE, CARBON DIOXIDE, DIFFUSION,
RIAL), *PHOSPHORUS SOURCES, *NU/
, *ADSORPTION, *WATER TREATMENT,
N LAGOONS, *ALGAE, LIGHT INTENS/
NT, WASTE IDENTIFICATION, ALGAE,
N CRITERIA, ODORS, EFFICIENCIES,
OTOSYNTHESI/ *OXIDATION LAGOONS,
, BACTERICIDES, CHLORINE, COSTS,
SUBSURFACE WATERS, *AGRICULTURE,
NUTRIENTS, */ *DOMESTIC ANIMALS,
ICAL TECHNIQUES, *AQUATIC WEEDS,
ODUCING ALGAE, EUTROPHICATION, /
WATER POLLUTION, WATER SAMPLING,
*WATER QUALITY, *BIOINDICATORS,
TIONAL USE.: *DEFINITIONS,
N EFFECTS, *PHOSPHATES, *ANALYT/
*PRODUCTIVITY, *EUTROPHICATION,
OLLUTION CONTROL, *LAKE ONTARIO,
*PHYTOPLANKTON, *LAKE MICHIGAN,
*GREAT LAKES, *WATER POLLUTION,
ATER, *NITRATES, EUTROPHICATION,
UTRIENTS, *PHOSPHORUS COMPOUNDS,
N CONTROL, WATER / *GREAT LAKES,
HORUS COMPOUNDS, *PHYTOPLANKTON,
ION ABATEMEN/ *OHIO, *LAKE ERIE,
NUTRIENTS, FE/ *EUTROPHICATION,
ERTILIZATION, *INLAND WATERWAYS,

*UPTAKE, CELLS, CONCENTRATION, CO   W70-06600
*UPTAKE, FIXATION, CELL COUNT, OP   W70-03983
*UPTAKE, SPECIFIC ACTIVITY, CHLOR   W70-02786
*UPWARD FLOW, PUTRESCIBILITY TEST   W70-05821
*URANIUM RADIOISOTOPES, RIVERS, M   W69-07846
*URBAN DRAINAGE, *AGRICULTURAL DR   W70-02787
*URBAN WASTES, MADISON(WIS), LAKE   W70-04506
*URBANIZATION, *WATER QUALITY, *S   W70-04727
*VASCULAR TISSUES, *ALGAE, ECOSYS   W70-10175
*VIETNAM, LITTLE LAKE(HANOI), RED   W70-03969
*VIRGINIA, NITROGEN COMPOUNDS, PH   W69-03695
*VISCOSE RAYON, ZINC COMPOUNDS, C   W70-06925
*VISCOSE RAYON PLANT WASTES, *MAN   W70-06925
*VITAMIN B-12, ALGAL GROWTH, CYAN   W69-06273
*VITAMINS, ALGAE, BIOASSAY, BIOIN   W69-06273
*VITAMINS, VITAMIN B, ALGAE, CULT   W70-05652
*VOLUME, *CARBON DIOXIDE, *AERATI   W70-03334
*WASHINGTON, URBANIZATION, WATER   W70-05663
*WASHINGTON, *SEWAGE TREATMENT, *   W69-03683
*WASHINGTON, *OXYGEN, *FLUCTUATIO   W69-10182
*WASHINGTON, *LAKES, *EUTROPHICAT   W69-10169
*WASTE DISPOSAL, *AQUACULTURE, *U   W70-09905
*WASTE DISPOSAL, *WASTE STORAGE,   W72-00941
*WASTE DISPOSAL, *WATER POLLUTION   W72-01801
*WASTE DISCHARGES, FLORA, FAUNA,   W68-00010
*WASTE DISPOSAL, DOMESTIC WASTES,   W70-03344
*WASTE STABILIZATION PONDS, *ALGA   W70-04786
*WASTE STORAGE, *WATER POLLUTION,   W72-00941
*WASTE TREATMENT, *WATERS, *ORGAN   W70-05750
*WASTE TREATMENT, WASTE WATER TRE   W71-03775
*WASTE TREATMENT, ALGAE, DISINFEC   W70-04788
*WASTE TREATMENT, *STABILIZATION,   W70-04786
*WASTE TREATMENT, *STABILIZATION,   W70-04787
*WASTE TREATMENT, FOOD CHAINS, CH   W70-04381
*WASTE TREATMENT, *AQUICULTURE, A   W69-05891
*WASTE WATER TREATMENT, *WATER RE   W69-08053
*WASTE WATER TREATMENT, *NUTRIENT   W70-05655
*WASTE WATER(POLLUTION), *WATER P   W70-05092
*WASTE WATER TREATMENT, ELECTROLY   W70-05267
*WASTE WATER TREATMENT, *BIOLOGIC   W70-10437
*WASTE WATER DISPOSAL, AIR POLLUT   W71-00940
*WASTE WATER TREATMENT.: /N, PHOT   W70-06600
*WASTE WATER, *ALGAE, *SLUDGE, ST   W70-05819
*WASTE WATER TREATMENT, LIGHT, MI   W70-06604
*WASTE WATER TREATMENT.: /ISSOLVE   W70-06601
*WASTE WATER TREATMENT, *POULTRY,   W70-09320
*WASTE WATER TREATMENT, *WATER PO   W70-09186
*WASTE WATER TREATMENT, *OXIDATIO   W70-07034
*WASTE WATER TREATMENT, *OXIDATIO   W70-07031
*WASTE WATER TREATMENT.: /Y, PHOT   W70-08416
*WASTE WATER DISPOSAL, *INDUSTRIA   W72-01786
*WASTE WATER TREATMENT, *WASTE WA   W72-01786
*WASTE WATER TREATMENT, *WATER PO   W71-04557
*WASTE WATER TREATMENT.: /A, BIOC   W71-05013
*WASTE WATER TREATMENT, *WATER PO   W71-04555
*WASTE WATER TREATMENT, *WATER PO   W71-04554
*WASTE WATER TREATMENT, *WATER PO   W71-04556
*WASTE WATER TREATMENT.: /, AQUAT   W71-09546
*WASTE WATER TREATMENT, *OXIDATIO   W71-09548
*WASTE WATER TREATMENT.: /NS, ALG   W71-08970
*WASTE WATER TREATMENT, *INDUSTRI   W72-00027
*WASTE WATER TREATMENT, TREATMENT   W71-13313
*WASTE WATER TREATMENT.: /L, DEGR   W71-13335
*WASTE WATER TREATMENT.: /NG, MET   W71-11377
*WASTE WATER(MUNICIPAL AND INDUST   W69-09340
*WASTE WATER TREATMENT, ALGAE, FI   W72-03641
*WASTE WATER TREATMENT, *OXIDATIO   W72-06022
*WASTE WATER TREATMENT.: /VELOPME   W72-03299
*WASTE WATER TREATMENT.: /, DESIG   W72-06854
*WASTE WATER TREATMENT, ALGAE, PH   W72-10573
*WASTE WATER TREATMENT.: /ACTERIA   W72-12987
*WASTES, ALGAE, ASSAY, DISSOLVED   W70-03334
*WASTES, FERTILIZERS, *ESSENTIAL   W69-05323
*WATER ANALYSIS, *SAMPLING, PLANT   W70-04382
*WATER ANALYSIS, *ODORS, *ODOR PR   W69-00387
*WATER ANALYSIS, *CHEMICAL ANALYS   W69-00659
*WATER ANALYSIS, *BIOASSAY, *COMP   W72-12240
*WATER BLOOM, TOXIC ALGAE, RECREA   W69-08279
*WATER CHEMISTRY, *WATER POLLUTIO   W70-05269
*WATER CHEMISTRY, POLLUTANTS, ALG   W70-02646
*WATER CIRCULATION, PATH OF POLLU   W71-05878
*WATER CIRCULATION, NUISANCE ALGA   W69-05763
*WATER POLLUTION CONTROL, WATER P   W69-03948
*WATER POLLUTION SOURCES, AGRICUL   W69-05323
*WATER POLLUTION EFFECTS, ALGAE,   W69-04805
*WATER POLLUTION, *WATER POLLUTIO   W69-03948
*WATER POLLUTION SOURCES, *CYCLIN   W69-04800
*WATER POLLUTION CONTROL, *POLLUT   W69-06305
*WATER POLLUTION SOURCES, *LAKES,   W69-07084
*WATER POLLUTION EFFECTS, *WATER   W69-07833

Y, *MINERALOGY, *MICROORGANISMS,
LGAE, *RIVERS, *WATER POLLUTION,
 *WISCONSIN, *SEWAGE EFFLUENTS,
N SO/ *ECOLOGY, *ALGAE, *RIVERS,
ATES, *ANALYT/ *WATER CHEMISTRY,
YNTHESIS, *PHYTOPLANKTON, *DDT,/
POUNDS, *NUTRIENTS, *PESTICIDES,
ND, PEN/ *ALGAE, *BIOINDICATORS,
ATION, *WATER POLLUTION SOURCES,
YFISH, POLLUTANT IDENTIFICATION,
RADIOACTIVE WASTES, *MUNICIPAL /
POLLUTION EFFE/ *EUTROPHICATION,
UALITY CONTROL, WATER POLLUTION,
, WEED CONTROL, SOLAR RADIATION,
TROPHY, PHYTOPLANKTON, *REVIEWS,
T REQUIREMENTS, *EUTROPHICATION,
*WASHINGTON, *SEWAGE TREATMENT,
HICATION, *PHOSPHORUS COMPOUNDS,
ICATION, *PHYSIOLOGICAL ECOLOGY,
G NUTRIENTS, *EUTROPH/ *STREAMS,
ATION, *WATER POLLUTION CONTROL,
, SEWAGE, DRAINAGE DIST/ *ALGAE,
, *GROWTH RATES, *BIOINDICATORS,
*MUNICIPAL WASTES, *GREAT LAKES,
CHEMISTRY, WATE/ *WATER QUALITY,
ATION, *WATER POLLUTION SOURCES,
URCES, *WATER POLLUTION EFFECTS,
TS, *EUTROPHICATION, *MEROMIXIS,
ATION, *WATER POLLUTION SOURCES,
POLLUTION EFFE/ *EUTROPHICATION,
POLLUTION EFFE/ *EUTROPHICATION,
URCES, *WATER POLLUTION EFFECTS,
POLLUTION EFFECTS, *EUTROPHICAT/
HICAT/ *WATER POLLUTION SOURCES,
TRIENTS, ALGAE, FISH POPULATION,
*ALGAE, *WASTE WATER TREATMENT,
*ALGAE, *WASTE WATER TREATMENT,
*ALGAE, *WASTE WATER TREATMENT,
NTARIO, *WA/ *SELF-PURIFICATION,
*ALGAE, *WASTE WATER TREATMENT,
Y, *BIOINDICATORS, *BIOASSAY, T/
LUTION EFFECTS, *CLASSIFICATION,
POLLUTION EFFECTS, *CLASSIFICAT/
FICAT/ *WATER POLLUTION SOURCES,
ON,/ *COPPER, *TOXICITY, *ALGAE,
ARINE BIOLOGY, STREAM POLLUTION,
QUALITY, *RECREATION, *SWIMMING,
T, *ECOSYSTEMS, GREAT LAKES, TR/
SUPPLY, *BIOLOGICAL PROPERTIES,
*TASTE, *ODOR, *AMINO ACIDS, P/
ER DISPOSAL, *INDUSTRIAL WASTES,
*WASTE DISPOSAL, *WASTE STORAGE,
*AQUATIC LIFE, *WASTE DISPOSAL,
NEVADA, *EUTROPHICATION, *LAKES,
*SEWAGE TREATMENT, *WASTE WATE/
ALGAE, TROPH/ *SEWAGE, *STREAMS,
NG, *SEWAGE TREATMENT, *TRACERS,
GENTS, *INSECTICIDES, *TOXICITY,
RATION, *WATER POLLUTION CONTRO/
AT/ *EUTROPHICATION, *NUTRIENTS,
ES, *MARINE BACTERIA, *TOXICITY,
*OIL, *OIL WASTES, *SURFACTANTS,
OLLUTION TREATMENT, *FILTRATION,
, *GREAT LAKES, *TRACE ELEMENTS,
ICATORS, *WATER QUALITY, *DISSO/
S, ALG/ *EUTROPHICATION, *LAKES,
S, *WAT/ *ALGAE, *BIOINDICATORS,
OLLUTION SOURCES, *THERMAL POLL/
CH EQUIPMENT,/ *WATER RESOURCES,
TION, *PARTICLE SIZE, *CLAYS, */
DEMAND, *WASTE WATER TREATMENT,

IDES, RESIDUES, *PESTICIDE KINE/
*LIMNOLOGY, *INDUSTRIAL WASTES,
ATION, *WATER POLLUTION CONTROL,
POLLUT/ *LAKES, *EUTROPHICATION,
NTROL, *WATER POLLUTION SOURCES,
REATMENT, *BIOLOGICAL TREATMENT,
NT REMOVAL, SOIL EROSION, ALGAE,
POLLUTION EFFECTS, *WATER SOURC/
SOURC/ *WATER POLLUTION SOURCES,
IRGINIA, NITROGEN COMPOUNDS, PH/
CATION, *WASTE WATER(POLLUTION),
ICROORGANISMS, *WATER POLLUTION,
TATION, *LAGOONS, ACTIVATED SLU/
OL, *WATER SUPPLY, *BASINS, WAT/
, D/ EFFECTS, *IMPOUNDED WATERS,
MMING, *WATER POLLUTION EFFECTS/
ALS, *DELTAS, *DRAINAGE EFFECTS,
POLLUTION, *NUCLEAR POWERPLANTS,

*WATER POLLUTION, *WATER PURIFICA   W69-07428
*WATER POLLUTION SOURCES, IRRIGAT   W69-07096
*WATER POLLUTION CONTROL, *LOCAL   W69-06909
*WATER POLLUTION, *WATER POLLUTIO   W69-07096
*WATER POLLUTION EFFECTS, *PHOSPH   W70-05269
*WATER POLLUTION EFFECTS, *PHOTOS   W70-05272
*WATER POLLUTION CONTROL, EUTROPH   W70-05470
*WATER POLLUTION, DIATOMS, MARYLA   W70-04510
*WATER POLLUTION EFFECTS, *PRIMAR   W69-02959
*WATER POLLUTION EFFECTS, GAS CHR   W69-02782
*WATER POLLUTION, *MINE WASTES, *   W69-01165
*WATER POLLUTION SOURCES, *WATER   W69-02959
*WATER POLLUTION SOURCES, SEPARAT   W69-01273
*WATER POLLUTION CONTROL, ALGICID   W68-00478
*WATER POLLUTION EFFECTS.: /OLIGO   W68-00680
*WATER POLLUTION CONTROL, *WATER   W69-03374
*WATER POLLUTION CONTROL, PACIFIC   W69-03683
*WATER POLLUTION SOURCES, ALGAE,   W69-03370
*WATER POLLUTION EFFECTS, ALGAE,   W69-03513
*WATER POLLUTION EFFECTS, *CYCLIN   W69-03369
*WATER POLLUTION TREATMENT, ALGAE   W69-03374
*WATER POLLUTION TREATMENT, LAKES   W70-04455
*WATER POLLUTION, *WASTE TREATMEN   W70-04381
*WATER POLLUTION, SEWAGE, AESTHET   W70-04430
*WATER POLLUTION, *RIVERS, WATER   W70-03068
*WATER POLLUTION EFFECTS, *WATER   W70-03975
*WATER POLLUTION CONTROL, *CONFER   W70-03975
*WATER POLLUTION EFFECTS, BIOINDI   W70-03974
*WATER POLLUTION EFFECTS, *WATER   W70-03975
*WATER POLLUTION SOURCES, *WATER   W70-03975
*WATER POLLUTION SOURCES, *WATER   W70-03975
*WATER POLLUTION CONTROL, *CONFER   W70-03975
*WATER POLLUTION SOURCES, *WATER   W70-04266
*WATER POLLUTION EFFECTS, *EUTROP   W70-04266
*WATER POLLUTION CONTROL, FARM WA   W70-04266
*WATER POLLUTION CONTROL, *DENITR   W71-04557
*WATER POLLUTION CONTROL, *DENITR   W71-04554
*WATER POLLUTION CONTROL, *DENITR   W71-04555
*WATER POLLUTION CONTROL, *LAKE O   W71-05878
*WATER POLLUTION CONTROL, *DENITR   W71-04556
*WATER POLLUTION, *AQUATIC BIOLOG   W71-10786
*WATER POLLUTION CONTROL, WATERCO   W72-00462
*WATER POLLUTION SOURCES, *WATER   W72-00462
*WATER POLLUTION EFFECTS, *CLASSI   W72-00462
*WATER POLLUTION EFFECTS, CHELATI   W71-11561
*WATER POLLUTION DETECTION.: /, M   W71-10786
*WATER POLLUTION EFFECTS, BACTERI   W71-09880
*WATER POLLUTION EFFECTS, *FALLOU   W71-09013
*WATER POLLUTION SOURCES, TASTE,   W72-01813
*WATER POLLUTION SOURCES, *ALGAE,   W72-01812
*WATER POLLUTION CONTROL, FISH CO   W72-01786
*WATER POLLUTION, WATER POLLUTION   W72-00941
*WATER POLLUTION EFFECTS, FOOD CH   W72-01801
*WATER POLLUTION, DEGRADATION, SE   W72-02817
*WATER POLLUTION, WASTE DISPOSAL,   W72-01786
*WATER POLLUTION EFFECTS, *FISH,   W72-01788
*WATER POLLUTION, DESIGN, DESIGN   W70-02609
*WATER POLLUTION, FOOD CHAINS, AQ   W70-01466
*WATER POLLUTION TREATMENT, *FILT   W70-01027
*WATER POLLUTION SOURCES, *PHOSPH   W70-01031
*WATER POLLUTION, AQUATIC ALGAE,   W70-01467
*WATER POLLUTION, CLEANING, DISAS   W70-01470
*WATER POLLUTION CONTROL, *FILTER   W70-01027
*WATER POLLUTION EFFECTS, CHLORID   W69-08562
*WATER POLLUTION EFFECTS, *BIOIND   W70-00475
*WATER POLLUTION CONTROL, *REVIEW   W70-07393
*WATER POLLUTION EFFECTS, *REVIEW   W70-07027
*WATER POLLUTION EFFECTS, WATER P   W70-07847
*WATER POLLUTION CONTROL, *RESEAR   W70-08627
*WATER POLLUTION CONTROL, *FILTRA   W70-08628
*WATER POLLUTION CONTROL, *EUTROP   W70-09186
*WATER POLLUTION RESEARCH.:   W70-08627
*WATER POLLUTION EFFECTS, *PESTIC   W70-09768
*WATER POLLUTION EFFECTS, *EUTROP   W70-06975
*WATER POLLUTION SOURCES, *WATER   W71-03048
*WATER POLLUTION CONTROL, *WATER   W71-03048
*WATER POLLUTION EFFECTS, WATER U   W71-03048
*WATER POLLUTION CONTROL, FISH, T   W70-10437
*WATER POLLUTION CONTROL, WATER Q   W73-01122
*WATER POLLUTION SOURCES, *WATER   W73-00758
*WATER POLLUTION EFFECTS, *WATER   W73-00758
*WATER PROPERTIES, *NUTRIENTS, *V   W69-03695
*WATER PURIFICATION, ALGAE, SEWAG   W70-05092
*WATER PURIFICATION, *BACTERIA, *   W69-07428
*WATER PURIFICATION, *FLOW AUGMEN   W72-06017
*WATER PURIFICATION, *ALGAL CONTR   W72-01814
*WATER QUALITY, *DISSOLVED SOLIDS   W72-01773
*WATER QUALITY, *RECREATION, *SWI   W71-09880
*WATER QUALITY, *FARM WASTES, *IN   W71-09577
*WATER QUALITY CONTROL, *BIODEGRA   W71-11881

CHLOROPHYLL, *OHIO RIVER, *ANAL/   *WATER QUALITY, *BIOINDICATORS, *   W71-04206
FILTRATION, CHL/ *WATER SUPPLY,   *WATER QUALITY, *WATER TREATMENT,   W71-05943
ERIA, WATER CHEM/ *LAKE ONTARIO,   *WATER QUALITY, *MONITORING, BACT   W71-05879
OPPER SULFATE, / *POTABLE WATER,   *WATER QUALITY CONTROL, *ALGAE, C   W71-00110
PERIOR, *PERIPHYTON, *LIMNOLOGY,   *WATER QUALITY, BIOINDICATORS, AQ   W70-05958
NCE ALGAE, *ALGICIDES, *PHENOLS,   *WATER QUALITY CONTROL, CRUSTACEA   W70-10161
N DEMAND, SU/ *DATA COLLECTIONS,   *WATER QUALITY, BIOCHEMICAL OXYGE   W70-09557
TER POLLUTION EFFECTS, *REVIEWS,   *WATER QUALITY, WATER POLLUTION,   W70-07027
LLUTION EFFECTS, *BIOINDICATORS,   *WATER QUALITY, *DISSOLVED OXYGEN   W70-00475
ACE ELEMENTS, *WATER POLLUTION /   *WATER QUALITY, *GREAT LAKES, *TR   W69-08562
WASHINGTON/ *EFFLUENT DIVERSION,   *WATER QUALITY MANAGEMENT, *LAKE   W69-09349
ES, *FISH, *FISH KILLS, *ROTIFE/   *WATER QUALITY, *ALGAE, *PESTICID   W72-04506
CUT, *LAKES, *LOCAL GOVERNMENTS,   *WATER QUALITY CONTROL, ALGAE, AQ   W72-05774
OPHICATION, NITROGEN, PHOSPHORU/   *WATER QUALITY, *NUTRIENTS, *EUTR   W73-01122
WATER ANALYSIS, *BIOASSAY, *COM/   *WATER QUALITY, *BIOINDICATORS, *   W72-12240
DROGEN ION CO/ *WATER TREATMENT,   *WATER QUALITY, *ALGAE, *ODOR, HY   W72-09694
TREATMENT, *FILTRATION, *ALGAE,   *WATER QUALITY, *ACTIVATED CARBON   W72-09693
RGE, *DATA COLLECTIONS, PIT REC/   *WATER QUALITY, *ARTIFICIAL RECHA   W69-05894
AE, *ALGICIDES, *COPPER SULFATE,   *WATER QUALITY CONTROL, ALGAL CON   W69-05704
COLOR, TASTE, ODOR, TOX/ *ALGAE,   *WATER QUALITY, *EUTROPHICATION,   W69-05697
*RESERVOIR, MASSACHUSETTS, POT/   *WATER QUALITY CONTROL, *ECOLOGY,   W70-05264
OUNDMENTS, *RESERVOIRS, *MIXING,   *WATER QUALITY CONTROL, WATER QUA   W70-04484
*TE/ *RESERVOIRS, *URBANIZATION,   *WATER QUALITY, *STRATIFICATION,   W70-04727
MANAGEMENT, STREAM IMPROVEMENT,   *WATER QUALITY CONTROL, *MIDGES,   W70-05428
GE INPUT, BIOLOGICAL PROPERTIES,   *WATER QUALITY, TURBIDITY, PLANT   W68-00475
*RIVERS, WATER CHEMISTRY, WATE/   *WATER QUALITY, *WATER POLLUTION,   W70-03068
*DETERGENTS, *PHOSPHATES, CHELA/   *WATER QUALITY, *EUTROPHICATION,   W70-04373
N CONTROL, *RESEARCH EQUIPMENT,/   *WATER RESOURCES, *WATER POLLUTIO   W70-08627
RGANIC COMPOUNDS, CHEMICAL WAST/   *WATER REUSE, *RECLAIMED WATER, O   W70-10433
G WATER, *ELECTRIC POWER PLANTS,   *WATER REUSE, COOLING TOWERS, DAM   W71-11393
ENT, *INDUSTRIAL WASTES, OXIDAT/   *WATER REUSE, *WASTE WATER TREATM   W72-00027
CONCENTRATION, ORGANIC LOADING,   *WATER REUSE, RESEARCH AND DEVELO   W72-03299
*CORROSION, *CHEMICAL REACTIONS,   *WATER REUSE, DISSOLVED OXYGEN, P   W70-02294
CALIFORNIA, WATER QUALITY, AEST/   *WATER REUSE, *RECLAIMED WATER, *   W70-05645
COMPOU/ *WASTE WATER TREATMENT,   *WATER REUSE, *NITROGEN, NITROGEN   W69-08053
ENITRIFICATION, *BIODEGRADATION,   *WATER REUSE, *RECLAIMED WATER, N   W69-08054
NDWATER, RECH/ *WATER TREATMENT,   *WATER REUSE, *RECLAMATION, *GROU   W72-04318
TEMPERATURE, NITRATES, AMMONIA,   *WATER REUSE, WASTE WATER TREATME   W72-06017
GE TREATMENT, *NUTRIENT REMOVAL,   *WATER REUSE, SEASONAL, WATER TRE   W72-08133
URCES, *WATER POLLUTION EFFECTS,   *WATER SOURCES, *GROUNDWATER, DDT   W73-00758
*GEOCHEMISTRY, *EUTROPHICATION,   *WATER STRUCTURE, ATMOSPHERE, SEA   W70-04385
D LOSSES, OOCISTUS, FILTRATION /   *WATER SUPPLIES, GREEN ALGAE, HEA   W70-04199
GAE, ODOR, TAST/ *ALGAE CONTROL,   *WATER SUPPLY, *NEW HAMPSHIRE, AL   W69-08282
RWAYS, *WATER POLLUTION EFFECTS,   *WATER SUPPLY, BIOINDICATORS, CAL   W69-07833
DOR, FILTERS, PLANKTON,/ *ALGAE,   *WATER SUPPLY, *CONTROL, TASTE, O   W70-05389
TIES, *W/ *SANITARY ENGINEERING,   *WATER SUPPLY, *BIOLOGICAL PROPER   W72-01813
ER PURIFICATION, *ALGAL CONTROL,   *WATER SUPPLY, *BASINS, WATER POL   W72-01814
ATER TREATMENT, FILTRATION, CHL/   *WATER SUPPLY, *WATER QUALITY, *W   W71-05943
SERVOIRS, WATER/ *ALGAE, *MAINE,   *WATER SUPPLY, COPPER SULFATE, RE   W70-08096
, QUALITY CONTR/ *LAKE MICHIGAN,   *WATER SUPPLY, INTAKES, NUTRIENTS   W70-00263
ACTERIA, *ESTUARINE ENVIRONMENT,   *WATER TEMPERATURE, *SEASONAL, RH   W70-00713
S, RIVERS, LAKES, *GROWTH RATES,   *WATER TEMPERATURE, SEASONAL, TES   W68-00472
*ALGAE, *BIOLOGICAL TREATMENT,   *WATER TREATMENT, *POLLUTANT IDEN   W69-07389
*WATER SUPPLY, *WATER QUALITY,   *WATER TREATMENT, FILTRATION, CHL   W71-05943
*CORROSION, *CHEMICAL REACTION/   *WATER TREATMENT, *COOLING WATER,   W70-02294
*ALGAE, *TASTE, *ODOR, WATER Q/   *WATER TREATMENT, *ACTINOMYCETES,   W72-13605
*ALGAE, *ODOR, HYDROGEN ION CO/   *WATER TREATMENT, *WATER QUALITY,   W72-09694
ERING, *FILTRATION, DESIGN, *AL/   *WATER TREATMENT, SANITARY ENGINE   W72-08859
ERING, *PATHOGENIC BACTERIA, AL/   *WATER TREATMENT, SANITARY ENGINE   W72-13601
LGAE, *WATER QUALITY, *ACTIVATE/   *WATER TREATMENT, *FILTRATION, *A   W72-09693
S, SANDS, LAKE MICHIGAN, WATER /   *WATER TREATMENT, *ALGAE, *FILTER   W73-02426
ST ANALYSIS, WATER PURIFICATION,   *WATER TREATMENT.: / CLEANING, CO   W72-08688
ND MAINTENANCE, DESIGN CRITERIA,   *WATER TREATMENT, *NEW YORK, WATE   W72-08686
CONTROL, *TASTE, *ODOR, *NATURA/   *WATER TREATMENT, *ALGAE, *ALGAL   W72-03562
*RESERVOIRS, *DECIDUOUS TREES, /   *WATER TREATMENT, *TASTE, *ODOR,   W72-03475
TION, HEAD LOSS, FILTERS, *SCEN/   *WATER TREATMENT, *ALGAE, *FILTRA   W72-03703
*ORGANIC MATTER, ALGAE, COPPER /   *WATER TREATMENT, WATER QUALITY,   W72-04495
NITARY ENGINEERING, *FILTERS, */   *WATER TREATMENT, *FILTRATION, SA   W72-07334
RECLAMATION, *GROUNDWATER, RECH/   *WATER TREATMENT, *WATER REUSE, *   W72-04318
ON, *MANGANESE, *COLOR, *ALGAE,/   *WATER TREATMENT, *TURBIDITY, *IR   W72-04145
ION, *SLIME, *ULTRAVIOLET RADIA/   *WATER TREATMENT, *ALGAE, FILTRAT   W72-03824
*ODOR, *ALGICIDES, ACTIVATED C/   *WATER TREATMENT, *ALGAE, *TASTE,   W72-03746
DECOMPOSING ORGANIC MATTER, *PO/   *WATER TREATMENT, ODOR, *ALGAE, *   W72-04242
*ACTIVATED CARBON, *ADSORPTION,   *WATER TREATMENT, *WASTE WATER TR   W72-03641
MYCETES, *ODOR, ACTIVATED CARBO/   *WATER TREATMENT, *ALGAE, *ACTINO   W72-04155
NS, HISTORY, WATER PURIFICATION,   *WATER UTILIZATION, DECOMPOSING O   W68-00483
ONSIN, *ADMINISTRATIVE AGENCIES,   *WATER WORKS, *WASTE DISPOSAL, DO   W70-03344
DISON(WIS), *'WORKING' OF LAKES,   *WATER-BLOOM, 'BREAKING' OF LAKES   W69-08283
CATION, LAKES, STREAMS, ALGAE,/   *WATER, *FERTILIZATION, *EUTROPHI   W69-10178
E ALGAE, INTAKES, SANDS, FLOW R/   *WATER, *FILTERS, *ALGAE, NUISANC   W70-04199
LIGHT, TEMPERATURE,/ *LIMNOLOGY,   *WATER, *RIVERS, *LAKES, *PONDS,   W70-05404
ER POLLUTION,/ *WASTE TREATMENT,   *WATERS, *ORGANISMS, STREAMS, WAT   W70-05750
*ECOLOGY, *MATHEMATICAL MODELS,   *WATERSHED MANAGEMENT, GEOCHEMIST   W71-09012
CONTROL, ALGICIDES, BIOCONTROL,   *WATERSHED MANAGEMENT, *ENVIRONME   W68-00478
KES(WIS), *AQUATIC WEED CUTTING,   *WEED EXTRACTING STEEL CABLES, AL   W70-04494
IRA STAGNALIS, NE/ *NEW SPECIES,   *WESTERN LAKE ERIE(OHIO), GONGROS   W70-04468
S, WATER POLLUT/ *NEWFOUND LAKE,   *WINNISQUAM LAKE, WATER AND PLANT   W70-02764
, *COLORADO, *LAKES, *MOUNTAINS,   *WINTER, CYANOPHYTA, PLANKTON, CO   W69-10154
LYSIS, P/ *ALGAE, *PLANT GROWTH,   *WISCONSIN, ECOLOGY, CHEMICAL ANA   W70-06229
RODUCING ALGAE, *NUISANCE ALGAE,   *WISCONSIN, *LAKES, PRODUCTIVITY,   W70-05112

IES, *WATER WORKS, *WASTE DISPO/
ATER POLLUTION CONTROL, *LOCAL /
TION, *IRON, *LAKES, *MANGANESE,
, SCREENS, TREATMENT FACILITIES,
FLOS AQUAE, OXYGEN DEPLETION.:
S, DINOFLAGELLATES, OCEANS, BIO/
IGOCLADUS LAMINOSUS, PHORMIDIUM/
AND HOT SPRINGS, SINTER, NUCLEI/
ECTS, *AQUATIC PLANTS, AQUATIC /
AL EFFECTS, SALINITY, TEMPERATU/
E, *RESINS, SEA WATER, PHYSICOC/
RADIOISOTOPES, *SALT MARSHES, */
*PERIPHYTON, *CYCLING NUTRIENTS,

NDALIA(OHIO), RHIZOPODS, SPIDER/
LGAE, LAKES, OXYGEN, EUTROPHICA/
EROMIXIS, *PRIMARY PRODUCTIVITY,
OPLANKTON, *STATISTICAL METHODS,
LEUCOCYTES, *MUSCULAR EXERTION,
ON, *PHYTOTOXICITY, ALGAL POISO/
OCYSTIS, APHANIZOMENON, ANABAEN/

*WISCONSIN, *ADMINISTRATIVE AGENC	W70-03344
*WISCONSIN, *SEWAGE EFFLUENTS, *W	W69-06909
*WISCONSIN, ALGAE, CALCIUM, CHELA	W69-05867
*WISCONSIN.: /FICATION, MEMBRANES	W73-02426
*YAHARA RIVER(WIS), APHANIZOMENON	W72-01797
*YEASTS, POPULATION, MARINE PLANT	W70-04368
*YELLOWSTONE PARK, *ICELAND, MAST	W70-05270
*YELLOWSTONE NATIONAL PARK, *ICEL	W69-10160
*ZINC RADIOISOTOPES, *AQUATIC INS	W71-09005
*ZINC RADIOISOTOPES, *ENVIRONMENT	W69-08267
*ZINC RADIOISOTOPES, *MARINE ALGA	W69-08275
*ZINC RADIOISOTOPES, *PHOSPHORUS	W69-08274
*ZINC RADIOISOTOPES, STREAMFLOW,	W69-07862
*ZINC.:	W71-09005
*ZOOGLEAL SLIME, DAYTON(OHIO), VA	W72-01819
*ZOOPLANKTON, *PHOTOSYNTHESIS, *A	W71-10098
*ZOOPLANKTON, SULFUR BACTERIA, DI	W70-05760
*ZOOPLANKTON, *BIOMASS, *SUCESSIO	W73-01446
*2-METHYL-5-VINYLPYRIDINE, *GLUE	W70-01788
*2-4 D, *HERBICIDES, *PHYTOPLANKT	W69-00994
*2,3-DICHLORONAPHTHOQUINONE, MICR	W70-03310

GENERAL INDEX

The middle rank of this index lists in alphabetical order each word of each descriptor except those included in the *Significant Descriptor Index*, which should be consulted first. The numbers at the right are the accession numbers appearing at the end of the abstract.

KE ERIE, LAKE ONTARIO, POLLUTION
NUTRIENTS, POLLUTANTS, POLLUTION
RA/ *EUTROPHICATION, *POLLUTION
D), OYSTERS, CHANNELS, POLLUTION
ON SOURCES, DREDGING, *POLLUTION
S, CONTROL, POLLUTION ABATEMENT,
LGAE, PLANTS, CONTROL, POLLUTION
UTANT IDENTIFICATION, *POLLUTION
ODOR-PRODUCING ALGAE, POLLUTION
UTION IDENTIFICATION, *POLLUTION
ER POLLUTION CONTROL, *POLLUTION
OSAL, WATER POLLUTION, POLLUTION
YTOPLANKTON, PLANKTON, POLLUTION
OL, FISH CONSERVATION, POLLUTION
SOTA, *EUTROPHICATION, POLLUTION
AMPHORA OVALIS, OPEPHORA MARTYI,
L COUNTS, SPECIES DETERMINATION,
ABILIZATION, PONDS, PENETRATION,
NIQUE, OXYGEN ELECTRODES, ATOMIC
ROG/ *REQUIREMENTS, SCENEDESMUS,
TREATMENT, LAGOONS, OXYGENATION,
NGE, WATER CIRCULATION, NUTRIENT
OTTOM SEDIMENTS, BIBLIOGRAPHIES,
LABORATORY TESTS, PRODUCTIVITY,
IXING, DIFFUSION, SEDIMENTATION,
*MARINE ANIMALS, *RADIOISOTOPES,
BACTERIA, BIOASSAY, METABOLISM,
T RELATIONSHIPS, NUCLEAR WASTES,
N ADSORPTION, COLLOIDS, AQUARIA,
ADIOISOTOPES, *ALGAE, *SORPTION,
AGE, CESIUM, NUCLEAR EXPLOSIONS,
GINOSA, NOSTOC, NOSTOC MUSCORUM,
E WASTE DISPOSAL, OCEANS, FOODS,
                        *PHOSPHATE
ALITY CONTROL, SEWAGE EFFLUENTS,
MAINTENANCE, *OXIDATION LAGOONS,
ASSAY, RADIOACTIVITY TECHNIQUES,
, EXPORT LOSSES, CHEMOSYNTHESIS,
N / *LEWIS AND CLARK LAKE (S D),
ITY OF WISCONSIN, ASIA/ NATIONAL
ITY OF WISCONSIN, ASIA/ NATIONAL
FUNGIA FUNGITES, ZOOXANTHELLAE,

L POLLUTION, LABORATORY ANIMALS,
, *NITROGEN LIMITATION, BA/ *FAT
DIONUCLIDE UPTAKE, *RADIONUCLIDE
EAE, CHLAMYDOMONAS MOEWUSSI, FAT
PESTICIDE METABOLISM, *PESTICIDE
FICIAL SUBSTRATES, *RADIONUCLIDE
COLOGY, FLUORESCENCE MICROSCOPY,
COSE, MECHAELIS-MENTEN EQUATION,
FLAGELLATES, LAKE ERKEN(SWEDEN),
CHAMYDOMONADS, LAPPLAND(SWEDEN),
E DEGRADATION, SEVIN, MALATHION,
URE, CHEMOSTATS, BATCH CULTURES,
ATP, CMU, NITROGEN-15, PYRUVATE,
*BIOCHEMISTRY METHODS, SEASONS,
NOIDS, SPECIES, DIVERSITY INDEX,
ATION, TAXONOMY, FLAVOBACTERIUM,
RICHIA, SARCINA, STAPHYLOCOCCUS,
VOBACTERIUM-CYTOPHAGA', VIBRIOS,
M, MELOSIRA ISLANDICA, NITZSCHIA
ICIPAL WASTES, METHANE BACTERIA,
DRICA, GLYCOLYSIS, TRICARBOXYLIC
LAX POLYEDRA, 3,5-DIAMINOBENZOIC
PHORUS/ *EUTROPHICATION, *LAKES,
INHIBITION, LIMNOLOGY, ACIDITY,
LUTION, WATER POLLUTION EFFECTS,
GEOGRAPHICAL REGIONS, KENTUCKY,

ABATEMENT, OLIGOTROPHY, DEPTH, PH	W70-03964
ABATEMENT, NUISANCE ALGAE, PHOSPH	W69-01453
ABATEMENT, DRAINAGE SYSTEMS, FLOW	W68-00479
ABATEMENT, HUMAN POPULATION.: /IL	W68-00461
ABATEMENT, DIKES, SOIL CONSE: /TI	W68-00172
ABATEMENT, PACIFIC COAST REGION,	W69-03683
ABATEMENT, ABATEMENT, PACIFIC COA	W69-03683
ABATEMENT, EFFLUENTS, STREAMFLOW,	W69-07389
ABATEMENT, SEWAGE TREATMENT, SEWA	W69-06909
ABATEMENT, TIDAL WATERS, MIXING,	W69-06203
ABATEMENT, POLLUTANTS, WASTES, TH	W69-06305
ABATEMENT, COLIFORMS, WATER QUALI	W70-04810
ABATEMENT, POLLUTANT IDENTIFICATI	W70-02792
ABATEMENT, BIBLIOGRAPHIES, ECOLOG	W72-01786
ABATEMENT, GROWTH RATES, CHLOROPH	W69-10167
ABBOT'S POND(SOMERSET, ENGLAND),	W71-12870
ABSORBANCE, COENOBIA, SWIRLING, V	W70-03334
ABSORPTION.: /LIGHT INTENSITY, ST	W70-03923
ABSORPTION SPECTROPHOTOMETRY, TOT	W70-03309
ABSORPTION, CELL CONCENTRATION, U	W70-05409
ABSORPTION, ECOLOGY, WATER PURIFI	W69-07389
ABSORPTION, *CHEMICAL PROPERTIES,	W68-00857
ABSORPTION, FISHES, ALGAE.: /S, B	W68-00858
ABSORPTION, SEDIMENTS, ORGANIC MA	W72-00845
ABSORPTION, BIOCHEMICAL OXYGEN DE	W71-11381
ABSORPTION, MARINE ALGAE, MARINE	W71-09850
ABSORPTION, COMPETITION, OLIGOTRO	W71-10079
ABSORPTION, CHELATION, METABOLISM	W71-09005
ABSORPTION, WATER POLLUTION EFFEC	W71-09863
ABSORPTION, GULF OF MEXICO, ATLAN	W69-08524
ABSORPTION, ADSORPTION, ATLANTIC	W70-00707
ABSORPTION, SPECTROMETRY.: / AERU	W70-02245
ABSORPTION, HAZARDS, CESIUM, HYDR	W70-02786
ABSORPTION.:	W70-01031
ABSORPTION, LAKES.: /NT, WATER QU	W70-01031
ABSORPTION, BIODEGRADATION, CHEMI	W70-08392
ABSORPTION.: / MODEL STUDIES, BIO	W73-01454
ABUNDANCE.: /TECOLOGY, MORPHOLOGY	W70-02780
ABUNDANCE, BLUE-GREEN ALGAE, GREE	W70-05375
ACADEMY OF SCIENCES(USA), UNIVERS	W70-03975
ACADEMY OF SCIENCES(USA), UNIVERS	W70-03975
ACANTHURUS TRIOSTEGUS, CONUS EBRA	W69-08526
ACCELERATED EUTROPHICATION.:	W68-01244
ACCLIMATIZATION, MARINE ANIMALS,	W71-02494
ACCUMULATION, *LABORATORY STUDIES	W70-07957
ACCUMULATION, *RADIONUCLIDE EXCHA	W70-00707
ACCUMULATION.: /A PALEA, MYXOPHYC	W70-02965
ACCUMULATION.: /LYOPHILIZATION, *	W69-02782
ACCUMULATION, RUTHENIUM-106, WHIT	W70-04371
ACETATE, CARBON-14, AUTORADIOGRAP	W70-04811
ACETATE, GLYCOLLIC ACID.: /S, GLU	W70-04284
ACETATE, LAKE NORRVIKEN(SWEDEN),	W70-04194
ACETATE, TURNOVER TIMES: /EASES,	W70-04287
ACETONE, ETHANOL.: /OSA, PESTICID	W70-02968
ACETYLENE REDUCTION TECHNIQUE, HE	W70-02775
ACETYLENE REDUCTION, REDUCTANT.: /	W70-04249
ACETYLENE REDUCTION TEST.: /HODS,	W70-03429
ACHNANTHES, ACTINOPTYCHUS, AMPHIP	W70-03325
ACHROMOBACTER, GULF OF MARSEILLES	W70-04280
ACHROMOBACTER(ALKALIGENES), RHODO	W70-03952
ACHROMOBACTERS, GRAM-POSITIVE BAC	W70-00713
ACICULARIS, NITZSCHIA DISSIPATA,	W69-10158
ACID BACTERIA, SULFUR BACTERIA, A	W72-06854
ACID CYCLE, AUTORADIOGRAPHY, MONO	W70-02965
ACID DIHYDROCHLORIDE, INFRARED GA	W69-08278
ACID LAKES, BIBLIOGRAPHIES, *PHOS	W68-00857
ACID SOILS, ACIDIC WATER, ACIDS,	W70-02792
ACID STREAMS, MINE DRAINAGE, CHLO	W70-02770
ACID STREAMS, MICROFLORA.: /HIPS,	W68-00891

CULES, NUCLEIC ACID, RIBONUCLEIC
OUNTY, KENTUCKY.:
TY CRITERIA, CYANIDE, NAPHTHENIC
TEN EQUATION, ACETATE, GLYCOLLIC
MIYA'S MINERAL MEDIUM, GLYOXYLIC
DE-COATED SAND, CRUSHED BAUXITE,
　　　　　　　　　　　　　　　*NITRIC
PRODUCTS, PHOTIC ZONE, GLYCOLLIC
TS, ASCOPHYLLUM NODOSUM, ALGINIC
SUBSTRATE COMPLEXES, RIBONUCLEIC
LATION, *MACROMOLECULES, NUCLEIC
AND HOT SPRINGS, SINTER, NUCLEIC
RINGS, HABITATS, ALKALINE WATER,
　LIMNOLOGY, ACIDITY, ACID SOILS,
TRATION, WATE/ *ACID MINE WATER,
LLUTANTS, INHIBITION, LIMNOLOGY,
NDS, HYDROGEN ION CONCENTRATION,
LANTS, AMMONIA, HARDNESS(WATER),
OLLUSKS, MINE ACIDS, COAL MINES,
RBOXYLIC ACIDS, AMINO CARBOXYLIC
DETERGENTS, RADIOACTIVITY, AMINO
*SEAWATER, *FRESH WATER, *HUMIC
HLORELLA PYRENOIDOSA, CARBOXYLIC
, / *ALGAE, *CHLORELLA, *ORGANIC
　TRACE ELEMENTS, VITAMINS, AMINO
HORUS RADIOISOTOPES, IONS, HUMIC
ERVOIRS, NITROGEN, GROWTH, AMINO
, DIATOMS, CORAL, MOLLUSKS, MINE
PSEUDOMONAS, TEMPERATURE, AMINO
SA, VOLATILE ACIDS, NON-VOLATILE
ROTEINS, WALLS, STRUCTURE, AMINO
　CHLORELLA PYRENOIDOSA, VOLATILE
LGAE, CONTINUOUS CULTURES, FATTY
IDITY, ACID SOILS, ACIDIC WATER,
S, *ALGAE, *TASTE, *ODOR, *AMINO
TLAND), SCOTLAND, PHAEOPIGMENTS,
LIFE, PTYCHOCHEILUS OREGONENSIS,
　HOLE OCEANOGRAPHIC INSTITUTION,
TTLE(WASH), REVENUE BONDS, METRO
RADIOACTIVE WASTE, WATER QUALITY
FUNGI, VIRUSES, ALGAE, LICHENS,
ZOMENON, OSCILLATORIA, MELOSIRA,
Y(KY), THORIUM-232, BISMUTH-214,
SMUS, EUGLENA, BACTERIA, YEASTS,
TIONS, E. COLI, BACTERIA, FUNGI,
OR, *AMINO ACIDS, POTABLE WATER,
BACTERIA, ULTRAVIOLET RADIATION,
ALGAE, EUTROPHICATION, ALCOHOLS,
UM COMPOUNDS, *ACTIVATED CARBON,
ES, DIVERSITY INDEX, ACHNANTHES,
CHARIS CANADENSIS/ *ANTAGONISTIC
L CONTROL, DISINFECTION, MODE OF
Y, BIOASSAY, INHIBITION, MODE OF
IPAL WASTES, SEWAGE, ADSORPTION,
, ELECTRODIALYSIS, ION EXCHANGE,
EMICAL OXYGEN DEMAND, *BACTERIA,
ENT, SOLID WASTES, MICROBIOLOGY,
IARY TREATMENT, ALGAE, BACTERIA,
N, SORPTION, FLOCCULATION, COAL,
ONTROL, ECOLOGICAL DISTRIBUTION,
TION, BIOCHEMICAL OXYGEN DEMAND,
　SEDIMENTATION, ORGANIC LOADING,
ANAEROBIC CONDITIONS, CANNERIES,
ION DITCH.:　　　　　　　　*FWPCA,
, CHEMICAL OXYGEN DEMAND, COLOR,
AE, *CATTLE, SLURRIES, SPRAYING,
T, SUSPENDED LOAD, PILOT PLANTS,
TIC MICROORGANISMS, TEMPERATURE,
S, COALS, DISTRIBUTION PATTERNS,
ATION, ODOR, AMMONIUM COMPOUNDS,
RS, *DESIGN, COAGULATION, ALGAE,
LGAE, *TASTE, *ODOR, *ALGICIDES,
S, CHLORINATION, COPPER SULFATE,
, *ALGAE, *ACTINOMYCETES, *ODOR,
N, *FLOW AUGMENTATION, *LAGOONS,
, COPPER SULFATE, SEDIMENTATION,
ROWTH, OXYGEN DEPLETION, NUCLEAR
COLOGY, MAXIMUM UPTAKE VELOCITY,
OGY, SWEDEN, LAKE ERKEN(SWEDEN),
ENIUM-106, WHITE OAK LAKE(TENN),
　　　　　　　　　　　　　　　　*ALGAL
BOUND RADIOSTRONTIUM, *MICROBIAL
TITUTED IMIDAZOLINES, BIOLOGICAL
BIOLOGICAL HA/ *UPTAKE, SPECIFIC
IS, NITZSCHIA DISSIPATA, SYNEDRA
YPTOPHYTA, PYROPHYTA, CHROOMONAS
OBSERVATIONS, CHRONIC EXPOSURE,
G CROP, CHLOROPHYLL, PHOSPHATES,
LETONEMA COSTATUM, PHYSIOLOGICAL
LUDGE, OXIDATION LAGOONS, ALGAE,

ACID SYNTHESIS.: /ION, *MACROMOLE　W69-10151
ACID WASTE MICROFLORA, MCCREARY C　W69-00096
ACID.: /BIAL ECOLOGY, WATER QUALI　W70-04510
ACID.: /S, GLUCOSE, MECHAELIS-MEN　W70-04284
ACID.: /TILE ACIDS, KETOACIDS, TA　W70-04195
ACID-WASHED BAUXITE, KANPUR(INDIA　W70-08628
ACID, *LANGELIER INDEX.:　　　　　W71-11393
ACID, CHEMOTROPHY, PARTICULATE MA　W70-02504
ACID, FUCUS VESICULOSUS, GROWTH F　W69-07826
ACID, ORCINOL METHOD.: /, ENZYME-　W69-10160
ACID, RIBONUCLEIC ACID SYNTHESIS.　W69-10151
ACID, SISJOTHANDI(ICELAND), ENZYM　W69-10160
ACIDIC WATER, TRAVERTINE, CALCIUM　W69-10160
ACIDIC WATER, ACIDS, PHOTOSYNTHES　W70-02792
ACIDIC WATER, HYDROGEN ION CONCEN　W70-05424
ACIDITY, ACID SOILS, ACIDIC WATER　W70-02792
ACIDITY, MUD, HYDROGEN SULFIDE.: /　W70-05399
ACIDITY, OXYGEN, ALGAE, SELF-PURI　W69-07838
ACIDITY, PYRITE, SULFUR COMPOUNDS　W69-07428
ACIDS.: /HLORELLA PYRENOIDOSA, CA　W70-02248
ACIDS.: /ISTRIBUTION, CHELATION,　W69-08275
ACIDS, *COLOR, RIVERS, BOGS, PHEN　W70-01074
ACIDS, AMINO CARBOXYLIC ACIDS.: /　W70-02248
ACIDS, BIOMASS, ORGANIC COMPOUNDS　W70-04195
ACIDS, CALCIUM, AMMONIA, NITROGEN　W70-02255
ACIDS, CARP, ICE, IRON COMPOUNDS,　W70-05399
ACIDS, CHLAMYDOMONAS, CHLORELLA,　W69-10180
ACIDS, COAL MINES, ACIDITY, PYRIT　W69-07428
ACIDS, INTERTIDAL AREAS, CARBOHYD　W70-03952
ACIDS, KETOACIDS, TAMIYA'S MINERA　W70-04195
ACIDS, MECHANICAL EQUIPMENT, ENZY　W70-04184
ACIDS, NON-VOLATILE ACIDS, KETOAC　W70-04195
ACIDS, OOCYSTIS POLYMORPHA, UNICE　W70-07957
ACIDS, PHOTOSYNTHESIS, PHOTOSYNTH　W70-02792
ACIDS, POTABLE WATER, ACTINOMYCET　W72-01812
ACRIDINE ORANGE, LINEWEAVER-BURKE　W70-04811
ACROCHEILUS ALUTACEUS, ULOTHRIX,　W69-07853
ACROPORA HYACINTHUS, POCILLOPORA　W69-08526
ACT(SEATTLE).: /INGTON(WASH), SEA　W70-04455
ACT, PATH OF POLLUTANT, FOOD CHAI　W72-00941
ACTIMOMYCETES, CYTOLOGICAL STUDIE　W71-11823
ACTINASTRUM, COELASTRUM, PEDIASTR　W70-08107
ACTINIUM-228, SWIFT RIFFLES, ZIRC　W69-09742
ACTINOMYCETES, PONDING, WASTE TRE　W70-08097
ACTINOMYCETES, PROTOZOA, ALGAE.: /　W71-02009
ACTINOMYCETES, PLANT GROWTH, DECO　W72-01812
ACTINOMYCETES, HYDROGEN ION CONCE　W70-00719
ACTINOMYCETES, CYANOPHYTA, SPECTR　W69-00387
ACTINOMYCETES, IOWA.: /, *POTASSI　W72-04242
ACTINOPTYCHUS, AMPHIPLEURA RUTILA　W70-03325
ACTION, *AQUATIC MACROPHYTES, ANA　W70-02249
ACTION, MORTALITY, INHIBITION, AL　W70-02370
ACTION, PHOTOSYNTHESIS.: /ORTALIT　W70-03519
ACTIVATED CARBON, FILTERS, SEASON　W70-04001
ACTIVATED SLUDGE, DENITRIFICATION　W69-01453
ACTIVATED SLUDGE, IRON COMPOUNDS,　W68-00256
ACTIVATED SLUDGE, RATES, CHLORELL　W68-00248
ACTIVATED SLUDGE.: / DEMAND, TERT　W69-08054
ACTIVATED CARBON.: /AE, FILTRATIO　W70-05470
ACTIVATED SLUDGE, MAINTENANCE COS　W70-04786
ACTIVATED SLUDGE, COST COMPARISON　W71-09548
ACTIVATED SLUDGE, AERATION, DISSO　W71-08970
ACTIVATED SLUDGE, BIOCHEMICAL OXY　W71-09546
ACTIVATED ALGAE, FEEDLOTS, OXIDAT　W71-06821
ACTIVATED SLUDGE, ALGAE, WASTE WA　W70-10433
ACTIVATED SLUDGE, FARM WASTES.: /　W70-07491
ACTIVATED SLUDGE, TREAT: /REATMEN　W70-09186
ACTIVATED SLUDGE, FISHERIES, CARP　W70-09905
ACTIVATED CARBON, CLAYS, ALGAE, W　W69-09884
ACTIVATED CARBON, CHLORINE, PHENO　W70-00263
ACTIVATED CARBON, *FLOCCULATION,　W72-07334
ACTIVATED CARBON, FILTRATION, *TO　W72-03746
ACTIVATED CARBON, WATER SUPPLY.: /　W72-03562
ACTIVATED CARBON, AMMONIA, CHLORI　W72-04155
ACTIVATED SLUDGE, ALGAE, NUTRIENT　W72-06017
ACTIVATED CARBON, CHLORINE, COSTS　W72-08859
ACTIVATION TECHNIQUES, ROCK LAKE(　W70-00264
ACTIVE TRANSPORT, BACTERIAL POPUL　W70-04194
ACTIVE TRANSPORT, CARBON-14, ENZY　W70-04284
ACTIVITY-DENSITY, VERTICAL, MICRO　W70-04371
ACTIVITY, *NUTRIENT REMOVAL.:　　　W70-08980
ACTIVITY, *RADIOSTRONTIUM COMPLEX　W71-04067
ACTIVITY, QUARTERNARY, AMMONIUM C　W70-02982
ACTIVITY, CHLORELLA PYRENOIDOSA,　W70-02786
ACUS, SINKING RATES, GRAZING PRES　W69-10158
ACUTA, SCHROEDERIA SETIGERA, MELO　W70-05387
ACUTE EXPOSURE.: /, PHYSIOLOGICAL　W70-01779
ADD: /CIUM, PRODUCTIVITY, STANDIN　W70-03512
ADJUSTMENT, ASTERIONELLA FORMOSA,　W70-04809
ADMINISTRATION, MAINTENANCE, *WAS　W71-08970

ION, SUPERVISORY CONTROL(POWER),
Y CONTROL, ALGAE, AQUATIC WEEDS,
EFFECTS, CATION EXCHANGE, CATION
HYSICOCHEMICAL PROPERTIES, ANION
QUARIA, WATER POLLUTION EFFECTS,
OMPOUNDS, *PHOSPHORUS COMPOUNDS,
IUM, PHYSICOCHEMICAL PROPERTIES,
  NUCLEAR EXPLOSIONS, ABSORPTION,
CE, EVAPORATION, BIODEGRADATION,
HORESIS, CHLORELLA, SCENEDESMUS,
ASTES, MUNICIPAL WASTES, SEWAGE,
LGAE, PERIPHYTON, INVERTEBRATES,
MUD-WATER INTERFACE, PHOSPHORUS,
RYING, SOLIDS-SETTLING, ECONOMIC
NSEE(SWITZ), PAFFIKERSEE(SWITZ),
TRATION FACTOR, LIMNODRILUS SPP,
CHLORELLA PYRENOIDOSA, TRIBONEMA
AND(N C), DETE/ *CONCENTRATIONS,
      *MANURE, ANAEROBIC LAGOON.:
ISMS, PHOSPHATES, HYDROGENATION,
E, CLAYS, ALDRIN, DIELDRIN, DDT,
TION, BIOCHEMICAL OXYGEN DEMAND,
GANIC LOADING, ACTIVATED SLUDGE,
AERATION, NUTRIENTS, OXYGEN SAG,
TREATMENT, TREATMENT FACILITIES,
  BIOCHEMICAL OXYGEN DEMAND(BOD),
LINITY, TEMPERATURE, MONITORING,
RIO).:
R TEMPERATURE, SEWAGE TREATMENT,
UD, BENTHOS, HYDROGEN, BACTERIA,
ORGANIC MATTER, AIR TEMPERATURE,
CRITERIA, FACILITIES, EQUATIONS,
E, *LIGHT, *NITRATES, *NITRITES,
R POLLUTION, SEWAGE, AESTHETICS,
MPERATURE, FLOW, ALGAE, BENTHOS,
DE, PHOTOSYNTHESIS, GROWTH RATE,
RIA, ODOR, ANAEROBIC CONDITIONS,
S.:   *POTATO PROCESSING WASTES,
, *ORGANIC WASTES, *OXYGENATION,
TERIA, EFFLUENTS, OXYGEN DEMAND,
GEN DEMAND, ANAEROBIC DIGESTION,
NITROGEN, BACTERIOLOGICAL STUDY,
S, PROTEUS VULGARIS, PSEUDOMONAS
                       *MICROCYSTIS
                        PSEUDOMONAS
ISTS, VETERINARIANS, MICROCYSTIS
ZOMENON FLOS AQUAE, *MICROCYSTIS
EAE, NOSTOCHINEAE, CLATHROCYSTIS
TION, CULTURE MEDIA, MICROCYSTIS
SULFATE, EFFLUENTS, MICROCYSTIS
HLORONAPHTHOQUINONE, MICROCYSTIS
STIGEOCLONIUM TENUE, MICROCYSTIS
LORELLA PYRENOIDOSA, MICROCYSTIS
ANABAENA CYLINDRICA, MICROCYSTIS
CHU NO 10 SOLUTION/ *MICROCYSTIS
  ANABAENA SPIROIDES, MICROCYSTIS
ANABAENA VARIABILIS, MICROCYSTIS
OGICALLY AVAILABLE), MICROCYSTIS
EDOGONIA, SPIROGYRA, MICROCYSTIS
MS, ALGAE NUTRITION, MICROCYSTIS
CHLORIS PENIOCYSTIS, DIPLOCYSTIS
ANABAENA FLOS-AQUAE, MICROCYSTIS
EMICAL COMPOSITION, *MICROCYSTIS
ITY, *ALGAL GROWTH, *MICROCYSTIS
E, PHOTOMICROGRAPHS, MICROCYSTIS
TER, *CALIFORNIA, WATER QUALITY,
LAKES, *WATER POLLUTION, SEWAGE,
IGAN, *LAKE ONTARIO, LAKE HURON,
EGETAT/ *HARTBEESPOORT DAM(SOUTH
IUM BREVE, CANADA, EUROPE, SOUTH
, INDUSTRIAL PLANTS, STRUCTURES,
DS, MACROMOLECULES, ANTIBIOTICS,
ERLAND, PERIDINIUM, OSCILLATORIA
M, LAMINARIA DIGITATA, LAMINARIA
S CASEI, BACILLUS THURINGIENSIS,
SILVER NITRATE, SIMAZIN, NUPHAR,
A, CHRYSOPHYTA, GRAZING, SNAILS,
Y CONTROL(POWER), ADMINISTRATIVE
SPO/ *WISCONSIN, *ADMINISTRATIVE
      *UPPER KLAMATH LAKE(ORE.),
STES, DYEING WASTES./ *ANTI-FOAM
OTOCARPUS, POLYSIPHONIA, ORGANIC
OMS, WATER BLOOM, AQUATIC ALGAE,
TROPHY, *NUTRIENT BALANCE, *LAKE
AQUATIC NUISANCES, TRANSPARENCY,
NUTRIENTS, NITROGEN, PHOSPHORUS,
OPHYLL, NUTRIENTS, CONDUCTIVITY,
WISCONSIN, MICHIGAN, PHOSPHORUS,
TION ASSESSMENT, URBAN DRAINAGE,
STES, SEWAGE, INDUSTRIAL WASTES,

ADMINISTRATIVE AGENCIES.: /ISTRAT   W72-05774
ADMINISTRATION; SUPERVISORY CONTR   W72-05774
ADSORPTION, HYDROLYSIS, MARINE AL   W71-09863
ADSORPTION, COLLOIDS, AQUARIA, AB   W71-09863
ADSORPTION.: /TION, METABOLISM, A   W71-09005
ADSORPTION, REMOVAL, DEGRADATION,   W71-13335
ADSORPTION, COLLOIDS, ZINC RADIOI   W70-00708
ADSORPTION, ATLANTIC OCEAN, DIATO   W70-00707
ADSORPTION, ALGAE, CLAYS, ALDRIN,   W70-09768
ADSORPTION, SEWAGE TREATMENT, PAR   W71-03035
ADSORPTION, ACTIVATED CARBON, FIL   W70-04001
ADSORPTION, PHYTOPLANKTON, SUCKER   W69-07853
ADSORPTION, EUTROPHICATION, ALGAE   W69-06859
ADVANTAGE, TEMPERATURE-EFFECT.: /   W70-05819
AEGERISEE(SWITZ), FURES: / GREIFE   W70-00270
AEOLOSOMA HEMPRICHI, ESCHERICHIA   W69-07861
AEQUALE, NAVICULA PELLICULOSA, LA   W69-10180
AEQUIPECTEN IRRADIANS, PIVERS ISL   W69-08267
AERATED LAGOON.:   W70-07491
AERATION, FLOCCULATION, FILTRATIO   W70-08627
AERATION, SILTS.: /SORPTION, ALGA   W70-09768
AERATION, TEMPERATURE, FLOW, ALGA   W71-11381
AERATION, DISSOLVED OXYGEN, SLUDG   W71-08970
AERATION, BIOCHEMICAL OXYGEN DEMA   W69-07520
AERATION, LAGOONS, DOMESTIC WASTE   W70-04060
AERATION, COLIFORMS, ALGAL CONTRO   W70-04788
AERIAL PHOTOGRAPHY, INFRARED RADI   W68-00010
AERO-HYDRAULIC GUN, ALLISTON(ONTA   W71-09546
AEROBIC BACTERIA, AQUATIC PLANTS,   W71-08221
AEROBIC BACTERIA, METABOLISM, CAN   W69-07838
AEROBIC BACTERIA, WATER TEMPERATU   W69-07096
AEROBIC CONDITIONS, ANAEROBIC CON   W70-04786
AEROBIC CONDITIONS, ANAEROBIC CON   W70-05261
AEROBIC CONDITIONS, ANAEROBIC CON   W70-04430
AEROBIC CONDITIONS, ANAEROBIC CON   W71-11381
AEROBIC CONDITIONS, ANAEROBIC CON   W70-05750
AEROBIC CONDITIONS, E. COLI, BACT   W71-02009
AEROBIC LAGOONS, ANAEROBIC LAGOON   W71-08970
AEROBIC TREATMENT, ANAEROBIC COND   W71-09546
AEROBIC TREATMENT, PHOTOSYNTHESIS   W72-00462
AEROBIC TREATMENT, CHEMICAL OXYGE   W71-00936
AEROBICITY.:   ORGANIC   W71-05742
AERUGINOSA, STAPHYLOCOCCUS AUREUS   W71-11823
AERUGINOSA, *LAKE WATERS.:   W70-08637
AERUGINOSA.:   W69-10171
AERUGINOSA, ENDOTOXIN, FAST-DEATH   W70-00273
AERUGINOSA, BOHEMIA, ELBE RIVER,   W70-00274
AERUGINOSA, COELOSPHAERIUM KUTZIN   W69-08283
AERUGINOSA, APHANIZOMENON FLOS-AQ   W70-00719
AERUGINOSA.: /ORUS, ALGAE, COPPER   W70-02787
AERUGINOSA, APHANIZOMENON FLOS-AQ   W70-02982
AERUGINOSA, NOSTOC, NOSTOC MUSCOR   W70-02245
AERUGINOSA, BURNTSIDE RIVER(MINN)   W70-03512
AERUGINOSA, OEDOGNIUM.: /RATION,   W70-03520
AERUGINOSA(KUTZ), CULTURE MEDIA,   W70-03507
AERUGINOSA, CRYPTOMONADIDA, MASTI   W70-05112
AERUGINOSA, APHANIZOMENON FLOS-AQ   W70-05091
AERUGINOSA, LAKE COMO(MINN), LAKE   W70-05269
AERUGINOSA, ANABAENOPSIS, RACIBOR   W70-05404
AERUGINOSA.:   ALGAL BLOO   W70-05565
AERUGINOSA, NOSTOC MUSCORUM, VANA   W69-06277
AERUGINOSA, ALGAL CULTURES, CHEMO   W69-06864
AERUGINOSA, *NUTRIENT AVAILABILIT   W69-05868
AERUGINOSA, ANTAGONISMS, ALGAL NU   W69-05867
AERUGINOSA, ANABAENA CIRCINALIS,   W69-05704
AESTHETICS, RECREATION, FISHING,   W70-05645
AESTHETICS, AEROBIC CONDITIONS, A   W70-04430
AESTHETICS, ECOLOGY, ECONOMICS, O   W70-01943
AFRICA), TRANSPARENCY, MARGINAL V   W70-06975
AFRICA, ISRAEL, MICROCYSTIS, PRYM   W70-05372
AFTERBAYS, WATER TYPES, INTAKES.:   W69-05023
AGARBACTERIUM, XANT: /ARY COMPOUN   W70-04280
AGARDHI, PHORMIDIUM, APHANIZOMENO   W69-10169
AGARDHII, DESICCATION, IMMERSION,   W70-01073
AGARS, STAPHYLOCOCCUS EPIDERMIDIS   W71-11823
AGASICLES, MANATEE, MARISA.: /C,   W70-10175
AGE, TURBIDITY, DIPTERA, MAYFL: /   W71-00668
AGENCIES.: /ISTRATION, SUPERVISOR   W72-05774
AGENCIES, *WATER WORKS, *WASTE DI   W70-03344
AGENCY LAKE(ORE.), *OSCILLATORA.:   W72-06051
AGENTS, *TEXTILE WASTES, FIBER WA   W70-04060
AGGREGATES, SEASONAL EFFECTS.: /C   W70-01068
AGING(BROLOGICAL), *AQUATIC PRODU   W68-00912
AGING, NUTRIENT SOURCES, WATER SA   W70-00264
AGRICULTURAL RUNOFF, NUTRIENT BUD   W69-09349
AGRICULTURE, WATER MANAGEMENT(APP   W71-03048
AGRICULTURAL WATERSHEDS, SEWAGE,   W70-05752
AGRICULTURAL WATERSHEDS, ORGANIC   W70-04193
AGRICULTURAL DRAINAGE, NORTH AMER   W70-03975
AGRICULTURE, ST LAWRENCE RIVER, A   W70-03964

379

TION ASSESSMENT, URBAN DRAINAGE,   AGRICULTURAL DRAINAGE, NORTH AMER   W70-03975
ATION, *WATER POLLUTION SOURCES,   AGRICULTURAL WATERSHEDS.: /ROPHIC   W69-05323
TE TREATMENT, *NUTRIENT REMOVAL,   AGRICULTURAL DRAINAGE, APPALACHIA   W69-07818
LUENTS, *SEWAGE DISPOSAL, ALGAE,   AGRICULTURAL WATERSHEDS, NUTRIENT   W70-05565
ALGAE COUNTS, PLANKTON ANALYSES,   AGRICULTURAL RUNOFF, ARKANSAS RIV   W70-04507
ILLINOIS, PHOSPHORUS, DRAINAGE,   AGRICULTURAL WATERSHEDS, NITROGEN   W70-04506
      *DIATOMITE FILTERS, FILTER   AIDS: W72-04145
                           FILTER   AIDS, DIATOMITE FILTERS.:   W72-04318
C WASTES, *WASTE WATER DISPOSAL,   AIR POLLUTION, WATER POLLUTION, O   W71-00940
AND DEVELOPMENT, GRANTS, ALGAE,   AIR POLLUTION, WATER POLLUTION, S   W71-06821
LING, PONDS, RIVERS, RESERVOIRS,   AIR TEMPERATURE, RAIN, GEOLOGY, M   W70-02646
YGEN, SILICATES, ORGANIC MATTER,   AIR TEMPERATURE, AEROBIC BACTERIA   W69-07096
POLLUTION, NITROGEN, PHOSPHORUS,   AIR, PHYTOPLANKTON, ALGAE, EFFICI   W70-04484
GS, TERRESTRIAL HABITATS, DEPTH,   AIR, SEDIMENTS, GEOLOGIC FORMATIO   W69-08272
DISCRIMINATOR, RHODAMINE WT DYE,   AIRBORNE TESTS, OIL SPILLS, GEOLO   W70-05377
SHOLD ODOR NUMBERS, CROP-DUSTING   AIRCRAFT, WATER SAMPLING, ONTARIO   W69-10155
AMAUGA RESERVOIR(TENN), FLORENCE(ALA).: / BROAD RIVER(TENN), CHICK   W71-07698
L COMMUNITIES, BACKWATER, PONDS,   ALASKA.: /YTA, DIATOMS, BIOLOGICA   W70-03969
                      SMITH LAKE(ALASKA).:   W69-03515
GRAPHITE LAKE, OLD J/ DETECTOR,   ALASKA, HUDEUC LAKE, PULLIN LAKE,   W69-08272
FIBERS, MAN-MADE FIBERS, METHYL   ALCOHOL, ETHYLENE GLYCOL, POLYMER   W70-10433
*TEXTILE MILL WASTES, ALFOL   ALCOHOLS.:   W70-08740
PRODUCING ALGAE, EUTROPHICATION,   ALCOHOLS, ACTINOMYCETES, CYANOPHY   W69-00387
ATION, ADSORPTION, ALGAE, CLAYS,   ALDRIN, DIELDRIN, DDT, AERATION,   W70-09768
UTANTS, WATER POLLUTION EFFECTS,   ALDRIN, DIELDRIN, ENDRIN, DDT.: /   W69-08565
ORINATED HYDROCARBON PESTICIDES,   ALDRIN, ENDRIN, DIELDRIN, SCENEDE   W70-03520
S, FINGERNAIL CLAMS, WHITE FISH,   ALEWIFE, GIZZARD SHAD, SEA LAMPHR   W70-01943
LANCE, OVERFISHING, COHO SALMON,   ALEWIFE, NUTRIENT REMOVAL.: /L BA   W70-04465
         *TEXTILE MILL WASTES,   ALFOL ALCOHOLS.:   W70-08740
OPLANKTON, *DIATOMS, *BLUE-GREEN   ALGA, COPPER SULFATE.: /NG, *PHYT   W70-02754
ROSYSTIS, ALGASTATIC SUBSTANCES,   ALGADYNAMIC SUBSTANCES, PHORMIDIN   W70-02510
ATIVE TREATMENT, ALGAE NUISANCE,   ALGAE BLOOM, ALGAE GROWTH, WATER   W69-08282
E MINNETONKA(MINN), MORPHOMETRY,   ALGAE BLOOMS, CLEARWATER LAKE(MIN   W70-05387
RAFTER METHOD, TREATMENT TIMING,   ALGAE CONTROL REQUIREMENTS, THRES   W69-10157
(ONTARIO), LAKE RAMSEY(ONTARIO),   ALGAE COUNTS, THRESHOLD ODOR NUMB   W69-10155
N METHODS, ALGAE IDENTIFICATION,   ALGAE COUNTS, ALGAL GROWTH, SEDGW   W69-10157
*NATIONAL WATER QUALITY NETWORK,   ALGAE COUNTS, PLANKTON ANALYSES,   W70-04507
NT, ALGAE NUISANCE, ALGAE BLOOM,   ALGAE GROWTH, WATER DENSITY, CONV   W69-08282
AL TREATMENT, AMMONIA STRIPPING,   ALGAE HARVESTING.: /ELECTROCHEMIC   W70-01981
GE, *COPPER APPLICATION METHODS,   ALGAE IDENTIFICATION, ALGAE COUNT   W69-10157
                  ALGAL TAXONOMY,   ALGAE IDENTIFICATION.:   W69-03514
*PIPETTE, UNIALGAL, NITZSCHIA,   ALGAE ISOLATION, CAPILLARITY, CAP   W70-04469
ATMENT, *PREVENTATIVE TREATMENT,   ALGAE NUISANCE, ALGAE BLOOM, ALGA   W69-08282
GINOSA.:   ALGAL BLOOMS,   ALGAE NUTRITION, MICROCYSTIS AERU   W70-05565
*BLUE-GREEN ALGAE, *ALGA BLOOMS,   ALGAE PHYSIOLOGY.:   W69-03512
TS, DRAINAGE EFFECTS, *SAMPLING,   ALGAE TOXINS, SALINE WATER, EQUIL   W69-00360
ER DENSITY, CONVECTION CURRENTS,   ALGAE TREATMENT COSTS, DINOBRYON,   W69-08282
       WATER PURIFICATION BY   ALGAE.:   W69-07389
W ZEALAND, SEWAGE TREATMENT POND   ALGAE.:   *NF   W70-02304
POLLUTION-TOLERANT   ALGAE.:   W70-07027
OOCYSTIS POLYMORPHA, UNICELLULAR   ALGAE.: / CULTURES, FATTY ACIDS,   W70-07993
TON, HYDROGEN ION CONCENTRATION,   ALGAE.: / PHYTOPLANKTON, ZOOPLANK   W70-04368
ENVIRONMENTAL EFFECTS, CHEMTROL,   ALGAE.: /*WATERSHED MANAGEMENT, *   W68-00478
STIC ANIMALS, SAMPLING, NUISANCE   ALGAE.: /, NUTRIENTS, DEPTH, DOME   W70-00274
EFFLUENTS, FLOW RATES, FILTERS,   ALGAE.: /ANTS, HUMUS, REAERATION,   W70-05821
S, CRABS, SHRIMP, MULLETS, FISH,   ALGAE.: /COSYSTEMS, SINKS, TRACER   W69-08274
SOTHERM, BLUE-GREEN ALGAE, GREEN   ALGAE.: /DE, UPTAKE, FREUNDLICH I   W70-08111
, MICROCYSTIS SCUMS, PITHIOPHORA   ALGAE.: /E FERTILIZERS, BLUEGILLS   W70-05417
FUNGI, ACTINOMYCETES, PROTOZOA,   ALGAE.: /IONS, E. COLI, BACTERIA,   W71-02009
ICIDES, WATER POLLUTION EFFECTS,   ALGAE.: /LABORATORY TESTS, INSECT   W71-03183
TREATMENT, PHOSPHATES, NITRATES,   ALGAE.: /NG FILTERS, WASTE WATER   W70-06792
ION, CARBON DIOXIDE, ALKALINITY,   ALGAE.: /NITROGEN, OXYGEN, OXIDAT   W70-05550
PHYCOLOGY, SPACE RESEARCH, TOXIC   ALGAE.: /OALGAE, MICRONUTRIENTS,   W69-07832
F ACTION, MORTALITY, INHIBITION,   ALGAE.: /OL, DISINFECTION, MODE O   W70-02370
FFECTS, EUTROPHICATION, NUISANCE   ALGAE.: /REAMS, WATER POLLUTION E   W70-00719
TION EFFECTS, YEASTS, CRUSTACEA,   ALGAE.: /RODUCTIVITY, WATER POLLU   W69-10152
SOLVED OXYGEN, AQUATIC BACTERIA,   ALGAE.: /R POLLUTION EFFECTS, DIS   W70-04727
ROSCOPY, BLUE-GREEN ALGAE, GREEN   ALGAE.: /RACTIONS, INFRARED SPECT   W70-03928
LIOGRAPHIES, ABSORPTION, FISHES,   ALGAE.: /S, BOTTOM SEDIMENTS, BIB   W68-00858
       NUTRIENT REQUIREMENTS(ALGAE).:   W70-04074
LOGY.:   *BLUE-GREEN   ALGAE, *ALGA BLOOMS, ALGAE PHYSIO   W69-03512
ONSIN, COPPER SULPHATE, NUISANCE   ALGAE, *ALGAL CONTROL, WEED CONTR   W68-00468
OOD WEBS, FORAGE FISH, PLANKTON,   ALGAE, *ALGAL CONTROL, PREDATION,   W70-05428
E, *WATER QUALITY CON/ *NUISANCE   ALGAE, *ALGICIDES, *COPPER SULFAT   W69-05704
HIRE, *COPPER SULFATE/ *NUISANCE   ALGAE, *ALGAL CONTROL, *NEW HAMPS   W69-08674
, FISH TOXINS, BIOASSAY, AQUATIC   ALGAE, *ALGAL TOXINS, KILLIFISHES   W70-08372
ER Q/ *AQUATIC PLANTS, *NUISANCE   ALGAE, *ALGICIDES, *PHENOLS, *WAT   W70-10161
UTANTS, BIOLOG/ *ALGAE, *AQUATIC   ALGAE, *AQUATIC ENVIRONMENT, POLL   W70-01233
D CLIMATES, SEMIARID CLIM/ *SOIL   ALGAE, *AQUATIC ALGAE, ALGAE, ARI   W69-03491
*ANALYTICAL TECHNIQUES, *AQUATIC   ALGAE, *AQUATIC MICROBIOLOGY, *AQ   W70-04283
MUD(LAKE), *PHOSPHORUS, *AQUATIC   ALGAE, *BIOASSAY, SCENEDESMUS, PH   W70-03955
ION-REDUCTION POTENTIAL, AQUATIC   ALGAE, *BIOLOGICAL TREATMENT, *WA   W71-09546
A, PIGMENTE/ *POLLUTION-TOLERANT   ALGAE, *BLUE GREEN ALGAE, *EUGLEN   W70-01233
TABOLISM, KI/ *BACTERIA, *MARINE   ALGAE, *CULTURES, PSEUDOMONAS, ME   W70-04280
TEM/ *PLANKTON, *ALGAE, *SESSILE   ALGAE, *CYCLING NUTRIENTS, ECOSYS   W73-01454
LIGHT INTENSITY, BIOMAS/ *MARINE   ALGAE, *DIATOMS, *BENTHIC FLORA,   W70-03325
ION EFFECT/ *OILY WATER, *MARINE   ALGAE, *EMULSIFIERS, WATER POLLUT   W70-01780
OCIATIONS, SHEAR WA/ *EPIPSAMMIC   ALGAE, *EPIPELIC ALGAE, ALGAL ASS   W71-12870
TION-TOLERANT ALGAE, *BLUE GREEN   ALGAE, *EUGLENA, PIGMENTED FLAGEL   W70-01233
A, LAKES, RESERVOIRS,/ *NUISANCE   ALGAE, *EUTROPHICATION, CYANOPHYT   W70-02803

380

AL RECHARGE, COPPER, *ALGICIDES,   ALGAE, *EVALUATION, FILTRATION.: /   W72-05143
*TEMPERATURE TOLERA/ *BLUE-GREEN   ALGAE, *GREEN ALGAE, *RED ALGAE,   W71-02496
  *EUTROPHICATION, *PHOSPHOROUS,   ALGAE, *HARVESTING OF ALGAE, EFFL   W71-13313
ESMIDS/ *EUTROPHICATION, AQUATIC   ALGAE, *LAKES, *CLASSIFICATION, D   W68-00255
, EXUDATION, T/ *OCEANS, *MARINE   ALGAE, *LITTORAL, *ORGANIC MATTER   W70-01073
EMULSIFIERS, *MORTALITY, *MARINE   ALGAE, *MARSH PLANTS, *MARINE ANI   W70-01777
OASSAY, EUTROPHICATION, OLIGOTR/   ALGAE, *NUTRIENT REQUIREMENTS, BI   W72-03183
YTA, *EUTROPHICATION, *PLANKTON,   ALGAE, *NUTRIENT REQUIREMENTS.: /   W69-03512
TROPHICATION, *LAKES, WISCONSIN,   ALGAE, *NUISANCE ALGAE, SPRAYING,   W68-00488
N, *LAKES, PROD/ *ODOR-PRODUCING   ALGAE, *NUISANCE ALGAE, *WISCONSI   W70-05112
PRIMARY PRODUCTIVITY, *PLANKTON,   ALGAE, *PHOTOSYNTHESIS, PRODUCTIV   W69-00632
OPHICATION, AQUATIC MI/ *AQUATIC   ALGAE, *PROTOZOA, *FLORIDA, *EUTR   W72-01993
AUNA, CHEMICAL ANALYSIS, SESSILE   ALGAE, *PRIMARY PRODUCTIVITY, ORE   W72-06052
OLERA/ *BLUE-GREEN ALGAE, *GREEN   ALGAE, *RED ALGAE, *TEMPERATURE T   W71-02496
OC/ *ZINC RADIOISOTOPES, *MARINE   ALGAE, *RESINS, SEA WATER, PHYSIC   W69-08275
HICATION, *CLADOPHORA, *NUISANCE   ALGAE, *SEWAGE, ALGICIDES, AQUATI   W69-04798
OURC/ *EUTROPHICATION, *NUISANCE   ALGAE, *SEWAGE, WATER POLLUTION S   W70-07261
HUMIC ACIDS, */ *OCEANS, *MARINE   ALGAE, *SEAWATER, *FRESH WATER, *   W70-01074
GAE, CHLORINATI/ *ODOR-PRODUCING   ALGAE, *TASTE-PRODUCING ALGAE, AL   W70-02373
-GREEN ALGAE, *GREEN ALGAE, *RED   ALGAE, *TEMPERATURE TOLERANCE, OP   W71-02496
VELOPMENT, WASTE IDENTIFICATION,   ALGAE, *WASTE WATER TREATMENT.: /   W72-03299
*NUTRIENT REMOVAL, SOIL EROSION,   ALGAE, *WATER POLLUTION CONTROL,   W73-01122
*ODOR-PRODUCING ALGAE, *NUISANCE   ALGAE, *WISCONSIN, *LAKES, PRODUC   W70-05112
 *FILTERS, *DESIGN, COAGULATION,   ALGAE, ACTIVATED CARBON, *FLOCCUL   W72-07334
YGEN, SLUDGE, OXIDATION LAGOONS,   ALGAE, ADMINISTRATION, MAINTENANC   W71-08970
GAL BLOOMS, WATER BLOOM, AQUATIC   ALGAE, AGING(BROLOGICAL), *AQUATI   W68-00912
AGE EFFLUENTS, *SEWAGE DISPOSAL,   ALGAE, AGRICULTURAL WATERSHEDS, N   W70-05565
ESEARCH AND DEVELOPMENT, GRANTS,   ALGAE, AIR POLLUTION, WATER POLLU   W71-06821
WA/ *EPIPSAMMIC ALGAE, *EPIPELIC   ALGAE, ALGAL ASSOCIATIONS, SHEAR   W71-12870
RODUCING ALGAE, *TASTE-PRODUCING   ALGAE, ALGAE, CHLORINATION, WATER   W70-02373
TS, COST COMPARISONS, ECONOMICS,   ALGAE, ALGAL CONTROL, ALGICIDES,   W69-08518
ONDITIONS, ANAEROBIC CONDITIONS,   ALGAE, ALGAL CONTROL, ECOLOGICAL   W70-04786
INTERKILLING, FISHKILL, NUISANCE   ALGAE, ALGAE, SEASONAL, LIMNOLOGY   W68-00487
OUNDS, *WATER POLLUTION SOURCES,   ALGAE, ALGAL CONTROL, ALGAL POISO   W69-03370
ARID CLIM/ *SOIL ALGAE, *AQUATIC   ALGAE, ALGAE, ARID CLIMATES, SEMI   W69-03491
ROL, *WATER POLLUTION TREATMENT,   ALGAE, ALGAL CONTROL, ANALYTICAL   W69-03374
OLOGY, *WATER POLLUTION EFFECTS,   ALGAE, ALGAL CONTROL, HARVESTING,   W69-03513
C WASTES, NUTRIENTS, PHOSPHORUS,   ALGAE, ALGICIDES, HARVESTING OF A   W69-07818
, NUTRIENTS, TERTIARY TREATMENT,   ALGAE, AMMONIA.: /DENITRIFICATION   W69-08053
ATION LAGOONS, *ORGANIC LOADING,   ALGAE, AMMONIA, SEDIMENTATION, FL   W71-05013
NT, IMPOUNDED WATERS, FRESHWATER   ALGAE, ANADROMOUS FISHES.: /RONME   W71-02491
ION SOURCES, *CYCLING NUTRIENTS,   ALGAE, ANALYTICAL TECHNIQUES, CLA   W69-04800
OTOSYNTHETIC OXYGEN, *CHLORELLA,   ALGAE, ANALYTICAL TECHNIQUES, AQU   W69-03371
REMENTS, *PHYSIOLOGICAL ECOLOGY,   ALGAE, ANALYTICAL TECHNIQUES, AQU   W69-03515
SIOLOGICAL ECOLOGY, *CYANOPHYTA,   ALGAE, ANALYTICAL TECHNIQUES, AQU   W69-03518
NOPHYTA, *PHYSIOLOGICAL ECOLOGY,   ALGAE, ANALYTICAL TECHNIQUES, AQU   W69-03517
N CYCLE, *NUTRIENT REQUIREMENTS,   ALGAE, ANALYTICAL TECHNIQUES, AQU   W69-03186
LORELLA, *PHOTOSYNTHETIC OXYGEN,   ALGAE, ANALYTICAL TECHNIQUES, AQU   W69-03362
OUNDS, *WATER POLLUTION EFFECTS,   ALGAE, ANIMALS, AQUATIC PLANTS, B   W69-04805
GAE, *SEWAGE, ALGICIDES, AQUATIC   ALGAE, AQUATIC PRODUCTIVITY, BIOA   W69-04798
FISH, *FISHERIES, ALGAE, AQUATIC   ALGAE, AQUATIC MICROBIOLOGY, AQUA   W69-04804
REMENTS, *PHYTOPLANKTON, AQUATIC   ALGAE, AQUATIC MICROBIOLOGY, AQUA   W69-04802
AL POISONING, ALGICIDES, AQUATIC   ALGAE, AQUATIC ENVIRONMENT, AQUAT   W69-05844
IFICATION, PHOSPHORUS COMPOUNDS,   ALGAE, AQUATIC PLANTS.: / *DENITR   W69-05323
AL NUTRIENTS, *FISH, *FISHERIES,   ALGAE, AQUATIC ALGAE, AQUATIC MIC   W69-04804
FISHK/ *EUTROPHICATION, *LAKES,   ALGAE, AQUATIC WEEDS, CYANOPHYTA,   W69-06535
, ANALYTICAL TECHNIQUES, AQUATIC   ALGAE, AQUATIC MICROBIOLOGY, AQUA   W69-03186
, ANALYTICAL TECHNIQUES, AQUATIC   ALGAE, AQUATIC MICROBIOLOGY, AQUA   W69-03185
, ANALYTICAL TECHNIQUES, AQUATIC   ALGAE, AQUATIC BACTERIA, AQUATIC   W69-03362
, ANALYTICAL TECHNIQUES, AQUATIC   ALGAE, AQUATIC BACTERIA, AQUATIC   W69-03188
TS, *NITROGEN COMPOUNDS, AQUATIC   ALGAE, AQUATIC MICROBIOLOGY, CHLA   W69-03364
 *PHYSIOLOGICAL ECOLOGY, AQUATIC   ALGAE, AQUATIC MICROBIOLOGY, CHLO   W69-03358
, ANALYTICAL TECHNIQUES, AQUATIC   ALGAE, AQUATIC MICROBIOLOGY, BALA   W69-03374
*EUTROPHICATION, ALGAE, AQUATIC   ALGAE, AQUATIC BACTERIA, AQUATIC   W69-03369
, ANALYTICAL TECHNIQUES, AQUATIC   ALGAE, AQUATIC MICROBIOLOGY, BALA   W69-03517
LING NUTRIENTS, *EUTROPHICATION,   ALGAE, AQUATIC ALGAE, AQUATIC BAC   W69-03369
, ANALYTICAL TECHNIQUES, AQUATIC   ALGAE, AQUATIC BACTERIA, AQUATIC   W69-03371
, *SALINE WATER, *THERMAL WATER,   ALGAE, AQUATIC ALGAE, AQUATIC MIC   W69-03516
, *THERMAL WATER, ALGAE, AQUATIC   ALGAE, AQUATIC MICROBIOLOGY, BALA   W69-03516
LOGY, *PHYTOPLANKTON, *CULTURES,   ALGAE, AQUATIC PRODUCTIVITY, CYCL   W69-03514
GAL CONTROL, HARVESTING, AQUATIC   ALGAE, AQUATIC ANIMALS, AQUATIC M   W69-03513
IREMENTS, *PHOSPHORUS COMPOUNDS,   ALGAE, AQUATIC ALGAE, AQUATIC MIC   W69-03373
, ANALYTICAL TECHNIQUES, AQUATIC   ALGAE, AQUATIC MICROBIOLOGY, AQUA   W69-03515
PHORUS COMPOUNDS, ALGAE, AQUATIC   ALGAE, AQUATIC MICROBIOLOGY, AQUA   W69-03373
IONS, AQUATIC PLANTS, WATERSHED,   ALGAE, AQUATIC VEGETATION, NITROG   W68-00511
HIES, CYANOPHY/ *EUTROPHICATION,   ALGAE, AQUATIC PLANTS, BIBLIOGRAP   W68-00680
S, FLOW RATES, CONN STREAM FLOW,   ALGAE, AQUATIC ALGAE, CURRENTS(WA   W68-00479
S, EUTROPHICATION, PRODUCTIVITY,   ALGAE, AQUATIC LIFE.: /N COMPOUND   W70-03068
CYCLING NUTRIENTS, ZOOPLANKTON,   ALGAE, AQUATIC PLANTS, PALEOLIMNO   W70-03975
CYCLING NUTRIENTS, ZOOPLANKTON,   ALGAE, AQUATIC PLANTS, PALEOLIMNO   W70-03975
LATION, WASTE TREATMENT, AQUATIC   ALGAE, AQUATIC PLANTS, TESTING, B   W70-04373
OGEN ION CONCENTRATION, NUISANCE   ALGAE, AQUATIC PLANTS, BOTTOM SED   W70-06225
, FISH, PLANKTON, WATER QUALITY,   ALGAE, AQUATIC LIFE, STREAM POLLU   W72-01786
HAINS, SAMPLING, SURVEYS, MARINE   ALGAE, AQUATIC LIFE, MILK, ATLANT   W72-00941
LGAL CONTROL, ALGICIDES, AQUATIC   ALGAE, AQUATIC MICROBIOLOGY, BIOA   W70-02248
*ANALYTICAL TECHNIQUES, *AQUATIC   ALGAE, AQUATIC MICROBIOLOGY, AQUA   W70-02792
, ANALYTICAL TECHNIQUES, *AQUATIC   ALGAE, AQUATIC ENVIRONMENT, AQUAT   W70-02968
TIC/ *CLADOPHORA, ALGAE, AQUATIC   ALGAE, AQUATIC MICROBIOLOGY, AQUA   W70-02784
ROBIOLOGY, AQUATIC/ *CLADOPHORA,   ALGAE, AQUATIC ALGAE, AQUATIC MIC   W70-02784
RNMENTS, *WATER QUALITY CONTROL,   ALGAE, AQUATIC WEEDS, ADMINISTRAT   W72-05774

381

IM/ *SOIL ALGAE, *AQUATIC ALGAE,
E WATERS, *AGRICULTURE, *WASTES,
DISSOLVED ORGANIC MATTER, ARCTIC
ICAL OXYGEN DEMAND, *ANAEROBICS,
S, *THERMAL POLLUTION, *ECOLOGY,
ES, *NUTRIENTS, *BIODEGRADATION,
FECTS, PHYTOPLANKTON, DIFFUSION,
BIBLIOGRAPHIES, LAKES, NUISANCE
YGEN DEMAND, TERTIARY TREATMENT,
STE WATER TREATMENT, *NUTRIENTS,
ATIC LIFE, *AQUATIC POPULATIONS,
IC FAUNA, AQUATIC PLANTS, MARINE
USTRIAL WASTES, OXIDATION PONDS,
ND, AERATION, TEMPERATURE, FLOW,
ENTS, ALGAL CONTROL, *NUTRIENTS,
N, *LAKES, LIMNOLOGY, *PLANKTON,
EFFECTS, *PRIMARY PRODUCTIVITY,
, ANALYTICAL TECHNIQUES, AQUATIC
AL POISONING, ALGICIDES, AQUATIC
TRATIFICATION, DISSOLVED OXYGEN,
HLAMYDOMONAS,/ *RAIN, *VITAMINS,
ANKTON, *PONDS, *EUTROPHICATION,
MARY PRODUCTIVITY, HARVESTING OF
ATIC PLANTS, ANAEROBIC BACTERIA,
*FARM LAGOONS, *BIODEGRADATION,
FFLICIENCIES, *LAGOONS, *SLUDGE,
IENCIES, *NITROGEN, *PHOSPHORUS,
ROGEN COMPOUNDS, NITROGEN CYCLE,
CONTROL, MORTALITY, CHLOROPHYLL,
TOPES, PHOSPHORUS RADIOISOTOPES,
TOLOGY, ENVIRONMENTAL C/ BENTHIC
TH.:   GREEN
URE EXPERIMENTS, RECYCLED/ GREEN
ON, SUPERSATURATION, RESERVOIRS,
ION, EPILIMNION, PHOTOSYNTHESIS,
SEWAGE TREATMENT, HARVESTING OF
ITY, LAKES, NUTRIENTS, BACTERIA,
A, FLORA, LABORATORY, MINERALS,/
HYL-5-VINYLPYRIDINE, *GLUE GREEN
ICITY, *WATER POLLUTION, AQUATIC
POLLUTION, FOOD CHAINS, AQUATIC
TENNESSEE RIVER, TROPHIC LEVEL,
*LAKES, *MANGANESE, *WISCONSIN,
S, CHLOROPHYTA, DINOFLAGELLATES,
N, SLUDGE, DETERGENTS, SPRAYING,
LLA PYRENOIDOSA, VOLATIL/ *GREEN
, CORES, DIATOMS, ODOR-PRODUCING
G ALGAE, *TASTE-PRODUCING ALGAE,
DENSITY, ROOTED AQUATIC PLANTS,
ION, BIODEGRADATION, ADSORPTION,
ES, *SEDIMENTS, *EUTROPHICATION,
AGRICULTURE, ST LAWRENCE RIVER,
ED SOILS, PHOSPHATES, NITROGEN,
S, *SUSPENDED LOAD, *FILTRATION,
LANKTON, LIMNOLOGY, ZOOPLANKTON,
ER POLLUTION EFFECTS, BIOASSAYS,
ELLA SOROKINIANA, COMPOSITION OF
BACTERIA, NITROGEN, PHOSPHORUS,
WATER QUALITY, *ORGANIC MATTER,
S, *OCEANOGRAPHY, *TRANSLATIONS,
SEA WATER, *VITAMINS, VITAMIN B,
NOPHYTA, BRACK/ *EUTROPHICATION,
CONN STREAM FLOW, ALGAE, AQUATIC
SODIUM ARSENITE, FISH, ANIMALS,
NTRATION, FUNGI, YEASTS, PYRITE,
NALYSIS, *NITROGEN, *PHOSPHORUS,
ERATURE, DEPTH, BACTERIA, SLIME,
IDES, ORGANIC MATTER,/ *NUISANCE
ATER INTERFACES, OREGON, SESSILE
AQUATIC ENVIRONMENT, PHOSPHATES,
NOWS, NUISANCE ALGAE, SUNFISHES,
EMPERATURE, WEATHER, EPILIMNION,
MAGNESIUM, NITRATES, PHOSPHATES,
-PRODUCING ALGAE, ODOR-PRODUCING
, SNAILS, PHOSPHORUS, NUTRIENTS,
AY, NEW YORK, LAKES, INHIBITION,
TRESS, THERMAL WATER, *PROTOZOA,
GE BACTERIA, SYSTEMATICS, FUNGI,
TION, SEDIMENTS, ORGANIC MATTER,
TUS, LIGHT INTENSITY, TURBIDITY,
CALCIUM, POTASSIUM, CLAMS, SIZE,
RIA, STABILIZATION, WASTE WATER,
S, ZOOPLANKTON, DINOFLAGELLATES,
NATION, *ODOR, *WASTE TREATMENT,
ASS, CULTURES, PLANT PHYSIOLOGY,
MPOUNDMENTS, TERTIARY TREATMENT,
KES, ODOR, TOX/ *EUTROPHICATION,
S, AQUATIC POPULATIONS, NUISANCE
OISOTOPES, STREAMFLOW, SAMPLING,
NIQUES, WATER POLLUTION EFFECTS,

ALGAE, ARID CLIMATES, SEMIARID CL   W69-03491
ALGAE, ASSAY, DISSOLVED SOLIDS, C   W70-03334
ALGAE, AUTOTROPHIC METABOLISM, HE   W71-10079
ALGAE, BACTERIA, WATER POLLUTION   W71-05742
ALGAE, BACTERIA, COLIFORMS, FISH,   W71-02479
ALGAE, BACTERIA, AQUATIC MICROORG   W71-00221
ALGAE, BACTERIA, EUTROPHICATION,   W70-04194
ALGAE, BACTERIA, WATER POLLUTION   W69-08283
ALGAE, BACTERIA, ACTIVATED SLUDGE   W69-08054
ALGAE, BACTERIA, CULTURES, CHEMIC   W70-05655
ALGAE, BENTHOS, ZOOPLANKTON, FRES   W71-02491
ALGAE, BENTHIC FLORA, ESTUARIES,   W71-02494
ALGAE, BENTHIC FAUNA, WATERSHEDS,   W72-00027
ALGAE, BENTHOS, AEROBIC CONDITION   W71-11381
ALGAE, BENTHIC FAUNA, CHEMICAL AN   W72-06052
ALGAE, BIBLIOGRAPHIES, CHEMICAL P   W69-00446
ALGAE, BIOMASS, CALCIUM COMPOUNDS   W69-02959
ALGAE, BIOASSAY, CYCLING NUTRIENT   W69-03518
ALGAE, BIOASSAY, CHLOROPHYTA, CYA   W69-03370
ALGAE, BIOCHEMICAL OXYGEN DEMAND(W70-04790
ALGAE, BIOASSAY, BIOINDICATORS, C   W69-06273
ALGAE, BIOMASS, CHLOROPHYLL, NEW   W69-06540
ALGAE, BIOMASS.: /NAL, ALGAE, PRI   W70-04290
ALGAE, BIOCHEMICAL OXYGEN DEMAND,   W71-08221
ALGAE, BIOCHEMICAL OXYGEN DEMAND,   W71-00936
ALGAE, BIOCHEMICAL OXYGEN DEMAND,   W70-06899
ALGAE, BIOLOGICAL TREATMENT, WAST   W70-07469
ALGAE, BIOLOGICAL TREATMENT, BOTT   W70-07031
ALGAE, BIOASSAY, VIABILITY.: /AL   W70-02364
ALGAE, BIOCHEMISTRY, ECOLOGY, SAM   W69-10151
ALGAE, BLOOMS, CLOSED SYSTEMS, CY   W69-07832
ALGAE, BLUE-GREEN ALGAE, HEAT DEA   W70-07313
ALGAE, BLUE-GREEN ALGAE, TEMPERAT   W70-07257
ALGAE, BODIES OF WATER, DISSOLVED   W70-00683
ALGAE, BOTTOM SEDIMENTS, THERMOCL   W70-06975
ALGAE, BOTTOM SEDIMENTS, WEED CON   W68-00478
ALGAE, BRACKISH WATER, ALKALINE W   W71-03031
ALGAE, BRACKISH, ECOSYSTEMS, FAUN   W72-02203
ALGAE, BREAM, WHITE BREAM, LEUKOC   W70-01788
ALGAE, BRINE SHRIMP, FISH, FOOD C   W70-01467
ALGAE, BRINE SHRIMP, FISH, MUSSEL   W70-01466
ALGAE, CADDISFLIES, FOOD CHAINS,   W70-02786
ALGAE, CALCIUM, CHELATION, CYANOP   W69-05867
ALGAE, CALIFORNIA, FLORIDA, PESTI   W70-05272
ALGAE, CENTRIFUGATION, FILTRATION   W70-04060
ALGAE, CHLORELLA VULGARIS, CHLORE   W70-04195
ALGAE, CHLOROPHYLL, CHEMTROL, COP   W68-00470
ALGAE, CHLORINATION, WATER QUALIT   W70-02373
ALGAE, CHLOROPHYLL, HYPOLIMNION,   W71-07698
ALGAE, CLAYS, ALDRIN, DIELDRIN, D   W70-09768
ALGAE, CLAYS, SILICATES.: /, *LAK   W69-03075
ALGAE, COLIFORMS, DISSOLVED: /ES,   W70-03964
ALGAE, COLIFORMS.: /IRON, DISSOLV   W69-05894
ALGAE, COLLOIDS, FLOCCULATION, MI   W71-10654
ALGAE, COLLOIDS, TRACE ELEMENTS,   W70-02510
ALGAE, COMPARATIVE PRODUCTIVITY,   W70-05424
ALGAE, CONTINUOUS CULTURES, FATTY   W70-07957
ALGAE, COPPER SULFATE, EFFLUENTS,   W70-02787
ALGAE, COPPER SULFATE, *CHEMICAL   W72-04495
ALGAE, CRUSTACEANS, BENTHOS, BRIN   W69-07440
ALGAE, CULTURES, CHEMICAL ANALYSI   W70-05652
ALGAE, CULTURES, CHLOROPHYTA, CYA   W70-05752
ALGAE, CURRENTS(WATER), LIMNOLOGY   W68-00479
ALGAE, CUTTING MANAGEMENT, POTABL   W70-05263
ALGAE, CYANOPHYTA, DOMINANT ORGAN   W68-00891
ALGAE, CYANOPHYTA, NITRATES, NITR   W69-05868
ALGAE, CYANOPHYTA, DIATOMS, CHLOR   W70-04371
ALGAE, CYANOPHYTA, VITAMINS, PEPT   W70-04369
ALGAE, CYANOPHYTA.: /, SEDIMENT-W   W72-06051
ALGAE, DAPHNIA, GROWTH RATES, PRE   W69-06536
ALGAE, DAPHNIA, LAKES, CHLOROPHYT   W70-02982
ALGAE, DENSITY, SEASONAL, MARINE   W70-02504
ALGAE, DETRITUS, PULP WASTES, LIV   W70-03501
ALGAE, DIATOMS, CYANOPHYTA, CHLOR   W69-05704
ALGAE, DIATOMS, CHLOROPHYTA, DISS   W70-01943
ALGAE, DIATOMS, EUTROPHICATION, C   W70-01579
ALGAE, DIATOMS, ANIMAL BEHAVIOR,   W70-07847
ALGAE, DIATOMS, PROTOZOA, PLANKTO   W72-01819
ALGAE, DIATOMS, CYANOPHYTA, CHLOR   W72-00845
ALGAE, DIATOMS, CYCLING NUTRIENTS   W70-00713
ALGAE, DIATOMS, TRACE ELEMENTS, M   W70-00708
ALGAE, DIFFUSION, SULFUR.: /BACTE   W70-01971
ALGAE, DISSOLVED OXYGEN, HYPOLIMN   W70-04253
ALGAE, DISINFECTION, BIOCHEMICAL   W70-04788
ALGAE, DIURNAL DISTRIBUTION, OXYG   W70-05094
ALGAE, DOMESTIC WASTES.: /KES, *I   W70-04765
ALGAE, DUCKS, COLOR, FISHKILL, LA   W69-08279
ALGAE, DUCKS(WILD), OYSTERS, CHAN   W68-00461
ALGAE, ECOSYSTEMS, ENVIRONMENT, R   W69-07862
ALGAE, ECOSYSTEMS, PHYSIOLOGICAL   W70-04287

*HABITATS, EPIPHYTOLOGY, MARINE	ALGAE, ECOLOGY, CULTURES, WASHING	W69-10161
, *TRITIUM, EPIPHYTOLOGY, MARINE	ALGAE, ECOLOGY, NUTRIENTS, SEAWAT	W69-09755
OSPHOROUS, ALGAE, *HARVESTING OF	ALGAE, EFFLUENT, QUALITY CONTROL,	W71-13313
PHOSPHORUS, AIR, PHYTOPLANKTON,	ALGAE, EFFICIENCIES, DISSOLVED OX	W70-04484
TOPES, PHOSPHORUS RADIOISOTOPES,	ALGAE, EMULSIONS, BACKGROUND RADI	W69-10163
THESIS, HYDROGEN SULFIDE, LAKES,	ALGAE, ENTERIC BACTERIA, METHANE	W70-03312
ALGAE, ALGICIDES, HARVESTING OF	ALGAE, ENVIRONMENT, CYANOPHYTA.: /	W69-07818
HLAMYDOMONAS, FRESHWATER, MARINE	ALGAE, ENZYMES.: / SCENEDESMUS, C	W70-05381
TIAL NUTRIENTS, *TRACE ELEMENTS,	ALGAE, EUTROPHICATION, MOLYBDENUM	W69-06277
ROGEN, HYDRO/ *ACTIVATED SLUDGE,	ALGAE, EUTROPHICATION, LIME, *NIT	W68-00012
NALYSIS, *ODORS, *ODOR PRODUCING	ALGAE, EUTROPHICATION, ALCOHOLS,	W69-00387
ROORGANISMS, NITROGEN, NUTRIENT/	ALGAE, EUTROPHICATION, LAKES, MIC	W68-00481
: *FWPCA, ACTIVATED	ALGAE, FEEDLOTS, OXIDATION DITCH.	W71-06821
YNECHOCOCCUS LIVIDUS, BLUE-GREEN	ALGAE, FILAMENTS.: /RMIS, MATS, S	W70-05270
LLUTION CONTROL, EUTROPHICATION,	ALGAE, FILTRATION, SORPTION, FLOC	W70-05470
UBESCENS, ZURICHSEE, FILAMENTOUS	ALGAE, FILAMENTOUS BLUE-GREEN ALG	W70-04253
EATMENT, *WASTE WATER TREATMENT,	ALGAE, FILTRATION, NEW HAMPSHIRE,	W72-03641
TS, *EUTROPHICATION, *NUTRIENTS,	ALGAE, FISH POPULATION, *WATER PO	W70-04266
N SOURCES, CHLOROPHYTA, NUISANCE	ALGAE, FISH, ZOOPLANKTON, SNAILS,	W70-03310
OCKING, COMMERCIAL FISH, SALMON,	ALGAE, FISH.: /HOSPHORUS, FISH ST	W70-04465
SSAY, TOXICITY, AQUATIC MAMMALS,	ALGAE, FISH, DIATOMS, FUNGI, INSE	W71-10786
TER POLLUTION EFFECTS, BACTERIA,	ALGAE, FISH, SEWERS, RUNOFF, WAST	W71-09880
*EUTROPHICATION, LAKES, STREAMS,	ALGAE, FISH, WISCONSIN, SEWAGE, T	W69-10178
DIMENTS, DREDGING, HARVESTING OF	ALGAE, FISH HARVEST, VITAMINS.: /	W69-09340
EN ION CONCENTRATION, NUTRIENTS,	ALGAE, FISH, STRATIFICATION, FLOC	W70-08834
KE(COLO), PASS LAKE(COLO), MICRO-	ALGAE, FLAGELLATES, CILIATES, COP	W69-10154
UTION, ANIMALS, BACTERIA, FUNGI,	ALGAE, FLOATING PLANTS, ROOTED AQ	W70-05750
BIOCHEMICAL OXYGEN DEMAND, FISH,	ALGAE, FOAMING, SEWAGE TREATMENT.	W70-08740
UNDANCE, BLUE-GREEN ALGAE, GREEN	ALGAE, FRAGILARIA, ASTERIONELLA,	W70-05375
ERIE, STATISTICS, GROWTH RATES,	ALGAE, FRESH WATER FISH.: /, LAKE	W68-00253
ALGAE, PLANKTON, DIATOMS, MARINE	ALGAE, FRESH WATER, FLOW RATES, P	W70-09904
A COLLECTIONS, FISHES, PLANKTON,	ALGAE, FROGS, MOSSES, HYPOLIMNION	W68-00857
TURE, *EPILIMNION, *HYPOLIMNION,	ALGAE, FUNGI, BACTERIA, FISH, LIM	W70-04475
M, GROWTH RATES, MICROORGANISMS,	ALGAE, FUNGI, LABORATORY TESTS, B	W72-01817
C MATTER, IRON OXIDES, BACTERIA,	ALGAE, FUNGI, SULFUR COMPOUNDS, C	W70-02294
COOLING WATER, *BIOMASS, *SLIME,	ALGAE, FUNGI, BACTERIA, BACTERICI	W72-12987
C LOADING, *SULFATES, *SULFIDES,	ALGAE, GASES, ODOR, OXIDATION-RED	W70-07034
IS, LIMNOLOGY,/ *EUTROPHICATION,	ALGAE, GERMANY, VOLUMETRIC ANALYS	W68-00475
AKE (S D), ABUNDANCE, BLUE-GREEN	ALGAE, GREEN ALGAE, FRAGILARIA, A	W70-05375
NFRARED SPECTROSCOPY, BLUE-GREEN	ALGAE, GREEN ALGAE.: /RACTIONS, I	W70-03928
RTILIZERS, PHOSPHATES, NITRATES,	ALGAE, GREAT LAKES.: /TRIENTS, FE	W69-07084
FREUNDLICH ISOTHERM, BLUE-GREEN	ALGAE, GREEN ALGAE.: /DE, UPTAKE,	W70-08111
CHLORELLA, / PESTICIDE RESIDUES,	ALGAE, GREAT LAKES, METABOLISM, *	W70-08652
T REQUIREMENTS, CULTURES, LAKES,	ALGAE, GROWTH RATES, BACTERIA, UL	W70-00719
HOSPHORUS COMPOUNDS, RESERVOIRS,	ALGAE, GROUND WATER, NITRATES, PH	W72-02817
G, BIODEGRADATION, ALGAE, MARINE	ALGAE, GROWTH RATES, ANALYTICAL T	W71-07339
PHOSPHORUS, SULFATES, VITAMINS,	ALGAE, GROWTH RATES, COPPER SULFA	W69-09349
IL POLLUTION, *FILAMENTOUS GREEN	ALGAE, HALIMONE PORTULACOIDES, SU	W70-09976
BIOLOGICAL COMMUNITIES, *AQUATIC	ALGAE, HALOPHYTES.: /QUILIBRIUM,	W69-00360
TRATION / *WATER SUPPLIES, GREEN	ALGAE, HEAD LOSSES, OOCISTUS, FIL	W70-04199
GREEN ALGAE, BLUE-GREEN	ALGAE, HEAT DEATH.:	W70-07313
KES, TEMPERATURE, PHYTOPLANKTON,	ALGAE, HEATED WATER, POPULATION,	W71-09767
GE, COST COMPARISONS, SEDIMENTS,	ALGAE, HYDROGEN ION CONCENTRATION	W71-09548
TER REUSE, COOLING TOWERS, DAMS,	ALGAE, HYDROGEN ION CONCENTRATION	W71-11393
, SOIL ANALYSIS, TASTE-PRODUCING	ALGAE, HYDROGEN ION CONCENTRATION	W72-11906
L, PRODUCTIVITY, PHOTOSYNTHESIS,	ALGAE, HYPOLIMNION, EPILIMNION, P	W69-10182
NKTON, DIATOMS, DINOFLAGELLATES,	ALGAE, HYPOLIMNION, OXYGEN, PRODU	W69-10169
SENTIAL NUTRIENTS, HARVESTING OF	ALGAE, INHIBITION, LIGHT PENETRAT	W69-03369
NVIRONMENTAL EFFECTS, HARVESTING	ALGAE, INHIBITION, LIGHT PENETRAT	W69-04798
S, SEDIMENTS, BIOTA, FISH, UTAH,	ALGAE, INSECTS, INVERTEBRATES, BA	W69-07846
EFFECTS, OLIGOCHAETES, DIPTERA,	ALGAE, INSECTS.: /WATER POLLUTION	W71-12077
STRY, LAKES, CONIFEROUS FORESTS,	ALGAE, INSECTS, PHOSPHORUS RADIOI	W71-09012
ATER, *FILTERS, *ALGAE, NUISANCE	ALGAE, INTAKES, SANDS, FLOW RATES	W70-04199
US, CYCLING NUTRIENTS, PLANKTON,	ALGAE, INVERTEBRATES, TURBIDITY,	W69-07861
N ADSORPTION, HYDROLYSIS, MARINE	ALGAE, INVERTEBRATES, NUCLEAR WAS	W71-09863
ATLANTIC OCEAN, DIATOMS, NERITIC	ALGAE, IONS, WATER POLLUTION EFFE	W70-00707
EMICAL ANALYSIS, ORGANIC MATTER,	ALGAE, IRON, HY: /LVED OXYGEN, CH	W70-05760
ONTIUM RADIOISOTOPES, MANGANESE,	ALGAE, IRON, BIRDS, RADIOACTIVITY	W69-08269
W AUGMENTATION, TENNESSEE RIVER,	ALGAE, IRON, ALKALINITY, HARDNESS	W73-00750
CLADOPHORA, BROMINE, FILAMENTOUS	ALGAE, KILL RATES, PHOTO REACTIVI	W70-02364
ORINATED HYDROCARBON PESTICIDES,	ALGAE, LABORATORY TESTS, WATER PO	W72-11105
TOZOA, BACTERIA, FUNGI, VIRUSES,	ALGAE, LICHENS, ACTIMOMYCETES, CY	W71-11823
NESIUM, POTASSIUM, HARVESTING OF	ALGAE, LIMNOLOGY, NUTRIENTS, PHYS	W70-02245
ODUCTIVITY, CYANOPHYTA, NUISANCE	ALGAE, LIMNOLOGY, CHLOROPHYTA, PL	W70-03959
RTILIZERS, NITROGEN, PHOSPHORUS,	ALGAE, LIVESTOCK, WATER POLLUTION	W71-10376
RUS, ADSORPTION, EUTROPHICATION,	ALGAE, MACROAQUATIC PLANTS, FERTI	W69-06859
, *MARINE MICROORGANISMS, MARINE	ALGAE, MARINE ANIMALS, MARINE BAC	W69-03752
ADIOISOTOPES, ABSORPTION, MARINE	ALGAE, MARINE FISH, INVERTEBRATES	W71-09850
ERIALS, TESTING, BIODEGRADATION,	ALGAE, MARINE ALGAE, GROWTH RATES	W71-07339
RSH PLANTS, *OILY WATER, AQUATIC	ALGAE, MARINE ALGAE, OIL, MARSH,	W70-01231
S, BLOODWORMS, AQUATIC BACTERIA,	ALGAE, MAYFLIES, CADDISFLIES.: /M	W70-00475
, TEMPERATURE, OXYGEN, SALINITY,	ALGAE, MECHANICAL EQUIPMENT.: /TS	W69-06203
RAYFISH, GROWTH STAGES, DIATOMS,	ALGAE, METABOLISM, ISOPODS, AMPHI	W69-09742
SUSPENDED LOAD, WASTES, SEWAGE,	ALGAE, MICROORGANISMS, PHOSPHATES	W70-08627
ACTERIA, *PHYTOPLANKTON, *LAKES,	ALGAE, MICROORGANISMS, BIOMASS, D	W70-08654
IC COMPOUNDS, SPECTROPHOTOMETRY,	ALGAE, MINNESOTA, SODIUM ARSENITE	W70-05269
ACID BACTERIA, SULFUR BACTERIA,	ALGAE, MIXING, HYDROGEN ION CONCE	W72-06854
DATION, WATER UTILIZATION, FISH,	ALGAE, MODEL STUDIES, BIOASSAY, A	W71-11881
LLUTION EFFECTS, MARINE ANIMALS,	ALGAE, MOLLUSKS, FISH, RADIOACTIV	W71-05531
TS, *CALIFORNIA, *NEVADA, DEPTH,	ALGAE, MOSSES, CHARA, FISH, LIMNO	W70-00711

383

I RIVER, SOUTH DAKOTA, NEBRASKA,
NIZATION OF TORONTO, FILAMENTOUS
ION, *RESERVOIRS, *GREAT PLAINS,
EFFECTS, BIOINDICATORS, NUISANCE
OPHYTA, EUTROPHICATION, NUISANCE
ROGEN UTILIZATION, *FRESH WATER,
IOCONTROL, INFECTION, NUTRIENTS,
NTS, *LAKES, PONDS, FERTILIZERS,
IES, ALGAL IDENTIFIC/ BLUE-GREEN
E, CURRENTS(WATER), HYPOLIMNION,
INSECTS, MITES, BACTERIS, FUNGI
FACES WATERS, CARBON, FERTILITY,
SENTIAL NUTRIENTS, HARVESTING OF
*EUTROPHICATION, *LAKE MICHIGAN,
FREEZE-THAW TESTS, HARVESTING OF
S, HARVESTING OF ALGAE, NUISANCE
S, HARVESTING OF ALGAE, NUISANCE
CHELATION, CYANOPHYTA, NUISANCE
NDS, NITROGEN FIXATION, NUISANCE
ES, NITROGEN COMPOUNDS, NUISANCE
NUM, NITROGEN FIXATION, NUISANCE
N EFFECTS, NITROGEN, PHOSPHORUS,
GY, NITROGEN COMPOUNDS, NUISANCE
AESTHETICS, RECREATION, FISHING,
ATER, WATER POLLUTION, BACTERIA,
AL, COLIFORMS, DISSOLVED OXYGEN,
, PHOSPHORUS, NITROGEN, AMMONIA,
ION, *LAGOONS, ACTIVATED SLUDGE,
US ALGAE, FILAMENTOUS BLUE-GREEN
E, FERTILIZATION, FISH, NUISANCE
CONTROL, SLIME, TASTE-PRODUCING
, *WATER SUPPLY, *NEW HAMPSHIRE,
ILY WATER, AQUATIC ALGAE, MARINE
FACTORS, BIOMASS, GROWTH RATES,
, MAGNESIUM, NUTRIENTS, NUISANCE
ES, STRATIGRAPHY, BIOINDICATORS,
IPAL WASTES, NUTRIENTS, NUISANCE
EQUIREMENTS, ECOLOGY, CHLORELLA,
S, *HYDROGEN SULFIDE, *SULFIDES,
ER TREATMENT, / *HOGS, *LAGOONS,
GINEERING, *PATHOGENIC BACTERIA,
TRIENTS, PESTICIDES, RECREATION,
RIA, DEGRADATION(DECOMPOSITION),
FECTS, EUTROPHICATION, FISHKILL,
NG, SEA WATER, CHLOROPHYLL, OIL,
                    MASS CULTURE OF
IQUES, BIOMASS, WATER POLLUTION,
EFFLUENTS, ECOSYSTEMS, PLANKTON,
URES, DIATOMS, ECOLOGY, NUISANCE
ALGISTATIC TESTS, SLIMEPRODUCING
LEGISLATION, NITROGEN, NUISANCE
S, POLLUTION ABATEMENT, NUISANCE
UTRIENTS, *PHOSPHORUS COMPOUNDS,
, FRESH WATER, SORPTION, SODIUM,
, PESTICIDES, OXIDATION LAGOONS,
LAGOONS, *WASTE WATER TREATMENT,
CONCENTRATION, DISSOLVED OXYGEN,
LLULOSE, DI/ *ECOLOGY, BACTERIA,
ENCE, PHOTOSYNTHESIS, BLUE-GREEN
UES, KINETICS, AQUATIC BACTERIA,
ES, TURBIDITY, SUSPENDED SOLIDS,
*FISH MANAGEMENT, FERTILIZATION,
SALINITY, BENTHOS, AQUATIC LIFE,
CLING NUTRIENTS, EUTROPHICATION,
G, WASTE TREATMENT, ENGINEERING,
*LAKE MICHIGAN, ALGAE, NUISANCE
RIENTS, FERTILIZATION, BACTERIA,
CLE SIZE, *CLAYS, *SILTS, SANDS,
, ECOLOGY, PRIMARY PRODUCTIVITY,
, PHYTOPLANKTON, IRON COMPOUNDS,
ALITY, CHLOROPHYLL, ZOOPLANKTON,
AL OXYGEN DEMAND, ODOR-PRODUCING
IOISOTOPES, REGRESSION ANALYSIS,
, HERBICIDES, FISH, FRESH WATER,
OAGULATION, WATERSHEDS, SEWERAGE
LAKES, ORGANIC MATTER, SEASONAL,
), HYDROLYSIS, CHELATION, FUNGI,
, *AQUATIC DRIFT, FISH FOOD ORG/
, *AQUATIC POPULATIONS, *RIVERS,
HICATION, *CONFERENCES, NUISANCE
OLOGICAL COMMUNITIES, INTERTIDAL
DEFINITIONS, *WATER BLOOM, TOXIC
ULFATE, *ALGAL CONTROL, *TOXINS,
POLLUTION SOURCES, TASTE, ODOR,
LAKES, *PILOT PLANTS, *NUISANCE
, DIATOMS, ANAEROBIC CONDITIONS,
S, ESTUARINE ENVIRONMENT, MARINE
COMPOUNDS, PHOSPHORUS, NUISANCE
BIOTA, SPECTROMETERS, BACTERIA,
OPHICATION, OLIGOTROPHY, SEWAGE,

ALGAE, NANNOPLANKTON, DIATOMS.: /   W70-05375
ALGAE, NAVICULA, APHANIZOMENON, O   W69-10155
ALGAE, NEBRASKA, TURBIDITY, STRAT   W72-04761
ALGAE, NEW YORK, SEWAGE EFFLUENTS   W70-03974
ALGAE, NITROGEN, PHOSPHATES, HYDR   W70-04381
ALGAE, NITRATE AMMONIUM, NITRITE,   W69-04521
ALGAE, NITROGEN, PHOSPHORUS, BICA   W68-00468
ALGAE, NITROGEN, PHOSPHORUS, WATE   W69-08668
ALGAE, NITZSCHIA, BIOLOGICAL STUD   W70-10173
ALGAE, NITROGEN, PHOSPHORUS, OXYG   W70-09889
ALGAE, NITRITES, NITRATES, DETERG   W71-00940
ALGAE, NITROGEN, PHOSPHORUS, WATE   W70-02973
ALGAE, NUISANCE ALGAE, NUTRIENT R   W69-03371
ALGAE, NUISANCE ALGAE, PLANKTON,   W68-01244
ALGAE, NUISANCE ALGAE, NUTRIENT R   W69-03362
ALGAE, NUTRIENT REQUIREMENTS, PHY   W69-03362
ALGAE, NUTRIENT REQUIREMENTS, PHY   W69-03371
ALGAE, NUTRIENT REQUIREMENTS, NUT   W69-05867
ALGAE, NUTRIENTS, PHOSPHORUS COMP   W69-04802
ALGAE, NUTRIENT REQUIREMENTS, NUT   W69-05868
ALGAE, NUTRIENTS, VITAMINS, WATER   W69-06277
ALGAE, NUTRIENTS, WATER PROPERTIE   W70-04382
ALGAE, NUTRIENT REQUIREMENTS, NUT   W70-04283
ALGAE, NUTRIENTS, TURBIDITY, TERT   W70-05645
ALGAE, NUTRIENTS, INDICATORS, ODO   W70-05264
ALGAE, NUTRIENTS, SOLUTES, CHLORI   W70-09557
ALGAE, NUTRIENTS, SEWAGE, WASTE W   W71-05879
ALGAE, NUTRIENTS, VIRUSES, BACTER   W72-06017
ALGAE, OCULAR MICROMETER, SUCROSE   W70-04253
ALGAE, ODOR, TASTE, DEPTH, CIRCUL   W70-05404
ALGAE, ODOR-PRODUCING ALGAE, DIAT   W69-05704
ALGAE, ODOR, TASTE, LIGHT TEMPERA   W69-08282
ALGAE, OIL, MARSH, POLLUTANTS.: /   W70-01231
ALGAE, OLIGOTROPHY, LIMITING FACT   W70-02779
ALGAE, OLIGOTROPHY, ODOR, PHYTOPL   W69-07833
ALGAE, OLIGOTROPHY, DIVERSION, CO   W70-05663
ALGAE, OLIGOTROPHY, PHYTOPLANKTON   W68-00680
ALGAE, ORGANIC MATTER, CULTURES,   W70-02804
ALGAE, ORGANIC LOADING, BIOCHEMIC   W71-11377
ALGAE, ORGANIC LOADING, WASTE WAT   W70-07508
ALGAE, ORGANIC MATTER, RADIOACTIV   W72-13601
ALGAE, OXYGEN DEMAND, WATER POLLU   W71-09577
ALGAE, OXYGENATION, SELF-PURIFICA   W72-01801
ALGAE, OXYGEN SAG, WATER QUALITY,   W69-03948
ALGAE, PACIFIC OCEAN.: /OTE SENSI   W70-05377
ALGAE, PACKED CELL VOLUME.:   W70-06053
ALGAE, PERIPHYTON, PHOSPHORUS, HY   W71-04206
ALGAE, PERIPHYTON, INVERTEBRATES,   W69-07853
ALGAE, PHAEOPHYTA, PHOTOSYNTHESIS   W69-07832
ALGAE, PHOTOMICROGRAPHS, MICROCYS   W69-05704
ALGAE, PHOSPHORUS, WASHINGTON, WI   W69-06858
ALGAE, PHOSPHATES, NITRATES, RIVE   W69-01453
ALGAE, PHOSPHATES, NITRATES, SEWA   W70-04074
ALGAE, PHOSPHORUS, TRACE ELEMENTS   W70-04385
ALGAE, PHOTOSYNTHESIS, DISSOLVED   W71-12183
ALGAE, PHOTOSYNTHESIS, DEPTH, THE   W72-10573
ALGAE, PHYTOPLANKTON, ALKALINE WA   W71-07731
ALGAE, PHYTOPLANKTON, *LAKES, *CE   W69-00976
ALGAE, PHYTOPLANKTON, OHIO RIVER,   W69-05697
ALGAE, PHYSIOLOGICAL ECOLOGY, CAR   W70-05109
ALGAE, PILOT PLANTS, HEAD LOSS, P   W72-08686
ALGAE, PLANKTON, NITROGEN, PHOSPH   W70-05399
ALGAE, PLANTS, WATER PROPERTIES,   W69-05023
ALGAE, PLANT GROWTH SUBSTANCES, P   W69-07826
ALGAE, PLANTS, CONTROL, POLLUTION   W69-03683
ALGAE, PLANKTON, BENTHIC FAUNA, S   W68-01244
ALGAE, PLANKTON, TUBIFICIDS, CALC   W70-05761
ALGAE, PLANKTON, MICROORGANISMS,   W70-08628
ALGAE, PLANKTON, SUBMERGED PLANTS   W71-00665
ALGAE, PLANTS, WATER POLLUTION EF   W69-10170
ALGAE, POLYCHAETA, CRUSTACEA, MOL   W69-08526
ALGAE, POLLUTION ABATEMENT, SEWAG   W69-06909
ALGAE, PONDS, PREDATION.: /US RAD   W70-03957
ALGAE, PONDS, WATER HYACINTH, CHE   W70-00269
ALGAE, POTABLE WATER.: /NATION, C   W71-05943
ALGAE, PRIMARY PRODUCTIVITY, HARV   W70-04290
ALGAE, PROTOZOA, LICHENS, PLANTS,   W69-07428
ALGAE, PROTOZOA, *AQUATIC INSECTS   W69-02782
ALGAE, PROTOZOA, AQUATIC INSECTS,   W73-00748
ALGAE, PUBLICATIONS, WATER POLLUT   W70-03973
ALGAE, PUGET SOUND, FLOURESCEIN D   W68-00010
ALGAE, RECREATIONAL USE.:   *   W69-08279
ALGAE, RESISTANCE, FISH, AMPHIBIA   W68-00859
ALGAE, RESERVOIRS, WATERSHEDS(BAS   W72-01813
ALGAE, ROOTED AQUATIC PLANTS, *RE   W68-00479
ALGAE, ROTIFERS, SESTON, RESPIRAT   W69-10154
ALGAE, SALINITY, TEMPERATURE, MON   W68-00010
ALGAE, SALTS, TIDES, SOUTHEAST U.   W69-03695
ALGAE, SANDS, HYDROGEN ION CONCEN   W69-08267
ALGAE, SCUM, WEEDS, NUTRIENTS, SE   W70-04193

384

385

CENEDESMUS, CHLORELLA, *EUGLENA,
, *MUD, *COPPER, COPPER SULFATE,
LOGICAL ECOLOGY, *PHYTOTOXICITY,
*WATER POLLUTION EFFECTS, ALGAE,
ATER POLLUTION TREATMENT, ALGAE,
*WATER POLLUTION SOURCES, ALGAE,
T COMPARISONS, ECONOMICS, ALGAE,
LUTION, WATER POLLUTION EFFECTS,
HKILLS, LIMNOLOGY, ALGAL TOXINS,
, LAKES, OXYGEN, EUTROPHICATION,
N, *WATER CHEMISTRY, POLLUTANTS,
RENCES, WATER POLLUTION EFFECTS,
TREATMENT, BIOLOGICAL TREATMENT,
UDIES, CHLORINE, COPPER SULFATE,
*HALOGENS, *ALGICIDES, CHLORINE,
LUTION, WATER POLLUTION SOURCES,
 ALGAE, AQUATIC MICROBI/ *ALGAE,
*CHLORELLA, CHLORINE, ALGICIDES,
TES, TOXICITY, FISH, BLOODWORMS,
RIES, WATER QUALITY, DETERGENTS,
YGEN, BIOCHEMICAL OXYGEN DEMAND,
, PHYTOPLANKTON, WATER ANALYSIS,
HICATION, LAKES, LAKE SEDIMENTS,
TON, HOWARD A. HANSON RES/ WOOD,
                             *PURE
S-AQUAE, MICROCYSTIS AERUGINOSA,
(MINN), LAKE MANAGEMENT PROGRAM,
, PEDIASTRUM, SELF-PURIFICATION,
MY, ALGAL PHYLOGENY, CLADOPHORA,
ALOUSE RIV/ TREE BARK, CHARCOAL,
 LAKE WAUBESA(WIS), MACROPHYTES,
UGINOSA, *NUTRIENT AVAILABILITY,
L GROWTH, NUTRIENT AVAILABILITY,
LENA GRACILIS Z,/ *VITAMIN B-12,
S CULTURE, *PILOT-PLANT STUDIES,
GAL PHYSIOLOGY/ *CYANOCOBALAMIN,
ILLARIOPHYCEAE, CHLORO/ *EUROPE,
TH EQUATION, INCIDENT LIGHT UTI/
UTRIENT SOURCES, WATER SAMPLING,
AE IDENTIFICATION, ALGAE COUNTS,
ESSMENT, SANIT/ *BIOSTIMULATION,
 GROWTH LIMITATION, SELENASTRUM,
IR EFFECTS, LAKE MEAD, DILUTION,
, NITZSCHIA, BIOLOGICAL STUDIES,
*IDENTIFICATION, *SIGNIFICANCE,
OMINE, IODINE, PHOTO REACTIVITY,
AKE(ORE), LAKE WASHINGTON(WASH),
IENT AVAILABILITY, ALGAL GROWTH,
LOT-PLANT STUDIES, ALGAL GROWTH,
 *CYANOCOBALAMIN, ALGAL GROWTH,
OCYSTIS AERUGINOSA, ANTAGONISMS,

AL BIOCHEMISTRY, ALGAL TAXONOMY,
MENTAL STUDIES, APPLIED STUDIES,
A, ANTAGONISMS, ALGAL NUTRITION,
, ALGAL GROWTH, ALGAL NUTRITION,
, ALGAL GROWTH, ALGAL NUTRITION,
, ALGAL GROWTH, ALGAL NUTRITION,
SHINGTON(WASH), ALGAL NUTRITION,
TRIENTS, *ALGAE, *ALGAL CONTROL,
N SOURCES, ALGAE, ALGAL CONTROL,
 *PHYTOPLANKTON, *PHYTOTOXICITY,
, *PHYTOTOXICITY, ALGAL CONTROL,
SIPPI RIVER BASIN, URBANIZATION,
RANSPORT, BACTERIAL POPULATIONS,
               *CYTOLOGICAL EFFECTS,
ONS.:
, *WEED EXTRACTING STEEL CABLES,
TROLYTES, CATIONIC ELECTROLYTES,
TISZA(HUNGARY), ZIMONA(HUNGARY),
 PHYSIOLOGY, ALGAL BIOCHEMISTRY,
ION.:
LUTA/ BIOCHEMICAL OXYGEN DEMAND,
MS, PHYTOPLANKTON, BACTERICIDES,
ED OXYGEN, FISHKILLS, LIMNOLOGY,
TS, CHLOROPHYLL A, PHOTOBIOLOGY,
GYMNODINIUM BREVIS, MYCROSYSTIS,
, SLIM/ *POTASSIUM PERMANGANATE,
PHORA, *NUISANCE ALGAE, *SEWAGE,
*ALGAL CONTROL, ALGAL POISONING,
S, NUTRIENTS, PHOSPHORUS, ALGAE,
, COPPER SULFATE, ALGAL CONTROL,
N, WATER QUALITY, ALGAL CONTROL,
 ALGAL CONTROL, ALGAL POISONING,
ATION, *WATER POLLUTION CONTROL,
, LAKES, WATER QUALITY, VIRUSES,
 PHOSPHORUS, MIGRATION, DIURNAL,
NESOTA, WATER POLLUTION EFFECTS,
, *ALGAE, *AQUATIC WEEDS, LAKES,
WTH RATES, *CHLORELLA, CHLORINE,

ALGAL CONTROL, BACTERICIDES, GROW   W69-00994
ALGAL CONTROL, ALGICIDES, TRACE E   W69-03366
ALGAL CONTROL, ALGAL POISONING, A   W69-03188
ALGAL CONTROL, HARVESTING, AQUATI   W69-03513
ALGAL CONTROL, ANALYTICAL TECHNIQ   W69-03374
ALGAL CONTROL, ALGAL POISONING, A   W69-03370
ALGAL CONTROL, ALGICIDES, SEDIMEN   W69-08518
ALGAL CONTRO: /EATMENT, WATER POL   W69-10171
ALGAL CONTROL, TRACE ELEMENTS, WA   W70-00264
ALGAL CONTROL, BIOCONTROL, TURBUL   W71-10098
ALGAL CONTROL, NITRATES, PHOSPHAT   W70-02646
ALGAL CONTROL, WATER POLLUTION, T   W70-02775
ALGAL CONTROL, AMMONIA, ION EXCHA   W70-01981
ALGAL CONTROL, MORTALITY, CHLOROP   W70-02364
ALGAL CONTROL, CHLORELLA, DISINFE   W70-02363
ALGAL CONTROL, BIOINDICATORS.: /L   W70-02304
ALGAL CONTROL, ALGICIDES, AQUATIC   W70-02248
ALGAL CONTROL, DISINFECTION, MODE   W70-02370
ALGAL CONTROL.: /DEPTH, OLIGOCHAE   W70-06217
ALGAL CONTROL, INDUSTRIAL WASTES,   W70-10437
ALGAL CONTROL, *COLOR, *ALKALINIT   W72-11906
ALGAL CONTROL, ALGAE, WATER QUALI   W72-06051
ALGAL CONTROL, *NUTRIENTS, ALGAE,   W72-06052
ALGAL COUNT, GREEN RIVER, WASHING   W72-11906
ALGAL CULTURES.:   W71-06002
ALGAL CULTURES, CHEMOSTATS, CARBO   W69-06864
ALGAL DENSITY, APHANIZOMENON, MEL   W69-10167
ALGAL ENUMERATION, SCENEDESMUS, G   W70-04381
ALGAL GENETICS, RUSSIA, BAVARIA,   W70-03973
ALGAL GROWTH, STREAM CONCOURSE, P   W70-03501
ALGAL GROWTH.: /NAJAS, ANACHARIS,   W70-03310
ALGAL GROWTH, ALGAL NUTRITION, AL   W69-05868
ALGAL GROWTH POTENTIAL, SELENASTR   W69-06864
ALGAL GROWTH, CYANOCOBALAMIN, EUG   W69-06273
ALGAL GROWTH, ALGAL NUTRITION, AL   W69-06865
ALGAL GROWTH, ALGAL NUTRITION, AL   W69-06277
ALGAL GROWTH, ANKISTRODESMUS, BAC   W69-07833
ALGAL GROWTH KINETICS, MONOD GROW   W69-03730
ALGAL GROWTH, OXYGEN DEPLETION, N   W70-00264
ALGAL GROWTH, SEDGWICK-RAFTER MET   W69-10157
ALGAL GROWTH, WATER POLLUTION ASS   W70-02775
ALGAL GROWTH, ANABAENA, BIOSTIMUL   W70-02779
ALGAL GROWTHS, OXYGEN DEPLETION.:   W73-00750
ALGAL IDENTIFICATION.: /EEN ALGAE   W70-10173
ALGAL KEY, CLUMP COUNT, DESMIDS,   W70-05389
ALGAL METABOLISM.: /LL RATES, *BR   W70-02363
ALGAL NUTRITION, ALGAL PHYSIOLOGY   W70-04506
ALGAL NUTRITION, ALGAL PHYSIOLOGY   W69-05868
ALGAL NUTRITION, ALGAL PHYSIOLOGY   W69-06865
ALGAL NUTRITION, ALGAL PHYSIOLOGY   W69-06277
ALGAL NUTRITION, ALGAL PHYSIOLOGY   W69-05867
ALGAL NUTRIENTS.:   W69-05844
ALGAL PHYLOGENY, CLADOPHORA, ALGA   W70-03973
ALGAL PHYSIOLOGY, ALGAL BIOCHEMIS   W70-03973
ALGAL PHYSIOLOGY, LAKE MENDOTA(WI   W69-05867
ALGAL PHYSIOLOGY, KJELDAHL PROCED   W69-05868
ALGAL PHYSIOLOGY, AMERICAN RESEAR   W69-06865
ALGAL PHYSIOLOGY, CALOTHRIX PARIE   W69-06277
ALGAL PHYSIOLOGY, HETEROTROPHY, M   W70-04506
ALGAL POISONING, ALGICIDES, AQUAT   W69-05844
ALGAL POISONING, ALGICIDES, AQUAT   W69-03370
ALGAL POISONING, SCENEDESMUS, CHL   W69-00994
ALGAL POISONING, ANALYTICAL TECHN   W69-03188
ALGAL POISONING, GASOLINE, FLOODS   W73-00758
ALGAL POPULATIONS, HETEROTROPHIC   W70-04194
ALGAL POPULATION.:   W70-02770
ALGAL SENESCENCE, GROWTH INHIBITI   W72-02417
ALGAL SUCCESSION, ANABAENA, CLADO   W70-04494
ALGAL SURFACE CHARGE, ELECTROSTAT   W70-05267
ALGAL TAXONOMY, SPIRULINA, CHODAT   W70-03969
ALGAL TAXONOMY, ALGAL PHYLOGENY,   W70-03973
ALGAL TAXONOMY, ALGAE IDENTIFICAT   W69-03514
ALGAL TOXINS, ORGANIC WASTES, POL   W69-01273
ALGAL TOXINS, LARVAE, PHENOLS, SH   W70-01068
ALGAL TOXINS, ALGAL CONTROL, TRAC   W70-00264
ALGAL: / BATCH CULTURES, CHEMOSTA   W70-02777
ALGASTATIC SUBSTANCES, ALGADYNAMI   W70-02510
ALGICIDAL TESTS, ALGISTATIC TESTS   W69-05704
ALGICIDES, AQUATIC ALGAE, AQUATIC   W69-04798
ALGICIDES, AQUATIC ALGAE, AQUATIC   W69-05844
ALGICIDES, HARVESTING OF ALGAE, E   W69-07818
ALGICIDES, TRACE ELEMENTS, CORES,   W69-03366
ALGICIDES, DOMINANT ORGANISMS, GR   W69-01977
ALGICIDES, AQUATIC ALGAE, BIOASSA   W69-03370
ALGICIDES, BIOCONTROL, *WATERSHED   W68-00478
ALGICIDES, FISH, STRE: /TREATMENT   W70-03973
ALGICIDES, CLOSTRIDIUM, FISH, AMP   W70-05372
ALGICIDES, WATER QUALITY.: /, MIN   W70-04506
ALGICIDES, AQUATIC WEED CONTROL,   W70-04494
ALGICIDES, ALGAL CONTROL, DISINFE   W70-02370

386

MICROBI/ *ALGAE, ALGAL CONTROL,　ALGICIDES, AQUATIC ALGAE, AQUATIC　W70-02248
HLORINE, COPPER SULF/ *HALOGENS,　ALGICIDES, CYTOLOGICAL STUDIES, C　W70-02364
ES, SEWAGE EFFLUENTS, WISCONSIN,　ALGICIDES, HYDROLYSIS, COPPER SUL　W70-02982
BACTERIA, *AQUATIC MICROBIOLOGY,　ALGICIDES, BIOASSAY, DISINFECTION　W69-10171
ECONOMICS, ALGAE, ALGAL CONTROL,　ALGICIDES, SEDIMENTS, DISSOLVED O　W69-08518
MASS, DIATOMS, CYANOPHYTA, IRON,　ALGICIDES, ZOOPLANKTON, FISH.: /O　W70-08654
D EXTRACTS, ASCOPHYLLUM NODOSUM,　ALGINIC ACID, FUCUS VESICULOSUS,　W69-07826
M PERMANGANATE, ALGICIDAL TESTS,　ALGISTATIC TESTS, SLIMEPRODUCING　W69-05704
DOSA, FUCALES, ALKANES, ALKENES,　ALIPHATIC HYDROCARBONS, PETROLEUM　W69-08284
A, STAPHYLOCOCCUS, ACHROMOBACTER(　ALKALIGENES), RHODOTORULA, GROWTH　W70-03952
ERATURE, *HOT SPRINGS, HABITATS,　ALKALINE WATER, ACIDIC WATER, TRA　W69-10160
BACTERIA, ALGAE, BRACKISH WATER,　ALKALINE WATER, SODIUM.: /IENTS,　W71-03031
ED OXYGEN, ALGAE, PHYTOPLANKTON,　ALKALINE WATER, ALKALINITY, FISH　W71-07731
, PHYTOPLANKTON, ALKALINE WATER,　ALKALINITY, FISH FARMING, FISH FO　W71-07731
RUS, HYDROGEN ION CONCENTRATION,　ALKALINITY, HARDNESS(WATER), CHLO　W71-04206
ION CONCENTRATION, TEMPERATURE,　ALKALINITY, BICARBONATES, BACTERI　W71-05267
ION CONCENTRATION, TEMPERATURE,　ALKALINITY, NITRIFICATION, *CORRO　W71-11393
ON, DISTRIBUTION, DEPTH, OXYGEN,　ALKALINITY, HYDROGEN ION CONCENTR　W71-00665
DEGRADATION(DECOMPOSITION), DDT,　ALKALINITY, STABILIZATION, WASTE　W70-08097
EFFLUENTS, NITROGEN, PHOSPHATES,　ALKALINITY, FISH, STREAMS, WATER　W69-10159
URE, HYDROGEN ION CONCENTRATION,　ALKALINITY,: /ARBONATES, TEMPERAT　W70-02804
POPULATION, INDUSTRIES, METHANE,　ALKALINITY, HYDROGEN ION CONCENTR　W70-02803
LLA, HYDROGEN ION CONCENTRATION,　ALKALINITY, DIATOMS, TEMPERATURE,　W70-02504
N CONCENTRATION, PHOTOSYNTHESIS,　ALKALINITY, NITRATES, COND: /N IO　W70-03309
NDS, HYDROGEN ION CONCENTRATION,　ALKALINITY, CALIFORNIA, LIGHT INT　W70-03334
AMPLING, EFFLUENTS, TRIBUTARIES,　ALKALINITY, HYDROGEN ION CONCENTR　W70-03983
ONDUCTIVITY, CALCIUM, MAGNESIUM,　ALKALINITY, SULFATES, CHLORIDES,　W69-05894
YGEN, OXIDATION, CARBON DIOXIDE,　ALKALINITY, ALGAE.: /NITROGEN, OX　W70-05550
OTABLE WATER, HERBICIDES, CHARA,　ALKALINITY, BULLHEADS, EELS, KILL　W70-05263
R PROPERTIES, WATER TEMPERATURE,　ALKALINITY, *PHOTOSYNTHESIS, LIGH　W68-00248
N, TENNESSEE RIVER, ALGAE, IRON,　ALKALINITY, HARDNESS, RADIOACTIVI　W73-00750
*PERIDINIUM SP., *GLENODININE,　ALKALOID TOXINS, IBOGAALKALOID, O　W71-07731
CHLORELLA PYRENOIDOSA, FUCALES,　ALKANES, ALKENES, ALIPHATIC HYDRO　W69-08284
A PYRENOIDOSA, FUCALES, ALKANES,　ALKENES, ALIPHATIC HYDROCARBONS,　W69-08284
EN ION CONCENTRATION, FLUORIDES,　ALKYL BENZENE SULFONATES, WIND VE　W70-00263
HU'S MEDIA, QUINONE DERIVATIVES,　ALKYL SUBSTITUTED IMIDAZOLINES, B　W70-02982
S, *ALKYL POLYETHOXYLAT/ *LINEAR　ALKYL SULFONATES, *AXENIC CULTURE　W70-03928
ALKYLBENZENE SULFONATES, LINEAR　ALKYLATE SULFONATES, WATER POLLUT　W69-09454
, WATER POLLUTION SOURC/ *LINEAR　ALKYLATE SULFONATES, *SURFACTANTS　W70-09438
CTERIA, DETERGENTS, SURFACTANTS,　ALKYLBENZENE SULFONATES, LINEAR A　W69-09454
ROTIFERS, WEEDS, WATER HYACINTH,　ALLIGATORWEED, CHEMCONTROL, MECHA　W70-10175
AERO-HYDRAULIC GUN,　ALLISTON(ONTARIO).:　W71-09546
YTARSUS, LIMNODRILUS, EUCYCLOPS,　ALONA, CHAOBORUS, / TANYPUS, TAN　W70-06975
OLOGY, CLADOCERAN, MONTANE ZONE,　ALPINE ZONE, BOULDER(COLO), TEA L　W69-10154
IONELLA FORMOSA, RECOLONIZATION,　ALSTER LAKE(GERMANY), GERMANY.: /　W70-00268
TRUN/ DUPLIN RIVER(GA), SPARTINA　ALTEMIFLORA, ORGANISMS, TURSIOPS　W69-08274
NIA, POTAMOGETON, TRAPA, PISTIA,　ALTERNANTHERA, HETERANTHERA, NAJA　W70-10175
E BREAM, LEUKOCYTIC COMPOSITION,　ALTERNATING CURRENT.: /REAM, WHIT　W70-01788
, ROTIFERS, SESTON, RESPIRATION,　ALTITUDE, SAMPLING, ZOOPLANKTON,　W69-10154
US, ANKISTRODESMUS, CHLOROMONAD,　ALUM.: /M, WESTELLA, STEPHANODISC　W70-08107
:　ALUM, *OSHKOSH(WIS), BACKWASHING.　W72-08859
-COATED SAND, CR/ *FILTER MEDIA,　ALUMINA-COATED SAND, FERRIC OXIDE　W70-08628
IRON SULFATE, CLADOPHORA BLANKE/　ALUMINUM SULFATE, IRON CHLORIDE,　W69-10170
IRON S/ RING TUBE, CANALIZATION,　ALUMINUM SULFATE, FERRICHLORIDE,　W70-05668
CHEILUS OREGONENSIS, ACROCHEILUS　ALUTACEUS, ULOTHRIX, CLADOPHORA,　W69-07853
IAL MEDIUM, ARTHROBACTER, FLORA,　AMBIENT TEMPERATURE, THERMAL SENS　W70-00713
, SCHROEDERIA SETIGERA, MELOSIRA　AMBIGUA, CERATIUM HIRUDINELLA, AN　W70-05387
TEMPERATURE COEFFICIENTS, NORTH　AMERICA, CENTRAL AMERICA, RATE EQ　W70-03957
GE, AGRICULTURAL DRAINAGE, NORTH　AMERICA, FOREST DRAINAGE.: /RAINA　W70-03975
GE, AGRICULTURAL DRAINAGE, NORTH　AMERICA, FOREST DRAINAGE.: /RAINA　W70-03975
FICIENTS, NORTH AMERICA, CENTRAL　AMERICA, RATE EQUATIONS, ANALYSIS　W70-03957
GAL NUTRITION, ALGAL PHYSIOLOGY,　AMERICAN RESEARCH AND DEVELOPMENT　W69-06865
IF), WATER POLLUTION ASSESSMENT,　AMERICAN RIVER(CALIF), CLEAR LAKE　W70-02777
A RUBESCENS, DINOBRYON, UROGLENA　AMERICANA, CYCLOTELLA, CERATIUM H　W70-05405
UNDS, ORGANO-METALLIC COMPOUNDS,　AMINES, ENGLAND.: /PHENOLIC COMPO　W72-12987
TION, DETERGENTS, RADIOACTIVITY,　AMINO ACIDS.: /ISTRIBUTION, CHELA　W69-08275
SONAL, PSEUDOMONAS, TEMPERATURE,　AMINO ACIDS, INTERTIDAL AREAS, CA　W70-03952
MUS, PROTEINS, WALLS, STRUCTURE,　AMINO ACIDS, MECHANICAL EQUIPMENT　W70-04184
TERIA, TRACE ELEMENTS, VITAMINS,　AMINO ACIDS, CALCIUM, AMMONIA, NI　W70-02255
S, RESERVOIRS, NITROGEN, GROWTH,　AMINO ACIDS, CHLAMYDOMONAS, CHLOR　W69-10180
A PYRENOIDOSA, CARBOXYLIC ACIDS,　AMINO CARBOXYLIC ACIDS.: /HLORELL　W70-02248
S, *GROWTH RATES, *ALGAE, 2-4-D,　AMINOTRIAZOLE, 2-4-5-T, SCENEDESM　W70-03519
*LITERATURE, ELEOCHARIS, 2,4-D,　AMMATE, CHLORAMINE, DE-K-PRUF-21,　W70-05263
TION, ELECTROCHEMICAL TREATMENT,　AMMONIA STRIPPING, ALGAE HARVESTI　W70-01981
ENTS, TERTIARY TREATMENT, ALGAE,　AMMONIA.: /DENITRIFICATION, NUTRI　W69-08053
BACTERIA, TEMPERATURE, NITRATES,　AMMONIA, *WATER REUSE, WASTE WATE　W72-06017
UTRIENTS, ENVIRONMENTAL EFFECTS,　AMMONIA, AMMONIUM COMPOUNDS, NITR　W69-03364
CHEMISTRY, PHOSPHORUS, NITROGEN,　AMMONIA, ALGAE, NUTRIENTS, SEWAGE　W71-05879
HYDROGEN SULFIDE, LEAD, COPPER,　AMMONIA, CONDUCTIVITY, CHR: /NDS,　W70-04510
EUGLENA, CHLOROPHYTA, NITRATES,　AMMONIA, CARBON, RECREATION, WATE　W72-04761
YCETES, *ODOR, ACTIVATED CARBON,　AMMONIA, CHLORINE.: /AE, *ACTINOM　W72-04155
ENITE, COPPER SULPHATE, MONURON,　AMMONIA, DALAPON, 2-4-: /DIUM ARS　W70-00269
RHODE ISLAND, SAMPLING, ECOLOGY,　AMMONIA, ENZYMES, CLASSIFICATION,　W70-00713
, TEMPERATURE, NITRATES, PLANTS,　AMMONIA, HARDNESS(WATER), ACIDITY　W69-07838
OGICAL TREATMENT, ALGAL CONTROL,　AMMONIA, ION EXCHANGE, REVERSE OS　W70-01981
VITAMINS, AMINO ACIDS, CALCIUM,　AMMONIA, NITROGEN FIXATION, IRON　W70-02255
L NUTRIENTS, NITROGEN COMPOUNDS,　AMMONIA, NITRATES, NITRITES, NUTR　W69-10177
TRATION, OXYGEN, CARBON DIOXIDE,　AMMONIA, NITRITES, HYDROGEN SULFI　W71-03031
ITRIFICATION, ORGANIC COMPOUNDS,　AMMONIA, NITRITES, NITRATES, YEAS　W69-06970
R, NITROGEN COMPOUNDS, NITROGEN,　AMMONIA, NUTRIENTS, BIOCHEMICAL O　W69-08054

PRODUCTIVITY, CALCIUM, NITRATES,
, NITRATES, ELECTRON MICROSCOPY,
AGOONS, *ORGANIC LOADING, ALGAE,
DIATOMS, MICROBIOLOGY, NITRATES,
MICAL PRECIPITATION, FILTRATION,
HERMAL STRATIFICATION, SULFIDES,
COLIFORMS, EUTROPHICATION, ODOR,
IOLOGICAL ACTIVITY, QUARTERNARY
CARBONATES, NITRATES, CHLORIDES,
ENVIRONMENTAL EFFECTS, AMMONIA,
DES, PRODUCTION LOSS, QUATERNARY
SITY, NITRATE, EUGLENA, CALCIUM,
ON, *FRESH WATER, ALGAE, NITRATE
TOXINS, ALGAE, RESISTANCE, FISH,
L, ALGICIDES, CLOSTRIDIUM, FISH,
NOALGAL, PORTUGAL, TAPES, JAPAN,
OSTATUM, DUNALIELLA TERTIOLECTA,
-106, PLATYMONAS, THALASSIOSIRA,
NDEX, ACHNANTHES, ACTINOPTYCHUS,
OMS, ALGAE, METABOLISM, ISOPODS,
OCIATIONS, SHEAR WATER(ENGLAND),
NOPTYCHUS, AMPHIPLEURA RUTILANS,
NI/ *ONONDAGA LAKE(NY), EUNOTIA,
ATION, CLAY LENSES, RHINE RIVER,
AGUE, MINERAL/ *TRANSFORMATIONS,
OGRAPHS, MICROCYSTIS AERUGINOSA,
AERU/ *BIOLOGICAL CONCENTRATION,
OMENON FLOS-AQUAE, OSCILLATORIA,
PHYCEA/ *EXTRACELLULAR PRODUCTS,
MONONA(WIS), LAKE WAUBESA(WIS),
ANADIUM, ZINC, EUGLENA GRACILIS,
CCUS BRAUNII, METABOLIC PATTERN,
A, FIXATION, TRINGFORD(ENGLAND),
A B629, ANABAENA INAEQUALIS 381,
S LIMNETICUS, LYNGBYA LIMNETICA,
N / *CONCEPTS, *TROPHIC BIOLOGY,
SA, COELOSPHAERIUM KUTZINGIANUM,
TIAL, SELENASTRUM CAPRICORNUTUM,
A FLOS AQUAE, ANABAENA MENDOTAE,
TORIA, ANABAENA CYLINDRICA B629,
AQUAE, SECCHI DISC TRANSPARENCY,
DOSAGE, SPECIES SUSCEPTIBILITY,
TZINGIANUM, ANABAENA FLOS AQUAE,
MENDOTA(WIS), LAKE WINGRA(WIS),
AERUGINOSA, APHAN/ *HETEROCYSTS,
DEA, CERATOPHYLLUM, MICROCYSTIS,
MICROCYSTIS,
BYA, MICROCYSTIS, APHANIZOMENON,
CS, MICROCYSTIS, DREDGE SAMPLES,
PSIS, RACIBORSKII, RAPHIDIOPSIS,
, CLOSTRIDIUM, FISH, AMPHIBIANS,
ROOCOCCUS, OSCILLATORIA, NOSTOC,
STEEL CABLES, ALGAL SUCCESSION,
STRUM, SCENEDESMUS, LEPOCINCLIS,
ONE, MICROCYSTIS, APHANIZOMENON,
S, COELOSPHAERIUM, OSCILLATORIA,
RENCY, MYXOPHYCEA, OSCILLATORIA,
H YIELDS, UNDERWATER SHOCK WAVE,
RIONELLA, APHANIZOMENON, SYNURA,
IA ECHINULATA, D/ HYDROXYLAMINE,
TION, SELENASTRUM, ALGAL GROWTH,
CESSES, GASTROENTERITIS, SYNURA,
ARGINAL VEGETATION, MICROCYSTIS,
MONAS, MELOSIRA, STEPHANODISCUS,
IROGYRA, MICROCYSTIS AERUGINOSA,
IC ACTION, *AQUATIC MACROPHYTES,
METRY, PACIFASTACUS LENIUSCULUS,
AMOGETON SPP, LAKE MENDOTA(WIS),
POND(WIS), HYDRODICTYON, NAJAS,
NEDESMUS, LEPOCINCLIS, ANABAENA,
OSCILLATORIA, NOSTOC, ANABAENA,
OUNDED WATERS, FRESHWATER ALGAE,
ADATION, *SEWAGE BACTERIA, ODOR,
LGAE, BIOCHEMICAL OXYGEN DEMAND,
: *MANURE,
HOSPHATES, METHANE, MUD, OXYGEN,
GROWTH RATE, AEROBIC CONDITIONS,
NTHESIS, RAINBOW TROUT, DIATOMS,
TRIENT REMOVAL, *EUTROPHICATION,
AE, BENTHOS, AEROBIC CONDITIONS,
EDIMENTATION, OXIDATION LAGOONS,
GE PROGRAMS, ANAEROBIC BACTERIA,
GE PROGRAMS, ANAEROBIC BACTERIA,
, CALIFORNIA, DRAINAGE PROGRAMS,
, CALIFORNIA, DRAINAGE PROGRAMS,
, CALIFORNIA, DRAINAGE PROGRAMS,
, CALIFORNIA, DRAINAGE PROGRAMS,
GE PROGRAMS, ANAEROBIC BACTERIA,
CESSING WASTES, AEROBIC LAGOONS,
EROBIC BACTERIA, AQUATIC PLANTS,

AMMONIA, PHOSPHATES, HYDROGEN ION   W70-05405
AMMONIA, SPECTROMETERS, DIFFUSION   W70-04369
AMMONIA, SEDIMENTATION, FLOTATION   W71-05013
AMMONIA, WATER POLLUTION EFFECTS,   W71-05026
AMMONIA, WATER PURIFICATION, WAST   W71-11393
AMMONIUM COMPOUNDS, OXIDATION RED   W70-05760
AMMONIUM COMPOUNDS, ACTIVATED CAR   W70-00263
AMMONIUM COMPOUNDS, MINIMUM LETHA   W70-02982
AMMONIUM COMPOUNDS, OXYGEN, SILIC   W69-07096
AMMONIUM COMPOUNDS, NITROGEN FIXA   W69-03364
AMMONIUM COMPOUNDS, PHENOLIC COMP   W72-12987
AMMONIUM SALTS, NITROGEN FIXATION   W70-02964
AMMONIUM, NITRITE, OXIDATION-REDU   W69-04521
AMPHIBIANS, INVERTEBRATES, AQUATI   W68-00859
AMPHIBIANS, ANABAENA, CYTOLOGICAL   W70-05372
AMPHIDINIUM, BLOOMS, MYTILUS, TRI   W70-05372
AMPHIDINIUM CARTERI, SYRACOSPHAER   W69-08278
AMPHIDINIUM KLEBSI, PORPHYRIDIUM   W70-00708
AMPHIPLEURA RUTILANS, AMPHORA, AU   W70-03325
AMPHIPODA, FISH, GROWTH RATES.: /   W69-09742
AMPHORA OVALIS, OPEPHORA MARTYI,   W71-12870
AMPHORA, AULACODISCUS, BACILLARIA   W70-03325
AMPHORA, GOMPHONEMA, PINNULARIA,   W70-03974
AMSTERDAM RHINE, POLDERS, COMPOSI   W69-07838
AMSTERDAM, HAARLEM, LEYDEN, THE H   W69-07838
ANABAENA CIRCINALIS, GLOEOTRICHIA   W69-05704
ANABAENA CYLINDRICA, MICROCYSTIS   W70-03520
ANABAENA CYLINDRICA B629, ANABAEN   W70-05091
ANABAENA CYLINDRICA, ZINC, XANTHO   W69-10180
ANABAENA CIR: /MENDOTA(WIS), LAKE   W69-08283
ANABAENA CYLINDRICA, BACILLUS SUB   W70-02964
ANABAENA CYLINDRICA, GLYCOLYSIS,   W70-02965
ANABAENA CYLINDRICA, NORTH SEA, F   W70-02504
ANABAENA FLOS AQUAE, TOLYPOTHRIX   W70-05091
ANABAENA FLOS AQUAE, TABELLARIA,   W70-05405
ANABAENA FLOS AQUAE, APHANIZOMENO   W70-03959
ANABAENA FLOS AQUAE, ANABAENA MEN   W69-08283
ANABAENA FLOS-AQUAE, MICROCYSTIS   W69-06864
ANABAENA HASSALI, LYNGBYA NOLLEI,   W69-08283
ANABAENA INAEQUALIS 381, ANABAENA   W70-05091
ANABAENA LEMMERMANNI, APHANOCAPSA   W69-10169
ANABAENA LEMMERMANNII, OTTAWA(ONT   W70-00273
ANABAENA MENDOTAE, ANABAENA HASSA   W69-08283
ANABAENA SPIROIDES, MICROCYSTIS A   W70-05112
ANABAENA VARIABILIS, MICROCYSTIS   W70-05091
ANABAENA.:   HYDRODICTYON, ELO   W68-00479
ANABAENA.:   W68-00470
ANABAENA, MELOSIRA, CERATIUM, OSC   W69-01977
ANABAENA, SEASONAL EFFECTS, DECOM   W70-05663
ANABAENA, WOLLEA, OSCILLATORIA, S   W70-05404
ANABAENA, CYTOLOGICAL STUDIES, PH   W70-05372
ANABAENA, ANACYSTIS, MICROCYSTIS,   W70-04503
ANABAENA, CLADOPHORA, PEDIASTRUM,   W70-04494
ANABAENA, ANACYSTIS, GLOEOTRICHIA   W70-03974
ANABAENA, GLOEOTRICHIA, JANESVILL   W70-03310
ANABAENA, APHANIZOMENON, GLOEOTRI   W70-04369
ANABAENA, DIAPTOMUS ASHLANDI, EPI   W69-10182
ANABAENA, PHYGON XL(DICHLONE), NE   W69-08674
ANABAENA, MELOSIRA, MALLOMONAS, U   W69-10157
ANABAENA, MICROCYSTIS, GLOEOTRICH   W70-02803
ANABAENA, BIOSTIMULATION, WATER P   W70-02779
ANABAENA, IOWA RIVER, CYCLOTELLA,   W70-08107
ANABAENA, EUCYCLOPS, NAIS, CRICOT   W70-06975
ANABAENA, APHANIZOMENON, MICROCYS   W72-04761
ANABAENOPSIS, RACIBORSKII, RAPHID   W70-05404
ANACHARIS CANADENSIS, MACROPHYTES   W70-02249
ANACHARIS CANADENSIS, MYRIOPHYLLU   W70-00711
ANACHARIS, NAIAS, SCIRPUS, NYMPHA   W70-05263
ANACHARIS, LAKE WAUBESA(WIS), MAC   W70-03310
ANACYSTIS, GLOEOTRICHIA, GLOEOCAP   W70-03974
ANACYSTIS, MICROCYSTIS, BOSMINA,   W70-04503
ANADROMOUS FISHES.: /RONMENT, IMP   W71-02491
ANAEROBIC CONDITIONS, AEROBIC CON   W71-02009
ANAEROBIC DIGESTION, AEROBIC TREA   W71-00936
ANAEROBIC LAGOON, AERATED LAGOON.   W70-07491
ANAEROBIC CONDITIONS, HYDROGEN SU   W70-05761
ANAEROBIC CONDITIONS, BIO: /SIS,   W70-05750
ANAEROBIC CONDITIONS, ALGAE, ROTI   W69-10154
ANAEROBIC DENITRIFICATION, LAND A   W70-01981
ANAEROBIC CONDITIONS, WATER POLLU   W71-11381
ANAEROBIC DIGESTION, ALGAE, SLUDG   W71-11375
ANAEROBIC CONDITIONS, LAGOONS.: /   W71-04555
ANAEROBIC CONDITIONS, LAGOONS.: /   W71-04556
ANAEROBIC BACTERIA, ANAEROBIC CON   W71-04556
ANAEROBIC BACTERIA, ANEROBIC COND   W71-04554
ANAEROBIC BACTERIA, ANAEROBIC CON   W71-04555
ANAEROBIC BACTERIA, ANAEROBIC CON   W71-04557
ANAEROBIC CONDITIONS, LAGOONS.: /   W71-04557
ANAEROBIC LAGOONS.:   *POTATO PRO   W71-08970
ANAEROBIC BACTERIA, ALGAE, BIOCHE   W71-08221

DUSTRIAL WASTES, WATER ANALYSIS, ANALYTICAL TECHNIQUES, SAMPLING, W72-03299
GAE, MARINE ALGAE, GROWTH RATES, ANALYTICAL TECHNIQUES, LABORATORY W71-07339
MENTS, IMPOUNDMENTS, RESERVOIRS, ANALYTICAL TECHNIQUES, ON-SITE TE W71-10098
SION, OXIDATION, PHOTOSYNTHESIS, ANALYTICAL TECHNIQUES, SPECTROSCO W71-11377
NTS, FILTRATION, MICROORGANISMS, ANALYTICAL TECHNIQUES, *WASTE WAT W71-13313
TEMPERATURE, CHEMICAL ANALYSIS, ANALYTICAL TECHNIQUES, CLOUD COVE W69-09676
TION EFFECTS, WATER UTILIZATION, ANALYTICAL TECHNIQUES, NUTRIENTS, W71-03048
KTON, MICROORGANISMS, TURBIDITY, ANALYTICAL TECHNIQUES, LABORATORY W70-08628
TEMPERATURE, SALINITY, BIOASSAY, ANALYTICAL TECHNIQUES, LETHAL LIM W70-08111
TERIONELLA, DINOBRYON, MELOSIRA, ANEBODA(SWEDEN), UPPSALA(SWEDEN), W70-05409
GE PROGRAMS, ANAEROBIC BACTERIA, ANEROBIC CONDITIONS, LAGOONS.: /A W71-04554
GAE, CENTRIFUGATION, FILTRATION, ANEROBIC DIGESTION, INCINERATION, W70-04060
           LOS ANGELES COUNTY(CALIF).: W70-05645
ATER, *PROTOZOA, ALGAE, DIATOMS, ANIMAL BEHAVIOR, *THERMODYNAMIC B W70-07847
INN), SOIL MINERALS, / *KILLING, ANIMAL MANURES, LAKE MINNETONKA(M W70-04193
EWAGE LAGOONS, PLANT PHYSIOLOGY, ANIMAL PHYSIOLOGY, NORTH DAKOTA, W72-01818
ON, MARINE ALGAE, MARIN/ *MARINE ANIMALS, *RADIOISOTOPES, ABSORPTI W71-09850
TOPES, PHYSICOCHEMICAL / *MARINE ANIMALS, *MARINE PLANTS, RADIOISO W71-09863
NE ALGAE, *MARSH PLANTS, *MARINE ANIMALS, *OILY WATER, LICHENS, TI W70-01777
SSENTIAL NUTRIENTS, */ *DOMESTIC ANIMALS, *WASTES, FERTILIZERS, *E W69-05323
*WATER POLLUTION EFFECTS, ALGAE, ANIMALS, AQUATIC PLANTS, BIOMASS, W69-04805
ONDWEEDS, SODIUM ARSENITE, FISH, ANIMALS, ALGAE, CUTTING MANAGEMEN W70-05263
RVESTING, AQUATIC ALGAE, AQUATIC ANIMALS, AQUATIC MICROBIOLOGY, AQ W69-03513
E, ZOOPLANKTON, AQUATIC ANIMALS, ANIMALS, APPALACHIAN MOUNTAIN REG W69-03695
UATIC LIFE, ZOOPLANKTON, AQUATIC ANIMALS, ANIMALS, APPALACHIAN MOU W69-03695
WATER POLLUTION EFFECTS, MARINE ANIMALS, ALGAE, MOLLUSKS, FISH, R W71-05531
, *THERMAL POLLUTION, LABORATORY ANIMALS, ACCLIMATIZATION, MARINE W71-02494
ANIMALS, ACCLIMATIZATION, MARINE ANIMALS, AQUATIC LIFE, AQUATIC AN W71-02494
E ANIMALS, AQUATIC LIFE, AQUATIC ANIMALS, BENTHIC FAUNA, AQUATIC P W71-02494
NISMS, STREAMS, WATER POLLUTION, ANIMALS, BACTERIA, FUNGI, ALGAE, W70-05750
MENTS, THERMOCLINE, HYPOLIMNION, ANIMALS, DENSITY, IRON, PHOSPHORU W70-06975
SEDIMENTS, RADIOACTIVITY, MARINE ANIMALS, ESTUARINE ENVIRONMENT, F W69-08267
QUATIC PLANTS, BOTTOM SEDIMENTS, ANIMALS, FISH, INFLOW, OUTLETS, N W70-06225
UATIC INSECTS, FISH, GAME BIRDS, ANIMALS, FOOD CHAINS, OXYGEN, ROT W70-10175
SPHORUS, *CORAL, *REEFS, *MARINE ANIMALS, FISH, ECOSYSTEMS, CYCLIN W69-08526
OORGANISMS, MARINE ALGAE, MARINE ANIMALS, MARINE BACTERIA, MARINE W69-03752
*RADIOISOTOPES, *DISTRIBUTION, ANIMALS, PLANTS, BOTTOM SEDIMENTS W69-08269
TUDIES, DATA COLLECTIONS, MARINE ANIMALS, PRIMARY PRODUCTIVITY, CH W73-01446
FE(RADIONUCLIDE), FILTER-FEEDING ANIMALS, RUTHENIUM-106, PLATYMONA W70-00708
KTON, NUTRIENTS, DEPTH, DOMESTIC ANIMALS, SAMPLING, NUISANCE ALGAE W70-00274
RIVER, DOLORES RIVER, URAVAN, / ANIMAS RIVER, DURANGO, SAN MIGUEL W69-07846
PES, PHYSICOCHEMICAL PROPERTIES, ANION ADSORPTION, COLLOIDS, AQUAR W71-09863
E, *ALKYL PHENOL POLYETHOXYLATE, ANIONIC SURFACTANTS, NONIONIC SUR W70-09438
STRUM, WESTELLA, STEPHANODISCUS, ANKISTRODESMUS, CHLOROMONAD, ALUM W70-08107
ILARIA, CYCLOTELLA, FLAGELLATES, ANKISTRODESMUS, GOMPHENEMA, SPIRO W70-02249
IC GROWTH, HETEROTROPHIC GROWTH, ANKISTRODESMUS, NOSTOC MUSCORUM, W70-02964
, CHLORO/ *EUROPE, ALGAL GROWTH, ANKISTRODESMUS, BACILLARIOPHYCEAE W69-07833
DIOXIDE, CHLORELLA PYRENOIDOSA, ANKISTRODESMUS BRAUNII, DUNALIELL W70-05261
N, CELL CONCENTRATION, UROGLENA, ANKISTRODESMUS, CHLOROCOCCALES, V W70-05409
RIDINIUM WILLEI, CHLOROCOCCALES, ANKISTRODESMUS, OOCYSTIS, CRYOMON W70-05405
VULGARIS, MONODUS SUBTERRANEUS, ANKISTRODESMUS FALCATUS, CHLAMYDO W70-05381
A AMBIGUA, CERATIUM HIRUDINELLA, ANKISTRODESMUS, GLENODINIUM PULVI W70-05387
, SCENEDESMUS, GROWTH EQUATIONS, ANKISTRODESMUS, HETEROTROPHY, CON W70-04381
YRA, CHAETOPHORA, STIGEOCLONIUM, ANKISTRODESMUS, APHANOCAPSA, CHRO W70-04371
SYNECHOCOCCUS, CARTERIA, GONIUM, ANKISTRODESMUS, MICROSPORA, PEDIA W70-03974
APSA, MERISMOPEDIA, MICROCYSTIS, ANKISTRODESMUS, PEDIASTRUM, TETRA W70-03969
ENSIONS, ELECTROKINETIC EFFECTS, ANN ARBOR(MICH), SAND BEDS.: /USP W71-03035
E WASHINGTON, LAKE MALAREN, LAKE ANNECY, LAKE VANERN, LAKE CONSTAN W70-03964
TROGEN ANNUAL BUDGET, PHOSPHORUS ANNUAL BUDGET.: / TROPHOLYTIC, NI W69-05142
, TROPHIC, TROPHOLYTIC, NITROGEN ANNUAL BUDGET, PHOSPHORUS ANNUAL W69-05142
TES, SWEDEN, LAKE ERKEN(SWEDEN), ANNUAL CYCLES, MICROBIAL ECOLOGY, W70-05109
IMNION, ALGAL CONTROL, SEASONAL, ANNUAL TURNOVER.: /CTIVITY, HYPOL W69-05142
ILATIVE CAPACITY, STANDING CROP, ANNUAL TURNOVER, FISH POPULATIONS W68-00912
OCKERELLI, BODY BURDENS, POMOXIS ANNULARIS, GASTROINTESTINAL TRACT W70-02786
OSOME CERCARIAE, TVA RESERVOIRS, ANOPHELES QUADRIMACULATUS, MADISO W70-06225
GROWTH, *MICROCYSTIS AERUGINOSA, ANTAGONISMS, ALGAL NUTRITION, ALG W69-05867
SES, OOCISTUS, FILTRATION MEDIA, ANTHRACITE, GRAVITY FILTERS.: /OS W70-04199
RMARY COMPOUNDS, MACROMOLECULES, ANTIBIOTICS, AGARBACTERIUM, XANT: W70-04280
T GROWTH, ANALYTICAL TECHNIQUES, ANTIBIOTICS, BIOASSAY, NORON, CAL W69-04801
RACIL, REDUCED SULFUR COMPOUNDS, ANTIBIOTIC EFFECTS, STIMULANTS, G W70-02510
IDES, *PAINTS, *TEST PROCEDURES, ANTIFOULING MATERIALS, TESTING, B W71-07339
ARBON RADIOISOTOPES, DETERGENTS, ANTIFREEZ: /ICAL OXYGEN DEMAND, C W70-04001
L, *BIKINI ATOLL, RUTHENIUM-106, ANTIMONY-125, GROUNDWATER, BISMUT W69-08269
L BALANCE, OVERFISHING, COHO SA/ ANTIPOLLUTION PROGRAMS, BIOLOGICA W70-04465
ORADIOGRAPHY), LEUCOTHRIX MUCOR, ANTITHAMNION SARNIENSE, QUANTITAT W69-10163
MORPHOLOGY, FRIDAY HARBOR(WASH), ANTITHAMNION SARNIENSE, RHODOCHOR W69-10161
H, *FILAMENTS, *AUTORADIOGRAPHY, ANTITHAMNION SARNIENSE, TRITIATED W69-09755
ESHOLD ODOR TESTS, ASTERIONELLA, APHANIZOMENON, SYNURA, ANABAENA, W69-10157
TO, FILAMENTOUS ALGAE, NAVICULA, APHANIZOMENON, ONTARIO(CANADA).: / W69-10155
SCILLATORIA AGARDHI, PHORMIDIUM, APHANIZOMENON FLOS-AQUAE, SECCHI W69-10169
NAGEMENT PROGRAM, ALGAL DENSITY, APHANIZOMENON, MELOSIRA, LYNGBYA. W69-10167
ERMANY), CHIRONOMIDS, CLADOCERA, APHANIZOMENON FLOS-AQUAE, STEPHAN W70-00268
DEPLETION.: *YAHARA RIVER(WIS), APHANIZOMENON FLOS AQUAE, OXYGEN W72-01797
QUINONE, MICROCYSTIS AERUGINOSA, APHANIZOMENON FLOS-AQUAE, MINIMUM W70-02982
CYCLOPS, GOMPHONEMA, PHORMIDIUM, APHANIZOMENON FLOS-AQUAE, LYNGBYA W70-02803
E MEDIA, MICROCYSTIS AERUGINOSA, APHANIZOMENON FLOS-AQUAE, GLOEOTR W70-00719
, FRAGILARIA, SYNEDRA, NAVICULA, APHANIZOMENON, OSCILLATORIA, MELO W70-08107
LOGY, MICROCYSTIS, GLOEOTRICHIA, APHANIZOMENON, MICRONUTRIENTS, DI W70-04381
HAERIUM, OSCILLATORIA, ANABAENA, APHANIZOMENON, GLOEOTRICHIA, POLY W70-04369
IC BIOLOGY, ANABAENA FLOS AQUAE, APHANIZOMENON FLOS AQUAE, GLOEOTR W70-03959

ASONAL EFFECTS, THERMAL EFFECTS,   APHANIZOMENON FLOS AQUAE, METOLIU   W70-03978
L ORONAPHTHOQUINONE, MICROCYSTIS,   APHANIZOMENON, ANABAENA, GLOEOTRI   W70-03310
E WAUBESA, LYNGBYA, MICROCYSTIS,   APHANIZOMENON, ANABAENA, MELOSIRA   W69-01977
IABILIS, MICROCYSTIS AERUGINOSA,   APHANIZOMENON FLOS-AQUAE, OSCILLA   W70-05091
SARCODINA, CILIATES, CYCLOTELLA,   APHANIZOMENON FLOS AQUAE, MELOSIR   W70-05112
ODISCUS AS/ *SHAGAWA LAKE(MINN),   APHANIZOMENON FLOS AQUAE, STEPHAN   W70-05387
OSIRA, STEPHANODISCUS, ANABAENA,   APHANIZOMENON, MICROCYSTIS, POTAM   W72-04761
IENT SOURCES, PHOSPHORUS BUDGET,   APHANIZOMINON FLOS AQUAE.: / NUTR   W70-04268
, STIGEOCLONIUM, ANKISTRODESMUS,   APHANOCAPSA, CHROOCOCCUS, CLOSTER   W70-04371
NSPARENCY, ANABAENA LEMMERMANNI,   APHANOCAPSA, BOSMINA LONGISPINA,   W69-10169
NKTON, AQUATIC ANIMALS, ANIMALS,   APPALACHIAN MOUNTAIN REGION.: /LA   W69-03695
REMOVAL, AGRICULTURAL DRAINAGE,   APPALACHIAN MOUNTAIN CLUB HUTS, T   W69-07818
R LINES, ATTENUATION MEASUREMENT   APPARATUS, HYDROGEN F LINE.: /OFE   W70-05377
ROOTED AQUATIC PLANTS, RATES OF   APPLICATION, APPLICATION TECHNIQU   W70-05568
IC PLANTS, RATES OF APPLICATION,   APPLICATION TECHNIQUES, SWIMMING,   W70-05568
LGAE, FISH, ZOOPLANKTON, SNAILS,   APPLICATION METHODS.: /NUISANCE A   W70-03310
RATURES, *COPPER DOSAGE, *COPPER   APPLICATION METHODS, ALGAE IDENTI   W69-10157
ODELS, BOD REMOVAL RATES, DESIGN   APPLICATIONS, FACULTATIVE PONDS,   W70-07034
ANAEROBIC DENITRIFICATION, LAND   APPLICATION, ELECTROCHEMICAL TREA   W70-01981
UNIVERSITY, FUNDAMENTAL STUDIES,   APPLIED STUDIES, ALGAL PHYSIOLOGY   W70-03973
EATMENT, ALGAE, WATER MANAGEMENT(APPLIED), LIMNOLOGY.: /LLUTION TR   W70-04721
S, AGRICULTURE, WATER MANAGEMENT(APPLIED), GEOCHEMISTRY, SOIL TYPE   W71-03048
ANCE, FOAMING DETERGENTS, PUBLIC   APPREHENSION.: /TES, NUTRIENT BAL   W70-05084
ES, *INDUSTRIAL WASTES, PEACHES,   APRICOTS, PEARS, CITRUS FRUITS, T   W71-09548
M WASTES, MERCENARIA MERCENARIA,   AQUACULTURE, RECYCLING WASTES.: /   W70-09905
HORUS BUDGET, APHANIZOMINON FLOS   AQUAE.: / NUTRIENT SOURCES, PHOSP   W70-04268
LAKE(ORE.), *APHANIZOMENON FLOS-   AQUAE, *CARBON-14 MEASUREMENTS.: /   W72-06052
OHEMIA, ELB/ *APHANIZOMENON FLOS-   AQUAE, *MICROCYSTIS AERUGINOSA, B   W70-00274
TERINARIANS, MIC/ *ANABAENA FLOS-   AQUAE, *STRAINS, PHYCOLOGISTS, VE   W70-00273
RIUM KUTZINGIANUM, ANABAENA FLOS   AQUAE, ANABAENA MENDOTAE, ANABAEN   W69-08283
*TROPHIC BIOLOGY, ANABAENA FLOS   AQUAE, APHANIZOMENON FLOS AQUAE,   W70-03959
A FLOS AQUAE, APHANIZOMENON FLOS   AQUAE, GLOEOTRICHIA ECHINULATA, B   W70-03959
S AERUGINOSA, APHANIZOMENON FLOS-   AQUAE, GLOEOTRICHIA ECHINULATA, C   W70-00719
, PHORMIDIUM, APHANIZOMENON FLOS-   AQUAE, LYNGBYA, STORM LAKE(IOWA),   W70-02803
, CYCLOTELLA, APHANIZOMENON FLOS   AQUAE, MELOSIRA, ASTERIONELLA FOR   W70-05112
RMAL EFFECTS, APHANIZOMENON FLOS   AQUAE, METOLIUS RIVER(ORE), COMMU   W70-03978
RUM CAPRICORNUTUM, ANABAENA FLOS-   AQUAE, MICROCYSTIS AERUGINOSA, AL   W69-06864
S AERUGINOSA, APHANIZOMENON FLOS-   AQUAE, MINIMUM LETHAL DOSES, CHU'   W70-02982
S AERUGINOSA, APHANIZOMENON FLOS-   AQUAE, OSCILLATORIA, ANABAENA CYL   W70-05091
A RIVER(WIS), APHANIZOMENON FLOS   AQUAE, OXYGEN DEPLETION.: *YAHAR   W72-01797
, PHORMIDIUM, APHANIZOMENON FLOS-   AQUAE, SECCHI DISC TRANSPARENCY,   W69-10169
S, CLADOCERA, APHANIZOMENON FLOS-   AQUAE, STEPHANODISCUS HANTZSCHII,   W70-00268
A LAKE(MINN), APHANIZOMENON FLOS   AQUAE, STEPHANODISCUS ASTREA, LAK   W70-05387
LYNGBYA LIMNETICA, ANABAENA FLOS   AQUAE, TABELLARIA, MELO: /TICUS,   W70-05405
NA INAEQUALIS 381, ANABAENA FLOS   AQUAE, TOLYPOTHRIX DISTORTA, GLOE   W70-05091
IES, ANION ADSORPTION, COLLOIDS,   AQUARIA, ABSORPTION, WATER POLLUT   W71-09863
SORPTION, CHELATION, METABOLISM,   AQUARIA, WATER POLLUTION EFFECTS,   W71-09005
, OXIDATION-REDUCTION POTENTIAL,   AQUATIC ALGAE, *BIOLOGICAL TREATM   W71-09546
UTRIENTS, *PRIMARY PRODUCTIVITY,   AQUATIC ALGAE, TROPHIC LEVEL, WAT   W72-01990
LGICIDES, ANALYTICAL TECHNIQUES,   AQUATIC ALGAE, AQUATIC ENVIRONMEN   W70-02968
IA, *TOXICITY, *WATER POLLUTION,   AQUATIC ALGAE, BRINE SHRIMP, FISH   W70-01467
ING, *MARSH PLANTS, *OILY WATER,   AQUATIC ALGAE, MARINE ALGAE, OIL,   W70-01231
, *WATER POLLUTION, FOOD CHAINS,   AQUATIC ALGAE, BRINE SHRIMP, FISH   W70-01466
ALGAE, ALGAL CONTROL, ALGICIDES,   AQUATIC ALGAE, AQUATIC MICROBIOLO   W70-02248
GY, AQUATIC/ *CLADOPHORA, ALGAE,   AQUATIC ALGAE, AQUATIC MICROBIOLO   W70-02784
FISHKILL, FISH TOXINS, BIOASSAY,   AQUATIC ALGAE, *ALGAL TOXINS, KIL   W70-08372
TY, BIOINDICATORS, AQUATIC LIFE,   AQUATIC ALGAE, WATER POLLUTION EF   W70-05958
R POLLUTION SOURCES, IRRIGATION,   AQUATIC ALGAE, WATER ANALYSIS, CA   W69-07096
SANCE ALGAE, *SEWAGE, ALGICIDES,   AQUATIC ALGAE, AQUATIC PRODUCTIVI   W69-04798
TOXICITY, BIOLOGICAL MEMBRANES,/   AQUATIC ALGAE, WATER POLLUTION, *   W69-05306
ROL, ALGAL POISONING, ALGICIDES,   AQUATIC ALGAE, AQUATIC ENVIRONMEN   W69-05844
IENTS, *FISH, *FISHERIES, ALGAE,   AQUATIC ALGAE, AQUATIC MICROBIOLO   W69-04804
NT REQUIREMENTS, *PHYTOPLANKTON,   AQUATIC ALGAE, AQUATIC MICROBIOLO   W69-04802
TES, CHELATION, WASTE TREATMENT,   AQUATIC ALGAE, AQUATIC PLANTS, TE   W70-04373
A, ALGAE, ANALYTICAL TECHNIQUES,   AQUATIC ALGAE, AQUATIC BACTERIA,   W69-03371
Y, ALGAE, ANALYTICAL TECHNIQUES,   AQUATIC ALGAE, AQUATIC MICROBIOLO   W69-03517
LGAE, ALGAL CONTROL, HARVESTING,   AQUATIC ALGAE, AQUATIC ANIMALS, A   W69-03513
NE WATER, *THERMAL WATER, ALGAE,   AQUATIC ALGAE, AQUATIC MICROBIOLO   W69-03516
Y, ALGAE, ANALYTICAL TECHNIQUES,   AQUATIC ALGAE, AQUATIC MICROBIOLO   W69-03515
ROL, ALGAL POISONING, ALGICIDES,   AQUATIC ALGAE, BIOASSAY, CHLOROPH   W69-03370
A, ALGAE, ANALYTICAL TECHNIQUES,   AQUATIC ALGAE, BIOASSAY, CYCLING   W69-03518
CONTROL, ANALYTICAL TECHNIQUES,   AQUATIC ALGAE, AQUATIC MICROBIOLO   W69-03374
S, *PHOSPHORUS COMPOUNDS, ALGAE,   AQUATIC ALGAE, AQUATIC MICROBIOLO   W69-03373
CTIVE WASTES, *MUNICIPAL WASTES,   AQUATIC ALGAE, URANIUM RADIOISOTO   W69-01165
TRIENTS, *EUTROPHICATION, ALGAE,   AQUATIC ALGAE, AQUATIC BACTERIA,   W69-03369
SPHATES, *PHYSIOLOGICAL ECOLOGY,   AQUATIC ALGAE, AQUATIC MICROBIOLO   W69-03358
EN CYCLE, ANALYTICAL TECHNIQUES,   AQUATIC ALGAE, AQUATIC MICROBIOLO   W69-03185
S, ALGAE, ANALYTICAL TECHNIQUES,   AQUATIC ALGAE, AQUATIC MICROBIOLO   W69-03186
N, ALGAE, ANALYTICAL TECHNIQUES,   AQUATIC ALGAE, AQUATIC BACTERIA,   W69-03362
NUTRIENTS, *NITROGEN COMPOUNDS,   AQUATIC ALGAE, AQUATIC MICROBIOLO   W69-03364
OISONING, ANALYTICAL TECHNIQUES,   AQUATIC ALGAE, AQUATIC BACTERIA,   W69-03188
RATES, CONN STREAM FLOW, ALGAE,   AQUATIC ALGAE, CURRENTS(WATER), L   W68-00479
TION, ALGAL BLOOMS, WATER BLOOM,   AQUATIC ALGAE, AGING(BROLOGICAL),   W68-00912
ATION, DESMIDS/ *EUTROPHICATION,   AQUATIC ALGAE, *LAKES, *CLASSIFIC   W68-00255
TROL, HARVESTING, AQUATIC ALGAE,   AQUATIC ANIMALS, AQUATIC MICROBIO   W69-03513
KTON, AQUATIC LIFE, ZOOPLANKTON,   AQUATIC ANIMALS, ANIMALS, APPALAC   W69-03695
N, MARINE ANIMALS, AQUATIC LIFE,   AQUATIC ANIMALS, BENTHIC FAUNA, A   W71-02494
ATURE, SLUDGE WORMS, BLOODWORMS,   AQUATIC BACTERIA, ALGAE, MAYFLIES   W70-00475
TICAL TECHNIQUES, AQUATIC ALGAE,   AQUATIC BACTERIA, AQUATIC MICROBI   W69-03371
PHICATION, ALGAE, AQUATIC ALGAE,   AQUATIC BACTERIA, AQUATIC MICROBI   W69-03369

TICAL TECHNIQUES, AQUATIC ALGAE,   AQUATIC BACTERIA, AQUATIC MICROBI   W69-03188
TICAL TECHNIQUES, AQUATIC ALGAE,   AQUATIC BACTERIA, AQUATIC MICROBI   W69-03362
OSYSTEMS, PHYSIOLOGICAL ECOLOGY,   AQUATIC BACTERIA, CARBON RADIOISO   W70-04287
TES, NITRATES, YEASTS, BACTERIA,   AQUATIC BACTERIA, WATER POLLUTION   W69-06970
ANALYTICAL TECHNIQUES, KINETICS,   AQUATIC BACTERIA, ALGAE, PHYSIOLO   W70-05109
UTION EFFECTS, DISSOLVED OXYGEN,   AQUATIC BACTERIA, ALGAE.: /R POLL   W70-04727
LAKE, BRITISH COLUMBIA, CANADA,   AQUATIC ECOSYSTEM MODEL, SIMULATI   W69-06536
CONCENTRATIONS, MICRONUTRIENTS,   AQUATIC ECOSYSTEM, MIDGE LARVAE,   W69-09340
TIC LIFE, ALGAE, SILTS, ECOLOGY,   AQUATIC ENVIRONMENT.: /ANTS, AQUA   W69-10080
DATION, WATER POLLUTION EFFECTS,   AQUATIC ENVIRONMENT, CHLORINATED   W70-08652
Y, NUT/ *ALGAE, *TROPHIC LEVELS,   AQUATIC ENVIRONMENT, WATER QUALIT   W70-02764
AE, WATER QUALITY, *TRO/ *ALGAE,   AQUATIC ENVIRONMENT, *AQUATIC ALG   W70-02772
TICAL TECHNIQUES, AQUATIC ALGAE,   AQUATIC ENVIRONMENT, AQUATIC MICR   W70-02968
QUATIC INSECTS, *AQUATIC PLANTS,   AQUATIC ENVIRONMENT, ECOSYSTEMS,   W71-09005
UCTIVITY, MODEL STUDIES, SALMON,   AQUATIC ENVIRONMENT, PHOSPHATES,   W69-06536
ONING, ALGICIDES, AQUATIC ALGAE,   AQUATIC ENVIRONMENT, AQUATIC PLAN   W69-05844
UTROPHICATION, LAKES, LIMNOLOGY,   AQUATIC ENVIRONMENT, AQUATIC PROD   W69-05142
, LIMNOLOGY, HYDROLOGIC ASPECTS,   AQUATIC ENVIRONMENT, WATER PROPER   W70-04475
TION, WATER SPORTS, ENVIRONMENT,   AQUATIC ENVIRONMENT, ESTUARINE EN   W69-04276
TOSYNTHESIS, *CALCIUM CARBONATE,   AQUATIC ENVIRONMENT, AQUATIC PROD   W69-00446
ION, *TOXICITY, *BIOASSAY, FISH,   AQUATIC ENVIRONMENT, ORGANIC PEST   W72-11105
WATER POLLUTION EFFECTS, FUNGI,   AQUATIC FUNGI.: /LLUTION CONTROL,   W69-05706
TOCHASTIC PROCESSES, ECOSYSTEMS,   AQUATIC HABITATS, WATER POLLUTION   W71-09012
AQUATIC ENVIRONMENT, ECOSYSTEMS,   AQUATIC HABITATS, DRAGONFLIES, AL   W71-09005
GLENOPHYTA, CYANOPHYTA, DIATOMS,   AQUATIC HABITAT, POPULATION.: /EU   W70-02770
FLOATING PLANTS, PHYTOPLANKTON,   AQUATIC INSECTS, FISH, GAME BIRDS   W70-10175
TIONS, *RIVERS, ALGAE, PROTOZOA,   AQUATIC INSECTS, FISH REPRODUCTIO   W73-00748
ER TEMPERATURE, *EUTROPHICATION,   AQUATIC LIFE, SESTON, SCENEDESMUS   W73-01446
ACCLIMATIZATION, MARINE ANIMALS,   AQUATIC LIFE, AQUATIC ANIMALS, BE   W71-02494
, *WATER QUALITY, BIOINDICATORS,   AQUATIC LIFE, AQUATIC ALGAE, WATE   W70-05958
OPHICATION, PRODUCTIVITY, ALGAE,   AQUATIC LIFE.: /N COMPOUNDS, EUTR   W70-03068
PLANKTON, WATER QUALITY, ALGAE,   AQUATIC LIFE, STREAM POLLUTION, B   W72-01786
SAMPLING, SURVEYS, MARINE ALGAE,   AQUATIC LIFE, MILK, ATLANTIC OCEA   W72-00941
ALGAE, MODEL STUDIES, BIOASSAY,   AQUATIC LIFE, COOLING WATER, SITE   W71-11881
ION SOURCES, PATH OF POLLUTANTS,   AQUATIC LIFE, ALGAE, SILTS, ECOLO   W69-10080
IES, ECOLOGY, SALINITY, BENTHOS,   AQUATIC LIFE, ALGAE, PLANTS, WATE   W69-05023
STANDING CROP, MICROENVIRONMENT,   AQUATIC LIFE, ENVIRONMENT, MICROO   W69-03611
BACTERIA, ENVIRONMENTAL EFFECTS,   AQUATIC LIFE, AQUATIC MICROORGANI   W69-04276
NIC COMPOUNDS, METALS, PLANKTON,   AQUATIC LIFE, ZOOPLANKTON, AQUATI   W69-03695
SURFACE DRAINAGE, ALGAL CONTROL,   AQUATIC LIFE, WATER CONSERVATION,   W70-03344
INDICATORS, *BIOASSAY, TOXICITY,   AQUATIC MAMMALS, ALGAE, FISH, DIA   W71-10786
OZOA, *FLORIDA, *EUTROPHICATION,   AQUATIC MICROORGANISMS, TROPHIC L   W72-01993
AL EFFECTS, *MORPHOLOGY, *ALGAE,   AQUATIC MICROBIOLOGY, CYANOPHYTA,   W69-10177
ICAL TECHNIQUES, *AQUATIC ALGAE,   AQUATIC MICROBIOLOGY, AQUATIC MIC   W70-02792
ATIC ALGAE, AQUATIC ENVIRONMENT,   AQUATIC MICROBIOLOGY, BALANCE OF   W70-02968
TIC ALGAE, AQUATIC MICROBIOLOGY,   AQUATIC MICROORGANISMS, AQUATIC P   W70-02792
LADOPHORA, ALGAE, AQUATIC ALGAE,   AQUATIC MICROBIOLOGY, AQUATIC PRO   W70-02784
FFECTS, PHYTOPLANKTON, PLANKTON,   AQUATIC MICROBIOLOGY, CYCLING NUT   W70-02245
NTROL, ALGICIDES, AQUATIC ALGAE,   AQUATIC MICROBIOLOGY, BIOASSAY, C   W70-02248
BIODEGRADATION, ALGAE, BACTERIA,   AQUATIC MICROORGANISMS, SAMPLING,   W71-00221
ORS, ECOSYSTEMS, MICROORGANISMS,   AQUATIC MICROORGANISMS, ENZYMES.:   W71-02479
C POWER, HEATED WATER, PROTEINS,   AQUATIC MICROORGANISMS, TEMPERATU   W70-09905
TICAL TECHNIQUES, AQUATIC ALGAE,   AQUATIC MICROBIOLOGY, AQUATIC WEE   W69-03515
RONMENTAL EFFECTS, AQUATIC LIFE,   AQUATIC MICROORGANISMS, MICROORGA   W69-04276
 AQUATIC ALGAE, AQUATIC ANIMALS,   AQUATIC MICROBIOLOGY, AQUATIC PRO   W69-03513
AQUATIC ALGAE, AQUATIC BACTERIA,   AQUATIC MICROBIOLOGY, BIOASSAY, C   W69-03371
TICAL TECHNIQUES, AQUATIC ALGAE,   AQUATIC MICROBIOLOGY, BALANCE OF   W69-03374
TICAL TECHNIQUES, AQUATIC ALGAE,   AQUATIC MICROBIOLOGY, BALANCE OF   W69-03517
COMPOUNDS, ALGAE, AQUATIC ALGAE,   AQUATIC MICROBIOLOGY, AQUATIC PRO   W69-03373
MAL WATER, ALGAE, AQUATIC ALGAE,   AQUATIC MICROBIOLOGY, BALANCE OF   W69-03516
AQUATIC ALGAE, AQUATIC BACTERIA,   AQUATIC MICROBIOLOGY, BALANCE OF   W69-03369
OLOGICAL ECOLOGY, AQUATIC ALGAE,   AQUATIC MICROBIOLOGY, CHLOROPHYTA   W69-03358
AQUATIC ALGAE, AQUATIC BACTERIA,   AQUATIC MICROBIOLOGY, BALANCE OF   W69-03188
TROGEN COMPOUNDS, AQUATIC ALGAE,   AQUATIC MICROBIOLOGY, CHLAMYDOMON   W69-03364
AQUATIC ALGAE, AQUATIC BACTERIA,   AQUATIC MICROBIOLOGY, BIOASSAY, C   W69-03362
TICAL TECHNIQUES, AQUATIC ALGAE,   AQUATIC MICROBIOLOGY, AQUATIC PRO   W69-03185
TICAL TECHNIQUES, AQUATIC ALGAE,   AQUATIC MICROBIOLOGY, AQUATIC PRO   W69-03186
, *PHYTOPLANKTON, AQUATIC ALGAE,   AQUATIC MICROBIOLOGY, AQUATIC PRO   W69-04802
FISHERIES, ALGAE, AQUATIC ALGAE,   AQUATIC MICROBIOLOGY, AQUATIC WEE   W69-04804
SSAY, *COMPARATIVE PRODUCTIVITY,   AQUATIC MICROBIOLOGY, EUTROPHICAT   W72-12240
FECTS, *BIOLOGICAL PRODUCTIVITY,   AQUATIC NUISANCES, TRANSPARENCY,   W69-09349
GRAPH, ARTHROPODA, CHIRONOMIDAE,   AQUATIC OLIGOCHAETES.: /THYTHERMO   W70-04253
WASTE TREATMENT, AQUATIC ALGAE,   AQUATIC PLANTS, TESTING, BIODEGRA   W70-04373
G NUTRIENTS, ZOOPLANKTON, ALGAE,   AQUATIC PLANTS, PALEOLIMNOLOG, M   W70-03975
G NUTRIENTS, ZOOPLANKTON, ALGAE,   AQUATIC PLANTS, PALEOLIMNOLOGY, M   W70-03975
ATIC ALGAE, AQUATIC ENVIRONMENT,   AQUATIC PLANTS, AQUATIC POPULATIO   W69-05844
ON, PHOSPHORUS COMPOUNDS, ALGAE,   AQUATIC PLANTS.: / *DENITRIFICATI   W69-05323
C WEED CONTROL, *EUTROPHICATION,   AQUATIC PLANTS, HERBICIDES, PESTI   W69-05705
C WEED CONTROL, *EUTROPHICATION,   AQUATIC PLANTS, HERBICIDES, PESTI   W69-05706
LLUTION EFFECTS, ALGAE, ANIMALS,   AQUATIC PLANTS, BIOMASS, FERTILIZ   W69-04805
L, *EUTROPHICATION, *HARVESTING,   AQUATIC PLANTS, AQUATIC WEEDS, EC   W69-06276
*ALGAL CONTROL, FISHKILL, ROOTED   AQUATIC PLANTS, GLACIAL LAKES, PL   W68-00488
YANOPHY/ *EUTROPHICATION, ALGAE,   AQUATIC PLANTS, BIBLIOGRAPHIES, C   W68-00680
PLANTS, *NUISANCE ALGAE, ROOTED   AQUATIC PLANTS, *RESERVOIRS, HYDR   W68-00479
MENTS, ON-SITE DATA COLLECTIONS,   AQUATIC PLANTS, WATERSHED, ALGAE,   W68-00511
, CHEMCONTROL, PONDWEEDS, ROOTED   AQUATIC PLANTS, RATES OF APPLICAT   W70-05568
, TRACERS, SAMPLING, PERIPHYTON,   AQUATIC PLANTS, BACKGROUND RADIAT   W69-09334
, *TERTIARY TREATMENT, NITROGEN,   AQUATIC PLANTS, BOTTOM SEDIMENTS,   W69-09340
AQUATIC ANIMALS, BENTHIC FAUNA,   AQUATIC PLANTS, MARINE ALGAE, BEN   W71-02494
N CONCENTRATION, NUISANCE ALGAE,   AQUATIC PLANTS, BOTTOM SEDIMENTS,   W70-06225

392

, ALGAE, FLOATING PLANTS, ROOTED
ON, OCHROMONAS, BACTERIA, FUNGI,
BIOLOGY, AQUATIC MICROORGANISMS,
AGE TREATMENT, AEROBIC BACTERIA,
ISSOLVED OXYGEN, DENSITY, ROOTED
RATURE, CHLOROPHYLL, CYANOPHYTA,
VANIA, RIVERS, STREAMS, EUGLENA,
FISH, AMPHIBIANS, INVERTEBRATES,
UTROPHICATION, MAINE, NUTRIENTS,
AN POPULATION, SEWAGE EFFLUENTS,
NG, RECREATION, SILT, NUTRIENTS,
TIC ENVIRONMENT, AQUATIC PLANTS,
   LIMNOLOGY, AQUATIC ENVIRONMENT,
TIC ALGAE, AQUATIC MICROBIOLOGY,
EWAGE, ALGICIDES, AQUATIC ALGAE,
   CARBONATE, AQUATIC ENVIRONMENT,
TIC ALGAE, AQUATIC MICROBIOLOGY,
TIC ALGAE, AQUATIC MICROBIOLOGY,
PHYTOPLANKTON, *CULTURES, ALGAE,
TIC ALGAE, AQUATIC MICROBIOLOGY,
C ANIMALS, AQUATIC MICROBIOLOGY,
ALGAE, COMPARATIVE PRODUCTIVITY,
   MICROORGANISMS, AQUATIC PLANTS,
TIC ALGAE, AQUATIC MICROBIOLOGY,
FFECTS, WATER POLLUTION SOURCES,
ICATORS, SECONDARY PRODUCTIVITY,
QUATIC PLANTS, WATERSHED, ALGAE,
ROBIOLOGY, AQUATIC PRODUCTIVITY,
TIC ALGAE, AQUATIC MICROBIOLOGY,
ROBIOLOGY, AQUATIC PRODUCTIVITY,
ROBIOLOGY, AQUATIC PRODUCTIVITY,
ION, *PHOSPHATE, FISH, PLANKTON,
AQUATIC WEEDS, LAKES, ALGICIDES,
TIC ALGAE, AQUATIC MICROBIOLOGY,
   *EUTROPHICATION, *LAKES, ALGAE,
ON, *HARVESTING, AQUATIC PLANTS,
AUNA, PHOTOSYNTHESIS, BIOASSAYS,
, *WATER QUALITY CONTROL, ALGAE,
   *OIL SPILLS, PROTEIN
GAL POISONING, GASOLINE, FLOODS,
GILARIA, MELOSIRA, ASTERIONELLA,
ONS, ELECTROKINETIC EFFECTS, ANN
LAKES, DISSOLVED ORGANIC MATTER,
VER, NEBESNA RIVER, MOOSE CREEK,
   PRIMARY PRODUCTIVITY, SURFACES,
   SEDIMENTS, GEOLOGIC FORMATIONS,
ITY, BIOLOGICAL MEMBRANES, HUMID
AE, LAKE STRATIFICATION, SURFACE
OTENTIAL NUTRIENT CONCENTRATION,
ERSTATE POLLUTION, *METROPOLITAN
ERATURE, AMINO ACIDS, INTERTIDAL
NOU/ *LAKE MENDOTA(WIS), PELAGIC
, POPULATION, OREGON, INTERTIDAL
EMPERATURE, PIPTIDES, INTERTIDAL
IL ALGAE, *AQUATIC ALGAE, ALGAE,
ING, WATER DEMAND, WATER SUPPLY,
, SOIL WATER MOVEMENT, LEACHING,
N ANALYSES, AGRICULTURAL RUNOFF,
, POLLUTION, FOSSIL, CIMARRON R/
OL, *OIL SPILLS, *TORREY CANYON,
INORGANIC PHOSPHATES, SOLUBLE R/
POUNDS, LAKES, SEWAGE EFFLUENTS,
   POLLUTION CONTROL, CHEMCONTROL,
OMETRY, ALGAE, MINNESOTA, SODIUM
ICIDES, *COPPER SULFATE, *SODIUM
PPER SULPHATE, PONDWEEDS, SODIUM
NTROLS, PARAQUAT, DIQUAT, SODIUM
OCONTROL, COPPER SULFATE, SODIUM
TIONS, BLACK SEA, CONTACT ZONES,
CERIUM-144, NITZSCHIA CLOSERIUM,
   HETERETROPHY, BACTERIAL MEDIUM,
TE, CHEMOSYNTHESIS, CONCRETIONS,
, METALIMNION, BATHYTHERMOGRAPH,
UM BOTULINUS, VENERUPIS, OSTREA,
   WASTE TREATMENT, COLONY COUNTS,
ARCH AND DEVELOPMENT CORP(MASS),
IAL FERTILIZERS.:
, WATER CONVEYANCE, WATER COSTS,
S, MILWAUKEE(WIS), CHICAGO(ILL),
ADISON(WIS), PSOROPHORA CILIATA,
ECHNIQUES, LABORATORY EQUIPMENT,
S VESICULOSUS, CHONDRUS CRISPUS,
OYA(NORWAY), FUCUS VESCICULOSUS,
, FUCUS VESI/ *SEAWEED EXTRACTS,
ORTALITY, CELL WALL, HEXOSAMINE,
POROBOLOMYCES ROSEUS, AUTOLYSIS,
NCENTRATION, 'EXTRA-DEEP PONDS',
SCILLATORIA, ANABAENA, DIAPTOMUS
                          LAKE
S(USA), UNIVERSITY OF WISCONSIN,

AQUATIC PLANTS, SELF-PURIFICATION   W70-05750
AQUATIC PLANTS, ASSAY, DISTRIBUTI   W70-02510
AQUATIC PLANTS, AQUATIC PRODUCTIV   W70-02792
AQUATIC PLANTS, ANAEROBIC BACTERI   W71-08221
AQUATIC PLANTS, ALGAE, CHLOROPHYL   W71-07698
AQUATIC PLANTS, PHOSPHORUS, DENSI   W72-04761
AQUATIC POPULATIONS, EUTROPHICATI   W70-04510
AQUATIC POPULATIONS, BIBLIOGRAPHI   W68-00859
AQUATIC POPULATIONS, WATER QUALIT   W68-00470
AQUATIC POPULATIONS, NUISANCE ALG   W68-00461
AQUATIC POPULATIONS, OLIGOTROPHY,   W68-00172
AQUATIC POPULATIONS, CONSERVATION   W69-05844
AQUATIC PRODUCTIVITY, HYPOLIMNION   W69-05142
AQUATIC PRODUCTIVITY, CYANOPHYTA,   W69-04802
AQUATIC PRODUCTIVITY, BIOASSAY, C   W69-04798
AQUATIC PRODUCTIVITY, EUTROPHICAT   W69-00446
AQUATIC PRODUCTIVITY, AQUATIC WEE   W69-03185
AQUATIC PRODUCTIVITY, AQUATIC WEE   W69-03186
AQUATIC PRODUCTIVITY, CYCLING NUT   W69-03514
AQUATIC PRODUCTIVITY, AQUATIC WEE   W69-03373
AQUATIC PRODUCTIVITY, BALANCE OF   W69-03513
AQUATIC PRODUCTIVITY, DISTRIBUTIO   W70-05424
AQUATIC PRODUCTIVITY, BALANCE OF   W70-02792
AQUATIC PRODUCTIVITY, CHLOROPHYTA   W70-02784
AQUATIC PRODUCTIVITY, LIMNOLOGY,   W69-09135
AQUATIC PRODUCTIVITY, WATER POLLU   W69-10152
AQUATIC VEGETATION, NITROGEN, PHO   W68-00511
AQUATIC WEEDS, CHLOROPHYTA, CYANO   W69-03373
AQUATIC WEEDS, BALANCE OF NATURE,   W69-03515
AQUATIC WEEDS, BALANCE OF NATURE,   W69-03185
AQUATIC WEEDS, BIOASSAY, CYANOPHY   W69-03186
AQUATIC WEED CONTROL, INSECT CONT   W70-05417
AQUATIC WEED CONTROL, MECHANICAL   W70-04494
AQUATIC WEEDS, BALANCE OF NATURE,   W69-04804
AQUATIC WEEDS, CYANOPHYTA, FISHKI   W69-06535
AQUATIC WEEDS, ECONOMIC FEASIBILI   W69-06229
AQUATIC WEEDS, THERMAL STRATIFICA   W70-02251
AQUATIC WEEDS, ADMINISTRATION, SU   W72-05774
AQUICULTURE.:   W69-05891
AQUIFERS, WATER LAW, NITRATES, TU   W73-00758
ARAPHIDINEAE, CENTRIC DI: /R, FRA   W70-05663
ARBOR(MICH), SAND BEDS.: /USPENSI   W71-03035
ARCTIC ALGAE, AUTOTROPHIC METABOL   W71-10079
ARCTIC CIRCLE HOT SPRINGS, RADON,   W69-08272
ARCTIC, ENZYMES, CHLOROPHYLL, PHO   W70-04809
ARCTIC, TUNDRA, SCALING, GRAVELS,   W69-08272
AREA.: /, WATER POLLUTION, *TOXIC   W69-05306
AREA(LAKES), PHAEDACTYLUM, WASTE   W69-09340
AREAL INCOME(NUTRIENTS), LAKE VOL   W70-00270
AREAS, BLONDIN.: /EVOLUTION, *INT   W70-04430
AREAS, CARBOHYDRATES, YEASTS, ENZ   W70-03952
AREAS, HORIZONTAL, BLOOMS, GELATI   W70-06229
AREAS, PHYSIOLOGICAL ECOLOGY, LIT   W70-03325
AREAS, SOIL BACTERIA, LABORATORY   W70-04365
ARID CLIMATES, SEMIARID CLIMATES,   W69-03491
ARID LANDS, LIMNOLOGY.: / MONITOR   W70-05645
ARID LANDS, DEPTH, CYANOPHYTA, CH   W71-04067
ARKANSAS RIVER, MERRIMACK RIVER,   W70-04507
ARKANSAS RIVER(OKLA), TULSA(OKLA)   W70-04001
AROMATIC SOLVENTS, PHAEODACTYLUM   W70-01470
ARSENATE INTERFERENCE, DISSOLVED   W70-05269
ARSENIC COMPOUNDS, SPECTROPHOTOME   W70-05269
ARSENIC COMPOUNDS, NITROGEN, PHOS   W70-04494
ARSENITE.: /OMPOUNDS, SPECTROPHOT   W70-05269
ARSENITE, NUISANCE ALGAE, SNAILS,   W70-05568
ARSENITE, FISH, ANIMALS, ALGAE, C   W70-05263
ARSENITE, COPPER SULPHATE, MONURO   W70-00269
ARSENITE,: /ECHANICAL CONTROL, BI   W70-10175
ARTEMIA, BALANUS, DESORPTION, EMB   W69-07440
ARTEMIA, CARTERIA, NORTH TEMPERAT   W70-00708
ARTHROBACTER, FLORA, AMBIENT TEMP   W70-00713
ARTHROBACTER, PYROPHOSPHATE, ORTH   W69-07428
ARTHROPODA, CHIRONOMIDAE, AQUATIC   W70-04253
ARTHROPODS.: /AXIDOMUS, CLOSTRIDI   W70-05372
ARTHROPODS.: /PODS, SPIDERS, MILK   W72-01819
ARTHUR D LITTLE INC(MASS), CARNEG   W69-06865
ARTIFICIAL FERTILIZATION, COMMERC   W68-00487
ARTIFICIAL WATER COURSES, CALIFOR   W71-09577
ARTIFICIAL CIRCULATION, NUISANCES   W69-10178
ARTIFICIAL KEY.: /DRIMACULATUS, M   W70-06225
ASBESTOS CEMENT, CHLORELLA, CARBO   W71-07339
ASCOPHYLLUM NODOSUM, LAMINARIA DI   W70-01073
ASCOPHYLLUM NODOSUM, ECTOCARPUS C   W70-01074
ASCOPHYLLUM NODOSUM, ALGINIC ACID   W69-07826
ASCOPHYLLUM NODOSUM, POLYSIPHONIA   W70-04365
ASCOSPOROGENOUS.: /BE(GERMANY), S   W70-04368
ASHKELON(ISRAEL), TEMPERATURE PRO   W70-04790
ASHLANDI, EPISCHURA NEVADENSIS, D   W69-10182
ASHTABULA.:   W70-09093
ASIA, NATIONAL RESEARCH COUNCIL(U   W70-03975

393

S(USA), UNIVERSITY OF WISCONSIN,　　ASIA, NATIONAL RESEARCH COUNCIL(U　W70-03975
EDIES, JUDICIAL DECISIONS, LEGAL　　ASPECTS.: /, STREAMS, SEWAGE, REM　W69-06909
EFFICIENCY, HEAT BUDGET, SOCIAL　　ASPECTS.: /ON, TURBINES, ECONOMIC　W70-08834
RIA, FISH, LIMNOLOGY, HYDROLOGIC　　ASPECTS, AQUATIC ENVIRONMENT, WAT　W70-04475
T-OF-WAY, ESTIMATED COSTS, LEGAL　　ASPECTS, LEGISLATION, SEWAGE DIST　W70-03344
A CYLINDRICA, BACILLUS SUBTILIS,　　ASPERGILLUS ORYZAE NIGER, S: /AEN　W70-02964
VULGARIS, KARATELLA COHCLEARIS,　　ASPLANCHA, LECANE, BOSMINA, NAVIC　W70-02249
ROCYSTIS, KERATELLA, POLYARTHRA,　　ASPLANCHNA, BRANCHIONUS.: /A, MIC　W70-05375
UTROPHICATION, PROVISIONAL ALGAL　　ASSAY PROCEDURE, BATCH CULTURES,　W70-02777
BACTERIA, FUNGI, AQUATIC PLANTS,　　ASSAY, DISTRIBUTION, DENSITY, CHR　W70-02510
S, *AGRICULTURE, *WASTES, ALGAE,　　ASSAY, DISSOLVED SOLIDS, CHEMICAL　W70-03334
Y CONSUMPTION, PROVISIONAL ALGAL　　ASSAY, PROCEDURES, SELENASTRUM CA　W70-02779
LENA GRACILIS Z, MICROBIOLOGICAL　　ASSAY, SPECIES COMPOSITION.: /EUG　W69-06273
ON(WIS), EUROPE, WATER POLLUTION　　ASSESSMENT, URBAN DRAINAGE, AGRIC　W70-03975
ON(WIS), EUROPE, WATER POLLUTION　　ASSESSMENT, URBAN DRAINAGE, AGRIC　W70-03975
N, ALGAL GROWTH, WATER POLLUTION　　ASSESSMENT, SANITARY ENGINEERING　W70-02775
BIOSTIMULATION, WATER POLLUTION　　ASSESSMENT.: /L GROWTH, ANABAENA,　W70-02779
US RIVER(CALIF), WATER POLLUTION　　ASSESSMENT, AMERICAN RIVER(CALIF)　W70-02777
OSA, SELENASTRUM GRACILE, CARBON　　ASSIMILATION, BACILLARIOPHYCEAE,　W70-03983
NITROGEN　　ASSIMILATION.:　W69-06970
(WASH), *REACTOR, RADIONUCLIDES,　　ASSIMILATION, BA-LA, HALF-LIFE, P　W69-07853
OPIC LEVEL, WATER QUALITY, WASTE　　ASSIMILATIVE CAPACITY, STANDING C　W68-00912
LUTION, MARINE ORGANISMS, MARINE　　ASSOCIATIONS, MOSCOW, PETROLEUM P　W69-07440
DOR, TAST/ *AMERICAN WATER WORKS　　ASSOCIATION, CHEMICAL INDUSTRY, O　W69-05943
IC ALGAE, *EPIPELIC ALGAE, ALGAL　　ASSOCIATIONS, SHEAR WATER(ENGLAND　W71-12870
PORTULACOIDES, SUAEDA MARITIMA,　　ASTER TRIPOLIUM, PEMBROKE, SOUTH-　W70-09976
), CARLETON POND(MAINE), SYNURA,　　ASTERIONELLA, TYPHOID FEVER, KENN　W70-08096
NS, WESTHAMPTON LAKE(VA).:　　ASTERIONELLA SPIROIDES, CLADOCERA　W71-10098
AGILARIA CROTONENSIS, NITZSCHIA,　　ASTERIONELLA GRACILLIMA, GROWTH F　W70-02804
ENT LIMITATION, CAYUGA LAKE(NY),　　ASTERIONELLA FORMOSA, SYNEDRA, ME　W70-01579
RECTION, TABELLARIA, FRAGILARIA,　　ASTERIONELLA, FILTER CLOGGING, ST　W70-00263
ENODINIUM, CRYPTOMONAS, CILIATA,　　ASTERIONELLA FORMOSA, RECOLONIZAT　W70-00268
UIREMENTS, THRESHOLD ODOR TESTS,　　ASTERIONELLA, APHANIZOMENON, SYNU　W69-10157
XCRETION, STEPHANODISCUS TENUIS,　　ASTERIONELLA FORMOSA, MELOSIRA BI　W69-10158
HROMOBACTER, GULF OF MARSEILLES,　　ASTERIONELLA JAPONICA, CHAETOCERO　W70-04280
LTRA-VIOLET LIGHT STERILIZATION,　　ASTERIONELLA, SCENEDESMUS OBLIQUU　W70-03955
ALGAE, GREEN ALGAE, FRAGILARIA,　　ASTERIONELLA, MELOSIRA, MICROCYST　W70-05375
ANACYSTIS, MICROCYSTIS, BOSMINA,　　ASTERIONELLA, MELOSIRA, SYNEDRA,　W70-04503
TATUM, PHYSIOLOGICAL ADJUSTMENT,　　ASTERIONELLA FORMOSA, CYCLOTELLA　W70-04809
ANIZOMENON FLOS AQUAE, MELOSIRA,　　ASTERIONELLA FORMOSA, FRAGILARIA,　W70-05112
TAE, HETEROKONTAE, GLOEOTRICHIA,　　ASTERIONELLA, DINOBRYON, MELOSIRA　W70-05409
TON CORER, FRAGILARIA, MELOSIRA,　　ASTERIONELLA, ARAPHIDINEAE, CENTR　W70-05663
MENON FLOS AQUAE, STEPHANODISCUS　　ASTREA, LAKE MINNETONKA(MINN), MO　W70-05387
APHICAL REGIONS, COASTAL PLAINS,　　ATLANTIC COASTAL PLAIN, INORGANIC　W69-03695
LOSIONS, ABSORPTION, ADSORPTION,　　ATLANTIC OCEAN, DIATOMS, NERITIC　W70-00707
COPPER, IRON, COBALT, MANGANESE,　　ATLANTIC OCEAN, PUERTO RICO, SPEC　W69-08525
ION, ABSORPTION, GULF OF MEXICO,　　ATLANTIC OCEAN, RADIOACTIVITY, GA　W69-08524
ARINE ALGAE, AQUATIC LIFE, MILK,　　ATLANTIC OCEAN, LANDFILLS, DISPOS　W72-00941
UTROPHICATION, *WATER STRUCTURE,　　ATMOSPHERE, SEA WATER, FRESH WATE　W70-04385
RUNOFF, UNDERFLOW, PRECIPITATION(　　ATMOSPHERIC), DRAINAGE, FERTILIZE　W70-04810
ION, ROADS, ROOFS, PRECIPITATION(　　ATMOSPHERIC), WETLANDS, SEEPAGE,　W69-10178
06, ANTIMONY-125, GRO/ *ENIWETOK　　ATOLL, *BIKINI ATOLL, RUTHENIUM-1　W69-08269
5, GRO/ *ENIWETOK ATOLL, *BIKINI　　ATOLL, RUTHENIUM-106, ANTIMONY-12　W69-08269
OCEA, WOODS HOLE OCEAN/ ENIWETOK　　ATOLL, TURNOVER TIME, TRIDACNA CR　W69-08526
ER TECHNIQUE, OXYGEN ELECTRODES,　　ATOMIC ABSORPTION SPECTROPHOTOMET　W70-03309
TIONAL RESEARCH COUNCIL(USA), US　　ATOMIC ENERGY COMMISSION, NATIONA　W70-03975
TIONAL RESEARCH COUNCIL(USA), US　　ATOMIC ENERGY COMMISSION, NATIONA　W70-03975
ACETYLENE/ *ANABAENA CYLINDRICA,　　ATP, CMU, NITROGEN-15, PYRUVATE,　W70-04249
FRONDS, HYDROCARBONS, BACTERIAL　　ATTACHMENT, ECTOCARPUS, POLYSIPH　W70-01068
FILTERS, SOLAR FRAUNHOFER LINES,　　ATTENUATION MEASUREMENT APPARATUS　W70-05377
ATION RATES, MANGANESE, LOW-FLOW　　AUGMENTATION, TENNESSEE RIVER, AL　W73-00750
SLU/ *WATER PURIFICATION, *FLOW　　AUGMENTATION, *LAGOONS, ACTIVATED　W72-06017
, AMPHIPLEURA RUTILANS, AMPHORA,　　AULACODISCUS, BACILLARIA, BIDDULP　W70-03325
CA MILIARIS, HELGOLAND(GERMANY),　　AUREOBASIDIUM PULLULANS, HANSENIA　W70-04368
MONAS AERUGINOSA, STAPHYLOCOCCUS　　AUREUS, LACTOBACILLUS CASEI, BACI　W71-11823
:　　AUSTIN(TEX), COLORADO RIVER(TEX).　W70-04727
UP, FOOD WASTES.:　　SHEPPARTION(　　AUSTRALIA), OXIDATION DITCHES, SO　W71-09548
SK(USSR), ENGLAND, OSCILLATORIA,　　AUSTRALIA, SANTA ROSA(CALIF), LYU　W70-05092
KLOPEINER(AUSTRIA), LAKE WORTHER(　　AUSTRIA), VERTICAL, CHALK, METALI　W70-05405
VERTICAL, CHALK/ *LAKE KLOPEINER(　　AUSTRIA), LAKE WORTHER(AUSTRIA),　W70-05405
LAKE SCHLIER(BAVARIA), LAKE ZELL(　　AUSTRIA), LAKE HALL WILER(SWITZER　W70-05668
NA, EUCYCLOPS, NAIS, CRICOTOPUS,　　AUSTROCLOEON, ENALLAGMA, PSEUDAGR　W70-06975
SSES, CHEMOS/ SPECIES DIVERSITY,　　AUTECOLOGY, MORPHOLOGY, EXPORT LO　W70-02780
UTROPHICATION, *TENNESSEE VALLEY　　AUTHORITY PROJECT, *RESERVOIRS, T　W71-07698
LATES, SAPROBIEN SYSTEM, CILIATE　　AUTOECOLOGY.:　　*CILIATES, FLAGEL　W71-03031
LLA PYRENOIDOSA(EMERSON STRAIN),　　AUTOINHIBITORS.: / GROWTH, CHLORE　W69-07442
DUCTS, HORMONES, AUXINS, METABO/　　AUTOINHIBITORS, EXTRACELLULAR PRO　W69-04801
GERMANY), SPOROBOLOMYCES ROSEUS,　　AUTOLYSIS, ASCOSPOROGENOUS.: /BE(　W70-04368
PHYCOCOLLOIDS, TA/ *ANTIBIOSIS,　　AUTOLYSIS, EXCRETION, DEPURATION,　W70-01068
E, FLOW RATES, SLUDGE, SLURRIES,　　AUTOMATIC CONTROL, PIPING SYSTEMS　W72-08688
LYSIS, TRICARBOXYLIC ACID CYCLE,　　AUTORADIOGRAPHY, MONODUS SUBTERRA　W70-02965
HEMICAL EXTRACTION, MILL/ *MICRO-　　AUTORADIOGRAPHY, PHYSIOLOGY, BIOC　W69-10163
FILM, SYNECHOCOCCUS, GRAIN COUNT(　　AUTORADIOGRAPHY), LEUCOTHRIX MUCO　W69-10163
MICROSCOPY, ACETATE, CARBON-14,　　AUTORADIOGRAPHY, MICHAELIS-MENTEN　W70-04811
BO/ HETEROTROPHIC, CHEMOTROPHIC,　　AUTOTROPHIC, SILICON, PEDIASTRUM　W70-02804
UTRIENTS, CHLORELLA PYRENOIDOSA,　　AUTOTROPHIC GROWTH, HETEROTROPHIC　W70-02964
MATHEMATICAL STUDIES,　　AUTOTROPHIC, HETEROTROPHIC.:　W70-05750
ED ORGANIC MATTER, ARCTIC ALGAE,　　AUTOTROPHIC METABOLISM, HETEROTRO　W71-10079
N UPTAKE TECHNIQ/ *HETEROTROPHY,　　AUTOTROPHY, CARBON-14, RADIOCARBO　W70-04287
XTRACELLULAR PRODUCTS, HORMONES,　　AUXINS, METABOLITES, INHIBITION,　W69-04801

ROCYSTIS AERUGINOSA, / *NUTRIENT   AVAILABILITY, *ALGAL GROWTH, *MIC   W69-05867
ICROCYSTIS AERUGINOSA, *NUTRIENT   AVAILABILITY, ALGAL GROWTH, ALGAL   W69-05868
IAL, SE/ *ALGAL GROWTH, NUTRIENT   AVAILABILITY, ALGAL GROWTH POTENT   W69-06864
GETATION, NITROGEN, PHOSPHOROUS,   AVAILABLE NUTRIENTS, SEDIMENT COR   W68-00511
OSPHORUS, PHOSPHATE(BIOLOGICALLY   AVAILABLE), MICROCYSTIS AERUGINOS   W70-05269
, *FIXED-LIGHT TESTS, *EFFECTIVE-   AVERAGE-LIGHT-INTENSITY, PHOTO-EL   W70-03923
IUM PARVUM, SYNERGISTIC EFFECTS,   AXENIC CULTURES, MONOALGAL, PORTU   W70-03923
SALTS, NITROGEN FIXATION, BORON,   AZOTOBACTER, MOLYBDENUM, COBALT,   W70-02964
OIDOSA, BICHOROMATE, DRY WEIGHT,   B COLI COMMUNAE, EUGLENE GRACILIS   W69-05092
WATER POLLUTION SOURCES, VITAMIN   B.: /S, WATER POLLUTION EFFECTS,   W69-06277
CATION, ORGANIC SOLUTES, VITAMIN   B-1, VITAMIN B-12, THIAMINE, SERR   W69-05652
STOC MUSCORUM, VANADIUM, VITAMIN   B-12.: /IPLOCYSTIS AERUGINOSA, NO   W69-06277
N, EUGLENA GRACILIS Z,/ *VITAMIN   B-12, ALGAL GROWTH, CYANOCOBALAMI   W69-06273
IC SOLUTES, VITAMIN B-1, VITAMIN   B-12, THIAMINE, SERRATIA MARINORU   W70-05652
, *SEA WATER, *VITAMINS, VITAMIN   B, ALGAE, CULTURES, CHEMICAL ANAL   W70-05652
IES, BACTERIA, MOLLUSKS, VITAMIN   B, RED TIDE, SALINITY, FLORIDA, T   W70-05372
OR, RADIONUCLIDES, ASSIMILATION,   BA-LA, HALF-LIFE, PTYCHOCHEILUS O   W69-07853
RUTILANS, AMPHORA, AULACODISCUS,   BACILLARIA, BIDDULPHIA, CHAETOCER   W70-03325
UM GRACILE, CARBON ASSIMILATION,   BACILLARIOPHYCEAE, DINOPHYCEAE.: /   W70-03983
CONJUGATOPHYCAEA, PLANCTOMYCES,   BACILLARIOPHYTA, GLOECAPSA, MERIS   W70-03969
E, ALGAL GROWTH, ANKISTRODESMUS,   BACILLARIOPHYCEAE, CHLOROPHYCEAE,   W69-07833
CULA PELLICULOSA, XANTHOPHYCEAE,   BACILLARIOPHYCEAE, NITZSCHIA PALE   W70-02965
CHIZOMYCETES, CYANOPHYTA, FUNGI,   BACILLARIOPHYCEAE, CHAROPHYTA, RH   W70-02784
PRICORNUTUM PRINTZ, OSCILLATORI/   BACILLARIOPHYCEAE, SELENASTRUM CA   W70-05752
CYLINDRICA, ZINC, XANTHOPHYCEAE,   BACILLARIOPHYCEAE, MYXOPHYCEAE, C   W69-10180
YCOBACTERIUM, BACILLUS SUBTILIS,   BACILLUS MYCOIDES, BACILLUS MESEN   W69-07838
LUS SUBTILIS, BACILLUS MYCOIDES,   BACILLUS MESENTERICUS, DIATOMEA,   W69-07838
ION, PSEUDOMONAS, MYCOBACTERIUM,   BACILLUS SUBTILIS, BACILLUS MYCOI   W69-07838
A GRACILIS, ANABAENA CYLINDRICA,   BACILLUS SUBTILIS, ASPERGILLUS OR   W70-02964
OTH/ RICKETTSIAE, STERILIZATION,   BACILLUS SUBTILIS, BACILLUS STEAR   W71-11823
TERILIZATION, BACILLUS SUBTILIS,   BACILLUS STEAROTHERMOPHILUS, PROT   W71-11823
CUS AUREUS, LACTOBACILLUS CASEI,   BACILLUS THURINGIENSIS, AGARS, ST   W71-11823
POSITIVE BACTERIA, SARCINE-LIKE,   BACILLUS-LIKE, MICROCOCCUS-LIKE,   W70-00713
IDLEY CREEK(PA), NOBS CREEK(MD),   BACK RIVER(MD), OSCILLATORIA, CAT   W70-04510
LITY TEST, BODY IMMEDIUM FILTER,   BACK-WASHING.: / FLOW, PUTRESCIBI   W70-05821
ING, PERIPHYTON, AQUATIC PLANTS,   BACKGROUND RADIATION, GEOCHEMISTR   W69-09334
RADIOISOTOPES, ALGAE, EMULSIONS,   BACKGROUND RADIATION, IODINE RADI   W69-10163
, ALGAE, INSECTS, INVERTEBRATES,   BACKGROUND RADIATION, FLOCCULATIO   W69-07846
COMBINED SEWERS, STORMWATER OVE/   BACKWASHING, *SUSPENDED SOLIDS, *   W71-10654
ALUM, *OSHKOSH(WIS),   BACKWASHING.:   W72-08859
TROPHICATION, PONDS, RESERVOIRS,   BACKWATER, FERTILIZATION, NITROGE   W70-00274
DIATOMS, BIOLOGICAL COMMUNITIES,   BACKWATER, PONDS, ALASKA.: /YTA,   W70-03969
ER, CHEMICAL ANALYSIS, PROTEINS,   BACTERIA.: /OMPOSING ORGANIC MATT   W72-01812
AQUATIC LIFE, STREAM POLLUTION,   BACTERIA, IMPURITIES, WASTE TREAT   W72-01786
RICKLING FILTERS, *BIOTA, SEWAGE   BACTERIA, SYSTEMATICS, FUNGI, ALG   W72-01819
POLLUTION EFFECTS, FOOD CHAINS,   BACTERIA, DEGRADATION(DECOMPOSITI   W72-01801
TERIA, AQUATIC PLANTS, ANAEROBIC   BACTERIA, ALGAE, BIOCHEMICAL OXYG   W71-08221
, FISH, DIATOMS, FUNGI, INSECTS,   BACTERIA, PROTOZOA, SEWAGE, LAKES   W71-10786
MMING, *WATER POLLUTION EFFECTS,   BACTERIA, ALGAE, FISH, SEWERS, RU   W71-09880
ATURE, SEWAGE TREATMENT, AEROBIC   BACTERIA, AQUATIC PLANTS, ANAEROB   W71-08221
RIENTS, LAKES, WATER PROPERTIES,   BACTERIA, BIOASSAY, METABOLISM, A   W71-10079
HEATED WATER, THERMAL POLLUTION,   BACTERIA, EFFLUENTS, OXYGEN DEMAN   W72-00462
EN, PHOSPHOROUS, NITROGEN FIXING   BACTERIA, *ACTIVATED SLUDGE, STOR   W71-13335
*CULTURES, *BIOASSAYS, PROTOZOA,   BACTERIA, FUNGI, VIRUSES, ALGAE,   W71-11823
IA, DRAINAGE PROGRAMS, ANAEROBIC   BACTERIA, ANAEROBIC CONDITIONS, L   W71-04556
IO, *WATER QUALITY, *MONITORING,   BACTERIA, WATER CHEMISTRY, PHOSPH   W71-05879
ADATION, CHLORINATION, COLIFORM,   BACTERIA, BIOCHEMICAL OXYGEN DEMA   W71-05013
YGEN DEMAND, *ANAEROBICS, ALGAE,   BACTERIA, WATER POLLUTION SOURCES   W71-05742
IA, DRAINAGE PROGRAMS, ANAEROBIC   BACTERIA, ANEROBIC CONDITIONS, LA   W71-04554
CARBON, WATER POLLUTION EFFECTS,   BACTERIA, PERIPHYTON, OHIO.: /R,   W71-04206
IA, DRAINAGE PROGRAMS, ANAEROBIC   BACTERIA, ANAEROBIC CONDITIONS, L   W71-04557
T, INDUSTRIAL WASTES, NUTRIENTS,   BACTERIA, ALGAE, TREATMENT FACILI   W71-06737
ERIE, BIOCHEMICAL OXYGEN DEMAND,   BACTERIA, ECOLOGY, WATER POLLUTIO   W71-04758
IA, DRAINAGE PROGRAMS, ANAEROBIC   BACTERIA, ANAEROBIC CONDITIONS, L   W71-04555
S, ACHROMOBACTERS, GRAM-POSITIVE   BACTERIA, SARCINE-LIKE, BACILLUS-   W70-00713
COCHEMICAL PROPERTIES, GENETICS,   BACTERIA, PEPTIDES, ENVIRONMENTAL   W70-00273
OXIDE, *REVIEWS, BIBLIOGRAPHIES,   BACTERIA, DECOMPOSING ORGANIC MAT   W70-00664
LUDGE WORMS, BLOODWORMS, AQUATIC   BACTERIA, ALGAE, MAYFLIES, CADDIS   W70-00475
ZOOPLANKTON, PHYTOPLANKTON, ICE,   BACTERIA, TRIPTON, BOTTOM SEDIMEN   W69-10154
, TRAVERTINE, CALCIUM CARBONATE,   BACTERIA, BIOMASS, CHLOROPHYLL, P   W69-10160
IMNOLOGY, OCEANOGRAPHY, TRACERS,   BACTERIA, GROWTH RATES, TRITIUM,   W69-10163
BIODEGRADATION, *ALGAE, *AQUATIC   BACTERIA, DETERGENTS, SURFACTANTS   W69-09454
RUS, ORGANOPHOSPHORUS COMPOUNDS,   BACTERIA, DIATOMS, DETRITUS, CYCL   W69-09334
ERMOCLINE, ZOOPLANKTON, TRIPTON,   BACTERIA, SEDIMENTATION, COPEPODS   W69-10169
OTOSYNTHESIS, CHEMICAL ANALYSIS,   BACTERIA, NUTRIENTS, TEMPERATURE,   W69-10167
RICIDES, *CHLORINATION, *AQUATIC   BACTERIA, *AQUATIC MICROBIOLOGY,   W69-10171
TROGEN COMPOUNDS, SEWAGE, SEWAGE   BACTERIA, SEWAGE TREATMENT, WATER   W69-10171
ODUCTIVITY, *ZOOPLANKTON, SULFUR   BACTERIA, DIEL MIGRATION, CARBON   W70-05760
S, PYCNOCLINE, PHOTOLITHOTROPHIC   BACTERIA, DESULFOVIBRIO, MESOTHER   W70-05760
TORMS, TURBULENCE, PILOT PLANTS,   BACTERIA, SEDIMENTATION, TOXICITY   W70-05819
LAKES, NUTRIENTS, FERTILIZATION,   BACTERIA, ALGAE, PLANKTON, TUBIFI   W70-05761
AGOONS, *BIODEGRADATION, *SEWAGE   BACTERIA, ODOR, ANAEROBIC CONDITI   W71-02009
IDE, SALINITY, LAKES, NUTRIENTS,   BACTERIA, ALGAE, BRACKISH WATER,   W71-03031
NS, AEROBIC CONDITIONS, E. COLI,   BACTERIA, FUNGI, ACTINOMYCETES, P   W71-02009
RMAL POLLUTION, *ECOLOGY, ALGAE,   BACTERIA, COLIFORMS, FISH, FUNGI,   W71-02479
TRIENTS, *BIODEGRADATION, ALGAE,   BACTERIA, AQUATIC MICROORGANISMS,   W71-00221
ENTRATION, SCENEDESMUS, EUGLENA,   BACTERIA, YEASTS, ACTINOMYCETES,   W70-08097
OTABLE WATER, CYANOPHYTA, SULFUR   BACTERIA, COLOR, IRON BACTERIA, T   W70-08107
TION-REDUCTION POTENTIAL, SULFUR   BACTERIA, *WASTE WATER TREATMENT,   W70-07034
TA, SULFUR BACTERIA, COLOR, IRON   BACTERIA, TASTE, FILTERS, IOWA, D   W70-08107

HLORELLA, CHELATION, OCHROMONAS,
  LAKES, SEWAGE TREATMENT, SEWAGE
POLYSIPHONIA HARVEYI, EPIPHYTIC
  RES, LAKES, ALGAE, GROWTH RATES,
, INDUSTRIAL WASTES, NEUTRIENTS,
BOLISM, WIND VELOCITY, HUMIDITY,
S, ELECTROLYTES, ZETA POTENTIAL,
  OYSTERS, INDUSTRIES, CHELATION,
TERGENTS, *INSECTICIDES, *MARINE
TES, DISSOLVED OXYGEN, DETRITUS,
EN, SOLAR RADIATION, PERIPHYTON,
TS, ORGANIC MATTER, IRON OXIDES,
PHYTA, CHLOROPHYTA, CHLOROPHYLL,
  CROP, ALGAE, ZOOPLANKTON, FISH,
DATION LAGOONS, *DESIGN, *SULFUR
OLIMNOLOGY, MATHEMATICAL MODELS,
OLIMNOLOGY, MATHEMATICAL MODELS,
, PHYSIOLOGICAL ECOLOGY, AQUATIC
PHYTOPLANKTON, DIFFUSION, ALGAE,
TITION, *PHYTOPLANKTON, *AQUATIC
IDE, INHIBITION, NITROGEN FIXING
CELLULOSE, CHLORELLA, PEPTIDES,
ENT REQUIREMENTS, *DISTRIBUTION,
MICROBIAL ECOLOGY, PURPLE SULFUR
ALGAE, ENTERIC BACTERIA, METHANE
N SULFIDE, LAKES, ALGAE, ENTERIC
, DIFFUSION, TURBULENCE, TOXINS,
BIOINDICATORS, FUNGI, ESTUARIES,
PIPTIDES, INTERTIDAL AREAS, SOIL
TESTING, CAPILLARY CONDUCTIVITY,
LYTICAL TECHNIQUES, TEM/ *MARINE
DIOISOTOPES, TEMPERATURE, DEPTH,
HOS, HYDROGEN, BACTERIA, AEROBIC
SUPPLY, MUD, BENTHOS, HYDROGEN,
NIA, NITRITES, NITRATES, YEASTS,
MATTER, AIR TEMPERATURE, AEROBIC
RATES, YEASTS, BACTERIA, AQUATIC
XPLOSIONS, BIOTA, SPECTROMETERS,
MAND, TERTIARY TREATMENT, ALGAE,
RADIOISOTOPES, *TRANSFER, WATER,
G NUTRIENTS, GEORGIA, SEDIMENTS,
GRAPHIES, LAKES, NUISANCE ALGAE,
NT, *AQUICULTURE, ALGAE, YEASTS,
NTATION, METHANE, ALGAL CONTROL,
FECTS, DISSOLVED OXYGEN, AQUATIC
ION, *HYPOLIMNION, ALGAE, FUNGI,
ATES, BRACKISH WATER, ESTUARIES,
L SPRINGS, *OREGON, THERMOPHILIC
ER TREATMENT, *NUTRIENTS, ALGAE,
HIBITION, CYANOCOBALAMIN, MUTANT
HNIQUES, DINOFLAGELLATES, MARINE
REAMS, WATER POLLUTION, ANIMALS,
TH, FROGS, CRAYFISH, MOSQUITOES,
POTABLE WATER, WATER POLLUTION,
AL TECHNIQUES, KINETICS, AQUATIC
HNIQUES, *MARINE MICRO/ *AQUATIC
MICROORGANISMS, *ALGAE, *AQUATIC
ON, *ALGAE, CHEMICAL PROPERTIES,
VIRONMENT, WATER QUALITY, SEWAGE
CHNIQUES, AQUATIC ALGAE, AQUATIC
NE ALGAE, MARINE ANIMALS, MARINE
CHNIQUES, AQUATIC ALGAE, AQUATIC
LAKES, *CELLULOSE, DI/ *ECOLOGY,
CHNIQUES, AQUATIC ALGAE, AQUATIC
N, ALGAE, AQUATIC ALGAE, AQUATIC
TTER, CONDUCTIVITY, TEMPERATURE,
ANITARY ENGINEERING, *PATHOGENIC
HORUS, SCENEDESMUS, CHLOROPHYTA,
*BIOMASS, *SLIME, ALGAE, FUNGI,
COLI, PUBLIC HEALTH, PATHOGENIC
UDGE, ALGAE, NUTRIENTS, VIRUSES,
BACTERIA, ACID BACTERIA, SULFUR
L WASTES, METHANE BACTERIA, ACID
TION, *MUNICIPAL WASTES, METHANE
                GREAT LAKES BASIN,
TAKE VELOCITY, ACTIVE TRANSPORT,
SARGASSUM, FRONDS, HYDROCARBONS,
, RADIOCARBON UPTAKE TECHNIQUES,
AKE TECHNIQUES, BACTERIAL PLATE,
, TAXONOMIC TYPES, HETERETROPHY,
BLING TIME, PACKED CELL VOLUMES,
ATURE, ALKALINITY, BICARBONATES,
E MICROORGANISMS, PHYTOPLANKTON,
ORELLA, *EUGLENA, ALGAL CONTROL,
*SLIME, ALGAE, FUNGI, BACTERIA,
.:   ORGANIC NITROGEN,
XYGEN, ALGAE, SELF-PURIFICATION,
RA, THALASSIOSIRA, COSCINDISCUS,
ORGANIC MATTER, INSECTS, MITES,
Y, HETEROTROPHY, MICRONUTRIENTS,

BACTERIA, FUNGI, AQUATIC PLANTS,	W70-02510
BACTERIA, NITROGEN, PHOSPHORUS, A	W70-02787
BACTERIA, SUBLITTORAL, YELLOW MAT	W70-01073
BACTERIA, ULTRAVIOLET RADIATION,	W70-00719
BACTERIA, ALGAE, TREATMENT FACILI	W70-01519
BACTERIA, HYDROGEN ION CONCENTRAT	W70-01073
BACTERIA, PROTOZOA, ALGAE, SOLID	W70-01027
BACTERIA, STREPTOCOCCUS, SECONDAR	W70-01068
BACTERIA, *TOXICITY, *WATER POLLU	W70-01467
BACTERIA, WATER POLLUTION EFFECTS	W70-02249
BACTERIA, SPECT: /EFFLUENTS, OXYG	W70-01933
BACTERIA, ALGAE, FUNGI, SULFUR CO	W70-02294
BACTERIA, TRACE ELEMENTS, VITAMIN	W70-02255
BACTERIA, BENTHIC FAUNA, PHOTOSYN	W70-02251
BACTERIA, STABILIZATION, WASTE WA	W70-01971
BACTERIA, PHYTOPLANKTON, GEOCHEMI	W70-03975
BACTERIA, PHYTOPLANKTON, GEOCHEMI	W70-03975
BACTERIA, CARBON RADIOISOTOPES, E	W70-04287
BACTERIA, EUTROPHICATION, GLUCOSE	W70-04194
BACTERIA, *ALGAE, ORGANIC MATTER,	W70-04284
BACTERIA, RADIOACTIVITY TECHNIQUE	W70-04249
BACTERIA, SNAILS, WEIGHT, VACUUM	W70-04184
BACTERIA, SAMPLING, SEA WATER, SE	W70-03952
BACTERIA, POPULATION EQUIVALENTS(	W70-03312
BACTERIA, CHEMICAL ANALYSIS, WATE	W70-03312
BACTERIA, METHANE BACTERIA, CHEMI	W70-03312
BACTERIA, COPPER SULFATE, CIRCULA	W70-04369
BACTERIA, SAMPLING, SEASONAL, HYD	W70-04368
BACTERIA, LABORATORY TESTS.: /E,	W70-04365
BACTERIA, DIATOMS, TEST PROCEDURE	W70-04469
BACTERIA, *ALGAE, *ANALYSIS, *ANA	W70-04365
BACTERIA, SLIME, ALGAE, CYANOPHYT	W70-04371
BACTERIA, METABOLISM, CANALS, SAN	W69-07838
BACTERIA, AEROBIC BACTERIA, METAB	W69-07838
BACTERIA, AQUATIC BACTERIA, WATER	W69-06970
BACTERIA, WATER TEMPERATURE, IRON	W69-07096
BACTERIA, WATER POLLUTION EFFECTS	W69-06970
BACTERIA, ALGAE, SANDS, HYDROGEN	W69-08267
BACTERIA, ACTIVATED SLUDGE.: / DE	W69-08054
BACTERIA, SEDIMENTS, MINNOWS, DET	W69-07861
BACTERIA, DETRITUS, ESTUARIES, PL	W69-08274
BACTERIA, WATER POLLUTION EFFECTS	W69-08283
BACTERIA, FISH, PROTEINS, OIL WAS	W69-05891
BACTERIA, COLIFORMS.: /ION, FERME	W70-04787
BACTERIA, ALGAE.: /R POLLUTION EF	W70-04727
BACTERIA, FISH, LIMNOLOGY, HYDROL	W70-04475
BACTERIA, MOLLUSKS, VITAMIN B, RE	W70-05372
BACTERIA, THERMAO POLLUTION, NUTR	W70-05270
BACTERIA, CULTURES, CHEMICAL OXYG	W70-05655
BACTERIA, CYCLOTELLA NANA, GEIGER	W70-05652
BACTERIA, WATER CHEMISTRY, CARBON	W70-05652
BACTERIA, FUNGI, ALGAE, FLOATING	W70-05750
BACTERIA, IRRIGATION, WILLOW TREE	W70-05263
BACTERIA, ALGAE, NUTRIENTS, INDIC	W70-05264
BACTERIA, ALGAE, PHYSIOLOGICAL EC	W70-05109
BACTERIA, *ALGAE, *ANALYTICAL TEC	W70-04811
BACTERIA, *AQUATIC FUNGI, YEASTS,	W69-00096
BACTERIA, ORGANIC POLLUTANTS, *ST	W68-00483
BACTERIA, ENVIRONMENTAL EFFECTS,	W69-04276
BACTERIA, AQUATIC MICROBIOLOGY, B	W69-03371
BACTERIA, MARINE FUNGI.: /S, MARI	W69-03752
BACTERIA, AQUATIC MICROBIOLOGY, B	W69-03362
BACTERIA, ALGAE, PHYTOPLANKTON, *	W69-00976
BACTERIA, AQUATIC MICROBIOLOGY, B	W69-03188
BACTERIA, AQUATIC MICROBIOLOGY, B	W69-03369
BACTERIA, NUTRIENTS, ESTUARINE EN	W73-00748
BACTERIA, ALGAE, ORGANIC MATTER,	W72-13601
BACTERIA, PHOSPHORUS RADIOISOTOPE	W73-01454
BACTERIA, BACTERICIDES, CHLORINE,	W72-12987
BACTERIA, TEMPERATURE, NITRATES,	W72-06017
BACTERIA, E. COLI, PUBLIC HEALTH,	W72-06017
BACTERIA, ALGAE, MIXING, HYDROGEN	W72-06854
BACTERIA, SULFUR BACTERIA, ALGAE,	W72-06854
BACTERIA, ACID BACTERIA, SULFUR B	W72-06854
BACTERIAL POLLUTION.:	W69-03948
BACTERIAL POPULATIONS, ALGAL POPU	W70-04194
BACTERIAL ATTACHMENT, ECOTOCARPUS	W70-01068
BACTERIAL PLATE, BACTERIAL PIGMEN	W70-05760
BACTERIAL PIGMENTS, DIAPTOMUS, EP	W70-05760
BACTERIAL MEDIUM, ARTHROBACTER, F	W70-00713
BACTERIAL PLATE COUNTS, LANCASTER	W71-05267
BACTERICIDE, KINETICS, HARDNESS,	W71-05267
BACTERICIDES, ALGAL TOXINS, LARVA	W70-01068
BACTERICIDES, GROWTH RATES.: /CHL	W69-00994
BACTERICIDES, CHLORINE, COSTS, *W	W72-12987
BACTERIOLOGICAL STUDY, AEROBICITY	W71-05742
BACTERIOPHAGE, RESERVOIRS, SPORES	W69-07838
BACTERIOSTRUM, LEPTOCYLINDRUS, CO	W73-01446
BACTERIS, FUNGI, ALGAE, NITRITES,	W71-00940
BADFISH RIVER(WIS), DETROIT LAKES	W70-04506

ELOSIRA BAICALENSIS, GYMNODINIUM	BAICALENSE, CYCLOTELLA MINUTA.: / W70-04290
, PHOTOSYNTHETIC LAYER, MELOSIRA	BAICALENSIS, GYMNODINIUM BAICALEN W70-04290
OSI/ SEASONAL DISTRIBUTION, LAKE	BAIKAL, PHOTOSYNTHETIC LAYER, MEL W70-04290
TIC MICROBIOLOGY, AQUATIC WEEDS,	BALANCE OF NATURE, EUTROPHICATION W69-04804
BACTERIA, AQUATIC MICROBIOLOGY,	BALANCE OF NATURE, BIOASSAY, CHLO W69-03188
TIC PRODUCTIVITY, AQUATIC WEEDS,	BALANCE OF NATURE, BIOASSAY, CHLO W69-03185
BACTERIA, AQUATIC MICROBIOLOGY,	BALANCE OF NATURE, WISCONSIN, BIO W69-03369
TIC ALGAE, AQUATIC MICROBIOLOGY,	BALANCE OF NATURE, BIOASSAY, CHLO W69-03516
TIC ALGAE, AQUATIC MICROBIOLOGY,	BALANCE OF NATURE, BIOASSAY, BIOC W69-03374
ROBIOLOGY, AQUATIC PRODUCTIVITY,	BALANCE OF NATURE, CYCLING NUTRIE W69-03513
TIC ALGAE, AQUATIC MICROBIOLOGY,	BALANCE OF NATURE, CYCLING NUTRIE W69-03517
TIC MICROBIOLOGY, AQUATIC WEEDS,	BALANCE OF NATURE, BIOASSAY, CHLO W69-03515
VIRONMENT, AQUATIC MICROBIOLOGY,	BALANCE OF NATURE, BIOASSAY, CHLO W70-02968
IC PLANTS, AQUATIC PRODUCTIVITY,	BALANCE OF NATURE, BIOASSAY, CHLO W70-02792
WEED BEDS, PHOSPHORUS, NITROGEN	BALANCE.: /S), LAKE KEGONSA(WIS), W70-02969
U/ *CULTURAL EUTROPHY, *NUTRIENT	BALANCE, *LAKE AGING, NUTRIENT SO W70-00264
ITY, ALGAE, STRATIFICATION, SALT	BALANCE, BIBLIOGRAPHIES.: / SALIN W72-01773
IC APP/ POLYPHOSPHATES, NUTRIENT	BALANCE, FOAMING DETERGENTS, PUBL W70-05084
TIPOLLUTION PROGRAMS, BIOLOGICAL	BALANCE, OVERFISHING, COHO SALMON W70-04465
WATER POLLUTION CONTROL, OXYGEN	BALANCE, SEWAGE, RUNOFF.: /URCES, W69-08668
ENTRATION, COMPOSITION, MATERIAL	BALANCES, ORGANISMS.: /ELLS, CONC W70-06600
ACK SEA, CONTACT ZONES, ARTEMIA,	BALANUS, DESORPTION, EMBRYOLOGY, W69-07440
STANCE, PFAFFIKERSEE, TURLERSEE,	BALDEGGERSEE, GREIFENSEE, ZURICHS W70-03964
GESTIBILITY, *IN VITRO, IN VIVO,	BALL-MILL, MEICELASE, HYDROGEN PE W70-04184
KE(ONTARIO), GULL LAKE(ONTARIO),	BALSAM LAKE(ONTARIO), STURGEON LA W70-02795
HAMNION SARNIENSE, RHODOCHORTON,	BANGIA FUSCO-PURPUREA, SPHACELARI W69-10161
*THERMAL	BAR.: W69-05763
E), COBBOSSEECONTEE LAKE(MAINE),	BARGE.: /VER, KENNEBEC RIVER(MAIN W70-08096
EAM CONCOURSE, PALOUSE RIV/ TREE	BARK, CHARCOAL, ALGAL GROWTH, STR W70-03501
E(ENGLAND), LOWESWATER(ENGLAND),	BARNES SOUTH RES: /AND), ESTHWAIT W69-10180
ES, FAXAFLOI FJORD, ICELAND, DNA	BASE COMPOSITION.: / CAPE REYKJAN W69-10161
POTOMAC RIVER	BASIN.: W70-09557
*OHIO RIVER	BASIN.: W72-01788
FEEDLOTS, WABASH RIVER	BASIN.: W71-10376
GREAT LAKES	BASIN, BACTERIAL POLLUTION.: W69-03948
UNDWATER, DDT, MISSISSIPPI RIVER	BASIN, URBANIZATION, ALGAL POISON W73-00758
VICE, GREAT LAKES-ILLINOIS RIVER	BASINS PROJECT, INDIANA HARBOR SH W70-00263
, ALGAE, SEDIMENTATION, SETTLING	BASINS.: /OCHEMICAL OXYGEN DEMAND W70-09320
R, ALGAE, RESERVOIRS, WATERSHEDS(	BASINS), INSECTICIDES, HARDNESS(W W72-01813
EPOSITION(SEDIMENTS), WATERSHEDS(	BASINS), WATER POLLUTION EFFECTS, W69-10182
TION, *PONDS, *SOIL, GROUNDWATER	BASINS, WATER SUPPLY, MUD, BENTHO W69-07838
NTROL, MOSQUITOS, ALGAL CONTROL,	BASS, CARP, CATFISHES, NITROGEN, W70-05417
TS, *ALGAE, *ROTIFERS, PLANKTON,	BASS, PHYTOPLANKTON, PONDS, ZOOPL W70-02249
A, *CHEMCONTROL, EUTROPHICATION,	BASS, WATER POLLUTION CONTROL, MI W70-02982
LS, KILLIFISHES, TROUT, SUCKERS,	BASS, WATER HYACINTH, FROGS, CRAY W70-05263
REITH/ *CARBON FILTERS, LEBANON,	BATAVIA, CLARKSBURG, BUFFALO, DUN W72-03641
TINUOUS CULTURES, CHLAMYDOMONAS,	BATCH CULTURES, OPTICAL DENSITY, W70-04381
CONTINUOUS CULTURE, CHEMOSTATS,	BATCH CULTURES, ACETYLENE REDUCTI W70-02775
OVISIONAL ALGAL ASSAY PROCEDURE,	BATCH CULTURES, CHEMOSTATS, CHLOR W70-02777
Y STUDIES, *NITROGEN LIMITATION,	BATCH CULTURES, CHLORELLA SOROKIN W70-07957
Y, WATER CHEMISTRY, RESPIRATION,	BATHYMETRY, SALINITY, PHOSPHORU: / W70-05387
LOTATION TECHNIQUE, METALIMNION,	BATHYTHERMOGRAPH, ARTHROPODA, CHI W70-04253
POLYELECTROLYTE-COATED FILTERS,	BAUXITE.: W70-01027
ERRIC OXIDE-COATED SAND, CRUSHED	BAUXITE, ACID-WASHED BAUXITE, KAN W70-08628
ND, CRUSHED BAUXITE, ACID-WASHED	BAUXITE, KANPUR(INDIA).: /ATED SA W70-08628
KE TEGERN(BAVARIA), LAKE SCHLIER(	BAVARIA), LAKE ZELL(AUSTRIA), LAK W70-05668
LLIKON(SWITZERLAND)), LAKE TEGERN(	BAVARIA), LAKE SCHLIER(BAVARIA), W70-05668
DOPHORA, ALGAL GENETICS, RUSSIA,	BAVARIA, MORICHES BAY(NY), ONONDA W70-03973
N, CYTOPHAGA, PROTEUS, SERRATIA,	BAY OF FUNDY(CANADA), RHYZOIDS.: / W70-03952
S, NEVADA, *LAKE MEAD, LAS VEGAS	BAY.: /D ORGANIC MATTER, LAS VEGA W72-02817
, POTOMAC ESTUARY, SAN FRANCISCO	BAY.: /IATIONS, SAN JOAQUIN RIVER W69-07520
NITRATES, PHOSPHATES, LAS VEGAS	BAY.: /OIRS, ALGAE, GROUND WATER, W72-02817
N), LAKE JOSEPHINE(MINN), GRAY'S	BAY(MINN), SMITH'S BAY(MINN), PHO W70-05269
MINN), GRAY'S BAY(MINN), SMITH'S	BAY(MINN), PHOSPHATE DETERMINATIO W70-05269
ETICS, RUSSIA, BAVARIA, MORICHES	BAY(NY), ONONDAGA LAKE(NY).: /GEN W70-03973
IVERSITY I/ DESICCATION, YAQUINA	BAY(ORE), CAROTENOIDS, SPECIES, D W70-03325
S, HETERETROPHY, B/ NARRAGANSETT	BAY(RHODE ISLAND), TAXONOMIC TYPE W70-00713
RITIATED THYMIDINE, NARRAGANSETT	BAY(RI), CAPE REYKJANES(ICELAND), W69-09755
LLEY MASTER DRAIN, SAN FRANCISCO	BAY-DELTA.: /RAIN, SAN JOAQUIN VA W71-09577
LONG ISLAND SOUND, NARRAGANSETT	BAY, CAPE REYKJANES, FAXAFLOI FJO W69-10161
TUARIES, *ESTUARINE ENVIRONMENT,	BAYS, *FLUCTUATION, RESISTANCE, L W71-02494
, REGIONS, GEOGRAPHICAL REGIONS,	BAYS, BODIES OF WATER.: /T REGION W69-03683
THER, WATER, MOVEMENT, CULTURES,	BAYS, STREAMS, WEEDS, LITTORAL, W W70-06229
NFRARED RADIATION, DYE RELEASES,	BEACHES, SANDS, BIOTA, WASTE WATE W68-00010
OLLER DRYING, SPRAY-DRIED, GLASS	BEADS, TRICHODERMA, DISINTEGRATIO W70-04184
S, PIVERS ISLAND(N C), DETECTOR,	BEAUFORT(N C), CRASSOSTREA VIRGIN W69-08267
UM IDAHOENSE, TENDIPES PLUMOSUS,	BEAVER DAM LAKE(WIS), DELAVAN LAK W70-06217
C EFFECTS, ANN ARBOR(MICH), SAND	BEDS.: /USPENSIONS, ELECTROKINETI W71-03035
SA(WIS), LAKE KEGONSA(WIS), WEED	BEDS, PHOSPHORUS, NITROGEN BALANC W70-02969
ANIMAL BEHAVIOR, *THERMODYNAMIC	BEHAVIOR.: /OZOA, ALGAE, DIATOMS, W70-07847
PROTOZOA, ALGAE, DIATOMS, ANIMAL	BEHAVIOR, *THERMODYNAMIC BEHAVIOR W70-07847
, RAJGHAT(INDIA), BOMBAY(INDIA),	BENARES(INDIA), MACROPHYTES, CLAD W70-05404
A, SEDIMENTATION, TOXICITY, COST-	BENEFIT ANALYSIS.: /ANTS, BACTERI W70-05819
MMATE, CHLORAMINE, DE-K-PRUF-21,	BENOCLORS, NIGROSINE DYE, SODIUM W70-05263
TEMS, CYTOLOGY, ENVIRONMENTAL C/	BENTHIC ALGAE, BLOOMS, CLOSED SYS W69-07832
RES, WISCONSIN, BOTTOM SEDIMENT,	BENTHIC FAUNA.: /ACE ELEMENTS, CO W69-03366
ALGAE, NUISANCE ALGAE, PLANKTON,	BENTHIC FAUNA, SLUDGE WORMS, BLOO W68-01244
, AQUATIC LIFE, AQUATIC ANIMALS,	BENTHIC FAUNA, AQUATIC PLANTS, MA W71-02494
AE, ZOOPLANKTON, FISH, BACTERIA,	BENTHIC FAUNA, PHOTOSYNTHESIS, BI W70-02251
WASTES, OXIDATION PONDS, ALGAE,	BENTHIC FAUNA, WATERSHEDS, WATER W72-00027

LGAL CONTROL, *NUTRIENTS, ALGAE,
GMENTS, ORGANIC MATTER, BIOMASS,
NTAL EFFECTS, *LIMITING FACTORS,
A, AQUATIC PLANTS, MARINE ALGAE,
SUES, *ALGAE, ECOSYSTEMS, CHARA,
RADIOISOTOPES, IRON, MANGANESE,
PENDED SOLIDS, SALTWATER TONGUE,
L PROPERTIES, ECOLOGY, SALINITY,
ATION, TEMPERATURE, FLOW, ALGAE,
RANSLATIONS, ALGAE, CRUSTACEANS,
PHYTA, CHLOROPHYTA, SCENEDESMUS,
WATER BASINS, WATER SUPPLY, MUD,
LIMITING FACTORS, BENTHIC FLORA,
ECOLOGY, KINETICS, FARM WASTES,
ALGAE, WATER CHEMISTRY, BIOMASS,
*ECOLOGY, BOTTOM SEDIMENTS, MUD,
FE, *AQUATIC POPULATIONS, ALGAE,
Y FROM POLLUTION, *KUWAIT CRUDE,
LUEGILLS, CHLORELLA PYRENOIDOSA,
CONCENTRATION, FLUORIDES, ALKYL
POLYPEPTIDES, PHOTOASSIMILATION,

IVERSITY, EXPERIMENTAL STREAMS,/
UM-106, RHODIUM-106, CERIUM-144,
N, CENTRAL CITY(COLO), MUTATION,
*POTASSIUM PERMANGANATE,
WATER POLLUTION, ORGANIC MATTER,
IENTS, LAKES, CHANNELS, REVIEWS,
ENTS, *CARBON DIOXIDE, *REVIEWS,
CATION, *REVIEWS, DOCUMENTATION,
NSERVATION, POLLUTION ABATEMENT,
E, STRATIFICATION, SALT BALANCE,
ODOR, SCUM, FISHING, WISCONSIN,
LGAE, *RESEARCH AND DEVELOPMENT,
S, *DIATOMS, *ALGAE, ECOLOGY, S/
HICATION, ALGAE, AQUATIC PLANTS,
ROPHICATION, *LAKES, ACID LAKES,
ERTEBRATES, AQUATIC POPULATIONS,
A COLLECTIONS, BOTTOM SEDIMENTS,
INORGANIC COMPOUNDS, PHOSPHORUS,
*EUTROPHICATION, PHYTOPLANKTON,
ES, LIMNOLOGY, *PLANKTON, ALGAE,
ORY TESTS, BIOLOGICAL TREATMENT,
ACE WATERS, RESERVOIRS, CONTROL,
TS, ALGAE, NITROGEN, PHOSPHORUS,
OGY, WATER PURIFICATION, CARBON,
ES, PHOSPHATES, POTASSIUM, IRON,
N, OLIGOTROPHY, LAKES, MICHIGAN,
RATION, TEMPERATURE, ALKALINITY,
T PENETRATION, DISSOLVED OXYGEN,
*PHYTOPLANKTON, *PHOTOSYNTHESIS,
PHOTOSYNTHESIS, CARBON DIOXIDE,
NTIAL NUTRIENTS, EUTROPHICATION,
GERMANY), CHLORELLA PYRENOIDOSA,
SOMA LEUCHTENBER GIANUM, CYCLOPS
PHORA, AULACODISCUS, BACILLARIA,
, ASTERIONELLA FORMOSA, MELOSIRA
ISCUS HANTZSCHII, STEPHANODISCUS
UCTIVITY, LIMNOLOGY, *NUTRIENTS,
ONDITIONS, ANAEROBIC CONDITIONS,
PESTICIDE
, / *PESTICIDE RESIDUES, *ALGAE,
AQUATIC MICROBIOLOGY, ALGICIDES,
EA WATER, TEMPERATURE, SALINITY,
GELLATES, FISHKILL, FISH TOXINS,
PRODUCTIVITY, BALANCE OF NATURE,
MICROBIOLOGY, BALANCE OF NATURE,
, MORTALITY, CHLOROPHYLL, ALGAE,
TIC ALGAE, AQUATIC MICROBIOLOGY,
PHOSPHATES, SILICATES, CITRATES,
NT-WATER INTERFACES, CALIFORNIA,
, CULTURES, DIATOMS, CYANOPHYTA,
ALGAE, *NUTRIENT REQUIREMENTS,
MARSHES, *LAKES, *PHYTOPLANKTON,
ION, FISH, ALGAE, MODEL STUDIES,
KES, WATER PROPERTIES, BACTERIA,
TIC PRODUCTIVITY, AQUATIC WEEDS,
QUATIC WEEDS, BALANCE OF NATURE,
BACTERIA, AQUATIC MICROBIOLOGY,
ONING, ALGICIDES, AQUATIC ALGAE,
Y, BALANCE OF NATURE, WISCONSIN,
MICROBIOLOGY, BALANCE OF NATURE,
QUATIC WEEDS, BALANCE OF NATURE,
TIC ALGAE, AQUATIC PRODUCTIVITY,
MICROBIOLOGY, BALANCE OF NATURE,
MICROBIOLOGY, BALANCE OF NATURE,
TICAL TECHNIQUES, AQUATIC ALGAE,
BACTERIA, AQUATIC MICROBIOLOGY,
GAS CHROMATOGRAPHY, RESISTANCE,

BENTHIC FAUNA, CHEMICAL ANALYSIS,   W72-06052
BENTHIC FLORA, PHAEOPHYTA, POLAR   W70-01074
BENTHIC FLORA, BENTHOS, MINE WATE   W70-02770
BENTHIC FLORA, ESTUARIES, *ESTUAR   W71-02494
BENTHIC FLORA, FLOATING PLANTS, P   W70-10175
BENTHIC FLORA, STREAMS, CULTURES,   W69-07845
BENTHIC FORMS, EPIPHYTIC FORMS, C   W73-00748
BENTHOS, AQUATIC LIFE, ALGAE, PLA   W69-05023
BENTHOS, AEROBIC CONDITIONS, ANAE   W71-11381
BENTHOS, BRINE SHRIMP, ECOSYSTEMS   W69-07440
BENTHOS, CONVECTION, WAVES(WATER)   W70-03969
BENTHOS, HYDROGEN, BACTERIA, AERO   W69-07838
BENTHOS, MINE WATER, WATER POLLUT   W70-02770
BENTHOS, NITRATES, FLUORESCENCE,   W70-02775
BENTHOS, SALINITY, SNAILS, SULFAT   W70-03978
BENTHOS, WISCONSIN, SAMPLING, CON   W70-06217
BENTHOS, ZOOPLANKTON, FRESHWATER   W71-02491
BENTLASS SALT MARSH.: /, *RECOVER   W70-01231
BENZENE COMPOUNDS.: /NCE LIMIT, B   W72-11105
BENZENE SULFONATES, WIND VELOCITY   W70-00263
BERKELSE LAKE(NETHERLANDS), HETER   W70-04369
BERLIN(WEST), GERMANY: W68-00472
BERLIN, *ALGAE-TITRE.:   W68-00475
BERRY CREEK(ORE), OREGON STATE UN   W70-03978
BETA RADIOACTIVITY, GONIOBASIS, D   W69-09742
BETA RAYS, COSMIC RADIATION, S: /   W69-08272
BEVERLY HILLS(CALIF).:   W71-00110
BIBLIOGRAPHIES.: /WATER QUALITY,   W70-07027
BIBLIOGRAPHIES, HERBICIDES, FISH,   W70-00269
BIBLIOGRAPHIES, BACTERIA, DECOMPO   W70-00664
BIBLIOGRAPHIES, PHOSPHORUS COMPOU   W69-08518
BIBLIOGRAPHIES, ECOLOGY, BIOLOGY,   W72-01786
BIBLIOGRAPHIES.: / SALINITY, ALGA   W72-01773
BIBLIOGRAPHIES, LAKES, NUISANCE A   W69-08283
BIBLIOGRAPHIES, CHLORELLA, CHLORO   W69-06865
BIBLIOGRAPHIES, *MINNESOTA, *LAKE   W69-00659
BIBLIOGRAPHIES, CYANOPHYTA, FRESH   W68-00680
BIBLIOGRAPHIES, *PHOSPHORUS RADIO   W68-00857
BIBLIOGRAPHIES, BOTTOM SEDIMENTS,   W68-00859
BIBLIOGRAPHIES, ABSORPTION, FISHE   W68-00858
BIBLIOGRAPHIES, NITRATES, NUTRIEN   W68-00860
BIBLIOGRAPHIES, *ALGAE, *CHEMICAL   W68-00856
BIBLIOGRAPHIES, CHEMICAL PRECIPIT   W69-00446
BIBLIOGRAPHIES, CHLORELLA, *SEWAG   W68-00855
BIBLIOGRAPHIES, ANALYTICAL TECHNI   W70-05389
BICARBONATES, *AQUATIC POPULATION   W68-00468
BICARBONATES, PHOTOSYNTHESIS, NUT   W69-07389
BICARBONATE, ZINC, COBALT, CALCIU   W70-03978
BICARBONATES, IRON, PHOSPHORUS, N   W72-03183
BICARBONATES, BACTERICIDE, KINETI   W71-05267
BICARBONATES, HYDROGEN ION CONCEN   W70-06225
BICARBONATES, RADIOACTIVITY, CARB   W70-02504
BICARBONATES, CARBONATES, TEMPERA   W70-02804
BICARBONATES, CARBON, CARBONATES,   W70-02792
BICHOROMATE, DRY WEIGHT, B COLI C   W70-05092
BICUSPIDATUS, REPRODUCTIVE RATE.:   W69-10182
BIDDULPHIA, CHAETOCEROS, COCCONEI   W70-03325
BINDERANA, GENERATION TIME, CHLOR   W69-10158
BINDERANUS, CLADOCERA, WATER MASS   W70-00263
BIO MASS, OLIGOTROPHY, *PRIMARY P   W68-00912
BIO: /SIS, GROWTH RATE, AEROBIC C   W70-05750
BIOACCUMULATION.:   W69-08565
BIOASSAY, BIOINDICATORS, FISHKILL   W69-08565
BIOASSAY, DISINFECTION, ENVIRONME   W69-10171
BIOASSAY, ANALYTICAL TECHNIQUES,   W70-08111
BIOASSAY, AQUATIC ALGAE, *ALGAL T   W70-08372
BIOASSAY, CHLOROPHYTA, CYANOPHYTA   W70-02792
BIOASSAY, CHLOROPHYTA, ENVIRONMEN   W70-02968
BIOASSAY, VIABILITY.: /AL CONTROL   W70-02364
BIOASSAY, CHLOROPHYTA, CHLORELLA,   W70-02248
BIOASSAY, NEW YORK, LAKES, INHIBI   W70-01579
BIOASSAY, SCENEDESMUS, PHYSIOLOGI   W70-02775
BIOASSAY, TEST PROCEDURES.: /LGAE   W71-06002
BIOASSAY, EUTROPHICATION, OLIGOTR   W72-03183
BIOASSAY, LABORATORY TESTS, PRODU   W72-00845
BIOASSAY, AQUATIC LIFE, COOLING W   W71-11881
BIOASSAY, METABOLISM, ABSORPTION,   W71-10079
BIOASSAY, CYANOPHYTA, ENVIRONMENT   W69-03186
BIOASSAY, CHLOROPHYTA, CYANOPHYTA   W69-03185
BIOASSAY, CHLOROPHYTA, CYANOPHYTA   W69-03362
BIOASSAY, CHLOROPHYTA, CYANOPHYTA   W69-03370
BIOASSAY, BIOCHEMICAL OXYGEN DEMA   W69-03369
BIOASSAY, CHLOROPHYTA, CYANOPHYTA   W69-03188
BIOASSAY, CHLOROPHYTA, DIATOMS, C   W69-03515
BIOASSAY, CYCLING NUTRIENTS, ENVI   W69-04798
BIOASSAY, CHLOROPHYTA, DIATOMS, E   W69-03516
BIOASSAY, BIOCHEMICAL OXYGEN DEMA   W69-03374
BIOASSAY, CYCLING NUTRIENTS, ENVI   W69-03518
BIOASSAY, CHLOROPHYTA, CYANOPHYTA   W69-03371
BIOASSAY, GROWTH RATES, PESTICIDE   W70-03520

398

MYDOMONAS, CHLORELLA, MORTALITY,
    RADIOISOTOPES, DINOFLAGELLATES,
    RIENT INTERACTIONS, *RADIOCARBON
    CULTU/ *EUTROPHICATION, *ALGAE,
    MONAS,/ *RAIN, *VITAMINS, ALGAE,
    LYTICAL TECHNIQUES, ANTIBIOTICS,
    TORAL, NUTRIENTS, MODEL STUDIES,
    OISOTOPES, LAKES, PHYTOPLANKTON,
    UNDS, METABOLISM, BIOINDICATORS,
    ERFACES, EUGLENOPHYTA, TOXICITY,
    ON RADIOISOTOPES, CHLAMYDOMONAS,
    OPHYTA, CYANOPHYTA, CHRYSOPHYTA,
    CHAINS, BIOLOGICAL COMMUNITIES,
    TIVITY, WATER POLLUTION EFFECTS,
    , BENTHIC FAUNA, PHOTOSYNTHESIS,
    IDE, HYDROGEN ION CONCENTRATION,
    NS, CANNERIES, ACTIVATED SLUDGE,
    ANTS, ANAEROBIC BACTERIA, ALGAE,
    N, STORM RUN-OFF, SEDIMENTATION,
    TREATMENT, *ANAEROBIC DIGESTION,
    SION, SEDIMENTATION, ABSORPTION,
    ULFIDES, ALGAE, ORGANIC LOADING,
    IGHT PENETRATION, ON-SITE TESTS,
    PHENOLS, ODOR, COLOR, TURBIDITY,
    ALGAE, FUNGI, LABORATORY TESTS,
    ONS, *ALGAE CONTROL, *LAKE ERIE,
    HLORINATION, COLIFORM, BACTERIA,
    WATER QUALITY, DISSOLVED OXYGEN,
    E, *LAGOONS, *OXIDATION LAGOONS,
    EROBIC TREATMENT, OXYGEN DEMAND,
    RON COMPOUNDS, HYDROGEN SULFIDE,
    NCIES, *LAGOONS, *SLUDGE, ALGAE,
    LAGOONS, *BIODEGRADATION, ALGAE,
    , WATER POLLUTION, WASTE WATERS,
    OTOSYNTHESIS, COSTS, EFFICIENCY,
    , CHEMICAL WASTES, ION EXCHANGE,
    ATA COLLECTIONS, *WATER QUALITY,
    CRO-AUTORADIOGRAPHY, PHYSIOLOGY,
    *DISSOLVED OXYGEN, PENNSYLVANIA,
    UATIC PLANTS, SELF-PURIFICATION,
    YCLES, UNITED STATES, CHLORELLA,
    CATION, DISSOLVED OXYGEN, ALGAE,
    TREATMENT, ALGAE, DISINFECTION,
    , DISSOLVED OXYGEN, TEMPERATURE,
    OBIOLOGY, *AQUATIC PRODUCTIVITY,
    LTERS, SEASONAL, STRATIFICATION,
    A, COLIFORMS, OXIDATION LAGOONS,
    GOVERNMENTS, STATE GOVERNMENTS,
    NUTRIENTS, OXYGEN SAG, AERATION,
    S, NITROGEN, AMMONIA, NUTRIENTS,
    GY, BALANCE OF NATURE, BIOASSAY,
    OF NATURE, WISCONSIN, BIOASSAY,
    TOXINS, ORGANIC WASTES, POLLUTA/
    S, CHEMICAL PRECIPITATION, IONS,
    ASTES, *ALGAE, WATER PROPERTIES,
    IMPOUNDMENTS, *DISSOLVED OXYGEN,
    GEN, HYDROGEN ION CONCENTRATION,
    NG, CENTRIFUGATION, RESPIRATION,
    EN SULFIDE, ELECTRON MICROSCOPY,
    STUDIES, ALGAL PHYSIOLOGY, ALGAL
    PHOSPHORUS RADIOISOTOPES, ALGAE,
    CHLOROPHYLL, ENZUMES, PEPTIDES,

    NARY AMMONIUM COMPOUNDS, PHENOL/
    GATIONS, ZOO-PLANKTON, SPRAYING,
    ER POLLUTION CONTROL, ALGICIDES,
    CHEMCONTROL, MECHANICAL CONTROL,
    OLOGY, ENVIRONMENT, CHEMCONTROL,
    , EUTROPHICATION, ALGAL CONTROL,
    MONIA, SEDIMENTATION, FLOTATION,
    ANTIFOULING MATERIALS, TESTING,
    *OXIDATION LAGOONS, ABSORPTION,
    TICS, *PERSISTENCE, EVAPORATION,
    AND, MINNESOTA, WATER CHEMISTRY,
    , NITROGEN COMPOUNDS, OXIDATION,
    RIENTS, *ALGAE, *MICROORGANISMS,
    ALGAE, AQUATIC PLANTS, TESTING,
    PLANTS, DINOFLAGELLATES, OCEANS,
    WATER POLLUTION EFFECTS, OREGON,
    MIXIS, *WATER POLLUTION EFFECTS,
    , ORGANIC COMPOUNDS, METABOLISM,
    OPES, DINOFLAGELLATES, BIOASSAY,
    AIN, *VITAMINS, ALGAE, BIOASSAY,
    FFECTS, WATER POLLUTION SOURCES,
    OLLUTION EFFECTS, *WATER SUPPLY,
    , *PHYTOPLANKTON, BRITISH ISLES,
    PHYTA, CHLOROPHYTA, GREAT LAKES,
    DIMENTATION RATES, STRATIGRAPHY,
    TON, *LIMNOLOGY, *WATER QUALITY,

BIOASSAY, INHIBITION, MODE OF ACT	W70-03519
BIOASSAY, BIOINDICATORS.: /CARBON	W70-04194
BIOASSAY, *NUTRIENT TYPES, MICROQ	W70-04765
BIOASSAY, BIOLOGICAL COMMUNITIES,	W69-07832
BIOASSAY, BIOINDICATORS, CHLAMYDO	W69-06273
BIOASSAY, NORON, CALCIUM, CHLORIN	W69-04801
BIOASSAY, RADIOACTIVITY TECHNIQUE	W73-01454
BIOASSAYS.: /ECOLOGY, CARBON RADI	W70-05109
BIOASSAYS.: /ATTER, ORGANIC COMPO	W70-04287
BIOASSAYS, DIATOMS.: /D-WATER INT	W70-04381
BIOASSAYS, WATER POLLUTION EFFECT	W70-04284
BIOASSAYS, SALINE WATER, CHLORELL	W70-03974
BIOASSAYS, EUTROPHICATION.: /FOOD	W70-05272
BIOASSAYS, ALGAE, COMPARATIVE PRO	W70-05424
BIOASSAYS, AQUATIC WEEDS, THERMAL	W70-02251
BIOCARBONATES, DISSOLVED OXYGEN,	W70-02249
BIOCHEMICAL OXYGEN DEMAND, OXIDAT	W71-09546
BIOCHEMICAL OXYGEN DEMAND, OXIDAT	W71-08221
BIOCHEMICAL OXYGEN DEMAND, WASTE	W71-10654
BIOCHEMICAL OXYGEN DEMAND, ACTIVA	W71-09548
BIOCHEMICAL OXYGEN DEMAND, AERATI	W71-11381
BIOCHEMICAL OXYGEN DEMAND, OXIDAT	W71-11377
BIOCHEMICAL OXYGEN DEMAND, DIURNA	W72-01818
BIOCHEMICAL OXYGEN DEMAND, HYDROG	W72-03299
BIOCHEMICAL OXYGEN DEMAND, SEWAGE	W72-01817
BIOCHEMICAL OXYGEN DEMAND, BACTER	W71-04758
BIOCHEMICAL OXYGEN DEMAND, DELAWA	W71-05013
BIOCHEMICAL OXYGEN DEMAND, SUSPEN	W70-08627
BIOCHEMICAL OXYGEN DEMAND, CHEMIC	W70-06601
BIOCHEMICAL OXYGEN DEMAND, SUSPEN	W70-06792
BIOCHEMICAL OXYGEN DEMAND, NUTRIE	W70-06925
BIOCHEMICAL OXYGEN DEMAND, MIXING	W70-06899
BIOCHEMICAL OXYGEN DEMAND, ANAERO	W71-00936
BIOCHEMICAL OXYGEN DEMAND, FISH,	W70-08740
BIOCHEMICAL OXYGEN DEMAND, ALGAE,	W70-09320
BIOCHEMICAL OXYGEN DEMAND, CHEMIC	W70-10433
BIOCHEMICAL OXYGEN DEMAND, SUSPEN	W70-09557
BIOCHEMICAL EXTRACTION, MILLIPORE	W69-10163
BIOCHEMICAL OXYGEN DEMAND, TEMPER	W69-10159
BIOCHEMICAL OXYGEN DEMAND, TOXICI	W70-05750
BIOCHEMICAL OXYGEN DEMAND, OXIDAT	W70-05092
BIOCHEMICAL OXYGEN DEMAND(BOD), M	W70-04790
BIOCHEMICAL OXYGEN DEMAND(BOD), A	W70-04788
BIOCHEMICAL OXYGEN DEMAND(BOD), E	W70-04787
BIOCHEMICAL OXYGEN DEMAND, CYCLIN	W70-04283
BIOCHEMICAL OXYGEN DEMAND, CARBON	W70-04001
BIOCHEMICAL OXYGEN DEMAND, PHOTOS	W70-03312
BIOCHEMICAL OXYGEN DEMAND, ODOR-P	W69-06909
BIOCHEMICAL OXYGEN DEMAND, PHOTOS	W69-07520
BIOCHEMICAL OXYGEN DEMAND, TERTIA	W69-08054
BIOCHEMICAL OXYGEN DEMAND, CHLORO	W69-03374
BIOCHEMICAL OXYGEN DEMAND, ENVIRO	W69-03369
BIOCHEMICAL OXYGEN DEMAND, ALGAL	W69-01273
BIOCHEMICAL OXYGEN DEMAND, MINNES	W69-00446
BIOCHEMICAL OXYGEN DEMAND, *BACTE	W68-00256
BIOCHEMICAL OXYGEN DEMAND, ALGAL	W72-11906
BIOCHEMICAL OXYGEN DEMAND, VIRGIN	W72-10573
BIOCHEMICAL OXYGEN DEMAND, NITROG	W72-06022
BIOCHEMISTRY, VITAMINS, EPIPHYTOL	W70-03952
BIOCHEMISTRY, ALGAL TAXONOMY, ALG	W70-03973
BIOCHEMISTRY, ECOLOGY, SAMPLING,	W69-10151
BIOCHEMISTRY, HYDROGEN ION CONCEN	W70-01068
BIOCIDES.:	W69-05706
BIOCIDES.:	W69-05705
BIOCIDES, PRODUCTION LOSS, QUATER	W72-12987
BIOCONTROL, INFECTION, NUTRIENTS,	W68-00468
BIOCONTROL, *WATERSHED MANAGEMENT	W68-00478
BIOCONTROL, COPPER SULFATE, SODIU	W70-10175
BIOCONTROL, INTEGRATED CONTROL ME	W70-07393
BIOCONTROL, TURBULENCE, COPEPODS,	W71-10098
BIODEGRADATION, CHLORINATION, COL	W71-05013
BIODEGRADATION, ALGAE, MARINE ALG	W71-07339
BIODEGRADATION, CHEMICAL OXYGEN D	W70-08392
BIODEGRADATION, ADSORPTION, ALGAE	W70-09768
BIODEGRADATION.: /ICAL OXYGEN DEM	W69-00446
BIODEGRADATION, NITRIFICATION, DE	W69-08053
BIODEGRADATION, NITRIFICATION, DE	W69-06970
BIODEGRADATION.: /ATMENT, AQUATIC	W70-04373
BIOINDICATORS, FUNGI, ESTUARIES,	W70-04368
BIOINDICATORS, LIGHT INTENSITY, E	W70-03978
BIOINDICATORS, NUISANCE ALGAE, NE	W70-03974
BIOINDICATORS, BIOASSAYS.: /ATTER	W70-04287
BIOINDICATORS.: /CARBON RADIOISOT	W70-04194
BIOINDICATORS, CHLAMYDOMONAS, CAR	W69-06273
BIOINDICATORS, NITROGEN FIXATION.	W69-06864
BIOINDICATORS, CALCIUM, CHLORIDES	W69-07833
BIOINDICATORS, *ESTIMATING EQUATI	W68-00255
BIOINDICATORS, FISH, WASHINGTON,	W70-04503
BIOINDICATORS, ALGAE, OLIGOTROPHY	W70-05663
BIOINDICATORS, AQUATIC LIFE, AQUA	W70-05958

SURVEYS, BIOLOGICAL COMMUNITIES,
OLLUTION SOURCES, ALGAL CONTROL,
ORING, *RADIOISOTOPES, *STREAMS,
H RATES, *DAPHNIA, PRODUCTIVITY,
CIDE RESIDUES, *ALGAE, BIOASSAY,
S COMPOUNDS, NITROGEN COMPOUNDS,
RRANEAN DRAINAGE, MECHANICAL AND
UTROPHICATION, *CANALS, *CITIES,
CHLORIS, THORACOMONAS, YTTRIUM, '
OMS, CARP, PLANKTON, METABOLISM,
ES, COST COMPARISONS, TREATMENT,
AQUATIC ENVIRONMENT, POLLUTANTS,
RAINAGE, *AGRICULTURAL DRAINAGE,
ST U S, ECOLOGICAL DISTRIBUTION,
ACTIVITY, CHLORELLA PYRENOIDOSA,
QUENCY DISTRIBUTION, POPULATION,
ALKYL SUBSTITUTED IMIDAZOLINES,
AUNA, *TROPHIC LEVELS, *SURVEYS,
MARINE BIOLOGY, *PRIMARY
XYGEN DEMAND, *ACTIVATED SLUDGE,
IC/ BLUE-GREEN ALGAE, NITZSCHIA,
*NUTRIENTS, *TERTIARY TREATMENT,
, *NITROGEN, *PHOSPHORUS, ALGAE,
OMPOUNDS, NITROGEN CYCLE, ALGAE,
C ENVIRONMENT, WATER PROPERTIES,
SIDUES, ECOSYSTEMS, FOOD CHAINS,
E REACTIVE PHOSPHORUS, PHOSPHATE(
HANNEL MORPHOLOGY, SEWAGE INPUT,
STES, *SEWAGE TREATMENT, SEWAGE,
XINS, SALINE WATER, EQUILIBRIUM,
UTROPHICATION, LABORATORY TESTS,
L STUDIES, WASTE IDENTIFICATION,
RIVERS, MILLS, COLORADO, WASTES,
TURE, CARBON DIOXIDE, NUTRIENTS,
UTROPHICATION, *ALGAE, BIOASSAY,
GAE, WATER POLLUTION, *TOXICITY,
, SEWAGE, EUGLENOPHYTA, DIATOMS,
COHO SA/ ANTIPOLLUTION PROGRAMS,
SLANDS, ECOLOGICAL DISTRIBUTION,
UCTIVITY.: MARINE
Y, T/ *WATER POLLUTION, *AQUATIC
ANIZOMENON / *CONCEPTS, *TROPHIC
NC(MASS), CARNEGIE DEPT OF PLANT
TEMENT, BIBLIOGRAPHIES, ECOLOGY,
TOSYNTHETIC PIGMENTS, PRODUCTION
OGY, SUSPENDED SEDIMENTS, MARINE
IES, MICROORGANISMS, FRESH WATER
PONDS, POLLUTANT IDENTIFICATION,
ODUCTIVITY, HARVESTING OF ALGAE,
S, CHLORELLA PYRENOIDOSA, *ALGAE
EUTHROPHICATION, MODEL STUDIES,
GRAZING, ALGAE, WATER CHEMISTRY,
ENOLS, PIGMENTS, ORGANIC MATTER,
NE, CALCIUM CARBONATE, BACTERIA,
*PONDS, *EUTROPHICATION, ALGAE,
S, *PRIMARY PRODUCTIVITY, ALGAE,
, PHYTOPLANKTON, PHOTOSYNTHESIS,
*BENTHIC FLORA, LIGHT INTENSITY,
, *LAKES, ALGAE, MICROORGANISMS,
ITY, ECOLOGY, LOTIC ENVIRONMENT,
ALGAE, ANIMALS, AQUATIC PLANTS,
C THEORY, ENVIRONMENTAL FACTORS,
MPERATURE, VELOCITY, CYANOPHYTA,
STUDIES, CULTURES, FLUOROMETRY,
GAE, *CHLORELLA, *ORGANIC ACIDS,
R), VELOCITY, DIATOMS, SEASONAL,
GAE, *THERMAL WATER, PERIPHYTON,
ENTRATION, STANDING CROP, LIGHT,
SCENCE, LAKES, RIVERS, NITROGEN,
EN DEMAND, NITROGEN, PHOSPHATES,
O RIVER, *ANALYTICAL TECHNIQUES,
NASTRUM, ALGAL GROWTH, ANABAENA,
, TRITIUM, CARBON RADIOISOTOPES,
AMS, *ECOSYSTEMS, RADIOACTIVITY,
OLOGICAL COMMUNITIES, SEDIMENTS,
*LAKES, *EUTROPHICATION, SEWAGE,
NT, FALLOUT, NUCLEAR EXPLOSIONS,
N, DYE RELEASES, BEACHES, SANDS,
MIN B1, VITAMIN B6, VITAMIN B12,
TON, AQUATIC INSECTS, FISH, GAME
TECHNIQUES, SWIMMING, WATERFOWL,
SOTOPES, MANGANESE, ALGAE, IRON,
-106, ANTIMONY-125, GROUNDWATER,
E, MEAD COUNTY(KY), THORIUM-232,
N, FOSSIL, CIMARRON RIVER(OKLA),
ATER EFFECTS, LAKE MENDOTA(WIS),
, BOULDER(COLO), TEA LAKE(COLO),
CAL OCEANOGRAPHY, *TRANSACTIONS,
LORIDE, IRON SULFATE, CLADOPHORA
VER, CELAKOVICE(CZECHOSLOVAKIA),

BIOINDICATORS, LAKE MORPHOLOGY, E   W71-12077
BIOINDICATORS.: /LLUTION, WATER P   W70-02304
BIOINDICATORS, *NEW YORK, NUCLEAR   W69-10080
BIOINDICATORS, SECONDARY PRODUCTI   W69-10152
BIOINDICATORS, FISHKILL, HAZARDS,   W69-08565
BIOINDICATORS, PRODUCTIVITY, WAST   W69-08518
BIOLOGICAL PURIFICATION, SWITZERL   W69-08668
BIOLOGICAL COMMUNITIES, SEWAGE, D   W70-00268
BIOLOGICAL DILUTION', PHYSIOLOGIC   W70-00708
BIOLOGICAL COMMUNITIES, DAPHNIA,   W70-00274
BIOLOGICAL TREATMENT, ALGAL CONTR   W70-01981
BIOLOGICAL COMMUNITIES, WATER POL   W70-01233
BIOLOGICAL RESPONSE, FERTILIZATIO   W70-02787
BIOLOGICAL INDICATORS, NEW HAMPSH   W70-02764
BIOLOGICAL HALF-LIFE, CLINCH RIVE   W70-02786
BIOLOGICAL INDICATORS, NEW HAMPSH   W70-02772
BIOLOGICAL ACTIVITY, QUARTERNARY,   W70-02982
BIOLOGICAL COMMUNITIES, BIOINDICA   W71-12077
BIOLOGICAL PRODUCTIVITY.:   W71-10788
BIOLOGICAL TREATMENT, PILOT PLANT   W70-05821
BIOLOGICAL STUDIES, ALGAL IDENTIF   W70-10173
BIOLOGICAL TREATMENT, NITROGEN, P   W70-08980
BIOLOGICAL TREATMENT, WASTE WATER   W70-07469
BIOLOGICAL TREATMENT, BOTTOM SEDI   W70-07031
BIOLOGICAL PROPERTIES, ECOLOGY, S   W70-04475
BIOLOGICAL COMMUNITIES, BIOASSAYS   W70-05272
BIOLOGICALLY AVAILABLE), MICROCYS   W70-05269
BIOLOGICAL PROPERTIES, *WATER QUA   W68-00475
BIOLOGICAL TREATMENT, SOLID WASTE   W68-00248
BIOLOGICAL COMMUNITIES, *AQUATIC   W69-00360
BIOLOGICAL TREATMENT, BIBLIOGRAPH   W68-00855
BIOLOGICAL PROPERTIES, PHYSICAL P   W69-01273
BIOLOGICAL COMMUNITIES, SEDIMENTS   W69-07846
BIOLOGICAL PROPERTIES, WATER POLL   W69-07442
BIOLOGICAL COMMUNITIES, CULTURES,   W69-07832
BIOLOGICAL MEMBRANES, HUMID AREA.   W69-05306
BIOLOGICAL COMMUNITIES, BACKWATER   W70-03969
BIOLOGICAL BALANCE, OVERFISHING,   W70-04465
BIOLOGICAL COMMUNITIES, SYSTEMATI   W72-10623
BIOLOGY, *PRIMARY BIOLOGICAL PROD   W71-10788
BIOLOGY, *BIOINDICATORS, *BIOASSA   W71-10786
BIOLOGY, ANABAENA FLOS AQUAE, APH   W70-03959
BIOLOGY, CARNEGIE INSTITUTION OF   W69-06865
BIOLOGY, FISH, PLANKTON, WATER QU   W72-01786
BIOLOGY, RUSSIA, SECOND INTERNATI   W69-07440
BIOLOGY, STREAM POLLUTION, *WATER   W71-10786
BIOLOGY, SUSPENDED SEDIMENTS, MAR   W71-10786
BIOLOGY, WATER POLLUTION, WATER P   W70-02304
BIOMASS.: /NAL, ALGAE, PRIMARY PR   W70-04290
BIOMASS.:   W73-02218
BIOMASS, *LIGHT INTENSITY.: /ENT,   W73-02218
BIOMASS, BENTHOS, SALINITY, SNAIL   W70-03978
BIOMASS, BENTHIC FLORA, PHAEOPHYT   W70-01074
BIOMASS, CHLOROPHYLL, PROTEINS, L   W69-10160
BIOMASS, CHLOROPHYLL, NEW YORK.: /   W69-06540
BIOMASS, CALCIUM COMPOUNDS, CHLOR   W69-02959
BIOMASS, CULTURES, PLANT PHYSIOLO   W70-05094
BIOMASS, DISTRIBUTION, ECOSYSTEMS   W70-03325
BIOMASS, DIATOMS, CYANOPHYTA, IRO   W70-08654
BIOMASS, FLOW RATES, DEPTH, DISSO   W70-00265
BIOMASS, FERTILIZATION, LAKES, LI   W69-04805
BIOMASS, GROWTH RATES, ALGAE, OLI   W70-02779
BIOMASS, INSECTS, DIATOMS, CHLORO   W71-00668
BIOMASS, OCEANOGRAPHY.: /OLOGICAL   W69-08278
BIOMASS, ORGANIC COMPOUNDS, CARBO   W70-04195
BIOMASS, PIGMENTS, PRODUCTIVITY,   W70-02780
BIOMASS, PHYSIOLOGICAL ECOLOGY, C   W70-03309
BIOMASS, RESPIRATION, LIFE CYCLES   W71-00665
BIOMASS, SEWAGE, FARM WASTES, CAR   W70-02777
BIOMASS, SYMBIOSIS, OXYDATION LAG   W70-05655
BIOMASS, WATER POLLUTION, ALGAE,   W71-04206
BIOSTIMULATION, WATER POLLUTION A   W70-02779
BIOTA, CLAMS, FALLOUT, PACIFIC OC   W69-08269
BIOTA, CYCLES, MICHIGAN, INVERTEB   W69-09334
BIOTA, FISH, UTAH, ALGAE, INSECTS   W69-07846
BIOTA, PHYTOPLANKTON, DIATOMS, DI   W69-10169
BIOTA, SPECTROMETERS, BACTERIA, A   W69-08267
BIOTA, WASTE WATER(POLLUTION), CU   W68-00010
BIOTIN, HISTIDINE, URACIL, REDUCE   W70-02510
BIRDS, ANIMALS, FOOD CHAINS, OXYG   W70-10175
BIRDS, COSTS, COST-SHARING.: /ON   W70-05568
BIRDS, RADIOACTIVITY, INVERTEBRAT   W69-08269
BISMUTH-207, VERTEBRATES, PLUTONI   W69-08269
BISMUTH-214, ACTINIUM-228, SWIFT   W69-09742
BIXBY(OKLA), KEYSTONE RESERVOIR(O   W70-04001
BLACK EARTH CREEK(WIS).: /WASTE W   W70-04382
BLACK LAKE(COLO), PASS LAKE(COLO)   W69-10154
BLACK SEA, CONTACT ZONES, ARTEMIA   W69-07440
BLANKETS, HYDRODICTYON, MACROPHYT   W69-10170
BLATNA(CZECHOSLOVAKIA), CYPRINID   W70-00274

R, PLECTONEMA, NOSTOC PALUDOSUM, CHEMOTROPHY, PARTICULATE MATTER, IZATION, *BACTERIA COUNT, WARING RAS CHODATI, DINOPHYCEAE, FILTER POLLUTION, *METROPOLITAN AREAS, ON, BENTHIC FAUNA, SLUDGE WORMS, WATER TEMPERATURE, SLUDGE WORMS, H, OLIGOCHAETES, TOXICITY, FISH, IS), *'WORKING' OF LAKES, *WATER-TREATMENT, ALGAE NUISANCE, ALGAE ROPHICATION, ALGAL BLOOMS, WATER USE.: *DEFINITIONS, *WATER WAUBESA(WIS), LAKE KEGONSA(WIS), INDIA, METHYL ORANGE, ALGAL
ALGAL
*PERIDINIUM POLONICUM, PLANKTON LYMERS, *BIOLOGICAL LIFE, ALGAE YSILANES, SCENEDESMUS QU/ *ALGAL YSILANES, SCENEDESMUS QU/ *ALGAL TIS AERUGINOSA.: ALGAL
*BLUE-GREEN ALGAE, *ALGA REGULATORS, MYRIOPHYLLUM, WATER *SHAGAWA LAKE(MINN), ELY(MINN), ETONKA(MINN), MORPHOMETRY, ALGAE ENVIRONMENTAL C/ BENTHIC ALGAE, OLOGY, WATER TEMPERATURE, *ALGAE WIS), PELAGIC AREAS, HORIZONTAL, RESPIROMETER, PROTOPLASM, WATER-ANTITIES, MACROQUANTITIES, ALGAL UGAL, TAPES, JAPAN, AMPHIDINIUM, HOE(CALIF)/ *COMPLEXONES, *ALGAL ATER POLLUTION COMMISSION, ALGAL QUALITY, *EUTROPHICATION, ALGAL E, AGING/ *EUTROPHICATION, ALGAL N TECHNIQUE, PARTICULATE MATTER, OGICAL STUDIES, ALGAL IDENTIFIC/ ERIMENTS, RECYCLED/ GREEN ALGAE, DE, UPTAKE, FREUNDLICH ISOTHERM, GREEN ALGAE, RACTIONS, INFRARED SPECTROSCOPY, , FILAMENTOUS ALGAE, FILAMENTOUS IS, MATS, SYNECHOCOCCUS LIVIDUS, AND CLARK LAKE (S D), ABUNDANCE, N, FLUORESCENCE, PHOTOSYNTHESIS, HIOPHORA/ PHOSPHATE FERTILIZERS, COMPOUNDS, MEAN TOLERANCE LIMIT, CEDURES, STATISTICAL PROCEDURES, EAMS, LAKES, RIVERS, RESERVOIRS, TENTION TIME, LABORATORY MODELS, ONMENTAL CONTROL, RECIRCULATION, CTION, BIOCHEMICAL OXYGEN DEMAND(ATURE, BIOCHEMICAL OXYGEN DEMAND(ALGAE, BIOCHEMICAL OXYGEN DEMAND(ON, *DESULFOVIBRIO, FACULTATIVE, IENTS, FACULTATIVE POND DESIGNS, ONS, GEOGRAPHICAL REGIONS, BAYS, ERSATURATION, RESERVOIRS, ALGAE, ISLAND), DILUTION RATE, FRUITING LF-LIFE, HYDROPSYCHE COCKERELLI, PWARD FLOW, PUTRESCIBILITY TEST, R, *HUMIC ACIDS, *COLOR, RIVERS, AQUAE, *MICROCYSTIS AERUGINOSA, , SNAILS, WEIGHT, VACUUM DRYING, ED, ZIRCONIUM-95-NIOBIUM-95, PRE-R GANGES(INDIA), RAJGHAT(INDIA), ON(WASH), SEATTLE(WASH), REVENUE MOEA, PISONIA, PANDANUS, MUSCLE, , CHAETOCEROS LAUNDERI, LAUDERIA LLERITE, ORPIMENT, TETRAHEDRITE, MONIUM SALTS, NITROGEN FIXATION, CKING, HETERETROPHS, ITALY, LAKE RIA, PANDORINA MORUM, PEDIASTRUM NAH RIVER LABORATORY, PLECTONEMA AUTOTROPHIC, SILICON, PEDIASTRUM WITZERLAND), LINSLEY POND(CONN), ABAENA LEMMERMANNI, APHANOCAPSA, NABAENA, ANACYSTIS, MICROCYSTIS, A COHCLEARIS, ASPLANCHA, LECANE, WINNISQUAM, LONG POND, CONCORD, QUABBIN RESERVOIR(MASS), BORYANUM, STAURASTRUM PARADOXUM, ATTERN, ANABAENA CYLINDRICA, GL/ AQUAE, GLOEOTRICHIA ECHINULATA, R DIVERSION, INSTREAM TREATMENT, , SMELTS, FISHKILL, CRUSTACEANS, TMENT, NITROGEN, AQUATIC PLANTS, ESIUM, FALLOUT, SILTS, DETRITUS, E, GROWTH RATES, COPPER SULFATE, PSEUDOMONAS, SUBSURFACE-WATERS, LANKTON, ICE, BACTERIA, TRIPTON,

BLEACHING, WASHINGTON(DC).: /RIVE W70-02255
BLELHAM TARN(ENGLAND), CHLORELLA W70-02504
BLENDOR MORTALITY, CELL WALL, HEX W70-04365
BLOCKING, HETERETROPHS, ITALY, LA W69-07833
BLONDIN.: /EVOLUTION, *INTERSTATE W70-04430
BLOODWORMS, WATER POLLUTION.: /KT W68-01244
BLOODWORMS, AQUATIC BACTERIA, ALG W70-00475
BLOODWORMS, ALGAL CONTROL.: /DEPT W70-06217
BLOOM, 'BREAKING' OF LAKES, CHROO W69-08283
BLOOM, ALGAE GROWTH, WATER DENSIT W69-08282
BLOOM, AQUATIC ALGAE, AGING(BROLO W68-00912
BLOOM, TOXIC ALGAE, RECREATIONAL W69-08279
BLOOMING, LAKE MENDOTA(WIS), LAKE W70-05112
BLOOMS.: W70-05550
BLOOMS.: W70-00664
BLOOMS.: /I, JAPAN, *GLENODININE, W70-08372
BLOOMS, *FISH TOXICITY TEST, SLUD W70-05819
BLOOMS, *ORGANO-SILICATES, *ALKOX W71-03056
BLOOMS, *ORGANO-SILICATES, *ALKOX W71-05719
BLOOMS, ALGAE NUTRITION, MICROCYS W70-05565
BLOOMS, ALGAE PHYSIOLOGY.: W69-03512
BLOOMS, CHAINING.: /SALIX, GROWTH W70-05263
BLOOMS, CHLORELLA PYRENOIDOSA, MI W70-03512
BLOOMS, CLEARWATER LAKE(MINN), TR W70-05387
BLOOMS, CLOSED SYSTEMS, CYTOLOGY, W69-07832
BLOOMS, CYANOPHYTA, NORTH DAKOTA. W70-09093
BLOOMS, GELATINOUS MATTER.: /OTA(W70-06229
BLOOMS, LAKE GEORGE(UGANDA), RHIN W70-04369
BLOOMS, LIEBIG'S LAW OF THE MINIM W70-04765
BLOOMS, MYTILUS, TRICHODESMIUM ER W70-05372
BLOOMS, NTA, EDTA, HEIDA, LAKE TA W70-04373
BLOOMS, NUTRIENT LOAD, SYNTHETIC W69-08674
BLOOMS, NUTRIENTS, LIMNOLOGY, DYS W69-09723
BLOOMS, WATER BLOOM, AQUATIC ALGA W68-00912
BLUE FLUORESCING MATE: /LON COLUM W70-01074
BLUE-GREEN ALGAE, NITZSCHIA, BIOL W70-10173
BLUE-GREEN ALGAE, TEMPERATURE EXP W70-07257
BLUE-GREEN ALGAE, GREEN ALGAE.: / W70-08111
BLUE-GREEN ALGAE, HEAT DEATH.: W70-07313
BLUE-GREEN ALGAE, GREEN ALGAE.: / W70-03928
BLUE-GREEN ALGAE, OCULAR MICROMET W70-04253
BLUE-GREEN ALGAE, FILAMENTS.: /RM W70-05270
BLUE-GREEN ALGAE, GREEN ALGAE, FR W70-05375
BLUE-GREEN ALGAE, PHYTOPLANKTON, W69-05697
BLUEGILLS, MICROCYSTIS SCUMS, PIT W70-05417
BLUEGILLS, CHLORELLA PYRENOIDOSA, W72-11105
BMT.: /HOSLOVAKIA, LABORATORY PRO W72-12240
BOATING, EUTHROPHICATION.: /, STR W71-09880
BOD REMOVAL RATES, DESIGN APPLICA W70-07034
BOD REMOVAL, GAS PRODUCTION.: /IR W70-04787
BOD), AERATION, COLIFORMS, ALGAL W70-04788
BOD), EVAPORATION, FERMENTATION, W70-04787
BOD), MIXING, WINDS.: /D OXYGEN, W70-04790
BOD, DETENTION TIME.: *SULFATE I W70-01971
BOD, SLUDGE.: / TEMPERATURE, NUTR W70-04786
BODIES OF WATER.: /T REGION, REGI W69-03683
BODIES OF WATER, DISSOLVED OXYGEN W70-00683
BODIES, ULVA LACTUCA VAR LATISSIM W70-01073
BODY BURDENS, POMOXIS ANNULARIS, W70-02786
BODY IMMEDIUM FILTER, BACK-WASHIN W70-05821
BOGS, PHENOLS, PIGMENTS, ORGANIC W70-01074
BOHEMIA, ELBE RIVER, CELAKOVICE(C W70-00274
BOILING.: /LA, PEPTIDES, BACTERIA W70-04184
BOMB MATERIAL, RUTHENIUM-106-RHOD W69-08524
BOMBAY(INDIA), BENARES(INDIA), MA W70-05404
BONDS, METRO ACT(SEATTLE).: /INGT W70-04455
BONE, LEAVES, SHELL, RHODIUM-102. W69-08269
BOREALIS, LEPTOCYLINDRUS DANICUS, W70-04280
BORNITE, CHALCOPYRITE, CHEMOSYNTH W69-07428
BORON, AZOTOBACTER, MOLYBDENUM, C W70-02964
BORREVANNET NORWAY, LAKE LUGANO I W69-07833
BORYA: / PUSILLA, OOCYSTIS SOLITA W69-06540
BORYANUM, REACTOR.: SAVAN W69-07845
BORYANUM, STAURASTRUM PARADOXUM, W70-02804
BOSMINA COREGONI LONGISPINA.: /(S W69-10169
BOSMINA LONGISPINA, ZURICHSEE(SWI W69-10169
BOSMINA, ASTERIONELLA, MELOSIRA, W70-04503
BOSMINA, NAVICULA, FRAGILARIA, CY W70-02249
BOSTON LOT RESERVOIR, LEBANON, WA W69-08674
BOSTON(MASS).: W70-05264
BOTRYOCCUS BRAUNII, FRAGILARIA CR W70-02804
BOTRYOCOCCUS BRAUNII, METABOLIC P W70-02965
BOTRYOCOCCUS, PLANKTON DISTRIBUTI W70-03959
BOTTOM SEALING, LAKE DILUTION.: / W69-09340
BOTTOM SEDIMENTS.: /AGE EFFLUENTS W69-08674
BOTTOM SEDIMENTS, DREDGING, HARVE W69-09340
BOTTOM SEDIMENTS, CHUTES, EDDIES, W69-09742
BOTTOM SEDIMENTS, LAKE MICHIGAN, W69-09349
BOTTOM SEDIMENTS, SURFACE WATERS, W70-00713
BOTTOM SEDIMENTS, SOLAR RADIATION W69-10154

ILIMNION, PHOTOSYNTHESIS, ALGAE,
NUISANCE ALGAE, AQUATIC PLANTS,
E, *TREATMENT, *LAKES, *ECOLOGY,
LE, ALGAE, BIOLOGICAL TREATMENT,
YGEN, CARBON, ENZYMES, SAMPLING,
*DISTRIBUTION, ANIMALS, PLANTS,
NG, GRAVELS, MUD, SANDS, SHALES,
*CHEMICAL PROPERTIES, EPILMNION,
ODELS, ON-SITE DATA COLLECTIONS,
TIC POPULATIONS, BIBLIOGRAPHIES,
TREATMENT, HARVESTING OF ALGAE,
RACE ELEMENTS, CORES, WISCONSIN,
MARGINAL VEGETATION FAUNA, SANDY
RYTHREUM, SAXIDOMUS, CLOSTRIDIUM
ERAN, MONTANE ZONE, ALPINE ZONE,
AL-BOUND RADIOSTRONTIUM, *FUNGAL-
ND RADI/ *RADIOSTRONTIUM, *ALGAL-
OLIOSUS, EXPERIMENTAL LIMNOLOGY,
OCLOEON, ENALLAGMA, PSEUDAGRION,
GEN SULFIDE, IRON, FUNGI, CLAYS,
LTURES, CHLOROPHYTA, CYANOPHYTA,
KES, NUTRIENTS, BACTERIA, ALGAE,
DS, RUNNING WATERS, FRESH WATER,
HYTA, CALIFORNIA, UNITED STATES,
A, LABORATORY, MINERALS,/ ALGAE,
RATELLA, POLYARTHRA, ASPLANCHNA,
SCRIPTIONS, GONGROSIRA LACUSTRIS
E, PINGO LAKE, TWELVE MILE LAKE,
AEA, CRASSIPES EICHORNIA, TYPHA,
ROCARBONS, *OLEFI/ *BOTRYOCOCCUS
ELLA PYRENOIDOSA, ANKISTRODESMUS
TAURASTRUM PARADOXUM, BOTRYOCCUS
ENA CYLINDRICA, GL/ BOTRYOCOCCUS
ORKING' OF LAKES, *WATER-BLOOM, '
*GLUE GREEN ALGAE, BREAM, WHITE
INYLPYRIDINE, *GLUE GREEN ALGAE,
S, ESTUARINE ENVIRONMENT, PONDS,
NIFICUM, PYRODINIUM, GYMNODINIUM
DIS, MICROCOCCUS, STREPTOCOCCUS,
ONYAULAX TAMERENSIS, GYMNODINIUM
, ORGANISMS, TURSIOPS TRUNCATUS,
NS, ALGAE, CRUSTACEANS, BENTHOS,
ION, FOOD CHAINS, AQUATIC ALGAE,
*WATER POLLUTION, AQUATIC ALGAE,
ECOSYSTEM MODEL/ KOOTENAY LAKE,
S, OLIGOTROPHIC, *PHYTOPLANKTON,
, NORRIS RESERVOIR(TENN), FRENCH
ATER BLOOM, AQUATIC ALGAE, AGING(
BROMINE,

RATES, PHOTO REACT/ *CLADOPHORA,

NOSA, POINT LEPREAU(CANADA), NEW
NTARIO), STURGEON LAKE(ONTARIO),
ON, TROPHIC INDICATORS, NUTRIENT
ANNUAL BUDGET, PHOSPHORUS ANNUAL
UX, NUTRIENT SOURCES, PHOSPHORUS
*SEDIMENTS, *NUTRIENT
IC, TROPHOLYTIC, NITROGEN ANNUAL
UTRIENT CONCENTRATION, *NUTRIENT
BINES, ECONOMIC EFFICIENCY, HEAT
EMISTRY, WATER TEMPERATURE, HEAT
YGEN, EVAPORATION CONTROL, WATER
Y, AGRICULTURAL RUNOFF, NUTRIENT
KE WINNISQUAM(N H), MA/ NUTRIENT
S, LEBANON, BATAVIA, CLARKSBURG,
, HERBICIDES, CHARA, ALKALINITY,
TROL, *BIBLIOGRAPHIES, CATTAILS,
IATOMS, CYANOPHYTA, CHLOROPHYTA,
R MILFOIL COONTAIL, WILD CELERY,
FE, HYDROPSYCHE COCKERELLI, BODY
EROTROPHIC POTENTIAL, LINEWEAVER-
DEN, ENZYME KINETICS, LINEWEAVER-
, LAKE ERKEN(SWEDEN), LINEWEAVER-
NTS, ACRIDINE ORANGE, LINEWEAVER-
NOIDOSA, MICROCYSTIS AERUGINOSA,
A LEMMERMANNII, OTTAWA(ONTARIO),
TATUM(GREV) CL, CHLORELLA OVALIS
MUL/ *SCIRPUS FLUVIATILIS, *LAKE
RA FRACTA, CLADOPHORA GLOMERATA,
NE.:
NDUSTRIAL WASTES, SEWAGE WASTES,
ORGANIC GROWTH FACTORS, VITAMIN
VITAMIN B1, VITAMIN B6, VITAMIN
WTH FACTORS, VITAMIN B1, VITAMIN
SCILLATORIA, ANABAENA CYLINDRICA
SLAND(N C), DETECTOR, BEAUFORT(N
ECTEN IRRADIANS, PIVERS ISLAND(N
GE, ELECTROSTATIC REPULSION, DOW
LLUSKS, STRATIFICATION, DIATOMS,

BOTTOM SEDIMENTS, THERMOCLINE, HY   W70-06975
BOTTOM SEDIMENTS, ANIMALS, FISH,   W70-06225
BOTTOM SEDIMENTS, MUD, BENTHOS, W   W70-06217
BOTTOM SEDIMENTS, NUTRIENTS, *WAS   W70-07031
BOTTOM SEDIMENTS, SODIUM, POTASSI   W70-09889
BOTTOM SEDIMENTS, SOILS, NUCLEAR   W69-08269
BOTTOM SEDIMENTS, CATTAILS, RADIO   W69-08272
BOTTOM SEDIMENTS.: / ABSORPTION,   W68-00857
BOTTOM SEDIMENTS, BIBLIOGRAPHIES,   W68-00858
BOTTOM SEDIMENTS, WATERSHED MANAG   W68-00859
BOTTOM SEDIMENTS, WEED CONTROL, S   W68-00478
BOTTOM SEDIMENT, BENTHIC FAUNA.: /   W69-03366
BOTTOMS FAUNA, MARGINAL VEGETATIO   W70-06975
BOTULINUS, VENERUPIS, OSTREA, ART   W70-05372
BOULDER(COLO), TEA LAKE(COLO), BL   W69-10154
BOUND RADIOSTRONTIUM, *MICROBIAL   W71-04067
BOUND RADIOSTRONTIUM, *FUNGAL-BOU   W71-04067
BRACHIONUS, SYNCHAETA PECTINATA,   W70-02249
BRACHYTHEMIS, CHYDORUS, PLEUROXUS   W70-06975
BRACKISH WATER, FISHERIES, LIMNOL   W70-05761
BRACKISH WATER, OLIGOTROPHY, TEMP   W70-05752
BRACKISH WATER, ALKALINE WATER, S   W71-03031
BRACKISH WA: /S, CARP, CLAMS, PON   W70-09905
BRACKISH WATER, ESTUARIES, BACTER   W70-05372
BRACKISH, ECOSYSTEMS, FAUNA, FLOR   W72-02203
BRANCHIONUS.: /A, MICROCYSTIS, KE   W70-05375
BRAND, GONGROSIRA DEBARYANA RAB,   W70-04468
BRANT LAKE, IMIKPUK LAKE, NORTH M   W69-08272
BRASENIA, DOW CONTACT, SPARGANIUM   W70-05263
BRAUNII, *ANACYSTIS MONTANA, *HYD   W69-08284
BRAUNII, DUNALIELLA TERTIOLECTA, '   W70-05261
BRAUNII, FRAGILARIA CROTONENSIS,   W70-02804
BRAUNII, METABOLIC PATTERN, ANABA   W70-02965
BREAKING' OF LAKES, CHROOCOCCACEA   W69-08283
BREAM, LEUKOCYTIC COMPOSITION, AL   W70-01788
BREAM, WHITE BREAM, LEUKOCYTIC CO   W70-01788
BREEDINGS, SALINE WATER-FRESHWATE   W73-00748
BREVE, CANADA, EUROPE, SOUTH AFRI   W70-05372
BREVIBACTERIUM, CORYNEBACTERIUM,   W71-11823
BREVIS, MYCROSYSTIS, ALGASTATIC S   W70-02510
BREVOORTIA TYRRANUS, CONCENTRATIO   W69-08274
BRINE SHRIMP, ECOSYSTEMS, FISH, P   W69-07440
BRINE SHRIMP, FISH, MUSSELS.: /UT   W70-01466
BRINE SHRIMP, FISH, FOOD CHAINS,   W70-01467
BRITISH COLUMBIA, CANADA, AQUATIC   W69-06536
BRITISH ISLES, BIOINDICATORS, *ES   W68-00255
BROAD RIVER(TENN), CHICKAMAUGA RE   W71-07698
BROLOGICAL), *AQUATIC PRODUCTIVIT   W68-00912
BROMAMINE, CHLORAMINE.:   W70-02370
BROMINE, BROMAMINE, CHLORAMINE.:   W70-02370
BROMINE, FILAMENTOUS ALGAE, KILL   W70-02364
BROMINE, IODINE.:   W70-05178
BRUNSWICK(CANADA), SERVALL OMNI-M   W70-04365
BUCKHORN LAKE(ONTARIO), CLEAR LAK   W70-02795
BUDGET.:   *CULTURAL EUTROPHICATI   W71-03048
BUDGET.: / TROPHOLYTIC, NITROGEN   W69-05142
BUDGET, APHANIZOMINON FLOS AQUAE.   W70-04268
BUDGET, CYCLING NUTRIENTS.:   W68-00511
BUDGET, PHOSPHORUS ANNUAL BUDGET.   W69-05142
BUDGET, POTENTIAL NUTRIENT CONCEN   W70-00270
BUDGET, SOCIAL ASPECTS.: /ON, TUR   W70-08834
BUDGET, TURBIDITY, *RESERVOIR EVA   W72-01773
BUDGET, WATER CHEMISTRY, WATER TE   W72-01773
BUDGETS, COMPARATIVE STUDIES, NUT   W69-09349
BUDGETS, LAKE SEBASTICOOK(ME), LA   W70-07261
BUFFALO, DUNREITH, OIL SPILLS.: /   W72-03641
BULLHEADS, EELS, KILLIFISHES, TRO   W70-05263
BULRUSH, DREDGING, COPPER SULPHAT   W70-05263
BULRUSH, PLANT GROWTH SUBSTANCES,   W72-00845
BUR REED, DUCKWEED, WHITE WATER L   W70-05568
BURDENS, POMOXIS ANNULARIS, GASTR   W70-02786
BURK EQUATION, UPTAKE VELOCITY, M   W70-05109
BURK EQUATION, MICHAELIS KINETICS   W70-04194
BURK EQUATION, MAXIMUM UPTAKE VEL   W70-04287
BURKE EQUATION.: /AND, PHAEOPIGME   W70-04811
BURNTSIDE RIVER(MINN).: /LLA PYRE   W70-03512
BURTON LAKE(SASKATCHEWAN), HUMBOL   W70-00273
BUTCHER, NJOER LAKE(NORWAY), HOBO   W70-05752
BUTTE DES MORTS(WISC), GROWTH STI   W72-00845
BUTTERNUT CREEK(NY).: *CLADOPHO   W72-05453
BYELORUSSIA(USSR), POLYCHLOROPINE   W70-08654
BYPRODUCTS, FERTILIZERS, ALGAE, Z   W70-09905
B1, VITAMIN B6, VITAMIN B12, BIOT   W70-02510
B12, BIOTIN, HISTIDINE, URACIL, R   W70-02510
B6, VITAMIN B12, BIOTIN, HISTIDIN   W70-02510
B629, ANABAENA INAEQUALIS 381, AN   W70-05091
C), CRASSOSTREA VIRGINICA, MERCEN   W69-08267
C), DETECTOR, BEAUFORT(N C), CRAS   W69-08267
C-31, DOW A-21, NALCO N-670.: /AR   W70-05267
C: /LIFE CYCLES, OLIGOCHAETES, MO   W71-00665

INC, COBALT, CALCIUM, MAGNESIUM,	C: /TASSIUM, IRON, BICARBONATE, Z	W70-03978
CUTTING, *WEED EXTRACTING STEEL	CABLES, ALGAL SUCCESSION, ANABAEN	W70-04494
ULTER COUNTER, EUGLENA GRACILIS,	CACHONINA NIEI, SKELETONEMA COSTA	W69-08278
PHYTOPLANKTON, SUCKERS, SHINERS,	CADDISFLIES, TEMPERATURE, FLOW, D	W69-07853
SEE RIVER, TROPHIC LEVEL, ALGAE,	CADDISFLIES, FOOD CHAINS, STABLE	W70-02786
UATIC BACTERIA, ALGAE, MAYFLIES,	CADDISFLIES.: /MS, BLOODWORMS, AQ	W70-00475
IOISOTOPE/ CESIUM RADIOISOTOPES,	CADMIUM RADIOISOTOPES, CERIUM RAD	W71-09850
WATER, ACIDIC WATER, TRAVERTINE,	CALCIUM CARBONATE, BACTERIA, BIOM	W69-10160
LOGY, MINNESOTA, WATER ANALYSIS,	CALCIUM CARBONATE, HYPOLIMNION.: /	W69-00632
NUTRIENTS, PHOSPHORUS, NITROGEN,	CALCIUM CHLORIDE, TEMPERATURE, ME	W72-05453
RY PRODUCTIVITY, ALGAE, BIOMASS,	CALCIUM COMPOUNDS, CHLORIDES, SAL	W69-02959
EN ION CONCENTRATION, MAGNESIUM,	CALCIUM.: /S, CHLOROPHYLL, HYDROG	W70-01579
GHT INTENSITY, NITRATE, EUGLENA,	CALCIUM, AMMONIUM SALTS, NITROGEN	W70-02964
ELEMENTS, VITAMINS, AMINO ACIDS,	CALCIUM, AMMONIA, NITROGEN FIXATI	W70-02255
, NUTRIENTS, SOLUTES, CHLORIDES,	CALCIUM, CARBONATES, PHOSPHATES.:	W70-09557
, *MANGANESE, *WISCONSIN, ALGAE,	CALCIUM, CHELATION, CYANOPHYTA, N	W69-05867
S, ANTIBIOTICS, BIOASSAY, NORON,	CALCIUM, CHLORINE, COBALT, COPPER	W69-04801
S, *WATER SUPPLY, BIOINDICATORS,	CALCIUM, CHLORIDES, CONDUCTIVITY,	W69-07833
ER, SEDIMENT LOAD, CONDUCTIVITY,	CALCIUM, MAGNESIUM, ALKALINITY, S	W69-05894
ES, TESTS, NITROGEN, PHOSPHORUS,	CALCIUM, MAGNESIUM, SODIUM, SILIC	W68-00856
IRON, BICARBONATE, ZINC, COBALT,	CALCIUM, MAGNESIUM, C: /TASSIUM,	W70-03978
NUTRIENTS, ESSENTIAL NUTRIENTS,	CALCIUM, MAGNESIUM, POTASSIUM, HA	W70-02245
M, NITROGEN, PHOSPHORUS, SULFUR,	CALCIUM, MAGNESIUM, POTASSIUM, SO	W70-02804
CENTRATION, NITRATES, POTASSIUM,	CALCIUM, MAGNESIUM, IRON, ORGANIC	W70-00719
SEASONAL, PRIMARY PRODUCTIVITY,	CALCIUM, NITRATES, AMMONIA, PHOSP	W70-05405
UM, SILICATES, DISSOLVED SOLIDS,	CALCIUM, PRODUCTIVITY, STANDING C	W70-03512
N, COLLOIDS, ZINC RADIOISOTOPES,	CALCIUM, POTASSIUM, CLAMS, SIZE,	W70-00708
IUM, POTASSIUM, IRON, MANGANESE,	CALCIUM, PIGMENTS.: /DIMENTS, SOD	W70-09889
IA, ALGAE, PLANKTON, TUBIFICIDS,	CALCIUM, SILICATES, DIATOMS, WIND	W70-05761
ICO, SPECTROSCOPY, DISTRIBUTION,	CALCIUM, TROPHIC LEVELS, FOOD WEB	W69-08525
HYLL CONTENT, CENTRIFUGE METHOD,	CALCULATED VOLUME METHOD, CYCLOTE	W70-03959
L PHYSIOLOGY, *DESIGN THEORY AND	CALCULATIONS, PHOTOSYNTHESIS, ILL	W70-04786
LGAE HARVESTING, *CENTRAL VALLEY(| CALIF.).: *A	W71-04556
LGAE HARVESTING, *CENTRAL VALLEY(| CALIF.).: *A	W71-04557
SIUM PERMANGANATE, BEVERLY HILLS(| CALIF).: *POTAS	W71-00110
LOS ANGELES COUNTY(| CALIF).:	W70-05645
MS, NTA, EDTA, HEIDA, LAKE TAHOE(| CALIF).: /OMPLEXONES, *ALGAL BLOO	W70-04373
ACTERIAL PLATE COUNTS, LANCASTER(| CALIF).: / PACKED CELL VOLUMES, B	W71-05267
ULATE ORGANIC CARBON, LAKE TAHOE(| CALIF), CHLORELLA PYRENOIDOSA, SE	W70-03983
UTION ASSESSMENT, AMERICAN RIVER(| CALIF), CLEAR LAKE(CALIF), EEL RI	W70-02777
MERICAN RIVER(CALIF), CLEAR LAKE(| CALIF), EEL RIVER(CALIF), JOINT I	W70-02777
F), CLEAR LAKE(CALIF), EEL RIVER(| CALIF), JOINT INDUSTRY/GOVERNMENT	W70-02777
ON, LAKE MONONA(WIS), LAKE TAHOE(| CALIF), LAKE WASHINGTON(WASH), ZU	W69-06858
ILLATORIA, AUSTRALIA, SANTA ROSA(| CALIF), LYUBLIN(USSR), LYUBERETS(	W70-05092
UMNE RIVER(CA/ SAN JOAQUIN RIVER(| CALIF), MERCED RIVER(CALIF), TUOL	W70-02777
ENGINEERING RESEARCH LABORATORY(| CALIF), NATIONAL EUTROPHICATION R	W70-02775
T, *CHAOBORID MIDGES, CLEAR LAKE(| CALIF), NUTRIENT ENERGY PATHWAYS,	W70-05428
TER PLAN, DELTA WATER FACILITIES(| CALIF), OXYGEN DYNAMICS, SACRAMEN	W70-05094
CEL/ STRIPPING, FIREBAUGH CENTER(| CALIF), SCENEDESMUS QUADRICAUDA,	W70-03334
CED RIVER(CALIF), TUOLUMNE RIVER(| CALIF), STANISLAUS RIVER(CALIF),	W70-02777
AQUIN RIVER(CALIF), MERCED RIVER(| CALIF), TUOLUMNE RIVER(CALIF), ST	W70-02777
E RIVER(CALIF), STANISLAUS RIVER(| CALIF), WATER POLLUTION ASSESSMEN	W70-02777
EASUREMENTS, EUPHOT/ *LAKE TAHOE(| CALIF-NEV), TRANSPARENCY, LIGHT M	W70-01933
TS, MORPHOMETRY, PA/ *LAKE TAHOE(| CALIF-NEV), HYDROPHYTES, LIVERWOR	W70-00711
XYGEN DEMAND, OXIDATION LAGOONS,	CALIF: / CHLORELLA, BIOCHEMICAL O	W70-05092
R/ SACRAMENTO-SAN JOAQUIN DELTA,	CALIFORNIA WATER PLAN, DELTA WATE	W70-05094
ROPHYTA, DINOFLAGELLATES, ALGAE,	CALIFORNIA, FLORIDA, PESTICIDE RE	W70-05272
LLATES, CYANOPHYTA, CHRYSOPHYTA,	CALIFORNIA, UNITED STATES, BRACKI	W70-05372
N ION CONCENTRATION, ALKALINITY,	CALIFORNIA, LIGHT INTENSITY, TEMP	W70-03334
OSPHORUS, WASHINGTON, WISCONSIN,	CALIFORNIA.: / NUISANCE ALGAE, PH	W69-06858
SOTA, SEDIMENT-WATER INTERFACES,	CALIFORNIA, BIOASSAY, SCENEDESMUS	W70-02775
ROWTH, *BIOASSAYS, *FLUOROMETRY,	CALIFORNIA, NUTRIENTS, EUTROPHICA	W70-02777
BIODEGRADATION, *DRAINAGE WATER,	CALIFORNIA, DRAINAGE PROGRAMS, AN	W71-04555
ES, EVAPORATION, MICROORGANISMS,	CALIFORNIA, *PHOTOSYNTHESIS, WAST	W71-05267
CATION.: *CENTRAL VALLEY(| CALIFORNIA), *BACTERIAL DENITRIFI	W71-04554
BIODEGRADATION, *DRAINAGE WATER,	CALIFORNIA, DRAINAGE PROGRAMS, AN	W71-04554
BIODEGRADATION, *DRAINAGE WATER,	CALIFORNIA, DRAINAGE PROGRAMS, AN	W71-04556
BIODEGRADATION, *DRAINAGE WATER,	CALIFORNIA, DRAINAGE PROGRAMS, AN	W71-04557
CATION.: *CENTRAL VALLEY(| CALIFORNIA), *BACTERIAL DENITRIFI	W71-04555
COSTS, ARTIFICIAL WATER COURSES,	CALIFORNIA.: / CONVEYANCE, WATER	W71-09577
COPPER SULFATE, CHLORINE, WELLS,	CALIFORNIA.: /Y CONTROL, *ALGAE,	W71-00110
-PURPUREA, SPHACELARIA, GONIDIA,	CALLOPHYLLIS HAENOPHYLLA, TIDAL C	W69-10161
DISSOLVED SOLIDS, HEDRIODISCUS,	CALOPARYP: /CTROPHOTOMETRY, TOTAL	W70-03309
BIOLOGY, *DIURNAL OXYGEN CURVES,	CALORIMETRY, SOLARIMETRY, YELLOWS	W70-03309
ODGETTS, CHROOCOCCUS PRESCOTTII,	CALOTHRIX FUSCA, MICROCOLEUS LACU	W70-04468
GAL NUTRITION, ALGAL PHYSIOLOGY,	CALOTHRIX PARIETINA, COCCOCHLORIS	W69-06277
*CHICAGO(ILL), COOK COUNTY(ILL),	CALUMET RIVER(ILL), POOLS, WIND D	W70-00263
NAVICULA, APHANIZOMENON, ONTARIO(| CANADA): /O, FILAMENTOUS ALGAE,	W69-10155
LAND), WHITEFISH, PUNCHBOWL LAKE(| CANADA).: /S, LAKE ZURICH(SWITZER	W70-05399
ENT LEVELS, S/ *TWELVE MILE LAKE(| CANADA), *ONTARIO, *CANADA, NUTRI	W70-02973
SKAHA LAKE(CANADA), OSOYOOS LAKE(| CANADA), DEFORMED CHIRONOMIDS.: /	W71-12077
GUINS, LAK/ PRINCE EDWARD ISLAND(| CANADA), NOVA SCOTIA(CANADA), PEN	W70-05399
LYSIPHONIA LANOSA, POINT LEPREAU(| CANADA), NEW BRUNSWICK(CANADA), S	W70-04365
KANAGAN LAKE(CANADA), SKAHA LAKE(| CANADA), OSOYOOS LAKE(CANADA), DE	W71-12077
WARD ISLAND(CANADA), NOVA SCOTIA(| CANADA), PENGUINS, LAKE ZURICH(SW	W70-05399
PROTEUS, SERRATIA, BAY OF FUNDY(| CANADA), RHYZOIDS.: /, CYTOPHAGA,	W70-03952
T LEPREAU(CANADA), NEW BRUNSWICK(| CANADA), SERVALL OMNI-MIXER HOMOG	W70-04365
OPHIC COMMUNITIES, OKANAGAN LAKE(| CANADA), SKAHA LAKE(CANADA), OSOY	W71-12077
KOOTENAY LAKE, BRITISH COLUMBIA,	CANADA, AQUATIC ECOSYSTEM MODEL,	W69-06536
, PYRODINIUM, GYMNODINIUM BREVE,	CANADA, EUROPE, SOUTH AFRICA, ISR	W70-05372

N, LAKE MENDOTA, LAKE FURES, LA/
*AQUATIC MACROPHYTES, ANACHARIS
IFASTACUS LENIUSCULUS, ANACHARIS
INS PROJECT, INDIANA HARBOR SHIP
ERRICHLORIDE, IRON S/ RING TUBE,
A, AEROBIC BACTERIA, METABOLISM,
CES HANSENII, RHODOTORULA RUBRA,
DIDDENSII, CANDIDA ZEYLANOIDES,
IDA ZEYLANOIDES, CANDIDA KRUSEI,
CANDIDA KRUSEI, CANDIDA OBTUSA,
TORULA RUBRA, CANDIDA DIDDENSII,
TREATMENT, ANAEROBIC CONDITIONS,
, SURVIVAL, SUSCEPTIBIL/ *TORREY
, *GAMOSOL, *OIL SPILLS, *TORREY
ITY.:   *TORREY
ION, CLIFF VEGETATION, / *TORREY
CHEMICAL OXYGEN DEMAND, STORAGE
ATER QUALITY, WASTE ASSIMILATIVE
THYMIDINE, NARRAGANSETT BAY(RI),
ISLAND SOUND, NARRAGANSETT BAY,
A, ALGAE ISOLATION, CAPILLARITY,
GAL, NITZSCHIA, ALGAE ISOLATION,
METHODOLOGY, CULTURES, TESTING,
AL GROWTH POTENTIAL, SELENASTRUM
BACILLARIOPHYCEAE, SELENASTRUM
L ASSAY, PROCEDURES, SELENASTRUM
ION CONCENTRATION, FLUORESCENCE,
NFALL, TEMPERATURE, COLORIMETRY,
NTRATION, TEMPERATURE, PROTEINS,
GROWTH SUBSTANCES, PLANT GROWTH,
, AMINO ACIDS, INTERTIDAL AREAS,
IDS, BIOMASS, ORGANIC COMPOUNDS,
YRENOIDOSA, SELENASTRUM GRACILE,
N, BIOMASS, SEWAGE, FARM WASTES,
ICARBONATES, CARBON, CARBONATES,
ITROGEN FIXATION, PHOTOCHEMICAL,
ENTS, CHELATION, PHOTOSYNTHESIS,
EWAGE, SULFATE, MANGANESE, IRON,
NGANESE, PHOTOSYNTHESIS, OXYGEN,
MUS, CHLAMYDOMONAS, TEMPERATURE,
NITY, NITRIFICATION, *CORROSION,
COLORIMETRY, SAMPLING, METHANE,
ENT, ASBESTOS CEMENT, CHLORELLA,
ROL, DECOMPOSING ORGANIC MATTER,
SULFUR BACTERIA, DIEL MIGRATION,
MICROORGANISMS, LIGHT INTENSITY,
ROGEN ION CONCENTRATION, OXYGEN,
NUTRIENTS, HYDROSTATIC PRESSURE,
ENSITY, CULTURES, RADIOACTIVITY,
ATURE, ORGANIC MATTER, HARDNESS,
NT, TEMPERATURE, PHOTOSYNTHESIS,
ODIUM COMPOUNDS, PHOTOSYNTHESIS,
ITROGEN FIXATION, *ALGAE, LIGHT,
LAMYDOMONAS, CHLORELLA, ENZYMES,
ROPHICATION, LIGHT, TEMPERATURE,
, PHYSIOLOGICAL ECOLOGY, OXYGEN,
, CHLORELLA, *SEWAGE, EFFLUENTS,
WATER CHEMISTRY, *PHYTOPLANKTON,
ON RADIOISOTOPES, PACIFIC OCEAN,
ENTIAL, SOLAR RADIATION, ENERGY,
ON, NITROGEN, OXYGEN, OXIDATION,
PIDS, LIGHT, ALGAE, TEMPERATURE,
DOSA, ANKISTRODESMUS BRAUNII, D/
RIENTS, EFFLUENTS, ALGAE, WEEDS,
AGOONS, *ALGAE, LIGHT INTENSITY,
*VISCOSE RAYON, ZINC COMPOUNDS,
UPHOTIC ZONE, PLANT CAROTENOIDS,
ALGAE, DIATOMS, EUTROPHICATION,
IS, BICARBONATES, RADIOACTIVITY,
ORELLA, TRACERS, PHOTOSYNTHESIS,
LORELLA, TRACERS, RADIOACTIVITY,
REMENT, PROTEINS, RADIOACTIVITY,
TOSYNTHESIS, DIATOMS, CHLORELLA,
WTH RATES, TRITIUM, PHOTOGRAPHY,
MMUNITIES, *ALGAE, *METHODOLOGY,
ATER POLLUTION EFFECTS, DIATOMS,
A, ALGAE, PHYSIOLOGICAL ECOLOGY,
ARINE BACTERIA, WATER CHEMISTRY,
AE, SEWAGE, DIATOMS, CYANOPHYTA,
RADIOCHEMICAL ANALYSIS, TRITIUM,
ECOLOGY, OXYGEN, CARBON DIOXIDE,
CTERIA, EUTROPHICATION, GLUCOSE,
GICAL ECOLOGY, AQUATIC BACTERIA,
TION, BIOCHEMICAL OXYGEN DEMAND,
RGANIC MATTER, LAKES, DIFFUSION,
TOTAL ORGANIC
N, FLOCCULATION, COAL, ACTIVATED
RS LAKE, DUNLOP LAKE, *INORGANIC
OSA, ALGAL CULTURES, CHEMOSTATS,
GLAND, LAKE WINDERMERE(ENGLAND),

CANADA, MESOTROPHY, LAKE NORRVIKE   W70-03964
CANADENSIS, MACROPHYTES, POTAMOGE   W70-02249
CANADENSIS, MYRIOPHYLLUM, POTAMOG   W70-00711
CANAL.: /LAKES-ILLINOIS RIVER BAS   W70-00263
CANALIZATION, ALUMINUM SULFATE, F   W70-05668
CANALS, SANDS, CHLORINATION, MICR   W69-07838
CANDIDA DIDDENSII, CANDIDA ZEYLAN   W70-04368
CANDIDA KRUSEI, CANDIDA OBTUSA, C   W70-04368
CANDIDA OBTUSA, CANDIDA TROPICALI   W70-04368
CANDIDA TROPICALIS, NOCTILUCA MIL   W70-04368
CANDIDA ZEYLANOIDES, CANDIDA KRUS   W70-04368
CANNERIES, ACTIVATED SLUDGE, BIOC   W71-09546
CANYON, *DETERGENT CONCENTRATIONS   W70-01779
CANYON, AROMATIC SOLVENTS, PHAEOD   W70-01470
CANYON, REPOPULATION, SUSCEPTIBIL   W70-01780
CANYON, SUSCEPTIBILITY, REGENERAT   W70-01777
CAPACITY, EFFLUENTS, CONSTRUCTION   W71-00936
CAPACITY, STANDING CROP, ANNUAL T   W68-00912
CAPE REYKJANES(ICELAND), SUDURNES   W69-09755
CAPE REYKJANES, FAXAFLOI FJORD, I   W69-10161
CAPILLARITY CONTROL DEVICE.: /CHI   W70-04469
CAPILLARITY, CAPILLARITY CONTROL   W70-04469
CAPILLARY CONDUCTIVITY, BACTERIA,   W70-04469
CAPRICORNUTUM, ANABAENA FLOS-AQUA   W69-06864
CAPRICORNUTUM PRINTZ, OSCILLATORI   W70-05752
CAPRICORNUTUM, SELENASTRUM GRACIL   W70-02779
CARBOHYDRATES, TEMPERATURE, SALIN   W70-01074
CARBOHYDRATES, PHENOLS, NITROGEN,   W70-01073
CARBOHYDRATES, LIPIDS, KINETICS,   W71-13413
CARBOHYDRATES.: /N, ALGAE, PLANT   W69-07826
CARBOHYDRATES, YEASTS, ENZYMES, C   W70-03952
CARBOHYDRATES.: /LLA, *ORGANIC AC   W70-04195
CARBON ASSIMILATION, BACILLARIOPH   W70-03983
CARBON CYCLE, PHOSPHORUS, PHOSPHA   W70-02777
CARBON CYCLE.: /EUTROPHICATION, B   W70-02792
CARBON DIOXIDE, SYNTHESIS, ENZYME   W70-02965
CARBON DIOXIDE, BICARBONATES, CAR   W70-02804
CARBON DIOXIDE.: /H SUBSTANCES, S   W70-02969
CARBON DIOXIDE, SCENEDESMUS, CHLO   W70-02964
CARBON DIOXIDE, HYDROGEN ION CONC   W70-02249
CARBON DIOXIDE, SOLUBILITY, *PHOS   W71-11393
CARBON DIOXIDE, DIFFUSION, *WASTE   W71-11377
CARBON DIOXIDE.: /BORATORY EQUIPM   W71-07339
CARBON DIOXIDE.: / POLLUTION CONT   W72-01788
CARBON DIOXIDE, HYDROGEN ION CONC   W70-05760
CARBON DIOXIDE, DISSOLVED OXYGEN,   W71-00668
CARBON DIOXIDE, AMMONIA, NITRITES   W71-03031
CARBON DIOXIDE, CYANOPHYTA, PHYSI   W69-10160
CARBON DIOXIDE, WATER POLLUTION E   W69-10158
CARBON DIOXIDE, FISH.: /Y, TEMPER   W69-10157
CARBON DIOXIDE, DEPTH, THERMOCLIN   W69-10169
CARBON DIOXIDE, NITRATES, PHOSPHA   W70-00161
CARBON DIOXIDE, INHIBITION, NITRO   W70-04249
CARBON DIOXIDE, GROWTH RATES, PLA   W70-04369
CARBON DIOXIDE, NUTRIENTS, BIOLOG   W69-07442
CARBON DIOXIDE, CARBON RADIOISOTO   W69-07862
CARBON DIOXIDE, *ALGAE, *HYDROGEN   W68-00855
CARBON DIOXIDE, NUTRIENT REQUIREM   W70-05424
CARBON DIOXIDE, LIGHT, TEMPERATUR   W70-05652
CARBON DIOXIDE, PHOTOSYNTHESIS, G   W70-05750
CARBON DIOXIDE, ALKALINITY, ALGAE   W70-05550
CARBON DIOXIDE, OXIDATION, CHEMIC   W70-05651
CARBON DIOXIDE, CHLORELLA PYRENOI   W70-05261
CARBON DIOXIDE, VITAMINS, NITRATE   W70-04810
CARBON DIOXIDE, GROWTH RATES, TEM   W72-06022
CARBON DISULFIDE.: /NTS, NYLON-6,   W70-06925
CARBON FIXATION, PELAGIC REGION,   W70-01933
CARBON RADIOISOTOPES, CHLOROPHYLL   W70-01579
CARBON RADIOISOTOPES, LAKES, ORGA   W70-02504
CARBON RADIOISOTOPES, FISH, CLAMS   W70-02786
CARBON RADIOISOTOPES, HYDROGEN IO   W70-08097
CARBON RADIOISOTOPES, PHOSPHORUS   W69-10151
CARBON RADIOISOTOPES, LIGHT INTEN   W69-10158
CARBON RADIOISOTOPES, PHOSPHORUS   W69-10163
CARBON RADIOISOTOPES, DIATOMS, ME   W71-12870
CARBON RADIOISOTOPES, KINETICS, T   W70-04811
CARBON RADIOISOTOPES, LAKES, PHYT   W70-05109
CARBON RADIOISOTOPES, PACIFIC OCE   W70-05652
CARBON RADIOISOTOPES, POTASSIUM,   W70-05405
CARBON RADIOISOTOPES, BIOTA, CLAM   W69-08269
CARBON RADIOISOTOPES, DETRITUS, C   W69-07862
CARBON RADIOISOTOPES, DINOFLAGELL   W70-04194
CARBON RADIOISOTOPES, ENZYMES, KI   W70-04287
CARBON RADIOISOTOPES, DETERGENTS,   W70-04001
CARBON RADIOISOTOPES, CHLAMYDOMON   W70-04284
CARBON.:   W72-04495
CARBON.: /AE, FILTRATION, SORPTIO   W70-05470
CARBON.: /AKES, QUIRKE LAKE, PECO   W70-05424
CARBON-14.: / MICROCYSTIS AERUGIN   W69-06864
CARBON-14.: /TLAND), SCOTLAND, EN   W70-02504

UTION VIII, NUTRIENT LIMITATION,
:   *ALGAE TOXICITY,
LUORESCENCE MICROSCOPY, ACETATE,
HNIQ/ *HETEROTROPHY, AUTOTROPHY,
ERKEN(SWEDEN), ACTIVE TRANSPORT,
MENT, *WASTE WATER T/ *ACTIVATED
ONOCHRYSIS LUTHE/ *DNA, *ORGANIC
LGAE, *WATER QUALITY, *ACTIVATED
N, COAGULATION, ALGAE, ACTIVATED
*POTASSIUM COMPOUNDS, *ACTIVATED
*ACTINOMYCETES, *ODOR, ACTIVATED
ON, ECOLOGY, WATER PURIFICATION,
S, EUTROPHICATION, BICARBONATES,
R, AMMONIUM COMPOUNDS, ACTIVATED
LOROFORM-METHANE EXTRACTS, LIPID
ULFATE, SEDIMENTATION, ACTIVATED
OROPHYLL A, SHAYLER RUN, ORGANIC
DISTRIBUTION PATTERNS, ACTIVATED
Y, BIOINDICATORS, CHLAMYDOMONAS,
E, NITROGEN, PHOSPHORUS, OXYGEN,
TION, SAMPLING, SURFACES WATERS,
S, SEWAGE, ADSORPTION, ACTIVATED
TE, *ODOR, *ALGICIDES, ACTIVATED
N DIOXIDE, OXIDATION, CHEMICALS,
CAL DENSITY, PARTICULATE ORGANIC
ALGAE, ORGANIC MATTER, CULTURES,
   PHENOLS, NITROGEN, RESPIRATION,
ALGAE, *CULTURES, PHYTOPLANKTON,
   CHLOROPHYTA, NITRATES, AMMONIA,
SONAL, DIURNAL, LIGHT INTENSITY,
ONMENTS, VARIABILITY, CHLORELLA,
TER), CHLORIDES, ORGANIC MATTER,
ATION, COPPER SULFATE, ACTIVATED
CIDIC WATER, TRAVERTINE, CALCIUM
   SCHISTOSOME DERMATITIS, COPPER
QUATI/ *PHOTOSYNTHESIS, *CALCIUM
NNESOTA, WATER ANALYSIS, CALCIUM
, AQUATIC ALGAE, WATER ANALYSIS,
TS, SOLUTES, CHLORIDES, CALCIUM,
PHICATION, BICARBONATES, CARBON,
IRS, SCUM, NITROGEN, PHOSPHORUS,
S, CARBON DIOXIDE, BICARBONATES,
   ALGAE, FUNGI, SULFUR COMPOUNDS,
NOIDOSA, CARBOXYLIC ACIDS, AMINO
ATE, NTA, CHLORELLA PYRENOIDOSA,
ERIONELLA, TYP/ *AUGUSTA(MAINE),
ASS), ARTHUR D LITTLE INC(MASS),
   CARNEGIE DEPT OF PLANT BIOLOGY,
F WATER, DISSOLVED OXYGEN, NORTH
, *SORPTION, *TEMPERATURE, SOUTH
ER, OYSTERS, CLAMS, CRABS, NORTH
IDE METABOLISM, *PESTICID/ SOUTH
UM-232, BISMUTH-214, AC/ *COTTUS
ASUREMENTS, EUPHOTIC ZONE, PLANT
OKLA), KEYSTONE RESERVOIR(OKLA),
   DESICCATION, YAQUINA BAY(ORE),
MOSQUITOS, ALGAL CONTROL, BASS,
RE, ACTIVATED SLUDGE, FISHERIES,
ROPHYTA, DISSOLVED OXYGEN, PIKE,
SYNTHESIS, PHYTOPLANKTON, CISCO,
ADIOISOTOPES, IONS, HUMIC ACIDS,
OMPOUND, PERCHES, FISH, DIATOMS,
L ANALYSIS, PHYSICAL PROPERTIES,
R(POLAND), MONTANE RIVER COURSE,
U/ *OSCILLATORIA, NATIONAL WATER
NIQUES, W/ *BIOTIN, *AMPHIDINIUM
ALIELLA TERTIOLECTA, AMPHIDINIUM
E, CHLOROPHYCEAE, CHRYSOPHYCEAE,
HALOCHLOROCOCCUM, SYNECHOCOCCUS,
4, NITZSCHIA CLOSERIUM, ARTEMIA,
HYLOCOCCUS AUREUS, LACTOBACILLUS
C EFFECTS, STIMULANTS, GONYAULAX
ITOS, ALGAL CONTROL, BASS, CARP,
), BACK RIVER(MD), OSCILLATORIA,
LUTION EFFECTS, CATION EXCHANGE,
RPTION, WATER POLLUTION EFFECTS,
THETIC ORGANIC POLYELECTROLYTES,
C WEED CONTROL, *BIBLIOGRAPHIES,
SANDS, SHALES, BOTTOM SEDIMENTS,

POSTING ORGANIC MATTER, LAGOONS,
OLLUTION EFFECTS, WATER QUALITY,
COCCUS, SPINES, SCHIZOMYCOPHYTA,
PERIMENTS, *NUTRIENT LIMITATION,
A RUBESCENS, MADISON LAKES(WIS),
AERUGINOSA, BOHEMIA, ELBE RIVER,
AE, WATER MILFOIL COONTAIL, WILD
E, KILL RATES, PHOTO REACTIVITY,
EMENTS, SCENEDESMUS, ABSORPTION,
ICU/ *GROWTH, *UPTAKE, FIXATION,

CARBON-14.: /KES(NY), RODHE'S SOL   W70-01579
CARBON-14, DDT, MALATHION, SEVIN.   W70-02198
CARBON-14, AUTORADIOGRAPHY, MICHA   W70-04811
CARBON-14, RADIOCARBON UPTAKE TEC   W70-04287
CARBON-14, ENZYME KINETICS, GLUCO   W70-04284
CARBON, *ADSORPTION, *WATER TREAT   W72-03641
CARBON, *CELL SIZE, EUKARYOTES, M   W69-08278
CARBON, *ELECTROLYTES, FILTERS, *   W72-09693
CARBON, *FLOCCULATION, *HEAD LOSS   W72-07334
CARBON, ACTINOMYCETES, IOWA.: /,   W72-04242
CARBON, AMMONIA, CHLORINE.: /AE,   W72-04155
CARBON, BICARBONATES, PHOTOSYNTHE   W69-07389
CARBON, CARBONATES, CARBON CYCLE.   W70-02792
CARBON, CHLORINE, PHENOLS, OIL WA   W70-00263
CARBON, CHLOROPLASTS, INFRARED AN   W70-05651
CARBON, CHLORINE, COSTS, TREATMEN   W72-08859
CARBON, CINCINNATI(OHIO).: /, CHL   W71-04206
CARBON, CLAYS, ALGAE, WATER PURIF   W69-09884
CARBON, ECOSYSTEMS, EUGLENA, EUTR   W69-06273
CARBON, ENZYMES, SAMPLING, BOTTOM   W69-09889
CARBON, FERTILITY, ALGAE, NITROGE   W70-02973
CARBON, FILTERS, SEASONAL, STRATI   W70-04001
CARBON, FILTRATION, *TOXICITY, *D   W72-03746
CARBON, GRAVIMETRY.: /TURE, CARBO   W70-05651
CARBON, LAKE TAHOE(CALIF), CHLORE   W70-03983
CARBON, METABOLISM, NITROGEN, PHO   W70-02804
CARBON, PRIMARY PRODUCTIVITY, OXY   W70-01073
CARBON, RADIOACTIVITY, CHLORELLA,   W70-03983
CARBON, RECREATION, WATER POLLUTI   W72-04761
CARBON, SEWAGE TREATMENT, TERTIAR   W70-05387
CARBON, TEMPERATURE, LIGHT INTENS   W70-02965
CARBON, WATER POLLUTION EFFECTS,   W71-04206
CARBON, WATER SUPPLY.: /, CHLORIN   W72-03562
CARBONATE, BACTERIA, BIOMASS, CHL   W69-10160
CARBONATE, CERCARIAE, WATER MILFO   W70-05568
CARBONATE, AQUATIC ENVIRONMENT, A   W69-00446
CARBONATE, HYPOLIMNION.: /OGY, MI   W69-00632
CARBONATES, NITRATES, CHLORIDES,   W69-07096
CARBONATES, PHOSPHATES.: /NUTRIEN   W70-09557
CARBONATES, CARBON CYCLE.: /EUTRO   W70-02792
CARBONATES, TEMPERATURE, SHALLOW   W70-02803
CARBONATES, TEMPERATURE, HYDROGEN   W70-02804
CARBONATES.: /N OXIDES, BACTERIA,   W70-02294
CARBOXYLIC ACIDS.: /HLORELLA PYRE   W70-02248
CARBOXYLIC ACIDS, AMINO CARBOXYLI   W70-02248
CARLETON POND(MAINE), SYNURA, AST   W70-08096
CARNEGIE DEPT OF PLANT BIOLOGY, C   W69-06865
CARNEGIE INSTITUTION OF WASHINGTO   W69-06865
CAROLINA.: /OIRS, ALGAE, BODIES O   W70-00683
CAROLINA, CESIUM, STRONTIUM RADIO   W69-07845
CAROLINA, SEDIMENTS, RADIOACTIVIT   W69-08267
CAROLINA, LYOPHILIZATION, *PESTIC   W69-02782
CAROLINAE, MEAD COUNTY(KY), THORI   W69-09742
CAROTENOIDS, CARBON FIXATION, PEL   W70-01933
CAROTENOIDS.: /IVER(OKLA), BIXBY(W70-04001
CAROTENOIDS, SPECIES, DIVERSITY I   W70-03325
CARP, CATFISHES, NITROGEN, POTASS   W70-05417
CARP, CLAMS, PONDS, RUNNING WATER   W70-09905
CARP, D: /S, ALGAE, DIATOMS, CHLO   W70-01943
CARP, DIATOMS, CHLOROPHYTA, CHLAM   W69-02959
CARP, ICE, IRON COMPOUNDS, HYDROG   W70-05399
CARP, PLANKTON, METABOLISM, BIOLO   W70-00274
CARP, TILAPIA, TEMPERATURE, NU: /   W70-06975
CARPATHIAN SUBMONTANE REGION, GRA   W70-02784
CARRIER(ISRAEL), THRESHOLD ODOR N   W70-02373
CARTERI, *RADIOCARBON UPTAKE TECH   W70-05652
CARTERI, SYRACOSPHAERA ELONGATA,   W69-08278
CARTERIA, COMPARATIVE STUDIES, CR   W69-07833
CARTERIA, GONIUM, ANKISTRODESMUS,   W70-03974
CARTERIA, NORTH TEMPERATE ZONE, P   W70-00708
CASEI, BACILLUS THURINGIENSIS, AG   W71-11823
CATENELLA, GONYAULAX TAMERENSIS,   W70-02510
CATFISHES, NITROGEN, POTASSIUM.: /   W70-05417
CATHERWOOD DIATOMETER, SAVANNAH R   W70-04510
CATION ADSORPTION, HYDROLYSIS, MA   W71-09863
CATION EXCHANGE, CATION ADSORPTIO   W71-09863
CATIONIC ELECTROLYTES, ALGAL SURF   W70-05267
CATTAILS, BULRUSH, DREDGING, COPP   W70-05263
CATTAILS, RADIOACTIVITY, INTERFAC   W69-08272
CATTARAUGUS CREEK(NY).:
                                        W69-10080
CATTLE.: /GE, SOLID WASTES, DECOM   W71-05742
CATTLE.: /ODOR, TOXICITY, WATER P   W69-08279
CAULOBACTERIALES, CHLOROCOCCALES,   W70-03969
CAYUGA LAKE(NY), ASTERIONELLA FOR   W70-01579
CEDAR RIVER(WASH), ZURICHSEE(SWIT   W70-00270
CELAKOVICE(CZECHOSLOVAKIA), BLATN   W70-00274
CELERY, BUR REED, DUCKWEED, WHITE   W70-05568
CELL COLOR, CELL DAMAGE.: /S ALGA   W70-02364
CELL CONCENTRATION, UROGLENA, ANK   W70-05409
CELL COUNT, OPTICAL DENSITY, PART   W70-03983

CALIF), SCENEDESMUS QUADRICAUDA,
S, PHOTO REACTIVITY, CELL COLOR,
HT EXTINCTION COEFFICIENT, ALGAL
VOLATILE SUSPENDED SOLIDS, MULTI-
EGHINIANA, LIGHT-SATURATED RATE,
ESIS, MEAN DOUBLING TIME, PACKED
     MASS CULTURE OF ALGAE, PACKED
COUNT, WARING BLENDOR MORTALITY,
S, EXTRACELLULAR PRODUCTS, ALGAL
, MATERIAL BALANCES, O/ *UPTAKE,
TH FACTORS, COLONIAL MORPHOLOGY,
R COUNTER, *POPULATION DYNAMICS,
CHEMICAL DEGRADATION, VITAMINS,
LEAVES, FERTILIZERS, PESTICIDES,
ULTRAVIOLET RADIATION, SULFATES,
, LABORATORY EQUIPMENT, ASBESTOS
QUAE, LYNGBYA, STORM LAKE(IOWA),
CAUDA, CEL/ STRIPPING, FIREBAUGH
URE COEFFICIENTS, NORTH AMERICA,
RCTIC CIRCLE HOT SPRINGS, RADON,
                          SOUTH -
IRA, ASTERIONELLA, ARAPHIDINEAE,
ING, PHYSICOCHEMICAL PROPERTIES,
GE, DETERGENTS, SPRAYING, ALGAE,
ION CONCENTRATION, TEMPERATURE,
ASONICS, DISINFECTION, SAMPLING,
CONCENTRATION, ORGANIC LOADING,
T PLANKTON, CHLOROPHYLL CONTENT,
LEGANS, EUGLENA GRACILIS, FOREST
, MELOSIRA, SYNEDRA, FRAGILARIA,
ERIA SETIGERA, MELOSIRA AMBIGUA,
UROGLENA AMERICANA, CYCLOTELLA,
HANIZOMENON, ANABAENA, MELOSIRA,
ENA.:   HYDRODICTYON, ELODEA,
HORNIA, POTAMOGETON, TRAPA, PIS/
LES QUADRI/ LEECHES, SCHISTOSOME
ME DERMATITIS, COPPER CARBONATE,
ISOTOPES, CADMIUM RADIOISOTOPES,
ARTEMIA, C/ *MARINE ENVIRONMENT,
-95, RUTHENIUM-106, RHODIUM-106,
-106, RUTHENIUM-103-RHODIUM-103,
BIUM, EUROPIUM-155, SILVER-110M,
IOISOTOPES, CERIUM RADIOISOTOPE/
SUCCESSION, ZINC RADIOISOTOPES,
OTOPES, STRONTIUM RADIOISOTOPES,
ANS, FOODS, ABSORPTION, HAZARDS,
, ION EXCHANGE, RESINS, LEAKAGE,
OSIONS, STRONTIUM RADIOISOTOPES,
ASSIUM RADIOISOTOPES, MANGANESE,
XPLOSIONS, COBALT RADIOISOTOPES,
N, *TEMPERATURE, SOUTH CAROLINA,
NTS, DINOPHYCEAE, CRYPTOPHYCEAE,
RSEILLES, ASTERIONELLA JAPONICA,
ODISCUS, BACILLARIA, BIDDULPHIA,
IOSI/ *ALGAL NUTRITION, DITYLUM,
AGILARIA, OEDOGONIUM, SPIROGYRA,
ORS, MYRIOPHYLLUM, WATER BLOOMS,
, MONITORING, MICROBIOLOGY, FOOD
TOXICITY, *WATER POLLUTION, FOOD
, *WATER POLLUTION EFFECTS, FOOD
ICIDE RESIDUES, ECOSYSTEMS, FOOD
OLLUTION, *WASTE TREATMENT, FOOD
UTRIENTS, ENERGY TRANSFER, *FOOD
ALGAE, BRINE SHRIMP, FISH, FOOD
FISH, GAME BIRDS, ANIMALS, FOOD
*RADIOISOTOPES, SEA WATER, FOOD
ITY ACT, PATH OF POLLUTANT, FOOD
LEVEL, ALGAE, CADDISFLIES, FOOD
ROPHIC LEVEL, PRODUCTIVITY, FOOD
RGINIA, PENNSYLVANIA, MARCASITE,
ORPIMENT, TETRAHEDRITE, BORNITE,
LAKE WORTHER(AUSTRIA), VERTICAL,
LLATORIA RUBESCENS, OSCILLATORIA
                     *RESPIROMETER
LAKE LOTSJON(SWEDEN), PERMEASES,
/ *LAKE WASHINGTON(WASH), *LAKE
EARTH, MUD, LAKES, CULTIVATION,
LAKE(WASH), WILLIAMS LAKE(WASH),
WAGE EFFLUENT, NUTRIENTS, LAKES,
NCE ALGAE, DUCKS(WILD), OYSTERS,
, LIMNODRILUS, EUCYCLOPS, ALONA,
MENT, POTABLE WATER, HERBICIDES,
LAR TISSUES, *ALGAE, ECOSYSTEMS,
, *NEVADA, DEPTH, ALGAE, MOSSES,
EFFECTS, ECOLOGY, WATER QUALITY
IS, VIBRIO, MORPHOLOGY, CULTURAL
SICAL CHARACTERISTICS, *CHEMICAL
TERIS/ *RIVER REGIONS, *PHYSICAL
SKATCHEWAN), MICE, MORPHOLOGICAL
NCOURSE, PALOUSE RIV/ TREE BARK,

CELL COUNTS, SPECIES DETERMINATIO   W70-03334
CELL DAMAGE.: /S ALGAE, KILL RATE   W70-02364
CELL DENSITY.: / TURBIDOSTAT, LIG   W70-03923
CELL SERIAL SYSTEM, PH, FACULTATI   W70-04790
CELL VOLUME, NITZSCHIA CLOSTERIUM   W70-05381
CELL VOLUMES, BACTERIAL PLATE COU   W71-05267
CELL VOLUME.:   W70-06053
CELL WALL, HEXOSAMINE, ASCOPHYLLU   W70-04365
CELLS.: /E, PHYTOPLANKTON DYNAMIC   W69-05697
CELLS, CONCENTRATION, COMPOSITION   W70-06600
CELLULAR MORPHOLOGY, FLAGELLATION   W70-03952
CELLULAR SIZE, CHLORELLA VULGARIS   W69-06540
CELLULOSE, CHLORELLA, PEPTIDES, B   W70-04184
CELLULOSE, CYCLES, HARVESTING, DR   W70-05264
CELLULOSE, INDUSTR: /, SALINITY,   W70-01074
CEMENT, CHLORELLA, CARBON DIOXIDE   W71-07339
CENTER LAKE(IOWA), DIAMOND LAKE(I   W70-02803
CENTER(CALIF), SCENEDESMUS QUADRI   W70-03334
CENTRAL AMERICA, RATE EQUATIONS,   W70-03957
CENTRAL CITY(COLO), MUTATION, BET   W69-08272
CENTRAL STATES.:   W72-03562
CENTRIC DI: /R, FRAGILARIA, MELOS   W70-05663
CENTRIFUGATION, MICROSCOPY, NETS,   W70-05389
CENTRIFUGATION, FILTRATION, ANERO   W70-04060
CENTRIFUGATION, COAGULATION, DEWA   W71-11375
CENTRIFUGATION, SEDIMENTATION, CH   W71-11823
CENTRIFUGATION, RESPIRATION, BIOC   W72-06022
CENTRIFUGE METHOD, CALCULATED VOL   W70-03959
CENTRIFUGE, GLENODINIUM PULVISCUL   W69-06540
CER: /STIS, BOSMINA, ASTERIONELLA   W70-04503
CERATIUM HIRUDINELLA, ANKISTRODES   W70-05387
CERATIUM HIRUNDINELLA, PERIDINIUM   W70-05405
CERATIUM, OSCILLATORIA, EUDORINA.   W69-01977
CERATOPHYLLUM, MICROCYSTIS, ANABA   W68-00479
CERATOPHYLLUM, MYRIOPHYLLUM, EICH   W70-10175
CERCARIAE, TVA RESERVOIRS, ANOPHE   W70-05568
CERCARIAE, WATER MILFOIL COONTAIL   W70-05568
CERIUM RADIOISOTOPES, ECHINODERMS   W71-09850
CERIUM-144, NITZSCHIA CLOSERIUM,   W70-00708
CERIUM-144, BETA RADIOACTIVITY, G   W69-09742
CERIUM-144-PRASEODYMIUM-144.: /UM   W69-08524
CERIUM, GUETTARDA, IPOMOEA, PISON   W69-08269
CESIUM RADIOISOTOPES, CADMIUM RAD   W71-09850
CESIUM, COBALT RADIOISOTOPES, SEA   W70-04371
CESIUM, FALLOUT, SILTS, DETRITUS,   W69-09742
CESIUM, HYDROGEN, DEUTERIUM, TRIT   W70-02786
CESIUM, NUCLEAR EXPLOSIONS, ABSOR   W70-00707
CESIUM, PHYSICOCHEMICAL PROPERTIE   W70-00708
CESIUM, RADIUM RADIOISOTOPES, URA   W69-08524
CESIUM, STRONTIUM RADIOISOTOPES,   W69-08269
CESIUM, STRONTIUM RADIOISOTOPES,   W69-07845
CH: /, APHANIZOMENON, MICRONUTRIE   W70-04381
CHAETOCEROS LAUNDERI, LAUDERIA BO   W70-04280
CHAETOCEROS, COCCONEIS, COSCINODI   W70-03325
CHAETOCEROS, SKELETONEMA, THALASS   W70-04503
CHAETOPHORA, STIGEOCLONIUM, ANKIS   W70-04371
CHAINING.: /SALIX, GROWTH REGULAT   W70-05263
CHAINS.: /S, HARVESTING, DREDGING   W70-05264
CHAINS, AQUATIC ALGAE, BRINE SHRI   W70-01466
CHAINS, BACTERIA, DEGRADATION(DEC   W72-01801
CHAINS, BIOLOGICAL COMMUNITIES, B   W70-05272
CHAINS, CHLORELLA, CHLOROPHYLL, P   W70-04381
CHAINS, FOOD WEBS, FORAGE FISH, P   W70-05428
CHAINS, MUSSELS.: /UTION, AQUATIC   W70-01467
CHAINS, OXYGEN, ROTIFERS, WEEDS,   W70-10175
CHAINS, PHOSPHORUS RADIOISOTOPES,   W70-00708
CHAINS, SAMPLING, SURVEYS, MARINE   W72-00941
CHAINS, STABLE ISOTOPES, RADIOACT   W70-02786
CHAINS, TOXICITY, WATER POLLUTION   W72-01788
CHALCOCITE, COVELLITE, SPHALERITE   W69-07428
CHALCOPYRITE, CHEMOSYNTHESIS, CON   W69-07428
CHALK, METALIMNION, OSCILLATORIA   W70-05405
CHALYBIA, HYDRODICTYON RETICULATU   W69-05704
CHAMBER.:   W71-00668
CHAMYDOMONADS, LAPPLAND(SWEDEN),   W70-04287
CHANGES, *NUTRIENT CONCENTRATION,   W70-00270
CHANNEL MORPHOLOGY, SEWAGE INPUT,   W68-00475
CHANNELED SCABLANDS(WASH), WASHIN   W70-00264
CHANNELS, REVIEWS, BIBLIOGRAPHIES   W70-00269
CHANNELS, POLLUTION ABATEMENT, HU   W68-00461
CHAOBORUS.: / TANYPUS, TANYTARSUS   W70-06975
CHARA, ALKALINITY, BULLHEADS, EEL   W70-05263
CHARA, BENTHIC FLORA, FLOATING PL   W70-10175
CHARA, FISH, LIMNOLOGY, LAKE TROU   W70-00711
CHARACTERIZATION.: /TER POLLUTION   W70-02764
CHARACTERISTICS, TERMARY COMPOUND   W70-04280
CHARACTERISTICS, ORGANISMS, SUSPE   W73-00748
CHARACTERISTICS, *CHEMICAL CHARAC   W73-00748
CHARACTERS, TRIS.: /, HUMBOLDT(SA   W70-00273
CHARCOAL, ALGAL GROWTH, STREAM CO   W70-03501

ONIC ELECTROLYTES, ALGAL SURFACE
PHYTA, FUNGI, BACILLARIOPHYCEAE,
*SUSPENDED SOLIDS,
FLUORESCEIN, MACROPH/ MICHIGAN,
NISMS, PAPERMILL WASTE EFFLUENT,
CTION, POTATO PROCESSING WASTES,
ATION, *DETERGENTS, *PHOSPHATES,
ESE, *WISCONSIN, ALGAE, CALCIUM,
REDUCTION(CHEMICAL), HYDROLYSIS,
OPERTIES, TRACERS, DISTRIBUTION,
SHELLFISH, OYSTERS, INDUSTRIES,
SCENEDESMUS, CHLOROPHYLL, IRON,
IRON, MANGANESE, TRACE ELEMENTS,
HIBITORS, SUCCESSION, CHLORELLA,
ALGAE, *WATER POLLUTION EFFECTS,
IPS, NUCLEAR WASTES, ABSORPTION,
NTATION, SEDIMENTS, TASTE, WATER
SNAILS, LEGISLATION, WISCONSIN,
ION SOURCES, SEWAGE, PHOSPHORUS,
ONTROL, WATER POLLUTION CONTROL,
, WATER HYACINTH, ALLIGATORWEED,
AE, WEEDS, ECOLOGY, ENVIRONMENT,
ANT GROWTH, *WISCONSIN, ECOLOGY,
ION POTENTIAL, DISSOLVED OXYGEN,
N, PHOSPHORUS, SILICA, SAMPLING,
WTH, DECOMPOSING ORGANIC MATTER,
, RADIOACTIVITY, PHOTOSYNTHESIS,
ER CHEMISTRY, WATER TEMPERATURE,
ES, CHLOROPHYLL, PHOTOSYNTHESIS,
ITY, PHOSPHORUS, NITROGEN, FLOW,
INS, VITAMIN B, ALGAE, CULTURES,
RSHEDS, SEWAGE, WATER POLLUTION,
S, CHLORINATION, MICROORGANISMS,
NTENSITY, VELOCITY, TEMPERATURE,
EUTROPHICATION, DATA COLLECTION,
ERIC BACTERIA, METHANE BACTERIA,
ALGAE, ASSAY, DISSOLVED SOLIDS,
L, WATER PROPERTIES, PHOSPHORUS,
NUTRIENTS, ALGAE, BENTHIC FAUNA,
NEGIE INSTITUTION OF WASHINGTON,
R, ALGAE, PONDS, WATER HYACINTH,
, MECHANICAL EQUIPMENT, ENZYMES,
*NICKEL, DETECTION LIMITS,
MERICAN WATER WORKS ASSOCIATION,
RIENTS, *FERMENTATION, TOXICITY,
DSORPTION, REMOVAL, DEGRADATION,
OONS, BIOCHEMICAL OXYGEN DEMAND,
ONS, ABSORPTION, BIODEGRADATION,
ANGE, BIOCHEMICAL OXYGEN DEMAND,
IC DIGESTION, AEROBIC TREATMENT,
ENTS, ALGAE, BACTERIA, CULTURES,
ES, *EUTROPHICATION, *HYDROLOGY,
N, *TEMPERATURE, *COOLING WATER,
PLANKTON, ALGAE, BIBLIOGRAPHIES,
UTROPHICATION, PLANKTON, *ALGAE,
THERMAL POLLUTION, TEMPERATURE,
IOXIDE, SOLUBILITY, *PHOSPHATES,
, CENTRIFUGATION, SEDIMENTATION,
TROPHICATION, SPECIES DIVERSITY,
RIAL WASTES, TERTIARY TREATMENT,
TION, ELECTRODIALYSIS, SORPTION,
UTICALS, EXTRACELLULAR PRODUCTS,
, PONDS, HYDROLYSIS, METABOLISM,
APHICAL REGIONS, PHOTOSYNTHESIS,
E WATER TREATMENT, ELECTROLYTES,
RGASSO SEA, SYNURA, GYMNODINIUM,
APHTHOQUIN/ *SELECTIVE TOXICITY,
TION, PHYSI/ *REMOVAL PROCESSES,
LAIMED WATER, ORGANIC COMPOUNDS,
*PLANT GROWTH, *ALGAE, ELEMENTS(
CORROSION, OXIDATION, REDUCTION(
TURE, CARBON DIOXIDE, OXIDATION,
EUTROPHICATION, SEQUENCE, LAKES,
ECTS, *WATER QUAL/ *AGRICULTURAL
RIFICATION, WATER QUALITY, WATER
T INTENSITY, SPRING WATER, WATER
, WATER POLLUTION SOURCES, WATER
TY, *MONITORING, BACTERIA, WATER
ION CONTROL, WATER BUDGET, WATER
RIENTS, LIGHT, WASHINGTON, WATER
*WATER POLLUTION, *RIVERS, *WATER
CTIVITY, *EUTROPHICATION, *WATER
VIRONMENT, MICROORGANISMS, WATER
DISPOSAL, SEWAGE DISPOSAL, WATER
OXYGEN DEMAND, MINNESOTA, WATER
GELLATES, MARINE BACTERIA, WATER
YDROGEN ION CONCENTRATION, WATER
TS, *PHOSPHATES, *ANALYT/ *WATER
ENT, EFFLUENTS, LIMNOLOGY, WATER
SODIUM, VIRUSES, VITAMINS, WATER

CHARGE, ELECTROSTATIC REPULSION, W70-05267
CHAROPHYTA, RHODOPHYTA.: /, CYANO W70-02784
CHASE CITY(VA).: W72-10573
CHEBOYGAN COUNTY, STURGEON RIVER, W69-09334
CHEESE FACTORY WASTE WATERS.: /GA W71-00940
CHEESE MANUFACTURING WASTES, STAR W70-03312
CHELATION, WASTE TREATMENT, AQUAT W70-04373
CHELATION, CYANOPHYTA, NUISANCE A W69-05867
CHELATION, FUNGI, ALGAE, PROTOZOA W69-07428
CHELATION, DETERGENTS, RADIOACTIV W69-08275
CHELATION, BACTERIA, STREPTOCOCCU W70-01068
CHELATION, ENZYMES, CHLORINE, LIG W70-02964
CHELATION, PHOTOSYNTHESIS, CARBON W70-02804
CHELATION, OCHROMONAS, BACTERIA, W70-02510
CHELATION, SPECTROPHOTOMETRY, NUT W71-11561
CHELATION, METABOLISM, AQUARIA, W W71-09005
CHEM: /PLANKTON, SULFATES, SEDIME W69-07833
CHEMCONTROL, PONDWEEDS, ROOTED AQ W70-05568
CHEMCONTROL, INDUSTRIAL WASTES, S W70-05668
CHEMCONTROL, ARSENIC COMPOUNDS, N W70-04494
CHEMCONTROL, MECHANICAL CONTROL, W70-10175
CHEMCONTROL, BIOCONTROL, INTEGRAT W70-07393
CHEMICAL ANALYSIS, PHYTOPLANKTON, W70-06229
CHEMICAL ANALYSIS, ORGANIC MATTER W70-05760
CHEMICAL ANALYSIS, PHYSICAL PROPE W70-06975
CHEMICAL ANALYSIS, PROTEINS, BACT W72-01812
CHEMICAL ANALYSIS, TRACE ELEMENTS W70-01933
CHEMICAL ANALYSIS, ANALYTICAL TEC W69-09676
CHEMICAL ANALYSIS, BACTERIA, NUTR W69-10167
CHEMICAL ANALYSIS, NUTRIENTS, EFF W70-04810
CHEMICAL ANALYSIS, ANALYTICAL TEC W70-05652
CHEMICAL ANALYSIS, PHYSICAL PROPE W70-05752
CHEMICAL ANALYSIS, TEMPERATURE, N W69-07838
CHEMICAL ANALYSIS, PHYSIOLOGICAL W69-07862
CHEMICAL ANALYSIS, RELIABILITY, W W70-04382
CHEMICAL ANALYSIS, WATER POLLUTIO W70-03312
CHEMICAL ANALYSIS, PHOSPHATES, IR W70-03334
CHEMICAL ANALYSIS.: / WEED CONTRO W68-00479
CHEMICAL ANALYSIS, SESSILE ALGAE, W72-06052
CHEMICAL COMPOSITION(ENGLAND), FO W69-06865
CHEMICAL CONTROLS, PARAQUAT, DIQU W70-00269
CHEMICAL DEGRADATION, VITAMINS, C W70-04184
CHEMICAL INTERFERENCES.: W71-11687
CHEMICAL INDUSTRY, ODOR, TASTE, C W71-05943
CHEMICAL OXYGEN DEMAND, HYDROGEN W71-13413
CHEMICAL OXYGEN DEMAND, *WASTE WA W71-13335
CHEMICAL OXYGEN DEMAND, DISSOLVED W70-06601
CHEMICAL OXYGEN DEMAND, CHLOROPHY W70-08392
CHEMICAL OXYGEN DEMAND, COLOR, AC W70-10433
CHEMICAL OXYGEN DEMAND, STORAGE C W71-00936
CHEMICAL OXYGEN DEMAND, NITROGEN, W70-05655
CHEMICAL PROPERTIES, WISCONSIN, S W70-05113
CHEMICAL PROPERTIES, ECOLOGY, SAL W69-05023
CHEMICAL PRECIPITATION, IONS, BIO W69-00446
CHEMICAL PROPERTIES, BACTERIA, OR W68-00483
CHEMICAL PROPERTIES, ECOLOGY, ECO W70-07313
CHEMICAL PRECIPITATION, FILTRATIO W71-11393
CHEMICAL PRECIPITATION, LIFE CYCL W71-11823
CHEMICAL PRECIPITATION.: /URAL EU W69-08518
CHEMICAL PRECIPITATION, HARVESTIN W69-10178
CHEMICAL PRECIPITATION, ULTIMATE W70-01981
CHEMICAL RESISTANCE, EUPHAUSIA, P W70-01068
CHEMICAL REACTIONS, CHLORELLA, TR W70-08097
CHEMICAL REACTIONS, ZOOPLANKTON, W69-03611
CHEMICAL REACTIONS, ELECTROCHEMIS W70-05267
CHEMICAL SYMBIOSIS, MICROBIAL ECO W70-04503
CHEMICAL STRUCTURE, 2,3-DICHLORON W70-02982
CHEMICAL TREATMENT, OVERFERTILIZA W70-07469
CHEMICAL WASTES, ION EXCHANGE, BI W70-10433
CHEMICAL), CHLOROPHYTA, CYANOPHYT W70-02964
CHEMICAL), HYDROLYSIS, CHELATION, W69-07428
CHEMICALS, CARBON, GRAVIMETRY.: / W70-05651
CHEMICALS, FISHING, SPORT FISHING W68-00172
CHEMICALS, *DELTAS, *DRAINAGE EFF W71-09577
CHEMISTRY.: /AYS, ALGAE, WATER PU W69-09884
CHEMISTRY, WATER TEMPERATURE, CHE W69-09676
CHEMISTRY, WATER ANALYSIS.: /LGAE W69-08562
CHEMISTRY, PHOSPHORUS, NITROGEN, W71-05879
CHEMISTRY, WATER TEMPERATURE, HEA W72-01773
CHEMISTRY, OXYGEN, HYDROGEN ION C W70-03309
CHEMISTRY, WATER POLLUTION SOURCE W70-03068
CHEMISTRY, POLLUTANTS, ALGAL CONT W70-02646
CHEMISTRY, PLANTS, SOUTHWEST U. S W69-03611
CHEMISTRY, ECOLOGY, WAT: / WASTE W69-04276
CHEMISTRY, BIODEGRADATION.: /ICAL W69-00446
CHEMISTRY, CARBON RADIOISOTOPES, W70-05652
CHEMISTRY, *PHYTOPLANKTON, CARBON W70-05424
CHEMISTRY, *WATER POLLUTION EFFEC W70-05269
CHEMISTRY, RESPIRATION, BATHYMETR W70-05387
CHEMISTRY, WATER POLLUT: /MENTS, W69-04801

ENTS, NUTRIENTS, TOXICITY, WATER
NTS, PHOSPHORUS COMPOUNDS, WATER
OUNDS, *CYCLING NUTRIENTS, WATER
Y, STANDING CROP, STREAMS, WATER
DUCTIVITY, SOIL CHEMISTRY, WATER
KTON, PRIMARY PRODUCTIVITY, SOIL
, WATER CONSTITUENTS, ANALYTICAL
ES, PHYSIOLOGICAL ECOLOGY, WATER
ES, PHYSIOLOGICAL ECOLOGY, WATER
ATMENT, INDUSTRIAL WASTES, WATER
PHICATION, GRAZING, ALGAE, WATER
OROBIUM, FILINIA, MONIMOLIMNION,
ASSAY PROCEDURE, BATCH CULTURES,
RCH PROGRAM, CONTINUOUS CULTURE,
STIS AERUGINOSA, ALGAL CULTURES,
AHEDRITE, BORNITE, CHALCOPYRITE,
ONOMIC GROUPS, MOLECULAR FILTER,
LOGY, MORPHOLOGY, EXPORT LOSSES,
N, PEDIASTRUM BO/ HETEROTROPHIC,
TS, PHOTIC ZONE, GLYCOLLIC ACID,
AGEMENT, *ENVIRONMENTAL EFFECTS,
OR-PRODUCING ALGAE, CHLOROPHYLL,
                USBR RESEARCH CONTRACTS,
RIVER(TENN), POTOMOGETON PECTI/
LORIS PENIOCYSTIS, MADISON(WIS),
SHING WATERWAYS, MILWAUKEE(WIS),
TENN), FRENCH BROAD RIVER(TENN),
, OSOYOOS LAKE(CANADA), DEFORMED
PTERA, EPHEMEROPTERA, TIPULIDAE,
ROPHIC WATERS, HAMBURG(GERMANY),
MUS PROMOSUS, TRICHO/ HEXAGENIA,
EFISH, WURTTEMBERG(GERMANY), PL/
N, BATHYTHERMOGRAPH, ARTHROPODA,
AWICKA(WIS), LAKE PEWAUKEE(WIS),
GENIA, CHIRONOMIDAE, PROCLADIUS,
ATER(POLLUTION), PHOTOSYNTHESIS,
TRACE ELEMENTS, LIGHT INTENSITY,
TEROTROPHY, CONTINUOUS CULTURES,
DIFFUSION, CARBON RADIOISOTOPES,
ASSAYS, SALINE WATER, CHLORELLA,
PILIMNION, CONNECTICUT, EUGLENA,
OTRIAZOLE, 2-4-5-T, SCENEDESMUS,
SCO, CARP, DIATOMS, CHLOROPHYTA,
NTS, INDICATOR SPECIES, EUGLENA,
TIC ALGAE, AQUATIC MICROBIOLOGY,
ALGAE, BIOASSAY, BIOINDICATORS,
RANEUS, ANKISTRODESMUS FALCATUS,
TE, ORGANIC MATTER, SCENEDESMUS,
ELLA, NUTRIENTS, TRACE ELEMENTS,
MINNOWS, PEDALFERS, SCENEDESMUS,
E, NITZSCHIA PALEA, MYXOPHYCEAE,
OPHYTA, RHODOPHYTA, SCENEDESMUS,
CEANS, ROTIFERS, SWIMMING POOLS,
YTA, DIATOMS, SEASONAL, EUGLENA,
FILTERS, IOWA, DIATOMS, EUGLENA,
ON, GROWTH-PROMOTING PROPERTIES,
, NITROGEN, GROWTH, AMINO ACIDS,
, CHLOROPHYTA, COPEPODS, CYCLES,
NT GROWTH SUBSTANCES, CHLORELLA,
ONTROL, EUTROPHICATION, DAPHNIA,
                BROMINE, BROMAMINE,
TURE, ELEOCHARIS, 2,4-D, AMMATE,
BENOCLORS, NIGROSINE DYE, SODIUM

OGEN LIMITATION, BATCH CULTURES,
, THIMET, TRI/ SEVIN, MALATHION,
, SKELETONEMA COSTATUM(GREV) CL,
ILANES, SCENEDESMUS QUADRICAUDA,
RISODIUM NITRILOTRIACETATE, NTA,
E MATTER, BLELHAM TARN(ENGLAND),
DEGRADATION, SEVIN, MALATHION, /
MACRONUTRIENTS, MICRONUTRIENTS,
HA/ *UPTAKE, SPECIFIC ACTIVITY,
ILANES, SCENEDESMUS QUADRICAUDA,
PERTIES, CHLAMYDOMONAS MOEWUSII,
SIRA BINDERANA, GENERATION TIME,
SMUS BRAUNII, D/ CARBON DIOXIDE,
YUBERETS(USSR), MUNICH(GERMANY),
LIQUUS, SCENEDESMUS QUADRICAUDA,
DRICAUDA, CHLORELLA PYRENOIDOSA,
OLUMBIA), CHLORELLA PYRENOIDOSA,
AROPHILA, CHLORELLA ELLIPSOIDEA,
TOOXIDATION, CHLORELLA VULGARIS,
GARIS(COLUMBIA)/ PHOTOOXIDATION,
OIDOSA, CHLORELLA SACCHAROPHILA,
S, CHLORELLA VULGARIS(COLUMBIA),
ULATION DYNAMICS, CELLULAR SIZE,
A/ *MASS CULTURE, *ALGAL GROWTH,
EDIMENTS, PRECAMBRIAN SEDIMENTS,
A LAKE(MINN), ELY(MINN), BLOOMS,

CHEMISTRY, WATER POLLUTION EFFECT   W69-05867
CHEMISTRY, WATER POLLUTION EFFECT   W69-05868
CHEMISTRY, WATER ANALYSIS, PHOSPH   W69-05142
CHEMISTRY, WATER: /S, PRODUCTIVIT   W69-04805
CHEMISTRY, WATER POLLUTION EFFECT   W69-06273
CHEMISTRY, WATER CHEMISTRY, WATER   W69-06273
CHEMISTRY, WASTE WATER EFFECTS, L   W70-04382
CHEMISTRY, GEOGRAPHICAL REGIONS,   W70-03975
CHEMISTRY, GEOGRAPHICAL REGIONS,   W70-03975
CHEMISTRY, DIATOMS, CHLOROPHYTA,   W70-03974
CHEMISTRY, BIOMASS, BENTHOS, SALI   W70-03978
CHEMOCLINE, MIXOLIMNION, RADIOCAR   W70-05760
CHEMOSTATS, CHLOROPHYLL A, PHOTOB   W70-02777
CHEMOSTATS, BATCH CULTURES, ACETY   W70-02775
CHEMOSTATS, CARBON-14.: / MICROCY   W69-06864
CHEMOSYNTHESIS, CONCRETIONS, ARTH   W69-07428
CHEMOSYNTHETIC, SAPROPHYTIC, PLAT   W68-00891
CHEMOSYNTHESIS, ABUNDANCE.: /TECO   W70-02780
CHEMOTROPHIC, AUTOTROPHIC, SILICO   W70-02804
CHEMOTROPHY, PARTICULATE MATTER,   W70-02504
CHEMTROL, ALGAE.: /*WATERSHED MAN   W68-00478
CHEMTROL, COPPER SULFATE, LAKES,   W68-00470
CHENEY DAM, KANS.:   W72-01773
CHEROKEE RESERVOIR(TENN), HOLSTON   W71-07698
CHICAGO NATURAL HISTORY MUSEUM, N   W70-00719
CHICAGO(ILL), ARTIFICIAL CIRCULAT   W69-10178
CHICKAMAUGA RESERVOIR(TENN), FLOR   W71-07698
CHIRONOMIDS.: /SKAHA LAKE(CANADA)   W71-12077
CHIRONOMIDAE, HALF-LIFE.: / PLECO   W69-09742
CHIRONOMIDS, CLADOCERA, APHANIZOM   W70-00268
CHIRONOMIDAE, PROCLADIUS, CHIRONO   W70-01943
CHIRONOMIDS, LAKE CONSTANCE, WHIT   W70-05761
CHIRONOMIDAE, AQUATIC OLIGOCHAETE   W70-04253
CHIRONOMUS, PROCLADIUS, CORETHRA   W70-06217
CHIRONOMUS PROMOSUS, TRICHOPTERA,   W70-01943
CHL: /NTS, PHYTOPLANKTON, WASTE W   W70-04371
CHLAMYDOMONAS, CHLORELLA, ENZYMES   W70-04369
CHLAMYDOMONAS, BATCH CULTURES, OP   W70-04381
CHLAMYDOMONAS, BIOASSAYS, WATER P   W70-04284
CHLAMYDOMONAS, EUGLENA.: /TA, BIO   W70-03974
CHLAMYDOMONAS, PHOSPHORUS RADIOIS   W70-03957
CHLAMYDOMONAS, CHLORELLA, MORTALI   W70-03519
CHLAMYDOMONAS, CHLORELLA.: /N, CI   W69-02959
CHLAMYDOMONAS.: /ISCONSIN, NUTRIE   W69-01977
CHLAMYDOMONAS, CHLOROPHYTA, CYANO   W69-03364
CHLAMYDOMONAS, CARBON, ECOSYSTEMS   W69-06273
CHLAMYDOMONAS MOEWUSII, SKELETONE   W70-05381
CHLAMYDOMONAS, FRESHWATER, MARINE   W70-05381
CHLAMYDOMONAS, CHRYSOPHYTA,: /LOR   W70-05409
CHLAMYDOMONAS, TEMPERATURE, CARBO   W70-02249
CHLAMYDOMONAS MOEWUSSI, FAT ACCUM   W70-02965
CHLAMYDOMONAS, ECOLOGY, PHOSPHORU   W70-02965
CHLAMYDOMONAS, DAPHNIA, DIATOMS,   W70-10161
CHLAMYDOMONAS, CHLORELLA, SCENEDE   W70-10173
CHLAMYDOMONAS, SAMPLING, SCENEDES   W70-08107
CHLAMYDOMONAS MOEWUSII, CHLORELLA   W69-10180
CHLAMYDOMONAS, CHLORELLA, WISCONS   W69-10180
CHLAMYDOMONAS, ALGAE, ZOOPLANKTON   W70-00268
CHLAMYDOMONAS, SCENEDESMUS, WISCO   W72-00845
CHLAMYDOMONAS, LABORATORY TESTS,   W72-02417
CHLORAMINE.:   W70-02370
CHLORAMINE, DE-K-PRUF-21, BENOCLO   W70-05263
CHLORATE, SANTOBRITE, CLADOPHORA,   W70-05263
CHLORDANE.:   W70-09768
CHLORELLA SOROKINIANA, COMPOSITIO   W70-07957
CHLORELLA PYRENOIDOSA, SCHRODERIA   W70-08097
CHLORELLA OVALIS BUTCHER, NJOER L   W70-05752
CHLORELLA VULGARIS, DAPHNIA MAGNA   W71-03056
CHLORELLA PYRENOIDOSA, CARBOXYLIC   W70-02248
CHLORELLA PYRENOIDOSA, FIXATION,   W70-02504
CHLORELLA PYRENOIDOSA, PESTICIDE   W70-02968
CHLORELLA PYRENOIDOSA, AUTOTROPHI   W70-02964
CHLORELLA PYRENOIDOSA, BIOLOGICAL   W70-02786
CHLORELLA VULGARIS, DAPHNIA MAGNA   W71-05719
CHLORELLA PYRENOIDOSA, TRIBONEMA   W69-10180
CHLORELLA PYRENOIDOSA, DIATOMA EL   W69-10158
CHLORELLA PYRENOIDOSA, ANKISTRODE   W70-05261
CHLORELLA PYRENOIDOSA, BICHROMAT   W70-05092
CHLORELLA PYRENOIDOSA, CHLORELLA   W70-05381
CHLORELLA VULGARIS, MONODUS SUBTE   W70-05381
CHLORELLA SACCHAROPHILA, CHLORELL   W70-04809
CHLORELLA LUTEOVIRIDIS, SKELETONE   W70-04809
CHLORELLA VULGARIS(COLUMBIA), CHL   W70-04809
CHLORELLA VULGARIS, CHLORELLA VUL   W70-04809
CHLORELLA ELLIPSOIDEA, CHLORELLA   W70-04809
CHLORELLA PYRENOIDOSA, CHLORELLA   W70-04809
CHLORELLA VULGARIS, CORNELL UNIVE   W69-06540
CHLORELLA PYRENOIDOSA(EMERSON STR   W69-07442
CHLORELLA PYRENOIDOSA, FUCALES, A   W69-08284
CHLORELLA PYRENOIDOSA, MICROCYSTI   W70-03512

GANIC CARBON, LAKE TAHOE(CALIF),
EDESMUS OBLIQUUS, HELIX POMATIA,
GREEN ALGAE, CHLORELLA VULGARIS,
ENOIDOSA, VOLATIL/ *GREEN ALGAE,
MASS.:   KINETIC MODELS,
MEAN TOLERANCE LIMIT, BLUEGILLS,
ONDITIONS, ANAEROBIC CONDITIONS,
OMS, CHLOROPHYTA, CHLAMYDOMONAS,
Y, ALGAL POISONING, SCENEDESMUS,
IOLOGY, ACTIVATED SLUDGE, RATES,
GICAL TREATMENT, BIBLIOGRAPHIES,
OROPHYLL, CYCLES, UNITED STATES,
, ELECTROCHEMISTRY, RESISTIVITY,
ES, CHLOROPHYLL, PHOTOSYNTHESIS,
NKTON, *ALGAE, *LIGHT INTENSITY,
OTYPES, CHLOROPHYTA, PHOTOMETRY,
EGRADATION, VITAMINS, CELLULOSE,
S, SANDS, FLOW RATES, TURBIDITY,
OPHYTA, BIOASSAYS, SALINE WATER,
PLANKTON, CARBON, RADIOACTIVITY,
OMS, WEIGHT, PILOT PLANTS, IRON,
MICAL ANALYSIS, WATER POLLUTION,
5-T, SCENEDESMUS, CHLAMYDOMONAS,
, *WASTE TREATMENT, FOOD CHAINS,
LIGHT INTENSITY, CHLAMYDOMONAS,
AND DEVELOPMENT, BIBLIOGRAPHIES,
FFECTS, PHOTOSYNTHESIS, DIATOMS,
WTH, AMINO ACIDS, CHLAMYDOMONAS,
, *ALGAE, TOXICITY, SCENEDESMUS,
TORY EQUIPMENT, ASBESTOS CEMENT,
ULRUSH, PLANT GROWTH SUBSTANCES,
, OLIGOTROPHY, LIMITING FACTORS,
OISOTOPES, *AQUATIC ENVIRONMENT,
*NUTRIENT REQUIREMENTS, ECOLOGY,
MICAL), CHLOROPHYTA, CYANOPHYTA,
ISM, *ENVIRONMENTS, VARIABILITY,
UTRIENTS, PHYSIOLOGICAL ECOLOGY,
, SAMPLING TEMPERATURE, TRACERS,
ICIDES, CHLORINE, ALGAL CONTROL,
OLISM, RADIOACTIVITY TECHNIQUES,
TOXINS, INHIBITORS, SUCCESSION,
OBIOLOGY, BIOASSAY, CHLOROPHYTA,
ATION, FILTERS, ELECTROPHORESIS,
, *ALGAE, TOXICITY, SCENEDESMUS,
, *OXYGENATION, *PHOTOSYNTHESIS,
OWTH, *ALGAE, LIPIDS, NUTRIENTS,
PLING, SCENEDESMUS, CHLOROPHYTA,
METABOLISM, CHEMICAL REACTIONS,
EASONAL, EUGLENA, CHLAMYDOMONAS,
WASTE WATER TREATMENT, CULTURES,
CHLAMYDOMONAS, DAPHNIA, DIATOMS,
HERM, BLUE-GREEN ALGAE,/ *COPPER
A BLANKE/ ALUMINUM SULFATE, IRON
S, PHOSPHORUS, NITROGEN, CALCIUM
MENTS, *WATER POLLUTION EFFECTS,
YGEN, ALGAE, NUTRIENTS, SOLUTES,
ON, ALKALINITY, HARDNESS(WATER),
MAGNESIUM, ALKALINITY, SULFATES,
ANALYSIS, CARBONATES, NITRATES,
SUPPLY, BIOINDICATORS, CALCIUM,
GAE, BIOMASS, CALCIUM COMPOUNDS,
LGAE, OXYGEN SAG, WATER QUALITY,
TICIDE REMOVAL, *FOOD WEBS, DDT,
ON EFFECTS, AQUATIC ENVIRONMENT,
ION CONCENTRATION, TEMPERATURE,
HOTOSYN/ *MARINE PLANTS, *ALGAE,
, *TASTE-PRODUCING ALGAE, ALGAE,
Y, *WATER TREATMENT, FILTRATION,
TION, FLOTATION, BIODEGRADATION,
STREAMS, *DISTRIBUTION SYSTEMS,
DOMESTIC WASTES, SEDIMENTATION,
ERIA, METABOLISM, CANALS, SANDS,
ODOR, ACTIVATED CARBON, AMMONIA,
SEDIMENTATION, ACTIVATED CARBON,
, FUNGI, BACTERIA, BACTERICIDES,
OTICS, BIOASSAY, NORON, CALCIUM,
GENS, *GROWTH RATES, *CHLORELLA,
A, DISIN/ *HALOGENS, *ALGICIDES,
ALGICIDES, CYTOLOGICAL STUDIES,
PHYLL, IRON, CHELATION, ENZYMES,
ONMENTAL SANITATION, INHIBITION,
IUM COMPOUNDS, ACTIVATED CARBON,
CONTROL, *ALGAE, COPPER SULFATE,
EDESMUS, CHLOROPHYTA, CHLORELLA,
LLE GREEN LAKE(NY), *MARL LAKES,
LAKE(HANOI), RED RIVER(VIETNAM),
HIZOMYCOPHYTA, CAULOBACTERIALES,
ATION, UROGLENA, ANKISTRODESMUS,
HIRUNDINELLA, PERIDINIUM WILLEI,
D CARBON, CHLOROPLASTS, INFRARE/

CHLORELLA PYRENOIDOSA, SELENASTRU   W70-03983
CHLORELLA PYRENOIDOSA, NITROGEN U   W70-04184
CHLORELLA PYRENOIDOSA, VOLATILE A   W70-04195
CHLORELLA VULGARIS, CHLORELLA PYR   W70-04195
CHLORELLA PYRENOIDOSA, *ALGAE BIO   W73-02218
CHLORELLA PYRENOIDOSA, BENZENE CO   W72-11105
CHLORELLA.: /*NITRITES, AEROBIC C   W70-05261
CHLORELLA.: /N, CISCO, CARP, DIAT   W69-02959
CHLORELLA, *EUGLENA, ALGAL CONTRO   W69-00994
CHLORELLA, HYDROGEN ION CONCENTRA   W68-00248
CHLORELLA, *SEWAGE, EFFLUENTS, CA   W68-00855
CHLORELLA, BIOCHEMICAL OXYGEN DEM   W70-05092
CHLORELLA, SCENEDESMUS.: /ACTIONS   W70-05267
CHLORELLA, GRAZING, ZOOPLANKTON,   W70-04809
CHLORELLA, CHLOROPHYTA, DIATOMS,   W70-05381
CHLORELLA, NUTRIENTS, TRACE ELEME   W70-05409
CHLORELLA, PEPTIDES, BACTERIA, SN   W70-04184
CHLORELLA, SCENEDESMUS, EUGLENA.:   W70-04199
CHLORELLA, CHLAMYDOMONAS, EUGLENA   W70-03974
CHLORELLA, SAMPLING, EFFLUENTS, T   W70-03983
CHLORELLA, SODIUM, POTASSIUM, SIL   W70-03512
CHLORELLA, SCENEDESMUS.: /IA, CHE   W70-03312
CHLORELLA, MORTALITY, BIOASSAY, I   W70-03519
CHLORELLA, CHLOROPHYLL, PHYSIOLOG   W70-04381
CHLORELLA, ENZYMES, CARBON DIOXID   W70-04369
CHLORELLA, CHLOROPHYTA, DIATOMS,   W69-06865
CHLORELLA, CARBON RADIOISOTOPES,   W69-10158
CHLORELLA, WISCONSIN, LAKE MICHIG   W69-10180
CHLORELLA, DAPHNIA, DETERIORATION   W71-05719
CHLORELLA, CARBON DIOXIDE.: /BORA   W71-07339
CHLORELLA, CHLAMYDOMONAS, SCENEDE   W72-00845
CHLORELLA, ANALYTICAL TECHNIQUES,   W70-02779
CHLORELLA, TRACERS, PHOTOSYNTHESI   W70-02786
CHLORELLA, ALGAE, ORGANIC MATTER,   W70-02804
CHLORELLA, MANGANESE, PHOTOSYNTHE   W70-02964
CHLORELLA, CARBON, TEMPERATURE, L   W70-02965
CHLORELLA, SCENEDESMUS.: /LOGY, N   W70-02245
CHLORELLA, HYDROGEN ION CONCENTRA   W70-02504
CHLORELLA, DISINFECTION, MORTALIT   W70-02363
CHLORELLA, OXIDATION LAGOONS.: /B   W70-02198
CHLORELLA, CHELATION, OCHROMONAS,   W70-02510
CHLORELLA, CYCLING NUTRIENTS, ESS   W70-02248
CHLORELLA, SCENEDESMUS, ADSORPTIO   W71-03035
CHLORELLA, DAPHNIA, DETERIORATION   W71-03056
CHLORELLA, PESTICIDES, PESTICIDE   W70-08416
CHLORELLA, CULTURES.:   *GR   W70-07957
CHLORELLA, CHLORINE, LIME, COAGUL   W70-08107
CHLORELLA, TRACERS, RADIOACTIVITY   W70-08097
CHLORELLA, SCENEDESMUS, SAMPLING,   W70-10173
CHLORELLA, WATER POLLUTION EFFECT   W70-09438
CHLORELLA, SCENEDESMUS, OXYGEN, C   W70-10161
CHLORIDE, UPTAKE, FREUNDLICH ISOT   W70-08111
CHLORIDE, IRON SULFATE, CLADOPHOR   W69-10170
CHLORIDE, TEMPERATURE, METABOLISM   W72-05453
CHLORIDES, HALOGENS, ALGAE, WATER   W69-08562
CHLORIDES, CALCIUM, CARBONATES, P   W70-09557
CHLORIDES, ORGANIC MATTER, CARBON   W71-04206
CHLORIDES, SODIUM, IRON, DISSOLVE   W69-05894
CHLORIDES, AMMONIUM COMPOUNDS, OX   W69-07096
CHLORIDES, CONDUCTIVITY, CULTURES   W69-07833
CHLORIDES, SALINE LAKES, HEAVY ME   W69-02959
CHLORIDES, PHOSPHORUS, SULFATES,   W69-03948
CHLORINATED HYDROCARBON PESTICIDE   W70-03520
CHLORINATED HYDROCARBON PESTICIDE   W70-08652
CHLORINATED HYDROCARBON PESTICIDE   W72-11105
CHLORINATION, SODIUM COMPOUNDS, P   W70-00161
CHLORINATION, WATER QUALITY CONTR   W70-02373
CHLORINATION, COAGULATION, WATERS   W71-05943
CHLORINATION, COLIFORM, BACTERIA,   W71-05013
CHLORINATION, COPPER SULFATE, ACT   W72-03562
CHLORINATION, SLUDGE, DETERGENTS,   W70-04060
CHLORINATION, MICROORGANISMS, CHE   W69-07838
CHLORINE.: /AE, *ACTINOMYCETES, *   W72-04155
CHLORINE, COSTS, TREATMENT FACILI   W72-08859
CHLORINE, COSTS, *WASTE WATER TRE   W72-12987
CHLORINE, COBALT, COPPER, ECOLOGY   W69-04801
CHLORINE, ALGICIDES, ALGAL CONTRO   W70-02370
CHLORINE, ALGAL CONTROL, CHLORELL   W70-02363
CHLORINE, COPPER SULFATE, ALGAL C   W70-02364
CHLORINE, LIGHT INTENSITY, NITRAT   W70-02964
CHLORINE, NITROGEN COMPOUNDS, SEW   W69-10171
CHLORINE, PHENOLS, OIL WASTES, HY   W70-00263
CHLORINE, WELLS, CALIFORNIA.: /Y   W71-00110
CHLORINE, LIME, COAGULATION.: /EN   W70-08107
CHLOROBIUM, FILINIA, MONIMOLIMNIO   W70-05760
CHLOROCOCCUS, SPINES, SCHIZOMYCOP   W70-03969
CHLOROCOCCALES, CONJUGATOPHYCAEA,   W70-03969
CHLOROCOCCALES, VOLVOCALES, CONJU   W70-05409
CHLOROCOCCALES, ANKISTRODESMUS, O   W70-05405
CHLOROFORM-METHANE EXTRACTS, LIPI   W70-05651

HORA, ENTEROMORPHA, LEPOCINCLIS,
STEPHANODISCUS, ANKISTRODESMUS,
KISTRODESMUS, BACILLARIOPHYCEAE,
*EUTROPHICATION, ALGAE, BIOMASS,
*OLEFINS, GREEN RIVER FORMATION,
, DIATOMS, ODOR-PRODUCING ALGAE,
HORUS, ORGANIC MATTER, NITROGEN,
DIATOMS, CYANOPHYTA, PHOSPHATES,
Y, TEMPERATURE, LIGHT INTENSITY,
CHLOROPHYTA, DIATOMS, PIGMENTS,
ON, TEMPERATURE, DEPTH, DENSITY,
L SPILLS, GEOLOGICAL SS POLARIS,
TER), REMOTE SENSING, SEA WATER,
VITY, SURFACES, ARCTIC, ENZYMES,
ADIOISOTOPES, KINETICS, TRITIUM,
OIRS, LAKES, REAERATION, SESTON,
NTAL RELATIONSHIPS, FLUOROMETRIC
LAKE MENDOTA(WIS), NET PLANKTON,
NITROGEN, *ALGAE, SEWAGE PLANTS,
UM, PRODUCTIVITY, STANDING CROP,
YNTHESIS, RESPIRATION, PIGMENTS,
EATMENT, FOOD CHAINS, CHLORELLA,
ADATION, CHEMICAL OXYGEN DEMAND,
N EFFECTS, PHOSPHORUS, NITROGEN,
IC CARBON, / *AUTOTROPHIC INDEX,
XYGEN, PRODUCTIVITY, PHOSPHATES,
WAGE TREATMENT, SOLAR RADIATION,
IOLET RADIATION, TRANSMISSIVITY,
NTS, TEMPERATURE, LIGHT QUALITY,
LLUTION ABATEMENT, GROWTH RATES,
UM CARBONATE, BACTERIA, BIOMASS,
EN, CARBON DIOXIDE, SCENEDESMUS,
LFATE, ALGAL CONTROL, MORTALITY,
DESMUS, CYANOPHYTA, CHLOROPHYTA,
URE, BATCH CULTURES, CHEMOSTATS,
RNIA, NUTRIENTS, EUTROPHICATION,
, PHYTOPLANKTON, ORGANIC MATTER,
LIMNOLOGY, PRIMARY PRODUCTIVITY,
PHICATION, CARBON RADIOISOTOPES,
OPHYTA, CRUSTACEANS, HERBIVORES,
CHEMICAL OXYGEN DEMAND, DIURNAL,
Y, ROOTED AQUATIC PLANTS, ALGAE,
TURBULENCE, COPEPODS, ROTIFERS,
E ANIMALS, PRIMARY PRODUCTIVITY,
TY, STRATIFICATION, TEMPERATURE,
CTIVITY, PHYTOPLANKTON, EUGLENA,
YSTEMS, PHOSPHORUS, SCENEDESMUS,
TER, ALGAE, DIATOMS, CYANOPHYTA,
TY, BALANCE OF NATURE, BIOASSAY,
TS, ACID STREAMS, MINE DRAINAGE,
MENTS, PRODUCTIVITY, PERIPHYTON,
ROBIOLOGY, AQUATIC PRODUCTIVITY,
ORUS, NUTRIENTS, ALGAE, DIATOMS,
AQUATIC MICROBIOLOGY, BIOASSAY,
UTION), SCENEDESMUS, CYANOPHYTA,
LAKES, WATER POLLUTION SOURCES,
YDROGEN SULFIDE, COPPER SULFATE,
UNFISHES, ALGAE, DAPHNIA, LAKES,
WTH, *ALGAE, ELEMENTS(CHEMICAL),
BON DIOXIDE, SYNTHESIS, ENZYMES,
GY, BALANCE OF NATURE, BIOASSAY,
YSIOLOGICAL ECOLOGY, RHODOPHYTA,
OPHYTA, PHYTOPLANKTON, ROTIFERS,
N, SESTON, CRUSTACEANS, DIATOMS,
ES, ORGANIC COMPOUNDS, TOXICITY,
DOMESTIC WASTES, SHALLOW WATER,
, ARID LANDS, DEPTH, CYANOPHYTA,
HYTA, BIOMASS, INSECTS, DIATOMS,
YDOMONAS, SAMPLING, SCENEDESMUS,
TION(DECOMPOSITION), CYANOPHYTA,
AE, *SEWAGE TREATMENT, *INDIANA,
IME, ALGAE, CYANOPHYTA, DIATOMS,
TATES, HABI/ *ALGAE, *LAKE ERIE,
HOSPHORUS, NITROGEN, CYANOPHYTA,
*SURFACTANTS, *SORPTION, *ALGAE,
TER POLLUTION EFFECTS, NEW YORK,
*ALGAE, TEMPERATURE, CYANOPHYTA,
S, NITROGEN, IRON, GROWTH RATES,
HYTA, NUISANCE ALGAE, LIMNOLOGY,
ASTES, WATER CHEMISTRY, DIATOMS,
N, CYANOPHYTA, SEWAGE EFFLUENTS,
IATOMS, OLIGOTROPHY, CYANOPHYTA,
, NETS, NANNOPLANKTON, PIGMENTS,
GE, PHYTOPLANKTON, PRODUCTIVITY,
AE, *LIGHT INTENSITY, CHLORELLA,
OLOGICAL ECOLOGY, MASSACHUSETTS,
PHYSIOLOGICAL ECOLOGY, ECOTYPES,
EUTROPHICATION, ALGAE, CULTURES,
EN, PHOSPHORUS, DIATOMS, *ALGAE,
TIC MICROBIOLOGY, CHLAMYDOMONAS,

CHLOROGONIUM, SCENEDESMUS, CYCLOT   W69-02959
CHLOROMONAD, ALUM.: /M, WESTELLA,   W70-08107
CHLOROPHYCEAE, CHRYSOPHYCEAE, CAR   W69-07833
CHLOROPHYLL, NEW YORK.: /*PONDS,   W69-06540
CHLOROPHYLL DERIVATIVES, RECENT S   W69-08284
CHLOROPHYLL, CHEMTROL, COPPER SUL   W68-00470
CHLOROPHYLL.: /TS, DIATOMS, PHOSP   W70-05663
CHLOROPHYLL, MAGNESIUM, POTASSIUM   W70-05409
CHLOROPHYLL, NUTRIENTS, CONDUCTIV   W70-05752
CHLOROPHYLL, PHOTOSYNTHESIS, GROW   W70-05381
CHLOROPHYLL, ICE, SEASONAL, DIURN   W70-05387
CHLOROPHYLL FLUORESCENCE, FABRY-P   W70-05377
CHLOROPHYLL, OIL, ALGAE, PACIFIC   W70-05377
CHLOROPHYLL, PHOTOSYNTHESIS, CHLO   W70-04809
CHLOROPHYLL.: / DIATOMS, CARBON R   W70-04811
CHLOROPHYLL, CYCLES, UNITED STATE   W70-05092
CHLOROPHYLL TECHNIQUE.: /NVIRONME   W70-05094
CHLOROPHYLL CONTENT, CENTRIFUGE M   W70-03959
CHLOROPHYLL, PHOTOSYNTHESIS, TRIB   W70-04268
CHLOROPHYLL, PHOSPHATES, ADD: /CI   W70-03512
CHLOROPHYLL, TEMPERATURE, OXYGEN,   W70-03325
CHLOROPHYLL, PHYSIOLOGICAL ECOLOG   W70-04381
CHLOROPHYLL, LAGOONS, PONDS, OXID   W70-08392
CHLOROPHYLL, PRODUCTIVITY, REGION   W70-07261
CHLOROPHYLL A, SHAYLER RUN, ORGAN   W71-04206
CHLOROPHYLL, EPILIMNION, EFFLUENT   W69-10169
CHLOROPHYLL, PRODUCTIVITY, PHOTOS   W69-10182
CHLOROPHYLL.: /ECTROSCOPY, ULTRAV   W69-09676
CHLOROPHYLL, ZOOPLANKTON, ALGAE,   W69-08526
CHLOROPHYLL, PHOTOSYNTHESIS, CHEM   W69-10167
CHLOROPHYLL, PROTEINS, LIGHT, NUT   W69-10160
CHLOROPHYLL, IRON, CHELATION, ENZ   W70-02964
CHLOROPHYLL, ALGAE, BIOASSAY, VIA   W70-02364
CHLOROPHYLL, BACTERIA, TRACE ELEM   W70-02255
CHLOROPHYLL A, PHOTOBIOLOGY, ALGA   W70-02777
CHLOROPHYLL, FLUORESCENCE, LAKES,   W70-02777
CHLOROPHYLL, DYNAMICS, TEMPERATUR   W70-02780
CHLOROPHYLL, SAMPLING, PHYTOPLANK   W70-01933
CHLOROPHYLL, HYDROGEN ION CONCENT   W70-01579
CHLOROPHYLL, ENZUMES, PEPTIDES, B   W70-01068
CHLOROPHYLL.: /ON-SITE TESTS, BIO   W72-01818
CHLOROPHYLL, HYPOLIMNION, PHOSPHO   W71-07698
CHLOROPHYLL, PIGMENTS, IMPOUNDMEN   W71-10098
CHLOROPHYLL, PIGMENTS, MAGN: /RIN   W73-01446
CHLOROPHYLL, CYANOPHYTA, AQUATIC   W72-04761
CHLOROPHYTA, NITRATES, AMMONIA, C   W72-04761
CHLOROPHYTA, BACTERIA, PHOSPHORUS   W73-01454
CHLOROPHYTA, BULRUSH, PLANT GROWT   W72-00845
CHLOROPHYTA, CYANOPHYTA, DIATOMS,   W70-02792
CHLOROPHYTA, CHRYSOPHYTA, EUGLENO   W70-02770
CHLOROPHYTA, CHRYSOPHYTA, CYANOPH   W70-02780
CHLOROPHYTA, GONOPHYTA, DIATOMS,   W70-02784
CHLOROPHYTA, DISSOLVED OXYGEN, PI   W70-01943
CHLOROPHYTA, CHLORELLA, CYCLING N   W70-02248
CHLOROPHYTA, CHLOROPHYLL, BACTERI   W70-02255
CHLOROPHYTA, NUISANCE ALGAE, FISH   W70-03310
CHLOROPHYTA, CRUSTACEANS, SOIL ER   W70-02803
CHLOROPHYTA, STREAMS, DIATOMS, RE   W70-02982
CHLOROPHYTA, CYANOPHYTA, CHLORELL   W70-02964
CHLOROPHYTA, RHODOPHYTA, SCENEDES   W70-02965
CHLOROPHYTA, ENVIRONMENTAL EFFECT   W70-02968
CHLOROPHYTA.: /CAL PROPERTIES, PH   W69-10161
CHLOROPHYTA, COPEPODS, CYCLES, CH   W70-00268
CHLOROPHYTA, DEPOSITION(SEDIMENTS   W69-10182
CHLOROPHYTA, LAKES, RESERVOIRS, N   W69-10180
CHLOROPHYTA, CHRYSOPHYTA, CYANOPH   W70-00711
CHLOROPHYTA, SIEROZEMS, SOIL MICR   W71-04067
CHLOROPHYTA, CHRYSOPHYTA, GRAZING   W71-00668
CHLOROPHYTA, CHLORELLA, CHLORINE,   W70-08107
CHLOROPHYTA, *ALGAE, SORPTION, WA   W70-09438
CHLOROPHYTA, CYANOPHYTA, DIATOMS,   W70-10173
CHLOROPHYTA, LIGHT INTENSITY, SUC   W70-04371
CHLOROPHYTA, CYANOPHYTA, UNITED S   W70-04468
CHLOROPHYTA, DIATOMS, WEIGHT, PIL   W70-03512
CHLOROPHYTA, CYANOPHYTA, DETERGEN   W70-03928
CHLOROPHYTA, NUTRIENTS, ECOSYSTEM   W70-03973
CHLOROPHYTA, SCENEDESMUS, BENTHOS   W70-03969
CHLOROPHYTA, CYANOPHYTA, CHRYSOPH   W70-03983
CHLOROPHYTA, PLANKTON, PHYTOPLANK   W70-03959
CHLOROPHYTA, EUGLENOPHRYTA, CYANOP   W70-03974
CHLOROPHYTA, RHODOPHYTA, PHYSIOLO   W70-04510
CHLOROPHYTA, GREAT LAKES, BIOINDI   W70-04503
CHLOROPHYTA, CYANOPHYTA, CHRYSOPH   W70-05389
CHLOROPHYTA, HYDROGEN ION CONCENT   W70-05404
CHLOROPHYTA, DIATOMS, PIGMENTS, C   W70-05381
CHLOROPHYTA, DINOFLAGELLATES, ALG   W70-05272
CHLOROPHYTA, PHOTOMETRY, CHLORELL   W70-05409
CHLOROPHYTA, CYANOPHYTA, BRACKISH   W70-05752
CHLOROPHYTA, CYANOPHYTA, LAND MAN   W68-00172
CHLOROPHYTA, CYANOPHYTA, DIATOMS,   W69-03364

AQUATIC MICROBIOLOGY, BIOASSAY,
ATER POLLUTION EFFECTS, DIATOMS,
TIC ALGAE, AQUATIC MICROBIOLOGY,
GY, BALANCE OF NATURE, BIOASSAY,
DS, BALANCE OF NATURE, BIOASSAY,
OPLANKTON, CISCO, CARP, DIATOMS,
  AQUATIC MICROBIOLOGY, BIOASSAY,
ICIDES, AQUATIC ALGAE, BIOASSAY,
DS, BALANCE OF NATURE, BIOASSAY,
GY, BALANCE OF NATURE, BIOASSAY,
AL EFFECTS, ESSENTIAL NUTRIENTS,
TIC PRODUCTIVITY, AQUATIC WEEDS,
SSAY, BIOCHEMICAL OXYGEN DEMAND,
ITY, WATER TEMPERATURE, DIATOMS,
MENT, BIBLIOGRAPHIES, CHLORELLA,
CING ALGAE, DIATOMS, CYANOPHYTA,
L POLLUTION, DIFFUSION, DIATOMS,
-METHANE EXTRACTS, LIPID CARBON,
HIA CLOSTERIUM, NITZSCHIA PALEA,
SIS, *GROWTH RATES, *TURBULENCE,
ARY), ALGAL TAXONOMY, SPIRULINA,
RUCIGENIA RECTANGULARIS, DICERAS
XTRACELLULAR, FUCUS VESICULOSUS,
, COPPER, AMMONIA, CONDUCTIVITY,
TIS, CRYOMONAS EROSA, COSMARIUM,
TON(ND), LAKOTA(ND), HARVEY(ND),
DRIN, DIELDRIN, SCENEDESMUS, GAS
URES, DIAGENESIS, SEDIMENTS, GAS
S, CYANOPHYTA, SPECTROSCOPY, GAS
N, *WATER POLLUTION EFFECTS, GAS
OLAR REGIONS, SPECTROPHOTOMETRY,
S, ASSAY, DISTRIBUTION, DENSITY,
AMYDOMONAS LABORATORY TESTS, GAS
S, TRO/ *BIOTA, *COLUMBIA RIVER,
TECHNIQUES, *PHOSPHATES, *ALGAE,
ITY, PHYSIOLOGICAL OBSERVATIONS,
AL EFFECTS, DECOMPOSITION RATES,
ATER-BLOOM, 'BREAKING' OF LAKES,
ORINA, SPIROGYRA, SYNECHOCOCCUS,
TUS, URONEMA ELONGATUM HODGETTS,
UM, ANKISTRODESMUS, APHANOCAPSA,
E(MINN), CRYPTOPHYTA, PYROPHYTA,
DA, MASTIGOPHORA, CRYPTOPHYCEAE,
ACILLARIOPHYCEAE, CHLOROPHYCEAE,
, SEDIMENTS/ *ALGAE, CYANOPHYTA,
DIFFUSION, DIATOMS, CHLOROPHYTA,
SH, DINOFLAGELLATES, CYANOPHYTA,
GMENTS, CHLOROPHYTA, CYANOPHYTA,
, TRACE ELEMENTS, CHLAMYDOMONAS,
RATES, CHLOROPHYTA, CYANOPHYTA,
PHYTA, EUGLENOPHYTA, CYANOPHYTA,
AMS, MINE DRAINAGE, CHLOROPHYTA,
TIVITY, PERIPHYTON, CHLOROPHYTA,
, INSECTS, DIATOMS, CHLOROPHYTA,
TES, SHALLOW WATER, CHLOROPHYTA,
MONAS, ALGAE, ZOOPLANKTON, FISH,
AERUGINOSA(KUTZ), CULTURE MEDIA,
LOS-AQUAE, MINIMUM LETHAL DOSES,
LTS, DETRITUS, BOTTOM SEDIMENTS,
  RATES, CHRONOLOGY, CLADOCERANS,
AGMA, PSEUDAGRION, BRACHYTHEMIS,
OPHORA, HOLOPHYTIC MASTIGOPHORA,
ULATUS, MADISON(WIS), PSOROPHORA
HACUS, GLENODINIUM, CRYPTOMONAS,
, FLAGELLATES, SAPROBIEN SYSTEM,
COLO), MICRO-ALGAE, FLAGELLATES,
OPHYCEAE, VOLVOCALES, SARCODINA,
IATOMEA, CLADOPHORA, FLAGELLATA,
TULSA(OKLA), POLLUTION, FOSSIL,
A, SHAYLER RUN, ORGANIC CARBON,
IS), LAKE WAUBESA(WIS), ANABAENA
MICROCYSTIS AERUGINOSA, ANABAENA
FLAGELLATA, CILIATES, RANUNCULUS
BESNA RIVER, MOOSE CREEK, ARCTIC
PLANKTON, *LAKE MICHIGAN, *WATER
ATER, FRESH WATER, SHORES, WATER
OXINS, BACTERIA, COPPER SULFATE,
SANCE ALGAE, ODOR, TASTE, DEPTH,
HERMOCLINE, *ION EXCHANGE, WATER
E(WIS), CHICAGO(ILL), ARTIFICIAL
ATER STORAGE, EVAPORATION, WATER
N CONTROL, *LAKE ONTARIO, *WATER
, PHOTOSYNTHESIS, PHYTOPLANKTON,
DRAINAGE DISTRICTS, DEBT, COSTS,
ANTS, MAPPING, HYDROGEN SULFIDE,
UCTIVITY, PHOSPHATES, SILICATES,
ASTES, PEACHES, APRICOTS, PEARS,
RCLE HOT SPRINGS, RADON, CENTRAL
     *SUSPENDED SOLIDS, CHASE
ONTROL, PACIFIC NORTHWEST U. S.,

CHLOROPHYTA, CYANOPHYTA, DIATOMS,   W69-03371
CHLOROPHYTA, CYANOPHYTA, NUISANCE   W69-01977
CHLOROPHYTA, CYANOPHYTA, ESSENTIA   W69-03358
CHLOROPHYTA, CYANOPHYTA, DIATOMS,   W69-03188
CHLOROPHYTA, CYANOPHYTA, ENVIRONM   W69-03185
CHLOROPHYTA, CHLAMYDOMONAS, CHLOR   W69-02959
CHLOROPHYTA, CYANOPHYTA, ENVIRONM   W69-03362
CHLOROPHYTA, CYANOPHYTA, DIATOMS,   W69-03370
CHLOROPHYTA, DIATOMS, CYCLING NUT   W69-03515
CHLOROPHYTA, DIATOMS, ENVIRONMENT   W69-03516
CHLOROPHYTA, DIATOMS, INHIBITION,   W69-03517
CHLOROPHYTA, CYANOPHYTA, DIATOMS,   W69-03373
CHLOROPHYTA, CYCLING NUTRIENTS, E   W69-03374
CHLOROPHYTA.: /IEWS, LIGHT INTENS   W68-00860
CHLOROPHYTA, DIATOMS, LIGHT, LIPI   W69-06865
CHLOROPHYTA, IRON, MANGANESE, EUT   W69-05704
CHLOROPHYTA, CHRYSOPHYTA, DINOFLA   W69-05763
CHLOROPLASTS, INFRARED ANALYSIS.:   W70-05651
CHLOROPLASTS, SCENEDESMUS OBLIQUU   W70-05381
CHLORRELA, SCENEDESMUS, KINETICS,   W69-03730
CHODATE: /A(HUNGARY), ZIMONA(HUNG   W70-03969
CHODATI, DINOPHYCEAE, FILTER BLOC   W69-07833
CHONDRUS CRISPUS, ASCOPHYLLUM NOD   W70-01073
CHR: /NDS, HYDROGEN SULFIDE, LEAD   W70-04510
CHROCOCCUS LIMNETICUS, LYNGBYA LI   W70-05405
CHROMATIUM VINOSUM, THIOCAPSA FLO   W70-03312
CHROMATOGRAPHY, RESISTANCE, BIOAS   W70-03520
CHROMATOGRAPHY, LIPIDS, OIL, OIL   W69-08284
CHROMATOGRAPHY, ANALYTICAL TECHNI   W69-00387
CHROMATOGRAPHY.: /T IDENTIFICATIO   W69-02782
CHROMATOGRAPHY, HYDROGEN ION CONC   W70-01074
CHROMATOGRAPHY, RADIOACTIVITY, TR   W70-02510
CHROMATOGRAPHY, DEGRADATION, WATE   W70-08652
CHROMIUM, PHOSPHORUS RADIOISOTOPE   W69-07853
CHROMOTOGRAPHY, RADIOCHEMICAL ANA   W71-03033
CHRONIC EXPOSURE, ACUTE EXPOSURE.   W70-01779
CHRONOLOGY, CLADOCERANS, CHYDORID   W70-05663
CHROOCOCCACEAE, NOSTOCHINEAE, CLA   W69-08283
CHROOCOCCUS, OSCILLATORIA, NOSTOC   W70-04503
CHROOCOCCUS PRESCOTTII, CALOTHRIX   W70-04468
CHROOCOCCUS, CLOSTERIUM, GLEO: /I   W70-04371
CHROOMONAS ACUTA, SCHROEDERIA SET   W70-05387
CHRYSOPHYCEAE, VOLVOCALES, SARCOD   W70-05112
CHRYSOPHYCEAE, CARTERIA, COMPARAT   W69-07833
CHRYSOPHYTA, CULTURES, DIAGENESIS   W69-08284
CHRYSOPHYTA, DINOFLAGELLATES, CYA   W69-05763
CHRYSOPHYTA, CALIFORNIA, UNITED S   W70-05372
CHRYSOPHYTA, RHODOPHYTA, EUGLENOP   W70-05389
CHRYSOPHYTA,: /LORELLA, NUTRIENTS   W70-05409
CHRYSOPHYTA, PHOTOSYNTHESIS.: /TH   W70-03983
CHRYSOPHYTA, BIOASSAYS, SALINE WA   W70-03974
CHRYSOPHYTA, EUGLENOPHYTA, CYANOP   W70-02770
CHRYSOPHYTA, CYANOPHYTA, SAMPLING   W70-02780
CHRYSOPHYTA, GRAZING, SNAILS, AGE   W71-00668
CHRYSOPHYTA, CYANOPHYTA, SAMPLING   W70-00711
CHRYSOPHYTA, SESTON, SELF-PURIFIC   W70-00268
CHU NO 10 SOLUTION, MINIMUM NUTRI   W70-03507
CHU'S MEDIA, QUINONE DERIVATIVES,   W70-02982
CHUTES, EDDIES, HABITATS, CRAYFIS   W69-09742
CHYDORID CLADOCERANS, SPECIES COM   W70-05663
CHYDORUS, PLEUROXUS, TANYPUS, TAN   W70-06975
CILIATA.: /PHORA, HOLOZOIC MASTIG   W72-01817
CILIATA, ARTIFICIAL KEY.: /DRIMAC   W70-06225
CILIATA, ASTERIONELLA FORMOSA, RE   W70-00268
CILIATE AUTOECOLOGY.:   *CILIATES   W71-03031
CILIATES, COPEPODIDS, NAUPLII, TE   W69-10154
CILIATES, CYCLOTELLA, APHANIZOMEN   W70-05112
CILIATES, RANUNCULUS CIRCINATUS,   W69-07838
CIMARRON RIVER(OKLA), BIXBY(OKLA)   W70-04001
CINCINNATI(OHIO).: /, CHLOROPHYLL   W71-04206
CIR: /MENDOTA(WIS), LAKE MONONA(W   W69-08283
CIRCINALIS, GLOEOTRICHIA ECHINULA   W69-05704
CIRCINATUS, POTAMOGETON PUSILLUS.   W69-07838
CIRCLE HOT SPRINGS, RADON, CENTRA   W69-08272
CIRCULATION, NUISANCE ALGAE, THER   W69-05763
CIRCULATION, FLOODS, DENSITY DURR   W69-06203
CIRCULATION.: /ION, TURBULENCE, T   W70-04369
CIRCULATION, DIURNAL, WINDS, CURR   W70-05404
CIRCULATION, NUTRIENT ABSORPTION,   W68-00857
CIRCULATION, NUISANCES.: /ILWAUKE   W69-10178
CIRCULATION, SEDIMENTS.: /EVEL, W   W69-09723
CIRCULATION, PATH OF POLLUTANTS,   W71-05878
CISCO, CARP, DIATOMS, CHLOROPHYTA   W69-02959
CITIES.: /ATMENT, LAKES, SEWAGE,   W70-04455
CITIES, LAKE HURON, LAKE ERIE, LA   W70-00667
CITRATES, BIOASSAY, NEW YORK, LAK   W71-01579
CITRUS FRUITS, TOMATOES, *AEROBIC   W71-09548
CITY(COLO), MUTATION, BETA RAYS,   W69-08272
CITY(VA).:   W72-10573
CIVIL ENGINEERING, WASTE TREATMEN   W69-03683

CHER, SKELETONEMA COSTATUM(GREV)
LATION MODEL, PHOSPHORUS EFFECT,
L EFFICIENCIES, TROPHIC ECOLOGY,
, HAMBURG(GERMANY), CHIRONOMIDS,
CHII, STEPHANODISCUS BINDERANUS,
TUDIES, *EXPERIMENTAL LIMNOLOGY,
.: ASTERIONELLA SPIROIDES,
HRONOLOGY, CLADOCERANS, CHYDORID
DECOMPOSITION RATES, CHRONOLOGY,
A), BENARES(INDIA), MACROPHYTES,
YE, SODIUM CHLORATE, SANTOBRITE,
LES, ALGAL SUCCESSION, ANABAENA,
ACROCHEILUS ALUTACEUS, ULOTHRIX,
BACILLUS MESENTERICUS, DIATOMEA,
SYRACUSE, WARBURG RESPIROMETER,
ROCOLEUS LACUSTRIS (RAB) FARLOW,
ALGAL TAXONOMY, ALGAL PHYLOGENY,
(NY), ENTEROMORPHA INTESTINALIS,
TE, IRON CHLORIDE, IRON SULFATE,

EDRA, COELOSPHAERIUM, SPIROGYRA,
REEK(NY).: *CLADOPHORA FRACTA,
RAISIN RIVER, DETROIT R/ LAKE ST
TEMPERATURE, SEA WATER, OYSTERS,
UM, CARBON RADIOISOTOPES, BIOTA,
ITY, *ALGAE, *REVIEWS, HABITATS,
TIVATED SLUDGE, FISHERIES, CARP,
DIOISOTOPES, CALCIUM, POTASSIUM,
SIS, CARBON RADIOISOTOPES, FISH,
TRICHOPTERA, LEECHES, FINGERNAIL
-GREEN ALGAE, GREEN / *LEWIS AND
ARBON FILTERS, LEBANON, BATAVIA,
HETEROTROPHY, FLAGELLATES, LAKE
LING, ECOLOGY, AMMONIA, ENZYMES,
HNIQUES, *ANALYTICAL TECHNIQUES,
S, CHROOCOCCACEAE, NOSTOCHINEAE,
YDEN, THE HAGUE, MINERALIZATION,
S, ALGAE, ANALYTICAL TECHNIQUES,
ODEGRADATION, ADSORPTION, ALGAE,
TION PATTERNS, ACTIVATED CARBON,
, HYDROGEN SULFIDE, IRON, FUNGI,
NALYTICAL TECHNIQUES, CLAY LOAM,
DIMENTS, *EUTROPHICATION, ALGAE,
AE, VOLVOCALES, INDUSTRIAL USES,
*SURFACTANTS, *WATER POLLUTION,
ROL, PIPING SYSTEMS(MECHANICAL),
NE, COSTS, TREATMENT FACILITIES,
SESSMENT, AMERICAN RIVER(CALIF),
NTARIO), BUCKHORN LAKE(ONTARIO),
Y MANAGEMENT, *CHAOBORID MIDGES,
INN), MORPHOMETRY, ALGAE BLOOMS,
N, SUSCEPTIBILITY, REGENERATION,
TILIZATION, PHYSICAL PROPERTIES,
NITRIFICATION, COLIFORMS, PONDS,
GAE, *AQUATIC ALGAE, ALGAE, ARID
, ALGAE, ARID CLIMATES, SEMIARID
HEMISTRY, SOIL TYPES, HYDROLOGY,
EASONAL, *SULFIDES, TEMPERATURE,
TOSYNTHESIS, DEPTH, THERMOCLINE,
RENOIDOSA, BIOLOGICAL HALF-LIFE,
ZONE, HELSINGOR(DENMARK), SALINO-
FRAGILARIA, ASTERIONELLA, FILTER
ZECHOSLOVAKIA), CYPRINID FISHES,
MENTAL C/ BENTHIC ALGAE, BLOOMS,
VIRONMENT, CERIUM-144, NITZSCHIA
TED RATE, CELL VOLUME, NITZSCHIA
ESMUS, APHANOCAPSA, CHROOCOCCUS,
CHODESMIUM ERYTHREUM, SAXIDOMUS,
, MIGRATION, DIURNAL, ALGICIDES,
ANALYSIS, ANALYTICAL TECHNIQUES,
L DRAINAGE, APPALACHIAN MOUNTAIN
ATION, *SIGNIFICANCE, ALGAL KEY,
RIC TESTS, ONE-SAMPLE RUNS TEST,
OIKOPLEURA, FRITILLARIA, OBELIA,
CYLINDRICA, NORTH SEA, FIRTH OF
LENE/ *ANABAENA CYLINDRICA, ATP,
LOGICAL DILUTION', PHYSIOLOGICAL
AZOTOBACTER, MOLYBDENUM, COBALT,
RHODOPHYTA, EUGLENOPHYTA, SLIME,
ULFONATES, WIND VELOCITY, TASTE,
HYTA, CHLORELLA, CHLORINE, LIME,
ON, TEMPERATURE, CENTRIFUGATION,
TMENT, FILTRATION, CHLORINATION,
TIVITY, PESTICIDES, TASTE, ODOR,
ENGINEERING, *FILTERS, *DESIGN,
REMOVAL.:
MS, CORAL, MOLLUSKS, MINE ACIDS,
TRATION, SORPTION, FLOCCULATION,
IDE REMOVAL, *ADSORPTION, SOILS,
ON ABATEMENT, ABATEMENT, PACIFIC

CL, CHLORELLA OVALIS BUTCHER, NJO   W70-05752
CLADECERA.: /COSYSTEM MODEL, SIMU   W69-06536
CLADOCERA.: /ADRICAUDA, ECOLOGICA   W69-10152
CLADOCERA, APHANIZOMENON FLOS-AQU   W70-00268
CLADOCERA, WATER MASSES, U S PUBL   W70-00263
CLADOCERAN, MONTANE ZONE, ALPINE   W69-10154
CLADOCERANS, WESTHAMPTON LAKE(VA)   W71-10098
CLADOCERANS, SPECIES COMPOSITION,   W70-05663
CLADOCERANS, CHYDORID CLADOCERANS   W70-05663
CLADOPHORA, OEDOGONIA, SPIROGYRA,   W70-05404
CLADOPHORA, POTAMOGETON SPP, LAKE   W70-05263
CLADOPHORA, PEDIASTRUM, HYDRODICT   W70-04494
CLADOPHORA, PROSOPIUM WILLIAMSONI   W69-07853
CLADOPHORA, FLAGELLATA, CILIATES,   W69-07838
CLADOPHORA, ENTEROMORPHA, LEPOCIN   W69-02959
CLADOPHORA, TAXONOMIC DESCRIPTION   W70-04468
CLADOPHORA, ALGAL GENETICS, RUSSI   W70-03973
CLADOPHORA, CYCLOTELLA, HALOCHLOR   W70-03974
CLADOPHORA BLANKETS, HYDRODICTYON   W69-10170
CLADOPHORA GLOMERATA.:   W70-00667
CLADOPHORA, DIAPTOMUS, CYCLOPS, G   W70-02803
CLADOPHORA GLOMERATA, BUTTERNUT C   W72-05453
CLAIR, ROUGE RIVER, HURON RIVER,   W69-01445
CLAMS, CRABS, NORTH CAROLINA, SED   W69-08267
CLAMS, FALLOUT, PACIFIC OCE: /ITI   W69-08269
CLAMS, MUSSELS, SHELLFISH, DINOFL   W70-05372
CLAMS, PONDS, RUNNING WATERS, FRE   W70-09905
CLAMS, SIZE, ALGAE, DIATOMS, TRAC   W70-00708
CLAMS, STRONTIUM RADIOISOTOPES, T   W70-02786
CLAMS, WHITE FISH, ALEWIFE, GIZZA   W70-01943
CLARK LAKE (S D), ABUNDANCE, BLUE   W70-05375
CLARKSBURG, BUFFALO, DUNREITH, OI   W72-03641
CLASSIFICATION, SHAGAWA LAKE(MINN   W70-02775
CLASSIFICATION, GROWTH RATES, SEA   W70-00713
CLASSIFICATION, *MODEL STUDIES, W   W69-01273
CLATHROCYSTIS AERUGINOSA, COELOSP   W69-08283
CLAY LENSES, RHINE RIVER, AMSTERD   W69-07838
CLAY LOAM, CLAYS, COLLOIDS, EUTRO   W69-04800
CLAYS, ALDRIN, DIELDRIN, DDT, AER   W70-09768
CLAYS, ALGAE, WATER PURIFICATION,   W69-09884
CLAYS, BRACKISH WATER, FISHERIES,   W70-05761
CLAYS, COLLOIDS, EUTROPHICATION,   W69-04800
CLAYS, SILICATES.: /, *LAKES, *SE   W69-03075
CLEAN WATER, GENUS, SPECIES, TOUC   W70-05389
CLEANING, DISASTERS, OILY WATER,   W70-01470
CLEANING, COST ANALYSIS, WATER PU   W72-08688
CLEANING, *ALGAL CONTROL, WISCONS   W72-08859
CLEAR LAKE(CALIF), EEL RIVER(CALI   W70-02777
CLEAR LAKE(ONTARIO), RICE LAKE(ON   W70-02795
CLEAR LAKE(CALIF), NUTRIENT ENERG   W70-05428
CLEARWATER LAKE(MINN), TROUT LAKE   W70-05387
CLIFF VEGETATION, SAND DUNES.: /O   W70-01777
CLIMATE, NITROGEN, PHOSPHORUS.: /   W70-05565
CLIMATES, PHOTOSYNTHESIS, PRODUCT   W70-05092
CLIMATES, SEMIARID CLIMATES, *ABS   W69-03491
CLIMATES, *ABSORPTION, *CALCIUM,   W69-03491
CLIMATES, TROPHIC LEVEL, NITROGEN   W71-03048
CLIMATIC DATA, STRATIFICATION, DI   W70-04790
CLIMATIC ZONES, ORGANIC LOADING,   W72-10573
CLINCH RIVER(TENNESSEE), PHYSICAL   W70-02786
CLINE.: /LA MENEGHINIANA, PHOTIC   W70-04809
CLOGGING, STEPHANODISCUS HANTZSCH   W70-00263
CLOROCOCCALES, DAPHNIA PULICARIS,   W70-00274
CLOSED SYSTEMS, CYTOLOGY, ENVIRON   W69-07832
CLOSERIUM, ARTEMIA, CARTERIA, NOR   W70-00708
CLOSTERIUM, NITZSCHIA PALEA, CHLO   W70-05381
CLOSTERIUM, GLEO: /IUM, ANKISTROD   W70-04371
CLOSTRIDIUM BOTULINUS, VENERUPIS,   W70-05372
CLOSTRIDIUM, FISH, AMPHIBIANS, AN   W70-05372
CLOUD COVER, SPECTROSCOPY, ULTRAV   W69-09676
CLUB HUTS, TEGERNSEE, GERMANY, EX   W69-07818
CLUMP COUNT, DESMIDS, FLAGELLATES   W70-05389
CLUSTER ANALYSIS, REYKJAVIK(ICELA   W69-09755
CLUSTER ANALYSIS, MULTIVARIATE AN   W73-01446
CLYDE(SCOTLAND), SCOTLAND, ENGLAN   W70-02504
CMU, NITROGEN-15, PYRUVATE, ACETY   W70-04249
CO: / THORACOMONAS, YTTRIUM, 'BIO   W70-00708
CO: /, NITROGEN FIXATION, BORON,   W70-02964
CO: /A, CYANOPHYTA, CHRYSOPHYTA,   W70-05389
COAGULATION, PLANKTON, MONITORING   W70-00263
COAGULATION.: /ENEDESMUS, CHLOROP   W70-08107
COAGULATION, DEWATERING, *COST AN   W71-11375
COAGULATION, WATERSHEDS, SEWERAGE   W71-05943
COAGULATION, ELECTROLYTES, POTABL   W72-13601
COAGULATION, ALGAE, ACTIVATED CAR   W72-07334
COAL CONTACT PROCESSES, PESTICIDE   W70-05470
COAL MINES, ACIDITY, PYRITE, SULF   W69-07428
COAL, ACTIVATED CARBON.: /AE, FIL   W70-05470
COALS, DISTRIBUTION PATTERNS, ACT   W69-09884
COAST REGION, REGIONS, GEOGRAPHIC   W69-03683

OUTHWEST U. S., REGIONS, PACIFIC   COAST REGIONS, GEOGRAPHICAL REGIO   W69-03611
TIPLE-USES, SAVANNAH RIVER, GULF   COAST.: / COLLOIDAL FRACTION, MUL   W73-00748
EGIONS, COASTAL PLAINS, ATLANTIC   COASTAL PLAIN, INORGANIC COMPOUND   W69-03695
, REGIONS, GEOGRAPHICAL REGIONS,   COASTAL PLAINS, ATLANTIC COASTAL   W69-03695
              POLYELECTROLYTE-   COATED FILTERS, BAUXITE.:   W70-01027
SAND, CR/ *FILTER MEDIA, ALUMINA-   COATED SAND, FERRIC OXIDE-COATED   W70-08628
LUMINA-COATED SAND, FERRIC OXIDE-   COATED SAND, CRUSHED BAUXITE, ACI   W70-08628
HAINS, PHOSPHORUS RADIOISOTOPES,   COBALT RADIOISOTOPES, URANIUM RAD   W70-00708
OISOTOPES, IODINE RADIOISOTOPES,   COBALT RADIOISOTOPES, WATER POLLU   W71-09850
ENTHIC FLORA, STREAMS, CULTURES,   COBALT RADIOISOTOPES.: /GANESE, B   W69-07845
ENTS, SOILS, NUCLEAR EXPLOSIONS,   COBALT RADIOISOTOPES, CESIUM, STR   W69-08269
ION, ZINC RADIOISOTOPES, CESIUM,   COBALT RADIOISOTOPES, SEASONAL, R   W70-04371
ASSIUM, IRON, BICARBONATE, ZINC,   COBALT, CALCIUM, MAGNESIUM, C: /T   W70-03978
ASSAY, NORON, CALCIUM, CHLORINE,   COBALT, COPPER, ECOLOGY, ENVIRONM   W69-04801
 BORON, AZOTOBACTER, MOLYBDENUM,   COBALT, CO: /, NITROGEN FIXATION,   W70-02964
*ALGAE, *HERRINGS, COPPER, IRON,   COBALT, MANGANESE, ATLANTIC OCEAN   W69-08525
MEADOW LAKE, GYTTJA, DUFF, OOZE,   COBBLES, EQUISETUM, THORIUM-232,   W69-08272
ID FEVER, KENNEBEC RIVER(MAINE),   COBBOSSEECONTEE LAKE(MAINE), BARG   W70-08096
PHIC, MYCELOID, CORYNEFORM RODS,   COCCI,: /E, PSYCHROPHILES, PLEMOR   W70-00713
-AQUAE, GLOEOTRICHIA ECHINULATA,   COCCOCHLORIS PENIOCYSTIS, MADISON   W70-00719
PHYSIOLOGY, CALOTHRIX PARIETINA,   COCCOCHLORIS PENIOCYSTIS, DIPLOCY   W69-06277
E TECHNIQUES, MICROBIAL ECOLOGY,   COCCOLITHOPHORES, SEAWATER MEDIUM   W70-05272
 PERIDINI/ SKELETONEMA COSTATUM,   COCCOLITHUS HUXLEYI, PYRAMIMONAS,   W70-05272
ROS, SKELETONEMA, THALASSIOSIRA,   COCCOLITHUS, SARGASSO SEA, SYNURA   W70-04503
LLARIA, BIDDULPHIA, CHAETOCEROS,   COCCONEIS, COSCINODISCUS, DIMEROG   W70-03325
PHYSICAL HALF-LIFE, HYDROPSYCHE   COCKERELLI, BODY BURDENS, POMOXIS   W70-02786
ATIVE LAGOONS, LABORATORY MODEL,   COD REMOVAL, MILK SUBSTRATE, NITR   W70-07031
L, PLUMBING CODES, WELL DRILLING   CODES.: /NIDE, NEW PRODUCT CONTRO   W73-00758
E, NEW PRODUCT CONTROL, PLUMBING   CODES, WELL DRILLING CODES.: /NID   W73-00758
IC TURBIDOSTAT, LIGHT EXTINCTION   COEFFICIENT, ALGAL CELL DENSITY.:   W70-03923
ASTRUM, RHODOTORULA, TEMPERATURE   COEFFICIENTS, NORTH AMERICA, CENT   W70-03957
ALUES, EUDORINA, CULTURE MEDIUM,   COELASTRUM, PEDIASTRUM.: /CTION V   W70-05409
LLATORIA, MELOSIRA, ACTINASTRUM,   COELASTRUM, PEDIASTRUM, WESTELLA,   W70-08107
, TABELLARIA, MELOSIRA, SYNEDRA,   COELOSPHAERIUM, SPIROGYRA, CLADOP   W70-02803
OGYRA, MYXOPHYCEAE, MICROCYSTIS,   COELOSPHAERIUM, OSCILLATORIA, ANA   W70-04369
INEAE, CLATHROCYSTIS AERUGINOSA,   COELOSPHAERIUM KUTZINGIANUM, ANAB   W69-08283
ECIES DETERMINATION, ABSORBANCE,   COENOBIA, SWIRLING, VOLATILE SOLI   W70-03334
, POLYARTHRA VULGARIS, KARATELLA   COHCLEARIS, ASPLANCHA, LECANE, BO   W70-02249
SERRATIA MARINORUBRA, GYRODINIUM   COHNII, MICROBIAL ECOLOGY, COULTE   W70-05652
BIOLOGICAL BALANCE, OVERFISHING,   COHO SALMON, ALEWIFE, NUTRIENT RE   W70-04465
IFE, GIZZARD SHAD, SEA LAMPHREY,   COHO SALMON.: /, WHITE FISH, ALEW   W70-01943
EMEROPTERA, PLECOPTERA, DIPTERA,   COLEOPTERA, HEMIPTERA, MOLLUSCA,   W69-07846
DOSA, BICHOROMATE, DRY WEIGHT, B   COLI COMMUNAE, EUGLENE GRACILIS,   W70-05092
NDITIONS, AEROBIC CONDITIONS, E.   COLI, BACTERIA, FUNGI, ACTINOMYCE   W71-02009
ATION, LIFE CYCLES, GENETICS, E.   COLI, ISOLATION.: /MICAL PRECIPIT   W71-11823
AEOLOSOMA HEMPRICHI, ESCHERICHIA   COLI, MYRIOPHYLLUM, LEPOMIS MACRO   W69-07861
QUUS, DACTYLOCOCCUS, ESCHERICHIA   COLI, OOCYSTIS, LEMNA, SPIRODELA,   W70-10161
NUTRIENTS, VIRUSES, BACTERIA, E.   COLI, PUBLIC HEALTH, PATHOGENIC B   W72-06017
N, BIODEGRADATION, CHLORINATION,   COLIFORM, BACTERIA, BIOCHEMICAL O   W71-05013
DS, PHOSPHATES, NITROGEN, ALGAE,   COLIFORMS.: /IRON, DISSOLVED SOIL   W69-05894
ETHANE, ALGAL CONTROL, BACTERIA,   COLIFORMS.: /ION, FERMENTATION, M   W70-04787
POLLUTION, POLLUTION ABATEMENT,   COLIFORMS, WATER QUALITY, PHOSPHO   W70-04810
AL OXYGEN DEMAND(BOD), AERATION,   COLIFORMS, ALGAL CONTROL, DISSOLV   W70-04788
NAEROBIC, OXYGEN, NITRIFICATION,   COLIFORMS, PONDS, CLIMATES, PHOTO   W70-05092
SEWAGE TREATMENT, NORTH DAKOTA,   COLIFORMS, OXIDATION LAGOONS, BIO   W70-03312
LTURE, ST LAWRENCE RIVER, ALGAE,   COLIFORMS, DISSOLVED: /ES, AGRICU   W70-03964
TER POLLUTION EFFECTS, NITROGEN,   COLIFORMS, WATER QUALITY.: /S, WA   W71-05407
ENTS, *DISSOLVED OXYGEN, *ALGAE,   COLIFORMS, EUTROPHICATION(PHYTOPL   W71-05407
BACTERICIDE, KINETICS, HARDNESS,   COLIFORMS, CYTOLOGICAL STUDIES, E   W71-05267
TREATMENT, COPPER SULFATE, ODOR,   COLIFORMS, ECONOMICS.: /OLLUTION   W72-01814
 CHLORELLA, SCENEDESMUS, OXYGEN,   COLIFORMS, SEDIMENTS, MUD, SANDS,   W70-10161
ITY, DATA STORAGE AND RETRIEVAL,   COLIFORMS, DISSOLVED OXYGEN, ALGA   W70-09557
TION, *ECOLOGY, ALGAE, BACTERIA,   COLIFORMS, FISH, FUNGI, HEAT, WAT   W71-02479
FACILITIES, FILTRATION, SEWAGE,   COLIFORMS, EUTROPHICATION, ODOR,   W70-00263
HEMICAL OXYGEN DEMAND, SU/ *DATA   COLLECTIONS, *WATER QUALITY, BIOC   W70-09557
POLLUTION SOURCES, ON-SITE DATA   COLLECTIONS, *ON-SITE INVESTIGATI   W70-09976
E ERIE,/ *SYSTEMS ANALYSIS, DATA   COLLECTIONS, *ALGAE CONTROL, *LAK   W71-04758
ODS, MARINE PLANTS, ON-SITE DATA   COLLECTIONS.: /MORTALITY, GASTROP   W70-01780
PHYSIOLOGY, EUTROPHICATION, DATA   COLLECTION, CHEMICAL ANALYSIS, RE   W70-04382
ITY, *ARTIFICIAL RECHARGE, *DATA   COLLECTIONS, PIT RECHARGE, WATER   W69-05894
HORUS RADIOISOTOPE, ON-SITE DATA   COLLECTIONS, FISHES, PLANKTON, AL   W68-00857
*BOTTOM SEDIMENTS, ON-SITE DATA   COLLECTIONS, AQUATIC PLANTS, WATE   W68-00511
YSIS, PH, SAMPLING, ON-SITE DATA   COLLECTIONS, LIMNOLOGY.: /AL ANAL   W69-00659
ATHEMATICAL MODELS, ON-SITE DATA   COLLECTIONS, BOTTOM SEDIMENTS, BI   W68-00858
SMUS, MATHEMATICAL STUDIES, DATA   COLLECTIONS, MARINE ANIMALS, PRIM   W73-01446
          SPRING CREEK(PA), STATE   COLLEGE(PA), FLORA.:   W69-10159
BENTHIC FORMS, EPIPHYTIC FORMS,   COLLOIDAL FRACTION, MULTIPLE-USES   W73-00748
INE WATER-FRESHWATER INTERFACES,   COLLOIDS.: /PONDS, BREEDINGS, SAL   W73-00748
CHEMICAL PROPERTIES, ADSORPTION,   COLLOIDS, ZINC RADIOISOTOPES, CAL   W70-00708
, LIMNOLOGY, ZOOPLANKTON, ALGAE,   COLLOIDS, TRACE ELEMENTS, DISSOLV   W70-02510
PENDED LOAD, *FILTRATION, ALGAE,   COLLOIDS, FLOCCULATION, MIXING, E   W71-10654
AL PROPERTIES, ANION ADSORPTION,   COLLOIDS, AQUARIA, ABSORPTION, WA   W71-09863
AL TECHNIQUES, CLAY LOAM, CLAYS,   COLLOIDS, EUTROPHICATION, IMPOUND   W69-04800
NE ZONE, BOULDER(COLO), TEA LAKE(COLO), BLACK LAKE(COLO), PASS LAK   W69-10154
LO), BLACK LAKE(COLO), PASS LAKE(COLO), MICRO-ALGAE, FLAGELLATES,   W69-10154
HOT SPRINGS, RADON, CENTRAL CITY(COLO), MUTATION, BETA RAYS, COSMI   W69-08272
OLO), TEA LAKE(COLO), BLACK LAKE(COLO), PASS LAKE(COLO), MICRO-ALG   W69-10154
NTANE ZONE, ALPINE ZONE, BOULDER(COLO), TEA LAKE(COLO), BLACK LAKE   W69-10154
S), RHODOTORULA, GROWTH FACTORS,   COLONIAL MORPHOLOGY, CELLULAR MOR   W70-03952
TE/ *GROWTH DYNAMICS, *BACTERIAL   COLONIZATION, *ARTIFICIAL SUBSTRA   W70-04371

```
SEDIMENTS, *NUTRIENTS, NITROGEN          COMPOUNDS, PHOSPHOROUS COMPOUNDS,       W72-06051
OSING ORGANIC MATTER, *POTASSIUM         COMPOUNDS, *ACTIVATED CARBON, ACT       W72-04242
ICATION, *NUTRIENTS, *PHOSPHORUS         COMPOUNDS, ALGAE, PHOSPHATES, NIT       W70-04074
NTHESIS, ORGANIC MATTER, ORGANIC         COMPOUNDS, METABOLISM, BIOINDICAT       W70-04287
ULTURAL CHARACTERISTICS, TERMARY         COMPOUNDS, MACROMOLECULES, ANTIBI       W70-04280
*ORGANIC ACIDS, BIOMASS, ORGANIC         COMPOUNDS, CARBOHYDRATES.: /LLA,        W70-04195
IC DIGESTION, INCINERATION, IRON         COMPOUNDS, SULFATES.: /ON, ANEROB       W70-04060
PENETRATION, LIMNOLOGY, NITROGEN         COMPOUNDS, NUISANCE ALGAE, NUTRIE       W70-04283
BORATORY TESTS, CULTURES, SODIUM         COMPOUNDS, NITROGEN.: / RATES, LA       W70-03507
MICAL ANALYSIS, PHOSPHATES, IRON         COMPOUNDS, HYDROGEN ION CONCENTRA       W70-03334
ROPHICATION, LIMNOLOGY, NITROGEN         COMPOUNDS, NITROGEN FIXATION, NUI       W69-04802
MANAGEMENT, LIMNOLOGY, NITROGEN          COMPOUNDS, NITROGEN CYCLE, NUTRIE       W69-04804
, CYANOPHYTA, NITRATES, NITROGEN         COMPOUNDS, NUISANCE ALGAE, NUTRIE       W69-05868
*NITROGEN COMPOUNDS, *PHOSPHORUS         COMPOUNDS, *CYCLING NUTRIENTS, WA       W69-05142
NCE ALGAE, NUTRIENTS, PHOSPHORUS         COMPOUNDS, PHYSIOLOGICAL ECOLOGY,       W69-04802
UIREMENTS, NUTRIENTS, PHOSPHORUS         COMPOUNDS, WATER CHEMISTRY, WATER       W69-05868
HIES, *EUTROPHICATION, *NITROGEN         COMPOUNDS, *NUTRIENTS, *PHOSPHORU       W69-04805
ON, *DENITRIFICATION, PHOSPHORUS         COMPOUNDS, ALGAE, AQUATIC PLANTS.       W69-05323
*CYCLING NUTRIENTS, / *NITROGEN          COMPOUNDS, *PHOSPHORUS COMPOUNDS,       W69-05142
MPOUNDS, *NUTRIENTS, *PHOSPHORUS         COMPOUNDS, *WATER POLLUTION EFFEC       W69-04805
XYGEN, OXYGEN DEMAND, PHOSPHORUS         COMPOUNDS, PLANT GROWTH, PRIMARY        W69-06535
ICROORGANISMS, BIODEG/ *NITROGEN         COMPOUNDS, *NUTRIENTS, *ALGAE, *M       W69-06970
TION, LIMITING FACTORS, NITROGEN         COMPOUNDS, NUTRIENTS, ODOR, OLIGO       W69-06535
CATION, DENITRIFICATION, ORGANIC         COMPOUNDS, AMMONIA, NITRITES, NIT       W69-06970
S, NITRATES, CHLORIDES, AMMONIUM         COMPOUNDS, OXYGEN, SILICATES, ORG       W69-07096
L MINES, ACIDITY, PYRITE, SULFUR         COMPOUNDS, MOLYBDENUM, ROTIFERS,        W69-07428
TION, BIBLIOGRAPHIES, PHOSPHORUS         COMPOUNDS, NITROGEN COMPOUNDS, BI       W69-08518
EUSE, *RECLAIMED WATER, NITROGEN         COMPOUNDS, NITROGEN, AMMONIA, NUT       W69-08054
WATER REUSE, *NITROGEN, NITROGEN         COMPOUNDS, OXIDATION, BIODEGRADAT       W69-08053
, PHOSPHORUS COMPOUNDS, NITROGEN         COMPOUNDS, BIOINDICATORS, PRODUCT       W69-08518
*TERTIARY TREATMENT, *PHOSPHORUS         COMPOUNDS, *NUTRIENTS, *PESTICIDE       W70-05470
ENEDESMUS, LIGHT, CULTURES, IRON         COMPOUNDS, DIATOMS, CYANOPHYTA, P       W70-05409
SOLVENT EXTRACTIONS, PHOSPHORUS          COMPOUNDS, LAKES, SEWAGE EFFLUENT       W70-05269
LAKES, SEWAGE EFFLUENTS, ARSENIC         COMPOUNDS, SPECTROPHOTOMETRY, ALG       W70-05269
ES, *BIOASSAYS, ORGANOPHOSPHORUS         COMPOUNDS, SOLVENT EXTRACTIONS, P       W70-05269
*NU/ *EUTROPHICATION, *NITROGEN          COMPOUNDS, *PHOSPHORUS COMPOUNDS,       W70-04765
HYDROGEN ION CONCENTRATION, IRON         COMPOUNDS, HYDROGEN SULFIDE, LEAD       W70-04510
*NITROGEN COMPOUNDS, *PHOSPHORUS         COMPOUNDS, *NUTRIENTS, *BIOASSAY,       W70-04765
ON CONTROL, CHEMCONTROL, ARSENIC         COMPOUNDS, NITROGEN, PHOSPHORUS.:       W70-04494
E, VITAMINS, NITRATES, INORGANIC         COMPOUNDS, SURFACE RUNOFF, UNDERF       W70-04810
NS, HUMIC ACIDS, CARP, ICE, IRON         COMPOUNDS, HYDROGEN ION CONCENTRA       W70-05399
BACTERIA, ACTIVATED SLUDGE, IRON         COMPOUNDS, LIME, OXYGEN, TERTIARY       W68-00256
ENTAL EFFECTS, AMMONIA, AMMONIUM         COMPOUNDS, NITROGEN FIXATION, NUT       W69-03364
ROPHICATION, LIMNOLOGY, NITROGEN         COMPOUNDS, NITROGEN-FIXATION, PHO       W69-03186
CTIVITY, ALGAE, BIOMASS, CALCIUM         COMPOUNDS, CHLORIDES, SALINE LAKE       W69-02959
UTROPH/ *ADSORPTION, *PHOSPHORUS         COMPOUNDS, *LAKES, *SEDIMENTS, *E       W69-03075
ATER POLLUTION SOURCES, NITROGEN         COMPOUNDS, PHOSPHATES, GREAT LAKE       W69-01445
*ESSENTIAL NUTRIENTS, *NITROGEN          COMPOUNDS, AQUATIC ALGAE, AQUATIC       W69-03364
ROPHICATION, LIMNOLOGY, NITROGEN         COMPOUNDS, NITROGEN FIXATION, NUT       W69-03185
TS, *EUTROPHICATION, *PHOSPHORUS         COMPOUNDS, *WATER POLLUTION SOURC       W69-03370
TRIENT REQUIREMENTS, *PHOSPHORUS         COMPOUNDS, ALGAE, AQUATIC ALGAE,        W69-03373
UTRIENT REQUIREMENTS, PHOSPHORUS         COMPOUNDS, PHOSPHORUS, PHYTOPLANK       W69-03358
ENVIRONMENTAL EFFECTS, NITROGEN          COMPOUNDS, NUTRIENT REQUIREMENTS,       W69-03358
SEASONAL, CYANOPHYTA, *INORGANIC         COMPOUNDS, PHOSPHORUS, BIBLIOGRAP       W68-00860
TLANTIC COASTAL PLAIN, INORGANIC         COMPOUNDS, METALS, PLANKTON, AQUA       W69-03695
POLLUTION SOURCES,/ *PHOSPHORUS          COMPOUNDS, *PHYTOPLANKTON, *WATER       W69-04800
*FRESH WATER, ALGAE,/ *NITROGEN          COMPOUNDS, *NITROGEN UTILIZATION,       W69-04521
, *PHYSI/ *CYANOPHYTA, *NITROGEN         COMPOUNDS, *NUTRIENT REQUIREMENTS       W69-03515
*NUTRIENTS, *VIRGINIA, NITROGEN          COMPOUNDS, PHOSPHORUS, NUISANCE A       W69-03695
PENETRATION, LIMNOLOGY, NITROGEN         COMPOUNDS, NUTRIENT REQUIREMENTS,       W69-04798
UIREMENTS, NUTRIENTS, PHOSPHORUS         COMPOUNDS,: /POUNDS, NUTRIENT REQ       W69-04798
PENETRATION, LIMNOLOGY, NITROGEN         COMPOUNDS, NUTRIENT REQUIREMENTS,       W69-03514
PENETRATION, LIMNOLOGY, NITROGEN         COMPOUNDS, NITROGEN FIXATION, NUT       W69-03518
CKWATER, FERTILIZATION, NITROGEN         COMPOUNDS, PHOSPHORUS COMPOUND, P       W70-00274
EMPERATURE, EPILIMNION, NITROGEN         COMPOUNDS, SEWAGE EFFLUENTS, LIGH       W69-10167
GY, NUTRIENTS, SEAWATER, ORGANIC         COMPOUNDS, HABITATS, STATISTICAL        W69-09755
IRON, IONS, PHOSPHATES, ORGANIC          COMPOUNDS, TOXICITY, CHLOROPHYTA,       W69-10180
PHOSPHATES, PHYTOPLANKTON, IRON          COMPOUNDS, ALGAE, PLANTS, WATER P       W69-10170
A, ESSENTIAL NUTRIENTS, NITROGEN         COMPOUNDS, AMMONIA, NITRATES, NIT       W69-10177
, EUTROPHICATION, ODOR, AMMONIUM         COMPOUNDS, ACTIVATED CARBON, CHLO       W70-00263
, INHIBITION, CHLORINE, NITROGEN         COMPOUNDS, SEWAGE, SEWAGE BACTERI       W69-10171
TS, *ALGAE, CHLORINATION, SODIUM         COMPOUNDS, PHOTOSYNTHESIS, CARBON       W70-00161
RS, PHOSPHORUS, ORGANOPHOSPHORUS         COMPOUNDS, BACTERIA, DIATOMS, DET       W69-09334
, BACTERIA, ALGAE, FUNGI, SULFUR         COMPOUNDS, CARBONATES.: /N OXIDES       W70-02294
NUTRIENTS, INHIBITION, NITROGEN          COMPOUNDS, NUTRIENT REQUIREMENTS,       W70-02248
AMMONIA, NITROGEN FIXATION, IRON         COMPOUNDS, LIGHT PENETRATION, WAT       W70-02255
ISTIDINE, URACIL, REDUCED SULFUR         COMPOUNDS, ANTIBIOTIC EFFECTS, ST       W70-02510
ACTIVITY, QUARTERNARY, AMMONIUM          COMPOUNDS, MINIMUM LETHAL EXPOSUR       W70-02982
ON SOURCES, PHOSPHORUS, NITROGEN         COMPOUNDS, EUTROPHICATION, PRODUC       W70-03068
ENCIES, *NITRIFICATION, NITROGEN         COMPOUNDS, NITROGEN CYCLE, ALGAE,       W70-07031
S, NYLON-6, *VISCOSE RAYON, ZINC         COMPOUNDS, CARBON DISULFIDE.: /NT       W70-06925
HEMICAL OXYGEN DEM/ *ALGAE, IRON         COMPOUNDS, HYDROGEN SULFIDE, BIOC       W70-06925
RATIFICATION, SULFIDES, AMMONIUM         COMPOUNDS, OXIDATION REDUCTION PO       W70-05760
S, INFRARED SPECTROSCOPY, SULFUR         COMPOUNDS, WASTE STABILIZATION PO       W70-07034
ATER POLLUTION EFFECTS, *ORGANIC         COMPOUNDS, *OXYGENATION, *PHOTOSY       W70-08416
REUSE, *RECLAIMED WATER, ORGANIC         COMPOUNDS, CHEMICAL WASTES, ION E       W70-10433
PHORUS, *PHOSPHATES, *PHOSPHORUS         COMPOUNDS, *CHEMICAL PRECIPITATIO       W70-09186
SH PHYSIOLOGY, RED TIDE, ORGANIC         COMPOUNDS, DINOFLAGELLATES.: / FI       W71-07731
, NITROGEN COMPOUNDS, PHOSPHORUS         COMPOUNDS, RESERVOIRS, ALGAE, GRO       W72-02817
TION, SEWAGE TREATMENT, NITROGEN         COMPOUNDS, PHOSPHORUS COMPOUNDS,        W72-02817
```

N CYCLE, *PHOSPHORUS, PHOSPHORUS
IRE, NEW YORK, VIRGINIA, ORGANIC
RATURE, MICROORGANISMS, NITROGEN
NITROGEN COMPOUNDS, *PHOSPHORUS
*OXYGENATION, *TOXICITY, ORGANIC
TION, FILTRATION, MODEL STUDIES,
RBON RADIOISOTOPES, HYDROGEN ION
ARY TREAT/ *ALGAE, *HYDROGEN ION
TION, *TEMPERATURE, HYDROGEN ION
, DISSOLVED OXYGEN, HYDROGEN ION
IAL BALANCES, O/ *UPTAKE, CELLS,
YGEN, BICARBONATES, HYDROGEN ION
PHYTON, PHOSPHORUS, HYDROGEN ION
OXYGEN, ALKALINITY, HYDROGEN ION
ALITY, TEMPERATURE, HYDROGEN ION
OWERS, DAMS, ALGAE, HYDROGEN ION
TORING.:
ICAL OXYGEN DEMAND, HYDROGEN ION
ICAL OXYGEN DEMAND, HYDROGEN ION
, SEDIMENTS, ALGAE, HYDROGEN ION
E, ORGANIC LOADING, HYDROGEN ION
H TOXINS, TOXICITY, HYDROGEN ION
*FUNGI, *BACTERIA, HYDROGEN ION
HOSPHORUS, SULPHUR, HYDROGEN ION
CHEMISTRY, OXYGEN, HYDROGEN ION
NATES, TEMPERATURE, HYDROGEN ION
ETHANE, ALKALINITY, HYDROGEN ION
RE, CARBON DIOXIDE, HYDROGEN ION
TRACERS, CHLORELLA, HYDROGEN ION
IDES, BIOCHEMISTRY, HYDROGEN ION
TOPES, CHLOROPHYLL, HYDROGEN ION
IBIL/ *TORREY CANYON, *DETERGENT
RY, CHROMATOGRAPHY, HYDROGEN ION
HUMIDITY, BACTERIA, HYDROGEN ION
MONOMICTIC, DEPTH EFFECTS, IONIC
, *NUTRIENT CONTROL, *PHOSPHORUS
HENOLS, OIL WASTES, HYDROGEN ION
, DISSOLVED SOLIDS, HYDROGEN ION
USEUM, NOSTOC MUSCORUN, NUTRIENT
ION, ACTINOMYCETES, HYDROGEN ION
RIENT BUDGET, POTENTIAL NUTRIENT
(WASH), *LAKE CHANGES, *NUTRIENT
ATIC FUNGI, YEASTS, HYDROGEN ION
WATER TEMPERATURE, HYDROGEN ION
N DIOXIDE, *ALGAE, *HYDROGEN ION
N, LIME, *NITROGEN, HYDROGEN ION
, RATES, CHLORELLA, HYDROGEN ION
MMONIA, PHOSPHATES, HYDROGEN ION
IVITY, CHLOROPHYTA, HYDROGEN ION
TIFICATION, OXYGEN, HYDROGEN ION
CE, IRON COMPOUNDS, HYDROGEN ION
T, *BOD REMOVAL, *ISRAEL, *ALGAL
INDUSTRIAL WASTES, HYDROGEN ION
XIMUM UPTAKE VELOCITY, SUBSTRATE
S, SCENEDESMUS, ABSORPTION, CELL
ATER, ACIDIC WATER, HYDROGEN ION
ON, CARBON DIOXIDE, HYDROGEN ION
ERIA, ALGAE, SANDS, HYDROGEN ION
SP/ RADIOAUTOGRAPHS, HALF-LIFE,
TRUNCATUS, BREVOORTIA TYRRANUS,
UCTIVITY, CULTURES, HYDROGEN ION
A, MICROCYSTIS AERU/ *BIOLOGICAL
CE WATERS, STREAMS, HYDROGEN ION
ES, IRON COMPOUNDS, HYDROGEN ION
IONS, ANALYSIS OF VARIANCE, FOOD
TARIES, ALKALINITY, HYDROGEN ION
NKTON, ZOOPLANKTON, HYDROGEN ION
TROGEN, PHOSPHATES, HYDROGEN ION
TURE, PILOT PLANTS, HYDROGEN ION
SPHOROUS COMPOUNDS, HYDROGEN ION
ITY, *ALGAE, *ODOR, HYDROGEN ION
RIA, ALGAE, MIXING, HYDROGEN ION
, DISSOLVED OXYGEN, HYDROGEN ION
ORGANIC PESTICIDES, HYDROGEN ION
TE-PRODUCING ALGAE, HYDROGEN ION
LGAE TREATMENT COSTS, DINOBRYON,
IRE, LAKE WINNISQUAM, LONG POND,
, CHARCOAL, ALGAL GROWTH, STREAM
E, CHALCOPYRITE, CHEMOSYNTHESIS,
SYNTHESIS, ALKALINITY, NITRATES,
NCE, OPTIMUM GROWTH TEMPERATURE,
R, ANAEROBIC CONDITIONS, AEROBIC
SEWAGE BACTERIA, ODOR, ANAEROBIC
MS, ANAEROBIC BACTERIA, ANEROBIC
METHANE, MUD, OXYGEN, ANAEROBIC
ONS, *SLUDGE, ALGAE,/ *ANAEROBIC
LAKE WASHINGTON(WASH), NUISANCE
AINBOW TROUT, DIATOMS, ANAEROBIC
S, ANAEROBIC BACTERIA, ANAEROBIC
S, ANAEROBIC BACTERIA, ANAEROBIC

COMPOUNDS, WATER SUPPLY, WATER MA   W72-01867
COMPOUNDS, LAGOON, FILTERS, PESTI   W72-03641
COMPOUNDS, *PHOSPHORUS COMPOUNDS,   W71-13335
COMPOUNDS, ADSORPTION, REMOVAL, D   W71-13335
COMPOUNDS, *PHENOLS, PESTICIDES,   W71-12183
COMPUTERS, NUTRIENTS, EUTROPHICAT   W70-08627
CONCENTRATION, SCENEDESMUS, EUGLE   W70-08097
CONCENTRATION, *NUTRIENTS, *TERTI   W70-08980
CONCENTRATION, NUTRIENTS, ALGAE,   W70-08834
CONCENTRATION, EPILIMNION, PHOTOS   W70-06975
CONCENTRATION, COMPOSITION, MATER   W70-06600
CONCENTRATION, NUISANCE ALGAE, AQ   W70-06225
CONCENTRATION, ALKALINITY, HARDNE   W71-04206
CONCENTRATION, STANDING CROP, LIG   W71-00665
CONCENTRATION, OXYGEN, CARBON DIO   W71-03031
CONCENTRATION, TEMPERATURE, ALKAL   W71-11393
CONCENTRATION, ENVIRONMENTAL MONI   W72-00941
CONCENTRATION, TEMPERATURE, PROTE   W71-13413
CONCENTRATION, ORGANIC LOADING, *   W72-03299
CONCENTRATION, ELECTRIC POWER, SE   W71-09548
CONCENTRATION, TEMPERATURE, CENTR   W71-11375
CONCENTRATION, DISSOLVED OXYGEN,   W71-07731
CONCENTRATION, TEMPERATURE, ALKAL   W71-05267
CONCENTRATION, TIME.: /ECOLOGY, P   W70-02965
CONCENTRATION, PHOTOSYNTHESIS, AL   W70-03309
CONCENTRATION, ALKALINITY,: /ARBO   W70-02804
CONCENTRATION, TOXINS, OXYGEN, DE   W70-02803
CONCENTRATION, BIOCARBONATES, DIS   W70-02249
CONCENTRATION, ALKALINITY, DIATOM   W70-02504
CONCENTRATION, OXIDATION,: / PEPT   W70-01068
CONCENTRATION, MAGNESIUM, CALCIUM   W70-01579
CONCENTRATIONS, SURVIVAL, SUSCEPT   W70-01779
CONCENTRATION, FLUORESCENCE, CARB   W70-01074
CONCENTRATION, SULFATES, RESPIRAT   W70-01073
CONCENTRATION, SEASONAL EFFECTS.:   W70-01933
CONCENTRATIONS, MICRONUTRIENTS, A   W69-09340
CONCENTRATION, FLUORIDES, ALKYL B   W70-00263
CONCENTRATION, TEMPERATURE, NITRO   W70-00265
CONCENTRATION, WISCONSIN, SULFUR,   W70-00719
CONCENTRATION, NITRATES, POTASSIU   W70-00719
CONCENTRATION, AREAL INCOME(NUTRI   W70-00270
CONCENTRATION, *NUTRIENT BUDGET,   W70-00270
CONCENTRATION, IRON, SULFATE, TEM   W69-00096
CONCENTRATION, FUNGI, YEASTS, PYR   W68-00891
CONCENTRATION, PHOSPHORUS, NITROG   W68-00855
CONCENTRATION, *PHOSPHORUS, SEWAG   W68-00012
CONCENTRATION, PLANT TISSUES, SCE   W68-00248
CONCENTRATION, PHYTOPLANKTON, ODO   W70-05405
CONCENTRATION, NITRATES, PHOSPHAT   W70-05404
CONCENTRATION, TEMPERATURE, DEPTH   W70-05387
CONCENTRATION, ACIDITY, MUD, HYDR   W70-05399
CONCENTRATION, 'EXTRA-DEEP PONDS'   W70-04790
CONCENTRATION, IRON COMPOUNDS, HY   W70-04510
CONCENTRAT: / UPTAKE VELOCITY, MA   W70-05109
CONCENTRATION, UROGLENA, ANKISTRO   W70-05409
CONCENTRATION, WATER CHEMISTRY, *   W70-05424
CONCENTRATION, PHOSPHATES, NITRAT   W70-05760
CONCENTRATION.: /CTROMETERS, BACT   W69-08267
CONCENTRATION FACTOR, LIMNODRILUS   W69-07861
CONCENTRATION PROCESSES.: /RSIOPS   W69-08274
CONCENTRATION, IRON, LAKES, MAGNE   W69-07833
CONCENTRATION, ANABAENA CYLINDRIC   W70-03501
CONCENTRATION, IRON, MAGNESIUM, N   W70-03501
CONCENTRATION, ALKALINITY, CALIFO   W70-03334
CONCENTRATION, MEXICO, PREDICTIVE   W70-03957
CONCENTRATION, NUTRIENTS, PHOSPHO   W70-03983
CONCENTRATION, ALGAE.: / PHYTOPLA   W70-04368
CONCENTRATION, CYCLING NUTRIENTS,   W70-04381
CONCENTRATION, ORGANIC LOADING, C   W72-06022
CONCENTRATION, CONDUCTIVITY, TEMP   W72-06051
CONCENTRATION.: /ENT, *WATER QUAL   W72-09694
CONCENTRATION, TEMPERATURE, TOXIC   W72-06854
CONCENTRATION, BIOCHEMICAL OXYGEN   W72-10573
CONCENTRATION, TEMPERATURE, CHLOR   W72-11105
CONCENTRATION, *ORGANI: /SIS, TAS   W72-11906
CONCORD(N H), PEANACOOK LAKE(N H)   W69-08282
CONCORD, BOSTON LOT RESERVOIR, LE   W69-08674
CONCOURSE, PALOUSE RIVER(IDAHO).:   W70-03501
CONCRETIONS, ARTHROBACTER, PYROPH   W69-07428
COND: /N ION CONCENTRATION, PHOTO   W70-03309
CONDENSER PAGGAGE.: /ATURE TOLERA   W71-02496
CONDITIONS, E. COLI, BACTERIA, FU   W71-02009
CONDITIONS, AEROBIC CONDITIONS, E   W71-02009
CONDITIONS, LAGOONS.: /AGE PROGRA   W71-04554
CONDITIONS, HYDROGEN SULFIDE, IRO   W70-05761
CONDITIONS.: /EFFLICIENCIES, *LAGO   W70-06899
CONDITIONS.: /MADISON LAKES(WIS),   W70-07261
CONDITIONS, ALGAE, ROTIFERS, SEST   W69-10154
CONDITIONS, LAGOONS.: /GE PROGRAM   W71-04557
CONDITIONS, LAGOONS.: /GE PROGRAM   W71-04556

S, ANAEROBIC BACTERIA, ANAEROBIC
ON, AEROBIC TREATMENT, ANAEROBIC
E, *LABORATORY TESTS, *ANAEROBIC
S, AEROBIC CONDITIONS, ANAEROBIC
E, FLOW, ALGAE, BENTHOS, AEROBIC
SYSTEMS, CYTOLOGY, ENVIRONMENTAL
ION, SEWAGE, AESTHETICS, AEROBIC
S, AEROBIC CONDITIONS, ANAEROBIC
E, AEROBIC CONDITIONS, ANAEROBIC
OSYNTHESIS, GROWTH RATE, AEROBIC
S, AEROBIC CONDITIONS, ANAEROBIC
T, *NITRATES, *NITRITES, AEROBIC
, FACILITIES, EQUATIONS, AEROBIC
S, AEROBIC CONDITIONS, ANAEROBIC
RIANS, SYNE/ COMMUNITY, SPECIFIC
SULFIDE, LEAD, COPPER, AMMONIA,
TENSITY, CHLOROPHYLL, NUTRIENTS,
N, HYPOLIMNION, LIGHT INTENSITY,
GY, CULTURES, TESTING, CAPILLARY
OINDICATORS, CALCIUM, CHLORIDES,
MUNICIPAL WATER, SEDIMENT LOAD,
S, TIDAL WATERS, ORGANIC MATTER,
NDS, HYDROGEN ION CONCENTRATION,
ASCOPHYLLUM NODOSUM, ECTOCARPUS
COND INTERNATIONAL OCEANOGRAPHIC
TREATMENT, ALGAE, WATER QUALITY,
MANAGEMENT, GEOCHEMISTRY, LAKES,
MUS, CHLOROCOCCALES, VOLVOCALES,
AULOBACTERIALES, CHLOROCOCCALES,
T, DRAINAGE SYSTEMS, FLOW RATES,
CHSEE(SWITZERLAND), LINSLEY POND(
WASTES, MADISON(WIS), LAKE ZOAR(
NOLOGY, WISCONSIN, URBANIZATION,
TRIENT/ *EUTROPHICATION, *LAKES,
ATURE, LIGHT INTENSITY, DAPHNIA,
R POLLUTION EFFECTS, EPILIMNION,
TURES, WASHINGTON, WAVES(WATER),
OCYCLOPS, KERATELLA, POLYARTHRA,
POLLUTION ABATEMENT, DIKES, SOIL
GAL CONTROL, AQUATIC LIFE, WATER
TIC PLANTS, AQUATIC POPULATIONS,
TS, *AQUATIC WEED CONTROL, WATER
, *WATER POLLUTION CONTROL, FISH
(GERMANY), PL/ CHIRONOMIDS, LAKE
, LAKE ANNECY, LAKE VANERN, LAKE
XIMUM UPTAKE VELOCITY, MICHAELIS
LA, SCENEDESMUS OBLIQUUS, GROWTH
SERVATION, WATER DYNAMICS, WATER
VATED SLUDGE, MAINTENANCE COSTS,
ND, STORAGE CAPACITY, EFFLUENTS,
ROPHYTES, GROUND SOLIDS, UPTAKE,
CHEMICAL INDUSTRY, ODOR, TASTE,
NUTRIENT LIMITED GROWTH, LUXURY
LOGY, KJELDAHL PROCEDURE, LUXURY
VAL.:   COAL
N DEMAND, SUSPENDED LOAD, SOLIDS
RAPHY, *TRANSACTIONS, BLACK SEA,
EICHORNIA, TYPHA, BRASENIA, DOW
TES, *ABSORPTION, *CALCIUM, SOIL
POLLUTION, WATER POLLUTION, SOIL
CT CO/ *MASTER PLANNING, *SOURCE
(WIS), NET PLANKTON, CHLOROPHYLL
WISCONSIN LAKE SURVEY, NITROGEN
S, ANKISTRODESMUS, HETEROTROPHY,
            *NUTRIENT REMOVAL,
ROKINIANA, COMPOSITION OF ALGAE,
EUTROPHICATION RESEARCH PROGRAM,
D, BENTHOS, WISCONSIN, SAMPLING,
            USBR RESEARCH
PUBLIC HEALTH, WATER POLLUTION,
, WATER POLLUTION EFFECTS, ALGAL
LATION, CAPILLARITY, CAPILLARITY
SECCHI DISC, VISIBILITY TESTS,
MCONTROL, BIOCONTROL, INTEGRATED
WATER RESOURCES COMMISSION, PEST
METHOD, TREATMENT TIMING, ALGAE
            *EUTROPHICATION
OXICITY, FISH, BLOODWORMS, ALGAL
N, GROWTH RATES, WATER POLLUTION
GAE, CHLORINATION, WATER QUALITY
PREDATION, COMPETITION, *INSECT
EUTROPHICATION, WATER POLLUTION
POLLUTION CONTROL, WATER QUALITY
EDS, ADMINISTRATION, SUPERVISORY
FATE, *ALKALINITY, WATER QUALITY
ON, LAKES, LAKE SEDIMENTS, ALGAL
BIOCHEMICAL OXYGEN DEMAND, ALGAL
EUTROPH/ *BIBLIOGRAPHIES, *ALGAL
S, *ALGAL CONTROL, *AQUATIC WEED
S, *ALGAL CONTROL, *AQUATIC WEED

CONDITIONS, LAGOONS.: /GE PROGRAM   W71-04555
CONDITIONS, CANNERIES, ACTIVATED   W71-09546
CONDITIONS, LIGHT INTENSITY, EUTR   W71-13413
CONDITIONS, WATER POLLUTION SOURC   W71-11381
CONDITIONS, ANAEROBIC CONDITIONS,   W71-11381
CONDITIONS, EXTRACELLULAR PRODUCT   W69-07832
CONDITIONS, ANAEROBIC CONDITIONS,   W70-04430
CONDITIONS, WATER SUPPLY.: /HETIC   W70-04430
CONDITIONS, BIO: /SIS, GROWTH RAT   W70-05750
CONDITIONS, ANAEROBIC CONDITIONS,   W70-05750
CONDITIONS, CHLORELLA.: /*NITRITE   W70-05261
CONDITIONS, ANAEROBIC CONDITIONS,   W70-05261
CONDITIONS, ANAEROBIC CONDITIONS,   W70-04786
CONDITIONS, ALGAE, ALGAL CONTROL,   W70-04786
CONDUCTANCE, SPECIES, MELOSIRA VA   W70-00265
CONDUCTIVITY, CHR: /NDS, HYDROGEN   W70-04510
CONDUCTIVITY, AGRICULTURAL WATERS   W70-05752
CONDUCTIVITY, ALGAE, SEWAGE, DIAT   W70-05405
CONDUCTIVITY, BACTERIA, DIATOMS,   W70-04469
CONDUCTIVITY, CULTURES, HYDROGEN   W69-07833
CONDUCTIVITY, CALCIUM, MAGNESIUM,   W69-05894
CONDUCTIVITY, TEMPERATURE, BACTER   W73-00748
CONDUCTIVITY, TEMPERATURE, DISSOL   W72-06051
CONFERVOIDES, NYLON COLUMN TECHNI   W70-01074
CONGRESS, HYPON: /OGY, RUSSIA, SE   W69-07440
CONIFEROUS FORESTS, DECIDUOUS FOR   W70-03512
CONIFEROUS FORESTS, ALGAE, INSECT   W71-09012
CONJUGATAE, HETEROKONTAE, GLOEOTR   W70-05409
CONJUGATOPHYCAEA, PLANCTOMYCES, B   W70-03969
CONN STREAM FLOW, ALGAE, AQUATIC   W68-00479
CONN), BOSMINA COREGONI LONGISPIN   W69-10169
CONN), KLAMATH LAKE(ORE), LAKE WA   W70-04506
CONNECTICUT, OREGON, WASHINGTON,   W70-04506
CONNECTICUT, WATER POLLUTION, *NU   W68-00511
CONNECTICUT, ROTIFERS.: /, TEMPER   W70-03959
CONNECTICUT, EUGLENA, CHLAMYDOMON   W70-03957
CONNECTICUT, RHODE ISLAND, TEMPER   W69-10161
CONOCHILUS, PYCNOCLINE, PHOTOLITH   W70-05760
CONSE: /TION SOURCES, DREDGING, *   W68-00172
CONSERVATION, PUBLIC HEALTH, WATE   W70-03344
CONSERVATION, WATER QUALITY, WATE   W69-05844
CONSERVATION, WATER CONTROL, FERT   W70-00269
CONSERVATION, POLLUTION ABATEMENT   W72-01786
CONSTANCE, WHITEFISH, WURTTEMBERG   W70-05761
CONSTANCE, PFAFFIKERSEE, TURLERSE   W70-03964
CONSTANT, LAKE NORRVIKEN(SWEDEN),   W70-04287
CONSTATS, RIVER RHINE.: /TERIONEL   W70-03955
CONSTITUENTS, ANALYTICAL CHEMISTR   W70-04382
CONSTRUCTION COSTS.: /UTION, ACTI   W70-04786
CONSTRUCTION.: /MICAL OXYGEN DEMA   W71-00936
CONSUMER ORGANISMS, SIMULIUM, PHY   W69-09334
CONSUMER.: /ER WORKS ASSOCIATION,   W71-05943
CONSUMPTION, PROVISIONAL ALGAL AS   W70-02779
CONSUMPTION, NUTRIENT REMOVAL, SP   W69-05868
CONTACT PROCESSES, PESTICIDE REMO   W70-05470
CONTACT PROCESS, SEWAGE TREATMENT   W70-06792
CONTACT ZONES, ARTEMIA, BALANUS,   W69-07440
CONTACT, SPARGANIUM, SALIX, GROWT   W70-05263
CONTAMINATION, *STRONTIUM RADIOSO   W69-03491
CONTAMINATION, NUTRIENTS.: / AIR   W71-06821
CONTAMINATION, CYANIDE, NEW PRODU   W73-00758
CONTENT, CENTRIFUGE METHOD, CALCU   W70-03959
CONTENT, LAKE ERKEN(SWEDEN), LAKE   W70-03959
CONTINUOUS CULTURES, CHLAMYDOMONA   W70-04381
CONTINUOUS FLOW SYSTEM.:   W72-00917
CONTINUOUS CULTURES, FATTY ACIDS,   W70-07957
CONTINUOUS CULTURE, CHEMOSTATS, B   W70-02775
CONTOURS, DEPTH, OLIGOCHAETES, TO   W70-06217
CONTRACTS, CHENEY DAM, KANS.:   W72-01773
CONTRACTS, REGULATION, RIGHT-OF-W   W70-03344
CONTRO: /EATMENT, WATER POLLUTION   W69-10171
CONTROL DEVICE.: /CHIA, ALGAE ISO   W70-04469
CONTROL METHODS.:   W68-00488
CONTROL MEASURES, PHYSICAL CONTRO   W70-07393
CONTROL ORGANIZATION OF TORONTO,   W69-10155
CONTROL REQUIREMENTS, THRESHOLD O   W69-10157
CONTROL.:   W70-07393
CONTROL.: /DEPTH, OLIGOCHAETES, T   W70-06217
CONTROL.: /RTICLE SIZE, POPULATIO   W72-02417
CONTROL.: /TE-PRODUCING ALGAE, AL   W70-02373
CONTROL.: /ALGAE, *ALGAL CONTROL,   W70-05428
CONTROL.: /HYTA, IRON, MANGANESE,   W69-05704
CONTROL.: /ROSION, ALGAE, *WATER   W73-01122
CONTROL(POWER), ADMINISTRATIVE AG   W72-05774
CONTROL, *CONTROL, *ALGAL CONTROL   W72-07442
CONTROL, *NUTRIENTS, ALGAE, BENTH   W72-06052
CONTROL, *COLOR, *ALKALINITY, *NI   W72-11906
CONTROL, *AQUATIC WEED CONTROL, *   W69-05706
CONTROL, *EUTROPHICATION, AQUATIC   W69-05706
CONTROL, *EUTROPHICATION, AQUATIC   W69-05705

417

TER CHEMISTRY, POLLUTANTS, ALGAL
LLUTION EFFECTS, WATER POLLUTION
, WATER QUALITY, WATER POLLUTION
LLUTION SOURCES, WATER POLLUTION
TING OF ALGAE, EFFLUENT, QUALITY
GE FISH, PLANKTON, ALGAE, *ALGAL
WAGE TREATMENT, *WATER POLLUTION
ENT, ENGINEERING, ALGAE, PLANTS,
TAMINATION, CYANIDE, NEW PRODUCT
AE, *TASTE, *ODOR, WATER QUALITY
TES, SLUDGE, SLURRIES, AUTOMATIC
PONDS, DESIGN FOR ENVIRONMENTAL
LLUTION SOURCES, WATER POLLUTION
LLUTION TREATMENT, WATER QUALITY
OF ALGAE, BOTTOM SEDIMENTS, WEED
E, *WATER QUALITY CONTROL, ALGAL
PRODUCTIVITY, HYPOLIMNION, ALGAL
PLY, INTAKES, NUTRIENTS, QUALITY
, LIMNOLOGY, ALGAL TOXINS, ALGAL
UALITY CONTROL, *CONTROL, *ALGAL
ENT FACILITIES, CLEANING, *ALGAL
EROSION, ALGAE, *WATER POLLUTION
, *AQUATIC PLANTS, *AQUATIC WEED
ITY, WA/ *COPPER SULFATE, *ALGAL
, WATER POLLUTION EFFECTS, ALGAL
ECTS OF POLLUTION, WATER QUALITY
EWAGE TREATMENT, WATER POLLUTION
IDENTIFICATION, WATER POLLUTION
CLASSIFICATION, *WATER POLLUTION
S, DISSOLVED OXYGEN, EVAPORATION
RATED CONTROL MEASURES, PHYSICAL
CIDES, HARVESTING, WATER QUALITY
QUALITY CONTROL, WATER POLLUTION
QUALITY CONTROL, WATER POLLUTION
CIDES, HARVESTING, WATER QUALITY
, WATER QUALITY, WATER POLLUTION
SULPHATE, NUISANCE ALGAE, *ALGAL
ATER POLLUTION, *WATER POLLUTION
OIRS, HYDROELECTRIC PLANTS, WEED
NTIFICATION, POLLUTANTS, QUALITY
*RIVERS, *ALGAE, WATER POLLUTION
AQUATIC WEED CONTROL, MECHANICAL
ERVOIRS, *MIXING, *WATER QUALITY
PONDS, WATER HYACINTH, CHEMICAL
NTHELLAE, ACANTHURUS TRIOSTEGUS,
LOROPHYTA, SCENEDESMUS, BENTHOS,
OM, ALGAE GROWTH, WATER DENSITY,
WATER, WATER REQUIREMENTS, WATER
LL),/ *POLLUTION, *CHICAGO(ILL),
TRIC POWER PLANTS, *WATER REUSE,
SHORTAGE FILTERS, SNAILS, ODOR,
STUDIES, BIOASSAY, AQUATIC LIFE,
M DESLUDGING./ *METAL CORROSION,
DROGEN SULFIDE, IRON, MAGNESIUM,
BONATE, CERCARIAE, WATER MILFOIL
RO-ALGAE, FLAGELLATES, CILIATES,
, *WINTER, CYANOPHYTA, PLANKTON,
RIPTON, BACTERIA, SEDIMENTATION,
PLANKTON, ROTIFERS, CHLOROPHYTA,
CONTROL, BIOCONTROL, TURBULENCE,
R MILFO/ SCHISTOSOME DERMATITIS,
ES, CATTAILS, BULRUSH, DREDGING,
E, LIGHT TEMPERATURE, NUTRIENTS,
N, TURBULENCE, TOXINS, BACTERIA,
GICIDES,/ *LAKES, *MUD, *COPPER,
TRIPHICATION, *LAKES, WISCONSIN,
NG ALGAE, CHLOROPHYLL, CHEMTROL,
SINS, WATER POLLUTION TREATMENT,
TRIBUTION SYSTEMS, CHLORINATION,
, VITAMINS, ALGAE, GROWTH RATES,
RAQUAT, DIQUAT, SODIUM ARSENITE,
, CYTOLOGICAL STUDIES, CHLORINE,
ISCONSIN, ALGICIDES, HYDROLYSIS,
SHALLOW WATER, HYDROGEN SULFIDE,
IA, NITROGEN, PHOSPHORUS, ALGAE,
TON, *DIATOMS, *BLUE-GREEN ALGA,
*ALGAE, *MAINE, *WATER SUPPLY,
MECHANICAL CONTROL, BIOCONTROL,
*WATER QUALITY CONTROL, *ALGAE,
QUALITY, *ORGANIC MATTER, ALGAE,
NG, *FILTRATION, DESIGN, *ALGAE,
ATION, FI/ *ARTIFICIAL RECHARGE,
MPOUNDS, HYDROGEN SULFIDE, LEAD,
ORON, CALCIUM, CHLORINE, COBALT,
ATLANTIC OCE/ *ALGAE, *HERRINGS,
*ALGAE, *FRESH WATER, *PEPTIDES,
TOZOA, LICHENS, PLANTS, DIATOMS,
NTOSPHENUS LAMOTTENII, GAMMARUS,
ND), LINSLEY POND(CONN), BOSMINA
, OSCILLATORIA RUBESCENS, PISTON

CONTROL, NITRATES, PHOSPHATES, SU   W70-02646
CONTROL, NITRATES,. NITROGEN FI: /   W69-10167
CONTROL, NUISANCE ALGAE, TASTE, O   W69-10155
CONTROL, OXYGEN BALANCE, SEWAGE,   W69-08668
CONTROL, OXIDATION LAGOON, NITROG   W71-13313
CONTROL, PREDATION, COMPETITION,   W70-05428
CONTROL, PACIFIC NORTHWEST U. S.,   W69-03683
CONTROL, POLLUTION ABATEMENT, ABA   W69-03683
CONTROL, PLUMBING CODES, WELL DRI   W73-00758
CONTROL, POLLUTANT IDENTIFICATION   W72-13605
CONTROL, PIPING SYSTEMS(MECHANICA   W72-08688
CONTROL, RECIRCULATION, BOD REMOV   W70-04787
CONTROL, RADIOACTIVE WASTE, WATER   W72-00941
CONTROL, SEWAGE EFFLUENTS, ABSORP   W70-01031
CONTROL, SOLAR RADIATION, *WATER   W68-00478
CONTROL, SLIME, TASTE-PRODUCING A   W69-05704
CONTROL, SEASONAL, ANNUAL TURNOVE   W69-05142
CONTROL, TREATMENT FACILITIES, FI   W70-00263
CONTROL, TRACE ELEMENTS, WASHINGT   W70-00264
CONTROL, TOXICITY.: /ITY, WATER Q   W72-07442
CONTROL, WISCONSIN.: /STS, TREATM   W72-08859
CONTROL, WATER QUALITY CONTROL.: /   W73-01122
CONTROL, WATER CONSERVATION, WATE   W70-00269
CONTROL, WATER SUPPLY, WATER QUAL   W69-10155
CONTROL, WATER POLLUTION, TOXICIT   W70-02775
CONTROL, WINNISQUAM LAKE, NEWFOUN   W70-02772
CONTROL, WATER POLLUTION EFFECTS,   W70-02248
CONTROL, WATER POLLUTION EFFECTS,   W70-02792
CONTROL, WATERCOURSES(LEGAL), ORG   W72-00462
CONTROL, WATER BUDGET, WATER CHEM   W72-01773
CONTROL, WATER QUALITY.: /, INTEG   W70-07393
CONTROL, WATER POLLUTION CONTROL,   W69-05706
CONTROL, WATER POLLUTION EFFECTS,   W69-05706
CONTROL, WATER POLLUTION EFFECTS.   W69-05705
CONTROL, WATER POLLUTION CONTROL,   W69-05705
CONTROL, WATER PO: /STE TREATMENT   W69-06535
CONTROL, WEED CONTROL, *INVESTIGA   W68-00468
CONTROL, WATER POLLUTION EFFECTS,   W69-03948
CONTROL, WATER PROPERTIES, PHOSPH   W68-00479
CONTROL, WATER POLLUTION, *WATER   W69-01273
CONTROL, WATER POLLUTION EFFECTS,   W70-05668
CONTROL, WATER POLLUTION CONTROL,   W70-04494
CONTROL, WATER QUALITY, WATER POL   W70-04484
CONTROLS, PARAQUAT, DIQUAT, SODIU   W70-00269
CONUS EBRAEUS.: / FUNGITES, ZOOXA   W69-08526
CONVECTION, WAVES(WATER), FISH, S   W70-03969
CONVECTION CURRENTS, ALGAE TREATM   W69-08282
CONVEYANCE, WATER COSTS, ARTIFICI   W71-09577
COOK COUNTY(ILL), CALUMET RIVER(I   W70-00263
COOLING TOWERS, DAMS, ALGAE, HYDR   W71-11393
COOLING TOWERS, FILTERS, EFFLUENT   W72-00027
COOLING WATER, SITES.: /E, MODEL   W71-11881
COOLING WATER TREATMENT, ON-STREA   W70-02294
COOLING, PROPERTY VALUES, RECREAT   W70-02251
COONTAIL, WILD CELERY, BUR REED,   W70-05568
COPEPODIDS, NAUPLII, TERRAMYCIN,   W69-10154
COPEPODS, EUTROPHICATION, OLIGOTR   W69-10154
COPEPODS, WATER POLLUTION SOURCES   W69-10169
COPEPODS, CYCLES, CHLAMYDOMONAS,   W70-00268
COPEPODS, ROTIFERS, CHLOROPHYLL,   W71-10098
COPPER CARBONATE, CERCARIAE, WATE   W70-05568
COPPER SULPHATE, PONDWEEDS, SODIU   W70-05263
COPPER SULFATE, STRATIFICATION.: /   W69-08282
COPPER SULFATE, CIRCULATION.: /IO   W70-04369
COPPER SULFATE, ALGAL CONTROL, AL   W69-03366
COPPER SULPHATE, NUISANCE ALGAE,   W68-00468
COPPER SULFATE, LAKES, NITROGEN.:   W68-00470
COPPER SULFATE, ODOR, COLIFORMS,   W72-01814
COPPER SULFATE, ACTIVATED CARBON,   W72-03562
COPPER SULPHATE, BOTTOM SEDIMENTS,   W69-09349
COPPER SULPHATE, MONURON, AMMONIA   W70-00269
COPPER SULFATE, ALGAL CONTROL, MO   W70-02364
COPPER SULFATE, FISHES, HERBICIDE   W70-02982
COPPER SULFATE, CHLOROPHYTA, CRUS   W70-02803
COPPER SULFATE, EFFLUENTS, MICROC   W70-02787
COPPER SULFATE.: /NG, *PHYTOPLANK   W70-02754
COPPER SULFATE, RESERVOIRS, WATER   W70-08096
COPPER SULFATE, SODIUM ARSENITE,:   W70-10175
COPPER SULFATE, CHLORINE, WELLS,   W71-00110
COPPER SULFATE, *CHEMICAL OXYGEN   W72-04495
COPPER SULFATE, SEDIMENTATION, AC   W72-08859
COPPER, *ALGICIDES, ALGAE, *EVALU   W72-05143
COPPER, AMMONIA, CONDUCTIVITY, CH   W70-04510
COPPER, ECOLOGY, ENVIRONMENTAL EF   W69-04801
COPPER, IRON, COBALT, MANGANESE,   W69-08525
COPPER, IRON, IONS, PHOSPHATES, O   W69-10180
CORAL, MOLLUSKS, MINE ACIDS, COAL   W69-07428
CORDULEGASTER.: / SALMO TRUTTA, E   W69-09334
COREGONI LONGISPINA.: /(SWITZERLA   W69-10169
CORER, FRAGILARIA, MELOSIRA, ASTE   W70-05663

S, AVAILABLE NUTRIENTS, SEDIMENT
RRENTS(WATER), INDUSTRIAL WASTE,
, ALGAE, OLIGOTROPHY, DIVERSION,
TROL, ALGICIDES, TRACE ELEMENTS,
EE(WIS), CHIRONOMUS, PROCLADIUS,
LLULAR SIZE, CHLORELLA VULGARIS,
MERICAN RESEARCH AND DEVELOPMENT
INCIPAL-COMPONENT ANALYSIS, PART
REUSE, DISSOLVED OXYGEN, PITTING(
T, ON-STREAM DESLUDGING./ *METAL
ENTIAL, DISSOLVED OXYGEN, ODORS,
RIFICATION, *BACTERIA, *ENZYMES,
, BACTERIOSTRUM, LEPTOCYLINDRUS,
, STREPTOCOCCUS, BREVIBACTERIUM,
HROPHILES, PLEMORPHIC, MYCELOID,
VICULA, MELOSIRA, THALASSIOSIRA,
DULPHIA, CHAETOCEROS, COCCONEIS,
SMUS, OOCYSTIS, CRYOMONAS EROSA,
CITY(COLO), MUTATION, BETA RAYS,
ORUS, FISH/ *NUTRIENTS, *BENEFIT-
G SYSTEMS(MECHANICAL), CLEANING,
SIBILITY, TECHNICAL FEASIBILITY,
OXYGEN DEMAND, ACTIVATED SLUDGE,
OGEN, *PHOSPHORUS, EFFICIENCIES,
TION SOURCES, CYCLING NUTRIENTS,
IME, OXYGEN, TERTIARY TREATMENT,
CTERIA, SEDIMENTATION, TOXICITY,
IMMING, WATERFOWL, BIRDS, COSTS,
, DRAIN, SAN JOAQUIN VA/ *CONTRA
SCILLATORIA VAUCHER, SKELETONEMA
RAMIMONAS, PERIDINI/ SKELETONEMA
AMYDOMONAS MOEWUSII, SKELETONEMA
ORELLA LUTEOVIRIDIS, SKELETONEMA
LIS, CACHONINA NIEI, SKELETONEMA
MAINTENANCE COSTS, CONSTRUCTION
ACTERIA, BACTERICIDES, CHLORINE,
REMENTS, WATER CONVEYANCE, WATER
EWAGE, DRAINAGE DISTRICTS, DEBT,
N, ACTIVATED SLUDGE, MAINTENANCE
UES, SWIMMING, WATERFOWL, BIRDS,
ECTION CURRENTS, ALGAE TREATMENT
LAGOONS, PONDS, PHOTOSYNTHESIS,
REATMENT, *GREAT LAKES, CONTROL,
ULATION, RIGHT-OF-WAY, ESTIMATED
ION, ACTIVATED CARBON, CHLORINE,
EL RIVER, DOLORES RIVER, URAVAN,
CHLORIDE, INFRARED GAS ANALYSIS,
INIUM COHNII, MICROBIAL ECOLOGY,
SCONSIN, ASIA, NATIONAL RESEARCH
SCONSIN, ASIA, NATIONAL RESEARCH
PPING FILM, SYNECHOCOCCUS, GRAIN
*SIGNIFICANCE, ALGAL KEY, CLUMP
EMOSYNTHETIC, SAPROPHYTIC, PLATE
OWARD A. HANSON RES/ WOOD, ALGAL
*GROWTH, *UPTAKE, FIXATION, CELL
*ALGAE HOMOGENIZATION, *BACTERIA
ACTERIA, CYCLOTELLA NANA, GEIGER
LLULAR SIZE, CHLORELLA/ *COULTER
, INFRARED GAS ANALYSIS, COULTER
HNII, MICROBIAL ECOLOGY, COULTER
UNIVERSITY, ELECTRONIC PARTICLE
NIUM PULVISCULUS, HAEMOCYTOMETER
ODS, ALGAE IDENTIFICATION, ALGAE
RS, MILK WASTE TREATMENT, COLONY
ED CELL VOLUMES, BACTERIAL PLATE
NAL WATER QUALITY NETWORK, ALGAE
), SCENEDESMUS QUADRICAUDA, CELL
IO), LAKE RAMSEY(ONTARIO), ALGAE
                            LOS ANGELES
*POLLUTION, *CHICAGO(ILL), COOK
214, AC/ *COTTUS CAROLINAE, MEAD
ACID WASTE MICROFLORA, MCCREARY
IN, MACROPH/ MICHIGAN, CHEBOYGAN
METER,/ *ONONDAGA LAKE, ONONDAGA
NIC PHOSPHATES.:   HARRIS
N, SAN JOAQUIN VA/ *CONTRA COSTA
AWA RIVER(POLAND), MONTANE RIVER
E, WATER COSTS, ARTIFICIAL WATER
SYLVANIA, MARCASITE, CHALCOCITE,
IS, ANALYTICAL TECHNIQUES, CLOUD
TURE, SEA WATER, OYSTERS, CLAMS,
TON, ECOSYSTEMS, SINKS, TRACERS,
HARIS, NAIAS, SCIRPUS, NYMPHAEA,
D(N C), DETECTOR, BEAUFORT(N C),
ATER PROPERTIES.:
S, *PESTICIDE RESIDUES, DIATOMS,
RS, *BASS, WATER HYACINTH, FROGS,
MENTS, CHUTES, EDDIES, HABITATS,
RA, FISH, LIMNOLOGY, LAKE TROUT,
TRUCTURE, RIDLEY CREEK(PA), NOBS

CORES.: /ON, NITROGEN, PHOSPHOROU   W68-00511
CORES, DIATOMS, ODOR-PRODUCING AL   W68-00470
CORES, SEWAGE EFFLUENTS, DIATOMS,   W70-05663
CORES, WISCONSIN, BOTTOM SEDIMENT   W69-03366
CORETHRA PUNCTIPENNIS, PISIDIUM I   W70-06217
CORNELL UNIVERSITY, ELECTRONIC PA   W69-06540
CORP(MASS), ARTHUR D LITTLE INC(M   W69-06865
CORRELATION, DECAPODS, SPECIES DI   W73-01446
CORROSION), RUSTING, METALS, INHI   W70-02294
CORROSION, COOLING WATER TREATMEN   W70-02294
CORROSION, OXIDATION, PHOTOSYNTHE   W71-11377
CORROSION, OXIDATION, REDUCTION(C   W69-07428
CORYCAEUS, OIKOPLEURA, FRITILLARI   W73-01446
CORYNEBACTERIUM, ELECTROSTATIC PR   W71-11823
CORYNEFORM RODS, COCCI,: /E, PSYC   W70-00713
COSCINDISCUS, BACTERIOSTRUM, LEPT   W73-01446
COSCINODISCUS, DIMEROGRAMMA, DIPL   W70-03325
COSMARIUM, CHROCOCCUS LIMNETICUS,   W70-05405
COSMIC RADIATION, S: /N, CENTRAL   W69-08272
COST ANALYSIS, *LAKE ERIE, PHOSPH   W70-04465
COST ANALYSIS, WATER PURIFICATION   W72-08688
COST ANALYSIS, SLUDGE, INCINERATI   W71-10654
COST COMPARISONS, SEDIMENTS, ALGA   W71-09548
COST COMPARISONS, TREATMENT, BIOL   W70-01981
COST COMPARISONS, ECONOMICS, ALGA   W69-08518
COST COMPARISONS.: / COMPOUNDS, L   W68-00256
COST-BENEFIT ANALYSIS.: /ANTS, BA   W70-05819
COST-SHARING.: /ON TECHNIQUES, SW   W70-05568
COSTA COUNTY, WATER DETERIORATION   W71-09577
COSTATUM(GREV) CL, CHLORELLA OVAL   W70-05752
COSTATUM, COCCOLITHUS HUXLEYI, PY   W70-05272
COSTATUM, SYNECHOCOCCUS ELON: /HL   W70-05381
COSTATUM, PHYSIOLOGICAL ADJUSTMEN   W70-04809
COSTATUM, DUNALIELLA TERTIOLECTA,   W69-08278
COSTS.: /UTION, ACTIVATED SLUDGE,   W70-04786
COSTS, *WASTE WATER TREATMENT.: /   W72-12987
COSTS, ARTIFICIAL WATER COURSES,   W71-09577
COSTS, CITIES.: /ATMENT, LAKES, S   W70-04455
COSTS, CONSTRUCTION COSTS.: /UTIO   W70-04786
COSTS, COST-SHARING.: /ON TECHNIQ   W70-05568
COSTS, DINOBRYON, CONCORD(N H), P   W69-08282
COSTS, EFFICIENCY, BIOCHEMICAL OX   W70-09320
COSTS, EUTROPHICATION, NITRATES,   W70-03964
COSTS, LEGAL ASPECTS, LEGISLATION   W70-03344
COSTS, TREATMENT FACILITIES, CLEA   W72-08859
COTTUS SPP, RHINICHTHYS OSCULUS,   W69-07846
COULTER COUNTER, EUGLENA GRACILIS   W69-08278
COULTER COUNTER, GROWTH INHIBITIO   W70-05652
COUNCIL(USA), US ATOMIC ENERGY CO   W70-03975
COUNCIL(USA), US ATOMIC ENERGY CO   W70-03975
COUNT(AUTORADIOGRAPHY), LEUCOTHRI   W69-10163
COUNT, DESMIDS, FLAGELLATES, MESO   W70-05389
COUNT, FILAMENTOUS, UNICELLULAR,   W68-00891
COUNT, GREEN RIVER, WASHINGTON, H   W72-11906
COUNT, OPTICAL DENSITY, PARTICULA   W70-03983
COUNT, WARING BLENDOR MORTALITY,   W70-04365
COUNTER.: /ANOCOBALAMIN, MUTANT B   W70-05652
COUNTER, *POPULATION DYNAMICS, CE   W69-06540
COUNTER, EUGLENA GRACILIS, CACHON   W69-08278
COUNTER, GROWTH INHIBITION, CYANO   W70-05652
COUNTING, EUDORINA ELEGANS, EUGLE   W69-06540
COUNTING, ITHACA(NY), GLENODINIUM   W69-06540
COUNTS, ALGAL GROWTH, SEDGWICK-RA   W69-10157
COUNTS, ARTHROPODS.: /PODS, SPIDE   W72-01819
COUNTS, LANCASTER(CALIF).: / PACK   W71-05267
COUNTS, PLANKTON ANALYSES, AGRICU   W70-04507
COUNTS, SPECIES DETERMINATION, AB   W70-03334
COUNTS, THRESHOLD ODOR NUMBERS, C   W69-10155
COUNTY(CALIF).:   W70-05645
COUNTY(ILL), CALUMET RIVER(ILL),   W70-00263
COUNTY(KY), THORIUM-232, BISMUTH-   W69-09742
COUNTY, KENTUCKY.:   W69-00096
COUNTY, STURGEON RIVER, FLUORESCE   W69-09334
COUNTY, SYRACUSE, WARBURG RESPIRO   W69-02959
COUNTY, TEX, POLYPHOSPHATES, ORGA   W69-04800
COUNTY, WATER DETERIORATION, DRAI   W71-09577
COURSE, CARPATHIAN SUBMONTANE REG   W70-02784
COURSES, CALIFORNIA.: / CONVEYANC   W71-09577
COVELLITE, SPHALERITE, MILLERITE,   W69-07428
COVER, SPECTROSCOPY, ULTRAVIOLET   W69-09676
CRABS, NORTH CAROLINA, SEDIMENTS,   W69-08267
CRABS, SHRIMP, MULLETS, FISH, ALG   W69-08274
CRASSIPES EICHORNIA, TYPHA, BRASE   W70-05263
CRASSOSTREA VIRGINICA, MERCENARIA   W69-08267
CRATER LAKE, ORE, GROWTH RATES, W   W68-00478
CRAYFISH, POLLUTANT IDENTIFICATIO   W69-02782
CRAYFISH, MOSQUITOES, BACTERIA, I   W70-05263
CRAYFISH, GROWTH STAGES, DIATOMS,   W69-09742
CRAYFISH, STONEFLIES, SNAILS, ECO   W70-00711
CREEK(MD), BACK RIVER(MD), OSCILL   W70-04510

```
                  CATTARAUGUS      CREEK(NY).:                          W69-10080
  CLADOPHORA GLOMERATA, BUTTERNUT   CREEK(NY).:       *CLADOPHORA FRACTA,  W72-05453
TY, EXPERIMENTAL STREAMS,/ BERRY    CREEK(ORE), OREGON STATE UNIVERSI   W70-03978
IVE/ COMMUNITY STRUCTURE, RIDLEY    CREEK(PA), NOBS CREEK(MD), BACK R   W70-04510
RA.:                      SPRING    CREEK(PA), STATE COLLEGE(PA), FLO   W69-10159
), NINE SPRINGS CREEK(WIS), DOOR    CREEK(WIS), DUTCH MILL DRAINAGE D   W70-05113
WEATHER CREEK(WIS), NINE SPRINGS    CREEK(WIS), DOOR CREEK(WIS), DUTC   W70-05113
YAHARA RIVER(WIS), STORKWEATHER     CREEK(WIS), NINE SPRINGS CREEK(WI   W70-05113
, LAKE MENDOTA(WIS), BLACK EARTH    CREEK(WIS).: /WASTE WATER EFFECTS,  W70-04382
UKON RIVER, NEBESNA RIVER, MOOSE    CREEK, ARCTIC CIRCLE HOT SPRINGS,   W69-08272
       MILLSBORO(DELAWARE), SWAN    CREEK, INDIAN RIVER.:               W71-05013
STIS, ANABAENA, EUCYCLOPS, NAIS,    CRICOTOPUS, AUSTROCLOEON, ENALLAG   W69-06975
LAR, FUCUS VESICULOSUS, CHONDRUS    CRISPUS, ASCOPHYLLUM NODOSUM, LAM   W70-01073
OMOGETON PECTINATUS, POTOMOGETON    CRISPUS, HETERAMTHERA, PICKWICK R   W71-07698
ENSIS, MYRIOPHYLLUM, POTAMOGETON    CRISPUS, MICROSPORA WITTROCKII, S   W70-00711
GE TREAT/ *FARM LAGOONS, *DESIGN    CRITERIA, WATER TEMPERATURE, SEWA   W71-08221
LLUTION, TOXICITY, LAKES, DESIGN    CRITERIA, REMOTE SENSING, MINNESO   W70-02775
*WATER POLLUTION, DESIGN, DESIGN    CRITERIA, OXIDATION LAGOONS PHOTO   W70-02609
EST(US), *SOUTHWEST(US), *DESIGN    CRITERIA, POND EFFLUENT, COLORADO   W70-04788
, *BIOLOGICAL TREATMENT, *DESIGN    CRITERIA, DISSOLVED OXYGEN, TEMPE   W70-04787
*STABILIZATION, *PONDS, *DESIGN     CRITERIA, FACILITIES, EQUATIONS,   W70-04786
MICROBIAL ECOLOGY, WATER QUALITY    CRITERIA, CYANIDE, NAPHTHENIC ACI   W70-04510
PERATION AND MAINTENANCE, DESIGN    CRITERIA, *WATER TREATMENT, *NEW    W72-08686
, DEPTH, ORGANIC LOADING, DESIGN    CRITERIA, ODORS, EFFICIENCIES, *W   W72-06854
K ATOLL, TURNOVER TIME, TRIDACNA    CROCEA, WOODS HOLE OCEANOGRAPHIC    W69-08526
ROPHIC METABOLISM, HETEROTROPHY,    CROOKED LAKE(IND), LITTLE CROOKED   W71-10079
ROPHY, CROOKED LAKE(IND), LITTLE    CROOKED LAKE(IND).: /ISM, HETEROT   W71-10079
COUNTS, THRESHOLD ODOR NUMBERS,     CROP-DUSTING AIRCRAFT, WATER SAMP   W69-10155
ORGANIC WASTES, OXYGEN, STANDING    CROP, ALGAE, ZOOPLANKTON, FISH, B   W70-02251
ASSIMILATIVE CAPACITY, STANDING     CROP, ANNUAL TURNOVER, FISH POPUL   W68-00912
CALCIUM, PRODUCTIVITY, STANDING     CROP, CHLOROPHYLL, PHOSPHATES, AD   W70-03512
CS, TEMPERATURE, SILTS, STANDING    CROP, ECOLOGY, PHOTOSYNT: /DYNAMI   W70-02780
INE, LIGHT PENETRATION, STANDING    CROP, INDUSTRIAL WASTE: /THERMOCL   W71-07698
OGEN ION CONCENTRATION, STANDING    CROP, LIGHT, BIOMASS, RESPIRATION   W71-00665
EQUATIONS, OXYGENATION, STANDING    CROP, MICROENVIRONMENT, AQUATIC L   W69-03611
OSPHORUS, PRODUCTIVITY, STANDING    CROP, STREAMS, WATER CHEMISTRY, W   W69-04805
NTIAL NUTRIENTS, *NITRIFICATION,    CROPS, TOXICITY, SOIL POROSITY, I   W69-05323
LAKE MENDOTA(WIS), PHYTOPLANKTON    CROPS, TROPHIC, TROPHOLYTIC, NITR   W69-05142
, BOTRYOCCUS BRAUNII, FRAGILARIA    CROTONENSIS, NITZSCHIA, ASTERIONE   W70-02804
, CARTERIA, COMPARATIVE STUDIES,    CRUCIGENIA RECTANGULARIS, DICERAS   W69-07833
RECOVERY FROM POLLUTION, *KUWAIT    CRUDE, BENTLASS SALT MARSH.: /, *   W70-01231
LIPHATIC HYDROCARBONS, PETROLEUM    CRUDES, MASS SPECTROMETRY, N-HEPT   W69-08284
DESMUS QUADRICAUDA, PORPHYRIDIUM    CRUENTUM, LEMNA MINOR, PHOTOPHOSP   W70-02964
AMPHIDINIUM KLEBSI, PORPHYRIDIUM    CRUENTUM, RADIONUCLIDE UPTAKE, NA   W70-00708
SAND, FERRIC OXIDE-COATED SAND,     CRUSHED BAUXITE, ACID-WASHED BAUX   W70-08628
WATER POLLUTION EFFECTS, YEASTS,    CRUSTACEA, ALGAE.: /RODUCTIVITY,    W69-10152
ZOOPLANKTON, ALGAE, POLYCHAETA,     CRUSTACEA, MOLLUSCA, ECHINODERMAT   W69-08526
ON, LAKE ERIE, INSECTS, DIPTERA,    CRUSTACEA, MOLLUSCS, OLIGOCHAETES   W70-04253
ANOGRAPHY, *TRANSLATIONS, ALGAE,    CRUSTACEANS, BENTHOS, BRINE SHRIM   W69-07440
AGE EFFLUENTS, SMELTS, FISHKILL,    CRUSTACEANS, BOTTOM SEDIMENTS.: /   W69-08674
ILIMNION, PHYTOPLANKTON, SESTON,    CRUSTACEANS, DIATOMS, CHLOROPHYTA   W69-10182
PHENOLS, *WATER QUALITY CONTROL,    CRUSTACEANS, ROTIFERS, SWIMMING P   W70-10161
DE, COPPER SULFATE, CHLOROPHYTA,    CRUSTACEANS, SOIL EROSION, SEWAGE   W70-02803
SULFIDES, SEA WATER, PHAWOPHYTA,    CRUSTACEANS, HERBIVORES, CHLOROPH   W70-01068
GAE, MARINE FISH, INVERTEBRATES,    CRUSTACEANS, MOLLUSKS, STRONTIUM    W71-09850
CALES, ANKISTRODESMUS, OOCYSTIS,    CRYOMONAS EROSA, COSMARIUM, CHROC   W70-05405
IROIDES, MICROCYSTIS AERUGINOSA,    CRYPTOMONADIDA, MASTIGOPHORA, CRY   W70-05112
PARAMECIUM, PHACUS, GLENODINIUM,    CRYPTOMONAS, CILIATA, ASTERIONELL   W70-00268
A, CRYPTOMONADIDA, MASTIGOPHORA,    CRYPTOPHYCEAE, CHRYSOPHYCEAE, VOL   W70-05112
, MICROPLANKTON, PSEUDOPLANKTON,    CRYPTOPHYCEAE, VOLVOCALES, INDUST   W70-05389
ON, MICRONUTRIENTS, DINOPHYCEAE,    CRYPTOPHYCEAE, CH: /, APHANIZOMEN   W70-04381
LAKE(MINN), KIMBALL LAKE(MINN),     CRYPTOPHYTA, PYROPHYTA, CHROOMONA   W70-05387
ALES, DAPHNIA PULICARIS, DAPHNIA    CUCULLATA, DAPHNIA HYALINA, DAPHN   W70-00274
LORIDA).:                           CUE LAKE(FLORIDA), MCCLOUD LAKE(F   W72-01993
              *ANDERSON-            CUE LAKE, MELROSE(FLORIDA).:        W72-01990
DIATOMACEOUS EARTH, MUD, LAKES,     CULTIVATION, CHANNEL MORPHOLOGY,    W68-00475
S IMMOBILIS, VIBRIO, MORPHOLOGY,    CULTURAL CHARACTERISTICS, TERNARY   W70-04280
DIVERSITY, CHEMICAL PRECIPITATI/    CULTURAL EUTROPHICATION, SPECIES    W69-08518
  *MICROCYSTIS AERUGINOSA(KUTZ),    CULTURE MEDIA, CHU NO 10 SOLUTION   W70-03507
TIGLEOCLONIUS, INSOLUBLE WASTES,    CULTURE MEDIA.:                  S  W68-00248
N), EXTINCTION VALUES, EUDORINA,    CULTURE MEDIUM, COELASTRUM, PEDIA   W70-05409
NOSA, APHANIZOME/ *PURIFICATION,    CULTURE MEDIA, MICROCYSTIS AERUGI   W70-00719
UME.:                       MASS    CULTURE OF ALGAE, PACKED CELL VOL   W70-06053
S, NUTRIENT LIMITED/ *CONTINUOUS    CULTURE, *CHEMOSTATS, TURBIDOSTAT   W70-02779
PYRENOIDOSA(EMERSON STRA/ *MASS     CULTURE, *ALGAL GROWTH, CHLORELLA   W69-07442
TURE, *LABORATORY STUDIES, *MASS    CULTURE, *PILOT-PLANT STUDIES, AL   W69-06865
SS CULTURE, *PILOT-PLANT/ *ALGAL    CULTURE, *LABORATORY STUDIES, *MA   W69-06865
CAL INDEX, *SAPROBITY, P/ *ALGAL    CULTURE, *BIOMASS TITER, *BIOLOGI   W72-12240
ION RESEARCH PROGRAM, CONTINUOUS    CULTURE, CHEMOSTATS, BATCH CULTUR   W70-02775
GAS EXCHANGE, GREEN PLANTS, MASS    CULTURE, MEDICINE, METABOLIC PATT   W69-07832
FLOW RATES, PUMPS, ELECTROLYSIS,    CULTURES.: / ALGAE, FRESH WATER,    W70-09904
E, LIPIDS, NUTRIENTS, CHLORELLA,    CULTURES.:              *GROWTH, *ALGA  W70-07957
            *PURE ALGAL             CULTURES.:                          W71-06002
OASS/ *LABORATORY TESTS, *ALGAE,    CULTURES, DIATOMS, CYANOPHYTA, BI   W71-06002
COMPOSITION OF ALGAE, CONTINUOUS    CULTURES, FATTY ACIDS, OOCYSTIS P   W70-07957
DIATOMS, TEMPERATURE, MANGANESE,    CULTURES, RECIRCULATED WATER, NIT   W70-07257
IES, *NITROGEN LIMITATION, BATCH    CULTURES, CHLORELLA SOROKINIANA,    W70-07957
SORPTION, WASTE WATER TREATMENT,    CULTURES, CHLORELLA, WATER POLLUT   W70-09438
TIONS, WEATHER, WATER, MOVEMENT,    CULTURES, BAYS, STREAMS, WEEDS, L   W70-06229
NUOUS CULTURE, CHEMOSTATS, BATCH    CULTURES, ACETYLENE REDUCTION TEC   W70-02775
```

421

NAL ALGAL ASSAY PROCEDURE, BATCH
A, STREAMS, DIATOMS, RESERVOIRS,
HLORELLA, ALGAE, ORGANIC MATTER,
OPHYTES, *NUTRIENT REQUIREMENTS,
LANKTON, DIATOMS, ECOLOGY, SOIL,
HYTOLOGY, MARINE ALGAE, ECOLOGY,
RADIOISOTOPES, LIGHT INTENSITY,
NGANESE, BENTHIC FLORA, STREAMS,
ALCIUM, CHLORIDES, CONDUCTIVITY,
IOASSAY, BIOLOGICAL COMMUNITIES,
E, MICROCYSTIS AERUGINOSA, ALGAL
*ALGAE, CYANOPHYTA, CHRYSOPHYTA,
BON CYCLE, *CYTOLOGICAL STUDIES,
, BRACK/ *EUTROPHICATION, ALGAE,
NT, *NUTRIENTS, ALGAE, BACTERIA,
ER, *VITAMINS, VITAMIN B, ALGAE,
VUM, SYNERGISTIC EFFECTS, AXENIC
TON, *ALGAE, SCENEDESMUS, LIGHT,
ANKTON, PHOTOSYNTHESIS, BIOMASS,
CARBOHYDRATES, YEASTS, ENZYMES,
GROWTH RATES, LABORATORY TESTS,
LINEAR ALKYL SULFONATES, *AXENIC
DESMUS, HETEROTROPHY, CONTINUOUS
SOLATION, *CONTROL, METHODOLOGY,
S CULTURES, CHLAMYDOMONAS, BATCH
CATION/ *ALGAL CULTURES, *AXENIC
STRUM, SELF-PURIFICATION/ *ALGAL
KOCYTIC COMPOSITION, ALTERNATING
CALLOPHYLLIS ·HAENOPHYLLA, TIDAL
ZCHIA SERIATA, NAVICULA, FLORIDA
RIES, *STREAMS, LIGHT INTENSITY,
WINDS, THERMOCLINE, TEMPERATURE,
DROGRAPHY, FLORIDA, THERMOCLINE,
TON, *SEDIMENTS, ORGANIC MATTER,
ENT, *FLUOROMETRY, DYE RELEASES,
TH, CIRCULATION, DIURNAL, WINDS,
ANOPHYTA, ENVIRONMENTAL EFFECTS,
, BIOTA, WASTE WATER(POLLUTION),
REAM FLOW, ALGAE, AQUATIC ALGAE,
ROWTH, WATER DENSITY, CONVECTION
ORESIS, MICROORGANISMS, ELECTRIC
*THERMOBIOLOGY, *DIURNAL OXYGEN
ARSENITE, FISH, ANIMALS, ALGAE,
ADISON LAKES(WIS), *AQUATIC WEED
TOXINS MEMBRANE, NORTHEAST U.S/
ECOLOGY, WATER QUALITY CRITERIA,
PLANNING, *SOURCE CONTAMINATION,
LTER COUNTER, GROWTH INHIBITION,
Z,/ *VITAMIN B-12, ALGAL GROWTH,
N, *LAKES, ALGAE, AQUATIC WEEDS,
A, CHRYSOPHYTA, DINOFLAGELLATES,
, ODOR-PRODUCING ALGAE, DIATOMS,
ROBIOLOGY, AQUATIC PRODUCTIVITY,
NSIN, ALGAE, CALCIUM, CHELATION,
, *NITROGEN, *PHOSPHORUS, ALGAE,
, DIAGENESIS, SEDIMENTS/ *ALGAE,
WISCONSIN, BIB/ *EUTROPHICATION,
DISSOLVED OXYGEN, OXYGEN DEMAND,
ARVESTING OF ALGAE, ENVIRONMENT,
N, ALGAE, CULTURES, CHLOROPHYTA,
L ECOLOGY, DIATOMS, OLIGOTROPHY,
TIC POPULATIONS, EUTROPHICATION,
L WATERSHEDS, NITROGEN FIXATION,
HORUS, DRAINAGE, MARSHES, FARMS,
ELS, SHELLFISH, DINOFLAGELLATES,
LTURES, IRON COMPOUNDS, DIATOMS,
ESIS, *MINNESOTA, PHYTOPLANKTON,
ENTRATION, NITRATES, PHOSPHATES,
CTIVITY, ALGAE, SEWAGE, DIATOMS,
PLANKTON, PIGMENTS, CHLOROPHYTA,
, DEPTH, BACTERIA, SLIME, ALGAE,
*ALGAE, *LAKE ERIE, CHLOROPHYTA,
OROPHYLL, PHYSIOLOGICAL ECOLOGY,
RGANIC MATTER,/ *NUISANCE ALGAE,
*SORPTION, *ALGAE, CHLOROPHYTA,
CATION, *BIOMASS, *PRODUCTIVITY,
S FORESTS, PHOSPHORUS, NITROGEN,
MUS, BENTH/ *ALGAE, TEMPERATURE,
TOMS, CHLOROPHYTA, EUGLENOPHYTA,
, PHYTOPLANKTON, PHOTOSYNTHESIS,
IRON, GROWTH RATES, CHLOROPHYTA,
S, DIATOMS, *ALGAE, CHLOROPHYTA,
NOLOGY, PLANT GROWTH REGULATORS,
AQUATIC PLANTS, BIBLIOGRAPHIES,
CATION, ALCOHOLS, ACTINOMYCETES,
N, FUNGI, YEASTS, PYRITE, ALGAE,
PHYTOPLANKTON, ALGAE, SEASONAL,
F NATURE, BIOASSAY, CHLOROPHYTA,
IC ALGAE, BIOASSAY, CHLOROPHYTA,
N EFFECTS, DIATOMS, CHLOROPHYTA,

CULTURES, CHEMOSTATS, CHLOROPHYLL   W70-02777
CULTURES, SEWAGE EFFLUENTS, WISCO   W70-02982
CULTURES, CARBON, METABOLISM, NIT   W70-02804
CULTURES, LAKES, ALGAE, GROWTH RA   W70-00719
CULTURES, CYANOPHYTA, WATER POLLU   W69-10180
CULTURES, WASHINGTON, WAVES(WATER   W69-10161
CULTURES, RADIOACTIVITY, CARBON D   W69-10158
CULTURES, COBALT RADIOISOTOPES.: /   W69-07845
CULTURES, HYDROGEN ION CONCENTRAT   W69-07833
CULTURES, DIATOMS, ECOLOGY, NUISA   W69-07832
CULTURES, CHEMOSTATS, CARBON-14.:   W69-06864
CULTURES, DIAGENESIS, SEDIMENTS,   W69-08284
CULTURES, FLUOROMETRY, BIOMASS, O   W69-08278
CULTURES, CHLOROPHYTA, CYANOPHYTA   W70-05752
CULTURES, CHEMICAL OXYGEN DEMAND,   W70-05655
CULTURES, CHEMICAL ANALYSIS, ANAL   W70-05652
CULTURES, MONOALGAL, PORTUGAL, TA   W70-05372
CULTURES, IRON COMPOUNDS, DIATOMS   W70-05409
CULTURES, PLANT PHYSIOLOGY, ALGAE   W70-05094
CULTURES, HYDROGEN SULFIDE, ELECT   W70-03952
CULTURES, SODIUM COMPOUNDS, NITRO   W70-03507
CULTURES, *ALKYL POLYETHOXYLATE,   W70-03928
CULTURES, CHLAMYDOMONAS, BATCH CU   W70-04381
CULTURES, TESTING, CAPILLARY COND   W70-04469
CULTURES, OPTICAL DENSITY, SOIL-W   W70-04306
CULTURES, PEDIASTRUM, SELF-PURIFI   W70-04381
CULTURES, *AXENIC CULTURES, PEDIA   W70-04381
CURRENT.: /REAM, WHITE BREAM, LEU   W70-01788
CURRENT, PUGET SOUND(WASH), LONG   W69-10161
CURRENT, YTTRIUM-90, KATODINIUM· R   W70-00707
CURRENTS(WATER), VELOCITY, DIATOM   W70-02780
CURRENTS(WATER), HYPOLIMNION, ALG   W70-09889
CURRENTS(WATER), PHYTOPLANKTON, Z   W70-04368
CURRENTS(WATER), WATER POLLUTION   W70-03501
CURRENTS(WATER), REMOTE SENSING,   W70-05377
CURRENTS(WATER), WATER POLLUTION,   W70-05404
CURRENTS(WATER), INDUSTRIAL WASTE   W68-00470
CURRENTS(WATER).: /BEACHES, SANDS   W68-00010
CURRENTS(WATER), LIMNOLOGY, LAKES   W68-00479
CURRENTS, ALGAE TREATMENT COSTS,   W69-08282
CURRENTS, *WASTE TREATMENT, WASTE   W71-03775
CURVES, CALORIMETRY, SOLARIMETRY,   W70-03309
CUTTING MANAGEMENT, POTABLE WATER   W70-05263
CUTTING, *WEED EXTRACTING STEEL C   W70-04494
CYANAPHYTA, ALGAL CONTROL, *ALGAL   W69-05306
CYANIDE, NAPHTHENIC ACID.: /BIAL   W70-04510
CYANIDE, NEW PRODUCT CONTROL, PLU   W73-00758
CYANOCORALAMIN, MUTANT BACTERIA,   W70-05652
CYANOCOBALAMIN, EUGLENA GRACILIS   W69-06273
CYANOPHYTA, FISHKILL, FERTILIZATI   W69-06535
CYANOPHYTA.: /DIATOMS, CHLOROPHYT   W69-05763
CYANOPHYTA, CHLOROPHYTA, IRON, MA   W69-05704
CYANOPHYTA, ENVIRONMENTAL EFFECTS   W69-04802
CYANOPHYTA, NUISANCE ALGAE, NUTRI   W69-05867
CYANOPHYTA, NITRATES, NITROGEN CO   W69-05868
CYANOPHYTA, CHRYSOPHYTA, CULTURES   W69-08284
CYANOPHYTA, ODOR, SCUM, FISHING,   W69-08283
CYANOPHYTA, DIAT: /S, SEDIMENTS,   W69-08518
CYANOPHYTA.: /ALGAE, ALGICIDES, H   W69-07818
CYANOPHYTA, BRACKISH WATER, OLIGO   W70-05752
CYANOPHYTA, CHLOROPHYTA, GREAT LA   W70-04503
CYANOPHYTA, SEWAGE EFFLUENTS, CHL   W70-04510
CYANOPHYTA, SEWAGE EFFLUENTS, MIN   W70-04506
CYANOPHYTA, PLANKTON, ZOOPLANKTON   W70-05112
CYANOPHYTA, CHRYSOPHYTA, CALIFORN   W70-05372
CYANOPHYTA, PHOSPHATES, CHLOROPHY   W70-05409
CYANOPHYTA, DIATOMS, STRATIFICATI   W70-05387
CYANOPHYTA, OXIDATION-REDU: /CONC   W70-05404
CYANOPHYTA, CARBON RADIOISOTOPES,   W70-05405
CYANOPHYTA, CHRYSOPHYTA, RHODOPHY   W70-05389
CYANOPHYTA, DIATOMS, CHLOROPHYTA,   W70-04371
CYANOPHYTA, UNITED STATES, HABITA   W70-04468
CYANOPHYTA, EUTROPHICATION, NUISA   W70-04381
CYANOPHYTA, VITAMINS, PEPTIDES, O   W70-04369
CYANOPHYTA, DETERGENTS, SPECTROSC   W70-03928
CYANOPHYTA, NUISANCE ALGAE, LIMNO   W70-03959
CYANOPHYTA, CHLOROPHYTA, DIATOMS,   W70-03512
CYANOPHYTA, CHLOROPHYTA, SCENEDES   W70-03969
CYANOPHYTA, CHRYSOPHYTA, BIOASSAY   W70-03974
CYANOPHYTA, CYTOLOGICAL STUDIES,   W70-03973
CYANOPHYTA, CHRYSOPHYTA, PHOTOSYN   W70-03983
CYANOPHYTA, LAND MANAGEMENT, WATE   W68-00172
CYANOPHYTA, ENVIRONMENTAL EFFECTS   W68-00470
CYANOPHYTA, FRESH WATER FISH, LAK   W68-00680
CYANOPHYTA, SPECTROSCOPY, GAS CHR   W69-00387
CYANOPHYTA, DOMINANT ORGANISM, NU   W68-00891
CYANOPHYTA, *INORGANIC COMPOUNDS,   W68-00860
CYANOPHYTA, ENVIRONMENTAL EFFECTS   W69-03185
CYANOPHYTA, DIATOMS, DISINFECTION   W69-03370
CYANOPHYTA, NUISANCE ALGAE, WISCO   W69-01977

ITY, AQUATIC WEEDS, CHLOROPHYTA, CYANOPHYTA, DIATOMS, CYCLING NUTR W69-03373
F NATURE, BIOASSAY, CHLOROPHYTA, CYANOPHYTA, DIATOMS, CYCLING NUTR W69-03188
OGY, CHLAMYDOMONAS, CHLOROPHYTA, CYANOPHYTA, DIATOMS, CYCLING NUTR W69-03364
OBIOLOGY, BIOASSAY, CHLOROPHYTA, CYANOPHYTA, DIATOMS, ENVIRONMENTA W69-03371
UATIC MICROBIOLOGY, CHLOROPHYTA, CYANOPHYTA, ESSENTIAL NUTRIENTS, W69-03358
OBIOLOGY, BIOASSAY, CHLOROPHYTA, CYANOPHYTA, ENVIRONMENTAL EFFECTS W69-03362
TIVITY, AQUATIC WEEDS, BIOASSAY, CYANOPHYTA, ENVIRONMENTAL EFFECTS W69-03186
HOSPHORUS RADIOISOTOPES, INFLOW, CYANOPHYTA, LITTORAL, NUTRIENTS, W73-01454
ATION, TEMPERATURE, CHLOROPHYLL, CYANOPHYTA, AQUATIC PLANTS, PHOSP W72-04761
, *PRIMARY PRODUCTIVITY, OREGON, CYANOPHYTA.: /YSIS, SESSILE ALGAE W72-06052
TERFACES, OREGON, SESSILE ALGAE, CYANOPHYTA.: /, SEDIMENT-WATER IN W72-06051
BIOMASS, PHYSIOLOGICAL ECOLOGY, CYANOPHYTA, DIFFUSION, RESPIRATIO W70-03309
ELEMENTS(CHEMICAL), CHLOROPHYTA, CYANOPHYTA, CHLORELLA, MANGANESE, W70-02964
NUISANCE ALGAE, *EUTROPHICATION, CYANOPHYTA, LAKES, RESERVOIRS, SC W70-02803
S(RIVER), THALLI, SCHIZOMYCETES, CYANOPHYTA, FUNGI, BACILLARIOPHYC W70-02784
F NATURE, BIOASSAY, CHLOROPHYTA, CYANOPHYTA, DIATOMS, ENVIRONMENTA W70-02792
HLORELLA, ANALYTICAL TECHNIQUES, CYANOPHYTA, DIATOMS, EUTROPHICATI W70-02779
HYTON, CHLOROPHYTA, CHRYSOPHYTA, CYANOPHYTA, SAMPLING, PHYTOPLANKT W70-02780
HYTA, CHRYSOPHYTA, EUGLENOPHYTA, CYANOPHYTA, DIATOMS, AQUATIC HABI W70-02770
E WATER(POLLUTION), SCENEDESMUS, CYANOPHYTA, CHLOROPHYTA, CHLOROPH W70-02255
ER POLLUTION EFFECTS, LIMNOLOGY, CYANOPHYTA.: /ITUS, BACTERIA, WAT W70-02249
D OXYGEN, TEMPERATURE, VELOCITY, CYANOPHYTA, BIOMASS, INSECTS, DIA W71-00668
NT, LEACHING, ARID LANDS, DEPTH, CYANOPHYTA, CHLOROPHYTA, SIEROZEM W71-04067
REATMENT, *INDIANA, CHLOROPHYTA, CYANOPHYTA, DIATOMS, SEASONAL, EU W70-10173
ATER TEMPERATURE, *ALGAE BLOOMS, CYANOPHYTA, NORTH DAKOTA.: /GY, W W70-09093
TS, *DEGRADATION(DECOMPOSITION), CYANOPHYTA, CHLOROPHYTA, *ALGAE, W70-09438
HEMICAL ANALYSIS, PHYTOPLANKTON, CYANOPHYTA, SEDIMENTS, PLANT POPU W70-06229
ICROORGANISMS, BIOMASS, DIATOMS, CYANOPHYTA, IRON, ALGICIDES, ZOOP W70-08654
*ALGAL CONTROL, *POTABLE WATER, CYANOPHYTA, SULFUR BACTERIA, COLO W70-08107
WATER, CHLOROPHYTA, CHRYSOPHYTA, CYANOPHYTA, SAMPLING, DENSITY, LI W70-00711
ITY, IRON, MANGANESE, WATERFOWL, CYANOPHYTA.: /EFFECTS, HETEROGENE W70-00273
STATIC PRESSURE, CARBON DIOXIDE, CYANOPHYTA, PHYSIOLOGICAL ECOLOGY W69-10160
DO, *LAKES, *MOUNTAINS, *WINTER, CYANOPHYTA, PLANKTON, COPEPODS, E W69-10154
ING, THERMAL SPRINGS, EFFLUENTS, CYANOPHYTA.: /TRY, ECOLOGY, SAMPL W69-10151
Y, *ALGAE, AQUATIC MICROBIOLOGY, CYANOPHYTA, ESSENTIAL NUTRIENTS, W69-10177
, TUBIFICIDS, PLANKTON, DIATOMS, CYANOPHYTA, PHYTOPLANKTON, ROTIFE W70-00268
ASINS), WATER POLLUTION EFFECTS, CYANOPHYTA, EURTROPHICATION.: /(B W69-10182
IATOMS, ECOLOGY, SOIL, CULTURES, CYANOPHYTA, WATER POLLUTION SOURC W69-10180
DIES, GAMMA RAYS, SPECTROMETERS, CYANOPHYTA, URANIUM RADIOISOTOPES W69-09742
ESTS, *ALGAE, CULTURES, DIATOMS, CYANOPHYTA, BIOASSAY, TEST PROCED W71-06002
ORGANIC MATTER, ALGAE, DIATOMS, CYANOPHYTA, CHLOROPHYTA, BULRUSH, W72-00845
ATES, CARBON, CARBONATES, CARBON CYCLE.: /EUTROPHICATION, BICARBON W70-02792
URES, / *ALGAE, *CARBON, *CARBON CYCLE, *CYTOLOGICAL STUDIES, CULT W69-08278
*NITRATES, EUTROPHIC/ *NITROGEN CYCLE, *FARM WASTE, GROUND WATER, W69-05323
*ESSENTIAL NUTRIENTS, *NITROGEN CYCLE, *NUTRIENT REQUIREMENTS, AL W69-03186
ROPHICATION, *NITROGEN, NITROGEN CYCLE, *PHOSPHORUS, PHOSPHORUS CO W72-01867
ON, NITROGEN COMPOUNDS, NITROGEN CYCLE, ALGAE, BIOLOGICAL TREATMEN W70-07031
*ESSENTIAL NUTRIENTS, *NITROGEN CYCLE, ANALYTICAL TECHNIQUES, AQU W69-03185
, GLYCOLYSIS, TRICARBOXYLIC ACID CYCLE, AUTORADIOGRAPHY, MONODUS S W70-02965
*PHOSPHORUS CYCLE, GLOEOTRICHA, P-32.: W73-01454
ES, NITRATES, NITROGEN, NITROGEN CYCLE, NITROGEN FIXATION, PHOSPHA W69-04805
GY, NITROGEN COMPOUNDS, NITROGEN CYCLE, NUTRIENT REQUIREME: /MNOLO W69-04804
ASS, SEWAGE, FARM WASTES, CARBON CYCLE, PHOSPHORUS, PHOSPHATE, NIT W70-02777
ET, ENGLAND), SAND GRAINS, TIDAL CYCLES.: /YI, ABBOT'S POND(SOMERS W71-12870
ROTIFERS, CHLOROPHYTA, COPEPODS, CYCLES, CHLAMYDOMONAS, ALGAE, ZOO W70-00268
ON, CHEMICAL PRECIPITATION, LIFE CYCLES, GENETICS, E. COLI, ISOLAT W71-11823
, *FLUCTUATION, RESISTANCE, LIFE CYCLES, GROWTH STAGES.: /NT, BAYS W71-02494
TILIZERS, PESTICIDES, CELLULOSE, CYCLES, HARVESTING, DREDGING, MON W70-05264
EDEN, LAKE ERKEN(SWEDEN), ANNUAL CYCLES, MICROBIAL ECOLOGY, RADIOC W70-05109
COSYSTEMS, RADIOACTIVITY, BIOTA, CYCLES, MICHIGAN, INVERTEBRATES, W69-09334
IGHT, BIOMASS, RESPIRATION, LIFE CYCLES, OLIGOCHAETES, MOLLUSKS, S W71-00665
REAERATION, SESTON, CHLOROPHYLL, CYCLES, UNITED STATES, CHLORELLA, W70-05092
FFECTS, WATER POLLUTION SOURCES, CYCLING NUTRIENTS, COST COMPARISO W69-08518
A, SEDIMENTS, MINNOWS, DETRITUS, CYCLING NUTRIENTS, PLANKTON, ALGA W69-07861
HLOROPHYTA, CYANOPHYTA, DIATOMS, CYCLING NUTRIENTS, ENVIRONMENTAL W69-03364
HLOROPHYTA, CYANOPHYTA, DIATOMS, CYCLING NUTRIENTS, ENVIRONMENTAL W69-03188
HLOROPHYTA, CYANOPHYTA, DIATOMS, CYCLING NUTRIENTS, ENVIRONMENTAL W69-03373
*SEDIMENTS, *NUTRIENT BUDGET, CYCLING NUTRIENTS.: W68-00511
, OXIDATION-REDUCTION POTENTIAL, CYCLING NUTRIENTS, EUTROPHICATION W69-04521
ICAL OXYGEN DEMAND, CHLOROPHYTA, CYCLING NUTRIENTS, ENVIRONMENTAL W69-03374
AQUATIC PRODUCTIVITY, BIOASSAY, CYCLING NUTRIENTS, ENVIRONMENTAL W69-04798
BIOASSAY, CHLOROPHYTA, DIATOMS, CYCLING NUTRIENTS, ENVIRONMENTAL W69-03515
ES, ALGAE, AQUATIC PRODUCTIVITY, CYCLING NUTRIENTS, ENVIRONMENTAL W69-03514
PRODUCTIVITY, BALANCE OF NATURE, CYCLING NUTRIENTS, ENVIRONMENTAL W69-03513
MICROBIOLOGY, BALANCE OF NATURE, CYCLING NUTRIENTS, ENVIRONMENTAL W69-03517
NIQUES, AQUATIC ALGAE, BIOASSAY, CYCLING NUTRIENTS, ENVIRONMENTAL W69-03518
CHEMISTRY, GEOGRAPHICAL REGIONS, CYCLING NUTRIENTS, ZOOPLANKTON, A W70-03975
CHEMISTRY, GEOGRAPHICAL REGIONS, CYCLING NUTRIENTS, ZOOPLANKTON, A W70-03975
TES, HYDROGEN ION CONCENTRATION, CYCLING NUTRIENTS, MUD-WATER INTE W70-04381
VITY, BIOCHEMICAL OXYGEN DEMAND, CYCLING NUTRIENTS, DIATOMS, ENVIR W70-04283
ARINE ANIMALS, FISH, ECOSYSTEMS, CYCLING NUTRIENTS, TEMPERATURE, L W69-08526
DS, BACTERIA, DIATOMS, DETRITUS, CYCLING NUTRIENTS, LIG: / COMPOUN W69-09334
SITY, TURBIDITY, ALGAE, DIATOMS, CYCLING NUTRIENTS, PHYSIOLOGICAL W70-00713
NALYSIS, SUCCESSION, ECOSYSTEMS, CYCLING NUTRIENTS.: /S, SYSTEMS A W70-02779
CATION, EPILIMNION, HYPOLIMNION, CYCLING NUTRIENTS, TASTE, ODOR, C W70-02251
IOASSAY, CHLOROPHYTA, CHLORELLA, CYCLING NUTRIENTS, ESSENTIAL NUTR W70-02248
PLANKTON, AQUATIC MICROBIOLOGY, CYCLING NUTRIENTS, ESSENTIAL NUTR W70-02245
DIAPHANOSOMA LEUCHTENBER GIANUM, CYCLOPS BICUSPIDATUS, REPRODUCTIV W69-10182
PIROGYRA, CLADOPHORA, DIAPTOMUS, CYCLOPS, GOMPHONEMA, PHORMIDIUM, W70-02803
GELLATES, MONURON, TRANSPARENCY, CYCLOPS, SCENEDESMUS OBLIQUUS, DA W70-10161

423

S, SYNURA, ANABAENA, IOWA RIVER,
, BOSMINA, NAVICULA, FRAGILARIA,
ALENSIS, GYMNODINIUM BAICALENSE,
ETHOD, CALCULATED VOLUME METHOD,
MORPHA INTESTINALIS, CLADOPHORA,
CLIS, CHLOROGONIUM, SCENEDESMUS,
VOLVOCALES, SARCODINA, CILIATES,
, DINOBRYON, UROGLENA AMERICANA,
TURATED / *ADAPTATION, *SPECIES,
DJUSTMENT, ASTERIONELLA FORMOSA,
CYANOCOBALAMIN, MUTANT BACTERIA,
RESERVOIRS, *TURBID RESERVOIRS,
OS-AQUAE, OSCILLATORIA, ANABAENA
, PYRUVATE, ACETYLENE/ *ANABAENA
OLOGICAL CONCENTRATION, ANABAENA
ON, TRINGFORD(ENGLAND), ANABAENA
NII, METABOLIC PATTERN, ANABAENA
ZINC, EUGLENA GRACILIS, ANABAENA
EXTRACELLULAR PRODUCTS, ANABAENA
OCYSTIS, OSCILLATORIA, NAVICULA,
OVAKIA), BLATNA(CZECHOSLOVAKIA),
PPER SULF/ *HALOGENS, ALGICIDES,
, ALGAE, LICHENS, ACTIMOMYCETES,
, KINETICS, HARDNESS, COLIFORMS,
TON, PHOTOSYNTHESIS, CYANOPHYTA,
IUM, FISH, AMPHIBIANS, ANABAENA,
NATION, *STRONTIUM RADIOISOTOPES,
C ALGAE, BLOOMS, CLOSED SYSTEMS,
ION, IODINE TEST, GRAM REACTION,
ITIVITY, GENERA, !FLAVOBACTERIUM-
DAPHNIA HYALINA, DAPHNIA MAGNA,
BOHEMIA, ELBE RIVER, CELAKOVICE(
LAKOVICE(CZECHOSLOVAKIA), BLATNA(
GICAL INDEX, *SAPROBITY, PRAGUE,
D DEVELOPMENT CORP(MASS), ARTHUR
GREEN / *LEWIS AND CLARK LAKE (S
PHYTOTOXICITY, ALGAL POISO/ *2-4
IDES, *GROWTH RATES, *ALGAE, 2-4-
21/ *LITERATURE, ELEOCHARIS, 2,4-
A, DISSOLVED OXYGEN, PIKE, CARP,
SIMAZINE 80 W,
, CYCLOPS, SCENEDESMUS OBLIQUUS,
*ALGAE BLOOMS, CYANOPHYTA, NORTH
(WASTES), RED RIVER VALLEY(NORTH
WASTES), SEWAGE TREATMENT, NORTH
*DIATOMS, *ALGAE, ECOLOGY, SOUTH
TRIBUTION, MISSOURI RIVER, SOUTH
PHYSIOLOGY, NORTH DAKOTA, SOUTH
IOLOGY, ANIMAL PHYSIOLOGY, NORTH
PPER SULPHATE, MONURON, AMMONIA,
OENSE, TENDIPES PLUMOSUS, BEAVER
MARGINAL VEGETAT/ *HARTBEESPOORT
USBR RESEARCH CONTRACTS, CHENEY
OTO REACTIVITY, CELL COLOR, CELL
S, *WATER REUSE, COOLING TOWERS,
AUDERIA BOREALIS, LEPTOCYLINDRUS
LOROCOCCALES, DAPHNIA PULICARIS,
APHNIA SCHODLERI, DAPHNIA PULEX,
IA PULICARIS, DAPHNIA CUCULLATA,
HNIA CUCULLATA, DAPHNIA HYALINA,
QUADRICAUDA, CHLORELLA VULGARIS,
QUADRICAUDA, CHLORELLA VULGARIS,
PULEX, DAPHNIA GALEATA MENDOTAE,
TIVE STUDIES, DAPHNIA SCHODLERI,
CYPRINID FISHES, CLOROCOCCALES,
BODY SIZE, *COMPARATIVE STUDIES,
Y, TEMPERATURE, LIGHT INTENSITY,
*ALGAL CONTROL, EUTROPHICATION,
OXICITY, SCENEDESMUS, CHLORELLA,
OXICITY, SCENEDESMUS, CHLORELLA,
, SWIMMING POOLS, CHLAMYDOMONAS,
ENVIRONMENT, PHOSPHATES, ALGAE,
UISANCE ALGAE, SUNFISHES, ALGAE,
ABOLISM, BIOLOGICAL COMMUNITIES,
ENTRATION, PHOSPHATES, NITRATES,
LANT PHYSIOLOGY, EUTROPHICATION,
ANALYSIS, PH, SAMPLING, ON-SITE
ENTS, *BOTTOM SEDIMENTS, ON-SITE
UM, MATHEMATICAL MODELS, ON-SITE
PHOSPHORUS RADIOISOTOPE, ON-SITE
STROPODS, MARINE PLANTS, ON-SITE
WATER POLLUTION SOURCES, ON-SITE
*LAKE ERIE,/ *SYSTEMS ANALYSIS,
ENEDESMUS, MATHEMATICAL STUDIES,
, NAVICULA, MEL/ *CHLOROPHYLL A,
MAND, SUSPENDED LOAD, TURBIDITY,
*BIOGEOCHEMICAL
*SULFIDES, TEMPERATURE, CLIMATIC
MELOSIRA, MALLOMONAS, ULOTHRIX,
ZOPODS, SPIDER/ *ZOOGLEAL SLIME,

CYCLOTELLA, FRAGILARIA, SYNEDRA,	W70-08107
CYCLOTELLA, FLAGELLATES, ANKISTRO	W70-02249
CYCLOTELLA MINUTA.: /ELOSIRA BAIC	W70-04290
CYCLOTELLA COMENSIS, RHODOMONAS L	W70-03959
CYCLOTELLA, HALOCHLOROCOCCUM, SYN	W70-03974
CYCLOTELLA, STEPHANODSICUS.: /CIN	W69-02959
CYCLOTELLA, APHANIZOMENON FLOS AQ	W70-05112
CYCLOTELLA, CERATIUM HIRUNDINELLA	W70-05405
CYCLOTELLA MENEGHINIANA, LIGHT-SA	W70-05381
CYCLOTELLA MENEGHINIANA, PHOTIC Z	W70-04809
CYCLOTELLA NANA, GEIGER COUNTER.:	W70-05652
CYCLOTELLA, TRACHELOMONAS, MELOSI	W72-04761
CYLINDRICA B629, ANABAENA INAEQUA	W70-05091
CYLINDRICA, ATP, CMU, NITROGEN-15	W70-04249
CYLINDRICA, MICROCYSTIS AERUGINOS	W70-03520
CYLINDRICA, NORTH SEA, FIRTH OF C	W70-02504
CYLINDRICA, GLYCOLYSIS, TRICARBOX	W70-02965
CYLINDRICA, BACILLUS SUBTILIS, AS	W70-02964
CYLINDRICA, ZINC, XANTHOPHYCEAE,	W69-10180
CYMBELLA, LYNGBYA, FRAGILARIA, OE	W70-04371
CYPRINID FISHES, CLOROCOCCALES, D	W70-00274
CYTOLOGICAL STUDIES, CHLORINE, CO	W70-02364
CYTOLOGICAL STUDIES, FILTRATION,	W71-11823
CYTOLOGICAL STUDIES, EVAPORATION,	W71-05267
CYTOLOGICAL STUDIES, ELECTRON MIC	W70-03973
CYTOLOGICAL STUDIES, PHY: /OSTRID	W70-05372
CYTOLOGICAL STUDIES.: /IL CONTAMI	W69-03491
CYTOLOGY, ENVIRONMENTAL CONDITION	W69-07832
CYTOPHAGA, PROTEUS, SERRATIA, BAY	W70-03952
CYTOPHAGA', VIBRIOS, ACHROMOBACTE	W70-00713
CZECHOSLOVAKIA.: /HNIA CUCULLATA,	W70-00274
CZECHOSLOVAKIA), BLATNA(CZECHOSLO	W70-00274
CZECHOSLOVAKIA), CYPRINID FISHES,	W70-00274
CZECHOSLOVAKIA, LABORATORY PROCED	W72-12240
D LITTLE INC(MASS), CARNEGIE DEPT	W69-06865
D), ABUNDANCE, BLUE-GREEN ALGAE,	W70-05375
D, *HERBICIDES, *PHYTOPLANKTON, *	W69-00994
D, AMINOTRIAZOLE, 2-4-5-T, SCENED	W70-03519
D, AMMATE, CHLORAMINE, DE-K-PRUF-	W70-05263
D: /S, ALGAE, DIATOMS, CHLOROPHYT	W70-01943
DACTHAL.:	W70-03519
DACTYLOCOCCUS, ESCHERICHIA COLI,	W70-10161
DAKOTA.: /GY, WATER TEMPERATURE,	W70-09093
DAKOTA), SULFATE REDUCTION, POTAT	W70-03312
DAKOTA, COLIFORMS, OXIDATION LAGO	W70-03312
DAKOTA, FIELD INVESTIGATIONS, WAT	W69-00659
DAKOTA, NEBRASKA, ALGAE, NANNOPLA	W70-05375
DAKOTA, PHOTOSYNTHESIS, LIGHT PEN	W72-01818
DAKOTA, SOUTH DAKOTA, PHOTOSYNTHE	W72-01818
DALAPON, 2-4-: /DIUM ARSENITE, CO	W70-00269
DAM LAKE(WIS), DELAVAN LAKE(WIS).	W70-06217
DAM(SOUTH AFRICA), TRANSPARENCY,	W70-06975
DAM, KANS.:	W72-01773
DAMAGE.: /S ALGAE, KILL RATES, PH	W70-02364
DAMS, ALGAE, HYDROGEN ION CONCENT	W71-11393
DANICUS, PHAEODACTYLUM TRICORNUTU	W70-04280
DAPHNIA CUCULLATA, DAPHNIA HYALIN	W70-00274
DAPHNIA GALEATA MENDOTAE, DAPHNIA	W70-03957
DAPHNIA HYALINA, DAPHNIA MAGNA, C	W70-00274
DAPHNIA MAGNA, CZECHOSLOVAKIA.: /	W70-00274
DAPHNIA MAGNA, LIMNAEA STAGNALIS,	W71-05719
DAPHNIA MAGNA, LIMNAEA STAGNALIS,	W71-03056
DAPHNIA MAGNA, SELENASTRUM, RHODO	W70-03957
DAPHNIA PULEX, DAPHNIA GALEATA ME	W70-03957
DAPHNIA PULICARIS, DAPHNIA CUCULL	W70-00274
DAPHNIA SCHODLERI, DAPHNIA PULEX,	W70-00274
DAPHNIA, CONNECTICUT, ROTIFERS.: /	W70-03959
DAPHNIA, CHLAMYDOMONAS, LABORATOR	W72-02417
DAPHNIA, DETERIORATION, WATER TRE	W71-05719
DAPHNIA, DETERIORATION, WATER TRE	W71-03056
DAPHNIA, DIATOMS, CHLORELLA, SCEN	W70-10161
DAPHNIA, GROWTH RATES, PREDATION.	W69-06536
DAPHNIA, LAKES, CHLOROPHYTA, STRE	W70-02982
DAPHNIA, NANNOPLANKTON, NUTRIENTS	W70-00274
DAPHNIA, SILICA, THERMAL STRATIFI	W70-05760
DATA COLLECTION, CHEMICAL ANALYSI	W70-04382
DATA COLLECTIONS, LIMNOLOGY.: /AL	W69-00659
DATA COLLECTIONS, AQUATIC PLANTS,	W68-00511
DATA COLLECTIONS, BOTTOM SEDIMENT	W68-00858
DATA COLLECTIONS, FISHES, PLANKTO	W68-00857
DATA COLLECTIONS.: /MORTALITY, GA	W70-01780
DATA COLLECTIONS, *ON-SITE INVEST	W70-09976
DATA COLLECTIONS, *ALGAE CONTROL,	W71-04758
DATA COLLECTIONS, MARINE ANIMALS,	W73-01446
DATA INTERPRETATION, *RHONE RIVER	W73-01446
DATA STORAGE AND RETRIEVAL, COLIF	W70-09557
DATA, *ABUNDANCE RATIOS.:	W71-05531
DATA, STRATIFICATION, DISSOLVED O	W70-04790
DAVID MONIE TEST, SUM: /ANABAENA,	W69-10157
DAYTON(OHIO), VANDALIA(OHIO), RHI	W72-01819

PALUDOSUM, BLEACHING, WASHINGTON(DC).: /RIVER, PLECTONEMA, NOSTOC   W70-02255
ANKTON, *CYANOPHYTA, WASHINGTON, DC, *ESTUARIES, HUMAN POPULATION,   W68-00461
BRAUNII, DUNALIELLA TERTIOLECTA, DCMU INHIBITOR.: /ANKISTRODESMUS   W70-05261
FECTS, ALDRIN, DIELDRIN, ENDRIN, DDT.: /UTANTS, WATER POLLUTION EF   W69-08565
ALGAE, CLAYS, ALDRIN, DIELDRIN, DDT, AERATION, SILTS.: /SORPTION,   W70-09768
ITY, DEGRADATION(DECOMPOSITION), DDT, ALKALINITY, STABILIZATION, W   W70-08097
*PESTICIDE REMOVAL, *FOOD WEBS, DDT, CHLORINATED HYDROCARBON PEST   W70-03520
    *ALGAE TOXICITY, CARBON-14, DDT, MALATHION, SEVIN.:   W70-02198
S, *WATER SOURCES, *GROUNDWATER, DDT, MISSISSIPPI RIVER BASIN, URB   W73-00758
FISH PHYSIOLOGY, FISH TAXONOMY, DDT, PHOSPHOTHIOATE PESTICIDES, H   W69-02782
ARIS, 2,4-D, AMMATE, CHLORAMINE, DE-K-PRUF-21, BENOCLORS, NIGROSIN   W70-05263
STIS AERUGINOSA, ENDOTOXIN, FAST- DEATH FACTOR, DOSAGE, SPECIES SUS   W70-00273
EN ALGAE, BLUE-GREEN ALGAE, HEAT DEATH.:   GRE   W70-07313
SIRA LACUSTRIS BRAND, GONGROSIRA DEBARYANA RAB, OOCYSTIS, OSCILLAT   W70-04468
A RUBRA, CA/ *NORTH SEA, MYCOTA, DEBARYOMYCES HANSENII, RHODOTORUL   W70-04368
ISSOLVED ORGANIC MATTER, ORGANIC DEBRIS, ORGANIC GROWTH FACTORS, V   W70-02510
KES, SEWAGE, DRAINAGE DISTRICTS, DEBT, COSTS, CITIES.: /ATMENT, LA   W70-04455
BETA RADIOACTIVITY, GONIOBASIS, DECAPODA, PLECOPTERA, EPHEMEROPTE   W69-09742
NENT ANALYSIS, PART CORRELATION, DECAPODS, SPECIES DIVERSITY INDEX   W73-01446
TER QUALITY, CONIFEROUS FORESTS, DECIDUOUS FORESTS, PHOSPHORUS, NI   W70-03512
EAMS, SEWAGE, REMEDIES, JUDICIAL DECISIONS, LEGAL ASPECTS.: /, STR   W69-06909
, EUTROPHICATION, PHYTOPLANKTON, DECOMPOSING ORGANIC MATTER, OKLAH   W70-04001
URIFICATION, *WATER UTILIZATION, DECOMPOSING ORGANIC MATTER, *STRE   W68-00483
VIEWS, BIBLIOGRAPHIES, BACTERIA, DECOMPOSING ORGANIC MATTER, WASTE   W70-00664
N CONCENTRATION, TOXINS, OXYGEN, DECOMPOSIN: /ALINITY, HYDROGEN IO   W70-02803
NTS, SEWAGE EFFLUENTS, DILUTION, DECOMPOSING ORGANIC MATTER, INSEC   W71-00940
XICITY, WATER POLLUTION CONTROL, DECOMPOSING ORGANIC MATTER, CARBO   W72-01788
ER, ACTINOMYCETES, PLANT GROWTH, DECOMPOSING ORGANIC MATTER, CHEMI   W72-01812
OD CHAINS, BACTERIA, DEGRADATION(DECOMPOSITION), ALGAE, OXYGENATIO   W72-01801
ON, WATER TREATMENT, DEGRADATION(DECOMPOSITION).: /IA, DETERIORATI   W71-05719
LS, NEUTRALIZATION, *DEGRADATION(DECOMPOSITION), METHANE, WATER PO   W71-13413
ON, WATER TREATMENT, DEGRADATION(DECOMPOSITION).: /IA, DETERIORATI   W71-03056
ECTICIDES, TOXICITY, DEGRADATION(DECOMPOSITION), DDT, ALKALINITY,   W70-08097
URCES, *DETERGENTS, *DEGRADATION(DECOMPOSITION), CYANOPHYTA, CHLOR   W70-09438
LES, ANABAENA, SEASONAL EFFECTS, DECOMPOSITION RATES, CHRONOLOGY,   W70-05663
ARM WASTE, SLUDGE, SOLID WASTES, DECOMPOSING ORGANIC MATTER, LAGO   W71-05742
EL, *ALGAL CONCENTRATION, 'EXTRA- DEEP PONDS', ASHKELON(ISRAEL), TE   W70-04790
ORGANISMS, HISTORY, TASTE, ODOR, DEEP WELLS, WATER TREATMENT, MECH   W70-08096
*LAKE WASHINGTON(WASH), *OXYGEN DEFICIT, SECCHI DISC TRANSPARENCY   W69-10182
E(CANADA), OSOYOOS LAKE(CANADA), DEFORMED CHIRONOMIDS.: /SKAHA LAK   W71-12077
TAL SOLIDS, VOLITILE SOLIDS, NON- DEGRADABLE SOLID.: /RY MANURE, TO   W71-00936
DETERIORATION, WATER TREATMENT, DEGRADATION(DECOMPOSITION).: /IA,   W71-03056
ATORY TESTS, GAS CHROMATOGRAPHY, DEGRADATION, WATER POLLUTION EFFE   W70-08652
*ALGAE, *INSECTICIDES, TOXICITY, DEGRADATION(DECOMPOSITION), DDT,   W70-08097
HEMICAL PRECIPITATION, *CHEMICAL DEGRADATION, *BIOLOGICAL TREATMEN   W70-09186
COMPOUNDS, ADSORPTION, REMOVAL, DEGRADATION, CHEMICAL OXYGEN DEMA   W71-13335
DETERIORATION, WATER TREATMENT, DEGRADATION(DECOMPOSITION).: /IA,   W71-05719
ATION, *LAKES, *WATER POLLUTION, DEGRADATION, SEWAGE TREATMENT, NI   W72-02817
EFFECTS, FOOD CHAINS, BACTERIA, DEGRADATION(DECOMPOSITION), ALGAE   W72-01801
CHLORELLA PYRENOIDOSA, PESTICIDE DEGRADATION, SEVIN, MALATHION, AC   W70-02968
CAL EQUIPMENT, ENZYMES, CHEMICAL DEGRADATION, VITAMINS, CELLULOSE,   W70-04184
*POTENTIATION, *ORGANIC DEGRADATION.:   W72-04506
PLUMOSUS, BEAVER DAM LAKE(WIS), DELAVAN LAKE(WIS).: /SE, TENDIPES   W70-06217
(WIS), SPAULDING POND(WIS), LAKE DELAVAN(WIS), PEWAUKEE LAKE(WIS),   W69-05867
, TENNESSEE RIVER, HUDSON RIVER, DELAWARE RIVER, WATER POLLUTION E   W70-04507
ER.:   MILLSBORO(DELAWARE), SWAN CREEK, INDIAN RIV   W71-05013
ERIA, BIOCHEMICAL OXYGEN DEMAND, DELAWARE, *WASTE WATER TREATMENT.   W71-05013
IN DELTA, CALIFORNIA WATER PLAN, DELTA WATER FACILITIES(CALIF), OX   W70-05094
MASTER DRAIN, SAN FRANCISCO BAY- DELTA.: /RAIN, SAN JOAQUIN VALLEY   W71-09577
AKE GEORGE(UGANDA), RHINE-MEEUSE DELTA(NETHERLANDS),: /R-BLOOMS, L   W70-04369
TA WATER/ SACRAMENTO-SAN JOAQUIN DELTA, CALIFORNIA WATER PLAN, DEL   W70-05094
COPPER SULFATE, *CHEMICAL OXYGEN DEMAND.: /ORGANIC MATTER, ALGAE,   W72-04495
DISINFECTION, BIOCHEMICAL OXYGEN DEMAND(BOD), AERATION, COLIFORMS,   W70-04788
XYGEN, ALGAE, BIOCHEMICAL OXYGEN DEMAND(BOD), MIXING, WINDS.: /D O   W70-04790
TEMPERATURE, BIOCHEMICAL OXYGEN DEMAND(BOD), EVAPORATION, FERMENT   W70-04787
ION LAGOONS, *BIOCHEMICAL OXYGEN DEMAND, *ANAEROBICS, ALGAE, BACTE   W71-05742
*FILTRATION, *BIOCHEMICAL OXYGEN DEMAND, *ACTIVATED SLUDGE, BIOLOG   W70-05821
R PROPERTIES, BIOCHEMICAL OXYGEN DEMAND, *BACTERIA, ACTIVATED SLUD   W68-00256
IC OXYGEN, / *BIOCHEMICAL OXYGEN DEMAND, *CHLORELLA, *PHOTOSYNTHET   W69-03362
ATED SLUDGE, *BIOCHEMICAL OXYGEN DEMAND, *COLOR, WASTE WATER TREAT   W70-04060
YGEN, *ALGA/ *BIOCHEMICAL OXYGEN DEMAND, *NUTRIENTS, *DISSOLVED OX   W71-05407
COLIFORMS, / *BIOCHEMICAL OXYGEN DEMAND, *PHOSPHORUS, *NITROGEN, *   W71-05409
CHLORELLA, / *BIOCHEMICAL OXYGEN DEMAND, *PHOTOSYNTHETIC OXYGEN, *   W69-03371
AL, DEGRADATION, CHEMICAL OXYGEN DEMAND, *WASTE WATER TREATMENT.: /   W71-13335
IOCHEMISTRY, *BIOCHEMICAL OXYGEN DEMAND, *WASTE WATER TREATMENT, *   W70-09186
IC DIGESTION, BIOCHEMICAL OXYGEN DEMAND, ACTIVATED SLUDGE, COST CO   W71-09548
ION, BACTERIA, EFFLUENTS, OXYGEN DEMAND, AEROBIC TREATMENT, PHOTOS   W72-00462
, ABSORPTION, BIOCHEMICAL OXYGEN DEMAND, AERATION, TEMPERATURE, FL   W71-11381
, EFFICIENCY, BIOCHEMICAL OXYGEN DEMAND, ALGAE, SEDIMENTATION, SET   W70-09320
TES, POLLUTA/ BIOCHEMICAL OXYGEN DEMAND, ALGAL TOXINS, ORGANIC WAS   W69-01273
OLVED OXYGEN, BIOCHEMICAL OXYGEN DEMAND, ALGAL CONTROL, *COLOR, *A   W72-11906
ATION, ALGAE, BIOCHEMICAL OXYGEN DEMAND, ANAEROBIC DIGESTION, AERO   W71-00936
, *LAKE ERIE, BIOCHEMICAL OXYGEN DEMAND, BACTERIA, ECOLOGY, WATER   W71-04758
MENT, *AEROBIC TREATMENT, OXYGEN DEMAND, BIOCHEMICAL OXYGEN DEMAND   W70-06792
RATIFICATION, BIOCHEMICAL OXYGEN DEMAND, CARBON RADIOISOTOPES, DET   W70-04001
RE, BIOASSAY, BIOCHEMICAL OXYGEN DEMAND, CHLOROPHYTA, CYCLING NUTR   W69-03374
TION LAGOONS, BIOCHEMICAL OXYGEN DEMAND, CHEMICAL OXYGEN DEMAND, D   W70-06601
ION EXCHANGE, BIOCHEMICAL OXYGEN DEMAND, CHEMICAL OXYGEN DEMAND, C   W70-10433
BIODEGRADATION, CHEMICAL OXYGEN DEMAND, CHLOROPHYLL, LAGOONS, PON   W70-08392
L OXYGEN DEMAND, CHEMICAL OXYGEN DEMAND, COLOR, ACTIVATED SLUDGE,   W70-10433

PRODUCTIVITY, BIOCHEMICAL OXYGEN DEMAND, CYCLING NUTRIENTS, DIATOM W70-04283
IMENTS, DISSOLVED OXYGEN, OXYGEN DEMAND, CYANOPHYTA, DIAT: /S, SED W69-08518
RM, BACTERIA, BIOCHEMICAL OXYGEN DEMAND, DELAWARE, *WASTE WATER TR W71-05013
N-SITE TESTS, BIOCHEMICAL OXYGEN DEMAND, DIURNAL, CHLOROPHYLL.: /O W72-01818
L OXYGEN DEMAND, CHEMICAL OXYGEN DEMAND, DISSOLVED OXYGEN, ODOR, O W70-06601
IN, BIOASSAY, BIOCHEMICAL OXYGEN DEMAND, ENVIRONMENTAL EFFECTS, EN W69-03369
WASTE WATERS, BIOCHEMICAL OXYGEN DEMAND, FISH, ALGAE, FOAMING, SEW W70-08740
R, TURBIDITY, BIOCHEMICAL OXYGEN DEMAND, HYDROGEN ION CONCENTRATIO W72-03299
ATION, TOXICITY, CHEMICAL OXYGEN DEMAND, HYDROGEN ION CONCENTRATIO W71-13413
ING, PROPERTY VALUES, RECREATION DEMAND, IR: /RON, MAGNESIUM, COOL W70-02251
*ALGAE, *BIODEGRADATION, *OXYGEN DEMAND, LABORATORY TESTS, LAGOONS W70-06600
LUDGE, ALGAE, BIOCHEMICAL OXYGEN DEMAND, MIXING, SEWAGE TREATMENT, W70-06899
TATION, IONS, BIOCHEMICAL OXYGEN DEMAND, MINNESOTA, WATER CHEMISTR W69-00446
TERIA, CULTURES, CHEMICAL OXYGEN DEMAND, NITROGEN, PHOSPHATES, BIO W70-05655
RESPIRATION, BIOCHEMICAL OXYGEN DEMAND, NITROGEN FIXATION.: /ION, W72-06022
OGEN SULFIDE, BIOCHEMICAL OXYGEN DEMAND, NUTRIENTS, WASTE WATER TR W70-06925
GOVERNMENTS, BIOCHEMICAL OXYGEN DEMAND, ODOR-PRODUCING ALGAE, POL W69-06909
S, CHLORELLA, BIOCHEMICAL OXYGEN DEMAND, OXIDATION LAGOONS, CALIF: W70-05092
VATED SLUDGE, BIOCHEMICAL OXYGEN DEMAND, OXIDATION-REDUCTION POTEN W71-09546
ANIC LOADING, BIOCHEMICAL OXYGEN DEMAND, OXIDATION-REDUCTION POTEN W71-11377
TERIA, ALGAE, BIOCHEMICAL OXYGEN DEMAND, OXIDATION LAGOONS, SLUDGE W71-08221
DOR, OLIGOTROPHY, OXYGEN, OXYGEN DEMAND, PHOSPHORUS COMPOUNDS, PLA W69-06535
AG, AERATION, BIOCHEMICAL OXYGEN DEMAND, PHOTOSYNTHETIC OXYGEN.: / W69-07520
TION LAGOONS, BIOCHEMICAL OXYGEN DEMAND, PHOTOSYNTHESIS, HYDROGEN W70-03312
RATORY TESTS, BIOCHEMICAL OXYGEN DEMAND, SEWAGE EFFLUENTS, SANITAR W72-01817
ROBIC TREATMENT, CHEMICAL OXYGEN DEMAND, STORAGE CAPACITY, EFFLUEN W71-00936
XYGEN DEMAND, BIOCHEMICAL OXYGEN DEMAND, SUSPENDED LOAD, SOLIDS CO W70-06792
OLVED OXYGEN, BIOCHEMICAL OXYGEN DEMAND, SUSPENDED LOAD, WASTES, S W70-08627
ATER QUALITY, BIOCHEMICAL OXYGEN DEMAND, SUSPENDED LOAD, TURBIDITY W70-09557
PENNSYLVANIA, BIOCHEMICAL OXYGEN DEMAND, TEMPERATURE, FLOW, SAMPLI W69-10159
A, NUTRIENTS, BIOCHEMICAL OXYGEN DEMAND, TERTIARY TREATMENT, ALGAE W69-08054
PURIFICATION, BIOCHEMICAL OXYGEN DEMAND, TOXICITY, OXYGEN-REDUCTIO W70-05750
ONCENTRATION, BIOCHEMICAL OXYGEN DEMAND, VIRGINIA.: /YDROGEN ION C W72-10573
ENT, SAMPLING, MONITORING, WATER DEMAND, WATER SUPPLY, ARID LANDS, W70-05645
CIDES, RECREATION, ALGAE, OXYGEN DEMAND, WATER POLLUTION, MUNICIPA W71-09577
EDIMENTATION, BIOCHEMICAL OXYGEN DEMAND, WASTE WATER TREATMENT, OV W71-10654
VADNAIS(MINN), MILLE LAC(MINN), DEMING LAKE(MINN), ORCHARD LAKE(M W70-05269
, BIODEGRADATION, NITRIFICATION, DENITRIFICATION, NUTRIENTS, TERTI W69-08053
LLUS-FERROBACILLUS, THIOBACILLUS DENITRIFICANS, W VIRGINIA, PENNSY W69-07428
, BIODEGRADATION, NITRIFICATION, DENITRIFICATION, ORGANIC COMPOUND W69-06970
ION EXCHANGE, ACTIVATED SLUDGE, DENITRIFICATION.: /ECTRODIALYSIS, W69-01453
L VALLEY(CALIFORNIA), *BACTERIAL DENITRIFICATION.:   *CENTRA W71-04555
L VALLEY(CALIFORNIA), *BACTERIAL DENITRIFICATION.:   *CENTRA W71-04554
OVAL, *EUTROPHICATION, ANAEROBIC DENITRIFICATION, LAND APPLICATION W70-01981
WASTES, WATER POLLUTION EFFECTS, DENITRIFICATION, SPECTROPHOTOMETR W70-02777
                 LAKE ERSOM(DENMARK).:   W71-00665
GHINIANA, PHOTIC ZONE, HELSINGOR(DENMARK), SALINO-CLINE.: /LA MENE W70-04809
ORES, WATER CIRCULATION, FLOODS, DENSITY DURRENTS, MEASUREMENTS, D W69-06203
TS, */ *BEER-LAMBERT LAW, *FIXED- DENSITY STUDIES, *FIXED-LIGHT TES W70-03923
TINCTION COEFFICIENT, ALGAL CELL DENSITY.: / TURBIDOSTAT, LIGHT EX W70-03923
ES, GRAZING PRESSURE, POPULATION DENSITY.: /EDRA ACUS, SINKING RAT W69-10158
, LAKE MANAGEMENT PROGRAM, ALGAL DENSITY, APHANIZOMENON, MELOSIRA, W69-10167
TIC PLANTS, ASSAY, DISTRIBUTION, DENSITY, CHROMATOGRAPHY, RADIOACT W70-02510
ALGAE BLOOM, ALGAE GROWTH, WATER DENSITY, CONVECTION CURRENTS, ALG W69-08282
NCENTRATION, TEMPERATURE, DEPTH, DENSITY, CHLOROPHYLL, ICE, SEASON W70-05387
ERMOCLINE, HYPOLIMNION, ANIMALS, DENSITY, IRON, PHOSPHORUS, SILICA W70-06975
RYSOPHYTA, CYANOPHYTA, SAMPLING, DENSITY, LIGHT, MUD, SILTS, SANDS W70-00711
E, FIXATION, CELL COUNT, OPTICAL DENSITY, PARTICULATE ORGANIC CARB W70-03983
, TEMPERATURE, DISSOLVED OXYGEN, DENSITY, ROOTED AQUATIC PLANTS, A W71-07698
URE, WEATHER, EPILIMNION, ALGAE, DENSITY, SEASONAL, MARINE PLANTS, W70-02504
DOMONAS, BATCH CULTURES, OPTICAL DENSITY, SOIL-WATER TECHNIQUES, P W70-04381
, WHITE OAK LAKE(TENN), ACTIVITY- DENSITY, VERTICAL, MICROCYSTIS, O W70-04371
YTA, AQUATIC PLANTS, PHOSPHORUS, DENSITY, WATER POLLUTION SOURCES, W72-04761
ON, OFFICE OF NAVAL RESEARCH, US DEPARTMENT OF INTERIOR, MADISON(W W70-03975
ON, OFFICE OF NAVAL RESEARCH, US DEPARTMENT OF INTERIOR, MADISON(W W70-03975
DILUTION, ALGAL GROWTHS, OXYGEN DEPLETION.: / EFFECTS, LAKE MEAD, W73-00750
APHANIZOMENON FLOS AQUAE, OXYGEN DEPLETION.: *YAHARA RIVER(WIS), W72-01797
R SAMPLING, ALGAL GROWTH, OXYGEN DEPLETION, NUCLEAR ACTIVATION TEC W70-00264
USTACEANS, DIATOMS, CHLOROPHYTA, DEPOSITION(SEDIMENTS), WATERSHEDS W69-10182
HUR D LITTLE INC(MASS), CARNEGIE DEPT OF PLANT BIOLOGY, CARNEGIE I W69-06865
ION, PELAGIC REGION, MONOMICTIC, DEPTH EFFECTS, IONIC CONCENTRATIO W71-01933
T SPRINGS, TERRESTRIAL HABITATS, DEPTH, AIR, SEDIMENTS, GEOLOGIC F W69-08272
, *PLANTS, *CALIFORNIA, *NEVADA, DEPTH, ALGAE, MOSSES, CHARA, FISH W70-00711
ITY, RADIOISOTOPES, TEMPERATURE, DEPTH, BACTERIA, SLIME, ALGAE, CY W70-04371
SH, NUISANCE ALGAE, ODOR, TASTE, DEPTH, CIRCULATION, DIURNAL, WIND W70-05404
MOVEMENT, LEACHING, ARID LANDS, DEPTH, CYANOPHYTA, CHLOROPHYTA, S W71-04067
ION CONCENTRATION, TEMPERATURE, DEPTH, DENSITY, CHLOROPHYLL, ICE, W70-05387
ME(NUTRIENTS), LAKE VOLUME, MEAN DEPTH, DILUTION, OSCILLATORIA RUB W70-00270
NVIRONMENT, BIOMASS, FLOW RATES, DEPTH, DISSOLVED OXYGEN, ALGAE, V W70-00265
ODUCTIVITY, TURBIDITY, SURFACES, DEPTH, DISTRIBUTION, TIME, EPILIM W71-07698
PHNIA, NANNOPLANKTON, NUTRIENTS, DEPTH, DOMESTIC ANIMALS, SAMPLING W70-00274
, WISCONSIN, SAMPLING, CONTOURS, DEPTH, OLIGOCHAETES, TOXICITY, FI W70-06217
OXYGEN, LIGHT INTENSITY, SLUDGE, DEPTH, ORGANIC LOADING, DESIGN CR W72-06854
ES, PHYTOPLANKTON, DISTRIBUTION, DEPTH, OXYGEN, ALKALINITY, HYDROG W71-00665
OLLUTION ABATEMENT, OLIGOTROPHY, DEPTH, PHYTOPLANKTON, ZOOPLANKTON W70-03964
ION, TEMPERATURE STRATIFICATION, DEPTH, PLANKTON, LIGHT PENETRATIO W70-05405
EASUREMENT, SEDIMENTS, SAMPLING, DEPTH, SEA WATER, FRES WATER, LIT W71-12870
PHOTOSYNTHESIS, CARBON DIOXIDE, DEPTH, THERMOCLINE, ZOOPLANKTON, W69-10169
REATMENT, ALGAE, PHOTOSYNTHESIS, DEPTH, THERMOCLINE, CLIMATIC ZONE W72-10573
NTIBIOSIS, AUTOLYSIS, EXCRETION, DEPURATION, PHYCOCOLLOIDS, TANNIN W70-01068

THAL DOSES, CHU'S MEDIA, QUINONE
EEN RIVER FORMATION, CHLOROPHYLL
CARIAE, WATER MILFO/ SCHISTOSOME
SCIRPUS FLUVIATILIS, *LAKE BUTTE
BATEMENT, EFFLUENTS, STREAMFLOW,
B) FARLOW, CLADOPHORA, TAXONOMIC
ROTENOIDS, SPECIES, DIVERSITY I/
IA DIGITATA, LAMINARIA AGARDHII,
ATORY MODELS, BOD REMOVAL RATES,
ACERS, *WATER POLLUTION, DESIGN,
ATER POLLUTION, TOXICITY, LAKES,
 SLUDGE, DEPTH, ORGANIC LOADING,
TURE, OPERATION AND MAINTENANCE,
AEROBIC PONDS, *ANAEROBIC PONDS,
NITARY ENGINEERING, *FILTRATION,
ENT, *TRACERS, *WATER POLLUTION,
ES, EXPERIMENTAL STUDIES, LAGOON
URE, NUTRIENTS, FACULTATIVE POND
OLING WATER TREATMENT, ON-STREAM

 FICANCE, ALGAL KEY, CLUMP COUNT,
 ALGAE, *LAKES, *CLASSIFICATION,

CONTACT ZONES, ARTEMIA, BALANUS,
INE, PHOTOLITHOTROPHIC BACTERIA,
ERENCES.:   *NICKEL,
REAM POLLUTION, *WATER POLLUTION
DE, ULOTHRIX, UPTAKE, HALF-LIFE,
N IRRADIANS, PIVERS ISLAND(N C),
LLIN LAKE, GRAPHITE LAKE, OLD J/
, BOD REMOVAL RATES, DESIGN APP/
DESULFOVIBRIO, FACULTATIVE, BOD,
ENE SULFONATES, *BIODEGRADATION,
CIDES, ESTUARIES, WATER QUALITY,
HOROUS, RUNOFF, DOMESTIC WASTES,
UNGI, ALGAE, NITRITES, NITRATES,
GE, TROUT, NITROGEN, PHOSPHORUS,
BLOOMS, NUTRIENT LOAD, SYNTHETIC
TION, *ALGAE, *AQUATIC BACTERIA,
RACERS, DISTRIBUTION, CHELATION,
HATES, NUTRIENT BALANCE, FOAMING
*ALGAE, CHLOROPHYTA, CYANOPHYTA,
MENTATION, CHLORINATION, SLUDGE,
EN DEMAND, CARBON RADIOISOTOPES,
ROWTH RATES, PESTICIDE RESIDUES,
EATMENT, *NUTRIENT R/ *AESTHETIC
SCENEDESMUS, CHLORELLA, DAPHNIA,
VA/ *CONTRA COSTA COUNTY, WATER
SCENEDESMUS, CHLORELLA, DAPHNIA,
UADRICAUDA, CELL COUNTS, SPECIES
TERIUM, ACHROMOBACTER, / SPECIES
N), SMITH'S BAY(MINN), PHOSPHATE
M./ *LINDANE, *ALGAE METABOLISM,
US COMPOUNDS, BACTERIA, DIATOMS,
SOTOPES, CESIUM, FALLOUT, SILTS,
 BIOLOGICAL COMMUNITIES, SEWAGE,
OTTOM SEDIMENTS, SURFACE WATERS,
BIOCARBONATES, DISSOLVED OXYGEN,
UM, NITRATES, PHOSPHATES, ALGAE,
R, BACTERIA, SEDIMENTS, MINNOWS,
S, GEORGIA, SEDIMENTS, BACTERIA,
N DIOXIDE, CARBON RADIOISOTOPES,
RONUTRIENTS, BADFISH RIVER(WIS),
IVER, HURON RIVER, RAISIN RIVER,
TION, HAZARDS, CESIUM, HYDROGEN,
IDIUM SP.:   *TEST
OLL/ *FARM WASTES, *RESEARCH AND
DING, *WATER REUSE, RESEARCH AND
RELLA, CH/ *ALGAE, *RESEARCH AND
HYSIOLOGY, AMERICAN RESEARCH AND
RESERVOIR SITES, WATER RESOURCES
CAPILLARITY, CAPILLARITY CONTROL
RE, CENTRIFUGATION, COAGULATION,
PHOSPHORUS, *PHOSPHATES, ENERGY,
ERIONELLA, ARAPHIDINEAE, CENTRIC
ANOPHYTA, CHRYSOPHYTA, CULTURES,
ICULOSA, GONYAULAX POLYEDRA, 3,5-
M LAKE(IOWA), CENTER LAKE(IOWA),
ASHLANDI, EPISCHURA NEVADENSIS,
OPHYCEA, OSCILLATORIA, ANABAENA,
PHAERIUM, SPIROGYRA, CLADOPHORA,
ERIAL PLATE, BACTERIAL PIGMENTS,
YGEN, OXYGEN DEMAND, CYANOPHYTA,
ION TIME, CHLORELLA PYRENOIDOSA,
*LABORATORY TESTS, *SCENEDESMUS,
E, *ORGANIC MATTER, *FILTRATION,
MYCOIDES, BACILLUS MESENTERICUS,
ER(MD), OSCILLATORIA, CATHERWOOD
            FILTER AIDS,
NEBRASKA, ALGAE, NANNOPLANKTON,

DERIVATIVES, ALKYL SUBSTITUTED IM   W70-02982
DERIVATIVES, RECENT SEDIMENTS, PR   W69-08284
DERMATITIS, COPPER CARBONATE, CER   W70-05568
DES MORTS(WISC), GROWTH STIMULATI   W72-00845
DESALINATION, TERTIARY TREATMENT,   W69-07389
DESCRIPTIONS, GONGROSIRA LACUSTRI   W70-04468
DESICCATION, YAQUINA BAY(ORE), CA   W70-03325
DESICCATION, IMMERSION, NARRAGANS   W70-01073
DESIGN APPLICATIONS, FACULTATIVE   W70-07034
DESIGN CRITERIA, OXIDATION LAGOON   W70-02609
DESIGN CRITERIA, REMOTE SENSING,   W70-02775
DESIGN CRITERIA, ODORS, EFFICIENC   W72-06854
DESIGN CRITERIA, *WATER TREATMENT   W72-08686
DESIGN FOR ENVIRONMENTAL CONTROL,   W70-04787
DESIGN, *ALGAE, COPPER SULFATE, S   W72-08859
DESIGN, DESIGN CRITERIA, OXIDATIO   W70-02609
DESIGN, SCALE-UP FACTORS, WASTE C   W70-07508
DESIGNS, BOD, SLUDGE.: / TEMPERAT   W70-04786
DESLUDGING.: /METAL CORROSION, CO   W70-02294
DESMIDS.:   W68-00860
DESMIDS, FLAGELLATES, MESOPLANKTO   W70-05389
DESMIDS, OLIGOTROPHIC, *PHYTOPLAN   W68-00255
DESMINS, FLAGELLATES.:   W68-00172
DESORPTION, EMBRYOLOGY, ENERGY TR   W69-07440
DESULFOVIBRIO, MESOTHERMY, MELOSI   W70-05760
DETECTION LIMITS, CHEMICAL INTERF   W71-11687
DETECTION.: /, MARINE BIOLOGY, ST   W71-10786
DETECTOR.:   *RADIONUCLI   W69-07862
DETECTOR, BEAUFORT(N C), CRASSOST   W69-08267
DETECTOR, ALASKA, HUDEUC LAKE, PU   W69-08272
DETENTION TIME, LABORATORY MODELS   W70-07034
DETENTION TIME.: *SULFATE ION, *   W70-01971
DETERGENTS, WATER POLLUTION, WAST   W70-08740
DETERGENTS, ALGAL CONTROL, INDUST   W70-10437
DETERGENTS, INDUSTRIAL WASTES, *C   W71-04072
DETERGENTS, PHOSPHATES.: /ERIS, F   W71-00940
DETERGENTS, RUNOFF, PERCOLATION,   W69-10178
DETERGENTS, FISH MANAGERS, FISH Y   W69-08674
DETERGENTS, SURFACTANTS, ALKYLBEN   W69-09454
DETERGENTS, RADIOACTIVITY, AMINO   W69-08275
DETERGENTS, PUBLIC APPREHENSION.:   W70-05084
DETERGENTS, SPECTROSCOPY.: /ION,   W70-03928
DETERGENTS, SPRAYING, ALGAE, CENT   W70-04060
DETERGENTS, ANTIFREEZ: /ICAL OXYG   W70-04001
DETERIORATION.: /NCE, BIOASSAY, G   W70-03520
DETERIORATION, *ADVANCED WASTE TR   W69-07818
DETERIORATION, WATER TREATMENT, D   W71-03056
DETERIORATION, DRAIN, SAN JOAQUIN   W71-09577
DETERIORATION, WATER TREATMENT, D   W71-05719
DETERMINATION, ABSORBANCE, COENOB   W70-03334
DETERMINATION, TAXONOMY, FLAVOBAC   W70-04280
DETERMINATION: /), GRAY'S BAY(MIN   W70-05269
DETOXIFICATION, LINDANE METABOLIS   W70-08652
DETRITUS, CYCLING NUTRIENTS, LIG: W69-09334
DETRITUS, BOTTOM SEDIMENTS, CHUTE   W69-09742
DETRITUS, TUBIFICIDS, PLANKTON, D   W70-00268
DETRITUS, LIGHT INTENSITY, TURBID   W70-00713
DETRITUS, BACTERIA, WATER POLLUTI   W70-02249
DETRITUS, PULP WASTES, LIVESTOCK,   W70-03501
DETRITUS, CYCLING NUTRIENTS, PLAN   W69-07861
DETRITUS, ESTUARIES, PLANKTON, EC   W69-08274
DETRITUS, COLUMBIA RIVER, ELECTRO   W69-07862
DETROIT LAKES(MINN), NUTRIENT REM   W70-04506
DETROIT RIVER, MAUMEE RIVER.: / R   W69-01445
DEUTERIUM, TRITIUM.: /ODS, ABSORP   W70-02786
DEVELOPMENT, TEST PROBLEMS, HORM   W71-07339
DEVELOPMENT, GRANTS, ALGAE, AIR P   W71-06821
DEVELOPMENT, WASTE IDENTIFICATION   W72-03299
DEVELOPMENT, BIBLIOGRAPHIES, CHLO   W69-06865
DEVELOPMENT CORP(MASS), ARTHUR D   W69-06865
DEVELOPMENT, WATER QUALITY, SOIL   W72-11906
DEVICE.: /CHIA, ALGAE ISOLATION,   W70-04469
DEWATERING, *COST ANALYSIS.: /ATU   W71-11375
DI-URNAL DISTRIBUTION, TEMPERATUR   W70-06604
DI: /R, FRAGILARIA, MELOSIRA, AST   W70-05663
DIAGENESIS, SEDIMENTS, GAS CHROMA   W69-08284
DIAMINOBENZOIC ACID DIHYDROCHLORI   W69-08278
DIAMOND LAKE(IOWA).: /NGBYA, STOR   W70-02803
DIAPHANOSOMA LEUCHTENBER GIANUM,   W69-10182
DIAPTOMUS ASHLANDI, EPISCHURA NEV   W69-10182
DIAPTOMUS, CYCLOPS, GOMPHONEMA, P   W70-02803
DIAPTOMUS, EPISCHURA, ORTHOCYCLOP   W70-05760
DIAT: /S, SEDIMENTS, DISSOLVED OX   W69-08518
DIATOMA ELONGATUM, MELOSIRA ISLAN   W69-10158
DIATOMACEOUS EARTH, MUD, LAKES, C   W68-00475
DIATOMACEOUS EARTH, ALGAE, TREATM   W72-04318
DIATOMEA, CLADOPHORA, FLAGELLATA,   W69-07838
DIATOMETER, SAVANNAH RIVER, POPUL   W70-04510
DIATOMITE FILTERS.:   W72-04318
DIATOMS.: /I RIVER, SOUTH DAKOTA,   W70-05375

427

NITIES, WATER POLLUTION EFFECTS,
LETHAL LIMITS, DINOFLAGELLATES,
OUGH FISH, NITROGEN, PHOSPHORUS,
THERMAL WATER, *PROTOZOA, ALGAE,
PHYTA, EUGLENOPHYTA, CYANOPHYTA,
, PHOTOSYNTHESIS, RAINBOW TROUT,
BITATS, CRAYFISH, GROWTH STAGES,
ER), FISH, SEWAGE, EUGLENOPHYTA,
USTRIAL WASTES, WATER CHEMISTRY,
TERIA, SLIME, ALGAE, CYANOPHYTA,
HERBICIDES, *PESTICIDE RESIDUES,
OASSAY, CHLOROPHYTA, CYANOPHYTA,
SIS, PHYTOPLANKTON, CISCO, CARP,
OMONAS, CHLOROPHYTA, CYANOPHYTA,
RAZING, WATER POLLUTION EFFECTS,
WEEDS, CHLOROPHYTA, CYANOPHYTA,
HT INTENSITY, WATER TEMPERATURE,
F NATURE, BIOASSAY, CHLOROPHYTA,
TY, CONDUCTIVITY, ALGAE, SEWAGE,
LIGHT, CULTURES, IRON COMPOUNDS,
MATTER, WATER POLLUTION EFFECTS,
LGAE, PROTOZOA, LICHENS, PLANTS,
E, THERMAL POLLUTION, DIFFUSION,
ING ALGAE, ODOR-PRODUCING ALGAE,
NMENTAL EFFECTS, PHOTOSYNTHESIS,
DETRITUS, TUBIFICIDS, PLANKTON,
TOPLANKTON, SESTON, CRUSTACEANS,
SPHORUS COMPOUND, PERCHES, FISH,
GHT INTENSITY, TURBIDITY, ALGAE,
S, PHOSPHORUS, NUTRIENTS, ALGAE,
ALGAE, MICROORGANISMS, BIOMASS,
Y, CYANOPHYTA, BIOMASS, INSECTS,
AETES, MOLLUSKS, STRATIFICATION,
G POOLS, CHLAMYDOMONAS, DAPHNIA,
ORATORY TESTS, *ALGAE, CULTURES,
EDIMENTS, ORGANIC MATTER, ALGAE,
N, SEWAGE, BIOTA, PHYTOPLANKTON,
OPHOSPHORUS COMPOUNDS, BACTERIA,
OASSAY, CHLOROPHYTA, CYANOPHYTA,
OASSAY, CHLOROPHYTA, CYANOPHYTA,
F NATURE, BIOASSAY, CHLOROPHYTA,
IOLOGICAL COMMUNITIES, CULTURES,
, *DDT, *MARINE ALGAE, TOXICITY,
XYGEN DEMAND, CYCLING NUTRIENTS,
N, LAKE MICHIGAN, PHYTOPLANKTON,
BACTERIA, TASTE, FILTERS, IOWA,
YORK, LAKES, INHIBITION, ALGAE,
OASSAY, CHLOROPHYTA, CYANOPHYTA,
CTIVITY, CHLOROPHYTA, GONOPIIYTA,
ALYTICAL TECHNIQUES, CYANOPHYTA,
BIOMASS, *SUCESSION, *ISOLATION,
Y, AQUATIC MAMMALS, ALGAE, FISH,
, INSECTICIDES, HARDNESS(WATER),
SSENTIAL NUTRIENTS, CHLOROPHYTA,
RAPHIES, CHLORELLA, CHLOROPHYTA,
CADDISFLIES, TEMPERATURE, FLOW,
BIOINDICATORS, *WATER POLLUTION,
WATER QUALITY, SEWAGE EFFLUENTS,
ITAMINS, HEAVY METALS, TOXICITY,
THODOLOGY, CARBON RADIOISOTOPES,
PHATES, *ALGAE, *EUTROPHICATION,
AL TECHNIQUES, *ALGAE, PLANKTON,
ION, ADSORPTION, ATLANTIC OCEAN,
OLLUTION, PHYSIOLOGICAL ECOLOGY,
WATER), INDUSTRIAL WASTE, CORES,
TENSITY, CHLORELLA, CHLOROPHYTA,
ERSION, CORES, SEWAGE EFFLUENTS,
*ALGAE, *TEMPERATURE,
ERIA, SYSTEMATICS, FUNGI, ALGAE,
IA, LAKES, CHLOROPHYTA, STREAMS,
CHLORELLA, GRAZING, ZOOPLANKTON,
CIUM, MAGNESIUM, SODIUM, SILICA,
SOTA, PHYTOPLANKTON, CYANOPHYTA,
GRAPHIES, ANALYTICAL TECHNIQUES,
SITY, CURRENTS(WATER), VELOCITY,
NDIANA, CHLOROPHYTA, CYANOPHYTA,
ANT ORGANISMS, *ACCLIMATIZATION,
N ION CONCENTRATION, ALKALINITY,
, POTASSIUM, CLAMS, SIZE, ALGAE,
APILLARY CONDUCTIVITY, BACTERIA,
TROGEN, CYANOPHYTA, CHLOROPHYTA,
TUBIFICIDS, CALCIUM, SILICATES,
OGEN, PHOSPHORUS, PHYTOPLANKTON,
GLENOPHYTA, TOXICITY, BIOASSAYS,
UDIES, CRUCIGENIA RECTANGULARIS,
TRIAZINE,
SHOCK WAVE, ANABAENA, PHYGON XL(
OXICITY, CHEMICAL STRUCTURE, 2,3-
IS, APHANIZOMENON, ANABAEN/ *2,3-
LYBIA, HYDRODICTYON RETICULATUM,

DIATOMS.: /ANTS, BIOLOGICAL COMMU   W70-01233
DIATOMS.: /ANALYTICAL TECHNIQUES,   W70-08111
DIATOMS, *ALGAE, CHLOROPHYTA, CYA   W68-00172
DIATOMS, ANIMAL BEHAVIOR, *THERMO   W70-07847
DIATOMS, AQUATIC HABITAT, POPULAT   W70-02770
DIATOMS, ANAEROBIC CONDITIONS, AL   W69-10154
DIATOMS, ALGAE, METABOLISM, ISOPO   W69-09742
DIATOMS, BIOLOGICAL COMMUNITIES,   W70-03969
DIATOMS, CHLOROPHYTA, EUGLENOPHYT   W70-03974
DIATOMS, CHLOROPHYTA, LIGHT INTEN   W70-04371
DIATOMS, CRAYFISH, POLLUTANT IDEN   W69-02782
DIATOMS, CYCLING NUTRIENTS, ENVIR   W69-03188
DIATOMS, CHLOROPHYTA, CHLAMYDOMON   W69-02959
DIATOMS, CYCLING NUTRIENTS, ENVIR   W69-03364
DIATOMS, CHLOROPHYTA, CYANOPHYTA,   W69-01977
DIATOMS, CYCLING NUTRIENTS, ENVIR   W69-03373
DIATOMS, CHLOROPHYTA.: /IEWS, LIG   W68-00860
DIATOMS, CYCLING NUTRIENTS, ENVIR   W69-03515
DIATOMS, CYANOPHYTA, CARBON RADIO   W70-05405
DIATOMS, CYANOPHYTA, PHOSPHATES,   W70-05409
DIATOMS, CARBON RADIOISOTOPES, KI   W70-04811
DIATOMS, CORAL, MOLLUSKS, MINE AC   W69-07428
DIATOMS, CHLOROPHYTA, CHRYSOPHYTA   W69-05763
DIATOMS, CYANOPHYTA, CHLOROPHYTA,   W69-05704
DIATOMS, CHLORELLA, CARBON RADIOI   W69-10158
DIATOMS, CYANOPHYTA, PHYTOPLANKTO   W70-00268
DIATOMS, CHLOROPHYTA, DEPOSITION(W69-10182
DIATOMS, CARP, PLANKTON, METABOLI   W70-00274
DIATOMS, CYCLING NUTRIENTS, PHYSI   W70-00713
DIATOMS, CHLOROPHYTA, DISSOLVED O   W70-01943
DIATOMS, CYANOPHYTA, IRON, ALGICI   W70-08654
DIATOMS, CHLOROPHYTA, CHRYSOPHYTA   W71-00668
DIATOMS, C: /LIFE CYCLES, OLIGOCH   W71-00665
DIATOMS, CHLORELLA, SCENEDESMUS,   W70-10161
DIATOMS, CYANOPHYTA, BIOASSAY, TE   W71-06002
DIATOMS, CYANOPHYTA, CHLOROPHYTA,   W72-00845
DIATOMS, DINOFLAGELLATES, ALGAE,   W69-10169
DIATOMS, DETRITUS, CYCLING NUTRIE   W69-09334
DIATOMS, DISINFECTION, ENVIRONMEN   W69-03371
DIATOMS, ENVIRONMENTAL EFFECTS, E   W69-03516
DIATOMS, ENVIRONMENTAL EFFECTS, I   W69-07832
DIATOMS, ECOLOGY, NUISANCE ALGAE,   W70-05272
DIATOMS, ESTUARIES, PHYSIOLOGICAL   W70-04283
DIATOMS, ENVIRONMENTAL EFFECTS, E   W69-10180
DIATOMS, ECOLOGY, SOIL, CULTURES,   W70-08107
DIATOMS, EUGLENA, CHLAMYDOMONAS,   W70-01579
DIATOMS, EUTROPHICATION, CARBON R   W70-02792
DIATOMS, ENVIRONMENTAL EFFECTS, E   W70-02784
DIATOMS, ENVIRONMENTAL EFFECTS, L   W70-02779
DIATOMS, EUTROPHICATION, WATER PO   W73-01446
DIATOMS, ECOLOGY, SALINITY, ORGAN   W71-10786
DIATOMS, FUNGI, INSECTS, BACTERIA   W72-01813
DIATOMS, FOULING, PLANKTON, SEWAG   W69-03517
DIATOMS, INHIBITION, LIGHT, LIGHT   W69-06865
DIATOMS, LIGHT, LIPIDS, NUTRIENT   W69-07853
DIATOMS, MANGANESE.: /S, SHINERS,   W70-04510
DIATOMS, MARYLAND, PENNSYLVANIA,   W70-04507
DIATOMS, MISSOURI RIVER, COLUMBIA   W71-10079
DIATOMS, MARL.: /, OLIGOTROPHY, V   W71-12870
DIATOMS, MEASUREMENT, SEDIMENTS,   W71-05026
DIATOMS, MICROBIOLOGY, NITRATES,   W70-09904
DIATOMS, MARINE ALGAE, FRESH WATE   W70-00707
DIATOMS, NERITIC ALGAE, IONS, WAT   W70-04503
DIATOMS, OLIGOTROPHY, CYANOPHYTA,   W68-00470
DIATOMS, ODOR-PRODUCING ALGAE, CH   W70-05381
DIATOMS, PIGMENTS, CHLOROPHYLL, P   W70-05663
DIATOMS, PHOSPHORUS, ORGANIC MATT   W71-02496
DIATOMS, PHOTOSYNTHESIS.:   W72-01819
DIATOMS, PROTOZOA, PLANKTON, WORM   W70-02982
DIATOMS, RESERVOIRS, CULTURES, SE   W70-04809
DIATOMS, RESPIRATION, PIGMENTS, O   W68-00856
DIATOMS, RESISTANCE, *PLANT GROWT   W70-05387
DIATOMS, STRATIFICATION, OXYGEN,   W70-05389
DIATOMS, SAMPLING, PHYSICOCHEMICA   W70-02780
DIATOMS, SEASONAL, BIOMASS, PIGME   W70-10173
DIATOMS, SEASONAL, EUGLENA, CHLAM   W70-07313
DIATOMS, THERMAL POLLUTION, TEMPE   W70-02504
DIATOMS, TEMPERATURE, WEATHER, EP   W70-00708
DIATOMS, TRACE ELEMENTS, METABOLI   W70-04469
DIATOMS, TEST PROCEDURES.: /NG, C   W70-03512
DIATOMS, WEIGHT, PILOT PLANTS, IR   W70-05761
DIATOMS, WINDS, NITROGEN, PHOSPHA   W70-04253
DIATOMS, ZOOPLANKTON, DINOFLAGELL   W70-04381
DIATOMS,: /D-WATER INTERFACES, EU   W69-07833
DICERAS CHODATI, DINOPHYCEAE, FIL   W70-00269
DICHLOBENIL.:
DICHLONE), NEW HAMPSHIRE, LAKE WI   W69-08674
DICHLORONAPHTHOQUINONE, MICROCYST   W70-02982
DICHLORONAPHTHOQUINONE, MICROCYST   W70-03310
DICTYOSPHAERIUM PULCHELLUM, GOMPH   W69-05704

ENII, RHODOTORULA RUBRA, CANDIDA   DIDDENSII, CANDIDA ZEYLANOIDES, C   W70-04368
QUAE, STEPHANODISCUS HANTZSCHII,   DIDINIUM, PARAMECIUM, PHACUS, GLE   W70-00268
, *ZOOPLANKTON, SULFUR BACTERIA,   DIEL MIGRATION, CARBON DIOXIDE, H   W70-05760
RBON PESTICIDES, ALDRIN, ENDRIN,   DIELDRIN, SCENEDESMUS, GAS CHROMA   W70-03520
WATER POLLUTION EFFECTS, ALDRIN,   DIELDRIN, ENDRIN, DDT.: /UTANTS,   W69-08565
DSORPTION, ALGAE, CLAYS, ALDRIN,   DIELDRIN, DDT, AERATION, SILTS.: /   W70-09768
OMETRY, N-HEPTADECANE, MONOENES,   DIENES, TRIENES.: /S, MASS SPECTR   W69-08284
NIUM, PHORMIDIUM RETZII, SURFACE   DIFFUSION RATE.: /RA ULNA, OEDOGO   W70-00265
YSIOLOGICAL ECOLOGY, CYANOPHYTA,   DIFFUSION, RESPIRATION, PHOSPHATE   W70-03309
ABILIZATION, WASTE WATER, ALGAE,   DIFFUSION, SULFUR.: /BACTERIA, ST   W70-01971
TION, SCOUR, SATURATION, MIXING,   DIFFUSION, SEDIMENTATION, ABSORPT   W71-11381
MPLING, METHANE, CARBON DIOXIDE,   DIFFUSION, *WASTE WATER TREATMENT   W71-11377
ISANCE ALGAE, THERMAL POLLUTION,   DIFFUSION, DIATOMS, CHLOROPHYTA,   W69-05763
ROSCOPY, AMMONIA, SPECTROMETERS,   DIFFUSION, TURBULENCE, TOXINS, BA   W70-04369
, *ALGAE, ORGANIC MATTER, LAKES,   DIFFUSION, CARBON RADIOISOTOPES,   W70-04284
OLLUTION EFFECTS, PHYTOPLANKTON,   DIFFUSION, ALGAE, BACTERIA, EUTRO   W70-04194
RIFUGATION, FILTRATION, ANEROBIC   DIGESTION, INCINERATION, IRON COM   W70-04060
, *AEROBIC TREATMENT, *ANAEROBIC   DIGESTION, BIOCHEMICAL OXYGEN DEM   W71-09548
ON, OXIDATION LAGOONS, ANAEROBIC   DIGESTION, ALGAE, SLUDGE, METHANE   W71-11375
ON, FILTERS, PESTICIDES, SLUDGE,   DIGESTION, INDIANA.: /OUNDS, LAGO   W72-03641
HEMICAL OXYGEN DEMAND, ANAEROBIC   DIGESTION, AEROBIC TREATMENT, CHE   W71-00936
    *OXIDATION LAGOONS, *ANAEROBIC   DIGESTION, *MUNICIPAL WASTES, MET   W72-06854
, ASCOPHYLLUM NODOSUM, LAMINARIA   DIGITATA, LAMINARIA AGARDHII, DES   W70-01073
OLYEDRA, 3,5-DIAMINOBENZOIC ACID   DIHYDROCHLORIDE, INFRARED GAS ANA   W69-08278
 DREDGING, *POLLUTION ABATEMENT,   DIKES, SOIL CONSE: /TION SOURCES,   W68-00172
 MARINE LABORATORY(RHODE ISLAND),   DILUTION RATE, FRUITING BODIES, U   W70-01073
  TREATMENT, BOTTOM SEALING, LAKE   DILUTION.: /R DIVERSION, INSTREAM   W69-09340
IENTS), LAKE VOLUME, MEAN DEPTH,   DILUTION, OSCILLATORIA RUBESCENS,   W70-00270
OR, EFFLUENTS, SEWAGE EFFLUENTS,   DILUTION, DECOMPOSING ORGANIC MAT   W71-00940
   *RESERVOIR EFFECTS, LAKE MEAD,   DILUTION, ALGAL GROWTHS, OXYGEN D   W73-00750
ORACOMONAS, YTTRIUM, 'BIOLOGICAL   DILUTION', PHYSIOLOGICAL CO: / TH   W70-00708
CEROS, COCCONEIS, COSCINODISCUS,   DIMEROGRAMMA, DIPLONEIS, EUNOTOGR   W70-03325
OHOL, ETHYLENE GLYCOL, POLYMERS,   DIMETHYL TEREPHTHALATE.: /HYL ALC   W70-10433
PHAERIUM PULCHELLUM, GOMPHONEMA,   DINOBRYON.: /RETICULATUM, DICTYOS   W69-05704
CURRENTS, ALGAE TREATMENT COSTS,   DINOBRYON, CONCORD(N H), PEANACOO   W69-08282
TAE, GLOEOTRICHIA, ASTERIONELLA,   DINOBRYON, MELOSIRA, ANEBODA(SWED   W70-05409
LIMNION, OSCILLATORIA RUBESCENS,   DINOBRYON, UROGLENA AMERICANA, CY   W70-05405
AL SYMBIOSIS, MICROBIAL ECOLOGY,   DINOBRYON, VOLVOX, PANDORINA, SPI   W70-04503
ELLA FORMOSA, SYNEDRA, MELOSIRA,   DINOBRYON, FRAGILARIA, RADIOCARBO   W70-01579
CYSTIS, GLOEOTRICHIA ECHINULATA,   DINOBRYON, SYNURA UVELLA, FRAGILA   W70-02803
TICAL TECHNIQUES, LETHAL LIMITS,   DINOFLAGELLATES, DIATOMS.: /ANALY   W70-08111
OURCES, WATER POLLUTION EFFECTS,   DINOFLAGELLATES, FISHKILL, FISH T   W70-08372
, BIOTA, PHYTOPLANKTON, DIATOMS,   DINOFLAGELLATES, ALGAE, HYPOLIMNI   W69-10169
GY, RED TIDE, ORGANIC COMPOUNDS,   DINOFLAGELLATES.: / FISH PHYSIOLO   W71-07731
TATS, CLAMS, MUSSELS, SHELLFISH,   DINOFLAGELLATES, CYANOPHYTA, CHRY   W70-05372
OGY, MASSACHUSETTS, CHLOROPHYTA,   DINOFLAGELLATES, ALGAE, CALIFORNI   W70-05272
ANALYSIS, ANALYTICAL TECHNIQUES,   DINOFLAGELLATES, MARINE BACTERIA,   W70-05652
ATOMS, CHLOROPHYTA, CHRYSOPHYTA,   DINOFLAGELLATES, CYANOPHYTA.: /DI   W69-05763
, GLUCOSE, CARBON RADIOISOTOPES,   DINOFLAGELLATES, BIOASSAY, BIOIND   W70-04194
OPLANKTON, DIATOMS, ZOOPLANKTON,   DINOFLAGELLATES, ALGAE, DISSOLVED   W70-04253
ASTS, POPULATION, MARINE PLANTS,   DINOFLAGELLATES, OCEANS, BIOINDIC   W70-04368
, APHANIZOMENON, MICRONUTRIENTS,   DINOPHYCEAE, CRYPTOPHYCEAE, CH: /   W70-04381
ASSIMILATION, BACILLARIOPHYCEAE,   DINOPHYCEAE.: /M GRACILE, CARBON   W70-03983
 RECTANGULARIS, DICERAS CHODATI,   DINOPHYCEAE, FILTER BLOCKING, HET   W69-07833
BESTOS CEMENT, CHLORELLA, CARBON   DIOXIDE.: /BORATORY EQUIPMENT, AS   W71-07339
COMPOSING ORGANIC MATTER, CARBON   DIOXIDE.: / POLLUTION CONTROL, DE   W72-01788
SULFATE, MANGANESE, IRON, CARBON   DIOXIDE.: /H SUBSTANCES, SEWAGE,   W70-02969
ION, *ALGAE, *NUTRIENTS, *CARBON   DIOXIDE, *REVIEWS, BIBLIOGRAPHIES   W70-00664
ULTU/ *SURFACE, *VOLUME, *CARBON   DIOXIDE, *AERATION, *NITRATES, *C   W70-03334
ELLA, *SEWAGE, EFFLUENTS, CARBON   DIOXIDE, *ALGAE, *HYDROGEN ION CO   W68-00855
ROGEN, OXYGEN, OXIDATION, CARBON   DIOXIDE, ALKALINITY, ALGAE.: /NIT   W70-05550
ON CONCENTRATION, OXYGEN, CARBON   DIOXIDE, AMMONIA, NITRITES, HYDRO   W71-03031
HELATION, PHOTOSYNTHESIS, CARBON   DIOXIDE, BICARBONATES, CARBONATES   W70-02804
TS, HYDROSTATIC PRESSURE, CARBON   DIOXIDE, CYANOPHYTA, PHYSIOLOGICA   W69-10160
NKISTRODESMUS BRAUNII, D/ CARBON   DIOXIDE, CHLORELLA PYRENOIDOSA, A   W70-05261
OLOGICAL ECOLOGY, OXYGEN, CARBON   DIOXIDE, CARBON RADIOISOTOPES, DE   W69-07862
PERATURE, PHOTOSYNTHESIS, CARBON   DIOXIDE, DEPTH, THERMOCLINE, ZOOP   W69-10169
GANISMS, LIGHT INTENSITY, CARBON   DIOXIDE, DISSOLVED OXYGEN, TEMPER   W71-00668
METRY, SAMPLING, METHANE, CARBON   DIOXIDE, DIFFUSION, *WASTE WATER   W71-11377
ORGANIC MATTER, HARDNESS, CARBON   DIOXIDE, FISH.: /Y, TEMPERATURE,   W69-10157
ONAS, CHLORELLA, ENZYMES, CARBON   DIOXIDE, GROWTH RATES, PLANKTON,   W70-04369
 *ALGAE, LIGHT INTENSITY, CARBON   DIOXIDE, GROWTH RATES, TEMPERATUR   W72-06022
BACTERIA, DIEL MIGRATION, CARBON   DIOXIDE, HYDROGEN ION CONCENTRATI   W70-05760
LAMYDOMONAS, TEMPERATURE, CARBON   DIOXIDE, HYDROGEN ION CONCENTRATI   W70-02249
FIXATION, *ALGAE, LIGHT, CARBON   DIOXIDE, INHIBITION, NITROGEN FIX   W70-04249
OISOTOPES, PACIFIC OCEAN, CARBON   DIOXIDE, LIGHT, TEMPERATURE.: /DI   W70-05652
HEMISTRY, *PHYTOPLANKTON, CARBON   DIOXIDE, NUTRIENT REQUIREMENTS, P   W70-05424
TION, LIGHT, TEMPERATURE, CARBON   DIOXIDE, NUTRIENTS, BIOLOGICAL PR   W70-07442
OMPOUNDS, PHOTOSYNTHESIS, CARBON   DIOXIDE, NITRATES, PHOSPHATES.: /   W70-00161
IGHT, ALGAE, TEMPERATURE, CARBON   DIOXIDE, OXIDATION, CHEMICALS, CA   W70-05651
 SOLAR RADIATION, ENERGY, CARBON   DIOXIDE, PHOTOSYNTHESIS, GROWTH R   W70-05750
, PHOTOSYNTHESIS, OXYGEN, CARBON   DIOXIDE, SCENEDESMUS, CHLOROPHYLL   W70-02964
 FIXATION, PHOTOCHEMICAL, CARBON   DIOXIDE, SYNTHESIS, ENZYMES, CHLO   W70-02965
ITRIFICATION, *CORROSION, CARBON   DIOXIDE, SOLUBILITY, *PHOSPHATES,   W71-11393
 EFFLUENTS, ALGAE, WEEDS, CARBON   DIOXIDE, VITAMINS, NITRATES, INOR   W70-04810
 CULTURES, RADIOACTIVITY, CARBON   DIOXIDE, WATER POLLUTION EFFECTS,   W69-10158
ETINA, COCCOCHLORIS PENIOCYSTIS,   DIPLOCYSTIS AERUGINOSA, NOSTOC MU   W69-06277
IS, COSCINODISCUS, DIMEROGRAMMA,   DIPLONEIS, EUNOTOGRAMMA, FRAGILAR   W70-03325
POLLUTION EFFECTS, OLIGOCHAETES,   DIPTERA, ALGAE, INSECTS.: /WATER   W71-12077
EPILIMNION, LAKE ERIE, INSECTS,   DIPTERA, CRUSTACEA, MOLLUSCS, OLI   W70-04253

NATA, EPHEMEROPTERA, PLECOPTERA,
GRAZING, SNAILS, AGE, TURBIDITY,
TH, CHEMICAL CONTROLS, PARAQUAT,
 CALUMET RIVER(ILL), POOLS, WIND
NTS, *WATER POLLUTION, CLEANING,
APHANIZOMENON FLOS-AQUAE, SECCHI
N(WASH), *OXYGEN DEFICIT, SECCHI
ETHODS.:   SECCHI
 *INTERTIDAL COMMUNITIES, *WASTE
AIRBORNE TESTS/ *FRAUNHOFER LINE
AL TOXINS, *PHYTOTOXICITY, *FISH
TUDIES, FILTRATION, ULTRASONICS,
CROBIOLOGY, ALGICIDES, BIOASSAY,
ORINE, ALGICIDES, ALGAL CONTROL,
ORINE, ALGAL CONTROL, CHLORELLA,
 *ODOR, *WASTE TREATMENT, ALGAE,
HLOROPHYTA, CYANOPHYTA, DIATOMS,
DRIED, GLASS BEADS, TRICHODERMA,
, PYROPHOSPHATE, ORTHOPHOSPHATE,
Y WATER, LICHENS, TIDAL EFFECTS,
CHEMICAL PRECIPITATION, ULTIMATE
STE WATER TREATMENT, WASTE WATER
TES, STABILIZATION PONDS, SEWAGE
HELLFISH, WASTE DISPOSAL, SEWAGE
INVERTEBRATES, SHELLFISH, WASTE
REATMENT, SEWAGE DISPOSAL, WASTE
TEMENT, SEWAGE TREATMENT, SEWAGE
WATER, SULFITE LIQUORS, ULTIMATE
SEWERS, SEWAGE TREATMENT, SEWAGE
E AGENCIES, *WATER WORKS, *WASTE
, *MINNESOTA, RECREATION, SEWAGE
AKES, *SEWAGE EFFLUENTS, SEWAGE
URE, FLOW, SAMPLING, WASTE WATER
TOPLANKTON, RADIOACTIVITY, WASTE
ABLE ISOTOPES, RADIOACTIVE WASTE
S, *RUNOFF, *FERTILIZAT/ *SEWAGE
TE WATER TREATMENT, *WASTE WATER
TE WATE/ *WATER POLLUTION, WASTE
MILK, ATLANTIC OCEAN, LANDFILLS,
FFLUENTS, *RADIOISOTOPES, *WASTE
NTRATION, ELECTRIC POWER, SEWAGE
*BIOASSAY, *AQUATIC LIFE, *WASTE
N, *ORGANIC WASTES, *WASTE WATER
DISPOSAL, DOMESTIC WAST/ *WASTE
ISPOSAL, *AQUACULTURE, *ULTIMATE
, PAPERMILL WASTE EF/ *BIOLOGIC '
NITZSCHIA ACICULARIS, NITZSCHIA
FRVOIRS, ALGAE, BODIES OF WATER,
ENTS, LAKE MICHIGAN, WASHINGTON,
Y, NUTRIENTS, WATER TEMPERATURE,
ENT, BIOMASS, FLOW RATES, DEPTH,
ALGAE, VELOCITY, WATER QUALITY,
LIGHT INTENSITY, CARBON DIOXIDE,
TORAGE AND RETRIEVAL, COLIFORMS,
TICAL TECHNIQUES, WATER QUALITY,
LUTION EFFECTS, *EUTROPHICATION,
 TEMPERATURE, LIGHT PENETRATION,
DEMAND, CHEMICAL OXYGEN DEMAND,
POSING ORGANIC MATTER, TOXICITY,
AS, NEVADA, *LAKE MEAD, LAS VEG/
GE TREATMENT, SELF-PURIFICATION,
ALGAE, AUTOTROPHIC/ *MARL LAKES,
ING, ACTIVATED SLUDGE, AERATION,
, OXIDATION-REDUCTION POTENTIAL,
LAGOONS, ALGAE, PHOTOSYNTHESIS,
ATER QUALITY, *DISSOLVED SOLIDS,
SEA WATER, FRES WATER, LITTORAL,
ITY, HYDROGEN ION CONCENTRATION,
OJECT, *RESERVOIRS, TEMPERATURE,
ON CONCENTRATION, BICARBONATES,
ALGAE, COLLOIDS, TRACE ELEMENTS,
HEMICAL REACTIONS, *WATER REUSE,
TS, ALGAE, DIATOMS, CHLOROPHYTA,
ORPTION SPECTROPHOTOMETRY, TOTAL
, OXIDATION REDUCTION POTENTIAL,
CAL TREATMENT, *DESIGN CRITERIA,
, CLIMATIC DATA, STRATIFICATION,
OURCES, WATER POLLUTION EFFECTS,
ATION, COLIFORMS, ALGAL CONTROL,
TASTE, FILTERS, WATER POLLUTION,
OLUBLE R/ ARSENATE INTERFERENCE,
E, DIURNAL DISTRIBUTION, OXYGEN,
ICULTURE, *WASTES, ALGAE, ASSAY,
A, SODIUM, POTASSIUM, SILICATES,
LANKTON, DINOFLAGELLATES, ALGAE,
TOPLANKTON, ALGAE, EFFICIENCIES,
LFATES, CHLORIDES, SODIUM, IRON,
ION, COLOR, TASTE, ODOR, TOXINS,
JOAQU/ STREETER-PHELPS EQUATION,
L CONTROL, ALGICIDES, SEDIMENTS,

DIPTERA, COLEOPTERA, HEMIPTERA, M   W69-07846
DIPTERA, MAYFL: /A, CHRYSOPHYTA,   W71-00668
DIQUAT, SODIUM ARSENITE, COPPER S   W70-00269
DIRECTION, TABELLARIA, FRAGILARIA   W70-00263
DISASTERS, OILY WATER, MARINE ALG   W70-01470
DISC TRANSPARENCY, ANABAENA LEMME   W69-10169
DISC TRANSPARENCY, MYXOPHYCEA, OS   W69-10182
DISC, VISIBILITY TESTS, CONTROL M   W68-00488
DISCHARGES, FLORA, FAUNA, *BIOLOG   W68-00010
DISCRIMINATOR, RHODAMINE WT DYE,   W70-05377
DISEASES, *FISH TOXINS, TOXICITY,   W71-07731
DISINFECTION, SAMPLING, CENTRIFUG   W71-11823
DISINFECTION, ENVIRONMENTAL EFFEC   W69-10171
DISINFECTION, MODE OF ACTION, MOR   W70-02370
DISINFECTION, MORTALITY.: /S, CHL   W70-02363
DISINFECTION, BIOCHEMICAL OXYGEN   W70-04788
DISINFECTION, ENVIRONMENTAL EFFEC   W69-03370
DISINTEGRATION.: / DRYING, SPRAY-   W70-04184
DISMUTATION, TANN: / ARTHROBACTER   W69-07428
DISPERSION, OIL-WATER INTERFACES,   W70-01777
DISPOSAL.: /ODIALYSIS, SORPTION,   W70-01981
DISPOSAL.: /LAGOONS, TOXICITY, WA   W69-09454
DISPOSAL.: /, NITROGEN, GROWTH RA   W68-00855
DISPOSAL, WATER CHEMISTRY, ECOLOG   W69-04276
DISPOSAL, SEWAGE DISPOSAL, WATER   W69-04276
DISPOSAL, WATER POLLUTION, LAKES,   W69-06909
DISPOSAL, WASTE DISPOSAL, WATER P   W69-06909
DISPOSAL, WATER REUSE, FISH FARMI   W69-05891
DISPOSAL, STORM DRAINS, SURFACE D   W70-03344
DISPOSAL, DOMESTIC WASTES, SEWERS   W70-03344
DISPOSAL, WATER POLLUTION, POLLUT   W70-04810
DISPOSAL, ALGAE, AGRICULTURAL WAT   W70-05565
DISPOSAL, DIURNAL, EFFLUENTS, NIT   W69-10159
DISPOSAL, NUCLEAR REACTORS, POWER   W70-00707
DISPOSAL, OCEANS, FOODS, ABSORPTI   W70-02786
DISPOSAL, *AGRICULTURAL WATERSHED   W70-02787
DISPOSAL, *INDUSTRIAL WASTES, *WA   W72-01786
DISPOSAL, *SEWAGE TREATMENT, *WAS   W72-01786
DISPOSAL, REGULATION,: /IC LIFE,   W72-00941
DISPOSAL, *WASTE STORAGE, *WATER   W72-00941
DISPOSAL, *WASTE WATER TREATMENT,   W71-09548
DISPOSAL, *WATER POLLUTION EFFECT   W72-01801
DISPOSAL, AIR POLLUTION, WATER PO   W71-00940
DISPOSAL, *AQUACULTURE, *ULTIMATE   W70-09905
DISPOSAL, DOMESTIC WASTES, INDUST   W70-09905
DISPOSERS', MICROSCOPIC ORGANISMS   W71-00940
DISSIPATA, SYNEDRA ACUS, SINKING   W69-10158
DISSOLVED OXYGEN, NORTH CAROLINA.   W70-00683
DISSOLVED OXYGEN.: / BOTTOM SEDIM   W69-09349
DISSOLVED OXYGEN, FISHKILLS, LIMN   W70-00264
DISSOLVED OXYGEN, ALGAE, VELOCITY   W70-00265
DISSOLVED SOLIDS, HYDROGEN ION CO   W70-00265
DISSOLVED OXYGEN, TEMPERATURE, VE   W71-00668
DISSOLVED OXYGEN, ALGAE, NUTRIENT   W70-09557
DISSOLVED OXYGEN, BIOCHEMICAL OXY   W70-08627
DISSOLVED OXYGEN, HYDROGEN ION CO   W70-06975
DISSOLVED OXYGEN, BICARBONATES, H   W70-06225
DISSOLVED OXYGEN, ODOR, ORGANIC L   W70-06601
DISSOLVED OXYGEN, ALGAE, WISCONSI   W72-01797
DISSOLVED ORGANIC MATTER, LAS VEG   W72-02817
DISSOLVED OXYGEN, PHOTOSYNTHESIS.   W72-01798
DISSOLVED ORGANIC MATTER, ARCTIC   W71-10079
DISSOLVED OXYGEN, SLUDGE, OXIDATI   W71-08970
DISSOLVED OXYGEN, ODORS, CORROSIO   W71-11377
DISSOLVED OXYGEN, LABORATORY TEST   W71-12183
DISSOLVED OXYGEN, EVAPORATION CON   W72-01773
DISSOLVED OXYGEN, DIURNAL, MIGRAT   W71-12870
DISSOLVED OXYGEN, ALGAE, PHYTOPLA   W71-07731
DISSOLVED OXYGEN, DENSITY, ROOTED   W71-07698
DISSOLVED OXYGEN, DETRITUS, BACTE   W70-02249
DISSOLVED SOLIDS, NUTRIENTS, VITA   W70-02510
DISSOLVED OXYGEN, PITTING(CORROSI   W70-02294
DISSOLVED OXYGEN, PIKE, CARP, D: /   W70-01943
DISSOLVED SOLIDS, HEDRIODISCUS, C   W70-03309
DISSOLVED OXYGEN, CHEMICAL ANALYS   W70-05760
DISSOLVED OXYGEN, TEMPERATURE, BI   W70-04787
DISSOLVED OXYGEN, ALGAE, BIOCHEMI   W70-04790
DISSOLVED OXYGEN, AQUATIC BACTERI   W70-04727
DISSOLVED OXYGEN.: /AND(BOD), AER   W70-04788
DISSOLVED OXYGEN, EUTROPHICATION,   W70-05264
DISSOLVED INORGANIC PHOSPHATES, S   W70-05269
DISSOLVED OXYGEN, MEASUREMENT, ES   W70-05094
DISSOLVED SOLIDS, CHEMICAL ANALYS   W70-03334
DISSOLVED SOLIDS, CALCIUM, PRODUC   W70-03512
DISSOLVED OXYGEN, HYPOLIMNION, EP   W70-04253
DISSOLVED OXYGEN, OXYGEN, LAKES.:   W70-04484
DISSOLVED SOLIDS, PHOSPHATES, NIT   W69-05894
DISSOLVED OXYGEN, LAKES, RIVERS,   W69-05697
DISSOLVED OXYGEN VARIATIONS, SAN   W69-07520
DISSOLVED OXYGEN, OXYGEN DEMAND,   W69-08518

NITROGEN, PHOSPHORUS, PLANKTON,
NTRATION, TEMPERATURE, TOXICITY,
CLIMATIC ZONES, ORGANIC LOADING,
TION, CONDUCTIVITY, TEMPERATURE,
IR OPERATIONS, DISSOLVED OXYGEN,
S, SEDIM/ *RESERVOIR OPERATIONS,
ION, INDUSTRIAL WASTES, ECOLOGY,
AWRENCE RIVER, ALGAE, COLIFORMS,
, ION EXCHANGE, REVERSE OSMOSIS,
ANABAENA FLOS AQUAE, TOLYPOTHRIX
GAL ASPECTS, LEGISLATION, SEWAGE
FLORA, LIGHT INTENSITY, BIOMASS,
HINULATA, BOTRYOCOCCUS, PLANKTON
YNTHETIC LAYER, MELOSI/ SEASONAL
PLANT PHYSIOLOGY, ALGAE, DIURNAL
ALGAE, ALGAL CONTROL, ECOLOGICAL
DUCTIVITY, AQUATIC PRODUCTIVITY,
SSOLVED OXYGEN, *ALGAE, *DIURNAL
YTOPLANKTON, *LAKES, *CELLULOSE,
ICOCHEMICAL PROPERTIES, TRACERS,
A, FUNGI, AQUATIC PLANTS, ASSAY,
ENTS, NORTHEAST U.S., ECOLOGICAL
IENTS, NORTHEAST U S, ECOLOGICAL
OLOGICAL DISTRIBUTION, FREQUENCY
, LIMNOLOGY, ECOLOGY, ECOLOGICAL
ITY, TURBIDITY, SURFACES, DEPTH,
ORS, LAKE MORPHOLOGY, ECOLOGICAL
Y, NATURAL RESOURCES, ECOLOGICAL
S, *PHOSPHATES, ENERGY, DI-URNAL
PROPERTIES, ECOLOGY, ECOLOGICAL
MICAL PROPERTIES, PHYTOPLANKTON,
A, SIEROZEMS, SOIL MICROBIOLOGY,
CEAN, PUERTO RICO, SPECTROSCOPY,
OVAL, *ADSORPTION, SOILS, COALS,
ATIC PLANTS, ISLANDS, ECOLOGICAL
EATMENT, LAKES, SEWAGE, DRAINAGE
SE RAYON, ZINC COMPOUNDS, CARBON
VATED ALGAE, FEEDLOTS, OXIDATION
CREEK(WIS), DUTCH MILL DRAINAGE
HEPPARTION(AUSTRALIA), OXIDATION
, THALASSIOSI/ *ALGAL NUTRITION,
LTURES, PLANT PHYSIOLOGY, ALGAE,
SEEPAGE LOSSES,
A, TEXAS, PHOSPHORUS, MIGRATION,
ESTS, BIOCHEMICAL OXYGEN DEMAND,
SAMPLING, WASTE WATER DISPOSAL,
S, INTERNAL WATER, WAVES(WATER),
ITY, CHLOROPHYLL, ICE, SEASONAL,
IGMENTS, OCEANS, ORGANIC MATTER,
TER, LITTORAL, DISSOLVED OXYGEN,
ODOR, TASTE, DEPTH, CIRCULATION,
NSPARENCY, CYCLOP/ *PHENYLUREAS,
ENT, *LAKE WASHINGTON/ *EFFLUENT
AKES), PHAEDACTYLUM, WASTE WATER
ECIPITATION, HARVESTING, ZONING,
ER POLLUTION CONTROL, LIMNOLOGY,
OINDICATORS, ALGAE, OLIGOTROPHY,
BAY(ORE), CAROTENOIDS, SPECIES,
T CORRELATION, DECAPODS, SPECIES
DUCTIVITY, DISTRIBUTION, SPECIES
CULTURAL EUTROPHICATION, SPECIES
BLOOMS, *BIOTIC FACTORS, SPECIES
, EXPORT LOSSES, CHEMOS/ SPECIES
KJANES, FAXAFLOI FJORD, ICELAND,
OSPH/ *EUTROPHICATION, *REVIEWS,
*ANABAENA
IVER, DURANGO, SAN MIGUEL RIVER,
NANNOPLANKTON, NUTRIENTS, DEPTH,
SNAILS, ECOLOGY, EUTROPHICATION,
AQUACULTURE, *ULTIMATE DISPOSAL,
TREATMENT, PHOSPHOROUS, RUNOFF,
TOMS, FOULING, PLANKTON, SEWAGE,
LAKES, GLACIATION, URBANIZATION,
WATER QUALITY, LAKE MORPHOMETRY,
TROPHICATION, *ALGAE, NUTRIENTS,
, PHOSPHATES, *SEWAGE TREATMENT,
OGEN, SEWAGE, INDUSTRIAL WASTES,
ENTS, TERTIARY TREATMENT, ALGAE,
CATION, PHOSPHORUS, *DETERGENTS,
, *WATER WORKS, *WASTE DISPOSAL,
T FACILITIES, AERATION, LAGOONS,
TON, PHYSICOCHEMICAL PROPERTIES,
ALGAE, WATER POLLUTION EFFECTS,
ALITY, ALGAL CONTROL, ALGICIDES,
ASTS, PYRITE, ALGAE, CYANOPHYTA,
K(WIS), NINE SPRINGS CREEK(WIS),
*CRITICAL TEMPERATURES, *COPPER
A, ENDOTOXIN, FAST-DEATH FACTOR,
MENON FLOS-AQUAE, MINIMUM LETHAL
INDUSTRIAL PHOTOSYNTHESIS, MEAN

DISSOLVED OXYGEN, HYPOLIMNION, GR   W68-00253
DISSOLVED OXYGEN, LIGHT INTENSITY   W72-06854
DISSOLVED OXYGEN, HYDROGEN ION CO   W72-10573
DISSOLVED OXYGEN, PHYTOPLANKTON,   W72-06051
DISSOLVED SOLIDS, SEDIMENTATION R   W73-00750
DISSOLVED OXYGEN, DISSOLVED SOLID   W73-00750
DISSOLVED OXYGEN, HARDNESS, TIDAL   W73-00478
DISSOLVED: /ES, AGRICULTURE, ST L   W70-03964
DISTILLATION, ELECTRODIALYSIS, SO   W70-01981
DISTORTA, GLOEOTRICHIA ECHINULATA   W70-05091
DISTRI: /WAY, ESTIMATED COSTS, LE   W70-03344
DISTRIBUTION, ECOSYSTEMS, POPULAT   W70-03325
DISTRIBUTION, WISCONSIN LAKE SURV   W70-03959
DISTRIBUTION, LAKE BAIKAL, PHOTOS   W70-04290
DISTRIBUTION, OXYGEN, DISSOLVED O   W70-05094
DISTRIBUTION, ACTIVATED SLUDGE, M   W70-04786
DISTRIBUTION, SPECIES DIVERSITY.:   W70-05424
DISTRIBUTION, *CALIFORNIA, ESTIMA   W69-03611
DISTRIBUTION PATTERNS, MEMBRANE P   W69-00976
DISTRIBUTION, CHELATION, DETERGEN   W69-08275
DISTRIBUTION, DENSITY, CHROMATOGR   W70-02510
DISTRIBUTION, FREQUENCY DISTRIBUT   W70-02772
DISTRIBUTION, BIOLOGICAL INDICATO   W70-02764
DISTRIBUTION, POPULATION, BIOLOGI   W70-02772
DISTRIBUTION.: /RONMENTAL EFFECTS   W70-02784
DISTRIBUTION, TIME, EPILIMNION, T   W71-07698
DISTRIBUTION, MIDGES, WATER POLLU   W71-12077
DISTRIBUTION, STOCHASTIC PROCESSE   W71-09012
DISTRIBUTION, TEMPERATURE, METABO   W70-06604
DISTRIBUTION, THERMAL EFFECTS.: /   W70-07313
DISTRIBUTION, DEPTH, OXYGEN, ALKA   W71-00665
DISTRIBUTION PATTERNS, SOIL PROFI   W71-04067
DISTRIBUTION, CALCIUM, TROPHIC LE   W69-08525
DISTRIBUTION PATTERNS, ACTIVATED   W69-09884
DISTRIBUTION, BIOLOGICAL COMMUNIT   W72-10623
DISTRICTS, DEBT, COSTS, CITIES.: /   W70-04455
DISULFIDE.: /NTS, NYLON-6, *VISCO   W70-06925
DITCH.:   *FWPCA, ACTI   W71-06821
DITCH(WIS), LAKE MONONA(WIS), LAK   W70-05113
DITCHES, SOUP, FOOD WASTES.:   S   W71-09548
DITYLUM, CHAETOCEROS, SKELETONEMA   W70-04503
DIURNAL DISTRIBUTION, OXYGEN, DIS   W70-05094
DIURNAL VARIATIONS.:   W69-09723
DIURNAL, ALGICIDES, CLOSTRIDIUM,   W70-05372
DIURNAL, CHLOROPHYLL.: /ON-SITE T   W72-01818
DIURNAL, EFFLUENTS, NITROGEN, PHO   W69-10159
DIURNAL, EPILIMNION, WINDS, THERM   W70-09889
DIURNAL, LIGHT INTENSITY, CARBON,   W70-05387
DIURNAL, LATITUDINAL STUDIES.: /P   W70-04809
DIURNAL, MIGRATION PATTERNS.: /WA   W71-12870
DIURNAL, WINDS, CURRENTS(WATER),   W70-05404
DIURON, FLAGELLATES, MONURON, TRA   W70-10161
DIVERSION, *WATER QUALITY MANAGEM   W69-09349
DIVERSION, INSTREAM TREATMENT, BO   W69-09340
DIVERSION, DREDGING, IMPOUNDMENTS   W69-10178
DIVERSION, STREAMS.: /OURCES, WAT   W69-10170
DIVERSION, CORES, SEWAGE EFFLUENT   W70-05663
DIVERSITY INDEX, ACHNANTHES, ACTI   W70-03325
DIVERSITY INDEX.: / ANALYSIS, PAR   W73-01446
DIVERSITY.: /CTIVITY, AQUATIC PRO   W70-05424
DIVERSITY, CHEMICAL PRECIPITATION   W69-08518
DIVERSITY, LAKE WAUBESA, LYNGBYA,   W69-01977
DIVERSITY, AUTECOLOGY, MORPHOLOGY   W70-02780
DNA BASE COMPOSITION.: / CAPE REY   W69-10161
DOCUMENTATION, BIBLIOGRAPHIES, PH   W69-08518
DOLIOLUM.:   W69-10177
DOLORES RIVER, URAVAN, COTTUS SPP   W69-07846
DOMESTIC ANIMALS, SAMPLING, NUISA   W70-00274
DOMESTIC WASTES, SHALLOW WATER, C   W70-00711
DOMESTIC WASTES, INDUSTRIAL WASTE   W70-09905
DOMESTIC WASTES, DETERGENTS, INDU   W71-04072
DOMESTIC WASTE, INDUSTRIAL WASTES   W72-01813
DOMESTIC WASTES, NUTRIENTS, PHOSP   W69-07818
DOMESTIC WASTES, PLANT GROWTH SUB   W68-00470
DOMESTIC WASTES, *SEWAGE TREATMEN   W68-00248
DOMESTIC WASTES, *ALGAE, WATER PR   W68-00256
DOMESTIC WASTES, SEWAGE TREATMENT   W68-00478
DOMESTIC WASTES.: /KES, *IMPOUNDM   W70-04765
DOMESTIC WASTES, POLLUTANTS, WATE   W70-05084
DOMESTIC WASTES, SEWERS, SEWAGE T   W70-03344
DOMESTIC WASTES, SEDIMENTATION, C   W70-04060
DOMESTIC WASTES, SEWAGE, INDUSTRI   W70-03964
DOMESTIC WASTES, INDUSTRIAL WASTE   W72-12240
DOMINANT ORGANISMS, GRAZING, WATE   W69-01977
DOMINANT ORGANISM, NUTRIENTS, RAI   W68-00891
DOOR CREEK(WIS), DUTCH MILL DRAIN   W70-05113
DOSAGE, *COPPER APPLICATION METHO   W69-10157
DOSAGE, SPECIES SUSCEPTIBILITY, A   W70-00273
DOSES, CHU'S MEDIA, QUINONE DERIV   W70-02982
DOUBLING TIME, PACKED CELL VOLUME   W71-05267

431

RVOIR(TENN), NAJAS, PHOTIC ZONE,
ECTROSTATIC REPULSION, DOW C-31,
CHARGE, ELECTROSTATIC REPULSION,
IPES EICHORNIA, TYPHA, BRASENIA,
T, ECOSYSTEMS, AQUATIC HABITATS,
STA COUNTY, WATER DETERIORATION,
DRAIN, SAN JOAQUIN VALLEY MASTER
IS), DOOR CREEK(WIS), DUTCH MILL
LUTION TREATMENT, LAKES, SEWAGE,
, *SALINITY, IRRIGATION EFFECTS,
ON, *DRAINAGE WATER, CALIFORNIA,
ON, *DRAINAGE WATER, CALIFORNIA,
ON, *DRAINAGE WATER, CALIFORNIA,
ON, *DRAINAGE WATER, CALIFORNIA,
PHICATION, *POLLUTION ABATEMENT,
, WATER POLLUTION, SNOW, RUNOFF,
DRAINAGE, NORTH AMERICA, FOREST
DRAINAGE, NORTH AMERICA, FOREST
HORELINE VEGETATION, STORMWATER,
TION EFFECTS, ACID STREAMS, MINE
*URBAN DRAINAGE, *AGRICULTURAL
BIOLOGICAL RESPONSE, FE/ *URBAN
FERTILIZED FIELDS, SUBTERRANEAN
ATER POLLUTION ASSESSMENT, URBAN
NT, URBAN DRAINAGE, AGRICULTURAL
ATER POLLUTION ASSESSMENT, URBAN
NT, URBAN DRAINAGE, AGRICULTURAL
DISPOSAL, STORM DRAINS, SURFACE
MPERATURE, HYDROGE/ STRIP MINES,
DUCTIVITY, NITROGEN, PHOSPHORUS,
LOW, PRECIPITATION(ATMOSPHERIC),
COSYSTEMS, ILLINOIS, PHOSPHORUS,
URRENTS(WATER), WATER POLLUTION,
*NUTRIENT REMOVAL, AGRICULTURAL
OPLANKTON, EUTROPHICATION, STORM
EATMENT, OVERFLOW, SEWERS, STORM
REATMENT, SEWAGE DISPOSAL, STORM
UM, MIC/ *CATIONS, DRAPARNALDIA,
A, STIGEOCLONIUM, MIC/ *CATIONS,
ULTURAL INFLUENCES, MICROCYSTIS,
BLIOGRAPHIES, CATTAILS, BULRUSH,
, CELLULOSE, CYCLES, HARVESTING,
GEMENT, WATER POLLUTION SOURCES,
QUATIC PLANTS, BOTTOM SEDIMENTS,
, HARVESTING, ZONING, DIVERSION,
TILIZATION, ROLLER DRYING, SPRAY-
OZOA, *AQUATIC INSECTS, *AQUATIC
CT CONTROL, PLUMBING CODES, WELL
URELLA PYRENOIDOSA, BICHOROMATE,
BACTERIA, SNAILS, WEIGHT, VACUUM
CTS, ESSENTIAL NUTRIENTS, FREEZE
CTS, ESSENTIAL NUTRIENTS, FREEZE
UTRIENTS, EUTROPHICATION, FREEZE
OMS, *FISH TOXICITY TEST, SLUDGE
SA, NITROGEN UTILIZATION, ROLLER
VITY, *RADIOSTRONTIUM COMPLEXES,
A, SUBLITTORAL, YELLOW MATERIAL,
TIC POPULATIONS, NUISANCE ALGAE,
OR, TOX/ *EUTROPHICATION, ALGAE,

COONTAIL, WILD CELERY, BUR REED,
LAKE, NORTH MEADOW LAKE, GYTTJA,
TION, INDUSTRIAL WASTES, GARBAGE
NOIDOSA, ANKISTRODESMUS BRAUNII,
NINA NIEI, SKELETONEMA COSTATUM,
NERATION, CLIFF VEGETATION, SAND
LAKES, QUIRKE LAKE, PECORS LAKE,
N, BATAVIA, CLARKSBURG, BUFFALO,
FLORA, ORGANISMS, TURSIOPS TRUN/
S RIVER, URAVAN, / ANIMAS RIVER,
TER CIRCULATION, FLOODS, DENSITY
TS, THRESHOLD ODOR NUMBERS, CROP-
NGS CREEK(WIS), DOOR CREEK(WIS),
E(WASH), RENTON TREATMENT PLANT,
ING, *MEASUREMENT, *FLUOROMETRY,
PHOTOGRAPHY, INFRARED RADIATION,
ALGAE, PUGET SOUND, FLOURESCEIN
LINE DISCRIMINATOR, RHODAMINE WT
-K-PRUF-21, BENOCLORS, NIGROSINE
, *TEXTILE WASTES, FIBER WASTES,
*MODEL TESTS,
ON, ORGANIC MATTER, CHLOROPHYLL,
EASE, SAMPLE PRESERVATION, WATER
, *ARTIFICIAL SUBSTRATE/ *GROWTH
WATER FACILITIES(CALIF), OXYGEN
ETER, SAVANNAH RIVER, POPULATION
A/ *COULTER COUNTER, *POPULATION
OTOSYNTHETIC RATE, PHYTOPLANKTON
AL BLOOMS, NUTRIENTS, LIMNOLOGY,
CONDITIONS, AEROBIC CONDITIONS,

DOUGLAS RESERVOIR(TENN), NORRIS R W71-07698
DOW A-21, NALCO N-670.: /ARGE, EL W70-05267
DOW C-31, DOW A-21, NALCO N-670.: W70-05267
DOW CONTACT, SPARGANIUM, SALIX, G W70-05263
DRAGONFLIES, ALGAE, SEDIMENTS, SO W71-09005
DRAIN, SAN JOAQUIN VALLEY MASTER   W71-09577
DRAIN, SAN FRANCISCO BAY-DELTA.: / W71-09577
DRAINAGE DITCH(WIS), LAKE MONONA(W70-05113
DRAINAGE DISTRICTS, DEBT, COSTS,   W70-04455
DRAINAGE EFFECTS, *SAMPLING, ALGA W69-00360
DRAINAGE PROGRAMS, ANAEROBIC BACT W71-04557
DRAINAGE PROGRAMS, ANAEROBIC BACT W71-04554
DRAINAGE PROGRAMS, ANAEROBIC BACT W71-04555
DRAINAGE PROGRAMS, ANAEROBIC BACT W71-04556
DRAINAGE SYSTEMS, FLOW RATES, CON W68-00479
DRAINAGE WATER, GRASSLANDS, SOIL   W70-04193
DRAINAGE.: /RAINAGE, AGRICULTURAL W70-03975
DRAINAGE.: /RAINAGE, AGRICULTURAL W70-03975
DRAINAGE.:   S W70-02251
DRAINAGE, CHLOROPHYTA, CHRYSOPHYT W70-02770
DRAINAGE, BIOLOGICAL RESPONSE, FE W70-02787
DRAINAGE, *AGRICULTURAL DRAINAGE, W70-02787
DRAINAGE, MECHANICAL AND BIOLOGIC W69-08668
DRAINAGE, AGRICULTURAL DRAINAGE,   W70-03975
DRAINAGE, NORTH AMERICA, FOREST D W70-03975
DRAINAGE, AGRICULTURAL DRAINAGE,   W70-03975
DRAINAGE, NORTH AMERICA, FOREST D W70-03975
DRAINAGE, ALGAL CONTROL, AQUATIC   W70-03344
DRAINAGE, FERROBACILLUS, WATER TE W68-00891
DRAINAGE, MARSHES, FARMS, CYANOPH W70-05112
DRAINAGE, FERTILIZER: /FF, UNDERF W70-04810
DRAINAGE, AGRICULTURAL WATERSHEDS W70-04506
DRAINAGE, ORGANIC MATTER, SEWAGE, W70-05404
DRAINAGE, APPALACHIAN MOUNTAIN CL W69-07818
DRAINS, SEWAGE EFFLUENTS, STREAMF W70-05112
DRAINS, SEWAGE EFFLUENTS, *SEWAGE W71-10654
DRAINS, SURFACE DRAINAGE, ALGAL C W70-03344
DRAPARNALDIA PLUMOSA, STIGEOCLONI W70-02245
DRAPARNALDIA, DRAPARNALDIA PLUMOS W70-02245
DREDGE SAMPLES, ANABAENA, SEASONA W70-05663
DREDGING, COPPER SULPHATE, PONDWE W70-05263
DREDGING, MONITORING, MICROBIOLOG W70-05264
DREDGING, *POLLUTION ABATEMENT, D W68-00172
DREDGING, HARVESTING OF ALGAE, FI W69-09340
DREDGING, IMPOUNDMENTS, WEEDS,: / W69-10178
DRIED, GLASS BEADS, TRICHODERMA,   W70-04184
DRIFT, FISH FOOD ORGANISMS, FISH   W69-02782
DRILLING CODES.: /NIDE, NEW PRODU W73-00758
DRY WEIGHT, B COLI COMMUNAC, EUGL W70-05092
DRYING, BOILING.: /LA, PEPTIDES,   W70-04184
DRYING, FREEZE-THAW TESTS, HARVES W69-03362
DRYING, FREEZE-THAW TESTS, INHIBI W69-03370
DRYING, FREEZE-THAW TESTS, FREEZI W69-03518
DRYING, SOLIDS-SETTLING, ECONOMIC W70-05819
DRYING, SPRAY-DRIED, GLASS BEADS, W70-04184
DTPA, SOIL WATER PROFILE.: / ACTI W71-04067
DU: /A HARVEYI, EPIPHYTIC BACTERI W70-01073
DUCKS(WILD), OYSTERS, CHANNELS, P W68-00461
DUCKS, COLOR, FISHKILL, LAKES, OD W69-08279
DUCKS, NITRATE POISONING.:   W69-04805
DUCKWEED, WHITE WATER LILY, SWIMM W70-05568
DUFF, OOZE, COBBLES, EQUISETUM, T W69-08272
DUMPS, OIL WASTES, INLETS(WATERWA W70-05112
DUNALIELLA TERTIOLECTA, DCMU INHI W70-05261
DUNALIELLA TERTIOLECTA, AMPHIDINI W69-08278
DUNES.: /ON, SUSCEPTIBILITY, REGE W70-01777
DUNLOP LAKE, *INORGANIC CARBON.: / W70-05424
DUNREITH, OIL SPILLS.: /S, LEBANO W72-03641
DUPLIN RIVER(GA), SPARTINA ALTEMI W69-08274
DURANGO, SAN MIGUEL RIVER, DOLORE W69-07846
DURRENTS, MEASUREMENTS, SEWAGE EF W69-06203
DUSTING AIRCRAFT, WATER SAMPLING, W69-10155
DUTCH MILL DRAINAGE DITCH(WIS), L W70-05113
DUWAMISH ESTUARY(WASH).: *SEATTL   W70-04283
DYE RELEASES, CURRENTS(WATER), RE W70-05377
DYE RELEASES, BEACHES, SANDS, BIO W68-00010
DYE.: /AL COMMUNITIES, INTERTIDAL W68-00010
DYE, AIRBORNE TESTS, OIL SPILLS,   W70-05377
DYE, SODIUM CHLORATE, SANTOBRITE, W70-05263
DYEING WASTES.: /ANTI-FOAM AGENTS W70-04060
DYES.:   W70-02609
DYNAMICS, TEMPERATURE, SILTS, STA W70-02780
DYNAMICS, WATER CONSTITUENTS, ANA W70-04382
DYNAMICS, *BACTERIAL COLONIZATION W70-04371
DYNAMICS, SACRAMENTO-SAN JOAQUIN   W70-05094
DYNAMICS, ALGAL COMMUNITIES, MICR W70-04510
DYNAMICS, CELLULAR SIZE, CHLORELL W69-06540
DYNAMICS, EXTRACELLULAR PRODUCTS, W69-05697
DYSTROPHY, OLIGOTROPHY, *PRIMARY   W69-09723
E. COLI, BACTERIA, FUNGI, ACTINOM W71-02009

PITATION, LIFE CYCLES, GENETICS,   E. COLI, ISOLATION.: /MICAL PRECI   W71-11823
E, NUTRIENTS, VIRUSES, BACTERIA,   E. COLI, PUBLIC HEALTH, PATHOGENI   W72-06017
IENT REQUIREMENTS, PHYSIOLOGICAL   E: /F ALGAE, NUISANCE ALGAE, NUTR   W69-03362
FFECTS, LAKE MENDOTA(WIS), BLACK   EARTH CREEK(WIS).: /WASTE WATER E   W70-04382
ECHNIQUES, TURBID/ *DIATOMACEOUS   EARTH, *FILTRATION, *SEPARATION T   W72-08686
ATTER, *FILTRATION, DIATOMACEOUS   EARTH, ALGAE, TREATMENT FACILITIE   W72-04318
ESTS, *SCENEDESMUS, DIATOMACEOUS   EARTH, MUD, LAKES, CULTIVATION, C   W68-00475
LGAE, *FILTRATION, DIATOMACEOUS   EARTH, SEPARATION TECHNIQUES, PRE   W72-08688
AE, ACANTHURUS TRIOSTEGUS, CONUS   EBRAEUS.: / FUNGITES, ZOOXANTHELL   W69-08526
POLYCHAETA, CRUSTACEA, MOLLUSCA,   ECHINODERMATA, PACIFIC OCEAN.: /   W69-08526
OISOTOPES, CERIUM RADIOISOTOPES,   ECHINODERMS.: /OPES, CADMIUM RADI   W71-09850
ZOMENON FLOS-AQUAE, GLOEOTRICHIA   ECHINULATA, COCCOCHLORIS PENIOCYS   W70-00719
BAENA, MICROCYSTIS, GLOEOTRICHIA   ECHINULATA, DINOBRYON, SYNURA UVE   W70-02803
ZOMENON FLOS AQUAE, GLOEOTRICHIA   ECHINULATA, BOTRYOCOCCUS, PLANKTO   W70-03959
LYPOTHRIX DISTORTA, GLOEOTRICHIA   ECHINULATA LB 1303.: /S AQUAE, TO   W70-05091
NABAENA CIRCINALIS, GLOEOTRICHIA   ECHINULATE, OSCILLATORIA RUBESCEN   W69-05704
         EUTROPHICATION PREVENTION,   ECOLOGIC IMBALANCES.:   W69-07084
                 *INDIA, MOOSI RIVER   ECOLOGICAL STUDY.:   W69-07096
ONDITIONS, ALGAE, ALGAL CONTROL,   ECOLOGICAL DISTRIBUTION, ACTIVATE   W70-04786
ALITY, NUTRIENTS, NORTHEAST U S,   ECOLOGICAL DISTRIBUTION, BIOLOGIC   W70-02764
VELS, NUTRIENTS, NORTHEAST U.S.,   ECOLOGICAL DISTRIBUTION, FREQUENC   W70-02772
TAL EFFECTS, LIMNOLOGY, ECOLOGY,   ECOLOGICAL DISTRIBUTION.: /RONMEN   W70-02784
OBTUSA, SCENEDESMUS QUADRICAUDA,   ECOLOGICAL EFFICIENCIES, TROPHIC   W69-10152
, TECHNOLOGY, NATURAL RESOURCES,   ECOLOGICAL DISTRIBUTION, STOCHAST   W71-09012
  BIOINDICATORS, LAKE MORPHOLOGY,   ECOLOGICAL DISTRIBUTION, MIDGES,   W71-12077
E, CHEMICAL PROPERTIES, ECOLOGY,   ECOLOGICAL DISTRIBUTION, THERMAL   W70-07313
TORAL, *AQUATIC PLANTS, ISLANDS,   ECOLOGICAL DISTRIBUTION, BIOLOGIC   W72-10623
*HYDRAULIC MODEL/ *PHYSIOLOGICAL   ECOLOGY, *BENTHIC FLORA, *ALGAE,   W71-00668
*MORPHOLOGY, *AL/ *PHYSIOLOGICAL   ECOLOGY, *ENVIRONMENTAL EFFECTS,   W69-10177
IVERS, ENVIRONME/ *PHYSIOLOGICAL   ECOLOGY, *AQUATIC ENVIRONMENTS, R   W70-03978
EDS, *INHIBITION, *PHYSIOLOGICAL   ECOLOGY, *PHYTOTOXICITY, ALGAL CO   W69-03188
HAT/ *CYANOPHYTA, *PHYSIOLOGICAL   ECOLOGY, *SALINE WATER, *THERMAL   W69-03516
TIC MICROBIOLOGY, *PHYSIOLOGICAL   ECOLOGY, *CYANOPHYTA, ALGAE, ANAL   W69-03518
S, / *CYANOPHYTA, *PHYSIOLOGICAL   ECOLOGY, *PHYTOPLANKTON, *CULTURE   W69-03514
  *EUTROPHICATION, *PHYSIOLOGICAL   ECOLOGY, *WATER POLLUTION EFFECTS   W69-03513
ION, *CYANOPHYTA, *PHYSIOLOGICAL   ECOLOGY, ALGAE, ANALYTICAL TECHNI   W69-03517
ENT REQUIREMENTS, *PHYSIOLOGICAL   ECOLOGY, ALGAE, ANALYTICAL TECHNI   W69-03515
ION, *PHOSPHATES, *PHYSIOLOGICAL   ECOLOGY, AQUATIC ALGAE, AQUATIC M   W69-03358
ALGAE, ECOSYSTEMS, PHYSIOLOGICAL   ECOLOGY, AQUATIC BACTERIA, CARBON   W70-04287
NTS, AQUATIC LIFE, ALGAE, SILTS,   ECOLOGY, AQUATIC ENVIRONMENT.: /A   W69-10080
EASONAL, RHODE ISLAND, SAMPLING,   ECOLOGY, AMMONIA, ENZYMES, CLASSI   W70-00713
UTION ABATEMENT, BIBLIOGRAPHIES,   ECOLOGY, BIOLOGY, FISH, PLANKTON,   W72-01786
ECOLOGICAL EFFICIENCIES, TROPHIC   ECOLOGY, CLADOCERA.: /ADRICAUDA,   W69-10152
ATS, EPIPHYTOLOGY, MARINE ALGAE,   ECOLOGY, CULTURES, WASHINGTON, WA   W69-10161
LGAE, *PLANT GROWTH, *WISCONSIN,   ECOLOGY, CHEMICAL ANALYSIS, PHYTO   W70-06229
RIPHYTON, BIOMASS, PHYSIOLOGICAL   ECOLOGY, CYANOPHYTA, DIFFUSION, R   W70-03309
LANKTON, *NUTRIENT REQUIREMENTS,   ECOLOGY, CHLORELLA, ALGAE, ORGANI   W70-02804
NOLOGY, NUTRIENTS, PHYSIOLOGICAL   ECOLOGY, CHLORELLA, SCENEDESMUS.:   W70-02245
ELLA, CHLOROPHYLL, PHYSIOLOGICAL   ECOLOGY, CYANOPHYTA, EUTROPHICATI   W70-04381
C BACTERIA, ALGAE, PHYSIOLOGICAL   ECOLOGY, CARBON RADIOISOTOPES, LA   W70-05109
BON UPTAKE TECHNIQUES, MICROBIAL   ECOLOGY, COCCOLITHOPHORES, SEAWAT   W70-05272
RA, GYRODINIUM COHNII, MICROBIAL   ECOLOGY, COULTER COUNTER, GROWTH   W70-05652
M, CHEMICAL SYMBIOSIS, MICROBIAL   ECOLOGY, DINOBRYON, VOLVOX, PANDO   W70-04503
, WATER POLLUTION, PHYSIOLOGICAL   ECOLOGY, DIATOMS, OLIGOTROPHY, CY   W70-04503
REPRODUCTION, INDUSTRIAL WASTES,   ECOLOGY, DISSOLVED OXYGEN, HARDNE   W73-00748
TENSITY, SEASONAL, PHYSIOLOGICAL   ECOLOGY, ECOTYPES, CHLOROPHYTA, P   W70-05409
LCIUM, CHLORINE, COBALT, COPPER,   ECOLOGY, ENVIRONMENTAL EFFECTS, G   W69-04801
PLANTS, WATER POLLUTION EFFECTS,   ECOLOGY, EFFECTS OF POLLUTION, WA   W70-02772
NVIRONMENTAL EFFECTS, LIMNOLOGY,   ECOLOGY, ECOLOGICAL DISTRIBUTION.   W70-02784
ONTARIO, LAKE HURON, AESTHETICS,   ECOLOGY, ECONOMICS, OHIO, COMMERC   W70-01943
EMPERATURE, CHEMICAL PROPERTIES,   ECOLOGY, ECOLOGICAL DISTRIBUTION,   W70-07313
CONTROL, *REVIEWS, ALGAE, WEEDS,   ECOLOGY, ENVIRONMENT, CHEMCONTROL   W70-07393
T, CRAYFISH, STONEFLIES, SNAILS,   ECOLOGY, EUTROPHICATION, DOMESTIC   W70-00711
*MARINE ENVIRONMENTS, MICROBIAL   ECOLOGY, FLUORESCENCE MICROSCOPY,   W70-04811
PHYTA, RHODOPHYTA, PHYSIOLOGICAL   ECOLOGY, INDUSTRIAL WASTES, HYDRO   W70-04510
SSAY, SCENEDESMUS, PHYSIOLOGICAL   ECOLOGY, KINETICS, FARM WASTES, B   W70-02775
TROPHIC LEVEL, LIGHT INTENSITY,   ECOLOGY, LOTIC ENVIRONMENT, BIOMA   W70-00265
MS, RIVERS, LAKES, PHYSIOLOGICAL   ECOLOGY, LIMITING FACTORS, ECOSYS   W71-02479
  INTERTIDAL AREAS, PHYSIOLOGICAL   ECOLOGY, LITTORAL, ESTUARIES, PHO   W70-03325
ERATURE, *DAPHNIA, PHYSIOLOGICAL   ECOLOGY, LAKES, ENVIRONMENTAL EFF   W70-03957
, *BIOLOGICAL UPTAKE, *MICROBIAL   ECOLOGY, MAXIMUM UPTAKE VELOCITY,   W70-04194
IATOMS, ESTUARIES, PHYSIOLOGICAL   ECOLOGY, MASSACHUSETTS, CHLOROPHY   W70-05272
  COMMUNITIES, CULTURES, DIATOMS,   ECOLOGY, NUISANCE ALGAE, PHAEOPHY   W69-07832
IUM, EPIPHYTOLOGY, MARINE ALGAE,   ECOLOGY, NUTRIENTS, SEAWATER, ORG   W69-09755
CHEMICAL ANALYSIS, PHYSIOLOGICAL   ECOLOGY, OXYGEN, CARBON DIOXIDE,   W69-07862
NE MICROORGANISMS, PHYSIOLOGICAL   ECOLOGY, ORGANIC MATTER, WATER PO   W70-04811
IENT REQUIREMENTS, PHYSIOLOGICAL   ECOLOGY, PHYTOPLANKTON, SE: /NUTR   W69-03371
EMENTS, NUTRIENTS, PHYSIOLOGICAL   ECOLOGY, PH: /ON, NUTRIENT REQUIR   W69-03364
HORUS, PHOSPHATES, PHYSIOLOGICAL   ECOLOGY, PHYTOPLAN: /IENTS, PHOSP   W69-03373
TROPHICATION, *NUTRIENTS, LAKES,   ECOLOGY, PRIMARY PRODUCTIVITY, AL   W71-00665
MPERATURE, SILTS, STANDING CROP,   ECOLOGY, PHOTOSYNT: /DYNAMICS, TE   W70-02780
OSYNTHETIC OXYGEN, PHYSIOLOGICAL   ECOLOGY, PHYTOPLANKTON, PLANKTON,   W70-02792
ESTICIDE RESIDUES, PHYSIOLOGICAL   ECOLOGY, PHYTOPLANKTON, PHYTOTOXI   W70-02968
POPULATION EQUIVALEN/ *MICROBIAL   ECOLOGY, PURPLE SULFUR BACTERIA,   W70-03312
YTA, SCENEDESMUS, CHLAMYDOMONAS,   ECOLOGY, PHOSPHORUS, SULPHUR, HYD   W70-02965
YTICAL TECHNIQUES, PHYSIOLOGICAL   ECOLOGY, PLANT PHYSIOLOGY.: /ANAL   W72-12240
EMICAL PROPERTIES, PHYSIOLOGICAL   ECOLOGY, RHODOPHYTA, CHLOROPHYTA.   W69-10161
WEDEN), ANNUAL CYCLES, MICROBIAL   ECOLOGY, RADIOCARBON UPTAKE TECHN   W70-05109
*ORGANIC MATTER, *PHYSIOLOGICAL   ECOLOGY, RADIOACTIVITY TECHNIQUES   W70-04194
OLUTES, *HETEROTROPHY, MICROBIAL   ECOLOGY, SWEDEN, LAKE ERKEN(SWEDE   W70-04284

433

OPERTIES, BIOLOGICAL PROPERTIES, ECOLOGY, STRATIFICATION.: /TER PR W70-04475
ESOTA, *LAKES, *DIATOMS, *ALGAE, ECOLOGY, SOUTH DAKOTA, FIELD INVE W69-00659
LING WATER, CHEMICAL PROPERTIES, ECOLOGY, SALINITY, BENTHOS, AQUAT W69-05023
IOISOTOPES, ALGAE, BIOCHEMISTRY, ECOLOGY, SAMPLING, THERMAL SPRING W69-10151
ICHIGAN, PHYTOPLANKTON, DIATOMS, ECOLOGY, SOIL, CULTURES, CYANOPHY W69-10180
*SUCCESSION, *ISOLATION, DIATOMS, ECOLOGY, SALINITY, ORGANIC MATTER W73-01446
OXIDE, CYANOPHYTA, PHYSIOLOGICAL ECOLOGY, THERMAL WATER.: /RBON DI W69-10160
CYCLING NUTRIENTS, PHYSIOLOGICAL ECOLOGY, WATER: /ALGAE, DIATOMS, W70-00713
PLANTS, WATER POLLUTION EFFECTS, ECOLOGY, WATER QUALITY CHARACTERI W70-02764
HEMICAL OXYGEN DEMAND, BACTERIA, ECOLOGY, WATER POLLUTION EFFECTS, W71-04758
AGOONS, OXYGENATION, ABSORPTION, ECOLOGY, WATER PURIFICATION, CARB W69-07389
EWAGE DISPOSAL, WATER CHEMISTRY, ECOLOGY, WAT: / WASTE DISPOSAL, S W69-04276
AKES, GREAT LAKES, PHYSIOLOGICAL ECOLOGY, WATER CHEMISTRY, GEOGRAP W70-03975
AKES, GREAT LAKES, PHYSIOLOGICAL ECOLOGY, WATER CHEMISTRY, GEOGRAP W70-03975
CS, ALGAL COMMUNITIES, MICROBIAL ECOLOGY, WATER QUALITY CRITERIA, W70-04510
SPHORUS COMPOUNDS, PHYSIOLOGICAL ECOLOGY,: / ALGAE, NUTRIENTS, PHO W69-04802
SLUDGE DRYING, SOLIDS-SETTLING, ECONOMIC ADVANTAGE, TEMPERATURE-E W70-05819
ICATION, FLOCCULATION, TURBINES, ECONOMIC EFFICIENCY, HEAT BUDGET, W70-08834
COLLOIDS, FLOCCULATION, MIXING, ECONOMIC FEASIBILITY, TECHNICAL F W71-10654
, AQUATIC PLANTS, AQUATIC WEEDS, ECONOMIC FEASIBILITY, LAKES, NUIS W69-06276
COPPER SULFATE, ODOR, COLIFORMS, ECONOMICS.: /OLLUTION TREATMENT W72-01814
LAKE HURON, AESTHETICS, ECOLOGY, ECONOMICS, OHIO, COMMERCIAL FISHI W70-01943
ING NUTRIENTS, COST COMPARISONS, ECONOMICS, ALGAE, ALGAL CONTROL, W69-08518
RITISH COLUMBIA, CANADA, AQUATIC ECOSYSTEM MODEL, SIMULATION MODEL W69-06536
RATIONS, MICRONUTRIENTS, AQUATIC ECOSYSTEM, MIDGE LARVAE, LAKE STR W69-09340
, *REEFS, *MARINE ANIMALS, FISH, ECOSYSTEMS, CYCLING NUTRIENTS, TE W69-08526
S, SYSTEMS ANALYSIS, SUCCESSION, ECOSYSTEMS, CYCLING NUTRIENTS.: / W70-02779
FUSION, RESPIRATION, PHOSPHATES, ECOSYSTEMS, ORGANIC MATTER, WATER W70-03309
ORY, MINERALS,/ ALGAE, BRACKISH, ECOSYSTEMS, FAUNA, FLORA, LABORAT W72-02203
TIC PLANTS, AQUATIC ENVIRONMENT, ECOSYSTEMS, AQUATIC HABITATS, DRA W71-09005
TRIBUTION, STOCHASTIC PROCESSES, ECOSYSTEMS, AQUATIC HABITATS, WAT W71-09012
IDES, *VASCULAR TISSUES, *ALGAE, ECOSYSTEMS, CHARA, BENTHIC FLORA, W70-10175
GICAL ECOLOGY, LIMITING FACTORS, ECOSYSTEMS, MICROORGANISMS, AQUAT W71-02479
DICATORS, CHLAMYDOMONAS, CARBON, ECOSYSTEMS, EUGLENA, EUTROPHICATI W69-06273
, DETRITUS, ESTUARIES, PLANKTON, ECOSYSTEMS, SINKS, TRACERS, CRABS W69-08274
ES, STREAMFLOW, SAMPLING, ALGAE, ECOSYSTEMS, ENVIRONMENT, RETENTIO W69-07862
STACEANS, BENTHOS, BRINE SHRIMP, ECOSYSTEMS, FISH, PHYTOPLANKTON, W69-07440
COMPOUNDS, MOLYBDENUM, ROTIFERS, ECOSYSTEMS, PHOTOS: /ITE, SULFUR W69-07428
, FISH, RADIOECOLOGY, EFFLUENTS, ECOSYSTEMS, PLANKTON, ALGAE, PERI W69-07853
ASHINGTON, PRIMARY PRODUCTIVITY, ECOSYSTEMS, ILLINOIS, PHOSPHORUS, W70-04506
IA, FLORIDA, PESTICIDE RESIDUES, ECOSYSTEMS, FOOD CHAINS, BIOLOGIC W70-05272
EW YORK, CHLOROPHYTA, NUTRIENTS, ECOSYSTEMS, PHYTOPLANKTON, PHOTOS W70-03973
WATER POLLUTION EFFECTS, ALGAE, ECOSYSTEMS, PHYSIOLOGICAL ECOLOGY W70-04287
TR/ *LAKES, STREAMS, FISH, LAND, ECOSYSTEMS, EUTROPHICATION, OLIGO W70-04193
NTENSITY, BIOMASS, DISTRIBUTION, ECOSYSTEMS, POPULATION, OREGON, I W70-03325
HLORRELA, SCENEDESMUS, KINETICS, ECOSYSTEMS.: /TES, *TURBULENCE, C W69-03730
ANTS, *ESTUARIES, *TIDAL WATERS, ECOSYSTEMS, ESTUARINE ENVIRONMENT W68-00010
SSILE ALGAE, *CYCLING NUTRIENTS, ECOSYSTEMS, PHOSPHORUS, SCENEDESM W73-01454
ROCARBONS, BACTERIAL ATTACHMENT, ECOTOCARPUS, POLYSIPHONIA, ORGANI W70-01068
SEASONAL, PHYSIOLOGICAL ECOLOGY, ECOTYPES, CHLOROPHYTA, PHOTOMETRY W70-05409
SCICULOSUS, ASCOPHYLLUM NODOSUM, ECTOCARPUS CONFERVOIDES, NYLON CO W70-01074
RITUS, BOTTOM SEDIMENTS, CHUTES, EDDIES, HABITATS, CRAYFISH, GROWT W69-09742
COMPLEXONES, *ALGAL BLOOMS, NTA, EDTA, HEIDA, LAKE TAHOE(CALIF).: / W70-04373
AKE, *PHAEODACTYLUM TRICORNUTUM, EDTA, LIGURIAN SEA, EUROPEAN MEDI W69-08275
A(CANADA), PENGUINS, LAK/ PRINCE EDWARD ISLAND(CANADA), NOVA SCOTI W70-05399
RIVER(CALIF), CLEAR LAKE(CALIF)/ EEL RIVER(CALIF), JOINT INDUSTRY/ W70-02777
S, CHARA, ALKALINITY, BULLHEADS, EELS, KILLIFISHES, TROUT, SUCKERS W70-05263
ECONOMIC ADVANTAGE, TEMPERATURE- EFFECT.: /RYING, SOLIDS-SETTLING, W70-05819
EL, SIMULATION MODEL, PHOSPHORUS EFFECT, CLADECERA.: /COSYSTEM MOD W69-06536
ATER POLLUTION EFFECTS, ECOLOGY, EFFECTS OF POLLUTION, WATER QUALI W70-02772
UTION TREATMENT, WATER POLLUTION EFFECTS.: /ERSISTENCE, WATER POLL W70-01777
IA, ORGANIC AGGREGATES, SEASONAL EFFECTS.: /COTOCARPUS, POLYSIPHON W70-01068
S, IONIC CONCENTRATION, SEASONAL EFFECTS.: /NOMICTIC, DEPTH EFFECT W70-01933
GEN, PHOSPHORUS, WATER POLLUTION EFFECTS.: /ERTILITY, ALGAE, NITRO W70-02973
, AQUATIC ALGAE, WATER POLLUTION EFFECTS.: /DICATORS, AQUATIC LIFE W70-05958
URES, CHLORELLA, WATER POLLUTION EFFECTS.: / WATER TREATMENT, CULT W70-09438
SOTA, NUTRIENTS, WATER POLLUTION EFFECTS.: /ABORATORY TESTS, MINNE W70-08637
MAC RIVER, SAVANNAH RIVER, SHOCK EFFECTS.: /ZOAN COMMUNITIES, POTO W70-07847
ECOLOGICAL DISTRIBUTION, THERMAL EFFECTS.: / PROPERTIES, ECOLOGY, W70-07313
LLUTION SOURCES, WATER POLLUTION EFFECTS.: /, CYANOPHYTA, WATER PO W69-10180
SHINGTON, LAKES, WATER POLLUTION EFFECTS.: /US, EUTROPHICATION, WA W70-00270
T RADIOISOTOPES, WATER POLLUTION EFFECTS.: /E RADIOISOTOPES, COBAL W71-09850
LGAE, LIVESTOCK, WATER POLLUTION EFFECTS.: /ITROGEN, PHOSPHORUS, A W71-10376
TER, POPULATION, WATER POLLUTION EFFECTS.: /KTON, ALGAE, HEATED WA W71-09767
QUATIC HABITATS, WATER POLLUTION EFFECTS.: /OCESSES, ECOSYSTEMS, A W71-09012
ITION), METHANE, WATER POLLUTION EFFECTS.: / *DEGRADATION(DECOMPOS W71-13413
QUATIC BACTERIA, WATER POLLUTION EFFECTS.: /S, YEASTS, BACTERIA, A W69-06970
ALGAE, BACTERIA, WATER POLLUTION EFFECTS.: /HIES, LAKES, NUISANCE W69-08283
ICAL PROPERTIES, WATER POLLUTION EFFECTS.: /IDE, NUTRIENTS, BIOLOG W69-07442
WATER CHEMISTRY, WATER POLLUTION EFFECTS.: / NUTRIENTS, TOXICITY, W69-05867
LLUTION CONTROL, WATER POLLUTION EFFECTS.: /LITY CONTROL, WATER PO W69-05705
DELAWARE RIVER, WATER POLLUTION EFFECTS.: /E RIVER, HUDSON RIVER, W70-04507
XYGEN, PLANKTON, WATER POLLUTION EFFECTS.: /NTS, PHOTOSYNTHETIC, O W69-03514
KTON, *REVIEWS, *WATER POLLUTION EFFECTS.: /OLIGOTROPHY, PHYTOPLAN W68-00680
PHYSIOLOGICAL EFFECTS.: W68-00859
ABORATORY TESTS, WATER POLLUTION EFFECTS.: /N PESTICIDES, ALGAE, L W72-11105
LUTION SOURCES, *WATER POLLUTION EFFECTS, *WATER SOURCES, *GROUNDW W73-00758
TY, IRRIGATION EFFECTS, DRAINAGE EFFECTS, *SAMPLING, ALGAE TOXINS, W69-00360
ROPH/ *STREAMS, *WATER POLLUTION EFFECTS, *CYCLING NUTRIENTS, *EUT W69-03369
LUTION SOURCES, *WATER POLLUTION EFFECTS, *PRIMARY PRODUCTIVITY, A W69-02959

```
      *ANAEROBIC WASTE PONDS, *DEPTH      EFFECTS, *HYDRAULIC LOAD EFFECTS,      W70-04787
IC LOAD EFFECTS, *DETENTION TIME          EFFECTS, *FACULATIVE PONDS, *AERO      W70-04787
ULIC LOAD EFFECTS, *ORGANIC LOAD          EFFECTS, *DETENTION TIME EFFECTS,      W70-04787
      *DEPTH EFFECTS, *HYDRAULIC LOAD      EFFECTS, *ORGANIC LOAD EFFECTS, *      W70-04787
LIZATION, *PONDS, *ENVIRONMENTAL          EFFECTS, *BIOLOGICAL TREATMENT, *      W70-04787
LANKTON, *DDT,/ *WATER POLLUTION          EFFECTS, *PHOTOSYNTHESIS, *PHYTOP      W70-05272
ATER CHEMISTRY, *WATER POLLUTION          EFFECTS, *PHOSPHATES, *ANALYTICAL      W70-05269
ENTS, POPULATIONS, ENVIRONMENTAL          EFFECTS, *PRIMARY PRODUCTIVITY, W      W70-05424
YCLING NUTRIENTS, *ENVIRONMENTAL          EFFECTS, *ESSENTIAL NUTRIENTS, *F      W69-04804
LAND WATERWAYS, *WATER POLLUTION          EFFECTS, *WATER SUPPLY, BIOINDICA      W69-07833
LUTION SOURCES, *WATER POLLUTION          EFFECTS, *EUTROPHICATION, *NUTRIE      W70-04266
LUTION SOURCES, *WATER POLLUTION          EFFECTS, *WATER POLLUTION CONTROL      W70-03975
OPHICATI/ *ALGAE, *ENVIRONMENTAL          EFFECTS, *HUMAN POPULATION, *EUTR      W70-03973
LUTION SOURCES, *WATER POLLUTION          EFFECTS, *WATER POLLUTION CONTROL      W70-03975
LUTION SOURCES, *WATER POLLUTION          EFFECTS, *CLASSIFICATION, *WATER       W72-00462
R QUALITY, *DISSOLVED SOLIDS, D/          EFFECTS, *IMPOUNDED WATERS, *WATE      W72-01773
REAT LAKES, TR/ *WATER POLLUTION          EFFECTS, *FALLOUT, *ECOSYSTEMS, G      W71-09013
AL CHEMICALS, *DELTAS, *DRAINAGE          EFFECTS, *WATER QUALITY, *FARM WA      W71-09577
WAGE, *STREAMS, *WATER POLLUTION          EFFECTS, *FISH, ALGAE, TROPHIC LE      W72-01788
ARBORS, *SEWAGE, WATER POLLUTION          EFFECTS, *ALGAE.: *ESTUARIES, *H       W71-05500
OLOGICAL ECOLOGY, *ENVIRONMENTAL          EFFECTS, *MORPHOLOGY, *ALGAE, AQU      W69-10177
ASHINGTON, *SEWAGE FERTILIZATION          EFFECTS, *BIOLOGICAL PRODUCTIVITY      W69-09349
UALITY, *DISSO/ *WATER POLLUTION          EFFECTS, *BIOINDICATORS, *WATER Q      W70-00475
E, *CHLOROPHYLL, WATER POLLUTION          EFFECTS, *ORGANIC COMPOUNDS, *OXY      W70-08416
PESTICIDE KINE/ *WATER POLLUTION          EFFECTS, *PESTICIDES, RESIDUES, *      W70-09768
R, / RESISTANCE, WATER POLLUTION          EFFECTS, *SALT MARSHES, OILY WATE      W70-09976
*BIOINDICATORS, *WATER POLLUTION          EFFECTS, *REVIEWS, *WATER QUALITY      W70-07027
USTRIAL WASTES, *WATER POLLUTION          EFFECTS, *EUTROPHICATION, DISSOLV      W70-06975
LOGY, ALGAE, BACTE/ *TEMPERATURE          EFFECTS, *THERMAL POLLUTION, *ECO      W71-02479
WATER, *HABITATS, *ENVIRONMENTAL          EFFECTS, *LIMITING FACTORS, BENTH      W70-02770
                   *CYTOLOGICAL          EFFECTS, ALGAL POPULATION.:            W70-02770
WATER POLLUTION, WATER POLLUTION          EFFECTS, ACID STREAMS, MINE DRAIN      W70-02770
N, *CONFERENCES, WATER POLLUTION          EFFECTS, ALGAL CONTROL, WATER POL      W70-02775
LGAL SUSPENSIONS, ELECTROKINETIC          EFFECTS, ANN ARBOR(MICH), SAND BE      W71-03035
S, INSECTICIDES, WATER POLLUTION          EFFECTS, ALGAE.: /LABORATORY TEST      W71-03183
HY, DEGRADATION, WATER POLLUTION          EFFECTS, AQUATIC ENVIRONMENT, CHL      W70-08652
ITY, POLLUTANTS, WATER POLLUTION          EFFECTS, ALDRIN, DIELDRIN, ENDRIN      W69-08565
WATER POLLUTION, WATER POLLUTION          EFFECTS, ALGAL CONTRO: /EATMENT,       W69-10171
BOLISM, AQUARIA, WATER POLLUTION          EFFECTS, ADSORPTION.: /TION, META      W71-09005
REAMS, SEASONAL EFFECTS, THERMAL          EFFECTS, APHANIZOMENON FLOS AQUAE      W70-03978
ONAS, BIOASSAYS, WATER POLLUTION          EFFECTS, ANALYTICAL TECHNIQUES.: /      W70-04284
ICAL TECHNIQUES, WATER POLLUTION          EFFECTS, ALGAE, ECOSYSTEMS, PHYSI      W70-04287
ORUS COMPOUNDS, *WATER POLLUTION          EFFECTS, ALGAE, ANIMALS, AQUATIC       W69-04805
 ORGANIC MATTER, WATER POLLUTION          EFFECTS, ANALYTICAL TECHNIQUES, K      W70-05109
ENTS, MINNESOTA, WATER POLLUTION          EFFECTS, ALGICIDES, WATER QUALITY      W70-04506
, PRYMNESIUM PARVUM, SYNERGISTIC          EFFECTS, AXENIC CULTURES, MONOALG      W70-05372
CYCLING NUTRIENTS, ENVIRONMENTAL          EFFECTS, AMMONIA, AMMONIUM COMPOU      W69-03364
OGICAL ECOLOGY, *WATER POLLUTION          EFFECTS, ALGAE, ALGAL CONTROL, HA      W69-03513
, SEWAGE BACTERIA, ENVIRONMENTAL          EFFECTS, AQUATIC LIFE, AQUATIC MI      W69-04276
INDUSTRIAL WASTES, ENVIRONMENTAL          EFFECTS, ANALYTICAL TECHNIQUES, P      W72-12240
RY PRODUCTIVITY, WATER POLLUTION          EFFECTS, BIOASSAYS, ALGAE, COMPAR      W70-05424
ON, *MEROMIXIS, *WATER POLLUTION          EFFECTS, BIOINDICATORS, NUISANCE       W70-03974
ION, *SWIMMING, *WATER POLLUTION          EFFECTS, BACTERIA, ALGAE, FISH, S      W71-09880
 MATTER, CARBON, WATER POLLUTION          EFFECTS, BACTERIA, PERIPHYTON, OH      W71-04206
RIA, ABSORPTION, WATER POLLUTION          EFFECTS, CATION EXCHANGE, CATION       W71-09863
XICITY, *ALGAE, *WATER POLLUTION          EFFECTS, CHELATION, SPECTROPHOTOM      W71-11561
ERSHEDS(BASINS), WATER POLLUTION          EFFECTS, CYANOPHYTA, EURTROPHICAT      W69-10182
TRACE ELEMENTS, *WATER POLLUTION          EFFECTS, CHLORIDES, HALOGENS, ALG      W69-08562
ATORS, CYANOPHYTA, ENVIRONMENTAL          EFFECTS, CURRENTS(WATER), INDUSTR      W68-00470
RSHED MANAGEMENT, *ENVIRONMENTAL          EFFECTS, CHEMTROL, ALGAE.: /*WATE      W68-00478
ANISMS, GRAZING, WATER POLLUTION          EFFECTS, DIATOMS, CHLOROPHYTA, CY      W69-01977
S, *ALGAE, *SALINITY, IRRIGATION          EFFECTS, DRAINAGE EFFECTS, *SAMPL      W69-00360
EDGE SAMPLES, ANABAENA, SEASONAL          EFFECTS, DECOMPOSITION RATES, CHR      W70-05663
LLUTION SOURCES, WATER POLLUTION          EFFECTS, DISSOLVED OXYGEN, AQUATI      W70-04727
 ORGANIC MATTER, WATER POLLUTION          EFFECTS, DIATOMS, CARBON RADIOISO      W70-04811
LLUTION SOURCES, WATER POLLUTION          EFFECTS, DINOFLAGELLATES, FISHKIL      W70-08372
DUSTRIAL WASTES, WATER POLLUTION          EFFECTS, DENITRIFICATION, SPECTRO      W70-02777
CAL COMMUNITIES, WATER POLLUTION          EFFECTS, DIATOMS.: /ANTS, BIOLOGI      W70-01233
ALS, *OILY WATER, LICHENS, TIDAL          EFFECTS, DISPERSION, OIL-WATER IN      W70-01777
RESHWATER ALGAE, WATER POLLUTION          EFFECTS, DOMESTIC WASTES, INDUSTR      W72-12240
ATER AND PLANTS, WATER POLLUTION          EFFECTS, ECOLOGY, EFFECTS OF POLL      W70-02772
ATER AND PLANTS, WATER POLLUTION          EFFECTS, ECOLOGY, WATER QUALITY C      W70-02764
ANOPHYTA, DIATOMS, ENVIRONMENTAL          EFFECTS, ESSENTIAL NUTRIENTS, EUT      W70-02792
GHT PENETRATION, WATER POLLUTION          EFFECTS, EUTROPHICATION.: /DS, LI      W70-02255
SAY, DISINFECTION, ENVIRONMENTAL          EFFECTS, ENVIRONMENTAL SANITATION      W69-10171
VOLUME, STREAMS, WATER POLLUTION          EFFECTS, EUTROPHICATION, NUISANCE      W70-00719
PHYTA, CYANOPHYTA, ENVIRONMENTAL          EFFECTS, ESSENTIAL NUTRIENTS, FRE      W69-03362
PHYTA, CYANOPHYTA, ENVIRONMENTAL          EFFECTS, EUTROPHICATION, LIMNOLOG      W69-03185
CAL OXYGEN DEMAND, ENVIRONMENTAL          EFFECTS, ENVIRONMENTAL SANITATION      W69-03369
CYCLING NUTRIENTS, ENVIRONMENTAL          EFFECTS, ESSENTIAL NUTRIENTS, ENV      W69-03374
OMS, DISINFECTION, ENVIRONMENTAL          EFFECTS, ESSENTIAL NUTRIENTS, FRE      W69-03370
CYCLING NUTRIENTS, ENVIRONMENTAL          EFFECTS, ESSENTIAL NUTRIENTS, NUT      W69-03373
CYCLING NUTRIENTS, ENVIRONMENTAL          EFFECTS, ENZYMES, ESSENTIAL N: /       W69-03188
ASSAY, CYANOPHYTA, ENVIRONMENTAL          EFFECTS, EUTROPHICATION, LIMNOLOG      W69-03186
ANOPHYTA, DIATOMS, ENVIRONMENTAL          EFFECTS, ESSENTIAL NUTRIENTS, HAR      W69-03371
CYCLING NUTRIENTS, ENVIRONMENTAL          EFFECTS, ESSENTIAL NUTRIENTS, EUT      W69-03518
CYCLING NUTRIENTS, ENVIRONMENTAL          EFFECTS, ESSENTIAL NUTRIENTS, CHL      W69-03517
LLUTION CONTROL, WATER POLLUTION          EFFECTS, EUTROPHICATION, FISHKILL      W69-03948
CYCLING NUTRIENTS, ENVIRONMENTAL          EFFECTS, ESSENTIAL NUTRIENTS, EUT      W69-03514
CYCLING NUTRIENTS, ENVIRONMENTAL          EFFECTS, ENVIRONMENTAL SANITATION      W69-03513
```

CYCLING NUTRIENTS, ENVIRONMENTAL   EFFECTS, ESSENTIAL NUTRIENTS, EUT   W69-03515
UTRIENTS, DIATOMS, ENVIRONMENTAL   EFFECTS, ESSENTIAL NUTRIENTS, EUT   W70-04283
NMENTAL EFFECTS, WATER POLLUTION   EFFECTS, EPILIMNION, CONNECTICUT,   W70-03957
IVITY, CYANOPHYTA, ENVIRONMENTAL   EFFECTS, EUTROPHICATION, LIMNOLOG   W69-04802
LLUTION CONTROL, WATER POLLUTION   EFFECTS, FUNGI, AQUATIC FUNGI.: /   W69-05706
WASTE DISPOSAL, *WATER POLLUTION   EFFECTS, FOOD CHAINS, BACTERIA, D   W72-01801
, COPPER, ECOLOGY, ENVIRONMENTAL   EFFECTS, GRAZING, IRON, MANGANESE   W69-04801
IDENTIFICATION, *WATER POLLUTION   EFFECTS, GAS CHROMATOGRAPHY.: /T   W69-02782
CYCLING NUTRIENTS, ENVIRONMENTAL   EFFECTS, HARVESTING ALGAE, INHIBI   W69-04798
ACTERIA, PEPTIDES, ENVIRONMENTAL   EFFECTS, HETEROGENEITY, IRON, MAN   W70-00273
ELAGIC REGION, MONOMICTIC, DEPTH   EFFECTS, IONIC CONCENTRATION, SEA   W70-01933
SSAY, CHLOROPHYTA, ENVIRONMENTAL   EFFECTS, INHIBITION, LIMNOLOGY, P   W70-02968
OROPHYTA, DIATOMS, ENVIRONMENTAL   EFFECTS, INHIBITION, LIGHT, LIGHT   W69-03516
LATION ANALYSIS, WATER POLLUTION   EFFECTS, LAKES, PHOSPHORUS, TROPH   W69-06536
LLUTION SOURCES, WATER POLLUTION   EFFECTS, LABORATORY TESTS, MODEL   W70-04074
NALYTICAL CHEMISTRY, WASTE WATER   EFFECTS, LAKE MENDOTA(WIS), BLACK   W70-04382
, *MARINE ALGAE, WATER POLLUTION   EFFECTS, LETHAL LIMIT, GROWTH RAT   W70-01779
RITUS, BACTERIA, WATER POLLUTION   EFFECTS, LIMNOLOGY, CYANOPHYTA.: /   W70-02249
ONOPHYTA, DIATOMS, ENVIRONMENTAL   EFFECTS, LIMNOLOGY, ECOLOGY, ECOL   W70-02784
CTERIA, ECOLOGY, WATER POLLUTION   EFFECTS, LAKES, *CYANOPHYTA, *EUT   W71-04758
SPHATES, SODIUM, WATER POLLUTION   EFFECTS, LABORATORY TESTS.: / PHO   W70-07257
AL GROWTHS, OXYGEN D/ *RESERVOIR   EFFECTS, LAKE MEAD, DILUTION, ALG   W73-00750
, *AQUATIC LIFE, WATER POLLUTION   EFFECTS, MARINE ANIMALS, ALGAE, M   W71-05531
E, *EMULSIFIERS, WATER POLLUTION   EFFECTS, MORTALITY, GASTROPODS, M   W70-01780
NITROGEN FIXATION, ENVIRONMENTAL   EFFECTS, NUTRIE: / FLUORESCENCE,   W70-02775
LLUTION SOURCES, WATER POLLUTION   EFFECTS, NITROGEN, COLIFORMS, WAT   W71-05407
WATER POLLUTION, WATER POLLUTION   EFFECTS, NITROGEN, PHOSPHORUS, AL   W70-04382
E, PUBLICATIONS, WATER POLLUTION   EFFECTS, NEW YORK, CHLOROPHYTA, N   W70-03973
SENTIAL NUTRIENTS, ENVIRONMENTAL   EFFECTS, NITROGEN COMPOUNDS, NUTR   W69-03358
CTS, PERIPHYTON, WATER POLLUTION   EFFECTS, OREGON, BIOINDICATORS, L   W70-03978
IBUTION, MIDGES, WATER POLLUTION   EFFECTS, OLIGOCHAETES, DIPTERA, A   W71-12077
TIVITY, TRACERS, WATER POLLUTION   EFFECTS, PHYSIOLOGICAL: / RADIOAC   W70-02510
ICAL TECHNIQUES, WATER POLLUTION   EFFECTS, PHYTOPLANKTON, PLANKTON,   W70-02245
LLUTION SOURCES, WATER POLLUTION   EFFECTS, PHOSPHORUS, NITROGEN, CH   W70-07261
EAR POWERPLANTS, WATER POLLUTION   EFFECTS, PHOSPHORUS, NITROGEN,: /   W70-00264
IO, *GROWTH RATES, ENVIRONMENTAL   EFFECTS, PHOTOSYNTHESIS, DIATOMS,   W69-10158
IRONMENTS, RIVERS, ENVIRONMENTAL   EFFECTS, PERIPHYTON, WATER POLLUT   W70-03978
ICAL TECHNIQUES, WATER POLLUTION   EFFECTS, PHYTOPLANKTON, DIFFUSION   W70-04194
AKES, NUTRIENTS, WATER POLLUTION   EFFECTS, PRODUCTIVITY, SEDIMENTS,   W70-04721
ON, RADIOACTIVITY, RADIOACTIVITY   EFFECTS, RADIOECOLOGY, SORPTION,   W69-07440
NC RADIOISOTOPES, *ENVIRONMENTAL   EFFECTS, SALINITY, TEMPERATURE, S   W69-08267
LATE SULFONATES, WATER POLLUTION   EFFECTS, STABILIZATION, LAGOONS,   W69-09454
CED SULFUR COMPOUNDS, ANTIBIOTIC   EFFECTS, STIMULANTS, GONYAULAX CA   W70-02510
EUTROPHICATION, WATER POLLUTION   EFFECTS, SYSTEMS ANALYSIS, SUCCES   W70-02779
, EXPERIMENTAL STREAMS, SEASONAL   EFFECTS, THERMAL EFFECTS, APHANIZ   W70-03978
AL ECOLOGY, LAKES, ENVIRONMENTAL   EFFECTS, WATER POLLUTION EFFECTS,   W70-03957
, PHYTOPLANKTON, WATER POLLUTION   EFFECTS, WISCONSIN ZOOPLANKTON, T   W70-03959
WASTE TREATMENT, WATER POLLUTION   EFFECTS, WATER POLLUTION SOURCES,   W69-08518
ODOR, TOXICITY, WATER POLLUTION   EFFECTS, WATER QUALITY, CATTLE.: /   W69-08279
OLOGY, SORPTION, WATER POLLUTION   EFFECTS, WATER POLLUTION SOURCES.   W69-07440
TY, SYSTEMATICS, WATER POLLUTION   EFFECTS, WATER POLLUTION SOURCES,   W69-07832
TION, WATER LAW, WATER POLLUTION   EFFECTS, WATER POLLUTION SOURCES,   W69-06305
EUTROPHICATION, WATER POLLUTION   EFFECTS, WATER POLLUTION SOURCES,   W69-06864
TROL, NUTRIENTS, WATER POLLUTION   EFFECTS, WATER POLLUTION SOURCES,   W69-06276
WATER CHEMISTRY, WATER POLLUTION   EFFECTS, WATER POLLUTION SOURCES.   W69-06273
IENTS, VITAMINS, WATER POLLUTION   EFFECTS, WATER POLLUTION SOURCES,   W69-06277
WATER CHEMISTRY, WATER POLLUTION   EFFECTS, WATER QUALITY, WISCONSIN   W69-05868
LLUTION CONTROL, WATER POLLUTION   EFFECTS, WATER POLLUTION SOURCES,   W70-05668
LLUTION CONTROL, WATER POLLUTION   EFFECTS, WATER POLLUTION SOURCES.   W70-02792
LLUTION CONTROL, WATER POLLUTION   EFFECTS, WATER: /ATMENT, WATER PO   W70-02248
UTRIENTS, ALGAE, WATER POLLUTION   EFFECTS, WATER POLLUTION SOURCES,   W69-09135
GEN, PHOSPHORUS, WATER POLLUTION   EFFECTS, WATER POLLUTION SOURCES,   W69-08668
LLUTION SOURCES, WATER POLLUTION   EFFECTS, WATER POLLUTION CONTROL,   W69-10167
, FISH, STREAMS, WATER POLLUTION   EFFECTS, WATER POLLUTION SOURCES.   W69-10159
CARBON DIOXIDE, WATER POLLUTION   EFFECTS, WATER POLLUTION SOURCES.   W69-10158
, ALGAE, PLANTS, WATER POLLUTION   EFFECTS, WATER POLLUTION SOURCES,   W69-10170
TIC ALGAE, IONS, WATER POLLUTION   EFFECTS, WATER POLLUTION SOURCES,   W70-00707
*THERMAL POLL/ *WATER POLLUTION   EFFECTS, WATER POLLUTION SOURCES,   W70-07847
LUTION SOURCES, *WATER POLLUTION   EFFECTS, WATER UTILIZATION, ANALY   W71-03048
GHT PENETRATION, WATER POLLUTION   EFFECTS, WATER POLLUTION CONTROL,   W72-00917
TRATES, AMMONIA, WATER POLLUTION   EFFECTS, WASTE WATER TREATMENT, P   W71-05026
LLUTION SOURCES, WATER POLLUTION   EFFECTS, WATER QUALITY.: /ATER PO   W71-05409
IC PRODUCTIVITY, WATER POLLUTION   EFFECTS, YEASTS, CRUSTACEA, ALGAE   W69-10152
NEDESMUS QUADRICAUDA, ECOLOGICAL   EFFICIENCIES, TROPHIC ECOLOGY, CL   W69-10152
ROGEN COMPOUN/ *DENITRIFICATION,   EFFICIENCIES, *NITRIFICATION, NIT   W70-07031
TRIENTS, *NITROGEN, *PHOSPHORUS,   EFFICIENCIES, COST COMPARISONS, T   W70-01981
ORUS, AIR, PHYTOPLANKTON, ALGAE,   EFFICIENCIES, DISSOLVED OXYGEN, O   W70-04484
LOADING, DESIGN CRITERIA, ODORS,   EFFICIENCIES, *WASTE WATER TREATM   W72-06854
FLOCCULATION, TURBINES, ECONOMIC   EFFICIENCY, HEAT BUDGET, SOCIAL A   W70-08834
S, PONDS, PHOTOSYNTHESIS, COSTS,   EFFICIENCY, BIOCHEMICAL OXYGEN DE   W70-09320
, ODOR, COOLING TOWERS, FILTERS,   EFFLUENT SAMPLING.: /TERS, SNAILS   W72-00027
US, ALGAE, *HARVESTING OF ALGAE,   EFFLUENT, QUALITY CONTROL, OXIDAT   W71-13313
COPIC ORGANISMS, PAPERMILL WASTE   EFFLUENT, CHEESE FACTORY WASTE WA   W71-00940
PHATES, CHLOROPHYLL, EPILIMNION,   EFFLUENT, TEMPERATURE, PHOTOSYNTH   W69-10169
MPTON UNIV/ *PHAEODACTYLUM, *GAS   EFFLUENT, FOOD PRODUCTION, SOUTHA   W70-00161
TER CONTROL, FERTILIZERS, SEWAGE   EFFLUENT, NUTRIENTS, LAKES, CHANN   W70-00269
*NITROGEN, *PHOSPHORUS, *SEWAGE   EFFLUENT, *SEWAGE TREATMENT, *NUT   W72-08133
WEST(US), *DESIGN CRITERIA, POND   EFFLUENT, COLORADO, WEST(US), STA   W70-04788
ES, WATER QUALITY, ALGAE, SEWAGE   EFFLUENTS.: /ATER POLLUTION SOURC   W70-05084
ICATION, RECREATIONAL USE, WASTE   EFFLUENTS.:   *CULTURAL EUTROPH   W69-06535

436

NT, NITROGEN, PHOSPHORUS, SEWAGE
OLLUTION, WATER POLLUTION, ODOR,
LLUTION, ODOR, EFFLUENTS, SEWAGE
OXYGEN DEMAND, STORAGE CAPACITY,
PILOT PLANTS, HUMUS, REAERATION,
UCTIVITY, NUISANCE ALGAE, SEWAGE
LOGY, SAMPLING, THERMAL SPRINGS,
, WASTE WATER DISPOSAL, DIURNAL,
NION, NITROGEN COMPOUNDS, SEWAGE
T, WATER QUALITY CONTROL, SEWAGE
LUTION CONTROL, *FILTERS, SEWAGE
 POWERPLANTS, SHIPS, SUBMARINES,
MPSHIRE, *COPPER SULFATE, SEWAGE
*LAKES, *EUTROPHICATION, *SEWAGE
ER, THERMAL POLLUTION, BACTERIA,
IOCHEMICAL OXYGEN DEMAND, SEWAGE
OW, SEWERS, STORM DRAINS, SEWAGE
S, RUNOFF, WASTES, WATER SUPPLY,
R, LAKES, SILTS/ *ALGAE, *SEWAGE
N IDENTIFICATION, SEWAGE, SEWAGE
OSPHORUS, ALGAE, COPPER SULFATE,
IENTS, LIGHT PENETRATION, SEWAGE
MS, RESERVOIRS, CULTURES, SEWAGE
RUSTACEANS, SOIL EROSION, SEWAGE
OL, *LOCAL / *WISCONSIN, *SEWAGE
IFICATION, *POLLUTANT ABATEMENT,
Y DURRENTS, MEASUREMENTS, SEWAGE
, SEDIMENTS, FISH, RADIOECOLOGY,
TROPHICATION, CYANOPHYTA, SEWAGE
GEN FIXATION, CYANOPHYTA, SEWAGE
W, CHEMICAL ANALYSIS, NUTRIENTS,
NCE ALGAE, WATER QUALITY, SEWAGE
E, AGRICULTURAL/ *LAKES, *SEWAGE
TROPHY, DIVERSION, CORES, SEWAGE
E TREATMENT, TERTIARY TREATMENT,
OPHICATION, STORM DRAINS, SEWAGE
AL PROPERTIES, WISCONSIN, SEWAGE
 RIVER(WI/ *TRUAX FIELD(MADISON)
SPHORUS COMPOUNDS, LAKES, SEWAGE
NNESOTA, *EUTROPHICATION, SEWAGE
AE, PHOSPHATES, NITRATES, SEWAGE
LATION, *BIOLOGICAL COMMUNITIES,
IOACTIVITY, CHLORELLA, SAMPLING,
NUISANCE ALGAE, NEW YORK, SEWAGE
 WEEDS, NUTRIENTS, SEPTIC TANKS,
LIOGRAPHIES, CHLORELLA, *SEWAGE,
UARIES, HUMAN POPULATION, SEWAGE
IS/ CERATOPHYLLUM, MYRIOPHYLLUM,
AS, SCIRPUS, NYMPHAEA, CRASSIPES
NT, LAKE NORRVIKEN(SWEDEN), LAKE
MICROCYSTIS AERUGINOSA, BOHEMIA,
ARUM, RADULESPORES, ZYMOLOGICAL,
ELECTROPHORESIS, MICROORGANISMS,
LANKTON, FISH, SHELLFISH, STEAM,
GAE, HYDROGEN ION CONCENTRATION,
, OUTLETS, HYDROELECTRIC PLANTS,
E-AVERAGE-LIGHT-INTENSITY, PHOTO-
LECTROLYTES, CHEMICAL REACTIONS,
NITRIFICATION, LAND APPLICATION,
IDIUM, WINKLER TECHNIQUE, OXYGEN
, REVERSE OSMOSIS, DISTILLATION,
Y TREATMENT, TERTIARY TREATMENT,
(MICH) SAN/ *ALGAL SUSPENSIONS,
 FRESH WATER, FLOW RATES, PUMPS,
ROL, *FILTERS, SEWAGE EFFLUENTS,
GANIC POLYELECTROLYTES, CATIONIC
ULATION, *WASTE WATER TREATMENT,
CIDES, TASTE, ODOR, COAGULATION,
MES, CULTURES, HYDROGEN SULFIDE,
EN, NITROGEN FIXATION, NITRATES,
CYANOPHYTA, CYTOLOGICAL STUDIES,
LA VULGARIS, CORNELL UNIVERSITY,
TOPES, DETRITUS, COLUMBIA RIVER,
*PROTOZOA, *ALGAE, LETHAL LIMIT,
E, *SANDS, *FILTRATION, FILTERS,
BREVIBACTERIUM, CORYNEBACTERIUM,
CTROLYTES, ALGAL SURFACE CHARGE,
ONIC PARTICLE COUNTING, EUDORINA
IREMENTS, *PLANT GROWTH, *ALGAE,
, SODIUM, IRON, MANGANESE, TRACE
PLANKTON, ALGAE, COLLOIDS, TRACE
TA, CHLOROPHYLL, BACTERIA, TRACE
THESIS, CHEMICAL ANALYSIS, TRACE
, *COBALT, *COPPER, *IRON, TRACE
 *ECOSYSTEMS, GREAT LAKES, TRACE
LLUTION / *RADIOACTIVITY, *TRACE
AMS, SIZE, ALGAE, DIATOMS, TRACE
ER QUALITY, *GREAT LAKES, *TRACE
OPISTHONEMA OGLIMUN, UDO/ *TRACE
GAL TOXINS, ALGAL CONTROL, TRACE

EFFLUENTS.: /, BIOLOGICAL TREATME   W70-08980
EFFLUENTS, SEWAGE EFFLUENTS, DILU   W71-00940
EFFLUENTS, DILUTION, DECOMPOSING   W71-00940
EFFLUENTS, CONSTRUCTION.: /MICAL   W71-00936
EFFLUENTS, FLOW RATES, FILTERS, A   W70-05821
EFFLUENTS, PHOSPHORUS, EUTROPHICA   W70-00270
EFFLUENTS, CYANOPHYTA.: /TRY, ECO   W69-10151
EFFLUENTS, NITROGEN, PHOSPHATES,   W69-10159
EFFLUENTS, LIGHT INTENSITY, LAKES   W69-10167
EFFLUENTS, ABSORPTION, LAKES.: /N   W70-01031
EFFLUENTS, ELECTROLYTES, ZETA POT   W70-01027
EFFLUENTS, ION EXCHANGE, RESINS,   W70-00707
EFFLUENTS, SMELTS, FISHKILL, CRUS   W69-08674
EFFLUENTS, *NUTRIENTS, *NITROGEN,   W69-09349
EFFLUENTS, OXYGEN DEMAND, AEROBIC   W72-00462
EFFLUENTS, SANITARY ENGINEERING.:   W72-01817
EFFLUENTS, *SEWAGE TREATMENT, STO   W71-10654
EFFLUENTS, STREAMS, LAKES, RIVERS   W71-09880
EFFLUENTS, *ESTUARIES, FRESH WATE   W70-02255
EFFLUENTS, SEWAGE TREATMENT, WATE   W70-02248
EFFLUENTS, MICROCYSTIS AERUGINOSA   W70-02787
EFFLUENTS, OXYGEN, SOLAR RADIATIO   W70-01933
EFFLUENTS, WISCONSIN, ALGICIDES,   W70-02982
EFFLUENTS, FARM WASTES. POPULATIO   W70-02803
EFFLUENTS, *WATER POLLUTION CONTR   W69-06909
EFFLUENTS, STREAMFLOW, DESALINATI   W69-07389
EFFLUENTS, TEMPERATURE, OXYGEN, S   W69-06203
EFFLUENTS, ECOSYSTEMS, PLANKTON,   W69-07853
EFFLUENTS, CHLOROPHYTA, RHODOPHYT   W70-04510
EFFLUENTS, MINNESOTA, WATER POLLU   W70-04506
EFFLUENTS, ALGAE, WEEDS, CARBON D   W70-04810
EFFLUENTS, DIATOMS, MISSOURI RIVE   W70-04507
EFFLUENTS, *SEWAGE DISPOSAL, ALGA   W70-05565
EFFLUENTS, DIATOMS, PHOSPHORUS, O   W70-05663
EFFLUENTS, LIMNOLOGY, WATER CHEMI   W70-05387
EFFLUENTS, STREAMFLOW, MEASUREMEN   W70-05112
EFFLUENTS, NITROGEN, PHOSPHORUS.:   W70-05113
EFFLUENTS, *MADISON LAKES, YAHARA   W70-05113
EFFLUENTS, ARSENIC COMPOUNDS, SPE   W70-05269
EFFLUENTS, TERTIARY TREATMENT, AL   W70-03512
EFFLUENTS, TERTIARY TREATMENT, WA   W70-04074
EFFLUENTS, EUTROPHICATION, PHYTOP   W70-04001
EFFLUENTS, TRIBUTARIES, ALKALINIT   W70-03983
EFFLUENTS, SEWAGE TREATMENT, INDU   W70-03974
EFFLUENTS, SWAMPS, FOREST SOILS,   W70-04193
EFFLUENTS, CARBON DIOXIDE, *ALGAE   W68-00855
EFFLUENTS, AQUATIC POPULATIONS, N   W68-00461
EICHHORNIA, POTAMOGETON, TRAPA, P   W70-10175
EICHORNIA, TYPHA, BRASENIA, DOW C   W70-05263
EKOLN(SWEDEN), LAKE LOTSJON(SWEDE   W70-04287
ELBE RIVER, CELAKOVICE(CZECHOSLOV   W70-00274
ELBE(GERMANY), SPOROBOLOMYCES ROS   W70-04368
ELECTRIC CURRENTS, *WASTE TREATME   W71-03775
ELECTRIC POWER, HEATED WATER, PRO   W70-09905
ELECTRIC POWER, SEWAGE DISPOSAL,   W71-09548
ELECTRIC POWERPLANTS, ENGINEERING   W69-05023
ELECTRIC TURBIDOSTAT, LIGHT EXTIN   W70-03923
ELECTROCHEMISTRY, RESISTIVITY, CH   W70-05267
ELECTROCHEMICAL TREATMENT, AMMONI   W70-01981
ELECTRODES, ATOMIC ABSORPTION SPE   W70-03309
ELECTRODIALYSIS, SORPTION, CHEMIC   W70-01981
ELECTRODIALYSIS, ION EXCHANGE, AC   W69-01453
ELECTROKINETIC EFFECTS, ANN ARBOR   W71-03035
ELECTROLYSIS, CULTURES.: / ALGAE,   W70-09904
ELECTROLYTES, ZETA POTENTIAL, BAC   W70-01027
ELECTROLYTES, ALGAL SURFACE CHARG   W70-05267
ELECTROLYTES, CHEMICAL REACTIONS,   W70-05267
ELECTROLYTES, POTABLE WATER.: /TI   W72-13601
ELECTRON MICROSCOPY, BIOCHEMISTRY   W70-03952
ELECTRON MICROSCOPY, AMMONIA, SPE   W70-04369
ELECTRON MICROSCOPY, WASTE TREATM   W70-03973
ELECTRONIC PARTICLE COUNTING, EUD   W69-06540
ELECTRONICS, SCALING.: / RADIOISO   W69-07862
ELECTROPHORESIS, MICROORGANISMS,   W71-03775
ELECTROPHORESIS, CHLORELLA, SCENE   W71-03035
ELECTROSTATIC PRECIPITATION, THER   W71-11823
ELECTROSTATIC REPULSION, DOW C-31   W70-05267
ELEGANS, EUGLENA GRACILIS, FOREST   W69-06540
ELEMENTS(CHEMICAL), CHLOROPHYTA,   W70-02964
ELEMENTS, CHELATION, PHOTOSYNTHES   W70-02804
ELEMENTS, DISSOLVED SOLIDS, NUTRI   W70-02510
ELEMENTS, VITAMINS, AMINO ACIDS,   W70-02255
ELEMENTS, TEMPERATURE, IRON, NUTR   W70-01933
ELEMENTS, ALGAE, SEA WATER, FRESH   W71-11687
ELEMENTS, RADIUM RADIOISOTOPES, T   W71-09013
ELEMENTS, *AQUATIC LIFE, WATER PO   W71-05531
ELEMENTS, METABOLISM.: /SSIUM, CL   W70-00708
ELEMENTS, *WATER POLLUTION EFFECT   W69-08562
ELEMENTS, *ANCHOA LAMPROTAENIA, *   W69-08525
ELEMENTS, WASHINGTON, NUCLEAR POW   W70-00264

TA, *ESSENTIAL NUTRIENTS, *TRACE
TRY, CHLORELLA, NUTRIENTS, TRACE
PLE PRESERVATION, WAT/ *NUTRIENT
SODIUM, ALGAE, PHOSPHORUS, TRACE
PEPTIDES, ORGANIC MATTER, TRACE
ALGAL CONTROL, ALGICIDES, TRACE
, PREDATION, PRODUCTIVITY, TRACE
MINE, DE-K-PRUF-21/ *LITERATURE,
YMNODINIUM SIMPLEX, FLAGELLATES,

LORELLA SACCHAROPHILA, CHLORELLA
OLI, OOCYSTIS, LEMNA, SPIRODELA,
S, ANABAENA.:   HYDRODICTYON,
PHENANTHROQUINONE, MADISON(WIS),
LETONEMA COSTATUM, SYNECHOCOCCUS
PHIDINIUM CARTERI, SYRACOSPHAERA
T, RADIOCOCCUS NIMBATUS, URONEMA
, CHLORELLA PYRENOIDOSA, DIATOMA
NOIDOSA, M/ *SHAGAWA LAKE(MINN),
S, ARTEMIA, BALANUS, DESORPTION,
AL GROWTH, CHLORELLA PYRENOIDOSA(
PHOSPHORUS RADIOISOTOPES, ALGAE,
NAIS, CRICOTOPUS, AUSTROCLOEON,
RA, HETERANTHERA, NAJAS, SILVEX,
NARIANS, MICROCYSTIS AERUGINOSA,
UTION EFFECTS, ALDRIN, DIELDRIN,
HYDROCARBON PESTICIDES, ALDRIN,
RESEARCH COUNCIL(USA), US ATOMIC
RESEARCH COUNCIL(USA), US ATOMIC
GES, CLEAR LAKE(CALIF), NUTRIENT
FISH FOOD ORGANISMS, NUTRIENTS,
BALANUS, DESORPTION, EMBRYOLOGY,
TION POTENTIAL, SOLAR RADIATION,
ALGAE, *PHOSPHORUS, *PHOSPHATES,
RONMENTAL ENGINEERING, *SANITARY
AE, *DISPERSIONS, *ENVIRONMENTAL
R POLLUTION ASSESSMENT, SANITARY
ROTOZOA, *SYSTEMATICS, *SANITARY
OGICAL PROPERTIES, *W/ *SANITARY
MAND, SEWAGE EFFLUENTS, SANITARY
MEASUREMENT, ESTUARIES, SANITARY
IC PLANTS, ELECTRIC POWERPLANTS,
IL ENGINEERING, WASTE TREATMENT,
, PACIFIC NORTHWEST U. S., CIVIL
SENTIAL NUTRIENTS, ENVIRONMENTAL
, AL/ *WATER TREATMENT, SANITARY
*AL/ *WATER TREATMENT, SANITARY
TREATMENT, *FILTRATION, SANITARY
GANO-METALLIC COMPOUNDS, AMINES,
WASHINGTON, CHEMICAL COMPOSITION(
A MARTYI, ABBOT'S POND(SOMERSET,
ALGAL ASSOCIATIONS, SHEAR WATER(
OTLAND, ENGLAND, LAKE WINDERMERE(
PYRENOIDOSA, FIXATION, TRINGFORD(
PARTICULATE MATTER, BLELHAM TARN(
KE MENDOTA(WIS), ENGLAND, LONDON(
LAND, LONDON(ENGLAND), ESTHWAITE(
, ESTHWAITE(ENGLAND), LOWESWATER(
PELLICULOSA, LAKE MENDOTA(WIS),
TH OF CLYDE(SCOTLAND), SCOTLAND,
SLOCH' RIVER(USSR), MINSK(USSR),
IDACNA CROCEA, WOODS HOLE OCEAN/
COLORADO RIVER, TEXAS, *NUTRIENT
ITY STRUCTURE, OXYTREMA, ORGANIC
HYDROGEN SULFIDE, LAKES, ALGAE,
UM VINOSUM, THIOCAPSA FLORIDANA,
TZSCHIA, NAVICULA, SYRACUSE(NY),
ARBURG RESPIROMETER, CLADOPHORA,
, SIMULIUM, PHYSA, SALMO TRUTTA,
ASTRUM, SELF-PURIFICATION, ALGAL
OGY, HYDROLOGIC ASPECTS, AQUATIC
EROMIXIS, *WAT/ *ALGAE, *AQUATIC
*PHYSIOLOGICAL ECOLOGY, *AQUATIC
, *AQUATIC ENVIRONMENTS, RIVERS,
A, PHYSIOLOGICAL ECOLOGY, LAKES,
AND, CYCLING NUTRIENTS, DIATOMS,
AL EFFECTS, ESSENTIAL NUTRIENTS,
OASSAY, CHLOROPHYTA, CYANOPHYTA,
YTA, DIATOMS, CYCLING NUTRIENTS,
YTA, DIATOMS, CYCLING NUTRIENTS,
N DEMAND, ENVIRONMENTAL EFFECTS,
CHLOROPHYTA, CYCLING NUTRIENTS,
SSAY, BIOCHEMICAL OXYGEN DEMAND,
ANOPHYTA, DIATOMS, DISINFECTION,
HLOROPHYTA, CYANOPHYTA, DIATOMS,
TIC WEEDS, BIOASSAY, CYANOPHYTA,
CYANOPHYTA, ESSENTIAL NUTRIENTS,
OASSAY, CHLOROPHYTA, CYANOPHYTA,
YTA, DIATOMS, CYCLING NUTRIENTS,
TER SPORTS, ENVIRONMENT, AQUATIC

ELEMENTS, ALGAE, EUTROPHICATION,   W69-06277
ELEMENTS, CHLAMYDOMONAS, CHRYSOPH   W70-05409
ELEMENTS, PHOSPHORUS RELEASE, SAM   W70-04382
ELEMENTS, PLANKTON, LAKES, RIVERS   W70-04385
ELEMENTS, LIGHT INTENSITY, CHLAMY   W70-04369
ELEMENTS, CORES, WISCONSIN, BOTTO   W69-03366
ELEMENTS, SODIUM, VIRUSES, VITAMI   W69-04801
ELEOCHARIS, 2,4-D, AMMATE, CHLORA   W70-05263
ELEUTHERA ISLAND, NITZCHIA SERIAT   W70-00707
ELLIOT LAKE, ONTARIO.:   W69-01165
ELLIPSOIDEA, CHLORELLA LUTEOVIRID   W70-04809
ELODEA.: /LOCOCCUS, ESCHERICHIA C   W70-10161
ELODEA, CERATOPHYLLUM, MICROCYSTI   W68-00479
ELODEA, REDOX COMP: /L EXPOSURE,   W70-02982
ELON: /HLAMYDOMONAS MOEWUSII, SKE   W70-05381
ELONGATA, THALASSIOSIRA FLUVIAT: /   W69-08278
ELONGATUM HODGETTS, CHROOCOCCUS P   W70-04468
ELONGATUM, MELOSIRA ISLANDICA, NI   W69-10158
ELY(MINN), BLOOMS, CHLORELLA PYRE   W70-03512
EMBRYOLOGY, ENERGY TRANSFORMATION   W69-07440
EMERSON STRAIN), AUTOINHIBITORS.:   W69-07442
EMULSIONS, BACKGROUND RADIATION,   W69-10163
ENALLAGMA, PSEUDAGRION, BRACHYTHE   W70-06975
ENDOTHAL, POTASSIUM PERMANGANATE,   W70-10175
ENDOTOXIN, FAST-DEATH FACTOR, DOS   W70-00273
ENDRIN, DDT.: /UTANTS, WATER POLL   W69-08565
ENDRIN, DIELDRIN, SCENEDESMUS, GA   W70-03520
ENERGY COMMISSION, NATIONAL SCIEN   W70-03975
ENERGY COMMISSION, NATIONAL SCIEN   W70-03975
ENERGY PATHWAYS, TROPHIC RELATION   W70-05428
ENERGY TRANSFER, *FOOD CHAINS, FO   W70-05428
ENERGY TRANSFORMATIONS, HETEROTRO   W69-07440
ENERGY, CARBON DIOXIDE, PHOTOSYNT   W70-05750
ENERGY, DI-URNAL DISTRIBUTION, TE   W70-06604
ENGINEERING, *SEWAGE TREATMENT, *   W70-02609
ENGINEERING, *SANITARY ENGINEERIN   W70-02609
ENGINEERING RESEARCH LABORATORY(C   W70-02775
ENGINEERING, MICROORGANISMS, NUIS   W72-01798
ENGINEERING, *WATER SUPPLY, *BIOL   W72-01813
ENGINEERING.: /CHEMICAL OXYGEN DE   W72-01817
ENGINEERING, WATER TREATMENT, WAS   W70-05094
ENGINEERING STRUCTURES, INDUSTRIA   W69-05023
ENGINEERING, ALGAE, PLANTS, CONTR   W69-03683
ENGINEERING, WASTE TREATMENT, ENG   W69-03683
ENGINEERING, H: /NTAL EFFECTS, ES   W69-03374
ENGINEERING, *PATHOGENIC BACTERIA   W72-13601
ENGINEERING, *FILTRATION, DESIGN,   W72-08859
ENGINEERING, *FILTERS, *DESIGN, C   W72-07334
ENGLAND.: /PHENOLIC COMPOUNDS, OR   W72-12987
ENGLAND), FOOD SOURCES, GREENHOUS   W69-06865
ENGLAND), SAND GRAINS, TIDAL CYCL   W71-12870
ENGLAND), AMPHORA OVALIS, OPEPHOR   W71-12870
ENGLAND), CARBON-14.: /TLAND), SC   W70-02504
ENGLAND), ANABAENA CYLINDRICA, NO   W70-02504
ENGLAND), CHLORELLA PYRENOIDOSA,   W70-02504
ENGLAND), ESTHWAITE(ENGLAND), LOW   W69-10180
ENGLAND), LOWESWATER(ENGLAND), BA   W69-10180
ENGLAND), BARNES SOUTH RES: /AND)   W69-10180
ENGLAND, LONDON(ENGLAND), ESTHWAI   W69-10180
ENGLAND, LAKE WINDERMERE(ENGLAND)   W70-02504
ENGLAND, OSCILLATORIA, AUSTRALIA,   W70-05092
ENIWETOK ATOLL, TURNOVER TIME, TR   W69-08526
ENRICHMENT TESTS, *NITROGEN, *PHO   W69-06004
ENRICHMENT, INTERACTIONS.: /OMMUN   W70-03978
ENTERIC BACTERIA, METHANE BACTERI   W70-03312
ENTEROCOCCI, PARK RIVER: /HROMATI   W70-03312
ENTEROMORPHA INTESTINALIS, CLADOP   W70-03974
ENTEROMORPHA, LEPOCINCLIS, CHLORO   W69-02959
ENTOSPHENUS LAMOTTENII, GAMMARUS,   W69-09334
ENUMERATION, SCENEDESMUS, GROWTH   W70-04381
ENVIRONMENT, WATER PROPERTIES, BI   W70-04475
ENVIRONMENTS, *EUTROPHICATION, *M   W70-03974
ENVIRONMENTS, RIVERS, ENVIRONMENT   W70-03978
ENVIRONMENTAL EFFECTS, PERIPHYTON   W70-03978
ENVIRONMENTAL EFFECTS, WATER POLL   W70-03957
ENVIRONMENTAL EFFECTS, ESSENTIAL   W70-04283
ENVIRONMENTAL ENGINEERING, H: /NT   W69-03374
ENVIRONMENTAL EFFECTS, EUTROPHICA   W69-03185
ENVIRONMENTAL EFFECTS, ESSENTIAL   W69-03373
ENVIRONMENTAL EFFECTS, ENZYMES, E   W69-03188
ENVIRONMENTAL SANITATION, ESSENTI   W69-03369
ENVIRONMENTAL EFFECTS, ESSENTIAL   W69-03374
ENVIRONMENTAL EFFECTS, ENVIRONMEN   W69-03369
ENVIRONMENTAL EFFECTS, ESSENTIAL   W69-03370
ENVIRONMENTAL EFFECTS, ESSENTIAL   W69-03371
ENVIRONMENTAL EFFECTS, EUTROPHICA   W69-03186
ENVIRONMENTAL EFFECTS, NITROGEN C   W69-03358
ENVIRONMENTAL EFFECTS, ESSENTIAL   W69-03362
ENVIRONMENTAL EFFECTS, AMMONIA, A   W69-03364
ENVIRONMENT, ESTUARINE ENVIRONMEN   W69-04276

438

WATER QUALITY, SEWAGE BACTERIA,	ENVIRONMENTAL EFFECTS, AQUATIC LI W69-04276
CE OF NATURE, CYCLING NUTRIENTS,	ENVIRONMENTAL EFFECTS, ESSENTIAL W69-03517
UTRIENTS, ENVIRONMENTAL EFFECTS,	ENVIRONMENTAL SANITATION, ESSENTI W69-03513
YTA, DIATOMS, CYCLING NUTRIENTS,	ENVIRONMENTAL EFFECTS, ESSENTIAL W69-03515
, AQUATIC ENVIRONMENT, ESTUARINE	ENVIRONMENT, WATER QUALITY, SEWAG W69-04276
TANTS, RECREATION, WATER SPORTS,	ENVIRONMENT, AQUATIC ENVIRONMENT, W69-04276
AE, BIOASSAY, CYCLING NUTRIENTS,	ENVIRONMENTAL EFFECTS, ESSENTIAL W69-03518
TY, BIOASSAY, CYCLING NUTRIENTS,	ENVIRONMENTAL EFFECTS, HARVESTING W69-04798
CE OF NATURE, CYCLING NUTRIENTS,	ENVIRONMENTAL EFFECTS, ENVIRONMEN W69-03513
MICROENVIRONMENT, AQUATIC LIFE,	ENVIRONMENT, MICROORGANISMS, WATE W69-03611
BIOASSAY, CHLOROPHYTA, DIATOMS,	ENVIRONMENTAL EFFECTS, INHIBITION W69-03516
LORINE, COBALT, COPPER, ECOLOGY,	ENVIRONMENTAL EFFECTS, GRAZING, I W69-04801
PRODUCTIVITY, CYCLING NUTRIENTS,	ENVIRONMENTAL EFFECTS, ESSENTIAL W69-03514
T GROWTH REGULATORS, CYANOPHYTA,	ENVIRONMENTAL EFFECTS, CURRENTS(W W68-00470
AL WATERS, ECOSYSTEMS, ESTUARINE	ENVIRONMENT, MARINE ALGAE, SALINI W68-00010
SIS, *CALCIUM CARBONATE, AQUATIC	ENVIRONMENT, AQUATIC PRODUCTIVITY W69-00446
OLLUTION/ *ESTUARIES, *ESTUARINE	ENVIRONMENT, *INSTRUMENTATION, *P W69-06203
ATION, LAKES, LIMNOLOGY, AQUATIC	ENVIRONMENT, AQUATIC PRODUCTIVITY W69-05142
QUATIC PRODUCTIVITY, CYANOPHYTA,	ENVIRONMENTAL EFFECTS, EUTROPHICA W69-04802
LGICIDES, AQUATIC ALGAE, AQUATIC	ENVIRONMENT, AQUATIC PLANTS, AQUA W69-05844
, MODEL STUDIES, SALMON, AQUATIC	ENVIRONMENT, PHOSPHATES, ALGAE, D W69-06536
OW, SAMPLING, ALGAE, ECOSYSTEMS,	ENVIRONMENT, RETENTION, PHOTOPERI W69-07862
LOOMS, CLOSED SYSTEMS, CYTOLOGY,	ENVIRONMENTAL CONDITIONS, EXTRACE W69-07832
ALGICIDES, HARVESTING OF ALGAE,	ENVIRONMENT, CYANOPHYTA.: /ALGAE, W69-07818
M RADIOISOTOPES, *URAN/ *AQUATIC	ENVIRONMENTS, *ECOSYSTEMS, *RADIU W69-07846
IVITY, MARINE ANIMALS, ESTUARINE	ENVIRONMENT, FALLOUT, NUCLEAR EXP W69-08267
*GLUCOSE, *ACETATE, *FRESHWATER	ENVIRONMENTS, *TURNOVER RATES, *H W70-05109
SACRAMENTO-SAN JOAQUIN ESTUARY,	ENVIRONMENTAL RELATIONSHIPS, FLUO W70-05094
RIENT REQUIREMENTS, POPULATIONS,	ENVIRONMENTAL EFFECTS, *PRIMARY P W70-05424
DS, *ANAEROBIC PONDS, DESIGN FOR	ENVIRONMENTAL CONTROL, RECIRCULAT W70-04787
ROPHY, *ORGANIC SOLUTES, *MARINE	ENVIRONMENTS, MICROBIAL ECOLOGY, W70-04811
*LABORATORY TESTS, RESPIRATION,	ENVIRONMENT, PHOTOSYNTHESIS, OXYG W70-00265
NFECTION, ENVIRONMENTAL EFFECTS,	ENVIRONMENTAL SANITATION, INHIBIT W69-10171
GICIDES, BIOASSAY, DISINFECTION,	ENVIRONMENTAL EFFECTS, ENVIRONMEN W69-10171
LIGHT INTENSITY, ECOLOGY, LOTIC	ENVIRONMENT, BIOMASS, FLOW RATES, W70-00265
S, GENETICS, BACTERIA, PEPTIDES,	ENVIRONMENTAL EFFECTS, HETEROGENE W70-00273
*SEASONAL/ *BACTERIA, *ESTUARINE	ENVIRONMENT, *WATER TEMPERATURE, W70-00713
A CLOSERIUM, ARTEMIA, C/ *MARINE	ENVIRONMENT, CERIUM-144, NITZSCHI W70-00708
, ALGAE, SILTS, ECOLOGY, AQUATIC	ENVIRONMENT.: /ANTS, AQUATIC LIFE W69-10080
N, *LAKE ONTARIO, *GROWTH RATES,	ENVIRONMENTAL EFFECTS, PHOTOSYNTH W69-10158
CHNIQUES, AQUATIC ALGAE, AQUATIC	ENVIRONMENT, AQUATIC MICROBIOLOGY W70-02968
F NATURE, BIOASSAY, CHLOROPHYTA,	ENVIRONMENTAL EFFECTS, INHIBITION W70-02968
*ALGAE, *TROPHIC LEVELS, AQUATIC	ENVIRONMENT, WATER QUALITY, NUTRI W70-02764
FLUORESCENCE, NITROGEN FIXATION,	ENVIRONMENTAL EFFECTS, NUTRIE: / W70-02775
NTS, PHOSPHORUS, KINETIC THEORY,	ENVIRONMENTAL FACTORS, BIOMASS, G W70-02779
R QUALITY, *TRO/ *ALGAE, AQUATIC	ENVIRONMENT, *AQUATIC ALGAE, WATE W70-02772
CHLOROPHYTA, GONOPHYTA, DIATOMS,	ENVIRONMENTAL EFFECTS, LIMNOLOGY, W70-02784
PHOTOS/ *RADIOISOTOPES, *AQUATIC	ENVIRONMENT, CHLORELLA, TRACERS, W70-02786
HLOROPHYTA, CYANOPHYTA, DIATOMS,	ENVIRONMENTAL EFFECTS, ESSENTIAL W70-02792
*ALGAE, *AQUATIC ALGAE, *AQUATIC	ENVIRONMENT, POLLUTANTS, BIOLOGIC W70-01233
CONCENTRATION,	ENVIRONMENTAL MONITORING.: W72-00941
NSECTS, *AQUATIC PLANTS, AQUATIC	ENVIRONMENT, ECOSYSTEMS, AQUATIC W71-09005
HIC FLORA, ESTUARIES, *ESTUARINE	ENVIRONMENT, BAYS, *FLUCTUATION, W71-02494
KTON, FRESHWATER FISH, ESTUARINE	ENVIRONMENT, IMPOUNDED WATERS, FR W71-02491
*CHEMICAL ENVIRONMENT, *PHYSICAL	ENVIRONMENT, PHAGE, FERMENTOR, FA W71-02009
, *BIOTA, *TEMPERATURE, *AQUATIC	ENVIRONMENT, *AQUATIC LIFE, *AQUA W71-02491
T, PHAGE, FERMENTOR, / *CHEMICAL	ENVIRONMENT, *PHYSICAL ENVIRONMEN W71-02009
WATER POLLUTION EFFECTS, AQUATIC	ENVIRONMENT, CHLORINATED HYDROCAR W70-08652
*REVIEWS, ALGAE, WEEDS, ECOLOGY,	ENVIRONMENT, CHEMCONTROL, BIOCONT W70-07393
ESTIC WASTES, INDUSTRIAL WASTES,	ENVIRONMENTAL EFFECTS, ANALYTICAL W72-12240
, BACTERIA, NUTRIENTS, ESTUARINE	ENVIRONMENT, PONDS, BREEDINGS, SA W73-00748
RGANIC MATTER, RIVERS, ESTUARINE	ENVIRONMENT, WATER TEMPERATURE, * W73-01446
E TESTS, LABORATORY TESTS, LOTIC	ENVIRONMENTS, LENTIC ENVIRONMENTS W72-05453
ESTS, LOTIC ENVIRONMENTS, LENTIC	ENVIRONMENTS.: /STS, LABORATORY T W72-05453
XICITY, *BIOASSAY, FISH, AQUATIC	ENVIRONMENT, ORGANIC PESTICIDES, W72-11105
ACEANS, HERBIVORES, CHLOROPHYLL,	ENZUMES, PEPTIDES, BIOCHEMISTRY, W70-01068
HETEROTROPHIC POTENTIAL, SWEDEN,	ENZYME KINETICS, LINEWEAVER-BURK W70-04194
N), ACTIVE TRANSPORT, CARBON-14,	ENZYME KINETICS, GLUCOSE, MECHAEL W70-04284
LEIC ACID, SISJOTHANDI(ICELAND),	ENZYME-SUBSTRATE COMPLEXES, RIBON W69-10160
GANISMS, AQUATIC MICROORGANISMS,	ENZYMES.: /S, ECOSYSTEMS, MICROOR W71-02479
MONAS, FRESHWATER, MARINE ALGAE,	ENZYMES.: / SCENEDESMUS, CHLAMYDO W70-05381
PRODUCTIVITY, SURFACES, ARCTIC,	ENZYMES, CHLOROPHYLL, PHOTOSYNTHE W70-04809
INO ACIDS, MECHANICAL EQUIPMENT,	ENZYMES, CHEMICAL DEGRADATION, VI W70-04184
AL AREAS, CARBOHYDRATES, YEASTS,	ENZYMES, CULTURES, HYDROGEN SULFI W70-03952
NSITY, CHLAMYDOMONAS, CHLORELLA,	ENZYMES, CARBON DIOXIDE, GROWTH R W70-04369
AND, SAMPLING, ECOLOGY, AMMONIA,	ENZYMES, CLASSIFICATION, GROWTH R W70-00713
ICAL, CARBON DIOXIDE, SYNTHESIS,	ENZYMES, CHLOROPHYTA, RHODOPHYTA, W70-02965
S, CHLOROPHYLL, IRON, CHELATION,	ENZYMES, CHLORINE, LIGHT INTENSIT W70-02964
UTRIENTS, ENVIRONMENTAL EFFECTS,	ENZYMES, ESSENTIAL N: / CYCLING N W69-03188
BACTERIA, CARBON RADIOISOTOPES,	ENZYMES, KINETICS, PHOTOSYNTHESIS W70-04287
GEN, PHOSPHORUS, OXYGEN, CARBON,	ENZYMES, SAMPLING, BOTTOM SEDIMEN W70-09889
ONIOBASIS, DECAPODA, PLECOPTERA,	EPHEMEROPTERA, TIPULIDAE, CHIRONO W69-09742
PUNCTATUS, TRICHOPTERA, ODONATA,	EPHEMEROPTERA, PLECOPTERA, DIPTER W69-07846
PLANTS, GLACIAL LAKES, PLANKTON,	EPIDEMICS.: /ILL, ROOTED AQUATIC W68-00488
INGIENSIS, AGARS, STAPHYLOCOCCUS	EPIDERMIDIS, MICROCOCCUS, STREPTO W71-11823
ACES, DEPTH, DISTRIBUTION, TIME,	EPILIMNION, THERMOCLINE, LIGHT PE W71-07698
ACTERIA, NUTRIENTS, TEMPERATURE,	EPILIMNION, NITROGEN COMPOUNDS, S W69-10167
TIVITY, PHOSPHATES, CHLOROPHYLL,	EPILIMNION, EFFLUENT, TEMPERATURE W69-10169
TOSYNTHESIS, ALGAE, HYPOLIMNION,	EPILIMNION, PHYTOPLANKTON, SESTON W69-10182
AL WATER, WAVES(WATER), DIURNAL,	EPILIMNION, WINDS, THERMOCLINE, T W70-09889

GEN, HYDROGEN ION CONCENTRATION,
, DIATOMS, TEMPERATURE, WEATHER,
C WEEDS, THERMAL STRATIFICATION,
, DISSOLVED OXYGEN, HYPOLIMNION,
FFECTS, WATER POLLUTION EFFECTS,
GNESIUM, POTASSIUM, TEMPERATURE,
ION, PHYTOPLANKTON, ODOR, WINDS,
BSORPTION, *CHEMICAL PROPERTIES,
LATISSIMA, POLYSIPHONIA HARVEYI,
SALTWATER TONGUE, BENTHIC FORMS,
LORIDE, TEMPERATURE, METABOLISM,
ARINE MICROORGANISMS, *HABITATS,
*BACTERIA, *CULTURES, *TRITIUM,
ROSCOPY, BIOCHEMISTRY, VITAMINS,
A, ANABAENA, DIAPTOMUS ASHLANDI,
, BACTERIAL PIGMENTS, DIAPTOMUS,
CRIDINE ORANGE, LINEWEAVER-BURKE
UTORADIOGRAPHY, MICHAELIS-MENTEN
OPHIC POTENTIAL, LINEWEAVER-BURK
AKE TECHNIQUES, MICHAELIS-MENTEN
ETICS, GLUCOSE, MECHAELIS-MENTEN
ENZYME KINETICS, LINEWEAVER-BURK
IQUES, GLUCOSE, MICHAELIS-MENTEN
E ERKEN(SWEDEN), LINEWEAVER-BURK
AL GROWTH KINETICS, MONOD GROWTH
IONS, SAN JOAQU/ STREETER-PHELPS
IBUTION, *CALIFORNIA, ESTIMATING
SLES, BIOINDICATORS, *ESTIMATING
H AMERICA, CENTRAL AMERICA, RATE
ONCENTRATION, MEXICO, PREDICTIVE
ENUMERATION, SCENEDESMUS, GROWTH
S, *DESIGN CRITERIA, FACILITIES,
, SORPTION REACTIONS, SOLUBILITY
ING, ALGAE TOXINS, SALINE WATER,
TS, PHOSPHORUS, INFLUENT SEWAGE,
GEN, SALINITY, ALGAE, MECHANICAL
RUCTURE, AMINO ACIDS, MECHANICAL
TER POLLUTION CONTROL, *RESEARCH
LLS, WATER TREATMENT, MECHANICAL
NALYTICAL TECHNIQUES, LABORATORY
KE, GYTTJA, DUFF, OOZE, COBBLES,
RPLE SULFUR BACTERIA, POPULATION
TROGEN FIXATION, NUTRIENTS, LAKE
NE/ *NEW SPECIES, *WESTERN LAKE
I/ *BIOLOGY, *GREAT LAKES, *LAKE
LLUTION / *EUTROPHICATION, *LAKE
POLLUTION ABATEMEN/ *OHIO, *LAKE
LLECTIONS, *ALGAE CONTROL, *LAKE
ITED STATES, HABI/ *ALGAE, *LAKE
N, HYPOLIMNION, EPILIMNION, LAKE
, EUTROPHICATION, NITRATES, LAKE
ULFIDE, CITIES, LAKE HURON, LAKE
S, *BENEFIT-COST ANALYSIS, *LAKE
, HYPOLIMNION, GREAT LAKES, LAKE
IS-MENTEN EQUATION, SWEDEN, LAKE
ELIS KINETICS, FLAGELLATES, LAKE
MICROBIAL ECOLOGY, SWEDEN, LAKE
E SURVEY, NITROGEN CONTENT, LAKE
Y, ORGANIC SOLUTES, SWEDEN, LAKE
A(SWEDEN), UPPSALA(SWEDEN), LAKE
ISTRODESMUS, OOCYSTIS, CRYOMONAS
SPHORUS, *NUTRIENT REMOVAL, SOIL
*PESTICIDES, FARM WASTES, SILT,
, CHLOROPHYTA, CRUSTACEANS, SOIL
TER, RUNOFF, FROZEN GROUND, SOIL
DRAINAGE WATER, GRASSLANDS, SOIL
                                  LAKE
, BLOOMS, MYTILUS, TRICHODESMIUM
UM NODOSUM, VIBRIO, FLAVOBACTER,
DRILUS SPP, AEOLOSOMA HEMPRICHI,
EDESMUS OBLIQUUS, DACTYLOCOCCUS,
DIATOMS, ENVIRONMENTAL EFFECTS,
MICROBIOLOGY, CYCLING NUTRIENTS,
A, CHLORELLA, CYCLING NUTRIENTS,
QUATIC MICROBIOLOGY, CYANOPHYTA,
DIATOMS, ENVIRONMENTAL EFFECTS,
IOLOGY, CHLOROPHYTA, CYANOPHYTA,
NFECTION, ENVIRONMENTAL EFFECTS,
ENVIRONMENTAL EFFECTS, ENZYMES,
UTRIENTS, ENVIRONMENTAL EFFECTS,
DIATOMS, ENVIRONMENTAL EFFECTS,
FECTS, ENVIRONMENTAL SANITATION,
ANOPHYTA, ENVIRONMENTAL EFFECTS,
UTRIENTS, ENVIRONMENTAL EFFECTS,
UTRIENTS, ENVIRONMENTAL EFFECTS,
UTRIENTS, ENVIRONMENTAL EFFECTS,
FECTS, ENVIRONMENTAL SANITATION,
UTRIENTS, ENVIRONMENTAL EFFECTS,
UTRIENTS, ENVIRONMENTAL EFFECTS,
(WIS), ENGLAND, LONDON(ENGLAND),

EPILIMNION, PHOTOSYNTHESIS, ALGAE   W70-06975
EPILIMNION, ALGAE, DENSITY, SEASO   W70-02504
EPILIMNION, HYPOLIMNION, CYCLING   W70-02251
EPILIMNION, LAKE ERIE, INSECTS, D   W70-04253
EPILIMNION, CONNECTICUT, EUGLENA,   W70-03957
EPILIMNION, LIGHT INTENSITY, SEAS   W70-05409
EPILIMNION, HYPOLIMNION, LIGHT IN   W70-05405
EPILMNION, BOTTOM SEDIMENTS.: / A   W68-00857
EPIPHYTIC BACTERIA, SUBLITTORAL,   W70-01073
EPIPHYTIC FORMS, COLLOIDAL FRACTI   W73-00748
EPIPHYTOLOGY, ON-SITE TESTS, LABO   W72-05453
EPIPHYTOLOGY, MARINE ALGAE, ECOLO   W69-10161
EPIPHYTOLOGY, MARINE ALGAE, ECOLO   W69-09755
EPIPHYTOLOGY.: /IDE, ELECTRON MIC   W70-03952
EPISCHURA NEVADENSIS, DIAPHANOSOM   W69-10182
EPISCHURA, ORTHOCYCLOPS, KERATELL   W70-05760
EQUATION.: /AND, PHAEOPIGMENTS, A   W70-04811
EQUATION, LOCH EWE(SCOTLAND), SCO   W70-04811
EQUATION, UPTAKE VELOCITY, MAXIMU   W70-05109
EQUATION, RELATIVE HETEROTROPHIC   W70-05109
EQUATION, ACETATE, GLYCOLLIC ACID   W70-04284
EQUATION, MICHAELIS KINETICS, FLA   W70-04194
EQUATION, SWEDEN, LAKE ERKEN(SWED   W70-04287
EQUATION, MAXIMUM UPTAKE VELOCITY   W70-04287
EQUATION, INCIDENT LIGHT UTILIZAT   W69-03730
EQUATION, DISSOLVED OXYGEN VARIAT   W69-07520
EQUATIONS, OXYGENATION, STANDING   W69-03611
EQUATIONS, PERIOD OF GROWTH.: / I   W68-00255
EQUATIONS, ANALYSIS OF VARIANCE,   W70-03957
EQUATIONS, ZOOGEOGRAPHY, THERMOBI   W70-03957
EQUATIONS, ANKISTRODESMUS, HETERO   W70-04381
EQUATIONS, AEROBIC CONDITIONS, AN   W70-04786
EQUILIBRIUM THEORY.: /LITHOSPHERE   W70-04385
EQUILIBRIUM, BIOLOGICAL COMMUNITI   W69-00360
EQUILIBRIUM, MATHEMATICAL MODELS,   W68-00858
EQUIPMENT.: /TS, TEMPERATURE, OXY   W69-06203
EQUIPMENT, ENZYMES, CHEMICAL DEGR   W70-04184
EQUIPMENT, *ANALYTICAL TECHNIQUES   W70-08627
EQUIPMENT, PUMPING PLANTS, MAINE.   W70-08096
EQUIPMENT, ASBESTOS CEMENT, CHLOR   W71-07339
EQUISETUM, THORIUM-232, YUKON RIV   W69-08272
EQUIVALENTS(WASTES), RED RIVER VA   W70-03312
ERIE.:   *ALGAE, *CYANOPHYTA, *NI   W70-05091
ERIE(OHIO), GONGROSIRA STAGNALIS,   W70-04468
ERIE, *LAKE MICHIGAN, *LAKE ONTAR   W70-01943
ERIE, *NUTRIENTS, ALGAE, WATER PO   W69-01445
ERIE, *WATER POLLUTION CONTROL, *   W69-06305
ERIE, BIOCHEMICAL OXYGEN DEMAND,   W71-04758
ERIE, CHLOROPHYTA, CYANOPHYTA, UN   W70-04468
ERIE, INSECTS, DIPTERA, CRUSTACEA   W70-04253
ERIE, LAKE ONTARIO, POLLUTION ABA   W70-03964
ERIE, LAKE ONTARIO, LAKE MICHIGAN   W70-00667
ERIE, PHOSPHORUS, FISH STOCKING,   W70-04465
ERIE, STATISTICS, GROWTH RATES, A   W68-00253
ERKEN(SWEDEN), LINEWEAVER-BURK EQ   W70-04287
ERKEN(SWEDEN), ACETATE, LAKE NORR   W70-04194
ERKEN(SWEDEN), ACTIVE TRANSPORT,   W70-04284
ERKEN(SWEDEN), LAKE MENDOTA(WIS),   W70-03959
ERKEN(SWEDEN), ANNUAL CYCLES, MIC   W70-05109
ERKEN(SWEDEN), EXTINCTION VALUES,   W70-05409
EROSA, COSMARIUM, CHROCOCCUS LIMN   W70-05405
EROSION, ALGAE, *WATER POLLUTION   W73-01122
EROSION, SEDIMENTATION, RUNOFF, F   W71-10376
EROSION, SEWAGE EFFLUENTS, FARM W   W70-02803
EROSION, WATER POLLUTION, *DENITR   W69-05323
EROSION.: /LUTION, SNOW, RUNOFF,   W70-04193
ERSOM(DENMARK).:   W71-00665
ERYTHREUM, SAXIDOMUS, CLOSTRIDIUM   W70-05372
ESCHERICHIA, SARCINA, STAPHYLOCOC   W70-03952
ESCHERICHIA COLI, MYRIOPHYLLUM, L   W69-07861
ESCHERICHIA COLI, OOCYSTIS, LEMNA   W70-10161
ESSENTIAL NUTRIENTS, EUTROPHICATI   W70-02792
ESSENTIAL NUTRIENTS, CALCIUM, MAG   W70-02245
ESSENTIAL NUTRIENTS, INHIBITION,   W70-02248
ESSENTIAL NUTRIENTS, NITROGEN COM   W69-10177
ESSENTIAL NUTRIENTS, EUTROPHICATI   W70-04283
ESSENTIAL NUTRIENTS, ENVIRONMENTA   W69-03358
ESSENTIAL NUTRIENTS, FREEZE DRYIN   W69-03370
ESSENTIAL N: / CYCLING NUTRIENTS,   W69-03188
ESSENTIAL NUTRIENTS, NUTRIENTS, P   W69-03373
ESSENTIAL NUTRIENTS, HARVESTING O   W69-03371
ESSENTIAL NUTRIENTS, HARVESTING O   W69-03369
ESSENTIAL NUTRIENTS, FREEZE DRYIN   W69-03362
ESSENTIAL NUTRIENTS, ENVIRONMENTA   W69-03374
ESSENTIAL NUTRIENTS, EUTROPHICATI   W69-03515
ESSENTIAL NUTRIENTS, EUTROPHICATI   W69-03514
ESSENTIAL NUTRIENTS, INHIBITION,   W69-03513
ESSENTIAL NUTRIENTS, EUTROPHICATI   W69-03518
ESSENTIAL NUTRIENTS, CHLOROPHYTA,   W69-03517
ESTHWAITE(ENGLAND), LOWESWATER(EN   W69-10180

CITY, *CYANOPHYTA, *CHEMCONTROL,
RATION, WATER POLLUTION EFFECTS,
L STUDIES, COMPUTERS, NUTRIENTS,
S, PRODUCTIVITY, PHOTOSYNTHESIS,
S, NUTRIENT BUDGET.:   *CULTURAL
, CHEMICAL PRECIPITATI/ CULTURAL
WATER QUALITY, WISCONSIN, LAKES,
 WASTE, GROUND WATER, *NITRATES,
HOSPHORUS, PONDS, PHYTOPLANKTON,
QUATIC WEEDS, BALANCE OF NATURE,
A, CHLOROPHYTA, IRON, MANGANESE,
TERFACE, PHOSPHORUS, ADSORPTION,
 WASTE EFFLUENTS.:   *CULTURAL
 LAKE TAHOE(CALIF), L/ *CULTURAL
AS, CARBON, ECOSYSTEMS, EUGLENA,
F/ *ALGAE, *BIOASSAY, NUTRIENTS,
IC IMBALANCES.:
TRIENTS, *TRACE ELEMENTS, ALGAE,
FERTILIZERS, *CYCLING NUTRIENTS,
H RATES, *ALGAE, PHOTOSYNTHESIS,
 BIOINDICATORS, LIGHT INTENSITY,
T, *GREAT LAKES, CONTROL, COSTS,
OLOGICAL COMMUNITIES, EFFLUENTS,
IS, *SAMPLING, PLANT PHYSIOLOGY,
YSIOLOGICAL ECOLOGY, CYANOPHYTA,
AL EFFECTS, ESSENTIAL NUTRIENTS,
STREAMS, FISH, LAND, ECOSYSTEMS,
TON, DIFFUSION, ALGAE, BACTERIA,
TER TREATMENT, DISSOLVED OXYGEN,
ANOPHYTA, PLANKTON, ZOOPLANKTON,
OLOGICAL COMMUNITIES, BIOASSAYS,

S, EUGLENA, AQUATIC POPULATIONS,
CIDES, *WATER POLLUTION CONTROL,
, *ODORS, *ODOR PRODUCING ALGAE,
VIRONMENT, AQUATIC PRODUCTIVITY,
              ACCELERATED
ANOPHYTA, ENVIRONMENTAL EFFECTS,
ANOPHYTA, ENVIRONMENTAL EFFECTS,
ISMS, NITROGEN, NUTRIENT/ ALGAE,
HYDRO/ *ACTIVATED SLUDGE, ALGAE,
UES, CLAY LOAM, CLAYS, COLLOIDS,
ONTROL, WATER POLLUTION EFFECTS,
AL EFFECTS, ESSENTIAL NUTRIENTS,
ON POTENTIAL, CYCLING NUTRIENTS,
ANOPHYTA, ENVIRONMENTAL EFFECTS,
AL EFFECTS, ESSENTIAL NUTRIENTS,
AL EFFECTS, ESSENTIAL NUTRIENTS,
E AGING, NUTRIENT SOU/ *CULTURAL
, TEMPERATURE, PRESSURE, MIXING,
, *TROPHIC LEVEL, WATER STORAGE,
ESTICIDE KINETICS, *PERSISTENCE,
SOLVED SOLIDS, DISSOLVED OXYGEN,
AT BUDGET, TURBIDITY, *RESERVOIR
 COLIFORMS, CYTOLOGICAL STUDIES,
BIOCHEMICAL OXYGEN DEMAND(BOD),
MICHAELIS-MENTEN EQUATION, LOCH
ES, *COPPER DOSAGE/ *MICROSCOPIC
LIDE ACCUMULATION, *RADIONUCLIDE
HIPS, SUBMARINES, EFFLUENTS, ION
WATER POLLUTION EFFECTS, CATION
COMPOUNDS, CHEMICAL WASTES, ION
ENT, ALGAL CONTROL, AMMONIA, ION
TREATMENT, ELECTRODIALYSIS, ION
, HYPOLIMNION, THERMOCLINE, *ION
ONS, EXTRACELLULAR PRODUCTS, GAS
ON, *GROUNDWATER, RECHARGE, *ION
ITION, MICROFOSSILS, COMPETITIVE
IDS, TA/ *ANTIBIOSIS, AUTOLYSIS,
*GRANULAR LEUCOCYTES, *MUSCULAR
N CLUB HUTS, TEGERNSEE, GERMANY,
K(ORE), OREGON STATE UNIVERSITY,
, CAYUGA LAKE(NY), / *ENRICHMENT
CROPHYTES, POTAMOGETON FOLIOSUS,
E, BLUE-GREEN ALGAE, TEMPERATURE
*JAPAN, *SLAUGHTERHOUSE WASTES,
, URANIUM RADIOISOTOPES, NUCLEAR
RESINS, LEAKAGE, CESIUM, NUCLEAR
BOTTOM SEDIMENTS, SOILS, NUCLEAR
NE ENVIRONMENT, FALLOUT, NUCLEAR
VERSITY, AUTECOLOGY, MORPHOLOGY,
VATIONS, CHRONIC EXPOSURE, ACUTE
SIOLOGICAL OBSERVATIONS, CHRONIC
MONIUM COMPOUNDS, MINIMUM LETHAL
HOTO-ELECTRIC TURBIDOSTAT, LIGHT
ALA(SWEDEN), LAKE ERKEN(SWEDEN),
*ISRAEL, *ALGAL CONCENTRATION, '
OLOGY, ENVIRONMENTAL CONDITIONS,
IC RATE, PHYTOPLANKTON DYNAMICS,
AUXINS, METABO/ AUTOINHIBITORS,

EUTROPHICATION, BASS, WATER POLLU   W70-02982
EUTROPHICATION.: /DS, LIGHT PENET   W70-02255
EUTROPHICATION.: /ILTRATION, MODE   W70-08627
EUTROPHICATION, TOXICITY, MIDGES,   W70-06225
EUTROPHICATION, TROPHIC INDICATOR   W71-03048
EUTROPHICATION, SPECIES DIVERSITY   W69-08518
EUTROPHICATION.: /UTION EFFECTS,   W69-05868
EUTROPHICATION, *WATER POLLUTION   W69-05323
EUTROPHICATION, LAKES, LIMNOLOGY,   W69-05142
EUTROPHICATION, FISH FARMING, FIS   W69-04804
EUTROPHICATION, WATER POLLUTION C   W69-05704
EUTROPHICATION, ALGAE, MACROAQUAT   W69-06859
EUTROPHICATION, RECREATIONAL USE,   W69-06535
EUTROPHICATION, LAKE MONONA(WIS),   W69-06858
EUTROPHICATION, ORGANIC MATTER, P   W69-06273
EUTROPHICATION, WATER POLLUTION E   W69-06864
EUTROPHICATION PREVENTION, ECOLOG   W69-07084
EUTROPHICATION, MOLYBDENUM, NITRO   W69-06277
EUTROPHICATION, ALGAE, PLANT GROW   W69-07826
EUTROPHICATION, LIGHT, TEMPERATUR   W69-07442
EUTROPHICATION, GRAZING, ALGAE, W   W70-03978
EUTROPHICATION, NITRATES, LAKE ER   W70-03964
EUTROPHICATION, PHYTOPLANKTON, DE   W70-04001
EUTROPHICATION, DATA COLLECTION,   W70-04382
EUTROPHICATION, NUISANCE ALGAE, N   W70-04381
EUTROPHICATION, LIGHT INTENSITY,   W70-04283
EUTROPHICATION, OLIGOTROPHY, SEWA   W70-04193
EUTROPHICATION, GLUCOSE, CARBON R   W70-04194
EUTROPHICATION, FISH, LAND, VALUE   W70-05264
EUTROPHICATION, STORM DRAINS, SEW   W70-05112
EUTROPHICATION.: /FOOD CHAINS, BI   W70-05272
EUTROPHICATION TREATMENT.:   W70-04721
EUTROPHICATION, CYANOPHYTA, SEWAG   W70-04510
EUTROPHICATION, ALGAE, FILTRATION   W70-05470
EUTROPHICATION, ALCOHOLS, ACTINOM   W69-00387
EUTROPHICATION, *LAKES, LIMNOLOGY   W69-00446
EUTROPHICATION.:   W68-01244
EUTROPHICATION, LIMNOLOGY, NITROG   W69-03186
EUTROPHICATION, LIMNOLOGY, NITROG   W69-03185
EUTROPHICATION, LAKES, MICROORGAN   W68-00481
EUTROPHICATION, LIME, *NITROGEN,   W68-00012
EUTROPHICATION, IMPOUNDMENTS, LAK   W69-04800
EUTROPHICATION, FISHKILL, ALGAE,   W69-03948
EUTROPHICATION, INHIBITION, LIGHT   W69-03514
EUTROPHICATION.: /IDATION-REDUCTI   W69-04521
EUTROPHICATION, LIMNOLOGY, NITROG   W69-04802
EUTROPHICATION, FREEZE DRYING, FR   W69-03518
EUTROPHICATION, INHIBITION, LIGHT   W69-03515
EUTROPHY, *NUTRIENT BALANCE, *LAK   W70-00264
EVAPORATION, SUPERSATURATION, RES   W70-00683
EVAPORATION, WATER CIRCULATION, S   W69-09723
EVAPORATION, BIODEGRADATION, ADSO   W70-09768
EVAPORATION CONTROL, WATER BUDGET   W72-01773
EVAPORATION, SALINITY, ALGAE, STR   W72-01773
EVAPORATION, MICROORGANISMS, CALI   W71-05267
EVAPORATION, FERMENTATION, METHAN   W70-04787
EWE(SCOTLAND), SCOTLAND, PHAEOPIG   W70-04811
EXAMINATION, *CRITICAL TEMPERATUR   W69-10157
EXCHANGE, *OPEN SEA, WEAPONS TEST   W70-00707
EXCHANGE, RESINS, LEAKAGE, CESIUM   W70-00707
EXCHANGE, CATION ADSORPTION, HYDR   W71-09863
EXCHANGE, BIOCHEMICAL OXYGEN DEMA   W70-10433
EXCHANGE, REVERSE OSMOSIS, DISTIL   W70-01981
EXCHANGE, ACTIVATED SLUDGE, DENIT   W69-01453
EXCHANGE, WATER CIRCULATION, NUTR   W68-00857
EXCHANGE, GREEN PLANTS, MASS CULT   W69-07832
EXCHANGE, *ORGANIC MATTER, *FILTR   W72-04318
EXCLUSION PRINCIPLE, OSCILLATORIA   W70-05663
EXCRETION, DEPURATION, PHYCOCOLLO   W70-01068
EXERTION, *2-METHYL-5-VINYLPYRIDI   W70-01788
EXPANDING TECHNOLOGY, OVERPOPULAT   W69-07818
EXPERIMENTAL STREAMS, SEASONAL EF   W70-03978
EXPERIMENTS, *NUTRIENT LIMITATION   W70-01579
EXPERIMENTAL LIMNOLOGY, BRACHIONU   W70-02249
EXPERIMENTS, RECYCLED WATER.: /GA   W70-07257
EXPERIMENTAL STUDIES, LAGOON DESI   W70-07508
EXPLOSIONS, STRONTIUM RADIOISOTOP   W70-00708
EXPLOSIONS, ABSORPTION, ADSORPTIO   W70-00707
EXPLOSIONS, COBALT RADIOISOTOPES,   W69-08269
EXPLOSIONS, BIOTA, SPECTROMETERS,   W69-08267
EXPORT LOSSES, CHEMOSYNTHESIS, AB   W70-02780
EXPOSURE.: /, PHYSIOLOGICAL OBSER   W70-01779
EXPOSURE, ACUTE EXPOSURE.: /, PHY   W70-01779
EXPOSURE, PHENANTHROQUINONE, MADI   W70-02982
EXTINCTION COEFFICIENT, ALGAL CEL   W70-03923
EXTINCTION VALUES, EUDORINA, CULT   W70-05409
EXTRA-DEEP PONDS', ASHKELON(ISRAE   W70-04790
EXTRACELLULAR PRODUCTS, GAS EXCHA   W69-07832
EXTRACELLULAR PRODUCTS, ALGAL CEL   W69-05697
EXTRACELLULAR PRODUCTS, HORMONES,   W69-04801

S, FLAGELLATES, PHARMACEUTICALS,

S), *AQUATIC WEED CUTTING, *WEED
L PHENOL POLYETHOXYLATE, ORGANIC
ANOPHOSPHORUS COMPOUNDS, SOLVENT
OGRAPHY, PHYSIOLOGY, BIOCHEMICAL
.: *BIOLOGICAL
STS, INFRARE/ CHLOROFORM-METHANE
GINIC ACID, FUCUS VESI/ *SEAWEED
GAE, *LITTORAL, *ORGANIC MATTER,
ACROPORA HYACINTHUS, POCILLOPORA
MEASUREMENT APPARATUS, HYDROGEN
OCCOLITHOPHORES, SEAWATER MEDIUM
LARIS, CHLOROPHYLL FLUORESCENCE,
LIFORNIA WATER PLAN, DELTA WATER
ATION, *PONDS, *DESIGN CRITERIA,
MS, FISH GENETICS, FISH HANDLING
WASTE WATER TREATMENT, TREATMENT
ENTS, QUALITY CONTROL, TREATMENT
ENTS, BACTERIA, ALGAE, TREATMENT
ENTS, BACTERIA, ALGAE, TREATMENT
WASTE WATER TREATMENT, TREATMENT
, *FILTERS, *NEW YORK, TREATMENT
LTRAVIOLET RADIATION, *TREATMENT
RBON, CHLORINE, COSTS, TREATMENT
OMACEOUS EARTH, ALGAE, TREATMENT
N, MEMBRANES, SCREENS, TREATMENT
ERUGINOSA, ENDOTOXIN, FAST-DEATH
GRAPHS, HALF-LIFE, CONCENTRATION
ASTERIONELLA GRACILLIMA, GROWTH
DUCTI/ *PHYTOPLANKTON, *LIMITING
ENVIRONMENTAL EFFECTS, *LIMITING
S, KINETIC THEORY, ENVIRONMENTAL
ES, ALGAE, OLIGOTROPHY, LIMITING
LKALIGENES), RHODOTORULA, GROWTH
PHYSIOLOGICAL ECOLOGY, LIMITING
ACID, FUCUS VESICULOSUS, GROWTH
ISHKILL, FERTILIZATION, LIMITING
TARD GROWTH TEST, ORGANIC GROWTH
WAUBESA, LYNGB/ *BLOOMS, *BIOTIC
, ORGANIC DEBRIS, ORGANIC GROWTH
STUDIES, LAGOON DESIGN, SCALE-UP
PAPERMILL WASTE EFFLUENT, CHEESE
L ENVIRONMENT, PHAGE, FERMENTOR,
OVAL RATES, DESIGN APPLICATIONS,
: *SULFATE ION, *DESULFOVIBRIO,
INATION, TEMPERATURE, NUTRIENTS,
S, MULTI-CELL SERIAL SYSTEM, PH,
DUS SUBTERRANEUS, ANKISTRODESMUS
ANIMALS, ESTUARINE ENVIRONMENT,
BON RADIOISOTOPES, BIOTA, CLAMS,
STRONTIUM RADIOISOTOPES, CESIUM,
SCA, MICROCOLEUS LACUSTRIS (RAB)
ATION, *WATER POLLUTION CONTROL,
PHYSIOLOGICAL ECOLOGY, KINETICS,
SOIL EROSION, SEWAGE EFFLUENTS,
VERS, NITROGEN, BIOMASS, SEWAGE,
IES, SPRAYING, ACTIVATED SLUDGE,
MESTIC WASTE, INDUSTRIAL WASTES,
, WATER POLLUTION SOURCES, ODOR,
, SLUDGE, PHOTOSYNTHETIC OXYGEN,
ONTROL, *NUTRIENTS, *PESTICIDES,
MATE DISPOSAL, WATER REUSE, FISH
OF NATURE, EUTROPHICATION, FISH
ALKALINE WATER, ALKALINITY, FISH
, PHOSPHORUS, DRAINAGE, MARSHES,
CROCYSTIS AERUGINOSA, ENDOTOXIN,
PHYCEAE, CHLAMYDOMONAS MOEWUSSI,
N OF ALGAE, CONTINUOUS CULTURES,
CONSIN, BOTTOM SEDIMENT, BENTHIC
ITIES, *WASTE DISCHARGES, FLORA,
S, LAKES, ECOLOGY, PRI/ *BENTHIC
BIOLOGICAL CO/ *LAKES, *BENTHIC
C LIFE, AQUATIC ANIMALS, BENTHIC
TROL, *NUTRIENTS, ALGAE, BENTHIC
RICE, SILTS, PLANKTON, MAYFLIES,
S,/ ALGAE, BRACKISH, ECOSYSTEMS,
VEGETATION FAUNA, SANDY BOTTOMS
OPLANKTON, DIATOMS, ZO/ *BENTHIC
LANKTON, FISH, BACTERIA, BENTHIC
RANSPARENCY, MARGINAL VEGETATION
UISANCE ALGAE, PLANKTON, BENTHIC
OXIDATION PONDS, ALGAE, BENTHIC
ARRAGANSETT BAY, CAPE REYKJANES,
TERIA, IRRIGATION, WILLOW TREES,
PLANTS, AQUATIC WEEDS, ECONOMIC
, FLOCCULATION, MIXING, ECONOMIC
ECONOMIC FEASIBILITY, TECHNICAL
ON TREATMENT, FINANCING, GRANTS,
LA PYRENOIDOSA, SYNTHETIC SEWAGE

EXTRACELLULAR PRODUCTS, CHEMICAL   W70-01068
EXTRACTABLE PHOSPHORUS.:   W69-03075
EXTRACTING STEEL CABLES, ALGAL SU   W70-04494
EXTRACTIONS, INFRARED SPECTROSCOP   W70-03928
EXTRACTIONS, PHOSPHORUS COMPOUNDS   W70-05269
EXTRACTION, MILLIPORE FILTERS, PH   W69-10163
EXTRACTION, GROWTH, SHOCK LOADING   W70-06604
EXTRACTS, LIPID CARBON, CHLOROPLA   W70-05651
EXTRACTS, ASCOPHYLLUM NODOSUM, AL   W69-07826
EXUDATION, TIDES, PHOTOSYNTHESIS,   W70-01073
EYDOUXI, PORITES LOBATA, LEPTASTR   W69-08526
F LINE.: /OFER LINES, ATTENUATION   W70-05377
F.: /NIQUES, MICROBIAL ECOLOGY, C   W70-05272
FABRY-PEROT FILTERS, SOLAR FRAUNH   W70-05377
FACILITIES(CALIF), OXYGEN DYNAMIC   W70-05094
FACILITIES, EQUATIONS, AEROBIC CO   W70-04786
FACILITIES, FISH MANAGEMENT, LIMN   W69-04804
FACILITIES, AERATION, LAGOONS, DO   W70-04060
FACILITIES, FILTRATION, SEWAGE, C   W70-00263
FACILITIES, TEST PROCEDURES, META   W70-01519
FACILITIES, TEST PROCEDURES, META   W71-06737
FACILITIES, MISSISSIPPI.: /UES, *   W71-13313
FACILITIES.: /OLLOIDS, FLOW RATES   W72-04145
FACILITIES, FILTERS.: /*SLIME, *U   W72-03824
FACILITIES, CLEANING, *ALGAL CONT   W72-08859
FACILITIES.: /, *FILTRATION, DIAT   W72-04318
FACILITIES, *WISCONSIN.: /FICATIO   W73-02426
FACTOR, DOSAGE, SPECIES SUSCEPTIB   W70-00273
FACTOR, LIMNODRILUS SPP, AEOLOSOM   W69-07861
FACTORS.: /ROTONENSIS, NITZSCHIA.   W70-02804
FACTORS, *NUTRIENTS, *PRIMARY PRO   W70-01579
FACTORS, BENTHIC FLORA, BENTHOS,   W70-02770
FACTORS, BIOMASS, GROWTH RATES, A   W70-02779
FACTORS, CHLORELLA, ANALYTICAL TE   W70-02779
FACTORS, COLONIAL MORPHOLOGY, CEL   W70-03952
FACTORS, ECOSYSTEMS, MICROORGANIS   W71-02479
FACTORS, LAMINARIA SACCHARINA, MA   W69-07826
FACTORS, NITROGEN COMPOUNDS, NUTR   W69-06535
FACTORS, POLYSACCHARIDES.: /, MUS   W69-07826
FACTORS, SPECIES DIVERSITY, LAKE   W69-01977
FACTORS, VITAMIN B1, VITAMIN B6,   W70-02510
FACTORS, WASTE COMPOSITION.: /AL   W70-07508
FACTORY WASTE WATERS.: /GANISMS,   W71-00940
FACULTATIVE, MICROAEROPHILIC, INO   W71-02009
FACULTATIVE PONDS, INFRARED SPECT   W70-07034
FACULTATIVE, BOD, DETENTION TIME.   W70-01971
FACULTATIVE POND DESIGNS, BOD, SL   W70-04786
FACULTATIVE POND, ANAEROBIC POND.   W70-04790
FALCATUS, CHLAMYDOMONAS MOEWUSII,   W70-05381
FALLOUT, NUCLEAR EXPLOSIONS, BIOT   W69-08267
FALLOUT, PACIFIC OCE: /ITIUM, CAR   W69-08269
FALLOUT, SILTS, DETRITUS, BOTTOM   W69-09742
FARLOW, CLADOPHORA, TAXONOMIC DES   W70-04468
FARM WASTES, MUNICIPAL WASTES, SE   W70-04266
FARM WASTES, BENTHOS, NITRATES, F   W70-02775
FARM WASTES. POPULATION, INDUSTRI   W70-02803
FARM WASTES, CARBON CYCLE, PHOSPH   W70-02777
FARM WASTES.: /AE, *CATTLE, SLURR   W70-07491
FARM WASTES.: /ANKTON, SEWAGE, DO   W72-01813
FARM WASTE, SLUDGE, SOLID WASTES,   W71-05742
FARM WASTES, WASTE WATER TREATMEN   W71-08221
FARM WASTES, SILT, EROSION, SEDIM   W71-10376
FARMING.: / SULFITE LIQUORS, ULTI   W69-05891
FARMING, FISH ROOD ORGANISMS, FIS   W69-04804
FARMING, FISH FOOD ORGANISMS, FIS   W71-07731
FARMS, CYANOPHYTA, PLANKTON, ZOOP   W70-05112
FAST-DEATH FACTOR, DOSAGE, SPECIE   W70-00273
FAT ACCUMULATION.: /A PALEA, MYXO   W70-02965
FATTY ACIDS, OOCYSTIS POLYMORPHA,   W70-07957
FAUNA.: /ACE ELEMENTS, CORES, WIS   W69-03366
FAUNA, *BIOLOGICAL COMMUNITIES, I   W68-00010
FAUNA, *EUTROPHICATION, *NUTRIENT   W71-00665
FAUNA, *TROPHIC LEVELS, *SURVEYS,   W71-12077
FAUNA, AQUATIC PLANTS, MARINE ALG   W71-02494
FAUNA, CHEMICAL ANALYSIS, SESSILE   W72-06052
FAUNA, FISH, OLIGOCHAETES, MIDGES   W70-01943
FAUNA, FLORA, LABORATORY, MINERAL   W72-02203
FAUNA, MARGINAL VEGETATION, MICRO   W70-06975
FAUNA, NITROGEN, PHOSPHORUS, PHYT   W70-04253
FAUNA, PHOTOSYNTHESIS, BIOASSAYS,   W70-02251
FAUNA, SANDY BOTTOMS FAUNA, MARGI   W70-06975
FAUNA, SLUDGE WORMS, BLOODWORMS,   W68-01244
FAUNA, WATERSHEDS, WATER SHORTAGE   W72-00027
FAXAFLOI FJORD, ICELAND, DNA BASE   W69-10161
FE: /S, CRAYFISH, MOSQUITOES, BAC   W70-05263
FEASIBILITY, LAKES, NUISANCE ALGA   W69-06276
FEASIBILITY, TECHNICAL FEASIBILIT   W71-10654
FEASIBILITY, COST ANALYSIS, SLUDG   W71-10654
FEDERAL GOVERNMENT, STATE GOVERNM   W69-06305
FEED, SYMBIOTIC RELATIONSHIPS.: /   W70-05655

HALF-LIFE(RADIONUCLIDE), FILTER-

ION LAGOON, NITROGEN, NUTRIENTS, IMOMYCETES, CYTOLOGICAL STUDIES, RADIA/ *WATER TREATMENT, *ALGAE, , *WASTE WATER TREATMENT, ALGAE, R, *ALGICIDES, ACTIVATED CARBON, WATER QUALITY, *WATER TREATMENT, CYCLINE, STREPTOMYCIN, MILLIPORE Y CONTROL, TREATMENT FACILITIES, *ALGAE HARVESTING, *DIATOMITE *ALGICIDES, ALGAE, *EVALUATION, MICHIGAN, WATER QUALITY CONTROL, RCES, WATER POLLUTION TREATMENT, FRAGILARIA, RADIOCARBON METHOD, PROMOSUS, TRICHOPTERA, LEECHES, , LAKE TEGERN(GERMANY), GERMANY, MUS QUADRICAUDA, CEL/ STRIPPING, ANABAENA CYLINDRICA, NORTH SEA, ASTES, *WATER POLLUTION CONTROL, TON, ALKALINE WATER, ALKALINITY, ULTIMATE DISPOSAL, WATER REUSE, LANCE OF NATURE, EUTROPHICATION, *WATER QUALITY CONTROL, *MIDGES, AQUATIC INSECTS, *AQUATIC DRIFT, WATER, ALKALINITY, FISH FARMING, SH FARMING, FISH ROOD ORGANISMS, H ROOD ORGANISMS, FISH GENETICS, , DREDGING, HARVESTING OF ALGAE, IENT LOAD, SYNTHETIC DETERGENTS, ETICS, FISH HANDLING FACILITIES, NT, *WATER QUALITY CONTROL, *MI/ ATIC DRIFT, FISH FOOD ORGANISMS, SH FARMING, FISH FOOD ORGANISMS, STANDING CROP, ANNUAL TURNOVER, OPHYTA/ *EUTROPHICATION, SEWAGE, TROPHICATION, *NUTRIENTS, ALGAE, LGAE, PROTOZOA, AQUATIC INSECTS, E, EUTROPHICATION, FISH FARMING, NALYSIS, *LAKE ERIE, PHOSPHORUS, FOOD ORGANISMS, FISH PHYSIOLOGY, ECTS, DINOFLAGELLATES, FISHKILL, HETIC DETERGENTS, FISH MANAGERS, GROWTH RATES, ALGAE, FRESH WATER COMMERCIAL FISH, SALMON, ALGAE, A, IRON, ALGICIDES, ZOOPLANKTON, ATTER, HARDNESS, CARBON DIOXIDE, *EUTROPHICATION, FERTILIZATION, LEECHES, FINGERNAIL CLAMS, WHITE TERS, BIOCHEMICAL OXYGEN DEMAND, ODEGRADATION, WATER UTILIZATION, TRACERS, CRABS, SHRIMP, MULLETS, ROL, *TOXINS, ALGAE, RESISTANCE, DIURNAL, ALGICIDES, CLOSTRIDIUM, ATE, PONDWEEDS, SODIUM ARSENITE, IFICATION, *TOXICITY, *BIOASSAY, ANDING CROP, ALGAE, ZOOPLANKTON, , DEPTH, OLIGOCHAETES, TOXICITY, LAMYDOMONAS, ALGAE, ZOOPLANKTON, SYNTHESIS, CARBON RADIOISOTOPES, S, PHOSPHORUS COMPOUND, PERCHES, OXICITY, AQUATIC MAMMALS, ALGAE, *CORAL, *REEFS, *MARINE ANIMALS, BENTHOS, ZOOPLANKTON, FRESHWATER ON, AQUATIC ALGAE, BRINE SHRIMP, EWS, BIBLIOGRAPHIES, HERBICIDES, OGY, ALGAE, BACTERIA, COLIFORMS, PHYTOPLANKTON, AQUATIC INSECTS, ION, *LIMNOLOGY, *PHYTOPLANKTON, METABOLISM, ISOPODS, AMPHIPODA, ANTS, BOTTOM SEDIMENTS, ANIMALS, ABSORPTION, MARINE ALGAE, MARINE RAPHIES, CYANOPHYTA, FRESH WATER ISSOLVED OXYGEN, EUTROPHICATION, ATION, OLIGOTR/ *LAKES, STREAMS, LIMNION, ALGAE, FUNGI, BACTERIA, DA, DEPTH, ALGAE, MOSSES, CHARA, RVAE, GROWTH RATES, *FRESH WATER NS, AQUATIC ALGAE, BRINE SHRIMP, ATIONS, PLANT POPULATIONS, ROUGH GHT, TEMPERATURE, FERTILIZATION, ILTS, PLANKTON, MAYFLIES, FAUNA, NTHOS, BRINE SHRIMP, ECOSYSTEMS, NDS, *FERTILIZATION, *PHOSPHATE, *FOOD CHAINS, FOOD WEBS, FORAGE IBLIOGRAPHIES, ECOLOGY, BIOLOGY, ULTURE, ALGAE, YEASTS, BACTERIA, RADIOISOTOPES, IRON, SEDIMENTS, MARINE ANIMALS, ALGAE, MOLLUSKS, HORUS, FISH STOCKING, COMMERCIAL TIVITY, INVERTEBRATES, PLANKTON, NTHOS, CONVECTION, WAVES(WATER),

FILTRATION, MICROORGANISMS, ANALY W71-13313
FILTRATION, ULTRASONICS, DISINFEC W71-11823
FILTRATION, *SLIME, *ULTRAVIOLET W72-03824
FILTRATION, NEW HAMPSHIRE, NEW YO W72-03641
FILTRATION, *TOXICITY, *DISTRIBUT W72-03746
FILTRATION, CHLORINATION, COAGULA W71-05943
FILTRATION, MESOTROPHY.: /, TETRA W69-10154
FILTRATION, SEWAGE, COLIFORMS, EU W70-00263
FILTRATION.: W72-08688
FILTRATION.: /L RECHARGE, COPPER, W72-05143
FILTRATION, WATER PURIFICATION, M W73-02426
FINANCING, GRANTS, FEDERAL GOVERN W69-06305
FINGER LAKES(NY), RODHE'S SOLUTIO W70-01579
FINGERNAIL CLAMS, WHITE FISH, ALE W70-01943
FINLAND, SEWAGE RING TUBES.: /NY) W70-05761
FIREBAUGH CENTER(CALIF), SCENEDES W70-03334
FIRTH OF CLYDE(SCOTLAND), SCOTLAN W70-02504
FISH CONSERVATION, POLLUTION ABAT W72-01786
FISH FARMING, FISH FOOD ORGANISMS W71-07731
FISH FARMING.: / SULFITE LIQUORS, W69-05891
FISH FARMING, FISH ROOD ORGANISMS W69-04804
FISH FOOD ORGANISMS, NUTRIENTS, E W70-05428
FISH FOOD ORGANISMS, FISH PHYSIOL W69-02782
FISH FOOD ORGANISMS, FISH PHYSIOL W71-07731
FISH GENETICS, FISH HANDLING FACI W69-04804
FISH HANDLING FACILITIES, FISH MA W69-04804
FISH HARVEST, VITAMINS.: /DIMENTS W69-09340
FISH MANAGERS, FISH YIELDS, UNDER W69-08674
FISH MANAGEMENT, LIMNOLOGY, NITRO W69-04804
FISH MANAGEMENT, STREAM IMPROVEME W70-05428
FISH PHYSIOLOGY, FISH TAXONOMY, D W69-02782
FISH PHYSIOLOGY, RED TIDE, ORGANI W71-07731
FISH POPULATIONS.: /IVE CAPACITY, W68-00912
FISH POPULATIONS, PLANKTON, *CYAN W68-00461
FISH POPULATION, *WATER POLLUTION W70-04266
FISH REPRODUCTION, INDUSTRIAL WAS W73-00748
FISH ROOD ORGANISMS, FISH GENETIC W69-04804
FISH STOCKING, COMMERCIAL FISH, S W70-04465
FISH TAXONOMY, DDT, PHOSPHOTHIOAT W69-02782
FISH TOXINS, BIOASSAY, AQUATIC AL W70-08372
FISH YIELDS, UNDERWATER SHOCK WAV W69-08674
FISH.: /, LAKE ERIE, STATISTICS, W68-00253
FISH.: /HOSPHORUS, FISH STOCKING, W70-04465
FISH.: /OMASS, DIATOMS, CYANOPHYT W70-08654
FISH.: /Y, TEMPERATURE, ORGANIC M W69-10157
FISH, *LIMNOLOGY, FISHKILL, NUTRI W69-05844
FISH, ALEWIFE, GIZZARD SHAD, SEA W70-01943
FISH, ALGAE, FOAMING, SEWAGE TREA W70-08740
FISH, ALGAE, MODEL STUDIES, BIOAS W71-11881
FISH, ALGAE.: /COSYSTEMS, SINKS, W69-08274
FISH, AMPHIBIANS, INVERTEBRATES, W68-00859
FISH, AMPHIBIANS, ANABAENA, CYTOL W70-05372
FISH, ANIMALS, ALGAE, CUTTING MAN W70-05263
FISH, AQUATIC ENVIRONMENT, ORGANI W72-11105
FISH, BACTERIA, BENTHIC FAUNA, PH W70-02251
FISH, BLOODWORMS, ALGAL CONTROL.: W70-06217
FISH, CHRYSOPHYTA, SESTON, SELF-P W70-00268
FISH, CLAMS, STRONTIUM RADIOISOTO W70-02786
FISH, DIATOMS, CARP, PLANKTON, ME W70-00274
FISH, DIATOMS, FUNGI, INSECTS, BA W71-10786
FISH, ECOSYSTEMS, CYCLING NUTRIEN W69-08526
FISH, ESTUARINE ENVIRONMENT, IMPO W71-02491
FISH, FOOD CHAINS, MUSSELS.: /UTI W70-01467
FISH, FRESH WATER, ALGAE, PONDS, W70-00269
FISH, FUNGI, HEAT, WATER POLLUTIO W71-02479
FISH, GAME BIRDS, ANIMALS, FOOD C W70-10175
FISH, GREAT LAKES, LEGISLATION, N W69-06858
FISH, GROWTH RATES.: /OMS, ALGAE, W69-09742
FISH, INFLOW, OUTLETS, NUTRIENTS, W70-06225
FISH, INVERTEBRATES, CRUSTACEANS, W71-09850
FISH, LAKES, *LAKE ZURICH, *MUNIC W68-00680
FISH, LAND, VALUE, LEAVES, FERTIL W70-05264
FISH, LAND, ECOSYSTEMS, EUTROPHIC W70-04193
FISH, LIMNOLOGY, HYDROLOGIC ASPEC W70-04475
FISH, LIMNOLOGY, LAKE TROUT, CRAY W70-00711
FISH, MICHIGAN, TROUT, WINTERKILL W68-00487
FISH, MUSSELS.: /UTION, FOOD CHAI W70-01466
FISH, NITROGEN, PHOSPHORUS, DIATO W68-00172
FISH, NUISANCE ALGAE, ODOR, TASTE W70-05404
FISH, OLIGOCHAETES, MIDGES, SNAIL W70-01943
FISH, PHYTOPLANKTON, PLANKTON, RA W69-07440
FISH, PLANKTON, AQUATIC WEED CONT W70-05417
FISH, PLANKTON, ALGAE, *ALGAL CON W70-05428
FISH, PLANKTON, WATER QUALITY, AL W72-01786
FISH, PROTEINS, OIL WASTES, OILY W69-05891
FISH, RADIOECOLOGY, EFFLUENTS, EC W69-07853
FISH, RADIOACTIVE WASTES.: /CTS, W71-05531
FISH, SALMON, ALGAE, FISH.: /HOSP W70-04465
FISH, SEA WATER, GAMMA RAYS, RADI W69-08269
FISH, SEWAGE, EUGLENOPHYTA, DIATO W70-03969

LUTION EFFECTS, BACTERIA, ALGAE,
FERTILIZERS, ALGAE, ZOOPLANKTON,
CONCENTRATION, NUTRIENTS, ALGAE,
ITROGEN, PHOSPHATES, ALKALINITY,
TER QUALITY, VIRUSES, ALGICIDES,
MPERATURE, SALINITY, REAERATION,
CYCLES, MICHIGAN, INVERTEBRATES,
TMENT, *WATER POLLUTION CONTROL,
L COMMUNITIES, SEDIMENTS, BIOTA,
YTA, GREAT LAKES, BIOINDICATORS,
LABORATORY TESTS, INVERTEBRATES,
HICATION, LAKES, STREAMS, ALGAE,
ES, CHLOROPHYTA, NUISANCE ALGAE,
, TEMPERATURE, ACTIVATED SLUDGE,
N, FUNGI, CLAYS, BRACKISH WATER,
RS, FRESHWATER ALGAE, ANADROMOUS
NTS, BIBLIOGRAPHIES, ABSORPTION,
BLATNA(CZECHOSLOVAKIA), CYPRINID
DES, HYDROLYSIS, COPPER SULFATE,
OTOPE, ON-SITE DATA COLLECTIONS,
QUALITY, AESTHETICS, RECREATION,
LIGHT PENETRATION, OXYGEN, SPORT
OGY, ECONOMICS, OHIO, COMMERCIAL
LAKES, CHEMICALS, FISHING, SPORT
ION, SEQUENCE, LAKES, CHEMICALS,
ICATION, CYANOPHYTA, ODOR, SCUM,
OPHICATION, ALGAE, DUCKS, COLOR,
FERTILIZATION, FISH, *LIMNOLOGY,
LGAE, AQUATIC WEEDS, CYANOPHYTA,
AYING, SEASONAL, *ALGAL CONTROL,
MICHIGAN, TROUT, WINTERKILLING,
LLUTION EFFECTS, EUTROPHICATION,
*ALGAE, BIOASSAY, BIOINDICATORS,
LFATE, SEWAGE EFFLUENTS, SMELTS,
LUTION EFFECTS, DINOFLAGELLATES,
R TEMPERATURE, DISSOLVED OXYGEN,
SOURCES, BIOINDICATORS, NITROGEN
CHEMICAL OXYGEN DEMAND, NITROGEN
TER, *GRASSLANDS, MOU/ *NITROGEN
OPHICATION, MOLYBDENUM, NITROGEN
TROGEN, NITROGEN CYCLE, NITROGEN
GY, NITROGEN COMPOUNDS, NITROGEN
GY, NITROGEN COMPOUNDS, NITROGEN
GY, NITROGEN COMPOUNDS, NITROGEN-
GY, NITROGEN COMPOUNDS, NITROGEN
IA, AMMONIUM COMPOUNDS, NITROGEN
*ALGAE, *CYANOPHYTA, *NITROGEN
GRICULTURAL WATERSHEDS, NITROGEN
ATES, PLANKTON, OXYGEN, NITROGEN
SITY, PARTICU/ *GROWTH, *UPTAKE,
IOXIDE, INHIBITION, N/ *NITROGEN
GAE, *NITROGEN, WISCO/ *NITROGEN-
, IRON, ORGANIC MATTER, NITROGEN
XTRACELLULAR PRODUCTION, *CARBON
LIMATES, TROPHIC LEVEL, NITROGEN
ZONE, PLANT CAROTENOIDS, CARBON
REQUIREMENTS, NITROGEN, NITROGEN
ALCIUM, AMMONIUM SALTS, NITROGEN
NITRATES, FLUORESCENCE, NITROGEN
CIDS, CALCIUM, AMMONIA, NITROGEN
ENGLAND), CHLORELLA PYRENOIDOSA,
NITROGEN, PHOSPHOROUS, NITROGEN
ON DIOXIDE, INHIBITION, NITROGEN
SOD MATS, NON-LEGUMES, *NITROGEN
TT BAY, CAPE REYKJANES, FAXAFLOI
IA, *OPISTHONEMA OGLIMUN, UDOTEA
ENTERICUS, DIATOMEA, CLADOPHORA,
RK EQUATION, MICHAELIS KINETICS,
ALGAL KEY, CLUMP COUNT, DESMIDS,
DESMINS,
), PASS LAKE(COLO), MICRO-ALGAE,
NIOBIUM-95, GYMNODINIUM SIMPLEX,
URATION, PHYCOCOLLOIDS, TANNINS,
NTED FLAGELLATES, *NON-PIGMENTED
LLATES.:   *PIGMENTED
AVICULA, FRAGILARIA, CYCLOTELLA,
DUCTION TECHNIQUE, HETEROTROPHY,
GREEN ALGAE, *EUGLENA, PIGMENTED
LIATE AUTOECOLOGY.:   *CILIATES,
*CHEMOSTAT, PHYSIOLOGY,
Y, CYCLOP/ *PHENYLUREAS, DIURON,
MORPHOLOGY, CELLULAR MORPHOLOGY,
ER(NORWAY), GAULA RIVER(NORWAY),
E, THERMAL SENSITIVITY, GENERA, '
A, *ASCOPHYLLUM NODOSUM, VIBRIO,
SPECIES DETERMINATION, TAXONOMY,
CED LAKES, *FERTILIZATION, WATER
ANIMALS, BACTERIA, FUNGI, ALGAE,
COSYSTEMS, CHARA, BENTHIC FLORA,
PHATES, HYDROGENATION, AERATION,

FISH, SEWERS, RUNOFF, WASTES, WAT   W71-09880
FISH, SHELLFISH, STEAM, ELECTRIC   W70-09905
FISH, STRATIFICATION, FLOCCULATIO   W70-08834
FISH, STREAMS, WATER POLLUTION EF   W69-10159
FISH, STRE: /TREATMENT, LAKES, WA   W70-03973
FISH, TASTE, ODOR, OXYGEN SAG, PO   W73-00750
FISH, TEMPERATURE, TRACERS, SAMPL   W69-09334
FISH, TOXICITY, FRESH WATER, PULP   W70-10437
FISH, UTAH, ALGAE, INSECTS, INVER   W69-07846
FISH, WASHINGTON, LAKES, INDIANA,   W70-04503
FISH, WATER ANALYSIS, INSECTS, SN   W72-01801
FISH, WISCONSIN, SEWAGE, TROUT, N   W69-10178
FISH, ZOOPLANKTON, SNAILS, APPLIC   W70-03310
FISHERIES, CARP, CLAMS, PONDS, RU   W70-09905
FISHERIES, LIMNOLOGY.: /FIDE, IRO   W70-05761
FISHES.: /RONMENT, IMPOUNDED WATE   W71-02491
FISHES, ALGAE.: /S, BOTTOM SEDIME   W68-00858
FISHES, CLOROCOCCALES, DAPHNIA PU   W70-00274
FISHES, HERBICIDES, MIDGES, PESTI   W70-02982
FISHES, PLANKTON, ALGAE, FROGS, M   W68-00857
FISHING, ALGAE, NUTRIENTS, TURBID   W70-05645
FISHING, ON-SITE INVESTIGATIONS.:   W68-00487
FISHING, PUBLIC HEALTH, WALLEYE,   W70-01943
FISHING, RECREATION, SILT, NUTRIE   W68-00172
FISHING, SPORT FISHING, RECREATIO   W68-00172
FISHING, WISCONSIN, BIBLIOGRAPHIE   W69-08283
FISHKILL, LAKES, ODOR, TOXICITY,   W69-08279
FISHKILL, NUTRIENTS, *ALGAE, *ALG   W69-05844
FISHKILL, FERTILIZATION, LIMITING   W69-06535
FISHKILL, ROOTED AQUATIC PLANTS,   W68-00488
FISHKILL, NUISANCE ALGAE, ALGAE,   W68-00487
FISHKILL, ALGAE, OXYGEN SAG, WATE   W69-03948
FISHKILL, HAZARDS, PESTICIDE KINE   W69-08565
FISHKILL, CRUSTACEANS, BOTTOM SED   W69-08674
FISHKILL, FISH TOXINS, BIOASSAY,   W70-08372
FISHKILLS, LIMNOLOGY, ALGAL TOXIN   W70-00264
FIXATION.: /CTS, WATER POLLUTION   W69-06864
FIXATION.: /ION, RESPIRATION, BIO   W72-06022
FIXATION, IRRIGATION, ORGANIC MAT   W72-08110
FIXATION, NUISANCE ALGAE, NUTRIEN   W69-06277
FIXATION, PHOSPHATES, PHOSPHORUS,   W69-04805
FIXATION, NUTRIENTS, NU: /LIMNOLO   W69-03518
FIXATION, NUISANCE ALGAE, NUTRIEN   W69-04802
FIXATION, PHOSPHORUS COMP: /MNOLO   W69-03186
FIXATION, NUTRIENT REQUIREMENT: /   W69-03185
FIXATION, NUTRIENT REQUIREMENTS,   W69-03364
FIXATION, NUTRIENTS, LAKE ERIE.:   W70-05091
FIXATION, CYANOPHYTA, SEWAGE EFFL   W70-04506
FIXATION, NITRATES, ELECTRON MICR   W70-04369
FIXATION, CELL COUNT, OPTICAL DEN   W70-03983
FIXATION, *ALGAE, LIGHT, CARBON D   W70-04249
FIXATION, *NUTRIENTS, *LAKES, *AL   W70-03429
FIXATION, VOLUME, STREAMS, WATER   W70-00719
FIXATION, *EXCRETION, STEPHANODIS   W69-10158
FIXATION, FOREST, WINDS, LIGHT, T   W71-03048
FIXATION, PELAGIC REGION, MONOMIC   W70-01933
FIXATION, PHOTOCHEMICAL, CARBON D   W70-02965
FIXATION, BORON, AZOTOBACTER, MOL   W70-02964
FIXATION, ENVIRONMENTAL EFFECTS,   W70-02775
FIXATION, IRON COMPOUNDS, LIGHT P   W70-02255
FIXATION, TRINGFORD(ENGLAND), ANA   W70-02504
FIXING BACTERIA, *ACTIVATED SLUDG   W71-13335
FIXING BACTERIA, RADIOACTIVITY TE   W70-04249
FIXING ORGANISMS.: /NITROGEN-15,   W72-08110
FJORD, ICELAND, DNA BASE COMPOSIT   W69-10161
FLABELLUM, ZINC, NICKEL, PUERTO R   W69-08525
FLAGELLATA, CILIATES, RANUNCULUS   W69-07838
FLAGELLATES, LAKE ERKEN(SWEDEN),   W70-04194
FLAGELLATES, MESOPLANKTON, MICROP   W70-05389
FLAGELLATES.:   W68-00172
FLAGELLATES, CILIATES, COPEPODIDS   W69-10154
FLAGELLATES, ELEUTHERA ISLAND, NI   W70-00707
FLAGELLATES, PHARMACEUTICALS, EXT   W70-01068
FLAGELLATES.:   *PIGME   W72-01798
FLAGELLATES, *NON-PIGMENTED FLAGE   W72-01798
FLAGELLATES, ANKISTRODESMUS, GOMP   W70-02249
FLAGELLATES, LAKE CLASSIFICATION,   W70-02775
FLAGELLATES.: /RANT ALGAE, *BLUE   W70-01233
FLAGELLATES, SAPROBIEN SYSTEM, CI   W71-03031
FLAGELLATES.:   W70-09904
FLAGELLATES, MONURON, TRANSPARENC   W70-10161
FLAGELLATION, IODINE TEST, GRAM R   W70-03952
FLAKK(NORWAY), RANHEIM(NORWAY), T   W70-01074
FLAVOBACTERIUM-CYTOPHAGA', VIBRIO   W70-00713
FLAVOBACTER, ESCHERICHIA, SARCINA   W70-03952
FLAVOBACTERIUM, ACHROMOBACTER, GU   W70-04280
FLEAS, LARVAE, GROWTH RATES, *FRE   W68-00487
FLOATING PLANTS, ROOTED AQUATIC P   W70-05750
FLOATING PLANTS, PHYTOPLANKTON, A   W70-10175
FLOCCULATION, FILTRATION, MODEL S   W70-08627

TS, ALGAE, FISH, STRATIFICATION,   FLOCCULATION, TURBINES, ECONOMIC   W70-08834
EAD LOSS, FILTERS, *SCENEDESMUS,   FLOCCULATION, WATER POLLUTION TRE   W72-03703
D, *FILTRATION, ALGAE, COLLOIDS,   FLOCCULATION, MIXING, ECONOMIC FE   W71-10654
ON, ALGAE, FILTRATION, SORPTION,   FLOCCULATION, COAL, ACTIVATED CAR   W70-05470
RTEBRATES, BACKGROUND RADIATION,   FLOCCULATION, POLLUTANTS, SAMPLIN   W69-07846
TION, ALGAL POISONING, GASOLINE,   FLOODS, AQUIFERS, WATER LAW, NITR   W73-00758
ATER, SHORES, WATER CIRCULATION,   FLOODS, DENSITY DURRENTS, MEASURE   W69-06203
   WATER LAW, NITRATES, TURBIDITY,   FLOODWATER, REGULATION, WELL REGU   W73-00758
NG CREEK(PA), STATE COLLEGE(PA),   FLORA.:   SPRI W69-10159
*PHYSIOLOGICAL ECOLOGY, *BENTHIC   FLORA, *ALGAE, *HYDRAULIC MODELS,   W71-00668
ITIES, *LABORATORIES, / *BENTHIC   FLORA, *ALGAE, BIOLOGICAL COMMUN   W70-02780
DUCTIVITY, *MEASUREMEN/ *BENTHIC   FLORA, *CHLOROPHYLL, *PRIMARY PRO   W69-10151
   BACTERIAL MEDIUM, ARTHROBACTER,   FLORA, AMBIENT TEMPERATURE, THERM   W70-00713
ECTS, *LIMITING FACTORS, BENTHIC   FLORA, BENTHOS, MINE WATER, WATER   W70-02770
IC PLANTS, MARINE ALGAE, BENTHIC   FLORA, ESTUARIES, *ESTUARINE ENVI   W71-02494
   COMMUNITIES, *WASTE DISCHARGES,   FLORA, FAUNA, *BIOLOGICAL COMMUNI   W68-00010
LGAE, ECOSYSTEMS, CHARA, BENTHIC   FLORA, FLOATING PLANTS, PHYTOPLAN   W70-10175
AE, BRACKISH, ECOSYSTEMS, FAUNA,   FLORA, LABORATORY, MINERALS, POLL   W72-02203
MARINE ALGAE, *DIATOMS, *BENTHIC   FLORA, LIGHT INTENSITY, BIOMASS,   W70-03325
ORGANIC MATTER, BIOMASS, BENTHIC   FLORA, PHAEOPHYTA, POLAR REGIONS,   W70-01074
OTOPES, IRON, MANGANESE, BENTHIC   FLORA, STREAMS, CULTURES, COBALT   W69-07845
N), CHICKAMAUGA RESERVOIR(TENN),   FLORENCE(ALA).: / BROAD RIVER(TEN   W71-07698
AND, NITZCHIA SERIATA, NAVICULA,   FLORIDA CURRENT, YTTRIUM-90, KATO   W70-00707
E, TROPHIC LEVEL, WATER QUALITY,   FLORIDA).: /UCTIVITY, AQUATIC ALGA   W72-01990
      *ANDERSON-CUE LAKE, MELROSE(FLORIDA).:   W72-01990
   CUE LAKE(FLORIDA), MCCLOUD LAKE(FLORIDA).:   W72-01993
                        CUE LAKE(FLORIDA), MCCLOUD LAKE(FLORIDA).:   W72-01993
OFLAGELLATES, ALGAE, CALIFORNIA,   FLORIDA, PESTICIDE RESIDUES, ECOS   W70-05272
, VITAMIN B, RED TIDE, SALINITY,   FLORIDA, TEXAS, PHOSPHORUS, MIGRA   W70-05372
SAMPLING, SEASONAL, HYDROGRAPHY,   FLORIDA, THERMOCLINE, CURRENTS(WA   W70-04368
), CHROMATIUM VINOSUM, THIOCAPSA   FLORIDANA, ENTEROCOCCI, PARK RIVE   W70-03312
EPTS, *TROPHIC BIOLOGY, ANABAENA   FLOS AQUAE, APHANIZOMENON FLOS AQ   W70-03959
ABAENA FLOS AQUAE, APHANIZOMENON   FLOS AQUAE, GLOEOTRICHIA ECHINULA   W70-03959
PHOSPHORUS BUDGET, APHANIZOMINON   FLOS AQUAE.: / NUTRIENT SOURCES,   W70-04268
, THERMAL EFFECTS, APHANIZOMENON   FLOS AQUAE, METOLIUS RIVER(ORE),   W70-03978
HAGAWA LAKE(MINN), APHANIZOMENON   FLOS AQUAE, STEPHANODISCUS ASTREA   W70-05387
CUS, LYNGBYA LIMNETICA, ANABAENA   FLOS AQUAE, TABELLARIA, MELO: /TI   W70-05405
IATES, CYCLOTELLA, APHANIZOMENON   FLOS AQUAE, MELOSIRA, ASTERIONELL   W70-05112
NABAENA INAEQUALIS 381, ANABAENA   FLOS AQUAE, TOLYPOTHRIX DISTORTA,   W70-05091
SPHAERIUM KUTZINGIANUM, ANABAENA   FLOS AQUAE, ANABAENA MENDOTAE, AN   W69-08283
YAHARA RIVER(WIS), APHANIZOMENON   FLOS AQUAE, OXYGEN DEPLETION.: *   W72-01797
SA, BOHEMIA, ELB/ *APHANIZOMENON   FLOS AQUAE, *MICROCYSTIS AERUGINO   W70-00274
S, VETERINARIANS, MIC/ *ANABAENA   FLOS-AQUAE, *STRAINS, PHYCOLOGIST   W70-00273
CYSTIS AERUGINOSA, APHANIZOMENON   FLOS-AQUAE, GLOEOTRICHIA ECHINULA   W70-00719
ARDHI, PHORMIDIUM, APHANIZOMENON   FLOS-AQUAE, SECCHI DISC TRANSPARE   W69-10169
NOMIDS, CLADOCERA, APHANIZOMENON   FLOS-AQUAE, STEPHANODISCUS HANTZS   W70-00268
CYSTIS AERUGINOSA, APHANIZOMENON   FLOS-AQUAE, MINIMUM LETHAL DOSES,   W70-02982
ONEMA, PHORMIDIUM, APHANIZOMENON   FLOS-AQUAE, LYNGBYA, STORM LAKE(I   W70-02803
ENASTRUM CAPRICORNUTUM, ANABAENA   FLOS-AQUAE, MICROCYSTIS AERUGINOS   W69-06864
CYSTIS AERUGINOSA, APHANIZOMENON   FLOS-AQUAE, OSCILLATORIA, ANABAEN   W70-05091
AMATH LAKE(ORE.), *APHANIZOMENON   FLOS-AQUAE, *CARBON-14 MEASUREMEN   W72-06052
LGAE, OCULAR MICROMETER, SUCROSE-   FLOTATION TECHNIQUE, METALIMNION,   W70-04253
, ALGAE, AMMONIA, SEDIMENTATION,   FLOTATION, BIODEGRADATION, CHLORI   W71-05013
, INTERTIDAL ALGAE, PUGET SOUND,   FLOURESCEIN DYE.: /AL COMMUNITIES   W68-00010
IMENTATION RATES, MANGANESE, LOW-   FLOW AUGMENTATION, TENNESSEE RIVE   W73-00750
SEPARATION TECHNIQUES, PRESSURE,   FLOW RATES, SLUDGE, SLURRIES, AUT   W72-08688
ION ABATEMENT, DRAINAGE SYSTEMS,   FLOW RATES, CONN STREAM FLOW, ALG   W68-00479
NUISANCE ALGAE, INTAKES, SANDS,   FLOW RATES, TURBIDITY, CHLORELLA,   W70-04199
NESE, *COLOR, *ALGAE, *COLLOIDS,   FLOW RATES, *FILTERS, *NEW YORK,   W72-04145
OGY, LOTIC ENVIRONMENT, BIOMASS,   FLOW RATES, DEPTH, DISSOLVED OXYG   W70-00265
TOMS, MARINE ALGAE, FRESH WATER,   FLOW RATES, PUMPS, ELECTROLYSIS,   W70-09904
S, HUMUS, REAERATION, EFFLUENTS,   FLOW RATES, FILTERS, ALGAE.: /ANT   W70-05821
   *NUTRIENT REMOVAL, CONTINUOUS   FLOW SYSTEM.:   W72-00917
N DEMAND, AERATION, TEMPERATURE,   FLOW, ALGAE, BENTHOS, AEROBIC CON   W71-11381
SYSTEMS, FLOW RATES, CONN STREAM   FLOW, ALGAE, AQUATIC ALGAE, CURRE   W68-00479
*LAKES, *TRIBUTARIES, SAMPLING,   FLOW, ANALYSIS, FERTILIZATION, PL   W70-02969
R QUALITY, PHOSPHORUS, NITROGEN,   FLOW, CHEMICAL ANALYSIS, NUTRIENT   W70-04810
INERS, CADDISFLIES, TEMPERATURE,   FLOW, DIATOMS, MANGANESE.: /S, SH   W69-07853
TS, NUTRIENTS, ALGAE, STRATIFIED   FLOW, MIXING, FLOW, MODEL STUDIES   W71-05878
ALGAE, STRATIFIED FLOW, MIXING,   FLOW, MODEL STUDIES, MATHEMATICAL   W71-05878
MMEDIUM FILTER, BACK-WA/ *UPWARD   FLOW, PUTRESCIBILITY TEST, BODY I   W70-05821
ICAL OXYGEN DEMAND, TEMPERATURE,   FLOW, SAMPLING, WASTE WATER DISPO   W69-10159
EED, ZIRCONIUM-95-NI/ *SARGASSUM   FLUITANS, *SARGASSUM NATANS, SEAW   W69-08524
ECTROPHOTOMETRY, SUSPENDED LOAD,   FLUORESCE: /TEMPERATURE, IONS, SP   W70-03334
HEBOYGAN COUNTY, STURGEON RIVER,   FLUORESCEIN, MACROPHYTES, GROUND   W69-09334
PHY, HYDROGEN ION CONCENTRATION,   FLUORESCENCE, CARBOHYDRATES, TEMP   W70-01074
   FARM WASTES, BENTHOS, NITRATES,   FLUORESCENCE, NITROGEN FIXATION,   W70-02775
TS, EUTROPHICATION, CHLOROPHYLL,   FLUORESCENCE, LAKES, RIVERS, NITR   W70-02777
AKES, RIVERS, LIGHT PENETRATION,   FLUORESCENCE, PHOTOSYNTHESIS, BLU   W69-05697
ENVIRONMENTS, MICROBIAL ECOLOGY,   FLUORESCENCE MICROSCOPY, ACETATE,   W70-04811
OLOGICAL SS POLARIS, CHLOROPHYLL   FLUORESCENCE, FABRY-PEROT FILTERS   W70-05377
HNIQUE, PARTICULATE MATTER, BLUE   FLUORESCING MATE: /LON COLUMN TEC   W70-01074
TES, HYDROGEN ION CONCENTRATION,   FLUORIDES, ALKYL BENZENE SULFONAT   W70-00263
RY, ENVIRONMENTAL RELATIONSHIPS,   FLUOROMETRIC CHLOROPHYLL TECHNIQU   W70-05094
   *CYTOLOGICAL STUDIES, CULTURES,   FLUOROMETRY, BIOMASS, OCEANOGRAPH   W69-08278
), CHICAG/ MANURE, MADISON(WIS),   FLUSHING WATERWAYS, MILWAUKEE(WIS   W69-10178
OSPHAERA ELONGATA, THALASSIOSIRA   FLUVIAT: /HIDINIUM CARTERI, SYRAC   W69-08278
S(WISC), GROWTH STIMUL/ *SCIRPUS   FLUVIATILIS, *LAKE BUTTE DES MORT   W72-00845
ER WASTES, DYEING WASTES./ *ANTI-   FOAM AGENTS, *TEXTILE WASTES, FIB   W70-04060
OLYPHOSPHATES, NUTRIENT BALANCE,   FOAMING DETERGENTS, PUBLIC APPREH   W70-05084

447

ICAL OXYGEN DEMAND, FISH, ALGAE,
DENSIS, MACROPHYTES, POTAMOGETON
UATIC ALGAE, BRINE SHRIMP, FISH,
ES, *TOXICITY, *WATER POLLUTION,
OPHIC LEVEL, ALGAE, CADDISFLIES,
ECTS, FISH, GAME BIRDS, ANIMALS,
AE, TROPHIC LEVEL, PRODUCTIVITY,
QUALITY ACT, PATH OF POLLUTANT,
POSAL, *WATER POLLUTION EFFECTS,
LING, *RADIOISOTOPES, SEA WATER,
DGING, MONITORING, MICROBIOLOGY,
PESTICIDE RESIDUES, ECOSYSTEMS,
TER POLLUTION, *WASTE TREATMENT,
EQUATIONS, ANALYSIS OF VARIANCE,
R QUALITY CONTROL, *MIDGES, FISH
IC INSECTS, *AQUATIC DRIFT, FISH
, ALKALINITY, FISH FARMING, FISH
*PHAEODACTYLUM, *GAS EFFLUENT,
, CHEMICAL COMPOSITION(ENGLAND),
RALIA), OXIDATION DITCHES, SOUP,
                    *CHICKENS,
BUTION, CALCIUM, TROPHIC LEVELS,
, ENERGY TRANSFER, *FOOD CHAINS,
NTRATION, SULFATES, RESPIRATION,
IOACTIVE WASTE DISPOSAL, OCEANS,
ANSFER, *FOOD CHAINS, FOOD WEBS,
SFORMATION, METAZOA, RADIOLARIA,
, JOINT INDUSTRY/GOVERNMENT TASK
AL TECHNIQUES, LABORATORY TESTS,
ORINA ELEGANS, EUGLENA GRACILIS,
ULTURAL DRAINAGE, NORTH AMERICA,
ULTURAL DRAINAGE, NORTH AMERICA,
SEPTIC TANKS, EFFLUENTS, SWAMPS,
ROPHIC LEVEL, NITROGEN FIXATION,
GEOCHEMISTRY, LAKES, CONIFEROUS
ALGAE, WATER QUALITY, CONIFEROUS
Y, CONIFEROUS FORESTS, DECIDUOUS
ROCARBONS, *OLEFINS, GREEN RIVER
RIOPHYCEAE, MYXOPHYCEAE, COMPLEX
ENSE, TRITIATED GLUCOSE, GONIDIA
DEPTH, AIR, SEDIMENTS, GEOLOGIC
LOGICAL ADJUSTMENT, ASTERIONELLA
OS AQUAE, MELOSIRA, ASTERIONELLA
PHANODISCUS TENUIS, ASTERIONELLA
YPTOMONAS, CILIATA, ASTERIONELLA
N, CAYUGA LAKE(NY), ASTERIONELLA
TONGUE, BENTHIC FORMS, EPIPHYTIC
OLIDS, SALTWATER TONGUE, BENTHIC
R(OKLA), TULSA(OKLA), POLLUTION,
CIDES, HARDNESS(WATER), DIATOMS,
NY STATE SCIENCE AND TECHNOLOGY
RGY COMMISSION, NATIONAL SCIENCE
RGY COMMISSION, NATIONAL SCIENCE
TERNUT CREEK(NY).:   *CLADOPHORA
ORMS, EPIPHYTIC FORMS, COLLOIDAL
PULSE-LABELING, POLYPHOSPHATES,
NABAENA, IOWA RIVER, CYCLOTELLA,
A, SYNEDRA, MELOSIRA, DINOBRYON,
ULATA, DINOBRYON, SYNURA UVELLA,
NCHA, LECANE, BOSMINA, NAVICULA,
M PARADOXUM, BOTRYOCCUS BRAUNII,
OLS, WIND DIRECTION, TABELLARIA,
GRAMMA, DIPLONEIS, EUNOTOGRAMMA,
IA, NAVICULA, CYMBELLA, LYNGBYA,
MELOSIRA, ASTERIONELLA FORMOSA,
ASTERIONELLA, MELOSIRA, SYNEDRA,
LATORIA RUBESCENS, PISTON CORER,
, BLUE-GREEN ALGAE, GREEN ALGAE,
QUIN RIVER, POTOMAC ESTUARY, SAN
JOAQUIN VALLEY MASTER DRAIN, SAN
ENCE, FABRY-PEROT FILTERS, SOLAR
AL EFFECTS, ESSENTIAL NUTRIENTS,
AL EFFECTS, ESSENTIAL NUTRIENTS,
NTIAL NUTRIENTS, EUTROPHICATION,
, EUTROPHICATION, FREEZE DRYING,
ENTIAL NUTRIENTS, FREEZE DRYING,
ENTIAL NUTRIENTS, FREEZE DRYING,
REEZE DRYING, FREEZE-THAW TESTS,
R(TENN), NORRIS RESERVOIR(TENN),
T U.S., ECOLOGICAL DISTRIBUTION,
CTIONS, *ON-SITE INVESTIGATIONS,
NTS, SAMPLING, DEPTH, SEA WATER,
RACE ELEMENTS, ALGAE, SEA WATER,
DICATOR SPECIES, MICROORGANISMS,
PLANKTON, DIATOMS, MARINE ALGAE,
P, CLAMS, PONDS, RUNNING WATERS,
LLUTION CONTROL, FISH, TOXICITY,
, *SEWAGE EFFLUENTS, *ESTUARIES,
IBLIOGRAPHIES, HERBICIDES, FISH,
STATISTICS, GROWTH RATES, ALGAE,

FOAMING, SEWAGE TREATMENT.: /CHEM   W70-08740
FOLIOSUS, EXPERIMENTAL LIMNOLOGY,   W70-02249
FOOD CHAINS, MUSSELS.: /UTION, AQ   W70-01467
FOOD CHAINS, AQUATIC ALGAE, BRINE   W70-01466
FOOD CHAINS, STABLE ISOTOPES, RAD   W70-02786
FOOD CHAINS, OXYGEN, ROTIFERS, WE   W70-10175
FOOD CHAINS, TOXICITY, WATER POLL   W72-01788
FOOD CHAINS, SAMPLING, SURVEYS, M   W72-00941
FOOD CHAINS, BACTERIA, DEGRADATIO   W72-01801
FOOD CHAINS, PHOSPHORUS RADIOISOT   W70-00708
FOOD CHAINS.: /S, HARVESTING, DRE   W70-05264
FOOD CHAINS, BIOLOGICAL COMMUNITI   W70-05272
FOOD CHAINS, CHLORELLA, CHLOROPHY   W70-04381
FOOD CONCENTRATION, MEXICO, PREDI   W70-03957
FOOD ORGANISMS, NUTRIENTS, ENERGY   W70-05428
FOOD ORGANISMS, FISH PHYSIOLOGY,   W69-02782
FOOD ORGANISMS, FISH PHYSIOLOGY,   W71-07731
FOOD PRODUCTION, SOUTHAMPTON UNIV   W70-00161
FOOD SOURCES, GREENHOUSE STUDIES,   W69-06865
FOOD WASTES.:   SHEPPARTION(AUST   W71-09548
FOOD WASTES, POULTRY WASTES.:   W70-09320
FOOD WEBS.: /SPECTROSCOPY, DISTRI   W69-08525
FOOD WEBS, FORAGE FISH, PLANKTON,   W70-05428
FOOD: /CTERIA, HYDROGEN ION CONCE   W70-01073
FOODS, ABSORPTION, HAZARDS, CESIU   W70-02786
FORAGE FISH, PLANKTON, ALGAE, *AL   W70-05428
FORAMINIFERA, THIOBACILLUS-FERROB   W69-07428
FORCE ON EUTROPHICATION, PROVISIO   W70-02777
FOREIGN RESEARCH.: /ITY, ANALYTIC   W70-08628
FOREST CENTRIFUGE, GLENODINIUM PU   W69-06540
FOREST DRAINAGE.: /RAINAGE, AGRIC   W70-03975
FOREST DRAINAGE.: /RAINAGE, AGRIC   W70-03975
FOREST SOILS, MINNESOTA, WISCONSI   W70-04193
FOREST, WINDS, LIGHT, TEMPERATURE   W71-03048
FORESTS, ALGAE, INSECTS, PHOSPHOR   W71-09012
FORESTS, DECIDUOUS FORESTS, PHOSP   W70-03512
FORESTS, PHOSPHORUS, NITROGEN, CY   W70-03512
FORMATION, CHLOROPHYLL DERIVATIVE   W69-08284
FORMATION, GROWTH-PROMOTING PROPE   W69-10180
FORMATION, LONG ISLAND SOUND(NY),   W69-09755
FORMATIONS, ARCTIC, TUNDRA, SCALI   W69-08272
FORMOSA, CYCLOTELLA MENEGHINIANA,   W70-04809
FORMOSA, FRAGILARIA,: /ZOMENON FL   W70-05112
FORMOSA, MELOSIRA BINDERANA, GENE   W69-10158
FORMOSA, RECOLONIZATION, ALSTER L   W70-00268
FORMOSA, SYNEDRA, MELOSIRA, DINOB   W70-01579
FORMS, COLLOIDAL FRACTION, MULTIP   W73-00748
FORMS, EPIPHYTIC FORMS, COLLOIDAL   W73-00748
FOSSIL, CIMARRON RIVER(OKLA), BIX   W70-04001
FOULING, PLANKTON, SEWAGE, DOMEST   W72-01813
FOUND, SYRACUSE(NY), SYRACUSE UNI   W70-03973
FOUNDATION, OFFICE OF NAVAL RESEA   W70-03975
FOUNDATION, OFFICE OF NAVAL RESEA   W70-03975
FRACTA, CLADOPHORA GLOMERATA, BUT   W72-05453
FRACTION, MULTIPLE-USES, SAVANNAH   W73-00748
FRACTIONATION.: / METAPHOSPHATES,   W71-03033
FRAGILARIA, SYNEDRA, NAVICULA, AP   W70-08107
FRAGILARIA, RADIOCARBON METHOD, F   W70-01579
FRAGILARIA, TABELLARIA, MELOSIRA,   W70-02803
FRAGILARIA, CYCLOTELLA, FLAGELLAT   W70-02249
FRAGILARIA CROTONENSIS, NITZSCHIA   W70-02804
FRAGILARIA, ASTERIONELLA, FILTER   W70-00263
FRAGILARIA, GOMPHONEMA, GYROSIGMA   W70-03325
FRAGILARIA, OEDOGONIUM, SPIROGYRA   W70-04371
FRAGILARIA,: /ZOMENON FLOS AQUAE,   W70-05112
FRAGILARIA, CER: /STIS, BOSMINA,   W70-04503
FRAGILARIA, MELOSIRA, ASTERIONELL   W70-05663
FRAGILARIA, ASTERIONELLA, MELOSIR   W70-05375
FRANCISCO BAY.: /IATIONS, SAN JOA   W69-07520
FRANCISCO BAY-DELTA.: /RAIN, SAN   W71-09577
FRAUNHOFER LINES, ATTENUATION MEA   W70-05377
FREEZE DRYING, FREEZE-THAW TESTS,   W69-03362
FREEZE DRYING, FREEZE-THAW TESTS,   W69-03370
FREEZE DRYING, FREEZE-THAW TESTS,   W69-03518
FREEZE-THAW TESTS, FREEZING, LIGH   W69-03518
FREEZE-THAW TESTS, HARVESTING OF   W69-03362
FREEZE-THAW TESTS, INHIBITION, NU   W69-03370
FREEZING, LIGHT INTENSITY, LIGHT   W69-03518
FRENCH BROAD RIVER(TENN), CHICKAM   W71-07698
FREQUENCY DISTRIBUTION, POPULATIO   W70-02772
FREQUENCY, *FREQUENCY ANALYSIS.: /   W70-09976
FRES WATER, LITTORAL, DISSOLVED O   W71-12870
FRESH WATER.: / *COPPER, *IRON, T   W71-11687
FRESH WATER BIOLOGY, SUSPENDED SE   W71-10786
FRESH WATER, FLOW RATES, PUMPS, E   W70-09904
FRESH WATER, BRACKISH WA: /S, CAR   W70-09905
FRESH WATER, PULP WASTES, PESTICI   W70-10437
FRESH WATER, LAKES, SILTS, PHOSPH   W70-02255
FRESH WATER, ALGAE, PONDS, WATER   W70-00269
FRESH WATER FISH.: /, LAKE ERIE,   W68-00253

TANCE, *PLANT GROWTH REGULATORS,   FRESH WATER LAKES, LABORATORY TES   W68-00856
NTS, BIBLIOGRAPHIES, CYANOPHYTA,   FRESH WATER FISH, LAKES, *LAKE ZU   W68-00680
TIDAL WATERS, MIXING, SEA WATER,   FRESH WATER, SHORES, WATER CIRCUL   W69-06203
TRUCTURE, ATMOSPHERE, SEA WATER,   FRESH WATER, SORPTION, SODIUM, AL   W70-04385
TER, SCENEDESMUS, CHLAMYDOMONAS,   FRESHWATER, MARINE ALGAE, ENZYMES   W70-05381
NS, ALGAE, BENTHOS, ZOOPLANKTON,   FRESHWATER FISH, ESTUARINE ENVIRO   W71-02491
E ENVIRONMENT, IMPOUNDED WATERS,   FRESHWATER ALGAE, ANADROMOUS FISH   W71-02491
, PONDS, BREEDINGS, SALINE WATER-   FRESHWATER INTERFACES, COLLOIDS.:   W73-00748
UTROPHICATION, MUNICIPAL WASTES,   FRESHWATER ALGAE, WATER POLLUTION   W72-12240
LGAE,/ *COPPER CHLORIDE, UPTAKE,   FREUNDLICH ISOTHERM, BLUE-GREEN A   W70-08111
  *LEUCOTHRIX MUCOR, MORPHOLOGY,   FRIDAY HARBOR(WASH), ANTITHAMNION   W69-10161
YLINDRUS, CORYCAEUS, OIKOPLEURA,   FRITILLARIA, OBELIA, CLUSTER ANAL   W73-01446
, SUCKERS, BASS, WATER HYACINTH,   FROGS, CRAYFISH, MOSQUITOES, BACT   W70-05263
CTIONS, FISHES, PLANKTON, ALGAE,   FROGS, MOSSES, HYPOLIMNION, THERM   W68-00857
PHAUSIA, PHAEOCYSTIS, SARGASSUM,   FRONDS, HYDROCARBONS, BACTERIAL A   W70-01068
), SMOLA(NORWAY), HITRA(NORWAY),   FROYA(NORWAY), FUCUS VESCICULOSUS   W70-01074
RRIGATION, GROUND WATER, RUNOFF,   FROZEN GROUND, SOIL EROSION, WATE   W69-05323
RY(RHODE ISLAND), DILUTION RATE,   FRUITING BODIES, ULVA LACTUCA VAR   W70-01073
PEACHES, APRICOTS, PEARS, CITRUS   FRUITS, TOMATOES, *AEROBIC TREATM   W71-09548
EDIMENTS, CHLORELLA PYRENOIDOSA,   FUCALES, ALKANES, ALKENES, ALIPHA   W69-08284
), HITRA(NORWAY), FROYA(NORWAY),   FUCUS VESCICULOSUS, ASCOPHYLLUM N   W70-01074
LGAL SUBSTANCES, *EXTRACELLULAR,   FUCUS VESICULOSUS, CHONDRUS CRISP   W70-01073
COPHYLLUM NODOSUM, ALGINIC ACID,   FUCUS VESICULOSUS, GROWTH FACTORS   W69-07826
RACUSE(NY), SYRACUSE UNIVERSITY,   FUNDAMENTAL STUDIES, APPLIED STUD   W70-03973
PHAGA, PROTEUS, SERRATIA, BAY OF   FUNDY(CANADA), RHYZOIDS.: /, CYTO   W70-03952
OLLUTION EFFECTS, FUNGI, AQUATIC   FUNGI.: /LLUTION CONTROL, WATER P   W69-05706
ANIMALS, MARINE BACTERIA, MARINE   FUNGI.: /S, MARINE ALGAE, MARINE   W69-03752
C CONDITIONS, E. COLI, BACTERIA,   FUNGI, ACTINOMYCETES, PROTOZOA, A   W71-02009
ATTER, INSECTS, MITES, BACTERIS,   FUNGI, ALGAE, NITRITES, NITRATES,   W71-00940
A, SEWAGE BACTERIA, SYSTEMATICS,   FUNGI, ALGAE, DIATOMS, PROTOZOA,   W72-01819
HEMICAL), HYDROLYSIS, CHELATION,   FUNGI, ALGAE, PROTOZOA, LICHENS,   W69-07428
ER POLLUTION, ANIMALS, BACTERIA,   FUNGI, ALGAE, FLOATING PLANTS, RO   W70-05750
ONTROL, WATER POLLUTION EFFECTS,   FUNGI, AQUATIC FUNGI.: /LLUTION C   W69-05706
CHELATION, OCHROMONAS, BACTERIA,   FUNGI, AQUATIC PLANTS, ASSAY, DIS   W70-02510
ALLI, SCHIZOMYCETES, CYANOPHYTA,   FUNGI, BACILLARIOPHYCEAE, CHAROPH   W70-02784
EPILIMNION, *HYPOLIMNION, ALGAE,   FUNGI, BACTERIA, FISH, LIMNOLOGY,   W70-04475
WATER, *BIOMASS, *SLIME, ALGAE,   FUNGI, BACTERIA, BACTERICIDES, CH   W72-12987
DITIONS, HYDROGEN SULFIDE, IRON,   FUNGI, CLAYS, BRACKISH WATER, FIS   W70-05761
GELLATES, OCEANS, BIOINDICATORS,   FUNGI, ESTUARIES, BACTERIA, SAMPL   W70-04368
LGAE, BACTERIA, COLIFORMS, FISH,   FUNGI, HEAT, WATER POLLUTION, STR   W71-02479
C MAMMALS, ALGAE, FISH, DIATOMS,   FUNGI, INSECTS, BACTERIA, PROTOZO   W71-10786
TH RATES, MICROORGANISMS, ALGAE,   FUNGI, LABORATORY TESTS, BIOCHEMI   W72-01817
R, IRON OXIDES, BACTERIA, ALGAE,   FUNGI, SULFUR COMPOUNDS, CARBONAT   W70-02294
*BIOASSAYS, PROTOZOA, BACTERIA,   FUNGI, VIRUSES, ALGAE, LICHENS, A   W71-11823
GAE, *AQUATIC BACTERIA, *AQUATIC   FUNGI, YEASTS, HYDROGEN ION CONCE   W69-00096
URE, HYDROGEN ION CONCENTRATION,   FUNGI, YEASTS, PYRITE, ALGAE, CYA   W68-00891
TES LOBATA, LEPTASTREA PURPUREA,   FUNGIA FUNGITES, ZOOXANTHELLAE, A   W69-08526
ATA, LEPTASTREA PURPUREA, FUNGIA   FUNGITES, ZOOXANTHELLAE, ACANTHUR   W69-08526
KE NORRVIKEN, LAKE MENDOTA, LAKE   FURES, LAKE SEBASTICOOK, LAKE WAS   W70-03964
KERSEE(SWITZ), AEGERISEE(SWITZ),   FURES: / GREIFENSEE(SWITZ), PAFFI   W70-00270
HROOCOCCUS PRESCOTTII, CALOTHRIX   FUSCA, MICROCOLEUS LACUSTRIS (RAB   W70-04468
SARNIENSE, RHODOCHORTON, BANGIA   FUSCO-PURPUREA, SPHACELARIA, GONI   W69-10161
NALIS, NEPHROCYTIUM OBESUM W AND   G S WEST, RADIOCOCCUS NIMBATUS, U   W70-04468
SMS, TURSIOPS TRUN/ DUPLIN RIVER(GA), SPARTINA ALTEMIFLORA, ORGANI   W69-08274
CHODLERI, DAPHNIA PULEX, DAPHNIA   GALEATA MENDOTAE, DAPHNIA MAGNA,   W70-03957
BITION, SILICON, VANADIUM, ZINC,   GALLIUM.: /INS, METABOLITES, INHI   W69-04801
TE, FERRICHLORIDE, IRON SULFATE,   GALVANIZING WASTES, LAKE ZURICH(S   W70-05668
PLANKTON, AQUATIC INSECTS, FISH,   GAME BIRDS, ANIMALS, FOOD CHAINS,   W70-10175
KENTUCKY, LIFE HISTORY STUDIES,   GAMMA RAYS, SPECTROMETERS, CYANOP   W69-09742
, ATLANTIC OCEAN, RADIOACTIVITY,   GAMMA RAYS, POTASSIUM RADIOISOTOP   W69-08524
ATES, PLANKTON, FISH, SEA WATER,   GAMMA RAYS, RADIOCHEMICAL ANALYSI   W69-08269
, ORGANIC MATTER, RADIOISOTOPES,   GAMMA RAYS, URANIUM RADIOISOTOPES   W69-08272
TRUTTA, ENTOSPHENUS LAMOTTENII,   GAMMARUS, CORDULEGASTER.: / SALMO   W69-09334
MBAY(IND)/ *INDIA, *BLOOMS, RIVER   GANGES(INDIA), RAJGHAT(INDIA), BO   W70-05404
ERTILIZATION, INDUSTRIAL WASTES,   GARBAGE DUMPS, OIL WASTES, INLETS   W70-05112
C ACID DIHYDROCHLORIDE, INFRARED   GAS ANALYSIS, COULTER COUNTER, EU   W69-08278
CULTURES, DIAGENESIS, SEDIMENTS,   GAS CHROMATOGRAPHY, LIPIDS, OIL,   W69-08284
CETES, CYANOPHYTA, SPECTROSCOPY,   GAS CHROMATOGRAPHY, ANALYTICAL TE   W69-00387
ATION, *WATER POLLUTION EFFECTS,   GAS CHROMATOGRAPHY.: /T IDENTIFIC   W69-02782
, ENDRIN, DIELDRIN, SCENEDESMUS,   GAS CHROMATOGRAPHY, RESISTANCE, B   W70-03520
*CHLAMYDOMONAS LABORATORY TESTS,   GAS CHROMATOGRAPHY, DEGRADATION,   W70-08652
DITIONS, EXTRACELLULAR PRODUCTS,   GAS EXCHANGE, GREEN PLANTS, MASS   W69-07832
ROL, RECIRCULATION, BOD REMOVAL,   GAS PRODUCTION.: /IRONMENTAL CONT   W70-04787
LAKE(NETHERLANDS), HETEROCYSTS,   GAS-VACUOLES, LLANGORS LAKE(WALES   W70-04369
NG, *SULFATES, *SULFIDES, ALGAE,   GASES, ODOR, OXIDATION-REDUCTION   W70-07034
, URBANIZATION, ALGAL POISONING,   GASOLINE, FLOODS, AQUIFERS, WATER   W73-00758
, IOWA RIVER, / *UNIT PROCESSES,   GASTROENTERITIS, SYNURA, ANABAENA   W70-08107
BODY BURDENS, POMOXIS ANNULARIS,   GASTROINTESTINAL TRACT.: /RELLI,   W70-02786
ER POLLUTION EFFECTS, MORTALITY,   GASTROPODS, MARINE PLANTS, ON-SIT   W70-01780
USTACEA, MOLLUSCS, OLIGOCHAETES,   GASTROPODS.: /NSECTS, DIPTERA, CR   W70-04253
NSET(NORWAY), NID RIVER(NORWAY),   GAULA RIVER(NORWAY), FLAKK(NORWAY   W70-01074
UTANT BACTERIA, CYCLOTELLA NANA,   GEIGER COUNTER.: /ANOCOBALAMIN, M   W70-05652
LAGIC AREAS, HORIZONTAL, BLOOMS,   GELATINOUS MATTER.: /OTA(WIS), PE   W70-06229
EMPERATURE, THERMAL SENSITIVITY,   GENERA, 'FLAVOBACTERIUM-CYTOPHAGA   W70-00713
LLA FORMOSA, MELOSIRA BINDERANA,   GENERATION TIME, CHLORELLA PYRENO   W69-10158
LTH, PHYSICOCHEMICAL PROPERTIES,   GENETICS, BACTERIA, PEPTIDES, ENV   W70-00273
ICAL PRECIPITATION, LIFE CYCLES,   GENETICS, E. COLI, ISOLATION.: /M   W71-11823
GAL PHYLOGENY, CLADOPHORA, ALGAL   GENETICS, RUSSIA, BAVARIA, MORICH   W70-03973
RMING, FISH ROOD ORGANISMS, FISH   GENETICS, FISH HANDLING FACILITIE   W69-04804
S, INDUSTRIAL USES, CLEAN WATER,   GENUS, SPECIES, TOUCH SENSATION.:   W70-05389

449

MODELS, BACTERIA, PHYTOPLANKTON,
MODELS, BACTERIA, PHYTOPLANKTON,
L MODELS, *WATERSHED MANAGEMENT,
IC PLANTS, BACKGROUND RADIATION,
TURE, WATER MANAGEMENT(APPLIED),
OGICAL ECOLOGY, WATER CHEMISTRY,
OGICAL ECOLOGY, WATER CHEMISTRY,
, RAINFALL-RUNOFF RELATIONSHIPS,
, PACIFIC COAST REGION, REGIONS,
REGIONS, PACIFIC COAST REGIONS,
TIDES, SOUTHEAST U. S., REGIONS,
HABITATS, DEPTH, AIR, SEDIMENTS,
DYE, AIRBORNE TESTS, OIL SPILLS,
SERVOIRS, AIR TEMPERATURE, RAIN,
, PROTOPLASM, WATER-BLOOMS, LAKE
ALT MARSHES, *CYCLING NUTRIENTS,
                        BERLIN(WEST),
ONIZATION, ALSTER LAKE(GERMANY),
/ *HYPERTROPHIC WATERS, HAMBURG(
OSA, RECOLONIZATION, ALSTER LAKE(
RADULESPORES, ZYMOLOGICAL, ELBE(
S, NOCTILUCA MILIARIS, HELGOLAND(
KE SCHLIER(GERMANY), LAKE TEGERN(
ONSTANCE, WHITEFISH, WURTTEMBERG(
NY), PLON(GERMANY), LAKE SCHLIER(
FISH, WURTTEMBERG(GERMANY), PLON(
N(USSR), LYUBERETS(USSR), MUNICH(
N MOUNTAIN CLUB HUTS, TEGERNSEE,
(GERMANY), LAKE TEGERN(GERMANY),
NOLOGY,/ *EUTROPHICATION, ALGAE,
DENSIS, DIAPHANOSOMA LEUCHTENBER
NAIL CLAMS, WHITE FISH, ALEWIFE,
FISHKILL, ROOTED AQUATIC PLANTS,
C WAST/ *EUTROPHICATION, *LAKES,
ION, ROLLER DRYING, SPRAY-DRIED,
, LAKE SHATUTAKEE, MASCOMA LAKE,
I, DIDINIUM, PARAMECIUM, PHACUS,
OCYTOMETER COUNTING, ITHACA(NY),
ENA GRACILIS, FOREST CENTRIFUGE,
IUM HIRUDINELLA, ANKISTRODESMUS,
LLATORIA, SPIRULINA, VOLVOCALES,
OCAPSA, CHROOCOCCUS, CLOSTERIUM,
NABAENA HASSALI, LYNGBYA NOLLEI,
, PLANCTOMYCES, BACILLARIOPHYTA,
ABAENA, ANACYSTIS, GLOEOTRICHIA,
EPOC INCLIS, ANABAENA, ANACYSTIS,
ATORIA, ANABAENA, APHANIZOMENON,
HNIQUES, PHYCOLOGY, MICROCYSTIS,
AQUAE, APHANIZOMENON FLOS AQUAE,
AERUGINOSA, ANABAENA CIRCINALIS,
CALES, CONJUGATAE, HETEROKONTAE,
LOS AQUAE, TOLYPOTHRIX DISTORTA,
INOSA, APHANIZOMENON FLOS-AQUAE,
XYLAMINE, ANABAENA, MICROCYSTIS,
CYSTIS, APHANIZOMENON, ANABAENA,
                *PHOSPHORUS CYCLE,
                        CLADOPHORA
*CLADOPHORA FRACTA, CLADOPHORA
ALGAE, BACTERIA, EUTROPHICATION,
NTITHAMNION SARNIENSE, TRITIATED
ORT, CARBON-14, ENZYME KINETICS,
, RADIOCARBON UPTAKE TECHNIQUES,
FIBERS, METHYL ALCOHOL, ETHYLENE
ACELLULAR PRODUCTS, PHOTIC ZONE,
HAELIS-MENTEN EQUATION, ACETATE,
IC PATTERN, ANABAENA CYLINDRICA,
OACIDS, TAMIYA'S MINERAL MEDIUM,
LA, FLAGELLATES, ANKISTRODESMUS,
CLADOPHORA, DIAPTOMUS, CYCLOPS,
ONEIS, EUNOTOGRAMMA, FRAGILARIA,
DAGA LAKE(NY), EUNOTIA, AMPHORA,
TUM, DICTYOSPHAERIUM PULCHELLUM,
DOPHORA, TAXONOMIC DESCRIPTIONS,
ONS, GONGROSIRA LACUSTRIS BRAND,
ECIES, *WESTERN LAKE ERIE(OHIO),
ON SARNIENSE, TRITIATED GLUCOSE,
GIA FUSCO-PURPUREA, SPHACELARIA,
CERIUM-144, BETA RADIOACTIVITY,
COCCUM, SYNECHOCOCCUS, CARTERIA,
UATIC PRODUCTIVITY, CHLOROPHYTA,
ANTIBIOTIC EFFECTS, STIMULANTS,
STIMULANTS, GONYAULAX CATENELLA,
S LUTHERI, NAVICULA PELLICULOSA,
GYMNODINIUM BREVE, C/ GONYAULAX,
YRODINIUM, GYMNODINIUM BREVE, C/
RANTS, FEDERAL GOVERNMENT, STATE
NTROL, *LOCAL GOVERNMENTS, STATE
*WATER POLLUTION CONTROL, *LOCAL
MENT, FINANCING, GRANTS, FEDERAL
OL/ *CONNECTICUT, *LAKES, *LOCAL

GEOCHEMIS: /NOLOGY, MATHEMATICAL   W70-03975
GEOCHEMIS: /NOLOGY, MATHEMATICAL   W70-03975
GEOCHEMISTRY, LAKES, CONIFEROUS F   W71-09012
GEOCHEMISTRY, SPRING WATERS, PHOS   W69-09334
GEOCHEMISTRY, SOIL TYPES, HYDROLO   W71-03048
GEOGRAPHICAL REGIONS, CYCLING NUT   W70-03975
GEOGRAPHICAL REGIONS, CYCLING NUT   W70-03975
GEOGRAPHICAL REGIONS, KENTUCKY, A   W68-00891
GEOGRAPHICAL REGIONS, BAYS, BODIE   W69-03683
GEOGRAPHICAL REGIONS, PHOTOSYNTHE   W69-03611
GEOGRAPHICAL REGIONS, COASTAL PLA   W69-03695
GEOLOGIC FORMATIONS, ARCTIC, TUND   W69-08272
GEOLOGICAL SS POLARIS, CHLOROPHYL   W70-05377
GEOLOGY, MAPPING, NUTRIENTS.: /RE   W70-02646
GEORGE(UGANDA), RHINE-MEEUSE DELT   W70-04369
GEORGIA, SEDIMENTS, BACTERIA, DET   W69-08274
GERMANY.:   W68-00472
GERMANY.: /IONELLA FORMOSA, RECOL   W70-00268
GERMANY), CHIRONOMIDS, CLADOCERA,   W70-00268
GERMANY), GERMANY.: /IONELLA FORM   W70-00268
GERMANY), SPOROBOLOMYCES ROSEUS,   W70-04368
GERMANY), AUREOBASIDIUM PULLULANS   W70-04368
GERMANY), GERMANY, FINLAND, SEWAG   W70-05761
GERMANY), PLON(GERMANY), LAKE SCH   W70-05761
GERMANY), LAKE TEGERN(GERMANY), G   W70-05761
GERMANY), LAKE SCHLIER(GERMANY),   W70-05761
GERMANY), CHLORELLA PYRENOIDOSA,   W70-05092
GERMANY, EXPANDING TECHNOLOGY, OV   W69-07818
GERMANY, FINLAND, SEWAGE RING TUB   W70-05761
GERMANY, VOLUMETRIC ANALYSIS, LIM   W68-00475
GIANUM, CYCLOPS BICUSPIDATUS, REP   W69-10182
GIZZARD SHAD, SEA LAMPHREY, COHO   W70-01943
GLACIAL LAKES, PLANKTON, EPIDEMIC   W68-00488
GLACIATION, URBANIZATION, DOMESTI   W69-07818
GLASS BEADS, TRICHODERMA, DISINTE   W70-04184
GLEN LAKE.: / WADLEIGH STATE PARK   W69-08674
GLENODINIUM, CRYPTOMONAS, CILIATA   W70-00268
GLENODINIUM QUADRIDENS, OOCYSTIS   W69-06540
GLENODINIUM PULVISCULUS, HAEMOCYT   W69-06540
GLENODINIUM PULVISCULUS, SCENEDES   W70-05387
GLENODINIUM, TRACHELOMONAS.: /SCI   W70-05404
GLEO: /IUM, ANKISTRODESMUS, APHAN   W70-04371
GLOCOTRICHIA PISUM, NOSTOC VERRUC   W69-08283
GLOECAPSA, MERISMOPEDIA, MICROCYS   W70-03969
GLOEOCAPSA, STAURONEIS.: /LIS, AN   W70-03974
GLOEOTRICHIA, GLOEOCAPSA, STAURON   W70-03974
GLOEOTRICHIA, POLYPEPTIDES, PHOTO   W70-04369
GLOEOTRICHIA, APHANIZOMENON, MICR   W70-04381
GLOEOTRICHIA ECHINULATA, BOTRYOCO   W70-03959
GLOEOTRICHIA ECHINULATE, OSCILLAT   W69-05704
GLOEOTRICHIA, ASTERIONELLA, DINOB   W70-05409
GLOEOTRICHIA ECHINULATE LB 1303.:   W70-05091
GLOEOTRICHIA ECHINULATA, COCCOCHL   W70-00719
GLOEOTRICHIA ECHINULATA, DINOBRYO   W70-02803
GLOEOTRICHIA, JANESVILLE(WIS), SP   W70-03310
GLOEOTRICHA, P-32.:   W73-01454
GLOMERATA.:   W70-00667
GLOMERATA, BUTTERNUT CREEK(NY).:   W72-05453
GLUCOSE, CARBON RADIOISOTOPES, DI   W70-04194
GLUCOSE, GONIDIA FORMATION, LONG   W69-09755
GLUCOSE, MECHAELIS-MENTEN EQUATIO   W70-04284
GLUCOSE, MICHAELIS-MENTEN EQUATIO   W70-04287
GLYCOL, POLYMERS, DIMETHYL TEREPH   W70-10433
GLYCOLLIC ACID, CHEMOTROPHY, PART   W70-02504
GLYCOLLIC ACID.: /S, GLUCOSE, MEC   W70-04284
GLYCOLYSIS, TRICARBOXYLIC ACID CY   W70-02965
GLYOXYLIC ACID.: /TILE ACIDS, KET   W70-04195
GOMPHENEMA, SPIROGYRA.: /CYCLOTEL   W70-02249
GOMPHONEMA, PHORMIDIUM, APHANIZOM   W70-02803
GOMPHONEMA, GYROSIGMA, MELOSIRA,   W70-03325
GOMPHONEMA, PINNULARIA, NITZSCHIA   W70-03974
GOMPHONEMA, DINOBRYON.: /RETICULA   W69-05704
GONGROSIRA LACUSTRIS BRAND, GONGR   W70-04468
GONGROSIRA DEBARYANA RAB, OOCYSTI   W70-04468
GONGROSIRA STAGNALIS, NEPHROCYTIU   W70-04468
GONIDIA FORMATION, LONG ISLAND SO   W69-09755
GONIDIA, CALLOPHYLLIS HAENOPHYLLA   W69-10161
GONIOBASIS, DECAPODA, PLECOPTERA,   W69-09742
GONIUM, ANKISTRODESMUS, MICROSPOR   W70-03974
GONOPHYTA, DIATOMS, ENVIRONMENTAL   W70-02784
GONYAULAX CATENELLA, GONYAULAX TA   W70-02510
GONYAULAX TAMERENSIS, GYMNODINIUM   W70-02510
GONYAULAX POLYEDRA, 3,5-DIAMINOBE   W69-08278
GONYAULAX VENIFICUM, PYRODINIUM,   W70-05372
GONYAULAX, GONYAULAX VENIFICUM, P   W70-05372
GOVERNMENTS, WATER QUALITY: /G, G   W69-06305
GOVERNMENTS, BIOCHEMICAL OXYGEN D   W69-06909
GOVERNMENTS, STATE GOVERNMENTS, B   W69-06909
GOVERNMENT, STATE GOVERNMENTS, WA   W69-06305
GOVERNMENTS, *WATER QUALITY CONTR   W72-05774

OL, MUSTARD GROWTH TEST, ORGANIC
ALGINIC ACID, FUCUS VESICULOSUS,
MATTER, ORGANIC DEBRIS, ORGANIC
ZSCHIA, ASTERIONELLA GRACILLIMA,
ALGAL SENESCENCE,
ROBIAL ECOLOGY, COULTER COUNTER,
OOD SOURCES, GREENHOUSE STUDIES,
ATION, INCIDENT LIGHT UTI/ ALGAL
RICORNUTUM, SELENASTRUM GRACILE,
HAMNION SARNIENSE, QUANTITATION,
TH, NUTRIENT AVAILABILITY, ALGAL
ENT, PHOSPHATES, ALGAE, DAPHNIA,
ENTRATION, PHOSPHORUS, NITROGEN,
NA, ALGAL CONTROL, BACTERICIDES,
TILIZATION, WATER FLEAS, LARVAE,
AT LAKES, LAKE ERIE, STATISTICS,
CRATER LAKE, ORE,
VOLUMETRIC ANALYSIS, LIMNOLOGY,
CARBON DIOXIDE, PHOTOSYNTHESIS,
TS, CHLOROPHYLL, PHOTOSYNTHESIS,
TRIENTS, *NUTRIENT REQUIREMENTS,
ATOGRAPHY, RESISTANCE, BIOASSAY,
ORELLA, ENZYMES, CARBON DIOXIDE,
NTS, PHOSPHORUS, NITROGEN, IRON,
OCEANOGRAPHY, TRACERS, BACTERIA,
OPHICATION, POLLUTION ABATEMENT,
OLISM, ISOPODS, AMPHIPODA, FISH,
ORUS, SULFATES, VITAMINS, ALGAE,
MMONIA, ENZYMES, CLASSIFICATION,
REMENTS, CULTURES, LAKES, REMENTS,
ENVIRONMENTAL FACTORS, BIOMASS,
POLLUTION EFFECTS, LETHAL LIMIT,
ESTS, PARTICLE SIZE, POPULATION,
S, LARVAE, SNAILS, MITES, MOLDS,
D SLUDGE, *BACTERIA, METABOLISM,
EGRADATION, ALGAE, MARINE ALGAE,
LIGHT INTENSITY, CARBON DIOXIDE,
DOW CONTACT, SPARGANIUM, SALIX,
WTH SUBSTANCES, LIMNOLOGY, PLANT
ICA, DIATOMS, RESISTANCE, *PLANT
ON, MANGANESE, MOLYBDENUM, PLANT
TES, EDDIES, HABITATS, CRAYFISH,
UATION, RESISTANCE, LIFE CYCLES,
IS, *LAKE BUTTE DES MORTS(WISC),
YTA, CHLOROPHYTA, BULRUSH, PLANT
, ANALYSIS, FERTILIZATION, PLANT
PHOMETRY, DOMESTIC WASTES, PLANT
TS, EUTROPHICATION, ALGAE, PLANT
*TEMPERATURE TOLERANCE, OPTIMUM
IA SACCHARINA, MANNITOL, MUSTARD
*ESTIMATING EQUATIONS, PERIOD OF
WAUBESA(WIS), MACROPHYTES, ALGAL
MYXOPHYCEAE, COMPLEX FORMATION,
*NUTRIENT REQUIREMENTS, *PLANT
CALIFORNIA, NUTRIENTS, / *ALGAL
*NUTRIENT AVAILABILITY, *ALGAL
A, NOSTOC PALUDOSUM, BLE/ *ALGAL
OXYGEN CURVES, CALORIMET/ *ALGAL
ICAL ANALYSIS, P/ *ALGAE, *PLANT
A, *NUTRIENT AVAILABILITY, ALGAL
URE, *PILOT-PLANT STUDIES, ALGAL
YSIOLOGY/ *CYANOCOBALAMIN, ALGAL
TA, LAKES, RESERVOIRS, NITROGEN,
UTOTROPHIC GROWTH, HETEROTROPHIC
H LIMITATION, SELENASTRUM, ALGAL
OPHYCEAE, CHLORO/ *EUROPE, ALGAL
RAPHIES, *EUTROPHICATION, *PLANT
, PLANT GROWTH SUBSTANCES, PLANT
RSON STRA/ *MASS CULTURE, *ALGAL
RACILIS Z,/ *VITAMIN B-12, ALGAL
ABLE WATER, ACTINOMYCETES, PLANT
LORELLA PYRENOIDOSA, AUTOTROPHIC
, TURBIDOSTATS, NUTRIENT LIMITED
GAL GROWTH POTENTIAL, SE/ *ALGAL
T SOURCES, WATER SAMPLING, ALGAL
AND, PHOSPHORUS COMPOUNDS, PLANT
NTIFICATION, ALGAE COUNTS, ALGAL
*BIOLOGICAL EXTRACTION,
RIV/ TREE BARK, CHARCOAL, ALGAL
GAE NUISANCE, ALGAE BLOOM, ALGAE
OLLUTION EFFECTS, ECOLOGY/ PLANT
T, SANIT/ *BIOSTIMULATION, ALGAL
SPHORUS SOURCES, *NUISANCE PLANT
ECTS, LAKE MEAD, DILUTION, ALGAL
ROPIUM-155, SILVER-110M, CERIUM,
, MULTIPLE-USES, SAVANNAH RIVER,
, FLAVOBACTERIUM, ACHROMOBACTER,
, *ALGAE, *SORPTION, ABSORPTION,
RIO), TWELVE MILE LAKE(ONTARIO),
AERO-HYDRAULIC

GROWTH FACTORS, POLYSACCHARIDES.:   W69-07826
GROWTH FACTORS, LAMINARIA SACCHAR   W69-07826
GROWTH FACTORS, VITAMIN B1, VITAM   W70-02510
GROWTH FACTORS.: /ROTONENSIS, NIT   W70-02804
GROWTH INHIBITIONS.:   W72-02417
GROWTH INHIBITION, CYANOCOBALAMIN   W70-05652
GROWTH K: /OMPOSITION(ENGLAND), F   W69-06865
GROWTH KINETICS, MONOD GROWTH EQU   W69-03730
GROWTH LIMITATION, SELENASTRUM, A   W70-02779
GROWTH MODES.: /HRIX MUCOR, ANTIT   W69-10163
GROWTH POTENTIAL, SELENASTRUM CAP   W69-06864
GROWTH RATES, PREDATION.: /VIRONM   W69-06536
GROWTH RATES, STABILIZATION PONDS   W68-00855
GROWTH RATES.: /CHLORELLA, *EUGLE   W69-00994
GROWTH RATES, *FRESH WATER FISH   W68-00487
GROWTH RATES, ALGAE, FRESH WATER   W68-00253
GROWTH RATES, WATER PROPERTIES.:   W68-00478
GROWTH RATES, *LABORATORY TESTS,   W68-00475
GROWTH RATE, AEROBIC CONDITIONS,   W70-05750
GROWTH RATE, ORGANIC MATTER, SCEN   W70-05381
GROWTH RATES, LABORATORY TESTS, C   W70-03507
GROWTH RATES, PESTICIDE RESIDUES,   W70-03520
GROWTH RATES, PLANKTON, OXYGEN, N   W70-04369
GROWTH RATES, CHLOROPHYTA, CYANOP   W70-03983
GROWTH RATES, TRITIUM, PHOTOGRAPH   W69-10163
GROWTH RATES, CHLOROPHYLL, PHOTOS   W69-10167
GROWTH RATES.: /OMS, ALGAE, METAB   W69-09742
GROWTH RATES, COPPER SULFATE, BOT   W69-09349
GROWTH RATES, SEAWATER, INCUBATIO   W70-00713
GROWTH RATES, BACTERIA, ULTRAVIOL   W70-00719
GROWTH RATES, ALGAE, OLIGOTROPHY,   W70-02779
GROWTH RATES, SPORES.: /E, WATER   W70-01779
GROWTH RATES, WATER POLLUTION CON   W72-02417
GROWTH RATES, HABITATS.: /ROTIFER   W72-01819
GROWTH RATES, MICROORGANISMS, ALG   W72-01817
GROWTH RATES, ANALYTICAL TECHNIQU   W71-07339
GROWTH RATES, TEMPERATURE, PILOT   W72-06022
GROWTH REGULATORS, MYRIOPHYLLUM,   W70-05263
GROWTH REGULATORS, CYANOPHYTA, EN   W68-00470
GROWTH REGULATORS, FRESH WATER LA   W68-00856
GROWTH REGULATORS, PREDATION, PRO   W69-04801
GROWTH STAGES, DIATOMS, ALGAE, ME   W69-09742
GROWTH STAGES.: /NT, BAYS, *FLUCT   W71-02494
GROWTH STIMULATION, SURFACE LITTE   W72-00845
GROWTH SUBSTANCES, CHLORELLA, CHL   W72-00845
GROWTH SUBSTANCES, SEWAGE, SULFAT   W70-02969
GROWTH SUBSTANCES, LIMNOLOGY, PLA   W68-00470
GROWTH SUBSTANCES, PLANT GROWTH,   W69-07826
GROWTH TEMPERATURE, CONDENSER PAG   W71-02496
GROWTH TEST, ORGANIC GROWTH FACTO   W69-07826
GROWTH.: / ISLES, BIOINDICATORS,   W68-00255
GROWTH.: /NAJAS, ANACHARIS, LAKE   W70-03310
GROWTH-PROMOTING PROPERTIES, CHLA   W69-10180
GROWTH, *ALGAE, ELEMENTS(CHEMICAL   W70-02964
GROWTH, *BIOASSAYS, *FLUOROMETRY,   W70-02777
GROWTH, *MICROCYSTIS AERUGINOSA,   W69-05867
GROWTH, *POTOMAC RIVER, PLECTONEM   W70-02255
GROWTH, *THERMOBIOLOGY, *DIURNAL   W70-03309
GROWTH, *WISCONSIN, ECOLOGY, CHEM   W70-06229
GROWTH, ALGAL NUTRITION, ALGAL PH   W69-05868
GROWTH, ALGAL NUTRITION, ALGAL PH   W69-06865
GROWTH, ALGAL NUTRITION, ALGAL PH   W69-06277
GROWTH, AMINO ACIDS, CHLAMYDOMONA   W69-10180
GROWTH, ANKISTRODESMUS, NOSTOC MU   W70-02964
GROWTH, ANABAENA, BIOSTIMULATION,   W70-02779
GROWTH, ANKISTRODESMUS, BACILLARI   W69-07833
GROWTH, ANALYTICAL TECHNIQUES, AN   W69-04801
GROWTH, CARBOHYDRATES.: /N, ALGAE   W69-07826
GROWTH, CHLORELLA PYRENOIDOSA(EME   W69-07442
GROWTH, CYANOCOBALAMIN, EUGLENA G   W69-06273
GROWTH, DECOMPOSING ORGANIC MATTE   W72-01812
GROWTH, HETEROTROPHIC GROWTH, ANK   W70-02964
GROWTH, LUXURY CONSUMPTION, PROVI   W70-02779
GROWTH, NUTRIENT AVAILABILITY, AL   W69-06864
GROWTH, OXYGEN DEPLETION, NUCLEAR   W70-00264
GROWTH, PRIMARY PRODUCTIVITY, RIV   W69-06535
GROWTH, SEDGWICK-RAFTER METHOD, T   W69-10157
GROWTH, SHOCK LOADING.:   W70-06604
GROWTH, STREAM CONCOURSE, PALOUSE   W70-03501
GROWTH, WATER DENSITY, CONVECTION   W69-08282
GROWTH, WATER AND PLANTS, WATER P   W70-02772
GROWTH, WATER POLLUTION ASSESSMEN   W70-02775
GROWTHS, *NUTRIENT CONTROL, *PHOS   W69-09340
GROWTHS, OXYGEN DEPLETION.: / EFF   W73-00750
GUETTARDA, IPOMOEA, PISONIA, PAND   W69-08269
GULF COAST.: / COLLOIDAL FRACTION   W73-00748
GULF OF MARSEILLES, ASTERIONELLA   W70-04280
GULF OF MEXICO, ATLANTIC OCEAN, R   W69-08524
GULL LAKE(ONTARIO), BALSAM LAKE(O   W70-02795
GUN, ALLISTON(ONTARIO).:   W71-09546

CATENELLA, GONYAULAX TAMERENSIS,
UM-95, RUBIDIUM-106, NIOBIUM-95,
TIC LAYER, MELOSIRA BAICALENSIS,
CCOLITHUS, SARGASSO SEA, SYNURA,
GONYAULAX VENIFICUM, PYRODINIUM,
 THIAMINE, SERRATIA MARINORUBRA,
OGRAMMA, FRAGILARIA, GOMPHONEMA,
IMIKPUK LAKE, NORTH MEADOW LAKE,
KE WASHINGTON(WASH), MACROFAUNA,
, CONCORD(N H), PEANACOOK LAKE(N
BASTICOOK(ME), LAKE WINNISQUAM(N
MENT COSTS, DINOBRYON, CONCORD(N
ENTS, ENVIRONMENTAL ENGINEERING,
AL/ *TRANSFORMATIONS, AMSTERDAM,
TA, CYANOPHYTA, DIATOMS, AQUATIC
ILS, MITES, MOLDS, GROWTH RATES,
HYTA, CYANOPHYTA, UNITED STATES,
RIVERS, HOT SPRINGS, TERRESTRIAL
 *BACKGROUND RADIATION, *AQUATIC
SH/ *TOXICITY, *ALGAE, *REVIEWS,
C PROCESSES, ECOSYSTEMS, AQUATIC
ENVIRONMENT, ECOSYSTEMS, AQUATIC
OTTOM SEDIMENTS, CHUTES, EDDIES,
TS, SEAWATER, ORGANIC COMPOUNDS,
GAE, *TEMPERATURE, *HOT SPRINGS,
RIFUGE, GLENODINIUM PULVISCULUS,
HACELARIA, GONIDIA, CALLOPHYLLIS
 AMSTERDAM, HAARLEM, LEYDEN, THE
OPTERA, TIPULIDAE, CHIRONOMIDAE,
NORTH TEMPERATE ZONE, PLUTONIUM,
HLORELLA PYRENOIDOSA, BIOLOGICAL
LINCH RIVER(TENNESSEE), PHYSICAL
 *RADIONUCLIDE, ULOTHRIX, UPTAKE,
IONUCLIDES, ASSIMILATION, BA-LA,
LIMNODRILUS SP/ RADIOAUTOGRAPHS,
UTION, *FILAMENTOUS GREEN ALGAE,
VARIA), LAKE ZELL(AUSTRIA), LAKE
EIFENSEE, ZURICHSEE, MOSES LAKE,
-VACUOLES, LLANGORS LAKE(WALES),
TINALIS, CLADOPHORA, CYCLOTELLA,
AROGRAPHIC ANALYSIS, PHOTOMETRY,
ER POLLUTION EFFECTS, CHLORIDES,
CAL COMMUNITIES, *AQUATIC ALGAE,
ADOCERA, / *HYPERTROPHIC WATERS,
ION, ALGAL BLOOMS, NUTRIEN/ *NEW
TION, BIOLOGICAL INDICATORS, NEW
TION, BIOLOGICAL INDICATORS, NEW
ANCE ALGAE, *ALGAL CONTROL, *NEW
ABAENA, PHYGON XL(DICHLONE), NEW
REATMENT, ALGAE, FILTRATION, NEW
GAE CONTROL, *WATER SUPPLY, *NEW
D ORGANISMS, FISH GENETICS, FISH
O/ *BLOOM, *VIETNAM, LITTLE LAKE(
FATE, *ACTIVATED SLUDGE PROCESS,
RMANY), AUREOBASIDIUM PULLULANS,
 *NORTH SEA, MYCOTA, DEBARYOMYCES
EEN RIVER, WASHINGTON, HOWARD A.
MENON FLOS-AQUAE, STEPHANODISCUS
 FILTER CLOGGING, STEPHANODISCUS
IS RIVER BASINS PROJECT, INDIANA
OTHRIX MUCOR, MORPHOLOGY, FRIDAY
N ION CONCENTRATION, ALKALINITY,
ATERSHEDS(BASINS), INSECTICIDES,
TURE, NITRATES, PLANTS, AMMONIA,
CHLORIDES, PHOSPHORUS, SULFATES,
RBONATES, BACTERICIDE, KINETICS,
EY, TEMPERATURE, ORGANIC MATTER,
STES, ECOLOGY, DISSOLVED OXYGEN,
 RIVER, ALGAE, IRON, ALKALINITY,
S, ORGANIC PHOSPHATES.:
DGING, HARVESTING OF ALGAE, FISH
NTS, BOTTOM SEDIMENTS, DREDGING,
EATMENT, CHEMICAL PRECIPITATION,
.).:   *ALGAE
.).:   *ALGAE
, CALCIUM, MAGNESIUM, POTASSIUM,
ATMENT, AMMONIA STRIPPING, ALGAE
UTRIENTS, ENVIRONMENTAL EFFECTS,
SEDIMENTS, WATERSHED MANAGEMENT,
AL EFFECTS, ESSENTIAL NUTRIENTS,
N EFFECTS, ALGAE, ALGAL CONTROL,
SANITATION, ESSENTIAL NUTRIENTS,
REEZE DRYING, FREEZE-THAW TESTS,
MESTIC WASTES, SEWAGE TREATMENT,
S, PHOSPHORUS, ALGAE, ALGICIDES,
 PLANTS, HERBICIDES, PESTICIDES,
 PLANTS, HERBICIDES, PESTICIDES,
AL, ALGAE, PRIMARY PRODUCTIVITY,
, PESTICIDES, CELLULOSE, CYCLES,
.:   *ALGAE

GYMNODINIUM BREVIS, MYCROSYSTIS,   W70-02510
GYMNODINIUM SIMPLEX, FLAGELLATES,   W70-00707
GYMNODINIUM BAICALENSE, CYCLOTELL   W70-04290
GYMNODINIUM, CHEMICAL SYMBIOSIS,   W70-04503
GYMNODINIUM BREVE, CANADA, EUROPE   W70-05372
GYRODINIUM COHNII, MICROBIAL ECOL   W70-05652
GYROSIGMA, MELOSIRA, NAVICULA, NI   W70-03325
GYTTJA, DUFF, OOZE, COBBLES, EQUI   W69-08282
GYTTJA, OSCILLATORIA RUBESCENS, Z   W70-04253
H).: / TREATMENT COSTS, DINOBRYON   W69-08282
H), MADISON LAKES(WIS), LAKE WASH   W70-07261
H), PEANACOOK LAKE(N H).: / TREAT   W69-08282
H: /NTAL EFFECTS, ESSENTIAL NUTRI   W69-03374
HAARLEM, LEYDEN, THE HAGUE, MINER   W69-07838
HABITAT, POPULATION.: /EUGLENOPHY   W70-02770
HABITATS.: /ROTIFERS, LARVAE, SNA   W72-01819
HABITATS.: /, *LAKE ERIE, CHLOROP   W70-04468
HABITATS, DEPTH, AIR, SEDIMENTS,   W69-08272
HABITATS, *ALASKA, LAKES, RIVERS,   W69-08272
HABITATS, CLAMS, MUSSELS, SHELLFI   W70-05372
HABITATS, WATER POLLUTION EFFECTS   W71-09012
HABITATS, DRAGONFLIES, ALGAE, SED   W71-09005
HABITATS, CRAYFISH, GROWTH STAGES   W69-09742
HABITATS, STATISTICAL METHODS.: /   W69-09755
HABITATS, ALKALINE WATER, ACIDIC   W69-10160
HAEMOCYTOMETER COUNTING, ITHACA(N   W69-06540
HAENOPHYLLA, TIDAL CURRENT, PUGET   W69-10161
HAGUE, MINERALIZATION, CLAY LENSE   W69-07838
HALF-LIFE.: / PLECOPTERA, EPHEMER   W69-09742
HALF-LIFE(RADIONUCLIDE), FILTER-F   W70-00708
HALF-LIFE, CLINCH RIVER(TENNESSEE   W70-02786
HALF-LIFE, HYDROPSYCHE COCKERELLI   W70-02786
HALF-LIFE, DETECTOR.:   W69-07862
HALF-LIFE, PTYCHOCHEILUS OREGONEN   W69-07853
HALF-LIFE, CONCENTRATION FACTOR,   W69-07861
HALIMONE PORTULACOIDES, SUAEDA MA   W70-09976
HALL WILER(SWITZERLAND).: /IER(BA   W70-05668
HALLWILLERSEE.: /BALDEGGERSEE, GR   W70-03964
HALOBACTERIUM, RESPIROMETER, PROT   W70-04369
HALOCHLOROCOCCUM, SYNECHOCOCCUS,   W70-03974
HALOGENS.: /ICAL TECHNIQUES, *POL   W70-05178
HALOGENS, ALGAE, WATER POLLUTION   W69-08562
HALOPHYTES.: /QUILIBRIUM, BIOLOGI   W69-08282
HAMBURG(GERMANY), CHIRONOMIDS, CL   W70-00268
HAMPSHIRE WATER POLLUTION COMMISS   W69-08674
HAMPSHIRE.: /DISTRIBUTION, POPULA   W70-02772
HAMPSHIRE.: / ECOLOGICAL DISTRIBU   W70-02764
HAMPSHIRE, *COPPER SULFATE, SEWAG   W69-08674
HAMPSHIRE, LAKE WINNISQUAM, LONG   W69-08674
HAMPSHIRE, NEW YORK, VIRGINIA, OR   W72-03641
HAMPSHIRE, ALGAE, ODOR, TASTE, LI   W69-08282
HANDLING FACILITIES, FISH MANAGEM   W69-04804
HANOI), RED RIVER(VIETNAM), CHLOR   W70-03969
HANOVER TREATMENT PLANT.: /UM SUL   W70-09186
HANSENIASPORA UVARUM, RADULOSPORE   W70-04368
HANSENII, RHODOTORULA RUBRA, CAND   W70-04368
HANSON RESERVOIR, TACOMA(WASH).: /   W72-11906
HANTZSCHII, DIDINIUM, PARAMECIUM,   W70-00268
HANTZSCHII, STEPHANODISCUS BINDER   W70-00263
HARBOR SHIP CANAL.: /LAKES-ILLINO   W70-00263
HARBOR(WASH), ANTITHAMNION SARNIE   W69-10161
HARDNESS(WATER), CHLORIDES, ORGAN   W71-04206
HARDNESS(WATER), DIATOMS, FOULING   W72-01813
HARDNESS(WATER), ACIDITY, OXYGEN,   W69-07838
HARDNESS(WATER).: /ATER QUALITY,   W69-03948
HARDNESS, COLIFORMS, CYTOLOGICAL   W71-05267
HARDNESS, CARBON DIOXIDE, FISH.: /   W69-10157
HARDNESS, TIDAL WATERS, ORGANIC M   W73-00748
HARDNESS, RADIOACTIVITY, TEMPERAT   W73-00750
HARRIS COUNTY, TEX, POLYPHOSPHATE   W69-04800
HARVEST, VITAMINS.: /DIMENTS, DRE   W69-09340
HARVESTING OF ALGAE, FISH HARVEST   W69-09340
HARVESTING, ZONING, DIVERSION, DR   W69-10178
HARVESTING, *CENTRAL VALLEY(CALIF   W71-04557
HARVESTING, *CENTRAL VALLEY(CALIF   W71-04556
HARVESTING OF ALGAE, LIMNOLOGY, N   W70-02245
HARVESTING.: /ELECTROCHEMICAL TRE   W70-01981
HARVESTING ALGAE, INHIBITION, LIG   W69-04798
HARVESTING OF ALGAE, WATER ZONING   W68-00859
HARVESTING OF ALGAE, NUISANCE ALG   W69-03371
HARVESTING, AQUATIC ALGAE, AQUATI   W69-03513
HARVESTING OF ALGAE, INHIBITION,   W69-03369
HARVESTING OF ALGAE, NUISANCE ALG   W69-03362
HARVESTING OF ALGAE, BOTTOM SEDIM   W68-00478
HARVESTING OF ALGAE, ENVIRONMENT,   W69-07818
HARVESTING, WATER QUALITY CONTROL   W69-05705
HARVESTING, WATER QUALITY CONTROL   W69-05706
HARVESTING OF ALGAE, BIOMASS.: /N   W70-04290
HARVESTING, DREDGING, MONITORING,   W70-05264
HARVESTING, *DIATOMITE FILTRATION   W72-08688

```
E LAKE, OLD J/ DETECTOR, ALASKA,        HUDEUC LAKE, PULLIN LAKE, GRAPHIT  W69-08272
COLORADO RIVER, TENNESSEE RIVER,        HUDSON RIVER, DELAWARE RIVER, WAT  W70-04507
YTA, WASHINGTON, DC, *ESTUARIES,        HUMAN POPULATION, SEWAGE EFFLUENT  W68-00461
, CHANNELS, POLLUTION ABATEMENT,        HUMAN POPULATION.: /ILD), OYSTERS  W68-00461
RIO), BURTON LAKE(SASKATCHEWAN),        HUMBOLDT(SASKATCHEWAN), MICE, MOR  W70-00273
 PHOSPHORUS RADIOISOTOPES, IONS,        HUMIC ACIDS, CARP, ICE, IRON COMP  W70-05399
*TOXICITY, BIOLOGICAL MEMBRANES,        HUMID AREA.: /, WATER POLLUTION,   W69-05306
YGEN, METABOLISM, WIND VELOCITY,        HUMIDITY, BACTERIA, HYDROGEN ION   W70-01073
LOGICAL TREATMENT, PILOT PLANTS,        HUMUS, REAERATION, EFFLUENTS, FLO  W70-05821
TRASTRUM, TISZA(HUNGARY), ZIMONA(       HUNGARY), ALGAL TAXONOMY, SPIRULI  W70-03969
M, TETRAEDRON, TETRASTRUM, TISZA(       HUNGARY), ZIMONA(HUNGARY), ALGAL   W70-03969
T R/ LAKE ST CLAIR, ROUGE RIVER,        HURON RIVER, RAISIN RIVER, DETROI  W69-01445
KE MICHIGAN, *LAKE ONTARIO, LAKE        HURON, AESTHETICS, ECOLOGY, ECONO  W70-01943
, HYDROGEN SULFIDE, CITIES, LAKE        HURON, LAKE ERIE, LAKE ONTARIO, L  W70-00667
*EFFECTS, NAVICULA SEMINULUM VAR.       HUSTEDTII.:                        W71-03183
INAGE, APPALACHIAN MOUNTAIN CLUB        HUTS, TEGERNSEE, GERMANY, EXPANDI  W69-07818
KELETONEMA COSTATUM, COCCOLITHUS        HUXLEYI, PYRAMIMONAS, PERIDINIUM   W70-05272
IS, ORGANIC MATTER, ALGAE, IRON,        HY: /LVED OXYGEN, CHEMICAL ANALYS  W70-05760
HES, TROUT, SUCKERS, BASS, WATER        HYACINTH, FROGS, CRAYFISH, MOSQUI  W70-05263
, OXYGEN, ROTIFERS, WEEDS, WATER        HYACINTH, ALLIGATORWEED, CHEMCONT  W70-10175
FRESH WATER, ALGAE, PONDS, WATER        HYACINTH, CHEMICAL CONTROLS, PARA  W70-00269
ANOGRAPHIC INSTITUTION, ACROPORA        HYACINTHUS, POCILLOPORA EYDOUXI,   W69-08526
ARIS, DAPHNIA CUCULLATA, DAPHNIA        HYALINA, DAPHNIA MAGNA, CZECHOSLO  W70-00274
REST, WINDS, LIGHT, TEMPERATURE,        HYD: /EVEL, NITROGEN FIXATION, FO  W71-03048
:                          AERO-        HYDRAULIC GUN, ALLISTON(ONTARIO).  W71-09546
AQUATIC ENVIRONMENT, CHLORINATED        HYDROCARBON PESTICIDES, PLANT PHY  W70-08652
 PHAEOCYSTIS, SARGASSUM, FRONDS,        HYDROCARBONS, BACTERIAL ATTACHMEN  W70-01068
LES, ALKANES, ALKENES, ALIPHATIC        HYDROCARBONS, PETROLEUM CRUDES, M  W69-08284
AL, *FOOD WEBS, DDT, CHLORINATED        HYDROCARBON PESTICIDES, ALDRIN, E  W70-03520
RATION, TEMPERATURE, CHLORINATED        HYDROCARBON PESTICIDES, ALGAE, LA  W72-11105
NABAENA, CLADOPHORA, PEDIASTRUM,        HYDRODICTYON, POTAMOGETON, RANUNC  W70-04494
UBESCENS, OSCILLATORIA CHALYBIA,        HYDRODICTYON RETICULATUM, DICTYOS  W69-05704
UM, MICROCYSTIS, ANABAENA.:             HYDRODICTYON, ELODEA, CERATOPHYLL  W68-00479
ON SULFATE, CLADOPHORA BLANKETS,        HYDRODICTYON, MACROPHYTES.: /, IR  W69-10170
LLE(WIS), SPAULDING'S POND(WIS),        HYDRODICTYON, NAJAS, ANACHARIS, L  W70-03310
TED AQUATIC PLANTS, *RESERVOIRS,        HYDROELECTRIC PLANTS, WEED CONTRO  W68-00479
IES, WATER TEMPERATURE, OUTLETS,        HYDROELECTRIC PLANTS, ELECTRIC PO  W69-05023
TENUATION MEASUREMENT APPARATUS,        HYDROGEN F LINE.: /OFER LINES, AT  W70-05377
DIATOMS, STRATIFICATION, OXYGEN,        HYDROGEN ION CONCENTRATION, TEMPE  W70-05387
 *ACID MINE WATER, ACIDIC WATER,        HYDROGEN ION CONCENTRATION, WATER  W70-05424
KTON, PRODUCTIVITY, CHLOROPHYTA,        HYDROGEN ION CONCENTRATION, NITRA  W70-05404
, NITRATES, AMMONIA, PHOSPHATES,        HYDROGEN ION CONCENTRATION, PHYTO  W70-05405
CIDS, CARP, ICE, IRON COMPOUNDS,        HYDROGEN ION CONCENTRATION, ACIDI  W70-05399
 DIEL MIGRATION, CARBON DIOXIDE,        HYDROGEN ION CONCENTRATION, PHOSP  W70-05760
ICAL ECOLOGY, INDUSTRIAL WASTES,        HYDROGEN ION CONCENTRATION, IRON   W70-04510
OMETERS, BACTERIA, ALGAE, SANDS,        HYDROGEN ION CONCENTRATION.: /CTR  W69-08267
LORIDES, CONDUCTIVITY, CULTURES,        HYDROGEN ION CONCENTRATION, IRON,  W69-07833
EUTROPHICATION, LIME, *NITROGEN,        HYDROGEN ION CONCENTRATION, *PHOS  W68-00012
IVATED SLUDGE, RATES, CHLORELLA,        HYDROGEN ION CONCENTRATION, PLANT  W68-00248
ERROBACILLUS, WATER TEMPERATURE,        HYDROGEN ION CONCENTRATION, FUNGI  W68-00891
ACTERIA, *AQUATIC FUNGI, YEASTS,        HYDROGEN ION CONCENTRATION, IRON,  W69-00096
NCE ALGAE, NITROGEN, PHOSPHATES,        HYDROGEN ION CONCENTRATION, CYCLI  W70-04381
SIS, PHOSPHATES, IRON COMPOUNDS,        HYDROGEN ION CONCENTRATION, ALKAL  W70-03334
OWMELT, SURFACE WATERS, STREAMS,        HYDROGEN ION CONCENTRATION, IRON,  W70-03501
LUENTS, TRIBUTARIES, ALKALINITY,        HYDROGEN ION CONCENTRATION, NUTRI  W70-03983
ER), PHYTOPLANKTON, ZOOPLANKTON,        HYDROGEN ION CONCENTRATION, ALGAE  W70-04368
ONATES, CARBONATES, TEMPERATURE,        HYDROGEN ION CONCENTRATION, ALKAL  W70-02804
S, ECOLOGY, PHOSPHORUS, SULPHUR,        HYDROGEN ION CONCENTRATION, TIME.  W70-02965
INGTON, WATER CHEMISTRY, OXYGEN,        HYDROGEN ION CONCENTRATION, PHOTO  W70-03309
ND VELOCITY, HUMIDITY, BACTERIA,        HYDROGEN ION CONCENTRATION, SULFA  W70-01073
ECTROPHOTOMETRY, CHROMATOGRAPHY,        HYDROGEN ION CONCENTRATION, FLUOR  W70-01074
RBON RADIOISOTOPES, CHLOROPHYLL,        HYDROGEN ION CONCENTRATION, MAGNE  W70-01579
TEMPERATURE, TRACERS, CHLORELLA,        HYDROGEN ION CONCENTRATION, ALKAL  W70-02504
AS, TEMPERATURE, CARBON DIOXIDE,        HYDROGEN ION CONCENTRATION, BIOCA  W70-02249
INDUSTRIES, METHANE, ALKALINITY,        HYDROGEN ION CONCENTRATION, TOXIN  W70-02803
WATER QUALITY, DISSOLVED SOLIDS,        HYDROGEN ION CONCENTRATION, TEMPE  W70-00265
, CHLORINE, PHENOLS, OIL WASTES,        HYDROGEN ION CONCENTRATION, FLUOR  W70-00263
VIOLET RADIATION, ACTINOMYCETES,        HYDROGEN ION CONCENTRATION, NITRA  W70-00719
ENZUMES, PEPTIDES, BIOCHEMISTRY,        HYDROGEN ION CONCENTRATION, OXIDA  W70-01068
OACTIVITY, CARBON RADIOISOTOPES,        HYDROGEN ION CONCENTRATION, SCENE  W70-08097
THERMAL POLLUTION, *TEMPERATURE,        HYDROGEN ION CONCENTRATION, NUTRI  W70-08834
TER, WATER QUALITY, TEMPERATURE,        HYDROGEN ION CONCENTRATION, OXYGE  W71-03031
, ALGAE, PERIPHYTON, PHOSPHORUS,        HYDROGEN ION CONCENTRATION, ALKAL  W71-04206
TION, DEPTH, OXYGEN, ALKALINITY,        HYDROGEN ION CONCENTRATION, STAND  W71-00665
UTROPHICATION, DISSOLVED OXYGEN,        HYDROGEN ION CONCENTRATION, EPILI  W70-06975
 DISSOLVED OXYGEN, BICARBONATES,        HYDROGEN ION CONCENTRATION, NUISA  W70-06225
T COMPARISONS, SEDIMENTS, ALGAE,        HYDROGEN ION CONCENTRATION, ELECT  W71-09548
LUDGE, METHANE, ORGANIC LOADING,        HYDROGEN ION CONCENTRATION, TEMPE  W71-11375
SE, COOLING TOWERS, DAMS, ALGAE,        HYDROGEN ION CONCENTRATION, TEMPE  W71-11393
N, *PROTOZOA, *FUNGI, *BACTERIA,        HYDROGEN ION CONCENTRATION, TEMPE  W71-05267
ISEASES, *FISH TOXINS, TOXICITY,        HYDROGEN ION CONCENTRATION, DISSO  W71-07731
DITY, BIOCHEMICAL OXYGEN DEMAND,        HYDROGEN ION CONCENTRATION, ORGAN  W72-03299
OXICITY, CHEMICAL OXYGEN DEMAND,        HYDROGEN ION CONCENTRATION, TEMPE  W71-13413
GANIC LOADING, DISSOLVED OXYGEN,        HYDROGEN ION CONCENTRATION, BIOCH  W72-10573
OMPOUNDS, PHOSPHOROUS COMPOUNDS,        HYDROGEN ION CONCENTRATION, CONDU  W72-06051
, *WATER QUALITY, *ALGAE, *ODOR,        HYDROGEN ION CONCENTRATION.: /ENT  W72-09694
ATES, TEMPERATURE, PILOT PLANTS,        HYDROGEN ION CONCENTRATION, ORGAN  W72-06022
ENVIRONMENT, ORGANIC PESTICIDES,        HYDROGEN ION CONCENTRATION, TEMPE  W72-11105
 SULFUR BACTERIA, ALGAE, MIXING,        HYDROGEN ION CONCENTRATION, TEMPE  W72-06854
ANALYSIS, TASTE-PRODUCING ALGAE,        HYDROGEN ION CONCENTRATION, *ORGA  W72-11906
```

, IN VIVO, BALL-MILL, MEICELASE,
ATES, YEASTS, ENZYMES, CULTURES,
N CONCENTRATION, IRON COMPOUNDS,
D, OXYGEN, ANAEROBIC CONDITIONS,
ION CONCENTRATION, ACIDITY, MUD,
RECIPITATES.:
GEN DEM/ *ALGAE, IRON COMPOUNDS,
RBON DIOXIDE, AMMONIA, NITRITES,
POPULATION, POLLUTANTS, MAPPING,
TES, TEMPERATURE, SHALLOW WATER,
, TASTE, ODOR, COLOR, TURBIDITY,
L OXYGEN DEMAND, PHOTOSYNTHESIS,
DS, ABSORPTION, HAZARDS, CESIUM,
INS, WATER SUPPLY, MUD, BENTHOS,
GAE, MICROORGANISMS, PHOSPHATES,
S, BACTERIA, SAMPLING, SEASONAL,
UNGI, BACTERIA, FISH, LIMNOLOGY,
LIED), GEOCHEMISTRY, SOIL TYPES,
ILIZATION, WASTE STORAGE, PONDS,
EFFLUENTS, WISCONSIN, ALGICIDES,
ION EXCHANGE, CATION ADSORPTION,
 OXIDATION, REDUCTION(CHEMICAL),
TRY, PA/ *LAKE TAHOE(CALIF-NEV),
(TENNESSEE), PHYSICAL HALF-LIFE,
YLL, PROTEINS, LIGHT, NUTRIENTS,
TIS, GLOEOTRICHIA ECHINULATA, D/
                    *MICROSTRAINERS,
ATIC PLANTS, ALGAE, CHLOROPHYLL,
RMAL STRATIFICATION, EPILIMNION,
DIATOMS, DINOFLAGELLATES, ALGAE,
UCTIVITY, PHOTOSYNTHESIS, ALGAE,
, BOTTOM SEDIMENTS, THERMOCLINE,
E, TEMPERATURE, CURRENTS(WATER),
VIRONMENT, AQUATIC PRODUCTIVITY,
LLATES, ALGAE, DISSOLVED OXYGEN,
ANKTON, ODOR, WINDS, EPILIMNION,
TER ANALYSIS, CALCIUM CARBONATE,
 PLANKTON, ALGAE, FROGS, MOSSES,
RUS, PLANKTON, DISSOLVED OXYGEN,
NATIONAL OCEANOGRAPHIC CONGRESS,
, *GLENODININE, ALKALOID TOXINS,
ING, ZOOPLANKTON, PHYTOPLANKTON,
OTOPES, IONS, HUMIC ACIDS, CARP,
RE, DEPTH, DENSITY, CHLOROPHYLL,
 NUTRIENTS, OXYGEN REQUIREMENTS,
EST, CLUSTER ANALYSIS, REYKJAVIK(
AGANSETT BAY(RI), CAPE REYKJANES(
APE REYKJANES(ICELAND), SUDURNES(
INTER, NUCLEIC ACID, SISJOTHANDI(
 CAPE REYKJANES, FAXAFLOI FJORD,
ORMATION, LONG ISLAND SOUND(NY),
P, RHINICHTHYS OSCULUS, SUCKERS,
ETRITUS, PULP WASTES, LIVESTOCK,
 STREAM CONCOURSE, PALOUSE RIVER(
CORETHRA PUNCTIPENNIS, PISIDIUM
YTOTOXICITY, PLANKTON, POLLUTANT
EUTROPHICATION, PONDS, POLLUTANT
QUIREMENTS, NUTRIENTS, POLLUTION
, POLLUTION ABATEMENT, POLLUTANT
SCHIA, BIOLOGICAL STUDIES, ALGAL
ALYSIS, WATER QUALITY, POLLUTANT
OPPER APPLICATION METHODS, ALGAE
WASTES, GREAT LAKES, *POLLUTANT
RESEARCH AND DEVELOPMENT, WASTE
NT, *INSTRUMENTATION, *POLLUTION
NT, *WATER TREATMENT, *POLLUTANT
IFICATION, *MODEL STUDIES, WASTE
OXINS, ORGANIC WASTES, POLLUTANT
        ALGAL TAXONOMY, ALGAE
ES, DIATOMS, CRAYFISH, POLLUTANT
TAKE TECHNIQUES, WATER POLLUTION
WATER QUALITY CONTROL, POLLUTANT
SAY, FISH, AQUATIC E/ *POLLUTANT
TERWAYS, MILWAUKEE(WIS), CHICAGO(
TION, *CHICAGO(ILL), COOK COUNTY(
IVER(ILL),/ *POLLUTION, *CHICAGO(
 COOK COUNTY(ILL), CALUMET RIVER(
BLIC HEALTH SERVICE, GREAT LAKES-
RIMARY PRODUCTIVITY, ECOSYSTEMS,
ND CALCULATIONS, PHOTOSYNTHESIS,
ROPHICATION PREVENTION, ECOLOGIC
            HYDROGEN SULFIDE,
E DERIVATIVES, ALKYL SUBSTITUTED
E, TWELVE MILE LAKE, BRANT LAKE,
 FLOW, PUTRESCIBILITY TEST, BODY
LAMINARIA AGARDHII, DESICCATION,
TYLUM TRICORNUTUM, STICHOCHRYSIS
TER FISH, ESTUARINE ENVIRONMENT,
ROTIFERS, CHLOROPHYLL, PIGMENTS,
NG, ZONING, DIVERSION, DREDGING,

HYDROGEN PEROXIDE, RATS, TRYPSIN,   W70-04184
HYDROGEN SULFIDE, ELECTRON MICROS   W70-03952
HYDROGEN SULFIDE, LEAD, COPPER, A   W70-04510
HYDROGEN SULFIDE, IRON, FUNGI, CL   W70-05761
HYDROGEN SULFIDE.: /DS, HYDROGEN   W70-05399
HYDROGEN SULFIDE, IMHOFF TANKS, P   W72-00027
HYDROGEN SULFIDE, BIOCHEMICAL OXY   W70-06925
HYDROGEN SULFIDE, SALINITY, LAKES   W71-03031
HYDROGEN SULFIDE, CITIES, LAKE HU   W70-00667
HYDROGEN SULFIDE, COPPER SULFATE,   W70-02803
HYDROGEN SULFIDE, IRON, MAGNESIUM   W70-02251
HYDROGEN SULFIDE, LAKES, ALGAE, E   W70-03312
HYDROGEN, DEUTERIUM, TRITIUM.: /O   W70-02786
HYDROGEN, BACTERIA, AEROBIC BACTE   W69-07838
HYDROGENATION, AERATION, FLOCCULA   W70-08627
HYDROGRAPHY, FLORIDA, THERMOCLINE   W70-04368
HYDROLOGIC ASPECTS, AQUATIC ENVIR   W70-04475
HYDROLOGY, CLIMATES, TROPHIC LEVE   W71-03048
HYDROLYSIS, METABOLISM, CHEMICAL   W70-08097
HYDROLYSIS, COPPER SULFATE, FISHE   W70-02982
HYDROLYSIS, MARINE ALGAE, INVERTE   W71-09863
HYDROLYSIS, CHELATION, FUNGI, ALG   W69-07428
HYDROPHYTES, LIVERWORTS, MORPHOME   W70-00711
HYDROPSYCHE COCKERELLI, BODY BURD   W70-02786
HYDROSTATIC PRESSURE, CARBON DIOX   W69-10160
HYDROXYLAMINE, ANABAENA, MICROCYS   W70-02803
HYPOCHLORITES.:   W72-03824
HYPOLIMNION, PHOSPHORUS, NITROGEN   W71-07698
HYPOLIMNION, CYCLING NUTRIENTS, T   W70-02251
HYPOLIMNION, OXYGEN, PRODUCTIVITY   W69-10169
HYPOLIMNION, EPILIMNION, PHYTOPLA   W70-10182
HYPOLIMNION, ANIMALS, DENSITY, IR   W70-06975
HYPOLIMNION, ALGAE, NITROGEN, PHO   W70-09889
HYPOLIMNION, ALGAL CONTROL, SEASO   W69-05142
HYPOLIMNION, EPILIMNION, LAKE ERI   W70-04253
HYPOLIMNION, LIGHT INTENSITY, CON   W70-05405
HYPOLIMNION.: /OGY, MINNESOTA, WA   W69-00632
HYPOLIMNION, THERMOCLINE, *ION EX   W68-00857
HYPOLIMNION, GREAT LAKES, LAKE ER   W68-00253
HYPON: /OGY, RUSSIA, SECOND INTER   W69-07440
IBOGAALKALOID, ORGANIC TOXINS.: /   W71-07731
ICE, BACTERIA, TRIPTON, BOTTOM SE   W69-10154
ICE, IRON COMPOUNDS, HYDROGEN ION   W70-05399
ICE, SEASONAL, DIURNAL, LIGHT INT   W70-05387
ICED LAKES, *FERTILIZATION, WATER   W68-00487
ICELAND).: /TS, ONE-SAMPLE RUNS T   W69-09755
ICELAND), SUDURNES(ICELAND), NON-   W69-09755
ICELAND), NON-PARAMETRIC TESTS, O   W69-09755
ICELAND), ENZYME-SUBSTRATE COMPLE   W69-10160
ICELAND, DNA BASE COMPOSITION.: /   W69-10161
ICELAND, TRITIATED THYMIDINE, NAR   W69-09755
ICTALURUS PUNCTATUS, TRICHOPTERA,   W69-07846
IDAHO.: /ES, PHOSPHATES, ALGAE, D   W70-03501
IDAHO).: /CHARCOAL, ALGAL GROWTH,   W70-03501
IDAHOENSE, TENDIPES PLUMOSUS, BEA   W70-06217
IDENTIFI: /OGY, PHYTOPLANKTON, PH   W70-02968
IDENTIFICATION, BIOLOGY, WATER PO   W70-02304
IDENTIFICATION, SEWAGE, SEWAGE EF   W70-02248
IDENTIFICATION, WATER POLLUTION C   W70-02792
IDENTIFICATION.: /EEN ALGAE, NITZ   W70-10173
IDENTIFICATION.: /PLING, WATER AN   W71-00221
IDENTIFICATION, ALGAE COUNTS, ALG   W69-10157
IDENTIFICATION.: /RCES, MUNICIPAL   W71-05879
IDENTIFICATION, ALGAE, *WASTE WAT   W72-03299
IDENTIFICATION, *POLLUTION ABATEM   W69-06203
IDENTIFICATION, *POLLUTANT ABATEM   W69-07389
IDENTIFICATION, BIOLOGICAL PROPER   W69-01273
IDENTIFICATION, POLLUTANTS, QUALI   W69-01273
IDENTIFICATION.:   W69-03514
IDENTIFICATION, *WATER POLLUTION   W69-02782
IDENTIFICATION, ORGANIC SOLUTES,   W70-05652
IDENTIFICATION, IOWA.: /, *ODOR,   W72-13605
IDENTIFICATION, *TOXICITY, *BIOAS   W72-11105
ILL), ARTIFICIAL CIRCULATION, NUI   W69-10178
ILL), CALUMET RIVER(ILL), POOLS,   W70-00263
ILL), COOK COUNTY(ILL), CALUMET R   W70-00263
ILL), POOLS, WIND DIRECTION, TABE   W70-00263
ILLINOIS RIVER BASINS PROJECT, IN   W70-00263
ILLINOIS, PHOSPHORUS, DRAINAGE, A   W70-04506
ILLUMINATION, TEMPERATURE, NUTRIE   W70-04786
IMBALANCES.:   EUT   W69-07084
IMHOFF TANKS, PRECIPITATES.:   W72-00027
IMIDAZOLINES, BIOLOGICAL ACTIVITY   W70-02982
IMIKPUK LAKE, NORTH MEADOW LAKE,   W69-08272
IMMEDIUM FILTER, BACK-WASHING.: /   W70-05821
IMMERSION, NARRAGANSETT MARINE LA   W70-01073
IMMOBILIS, VIBRIO, MORPHOLOGY, CU   W70-04280
IMPOUNDED WATERS, FRESHWATER ALGA   W71-02491
IMPOUNDMENTS, RESERVOIRS, ANALYTI   W71-10098
IMPOUNDMENTS, WEEDS,: /, HARVESTI   W69-10178

CLAYS, COLLOIDS, EUTROPHICATION,
VESTIGATIONS, *IMPOUNDED WATERS,
OL, *MI/ FISH MANAGEMENT, STREAM
IFE, STREAM POLLUTION, BACTERIA,
ABAENA CYLINDRICA B629, ANABAENA
MENT CORP(MASS), ARTHUR D LITTLE
KINETICS, MONOD GROWTH EQUATION,
FILTRATION, ANEROBIC DIGESTION,
SIBILITY, COST ANALYSIS, SLUDGE,
AL NUTRIENT CONCENTRATION, AREAL
ICATION, GROWTH RATES, SEAWATER,
D LAKE(IND), LITTLE CROOKED LAKE(
LISM, HETEROTROPHY, CROOKED LAKE(
  *NITRIC ACID, *LANGELIER
ION, DECAPODS, SPECIES DIVERSITY
URE, *BIOMASS TITER, *BIOLOGICAL
CAROTENOIDS, SPECIES, DIVERSITY
, ORGANIC CARBON, / *AUTOTROPHIC
ITE, ACID-WASHED BAUXITE, KANPUR(
S(INDIA), RAJGHAT(INDIA), BOMBAY(
MS, RIVER GANGES(INDIA), RAJGHAT(
T(INDIA), BOMBAY(INDIA), BENARES(
D/ *INDIA, *BLOOMS, RIVER GANGES(
S.:
MILLSBORO(DELAWARE), SWAN CREEK,
S-ILLINOIS RIVER BASINS PROJECT,
, PESTICIDES, SLUDGE, DIGESTION,
CATORS, FISH, WASHINGTON, LAKES,
NCE ALGAE, WISCONSIN, NUTRIENTS,
ION, BACTERIA, ALGAE, NUTRIENTS,
CULTURAL EUTROPHICATION, TROPHIC
RY(LAKE SUPERIOR), WATER QUALITY
RIBUTION, POPULATION, BIOLOGICAL
LOGICAL DISTRIBUTION, BIOLOGICAL
, MOSQUITOES, SAMPLING, CONTROL,
RADIATION, SULFATES, CELLULOSE,
C WASTES, WASTE WATER TREATMENT,
PHOSPHORUS, PHOSPHATE, NITRATES,
FF, DOMESTIC WASTES, DETERGENTS,
ER TREATMENT, WATER UTILIZATION,
WASTE WATER TREATMENT, *POULTRY,
LITY, DETERGENTS, ALGAL CONTROL,
IMATE DISPOSAL, DOMESTIC WASTES,
, *OIL WASTES, *CHEMICAL WASTES,
LANKTON, SEWAGE, DOMESTIC WASTE,
C WASTES, WASTE WATER TREATMENT,
OUBLING TIME, PACKED CELL VOLUM/
IGHT PENETRATION, STANDING CROP,
*LAKE ONTARIO, WATER TURBIDITY,
L), ORGANIC WASTES, PULP WASTES,
ATMOSPHERIC), WETLANDS, SEEPAGE,
*NU/ *WASTE WATER(MUNICIPAL AND
WATER POLLUTION, FERTILIZATION,
SEWAGE, PHOSPHORUS, CHEMCONTROL,
KTON, CRYPTOPHYCEAE, VOLVOCALES,
ODOPHYTA, PHYSIOLOGICAL ECOLOGY,
DES, SALINE LAKES, HEAVY METALS,
MENTAL EFFECTS, CURRENTS(WATER),
S, PHOSPHORUS, NITROGEN, SEWAGE,
ERTIES, DOMESTIC WASTES, SEWAGE,
OKLAHOMA, RESERVOIRS, SAMPLING,
AGE EFFLUENTS, SEWAGE TREATMENT,
RPLANTS, ENGINEERING STRUCTURES,
ATIC INSECTS, FISH REPRODUCTION,
LUTION EFFECTS, DOMESTIC WASTES,
AE, PHENOLS, SHELLFISH, OYSTERS,
LUENTS, FARM WASTES. POPULATION,
(CALIF), EEL RIVER(CALIF), JOINT
ATER WORKS ASSOCIATION, CHEMICAL
GEMENT, *FARM MANAGEMENT, *DAIRY
-PLANKTON, SPRAYING, BIOCONTROL,
CLING NUTRIENTS, *SEDIMENT-WATER
TERIA, PHOSPHORUS RADIOISOTOPES,
BOTTOM SEDIMENTS, ANIMALS, FISH,
LAKE WASHINGTON(WASH), *CULTURAL
URNOVERS, NUTRIENTS, PHOSPHORUS,
(MINN), NUTRIENT TRAPS, NUTRIENT
CTS, LIPID CARBON, CHLOROPLASTS,
INOBENZOIC ACID DIHYDROCHLORIDE,
MONITORING, AERIAL PHOTOGRAPHY,
ETHOXYLATE, ORGANIC EXTRACTIONS,
APPLICATIONS, FACULTATIVE PONDS,
RFACTANTS, NONIONIC SURFACTANTS,
  ALGAL SENESCENCE, GROWTH
RIENT REQUIREMENTS, *POLLUTANTS,
ATES, BIOASSAY, NEW YORK, LAKES,
TION, MODE OF ACTION, MORTALITY,
NUTRIENTS, ESSENTIAL NUTRIENTS,
OROPHYTA, ENVIRONMENTAL EFFECTS,
MENTS, NUTRIENTS, PHYTOPLANKTON,

IMPOUNDMENTS, LAKES, LOAMS, RESER W69-04800
IMPOUNDMENTS, *DISSOLVED OXYGEN, W72-11906
IMPROVEMENT, *WATER QUALITY CONTR W70-05428
IMPURITIES, WASTE TREATMENT.: / L W72-01786
INAEQUALIS 381, ANABAENA FLOS AQU W70-05091
INC(MASS), CARNEGIE DEPT OF PLANT W69-06865
INCIDENT LIGHT UTILIZATION.: /TH W69-03730
INCINERATION, IRON COMPOUNDS, SUL W70-04060
INCINERATION, STORM RUN-OFF, SEDI W71-10654
INCOME(NUTRIENTS), LAKE VOLUME, M W70-00270
INCUBATION, PSEUDOMONAS, SUBSURFA W70-00713
IND).: /ISM, HETEROTROPHY, CROOKE W71-10079
IND), LITTLE CROOKED LAKE(IND).: / W71-10079
INDEX.: W71-11393
INDEX.: / ANALYSIS, PART CORRELAT W73-01446
INDEX, *SAPROBITY, PRAGUE, CZECHO W72-12240
INDEX, ACHNANTHES, ACTINOPTYCHUS, W70-03325
INDEX, CHLOROPHYLL A, SHAYLER RUN W71-04206
INDIA).: /ATED SAND, CRUSHED BAUX W70-08628
INDIA), BENARES(INDIA), MACROPHYT W70-05404
INDIA), BOMBAY(INDIA), BENARES(IN W70-05404
INDIA), MACROPHYTES, CLADOPHORA, W70-05404
INDIA), RAJGHAT(INDIA), BOMBAY(IN W70-05404
INDIA, METHYL ORANGE, ALGAL BLOOM W70-05550
INDIAN RIVER.:' W71-05013
INDIANA HARBOR SHIP CANAL.: /LAKE W70-00263
INDIANA.: /OUNDS, LAGOON, FILTERS W72-03641
INDIANA, OCHROMONAS, PHYTOPLANKTO W70-04503
INDICATOR SPECIES, EUGLENA, CHLAM W69-01977
INDICATORS, ODOR, TASTE, FILTERS, W70-05264
INDICATORS, NUTRIENT BUDGET.: * W71-03048
INDICATORS.: PERIPHYTON INVENTO W70-05958
INDICATORS, NEW HAMPSHIRE.: /DIST W70-02772
INDICATORS, NEW HAMPSHIRE.: / ECO W70-02764
INDUS: /ICATION, TOXICITY, MIDGES W70-06225
INDUSTR: /, SALINITY, ULTRAVIOLET W70-01074
INDUSTRIAL WASTES, NEUTRIENTS, BA W70-01519
INDUSTRIAL WASTES, WATER POLLUTIO W70-02777
INDUSTRIAL WASTES, *CHELATION.: / W71-04072
INDUSTRIAL WASTES, OXIDATION LAGO W70-07508
INDUSTRIAL WASTES, LAGOONS, PONDS W70-09320
INDUSTRIAL WASTES, THERMAL POLLUT W70-10437
INDUSTRIAL WASTES, SEWAGE WASTES, W70-09905
INDUSTRIAL WASTES, WATER ANALYSIS W72-03299
INDUSTRIAL WASTES, FARM WASTES.: / W72-01813
INDUSTRIAL WASTES, NUTRIENTS, BAC W71-06737
INDUSTRIAL PHOTOSYNTHESIS, MEAN D W71-05267
INDUSTRIAL WASTE: /THERMOCLINE, L W71-07698
INDUSTRIAL PLANTS, WATER SPORTS.: W71-09880
INDUSTRIAL WASTES, TOXINS, SOLID W72-00462
INDUSTRIAL WASTES, TERTIARY TREAT W69-10178
INDUSTRIAL), *PHOSPHORUS SOURCES, W69-09340
INDUSTRIAL WASTES, GARBAGE DUMPS, W70-05112
INDUSTRIAL WASTES, SEWAGE TREATME W70-05668
INDUSTRIAL USES, CLEAN WATER, GEN W70-05389
INDUSTRIAL WASTES, HYDROGEN ION C W70-04510
INDUSTRIAL WASTES, LAKES, PHOTOSY W69-02959
INDUSTRIAL WASTE, CORES, DIATOMS, W68-00470
INDUSTRIAL WASTES, DOMESTIC WASTE W68-00478
INDUSTRIAL WASTES, AGRICULTURE, S W70-03964
INDUSTRIAL WASTES, MUNICIPAL WAST W70-04001
INDUSTRIAL WASTES, WATER CHEMISTR W70-03974
INDUSTRIAL PLANTS, STRUCTURES, AF W69-05023
INDUSTRIAL WASTES, ECOLOGY, DISSO W73-00748
INDUSTRIAL WASTES, ENVIRONMENTAL W72-12240
INDUSTRIES, CHELATION, BACTERIA, W70-01068
INDUSTRIES, METHANE, ALKALINITY, W70-02803
INDUSTRY/GOVERNMENT TASK FORCE ON W70-02777
INDUSTRY, ODOR, TASTE, CONSUMER.: W71-05943
INDUSTRY, *ALGAE, *CATTLE, SLURRI W70-07491
INFECTION, NUTRIENTS, ALGAE, NITR W68-00468
INFERFACE, MUD-WATER INTERFACE, P W69-06859
INFLOW, CYANOPHYTA, LITTORAL, NUT W73-01454
INFLOW, OUTLETS, NUTRIENTS, PRODU W70-06225
INFLUENCES, MICROCYSTIS, DREDGE S W70-05663
INFLUENT SEWAGE, EQUILIBRIUM, MAT W68-00858
INFLUX, NUTRIENT SOURCES, PHOSPHO W70-04268
INFRARED ANALYSIS.: /ETHANE EXTRA W70-05651
INFRARED GAS ANALYSIS, COULTER CO W69-08278
INFRARED RADIATION, DYE RELEASES, W68-00010
INFRARED SPECTROSCOPY, BLUE-GREEN W70-03928
INFRARED SPECTROSCOPY, SULFUR COM W70-07034
INFRARED SPECTRA.: /E, ANIONIC SU W70-09438
INHIBITIONS.: W72-02417
INHIBITION, LIMNOLOGY, ACIDITY, A W70-02792
INHIBITION, ALGAE, DIATOMS, EUTRO W70-01579
INHIBITION, ALGAE.: /OL, DISINFEC W70-02370
INHIBITION, NITROGEN COMPOUNDS, N W70-02248
INHIBITION, LIMNOLOGY, PESTICIDES W70-02968
INHIBITION.: /S, NUTRIENT REQUIRE W69-10177

457

FECTS, ENVIRONMENTAL SANITATION,
CHLORELLA, MORTALITY, BIOASSAY,
, *ALGAE, LIGHT, CARBON DIOXIDE,
REEZE DRYING, FREEZE-THAW TESTS,
SANITATION, ESSENTIAL NUTRIENTS,
NUTRIENTS, HARVESTING OF ALGAE,
, HORMONES, AUXINS, METABOLITES,
NTIAL NUTRIENTS, EUTROPHICATION,
DIATOMS, ENVIRONMENTAL EFFECTS,
NTIAL NUTRIENTS, EUTROPHICATION,
ENTAL EFFECTS, HARVESTING ALGAE,
NUTRIENTS, CHLOROPHYTA, DIATOMS,
ECOLOGY, GROWTH, COULTER COUNTER, GROWTH
II, DUNALIELLA TERTIOLECTA, DCMU
DS, NUTRIENTS, VITAMINS, TOXINS,
ING(CORROSION), RUSTING, METALS,
STES, GARBAGE DUMPS, OIL WASTES,
R, FACULTATIVE, MICROAEROPHILIC,
ARSENATE INTERFERENCE, DISSOLVED
BON DIOXIDE, VITAMINS, NITRATES,
PLAINS, ATLANTIC COASTAL PLAIN,
TION, CHANNEL MORPHOLOGY, SEWAGE
PLANKTON, AQUATIC WEED CONTROL,
RIN, *DIATOMS, LABORATORY TESTS,
RESERVOIRS, WATERSHEDS(BASINS),
S, OLIGOCHAETES, DIPTERA, ALGAE,
/ *ZINC RADIOISOTOPES, *AQUATIC
D ORG/ ALGAE, PROTOZOA, *AQUATIC
LS, ALGAE, FISH, DIATOMS, FUNGI,
, VELOCITY, CYANOPHYTA, BIOMASS,
OLIMNION, EPILIMNION, LAKE ERIE,
G PLANTS, PHYTOPLANKTON, AQUATIC
RIVERS, ALGAE, PROTOZOA, AQUATIC
MENTS, BIOTA, FISH, UTAH, ALGAE,
ION, DECOMPOSING ORGANIC MATTER,
AKES, CONIFEROUS FORESTS, ALGAE,
RTEBRATES, FISH, WATER ANALYSIS,
STIGLEOCLONIUS,
CROCEA, WOODS HOLE OCEANOGRAPHIC
DEPT OF PLANT BIOLOGY, CARNEGIE
DACTYLUM, WASTE WATER DIVERSION,
THERMAL POLLUTION, METHODOLOGY,
UCTURES, AFTERBAYS, WATER TYPES,
*LAKE MICHIGAN, *WATER SUPPLY,
FILTERS, *ALGAE, NUISANCE ALGAE,
ONMENT, CHEMCONTROL, BIOCONTROL,
MPERATURE, PHOTOSYNTHESIS, LIGHT
, MODEL STUDIES, BIOMASS, *LIGHT
OXIDATION LAGOONS, *ALGAE, LIGHT
OXICITY, DISSOLVED OXYGEN, LIGHT
L, ICE, SEASONAL, DIURNAL, LIGHT
, DIA/ *PLANKTON, *ALGAE, *LIGHT
, EPILIMNION, HYPOLIMNION, LIGHT
, TEMPERATURE, EPILIMNION, LIGHT
RODUCT/ *PLANKTON, *ALGAE, LIGHT
OLIGOTROPHY, TEMPERATURE, LIGHT
NUTRIENTS, EUTROPHICATION, LIGHT
HOSPHORUS, PHOTOSYNTHESIS, LIGHT
TESTS, *EFFECTIVE-AVERAGE-LIGHT-
N, ALKALINITY, CALIFORNIA, LIGHT
OLIGOTROPHY, TEMPERATURE, LIGHT
TS, OREGON, BIOINDICATORS, LIGHT
YTA, DIATOMS, CHLOROPHYTA, LIGHT
IC MATTER, TRACE ELEMENTS, LIGHT
RETENTION, PHOTOPERIODISM, LIGHT
IAL NUTRIENTS, INHIBITION, LIGHT
UTROPHICATION, INHIBITION, LIGHT
FFECTS, INHIBITION, LIGHT, LIGHT
IATOMS, INHIBITION, LIGHT, LIGHT
EEZE-THAW TESTS, FREEZING, LIGHT
UTROPHICATION, INHIBITION, LIGHT
LANT POPULATIONS, REVIEWS, LIGHT
SPIRATION, MICROORGANISMS, LIGHT
IS, OXYGEN, TROPHIC LEVEL, LIGHT
*ALGAE, *THERMAL SPRINGS, *LIGHT
SURFACE WATERS, DETRITUS, LIGHT
MPOUNDS, SEWAGE EFFLUENTS, LIGHT
LLA, CARBON RADIOISOTOPES, LIGHT
TS, *ANAEROBIC CONDITIONS, LIGHT
LAGOONS, LABORATORY TESTS, LIGHT
TER, PRIMARY PRODUCTIVITY, LIGHT
ELLA, CARBON, TEMPERATURE, LIGHT
LATION, ENZYMES, CHLORINE, LIGHT
*DIATOMS, *BENTHIC FLORA, LIGHT
, *LABORATORIES, *STREAMS, LIGHT
E, OXYTREMA, ORGANIC ENRICHMENT,
AY, / *MICRONUTRIENTS, *NUTRIENT
IMENT-WATER INFERFACE, MUD-WATER
IMENTS, CATTAILS, RADIOACTIVITY,
ON, CYCLING NUTRIENTS, MUD-WATER

INHIBITION, CHLORINE, NITROGEN CO W69-10171
INHIBITION, MODE OF ACTION, PHOTO W70-03519
INHIBITION, NITROGEN FIXING BACTE W70-04249
INHIBITION, NUTRIENT REQUI: /S, F W69-03370
INHIBITION, LIGHT INTENSITY, LIMN W69-03513
INHIBITION, LIGHT PENETRATION, NI W69-03369
INHIBITION, SILICON, VANADIUM, ZI W69-04801
INHIBITION, LIGHT INTENSITY, LIGH W69-03514
INHIBITION, LIGHT, LIGHT INTENSIT W69-03516
INHIBITION, LIGHT INTENSITY, LIGH W69-03515
INHIBITION, LIGHT PENETRATION, LI W69-04798
INHIBITION, LIGHT, LIGHT INTENSIT W69-03517
INHIBITION, CYANOCOBALAMIN, MUTAN W70-05652
INHIBITOR.: /ANKISTRODESMUS BRAUN W70-05261
INHIBITORS, SUCCESSION, CHLORELLA W70-02510
INHIBITORS, SILTS, ORGANIC MATTER W70-02294
INLETS(WATERWAYS), ORGANIC MA: /A W70-05112
INOCULATION.: /T, PHAGE, FERMENTO W71-02009
INORGANIC PHOSPHATES, SOLUBLE REA W70-05269
INORGANIC COMPOUNDS, SURFACE RUNO W70-04810
INORGANIC COMPOUNDS, METALS, PLAN W69-03695
INPUT, BIOLOGICAL PROPERTIES, *WA W68-00475
INSECT CONTROL, MOSQUITOS, ALGAL W70-05417
INSECTICIDES, WATER POLLUTION EFF W71-03183
INSECTICIDES, HARDNESS(WATER), DI W72-01813
INSECTS.: /WATER POLLUTION EFFECT W71-12077
INSECTS, *AQUATIC PLANTS, AQUATIC W71-09005
INSECTS, *AQUATIC DRIFT, FISH FOO W69-02782
INSECTS, BACTERIA, PROTOZOA, SEWA W71-10786
INSECTS, DIATOMS, CHLOROPHYTA, CH W71-00668
INSECTS, DIPTERA, CRUSTACEA, MOLL W70-04253
INSECTS, FISH, GAME BIRDS, ANIMAL W70-10175
INSECTS, FISH REPRODUCTION, INDUS W73-00748
INSECTS, INVERTEBRATES, BACKGROUN W69-07846
INSECTS, MITES, BACTERIS, FUNGI, W71-00940
INSECTS, PHOSPHORUS RADIOISOTOPES W71-09012
INSECTS, SNAILS, TOXICITY, RIVERS W72-01801
INSOLUBLE WASTES, CULTURE MEDIA.: W68-00248
INSTITUTION, ACROPORA HYACINTHUS, W69-08526
INSTITUTION OF WASHINGTON, CHEMIC W69-06865
INSTREAM TREATMENT, BOTTOM SEALIN W69-09340
INSTRUMENTATION, SEA WATER.: /ES, W70-10437
INTAKES.: /INDUSTRIAL PLANTS, STR W69-05023
INTAKES, NUTRIENTS, QUALITY CONTR W70-00263
INTAKES, SANDS, FLOW RATES, TURBI W70-04199
INTEGRATED CONTROL MEASURES, PHYS W70-07393
INTENSITY.: /UTION, NUTRIENTS, TE W70-05270
INTENSITY.: /ENT, EUTROPHICATION W73-02218
INTENSITY, CARBON DIOXIDE, GROWTH W72-06022
INTENSITY, SLUDGE, DEPTH, ORGANIC W72-06854
INTENSITY, CARBON, SEWAGE TREATME W70-05387
INTENSITY, CHLORELLA, CHLOROPHYTA W70-05381
INTENSITY, CONDUCTIVITY, ALGAE, S W70-05405
INTENSITY, SEASONAL, PHYSIOLOGICA W70-05409
INTENSITY, TEMPERATURE, PRIMARY P W70-04809
INTENSITY, CHLOROPHYLL, NUTRIENTS W70-05752
INTENSITY, LIGHT PENETRATION, LIM W70-04283
INTENSITY, STABILIZATION, PONDS, W70-03923
INTENSITY, PHOTO-ELECTRIC TURBIDO W70-03923
INTENSITY, TEMPERATURE, IONS, SPE W70-03334
INTENSITY, DAPHNIA, CONNECTICUT, W70-03959
INTENSITY, EUTROPHICATION, GRAZIN W70-03978
INTENSITY, SUCCESSION, ZINC RADIO W70-04371
INTENSITY, CHLAMYDOMONAS, CHLOREL W70-04369
INTENSITY, VELOCITY, TEMPERATURE, W69-07862
INTENSITY, LIMNOLOGY,: /N, ESSENT W69-03513
INTENSITY, LIGHT PENETRAT: /TS, E W69-03515
INTENSITY, LIGHT PENETRATION, LIM W69-03516
INTENSITY, LIGHT PENETRATION, LIM W69-03517
INTENSITY, LIGHT PENETRATION, LIM W69-03518
INTENSITY, LIGHT PENETRATION, LIM W69-03514
INTENSITY, WATER TEMPERATURE, DIA W68-00860
INTENSITY, CARBON DIOXIDE, DISSOL W71-00668
INTENSITY, ECOLOGY, LOTIC ENVIRON W70-00265
INTENSITY, SPRING WATER, WATER CH W69-09676
INTENSITY, TURBIDITY, ALGAE, DIAT W70-00713
INTENSITY, LAKES, WATER POLLUTION W69-10167
INTENSITY, CULTURES, RADIOACTIVIT W69-10158
INTENSITY, EUTROPHICATION, NUTRIE W71-13413
INTENSITY, PHOTOPERIODISM, LIGHT W72-00917
INTENSITY, TROPHIC LEVEL, SAMPLIN W70-02504
INTENSITY, NUTRIENT REQUIREMENTS, W70-02965
INTENSITY, NITRATE, EUGLENA, CALC W70-02964
INTENSITY, BIOMASS, DISTRIBUTION, W70-03325
INTENSITY, CURRENTS(WATER), VELOC W70-02780
INTERACTIONS.: /OMMUNITY STRUCTUR W70-03978
INTERACTIONS, *RADIOCARBON BIOASS W70-04765
INTERFACE, PHOSPHORUS, ADSORPTION W69-06859
INTERFACES, ORGANIC MATTER, RADIO W69-08272
INTERFACES, EUGLENOPHYTA, TOXICIT W70-04381

NSING, MINNESOTA, SEDIMENT-WATER
L EFFECTS, DISPERSION, OIL-WATER
E, WATER QUALITY, SEDIMENT-WATER
EEDINGS, SALINE WATER-FRESHWATER
CKEL, DETECTION LIMITS, CHEMICAL
 PHOSPHATES, SOLUBLE R/ ARSENATE
NAVAL RESEARCH, US DEPARTMENT OF
NAVAL RESEARCH, US DEPARTMENT OF
HICATION, *HYDRODYNAMICS, LAKES,
ODUCTION BIOLOGY, RUSSIA, SECOND
ICULA, MEL/ *CHLOROPHYLL A, DATA
MONAS, TEMPERATURE, AMINO ACIDS,
CHNIQUES, TEMPERATURE, PIPTIDES,
 FAUNA, *BIOLOGICAL COMMUNITIES,
ECOSYSTEMS, POPULATION, OREGON,
CULA, SYRACUSE(NY), ENTEROMORPHA
UALITY INDICATORS.:   PERIPHYTON
-PURIFICATION, LABORATORY TESTS,
PTION, HYDROLYSIS, MARINE ALGAE,
TION, MARINE ALGAE, MARINE FISH,
TIVITY, BIOTA, CYCLES, MICHIGAN,
E, RESISTANCE, FISH, AMPHIBIANS,
GANISMS, MICROORGANISMS, SESTON,
OTA, FISH, UTAH, ALGAE, INSECTS,
MS, PLANKTON, ALGAE, PERIPHYTON,
LING NUTRIENTS, PLANKTON, ALGAE,
GAE, IRON, BIRDS, RADIOACTIVITY,
AE, ECOLOGY, SOUTH DAKOTA, FIELD
, OXYGEN, SPORT FISHING, ON-SITE
-SITE DATA COLLECTIONS, *ON-SITE
DEVELOPMENT, WATER QUALITY, SOIL
EMULSIONS, BACKGROUND RADIATION,
DIOISOTOPES, ZINC RADIOISOTOPES,
LLULAR MORPHOLOGY, FLAGELLATION,
                          BROMINE,
ETABOLIS/ *KILL RATES, *BROMINE,
RE, TRACERS, CHLORELLA, HYDROGEN
RATURE, CARBON DIOXIDE, HYDROGEN
Y, PHOSPHORUS, SULPHUR, HYDROGEN
ATER CHEMISTRY, OXYGEN, HYDROGEN
ARBONATES, TEMPERATURE, HYDROGEN
TY, HUMIDITY, BACTERIA, HYDROGEN
OMETRY, CHROMATOGRAPHY, HYDROGEN
OISOTOPES, CHLOROPHYLL, HYDROGEN
S, METHANE, ALKALINITY, HYDROGEN
THANE, ORGANIC LOADING, HYDROGEN
SONS, SEDIMENTS, ALGAE, HYDROGEN
NG TOWERS, DAMS, ALGAE, HYDROGEN
CHEMICAL OXYGEN DEMAND, HYDROGEN
CHEMICAL OXYGEN DEMAND, HYDROGEN
*FISH TOXINS, TOXICITY, HYDROGEN
ZOA, *FUNGI, *BACTERIA, HYDROGEN
DIATION, ACTINOMYCETES, HYDROGEN
PEPTIDES, BIOCHEMISTRY, HYDROGEN
LITY, DISSOLVED SOLIDS, HYDROGEN
E, PHENOLS, OIL WASTES, HYDROGEN
TH, OXYGEN, ALKALINITY, HYDROGEN
TION, DISSOLVED OXYGEN, HYDROGEN
D OXYGEN, BICARBONATES, HYDROGEN
OLLUTION, *TEMPERATURE, HYDROGEN
ERTIARY TREAT/ *ALGAE, *HYDROGEN
, CARBON RADIOISOTOPES, HYDROGEN
PERIPHYTON, PHOSPHORUS, HYDROGEN
R QUALITY, TEMPERATURE, HYDROGEN
OGY, INDUSTRIAL WASTES, HYDROGEN
RATION, CARBON DIOXIDE, HYDROGEN
NE WATER, ACIDIC WATER, HYDROGEN
P, ICE, IRON COMPOUNDS, HYDROGEN
DUCTIVITY, CHLOROPHYTA, HYDROGEN
S, AMMONIA, PHOSPHATES, HYDROGEN
STRATIFICATION, OXYGEN, HYDROGEN
PHATES, IRON COMPOUNDS, HYDROGEN
URFACE WATERS, STREAMS, HYDROGEN
RIBUTARIES, ALKALINITY, HYDROGEN
OPLANKTON, ZOOPLANKTON, HYDROGEN
, NITROGEN, PHOSPHATES, HYDROGEN
LUS, WATER TEMPERATURE, HYDROGEN
*AQUATIC FUNGI, YEASTS, HYDROGEN
ARBON DIOXIDE, *ALGAE, *HYDROGEN
ATION, LIME, *NITROGEN, HYDROGEN
UDGE, RATES, CHLORELLA, HYDROGEN
BACTERIA, ALGAE, SANDS, HYDROGEN
CONDUCTIVITY, CULTURES, HYDROGEN
ACTERIA, ALGAE, MIXING, HYDROGEN
 PHOSPHOROUS COMPOUNDS, HYDROGEN
NT, ORGANIC PESTICIDES, HYDROGEN
TASTE-PRODUCING ALGAE, HYDROGEN
QUALITY, *ALGAE, *ODOR, HYDROGEN
DING, DISSOLVED OXYGEN, HYDROGEN
PERATURE, PILOT PLANTS, HYDROGEN

INTERFACES, CALIFORNIA, BIOASSAY,   W70-02775
INTERFACES, PERSISTENCE, WATER PO   W70-01777
INTERFACES, OREGON, SESSILE ALGAE   W72-06051
INTERFACES, COLLOIDS.: /PONDS, BR   W73-00748
INTERFERENCES.:   *NI   W71-11687
INTERFERENCE, DISSOLVED INORGANIC   W70-05269
INTERIOR, MADISON(WIS), EUROPE, W   W70-03975
INTERIOR, MADISON(WIS), EUROPE, W   W70-03975
INTERNAL WATER, WAVES(WATER), DIU   W70-09889
INTERNATIONAL OCEANOGRAPHIC CONGR   W69-07440
INTERPRETATION, *RHONE RIVER, NAV   W73-01446
INTERTIDAL AREAS, CARBOHYDRATES,   W70-03952
INTERTIDAL AREAS, SOIL BACTERIA,   W70-04365
INTERTIDAL ALGAE, PUGET SOUND, FL   W68-00010
INTERTIDAL AREAS, PHYSIOLOGICAL E   W70-03325
INTESTINALIS, CLADOPHORA, CYCLOTE   W70-03974
INVENTORY(LAKE SUPERIOR), WATER Q   W70-05958
INVERTEBRATES, FISH, WATER ANALYS   W72-01801
INVERTEBRATES, NUCLEAR WASTES, WA   W71-09863
INVERTEBRATES, CRUSTACEANS, MOLLU   W71-09850
INVERTEBRATES, FISH, TEMPERATURE,   W69-09334
INVERTEBRATES, AQUATIC POPULATION   W68-00859
INVERTEBRATES, SHELLFISH, WASTE D   W69-04276
INVERTEBRATES, BACKGROUND RADIATI   W69-07846
INVERTEBRATES, ADSORPTION, PHYTOP   W69-07853
INVERTEBRATES, TURBIDITY, WORMS,   W69-07861
INVERTEBRATES, PLANKTON, FISH, SE   W69-08269
INVESTIGATIONS, WATER POLLUTION,   W69-00659
INVESTIGATIONS.: /GHT PENETRATION   W68-00487
INVESTIGATIONS, FREQUENCY, *FREQU   W70-09976
INVESTIGATIONS, IMPOUNDED WATERS   W72-11906
IODINE RADIOISOTOPES, NUTRIENT RE   W69-10163
IODINE RADIOISOTOPES, COBALT RADI   W71-09850
IODINE TEST, GRAM REACTION, CYTOP   W70-03952
IODINE.:   W70-05178
IODINE, PHOTO REACTIVITY, ALGAL M   W70-02363
ION CONCENTRATION, ALKALINITY, DI   W70-02504
ION CONCENTRATION, BIOCARBONATES,   W70-02249
ION CONCENTRATION, TIME.: /ECOLOG   W70-02965
ION CONCENTRATION, PHOTOSYNTHESIS   W70-03309
ION CONCENTRATION, ALKALINITY,: /   W70-02804
ION CONCENTRATION, SULFATES, RESP   W70-01073
ION CONCENTRATION, FLUORESCENCE,   W70-01074
ION CONCENTRATION, MAGNESIUM, CAL   W70-01579
ION CONCENTRATION, TOXINS, OXYGEN   W70-02803
ION CONCENTRATION, TEMPERATURE, C   W71-11375
ION CONCENTRATION, ELECTRIC POWER   W71-09548
ION CONCENTRATION, TEMPERATURE, A   W71-11393
ION CONCENTRATION, ORGANIC LOADIN   W72-03299
ION CONCENTRATION, TEMPERATURE, P   W71-13413
ION CONCENTRATION, DISSOLVED OXYG   W71-07731
ION CONCENTRATION, TEMPERATURE, A   W71-05267
ION CONCENTRATION, NITRATES, POTA   W70-00719
ION CONCENTRATION, OXIDATION,: /   W70-01068
ION CONCENTRATION, TEMPERATURE, N   W70-00265
ION CONCENTRATION, FLUORIDES, ALK   W70-00263
ION CONCENTRATION, STANDING CROP,   W71-00665
ION CONCENTRATION, EPILIMNION, PH   W70-06975
ION CONCENTRATION, NUISANCE ALGAE   W70-06225
ION CONCENTRATION, NUTRIENTS, ALG   W70-08834
ION CONCENTRATION, *NUTRIENTS, *T   W70-08980
ION CONCENTRATION, SCENEDESMUS, E   W70-08097
ION CONCENTRATION, ALKALINITY, HA   W71-04206
ION CONCENTRATION, OXYGEN, CARBON   W71-03031
ION CONCENTRATION, IRON COMPOUNDS   W70-04510
ION CONCENTRATION, PHOSPHATES, NI   W70-05760
ION CONCENTRATION, WATER CHEMISTR   W70-05424
ION CONCENTRATION, ACIDITY, MUD,   W70-05399
ION CONCENTRATION, NITRATES, PHOS   W70-05404
ION CONCENTRATION, PHYTOPLANKTON,   W70-05405
ION CONCENTRATION, TEMPERATURE, D   W70-05387
ION CONCENTRATION, ALKALINITY, CA   W70-03334
ION CONCENTRATION, IRON, MAGNESIU   W70-03501
ION CONCENTRATION, NUTRIENTS, PHO   W70-03983
ION CONCENTRATION, ALGAE.: / PHYT   W70-04368
ION CONCENTRATION, CYCLING NUTRIE   W70-04381
ION CONCENTRATION, FUNGI, YEASTS,   W68-00891
ION CONCENTRATION, IRON, SULFATE,   W69-00096
ION CONCENTRATION, PHOSPHORUS, NI   W68-00855
ION CONCENTRATION, *PHOSPHORUS, S   W68-00012
ION CONCENTRATION, PLANT TISSUES,   W68-00248
ION CONCENTRATION.: /CTROMETERS,   W69-08267
ION CONCENTRATION, IRON, LAKES, M   W69-07833
ION CONCENTRATION, TEMPERATURE, T   W72-06854
ION CONCENTRATION, CONDUCTIVITY,   W72-06051
ION CONCENTRATION, TEMPERATURE, C   W72-11105
ION CONCENTRATION, *ORGANI: /SIS,   W72-11906
ION CONCENTRATION.: /ENT, *WATER   W72-09694
ION CONCENTRATION, BIOCHEMICAL OX   W72-10573
ION CONCENTRATION, ORGANIC LOADIN   W72-06022

459

IARY TREATMENT, ELECTRODIALYSIS,
ANIC COMPOUNDS, CHEMICAL WASTES,
S, SHIPS, SUBMARINES, EFFLUENTS,
EATMENT, ALGAL CONTROL, AMMONIA,
BOD, DETENTION TIME.: *SULFATE
GION, MONOMICTIC, DEPTH EFFECTS,
RAPHIES, CHEMICAL PRECIPITATION,
ACERS, PHOSPHORUS RADIOISOTOPES,
WATER, *PEPTIDES, COPPER, IRON,
A, LIGHT INTENSITY, TEMPERATURE,
C OCEAN, DIATOMS, NERITIC ALGAE,
STROENTERITIS, SYNURA, ANABAENA,
ACTIVATED CARBON, ACTINOMYCETES,
NTROL, POLLUTANT IDENTIFICATION,
        *METABOLITES, *CEDAR RIVER(
ASSIUM PERMANGANATE, *DES MOINES(
CENTER LAKE(IOWA), DIAMOND LAKE(
FLOS-AQUAE, LYNGBYA, STORM LAKE(
A, STORM LAKE(IOWA), CENTER LAKE(
, IRON BACTERIA, TASTE, FILTERS,
SILVER-110M, CERIUM, GUETTARDA,
PERTY VALUES, RECREATION DEMAND,
NOPHYTA, SULFUR BACTERIA, COLOR,
OPHORA BLANKE/ ALUMINUM SULFATE,
ATES, PHOSPHATES, PHYTOPLANKTON,
BIOCHEMICAL OXYGEN DEM/ *ALGAE,
IUM, AMMONIA, NITROGEN FIXATION,
, CHEMICAL ANALYSIS, PHOSPHATES,
NEROBIC DIGESTION, INCINERATION,
S, IONS, HUMIC ACIDS, CARP, ICE,
E, SCENEDESMUS, LIGHT, CULTURES,
TES, HYDROGEN ION CONCENTRATION,
ND, *BACTERIA, ACTIVATED SLUDGE,
UTRIENTS, PHOSPHATES, SILICATES,
HIBITORS, SILTS, ORGANIC MATTER,
ALUMINUM SULFATE, IRON CHLORIDE,
ALUMINUM SULFATE, FERRICHLORIDE,
BIC BACTERIA, WATER TEMPERATURE,
S, BIOMASS, DIATOMS, CYANOPHYTA,
NTATION, TENNESSEE RIVER, ALGAE,
NITRATES, PHOSPHATES, POTASSIUM,
RADIOISOTOPES, MANGANESE, ALGAE,
CES, SEWAGE, SULFATE, MANGANESE,
OXIDE, SCENEDESMUS, CHLOROPHYLL,
, DIATOMS, WEIGHT, PILOT PLANTS,
OCE/ *ALGAE, *HERRINGS, COPPER,
TY, SULFATES, CHLORIDES, SODIUM,
IC CONDITIONS, HYDROGEN SULFIDE,
NUTRIENTS, PHOSPHORUS, NITROGEN,
ANALYSIS, ORGANIC MATTER, ALGAE,
*FRESH WATER, *PEPTIDES, COPPER,
RES, HYDROGEN ION CONCENTRATION,
AMS, HYDROGEN ION CONCENTRATION,
OR, TURBIDITY, HYDROGEN SULFIDE,
M, MAGNESIUM, POTASSIUM, SODIUM,
ONMENTAL EFFECTS, HETEROGENEITY,
OM SEDIMENTS, SODIUM, POTASSIUM,
DIOISOTOPES, ZINC RADIOISOTOPES,
IATOMS, CYANOPHYTA, CHLOROPHYTA,
ENVIRONMENTAL EFFECTS, GRAZING,
IS, TRACE ELEMENTS, TEMPERATURE,
, POTASSIUM, CALCIUM, MAGNESIUM,
, HYPOLIMNION, ANIMALS, DENSITY,
, LAKES, MICHIGAN, BICARBONATES,
OPHIC LEVEL, ZINC RADIOISOTOPES,
STS, HYDROGEN ION CONCENTRATION,
ROGEN, *CONDUCTIVITY, *LEACHING,
TE/ *CONCENTRATIONS, AEQUIPECTEN
CROPS, TOXICITY, SOIL POROSITY,
UTION, *WATER POLLUTION SOURCES,
AKES, PLANTS, *ALGAE, *SALINITY,
CRAYFISH, MOSQUITOES, BACTERIA,
SLANDS, MOU/ *NITROGEN FIXATION,
TE WAT/ *FARM WASTES, *SPRINKLER
CURRENT, PUGET SOUND(WASH), LONG
GLUCOSE, GONIDIA FORMATION, LONG
A), PENGUINS, LAK/ PRINCE EDWARD
S, AEQUIPECTEN IRRADIANS, PIVERS
            *SAKHALIN
AGANSETT MARINE LABORATORY(RHODE
ROPHY, B/ NARRAGANSETT BAY(RHODE
SIMPLEX, FLAGELLATES, ELEUTHERA
ER TEMPERATURE, *SEASONAL, RHODE
WAVES(WATER), CONNECTICUT, RHODE
OSA, DIATOMA ELONGATUM, MELOSIRA
GAE, *LITTORAL, *AQUATIC PLANTS,
TROPHIC, *PHYTOPLANKTON, BRITISH
TOXINS, KILLIFISHES, *TOXICITY,
LIFE CYCLES, GENETICS, E. COLI,
ETTE, UNIALGAL, NITZSCHIA, ALGAE

ION EXCHANGE, ACTIVATED SLUDGE, D   W69-01453
ION EXCHANGE, BIOCHEMICAL OXYGEN   W70-10433
ION EXCHANGE, RESINS, LEAKAGE, CE   W70-00707
ION EXCHANGE, REVERSE OSMOSIS, DI   W70-01981
ION, *DESULFOVIBRIO, FACULTATIVE,   W70-01971
IONIC CONCENTRATION, SEASONAL EFF   W70-01933
IONS, BIOCHEMICAL OXYGEN DEMAND,   W69-00446
IONS, HUMIC ACIDS, CARP, ICE, IRO   W70-05399
IONS, PHOSPHATES, ORGANIC COMPOUN   W69-10180
IONS, SPECTROPHOTOMETRY, SUSPENDE   W70-03334
IONS, WATER POLLUTION EFFECTS, WA   W70-00707
IOWA RIVER, CYCLOTELLA, FRAGILARI   W70-08107
IOWA.: /, *POTASSIUM COMPOUNDS, *   W72-04242
IOWA.: /, *ODOR, WATER QUALITY CO   W72-13605
IOWA).:   W72-13605
IOWA).:   *POT W72-04242
IOWA).: /NGBYA, STORM LAKE(IOWA),   W70-02803
IOWA), CENTER LAKE(IOWA), DIAMOND   W70-02803
IOWA), DIAMOND LAKE(IOWA).: /NGBY   W70-02803
IOWA, DIATOMS, EUGLENA, CHLAMYDOM   W70-08107
IPOMOEA, PISONIA, PANDANUS, MUSCL   W69-08269
IR: /RON, MAGNESIUM, COOLING, PRO   W70-02251
IRON BACTERIA, TASTE, FILTERS, IO   W70-08107
IRON CHLORIDE, IRON SULFATE, CLAD   W69-10170
IRON COMPOUNDS, ALGAE, PLANTS, WA   W69-10170
IRON COMPOUNDS, HYDROGEN SULFIDE,   W70-06925
IRON COMPOUNDS, LIGHT PENETRATION   W70-02255
IRON COMPOUNDS, HYDROGEN ION CONC   W70-03334
IRON COMPOUNDS, SULFATES.: /ON, A   W70-04060
IRON COMPOUNDS, HYDROGEN ION CONC   W70-05399
IRON COMPOUNDS, DIATOMS, CYANOPHY   W70-05409
IRON COMPOUNDS, HYDROGEN SULFIDE,   W70-04510
IRON COMPOUNDS, LIME, OXYGEN, TER   W68-00256
IRON OXIDES, *PLANT POPULATIONS,   W68-00860
IRON OXIDES, BACTERIA, ALGAE, FUN   W70-02294
IRON SULFATE, CLADOPHORA BLANKETS   W69-10170
IRON SULFATE, GALVANIZING WASTES,   W70-05668
IRON.: /ER, AIR TEMPERATURE, AERO   W69-07096
IRON, ALGICIDES, ZOOPLANKTON, FIS   W70-08654
IRON, ALKALINITY, HARDNESS, RADIO   W73-00750
IRON, BICARBONATE, ZINC, COBALT,   W70-03978
IRON, BIRDS, RADIOACTIVITY, INVER   W69-08269
IRON, CARBON DIOXIDE.: /H SUBSTAN   W70-02969
IRON, CHELATION, ENZYMES, CHLORIN   W70-02964
IRON, CHLORELLA, SODIUM, POTASSIU   W70-03512
IRON, COBALT, MANGANESE, ATLANTIC   W69-08525
IRON, DISSOLVED SOILDS, PHOSPHATE   W69-05894
IRON, FUNGI, CLAYS, BRACKISH WATE   W70-05761
IRON, GROWTH RATES, CHLOROPHYTA,   W70-03983
IRON, HY: /LVED OXYGEN, CHEMICAL   W70-05760
IRON, IONS, PHOSPHATES, ORGANIC C   W69-10180
IRON, LAKES, MAGNESIUM, NUTRIENTS   W69-07833
IRON, MAGNESIUM, NITRATES, PHOSPH   W70-03501
IRON, MAGNESIUM, COOLING, PROPERT   W70-02251
IRON, MANGANESE, TRACE ELEMENTS,   W70-02804
IRON, MANGANESE, WATERFOWL, CYANO   W70-00273
IRON, MANGANESE, CALCIUM, PIGMENT   W70-09889
IRON, MANGANESE, BENTHIC FLORA, S   W69-07845
IRON, MANGANESE, EUTROPHICATION,   W69-05704
IRON, MANGANESE, MOLYBDENUM, PLAN   W69-04801
IRON, NUTRIENTS, LIGHT PENETRATIO   W70-01933
IRON, ORGANIC MATTER, NITROGEN FI   W70-00719
IRON, PHOSPHORUS, SILICA, SAMPLIN   W70-06975
IRON, PHOSPHORUS, NITROGEN, WATER   W72-03183
IRON, SEDIMENTS, FISH, RADIOECOLO   W69-07853
IRON, SULFATE, TEMPERATURE, KENTU   W69-00096
IRON, TEMPERATURE, SOIL ANALYSIS,   W72-11906
IRRADIANS, PIVERS ISLAND(N C), DE   W69-08267
IRRIGATION, GROUND WATER, RUNOFF,   W69-05323
IRRIGATION, AQUATIC ALGAE, WATER   W69-07096
IRRIGATION EFFECTS, DRAINAGE EFFE   W69-00360
IRRIGATION, WILLOW TREES, FE: /S,   W70-05263
IRRIGATION, ORGANIC MATTER, *GRAS   W72-08110
IRRIGATION, *ORGANIC WASTES, *WAS   W71-00940
ISLAND SOUND, NARRAGANSETT BAY, C   W69-10161
ISLAND SOUND(NY), ICELAND, TRITIA   W69-09755
ISLAND(CANADA), NOVA SCOTIA(CANAD   W70-05399
ISLAND(N C), DETECTOR, BEAUFORT(N   W69-08267
ISLAND(USSR).:   W72-10623
ISLAND), DILUTION RATE, FRUITING   W70-01073
ISLAND), TAXONOMIC TYPES, HETERET   W70-00713
ISLAND, NITZSCHIA SERIATA, NAVICU   W70-00707
ISLAND, SAMPLING, ECOLOGY, AMMONI   W70-00713
ISLAND, TEMPERATURE, NUTRIENT REQ   W69-10161
ISLANDICA, NITZSCHIA ACICULARIS,   W69-10158
ISLANDS, ECOLOGICAL DISTRIBUTION,   W72-10623
ISLES, BIOINDICATORS, *ESTIMATING   W68-00255
ISOLATION.: /QUATIC ALGAE, *ALGAL   W70-08372
ISOLATION.: /MICAL PRECIPITATION,   W71-11823
ISOLATION, CAPILLARITY, CAPILLARI   W70-04469

GES, DIATOMS, ALGAE, METABOLISM,
PER CHLORIDE, UPTAKE, FREUNDLICH
CADDISFLIES, FOOD CHAINS, STABLE
                 *LAKE KINNERET(
LLATORIA, NATIONAL WATER CARRIER(
ON, 'EXTRA-DEEP PONDS', ASHKELON(
E, CANADA, EUROPE, SOUTH AFRICA,
BORREVANNET NORWAY, LAKE LUGANO
, FILTER BLOCKING, HETERETROPHS,
EED, WHITE WATER LILY, SWIMMERS'
SCULUS, HAEMOCYTOMETER COUNTING,
ZOMENON, ANABAENA, GLOEOTRICHIA,
LANKTON CROPS, TR/ TAKASUKA POND(
POLONICUM, PLANKTO/ LAKE SAGAMI(
RES, MONOALGAL, PORTUGAL, TAPES,
GULF OF MARSEILLES, ASTERIONELLA
, RESERVOIRS, WATER QUALITY, NEW
LAN, DELTA WATER/ SACRAMENTO-SAN
OXYGEN DYNAMICS, SACRAMENTO-SAN
DISSOLVED OXYGEN VARIATIONS, SAN
R(CALIF), TUOLUMNE RIVER(CA/ SAN
WATER DETERIORATION, DRAIN, SAN
PULL IN LAKE, GRAPHITE LAKE, OLD
R LAKE(CALIF), EEL RIVER(CALIF),
E(MINN), LAKE OWASSO(MINN), LAKE
ICKEL, PUERTO RICO, LA PARGUERA,
AKES, STREAMS, SEWAGE, REMEDIES,
S, 2,4-D, AMMATE, CHLORAMINE, DE-
RCES, GREENHOUSE STUDIES, GROWTH
OTOSYNTHETIC OXYGEN, LEMMON(SD),
ED BAUXITE, ACID-WASHED BAUXITE,
RESEARCH CONTRACTS, CHENEY DAM,
PECTINATA, POLYARTHRA VULGARIS,
LA, FLORIDA CURRENT, YTTRIUM-90,
TA(WIS), LAKE WAUBESA(WIS), LAKE
NA(WIS), LAKE WAUBESA(WIS), LAKE
NA(WIS), LAKE WAUBESA(WIS), LAKE
OTA(WIS), LAKE MONONA(WIS), LAKE
NA(WIS), LAKE WAUBESA(WIS), LAKE
NA(WIS), LAKE WAUBESA(WIS), LAKE
ONA(WIS), LAKE WINGRA(WIS), LAKE
LAKE MENDOTA, LAKE WAUBESA, LAKE
RA, ASTERIONELLA, TYPHOID FEVER,
ION, IRON, SULFATE, TEMPERATURE,
STE MICROFLORA, MCCREARY COUNTY,
TIONSHIPS, GEOGRAPHICAL REGIONS,
INS, *RADIOACTIVITY, *FOOD WEBS,
RIONELLA, MELOSIRA, MICROCYSTIS,
PTOMUS, EPISCHURA, ORTHOCYCLOPS,
ATILE ACIDS, NON-VOLATILE ACIDS,
, PSOROPHORA CILIATA, ARTIFICIAL
TIFICATION, *SIGNIFICANCE, ALGAL
MARRON RIVER(OKLA), BIXBY(OKLA),

URICH(SWITZERLAND), SWITZERLAND,
ORA, BROMINE, FILAMENTOUS ALGAE,
Y, AQUATIC ALGAE, *ALGAL TOXINS,
RA, ALKALINITY, BULLHEADS, EELS,
ALGAE, *PESTICIDES, *FISH, *FISH
ER LAKE(MINN), TROUT LAKE(MINN),
DOSA, *ALGAE BIOMASS.:
IOASSAYS, NUTRIENTS, PHOSPHORUS,
NEDESMUS, PHYSIOLOGICAL ECOLOGY,
PESTICIDES, RESIDUES, *PESTICIDE
RS, FISHKILL, HAZARDS, PESTICIDE
SORPTION, SOILS, COA/ *PESTICIDE
PROTEINS, CARBOHYDRATES, LIPIDS,
NITY, BICARBONATES, BACTERICIDE,
 EFFECTS, ANALYTICAL TECHNIQUES,
, DIATOMS, CARBON RADIOISOTOPES,
, CARBON RADIOISOTOPES, ENZYMES,
LTURES, PSEUDOMONAS, METABOLISM,
ROPHIC POTENTIAL, SWEDEN, ENZYME
IVE TRANSPORT, CARBON-14, ENZYME
EWEAVER-BURK EQUATION, MICHAELIS
BULENCE, CHLORRELA, SCENEDESMUS,
INCIDENT LIGHT UTI/ ALGAL GROWTH
                           *LAKE
GAL NUTRITION, ALGAL PHYSIOLOGY,

, MADISON(WIS), LAKE ZOAR(CONN),
N FLOS-AQUAE, *CARBON-14/ *UPPER
RE.), *OSCILLATORA.:   *UPPER
ONAS, THALASSIOSIRA, AMPHIDINIUM
AUSTRIA), VERTICAL, CHALK/ *LAKE
CANADA, AQUATIC ECOSYSTEM MODEL/
II, CANDIDA ZEYLANOIDES, CANDIDA
LAKE(ONTARI/ *ONTARIO, *CANADA,
OLUTION/ *MICROCYSTIS AERUGINOSA(
YSTIS AERUGINOSA, COELOSPHAERIUM

ISOPODS, AMPHIPODA, FISH, GROWTH   W69-09742
ISOTHERM, BLUE-GREEN ALGAE, GREEN   W70-08111
ISOTOPES, RADIOACTIVE WASTE DISPO   W70-02786
ISRAEL).:   W70-09889
ISRAEL), THRESHOLD ODOR NUMBERS.:   W70-02373
ISRAEL), TEMPERATURE PROFILES, VO   W70-04790
ISRAEL, MICROCYSTIS, PRYMNESIUM P   W70-05372
ITALY, LAKE MARIDALSVANNET NORWAY   W69-07833
ITALY, LAKE BORREVANNET NORWAY, L   W69-07833
ITCH.: /D CELERY, BUR REED, DUCKW   W70-05568
ITHACA(NY), GLENODINIUM QUADRIDEN   W69-06540
JANESVILLE(WIS), SPAULDING'S POND   W70-03310
JAPAN), LAKE MENDOTA(WIS), PHYTOP   W69-05142
JAPAN, *GLENODININE, *PERIDINIUM   W70-08372
JAPAN, AMPHIDINIUM, BLOOMS, MYTIL   W70-05372
JAPONICA, CHAETOCEROS LAUNDERI, L   W70-04280
JERSEY, TEMPERATURE, ORGANIC MATT   W69-10157
JOAQUIN DELTA, CALIFORNIA WATER P   W70-05094
JOAQUIN ESTUARY, ENVIRONMENTAL RE   W70-05094
JOAQUIN RIVER, POTOMAC ESTUARY, S   W69-07520
JOAQUIN RIVER(CALIF), MERCED RIVE   W70-02777
JOAQUIN VALLEY MASTER DRAIN, SAN   W71-09577
JOHN LAKE, PINGO LAKE, TWELVE MIL   W69-08272
JOINT INDUSTRY/GOVERNMENT TASK FO   W70-02777
JOSEPHINE(MINN), GRAY'S BAY(MINN)   W70-05269
JOYUDA.: /OTEA FLABELLUM, ZINC, N   W69-08525
JUDICIAL DECISIONS, LEGAL ASPECTS   W69-06909
K-PRUF-21, BENOCLORS, NIGROSINE D   W70-05263
K: /OMPOSITION(ENGLAND), FOOD SOU   W69-06865
KADOKA(SD).:   *PH   W72-01818
KANPUR(INDIA).: /ATED SAND, CRUSH   W70-08628
KANS.:   USBR   W72-01773
KARATELLA COHCLEARIS, ASPLANCHA,   W70-02249
KATODINIUM ROTUNDATA.: /A, NAVICU   W70-00707
KEGONSA(WIS).: /(WIS), LAKE MENDO   W70-02787
KEGONSA(WIS), WEED BEDS, PHOSPHOR   W70-02969
KEGONSA(WIS).: /A(WIS), LAKE MONO   W72-01813
KEGONSA(WIS), LAKE WAUBESA(WIS),   W69-06276
KEGONSA(WIS).: /H(WIS), LAKE MONO   W70-05113
KEGONSA(WIS), BLOOMING, LAKE MEND   W70-05112
KEGONSA(WIS), LAKE WAUBESA(WIS),   W69-03366
KEGONSA, LAKE ZURICH, OSCILLATO: /   W69-09349
KENNEBEC RIVER(MAINE), COBBOSSEEC   W70-08096
KENTUCKY.: /DROGEN ION CONCENTRAT   W69-00096
KENTUCKY.:   ACID WA   W69-00096
KENTUCKY, ACID STREAMS, MICROFLOR   W68-00891
KENTUCKY, LIFE HISTORY STUDIES, G   W69-09742
KERATELLA, POLYARTHRA, ASPLANCHNA   W70-05375
KERATELLA, POLYARTHRA, CONOCHILUS   W70-05760
KETOACIDS, TAMIYA'S MINERAL MEDIU   W70-04195
KEY.: /DRIMACULATUS, MADISON(WIS)   W70-06225
KEY, CLUMP COUNT, DESMIDS, FLAGEL   W70-05389
KEYSTONE RESERVOIR(OKLA), CAROTEN   W70-04001
KIEV RESERVOIR(USSR).:   W71-00221
KILCHBERG(SWITZERLAND), ZOLLIKON(W70-05668
KILL RATES, PHOTO REACTIVITY, CEL   W70-02364
KILLIFISHES, *TOXICITY, ISOLATION   W70-08372
KILLIFISHES, TROUT, SUCKERS, BASS   W70-05263
KILLS, *ROTIFERS, *ECOLOGY, *ECOS   W72-04506
KIMBALL LAKE(MINN), CRYPTOPHYTA,   W70-05387
KINETIC MODELS, CHLORELLA PYRENOI   W73-02218
KINETIC THEORY, ENVIRONMENTAL FAC   W70-02779
KINETICS, FARM WASTES, BENTHOS, N   W70-02775
KINETICS, *PERSISTENCE, EVAPORATI   W70-09768
KINETICS, PESTICIDE TOXICITY, POL   W69-08565
KINETICS, *PESTICIDE REMOVAL, *AD   W69-09884
KINETICS, MATHEMATICAL MODELS, NE   W71-13413
KINETICS, HARDNESS, COLIFORMS, CY   W71-05267
KINETICS, AQUATIC BACTERIA, ALGAE   W70-05109
KINETICS, TRITIUM, CHLOROPHYLL.: /   W70-04811
KINETICS, PHOTOSYNTHESIS, ORGANIC   W70-04287
KINETICS, NITROGEN COMPOUNDS.: /U   W70-04280
KINETICS, LINEWEAVER-BURK EQUATIO   W70-04194
KINETICS, GLUCOSE, MECHAELIS-MENT   W70-04284
KINETICS, FLAGELLATES, LAKE ERKEN   W70-04194
KINETICS, ECOSYSTEMS.: /TES, *TUR   W69-03730
KINETICS, MONOD GROWTH EQUATION,   W69-03730
KINNERET(ISRAEL).:   W70-09889
KJELDAHL PROCEDURE, LUXURY CONSUM   W69-05868
KLAMATH LAKE.:   W69-04802
KLAMATH LAKE(ORE), LAKE WASHINGTO   W70-04506
KLAMATH LAKE(ORE.), *APHANIZOMENO   W72-06052
KLAMATH LAKE(ORE.), AGENCY LAKE(O   W72-06051
KLEBSI, PORPHYRIDIUM CRUENTUM, RA   W70-00708
KLOPEINER(AUSTRIA), LAKE WORTHER(W70-05405
KOOTENAY LAKE, BRITISH COLUMBIA,   W69-06536
KRUSEI, CANDIDA OBTUSA, CANDIDA T   W70-04368
KUSHOG LAKE(ONTARIO), TWELVE MILE   W70-02795
KUTZ), CULTURE MEDIA, CHU NO 10 S   W70-03507
KUTZINGIANUM, ANABAENA FLOS AQUAE   W69-08283

461

*COTTUS CAROLINAE, MEAD COUNTY(KY), THORIUM-232, BISMUTH-214, AC   W69-09742
LLUM, ZINC, NICKEL, PUERTO RICO,   LA PARGUERA, JOYUDA.: /OTEA FLABE   W69-08525
 RADIONUCLIDES, ASSIMILATION, BA- LA, HALF-LIFE, PTYCHOCHEILUS OREG   W69-07853
HOSPHATES, METAPHOSPHATES, PULSE- LABELING, POLYPHOSPHATES, FRACTIO   W71-03033
ATER POLLU/ *DIELDRIN, *DIATOMS,   LABORATORY TESTS, INSECTICIDES, W   W71-03183
TEMPERATURE, *THERMAL POLLUTION,   LABORATORY ANIMALS, ACCLIMATIZATI   W71-02494
LORA, *ALGAE, *HYDRAULIC MODELS,   LABORATORY TESTS, PERIPHYTON, STR   W71-00668
LISM, *CHLORELLA, *CHLAMYDOMONAS   LABORATORY TESTS, GAS CHROMATOGRA   W70-08652
URBIDITY, ANALYTICAL TECHNIQUES,   LABORATORY TESTS, FOREIGN RESEARC   W70-08628
GAE, *GROWTH RATES, *PHOSPHATES,   LABORATORY TESTS, MINNESOTA, NUTR   W70-08637
SODIUM, WATER POLLUTION EFFECTS,   LABORATORY TESTS.: / PHOSPHATES,   W70-07257
ANAEROBIC TREATMENT, *AUSTRALIA,   LABORATORY-SCALE STUDIES, PLANT-S   W70-06899
LK SUBSTR/ *FACULTATIVE LAGOONS,   LABORATORY MODEL, COD REMOVAL, MI   W70-07031
*BIODEGRADATION, *OXYGEN DEMAND,   LABORATORY TESTS, LAGOONS, OXIDAT   W70-06600
TES, DESIGN APP/ DETENTION TIME,   LABORATORY MODELS, BOD REMOVAL RA   W70-07034
T, SANITARY ENGINEERING RESEARCH   LABORATORY(CALIF), NATIONAL EUTRO   W70-02775
, IMMERSION, NARRAGANSETT MARINE   LABORATORY(RHODE ISLAND), DILUTIO   W70-01073
TH RATES, ANALYTICAL TECHNIQUES,   LABORATORY EQUIPMENT, ASBESTOS CE   W71-07339
HOTOSYNTHESIS, DISSOLVED OXYGEN,   LABORATORY TESTS, TEMPERATURE, WA   W71-12183
LAKES, *PHYTOPLANKTON, BIOASSAY,   LABORATORY TESTS, PRODUCTIVITY, A   W72-00845
CATION, PHOTOSYNTHESIS, LAGOONS,   LABORATORY TESTS, LIGHT INTENSITY   W72-00917
CKISH, ECOSYSTEMS, FAUNA, FLORA,   LABORATORY, MINERALS, POLLUTION.:   W72-02203
ICATION, DAPHNIA, CHLAMYDOMONAS,   LABORATORY TESTS, PARTICLE SIZE,   W72-02417
 OXYGENATION, SELF-PURIFICATION,   LABORATORY TESTS, INVERTEBRATES,   W72-01801
S, MICROORGANISMS, ALGAE, FUNGI,   LABORATORY TESTS, BIOCHEMICAL OXY   W72-01817
REACTOR.:   SAVANNAH RIVER   LABORATORY, PLECTONEMA BORYANUM,   W69-07845
IENT REQUIREMENTS, GROWTH RATES,   LABORATORY TESTS, CULTURES, SODIU   W70-03507
INTERTIDAL AREAS, SOIL BACTERIA,   LABORATORY TESTS.: /E, PIPTIDES,   W70-04365
OURCES, WATER POLLUTION EFFECTS,   LABORATORY TESTS, MODEL STUDIES,   W70-04074
H REGULATORS, FRESH WATER LAKES,   LABORATORY TESTS.: / *PLANT GROWT   W68-00856
TMENT, BIBLIOG/ *EUTROPHICATION,   LABORATORY TESTS, BIOLOGICAL TREA   W68-00855
D HYDROCARBON PESTICIDES, ALGAE,   LABORATORY TESTS, WATER POLLUTION   W72-11105
SM, EPIPHYTOLOGY, ON-SITE TESTS,   LABORATORY TESTS, LOTIC ENVIRONME   W72-05453
PROBITY, PRAGUE, CZECHOSLOVAKIA,   LABORATORY PROCEDURES, STATISTICA   W72-12240
MINN), LAKE VADNAIS(MINN), MILLE   LAC(MINN), DEMING LAKE(MINN), ORC   W70-05269
RUGINOSA, STAPHYLOCOCCUS AUREUS,   LACTOBACILLUS CASEI, BACILLUS THU   W71-11823
TION RATE, FRUITING BODIES, ULVA   LACTUCA VAR LATISSIMA, POLYSIPHON   W70-01073
CYCLOTELLA COMENSIS, RHODOMONAS   LACUSTR: /LCULATED VOLUME METHOD,   W70-03959
II, CALOTHRIX FUSCA, MICROCOLEUS   LACUSTRIS (RAB) FARLOW, CLADOPHOR   W70-04468
XONOMIC DESCRIPTIONS, GONGROSIRA   LACUSTRIS BRAND, GONGROSIRA DEBAR   W70-04468
SE WASTES, EXPERIMENTAL STUDIES,   LAGOON DESIGN, SCALE-UP FACTORS,   W70-07508
ANURE, ANAEROBIC LAGOON, AERATED   LAGOON.:   *M   W70-07491
          *MANURE, ANAEROBIC   LAGOON, AERATED LAGOON.:   W70-07491
RK, VIRGINIA, ORGANIC COMPOUNDS,   LAGOON, FILTERS, PESTICIDES, SLUD   W72-03641
UENT, QUALITY CONTROL, OXIDATION   LAGOON, NITROGEN, NUTRIENTS, FILT   W71-13313
SIGN, DESIGN CRITERIA, OXIDATION   LAGOONS PHOTOSYNTHESIS.: /ION, DE   W70-02609
TECHNIQUES, CHLORELLA, OXIDATION   LAGOONS.: /BOLISM, RADIOACTIVITY   W70-02198
STES, AEROBIC LAGOONS, ANAEROBIC   LAGOONS.:   *POTATO PROCESSING WA   W71-08970
ASTE WATER TREATMENT, *OXIDATION   LAGOONS.: /R, SEWAGE DISPOSAL, *W   W71-09548
ON, INDUSTRIAL WASTES, OXIDATION   LAGOONS.: /TMENT, WATER UTILIZATI   W70-07508
SCENEDESMUS, SAMPLING, OXYDATION   LAGOONS.: /MYDOMONAS, CHLORELLA,   W70-10173
BACTERIA, ANAEROBIC CONDITIONS,   LAGOONS.: /GE PROGRAMS, ANAEROBIC   W71-04557
C BACTERIA, ANAEROBIC CONDITIONS, LAGOONS.: /AGE PROGRAMS, ANAEROBI   W71-04554
 BACTERIA, ANAEROBIC CONDITIONS,   LAGOONS.: /GE PROGRAMS, ANAEROBIC   W71-04555
 BACTERIA, ANAEROBIC CONDITIONS,   LAGOONS.: /GE PROGRAMS, ANAEROBIC   W71-04556
S, BIOMASS, SYMBIOSIS, OXYDATION   LAGOONS.: /D, NITROGEN, PHOSPHATE   W70-05655
US, SEWAGE TREATMENT, *OXIDATION   LAGOONS.: /NCENTRATION, *PHOSPHOR   W68-00012
ENT, STABILIZATION PONDS, SEWAGE   LAGOONS.: /MENT, SECONDARY TREATM   W68-00012
(US), *DESIGN CRITERIA/ *AERATED   LAGOONS, *MIDWEST(US), *SOUTHWEST   W70-04788
AMMONIA, / *POULTRY, *OXIDATION   LAGOONS, *ORGANIC LOADING, ALGAE,   W71-05013
BIOCHEMICAL/ *FARM WASTES, *FARM   LAGOONS, *BIODEGRADATION, ALGAE,   W71-00936
IOLOGICAL TREATMENT, *OXIDATION,   LAGOONS, *BAFFLES, *WASTE WATER T   W70-09320
T, *INDIANA, CHLOROPHYT/ *SEWAGE   LAGOONS, *ALGAE, *SEWAGE TREATMEN   W70-10173
ASTE WATER TREATMENT, *OXIDATION   LAGOONS, *MODEL STUDIES.: /IA, *W   W70-07034
ATION, AEROBIC TREAT/ *OXIDATION   LAGOONS, *ORGANIC WASTES, *OXYGEN   W71-09546
ND, *ANAEROBICS, ALG/ *OXIDATION   LAGOONS, *BIOCHEMICAL OXYGEN DEMA   W71-05742
TEMPERATURE, SEWAGE TREAT/ *FARM   LAGOONS, *DESIGN CRITERIA, WATER   W71-08221
O/ *SEWAGE TREATMENT, *OXIDATION   LAGOONS, *ALGAE, *BACTERIA, *DISS   W72-01818
A, STABIL/ *SULFIDES, *OXIDATION   LAGOONS, *DESIGN, *SULFUR BACTERI   W70-01971
GE TREAT/ *MICROBIOLOGY, *SEWAGE   LAGOONS, *INDUSTRIAL WASTES, SEWA   W70-03312
ASTE WATER TREATMENT, *OXIDATION   LAGOONS, *ALGAE, LIGHT INTENSITY,   W72-06022
UNICIPAL WASTES, MET/ *OXIDATION   LAGOONS, *ANAEROBIC DIGESTION, *M   W72-06854
ALGAE, PHOTOSYNTHESI/ *OXIDATION   LAGOONS, *WASTE WATER TREATMENT,   W72-10573
SOLVED OXYGEN, SLUDGE, OXIDATION   LAGOONS, ALGAE, ADMINISTRATION, M   W71-08970
OTATO PROCESSING WASTES, AEROBIC   LAGOONS, ANAEROBIC LAGOONS.:   *P   W71-08970
ATMENT, SEDIMENTATION, OXIDATION   LAGOONS, ANAEROBIC DIGESTION, ALG   W71-11375
 *PHENOLS, PESTICIDES, OXIDATION   LAGOONS, ALGAE, PHOTOSYNTHESIS, D   W71-12183
TION AND MAINTENANCE, *OXIDATION   LAGOONS, ABSORPTION, BIODEGRADATI   W70-08392
ON, *ALGAE, *LAGOONS, *OXIDATION   LAGOONS, BIOCHEMICAL OXYGEN DEMAN   W70-06601
RTH DAKOTA, COLIFORMS, OXIDATION   LAGOONS, BIOCHEMICAL OXYGEN DEMAN   W70-03312
ES, DECOMPOSTING ORGANIC MATTER,   LAGOONS, CATTLE.: /GE, SOLID WAST   W71-05742
HEMICAL OXYGEN DEMAND, OXIDATION   LAGOONS, CALIF: / CHLORELLA, BIOC   W70-05092
TREATMENT FACILITIES, AERATION,   LAGOONS, DOMESTIC WASTES, SEDIMEN   W72-04060
*EUTROPHICATION, PHOTOSYNTHESIS,   LAGOONS, LABORATORY TESTS, LIGHT   W72-00917
MOVAL, MILK SUBSTR/ *FACULTATIVE   LAGOONS, LABORATORY MODEL, COD RE   W70-07031
ASTE WATER TREATMENT, *OXIDATION   LAGOONS, MILK, *MODEL STUDIES.: /   W70-07031
OXYGEN DEMAND, LABORATORY TESTS,   LAGOONS, OXIDATION, PHOTOSYNTHESE   W70-06600
ESALINATION, TERTIARY TREATMENT,   LAGOONS, OXYGENATION, ABSORPTION,   W69-01073
ICAL OXYGEN DEMAND, CHLOROPHYLL,   LAGOONS, PONDS, OXIDATION, WASTE   W70-08392
NT, *POULTRY, INDUSTRIAL WASTES,   LAGOONS, PONDS, PHOTOSYNTHESIS, C   W70-09320

462

TERIA, *DISSOLVED OXYGEN, SEWAGE
HEMICAL OXYGEN DEMAND, OXIDATION
OLLUTION EFFECTS, STABILIZATION,
, LIGHT PENETRATION, PHOSPHORUS,
ALGAE, GREEN / *LEWIS AND CLARK
, LAKE WASHINGTON, LAKE MALAREN,

, MELOSI/ SEASONAL DISTRIBUTION,
R BLOCKING, HETERETROPHS, ITALY,
IQUE, HETEROTROPHY, FLAGELLATES,
ILABLE), MICROCYSTIS AERUGINOSA,
MBERG(GERMANY), PL/ CHIRONOMIDS,
LAREN, LAKE ANNECY, LAKE VANERN,
UBESA(WIS), SPAULDING POND(WIS),
TREAM TREATMENT, BOTTOM SEALING,
ONSTANT, LAKE NORRVIKEN(SWEDEN),
, *NITROGEN FIXATION, NUTRIENTS,
ALIS, NE/ *NEW SPECIES, *WESTERN
OXYGEN, HYPOLIMNION, EPILIMNION,
COSTS, EUTROPHICATION, NITRATES,
XYGEN, HYPOLIMNION, GREAT LAKES,
GEN SULFIDE, CITIES, LAKE HURON,
N LAKE SURVEY, NITROGEN CONTENT,
OPHY, MICROBIAL ECOLOGY, SWEDEN,
CHAELIS-MENTEN EQUATION, SWEDEN,
MICHAELIS KINETICS, FLAGELLATES,
TROPHY, ORGANIC SOLUTES, SWEDEN,
NEBODA(SWEDEN), UPPSALA(SWEDEN),

Y, LAKE NORRVIKEN, LAKE MENDOTA,
METER, PROTOPLASM, WATER-BLOOMS,
ER(BAVARIA), LAKE ZELL(AUSTRIA),
PPING, HYDROGEN SULFIDE, CITIES,
, *LAKE MICHIGAN, *LAKE ONTARIO,
D LAKE(MINN), LAKE OWASSO(MINN),
MONONA(WIS), LAKE WAUBESA(WIS),
MONONA(WIS), LAKE WAUBESA(WIS),
E MONONA(WIS), LAKE WINGRA(WIS),
MENDOTA(WIS), LAKE MONONA(WIS),
MENDOTA(WIS), LAKE WAUBESA(WIS),
MONONA(WIS), LAKE WAUBESA(WIS),
ONA, LAKE MENDOTA, LAKE WAUBESA,
MONONA(WIS), LAKE WAUBESA(WIS),
KEN(SWEDEN), LAKE EKOLN(SWEDEN),
SUBTERRANEAN DRAINAGE, MECHANIC/
ITALY, LAKE BORREVANNET NORWAY,
KE SEBASTICOOK, LAKE WASHINGTON,
NSITY, / *LAKE MINNETONKA(MINN),
ANNET NORWAY, LAKE LUGANO ITALY,
S, OXYGEN D/ *RESERVOIR EFFECTS,
S), LAKE KEGONSA(WIS), LAKE WAU/
CROPS, TR/ TAKASUKA POND(JAPAN),
GAL NUTRITION, ALGAL PHYSIOLOGY,
RICHIA PISUM, NOSTOC VERRUCOSUM,
ADA, MESOTROPHY, LAKE NORRVIKEN,
GEN CONTENT, LAKE ERKEN(SWEDEN),
CHEMISTRY, WASTE WATER EFFECTS,

*MADISON(WIS), NUISANCE ODORS,
TE, CLADOPHORA, POTAMOGETON SPP,
S), LAKE KEGONSA(WIS), BLOOMING,
ECOVERY, WISCONSIN, LAKE MONONA,
A AEQUALE, NAVICULA PELLICULOSA,
AL PROBLEMS, SCIOTO RIVER(OHIO),
TAL SAMPLING, VERTICAL SAMPLING,
TION SURVEY, MADISON LAKES(WIS),
S), LAKE NAGAWICKA(WIS), LAKE P/
MYDOMONAS, CHLORELLA, WISCONSIN,
OPPER SULFATE, BOTTOM SEDIMENTS,
HURON, LAKE ERIE, LAKE ONTARIO,
ATMENT, *ALGAE, *FILTERS, SANDS,
OS AQUAE, STEPHANODISCUS ASTREA,
S-ST PAUL(MINN).:
RAPS, NUTRIENT INFLUX, NUTRIENT/
ALS, / *KILLING, ANIMAL MANURES,
DUTCH MILL DRAINAGE DITCH(WIS),
S), LAKE KEGONSA/ *MADISON(WIS),
ISANCE ODORS, LAKE MENDOTA(WIS),
C VERRUCOSUM, LAKE MENDOTA(WIS),
L PHYSIOLOGY, LAKE MENDOTA(WIS),
F), L/ *CULTURAL EUTROPHICATION,
S), LAKE WAU/ LAKE MENDOTA(WIS),
MOVAL, LAKE RECOVERY, WISCONSIN,
WIS), LAKE P/ LAKE MENDOTA(WIS),
CAL SAMPLING, LAKE MENDOTA(WIS),
RIVER(OHIO), LAKE MENDOTA(WIS),
ICAL COMMUNITIES, BIOINDICATORS,
ATIC POPULATIONS, WATER QUALITY,
UKEE LAKE(WIS), NORTH LAKE(WIS),
MENDOTA(WIS), LAKE MONONA(WIS),

LAGOONS, PLANT PHYSIOLOGY, ANIMAL   W72-01818
LAGOONS, SLUDGE, PHOTOSYNTHETIC O   W71-08221
LAGOONS, TOXICITY, WASTE WATER TR   W69-09454
LAGOONS: /LINITY, *PHOTOSYNTHESIS   W68-00248
LAKE (S D), ABUNDANCE, BLUE-GREEN   W70-05375
LAKE ANNECY, LAKE VANERN, LAKE CO   W70-03964
LAKE ASHTABULA.:   W70-09093
LAKE BAIKAL, PHOTOSYNTHETIC LAYER   W70-04290
LAKE BORREVANNET NORWAY, LAKE LUG   W69-07833
LAKE CLASSIFICATION, SHAGAWA LAKE   W70-02775
LAKE COMO(MINN), LAKE VADNAIS(MIN   W70-05269
LAKE CONSTANCE, WHITEFISH, WURTTE   W70-05761
LAKE CONSTANCE, PFAFFIKERSEE, TUR   W70-03964
LAKE DELAVAN(WIS), PEWAUKEE LAKE(W69-05867
LAKE DILUTION.: /R DIVERSION, INS   W69-09340
LAKE EKOLN(SWEDEN), LAKE LOTSJON(W70-04287
LAKE ERIE.:   *ALGAE, *CYANOPHYTA   W70-05091
LAKE ERIE(OHIO), GONGROSIRA STAGN   W70-04468
LAKE ERIE, INSECTS, DIPTERA, CRUS   W70-04253
LAKE ERIE, LAKE ONTARIO, POLLUTIO   W70-03964
LAKE ERIE, STATISTICS, GROWTH RAT   W68-00253
LAKE ERIE, LAKE ONTARIO, LAKE MIC   W70-00667
LAKE ERKEN(SWEDEN), LAKE MENDOTA(W70-03959
LAKE ERKEN(SWEDEN), ACTIVE TRANSP   W70-04284
LAKE ERKEN(SWEDEN), LINEWEAVER-BU   W70-04287
LAKE ERKEN(SWEDEN), ACETATE, LAKE   W70-04194
LAKE ERKEN(SWEDEN), ANNUAL CYCLES   W70-05109
LAKE ERKEN(SWEDEN), EXTINCTION VA   W70-05409
LAKE ERSOM(DENMARK).:   W71-00665
LAKE FURES, LAKE SEBASTICOOK, LAK   W70-03964
LAKE GEORGE(UGANDA), RHINE-MEEUSE   W70-04369
LAKE HALL WILER(SWITZERLAND).: /I   W70-05668
LAKE HURON, LAKE ERIE, LAKE ONTAR   W70-00667
LAKE HURON, AESTHETICS, ECOLOGY,   W70-01943
LAKE JOSEPHINE(MINN), GRAY'S BAY(W70-05269
LAKE KEGONSA(WIS).: /H(WIS), LAKE   W70-05113
LAKE KEGONSA(WIS), BLOOMING, LAKE   W70-05112
LAKE KEGONSA(WIS), LAKE WAUBESA(W   W69-03366
LAKE KEGONSA(WIS), LAKE WAUBESA(W   W69-06276
LAKE KEGONSA(WIS).: /(WIS), LAKE   W70-02787
LAKE KEGONSA(WIS), WEED BEDS, PHO   W70-02969
LAKE KEGONSA, LAKE ZURICH, OSCILL   W69-09349
LAKE KEGONSA(WIS).: /A(WIS), LAKE   W72-01813
LAKE LOTSJON(SWEDEN), PERMEASES,   W70-04287
LAKE LUCERNE, FERTILIZED FIELDS,   W69-08668
LAKE LUGANO ITALY, LAKE MARIDALSV   W69-07833
LAKE MALAREN, LAKE ANNECY, LAKE V   W70-03964
LAKE MANAGEMENT PROGRAM, ALGAL DE   W69-10167
LAKE MARIDALSVANNET NORWAY, LAKE   W69-07833
LAKE MEAD, DILUTION, ALGAL GROWTH   W73-00750
LAKE MENDOTA(WIS), LAKE MONONA(WI   W69-06276
LAKE MENDOTA(WIS), PHYTOPLANKTON   W69-05142
LAKE MENDOTA(WIS), LAKE MONONA(WI   W69-05867
LAKE MENDOTA(WIS), LAKE MONONA(WI   W69-08283
LAKE MENDOTA, LAKE FURES, LAKE SE   W70-03964
LAKE MENDOTA(WIS), NET PLANKTON,   W70-03959
LAKE MENDOTA(WIS), BLACK EARTH CR   W70-04382
LAKE MENDOTA(WIS).:   W69-03185
LAKE MENDOTA(WIS), LAKE MONONA(WI   W69-03366
LAKE MENDOTA(WIS), ANACHARIS, NAI   W70-05263
LAKE MENDOTA(WIS), LAKE WINGRA(WI   W70-05112
LAKE MENDOTA, LAKE WAUBESA, LAKE   W69-09349
LAKE MENDOTA(WIS), ENGLAND, LONDO   W69-10180
LAKE MENDOTA(WIS), LAKE MONONA(WI   W72-01813
LAKE MENDOTA(WIS), LAKE MONONA(WI   W70-02969
LAKE MENDOTA(WIS), LAKE WAUBESA(W   W70-02787
LAKE MENDOTA(WIS), LAKE MONONA(WI   W70-06217
LAKE MICHIGAN, PHYTOPLANKTON, DIA   W69-10180
LAKE MICHIGAN, WASHINGTON, DISSOL   W69-09349
LAKE MICHIGAN, WATER TEMPERATURE,   W70-00667
LAKE MICHIGAN, WATER QUALITY CONT   W73-02426
LAKE MINNETONKA(MINN), MORPHOMETR   W70-05387
LAKE MINNETONKA(MINN), MINNEAPOLI   W70-04810
LAKE MINNETONKA(MINN), NUTRIENT T   W70-04268
LAKE MINNETONKA(MINN), SOIL MINER   W70-04193
LAKE MONONA(WIS), LAKE WAUBESA(WI   W70-05113
LAKE MONONA(WIS), LAKE WAUBESA(WI   W70-05112
LAKE MONONA(WIS), LAKE WINGRA(WIS   W69-03366
LAKE MONONA(WIS), LAKE WAUBESA(WI   W69-08283
LAKE MONONA(WIS), LAKE WAUBESA(WI   W69-05867
LAKE MONONA(WIS), LAKE TAHOE(CALI   W69-06858
LAKE MONONA(WIS), LAKE KEGONSA(WI   W69-06276
LAKE MONONA, LAKE MENDOTA, LAKE W   W69-09349
LAKE MONONA(WIS), LAKE NAGAWICKA(W70-06217
LAKE MONONA(WIS), LAKE WAUBESA(WI   W70-02969
LAKE MONONA(WIS), LAKE WAUBESA(WI   W72-01813
LAKE MORPHOLOGY, ECOLOGICAL DISTR   W71-12077
LAKE MORPHOMETRY, DOMESTIC WASTES   W68-00470
LAKE NAGAWICKA(WIS).: /WIS), PEWA   W69-05867
LAKE NAGAWICKA(WIS), LAKE PEWAUKE   W70-06217

ES, LAKE ERKEN(SWEDEN), ACETATE,
KE VELOCITY, MICHAELIS CONSTANT,
E FURES, LA/ CANADA, MESOTROPHY,
OPHICATION, NITRATES, LAKE ERIE,
, CITIES, LAKE HURON, LAKE ERIE,
LAKE(MINN), ORCHARD LAKE(MINN),
ONONA(WIS), LAKE NAGAWICKA(WIS),
S/ *AIR SPRAY, SUDBURY(ONTARIO),
ATIVE STUDIES, NUTRIENT REMOVAL,
*PERIDINIUM POLONICUM, PLANKTO/
TEMBERG(GERMANY), PLON(GERMANY),
TZERLAND), LAKE TEGERN(BAVARIA),
VIKEN, LAKE MENDOTA, LAKE FURES,
ODIC VARIATIONS, SEASONAL VARIA/
QUAM(N H), MA/ NUTRIENT BUDGETS,
UTRIENT/ *EUTROPHICATION, LAKES,
R, LEBANON, WADLEIGH STATE PARK,
RIVER, RAISIN RIVER, DETROIT R/
AQUATIC ECOSYSTEM, MIDGE LARVAE,
ICATORS.:   PERIPHYTON INVENTORY(
PLANKTON DISTRIBUTION, WISCONSIN
*ALGAL BLOOMS, NTA, EDTA, HEIDA,
ITY, PARTICULATE ORGANIC CARBON,
UTROPHICATION, LAKE MONONA(WIS),

GERMANY), LAKE SCHLIER(GERMANY),
ZERLAND), ZOLLIKON(SWITZERLAND),
MOSSES, CHARA, FISH, LIMNOLOGY,
TIS AERUGINOSA, LAKE COMO(MINN),
GTON, LAKE MALAREN, LAKE ANNECY,
RATION, AREAL INCOME(NUTRIENTS),
ISQUAM(N H), MADISON LAKES(WIS),
A, LAKE FURES, LAKE SEBASTICOOK,
E ZOAR(CONN), KLAMATH LAKE(ORE),
MONONA(WIS), LAKE TAHOE(CALIF),

WINGRA(WIS), LAKE KEGONSA(WIS),
OTIC FACTORS, SPECIES DIVERSITY,
MONONA(WIS), LAKE KEGONSA(WIS),
MENDOTA(WIS), LAKE MONONA(WIS),
REMOVAL, SPAULDING'S POND(WIS),
MENDOTA(WIS), LAKE MONONA(WIS),
GE DITCH(WIS), LAKE MONONA(WIS),
*MADISON(WIS), LAKE MONONA(WIS),
NSIN, LAKE MONONA, LAKE MENDOTA,
MENDOTA(WIS), LAKE MONONA(WIS),
HYDRODICTYON, NAJAS, ANACHARIS,
MENDOTA(WIS), LAKE MONONA(WIS),
N LAKES(WIS), LAKE MENDOTA(WIS),
DE(SCOTLAND), SCOTLAND, ENGLAND,
S), BLOOMING, LAKE MENDOTA(WIS),
MENDOTA(WIS), LAKE MONONA(WIS),
.:
GON XL(DICHLONE), NEW HAMPSHIRE,
T BUDGETS, LAKE SEBASTICOOK(ME),
CHALK/ *LAKE KLOPEINER(AUSTRIA),
BAVARIA), LAKE SCHLIER(BAVARIA),
),/ *URBAN WASTES, MADISON(WIS),
RON SULFATE, GALVANIZING WASTES,
, NOVA SCOTIA(CANADA), PENGUINS,
ALY, LAKE MARIDALSVANNET NORWAY,
OTA, LAKE WAUBESA, LAKE KEGONSA,
                              KLAMATH
E SHATUTAKEE, MASCOMA LAKE, GLEN
NTROL, WINNISQUAM LAKE, NEWFOUND
                                SMITH
GEMENT, *CHAOBORID MIDGES, CLEAR
NT, AMERICAN RIVER(CALIF), CLEAR
NUTRIENT LEVELS, S/ *TWELVE MILE
ES, OKANAGAN LAKE(CANADA), SKAHA
DA), SKAHA LAKE(CANADA), OSOYOOS
/ TROPHIC COMMUNITIES, OKANAGAN
ITZERLAND), WHITEFISH, PUNCHBOWL
ALPINE ZONE, BOULDER(COLO), TEA
DER(COLO), TEA LAKE(COLO), BLACK
KE(COLO), BLACK LAKE(COLO), PASS
DA).:   CUE
     CUE LAKE(FLORIDA), MCCLOUD
FORMOSA, RECOLONIZATION, ALSTER
CHLORO/ *BLOOM, *VIETNAM, LITTLE
ROOKED LAKE(IND), LITTLE CROOKED
ETABOLISM, HETEROTROPHY, CROOKED
OWA), CENTER LAKE(IOWA), DIAMOND
YNGBYA, STORM LAKE(IOWA), CENTER
MENON FLOS-AQUAE, LYNGBYA, STORM
EC RIVER(MAINE), COBBOSSEECONTEE
                              LAWRENCE
ES, LAKE CLASSIFICATION, SHAGAWA
LORELLA PYRENOIDOSA, M/ *SHAGAWA
UAE, STEPHANODISCUS AS/ *SHAGAWA

LAKE NORRVIKEN(SWEDEN), A: /ELLAT   W70-04194
LAKE NORRVIKEN(SWEDEN), LAKE EKOL   W70-04287
LAKE NORRVIKEN, LAKE MENDOTA, LAK   W70-03964
LAKE ONTARIO, POLLUTION ABATEMENT   W70-03964
LAKE ONTARIO, LAKE MICHIGAN, WATE   W70-00667
LAKE OWASSO(MINN), LAKE JOSEPHINE   W70-05269
LAKE PEWAUKEE(WIS), CHIRONOMUS, P   W70-06217
LAKE RAMSEY(ONTARIO), ALGAE COUNT   W69-10155
LAKE RECOVERY, WISCONSIN, LAKE MO   W69-09349
LAKE SAGAMI, JAPAN, *GLENODININE,   W70-08372
LAKE SCHLIER(GERMANY), LAKE TEGER   W70-05761
LAKE SCHLIER(BAVARIA), LAKE ZELL(W70-05668
LAKE SEBASTICOOK, LAKE WASHINGTON   W70-03964
LAKE SEBASTICOOK, ME, MAINE, PERI   W68-00481
LAKE SEBASTICOOK(ME), LAKE WINNIS   W70-07261
LAKE SEDIMENTS, ALGAL CONTROL, *N   W72-06052
LAKE SHATUTAKEE, MASCOMA LAKE, GL   W69-08674
LAKE ST CLAIR, ROUGE RIVER, HURON   W69-01445
LAKE STRATIFICATION, SURFACE AREA   W69-09340
LAKE SUPERIOR, WATER QUALITY IND   W70-05958
LAKE SURVEY, NITROGEN CONTENT, LA   W70-03959
LAKE TAHOE(CALIF).: /OMPLEXONES,   W70-04373
LAKE TAHOE(CALIF), CHLORELLA PYRE   W70-03983
LAKE TAHOE(CALIF), LAKE WASHINGTO   W69-06858
LAKE TAHOE.:   W69-09135
LAKE TEGERN(GERMANY), GERMANY, FI   W70-05761
LAKE TEGERN(BAVARIA), LAKE SCHLIE   W70-05668
LAKE TROUT, CRAYFISH, STONEFLIES,   W70-00711
LAKE VADNAIS(MINN), MILLE LAC(MIN   W70-05269
LAKE VANERN, LAKE CONSTANCE, PFAF   W70-03964
LAKE VOLUME, MEAN DEPTH, DILUTION   W70-00270
LAKE WASHINGTON(WASH), NUISANCE C   W70-07261
LAKE WASHINGTON, LAKE MALAREN, LA   W70-03964
LAKE WASHINGTON(WASH), ALGAL NUTR   W70-04506
LAKE WASHINGTON(WASH), ZURICHSEE(W69-06858
LAKE WASHINGTON.:   W69-03513
LAKE WAUBESA(WIS), PRECIPITATION.   W69-03366
LAKE WAUBESA, LYNGBYA, MICROCYSTI   W69-01977
LAKE WAUBESA(WIS), MADISON(WIS),   W69-06276
LAKE WAUBESA(WIS), SPAULDING POND   W69-05867
LAKE WAUBESA(WIS), GREEN LAKE(WIS   W69-05868
LAKE WAUBESA(WIS), ANABAENA CIR: /   W69-08283
LAKE WAUBESA(WIS), LAKE KEGONSA(W   W70-05113
LAKE WAUBESA(WIS), LAKE KEGONSA(W   W70-05112
LAKE WAUBESA, LAKE KEGONSA, LAKE   W69-09349
LAKE WAUBESA(WIS), LAKE KEGONSA(W   W72-01813
LAKE WAUBESA(WIS), MACROPHYTES, A   W70-03310
LAKE WAUBESA(WIS), LAKE KEGONSA(W   W70-02969
LAKE WAUBESA(WIS), LAKE KEGONSA(W   W70-02787
LAKE WINDERMERE(ENGLAND), CARBON-   W70-02504
LAKE WINGRA(WIS), ANABAENA SPIROI   W70-05112
LAKE WINGRA(WIS), LAKE KEGONSA(WI   W69-03366
LAKE WINNEBAGO(WIS), MENASHA(WIS)   W72-01814
LAKE WINNISQUAM, LONG POND, CONCO   W69-08674
LAKE WINNISQUAM(N H), MADISON LAK   W70-07261
LAKE WORTHER(AUSTRIA), VERTICAL,   W70-05405
LAKE ZELL(AUSTRIA), LAKE HALL WIL   W70-05668
LAKE ZOAR(CONN), KLAMATH LAKE(ORE   W70-04506
LAKE ZURICH(SWITZERLAND), SWITZER   W70-05668
LAKE ZURICH(SWITZERLAND), WHITEFI   W70-05399
LAKE ZURICH, SWITZERLAND, NITELV   W69-07833
LAKE ZURICH, OSCILLATO: /AKE MEND   W69-09349
LAKE.:   W69-04802
LAKE.: / WADLEIGH STATE PARK, LAK   W69-08674
LAKE.: /LLUTION, WATER QUALITY CO   W70-02772
LAKE(ALASKA).:   W69-03515
LAKE(CALIF), NUTRIENT ENERGY PATH   W70-05428
LAKE(CALIF), EEL RIVER(CALIF), JO   W70-02777
LAKE(CANADA), *ONTARIO, *CANADA,   W70-02973
LAKE(CANADA), OSOYOOS LAKE(CANADA   W71-12077
LAKE(CANADA), DEFORMED CHIRONOMID   W71-12077
LAKE(CANADA), SKAHA LAKE(CANADA),   W71-12077
LAKE(CANADA).: /S, LAKE ZURICH(SW   W70-05399
LAKE(COLO), BLACK LAKE(COLO), PAS   W69-10154
LAKE(COLO), PASS LAKE(COLO), MICR   W69-10154
LAKE(COLO), MICRO-ALGAE, FLAGELLA   W69-10154
LAKE(FLORIDA), MCCLOUD LAKE(FLORI   W72-01993
LAKE(FLORIDA).:   W72-01993
LAKE(GERMANY), GERMANY.: /IONELLA   W70-00268
LAKE(HANOI), RED RIVER(VIETNAM),   W70-03969
LAKE(IND).: /ISM, HETEROTROPHY, C   W71-10079
LAKE(IND), LITTLE CROOKED LAKE(IN   W71-10079
LAKE(IOWA).: /NGBYA, STORM LAKE(I   W70-02803
LAKE(IOWA), DIAMOND LAKE(IOWA).: /   W70-02803
LAKE(IOWA), CENTER LAKE(IOWA), DI   W70-02803
LAKE(MAINE), BARGE.: /VER, KENNEB   W70-08096
LAKE(MICHIGAN), *MACROPHYTES.:   W71-00832
LAKE(MINN).: /ROTROPHY, FLAGELLAT   W70-02775
LAKE(MINN), ELY(MINN), BLOOMS, CH   W70-03512
LAKE(MINN), APHANIZOMENON FLOS AQ   W70-05387

OMETRY, ALGAE BLOOMS, CLEARWATER   LAKE(MINN), TROUT LAKE(MINN), KIM   W70-05387
MS, CLEARWATER LAKE(MINN), TROUT   LAKE(MINN), KIMBALL LAKE(MINN), C   W70-05387
MINN), TROUT LAKE(MINN), KIMBALL   LAKE(MINN), CRYPTOPHYTA, PYROPHYT   W70-05387
S(MINN), MILLE LAC(MINN), DEMING   LAKE(MINN), ORCHARD LAKE(MINN), L   W70-05269
INN), DEMING LAKE(MINN), ORCHARD   LAKE(MINN), LAKE OWASSO(MINN), LA   W70-05269
NOBRYON, CONCORD(N H), PEANACOOK   LAKE(N H).: / TREATMENT COSTS, DI   W69-08282
DES, PHOTOASSIMILATION, BERKELSE   LAKE(NETHERLANDS), HETEROCYSTS, G   W70-04369
 CHLORELLA OVALIS BUTCHER, NJOER   LAKE(NORWAY), HOBOL RIVER(NORWAY)   W70-05752
ORWAY), HOBOL RIVER(NORWAY), VAN   LAKE(NORWAY).: /HER, NJOER LAKE(N   W70-05752
ARIA, MORICHES BAY(NY), ONONDAGA   LAKE(NY).: /GENETICS, RUSSIA, BAV   W70-03973
ONEMA, PINNULARIA, NI/ *ONONDAGA   LAKE(NY), EUNOTIA, AMPHORA, GOMPH   W70-03974
, FILINIA, / *FAYETTEVILLE GREEN   LAKE(NY), *MARL LAKES, CHLOROBIUM   W70-03760
TS, *NUTRIENT LIMITATION, CAYUGA   LAKE(NY), ASTERIONELLA FORMOSA, S   W70-01579
), BUCKHORN LAKE(ONTARIO), CLEAR   LAKE(ONTARIO), RICE LAKE(ONTARIO)   W70-02795
, BALSAM LAKE(ONTARIO), STURGEON   LAKE(ONTARIO), BUCKHORN LAKE(ONTA   W70-02795
 TWELVE MILE LAKE(ONTARIO), GULL   LAKE(ONTARIO), BALSAM LAKE(ONTARI   W70-02795
NTARI/ *ONTARIO, *CANADA, KUSHOG   LAKE(ONTARIO), TWELVE MILE LAKE(O   W70-02795
USHOG LAKE(ONTARIO), TWELVE MILE   LAKE(ONTARIO), GULL LAKE(ONTARIO)   W70-02795
STURGEON LAKE(ONTARIO), BUCKHORN   LAKE(ONTARIO), CLEAR LAKE(ONTARIO   W70-02795
RIO), GULL LAKE(ONTARIO), BALSAM   LAKE(ONTARIO), STURGEON LAKE(ONTA   W70-02795
ARIO), CLEAR LAKE(ONTARIO), RICE   LAKE(ONTARIO).: /UCKHORN LAKE(ONT   W70-02795
SCILLATORA.:   *UPPER KLAMATH   LAKE(ORE.), AGENCY LAKE(ORE.), *O   W72-06051
UPPER KLAMATH LAKE(ORE.), AGENCY   LAKE(ORE.), *OSCILLATORA.:   *   W72-06051
QUAE, *CARBON-14/ *UPPER KLAMATH   LAKE(ORE.), *APHANIZOMENON FLOS-A   W72-06052
N(WIS), LAKE ZOAR(CONN), KLAMATH   LAKE(ORE), LAKE WASHINGTON(WASH),   W70-04506
RMANNII, OTTAWA(ONTARIO), BURTON   LAKE(SASKATCHEWAN), HUMBOLDT(SASK   W70-00273
LATION, RUTHENIUM-106, WHITE OAK   LAKE(TENN), ACTIVITY-DENSITY, VER   W70-04371
ROIDES, CLADOCERANS, WESTHAMPTON   LAKE(VA).:   ASTERIONELLA SPI   W71-10098
EROCYSTS, GAS-VACUOLES, LLANGORS   LAKE(WALES), HALOBACTERIUM, RESPI   W70-04369
LEAR ACTIVATION TECHNIQUES, ROCK   LAKE(WASH), WILLIAMS LAKE(WASH),   W70-00264
IQUES, ROCK LAKE(WASH), WILLIAMS   LAKE(WASH), CHANNELED SCABLANDS(W   W70-00264
S, BEAVER DAM LAKE(WIS), DELAVAN   LAKE(WIS).: /SE, TENDIPES PLUMOSU   W70-06217
D(WIS), LAKE WAUBESA(WIS), GREEN   LAKE(WIS).: /VAL, SPAULDING'S PON   W69-05868
IS), LAKE DELAVAN(WIS), PEWAUKEE   LAKE(WIS), NORTH LAKE(WIS), LAKE   W69-05867
(WIS), PEWAUKEE LAKE(WIS), NORTH   LAKE(WIS), LAKE NAGAWICKA(WIS).: /   W69-05867
E, TENDIPES PLUMOSUS, BEAVER DAM   LAKE(WIS), DELAVAN LAKE(WIS).: /S   W70-06217
E, *BIOAS/ *EUTROPHICATION, *MUD(LAKE), *PHOSPHORUS, *AQUATIC ALGA   W70-03955
QUIRKE LAKE, PECORS LAKE, DUNLOP   LAKE, *INORGANIC CARBON.: /AKES,   W70-05424
, NITROGEN COM/ *EUTROPHICATION,   LAKE, *LAKE SEDIMENTS, *NUTRIENTS   W72-06051
PLANTS, WATER POLLUT/ *NEWFOUND   LAKE, *WINNISQUAM LAKE, WATER AND   W70-02764
HN LAKE, PINGO LAKE, TWELVE MILE   LAKE, BRANT LAKE, IMIKPUK LAKE, N   W69-08272
QUATIC ECOSYSTEM MODEL/ KOOTENAY   LAKE, BRITISH COLUMBIA, CANADA, A   W69-06536
TARIO LAKES, QUIRKE LAKE, PECORS   LAKE, DUNLOP LAKE, *INORGANIC CAR   W70-05424
E PARK, LAKE SHATUTAKEE, MASCOMA   LAKE, GLEN LAKE.: / WADLEIGH STAT   W69-08674
TOR, ALASKA, HUDEUC LAKE, PULLIN   LAKE, GRAPHITE LAKE, OLD JOHN LAK   W69-08272
LAKE, IMIKPUK LAKE, NORTH MEADOW   LAKE, GYTTJA, DUFF, OOZE, COBBLES   W69-08272
EE, GREIFENSEE, ZURICHSEE, MOSES   LAKE, HALLWILLERSEE.: /BALDEGGERS   W70-03964
GO LAKE, TWELVE MILE LAKE, BRANT   LAKE, IMIKPUK LAKE, NORTH MEADOW   W69-08272
                    *ANDERSON-CUE   LAKE, MELROSE(FLORIDA).:   W72-01990
ATER QUALITY CONTROL, WINNISQUAM   LAKE, NEWFOUND LAKE.: /LLUTION, W   W70-02772
E MILE LAKE, BRANT LAKE, IMIKPUK   LAKE, NORTH MEADOW LAKE, GYTTJA,   W69-08272
DEUC LAKE, PULLIN LAKE, GRAPHITE   LAKE, OLD JOHN LAKE, PINGO LAKE,   W69-08272
WARBURG RESPIROMETER,/ *ONONDAGA   LAKE, ONONDAGA COUNTY, SYRACUSE,   W69-02959
                             ELLIOT   LAKE, ONTARIO.:   W69-01165
OPERTIES.:   CRATER   LAKE, ORE, GROWTH RATES, WATER PR   W68-00478
INORGANIC/ ONTARIO LAKES, QUIRKE   LAKE, PECORS LAKE, DUNLOP LAKE, *   W70-05424
IN LAKE, GRAPHITE LAKE, OLD JOHN   LAKE, PINGO LAKE, TWELVE MILE LAK   W69-08272
OLD J/ DETECTOR, ALASKA, HUDEUC   LAKE, PULLIN LAKE, GRAPHITE LAKE,   W69-08272
PHITE LAKE, OLD JOHN LAKE, PINGO   LAKE, TWELVE MILE LAKE, BRANT LAK   W69-08272
LUT/ *NEWFOUND LAKE, *WINNISQUAM   LAKE, WATER AND PLANTS, WATER POL   W70-02764
:   GREAT   LAKES BASIN, BACTERIAL POLLUTION.   W69-03948
L, SEWAGE EFFLUENTS, ABSORPTION,   LAKES.: /NT, WATER QUALITY CONTRO   W70-01031
, PLANKTON, LAKES, RIVERS, GREAT   LAKES.: /OSPHORUS, TRACE ELEMENTS   W70-04385
NCIES, DISSOLVED OXYGEN, OXYGEN,   LAKES.: /PLANKTON, ALGAE, EFFICIE   W70-04484
OSPHATES, NITRATES, ALGAE, GREAT   LAKES.: /TRIENTS, FERTILIZERS, PH   W69-07084
NTS, BADFISH RIVER(WIS), DETROIT   LAKES(MINN), NUTRIENT REMOVAL, MA   W70-04506
ARIA, RADIOCARBON METHOD, FINGER   LAKES(NY), RODHE'S SOLUTION VIII,   W70-01579
E, FERTILIZATION SURVEY, MADISON   LAKES(WIS), LAKE MENDOTA(WIS), LA   W70-02787
 OSCILLATORIA RUBESCENS, MADISON   LAKES(WIS), CEDAR RIVER(WASH), ZU   W70-00270
), LAKE WINNISQUAM(N H), MADISON   LAKES(WIS), LAKE WASHINGTON(WASH)   W70-07261
, *WEED EXTRACTING STE/ *MADISON   LAKES(WIS), *AQUATIC WEED CUTTING   W70-04494
AKE STRATIFICATION, SURFACE AREA(LAKES), PHAEDACTYLUM, WASTE WATER   W69-09340
U S PUBLIC HEALTH SERVICE, GREAT   LAKES-ILLINOIS RIVER BASINS PROJE   W70-00263
ECOLOGY, *EUTROPHICATION, *GREAT   LAKES, *ALGAE, NUTRIENTS, PHOSPHO   W70-00667
ULTRA-VIOLET LIGHT STER/ *DUTCH   LAKES, *ALLOCHTHONOUS PHOSPHATES,   W70-03955
COLOGY, WATER POLLUTION EFFECTS,   LAKES, *CYANOPHYTA, *EUTROPHICATI   W71-04758
IENTS, OXYGEN REQUIREMENTS, ICED   LAKES, *FERTILIZATION, WATER FLEA   W68-00487
 PROCEDURES, NURTRIENTS, RIVERS,   LAKES, *GROWTH RATES, *WATER TEMP   W68-00472
S, CYANOPHYTA, FRESH WATER FISH,   LAKES, *LAKE ZURICH, *MUNICIPAL W   W68-00680
, *LAKE ONTARI/ *BIOLOGY, *GREAT   LAKES, *LAKE ERIE, *LAKE MICHIGAN   W70-01943
PHOSPHORUS, N/ *EUTROPHICATION,   LAKES, *NUTRIENTS, UNITED STATES,   W68-00478
GAE, CURRENTS(WATER), LIMNOLOGY,   LAKES, *PILOT PLANTS, *NUISANCE A   W68-00479
SOURCES, MUNICIPAL WASTES, GREAT   LAKES, *POLLUTANT IDENTIFICATION.   W71-05879
LLUTION, *WATER QUALITY, *GREAT   LAKES, *TRACE ELEMENTS, *WATER PO   W69-08562
F / *MADISON(WIS), *'WORKING' OF   LAKES, *WATER-BLOOM, 'BREAKING' O   W69-08283
OLLUTION CONTROL, WATER / *GREAT   LAKES, *WATER POLLUTION, *WATER P   W69-03948
ASTES, *MUNICIPAL WASTES, *GREAT   LAKES, *WATER POLLUTION, SEWAGE,   W70-04430
ICATION, *ALGAE, *AQUATIC WEEDS,   LAKES, ALGICIDES, AQUATIC WEED CO   W70-04494
HOTOSYNTHESIS, HYDROGEN SULFIDE,   LAKES, ALGAE, ENTERIC BACTERIA, M   W70-03312
NUTRIENT REQUIREMENTS, CULTURES,   LAKES, ALGAE, GROWTH RATES, BACTE   W70-00719

S/ *EUTROPHICATION, *LAKES, ACID
, CYANOPHYTA, CHLOROPHYTA, GREAT
ERS, SEWAGE EFFLUENT, NUTRIENTS,
FISH/ *EUTROPHICATION, SEQUENCE,
YETTEVILLE GREEN LAKE(NY), *MARL
LGAE, SUNFISHES, ALGAE, DAPHNIA,
KES, *WATER-BLOOM, 'BREAKING' OF
TRIENT ENRICHMENT TEST/ HIGHLAND
NTS, *TERTIARY TREATMENT, *GREAT
ERSHED MANAGEMENT, GEOCHEMISTRY,
DESMUS, DIATOMACEOUS EARTH, MUD,
TROL, WATER POLLUTION, TOXICITY,
ACTERIA, *ALGAE, ORGANIC MATTER,
ARCTIC ALGAE, AUTOTROPHIC/ *MARL
NA, *EUTROPHICATION, *NUTRIENTS,
*DAPHNIA, PHYSIOLOGICAL ECOLOGY,
FECTS, WATER QUALITY, WISCONSIN,
SURFA/ *BIOASSAY, PHYTOPLANKTON,
CONFERENCES, STREAMS, ESTUARIES,
CONFERENCES, STREAMS, ESTUARIES,
IUM COMPOUNDS, CHLORIDES, SALINE
BIOINDICATORS, FISH, WASHINGTON,
S, CITRATES, BIOASSAY, NEW YORK,
*EUTROPHICATION, *HYDRODYNAMICS,
T GROWTH REGULATORS, FRESH WATER
OLVED OXYGEN, HYPOLIMNION, GREAT
ROL, *NUTRIENT/ *EUTROPHICATION,
OGY, *PHYTOPLANKTON, FISH, GREAT
, PHYTOPLANKTON, EUTROPHICATION,
PLANTS, BIOMASS, FERTILIZATION,
ATION, *ALGAE, PLANNING, RIVERS,
S, EUTROPHICATION, IMPOUNDMENTS,
YDROGEN ION CONCENTRATION, IRON,
PESTICIDE RESIDUES, ALGAE, GREAT
AY, EUTROPHICATION, OLIGOTROPHY,
NUTRIENT/ ALGAE, EUTROPHICATION,
GEN COMPOUNDS, PHOSPHATES, GREAT
PHYLL, CHEMTROL, COPPER SULFATE,
TIC WEEDS, ECONOMIC FEASIBILITY,
HING, WISCONSIN, BIBLIOGRAPHIES,
TOSYNTHESIS, *RESPIRATION, GREAT
CTS, BACTERIA, PROTOZOA, SEWAGE,
TES, HYDROGEN SULFIDE, SALINITY,
, ALGAE, DUCKS, COLOR, FISHKILL,
ANDING CROP, A/ *EUTROPHICATION,
OACTIVITY, CARBON RADIOISOTOPES,
ANKTON, *PHOTOSYNTHESIS, *ALGAE,
ALYSIS, WATER POLLUTION EFFECTS,
HEAVY METALS, INDUSTRIAL WASTES,
STREAMS, ESTUARIES, LAKES, GREAT
STREAMS, ESTUARIES, LAKES, GREAT
L ECOLOGY, CARBON RADIOISOTOPES,
ATER POLLUTION, STREAMS, RIVERS,
, ROOTED AQUATIC PLANTS, GLACIAL
SPHORUS, *NITROGEN, *ALKALINITY,
DUNLOP LAKE, *INORGANIC/ ONTARIO
UCTIVITY, SYMBIOSIS, RESERVOIRS,
AE, *EUTROPHICATION, CYANOPHYTA,
OMPOUNDS, TOXICITY, CHLOROPHYTA,
TION, CHLOROPHYLL, FLUORESCENCE,
ATER SUPPLY, EFFLUENTS, STREAMS,
HORUS, TRACE ELEMENTS, PLANKTON,
ION, *AQUATIC HABITATS, *ALASKA,
ODOR, TOXINS, DISSOLVED OXYGEN,
PHOSPHATES, SULFATES, SEASONAL,
RSHEDS, *RUNOFF, *FERTILIZATION,
GAE, *WATER POLLUTION TREATMENT,
TRACTIONS, PHOSPHORUS COMPOUNDS,
LUENTS, *ESTUARIES, FRESH WATER,
*FERTILIZATION, *EUTROPHICATION,
WASTE DISPOSAL, WATER POLLUTION,
*CHEMICAL PROPERTIES, NUTRIENTS,
DIES, MATHEMATICAL MODELS, GREAT
TS, *FALLOUT, *ECOSYSTEMS, GREAT
ITY, *ORGANIC MATTER, NUTRIENTS,
RUS, EUTROPHICATION, WASHINGTON,
WAGE EFFLUENTS, LIGHT INTENSITY,
ISCONSIN, WATER QUALITY CONTROL,
RON MICROSCOPY, WASTE TREATMENT,
ELD(MADISON) EFFLUENTS, *MADISON
QUALITY STANDARDS, GRAFTON(ND),
ES, *FIXED-LIGHT TESTS, */ *BEER-
CUS VESICULOSUS, GROWTH FACTORS,
US CRISPUS, ASCOPHYLLUM NODOSUM,
LUM NODOSUM, LAMINARIA DIGITATA,
TIGOCLADUS LAMINOSUS, PHORMIDIUM
NE PARK, *ICELAND, MASTIGOCLADUS
PHYSA, SALMO TRUTTA, ENTOSPHENUS
FISH, ALEWIFE, GIZZARD SHAD, SEA
N, UDO/ *TRACE ELEMENTS, *ANCHOA

LAKES, BIBLIOGRAPHIES, *PHOSPHORU   W68-00857
LAKES, BIOINDICATORS, FISH, WASHI   W70-04503
LAKES, CHANNELS, REVIEWS, BIBLIOG   W70-00269
LAKES, CHEMICALS, FISHING, SPORT   W68-00172
LAKES, CHLOROBIUM, FILINIA, MONIM   W70-05760
LAKES, CHLOROPHYTA, STREAMS, DIAT   W70-02982
LAKES, CHROOCOCCACEAE, NOSTOCHINE   W69-08283
LAKES, COLORADO RIVER, TEXAS, *NU   W69-06004
LAKES, CONTROL, COSTS, EUTROPHICA   W70-03964
LAKES, CONIFEROUS FORESTS, ALGAE,   W71-09012
LAKES, CULTIVATION, CHANNEL MORPH   W68-00475
LAKES, DESIGN CRITERIA, REMOTE SE   W70-02775
LAKES, DIFFUSION, CARBON RADIOISO   W70-04284
LAKES, DISSOLVED ORGANIC MATTER,   W71-10079
LAKES, ECOLOGY, PRIMARY PRODUCTIV   W71-00665
LAKES, ENVIRONMENTAL EFFECTS, WAT   W70-03957
LAKES, EUTROPHICATION.: /UTION EF   W69-05868
LAKES, EUTROPHICATION, SAMPLING,   W70-02973
LAKES, GREAT LAKES, PHYSIOLOGICAL   W70-03975
LAKES, GREAT LAKES, PHYSIOLOGICAL   W70-03975
LAKES, HEAVY METALS, INDUSTRIAL W   W69-02959
LAKES, INDIANA, OCHROMONAS, PHYTO   W70-04503
LAKES, INHIBITION, ALGAE, DIATOMS   W70-01579
LAKES, INTERNAL WATER, WAVES(WATE   W70-09889
LAKES, LABORATORY TESTS.: / *PLAN   W68-00856
LAKES, LAKE ERIE, STATISTICS, GRO   W68-00253
LAKES, LAKE SEDIMENTS, ALGAL CONT   W72-06052
LAKES, LEGISLATION, NITROGEN, NUI   W69-06858
LAKES, LIMNOLOGY, AQUATIC ENVIRON   W69-05142
LAKES, LIMNOLOGY, MIDGES, NITRATE   W69-04805
LAKES, LIMNOLOGY, WISCONSIN, URBA   W70-04506
LAKES, LOAMS, RESERVOIRS, TEXAS.:   W69-04800
LAKES, MAGNESIUM, NUTRIENTS, NUIS   W69-07833
LAKES, METABOLISM, *CHLORELLA, *C   W70-08652
LAKES, MICHIGAN, BICARBONATES, IR   W72-03183
LAKES, MICROORGANISMS, NITROGEN,   W68-00481
LAKES, MICHIGAN.: /SOURCES, NITRO   W69-01445
LAKES, NITROGEN.: / ALGAE, CHLORO   W68-00470
LAKES, NUISANCE ALGAE, WATER POLL   W69-06276
LAKES, NUISANCE ALGAE, BACTERIA,   W69-08283
LAKES, NUISANCE ALGAE, WATER QUAL   W72-05453
LAKES, NUTRIENTS.: /, FUNGI, INSE   W71-10786
LAKES, NUTRIENTS, BACTERIA, ALGAE   W71-03031
LAKES, ODOR, TOXICITY, WATER POLL   W69-08279
LAKES, ORGANIC WASTES, OXYGEN, ST   W70-02251
LAKES, ORGANIC MATTER, SEA WATER,   W70-02504
LAKES, OXYGEN, EUTROPHICATION, AL   W71-10098
LAKES, PHOSPHORUS, TROPHIC LEVEL,   W69-06536
LAKES, PHOTOSYNTHESIS, PHYTOPLANK   W69-02959
LAKES, PHYSIOLOGICAL ECOLOGY, WAT   W70-03975
LAKES, PHYSIOLOGICAL ECOLOGY, WAT   W70-03975
LAKES, PHYTOPLANKTON, BIOASSAYS.:   W70-05109
LAKES, PHYSIOLOGICAL ECOLOGY, LIM   W71-02479
LAKES, PLANKTON, EPIDEMICS.: /ILL   W68-00488
LAKES, PRODUCTIVITY, PHYTOPLANKTO   W70-02795
LAKES, QUIRKE LAKE, PECORS LAKE,   W70-05424
LAKES, REAERATION, SESTON, CHLORO   W70-05092
LAKES, RESERVOIRS, SCUM, NITROGEN   W70-02803
LAKES, RESERVOIRS, NITROGEN, GROW   W69-10180
LAKES, RIVERS, NITROGEN, BIOMASS,   W70-02777
LAKES, RIVERS, RESERVOIRS, BOATIN   W71-09880
LAKES, RIVERS, GREAT LAKES.: /OSP   W70-04385
LAKES, RIVERS, HOT SPRINGS, TERRE   W69-08272
LAKES, RIVERS, LIGHT PENETRATION,   W69-05697
LAKES, SAMPLING, PONDS, RIVERS, R   W70-02646
LAKES, SEWAGE TREATMENT, SEWAGE B   W70-02787
LAKES, SEWAGE, DRAINAGE DISTRICTS   W70-04455
LAKES, SEWAGE EFFLUENTS, ARSENIC   W70-05269
LAKES, SILTS, PHOSPHATES, NITROGE   W70-02255
LAKES, STREAMS, ALGAE, FISH, WISC   W69-10178
LAKES, STREAMS, SEWAGE, REMEDIES,   W69-06909
LAKES, TESTS, NITROGEN, PHOSPHORU   W68-00856
LAKES, THERMAL STRATIFICATION.: /   W71-05878
LAKES, TRACE ELEMENTS, RADIUM RAD   W71-09013
LAKES, WATER PROPERTIES, BACTERIA   W71-10079
LAKES, WATER POLLUTION EFFECTS.: /   W70-00270
LAKES, .WATER POLLUTION SOURCES, W   W69-10167
LAKES, WATER POLLUTION SOURCES, C   W70-03310
LAKES, WATER QUALITY, VIRUSES, AL   W70-03973
LAKES, YAHARA RIVER(WIS), STORKWE   W70-05113
LAKOTA(ND), HARVEY(ND), CHROMATIU   W70-03312
LAMBERT LAW, *FIXED-DENSITY STUDI   W70-03923
LAMINARIA SACCHARINA, MANNITOL, M   W69-07826
LAMINARIA DIGITATA, LAMINARIA AGA   W70-01073
LAMINARIA AGARDHII, DESICCATION,   W70-01073
LAMINOSUM, OSCILLATORIA TENUIS, O   W70-05270
LAMINOSUS, PHORMIDIUM LAMINOSUM,   W70-05270
LAMOTTENII, GAMMARUS, CORDULEGAST   W69-09334
LAMPHREY, COHO SALMON.: /, WHITE   W70-01943
LAMPROTAENIA, *OPISTHONEMA OGLIMU   W69-08525

VOLUMES, BACTERIAL PLATE COUNTS,
TION, ANAEROBIC DENITRIFICATION,
*ALGAE, CHLOROPHYTA, CYANOPHYTA,
OLIGOTR/ *LAKES, STREAMS, FISH,
ED OXYGEN, EUTROPHICATION, FISH,
ATIC LIFE, MILK, ATLANTIC OCEAN,
L WATER MOVEMENT, LEACHING, ARID
WATER DEMAND, WATER SUPPLY, ARID
RIO, FLAVOBACTER,/ *POLYSIPHONIA
SCOPHYLLUM NODOSUM, POLYSIPHONIA
EDEN), PERMEASES, CHAMYDOMONADS,
ES, *FERTILIZATION, WATER FLEAS,
RIENTS, AQUATIC ECOSYSTEM, MIDGE
TON, BACTERICIDES, ALGAL TOXINS,
TON, WORMS, NEMATODES, ROTIFERS,
UND WATER, NITRATES, PHOSPHATES,
, LAS VEGAS, NEVADA, *LAKE MEAD,
S VEG/ DISSOLVED ORGANIC MATTER,
*ABS,
Y, YELLOWSTONE NATIONAL PARK, MT
RUITING BODIES, ULVA LACTUCA VAR
*ULVA
OCEANS, ORGANIC MATTER, DIURNAL,
JAPONICA, CHAETOCEROS LAUNDERI,
TERIONELLA JAPONICA, CHAETOCEROS
ANTITIES, ALGAL BLOOMS, LIEBIG'S
G'S LAW OF THE MINIMUM, VOISIN'S
ED-LIGHT TESTS, */ *BEER-LAMBERT
ASOLINE, FLOODS, AQUIFERS, WATER
, SEWAGE, WATER POLLUTION, WATER
YTES.:
DUSTRIAL WASTES, AGRICULTURE, ST
ION, LAKE BAIKAL, PHOTOSYNTHETIC
ISTORTA, GLOEOTRICHIA ECHINULATA
SOIL WATER, SOIL WATER MOVEMENT,
RON COMPOUNDS, HYDROGEN SULFIDE,
EFFLUENTS, ION EXCHANGE, RESINS,
TROPHICATION, FISH, LAND, VALUE,
PISONIA, PANDANUS, MUSCLE, BONE,
FALO, DUNREITH/ *CARBON FILTERS,
, CONCORD, BOSTON LOT RESERVOIR,
KARATELLA COHCLEARIS, ASPLANCHA,
TUBES, *DIFFUSERS, *RAYTOWN(MO),
HIRONOMUS PROMOSUS, TRICHOPTERA,
VA RESERVOIRS, ANOPHELES QUADRI/
E, REMEDIES, JUDICIAL DECISIONS,
, RIGHT-OF-WAY, ESTIMATED COSTS,
POLLUTION CONTROL, WATERCOURSES(
ESTIMATED COSTS, LEGAL ASPECTS,
HYTOPLANKTON, FISH, GREAT LAKES,
RSENITE, NUISANCE ALGAE, SNAILS,
ADOW, NITROGEN-15, SOD MATS, NON-
CCHI DISC TRANSPARENCY, ANABAENA
SPECIES SUSCEPTIBILITY, ANABAENA
*PHOTOSYNTHETIC OXYGEN,
DRICAUDA, PORPHYRIDIUM CRUENTUM,
CUS, ESCHERICHIA COLI, OOCYSTIS,
WORTS, MORPHOMETRY, PACIFASTACUS
THE HAGUE, MINERALIZATION, CLAY
ATORY TESTS, LOTIC ENVIRONMENTS,
OSPORA, PEDIASTRUM, SCENEDESMUS,
METER, CLADOPHORA, ENTEROMORPHA,
ESCHERICHIA COLI, MYRIOPHYLLUM,
OSUM, POLYSIPHONIA LANOSA, POINT
LLOPORA EYDOUXI, PORITES LOBATA,
ROS LAUNDERI, LAUDERIA BOREALIS,
RA, COSCINDISCUS, BACTERIOSTRUM,
PHANIZOMENON FLOS-AQUAE, MINIMUM
ARY, AMMONIUM COMPOUNDS, MINIMUM
ALGAE, WATER POLLUTION EFFECTS,
N, *BACTERIA, *PROTOZOA, *ALGAE,
BIOASSAY, ANALYTICAL TECHNIQUES,
ISCHURA NEVADENSIS, DIAPHANOSOMA
2-M/ *FISH HEMATOLOGY, *GRANULAR
S, GRAIN COUNT(AUTORADIOGRAPHY),
GREEN ALGAE, BREAM, WHITE BREAM,
AQUATIC MICROORGANISMS, TROPHIC
N, PRIMARY PRODUCTIVITY, TROPHIC
OTOPES, TENNESSEE RIVER, TROPHIC
PHOTOSYNTHESIS, OXYGEN, TROPHIC
ES, HYDROLOGY, CLIMATES, TROPHIC
N EFFECTS, *FISH, ALGAE, TROPHIC
ECTS, LAKES, PHOSPHORUS, TROPHIC
TIVITY, LIGHT INTENSITY, TROPHIC
UCTIVITY, AQUATIC ALGAE, TROPHIC
*PRIMARY PRODUCTIVITY, *TROPHIC
Y, *PRIMARY PRODUCTIVITY, TROPIC
HOSPHORUS RADIOISOTOPES, TROPHIC
*LAKES, *BENTHIC FAUNA, *TROPHIC
R QUALITY, NUT/ *ALGAE, *TROPHIC

LANCASTER(CALIF).: / PACKED CELL   W71-05267
LAND APPLICATION, ELECTROCHEMICAL   W70-01981
LAND MANAGEMENT, WATER POLLUTION   W68-00172
LAND, ECOSYSTEMS, EUTROPHICATION,   W70-04193
LAND, VALUE, LEAVES, FERTILIZERS,   W70-05264
LANDFILLS, DISPOSAL, REGULATION,:   W72-00941
LANDS, DEPTH, CYANOPHYTA, CHLOROP   W71-04067
LANDS, LIMNOLOGY.: / MONITORING,   W70-05645
LANOSA, *ASCOPHYLLUM NODOSUM, VIB   W70-03952
LANOSA, POINT LEPREAU(CANADA), NE   W70-04365
LAPPLAND(SWEDEN), ACETATE, TURNOV   W70-04287
LARVAE, GROWTH RATES, *FRESH WATE   W68-00487
LARVAE, LAKE STRATIFICATION, SURF   W69-09340
LARVAE, PHENOLS, SHELLFISH, OYSTE   W70-01068
LARVAE, SNAILS, MITES, MOLDS, GRO   W72-01819
LAS VEGAS BAY.: /OIRS, ALGAE, GRO   W72-02817
LAS VEGAS BAY.: /D ORGANIC MATTER   W72-02817
LAS VEGAS, NEVADA, *LAKE MEAD, LA   W72-02817
LAS.:   W69-09454
LASSEN NATIONAL PARK, MT RAINIER   W70-03309
LATISSIMA, POLYSIPHONIA HARVEYI,   W70-01073
LATISSIMA, *AMMONIA NITROGEN.:   W71-05500
LATITUDINAL STUDIES.: /PIGMENTS,   W70-04809
LAUDERIA BOREALIS, LEPTOCYLINDRUS   W70-04280
LAUNDERI, LAUDERIA BOREALIS, LEPT   W70-04280
LAW OF THE MINIMUM, VOISIN'S LAW   W70-04765
LAW OF THE MAXIMUM.: /OOMS, LIEBI   W70-04765
LAW, *FIXED-DENSITY STUDIES, *FIX   W70-03923
LAW, NITRATES, TURBIDITY, FLOODWA   W73-00758
LAW, WATER POLLUTION EFFECTS, WAT   W69-06305
LAWRENCE LAKE(MICHIGAN), *MACROPH   W71-00832
LAWRENCE RIVER, ALGAE, COLIFORMS,   W70-03964
LAYER, MELOSIRA BAICALENSIS, GYMN   W70-04290
LB 1303.: /S AQUAE, TOLYPOTHRIX D   W70-05091
LEACHING, ARID LANDS, DEPTH, CYAN   W71-04067
LEAD, COPPER, AMMONIA, CONDUCTIVI   W70-04510
LEAKAGE, CESIUM, NUCLEAR EXPLOSIO   W70-00707
LEAVES, FERTILIZERS, PESTICIDES,   W70-05264
LEAVES, SHELL, RHODIUM-102.: /A,   W69-08269
LEBANON, BATAVIA, CLARKSBURG, BUF   W72-03641
LEBANON, WADLEIGH STATE PARK, LAK   W69-08674
LECANE, BOSMINA, NAVICULA, FRAGIL   W70-02249
LEE'S SUMMIT(MO), OXYGEN TRANSFER   W70-06601
LEECHES, FINGERNAIL CLAMS, WHITE   W70-01943
LEECHES, SCHISTOSOME CERCARIAE, T   W70-06225
LEGAL ASPECTS.: /, STREAMS, SEWAG   W69-06909
LEGAL ASPECTS, LEGISLATION, SEWAG   W70-03344
LEGAL), ORGANIC WASTES, PULP WAST   W72-00462
LEGISLATION, SEWAGE DISTRI: /WAY,   W70-03344
LEGISLATION, NITROGEN, NUISANCE A   W69-06858
LEGISLATION, WISCONSIN, CHEMCONTR   W70-05568
LEGUMES, *NITROGEN FIXING ORGANIS   W72-08110
LEMMERMANNI, APHANOCAPSA, BOSMINA   W69-10169
LEMMERMANNII, OTTAWA(ONTARIO), BU   W70-00273
LEMMON(SD), KADOKA(SD).:   W72-01818
LEMNA MINOR, PHOTOPHOSPHORYLATION   W70-02964
LEMNA, SPIRODELA, ELODEA.: /LOCOC   W70-10161
LENIUSCULUS, ANACHARIS CANADENSIS   W70-00711
LENSES, RHINE RIVER, AMSTERDAM RH   W69-07838
LENTIC ENVIRONMENTS.: /STS, LABOR   W72-05453
LEPOCINCLIS, ANABAENA, ANACYSTIS,   W70-03974
LEPOCINCLIS, CHLOROGONIUM, SCENED   W69-02959
LEPOMIS MACROCHIRUS, PIMEPHALES N   W69-07861
LEPREAU(CANADA), NEW BRUNSWICK(CA   W70-04365
LEPTASTREA PURPUREA, FUNGIA FUNGI   W69-08526
LEPTOCYLINDRUS DANICUS, PHAEODACT   W70-04280
LEPTOCYLINDRUS, CORYCAEUS, OIKOPL   W73-01446
LETHAL DOSES, CHU'S MEDIA, QUINON   W70-02982
LETHAL EXPOSURE, PHENANTHROQUINON   W70-02982
LETHAL LIMIT, GROWTH RATES, SPORE   W70-01779
LETHAL LIMIT, ELECTROPHORESIS, MI   W71-03775
LETHAL LIMITS, DINOFLAGELLATES, D   W70-08111
LEUCHTENBER GIANUM, CYCLOPS BICUS   W69-10182
LEUCOCYTES, *MUSCULAR EXERTION, *   W70-01788
LEUCOTHRIX MUCOR, ANTITHAMNION SA   W69-10163
LEUKOCYTIC COMPOSITION, ALTERNATI   W70-01788
LEVEL.: /LORIDA, *EUTROPHICATION,   W72-01993
LEVEL.: /VITY, LIMNOLOGY, PLANKTO   W69-09135
LEVEL, ALGAE, CADDISFLIES, FOOD C   W70-02786
LEVEL, LIGHT INTENSITY, ECOLOGY,   W70-00265
LEVEL, NITROGEN FIXATION, FOREST,   W71-03048
LEVEL, PRODUCTIVITY, FOOD CHAINS,   W72-01788
LEVEL, PRODUCTIVITY, MODEL STUDIE   W69-06536
LEVEL, SAMPLING TEMPERATURE, TRAC   W70-02504
LEVEL, WATER QUALITY, FLORIDA.: /   W72-01990
LEVEL, WATER STORAGE, EVAPORATION   W69-09723
LEVEL, WATER QUALITY, WASTE ASSIM   W68-00912
LEVEL, ZINC RADIOISOTOPES, IRON,   W69-07853
LEVELS, *SURVEYS, BIOLOGICAL COMM   W71-12077
LEVELS, AQUATIC ENVIRONMENT, WATE   W70-02764

467

, DISTRIBUTION, CALCIUM, TROPHIC
C ALGAE, WATER QUALITY, *TROPHIC
DA), *ONTARIO, *CANADA, NUTRIENT
SFORMATIONS, AMSTERDAM, HAARLEM,
BACTERIA, FUNGI, VIRUSES, ALGAE,
ELATION, FUNGI, ALGAE, PROTOZOA,
S, *MARINE ANIMALS, *OILY WATER,
, MACROQUANTITIES, ALGAL BLOOMS,
NTATION, CHEMICAL PRECIPITATION,
BAYS, *FLUCTUATION, RESISTANCE,
OP, LIGHT, BIOMASS, RESPIRATION,
OACTIVITY, *FOOD WEBS, KENTUCKY,
A, TIPULIDAE, CHIRONOMIDAE, HALF-
ON, PRODUCTIVITY, ALGAE, AQUATIC
TEMPERATE ZONE, PLUTONIUM, HALF-
TY TEST,/ *POLYMERS, *BIOLOGICAL
, *AQUATIC ENVIRONMENT, *AQUATIC
RS, ALGAE, PROTOZOA, A/ *AQUATIC
LUTION EFFE/ *BIOASSAY, *AQUATIC
CES, PATH OF POLLUTANTS, AQUATIC
LOGY, SALINITY, BENTHOS, AQUATIC
, ENVIRONMENTAL EFFECTS, AQUATIC
IZATION, MARINE ANIMALS, AQUATIC
QUALITY, BIOINDICATORS, AQUATIC
LLA PYRENOIDOSA, BIOLOGICAL HALF-
DRILUS SP/ RADIOAUTOGRAPHS, HALF-
MODEL STUDIES, BIOASSAY, AQUATIC
ONUCLIDE, ULOTHRIX, UPTAKE, HALF-
CROP, MICROENVIRONMENT, AQUATIC
RIVER(TENNESSEE), PHYSICAL HALF-
, SURVEYS, MARINE ALGAE, AQUATIC
LIDES, ASSIMILATION, BA-LA, HALF-
RATURE, *EUTROPHICATION, AQUATIC
N, WATER QUALITY, ALGAE, AQUATIC
IVITY, *TRACE ELEMENTS, *AQUATIC
DRAINAGE, ALGAL CONTROL, AQUATIC
OUNDS, METALS, PLANKTON, AQUATIC
MS, DETRITUS, CYCLING NUTRIENTS,
ITY, PHOTO-ELECTRIC TURBIDOSTAT,
GHT, PHOSPHORUS, PHOTOSYNTHESIS,
TRATION, ALKALINITY, CALIFORNIA,
IENTS, OLIGOTROPHY, TEMPERATURE,
ORGANIC MATTER, TRACE ELEMENTS,
NTIAL NUTRIENTS, EUTROPHICATION,
EFFECTS, OREGON, BIOINDICATORS,
YANOPHYTA, DIATOMS, CHLOROPHYTA,
NTS, EUTROPHICATION, INHIBITION,
NTAL EFFECTS, INHIBITION, LIGHT,
NG, FREEZE-THAW TESTS, FREEZING,
YTA, DIATOMS, INHIBITION, LIGHT,
ES, *PLANT POPULATIONS, REVIEWS,
NTS, EUTROPHICATION, INHIBITION,
ESSENTIAL NUTRIENTS, INHIBITION,
MENT, RETENTION, PHOTOPERIODISM,
WATER, OLIGOTROPHY, TEMPERATURE,
MARY PRODUCT/ *PLANKTON, *ALGAE,
TS, TEMPERATURE, PHOTOSYNTHESIS,
ASSIUM, TEMPERATURE, EPILIMNION,
ROPHYLL, ICE, SEASONAL, DIURNAL,
WINDS, EPILIMNION, HYPOLIMNION,
MENTS, SURFACE WATERS, DETRITUS,
YNTHESIS, OXYGEN, TROPHIC LEVEL,
CHLORELLA, CARBON RADIOISOTOPES,
GEN COMPOUNDS, SEWAGE EFFLUENTS,
ESIS, LAGOONS, LABORATORY TESTS,
RY TESTS, *ANAEROBIC CONDITIONS,
NITIES, *LABORATORIES, *STREAMS,
CHLORELLA, CARBON, TEMPERATURE,
N, CHELATION, ENZYMES, CHLORINE,
ALGAE, *DIATOMS, *BENTHIC FLORA,
SEA WATER, PRIMARY PRODUCTIVITY,
IS, RESPIRATION, MICROORGANISMS,
URE, TOXICITY, DISSOLVED OXYGEN,
ENT, *OXIDATION LAGOONS, *ALGAE,
TAHOE(CALIF-NEV), TRANSPARENCY,
S, TEMPERATURE, IRON, NUTRIENTS,
TROGEN FIXATION, IRON COMPOUNDS,
RESERVOIRS, PONDS, TEMPERATURE,
LIGHT INTENSITY, PHOTOPERIODISM,
, TIME, EPILIMNION, THERMOCLINE,
A, SOUTH DAKOTA, PHOTOSYNTHESIS,
WATER POLLUTION SOURCES, RUNOFF,
STRATIFICATION, DEPTH, PLANKTON,
DISSOLVED OXYGEN, LAKES, RIVERS,
HARVESTING OF ALGAE, INHIBITION,
ON, INHIBITION, LIGHT INTENSITY,
S, HARVESTING ALGAE, INHIBITION,
IBITION, LIGHT, LIGHT INTENSITY,
ON, INHIBITION, LIGHT INTENSITY,
ESTS, FREEZING, LIGHT INTENSITY,

LEVELS, FOOD WEBS.: /SPECTROSCOPY   W69-08525
LEVELS, NUTRIENTS, NORTHEAST U.S.   W70-02772
LEVELS, SOFT WATER.: /E LAKE(CANA   W70-02973
LEYDEN, THE HAGUE, MINERALIZATION   W69-07838
LICHENS, ACTIMOMYCETES, CYTOLOGIC   W71-11823
LICHENS, PLANTS, DIATOMS, CORAL,   W69-07428
LICHENS, TIDAL EFFECTS, DISPERSIO   W70-01777
LIEBIG'S LAW OF THE MINIMUM, VOIS   W70-04765
LIFE CYCLES, GENETICS, E. COLI, I   W71-11823
LIFE CYCLES, GROWTH STAGES.: /NT,   W71-02494
LIFE CYCLES, OLIGOCHAETES, MOLLUS   W71-00665
LIFE HISTORY STUDIES, GAMMA RAYS,   W69-09742
LIFE.: / PLECOPTERA, EPHEMEROPTER   W69-09742
LIFE.: /N COMPOUNDS, EUTROPHICATI   W70-03068
LIFE(RADIONUCLIDE), FILTER-FEEDIN   W70-00708
LIFE, *ALGAE BLOOMS, *FISH TOXICI   W70-05819
LIFE, *AQUATIC POPULATIONS, ALGAE   W71-02491
LIFE, *AQUATIC POPULATIONS, *RIVE   W73-00748
LIFE, *WASTE DISPOSAL, *WATER POL   W72-01801
LIFE, ALGAE, SILTS, ECOLOGY, AQUA   W69-10080
LIFE, ALGAE, PLANTS, WATER PROPER   W69-05023
LIFE, AQUATIC MICROORGANISMS, MIC   W69-04276
LIFE, AQUATIC ANIMALS, BENTHIC FA   W71-02494
LIFE, AQUATIC ALGAE, WATER POLLUT   W70-05958
LIFE, CLINCH RIVER(TENNESSEE), PH   W70-02786
LIFE, CONCENTRATION FACTOR, LIMNO   W69-07861
LIFE, COOLING WATER, SITES.: /E,   W71-11881
LIFE, DETECTOR.:   *RADI   W69-07862
LIFE, ENVIRONMENT, MICROORGANISMS   W69-03611
LIFE, HYDROPSYCHE COCKERELLI, BOD   W70-02786
LIFE, MILK, ATLANTIC OCEAN, LANDF   W72-00941
LIFE, PTYCHOCHEILUS OREGONENSIS,   W69-07853
LIFE, SESTON, SCENEDESMUS, MATHEM   W73-01446
LIFE, STREAM POLLUTION, BACTERIA,   W72-01786
LIFE, WATER POLLUTION EFFECTS, MA   W71-05531
LIFE, WATER CONSERVATION, PUBLIC   W70-03344
LIFE, ZOOPLANKTON, AQUATIC ANIMAL   W69-03695
LIG: / COMPOUNDS, BACTERIA, DIATO   W69-09334
LIGHT EXTINCTION COEFFICIENT, ALG   W70-03923
LIGHT INTENSITY, STABILIZATION, P   W70-03923
LIGHT INTENSITY, TEMPERATURE, ION   W70-03334
LIGHT INTENSITY, DAPHNIA, CONNECT   W70-03959
LIGHT INTENSITY, CHLAMYDOMONAS, C   W70-04369
LIGHT INTENSITY, LIGHT PENETRATIO   W70-04283
LIGHT INTENSITY, EUTROPHICATION,   W70-03978
LIGHT INTENSITY, SUCCESSION, ZINC   W70-04371
LIGHT INTENSITY, LIGHT PENETRAT: /   W69-03515
LIGHT INTENSITY, LIGHT PENETRATIO   W69-03516
LIGHT INTENSITY, LIGHT PENETRATIO   W69-03518
LIGHT INTENSITY, LIGHT PENETRATIO   W69-03517
LIGHT INTENSITY, WATER TEMPERATUR   W68-00860
LIGHT INTENSITY, LIGHT PENETRATIO   W69-03514
LIGHT INTENSITY, LIMNOLOGY,: /N,   W69-03513
LIGHT INTENSITY, VELOCITY, TEMPER   W69-07862
LIGHT INTENSITY, CHLOROPHYLL, NUT   W70-05752
LIGHT INTENSITY, TEMPERATURE, PRI   W70-04809
LIGHT INTENSITY.: /UTION, NUTRIEN   W70-05270
LIGHT INTENSITY, SEASONAL, PHYSIO   W70-05409
LIGHT INTENSITY, CARBON, SEWAGE T   W70-05387
LIGHT INTENSITY, CONDUCTIVITY, AL   W70-05405
LIGHT INTENSITY, TURBIDITY, ALGAE   W70-00713
LIGHT INTENSITY, ECOLOGY, LOTIC E   W70-00265
LIGHT INTENSITY, CULTURES, RADIOA   W69-10158
LIGHT INTENSITY, LAKES, WATER POL   W69-10167
LIGHT INTENSITY, PHOTOPERIODISM,   W72-00917
LIGHT INTENSITY, EUTROPHICATION,   W71-13413
LIGHT INTENSITY, CURRENTS(WATER),   W70-02780
LIGHT INTENSITY, NUTRIENT REQUIRE   W70-02965
LIGHT INTENSITY, NITRATE, EUGLENA   W70-02964
LIGHT INTENSITY, BIOMASS, DISTRIB   W70-03325
LIGHT INTENSITY, TROPHIC LEVEL, S   W70-02504
LIGHT INTENSITY, CARBON DIOXIDE,   W71-00668
LIGHT INTENSITY, SLUDGE, DEPTH, O   W72-06854
LIGHT INTENSITY, CARBON DIOXIDE,   W72-06022
LIGHT MEASUREMENTS, EUPHOTIC ZONE   W70-01933
LIGHT PENETRATION, SEWAGE EFFLUEN   W70-01933
LIGHT PENETRATION, WATER POLLUTIO   W70-02255
LIGHT PENETRATION, DISSOLVED OXYG   W70-06225
LIGHT PENETRATION, WATER POLLUTIO   W72-00917
LIGHT PENETRATION, STANDING CROP,   W71-07698
LIGHT PENETRATION, ON-SITE TESTS,   W72-01818
LIGHT PENETRATION, SAMPLING, PRIM   W72-04761
LIGHT PENETRATION, SEASONAL, PRIM   W70-05405
LIGHT PENETRATION, FLUORESCENCE,   W69-05697
LIGHT PENETRATION, NITROGEN COMP: W69-03369
LIGHT PENETRATION, LIMNOLOGY, NIT   W69-03514
LIGHT PENETRATION, LIMNOLOGY, NIT   W69-04798
LIGHT PENETRATION, LIMNOLOGY, PHO   W69-03516
LIGHT PENETRAT: /TS, EUTROPHICATI   W69-03515
LIGHT PENETRATION, LIMNOLOGY, NIT   W69-03518

IBITION, LIGHT, LIGHT INTENSITY,   LIGHT PENETRATION, LIMNOLOGY, NUT   W69-03517
RE, ALKALINITY, *PHOTOSYNTHESIS,   LIGHT PENETRATION, PHOSPHORUS, LA   W68-00248
GAE, ALGAE, SEASONAL, LIMNOLOGY,   LIGHT PENETRATION, OXYGEN, SPORT   W68-00487
EUTROPHICATION, LIGHT INTENSITY,   LIGHT PENETRATION, LIMNOLOGY, NIT   W70-04283
CYCLING NUTRIENTS, TEMPERATURE,   LIGHT QUALITY, CHLOROPHYLL, ZOOPL   W69-08526
THONOUS PHOSPHATES, ULTRA-VIOLET   LIGHT STERILIZATION, ASTERIONELLA   W70-03955
W HAMPSHIRE, ALGAE, ODOR, TASTE,   LIGHT TEMPERATURE, NUTRIENTS, COP   W69-08282
, *FIXED-DENSITY STUDIES, *FIXED-   LIGHT TESTS, *EFFECTIVE-AVERAGE-L   W70-03923
MONOD GROWTH EQUATION, INCIDENT   LIGHT UTILIZATION.: /TH KINETICS,   W69-03730
-LIGHT TESTS, *EFFECTIVE-AVERAGE-   LIGHT-INTENSITY, PHOTO-ELECTRIC T   W70-03923
PECIES, CYCLOTELLA MENEGHINIANA,   LIGHT-SATURATED RATE, CELL VOLUME   W70-05381
IQUES, *MICROORGANISMS, *LIPIDS,   LIGHT, ALGAE, TEMPERATURE, CARBON   W70-05651
ON CONCENTRATION, STANDING CROP,   LIGHT, BIOMASS, RESPIRATION, LIFE   W71-00665
, N/ *NITROGEN FIXATION, *ALGAE,   LIGHT, CARBON DIOXIDE, INHIBITION   W70-04249
*PLANKTON, *ALGAE, SCENEDESMUS,   LIGHT, CULTURES, IRON COMPOUNDS,   W70-05409
HLOROPHYTA, DIATOMS, INHIBITION,   LIGHT, LIGHT INTENSITY, LIGHT PEN   W69-03517
VIRONMENTAL EFFECTS, INHIBITION,   LIGHT, LIGHT INTENSITY, LIGHT PEN   W69-03516
CHLORELLA, CHLOROPHYTA, DIATOMS,   LIGHT, LIPIDS, NUTRIENT REQUIREME   W69-06865
ABOLISM, *WASTE WATER TREATMENT,   LIGHT, MICROORGANISMS, PLANTS.: /   W70-06604
, CYANOPHYTA, SAMPLING, DENSITY,   LIGHT, MUD, SILTS, SANDS, GRAVELS   W70-00711
BIOMASS, CHLOROPHYLL, PROTEINS,   LIGHT, NUTRIENTS, HYDROSTATIC PRE   W69-10160
ITROGEN FIXATION, FOREST, WINDS,   LIGHT, TEMPERATURE, HYD: /EVEL, N   W71-03048
PHOTOSYNTHESIS, EUTROPHICATION,   LIGHT, TEMPERATURE, CARBON DIOXID   W69-07442
*WATER, *RIVERS, *LAKES, *PONDS,   LIGHT, TEMPERATURE, FERTILIZATION   W70-05404
, PACIFIC OCEAN, CARBON DIOXIDE,   LIGHT, TEMPERATURE.: /DIOISOTOPES   W70-05652
L POLLUTION, STREAMS, NUTRIENTS,   LIGHT, WASHINGTON, WATER CHEMISTR   W70-03309
PHAEODACTYLUM TRICORNUTUM, EDTA,   LIGURIAN SEA, EUROPEAN MEDITERRAN   W69-08275
GRAM-POSITIVE BACTERIA, SARCINE-   LIKE, BACILLUS-LIKE, MICROCOCCUS-   W70-00713
BACTERIA, SARCINE-LIKE, BACILLUS-   LIKE, MICROCOCCUS-LIKE, PSYCHROPH   W70-00713
LIKE, BACILLUS-LIKE, MICROCOCCUS-   LIKE, PSYCHROPHILES, PLEMORPHIC,   W70-00713
BUR REED, DUCKWEED, WHITE WATER   LILY, SWIMMERS' ITCH.: /D CELERY,   W70-05568
D SLUDGE, ALGAE, EUTROPHICATION,   LIME, *NITROGEN, HYDROGEN ION CON   W68-00012
HLOROPHYTA, CHLORELLA, CHLORINE,   LIME, COAGULATION.: /ENEDESMUS, C   W70-08107
CTIVATED SLUDGE, IRON COMPOUNDS,   LIME, OXYGEN, TERTIARY TREATMENT,   W68-00256
EMICAL COMPOUNDS, MEAN TOLERANCE   LIMIT, BLUEGILLS, CHLORELLA PYREN   W72-11105
TERIA, *PROTOZOA, *ALGAE, LETHAL   LIMIT, ELECTROPHORESIS, MICROORGA   W71-03775
WATER POLLUTION EFFECTS, LETHAL   LIMIT, GROWTH RATES, SPORES.: /E,   W70-01779
NRICHMENT EXPERIMENTS, *NUTRIENT   LIMITATION, CAYUGA LAKE(NY), ASTE   W70-01579
RODHE'S SOLUTION VIII, NUTRIENT   LIMITATION, CARBON-14.: /KES(NY),   W70-01579
TUM, SELENASTRUM GRACILE, GROWTH   LIMITATION, SELENASTRUM, ALGAL GR   W70-02779
, *LABORATORY STUDIES, *NITROGEN   LIMITATION, BATCH CULTURES, CHLOR   W70-07957
EMOSTATS, TURBIDOSTATS, NUTRIENT   LIMITED GROWTH, LUXURY CONSUMPTIO   W70-02779
ROWTH RATES, ALGAE, OLIGOTROPHY,   LIMITING FACTORS, CHLORELLA, ANAL   W70-02779
S, LAKES, PHYSIOLOGICAL ECOLOGY,   LIMITING FACTORS, ECOSYSTEMS, MIC   W71-02479
OPHYTA, FISHKILL, FERTILIZATION,   LIMITING FACTORS, NITROGEN COMPOU   W69-06535
              *NICKEL, DETECTION   LIMITS, CHEMICAL INTERFERENCES.:   W71-11687
Y, ANALYTICAL TECHNIQUES, LETHAL   LIMITS, DINOFLAGELLATES, DIATOMS.   W70-08111
LORELLA VULGARIS, DAPHNIA MAGNA,   LIMNAEA STAGNALIS, OUSPENSKII'S M   W71-03056
LORELLA VULGARIS, DAPHNIA MAGNA,   LIMNAEA STAGNALIS, OUSPENSKII'S M   W71-05719
, CHROCOCCUS LIMNETICUS, LYNGBYA   LIMNETICA, ANABAENA FLOS AQUAE, T   W70-05405
NAS EROSA, COSMARIUM, CHROCOCCUS   LIMNETICUS, LYNGBYA LIMNETICA, AN   W70-05405
HALF-LIFE, CONCENTRATION FACTOR,   LIMNODRILUS SPP, AEOLOSOMA HEMPRI   W69-07861
PLEUROXUS, TANYPUS, TANYTARSUS,   LIMNODRILUS, EUCYCLOPS, ALONA, CH   W70-06975
EMAND, WATER SUPPLY, ARID LANDS,   LIMNOLOGY.: / MONITORING, WATER D   W70-05645
LAYS, BRACKISH WATER, FISHERIES,   LIMNOLOGY.: /FIDE, IRON, FUNGI, C   W70-05761
LGAE, WATER MANAGEMENT(APPLIED),   LIMNOLOGY.: /LLUTION TREATMENT, A   W70-04721
PLING, ON-SITE DATA COLLECTIONS,   LIMNOLOGY.: /AL ANALYSIS, PH, SAM   W69-00659
CTIVITY, EUTROPHICATION, *LAKES,   LIMNOLOGY, *PLANKTON, ALGAE, BIBL   W69-00446
LOGICAL), *AQUATIC PRODUCTIVITY,   LIMNOLOGY, *NUTRIENTS, BIO MASS,   W68-00912
NUISANCE ALGAE, ALGAE, SEASONAL,   LIMNOLOGY, LIGHT PENETRATION, OXY   W68-00487
WASTES, PLANT GROWTH SUBSTANCES,   LIMNOLOGY, PLANT GROWTH REGULATOR   W68-00470
E, GERMANY, VOLUMETRIC ANALYSIS,   LIMNOLOGY, GROWTH RATES, *LABORAT   W68-00475
ION, PLANT TISSUES, SCENEDESMUS,   LIMNOLOGY, WATER PROPERTIES, WATE   W68-00248
AQUATIC ALGAE, CURRENTS(WATER),   LIMNOLOGY, LAKES, *PILOT PLANTS,   W68-00479
HT INTENSITY, LIGHT PENETRATION,   LIMNOLOGY, NITROGEN COMPOUNDS, NI   W69-03518
HT INTENSITY, LIGHT PENETRATION,   LIMNOLOGY, NUTRIENT REQUIREMENTS,   W69-03517
, BIOMASS, FERTILIZATION, LAKES,   LIMNOLOGY, MIDGES, NITRATES, NITR   W69-04805
HT INTENSITY, LIGHT PENETRATION,   LIMNOLOGY, PHOTOSYNTHETIC OXYGEN,   W69-03516
, INHIBITION, LIGHT PENETRATION,   LIMNOLOGY, NITROGEN COMPOUNDS, NU   W69-04798
NMENTAL EFFECTS, EUTROPHICATION,   LIMNOLOGY, NITROGEN COMPOUNDS, NI   W69-04802
ING FACILITIES, FISH MANAGEMENT,   LIMNOLOGY, NITROGEN COMPOUNDS, NI   W69-04804
HT INTENSITY, LIGHT PENETRATION,   LIMNOLOGY, NITROGEN COMPOUNDS, NU   W69-03514
NMENTAL EFFECTS, EUTROPHICATION,   LIMNOLOGY, NITROGEN COMPOUNDS, NI   W69-03186
NMENTAL EFFECTS, EUTROPHICATION,   LIMNOLOGY, NITROGEN COMPOUNDS, NI   W69-03185
TS, INHIBITION, LIGHT INTENSITY,   LIMNOLOGY,: /N, ESSENTIAL NUTRIEN   W69-03513
, TERTIARY TREATMENT, EFFLUENTS,   LIMNOLOGY, WATER CHEMISTRY, RESPI   W70-05387
PLANKTON, EUTROPHICATION, LAKES,   LIMNOLOGY, AQUATIC ENVIRONMENT, A   W69-05142
HT INTENSITY, LIGHT PENETRATION,   LIMNOLOGY, NITROGEN COMPOUNDS, NU   W70-04283
ITY, CYANOPHYTA, NUISANCE ALGAE,   LIMNOLOGY, CHLOROPHYTA, PLANKTON,   W70-03959
N, ALGAE, FUNGI, BACTERIA, FISH,   LIMNOLOGY, HYDROLOGIC ASPECTS, AQ   W70-04475
*ALGAE, PLANNING, RIVERS, LAKES,   LIMNOLOGY, WISCONSIN, URBANIZATIO   W70-04506
TION, WATER QUALITY, *NUTRIENTS,   LIMNOLOGY, WATER TEMPERATURE, *AL   W70-09093
EMENTS, *POLLUTANTS, INHIBITION,   LIMNOLOGY, ACIDITY, ACID SOILS, A   W70-02792
DIATOMS, ENVIRONMENTAL EFFECTS,   LIMNOLOGY, ECOLOGY, ECOLOGICAL DI   W70-02784
PHICATION, *CALIFORNIA, *NEVADA,   LIMNOLOGY, PRIMARY PRODUCTIVITY,   W70-01933
POTASSIUM, HARVESTING OF ALGAE,   LIMNOLOGY, NUTRIENTS, PHYSIOLOGIC   W70-02245
VIRONMENTAL EFFECTS, INHIBITION,   LIMNOLOGY, PESTICIDES, PESTICIDE   W70-02968
*ORGANIC MATTER, *PHYTOPLANKTON,   LIMNOLOGY, ZOOPLANKTON, ALGAE, CO   W70-02510
CTERIA, WATER POLLUTION EFFECTS,   LIMNOLOGY, CYANOPHYTA.: /ITUS, BA   W70-02249
TAMOGETON FOLIOSUS, EXPERIMENTAL   LIMNOLOGY, BRACHIONUS, SYNCHAETA   W70-02249

469

PTH, ALGAE, MOSSES, CHARA, FISH,
OURCES, WATER POLLUTION CONTROL,
N/ *FIELD STUDIES, *EXPERIMENTAL
VITY TECHNIQUES, *RADIOISOTOPES,
TOXICITY, *CULTURES, *ISOLATION
RE, DISSOLVED OXYGEN, FISHKILLS,
CATION, ALGAL BLOOMS, NUTRIENTS,
N SOURCES, AQUATIC PRODUCTIVITY,
PLATTE RIVER,
LGAE METABOLISM, DETOXIFICATION,
DYE, AIRBORNE TESTS/ *FRAUNHOFER
EASUREMENT APPARATUS, HYDROGEN F
CTANTS, ALKYLBENZENE SULFONATES,
-PEROT FILTERS, SOLAR FRAUNHOFER
PHAEOPIGMENTS, ACRIDINE ORANGE,
ELATIVE HETEROTROPHIC POTENTIAL,
ION, SWEDEN, LAKE ERKEN(SWEDEN),
ENTIAL, SWEDEN, ENZYME KINETICS,
GISPINA, ZURICHSEE(SWITZERLAND),
RE/ CHLOROFORM-METHANE EXTRACTS,
RATURE, PROTEINS, CARBOHYDRATES,
TURES.:   *GROWTH, *ALGAE,
LA, CHLOROPHYTA, DIATOMS, LIGHT,
, SEDIMENTS, GAS CHROMATOGRAPHY,
OROUS UPTAKE RATE.:   MIXED
OIL WASTES, OILY WATER, SULFITE
SOLUBILITY / *EQUILIBRIUM MODEL,
SC), GROWTH STIMULATION, SURFACE
HETEROTROPHY, CROOKED LAKE(IND),
DEVELOPMENT CORP(MASS), ARTHUR D
TNAM), CHLORO/ *BLOOM, *VIETNAM,
G, DEPTH, SEA WATER, FRES WATER,
CULTURES, BAYS, STREAMS, WEEDS,
AL AREAS, PHYSIOLOGICAL ECOLOGY,
DIOISOTOPES, INFLOW, CYANOPHYTA,
E TAHOE(CALIF-NEV), HYDROPHYTES,
RS, NITROGEN, PHOSPHORUS, ALGAE,
S, ALGAE, DETRITUS, PULP WASTES,
REBRIFORMIS, MATS, SYNECHOCOCCUS
NDS), HETEROCYSTS, GAS-VACUOLES,
ONDS, *DEPTH EFFECTS, *HYDRAULIC
HYDRAULIC LOAD EFFECTS, *ORGANIC
S, FLO/ *ULTRASONICS, *SUSPENDED
UPPLY, MUNICIPAL WATER, SEDIMENT
NS, SPECTROPHOTOMETRY, SUSPENDED
, *TERTIARY TREATMENT, SUSPENDED
HEMICAL OXYGEN DEMAND, SUSPENDED
MMISSION, ALGAL BLOOMS, NUTRIENT
HEMICAL OXYGEN DEMAND, SUSPENDED
HEMICAL OXYGEN DEMAND, SUSPENDED
OGICAL EXTRACTION, GROWTH, SHOCK
                       *SITE SELECTION,
ROGEN ION CONCENTRATION, ORGANIC
GAE/ *HYDROGEN SULFIDE, *ORGANIC
DISSOLVED OXYGEN, ODOR, ORGANIC
RY, *OXIDATION LAGOONS, *ORGANIC
SCREENS, SEDIMENTATION, ORGANIC
LFIDE, *SULFIDES, ALGAE, ORGANIC
ROGEN ION CONCENTRATION, ORGANIC
MOCLINE, CLIMATIC ZONES, ORGANIC
NTENSITY, SLUDGE, DEPTH, ORGANIC
ALGAE, SLUDGE, METHANE, ORGANIC
*HOGS, *LAGOONS, ALGAE, ORGANIC
DO, WEST(US), STABILIZATION POND
GAE, ANALYTICAL TECHNIQUES, CLAY
OPHICATION, IMPOUNDMENTS, LAKES,
US, POCILLOPORA EYDOUXI, PORITES
APHY, MICHAELIS-MENTEN EQUATION,
OSA, LAKE MENDOTA(WIS), ENGLAND,
ATED GLUCOSE, GONIDIA FORMATION,
IDAL CURRENT, PUGET SOUND(WASH),
NEW HAMPSHIRE, LAKE WINNISQUAM,
EMMERMANNI, APHANOCAPSA, BOSMINA
LEY POND(CONN), BOSMINA COREGONI

TED CARBON, *FLOCCULATION, *HEAD
TMENT, *ALGAE, *FILTRATION, HEAD
OLIDS, ALGAE, PILOT PLANTS, HEAD
DS, PHENOL/ BIOCIDES, PRODUCTION
, AUTECOLOGY, MORPHOLOGY, EXPORT
                            SEEPAGE
ATER SUPPLIES, GREEN ALGAE, HEAD
QUAM, LONG POND, CONCORD, BOSTON
LEVEL, LIGHT INTENSITY, ECOLOGY,
ON-SITE TESTS, LABORATORY TESTS,
WEDEN), LAKE EKOLN(SWEDEN), LAKE
SEDIMENTATION RATES, MANGANESE,
ON(ENGLAND), ESTHWAITE(ENGLAND),
RRANEAN DRAINAGE, MECHANIC/ LAKE
Y, LAKE BORREVANNET NORWAY, LAKE

LIMNOLOGY, LAKE TROUT, CRAYFISH,   W70-00711
LIMNOLOGY, DIVERSION, STREAMS.: /   W69-10170
LIMNOLOGY, CLADOCERAN, MONTANE ZO   W69-10154
LIMNOLOGY, OCEANOGRAPHY, TRACERS,   W69-10163
LIMNOLOGY, PUBLIC HEALTH, PHYSICO   W70-00273
LIMNOLOGY, ALGAL TOXINS, ALGAL CO   W70-00264
LIMNOLOGY, DYSTROPHY, OLIGOTROPHY   W69-09723
LIMNOLOGY, PLANKTON, PRIMARY PROD   W69-09135
LINCOLN(NEBR).:   W69-05894
LINDANE METABOLISM.: /LINDANE, *A   W70-08652
LINE DISCRIMINATOR, RHODAMINE WT   W70-05377
LINE.: /OFER LINES, ATTENUATION M   W70-05377
LINEAR ALKYLATE SULFONATES, WATER   W69-09454
LINES, ATTENUATION MEASUREMENT AP   W70-05377
LINEWEAVER-BURKE EQUATION.: /AND,   W70-04811
LINEWEAVER-BURK EQUATION, UPTAKE   W70-05109
LINEWEAVER-BURK EQUATION, MAXIMUM   W70-04287
LINEWEAVER-BURK EQUATION, MICHAEL   W70-04194
LINSLEY POND(CONN), BOSMINA COREG   W69-10169
LIPID CARBON, CHLOROPLASTS, INFRA   W70-05651
LIPIDS, KINETICS, MATHEMATICAL MO   W71-13413
LIPIDS, NUTRIENTS, CHLORELLA, CUL   W70-07957
LIPIDS, NUTRIENT REQUIREMENTS, OR   W69-06865
LIPIDS, OIL, OIL SHALE.: /GENESIS   W69-08284
LIQUOR, SUSPENDED SOLIDS, *PHOSPH   W71-13335
LIQUORS, ULTIMATE DISPOSAL, WATER   W69-05891
LITHOSPHERE, SORPTION REACTIONS,   W70-04385
LITTER.: /LAKE BUTTE DES MORTS(WI   W72-00845
LITTLE CROOKED LAKE(IND).: /ISM,   W71-10079
LITTLE INC(MASS), CARNEGIE DEPT O   W69-06865
LITTLE LAKE(HANOI), RED RIVER(VIE   W70-03969
LITTORAL, DISSOLVED OXYGEN, DIURN   W71-12870
LITTORAL, WINDS.: /TER, MOVEMENT,   W70-06229
LITTORAL, ESTUARIES, PHOTOSYNTHES   W70-03325
LITTORAL, NUTRIENTS, MODEL STUDIE   W73-01454
LIVERWORTS, MORPHOMETRY, PACIFAST   W70-00711
LIVESTOCK, WATER POLLUTION EFFECT   W71-10376
LIVESTOCK, IDAHO.: /ES, PHOSPHATE   W70-03501
LIVIDUS, BLUE-GREEN ALGAE, FILAME   W70-05270
LLANGORS LAKE(WALES), HALOBACTERI   W70-04369
LOAD EFFECTS, *ORGANIC LOAD EFFEC   W70-04787
LOAD EFFECTS, *DETENTION TIME EFF   W70-04787
LOAD, *FILTRATION, ALGAE, COLLOID   W71-10654
LOAD, CONDUCTIVITY, CALCIUM, MAGN   W69-05894
LOAD, FLUORESCE: /TEMPERATURE, IO   W70-03334
LOAD, PILOT PLANTS, ACTIVATED SLU   W70-09186
LOAD, SOLIDS CONTACT PROCESS, SEW   W70-06792
LOAD, SYNTHETIC DETERGENTS, FISH   W69-08674
LOAD, TURBIDITY, DATA STORAGE AND   W70-09557
LOAD, WASTES, SEWAGE, ALGAE, MICR   W70-08627
LOADING.:   *BIOL   W70-06604
LOADING.:   W71-08221
LOADING, *WATER REUSE, RESEARCH A   W72-03299
LOADING, *SULFATES, *SULFIDES, AL   W70-07034
LOADING, *WASTE WATER TREATMENT.:   W70-06601
LOADING, ALGAE, AMMONIA, SEDIMENT   W71-05013
LOADING, ACTIVATED SLUDGE, AERATI   W71-08970
LOADING, BIOCHEMICAL OXYGEN DEMAN   W71-11377
LOADING, CENTRIFUGATION, RESPIRAT   W72-06022
LOADING, DISSOLVED OXYGEN, HYDROG   W72-10573
LOADING, DESIGN CRITERIA, ODORS,   W72-06854
LOADING, HYDROGEN ION CONCENTRATI   W71-11375
LOADING, WASTE WATER TREATMENT, W   W70-07508
LOADINGS, ODOR PERSISTENCE.: /ORA   W70-04788
LOAM, CLAYS, COLLOIDS, EUTROPHICA   W69-04800
LOAMS, RESERVOIRS, TEXAS.: / EUTR   W69-04800
LOBATA, LEPTASTREA PURPUREA, FUNG   W70-08526
LOCH EWE(SCOTLAND), SCOTLAND, PHA   W70-04811
LONDON(ENGLAND), ESTHWAITE(ENGLAN   W69-10180
LONG ISLAND SOUND(NY), ICELAND, T   W69-09755
LONG ISLAND SOUND, NARRAGANSETT B   W69-10161
LONG POND, CONCORD, BOSTON LOT RE   W69-08674
LONGISPINA, ZURICHSEE(SWITZERLAND   W69-10169
LONGISPINA.: /(SWITZERLAND), LINS   W69-10169
LOS ANGELES COUNTY(CALIF).:   W70-05645
LOSS.: /OAGULATION, ALGAE, ACTIVA   W72-07334
LOSS, FILTERS, *SCENEDESMUS, FLOC   W72-03703
LOSS, PERMEABILITY, TEMPERATURE,   W72-08686
LOSS, QUATERNARY AMMONIUM COMPOUN   W72-12987
LOSSES, CHEMOSYNTHESIS, ABUNDANCE   W70-02780
LOSSES, DIURNAL VARIATIONS.:   W69-09723
LOSSES, OOCISTUS, FILTRATION MEDI   W70-04199
LOT RESERVOIR, LEBANON, WADLEIGH   W69-08674
LOTIC ENVIRONMENT, BIOMASS, FLOW   W70-00265
LOTIC ENVIRONMENTS, LENTIC ENVIRO   W72-05453
LOTSJON(SWEDEN), PERMEASES, CHAMY   W70-04287
LOW-FLOW AUGMENTATION, TENNESSEE   W73-00750
LOWESWATER(ENGLAND), BARNES SOUTH   W69-10180
LUCERNE, FERTILIZED FIELDS, SUBTE   W69-08668
LUGANO ITALY, LAKE MARIDALSVANNET   W69-07833

CHLORELLA ELLIPSOIDEA, CHLORELLA
LL SIZE, EUKARYOTES, MONOCHRYSIS
 PHYSIOLOGY, KJELDAHL PROCEDURE,
OSTATS, NUTRIENT LIMITED GROWTH,
OSMARIUM, CHROCOCCUS LIMNETICUS,
AENA MENDOTAE, ANABAENA HASSALI,
ENSITY, APHANIZOMENON, MELOSIRA,
SCILLATORIA, NAVICULA, CYMBELLA,
SPECIES DIVERSITY, LAKE WAUBESA,
IDIUM, APHANIZOMENON FLOS-AQUAE,
LISM, *PESTICID/ SOUTH CAROLINA,
ANTA ROSA(CALIF), LYUBLIN(USSR),
A, AUSTRALIA, SANTA ROSA(CALIF),
STES, INLETS(WATERWAYS), ORGANIC
SORPTION, EUTROPHICATION, ALGAE,
CHIA COLI, MYRIOPHYLLUM, LEPOMIS
RUBESCE/ *LAKE WASHINGTON(WASH),
RACTERISTICS, TERMARY COMPOUNDS,
HLORELLA PYRENOIDOSA, AUTOTROPH/
S, ANACHARIS, LAKE WAUBESA(WIS),
CROPHYTES, ANACHARIS CANADENSIS,
 *ANTAGONISTIC ACTION, *AQUATIC
ADOPHORA BLANKETS, HYDRODICTYON,
TY, STURGEON RIVER, FLUORESCEIN,
, BOMBAY(INDIA), BENARES(INDIA),
NUTRIENT TYPES, MICROQUANTITIES,
VISCOSE RAYON PLANT WASTES, *MAN-
LENE GLY/ *POLYESTER FIBERS, MAN-
ICOOK(ME), LAKE WINNISQUAM(N H),
ILUTION, OSCILLATORIA RUBESCENS,
RESPONSE, FERTILIZATION SURVEY,
HAL EXPOSURE, PHENANTHROQUINONE,
MILWAUKEE(WIS), CHICAG/ MANURE,
ULATA, COCCOCHLORIS PENIOCYSTIS,
OIRS, ANOPHELES QUADRIMACULATUS,
AMATH LAKE(ORE),/ *URBAN WASTES,
ARCH, US DEPARTMENT OF INTERIOR,
ARCH, US DEPARTMENT OF INTERIOR,
KEGONSA(WIS), LAKE WAUBESA(WIS),

S, YAHARA RIVER(WI/ *TRUAX FIELD(
UCTIVITY, CHLOROPHYLL, PIGMENTS,
ULLATA, DAPHNIA HYALINA, DAPHNIA
UDA, CHLORELLA VULGARIS, DAPHNIA
UDA, CHLORELLA VULGARIS, DAPHNIA
APHNIA GALEATA MENDOTAE, DAPHNIA
YDROGEN ION CONCENTRATION, IRON,
ARBONATE, ZINC, COBALT, CALCIUM,
OPHYTA, PHOSPHATES, CHLOROPHYLL,
, NITROGEN, PHOSPHORUS, CALCIUM,
 ION CONCENTRATION, IRON, LAKES,
ENT LOAD, CONDUCTIVITY, CALCIUM,
N, NITRATES, POTASSIUM, CALCIUM,
EN, PHOSPHORUS, SULFUR, CALCIUM,
RBIDITY, HYDROGEN SULFIDE, IRON,
YLL, HYDROGEN ION CONCENTRATION,
S, ESSENTIAL NUTRIENTS, CALCIUM,
OSCILLATORIA TENUIS, RAM/ PROTON
NICAL EQUIPMENT, PUMPING PLANTS,
VER(MAINE), COBBOSSEECONTEE LAKE(
URA, ASTERIONELLA, TYP/ *AUGUSTA(
A, TYPHOID FEVER, KENNEBEC RIVER(
 *AUGUSTA(MAINE), CARLETON POND(
IONS, WATER QU/ *EUTROPHICATION,
NAL VARIA/ LAKE SEBASTICOOK, ME,
DISTRIBUTION, ACTIVATED SLUDGE,
PERTIES, CONTROL, *OPERATION AND
LAGOONS, ALGAE, ADMINISTRATION,
LITY, TEMPERATURE, OPERATION AND
BASTICOOK, LAKE WASHINGTON, LAKE
SCHRODERIA, THIMET, TRI/ SEVIN,
*ALGAE TOXICITY, CARBON-14, DDT,
A, PESTICIDE DEGRADATION, SEVIN,
                          *AFRICA,
NON, SYNURA, ANABAENA, MELOSIRA,
RS, *BIOASSAY, TOXICITY, AQUATIC
ETHYLENE GLY/ *POLYESTER FIBERS,
IRY INDUSTRY, */ *ANIMAL WASTES,
ANIMAL WASTES, MANAGEMENT, *FARM
, PHOSPHORUS, AGRICULTURE, WATER
*MATHEMATICAL MODELS, *WATERSHED
S COMPOUNDS, WATER SUPPLY, WATER
, / *LAKE MINNETONKA(MINN), LAKE
FLUENT DIVERSION, *WATER QUALITY
YNTHESIS, TRIBUTARIES, WATERSHED
OLLUTION TREATMENT, ALGAE, WATER
NUTRIENTS, *LAKES, *TROUT, *FISH
EAR LAKE(CALIF), NUTRI/ *FISHERY
WATER QUALITY CONTROL, *MI/ FISH
E, FISH, ANIMALS, ALGAE, CUTTING

LUTEOVIRIDIS, SKELETONEMA COSTATU   W70-04809
LUTHERI, NAVICULA PELLICULOSA, GO   W69-08278
LUXURY CONSUMPTION, NUTRIENT REMO   W69-05868
LUXURY CONSUMPTION, PROVISIONAL A   W70-02779
LYNGBYA LIMNETICA, ANABAENA FLOS   W70-05405
LYNGBYA NOLLEI, GLOCOTRICHIA PISU   W69-08283
LYNGBYA.: /EMENT PROGRAM, ALGAL D   W69-10167
LYNGBYA, FRAGILARIA, OEDOGONIUM,   W70-04371
LYNGBYA, MICROCYSTIS, APHANIZOMEN   W69-01977
LYNGBYA, STORM LAKE(IOWA), CENTER   W70-02803
LYOPHILIZATION, *PESTICIDE METABO   W69-02782
LYUBERETS(USSR), MUNICH(GERMANY),   W70-05092
LYUBLIN(USSR), LYUBERETS(USSR), M   W70-05092
MA: /ASTES, GARBAGE DUMPS, OIL WA   W70-05112
MACROAQUATIC PLANTS, FERTILIZATIO   W69-06859
MACROCHIRUS, PIMEPHALES NOTATUS.:   W69-07861
MACROFAUNA, GYTTJA, OSCILLATORIA   W70-04253
MACROMOLECULES, ANTIBIOTICS, AGAR   W70-04280
MACRONUTRIENTS, MICRONUTRIENTS, C   W70-02964
MACROPHYTES, ALGAL GROWTH.: /NAJA   W70-03310
MACROPHYTES, POTAMOGETON FOLIOSUS   W70-02249
MACROPHYTES, ANACHARIS CANADENSIS   W70-02249
MACROPHYTES.: /, IRON SULFATE, CL   W69-10170
MACROPHYTES, GROUND SOLIDS, UPTAK   W69-09334
MACROPHYTES, CLADOPHORA, OEDOGONI   W70-05404
MACROQUANTITIES, ALGAL BLOOMS, LI   W70-04765
MADE FIBER PLANTS, NYLON-6, *VISC   W70-06925
MADE FIBERS, METHYL ALCOHOL, ETHY   W70-10433
MADISON LAKES(WIS), LAKE WASHINGT   W70-07261
MADISON LAKES(WIS), CEDAR RIVER(W   W70-00270
MADISON LAKES(WIS), LAKE MENDOTA(W70-02787
MADISON(WIS), ELODEA, REDOX COMP: W70-02982
MADISON(WIS), FLUSHING WATERWAYS,   W69-10178
MADISON(WIS), CHICAGO NATURAL HIS   W70-00719
MADISON(WIS), PSOROPHORA CILIATA,   W70-06225
MADISON(WIS), LAKE ZOAR(CONN), KL   W70-04506
MADISON(WIS), EUROPE, WATER POLLU   W70-03975
MADISON(WIS), EUROPE, WATER POLLU   W70-03975
MADISON(WIS), UNIVERSITY OF WISCO   W69-06276
MADISON(WIS).:   W69-03369
MADISON) EFFLUENTS, *MADISON LAKE   W70-05113
MAGN: /RINE ANIMALS, PRIMARY PROD   W73-01446
MAGNA, CZECHOSLOVAKIA.: /HNIA CUC   W70-00274
MAGNA, LIMNAEA STAGNALIS, OUSPENS   W71-03056
MAGNA, LIMNAEA STAGNALIS, OUSPENS   W71-05719
MAGNA, SELENASTRUM, RHODOTORULA,   W70-03957
MAGNESIUM, NITRATES, PHOSPHATES,   W70-03501
MAGNESIUM, C: /TASSIUM, IRON, BIC   W70-03978
MAGNESIUM, POTASSIUM, TEMPERATURE   W70-05409
MAGNESIUM, SODIUM, SILICA, DIATOM   W68-00856
MAGNESIUM, NUTRIENTS, NUISANCE AL   W69-07833
MAGNESIUM, ALKALINITY, SULFATES,   W69-05894
MAGNESIUM, IRON, ORGANIC MATTER,   W70-00719
MAGNESIUM, POTASSIUM, SODIUM, IRO   W70-02804
MAGNESIUM, COOLING, PROPERTY VALU   W70-02251
MAGNESIUM, CALCIUM.: /S, CHLOROPH   W70-01579
MAGNESIUM, POTASSIUM, HARVESTING   W70-02245
MAGNETIC RESONANCE SPECTROMETRY,   W69-00387
MAINE.: /, WATER TREATMENT, MECHA   W70-08096
MAINE), BARGE.: /VER, KENNEBEC RI   W70-08096
MAINE), CARLETON POND(MAINE), SYN   W70-08096
MAINE), COBBOSSEECONTEE LAKE(MAIN   W70-08096
MAINE), SYNURA, ASTERIONELLA, TYP   W70-08096
MAINE, NUTRIENTS, AQUATIC POPULAT   W68-00470
MAINE, PERIODIC VARIATIONS, SEASO   W68-00481
MAINTENANCE COSTS, CONSTRUCTION C   W70-04786
MAINTENANCE, *OXIDATION LAGOONS,   W70-08392
MAINTENANCE, *WASTE WATER TREATME   W71-08970
MAINTENANCE, DESIGN CRITERIA, *WA   W72-08686
MALAREN, LAKE ANNECY, LAKE VANERN   W70-03964
MALATHION, CHLORELLA PYRENOIDOSA,   W70-08097
MALATHION, SEVIN.:   W70-02198
MALATHION, ACETONE, ETHANOL.: /OS   W70-02968
MALAWI.:   W70-02646
MALLOMONAS, ULOTHRIX, DAVID MONIE   W69-10157
MAMMALS, ALGAE, FISH, DIATOMS, FU   W71-10786
MAN-MADE FIBERS, METHYL ALCOHOL,   W70-10433
MANAGEMENT, *FARM MANAGEMENT, *DA   W70-07491
MANAGEMENT, *DAIRY INDUSTRY, *ALG   W70-07491
MANAGEMENT(APPLIED), GEOCHEMISTRY   W71-03048
MANAGEMENT, GEOCHEMISTRY, LAKES,   W71-09012
MANAGEMENT.: /OSPHORUS, PHOSPHORU   W72-01867
MANAGEMENT PROGRAM, ALGAL DENSITY   W69-10167
MANAGEMENT, *LAKE WASHINGTON, *SE   W69-09349
MANAGEMENT, SEDIMENTS.: /, PHOTOS   W70-04268
MANAGEMENT(APPLIED), LIMNOLOGY.: /   W70-04721
MANAGEMENT, FERTILIZATION, ALGAE,   W70-05399
MANAGEMENT, *CHAOBORID MIDGES, CL   W70-05428
MANAGEMENT, STREAM IMPROVEMENT, *   W70-05428
MANAGEMENT, POTABLE WATER, HERBIC   W70-05263

471

ON PATTERNS, MEMBRANE PROCESSES,
STUARIES, PHYSIOLOGICAL ECOLOGY,
Y CONTROL, *ECOLOGY, *RESERVOIR,
CUS BINDERANUS, CLADOCERA, WATER
ATION, DRAIN, SAN JOAQUIN VALLEY
UM/ *YELLOWSTONE PARK, *ICELAND,
STIS AERUGINOSA, CRYPTOMONADIDA,
ARCODINA, MASTIGOPHORA, HOLOZOIC
OLOZOIC MASTIGOPHORA, HOLOPHYTIC
RA, HOLOPHYTIC MASTI/ SARCODINA,
ICULATE MATTER, BLUE FLUORESCING
LLS, CONCENTRATION, COMPOSITION,
IC BACTERIA, SUBLITTORAL, YELLOW
IRCONIUM-95-NIOBIUM-95, PRE-BOMB
S, *TEST PROCEDURES, ANTIFOULING
OW, MIXING, FLOW, MODEL STUDIES,
CARBOHYDRATES, LIPIDS, KINETICS,
, HETEROTROPHIC.:
S, INFLUENT SEWAGE, EQUILIBRIUM,
AQUATIC PLANTS, PALEOLIMNOLOGY,
AQUATIC PLANTS, PALEOLIMNOLOGY,
UATIC LIFE, SESTON, SCENEDESMUS,
OUNTAIN MEADOW, NITROGEN-15, SOD
UIS, OSCILLATORIA TEREBRIFORMIS,
, HORIZONTAL, BLOOMS, GELATINOUS
UOUS TREES, *DECOMPOSING ORGANIC
ECHARGE, *ION EXCHANGE, *ORGANIC
EN FIXATION, IRRIGATION, ORGANIC
OR, *ALGAE, *DECOMPOSING ORGANIC
, ZOOPLANKTON, ALGAE, / *ORGANIC
ADIOACTIVITY T/ *LAKES, *ORGANIC
UTILIZATION, DECOMPOSING ORGANIC
UNDS, OXYGEN, SILICATES, ORGANIC
YGEN, CHEMICAL ANALYSIS, ORGANIC
EATMENT, WATER QUALITY, *ORGANIC
, ABSORPTION, SEDIMENTS, ORGANIC
*MARL LAKES, DISSOLVED ORGANIC
BOGS, PHENOLS, PIGMENTS, ORGANIC
UALITY, WATER POLLUTION, ORGANIC
ON COLUMN TECHNIQUE, PARTICULATE
C ACID, CHEMOTROPHY, PARTICULATE
DNESS(WATER), CHLORIDES, ORGANIC
ION CONTROL, DECOMPOSING ORGANIC
LANT GROWTH, DECOMPOSING ORGANIC
SAMPLING, PHYTOPLANKTON, ORGANIC
HARDNESS, TIDAL WATERS, ORGANIC
OLOGY, CHLORELLA, ALGAE, ORGANIC
ER, *SEDIMENTS, ORGANIC
ATION, PIGMENTS, OCEANS, ORGANIC
ARINE ALGAE, *LITTORAL, *ORGANIC
NEW JERSEY, TEMPERATURE, ORGANIC
S, DILUTION, DECOMPOSING ORGANIC
TALS, INHIBITORS, SILTS, ORGANIC
MEAD, LAS VEG/ DISSOLVED ORGANIC
LID WASTES, DECOMPOSTING ORGANIC
QUATIC BACTERIA, *ALGAE, ORGANIC
TS, DIATOMS, PHOSPHORUS, ORGANIC
ALCIUM, MAGNESIUM, IRON, ORGANIC
*PRIMARY PRODUCTIVITY, *ORGANIC
YTOPLANKTON, DECOMPOSING ORGANIC
INETICS, PHOTOSYNTHESIS, ORGANIC
ROWTH FACTOR/ *DISSOLVED ORGANIC
RE, NITROGEN, NUTRIENTS, ORGANIC
, NUTRIENT REQUIREMENTS, ORGANIC
EUGLENA, EUTROPHICATION, ORGANIC
DIOACTIVITY, INTERFACES, ORGANIC
HOGENIC BACTERIA, ALGAE, ORGANIC
TOMS, ECOLOGY, SALINITY, ORGANIC
OSYNTHESIS, GROWTH RATE, ORGANIC
TER POLLUTION, DRAINAGE, ORGANIC
*PHYTOPLANKTON, *LAKES, ORGANIC
AGRICULTURAL WATERSHEDS, ORGANIC
ON RADIOISOTOPES, LAKES, ORGANIC
AL TECHNIQUES, SAMPLING, ORGANIC
CYANOPHYTA, *DECOMPOSING ORGANIC
YTA, VITAMINS, PEPTIDES, ORGANIC
, PHYSIOLOGICAL ECOLOGY, ORGANIC
ANA/ *CHEMICAL ANALYSIS, ORGANIC
PHOSPHATES, ECOSYSTEMS, ORGANIC
S, BACTERIA, DECOMPOSING ORGANIC
PROTOZOA, *ENVIRONMENT, *ORGANIC
ER, RAISIN RIVER, DETROIT RIVER,
-BURK EQUATION, UPTAKE VELOCITY,
EDEN), LINEWEAVER-BURK EQUATION,
ICAL UPTAKE, *MICROBIAL ECOLOGY,
THE MINIMUM, VOISIN'S LAW OF THE
SNAILS, AGE, TURBIDITY, DIPTERA,
DWORMS, AQUATIC BACTERIA, ALGAE,
ION, WILD RICE, SILTS, PLANKTON,
CUE LAKE(FLORIDA),

MASSACHUSETTS.: /LOSE, DISTRIBUTI   W69-00976
MASSACHUSETTS, CHLOROPHYTA, DINOF   W70-05272
MASSACHUSETTS, POTABLE WATER, WAT   W70-05264
MASSES, U S PUBLIC HEALTH SERVICE   W70-00263
MASTER DRAIN, SAN FRANCISCO BAY-D   W71-09577
MASTIGOCLADUS LAMINOSUS, PHORMIDI   W70-05270
MASTIGOPHORA, CRYPTOPHYCEAE, CHRY   W70-05112
MASTIGOPHORA, HOLOPHYTIC MASTIGOP   W72-01817
MASTIGOPHORA, CILIATA.: /PHORA, H   W72-01817
MASTIGOPHORA, HOLOZOIC MASTIGOPHO   W72-01817
MATE: /LON COLUMN TECHNIQUE, PART   W70-01074
MATERIAL BALANCES, ORGANISMS.: /E   W70-06600
MATERIAL, DU: /A HARVEYI, EPIPHYT   W70-01073
MATERIAL, RUTHENIUM-106-RHODIUM-1   W69-08524
MATERIALS, TESTING, BIODEGRADATIO   W71-07339
MATHEMATICAL MODELS, GREAT LAKES,   W71-05878
MATHEMATICAL MODELS, NEUTRALIZATI   W71-13413
MATHEMATICAL STUDIES, AUTOTROPHIC   W70-05750
MATHEMATICAL MODELS, ON-SITE DATA   W68-00858
MATHEMATICAL MODELS, BACTERIA, PH   W70-03975
MATHEMATICAL MODELS, BACTERIA, PH   W70-03975
MATHEMATICAL STUDIES, DATA COLLEC   W73-01446
MATS, NON-LEGUMES, *NITROGEN FIXI   W72-08110
MATS, SYNECHOCOCCUS LIVIDUS, BLUE   W70-05270
MATTER.: /OTA(WIS), PELAGIC AREAS   W70-06229
MATTER, *ALGAE, *PHENOLS.: /DECID   W72-03475
MATTER, *FILTRATION, DIATOMACEOUS   W72-04318
MATTER, *GRASSLANDS, MOUNTAINS, *   W72-08110
MATTER, *POTASSIUM COMPOUNDS, *AC   W72-04242
MATTER, *PHYTOPLANKTON, LIMNOLOGY   W70-02510
MATTER, *PHYSIOLOGICAL ECOLOGY, R   W70-04194
MATTER, *STREAM POLLUTION.: /TER   W68-00483
MATTER, AIR TEMPERATURE, AEROBIC   W69-07096
MATTER, ALGAE, IRON, HY: /LVED OX   W70-05760
MATTER, ALGAE, COPPER SULFATE, *C   W72-04495
MATTER, ALGAE, DIATOMS, CYANOPHYT   W72-00845
MATTER, ARCTIC ALGAE, AUTOTROPHIC   W71-10079
MATTER, BIOMASS, BENTHIC FLORA, P   W70-01074
MATTER, BIBLIOGRAPHIES.: /WATER Q   W70-07027
MATTER, BLUE FLUORESCING MATE: /L   W70-01074
MATTER, BLELHAM TARN(ENGLAND), CH   W70-02504
MATTER, CARBON, WATER POLLUTION E   W71-04206
MATTER, CARBON DIOXIDE.: / POLLUT   W72-01788
MATTER, CHEMICAL ANALYSIS, PROTEI   W72-01812
MATTER, CHLOROPHYLL, DYNAMICS, TE   W70-02780
MATTER, CONDUCTIVITY, TEMPERATURE   W73-00748
MATTER, CULTURES, CARBON, METABOL   W70-02804
MATTER, CURRENTS(WATER), WATER PO   W70-03501
MATTER, DIURNAL, LATITUDINAL STUD   W70-04809
MATTER, EXUDATION, TIDES, PHOTOSY   W70-01073
MATTER, HARDNESS, CARBON DIOXIDE,   W69-10157
MATTER, INSECTS, MITES, BACTERIS,   W71-00940
MATTER, IRON OXIDES, BACTERIA, AL   W70-02294
MATTER, LAS VEGAS, NEVADA, *LAKE   W72-02817
MATTER, LAGOONS, CATTLE.: /GE, SO   W71-05742
MATTER, LAKES, DIFFUSION, CARBON   W70-04284
MATTER, NITROGEN, CHLOROPHYLL.: /   W70-05663
MATTER, NITROGEN FIXATION, VOLUME   W70-00719
MATTER, NUTRIENTS, LAKES, WATER P   W71-10079
MATTER, OKLAHOMA, RESERVOIRS, SAM   W70-04001
MATTER, ORGANIC COMPOUNDS, METABO   W70-04287
MATTER, ORGANIC DEBRIS, ORGANIC G   W70-02510
MATTER, P: /CENTRATION, TEMPERATU   W70-00265
MATTER, PHOTOSYNTHESIS, TEMPERATU   W69-06865
MATTER, PHYTOPLANKTON, PRIMARY PR   W69-06273
MATTER, RADIOISOTOPES, GAMMA RAYS   W69-08272
MATTER, RADIOACTIVITY, PESTICIDES   W72-13601
MATTER, RIVERS, ESTUARINE ENVIRON   W73-01446
MATTER, SCENEDESMUS, CHLAMYDOMONA   W70-05381
MATTER, SEWAGE, PHYTOPLANKTON, PR   W70-05404
MATTER, SEASONAL, ALGAE, PRIMARY   W70-04290
MATTER, SEDIMENTS, FERTILIZERS, W   W70-04193
MATTER, SEA WATER, PRIMARY PRODUC   W70-02504
MATTER, TOXICITY, PHENOLS, ODOR,   W72-03299
MATTER, TOXICITY, DISSOLVED OXYGE   W72-01797
MATTER, TRACE ELEMENTS, LIGHT INT   W70-04369
MATTER, WATER POLLUTION EFFECTS,   W70-04811
MATTER, WATER POLLUTION EFFECTS,   W70-05109
MATTER, WATER POLLUTION, THERMAL   W70-03309
MATTER, WASTE WATER(POLLUTION).: /   W70-00664
MATTER, WATER QUALITY, TEMPERATUR   W71-03031
MAUMEE RIVER.: / RIVER, HURON RIV   W69-01445
MAXIMUM UPTAKE VELOCITY, SUBSTRAT   W70-05109
MAXIMUM UPTAKE VELOCITY, MICHAELI   W70-04287
MAXIMUM UPTAKE VELOCITY, ACTIVE T   W70-04194
MAXIMUM.: /OOMS, LIEBIG'S LAW OF   W70-04765
MAYFL: /A, CHRYSOPHYTA, GRAZING,   W71-00668
MAYFLIES, CADDISFLIES.: /MS, BLOO   W70-00475
MAYFLIES, FAUNA, FISH, OLIGOCHAET   W70-01943
MCCLOUD LAKE(FLORIDA).:   W72-01993

ONTENT, LAKE ERKEN(SWEDEN), LAKE   MENDOTA(WIS), NET PLANKTON, CHLOR   W70-03959
MESOTROPHY, LAKE NORRVIKEN, LAKE   MENDOTA, LAKE FURES, LAKE SEBASTI   W70-03964
RY, WISCONSIN, LAKE MONONA, LAKE   MENDOTA, LAKE WAUBESA, LAKE KEGON   W69-09349
, DAPHNIA PULEX, DAPHNIA GALEATA   MENDOTAE, DAPHNIA MAGNA, SELENAST   W70-03957
M, ANABAENA FLOS AQUAE, ANABAENA   MENDOTAE, ANABAENA HASSALI, LYNGB   W69-08283
ADAPTATION, *SPECIES, CYCLOTELLA   MENEGHINIANA, LIGHT-SATURATED RAT   W70-05381
ASTERIONELLA FORMOSA, CYCLOTELLA   MENEGHINIANA, PHOTIC ZONE, HELSIN   W70-04809
N-14, AUTORADIOGRAPHY, MICHAELIS-   MENTEN EQUATION, LOCH EWE(SCOTLAN   W70-04811
BON UPTAKE TECHNIQUES, MICHAELIS-   MENTEN EQUATION, RELATIVE HETEROT   W70-05109
YME KINETICS, GLUCOSE, MECHAELIS-   MENTEN EQUATION, ACETATE, GLYCOLL   W70-04284
E TECHNIQUES, GLUCOSE, MICHAELIS-   MENTEN EQUATION, SWEDEN, LAKE ERK   W70-04287
ER(CA/ SAN JOAQUIN RIVER(CALIF),   MERCED RIVER(CALIF), TUOLUMNE RIV   W70-02777
E, RECYCLING / PETROLEUM WASTES,   MERCENARIA MERCENARIA, AQUACULTUR   W70-09905
G / PETROLEUM WASTES, MERCENARIA   MERCENARIA, AQUACULTURE, RECYCLIN   W70-09905
ORT(N C), CRASSOSTREA VIRGINICA,   MERCENARIA MERCENARIA, PANOPEUS H   W69-08267
RASSOSTREA VIRGINICA, MERCENARIA   MERCENARIA, PANOPEUS HERBSTII.: /   W69-08267
CES, BACILLARIOPHYTA, GLOECAPSA,   MERISMOPEDIA, MICROCYSTIS, ANKIST   W70-03969
CULTURAL RUNOFF, ARKANSAS RIVER,   MERRIMACK RIVER, RED RIVER, SAVAN   W70-04507
LIS, BACILLUS MYCOIDES, BACILLUS   MESENTERICUS, DIATOMEA, CLADOPHOR   W69-07838
UMP COUNT, DESMIDS, FLAGELLATES,   MESOPLANKTON, MICROPLANKTON, PSEU   W70-05389
TROPHIC BACTERIA, DESULFOVIBRIO,   MESOTHERMY, MELOSIRA, SYNEDRA, RH   W70-05760
MENDOTA, LAKE FURES, LA/ CANADA,   MESOTROPHY, LAKE NORRVIKEN, LAKE   W70-03964
EPTOMYCIN, MILLIPORE FILTRATION,   MESOTROPHY.: /, TETRACYCLINE, STR   W69-10154
DRICA, GL/ BOTRYOCOCCUS BRAUNII,   METABOLIC PATTERN, ANABAENA CYLIN   W70-02965
PLANTS, MASS CULTURE, MEDICINE,   METABOLIC PATTERNS, MICROALGAE, M   W69-07832
GEN, BACTERIA, AEROBIC BACTERIA,   METABOLISM, CANALS, SANDS, CHLORI   W69-07838
E ALGAE, *CULTURES, PSEUDOMONAS,   METABOLISM, KINETICS, NITROGEN CO   W70-04280
GANIC MATTER, ORGANIC COMPOUNDS,   METABOLISM, BIOINDICATORS, BIOASS   W70-04287
LINA, LYOPHILIZATION, *PESTICIDE   METABOLISM, *PESTICIDE ACCUMULATI   W69-02782
RGANIC MATTER, CULTURES, CARBON,   METABOLISM, NITROGEN, PHOSPHORUS,   W70-02804
ALGAE, *TOXICITY, *INSECTICIDES,   METABOLISM, RADIOACTIVITY TECHNIQ   W70-02198
IODINE, PHOTO REACTIVITY, ALGAL   METABOLISM.: /LL RATES, *BROMINE,   W70-02363
, GROWTH STAGES, DIATOMS, ALGAE,   METABOLISM, ISOPODS, AMPHIPODA, F   W69-09742
, FISH, DIATOMS, CARP, PLANKTON,   METABOLISM, BIOLOGICAL COMMUNITIE   W70-00274
ALGAE, DIATOMS, TRACE ELEMENTS,   METABOLISM.: /SSIUM, CLAMS, SIZE,   W70-00708
N, PRIMARY PRODUCTIVITY, OXYGEN,   METABOLISM, WIND VELOCITY, HUMIDI   W70-01073
URNAL DISTRIBUTION, TEMPERATURE,   METABOLISM, *WASTE WATER TREATMEN   W70-06604
ASTE STORAGE, PONDS, HYDROLYSIS,   METABOLISM, CHEMICAL REACTIONS, C   W70-08097
NE METABOLISM./ *LINDANE, *ALGAE   METABOLISM, DETOXIFICATION, LINDA   W70-08652
ABOLISM, DETOXIFICATION, LINDANE   METABOLISM.: /LINDANE, *ALGAE MET   W70-08652
DE RESIDUES, ALGAE, GREAT LAKES,   METABOLISM, *CHLORELLA, *CHLAMYDO   W70-08652
A, *ACTIVATED SLUDGE, *BACTERIA,   METABOLISM, GROWTH RATES, MICROOR   W72-01817
R WASTES, ABSORPTION, CHELATION,   METABOLISM, AQUARIA, WATER POLLUT   W71-09005
PROPERTIES, BACTERIA, BIOASSAY,   METABOLISM, ABSORPTION, COMPETITI   W71-10079
ATTER, ARCTIC ALGAE, AUTOTROPHIC   METABOLISM, HETEROTROPHY, CROOKED   W71-10079
, CALCIUM CHLORIDE, TEMPERATURE,   METABOLISM, EPIPHYTOLOGY, ON-SITE   W72-05453
ULAR PRODUCTS, HORMONES, AUXINS,   METABOLITES, INHIBITION, SILICON,   W69-04801
   METAL TOXICITY.:   W70-10437

ER, SUCROSE-FLOTATION TECHNIQUE,   METALIMNION, BATHYTHERMOGRAPH, AR   W70-04253
RTHER(AUSTRIA), VERTICAL, CHALK,   METALIMNION, OSCILLATORIA RUBESCE   W70-05405
UNDS, PHENOLIC COMPOUNDS, ORGANO-   METALLIC COMPOUNDS, AMINES, ENGLA   W72-12987
CTROPHOTOMETRY, NUTRIENTS, HEAVY   METALS.: /EFFECTS, CHELATION, SPE   W71-11561
ENT FACILITIES, TEST PROCEDURES,   METALS.: /BACTERIA, ALGAE, TREATM   W71-06737
ENT FACILITIES, TEST PROCEDURES,   METALS.: /BACTERIA, ALGAE, TREATM   W70-01519
EN, PITTING(CORROSION), RUSTING,   METALS, INHIBITORS, SILTS, ORGANI   W70-02294
, CHLORIDES, SALINE LAKES, HEAVY   METALS, INDUSTRIAL WASTES, LAKES,   W69-02959
STAL PLAIN, INORGANIC COMPOUNDS,   METALS, PLANKTON, AQUATIC LIFE, Z   W69-03695
ON, OLIGOTROPHY, VITAMINS, HEAVY   METALS, TOXICITY, DIATOMS, MARL.:   W71-10079
PHORUS, SODIUM, POTASSIUM, 'TRACE   METALS, VITAMINS, WATER POLLUTION   W70-04503
YSTIS NIDULANS, OLIGOPHOSPHATES,   METAPHOSPHATES, PULSE-LABELING, P   W71-03033
, THIO/ *MINERAL TRANSFORMATION,   METAZOA, RADIOLARIA, FORAMINIFERA   W69-07428
LAKES, ALGAE, ENTERIC BACTERIA,   METHANE BACTERIA, CHEMICAL ANALYS   W70-03312
IC DIGESTION, *MUNICIPAL WASTES,   METHANE BACTERIA, ACID BACTERIA,   W72-06854
HLOROPLASTS, INFRARE/ CHLOROFORM-   METHANE EXTRACTS, LIPID CARBON, C   W70-05651
BOD), EVAPORATION, FERMENTATION,   METHANE, ALGAL CONTROL, BACTERIA,   W70-04787
WASTES. POPULATION, INDUSTRIES,   METHANE, ALKALINITY, HYDROGEN ION   W70-02803
TROSCOPY, COLORIMETRY, SAMPLING,   METHANE, CARBON DIOXIDE, DIFFUSIO   W71-11377
MS, WINDS, NITROGEN, PHOSPHATES,   METHANE, MUD, OXYGEN, ANAEROBIC C   W70-05761
EROBIC DIGESTION, ALGAE, SLUDGE,   METHANE, ORGANIC LOADING, HYDROGE   W71-11375
ON, *DEGRADATION(DECOMPOSITION),   METHANE, WATER POLLUTION EFFECTS.   W71-13413
LEXES, RIBONUCLEIC ACID, ORCINOL   METHOD.: /, ENZYME-SUBSTRATE COMP   W69-10160
CHLOROPHYLL CONTENT, CENTRIFUGE   METHOD, CALCULATED VOLUME METHOD,   W70-03959
RIFUGE METHOD, CALCULATED VOLUME   METHOD, CYCLOTELLA COMENSIS, RHOD   W70-03959
NOBRYON, FRAGILARIA, RADIOCARBON   METHOD, FINGER LAKES(NY), RODHE'S   W70-01579
S, ALGAL GROWTH, SEDGWICK-RAFTER   METHOD, TREATMENT TIMING, ALGAE C   W69-10157
TRIAL WASTES, THERMAL POLLUTION,   METHODOLOGY, INSTRUMENTATION, SEA   W70-10437
ORY TESTS, *ISOLATION, *CONTROL,   METHODOLOGY, CULTURES, TESTING, C   W70-04469
DISC, VISIBILITY TESTS, CONTROL   METHODS.:   SECCHI   W68-00488
COMPOUNDS, HABITATS, STATISTICAL   METHODS.: /TS, SEAWATER, ORGANIC   W69-09755
ZOOPLANKTON, SNAILS, APPLICATION   METHODS.: /NUISANCE ALGAE, FISH,   W70-03310
EASONS, ACETYLENE R/ *ANALYTICAL   METHODS, *BIOCHEMISTRY METHODS, S   W70-03429
*S/ *PHYTOPLANKTON, *STATISTICAL   METHODS, *ZOOPLANKTON, *BIOMASS,   W73-01446
PPER DOSAGE, *COPPER APPLICATION   METHODS, ALGAE IDENTIFICATION, AL   W69-10157
NALYTICAL METHODS, *BIOCHEMISTRY   METHODS, SEASONS, ACETYLENE REDUC   W70-03429
LYESTER FIBERS, MAN-MADE FIBERS,   METHYL ALCOHOL, ETHYLENE GLYCOL,   W70-10433
INDIA,   METHYL ORANGE, ALGAL BLOOMS,   W70-05550
UCOCYTES, *MUSCULAR EXERTION, *2-   METHYL-5-VINYLPYRIDINE, *GLUE GRE   W70-01788
FECTS, APHANIZOMENON FLOS AQUAE,   METOLIUS RIVER(ORE), COMMUNITY ST   W70-03978
), SEATTLE(WASH), REVENUE BONDS,   METRO ACT(SEATTLE).: /INGTON(WASH   W70-04455
, *SORPTION, ABSORPTION, GULF OF   MEXICO, ATLANTIC OCEAN, RADIOACTI   W69-08524

OF VARIANCE, FOOD CONCENTRATION,   MEXICO, PREDICTIVE EQUATIONS, ZOO   W70-03957
                              GREAT   MIAMI RIVER(OHIO).:   W70-00475
CHEWAN), HUMBOLDT(SASKATCHEWAN),   MICE, MORPHOLOGICAL CHARACTERS, T   W70-00273
LECTROKINETIC EFFECTS, ANN ARBOR(MICH), SAND BEDS.: /USPENSIONS, E   W71-03035
UATION, MAXIMUM UPTAKE VELOCITY,   MICHAELIS CONSTANT, LAKE NORRVIKE   W70-04287
ETICS, LINEWEAVER-BURK EQUATION,   MICHAELIS KINETICS, FLAGELLATES,   W70-04194
RBON UPTAKE TECHNIQUES, GLUCOSE,   MICHAELIS-MENTEN EQUATION, SWEDEN   W70-04287
ATE, CARBON-14, AUTORADIOGRAPHY,   MICHAELIS-MENTEN EQUATION, LOCH E   W70-04811
, RADIOCARBON UPTAKE TECHNIQUES,   MICHAELIS-MENTEN EQUATION, RELATI   W70-05109
POUNDS, PHOSPHATES, GREAT LAKES,   MICHIGAN.: /SOURCES, NITROGEN COM   W69-01445
                  LAWRENCE LAKE(MICHIGAN), *MACROPHYTES.:   W71-00832
S, PLANT/ *LAKES, *PRODUCTIVITY,   MICHIGAN, *EPIPHYTOLOGY, NUTRIENT   W71-00832
NUTRIENTS, QUALITY CONTR/ *LAKE   MICHIGAN, *WATER SUPPLY, INTAKES,   W70-00263
ONAS, CHLORELLA, WISCONSIN, LAKE   MICHIGAN, PHYTOPLANKTON, DIATOMS,   W69-10180
N, LAKE ERIE, LAKE ONTARIO, LAKE   MICHIGAN, WATER TEMPERATURE, PHOS   W70-00667
EON RIVER, FLUORESCEIN, MACROPH/   MICHIGAN, CHEBOYGAN COUNTY, STURG   W69-09334
S, RADIOACTIVITY, BIOTA, CYCLES,   MICHIGAN, INVERTEBRATES, FISH, TE   W69-09334
SULFATE, BOTTOM SEDIMENTS, LAKE   MICHIGAN, WASHINGTON, DISSOLVED O   W69-09349
*GREAT LAKES, *LAKE ERIE, *LAKE   MICHIGAN, *LAKE ONTARIO, LAKE HUR   W70-01943
ROPHICATION, OLIGOTROPHY, LAKES,   MICHIGAN, BICARBONATES, IRON, PHO   W72-03183
PLANKTON/ *EUTROPHICATION, *LAKE   MICHIGAN, ALGAE, NUISANCE ALGAE,   W68-01244
GROWTH RATES, *FRESH WATER FISH,   MICHIGAN, TROUT, WINTERKILLING, F   W68-00487
EST SOILS, MINNESOTA, WISCONSIN,   MICHIGAN, PHOSPHORUS, AGRICULTURA   W70-04193
SANCE ALG/ *PHYTOPLANKTON, *LAKE   MICHIGAN, *WATER CIRCULATION, NUI   W69-05763
T, *ALGAE, *FILTERS, SANDS, LAKE   MICHIGAN, WATER QUALITY CONTROL,   W73-02426
ACK LAKE(COLO), PASS LAKE(COLO),   MICRO-ALGAE, FLAGELLATES, CILIATE   W69-10154
, PHAGE, FERMENTOR, FACULTATIVE   MICROAEROPHILIC, INOCULATION.: /T   W71-02009
E, MEDICINE, METABOLIC PATTERNS,   MICROALGAE, MICRONUTRIENTS, PHYCO   W69-07832
*ORGANIC SOLUTES, *HETEROTROPHY,   MICROBIAL ECOLOGY, SWEDEN, LAKE E   W70-04284
GYMNODINIUM, CHEMICAL SYMBIOSIS,   MICROBIAL ECOLOGY, DINOBRYON, VOL   W70-04503
. RADIOCARBON UPTAKE TECHNIQUES,   MICROBIAL ECOLOGY, COCCOLITHOPHOR   W70-05272
KE ERKEN(SWEDEN), ANNUAL CYCLES,   MICROBIAL ECOLOGY, RADIOCARBON UP   W70-05109
ION DYNAMICS, ALGAL COMMUNITIES,   MICROBIAL ECOLOGY, WATER QUALITY   W70-04510
C SOLUTES, *MARINE ENVIRONMENTS,   MICROBIAL ECOLOGY, FLUORESCENCE M   W70-04811
MARINORUBRA, GYRODINIUM COHNII,   MICROBIAL ECOLOGY, COULTER COUNTE   W70-05652
ARVESTING, DREDGING, MONITORING,   MICROBIOLOGY, FOOD CHAINS.: /S, H   W70-05264
NIQUES, *AQUATIC ALGAE, *AQUATIC   MICROBIOLOGY, *AQUATIC PRODUCTIVI   W70-04283
NOCOBALAMIN, EUGLENA GRACILIS Z,   MICROBIOLOGICAL ASSAY, SPECIES CO   W69-06273
ATES, PHOTOSYNTHESIS, NUTRIENTS,   MICROBIOLOGY.: / CARBON, BICARBON   W70-07389
LOGICAL TREATMENT, SOLID WASTES,   MICROBIOLOGY, ACTIVATED SLUDGE, R   W68-00248
CHNIQUES, AQUATIC ALGAE, AQUATIC   MICROBIOLOGY, BALANCE OF NATURE,   W69-03517
S, ALGAE, AQUATIC ALGAE, AQUATIC   MICROBIOLOGY, AQUATIC WEEDS, BALA   W69-04804
PLANKTON, AQUATIC ALGAE, AQUATIC   MICROBIOLOGY, AQUATIC PRODUCTIVIT   W69-04802
R, ALGAE, AQUATIC ALGAE, AQUATIC   MICROBIOLOGY, BALANCE OF NATURE,   W69-03516
MICROO/ *MICROBIOLOGY, *AQUATIC   MICROBIOLOGY, *ESTUARIES, *MARINE   W69-03752
                         ESTUARINE   MICROBIOLOGY.:   W69-03752
OGY, *CYANOPHYTA, ALGA/ *AQUATIC   MICROBIOLOGY, *PHYSIOLOGICAL ECOL   W69-03518
S, ALGAE, AQUATIC ALGAE, AQUATIC   MICROBIOLOGY, AQUATIC PRODUCTIVIT   W69-03373
CIINIQUES, AQUATIC ALGAE, AQUATIC   MICROBIOLOGY, AQUATIC PRODUCTIVIT   W69-03186
ALGAE, AQUATIC ANIMALS, AQUATIC   MICROBIOLOGY, AQUATIC PRODUCTIVIT   W69-03513
ALGAE, AQUATIC BACTERIA, AQUATIC   MICROBIOLOGY, BIOASSAY, CHLOROPHY   W69-03362
ALGAE, AQUATIC BACTERIA, AQUATIC   MICROBIOLOGY, BALANCE OF NATURE,   W69-03188
CHNIQUES, AQUATIC ALGAE, AQUATIC   MICROBIOLOGY, BALANCE OF NATURE,   W69-03374
ALGAE, AQUATIC BACTERIA, AQUATIC   MICROBIOLOGY, BALANCE OF NATURE,   W69-03369
OMPOUNDS, AQUATIC BACTERIA, AQUATIC   MICROBIOLOGY, CHLAMYDOMONAS, CHLO   W69-03364
ALGAE, AQUATIC BACTERIA, AQUATIC   MICROBIOLOGY, BIOASSAY, CHLOROPHY   W69-03371
CHNIQUES, AQUATIC ALGAE, AQUATIC   MICROBIOLOGY, AQUATIC PRODUCTIVIT   W69-03185
CHNIQUES, AQUATIC ALGAE, AQUATIC   MICROBIOLOGY, AQUATIC WEEDS, BALA   W69-03515
ECOLOGY, AQUATIC ALGAE, AQUATIC   MICROBIOLOGY, CHLOROPHYTA, CYANOP   W69-03358
ALGAE, *EUTROPHICATION, DIATOMS,   MICROBIOLOGY, NITRATES, AMMONIA,   W71-05026
TA, CHLOROPHYTA, SIEROZEMS, SOIL   MICROBIOLOGY, DISTRIBUTION PATTER   W71-04067
ION, *AQUATIC BACTERIA, *AQUATIC   MICROBIOLOGY, ALGICIDES, BIOASSAY   W69-10171
TS, *MORPHOLOGY, *ALGAE, AQUATIC   MICROBIOLOGY, CYANOPHYTA, ESSENTI   W69-10177
PHYTOPLANKTON, PLANKTON, AQUATIC   MICROBIOLOGY, CYCLING NUTRIENTS,   W70-02245
LGICIDES, AQUATIC ALGAE, AQUATIC   MICROBIOLOGY, BIOASSAY, CHLOROPHY   W70-02248
AE, AQUATIC ENVIRONMENT, AQUATIC   MICROBIOLOGY, BALANCE OF NATURE,   W70-02968
HNIQUES, *AQUATIC ALGAE, AQUATIC   MICROBIOLOGY, AQUATIC MICROORGANI   W70-02792
A, ALGAE, AQUATIC ALGAE, AQUATIC   MICROBIOLOGY, AQUATIC PRODUCTIVIT   W70-02784
RGANISMS, SOIL ALGAE, TURF, SOIL   MICROBIOLOGY, WATER POLLUTION SOU   W72-08110
OMPARATIVE PRODUCTIVITY, AQUATIC   MICROBIOLOGY, EUTROPHICATION, MUN   W72-12240
IA, SARCINE-LIKE, BACILLUS-LIKE,   MICROCOCCUS-LIKE, PSYCHROPHILES,   W70-00713
ARS, STAPHYLOCOCCUS EPIDERMIDIS,   MICROCOCCUS, STREPTOCOCCUS, BREVI   W71-11823
CUS PRESCOTTII, CALOTHRIX FUSCA,   MICROCOLEUS LACUSTRIS (RAB) FARLO   W70-04468
OIL-WATER TECHNIQUES, PHYCOLOGY,   MICROCYSTIS, GLOEOTRICHIA, APHANI   W70-04381
IA, NOSTOC, ANABAENA, ANACYSTIS,   MICROCYSTIS, BOSMINA, ASTERIONELL   W70-04503
NN), ACTIVITY-DENSITY, VERTICAL,   MICROCYSTIS, OSCILLATORIA, NAVICU   W70-04371
SIOLOGY, SPIROGYRA, MYXOPHYCEAE,   MICROCYSTIS, COELOSPHAERIUM, OSCI   W70-04369
CENTRATION, ANABAENA CYLINDRICA,   MICROCYSTIS AERUGINOSA, OEDOGNIUM   W70-03520
OPHYTA, GLOECAPSA, MERISMOPEDIA,   MICROCYSTIS, ANKISTRODESMUS, PEDI   W70-03969
, BLOOMS, CHLORELLA PYRENOIDOSA,   MICROCYSTIS AERUGINOSA, BURNTSIDE   W70-03512
IVERSITY, LAKE WAUBESA, LYNGBYA,   MICROCYSTIS, APHANIZOMENON, ANABA   W69-01977
ODICTYON, ELODEA, CERATOPHYLLUM,   MICROCYSTIS, ANABAENA.:   HYDR   W68-00479
                                                  MICROCYSTIS, ANABAENA.:   W68-00470
RICORNUTUM, ANABAENA FLOS-AQUAE,   MICROCYSTIS AERUGINOSA, ALGAL CUL   W69-06864
ODUCING ALGAE, PHOTOMICROGRAPHS,   MICROCYSTIS AERUGINOSA, ANABAENA   W69-05704
OSPHATE(BIOLOGICALLY AVAILABLE),   MICROCYSTIS AERUGINOSA, LAKE COMO   W70-05269
WINGRA(WIS), ANABAENA SPIROIDES,   MICROCYSTIS AERUGINOSA, CRYPTOMON   W70-05112
GILARIA, ASTERIONELLA, MELOSIRA,   MICROCYSTIS, KERATELLA, POLYARTHR   W70-05375
A, EUROPE, SOUTH AFRICA, ISRAEL,   MICROCYSTIS, PRYMNESIUM PARVUM, S   W70-05372
TON(WASH), *CULTURAL INFLUENCES,   MICROCYSTIS, DREDGE SAMPLES, ANAB   W70-05663

476

ALGAL BLOOMS, ALGAE NUTRITION,
ETEROCYSTS, ANABAENA VARIABILIS,
HOSPHATE FERTILIZERS, BLUEGILLS,
LADOPHORA, OEDOGONIA, SPIROGYRA,
E/ *PURIFICATION, CULTURE MEDIA,
NS, PHYCOLOGISTS, VETERINARIANS,
ATA, D/ HYDROXYLAMINE, ANABAENA,
LGAE, COPPER SULFATE, EFFLUENTS,
EN/ *2,3-DICHLORONAPHTHOQUINONE,
URE, 2,3-DICHLORONAPHTHOQUINONE,
ARNALDIA PLUMOSA, STIGEOCLONIUM,
ICROCYSTIS, STIGEOCLONIUM TENUE,
TOMS FAUNA, MARGINAL VEGETATION,
DISCUS, ANABAENA, APHANIZOMENON,
ONS, OXYGENATION, STANDING CROP,
REGIONS, KENTUCKY, ACID STREAMS,
UCKY.:   ACID WASTE
LADOCERANS, SPECIES COMPOSITION,
MENTOUS BLUE-GREEN ALGAE, OCULAR
IS, GLOEOTRICHIA, APHANIZOMENON,
 ALGAL PHYSIOLOGY, HETEROTROPHY,
METABOLIC PATTERNS, MICROALGAE,
DOSA, AUTOTROPH/ MACRONUTRIENTS,
ROL, *PHOSPHORUS CONCENTRATIONS,
CTERI/ *ALGAE, *ECOLOGY, *MARINE
TOLOGY, MARINE ALGAE, E/ *MARINE
E, AQUATIC MICROBIOLOGY, AQUATIC
, *WASTE WATER TREATMENT, LIGHT,
, LETHAL LIMIT, ELECTROPHORESIS,
SYSTEMS, MICROORGANISMS, AQUATIC
Y, LIMITING FACTORS, ECOSYSTEMS,
MS, PHOTOSYNTHESIS, RESPIRATION,
 HEATED WATER, PROTEINS, AQUATIC
DATION, ALGAE, BACTERIA, AQUATIC
, *PHYTOPLANKTON, *LAKES, ALGAE,
TE, RESERVOIRS, WATER POLLUTION,
DED LOAD, WASTES, SEWAGE, ALGAE,
 *SILTS, SANDS, ALGAE, PLANKTON,
ED SLUDGE, STORAGE, TEMPERATURE,
NITROGEN, NUTRIENTS, FILTRATION,
TEMATICS, *SANITARY ENGINEERING,
TERIA, METABOLISM, GROWTH RATES,
LORIDA, *EUTROPHICATION, AQUATIC
GY, SUSPEND/ *INDICATOR SPECIES,
YTOLOGICAL STUDIES, EVAPORATION,
SM, CANALS, SANDS, CHLORINATION,
 REQUIREMENTS, *DISTRIB/ *MARINE
 *ANALYTICAL TECHNIQUES, *MARINE
BAC/ *ACID MINE WATER, *AQUIATIC
IC LIFE, AQUATIC MICROORGANISMS,
L EFFECTS, AQUATIC LIFE, AQUATIC
ICROBIOLOGY, *ESTUARIES, *MARINE
EATION, WAT/ *NUTRIENTS, *MARINE
MENT, AQUATIC LIFE, ENVIRONMENT,
T/ ALGAE, EUTROPHICATION, LAKES,
R, *GRASSLANDS, MOUNTAINS, *SOIL
MIDS, FLAGELLATES, MESOPLANKTON,
ARBON BIOASSAY, *NUTRIENT TYPES,
WASTE EF/ *BIOLOGIC 'DISPOSERS',
ACTION, MILLIPORE FILTERS, PHASE
 MICROBIAL ECOLOGY, FLUORESCENCE
ICAL PROPERTIES, CENTRIFUGATION,
URES, HYDROGEN SULFIDE, ELECTRON
GEN FIXATION, NITRATES, ELECTRON
A, CYTOLOGICAL STUDIES, ELECTRON
ARTERIA, GONIUM, ANKISTRODESMUS,
RIOPHYLLUM, POTAMOGETON CRISPUS,
CRONUTRIENTS, AQUATIC ECOSYSTEM,
 *FISHERY MANAGEMENT, *CHAOBORID
HESIS, EUTROPHICATION, TOXICITY,
FERTILIZATION, LAKES, LIMNOLOGY,
PER SULFATE, FISHES, HERBICIDES,
LIES, FAUNA, FISH, OLIGOCHAETES,
HOLOGY, ECOLOGICAL DISTRIBUTION,
ORAL, DISSOLVED OXYGEN, DIURNAL,
OPLANKTON, SULFUR BACTERIA, DIEL
ITY, FLORIDA, TEXAS, PHOSPHORUS,
AN, / ANIMAS RIVER, DURANGO, SAN
ADA, NUTRIENT LEVELS, S/ *TWELVE
DA, KUSHOG LAKE(ONTARIO), TWELVE
LD JOHN LAKE, PINGO LAKE, TWELVE
PPER CARBONATE, CERCARIAE, WATER
A, CANDIDA TROPICALIS, NOCTILUCA
, LABORATORY MODEL, COD REMOVAL,
DALIA(OHIO), RHIZOPODS, SPIDERS,
R TREATMENT, *OXIDATION LAGOONS,
EYS, MARINE ALGAE, AQUATIC LIFE,
EEK(WIS), DOOR CREEK(WIS), DUTCH
                         *TEXTILE
BILITY, *IN VITRO, IN VIVO, BALL-

MICROCYSTIS AERUGINOSA.:	W70-05565
MICROCYSTIS AERUGINOSA, APHANIZOM	W70-05091
MICROCYSTIS SCUMS, PITHIOPHORA AL	W70-05417
MICROCYSTIS AERUGINOSA, ANABAENOP	W70-05404
MICROCYSTIS AERUGINOSA, APHANIZOM	W70-00719
MICROCYSTIS AERUGINOSA, ENDOTOXIN	W70-00273
MICROCYSTIS, GLOEOTRICHIA ECHINUL	W70-02803
MICROCYSTIS AERUGINOSA.: /ORUS, A	W70-02787
MICROCYSTIS, APHANIZOMENON, ANABA	W70-03310
MICROCYSTIS AERUGINOSA, APHANIZOM	W70-02982
MICROCYSTIS, STIGEOCLONIUM TENUE,	W70-02245
MICROCYSTIS AERUGINOSA, NOSTOC, N	W70-02245
MICROCYSTIS, ANABAENA, EUCYCLOPS,	W70-06975
MICROCYSTIS, POTAMOGETON, POLYGON	W72-04761
MICROENVIRONMENT, AQUATIC LIFE, E	W69-03611
MICROFLORA.: /HIPS, GEOGRAPHICAL	W68-00891
MICROFLORA, MCCREARY COUNTY, KENT	W69-00096
MICROFOSSILS, COMPETITIVE EXCLUSI	W70-05663
MICROMETER, SUCROSE-FLOTATION TEC	W70-04253
MICRONUTRIENTS, DINOPHYCEAE, CRYP	W70-04381
MICRONUTRIENTS, BADFISH RIVER(WIS	W70-04506
MICRONUTRIENTS, PHYCOLOGY, SPACE	W69-07832
MICRONUTRIENTS, CHLORELLA PYRENOI	W70-02964
MICRONUTRIENTS, AQUATIC ECOSYSTEM	W69-09340
MICROORGANISMS, PHYTOPLANKTON, BA	W70-01068
MICROORGANISMS, *HABITATS, EPIPHY	W69-10161
MICROORGANISMS, AQUATIC PLANTS, A	W70-02792
MICROORGANISMS, PLANTS.: /ABOLISM	W70-06604
MICROORGANISMS, ELECTRIC CURRENTS	W71-03775
MICROORGANISMS, ENZYMES.: /S, ECO	W71-02479
MICROORGANISMS, AQUATIC MICROORGA	W71-02479
MICROORGANISMS, LIGHT INTENSITY,	W71-00668
MICROORGANISMS, TEMPERATURE, ACTI	W70-09905
MICROORGANISMS, SAMPLING, WATER A	W71-00221
MICROORGANISMS, BIOMASS, DIATOMS,	W70-08654
MICROORGANISMS, HISTORY, TASTE, O	W70-08096
MICROORGANISMS, PHOSPHATES, HYDRO	W70-08627
MICROORGANISMS, TURBIDITY, ANALYT	W70-08628
MICROORGANISMS, NITROGEN COMPOUND	W71-13335
MICROORGANISMS, ANALYTICAL TECHNI	W71-13313
MICROORGANISMS, NUISANCE ALGAE, S	W72-01798
MICROORGANISMS, ALGAE, FUNGI, LAB	W72-01817
MICROORGANISMS, TROPHIC LEVEL.: /	W72-01993
MICROORGANISMS, FRESH WATER BIOLO	W71-10786
MICROORGANISMS, CALIFORNIA, *PHOT	W71-05267
MICROORGANISMS, CHEMICAL ANALYSIS	W69-07838
MICROORGANISMS, *ALGAE, *NUTRIENT	W70-03952
MICROORGANISMS, PHYSIOLOGICAL ECO	W70-04811
MICROORGANISMS, *ALGAE, *AQUATIC	W69-00096
MICROORGANISMS, SESTON, INVERTEBR	W69-04276
MICROORGANISMS, MICROORGANISMS, S	W69-04276
MICROORGANISMS, MARINE ALGAE, MAR	W69-03752
MICROORGANISMS, *POLLUTANTS, RECR	W69-04276
MICROORGANISMS, WATER CHEMISTRY,	W69-03611
MICROORGANISMS, NITROGEN, NUTRIEN	W68-00481
MICROORGANISMS, SOIL ALGAE, TURF,	W72-08110
MICROPLANKTON, PSEUDOPLANKTON, CR	W70-05389
MICROQUANTITIES, MACROQUANTITIES,	W70-04765
MICROSCOPIC ORGANISMS, PAPERMILL	W71-00940
MICROSCOPY, STAINING, STRIPPING F	W69-10163
MICROSCOPY, ACETATE, CARBON-14, W	W70-04811
MICROSCOPY, NETS, NANNOPLANKTON,	W70-05389
MICROSCOPY, BIOCHEMISTRY, VITAMIN	W70-03952
MICROSCOPY, AMMONIA, SPECTROMETER	W70-04369
MICROSCOPY, WASTE TREATMENT, LAKE	W70-03973
MICROSPORA, PEDIASTRUM, SCENEDESM	W70-03974
MICROSPORA WITTROCKII, SPIROGYRA,	W70-00711
MIDGE LARVAE, LAKE STRATIFICATION	W69-09340
MIDGES, CLEAR LAKE(CALIF), NUTRIE	W70-05428
MIDGES, MOSQUITOES, SAMPLING, CON	W70-06225
MIDGES, NITRATES, NITROGEN, NITRO	W69-04805
MIDGES, PESTICIDES.: /OLYSIS, COP	W70-02982
MIDGES, SNAILS, PHOSPHORUS, NUTRI	W70-01943
MIDGES, WATER POLLUTION EFFECTS,	W71-12077
MIGRATION PATTERNS.: /WATER, LITT	W71-12870
MIGRATION, CARBON DIOXIDE, HYDROG	W70-05760
MIGRATION, DIURNAL, ALGICIDES, CL	W70-05372
MIGUEL RIVER, DOLORES RIVER, URAV	W69-07846
MILE LAKE(CANADA), *ONTARIO, *CAN	W70-02973
MILE LAKE(ONTARIO), GULL LAKE(ONT	W70-02795
MILE LAKE, BRANT LAKE, IMIKPUK LA	W69-08272
MILFOIL COONTAIL, WILD CELERY, BU	W70-05568
MILIARIS, HELGOLAND(GERMANY), AUR	W70-04368
MILK SUBSTRATE, NITROGEN REMOVAL,	W70-07031
MILK WASTE TREATMENT, COLONY COUN	W72-01819
MILK, *MODEL STUDIES.: /ASTE WATE	W70-07031
MILK, ATLANTIC OCEAN, LANDFILLS,	W72-00941
MILL DRAINAGE DITCH(WIS), LAKE MO	W70-05113
MILL WASTES, ALFOL ALCOHOLS.:	W70-08740
MILL, MEICELASE, HYDROGEN PEROXID	W70-04184

477

COMO(MINN), LAKE VADNAIS(MINN),
ALCOCITE, COVELLITE, SPHALERITE,
SIOLOGY, BIOCHEMICAL EXTRACTION,
CIN, TETRACYCLINE, STREPTOMYCIN,
                *GLUCOSE, WATERLOO
*URANIUM RADIOISOTOPES, RIVERS,
INDIAN RIVER.:
ADISON(WIS), FLUSHING WATERWAYS,
LANTS, DIATOMS, CORAL, MOLLUSKS,
POLLUTION EFFECTS, ACID STREAMS,
NTAL EFFECTS, *LI/ *ALGAE, *ACID
FACTORS, BENTHIC FLORA, BENTHOS,
S, *POLLUTANTS, INHIBITIO/ *ACID
N ION CONCENTRATION, WATE/ *ACID
SMS, *ALGAE, *AQUATIC BAC/ *ACID
ATILE ACIDS, KETOACIDS, TAMIYA'S
DAM, HAARLEM, LEYDEN, THE HAGUE,
RES, LAKE MINNETONKA(MINN), SOIL
STEMS, FAUNA, FLORA, LABORATORY,
ORAL, MOLLUSKS, MINE ACIDS, COAL
ATER TEMPERATURE, HYDROGE/ STRIP
QUARTERNARY, AMMONIUM COMPOUNDS,
INOSA, APHANIZOMENON FLOS-AQUAE,
LTURE MEDIA, CHU NO 10 SOLUTION,
LGAL BLOOMS, LIEBIG'S LAW OF THE
TONKA(MINN), MINNEAPOLIS-ST PAUL(
STIS AERUGINOSA, BURNTSIDE RIVER(
AKE CLASSIFICATION, SHAGAWA LAKE(
STEPHANODISCUS AS/ *SHAGAWA LAKE(
OSA, M/ *SHAGAWA LAKE(MINN), ELY(
, TROUT LAKE(MINN), KIMBALL LAKE(
), LAKE VADNAIS(MINN), MILLE LAC(
LA PYRENOIDOSA, M/ *SHAGAWA LAKE(
AKE OWASSO(MINN), LAKE JOSEPHINE(
LEARWATER LAKE(MINN), TROUT LAKE(
ORCHARD LAKE(MINN), LAKE OWASSO(
DEMING LAKE(MINN), ORCHARD LAKE(
ICROCYSTIS AERUGINOSA, LAKE COMO(
LGAL DENSITY, / *LAKE MINNETONKA(
A, LAKE COMO(MINN), LAKE VADNAIS(
:   LAKE MINNETONKA(
NODISCUS ASTREA, LAKE MINNETONKA(
NFLUX, NUTRIENT/ LAKE MINNETONKA(
ADFISH RIVER(WIS), DETROIT LAKES(
N), MILLE LAC(MINN), DEMING LAKE(
), GRAY'S BAY(MINN), SMITH'S BAY(
LAKE JOSEPHINE(MINN), GRAY'S BAY(
MINERALS, PLANT RESIDUES, MORRIS(
ANIMAL MANURES, LAKE MINNETONKA(
Y, ALGAE BLOOMS, CLEARWATER LAKE(
        LAKE MINNETONKA(MINN),
OUNDS, SPECTROPHOTOMETRY, ALGAE,
EFFLUENTS, SWAMPS, FOREST SOILS,
N, CYANOPHYTA, SEWAGE EFFLUENTS,
VITY, WATER QUALITY, *LIMNOLOGY,
IONS, BIOCHEMICAL OXYGEN DEMAND,
DESIGN CRITERIA, REMOTE SENSING,
, *PHOSPHATES, LABORATORY TESTS,
PROGRAM, ALGAL DENSITY, / *LAKE
*KILLING, ANIMAL MANURES, LAKE
NUTRIENT INFLUX, NUTRIENT/ LAKE
PAUL(MINN).:   LAKE
UAE, STEPHANODISCUS ASTREA, LAKE
FER, WATER, BACTERIA, SEDIMENTS,
, BASS, WATER POLLUTION CONTROL,
YTOPLANKTON, PONDS, ZOOPLANKTON,
DA, PORPHYRIDIUM CRUENTUM, LEMNA
AL PONDS, SVISLOCH' RIVER(USSR),
MNODINIUM BAICALENSE, CYCLOTELLA
TREATMENT, TREATMENT FACILITIES,
ATER SOURCES, *GROUNDWATER, DDT,
LITY, SEWAGE EFFLUENTS, DIATOMS,
NKTON, *ROTIFERS, *DISTRIBUTION,
MPOSING ORGANIC MATTER, INSECTS,
TODES, ROTIFERS, LARVAE, SNAILS,
PHOSPHOROUS UPTAKE RATE.:
BRUNSWICK(CANADA), SERVALL OMNI-
RESPIRATION, SCOUR, SATURATION,
, ALGAE, COLLOIDS, FLOCCULATION,
TURATION, TEMPERATURE, PRESSURE,
TRIENTS, ALGAE, STRATIFIED FLOW,
ACTERIA, SULFUR BACTERIA, ALGAE,
OXYGEN SUPERSATURATION, VERTICAL
LGAE, BIOCHEMICAL OXYGEN DEMAND,
LLUTION ABATEMENT, TIDAL WATERS,
BIOCHEMICAL OXYGEN DEMAND(BOD),
INIA, MONIMOLIMNION, CHEMOCLINE,
TION TUBES, *DIFFUSERS, *RAYTOWN(
SERS, *RAYTOWN(MO), LEE'S SUMMIT(
ES, ALGAL CONTROL, DISINFECTION,

MILLE LAC(MINN), DEMING LAKE(MINN	W70-05269
MILLERITE, ORPIMENT, TETRAHEDRITE	W69-07428
MILLIPORE FILTERS, PHASE MICROSCO	W69-10163
MILLIPORE FILTRATION, MESOTROPHY.	W69-10154
MILLS(PA).:	W71-05026
MILLS, COLORADO, WASTES, BIOLOGIC	W69-07846
MILLSBORO(DELAWARE), SWAN CREEK,	W71-05013
MILWAUKEE(WIS), CHICAGO(ILL), ART	W69-10178
MINE ACIDS, COAL MINES, ACIDITY,	W69-07428
MINE DRAINAGE, CHLOROPHYTA, CHRYS	W70-02770
MINE WATER, *HABITATS, *ENVIRONME	W70-02770
MINE WATER, WATER POLLUTION, WATE	W70-02770
MINE WATER, *NUTRIENT REQUIREMENT	W70-02792
MINE WATER, ACIDIC WATER, HYDROGE	W70-05424
MINE WATER, *AQUATIC MICROORGANI	W69-00096
MINERAL MEDIUM, GLYOXYLIC ACID.: /	W70-04195
MINERALIZATION, CLAY LENSES, RHIN	W69-07838
MINERALS, PLANT RESIDUES, MORRIS(	W70-04193
MINERALS, POLLUTION.: /ISH, ECOSY	W72-02203
MINES, ACIDITY, PYRITE, SULFUR CO	W69-07428
MINES, DRAINAGE, FERROBACILLUS, W	W68-00891
MINIMUM LETHAL EXPOSURE, PHENANTH	W70-02982
MINIMUM LETHAL DOSES, CHU'S MEDIA	W70-02982
MINIMUM NUTRIENTS.: /SA(KUTZ), CU	W70-03507
MINIMUM, VOISIN'S LAW OF THE MAXI	W70-04765
MINN).: LAKE MINNE	W70-04810
MINN).: /LLA PYRENOIDOSA, MICROCY	W70-03512
MINN).: /ROTROPHY, FLAGELLATES, L	W70-02775
MINN), APHANIZOMENON FLOS AQUAE,	W70-05387
MINN), BLOOMS, CHLORELLA PYRENOID	W70-03512
MINN), CRYPTOPHYTA, PYROPHYTA, CH	W70-05387
MINN), DEMING LAKE(MINN), ORCHARD	W70-05269
MINN), ELY(MINN), BLOOMS, CHLOREL	W70-03512
MINN), GRAY'S BAY(MINN), SMITH'S	W70-05269
MINN), KIMBALL LAKE(MINN), CRYPTO	W70-05387
MINN), LAKE JOSEPHINE(MINN), GRAY	W70-05269
MINN), LAKE OWASSO(MINN), LAKE JO	W70-05269
MINN), LAKE VADNAIS(MINN), MILLE	W70-05269
MINN), LAKE MANAGEMENT PROGRAM, A	W69-10167
MINN), MILLE LAC(MINN), DEMING LA	W70-05269
MINN), MINNEAPOLIS-ST PAUL(MINN).	W70-04810
MINN), MORPHOMETRY, ALGAE BLOOMS,	W70-05387
MINN), NUTRIENT TRAPS, NUTRIENT I	W70-04268
MINN), NUTRIENT REMOVAL, MASS COM	W70-04506
MINN), ORCHARD LAKE(MINN), LAKE O	W70-05269
MINN), PHOSPHATE DETERMINATION: /	W70-05269
MINN), SMITH'S BAY(MINN), PHOSPHA	W70-05269
MINN), SOIL PERCOLATION.: / SOIL	W70-04193
MINN), SOIL MINERALS, PLANT RESID	W70-04193
MINN), TROUT LAKE(MINN), KIMBALL	W70-05387
MINNEAPOLIS-ST PAUL(MINN).:	W70-04810
MINNESOTA, SODIUM ARSENITE.: /OMP	W70-05269
MINNESOTA, WISCONSIN, MICHIGAN, P	W70-04193
MINNESOTA, WATER POLLUTION EFFECT	W70-04506
MINNESOTA, WATER ANALYSIS, CALCIU	W69-00632
MINNESOTA, WATER CHEMISTRY, BIODE	W69-00446
MINNESOTA, SEDIMENT-WATER INTERFA	W70-02775
MINNESOTA, NUTRIENTS, WATER POLLU	W70-08637
MINNETONKA(MINN), LAKE MANAGEMENT	W69-10167
MINNETONKA(MINN), SOIL MINERALS,	W70-04193
MINNETONKA(MINN), NUTRIENT TRAPS,	W70-04268
MINNETONKA(MINN), MINNEAPOLIS-ST	W70-04810
MINNETONKA(MINN), MORPHOMETRY, AL	W70-05387
MINNOWS, DETRITUS, CYCLING NUTRIE	W69-07861
MINNOWS, NUISANCE ALGAE, SUNFISHE	W70-02982
MINNOWS, PEDALFERS, SCENEDESMUS,	W70-02249
MINOR, PHOTOPHOSPHORYLATION, VANA	W70-02964
MINSK(USSR), ENGLAND, OSCILLATORI	W70-05092
MINUTA.: /ELOSIRA BAICALENSIS, GY	W70-04290
MISSISSIPPI.: /UES, *WASTE WATER	W71-13313
MISSISSIPPI RIVER BASIN, URBANIZA	W73-00758
MISSOURI RIVER, COLUMBIA RIVER, C	W70-04507
MISSOURI RIVER, SOUTH DAKOTA, NEB	W70-05375
MITES, BACTERIS, FUNGI, ALGAE, NI	W71-00940
MITES, MOLDS, GROWTH RATES, HABIT	W72-01819
MIXED LIQUOR, SUSPENDED SOLIDS, *	W71-13335
MIXER HOMOGENIZER.: /CANADA), NEW	W70-04365
MIXING, DIFFUSION, SEDIMENTATION,	W71-11381
MIXING, ECONOMIC FEASIBILITY, TEC	W71-10654
MIXING, EVAPORATION, SUPERSATURAT	W70-00683
MIXING, FLOW, MODEL STUDIES, MATH	W71-05878
MIXING, HYDROGEN ION CONCENTRATIO	W72-06854
MIXING, ROANOKE RAPIDS RESERVOIR(	W70-00683
MIXING, SEWAGE TREATMENT, WASTE W	W70-06899
MIXING, SEA WATER, FRESH WATER, S	W69-06203
MIXING, WINDS.: /D OXYGEN, ALGAE,	W70-04790
MIXOLIMNION, RADIOCARBON UPTAKE T	W70-05760
MO), LEE'S SUMMIT(MO), OXYGEN TRA	W70-06601
MO), OXYGEN TRANSFER, PH, SUSPEND	W70-06601
MODE OF ACTION, MORTALITY, INHIBI	W70-02370

MORTALITY, BIOASSAY, INHIBITION,
UTION EFFECTS, LABORATORY TESTS,
US, TROPHIC LEVEL, PRODUCTIVITY,
ATION, FLOCCULATION, FILTRATION,
, STRATIFIED FLOW, MIXING, FLOW,
WATER UTILIZATION, FISH, ALGAE,
CYANOPHYTA, LITTORAL, NUTRIENTS,
ASTE TREATMENT, EUTHROPHICATION,
*FACULTATIVE LAGOONS, LABORATORY
TIONS, SOLUBILITY / *EQUILIBRIUM
ATIC ECOSYSTEM MODEL, SIMULATION
UMBIA, CANADA, AQUATIC ECOSYSTEM
OCHEMIS/ *ECOLOGY, *MATHEMATICAL
TS, PALEOLIMNOLOGY, MATHEMATICAL
TS, PALEOLIMNOLOGY, MATHEMATICAL
APP/ DETENTION TIME, LABORATORY
LGAE BIOMASS.:   KINETIC
LOW, MODEL STUDIES, MATHEMATICAL
ENTHIC FLORA, *ALGAE, *HYDRAULIC
, LIPIDS, KINETICS, MATHEMATICAL
EWAGE, EQUILIBRIUM, MATHEMATICAL
SARNIENSE, QUANTITATION, GROWTH
MOTING PROPERTIES, CHLAMYDOMONAS
RODESMUS FALCATUS, CHLAMYDOMONAS
ALEA, MYXOPHYCEAE, CHLAMYDOMONAS
*POTASSIUM PERMANGANATE, *DES
ROTIFERS, LARVAE, SNAILS, MITES,
SAPROPHYTIC,/ TAXONOMIC GROUPS,
DIPTERA, COLEOPTERA, HEMIPTERA,
N, ALGAE, POLYCHAETA, CRUSTACEA,
IE, INSECTS, DIPTERA, CRUSTACEA,
LICHENS, PLANTS, DIATOMS, CORAL,
KISH WATER, ESTUARIES, BACTERIA,
EFFECTS, MARINE ANIMALS, ALGAE,
ISH, INVERTEBRATES, CRUSTACEANS,
TION, LIFE CYCLES, OLIGOCHAETES,
EN FIXATION, BORON, AZOTOBACTER,
IDITY, PYRITE, SULFUR COMPOUNDS,
ELEMENTS, ALGAE, EUTROPHICATION,
FECTS, GRAZING, IRON, MANGANESE,
IRA, MALLOMONAS, ULOTHRIX, DAVID
MARL LAKES, CHLOROBIUM, FILINIA,
Y, TERTIARY TREATMENT, SAMPLING,
E, CYCLES, HARVESTING, DREDGING,
NE ALGAE, SALINITY, TEMPERATURE,
EASONAL, RADIOCHEMICAL ANALYSIS,
Y, TASTE, COAGULATION, PLANKTON,
CONCENTRATION, ENVIRONMENTAL
GISTIC EFFECTS, AXENIC CULTURES,
CARBON, *CELL SIZE, EUKARYOTES,
IGHT UTI/ ALGAL GROWTH KINETICS,
PYRENOIDOSA, CHLORELLA VULGARIS,
LIC ACID CYCLE, AUTORADIOGRAPHY,
ASS SPECTROMETRY, N-HEPTADECANE,
CARBON FIXATION, PELAGIC REGION,
AMPLING, LAKE MENDOTA(WIS), LAKE
R(OHIO), LAKE MENDOTA(WIS), LAKE
LAKE P/ LAKE MENDOTA(WIS), LAKE
RUCOSUM, LAKE MENDOTA(WIS), LAKE
*CULTURAL EUTROPHICATION, LAKE
SIOLOGY, LAKE MENDOTA(WIS), LAKE
AKE WAU/ LAKE MENDOTA(WIS), LAKE
H MILL DRAINAGE DITCH(WIS), LAKE
AKE KEGONSA/ *MADISON(WIS), LAKE
E ODORS, LAKE MENDOTA(WIS), LAKE
, LAKE RECOVERY, WISCONSIN, LAKE
BOTRYOCOCCUS BRAUNII, *ANACYSTIS
SUBMONTAN/ *SKAWA RIVER(POLAND),
ERIMENTAL LIMNOLOGY, CLADOCERAN,
ODIUM ARSENITE, COPPER SULPHATE,
HENYLUREAS, DIURON, FLAGELLATES,
232, YUKON RIVER, NEBESNA RIVER,
*INDIA,
ALGAL GENETICS, RUSSIA, BAVARIA,
), HUMBOLDT(SASKATCHEWAN), MICE,
ANTITHAMNION/ *LEUCOTHRIX MUCOR,
SPECIES DIVERSITY, AUTECOLOGY,
COMMUNITIES, BIOINDICATORS, LAKE
STICHOCHRYSIS IMMOBILIS, VIBRIO,
S, COLONIAL MORPHOLOGY, CELLULAR
TORULA, GROWTH FACTORS, COLONIAL
MUD, LAKES, CULTIVATION, CHANNEL
POPULATIONS, WATER QUALITY, LAKE
S ASTREA, LAKE MINNETONKA(MINN),
F-NEV), HYDROPHYTES, LIVERWORTS,
, SOIL MINERALS, PLANT RESIDUES,
ONTROL, CHLORELLA, DISINFECTION,
, COPPER SULFATE, ALGAL CONTROL,
L, DISINFECTION, MODE OF ACTION,
IFIERS, WATER POLLUTION EFFECTS,

MODE OF ACTION, PHOTOSYNTHESIS.: / W70-03519
MODEL STUDIES, RADIOACTIVITY TECH   W70-04074
MODEL STUDIES, SALMON, AQUATIC EN   W69-06536
MODEL STUDIES, COMPUTERS, NUTRIEN   W70-08627
MODEL STUDIES, MATHEMATICAL MODEL   W71-05878
MODEL STUDIES, BIOASSAY, AQUATIC   W71-11881
MODEL STUDIES, BIOASSAY, RADIOACT   W73-01454
MODEL STUDIES, BIOMASS, *LIGHT IN   W73-02218
MODEL, COD REMOVAL, MILK SUBSTRAT   W70-07031
MODEL, LITHOSPHERE, SORPTION REAC   W70-04385
MODEL, PHOSPHORUS EFFECT, CLADECE   W69-06536
MODEL, SIMULATION MODEL, PHOSPHOR   W69-06536
MODELS, *WATERSHED MANAGEMENT, GE   W71-09012
MODELS, BACTERIA, PHYTOPLANKTON,   W70-03975
MODELS, BACTERIA, PHYTOPLANKTON,   W70-03975
MODELS, BOD REMOVAL RATES, DESIGN   W70-07034
MODELS, CHLORELLA PYRENOIDOSA, *A   W73-02218
MODELS, GREAT LAKES, THERMAL STRA   W71-05878
MODELS, LABORATORY TESTS, PERIPHY   W71-00668
MODELS, NEUTRALIZATION, *DEGRADAT   W71-13413
MODELS, ON-SITE DATA COLLECTIONS,   W68-00858
MODES.: /HRIX MUCOR, ANTITHAMNION   W69-10163
MOEWUSII, CHLORELLA PYRENOIDOSA,   W69-10180
MOEWUSII, SKELETONEMA COSTATUM, S   W70-05381
MOEWUSSI, FAT ACCUMULATION.: /A P   W70-02965
MOINES(IOWA).:   W72-04242
MOLDS, GROWTH RATES, HABITATS.: /   W72-01819
MOLECULAR FILTER, CHEMOSYNTHETIC,   W68-00891
MOLLUSCA, RAFFINATE, THORIUM, TAI   W69-07846
MOLLUSCA, ECHINODERMATA, PACIFIC   W69-08526
MOLLUSCS, OLIGOCHAETES, GASTROPOD   W70-04253
MOLLUSKS, MINE ACIDS, COAL MINES,   W69-07428
MOLLUSKS, VITAMIN B, RED TIDE, SA   W70-05372
MOLLUSKS, FISH, RADIOACTIVE WASTE   W71-05531
MOLLUSKS, STRONTIUM RADIOISOTOPES   W71-09850
MOLLUSKS, STRATIFICATION, DIATOMS   W71-00665
MOLYBDENUM, COBALT, CO: /, NITROG   W70-02964
MOLYBDENUM, ROTIFERS, ECOSYSTEMS,   W69-07428
MOLYBDENUM, NITROGEN FIXATION, NU   W69-06277
MOLYBDENUM, PLANT GROWTH REGULATO   W69-04801
MONIE TEST, SUM: /ANABAENA, MELOS   W69-10157
MONIMOLIMNION, CHEMOCLINE, MIXOLI   W70-05760
MONITORING, WATER DEMAND, WATER S   W70-05645
MONITORING, MICROBIOLOGY, FOOD CH   W70-05264
MONITORING, AERIAL PHOTOGRAPHY, I   W68-00010
MONITORING, PIGMENTS, PHYTOPLANKT   W70-04371
MONITORING, TEM: /S, WIND VELOCIT   W70-00263
MONITORING.:   W72-00941
MONOALGAL, PORTUGAL, TAPES, JAPAN   W70-05372
MONOCHRYSIS LUTHERI, NAVICULA PEL   W69-08278
MONOD GROWTH EQUATION, INCIDENT L   W69-03730
MONODUS SUBTERRANEUS, ANKISTRODES   W70-05381
MONODUS SUBTERRANEUS, NAVICULA PE   W70-02965
MONOENES, DIENES, TRIENES.: /S, M   W69-08284
MONOMICTIC, DEPTH EFFECTS, IONIC   W70-01933
MONONA(WIS), LAKE WAUBESA(WIS), L   W70-02969
MONONA(WIS), LAKE WAUBESA(WIS), L   W72-01813
MONONA(WIS), LAKE NAGAWICKA(WIS),   W70-06217
MONONA(WIS), LAKE WAUBESA(WIS), A   W69-08283
MONONA(WIS), LAKE TAHOE(CALIF), L   W69-06858
MONONA(WIS), LAKE WAUBESA(WIS), S   W69-05867
MONONA(WIS), LAKE KEGONSA(WIS),   W69-06276
MONONA(WIS), LAKE WAUBESA(WIS), L   W70-05113
MONONA(WIS), LAKE WAUBESA(WIS), L   W70-05112
MONONA(WIS), LAKE WINGRA(WIS), LA   W69-03366
MONONA, LAKE MENDOTA, LAKE WAUBES   W69-09349
MONTANA, *HYDROCARBONS, *OLEFINS,   W69-08284
MONTANE RIVER COURSE, CARPATHIAN   W70-02784
MONTANE ZONE, ALPINE ZONE, BOULDE   W69-10154
MONURON, AMMONIA, DALAPON, 2-4-: / W70-00269
MONURON, TRANSPARENCY, CYCLOPS, S   W70-10161
MOOSE CREEK, ARCTIC CIRCLE HOT SP   W69-08272
MOOSI RIVER ECOLOGICAL STUDY.:   W69-07096
MORICHES BAY(NY), ONONDAGA LAKE(N   W70-03973
MORPHOLOGICAL CHARACTERS, TRIS.: / W70-00273
MORPHOLOGY, FRIDAY HARBOR(WASH),   W69-10161
MORPHOLOGY, EXPORT LOSSES, CHEMOS   W70-02780
MORPHOLOGY, ECOLOGICAL DISTRIBUTI   W71-12077
MORPHOLOGY, CULTURAL CHARACTERIST   W70-04280
MORPHOLOGY, FLAGELLATION, IODINE   W70-03952
MORPHOLOGY, CELLULAR MORPHOLOGY,   W70-03952
MORPHOLOGY, SEWAGE INPUT, BIOLOGI   W68-00475
MORPHOMETRY, DOMESTIC WASTES, PLA   W68-00470
MORPHOMETRY, ALGAE BLOOMS, CLEARW   W70-05387
MORPHOMETRY, PACIFASTACUS LENIUSC   W70-00711
MORRIS(MINN), SOIL PERCOLATION.: / W70-04193
MORTALITY.: /S CHLORINE, ALGAL C   W70-02363
MORTALITY, CHLOROPHYLL, ALGAE, BI   W70-02364
MORTALITY, INHIBITION, ALGAE.: /O   W70-02370
MORTALITY, GASTROPODS, MARINE PLA   W70-01780

ESMUS, CHLAMYDOMONAS, CHLORELLA,
*BACTERIA COUNT, WARING BLENDOR
PUS FLUVIATILIS, *LAKE BUTTE DES
A, OOCYSTIS SOLITARIA, PANDORINA
ORGANISMS, MARINE ASSOCIATIONS,
EGGERSEE, GREIFENSEE, ZURICHSEE,
WATER HYACINTH, FROGS, CRAYFISH,
UTROPHICATION, TOXICITY, MIDGES,
IC WEED CONTROL, INSECT CONTROL,
LIFORNIA, *NEVADA, DEPTH, ALGAE,
FISHES, PLANKTON, ALGAE, FROGS,
RICULTURAL DRAINAGE, APPALACHIAN
IC ANIMALS, ANIMALS, APPALACHIAN
ON, ORGANIC MATTER, *GRASSLANDS,
ANT POPULATIONS, WEATHER, WATER,
OUS SOIL, SOIL WATER, SOIL WATER
ETRY, YELLOWSTONE NATIONAL PARK,
L PARK, MT LASSEN NATIONAL PARK,
RADIOGRAPHY, ANTITH/ *LEUCOTHRIX
UNT(AUTORADIOGRAPHY), LEUCOTHRIX
WASH), ANTITHAMNION/ *LEUCOTHRIX
ENTS, *SEDIMENT-WATER INFERFACE,
ONCENTRATION, CYCLING NUTRIENTS,
ROUNDWATER BASINS, WATER SUPPLY,
KES, *ECOLOGY, BOTTOM SEDIMENTS,
OGEN ION CONCENTRATION, ACIDITY,
SCENEDESMUS, DIATOMACEOUS EARTH,
, NITROGEN, PHOSPHATES, METHANE,
RCTIC, TUNDRA, SCALING, GRAVELS,
S, OXYGEN, COLIFORMS, SEDIMENTS,
PHYTA, SAMPLING, DENSITY, LIGHT,
, SINKS, TRACERS, CRABS, SHRIMP,
ILES, VOLATILE SUSPENDED SOLIDS,
HYTIC FORMS, COLLOIDAL FRACTION,
LARIA, OBELIA, CLUSTER ANALYSIS,
LYUBLIN(USSR), LYUBERETS(USSR),
ONS, PIT RECHARGE, WATER SUPPLY,
POLLUTION CONTROL, FARM WASTES,
RS, SAMPLING, INDUSTRIAL WASTES,
IC MICROBIOLOGY, EUTROPHICATION,
HORUS SOURCES, *NU/ *WASTE WATER(
, PHOTOSYNTHESIS, ALGAE, SEWAGE,
FARM WASTES, *INDUSTRIAL WASTES,
OXYGEN DEMAND, WATER POLLUTION,
UTION), WATER POLLUTION SOURCES,
RDA, IPOMOEA, PISONIA, PANDANUS,
OPY, MASS SPECTROMETRY, SYMPLOCA
, DIPLOCYSTIS AERUGINOSA, NOSTOC
C GROWTH, ANKISTRODESMUS, NOSTOC
YSTIS AERUGINOSA, NOSTOC, NOSTOC
O NATURAL HISTORY MUSEUM, NOSTOC
ON(WIS), CHICAGO NATURAL HISTORY
BRINE SHRIMP, FISH, FOOD CHAINS,
UATIC ALGAE, BRINE SHRIMP, FISH,
LGAE, *REVIEWS, HABITATS, CLAMS,
LAMINARIA SACCHARINA, MANNITOL,
OWTH INHIBITION, CYANOCOBALAMIN,
INGS, RADON, CENTRAL CITY(COLO),
LIKE, PSYCHROPHILES, PLEMORPHIC,
LDERS, COMPOSITION, PSEUDOMONAS,
IUM, BACILLUS SUBTILIS, BACILLUS
ODOTORULA RUBRA, CA/ *NORTH SEA,
TAMERENSIS, GYMNODINIUM BREVIS,
NIUSCULUS, ANACHARIS CANADENSIS,
ETON, TRAPA, PIS/ CERATOPHYLLUM,
OMA HEMPRICHI, ESCHERICHIA COLI,
ANIUM, SALIX, GROWTH REGULATORS,
PES, JAPAN, AMPHIDINIUM, BLOOMS,
HAERIUM/ *PHYSIOLOGY, SPIROGYRA,
OCKII, SPIROGYRA, XANTHOPHYCEAE,
ANTHOPHYCEAE, BACILLARIOPHYCEAE,
FICIT, SECCHI DISC TRANSPARENCY,
ILLARIOPHYCEAE, NITZSCHIA PALEA,
ISLAND(N C), DETECTOR, BEAUFORT(
IPECTEN IRRADIANS, PIVERS ISLAND(
ON, CONCORD(N H), PEANACOOK LAKE(
SEBASTICOOK(ME), LAKE WINNISQUAM(
ATMENT COSTS, DINOBRYON, CONCORD(
OLEUM CRUDES, MASS SPECTROMETRY,
LSION, DOW C-31, DOW A-21, NALCO
NTAL EFFECTS, ENZYMES, ESSENTIAL
LAKE(WIS), NORTH LAKE(WIS), LAKE
OTA(WIS), LAKE MONONA(WIS), LAKE
P, LAKE MENDOTA(WIS), ANACHARIS,
ICROCYSTIS, ANABAENA, EUCYCLOPS,
AMOGETON, POLYGONUM, SAGITTARIA,
LDING'S POND(WIS), HYDRODICTYON,
THERA, PICKWICK RESERVOIR(TENN),
IA, ALTERNANTHERA, HETERANTHERA,
C REPULSION, DOW C-31, DOW A-21,

MORTALITY, BIOASSAY, INHIBITION,   W70-03519
MORTALITY, CELL WALL, HEXOSAMINE,   W70-04365
MORTS(WISC), GROWTH STIMULATION,   W72-00845
MORUM, PEDIASTRUM BORYA: / PUSILL   W69-06540
MOSCOW, PETROLEUM POLLUTION, PHOT   W69-07440
MOSES LAKE, HALLWILLERSEE.: /BALD   W70-03964
MOSQUITOES, BACTERIA, IRRIGATION,   W70-05263
MOSQUITOES, SAMPLING, CONTROL, IN   W70-06225
MOSQUITOS, ALGAL CONTROL, BASS, C   W70-05417
MOSSES, CHARA, FISH, LIMNOLOGY, L   W70-00711
MOSSES, HYPOLIMNION, THERMOCLINE,   W68-00857
MOUNTAIN CLUB HUTS, TEGERNSEE, GE   W69-07818
MOUNTAIN REGION.: /LANKTON, AQUAT   W69-03695
MOUNTAINS, *SOIL MICROORGANISMS,   W72-08110
MOVEMENT, CULTURES, BAYS, STREAMS   W70-06229
MOVEMENT, LEACHING, ARID LANDS, D   W71-04067
MT LASSEN NATIONAL PARK, MT RAINI   W70-03309
MT RAINIER NATIONAL PARK, OHANAPE   W70-03309
MUCOR, *GROWTH, *FILAMENTS, *AUTO   W69-09755
MUCOR, ANTITHAMNION SARNIENSE, QU   W69-10163
MUCOR, MORPHOLOGY, FRIDAY HARBOR(W69-10161
MUD-WATER INTERFACE, PHOSPHORUS,   W69-06859
MUD-WATER INTERFACES, EUGLENOPHYT   W70-04381
MUD, BENTHOS, HYDROGEN, BACTERIA,   W69-07838
MUD, BENTHOS, WISCONSIN, SAMPLING   W70-06217
MUD, HYDROGEN SULFIDE.: /DS, HYDR   W70-05399
MUD, LAKES, CULTIVATION, CHANNEL   W68-00475
MUD, OXYGEN, ANAEROBIC CONDITIONS   W70-05761
MUD, SANDS, SHALES, BOTTOM SEDIME   W69-08272
MUD, SANDS, SLUDGE, RECREATION.: /   W70-10161
MUD, SILTS, SANDS, GRAVELS, SEDIM   W70-00711
MULLETS, FISH, ALGAE.: /COSYSTEMS   W69-08274
MULTI-CELL SERIAL SYSTEM, PH, FAC   W70-04790
MULTIPLE-USES, SAVANNAH RIVER, GU   W73-00748
MULTIVARIATE ANALYSIS, PRINCIPAL-   W73-01446
MUNICH(GERMANY), CHLORELLA PYRENO   W70-05092
MUNICIPAL WATER, SEDIMENT LOAD, C   W69-05894
MUNICIPAL WASTES, SEPTIC TANKS, R   W70-04266
MUNICIPAL WASTES, SEWAGE, ADSORPT   W70-04001
MUNICIPAL WASTES, FRESHWATER ALGA   W72-12240
MUNICIPAL AND INDUSTRIAL), *PHOSP   W69-09340
MUNICIPAL WA: / AEROBIC TREATMENT   W72-00462
MUNICIPAL WASTES, NUTRIENTS, PEST   W71-09577
MUNICIPAL WATER, WATER REQUIREMEN   W71-09577
MUNICIPAL WASTES, GREAT LAKES, *P   W71-05879
MUSCLE, BONE, LEAVES, SHELL, RHOD   W69-08269
MUSCORUM.: /UIS, RAMMAN SPECTROSC   W69-00387
MUSCORUM, VANADIUM, VITAMIN B-12.   W69-06277
MUSCORUM, SCENEDESMUS OBLIQUUES,   W70-02964
MUSCORUM, ABSORPTION, SPECTROMETR   W70-02245
MUSCORUN, NUTRIENT CONCENTRATION,   W70-00719
MUSEUM, NOSTOC MUSCORUN, NUTRIENT   W70-00719
MUSSELS.: /UTION, AQUATIC ALGAE,   W70-01467
MUSSELS.: /UTION, FOOD CHAINS, AQ   W70-01466
MUSSELS, SHELLFISH, DINOFLAGELLAT   W70-05372
MUSTARD GROWTH TEST, ORGANIC GROW   W69-07826
MUTANT BACTERIA, CYCLOTELLA NANA,   W70-05652
MUTATION, BETA RAYS, COSMIC RADIA   W69-08272
MYCELOID, CORYNEFORM RODS, COCCI,   W70-00713
MYCOBACTERIUM, BACILLUS SUBTILIS,   W69-07838
MYCOIDES, BACILLUS MESENTERICUS,   W69-07838
MYCOTA, DEBARYOMYCES HANSENII, RH   W70-04368
MYCROSYSTIS, ALGASTATIC SUBSTANCE   W70-02510
MYRIOPHYLLUM, POTAMOGETON CRISPUS   W70-00711
MYRIOPHYLLUM, EICHHORNIA, POTAMOG   W70-10175
MYRIOPHYLLUM, LEPOMIS MACROCHIRUS   W69-07861
MYRIOPHYLLUM, WATER BLOOMS, CHAIN   W70-05263
MYTILUS, TRICHODESMIUM ERYTHREUM,   W70-05372
MYXOPHYCEAE, MICROCYSTIS, COELOSP   W70-04369
MYXOPHYCEAE, SALVELINUS NAMAYCUSH   W70-00711
MYXOPHYCEAE, COMPLEX FORMATION, G   W69-10180
MYXOPHYCEA, OSCILLATORIA, ANABAEN   W69-10182
MYXOPHYCEAE, CHLAMYDOMONAS MOEWUS   W70-02965
N C), CRASSOSTREA VIRGINICA, MERC   W69-08267
N C), DETECTOR, BEAUFORT(N C), CR   W69-08267
N H).: / TREATMENT COSTS, DINOBRY   W69-08282
N H), MADISON LAKES(WIS), LAKE WA   W70-07261
N H), PEANACOOK LAKE(N H).: / TRE   W69-08282
N-HEPTADECANE, MONOENES, DIENES,   W69-08284
N-670.: /ARGE, ELECTROSTATIC REPU   W70-05267
N: / CYCLING NUTRIENTS, ENVIRONME   W69-03188
NAGAWICKA(WIS).: /WIS), PEWAUKEE   W69-05867
NAGAWICKA(WIS), LAKE PEWAUKEE(WIS   W70-06217
NAIAS, SCIRPUS, NYMPHAEA, CRASSIP   W70-05263
NAIS, CRICOTOPUS, AUSTROCLOEON, E   W70-06975
NAJAS.: /OMENON, MICROCYSTIS, POT   W72-04761
NAJAS, ANACHARIS, LAKE WAUBESA(WI   W70-03310
NAJAS, PHOTIC ZONE, DOUGLAS RESER   W71-07698
NAJAS, SILVEX, ENDOTHAL, POTASSIU   W70-10175
NALCO N-670.: /ARGE, ELECTROSTATI   W70-05267

PHYCEAE, MYXOPHYCEAE, SALVELINUS
MIN, MUTANT BACTERIA, CYCLOTELLA
M CRUENTUM, RADIONUCLIDE UPTAKE,
BIOLOGICAL COMMUNITIES, DAPHNIA,
, SOUTH DAKOTA, NEBRASKA, ALGAE,
ENTRIFUGATION, MICROSCOPY, NETS,
WISCONSIN ZOOPLANKTON, TRIPTON,
WATER QUALITY CRITERIA, CYANIDE,
AXONOMIC TYPES, HETERETROPHY, B/
GARDHII, DESICCATION, IMMERSION,
), ICELAND, TRITIATED THYMIDINE,
SOUND(WASH), LONG ISLAND SOUND,
*SARGASSUM FLUITANS, *SARGASSUM
, UNIVERSITY OF WISCONSIN, ASIA/
, UNIVERSITY OF WISCONSIN, ASIA/
RING RESEARCH LABORATORY(CALIF),
LASSEN NATIONAL PARK, MT RAINIER
IMETRY, SOLARIMETRY, YELLOWSTONE
OWSTONE NATIONAL PARK, MT LASSEN
GS, SINTER, NUCLEI/ *YELLOWSTONE
, UNIVERSITY OF WISCONSIN, ASIA,
, UNIVERSITY OF WISCONSIN, ASIA,
A), US ATOMIC ENERGY COMMISSION,
A), US ATOMIC ENERGY COMMISSION,
HRESHOLD ODOR NU/ *OSCILLATORIA,
NIOCYSTIS, MADISON(WIS), CHICAGO
HORUS RADIOISOTOPES, TECHNOLOGY,
COUNT, FILAMENTOUS, UNICELLULAR,
AQUATIC MICROBIOLOGY, BALANCE OF
AQUATIC MICROBIOLOGY, BALANCE OF
IVITY, AQUATIC WEEDS, BALANCE OF
OLOGY, AQUATIC WEEDS, BALANCE OF
AQUATIC MICROBIOLOGY, BALANCE OF
AQUATIC PRODUCTIVITY, BALANCE OF
AQUATIC MICROBIOLOGY, BALANCE OF
AQUATIC PRODUCTIVITY, BALANCE OF
OLOGY, AQUATIC WEEDS, BALANCE OF
AQUATIC MICROBIOLOGY, BALANCE OF
AGELLATES, CILIATES, COPEPODIDS,
AL SCIENCE FOUNDATION, OFFICE OF
AL SCIENCE FOUNDATION, OFFICE OF
EUKARYOTES, MONOCHRYSIS LUTHERI,
PYRENOIDOSA, TRIBONEMA AEQUALE,
DIOGRAPHY, MONODUS SUBTERRANEUS,
:   *EFFECTS,
CYCLOTELLA, FRAGILARIA, SYNEDRA,
GOMPHONEMA, GYROSIGMA, MELOSIRA,
RIS, ASPLANCHA, LECANE, BOSMINA,
N OF TORONTO, FILAMENTOUS ALGAE,
UTHERA ISLAND, NITZCHIA SERIATA,
MPHONEMA, PINNULARIA, NITZSCHIA,
ICAL, MICROCYSTIS, OSCILLATORIA,
TA INTERPRETATION, *RHONE RIVER,
MIXING, ROANOKE RAPIDS RESERVOIR(
GRAFTON(ND), LAKOTA(ND), HARVEY(
Y STANDARDS, GRAFTON(ND), LAKOTA(
WATER QUALITY STANDARDS, GRAFTON(
URBID R/ *SALT VALLEY RESERVOIRS(
SETUM, THORIUM-232, YUKON RIVER,
          PLATTE RIVER, LINCOLN(
N, MISSOURI RIVER, SOUTH DAKOTA,
ESERVOIRS, *GREAT PLAINS, ALGAE,
TOMS, PROTOZOA, PLANKTON, WORMS,
RIE(OHIO), GONGROSIRA STAGNALIS,
RPTION, ATLANTIC OCEAN, DIATOMS,
RKEN(SWEDEN), LAKE MENDOTA(WIS),
ORGE(UGANDA), RHINE-MEEUSE DELTA(
PHOTOASSIMILATION, BERKELSE LAKE(
IES, CENTRIFUGATION, MICROSCOPY,
STUDIES, *NATIONAL WATER QUALITY
, KINETICS, MATHEMATICAL MODELS,
ER TREATMENT, INDUSTRIAL WASTES,
RPHOMETRY, PA/ *LAKE TAHOE(CALIF-
MENTS, EUPHOT/ *LAKE TAHOE(CALIF-
OLVED ORGANIC MATTER, LAS VEGAS,
A, DIAPTOMUS ASHLANDI, EPISCHURA
A LANOSA, POINT LEPREAU(CANADA),
, ANABAENA, PHYGON XL(DICHLONE),
ER TREATMENT, ALGAE, FILTRATION,
RIBUTION, BIOLOGICAL INDICATORS,
PULATION, BIOLOGICAL INDICATORS,
PPLY, RESERVOIRS, WATER QUALITY,
*SOURCE CONTAMINATION, CYANIDE,
ON, ALGAE, BIOMASS, CHLOROPHYLL,
ATIONS, WATER POLLUTION EFFECTS,
, BIOINDICATORS, NUISANCE ALGAE,
, SILICATES, CITRATES, BIOASSAY,
LGAE, FILTRATION, NEW HAMPSHIRE,
UALITY CONTROL, WINNISQUAM LAKE,

NAMAYCUSH, SUBSTRATE TYPE.: /NTHO   W70-00711
NANA, GEIGER COUNTER.: /ANOCOBALA   W70-05652
NANNOCHLORIS, THORACOMONAS, YTTRI   W70-00708
NANNOPLANKTON, NUTRIENTS, DEPTH,   W70-00274
NANNOPLANKTON, DIATOMS.: /I RIVER   W70-05375
NANNOPLANKTON, PIGMENTS, CHLOROPH   W70-05389
NANNOPLANKTON, NUTRIENTS, OLIGOTR   W70-03959
NAPHTHENIC ACID.: /BIAL ECOLOGY,   W70-04510
NARRAGANSETT BAY(RHODE ISLAND), T   W70-00713
NARRAGANSETT MARINE LABORATORY(RH   W70-01073
NARRAGANSETT BAY(RI), CAPE REYKJA   W69-09755
NARRAGANSETT BAY, CAPE REYKJANES,   W69-10161
NATANS, SEAWEED, ZIRCONIUM-95-NIO   W69-08524
NATIONAL ACADEMY OF SCIENCES(USA)   W70-03975
NATIONAL ACADEMY OF SCIENCES(USA)   W70-03975
NATIONAL EUTROPHICATION RESEARCH   W70-02775
NATIONAL PARK, OHANAPECOSH HOT SP   W70-03309
NATIONAL PARK, MT LASSEN NATIONAL   W70-03309
NATIONAL PARK, MT RAINIER NATIONA   W70-03309
NATIONAL PARK, *ICELAND HOT SPRIN   W69-10160
NATIONAL RESEARCH COUNCIL(USA), U   W70-03975
NATIONAL RESEARCH COUNCIL(USA), U   W70-03975
NATIONAL SCIENCE FOUNDATION, OFFI   W70-03975
NATIONAL SCIENCE FOUNDATION, OFFI   W70-03975
NATIONAL WATER CARRIER(ISRAEL), T   W70-02373
NATURAL HISTORY MUSEUM, NOSTOC MU   W70-00719
NATURAL RESOURCES, ECOLOGICAL DIS   W71-09012
NATURAL RECOVERY.: /HYTIC, PLATE   W68-00891
NATURE, BIOASSAY, BIOCHEMICAL OXY   W69-03374
NATURE, BIOASSAY, CHLOROPHYTA, CY   W69-03188
NATURE, BIOASSAY, CHLOROPHYTA, CY   W69-03185
NATURE, BIOASSAY, CHLOROPHYTA, DI   W69-03515
NATURE, BIOASSAY, CHLOROPHYTA, DI   W69-03516
NATURE, BIOASSAY, CHLOROPHYTA, EN   W70-02968
NATURE, BIOASSAY, CHLOROPHYTA, CY   W70-02792
NATURE, CYCLING NUTRIENTS, ENVIRO   W69-03517
NATURE, CYCLING NUTRIENTS, ENVIRO   W69-03513
NATURE, EUTROPHICATION, FISH FARM   W69-04804
NATURE, WISCONSIN, BIOASSAY, BIOC   W69-03369
NAUPLII, TERRAMYCIN, TETRACYCLINE   W69-10154
NAVAL RESEARCH, US DEPARTMENT OF   W70-03975
NAVAL RESEARCH, US DEPARTMENT OF   W70-03975
NAVICULA PELLICULOSA, GONYAULAX P   W69-08278
NAVICULA PELLICULOSA, LAKE MENDOT   W69-10180
NAVICULA PELLICULOSA, XANTHOPHYCE   W70-02965
NAVICULA SEMINULUM VAR.HUSTEDTII.   W71-03183
NAVICULA, APHANIZOMENON, OSCILLAT   W70-08107
NAVICULA, NITZSCHIA, SKELETONEMA,   W70-03325
NAVICULA, FRAGILARIA, CYCLOTELLA,   W70-02249
NAVICULA, APHANIZOMENON, ONTARIO(W69-10155
NAVICULA, FLORIDA CURRENT, YTTRIU   W70-00707
NAVICULA, SYRACUSE(NY), ENTEROMOR   W70-03974
NAVICULA, CYMBELLA, LYNGBYA, FRAG   W70-04371
NAVICULA, MELOSIRA, THALASSIOSIRA   W73-01446
NC).: /SUPERSATURATION, VERTICAL   W70-00683
ND), CHROMATIUM VINOSUM, THIOCAPS   W70-03312
ND), HARVEY(ND), CHROMATIUM VINOS   W70-03312
ND), LAKOTA(ND), HARVEY(ND), CHRO   W70-03312
NEB), *CLEAR WATER RESERVOIRS, *T   W72-04761
NEBESNA RIVER, MOOSE CREEK, ARCTI   W69-08272
NEBR).:   W69-05894
NEBRASKA, ALGAE, NANNOPLANKTON, D   W70-05375
NEBRASKA, TURBIDITY, STRATIFICATI   W72-04761
NEMATODES, ROTIFERS, LARVAE, SNAI   W72-01819
NEPHROCYTIUM OBESUM W AND G S WES   W70-04468
NERITIC ALGAE, IONS, WATER POLLUT   W70-00707
NET PLANKTON, CHLOROPHYLL CONTENT   W70-03959
NETHERLANDS),: /R-BLOOMS, LAKE GE   W70-04369
NETHERLANDS), HETEROCYSTS, GAS-VA   W70-04369
NETS, NANNOPLANKTON, PIGMENTS, CH   W70-05389
NETWORK, ALGAE COUNTS, PLANKTON A   W70-04507
NEUTRALIZATION, *DEGRADATION(DECO   W71-13413
NEUTRIENTS, BACTERIA, ALGAE, TREA   W70-01519
NEV), HYDROPHYTES, LIVERWORTS, MO   W70-00711
NEV), TRANSPARENCY, LIGHT MEASURE   W70-01933
NEVADA, *LAKE MEAD, LAS VEGAS BAY   W72-02817
NEVADENSIS, DIAPHANOSOMA LEUCHTEN   W69-10182
NEW BRUNSWICK(CANADA), SERVALL OM   W70-04365
NEW HAMPSHIRE, LAKE WINNISQUAM, L   W69-08674
NEW HAMPSHIRE, NEW YORK, VIRGINIA   W72-03641
NEW HAMPSHIRE.: / ECOLOGICAL DIST   W70-02764
NEW HAMPSHIRE.: /DISTRIBUTION, PO   W70-02772
NEW JERSEY, TEMPERATURE, ORGANIC   W69-10157
NEW PRODUCT CONTROL, PLUMBING COD   W73-00758
NEW YORK.: /*PONDS, *EUTROPHICATI   W69-06540
NEW YORK, CHLOROPHYTA, NUTRIENTS,   W70-03973
NEW YORK, SEWAGE EFFLUENTS, SEWAG   W70-03974
NEW YORK, LAKES, INHIBITION, ALGA   W70-01579
NEW YORK, VIRGINIA, ORGANIC COMPO   W72-03641
NEWFOUND LAKE.: /LLUTION, WATER Q   W70-02772

ENTS(WATER), HYPOLIMNION, ALGAE, TREATMENT, BIOLOGICAL TREATMENT, OPHYTA, LAKES, RESERVOIRS, SCUM, R, CULTURES, CARBON, METABOLISM, WAGE TREATMENT, SEWAGE BACTERIA, LL, FLUORESCENCE, LAKES, RIVERS, WATER, LAKES, SILTS, PHOSPHATES, NTENSITY, NUTRIENT REQUIREMENTS, ATERS, CARBON, FERTILITY, ALGAE, , FLAGELLATES, ELEUTHERA ISLAND, *MARINE ENVIRONMENT, CERIUM-144, A ELONGATUM, MELOSIRA ISLANDICA, ISLANDICA, NITZSCHIA ACICULARIS, ANTHOPHYCEAE, BACILLARIOPHYCEAE, GHT-SATURATED RATE, CELL VOLUME, LL VOLUME, NITZSCHIA CLOSTERIUM, LARITY, CAP/ *PIPETTE, UNIALGAL, AMPHORA, GOMPHONEMA, PINNULARIA, , GYROSIGMA, MELOSIRA, NAVICULA, BRAUNII, FRAGILARIA CROTONENSIS, GAL IDENTIFIC/ BLUE-GREEN ALGAE, V) CL, CHLORELLA OVALIS BUTCHER, GINOSA(KUTZ), CULTURE MEDIA, CHU ITY STRUCTURE, RIDLEY CREEK(PA), DIDA OBTUSA, CANDIDA TROPICALIS, *SEAWEED EXTRACTS, ASCOPHYLLUM FUCUS VESCICULOSUS, ASCOPHYLLUM S, CHONDRUS CRISPUS, ASCOPHYLLUM LL WALL, HEXOSAMINE, ASCOPHYLLUM OLYSIPHONIA LANOSA, *ASCOPHYLLUM DOTAE, ANABAENA HASSALI, LYNGBYA , TOTAL SOLIDS, VOLITILE SOLIDS, N MEADOW, NITROGEN-15, SOD MATS, NES(ICELAND), SUDURNES(ICELAND), LLA PYRENOIDOSA, VOLATILE SOLIDS, ETHOXYLATE, ANIONIC SURFACTANTS, CHNIQUES, ANTIBIOTICS, BIOASSAY, C ZONE, DOUGLAS RESERVOIR(TENN), AKE ERKEN(SWEDEN), ACETATE, LAKE LOCITY, MICHAELIS CONSTANT, LAKE ES, LA/ CANADA, MESOTROPHY, LAKE ORULA, TEMPERATURE COEFFICIENTS, DRAINAGE, AGRICULTURAL DRAINAGE, DRAINAGE, AGRICULTURAL DRAINAGE, EA WATER, OYSTERS, CLAMS, CRABS, DIES OF WATER, DISSOLVED OXYGEN, T PHYSIOLOGY, ANIMAL PHYSIOLOGY, TURE, *ALGAE BLOOMS, CYANOPHYTA, STRIAL WASTES, SEWAGE TREATMENT, ALENTS(WASTES), RED RIVER VALLEY(ELAVAN(WIS), PEWAUKEE LAKE(WIS), LAKE, BRANT LAKE, IMIKPUK LAKE, D(ENGLAND), ANABAENA CYLINDRICA, IA CLOSERIUM, ARTEMIA, CARTERIA, ITY, *TROPHIC LEVELS, NUTRIENTS, NMENT, WATER QUALITY, NUTRIENTS, CONTROL, *ALGAL TOXINS MEMBRANE, WATER POLLUTION CONTROL, PACIFIC WITZERLAND, NITELV RIVER NORWAY,), HOBOL RIVER(NORWAY), VAN LAKE(, NID RIVER(NORWAY), GAULA RIVER(LA(NORWAY), HITRA(NORWAY), FROYA(RD(NORWAY), SMOLA(NORWAY), HITRA(ATERS, TYNSET(NORWAY), NID RIVER(, TRONDHEIMSFJORD(NORWAY), SMOLA(RELLA OVALIS BUTCHER, NJOER LAKE(STANCES, SUBPOLAR WATERS, TYNSET(WAY), GAULA RIVER(NORWAY), FLAKK(RANHEIM(NORWAY), TRONDHEIMSFJORD((NORWAY), FLAKK(NORWAY), RANHEIM(NJOER LAKE(NORWAY), HOBOL RIVER(ETROPHS, ITALY, LAKE BORREVANNET UGANO ITALY, LAKE MARIDALSVANNET URICH, SWITZERLAND, NITELV RIVER OCYSTIS, DIPLOCYSTIS AERUGINOSA, MICROCYSTIS AERUGINOSA, NOSTOC, OTROPHIC GROWTH, ANKISTRODESMUS, CHICAGO NATURAL HISTORY MUSEUM, WTH, *POTOMAC RIVER, PLECTONEMA, GBYA NOLLEI, GLOCOTRICHIA PISUM, CCUS, CHROOCOCCUS, OSCILLATORIA, M TENUE, MICROCYSTIS AERUGINOSA, AKING' OF LAKES, CHROOCOCCACEAE, LEPOMIS MACROCHIRUS, PIMEPHALES K/ PRINCE EDWARD ISLAND(CANADA), X/ *TRISODIUM NITRILOTRIACETATE, F)/ *COMPLEXONES, *ALGAL BLOOMS, IES, CARP, TILAPIA, TEMPERATURE, S, NITROGEN FIXATION, NUTRIENTS, ALGAL GROWTH, OXYGEN DEPLETION,

NITROGEN, PHOSPHORUS, OXYGEN, CAR W70-09889
NITROGEN, PHOSPHORUS, SEWAGE EFFL W70-08980
NITROGEN, PHOSPHORUS, CARBONATES, W70-02803
NITROGEN, PHOSPHORUS, SULFUR, CAL W70-02804
NITROGEN, PHOSPHORUS, ALGAE, COPP W70-02787
NITROGEN, BIOMASS, SEWAGE, FARM W W70-02777
NITROGEN, WASTE WATER(POLLUTION), W70-02255
NITROGEN, NITROGEN FIXATION, PHOT W70-02965
NITROGEN, PHOSPHORUS, WATER POLLU W70-02973
NITZCHIA SERIATA, NAVICULA, FLORI W70-00707
NITZSCHIA CLOSERIUM, ARTEMIA, CAR W70-00708
NITZSCHIA ACICULARIS, NITZSCHIA D W69-10158
NITZSCHIA DISSIPATA, SYNEDRA ACUS W69-10158
NITZSCHIA PALEA, MYXOPHYCEAE, CHL W70-02965
NITZSCHIA CLOSTERIUM, NITZSCHIA P W70-05381
NITZSCHIA PALEA, CHLOROPLASTS, SC W70-05381
NITZSCHIA, ALGAE ISOLATION, CAPIL W70-04469
NITZSCHIA, NAVICULA, SYRACUSE(NY) W70-03974
NITZSCHIA, SKELETONEMA, PLAGIOGRA W70-03325
NITZSCHIA, ASTERIONELLA GRACILLIM W70-02804
NITZSCHIA, BIOLOGICAL STUDIES, AL W70-10173
NJOER LAKE(NORWAY), HOBOL RIVER(N W70-05752
NO 10 SOLUTION, MINIMUM NUTRIENTS W70-03507
NOBS CREEK(MD), BACK RIVER(MD), O W70-04510
NOCTILUCA MILIARIS, HELGOLAND(GER W70-04368
NODOSUM, ALGINIC ACID, FUCUS VESI W69-07826
NODOSUM, ECTOCARPUS CONFERVOIDES, W70-01074
NODOSUM, LAMINARIA DIGITATA, LAMI W70-01073
NODOSUM, POLYSIPHONIA LANOSA, POI W70-04365
NODOSUM, VIBRIO, FLAVOBACTER, ESC W70-03952
NOLLEI, GLOCOTRICHIA PISUM, NOSTO W69-08283
NON-DEGRADABLE SOLID.: /RY MANURE W71-00936
NON-LEGUMES, *NITROGEN FIXING ORG W72-08110
NON-PARAMETRIC TESTS, ONE-SAMPLE W69-09755
NON-VOLATILE ACIDS, KETOACIDS, TA W70-04195
NONIONIC SURFACTANTS, INFRARED SP W70-09438
NORON, CALCIUM, CHLORINE, COBALT, W69-04801
NORRIS RESERVOIR(TENN), FRENCH BR W71-07698
NORRVIKEN(SWEDEN), A: /ELLATES, L W70-04194
NORRVIKEN(SWEDEN), LAKE EKOLN(SWE W70-04287
NORRVIKEN, LAKE MENDOTA, LAKE FUR W70-03964
NORTH AMERICA, CENTRAL AMERICA, R W70-03957
NORTH AMERICA, FOREST DRAINAGE.: / W70-03975
NORTH AMERICA, FOREST DRAINAGE.: / W70-03975
NORTH CAROLINA, SEDIMENTS, RADIOA W69-08267
NORTH CAROLINA.: /OIRS, ALGAE, BO W70-00683
NORTH DAKOTA, SOUTH DAKOTA, PHOTO W72-01818
NORTH DAKOTA.: /GY, WATER TEMPERA W70-09093
NORTH DAKOTA, COLIFORMS, OXIDATIO W70-03312
NORTH DAKOTA), SULFATE REDUCTION, W70-03312
NORTH LAKE(WIS), LAKE NAGAWICKA(W W69-05867
NORTH MEADOW LAKE, GYTTJA, DUFF, W69-08272
NORTH SEA, FIRTH OF CLYDE(SCOTLAN W70-02504
NORTH TEMPERATE ZONE, PLUTONIUM, W70-00708
NORTHEAST U.S., ECOLOGICAL DISTRI W70-02772
NORTHEAST U S, ECOLOGICAL DISTRIB W70-02764
NORTHEAST U.S.: /ANAPHYTA, ALGAL W69-05306
NORTHWEST U. S., CIVIL ENGINEERIN W69-03683
NORWA: /ET NORWAY, LAKE ZURICH, S W69-07833
NORWAY).: /HER, NJOER LAKE(NORWAY W70-05752
NORWAY), FLAKK(NORWAY), RANHEIM(N W70-01074
NORWAY), FUCUS VESCICULOSUS, ASCO W70-01074
NORWAY), FROYA(NORWAY), FUCUS VES W70-01074
NORWAY), GAULA RIVER(NORWAY), FLA W70-01074
NORWAY), HITRA(NORWAY), FROYA(NOR W70-01074
NORWAY), HOBOL RIVER(NORWAY), VAN W70-05752
NORWAY), NID RIVER(NORWAY), GAULA W70-01074
NORWAY), RANHEIM(NORWAY), TRONDHE W70-01074
NORWAY), SMOLA(NORWAY), HITRA(NOR W70-01074
NORWAY), TRONDHEIMSFJORD(NORWAY), W70-01074
NORWAY), VAN LAKE(NORWAY).: /HER, W70-05752
NORWAY, LAKE LUGANO ITALY, LAKE M W69-07833
NORWAY, LAKE ZURICH, SWITZERLAND, W69-07833
NORWAY, NORWA: /ET NORWAY, LAKE Z W69-07833
NOSTOC MUSCORUM, VANADIUM, VITAMI W69-06277
NOSTOC MUSCORUM, ABSORPTION, SPEC W70-02245
NOSTOC MUSCORUM, SCENEDESMUS OBLI W70-02964
NOSTOC MUSCORUM, NUTRIENT CONCENT W70-00719
NOSTOC PALUDOSUM, BLEACHING, WASH W70-02255
NOSTOC VERRUCOSUM, LAKE MENDOTA(W W69-08283
NOSTOC, ANABAENA, ANACYSTIS, MICR W70-04503
NOSTOC, NOSTOC MUSCORUM, ABSORPTI W69-08283
NOSTOCHINEAE, CLATHROCYSTIS AERUG W69-08283
NOTATUS.: /IA COLI, MYRIOPHYLLUM, W69-07861
NOVA SCOTIA(CANADA), PENGUINS, LA W70-05399
NTA, CHLORELLA PYRENOIDOSA, CARBO W70-02248
NTA, EDTA, HEIDA, LAKE TAHOE(CALI W70-04373
NU: /L ANALYSIS, PHYSICAL PROPERT W70-06975
NU: /LIMNOLOGY, NITROGEN COMPOUND W69-03518
NUCLEAR ACTIVATION TECHNIQUES, RO W70-00264

ISOTOPES, URANIUM RADIOISOTOPES,
CHANGE, RESINS, LEAKAGE, CESIUM,
ESTUARINE ENVIRONMENT, FALLOUT,
PLANTS, BOTTOM SEDIMENTS, SOILS,
ROL, TRACE ELEMENTS, WASHINGTON,
, RADIOACTIVITY, WASTE DISPOSAL,
NGE, *OPEN SEA, WEAPONS TESTING,
REAMS, BIOINDICATORS, *NEW YORK,
SOIL-WATER-PLANT RELATIONSHIPS,
IS, MARINE ALGAE, INVERTEBRATES,
HOSPHORYLATION, *MACROMOLECULES,
K, *ICELAND HOT SPRINGS, SINTER,
UALITY, WATER POLLUTION CONTROL,
LLUTION EFFECTS, EUTROPHICATION,
PTH, DOMESTIC ANIMALS, SAMPLING,
PHO/ *NUTRIENTS, *PRODUCTIVITY,
ARY ENGINEERING, MICROORGANISMS,
TES, HYDROGEN ION CONCENTRATION,
POLLUTION SOURCES, CHLOROPHYTA,
ATER POLLUTION CONTROL, MINNOWS,
ON, LAKES, MAGNESIUM, NUTRIENTS,
IES, CULTURES, DIATOMS, ECOLOGY,
ISCONSIN, BIBLIOGRAPHIES, LAKES,
AT LAKES, LEGISLATION, NITROGEN,
, MOLYBDENUM, NITROGEN FIXATION,
CALCIUM, CHELATION, CYANOPHYTA,
KE MICHIGAN, *WATER CIRCULATION,
DS, ECONOMIC FEASIBILITY, LAKES,
A, NITRATES, NITROGEN COMPOUNDS,
NITROGEN COMPOUNDS, PHOSPHORUS,
EN COMPOUNDS, NITROGEN FIXATION,
IATOMS, CHLOROPHYTA, CYANOPHYTA,
HICATION, *LAKE MICHIGAN, ALGAE,
H, *MUNICIPAL WASTES, NUTRIENTS,
POLLUTANTS, POLLUTION ABATEMENT,
EFFLUENTS, AQUATIC POPULATIONS,
KES, WISCONSIN, COPPER SULPHATE,
TROUT, WINTERKILLING, FISHKILL,
THAW TESTS, HARVESTING OF ALGAE,
NUTRIENTS, HARVESTING OF ALGAE,
*MISSISSIPPI RIVER, *OHIO RIVER,
OGY, CYANOPHYTA, EUTROPHICATION,
OLLUTION EFFECTS, BIOINDICATORS,
, *EUTROPHICATION, *CONFERENCES,
MASS, *PRODUCTIVITY, CYANOPHYTA,
, LIMNOLOGY, NITROGEN COMPOUNDS,
LOW R/ *WATER, *FILTERS, *ALGAE,
EMPERATURE, FERTILIZATION, FISH,
OPPER SULFATE, *SODIUM ARSENITE,
ESIS, *RESPIRATION, GREAT LAKES,
KES(WIS), LAKE WASHINGTON(WASH),
, LAKE MONONA(WI/ *MADISON(WIS),
, *PREVENTATIVE TREATMENT, ALGAE
GO(ILL), ARTIFICIAL CIRCULATION,
BIOLOGICAL PRODUCTIVITY, AQUATIC
CARRIER(ISRAEL), THRESHOLD ODOR
O), ALGAE COUNTS, THRESHOLD ODOR
FENAC, SILVER NITRATE, SIMAZIN,
BORATORY TESTS, TEST PROCEDURES,
FIXATION, ENVIRONMENTAL EFFECTS,
ION EXCHANGE, WATER CIRCULATION,
TH POTENTIAL, SE/ *ALGAL GROWTH,
NTS, PUBLIC APP/ POLYPHOSPHATES,
K(ME), LAKE WINNISQUAM(N H), MA/
ROPHICATION, TROPHIC INDICATORS,
ANSPARENCY, AGRICULTURAL RUNOFF,
ION, *NUTRIENT BUDGET, POTENTIAL
HISTORY MUSEUM, NOSTOC MUSCORUN,
BORID MIDGES, CLEAR LAKE(CALIF),
INNETONKA(MINN), NUTRIENT TRAPS,
LUTION COMMISSION, ALGAL BLOOMS,
LAKE(CANADA), *ONTARIO, *CANADA,
AKES(NY), RODHE'S SOLUTION VIII,
TURE, *CHEMOSTATS, TURBIDOSTATS,

ESTING OF ALGAE, NUISANCE ALGAE,
ESTING OF ALGAE, NUISANCE ALGAE,
TAL EFFECTS, NITROGEN COMPOUNDS,
, FREEZE-THAW TESTS, INHIBITION,
, LIMNOLOGY, NITROGEN COMPOUNDS,
UM COMPOUNDS, NITROGEN FIXATION,
EN COMPOUNDS, NITROGEN FIXATION,
Y, LIGHT PENETRATION, LIMNOLOGY,
ROGEN COMPOUNDS, NITROGEN CYCLE,
, LIMNOLOGY, NITROGEN COMPOUNDS,
ROGEN COMPOUNDS, NUISANCE ALGAE,

RIVER(WIS), DETROIT LAKES(MINN),
ERFISHING, COHO SALMON, ALEWIFE,
*PHYTOPLANKTON, CARBON DIOXIDE,

NUCLEAR EXPLOSIONS, STRONTIUM RAD   W70-00708
NUCLEAR EXPLOSIONS, ABSORPTION, A   W70-00707
NUCLEAR EXPLOSIONS, BIOTA, SPECTR   W69-08267
NUCLEAR EXPLOSIONS, COBALT RADIOI   W69-08269
NUCLEAR POWERPLANTS, WATER POLLUT   W70-00264
NUCLEAR REACTORS, POWERPLANTS, SH   W70-00707
NUCLEAR SHIP SAVANNAH, ZIRCONIUM-   W70-00707
NUCLEAR WASTES, WATER POLLUTION S   W69-10080
NUCLEAR WASTES, ABSORPTION, CHELA   W71-09005
NUCLEAR WASTES, WATER POLLUTION S   W71-09863
NUCLEIC ACID, RIBONUCLEIC ACID SY   W69-10151
NUCLEIC ACID, SISJOTHANDI(ICELAND   W69-10160
NUISANCE ALGAE, TASTE, ODOR.: / Q   W69-10155
NUISANCE ALGAE.: /REAMS, WATER PO   W70-00719
NUISANCE ALGAE.: /, NUTRIENTS, DE   W70-00274
NUISANCE ALGAE, SEWAGE EFFLUENTS,   W70-00270
NUISANCE ALGAE, SEWAGE TREATMENT,   W72-01798
NUISANCE ALGAE, AQUATIC PLANTS, B   W70-06225
NUISANCE ALGAE, FISH, ZOOPLANKTON   W70-03310
NUISANCE ALGAE, SUNFISHES, ALGAE,   W70-02982
NUISANCE ALGAE, OLIGOTROPHY, ODOR   W69-07833
NUISANCE ALGAE, PHAEOPHYTA, PHOTO   W69-07832
NUISANCE ALGAE, BACTERIA, WATER P   W69-08283
NUISANCE ALGAE, PHOSPHORUS, WASHI   W69-06858
NUISANCE ALGAE, NUTRIENTS, VITAMI   W69-06277
NUISANCE ALGAE, NUTRIENT REQUIREM   W69-05867
NUISANCE ALGAE, THERMAL POLLUTION   W69-05763
NUISANCE ALGAE, WATER POLLUTION C   W69-06276
NUISANCE ALGAE, NUTRIENT REQUIREM   W69-05868
NUISANCE ALGAE, SALTS, TIDES, SOU   W69-03695
NUISANCE ALGAE, NUTRIENTS, PHOSPH   W69-04802
NUISANCE ALGAE, WISCONSIN, NUTRIE   W69-01977
NUISANCE ALGAE, PLANKTON, BENTHIC   W68-01244
NUISANCE ALGAE, OLIGOTROPHY, PHYT   W68-00680
NUISANCE ALGAE, PHOSPHATES, NITRA   W69-01453
NUISANCE ALGAE, DUCKS(WILD), OYST   W68-00461
NUISANCE ALGAE, *ALGAL CONTROL, W   W68-00468
NUISANCE ALGAE, ALGAE, SEASONAL,   W68-00487
NUISANCE ALGAE, NUTRIENT REQUIREM   W69-03362
NUISANCE ALGAE, NUTRIENT REQUIREM   W69-03371
NUISANCE ALGAE, WATER QUALITY, SE   W70-04507
NUISANCE ALGAE, NITROGEN, PHOSPHA   W70-04381
NUISANCE ALGAE, NEW YORK, SEWAGE   W70-03974
NUISANCE ALGAE, PUBLICATIONS, WAT   W70-03973
NUISANCE ALGAE, LIMNOLOGY, CHLORO   W70-03959
NUISANCE ALGAE, NUTRIENT REQUIREM   W70-04283
NUISANCE ALGAE, INTAKES, SANDS, F   W70-04199
NUISANCE ALGAE, ODOR, TASTE, DEPT   W70-05404
NUISANCE ALGAE, SNAILS, LEGISLATI   W70-05568
NUISANCE ALGAE, WATER QUALITY, NU   W72-05453
NUISANCE CONDITIONS.: /MADISON LA   W70-07261
NUISANCE ODORS, LAKE MENDOTA(WIS)   W69-03366
NUISANCE, ALGAE BLOOM, ALGAE GROW   W69-08282
NUISANCES.: /ILWAUKEE(WIS), CHICA   W69-10178
NUISANCES, TRANSPARENCY, AGRICULT   W69-09349
NUMBERS.: /ATORIA, NATIONAL WATER   W70-02373
NUMBERS, CROP-DUSTING AIRCRAFT, W   W69-10155
NUPHAR, AGASICLES, MANATEE, MARIS   W70-10175
NURTRIENTS, RIVERS, LAKES, *GROWT   W68-00472
NUTRIE: / FLUORESCENCE, NITROGEN   W70-02775
NUTRIENT ABSORPTION, *CHEMICAL PR   W68-00857
NUTRIENT AVAILABILITY, ALGAL GROW   W69-06864
NUTRIENT BALANCE, FOAMING DETERGE   W70-05084
NUTRIENT BUDGETS, LAKE SEBASTICOO   W70-07261
NUTRIENT BUDGET.:   *CULTURAL EUT   W71-03048
NUTRIENT BUDGETS, COMPARATIVE STU   W69-09349
NUTRIENT CONCENTRATION, AREAL INC   W70-00270
NUTRIENT CONCENTRATION, WISCONSIN   W70-00719
NUTRIENT ENERGY PATHWAYS, TROPHIC   W70-05428
NUTRIENT INFLUX, NUTRIENT SOURCES   W70-04268
NUTRIENT LOAD, SYNTHETIC DETERGEN   W69-08674
NUTRIENT LEVELS, SOFT WATER.: /E   W70-02973
NUTRIENT LIMITATION, CARBON-14.: /   W70-01579
NUTRIENT LIMITED GROWTH, LUXURY C   W70-02779
NUTRIENT MEDIA.:   W69-03491
NUTRIENT REQUIREMENTS, PHYSIOLOGI   W69-03362
NUTRIENT REQUIREMENTS, PHYSIOLOGI   W69-03371
NUTRIENT REQUIREMENTS, PHOSPHORUS   W69-03358
NUTRIENT REQUI: /S, FREEZE DRYING   W69-03370
NUTRIENT REQUIREMENTS, PHOTOSYNTH   W69-03514
NUTRIENT REQUIREMENTS, NUTRIENTS,   W69-03364
NUTRIENT REQUIREMENT: /GY, NITROG   W69-03185
NUTRIENT REQUIREMENTS, NUTRIENTS,   W69-03517
NUTRIENT REQUIREME: /MNOLOGY, NIT   W69-04804
NUTRIENT REQUIREMENTS, NUTRIENTS,   W69-04798
NUTRIENT REQUIREMENTS, NUTRIENTS,   W70-04283
NUTRIENT REQUIREMENTS(ALGAE).:   W70-04074
NUTRIENT REMOVAL, MASS COMMUNICAT   W70-04506
NUTRIENT REMOVAL.: /L BALANCE, OV   W70-04465
NUTRIENT REQUIREMENTS, POPULATION   W70-05424

ROPHYTA, DIATOMS, LIGHT, LIPIDS,
L PROCEDURE, LUXURY CONSUMPTION,
ROGEN COMPOUNDS, NUISANCE ALGAE,
ION, CYANOPHYTA, NUISANCE ALGAE,
OF WISCONSIN, YAHARA RIVER(WIS),
  INHIBITION, NITROGEN COMPOUNDS,
N, TEMPERATURE, LIGHT INTENSITY,
NT BUDGETS, COMPARATIVE STUDIES,
RADIATION, IODINE RADIOISOTOPES,
ICUT, RHODE ISLAND, TEMPERATURE,
DS, AMMONIA, NITRATES, NITRITES,
                         *SEWAGE
*NUTRIENT BALANCE, *LAKE AGING,
NUTRIENT TRAPS, NUTRIENT INFLUX,
NUTRIENT/ LAKE MINNETONKA(MINN),
IPS.:   *POTOMAC RIVER ESTUARY,
R POLLUTION, SOIL CONTAMINATION,
CTERIA, PROTOZOA, SEWAGE, LAKES,
FFECTS, WATER POLLUTION SOURCES,
  SUCCESSION, ECOSYSTEMS, CYCLING
ERATURE, RAIN, GEOLOGY, MAPPING,
DIA, CHU NO 10 SOLUTION, MINIMUM
                           ALGAL
MENTS, *NUTRIENT BUDGET, CYCLING
IENT CONCENTRATION, AREAL INCOME(
L, FERTILIZERS, SEWAGE EFFLUENT,
E, *EUTROPHICATION, OLIGOTROPHY,
UNITIES, DAPHNIA, NANNOPLANKTON,
NTRATION, TEMPERATURE, NITROGEN,
ICHIGAN, *WATER SUPPLY, INTAKES,
ROBIOLOGY, CYANOPHYTA, ESSENTIAL
S, CHLOROPHYLL, PROTEINS, LIGHT,
IS, CHEMICAL ANALYSIS, BACTERIA,
NITRITES, NUTRIENT REQUIREMENTS,
OPHICATION, RUNOFF, FERTILIZERS,
HYTOLOGY, MARINE ALGAE, ECOLOGY,
, *EUTROPHICATION, ALGAL BLOOMS,
OPHICATION, *LAKES, *PHOSPHATES,
ERIA, DIATOMS, DETRITUS, CYCLING
RBIDITY, ALGAE, DIATOMS, CYCLING
PHICATION, *GREAT LAKES, *ALGAE,
EPILIMNION, HYPOLIMNION, CYCLING
ATIC ENVIRONMENT, WATER QUALITY,
RACE ELEMENTS, DISSOLVED SOLIDS,
  WATER QUALITY, *TROPHIC LEVELS,
ENVIRONMENTAL EFFECTS, ESSENTIAL
SSAYS, *FLUOROMETRY, CALIFORNIA,
EORY, ENVIRONMENTAL/ *BIOASSAYS,
ION, THERMAL POLLUTION, STREAMS,
N, AQUATIC MICROBIOLOGY, CYCLING
ACE ELEMENTS, TEMPERATURE, IRON,
OMPOUNDS, NUTRIENT REQUIREMENTS,
GY, CYCLING NUTRIENTS, ESSENTIAL
  HARVESTING OF ALGAE, LIMNOLOGY,
TES, MIDGES, SNAILS, PHOSPHORUS,
CHLOROPHYTA, CHLORELLA, CYCLING
LA, CYCLING NUTRIENTS, ESSENTIAL
S, CHELATION, SPECTROPHOTOMETRY,
Y PRODUCTIVITY, *ORGANIC MATTER,
STRIAL WASTES, MUNICIPAL WASTES,
ER TREATMENT, INDUSTRIAL WASTES,
HORUS, NITROGEN, AMMONIA, ALGAE,
CIRCULATION, PATH OF POLLUTANTS,
NITRO/ *EUTROPHICATION, *ALGAE,
ROL, OXIDATION LAGOON, NITROGEN,
LIGHT INTENSITY, EUTROPHICATION,
HICATION, *LIMNOLOGY, *ESSENTIAL
N, *PHOSPHATES, *EUTROPHICATION,
LIZATION, ANALYTICAL TECHNIQUES,
DROGEN SULFIDE, SALINITY, LAKES,
  ANIMALS, FISH, INFLOW, OUTLETS,
CAL TREATMENT, BOTTOM SEDIMENTS,
FIDE, BIOCHEMICAL OXYGEN DEMAND,
ODYNAMICS, LAKES, INTE/ *CYCLING
TIVITY, MICHIGAN, *EPIPHYTOLOGY,
IFORMS, DISSOLVED OXYGEN, ALGAE,
        *GROWTH, *ALGAE, LIPIDS,
ES, LABORATORY TESTS, MINNESOTA,
ATION, MODEL STUDIES, COMPUTERS,
URE, HYDROGEN ION CONCENTRATION,
*LAKE ZURICH, *MUNICIPAL WASTES,
PHORUS RADIOISOTOPE, *TURNOVERS,
, CYANOPHYTA, DOMINANT ORGANISM,
HYTA, NUISANCE ALGAE, WISCONSIN,
S, *ALGAE, *CHEMICAL PROPERTIES,
HORUS, BIBLIOGRAPHIES, NITRATES,
ECHNIQUES, *BIOASSAY, *ESSENTIAL
LGAE, *ALGAL CONTROL, *ESSENTIAL
ENVIRONMENTAL EFFECTS, ESSENTIAL
OMPOUNDS, NUTRIENT REQUIREMENTS,

PRODUCTIVITY, BIOASSAY, CYCLING
*PHOSPHORUS COMPOUNDS, *CYCLING
ION-REDUCTION POTENTIAL, CYCLING
ATER POLLUTION SOURCES, *CYCLING
IMNOLOGY, NUTRIENT REQUIREMENTS,
LOGY, BALANCE OF NATURE, CYCLING
TROGEN FIXATION, NUISANCE ALGAE,
EN COMPOUNDS, NITROGEN FIXATION,
ENVIRONMENTAL EFFECTS, ESSENTIAL
NVIRONMENTAL EFFECTS, *ESSENTIAL
AQUATIC ALGAE, BIOASSAY, CYCLING
, *ESSENTIAL NUTRIENTS/ *CYCLING
ENVIRONMENTAL EFFECTS, ESSENTIAL
TA, CYANOPHYTA, DIATOMS, CYCLING
FIXATION, NUTRIENT REQUIREMENTS,
Y, CHLOROPHYTA, DIATOMS, CYCLING
AL EFFECTS, ESSENTIAL NUTRIENTS,
E, AQUATIC PRODUCTIVITY, CYCLING
UTRIENTS/ *CHLOROPHYTA, *CYCLING
UTRIENTS, *NIT/ *ALGAE, *CYCLING
NUTRIENTS, *ENZYMES, *ESSENTIAL
ENVIRONMENTAL EFFECTS, ESSENTIAL
VITY, BALANCE OF NATURE, CYCLING
TA, CYANOPHYTA, DIATOMS, CYCLING
LOROPHYTA, CYANOPHYTA, ESSENTIAL
ENVIRONMENTAL EFFECTS, ESSENTIAL
ENVIRONMENTAL EFFECTS, ESSENTIAL
IRONMENTAL SANITATION, ESSENTIAL
ENVIRONMENTAL EFFECTS, ESSENTIAL
IRONMENTAL SANITATION, ESSENTIAL
NUTRIENTS, *ENZYMES, *ESSENTIAL
TIC WEEDS, *BIOASSAY, *ESSENTIAL
TA, CYANOPHYTA, DIATOMS, CYCLING
ENVIRONMENTAL EFFECTS, ESSENTIAL
*ANALYTICAL TECHNIQUES, *CYCLING
GEN DEMAND, CHLOROPHYTA, CYCLING
ATER POLLUTION EFFECTS, *CYCLING
*ANALYTICAL TECHNIQUES, *CYCLING
ENVIRONMENTAL EFFECTS, ESSENTIAL
ATER QU/ *EUTROPHICATION, MAINE,
SPORT FISHING, RECREATION, SILT,
GE TRE/ *EUTROPHICATION, *ALGAE,
C POLLUTANTS, *STREAMS, EUGLENA,
SPRAYING, BIOCONTROL, INFECTION,
CED LA/ *EUTROPHICATION, *LAKES,
PLANKTON, DISS/ *EUTROPHICATION,
LAKES, MICROORGANISMS, NITROGEN,
NITROGEN, PHOSPHOROUS, AVAILABLE
CE ALGAE, NUTRIENT REQUIREMENTS,
ION, FISH, *LIMNOLOGY, FISHKILL,
*WASTES, FERTILIZERS, *ESSENTIAL
CE ALGAE, NUTRIENT REQUIREMENTS,
*COBALT, *CYANOPHYTA, *ESSENTIAL
TROGEN FIXATION, NUISANCE ALGAE,
ALGAE, WATER POLLUTION CONTROL,
HT, TEMPERATURE, CARBON DIOXIDE,
ING FACTORS, NITROGEN COMPOUNDS,
AC/ *PHOSPHATE, *LAKES, *CYCLING
WATER POLLUTION SOURCES, *LAKES,
N, BICARBONATES, PHOTOSYNTHESIS,
POLLUTION EF/ *ALGAE, *BIOASSAY,
SOTOPES, *SALT MARSHES, *CYCLING
IMALS, FISH, ECOSYSTEMS, CYCLING
WATER POLLUTION SOURCES, CYCLING
ODOR, TASTE, LIGHT TEMPERATURE,
ENTS, MINNOWS, DETRITUS, CYCLING
PLANT G/ *FERTILIZERS, *CYCLING
TRATION, IRON, LAKES, MAGNESIUM,
TREAMFLOW/ *PERIPHYTON, *CYCLING
IA, WATER POLLUTION, REAERATION,
NITRIFICATION, DENITRIFICATION,
, URBANIZATION, DOMESTIC WASTES,
EN COMPOUNDS, NITROGEN, AMMONIA,
EFFECTS, NEW YORK, CHLOROPHYTA,
LANKTON, TRIPTON, NANNOPLANKTON,
ENVIRONMENTAL EFFECTS, ESSENTIAL
CE ALGAE, NUTRIENT REQUIREMENTS,
OCHEMICAL OXYGEN DEMAND, CYCLING
TS, NITROGEN, PHOSPHORUS, ALGAE,
ROGEN ION CONCENTRATION, CYCLING
Y, GEOGRAPHICAL REGIONS, CYCLING
ITY, HYDROGEN ION CONCENTRATION,
Y, GEOGRAPHICAL REGIONS, CYCLING
PHY, SEWAGE, ALGAE, SCUM, WEEDS,
OROPHYTA, PHOTOMETRY, CHLORELLA,
L, *MIDGES, FISH FOOD ORGANISMS,
*CYANOPHYTA, *NITROGEN FIXATION,
S, PRO/ *EUTROPHICATION, *LAKES,
ESIS, ILLUMINATION, TEMPERATURE,
TROGEN, FLOW, CHEMICAL ANALYSIS,

NUTRIENTS, ENVIRONMENTAL EFFECTS,   W69-04798
NUTRIENTS, WATER CHEMISTRY, WATER   W69-05142
NUTRIENTS, EUTROPHICATION.: /IDAT   W69-04521
NUTRIENTS, ALGAE, ANALYTICAL TECH   W69-04800
NUTRIENTS, PHOTOSYNTHET: /TION, L   W69-03517
NUTRIENTS, ENVIRONMENTAL EFFECTS,   W69-04802
NUTRIENTS, PHOSPHORUS COMPOUNDS,   W69-04802
NUTRIENTS, NU: /LIMNOLOGY, NITROG   W69-03518
NUTRIENTS, CHLOROPHYTA, DIATOMS,   W69-03517
NUTRIENTS, *FISH, *FISHERIES, ALG   W69-04804
NUTRIENTS, ENVIRONMENTAL EFFECTS,   W69-03518
NUTRIENTS, *ENVIRONMENTAL EFFECTS   W69-04804
NUTRIENTS, FREEZE DRYING, FREEZE-   W69-03370
NUTRIENTS, ENVIRONMENTAL EFFECTS,   W69-03188
NUTRIENTS, PHYSIOLOGICAL ECOLOGY,   W69-03364
NUTRIENTS, ENVIRONMENTAL EFFECTS,   W69-03515
NUTRIENTS, PHOSPHORUS, PHOSPHATES   W69-03373
NUTRIENTS, ENVIRONMENTAL EFFECTS,   W69-03514
NUTRIENTS, *ENZYMES, *ESSENTIAL N   W69-03186
NUTRIENTS, *ENZYMES, *ESSENTIAL N   W69-03185
NUTRIENTS, *NITROGEN CYCLE, *NUTR   W69-03186
NUTRIENTS, NUTRIENTS, PHOSPHORUS,   W69-03373
NUTRIENTS, ENVIRONMENTAL EFFECTS,   W69-03513
NUTRIENTS, ENVIRONMENTAL EFFECTS,   W69-03364
NUTRIENTS, ENVIRONMENTAL EFFECTS,   W69-03358
NUTRIENTS, EUTROPHICATION, INHIBI   W69-03515
NUTRIENTS, FREEZE DRYING, FREEZE-   W69-03362
NUTRIENTS, HARVESTING OF ALGAE, I   W69-03369
NUTRIENTS, EUTROPHICATION, INHIBI   W69-03514
NUTRIENTS, INHIBITION, LIGHT INTE   W69-03513
NUTRIENTS, *NITROGEN CYCLE, ANALY   W69-03185
NUTRIENTS, *NITROGEN COMPOUNDS, A   W69-03364
NUTRIENTS, ENVIRONMENTAL EFFECTS,   W69-03373
NUTRIENTS, HARVESTING OF ALGAE, N   W69-03371
NUTRIENTS, *EUTROPHICATION, *PHOS   W69-03358
NUTRIENTS, ENVIRONMENTAL EFFECTS,   W69-03374
NUTRIENTS, *EUTROPHICATION, ALGAE   W69-03369
NUTRIENTS, *EUTROPHICATION, *PHOS   W69-03370
NUTRIENTS, ENVIRONMENTAL ENGINEER   W69-03374
NUTRIENTS, AQUATIC POPULATIONS, W   W68-00470
NUTRIENTS, AQUATIC POPULATIONS, O   W68-00172
NUTRIENTS, DOMESTIC WASTES, *SEWA   W68-00248
NUTRIENTS, PLANT POPULATIONS, HIS   W68-00483
NUTRIENTS, ALGAE, NITROGEN, PHOSP   W68-00468
NUTRIENTS, OXYGEN REQUIREMENTS, I   W68-00487
NUTRIENTS, NITROGEN, PHOSPHORUS,   W68-00253
NUTRIENTS, PHOSPHORUS, PLANKTON,   W68-00481
NUTRIENTS, SEDIMENT CORES.: /ON,   W68-00511
NUTRIENTS, TOXICITY, WATER CHEMIS   W69-05867
NUTRIENTS, *ALGAE, *ALGAL CONTROL   W69-05844
NUTRIENTS, *NITRIFICATION, CROPS,   W69-05323
NUTRIENTS, PHOSPHORUS COMPOUNDS,   W69-05868
NUTRIENTS, *TRACE ELEMENTS, ALGAE   W69-06277
NUTRIENTS, VITAMINS, WATER POLLUT   W69-06277
NUTRIENTS, WATER POLLUTION EFFECT   W69-06276
NUTRIENTS, BIOLOGICAL PROPERTIES,   W69-07442
NUTRIENTS, ODOR, OLIGOTROPHY, OXY   W69-06535
NUTRIENTS, *SEDIMENT-WATER INFERF   W69-06859
NUTRIENTS, FERTILIZERS, PHOSPHATE   W69-07084
NUTRIENTS, MICROBIOLOGY.: / CARBO   W69-07389
NUTRIENTS, EUTROPHICATION, WATER   W69-06864
NUTRIENTS, GEORGIA, SEDIMENTS, BA   W69-08274
NUTRIENTS, TEMPERATURE, LIGHT QUA   W69-08526
NUTRIENTS, COST COMPARISONS, ECON   W69-08518
NUTRIENTS, COPPER SULFATE, STRATI   W69-08282
NUTRIENTS, PLANKTON, ALGAE, INVER   W69-07861
NUTRIENTS, EUTROPHICATION, ALGAE,   W69-07826
NUTRIENTS, NUISANCE ALGAE, OLIGOT   W69-07833
NUTRIENTS, *ZINC RADIOISOTOPES, S   W69-07862
NUTRIENTS, OXYGEN SAG, AERATION,   W69-07520
NUTRIENTS, TERTIARY TREATMENT, AL   W69-08053
NUTRIENTS, PHOSPHORUS, ALGAE, ALG   W69-07818
NUTRIENTS, BIOCHEMICAL OXYGEN DEM   W69-08054
NUTRIENTS, ECOSYSTEMS, PHYTOPLANK   W70-03973
NUTRIENTS, OLIGOTROPHY, TEMPERATU   W70-03959
NUTRIENTS, EUTROPHICATION, LIGHT   W70-04283
NUTRIENTS, PHOSPHORUS COMPOUND: /   W70-04283
NUTRIENTS, DIATOMS, ENVIRONMENTAL   W70-04283
NUTRIENTS, WATER PROPERTIES.: /EC   W70-04382
NUTRIENTS, MUD-WATER INTERFACES,   W70-04381
NUTRIENTS, ZOOPLANKTON, ALGAE, AQ   W70-03975
NUTRIENTS, PHOSPHORUS, NITROGEN,   W70-03983
NUTRIENTS, ZOOPLANKTON, ALGAE, AQ   W70-03975
NUTRIENTS, SEPTIC TANKS, EFFLUENT   W70-04193
NUTRIENTS, TRACE ELEMENTS, CHLAMY   W70-05409
NUTRIENTS, ENERGY TRANSFER, *FOOD   W70-05428
NUTRIENTS, LAKE ERIE.:   *ALGAE,   W70-05091
NUTRIENTS, WATER POLLUTION EFFECT   W70-04721
NUTRIENTS, FACULTATIVE POND DESIG   W70-04786
NUTRIENTS, EFFLUENTS, ALGAE, WEED   W70-04810

A, ALG/ *EUTROPHICATION, *LAKES,
TION, SEDIMENTS, PALEOLIMNOLOGY,
ICS, RECREATION, FISHING, ALGAE,
ALGAE, AGRICULTURAL WATERSHEDS,
E, LIGHT INTENSITY, CHLOROPHYLL,
LIC BACTERIA, THERMAO POLLUTION,
ATER POLLUTION, BACTERIA, ALGAE,
AGOONS, ACTIVATED SLUDGE, ALGAE,
, NUISANCE ALGAE, WATER QUALITY,
*ALGAE, *SESSILE ALGAE, *CYCLING
UCTIVITY, TEMPERATURE, BACTERIA,
S, INFLOW, CYANOPHYTA, LITTORAL,
.:   ALGAL BLOOMS, ALGAE
E), LAKE WASHINGTON(WASH), ALGAL
SKELETONEMA, THALASSIOSI/ *ALGAL
NOCOBALAMIN, ALGAL GROWTH, ALGAL
ANT STUDIES, ALGAL GROWTH, ALGAL
VAILABILITY, ALGAL GROWTH, ALGAL
S AERUGINOSA, ANTAGONISMS, ALGAL
OUND, SYRACUSE(NY),/ *PHYCOLOGY,
                    *MASSENA(
                 CATTARAUGUS CREEK(
PHORA GLOMERATA, BUTTERNUT CREEK(
MORICHES BAY(NY), ONONDAGA LAKE(
INIA, / *FAYETTEVILLE GREEN LAKE(
NUTRIENT LIMITATION, CAYUGA LAKE(
A, NITZSCHIA, NAVICULA, SYRACUSE(
, PINNULARIA, NI/ *ONONDAGA LAKE(
HAEMOCYTOMETER COUNTING, ITHACA(
DIA FORMATION, LONG ISLAND SOUND(
S, RUSSIA, BAVARIA, MORICHES BAY(
RADIOCARBON METHOD, FINGER LAKES(
E AND TECHNOLOGY FOUND, SYRACUSE(
ODOSUM, ECTOCARPUS CONFERVOIDES,
WASTES, *MAN-MADE FIBER PLANTS,
WIS), ANACHARIS, NAIAS, SCIRPUS,
CUMULATION, RUTHENIUM-106, WHITE
YCAEUS, OIKOPLEURA, FRITILLARIA,
NGROSIRA STAGNALIS, NEPHROCYTIUM
US, NOSTOC MUSCORUM, SCENEDESMUS
ANSPARENCY, CYCLOPS, SCENEDESMUS
EDESMUS QUADRICAUDA, SCENEDESMUS
ATION, ASTERIONELLA, SCENEDESMUS
PALEA, CHLOROPLASTS, SCENEDESMUS
L, SUSCEPTIBILITY, PHYSIOLOGICAL
ANOIDES, CANDIDA KRUSEI, CANDIDA
ECOLOGICAL EFFIC/ *DAPHNIA PULEX
, BIOTA, CLAMS, FALLOUT, PACIFIC
MOLLUSCA, ECHINODERMATA, PACIFIC
CHLOROPHYLL, OIL, ALGAE, PACIFIC
Y, CARBON RADIOISOTOPES, PACIFIC
ABSORPTION, ADSORPTION, ATLANTIC
AE, AQUATIC LIFE, MILK, ATLANTIC
RON, COBALT, MANGANESE, ATLANTIC
RPTION, GULF OF MEXICO, ATLANTIC
IME, TRIDACNA CROCEA, WOODS HOLE
CULTURES, FLUOROMETRY, BIOMASS,
GY, RUSSIA, SECOND INTERNATIONAL
K SEA, CONTACT ZONE/ *BIOLOGICAL
QUES, *RADIOISOTOPES, LIMNOLOGY,
MARINE PLANTS, DINOFLAGELLATES,
PES, RADIOACTIVE WASTE DISPOSAL,
DIATOMS, RESPIRATION, PIGMENTS,
ISH, WASHINGTON, LAKES, INDIANA,
UCCESSION, CHLORELLA, CHELATION,
E, FILAMENTOUS BLUE-GREEN ALGAE,
CTALURUS PUNCTATUS, TRICHOPTERA,
WATER CARRIER(ISRAEL), THRESHOLD
NTARIO), ALGAE COUNTS, THRESHOLD
S), STABILIZATION POND LOADINGS,
CONTROL REQUIREMENTS, THRESHOLD
CONTROL, NUISANCE ALGAE, TASTE,
ENTS, BIOCHEMICAL OXYGEN DEMAND,
L, SLIME, TASTE-PRODUCING ALGAE,
NDUSTRIAL WASTE, CORES, DIATOMS,
C MATTER, *PO/ *WATER TREATMENT,
*WATER POLLUTION SOURCES, TASTE,
WAGE, COLIFORMS, EUTROPHICATION,
IODEGRADATION, *SEWAGE BACTERIA,
ADIOACTIVITY, PESTICIDES, TASTE,
UTION TREATMENT, COPPER SULFATE,
GANIC MATTER, TOXICITY, PHENOLS,
MNION, CYCLING NUTRIENTS, TASTE,
WATER SHORTAGE FILTERS, SNAILS,
MICROORGANISMS, HISTORY, TASTE,
AIR POLLUTION, WATER POLLUTION,
CTERIA, WATER POLLUTION SOURCES,
*WATER SUPPLY, *CONTROL, TASTE,
, NITROGEN COMPOUNDS, NUTRIENTS,
OXYGEN DEMAND, DISSOLVED OXYGEN,

NUTRIENTS, FERTILIZATION, BACTERI   W70-05761
NUTRIENTS, PRODUCTIVITY, SEDIMENT   W70-05663
NUTRIENTS, TURBIDITY, TERTIARY TR   W70-05645
NUTRIENTS, FERTILIZATION, PHYSICA   W70-05565
NUTRIENTS, CONDUCTIVITY, AGRICULT   W70-05752
NUTRIENTS, TEMPERATURE, PHOTOSYNT   W70-05270
NUTRIENTS, INDICATORS, ODOR, TAST   W70-05264
NUTRIENTS, VIRUSES, BACTERIA, E.   W72-06017
NUTRIENTS, PHOSPHORUS, NITROGEN,   W72-05453
NUTRIENTS, ECOSYSTEMS, PHOSPHORUS   W73-01454
NUTRIENTS, ESTUARINE ENVIRONMENT,   W73-00748
NUTRIENTS, MODEL STUDIES, BIOASSA   W73-01454
NUTRITION, MICROCYSTIS AERUGINOSA   W70-05565
NUTRITION, ALGAL PHYSIOLOGY, HETE   W70-04506
NUTRITION, DITYLUM, CHAETOCEROS,   W70-04503
NUTRITION, ALGAL PHYSIOLOGY, CALO   W69-06277
NUTRITION, ALGAL PHYSIOLOGY, AMER   W69-06865
NUTRITION, ALGAL PHYSIOLOGY, KJEL   W69-05868
NUTRITION, ALGAL PHYSIOLOGY, LAKE   W69-05867
NY STATE SCIENCE AND TECHNOLOGY F   W70-03973
NY).:   W72-08686
NY).:   W69-10080
NY).:   *CLADOPHORA FRACTA, CLADO W72-05453
NY).: /GENETICS, RUSSIA, BAVARIA,   W70-03973
NY), *MARL LAKES, CHLOROBIUM, FIL   W70-05760
NY), ASTERIONELLA FORMOSA, SYNEDR   W70-01579
NY), ENTEROMORPHA INTESTINALIS, C   W70-03974
NY), EUNOTIA, AMPHORA, GOMPHONEMA   W70-03974
NY), GLENODINIUM QUADRIDENS, OOCY   W69-06540
NY), ICELAND, TRITIATED THYMIDINE   W69-09755
NY), ONONDAGA LAKE(NY).: /GENETIC   W70-03973
NY), RODHE'S SOLUTION VIII, NUTRI   W70-01579
NY), SYRACUSE UNIVERSITY, FUNDAME   W70-03973
NYLON COLUMN TECHNIQUE, PARTICULA   W70-01074
NYLON-6, *VISCOSE RAYON, ZINC COM   W70-06925
NYMPHAEA, CRASSIPES EICHORNIA, TY   W70-05263
OAK LAKE(TENN), ACTIVITY-DENSITY,   W70-04371
OBELIA, CLUSTER ANALYSIS, MULTIVA   W73-01446
OBESUM W AND G S WEST, RADIOCOCCU   W70-04468
OBLIQUUES, SCENEDESMUS QUADRICAUD   W70-02964
OBLIQUUS, DACTYLOCOCCUS, ESCHERIC   W70-10161
OBLIQUUS, HELIX POMATIA, CHLORELL   W70-04184
OBLIQUUS, GROWTH CONSTATS, RIVER   W70-03955
OBLIQUUS, SCENEDESMUS QUADRICAUDA   W70-05381
OBSERVATIONS, CHRONIC EXPOSURE, A   W70-01779
OBTUSA, CANDIDA TROPICALIS, NOCTI   W70-04368
OBTUSA, SCENEDESMUS QUADRICAUDA,   W69-10152
OCE: /ITIUM, CARBON RADIOISOTOPES   W69-08269
OCEAN.: / POLYCHAFTA, CRUSTACEA,   W69-08526
OCEAN.: /OTE SENSING, SEA WATER,   W70-05377
OCEAN, CARBON DIOXIDE, LIGHT, TEM   W70-05652
OCEAN, DIATOMS, NERITIC ALGAE, IO   W70-00707
OCEAN, LANDFILLS, DISPOSAL, REGUL   W72-00941
OCEAN, PUERTO RICO, SPECTROSCOPY,   W69-08525
OCEAN, RADIOACTIVITY, GAMMA RAYS,   W69-08524
OCEANOGRAPHIC INSTITUTION, ACROPO   W69-08526
OCEANOGRAPHY.: /OLOGICAL STUDIES,   W69-08278
OCEANOGRAPHIC CONGRESS, HYPON: /O   W69-07440
OCEANOGRAPHY, *TRANSACTIONS, BLAC   W69-07440
OCEANOGRAPHY, TRACERS, BACTERIA,   W69-10163
OCEANS, BIOINDICATORS, FUNGI, EST   W70-04368
OCEANS, FOODS, ABSORPTION, HAZARD   W70-02786
OCEANS, ORGANIC MATTER, DIURNAL,   W70-04809
OCHROMONAS, PHYTOPLANKTON, SEAWEE   W70-04503
OCHROMONAS, BACTERIA, FUNGI, AQUA   W70-02510
OCULAR MICROMETER, SUCROSE-FLOTAT   W70-04253
ODONATA, EPHEMEROPTERA, PLECOPTER   W69-07846
ODOR NUMBERS.: /ATORIA, NATIONAL   W70-02373
ODOR NUMBERS, CROP-DUSTING AIRCRA   W69-10155
ODOR PERSISTENCE.: /ORADO, WEST(U   W70-04788
ODOR TESTS, ASTERIONELLA, APHANIZ   W69-10157
ODOR.: / QUALITY, WATER POLLUTION   W69-10155
ODOR-PRODUCING ALGAE, POLLUTION A   W69-06909
ODOR-PRODUCING ALGAE, DIATOMS, CY   W69-05704
ODOR-PRODUCING ALGAE, CHLOROPHYLL   W68-00470
ODOR, *ALGAE, *DECOMPOSING ORGANI   W72-04242
ODOR, ALGAE, RESERVOIRS, WATERSHE   W72-01813
ODOR, AMMONIUM COMPOUNDS, ACTIVAT   W70-00263
ODOR, ANAEROBIC CONDITIONS, AEROB   W71-02009
ODOR, COAGULATION, ELECTROLYTES,   W72-13601
ODOR, COLIFORMS, ECONOMICS.: /OLL   W72-01814
ODOR, COLOR, TURBIDITY, BIOCHEMIC   W72-03299
ODOR, COLOR, TURBIDITY, HYDROGEN   W70-02251
ODOR, COOLING TOWERS, FILTERS, EF   W72-00027
ODOR, DEEP WELLS, WATER TREATMENT   W70-08096
ODOR, EFFLUENTS, SEWAGE EFFLUENTS   W71-00940
ODOR, FARM WASTE, SLUDGE, SOLID W   W71-05742
ODOR, FILTERS, PLANKTON, WATER PO   W70-05389
ODOR, OLIGOTROPHY, OXYGEN, OXYGEN   W69-06535
ODOR, ORGANIC LOADING, *WASTE WAT   W70-06601

MENT, ODOR, *ALGAE, *DECOMPOSING
, *LAKE MEAD, LAS VEG/ DISSOLVED
UDGE, SOLID WASTES, DECOMPOSING
R POLLUTION CONTROL, DECOMPOSING
CETES, PLANT GROWTH, DECOMPOSING
UCTIVITY, ABSORPTION, SEDIMENTS,
HKILL, *CYANOPHYTA, *DECOMPOSING
*WATER QUALITY, WATER POLLUTION,
EFFLUENTS, DILUTION, DECOMPOSING
ITY, HARDNESS(WATER), CHLORIDES,
TY, CARBON RADIOISOTOPES, LAKES,
RGANIC GROWTH FACTOR/ *DISSOLVED
TING, METALS, INHIBITORS, SILTS,
IRATION, PHOSPHATES, ECOSYSTEMS,
ENTS, ECOLOGY, CHLORELLA, ALGAE,
OPHYTA, SAMPLING, PHYTOPLANKTON,
EMPERATURE, NITROGEN, NUTRIENTS,
OGRAPHIES, BACTERIA, DECOMPOSING
SSIUM, CALCIUM, MAGNESIUM, IRON,
RIVERS, BOGS, PHENOLS, PIGMENTS,
UALITY, NEW JERSEY, TEMPERATURE,
, LIPIDS, NUTRIENT REQUIREMENTS,
UM COMPOUNDS, OXYGEN, SILICATES,
AILS, RADIOACTIVITY, INTERFACES,
YSTEMS, EUGLENA, EUTROPHICATION,
ROUNDWATER, *SESTON, *SEDIMENTS,
PHORUS, AGRICULTURAL WATERSHEDS,
TION, PHYTOPLANKTON, DECOMPOSING
CYANOPHYTA, VITAMINS, PEPTIDES,
ZYMES, KINETICS, PHOTOSYNTHESIS,
PRIMARY/ *PHYTOPLANKTON, *LAKES,
KTON, *AQUATIC BACTERIA, *ALGAE,
*WATER UTILIZATION, DECOMPOSING
, OIL WASTES, INLETS(WATERWAYS),
RGANISMS, PHYSIOLOGICAL ECOLOGY,
FFECTS, ANA/ *CHEMICAL ANALYSIS,
, RESPIRATION, PIGMENTS, OCEANS,
TER), WATER POLLUTION, DRAINAGE,
LL, PHOTOSYNTHESIS, GROWTH RATE,
OLVED OXYGEN, CHEMICAL ANALYSIS,
EFFLUENTS, DIATOMS, PHOSPHORUS,
STUDY, AEROBICITY.:
SSAY, FISH, AQUATIC ENVIRONMENT,
RIS COUNTY, TEX, POLYPHOSPHATES,
, CHEMICAL PROPERTIES, BACTERIA,
C/ *POLYELECTROLYTES, *SYNTHETIC
Y TRANSFORMATIONS, HETEROTROPHY,
*TURNOVER RATES, *HETEROTROPHY,
WATER POLLUTION IDENTIFICATION,
ALKALOID TOXINS, IBOGAALKALOID,
ON CONTROL, WATERCOURSES(LEGAL),
CROP, A/ *EUTROPHICATION, LAKES,
CAL OXYGEN DEMAND, ALGAL TOXINS,
ITE, ALGAE, CYANOPHYTA, DOMINANT
COMPOSITION, MATERIAL BALANCES,
S, NON-LEGUMES, *NITROGEN FIXING
TS, PHOSPHORUS, PLANKTON, PLANTS(
GAL CONTROL, ALGICIDES, DOMINANT
SECTS, *AQUATIC DRIFT, FISH FOOD
ICATION, FISH FARMING, FISH ROOD
LITY CONTROL, *MIDGES, FISH FOOD
ROPHY, ORGANIC POLLUTION, MARINE
RIVER(GA), SPARTINA ALTEMIFLORA,
TICS, *CHEMICAL CHARACTERISTICS,
ICROORGANISMS, *ALGAE, *DOMINANT
IOLOGIC 'DISPOSERS', MICROSCOPIC
ALINITY, FISH FARMING, FISH FOOD
GROUND SOLIDS, UPTAKE, CONSUMER
ODORIFEROUS PRODUCTS,
SOURCES COMMISSION, PEST CONTROL
M COMPOUNDS, PHENOLIC COMPOUNDS,
STRY, SPRING WATERS, PHOSPHORUS,
ALYTICAL TECHNIQUES, *BIOASSAYS,
OVELLITE, SPHALERITE, MILLERITE,
PIGMENTS, DIAPTOMUS, EPISCHURA,
NS, ARTHROBACTER, PYROPHOSPHATE,
, BACILLUS SUBTILIS, ASPERGILLUS
BESA, LAKE KEGONSA, LAKE ZURICH,
ND, PER/ *LAKE WASHINGTON(WASH),
ESCENS, SWITZERLAND, PERIDINIUM,
KE VOLUME, MEAN DEPTH, DILUTION,
I DISC TRANSPARENCY, MYXOPHYCEA,
YNEDRA, NAVICULA, APHANIZOMENON,
INULATE, OSCILLATORIA RUBESCENS,
INALIS, GLOEOTRICHIA ECHINULATE,
ELENASTRUM CAPRICORNUTUM PRINTZ,
COMPETITIVE EXCLUSION PRINCIPLE,
LAMINOSUM, OSCILLATORIA TENUIS,
LAMINOSUS, PHORMIDIUM LAMINOSUM,
), VERTICAL, CHALK, METALIMNION,

ORGANIC MATTER, *POTASSIUM COMPOU   W72-04242
ORGANIC MATTER, LAS VEGAS, NEVADA   W72-02817
ORGANIC MATTER, LAGOONS, CATTLE.:   W71-05742
ORGANIC MATTER, CARBON DIOXIDE.: / W72-01788
ORGANIC MATTER, CHEMICAL ANALYSIS   W72-01812
ORGANIC MATTER, ALGAE, DIATOMS, C   W72-00845
ORGANIC MATTER, TOXICITY, DISSOLV   W72-01797
ORGANIC MATTER, BIBLIOGRAPHIES.: / W70-07027
ORGANIC MATTER, INSECTS, MITES, B   W71-00940
ORGANIC MATTER, CARBON, WATER POL   W71-04206
ORGANIC MATTER, SEA WATER, PRIMAR   W70-02504
ORGANIC MATTER, ORGANIC DEBRIS, O   W70-02510
ORGANIC MATTER, IRON OXIDES, BACT   W70-02294
ORGANIC MATTER, WATER POLLUTION,   W70-03309
ORGANIC MATTER, CULTURES, CARBON,   W70-02804
ORGANIC MATTER, CHLOROPHYLL, DYNA   W70-02780
ORGANIC MATTER, P: /CENTRATION, T   W70-00265
ORGANIC MATTER, WASTE WATER(POLLU   W70-00664
ORGANIC MATTER, NITROGEN FIXATION   W70-00719
ORGANIC MATTER, BIOMASS, BENTHIC   W70-01074
ORGANIC MATTER, HARDNESS, CARBON   W69-10157
ORGANIC MATTER, PHOTOSYNTHESIS, T   W69-06865
ORGANIC MATTER, AIR TEMPERATURE,   W69-07096
ORGANIC MATTER, RADIOISOTOPES, GA   W69-08272
ORGANIC MATTER, PHYTOPLANKTON, PR   W69-06273
ORGANIC MATTER, CURRENTS(WATER),   W70-03501
ORGANIC MATTER, SEDIMENTS, FERTIL   W70-04193
ORGANIC MATTER, OKLAHOMA, RESERVO   W70-04001
ORGANIC MATTER, TRACE ELEMENTS, L   W70-04369
ORGANIC MATTER, ORGANIC COMPOUNDS   W70-04287
ORGANIC MATTER, SEASONAL, ALGAE,   W70-04290
ORGANIC MATTER, LAKES, DIFFUSION,   W70-04284
ORGANIC MATTER, *STREAM POLLUTION   W68-00483
ORGANIC MA: /ASTES, GARBAGE DUMPS   W70-05112
ORGANIC MATTER, WATER POLLUTION F   W70-04811
ORGANIC MATTER, WATER POLLUTION E   W70-05109
ORGANIC MATTER, DIURNAL, LATITUDI   W70-04809
ORGANIC MATTER, SEWAGE, PHYTOPLAN   W70-05404
ORGANIC MATTER, SCENEDESMUS, CHLA   W70-05381
ORGANIC MATTER, ALGAE, IRON, HY: / W70-05760
ORGANIC MATTER, NITROGEN, CHLOROP   W70-05663
ORGANIC NITROGEN, BACTERIOLOGICAL   W71-05742
ORGANIC PESTICIDES, HYDROGEN ION   W72-11105
ORGANIC PHOSPHATES.:   HAR   W69-04800
ORGANIC POLLUTANTS, *STREAMS, EUG   W68-00483
ORGANIC POLYELECTROLYTES, CATIONI   W70-05267
ORGANIC POLLUTION, MARINE ORGANIS   W69-07440
ORGANIC SOLUTES, SWEDEN, LAKE ERK   W70-05109
ORGANIC SOLUTES, VITAMIN B-1, VIT   W70-05652
ORGANIC TOXINS.: /, *GLENODININE,   W71-07731
ORGANIC WASTES, PULP WASTES, INDU   W72-00462
ORGANIC WASTES, OXYGEN, STANDING   W70-02251
ORGANIC WASTES, POLLUTANT IDENTIF   W69-01273
ORGANISM, NUTRIENTS, RAINFALL-RUN   W68-00891
ORGANISMS.: /ELLS, CONCENTRATION,   W70-06600
ORGANISMS.: /NITROGEN-15, SOD MAT   W72-08110
ORGANISMS).: /, NITROGEN, NUTRIEN   W68-00481
ORGANISMS, GRAZING, WATER POLLUTI   W69-01977
ORGANISMS, FISH PHYSIOLOGY, FISH   W69-02782
ORGANISMS, FISH GENETICS, FISH HA   W69-04804
ORGANISMS, NUTRIENTS, ENERGY TRAN   W70-05428
ORGANISMS, MARINE ASSOCIATIONS, M   W69-07440
ORGANISMS, TURSIOPS TRUNCATUS, BR   W69-08274
ORGANISMS, SUSPENDED SOLIDS, SALT   W73-00748
ORGANISMS, *ACCLIMATIZATION, DIAT   W70-07313
ORGANISMS, PAPERMILL WASTE EFFLUE   W71-00940
ORGANISMS, FISH PHYSIOLOGY, RED T   W71-07731
ORGANISMS, SIMULIUM, PHYSA, SALMO   W69-09334
ORGANISMS' ODORS.:   W72-01812
ORGANIZATION OF TORONTO, FILAMENT   W69-10155
ORGANO-METALLIC COMPOUNDS, AMINES   W72-12987
ORGANOPHOSPHORUS COMPOUNDS, BACTE   W69-09334
ORGANOPHOSPHORUS COMPOUNDS, SOLVE   W70-05269
ORPIMENT, TETRAHEDRITE, BORNITE,   W69-07428
ORTHOCYCLOPS, KERATELLA, POLYARTH   W70-05760
ORTHOPHOSPHATE, DISMUTATION, TANN   W69-07428
ORYZAE NIGER, S: /AENA CYLINDRICA   W70-02964
OSCILLATORIA: /AKE MENDOTA, LAKE WAU   W69-09349
OSCILLATORIA RUBESCENS, SWITZERLA   W69-10169
OSCILLATORIA AGARDHI, PHORMIDIUM,   W69-10169
OSCILLATORIA RUBESCENS, MADISON L   W70-00270
OSCILLATORIA, ANABAENA, DIATOMS   W69-10182
OSCILLATORIA, MELOSIRA, ACTINASTR   W70-08107
OSCILLATORIA CHALYBIA, HYDRODICTY   W69-05704
OSCILLATORIA RUBESCENS, OSCILLATO   W69-05704
OSCILLATORIA VAUCHER, SKELETONEMA   W70-05752
OSCILLATORIA RUBESCENS, PISTON CO   W70-05663
OSCILLATORIA TEREBRIFORMIS, MATS,   W70-05270
OSCILLATORIA TENUIS, OSCILLATORIA   W70-05270
OSCILLATORIA RUBESCENS, DINOBRYON   W70-05405

S, ACTIVATED SLUDGE, BIOCHEMICAL	OXYGEN DEMAND, OXIDATION-REDUCTIO	W71-09546
LVED OXYGEN, *ALGA/ *BIOCHEMICAL	OXYGEN DEMAND, *NUTRIENTS, *DISSO	W71-05407
*OXIDATION LAGOONS, *BIOCHEMICAL	OXYGEN DEMAND, *ANAEROBICS, ALGAE	W71-05742
BIC BACTERIA, ALGAE, BIOCHEMICAL	OXYGEN DEMAND, OXIDATION LAGOONS,	W71-08221
OGEN, *COLIFORMS, / *BIOCHEMICAL	OXYGEN DEMAND, *PHOSPHORUS, *NITR	W71-05409
POLLUTION, BACTERIA, EFFLUENTS,	OXYGEN DEMAND, AEROBIC TREATMENT,	W72-00462
FERMENTATION, TOXICITY, CHEMICAL	OXYGEN DEMAND, HYDROGEN ION CONCE	W71-13413
, REMOVAL, DEGRADATION, CHEMICAL	OXYGEN DEMAND, *WASTE WATER TREAT	W71-13335
R, COLOR, TURBIDITY, BIOCHEMICAL	OXYGEN DEMAND, HYDROGEN ION CONCE	W72-03299
TION, ON-SITE TESTS, BIOCHEMICAL	OXYGEN DEMAND, DIURNAL, CHLOROPHY	W72-01818
I, LABORATORY TESTS, BIOCHEMICAL	OXYGEN DEMAND, SEWAGE EFFLUENTS,	W72-01817
ALGAE, COPPER SULFATE, *CHEMICAL	OXYGEN DEMAND.: /ORGANIC MATTER,	W72-04495
ES, SEDIMENTS, DISSOLVED OXYGEN,	OXYGEN DEMAND, CYANOPHYTA, DIAT: /	W69-08518
ENTS, ODOR, OLIGOTROPHY, OXYGEN,	OXYGEN DEMAND, PHOSPHORUS COMPOUN	W69-06535
, STATE GOVERNMENTS, BIOCHEMICAL	OXYGEN DEMAND, ODOR-PRODUCING ALG	W69-06909
AMMONIA, NUTRIENTS, BIOCHEMICAL	OXYGEN DEMAND, TERTIARY TREATMENT	W69-08054
XYGEN SAG, AERATION, BIOCHEMICAL	OXYGEN DEMAND, PHOTOSYNTHETIC OXY	W69-07520
TMENT, *FILTRATION, *BIOCHEMICAL	OXYGEN DEMAND, *ACTIVATED SLUDGE,	W70-05821
AE, BACTERIA, CULTURES, CHEMICAL	OXYGEN DEMAND, NITROGEN, PHOSPHAT	W70-05655
, SELF-PURIFICATION, BIOCHEMICAL	OXYGEN DEMAND, TOXICITY, OXYGEN-R	W70-05750
OLVED OXYGEN, ALGAE, BIOCHEMICAL	OXYGEN DEMAND(BOD), MIXING, WINDS	W70-04790
ALGAE, DISINFECTION, BIOCHEMICAL	OXYGEN DEMAND(BOD), AERATION, COL	W70-04788
OXYGEN, TEMPERATURE, BIOCHEMICAL	OXYGEN DEMAND(BOD), EVAPORATION,	W70-04787
D STATES, CHLORELLA, BIOCHEMICAL	OXYGEN DEMAND, OXIDATION LAGOONS,	W70-05092
PRECIPITATION, IONS, BIOCHEMICAL	OXYGEN DEMAND, MINNESOTA, WATER C	W69-00446
NIC WASTES, POLLUTA/ BIOCHEMICAL	OXYGEN DEMAND, ALGAL TOXINS, ORGA	W69-01273
E, WATER PROPERTIES, BIOCHEMICAL	OXYGEN DEMAND, *BACTERIA, ACTIVAT	W68-00256
YGEN, *CHLORELLA, / *BIOCHEMICAL	OXYGEN DEMAND, *PHOTOSYNTHETIC OX	W69-03371
OF NATURE, BIOASSAY, BIOCHEMICAL	OXYGEN DEMAND, CHLOROPHYTA, CYCLI	W69-03374
WISCONSIN, BIOASSAY, BIOCHEMICAL	OXYGEN DEMAND, ENVIRONMENTAL EFFE	W69-03369
SYNTHETIC OXYGEN, / *BIOCHEMICAL	OXYGEN DEMAND, *CHLORELLA, *PHOTO	W69-03362
QUATIC PRODUCTIVITY, BIOCHEMICAL	OXYGEN DEMAND, CYCLING NUTRIENTS,	W70-04283
NAL, STRATIFICATION, BIOCHEMICAL	OXYGEN DEMAND, CARBON RADIOISOTOP	W70-04001
*ACTIVATED SLUDGE, *BIOCHEMICAL	OXYGEN DEMAND, *COLOR, WASTE WATE	W70-04060
GATION, RESPIRATION, BIOCHEMICAL	OXYGEN DEMAND, NITROGEN FIXATION.	W72-06022
, *DISSOLVED OXYGEN, BIOCHEMICAL	OXYGEN DEMAND, ALGAL CONTROL, *CO	W72-11906
N ION CONCENTRATION, BIOCHEMICAL	OXYGEN DEMAND, VIRGINIA.: /YDROGE	W72-10573
E MEAD, DILUTION, ALGAL GROWTHS,	OXYGEN DEPLETION.: / EFFECTS, LAK	W73-00750
(WIS), APHANIZOMENON FLOS AQUAE,	OXYGEN DEPLETION.: *YAHARA RIVER	W72-01797
S, WATER SAMPLING, ALGAL GROWTH,	OXYGEN DEPLETION, NUCLEAR ACTIVAT	W70-00264
, DELTA WATER FACILITIES(CALIF),	OXYGEN DYNAMICS, SACRAMENTO-SAN J	W70-05094
, PHORMIDIUM, WINKLER TECHNIQUE,	OXYGEN ELECTRODES, ATOMIC ABSORPT	W70-03309
TROPHICATION, *LAKES, NUTRIENTS,	OXYGEN REQUIREMENTS, ICED LAKES,	W68-00487
EUTROPHICATION, FISHKILL, ALGAE,	OXYGEN SAG, WATER QUALITY, CHLORI	W69-03948
OLLUTION, REAERATION, NUTRIENTS,	OXYGEN SAG, AERATION, BIOCHEMICAL	W69-07520
, REAERATION, FISH, TASTE, ODOR,	OXYGEN SAG, POTOMAC RIVER.: /NITY	W73-00750
*RAYTOWN(MO), LEE'S SUMMIT(MO),	OXYGEN TRANSFER, PH, SUSPENDED SO	W70-06601
EETER-PHELPS EQUATION, DISSOLVED	OXYGEN VARIATIONS, SAN JOAQUIN RI	W69-07520
AL OXYGEN DEMAND, PHOTOSYNTHETIC	OXYGEN.: /AG, AERATION, BIOCHEMIC	W69-07520
IFORMS, ALGAL CONTROL, DISSOLVED	OXYGEN.: /AND(BOD), AERATION, COL	W70-04788
MICHIGAN, WASHINGTON, DISSOLVED	OXYGEN.: / BOTTOM SEDIMENTS, LAKE	W69-09349
HEMICAL OXYGEN DEMAND, TOXICITY,	OXYGEN-REDUCTION POTENTIAL, SOLAR	W70-05750
TUARIES, *EUTROPHICA/ *DISSOLVED	OXYGEN, *ALGAE, *RESPIRATION, *ES	W69-07520
TION/ *PHYTOPLANKTON, *DISSOLVED	OXYGEN, *ALGAE, *DIURNAL DISTRIBU	W69-03611
N DEMAND, *NUTRIENTS, *DISSOLVED	OXYGEN, *ALGAE, COLIFORMS, EUTROP	W71-05407
L OXYGEN DEMAND, *PHOTOSYNTHETIC	OXYGEN, *CHLORELLA, ALGAE, ANALYT	W69-03371
TORS, *WATER QUALITY, *DISSOLVED	OXYGEN, *OHIO, WATER TEMPERATURE,	W70-00475
SYNTH/ *RADIONUCLIDE, *DISSOLVED	OXYGEN, *RIVERS, OXIDATION, PHOTO	W71-11381
GEN ION CONCENTRATION, DISSOLVED	OXYGEN, ALGAE, PHYTOPLANKTON, ALK	W71-07731
ANIC MATTER, TOXICITY, DISSOLVED	OXYGEN, ALGAE, WISCONSIN.: /G ORG	W72-01797
SS, FLOW RATES, DEPTH, DISSOLVED	OXYGEN, ALGAE, VELOCITY, WATER QU	W70-00265
TOPLANKTON, DISTRIBUTION, DEPTH,	OXYGEN, ALKALINITY, HYDROGEN ION	W71-00665
RETRIEVAL, COLIFORMS, DISSOLVED	OXYGEN, ALGAE, NUTRIENTS, SOLUTES	W70-09557
AND, *CHLORELLA, *PHOTOSYNTHETIC	OXYGEN, ALGAE, ANALYTICAL TECHNIQ	W69-03362
MONIA, HARDNESS(WATER), ACIDITY,	OXYGEN, ALGAE, SELF-PURIFICATION,	W69-07838
DATA, STRATIFICATION, DISSOLVED	OXYGEN, ALGAE, BIOCHEMICAL OXYGEN	W70-04790
ROGEN, PHOSPHATES, METHANE, MUD,	OXYGEN, ANAEROBIC CONDITIONS, HYD	W70-05761
TER POLLUTION EFFECTS, DISSOLVED	OXYGEN, AQUATIC BACTERIA, ALGAE.:	W70-04727
RE, LIGHT PENETRATION, DISSOLVED	OXYGEN, BICARBONATES, HYDROGEN IO	W70-06225
NIQUES, WATER QUALITY, DISSOLVED	OXYGEN, BIOCHEMICAL OXYGEN DEMAND	W70-08627
WATERS, IMPOUNDMENTS, *DISSOLVED	OXYGEN, BIOCHEMICAL OXYGEN DEMAND	W72-11906
ON, ALGAE, NITROGEN, PHOSPHORUS,	OXYGEN, CARBON, ENZYMES, SAMPLING	W70-09889
URE, HYDROGEN ION CONCENTRATION,	OXYGEN, CARBON DIOXIDE, AMMONIA,	W71-03031
ELLA, MANGANESE, PHOTOSYNTHESIS,	OXYGEN, CARBON DIOXIDE, SCENEDESM	W70-02964
ANALYSIS, PHYSIOLOGICAL ECOLOGY,	OXYGEN, CARBON DIOXIDE, CARBON RA	W69-07862
N REDUCTION POTENTIAL, DISSOLVED	OXYGEN, CHEMICAL ANALYSIS, ORGANI	W70-05760
DIATOMS, CHLORELLA, SCENEDESMUS,	OXYGEN, COLIFORMS, SEDIMENTS, MUD	W70-10161
ROGEN ION CONCENTRATION, TOXINS,	OXYGEN, DECOMPOSIN: /ALINITY, HYD	W70-02803
RATION, BIOCARBONATES, DISSOLVED	OXYGEN, DETRITUS, BACTERIA, WATER	W70-02249
SERVOIRS, TEMPERATURE, DISSOLVED	OXYGEN, DENSITY, ROOTED AQUATIC P	W71-07698
FRES WATER, LITTORAL, DISSOLVED	OXYGEN, DIURNAL, MIGRATION PATTER	W71-12870
GY, ALGAE, DIURNAL DISTRIBUTION,	OXYGEN, DISSOLVED OXYGEN, MEASURE	W70-05094
*RESERVOIR OPERATIONS, DISSOLVED	OXYGEN, DISSOLVED SOLIDS, SEDIMEN	W73-00750
NITROGEN, *COLIFORMS, *DISSOLVED	OXYGEN, ESTUARIES, ALGAE, WATER P	W71-05409
*PHOTOSYNTHESIS, *ALGAE, LAKES,	OXYGEN, EUTROPHICATION, ALGAL CON	W71-10098
TERS, WATER TREATMENT, DISSOLVED	OXYGEN, EUTROPHICATION, FISH, LAN	W70-05264
TY, *DISSOLVED SOLIDS, DISSOLVED	OXYGEN, EVAPORATION CONTROL, WATE	W72-01773
LAGOONS, SLUDGE, PHOTOSYNTHETIC	OXYGEN, FARM WASTES, WASTE WATER	W71-08221
TS, WATER TEMPERATURE, DISSOLVED	OXYGEN, FISHKILLS, LIMNOLOGY, ALG	W70-00264
TRIAL WASTES, ECOLOGY, DISSOLVED	OXYGEN, HARDNESS, TIDAL WATERS, O	W73-00748

ONES, ORGANIC LOADING, DISSOLVED   OXYGEN, HYDROGEN ION CONCENTRATIO W72-10573
HT, WASHINGTON, WATER CHEMISTRY,   OXYGEN, HYDROGEN ION CONCENTRATIO W70-03309
ECTS, *EUTROPHICATION, DISSOLVED   OXYGEN, HYDROGEN ION CONCENTRATIO W70-06975
OPHYTA, DIATOMS, STRATIFICATION,   OXYGEN, HYDROGEN ION CONCENTRATIO W70-05387
PHOSPHORUS, PLANKTON, DISSOLVED   OXYGEN, HYPOLIMNION, GREAT LAKES, W68-00253
INOFLAGELLATES, ALGAE, DISSOLVED   OXYGEN, HYPOLIMNION, EPILIMNION,   W70-04253
EFFICIENCIES, DISSOLVED OXYGEN,   OXYGEN, LAKES.: /PLANKTON, ALGAE, W70-04484
, TASTE, ODOR, TOXINS, DISSOLVED   OXYGEN, LAKES, RIVERS, LIGHT PENE W69-05697
ALGAE, PHOTOSYNTHESIS, DISSOLVED   OXYGEN, LABORATORY TESTS, TEMPERA W71-12183
                  *PHOTOSYNTHETIC   OXYGEN, LEMMON(SD), KADOKA(SD).:   W72-01818
TEMPERATURE, TOXICITY, DISSOLVED   OXYGEN, LIGHT INTENSITY, SLUDGE,   W72-06854
N, CARBON, PRIMARY PRODUCTIVITY,   OXYGEN, METABOLISM, WIND VELOCITY W70-01073
DISTRIBUTION, OXYGEN, DISSOLVED   OXYGEN, MEASUREMENT, ESTUARIES, S W70-05094
ATION, ALGAE, SEWAGE, ANAEROBIC,   OXYGEN, NITRIFICATION, COLIFORMS, W70-05092
DIOXIDE, GROWTH RATES, PLANKTON,   OXYGEN, NITROGEN FIXATION, NITRAT W70-04369
LGAE, BODIES OF WATER, DISSOLVED   OXYGEN, NORTH CAROLINA.: /OIRS, A W70-00683
N-REDUCTION POTENTIAL, DISSOLVED   OXYGEN, ODORS, CORROSION, OXIDATI W71-11377
HEMICAL OXYGEN DEMAND, DISSOLVED   OXYGEN, ODOR, ORGANIC LOADING, *W W70-06601
, ALGAE, EFFICIENCIES, DISSOLVED   OXYGEN, OXYGEN, LAKES.: /PLANKTON W70-04484
OIRS, *EUTROPHICATION, NITROGEN,   OXYGEN, OXIDATION, CARBON DIOXIDE W70-05550
S, NUTRIENTS, ODOR, OLIGOTROPHY,   OXYGEN, OXYGEN DEMAND, PHOSPHORUS W69-06535
ALGICIDES, SEDIMENTS, DISSOLVED   OXYGEN, OXYGEN DEMAND, CYANOPHYTA W69-08518
ION, *EUTROPHICATION, *DISSOLVED   OXYGEN, PENNSYLVANIA, BIOCHEMICAL W69-10159
NT, SELF-PURIFICATION, DISSOLVED   OXYGEN, PHOTOSYNTHESIS.: /TREATME W72-01798
, PHOTOSYNTHESIS, PHOTOSYNTHETIC   OXYGEN, PHYSIOLOGICAL ECOLOGY, PH W70-02792
UCTIVITY, TEMPERATURE, DISSOLVED   OXYGEN, PHYTOPLANKTON, WATER ANAL W72-06051
ACTIONS, *WATER REUSE, DISSOLVED   OXYGEN, PITTING(CORROSION), RUSTI W70-02294
DIATOMS, CHLOROPHYTA, DISSOLVED   OXYGEN, PIKE, CARP, D: /S, ALGAE, W70-01943
NT REQUIREMENTS, PHOTOSYNTHETIC,   OXYGEN, PLANKTON, WATER POLLUTION W69-03514
FLAGELLATES, ALGAE, HYPOLIMNION,   OXYGEN, PRODUCTIVITY, PHOSPHATES, W69-10169
STICIDE TOXICITY, PHOTOSYNTHETIC   OXYGEN, PRODUCTIVITY, *WASTE WATE W70-08416
AME BIRDS, ANIMALS, FOOD CHAINS,   OXYGEN, ROTIFERS, WEEDS, WATER HY W70-10175
MENTS, CHLOROPHYLL, TEMPERATURE,   OXYGEN, SALINITY.: /PIRATION, PIG W70-03325
, SEWAGE EFFLUENTS, TEMPERATURE,   OXYGEN, SALINITY, ALGAE, MECHANIC W69-06203
S, *ALGAE, *BACTERIA, *DISSOLVED   OXYGEN, SEWAGE LAGOONS, PLANT PHY W72-01818
, CHLORIDES, AMMONIUM COMPOUNDS,   OXYGEN, SILICATES, ORGANIC MATTER W69-07096
ATED SLUDGE, AERATION, DISSOLVED   OXYGEN, SLUDGE, OXIDATION LAGOONS W71-08970
T PENETRATION, SEWAGE EFFLUENTS,   OXYGEN, SOLAR RADIATION, PERIPHYT W70-01933
L, LIMNOLOGY, LIGHT PENETRATION,   OXYGEN, SPORT FISHING, ON-SITE IN W68-00487
HICATION, LAKES, ORGANIC WASTES,   OXYGEN, STANDING CROP, ALGAE, ZOO W70-02251
NSITY, CARBON DIOXIDE, DISSOLVED   OXYGEN, TEMPERATURE, VELOCITY, CY W71-00668
ED SLUDGE, IRON COMPOUNDS, LIME,   OXYGEN, TERTIARY TREATMENT, COST   W68-00256
ATION, LIMNOLOGY, PHOTOSYNTHETIC   OXYGEN, TEMPERATURE, THERMAL POLL W69-03516
ENT, *DESIGN CRITERIA, DISSOLVED   OXYGEN, TEMPERATURE, BIOCHEMICAL   W70-04787
POTASSIUM, SELECTIVITY, POISONS,   OXYGEN, TRACERS, PHOSPHORUS RADIO W70-05399
ON, ENVIRONMENT, PHOTOSYNTHESIS,   OXYGEN, TROPHIC LEVEL, LIGHT INTE W70-00265
GRADATION(DECOMPOSITION), ALGAE,   OXYGENATION, SELF-PURIFICATION, L W72-01801
ALIFORNIA, ESTIMATING EQUATIONS,   OXYGENATION, STANDING CROP, MICRO W69-03611
ON, TERTIARY TREATMENT, LAGOONS,   OXYGENATION, ABSORPTION, ECOLOGY, W69-07389
RIVER(ORE), COMMUNITY STRUCTURE,   OXYTREMA, ORGANIC ENRICHMENT, INT W70-03978
ALINITY, TEMPERATURE, SEA WATER,   OYSTERS, CLAMS, CRABS, NORTH CARO W69-08267
NS, NUISANCE ALGAE, DUCKS(WILD),   OYSTERS, CHANNELS, POLLUTION ABAT W68-00461
INS, LARVAE, PHENOLS, SHELLFISH,   OYSTERS, INDUSTRIES, CHELATION, B W70-01068
*PHOSPHORUS CYCLE, GLOEOTRICHA,   P-32.:   W73-01454
OGEN, NUTRIENTS, ORGANIC MATTER,   P: /CENTRATION, TEMPERATURE, NITR W70-00265
, WATER POLLUTION SOURCES, WATER   P: /ERIA, SEDIMENTATION, COPEPODS W69-10169
              *GLUCOSE, WATERLOO MILLS(PA).:   W71-05026
SPRING CREEK(PA), STATE COLLEGE(PA), FLORA.:   W69-10159
OMMUNITY STRUCTURE, RIDLEY CREEK(PA), NOBS CREEK(MD), BACK RIVER(M W70-04510
              SPRING CREEK(PA), STATE COLLEGE(PA), FLORA.:   W69-10159
PHYTES, LIVERWORTS, MORPHOMETRY,   PACIFASTACUS LENIUSCULUS, ANACHAR W70-00711
POLLUTION ABATEMENT, ABATEMENT,   PACIFIC COAST REGION, REGIONS, GE W69-03683
LANTS, SOUTHWEST U. S., REGIONS,   PACIFIC COAST REGIONS, GEOGRAPHIC W69-03611
TMENT, *WATER POLLUTION CONTROL,   PACIFIC NORTHWEST U. S., CIVIL EN W69-03683
WATER, CHLOROPHYLL, OIL, ALGAE,   PACIFIC OCEAN.: /OTE SENSING, SEA W70-05377
CHEMISTRY, CARBON RADIOISOTOPES,   PACIFIC OCEAN, CARBON DIOXIDE, LI W70-05652
ISOTOPES, BIOTA, CLAMS, FALLOUT,   PACIFIC OCE: /ITIUM, CARBON RADIO W69-08269
STACEA, MOLLUSCA, ECHINODERMATA,   PACIFIC OCEAN.: / POLYCHAETA, CRU W69-08526
          MASS CULTURE OF ALGAE,   PACKED CELL VOLUME.:   W70-06053
TOSYNTHESIS, MEAN DOUBLING TIME,   PACKED CELL VOLUMES, BACTERIAL PL W71-05267
ERSEE(SWITZ), GREIFENSEE(SWITZ),   PAFFIKERSEE(SWITZ), AEGERISEE(SWI W70-00270
UM GROWTH TEMPERATURE, CONDENSER   PAGGAGE.: /ATURE TOLERANCE, OPTIM W71-02496
NITZSCHIA CLOSTERIUM, NITZSCHIA   PALEA, CHLOROPLASTS, SCENEDESMUS   W70-05381
AE, BACILLARIOPHYCEAE, NITZSCHIA   PALEA, MYXOPHYCEAE, CHLAMYDOMONAS W70-02965
ION, WATER POLLUTION, SEDIMENTS,   PALEOLIMNOLOGY, NUTRIENTS, PRODUC W70-05663
PLANKTON, ALGAE, AQUATIC PLANTS,   PALEOLIMNOLOGY, MATHEMATICAL MODE W70-03975
PLANKTON, ALGAE, AQUATIC PLANTS,   PALEOLIMNOLOGY, MATHEMATICAL MODE W70-03975
ALGAL GROWTH, STREAM CONCOURSE,   PALOUSE RIVER(IDAHO).: /CHARCOAL, W70-03501
OTOMAC RIVER, PLECTONEMA, NOSTOC   PALUDOSUM, BLEACHING, WASHINGTON(W70-02255
UM, GUETTARDA, IPOMOEA, PISONIA,   PANDANUS, MUSCLE, BONE, LEAVES, S W69-08269
TIS PUSILLA, OOCYSTIS SOLITARIA,   PANDORINA MORUM, PEDIASTRUM BORYA W69-06540
BIAL ECOLOGY, DINOBRYON, VOLVOX,   PANDORINA, SPIROGYRA, SYNECHOCOCC W70-04503
ADYNAMIC SUBSTANCES, PHORMIDINE,   PANDORININE,: /IC SUBSTANCES, ALG W70-02510
IRGINICA, MERCENARIA MERCENARIA,   PANOPEUS HERBSTII.: /RASSOSTREA V W69-08267
SPOSERS', MICROSCOPIC ORGANISMS,   PAPERMILL WASTE EFFLUENT, CHEESE   W71-00940
PEDIASTRUM BORYANUM, STAURASTRUM   PARADOXUM, BOTRYOCCUS BRAUNII, FR W70-02804
HANODISCUS HANTZSCHII, DIDINIUM,   PARAMECIUM, PHACUS, GLENODINIUM,   W70-00268
ICELAND) AND SUDURNES(ICELAND), NON-   PARAMETRIC TESTS, ONE-SAMPLE RUNS W69-09755
TER HYACINTH, CHEMICAL CONTROLS,   PARAQUAT, DIQUAT, SODIUM ARSENITE W70-00269
M, ZINC, NICKEL, PUERTO RICO, LA   PARGUERA, JOYUDA.: /OTEA FLABELLU W69-08525

ION, ALGAL PHYSIOLOGY, CALOTHRIX   PARIETINA, COCCOCHLORIS PENIOCYST   W69-06277
HIOCAPSA FLORIDANA, ENTEROCOCCI/   PARK RIVER: /HROMATIUM VINOSUM, T   W70-03312
R, NUCLEI/ *YELLOWSTONE NATIONAL   PARK, *ICELAND HOT SPRINGS, SINTE   W69-10160
INOSUS, PHORMIDIUM/ *YELLOWSTONE   PARK, *ICELAND, MASTIGOCLADUS LAM   W70-05270
SERVOIR, LEBANON, WADLEIGH STATE   PARK, LAKE SHATUTAKEE, MASCOMA LA   W69-08674
OLARIMETRY, YELLOWSTONE NATIONAL   PARK, MT LASSEN NATIONAL PARK, MT   W70-03309
ATIONAL PARK, MT LASSEN NATIONAL   PARK, MT RAINIER NATIONAL PARK, O   W70-03309
TIONAL PARK, MT RAINIER NATIONAL   PARK, OHANAPECOSH HOT SPRINGS(WAS   W70-03309
S, PRINCIPAL-COMPONENT ANALYSIS,   PART CORRELATION, DECAPODS, SPECI   W73-01446
, CORNELL UNIVERSITY, ELECTRONIC   PARTICLE COUNTING, EUDORINA ELEGA   W69-06540
S, ADSORPTION, SEWAGE TREATMENT,   PARTICLE SIZE.: /ELLA, SCENEDESMU   W71-03035
CHLAMYDOMONAS, LABORATORY TESTS,   PARTICLE SIZE, POPULATION, GROWTH   W72-02417
NE, GLYCOLLIC ACID, CHEMOTROPHY,   PARTICULATE MATTER, BLELHAM TARN(W70-02504
RVOIDES, NYLON COLUMN TECHNIQUE,   PARTICULATE MATTER, BLUE FLUORESC   W70-01074
ON, CELL COUNT, OPTICAL DENSITY,   PARTICULATE ORGANIC CARBON, LAKE   W70-03983
ISRAEL, MICROCYSTIS, PRYMNESIUM   PARVUM, SYNERGISTIC EFFECTS, AXEN   W70-05372
EA LAKE(COLO), BLACK LAKE(COLO),   PASS LAKE(COLO), MICRO-ALGAE, FLA   W69-10154
WASTES, WATER POLLUTION SOURCES,   PATH OF POLLUTANTS, AQUATIC LIFE,   W69-10080
ACTIVE WASTE, WATER QUALITY ACT,   PATH OF POLLUTANT, FOOD CHAINS, S   W72-00941
AKE ONTARIO, *WATER CIRCULATION,   PATH OF POLLUTANTS, NUTRIENTS, AL   W71-05878
ITIONS, WATER POLLUTION SOURCES,   PATH OF POLLUTANT.: /AEROBIC COND   W71-11381
ACTERIA, E. COLI, PUBLIC HEALTH,   PATHOGENIC BACTERIA, TEMPERATURE,   W72-06017
*EPIPHYTOLOGY, NUTRIENTS, PLANT   PATHOLOGY, *ALGAE.: /Y, MICHIGAN,   W71-00832
EAR LAKE(CALIF), NUTRIENT ENERGY   PATHWAYS, TROPHIC RELATIONSHIPS.:   W70-05428
BOTRYOCOCCUS BRAUNII, METABOLIC   PATTERN, ANABAENA CYLINDRICA, GLY   W70-02965
OLVED OXYGEN, DIURNAL, MIGRATION   PATTERNS.: /WATER, LITTORAL, DISS   W71-12870
SOIL MICROBIOLOGY, DISTRIBUTION   PATTERNS, SOIL PROFILES.: /OZEMS,   W71-04067
TION, SOILS, COALS, DISTRIBUTION   PATTERNS, ACTIVATED CARBON, CLAYS   W69-09884
ASS CULTURE, MEDICINE, METABOLIC   PATTERNS, MICROALGAE, MICRONUTRIE   W69-07832
*LAKES, *CELLULOSE, DISTRIBUTION   PATTERNS, MEMBRANE PROCESSES, MAS   W69-00976
MINNETONKA(MINN), MINNEAPOLIS-ST   PAUL(MINN).:   LAKE   W70-04810
*CANNERIES, *INDUSTRIAL WASTES,   PEACHES, APRICOTS, PEARS, CITRUS   W71-09548
COSTS, DINOBRYON, CONCORD(N H),   PEANACOOK LAKE(N H).: / TREATMENT   W69-08282
TRIAL WASTES, PEACHES, APRICOTS,   PEARS, CITRUS FRUITS, TOMATOES, *   W71-09548
NIC/ ONTARIO LAKES, QUIRKE LAKE,   PECORS LAKE, DUNLOP LAKE, *INORGA   W70-05424
LIMNOLOGY, BRACHIONUS, SYNCHAETA   PECTINATA, POLYARTHRA VULGARIS, K   W70-02249
HOLSTON RIVER(TENN), POTOMOGETON   PECTINATUS, POTOMOGETON CRISPUS,   W71-07698
ON, PONDS, ZOOPLANKTON, MINNOWS,   PEDALFERS, SCENEDESMUS, CHLAMYDOM   W70-02249
MOTROPHIC, AUTOTROPHIC, SILICON,   PEDIASTRUM BORYANUM, STAURASTRUM   W70-02804
LOSIRA, ACTINASTRUM, COELASTRUM,   PEDIASTRUM, WESTELLA, STEPHANODIS   W70-08107
INA, CULTURE MEDIUM, COELASTRUM,   PEDIASTRUM.: /CTION VALUES, EUDOR   W70-05409
STIS SOLITARIA, PANDORINA MORUM,   PEDIASTRUM BORYA: / PUSILLA, OOCY   W69-06540
IUM, ANKISTRODESMUS, MICROSPORA,   PEDIASTRUM, SCENEDESMUS, LEPOCINC   W70-03974
UCCESSION, ANABAENA, CLADOPHORA,   PEDIASTRUM, HYDRODICTYON, POTAMOG   W70-04494
LGAL CULTURES, *AXENIC CULTURES,   PEDIASTRUM, SELF-PURIFICATION, AL   W70-04381
IA, MICROCYSTIS, ANKISTRODESMUS,   PEDIASTRUM, TETRAEDRON, TETRASTRU   W70-03969
, GELATINOU/ *LAKE MENDOTA(WIS),   PELAGIC AREAS, HORIZONTAL, BLOOMS   W70-06229
NT CAROTENOIDS, CARBON FIXATION,   PELAGIC REGION, MONOMICTIC, DEPTH   W70-01933
, MONODUS SUBTERRANEUS, NAVICULA   PELLICULOSA, XANTHOPHYCEAE, BACIL   W70-02965
OSA, TRIBONEMA AEQUALE, NAVICULA   PELLICULOSA, LAKE MENDOTA(WIS), E   W69-10180
S, MONOCHRYSIS LUTHERI, NAVICULA   PELLICULOSA, GONYAULAX POLYEDRA,   W69-08278
UAEDA MARITIMA, ASTER TRIPOLIUM,   PEMBROKE, SOUTH-WEST WALES.: /, S   W70-09976
HIBITION, LIGHT INTENSITY, LIGHT   PENETRAT: /TS, EUTROPHICATION, IN   W69-03515
N, LIGHT, LIGHT INTENSITY, LIGHT   PENETRATION, LIMNOLOGY, PHOTOSYNT   W69-03516
HIBITION, LIGHT INTENSITY, LIGHT   PENETRATION, LIMNOLOGY, NITROGEN   W69-03514
TING OF ALGAE, INHIBITION, LIGHT   PENETRATION, NITROGEN COMP: /RVES   W69-03369
FREEZING, LIGHT INTENSITY, LIGHT   PENETRATION, LIMNOLOGY, NITROGEN   W69-03518
VESTING ALGAE, INHIBITION, LIGHT   PENETRATION, LIMNOLOGY, NITROGEN   W69-04798
N, LIGHT, LIGHT INTENSITY, LIGHT   PENETRATION, LIMNOLOGY, NUTRIENT   W69-03517
KALINITY, *PHOTOSYNTHESIS, LIGHT   PENETRATION, PHOSPHORUS, LAGOONS: W68-00248
LGAE, SEASONAL, LIMNOLOGY, LIGHT   PENETRATION, OXYGEN, SPORT FISHIN   W68-00487
VED OXYGEN, LAKES, RIVERS, LIGHT   PENETRATION, FLUORESCENCE, PHOTOS   W69-05697
INTENSITY, STABILIZATION, PONDS,   PENETRATION, ABSORPTION.: /LIGHT   W70-03923
HICATION, LIGHT INTENSITY, LIGHT   PENETRATION, LIMNOLOGY, NITROGEN   W70-04283
FICATION, DEPTH, PLANKTON, LIGHT   PENETRATION, SEASONAL, PRIMARY PR   W70-05405
VOIRS, PONDS, TEMPERATURE, LIGHT   PENETRATION, DISSOLVED OXYGEN, BI   W70-06225
PERATURE, IRON, NUTRIENTS, LIGHT   PENETRATION, SEWAGE EFFLUENTS, OX   W70-01933
FIXATION, IRON COMPOUNDS, LIGHT   PENETRATION, WATER POLLUTION EFFE   W70-02255
, EPILIMNION, THERMOCLINE, LIGHT   PENETRATION, STANDING CROP, INDUS   W71-07698
INTENSITY, PHOTOPERIODISM, LIGHT   PENETRATION, WATER POLLUTION EFFE   W72-00917
POLLUTION SOURCES, RUNOFF, LIGHT   PENETRATION, SAMPLING, PRIMARY PR   W72-04761
TH DAKOTA, PHOTOSYNTHESIS, LIGHT   PENETRATION, ON-SITE TESTS, BIOCH   W72-01818
ND(CANADA), NOVA SCOTIA(CANADA),   PENGUINS, LAKE ZURICH(SWITZERLAND   W70-05399
ALOTHRIX PARIETINA, COCCOCHLORIS   PENIOCYSTIS, DIPLOCYSTIS AERUGINO   W69-06277
TRICHIA ECHINULATA, COCCOCHLORIS   PENIOCYSTIS, MADISON(WIS), CHICAG   W70-00719
TROPHICATION, *DISSOLVED OXYGEN,   PENNSYLVANIA, BIOCHEMICAL OXYGEN   W69-10159
EFFECTS, WASTE WATER TREATMENT,   PENNSYLVANIA.: /, WATER POLLUTION   W71-05026
ILLUS DENITRIFICANS, W VIRGINIA,   PENNSYLVANIA, MARCASITE, CHALCOCI   W69-07428
LGAE, PHYTOPLANKTON, OHIO RIVER,   PENNSYLVANIA, WATER SUPPLY.: /N A   W69-05697
ER POLLUTION, DIATOMS, MARYLAND,   PENNSYLVANIA, RIVERS, STREAMS, EU   W70-04510
NCE ALGAE, CYANOPHYTA, VITAMINS,   PEPTIDES, ORGANIC MATTER, TRACE E   W70-04369
VITAMINS, CELLULOSE, CHLORELLA,   PEPTIDES, BACTERIA, SNAILS, WEIGH   W70-04184
ERBIVORES, CHLOROPHYLL, ENZUMES,   PEPTIDES, BIOCHEMISTRY, HYDROGEN   W70-01068
PROPERTIES, GENETICS, BACTERIA,   PEPTIDES, ENVIRONMENTAL EFFECTS,   W70-00273
COMPOUNDS, PHOSPHORUS COMPOUND,   PERCHES, FISH, DIATOMS, CARP, PLA   W70-00274
PHOSPHORUS, DETERGENTS, RUNOFF,   PERCOLATION, ROADS, ROOFS, PRECIP   W69-10178
ANT RESIDUES, MORRIS(MINN), SOIL   PERCOLATION.: / SOIL MINERALS, PL   W70-04193
CLOTELLA, CERATIUM HIRUNDINELLA,   PERIDINIUM WILLEI, CHLOROCOCCALES   W70-05405
OCCOLITHUS HUXLEYI, PYRAMIMONAS,   PERIDINIUM TROCHOIDEUM, VINEYARD   W70-05272
LLATORIA RUBESCENS, SWITZERLAND,   PERIDINIUM, OSCILLATORIA AGARDHI,   W69-10169

DICATORS, *ESTIMATING EQUATIONS,
IA/ LAKE SEBASTICOOK, ME, MAINE,
R), WATER QUALITY INDICATORS.:
, RIVERS, ENVIRONMENTAL EFFECTS,
TS, ECOSYSTEMS, PLANKTON, ALGAE,
 TEMPERATURE, TRACERS, SAMPLING,
BIOMASS, WATER POLLUTION, ALGAE,
TER POLLUTION EFFECTS, BACTERIA,
RAULIC MODELS, LABORATORY TESTS,
LUENTS, OXYGEN, SOLAR RADIATION,
CTIVITY, *ALGAE, *THERMAL WATER,
BIOMASS, PIGMENTS, PRODUCTIVITY,
).:   *POTASSIUM
JAS, SILVEX, ENDOTHAL, POTASSIUM
                          *POTASSIUM
                          *POTASSIUM
GISTATIC TESTS, SLIM/ *POTASSIUM
 ALGAE, PILOT PLANTS, HEAD LOSS,
N(SWEDEN), LAKE LOTSJON(SWEDEN),
 CHLOROPHYLL FLUORESCENCE, FABRY-
, BALL-MILL, MEICELASE, HYDROGEN
TABILIZATION POND LOADINGS, ODOR
ISPERSION, OIL-WATER INTERFACES,
ARIO WATER RESOURCES COMMISSION,

ATHION, / CHLORELLA PYRENOIDOSA,
HIBITION, LIMNOLOGY, PESTICIDES,
, PESTICIDES, PESTICIDE REMOVAL,
ICITY, FRESH WATER, PULP WASTES,
LAKES, METABOLISM, *CHLORELLA, /
YNTHESIS, CHLORELLA, PESTICIDES,
TES, ALGAE, CALIFORNIA, FLORIDA,
       COAL CONTACT PROCESSES,
ISTANCE, BIOASSAY, GROWTH RATES,
STES, THERMAL POLLUTION, TOXINS,
IOINDICATORS, FISHKILL, HAZARDS,
LL, HAZARDS, PESTICIDE KINETICS,

ION, AQUATIC PLANTS, HERBICIDES,
ION, AQUATIC PLANTS, HERBICIDES,
BS, DDT, CHLORINATED HYDROCARBON
AND, VALUE, LEAVES, FERTILIZERS,
SH TAXONOMY, DDT, PHOSPHOTHIOATE
RONMENT, CHLORINATED HYDROCARBON
ION, *PHOTOSYNTHESIS, CHLORELLA,
ATE, FISHES, HERBICIDES, MIDGES,
 EFFECTS, INHIBITION, LIMNOLOGY,
ANIC COMPOUNDS, LAGOON, FILTERS,
TY, ORGANIC COMPOUNDS, *PHENOLS,
ES, MUNICIPAL WASTES, NUTRIENTS,
SH, AQUATIC ENVIRONMENT, ORGANIC
ERATURE, CHLORINATED HYDROCARBON
, ORGANIC MATTER, RADIOACTIVITY,
ENARIA, AQUACULTURE, RECYCLING /
ALKENES, ALIPHATIC HYDROCARBONS,
MS, MARINE ASSOCIATIONS, MOSCOW,
NG POND(WIS), LAKE DELAVAN(WIS),
(WIS), LAKE NAGAWICKA(WIS), LAKE
CY, LAKE VANERN, LAKE CONSTANCE,
OLIDS, MULTI-CELL SERIAL SYSTEM,
ER ANALYSIS, *CHEMICAL ANALYSIS,
E'S SUMMIT(MO), OXYGEN TRANSFER,
UTRIENTS, PHYSIOLOGICAL ECOLOGY,
ANTZSCHII, DIDINIUM, PARAMECIUM,
TIFICATION, SURFACE AREA(LAKES),
 CHEMICAL RESISTANCE, EUPHAUSIA,
ORREY CANYON, AROMATIC SOLVENTS,
OREALIS, LEPTOCYLINDRUS DANICUS,
IATOMS, ECOLOGY, NUISANCE ALGAE,
 MATTER, BIOMASS, BENTHIC FLORA,
N, LOCH EWE(SCOTLAND), SCOTLAND,
IRONMENT, *PHYSICAL ENVIRONMENT,
OCOLLOIDS, TANNINS, FLAGELLATES,
L EXTRACTION, MILLIPORE FILTERS,
ODUCTIVITY, SULFIDES, SEA WATER,
 VARIATIONS, SAN JOAQU/ STREETER-
POUNDS, MINIMUM LETHAL EXPOSURE,
 *ALKYL POLY ETHOXYLATE, *ALKYL
S, *ALKYL POLYETHOXYLATE, *ALKYL
, QUATERNARY AMMONIUM COMPOUNDS,
URE, COLORIMETRY, CARBOHYDRATES,
NDS, ACTIVATED CARBON, CHLORINE,
PLING, ORGANIC MATTER, TOXICITY,
MIC ACIDS, *COLOR, RIVERS, BOGS,
TERICIDES, ALGAL TOXINS, LARVAE,
STANCES, ALGADYNAMIC SUBSTANCES,
 HOT SPRINGS(WASH), SCHIZOTHRIX,
DIAPTOMUS, CYCLOPS, GOMPHONEMA,
RIANS, SYNEDRA ULNA, OEDOGONIUM,
ERIDINIUM, OSCILLATORIA AGARDHI,

PERIOD OF GROWTH.: / ISLES, BIOIN   W68-00255
PERIODIC VARIATIONS, SEASONAL VAR   W68-00481
PERIPHYTON INVENTORY(LAKE SUPERIO   W70-05958
PERIPHYTON, WATER POLLUTION EFFEC   W70-03978
PERIPHYTON, INVERTEBRATES, ADSORP   W69-07853
PERIPHYTON, AQUATIC PLANTS, BACKG   W69-09334
PERIPHYTON, PHOSPHORUS, HYDROGEN   W71-04206
PERIPHYTON, OHIO.: /R, CARBON, WA   W71-04206
PERIPHYTON, STREAMS, PHOTOSYNTHES   W71-00668
PERIPHYTON, BACTERIA, SPECT: /EFF   W70-01933
PERIPHYTON, BIOMASS, PHYSIOLOGICA   W70-03309
PERIPHYTON, CHLOROPHYTA, CHRYSOPH   W70-02780
PERMANGANATE, BEVERLY HILLS(CALIF   W71-00110
PERMANGANATE, FENAC, SILVER NITRA   W70-10175
PERMANGANATE, *DES MOINES(IOWA).:   W72-04242
PERMANGANATE.:   W72-04155
PERMANGANATE, ALGICIDAL TESTS, AL   W69-05704
PERMEABILITY, TEMPERATURE, OPERAT   W72-08686
PERMEASES, CHAMYDOMONADS, LAPPLAN   W70-04287
PEROT FILTERS, SOLAR FRAUNHOFER L   W70-05377
PEROXIDE, RATS, TRYPSIN, SCENEDES   W70-04184
PERSISTENCE.: /ORADO, WEST(US), S   W70-04788
PERSISTENCE, WATER POLLUTION TREA   W70-01777
PEST CONTROL ORGANIZATION OF TORO   W69-10155
PESTICIDE SORPTION.:   W69-09884
PESTICIDE DEGRADATION, SEVIN, MAL   W70-02968
PESTICIDE REMOVAL, PESTICIDE RESI   W70-02968
PESTICIDE RESIDUES, PHYSIOLOGICAL   W70-02968
PESTICIDE RESIDUES, HERBICIDES, E   W70-10437
PESTICIDE RESIDUES, ALGAE, GREAT   W70-08652
PESTICIDE TOXICITY, PHOTOSYNTHETI   W70-08416
PESTICIDE RESIDUES, ECOSYSTEMS, F   W70-05272
PESTICIDE REMOVAL.:   W70-05470
PESTICIDE RESIDUES, DETERIORATION   W70-03520
PESTICIDE RESIDUES, PHOSPHATES, A   W69-06305
PESTICIDE KINETICS, PESTICIDE TOX   W69-08565
PESTICIDE TOXICITY, POLLUTANTS, W   W69-08565
PESTICIDE BIOACCUMULATION.:   W69-08565
PESTICIDES, HARVESTING, WATER QUA   W69-05706
PESTICIDES, HARVESTING, WATER QUA   W69-05705
PESTICIDES, ALDRIN, ENDRIN, DIELD   W70-03520
PESTICIDES, CELLULOSE, CYCLES, HA   W70-05264
PESTICIDES, HERBICIDES, *PESTICID   W69-02782
PESTICIDES, PLANT PHYSIOLOGY.: /I   W70-08652
PESTICIDES, PESTICIDE TOXICITY, P   W70-08416
PESTICIDES.: /OLYSIS, COPPER SULF   W70-02982
PESTICIDES, PESTICIDE REMOVAL, PE   W70-02968
PESTICIDES, SLUDGE, DIGESTION, IN   W72-03641
PESTICIDES, OXIDATION LAGOONS, AL   W71-12183
PESTICIDES, RECREATION, ALGAE, OX   W71-09577
PESTICIDES, HYDROGEN ION CONCENTR   W72-11105
PESTICIDES, ALGAE, LABORATORY TES   W72-11105
PESTICIDES, TASTE, ODOR, COAGULAT   W72-13601
PETROLEUM WASTES, MERCENARIA MERC   W70-09905
PETROLEUM CRUDES, MASS SPECTROMET   W69-08284
PETROLEUM POLLUTION, PHOTOSYNTHET   W69-07440
PEWAUKEE LAKE(WIS), NORTH LAKE(WI   W69-05867
PEWAUKEE(WIS), CHIRONOMUS, PROCLA   W70-06217
PFAFFIKERSEE, TURLERSEE, BALDEGGE   W70-03964
PH, FACULTATIVE POND, ANAEROBIC P   W70-04790
PH, SAMPLING, ON-SITE DATA COLLEC   W69-00659
PH, SUSPENDED SOLIDS.: /N(MO), LE   W70-06601
PH: /ON, NUTRIENT REQUIREMENTS, N   W69-03364
PHACUS, GLENODINIUM, CRYPTOMONAS,   W70-00268
PHAEDACTYLUM, WASTE WATER DIVERSI   W69-09340
PHAEOCYSTIS, SARGASSUM, FRONDS, H   W70-01068
PHAEODACTYLUM TRICORNUTUM.: /, *T   W70-01470
PHAEODACTYLUM TRICORNUTUM, STICHO   W70-04280
PHAEOPHYTA, PHOTOSYNTHESIS, PHYTO   W69-07832
PHAEOPHYTA, POLAR REGIONS, SPECTR   W70-01074
PHAEOPIGMENTS, ACRIDINE ORANGE, L   W70-04811
PHAGE, FERMENTOR, FACULTATIVE, MI   W71-02009
PHARMACEUTICALS, EXTRACELLULAR PR   W70-01068
PHASE MICROSCOPY, STAINING, STRIP   W69-10163
PHAWOPHYTA, CRUSTACEANS, HERBIVOR   W70-01068
PHELPS EQUATION, DISSOLVED OXYGEN   W69-07520
PHENANTHROQUINONE, MADISON(WIS),   W70-02982
PHENOL POLYETHOXYLATE, ANIONIC SU   W70-09438
PHENOL POLYETHOXYLATE, ORGANIC EX   W70-03928
PHENOLIC COMPOUNDS, ORGANO-METALL   W72-12987
PHENOLS, NITROGEN, RESPIRATION, C   W70-01073
PHENOLS, OIL WASTES, HYDROGEN ION   W70-00263
PHENOLS, ODOR, COLOR, TURBIDITY,   W72-03299
PHENOLS, PIGMENTS, ORGANIC MATTER   W70-01074
PHENOLS, SHELLFISH, OYSTERS, INDU   W70-01068
PHORMIDINE, PANDORININE,: /IC SUB   W70-02510
PHORMIDIUM, WINKLER TECHNIQUE, OX   W70-03309
PHORMIDIUM, APHANIZOMENON FLOS-AQ   W70-02803
PHORMIDIUM RETZII, SURFACE DIFFUS   W70-00265
PHORMIDIUM, APHANIZOMENON FLOS-AQ   W69-10169

496

CELAND, MASTIGOCLADUS LAMINOSUS,
'S BAY(MINN), SMITH'S BAY(MINN),
MICROCYSTIS SCUMS, PITHIOPHORA/
ES, SOLUBLE REACTIVE PHOSPHORUS,
ASTES, CARBON CYCLE, PHOSPHORUS,
OLVED SOLIDS, CHEMICAL ANALYSIS,
NOPHYTA, DIFFUSION, RESPIRATION,
UTANTS, ALGAL CONTROL, NITRATES,
RIES, FRESH WATER, LAKES, SILTS,
UTRIENTS, *PRIMARY PRODUCTIVITY,
, *PEPTIDES, COPPER, IRON, IONS,
L, DIURNAL, EFFLUENTS, NITROGEN,
POLIMNION, OXYGEN, PRODUCTIVITY,
ERTILIZERS, NUTRIENTS, NITRATES,
HESIS, CARBON DIOXIDE, NITRATES,
AKE MICHIGAN, WATER TEMPERATURE,
, ALGAE, GROUND WATER, NITRATES,
, SEWAGE, ALGAE, MICROORGANISMS,
NITRITES, NITRATES, DETERGENTS,
CHLORIDES, CALCIUM, CARBONATES,
FILTERS, WASTE WATER TREATMENT,
S, RECIRCULATED WATER, NITRATES,
NTERFERENCE, DISSOLVED INORGANIC
ITY, CALCIUM, NITRATES, AMMONIA,
COMPOUNDS, DIATOMS, CYANOPHYTA,
GEN ION CONCENTRATION, NITRATES,
HEMICAL OXYGEN DEMAND, NITROGEN,
IDE, HYDROGEN ION CONCENTRATION,
CATES, DIATOMS, WINDS, NITROGEN,
ITY, STANDING CROP, CHLOROPHYLL,
TION, IRON, MAGNESIUM, NITRATES,
C ALGAE, *BIOASSAY, SCENEDESMUS,
ER/ *DUTCH LAKES, *ALLOCHTHONOUS
S, *PHOSPHORUS COMPOUNDS, ALGAE,
ITY, SNAILS, SULFATES, NITRATES,
ATION, NUISANCE ALGAE, NITROGEN,
ES, SALMON, AQUATIC ENVIRONMENT,
*LAKES, NUTRIENTS, FERTILIZERS,
ION, TOXINS, PESTICIDE RESIDUES,
SODIUM, IRON, DISSOLVED SOILDS,
UTRIENTS, NUTRIENTS, PHOSPHORUS,
IOGRAPHIES, NITRATES, NUTRIENTS,
ION SOURCES, NITROGEN COMPOUNDS,
UTION ABATEMENT, NUISANCE ALGAE,
OPHICATION, SEWAGE, *PHOSPHORUS,
TY, TEX, POLYPHOSPHATES, ORGANIC
TROGEN CYCLE, NITROGEN FIXATION,
E, AQUATIC VEGETATION, NITROGEN,
NITROGEN, WASTE WATER TREATMENT,
ON, *ALGAE, NUTRIENTS, NITROGEN,
*NUTRIENTS, NITROGEN COMPOUNDS,
SPIRATION, BATHYMETRY, SALINITY,
L PROPERTIES, CLIMATE, NITROGEN,
TION, ALGAE, PLANKTON, NITROGEN,
IVITY, POISONS, OXYGEN, TRACERS,
WATER POLLUTION SOURCES, SEWAGE,
ORES, SEWAGE EFFLUENTS, DIATOMS,
TIDE, SALINITY, FLORIDA, TEXAS,
COMPOUNDS, SOLVENT EXTRACTIONS,
SIN, SEWAGE EFFLUENTS, NITROGEN,
*LAKES, PRODUCTIVITY, NITROGEN,
NIC PHOSPHATES, SOLUBLE REACTIVE
WASTES, POLLU/ *EUTROPHICATION,
EMENT, COLIFORMS, WATER QUALITY,
WEED CONTROL, WATER PROPERTIES,
OTOSYNTHESIS, LIGHT PENETRATION,
OPHICATION, NUTRIENTS, NITROGEN,
AKES, *NUTRIENTS, UNITED STATES,
PULATIONS, ROUGH FISH, NITROGEN,
OORGANISMS, NITROGEN, NUTRIENTS,
ION, NUTRIENTS, ALGAE, NITROGEN,
HOLYTIC, NITROGEN ANNUAL BUDGET,
WATER CHEMISTRY, WATER ANALYSIS,
UTRIENT REQUIREMENTS, NUTRIENTS,
, NITROGEN FIXATION, PHOSPHATES,
TION, NUISANCE ALGAE, NUTRIENTS,
, *VIRGINIA, NITROGEN COMPOUNDS,
N SAG, WATER QUALITY, CHLORIDES,
YANOPHYTA, *INORGANIC COMPOUNDS,
OISOTOPE, *TURNOVERS, NUTRIENTS,
AE, *HYDROGEN ION CONCENTRATION,
TRIENTS, LAKES, TESTS, NITROGEN,
UIREMENTS, PHOSPHORUS COMPOUNDS,
ESSENTIAL NUTRIENTS, NUTRIENTS,
EN COMPOUNDS, NITROGEN-FIXATION,
OMPOUNDS, NUTRIENT REQUIREMENTS,
                          EXTRACTABLE
TER POLLUTION, *DENITRIFICATION,
UTRIENT REQUIREMENTS, NUTRIENTS,
ATION, NITROGEN, NUISANCE ALGAE,

PHORMIDIUM LAMINOSUM, OSCILLATORI	W70-05270
PHOSPHATE DETERMINATION: /), GRAY	W70-05269
PHOSPHATE FERTILIZERS, BLUEGILLS,	W70-05417
PHOSPHATE(BIOLOGICALLY AVAILABLE)	W70-05269
PHOSPHATE, NITRATES, INDUSTRIAL W	W70-02777
PHOSPHATES, IRON COMPOUNDS, HYDRO	W70-03334
PHOSPHATES, ECOSYSTEMS, ORGANIC M	W70-03309
PHOSPHATES, SULFATES, SEASONAL, L	W70-02646
PHOSPHATES, NITROGEN, WASTE WATER	W70-02255
PHOSPHATES, SILICATES, CITRATES,	W70-01579
PHOSPHATES, ORGANIC COMPOUNDS, TO	W69-10180
PHOSPHATES, ALKALINITY, FISH, STR	W69-10159
PHOSPHATES, CHLOROPHYLL, EPILIMNI	W69-10169
PHOSPHATES, PHYTOPLANKTON, IRON C	W69-10170
PHOSPHATES.: /OMPOUNDS, PHOTOSYNT	W70-00161
PHOSPHATES, TURBIDITY.: /TARIO, L	W70-00667
PHOSPHATES, LAS VEGAS BAY.: /OIRS	W72-02817
PHOSPHATES, HYDROGENATION, AERATI	W70-08627
PHOSPHATES.: /ERIS, FUNGI, ALGAE,	W71-00940
PHOSPHATES.: /NUTRIENTS, SOLUTES,	W70-09557
PHOSPHATES, NITRATES, ALGAE.: /NG	W70-06792
PHOSPHATES, SODIUM, WATER POLLUTI	W70-07257
PHOSPHATES, SOLUBLE REACTIVE PHOS	W70-05269
PHOSPHATES, HYDROGEN ION CONCENTR	W70-05405
PHOSPHATES, CHLOROPHYLL, MAGNESIU	W70-05409
PHOSPHATES, CYANOPHYTA, OXIDATION	W70-05404
PHOSPHATES, BIOMASS, SYMBIOSIS, O	W70-05655
PHOSPHATES, NITRATES, DAPHNIA, SI	W70-05760
PHOSPHATES, METHANE, MUD, OXYGEN,	W70-05761
PHOSPHATES, ADD: /CIUM, PRODUCTIV	W70-03512
PHOSPHATES, ALGAE, DETRITUS, PULP	W70-03501
PHOSPHATES, WATER POLLUTION SOURC	W70-03955
PHOSPHATES, ULTRA-VIOLET LIGHT ST	W70-03955
PHOSPHATES, NITRATES, SEWAGE EFFL	W70-04074
PHOSPHATES, POTASSIUM, IRON, BICA	W70-03978
PHOSPHATES, HYDROGEN ION CONCENTR	W70-04381
PHOSPHATES, ALGAE, DAPHNIA, GROWT	W69-06536
PHOSPHATES, NITRATES, ALGAE, GREA	W69-07084
PHOSPHATES, ALGAE, SEWAGE, WATER	W69-06305
PHOSPHATES, NITROGEN, ALGAE, COLI	W69-05894
PHOSPHATES, PHYSIOLOGICAL ECOLOGY	W69-03373
PHOSPHATES, SILICATES, IRON OXIDE	W68-00860
PHOSPHATES, GREAT LAKES, MICHIGAN	W69-01445
PHOSPHATES, NITRATES, RIVERS, SEW	W69-01453
PHOSPHATES, *SEWAGE TREATMENT, DO	W68-00256
PHOSPHATES.: HARRIS COUN	W69-04800
PHOSPHATES, PHOSPHORUS, PRODUCTIV	W69-04805
PHOSPHOROUS, AVAILABLE NUTRIENTS,	W68-00511
PHOSPHOROUS, RUNOFF, DOMESTIC WAS	W71-04072
PHOSPHOROUS, NITROGEN FIXING BACT	W71-13335
PHOSPHOROUS COMPOUNDS, HYDROGEN I	W72-06051
PHOSPHORU: /, WATER CHEMISTRY, RE	W70-05387
PHOSPHORUS.: /TILIZATION, PHYSICA	W70-05565
PHOSPHORUS, POTASSIUM, SELECTIVIT	W70-05399
PHOSPHORUS RADIOISOTOPES, IONS, H	W70-05399
PHOSPHORUS, CHEMCONTROL, INDUSTRI	W70-05668
PHOSPHORUS, ORGANIC MATTER, NITRO	W70-05663
PHOSPHORUS, MIGRATION, DIURNAL, A	W70-05372
PHOSPHORUS COMPOUNDS, LAKES, SEWA	W70-05269
PHOSPHORUS.: / PROPERTIES, WISCON	W70-05113
PHOSPHORUS, DRAINAGE, MARSHES, FA	W70-05112
PHOSPHORUS, PHOSPHATE(BIOLOGICALL	W70-05269
PHOSPHORUS, *DETERGENTS, DOMESTIC	W70-05084
PHOSPHORUS, NITROGEN, FLOW, CHEMI	W70-04810
PHOSPHORUS, CHEMICAL ANALYSIS.: /	W68-00479
PHOSPHORUS, LAGOONS: /LINITY, *PH	W68-00248
PHOSPHORUS, PLANKTON, DISSOLVED O	W68-00253
PHOSPHORUS, NITROGEN, SEWAGE, IND	W68-00478
PHOSPHORUS, DIATOMS, *ALGAE, CHLO	W68-00172
PHOSPHORUS, PLANKTON, PLANTS(ORGA	W68-00481
PHOSPHORUS, BICARBONATES, *AQUATI	W68-00468
PHOSPHORUS ANNUAL BUDGET.: / TROP	W69-05142
PHOSPHORUS, PONDS, PHYTOPLANKTON,	W69-05142
PHOSPHORUS COMPOUNDS,: /POUNDS, N	W69-04798
PHOSPHORUS, PRODUCTIVITY, STANDIN	W69-04805
PHOSPHORUS COMPOUNDS, PHYSIOLOGIC	W69-04802
PHOSPHORUS, NUISANCE ALGAE, SALTS	W69-03695
PHOSPHORUS, SULFATES, HARDNESS(WA	W69-03948
PHOSPHORUS, BIBLIOGRAPHIES, NITRA	W68-00860
PHOSPHORUS, INFLUENT SEWAGE, EQUI	W68-00858
PHOSPHORUS, NITROGEN, GROWTH RATE	W68-00855
PHOSPHORUS, CALCIUM, MAGNESIUM, S	W68-00856
PHOSPHORUS, PHYTOPLANKTON.: / REQ	W69-03358
PHOSPHORUS, PHOSPHATES, PHYSIOLOG	W69-03373
PHOSPHORUS COMP: /MNOLOGY, NITROG	W69-03186
PHOSPHORUS COMPOUNDS, PHOSPHORUS,	W69-03358
PHOSPHORUS.:	W69-03075
PHOSPHORUS COMPOUNDS, ALGAE, AQUA	W69-05323
PHOSPHORUS COMPOUNDS, WATER CHEMI	W69-05868
PHOSPHORUS, WASHINGTON, WISCONSIN	W69-06858

OSYSTEM MODEL, SIMULATION MODEL,
 INFERFACE, MUD–WATER INTERFACE,
GOTROPHY, OXYGEN, OXYGEN DEMAND,
 WATER POLLUTION EFFECTS, LAKES,
 ION, DOMESTIC WASTES, NUTRIENTS,
 IOTA, *COLUMBIA RIVER, CHROMIUM,
 , DOCUMENTATION, BIBLIOGRAPHIES,
 S, FERTILIZERS, ALGAE, NITROGEN,
 LITY, WATER POLLUTION, NITROGEN,
 WATER, SORPTION, SODIUM, ALGAE,
 OL, ARSENIC COMPOUNDS, NITROGEN,
 VATION, WAT/ *NUTRIENT ELEMENTS,
 N, *ALGAE, *NUTRIENTS, NITROGEN,
 DUCTIVITY, ECOSYSTEMS, ILLINOIS,
 TER POLLUTION EFFECTS, NITROGEN,
 NEFIT–COST ANALYSIS, *LAKE ERIE,
 EN ION CONCENTRATION, NUTRIENTS,
 MINNESOTA, WISCONSIN, MICHIGAN,
 ROUS FORESTS, DECIDUOUS FORESTS,
 ECTICUT, EUGLENA, CHLAMYDOMONAS,
 CHLORELLA, *SCENEDESMUS, *LIGHT,
 UTRIENT REQUIREMENTS, NUTRIENTS,
 TRIENT INFLUX, NUTRIENT SOURCES,
 S, ZO/ *BENTHIC FAUNA, NITROGEN,
 ENTS, *EUTROPHICATION, NITROGEN,
 *CYCLING NUTRIENTS, ECOSYSTEMS,
 NEDESMUS, CHLOROPHYTA, BACTERIA,
 OSYNTHESIS, L/ *ALGAE, NITROGEN,
 S, MICHIGAN, BICARBONATES, IRON,
 EN, NITROGEN CYCLE, *PHOSPHORUS,
 E TREATMENT, NITROGEN COMPOUNDS,
 YLL, CYANOPHYTA, AQUATIC PLANTS,
 ALGAE, WATER QUALITY, NUTRIENTS,
 , RUNOFF, FERTILIZERS, NITROGEN,
 IFEROUS FORESTS, ALGAE, INSECTS,
 RING, BACTERIA, WATER CHEMISTRY,
 ALGAE, CHLOROPHYLL, HYPOLIMNION,
 TECHNIQUES, NUTRIENTS, NITROGEN,
 ER POLLUTION, ALGAE, PERIPHYTON,
 LIMNION, ANIMALS, DENSITY, IRON,
 OURCES, WATER POLLUTION EFFECTS,
 ), HYPOLIMNION, ALGAE, NITROGEN,
 BIOLOGICAL TREATMENT, NITROGEN,
 SOTOPES, SEA WATER, FOOD CHAINS,
 ONCENTRATION, WISCONSIN, SULFUR,
 *GREAT LAKES, *ALGAE, NUTRIENTS,
 RTILIZATION, NITROGEN COMPOUNDS,
 UISANCE ALGAE, SEWAGE EFFLUENTS,
 PLANTS, WATER POLLUTION EFFECTS,
 OTOGRAPHY, CARBON RADIOISOTOPES,
 CONSIN, SEWAGE, TROUT, NITROGEN,
 OACTIVITY, CARBON RADIOISOTOPES,
 FFLUENTS, *NUTRIENTS, *NITROGEN,
 ON, GEOCHEMISTRY, SPRING WATERS,
 H, OLIGOCHAETES, MIDGES, SNAILS,
 BON, FERTILITY, ALGAE, NITROGEN,
 MISTRY, WATER POLLUTION SOURCES,
 ), LAKE KEGONSA(WIS), WEED BEDS,
 EDESMUS, CHLAMYDOMONAS, ECOLOGY,
 S, CARBON, METABOLISM, NITROGEN,
 WAGE, FARM WASTES, CARBON CYCLE,
 MENT, SEWAGE BACTERIA, NITROGEN,
 KES, RESERVOIRS, SCUM, NITROGEN,
 ONMENTAL/ *BIOASSAYS, NUTRIENTS,
 PHYSIOLOGY, FISH TAXONOMY, DDT,
 ORMOSA, CYCLOTELLA MENEGHINIANA,
 OTROPH/ *EXTRACELLULAR PRODUCTS,
 PICKWICK RESERVOIR(TENN), NAJAS,
 , FILAMENTOUS ALGAE, KILL RATES,
 *KILL RATES, *BROMINE, IODINE,
 FECTIVE–AVERAGE–LIGHT–INTENSITY,
 NON, GLOEOTRICHIA, POLYPEPTIDES,
 URES, CHEMOSTATS, CHLOROPHYLL A,
 TS, NITROGEN, NITROGEN FIXATION,
 BACTERIA, GROWTH RATES, TRITIUM,
 TEMPERATURE, MONITORING, AERIAL
 YARTHRA, CONOCHILUS, PYCNOCLINE,
 NIQUES, *POLAROGRAPHIC ANALYSIS,
 ECOLOGY, ECOTYPES, CHLOROPHYTA,
 TIC TESTS, SLIMEPRODUCING ALGAE,
 S, CHLORELLA VULGARIS(COLUMBIA)/
 SYSTEMS, ENVIRONMENT, RETENTION,
 BORATORY TESTS, LIGHT INTENSITY,
 PHYRIDIUM CRUENTUM, LEMNA MINOR,
 OLYBDENUM, ROTIFERS, ECOSYSTEMS,
 , SILTS, STANDING CROP, ECOLOGY,
 ACID SOILS, ACIDIC WATER, ACIDS,
 YANOPHYTA, CHLORELLA, MANGANESE,
 NESE, TRACE ELEMENTS, CHELATION,
 IC WATER, ACIDS, PHOTOSYNTHESIS,

PHOSPHORUS EFFECT, CLADECERA.: /C   W69-06536
PHOSPHORUS, ADSORPTION, EUTROPHIC   W69-06859
PHOSPHORUS COMPOUNDS, PLANT GROWT   W69-06535
PHOSPHORUS, TROPHIC LEVEL, PRODUC   W69-06536
PHOSPHORUS, ALGAE, ALGICIDES, HAR   W69-07818
PHOSPHORUS RADIOISOTOPES, TROPHIC   W69-07853
PHOSPHORUS COMPOUNDS, NITROGEN CO   W69-08518
PHOSPHORUS, WATER POLLUTION EFFEC   W69-08668
PHOSPHORUS, AIR, PHYTOPLANKTON, A   W70-04484
PHOSPHORUS, TRACE ELEMENTS, PLANK   W70-04385
PHOSPHORUS.: / CONTROL, CHEMCONTR   W70-04494
PHOSPHORUS RELEASE, SAMPLE PRESER   W70-04382
PHOSPHORUS, SODIUM, POTASSIUM, TR   W70-04503
PHOSPHORUS, DRAINAGE, AGRICULTURA   W70-04506
PHOSPHORUS, ALGAE, NUTRIENTS, WAT   W70-04382
PHOSPHORUS, FISH STOCKING, COMMER   W70-04465
PHOSPHORUS, NITROGEN, IRON, GROWT   W70-03983
PHOSPHORUS, AGRICULTURAL WATERSHE   W70-04193
PHOSPHORUS, NITROGEN, CYANOPHYTA,   W70-03512
PHOSPHORUS RADIOISOTOPES, REGRESS   W70-03957
PHOSPHORUS, PHOTOSYNTHESIS, LIGHT   W70-03923
PHOSPHORUS COMPOUND: /CE ALGAE, N   W70-04283
PHOSPHORUS BUDGET, APHANIZOMINON   W70-04268
PHOSPHORUS, PHYTOPLANKTON, DIATOM   W70-04253
PHOSPHORUS, *NUTRIENT REMOVAL, SO   W73-01122
PHOSPHORUS, SCENEDESMUS, CHLOROPH   W73-01454
PHOSPHORUS RADIOISOTOPES, INFLOW,   W73-01454
PHOSPHORUS, *EUTROPHICATION, PHOT   W72-00917
PHOSPHORUS, NITROGEN, WATER QUALI   W72-03183
PHOSPHORUS COMPOUNDS, WATER SUPPL   W72-01867
PHOSPHORUS COMPOUNDS, RESERVOIRS,   W72-02817
PHOSPHORUS, DENSITY, WATER POLLUT   W72-04761
PHOSPHORUS, NITROGEN, CALCIUM CHL   W72-05453
PHOSPHORUS, ALGAE, LIVESTOCK, WAT   W71-10376
PHOSPHORUS RADIOISOTOPES, TECHNOL   W71-09012
PHOSPHORUS, NITROGEN, AMMONIA, AL   W71-05879
PHOSPHORUS, NITROGEN, PHYTOPLANKT   W71-07698
PHOSPHORUS, AGRICULTURE, WATER MA   W71-03048
PHOSPHORUS, HYDROGEN ION CONCENTR   W71-04206
PHOSPHORUS, SILICA, SAMPLING, CHE   W70-06975
PHOSPHORUS, NITROGEN, CHLOROPHYLL   W70-07261
PHOSPHORUS, OXYGEN, CARBON, ENZYM   W70-09889
PHOSPHORUS, SEWAGE EFFLUENTS.: /,   W70-08980
PHOSPHORUS RADIOISOTOPES, COBALT   W70-00708
PHOSPHORUS.: /USCORUN, NUTRIENT C   W70-00719
PHOSPHORUS, POPULATION, POLLUTANT   W70-00667
PHOSPHORUS COMPOUND, PERCHES, FIS   W70-00274
PHOSPHORUS, EUTROPHICATION, WASHI   W70-00270
PHOSPHORUS, NITROGEN.: /EAR POWER   W70-00264
PHOSPHORUS RADIOISOTOPES, ALGAE,   W69-10163
PHOSPHORUS, DETERGENTS, RUNOFF, P   W69-10178
PHOSPHORUS RADIOISOTOPES, ALGAE,   W69-10151
PHOSPHORUS, SULFATES, VITAMINS, A   W69-09349
PHOSPHORUS, ORGANOPHOSPHORUS COMP   W69-09334
PHOSPHORUS, NUTRIENTS, ALGAE, DIA   W70-01943
PHOSPHORUS, WATER POLLUTION EFFEC   W70-02973
PHOSPHORUS, NITROGEN COMPOUNDS, E   W70-03068
PHOSPHORUS, NITROGEN BALANCE.: /S   W70-02969
PHOSPHORUS, SULPHUR, HYDROGEN ION   W70-02965
PHOSPHORUS, SULFUR, CALCIUM, MAGN   W70-02804
PHOSPHORUS, PHOSPHATE, NITRATES,   W70-02777
PHOSPHORUS, ALGAE, COPPER SULFATE   W70-02787
PHOSPHORUS, CARBONATES, TEMPERATU   W70-02803
PHOSPHORUS, KINETIC THEORY, ENVIR   W70-02779
PHOSPHOTHIOATE PESTICIDES, HERBIC   W69-02782
PHOTIC ZONE, HELSINGOR(DENMARK),   W70-04809
PHOTIC ZONE, GLYCOLLIC ACID, CHEM   W70-02504
PHOTIC ZONE, DOUGLAS RESERVOIR(TE   W71-07698
PHOTO REACTIVITY, CELL COLOR, CEL   W70-02364
PHOTO REACTIVITY, ALGAL METABOLIS   W70-02363
PHOTO–ELECTRIC TURBIDOSTAT, LIGHT   W70-03923
PHOTOASSIMILATION, BERKELSE LAKE(W70-04369
PHOTOBIOLOGY, ALGAL: / BATCH CULT   W70-02777
PHOTOCHEMICAL, CARBON DIOXIDE, SY   W70-02965
PHOTOGRAPHY, CARBON RADIOISOTOPES   W69-10163
PHOTOGRAPHY, INFRARED RADIATION,   W68-00010
PHOTOLITHOTROPHIC BACTERIA, DESUL   W70-05760
PHOTOMETRY, HALOGENS.: /ICAL TECH   W70-05178
PHOTOMETRY, CHLORELLA, NUTRIENTS,   W70-05409
PHOTOMICROGRAPHS, MICROCYSTIS AER   W69-05704
PHOTOOXIDATION, CHLORELLA VULGARI   W70-04809
PHOTOPERIODISM, LIGHT INTENSITY,   W69-07862
PHOTOPERIODISM, LIGHT PENETRATION   W72-00917
PHOTOPHOSPHORYLATION, VANADIUM, Z   W70-02964
PHOTOS: /ITE, SULFUR COMPOUNDS, M   W69-07428
PHOTOSYNT: /DYNAMICS, TEMPERATURE   W70-02780
PHOTOSYNTHESIS, PHOTOSYNTHETIC OX   W70-02792
PHOTOSYNTHESIS, OXYGEN, CARBON DI   W70-02964
PHOTOSYNTHESIS, CARBON DIOXIDE, B   W70-02804
PHOTOSYNTHETIC OXYGEN, PHYSIOLOGI   W70-02792

ENVIRONMENT, CHLORELLA, TRACERS,
SIGN CRITERIA, OXIDATION LAGOONS
, FISH, BACTERIA, BENTHIC FAUNA,
OONS, BIOCHEMICAL OXYGEN DEMAND,
AL ECOLOGY, LITTORAL, ESTUARIES,
GEN, HYDROGEN ION CONCENTRATION,
G, PHYTOPLANKTON, RADIOACTIVITY,
CIDES, OXIDATION LAGOONS, ALGAE,
-PURIFICATION, DISSOLVED OXYGEN,
EN, PHOSPHORUS, *EUTROPHICATION,
XYGEN DEMAND, AEROBIC TREATMENT,
MAND, OXIDATION LAGOONS, SLUDGE,
LVED OXYGEN, *RIVERS, OXIDATION,
EN, ODORS, CORROSION, OXIDATION,
OGY, NORTH DAKOTA, SOUTH DAKOTA,
TH RATES, ENVIRONMENTAL EFFECTS,
ILIMNION, EFFLUENT, TEMPERATURE,
MENT, GROWTH RATES, CHLOROPHYLL,
ATION, OLIGOTROPHY, TEMPERATURE,
  CHLORINATION, SODIUM COMPOUNDS,
TION, CHLOROPHYLL, PRODUCTIVITY,
TESTS, RESPIRATION, ENVIRONMENT,
RGANIC MATTER, EXUDATION, TIDES,
DUSTRIAL WASTES, LAGOONS, PONDS,
  PESTICIDES, PESTICIDE TOXICITY,
TORY TESTS, PERIPHYTON, STREAMS,
UTLETS, NUTRIENTS, PRODUCTIVITY,
N ION CONCENTRATION, EPILIMNION,
ATORY TESTS, LAGOONS, OXIDATION,
  *ALGAE, *TEMPERATURE, DIATOMS,
E, PACKED CELL VOLUM/ INDUSTRIAL
IFICATION, CARBON, BICARBONATES,
IGHT, TE/ *GROWTH RATES, *ALGAE,
TION, BIOCHEMICAL OXYGEN DEMAND,
NT REQUIREMENTS, ORGANIC MATTER,
NS, MOSCOW, PETROLEUM POLLUTION,
OGY, NUISANCE ALGAE, PHAEOPHYTA,
N DYNAMICS, EXTRACELLULAR PRODU/
LIGHT PENETRATION, FLUORESCENCE,
S, ARCTIC, ENZYMES, CHLOROPHYLL,
*DESIGN THEORY AND CALCULATIONS,
ARY PRODUCTIVITY, PHYTOPLANKTON,
ION, COLIFORMS, PONDS, CLIMATES,
  DIATOMS, PIGMENTS, CHLOROPHYLL,
LLUTION, NUTRIENTS, TEMPERATURE,
DIATION, ENERGY, CARBON DIOXIDE,
ETALS, INDUSTRIAL WASTES, LAKES,
Y, LIGHT PENETRATION, LIMNOLOGY,
OMPOUNDS, NUTRIENT REQUIREMENTS,
T REGIONS, GEOGRAPHICAL REGIONS,
UTRIENT REQUIREMENTS, NUTRIENTS,
SONAL DISTRIBUTION, LAKE BAIKAL,
GAE, SEWAGE PLANTS, CHLOROPHYLL,
ADIOISOTOPES, ENZYMES, KINETICS,
CTERIA, RADIOACTIVITY TECHNIQUE,
LANKTON, WASTE WATER(POLLUTION),
ENTS, ECOSYSTEMS, PHYTOPLANKTON,
SAY, INHIBITION, MODE OF ACTION,
SCENEDESMUS, *LIGHT, PHOSPHORUS,
OPHYTA, CYANOPHYTA, CHRYSOPHYTA,
, *WASTE WATER TREATMENT, ALGAE,
, ANABAENA, CYTOLOGICAL STUDIES,
UTOLYSIS, EXCRETION, DEPURATION,
  *ANABAENA FLOS-AQUAE, *STRAINS,
  DENSITY, SOIL-WATER TECHNIQUES,
RNS, MICROALGAE, MICRONUTRIENTS,
UNDERWATER SHOCK WAVE, ANABAENA,
CHEMISTRY, ALGAL TAXONOMY, ALGAL
E, CONSUMER ORGANISMS, SIMULIUM,
OL, INTEGRATED CONTROL MEASURES,
F-LIFE, CLINCH RIVER(TENNESSEE),
CA, SAMPLING, CHEMICAL ANALYSIS,
ER POLLUTION, CHEMICAL ANALYSIS,
SHEDS, NUTRIENTS, FERTILIZATION,
FICATION, BIOLOGICAL PROPERTIES,
AL TREATMENT, OVERFERTILIZATION,
GAE, PLANKTON, SUBMERGED PLANTS,
ATION, LIMNOLOGY, PUBLIC HEALTH,
STRONTIUM RADIOISOTOPES, CESIUM,
, *MARINE PLANTS, RADIOISOTOPES,
L TECHNIQUES, DIATOMS, SAMPLING,
PTH, PHYTOPLANKTON, ZOOPLANKTON,
ARINE ALGAE, *RESINS, SEA WATER,
PERATURE, NUTRIENT REQUIREMENTS,
TS, PHYSIOCOCHEMICAL PROPERTIES,
URE, CARBON DIOXIDE, CYANOPHYTA,
GAE, DIATOMS, CYCLING NUTRIENTS,
  YTTRIUM, 'BIOLOGICAL DILUTION',
LLUTION, STREAMS, RIVERS, LAKES,
YNTHESIS, PHOTOSYNTHETIC OXYGEN,

PHOTOSYNTHESIS, CARBON RADIOISOTO	W70-02786
PHOTOSYNTHESIS.: /ION, DESIGN, DE	W70-02609
PHOTOSYNTHESIS, BIOASSAYS, AQUATI	W70-02251
PHOTOSYNTHESIS, HYDROGEN SULFIDE,	W70-03312
PHOTOSYNTHESIS, RESPIRATION, PIGM	W70-03325
PHOTOSYNTHESIS, ALKALINITY, NITRA	W70-03309
PHOTOSYNTHESIS, CHEMICAL ANALYSIS	W70-01933
PHOTOSYNTHESIS, DISSOLVED OXYGEN,	W71-12183
PHOTOSYNTHESIS.: /TREATMENT, SELF	W72-01798
PHOTOSYNTHESIS, LAGOONS, LABORATO	W72-00917
PHOTOSYNTHESIS, ALGAE, SEWAGE, MU	W72-00462
PHOTOSYNTHETIC OXYGEN, FARM WASTE	W71-08221
PHOTOSYNTHESIS, RESPIRATION, SCOU	W71-11381
PHOTOSYNTHESIS, ANALYTICAL TECHNI	W71-11377
PHOTOSYNTHESIS, LIGHT PENETRATION	W72-01818
PHOTOSYNTHESIS, DIATOMS, CHLORELL	W69-10158
PHOTOSYNTHESIS, CARBON DIOXIDE, D	W69-10169
PHOTOSYNTHESIS, CHEMICAL ANALYSIS	W69-10167
PHOTOSYNTHESIS, RAINBOW TROUT, DI	W69-10154
PHOTOSYNTHESIS, CARBON DIOXIDE, N	W70-00161
PHOTOSYNTHESIS, ALGAE, HYPOLIMNIO	W69-10182
PHOTOSYNTHESIS, OXYGEN, TROPHIC L	W70-00265
PHOTOSYNTHESIS, SOLAR RADIATION,	W70-01073
PHOTOSYNTHESIS, COSTS, EFFICIENCY	W70-09320
PHOTOSYNTHETIC OXYGEN, PRODUCTIVI	W70-08416
PHOTOSYNTHESIS, RESPIRATION, MICR	W71-00668
PHOTOSYNTHESIS, EUTROPHICATION, T	W70-06225
PHOTOSYNTHESIS, ALGAE, BOTTOM SED	W70-06975
PHOTOSYNTHESES, PLANTS, TEMPERATU	W70-06600
PHOTOSYNTHESIS.:	W71-02496
PHOTOSYNTHESIS, MEAN DOUBLING TIM	W71-05267
PHOTOSYNTHESIS, NUTRIENTS, MICROB	W69-07389
PHOTOSYNTHESIS, EUTROPHICATION, L	W69-07442
PHOTOSYNTHETIC OXYGEN.: /AG, AERA	W69-07520
PHOTOSYNTHESIS, TEMPERATURE, TROP	W69-06865
PHOTOSYNTHETIC PIGMENTS, PRODUCTI	W69-07440
PHOTOSYNTHESIS, PHYTOPLANKTON, PR	W69-07832
PHOTOSYNTHETIC RATE, PHYTOPLANKTO	W69-05697
PHOTOSYNTHESIS, BLUE-GREEN ALGAE,	W69-05697
PHOTOSYNTHESIS, CHLORELLA, GRAZIN	W70-04809
PHOTOSYNTHESIS, ILLUMINATION, TEM	W70-04786
PHOTOSYNTHESIS, BIOMASS, CULTURES	W70-05094
PHOTOSYNTHESIS, PRODUCTIVITY, SYM	W70-05092
PHOTOSYNTHESIS, GROWTH RATE, ORGA	W70-05381
PHOTOSYNTHESIS, LIGHT INTENSITY.:	W70-05270
PHOTOSYNTHESIS, GROWTH RATE, AERO	W70-05750
PHOTOSYNTHESIS, PHYTOPLANKTON, CI	W69-02959
PHOTOSYNTHETIC OXYGEN, TEMPERATUR	W69-03516
PHOTOSYNTHETIC, OXYGEN, PLANKTON,	W69-03514
PHOTOSYNTHESIS, CHEMICAL REACTION	W69-03611
PHOTOSYNTHET: /TION, LIMNOLOGY, N	W69-03517
PHOTOSYNTHETIC LAYER, MELOSIRA BA	W70-04290
PHOTOSYNTHESIS, TRIBUTARIES, WATE	W70-04268
PHOTOSYNTHESIS, ORGANIC MATTER, O	W70-04287
PHOTOSYNTHESIS.: /ROGEN FIXING BA	W70-04249
PHOTOSYNTHESIS, CHL: /NTS, PHYTOP	W70-04371
PHOTOSYNTHESIS, CYANOPHYTA, CYTOL	W70-03973
PHOTOSYNTHESIS.: /ORTALITY, BIOAS	W70-03519
PHOTOSYNTHESIS, LIGHT INTENSITY,	W70-03923
PHOTOSYNTHESIS.: /TH RATES, CHLOR	W70-03983
PHOTOSYNTHESIS, DEPTH, THERMOCLIN	W72-10573
PHY: /OSTRIDIUM, FISH, AMPHIBIANS	W70-05372
PHYCOCOLLOIDS, TANNINS, FLAGELLAT	W70-01068
PHYCOLOGISTS, VETERINARIANS, MICR	W70-00273
PHYCOLOGY, MICROCYSTIS, GLOEOTRIC	W70-04381
PHYCOLOGY, SPACE RESEARCH, TOXIC	W69-07832
PHYGON XL(DICHLONE), NEW HAMPSHIR	W69-08674
PHYLOGENY, CLADOPHORA, ALGAL GENE	W70-03973
PHYSA, SALMO TRUTTA, ENTOSPHENUS	W69-09334
PHYSICAL CONTROL, WATER QUALITY.:	W70-07393
PHYSICAL HALF-LIFE, HYDROPSYCHE C	W70-02786
PHYSICAL PROPERTIES, CARP, TILAPI	W70-06975
PHYSICAL PROPERTIES.: /EWAGE, WAT	W70-05752
PHYSICAL PROPERTIES, CLIMATE, NIT	W70-05565
PHYSICAL PROPERTIES, WASTE WATER(	W69-01273
PHYSICAL TREATMENT, UNIT OPERATIO	W70-07469
PHYSICOCHEMICAL PROPERTIES, PHYTO	W71-00665
PHYSICOCHEMICAL PROPERTIES, GENET	W70-00273
PHYSICOCHEMICAL PROPERTIES, ADSOR	W70-00708
PHYSICOCHEMICAL PROPERTIES, ANION	W71-09863
PHYSICOCHEMICAL PROPERTIES, CENTR	W70-05389
PHYSICOCHEMICAL PROPERTIES, DOMES	W70-03964
PHYSICOCHEMICAL PROPERTIES, TRACE	W69-08275
PHYSIOCOCHEMICAL PROPERTIES, PHYS	W69-10161
PHYSIOLOGICAL ECOLOGY, RHODOPHYTA	W69-10161
PHYSIOLOGICAL ECOLOGY, THERMAL WA	W69-10160
PHYSIOLOGICAL ECOLOGY, WATER: /AL	W70-00713
PHYSIOLOGICAL CO: / THORACOMONAS,	W70-00708
PHYSIOLOGICAL ECOLOGY, LIMITING F	W71-02479
PHYSIOLOGICAL ECOLOGY, PHYTOPLANK	W70-02792

TIONS, SURVIVAL, SUSCEPTIBILITY,   PHYSIOLOGICAL OBSERVATIONS, CHRON   W70-01779
   OF ALGAE, LIMNOLOGY, NUTRIENTS,   PHYSIOLOGICAL ECOLOGY, CHLORELLA,   W70-02245
ATION, OREGON, INTERTIDAL AREAS,   PHYSIOLOGICAL ECOLOGY, LITTORAL,   W70-03325
IDE REMOVAL, PESTICIDE RESIDUES,   PHYSIOLOGICAL ECOLOGY, PHYTOPLANK   W70-02968
RMAL WATER, PERIPHYTON, BIOMASS,   PHYSIOLOGICAL ECOLOGY, CYANOPHYTA   W70-03309
LIFORNIA, BIOASSAY, SCENEDESMUS,   PHYSIOLOGICAL ECOLOGY, KINETICS,   W70-02775
RACERS, WATER POLLUTION EFFECTS,   PHYSIOLOGICAL: / RADIOACTIVITY, T   W70-02510
   TEMPERATURE, CHEMICAL ANALYSIS,   PHYSIOLOGICAL ECOLOGY, OXYGEN, CA   W69-07862
GRAZING, *TEMPERATURE, *DAPHNIA,   PHYSIOLOGICAL ECOLOGY, LAKES, ENV   W70-03957
, ESTUARIES, LAKES, GREAT LAKES,   PHYSIOLOGICAL ECOLOGY, WATER CHEM   W70-03975
, ESTUARIES, LAKES, GREAT LAKES,   PHYSIOLOGICAL ECOLOGY, WATER CHEM   W70-03975
TALS, VITAMINS, WATER POLLUTION,   PHYSIOLOGICAL ECOLOGY, DIATOMS, O   W70-04503
   CHAINS, CHLORELLA, CHLOROPHYLL,   PHYSIOLOGICAL ECOLOGY, CYANOPHYTA   W70-04381
TION EFFECTS, ALGAE, ECOSYSTEMS,   PHYSIOLOGICAL ECOLOGY, AQUATIC BA   W70-04287
NION, LIGHT INTENSITY, SEASONAL,   PHYSIOLOGICAL ECOLOGY, ECOTYPES,   W70-05409
E, TOXICITY, DIATOMS, ESTUARIES,   PHYSIOLOGICAL ECOLOGY, MASSACHUSE   W70-05272
NETICS, AQUATIC BACTERIA, ALGAE,   PHYSIOLOGICAL ECOLOGY, CARBON RAD   W70-05109
LUENTS, CHLOROPHYTA, RHODOPHYTA,   PHYSIOLOGICAL ECOLOGY, INDUSTRIAL   W70-04510
HNIQUES, *MARINE MICROORGANISMS,   PHYSIOLOGICAL ECOLOGY, ORGANIC MA   W70-04811
EOVIRIDIS, SKELETONEMA COSTATUM,   PHYSIOLOGICAL ADJUSTMENT, ASTERIO   W70-04809
                                     PHYSIOLOGICAL EFFECTS.:   W68-00859
NUTRIENTS, PHOSPHORUS COMPOUNDS,   PHYSIOLOGICAL ECOLOGY,: / ALGAE,   W69-04802
TRIENTS, PHOSPHORUS, PHOSPHATES,   PHYSIOLOGICAL ECOLOGY, PHYTOPLAN: W69-03373
UTRIENT REQUIREMENTS, NUTRIENTS,   PHYSIOLOGICAL ECOLOGY, PH: /ON, N   W69-03364
CE ALGAE, NUTRIENT REQUIREMENTS,   PHYSIOLOGICAL E: /F ALGAE, NUISAN   W69-03362
CE ALGAE, NUTRIENT REQUIREMENTS,   PHYSIOLOGICAL ECOLOGY, PHYTOPLANK   W69-03371
   EFFECTS, ANALYTICAL TECHNIQUES,   PHYSIOLOGICAL ECOLOGY, PLANT PHYS   W72-12240
ES, PHYSIOLOGICAL ECOLOGY, PLANT   PHYSIOLOGY.: /ANALYTICAL TECHNIQU   W72-12240
GREEN ALGAE, *ALGA BLOOMS, ALGAE   PHYSIOLOGY.:   *BLUE-   W69-03512
DRIFT, FISH FOOD ORGANISMS, FISH   PHYSIOLOGY, FISH TAXONOMY, DDT, P   W69-02782
THESIS, BIOMASS, CULTURES, PLANT   PHYSIOLOGY, ALGAE, DIURNAL DISTRI   W70-05094
ASTE STABILIZATION PONDS, *ALGAL   PHYSIOLOGY, *DESIGN THEORY AND CA   W70-04786
WATER ANALYSIS, *SAMPLING, PLANT   PHYSIOLOGY, EUTROPHICATION, DATA   W70-04382
ON(WASH), ALGAL NUTRITION, ALGAL   PHYSIOLOGY, HETEROTROPHY, MICRONU   W70-04506
   STUDIES, APPLIED STUDIES, ALGAL   PHYSIOLOGY, ALGAL BIOCHEMISTRY, A   W70-03973
AGONISMS, ALGAL NUTRITION, ALGAL   PHYSIOLOGY, LAKE MENDOTA(WIS), LA   W69-05867
L GROWTH, ALGAL NUTRITION, ALGAL   PHYSIOLOGY, CALOTHRIX PARIETINA,   W69-06277
L GROWTH, ALGAL NUTRITION, ALGAL   PHYSIOLOGY, KJELDAHL PROCEDURE, L   W69-05868
L GROWTH, ALGAL NUTRITION, ALGAL   PHYSIOLOGY, AMERICAN RESEARCH AND   W69-06865
UTION, *TOXICITY.:   *FISH   PHYSIOLOGY, *ALGAE, *THERMAL POLL   W70-01788
                 *CHEMOSTAT,   PHYSIOLOGY, FLAGELLATES.:   W70-09904
ED HYDROCARBON PESTICIDES, PLANT   PHYSIOLOGY.: /IRONMENT, CHLORINAT   W70-08652
N, MILL/ *MICRO-AUTORADIOGRAPHY,   PHYSIOLOGY, BIOCHEMICAL EXTRACTIO   W70-10163
AGOONS, PLANT PHYSIOLOGY, ANIMAL   PHYSIOLOGY, NORTH DAKOTA, SOUTH D   W72-01818
ED OXYGEN, SEWAGE LAGOONS, PLANT   PHYSIOLOGY, ANIMAL PHYSIOLOGY, NO   W72-01818
RMING, FISH FOOD ORGANISMS, FISH   PHYSIOLOGY, RED TIDE, ORGANIC COM   W71-07731
OSPHATES, PHYSIOLOGICAL ECOLOGY,   PHYTOPLAN: /IENTS, PHOSPHORUS, PH   W69-03373
IREMENTS, PHYSIOLOGICAL ECOLOGY,   PHYTOPLANKTON, SE: /NUTRIENT REQU   W69-03371
L WASTES, LAKES, PHOTOSYNTHESIS,   PHYTOPLANKTON, CISCO, CARP, DIATO   W69-02959
HOSPHORUS COMPOUNDS, PHOSPHORUS,   PHYTOPLANKTON.: / REQUIREMENTS, P   W69-03358
YANOPH/ *EUTROPHICATION, *LAKES,   PHYTOPLANKTON, ALGAE, SEASONAL, C   W68-00860
LGAE, *CHEMICA/ *EUTROPHICATION,   PHYTOPLANKTON, BIBLIOGRAPHIES, *A   W68-00856
, DI/ *ECOLOGY, BACTERIA, ALGAE,   PHYTOPLANKTON, *LAKES, *CELLULOSE   W69-00976
TER ANALYSIS, PHOSPHORUS, PONDS,   PHYTOPLANKTON, EUTROPHICATION, LA   W69-05142
POND(JAPAN), LAKE MENDOTA(WIS),   PHYTOPLANKTON CROPS, TROPHIC, TRO   W69-05142
TS, NUISANCE ALGAE, OLIGOTROPHY,   PHYTOPLANKTON, *REVIEWS, *WATER P   W68-00680
   BRINE SHRIMP, ECOSYSTEMS, FISH,   PHYTOPLANKTON, PLANKTON, RADIOACT   W69-07440
EUTROPHICATION, ORGANIC MATTER,   PHYTOPLANKTON, PRIMARY PRODUCTIVI   W69-06273
ULAR PRODU/ PHOTOSYNTHETIC RATE,   PHYTOPLANKTON DYNAMICS, EXTRACELL   W69-05697
HOTOSYNTHESIS, BLUE-GREEN ALGAE,   PHYTOPLANKTON, OHIO RIVER, PENNSY   W69-05697
ISANCE ALGAE, OLIGOTROPHY, ODOR,   PHYTOPLANKTON, SULFATES, SEDIMENT   W69-07833
YTON, INVERTEBRATES, ADSORPTION,   PHYTOPLANKTON, SUCKERS, SHINERS,   W69-07853
GAE, PHAEOPHYTA, PHOTOSYNTHESIS,   PHYTOPLANKTON, PRIMARY PRODUCTIVI   W69-07832
N ABATEMENT, OLIGOTROPHY, DEPTH,   PHYTOPLANKTON, ZOOPLANKTON, PHYSI   W70-03964
OROPHYTA, NUTRIENTS, ECOSYSTEMS,   PHYTOPLANKTON, PHOTOSYNTHESIS, CY   W70-03973
IMNOLOGY, CHLOROPHYLL, PLANKTON,   PHYTOPLANKTON, WATER POLLUTION EF   W70-03959
LAKES, *RESERVOIRS, ZOOPLANKTON,   PHYTOPLANKTON, TEMPERATURE, *EPIL   W70-04475
TON, LAKES, INDIANA, OCHROMONAS,   PHYTOPLANKTON, SEAWEEDS.: /ASHING   W70-04503
TION, NITROGEN, PHOSPHORUS, AIR,   PHYTOPLANKTON, ALGAE, EFFICIENCIE   W70-04484
HIC FAUNA, NITROGEN, PHOSPHORUS,   PHYTOPLANKTON, DIATOMS, ZOOPLANKT   W70-04253
   ANALYSIS, MONITORING, PIGMENTS,   PHYTOPLANKTON, WASTE WATER(POLLUT   W70-04371
A, THERMOCLINE, CURRENTS(WATER),   PHYTOPLANKTON, ZOOPLANKTON, HYDRO   W70-04368
ADIOISOTOPES, *ALGAE, *CULTURES,   PHYTOPLANKTON, CARBON, RADIOACTIV   W70-03983
, MATHEMATICAL MODELS, BACTERIA,   PHYTOPLANKTON, GEOCHEMIS: /NOLOGY   W70-03975
, MATHEMATICAL MODELS, BACTERIA,   PHYTOPLANKTON, GEOCHEMIS: /NOLOGY   W70-03975
TIES, EFFLUENTS, EUTROPHICATION,   PHYTOPLANKTON, DECOMPOSING ORGANI   W70-04001
NIQUES, WATER POLLUTION EFFECTS,   PHYTOPLANKTON, DIFFUSION, ALGAE,   W70-04194
PHICATION, PRIMARY PRODUCTIVITY,   PHYTOPLANKTON, PHOTOSYNTHESIS, BI   W70-05094
GY, CARBON RADIOISOTOPES, LAKES,   PHYTOPLANKTON, BIOASSAYS.: /ECOLO   W70-05109
RAINAGE, ORGANIC MATTER, SEWAGE,   PHYTOPLANKTON, PRODUCTIVITY, CHLO   W70-05404
TES, HYDROGEN ION CONCENTRATION,   PHYTOPLANKTON, ODOR, WINDS, EPILI   W70-05405
AE, *PHOTOSYNTHESIS, *MINNESOTA,   PHYTOPLANKTON, CYANOPHYTA, DIATOM   W70-05387
*POTOMAC RIVER ESTUARY, NUTRIENT-   PHYTOPLANKTON RELATIONSHIPS.:   W71-05407
POLIMNION, PHOSPHORUS, NITROGEN,   PHYTOPLANKTON, PRODUCTIVITY, TURB   W71-07698
ALGAE, COLIFORMS, EUTROPHICATION(PHYTOPLANKTON), ESTUARIES, WATER   W71-05407
RATION, DISSOLVED OXYGEN, ALGAE,   PHYTOPLANKTON, ALKALINE WATER, AL   W71-07731
   SAMPLING, PRIMARY PRODUCTIVITY,   PHYTOPLANKTON, EUGLENA, CHLOROPHY   W72-04761
, S/ *PLANKTON, *PHOTOSYNTHESIS,   PHYTOPLANKTON, ALGAE, TEMPERATURE   W71-10788
OWERPLANTS, *LAKES, TEMPERATURE,   PHYTOPLANKTON, ALGAE, HEATED WATE   W71-09767
*EUTROPHICATION, SEWAGE, BIOTA,   PHYTOPLANKTON, DIATOMS, DINOFLAGE   W69-10169

RELLA, WISCONSIN, LAKE MICHIGAN,
NUTRIENTS, NITRATES, PHOSPHATES,
UTRIENT REQUIREMENTS, NUTRIENTS,
ECOLOGY, *MARINE MICROORGANISMS,
ALGAE, HYPOLIMNION, EPILIMNION,
, PLANKTON, DIATOMS, CYANOPHYTA,
ALTITUDE, SAMPLING, ZOOPLANKTON,
BENTHIC FLORA, FLOATING PLANTS,
NTS, PHYSICOCHEMICAL PROPERTIES,
SIN, ECOLOGY, CHEMICAL ANALYSIS,
UCTIVITY, CHLOROPHYLL, SAMPLING,
NIQUES, WATER POLLUTION EFFECTS,
LGAE, *ROTIFERS, PLANKTON, BASS,
ION, SAMPLING, SURFA/ *BIOASSAY,
RESIDUES, PHYSIOLOGICAL ECOLOGY,
C OXYGEN, PHYSIOLOGICAL ECOLOGY,
RYSOPHYTA, CYANOPHYTA, SAMPLING,
ALKALINITY, LAKES, PRODUCTIVITY,
, TEMPERATURE, DISSOLVED OXYGEN,
OLOGICAL ECOLOGY, PHYTOPLANKTON,
TOMOGETON CRISPUS, HETERAMTHERA,
     *PIGMENTED FLAGELLATES, *NON-
AE, *BLUE GREEN ALGAE, *EUGLENA,
SSIUM, IRON, MANGANESE, CALCIUM,
ES, PHOTOSYNTHESIS, RESPIRATION,
ITY, DIATOMS, SEASONAL, BIOMASS,
COPEPODS, ROTIFERS, CHLOROPHYLL,
, *COLOR, RIVERS, BOGS, PHENOLS,
IMARY PRODUCTIVITY, CHLOROPHYLL,
MICROSCOPY, NETS, NANNOPLANKTON,
OPLANKTON, DIATOMS, RESPIRATION,
CHLORELLA, CHLOROPHYTA, DIATOMS,
QUES, BACTERIAL PLATE, BACTERIAL
IOCHEMICAL ANALYSIS, MONITORING,
ROLEUM POLLUTION, PHOTOSYNTHETIC
, CHLOROPHYTA, DISSOLVED OXYGEN,
TIARY TREATMENT, SUSPENDED LOAD,
A, CHLOROPHYTA, DIATOMS, WEIGHT,
ED SLUDGE, BIOLOGICAL TREATMENT,
AE, *SLUDGE, STORMS, TURBULENCE,
XIDE, GROWTH RATES, TEMPERATURE,
BIDITY, SUSPENDED SOLIDS, ALGAE,
RIOPHYLLUM, LEPOMIS MACROCHIRUS,
E, GRAPHITE LAKE, OLD JOHN LAKE,
), EUNOTIA, AMPHORA, GOMPHONEMA,
GE, SLURRIES, AUTOMATIC CONTROL,
LYTICAL TECHNIQUES, TEMPERATURE,
VER(USSR), MINSK(USSR), ENGLAND/
OCLADIUS, CORETHRA PUNCTIPENNIS,
IOM, CERIUM, GUETTARDA, IPOMOEA,
EICHHORNIA, POTAMOGETON, TRAPA,
INCIPLE, OSCILLATORIA RUBESCENS,
LI, LYNGBYA NOLLEI, GLOCOTRICHIA
IAL RECHARGE, *DATA COLLECTIONS,
S, BLUEGILLS, MICROCYSTIS SCUMS,
*WATER REUSE, DISSOLVED OXYGEN,
TRATIONS, AEQUIPECTEN IRRADIANS,
AVICULA, NITZSCHIA, SKELETONEMA,
COASTAL PLAINS, ATLANTIC COASTAL
ROPHICATION, *RESERVOIRS, *GREAT
S, GEOGRAPHICAL REGIONS, COASTAL
JOAQUIN DELTA, CALIFORNIA WATER
HLOROCOCCALES, CONJUGATOPHYCAEA,
R QUALITY NETWORK, ALGAE COUNTS,
NODININE, *PERIDINIUM POLONICUM,
RICHIA ECHINULATA, BOTRYOCOCCUS,
E ALGAE, LIMNOLOGY, CHLOROPHYTA,
(SWEDEN), LAKE MENDOTA(WIS), NET
GAE, PHOSPHORUS, TRACE ELEMENTS,
S, CARBON DIOXIDE, GROWTH RATES,
CHAINS, FOOD WEBS, FORAGE FISH,
FERTILIZATION, *PHOSPHATE, FISH,
MPERATURE STRATIFICATION, DEPTH,
ANAGEMENT, FERTILIZATION, ALGAE,
*CONTROL, TASTE, ODOR, FILTERS,
FERTILIZATION, BACTERIA, ALGAE,
AGE, MARSHES, FARMS, CYANOPHYTA,
IN, INORGANIC COMPOUNDS, METALS,
D AQUATIC PLANTS, GLACIAL LAKES,
TIES, BACTERIA/ *EUTROPHICATION,
NUTRIENTS, NITROGEN, PHOSPHORUS,
NITROGEN, NUTRIENTS, PHOSPHORUS,
ATION, SEWAGE, FISH POPULATIONS,
ED CONTROL, *INVESTIGATIONS, ZOO-
MICHIGAN, ALGAE, NUISANCE ALGAE,
N-SITE DATA COLLECTIONS, FISHES,
REMENTS, PHOTOSYNTHETIC OXYGEN,
S, RADIOACTIVITY, INVERTEBRATES,
OECOLOGY, EFFLUENTS, ECOSYSTEMS,
WS, DETRITUS, CYCLING NUTRIENTS,

PHYTOPLANKTON, ·DIATOMS, ECOLOGY,	W69-10180
PHYTOPLANKTON, IRON COMPOUNDS, AL	W69-10170
PHYTOPLANKTON, INHIBITION.: /S, N	W69-10177
PHYTOPLANKTON, BACTERICIDES, ALGA	W70-01068
PHYTOPLANKTON, SESTON, CRUSTACEAN	W69-10182
PHYTOPLANKTON, ROTIFERS, CHLOROPH	W70-00268
PHYTOPLANKTON, ICE, BACTERIA, TRI	W69-10154
PHYTOPLANKTON, AQUATIC INSECTS, F	W70-10175
PHYTOPLANKTON, DISTRIBUTION, DEPT	W71-00665
PHYTOPLANKTON, CYANOPHYTA, SEDIME	W70-06229
PHYTOPLANKTON, RADIOACTIVITY, PHO	W70-01933
PHYTOPLANKTON, PLANKTON, AQUATIC	W70-02245
PHYTOPLANKTON, PONDS, ZOOPLANKTON	W70-02249
PHYTOPLANKTON, LAKES, EUTROPHICAT	W70-02973
PHYTOPLANKTON, PHYTOTOXICITY, PLA	W70-02968
PHYTOPLANKTON, PLANKTON, POLLUTIO	W70-02792
PHYTOPLANKTON, ORGANIC MATTER, CH	W70-02780
PHYTOPLANKTON.: /US, *NITROGEN, *	W70-02795
PHYTOPLANKTON, WATER ANALYSIS, AL	W72-06051
PHYTOTOXICITY, PLANKTON, POLLUTAN	W70-02968
PICKWICK RESERVOIR(TENN), NAJAS,	W71-07698
PIGMENTED FLAGELLATES.:	W72-01798
PIGMENTED FLAGELLATES.: /RANT ALG	W70-01233
PIGMENTS.: /DIMENTS, SODIUM, POTA	W70-09889
PIGMENTS, CHLOROPHYLL, TEMPERATUR	W70-03325
PIGMENTS, PRODUCTIVITY, PERIPHYTO	W70-02780
PIGMENTS, IMPOUNDMENTS, RESERVOIR	W71-10098
PIGMENTS, ORGANIC MATTER, BIOMASS	W70-01074
PIGMENTS, MAGN: /RINE ANIMALS, PR	W73-01446
PIGMENTS, CHLOROPHYTA, CYANOPHYTA	W70-05389
PIGMENTS, OCEANS, ORGANIC MATTER,	W70-04809
PIGMENTS, CHLOROPHYLL, PHOTOSYNTH	W70-05381
PIGMENTS, DIAPTOMUS, EPISCHURA, O	W70-05760
PIGMENTS, PHYTOPLANKTON, WASTE WA	W70-04371
PIGMENTS, PRODUCTION BIOLOGY, RUS	W69-07440
PIKE, CARP, D: /S, ALGAE, DIATOMS	W70-01943
PILOT PLANTS, ACTIVATED SLUDGE, T	W70-09186
PILOT PLANTS, IRON, CHLORELLA, SO	W70-03512
PILOT PLANTS, HUMUS, REAERATION,	W70-05821
PILOT PLANTS, BACTERIA, SEDIMENTA	W70-05663
PILOT PLANTS, HYDROGEN ION CONCEN	W72-06022
PILOT PLANTS, HEAD LOSS, PERMEABI	W72-08686
PIMEPHALES NOTATUS.: /IA COLI, MY	W69-07861
PINGO LAKE, TWELVE MILE LAKE, BRA	W69-08272
PINNULARIA, NITZSCHIA, NAVICULA,	W70-03974
PIPING SYSTEMS(MECHANICAL), CLEAN	W72-08688
PIPTIDES, INTERTIDAL AREAS, SOIL	W70-04365
PISCICULTURAL PONDS, SVISLOCH' RI	W70-05092
PISIDIUM IDAHOENSE, TENDIPES PLUM	W70-06217
PISONIA, PANDANUS, MUSCLE, BONE,	W69-08269
PISTIA, ALTERNANTHERA, HETERANTHE	W70-10175
PISTON CORER, FRAGILARIA, MELOSIR	W70-05663
PISUM, NOSTOC VERRUCOSUM, LAKE ME	W69-08283
PIT RECHARGE, WATER SUPPLY, MUNIC	W69-05894
PITHIOPHORA ALGAE.: /E FERTILIZER	W70-05417
PITTING(CORROSION), RUSTING, META	W70-02294
PIVERS ISLAND(N C), DETECTOR, BEA	W69-08267
PLAGIOGRAMMA,: /IGMA, MELOSIRA, N	W70-03325
PLAIN, INORGANIC COMPOUNDS, METAL	W69-03695
PLAINS, ALGAE, NEBRASKA, TURBIDIT	W72-04761
PLAINS, ATLANTIC COASTAL PLAIN, I	W69-03695
PLAN, DELTA WATER FACILITIES(CALI	W70-05094
PLANCTOMYCES, BACILLARIOPHYTA, GL	W70-03969
PLANKTON ANALYSES, AGRICULTURAL R	W70-04507
PLANKTON BLOOMS.: /I, JAPAN, *GLE	W70-08372
PLANKTON DISTRIBUTION, WISCONSIN	W70-03959
PLANKTON, PHYTOPLANKTON, WATER PO	W70-03959
PLANKTON, CHLOROPHYLL CONTENT, CE	W70-03959
PLANKTON, LAKES, RIVERS, GREAT LA	W70-04385
PLANKTON, OXYGEN, NITROGEN FIXATI	W70-04369
PLANKTON, ALGAE, *ALGAL CONTROL,	W70-05428
PLANKTON, AQUATIC WEED CONTROL, I	W70-05417
PLANKTON, LIGHT PENETRATION, SEAS	W70-05405
PLANKTON, NITROGEN, PHOSPHORUS, P	W70-05399
PLANKTON, WATER POLLUTION, SURFAC	W70-05389
PLANKTON, TUBIFICIDS, CALCIUM, SI	W70-05761
PLANKTON, ZOOPLANKTON, EUTROPHICA	W70-05112
PLANKTON, AQUATIC LIFE, ZOOPLANKT	W69-03695
PLANKTON, EPIDEMICS.: /ILL, ROOTE	W68-00488
PLANKTON, *ALGAE, CHEMICAL PROPER	W68-00483
PLANKTON, DISSOLVED OXYGEN, HYPOL	W68-00253
PLANKTON, PLANTS(ORGANISMS).: /,	W68-00481
PLANKTON, *CYANOPHYTA, WASHINGTON	W68-00461
PLANKTON, SPRAYING, BIOCONTROL, I	W68-00468
PLANKTON, BENTHIC FAUNA, SLUDGE W	W68-01244
PLANKTON, ALGAE, FROGS, MOSSES, H	W68-00857
PLANKTON, WATER POLLUTION EFFECTS	W69-03514
PLANKTON, FISH, SEA WATER, GAMMA	W69-08269
PLANKTON, ALGAE, PERIPHYTON, INVE	W69-07853
PLANKTON, ALGAE, INVERTEBRATES, T	W69-07861

, BACTERIA, DETRITUS, ESTUARIES,   PLANKTON, ECOSYSTEMS, SINKS, TRAC   W69-08274
ECOSYSTEMS, FISH, PHYTOPLANKTON,   PLANKTON, RADIOACTIVITY, RADIOACT   W69-07440
E, *CLAYS, *SILTS, SANDS, ALGAE,   PLANKTON, MICROORGANISMS, TURBIDI   W70-08628
*ANALYTICAL TECHNIQUES, *ALGAE,   PLANKTON, DIATOMS, MARINE ALGAE,   W70-09904
GY, PRIMARY PRODUCTIVITY, ALGAE,   PLANKTON, SUBMERGED PLANTS, PHYSI   W71-00665
FUNGI, ALGAE, DIATOMS, PROTOZOA,   PLANKTON, WORMS, NEMATODES, ROTIF   W72-01819
RDNESS(WATER), DIATOMS, FOULING,   PLANKTON, SEWAGE, DOMESTIC WASTE,   W72-01813
RAPHIES, ECOLOGY, BIOLOGY, FISH,   PLANKTON, WATER QUALITY, ALGAE, A   W72-01786
Y, PHYTOPLANKTON, PHYTOTOXICITY,   PLANKTON, POLLUTANT IDENTIFI: /OG   W70-02968
OLLUTION EFFECTS, PHYTOPLANKTON,   PLANKTON, AQUATIC MICROBIOLOGY, C   W70-02245
UATIC PLANTS, *ALGAE, *ROTIFERS,   PLANKTON, BASS, PHYTOPLANKTON, PO   W70-02249
E, VEGETATION, WILD RICE, SILTS,   PLANKTON, MAYFLIES, FAUNA, FISH,   W70-01943
OLOGICAL ECOLOGY, PHYTOPLANKTON,   PLANKTON, POLLUTION ABATEMENT, PO   W70-02792
AQUATIC PRODUCTIVITY, LIMNOLOGY,   PLANKTON, PRIMARY PRODUCTIVITY, T   W69-09135
*MOUNTAINS, *WINTER, CYANOPHYTA,   PLANKTON, COPEPODS, EUTROPHICATIO   W69-10154
S, SEWAGE, DETRITUS, TUBIFICIDS,   PLANKTON, DIATOMS, CYANOPHYTA, PH   W70-00268
ND VELOCITY, TASTE, COAGULATION,   PLANKTON, MONITORING, TEM: /S, WI   W70-00263
D, PERCHES, FISH, DIATOMS, CARP,   PLANKTON, METABOLISM, BIOLOGICAL   W70-00274
RENCES, *EUTROPHICATION, *ALGAE,   PLANNING, RIVERS, LAKES, LIMNOLOG   W70-04506
CYANIDE, NEW PRODUCT CO/ *MASTER   PLANNING, *SOURCE CONTAMINATION,   W73-00758
TTLE INC(MASS), CARNEGIE DEPT OF   PLANT BIOLOGY, CARNEGIE INSTITUTI   W69-06865
GHT MEASUREMENTS, EUPHOTIC ZONE,   PLANT CAROTENOIDS, CARBON FIXATIO   W70-01933
, FLOW, ANALYSIS, FERTILIZATION,   PLANT GROWTH SUBSTANCES, SEWAGE,   W70-02969
ATER POLLUTION EFFECTS, ECOLOGY/   PLANT GROWTH, WATER AND PLANTS, W   W70-02772
, *PHOSPHORUS SOURCES, *NUISANCE   PLANT GROWTHS, *NUTRIENT CONTROL,   W69-09340
S, POTABLE WATER, ACTINOMYCETES,   PLANT GROWTH, DECOMPOSING ORGANIC   W72-01812
YANOPHYTA, CHLOROPHYTA, BULRUSH,   PLANT GROWTH SUBSTANCES, CHLORELL   W72-00845
EN DEMAND, PHOSPHORUS COMPOUNDS,   PLANT GROWTH, PRIMARY PRODUCTIVIT   W69-06535
ALGAE, PLANT GROWTH SUBSTANCES,   PLANT GROWTH, CARBOHYDRATES.: /N,   W69-07826
UTRIENTS, EUTROPHICATION, ALGAE,   PLANT GROWTH SUBSTANCES, PLANT 'GR   W69-07826
NT GROWTH SUBSTANCES, LIMNOLOGY,   PLANT GROWTH REGULATORS, CYANOPHY   W68-00470
KE MORPHOMETRY, DOMESTIC WASTES,   PLANT GROWTH SUBSTANCES, LIMNOLOG   W68-00470
NG, IRON, MANGANESE, MOLYBDENUM,   PLANT GROWTH REGULATORS, PREDATIO   W69-04801
HIGAN, *EPIPHYTOLOGY, NUTRIENTS,   PLANT PATHOLOGY, *ALGAE.: /Y, MIC   W71-00832
ORINATED HYDROCARBON PESTICIDES,   PLANT PHYSIOLOGY.: /IRONMENT, CHL   W70-08652
ISSOLVED OXYGEN, SEWAGE LAGOONS,   PLANT PHYSIOLOGY, ANIMAL PHYSIOLO   W72-01818
EDS, *WATER ANALYSIS, *SAMPLING,   PLANT PHYSIOLOGY, EUTROPHICATION,   W70-04382
OTOSYNTHESIS, BIOMASS, CULTURES,   PLANT PHYSIOLOGY, ALGAE, DIURNAL   W70-05094
CHNIQUES, PHYSIOLOGICAL ECOLOGY,   PLANT PHYSIOLOGY.: /ANALYTICAL TE   W72-12240
S, *STREAMS, EUGLENA, NUTRIENTS,   PLANT POPULATIONS, HISTORY, WATER   W68-00483
TIES, *WATER QUALITY, TURBIDITY,   PLANT POPULATIONS.: /GICAL PROPER   W68-00475
OLIGOTROPHY, *FISH POPULATIONS,   PLANT POPULATIONS, ROUGH FISH, NI   W68-00172
PLANKTON, CYANOPHYTA, SEDIMENTS,   PLANT POPULATIONS, WEATHER, WATER   W70-06229
ES, ALGAE, SEDIMENTS, SOIL-WATER-   PLANT RELATIONSHIPS, NUCLEAR WAST   W71-09005
MINNETONKA(MINN), SOIL MINERALS,   PLANT RESIDUES, MORRIS(MINN), SOI   W70-04193
Y STUDIES, *MASS CULTURE, *PILOT-   PLANT STUDIES, ALGAL GROWTH, ALGA   W69-06865
LLA, HYDROGEN ION CONCENTRATION,   PLANT TISSUES, SCENEDESMUS, LIMNO   W68-00248
NTS, NYLON-6, *V/ *VISCOSE RAYON   PLANT WASTES, *MAN-MADE FIBER PLA   W70-06925
LUDGE PROCESS, HANOVER TREATMENT   PLANT.: /UM SULFATE, *ACTIVATED S   W70-09186
S, WASTE WATER TREATMENT, *FIBER(PLANT).: /OXYGEN DEMAND, NUTRIENT   W70-06925
RALIA, LABORATORY-SCALE STUDIES,   PLANT-SCALE STUDIES, REMOVAL.: /T   W70-06899
*SEATTLE(WASH), RENTON TREATMENT   PLANT, DUWAMISH ESTUARY(WASH).:   W70-04283
PHORUS COMPOUNDS, ALGAE, AQUATIC   PLANTS.: / *DENITRIFICATION, PHOS   W69-05323
REATMENT, LIGHT, MICROORGANISMS,   PLANTS.: /ABOLISM, *WASTE WATER T   W70-06604
COMMUNITIES, SYSTEMATICS, MARINE   PLANTS.: /STRIBUTION, BIOLOGICAL   W72-10623
NUTRIENTS, PHOSPHORUS, PLANKTON,   PLANTS(ORGANISMS).: /, NITROGEN,   W68-00481
TION EFFECTS; DRAINAGE / *LAKES,   PLANTS, *ALGAE, *SALINITY, IRRIGA   W69-00360
IUM COMPOUNDS, PHOTOSYN/ *MARINE   PLANTS, *ALGAE, CHLORINATION, SOD   W70-00161
TER CO/ *AQUATIC WEEDS, *AQUATIC   PLANTS, *AQUATIC WEED CONTROL, WA   W70-00269
ON, BASS, PHYTOPLANKTO/ *AQUATIC   PLANTS, *ALGAE, *ROTIFERS, PLANKT   W70-02249
*AQUATIC WEED CONTROL, *AQUATIC   PLANTS, *HERBICIDES, *VASCULAR TI   W70-10175
AKES, *PHYTOPLANKTON, / *AQUATIC   PLANTS, *INHIBITORS, *MARSHES, *L   W72-00845
MORTALITY, *MARINE ALGAE, *MARSH   PLANTS, *MARINE ANIMALS, *OILY WA   W70-01777
ES, *PHENOLS, *WATER O/ *AQUATIC   PLANTS, *NUISANCE ALGAE, *ALGICID   W70-10161
WATER), LIMNOLOGY, LAKES, *PILOT   PLANTS, *NUISANCE ALGAE, ROOTED A   W68-00479
E, MAR/ *ALGAL POISONING, *MARSH   PLANTS, *OILY WATER, AQUATIC ALGA   W70-01231
*NUISANCE ALGAE, ROOTED AQUATIC   PLANTS, *RESERVOIRS, HYDROELECTRI   W68-00479
*COOLING WATER, *ELECTRIC POWER   PLANTS, *WATER REUSE, COOLING TOW   W71-11393
TREATMENT, SUSPENDED LOAD, PILOT   PLANTS, ACTIVATED SLUDGE, TREAT: /   W70-09186
OXYGEN, DENSITY, ROOTED AQUATIC   PLANTS, ALGAE, CHLOROPHYLL, HYPOL   W71-07698
ANALYSIS, TEMPERATURE, NITRATES,   PLANTS, AMMONIA, HARDNESS(WATER),   W69-07838
TMENT, AEROBIC BACTERIA, AQUATIC   PLANTS, ANAEROBIC BACTERIA, ALGAE   W71-08221
OPES, *AQUATIC INSECTS, *AQUATIC   PLANTS, AQUATIC ENVIRONMENT, ECOS   W71-09005
AQUATIC MICROORGANISMS, AQUATIC   PLANTS, AQUATIC PRODUCTIVITY, BAL   W70-02792
AE, AQUATIC ENVIRONMENT, AQUATIC   PLANTS, AQUATIC POPULATIONS, CONS   W69-05844
OPHICATION, *HARVESTING, AQUATIC   PLANTS, AQUATIC WEEDS, ECONOMIC F   W69-06276
OMONAS, BACTERIA, FUNGI, AQUATIC   PLANTS, ASSAY, DISTRIBUTION, DENS   W70-02510
S, SAMPLING, PERIPHYTON, AQUATIC   PLANTS, BACKGROUND RADIATION, GEO   W69-09334
LUDGE, STORMS, TURBULENCE, PILOT   PLANTS, BACTERIA, SEDIMENTATION,   W70-05819
*EUTROPHICATION, ALGAE, AQUATIC   PLANTS, BIBLIOGRAPHIES, CYANOPHYT   W68-00680
EFFECTS, ALGAE, ANIMALS, AQUATIC   PLANTS, BIOMASS, FERTILIZATION, L   W69-04805
SOTOPES, *DISTRIBUTION, ANIMALS,   PLANTS, BOTTOM SEDIMENTS, SOILS,   W69-08269
ARY TREATMENT, NITROGEN, AQUATIC   PLANTS, BOTTOM SEDIMENTS, DREDGIN   W69-09340
TRATION, NUISANCE ALGAE, AQUATIC   PLANTS, BOTTOM SEDIMENTS, ANIMALS   W70-06225
HORUS, *NITROGEN, *ALGAE, SEWAGE   PLANTS, CHLOROPHYLL, PHOTOSYNTHES   W70-04268
E TREATMENT, ENGINEERING, ALGAE,   PLANTS, CONTROL, POLLUTION ABATEM   W69-03683
BIO/ *YEASTS, POPULATION, MARINE   PLANTS, DINOFLAGELLATES, OCEANS,   W70-04368
FUNGI, ALGAE, PROTOZOA, LICHENS,   PLANTS, DIATOMS, CORAL, MOLLUSKS,   W69-07428
PERATURE, OUTLETS, HYDROELECTRIC   PLANTS, ELECTRIC POWERPLANTS, ENG   W69-05023
ROPHICATION, ALGAE, MACROAQUATIC   PLANTS, FERTILIZATION.: /ION, EUT   W69-06859

502

ONTROL, FISHKILL, ROOTED AQUATIC
ONTROL, *EUTROPHICATION, AQUATIC
ONTROL, *EUTROPHICATION, AQUATIC
, SUSPENDED SOLIDS, ALGAE, PILOT
DGE, BIOLOGICAL TREATMENT, PILOT
GROWTH RATES, TEMPERATURE, PILOT
OROPHYTA, DIATOMS, WEIGHT, PILOT
IBU/ *ALGAE, *LITTORAL, *AQUATIC
AR PRODUCTS, GAS EXCHANGE, GREEN
T, MECHANICAL EQUIPMENT, PUMPING
ANIMALS, BENTHIC FAUNA, AQUATIC
ON PLANT WASTES, *MAN-MADE FIBER
S, MORTALITY, GASTROPODS, MARINE
NTS, ZOOPLANKTON, ALGAE, AQUATIC
NTS, ZOOPLANKTON, ALGAE, AQUATIC
VITY, ALGAE, PLANKTON, SUBMERGED
, CHARA, BENTHIC FLORA, FLOATING
CHLOROPHYLL, CYANOPHYTA, AQUATIC
MICAL / *MARINE ANIMALS, *MARINE
NTROL, PONDWEEDS, ROOTED AQUATIC
BACTERIA, FUNGI, ALGAE, FLOATING
 FLOATING PLANTS, ROOTED AQUATIC
MICROORGANISMS, WATER CHEMISTRY,
GINEERING STRUCTURES, INDUSTRIAL
REATMENT, AQUATIC ALGAE, AQUATIC
OONS, OXIDATION, PHOTOSYNTHESES,
RIO, WATER TURBIDITY, INDUSTRIAL
ECOLOGY/ PLANT GROWTH, WATER AND
AKE, *WINNISQUAM LAKE, WATER AND
PLANKTON, IRON COMPOUNDS, ALGAE,
Y, BENTHOS, AQUATIC LIFE, ALGAE,
N-SITE DATA COLLECTIONS, AQUATIC
ANTS, *RESERVOIRS, HYDROELECTRIC
ALGAE, DENSITY, SEASONAL, MARINE
, PACKED CELL VOLUMES, BACTERIAL
ER, CHEMOSYNTHETIC, SAPROPHYTIC,
BON UPTAKE TECHNIQUES, BACTERIAL

-FEEDING ANIMALS, RUTHENIUM-106,
OACTIVITY, GONIOBASIS, DECAPODA,
HOPTERA, ODONATA, EPHEMEROPTERA,
    SAVANNAH RIVER LABORATORY,
*ALGAL GROWTH, *POTOMAC RIVER,
MICROCOCCUS-LIKE, PSYCHROPHILES,
DAGRION, BRACHYTHEMIS, CHYDORUS,
WHITEFISH, WURTTEMBERG(GERMANY),
N, CYANIDE, NEW PRODUCT CONTROL,
IONS, DRAPARNALDIA, DRAPARNALDIA
IS, PISIDIUM IDAHOENSE, TENDIPES
WATER, BISMUTH-207, VERTEBRATES,
CARTERIA, NORTH TEMPERATE ZONE,
, WATER POLLUTION CONTROL, WATER
NSTITUTION, ACROPORA HYACINTHUS,
UM NODOSUM, POLYSIPHONIA LANOSA,
            DUCKS, NITRATE
OPLANKTON, *PHYTOTOXICITY, ALGAL
CES, ALGAE, ALGAL CONTROL, ALGAL
TOTOXICITY, ALGAL CONTROL, ALGAL
S, *ALGAE, *ALGAL CONTROL, ALGAL
ATER, AQUATIC ALGAE, MAR/ *ALGAL
OXIC/ *FISHKILL, *TOXINS, *ALGAL
RIVER BASIN, URBANIZATION, ALGAL
SPHORUS, POTASSIUM, SELECTIVITY,
RPATHIAN SUBMONTAN/ *SKAWA RIVER(
MASS, BENTHIC FLORA, PHAEOPHYTA,
TESTS, OIL SPILLS, GEOLOGICAL SS
S, RHINE RIVER, AMSTERDAM RHINE,
VITAMINS, WATER CHEMISTRY, WATER
D, ALGAL TOXINS, ORGANIC WASTES,
IDE RESIDUES, DIATOMS, CRAYFISH,
, PLANKTON, POLLUTION ABATEMENT,
EATMENT, *EUTROPHICATION, PONDS,
ANKTON, PHYTOTOXICITY, PLANKTON,
, WATER ANALYSIS, WATER QUALITY,
E, *ODOR, WATER QUALITY CONTROL,
WATER POLLUTION SOURCES, PATH OF
ASTE, WATER QUALITY ACT, PATH OF
            REMOVAL TIME(
RIO, *WATER CIRCULATION, PATH OF
UTROPHICATION, *WATER CHEMISTRY,
ALGAE, MARINE ALGAE, OIL, MARSH,
TIC ALGAE, *AQUATIC ENVIRONMENT,
WATER POLLUTION SOURCES, PATH OF
TRIENTS, PHOSPHORUS, POPULATION,
NU/ *EUTROPHICATION, *NUTRIENTS,
ASTES, POLLUTANT IDENTIFICATION,
AL PROPERTIES, BACTERIA, ORGANIC
KGROUND RADIATION, FLOCCULATION,
N CONTROL, *POLLUTION ABATEMENT,
DE KINETICS, PESTICIDE TOXICITY,

PLANTS, GLACIAL LAKES, PLANKTON,   W68-00488
PLANTS, HERBICIDES, PESTICIDES, H   W69-05705
PLANTS, HERBICIDES, PESTICIDES, H   W69-05706
PLANTS, HEAD LOSS, PERMEABILITY,   W72-08686
PLANTS, HUMUS, REAERATION, EFFLUE   W70-05821
PLANTS, HYDROGEN ION CONCENTRATIO   W72-06022
PLANTS, IRON, CHLORELLA, SODIUM,   W70-03512
PLANTS, ISLANDS, ECOLOGICAL DISTR   W72-10623
PLANTS, MASS CULTURE, MEDICINE, M   W69-07832
PLANTS, MAINE.: /, WATER TREATMEN   W70-08096
PLANTS, MARINE ALGAE, BENTHIC FLO   W71-02494
PLANTS, NYLON-6, *VISCOSE RAYON,   W70-06925
PLANTS, ON-SITE DATA COLLECTIONS.   W70-01780
PLANTS, PALEOLIMNOLOGY, MATHEMATI   W70-03975
PLANTS, PALEOLIMNOLOGY, MATHEMATI   W70-03975
PLANTS, PHYSICOCHEMICAL PROPERTIE   W71-00665
PLANTS, PHYTOPLANKTON, AQUATIC IN   W70-10175
PLANTS, PHOSPHORUS, DENSITY, WATE   W72-04761
PLANTS, RADIOISOTOPES, PHYSICOCHE   W71-09863
PLANTS, RATES OF APPLICATION, APP   W70-05568
PLANTS, ROOTED AQUATIC PLANTS, SE   W70-05750
PLANTS, SELF-PURIFICATION, BIOCHE   W70-05750
PLANTS, SOUTHWEST U. S., REGIONS,   W69-03611
PLANTS, STRUCTURES, AFTERBAYS, WA   W69-05023
PLANTS, TESTING, BIODEGRADATION.:   W70-04373
PLANTS, TEMPERATURE, *WASTE WATER   W70-06600
PLANTS, WATER SPORTS.: /LAKE ONTA   W71-09880
PLANTS, WATER POLLUTION EFFECTS,   W70-02772
PLANTS, WATER POLLUTION EFFECTS,   W70-02764
PLANTS, WATER POLLUTION EFFECTS,   W69-10170
PLANTS, WATER PROPERTIES, WATER T   W69-05023
PLANTS, WATERSHED, ALGAE, AQUATIC   W68-00511
PLANTS, WEED CONTROL, WATER PROPE   W68-Q0479
PLANTS,: /, WEATHER, EPILIMNION,   W70-02504
PLATE COUNTS, LANCASTER(CALIF).: /   W71-05267
PLATE COUNT, FILAMENTOUS, UNICELL   W68-00891
PLATE, BACTERIAL PIGMENTS, DIAPTO   W70-05760
PLATTE RIVER, LINCOLN(NEBR).:   W69-05894
PLATYMONAS, THALASSIOSIRA, AMPHID   W70-00708
PLECOPTERA, EPHEMEROPTERA, TIPULI   W69-09742
PLECOPTERA, DIPTERA, COLEOPTERA,   W69-07846
PLECTONEMA BORYANUM, REACTOR.:   W69-07845
PLECTONEMA, NOSTOC PALUDOSUM, BLE   W70-02255
PLEMORPHIC, MYCELOID, CORYNEFORM   W70-00713
PLEUROXUS, TANYPUS, TANYTARSUS, L   W70-06975
PLON(GERMANY), LAKE SCHLIER(GERMA   W70-05761
PLUMBING CODES, WELL DRILLING COD   W73-00758
PLUMOSA, STIGEOCLONIUM, MICROCYST   W70-02245
PLUMOSUS, BEAVER DAM LAKE(WIS), D   W70-06217
PLUTONIUM-239, RATS, ZIRCONIUM-95   W69-08269
PLUTONIUM, HALF-LIFE(RADIONUCLIDE   W70-00708
PO: /STE TREATMENT, WATER QUALITY   W69-06535
POCILLOPORA EYDOUXI, PORITES LOBA   W69-08526
POINT LEPREAU(CANADA), NEW BRUNSW   W70-04365
POISONING.:   W69-04805
POISONING, SCENEDESMUS, CHLORELLA   W69-00994
POISONING, ALGICIDES, AQUATIC ALG   W69-03370
POISONING, ANALYTICAL TECHNIQUES,   W69-03188
POISONING, ALGICIDES, AQUATIC ALG   W69-05844
POISONING, *MARSH PLANTS, *OILY W   W70-01231
POISONING, *ALGAL TOXINS, *PHYTOT   W71-07731
POISONING, GASOLINE, FLOODS, AQUI   W73-00758
POISONS, OXYGEN, TRACERS, PHOSPHO   W70-05399
POLAND), MONTANE RIVER COURSE, CA   W70-02784
POLAR REGIONS, SPECTROPHOTOMETRY,   W70-01074
POLARIS, CHLOROPHYLL FLUORESCENCE   W70-05377
POLDERS, COMPOSITION, PSEUDOMONAS   W69-07838
POLLUT: /MENTS, SODIUM, VIRUSES,   W69-04801
POLLUTANT IDENTIFICATION, POLLUTA   W69-01273
POLLUTANT IDENTIFICATION, *WATER   W69-02782
POLLUTANT IDENTIFICATION, WATER P   W70-02792
POLLUTANT IDENTIFICATION, BIOLOGY   W70-02304
POLLUTANT IDENTIFI: /OGY, PHYTOPL   W70-02968
POLLUTANT IDENTIFICATION.: /PLING   W71-00221
POLLUTANT IDENTIFICATION, IOWA.: /   W72-13605
POLLUTANT.: /AEROBIC CONDITIONS,   W71-11381
POLLUTANT, FOOD CHAINS, SAMPLING,   W72-00941
POLLUTANTS).:   W71-05878
POLLUTANTS, NUTRIENTS, ALGAE, STR   W71-05878
POLLUTANTS, ALGAL CONTROL, NITRAT   W70-02646
POLLUTANTS.: /ILY WATER, AQUATIC   W70-01231
POLLUTANTS, BIOLOGICAL COMMUNITIE   W70-01233
POLLUTANTS, AQUATIC LIFE, ALGAE,   W69-10080
POLLUTANTS, MAPPING, HYDROGEN SUL   W70-00667
POLLUTANTS, POLLUTION ABATEMENT,   W69-01453
POLLUTANTS, QUALITY CONTROL, WATE   W69-01273
POLLUTANTS, *STREAMS, EUGLENA, NU   W68-00483
POLLUTANTS, SAMPLING.: /ATES, BAC   W69-07846
POLLUTANTS, WASTES, THERMAL POLLU   W69-06305
POLLUTANTS, WATER POLLUTION EFFEC   W69-08565

504

```
Y, RADIOACTIVITY, TRACERS, WATER    POLLUTION EFFECTS, PHYSIOLOGICAL:      W70-02510
AM LAKE, WATER AND PLANTS, WATER    POLLUTION EFFECTS, ECOLOGY, WATER      W70-02764
POUNDS, LIGHT PENETRATION, WATER    POLLUTION EFFECTS, EUTROPHICATION      W70-02255
ER QUALITY CONTROL, LAKES, WATER    POLLUTION SOURCES, CHLOROPHYTA, N      W70-03310
   *RIVERS, WATER CHEMISTRY, WATER  POLLUTION SOURCES, PHOSPHORUS, NI      W70-03068
GAE, NITROGEN, PHOSPHORUS, WATER    POLLUTION EFFECTS.: /ERTILITY, AL      W70-02973
ROL, EUTROPHICATION, BASS, WATER    POLLUTION CONTROL, MINNOWS, NUISA      W70-02982
C MATTER, CURRENTS(WATER), WATER    POLLUTION SOURCES, SURFACE RUNOFF      W70-03501
, STANISLAUS RIVER(CALIF), WATER    POLLUTION ASSESSMENT, AMERICAN RI      W70-02777
, DIATOMS, EUTROPHICATION, WATER    POLLUTION EFFECTS, SYSTEMS ANALYS      W70-02779
, WATER POLLUTION CONTROL, WATER    POLLUTION EFFECTS, WATER POLLUTIO      W70-02792
POLLUTANT IDENTIFICATION, WATER     POLLUTION CONTROL, WATER POLLUTIO      W70-02792
TRATES, INDUSTRIAL WASTES, WATER    POLLUTION EFFECTS, DENITRIFICATIO      W70-02777
COLOGY, PHYTOPLANKTON, PLANKTON,    POLLUTION ABATEMENT, POLLUTANT ID      W70-02792
, WATER POLLUTION EFFECTS, WATER    POLLUTION SOURCES.: /TION CONTROL      W70-02792
ANABAENA, BIOSTIMULATION, WATER     POLLUTION ASSESSMENT.: /L GROWTH,      W70-02779
T, *BIOLOGICAL TREATMENT, *WATER    POLLUTION CONTROL, FISH, TOXICITY      W70-10437
ATER, *OIL, *MARINE ALGAE, WATER    POLLUTION SOURCES, ON-SITE DATA C      W70-09976
OILY WATER, / RESISTANCE, WATER     POLLUTION EFFECTS, *SALT MARSHES,      W70-09976
ESIDUES, *PESTICIDE KINE/ *WATER    POLLUTION EFFECTS, *PESTICIDES, R      W70-09768
, WATER POLLUTION SOURCES, WATER    POLLUTION EFFECTS, PHOSPHORUS, NI      W70-07261
*EUTROPHICATION, *LAKES, *WATER     POLLUTION CONTROL, *REVIEWS, ALGA      W70-07393
N SOURCES, *THERMAL POLL/ *WATER    POLLUTION EFFECTS, WATER POLLUTIO      W70-07847
   *ALGAE, *BIOINDICATORS, *WATER   POLLUTION EFFECTS, *REVIEWS, *WAT      W70-07027
LOGY, *INDUSTRIAL WASTES, WATER     POLLUTION EFFECTS, *EUTROPHICATIO      W70-06975
*WATER POLLUTION EFFECTS, WATER     POLLUTION SOURCES, *THERMAL POLLU      W70-07847
RATES, PHOSPHATES, SODIUM, WATER    POLLUTION EFFECTS, LABORATORY TES      W70-07257
*NUISANCE ALGAE, *SEWAGE, WATER     POLLUTION SOURCES, WATER POLLUTIO      W70-07261
ATORY TESTS, INSECTICIDES, WATER    POLLUTION EFFECTS, ALGAE.: /LABOR      W71-03183
S, ORGANIC MATTER, CARBON, WATER    POLLUTION EFFECTS, BACTERIA, PERI      W71-04206
PHYTOPLANKTON), ESTUARIES, WATER    POLLUTION SOURCES, WATER POLLUTIO      W71-05407
*LAKES, *EUTROPHICATION, *WATER     POLLUTION CONTROL, *WATER POLLUTI      W71-03048
*WATER POLLUTION CONTROL, *WATER    POLLUTION SOURCES, *WATER POLLUTI      W71-03048
, WATER POLLUTION SOURCES, WATER    POLLUTION EFFECTS, NITROGEN, COLI      W71-05407
, *WASTE WATER TREATMENT, *WATER    POLLUTION CONTROL, *DENITRIFICATI      W71-04555
DEMAND, BACTERIA, ECOLOGY, WATER    POLLUTION EFFECTS, LAKES, *CYANOP      W71-04758
, *WASTE WATER TREATMENT, *WATER    POLLUTION CONTROL, *DENITRIFICATI      W71-04556
*WATER POLLUTION SOURCES, *WATER    POLLUTION EFFECTS, WATER UTILIZAT      W71-03048
IOLOGY, NITRATES, AMMONIA, WATER    POLLUTION EFFECTS, WASTE WATER TR      W71-05026
, *WASTE WATER TREATMENT, *WATER    POLLUTION CONTROL, *DENITRIFICATI      W71-04554
, *WASTE WATER TREATMENT, *WATER    POLLUTION CONTROL, *DENITRIFICATI      W71-04557
N EFFECTS, DINOFLAGELLATE/ WATER    POLLUTION SOURCES, WATER POLLUTIO      W70-08372
                         *WATER     POLLUTION RESEARCH.:                   W70-08627
PARTICLE SIZE, *CLAYS, */ *WATER    POLLUTION CONTROL, *FILTRATION, *      W70-08628
, *WASTE WATER TREATMENT, *WATER    POLLUTION CONTROL, *EUTROPHICATIO      W70-09186
   WATER POLLUTION SOURCES, WATER   POLLUTION EFFECTS, DINOFLAGELLATE      W70-08372
UND/ *ALGAE, *CHLOROPHYLL, WATER    POLLUTION EFFECTS, *ORGANIC COMPO      W70-08416
PMENT,/ *WATER RESOURCES, *WATER    POLLUTION CONTROL, *RESEARCH EQUI      W70-08627
ROMATOGRAPHY, DEGRADATION, WATER    POLLUTION EFFECTS, AQUATIC ENVIRO      W70-08652
MENT, CULTURES, CHLORELLA, WATER    POLLUTION EFFECTS.: / WATER TREAT      W70-09438
STS, MINNESOTA, NUTRIENTS, WATER    POLLUTION EFFECTS.: /ABORATORY TE      W70-08637
SULFONATES, *SURFACTANTS, WATER     POLLUTION SOURCES, *DETERGENTS, *      W70-09438
, WATER POLLUTION SOURCES, WATER    POLLUTION EFFECTS, DISSOLVED OXYG      W70-04727
, SEDIMENTS, URBANIZATION, WATER    POLLUTION TREATMENT, ALGAE, WATER      W70-04721
ANALYSIS, ORGANIC MATTER, WATER     POLLUTION EFFECTS, ANALYTICAL TEC      W70-05109
L ECOLOGY, ORGANIC MATTER, WATER    POLLUTION EFFECTS, DIATOMS, CARBO      W70-04811
CATION, *LAKES, NUTRIENTS, WATER    POLLUTION EFFECTS, PRODUCTIVITY,       W70-04721
MESTIC WASTES, POLLUTANTS, WATER    POLLUTION SOURCES, WATER QUALITY,      W70-05084
, *STRATIFICATION, *TEXAS, WATER    POLLUTION SOURCES, WATER POLLUTIO      W70-04727
EWAGE DISPOSAL, WATER POLLUTION,    POLLUTION ABATEMENT, COLIFORMS, W      W70-04810
S, *PHYTOPLANKTON, *DDT,/ *WATER    POLLUTION EFFECTS, *PHOTOSYNTHESI      W70-05272
ANALYT/ *WATER CHEMISTRY, *WATER    POLLUTION EFFECTS, *PHOSPHATES, *      W70-05269
   *NUTRIENTS, *PESTICIDES, *WATER  POLLUTION CONTROL, EUTROPHICATION      W70-05470
TS, *PRIMARY PRODUCTIVITY, WATER    POLLUTION EFFECTS, BIOASSAYS, ALG      W70-05424
, *LAKES, *RIVERS, *ALGAE, WATER    POLLUTION CONTROL, WATER POLLUTIO      W70-05668
, WATER POLLUTION CONTROL, WATER    POLLUTION EFFECTS, WATER POLLUTIO      W70-05668
, WATER POLLUTION EFFECTS, WATER    POLLUTION SOURCES, SEWAGE, PHOSPH      W70-05668
UATIC LIFE, AQUATIC ALGAE, WATER    POLLUTION EFFECTS.: /DICATORS, AQ      W70-05958
OCARBON UPTAKE TECHNIQUES, WATER    POLLUTION IDENTIFICATION, ORGANIC      W70-05652
UCTIVITY, WASTE TREATMENT, WATER    POLLUTION EFFECTS, WATER POLLUTIO      W69-08518
GAE, NITROGEN, PHOSPHORUS, WATER    POLLUTION EFFECTS, WATER POLLUTIO      W69-08668
LL, LAKES, ODOR, TOXICITY, WATER    POLLUTION EFFECTS, WATER QUALITY,      W69-08279
T LAKES, *TRACE ELEMENTS, *WATER    POLLUTION EFFECTS, CHLORIDES, HAL      W69-08562
CIDE TOXICITY, POLLUTANTS, WATER    POLLUTION EFFECTS, ALDRIN, DIELDR      W69-08565
HLORIDES, HALOGENS, ALGAE, WATER    POLLUTION SOURCES, WATER CHEMISTR      W69-08562
S, NUTRIEN/ *NEW HAMPSHIRE WATER    POLLUTION COMMISSION, ALGAL BLOOM      W69-08674
, WATER POLLUTION EFFECTS, WATER    POLLUTION SOURCES, CYCLING NUTRIE      W69-08518
NUISANCE ALGAE, BACTERIA, WATER     POLLUTION EFFECTS.: /HIES, LAKES,      W69-08283
, WATER POLLUTION SOURCES, WATER    POLLUTION CONTROL, OXYGEN BALANCE      W69-08668·
, WATER POLLUTION EFFECTS, WATER    POLLUTION SOURCES, WATER POLLUTIO      W69-08668
, WATER POLLUTION EFFECTS, WATER    POLLUTION SOURCES.: /ER CHEMISTRY      W69-06273
HEMISTRY, WATER CHEMISTRY, WATER    POLLUTION EFFECTS, WATER POLLUTIO      W69-06273
NITRATES, EUTROPHICATION, *WATER    POLLUTION SOURCES, AGRICULTURAL W      W69-05323
, WATER POLLUTION CONTROL, WATER    POLLUTION SOURCES, WISCONSIN.: /S      W69-06276
LUTION CONTROL, NUTRIENTS, WATER    POLLUTION EFFECTS, WATER POLLUTIO      W69-06276
TEMEN/ *OHIO, *LAKE ERIE, *WATER    POLLUTION CONTROL, *POLLUTION ABA      W69-06305
NG, WATER QUALITY CONTROL, WATER    POLLUTION CONTROL, WATER POLLUTIO      W69-05706
NG, WATER QUALITY CONTROL, WATER    POLLUTION CONTROL, WATER POLLUTIO      W69-05705
, WATER POLLUTION CONTROL, WATER    POLLUTION EFFECTS.: /LITY CONTROL      W69-05705
, WATER POLLUTION CONTROL, WATER    POLLUTION EFFECTS, FUNGI, AQUATIC      W69-05706
```

```
LGAE, NUTRIENTS, VITAMINS, WATER   POLLUTION EFFECTS, WATER POLLUTIO   W69-06277
ATER POLLUTION, WATER LAW, WATER   POLLUTION EFFECTS, WATER POLLUTIO   W69-06305
OMPOUNDS, WATER CHEMISTRY, WATER   POLLUTION EFFECTS, WATER QUALITY,   W69-05868
TY, LAKES, NUISANCE ALGAE, WATER   POLLUTION CONTROL, NUTRIENTS, WAT   W69-06276
, WATER POLLUTION EFFECTS, WATER   POLLUTION SOURCES, WATER POLLUTIO   W69-06305
MANGANESE, EUTROPHICATION, WATER   POLLUTION CONTROL.: /HYTA, IRON,    W69-05704
TOXICITY, WATER CHEMISTRY, WATER   POLLUTION EFFECTS.: / NUTRIENTS,    W69-05867
, WATER POLLUTION SOURCES, WATER   POLLUTION TREATMENT, FINANCING, G   W69-06305
, WATER POLLUTION EFFECTS, WATER   POLLUTION SOURCES, VITAMIN B.: /S   W69-06277
PRODUCTIVITY, SYSTEMATICS, WATER   POLLUTION EFFECTS, WATER POLLUTIO   W69-07832
, WATER POLLUTION EFFECTS, WATER   POLLUTION SOURCES, WATER SUPPLY.:   W69-07832
ATION, *INLAND WATERWAYS, *WATER   POLLUTION EFFECTS, *WATER SUPPLY,   W69-07833
TREATMENT, WATER QUALITY, WATER    POLLUTION CONTROL, WATER PO: /STE   W69-06535
EMS, *SIMULATION ANALYSIS, WATER   POLLUTION EFFECTS, LAKES, PHOSPHO   W69-06536
NUTRIENTS, EUTROPHICATION, WATER   POLLUTION EFFECTS, WATER POLLUTIO   W69-06864
TS, BIOLOGICAL PROPERTIES, WATER   POLLUTION EFFECTS.: /IDE, NUTRIEN   W69-07442
NTS, FE/ *EUTROPHICATION, *WATER   POLLUTION SOURCES, *LAKES, NUTRIE   W69-07084
, WATER POLLUTION EFFECTS, WATER   POLLUTION SOURCES, BIOINDICATORS,   W69-06864
ONSIN, *SEWAGE EFFLUENTS, *WATER   POLLUTION CONTROL, *LOCAL GOVERNM   W69-06909
ACTERIA, AQUATIC BACTERIA, WATER   POLLUTION EFFECTS.: /S, YEASTS, B   W69-06970
S, RADIOECOLOGY, SORPTION, WATER   POLLUTION EFFECTS, WATER POLLUTIO   W69-07440
, WATER POLLUTION EFFECTS, WATER   POLLUTION SOURCES.: /GY, SORPTION   W69-07440
EN DEMAND, ODOR-PRODUCING ALGAE,   POLLUTION ABATEMENT, SEWAGE TREAT   W69-06909
RIVERS, *WATER POLLUTION, *WATER   POLLUTION SOURCES, IRRIGATION, AQ   W69-07096
PHYTOPLANKTON, *REVIEWS, *WATER    POLLUTION EFFECTS.: /OLIGOTROPHY,   W68-00680
ANOPHYTA, LAND MANAGEMENT, WATER   POLLUTION SOURCES, DREDGING, *POL   W68-00172
CONTROL, SOLAR RADIATION, *WATER   POLLUTION CONTROL, ALGICIDES, BIO   W68-00478
DUCKS(WILD), OYSTERS, CHANNELS,    POLLUTION ABATEMENT, HUMAN POPULA   W68-00461
E ERIE, *NUTRIENTS, ALGAE, WATER   POLLUTION SOURCES, NITROGEN COMPO   W69-01445
CONTROL, WATER POLLUTION, *WATER   POLLUTION SOURCES, SEPARATION TEC   W69-01273
RIVERS, SEWAGE TREATMENT, WATER    POLLUTION SOURCES, PRIMARY TREATM   W69-01453
ICATION, *NUTRIENTS, POLLUTANTS,   POLLUTION ABATEMENT, NUISANCE ALG   W69-01453
*WATER POLLUTION SOURCES, *WATER   POLLUTION EFFECTS, *PRIMARY PRODU   W69-02959
POLLUTANT IDENTIFICATION, *WATER   POLLUTION EFFECTS, GAS CHROMATOGR   W69-02782
HY, ANALYTICAL TECHNIQUES, WATER   POLLUTION SOURCES.: /CHROMATOGRAP   W69-00387
ON EFFE/ *EUTROPHICATION, *WATER   POLLUTION SOURCES, *WATER POLLUTI   W69-02959
MINANT ORGANISMS, GRAZING, WATER   POLLUTION EFFECTS, DIATOMS, CHLOR   W69-01977
LAKES, *WATER POLLUTION, *WATER    POLLUTION CONTROL, WATER POLLUTIO   W69-03948
S, *PHOSPHORUS COMPOUNDS, *WATER   POLLUTION EFFECTS, ALGAE, ANIMALS   W69-04805
NEERING, ALGAE, PLANTS, CONTROL,   POLLUTION ABATEMENT, ABATEMENT, P   W69-03683
*WATER POLLUTION CONTROL, WATER    POLLUTION EFFECTS, EUTROPHICATION   W69-03948
OMPOUNDS, *PHYTOPLANKTON, *WATER   POLLUTION SOURCES, *CYCLING NUTRI   W69-04800
NGTON, *SEWAGE TREATMENT, *WATER   POLLUTION CONTROL, PACIFIC NORTHW   W69-03683
, *PHYSIOLOGICAL ECOLOGY, WATER    POLLUTION EFFECTS, ALGAE, ALGAL C   W69-03513
NTHETIC, OXYGEN, PLANKTON, WATER   POLLUTION EFFECTS.: /NTS, PHOTOSY   W69-03514
*WATER POLLUTION CONTROL, *WATER   POLLUTION TREATMENT, ALGAE, ALGAL   W69-03374
N, *PHOSPHORUS COMPOUNDS, *WATER   POLLUTION SOURCES, ALGAE, ALGAL C   W69-03370
REMENTS, *EUTROPHICATION, *WATER   POLLUTION CONTROL, *WATER POLLUTI   W69-03374
ENTS, *EUTROPH/ *STREAMS, *WATER   POLLUTION EFFECTS, *CYCLING NUTRI   W69-03369
ON EFFECTS, *EUTROPHICAT/ *WATER   POLLUTION SOURCES, *WATER POLLUTI   W70-04266
CHLAMYDOMONAS, BIOASSAYS, WATER    POLLUTION EFFECTS, ANALYTICAL TEC   W70-04284
A, *ANALYTICAL TECHNIQUES, WATER   POLLUTION EFFECTS, ALGAE, ECOSYST   W70-04287
*WATER POLLUTION SOURCES, *WATER   POLLUTION EFFECTS, *EUTROPHICATIO   W70-04266
, ALGAE, FISH POPULATION, *WATER   POLLUTION CONTROL, FARM WASTES, M   W70-04266
ES, ANALYTICAL TECHNIQUES, WATER   POLLUTION EFFECTS, PHYTOPLANKTON,   W70-04194
*WATER POLLUTION EFFECTS, WATER    POLLUTION CONTROL, *CONFERENCES,    W70-03975
ENTAL EFFECTS, PERIPHYTON, WATER   POLLUTION EFFECTS, OREGON, BIOIND   W70-03978
TROPHICATION, *MEROMIXIS, *WATER   POLLUTION EFFECTS, BIOINDICATORS,   W70-03974
*WATER POLLUTION SOURCES, *WATER   POLLUTION EFFECTS, *WATER POLLUTI   W70-03975
ON EFFE/ *EUTROPHICATION, *WATER   POLLUTION SOURCES, *WATER POLLUTI   W70-03975
, WATER POLLUTION SOURCES, WATER   POLLUTION EFFECTS, LABORATORY TES   W70-04074
IOR, MADISON(WIS), EUROPE, WATER   POLLUTION ASSESSMENT, URBAN DRAIN   W70-03975
*WATER POLLUTION EFFECTS, *WATER   POLLUTION CONTROL, *CONFERENCES,    W70-03975
ON EFFE/ *EUTROPHICATION, *WATER   POLLUTION SOURCES, *WATER POLLUTI   W70-03975
UENTS, TERTIARY TREATMENT, WATER   POLLUTION EFFECTS, *WATER POLLUTI   W70-03975
IOR, MADISON(WIS), EUROPE, WATER   POLLUTION SOURCES, WATER POLLUTIO   W70-04074
ES, ENVIRONMENTAL EFFECTS, WATER   POLLUTION ASSESSMENT, URBAN DRAIN   W70-03975
TRATES, LAKE ERIE, LAKE ONTARIO,   POLLUTION EFFECTS, EPILIMNION, CO   W70-03957
SANCE ALGAE, PUBLICATIONS, WATER   POLLUTION ABATEMENT, OLIGOTROPHY,   W70-03964
, PLANKTON, PHYTOPLANKTON, WATER   POLLUTION EFFECTS, NEW YORK, CHLO   W70-03973
, SCENEDESMUS, PHOSPHATES, WATER   POLLUTION EFFECTS, WISCONSIN ZOOP   W70-03959
IABILITY, WATER POLLUTION, WATER   POLLUTION SOURCES.: /E, *BIOASSAY   W70-03955
NTROL, MECHANICAL CONTROL, WATER   POLLUTION EFFECTS, NITROGEN, PHOS   W70-04382
SON RIVER, DELAWARE RIVER, WATER   POLLUTION CONTROL, CHEMCONTROL, A   W70-04494
E, DRAINAGE DIST/ *ALGAE, *WATER   POLLUTION EFFECTS.: /E RIVER, HUD   W70-04507
WAGE EFFLUENTS, MINNESOTA, WATER   POLLUTION TREATMENT, LAKES, SEWAG   W70-04455
ON EFFECTS, *WATER SOURC/ *WATER   POLLUTION EFFECTS, ALGICIDES, WAT   W70-04506
*WATER POLLUTION SOURCES, *WATER   POLLUTION SOURCES, *WATER POLLUTI   W73-00758
VAL, SOIL EROSION, ALGAE, *WATER   POLLUTION SOURCES, *WATER SOURCES   W73-00758
WASTES, FRESHWATER ALGAE, WATER    POLLUTION CONTROL, WATER QUALITY    W73-01122
, TURF, SOIL MICROBIOLOGY, WATER   POLLUTION EFFECTS, DOMESTIC WASTE   W72-12240
, ALGAE, LABORATORY TESTS, WATER   POLLUTION SOURCES.: /, SOIL ALGAE   W72-08110
GREAT LAKES BASIN, BACTERIAL       POLLUTION EFFECTS.: /N PESTICIDES   W72-11105
SLUDGE WORMS, BLOODWORMS, WATER    POLLUTION.:                         W69-03948
OMPOSING ORGANIC MATTER, *STREAM   POLLUTION.: /KTON, BENTHIC FAUNA,   W68-01244
                     *ESTUARINE    POLLUTION.: /TER UTILIZATION, DEC   W68-00483
NA, FLORA, LABORATORY, MINERALS,   POLLUTION.:                         W69-06203
, NUTRIENTS, SEWAGE, WASTE WATER(  POLLUTION.: /ISH, ECOSYSTEMS, FAU   W72-02203
HOSPHATES, NITROGEN, WASTE WATER(  POLLUTION), WATER POLLUTION SOURC   W71-05879
                                   POLLUTION), SCENEDESMUS, CYANOPHY   W70-02255
```

```
SING ORGANIC MATTER, WASTE WATER( POLLUTION).: /, BACTERIA, DECOMPO      W70-00664
ACHES, SANDS, BIOTA, WASTE WATER( POLLUTION), CURRENTS(WATER).:/BE       W68-00010
PHYSICAL PROPERTIES, WASTE WATER( POLLUTION), WATER QUALITY: /IES,       W69-01273
ENTS, PHYTOPLANKTON, WASTE WATER( POLLUTION), PHOTOSYNTHESIS, CHL: /     W70-04371
*SELF-PURIFICATION, *WASTE WATER( POLLUTION), *WATER PURIFICATION,       W70-05092
                                   POLLUTION-TOLERANT ALGAE.:            W70-07027
ATER POLLUTION SOURCES, *THERMAL   POLLUTION, THERMAL POWERPLANTS, *     W70-07847
CCLIMATIZATION, DIATOMS, THERMAL   POLLUTION, TEMPERATURE, CHEMICAL      W70-07313
*REVIEWS, *WATER QUALITY, WATER    POLLUTION, ORGANIC MATTER, BIBLIO     W70-07027
IODEGRADATION, DETERGENTS, WATER   POLLUTION, WASTE WATERS, BIOCHEMI     W70-08740
ION CONCENTRATION, NU/ *THERMAL    POLLUTION, *TEMPERATURE, HYDROGEN     W70-08834
OPPER SULFATE, RESERVOIRS, WATER   POLLUTION, MICROORGANISMS, HISTOR     W70-08096
TICAL TECHNIQUES, BIOMASS, WATER   POLLUTION, ALGAE, PERIPHYTON, PHO     W71-04206
*BENTHOS, *TEMPERATURE, *THERMAL   POLLUTION, LABORATORY ANIMALS, AC     W71-02494
AE, / *RECOVERY, OIL SPILL, *OIL   POLLUTION, *FILAMENTOUS GREEN ALG     W70-09976
TROL, INDUSTRIAL WASTES, THERMAL   POLLUTION, METHODOLOGY, INSTRUMEN     W70-10437
STES, *WASTE WATER DISPOSAL, AIR   POLLUTION, WATER POLLUTION, ODOR,     W71-00940
R DISPOSAL, AIR POLLUTION, WATER   POLLUTION, ODOR, EFFLUENTS, SEWAG     W71-00940
*TEMPERATURE EFFECTS, *THERMAL     POLLUTION, *ECOLOGY, ALGAE, BACTE     W71-02479
*ECOLOGY, *BIOTA, *TEM/ *THERMAL   POLLUTION, *THERMAL POWERPLANTS,      W71-02491
IFORMS, FISH, FUNGI, HEAT, WATER   POLLUTION, STREAMS, RIVERS, LAKES     W71-02479
ACTERIA, SEWAGE TREATMENT, WATER   POLLUTION, WATER POLLUTION EFFECT     W69-10171
ON EFFECTS, ALGAL CONTROL, WATER   POLLUTION, TOXICITY, LAKES, DESIG     W70-02775
WAGE TREATMENT, *TRACERS, *WATER   POLLUTION, DESIGN, DESIGN CRITERI     W70-02609
T IDENTIFICATION, BIOLOGY, WATER   POLLUTION, WATER POLLUTION SOURCE     W70-02304
ION EFFECTS, ECOLOGY, EFFECTS OF   POLLUTION, WATER QUALITY CONTROL,     W70-02772
LORA, BENTHOS, MINE WATER, WATER   POLLUTION, WATER POLLUTION EFFECT     W70-02770
RY, WATE/ *WATER QUALITY, *WATER   POLLUTION, *RIVERS, WATER CHEMIST     W70-03068
MATTER, WATER POLLUTION, THERMAL   POLLUTION, STREAMS, NUTRIENTS, LI     W70-03309
CTERIA, CHEMICAL ANALYSIS, WATER   POLLUTION, CHLORELLA, SCENEDESMUS     W70-03312
NSERVATION, PUBLIC HEALTH, WATER   POLLUTION, CONTRACTS, REGULATION,     W70-03344
COSYSTEMS, ORGANIC MATTER, WATER   POLLUTION, THERMAL POLLUTION, STR     W70-03309
RINE BACTERIA, *TOXICITY, *WATER   POLLUTION, AQUATIC ALGAE, BRINE S     W70-01467
OIL WASTES, *SURFACTANTS, *WATER   POLLUTION, CLEANING, DISASTERS, O     W70-01470
ON, *KUWAIT C/ *SALT MARSH, *OIL   POLLUTION, *RECOVERY FROM POLLUTI     W70-01231
*INSECTICIDES, *TOXICITY, *WATER   POLLUTION, FOOD CHAINS, AQUATIC A     W70-01466
ISH PHYSIOLOGY, *ALGAE, *THERMAL   POLLUTION, *TOXICITY.:          *F   W70-01788
, *OIL POLLUTION, *RECOVERY FROM   POLLUTION, *KUWAIT CRUDE, BENTLAS     W70-01231
DEVELOPMENT, GRANTS, ALGAE, AIR    POLLUTION, WATER POLLUTION, SOIL      W71-06821
NTS, ALGAE, AIR POLLUTION, WATER   POLLUTION, SOIL CONTAMINATION, NU     W71-06821
TY, *DISTRIBUTION SYSTEMS, WATER   POLLUTION, WATER QUALITY.: /OXICI     W72-03746
*EUTROPHICATION, *LAKES, *WATER    POLLUTION, DEGRADATION, SEWAGE TR     W72-02817
*WATER QUALITY CONTROL/ *THERMAL   POLLUTION, *NUCLEAR POWERPLANTS,      W71-11881
INDICATORS, *BIOASSAY, T/ *WATER   POLLUTION, *AQUATIC BIOLOGY, *BIO     W71-10786
EDIMENTS, MARINE BIOLOGY, STREAM   POLLUTION, *WATER POLLUTION DETEC     W71-10786
ION, ALGAE, OXYGEN DEMAND, WATER   POLLUTION, MUNICIPAL WATER, WATER     W71-09577
ITY, ALGAE, AQUATIC LIFE, STREAM   POLLUTION, BACTERIA, IMPURITIES,      W72-01786
DISPOSAL, *WASTE STORAGE, *WATER   POLLUTION, WATER POLLUTION SOURCE     W72-00941
VE WASTES, HEATED WATER, THERMAL   POLLUTION, BACTERIA, EFFLUENTS, O     W72-00462
E TREATMENT, *WASTE WATE/ *WATER   POLLUTION, WASTE DISPOSAL, *SEWAG     W72-01786
CREATION, SEWAGE DISPOSAL, WATER   POLLUTION, POLLUTION ABATEMENT, C     W70-04810
*WASHINGTON, URBANIZATION, WATER   POLLUTION, SEDIMENTS, PALEOLIMNOL     W70-05663
TERS, *ORGANISMS, STREAMS, WATER   POLLUTION, ANIMALS, BACTERIA, FUN     W70-05750
LTURAL WATERSHEDS, SEWAGE, WATER   POLLUTION, CHEMICAL ANALYSIS, PHY     W70-05752
, ODOR, FILTERS, PLANKTON, WATER   POLLUTION, SURFACE WATERS, RESERV     W70-05389
L, WINDS, CURRENTS(WATER), WATER   POLLUTION, DRAINAGE, ORGANIC MATT     W70-05404
SACHUSETTS, POTABLE WATER, WATER   POLLUTION, BACTERIA, ALGAE, NUTRI     W70-05264
, STREAMFLOW, MEASUREMENT, WATER   POLLUTION, FERTILIZATION, INDUSTR     W70-05112
, THERMOPHILIC BACTERIA, THERMAO   POLLUTION, NUTRIENTS, TEMPERATURE     W70-05270
LONDIN/ *DEVOLUTION, *INTERSTATE   POLLUTION, *METROPOLITAN AREAS, B     W70-04430
CAL ANALYSIS, RELIABILITY, WATER   POLLUTION, WATER POLLUTION EFFECT     W70-04382
TY CONTROL, WATER QUALITY, WATER   POLLUTION, NITROGEN, PHOSPHORUS,      W70-04484
PAL WASTES, *GREAT LAKES, *WATER   POLLUTION, SEWAGE, AESTHETICS, AE     W70-04430
M, TRACE METALS, VITAMINS, WATER   POLLUTION, PHYSIOLOGICAL ECOLOGY,     W70-04503
*ALGAE, *BIOINDICATORS, *WATER     POLLUTION, DIATOMS, MARYLAND, PEN     W70-04510
TH RATES, *BIOINDICATORS, *WATER   POLLUTION, *WASTE TREATMENT, FOOD     W70-04381
KANSAS RIVER(OKLA), TULSA(OKLA),   POLLUTION, FOSSIL, CIMARRON RIVER     W70-04001
R, SEDIMENTS, FERTILIZERS, WATER   POLLUTION, SNOW, RUNOFF, DRAINAGE     W70-04193
LLUTANTS, QUALITY CONTROL, WATER   POLLUTION, *WATER POLLUTION SOURC     W69-01273
TIVE WASTES, *MUNICIPAL / *WATER   POLLUTION, *MINE WASTES, *RADIOAC     W69-01165
OTA, FIELD INVESTIGATIONS, WATER   POLLUTION, WATER SAMPLING, *WATER     W69-00659
TION, *LAKES, CONNECTICUT, WATER   POLLUTION, *NUTRIENT REQUIREMENTS     W68-00511
WATER, / *POWERPLANTS, *THERMAL    POLLUTION, *TEMPERATURE, *COOLING     W69-05023
OL, WATER / *GREAT LAKES, *WATER   POLLUTION, *WATER POLLUTION CONTR     W69-03948
MEMBRANES,/ AQUATIC ALGAE, WATER   POLLUTION, *TOXICITY, BIOLOGICAL      W69-05306
TIC OXYGEN, TEMPERATURE, THERMAL   POLLUTION, THERMAL SPRINGS, THERM     W69-03516
OZEN GROUND, SOIL EROSION, WATER   POLLUTION, *DENITRIFICATION, PHOS     W69-05323
NSERVATION, WATER QUALITY, WATER   POLLUTION, WATER RESOURCES.: / CO     W69-05844
PHOSPHATES, ALGAE, SEWAGE, WATER   POLLUTION, WATER LAW, WATER POLLU     W69-06305
ENT, POLLUTANTS, WASTES, THERMAL   POLLUTION, TOXINS, PESTICIDE RESI     W69-06305
ULATION, NUISANCE ALGAE, THERMAL   POLLUTION, DIFFUSION, DIATOMS, CH     W69-05763
ERALOGY, *MICROORGANISMS, *WATER   POLLUTION, *WATER PURIFICATION, *     W69-07428
ASSOCIATIONS, MOSCOW, PETROLEUM    POLLUTION, PHOTOSYNTHETIC PIGMENT     W69-07440
ECOLOGY, *ALGAE, *RIVERS, *WATER   POLLUTION, *WATER POLLUTION SOURC     W69-07096
TROPHICATION, *CALIFORNIA, WATER   POLLUTION, REAERATION, NUTRIENTS,     W69-07520
ORMATIONS, HETEROTROPHY, ORGANIC   POLLUTION, MARINE ORGANISMS, MARI     W69-07440
DISPOSAL, WASTE DISPOSAL, WATER    POLLUTION, LAKES, STREAMS, SEWAGE     W69-06909
JAPAN, *GLENODININE, *PERIDINIUM   POLONICUM, PLANKTON BLOOMS.: /I,      W70-08372
LYETHOXYLATE, ANIONIC SU/ *ALKYL   POLY ETHOXYLATE, *ALKYL PHENOL PO     W70-09438
BRACHIONUS, SYNCHAETA PECTINATA,   POLYARTHRA VULGARIS, KARATELLA CO     W70-02249
```

507

ELOSIRA, MICROCYSTIS, KERATELLA,
SCHURA, ORTHOCYCLOPS, KERATELLA,
CHLOROPHYLL, ZOOPLANKTON, ALGAE,
BYELORUSSIA(USSR),
NAVICULA PELLICULOSA, GONYAULAX
ELECTROLYTES, *SYNTHETIC ORGANIC
AUXITE.:
L POLY ETHOXYLATE, *ALKYL PHENOL
YL POLYETHOXYLATE, *ALKYL PHENOL
ONATES, *AXENIC CULTURES, *ALKYL
MENON, MICROCYSTIS, POTAMOGETON,
METHYL ALCOHOL, ETHYLENE GLYCOL,
CULTURES, FATTY ACIDS, OOCYSTIS
NA, APHANIZOMENON, GLOEOTRICHIA,
FOAMING DETERGENTS, PUBLIC APP/
S.:   HARRIS COUNTY, TEX,
METAPHOSPHATES, PULSE-LABELING,
TH TEST, ORGANIC GROWTH FACTORS,
HEXOSAMINE, ASCOPHYLLUM NODOSUM,
IES, ULVA LACTUCA VAR LATISSIMA,
CTERIAL ATTACHMENT, ECOTOCARPUS,
UDA, SCENEDESMUS OBLIQUUS, HELIX
PSYCHE COCKERELLI, BODY BURDENS,
*NEW ZEALAND, SEWAGE TREATMENT
PERATURE, NUTRIENTS, FACULTATIVE
SOUTHWEST(US), *DESIGN CRITERIA,
OLORADO, WEST(US)), STABILIZATION
PH, FACULTATIVE POND, ANAEROBIC
ZURICHSEE(SWITZERLAND), LINSLEY
HYTOPLANKTON CROPS, TR/ TAKASUKA
, TYP/ *AUGUSTA(MAINE), CARLETON
OVALIS, OPEPHORA MARTYI, ABBOT'S
IA, JANESVILLE(WIS), SPAULDING'S
S), LAKE WAUBESA(WIS), SPAULDING
N, NUTRIENT REMOVAL, SPAULDING'S
L SERIAL SYSTEM, PH, FACULTATIVE
HAMPSHIRE, LAKE WINNISQUAM, LONG
BACTERIA, YEASTS, ACTINOMYCETES,
R COMPOUNDS, WASTE STABILIZATION
ENTION TIME EFFECTS, *FACULATIVE
THEORY AN/ *WASTE STABILIZATION
CTS, *FACULATIVE PONDS, *AEROBIC
LOAD EFFECTS,/ *ANAEROBIC WASTE
ES, ALGAE, ORGANIC L/ *OXIDATION
OLOGICAL COMMUNITIES, BACKWATER,
T, *INDUSTRIAL WASTES, OXIDATION
UTRIENTS, ESTUARINE ENVIRONMENT,
XYGEN, NITRIFICATION, COLIFORMS,
ONDS, *ACRODIC PONDS, *ANAEROBIC
ROPHICATION, *NUTRIENTS, *LAKES,
Y, STABILIZATION, WASTE STORAGE,
DESIGN APPLICATIONS, FACULTATIVE
EN DEMAND, CHLOROPHYLL, LAGOONS,
LIGHT INTENSITY, STABILIZATION,
TRY, INDUSTRIAL WASTES, LAGOONS,
TRY, WATER ANALYSIS, PHOSPHORUS,
WAGE TREATMENT, *EUTROPHICATION,
PES, REGRESSION ANALYSIS, ALGAE,
RY TREATMENT, STABI/ *NUTRIENTS,
HYTA, *ECOLOGY, *EUTROPHICATION,
ATES, SEASONAL, LAKES, SAMPLING,
SLUDGE, FISHERIES, CARP, CLAMS,
CONDARY TREATMENT, STABILIZATION
GEN, GROWTH RATES, STABILIZATION
SK(USSR), ENGLAND/ PISCICULTURAL
*LAKES, *RECREATION, RESERVOIRS,
CIDES, FISH, FRESH WATER, ALGAE,
, PLANKTON, BASS, PHYTOPLANKTON,
ALGAL CONCENTRATION, 'EXTRA-DEEP
RUSH, DREDGING, COPPER SULPHATE,
SLATION, WISCONSIN, CHEMCONTROL,
*CHLORI/ *PSEUDOMONAS, *SWIMMING
CRUSTACEANS, ROTIFERS, SWIMMING
COUNTY(ILL), CALUMET RIVER(ILL),
, *ALGAE, NUTRIENTS, PHOSPHORUS,
SINKING RATES, GRAZING PRESSURE,
RONMENT, *AQUATIC LIFE, *AQUATIC
ON, CYANOPHYTA, SEDIMENTS, PLANT
PHYTA, DIATOMS, AQUATIC HABITAT,
*CYTOLOGICAL EFFECTS, ALGAL
IBUTION, FREQUENCY DISTRIBUTION,
OMASS, DISTRIBUTION, ECOSYSTEMS,
ECOLOGY, PURPLE SULFUR BACTERIA,
, SEWAGE EFFLUENTS, FARM WASTES.
TOPLANKTON, ALGAE, HEATED WATER,
LABORATORY TESTS, PARTICLE SIZE,
DIOXIDE, NUTRIENT REQUIREMENTS,
WOOD DIATOMETER, SAVANNAH RIVER,
TROPHY, *FISH POPULATIONS, PLANT
POPULATIONS, OLIGOTROPHY, *FISH

POLYARTHRA, ASPLANCHNA, BRANCHION	W70-05375
POLYARTHRA, CONOCHILUS, PYCNOCLIN	W70-05760
POLYCHAETA, CRUSTACEA, MOLLUSCA,	W69-08526
POLYCHLOROPINENE.:	W70-08654
POLYEDRA, 3,5-DIAMINOBENZOIC ACID	W69-08278
POLYELECTROLYTES, CATIONIC ELECTR	W70-05267
POLYELECTROLYTE-COATED FILTERS, B	W70-01027
POLYETHOXYLATE, ANIONIC SURFACTAN	W70-09438
POLYETHOXYLATE, ORGANIC EXTRACTIO	W70-03928
POLYETHOXYLATE, *ALKYL PHENOL POL	W70-03928
POLYGONUM, SAGITTARIA, NAJAS.: /O	W72-04761
POLYMERS, DIMETHYL TEREPHTHALATE.	W70-10433
POLYMORPHA, UNICELLULAR ALGAE.: /	W70-07957
POLYPEPTIDES, PHOTOASSIMILATION,	W70-04369
POLYPHOSPHATES, NUTRIENT BALANCE,	W70-05084
POLYPHOSPHATES, ORGANIC PHOSPHATE	W69-04800
POLYPHOSPHATES, FRACTIONATION.: /	W71-03033
POLYSACCHARIDES.: /, MUSTARD GROW	W69-07826
POLYSIPHONIA LANOSA, POINT LEPREA	W70-04365
POLYSIPHONIA HARVEYI, EPIPHYTIC B	W70-01073
POLYSIPHONIA, ORGANIC AGGREGATES,	W70-01068
POMATIA, CHLORELLA PYRENOIDOSA, N	W70-04184
POMOXIS ANNULARIS, GASTROINTESTIN	W70-02786
POND ALGAE.:	W70-02304
POND DESIGNS, BOD, SLUDGE.: / TEM	W70-04786
POND EFFLUENT, COLORADO, WEST(US)	W70-04788
POND LOADINGS, ODOR PERSISTENCE.:	W70-04788
POND.: /MULTI-CELL SERIAL SYSTEM,	W70-04790
POND(CONN), BOSMINA COREGONI LONG	W69-10169
POND(JAPAN), LAKE MENDOTA(WIS), P	W69-05142
POND(MAINE), SYNURA, ASTERIONELLA	W70-08096
POND(SOMERSET, ENGLAND), SAND GRA	W71-12870
POND(WIS), HYDRODICTYON, NAJAS, A	W70-03310
POND(WIS), LAKE DELAVAN(WIS), PEW	W69-05867
POND(WIS), LAKE WAUBESA(WIS), GRE	W69-05868
POND, ANAEROBIC POND.: /MULTI-CEL	W70-04790
POND, CONCORD, BOSTON LOT RESERVO	W69-08674
PONDING, WASTE TREATMENT.: /ENA,	W70-08097
PONDS.: /ARED SPECTROSCOPY, SULFU	W70-07034
PONDS, *AEROBIC PONDS, *ANAEROBIC	W70-04787
PONDS, *ALGAL PHYSIOLOGY, *DESIGN	W70-04786
PONDS, *ANAEROBIC PONDS, DESIGN F	W70-04787
PONDS, *DEPTH EFFECTS, *HYDRAULIC	W70-04787
PONDS, *HYDROGEN SULFIDE, *SULFID	W71-11377
PONDS, ALASKA.: /YTA, DIATOMS, BI	W70-03969
PONDS, ALGAE, BENTHIC FAUNA, WATE	W72-00027
PONDS, BREEDINGS, SALINE WATER-FR	W73-00748
PONDS, CLIMATES, PHOTOSYNTHESIS,	W70-05092
PONDS, DESIGN FOR ENVIRONMENTAL C	W70-04787
PONDS, FERTILIZERS, ALGAE, NITROG	W69-08668
PONDS, HYDROLYSIS, METABOLISM, CH	W70-08097
PONDS, INFRARED SPECTROSCOPY, SUL	W70-07034
PONDS, OXIDATION, WASTE WATER TRE	W70-08392
PONDS, PENETRATION, ABSORPTION.: /	W70-03923
PONDS, PHOTOSYNTHESIS, COSTS, EFF	W70-09320
PONDS, PHYTOPLANKTON, EUTROPHICAT	W69-05142
PONDS, POLLUTANT IDENTIFICATION,	W70-02304
PONDS, PREDATION.: /US RADIOISOTO	W70-03957
PONDS, PRIMARY TREATMENT, SECONDA	W68-00012
PONDS, RESERVOIRS, BACKWATER, FER	W70-00274
PONDS, RIVERS, RESERVOIRS, AIR TE	W70-02646
PONDS, RUNNING WATERS, FRESH WATE	W70-09905
PONDS, SEWAGE LAGOONS.: /MENT, SE	W68-00012
PONDS, SEWAGE DISPOSAL.: /, NITRO	W68-00855
PONDS, SVISLOCH' RIVER(USSR), MIN	W70-05092
PONDS, TEMPERATURE, LIGHT PENETRA	W70-06225
PONDS, WATER HYACINTH, CHEMICAL C	W70-00269
PONDS, ZOOPLANKTON, MINNOWS, PEDA	W70-02249
PONDS', ASHKELON(ISRAEL), TEMPERA	W70-04790
PONDWEEDS, SODIUM ARSENITE, FISH,	W70-05263
PONDWEEDS, ROOTED AQUATIC PLANTS,	W70-05568
POOLS, *BACTERIA, *BACTERICIDES,	W69-10171
POOLS, CHLAMYDOMONAS, DAPHNIA, DI	W70-10161
POOLS, WIND DIRECTION, TABELLARIA	W70-00263
POPULATION, POLLUTANTS, MAPPING,	W70-00667
POPULATION DENSITY.: /EDRA ACUS,	W69-10158
POPULATIONS, ALGAE, BENTHOS, ZOOP	W71-02491
POPULATIONS, WEATHER, WATER, MOVE	W70-06229
POPULATION.: /EUGLENOPHYTA, CYANO	W70-02770
POPULATION.:	W70-02770
POPULATION, BIOLOGICAL INDICATORS	W70-02772
POPULATION, OREGON, INTERTIDAL AR	W70-03325
POPULATION EQUIVALENTS(WASTES), R	W70-03312
POPULATION, INDUSTRIES, METHANE,	W70-02803
POPULATION, WATER POLLUTION EFFEC	W71-09767
POPULATION, GROWTH RATES, WATER P	W72-02417
POPULATIONS, ENVIRONMENTAL EFFECT	W70-05424
POPULATION DYNAMICS, ALGAL COMMUN	W70-04510
POPULATIONS, ROUGH FISH, NITROGEN	W68-00172
POPULATIONS, PLANT POPULATIONS, R	W68-00172

508

EATION, SILT, NUTRIENTS, AQUATIC   POPULATIONS, OLIGOTROPHY, *FISH P   W68-00172
*WATER QUALITY, TURBIDITY, PLANT   POPULATIONS.: /GICAL PROPERTIES,   W68-00475
ATION, MAINE, NUTRIENTS, AQUATIC   POPULATIONS, WATER QUALITY, LAKE   W68-00470
OSPHORUS, BICARBONATES, *AQUATIC   POPULATIONS.: /LGAE, NITROGEN, PH   W68-00468
REAMS, EUGLENA, NUTRIENTS, PLANT   POPULATIONS, HISTORY, WATER PURIF   W68-00483
ATION, SEWAGE EFFLUENTS, AQUATIC   POPULATIONS, NUISANCE ALGAE, DUCK   W68-00461
NELS, POLLUTION ABATEMENT, HUMAN   POPULATION.: /ILD), OYSTERS, CHAN   W68-00461
A/ *EUTROPHICATION, SEWAGE, FISH   POPULATIONS, PLANKTON, *CYANOPHYT   W68-00461
ASHINGTON, DC, *ESTUARIES, HUMAN   POPULATION, SEWAGE EFFLUENTS, AQU   W68-00461
, SILICATES, IRON OXIDES, *PLANT   POPULATIONS, REVIEWS, LIGHT INTEN   W68-00860
PHIBIANS, INVERTEBRATES, AQUATIC   POPULATIONS, BIBLIOGRAPHIES, BOTT   W68-00859
DING CROP, ANNUAL TURNOVER, FISH   POPULATIONS.: /IVE CAPACITY, STAN   W68-00912
, *ENVIRONMENTAL EFFECTS, *HUMAN   POPULATION, *EUTROPHICATION, *CON   W70-03973
RT, BACTERIAL POPULATIONS, ALGAL   POPULATIONS, HETEROTROPHIC POTENT   W70-04194
ITY, ACTIVE TRANSPORT, BACTERIAL   POPULATIONS, ALGAL POPULATIONS, H   W70-04194
ICATION, *NUTRIENTS, ALGAE, FISH   POPULATION, *WATER POLLUTION CONT   W70-04266
AGELLATES, OCEANS, BIO/ *YEASTS,   POPULATION, MARINE PLANTS, DINOFL   W70-04368
IVERS, STREAMS, EUGLENA, AQUATIC   POPULATIONS, EUTROPHICATION, CYAN   W70-04510
RONMENT, AQUATIC PLANTS, AQUATIC   POPULATIONS, CONSERVATION, WATER   W69-05844
OZOA, A/ *AQUATIC LIFE, *AQUATIC   POPULATIONS, *RIVERS, ALGAE, PROT   W73-00748
HYACINTHUS, POCILLOPORA EYDOUXI,   PORITES LOBATA, LEPTASTREA PURPUR   W69-08526
IFICATION, CROPS, TOXICITY, SOIL   POROSITY, IRRIGATION, GROUND WATE   W69-05323
IQUUES, SCENEDESMUS QUADRICAUDA,   PORPHYRIDIUM CRUENTUM, LEMNA MINO   W70-02964
ALASSIOSIRA, AMPHIDINIUM KLEBSI,   PORPHYRIDIUM CRUENTUM, RADIONUCLI   W70-00708
CTS, AXENIC CULTURES, MONOALGAL,   PORTUGAL, TAPES, JAPAN, AMPHIDINI   W70-05372
ILAMENTOUS GREEN ALGAE, HALIMONE   PORTULACOIDES, SUAEDA MARITIMA, A   W70-09976
', VIBRIOS, ACHROMOBACTERS, GRAM-   POSITIVE BACTERIA, SARCINE-LIKE,   W70-00713
AE, *TASTE, *ODOR, *AMINO ACIDS,   POTABLE WATER, ACTINOMYCETES, PLA   W72-01812
ION, WATERSHEDS, SEWERAGE ALGAE,   POTABLE WATER.: /NATION, COAGULAT   W71-05943
LOGY, *RESERVOIR, MASSACHUSETTS,   POTABLE WATER, WATER POLLUTION, B   W70-05264
MALS, ALGAE, CUTTING MANAGEMENT,   POTABLE WATER, HERBICIDES, CHARA,   W70-05263
ODOR, COAGULATION, ELECTROLYTES,   POTABLE WATER.: /TICIDES, TASTE,   W72-13601
HLORATE, SANTOBRITE, CLADOPHORA,   POTAMOGETON SPP, LAKE MENDOTA(WIS   W70-05263
CILIATES, RANUNCULUS CIRCINATUS,   POTAMOGETON PUSILLUS.: /GELLATA,   W69-07838
PHORA, PEDIASTRUM, HYDRODICTYON,   POTAMOGETON, RANUNCULUS.: / CLADO   W70-04494
ENA, APHANIZOMENON, MICROCYSTIS,   POTAMOGETON, POLYGONUM, SAGITTARI   W72-04761
CHARIS CANADENSIS, MYRIOPHYLLUM,   POTAMOGETON CRISPUS, MICROSPORA W   W70-00711
YLLUM, MYRIOPHYLLUM, EICHHORNIA,   POTAMOGETON, TRAPA, PISTIA, ALTER   W70-10175
ACHARIS CANADENSIS, MACROPHYTES,   POTAMOGETON FOLIOSUS, EXPERIMENTA   W70-02249
NTHERA, NAJAS, SILVEX, ENDOTHAL,   POTASSIUM PERMANGANATE, FENAC, SI   W70-10175
CEAN, RADIOACTIVITY, GAMMA RAYS,   POTASSIUM RADIOISOTOPES, MANGANES   W69-08524
BASS, CARP, CATFISHES, NITROGEN,   POTASSIUM.: /TOS, ALGAL CONTROL,   W70-05417
YANOPHYTA, CARBON RADIOISOTOPES,   POTASSIUM, OXY: /WAGE, DIATOMS, C   W70-05405
PLANKTON, NITROGEN, PHOSPHORUS,   POTASSIUM, SELECTIVITY, POISONS,   W70-05399
SPHATES, CHLOROPHYLL, MAGNESIUM,   POTASSIUM, TEMPERATURE, EPILIMNIO   W70-05409
S, NITROGEN, PHOSPHORUS, SODIUM,   POTASSIUM, TRACE METALS, VITAMINS   W70-04503
SULFATES, NITRATES, PHOSPHATES,   POTASSIUM, IRON, BICARBONATE, ZIN   W70-03978
PLANTS, IRON, CHLORELLA, SODIUM,   POTASSIUM, SILICATES, DISSOLVED S   W70-03512
PLING, BOTTOM SEDIMENTS, SODIUM,   POTASSIUM, IRON, MANGANESE, CALCI   W70-09889
L NUTRIENTS, CALCIUM, MAGNESIUM,   POTASSIUM, HARVESTING OF ALGAE, L   W70-02245
RUS, SULFUR, CALCIUM, MAGNESIUM,   POTASSIUM, SODIUM, IRON, MANGANES   W70-02804
GEN ION CONCENTRATION, NITRATES,   POTASSIUM, CALCIUM, MAGNESIUM, IR   W70-00719
DS, ZINC RADIOISOTOPES, CALCIUM,   POTASSIUM, CLAMS, SIZE, ALGAE, DI   W70-00708
ORTH DAKOTA), SULFATE REDUCTION,   POTATO PROCESSING WASTES, CHEESE   W70-03312
CONCENTRATION, *NUTRIENT BUDGET,   POTENTIAL NUTRIENT CONCENTRATION,   W70-00270
GE EFFLUENTS, ELECTROLYTES, ZETA   POTENTIAL, BACTERIA, PROTOZOA, AL   W70-01027
GASES, ODOR, OXIDATION-REDUCTION   POTENTIAL, SULFUR BACTERIA, *WAST   W70-07034
YGEN DEMAND, OXIDATION-REDUCTION   POTENTIAL, AQUATIC ALGAE, *BIOLOG   W71-09546
YGEN DEMAND, OXIDATION-REDUCTION   POTENTIAL, DISSOLVED OXYGEN, ODOR   W71-11377
ALGAL POPULATIONS, HETEROTROPHIC   POTENTIAL, RELATIVE HETEROTROPHIC   W70-04194
OTENTIAL, RELATIVE HETEROTROPHIC   POTENTIAL, SWEDEN, ENZYME KINETIC   W70-04194
EQUATION, RELATIVE HETEROTROPHIC   POTENTIAL, LINEWEAVER-BURK EQUATI   W70-05109
M COMPOUNDS, OXIDATION REDUCTION   POTENTIAL, DISSOLVED OXYGEN, CHEM   W70-05760
MAND, TOXICITY, OXYGEN-REDUCTION   POTENTIAL, SOLAR RADIATION, ENERG   W70-05750
RIENT AVAILABILITY, ALGAL GROWTH   POTENTIAL, SELENASTRUM CAPRICORNU   W69-06864
UM, NITRITE, OXIDATION-REDUCTION   POTENTIAL, CYCLING NUTRIENTS, EUT   W69-04521
N VARIATIONS, SAN JOAQUIN RIVER,   POTOMAC ESTUARY, SAN FRANCISCO BA   W69-07520
, SNAKE RIVER, RIO GRANDE RIVER,   POTOMAC RIVER.: /, SAVANNAH RIVER   W70-04507
                                    POTOMAC RIVER BASIN.:   W70-09557
E SHOCK, *PROTOZOAN COMMUNITIES,   POTOMAC RIVER, SAVANNAH RIVER, SH   W70-07847
, FISH, TASTE, ODOR, OXYGEN SAG,   POTOMAC RIVER.: /NITY, REAERATION   W73-00750
VOIR(TENN), HOLSTON RIVER(TENN),   POTOMOGETON PECTINATUS, POTOMOGET   W71-07698
R(TENN), POTOMOGETON PECTINATUS,   POTOMOGETON CRISPUS, HETERAMTHERA   W71-07698
*CHICKENS, FOOD WASTES,   POULTRY WASTES.:   W70-09320
NG TO/ *COOLING WATER, *ELECTRIC   POWER PLANTS, *WATER REUSE, COOLI   W71-11393
INISTRATION, SUPERVISORY CONTROL(POWER), ADMINISTRATIVE AGENCIES.:   W72-05774
FISH, SHELLFISH, STEAM, ELECTRIC   POWER, HEATED WATER, PROTEINS, AQ   W70-09905
OGEN ION CONCENTRATION, ELECTRIC   POWER, SEWAGE DISPOSAL, *WASTE WA   W71-09548
PHYTOPLANKTON, ALGAE,/ *THERMAL   POWERPLANTS, *LAKES, TEMPERATURE,   W71-09767
OL/ *THERMAL POLLUTION, *NUCLEAR   POWERPLANTS, *WATER QUALITY CONTR   W71-11881
EM/ *THERMAL POLLUTION, *THERMAL   POWERPLANTS, *ECOLOGY, *BIOTA, *T   W71-02491
CES, *THERMAL POLLUTION, THERMAL   POWERPLANTS, *THERMAL STRESS, THE   W70-07847
ASTE DISPOSAL, NUCLEAR REACTORS,   POWERPLANTS, SHIPS, SUBMARINES, E   W70-00707
CE ELEMENTS, WASHINGTON, NUCLEAR   POWERPLANTS, WATER POLLUTION EFFE   W70-00264
, HYDROELECTRIC PLANTS, ELECTRIC   POWERPLANTS, ENGINEERING STRUCTUR   W69-05023
, *BIOLOGICAL INDEX, *SAPROBITY,   PRAGUE, CZECHOSLOVAKIA, LABORATOR   W72-12240
NIUM-103-RHODIUM-103, CERIUM-144-   PRASEODYMIUM-144.: /UM-106, RUTHE   W69-08524
EAWEED, ZIRCONIUM-95-NIOBIUM-95,   PRE-BOMB MATERIAL, RUTHENIUM-106-   W69-08524
L DERIVATIVES, RECENT SEDIMENTS,   PRECAMBRIAN SEDIMENTS, CHLORELLA   W69-08284
ION, SPECIES DIVERSITY, CHEMICAL   PRECIPITATION.: /URAL EUTROPHICAT   W69-08518

ALGAE, BIBLIOGRAPHIES, CHEMICAL
KEGONSA(WIS), LAKE WAUBESA(WIS),
UNDS, SURFACE RUNOFF, UNDERFLOW,
NOFF, PERCOLATION, ROADS, ROOFS,
ES, TERTIARY TREATMENT, CHEMICAL
*PHOSPHORUS COMPOUNDS, *CHEMICAL
OLUBILITY, *PHOSPHATES, CHEMICAL
TROSTATIC PRECIPITATION, THERMAL
UGATION, SEDIMENTATION, CHEMICAL
, CORYNEBACTERIUM, ELECTROSTATIC
HYDROGEN SULFIDE, IMHOFF TANKS,
CTRODIALYSIS, SORPTION, CHEMICAL
S, ALGAE, DAPHNIA, GROWTH RATES,
GRESSION ANALYSIS, ALGAE, PONDS,
PLANKTON, ALGAE, *ALGAL CONTROL,
BDENUM, PLANT GROWTH REGULATORS,
NCE, FOOD CONCENTRATION, MEXICO,
ELONGATUM HODGETTS, CHROOCOCCUS
ENTS, PHOSPHORUS RELEASE, SAMPLE
S, LIGHT, NUTRIENTS, HYDROSTATIC
DRA ACUS, SINKING RATES, GRAZING
XYGEN, *SATURATION, TEMPERATURE,
US EARTH, SEPARATION TECHNIQUES,
EUTROPHICATION
SPHORUS COMPOUNDS, PLANT GROWTH,
, PHOTOSYNTHESIS, PHYTOPLANKTON,
, ORGANIC MATTER, PHYTOPLANKTON,
CONNECTICUT, OREGON, WASHINGTON,
ORGANIC MATTER, SEASONAL, ALGAE,
ON, LIGHT PENETRATION, SEASONAL,
, *CHLOROPHYLL, *EUTROPHICATION,
E, LIGHT INTENSITY, TEMPERATURE,
ATA COLLECTIONS, MARINE ANIMALS,
, NITROGEN, RESPIRATION, CARBON,
ODUCTIVITY, LIMNOLOGY, PLANKTON,
*CALIFORNIA, *NEVADA, LIMNOLOGY,
AKES, ORGANIC MATTER, SEA WATER,
COMMUNITIES, *ALGAE, *METHODOL/
FF, LIGHT PENETRATION, SAMPLING,
ION, *NUTRIENTS, LAKES, ECOLOGY,
ATMENT, WATER POLLUTION SOURCES,
TMENT, STABI/ *NUTRIENTS, PONDS,
A SCOTIA(CANADA), PENGUINS, LAK/
ANALYSIS, MULTIVARIATE ANALYSIS,
ROFOSSILS, COMPETITIVE EXCLUSION
YCEAE, SELENASTRUM CAPRICORNUTUM
E MENDOTA(WIS), LAK/ *BIOLOGICAL
*TEST DEVELOPMENT, *TEST
ICATION, PROVISIONAL ALGAL ASSAY
TION, ALGAL PHYSIOLOGY, KJELDAHL
ATER TEMPERATURE, SEASONAL, TEST
NALYSIS, *LABORATORY TESTS, TEST
CTIVITY, BACTERIA, DIATOMS, TEST
MPTION, PROVISIONAL ALGAL ASSAY,
LGAE, TREATMENT FACILITIES, TEST
TOMS, CYANOPHYTA, BIOASSAY, TEST
, TE/ *ALGICIDES, *PAINTS, *TEST
LGAE, TREATMENT FACILITIES, TEST
AGUE, CZECHOSLOVAKIA, LABORATORY
BORATORY PROCEDURES, STATISTICAL
MINUM SULFATE, *ACTIVATED SLUDGE
, SUSPENDED LOAD, SOLIDS CONTACT
TREATMENT, UNIT OPERATIONS, UNIT
EVOORTIA TYRRANUS, CONCENTRATION
DISTRIBUTION PATTERNS, MEMBRANE
COAL CONTACT
ERFERTILIZATION, PHYSI/ *REMOVAL
A, ANABAENA, IOWA RIVER, / *UNIT
LOGICAL DISTRIBUTION, STOCHASTIC
S, ANAEROBIC LAGOONS.:   *POTATO
KOTA), SULFATE REDUCTION, POTATO
TRICHO/ HEXAGENIA, CHIRONOMIDAE,
LAKE PEWAUKEE(WIS), CHIRONOMUS,
*WISCONSIN, *LAKES, PROD/ *ODOR-
*WATER ANALYSIS, *ODORS, *ODOR
RIAL WASTE, CORES, DIATOMS, ODOR-
ROL, ALGAL CONTROL, SLIME, TASTE-
IME, TASTE-PRODUCING ALGAE, ODOR-
BIOCHEMICAL OXYGEN DEMAND, ODOR-
ALGAE, ALGAE, CHLORINATI/ *ODOR-
I/ *ODOR-PRODUCING ALGAE, *TASTE-
EMPERATURE, SOIL ANALYSIS, TASTE-
URCE CONTAMINATION, CYANIDE, NEW
*OXYGEN
IUM COMPOUNDS, PHENOL/ BIOCIDES,
AEODACTYLUM, *GAS EFFLUENT, FOOD
CRETION, STEPHAN/ *EXTRACELLULAR
LUTION, PHOTOSYNTHETIC PIGMENTS,
RECIRCULATION, BOD REMOVAL, GAS
PONDS, CLIMATES, PHOTOSYNTHESIS,

PRECIPITATION, IONS, BIOCHEMICAL   W69-00446
PRECIPITATION.: /NGRA(WIS), LAKE   W69-03366
PRECIPITATION(ATMOSPHERIC), DRAIN   W70-04810
PRECIPITATION(ATMOSPHERIC), WETLA   W69-10178
PRECIPITATION, HARVESTING, ZONING   W69-10178
PRECIPITATION, *CHEMICAL DEGRADAT   W70-09186
PRECIPITATION, FILTRATION, AMMONI   W71-11393
PRECIPITATION.: /EBACTERIUM, ELEC   W71-11823
PRECIPITATION, LIFE CYCLES, GENET   W71-11823
PRECIPITATION, THERMAL PRECIPITAT   W71-11823
PRECIPITATES.:   W72-00027
PRECIPITATION, ULTIMATE DISPOSAL.   W70-01981
PREDATION.: /VIRONMENT, PHOSPHATE   W69-06536
PREDATION.: /US RADIOISOTOPES, RE   W70-03957
PREDATION, COMPETITION, *INSECT C   W70-05428
PREDATION, PRODUCTIVITY, TRACE EL   W69-04801
PREDICTIVE EQUATIONS, ZOOGEOGRAPH   W70-03957
PRESCOTTII, CALOTHRIX FUSCA, MICR   W70-04468
PRESERVATION, WATER DYNAMICS, WAT   W70-04382
PRESSURE, CARBON DIOXIDE, CYANOPH   W69-10160
PRESSURE, POPULATION DENSITY.: /E   W69-10158
PRESSURE, MIXING, EVAPORATION, SU   W70-00683
PRESSURE, FLOW RATES, SLUDGE, SLU   W72-08688
PREVENTION, ECOLOGIC IMBALANCES.:   W69-07084
PRIMARY PRODUCTIVITY, RIVERS, TAS   W69-06535
PRIMARY PRODUCTIVITY, SYSTEMATICS   W69-07832
PRIMARY PRODUCTIVITY, SOIL CHEMIS   W69-06273
PRIMARY PRODUCTIVITY, ECOSYSTEMS,   W70-04506
PRIMARY PRODUCTIVITY, HARVESTING   W70-04290
PRIMARY PRODUCTIVITY, CALCIUM, NI   W70-05405
PRIMARY PRODUCTIVITY, PHYTOPLANKT   W70-05094
PRIMARY PRODUCTIVITY, SURFACES, A   W70-04809
PRIMARY PRODUCTIVITY, CHLOROPHYLL   W73-01446
PRIMARY PRODUCTIVITY, OXYGEN, MET   W70-01073
PRIMARY PRODUCTIVITY, TROPHIC LEV   W69-09135
PRIMARY PRODUCTIVITY, CHLOROPHYLL   W70-01933
PRIMARY PRODUCTIVITY, LIGHT INTEN   W70-02504
PRIMARY PRODUCTIVITY, *BIOLOGICAL   W71-12870
PRIMARY PRODUCTIVITY, PHYTOPLANKT   W72-04761
PRIMARY PRODUCTIVITY, ALGAE, PLAN   W71-00665
PRIMARY TREATMENT, SECONDARY TREA   W69-01453
PRIMARY TREATMENT, SECONDARY TREA   W68-00012
PRINCE EDWARD ISLAND(CANADA), NOV   W70-05399
PRINCIPAL-COMPONENT ANALYSIS, PAR   W73-01446
PRINCIPLE, OSCILLATORIA RUBESCENS   W70-05663
PRINTZ, OSCILLATORIA VAUCHER, SKE   W70-05752
PROBLEMS, SCIOTO RIVER(OHIO), LAK   W72-01813
PROBLEMS, HORMIDIUM SP.:   W71-07339
PROCEDURE, BATCH CULTURES, CHEMOS   W70-02777
PROCEDURE, LUXURY CONSUMPTION, NU   W69-05868
PROCEDURES, SCENEDESMUS, *ALGAE.:   W68-00472
PROCEDURES, NURTRIENTS, RIVERS, L   W68-00472
PROCEDURES.: /NG, CAPILLARY CONDU   W70-04469
PROCEDURES, SELENASTRUM CAPRICORN   W70-02779
PROCEDURES, METALS.: /BACTERIA, A   W70-01519
PROCEDURES.: /LGAE, CULTURES, DIA   W71-06002
PROCEDURES, ANTIFOULING MATERIALS   W71-07339
PROCEDURES, METALS.: /BACTERIA, A   W71-06737
PROCEDURES, STATISTICAL PROCEDURE   W72-12240
PROCEDURES, BMT.: /HOSLOVAKIA, LA   W72-12240
PROCESS, HANOVER TREATMENT PLANT.   W70-09186
PROCESS, SEWAGE TREATMENT, TRICKL   W70-06792
PROCESSES.: /ILIZATION, PHYSICAL   W70-07469
PROCESSES.: /RSIOPS TRUNCATUS, BR   W69-08274
PROCESSES, MASSACHUSETTS.: /LOSE,   W69-00976
PROCESSES, PESTICIDE REMOVAL.:   W70-05470
PROCESSES, CHEMICAL TREATMENT, OV   W70-07469
PROCESSES, GASTROENTERITIS, SYNUR   W70-08107
PROCESSES, ECOSYSTEMS, AQUATIC HA   W71-09012
PROCESSING WASTES, AEROBIC LAGOON   W71-08970
PROCESSING WASTES, CHEESE MANUFAC   W70-03312
PROCLADIUS, CHIRONOMUS PROMOSUS,   W70-01943
PROCLADIUS, CORETHRA PUNCTIPENNIS   W70-06217
PRODUCING ALGAE, *NUISANCE ALGAE,   W70-05112
PRODUCING ALGAE, EUTROPHICATION,   W69-00387
PRODUCING ALGAE, CHLOROPHYLL, CHE   W68-00470
PRODUCING ALGAE, ODOR-PRODUCING A   W69-05704
PRODUCING ALGAE, DIATOMS, CYANOPH   W69-05704
PRODUCING ALGAE, POLLUTION ABATEM   W69-06909
PRODUCING ALGAE, *TASTE-PRODUCING   W70-02373
PRODUCING ALGAE, ALGAE, CHLORINAT   W70-02373
PRODUCING ALGAE, HYDROGEN ION CON   W72-11906
PRODUCT CONTROL, PLUMBING CODES,   W73-00758
PRODUCTION.:   W72-06022
PRODUCTION LOSS, QUATERNARY AMMON   W72-12987
PRODUCTION, SOUTHAMPTON UNIVERSIT   W70-00161
PRODUCTION, *CARBON FIXATION, *EX   W69-10158
PRODUCTION BIOLOGY, RUSSIA, SECON   W69-07440
PRODUCTION.: /IRONMENTAL CONTROL,   W70-04787
PRODUCTIVITY, SYMBIOSIS, RESERVOI   W70-05092

RIENTS, WATER POLLUTION EFFECTS,
OPHYLL, *EUTROPHICATION, PRIMARY
INTENSITY, TEMPERATURE, PRIMARY
SANCE ALGAE, *WISCONSIN, *LAKES,
*NEW YORK, *MEROMIXIS, *PRIMARY
ENTS, PALEOLIMNOLOGY, NUTRIENTS,
C MATTER, SEWAGE, PHYTOPLANKTON,
S, BIOASSAYS, ALGAE, COMPARATIVE
ENVIRONMENTAL EFFECTS, *PRIMARY
OMPARATIVE PRODUCTIVITY, AQUATIC
T PENETRATION, SEASONAL, PRIMARY
COMPOUNDS, PLANT GROWTH, PRIMARY
AKES, PHOSPHORUS, TROPHIC LEVEL,
C MATTER, PHYTOPLANKTON, PRIMARY
WATER POLLUTION SOURCES, AQUATIC
TROGEN COMPOUNDS, BIOINDICATORS,
YNTHESIS, PHYTOPLANKTON, PRIMARY
ATER POLLUTION EFFECTS, *PRIMARY
TE, AQUATIC ENVIRONMENT, AQUATIC
GAE, AGING(BROLOGICAL), *AQUATIC
EUTROPHICATION, *LAKES, *PRIMARY
LANKTON, ALGAE, *PHOTOSYNTHESIS,
BIO MASS, OLIGOTROPHY, *PRIMARY
E, AQUATIC MICROBIOLOGY, AQUATIC
GY, AQUATIC ENVIRONMENT, AQUATIC
LGICIDES, AQUATIC ALGAE, AQUATIC
IXATION, PHOSPHATES, PHOSPHORUS,
NT GROWTH REGULATORS, PREDATION,
E, AQUATIC MICROBIOLOGY, AQUATIC
E, AQUATIC MICROBIOLOGY, AQUATIC
E, AQUATIC MICROBIOLOGY, AQUATIC
NKTON, *CULTURES, ALGAE, AQUATIC
S, AQUATIC MICROBIOLOGY, AQUATIC
CUT, OREGON, WASHINGTON, PRIMARY
IPHYTON, *RADIOACTIVITY, *LAKES,
MATTER, SEASONAL, ALGAE, PRIMARY
*AQUATIC MICROBIOLOGY, *AQUATIC
ATES, DISSOLVED SOLIDS, CALCIUM,
TES, ALGAE, HYPOLIMNION, OXYGEN,
T, SOLAR RADIATION, CHLOROPHYLL,
RY TESTS, RESPIRATION,/ *PRIMARY
SECONDARY PRODUCTIVITY, AQUATIC
TY, LIMNOLOGY, PLANKTON, PRIMARY
IC FLORA, *CHLOROPHYLL, *PRIMARY
RATES, *GROWTH RATES, *DAPHNIA,
DYSTROPHY, OLIGOTROPHY, *PRIMARY
TIVITY, BIOINDICATORS, SECONDARY
RTILIZATION EFFECTS, *BIOLOGICAL
EN, RESPIRATION, CARBON, PRIMARY
CTERIA, STREPTOCOCCUS, SECONDARY
GANIC MATTER, SEA WATER, PRIMARY
NIA, *NEVADA, LIMNOLOGY, PRIMARY
NG FACTORS, *NUTRIENTS, *PRIMARY
ROGEN COMPOUNDS, EUTROPHICATION,
TER, PERIPHYTON, BIOMA/ *PRIMARY
, *NITROGEN, *ALKALINITY, LAKES,
GANISMS, AQUATIC PLANTS, AQUATIC
E, AQUATIC MICROBIOLOGY, AQUATIC
MS, SEASONAL, BIOMASS, PIGMENTS,
PHORUS, NITROGEN, PHYTOPLANKTON,
TS, *FISH, ALGAE, TROPHIC LEVEL,
TON, BIOASSAY, LABORATORY TESTS,
TIES, *ALGAE, *METHODOL/ PRIMARY
, *ESSENTIAL NUTRIENTS, *PRIMARY
T PENETRATION, SAMPLING, PRIMARY
*PHYTOPLANKTON, *ALGAE, *PRIMARY
INE BIOLOGY, *PRIMARY BIOLOGICAL
TOXICITY, PHOTOSYNTHETIC OXYGEN,
TER QUALITY/ RESERVOIR, *PRIMARY
OSPHORUS, NITROGEN, CHLOROPHYLL,
ISH, INFLOW, OUTLETS, NUTRIENTS,
TRIENTS, LAKES, ECOLOGY, PRIMARY
NALYSIS, *BIOASSAY, *COMPARATIVE
NALYSIS, SESSILE ALGAE, *PRIMARY
ECTIONS, MARINE ANIMALS, PRIMARY
ODORIFEROUS
ACID, CHEMOTROPH/ *EXTRACELLULAR
, PHARMACEUTICALS, EXTRACELLULAR
NC, XANTHOPHYCEA/ *EXTRACELLULAR
O/ AUTOINHIBITORS, EXTRACELLULAR
MENTAL CONDITIONS, EXTRACELLULAR
PLANKTON DYNAMICS, EXTRACELLULAR
TIUM COMPLEXES, DTPA, SOIL WATER
OGY, DISTRIBUTION PATTERNS, SOIL
', ASHKELON(ISRAEL), TEMPERATURE
INNETONKA(MINN), LAKE MANAGEMENT
NATIONAL EUTROPHICATION RESEARCH
NAGE WATER, CALIFORNIA, DRAINAGE
NAGE WATER, CALIFORNIA, DRAINAGE
NAGE WATER, CALIFORNIA, DRAINAGE

PRODUCTIVITY, SEDIMENTS, URBANIZA   W70-04721
PRODUCTIVITY, PHYTOPLANKTON, PHOT   W70-05094
PRODUCTIVITY, SURFACES, ARCTIC, E   W70-04809
PRODUCTIVITY, NITROGEN, PHOSPHORU   W70-05112
PRODUCTIVITY, *ZOOPLANKTON, SULFU   W70-05760
PRODUCTIVITY, SEDIMENTATION RATES   W70-05663
PRODUCTIVITY, CHLOROPHYTA, HYDROG   W70-05404
PRODUCTIVITY, AQUATIC PRODUCTIVIT   W70-05424
PRODUCTIVITY, WATER POLLUTION EFF   W70-05424
PRODUCTIVITY, DISTRIBUTION, SPECI   W70-05424
PRODUCTIVITY, CALCIUM, NITRATES,   W70-05405
PRODUCTIVITY, RIVERS, TASTE, THER   W69-06535
PRODUCTIVITY, MODEL STUDIES, SALM   W69-06536
PRODUCTIVITY, SOIL CHEMISTRY, WAT   W69-06273
PRODUCTIVITY, LIMNOLOGY, PLANKTON   W69-09135
PRODUCTIVITY, WASTE TREATMENT, WA   W69-08518
PRODUCTIVITY, SYSTEMATICS, WATER   W69-07832
PRODUCTIVITY, ALGAE, BIOMASS, CAL   W69-02959
PRODUCTIVITY, EUTROPHICATION, *LA   W69-00446
PRODUCTIVITY, LIMNOLOGY, *NUTRIEN   W68-00912
PRODUCTIVITY, *PLANKTON, ALGAE, *   W69-00632
PRODUCTIVITY, WATER QUALITY, *LIM   W69-00632
PRODUCTIVITY, TROPIC LEVEL, WATER   W68-00912
PRODUCTIVITY, CYANOPHYTA, ENVIRON   W69-04802
PRODUCTIVITY, HYPOLIMNION, ALGAL   W69-05142
PRODUCTIVITY, BIOASSAY, CYCLING N   W69-04798
PRODUCTIVITY, STANDING CROP, STRE   W69-04805
PRODUCTIVITY, TRACE ELEMENTS, SOD   W69-04801
PRODUCTIVITY, AQUATIC WEEDS, BIOA   W69-03186
PRODUCTIVITY, AQUATIC WEEDS, BALA   W69-03185
PRODUCTIVITY, AQUATIC WEEDS, CHLO   W69-03373
PRODUCTIVITY, CYCLING NUTRIENTS,   W69-03514
PRODUCTIVITY, BALANCE OF NATURE,   W69-03513
PRODUCTIVITY, ECOSYSTEMS, ILLINOI   W70-04506
PRODUCTIVITY, RADIOISOTOPES, TEMP   W70-04371
PRODUCTIVITY, HARVESTING OF ALGAE   W70-04290
PRODUCTIVITY, BIOCHEMICAL OXYGEN   W70-04283
PRODUCTIVITY, STANDING CROP, CHLO   W70-03512
PRODUCTIVITY, PHOSPHATES, CHLOROP   W69-10169
PRODUCTIVITY, PHOTOSYNTHESIS, ALG   W69-10182
PRODUCTIVITY, *STREAMS, *LABORATO   W70-00265
PRODUCTIVITY, WATER POLLUTION EFF   W69-10152
PRODUCTIVITY, TROPHIC LEVEL.: /VI   W69-09135
PRODUCTIVITY, *MEASUREMENT, PROTE   W69-10151
PRODUCTIVITY, BIOINDICATORS, SECO   W69-10152
PRODUCTIVITY, *TROPHIC LEVEL, WAT   W69-09723
PRODUCTIVITY, AQUATIC PRODUCTIVIT   W69-10152
PRODUCTIVITY, AQUATIC NUISANCES,   W69-09349
PRODUCTIVITY, OXYGEN, METABOLISM,   W70-01073
PRODUCTIVITY, SULFIDES, SEA WATER   W70-01068
PRODUCTIVITY, LIGHT INTENSITY, TR   W70-02504
PRODUCTIVITY, CHLOROPHYLL, SAMPLI   W70-01933
PRODUCTIVITY, PHOSPHATES, SILICAT   W70-01579
PRODUCTIVITY, ALGAE, AQUATIC LIFE   W70-03068
PRODUCTIVITY, *ALGAE, *THERMAL WA   W70-03309
PRODUCTIVITY, PHYTOPLANKTON.: /US   W70-02795
PRODUCTIVITY, BALANCE OF NATURE,   W70-02792
PRODUCTIVITY, CHLOROPHYTA, GONOPH   W70-02784
PRODUCTIVITY, PERIPHYTON, CHLOROP   W70-02780
.PRODUCTIVITY, TURBIDITY, SURFACES   W71-07698
PRODUCTIVITY, FOOD CHAINS, TOXICI   W72-01788
PRODUCTIVITY, ABSORPTION, SEDIMEN   W72-00845
PRODUCTIVITY, *BIOLOGICAL COMMUNI   W71-12870
PRODUCTIVITY, AQUATIC ALGAE, TROP   W72-01990
PRODUCTIVITY, PHYTOPLANKTON, EUGL   W72-04761
PRODUCTIVITY, *ORGANIC MATTER, NU   W71-10079
PRODUCTIVITY.:   MAR   W71-10788
PRODUCTIVITY, *WASTE WATER TREATM   W70-08416
PRODUCTIVITY, *EUTROPHICATION, WA   W70-09093
PRODUCTIVITY, REGIONAL ANALYSIS,   W70-07261
PRODUCTIVITY, PHOTOSYNTHESIS, EUT   W70-06225
PRODUCTIVITY, ALGAE, PLANKTON, SU   W71-00665
PRODUCTIVITY, AQUATIC MICROBIOLOG   W72-12240
PRODUCTIVITY, OREGON, CYANOPHYTA.   W72-06052
PRODUCTIVITY, CHLOROPHYLL, PIGMEN   W73-01446
PRODUCTS, ORGANISMS' ODORS.:   W72-01812
PRODUCTS, PHOTIC ZONE, GLYCOLLIC   W70-02504
PRODUCTS, CHEMICAL RESISTANCE, EU   W70-01068
PRODUCTS, ANABAENA CYLINDRICA, ZI   W69-10180
PRODUCTS, HORMONES, AUXINS, METAB   W69-04801
PRODUCTS, GAS EXCHANGE, GREEN PLA   W69-07832
PRODUCTS, ALGAL CELLS.: /E, PHYTO   W69-05697
PROFILE.: / ACTIVITY, *RADIOSTRON   W71-04067
PROFILES.: /OZEMS, SOIL MICROBIOL   W71-04067
PROFILES, VOLATILE SUSPENDED SOLI   W70-04790
PROGRAM, ALGAL DENSITY, APHANIZOM   W69-10167
PROGRAM, CONTINUOUS CULTURE, CHEM   W70-02775
PROGRAMS, ANAEROBIC BACTERIA, ANA   W71-04556
PROGRAMS, ANAEROBIC BACTERIA, ANE   W71-04554
PROGRAMS, ANAEROBIC BACTERIA, ANA   W71-04555

NAGE WATER, CALIFORNIA, DRAINAGE
RFISHING, COHO SA/ ANTIPOLLUTION
ION, *TENNESSEE VALLEY AUTHORITY
REAT LAKES-ILLINOIS RIVER BASINS
ONOMIDAE, PROCLADIUS, CHIRONOMUS
YCEAE, COMPLEX FORMATION, GROWTH-
T REQUIREMENTS, PHYSIOCOCHEMICAL
PLEX FORMATION, GROWTH-PROMOTING
, PUBLIC HEALTH, PHYSICOCHEMICAL
SOTOPES, CESIUM, PHYSICOCHEMICAL
RING, *WATER SUPPLY, *BIOLOGICAL
MATTER, NUTRIENTS, LAKES, WATER
, RADIOISOTOPES, PHYSICOCHEMICAL
UBMERGED PLANTS, PHYSICOCHEMICAL
ING, CHEMICAL ANALYSIS, PHYSICAL
POLLUTION, TEMPERATURE, CHEMICAL
ND MAINTENA/ *ALGAE, *BIOLOGICAL
SPHORUS, ALGAE, NUTRIENTS, WATER
ECTS, AQUATIC ENVIRONMENT, WATER
NT, WATER PROPERTIES, BIOLOGICAL
                        WATER QUALITY
ON, ZOOPLANKTON, PHYSICOCHEMICAL
ATOMS, SAMPLING, PHYSICOCHEMICAL
TRIENTS, FERTILIZATION, PHYSICAL
ION, CHEMICAL ANALYSIS, PHYSICAL
OPHICATION, *HYDROLOGY, CHEMICAL
SINS, SEA WATER, PHYSICOCHEMICAL
N DIOXIDE, NUTRIENTS, BIOLOGICAL
UATIC LIFE, ALGAE, PLANTS, WATER
RATURE, *COOLING WATER, CHEMICAL
, NITROGEN COMPOUNDS, PH/ *WATER
, NUTRIENT ABSORPTION, *CHEMICAL
WASTE IDENTIFICATION, BIOLOGICAL
BIOLOGICAL PROPERTIES, PHYSICAL
TION, PLANKTON, *ALGAE, CHEMICAL
TRIC PLANTS, WEED CONTROL, WATER
R LAKE, ORE, GROWTH RATES, WATER
, DOMESTIC WASTES, *ALGAE, WATER
IBLIOGRAPHIES, *ALGAE, *CHEMICAL
S, SCENEDESMUS, LIMNOLOGY, WATER
HOLOGY, SEWAGE INPUT, BIOLOGICAL
LFIDE, IRON, MAGNESIUM, COOLING,
ALUTACEUS, ULOTHRIX, CLADOPHORA,
                        *OIL SPILLS,
, ALGAE, YEASTS, BACTERIA, FISH,
E, *MICROORGANISMS, SCENEDESMUS,
M, ELECTRIC POWER, HEATED WATER,
ION CONCENTRATION, TEMPERATURE,
GANIC MATTER, CHEMICAL ANALYSIS,
BACTERIA, BIOMASS, CHLOROPHYLL,
MARY PRODUCTIVITY, *MEASUREMENT,
IS, BACILLUS STEAROTHERMOPHILUS,
TEST, GRAM REACTION, CYTOPHAGA,
COLI COMMUNAE, EUGLENE GRACILIS,
UDA.:
METRY, OSCILLATORIA TENUIS, RAM/
S), HALOBACTERIUM, RESPIROMETER,
TIC DRIFT, FISH FOOD ORG/ ALGAE,
OLYSIS, CHELATION, FUNGI, ALGAE,
ATOMS, FUNGI, INSECTS, BACTERIA,
FICATION, *CULTURES, *BIOASSAYS,
TEMATICS, FUNGI, ALGAE, DIATOMS,
LYTES, ZETA POTENTIAL, BACTERIA,
BACTERIA, FUNGI, ACTINOMYCETES,
TIC POPULATIONS, *RIVERS, ALGAE,
ITED GROWTH, LUXURY CONSUMPTION,
NT TASK FORCE ON EUTROPHICATION,
2,4-D, AMMATE, CHLORAMINE, DE-K-
UTH AFRICA, ISRAEL, MICROCYSTIS,
OTOPUS, AUSTROCLOEON, ENALLAGMA,
WTH RATES, SEAWATER, INCUBATION,

OTHERMOPHILUS, PROTEUS VULGARIS,
DAM RHINE, POLDERS, COMPOSITION,
TERIA, *MARINE ALGAE, *CULTURES,
, SAMPLING, SEA WATER, SEASONAL,
ES, MESOPLANKTON, MICROPLANKTON,
S QUADRIMACULATUS, MADISON(WIS),
BACILLUS-LIKE, MICROCOCCUS-LIKE,
ASSIMILATION, BA-LA, HALF-LIFE,
ENT BALANCE, FOAMING DETERGENTS,
CULTURES, *ISOLATION, LIMNOLOGY,
US, CLADOCERA, WATER MASSES, U S
OMICS, OHIO, COMMERCIAL FISHING,
QUATIC LIFE, WATER CONSERVATION,
NTS, VIRUSES, BACTERIA, E. COLI,
N, *CONFERENCES, NUISANCE ALGAE,
UDOTEA FLABELLUM, ZINC, NICKEL,
BALT, MANGANESE, ATLANTIC OCEAN,
L COMMUNITIES, INTERTIDAL ALGAE,

PROGRAMS, ANAEROBIC BACTERIA, ANA   W71-04557
PROGRAMS, BIOLOGICAL BALANCE, OVE   W70-04465
PROJECT, *RESERVOIRS, TEMPERATURE   W71-07698
PROJECT, INDIANA HARBOR SHIP CANA   W70-00263
PROMOSUS, TRICHOPTERA, LEECHES, F   W70-01943
PROMOTING PROPERTIES, CHLAMYDOMON   W69-10180
PROPERTIES, PHYSIOLOGICAL ECOLOGY   W69-10161
PROPERTIES, CHLAMYDOMONAS MOEWUSI   W69-10180
PROPERTIES, GENETICS, BACTERIA, P   W70-00273
PROPERTIES, ADSORPTION, COLLOIDS,   W70-00708
PROPERTIES, *WATER POLLUTION SOUR   W72-01813
PROPERTIES, BACTERIA, BIOASSAY, M   W71-10079
PROPERTIES, ANION ADSORPTION, COL   W71-09863
PROPERTIES, PHYTOPLANKTON, DISTRI   W71-00665
PROPERTIES, CARP, TILAPIA, TEMPER   W70-06975
PROPERTIES, ECOLOGY, ECOLOGICAL D   W70-07313
PROPERTIES, CONTROL, *OPERATION A   W70-08392
PROPERTIES.: /ECTS, NITROGEN, PHO   W70-04382
PROPERTIES, BIOLOGICAL PROPERTIES   W70-04475
PROPERTIES, ECOLOGY, STRATIFICATI   W70-04475
PROPERTIES.:   W70-04475
PROPERTIES, DOMESTIC WASTES, SEWA   W70-03964
PROPERTIES, CENTRIFUGATION, MICRO   W70-05389
PROPERTIES, CLIMATE, NITROGEN, PH   W70-05565
PROPERTIES.: /EWAGE, WATER POLLUT   W70-05752
PROPERTIES, WISCONSIN, SEWAGE EFF   W70-05113
PROPERTIES, TRACERS, DISTRIBUTION   W69-08275
PROPERTIES, WATER POLLUTION EFFEC   W69-07442
PROPERTIES, WATER TEMPERATURE, OU   W69-05023
PROPERTIES, ECOLOGY, SALINITY, BE   W69-05023
PROPERTIES, *NUTRIENTS, *VIRGINIA   W69-03695
PROPERTIES, EPILMNION, BOTTOM SED   W68-00857
PROPERTIES, PHYSICAL PROPERTIES,   W69-01273
PROPERTIES, WASTE WATER(POLLUTION   W69-01273
PROPERTIES, BACTERIA, ORGANIC POL   W68-00483
PROPERTIES, PHOSPHORUS, CHEMICAL   W68-00479
PROPERTIES.:   CRATE   W68-00478
PROPERTIES, BIOCHEMICAL OXYGEN DE   W68-00256
PROPERTIES, NUTRIENTS, LAKES, TES   W68-00856
PROPERTIES, WATER TEMPERATURE, AL   W68-00248
PROPERTIES, *WATER QUALITY, TURBI   W68-00475
PROPERTY VALUES, RECREATION DEMAN   W70-02251
PROSOPIUM WILLIAMSONI.: /CHEILUS   W69-07853
PROTEIN AQUICULTURE.:   W69-05891
PROTEINS, OIL WASTES, OILY WATER,   W69-05891
PROTEINS, WALLS, STRUCTURE, AMINO   W70-04184
PROTEINS, AQUATIC MICROORGANISMS,   W70-09905
PROTEINS, CARBOHYDRATES, LIPIDS,   W71-13413
PROTEINS, BACTERIA.: /OMPOSING OR   W72-01812
PROTEINS, LIGHT, NUTRIENTS, HYDRO   W69-10160
PROTEINS, RADIOACTIVITY, CARBON R   W69-10151
PROTEUS VULGARIS, PSEUDOMONAS AER   W71-11823
PROTEUS, SERRATIA, BAY OF FUNDY(C   W70-03952
PROTOCOCCALES.: /, DRY WEIGHT, B   W70-05092
PROTOCOCCUS, SCENEDESMUS QUADRICA   W71-11561
PROTON MAGNETIC RESONANCE SPECTRO   W69-00387
PROTOPLASM, WATER-BLOOMS, LAKE GE   W70-04369
PROTOZOA, *AQUATIC INSECTS, *AQUA   W69-02782
PROTOZOA, LICHENS, PLANTS, DIATOM   W69-07428
PROTOZOA, SEWAGE, LAKES, NUTRIENT   W71-10786
PROTOZOA, BACTERIA, FUNGI, VIRUSE   W71-11823
PROTOZOA, PLANKTON, WORMS, NEMATO   W72-01819
PROTOZOA, ALGAE, SOLID WASTES, TE   W70-01027
PROTOZOA, ALGAE.: /IONS, E. COLI,   W71-02009
PROTOZOA, AQUATIC INSECTS, FISH R   W73-00748
PROVISIONAL ALGAL ASSAY, PROCEDUR   W70-02779
PROVISIONAL ALGAL ASSAY PROCEDURE   W70-02777
PRUF-21, BENOCLORS, NIGROSINE DYE   W70-05263
PRYMNESIUM PARVUM, SYNERGISTIC EF   W70-05372
PSEUDAGRION, BRACHYTHEMIS, CHYDOR   W70-06975
PSEUDOMONAS, SUBSURFACE-WATERS, B   W70-00713
PSEUDOMONAS AERUGINOSA.:   W69-10171
PSEUDOMONAS AERUGINOSA, STAPHYLOC   W71-11823
PSEUDOMONAS, MYCOBACTERIUM, BACIL   W69-07838
PSEUDOMONAS, METABOLISM, KINETICS   W70-04280
PSEUDOMONAS, TEMPERATURE, AMINO A   W70-03952
PSEUDOPLANKTON, CRYPTOPHYCEAE, VO   W70-05389
PSOROPHORA CILIATA, ARTIFICIAL KE   W70-00225
PSYCHROPHILES, PLEMORPHIC, MYCELO   W70-00713
PTYCHOCHEILUS OREGONENSIS, ACROCH   W69-07853
PUBLIC APPREHENSION.: /TES, NUTRI   W70-05084
PUBLIC HEALTH, PHYSICOCHEMICAL PR   W70-00273
PUBLIC HEALTH SERVICE, GREAT LAKE   W70-00263
PUBLIC HEALTH, WALLEYE, VEGETATIO   W70-01943
PUBLIC HEALTH, WATER POLLUTION, C   W70-03344
PUBLIC HEALTH, PATHOGENIC BACTERI   W72-06017
PUBLICATIONS, WATER POLLUTION EFF   W70-03973
PUERTO RICO, LA PARGUERA, JOYUDA.   W69-08525
PUERTO RICO, SPECTROSCOPY, DISTRI   W69-08525
PUGET SOUND, FLOURESCEIN DYE.: /A   W68-00010

LLIS HAENOPHYLLA, TIDAL CURRENT,
YON RETICULATUM, DICTYOSPHAERIUM
AUDA, ECOLOGICAL EFFIC/ *DAPHNIA
DIES, DAPHNIA SCHODLERI, DAPHNIA
D FISHES, CLOROCOCCALES, DAPHNIA
DETECTOR, ALASKA, HUDEUC LAKE,
ELGOLAND(GERMANY)), AUREOBASIDIUM
ES, PHOSPHATES, ALGAE, DETRITUS,
OL, FISH, TOXICITY, FRESH WATER,
RCOURSES(LEGAL), ORGANIC WASTES,
OLIGOPHOSPHATES, METAPHOSPHATES,
, FOREST CENTRIFUGE, GLENODINIUM
LLA, ANKISTRODESMUS, GLENODINIUM
TREATMENT, MECHANICAL EQUIPMENT,
ALGAE, FRESH WATER, FLOW RATES,
ZURICH(SWITZERLAND), WHITEFISH,
THYS OSCULUS, SUCKERS, ICTALURUS
CHIRONOMUS, PROCLADIUS, CORETHRA
NTS, ROOTED AQUATIC PLANTS, SELF-
*WASTE WATER(POLLUTION), *WATER
ION), *WA/ *PHYTOPLANKTON, *SELF-
R), ACIDITY, OXYGEN, ALGAE, SELF-
TION, ABSORPTION, ECOLOGY, WATER
                                WATER
ANISMS, *WATER POLLUTION, *WATER
INAGE, MECHANICAL AND BIOLOGICAL
XENIC CULTURES, PEDIASTRUM, SELF-
LANT POPULATIONS, HISTORY, WATER
*BACTER/ *ALGAE, *BIOMASS, WATER
CE ALGAE, SEWAGE TREATMENT, SELF-
ITION), ALGAE, OXYGENATION, SELF-
TION, FILTRATION, AMMONIA, WATER
NTROL, *LAKE ONTARIO, *WA/ *SELF-
FISH, CHRYSOPHYTA, SESTON, SELF-
ATED CARBON, CLAYS, ALGAE, WATER
, CLEANING, COST ANALYSIS, WATER
ATER TREATMENT, *NEW YORK, WATER
*LAGOONS, ACTIVATED SLU/ *WATER
ALITY CONTROL, FILTRATION, WATER
N EQUIVALEN/ *MICROBIAL ECOLOGY,
ENSE, RHODOCHORTON, BANGIA FUSCO-
OUXI, PORITES LOBATA, LEPTASTREA
GLENODINIUM QUADRIDENS, OOCYSTIS
NUNCULUS CIRCINATUS, POTAMOGETON
M FILTER, BACK-WA/ *UPWARD FLOW,
RATELLA, POLYARTHRA, CONOCHILUS,
A COSTATUM, COCCOLITHUS HUXLEYI,
II, D/ CARBON DIOXIDE, CHLORELLA
ENEDESMUS QUADRICAUDA, CHLORELLA
D, SYMBIOTIC RELATIO/ *CHLORELLA
SSR), MUNICH(GERMANY), CHLORELLA
LA VULGARIS(COLUMBIA), CHLORELLA
ULTURE, *ALGAL GROWTH, CHLORELLA
PRECAMBRIAN SEDIMENTS, CHLORELLA
N), ELY(MINN), BLOOMS, CHLORELLA
E, CHLORELLA VULGARIS, CHLORELLA
LIQUUS, HELIX POMATIA, CHLORELLA
ON, LAKE TAHOE(CALIF), CHLORELLA
RANA, GENERATION TIME, CHLORELLA
HLAMYDOMONAS MOEWUSII, CHLORELLA
N, SEVIN, MALATHION, / CHLORELLA
ITRILOTRIACETATE, NTA, CHLORELLA
KE, SPECIFIC ACTIVITY, CHLORELLA
IENTS, MICRONUTRIENTS, CHLORELLA
BLELHAM TARN(ENGLAND), CHLORELLA
                            *CHLORELLA
TRI/ SEVIN, MALATHION, CHLORELLA
KINETIC MODELS, CHLORELLA
ANCE LIMIT, BLUEGILLS, CHLORELLA
ON CONCENTRATION, FUNGI, YEASTS,
MINE ACIDS, COAL MINES, ACIDITY,
GONYAULAX, GONYAULAX VENIFICUM,
ESIS, CONCRETIONS, ARTHROBACTER,
KIMBALL LAKE(MINN), CRYPTOPHYTA,
LINDRICA, ATP, CMU, NITROGEN-15,
ASS).:
CENEDESMUS OBLIQUUS, SCENEDESMUS
XIDE, RATS, TRYPSIN, SCENEDESMUS
TES, *ALKOXYSILANES, SCENEDESMUS
ENEDESMUS OBLIQUUES, SCENEDESMUS
BAUGH CENTER(CALIF), SCENEDESMUS
APHNIA PULEX OBTUSA, SCENEDESMUS
TES, *ALKOXYSILANES, SCENEDESMUS
PROTOCOCCUS, SCENEDESMUS
OUNTING, ITHACA(NY), GLENODINIUM
ARIAE, TVA RESERVOIRS, ANOPHELES
ONTROL, RADIOACTIVE WASTE, WATER
OLLUTION EFFECTS, ECOLOGY, WATER
LGAE, ALGAE, CHLORINATION, WATER

PUGET SOUND(WASH), LONG ISLAND SO   W69-10161
PULCHELLUM, GOMPHONEMA, DINOBRYON   W69-05704
PULEX OBTUSA, SCENEDESMUS QUADRIC   W69-10152
PULEX, DAPHNIA GALEATA MENDOTAE,   W70-03957
PULICARIS, DAPHNIA CUCULLATA, DAP   W70-00274
PULLIN LAKE, GRAPHITE LAKE, OLD J   W69-08272
PULLULANS, HANSENIASPORA UVARUM,   W70-04368
PULP WASTES, LIVESTOCK, IDAHO.: /   W70-03501
PULP WASTES, PESTICIDE RESIDUES,   W70-10437
PULP WASTES, INDUSTRIAL WASTES, T   W72-00462
PULSE-LABELING, POLYPHOSPHATES, F   W71-03033
PULVISCULUS, HAEMOCYTOMETER COUNT   W69-06540
PULVISCULUS, SCENEDESMUS: /RUDINE   W70-05387
PUMPING PLANTS, MAINE.: /, WATER   W70-08096
PUMPS, ELECTROLYSIS, CULTURES.: /   W70-09904
PUNCHBOWL LAKE(CANADA).: /S, LAKE   W70-05399
PUNCTATUS, TRICHOPTERA, ODONATA,   W69-07846
PUNCTIPENNIS, PISIDIUM IDAHOENSE,   W70-06217
PURIFICATION, BIOCHEMICAL OXYGEN   W70-05750
PURIFICATION, ALGAE, SEWAGE, ANAE   W70-05092
PURIFICATION, *WASTE WATER(POLLUT   W70-05092
PURIFICATION, BACTERIOPHAGE, RESE   W69-07838
PURIFICATION, CARBON, BICARBONATE   W69-07389
PURIFICATION BY ALGAE.:   W69-07389
PURIFICATION, *BACTERIA, *ENZYMES   W69-07428
PURIFICATION, SWITZERLAND.: / DRA   W69-08668
PURIFICATION, ALGAL ENUMERATION,   W70-04381
PURIFICATION, *WATER UTILIZATION,   W68-00483
PURIFICATION, *PROTOZOA, *FUNGI,   W71-05267
PURIFICATION, DISSOLVED OXYGEN, P   W72-01798
PURIFICATION, LABORATORY TESTS, I   W72-01801
PURIFICATION, WASTE WATER TREATME   W71-11393
PURIFICATION, *WATER POLLUTION CO   W71-05878
PURIFICATION.: /GAE, ZOOPLANKTON,   W70-00268
PURIFICATION, WATER QUALITY, WATE   W69-09884
PURIFICATION, *WATER TREATMENT.: /   W72-08688
PURIFICATION.: /SIGN CRITERIA, *W   W72-08686
PURIFICATION, *FLOW AUGMENTATION,   W72-06017
PURIFICATION, MEMBRANES, SCREENS,   W73-02426
PURPLE SULFUR BACTERIA, POPULATIO   W70-03312
PURPUREA, SPHACELARIA, GONIDIA, C   W69-10161
PURPUREA, FUNGIA FUNGITES, ZOOXAN   W69-08526
PUSILLA, OOCYSTIS SOLITARIA, PAND   W69-06540
PUSILLUS.: /GELLATA, CILIATES, RA   W69-07838
PUTRESCIBILITY TEST, BODY IMMEDIU   W70-05821
PYCNOCLINE, PHOTOLITHOTROPHIC BAC   W70-05760
PYRAMIMONAS, PERIDINIUM TROCHOIDE   W70-05272
PYRENOIDOSA, ANKISTRODESMUS BRAUN   W70-05261
PYRENOIDOSA, CHLORELLA VULGARIS,   W70-05381
PYRENOIDOSA, SYNTHETIC SEWAGE FEE   W70-05655
PYRENOIDOSA, BICHOROMATE, DRY WEI   W70-05092
PYRENOIDOSA, CHLORELLA SACCHAROPH   W70-04809
PYRENOIDOSA(EMERSON STRAIN), AUTO   W69-07442
PYRENOIDOSA, FUCALES, ALKANES, AL   W69-08284
PYRENOIDOSA, MICROCYSTIS AERUGINO   W70-03512
PYRENOIDOSA, VOLATILE ACIDS, NON-   W70-04195
PYRENOIDOSA, NITROGEN UTILIZATION   W70-04184
PYRENOIDOSA, SELENASTRUM GRACILE,   W70-03983
PYRENOIDOSA, DIATOMA ELONGATUM, M   W69-10158
PYRENOIDOSA, TRIBONEMA AEQUALE, N   W69-10180
PYRENOIDOSA, PESTICIDE DEGRADATIO   W70-02968
PYRENOIDOSA, CARBOXYLIC ACIDS, AM   W70-02248
PYRENOIDOSA, BIOLOGICAL HALF-LIFE   W70-02786
PYRENOIDOSA, AUTOTROPHIC GROWTH,   W70-02504
PYRENOIDOSA, FIXATION, TRINGFORD(W70-02504
PYRENOIDOSA.:   W70-08416
PYRENOIDOSA, SCHRODERIA, THIMET,   W70-08097
PYRENOIDOSA, *ALGAE BIOMASS.:   W73-02218
PYRENOIDOSA, BENZENE COMPOUNDS.: /   W72-11105
PYRITE, ALGAE, CYANOPHYTA, DOMINA   W68-00891
PYRITE, SULFUR COMPOUNDS, MOLYBDE   W69-07428
PYRODINIUM, GYMNODINIUM BREVE, CA   W70-05372
PYROPHOSPHATE, ORTHOPHOSPHATE, DI   W69-07428
PYROPHYTA, CHROOMONAS ACUTA, SCHR   W70-05387
PYRUVATE, ACETYLENE REDUCTION, RE   W70-04249
QUABBIN RESERVOIR(MASS), BOSTON(M   W70-05264
QUADRICAUDA, CHLORELLA PYRENOIDOS   W70-05381
QUADRICAUDA, SCENEDESMUS OBLIQUUS   W70-04184
QUADRICAUDA, CHLORELLA VULGARIS,   W71-03056
QUADRICAUDA, PORPHYRIDIUM CRUENTU   W70-02964
QUADRICAUDA, CELL COUNTS, SPECIES   W70-03334
QUADRICAUDA, ECOLOGICAL EFFICIENC   W69-10152
QUADRICAUDA, CHLORELLA VULGARIS,   W71-05719
QUADRICAUDA.:   W71-11561
QUADRIDENS, OOCYSTIS PUSILLA, OOC   W69-06540
QUADRIMACULATUS, MADISON(WIS), PS   W70-06225
QUALITY ACT, PATH OF POLLUTANT, F   W72-00941
QUALITY CHARACTERIZATION.: /TER P   W70-02764
QUALITY CONTROL.: /TE-PRODUCING A   W70-02373

OGY, EFFECTS OF POLLUTION, WATER QUALITY CONTROL, WINNISQUAM LAKE, W70-02772
EUTROPHICATION, WISCONSIN, WATER QUALITY CONTROL, LAKES, WATER POL W70-03310
*HARVESTING OF ALGAE, EFFLUENT, QUALITY CONTROL, OXIDATION LAGOON W71-13313
ON, *NUCLEAR POWERPLANTS, *WATER QUALITY CONTROL, *BIODEGRADATION, W71-11881
TICIDES, FARM WASTES, SIL/ WATER QUALITY CONTROL, *NUTRIENTS, *PES W71-10376
AKES, *LOCAL GOVERNMENTS, *WATER QUALITY CONTROL, ALGAE, AQUATIC W W72-05774
ULFATE, / *POTABLE WATER, *WATER QUALITY CONTROL, *ALGAE, COPPER S W71-00110
AE, *ALGICIDES, *PHENOLS, *WATER QUALITY CONTROL, CRUSTACEANS, ROT W70-10161
ATER SUPPLY, INTAKES, NUTRIENTS, QUALITY CONTROL, TREATMENT FACILI W70-00263
WATER POLLUTION TREATMENT, WATER QUALITY CONTROL, SEWAGE EFFLUENTS W70-01031
S, PESTICIDES, HARVESTING, WATER QUALITY CONTROL, WATER POLLUTION W69-05706
GICIDES, *COPPER SULFATE, *WATER QUALITY CONTROL, ALGAL CONTROL, S W69-05704
S, PESTICIDES, HARVESTING, WATER QUALITY CONTROL, WATER POLLUTION W69-05705
TS, *RESERVOIRS, *MIXING, *WATER QUALITY CONTROL, WATER QUALITY, W W70-04484
VOIR, MASSACHUSETTS, POT/ *WATER QUALITY CONTROL, *ECOLOGY, *RESER W70-05264
MENT, STREAM IMPROVEMENT, *WATER QUALITY CONTROL, *MIDGES, FISH FO W70-05428
TANT IDENTIFICATION, POLLUTANTS, QUALITY CONTROL, WATER POLLUTION, W69-01273
ES, *ALGAE, *TASTE, *ODOR, WATER QUALITY CONTROL, POLLUTANT IDENTI W72-13605
PPER SULFATE, *ALKALINITY, WATER QUALITY CONTROL, *CONTROL, *ALGAL W72-07442
*WATER POLLUTION CONTROL, WATER QUALITY CONTROL.: /ROSION, ALGAE, W73-01122
ERS, SANDS, LAKE MICHIGAN, WATER QUALITY CONTROL, FILTRATION, WATE W73-02426
NITIES, MICROBIAL ECOLOGY, WATER QUALITY CRITERIA, CYANIDE, NAPHTH W70-04510
INVENTORY(LAKE SUPERIOR), WATER QUALITY INDICATORS.:   PERIPHYTON W70-05958
TON/ *EFFLUENT DIVERSION, *WATER QUALITY MANAGEMENT, *LAKE WASHING W69-09349
*RIVER STUDIES, *NATIONAL WATER QUALITY NETWORK, ALGAE COUNTS, PL W70-04507
                          WATER QUALITY PROPERTIES.: W70-04475
ARCH MANUFACTURING WASTES, WATER QUALITY STANDARDS, GRAFTON(ND), L W70-03312
EASURES, PHYSICAL CONTROL, WATER QUALITY.: /, INTEGRATED CONTROL M W70-07393
ANOPHYTA, *EUTROPHICATION, WATER QUALITY.: /ON EFFECTS, LAKES, *CY W71-04758
ECTS, NITROGEN, COLIFORMS, WATER QUALITY.: /S, WATER POLLUTION EFF W71-05407
RON, PHOSPHORUS, NITROGEN, WATER QUALITY.: /HIGAN, BICARBONATES, I W72-03183
SYSTEMS, WATER POLLUTION, WATER QUALITY.: /OXICITY, *DISTRIBUTION W72-03746
, WATER POLLUTION EFFECTS, WATER QUALITY.: /ATER POLLUTION SOURCES W71-05409
LUTION EFFECTS, ALGICIDES, WATER QUALITY.: /, MINNESOTA, WATER POL W70-04506
TANKS, RECREATION WASTES, WATER QUALITY.: /NICIPAL WASTES, SEPTIC W70-04266
ENTS, *PHOSPHATES, CHELA/ *WATER QUALITY, *EUTROPHICATION, *DETERG W70-04373
ESERVOIRS, *URBANIZATION, *WATER QUALITY, *STRATIFICATION, *TEXAS, W70-04727
TOSYNTHESIS, PRODUCTIVITY, WATER QUALITY, *LIMNOLOGY, MINNESOTA, W W69-00632
ATA COLLECTIONS, PIT REC/ *WATER QUALITY, *ARTIFICIAL RECHARGE, *D W69-05894
TASTE, ODOR, TOX/ *ALGAE, *WATER QUALITY, *EUTROPHICATION, COLOR, W69-05697
MENTS, *WATER POLLUTION / *WATER QUALITY, *GREAT LAKES, *TRACE ELE W69-08562
ELTAS, *DRAINAGE EFFECTS, *WATER QUALITY, *FARM WASTES, *INDUSTRIA W71-09577
ATER CHEM/ *LAKE ONTARIO, *WATER QUALITY, *MONITORING, BACTERIA, W W71-05879
TION, CHL/ *WATER SUPPLY, *WATER QUALITY, *WATER TREATMENT, FILTRA W71-05943
SH, *FISH KILLS, *ROTIFE/ *WATER QUALITY, *ALGAE, *PESTICIDES, *FI W72-04506
COPPER / *WATER TREATMENT, WATER QUALITY, *ORGANIC MATTER, ALGAE, W72-04495
CATION, DAPHNIA, CHLAMYDO/ WATER QUALITY, *ALGAL CONTROL, EUTROPHI W72-02417
*WATER POLLUTION EFFECTS/ *WATER QUALITY, *RECREATION, *SWIMMING, W71-09880
FECTS, *IMPOUNDED WATERS, *WATER QUALITY, *DISSOLVED SOLIDS, DISSO W72-01773
HYLL, *OHIO RIVER, *ANAL/ *WATER QUALITY, *BIOINDICATORS, *CHLOROP W71-04206
UCTIVITY, *EUTROPHICATION, WATER QUALITY, *NUTRIENTS, LIMNOLOGY, W W70-09093
S, WATER CHEMISTRY, WATE/ WATER QUALITY, *WATER POLLUTION, *RIVER W70-03068
VIRONMENT, *AQUATIC ALGAE, WATER QUALITY, *TROPHIC LEVELS, NUTRIEN W70-02772
LOOMS, NUTRIENTS, LIMNOLO/ WATER QUALITY, *EUTROPHICATION, ALGAL B W69-09723
EFFECTS, *BIOINDICATORS, *WATER QUALITY, *DISSOLVED OXYGEN, *OHIO W70-00475
ION, NITROGEN, PHOSPHORU/ *WATER QUALITY, *NUTRIENTS, *EUTROPHICAT W73-01122
ION CO/ *WATER TREATMENT, *WATER QUALITY, *ALGAE, *ODOR, HYDROGEN W72-09694
ENT, *FILTRATION, *ALGAE, *WATER QUALITY, *ACTIVATED CARBON, *ELEC W72-09693
NALYSIS, *BIOASSAY, *COM/ *WATER QUALITY, *BIOINDICATORS, *WATER A W72-12240
, BIOLOGY, FISH, PLANKTON, WATER QUALITY, ALGAE, AQUATIC LIFE, STR W72-01786
UTROPHICATION, SUCCESSION, WATER QUALITY, ALGAL CONTROL, ALGICIDES W69-01977
, WATER POLLUTION SOURCES, WATER QUALITY, ALGAE, SEWAGE EFFLUENTS. W70-05084
LAIMED WATER, *CALIFORNIA, WATER QUALITY, AESTHETICS, RECREATION, W70-05645
*PERIPHYTON, *LIMNOLOGY, *WATER QUALITY, BIOINDICATORS, AQUATIC L W70-05958
D, SU/ *DATA COLLECTIONS, *WATER QUALITY, BIOCHEMICAL OXYGEN DEMAN W70-09557
SHKILL, ALGAE, OXYGEN SAG, WATER QUALITY, CHLORIDES, PHOSPHORUS, S W69-03948
NG NUTRIENTS, TEMPERATURE, LIGHT QUALITY, CHLOROPHYLL, ZOOPLANKTON W69-08526
, WATER POLLUTION EFFECTS, WATER QUALITY, CATTLE.: /ODOR, TOXICITY W69-08279
TERTIARY TREATMENT, ALGAE, WATER QUALITY, CONIFEROUS FORESTS, DECI W70-03512
T, *ANALYTICAL TECHNIQUES, WATER QUALITY, DISSOLVED OXYGEN, BIOCHE W70-08627
ES, HERBICIDES, ESTUARIES, WATER QUALITY, DETERGENTS, ALGAL CONTRO W70-10437
D OXYGEN, ALGAE, VELOCITY, WATER QUALITY, DISSOLVED SOLIDS, HYDROG W70-00265
ATIC ALGAE, TROPHIC LEVEL, WATER QUALITY, FLORIDA.: /UCTIVITY, AQU W72-01990
ENTS, AQUATIC POPULATIONS, WATER QUALITY, LAKE MORPHOMETRY, DOMEST W68-00470
EAT LAKES, NUISANCE ALGAE, WATER QUALITY, NUTRIENTS, PHOSPHORUS, N W72-05453
WATER SUPPLY, RESERVOIRS, WATER QUALITY, NEW JERSEY, TEMPERATURE, W69-10157
VELS, AQUATIC ENVIRONMENT, WATER QUALITY, NUTRIENTS, NORTHEAST U S W70-02764
SAMPLING, WATER ANALYSIS, WATER QUALITY, POLLUTANT IDENTIFICATION W71-00221
TION ABATEMENT, COLIFORMS, WATER QUALITY, PHOSPHORUS, NITROGEN, FL W70-04810
NT, ESTUARINE ENVIRONMENT, WATER QUALITY, SEWAGE BACTERIA, ENVIRON W69-04276
HIO RIVER, NUISANCE ALGAE, WATER QUALITY, SEWAGE EFFLUENTS, DIATOM W70-04507
SIS, ALGAL CONTROL, ALGAE, WATER QUALITY, SEDIMENT-WATER INTERFACE W72-06051
TER RESOURCES DEVELOPMENT, WATER QUALITY, SOIL INVESTIGATIONS, *IM W72-11906
T, BIOLOGICAL PROPERTIES, *WATER QUALITY, TURBIDITY, PLANT POPULAT W68-00475
IRONMENT, *ORGANIC MATTER, WATER QUALITY, TEMPERATURE, HYDROGEN IO W71-03031
Y, WASTE TREATMENT, LAKES, WATER QUALITY, VIRUSES, ALGICIDES, FISH W70-03973
G, *WATER QUALITY CONTROL, WATER QUALITY, WATER POLLUTION, NITROGE W70-04484
RODUCTIVITY, TROPIC LEVEL, WATER QUALITY, WASTE ASSIMILATIVE CAPAC W68-00912
, WATER POLLUTION EFFECTS, WATER QUALITY, WISCONSIN, LAKES, EUTROP W69-05868

POPULATIONS, CONSERVATION, WATER QUALITY, WATER POLLUTION, WATER R W69-05844

FICATION, WASTE TREATMENT, WATER   QUALITY, WATER POLLUTION CONTROL,   W69-06535
LUTION EFFECTS, *REVIEWS, *WATER   QUALITY, WATER POLLUTION, ORGANIC   W70-07027
GAL CONTROL, WATER SUPPLY, WATER   QUALITY, WATER POLLUTION,   W69-10155
ALGAE, WATER PURIFICATION, WATER   QUALITY, WATER POLLUTION CONTROL,   W69-09884
RNMENT, STATE GOVERNMENTS, WATER   QUALITY, WATER CHEMISTRY.: /AYS,   W69-06305
S, WASTE WATER(POLLUTION), WATER   QUALITY: /G, GRANTS, FEDERAL GOVE   W69-01273
X MUCOR, ANTITHAMNION SARNIENSE,   QUALITY: /IES, PHYSICAL PROPERTIE   W69-10163
IDAZOLINES, BIOLOGICAL ACTIVITY,   QUANTITATION, GROWTH MODES.: /HRI   W70-02982
ENOL/ BIOCIDES, PRODUCTION LOSS,   QUARTERNARY, AMMONIUM COMPOUNDS,   W72-12987
NIMUM LETHAL DOSES, CHU'S MEDIA,   QUATERNARY AMMONIUM COMPOUNDS, PH   W70-02982
LAKE, *INORGANIC/ ONTARIO LAKES,   QUINONE DERIVATIVES, ALKYL SUBSTI   W70-05424
IX FUSCA, MICROCOLEUS LACUSTRIS (QUIRKE LAKE, PECORS LAKE, DUNLOP   W70-04468
TRIS BRAND, GONGROSIRA DEBARYANA   RAB) FARLOW, CLADOPHORA, TAXONOMI   W70-04468
CYSTIS AERUGINOSA, ANABAENOPSIS,   RAB, OOCYSTIS, OSCILLATORIA: /CUS   W70-05404
NKTON, ALGAE, TEMPERATURE, SOLAR   RACIBORSKII, RAPHIDIOPSIS, ANABAE   W71-10788
FILTRATION, *SLIME, *ULTRAVIOLET   RADIATION.: /OSYNTHESIS, PHYTOPLA   W72-03824
SEWAGE EFFLUENTS, OXYGEN, SOLAR   RADIATION, *TREATMENT FACILITIES,   W70-01933
ES, ALGAE, EMULSIONS, BACKGROUND   RADIATION, PERIPHYTON, BACTERIA,   W69-10163
COVER, SPECTROSCOPY, ULTRAVIOLET   RADIATION, IODINE RADIOISOTOPES,   W69-09676
TRIPTON, BOTTOM SEDIMENTS, SOLAR   RADIATION, TRANSMISSIVITY, CHLORO   W69-10154
YTON, AQUATIC PLANTS, BACKGROUND   RADIATION, OXYG: /ICE, BACTERIA,   W69-09334
TUATION, SEWAGE TREATMENT, SOLAR   RADIATION, GEOCHEMISTRY, SPRING W   W69-10182
ON, TIDES, PHOTOSYNTHESIS, SOLAR   RADIATION, CHLOROPHYLL, PRODUCTIV   W70-01073
MPERATURE, SALINITY, ULTRAVIOLET   RADIATION, SALINITY, RAINFALL, TE   W70-01074
WTH RATES, BACTERIA, ULTRAVIOLET   RADIATION, SULFATES, CELLULOSE, I   W70-00719
XYGEN-REDUCTION POTENTIAL, SOLAR   RADIATION, ACTINOMYCETES, HYDROGE   W70-05750
NG, AERIAL PHOTOGRAPHY, INFRARED   RADIATION, ENERGY, CARBON DIOXIDE   W68-00010
M SEDIMENTS, WEED CONTROL, SOLAR   RADIATION, DYE RELEASES, BEACHES,   W68-00478
LO), MUTATION, BETA RAYS, COSMIC   RADIATION, *WATER POLLUTION CONTR   W69-08272
ASKA, LAKES, RIVERS/ *BACKGROUND   RADIATION, S: /N, CENTRAL CITY(CO   W69-08272
SECTS, INVERTEBRATES, BACKGROUND   RADIATION, *AQUATIC HABITATS, *AL   W69-07846
, MANGANESE, ALGAE, IRON, BIRDS,   RADIATION, FLOCCULATION, POLLUTAN   W69-08269
LES, BOTTOM SEDIMENTS, CATTAILS,   RADIOACTIVITY, INVERTEBRATES, PLA   W69-08272
RABS, NORTH CAROLINA, SEDIMENTS,   RADIOACTIVITY, INTERFACES, ORGANI   W69-08267
RIBUTION, CHELATION, DETERGENTS,   RADIOACTIVITY, MARINE ANIMALS, ES   W69-08275
, *TROUT, *STREAMS, *ECOSYSTEMS,   RADIOACTIVITY, AMINO ACIDS.: /IST   W69-09334
GULF OF MEXICO, ATLANTIC OCEAN,   RADIOACTIVITY, BIOTA, CYCLES, MIC   W69-08524
, FISH, PHYTOPLANKTON, PLANKTON,   RADIOACTIVITY, GAMMA RAYS, POTASS   W69-07440
ANKTON, PLANKTON, RADIOACTIVITY,   RADIOACTIVITY, RADIOACTIVITY EFFE   W69-07440
ITION, NITROGEN FIXING BACTERIA,   RADIOACTIVITY EFFECTS, RADIOECOLO   W70-04249
LABORATORY TESTS, MODEL STUDIES,   RADIOACTIVITY TECHNIQUE, PHOTOSYN   W70-04074
MATTER, *PHYSIOLOGICAL ECOLOGY,   RADIOACTIVITY TECHNIQUES, TAGGING   W70-04194
CULTURES, PHYTOPLANKTON, CARBON   RADIOACTIVITY TECHNIQUES, ANALYTI   W70-03983
M RADIOISOTOPES, *PHYTOPLANKTON,   RADIOACTIVITY, CHLORELLA, SAMPLIN   W70-00707
CTIVITY, *MEASUREMENT, PROTEINS,   RADIOACTIVITY, WASTE DISPOSAL, NU   W69-10151
6, RHODIUM-106, CERIUM-144, BETA   RADIOACTIVITY, CARBON RADIOISOTOP   W69-09742
OPES, LIGHT INTENSITY, CULTURES,   RADIOACTIVITY, GONIOBASIS, DECAPO   W69-10158
OPHYLL, SAMPLING, PHYTOPLANKTON,   RADIOACTIVITY, CARBON DIOXIDE, WA   W70-01933
CITY, *INSECTICIDES, METABOLISM,   RADIOACTIVITY, PHOTOSYNTHESIS, CH   W70-02198
BUTION, DENSITY, CHROMATOGRAPHY,   RADIOACTIVITY TECHNIQUES, CHLOREL   W70-02510
, *PHOTOSYNTHESIS, BICARBONATES,   RADIOACTIVITY, TRACERS, WATER POL   W70-02504
S, FOOD CHAINS, STABLE ISOTOPES,   RADIOACTIVITY, CARBON RADIOISOTOP   W70-02786
AL WASTES, TOXINS, SOLID WASTES,   RADIOACTIVE WASTE DISPOSAL, OCEAN   W72-00462
OURCES, WATER POLLUTION CONTROL,   RADIOACTIVE WASTES, HEATED WATER,   W72-00941
ANIMALS, ALGAE, MOLLUSKS, FISH,   RADIOACTIVE WASTE, WATER QUALITY   W71-05531
L REACTIONS, CHLORELLA, TRACERS,   RADIOACTIVE WASTES.: /CTS, MARINE   W70-08097
BACTERIA, ALGAE, ORGANIC MATTER,   RADIOACTIVITY, CARBON RADIOISOTOP   W72-13601
RIENTS, MODEL STUDIES, BIOASSAY,   RADIOACTIVITY, PESTICIDES, TASTE,   W73-01454
GAE, IRON, ALKALINITY, HARDNESS,   RADIOACTIVITY TECHNIQUES, ABSORPT   W73-00750
NTRATION FACTOR, LIMNODRILUS SP/   RADIOACTIVITY, TEMPERATURE, SALIN   W69-07861
ROTROPHY, AUTOTROPHY, CARBON-14,   RADIOAUTOGRAPHS, HALF-LIFE, CONCE   W70-04287
IMNION, CHEMOCLINE, MIXOLIMNION,   RADIOCARBON UPTAKE TECHNIQUES, GL   W70-05760
NNUAL CYCLES, MICROBIAL ECOLOGY,   RADIOCARBON UPTAKE TECHNIQUES, BA   W70-05109
D SOUND(MASS), WOODS HOLE(MASS).   RADIOCARBON UPTAKE TECHNIQUES, MI   W70-05272
MELOSIRA, DINOBRYON, FRAGILARIA,   RADIOCARBON UPTAKE TECHNIQUES, MI   W70-01579
SPHATES, *ALGAE, CHROMOTOGRAPHY,   RADIOCARBON METHOD, FINGER LAKES(W71-03033
COBALT RADIOISOTOPES, SEASONAL,   RADIOCHEMICAL ANALYSIS, TRACERS.:   W70-04371
ON, FISH, SEA WATER, GAMMA RAYS,   RADIOCHEMICAL ANALYSIS, MONITORIN   W69-08269
HROCYTIUM OBESUM W AND G S WEST,   RADIOCHEMICAL ANALYSIS, TRITIUM,   W70-04468
ISOTOPES, IRON, SEDIMENTS, FISH,   RADIOCOCCUS NIMBATUS, URONEMA ELO   W69-07853
ACTIVITY, RADIOACTIVITY EFFECTS,   RADIOECOLOGY, EFFLUENTS, ECOSYSTE   W69-07440
FFECTS, WATER POLLUTION SOURCES,   RADIOECOLOGY, SORPTION, WATER POL   W70-00707
A WATER, FOOD CHAINS, PHOSPHORUS   RADIOECOLOGY.: /WATER POLLUTION E   W70-00708
S, COBALT RADIOISOTOPES, URANIUM   RADIOISOTOPES, COBALT RADIOISOTOP   W70-00708
TIES, ADSORPTION, COLLOIDS, ZINC   RADIOISOTOPES, NUCLEAR EXPLOSIONS   W70-00708
PHOSPHORUS RADIOISOTOPES, COBALT   RADIOISOTOPES, CALCIUM, POTASSIUM   W70-00708
S, NUCLEAR EXPLOSIONS, STRONTIUM   RADIOISOTOPES, URANIUM RADIOISOTO   W70-00708
ESIS, DIATOMS, CHLORELLA, CARBON   RADIOISOTOPES, CESIUM, PHYSICOCHE   W69-10158
NS, BACKGROUND RADIATION, IODINE   RADIOISOTOPES, LIGHT INTENSITY, C   W69-10163
ES, TRITIUM, PHOTOGRAPHY, CARBON   RADIOISOTOPES, NUTRIENT REQUIREME   W69-10163
CARBON RADIOISOTOPES, PHOSPHORUS   RADIOISOTOPES, PHOSPHORUS RADIOIS   W69-10163
PROTEINS, RADIOACTIVITY, CARBON   RADIOISOTOPES, ALGAE, EMULSIONS,   W69-10151
CARBON RADIOISOTOPES, PHOSPHORUS   RADIOISOTOPES, PHOSPHORUS RADIOIS   W69-10151
ECTROMETERS, CYANOPHYTA, URANIUM   RADIOISOTOPES, ALGAE, BIOCHEMISTR   W69-09742
URANIUM RADIOISOTOPES, STRONTIUM   RADIOISOTOPES, STRONTIUM RADIOISO   W69-09742
DIOACTIVITY, WASTE D/ *STRONTIUM   RADIOISOTOPES, CESIUM, FALLOUT, S   W70-00707
ALCAREOU/ *CHELATION, *STRONTIUM   RADIOISOTOPES, *PHYTOPLANKTON, RA   W71-04067
, TRACERS, RADIOACTIVITY, CARBON   RADIOISOTOPES, *ALGAE, *FUNGI, *C   W70-08097
DIATOMS, EUTROPHICATION, CARBON   RADIOISOTOPES, HYDROGEN ION CONCE   W70-01579
HICATION, *CALI/ *LAKES, *CARBON   RADIOISOTOPES, CHLOROPHYLL, HYDRO   W70-01933
                                          RADIOISOTOPES, *BIOASSAY, *EUTROP

515

ISOTOPES, FISH, CLAMS, STRONTIUM
TRACERS, PHOTOSYNTHESIS, CARBON
ARBONATES, RADIOACTIVITY, CARBON
*CESIUM
S, RADIUM RADIOISOTOPES, THORIUM
*AQUATIC PLANTS, AQUATIC / *ZINC
, THORIUM RADIOISOTOPES, URANIUM
ESTS, ALGAE, INSECTS, PHOSPHORUS
AT LAKES, TRACE ELEMENTS, RADIUM
ES, *ALGAE, *METHODOLOGY, CARBON
S, CADMIUM RADIOISOTOPES, CERIUM
PES, CERIUM RADIOISOTOPE/ CESIUM
S, STRONTIUM RADIOISOTOPES, ZINC
CRUSTACEANS, MOLLUSKS, STRONTIUM
ES, IODINE RADIOISOTOPES, COBALT
*RUTHENIUM
OPES, ZINC RADIOISOTOPES, IODINE
*MARINE ANIMALS, *MARINE PLANTS,
E/ CESIUM RADIOISOTOPES, CADMIUM
DIOISOTOPES, TROPHIC LEVEL, ZINC
S, URANIUM RADIOISOTOPES, RADIUM
VIRONMENTS, *ECOSYSTEMS, *RADIUM
OILS, NUCLEAR EXPLOSIONS, COBALT
HYTON, *CYCLING NUTRIENTS, *ZINC
DIOISOTOPES, GAMMA RAYS, URANIUM
, OXYGEN, CARBON DIOXIDE, CARBON
ECTS, SALINITY, TEMPERATU/ *ZINC
ITY, INTERFACES, ORGANIC MATTER,
EMICAL ANALYSIS, TRITIUM, CARBON
M, STRONTIUM RADIOISOTOPES, ZINC
FICIDS, *ECOSYSTEMS, *PHOSPHORUS
OUTH CAROLINA, CESIUM, STRONTIUM
RADIOISOTOPES, CESIUM, STRONTIUM
*RADIUM RADIOISOTOPES, *URANIUM
FLORA, STREAMS, CULTURES, COBALT
MBIA RIVER, CHROMIUM, PHOSPHORUS
*EC/ *TRANSLOCATION, *PHOSPHORUS
*ZINC RADIOISOTOPES, *PHOSPHORUS
M, RADIUM RADIOISOTOPES, URANIUM
SOTOPES, *SALT MARSHES, */ *ZINC
OACTIVITY, GAMMA RAYS, POTASSIUM
TOPES, MANGANESE, CESIUM, RADIUM
SINS, SEA WATER, PHYSICOC/ *ZINC
NC RADIOISOTOPES, CESIUM, COBALT
OACTIVITY, *LAKES, PRODUCTIVITY,
COLOGY, AQUATIC BACTERIA, CARBON
IGHT INTENSITY, SUCCESSION, ZINC
MATTER, LAKES, DIFFUSION, CARBON
EUTROPHICATION, GLUCOSE, CARBON
PHYTOPLANKTON, CARBON,/ *CARBON
IOCHEMICAL OXYGEN DEMAND, CARBON
GLENA, CHLAMYDOMONAS, PHOSPHORUS
LLUTION EFFECTS, DIATOMS, CARBON
E, PHYSIOLOGICAL ECOLOGY, CARBON
ACTERIA, WATER CHEMISTRY, CARBON
AGE, DIATOMS, CYANOPHYTA, CARBON
ONS, OXYGEN, TRACERS, PHOSPHORUS
KES, BIBLIOGRAPHIES, *PHOSPHORUS
E, URANIUM RADIOISOTOPES, RADIUM
LIGOTROPHIC, *LAKES, *PHOSPHORUS
L WASTES, AQUATIC ALGAE, URANIUM
HLOROPHYTA, BACTERIA, PHOSPHORUS
MINERAL TRANSFORMATION, METAZOA,
A, HA/ *HANFORD(WASH), *REACTOR,
ERATE ZONE, PLUTONIUM, HALF-LIFE(
M KLEBSI, PORPHYRIDIUM CRUENTUM,
, SOIL CONTAMINATION, *STRONTIUM
I/ *RADIOSTRONTIUM, *ALGAL-BOUND
ND RADIOSTRONTIUM, *FUNGAL-BOUND
MS, GREAT LAKES, TRACE ELEMENTS,
IC ALGAE, URANIUM RADIOISOTOPES,
MMA RAYS, URANIUM RADIOISOTOPES,
ADIOISOTOPES, MANGANESE, CESIUM,
REEK, ARCTIC CIRCLE HOT SPRINGS,
PULLULANS, HANSENIASPORA UVARUM,
COLEOPTERA, HEMIPTERA, MOLLUSCA,
E COUNTS, ALGAL GROWTH, SEDGWICK-
RS, RESERVOIRS, AIR TEMPERATURE,
HY, TEMPERATURE, PHOTOSYNTHESIS,
A, DOMINANT ORGANISM, NUTRIENTS,
ESIS, SOLAR RADIATION, SALINITY,
LLUTION SOURCES, SURFACE RUNOFF,
ARK, MT LASSEN NATIONAL PARK, MT
CLAIR, ROUGE RIVER, HURON RIVER,
A, *BLOOMS, RIVER GANGES(INDIA),
ECTROMETRY, OSCILLATORIA TENUIS,
IR SPRAY, SUDBURY(ONTARIO), LAKE
:   TEMPERATURE
LA RIVER(NORWAY), FLAKK(NORWAY),
LADOPHORA, FLAGELLATA, CILIATES,

RADIOISOTOPES, TENNESSEE RIVER, T   W70-02786
RADIOISOTOPES, FISH, CLAMS, STRON   W70-02786
RADIOISOTOPES, LAKES, ORGANIC MAT   W70-02504
RADIOISOTOPES.:   W71-09013
RADIOISOTOPES, URANIUM RADIOISOTO   W71-09013
RADIOISOTOPES, *AQUATIC INSECTS,   W71-09005
RADIOISOTOPES, MARINE ALGAE, WATE   W71-09013
RADIOISOTOPES, TECHNOLOGY, NATURA   W71-09012
RADIOISOTOPES, THORIUM RADIOISOTO   W71-09013
RADIOISOTOPES, DIATOMS, MEASUREME   W71-12870
RADIOISOTOPES, ECHINODERMS.: /OPE   W71-09850
RADIOISOTOPES, CADMIUM RADIOISOTO   W71-09850
RADIOISOTOPES, IODINE RADIOISOTOP   W71-09850
RADIOISOTOPES, ZINC RADIOISOTOPES   W71-09850
RADIOISOTOPES, WATER POLLUTION EF   W71-09850
RADIOISOTOPES.:   W71-09863
RADIOISOTOPES, COBALT RADIOISOTOP   W71-09850
RADIOISOTOPES, PHYSICOCHEMICAL PR   W71-09863
RADIOISOTOPES, CERIUM RADIOISOTOP   W71-09850
RADIOISOTOPES, IRON, SEDIMENTS, F   W69-07853
RADIOISOTO: /OISOTOPES, GAMMA RAY   W69-08272
RADIOISOTOPES, *URANIUM RADIOISOT   W69-07846
RADIOISOTOPES, CESIUM, STRONTIUM   W69-08269
RADIOISOTOPES, STREAMFLOW, SAMPLI   W69-07862
RADIOISOTOPES, RADIUM RADIOISOTO: W69-08272
RADIOISOTOPES, DETRITUS, COLUMBIA   W69-07862
RADIOISOTOPES, *ENVIRONMENTAL EFF   W69-08267
RADIOISOTOPES, GAMMA RAYS, URANIU   W69-08272
RADIOISOTOPES, BIOTA, CLAMS, FALL   W69-08269
RADIOISOTOPES, IRON, MANGANESE, B   W69-07845
RADIOISOTOPES, *TRANSFER, WATER,   W69-07861
RADIOISOTOPES, ZINC RADIOISOTOPES   W69-07845
RADIOISOTOPES, MANGANESE, ALGAE,   W69-08269
RADIOISOTOPES, RIVERS, MILLS, COL   W69-07846
RADIOISOTOPES.: /GANESE, BENTHIC   W69-07845
RADIOISOTOPES, TROPHIC LEVEL, ZIN   W69-07853
RADIOISOTOPES, *TROUT, *STREAMS,   W69-09334
RADIOISOTOPES, *SALT MARSHES, *CY   W69-08274
RADIOISOTOPES.: /MANGANESE, CESIU   W69-08524
RADIOISOTOPES, *PHOSPHORUS RADIOI   W69-08274
RADIOISOTOPES, MANGANESE, CESIUM,   W69-08524
RADIOISOTOPES, URANIUM RADIOISOTO   W69-08524
RADIOISOTOPES, *MARINE ALGAE, *RE   W69-08275
RADIOISOTOPES, SEASONAL, RADIOCHE   W70-04371
RADIOISOTOPES, TEMPERATURE, DEPTH   W70-04371
RADIOISOTOPES, ENZYMES, KINETICS,   W70-04287
RADIOISOTOPES, CESIUM, COBALT RAD   W70-04371
RADIOISOTOPES, CHLAMYDOMONAS, BIO   W70-04284
RADIOISOTOPES, DINOFLAGELLATES, B   W70-04194
RADIOISOTOPES, *ALGAE, *CULTURES,   W70-03983
RADIOISOTOPES, DETERGENTS, ANTIFR   W70-04001
RADIOISOTOPES, REGRESSION ANALYSI   W70-03957
RADIOISOTOPES, KINETICS, TRITIUM,   W70-04811
RADIOISOTOPES, LAKES, PHYTOPLANKT   W70-05109
RADIOISOTOPES, PACIFIC OCEAN, CAR   W70-05652
RADIOISOTOPES, POTASSIUM, OXY: /W   W70-05405
RADIOISOTOPES, IONS, HUMIC ACIDS,   W70-05399
RADIOISOTOPE, ON-SITE DATA COLLEC   W68-00857
RADIOISOTOPES.: /ES, AQUATIC ALGA   W69-01165
RADIOISOTOPE, *TURNOVERS, NUTRIEN   W68-00858
RADIOISOTOPE, RADIUM RADIOISOTOP   W69-01165
RADIOISOTOPES, INFLOW, CYANOPHYTA   W73-01454
RADIOLARIA, FORAMINIFERA, THIOBAC   W69-07428
RADIONUCLIDES, ASSIMILATION, BA-L   W69-07853
RADIONUCLIDE), FILTER-FEEDING ANI   W70-00708
RADIONUCLIDE UPTAKE, NANNOCHLORIS   W70-00708
RADIOSOTOPES, CYTOLOGICAL STUDIES   W69-03491
RADIOSTRONTIUM, *FUNGAL-BOUND RAD   W71-04067
RADIOSTRONTIUM, *MICROBIAL ACTIVI   W71-04067
RADIUM RADIOISOTOPES, THORIUM RAD   W71-09013
RADIUM RADIOISOTOPES.: /ES, AQUAT   W69-01165
RADIUM RADIOISOTO: /OISOTOPES, GA   W69-08272
RADIUM RADIOISOTOPES, URANIUM RAD   W69-08524
RADON, CENTRAL CITY(COLO), MUTATI   W69-08272
RADULESPORES, ZYMOLOGICAL, ELBE(G   W70-04368
RAFFINATE, THORIUM, TAILINGS.: /   W69-07846
RAFTER METHOD, TREATMENT TIMING,   W69-10157
RAIN, GEOLOGY, MAPPING, NUTRIENTS   W70-02646
RAINBOW TROUT, DIATOMS, ANAEROBIC   W69-10154
RAINFALL-RUNOFF RELATIONSHIPS, GE   W68-00891
RAINFALL, TEMPERATURE, COLORIMETR   W70-01073
RAINFALL, SNOWMELT, SURFACE WATER   W70-03501
RAINIER NATIONAL PARK, OHANAPECOS   W70-03309
RAISIN RIVER, DETROIT RIVER, MAUM   W69-01445
RAJGHAT(INDIA), BOMBAY(INDIA), BE   W70-05404
RAMMAN SPECTROSCOPY, MASS SPECTRO   W69-00387
RAMSEY(ONTARIO), ALGAE COUNTS, TH   W69-10155
RANGES, TEMPERATURE REQUIREMENTS.   W71-02494
RANHEIM(NORWAY), TRONDHEIMSFJORD(W70-01074
RANUNCULUS CIRCINATUS, POTAMOGETO   W69-07838

TRUM, HYDRODICTYON, POTAMOGETON,
NOSA, ANABAENOPSIS, RACIBORSKII,
RATION, VERTICAL MIXING, ROANOKE
NORTH AMERICA, CENTRAL AMERICA,
NDED SOLIDS, *PHOSPHOROUS UPTAKE
CLOPS BICUSPIDATUS, REPRODUCTIVE
MIDIUM RETZII, SURFACE DIFFUSION
UDIES, DAPHNIA SCHOD/ *FILTERING
GE, STORMS, T/ *ADDITIVES, *FLOW
DIOXIDE, PHOTOSYNTHESIS, GROWTH
LA MENEGHINIANA, LIGHT-SATURATED
BORATORY(RHODE ISLAND), DILUTION
OROPHYLL, PHOTOSYNTHESIS, GROWTH
RACELLULAR PRODU/ PHOTOSYNTHETIC
ONDWEEDS, ROOTED AQUATIC PLANTS,
AL CONTROL, BACTERICIDES, GROWTH
ISOPODS, AMPHIPODA, FISH, GROWTH
TROPHICATION, LIGHT, TE/ *GROWTH
LE, 2-4-5-/ *HERBICIDES, *GROWTH
ICAL TECHNIQUES, *ALGAE, *GROWTH
ACTIVITY, ALGAL METABOLIS/ *KILL
C/ *BIOASSAY, *HALOGENS, *GROWTH
OINDICA/ *FEEDING RATES, *GROWTH
PRIMARY PRODUCTIV/ SEDIMENTATION
*COLOR, *ALGAE, *COLLOIDS, FLOW
ION, WATER FLEAS, LARVAE, GROWTH
RODUCTIVITY, BIOINDICA/ *FEEDING
ESHWATER ENVIRONMENTS, *TURNOVER
TRIC ANALYSIS, LIMNOLOGY, GROWTH
ICATION, *LAKES, *ALGAE, *GROWTH
*ALGAE, *PHOTOSYNTHESIS, *GROWTH
RTRIENTS, RIVERS, LAKES, *GROWTH
S, LAKE ERIE, STATISTICS, GROWTH
NMENTAL FACTORS, BIOMASS, GROWTH
ION, ALGAE, MARINE ALGAE, GROWTH
, CULTURES, LAKES, ALGAE, GROWTH
ION, POLLUTION ABATEMENT, GROWTH
MICROBIOLOGY, ACTIVATED SLUDGE,
OSPHORUS, NITROGEN, IRON, GROWTH
SEASONAL EFFECTS, DECOMPOSITION
BATEMENT, DRAINAGE SYSTEMS, FLOW
ULFATES, VITAMINS, ALGAE, GROWTH
LOTIC ENVIRONMENT, BIOMASS, FLOW
, LABORATORY MODELS, BOD REMOVAL
PLANKTON, *LAKE ONTARIO, *GROWTH
MUS, REAERATION, EFFLUENTS, FLOW
DISSIPATA, SYNEDRA ACUS, SINKING
AE, SNAILS, MITES, MOLDS, GROWTH
, *NUTRIENT REQUIREMENTS, GROWTH
DISSOLVED SOLIDS, SEDIMENTATION
E, *BACTERIA, METABOLISM, GROWTH
HY, RESISTANCE, BIOASSAY, GROWTH
BROMINE, FILAMENTOUS ALGAE, KILL
ENZYMES, CARBON DIOXIDE, GROWTH
OSPHATES, ALGAE, DAPHNIA, GROWTH
MARINE ALGAE, FRESH WATER, FLOW
ENZYMES, CLASSIFICATION, GROWTH
ATION TECHNIQUES, PRESSURE, FLOW
ON EFFECTS, LETHAL LIMIT, GROWTH
ON, PHOSPHORUS, NITROGEN, GROWTH
NTS, PRODUCTIVITY, SEDIMENTATION
NTENSITY, CARBON DIOXIDE, GROWTH
RAPHY, TRACERS, BACTERIA, GROWTH
ANCE ALGAE, INTAKES, SANDS, FLOW
        CRATER LAKE, ORE, GROWTH
ARTICLE SIZE, POPULATION, GROWTH
*BIOGEOCHEMICAL DATA, *ABUNDANCE
L, MEICELASE, HYDROGEN PEROXIDE,
207, VERTEBRATES, PLUTONIUM-239,
ER PLANTS, NYLON-6, *V/ *VISCOSE
FIBER PLANTS, NYLON-6, *VISCOSE
NTRAL CITY(COLO), MUTATION, BETA
NTIC OCEAN, RADIOACTIVITY, GAMMA
PLANKTON, FISH, SEA WATER, GAMMA
CKY, LIFE HISTORY STUDIES, GAMMA
NIC MATTER, RADIOISOTOPES, GAMMA
FLAGELLATION, IODINE TEST, GRAM
IUM MODEL, LITHOSPHERE, SORPTION
EGIONS, PHOTOSYNTHESIS, CHEMICAL
REATMENT, ELECTROLYTES, CHEMICAL
HYDROLYSIS, METABOLISM, CHEMICAL
ING WATER, *CORROSION, *CHEMICAL
ED INORGANIC PHOSPHATES, SOLUBLE
MENTOUS ALGAE, KILL RATES, PHOTO
L RATES, *BROMINE, IODINE, PHOTO
LABORATORY, PLECTONEMA BORYANUM,
CTIVITY, WASTE DISPOSAL, NUCLEAR
N, *CALIFORNIA, WATER POLLUTION,
TREATMENT, PILOT PLANTS, HUMUS,
Y, SYMBIOSIS, RESERVOIRS, LAKES,

RANUNCULUS.: / CLADOPHORA, PEDIAS   W70-04494
RAPHIDIOPSIS, ANABAENA, WOLLEA, O   W70-05404
RAPIDS RESERVOIR(NC).: /SUPERSATU   W70-00683
RATE EQUATIONS, ANALYSIS OF VARIA   W70-03957
RATE.:   MIXED LIQUOR, SUSPE   W71-13335
RATE.: /MA LEUCHTENBER GIANUM, CY   W69-10182
RATE.: /RA ULNA, OEDOGONIUM, PHOR   W70-00265
RATE, *BODY SIZE, *COMPARATIVE ST   W70-03957
RATE, *WASTE WATER, *ALGAE, *SLUD   W70-05819
RATE, AEROBIC CONDITIONS, ANAEROB   W70-05750
RATE, CELL VOLUME, NITZSCHIA CLOS   W70-05381
RATE, FRUITING BODIES, ULVA LACTU   W70-01073
RATE, ORGANIC MATTER, SCENEDESMUS   W70-05381
RATE, PHYTOPLANKTON DYNAMICS, EXT   W69-05697
RATES OF APPLICATION, APPLICATION   W70-05568
RATES.: /CHLORELLA, *EUGLENA, ALG   W69-00994
RATES.: /OMS, ALGAE, METABOLISM,   W69-09742
RATES, *ALGAE, PHOTOSYNTHESIS, EU   W69-07442
RATES, *ALGAE, 2-4-D, AMINOTRIAZO   W70-03519
RATES, *BIOINDICATORS, *WATER POL   W70-04381
RATES, *BROMINE, IODINE, PHOTO RE   W70-02363
RATES, *CHLORELLA, CHLORINE, ALGI   W70-02370
RATES, *DAPHNIA, PRODUCTIVITY, BI   W69-10152
RATES, *EUTROPHICATION, *LAKES, *   W69-00632
RATES, *FILTERS, *NEW YORK, TREAT   W72-04145
RATES, *FRESH WATER FISH, MICHIGA   W68-00487
RATES, *GROWTH RATES, *DAPHNIA, P   W69-10152
RATES, *HETEROTROPHY, ORGANIC SOL   W70-05109
RATES, *LABORATORY TESTS, *SCENED   W68-00475
RATES, *PHOSPHATES, LABORATORY TE   W70-08637
RATES, *TURBULENCE, CHLORRELA, SC   W69-03730
RATES, *WATER TEMPERATURE, SEASON   W68-00472
RATES, ALGAE, FRESH WATER FISH.: /   W68-00253
RATES, ALGAE, OLIGOTROPHY, LIMITI   W70-02779
RATES, ANALYTICAL TECHNIQUES, LAB   W71-07339
RATES, BACTERIA, ULTRAVIOLET RADI   W70-00719
RATES, CHLOROPHYLL, PHOTOSYNTHESI   W69-10167
RATES, CHLORELLA, HYDROGEN ION CO   W68-00248
RATES, CHLOROPHYTA, CYANOPHYTA, C   W70-03983
RATES, CHRONOLOGY, CLADOCERANS, C   W70-05663
RATES, CONN STREAM FLOW, ALGAE, A   W68-00479
RATES, COPPER SULFATE, BOTTOM SED   W69-09349
RATES, DEPTH, DISSOLVED OXYGEN, A   W70-00265
RATES, DESIGN APPLICATIONS, FACUL   W70-07034
RATES, ENVIRONMENTAL EFFECTS, PHO   W69-10158
RATES, FILTERS, ALGAE.: /ANTS, HU   W70-05821
RATES, GRAZING PRESSURE, POPULATI   W69-10158
RATES, HABITATS.: /ROTIFERS, LARV   W72-01819
RATES, LABORATORY TESTS, CULTURES   W70-03507
RATES, MANGANESE, LOW-FLOW AUGMEN   W73-00750
RATES, MICROORGANISMS, ALGAE, FUN   W72-01817
RATES, PESTICIDE RESIDUES, DETERI   W70-03520
RATES, PHOTO REACTIVITY, CELL COL   W70-02364
RATES, PLANKTON, OXYGEN, NITROGEN   W70-04369
RATES, PREDATION.: /VIRONMENT, PH   W69-06536
RATES, PUMPS, ELECTROLYSIS, CULTU   W70-09904
RATES, SEAWATER, INCUBATION, PSEU   W70-00713
RATES, SLUDGE, SLURRIES, AUTOMATI   W72-08688
RATES, SPORES.: /E, WATER POLLUTI   W70-01779
RATES, STABILIZATION PONDS, SEWAG   W68-00855
RATES, STRATIGRAPHY, BIOINDICATOR   W70-05663
RATES, TEMPERATURE, PILOT PLANTS,   W72-06022
RATES, TRITIUM, PHOTOGRAPHY, CARB   W69-10163
RATES, TURBIDITY, CHLORELLA, SCEN   W70-04199
RATES, WATER PROPERTIES.:   W68-00478
RATES, WATER POLLUTION CONTROL.: /   W72-02417
RATIOS.:   W71-05531
RATS, TRYPSIN, SCENEDESMUS QUADRI   W70-04184
RATS, ZIRCONIUM-95-NIOBIUM, EUROP   W69-08269
RAYON PLANT WASTES, *MAN-MADE FIB   W70-06925
RAYON, ZINC COMPOUNDS, CARBON DIS   W70-06925
RAYS, COSMIC RADIATION, S: /N, CE   W69-08272
RAYS, POTASSIUM RADIOISOTOPES, MA   W69-08524
RAYS, RADIOCHEMICAL ANALYSIS, TRI   W69-08269
RAYS, SPECTROMETERS, CYANOPHYTA,   W69-09742
RAYS, URANIUM RADIOISOTOPES, RADI   W69-08272
REACTION, CYTOPHAGA, PROTEUS, SER   W70-03952
REACTIONS, SOLUBILITY EQUILIBRIUM   W70-04385
REACTIONS, ZOOPLANKTON, AN: /AL R   W69-03611
REACTIONS, ELECTROCHEMISTRY, RESI   W70-05267
REACTIONS, CHLORELLA, TRACERS, RA   W70-08097
REACTIONS, *WATER REUSE, DISSOLVE   W70-02294
REACTIVE PHOSPHORUS, PHOSPHATE(BI   W70-05269
REACTIVITY, CELL COLOR, CELL DAMA   W70-02364
REACTIVITY, ALGAL METABOLISM.: /L   W70-02363
REACTOR.:   SAVANNAH RIVER   W69-07845
REACTORS, POWERPLANTS, SHIPS, SUB   W70-00707
REAERATION, NUTRIENTS, OXYGEN SAG   W69-07520
REAERATION, EFFLUENTS, FLOW RATES   W70-05821
REAERATION, SESTON, CHLOROPHYLL,   W70-05092

ACTIVITY, TEMPERATURE, SALINITY,
MATION, CHLOROPHYLL DERIVATIVES,
REC/ *WATER QUALITY, *ARTIFICIAL
RECHARGE, *DATA COLLECTIONS, PIT
AE, *EVALUATION, FI/ *ARTIFICIAL
USE, *RECLAMATION, *GROUNDWATER,
EMPERATURE, MANGANESE, CULTURES,
ESIGN FOR ENVIRONMENTAL CONTROL,
SLUDGE, *EFFLUENTS, WASTE WATER
, CILIATA, ASTERIONELLA FORMOSA,
ILAMENTOUS, UNICELLULAR, NATURAL
STUDIES, NUTRIENT REMOVAL, LAKE
, SEDIMENTS, MUD, SANDS, SLUDGE,
HYTA, NITRATES, AMMONIA, CARBON,
L WASTES, NUTRIENTS, PESTICIDES,
ESIUM, COOLING, PROPERTY VALUES,
INE MICROORGANISMS, *POLLUTANTS,
EMICALS, FISHING, SPORT FISHING,
R / *EUTROPHICATION, *MINNESOTA,
RNIA, WATER QUALITY, AESTHETICS,
IONS, *WATER BLOOM, TOXIC ALGAE,
.:   *CULTURAL EUTROPHICATION,
MUNICIPAL WASTES, SEPTIC TANKS,
COMPARATIVE STUDIES, CRUCIGENIA
ALGAE, TEMPERATURE EXPERIMENTS,
CENARIA MERCENARIA, AQUACULTURE,
POPULATION EQUIVALENTS(WASTES),
M, *VIETNAM, LITTLE LAKE(HANOI),
ARKANSAS RIVER, MERRIMACK RIVER,
, BACTERIA, MOLLUSKS, VITAMIN B,
FOOD ORGANISMS, FISH PHYSIOLOGY,
ROQUINONE, MADISON(WIS), ELODEA,
HOSPHATES, CYANOPHYTA, OXIDATION-
B12, BIOTIN, HISTIDINE, URACIL,
, PYRUVATE, ACETYLENE REDUCTION,
OXYGEN DEMAND, TOXICITY, OXYGEN-
S, AMMONIUM COMPOUNDS, OXIDATION
ATE AMMONIUM, NITRITE, OXIDATION-
STATS, BATCH CULTURES, ACETYLENE
STRY METHODS, SEASONS, ACETYLENE
HEMICAL OXYGEN DEMAND, OXIDATION-
HEMICAL OXYGEN DEMAND, OXIDATION-
S, ALGAE, GASES, ODOR, OXIDATION-
*ENZYMES, CORROSION, OXIDATION,
NITROGEN-15, PYRUVATE, ACETYLENE
ER VALLEY(NORTH DAKOTA), SULFATE
LFOIL COONTAIL, WILD CELERY, BUR
*TORREY CANYON, SUSCEPTIBILITY,
S, ANIMALS, APPALACHIAN MOUNTAIN
ER COURSE, CARPATHIAN SUBMONTANE
ENOIDS, CARBON FIXATION, PELAGIC
TEMENT, ABATEMENT, PACIFIC COAST
OGEN, CHLOROPHYLL, PRODUCTIVITY,
SYNTHESIS, TEMPERATURE, TROPICAL
S, *CHEMICAL CHARACTERIS/ *RIVER
ST REGION, REGIONS, GEOGRAPHICAL
AST U. S., REGIONS, GEOGRAPHICAL
Y, WATER CHEMISTRY, GEOGRAPHICAL
Y, WATER CHEMISTRY, GEOGRAPHICAL
, SALTS, TIDES, SOUTHEAST U. S.,
ABATEMENT, PACIFIC COAST REGION,
ST U. S., REGIONS, PACIFIC COAST
NOFF RELATIONSHIPS, GEOGRAPHICAL
IFIC COAST REGIONS, GEOGRAPHICAL
MISTRY, PLANTS, SOUTHWEST U. S.,
BENTHIC FLORA, PHAEOPHYTA, POLAR
MONAS, PHOSPHORUS RADIOISOTOPES,
LTH, WATER POLLUTION, CONTRACTS,
NTIC OCEAN, LANDFILLS, DISPOSAL,
TY, FLOODWATER, REGULATION, WELL
NITRATES, TURBIDITY, FLOODWATER,
GANESE, MOLYBDENUM, PLANT GROWTH
ATOMS, RESISTANCE, *PLANT GROWTH
STANCES, LIMNOLOGY, PLANT GROWTH
NTACT, SPARGANIUM, SALIX, GROWTH
UTRIENT ENERGY PATHWAYS, TROPHIC
SYNTHETIC SEWAGE FEED, SYMBIOTIC
N JOAQUIN ESTUARY, ENVIRONMENTAL
NISM, NUTRIENTS, RAINFALL-RUNOFF
GAE, SEDIMENTS, SOIL-WATER-PLANT
ESTUARY, NUTRIENT-PHYTOPLANKTON
QUES, MICHAELIS-MENTEN EQUATION,
ATIONS, HETEROTROPHIC POTENTIAL,
*NUTRIENT ELEMENTS, PHOSPHORUS
*MEASUREMENT, *FLUOROMETRY, DYE
OGRAPHY, INFRARED RADIATION, DYE
A COLLECTION, CHEMICAL ANALYSIS,
LLUTION, LAKES, STREAMS, SEWAGE,
, DYE RELEASES, CURRENTS(WATER),
OXICITY, LAKES, DESIGN CRITERIA,

REAERATION, FISH, TASTE, ODOR, OX   W73-00750
RECENT SEDIMENTS, PRECAMBRIAN SED   W69-08284
RECHARGE, *DATA COLLECTIONS, PIT   W69-05894
RECHARGE, WATER SUPPLY, MUNICIPAL   W69-05894
RECHARGE, COPPER, *ALGICIDES, ALG   W72-05143
RECHARGE, *ION EXCHANGE, *ORGANIC   W72-04318
RECIRCULATED WATER, NITRATES, PHO   W70-07257
RECIRCULATION, BOD REMOVAL, GAS P   W70-04787
RECLAMATION, WASTE WATER TREATMEN   W70-06053
RECOLONIZATION, ALSTER LAKE(GERMA   W70-00268
RECOVERY.: /HYTIC, PLATE COUNT, F   W68-00891
RECOVERY, WISCONSIN, LAKE MONONA,   W69-09349
RECREATION, /, OXYGEN, COLIFORMS   W70-10161
RECREATION, WATER POLLUTI: /LOROP   W72-04761
RECREATION, ALGAE, OXYGEN DEMAND,   W71-09577
RECREATION DEMAND, IR: /RON, MAGN   W70-02251
RECREATION, WATER SPORTS, ENVIRON   W69-04276
RECREATION, SILT, NUTRIENTS, AQUA   W68-00172
RECREATION, SEWAGE DISPOSAL, WATE   W70-04810
RECREATION, FISHING, ALGAE, NUTRI   W70-05645
RECREATIONAL USE.:   *DEFINIT   W69-08279
RECREATIONAL USE, WASTE EFFLUENTS   W69-06535
RECREATION WASTES, WATER QUALITY.   W70-04266
RECTANGULARIS, DICERAS CHODATI, D   W69-07833
RECYCLED WATER.: /GAE, BLUE-GREEN   W70-07257
RECYCLING WASTES.: /M WASTES, MER   W70-09905
RED RIVER VALLEY(NORTH DAKOTA), S   W70-03312
RED RIVER(VIETNAM), CHLOROCOCCUS,   W70-03969
RED RIVER, SAVANNAH RIVER, SNAKE   W70-04507
RED TIDE, SALINITY, FLORIDA, TEXA   W70-05372
RED TIDE, ORGANIC COMPOUNDS, DINO   W71-07731
REDOX COMP: /L EXPOSURE, PHENANTH   W70-02982
REDU: /CONCENTRATION, NITRATES, P   W70-05404
REDUCED SULFUR COMPOUNDS, ANTIBIO   W70-02510
REDUCTANT.: /TP, CMU, NITROGEN-15   W70-04249
REDUCTION POTENTIAL, SOLAR RADIAT   W70-05750
REDUCTION POTENTIAL, DISSOLVED OX   W70-05760
REDUCTION POTENTIAL, CYCLING NUTR   W69-04521
REDUCTION TECHNIQUE, HETEROTROPHY   W70-02775
REDUCTION TEST.: /HODS, *BIOCHEMI   W70-03429
REDUCTION POTENTIAL, AQUATIC ALGA   W71-09546
REDUCTION POTENTIAL, DISSOLVED OX   W71-11377
REDUCTION POTENTIAL, SULFUR BACTE   W70-07034
REDUCTION(CHEMICAL), HYDROLYSIS,   W69-07428
REDUCTION, REDUCTANT.: /TP, CMU,   W70-04249
REDUCTION, POTATO PROCESSING WAST   W70-03312
REED, DUCKWEED, WHITE WATER LILY,   W70-05568
REGENERATION, CLIFF VEGETATION, S   W70-01777
REGION.: /LANKTON, AQUATIC ANIMAL   W69-03695
REGION, GRADIENTS(RIVER), THALLI,   W70-02784
REGION, MONOMICTIC, DEPTH EFFECTS   W70-01933
REGION, REGIONS, GEOGRAPHICAL REG   W69-03683
REGIONAL ANALYSIS, RESERVOIRS.: /   W70-07261
REGIONS.: / ORGANIC MATTER, PHOTO   W69-06865
REGIONS, *PHYSICAL CHARACTERISTIC   W73-00748
REGIONS, BAYS, BODIES OF WATER.: /   W69-03683
REGIONS, COASTAL PLAINS, ATLANTIC   W69-03695
REGIONS, CYCLING NUTRIENTS, ZOOPL   W70-03975
REGIONS, CYCLING NUTRIENTS, ZOOPL   W70-03975
REGIONS, GEOGRAPHICAL REGIONS, CO   W69-03695
REGIONS, GEOGRAPHICAL REGIONS, BA   W69-03683
REGIONS, GEOGRAPHICAL REGIONS, PH   W69-03611
REGIONS, KENTUCKY, ACID STREAMS,   W68-00891
REGIONS, PHOTOSYNTHESIS, CHEMICAL   W69-03611
REGIONS, PACIFIC COAST REGIONS, G   W69-03611
REGIONS, SPECTROPHOTOMETRY, CHROM   W70-01074
REGRESSION ANALYSIS, ALGAE, PONDS   W70-03957
REGULATION, RIGHT-OF-WAY, ESTIMAT   W70-03344
REGULATION,: /IC LIFE, MILK, ATLA   W72-00941
REGULATIONS.: / NITRATES, TURBIDI   W73-00758
REGULATION, WELL REGULATIONS.: /   W73-00758
REGULATORS, PREDATION, PRODUCTIVI   W69-04801
REGULATORS, FRESH WATER LAKES, LA   W68-00856
REGULATORS, CYANOPHYTA, ENVIRONME   W68-00470
REGULATORS, MYRIOPHYLLUM, WATER B   W70-05263
RELATIONSHIPS.: /R LAKE(CALIF), N   W70-05428
RELATIONSHIPS.: /LA PYRENOIDOSA,   W70-05655
RELATIONSHIPS, FLUOROMETRIC CHLOR   W70-05094
RELATIONSHIPS, GEOGRAPHICAL REGIO   W68-00891
RELATIONSHIPS, NUCLEAR WASTES, AB   W71-09005
RELATIONSHIPS.:   *POTOMAC RIVER   W71-05407
RELATIVE HETEROTROPHIC POTENTIAL,   W70-05109
RELATIVE HETEROTROPHIC POTENTIAL,   W70-04194
RELEASE, SAMPLE PRESERVATION, WAT   W70-04382
RELEASES, CURRENTS(WATER), REMOTE   W70-05377
RELEASES, BEACHES, SANDS, BIOTA,   W68-00010
RELIABILITY, WATER POLLUTION, WAT   W70-04382
REMEDIES, JUDICIAL DECISIONS, LEG   W69-06909
REMOTE SENSING, SEA WATER, CHLORO   W70-05377
REMOTE SENSING, MINNESOTA, SEDIME   W70-02775

519

H BROAD RIVER(TENN), CHICKAMAUGA
CRISPUS, HETERAMTHERA, PICKWICK
NN), POTOMOGETON PECTI/ CHEROKEE
DOUGLAS RESERVOIR(TENN), NORRIS
VERTICAL MIXING, ROANOKE RAPIDS
*EUTROPHICATION, WATER QUALITY/
, LONG POND, CONCORD, BOSTON LOT
ER, WASHINGTON, HOWARD A. HANSON
ELF-PURIFICATION, BACTERIOPHAGE,
WATER POLLUTION, SURFACE WATERS,
THESIS, PRODUCTIVITY, SYMBIOSIS,
*LIMNOLOGY, *LAKES, *RECREATION,
POSING ORGANIC MATTER, OKLAHOMA,
ION, IMPOUNDMENTS, LAKES, LOAMS,
, *WATER SUPPLY, COPPER SULFATE,
PRODUCTIVITY, REGIONAL ANALYSIS,
CHES, SCHISTOSOME CERCARIAE, TVA
ECOLOGY, *EUTROPHICATION, PONDS,
G, EVAPORATION, SUPERSATURATION,
, *COPPER SULFATE, WATER SUPPLY,
S, TOXICITY, CHLOROPHYTA, LAKES,
ION SOURCES, TASTE, ODOR, ALGAE,
EY RESERVOIRS(NEB), *CLEAR WATER
ERVOIRS, *TURBID R/ *SALT VALLEY
*CLEAR WATER RESERVOIRS, *TURBID
COMPOUNDS, PHOSPHORUS COMPOUNDS,
ROPHYLL, PIGMENTS, IMPOUNDMENTS,
FLUENTS, STREAMS, LAKES, RIVERS,
LAKES, SAMPLING, PONDS, RIVERS,
TROPHICATION, CYANOPHYTA, LAKES,
, CHLOROPHYTA, STREAMS, DIATOMS,
ES, PESTICIDE REMOVAL, PESTICIDE
POLLUTION EFFECTS, *PESTICIDES,
ABOLISM, *CHLORELLA, / PESTICIDE
SH WATER, PULP WASTES, PESTICIDE
STICIDES, HERBICIDES, *PESTICIDE
ONKA(MINN), SOIL MINERALS, PLANT
IOASSAY, GROWTH RATES, PESTICIDE
, CALIFORNIA, FLORIDA, PESTICIDE
DICATORS, FISHKILL, / *PESTICIDE
MAL POLLUTION, TOXINS, PESTICIDE
ARINES, EFFLUENTS, ION EXCHANGE,
EXTRACELLULAR PRODUCTS, CHEMICAL
TS, *SALT MARSHES, OILY WATER, /
ENVIRONMENT, BAYS, *FLUCTUATION,
SCENEDESMUS, GAS CHROMATOGRAPHY,
*ALGAL CONTROL, *TOXINS, ALGAE,
NESIUM, SODIUM, SILICA, DIATOMS,
CAL REACTIONS, ELECTROCHEMISTRY,
RIA TENUIS, RAM/ PROTON MAGNETIC
T, WATER SAMPLING, ONTARIO WATER
ITY, SO/ *RESERVOIR SITES, WATER
QUALITY, WATER POLLUTION, WATER
OL, *RESEARCH EQUIPMENT,/ *WATER
DIOISOTOPES, TECHNOLOGY, NATURAL
VERS, OXIDATION, PHOTOSYNTHESIS,
, STANDING CROP, LIGHT, BIOMASS,
PHYTON, STREAMS, PHOTOSYNTHESIS,
ITIONS, ALGAE, ROTIFERS, SESTON,
GEN ION CONCENTRATION, SULFATES,
ARBOHYDRATES, PHENOLS, NITROGEN,
TY, *STREAMS, *LABORATORY TESTS,
ORAL, ESTUARIES, PHOTOSYNTHESIS,
ECOLOGY, CYANOPHYTA, DIFFUSION,
, GRAZING, ZOOPLANKTON, DIATOMS,
NTS, LIMNOLOGY, WATER CHEMISTRY,
ORGANIC LOADING, CENTRIFUGATION,
ONDAGA COUNTY, SYRACUSE, WARBURG
GORS LAKE(WALES), HALOBACTERIUM,
GRICULTURAL DRAINAGE, BIOLOGICAL
ALGAE, ECOSYSTEMS, ENVIRONMENT,
ILLATORIA CHALYBIA, HYDRODICTYON
OAD, TURBIDITY, DATA STORAGE AND
DRA ULNA, OEDOGONIUM, PHORMIDIUM
*WASTE WATER TREATMENT, *WATER
ICATION, *BIODEGRADATION, *WATER
NIA, WATER QUALITY, AEST/ *WATER
COMPOUNDS, CHEMICAL WAST/ *WATER
, RECH/ *WATER TREATMENT, *WATER
NDUSTRIAL WASTES, OXIDAT/ *WATER
, *ELECTRIC POWER PLANTS, *WATER
ION, *CHEMICAL REACTIONS, *WATER
IQUORS, ULTIMATE DISPOSAL, WATER
TRATION, ORGANIC LOADING, *WATER
TMENT, *NUTRIENT REMOVAL, *WATER
ATURE, NITRATES, AMMONIA, *WATER
WASHINGTON(WASH), SEATTLE(WASH),
CONTROL, AMMONIA, ION EXCHANGE,
ENT, NUTRIENTS, LAKES, CHANNELS,
IRON OXIDES, *PLANT POPULATIONS,

RESERVOIR(TENN), FLORENCE(ALA).: / W71-07698
RESERVOIR(TENN), NAJAS, PHOTIC ZO   W71-07698
RESERVOIR(TENN), HOLSTON RIVER(TE   W71-07698
RESERVOIR(TENN), FRENCH BROAD RIV   W71-07698
RESERVOIR(NC).: /SUPERSATURATION,   W70-00683
RESERVOIR, *PRIMARY PRODUCTIVITY,   W70-09093
RESERVOIR, LEBANON, WADLEIGH STAT   W69-08674
RESERVOIR, TACOMA(WASH).: /EN RIV   W72-11906
RESERVOIRS, SPORES, NITRITES,: /S   W69-07838
RESERVOIRS, CONTROL, BIBLIOGRAPHI   W70-05389
RESERVOIRS, LAKES, REAERATION, SE   W70-05092
RESERVOIRS, PONDS, TEMPERATURE, L   W70-06225
RESERVOIRS, SAMPLING, INDUSTRIAL   W70-04001
RESERVOIRS, TEXAS.: / EUTROPHICAT   W69-04800
RESERVOIRS, WATER POLLUTION, MICR   W70-08096
RESERVOIRS.: /OGEN, CHLOROPHYLL,   W70-07261
RESERVOIRS, ANOPHELES QUADRIMACUL   W70-06225
RESERVOIRS, BACKWATER, FERTILIZAT   W70-00274
RESERVOIRS, ALGAE, BODIES OF WATE   W70-00683
RESERVOIRS, WATER QUALITY, NEW JE   W69-10157
RESERVOIRS, NITROGEN, GROWTH, AMI   W69-10180
RESERVOIRS, WATERSHEDS(BASINS), I   W72-01813
RESERVOIRS, *TURBID RESERVOIRS, C   W72-04761
RESERVOIRS(NEB), *CLEAR WATER RES   W72-04761
RESERVOIRS, CYCLOTELLA, TRACHELOM   W72-04761
RESERVOIRS, ALGAE, GROUND WATER,   W72-02817
RESERVOIRS, ANALYTICAL TECHNIQUES   W71-10098
RESERVOIRS, BOATING, EUTHROPHICAT   W71-09880
RESERVOIRS, AIR TEMPERATURE, RAIN   W70-02646
RESERVOIRS, SCUM, NITROGEN, PHOSP   W70-02803
RESERVOIRS, CULTURES, SEWAGE EFFL   W70-02982
RESIDUES, PHYSIOLOGICAL ECOLOGY,   W70-02968
RESIDUES, *PESTICIDE KINETICS, *P   W70-09768
RESIDUES, ALGAE, GREAT LAKES, MET   W70-08652
RESIDUES, HERBICIDES, ESTUARIES,   W70-10437
RESIDUES, DIATOMS, CRAYFISH, POLL   W69-02782
RESIDUES, MORRIS(MINN), SOIL PERC   W70-04193
RESIDUES, DETERIORATION.: /NCE, B   W70-03520
RESIDUES, ECOSYSTEMS, FOOD CHAINS   W70-05272
RESIDUES, *ALGAE, BIOASSAY, BIOIN   W69-08565
RESIDUES, PHOSPHATES, ALGAE, SEWA   W69-06305
RESINS, LEAKAGE, CESIUM, NUCLEAR   W70-00707
RESISTANCE, EUPHAUSIA, PHAEOCYSTI   W70-01068
RESISTANCE, WATER POLLUTION EFFEC   W70-09976
RESISTANCE, LIFE CYCLES, GROWTH S   W71-02494
RESISTANCE, BIOASSAY, GROWTH RATE   W70-03520
RESISTANCE, FISH, AMPHIBIANS, INV   W68-00859
RESISTANCE, *PLANT GROWTH REGULAT   W68-00856
RESISTIVITY, CHLORELLA, SCENEDESM   W70-05267
RESONANCE SPECTROMETRY, OSCILLATO   W69-00387
RESOURCES COMMISSION, PEST CONTRO   W69-10155
RESOURCES DEVELOPMENT, WATER QUAL   W72-11906
RESOURCES.: / CONSERVATION, WATER   W69-05844
RESOURCES, *WATER POLLUTION CONTR   W70-08627
RESOURCES, ECOLOGICAL DISTRIBUTIO   W71-09012
RESPIRATION, SCOUR, SATURATION, M   W71-11381
RESPIRATION, LIFE CYCLES, OLIGOCH   W71-00665
RESPIRATION, MICROORGANISMS, LIGH   W71-00668
RESPIRATION, ALTITUDE, SAMPLING,   W69-10154
RESPIRATION, FOOD: /CTERIA, HYDRO   W70-01073
RESPIRATION, CARBON, PRIMARY PROD   W70-01073
RESPIRATION, ENVIRONMENT, PHOTOSY   W70-00265
RESPIRATION, PIGMENTS, CHLOROPHYL   W70-03325
RESPIRATION, PHOSPHATES, ECOSYSTE   W70-03309
RESPIRATION, PIGMENTS, OCEANS, OR   W70-04809
RESPIRATION, BATHYMETRY, SALINITY   W70-05387
RESPIRATION, BIOCHEMICAL OXYGEN D   W72-06022
RESPIROMETER, CLADOPHORA, ENTEROM   W69-02959
RESPIROMETER, PROTOPLASM, WATER-B   W70-04369
RESPONSE, FERTILIZATION SURVEY, M   W72-02787
RETENTION, PHOTOPERIODISM, LIGHT   W69-07862
RETICULATUM, DICTYOSPHAERIUM PULC   W69-05704
RETRIEVAL, COLIFORMS, DISSOLVED O   W70-09557
RETZII, SURFACE DIFFUSION RATE.:   W70-00265
REUSE, *NITROGEN, NITROGEN COMPOU   W69-08053
REUSE, *RECLAIMED WATER, NITROGEN   W69-08054
REUSE, *RECLAIMED WATER, *CALIFOR   W70-05645
REUSE, *RECLAIMED WATER, ORGANIC   W70-10433
REUSE, *RECLAMATION, *GROUNDWATER   W72-04318
REUSE, *WASTE WATER TREATMENT, *I   W72-00027
REUSE, COOLING TOWERS, DAMS, ALGA   W71-11393
REUSE, DISSOLVED OXYGEN, PITTING(W70-02294
REUSE, FISH FARMING.: / SULFITE L   W69-05891
REUSE, RESEARCH AND DEVELOPMENT,   W72-03299
REUSE, SEASONAL, WATER TREATMENT.   W72-08133
REUSE, WASTE WATER TREATMENT.: /R   W72-06017
REVENUE BONDS, METRO ACT(SEATTLE)   W70-04455
REVERSE OSMOSIS, DISTILLATION, EL   W70-01981
REVIEWS, BIBLIOGRAPHIES, HERBICID   W70-00269
REVIEWS, LIGHT INTENSITY, WATER T   W68-00860

DINE, NARRAGANSETT BAY(RI), CAPE
ND SOUND, NARRAGANSETT BAY, CAPE
PLE RUNS TEST, CLUSTER ANALYSIS,
UE, MINERALIZATION, CLAY LENSES,
OBLIQUUS, GROWTH CONSTATS, RIVER
TER-BLOOMS, LAKE GEORGE(UGANDA),
Y LENSES, RHINE RIVER, AMSTERDAM
LORES RIVER, URAVAN, COTTUS SPP,
E, DAYTON(OHIO), VANDALIA(OHIO),
  *FRAUNHOFER LINE DISCRIMINATOR,
GTON, WAVES(WATER), CONNECTICUT,
, NARRAGANSETT MARINE LABORATORY(
, *WATER TEMPERATURE, *SEASONAL,
ETERETROPHY, B/ NARRAGANSETT BAY(
M-95, NIOBIUM-95, RUTHENIUM-106,
US, MUSCLE, BONE, LEAVES, SHELL,
PRE-BOMB MATERIAL, RUTHENIUM-106-
M-106-RHODIUM-106, RUTHENIUM-103-
R(WASH), ANTITHAMNION SARNIENSE,
, MESOTHERMY, MELOSIRA, SYNEDRA,
UME METHOD, CYCLOTELLA COMENSIS,
, SEWAGE EFFLUENTS, CHLOROPHYTA,
OPHYTA, CYANOPHYTA, CHRYSOPHYTA,
OPERTIES, PHYSIOLOGICAL ECOLOGY,
, BACILLARIOPHYCEAE, CHAROPHYTA,
SYNTHESIS, ENZYMES, CHLOROPHYTA,
CUS, ACHROMOBACTER(ALKALIGENES),
TAE, DAPHNIA MAGNA, SELENASTRUM,
, MYCOTA, DEBARYOMYCES HANSENII,
SERRATIA, BAY OF FUNDY(CANADA),
ATED THYMIDINE, NARRAGANSETT BAY(
, *MACROMOLECULES, NUCLEIC ACID,
ND), ENZYME-SUBSTRATE COMPLEXES,
E(ONTARIO), CLEAR LAKE(ONTARIO),
EALTH, WALLEYE, VEGETATION, WILD
LUS SUBTILIS, BACILLUS STEAROTH/
FLABELLUM, ZINC, NICKEL, PUERTO
ANGANESE, ATLANTIC OCEAN, PUERTO
BACK RIVE/ COMMUNITY STRUCTURE,
BISMUTH-214, ACTINIUM-228, SWIFT
OLLUTION, CONTRACTS, REGULATION,
SULFATE, FERRICHLORIDE, IRON S/
RMANY), GERMANY, FINLAND, SEWAGE
ER, SAVANNAH RIVER, SNAKE RIVER,
TH SERVICE, GREAT LAKES-ILLINOIS
                    FEEDLOTS, WABASH
                             *OHIO
                           POTOMAC
, *GROUNDWATER, DDT, MISSISSIPPI
N/ *SKAWA RIVER(POLAND), MONTANE
                  *INDIA, MOOSI
KTON RELATIONSHIPS.:   *POTOMAC
                        *POTOMAC
, *HYDROCARBONS, *OLEFINS, GREEN
A), BOMBAY(IND/ *INDIA, *BLOOMS,
ANUM, REACTOR.:   SAVANNAH
LAKE ZURICH, SWITZERLAND, NITELV'
ESMUS OBLIQUUS, GROWTH CONSTATS,
ULATION EQUIVALENTS(WASTES), RED
                          SHENANDOAH
RO(DELAWARE), SWAN CREEK, INDIAN
SIN RIVER, DETROIT RIVER, MAUMEE
RIVER, RIO GRANDE RIVER, POTOMAC
TASTE, ODOR, OXYGEN SAG, POTOMAC
UOLUMNE RIVER(CALIF), STANISLAUS
SAN JOAQUIN RIVER(CALIF), MERCED
R(CALIF), CLEAR LAKE(CALIF), EEL
), MERCED RIVER(CALIF), TUOLUMNE
, TUOLUMNE RIVER(CA/ SAN JOAQUIN
R POLLUTION ASSESSMENT, AMERICAN
ORGANISMS, TURSIOPS TRUN/ DUPLIN
ROWTH, STREAM CONCOURSE, PALOUSE
(ILL), COOK COUNTY(ILL), CALUMET
              *METABOLITES, *CEDAR
IONELLA, TYPHOID FEVER, KENNEBEC
CREEK(PA), NOBS CREEK(MD), BACK
ICROCYSTIS AERUGINOSA, BURNTSIDE
TCHER, NJOER LAKE(NORWAY), HOBOL
OLAR WATERS, TYNSET(NORWAY), NID
ORWAY), NID RIVER(NORWAY), GAULA
                        GREAT MIAMI
AK/ *BIOLOGICAL PROBLEMS, SCIOTO
ON, FOSSIL, CIMARRON R/ ARKANSAS
LA), POLLUTION, FOSSIL, CIMARRON
HANIZOMENON FLOS AQUAE, METOLIUS
SE, CARPATHIAN SUBMONTAN/ *SKAWA
SA, BIOLOGICAL HALF-LIFE, CLINCH
HEROKEE RESERVOIR(TENN), HOLSTON
IS RESERVOIR(TENN), FRENCH BROAD
        AUSTIN(TEX), COLORADO

REYKJANES(ICELAND), SUDURNES(ICEL	W69-09755
REYKJANES, FAXAFLOI FJORD, ICELAN	W69-10161
REYKJAVIK(ICELAND).: /TS, ONE-SAM	W69-09755
RHINE RIVER, AMSTERDAM RHINE, POL	W69-07838
RHINE.: /TERIONELLA, SCENEDESMUS	W70-03955
RHINE-MEEUSE DELTA(NETHERLANDS),:	W70-04369
RHINE, POLDERS, COMPOSITION, PSEU	W69-07838
RHINICHTHYS OSCULUS, SUCKERS, ICT	W69-07846
RHIZOPODS, SPIDERS, MILK WASTE TR	W72-01819
RHODAMINE WT DYE, AIRBORNE TESTS,	W70-05377
RHODE ISLAND, TEMPERATURE, NUTRIE	W69-10161
RHODE ISLAND), DILUTION RATE, FRU	W70-01073
RHODE ISLAND), SAMPLING, ECOLOGY,	W70-00713
RHODE ISLAND), TAXONOMIC TYPES, H	W70-00713
RHODIUM-106, CERIUM-144, BETA RAD	W69-09742
RHODIUM-102.: /A, PISONIA, PANDAN	W69-08269
RHODIUM-106, RUTHENIUM-103-RHODIU	W69-08524
RHODIUM-103, CERIUM-144-PRASEODYM	W69-08524
RHODOCHORTON, BANGIA FUSCO-PURPUR	W69-10161
RHODOMON: /ACTERIA, DESULFOVIBRIO	W70-05760
RHODOMONAS LACUSTR: /LCULATED VOL	W70-03959
RHODOPHYTA, PHYSIOLOGICAL ECOLOGY	W70-04510
RHODOPHYTA, EUGLENOPHYTA, SLIME,	W70-05389
RHODOPHYTA, CHLOROPHYTA.: /CAL PR	W69-10161
RHODOPHYTA.: /, CYANOPHYTA, FUNGI	W70-02784
RHODOPHYTA, SCENEDESMUS, CHLAMYDO	W70-02965
RHODOTORULA, GROWTH FACTORS, COLO	W70-03952
RHODOTORULA, TEMPERATURE COEFFICI	W70-03957
RHODOTORULA RUBRA, CANDIDA DIDDEN	W70-04368
RHYZOIDS.: /, CYTOPHAGA, PROTEUS,	W70-03952
RI), CAPE REYKJANES(ICELAND), SUD	W69-09755
RIBONUCLEIC ACID SYNTHESIS.: /ION	W69-10151
RIBONUCLEIC ACID, ORCINOL METHOD.	W69-10160
RICE LAKE(ONTARIO).: /UCKHORN LAK	W70-02795
RICE, SILTS, PLANKTON, MAYFLIES,	W70-01943
RICKETTSIAE, STERILIZATION, BACIL	W71-11823
RICO, LA PARGUERA, JOYUDA.: /OTEA	W69-08525
RICO, SPECTROSCOPY, DISTRIBUTION,	W69-08525
RIDLEY CREEK(PA), NOBS CREEK(MD),	W70-04510
RIFFLES, ZIRCONIUM-95, NIOBIUM-95	W69-09742
RIGHT-OF-WAY, ESTIMATED COSTS, LE	W70-03344
RING TUBE, CANALIZATION, ALUMINUM	W70-05668
RING TUBES.: /NY), LAKE TEGERN(GE	W70-05761
RIO GRANDE RIVER, POTOMAC RIVER.:	W70-04507
RIVER BASINS PROJECT, INDIANA HAR	W70-00263
RIVER BASIN.:	W71-10376
RIVER BASIN.:	W72-01788
RIVER BASIN.:	W70-09557
RIVER BASIN, URBANIZATION, ALGAL	W73-00758
RIVER COURSE, CARPATHIAN SUBMONTA	W70-02784
RIVER ECOLOGICAL STUDY.:	W69-07096
RIVER ESTUARY, NUTRIENT-PHYTOPLAN	W71-05407
RIVER ESTUARY.:	W71-05409
RIVER FORMATION, CHLOROPHYLL DERI	W69-08284
RIVER GANGES(INDIA), RAJGHAT(INDI	W70-05404
RIVER LABORATORY, PLECTONEMA BORY	W69-07845
RIVER NORWAY, NORWA: /ET NORWAY,	W69-07833
RIVER RHINE.: /TERIONELLA, SCENED	W70-03955
RIVER VALLEY(NORTH DAKOTA), SULFA	W70-03312
RIVER.:	W70-03068
RIVER.: MILLSBO	W71-05013
RIVER.: / RIVER, HURON RIVER, RAI	W69-01445
RIVER.: /, SAVANNAH RIVER, SNAKE	W70-04507
RIVER.: /NITY, REAERATION, FISH,	W73-00750
RIVER(CALIF), WATER POLLUTION ASS	W70-02777
RIVER(CALIF), TUOLUMNE RIVER(CALI	W70-02777
RIVER(CALIF), JOINT INDUSTRY/GOVE	W70-02777
RIVER(CALIF), STANISLAUS RIVER(CA	W70-02777
RIVER(CALIF), MERCED RIVER(CALIF)	W70-02777
RIVER(CALIF), CLEAR LAKE(CALIF),	W70-02777
RIVER(GA), SPARTINA ALTEMIFLORA,	W69-08274
RIVER(IDAHO).: /CHARCOAL, ALGAL G	W70-03501
RIVER(ILL), POOLS, WIND DIRECTION	W70-00263
RIVER(IOWA).:	W72-13605
RIVER(MAINE), COBBOSSEECONTEE LAK	W70-08096
RIVER(MD), OSCILLATORIA, CATHERWO	W70-04510
RIVER(MINN).: /LLA PYRENOIDOSA, M	W70-03512
RIVER(NORWAY), VAN LAKE(NORWAY).:	W70-05752
RIVER(NORWAY), GAULA RIVER(NORWAY	W70-01074
RIVER(NORWAY), FLAKK(NORWAY), RAN	W70-01074
RIVER(OHIO).:	W70-00475
RIVER(OHIO), LAKE MENDOTA(WIS), L	W72-01813
RIVER(OKLA), TULSA(OKLA), POLLUTI	W70-04001
RIVER(OKLA), BIXBY(OKLA), KEYSTON	W70-04001
RIVER(ORE), COMMUNITY STRUCTURE,	W70-03978
RIVER(POLAND), MONTANE RIVER COUR	W70-02784
RIVER(TENNESSEE), PHYSICAL HALF-L	W70-02786
RIVER(TENN), POTOMOGETON PECTINAT	W71-07698
RIVER(TENN), CHICKAMAUGA RESERVOI	W71-07698
RIVER(TEX).:	W70-04727

521

PISCICULTURAL PONDS, SVISLOCH'
VIETNAM, LITTLE LAKE(HANOI), RED
SCENS, MADISON LAKES(WIS), CEDAR
UAE, OXYGEN DEPLETION.: *YAHARA
OTROPHY, MICRONUTRIENTS, BADFISH
FFLUENTS, *MADISON LAKES, YAHARA
UNIVERSITY OF WISCONSIN, YAHARA
IAN SUBMONTANE REGION, GRADIENTS(
OINDICATORS, *CHLOROPHYLL, *OHIO
*ALGAE, *SAMPLING, *MISSISSIPPI
WASTES, AGRICULTURE, ST LAWRENCE
LOW-FLOW AUGMENTATION, TENNESSEE
NERALIZATION, CLAY LENSES, RHINE
CYSTIS AERUGINOSA, BOHEMIA, ELBE
ISOTOPES, TRO/ *BIOTA, *COLUMBIA
AGE EFFLUENTS, DIATOMS, MISSOURI
IATOMS, MISSOURI RIVER, COLUMBIA
NTERITIS, SYNURA, ANABAENA, IOWA
O RIVER, TENNESSEE RIVER, HUDSON
ROUGE RIVER, HURON RIVER, RAISIN
NIMAS RIVER, DURANGO, SAN MIGUEL
DOLORES RIVER, URAVAN, / ANIMAS
ADIOISOTOPES, DETRITUS, COLUMBIA
IGAN, CHEBOYGAN COUNTY, STURGEON
RACTION, MULTIPLE-USES, SAVANNAH
RIVER, COLORADO RIVER, TENNESSEE
DETROIT R/ LAKE ST CLAIR, ROUGE
                                PLATTE
RON RIVER, RAISIN RIVER, DETROIT
S, AGRICULTURAL RUNOFF, ARKANSAS
HORIUM-232, YUKON RIVER, NEBESNA
L A, DATA INTERPRETATION, *RHONE
S, EQUISETUM, THORIUM-232, YUKON
PLING, *MISSISSIPPI RIVER, *OHIO
GREEN ALGAE, PHYTOPLANKTON, OHIO
UM, BLE/ *ALGAL GROWTH, *POTOMAC
CATHERWOOD DIATOMETER, SAVANNAH
H RIVER, SNAKE RIVER, RIO GRANDE
D OXYGEN VARIATIONS, SAN JOAQUIN
AKE ST CLAIR, ROUGE RIVER, HURON
UNOFF, ARKANSAS RIVER, MERRIMACK
RED RIVER, SAVANNAH RIVER, SNAKE
NSAS RIVER, MERRIMACK RIVER, RED
*PROTOZOAN COMMUNITIES, POTOMAC
UNITIES, POTOMAC RIVER, SAVANNAH
IMACK RIVER, RED RIVER, SAVANNAH
OTIFERS, *DISTRIBUTION, MISSOURI
RIVER, COLUMBIA RIVER, COLORADO
T TEST/ HIGHLAND LAKES, COLORADO
RONTIUM RADIOISOTOPES, TENNESSEE
RANGO, SAN MIGUEL RIVER, DOLORES
ON RES/ WOOD, ALGAL COUNT, GREEN
EE RIVER, HUDSON RIVER, DELAWARE
PSA FLORIDANA, ENTEROCOCCI, PARK
ESH WATER, *HUMIC ACIDS, *COLOR,
ECOLOGY, *AQUATIC ENVIRONMENTS,
OLOGY, SALINITY, ORGANIC MATTER,
TRACE ELEMENTS, PLANKTON, LAKES,
YSIS, INSECTS, SNAILS, TOXICITY,
QUATIC HABITATS, *ALASKA, LAKES,
UTROPHICATION, *ALGAE, PLANNING,
TS, TEST PROCEDURES, NURTRIENTS,
HEAT, WATER POLLUTION, STREAMS,
TOXINS, DISSOLVED OXYGEN, LAKES,
SOTOPES, *URANIUM RADIOISOTOPES,
HLOROPHYLL, FLUORESCENCE, LAKES,
EASONAL, LAKES, SAMPLING, PONDS,
PPLY, EFFLUENTS, STREAMS, LAKES,
NCE ALGAE, PHOSPHATES, NITRATES,
DIATOMS, MARYLAND, PENNSYLVANIA,
NT GROWTH, PRIMARY PRODUCTIVITY,
DETERGENTS, RUNOFF, PERCOLATION,
UPERSATURATION, VERTICAL MIXING,
, NUCLEAR ACTIVATION TECHNIQUES,
CARBON METHOD, FINGER LAKES(NY),
PLEMORPHIC, MYCELOID, CORYNEFORM
RENOIDOSA, NITROGEN UTILIZATION,
TROPHICATION, FISH FARMING, FISH
NTS, RUNOFF, PERCOLATION, ROADS,
TURE, DISSOLVED OXYGEN, DENSITY,
*PILOT PLANTS, *NUISANCE ALGAE,
SONAL, *ALGAL CONTROL, FISHKILL,
, FUNGI, ALGAE, FLOATING PLANTS,
SCONSIN, CHEMCONTROL, PONDWEEDS,
, OSCILLATORIA, AUSTRALIA, SANTA
L, ELBE(GERMANY), SPOROBOLOMYCES
INTENSITY, DAPHNIA, CONNECTICUT,
E, SULFUR COMPOUNDS, MOLYBDENUM,
IOCONTROL, TURBULENCE, COPEPODS,
ZOA, PLANKTON, WORMS, NEMATODES,

RIVER(USSR), MINSK(USSR), ENGLAND   W70-05092
RIVER(VIETNAM), CHLOROCOCCUS, SPI   W70-03969
RIVER(WASH), ZURICHSEE(SWITZ), TU   W70-00270
RIVER(WIS), APHANIZOMENON FLOS AQ   W72-01797
RIVER(WIS), DETROIT LAKES(MINN),   W70-04506
RIVER(WIS), STORKWEATHER CREEK(WI   W70-05113
RIVER(WIS), NUTRIENT REMOVAL.: /,   W69-06276
RIVER), THALLI, SCHIZOMYCETES, CY   W70-02784
RIVER, *ANALYTICAL TECHNIQUES, BI   W71-04206
RIVER, *OHIO RIVER, NUISANCE ALGA   W70-04507
RIVER, ALGAE, COLIFORMS, DISSOLVE   W70-03964
RIVER, ALGAE, IRON, ALKALINITY, H   W73-00750
RIVER, AMSTERDAM RHINE, POLDERS,   W69-07838
RIVER, CELAKOVICE(CZECHOSLOVAKIA)   W70-00274
RIVER, CHROMIUM, PHOSPHORUS RADIO   W69-07853
RIVER, COLUMBIA RIVER, COLORADO R   W70-04507
RIVER, COLORADO RIVER, TENNESSEE   W70-04507
RIVER, CYCLOTELLA, FRAGILARIA, SY   W70-08107
RIVER, DELAWARE RIVER, WATER POLL   W70-04507
RIVER, DETROIT RIVER, MAUMEE RIVE   W69-01445
RIVER, DOLORES RIVER, URAVAN, COT   W69-07846
RIVER, DURANGO, SAN MIGUEL RIVER,   W69-07846
RIVER, ELECTRONICS, SCALING.: / R   W69-07862
RIVER, FLUORESCEIN, MACROPHYTES,   W69-09334
RIVER, GULF COAST.: / COLLOIDAL F   W73-00748
RIVER, HUDSON RIVER, DELAWARE RIV   W70-04507
RIVER, HURON RIVER, RAISIN RIVER,   W69-01445
RIVER, LINCOLN(NEBR).:   W69-05894
RIVER, MAUMEE RIVER.: / RIVER, HU   W69-01445
RIVER, MERRIMACK RIVER, RED RIVER   W70-04507
RIVER, MOOSE CREEK, ARCTIC CIRCLE   W69-08272
RIVER, NAVICULA, MELOSIRA, THALÁS   W73-01446
RIVER, NEBESNA RIVER, MOOSE CREEK   W69-08272
RIVER, NUISANCE ALGAE, WATER QUAL   W70-04507
RIVER, PENNSYLVANIA, WATER SUPPLY   W69-05697
RIVER, PLECTONEMA, NOSTOC PALUDOS   W70-02255
RIVER, POPULATION DYNAMICS, ALGAL   W70-04510
RIVER, POTOMAC RIVER.: /, SAVANNA   W70-04507
RIVER, POTOMAC ESTUARY, SAN FRANC   W69-07520
RIVER, RAISIN RIVER, DETROIT RIVE   W69-01445
RIVER, RED RIVER, SAVANNAH RIVER,   W70-04507
RIVER, RIO GRANDE RIVER, POTOMAC   W70-04507
RIVER, SAVANNAH RIVER, SNAKE RIVE   W70-04507
RIVER, SAVANNAH RIVER, SHOCK EFFE   W70-07847
RIVER, SHOCK EFFECTS.: /ZOAN COMM   W70-07847
RIVER, SNAKE RIVER, RIO GRANDE RI   W70-04507
RIVER, SOUTH DAKOTA, NEBRASKA, AL   W70-05375
RIVER, TENNESSEE RIVER, HUDSON RI   W70-04507
RIVER, TEXAS, *NUTRIENT ENRICHMEN   W69-06004
RIVER, TROPHIC LEVEL, ALGAE, CADD   W70-02786
RIVER, URAVAN, COTTUS SPP, RHINIC   W69-07846
RIVER, WASHINGTON, HOWARD A. HANS   W72-11906
RIVER, WATER POLLUTION EFFECTS.: /   W70-04507
RIVER: /HROMATIUM VINOSUM, THIOCA   W70-03312
RIVERS, BOGS, PHENOLS, PIGMENTS,   W70-01074
RIVERS, ENVIRONMENTAL EFFECTS, PE   W70-03978
RIVERS, ESTUARINE ENVIRONMENT, WA   W73-01446
RIVERS, GREAT LAKES.: /OSPHORUS,   W70-04385
RIVERS, HISTOGRAMS.: / WATER ANAL   W72-01801
RIVERS, HOT SPRINGS, TERRESTRIAL   W69-08272
RIVERS, LAKES, LIMNOLOGY, WISCONS   W70-04506
RIVERS, LAKES, *GROWTH RATES, *WA   W68-00472
RIVERS, LAKES, PHYSIOLOGICAL ECOL   W71-02479
RIVERS, LIGHT PENETRATION, FLUORE   W69-05697
RIVERS, MILLS, COLORADO, WASTES,   W69-07846
RIVERS, NITROGEN, BIOMASS, SEWAGE   W70-02777
RIVERS, RESERVOIRS, AIR TEMPERATU   W70-02646
RIVERS, RESERVOIRS, BOATING, EUTH   W71-09880
RIVERS, SEWAGE TREATMENT, WATER P   W69-01453
RIVERS, STREAMS, EUGLENA, AQUATIC   W70-04510
RIVERS, TASTE, THERMAL STRATIFICA   W69-06535
ROADS, ROOFS, PRECIPITATION(ATMOS   W69-10178
ROANOKE RAPIDS RESERVOIR(NC).: /S   W70-00683
ROCK LAKE(WASH), WILLIAMS LAKE(WA   W70-00264
RODHE'S SOLUTION VIII, NUTRIENT L   W70-01579
RODS, COCCI,: /E, PSYCHROPHILES,   W70-00713
ROLLER DRYING, SPRAY-DRIED, GLASS   W70-04184
ROOD ORGANISMS, FISH GENETICS, FI   W69-04804
ROOFS, PRECIPITATION(ATMOSPHERIC)   W69-10178
ROOTED AQUATIC PLANTS, ALGAE, CHL   W71-07698
ROOTED AQUATIC PLANTS, *RESERVOIR   W68-00479
ROOTED AQUATIC PLANTS, GLACIAL LA   W68-00488
ROOTED AQUATIC PLANTS, SELF-PURIF   W70-05750
ROOTED AQUATIC PLANTS, RATES OF A   W70-05568
ROSA(CALIF), LYUBLIN(USSR), LYUBE   W70-05092
ROSEUS, AUTOLYSIS, ASCOSPOROGENOU   W70-04368
ROTIFERS.: /, TEMPERATURE, LIGHT   W73-03959
ROTIFERS, ECOSYSTEMS, PHOTOS: /IT   W69-07428
ROTIFERS, CHLOROPHYLL, PIGMENTS,   W71-10098
ROTIFERS, LARVAE, SNAILS, MITES,   W72-01819

522

TER CHEMISTRY, BIOMASS, BENTHOS,
R, CHEMICAL PROPERTIES, ECOLOGY,
ARINE ENVIRONMENT, MARINE ALGAE,
EFFLUENTS, TEMPERATURE, OXYGEN,
SOTOPES, *ENVIRONMENTAL EFFECTS,
, MOLLUSKS, VITAMIN B, RED TIDE,
MISTRY, RESPIRATION, BATHYMETRY,
PHOTIC ZONE, HELSINGOR(DENMARK),
ASENIA, DOW CONTACT, SPARGANIUM,
UMER ORGANISMS, SIMULIUM, PHYSA,
GIZZARD SHAD, SEA LAMPHREY, COHO
GICAL BALANCE, OVERFISHING, COHO
FISH STOCKING, COMMERCIAL FISH,
EL, PRODUCTIVITY, MODEL STUDIES,
SALINITY, ALGAE, STRATIFICATION,
LLUTION, *KUWAIT CRUDE, BENTLASS
RATE, EUGLENA, CALCIUM, AMMONIUM
NDS, PHOSPHORUS, NUISANCE ALGAE,
CS, ORGANISMS, SUSPENDED SOLIDS,
YRA, XANTHOPHYCEAE, MYXOPHYCEAE,
NT ELEMENTS, PHOSPHORUS RELEASE,
LAND), NON-PARAMETRIC TESTS, ONE-
INFLUENCES, MICROCYSTIS, DREDGE
LIGHT INTENSITY, TROPHIC LEVEL,
OOLING TOWERS, FILTERS, EFFLUENT
ATION, FLOCCULATION, POLLUTANTS,
*ZINC RADIOISOTOPES, STREAMFLOW,
TES, FISH, TEMPERATURE, TRACERS,
N, TOXICITY, MIDGES, MOSQUITOES,
IMENTS, MUD, BENTHOS, WISCONSIN,
, TURBIDITY, TERTIARY TREATMENT,
ANALYTICAL TECHNIQUES, DIATOMS,
IC MATTER, OKLAHOMA, RESERVOIRS,
ARBON, RADIOACTIVITY, CHLORELLA,
EMENTS, *DISTRIBUTION, BACTERIA,
ORS, FUNGI, ESTUARIES, BACTERIA,
IGATIONS, WATER POLLUTION, WATER
NALYSIS, *CHEMICAL ANALYSIS, PH,
PATH OF POLLUTANT, FOOD CHAINS,
RCES, RUNOFF, LIGHT PENETRATION,
ANALYSIS, ANALYTICAL TECHNIQUES,
DIATOMS, MEASUREMENT, SEDIMENTS,
TION, ULTRASONICS, DISINFECTION,
QUES, SPECTROSCOPY, COLORIMETRY,
ATES, SULFATES, SEASONAL, LAKES,
OPHYTA, CHRYSOPHYTA, CYANOPHYTA,
IMARY PRODUCTIVITY, CHLOROPHYLL,
S, *ALGAE, *LAKES, *TRIBUTARIES,
MCNDOTA(WI/ *INLAND, HORIZONTAL
D, HORIZONTAL SAMPLING, VERTICAL
PLANKTON, LAKES, EUTROPHICATION,
, SESTON, RESPIRATION, ALTITUDE,
RS, CROP-DUSTING AIRCRAFT, WATER
S, ALGAE, BIOCHEMISTRY, ECOLOGY,
RATURE, *SEASONAL, RHODE ISLAND,
OPHYTA, CHRYSOPHYTA, CYANOPHYTA,
E AGING, NUTRIENT SOURCES, WATER
RIENTS, DEPTH, DOMESTIC ANIMALS,
XYGEN DEMAND, TEMPERATURE, FLOW,
PHORUS, OXYGEN, CARBON, ENZYMES,
DIATOMS, EUGLENA, CHLAMYDOMONAS,
NSITY, IRON, PHOSPHORUS, SILICA,
ACTERIA, AQUATIC MICROORGANISMS,
DOMONAS, CHLORELLA, SCENEDESMUS,
SAN JOAQUIN VALLEY MASTER DRAIN,
JOAQUIN RIVER, POTOMAC ESTUARY,
ON, DISSOLVED OXYGEN VARIATIONS,
ER PLAN, DELTA WATER/ SACRAMENTO-
IF), OXYGEN DYNAMICS, SACRAMENTO-
NTY, WATER DETERIORATION, DRAIN,
RIVER(CALIF), TUOLUMNE RIVER(CA/
URAVAN, / ANIMAS RIVER, DURANGO,
INETIC EFFECTS, ANN ARBOR(MICH),
REGENERATION, CLIFF VEGETATION,
ABBOT'S POND(SOMERSET, ENGLAND),
COATED SAND, FERRIC OXIDE-COATED
R/ *FILTER MEDIA, ALUMINA-COATED
*PARTICLE SIZE, *CLAYS, *SILTS,
ADIATION, DYE RELEASES, BEACHES,
IC BACTERIA, METABOLISM, CANALS,
*ALGAE, NUISANCE ALGAE, INTAKES,
ING, DENSITY, LIGHT, MUD, SILTS,
SPECTROMETERS, BACTERIA, ALGAE,
TER TREATMENT, *ALGAE, *FILTERS,
, TUNDRA, SCALING, GRAVELS, MUD,
YGEN, COLIFORMS, SEDIMENTS, MUD,
ENCY, MARGINAL VEGETATION FAUNA,
OXYGEN DEMAND, SEWAGE EFFLUENTS,
WTH, WATER POLLUTION ASSESSMENT,
OXYGEN, MEASUREMENT, ESTUARIES,

SALINITY, SNAILS, SULFATES, NITRA   W70-03978
SALINITY, BENTHOS, AQUATIC LIFE,   W69-05023
SALINITY, TEMPERATURE, MONITORING   W68-00010
SALINITY, ALGAE, MECHANICAL EQUIP   W69-06203
SALINITY, TEMPERATURE, SEA WATER,   W69-08267
SALINITY, FLORIDA, TEXAS, PHOSPHO   W70-05372
SALINITY, PHOSPHORU: /, WATER CHE   W70-05387
SALINO-CLINE.: /LA MENEGHINIANA,   W70-04809
SALIX, GROWTH REGULATORS, MYRIOPH   W70-05263
SALMO TRUTTA, ENTOSPHENUS LAMOTTE   W69-09334
SALMON.: /, WHITE FISH, ALEWIFE,   W70-01943
SALMON, ALEWIFE, NUTRIENT REMOVAL   W70-04465
SALMON, ALGAE, FISH.: /HOSPHORUS,   W70-04465
SALMON, AQUATIC ENVIRONMENT, PHOS   W69-06536
SALT BALANCE, BIBLIOGRAPHIES.: /   W72-01773
SALT MARSH.: /, *RECOVERY FROM PO   W70-01231
SALTS, NITROGEN FIXATION, BORON,   W70-02964
SALTS, TIDES, SOUTHEAST U. S., RE   W69-03695
SALTWATER TONGUE, BENTHIC FORMS,   W73-00748
SALVELINUS NAMAYCUSH, SUBSTRATE T   W70-00711
SAMPLE PRESERVATION, WATER DYNAMI   W70-04382
SAMPLE RUNS TEST, CLUSTER ANALYSI   W69-09755
SAMPLES, ANABAENA, SEASONAL EFFEC   W70-05663
SAMPLING TEMPERATURE, TRACERS, CH   W70-02504
SAMPLING.: /TERS, SNAILS, ODOR, C   W72-00027
SAMPLING.: /ATES, BACKGROUND RADI   W69-07846
SAMPLING, ALGAE, ECOSYSTEMS, ENVI   W69-07862
SAMPLING, PERIPHYTON, AQUATIC PLA   W69-09334
SAMPLING, CONTROL, INDUS: /ICATIO   W70-06225
SAMPLING, CONTOURS, DEPTH, OLIGOC   W70-06217
SAMPLING, MONITORING, WATER DEMAN   W70-05645
SAMPLING, PHYSICOCHEMICAL PROPERT   W70-05389
SAMPLING, INDUSTRIAL WASTES, MUNI   W70-04001
SAMPLING, EFFLUENTS, TRIBUTARIES,   W70-03983
SAMPLING, SEA WATER, SEASONAL, PS   W70-03952
SAMPLING, SEASONAL, HYDROGRAPHY,   W70-04368
SAMPLING, *WATER ANALYSIS, *CHEMI   W69-00659
SAMPLING, ON-SITE DATA COLLECTION   W69-00659
SAMPLING, SURVEYS, MARINE ALGAE,   W72-00941
SAMPLING, PRIMARY PRODUCTIVITY, P   W72-04761
SAMPLING, ORGANIC MATTER, TOXICIT   W72-03299
SAMPLING, DEPTH, SEA WATER, FRES   W71-12870
SAMPLING, CENTRIFUGATION, SEDIMEN   W71-11823
SAMPLING, METHANE, CARBON DIOXIDE   W71-11377
SAMPLING, PONDS, RIVERS, RESERVOI   W70-02646
SAMPLING, PHYTOPLANKTON, ORGANIC   W70-02780
SAMPLING, PHYTOPLANKTON, RADIOACT   W70-01933
SAMPLING, FLOW, ANALYSIS, FERTILI   W70-02969
SAMPLINC, VERTICAL SAMPLING, LAKE   W70-02969
SAMPLING, LAKE MENDOTA(WIS), LAKE   W70-02969
SAMPLING, SURFACES WATERS, CARBON   W70-02973
SAMPLING, ZOOPLANKTON, PHYTOPLANK   W69-10154
SAMPLING, ONTARIO WATER RESOURCES   W69-10155
SAMPLING, THERMAL SPRINGS, EFFLUE   W69-10151
SAMPLING, ECOLOGY, AMMONIA, ENZYM   W70-00713
SAMPLING, DENSITY, LIGHT, MUD, SI   W70-00711
SAMPLING, ALGAL GROWTH, OXYGEN DE   W70-00264
SAMPLING, NUISANCE ALGAE.: /, NUT   W70-00274
SAMPLING, WASTE WATER DISPOSAL, D   W69-10159
SAMPLING, BOTTOM SEDIMENTS, SODIU   W70-09889
SAMPLING, SCENEDESMUS, CHLOROPHYT   W70-08107
SAMPLING, CHEMICAL ANALYSIS, PHYS   W70-05094
SAMPLING, WATER ANALYSIS, WATER Q   W71-00221
SAMPLING, OXYDATION LAGOONS.: /MY   W70-10173
SAN FRANCISCO BAY-DELTA.: /RAIN,   W71-09577
SAN FRANCISCO BAY.: /IATIONS, SAN   W69-07520
SAN JOAQUIN RIVER, POTOMAC ESTUAR   W69-07520
SAN JOAQUIN DELTA, CALIFORNIA WAT   W70-05094
SAN JOAQUIN ESTUARY, ENVIRONMENTA   W70-05094
SAN JOAQUIN VALLEY MASTER DRAIN,   W71-09577
SAN JOAQUIN RIVER(CALIF), MERCED   W70-02777
SAN MIGUEL RIVER, DOLORES RIVER,   W69-07846
SAND BEDS.: /USPENSIONS, ELECTROK   W71-03035
SAND DUNES.: /ON, SUSCEPTIBILITY,   W70-01777
SAND GRAINS, TIDAL CYCLES.: /YI,   W71-12870
SAND, CRUSHED BAUXITE, ACID-WASHE   W70-08628
SAND, FERRIC OXIDE-COATED SAND, C   W70-08628
SANDS, ALGAE, PLANKTON, MICROORGA   W70-08628
SANDS, BIOTA, WASTE WATER(POLLUTI   W68-00010
SANDS, CHLORINATION, MICROORGANIS   W69-07838
SANDS, FLOW RATES, TURBIDITY, CHL   W70-04199
SANDS, GRAVELS, SEDIMENTS, WATER   W70-00711
SANDS, HYDROGEN ION CONCENTRATION   W69-08267
SANDS, LAKE MICHIGAN, WATER QUALI   W73-02426
SANDS, SHALES, BOTTOM SEDIMENTS,   W69-08272
SANDS, SLUDGE, RECREATION.: /, OX   W70-10161
SANDY BOTTOMS FAUNA, MARGINAL VEG   W70-06975
SANITARY ENGINEERING.: /CHEMICAL   W72-01817
SANITARY ENGINEERING RESEARCH LAB   W70-02775
SANITARY ENGINEERING, WATER TREAT   W70-05094

BACTERIA, AL/ *WATER TREATMENT,   SANITARY ENGINEERING, *PATHOGENIC W72-13601
*WATER TREATMENT, *FILTRATION,   SANITARY ENGINEERING, *FILTERS, * W72-07334
, DESIGN, *AL/ *WATER TREATMENT,   SANITARY ENGINEERING, *FILTRATION W72-08859
RONMENTAL EFFECTS, ENVIRONMENTAL   SANITATION, ESSENTIAL NUTRIENTS, W69-03513
RONMENTAL EFFECTS, ENVIRONMENTAL   SANITATION, ESSENTIAL NUTRIENTS, W69-03369
RONMENTAL EFFECTS, ENVIRONMENTAL   SANITATION, INHIBITION, CHLORINE, W69-10171
NGLAND, OSCILLATORIA, AUSTRALIA,   SANTA ROSA(CALIF), LYUBLIN(USSR), W70-05092
NIGROSINE DYE, SODIUM CHLORATE,   SANTOBRITE, CLADOPHORA, POTAMOGET W70-05263
LOGY.:   *CILIATES, FLAGELLATES,   SAPROBIEN SYSTEM, CILIATE AUTOECO W71-03031
                      SAPROLEGNIA.: W68-00468
OLECULAR FILTER, CHEMOSYNTHETIC,   SAPROPHYTIC, PLATE COUNT, FILAMEN W68-00891
IBRIO, FLAVOBACTER, ESCHERICHIA,   SARCINA, STAPHYLOCOCCUS, ACHROMOB W70-03952
BACTERS, GRAM-POSITIVE BACTERIA,   SARCINE-LIKE, BACILLUS-LIKE, MICR W70-00713
MASTIGOPHORA, HOLOPHYTIC MASTI/   SARCODINA, MASTIGOPHORA, HOLOZOIC W72-01817
CEAE, CHRYSOPHYCEAE, VOLVOCALES,   SARCODINA, CILIATES, CYCLOTELLA, W70-05112
EMA, THALASSIOSIRA, COCCOLITHUS,   SARGASSO SEA, SYNURA, GYMNODINIUM W70-04503
ISTANCE, EUPHAUSIA, PHAEOCYSTIS,   SARGASSUM, FRONDS, HYDROCARBONS, W70-01068
RIDAY HARBOR(WASH), ANTITHAMNION   SARNIENSE, RHODOCHORTON, BANGIA F W69-10161
, LEUCOTHRIX MUCOR, ANTITHAMNION   SARNIENSE, QUANTITATION, GROWTH M W69-10163
, *AUTORADIOGRAPHY, ANTITHAMNION   SARNIENSE, TRITIATED GLUCOSE, GON W69-09755
TON LAKE(SASKATCHEWAN), HUMBOLDT(SASKATCHEWAN), MICE, MORPHOLOGICA W70-00273
II, OTTAWA(ONTARIO), BURTON LAKE(SASKATCHEWAN), HUMBOLDT(SASKATCHE W70-00273
, CYCLOTELLA MENEGHINIANA, LIGHT-   SATURATED RATE, CELL VOLUME, NITZ W70-05381
TOSYNTHESIS, RESPIRATION, SCOUR,   SATURATION, MIXING, DIFFUSION, SE W71-11381
ZOAN COMMUNITIES, POTOMAC RIVER,   SAVANNAH RIVER, SHOCK EFFECTS.: / W70-07847
LLATORIA, CATHERWOOD DIATOMETER,   SAVANNAH RIVER, POPULATION DYNAMI W70-04510
VER, MERRIMACK RIVER, RED RIVER,   SAVANNAH RIVER, SNAKE RIVER, RIO W70-04507
NEMA BORYANUM, REACTOR.:   SAVANNAH RIVER LABORATORY, PLECTO W69-07845
LLOIDAL FRACTION, MULTIPLE-USES,   SAVANNAH RIVER, GULF COAST.: / CO W73-00748
A, WEAPONS TESTING, NUCLEAR SHIP   SAVANNAH, ZIRCONIUM-95, RUBIDIUM- W70-00707
YTILUS, TRICHODESMIUM ERYTHREUM,   SAXIDOMUS, CLOSTRIDIUM BOTULINUS, W70-05372
, WILLIAMS LAKE(WASH), CHANNELED   SCABLANDS(WASH), WASHINGTON STATE W70-00264
REATMENT, *AUSTRALIA, LABORATORY-   SCALE STUDIES, PLANT-SCALE STUDIE W70-06899
LABORATORY-SCALE STUDIES, PLANT-   SCALE STUDIES, REMOVAL.: /TRALIA, W70-06899
RIMENTAL STUDIES, LAGOON DESIGN,   SCALE-UP FACTORS, WASTE COMPOSITI W70-07508
US, COLUMBIA RIVER, ELECTRONICS,   SCALING.: / RADIOISOTOPES, DETRIT W69-07862
OGIC FORMATIONS, ARCTIC, TUNDRA,   SCALING, GRAVELS, MUD, SANDS, SHA W69-08272
EMISTRY, RESISTIVITY, CHLORELLA,   SCENEDESMUS.: /ACTIONS, ELECTROCH W70-05267
IS, GROWTH RATE, ORGANIC MATTER,   SCENEDESMUS, CHLAMYDOMONAS, FRESH W70-05381
OROPLASTS, SCENEDESMUS OBLIQUUS,   SCENEDESMUS QUADRICAUDA, CHLORELL W70-05381
, NITZSCHIA PALEA, CHLOROPLASTS,   SCENEDESMUS OBLIQUUS, SCENEDESMUS W70-05381
CENTRATION, UROG/ *REQUIREMENTS,   SCENEDESMUS, ABSORPTION, CELL CON W70-05409
*FRESH WATER, *PLANKTON, *ALGAE,   SCENEDESMUS, LIGHT, CULTURES, IRO W70-05409
DESMUS, GLENODINIUM PULVISCULUS,   SCENEDESMUS: /RUDINELLA, ANKISTRO W70-05387
PURIFICATION, ALGAL ENUMERATION,   SCENEDESMUS, GROWTH EQUATIONS, AN W70-04381
ORUS, *AQUATIC ALGAE, *BIOASSAY,   SCENEDESMUS, PHOSPHATES, WATER PO W70-03955
, 2-4-D, AMINOTRIAZOLE, 2-4-5-T,   SCENEDESMUS, CHLAMYDOMONAS, CHLOR W70-03519
GHT STERILIZATION, ASTERIONELLA,   SCENEDESMUS OBLIQUUS, GROWTH CONS W70-03955
ODESMUS, MICROSPORA, PEDIASTRUM,   SCENEDESMUS, LEPOCINCLIS, ANABAEN W70-03974
CIDES, ALDRIN, ENDRIN, DIELDRIN,   SCENEDESMUS, GAS CHROMATOGRAPHY, W70-03520
RATURE, CYANOPHYTA, CHLOROPHYTA,   SCENEDESMUS, BENTHOS, CONVECTION, W70-03969
UCTURE/ *ALGAE, *MICROORGANISMS,   SCENEDESMUS, PROTEINS, WALLS, STR W70-04184
YDROGEN PEROXIDE, RATS, TRYPSIN,   SCENEDESMUS QUADRICAUDA, SCENEDES W70-04184
RYPSIN, SCENEDESMUS QUADRICAUDA,   SCENEDESMUS OBLIQUUS, HELIX POMAT W70-04184
LOW RATES, TURBIDITY, CHLORELLA,   SCENEDESMUS, EUGLENA.: / SANDS, F W70-04199
RPHA, LEPOCINCLIS, CHLOROGONIUM,   SCENEDESMUS, CYCLOTELLA, STEPHANO W69-02959
*PHYTOTOXICITY, ALGAL POISONING,   SCENEDESMUS, CHLORELLA, *EUGLENA, W69-00994
TURE, SEASONAL, TEST PROCEDURES,   SCENEDESMUS, *ALGAE.: /ER TEMPERA W68-00472
ON CONCENTRATION, PLANT TISSUES,   SCENEDESMUS, LIMNOLOGY, WATER PRO W68-00248
H RATES, *TURBULENCE, CHLORRELA,   SCENEDESMUS, KINETICS, ECOSYSTEMS W69-03730
*HERBICIDES, *ALGAE, TOXICITY,   SCENEDESMUS, CHLORELLA, DAPHNIA, W71-03056
RGANO-SILICATES, *ALKOXYSILANES,   SCENEDESMUS QUADRICAUDA, CHLORELL W71-03056
ERS, ELECTROPHORESIS, CHLORELLA,   SCENEDESMUS, ADSORPTION, SEWAGE T W71-03035
MONURON, TRANSPARENCY, CYCLOPS,   SCENEDESMUS OBLIQUUS, DACTYLOCOCC W70-10161
GLENA, CHLAMYDOMONAS, CHLORELLA,   SCENEDESMUS, SAMPLING, OXYDATION W70-10173
AS, DAPHNIA, DIATOMS, CHLORELLA,   SCENEDESMUS, OXYGEN, COLIFORMS, S W70-10161
UGLENA, CHLAMYDOMONAS, SAMPLING,   SCENEDESMUS, CHLOROPHYTA, CHLOREL W70-08107
PES, HYDROGEN ION CONCENTRATION,   SCENEDESMUS, EUGLENA, BACTERIA, Y W70-08097
AL EFFIC/ *DAPHNIA PULEX OBTUSA,   SCENEDESMUS QUADRICAUDA, ECOLOGIC W69-10152
               PROTOCOCCUS,   SCENEDESMUS QUADRICAUDA.: W71-11561
ANCES, CHLORELLA, CHLAMYDOMONAS,   SCENEDESMUS, WISCONSIN.: /H SUBST W72-00845
RGANO-SILICATES, *ALKOXYSILANES,   SCENEDESMUS QUADRICAUDA, CHLORELL W71-05719
*HERBICIDES, *ALGAE, TOXICITY,   SCENEDESMUS, CHLORELLA, DAPHNIA, W71-05719
NTERFACES, CALIFORNIA, BIOASSAY,   SCENEDESMUS, PHYSIOLOGICAL ECOLOG W70-02775
ITROGEN, WASTE WATER(POLLUTION),   SCENEDESMUS, CYANOPHYTA, CHLOROPH W70-02255
HYSIOLOGICAL ECOLOGY, CHLORELLA,   SCENEDESMUS.: /LOGY, NUTRIENTS, P W70-02245
ZOOPLANKTON, MINNOWS, PEDALFERS,   SCENEDESMUS, CHLAMYDOMONAS, TEMPE W70-02249
IPPING, FIREBAUGH CENTER(CALIF),   SCENEDESMUS QUADRICAUDA, CELL COU W70-03334
SIS, WATER POLLUTION, CHLORELLA,   SCENEDESMUS.: /IA, CHEMICAL ANALY W70-03312
ANKISTRODESMUS, NOSTOC MUSCORUM,   SCENEDESMUS OBLIQUUES, SCENEDESMU W70-02964
MUSCORUM, SCENEDESMUS OBLIQUUES,   SCENEDESMUS QUADRICAUDA, PORPHYRI W70-02964
NTHESIS, OXYGEN, CARBON DIOXIDE,   SCENEDESMUS, CHLOROPHYLL, IRON, C W70-02964
NZYMES, CHLOROPHYTA, RHODOPHYTA,   SCENEDESMUS, CHLAMYDOMONAS, ECOLO W70-02965
PHICATION, AQUATIC LIFE, SESTON,   SCENEDESMUS, MATHEMATICAL STUDIES W73-01446
TRIENTS, ECOSYSTEMS, PHOSPHORUS,   SCENEDESMUS, CHLOROPHYTA, BACTERI W73-01454
OIRS, ANOPHELES QUADRI/ LEECHES,   SCHISTOSOME CERCARIAE, TVA RESERV W70-06225
RBONATE, CERCARIAE, WATER MILFO/   SCHISTOSOME DERMATITIS, COPPER CA W70-05568
EGION, GRADIENTS(RIVER), THALLI,   SCHIZOMYCETES, CYANOPHYTA, FUNGI, W70-02784
(VIETNAM), CHLOROCOCCUS, SPINES,   SCHIZOMYCOPHYTA, CAULOBACTERIALES W70-03969
, OHANAPECOSH HOT SPRINGS(WASH),   SCHIZOTHRIX, PHORMIDIUM, WINKLER W70-03309

AND), LAKE TEGERN(BAVARIA), LAKE
RG(GERMANY), PLON(GERMANY), LAKE
E, *COMPARATIVE STUDIES, DAPHNIA
ALATHION, CHLORELLA PYRENOIDOSA,
TA, PYROPHYTA, CHROOMONAS ACUTA,
ACUSE(NY),/ *PHYCOLOGY, NY STATE
OMIC ENERGY COMMISSION, NATIONAL
OMIC ENERGY COMMISSION, NATIONAL
ONSIN, ASIA/ NATIONAL ACADEMY OF
ONSIN, ASIA/ NATIONAL ACADEMY OF
WIS), LAK/ *BIOLOGICAL PROBLEMS,
 MENDOTA(WIS), ANACHARIS, NAIAS,
INCE EDWARD ISLAND(CANADA), NOVA
HAELIS-MENTEN EQUATION, LOCH EWE(
DRICA, NORTH SEA, FIRTH OF CLYDE(
H SEA, FIRTH OF CLYDE(SCOTLAND),
EN EQUATION, LOCH EWE(SCOTLAND),
ON, PHOTOSYNTHESIS, RESPIRATION,
 *POTATOES, *INDUSTRIAL WASTES,
, WATER PURIFICATION, MEMBRANES,
UTROPHICATION, CYANOPHYTA, ODOR,
, CYANOPHYTA, LAKES, RESERVOIRS,
ION, OLIGOTROPHY, SEWAGE, ALGAE,
TILIZERS, BLUEGILLS, MICROCYSTIS
HETIC OXYGEN, LEMMON(SD), KADOKA(
 *PHOTOSYNTHETIC OXYGEN, LEMMON(
OLOGICAL ECOLOGY, PHYTOPLANKTON,
ITE FISH, ALEWIFE, GIZZARD SHAD,
N, METHODOLOGY, INSTRUMENTATION,
*MARINE WATER, *COPPER, *ALGAE,
ISOTOPES, LAKES, ORGANIC MATTER,
ENT, SEDIMENTS, SAMPLING, DEPTH,
R, *IRON, TRACE ELEMENTS, ALGAE,
ECONDARY PRODUCTIVITY, SULFIDES,
NKTON, *CYCLING, *RADIOISOTOPES,
CURRENTS(WATER), REMOTE SENSING,
ISTRIBUTION, BACTERIA, SAMPLING,
N, *WATER STRUCTURE, ATMOSPHERE,
SOTOPES, *MARINE ALGAE, *RESINS,
EFFECTS, SALINITY, TEMPERATURE,
, INVERTEBRATES, PLANKTON, FISH,
ABATEMENT, TIDAL WATERS, MIXING,
EANOGRAPHY, *TRANSACTIONS, BLACK
YLUM TRICORNUTUM, EDTA, LIGURIAN
AND), ANABAENA CYLINDRICA, NORTH
I, RHODOTORULA RUBRA, CA/ *NORTH
ASSIOSIRA, COCCOLITHUS, SARGASSO
N, *RADIONUCLIDE EXCHANGE, *OPEN
SION, INSTREAM TREATMENT, BOTTOM
L, PHOTOSYNTHETIC LAYER, MELOSI/
NIVERSITY, EXPERIMENTAL STREAMS,
YSTIS, DREDGE SAMPLES, ANABAENA,
OLYSIPHONIA, ORGANIC AGGREGATES,
TH EFFECTS, IONIC CONCENTRATION,
ME, MAINE, PERIODIC VARIATIONS,
LGAE, *NUISANCE ALGAE, SPRAYING,
ROWTH RATES, *WATER TEMPERATURE,
FISHKILL, NUISANCE ALGAE, ALGAE,
ITY, HYPOLIMNION, ALGAL CONTROL,
N, *LAKES, PHYTOPLANKTON, ALGAE,
EPTH, DENSITY, CHLOROPHYLL, ICE,
TH, PLANKTON, LIGHT PENETRATION,
RE, EPILIMNION, LIGHT INTENSITY,
TION, ACTIVATED CARBON, FILTERS,
S, CESIUM, COBALT RADIOISOTOPES,
LANKTON, *LAKES, ORGANIC MATTER,
, ESTUARIES, BACTERIA, SAMPLING,
, BACTERIA, SAMPLING, SEA WATER,
NITRATES, PHOSPHATES, SULFATES,
HER, EPILIMNION, ALGAE, DENSITY,
RENTS(WATER), VELOCITY, DIATOMS,
HLOROPHYTA, CYANOPHYTA, DIATOMS,
*NUTRIENT REMOVAL, *WATER REUSE,
METHODS, *BIOCHEMISTRY METHODS,
RO ACT(/ *LAKE WASHINGTON(WASH),
(WASH), REVENUE BONDS, METRO ACT(
OBIAL ECOLOGY, COCCOLITHOPHORES,
S, CLASSIFICATION, GROWTH RATES,
ARINE ALGAE, ECOLOGY, NUTRIENTS,
SUM FLUITANS, *SARGASSUM NATANS,
IANA, OCHROMONAS, PHYTOPLANKTON,
, LAKE MENDOTA, LAKE FURES, LAKE
VARIATIONS, SEASONAL VARIA/ LAKE
N H), MA/ NUTRIENT BUDGETS, LAKE
SHINGTON(WASH), *OXYGEN DEFICIT,
IDIUM, APHANIZOMENON FLOS-AQUAE,
NTROL METHODS.:
NTS, PRODUCTION BIOLOGY, RUSSIA,
IENTS, PONDS, PRIMARY TREATMENT,
TION SOURCES, PRIMARY TREATMENT,

SCHLIER(BAVARIA), LAKE ZELL(AUSTR   W70-05668
SCHLIER(GERMANY), LAKE TEGERN(GER   W70-05761
SCHODLERI, DAPHNIA PULEX, DAPHNIA   W70-03957
SCHRODERIA, THIMET, TRICHODERMA.:   W70-08097
SCHROEDERIA SETIGERA, MELOSIRA AM   W70-05387
SCIENCE AND TECHNOLOGY FOUND, SYR   W70-03973
SCIENCE FOUNDATION, OFFICE OF NAV   W70-03975
SCIENCE FOUNDATION, OFFICE OF NAV   W70-03975
SCIENCES(USA), UNIVERSITY OF WISC   W70-03975
SCIENCES(USA), UNIVERSITY OF WISC   W70-03975
SCIOTO RIVER(OHIO), LAKE MENDOTA(W72-01813
SCIRPUS, NYMPHAEA, CRASSIPES EICH   W70-05263
SCOTIA(CANADA), PENGUINS, LAKE ZU   W70-05399
SCOTLAND), SCOTLAND, PHAEOPIGMENT   W70-04811
SCOTLAND), SCOTLAND, ENGLAND, LAK   W70-02504
SCOTLAND, ENGLAND, LAKE WINDERMER   W70-02504
SCOTLAND, PHAEOPIGMENTS, ACRIDINE   W70-04811
SCOUR, SATURATION, MIXING, DIFFUS   W71-11381
SCREENS, SEDIMENTATION, ORGANIC L   W71-08970
SCREENS, TREATMENT FACILITIES, *W   W73-02426
SCUM, FISHING, WISCONSIN, BIBLIOG   W69-08283
SCUM, NITROGEN, PHOSPHORUS, CARBO   W70-02803
SCUM, WEEDS, NUTRIENTS, SEPTIC TA   W70-04193
SCUMS, PITHIOPHORA ALGAE.: /E FER   W70-05417
SD).:   *PHOTOSYNT   W72-01818
SD), KADOKA(SD).:   W72-01818
SE: /NUTRIENT REQUIREMENTS, PHYSI   W69-03371
SEA LAMPHREY, COHO SALMON.: /, WH   W70-01923
SEA WATER.: /ES, THERMAL POLLUTIO   W70-10437
SEA WATER, TEMPERATURE, SALINITY,   W70-08111
SEA WATER, PRIMARY PRODUCTIVITY,   W70-02504
SEA WATER, FRES WATER, LITTORAL,   W71-12870
SEA WATER, FRESH WATER.: / *COPPE   W71-11687
SEA WATER, PHAWOPHYTA, CRUSTACEAN   W70-01068
SEA WATER, FOOD CHAINS, PHOSPHORU   W70-00708
SEA WATER, CHLOROPHYLL, OIL, ALGA   W70-05377
SEA WATER, SEASONAL, PSEUDOMONAS,   W70-03952
SEA WATER, FRESH WATER, SORPTION,   W70-04385
SEA WATER, PHYSICOCHEMICAL PROPER   W69-08275
SEA WATER, OYSTERS, CLAMS, CRABS,   W69-08267
SEA WATER, GAMMA RAYS, RADIOCHEMI   W69-08269
SEA WATER, FRESH WATER, SHORES, W   W69-06203
SEA, CONTACT ZONES, ARTEMIA, BALA   W69-07440
SEA, EUROPEAN MEDITERRANEAN.: /CT   W69-08275
SEA, FIRTH OF CLYDE(SCOTLAND), SC   W70-02504
SEA, MYCOTA, DEBARYOMYCES HANSENI   W70-04368
SEA, SYNURA, GYMNODINIUM, CHEMICA   W70-04503
SEA, WEAPONS TESTING, NUCLEAR SHI   W70-00707
SEALING, LAKE DILUTION.: /R DIVER   W69-09340
SEASONAL DISTRIBUTION, LAKE BAIKA   W70-04290
SEASONAL EFFECTS, THERMAL EFFECTS   W70-03978
SEASONAL EFFECTS, DECOMPOSITION R   W70-05663
SEASONAL EFFECTS.: /COTOCARPUS, P   W70-01068
SEASONAL EFFECTS.: /NOMICTIC, DEP   W70-01933
SEASONAL VARIATIONS, SOURCES.: /,   W68-00481
SEASONAL, *ALGAL CONTROL, FISHKIL   W68-00488
SEASONAL, TEST PROCEDURES, SCENED   W68-00472
SEASONAL, LIMNOLOGY, LIGHT PENETR   W68-00487
SEASONAL, ANNUAL TURNOVER.: /CTIV   W69-05142
SEASONAL, CYANOPHYTA, *INORGANIC   W68-00860
SEASONAL, DIURNAL, LIGHT INTENSIT   W70-05387
SEASONAL, PRIMARY PRODUCTIVITY, C   W70-05405
SEASONAL, PHYSIOLOGICAL ECOLOGY,   W70-05409
SEASONAL, STRATIFICATION, BIOCHEM   W70-04001
SEASONAL, RADIOCHEMICAL ANALYSIS,   W70-04371
SEASONAL, ALGAE, PRIMARY PRODUCTI   W70-04290
SEASONAL, HYDROGRAPHY, FLORIDA, T   W70-04368
SEASONAL, PSEUDOMONAS, TEMPERATUR   W70-03952
SEASONAL, LAKES, SAMPLING, PONDS,   W70-02646
SEASONAL, MARINE PLANTS.: /, WEAT   W70-02504
SEASONAL, BIOMASS, PIGMENTS, PROD   W70-02780
SEASONAL, EUGLENA, CHLAMYDOMONAS,   W70-10173
SEASONAL, WATER TREATMENT.: /NT,   W72-08133
SEASONS, ACETYLENE REDUCTION TEST   W70-03429
SEATTLE(WASH), REVENUE BONDS, MET   W70-04455
SEATTLE).: /INGTON(WASH), SEATTLE   W70-04455
SEAWATER MEDIUM F.: /NIQUES, MICR   W70-05272
SEAWATER, INCUBATION, PSEUDOMONAS   W70-00713
SEAWATER, ORGANIC COMPOUNDS, HABI   W69-09755
SEAWEED, ZIRCONIUM-95-NIOBIUM-95,   W69-08524
SEAWEEDS.: /ASHINGTON, LAKES, IND   W70-04503
SEBASTICOOK, LAKE WASHINGTON, LAK   W70-03964
SEBASTICOOK, ME, MAINE, PERIODIC   W68-00481
SEBASTICOOK(ME), LAKE WINNISQUAM(W70-07261
SECCHI DISC TRANSPARENCY, MYXOPHY   W69-10182
SECCHI DISC TRANSPARENCY, ANABAEN   W69-10169
SECCHI DISC, VISIBILITY TESTS, CO   W68-00488
SECOND INTERNATIONAL OCEANOGRAPHI   W69-07440
SECONDARY TREATMENT, STABILIZATIO   W68-00012
SECONDARY TREATMENT, TERTIARY TRE   W69-01453

526

IA, PRODUCTIVITY, BIOINDICATORS,
LATION, BACTERIA, STREPTOCOCCUS,
ION, ALGAE COUNTS, ALGAL GROWTH,
HOSPHOROUS, AVAILABLE NUTRIENTS,
, WATER SUPPLY, MUNICIPAL WATER,
ERIA, REMOTE SENSING, MINNESOTA,
L CONTROL, ALGAE, WATER QUALITY,
EMENTS, CORES, WISCONSIN, BOTTOM
ION, *LAKES, *PRIMARY PRODUCTIV/
, ODOR, PHYTOPLANKTON, SULFATES,
ATION, LAGOONS, DOMESTIC WASTES,
BULENCE, PILOT PLANTS, BACTERIA,
NOLOGY, NUTRIENTS, PRODUCTIVITY,
DESIGN, *ALGAE, COPPER SULFATE,
SOLVED OXYGEN, DISSOLVED SOLIDS,
ZOOPLANKTON, TRIPTON, BACTERIA,
IOCHEMICAL OXYGEN DEMAND, ALGAE,
ORGANIC LOADING, ALGAE, AMMONIA,
*POULTRY, WASTE WATER TREATMENT,
GE, INCINERATION, STORM RUN-OFF,
DES, FARM WASTES, SILT, EROSION,
CTION, SAMPLING, CENTRIFUGATION,
, SATURATION, MIXING, DIFFUSION,
ES, *INDUSTRIAL WASTES, SCREENS,
EVAPORATION, WATER CIRCULATION,
IBUTARIES, WATERSHED MANAGEMENT,
S, FISHKILL, CRUSTACEANS, BOTTOM
AL PROPERTIES, EPILMNION, BOTTOM
DIATOMS, CHLOROPHYTA, DEPOSITION(
, ICE, BACTERIA, TRIPTON, BOTTOM
FALLOUT, SILTS, DETRITUS, BOTTOM
NITROGEN, AQUATIC PLANTS, BOTTOM
TH RATES, COPPER SULFATE, BOTTOM
MONAS, SUBSURFACE-WATERS, BOTTOM
GHT, MUD, SILTS, SANDS, GRAVELS,
IC HABITATS, DRAGONFLIES, ALGAE,
IVATED SLUDGE, COST COMPARISONS,
OISOTOPES, DIATOMS, MEASUREMENT,
, FRESH WATER BIOLOGY, SUSPENDED
TESTS, PRODUCTIVITY, ABSORPTION,
ARBON, ENZYMES, SAMPLING, BOTTOM
N, PHOTOSYNTHESIS, ALGAE, BOTTOM
AE, BIOLOGICAL TREATMENT, BOTTOM
YSIS, PHYTOPLANKTON, CYANOPHYTA,
SCENEDESMUS, OXYGEN, COLIFORMS,
ENT, HARVESTING OF ALGAE, BOTTOM
*NUTRIENT REQUIREMENTS, *BOTTOM
ON-SITE DATA COLLECTIONS, BOTTOM
ULATIONS, BIBLIOGRAPHIES, BOTTOM
S, RECENT SEDIMENTS, PRECAMBRIAN
ALGAE, ALGAL CONTROL, ALGICIDES,
CHLOROPHYLL DERIVATIVES, RECENT
RYSOPHYTA, CULTURES, DIAGENESIS,
ANKTON, SULFATES, SEDIMENTATION,
WASTES, BIOLOGICAL COMMUNITIES,
LEVEL, ZINC RADIOISOTOPES, IRON,
ES, *CYCLING NUTRIENTS, GEORGIA,
IBUTION, ANIMALS, PLANTS, BOTTOM
ERRESTRIAL HABITATS, DEPTH, AIR,
PES, *TRANSFER, WATER, BACTERIA,
VELS, MUD, SANDS, SHALES, BOTTOM
S, CLAMS, CRABS, NORTH CAROLINA,
URAL WATERSHEDS, ORGANIC MATTER,
POLLUTION EFFECTS, PRODUCTIVITY,
CE ALGAE, AQUATIC PLANTS, BOTTOM
ATMENT, *LAKES, *ECOLOGY, BOTTOM
, URBANIZATION, WATER POLLUTION,
NT/ *EUTROPHICATION, LAKES, LAKE
OM/ *EUTROPHICATION, LAKE, *LAKE
S.:
PITATION(ATMOSPHERIC), WETLANDS,
*SITE
NITROGEN, PHOSPHORUS, POTASSIUM,
OSCILLATORI/ BACILLARIOPHYCEAE,
E(CALIF), CHLORELLA PYRENOIDOSA,
GALEATA MENDOTAE, DAPHNIA MAGNA,
ABILITY, ALGAL GROWTH POTENTIAL,
URES, SELENASTRUM CAPRICORNUTUM,
TRUM GRACILE, GROWTH LIMITATION,
ISIONAL ALGAL ASSAY, PROCEDURES,
UISANCE ALGAE, SEWAGE TREATMENT,
OMPOSITION), ALGAE, OXYGENATION,
KTON, FISH, CHRYSOPHYTA, SESTON,
(WATER), ACIDITY, OXYGEN, ALGAE,
S, *AXENIC CULTURES, PEDIASTRUM,
G PLANTS, ROOTED AQUATIC PLANTS,
TIC ALGAE, ALGAE, ARID CLIMATES,
*EFFECTS, NAVICULA
ALGAL
EAN WATER, GENUS, SPECIES, TOUCH

SECONDARY PRODUCTIVITY, AQUATIC P   W69-10152
SECONDARY PRODUCTIVITY, SULFIDES,   W70-01068
SEDGWICK-RAFTER METHOD, TREATMENT   W69-10157
SEDIMENT CORES.: /ON, NITROGEN, P   W68-00511
SEDIMENT LOAD, CONDUCTIVITY, CALC   W69-05894
SEDIMENT-WATER INTERFACES, CALIFO   W70-02775
SEDIMENT-WATER INTERFACES, OREGON   W72-06051
SEDIMENT, BENTHIC FAUNA.: /ACE EL   W69-03366
SEDIMENTATION RATES, *EUTROPHICAT   W69-00632
SEDIMENTATION, SEDIMENTS, TASTE,   W69-07833
SEDIMENTATION, CHLORINATION, SLUD   W70-04060
SEDIMENTATION, TOXICITY, COST-BEN   W70-05819
SEDIMENTATION RATES, STRATIGRAPHY   W70-05663
SEDIMENTATION, ACTIVATED CARBON,   W72-08859
SEDIMENTATION RATES, MANGANESE, L   W73-00750
SEDIMENTATION, COPEPODS, WATER PO   W69-10169
SEDIMENTATION, SETTLING BASINS.: /   W70-09320
SEDIMENTATION, FLOTATION, BIODEGR   W71-05013
SEDIMENTATION, OXIDATION LAGOONS,   W71-11375
SEDIMENTATION, BIOCHEMICAL OXYGEN   W71-10654
SEDIMENTATION, RUNOFF, FERTILIZER   W71-10376
SEDIMENTATION, CHEMICAL PRECIPITA   W71-11823
SEDIMENTATION, ABSORPTION, BIOCHE   W71-11381
SEDIMENTATION, ORGANIC LOADING, A   W71-08970
SEDIMENTS.: /EVEL, WATER STORAGE,   W69-09723
SEDIMENTS.: /, PHOTOSYNTHESIS, TR   W70-04268
SEDIMENTS.: /AGE EFFLUENTS, SMELT   W69-08674
SEDIMENTS.: / ABSORPTION, *CHEMIC   W68-00857
SEDIMENTS), WATERSHEDS(BASINS), W   W69-10182
SEDIMENTS, SOLAR RADIATION, OXYG: W69-10154
SEDIMENTS, CHUTES, EDDIES, HABITA   W69-09742
SEDIMENTS, DREDGING, HARVESTING O   W69-09340
SEDIMENTS, LAKE MICHIGAN, WASHING   W69-09349
SEDIMENTS, SURFACE WATERS, DETRIT   W70-00713
SEDIMENTS, WATER POLLUTION SOURCE   W70-00711
SEDIMENTS, SOIL-WATER-PLANT RELAT   W71-09005
SEDIMENTS, ALGAE, HYDROGEN ION CO   W71-09548
SEDIMENTS, SAMPLING, DEPTH, SEA W   W71-12870
SEDIMENTS, MARINE BIOLOGY, STREAM   W71-10786
SEDIMENTS, ORGANIC MATTER, ALGAE,   W72-00845
SEDIMENTS, SODIUM, POTASSIUM, IRO   W70-09889
SEDIMENTS, THERMOCLINE, HYPOLIMNI   W70-06975
SEDIMENTS, NUTRIENTS, *WASTE WATE   W70-07031
SEDIMENTS, PLANT POPULATIONS, WEA   W70-06229
SEDIMENTS, MUD, SANDS, SLUDGE, RE   W70-10161
SEDIMENTS, WEED CONTROL, SOLAR RA   W68-00478
SEDIMENTS, ON-SITE DATA COLLECTIO   W68-00511
SEDIMENTS, BIBLIOGRAPHIES, ABSORP   W68-00858
SEDIMENTS, WATERSHED MANAGEMENT,   W68-00859
SEDIMENTS, CHLORELLA PYRENOIDOSA,   W69-08284
SEDIMENTS, DISSOLVED OXYGEN, OXYG   W69-08518
SEDIMENTS, PRECAMBRIAN SEDIMENTS,   W69-08284
SEDIMENTS, GAS CHROMATOGRAPHY, LI   W69-08284
SEDIMENTS, TASTE, WATER CHEM: /PL   W69-07833
SEDIMENTS, BIOTA, FISH, UTAH, ALG   W69-07846
SEDIMENTS, FISH, RADIOECOLOGY, EF   W69-07853
SEDIMENTS, BACTERIA, DETRITUS, ES   W69-08274
SEDIMENTS, SOILS, NUCLEAR EXPLOSI   W69-08269
SEDIMENTS, GEOLOGIC FORMATIONS, A   W69-08272
SEDIMENTS, MINNOWS, DETRITUS, CYC   W69-07861
SEDIMENTS, CATTAILS, RADIOACTIVIT   W69-08272
SEDIMENTS, RADIOACTIVITY, MARINE   W69-08267
SEDIMENTS, FERTILIZERS, WATER POL   W70-04193
SEDIMENTS, URBANIZATION, WATER PO   W70-04721
SEDIMENTS, ANIMALS, FISH, INFLOW,   W70-06225
SEDIMENTS, MUD, BENTHOS, WISCONSI   W70-06217
SEDIMENTS, PALEOLIMNOLOGY, NUTRIE   W70-05663
SEDIMENTS, ALGAL CONTROL, *NUTRIE   W72-06052
SEDIMENTS, *NUTRIENTS, NITROGEN C   W72-06051
SEEPAGE LOSSES, DIURNAL VARIATION   W69-09723
SEEPAGE, INDUSTRIAL WASTES, TERTI   W69-10178
SELECTION, LOADING.:   W71-08221
SELECTIVITY, POISONS, OXYGEN, TRA   W70-05399
SELENASTRUM CAPRICORNUTUM PRINTZ,   W70-05752
SELENASTRUM GRACILE, CARBON ASSIM   W70-03983
SELENASTRUM, RHODOTORULA, TEMPERA   W70-03957
SELENASTRUM CAPRICORNUTUM, ANABAE   W69-06864
SELENASTRUM GRACILE, GROWTH LIMIT   W70-02779
SELENASTRUM, ALGAL GROWTH, ANABAE   W70-02779
SELENASTRUM CAPRICORNUTUM, SELENA   W70-02779
SELF-PURIFICATION, DISSOLVED OXYG   W72-01798
SELF-PURIFICATION, LABORATORY TES   W72-01801
SELF-PURIFICATION.: /GAE, ZOOPLAN   W70-00268
SELF-PURIFICATION, BACTERIOPHAGE,   W69-07838
SELF-PURIFICATION, ALGAL ENUMERAT   W70-04381
SELF-PURIFICATION, BIOCHEMICAL OX   W70-05750
SEMIARID CLIMATES, *ABSORPTION, *   W69-03491
SEMINULUM VAR.HUSTEDTII.:   W71-03183
SENESCENCE, GROWTH INHIBITIONS.:   W72-02417
SENSATION.: / INDUSTRIAL USES, CL   W70-05389

, LAKES, DESIGN CRITERIA, REMOTE
ELEASES, CURRENTS(WATER), REMOTE
RA, AMBIENT TEMPERATURE, THERMAL
UTION, *WATER POLLUTION SOURCES,
FILTRATION, *DIATOMACEOUS EARTH,
, FARM WASTES, MUNICIPAL WASTES,
, ALGAE, SCUM, WEEDS, NUTRIENTS,
NG, SPORT FISH/ *EUTROPHICATION,
ILE SUSPENDED SOLIDS, MULTI-CELL
ATES, ELEUTHERA ISLAND, NITZCHIA
MIN B-1, VITAMIN B-12, THIAMINE,
AM REACTION, CYTOPHAGA, PROTEUS,
(CANADA), NEW BRUNSWICK(CANADA),
WATER MASSES, U S PUBLIC HEALTH
ENTHIC FAUNA, CHEMICAL ANALYSIS,
DIMENT-WATER INTERFACES, OREGON,
, RESERVOIRS, LAKES, REAERATION,
NION, EPILIMNION, PHYTOPLANKTON,
MICROORGANISMS, MICROORGANISMS,
BIC CONDITIONS, ALGAE, ROTIFERS,
, *EUTROPHICATION, AQUATIC LIFE,
ZOOPLANKTON, FISH, CHRYSOPHYTA,
A, CHROOMONAS ACUTA, SCHROEDERIA
EN DEMAND, ALGAE, SEDIMENTATION,
CITY TEST, SLUDGE DRYING, SOLIDS-
CITY, CARBON-14, DDT, MALATHION,
ENOIDOSA, PESTICIDE DEGRADATION,
OIDOSA, SCHRODERIA, THIMET, TRI/
ZATION, LAKES, SEWAGE TREATMENT,
INE, NITROGEN COMPOUNDS, SEWAGE,
ENT, *TRICKLING FILTERS, *BIOTA,
RINE ENVIRONMENT, WATER QUALITY,
ATES, SHELLFISH, WASTE DISPOSAL,
OWTH RATES, STABILIZATION PONDS,
ICATION, *MINNESOTA, RECREATION,
ION ABATEMENT, SEWAGE TREATMENT,
N CONCENTRATION, ELECTRIC POWER,
STS, LEGAL ASPECTS, LEGISLATION,
ASTES, SEWERS, SEWAGE TREATMENT,
, DIATOMS, RESERVOIRS, CULTURES,
HYTA, CRUSTACEANS, SOIL EROSION,
OLLUTION IDENTIFICATION, SEWAGE,
N, NUTRIENTS, LIGHT PENETRATION,
ESTS, BIOCHEMICAL OXYGEN DEMAND,
OVERFLOW, SEWERS, STORM DRAINS,
EPILIMNION, NITROGEN COMPOUNDS,
, *PRODUCTIVITY, NUISANCE ALGAE,
ION, WATER CONTROL, FERTILIZERS,
TCR POLLUTION CONTROL, *FILTERS,
REATMENT, WATER QUALITY CONTROL,
TREATMENT, NITROGEN, PHOSPHORUS,
ATER POLLUTION, ODOR, EFFLUENTS,
*NEW HAMPSHIRE, *COPPER SULFATE,
DENSITY DURRENTS, MEASUREMENTS,
N, EUTROPHICATION, STORM DRAINS,
N SOURCES, WATER QUALITY, ALGAE,
, OLIGOTROPHY, DIVERSION, CORES,
NS, PHOSPHORUS COMPOUNDS, LAKES,
CHEMICAL PROPERTIES, WISCONSIN,
C, *ESTUARIES, HUMAN POPULATION,
ATORS, NUISANCE ALGAE, NEW YORK,
EN/ *MINNESOTA, *EUTROPHICATION,
DS, ALGAE, PHOSPHATES, NITRATES,
, NUISANCE ALGAE, WATER QUALITY,
ONS, EUTROPHICATION, CYANOPHYTA,
, NITROGEN FIXATION, CYANOPHYTA,
CHLORELLA PYRENOIDOSA, SYNTHETIC
CULTIVATION, CHANNEL MORPHOLOGY,
TREATMENT, STABILIZATION PONDS,
E, *BACTERIA, *DISSOLVED OXYGEN,
*PHOSPHORUS, *NITROGEN, *ALGAE,
GERN(GERMANY), GERMANY, FINLAND,
CHEMCONTROL, INDUSTRIAL WASTES,
IURNAL, LIGHT INTENSITY, CARBON,
GAE, NEW YORK, SEWAGE EFFLUENTS,
USTRIAL WASTES, DOMESTIC WASTES,
ION CONCENTRATION, *PHOSPHORUS,
E, PHOSPHATES, NITRATES, RIVERS,
CING ALGAE, POLLUTION ABATEMENT,
MICROORGANISMS, NUISANCE ALGAE,
IGN CRITERIA, WATER TEMPERATURE,
, *WATER POLLUTION, DEGRADATION,
EN DEMAND, FISH, ALGAE, FOAMING,
ORELLA, SCENEDESMUS, ADSORPTION,
OCHEMICAL OXYGEN DEMAND, MIXING,
ED LOAD, SOLIDS CONTACT PROCESS,
POUNDS, SEWAGE, SEWAGE BACTERIA,
SHINGTON, *OXYGEN, *FLUCTUATION,
ATION, SEWAGE, SEWAGE EFFLUENTS,
*RUNOFF, *FERTILIZATION, LAKES,

SENSING, MINNESOTA, SEDIMENT-WATE   W70-02775
SENSING, SEA WATER, CHLOROPHYLL,   W70-05377
SENSITIVITY, GENERA, 'FLAVOBACTER   W70-00713
SEPARATION TECHNIQUES, *ANALYTICA   W69-01273
SEPARATION TECHNIQUES, PRESSURE,   W72-08688
SEPTIC TANKS, RECREATION WASTES,   W70-04266
SEPTIC TANKS, EFFLUENTS, SWAMPS,   W70-04193
SEQUENCE, LAKES, CHEMICALS, FISHI   W68-00172
SERIAL SYSTEM, PH, FACULTATIVE PO   W70-04790
SERIATA, NAVICULA, FLORIDA CURREN   W70-00707
SERRATIA MARINORUBRA, GYRODINIUM   W70-05652
SERRATIA, BAY OF FUNDY(CANADA), R   W70-03952
SERVALL OMNI-MIXER HOMOGENIZER.: /   W70-04365
SERVICE, GREAT LAKES-ILLINOIS RIV   W70-00263
SESSILE ALGAE, *PRIMARY PRODUCTIV   W72-06052
SESSILE ALGAE, CYANOPHYTA.: /, SE   W72-06051
SESTON, CHLOROPHYLL, CYCLES, UNIT   W70-05092
SESTON, CRUSTACEANS, DIATOMS, CHL   W69-10182
SESTON, INVERTEBRATES, SHELLFISH,   W69-04276
SESTON, RESPIRATION, ALTITUDE, SA   W69-10154
SESTON, SCENEDESMUS, MATHEMATICAL   W73-01446
SESTON, SELF-PURIFICATION.: /GAE,   W70-00268
SETIGERA, MELOSIRA AMBIGUA, CERAT   W70-05387
SETTLING BASINS.: /OCHEMICAL OXYG   W70-09320
SETTLING, ECONOMIC ADVANTAGE, TEM   W70-05819
SEVIN.:   *ALGAE TOXI   W70-02198
SEVIN, MALATHION, ACETONE, ETHANO   W70-02968
SEVIN, MALATHION, CHLORELLA PYREN   W70-08097
SEWAGE BACTERIA, NITROGEN, PHOSPH   W70-02787
SEWAGE BACTERIA, SEWAGE TREATMENT   W69-10171
SEWAGE BACTERIA, SYSTEMATICS, FUN   W72-01819
SEWAGE BACTERIA, ENVIRONMENTAL EF   W69-04276
SEWAGE DISPOSAL, WATER CHEMISTRY,   W69-04276
SEWAGE DISPOSAL.: /, NITROGEN, GR   W68-00855
SEWAGE DISPOSAL, WATER POLLUTION,   W70-04810
SEWAGE DISPOSAL, WASTE DISPOSAL,   W69-06909
SEWAGE DISPOSAL, *WASTE WATER TRE   W71-09548
SEWAGE DISTRI: /WAY, ESTIMATED CO   W70-03344
SEWAGE DISPOSAL, STORM DRAINS, SU   W70-03344
SEWAGE EFFLUENTS, WISCONSIN, ALGI   W70-02982
SEWAGE EFFLUENTS, FARM WASTES. PO   W70-02803
SEWAGE EFFLUENTS, SEWAGE TREATMEN   W70-02248
SEWAGE EFFLUENTS, OXYGEN, SOLAR R   W70-01933
SEWAGE EFFLUENTS, SANITARY ENGINE   W72-01817
SEWAGE EFFLUENTS, *SEWAGE TREATME   W71-10654
SEWAGE EFFLUENTS, LIGHT INTENSITY   W69-10167
SEWAGE EFFLUENTS, PHOSPHORUS, EUT   W70-00270
SEWAGE EFFLUENT, NUTRIENTS, LAKES   W70-00269
SEWAGE EFFLUENTS, ELECTROLYTES, Z   W70-01027
SEWAGE EFFLUENTS, ABSORPTION, LAK   W70-01031
SEWAGE EFFLUENTS.: /, BIOLOGICAL   W70-08980
SEWAGE EFFLUENTS, DILUTION, DECOM   W71-00940
SEWAGE EFFLUENTS, SMELTS, FISHKIL   W69-08674
SEWAGE EFFLUENTS, TEMPERATURE, OX   W69-06203
SEWAGE EFFLUENTS, STREAMFLOW, MEA   W70-05112
SEWAGE EFFLUENTS.: /ATER POLLUTIO   W70-05084
SEWAGE EFFLUENTS, DIATOMS, PHOSPH   W70-05663
SEWAGE EFFLUENTS, ARSENIC COMPOUN   W70-05269
SEWAGE EFFLUENTS, NITROGEN, PHOSP   W70-05113
SEWAGE EFFLUENTS, AQUATIC POPULAT   W68-00461
SEWAGE EFFLUENTS, SEWAGE TREATMEN   W70-03974
SEWAGE EFFLUENTS, TERTIARY TREATM   W70-03512
SEWAGE EFFLUENTS, TERTIARY TREATM   W70-04074
SEWAGE EFFLUENTS, DIATOMS, MISSOU   W70-04507
SEWAGE EFFLUENTS, CHLOROPHYTA, RH   W70-04510
SEWAGE EFFLUENTS, MINNESOTA, WATE   W70-04506
SEWAGE FEED, SYMBIOTIC RELATIONSH   W70-05655
SEWAGE INPUT, BIOLOGICAL PROPERTI   W68-00475
SEWAGE LAGOONS.: /MENT, SECONDARY   W68-00012
SEWAGE LAGOONS, PLANT PHYSIOLOGY,   W72-01818
SEWAGE PLANTS, CHLOROPHYLL, PHOTO   W70-04268
SEWAGE RING TUBES.: /NY), LAKE TE   W70-05761
SEWAGE TREATMENT.: /, PHOSPHORUS,   W70-05668
SEWAGE TREATMENT, TERTIARY TREATM   W70-05387
SEWAGE TREATMENT, INDUSTRIAL WAST   W70-03974
SEWAGE TREATMENT, HARVESTING OF A   W68-00478
SEWAGE TREATMENT, *OXIDATION LAGO   W68-00012
SEWAGE TREATMENT, WATER POLLUTION   W69-01453
SEWAGE TREATMENT, SEWAGE DISPOSAL   W69-06909
SEWAGE TREATMENT, SELF-PURIFICATI   W72-01798
SEWAGE TREATMENT, AEROBIC BACTERI   W71-08221
SEWAGE TREATMENT, NITROGEN COMPOU   W72-02817
SEWAGE TREATMENT.: /CHEMICAL OXYG   W70-08740
SEWAGE TREATMENT, PARTICLE SIZE.:   W71-03035
SEWAGE TREATMENT, WASTE WATER TRE   W70-06899
SEWAGE TREATMENT, TRICKLING FILTE   W70-06792
SEWAGE TREATMENT, WATER POLLUTION   W69-10171
SEWAGE TREATMENT, SOLAR RADIATION   W69-10182
SEWAGE TREATMENT, WATER POLLUTION   W70-02248
SEWAGE TREATMENT, SEWAGE BACTERIA   W70-02787

AGE LAGOONS, *INDUSTRIAL WASTES,   SEWAGE TREATMENT, NORTH DAKOTA, C   W70-03312
SPOSAL, DOMESTIC WASTES, SEWERS,   SEWAGE TREATMENT, SEWAGE DISPOSAL   W70-03344
                     *NEW ZEALAND,   SEWAGE TREATMENT POND ALGAE.:   W70-02304
ESTIC WASTES, INDUSTRIAL WASTES,   SEWAGE WASTES, BYPRODUCTS, FERTIL   W70-09905
*SEWAGE TREATM/ *EUTROPHICATION,   SEWAGE, *PHOSPHORUS, PHOSPHATES,   W68-00256
STRIAL WASTES, MUNICIPAL WASTES,   SEWAGE, ADSORPTION, ACTIVATED CAR   W70-04001
*GREAT LAKES, *WATER POLLUTION,   SEWAGE, AESTHETICS, AEROBIC CONDI   W70-04430
MS, EUTROPHICATION, OLIGOTROPHY,   SEWAGE, ALGAE, SCUM, WEEDS, NUTRI   W70-04193
DEMAND, SUSPENDED LOAD, WASTES,   SEWAGE, ALGAE, MICROORGANISMS, PH   W70-08627
ON), *WATER PURIFICATION, ALGAE,   SEWAGE, ANAEROBIC, OXYGEN, NITRIF   W70-05092
ESTIC WASTES, *SEWAGE TREATMENT,   SEWAGE, BIOLOGICAL TREATMENT, SOL   W68-00248
INGTON, *LAKES, *EUTROPHICATION,   SEWAGE, BIOTA, PHYTOPLANKTON, DIA   W69-10169
REATMENT FACILITIES, FILTRATION,   SEWAGE, COLIFORMS, EUTROPHICATION   W70-00263
*CITIES, BIOLOGICAL COMMUNITIES,   SEWAGE, DETRITUS, TUBIFICIDS, PLA   W70-00268
INTENSITY, CONDUCTIVITY, ALGAE,   SEWAGE, DIATOMS, CYANOPHYTA, CARB   W70-05405
ER), DIATOMS, FOULING, PLANKTON,   SEWAGE, DOMESTIC WASTE, INDUSTRIA   W72-01813
ATER POLLUTION TREATMENT, LAKES,   SEWAGE, DRAINAGE DISTRICTS, DEBT,   W70-04455
NUTRIENTS, PHOSPHORUS, INFLUENT   SEWAGE, EQUILIBRIUM, MATHEMATICAL   W68-00858
CONVECTION, WAVES(WATER), FISH,   SEWAGE, EUGLENOPHYTA, DIATOMS, BI   W70-03969
AKES, RIVERS, NITROGEN, BIOMASS,   SEWAGE, FARM WASTES, CARBON CYCLE   W70-02777
N, *CYANOPHYTA/ *EUTROPHICATION,   SEWAGE, FISH POPULATIONS, PLANKTO   W68-00461
ED STATES, PHOSPHORUS, NITROGEN,   SEWAGE, INDUSTRIAL WASTES, DOMEST   W68-00478
CAL PROPERTIES, DOMESTIC WASTES,   SEWAGE, INDUSTRIAL WASTES, AGRICU   W70-03964
GI, INSECTS, BACTERIA, PROTOZOA,   SEWAGE, LAKES, NUTRIENTS.: /, FUN   W71-10786
REATMENT, PHOTOSYNTHESIS, ALGAE,   SEWAGE, MUNICIPAL WA: / AEROBIC T   W72-00462
UTION, DRAINAGE, ORGANIC MATTER,   SEWAGE, PHYTOPLANKTON, PRODUCTIVI   W70-05404
FFECTS, WATER POLLUTION SOURCES,   SEWAGE, PHOSPHORUS, CHEMCONTROL,   W70-05668
WATER POLLUTION, LAKES, STREAMS,   SEWAGE, REMEDIES, JUDICIAL DECISI   W69-06909
LLUTION CONTROL, OXYGEN BALANCE,   SEWAGE, RUNOFF.: /URCES, WATER PO   W69-08668
IENTS, POLLUTION IDENTIFICATION,   SEWAGE, SEWAGE EFFLUENTS, SEWAGE   W70-02248
N, CHLORINE, NITROGEN COMPOUNDS,   SEWAGE, SEWAGE BACTERIA, SEWAGE T   W69-10171
ZATION, PLANT GROWTH SUBSTANCES,   SEWAGE, SULFATE, MANGANESE, IRON,   W70-02969
STREAMS, ALGAE, FISH, WISCONSIN,   SEWAGE, TROUT, NITROGEN, PHOSPHOR   W69-10178
OGEN, AMMONIA, ALGAE, NUTRIENTS,   SEWAGE, WASTE WATER(POLLUTION), W   W71-05879
IDE RESIDUES, PHOSPHATES, ALGAE,   SEWAGE, WATER POLLUTION, WATER LA   W69-06305
TIVITY, AGRICULTURAL WATERSHEDS,   SEWAGE, WATER POLLUTION, CHEMICAL   W70-05752
NATION, COAGULATION, WATERSHEDS,   SEWERAGE ALGAE, POTABLE WATER.: /   W71-05943
EFFECTS, BACTERIA, ALGAE, FISH,   SEWERS, RUNOFF, WASTES, WATER SUP   W71-09880
WASTE DISPOSAL, DOMESTIC WASTES,   SEWERS, SEWAGE TREATMENT, SEWAGE   W70-03344
WASTE WATER TREATMENT, OVERFLOW,   SEWERS, STORM DRAINS, SEWAGE EFFL   W71-10654
NG, *SUSPENDED SOLIDS, *COMBINED   SEWERS, STORMWATER OVERFLOWS.: /I   W71-10654
MS, WHITE FISH, ALEWIFE, GIZZARD   SHAD, SEA LAMPHREY, COHO SALMON.:   W70-01943
LAGELLATES, LAKE CLASSIFICATION,   SHAGAWA LAKE(MINN).: /ROTROPHY, F   W70-02775
CHROMATOGRAPHY, LIPIDS, OIL, OIL   SHALE.: /GENESIS, SEDIMENTS, GAS   W69-08284
A, SCALING, GRAVELS, MUD, SANDS,   SHALES, BOTTOM SEDIMENTS, CATTAIL   W69-08272
PHORUS, CARBONATES, TEMPERATURE,   SHALLOW WATER, HYDROGEN SULFIDE,   W70-02803
EUTROPHICATION, DOMESTIC WASTES,   SHALLOW WATER, CHLOROPHYTA, CHRYS   W70-00711
G, WATERFOWL, BIRDS, COSTS, COST-   SHARING.: /ON TECHNIQUES, SWIMMIN   W70-05568
BANON, WADLEIGH STATE PARK, LAKE   SHATUTAKEE, MASCOMA LAKE, GLEN LA   W69-08674
UTOTROPHIC INDEX, CHLOROPHYLL A,   SHAYLER RUN, ORGANIC CARBON, CINC   W71-04206
PELIC ALGAE, ALGAL ASSOCIATIONS,   SHEAR WATER(ENGLAND), AMPHORA OVA   W71-12870
PANDANUS, MUSCLE, BONE, LEAVES,   SHELL, RHODIUM-102.: /A, PISONIA,   W69-08269
VIEWS, HABITATS, CLAMS, MUSSELS,   SHELLFISH, DINOFLAGELLATES, CYANO   W70-05372
RGANISMS, SESTON, INVERTEBRATES,   SHELLFISH, WASTE DISPOSAL, SEWAGE   W69-04276
IZERS, ALGAE, ZOOPLANKTON, FISH,   SHELLFISH, STEAM, ELECTRIC POWER,   W70-09905
, ALGAL TOXINS, LARVAE, PHENOLS,   SHELLFISH, OYSTERS, INDUSTRIES, C   W70-01068
                                                  SHENANDOAH RIVER.:   W70-03068
DITCHES, SOUP, FOOD WASTES.:   SHEPPARTION(AUSTRALIA), OXIDATION   W71-09548
ORPTION, PHYTOPLANKTON, SUCKERS,   SHINERS, CADDISFLIES, TEMPERATURE   W69-07853
R BASINS PROJECT, INDIANA HARBOR   SHIP CANAL.: /LAKES-ILLINOIS RIVE   W70-00263
EN SEA, WEAPONS TESTING, NUCLEAR   SHIP SAVANNAH, ZIRCONIUM-95, RUBI   W70-00707
STERS, OILY WATER, MARINE ALGAE,   SHIPS.: /OLLUTION, CLEANING, DISA   W70-01470
, NUCLEAR REACTORS, POWERPLANTS,   SHIPS, SUBMARINES, EFFLUENTS, ION   W70-00707
, POTOMAC RIVER, SAVANNAH RIVER,   SHOCK EFFECTS.: /ZOAN COMMUNITIES   W70-07847
*BIOLOGICAL EXTRACTION, GROWTH,   SHOCK LOADING.:   W70-06604
ANAGERS, FISH YIELDS, UNDERWATER   SHOCK WAVE, ANABAENA, PHYGON XL(D   W69-08674
TOMAC RIVER, SAVAN/ *TEMPERATURE   SHOCK, *PROTOZOAN COMMUNITIES, PO   W70-07847
DRAINAGE.:   SHORELINE VEGETATION, STORMWATER,   W70-02251
MIXING, SEA WATER, FRESH WATER,   SHORES, WATER CIRCULATION, FLOODS   W69-06203
BENTHIC FAUNA, WATERSHEDS, WATER   SHORTAGE FILTERS, SNAILS, ODOR, C   W72-00027
GAE, CRUSTACEANS, BENTHOS, BRINE   SHRIMP, ECOSYSTEMS, FISH, PHYTOPL   W69-07440
POLLUTION, AQUATIC ALGAE, BRINE   SHRIMP, FISH, FOOD CHAINS, MUSSEL   W70-01467
OOD CHAINS, AQUATIC ALGAE, BRINE   SHRIMP, FISH, MUSSELS.: /UTION, F   W70-01466
OSYSTEMS, SINKS, TRACERS, CRABS,   SHRIMP, MULLETS, FISH, ALGAE.: /C   W69-08274
DEPTH, CYANOPHYTA, CHLOROPHYTA,   SIEROZEMS, SOIL MICROBIOLOGY, DIS   W71-04067
RUS, CALCIUM, MAGNESIUM, SODIUM,   SILICA, DIATOMS, RESISTANCE, *PLA   W68-00856
MALS, DENSITY, IRON, PHOSPHORUS,   SILICA, SAMPLING, CHEMICAL ANALYS   W70-06975
, PHOSPHATES, NITRATES, DAPHNIA,   SILICA, THERMAL STRATIFICATION, S   W70-05760
, *EUTROPHICATION, ALGAE, CLAYS,   SILICATES.: /, *LAKES, *SEDIMENTS   W69-03075
NITRATES, NUTRIENTS, PHOSPHATES,   SILICATES, IRON OXIDES, *PLANT PO   W68-00860
, PLANKTON, TUBIFICIDS, CALCIUM,   SILICATES, DIATOMS, WINDS, NITROG   W70-05761
DES, AMMONIUM COMPOUNDS, OXYGEN,   SILICATES, ORGANIC MATTER, AIR TE   W69-07096
N, CHLORELLA, SODIUM, POTASSIUM,   SILICATES, DISSOLVED SOLIDS, CALC   W70-03512
ESMUS QU/ *ALGAL BLOOMS, *ORGANO-   SILICATES, *ALKOXYSILANES, SCENED   W71-03056
RIMARY PRODUCTIVITY, PHOSPHATES,   SILICATES, CITRATES, BIOASSAY, NE   W70-01579
ESMUS QU/ *ALGAL BLOOMS, *ORGANO-   SILICATES, *ALKOXYSILANES, SCENED   W71-05719
PHIC, CHEMOTROPHIC, AUTOTROPHIC,   SILICON, PEDIASTRUM BORYANUM, STA   W70-02804
AUXINS, METABOLITES, INHIBITION,   SILICON, VANADIUM, ZINC, GALLIUM.   W69-04801
IENTS, *PESTICIDES, FARM WASTES,   SILT, EROSION, SEDIMENTATION, RUN   W71-10376
HING, SPORT FISHING, RECREATION,   SILT, NUTRIENTS, AQUATIC POPULATI   W68-00172

ALDRIN, DIELDRIN, DDT, AERATION,   SILTS.: /SORPTION, ALGAE, CLAYS,   W70-09768
RADIOISOTOPES, CESIUM, FALLOUT,   SILTS, DETRITUS, BOTTOM SEDIMENTS   W69-09742
POLLUTANTS, AQUATIC LIFE, ALGAE,   SILTS, ECOLOGY, AQUATIC ENVIRONME   W69-10080
N), RUSTING, METALS, INHIBITORS,   SILTS, ORGANIC MATTER, IRON OXIDE   W70-02294
*ESTUARIES, FRESH WATER, LAKES,   SILTS, PHOSPHATES, NITROGEN, WAST   W70-02255
WALLEYE, VEGETATION, WILD RICE,   SILTS, PLANKTON, MAYFLIES, FAUNA,   W70-01943
, SAMPLING, DENSITY, LIGHT, MUD,   SILTS, SANDS, GRAVELS, SEDIMENTS,   W70-00711
OROPHYLL, DYNAMICS, TEMPERATURE,   SILTS, STANDING CROP, ECOLOGY, PH   W70-02780
, POTASSIUM PERMANGANATE, FENAC,   SILVER NITRATE, SIMAZIN, NUPHAR,   W70-10175
                                          SILVER.:   W72-05143
CONIUM-95-NIOBIUM, EUROPIUM-155,   SILVER-110M, CERIUM, GUETTARDA, I   W69-08269
ERNANTHERA, HETERANTHERA, NAJAS,   SILVEX, ENDOTHAL, POTASSIUM PERMA   W70-10175
ANGANATE, FENAC, SILVER NITRATE,   SIMAZIN, NUPHAR, AGASICLES, MANAT   W70-10175
                                          SIMAZINE 80 W, DACTHAL.:   W70-03519
IUM-106, NIOBIUM-95, GYMNODINIUM   SIMPLEX, FLAGELLATES, ELEUTHERA I   W70-00707
CANADA, AQUATIC ECOSYSTEM MODEL,   SIMULATION MODEL, PHOSPHORUS EFFE   W69-06536
IDS, UPTAKE, CONSUMER ORGANISMS,   SIMULIUM, PHYSA, SALMO TRUTTA, EN   W69-09334
TZSCHIA DISSIPATA, SYNEDRA ACUS,   SINKING RATES, GRAZING PRESSURE,   W69-10158
ESTUARIES, PLANKTON, ECOSYSTEMS,   SINKS, TRACERS, CRABS, SHRIMP, MU   W69-08274
ONAL PARK, *ICELAND HOT SPRINGS,   SINTER, NUCLEIC ACID, SISJOTHANDI   W69-10160
T SPRINGS, SINTER, NUCLEIC ACID,   SISJOTHANDI(ICELAND), ENZYME-SUBS   W69-10160
GAE, WATER POLLUTION SOURCES, ON-   SITE DATA COLLECTIONS, *ON-SITE I   W70-09976
Y, GASTROPODS, MARINE PLANTS, ON-   SITE DATA COLLECTIONS.: /MORTALIT   W70-01780
LIBRIUM, MATHEMATICAL MODELS, ON-   SITE DATA COLLECTIONS, BOTTOM SED   W68-00858
UIREMENTS, *BOTTOM SEDIMENTS, ON-   SITE DATA COLLECTIONS, AQUATIC PL   W68-00511
ES, *PHOSPHORUS RADIOISOTOPE, ON-   SITE DATA COLLECTIONS, FISHES, PL   W68-00857
MICAL ANALYSIS, PH, SAMPLING, ON-   SITE DATA COLLECTIONS, LIMNOLOGY.   W69-00659
ATION, OXYGEN, SPORT FISHING, ON-   SITE INVESTIGATIONS.: /GHT PENETR   W68-00487
S, ON-SITE DATA COLLECTIONS, *ON-   SITE INVESTIGATIONS, FREQUENCY,   W70-09976
RE, METABOLISM, EPIPHYTOLOGY, ON-   SITE TESTS, LABORATORY TESTS, LOT   W72-05453
SYNTHESIS, LIGHT PENETRATION, ON-   SITE TESTS, BIOCHEMICAL OXYGEN DE   W72-01818
VOIRS, ANALYTICAL TECHNIQUES, ON-   SITE TESTS.: /IMPOUNDMENTS, RESER   W71-10098
AY, AQUATIC LIFE, COOLING WATER,   SITES.: /E, MODEL STUDIES, BIOASS   W71-11881
T, WATER QUALITY, SO/ *RESERVOIR   SITES, WATER RESOURCES DEVELOPMEN   W72-11906
TION, SEWAGE TREATMENT, PARTICLE   SIZE.: /ELLA, SCENEDESMUS, ADSORP   W71-03035
CONTROL, *FILTRATION, *PARTICLE   SIZE, *CLAYS, *SILTS, SANDS, ALGA   W70-08628
IA SCHOD/ *FILTERING RATE, *BODY   SIZE, *COMPARATIVE STUDIES, DAPHN   W70-03957
OPES, CALCIUM, POTASSIUM, CLAMS,   SIZE, ALGAE, DIATOMS, TRACE ELEME   W70-00708
, *POPULATION DYNAMICS, CELLULAR   SIZE, CHLORELLA VULGARIS, CORNELL   W69-06540
HE/ *DNA, *ORGANIC CARBON, *CELL   SIZE, EUKARYOTES, MONOCHRYSIS LUT   W69-08278
ONAS, LABORATORY TESTS, PARTICLE   SIZE, POPULATION, GROWTH RATES, W   W72-02417
MUNITIES, OKANAGAN LAKE(CANADA),   SKAHA LAKE(CANADA), OSOYOOS LAKE(W71-12077
, MELOSIRA, NAVICULA, NITZSCHIA,   SKELETONEMA, PLAGIOGRAMMA,: /IGMA   W70-03325
UGLENA GRACILIS, CACHONINA NIEI,   SKELETONEMA COSTATUM, DUNALIELLA   W69-08278
NUTRITION, DITYLUM, CHAETOCEROS,   SKELETONEMA, THALASSIOSIRA, COCCO   W70-04503
UM PRINTZ, OSCILLATORIA VAUCHER,   SKELETONEMA COSTATUM(GREV) CL, CH   W70-05752
HUXLEYI, PYRAMIMONAS, PERIDINI/   SKELETONEMA COSTATUM, COCCOLITHUS   W70-05272
ALCATUS, CHLAMYDOMONAS MOEWUSII,   SKELETONEMA COSTATUM, SYNECHOCOCC   W70-05381
PSOIDEA, CHLORELLA LUTEOVIRIDIS,   SKELETONEMA COSTATUM, PHYSIOLOGIC   W70-04809
S, TEMPERATURE, DEPTH, BACTERIA,   SLIME, ALGAE, CYANOPHYTA, DIATOMS   W70-04371
PHYTA, RHODOPHYTA, EUGLENOPHYTA,   SLIME, CO: /A, CYANOPHYTA, CHRYSO   W70-05389
O), RHIZOPODS, SPIDER/ *ZOOGLEAL   SLIME, DAYTON(OHIO), VANDALIA(OHI   W72-01819
QUALITY CONTROL, ALGAL CONTROL,   SLIME, TASTE-PRODUCING ALGAE, ODO   W69-05704
GICIDAL TESTS, ALGISTATIC TESTS,   SLIMEPRODUCING ALGAE, PHOTOMICROG   W69-05704
GAE BLOOMS, *FISH TOXICITY TEST,   SLUDGE DRYING, SOLIDS-SETTLING, E   W70-05819
M, *ALUMINUM SULFATE, *ACTIVATED   SLUDGE PROCESS, HANOVER TREATMENT   W70-09186
XYGEN, *OHIO, WATER TEMPERATURE,   SLUDGE WORMS, BLOODWORMS, AQUATIC   W70-00475
ALGAE, PLANKTON, BENTHIC FAUNA,   SLUDGE WORMS, BLOODWORMS, WATER P   W68-01244
, FACULTATIVE POND DESIGNS, BOD,   SLUDGE.: / TEMPERATURE, NUTRIENTS   W70-04786
MENT, ALGAE, BACTERIA, ACTIVATED   SLUDGE.: / DEMAND, TERTIARY TREAT   W69-08054
D, *COLOR, WASTE WAT/ *ACTIVATED   SLUDGE, *BIOCHEMICAL OXYGEN DEMAN   W70-04060
TREATMENT, *PROTOZOA, *ACTIVATED   SLUDGE, *BACTERIA, METABOLISM, GR   W72-01817
ECLAMATION, / *ALGAE, *ACTIVATED   SLUDGE, *EFFLUENTS, WASTE WATER R   W70-06053
TION, ORGANIC LOADING, ACTIVATED   SLUDGE, AERATION, DISSOLVED OXYGE   W71-08970
UGMENTATION, *LAGOONS, ACTIVATED   SLUDGE, ALGAE, NUTRIENTS, VIRUSES   W72-06017
OXYGEN DEMAND, COLOR, ACTIVATED   SLUDGE, ALGAE, WASTE WATER TREATM   W70-10433
ME, *NITROGEN, HYDRO/ *ACTIVATED   SLUDGE, ALGAE, EUTROPHICATION, LI   W68-00012
EMICAL OXYGEN DEMAND, *ACTIVATED   SLUDGE, BIOLOGICAL TREATMENT, PIL   W70-05821
CONDITIONS, CANNERIES, ACTIVATED   SLUDGE, BIOCHEMICAL OXYGEN DEMAND   W71-09546
HEMICAL OXYGEN DEMAND, ACTIVATED   SLUDGE, COST COMPARISONS, SEDIMEN   W71-09548
IALYSIS, ION EXCHANGE, ACTIVATED   SLUDGE, DENITRIFICATION.: /ECTROD   W69-01453
ES, SEDIMENTATION, CHLORINATION,   SLUDGE, DETERGENTS, SPRAYING, ALG   W70-04060
SSOLVED OXYGEN, LIGHT INTENSITY,   SLUDGE, DEPTH, ORGANIC LOADING, D   W72-06854
DS, LAGOON, FILTERS, PESTICIDES,   SLUDGE, DIGESTION, INDIANA.: /OUN   W72-03641
E, SLURRIES, SPRAYING, ACTIVATED   SLUDGE, FARM WASTES.: /AE, *CATTL   W70-07491
RGANISMS, TEMPERATURE, ACTIVATED   SLUDGE, FISHERIES, CARP, CLAMS, P   W70-09905
ICAL FEASIBILITY, COST ANALYSIS,   SLUDGE, INCINERATION, STORM RUN-O   W71-10654
GEN DEMAND, *BACTERIA, ACTIVATED   SLUDGE, IRON COMPOUNDS, LIME, OXY   W68-00256
OLOGICAL DISTRIBUTION, ACTIVATED   SLUDGE, MAINTENANCE COSTS, CONSTR   W70-04786
ONS, ANAEROBIC DIGESTION, ALGAE,   SLUDGE, METHANE, ORGANIC LOADING,   W71-11375
DGE, AERATION, DISSOLVED OXYGEN,   SLUDGE, OXIDATION LAGOONS, ALGAE,   W71-08970
XYGEN DEMAND, OXIDATION LAGOONS,   SLUDGE, PHOTOSYNTHETIC OXYGEN, FA   W71-08221
WASTES, MICROBIOLOGY, ACTIVATED   SLUDGE, RATES, CHLORELLA, HYDROGE   W70-00248
OLIFORMS, SEDIMENTS, MUD, SANDS,   SLUDGE, RECREATION.: /, OXYGEN, C   W70-10161
ECHNIQUES, PRESSURE, FLOW RATES,   SLUDGE, SLURRIES, AUTOMATIC CONTR   W72-08688
UTION SOURCES, ODOR, FARM WASTE,   SLUDGE, SOLID WASTES, DECOMPOSTIN   W71-05742
OGEN FIXING BACTERIA, *ACTIVATED   SLUDGE, STORAGE, TEMPERATURE, MIC   W71-13335
ED LOAD, PILOT PLANTS, ACTIVATED   SLUDGE, TREAT: /REATMENT, SUSPEND   W70-09186
DAIRY INDUSTRY, *ALGAE, *CATTLE,   SLURRIES, SPRAYING, ACTIVATED SLU   W70-07491
S, PRESSURE, FLOW RATES, SLUDGE,   SLURRIES, AUTOMATIC CONTROL, PIPI   W72-08688

530

OPPER SULFATE, SEWAGE EFFLUENTS,

SEPHINE(MINN), GRAY'S BAY(MINN),
ORWAY), TRONDHEIMSFJORD(NORWAY),
LOROPHYTA, CHRYSOPHYTA, GRAZING,
ISANCE ALGAE, FISH, ZOOPLANKTON,
AKE TROUT, CRAYFISH, STONEFLIES,
SODIUM ARSENITE, NUISANCE ALGAE,
MS, NEMATODES, ROTIFERS, LARVAE,
ERSHEDS, WATER SHORTAGE FILTERS,
UNA, FISH, OLIGOCHAETES, MIDGES,
TRY, BIOMASS, BENTHOS, SALINITY,
, FISH, WATER ANALYSIS, INSECTS,
, CHLORELLA, PEPTIDES, BACTERIA,
IVER, RED RIVER, SAVANNAH RIVER,
S, FERTILIZERS, WATER POLLUTION,
URCES, SURFACE RUNOFF, RAINFALL,
CONOMIC EFFICIENCY, HEAT BUDGET,
  *MOUNTAIN MEADOW, NITROGEN-15,
ROL, BIOCONTROL, COPPER SULFATE,
ICAL CONTROLS, PARAQUAT, DIQUAT,
TROPHOTOMETRY, ALGAE, MINNESOTA,
ING, COPPER SULPHATE, PONDWEEDS,
UF-21, BENOCLORS, NIGROSINE DYE,
NE PLANTS, *ALGAE, CHLORINATION,
TES, LABORATORY TESTS, CULTURES,
BRACKISH WATER, ALKALINE WATER,
EA WATER, FRESH WATER, SORPTION,
ALKALINITY, SULFATES, CHLORIDES,
, CALCIUM, MAGNESIUM, POTASSIUM,
MES, SAMPLING, BOTTOM SEDIMENTS,
NUTRIENTS, NITROGEN, PHOSPHORUS,
, PILOT PLANTS, IRON, CHLORELLA,
PHOSPHORUS, CALCIUM, MAGNESIUM,
N, PRODUCTIVITY, TRACE ELEMENTS,
TED WATER, NITRATES, PHOSPHATES,
TARIO, *CANADA, NUTRIENT LEVELS,
MOUNTAINS, *SOIL MICROORGANISMS,
Y, *LEACHING, IRON, TEMPERATURE,
URE, PIPTIDES, INTERTIDAL AREAS,
OPLANKTON, PRIMARY PRODUCTIVITY,
NG, *POLLUTION ABATEMENT, DIKES,
CLIMATES, *ABSORPTION, *CALCIUM,
AIR POLLUTION, WATER POLLUTION,
LFATE, CHLOROPHYTA, CRUSTACEANS,
ND WATER, RUNOFF, FROZEN GROUND,
OFF, DRAINAGE WATER, GRASSLANDS,
, PHOSPHORUS, *NUTRIENT REMOVAL,
RCES DEVELOPMENT, WATER QUALITY,
ICROORGANISMS, SOIL ALGAE, TURF,
NOPHYTA, CHLOROPHYTA, SIEROZEMS,
MANURES, LAKE MINNETONKA(MINN),
S, PLANT RESIDUES, MORRIS(MINN),
*NITRIFICATION, CROPS, TOXICITY,
OBIOLOGY, DISTRIBUTION PATTERNS,
NAGEMENT(APPLIED), GEOCHEMISTRY,
ALGAE, *FUNGI, *CALCAREOUS SOIL,
I, *CALCAREOUS SOIL, SOIL WATER,
*RADIOSTRONTIUM COMPLEXES, DTPA,
, DRAGONFLIES, ALGAE, SEDIMENTS,
BATCH CULTURES, OPTICAL DENSITY,
PHYTOPLANKTON, DIATOMS, ECOLOGY,
PES, *ALGAE, *FUNGI, *CALCAREOUS
LORIDES, SODIUM, IRON, DISSOLVED
BITION, LIMNOLOGY, ACIDITY, ACID
*PESTICIDE REMOVAL, *ADSORPTION,
TANKS, EFFLUENTS, SWAMPS, FOREST
IMALS, PLANTS, BOTTOM SEDIMENTS,
UORESCENCE, FABRY-PEROT FILTERS,
ITY, OXYGEN-REDUCTION POTENTIAL,
BOTTOM SEDIMENTS, WEED CONTROL,
ERIA, TRIPTON, BOTTOM SEDIMENTS,
*FLUCTUATION, SEWAGE TREATMENT,
XUDATION, TIDES, PHOTOSYNTHESIS,
ATION, SEWAGE EFFLUENTS, OXYGEN,
YTOPLANKTON, ALGAE, TEMPERATURE,
RNAL OXYGEN CURVES, CALORIMETRY,
URCES, ODOR, FARM WASTE, SLUDGE,
STES, INDUSTRIAL WASTES, TOXINS,
TIAL, BACTERIA, PROTOZOA, ALGAE,
T, SEWAGE, BIOLOGICAL TREATMENT,
VOLITILE SOLIDS, NON-DEGRADABLE
L OXYGEN DEMAND, SUSPENDED LOAD,

, OXYGEN TRANSFER, PH, SUSPENDED
CE, COENOBIA, SWIRLING, VOLATILE
SH TOXICITY TEST, SLUDGE DRYING,
TER OVE/ BACKWASHING, *SUSPENDED
:   MIXED LIQUOR, SUSPENDED
TECHNIQUES, TURBIDITY, SUSPENDED

SMELTS, FISHKILL, CRUSTACEANS, BO   W69-08674
SMITH LAKE(ALASKA).:   W69-03515
SMITH'S BAY(MINN), PHOSPHATE DETE   W70-05269
SMOLA(NORWAY), HITRA(NORWAY), FRO   W70-01074
SNAILS, AGE, TURBIDITY, DIPTERA,   W71-00668
SNAILS, APPLICATION METHODS.: /NU   W70-03310
SNAILS, ECOLOGY, EUTROPHICATION,   W70-00711
SNAILS, LEGISLATION, WISCONSIN, C   W70-05568
SNAILS, MITES, MOLDS, GROWTH RATE   W72-01819
SNAILS, ODOR, COOLING TOWERS, FIL   W72-00027
SNAILS, PHOSPHORUS, NUTRIENTS, AL   W70-01943
SNAILS, SULFATES, NITRATES, PHOSP   W70-03978
SNAILS, TOXICITY, RIVERS, HISTOGR   W72-01801
SNAILS, WEIGHT, VACUUM DRYING, BO   W70-04184
SNAKE RIVER, RIO GRANDE RIVER, PO   W70-04507
SNOW, RUNOFF, DRAINAGE WATER, GRA   W70-04193
SNOWMELT, SURFACE WATERS, STREAMS   W70-03501
SOCIAL ASPECTS.: /ON, TURBINES, E   W70-08834
SOD MATS, NON-LEGUMES, *NITROGEN   W72-08110
SODIUM ARSENITE,: /ECHANICAL CONT   W70-10175
SODIUM ARSENITE, COPPER SULPHATE,   W70-00269
SODIUM ARSENITE.: /OMPOUNDS, SPEC   W70-05269
SODIUM ARSENITE, FISH, ANIMALS, A   W70-05263
SODIUM CHLORATE, SANTOBRITE, CLAD   W70-05263
SODIUM COMPOUNDS, PHOTOSYNTHESIS,   W70-00161
SODIUM COMPOUNDS, NITROGEN.: / RA   W70-03507
SODIUM.: /IENTS, BACTERIA, ALGAE,   W71-03031
SODIUM, ALGAE, PHOSPHORUS, TRACE   W70-04385
SODIUM, IRON, DISSOLVED SOILS, P   W69-05894
SODIUM, IRON, MANGANESE, TRACE EL   W70-02804
SODIUM, POTASSIUM, IRON, MANGANES   W70-09889
SODIUM, POTASSIUM, TRACE METALS,   W70-04503
SODIUM, POTASSIUM, SILICATES, DIS   W70-03512
SODIUM, SILICA, DIATOMS, RESISTAN   W68-00856
SODIUM, VIRUSES, VITAMINS, WATER   W69-04801
SODIUM, WATER POLLUTION EFFECTS,   W70-07257
SOFT WATER.: /E LAKE(CANADA), *ON   W70-02973
SOIL ALGAE, TURF, SOIL MICROBIOLO   W72-08110
SOIL ANALYSIS, TASTE-PRODUCING AL   W72-11906
SOIL BACTERIA, LABORATORY TESTS.:   W70-04365
SOIL CHEMISTRY, WATER CHEMISTRY,   W69-06273
SOIL CONSE: /TION SOURCES, DREDGI   W68-00172
SOIL CONTAMINATION, *STRONTIUM RA   W69-03491
SOIL CONTAMINATION, NUTRIENTS.: /   W71-06821
SOIL EROSION, SEWAGE EFFLUENTS, F   W70-02803
SOIL EROSION, WATER POLLUTION, *D   W69-05323
SOIL EROSION,: /LUTION, SNOW, RUN   W70-04193
SOIL EROSION, ALGAE, *WATER POLLU   W73-01122
SOIL INVESTIGATIONS, *IMPOUNDED W   W72-11906
SOIL MICROBIOLOGY, WATER POLLUTIO   W72-08110
SOIL MICROBIOLOGY, DISTRIBUTION P   W71-04067
SOIL MINERALS, PLANT RESIDUES, MO   W70-04193
SOIL PERCOLATION.: / SOIL MINERAL   W70-04193
SOIL POROSITY, IRRIGATION, GROUND   W69-05323
SOIL PROFILES.: /OZEMS, SOIL MICR   W71-04067
SOIL TYPES, HYDROLOGY, CLIMATES,   W71-03048
SOIL WATER, SOIL WATER MOVEMENT,   W71-04067
SOIL WATER MOVEMENT, LEACHING, AR   W71-04067
SOIL WATER PROFILE.: / ACTIVITY,   W71-04067
SOIL-WATER-PLANT RELATIONSHIPS, N   W71-09005
SOIL-WATER TECHNIQUES, PHYCOLOGY,   W70-04381
SOIL, CULTURES, CYANOPHYTA, WATER   W69-10180
SOIL, SOIL WATER, SOIL WATER MOVE   W71-04067
SOILDS, PHOSPHATES, NITROGEN, ALG   W69-05894
SOILS, ACIDIC WATER, ACIDS, PHOTO   W70-02792
SOILS, COALS, DISTRIBUTION PATTER   W69-09884
SOILS, MINNESOTA, WISCONSIN, MICH   W70-04193
SOILS, NUCLEAR EXPLOSIONS, COBALT   W69-08269
SOLAR FRAUNHOFER LINES, ATTENUATI   W70-05377
SOLAR RADIATION, ENERGY, CARBON D   W70-05750
SOLAR RADIATION, *WATER POLLUTION   W68-00478
SOLAR RADIATION, OXYG: /ICE, BACT   W69-10154
SOLAR RADIATION, CHLOROPHYLL, PRO   W69-10182
SOLAR RADIATION, SALINITY, RAINFA   W70-01073
SOLAR RADIATION, PERIPHYTON, BACT   W70-01933
SOLAR RADIATION.: /OSYNTHESIS, PH   W71-10788
SOLARIMETRY, YELLOWSTONE NATIONAL   W70-03309
SOLID WASTES, DECOMPOSTING ORGANI   W71-05742
SOLID WASTES, RADIOACTIVE WASTES,   W72-00462
SOLID WASTES, TERTIARY TREATMENT.   W70-01027
SOLID WASTES, MICROBIOLOGY, ACTIV   W68-00248
SOLID.: /RY MANURE, TOTAL SOLIDS,   W71-00936
SOLIDS CONTACT PROCESS, SEWAGE TR   W70-06792
SOLIDS REMOVAL.:   W70-06792
SOLIDS.: /N(MO), LEE'S SUMMIT(MO)   W70-06601
SOLIDS.: /DETERMINATION, ABSORBAN   W70-03334
SOLIDS-SETTLING, ECONOMIC ADVANTA   W70-05819
SOLIDS, *COMBINED SEWERS, STORMWA   W71-10654
SOLIDS, *PHOSPHOROUS UPTAKE RATE.   W71-13335
SOLIDS, ALGAE, PILOT PLANTS, HEAD   W72-08686

POTASSIUM, SILICATES, DISSOLVED   SOLIDS, CALCIUM, PRODUCTIVITY, ST   W70-03512
                *SUSPENDED   SOLIDS, CHASE CITY(VA).:   W72-10573
*WASTES, ALGAE, ASSAY, DISSOLVED   SOLIDS, CHEMICAL ANALYSIS, PHOSPH   W70-03334
TERS, *WATER QUALITY, *DISSOLVED   SOLIDS, DISSOLVED OXYGEN, EVAPORA   W72-01773
                VOLATILE   SOLIDS, GRIT.:   W71-11375
ECTROPHOTOMETRY, TOTAL DISSOLVED   SOLIDS, HEDRIODISCUS, CALOPARYP: /   W70-03309
LOCITY, WATER QUALITY, DISSOLVED   SOLIDS, HYDROGEN ION CONCENTRATIO   W70-00265
URE PROFILES, VOLATILE SUSPENDED   SOLIDS, MULTI-CELL SERIAL SYSTEM,   W70-04790
Y MANURE, TOTAL SOLIDS, VOLITILE   SOLIDS, NON-DEGRADABLE SOLID.: /R   W71-00936
LOIDS, TRACE ELEMENTS, DISSOLVED   SOLIDS, NUTRIENTS, VITAMINS, TOXI   W70-02510
CTERISTICS, ORGANISMS, SUSPENDED   SOLIDS, SALTWATER TONGUE, BENTHIC   W73-00748
ONS, DISSOLVED OXYGEN, DISSOLVED   SOLIDS, SEDIMENTATION RATES, MANG   W73-00750
FLUORESCEIN, MACROPHYTES, GROUND   SOLIDS, UPTAKE, CONSUMER ORGANISM   W69-09334
ADABLE SOL/ *DAIRY MANURE, TOTAL   SOLIDS, VOLITILE SOLIDS, NON-DEGR   W71-00936
DENS, OOCYSTIS PUSILLA, OOCYSTIS   SOLITARIA, PANDORINA MORUM, PEDIA   W69-06540
LITHOSPHERE, SORPTION REACTIONS,   SOLUBILITY EQUILIBRIUM THEORY.: /   W70-04385
ION, *CORROSION, CARBON DIOXIDE,   SOLUBILITY, *PHOSPHATES, CHEMICAL   W71-11393
DISSOLVED INORGANIC PHOSPHATES,   SOLUBLE REACTIVE PHOSPHORUS, PHOS   W70-05269
CROBIAL/ *HETEROTROPHY, *ORGANIC   SOLUTES, *MARINE ENVIRONMENTS, MI   W70-04811
ECOLOGY, SWEDEN, LAKE/ *ORGANIC   SOLUTES, *HETEROTROPHY, MICROBIAL   W70-04284
SOLVED OXYGEN, ALGAE, NUTRIENTS,   SOLUTES, CHLORIDES, CALCIUM, CARB   W70-09557
ER RATES, *HETEROTROPHY, ORGANIC   SOLUTES, SWEDEN, LAKE ERKEN(SWEDE   W70-05109
OLLUTION IDENTIFICATION, ORGANIC   SOLUTES, VITAMIN B-1, VITAMIN B-1   W70-05652
ETHOD, FINGER LAKES(NY), RODHE'S   SOLUTION VIII, NUTRIENT LIMITATIO   W70-01579
(KUTZ), CULTURE MEDIA, CHU NO 10   SOLUTION, MINIMUM NUTRIENTS.: /SA   W70-03507
AYS, ORGANOPHOSPHORUS COMPOUNDS,   SOLVENT EXTRACTIONS, PHOSPHORUS C   W70-05269
SPILLS, *TORREY CANYON, AROMATIC   SOLVENTS, PHAEODACTYLUM TRICORNUT   W70-01470
S, OPEPHORA MARTYI, ABBOT'S POND(SOMERSET, ENGLAND), SAND GRAINS,   W71-12870
ATION, BATCH CULTURES, CHLORELLA   SOROKINIANA, COMPOSITION OF ALGAE   W70-07957
*EQUILIBRIUM MODEL, LITHOSPHERE,   SORPTION REACTIONS, SOLUBILITY EQ   W70-04385
               PESTICIDE   SORPTION.:   W69-09884
CYANOPHYTA, CHLOROPHYTA, *ALGAE,   SORPTION, WASTE WATER TREATMENT,   W70-09438
, DISTILLATION, ELECTRODIALYSIS,   SORPTION, CHEMICAL PRECIPITATION,   W70-01981
OSPHERE, SEA WATER, FRESH WATER,   SORPTION, SODIUM, ALGAE, PHOSPHOR   W70-04385
TROPHICATION, ALGAE, FILTRATION,   SORPTION, FLOCCULATION, COAL, ACT   W70-05470
OACTIVITY EFFECTS, RADIOECOLOGY,   SORPTION, WATER POLLUTION EFFECTS   W69-07440
PERIDINIUM TROCHOIDEUM, VINEYARD   SOUND(MASS), WOODS HOLE(MASS). RA   W70-05272
, GONIDIA FORMATION, LONG ISLAND   SOUND(NY), ICELAND, TRITIATED THY   W69-09755
AENOPHYLLA, TIDAL CURRENT, PUGET   SOUND(WASH), LONG ISLAND SOUND, N   W69-10161
UNITIES, INTERTIDAL ALGAE, PUGET   SOUND, FLOURESCEIN DYE.: /AL COMM   W68-00010
, PUGET SOUND(WASH), LONG ISLAND   SOUND, NARRAGANSETT BAY, CAPE REY   W69-10161
N(AUSTRALIA), OXIDATION DITCHES,   SOUP, FOOD WASTES.:   SHEPPARTIO   W71-09548
S, MARINE ALGAE, WATER POLLUTION   SOURCES.: /, URANIUM RADIOISOTOPE   W71-09013
NUCLEAR WASTES, WATER POLLUTION   SOURCES.: / ALGAE, INVERTEBRATES,   W71-09863
LLUTION EFFECTS, WATER POLLUTION   SOURCES.: /RBON DIOXIDE, WATER PO   W69-10158
VELS, SEDIMENTS, WATER POLLUTION   SOURCES.: /MUD, SILTS, SANDS, GRA   W70-00711
LLUTION EFFECTS, WATER POLLUTION   SOURCES.: /TION CONTROL, WATER PO   W70-02792
VARIATIONS, SEASONAL VARIATIONS,   SOURCES.: /, ME, MAINE, PERIODIC   W68-00481
ICAL TECHNIQUES, WATER POLLUTION   SOURCES.: /CHROMATOGRAPHY, ANALYT   W69-00387
LLUTION EFFECTS, WATER POLLUTION   SOURCES.: /GY, SORPTION, WATER PO   W69-07440
LLUTION EFFECTS, WATER POLLUTION   SOURCES.: /ER CHEMISTRY, WATER PO   W69-06273
               *NUTRIENT   SOURCES.:   W70-04266
MUS, PHOSPHATES, WATER POLLUTION   SOURCES.: /E, *BIOASSAY, SCENEDES   W70-03955
IL MICROBIOLOGY, WATER POLLUTION   SOURCES.: /, SOIL ALGAE, TURF, SO   W72-08110
, *WATER SOURC/ *WATER POLLUTION   SOURCES, *WATER POLLUTION EFFECTS   W73-00758
*WATER POLLUTION EFFECTS, *WATER   SOURCES, *GROUNDWATER, DDT, MISSI   W73-00758
EUTROPHICATION, *WATER POLLUTION   SOURCES, *WATER POLLUTION EFFECTS   W70-03975
, *EUTROPHICAT/ *WATER POLLUTION   SOURCES, *WATER POLLUTION EFFECTS   W70-04266
EUTROPHICATION, *WATER POLLUTION   SOURCES, *WATER POLLUTION EFFECTS   W70-03975
EUTROPHICATION, *WATER POLLUTION   SOURCES, *LAKES, NUTRIENTS, FERTI   W69-07084
*PHYTOPLANKTON, *WATER POLLUTION   SOURCES, *WATER POLLUTION EFFECTS   W69-02959
ON, *NUTRIENTS, *WATER POLLUTION   SOURCES, *CYCLING NUTRIENTS, ALGA   W69-04800
PAL AND INDUSTRIAL), *PHOSPHORUS   SOURCES, *PHOSPHATES, ALGAE, WATE   W70-01031
, *CLASSIFICAT/ *WATER POLLUTION   SOURCES, *NUISANCE PLANT GROWTHS,   W69-09340
AMINO ACIDS, P/ *WATER POLLUTION   SOURCES, *WATER POLLUTION EFFECTS   W72-00462
S, *SURFACTANTS, WATER POLLUTION   SOURCES, *ALGAE, *TASTE, *ODOR, *   W72-01812
LLUTION EFFECTS, WATER POLLUTION   SOURCES, *DETERGENTS, *DEGRADATIO   W70-09438
LUTION CONTROL, *WATER POLLUTION   SOURCES, *THERMAL POLLUTION, THER   W70-07847
WATER POLLUTION, WATER POLLUTION   SOURCES, *WATER POLLUTION EFFECTS   W71-03048
EUTROPHICATION, *WATER POLLUTION   SOURCES, ALGAL CONTROL, BIOINDICA   W70-02304
ORUS COMPOUNDS, *WATER POLLUTION   SOURCES, AGRICULTURAL WATERSHEDS.   W69-05323
LLUTION EFFECTS, WATER POLLUTION   SOURCES, ALGAE, ALGAL CONTROL, AL   W69-03370
LLUTION EFFECTS, WATER POLLUTION   SOURCES, AQUATIC PRODUCTIVITY, LI   W69-09135
LLUTION EFFECTS, WATER POLLUTION   SOURCES, BIOINDICATORS, NITROGEN   W69-06864
CONTROL, LAKES, WATER POLLUTION   SOURCES, CYCLING NUTRIENTS, COST   W69-08518
LAND MANAGEMENT, WATER POLLUTION   SOURCES, CHLOROPHYTA, NUISANCE AL   W70-03310
MICAL COMPOSITION(ENGLAND), FOOD   SOURCES, DREDGING, *POLLUTION ABA   W68-00172
ATER POLLUTION   SOURCES, GREENHOUSE STUDIES, GROW   W69-06865
ATER(POLLUTION), WATER POLLUTION   SOURCES, IRRIGATION, AQUATIC ALGA   W69-07096
LLUTION EFFECTS, WATER POLLUTION   SOURCES, MUNICIPAL WASTES, GREAT   W71-05879
UTRIENTS, ALGAE, WATER POLLUTION   SOURCES, NUTRIENTS.: /S, WATER PO   W69-10159
ALGAE, BACTERIA, WATER POLLUTION   SOURCES, NITROGEN COMPOUNDS, PHOS   W69-01445
, *MARINE ALGAE, WATER POLLUTION   SOURCES, ODOR, FARM WASTE, SLUDGE   W71-05742
OBIC CONDITIONS, WATER POLLUTION   SOURCES, ON-SITE DATA COLLECTIONS   W70-09976
NUCLEAR WASTES, WATER POLLUTION   SOURCES, PATH OF POLLUTANT.: /AER   W71-11381
WATER CHEMISTRY, WATER POLLUTION   SOURCES, PATH OF POLLUTANTS, AQUA   W69-10080
EWAGE TREATMENT, WATER POLLUTION   SOURCES, PHOSPHORUS, NITROGEN COM   W70-03068
TRAPS, NUTRIENT INFLUX, NUTRIENT   SOURCES, PRIMARY TREATMENT, SECON   W69-01453
LLUTION EFFECTS, WATER POLLUTION   SOURCES, PHOSPHORUS BUDGET, APHAN   W70-04268
    SOURCES, RADIOECOLOGY.: /WATER PO   W70-00707

US GREEN ALGAE, / *RECOVERY, OIL
ARKSBURG, BUFFALO, DUNREITH, OIL
SOLV/ *CRUDE OIL, *GAMOSOL, *OIL
MINE WT DYE, AIRBORNE TESTS, OIL
*OIL
ED RIVER(VIETNAM), CHLOROCOCCUS,
CHERICHIA COLI, OOCYSTIS, LEMNA,
TES, ANKISTRODESMUS, GOMPHENEMA,
LOSIRA, SYNEDRA, COELOSPHAERIUM,
CRISPUS, MICROSPORA WITTROCKII,
IS, COELOSPHAERIUM/ *PHYSIOLOGY,
LYNGBYA, FRAGILARIA, OEDOGONIUM,
Y, DINOBRYON, VOLVOX, PANDORINA,
ROPHYTES, CLADOPHORA, OEDOGONIA,
WIS), LAKE WINGRA(WIS), ANABAENA
ON LAKE(VA).:   ASTERIONELLA
ANABAENA, WOLLEA, OSCILLATORIA,
ZIMONA(HUNGARY), ALGAL TAXONOMY,
CTS, LETHAL LIMIT, GROWTH RATES,
TION, BACTERIOPHAGE, RESERVOIRS,
RES, ZYMOLOGICAL, ELBE(GERMANY),
LOGY, LIGHT PENETRATION, OXYGEN,
ENCE, LAKES, CHEMICALS, FISHING,
BIDITY, INDUSTRIAL PLANTS, WATER
, *POLLUTANTS, RECREATION, WATER
ONCENTRATION FACTOR, LIMNODRILUS
TOBRITE, CLADOPHORA, POTAMOGETON
R, DOLORES RIVER, URAVAN, COTTUS
OGEN UTILIZATION, ROLLER DRYING,
SEY(ONTARIO), ALGAE COUNTS/ *AIR
STRY, *ALGAE, *CATTLE, SLURRIES,
HLORINATION, SLUDGE, DETERGENTS,
SCONSIN, ALGAE, *NUISANCE ALGAE,
, *INVESTIGATIONS, ZOO-PLANKTON,
A), FLORA.:
ERMAL SPRINGS, *LIGHT INTENSITY,
KGROUND RADIATION, GEOCHEMISTRY,
), STORKWEATHER CREEK(WIS), NINE
R NATIONAL PARK, OHANAPECOSH HOT
PHOTOSYNTHESIS, *ALGAE, *THERMAL
C/ *CYANOPHYTA, *ALGAE, *THERMAL
STRY, ECOLOGY, SAMPLING, THERMAL
, AC/ *ALGAE, *TEMPERATURE, *HOT
¬, MOOSE CREEK, ARCTIC CIRCLE HOT
TONE NATIONAL PARK, *ICELAND HOT
ATS, *ALASKA, LAKES, RIVERS, HOT
TURE, THERMAL POLLUTION, THERMAL
NE TESTS, OIL SPILLS, GEOLOGICAL
R, RAISIN RIVER, DETROIT R/ LAKE
INDUSTRIAL WASTES, AGRICULTURE,
KE MINNETONKA(MINN), MINNEAPOLIS-
ND EFFLUENT, COLORADO, WEST(US),
OLOGY, *DESIGN THEORY AN/ *WASTE
PHOTOSYNTHESIS, LIGHT INTENSITY,
SPHORUS, NITROGEN, GROWTH RATES,
TREATMENT, SECONDARY TREATMENT,
ONATES, WATER POLLUTION EFFECTS,
OONS, *DESIGN, *SULFUR BACTERIA,
ROSCOPY, SULFUR COMPOUNDS, WASTE
DECOMPOSITION), DDT, ALKALINITY,
ALGAE, CADDISFLIES, FOOD CHAINS,
RESISTANCE, LIFE CYCLES, GROWTH
DIES, HABITATS, CRAYFISH, GROWTH
VULGARIS, DAPHNIA MAGNA, LIMNAEA
VULGARIS, DAPHNIA MAGNA, LIMNAEA
TERN LAKE ERIE(OHIO), GONGROSIRA
IPORE FILTERS, PHASE MICROSCOPY,
UFACTURING WASTES, WATER QUALITY
L, DYNAMICS, TEMPERATURE, SILTS,
, LAKES, ORGANIC WASTES, OXYGEN,
THERMOCLINE, LIGHT PENETRATION,
ITY, HYDROGEN ION CONCENTRATION,
D SOLIDS, CALCIUM, PRODUCTIVITY,
TY, WASTE ASSIMILATIVE CAPACITY,
TIMATING EQUATIONS, OXYGENATION,
HATES, PHOSPHORUS, PRODUCTIVITY,
R(CALIF), TUOLUMNE RIVER(CALIF),
ULGARIS, PSEUDOMONAS AERUGINOSA,
, BACILLUS THURINGIENSIS, AGARS,
AVOBACTER, ESCHERICHIA, SARCINA,
ES, CHEESE MANUFACTURING WASTES,
              SPRING CREEK(PA),
ING, GRANTS, FEDERAL GOVERNMENTS,
ION CONTROL, *LOCAL GOVERNMENTS,
LOT RESERVOIR, LEBANON, WADLEIGH
D, SYRACUSE(NY),/ *PHYCOLOGY, NY
REAMS,/ BERRY CREEK(ORE), OREGON
ELED SCABLANDS(WASH), WASHINGTON
              SOUTH – CENTRAL
CHRYSOPHYTA, CALIFORNIA, UNITED

SPILL, *OIL POLLUTION, *FILAMENTO   W70-09976
SPILLS.: /S, LEBANON, BATAVIA, CL   W72-03641
SPILLS, *TORREY CANYON, AROMATIC   W70-01470
SPILLS, GEOLOGICAL SS POLARIS, CH   W70-05377
SPILLS, PROTEIN AQUICULTURE.:   W69-05891
SPINES, SCHIZOMYCOPHYTA, CAULOBAC   W70-03969
SPIRODELA, ELODEA.: /LOCOCCUS, ES   W70-10161
SPIROGYRA.: /CYCLOTELLA, FLAGELLA   W70-02249
SPIROGYRA, CLADOPHORA, DIAPTOMUS,   W70-02803
SPIROGYRA, XANTHOPHYCEAE, MYXOPHY   W70-00711
SPIROGYRA, MYXOPHYCEAE, MICROCYST   W70-04369
SPIROGYRA, CHAETOPHORA, STIGEOCLO   W70-04371
SPIROGYRA, SYNECHOCOCCUS, CHROOCO   W70-04503
SPIROGYRA, MICROCYSTIS AERUGINOSA   W70-05404
SPIROIDES, MICROCYSTIS AERUGINOSA   W70-05112
SPIROIDES, CLADOCERANS, WESTHAMPT   W71-10098
SPIRULINA, VOLVOCALES, GLENODINIU   W70-05404
SPIRULINA, CHODATE: /A(HUNGARY),   W70-03969
SPORES.: /E, WATER POLLUTION EFFE   W70-01779
SPORES, NITRITES, /SELF-PURIFICA   W69-07838
SPOROBOLOMYCES ROSEUS, AUTOLYSIS,   W70-04368
SPORT FISHING, ON-SITE INVESTIGAT   W68-00487
SPORT FISHING, RECREATION, SILT,   W68-00172
SPORTS.: /LAKE ONTARIO, WATER TUR   W71-09880
SPORTS, ENVIRONMENT, AQUATIC ENVI   W69-04276
SPP, AEOLOSOMA HEMPRICHI, ESCHERI   W69-07861
SPP, LAKE MENDOTA(WIS), ANACHARIS   W70-05263
SPP, RHINICHTHYS OSCULUS, SUCKERS   W69-07846
SPRAY-DRIED, GLASS BEADS, TRICHOD   W70-04184
SPRAY, SUDBURY(ONTARIO), LAKE RAM   W69-10155
SPRAYING, ACTIVATED SLUDGE, FARM   W70-07491
SPRAYING, ALGAE, CENTRIFUGATION,   W70-04060
SPRAYING, SEASONAL, *ALGAL CONTRO   W68-00488
SPRAYING, BIOCONTROL, INFECTION,   W68-00468
SPRING CREEK(PA), STATE COLLEGE(P   W69-10159
SPRING WATER, WATER CHEMISTRY, WA   W69-09676
SPRING WATERS, PHOSPHORUS, ORGANO   W69-09334
SPRINGS CREEK(WIS), DOOR CREEK(WI   W70-05113
SPRINGS(WASH), SCHIZOTHRIX, PHORM   W70-03309
SPRINGS, *LIGHT INTENSITY, SPRING   W69-09676
SPRINGS, *OREGON, THERMOPHILIC BA   W70-05270
SPRINGS, EFFLUENTS, CYANOPHYTA.: /   W69-10151
SPRINGS, HABITATS, ALKALINE WATER   W69-10160
SPRINGS, RADON, CENTRAL CITY(COLO   W69-08272
SPRINGS, SINTER, NUCLEIC ACID, SI   W69-10160
SPRINGS, TERRESTRIAL HABITATS, DE   W69-08272
SPRINGS, THERMAL STRES: / TEMPERA   W69-03516
SS POLARIS, CHLOROPHYLL FLUORESCE   W70-05377
ST CLAIR, ROUGE RIVER, HURON RIVE   W69-01445
ST LAWRENCE RIVER, ALGAE, COLIFOR   W70-03964
ST PAUL(MINN).:   LA   W70-04810
STABILIZATION POND LOADINGS, ODOR   W70-04788
STABILIZATION PONDS, *ALGAL PHYSI   W70-04786
STABILIZATION, PONDS, PENETRATION   W70-03923
STABILIZATION PONDS, SEWAGE DISPO   W68-00855
STABILIZATION PONDS, SEWAGE LAGOO   W68-00012
STABILIZATION, LAGOONS, TOXICITY,   W69-09454
STABILIZATION, WASTE WATER, ALGAE   W70-01971
STABILIZATION PONDS.: /ARED SPECT   W70-07034
STABILIZATION, WASTE STORAGE, PON   W70-08097
STABLE ISOTOPES, RADIOACTIVE WAST   W70-02786
STAGES.: /NT, BAYS, *FLUCTUATION,   W71-02494
STAGES, DIATOMS, ALGAE, METABOLIS   W69-09742
STAGNALIS, OUSPENSKII'S MEDIUM.: /   W71-03056
STAGNALIS, OUSPENSKII'S MEDIUM.: /   W71-05719
STAGNALIS, NEPHROCYTIUM OBESUM W   W70-04468
STAINING, STRIPPING FILM, SYNECHO   W69-10163
STANDARDS, GRAFTON(ND), LAKOTA(ND   W70-03312
STANDING CROP, ECOLOGY, PHOTOSYNT   W70-02780
STANDING CROP, ALGAE, ZOOPLANKTON   W70-02251
STANDING CROP, INDUSTRIAL WASTE: /   W71-07698
STANDING CROP, LIGHT, BIOMASS, RE   W71-00665
STANDING CROP, CHLOROPHYLL, PHOSP   W70-03512
STANDING CROP, ANNUAL TURNOVER, F   W68-00912
STANDING CROP, MICROENVIRONMENT,   W69-03611
STANDING CROP, STREAMS, WATER CHE   W69-04805
STANISLAUS RIVER(CALIF), WATER PO   W70-02777
STAPHYLOCOCCUS AUREUS, LACTOBACIL   W71-11823
STAPHYLOCOCCUS EPIDERMIDIS, MICRO   W71-11823
STAPHYLOCOCCUS, ACHROMOBACTER(ALK   W70-03952
STARCH MANUFACTURING WASTES, WATE   W70-03312
STATE COLLEGE(PA), FLORA.:   W69-10159
STATE GOVERNMENTS, WATER QUALITY: W69-06305
STATE GOVERNMENTS, BIOCHEMICAL OX   W69-06909
STATE PARK, LAKE SHATUTAKEE, MASC   W69-08674
STATE SCIENCE AND TECHNOLOGY FOUN   W70-03973
STATE UNIVERSITY, EXPERIMENTAL ST   W70-03978
STATE UNIVERSITY.: /(WASH), CHANN   W70-00264
STATES.:   W72-03562
STATES, BRACKISH WATER, ESTUARIES   W70-05372

TON, CHLOROPHYLL, CYCLES, UNITED   STATES, CHLORELLA, BIOCHEMICAL OX   W70-05092
CHLOROPHYTA, CYANOPHYTA, UNITED   STATES, HABITATS.: /, *LAKE ERIE,   W70-04468
ATION, LAKES, *NUTRIENTS, UNITED   STATES, PHOSPHORUS, NITROGEN, SEW   W68-00478
ER, ORGANIC COMPOUNDS, HABITATS,   STATISTICAL METHODS.: /TS, SEAWAT   W69-09755
SLOVAKIA, LABORATORY PROCEDURES,   STATISTICAL PROCEDURES, BMT.: /HO   W72-12240
LIMNION, GREAT LAKES, LAKE ERIE,   STATISTICS, GROWTH RATES, ALGAE,   W68-00253
C, SILICON, PEDIASTRUM BORYANUM,   STAURASTRUM PARADOXUM, BOTRYOCCUS   W70-02804
YSTIS, GLOEOTRICHIA, GLOEOCAPSA,   STAURONEIS.: /LIS, ANABAENA, ANAC   W70-03974
E, ZOOPLANKTON, FISH, SHELLFISH,   STEAM, ELECTRIC POWER, HEATED WAT   W70-09905
ION, BACILLUS SUBTILIS, BACILLUS   STEAROTHERMOPHILUS, PROTEUS VULGA   W71-11823
C WEED CUTTING, *WEED EXTRACTING   STEEL CABLES, ALGAL SUCCESSION, A   W70-04494
MINN), APHANIZOMENON FLOS AQUAE,   STEPHANODISCUS ASTREA, LAKE MINNE   W70-05387
OTELLA, TRACHELOMONAS, MELOSIRA,   STEPHANODISCUS, ANABAENA, APHANIZ   W72-04761
OELASTRUM, PEDIASTRUM, WESTELLA,   STEPHANODISCUS, ANKISTRODESMUS, C   W70-08107
GING, STEPHANODISCUS HANTZSCHII,   STEPHANODISCUS BINDERANUS, CLADOC   W70-00263
OCERA, APHANIZOMENON FLOS-AQUAE,   STEPHANODISCUS HANTZSCHII, DIDINI   W70-00268
, ASTERIONELLA, FILTER CLOGGING,   STEPHANODISCUS HANTZSCHII, STEPHA   W70-00263
N, *CARBON FIXATION, *EXCRETION,   STEPHANODISCUS TENUIS, ASTERIONEL   W69-10158
GONIUM, SCENEDESMUS, CYCLOTELLA,   STEPHANODSICUS.: /CINCLIS, CHLORO   W69-02959
S PHOSPHATES, ULTRA-VIOLET LIGHT   STERILIZATION, ASTERIONELLA, SCEN   W70-03955
BACILLUS STEAROTH/ RICKETTSIAE,   STERILIZATION, BACILLUS SUBTILIS,   W71-11823
ICUS, PHAEODACTYLUM TRICORNUTUM,   STICHOCHRYSIS IMMOBILIS, VIBRIO,   W70-04280
OGONIUM, SPIROGYRA, CHAETOPHORA,   STIGEOCLONIUM, ANKISTRODESMUS, AP   W70-04371
PARNALDIA, DRAPARNALDIA PLUMOSA,   STIGEOCLONIUM, MICROCYSTIS, STIGE   W70-02245
OSA, STIGEOCLONIUM, MICROCYSTIS,   STIGEOCLONIUM TENUE, MICROCYSTIS   W70-02245
CULTURE MEDIA.:   STIGLEOCLONIUS, INSOLUBLE WASTES,   W68-00248
R COMPOUNDS, ANTIBIOTIC EFFECTS,   STIMULANTS, GONYAULAX CATENELLA,   W70-02510
KE BUTTE DES MORTS(WISC), GROWTH   STIMULATION, SURFACE LITTER.: /LA   W72-00845
OURCES, ECOLOGICAL DISTRIBUTION,   STOCHASTIC PROCESSES, ECOSYSTEMS,   W71-09012
IS, *LAKE ERIE, PHOSPHORUS, FISH   STOCKING, COMMERCIAL FISH, SALMON   W70-04465
LIMNOLOGY, LAKE TROUT, CRAYFISH,   STONEFLIES, SNAILS, ECOLOGY, EUTR   W70-00711
SUSPENDED LOAD, TURBIDITY, DATA   STORAGE AND RETRIEVAL, COLIFORMS,   W70-09557
EATMENT, CHEMICAL OXYGEN DEMAND,   STORAGE CAPACITY, EFFLUENTS, CONS   W71-00936
SOTOPES, *WASTE DISPOSAL, *WASTE   STORAGE, *WATER POLLUTION, WATER   W72-00941
DUCTIVITY, *TROPHIC LEVEL, WATER   STORAGE, EVAPORATION, WATER CIRCU   W69-09723
ALKALINITY, STABILIZATION, WASTE   STORAGE, PONDS, HYDROLYSIS, METAB   W70-08097
ING BACTERIA, *ACTIVATED SLUDGE,   STORAGE, TEMPERATURE, MICROORGANI   W71-13335
ADISON LAKES, YAHARA RIVER(WIS),   STORKWEATHER CREEK(WIS), NINE SPR   W70-05113
ON, ZOOPLANKTON, EUTROPHICATION,   STORM DRAINS, SEWAGE EFFLUENTS, S   W70-05112
TER TREATMENT, OVERFLOW, SEWERS,   STORM DRAINS, SEWAGE EFFLUENTS, *   W71-10654
WAGE TREATMENT, SEWAGE DISPOSAL,   STORM DRAINS, SURFACE DRAINAGE, A   W70-03344
HANIZOMENON FLOS-AQUAE, LYNGBYA,   STORM LAKE(IOWA), CENTER LAKE(IOW   W70-02803
ANALYSIS, SLUDGE, INCINERATION,   STORM RUN-OFF, SEDIMENTATION, BIO   W71-10654
GE EFFLUENTS, *SEWAGE TREATMENT,   STORM RUNOFF.: /TORM DRAINS, SEWA   W71-10654
, *WASTE WATER, *ALGAE, *SLUDGE,   STORMS, TURBULENCE, PILOT PLANTS,   W70-05819
PENDED SOLIDS, *COMBINED SEWERS,   STORMWATER OVERFLOWS.: /ING, *SUS   W71-10654
SHORELINE VEGETATION,   STORMWATER, DRAINAGE.:   W70-02251
H, CHLORELLA PYRENOIDOSA(EMERSON   STRAIN), AUTOINHIBITORS.: / GROWT   W69-07442
TURE, NUTRIENTS, COPPER SULFATE,   STRATIFICATION.: /, LIGHT TEMPERA   W69-08282
UCTIVITY, RIVERS, TASTE, THERMAL   STRATIFICATION, WASTE TREATMENT,   W69-06535
TRATES, DAPHNIA, SILICA, THERMAL   STRATIFICATION, SULFIDES, AMMONIU   W70-05760
*PONDS, *STABILIZATION, *THERMAL   STRATIFICATION, *SEASONAL, *SULFI   W70-04790
DES, TEMPERATURE, CLIMATIC DATA,   STRATIFICATION, DISSOLVED OXYGEN,   W70-04790
TOPLANKTON, CYANOPHYTA, DIATOMS,   STRATIFICATION, OXYGEN, HYDROGEN   W70-05387
LI/ *EUTROPHICATION, TEMPERATURE   STRATIFICATION, DEPTH, PLANKTON,   W70-05405
BIOLOGICAL PROPERTIES, ECOLOGY,   STRATIFICATION.: /TER PROPERTIES,   W70-04475
NDMENTS, *RESERVOIRS, / *THERMAL   STRATIFICATION, *AERATION, *IMPOU   W70-04484
VATED CARBON, FILTERS, SEASONAL,   STRATIFICATION, BIOCHEMICAL OXYGE   W70-04001
IOASSAYS, AQUATIC WEEDS, THERMAL   STRATIFICATION, EPILIMNION, HYPOL   W70-02251
IR EVAPORATION, SALINITY, ALGAE,   STRATIFICATION, SALT BALANCE, BIB   W72-01773
CAL MODELS, GREAT LAKES, THERMAL   STRATIFICATION.: /DIES, MATHEMATI   W71-05878
INS, ALGAE, NEBRASKA, TURBIDITY,   STRATIFICATION, TEMPERATURE, CHLO   W72-04761
TRATION, NUTRIENTS, ALGAE, FISH,   STRATIFICATION, FLOCCULATION, TUR   W70-08834
CYCLES, OLIGOCHAETES, MOLLUSKS,   STRATIFICATION, DIATOMS, C: /LIFE   W71-00665
IC ECOSYSTEM, MIDGE LARVAE(LAKE   STRATIFICATION, SURFACE AREA(LAKE   W69-09340
OF POLLUTANTS, NUTRIENTS, ALGAE,   STRATIFIED FLOW, MIXING, FLOW, MO   W71-05878
ODUCTIVITY, SEDIMENTATION RATES,   STRATIGRAPHY, BIOINDICATORS, ALGA   W70-05663
ALITY, VIRUSES, ALGICIDES, FISH,   STRE: /TREATMENT, LAKES, WATER QU   W70-03973
EE BARK, CHARCOAL, ALGAL GROWTH,   STREAM CONCOURSE, PALOUSE RIVER(I   W70-03501
ION, COOLING WATER TREATMENT, ON-   STREAM DESLUDGING.: /METAL CORROS   W70-02294
AINAGE SYSTEMS, FLOW RATES, CONN   STREAM FLOW, ALGAE, AQUATIC ALGAE   W68-00479
Y CONTROL, *MI/ FISH MANAGEMENT,   STREAM IMPROVEMENT, *WATER QUALIT   W70-05428
ER QUALITY, ALGAE, AQUATIC LIFE,   STREAM POLLUTION, BACTERIA, IMPUR   W72-01786
ENDED SEDIMENTS, MARINE BIOLOGY,   STREAM POLLUTION, *WATER POLLUTIO   W71-10786
STORM DRAINS, SEWAGE EFFLUENTS,   STREAMFLOW, MEASUREMENT, WATER PO   W70-05112
*POLLUTANT ABATEMENT, EFFLUENTS,   STREAMFLOW, DESALINATION, TERTIAR   W69-07389
NUTRIENTS, *ZINC RADIOISOTOPES,   STREAMFLOW, SAMPLING, ALGAE, ECOS   W69-07862
N CONTROL, LIMNOLOGY, DIVERSION,   STREAMS.: /OURCES, WATER POLLUTIO   W69-10170
CONTROL, *TASTE, *ODOR, *NATURAL   STREAMS, *DISTRIBUTION SYSTEMS, C   W72-03562
IZATION, *EUTROPHICATION, LAKES,   STREAMS, ALGAE, FISH, WISCONSIN,   W69-10178
IRON, MANGANESE, BENTHIC FLORA,   STREAMS, CULTURES, COBALT RADIOIS   W69-07845
AE, DAPHNIA, LAKES, CHLOROPHYTA,   STREAMS, DIATOMS, RESERVOIRS, CUL   W70-02982
POLLUTION CONTROL, *CONFERENCES,   STREAMS, ESTUARIES, LAKES, GREAT   W70-03975
POLLUTION CONTROL, *CONFERENCES,   STREAMS, ESTUARIES, LAKES, GREAT   W70-03975
MARYLAND, PENNSYLVANIA, RIVERS,   STREAMS, EUGLENA, AQUATIC POPULAT   W70-04510
EUTROPHICATION, OLIGOTR/ *LAKES,   STREAMS, FISH, LAND, ECOSYSTEMS,   W70-04193
NFALL, SNOWMELT, SURFACE WATERS,   STREAMS, HYDROGEN ION CONCENTRATI   W70-03501
WASTES, WATER SUPPLY, EFFLUENTS,   STREAMS, LAKES, RIVERS, RESERVOIR   W71-09880
N, WATER POLLUTION EFFECTS, ACID   STREAMS, MINE DRAINAGE, CHLOROPHY   W70-02770
RAPHICAL REGIONS, KENTUCKY, ACID   STREAMS, MICROFLORA.: /HIPS, GEOG   W68-00891

535

ER POLLUTION, THERMAL POLLUTION,
S, LABORATORY TESTS, PERIPHYTON,
H, FUNGI, HEAT, WATER POLLUTION,
N STATE UNIVERSITY, EXPERIMENTAL
ISPOSAL, WATER POLLUTION, LAKES,
US, PRODUCTIVITY, STANDING CROP,
TREATMENT, *WATERS, *ORGANISMS,
WATER, MOVEMENT, CULTURES, BAYS,
N, PHOSPHATES, ALKALINITY, FISH,
TTER, NITROGEN FIXATION, VOLUME,
ED OXYGEN VARIATIONS, SAN JOAQU/
INDUSTRIES, CHELATION, BACTERIA,
COCCUS EPIDERMIDIS, MICROCOCCUS,
UPLII, TERRAMYCIN, TETRACYCLINE,
LUTION, THERMAL SPRINGS, THERMAL
N, THERMAL POWERPLANTS, *THERMAL
LUS, WATER TEMPERATURE, HYDROGE/
ERS, PHASE MICROSCOPY, STAINING,
), SCENEDESMUS QUADRICAUDA, CEL/
ECTROCHEMICAL TREATMENT, AMMONIA
RBON RADIOISOTOPES, FISH, CLAMS,
ANOPHYTA, URANIUM RADIOISOTOPES,
DIOISOTOPES, NUCLEAR EXPLOSIONS,
TEBRATES, CRUSTACEANS, MOLLUSKS,
S, COBALT RADIOISOTOPES, CESIUM,
ERATURE, SOUTH CAROLINA, CESIUM,
S, SCENEDESMUS, PROTEINS, WALLS,
, METOLIUS RIVER(ORE), COMMUNITY
EMISTRY, *EUTROPHICATION, *WATER
CREEK(MD), BACK RIVE/ COMMUNITY
N/ *SELECTIVE TOXICITY, CHEMICAL
G STRUCTURES, INDUSTRIAL PLANTS,
LECTRIC POWERPLANTS, ENGINEERING
ONTIUM RADIOSOTOPES, CYTOLOGICAL
NIC MATTER, DIURNAL, LATITUDINAL
MENT, *OXIDATION LAGOONS, *MODEL
*OXIDATION LAGOONS, MILK, *MODEL
*FAT ACCUMULATION, *LABORATORY
CLADOCERAN, MONTANE ZON/ *FIELD
NETWORK, ALGAE COUNTS, P/ *RIVER
BEER-LAMBERT LAW, *FIXED-DENSITY
ANT/ *ALGAL CULTURE, *LABORATORY
IES, *MASS CULTURE, *PILOT-PLANT
TY, FUNDAMENTAL STUDIES, APPLIED
SYRACUSE UNIVERSITY, FUNDAMENTAL
IC.: MATHEMATICAL
EEN ALGAE, NITZSCHIA, BIOLOGICAL
UTILIZATION, FISH, ALGAE, MODEL
REATMENT, EUTHROPHICATION, MODEL
HYTA, LITTORAL, NUTRIENTS, MODEL
FLOCCULATION, FILTRATION, MODEL
HALOGENS, ALGICIDES, CYTOLOGICAL
SOPHYCEAE, CARTERIA, COMPARATIVE
BON, *CARBON CYCLE, *CYTOLOGICAL
G RATE, *BODY SIZE, *COMPARATIVE
ESTON, SCENEDESMUS, MATHEMATICAL
NTHESIS, CYANOPHYTA, CYTOLOGICAL
HARDNESS, COLIFORMS, CYTOLOGICAL
HENS, ACTIMOYCETES, CYTOLOGICAL
OOD WEBS, KENTUCKY, LIFE HISTORY
GLAND), FOOD SOURCES, GREENHOUSE
UGHTERHOUSE WASTES, EXPERIMENTAL
TIFIED FLOW, MIXING, FLOW, MODEL
F, NUTRIENT BUDGETS, COMPARATIVE
NT, *AUSTRALIA, LABORATORY-SCALE
MPHIBIANS, ANABAENA, CYTOLOGICAL
EFFECTS, LABORATORY TESTS, MODEL
ATORY-SCALE STUDIES, PLANT-SCALE
OPHIC LEVEL, PRODUCTIVITY, MODEL
CHNIQUES, CLASSIFICATION, *MODEL
*INDIA, MOOSI RIVER ECOLOGICAL
RGANIC NITROGEN, BACTERIOLOGICAL
(ONTARIO), BALSAM LAKE(ONTARIO),
OPH/ MICHIGAN, CHEBOYGAN COUNTY,
N ALGAE, HALIMONE PORTULACOIDES,
NIA HARVEYI, EPIPHYTIC BACTERIA,
AR REACTORS, POWERPLANTS, SHIPS,
Y PRODUCTIVITY, ALGAE, PLANKTON,
MONTANE RIVER COURSE, CARPATHIAN
*GELBSTOFF, *ALGAL SUBSTANCES,
ET(NORWAY), / *GELBSTOFF, *ALGAL
VESICULOSUS, CHONDRUS C/ *ALGAL
SIS, FERTILIZATION, PLANT GROWTH
BREVIS, MYCROSYSTIS, ALGASTATIC
GASTATIC SUBSTANCES, ALGADYNAMIC
LOROPHYTA, BULRUSH, PLANT GROWTH
ROPHICATION, ALGAE, PLANT GROWTH
Y, DOMESTIC WASTES, PLANT GROWTH
EDIA, QUINONE DERIVATIVES, ALKYL
XOPHYCEAE, SALVELINUS NAMAYCUSH,

STREAMS, NUTRIENTS, LIGHT, WASHIN   W70-03309
STREAMS, PHOTOSYNTHESIS, RESPIRAT   W71-00668
STREAMS, RIVERS, LAKES, PHYSIOLOG   W71-02479
STREAMS, SEASONAL EFFECTS, THERMA   W70-03978
STREAMS, SEWAGE, REMEDIES, JUDICI   W69-06909
STREAMS, WATER CHEMISTRY, WATER: /   W69-04805
STREAMS, WATER POLLUTION, ANIMALS   W70-05750
STREAMS, WEEDS, LITTORAL, WINDS.:   W70-06229
STREAMS, WATER POLLUTION EFFECTS,   W69-10159
STREAMS, WATER POLLUTION EFFECTS,   W70-00719
STREETER-PHELPS EQUATION, DISSOLV   W69-07520
STREPTOCOCCUS, SECONDARY PRODUCTI   W70-01068
STREPTOCOCCUS, BREVIBACTERIUM, CO   W71-11823
STREPTOMYCIN, MILLIPORE FILTRATIO   W69-10154
STRES: / TEMPERATURE, THERMAL POL   W69-03516
STRESS, THERMAL WATER, *PROTOZOA,   W70-07847
STRIP MINES, DRAINAGE, FERROBACIL   W68-00891
STRIPPING FILM, SYNECHOCOCCUS, GR   W69-10163
STRIPPING, FIREBAUGH CENTER(CALIF   W70-03334
STRIPPING, ALGAE HARVESTING.: /EL   W70-01981
STRONTIUM RADIOISOTOPES, TENNESSE   W70-02786
STRONTIUM RADIOISOTOPES, CESIUM,   W69-09742
STRONTIUM RADIOISOTOPES, CESIUM,   W70-00708
STRONTIUM RADIOISOTOPES, ZINC RAD   W71-09850
STRONTIUM RADIOISOTOPES, MANGANES   W69-08269
STRONTIUM RADIOISOTOPES, ZINC· RAD   W69-07845
STRUCTURE, AMINO ACIDS, MECHANICA   W70-04184
STRUCTURE, OXYTREMA, ORGANIC ENRI   W70-03978
STRUCTURE, ATMOSPHERE, SEA WATER,   W70-04385
STRUCTURE, RIDLEY CREEK(PA), NOBS   W70-04510
STRUCTURE, 2,3-DICHLORONAPHTHOQUI   W70-02982
STRUCTURES, AFTERBAYS, WATER TYPE   W69-05023
STRUCTURES, INDUSTRIAL PLANTS, ST   W69-05023
STUDIES.: /IL CONTAMINATION, *STR   W69-03491
STUDIES.: /PIGMENTS, OCEANS, ORGA   W70-04809
STUDIES.: /IA, *WASTE WATER TREAT   W70-07034
STUDIES.: /ASTE WATER TREATMENT,   W70-07031
STUDIES, *NITROGEN LIMITATION, BA   W70-07957
STUDIES, *EXPERIMENTAL LIMNOLOGY,   W69-10154
STUDIES, *NATIONAL WATER QUALITY   W70-04507
STUDIES, *FIXED-LIGHT TESTS, *EFF   W70-03923
STUDIES, *MASS CULTURE, *PILOT-PL   W69-06865
STUDIES, ALGAL GROWTH, ALGAL NUTR   W69-06865
STUDIES, ALGAL PHYSIOLOGY, ALGAL   W70-03973
STUDIES, APPLIED STUDIES, ALGAL P   W70-03973
STUDIES, AUTOTROPHIC, HETEROTROPH   W70-05750
STUDIES, ALGAL IDENTIFICATION.: /   W70-10173
STUDIES, BIOASSAY, AQUATIC LIFE,   W71-11881
STUDIES, BIOMASS, *LIGHT INTENSIT   W73-02218
STUDIES, BIOASSAY, RADIOACTIVITY   W73-01454
STUDIES, COMPUTERS, NUTRIENTS, EU   W70-08627
STUDIES, CHLORINE, COPPER SULFATE   W70-02364
STUDIES, CRUCIGENIA RECTANGULARIS   W69-07833
STUDIES, CULTURES, FLUOROMETRY, B   W69-08278
STUDIES, DAPHNIA SCHODLERI, DAPHN   W70-03957
STUDIES, DATA COLLECTIONS, MARINE   W73-01446
STUDIES, ELECTRON MICROSCOPY, WAS   W70-03973
STUDIES, EVAPORATION, MICROORGANI   W71-05267
STUDIES, FILTRATION, ULTRASONICS,   W71-11823
STUDIES, GAMMA RAYS, SPECTROMETER   W69-09742
STUDIES, GROWTH K: /OMPOSITION(EN   W69-06865
STUDIES, LAGOON DESIGN, SCALE-UP   W77-07508
STUDIES, MATHEMATICAL MODELS, GRE   W71-05878
STUDIES, NUTRIENT REMOVAL, LAKE R   W69-09349
STUDIES, PLANT-SCALE STUDIES, REM   W70-06899
STUDIES, PHY: /OSTRIDIUM, FISH, A   W70-05372
STUDIES, RADIOACTIVITY TECHNIQUES   W70-04074
STUDIES, REMOVAL.: /TRALIA, LABOR   W70-06899
STUDIES, SALMON, AQUATIC ENVIRONM   W69-06536
STUDIES, WASTE IDENTIFICATION, BI   W69-01273
STUDY.:   W69-07096
STUDY, AEROBICITY.:   O   W71-05742
STURGEON LAKE(ONTARIO), BUCKHORN   W70-02795
STURGEON RIVER, FLUORESCEIN, MACR   W69-09334
SUAEDA MARITIMA, ASTER TRIPOLIUM,   W70-09976
SUBLITTORAL, YELLOW MATERIAL, DU: W70-01073
SUBMARINES, EFFLUENTS, ION EXCHAN   W70-00707
SUBMERGED PLANTS, PHYSICOCHEMICAL   W71-05267
SUBMONTANE REGION, GRADIENTS(RIVE   W70-02784
SUBPOLAR WATERS, TYNSET(NORWAY),   W70-01074
SUBSTANCES, SUBPOLAR WATERS, TYNS   W70-01074
SUBSTANCES, *EXTRACELLULAR, FUCUS   W70-01073
SUBSTANCES, SEWAGE, SULFATE, MANG   W70-02969
SUBSTANCES, ALGADYNAMIC SUBSTANCE   W70-02510
SUBSTANCES, PHORMIDINE, PANDORINI   W70-02510
SUBSTANCES, CHLORELLA, CHLAMYDOMO   W72-00845
SUBSTANCES, PLANT GROWTH, CARBOHY   W69-07826
SUBSTANCES, LIMNOLOGY, PLANT GROW   W68-00470
SUBSTITUTED IMIDAZOLINES, BIOLOGI   W70-02982
SUBSTRATE TYPE.: /NTHOPHYCEAE, MY   W70-00711

536

537

KE NORRVIKEN(SWEDEN), LAKE EKOLN(SWEDEN), LAKE LOTSJON(SWEDEN), PE W70-04287
CHAELIS CONSTANT, LAKE NORRVIKEN(SWEDEN), LAKE EKOLN(SWEDEN), LAKE W70-04287
EY, NITROGEN CONTENT, LAKE ERKEN(SWEDEN), LAKE MENDOTA(WIS), NET P W70-03959
LAKE EKOLN(SWEDEN), LAKE LOTSJON(SWEDEN), PERMEASES, CHAMYDOMONADS W70-04287
LA, DINOBRYON, MELOSIRA, ANEBODA(SWEDEN), UPPSALA(SWEDEN), LAKE ER W70-05409
ELATIVE HETEROTROPHIC POTENTIAL, SWEDEN, ENZYME KINETICS, LINEWEAV W70-04194
HETEROTROPHY, MICROBIAL ECOLOGY, SWEDEN, LAKE ERKEN(SWEDEN), ACTIV W70-04284
COSE, MICHAELIS-MENTEN EQUATION, SWEDEN, LAKE ERKEN(SWEDEN), LINEW W70-04287
 *HETEROTROPHY, ORGANIC SOLUTES, SWEDEN, LAKE ERKEN(SWEDEN), ANNUA W70-05109
-232, BISMUTH-214, ACTINIUM-228, SWIFT RIFFLES, ZIRCONIUM-95, NIOB W69-09742
EED, DUCKWEED, WHITE WATER LILY, SWIMMERS' ITCH.: /D CELERY, BUR R W70-05568
CONTROL, CRUSTACEANS, ROTIFERS, SWIMMING POOLS, CHLAMYDOMONAS, DA W70-10161
ICATION, APPLICATION TECHNIQUES, SWIMMING, WATERFOWL, BIRDS, COSTS W70-05568
RMINATION, ABSORBANCE, COENOBIA, SWIRLING, VOLATILE SOLIDS.: /DETE W70-03334
, GREIFENSEE(SWITZ), PAFFIKERSEE(SWITZ), AEGERISEE(SWITZ), FURES: / W70-00270
), PAFFIKERSEE(SWITZ), AEGERISEE(SWITZ), FURES: / GREIFENSEE(SWITZ W70-00270
SH), ZURICHSEE(SWITZ), TURLERSEE(SWITZ), GREIFENSEE(SWITZ), PAFFIK W70-00270
Z), TURLERSEE(SWITZ), GREIFENSEE(SWITZ), PAFFIKERSEE(SWITZ), AEGER W70-00270
S), CEDAR RIVER(WASH), ZURICHSEE(SWITZ), TURLERSEE(SWITZ), GREIFEN W70-00270
N(WASH), OSCILLATORIA RUBESCENS, SWITZERLAND, PERIDINIUM, OSCILLAT W69-10169
A, BOSMINA LONGISPINA, ZURICHSEE(SWITZERLAND), LINSLEY POND(CONN), W69-10169
A(CANADA), PENGUINS, LAKE ZURICH(SWITZERLAND), WHITEFISH, PUNCHBOW W70-05399
F ZELL(AUSTRIA), LAKE HALL WILER(SWITZERLAND).: /IER(BAVARIA), LAK W70-05668
ASTES, LAKE ZURICH(SWITZERLAND), SWITZERLAND, KILCHBERG(SWITZERLAN W70-05668
KILCHBERG(SWITZERLAND), ZOLLIKON(SWITZERLAND), LAKE TEGERN(BAVARIA W70-05668
 GALVANIZING WASTES, LAKE ZURICH(SWITZERLAND), SWITZERLAND, KILCHB W70-05668
ZERLAND), SWITZERLAND, KILCHBERG(SWITZERLAND), ZOLLIKON(SWITZERLAN W70-05668
IDALSVANNET NORWAY, LAKE ZURICH, SWITZERLAND, NITELV RIVER NORWAY, W69-07833
CAL AND BIOLOGICAL PURIFICATION, SWITZERLAND.: / DRAINAGE, MECHANI W69-08668
LAKE WASHINGTON(WASH), ZURICHSEE(SWITZERLAND).: /KE TAHOE(CALIF), W69-06858
, NITROGEN, PHOSPHATES, BIOMASS, SYMBIOSIS, OXYDATION LAGOONS.: /D W70-05655
S, PHOTOSYNTHESIS, PRODUCTIVITY, SYMBIOSIS, RESERVOIRS, LAKES, REA W70-05092
A, SYNURA, GYMNODINIUM, CHEMICAL SYMBIOSIS, MICROBIAL ECOLOGY, DIN W70-04503
ENOIDOSA, SYNTHETIC SEWAGE FEED, SYMBIOTIC RELATIONSHIPS.: /LA PYR W70-05655
SPECTROSCOPY, MASS SPECTROMETRY, SYMPLOCA MUSCORUM.: /UIS, RAMMAN W69-00387
ERIMENTAL LIMNOLOGY, BRACHIONUS, SYNCHAETA PECTINATA, POLYARTHRA V W70-02249
SCOPY, STAINING, STRIPPING FILM, SYNECHOCOCCUS, GRAIN COUNT(AUTORA W69-10163
SCILLATORIA TEREBRIFORMIS, MATS, SYNECHOCOCCUS LIVIDUS, BLUE-GREEN W70-05270
MOEWUSII, SKELETONEMA COSTATUM, SYNECHOCOCCUS ELON: /HLAMYDOMONAS W70-05381
N, VOLVOX, PANDORINA, SPIROGYRA, SYNECHOCOCCUS, CHROOCOCCUS, OSCIL W70-04503
A, CYCLOTELLA, HALOCHLOROCOCCUM, SYNECHOCOCCUS, CARTERIA, GONIUM, W70-03974
ACICULARIS, NITZSCHIA DISSIPATA, SYNEDRA ACUS, SINKING RATES, GRAZ W69-10158
ANCE, SPECIES, MELOSIRA VARIANS, SYNEDRA ULNA, OEDOGONIUM, PHORMID W70-00265
RAGILARIA, TABELLARIA, MELOSIRA, SYNEDRA, COELOSPHAERIUM, SPIROGYR W70-02803
BOSMINA, ASTERIONELLA, MELOSIRA, SYNEDRA, FRAGILARIA, CER: /STIS, W70-04503
LAKE(NY), ASTERIONELLA FORMOSA, SYNEDRA, MELOSIRA, DINOBRYON, FRA W70-01579
A RIVER, CYCLOTELLA, FRAGILARIA, SYNEDRA, NAVICULA, APHANIZOMENON, W70-08107
LFOVIBRIO, MESOTHERMY, MELOSIRA, SYNEDRA, RHODOMON: /ACTERIA, DESU W70-05760
MICROCYSTIS, PRYMNESIUM PARVUM, SYNERGISTIC EFFECTS, AXENIC CULTU W70-05372
, NUCLEIC ACID, RIBONUCLEIC ACID SYNTHESIS.: /ION, *MACROMOLECULES W69-10151
, PHOTOCHEMICAL, CARBON DIOXIDE, SYNTHESIS, ENZYMES, CHLOROPHYTA, W70-02965
RELATIO/ *CHLORELLA PYRENOIDOSA, SYNTHETIC SEWAGE FEED, SYMBIOTIC W70-05655
ON, ALGAL BLOOMS, NUTRIENT LOAD, SYNTHETIC DETERGENTS, FISH MANAGE W69-08674
EOTRICHIA ECHINULATA, DINOBRYON, SYNURA UVELLA, FRAGILARIA, TABELL W70-02803
TS, ASTERIONELLA, APHANIZOMENON, SYNURA, ANABAENA, MELOSIRA, MALLO W69-10157
UNIT PROCESSES, GASTROENTERITIS, SYNURA, ANABAENA, IOWA RIVER, CYC W70-08107
TA(MAINE), CARLETON POND(MAINE), SYNURA, ASTERIONELLA, TYPHOID FEV W70-08096
SIRA, COCCOLITHUS, SARGASSO SEA, SYNURA, GYMNODINIUM, CHEMICAL SYM W70-04503
ERTIOLECTA, AMPHIDINIUM CARTERI, SYRACOSPHAERA ELONGATA, THALASSIO W69-08278
 TECHNOLOGY FOUND, SYRACUSE(NY), SYRACUSE UNIVERSITY, FUNDAMENTAL W70-03973
TE SCIENCE AND TECHNOLOGY FOUND, SYRACUSE(NY), SYRACUSE UNIVERSITY W70-03973
PINNULARIA, NITZSCHIA, NAVICULA, SYRACUSE(NY), ENTEROMORPHA INTEST W70-03974
*ONONDAGA LAKE, ONONDAGA COUNTY, SYRACUSE, WARBURG RESPIROMETER, C W69-02959
UTRIENT REMOVAL, CONTINUOUS FLOW SYSTEM.:   *N W72-00917
CILIATES, FLAGELLATES, SAPROBIEN SYSTEM, CILIATE AUTOECOLOGY.:   * W71-03031
PENDED SOLIDS, MULTI-CELL SERIAL SYSTEM, PH, FACULTATIVE POND, ANA W70-04790
OPLANKTON, PRIMARY PRODUCTIVITY, SYSTEMATICS, WATER POLLUTION EFFE W69-07832
ILTERS, *BIOTA, SEWAGE BACTERIA, SYSTEMATICS, FUNGI, ALGAE, DIATOM W72-01819
IBUTION, BIOLOGICAL COMMUNITIES, SYSTEMATICS, MARINE PLANTS.: /STR W72-10623
CATION, WATER POLLUTION EFFECTS, SYSTEMS ANALYSIS, SUCCESSION, ECO W70-02779
RRIES, AUTOMATIC CONTROL, PIPING SYSTEMS(MECHANICAL), CLEANING, CO W72-08688
 *NATURAL STREAMS, *DISTRIBUTION SYSTEMS, CHLORINATION, COPPER SUL W72-03562
C/ BENTHIC ALGAE, BLOOMS, CLOSED SYSTEMS, CYTOLOGY, ENVIRONMENTAL W69-07832
, *POLLUTION ABATEMENT, DRAINAGE SYSTEMS, FLOW RATES, CONN STREAM W68-00479
RATION, *TOXICITY, *DISTRIBUTION SYSTEMS, WATER POLLUTION, WATER Q W72-03746
GAE, 2-4-D, AMINOTRIAZOLE, 2-4-5- T, SCENEDESMUS, CHLAMYDOMONAS, CH W70-03519
 LIMNETICA, ANABAENA FLOS AQUAE, TABELLARIA, MELO: /TICUS, LYNGBYA W70-05405
RYON, SYNURA UVELLA, FRAGILARIA, TABELLARIA, MELOSIRA, SYNEDRA, CO W70-02803
VER(ILL), POOLS, WIND DIRECTION, TABELLARIA, FRAGILARIA, ASTERIONE W70-00263
TON, HOWARD A. HANSON RESERVOIR, TACOMA(WASH).: /EN RIVER, WASHING W72-11906
UDIES, RADIOACTIVITY TECHNIQUES, TAGGING.: /RATORY TESTS, MODEL ST W70-04074
                           LAKE TAHOE.:   W70-09135
HICATION, LAKE MONONA(WIS), LAKE TAHOE(CALIF), LAKE WASHINGTON(WAS W69-06858
PARTICULATE ORGANIC CARBON, LAKE TAHOE(CALIF), CHLORELLA PYRENOIDO W70-03983
L BLOOMS, NTA, EDTA, HEIDA, LAKE TAHOE(CALIF).: /OMPLEXONES, *ALGA W70-04373
VERWORTS, MORPHOMETRY, PA/ *LAKE TAHOE(CALIF-NEV), HYDROPHYTES, LI W70-00711
IGHT MEASUREMENTS, EUPHOT/ *LAKE TAHOE(CALIF-NEV), TRANSPARENCY, L W70-01933
A, MOLLUSCA, RAFFINATE, THORIUM, TAILINGS.: / COLEOPTERA, HEMIPTER W69-07846
A(WIS), PHYTOPLANKTON CROPS, TR/ TAKASUKA POND(JAPAN), LAKE MENDOT W69-05142
, GONYAULAX CATENELLA, GONYAULAX TAMERENSIS, GYMNODINIUM BREVIS, M W70-02510

539

, NON-VOLATILE ACIDS, KETOACIDS,
, SCUM, WEEDS, NUTRIENTS, SEPTIC
        HYDROGEN SULFIDE, IMHOFF
WASTES, MUNICIPAL WASTES, SEPTIC
TE, ORTHOPHOSPHATE, DISMUTATION,
TION, DEPURATION, PHYCOCOLLOIDS,
ACHYTHEMIS, CHYDORUS, PLEUROXUS,
S, CHYDORUS, PLEUROXUS, TANYPUS,
C CULTURES, MONOALGAL, PORTUGAL,
PHY, PARTICULATE MATTER, BLELHAM
ALIF), JOINT INDUSTRY/GOVERNMENT
Y CONTROL, ALGAL CONTROL, SLIME,
RON, TEMPERATURE, SOIL ANALYSIS,
NZENE SULFONATES, WIND VELOCITY,
IATION, CHEMICAL INDUSTRY, ODOR,
ION, FISH, NUISANCE ALGAE, ODOR,
AE, NUTRIENTS, INDICATORS, ODOR,
 BACTERIA, COLOR, IRON BACTERIA,
LY, *NEW HAMPSHIRE, ALGAE, ODOR,
*ALGAE, *WATER SUPPLY, *CONTROL,
QUALITY, *EUTROPHICATION, COLOR,
LUTION, MICROORGANISMS, HISTORY,
RTIES, *WATER POLLUTION SOURCES,
LLUTION CONTROL, NUISANCE ALGAE,
HYPOLIMNION, CYCLING NUTRIENTS,
TTER, RADIOACTIVITY, PESTICIDES,
URE, SALINITY, REAERATION, FISH,
H, PRIMARY PRODUCTIVITY, RIVERS,
FATES, SEDIMENTATION, SEDIMENTS,
R, CHEMOSYNTHETIC, SAPROPHYTIC,/
USTRIS (RAB) FARLOW, CLADOPHORA,
NARRAGANSETT BAY(RHODE ISLAND),
RACTER, / SPECIES DETERMINATION,
OLOGY, ALGAL BIOCHEMISTRY, ALGAL
HUNGARY), ZIMONA(HUNGARY), ALGAL
ORGANISMS, FISH PHYSIOLOGY, FISH
                            ALGAL
                          SPECIES,
ONE, ALPINE ZONE, BOULDER(COLO),
N, MIXING, ECONOMIC FEASIBILITY,
NSHIPS, FLUOROMETRIC CHLOROPHYLL
N FIXING BACTERIA, RADIOACTIVITY
AR MICROMETER, SUCROSE-FLOTATION
ARPUS CONFERVOIDES, NYLON COLUMN
CH CULTURES, ACETYLENE REDUCTION
SCHIZOTHRIX, PHORMIDIUM, WINKLER
G FACTORS, CHLORELLA, ANALYTICAL
UTRIENT REQUIREMENTS, ANALYTICAL
CIDES, METABOLISM, RADIOACTIVITY
 CONTROL, *ALGICIDES, ANALYTICAL
IC MICROBIO/ *ALGAE, *ANALYTICAL
E, CHEMICAL ANALYSIS, ANALYTICAL
EN DEPLETION, NUCLEAR ACTIVATION
LOGY, / *ECOLOGY, *RADIOACTIVITY
UNDMENTS, RESERVOIRS, ANALYTICAL
TION, PHOTOSYNTHESIS, ANALYTICAL
TION, MICROORGANISMS, ANALYTICAL
ALGAE, GROWTH RATES, ANALYTICAL
STES, WATER ANALYSIS, ANALYTICAL
*RESEARCH EQUIPMENT, *ANALYTICAL
, SALINITY, BIOASSAY, ANALYTICAL
TOMS, MARINE ALGAE,/ *ANALYTICAL
ORGANISMS, TURBIDITY, ANALYTICAL
CHROMOTOGRAPHY, RAD/ *ANALYTICAL
S, WATER UTILIZATION, ANALYTICAL
OPHYLL, *OHIO RIVER, *ANALYTICAL
OACTIVITY TECHNIQUES, ANALYTICAL
IOLOGICAL ECOLOGY, RADIOACTIVITY
TS, MODEL STUDIES, RADIOACTIVITY
TIC MICROBIOLOGY, */ *ANALYTICAL
  *LAKES, *BACTERIA, *ANALYTICAL
, *BIOINDICATORS, */ *ANALYTICAL
ER POLLUTION EFFECTS, ANALYTICAL
R ANALYSIS, *SAMPLI/ *ANALYTICAL
Y, CARBON-14, RADIOCARBON UPTAKE
, *ALGAE, *ANALYSIS, *ANALYTICAL
RES, OPTICAL DENSITY, SOIL-WATER
Y TESTS, *I/ *ALGAE, *ANALYTICAL
BIAL ECOLOGY, RADIOCARBON UPTAKE
IC BACTERIA, *ALGAE, *ANALYTICAL
ER POLLUTION EFFECTS, ANALYTICAL
TROL, BIBLIOGRAPHIES, ANALYTICAL
IUM CARTERI, *RADIOCARBON UPTAKE
S, CHEMICAL ANALYSIS, ANALYTICAL
IDS, LIGHT, ALGAE, / APPLICATION
ATES OF APPLICATION, APPLICATION
ANALYSIS, *CHLORINE, *ANALYTICAL
FFECTS, *PHOSPHATES, *ANALYTICAL
S HOLE(MASS). RADIOCARBON UPTAKE
MIXOLIMNION, RADIOCARBON UPTAKE

TAMIYA'S MINERAL MEDIUM, GLYOXYLI   W70-04195
TANKS, EFFLUENTS, SWAMPS, FOREST   W70-04193
TANKS, PRECIPITATES.:   W72-00027
TANKS, RECREATION WASTES, WATER Q   W70-04266
TANN: / ARTHROBACTER, PYROPHOSPHA   W69-07428
TANNINS, FLAGELLATES, PHARMACEUTI   W70-01068
TANYPUS, TANYTARSUS, LIMNODRILUS,   W70-06975
TANYTARSUS, LIMNODRILUS, EUCYCLOP   W70-06975
TAPES, JAPAN, AMPHIDINIUM, BLOOMS   W70-05372
TARN(ENGLAND), CHLORELLA PYRENOID   W70-02504
TASK FORCE ON EUTROPHICATION, PRO   W70-02777
TASTE-PRODUCING ALGAE, ODOR-PRODU   W69-05704
TASTE-PRODUCING ALGAE, HYDROGEN I   W72-11906
TASTE, COAGULATION, PLANKTON, MON   W70-00263
TASTE, CONSUMER.: /ER WORKS ASSOC   W71-05943
TASTE, DEPTH, CIRCULATION, DIURNA   W70-05404
TASTE, FILTERS, WATER TREATMENT,   W70-05264
TASTE, FILTERS, IOWA, DIATOMS, EU   W70-08107
TASTE, LIGHT TEMPERATURE, NUTRIEN   W69-08282
TASTE, ODOR, FILTERS, PLANKTON, W   W70-05389
TASTE, ODOR, TOXINS, DISSOLVED OX   W69-05697
TASTE, ODOR, DEEP WELLS, WATER TR   W70-08096
TASTE, ODOR, ALGAE, RESERVOIRS, W   W72-01813
TASTE, ODOR.: / QUALITY, WATER PO   W69-10155
TASTE, ODOR, COLOR, TURBIDITY, HY   W70-02251
TASTE, ODOR, COAGULATION, ELECTRO   W72-13601
TASTE, ODOR, OXYGEN SAG, POTOMAC   W73-00750
TASTE, THERMAL STRATIFICATION, WA   W69-06535
TASTE, WATER CHEM: /PLANKTON, SUL   W69-07833
TAXONOMIC GROUPS, MOLECULAR FILTE   W68-00891
TAXONOMIC DESCRIPTIONS, GONGROSIR   W70-04468
TAXONOMIC TYPES, HETERETROPHY, BA   W70-00713
TAXONOMY, FLAVOBACTERIUM, ACHROMO   W70-04280
TAXONOMY, ALGAL PHYLOGENY, CLADOP   W70-03973
TAXONOMY, SPIRULINA, CHODATE: /A(W70-03969
TAXONOMY, DDT, PHOSPHOTHIOATE PES   W69-02782
TAXONOMY, ALGAE IDENTIFICATION.:   W69-03514
TAXONS.:   W71-09767
TEA LAKE(COLO), BLACK LAKE(COLO),   W69-10154
TECHNICAL FEASIBILITY, COST ANALY   W71-10654
TECHNIQUE.: /NVIRONMENTAL RELATIO   W70-05094
TECHNIQUE, PHOTOSYNTHESIS.: /ROGE   W70-04249
TECHNIQUE, METALIMNION, BATHYTHER   W70-04253
TECHNIQUE, PARTICULATE MATTER, BL   W70-01074
TECHNIQUE, HETEROTROPHY, FLAGELLA   W70-02775
TECHNIQUE, OXYGEN ELECTRODES, ATO   W70-03309
TECHNIQUES, CYANOPHYTA, DIATOMS,   W70-02779
TECHNIQUES, WATER POLLUTION EFFEC   W70-02245
TECHNIQUES, CHLORELLA, OXIDATION   W70-02198
TECHNIQUES, AQUATIC ALGAE, AQUATI   W70-02968
TECHNIQUES, *AQUATIC ALGAE, AQUAT   W70-02792
TECHNIQUES, CLOUD COVER, SPECTROS   W69-09676
TECHNIQUES, ROCK LAKE(WASH), WILL   W70-00264
TECHNIQUES, *RADIOISOTOPES, LIMNO   W69-10163
TECHNIQUES, ON-SITE TESTS.: /IMPO   W71-10098
TECHNIQUES, SPECTROSCOPY, COLORIM   W71-11377
TECHNIQUES, *WASTE WATER TREATMEN   W71-13313
TECHNIQUES, LABORATORY EQUIPMENT,   W71-07339
TECHNIQUES, SAMPLING, ORGANIC MAT   W72-03299
TECHNIQUES, WATER QUALITY, DISSOL   W70-08627
TECHNIQUES, LETHAL LIMITS, DINOFL   W70-08111
TECHNIQUES, *ALGAE, PLANKTON, DIA   W70-09904
TECHNIQUES, LABORATORY TESTS, FOR   W70-08628
TECHNIQUES, *PHOSPHATES, *ALGAE,   W71-03033
TECHNIQUES, NUTRIENTS, NITROGEN,   W71-03048
TECHNIQUES, BIOMASS, WATER POLLUT   W71-04206
TECHNIQUES, WATER POLLUTION EFFEC   W70-04194
TECHNIQUES, ANALYTICAL TECHNIQUES   W70-04194
TECHNIQUES, TAGGING.: /RATORY TES   W70-04074
TECHNIQUES, *AQUATIC ALGAE, *AQUA   W70-04283
TECHNIQUES, WATER POLLUTION EFFEC   W70-04287
TECHNIQUES, *ALGAE, *GROWTH RATES   W70-04381
TECHNIQUES.: /NAS, BIOASSAYS, WAT   W70-04284
TECHNIQUES, *AQUATIC WEEDS, *WATE   W70-04382
TECHNIQUES, GLUCOSE, MICHAELIS-ME   W70-04287
TECHNIQUES, TEMPERATURE, PIPTIDES   W70-04365
TECHNIQUES, PHYCOLOGY, MICROCYSTI   W70-04381
TECHNIQUES, *PLANKTON, *LABORATOR   W70-04469
TECHNIQUES, MICHAELIS-MENTEN EQUA   W70-05109
TECHNIQUES, *MARINE MICROORGANISM   W70-04811
TECHNIQUES, KINETICS, AQUATIC BAC   W70-05109
TECHNIQUES, DIATOMS, SAMPLING, PH   W70-05389
TECHNIQUES, WATER POLLUTION IDENT   W70-05652
TECHNIQUES, DINOFLAGELLATES, MARI   W70-05652
TECHNIQUES, *MICROORGANISMS, *LIP   W70-05651
TECHNIQUES, SWIMMING, WATERFOWL,   W70-05568
TECHNIQUES, *POLAROGRAPHIC ANALYS   W70-05178
TECHNIQUES, *BIOASSAYS, ORGANOPHO   W70-05269
TECHNIQUES, MICROBIAL ECOLOGY, CO   W70-05272
TECHNIQUES, BACTERIAL PLATE, BACT   W70-05760

540

SSAY, *ESSE/ *ALGAE, *ANALYTICAL
GICAL ECOLOGY, ALGAE, ANALYTICAL
*NUTRIENT REQUIREME/ *ANALYTICAL
N, *CHLORELLA, ALGAE, ANALYTICAL
ALGAE, ALGAL CONTROL, ANALYTICAL
THETIC OXYGEN, ALGAE, ANALYTICAL
, *CYANOPHYTA, ALGAE, ANALYTICAL
ROL, ALGAL POISONING, ANALYTICAL
GICAL ECOLOGY, ALGAE, ANALYTICAL
EUTROPHICAT/ *ALGAE, *ANALYTICAL
EUTROPHICATION, *PH/ *ANALYTICAL
REQUIREMENTS, ALGAE, ANALYTICAL
, GAS CHROMATOGRAPHY, ANALYTICAL
NTS, *NITROGEN CYCLE, ANALYTICAL
ER POLLUTION SOURCES, SEPARATION
PARATION TECHNIQUES, *ANALYTICAL
ING NUTRIENTS, ALGAE, ANALYTICAL
NUTRIENTS,/ *ALGAE, *ANALYTICAL
ATION, *PLANT GROWTH, ANALYTICAL
STUDIES, BIOASSAY, RADIOACTIVITY
*DIATOMACEOUS EARTH, SEPARATION
EARTH, *FILTRATION, *SEPARATION
ECTROLYTES, FILTERS, *ANALYTICAL
NVIRONMENTAL EFFECTS, ANALYTICAL
*PHYCOLOGY, NY STATE SCIENCE AND
S, TEGERNSEE, GERMANY, EXPANDING
SECTS, PHOSPHORUS RADIOISOTOPES,
ND), ZOLLIKON(SWITZERLAND), LAKE
NY), LAKE SCHLIER(GERMANY), LAKE
APPALACHIAN MOUNTAIN CLUB HUTS,
AGULATION, PLANKTON, MONITORING,
SERIUM, ARTEMIA, CARTERIA, NORTH
UM, ARTHROBACTER, FLORA, AMBIENT
, *ESTUARINE ENVIRONMENT, *WATER
ON, FLUORESCENCE, CARBOHYDRATES,
R RADIATION, SALINITY, RAINFALL,
N, OLIGOTROPHY, NUTRIENTS, WATER
KE ONTARIO, LAKE MICHIGAN, WATER
SYNTHESIS, *OXYGEN, *SATURATION,
*DISSOLVED OXYGEN, *OHIO, WATER
IDS, HYDROGEN ION CONCENTRATION,
TER), CONNECTICUT, RHODE ISLAND,
L ANALYSIS, BACTERIA, NUTRIENTS,
ANIA, BIOCHEMICAL OXYGEN DEMAND,
LOROPHYLL, EPILIMNION, EFFLUENT,
NG WATER, WATER CHEMISTRY, WATER
DS, EUTROPHICATION, OLIGOTROPHY,
OIRS, WATER QUALITY, NEW JERSEY,
CROSCOPIC EXAMINATION, *CRITICAL
AUTHORITY PROJECT, *RESERVOIRS,
,/ *THERMAL POWERPLANTS, *LAKES,
LAGOONS, *DESIGN CRITERIA, WATER
LIC HEALTH, PATHOGENIC BACTERIA,
Y, CARBON DIOXIDE, GROWTH RATES,
RUS, NITROGEN, CALCIUM CHLORIDE,
ASKA, TURBIDITY, STRATIFICATION,
SOLVED OXYGEN, LABORATORY TESTS,
GAE, HYDROGEN ION CONCENTRATION,
ING, HYDROGEN ION CONCENTRATION,
SYNTHESIS, PHYTOPLANKTON, ALGAE,
RIA, *ACTIVATED SLUDGE, STORAGE,
HEMICAL OXYGEN DEMAND, AERATION,
R BUDGET, WATER CHEMISTRY, WATER
AND, HYDROGEN ION CONCENTRATION,
FIXATION, FOREST, WINDS, LIGHT,
RIA, HYDROGEN ION CONCENTRATION,
ARBON DIOXIDE, DISSOLVED OXYGEN,
TEMPERATURE RANGES,
*ORGANIC MATTER, WATER QUALITY,
RATURE TOLERANCE, OPTIMUM GROWTH
EQUIREMENTS.:
ROTEINS, AQUATIC MICROORGANISMS,
EPILIMNION, WINDS, THERMOCLINE,
TER, *COPPER, *ALGAE, SEA WATER,
TY, *NUTRIENTS, LIMNOLOGY, WATER
SICAL PROPERTIES, CARP, TILAPIA,
, ENERGY, DI-URNAL DISTRIBUTION,
RE/ *STREAMS, *ALGAE, *DIATOMS,
ION, DIATOMS, THERMAL POLLUTION,
GREEN ALGAE, BLUE-GREEN ALGAE,
VARIABILITY, CHLORELLA, CARBON,
OXIDE, BICARBONATES, CARBONATES,
C MATTER, CHLOROPHYLL, DYNAMICS,
ITROGEN, PHOSPHORUS, CARBONATES,
ERS, SCENEDESMUS, CHLAMYDOMONAS,
EMICAL ANALYSIS, TRACE ELEMENTS,
TENSITY, TROPHIC LEVEL, SAMPLING
CENTRATION, ALKALINITY, DIATOMS,
, PONDS, RIVERS, RESERVOIRS, AIR
PIRATION, PIGMENTS, CHLOROPHYLL,

TECHNIQUES, *AQUATIC WEEDS, *BIOA   W69-03364
TECHNIQUES, AQUATIC ALGAE, AQUATI   W69-03517
TECHNIQUES, *BIOASSAY, *ENZYMES,   W69-03373
TECHNIQUES, AQUATIC ALGAE, AQUATI   W69-03371
TECHNIQUES, AQUATIC ALGAE, AQUATI   W69-03374
TECHNIQUES, AQUATIC ALGAE, AQUATI   W69-03362
TECHNIQUES, AQUATIC ALGAE, BIOASS   W69-03518
TECHNIQUES, AQUATIC ALGAE, AQUATI   W69-03188
TECHNIQUES, AQUATIC ALGAE, AQUATI   W69-03515
TECHNIQUES, *CYCLING NUTRIENTS, *   W69-03358
TECHNIQUES, *CYCLING NUTRIENTS, *   W69-03370
TECHNIQUES, AQUATIC ALGAE, AQUATI   W69-03186
TECHNIQUES, WATER POLLUTION SOURC   W69-00387
TECHNIQUES, AQUATIC ALGAE, AQUATI   W69-03185
TECHNIQUES, *ANALYTICAL TECHNIQUE   W69-01273
TECHNIQUES, CLASSIFICATION, *MODE   W69-01273
TECHNIQUES, CLAY LOAM, CLAYS, COL   W69-04800
TECHNIQUES, *BIOASSAY, *ESSENTIAL   W69-04802
TECHNIQUES, ANTIBIOTICS, BIOASSAY   W69-04801
TECHNIQUES, ABSORPTION.: / MODEL   W73-01454
TECHNIQUES, PRESSURE, FLOW RATES,   W72-08688
TECHNIQUES, TURBIDITY, SUSPENDED   W72-08686
TECHNIQUES.: /TIVATED CARBON, *EL   W72-09693
TECHNIQUES, PHYSIOLOGICAL ECOLOGY   W72-12240
TECHNOLOGY FOUND, SYRACUSE(NY), S   W70-03973
TECHNOLOGY, OVERPOPULATION.: /HUT   W69-07818
TECHNOLOGY, NATURAL RESOURCES, EC   W71-09012
TEGERN(BAVARIA), LAKE SCHLIER(BAV   W70-05668
TEGERN(GERMANY), GERMANY, FINLAND   W70-05761
TEGERNSEE, GERMANY, EXPANDING TEC   W69-07818
TEM: /S, WIND VELOCITY, TASTE, CO   W70-00263
TEMPERATE ZONE, PLUTONIUM, HALF-L   W70-00708
TEMPERATURE, THERMAL SENSITIVITY,   W70-00713
TEMPERATURE, *SEASONAL, RHODE ISL   W70-00713
TEMPERATURE, SALINITY, ULTRAVIOLE   W70-01074
TEMPERATURE, COLORIMETRY, CARBOHY   W70-01073
TEMPERATURE, DISSOLVED OXYGEN, FI   W70-00264
TEMPERATURE, PHOSPHATES, TURBIDIT   W70-00667
TEMPERATURE, PRESSURE, MIXING, EV   W70-00683
TEMPERATURE, SLUDGE WORMS, BLOODW   W70-00475
TEMPERATURE, NITROGEN, NUTRIENTS,   W70-00265
TEMPERATURE, NUTRIENT REQUIREMENT   W69-10161
TEMPERATURE, EPILIMNION, NITROGEN   W69-10167
TEMPERATURE, FLOW, SAMPLING, WAST   W69-10159
TEMPERATURE, PHOTOSYNTHESIS, CARB   W69-10169
TEMPERATURE, CHEMICAL ANALYSIS, A   W69-09676
TEMPERATURE, PHOTOSYNTHESIS, RAIN   W69-10154
TEMPERATURE, ORGANIC MATTER, HARD   W69-10157
TEMPERATURES, *COPPER DOSAGE, *CO   W69-10157
TEMPERATURE, DISSOLVED OXYGEN, DE   W71-07698
TEMPERATURE, PHYTOPLANKTON, ALGAE   W71-09767
TEMPERATURE, SEWAGE TREATMENT, AE   W71-08221
TEMPERATURE, NITRATES, AMMONIA, *   W72-06017
TEMPERATURE, PILOT PLANTS, HYDROG   W72-06022
TEMPERATURE, METABOLISM, EPIPHYTO   W72-05453
TEMPERATURE, CHLOROPHYLL, CYANOPH   W72-04761
TEMPERATURE, WASTE WATER TREATMEN   W71-12183
TEMPERATURE, ALKALINITY, NITRIFIC   W71-11393
TEMPERATURE, CENTRIFUGATION, COAG   W71-11375
TEMPERATURE, SOLAR RADIATION.: /O   W71-10788
TEMPERATURE, MICROORGANISMS, NITR   W71-13335
TEMPERATURE, FLOW, ALGAE, BENTHOS   W71-11381
TEMPERATURE, HEAT BUDGET, TURBIDI   W72-01773
TEMPERATURE, PROTEINS, CARBOHYDRA   W71-13413
TEMPERATURE, HYD: /EVEL, NITROGEN   W71-03048
TEMPERATURE, ALKALINITY, BICARBON   W71-05267
TEMPERATURE, VELOCITY, CYANOPHYTA   W71-00668
TEMPERATURE REQUIREMENTS.:   W71-02494
TEMPERATURE, HYDROGEN ION CONCENT   W71-03031
TEMPERATURE, CONDENSER PAGGAGE.: /   W71-02496
TEMPERATURE RANGES, TEMPERATURE R   W71-02494
TEMPERATURE, ACTIVATED SLUDGE, FI   W70-09905
TEMPERATURE, CURRENTS(WATER), HYP   W70-09889
TEMPERATURE, SALINITY, BIOASSAY,   W70-08111
TEMPERATURE, *ALGAE BLOOMS, CYANO   W70-09903
TEMPERATURE, NU: /L ANALYSIS, PHY   W70-06975
TEMPERATURE, METABOLISM, *WASTE W   W70-06604
TEMPERATURE, MANGANESE, CULTURES,   W70-07257
TEMPERATURE, CHEMICAL PROPERTIES,   W70-07313
TEMPERATURE EXPERIMENTS, RECYCLED   W70-07257
TEMPERATURE, LIGHT INTENSITY, NUT   W70-02965
TEMPERATURE, HYDROGEN ION CONCENT   W70-02804
TEMPERATURE, SILTS, STANDING CROP   W70-02780
TEMPERATURE, SHALLOW WATER, HYDRO   W70-02803
TEMPERATURE, CARBON DIOXIDE, HYDR   W70-02249
TEMPERATURE, IRON, NUTRIENTS, LIG   W70-01933
TEMPERATURE, TRACERS, CHLORELLA,   W70-02504
TEMPERATURE, WEATHER, EPILIMNION,   W70-02504
TEMPERATURE, RAIN, GEOLOGY, MAPPI   W70-02646
TEMPERATURE, OXYGEN, SALINITY.: /   W70-03325

TY, CALIFORNIA, LIGHT INTENSITY,
ERATURE, AEROBIC BACTERIA, WATER
ORGANIC MATTER, PHOTOSYNTHESIS,
YNTHESIS, EUTROPHICATION, LIGHT,
, SILICATES, ORGANIC MATTER, AIR
CROORGANISMS, CHEMICAL ANALYSIS,
ENVIRONMENTAL EFFECTS, SALINITY,
, SUCKERS, SHINERS, CADDISFLIES,
DISM, LIGHT INTENSITY, VELOCITY,
MEASUREMENTS, SEWAGE EFFLUENTS,
SHIRE, ALGAE, ODOR, TASTE, LIGHT
, MICHIGAN, INVERTEBRATES, FISH,
, ECOSYSTEMS, CYCLING NUTRIENTS,
DS-SETTLING, ECONOMIC ADVANTAGE,
IDATION, PHOTOSYNTHESES, PLANTS,
*RECREATION, RESERVOIRS, PONDS,
TA, BRACKISH WATER, OLIGOTROPHY,
GEN, HYDROGEN ION CONCENTRATION,
A, THERMAO POLLUTION, NUTRIENTS,
IC OCEAN, CARBON DIOXIDE, LIGHT,
RGANISMS, *LIPIDS, LIGHT, ALGAE,
LOROPHYLL, MAGNESIUM, POTASSIUM,
, PLANKTON, LI/ *EUTROPHICATION,
*RIVERS, *LAKES, *PONDS, LIGHT,
IFICATION, *SEASONAL, *SULFIDES,
ANKTON, *ALGAE, LIGHT INTENSITY,
SIGN CRITERIA, DISSOLVED OXYGEN,
A-DEEP PONDS', ASHKELON(ISRAEL),
S, PHOTOSYNTHESIS, ILLUMINATION,
LANKTON, NUTRIENTS, OLIGOTROPHY,
EA WATER, SEASONAL, PSEUDOMONAS,
MAGNA, SELENASTRUM, RHODOTORULA,
YTA, SCENEDESMUS, BENTH/ *ALGAE,
IRS, ZOOPLANKTON, PHYTOPLANKTON,
ES, PRODUCTIVITY, RADIOISOTOPES,
NALYSIS, *ANALYTICAL TECHNIQUES,
PLANTS, WATER PROPERTIES, WATER
ON CONCENTRATION, IRON, SULFATE,
REVIEWS, LIGHT INTENSITY, WATER
, DRAINAGE, FERROBACILLUS, WATER
IMNOLOGY, PHOTOSYNTHETIC OXYGEN,
RONMENT, MARINE ALGAE, SALINITY,
RS, LAKES, *GROWTH RATES, *WATER
MNOLOGY, WATER PROPERTIES, WATER
ING, HYDROGEN ION CONCENTRATION,
*CONDUCTIVITY, *LEACHING, IRON,
ION CONCENTRATION, CONDUCTIVITY,
LINITY, HARDNESS, RADIOACTIVITY,
S, ORGANIC MATTER, CONDUCTIVITY,
DES, HYDROGEN ION CONCENTRATION,
PLANTS, HEAD LOSS, PERMEABILITY,
RS, ESTUARINE ENVIRONMENT, WATER
UNCTIPENNIS, PISIDIUM IDAHOENSE,
N, RUTHENIUM-106, WHITE OAK LAKE(
ERVOIR(TENN), FRENCH BROAD RIVER(
VER(TENN), CHICKAMAUGA RESERVOIR(
ESERVOIR(TENN), NORRIS RESERVOIR(
OGETON PECTI/ CHEROKEE RESERVOIR(
HETERAMTHERA, PICKWICK RESERVOIR(
, PHOTIC ZONE, DOUGLAS RESERVOIR(
E RESERVOIR(TENN), HOLSTON RIVER(
CLAMS, STRONTIUM RADIOISOTOPES,
COLUMBIA RIVER, COLORADO RIVER,
ANGANESE, LOW-FLOW AUGMENTATION,
OLOGICAL HALF-LIFE, CLINCH RIVER(
NIUM, MICROCYSTIS, STIGEOCLONIUM
TION, *EXCRETION, STEPHANODISCUS
ORMIDIUM LAMINOSUM, OSCILLATORIA
NANCE SPECTROMETRY, OSCILLATORIA
SCILLATORIA TENUIS, OSCILLATORIA
YLENE GLYCOL, POLYMERS, DIMETHYL
OLOGY, CULTURAL CHARACTERISTICS,
, CILIATES, COPEPODIDS, NAUPLII,
SKA, LAKES, RIVERS, HOT SPRINGS,
ION, DENITRIFICATION, NUTRIENTS,
ENTS, BIOCHEMICAL OXYGEN DEMAND,
UENTS, STREAMFLOW, DESALINATION,
*RUNOFF, *LAKES, *IMPOUNDMENTS,
TES, NITRATES, SEWAGE EFFLUENTS,
NSITY, CARBON, SEWAGE TREATMENT,
NG, ALGAE, NUTRIENTS, TURBIDITY,
TREATMENT, SECONDARY TREATMENT,
E, IRON COMPOUNDS, LIME, OXYGEN,
NDS, SEEPAGE, INDUSTRIAL WASTES,
, PROTOZOA, ALGAE, SOLID WASTES,
UTROPHICATION, SEWAGE EFFLUENTS,
KISTRODESMUS BRAUNII, DUNALIELLA
SKELETONEMA COSTATUM, DUNALIELLA
TEST PROCEDUR/ *EUTROPHICATION,
EST ANALYSIS, *LABORATORY TESTS,

TEMPERATURE, IONS, SPECTROPHOTOME W70-03334
TEMPERATURE, IRON.: /ER, AIR TEMP W69-07096
TEMPERATURE, TROPICAL REGIONS.: / W69-06865
TEMPERATURE, CARBON DIOXIDE, NUTR W69-07442
TEMPERATURE, AEROBIC BACTERIA, WA W69-07096
TEMPERATURE, NITRATES, PLANTS, AM W69-07838
TEMPERATURE, SEA WATER, OYSTERS, W69-08267
TEMPERATURE, FLOW, DIATOMS, MANGA W69-07853
TEMPERATURE, CHEMICAL ANALYSIS, P W69-07862
TEMPERATURE, OXYGEN, SALINITY, AL W69-06203
TEMPERATURE, NUTRIENTS, COPPER SU W69-08282
TEMPERATURE, TRACERS, SAMPLING, P W69-09334
TEMPERATURE, LIGHT QUALITY, CHLOR W69-08526
TEMPERATURE-EFFECT.: /RYING, SOLI W70-05819
TEMPERATURE, *WASTE WATER TREATME W70-06600
TEMPERATURE, LIGHT PENETRATION, D W70-06225
TEMPERATURE, LIGHT INTENSITY, CHL W70-05752
TEMPERATURE, DEPTH, DENSITY, CHLO W70-05387
TEMPERATURE, PHOTOSYNTHESIS, LIGH W70-05270
TEMPERATURE.: /DIOISOTOPES, PACIF W70-05652
TEMPERATURE, CARBON DIOXIDE, OXID W70-05651
TEMPERATURE, EPILIMNION, LIGHT IN W70-05409
TEMPERATURE STRATIFICATION, DEPTH W70-05405
TEMPERATURE, FERTILIZATION, FISH, W70-05404
TEMPERATURE, CLIMATIC DATA, STRAT W70-04790
TEMPERATURE, PRIMARY PRODUCTIVITY W70-04809
TEMPERATURE, BIOCHEMICAL OXYGEN D W70-04787
TEMPERATURE PROFILES, VOLATILE SU W70-04790
TEMPERATURE, NUTRIENTS, FACULTATI W70-04786
TEMPERATURE, LIGHT INTENSITY, DAP W70-03959
TEMPERATURE, AMINO ACIDS, INTERTI W70-03952
TEMPERATURE COEFFICIENTS, NORTH A W70-03957
TEMPERATURE, CYANOPHYTA, CHLOROPH W70-03969
TEMPERATURE, *EPILIMNION, *HYPOLI W70-04475
TEMPERATURE, DEPTH, BACTERIA, SLI W70-04371
TEMPERATURE, PIPTIDES, INTERTIDAL W70-04365
TEMPERATURE, OUTLETS, HYDROELECTR W69-05023
TEMPERATURE, KENTUCKY.: /DROGEN I W69-00096
TEMPERATURE, DIATOMS, CHLOROPHYTA W68-00860
TEMPERATURE, HYDROGEN ION CONCENT W68-00891
TEMPERATURE, THERMAL POLLUTION, T W69-03516
TEMPERATURE, MONITORING, AERIAL P W68-00010
TEMPERATURE, SEASONAL, TEST PROCE W68-00472
TEMPERATURE, ALKALINITY, *PHOTOSY W68-00248
TEMPERATURE, TOXICITY, DISSOLVED W72-06854
TEMPERATURE, SOIL ANALYSIS, TASTE W72-11906
TEMPERATURE, DISSOLVED OXYGEN, PH W72-06051
TEMPERATURE, SALINITY, REAERATION W73-00750
TEMPERATURE, BACTERIA, NUTRIENTS, W73-00748
TEMPERATURE, CHLORINATED HYDROCAR W72-11105
TEMPERATURE, OPERATION AND MAINTE W72-08686
TEMPERATURE, *EUTROPHICATION, AQU W73-01446
TENDIPES PLUMOSUS, BEAVER DAM LAK W70-06217
TENN), ACTIVITY-DENSITY, VERTICAL W70-04371
TENN), CHICKAMAUGA RESERVOIR(TENN W71-07698
TENN), FLORENCE(ALA).: / BROAD RI W71-07698
TENN), FRENCH BROAD RIVER(TENN), W71-07698
TENN), HOLSTON RIVER(TENN), POTOM W71-07698
TENN), NAJAS, PHOTIC ZONE, DOUGLA W71-07698
TENN), NORRIS RESERVOIR(TENN), FR W71-07698
TENN), POTOMOGETON PECTINATUS, PO W71-07698
TENNESSEE RIVER, TROPHIC LEVEL, A W70-02786
TENNESSEE RIVER, HUDSON RIVER, DE W70-04507
TENNESSEE RIVER, ALGAE, IRON, ALK W73-00750
TENNESSEE), PHYSICAL HALF-LIFE, H W70-02786
TENUE, MICROCYSTIS AERUGINOSA, NO W70-02245
TENUIS, ASTERIONELLA FORMOSA, MEL W69-10158
TENUIS, OSCILLATORIA TEREBRIFORMI W70-05270
TENUIS, RAMMAN SPECTROSCOPY, MASS W69-00387
TEREBRIFORMIS, MATS, SYNECHOCOCCU W70-05270
TEREPHTHALATE.: /HYL ALCOHOL, ETH W70-10433
TERMARY COMPOUNDS, MACROMOLECULES W70-04280
TERRAMYCIN, TETRACYLINE, STREPTO W69-10154
TERRESTRIAL HABITATS, DEPTH, AIR, W69-08272
TERTIARY TREATMENT, ALGAE, AMMONI W69-08053
TERTIARY TREATMENT, ALGAE, BACTER W69-08054
TERTIARY TREATMENT, LAGOONS, OXYG W69-07389
TERTIARY TREATMENT, ALGAE, DOMEST W70-04765
TERTIARY TREATMENT, WATER POLLUTI W70-04074
TERTIARY TREATMENT, EFFLUENTS, LI W70-05387
TERTIARY TREATMENT, SAMPLING, MON W70-05645
TERTIARY TREATMENT, ELECTRODIALYS W69-01453
TERTIARY TREATMENT, COST COMPARIS W68-00256
TERTIARY TREATMENT, CHEMICAL PREC W69-10178
TERTIARY TREATMENT.: /L, BACTERIA W70-01027
TERTIARY TREATMENT, ALGAE, WATER W70-03512
TERTIOLECTA, DCMU INHIBITOR.: /AN W70-05261
TERTIOLECTA, AMPHIDINIUM CARTERI, W69-08278
TEST ANALYSIS, *LABORATORY TESTS, W68-00472
TEST PROCEDURES, NURTIENTS, RIVE W68-00472

WATER ANALYSIS, INSECTS, SNAILS,
TA, *DECOMPOSING ORGANIC MATTER,
EVEL, PRODUCTIVITY, FOOD CHAINS,
OTROPHY, VITAMINS, HEAVY METALS,
LOGY, *BIOINDICATORS, *BIOASSAY,
IQUES, SAMPLING, ORGANIC MATTER,
Y, *FISH DISEASES, *FISH TOXINS,
-DICHLORONAPHTHOQUIN/ *SELECTIVE
ON, SEVIN.:   *ALGAE
ALGAL CONTROL, WATER POLLUTION,
EFFECTS, STABILIZATION, LAGOONS,
, PHOSPHATES, ORGANIC COMPOUNDS,
TION, BIOCHEMICAL OXYGEN DEMAND,
PLANTS, BACTERIA, SEDIMENTATION,
, CONTOURS, DEPTH, OLIGOCHAETES,
PHOTOSYNTHESIS, EUTROPHICATION,
TOPLANKTON, *DDT, *MARINE ALGAE,
S, COLOR, FISHKILL, LAKES, ODOR,
S, PESTICIDE KINETICS, PESTICIDE
UTRIENT REQUIREMENTS, NUTRIENTS,
UTRIENTS, *NITRIFICATION, CROPS,
OD WEBS, DDT/ *ALGAE, *PESTICIDE
-WATER INTERFACES, EUGLENOPHYTA,
YANAPHYTA, ALGAL CONTROL, *ALGAL
D TOXINS, IBOGAALKALOID, ORGANIC
*INHIBITION, *P/ *ALGAE, *ALGAL
TOXINS, *ALGAL POISONING, *ALGAL
GEN, FISHKILLS, LIMNOLOGY, ALGAL
ROMETERS, DIFFUSION, TURBULENCE,
DINOFLAGELLATES, FISHKILL, FISH
ROPHICATION, COLOR, TASTE, ODOR,
NIUM SP., *GLENODININE, ALKALOID
VED SOLIDS, NUTRIENTS, VITAMINS,
BIOASSAY, AQUATIC ALGAE, *ALGAL
YTOPLANKTON, BACTERICIDES, ALGAL
BIOCHEMICAL OXYGEN DEMAND, ALGAL
ITY, HYDROGEN ION CONCENTRATION,
ANTS, WASTES, THERMAL POLLUTION,
AINAGE EFFECTS, *SAMPLING, ALGAE
PULP WASTES, INDUSTRIAL WASTES,
OTOXICITY, *FISH DISEASES, *FISH
LLOUT, *ECOSYSTEMS, GREAT LAKES,
GRAPHY, *COBALT, *COPPER, *IRON,
ASSIUM, SODIUM, IRON, MANGANESE,
Y, ZOOPLANKTON, ALGAE, COLLOIDS,
OROPHYTA, CHLOROPHYLL, BACTERIA,
OTOSYNTHESIS, CHEMICAL ANALYSIS,
UM, CLAMS, SIZE, ALGAE, DIATOMS,
GY, ALGAL TOXINS, ALGAL CONTROL,
LATORS, PREDATION, PRODUCTIVITY,
LFATE, ALGAL CONTROL, ALGICIDES,
AMINS, PEPTIDES, ORGANIC MATTER,
TION, SODIUM, ALGAE, PHOSPHORUS,
HOTOMETRY, CHLORELLA, NUTRIENTS,
, PHOSPHORUS, SODIUM, POTASSIUM,
OGRAPHY, RADIOCHEMICAL ANALYSIS,
OTOPES, LIMNOLOGY, OCEANOGRAPHY,
HIC LEVEL, SAMPLING TEMPERATURE,
ES, PLANKTON, ECOSYSTEMS, SINKS,
TER, PHYSICOCHEMICAL PROPERTIES,
M, SELECTIVITY, POISONS, OXYGEN,
*AQUATIC ENVIRONMENT, CHLORELLA,
, CHEMICAL REACTIONS, CHLORELLA,
NVERTEBRATES, FISH, TEMPERATURE,
, CHROMATOGRAPHY, RADIOACTIVITY,
*TURBID RESERVOIRS, CYCLOTELLA,
RULINA, VOLVOCALES, GLENODINIUM,
OXIS ANNULARIS, GASTROINTESTINAL
WN(MO), LEE'S SUMMIT(MO), OXYGEN
OOD ORGANISMS, NUTRIENTS, ENERGY
IA, FORAMINIFERA, THIO/ *MINERAL
, DESORPTION, EMBRYOLOGY, ENERGY
TROSCOPY, ULTRAVIOLET RADIATION,
PRODUCTIVITY, AQUATIC NUISANCES,
IZOMENON FLOS-AQUAE, SECCHI DISC
H), *OXYGEN DEFICIT, SECCHI DISC
HARTBEESPOORT DAM(SOUTH AFRICA),
S, DIURON, FLAGELLATES, MONURON,
EUPHOT/ *LAKE TAHOE(CALIF-NEV),
EDEN, LAKE ERKEN(SWEDEN), ACTIVE
MAXIMUM UPTAKE VELOCITY, ACTIVE
HYLLUM, EICHHORNIA, POTAMOGETON,
LAKE MINNETONKA(MINN), NUTRIENT
S, ALKALINE WATER, ACIDIC WATER,
PILOT PLANTS, ACTIVATED SLUDGE,
CTIVATED SLUDGE PROCESS, HANOVER
KES, NUTRIENTS, QUALITY CONTROL,
GROWTH, SEDGWICK-RAFTER METHOD,
ES, NEUTRIENTS, BACTERIA, ALGAE,
*NEW ZEALAND, SEWAGE

TOXICITY, RIVERS, HISTOGRAMS.: /   W72-01801
TOXICITY, DISSOLVED OXYGEN, ALGAE   W72-01797
TOXICITY, WATER POLLUTION CONTROL   W72-01788
TOXICITY, DIATOMS, MARL.: /, OLIG   W71-10079
TOXICITY, AQUATIC MAMMALS, ALGAE,   W71-10786
TOXICITY, PHENOLS, ODOR, COLOR, T   W72-03299
TOXICITY, HYDROGEN ION CONCENTRAT   W71-07731
TOXICITY, CHEMICAL STRUCTURE, 2,3   W70-02982
TOXICITY, CARBON-14, DDT, MALATHI   W70-02198
TOXICITY, LAKES, DESIGN CRITERIA,   W70-02775
TOXICITY, WASTE WATER TREATMENT,   W69-09454
TOXICITY, CHLOROPHYTA, LAKES, RES   W69-10180
TOXICITY, OXYGEN-REDUCTION POTENT   W70-05750
TOXICITY, COST-BENEFIT ANALYSIS.:   W70-05819
TOXICITY, FISH, BLOODWORMS, ALGAL   W70-06217
TOXICITY, MIDGES, MOSQUITOES, SAM   W70-06225
TOXICITY, DIATOMS, ESTUARIES, PHY   W70-05272
TOXICITY, WATER POLLUTION EFFECTS   W69-08279
TOXICITY, POLLUTANTS, WATER POLLU   W69-08565
TOXICITY, WATER CHEMISTRY, WATER   W69-05867
TOXICITY, SOIL POROSITY, IRRIGATI   W69-05323
TOXICITY, *PESTICIDE REMOVAL, *FO   W70-03520
TOXICITY, BIOASSAYS, DIATOMS,: /D   W70-04381
TOXINS MEMBRANE, NORTHEAST U.S.: /   W69-05306
TOXINS.: /, *GLENODININE, ALKALOI   W71-07731
TOXINS, *AQUARIA, *AQUATIC WEEDS,   W69-03188
TOXINS, *PHYTOTOXICITY, *FISH DIS   W71-07731
TOXINS, ALGAL CONTROL, TRACE ELEM   W70-00264
TOXINS, BACTERIA, COPPER SULFATE,   W70-04369
TOXINS, BIOASSAY, AQUATIC ALGAE,   W70-08372
TOXINS, DISSOLVED OXYGEN, LAKES,   W69-05697
TOXINS, IBOGAALKALOID, ORGANIC TO   W71-07731
TOXINS, INHIBITORS, SUCCESSION, C   W70-02510
TOXINS, KILLIFISHES, *TOXICITY, I   W70-08372
TOXINS, LARVAE, PHENOLS, SHELLFIS   W70-01068
TOXINS, ORGANIC WASTES, POLLUTANT   W69-01273
TOXINS, OXYGEN, DECOMPOSIN: /ALIN   W70-02803
TOXINS, PESTICIDE RESIDUES, PHOSP   W69-06305
TOXINS, SALINE WATER, EQUILIBRIUM   W69-00360
TOXINS, SOLID WASTES, RADIOACTIVE   W72-00462
TOXINS, TOXICITY, HYDROGEN ION CO   W71-07731
TRACE ELEMENTS, RADIUM RADIOISOTO   W71-09013
TRACE ELEMENTS, ALGAE, SEA WATER,   W71-11687
TRACE ELEMENTS, CHELATION, PHOTOS   W70-02804
TRACE ELEMENTS, DISSOLVED SOLIDS,   W70-02510
TRACE ELEMENTS, VITAMINS, AMINO A   W70-02255
TRACE ELEMENTS, TEMPERATURE, IRON   W70-01933
TRACE ELEMENTS, METABOLISM,: /SSI   W70-00708
TRACE ELEMENTS, WASHINGTON, NUCLE   W70-00264
TRACE ELEMENTS, SODIUM, VIRUSES,   W69-04801
TRACE ELEMENTS, CORES, WISCONSIN,   W69-03366
TRACE ELEMENTS, LIGHT INTENSITY,   W70-04369
TRACE ELEMENTS, PLANKTON, LAKES,   W70-04385
TRACE ELEMENTS, CHLAMYDOMONAS, CH   W70-05409
TRACE METALS, VITAMINS, WATER POL   W70-04503
TRACERS.: /HATES, *ALGAE, CHROMOT   W71-03033
TRACERS, BACTERIA, GROWTH RATES,   W69-10163
TRACERS, CHLORELLA, HYDROGEN ION   W70-02504
TRACERS, CRABS, SHRIMP, MULLETS,   W69-08274
TRACERS, DISTRIBUTION, CHELATION,   W69-08275
TRACERS, PHOSPHORUS RADIOISOTOPES   W70-05399
TRACERS, PHOTOSYNTHESIS, CARBON R   W70-02786
TRACERS, RADIOACTIVITY, CARBON RA   W70-08097
TRACERS, SAMPLING, PERIPHYTON, AQ   W69-09334
TRACERS, WATER POLLUTION EFFECTS,   W70-02510
TRACHELOMONAS, MELOSIRA, STEPHANO   W72-04761
TRACHELOMONAS.: /SCILLATORIA, SPI   W70-05404
TRACT.: /RELLI, BODY BURDENS, POM   W70-02786
TRANSFER, PH, SUSPENDED SOLIDS.: /   W70-06601
TRANSFER, *FOOD CHAINS, FOOD WEBS   W70-05428
TRANSFORMATION, METAZOA, RADIOLAR   W69-07428
TRANSFORMATIONS, HETEROTROPHY, OR   W69-07440
TRANSMISSIVITY, CHLOROPHYLL.: /EC   W69-09676
TRANSPARENCY, AGRICULTURAL RUNOFF   W69-09349
TRANSPARENCY, ANABAENA LEMMERMANN   W69-10169
TRANSPARENCY, MYXOPHYCEA, OSCILLA   W69-10182
TRANSPARENCY, MARGINAL VEGETATION   W70-06975
TRANSPARENCY, CYCLOPS, SCENEDESMU   W70-10161
TRANSPARENCY, LIGHT MEASUREMENTS,   W70-01933
TRANSPORT, CARBON-14, ENZYME KINE   W70-04284
TRANSPORT, BACTERIAL POPULATIONS,   W70-04194
TRAPA, PISTIA, ALTERNANTHERA, HET   W70-10175
TRAPS, NUTRIENT INFLUX, NUTRIENT   W70-04268
TRAVERTINE, CALCIUM CARBONATE, BA   W69-10160
TREAT: /REATMENT, SUSPENDED LOAD,   W70-09186
TREATMENT PLANT.: /UM SULFATE, *A   W70-09186
TREATMENT FACILITIES, FILTRATION,   W70-00263
TREATMENT TIMING, ALGAE CONTROL R   W69-10157
TREATMENT FACILITIES, TEST PROCED   W70-01519
TREATMENT POND ALGAE.:   W70-02304

FLOW RATES, *FILTERS, *NEW YORK,
TION, DIATOMACEOUS EARTH, ALGAE,
HNIQUES, *WASTE WATER TREATMENT,
TES, NUTRIENTS, BACTERIA, ALGAE,
(WASH).: *SEATTLE(WASH), RENTON
, *COLOR, WASTE WATER TREATMENT,
SITY, CONVECTION CURRENTS, ALGAE
TIVATED CARBON, CHLORINE, COSTS,
URIFICATION, MEMBRANES, SCREENS,
S, CHLORINE, COSTS, *WASTE WATER
DORS, EFFICIENCIES, *WASTE WATER
YSIS, WATER PURIFICATION, *WATER
L, *WATER REUSE, SEASONAL, WATER
EUTROPHICATION
LANTS, TEMPERATURE, *WASTE WATER
NTROL, INDUSTRIAL WASTES, SEWAGE
INEERING, WATER TREATMENT, WASTE
OXYGEN, FARM WASTES, WASTE WATER
ATION, MAINTENANCE, *WASTE WATER
OLOGICAL TREATMENT, *WASTE WATER
WATER PURIFICATION, WASTE WATER
DIOXIDE, DIFFUSION, *WASTE WATER
ICAL OXYGEN DEMAND, *WASTE WATER
TESTS, TEMPERATURE, WASTE WATER
NTIFICATION, ALGAE, *WASTE WATER
S, FLOCCULATION, WATER POLLUTION
MONIA, *WATER REUSE, WASTE WATER
ION, BACTERIA, IMPURITIES, WASTE
A, ALGAE, SOLID WASTES, TERTIARY
S, ACTINOMYCETES, PONDING, WASTE
YGEN, PRODUCTIVITY, *WASTE WATER
ND, FISH, ALGAE, FOAMING, SEWAGE
S, PONDS, OXIDATION, WASTE WATER
VATED SLUDGE, ALGAE, WASTE WATER
IOLOGICAL TREATMENT, WASTE WATER
G, SEWAGE TREATMENT, WASTE WATER
R, ORGANIC LOADING, *WASTE WATER
N DEMAND, DELAWARE, *WASTE WATER
IA, *PHOTOSYNTHESIS, WASTE WATER
S, *WASTE WATER TREATMENT, WASTE
NUTRIENTS, NITROGEN, WASTE WATER
OL, *DENIT/ *ALGAE, *WASTE WATER
NISMS, ELECTRIC CURRENTS, *WASTE
A, DAPHNIA, DETERIORATION, WATER
OL, *DENIT/ *ALGAE, *WASTE WATER
R POLLUTION EFFECTS, WASTE WATER
OL, *DENIT/ *ALGAE, *WASTE WATER
A, DAPHNIA, DETERIORATION, WASTE
OL, *DENIT/ *ALGAE, *WASTE WATER
, *PHOSPHORUS, ALGAE, BIOLOGICAL
AE, ORGANIC LOADING, WASTE WATER
RATURE, METABOLISM, WASTE WATER
, TASTE, ODOR, DEEP WELLS, WATER
DIMENTS, NUTRIENTS, *WASTE WATER
AL OXYGEN DEMAND, MIXING, SEWAGE
, SOLIDS CONTACT PROCESS, SEWAGE
YGEN DEMAND, BIOCHEMI/ *TERTIARY
, TRICKLING FILTERS, WASTE WATER
SI/ *REMOVAL PROCESSES, CHEMICAL
ITROGEN CYCLE, ALGAE, BIOLOGICAL
I/ *TERTIARY TREATMENT, *AEROBIC
-SCALE STUDIES, PLAN/ *ANAEROBIC
ENT, OVERFERTILIZATION, PHYSICAL
L, SULFUR BACTERIA, *WASTE WATER
N DEMAND, NUTRIENTS, WASTE WATER
STE WATER TREATMENT, *BIOLOGICAL
ND, ANAEROBIC DIGESTION, AEROBIC
SCENEDESMUS, ADSORPTION, SEWAGE
*WATER POLLUTION / *WASTE WATER
*SEWAGE LAGOONS, *ALGAE, *SEWAGE
ATION, *ALGAL CONTROL, *TERTIARY
*TERTIARY TREATMENT, BIOLOGICAL
ICAL OXYGEN DEMAND, *WASTE WATER
A, *ALGAE, SORPTION, WASTE WATER
BAFFLES, *WASTE WAT/ *BIOLOGICAL
ENTRATION, *NUTRIENTS, *TERTIARY
LAGOONS, *BAFFLES, *WASTE WATER
HEMICAL DEGRADATION, *BIOLOGICAL
LLUTION CONTRO/ *WATER POLLUTION
S, *DOMESTIC WASTES, WASTE WATER
OSPHATES, ALGAE, WATER POLLUTION
, LAGOONS, TOXICITY, WASTE WATER
WASTE WATER DIVERSION, INSTREAM
N, *OXYGEN, *FLUCTUATION, SEWAGE
SEWAGE, SEWAGE BACTERIA, SEWAGE
AGE, INDUSTRIAL WASTES, TERTIARY
*BIOLOGICAL COMMUNITIES, *SEWAGE
IDAT/ *WATER REUSE, *WASTE WATER
FFLUENTS, OXYGEN DEMAND, AEROBIC
RGANISMS, NUISANCE ALGAE, SEWAGE

TREATMENT FACILITIES.: /OLLOIDS,   W72-04145
TREATMENT FACILITIES.: /, *FILTRA   W72-04318
TREATMENT FACILITIES, MISSISSIPPI   W71-13313
TREATMENT FACILITIES, TEST PROCED   W71-06737
TREATMENT PLANT, DUWAMISH ESTUARY   W70-04283
TREATMENT FACILITIES, AERATION, L   W70-04060
TREATMENT COSTS, DINOBRYON, CONCO   W69-08282
TREATMENT FACILITIES, CLEANING, *   W72-08859
TREATMENT FACILITIES, *WISCONSIN.   W73-02426
TREATMENT.: /ACTERIA, BACTERICIDE   W72-12987
TREATMENT.: /, DESIGN CRITERIA, O   W72-06854
TREATMENT.: / CLEANING, COST ANAL   W72-08688
TREATMENT.: /NT, *NUTRIENT REMOVA   W72-08133
TREATMENT.:   W70-04721
TREATMENT.: /N, PHOTOSYNTHESES, P   W70-06600
TREATMENT.: /, PHOSPHORUS, CHEMCO   W70-05668
TREATMENT.: /UARIES, SANITARY ENG   W70-05094
TREATMENT.: /DGE, PHOTOSYNTHETIC   W71-08221
TREATMENT.: /NS, ALGAE, ADMINISTR   W71-08970
TREATMENT.: /, AQUATIC ALGAE, *BI   W71-09546
TREATMENT.: /FILTRATION, AMMONIA,   W71-11393
TREATMENT.: /NG, METHANE, CARBON   W71-11377
TREATMENT.: /L, DEGRADATION, CHEM   W71-13335
TREATMENT.: /D OXYGEN, LABORATORY   W71-12183
TREATMENT.: /VELOPMENT, WASTE IDE   W72-03299
TREATMENT.: /FILTERS, *SCENEDESMU   W72-03703
TREATMENT.: /RATURE, NITRATES, AM   W72-06017
TREATMENT.: / LIFE, STREAM POLLUT   W72-01786
TREATMENT.: /L, BACTERIA, PROTOZO   W70-01027
TREATMENT.: /ENA, BACTERIA, YEAST   W70-08097
TREATMENT.: /Y, PHOTOSYNTHETIC OX   W70-08416
TREATMENT.: /CHEMICAL OXYGEN DEMA   W70-08740
TREATMENT.: / CHLOROPHYLL, LAGOON   W70-08392
TREATMENT.: / DEMAND, COLOR, ACTI   W70-10433
TREATMENT.: /PHOSPHORUS, ALGAE, B   W70-07469
TREATMENT.: /OXYGEN DEMAND, MIXIN   W70-06899
TREATMENT.: /ISSOLVED OXYGEN, ODO   W70-06601
TREATMENT.: /A, BIOCHEMICAL OXYGE   W71-05013
TREATMENT.: /OORGANISMS, CALIFORN   W71-05267
TREATMENT.: /MS, ELECTRIC CURRENT   W71-03775
TREATMENT, PHOSPHOROUS, RUNOFF, D   W71-04072
TREATMENT, *WATER POLLUTION CONTR   W71-04557
TREATMENT, WASTE WATER TREATMENT.   W71-03775
TREATMENT, DEGRADATION(DECOMPOSIT   W71-05719
TREATMENT, *WATER POLLUTION CONTR   W71-04554
TREATMENT, PENNSYLVANIA.: /, WATE   W71-05026
TREATMENT, *WATER POLLUTION CONTR   W71-04555
TREATMENT, DEGRADATION(DECOMPOSIT   W71-03056
TREATMENT, *WATER POLLUTION CONTR   W71-04556
TREATMENT, WASTE WATER TREATMENT.   W70-07469
TREATMENT, WATER UTILIZATION, IND   W70-07508
TREATMENT, LIGHT, MICROORGANISMS,   W70-06604
TREATMENT, MECHANICAL EQUIPMENT,   W70-08096
TREATMENT, *OXIDATION LAGOONS, MI   W70-07031
TREATMENT, WASTE WATER TREATMENT.   W70-06899
TREATMENT, TRICKLING FILTERS, WAS   W70-06792
TREATMENT, *AEROBIC TREATMENT, OX   W70-06792
TREATMENT, PHOSPHATES, NITRATES,   W70-06792
TREATMENT, OVERFERTILIZATION, PHY   W70-07469
TREATMENT, BOTTOM SEDIMENTS, NUTR   W70-07031
TREATMENT, OXYGEN DEMAND, BIOCHEM   W70-06792
TREATMENT, *AUSTRALIA, LABORATORY   W70-06899
TREATMENT, UNIT OPERATIONS, UNIT   W70-07469
TREATMENT, *OXIDATION LAGOONS, *M   W70-07034
TREATMENT, *FIBER(PLANT).: /OXYGE   W70-06925
TREATMENT, *WATER POLLUTION CONTR   W70-10437
TREATMENT, CHEMICAL OXYGEN DEMAND   W71-00936
TREATMENT, PARTICLE SIZE.: /ELLA,   W71-03035
TREATMENT, *BIOLOGICAL TREATMENT,   W70-10437
TREATMENT, *INDIANA, CHLOROPHYTA,   W70-10173
TREATMENT, SUSPENDED LOAD, PILOT   W70-09186
TREATMENT, NITROGEN, PHOSPHORUS,   W70-08980
TREATMENT, *WATER POLLUTION CONTR   W70-09186
TREATMENT, CULTURES, CHLORELLA, W   W70-09438
TREATMENT, *OXIDATION, LAGOONS, *   W70-09320
TREATMENT, BIOLOGICAL TREATMENT,   W70-08980
TREATMENT, *POULTRY, INDUSTRIAL W   W70-09320
TREATMENT, *BIODEGRADATION, *BIOC   W70-09186
TREATMENT, *FILTRATION, *WATER PO   W70-01027
TREATMENT, INDUSTRIAL WASTES, NEU   W70-01519
TREATMENT, WATER QUALITY CONTROL,   W70-01031
TREATMENT, WASTE WATER DISPOSAL.:   W69-09454
TREATMENT, BOTTOM SEALING, LAKE D   W69-09340
TREATMENT, SOLAR RADIATION, CHLOR   W69-10182
TREATMENT, WATER POLLUTION, WATER   W69-10171
TREATMENT, CHEMICAL PRECIPITATION   W69-10178
TREATMENT, *TRICKLING FILTERS, *B   W72-01819
TREATMENT, *INDUSTRIAL WASTES, OX   W72-00027
TREATMENT, PHOTOSYNTHESIS, ALGAE,   W72-00462
TREATMENT, SELF-PURIFICATION, DIS   W72-01798

SLUDGE, *BACTERIA, META/	*SEWAGE TREATMENT, *PROTOZOA, *ACTIVATED	W72-01817
LLUTION, WASTE DISPOSAL, *SEWAGE	TREATMENT, *WASTE WATER TREATMENT	W72-01786
*SEWAGE TREATMENT, *WASTE WATER	TREATMENT, *WASTE WATER DISPOSAL,	W72-01786
, RHIZOPODS, SPIDERS, MILK WASTE	TREATMENT, COLONY COUNTS, ARTHROP	W72-01819
SUPPLY, *BASINS, WATER POLLUTION	TREATMENT, COPPER SULFATE, ODOR,	W72-01814
LGAE, *BACTERIA, *DISSO/	*SEWAGE TREATMENT, *OXIDATION LAGOONS, *A	W72-01818
EAD LOSS, FILTERS, *SCEN/	*WATER TREATMENT, *ALGAE, *FILTRATION, H	W72-03703
LGAE, LIGHT INTENS/	*WASTE WATER TREATMENT, *OXIDATION LAGOONS, *A	W72-06022
NGANESE, *COLOR, *ALGAE,/	*WATER TREATMENT, *TURBIDITY, *IRON, *MA	W72-04145
SING ORGANIC MATTER, *PO/	*WATER TREATMENT, ODOR, *ALGAE, *DECOMPO	W72-04242
E, *EUTROPHICATION,/	*WASTE WATER TREATMENT, WATER TREATMENT, *ALGA	W72-01867
N,/ WASTE WATER TREATMENT, WATER	TREATMENT, *ALGAE, *EUTROPHICATIO	W72-01867
ATED CARBON, *ADSORPTION, *WATER	TREATMENT, *WASTE WATER TREATMENT	W72-03641
C MATTER, ALGAE, COPPER /	*WATER TREATMENT, WATER QUALITY, *ORGANI	W72-04495
*ALGICIDES, ACTIVATED C/	*WATER TREATMENT, *ALGAE, *TASTE, *ODOR,	W72-03746
TION, *GROUNDWATER, RECH/	*WATER TREATMENT, *WATER REUSE, *RECLAMA	W72-04318
LIME, *ULTRAVIOLET RADIA/	*WATER TREATMENT, *ALGAE, FILTRATION, *S	W72-03824
, *WATER TREATMENT, *WASTE WATER	TREATMENT, ALGAE, FILTRATION, NEW	W72-03641
OIRS, *DECIDUOUS TREES, /	*WATER TREATMENT, *TASTE, *ODOR, *RESERV	W72-03475
, *TASTE, *ODOR, *NATURA/	*WATER TREATMENT, *ALGAE, ALGAL CONTROL	W72-03562
R POLLUTION, DEGRADATION, SEWAGE	TREATMENT, NITROGEN COMPOUNDS, PH	W72-02817
, *ODOR, ACTIVATED CARBO/	*WATER TREATMENT, *ALGAE, *ACTINOMYCETES	W72-04155
LYTICAL TECHNIQUES, *WASTE WATER	TREATMENT, TREATMENT FACILITIES,	W71-13313
RAINS, SEWAGE EFFLUENTS, *SEWAGE	TREATMENT, STORM RUNOFF.: /TORM D	W71-10654
RM WASTES, *POULTRY, WASTE WATER	TREATMENT, SEDIMENTATION, OXIDATI	W71-11375
MICAL OXYGEN DEMAND, WASTE WATER	TREATMENT, OVERFLOW, SEWERS, STOR	W71-10654
TIAL, AQUATIC ALGAE, *BIOLOGICAL	TREATMENT, *WASTE WATER TREATMENT	W71-09546
TERIA, WATER TEMPERATURE, SEWAGE	TREATMENT, AEROBIC BACTERIA, AQUA	W71-08221
IC WASTES, *OXYGENATION, AEROBIC	TREATMENT, ANAEROBIC CONDITIONS,	W71-09546
R, SEWAGE DISPOSAL, *WASTE WATER	TREATMENT, *OXIDATION LAGOONS.: /	W71-09548
S, *DOMESTIC WASTES, WASTE WATER	TREATMENT, INDUSTRIAL WASTES, NUT	W71-06737
ITRUS FRUITS, TOMATOES, *AEROBIC	TREATMENT, *ANAEROBIC DIGESTION,	W71-09548
R SUPPLY, *WATER QUALITY, *WATER	TREATMENT, FILTRATION, CHLORINATI	W71-05943
GAE, *PLANKTON, *SEWAGE, *SEWAGE	TREATMENT, *EUTROPHICATION, PONDS	W70-02304
*METAL CORROSION, COOLING WATER	TREATMENT, ON-STREAM DESLUDGING.:	W70-02294
, *SANITARY ENGINEERING, *SEWAGE	TREATMENT, *TRACERS, *WATER POLLU	W70-02609
MPARISONS, TREATMENT, BIOLOGICAL	TREATMENT, ALGAL CONTROL, AMMONIA	W70-01981
AND APPLICATION, ELECTROCHEMICAL	TREATMENT, AMMONIA STRIPPING, ALG	W70-01981
EFFICIENCIES, COST COMPARISONS,	TREATMENT, BIOLOGICAL TREATMENT,	W70-01981
SION, *CHEMICAL REACTION/	*WATER TREATMENT, *COOLING WATER, *CORRO	W70-02294
SEWAGE, SEWAGE EFFLUENTS, SEWAGE	TREATMENT, WATER POLLUTION CONTRO	W70-02248
ES, PERSISTENCE, WATER POLLUTION	TREATMENT, WATER POLLUTION EFFECT	W70-01777
F, *FERTILIZATION, LAKES, SEWAGE	TREATMENT, SEWAGE BACTERIA, NITRO	W70-02787
OONS, *INDUSTRIAL WASTES, SEWAGE	TREATMENT, NORTH DAKOTA, COLIFORM	W70-03312
DOMESTIC WASTES, SEWERS, SEWAGE	TREATMENT, SEWAGE DISPOSAL, STORM	W70-03344
TION, SEWAGE EFFLUENTS, TERTIARY	TREATMENT, ALGAE, WATER QUALITY,	W70-03512
VIRONMENTAL EFFECTS, *BIOLOGICAL	TREATMENT, *DESIGN CRITERIA, DISS	W70-04787
ES, *CHLORINATION, *ODOR, *WASTE	TREATMENT, ALGAE, DISINFECTION, B	W70-04788
IES, SANITARY ENGINEERING, WATER	TREATMENT, WASTE TREATMENT.: /UAR	W70-05094
, *ENVIRONMENTAL EFFECTS/	*WASTE TREATMENT, *STABILIZATION, *PONDS	W70-04787
TREAMS, WATER POLLUTION, *WASTE	TREATMENT, *WATERS, *ORGANISMS, S	W70-05750
CAL OXYGEN DEMAND, *ACT/	*SEWAGE TREATMENT, *FILTRATION, *BIOCHEMI	W70-05821
TERIA, CULTURES, C/	*WASTE WATER TREATMENT, *NUTRIENTS, ALGAE, BAC	W70-05655
E WATER RECLAMATION, WASTE WATER	TREATMENT, *NUTRIENTS.: /TS, WAST	W70-06053
D, *ACTIVATED SLUDGE, BIOLOGICAL	TREATMENT, PILOT PLANTS, HUMUS, R	W70-05821
, NUTRIENTS, TURBIDITY, TERTIARY	TREATMENT, SAMPLING, MONITORING,	W70-05645
*NUTRIENTS, *PESTICI/	*TERTIARY TREATMENT, *PHOSPHORUS COMPOUNDS,	W70-05470
LIGHT INTENSITY, CARBON, SEWAGE	TREATMENT, TERTIARY TREATMENT, EF	W70-05387
RBON, SEWAGE TREATMENT, TERTIARY	TREATMENT, EFFLUENTS, LIMNOLOGY,	W70-05387
ORS, ODOR, TASTE, FILTERS, WATER	TREATMENT, DISSOLVED OXYGEN, EUTR	W70-05264
GAE, *FLOCCULATION, *WASTE WATER	TREATMENT, ELECTROLYTES, CHEMICAL	W70-05267
, *DESIGN CRITERIA, FACI/	*WASTE TREATMENT, *STABILIZATION, *PONDS	W70-04786
LAKES, *IMPOUNDMENTS, TERTIARY	TREATMENT, ALGAE, DOMESTIC WASTES	W70-04765
E DIST/ *ALGAE, *WATER POLLUTION	TREATMENT, LAKES, SEWAGE, DRAINAG	W70-04455
S, URBANIZATION, WATER POLLUTION	TREATMENT, ALGAE, WATER MANAGEMEN	W70-04721
ATES, SEWAGE EFFLUENTS, TERTIARY	TREATMENT, WATER POLLUTION SOURCE	W70-04074
YGEN DEMAND, *COLOR, WASTE WATER	TREATMENT, TREATMENT FACILITIES,	W70-04060
CATORS, *WATER POLLUTION, *WASTE	TREATMENT, FOOD CHAINS, CHLORELLA	W70-04381
S, *PHOSPHATES, CHELATION, WASTE	TREATMENT, AQUATIC ALGAE, AQUATIC	W70-04373
, *LAKES, *DETERGENTS, *TERTIARY	TREATMENT, *GREAT LAKES, CONTROL,	W70-03964
W YORK, SEWAGE EFFLUENTS, SEWAGE	TREATMENT, INDUSTRIAL WASTES, WAT	W70-03974
DIES, ELECTRON MICROSCOPY, WASTE	TREATMENT, LAKES, WATER QUALITY,	W70-03973
RECTIVE TREATMENT, *PREVENTATIVE	TREATMENT, ALGAE NUISANCE, ALGAE	W69-08282
OINDICATORS, PRODUCTIVITY, WASTE	TREATMENT, WATER POLLUTION EFFECT	W69-08518
ENTS, *EUTROPHICATION, *TERTIARY	TREATMENT, NITROGEN, AQUATIC PLAN	W69-09340
T, ALGAE NUISANCE, /	*CORRECTIVE TREATMENT, *PREVENTATIVE TREATMEN	W69-08282
REAMFLOW, DESALINATION, TERTIARY	TREATMENT, LAGOONS, OXYGENATION,	W69-07389
E, *BIOLOGICAL TREATMENT, *WATER	TREATMENT, *POLLUTANT IDENTIFICAT	W69-07389
LUTANT IDEN/ *ALGAE, *BIOLOGICAL	TREATMENT, *WATER TREATMENT, *POL	W69-07389
C DETERIORATION, *ADVANCED WASTE	TREATMENT, *NUTRIENT REMOVAL, AGR	W69-07818
GAE, POLLUTION ABATEMENT, SEWAGE	TREATMENT, SEWAGE DISPOSAL, WASTE	W69-06909
TRIFICATION, NUTRIENTS, TERTIARY	TREATMENT, ALGAE, AMMONIA.: /DENI	W69-08053
N, NITROGEN COMPOU/	*WASTE WATER TREATMENT, *WATER REUSE, *NITROGE	W69-08053
CHEMICAL OXYGEN DEMAND, TERTIARY	TREATMENT, ALGAE, BACTERIA, ACTIV	W69-08054
LLUTION SOURCES, WATER POLLUTION	TREATMENT, FINANCING, GRANTS, FED	W69-06305
EASTS, BACTERIA, FISH, P/	*WASTE TREATMENT, *AQUICULTURE, ALGAE, Y	W69-05891
E, THERMAL STRATIFICATION, WASTE	TREATMENT, WATER QUALITY, WATER P	W69-06535
LUTION CONTROL, *WATER POLLUTION	TREATMENT, ALGAE, ALGAL CONTROL,	W69-03374
OL, PACIFI/ *WASHINGTON, *SEWAGE	TREATMENT, *WATER POLLUTION CONTR	W69-03683
U. S., CIVIL ENGINEERING, WASTE	TREATMENT, ENGINEERING, ALGAE, PL	W69-03683

WATER POLLUTION SOURCES, PRIMARY
PHATES, NITRATES, RIVERS, SEWAGE
ES, PRIMARY TREATMENT, SECONDARY
T, SECONDARY TREATMENT, TERTIARY
WASTES, DOMESTIC WASTES, SEWAGE
RIENTS, DOMESTIC WASTES, *SEWAGE
GE TREATMENT, SEWAGE, BIOLOGICAL
DS, PRIMARY TREATMENT, SECONDARY
NCENTRATION, *PHOSPHORUS, SEWAGE
TABI/ *NUTRIENTS, PONDS, PRIMARY
*PHOSPHORUS, PHOSPHATES, *SEWAGE
OMPOUNDS, LIME, OXYGEN, TERTIARY
ON, LABORATORY TESTS, BIOLOGICAL
E, *ALKALINITY, WATER QUA/ WATER
, *TASTE, *ODOR, WATER Q/ WATER
, *ODOR, HYDROGEN ION CO/ *WATER
TENANCE, DESIGN CRITERIA, *WATER
HORUS, *SEWAGE EFFLUENT, *SEWAGE
*FILTRATION, DESIGN, *AL/ *WATER
*PATHOGENIC BACTERIA, AL/ *WATER
ENGINEERING, *FILTERS, */ *WATER
WATER QUALITY, *ACTIVATE/ *WATER
*OXIDATION LAGOONS, *WASTE WATER
S, LAKE MICHIGAN, WATER / *WATER
GEN, *IRRADIATION, *ALGAE, WASTE
, STREAM CONCOURSE, PALOUSE RIV/
, *ODOR, *RESERVOIRS, *DECIDUOUS
ES, BACTERIA, IRRIGATION, WILLOW

MOEWUSII, CHLORELLA PYRENOIDOSA,
TS, CHLOROPHYLL, PHOTOSYNTHESIS,
CHLORELLA, SAMPLING, EFFLUENTS,
ANABAENA CYLINDRICA, GLYCOLYSIS,
PYRENOIDOSA, SCHRODERIA, THIMET,
RYING, SPRAY-DRIED, GLASS BEADS,
N, AMPHIDINIUM, BLOOMS, MYTILUS,
S, SUCKERS, ICTALURUS PUNCTATUS,
PROCLADIUS, CHIRONOMUS PROMOSUS,
NTACT PROCESS, SEWAGE TREATMENT,
AROMATIC SOLVENTS, PHAEODACTYLUM
IONUCLIDE UPTAKE, *PHAEODACTYLUM
CYLINDRUS DANICUS, PHAEODACTYLUM
ENIWETOK ATOLL, TURNOVER TIME,
N-HEPTADECANE, MONOENES, DIENES,
CHLORELLA PYRENOIDOSA, FIXATION,
GITES, ZOOXANTHELLAE, ACANTHURUS
LACOIDES, SUAEDA MARITIMA, ASTER
DEPTH, THERMOCLINE, ZOOPLANKTON,
N, PHYTOPLANKTON, ICE, BACTERIA,
EFFECTS, WISCONSIN ZOOPLANKTON,
MICE, MORPHOLOGICAL CHARACTERS,
LONG ISLAND SOUND(NY), ICELAND,
OGRAPHY, ANTITHAMNION SARNIENSE,
DS, CESIUM, HYDROGEN, DEUTERIUM,
MA RAYS, RADIOCHEMICAL ANALYSIS,
CARBON RADIOISOTOPES, KINETICS,
TRACERS, BACTERIA, GROWTH RATES,
HUXLEYI, PYRAMIMONAS, PERIDINIUM
FLAKK(NORWAY), RANHEIM(NORWAY),
E(CANADA), SKAHA LAKE(CANADA), /
ICAUDA, ECOLOGICAL EFFICIENCIES,
ET.: *CULTURAL EUTROPHICATION,
SOIL TYPES, HYDROLOGY, CLIMATES,
RONMENT, PHOTOSYNTHESIS, OXYGEN,
ICATION, AQUATIC MICROORGANISMS,
ARY PRODUCTIVITY, AQUATIC ALGAE,
POLLUTION EFFECTS, *FISH, ALGAE,
RADIOISOTOPES, TENNESSEE RIVER,
Y PRODUCTIVITY, LIGHT INTENSITY,
OMIUM, PHOSPHORUS RADIOISOTOPES,
PLANKTON, PRIMARY PRODUCTIVITY,
TROSCOPY, DISTRIBUTION, CALCIUM,
TION EFFECTS, LAKES, PHOSPHORUS,
ALIF), NUTRIENT ENERGY PATHWAYS,
NDOTA(WIS), PHYTOPLANKTON CROPS,
), PHYTOPLANKTON CROPS, TROPHIC,
GOTROPHY, *PRIMARY PRODUCTIVITY,
ER, PHOTOSYNTHESIS, TEMPERATURE,
KRUSEI, CANDIDA OBTUSA, CANDIDA
E BLOOMS, CLEARWATER LAKE(MINN),
ES, CHARA, FISH, LIMNOLOGY, LAKE
ERATURE, PHOTOSYNTHESIS, RAINBOW
ALGAE, FISH, WISCONSIN, SEWAGE,
Y, BULLHEADS, EELS, KILLIFISHES,
ES, *FRESH WATER FISH, MICHIGAN,
ALTEMIFLORA, ORGANISMS, TURSIOPS
RGANISMS, SIMULIUM, PHYSA, SALMO
CELASE, HYDROGEN PEROXIDE, RATS,
ATE, FERRICHLORIDE, IRON S/ RING
), GERMANY, FINLAND, SEWAGE RING

TREATMENT, SECONDARY TREATMENT, T   W69-01453
TREATMENT, WATER POLLUTION SOURCE   W69-01453
TREATMENT, TERTIARY TREATMENT, EL   W69-01453
TREATMENT, ELECTRODIALYSIS, ION E   W69-01453
TREATMENT, HARVESTING OF ALGAE, B   W68-00478
TREATMENT, SEWAGE, BIOLOGICAL TRE   W68-00248
TREATMENT, SOLID WASTES, MICROBIO   W68-00248
TREATMENT, STABILIZATION PONDS, S   W68-00012
TREATMENT, *OXIDATION LAGOONS.: /   W68-00012
TREATMENT, SECONDARY TREATMENT, S   W68-00012
TREATMENT, DOMESTIC WASTES, *ALGA   W68-00256
TREATMENT, COST COMPARISONS.: / C   W68-00256
TREATMENT, BIBLIOGRAPHIES, CHLORE   W68-00855
TREATMENT, *ALGAE, *COPPER SULFAT   W72-07442
TREATMENT, *ACTINOMYCETES, *ALGAE   W72-13605
TREATMENT, *WATER QUALITY, *ALGAE   W72-09694
TREATMENT, *NEW YORK, WATER PURIF   W72-08686
TREATMENT, *NUTRIENT REMOVAL, *WA   W72-08133
TREATMENT, SANITARY ENGINEERING,   W72-08859
TREATMENT, SANITARY ENGINEERING,   W72-13601
TREATMENT, *FILTRATION, SANITARY   W72-07334
TREATMENT, *FILTRATION, *ALGAE, *   W72-09693
TREATMENT, ALGAE, PHOTOSYNTHESIS,   W72-10573
TREATMENT, *ALGAE, *FILTERS, SAND   W73-02426
TREATMENT, EUTROPHICATION, MODEL   W73-02218
TREE BARK, CHARCOAL, ALGAL GROWTH   W70-03501
TREES, *DECOMPOSING ORGANIC MATTE   W72-03475
TREES, FE: /S, CRAYFISH, MOSQUITO   W70-05263
TRIAZINE, DICHLOBENIL.:   W70-00269
TRIBONEMA AEQUALE, NAVICULA PELLI   W69-10180
TRIBUTARIES, WATERSHED MANAGEMENT   W70-04268
TRIBUTARIES, ALKALINITY, HYDROGEN   W70-03983
TRICARBOXYLIC ACID CYCLE, AUTORAD   W70-02965
TRICHODERMA.: /ATHION, CHLORELLA   W70-08097
TRICHODERMA, DISINTEGRATION.: / D   W70-04184
TRICHODESMIUM ERYTHREUM, SAXIDOMU   W70-05372
TRICHOPTERA, ODONATA, EPHEMEROPTE   W69-07846
TRICHOPTERA, LEECHES, FINGERNAIL   W70-01943
TRICKLING FILTERS, WASTE WATER TR   W70-06792
TRICORNUTUM.: /, *TORREY CANYON,   W70-01470
TRICORNUTUM, EDTA, LIGURIAN SEA,   W69-08275
TRICORNUTUM, STICHOCHRYSIS IMMOBI   W70-04280
TRIDACNA CROCEA, WOODS HOLE OCEAN   W69-08526
TRIENES.: /S, MASS SPECTROMETRY,   W69-08284
TRINGFORD(ENGLAND), ANABAENA CYLI   W70-02504
TRIOSTEGUS, CONUS EBRAEUS.: / FUN   W69-08526
TRIPOLIUM, PEMBROKE, SOUTH-WEST W   W70-09976
TRIPTON, BACTERIA, SEDIMENTATION,   W69-10169
TRIPTON, BOTTOM SEDIMENTS, SOLAR   W69-10154
TRIPTON, NANNOPLANKTON, NUTRIENTS   W70-03959
TRIS.: /, HUMBOLDT(SASKATCHEWAN),   W70-00273
TRITIATED THYMIDINE, NARRAGANSETT   W69-09755
TRITIATED GLUCOSE, GONIDIA FORMAT   W69-09755
TRITIUM.: /ODS, ABSORPTION, HAZAR   W70-02786
TRITIUM, CARBON RADIOISOTOPES, BI   W69-08269
TRITIUM, CHLOROPHYLL.: / DIATOMS,   W70-04811
TRITIUM, PHOTOGRAPHY, CARBON RADI   W69-10163
TROCHOIDEUM, VINEYARD SOUND(MASS)   W70-05272
TRONDHEIMSFJORD(NORWAY), SMOLA(NO   W70-01074
TROPHIC COMMUNITIES, OKANAGAN LAK   W71-12077
TROPHIC ECOLOGY, CLADOCERA.: /ADR   W69-10152
TROPHIC INDICATORS, NUTRIENT BUDG   W71-03048
TROPHIC LEVEL, NITROGEN FIXATION,   W71-03048
TROPHIC LEVEL, LIGHT INTENSITY, E   W70-00265
TROPHIC LEVEL.: /LORIDA, *EUTROPH   W72-01993
TROPHIC LEVEL, WATER QUALITY, FLO   W72-01990
TROPHIC LEVEL, PRODUCTIVITY, FOOD   W72-01788
TROPHIC LEVEL, ALGAE, CADDISFLIES   W70-02786
TROPHIC LEVEL, SAMPLING TEMPERATU   W70-02504
TROPHIC LEVEL, ZINC RADIOISOTOPES   W69-07853
TROPHIC LEVEL.: /VITY, LIMNOLOGY,   W69-09135
TROPHIC LEVELS, FOOD WEBS.: /SPEC   W69-08525
TROPHIC LEVEL, PRODUCTIVITY, MODE   W69-06536
TROPHIC RELATIONSHIPS.: /R LAKE(C   W69-05428
TROPHIC, TROPHOLYTIC, NITROGEN AN   W69-05142
TROPHOLYTIC, NITROGEN ANNUAL BUDG   W69-05142
TROPIC LEVEL, WATER QUALITY, WAST   W68-00912
TROPICAL REGIONS.: / ORGANIC MATT   W69-06865
TROPICALIS, NOCTILUCA MILIARIS, H   W70-04368
TROUT LAKE(MINN), KIMBALL LAKE(MI   W70-05387
TROUT, CRAYFISH, STONEFLIES, SNAI   W70-00711
TROUT, DIATOMS, ANAEROBIC CONDITI   W69-10154
TROUT, NITROGEN, PHOSPHORUS, DETE   W69-10178
TROUT, SUCKERS, BASS, WATER HYACI   W70-05263
TROUT, WINTERKILLING, FISHKILL, N   W68-00487
TRUNCATUS, BREVOORTIA TYRRANUS, C   W68-08274
TRUTTA, ENTOSPHENUS LAMOTTENII, G   W69-09334
TRYPSIN, SCENEDESMUS QUADRICAUDA,   W70-04184
TUBE, CANALIZATION, ALUMINUM SULF   W70-05668
TUBES.: /NY), LAKE TEGERN(GERMANY   W70-05761

LEE'S SUMMIT(MO), OXY/ *AERATION
L COMMUNITIES, SEWAGE, DETRITUS,
TION, BACTERIA, ALGAE, PLANKTON,
IMARRON R/ ARKANSAS RIVER(OKLA),
TS, GEOLOGIC FORMATIONS, ARCTIC,
VER(CALIF), MERCED RIVER(CALIF),
, WATER TEMPERATURE, PHOSPHATES,
TERS, DETRITUS, LIGHT INTENSITY,
G NUTRIENTS, TASTE, ODOR, COLOR,
RYSOPHYTA, GRAZING, SNAILS, AGE,
ALGAE, PLANKTON, MICROORGANISMS,
L OXYGEN DEMAND, SUSPENDED LOAD,
WATER TEMPERATURE, HEAT BUDGET,
*GREAT PLAINS, ALGAE, NEBRASKA,
TOXICITY, PHENOLS, ODOR, COLOR,
ER SPORTS./ *LAKE ONTARIO, WATER
EN, PHYTOPLANKTON, PRODUCTIVITY,
PLANKTON, ALGAE, INVERTEBRATES,
GAE, INTAKES, SANDS, FLOW RATES,
TION, FISHING, ALGAE, NUTRIENTS,
ICAL PROPERTIES, *WATER QUALITY,
, AQUIFERS, WATER LAW, NITRATES,
TRATION, *SEPARATION TECHNIQUES,
-LIGHT-INTENSITY, PHOTO-ELECTRIC
CONTINUOUS CULTURE, *CHEMOSTATS,
H, STRATIFICATION, FLOCCULATION,
TION, ALGAL CONTROL, BIOCONTROL,
MONIA, SPECTROMETERS, DIFFUSION,
WATER, *ALGAE, *SLUDGE, STORMS,
SOIL MICROORGANISMS, SOIL ALGAE,
R RIVER(WASH), ZURICHSEE(SWITZ),
N, LAKE CONSTANCE, PFAFFIKERSEE,
NADS, LAPPLAND(SWEDEN), ACETATE,
OODS HOLE OCEAN/ ENIWETOK ATOLL,
ALGAL CONTROL, SEASONAL, ANNUAL
CAPACITY, STANDING CROP, ANNUAL
SPARTINA ALTEMIFLORA, ORGANISMS,
LEECHES, SCHISTOSOME CERCARIAE,
LAKE, OLD JOHN LAKE, PINGO LAKE,
, *CANADA, KUSHOG LAKE(ONTARIO),
GAL SUBSTANCES, SUBPOLAR WATERS,
SALVELINUS NAMAYCUSH, SUBSTRATE
ETT BAY(RHODE ISLAND), TAXONOMIC
ENT(APPLIED), GEOCHEMISTRY, SOIL
TS, STRUCTURES, AFTERBAYS, WATER
*RADIOCARBON BIOASSAY, *NUTRIENT
, NYMPHAEA, CRASSIPES EICHORNIA,
ND(MAINE), SYNURA, ASTERIONELLA,
, TURSIOPS TRUNCATUS, BREVOORTIA
ERANUS, CLADOCERA, WATER MASSES,
ER QUALITY, NUTRIENTS, NORTHEAST
UTION CONTROL, PACIFIC NORTHWEST
E ALGAE, SALTS, TIDES, SOUTHEAST
TER CHEMISTRY, PLANTS, SOUTHWEST
HIC LEVELS, NUTRIENTS, NORTHEAST
ALGAL TOXINS MEMBRANE, NORTHEAST
PROTAENIA, *OPISTHONEMA OGLIMUN,
PLASM, WATER-BLOOMS, LAKE GEORGE(
ECIES, MELOSIRA VARIANS, SYNEDRA
ANABAENA, MELOSIRA, MALLOMONAS,
CTOR.:   *RADIONUCLIDE,
GONENSIS, ACROCHEILUS ALUTACEUS,
ES, OILY WATER, SULFITE LIQUORS,
ORPTION, CHEMICAL PRECIPITATION,
AKES, *ALLOCHTHONOUS PHOSPHATES,
CYTOLOGICAL STUDIES, FILTRATION,
QUES, CLOUD COVER, SPECTROSCOPY,
HYDRATES, TEMPERATURE, SALINITY,
, ALGAE, GROWTH RATES, BACTERIA,
DILUTION RATE, FRUITING BODIES,
GANIC COMPOUNDS, SURFACE RUNOFF,
NTS, FISH MANAGERS, FISH YIELDS,
ION, CAPILLARITY, CAP/ *PIPETTE,
HYTIC, PLATE COUNT, FILAMENTOUS,
ATTY ACIDS, OOCYSTIS POLYMORPHA,
RTILIZATION, PHYSICAL TREATMENT,
ICAL TREATMENT, UNIT OPERATIONS,
TROPHICATION, LAKES, *NUTRIENTS,
E ERIE, CHLOROPHYTA, CYANOPHYTA,
ON, SESTON, CHLOROPHYLL, CYCLES,
OPHYTA, CHRYSOPHYTA, CALIFORNIA,
TIONAL ACADEMY OF SCIENCES(USA),
GY FOUND, SYRACUSE(NY), SYRACUSE
BERRY CREEK(ORE), OREGON STATE
TIONAL ACADEMY OF SCIENCES(USA),
LAKE WAUBESA(WIS), MADISON(WIS),
IZE, CHLORELLA VULGARIS, CORNELL
CABLANDS(WASH), WASHINGTON STATE
NT, FOOD PRODUCTION, SOUTHAMPTON
AL STUDIES, LAGOON DESIGN, SCALE-

TUBES, *DIFFUSERS, *RAYTOWN(MO),   W70-06601
TUBIFICIDS, PLANKTON, DIATOMS, CY   W70-00268
TUBIFICIDS, CALCIUM, SILICATES, D   W70-05761
TULSA(OKLA), POLLUTION, FOSSIL, C   W70-04001
TUNDRA, SCALING, GRAVELS, MUD, SA   W69-08272
TUOLUMNE RIVER(CALIF), STANISLAUS   W70-02777
TURBIDITY.: /TARIO, LAKE MICHIGAN   W70-00667
TURBIDITY, ALGAE, DIATOMS, CYCLIN   W70-00713
TURBIDITY, HYDROGEN SULFIDE, IRON   W70-02251
TURBIDITY, DIPTERA, MAYFL: /A, CH   W71-00668
TURBIDITY, ANALYTICAL TECHNIQUES,   W70-08628
TURBIDITY, DATA STORAGE AND RETRI   W70-09557
TURBIDITY, *RESERVOIR EVAPORATION   W72-01773
TURBIDITY, STRATIFICATION, TEMPER   W72-04761
TURBIDITY, BIOCHEMICAL OXYGEN DEM   W72-03299
TURBIDITY, INDUSTRIAL PLANTS, WAT   W71-09880
TURBIDITY, SURFACES, DEPTH, DISTR   W71-07698
TURBIDITY, WORMS, SUNFISHES.: /S,   W69-07861
TURBIDITY, CHLORELLA, SCENEDESMUS   W70-04199
TURBIDITY, TERTIARY TREATMENT, SA   W70-05645
TURBIDITY, PLANT POPULATIONS.: /G   W68-00475
TURBIDITY, FLOODWATER, REGULATION   W73-00758
TURBIDITY, SUSPENDED SOLIDS, ALGA   W72-08686
TURBIDOSTAT, LIGHT EXTINCTION COE   W70-03923
TURBIDOSTATS, NUTRIENT LIMITED GR   W70-02779
TURBINES, ECONOMIC EFFICIENCY, HE   W70-08834
TURBULENCE, COPEPODS, ROTIFERS, C   W71-10098
TURBULENCE, TOXINS, BACTERIA, COP   W70-04369
TURBULENCE, PILOT PLANTS, BACTERI   W70-05819
TURF, SOIL MICROBIOLOGY, WATER PO   W72-08110
TURLERSEE(SWITZ), GREIFENSEE(SWIT   W70-00270
TURLERSEE, BALDEGGERSEE, GREIFENS   W70-03964
TURNOVER TIMES: /EASES, CHAMYDOMO   W70-04287
TURNOVER TIME, TRIDACNA CROCEA, W   W69-08526
TURNOVER.: /CTIVITY, HYPOLIMNION,   W69-05142
TURNOVER, FISH POPULATIONS.: /IVE   W68-00912
TURSIOPS TRUNCATUS, BREVOORTIA TY   W69-08274
TVA RESERVOIRS, ANOPHELES QUADRIM   W70-06225
TWELVE MILE LAKE, BRANT LAKE, IMI   W69-08272
TWELVE MILE LAKE(ONTARIO), GULL L   W70-02795
TYNSET(NORWAY), NID RIVER(NORWAY)   W70-01074
TYPE.: /NTHOPHYCEAE, MYXOPHYCEAE,   W70-00711
TYPES, HETERETROPHY, BACTERIAL ME   W70-00713
TYPES, HYDROLOGY, CLIMATES, TROPH   W71-03048
TYPES, INTAKES.: /INDUSTRIAL PLAN   W69-05023
TYPES, MICROQUANTITIES, MACROQUAN   W70-04765
TYPHA, BRASENIA, DOW CONTACT, SPA   W70-05263
TYPHOID FEVER, KENNEBEC RIVER(MAI   W70-08096
TYRRANUS, CONCENTRATION PROCESSES   W69-08274
U S PUBLIC HEALTH SERVICE, GREAT   W70-00263
U S, ECOLOGICAL DISTRIBUTION, BIO   W70-02764
U. S., CIVIL ENGINEERING, WASTE T   W69-03683
U. S., REGIONS, GEOGRAPHICAL REGI   W69-03695
U. S., REGIONS, PACIFIC COAST REG   W69-03611
U.S., ECOLOGICAL DISTRIBUTION, FR   W70-02772
U.S.: /ANAPHYTA, ALGAL CONTROL, *   W69-05306
UDOTEA FLABELLUM, ZINC, NICKEL, P   W69-08525
UGANDA), RHINE-MEEUSE DELTA(NETHE   W70-04369
ULNA, OEDOGONIUM, PHORMIDIUM RETZ   W70-00265
ULOTHRIX, DAVID MONIE TEST, SUM: /   W69-10157
ULOTHRIX, UPTAKE, HALF-LIFE, DETE   W69-07862
ULOTHRIX, CLADOPHORA, PROSOPIUM 'W   W69-07853
ULTIMATE DISPOSAL, WATER REUSE, F   W69-05891
ULTIMATE DISPOSAL.: /ODIALYSIS, S   W70-01981
ULTRA-VIOLET LIGHT STERILIZATION,   W70-03955
ULTRASONICS, DISINFECTION, SAMPLI   W71-11823
ULTRAVIOLET RADIATION, TRANSMISSI   W69-09676
ULTRAVIOLET RADIATION, SULFATES,   W70-01074
ULTRAVIOLET RADIATION, ACTINOMYCE   W70-00719
ULVA LACTUCA VAR LATISSIMA, POLYS   W70-01073
UNDERFLOW, PRECIPITATION(ATMOSPHE   W70-04810
UNDERWATER SHOCK WAVE, ANABAENA,   W69-08674
UNIALGAL, NITZSCHIA, ALGAE ISOLAT   W70-04469
UNICELLULAR, NATURAL RECOVERY.: /   W68-00891
UNICELLULAR ALGAE.: / CULTURES, F   W70-07957
UNIT OPERATIONS, UNIT PROCESSES.:   W70-07469
UNIT PROCESSES.: /ILIZATION, PHYS   W70-07469
UNITED STATES, PHOSPHORUS, NITROG   W68-00478
UNITED STATES, HABITATS.: /, *LAK   W70-04468
UNITED STATES, CHLORELLA, BIOCHEM   W70-05092
UNITED STATES, BRACKISH WATER, ES   W70-05372
UNIVERSITY OF WISCONSIN, ASIA, NA   W70-03975
UNIVERSITY, FUNDAMENTAL STUDIES,   W70-03973
UNIVERSITY, EXPERIMENTAL STREAMS,   W70-03978
UNIVERSITY OF WISCONSIN, ASIA, NA   W70-03975
UNIVERSITY OF WISCONSIN, YAHARA R   W69-06276
UNIVERSITY, ELECTRONIC PARTICLE C   W69-06540
UNIVERSITY.: /(WASH), CHANNELED S   W70-00264
UNIVERSITY.: /CTYLUM, *GAS EFFLUE   W70-00161
UP FACTORS, WASTE COMPOSITION.: /   W70-07508

RYON, MELOSIRA, ANEBODA(SWEDEN), UPPSALA(SWEDEN), LAKE ERKEN(SWEDE W70-05409
, SUSPENDED SOLIDS, *PHOSPHOROUS UPTAKE RATE.: MIXED LIQUOR W71-13335
MPHIDINIUM CARTERI, *RADIOCARBON UPTAKE TECHNIQUES, WATER POLLUTIO W70-05652
), WOODS HOLE(MASS). RADIOCARBON UPTAKE TECHNIQUES, MICROBIAL ECOL W70-05272
, MICROBIAL ECOLOGY, RADIOCARBON UPTAKE TECHNIQUES, MICHAELIS-MENT W70-05109
OCLINE, MIXOLIMNION, RADIOCARBON UPTAKE TECHNIQUES, BACTERIAL PLAT W70-05760
TOTROPHY, CARBON-14, RADIOCARBON UPTAKE TECHNIQUES, GLUCOSE, MICHA W70-04287
INEWEAVER-BURK EQUATION, MAXIMUM UPTAKE VELOCITY, MICHAELIS CONSTA W70-04287
AKE, *MICROBIAL ECOLOGY, MAXIMUM UPTAKE VELOCITY, ACTIVE TRANSPORT W70-04194
UATION, UPTAKE VELOCITY, MAXIMUM UPTAKE VELOCITY, SUBSTRATE CONCEN W70-05109
NTIAL, LINEWEAVER-BURK EQUATION, UPTAKE VELOCITY, MAXIMUM UPTAKE V W70-05109
UM U/ *HETEROTROPHY, *BIOLOGICAL UPTAKE, *MICROBIAL ECOLOGY, MAXIM W70-04194
M, EDTA, LIGURIAN/ *RADIONUCLID UPTAKE, *PHAEODACTYLUM TRICORNUTU W69-08275
N, *RADIONUCLIDE / *RADIONUCLID UPTAKE, *RADIONUCLIDE ACCUMULATIO W70-00707
EIN, MACROPHYTES, GROUND SOLIDS, UPTAKE, CONSUMER ORGANISMS, SIMUL W69-09334
-GREEN ALGAE,/ *COPPER CHLORIDE, UPTAKE, FREUNDLICH ISOTHERM, BLUE W70-08111
*RADIONUCLIDE, ULOTHRIX, UPTAKE, HALF-LIFE, DETECTOR.: W69-07862
PHYRIDIUM CRUENTUM, RADIONUCLIDE UPTAKE, NANNOCHLORIS, THORACOMONA W70-00708
VITAMIN B12, BIOTIN, HISTIDINE, URACIL, REDUCED SULFUR COMPOUNDS, W70-02510
OISOTOPES, COBALT RADIOISOTOPES, URANIUM RADIOISOTOPES, NUCLEAR EX W70-00708
RAYS, SPECTROMETERS, CYANOPHYTA, URANIUM RADIOISOTOPES, STRONTIUM W69-09742
ISOTOPES, THORIUM RADIOISOTOPES, URANIUM RADIOISOTOPES, MARINE ALG W71-09013
TTER, RADIOISOTOPES, GAMMA RAYS, URANIUM RADIOISOTOPES, RADIUM RAD W69-08272
E, CESIUM, RADIUM RADIOISOTOPES, URANIUM RADIOISOTOPES.: /MANGANES W69-08524
MUNICIPAL WASTES, AQUATIC ALGAE, URANIUM RADIOISOTOPES, RADIUM RAD W69-01165
SAN MIGUEL RIVER, DOLORES RIVER, URAVAN, COTTUS SPP, RHINICHTHYS O W69-07846
OPE, WATER POLLUTION ASSESSMENT, URBAN DRAINAGE, AGRICULTURAL DRAI W70-03975
OPE, WATER POLLUTION ASSESSMENT, URBAN DRAINAGE, AGRICULTURAL DRAI W70-03975
URBAN RUNOFF.: W71-04072
FFECTS, PRODUCTIVITY, SEDIMENTS, URBANIZATION, WATER POLLUTION TRE W70-04721
RS, LAKES, LIMNOLOGY, WISCONSIN, URBANIZATION, CONNECTICUT, OREGON W70-04506
ROPHICATION, *LAKES, GLACIATION, URBANIZATION, DOMESTIC WASTES, NU W69-07818
S, *EUTROPHICATION, *WASHINGTON, URBANIZATION, WATER POLLUTION, SE W70-05663
R, DDT, MISSISSIPPI RIVER BASIN, URBANIZATION, ALGAL POISONING, GA W73-00758
SPHORUS, *PHOSPHATES, ENERGY, DI- URNAL DISTRIBUTION, TEMPERATURE, W70-06604
CILLATORIA RUBESCENS, DINOBRYON, UROGLENA AMERICANA, CYCLOTELLA, C W70-05405
ABSORPTION, CELL CONCENTRATION, UROGLENA, ANKISTRODESMUS, CHLOROC W70-05409
G S WEST, RADIOCOCCUS NIMBATUS, URONEMA ELONGATUM HODGETTS, CHROO W70-04468
NATIONAL RESEARCH COUNCIL(USA), US ATOMIC ENERGY COMMISSION, NATI W70-03975
NATIONAL RESEARCH COUNCIL(USA), US ATOMIC ENERGY COMMISSION, NATI W70-03975
ATION, OFFICE OF NAVAL RESEARCH, US DEPARTMENT OF INTERIOR, MADISO W70-03975
ATION, OFFICE OF NAVAL RESEARCH, US DEPARTMENT OF INTERIOR, MADISO W70-03975
AGOONS, *MIDWEST(US), *SOUTHWEST(US), *DESIGN CRITERIA, POND EFFLU W70-04788
ERIA/ *AERATED LAGOONS, *MIDWEST(US), *SOUTHWEST(US), *DESIGN CRIT W70-04788
A, POND EFFLUENT, COLORADO, WEST(US), STABILIZATION POND LOADINGS, W70-04788
IA/ NATIONAL ACADEMY OF SCIENCES(USA), UNIVERSITY OF WISCONSIN, AS W70-03975
IA/ NATIONAL ACADEMY OF SCIENCES(USA), UNIVERSITY OF WISCONSIN, AS W70-03975
ASIA, NATIONAL RESEARCH COUNCIL(USA), US ATOMIC ENERGY COMMISSION W70-03975
ASIA, NATIONAL RESEARCH COUNCIL(USA), US ATOMIC ENERGY COMMISSION W70-03975
AM, KANS.: USBR RESEARCH CONTRACTS, CHENEY D W72-01773
BLOOM, TOXIC ALGAE, RECREATIONAL USE.: *DEFINITIONS, *WATER W69-08279
*DOMESTIC WATER USE.: W68-00483
RAL EUTROPHICATION, RECREATIONAL USE, WASTE EFFLUENTS.: *CULTU W69-06535
OPHYCEAE, VOLVOCALES, INDUSTRIAL USES, CLEAN WATER, GENUS, SPECIES W70-05389
MS, COLLOIDAL FRACTION, MULTIPLE- USES, SAVANNAH RIVER, GULF COAST. W73-00748
*SAKHALIN ISLAND(USSR).: W72-10623
KIEV RESERVOIR(USSR).: W71-00221
DS, SVISLOCH' RIVER(USSR), MINSK(USSR), ENGLAND, OSCILLATORIA, AUS W70-05092
ALIA, SANTA ROSA(CALIF), LYUBLIN(USSR), LYUBERETS(USSR), MUNICH(GE W70-05092
ICULTURAL PONDS, SVISLOCH' RIVER(USSR), MINSK(USSR), ENGLAND, OSCI W70-05092
CALIF), LYUBLIN(USSR), LYUBERETS(USSR), MUNICH(GERMANY), CHLORELLA W70-05092
BYELORUSSIA(USSR), POLYCHLOROPINENE.: W70-08654
UNITIES, SEDIMENTS, BIOTA, FISH, UTAH, ALGAE, INSECTS, INVERTEBRAT W69-07846
TORY, WATER PURIFICATION, *WATER UTILIZATION, DECOMPOSING ORGANIC W68-00483
GROWTH EQUATION, INCIDENT LIGHT UTILIZATION.: /TH KINETICS, MONOD W69-03730
*NITROGEN COMPOUNDS, *NITROGEN UTILIZATION, *FRESH WATER, ALGAE, W69-04521
CHLORELLA PYRENOIDOSA, NITROGEN UTILIZATION, ROLLER DRYING, SPRAY W70-04184
NG, WASTE WATER TREATMENT, WATER UTILIZATION, INDUSTRIAL WASTES, O W70-07508
*WATER POLLUTION EFFECTS, WATER UTILIZATION, ANALYTICAL TECHNIQUE W71-03048
CONTROL, *BIODEGRADATION, WATER UTILIZATION, FISH, ALGAE, MODEL S W71-11881
ASIDIUM PULLULANS, HANSENIASPORA UVARUM, RADULESPORES, ZYMOLOGICAL W70-04368
IA ECHINULATA, DINOBRYON, SYNURA UVELLA, FRAGILARIA, TABELLARIA, M W70-02803
S, CLADOCERANS, WESTHAMPTON LAKE(VA).: ASTERIONELLA SPIROIDE W71-10098
*SUSPENDED SOLIDS, CHASE CITY(VA).: W72-10573
E(NETHERLANDS), HETEROCYSTS, GAS- VACUOLES, LLANGORS LAKE(WALES), H W70-04369
TIDES, BACTERIA, SNAILS, WEIGHT, VACUUM DRYING, BOILING.: /LA, PEP W70-04184
ERUGINOSA, LAKE COMO(MINN), LAKE VADNAIS(MINN), MILLE LAC(MINN), D W70-05269
OIR/ *EUTROPHICATION, *TENNESSEE VALLEY AUTHORITY PROJECT, *RESERV W71-07698
ETERIORATION, DRAIN, SAN JOAQUIN VALLEY MASTER DRAIN, SAN FRANCISC W71-09577
TER RESERVOIRS, *TURBID R/ *SALT VALLEY RESERVOIRS(NEB), *CLEAR WA W72-04761
NITRIFICATION.: *CENTRAL VALLEY(CALIFORNIA), *BACTERIAL DE W71-04555
*ALGAE HARVESTING, *CENTRAL VALLEY(CALIF.).: W71-04556
*ALGAE HARVESTING, *CENTRAL VALLEY(CALIF.).: W71-04557
NITRIFICATION.: *CENTRAL VALLEY(CALIFORNIA), *BACTERIAL DE W71-04554
N EQUIVALENTS(WASTES), RED RIVER VALLEY(NORTH DAKOTA), SULFATE RED W70-03312
GEN, EUTROPHICATION, FISH, LAND, VALUE, LEAVES, FERTILIZERS, PESTI W70-05264
, LAKE ERKEN(SWEDEN), EXTINCTION VALUES, EUDORINA, CULTURE MEDIUM, W70-05409
ON, MAGNESIUM, COOLING, PROPERTY VALUES, RECREATION DEMAND, IR: /R W70-02251
KE(NORWAY), HOBOL RIVER(NORWAY), VAN LAKE(NORWAY).: /HER, NJOER LA W70-05752
ETABOLITES, INHIBITION, SILICON, VANADIUM, ZINC, GALLIUM.: /INS, M W69-04801

TIS AERUGINOSA, NOSTOC MUSCORUM,
MNA MINOR, PHOTOPHOSPHORYLATION,
*ZOOGLEAL SLIME, DAYTON(OHIO),
LAKE MALAREN, LAKE ANNECY, LAKE
E, FRUITING BODIES, ULVA LACTUCA
    *EFFECTS, NAVICULA SEMINULUM
A, APHAN/ *HETEROCYSTS, ANABAENA
GAE, *METABOLISM, *ENVIRONMENTS,
ICA, RATE EQUATIONS, ANALYSIS OF
C CONDUCTANCE, SPECIES, MELOSIRA
    SEEPAGE LOSSES, DIURNAL
HELPS EQUATION, DISSOLVED OXYGEN
E, PERIODIC VARIATIONS, SEASONAL
SEBASTICOOK, ME, MAINE, PERIODIC
PRICORNUTUM PRINTZ, OSCILLATORIA
S VEGAS, NEVADA, *LAKE MEAD, LAS
WATER, NITRATES, PHOSPHATES, LAS
G/ DISSOLVED ORGANIC MATTER, LAS
CEPTIBILITY, REGENERATION, CLIFF
FISHING, PUBLIC HEALTH, WALLEYE,
:   SHORELINE
AFRICA), TRANSPARENCY, MARGINAL
A, SANDY BOTTOMS FAUNA, MARGINAL
LANTS, WATERSHED, ALGAE, AQUATIC
LINEWEAVER-BURK EQUATION, UPTAKE
UPTAKE VELOCITY, MAXIMUM UPTAKE
PHOTOPERIODISM, LIGHT INTENSITY,
ICROBIAL ECOLOGY, MAXIMUM UPTAKE
ER-BURK EQUATION, MAXIMUM UPTAKE
, DISSOLVED OXYGEN, TEMPERATURE,
IGHT INTENSITY, CURRENTS(WATER),
DEPTH, DISSOLVED OXYGEN, ALGAE,
TIVITY, OXYGEN, METABOLISM, WIND
, ALKYL BENZENE SULFONATES, WIND
AXIDOMUS, CLOSTRIDIUM BOTULINUS,
M BREVE, C/ GONYAULAX, GONYAULAX
LLEI, GLOCOTRICHIA PISUM, NOSTOC
Y-125, GROUNDWATER, BISMUTH-207,
ESERVO/ *OXYGEN SUPERSATURATION,
I/ *INLAND, HORIZONTAL SAMPLING,
AUSTRIA), LAKE WORTHER(AUSTRIA),
AK LAKE(TENN), ACTIVITY-DENSITY,
RA(NORWAY), FROYA(NORWAY), FUCUS
UBSTANCES, *EXTRACELLULAR, FUCUS
LUM NODOSUM, ALGINIC ACID, FUCUS
S-AQUAE, *STRAINS, PHYCOLOGISTS,
Y, CHLOROPHYLL, ALGAE, BIOASSAY,
IA LANOSA, *ASCOPHYLLUM NODOSUM,
RNUTUM, STICHOCHRYSIS IMMOBILIS,
ERA, 'FLAVOBACTERIUM-CYTOPHAGA',
M, LITTLE LAKE(HANOI), RED RIVER(
NGER LAKES(NY), RODHE'S SOLUTION
MIMONAS, PERIDINIUM TROCHOIDEUM,
KOTA(ND), HARVEY(ND), CHROMATIUM
 *MUSCULAR EXERTION, *2-METHYL-5-
*ALLOCHTHONOUS PHOSPHATES, ULTRA-
TION, BIOCHEMICAL OXYGEN DEMAND,
S, THIOBACILLUS DENITRIFICANS, W
RATION, NEW HAMPSHIRE, NEW YORK,
CTOR, BEAUFORT(N C), CRASSOSTREA
TREATMENT, LAKES, WATER QUALITY,
SAYS, PROTOZOA, BACTERIA, FUNGI,
IVATED SLUDGE, ALGAE, NUTRIENTS,
CTIVITY, TRACE ELEMENTS, SODIUM,
.:   SECCHI DISC,
FFECTS, WATER POLLUTION SOURCES,
NOSA, NOSTOC MUSCORUM, VANADIUM,
IDENTIFICATION, ORGANIC SOLUTES,
N, ORGANIC SOLUTES, VITAMIN B-1,
BIOASSAY, *SEA WATER, *VITAMINS,
, ESTUARIES, BACTERIA, MOLLUSKS,
FACTORS, VITAMIN B1, VITAMIN B6,
DEBRIS, ORGANIC GROWTH FACTORS,
ANIC GROWTH FACTORS, VITAMIN B1,
RVESTING OF ALGAE, FISH HARVEST,
TION, NUISANCE ALGAE, NUTRIENTS,
S, ALGAE, WEEDS, CARBON DIOXIDE,
TRACE ELEMENTS, SODIUM, VIRUSES,
ECTRON MICROSCOPY, BIOCHEMISTRY,
R,/ *NUISANCE ALGAE, CYANOPHYTA,
, ENZYMES, CHEMICAL DEGRADATION,
SODIUM, POTASSIUM, TRACE METALS,
TS, DISSOLVED SOLIDS, NUTRIENTS,
PHYLL, BACTERIA, TRACE ELEMENTS,
PTION, COMPETITION, OLIGOTROPHY,
*NITROGEN, PHOSPHORUS, SULFATES,
ASE, HYDROG/ *DIGESTIBILITY, *IN
G/ *DIGESTIBILITY, *IN VITRO, IN
MS, LIEBIG'S LAW OF THE MINIMUM,
PYRENOIDOSA, VOLATILE ACIDS, NON-

VANADIUM, VITAMIN B-12.: /IPLOCYS	W69-06277
VANADIUM, ZINC, EUGLENA GRACILIS,	W70-02964
VANDALIA(OHIO), RHIZOPODS, SPIDER	W72-01819
VANERN, LAKE CONSTANCE, PFAFFIKER	W70-03964
VAR LATISSIMA, POLYSIPHONIA HARVE	W70-01073
VAR.HUSTEDTII.:	W71-03183
VARIABILIS, MICROCYSTIS AERUGINOS	W70-05091
VARIABILITY, CHLORELLA, CARBON, T	W70-02965
VARIANCE, FOOD CONCENTRATION, MEX	W70-03957
VARIANS, SYNEDRA ULNA, OEDOGONIUM	W70-00265
VARIATIONS.:	W69-09723
VARIATIONS, SAN JOAQUIN RIVER, PO	W69-07520
VARIATIONS, SOURCES.: /, ME, MAIN	W68-00481
VARIATIONS, SEASONAL VARIATIONS,	W68-00481
VAUCHER, SKELETONEMA COSTATUM(GRE	W70-05752
VEGAS BAY.: /D ORGANIC MATTER, LA	W72-02817
VEGAS BAY.: /OIRS, ALGAE, GROUND	W72-02817
VEGAS, NEVADA, *LAKE MEAD, LAS VE	W72-02817
VEGETATION, SAND DUNES.: /ON, SUS	W70-01777
VEGETATION, WILD RICE, SILTS, PLA	W70-01943
VEGETATION, STORMWATER, DRAINAGE.	W70-02251
VEGETATION FAUNA, SANDY BOTTOMS F	W70-06975
VEGETATION, MICROCYSTIS, ANABAENA	W70-06975
VEGETATION, NITROGEN, PHOSPHOROUS	W68-00511
VELOCITY, MAXIMUM UPTAKE VELOCITY	W70-05109
VELOCITY, SUBSTRATE CONCENTRAT: /	W70-05109
VELOCITY, TEMPERATURE, CHEMICAL A	W69-07862
VELOCITY, ACTIVE TRANSPORT, BACTE	W70-04194
VELOCITY, MICHAELIS CONSTANT, LAK	W70-04287
VELOCITY, CYANOPHYTA, BIOMASS, IN	W71-00668
VELOCITY, DIATOMS, SEASONAL, BIOM	W70-02780
VELOCITY, WATER QUALITY, DISSOLVE	W70-00265
VELOCITY, HUMIDITY, BACTERIA, HYD	W70-01073
VELOCITY, TASTE, COAGULATION, PLA	W70-00263
VENERUPIS, OSTREA, ARTHROPODS.: /	W70-05372
VENIFICUM, PYRODINIUM, GYMNODINIU	W70-05372
VERRUCOSUM, LAKE MENDOTA(WIS), LA	W69-08283
VERTEBRATES, PLUTONIUM-239, RATS,	W69-08269
VERTICAL MIXING, ROANOKE RAPIDS R	W70-00683
VERTICAL SAMPLING, LAKE MENDOTA(W	W70-02969
VERTICAL, CHALK, METALIMNION, OSC	W70-05405
VERTICAL, MICROCYSTIS, OSCILLATOR	W70-04371
VESCICULOSUS, ASCOPHYLLUM NODOSUM	W70-01074
VESICULOSUS, CHONDRUS CRISPUS, AS	W70-01073
VESICULOSUS, GROWTH FACTORS, LAMI	W69-07826
VETERINARIANS, MICROCYSTIS AERUGI	W70-00273
VIABILITY.: /AL CONTROL, MORTALIT	W70-02364
VIBRIO, FLAVOBACTER, ESCHERICHIA,	W70-03952
VIBRIO, MORPHOLOGY, CULTURAL CHAR	W70-04280
VIBRIOS, ACHROMOBACTERS, GRAM-POS	W70-00713
VIETNAM), CHLOROCOCCUS, SPINES, S	W70-03969
VIII, NUTRIENT LIMITATION, CARBON	W70-01579
VINEYARD SOUND(MASS), WOODS HOLE(	W70-05272
VINOSUM, THIOCAPSA FLORIDANA, ENT	W70-03312
VINYLPYRIDINE, *GLUE GREEN ALGAE,	W70-01788
VIOLET LIGHT STERILIZATION, ASTER	W70-03955
VIRGINIA.: /YDROGEN ION CONCENTRA	W72-10573
VIRGINIA, PENNSYLVANIA, MARCASITE	W69-07428
VIRGINIA, ORGANIC COMPOUNDS, LAGO	W72-03641
VIRGINICA, MERCENARIA MERCENARIA,	W69-08267
VIRUSES, ALGICIDES, FISH, STRE: /	W70-03973
VIRUSES, ALGAE, LICHENS, ACTIMOMY	W71-11823
VIRUSES, BACTERIA, E. COLI, PUBLI	W72-06017
VIRUSES, VITAMINS, WATER CHEMISTR	W69-04801
VISIBILITY TESTS, CONTROL METHODS	W68-00488
VITAMIN B.: /S, WATER POLLUTION E	W69-06277
VITAMIN B-12.: /IPLOCYSTIS AERUGI	W69-06277
VITAMIN B-1, VITAMIN B-12, THIAMI	W70-05652
VITAMIN B-12, THIAMINE, SERRATIA	W70-05652
VITAMIN B, ALGAE, CULTURES, CHEMI	W70-05652
VITAMIN B, RED TIDE, SALINITY, FL	W70-05372
VITAMIN B12, BIOTIN, HISTIDINE, U	W70-02510
VITAMIN B1, VITAMIN B6, VITAMIN B	W70-02510
VITAMIN B6, VITAMIN B12, BIOTIN,	W70-02510
VITAMINS.: /DIMENTS, DREDGING, HA	W69-09340
VITAMINS, WATER POLLUTION EFFECTS	W69-06277
VITAMINS, NITRATES, INORGANIC COM	W70-04810
VITAMINS, WATER CHEMISTRY, WATER	W69-04801
VITAMINS, EPIPHYTOLOGY.: /IDE, EL	W70-03952
VITAMINS, PEPTIDES, ORGANIC MATTE	W70-04369
VITAMINS, CELLULOSE, CHLORELLA, P	W70-04184
VITAMINS, WATER POLLUTION, PHYSIO	W70-04503
VITAMINS, TOXINS, INHIBITORS, SUC	W70-02510
VITAMINS, AMINO ACIDS, CALCIUM, A	W70-02255
VITAMINS, HEAVY METALS, TOXICITY,	W71-10079
VITAMINS, ALGAE, GROWTH RATES, CO	W69-09349
VITRO, IN VIVO, BALL-MILL, MEICEL	W70-04184
VIVO, BALL-MILL, MEICELASE, HYDRO	W70-04184
VOISIN'S LAW OF THE MAXIMUM.: /OO	W70-04765
VOLATILE ACIDS, KETOACIDS, TAMIYA	W70-04195

VULGARIS, CHLORELLA PYRENOIDOSA,
N(ISRAEL), TEMPERATURE PROFILES,

ABSORBANCE, COENOBIA, SWIRLING,
OL/ *DAIRY MANURE, TOTAL SOLIDS,
T, CENTRIFUGE METHOD, CALCULATED
SS CULTURE OF ALGAE, PACKED CELL
N, AREAL INCOME(NUTRIENTS), LAKE
IANA, LIGHT-SATURATED RATE, CELL
GANIC MATTER, NITROGEN FIXATION,
MEAN DOUBLING TIME, PACKED CELL
*EUTROPHICATION, ALGAE, GERMANY,
A, CRYPTOPHYCEAE, CHRYSOPHYCEAE,
ANKISTRODESMUS, CHLOROCOCCALES,
WOLLEA, OSCILLATORIA, SPIRULINA,
, PSEUDOPLANKTON, CRYPTOPHYCEAE,
S, MICROBIAL ECOLOGY, DINOBRYON,
N, CHLORELLA VULGARIS, CHLORELLA
MBIA)/ PHOTOOXIDATION, CHLORELLA
CHLORELLA PYRENOIDOSA, CHLORELLA
VOLATIL/ *GREEN ALGAE, CHLORELLA
NAMICS, CELLULAR SIZE, CHLORELLA
ENEDESMUS QUADRICAUDA, CHLORELLA
ENEDESMUS QUADRICAUDA, CHLORELLA
SYNCHAETA PECTINATA, POLYARTHRA
LLUS STEAROTHERMOPHILUS, PROTEUS
A STAGNALIS, NEPHROCYTIUM OBESUM
LUS, THIOBACILLUS DENITRIFICANS,
                        SIMAZINE 80
THESIS, ALGAE, SEWAGE, MUNICIPAL
NG WATERS, FRESH WATER, BRACKISH
                            FEEDLOTS,
, BOSTON LOT RESERVOIR, LEBANON,
TRIPOLIUM, PEMBROKE, SOUTH-WEST
STS, GAS-VACUOLES, LLANGORS LAKE(
, WARING BLENDOR MORTALITY, CELL
MMERCIAL FISHING, PUBLIC HEALTH,
RGANISMS, SCENEDESMUS, PROTEINS,
LAKE, ONONDAGA COUNTY, SYRACUSE,
HOMOGENIZATION, *BACTERIA COUNT,
REATMENT PLANT, DUWAMISH ESTUARY(
WARD A. HANSON RESERVOIR, TACOMA(
OCYSTIS, DREDG/ *LAKE WASHINGTON(
ONCENTRATION, / *LAKE WASHINGTON(
SC TRANSPARENC/ *LAKE WASHINGTON(
SSIMILATION, BA-LA, HA/ *HANFORD(
AMATH LAKE(ORE), LAKE WASHINGTON(
MUCOR, MORPHOLOGY, FRIDAY HARBOR(
, ROCK LAKE(WASH), WILLIAMS LAKE(
YLLA, TIDAL CURRENT, PUGET SOUND(
ATORIA RUBESCE/ *LAKE WASHINGTON(
ISON LAKES(WIS), LAKE WASHINGTON(
ITZERLAND, PER/ *LAKE WASHINGTON(
WAMISH ESTUARY(WASH).: *SEATTLE(
*LAKE WASHINGTON(WASH), SEATTLE(
AL PARK, OHANAPECOSH HOT SPRINGS(
DS, METRO ACT(/ *LAKE WASHINGTON(
LAKE(WASH), CHANNELED SCABLANDS(
ACTIVATION TECHNIQUES, ROCK LAKE(
MADISON LAKES(WIS), CEDAR RIVER(
KE TAHOE(CALIF), LAKE WASHINGTON(
ATED SAND, CRUSHED BAUXITE, ACID-
TEST, BODY IMMEDIUM FILTER, BACK-
ENCES, MICROCYSTIS, DREDG/ *LAKE
NA(WIS), LAKE TAHOE(CALIF), LAKE
BIOLOGY, CARNEGIE INSTITUTION OF
GEN, NUISANCE ALGAE, PHOSPHORUS,
BANIZATION, CONNECTICUT, OREGON,
REVENUE BONDS, METRO ACT(/ *LAKE
R(CONN), KLAMATH LAKE(ORE), LAKE
REAT LAKES, BIOINDICATORS, FISH,
TJA, OSCILLATORIA RUBESCE/ *LAKE
KE FURES, LAKE SEBASTICOOK, LAKE
ULATIONS, PLANKTON, *CYANOPHYTA,
                                LAKE
M(N H), MADISON LAKES(WIS), LAKE
NTS, PHOSPHORUS, EUTROPHICATION,
*NUTRIENT CONCENTRATION, / *LAKE
ASH), CHANNELED SCABLANDS(WASH),
, ALGAL CONTROL, TRACE ELEMENTS,
, SECCHI DISC TRANSPARENC/ *LAKE
MARINE ALGAE, ECOLOGY, CULTURES,
BESCENS, SWITZERLAND, PER/ *LAKE
*WATER QUALITY MANAGEMENT, *LAKE
BOTTOM SEDIMENTS, LAKE MICHIGAN,
TION, STREAMS, NUTRIENTS, LIGHT,
MA, NOSTOC PALUDOSUM, BLEACHING,
WOOD, ALGAL COUNT, GREEN RIVER,
TY, TROPIC LEVEL, WATER QUALITY,
LAGOON DESIGN, SCALE-UP FACTORS,

VOLATILE ACIDS, NON-VOLATILE ACID   W70-04195
VOLATILE SUSPENDED SOLIDS, MULTI-   W70-04790
VOLATILE SOLIDS, GRIT.:   W71-11375
VOLATILE SOLIDS.: /DETERMINATION,   W70-03334
VOLITILE SOLIDS, NON-DEGRADABLE S   W71-00936
VOLUME METHOD, CYCLOTELLA COMENSI   W70-03959
VOLUME.:   MA   W70-06053
VOLUME, MEAN DEPTH, DILUTION, OSC   W70-00270
VOLUME, NITZSCHIA CLOSTERIUM, NIT   W70-05381
VOLUME, STREAMS, WATER POLLUTION   W70-00719
VOLUMES, BACTERIAL PLATE COUNTS,   W71-05267
VOLUMETRIC ANALYSIS, LIMNOLOGY, G   W68-00475
VOLVOCALES, SARCODINA, CILIATES,   W70-05112
VOLVOCALES, CONJUGATAE, HETEROKON   W70-05409
VOLVOCALES, GLENODINIUM, TRACHELO   W70-05404
VOLVOCALES, INDUSTRIAL USES, CLEA   W70-05389
VOLVOX, PANDORINA, SPIROGYRA, SYN   W70-04503
VULGARIS(COLUMBIA), CHLORELLA PYR   W70-04809
VULGARIS, CHLORELLA VULGARIS(COLU   W70-04809
VULGARIS, MONODUS SUBTERRANEUS, A   W70-05381
VULGARIS, CHLORELLA PYRENOIDOSA,   W70-04195
VULGARIS, CORNELL UNIVERSITY, ELE   W69-06540
VULGARIS, DAPHNIA MAGNA, LIMNAEA   W71-03056
VULGARIS, DAPHNIA MAGNA, LIMNAEA   W71-05719
VULGARIS, KARATELLA COHCLEARIS, A   W70-02249
VULGARIS, PSEUDOMONAS AERUGINOSA,   W71-11823
W AND G S WEST, RADIOCOCCUS NIMBA   W70-04468
W VIRGINIA, PENNSYLVANIA, MARCASI   W69-07428
W, DACTHAL.:   W70-03519
WA: / AEROBIC TREATMENT, PHOTOSYN   W72-00462
WA: /S, CARP, CLAMS, PONDS, RUNNI   W70-09905
WABASH RIVER BASIN.:   W71-10376
WADLEIGH STATE PARK, LAKE SHATUTA   W69-08674
WALES.: /, SUAEDA MARITIMA, ASTER   W70-09976
WALES), HALOBACTERIUM, RESPIROMET   W70-04369
WALL, HEXOSAMINE, ASCOPHYLLUM NOD   W70-04365
WALLEYE, VEGETATION, WILD RICE, S   W70-01943
WALLS, STRUCTURE, AMINO ACIDS, ME   W70-04184
WARBURG RESPIROMETER, CLADOPHORA,   W69-02959
WARING BLENDOR MORTALITY, CELL WA   W70-04365
WASH).: *SEATTLE(WASH), RENTON T   W70-04283
WASH).: /EN RIVER, WASHINGTON, HO   W72-11906
WASH), *CULTURAL INFLUENCES, MICR   W70-05663
WASH), *LAKE CHANGES, *NUTRIENT C   W70-00270
WASH), *OXYGEN DEFICIT, SECCHI DI   W69-10182
WASH), *REACTOR, RADIONUCLIDES,   W69-07853
WASH), ALGAL NUTRITION, ALGAL PHY   W70-04506
WASH), ANTITHAMNION SARNIENSE, RH   W69-10161
WASH), CHANNELED SCABLANDS(WASH),   W70-00264
WASH), LONG ISLAND SOUND, NARRAGA   W69-10161
WASH), MACROFAUNA, GYTTJA, OSCILL   W70-04253
WASH), NUISANCE CONDITIONS.: /MAD   W70-07261
WASH), OSCILLATORIA RUBESCENS, SW   W69-10169
WASH), RENTON TREATMENT PLANT, DU   W70-04283
WASH), REVENUE BONDS, METRO ACT(S   W70-04455
WASH), SCHIZOTHRIX, PHORMIDIUM, W   W70-03309
WASH), SEATTLE(WASH), REVENUE BON   W70-04455
WASH), WASHINGTON STATE UNIVERSIT   W70-00264
WASH), WILLIAMS LAKE(WASH), CHANN   W70-00264
WASH), ZURICHSEE(SWITZ), TURLERSE   W70-00270
WASH), ZURICHSEE(SWITZERLAND).: /   W69-06858
WASHED BAUXITE, KANPUR(INDIA).: /   W70-08628
WASHING.: / FLOW, PUTRESCIBILITY   W70-05821
WASHINGTON(WASH), *CULTURAL INFLU   W70-05663
WASHINGTON(WASH), ZURICHSEE(SWITZ   W69-06858
WASHINGTON, CHEMICAL COMPOSITION(W69-06865
WASHINGTON, WISCONSIN, CALIFORNIA   W69-06858
WASHINGTON, PRIMARY PRODUCTIVITY,   W70-04506
WASHINGTON(WASH), SEATTLE(WASH),   W70-04455
WASHINGTON(WASH), ALGAL NUTRITION   W70-04506
WASHINGTON, LAKES, INDIANA, OCHRO   W70-04503
WASHINGTON(WASH), MACROFAUNA, GYT   W70-04253
WASHINGTON, LAKE MALAREN, LAKE AN   W70-03964
WASHINGTON, DC, *ESTUARIES, HUMAN   W68-00461
WASHINGTON.:   W69-03513
WASHINGTON(WASH), NUISANCE CONDIT   W70-07261
WASHINGTON, LAKES, WATER POLLUTIO   W70-00270
WASHINGTON(WASH), *LAKE CHANGES,   W70-00270
WASHINGTON STATE UNIVERSITY.: /(W   W70-00264
WASHINGTON, NUCLEAR POWERPLANTS,   W70-00264
WASHINGTON(WASH), *OXYGEN DEFICIT   W69-10182
WASHINGTON, WAVES(WATER), CONNECT   W69-10161
WASHINGTON(WASH), OSCILLATORIA RU   W69-10169
WASHINGTON, *SEWAGE FERTILIZATION   W69-09349
WASHINGTON, DISSOLVED OXYGEN.: /   W69-09349
WASHINGTON, WATER CHEMISTRY, OXYG   W70-03309
WASHINGTON(DC).: /RIVER, PLECTONE   W70-02255
WASHINGTON, HOWARD A. HANSON RESE   W72-11906
WASTE ASSIMILATIVE CAPACITY, STAN   W68-00912
WASTE COMPOSITION.: /AL STUDIES,   W70-07508

NS, STABLE ISOTOPES, RADIOACTIVE
, *PHYTOPLANKTON, RADIOACTIVITY,
, *WASTE WATE/ *WATER POLLUTION,
ESTON, INVERTEBRATES, SHELLFISH,
WAGE TREATMENT, SEWAGE DISPOSAL,
UTROPHICATION, RECREATIONAL USE,
MICROSCOPIC ORGANISMS, PAPERMILL
REUSE, RESEARCH AND DEVELOPMENT,
CLASSIFICATION, *MODEL STUDIES,
, KENTUCKY.:   ACID
RAULIC LOAD EFFECTS,/ *ANAEROBIC
SPECTROSCOPY, SULFUR COMPOUNDS,
DDT, ALKALINITY, STABILIZATION,
YEASTS, ACTINOMYCETES, PONDING,
(OHIO), RHIZOPODS, SPIDERS, MILK
POLLUTION, BACTERIA, IMPURITIES,
RY ENGINEERING, WATER TREATMENT,
THWEST U. S., CIVIL ENGINEERING,
, TASTE, THERMAL STRATIFICATION,
STHETIC DETERIORATION, *ADVANCED
DS, BIOINDICATORS, PRODUCTIVITY,
AL STUDIES, ELECTRON MICROSCOPY,
ERGENTS, *PHOSPHATES, CHELATION,
*NITROGEN, *IRRADIATION, *ALGAE,
STITUENTS, ANALYTICAL CHEMISTRY,
TORING, PIGMENTS, PHYTOPLANKTON,
OCHEMICAL OXYGEN DEMAND, *COLOR,
RFACE AREA(LAKES), PHAEDACTYLUM,
PROPERTIES, PHYSICAL PROPERTIES,
RELEASES, BEACHES, SANDS, BIOTA,
, *ACTIVATED SLUDGE, *EFFLUENTS,
LUENTS, WASTE WATER RECLAMATION,
TMENT, *ALGAE, *EUTROPHICATION,/
NITRATES, AMMONIA, *WATER REUSE,
ON, AMMONIA, WATER PURIFICATION,
TION, BIOCHEMICAL OXYGEN DEMAND,
, LABORATORY TESTS, TEMPERATURE,
ION, OX/ *FARM WASTES, *POULTRY,
TOSYNTHETIC OXYGEN, FARM WASTES,
MONIA, ALGAE, NUTRIENTS, SEWAGE,
WA/ *LAGOONS, *DOMESTIC WASTES,
LAGOONS, ALGAE, ORGANIC LOADING,
EMICAL OXYGEN DEMAND, NUTRIENTS,
US, ALGAE, BIOLOGICAL TREATMENT,
EMAND, MIXING, SEWAGE TREATMENT,
GE TREATMENT, TRICKLING FILTERS,
COLOR, ACTIVATED SLUDGE, ALGAE,
L WASTE EFFLUENT, CHEESE FACTORY
HYLL, LAGOONS, PONDS, OXIDATION
ON, DETERGENTS, WATER POLLUTION,
, CHLOROPHYTA, *ALGAE, SORPTION,
MMONIA, WATER POLLUTION EFFECTS,
OPHICATION, NUTRIENTS, NITROGEN,
MS, CALIFORNIA, *PHOTOSYNTHESIS,
TRIC CURRENTS, *WASTE TREATMENT,
RIA, DECOMPOSING ORGANIC MATTER,
TOXICITY, WASTE WATER TREATMENT,
TABILIZATION, LAGOONS, TOXICITY,
ND, TEMPERATURE, FLOW, SAMPLING,
WA/ *LAGOONS, *DOMESTIC WASTES,
*SULFUR BACTERIA, STABILIZATION,
ES, SILTS, PHOSPHATES, NITROGEN,
CTS, CURRENTS(WATER), INDUSTRIAL
UTROPHIC/ *NITROGEN CYCLE, *FARM
LING, PLANKTON, SEWAGE, DOMESTIC
ER POLLUTION SOURCES, ODOR, FARM
R POLLUTION CONTROL, RADIOACTIVE
ATION, STANDING CROP, INDUSTRIAL
EROSION, SEWAGE EFFLUENTS, FARM
                      *FINISHING
), OXIDATION DITCHES, SOUP, FOOD
                      *FINISHING
*CHICKENS, FOOD WASTES, POULTRY
RCENARIA, AQUACULTURE, RECYCLING
GAE, MOLLUSKS, FISH, RADIOACTIVE
SPRAYING, ACTIVATED SLUDGE, FARM
BSTRATE, NITROGEN REMOVAL, *MILK
C WASTE, INDUSTRIAL WASTES, FARM
ILE WASTES, FIBER WASTES, DYEING
TIARY TREATMENT, ALGAE, DOMESTIC
BACTERIA, POPULATION EQUIVALENTS(
*EUTROPHICATION, *AGRICULTURAL
TES, *SEWAGE TREATMENT, DOMESTIC
IAL WASTES, WATER AN/ *OIL, *OIL
C WASTES, DETERGENTS, INDUSTRIAL
ATION, ALGAE, BIOCHEMICAL/ *FARM
, *INDUSTRIAL WASTES, *MUNICIPAL
E EFFECTS, *WATER QUALITY, *FARM
, *SEWAGE BACTERIA, ODOR,/ *FARM
YLON-6, *V/ *VISCOSE RAYON PLANT

WASTE DISPOSAL, OCEANS, FOODS, AB   W70-02786
WASTE DISPOSAL, NUCLEAR REACTORS,   W70-00707
WASTE DISPOSAL, *SEWAGE TREATMENT   W72-01786
WASTE DISPOSAL, SEWAGE DISPOSAL,   W69-04276
WASTE DISPOSAL, WATER POLLUTION,   W69-06909
WASTE EFFLUENTS.:   *CULTURAL E   W69-06535
WASTE EFFLUENT, CHEESE FACTORY WA   W71-00940
WASTE IDENTIFICATION, ALGAE, *WAS   W72-03299
WASTE IDENTIFICATION, BIOLOGICAL   W69-01273
WASTE MICROFLORA, MCCREARY COUNTY   W69-00096
WASTE PONDS, *DEPTH EFFECTS, *HYD   W70-04787
WASTE STABILIZATION PONDS.: /ARED   W70-07034
WASTE STORAGE, PONDS, HYDROLYSIS,   W70-08097
WASTE TREATMENT.: /ENA, BACTERIA,   W70-08097
WASTE TREATMENT, COLONY COUNTS, A   W72-01819
WASTE TREATMENT.: / LIFE, STREAM   W72-01786
WASTE TREATMENT.: /UARIES, SANITA   W70-05094
WASTE TREATMENT, ENGINEERING, ALG   W69-03683
WASTE TREATMENT, WATER QUALITY, W   W69-06535
WASTE TREATMENT, *NUTRIENT REMOVA   W69-07818
WASTE TREATMENT, WATER POLLUTION   W69-08518
WASTE TREATMENT, LAKES, WATER QUA   W70-03973
WASTE TREATMENT, AQUATIC ALGAE, A   W70-04373
WASTE TREATMENT, EUTROPHICATION,   W73-02218
WASTE WATER EFFECTS, LAKE MENDOTA   W70-04382
WASTE WATER(POLLUTION), PHOTOSYNT   W70-04371
WASTE WATER TREATMENT, TREATMENT   W70-04060
WASTE WATER DIVERSION, INSTREAM T   W69-09340
WASTE WATER(POLLUTION), WATER QUA   W69-01273
WASTE WATER(POLLUTION), CURRENTS(W68-00010
WASTE WATER RECLAMATION, WASTE WA   W70-06053
WASTE WATER TREATMENT, *NUTRIENTS   W70-06053
WASTE WATER TREATMENT, WATER TREA   W72-01867
WASTE WATER TREATMENT.: /RATURE,   W72-06017
WASTE WATER TREATMENT.: /FILTRATI   W71-11393
WASTE WATER TREATMENT, OVERFLOW,   W71-10654
WASTE WATER TREATMENT.: /D OXYGEN   W71-12183
WASTE WATER TREATMENT, SEDIMENTAT   W71-11375
WASTE WATER TREATMENT.: /DGE, PHO   W71-08221
WASTE WATER(POLLUTION), WATER POL   W71-05879
WASTE WATER TREATMENT, INDUSTRIAL   W71-06737
WASTE WATER TREATMENT, WATER UTIL   W70-07508
WASTE WATER TREATMENT, *FIBER(PLA   W70-06925
WASTE WATER TREATMENT.: /PHOSPHOR   W70-07469
WASTE WATER TREATMENT.: /OXYGEN D   W70-06899
WASTE WATER TREATMENT, PHOSPHATES   W70-06792
WASTE WATER TREATMENT.: / DEMAND,   W70-10433
WASTE WATERS.: /GANISMS, PAPERMIL   W71-00940
WASTE WATER TREATMENT.: / CHLOROP   W70-08392
WASTE WATERS, BIOCHEMICAL OXYGEN   W70-08740
WASTE WATER TREATMENT, CULTURES,   W70-09438
WASTE WATER TREATMENT, PENNSYLVAN   W71-05026
WASTE WATER TREATMENT, PHOSPHOROU   W71-04072
WASTE WATER TREATMENT.: /OORGANIS   W71-05267
WASTE WATER TREATMENT.: /MS, ELEC   W71-03775
WASTE WATER(POLLUTION).: /, BACTE   W70-00664
WASTE WATER DISPOSAL.: /LAGOONS,   W69-09454
WASTE WATER TREATMENT, WASTE WATE   W69-09454
WASTE WATER DISPOSAL, DIURNAL, EF   W69-10159
WASTE WATER TREATMENT, INDUSTRIAL   W70-01519
WASTE WATER, ALGAE, DIFFUSION, SU   W70-01971
WASTE WATER(POLLUTION), SCENEDESM   W70-02255
WASTE, CORES, DIATOMS, ODOR-PRODU   W68-00470
WASTE, GROUND WATER, *NITRATES, E   W69-05323
WASTE, INDUSTRIAL WASTES, FARM WA   W72-01813
WASTE, SLUDGE, SOLID WASTES, DECO   W71-05742
WASTE, WATER QUALITY ACT, PATH OF   W72-00941
WASTE: /THERMOCLINE, LIGHT PENETR   W71-07698
WASTES. POPULATION, INDUSTRIES, M   W70-02803
WASTES.:   W71-06737
WASTES.:   SHEPPARTION(AUSTRALIA   W71-09548
WASTES.:   W70-01519
WASTES.:   W70-09320
WASTES.: /M WASTES, MERCENARIA ME   W70-09905
WASTES.: /CTS, MARINE ANIMALS, AL   W71-05531
WASTES.: /AE, *CATTLE, SLURRIES,   W70-07491
WASTES.: /L, COD REMOVAL, MILK SU   W70-07031
WASTES.: /ANKTON, SEWAGE, DOMESTI   W72-01813
WASTES.: /ANTI-FOAM AGENTS, *TEXT   W70-04060
WASTES.: /KES, *IMPOUNDMENTS, TER   W70-04765
WASTES), RED RIVER VALLEY(NORTH D   W70-03312
WASTES, *ALGAE, *LAKES, *TRIBUTAR   W70-02969
WASTES, *ALGAE, WATER PROPERTIES,   W68-00256
WASTES, *CHEMICAL WASTES, INDUSTR   W72-03299
WASTES, *CHELATION.: /FF, DOMESTI   W71-04072
WASTES, *FARM LAGOONS, *BIODEGRAD   W71-00936
WASTES, *GREAT LAKES, *WATER POLL   W70-04430
WASTES, *INDUSTRIAL WASTES, MUNIC   W70-09577
WASTES, *LAGOONS, *BIODEGRADATION   W71-02009
WASTES, *MAN-MADE FIBER PLANTS, N   W70-06925

LAKES, *WA/ *ALGAE, *INDUSTRIAL
TION, *MINE WASTES, *RADIOACTIVE
AT/ *OXIDATION LAGOONS, *ORGANIC
ATMENT, SEDIMENTATION, OX/ *FARM
, GRANTS, ALGAE, AIR POLL/ *FARM
ICIPAL / *WATER POLLUTION, *MINE
ION, *ALGAE, NUTRIENTS, DOMESTIC
RGANIC WASTES, *WASTE WAT/ *FARM
DETERGENTS, *DIATOMS, *OIL, *OIL
*SPRINKLER IRRIGATION, *ORGANIC
ERVOIRS, *LIMNOLOGY, *INDUSTRIAL
ASTE WATER DISPOSAL, *INDUSTRIAL
TER-PLANT RELATIONSHIPS, NUCLEAR
C LAGOONS.: *POTATO PROCESSING
ESTIC WASTES, SEWAGE, INDUSTRIAL
*TEXTILE MILL

*RADIOACTIVE WASTES, *MUNICIPAL
OLOGICAL ECOLOGY, KINETICS, FARM
OTOPES, RIVERS, MILLS, COLORADO,
ASTES, INDUSTRIAL WASTES, SEWAGE
NITROGEN, BIOMASS, SEWAGE, FARM
ATE REDUCTION, POTATO PROCESSING
STIGLEOCLONIUS, INSOLUBLE
T, PHOSPHOROUS, RUNOFF, DOMESTIC
ODOR, FARM WASTE, SLUDGE, SOLID
US, NITROGEN, SEWAGE, INDUSTRIAL
M AGENTS, *TEXTILE WASTES, FIBER
S, FISH REPRODUCTION, INDUSTRIAL
CTS, DOMESTIC WASTES, INDUSTRIAL
OON DES/ *JAPAN, *SLAUGHTERHOUSE
MAGE, DOMESTIC WASTE, INDUSTRIAL
ES./ *ANTI-FOAM AGENTS, *TEXTILE
OLOGY, EUTROPHICATION, MUNICIPAL
UTION, FERTILIZATION, INDUSTRIAL
TER POLLUTION SOURCES, MUNICIPAL
OXINS, SOLID WASTES, RADIOACTIVE
D CARBON, CHLORINE, PHENOLS, OIL
HYSIOLOGICAL ECOLOGY, INDUSTRIAL
TRIAL WASTES, GARBAGE DUMPS, OIL
SES(LEGAL), ORGANIC WASTES, PULP
AN/ *OIL, *OIL WASTES, *CHEMICAL
RE, *ULTIMATE DISPOSAL, DOMESTIC
ATER POLLUTION EFFECTS, DOMESTIC
TER, ORGANIC COMPOUNDS, CHEMICAL
TREATMENT, *POULTRY, INDUSTRIAL
ORIDE, IRON SULFATE, GALVANIZING
LAKES, HEAVY METALS, INDUSTRIAL
HOSPHATES, ALGAE, DETRITUS, PULP
ENT, *DAIRY INDUSTRY, */ *ANIMAL
ONN), KLAMATH LAKE(ORE),/ *URBAN
UACULTURE, RECYCLING / PETROLEUM
*ANAEROBIC DIGESTION, *MUNICIPAL
AGE, BIOLOGICAL TREATMENT, SOLID
RESERVOIRS, SAMPLING, INDUSTRIAL
, *WATER POLLUTION CONTROL, FARM
ALITY, *FARM WASTES, *INDUSTRIAL
ASTE WATER TREATMENT, INDUSTRIAL
ASTE WATER TREATMENT, INDUSTRIAL
S, *INDUSTRIAL WASTES, MUNICIPAL
LAKES, *LAKE ZURICH, *MUNICIPAL
ACIATION, URBANIZATION, DOMESTIC
S, BACTERIA, FISH, PROTEINS, OIL
STE WATER TREATMENT, *INDUSTRIAL
T, WATER UTILIZATION, INDUSTRIAL
*EUTROPHICATION, LAKES, ORGANIC
ISH, TOXICITY, FRESH WATER, PULP
CITRUS/ *CANNERIES, *INDUSTRIAL
LITY, LAKE MORPHOMETRY, DOMESTIC
EN DEMAND, ALGAL TOXINS, ORGANIC
HOSPHORUS, *DETERGENTS, DOMESTIC
*CHICKENS, FOOD
OL, WATERCOURSES(LEGAL), ORGANIC
INDUSTRIAL WASTES, TOXINS, SOLID
RGANIC L/ *POTATOES, *INDUSTRIAL
SAL, DOMESTIC WASTES, INDUSTRIAL
L OXYGEN DEMAND, SUSPENDED LOAD,
GY, *SEWAGE LAGOONS, *INDUSTRIAL
WORKS, *WASTE DISPOSAL, DOMESTIC
SPHORUS, CHEMCONTROL, INDUSTRIAL
AGE, INDUSTRIAL WASTES, DOMESTIC
IES, AERATION, LAGOONS, DOMESTIC
NG, INDUSTRIAL WASTES, MUNICIPAL
CONTROL, FARM WASTES, MUNICIPAL
ICOCHEMICAL PROPERTIES, DOMESTIC
COLOGY, EUTROPHICATION, DOMESTIC
L, *NUTRIENTS, *PESTICIDES, FARM
ING WASTES, CHEESE MANUFACTURING
BACTERIA, PROTOZOA, ALGAE, SOLID
), WETLANDS, SEEPAGE, INDUSTRIAL
GENTS, ALGAL CONTROL, INDUSTRIAL

WASTES, *MUNICIPAL WASTES, *GREAT   W70-04430
WASTES, *MUNICIPAL WASTES, AQUATI   W69-01165
WASTES, *OXYGENATION, AEROBIC TRE   W71-09546
WASTES, *POULTRY, WASTE WATER TRE   W71-11375
WASTES, *RESEARCH AND DEVELOPMENT   W71-06821
WASTES, *RADIOACTIVE WASTES, *MUN   W69-01165
WASTES, *SEWAGE TREATMENT, SEWAGE   W68-00248
WASTES, *SPRINKLER IRRIGATION, *O   W71-00940
WASTES, *SURFACTANTS, *WATER POLL   W70-01470
WASTES, *WASTE WATER DISPOSAL, AI   W71-00940
WASTES, *WATER POLLUTION EFFECTS,   W70-06975
WASTES, *WATER POLLUTION CONTROL,   W72-01786
WASTES, ABSORPTION, CHELATION, ME   W71-09005
WASTES, AEROBIC LAGOONS, ANAEROBI   W71-08970
WASTES, AGRICULTURE, ST LAWRENCE   W70-03964
WASTES, ALFOL ALCOHOLS.:   W70-08740
WASTES, AQUATIC ALGAE, URANIUM RA   W69-01165
WASTES, BENTHOS, NITRATES, FLUORE   W70-02775
WASTES, BIOLOGICAL COMMUNITIES, S   W69-07846
WASTES, BYPRODUCTS, FERTILIZERS,   W70-09905
WASTES, CARBON CYCLE, PHOSPHORUS,   W70-02777
WASTES, CHEESE MANUFACTURING WAST   W70-03312
WASTES, CULTURE MEDIA.:   W68-00248
WASTES, DETERGENTS, INDUSTRIAL WA   W71-04072
WASTES, DECOMPOSTING ORGANIC MATT   W71-05742
WASTES, DOMESTIC WASTES, SEWAGE T   W68-00478
WASTES, DYEING WASTES.: /ANTI-FOA   W70-04060
WASTES, ECOLOGY, DISSOLVED OXYGEN   W73-00748
WASTES, ENVIRONMENTAL EFFECTS, AN   W72-12240
WASTES, EXPERIMENTAL STUDIES, LAG   W70-07508
WASTES, FARM WASTES.: /ANKTON, SE   W72-01813
WASTES, FIBER WASTES, DYEING WAST   W70-04060
WASTES, FRESHWATER ALGAE, WATER P   W72-12240
WASTES, GARBAGE DUMPS, OIL WASTES   W70-05112
WASTES, GREAT LAKES, *POLLUTANT I   W71-05879
WASTES, HEATED WATER, THERMAL POL   W72-00462
WASTES, HYDROGEN ION CONCENTRATIO   W70-00263
WASTES, HYDROGEN ION CONCENTRATIO   W70-04510
WASTES, INLETS(WATERWAYS), ORGANI   W70-05112
WASTES, INDUSTRIAL WASTES, TOXINS   W72-00462
WASTES, INDUSTRIAL WASTES, WATER   W72-03299
WASTES, INDUSTRIAL WASTES, SEWAGE   W70-09905
WASTES, INDUSTRIAL WASTES, ENVIRO   W72-12240
WASTES, ION EXCHANGE, BIOCHEMICAL   W70-10433
WASTES, LAGOONS, PONDS, PHOTOSYNT   W70-09320
WASTES, LAKE ZURICH(SWITZERLAND),   W70-05668
WASTES, LAKES, PHOTOSYNTHESIS, PH   W69-02959
WASTES, LIVESTOCK, IDAHO.: /ES, P   W70-03501
WASTES, MANAGEMENT, *FARM MANAGEM   W70-07491
WASTES, MADISON(WIS), LAKE ZOAR(C   W70-04506
WASTES, MERCENARIA MERCENARIA, AQ   W70-09905
WASTES, METHANE BACTERIA, ACID BA   W72-06854
WASTES, MICROBIOLOGY, ACTIVATED S   W68-00248
WASTES, MUNICIPAL WASTES, SEWAGE,   W70-04001
WASTES, MUNICIPAL WASTES, SEPTIC   W70-04266
WASTES, MUNICIPAL WASTES, NUTRIEN   W71-09577
WASTES, NEUTRIENTS, BACTERIA, ALG   W70-01519
WASTES, NUTRIENTS, BACTERIA, ALGA   W71-06737
WASTES, NUTRIENTS, PESTICIDES, RE   W71-09577
WASTES, NUTRIENTS, NUISANCE ALGAE   W68-00680
WASTES, NUTRIENTS, PHOSPHORUS, AL   W69-07818
WASTES, OILY WATER, SULFITE LIQUO   W69-05891
WASTES, OXIDATION PONDS, ALGAE, B   W72-00027
WASTES, OXIDATION LAGOONS.: /TMEN   W70-07508
WASTES, OXYGEN, STANDING CROP, AL   W70-02251
WASTES, PESTICIDE RESIDUES, HERBI   W70-10437
WASTES, PEACHES, APRICOTS, PEARS,   W71-09548
WASTES, PLANT GROWTH SUBSTANCES,   W68-00470
WASTES, POLLUTANT IDENTIFICATION,   W69-01273
WASTES, POLLUTANTS, WATER POLLUTI   W70-05084
WASTES, POULTRY WASTES.:   W70-09320
WASTES, PULP WASTES, INDUSTRIAL W   W72-00462
WASTES, RADIOACTIVE WASTES, HEATE   W72-00462
WASTES, SCREENS, SEDIMENTATION, O   W71-08970
WASTES, SEWAGE WASTES, BYPRODUCTS   W70-09905
WASTES, SEWAGE, ALGAE, MICROORGAN   W70-08627
WASTES, SEWAGE TREATMENT, NORTH D   W70-03312
WASTES, SEWERS, SEWAGE TREATMENT,   W70-03344
WASTES, SEWAGE TREATMENT.: /, PHO   W70-05668
WASTES, SEWAGE TREATMENT, HARVEST   W68-00478
WASTES, SEDIMENTATION, CHLORINATI   W70-04060
WASTES, SEWAGE, ADSORPTION, ACTIV   W70-04001
WASTES, SEPTIC TANKS, RECREATION   W70-04266
WASTES, SEWAGE, INDUSTRIAL WASTES   W70-03964
WASTES, SHALLOW WATER, CHLOROPHYT   W70-00711
WASTES, SILT, EROSION, SEDIMENTAT   W71-10376
WASTES, STARCH MANUFACTURING WAST   W70-03312
WASTES, TERTIARY TREATMENT.: /L,   W70-01027
WASTES, TERTIARY TREATMENT, CHEMI   W69-10178
WASTES, THERMAL POLLUTION, METHOD   W70-10437

POLLUTION ABATEMENT, POLLUTANTS,   WASTES, THERMAL POLLUTION, TOXINS   W69-06305
  WASTES, PULP WASTES, INDUSTRIAL   WASTES, TOXINS, SOLID WASTES, RAD   W72-00462
IA, ALGAE, FISH, SEWERS, RUNOFF,   WASTES, WATER SUPPLY, EFFLUENTS,   W71-09880
DGE, PHOTOSYNTHETIC OXYGEN, FARM   WASTES, WASTE WATER TREATMENT.: /   W71-08221
DUSTRIAL WA/ *LAGOONS, *DOMESTIC   WASTES, WASTE WATER TREATMENT, IN   W71-06737
NE ALGAE, INVERTEBRATES, NUCLEAR   WASTES, WATER POLLUTION SOURCES.:   W71-09863
ES, *CHEMICAL WASTES, INDUSTRIAL   WASTES, WATER ANALYSIS, ANALYTICA   W72-03299
DUSTRIAL WA/ *LAGOONS, *DOMESTIC   WASTES, WASTE WATER TREATMENT, IN   W70-01519
IOINDICATORS, *NEW YORK, NUCLEAR   WASTES, WATER POLLUTION SOURCES,   W69-10080
ING WASTES, STARCH MANUFACTURING   WASTES, WATER QUALITY STANDARDS,   W70-03312
PHOSPHATE, NITRATES, INDUSTRIAL   WASTES, WATER POLLUTION EFFECTS,   W70-02777
TS, SEWAGE TREATMENT, INDUSTRIAL   WASTES, WATER CHEMISTRY, DIATOMS,   W70-03974
WASTES, SEPTIC TANKS, RECREATION   WASTES, WATER QUALITY.: /NICIPAL   W70-04266
POSAL, WATER CHEMISTRY, ECOLOGY,   WAT: / WASTE DISPOSAL, SEWAGE DIS   W69-04276
LING NUTRIENTS, WATER CHEMISTRY,   WATER ANALYSIS, PHOSPHORUS, PONDS   W69-05142
QUALITY, *LIMNOLOGY, MINNESOTA,   WATER ANALYSIS, CALCIUM CARBONATE   W69-00632
RCES, IRRIGATION, AQUATIC ALGAE,   WATER ANALYSIS, CARBONATES, NITRA   W69-07096
LUTION SOURCES, WATER CHEMISTRY,   WATER ANALYSIS.: /LGAE, WATER POL   W69-08562
DISSOLVED OXYGEN, PHYTOPLANKTON,   WATER ANALYSIS, ALGAL CONTROL, AL   W72-06051
MICAL WASTES, INDUSTRIAL WASTES,   WATER ANALYSIS, ANALYTICAL TECHNI   W72-03299
TORY TESTS, INVERTEBRATES, FISH,   WATER ANALYSIS, INSECTS, SNAILS,   W72-01801
QUATIC MICROORGANISMS, SAMPLING,   WATER ANALYSIS, WATER QUALITY, PO   W71-00221
EFFECTS, ECOLOGY/ PLANT GROWTH,   WATER AND PLANTS, WATER POLLUTION   W70-02772
NEWFOUND LAKE, *WINNISQUAM LAKE,   WATER AND PLANTS, WATER POLLUTION   W70-02764
R SPECIES, MICROORGANISMS, FRESH   WATER BIOLOGY, SUSPENDED SEDIMENT   W71-10786
  *EUTROPHICATION, ALGAL BLOOMS,   WATER BLOOM, AQUATIC ALGAE, AGING   W68-00912
GROWTH REGULATORS, MYRIOPHYLLUM,   WATER BLOOMS, CHAINING.: /SALIX,   W70-05263
VED OXYGEN, EVAPORATION CONTROL,   WATER BUDGET, WATER CHEMISTRY, WA   W72-01773
ODOR NU/ *OSCILLATORIA, NATIONAL   WATER CARRIER(ISRAEL), THRESHOLD   W70-02373
S, NUTRIENTS, LIGHT, WASHINGTON,   WATER CHEMISTRY, OXYGEN, HYDROGEN   W70-03309
LITY, *WATER POLLUTION, *RIVERS,   WATER CHEMISTRY, WATER POLLUTION   W70-03068
APORATION CONTROL, WATER BUDGET,   WATER CHEMISTRY, WATER TEMPERATUR   W72-01773
QUALITY, *MONITORING, BACTERIA,   WATER CHEMISTRY, PHOSPHORUS, NITR   W71-05879
TER PURIFICATION, WATER QUALITY,   WATER CHEMISTRY.: /AYS, ALGAE, WA   W69-09884
  *LIGHT INTENSITY, SPRING WATER,   WATER CHEMISTRY, WATER TEMPERATUR   W69-09676
TREATMENT, EFFLUENTS, LIMNOLOGY,   WATER CHEMISTRY, RESPIRATION, BAT   W70-05387
TER, HYDROGEN ION CONCENTRATION,   WATER CHEMISTRY, *PHYTOPLANKTON,   W70-05424
INOFLAGELLATES, MARINE BACTERIA,   WATER CHEMISTRY, CARBON RADIOISOT   W70-05652
EMICAL OXYGEN DEMAND, MINNESOTA,   WATER CHEMISTRY, BIODEGRADATION.:   W69-00446
CTIVITY, STANDING CROP, STREAMS,   WATER CHEMISTRY, WATER: /S, PRODU   W69-04805
WASTE DISPOSAL, SEWAGE DISPOSAL,   WATER CHEMISTRY, ECOLOGY, WAT: /   W69-04276
S COMPOUNDS, *CYCLING NUTRIENTS,   WATER CHEMISTRY, WATER ANALYSIS,   W69-05142
ENTS, SODIUM, VIRUSES, VITAMINS,   WATER CHEMISTRY, WATER POLLUT: /M   W69-04801
FE, ENVIRONMENT, MICROORGANISMS,   WATER CHEMISTRY, PLANTS, SOUTHWES   W69-03611
  ALGAE, WATER POLLUTION SOURCES,   WATER CHEMISTRY, WATER ANALYSIS.:   W69-08562
SEDIMENTATION, SEDIMENTS, TASTE,   WATER CHEM: /PLANKTON, SULFATES,   W69-07833
NUTRIENTS, PHOSPHORUS COMPOUNDS,   WATER CHEMISTRY, WATER POLLUTION   W69-05868
QUIREMENTS, NUTRIENTS, TOXICITY,   WATER CHEMISTRY, WATER POLLUTION   W69-05867
RY PRODUCTIVITY, SOIL CHEMISTRY,   WATER CHEMISTRY, WATER POLLUTION   W69-06273
AT LAKES, PHYSIOLOGICAL ECOLOGY,   WATER CHEMISTRY, GEOGRAPHICAL REG   W70-03975
  EUTROPHICATION, GRAZING, ALGAE,   WATER CHEMISTRY, BIOMASS, BENTHOS   W70-03978
GE TREATMENT, INDUSTRIAL WASTES,   WATER CHEMISTRY, DIATOMS, CHLOROP   W70-03974
AT LAKES, PHYSIOLOGICAL ECOLOGY,   WATER CHEMISTRY, GEOGRAPHICAL REG   W70-03975
  SEA WATER, FRESH WATER, SHORES,   WATER CIRCULATION, FLOODS, DENSIT   W69-06203
ION, THERMOCLINE, *ION EXCHANGE,   WATER CIRCULATION, NUTRIENT ABSOR   W68-00857
VEL, WATER STORAGE, EVAPORATION,   WATER CIRCULATION, SEDIMENTS.: /E   W69-09723
C PLANTS, *AQUATIC WEED CONTROL,   WATER CONSERVATION, WATER CONTROL   W70-00269
GE, ALGAL CONTROL, AQUATIC LIFE,   WATER CONSERVATION, PUBLIC HEALTH   W70-03344
LE PRESERVATION, WATER DYNAMICS,   WATER CONSTITUENTS, ANALYTICAL CH   W70-04382
EED CONTROL, WATER CONSERVATION,   WATER CONTROL, FERTILIZERS, SEWAG   W70-00269
CIPAL WATER, WATER REQUIREMENTS,   WATER CONVEYANCE, WATER COSTS, AR   W71-09577
REQUIREMENTS, WATER CONVEYANCE,   WATER COSTS, ARTIFICIAL WATER COU   W71-09577
VEYANCE, WATER COSTS, ARTIFICIAL   WATER COURSES, CALIFORNIA.: / CON   W71-09577
TREATMENT, SAMPLING, MONITORING,   WATER DEMAND, WATER SUPPLY, ARID   W70-05645
ANCE, ALGAE BLOOM, ALGAE GROWTH,   WATER DENSITY, CONVECTION CURRENT   W69-08282
OAQUIN VA/ *CONTRA COSTA COUNTY,   WATER DETERIORATION, DRAIN, SAN J   W71-09577
, *WASTE WATER TREATMENT, *WASTE   WATER DISPOSAL, *INDUSTRIAL WASTE   W72-01786
TY, WASTE WATER TREATMENT, WASTE   WATER DISPOSAL.: /LAGOONS, TOXICI   W69-09454
MPERATURE, FLOW, SAMPLING, WASTE   WATER DISPOSAL, DIURNAL, EFFLUENT   W69-10159
IGATION, *ORGANIC WASTES, *WASTE   WATER DISPOSAL, AIR POLLUTION, WA   W71-00940
AREA(LAKES), PHAEDACTYLUM, WASTE   WATER DIVERSION, INSTREAM TREATME   W69-09340
US RELEASE, SAMPLE PRESERVATION,   WATER DYNAMICS, WATER CONSTITUENT   W70-04382
NTS, ANALYTICAL CHEMISTRY, WASTE   WATER EFFECTS, LAKE MENDOTA(WIS),   W70-04382
TA, CALIFORNIA WATER PLAN, DELTA   WATER FACILITIES(CALIF), OXYGEN D   W70-05094
AS, LARVAE, GROWTH RATES, *FRESH   WATER FISH, MICHIGAN, TROUT, WINT   W68-00487
IBLIOGRAPHIES, CYANOPHYTA, FRESH   WATER FISH, LAKES, *LAKE ZURICH,   W68-00680
TICS, GROWTH RATES, ALGAE, FRESH   WATER FISH.: /, LAKE ERIE, STATIS   W68-00253
NTS, ICED LAKES, *FERTILIZATION,   WATER FLEAS, LARVAE, GROWTH RATES   W68-00487
LLIFISHES, TROUT, SUCKERS, BASS,   WATER HYACINTH, FROGS, CRAYFISH,   W70-05263
CHAINS, OXYGEN, ROTIFERS, WEEDS,   WATER HYACINTH, ALLIGATORWEED, CH   W70-10175
FISH, FRESH WATER, ALGAE, PONDS,   WATER HYACINTH, CHEMICAL CONTROLS   W70-00269
S, *CYCLING NUTRIENTS, *SEDIMENT-   WATER INFERFACE, MUD-WATER INTERF   W69-06859
, *SEDIMENT-WATER INFERFACE, MUD-   WATER INTERFACE, PHOSPHORUS, ADSO   W69-06859
NTRATION, CYCLING NUTRIENTS, MUD-   WATER INTERFACES, EUGLENOPHYTA, T   W70-04381
, ALGAE, WATER QUALITY, SEDIMENT-   WATER INTERFACES, OREGON, SESSILE   W72-06051
OTE SENSING, MINNESOTA, SEDIMENT-   WATER INTERFACES, CALIFORNIA, BIO   W70-02775
, TIDAL EFFECTS, DISPERSION, OIL-   WATER INTERFACES, PERSISTENCE, WA   W70-01777
  *PLANT GROWTH REGULATORS, FRESH   WATER LAKES, LABORATORY TESTS.: /   W68-00856
ALGAE, SEWAGE, WATER POLLUTION,   WATER LAW, WATER POLLUTION EFFECT   W69-06305
ING, GASOLINE, FLOODS, AQUIFERS,   WATER LAW, NITRATES, TURBIDITY, F   W73-00758

ELERY, BUR REED, DUCKWEED, WHITE
ATER POLLUTION TREATMENT, ALGAE,
SPHORUS COMPOUNDS, WATER SUPPLY,
TROGEN, PHOSPHORUS, AGRICULTURE,
ANODISCUS BINDERANUS, CLADOCERA,
IS, COPPER CARBONATE, CERCARIAE,
ALCAREOUS SOIL, SOIL WATER, SOIL
PEPODS, WATER POLLUTION SOURCES,
TO-SAN JOAQUIN DELTA, CALIFORNIA
UALITY, WATER POLLUTION CONTROL,
ONTROL, WATER POLLUTION CONTROL,
AGE, WATER POLLUTION, WATER LAW,
COSYSTEMS, *SIMULATION ANALYSIS,
ONTROL, WATER POLLUTION CONTROL,
ER POLLUTION CONTROL, NUTRIENTS,
RVESTING, WATER QUALITY CONTROL,
RVESTING, WATER QUALITY CONTROL,
ENTS, TOXICITY, WATER CHEMISTRY,
NS, CONSERVATION, WATER QUALITY,
DUES, PHOSPHATES, ALGAE, SEWAGE,
TAMINS, WATER POLLUTION EFFECTS,
SIBILITY, LAKES, NUISANCE ALGAE,
 WASTE TREATMENT, WATER QUALITY,
SOIL CHEMISTRY, WATER CHEMISTRY,
ORUS COMPOUNDS, WATER CHEMISTRY,
RIENTS, WATER POLLUTION EFFECTS,
ANCE ALGAE, NUTRIENTS, VITAMINS,
ER LAW, WATER POLLUTION EFFECTS,
MISTRY, WATER POLLUTION EFFECTS,
FFECTS, WATER POLLUTION SOURCES,
UTRIENTS, BIOLOGICAL PROPERTIES,
EFFECTS, RADIOECOLOGY, SORPTION,
CATION, WATER POLLUTION EFFECTS,
RPTION, WATER POLLUTION EFFECTS,
SSAY, NUTRIENTS, EUTROPHICATION,
STS, BACTERIA, AQUATIC BACTERIA,
IMARY PRODUCTIVITY, SYSTEMATICS,
S, *EUTROPHICATION, *CALIFORNIA,
MATICS, WATER POLLUTION EFFECTS,
SEWAGE DISPOSAL, WASTE DISPOSAL,
FFECTS, WATER POLLUTION SOURCES,
ATMENT, WATER POLLUTION EFFECTS,
, PRODUCTIVITY, WASTE TREATMENT,
ALGAE, WATER POLLUTION EFFECTS,
RS, ALGAE, NITROGEN, PHOSPHORUS,
PESTICIDE TOXICITY, POLLUTANTS,
FISHKILL, LAKES, ODOR, TOXICITY,
BLOOMS, NUTRIEN/ *NEW HAMPSHIRE
PHORUS, WATER POLLUTION EFFECTS,
CTS, CHLORIDES, HALOGENS, ALGAE,
, *PHOSPHATES, NUTRIENTS, ALGAE,
LAKES, NUISANCE ALGAE, BACTERIA,
TA, RECREATION, SEWAGE DISPOSAL,
TS, DOMESTIC WASTES, POLLUTANTS,
LOGICAL ECOLOGY, ORGANIC MATTER,
EMICAL ANALYSIS, ORGANIC MATTER,
LUENTS, STREAMFLOW, MEASUREMENT,
EFFECTS, *PRIMARY PRODUCTIVITY,
DIURNAL, WINDS, CURRENTS(WATER),
*RADIOCARBON UPTAKE TECHNIQUES,
TASTE, ODOR, FILTERS, PLANKTON,
R, MASSACHUSETTS, POTABLE WATER,
TION, *WASHINGTON, URBANIZATION,
*ALGAE, WATER POLLUTION CONTROL,
AGRICULTURAL WATERSHEDS, SEWAGE,
T, *WATERS, *ORGANISMS, STREAMS,
RS, AQUATIC LIFE, AQUATIC ALGAE,
ONTROL, WATER POLLUTION EFFECTS,
CATION, *LAKES, *RIVERS, *ALGAE,
EED CONTROL, MECHANICAL CONTROL,
 QUALITY CONTROL, WATER QUALITY,
TA, SEWAGE EFFLUENTS, MINNESOTA,
TROPHICATION, *LAKES, NUTRIENTS,
TIVITY, SEDIMENTS, URBANIZATION,
TASSIUM, TRACE METALS, VITAMINS,
R, HUDSON RIVER, DELAWARE RIVER,
*TEXAS, WATER POLLUTION EFFECTS,
UALITY, *STRATIFICATION, *TEXAS,
S, RELIABILITY, WATER POLLUTION,
ACTERIA, *ANALYTICAL TECHNIQUES,
TOPES, CHLAMYDOMONAS, BIOASSAYS,
 CHEMICAL ANALYSIS, RELIABILITY,
OPHYTA, PLANKTON, PHYTOPLANKTON,
Y, LAKES, ENVIRONMENTAL EFFECTS,
S, NUISANCE ALGAE, PUBLICATIONS,
OASSAY, SCENEDESMUS, PHOSPHATES,
 INTERIOR, MADISON(WIS), EUROPE,
ATMENT, WATER POLLUTION SOURCES,
E EFFLUENTS, TERTIARY TREATMENT,
VIRONMENTAL EFFECTS, PERIPHYTON,

WATER LILY, SWIMMERS' ITCH.: /D C   W70-05568
WATER MANAGEMENT(APPLIED), LIMNOL   W70-04721
WATER MANAGEMENT.: /OSPHORUS, PHO   W72-01867
WATER MANAGEMENT(APPLIED), GEOCHE   W71-03048
WATER MASSES, U S PUBLIC HEALTH S   W70-00263
WATER MILFOIL COONTAIL, WILD CELE   W70-05568
WATER MOVEMENT, LEACHING, ARID LA   W71-04067
WATER P: /ERIA, SEDIMENTATION, CO   W69-10169
WATER PLAN, DELTA WATER FACILITIE   W70-05094
WATER PO: /STE TREATMENT, WATER Q   W69-06535
WATER POLLUTION EFFECTS, FUNGI, A   W69-05706
WATER POLLUTION EFFECTS, WATER PO   W69-06305
WATER POLLUTION EFFECTS, LAKES, P   W69-06536
WATER POLLUTION EFFECTS.: /LITY C   W69-05705
WATER POLLUTION EFFECTS, WATER PO   W69-06276
WATER POLLUTION CONTROL, WATER PO   W69-05706
WATER POLLUTION CONTROL, WATER PO   W69-05705
WATER POLLUTION EFFECTS.: / NUTRI   W69-05867
WATER POLLUTION, WATER RESOURCES.   W69-05844
WATER POLLUTION, WATER LAW, WATER   W69-06305
WATER POLLUTION SOURCES, VITAMIN   W69-06277
WATER POLLUTION CONTROL, NUTRIENT   W69-06276
WATER POLLUTION CONTROL, WATER PO   W69-06535
WATER POLLUTION EFFECTS, WATER PO   W69-06273
WATER POLLUTION EFFECTS, WATER QU   W69-05868
WATER POLLUTION SOURCES, WISCONSI   W69-06276
WATER POLLUTION EFFECTS, WATER PO   W69-06277
WATER POLLUTION SOURCES, WATER PO   W69-06305
WATER POLLUTION SOURCES.: /ER CHE   W69-06273
WATER POLLUTION TREATMENT, FINANC   W69-06305
WATER POLLUTION EFFECTS.: /IDE, N   W69-07442
WATER POLLUTION EFFECTS, WATER PO   W69-07440
WATER POLLUTION SOURCES, BIOINDIC   W69-06864
WATER POLLUTION SOURCES.: /GY, SO   W69-07440
WATER POLLUTION EFFECTS, WATER PO   W69-06864
WATER POLLUTION EFFECTS.: /S, YEA   W69-06970
WATER POLLUTION EFFECTS, WATER PO   W69-07832
WATER POLLUTION, REAERATION, NUTR   W69-07520
WATER POLLUTION SOURCES, WATER SU   W69-07832
WATER POLLUTION, LAKES, STREAMS,   W69-06909
WATER POLLUTION CONTROL, OXYGEN B   W69-08668
WATER POLLUTION SOURCES, CYCLING   W69-08518
WATER POLLUTION EFFECTS, WATER PO   W69-08518
WATER POLLUTION SOURCES, AQUATIC   W69-09135
WATER POLLUTION EFFECTS, WATER PO   W69-08668
WATER POLLUTION EFFECTS, ALDRIN,   W69-08565
WATER POLLUTION EFFECTS, WATER QU   W69-08279
WATER POLLUTION COMMISSION, ALGAL   W69-08674
WATER POLLUTION SOURCES, WATER PO   W69-08668
WATER POLLUTION SOURCES, WATER CH   W69-08562
WATER POLLUTION EFFECTS, WATER PO   W69-09135
WATER POLLUTION EFFECTS.: /HIES,   W69-08283
WATER POLLUTION, POLLUTION ABATEM   W70-04810
WATER POLLUTION SOURCES, WATER QU   W70-05084
WATER POLLUTION EFFECTS, DIATOMS,   W70-04811
WATER POLLUTION EFFECTS, ANALYTIC   W70-05109
WATER POLLUTION, FERTILIZATION, I   W70-05112
WATER POLLUTION EFFECTS, BIOASSAY   W70-05424
WATER POLLUTION, DRAINAGE, ORGANI   W70-05404
WATER POLLUTION IDENTIFICATION, O   W70-05652
WATER POLLUTION, SURFACE WATERS,   W70-05389
WATER POLLUTION, BACTERIA, ALGAE,   W70-05264
WATER POLLUTION, SEDIMENTS, PALEO   W70-05663
WATER POLLUTION EFFECTS, WATER PO   W70-05668
WATER POLLUTION, CHEMICAL ANALYSI   W70-05752
WATER POLLUTION, ANIMALS, BACTERI   W70-05750
WATER POLLUTION EFFECTS.: /DICATO   W70-05958
WATER POLLUTION SOURCES, SEWAGE,   W70-05668
WATER POLLUTION CONTROL, WATER PO   W70-05668
WATER POLLUTION CONTROL, CHEMCONT   W70-04494
WATER POLLUTION, NITROGEN, PHOSPH   W70-04484
WATER POLLUTION EFFECTS, ALGICIDE   W70-04506
WATER POLLUTION EFFECTS, PRODUCTI   W70-04721
WATER POLLUTION TREATMENT, ALGAE,   W70-04721
WATER POLLUTION, PHYSIOLOGICAL EC   W70-04503
WATER POLLUTION EFFECTS.: /E RIVE   W70-04507
WATER POLLUTION EFFECTS, DISSOLVE   W70-04727
WATER POLLUTION SOURCES, WATER PO   W70-04727
WATER POLLUTION EFFECTS, NITROGEN   W70-04382
WATER POLLUTION EFFECTS, ALGAE, E   W70-04287
WATER POLLUTION EFFECTS, ANALYTIC   W70-04284
WATER POLLUTION, WATER POLLUTION   W70-04382
WATER POLLUTION EFFECTS, WISCONSI   W70-03959
WATER POLLUTION EFFECTS, EPILIMNI   W70-03957
WATER POLLUTION EFFECTS, NEW YORK   W70-03973
WATER POLLUTION SOURCES.: /E, *BI   W70-03955
WATER POLLUTION ASSESSMENT, URBAN   W70-03975
WATER POLLUTION EFFECTS, LABORATO   W70-04074
WATER POLLUTION SOURCES, WATER PO   W70-04074
WATER POLLUTION EFFECTS, OREGON,   W70-03978

CHELATION, METABOLISM, AQUARIA,
ECREATION, ALGAE, OXYGEN DEMAND,
ALGAE, HEATED WATER, POPULATION,
IUM RADIOISOTOPES, MARINE ALGAE,
SEWAGE, WASTE WATER(POLLUTION),
, COLLOIDS, AQUARIA, ABSORPTION,
OISOTOPES, COBALT RADIOISOTOPES,
S, ECOSYSTEMS, AQUATIC HABITATS,
ECOLOGICAL DISTRIBUTION, MIDGES,
N, PHOSPHORUS, ALGAE, LIVESTOCK,
ONDITIONS, ANAEROBIC CONDITIONS,
ENCE, WATER POLLUTION TREATMENT,
LVED OXYGEN, DETRITUS, BACTERIA,
IREMENTS, ANALYTICAL TECHNIQUES,
AGE EFFLUENTS, SEWAGE TREATMENT,
ON COMPOUNDS, LIGHT PENETRATION,
ER, *MARINE ALGAE, *EMULSIFIERS,
LLUTANT IDENTIFICATION, BIOLOGY,
L-WATER INTERFACES, PERSISTENCE,
ATION, BIOLOGY, WATER POLLUTION,
ATMENT, WATER POLLUTION CONTROL,
IMI/ *DETERGENTS, *MARINE ALGAE,
NNISQUAM LAKE, WATER AND PLANTS,
TE, NITRATES, INDUSTRIAL WASTES,
OLLUTION EFFECTS, ALGAL CONTROL,
THIC FLORA, BENTHOS, MINE WATER,
N/ EUTROPHICATION, *CONFERENCES,
   *BIOSTIMULATION, ALGAL GROWTH,
PLANT GROWTH, WATER AND PLANTS,
ROWTH, ANABAENA, BIOSTIMULATION,
OGRAPHY, RADIOACTIVITY, TRACERS,
OS, MINE WATER, WATER POLLUTION,
CALIF), STANISLAUS RIVER(CALIF),
OPHYTA, DIATOMS, EUTROPHICATION,
TES, ECOSYSTEMS, ORGANIC MATTER,
ORGANIC MATTER, CURRENTS(WATER),
N, WATER QUALITY CONTROL, LAKES,
UTION, *RIVERS, WATER CHEMISTRY,
ANE BACTERIA, CHEMICAL ANALYSIS,
TER CONSERVATION, PUBLIC HEALTH,
EMCONTROL, EUTROPHICATION, BASS,
ONTROL, WATER POLLUTION EFFECTS,
TY, ALGAE, NITROGEN, PHOSPHORUS,
CATION, WATER POLLUTION CONTROL,
EMENT, POLLUTANT IDENTIFICATION,
ICIPAL WASTES, FRESHWATER ALGAE,
 ALGAE, TURF, SOIL MICROBIOLOGY,
ICIDES, ALGAE, LABORATORY TESTS,
OSTRONTIUM COMPLEXES, DTPA, SOIL
RGANIC MATTER, NUTRIENTS, LAKES,
OS, AQUATIC LIFE, ALGAE, PLANTS,
CRATER LAKE, ORE, GROWTH RATES,
ROELECTRIC PLANTS, WEED CONTROL,
TISSUES, SCENEDESMUS, LIMNOLOGY,
ATMENT, DOMESTIC WASTES, *ALGAE,
N, PHOSPHORUS, ALGAE, NUTRIENTS,
IC ASPECTS, AQUATIC ENVIRONMENT,
NTS, PLANT POPULATIONS, HISTORY,

XYGENATION, ABSORPTION, ECOLOGY,
CIPITATION, FILTRATION, AMMONIA,
UNGI, *BACTER/ *ALGAE, *BIOMASS,
ACTIVATED CARBON, CLAYS, ALGAE,
NICAL), CLEANING, COST ANALYSIS,
IA, *WATER TREATMENT, *NEW YORK,
TER QUALITY CONTROL, FILTRATION,
ALGAE, *WATER POLLUTION CONTROL,
*FILTERS, SANDS, LAKE MICHIGAN,
OMYCETES, *ALGAE, *TASTE, *ODOR,
ES, WATER RESOURCES DEVELOPMENT,
E, *COPPER SULFATE, *ALKALINITY,
LFATE, WATER SUPPLY, RESERVOIRS,
LAYS, ALGAE, WATER PURIFICATION,
LGAL BLOOMS, NUTRIENTS, LIMNOLO/
E, *ALGAL CONTROL, WATER SUPPLY,
LGAE, WATER POLLUTION TREATMENT,
SSOLVED OXYGEN, ALGAE, VELOCITY,
ON EFFECTS, NITROGEN, COLIFORMS,
OURCES, WATER POLLUTION EFFECTS,
S, *CYANOPHYTA, *EUTROPHICATION,
TROL MEASURES, PHYSICAL CONTROL,
Y PRODUCTIVITY, *EUTROPHICATION,
UIPMENT, *ANALYTICAL TECHNIQUES,
RESIDUES, HERBICIDES, ESTUARIES,
NISMS, SAMPLING, WATER ANALYSIS,
, *ENVIRONMENT, *ORGANIC MATTER,
, *PESTICIDES, FARM WASTES, SIL/
LGAE, COPPER / *WATER TREATMENT,
ON, GREAT LAKES, NUISANCE ALGAE,
TROPHICATION, DAPHNIA, CHLAMYDO/

WATER POLLUTION EFFECTS, ADSORPTI   W71-09005
WATER POLLUTION, MUNICIPAL WATER,   W71-09577
WATER POLLUTION EFFECTS.: /KTON,   W71-09767
WATER POLLUTION SOURCES.: /, URAN   W71-09013
WATER POLLUTION SOURCES, MUNICIPA   W71-05879
WATER POLLUTION EFFECTS, CATION E   W71-09863
WATER POLLUTION EFFECTS.: /E RADI   W71-09850
WATER POLLUTION EFFECTS.: /OCESSE   W71-09012
WATER POLLUTION EFFECTS, OLIGOCHA   W71-12077
WATER POLLUTION EFFECTS.: /ITROGE   W71-10376
WATER POLLUTION SOURCES, PATH OF   W71-11381
WATER POLLUTION EFFECTS.: /ERSIST   W70-01777
WATER POLLUTION EFFECTS, LIMNOLOG   W70-02249
WATER POLLUTION EFFECTS, PHYTOPLA   W70-02245
WATER POLLUTION CONTROL, WATER PO   W70-02248
WATER POLLUTION EFFECTS, EUTROPHI   W70-02255
WATER POLLUTION EFFECTS, MORTALIT   W70-01780
WATER POLLUTION, WATER POLLUTION   W70-02304
WATER POLLUTION TREATMENT, WATER   W70-01777
WATER POLLUTION SOURCES, ALGAL CO   W70-02304
WATER POLLUTION EFFECTS, WATER: /   W70-02248
WATER POLLUTION EFFECTS, LETHAL L   W70-01779
WATER POLLUTION EFFECTS, ECOLOGY,   W70-02764
WATER POLLUTION EFFECTS, DENITRIF   W70-02777
WATER POLLUTION, TOXICITY, LAKES,   W70-02775
WATER POLLUTION, WATER POLLUTION   W70-02770
WATER POLLUTION EFFECTS, ALGAL CO   W70-02775
WATER POLLUTION ASSESSMENT, SANIT   W70-02775
WATER POLLUTION EFFECTS, ECOLOGY,   W70-02772
WATER POLLUTION ASSESSMENT.: /L G   W70-02779
WATER POLLUTION EFFECTS, PHYSIOLO   W70-02510
WATER POLLUTION EFFECTS, ACID STR   W70-02770
WATER POLLUTION ASSESSMENT, AMERI   W70-02777
WATER POLLUTION EFFECTS, SYSTEMS   W70-02779
WATER POLLUTION, THERMAL POLLUTIO   W70-03309
WATER POLLUTION SOURCES, SURFACE   W70-03501
WATER POLLUTION SOURCES, CHLOROPH   W70-03310
WATER POLLUTION SOURCES, PHOSPHOR   W70-03068
WATER POLLUTION, CHLORELLA, SCENE   W70-03312
WATER POLLUTION, CONTRACTS, REGUL   W70-03344
WATER POLLUTION CONTROL, MINNOWS,   W70-02982
WATER POLLUTION SOURCES.: /TION C   W70-02792
WATER POLLUTION EFFECTS.: /ERTILI   W70-02973
WATER POLLUTION EFFECTS, WATER PO   W70-02792
WATER POLLUTION CONTROL, WATER PO   W70-02792
WATER POLLUTION EFFECTS, DOMESTIC   W72-12240
WATER POLLUTION SOURCES.: /, SOIL   W72-08110
WATER POLLUTION EFFECTS.: /N PEST   W72-11105
WATER PROFILE.: / ACTIVITY, *RADI   W71-04067
WATER PROPERTIES, BACTERIA, BIOAS   W71-10079
WATER PROPERTIES, WATER TEMPERATU   W69-05023
WATER PROPERTIES.:   W68-00478
WATER PROPERTIES, PHOSPHORUS, CHE   W68-00479
WATER PROPERTIES, WATER TEMPERATU   W68-00248
WATER PROPERTIES, BIOCHEMICAL OXY   W68-00256
WATER PROPERTIES.: /ECTS, NITROGE   W70-04382
WATER PROPERTIES, BIOLOGICAL PROP   W70-04475
WATER PURIFICATION, *WATER UTILIZ   W68-00483
WATER PURIFICATION BY ALGAE.:   W69-07389
WATER PURIFICATION, CARBON, BICAR   W69-07389
WATER PURIFICATION, WASTE WATER T   W71-11393
WATER PURIFICATION, *PROTOZOA, *F   W71-05267
WATER PURIFICATION, WATER QUALITY   W69-09884
WATER PURIFICATION, *WATER TREATM   W72-08688
WATER PURIFICATION.: /SIGN CRITER   W72-08686
WATER PURIFICATION, MEMBRANES, SC   W73-02426
WATER QUALITY CONTROL.: /ROSION,   W73-01122
WATER QUALITY CONTROL, FILTRATION   W73-02426
WATER QUALITY CONTROL, POLLUTANT   W72-13605
WATER QUALITY, SOIL INVESTIGATION   W72-11906
WATER QUALITY CONTROL, *CONTROL,   W72-07442
WATER QUALITY, NEW JERSEY, TEMPER   W69-10157
WATER QUALITY, WATER CHEMISTRY.: /   W69-09884
WATER QUALITY, *EUTROPHICATION, A   W69-09723
WATER QUALITY, WATER POLLUTION CO   W69-10155
WATER QUALITY CONTROL, SEWAGE EFF   W70-01031
WATER QUALITY, DISSOLVED SOLIDS,   W70-00265
WATER QUALITY.: /S, WATER POLLUTI   W71-05407
WATER QUALITY.: /ATER POLLUTION S   W71-05409
WATER QUALITY.: /ON EFFECTS, LAKE   W71-04758
WATER QUALITY.: /, INTEGRATED CON   W70-07393
WATER QUALITY, *NUTRIENTS, LIMNOL   W70-09093
WATER QUALITY, DISSOLVED OXYGEN,   W70-08627
WATER QUALITY, DETERGENTS, ALGAL   W70-10437
WATER QUALITY, POLLUTANT IDENTIFI   W71-00221
WATER QUALITY, TEMPERATURE, HYDRO   W71-03031
WATER QUALITY CONTROL, *NUTRIENTS   W71-10376
WATER QUALITY, *ORGANIC MATTER, A   W72-04495
WATER QUALITY, NUTRIENTS, PHOSPHO   W72-05453
WATER QUALITY, *ALGAL CONTROL, EU   W72-02417

TES, IRON, PHOSPHORUS, NITROGEN,
ANALYSIS, ALGAL CONTROL, ALGAE,
BUTION SYSTEMS, WATER POLLUTION,
TION CONTROL, RADIOACTIVE WASTE,
Y, AQUATIC ALGAE, TROPHIC LEVEL,
COLOGY, BIOLOGY, FISH, PLANKTON,
ENTS, TERTIARY TREATMENT, ALGAE,
YTA, *EUTROPHICATION, WISCONSIN,
ES, STARCH MANUFACTURING WASTES,
ATER POLLUTION EFFECTS, ECOLOGY,
CING ALGAE, ALGAE, CHLORINATION,
, ECOLOGY, EFFECTS OF POLLUTION,
HIC LEVELS, AQUATIC ENVIRONMENT,
TIC ENVIRONMENT, *AQUATIC ALGAE,
XICITY, WATER POLLUTION EFFECTS,
BICIDES, PESTICIDES, HARVESTING,
L GOVERNMENT, STATE GOVERNMENTS,
UATIC POPULATIONS, CONSERVATION,
STRATIFICATION, WASTE TREATMENT,
BICIDES, PESTICIDES, HARVESTING,
MISTRY, WATER POLLUTION EFFECTS,
NUTRIENTS, AQUATIC POPULATIONS,
ON, FISHKILL, ALGAE, OXYGEN SAG,
IRONMENT, ESTUARINE ENVIRONMENT,
, *PHOTOSYNTHESIS, PRODUCTIVITY,
PERTIES, WASTE WATER(POLLUTION),
MARY PRODUCTIVITY, TROPIC LEVEL,
IC/ *EUTROPHICATION, SUCCESSION,

TS, P/ *RIVER STUDIES, *NATIONAL
COMMUNITIES, MICROBIAL ECOLOGY,
ER, *OHIO RIVER, NUISANCE ALGAE,
ER POLLUTION EFFECTS, ALGICIDES,
*MIXING, *WATER QUALITY CONTROL,
SEPTIC TANKS, RECREATION WASTES,
ROSCOPY, WASTE TREATMENT, LAKES,
PHYTON INVENTORY(LAKE SUPERIOR),
, *RECLAIMED WATER, *CALIFORNIA,
POLLUTION ABATEMENT, COLIFORMS,
UTANTS, WATER POLLUTION SOURCES,
IVATED SLUDGE, *EFFLUENTS, WASTE
ATER POLLUTION, MUNICIPAL WATER,
T VALLEY RESERVOIRS(NEB), *CLEAR
IRCRAFT, WATER SAMPLING, ONTARIO
WATER QUALITY, WATER POLLUTION,
R QUALITY, SO/ *RESERVOIR SITES,
FITE LIQUORS, ULTIMATE DISPOSAL,
INVESTIGATIONS, WATER POLLUTION,
NUMBERS, CROP-DUSTING AIRCRAFT,
, *LAKE AGING, NUTRIENT SOURCES,
LGAE, BENTHIC FAUNA, WATERSHEDS,
ER TURBIDITY, INDUSTRIAL PLANTS,
ANISMS, *POLLUTANTS, RECREATION,
RY PRODUCTIVITY, *TROPHIC LEVEL,
*ALGAL CONTROL, *COPPER SULFATE,
*COPPER SULFATE, *ALGAL CONTROL,
E, FISH, SEWERS, RUNOFF, WASTES,
HOSPHORUS, PHOSPHORUS COMPOUNDS,
OPPER SULFATE, ACTIVATED CARBON,
NKTON, OHIO RIVER, PENNSYLVANIA,
*DATA COLLECTIONS, PIT RECHARGE,
FFECTS, WATER POLLUTION SOURCES,
ONDS, *SOIL, GROUNDWATER BASINS,
PLING, MONITORING, WATER DEMAND,
ONDITIONS, ANAEROBIC CONDITIONS,
CULTURES, OPTICAL DENSITY, SOIL-
R TEMPERATURE, AEROBIC BACTERIA,
ALGAE, PLANTS, WATER PROPERTIES,
TIONS, REVIEWS, LIGHT INTENSITY,
US, LIMNOLOGY, WATER PROPERTIES,
MINES, DRAINAGE, FERROBACILLUS,
, WATER BUDGET, WATER CHEMISTRY,
*FARM LAGOONS, *DESIGN CRITERIA,
, SPRING WATER, WATER CHEMISTRY,
ICATION, OLIGOTROPHY, NUTRIENTS,
IE, LAKE ONTARIO, LAKE MICHIGAN,
ALITY, *DISSOLVED OXYGEN, *OHIO,
QUALITY, *NUTRIENTS, LIMNOLOGY,
, RIVERS, ESTUARINE ENVIRONMENT,
REMOVAL, *WATER REUSE, SEASONAL,
HESI/ *OXIDATION LAGOONS, *WASTE
SULFATE, *ALKALINITY, WATER QUA/
RICIDES, CHLORINE, COSTS, *WASTE
RIA, ODORS, EFFICIENCIES, *WASTE
ATION, LAGOONS, *BAFFLES, *WASTE
TIC OXYGEN, PRODUCTIVITY, *WASTE
IOCHEMICAL OXYGEN DEMAND, *WASTE
ROPHYTA, *ALGAE, SORPTION, WASTE
LAGOONS, PONDS, OXIDATION, WASTE
TMENT, *WATER POLLUTION / *WASTE

WATER QUALITY.: /HIGAN, BICARBONA W72-03183
WATER QUALITY, SEDIMENT-WATER INT W72-06051
WATER QUALITY.: /OXICITY, *DISTRI W72-03746
WATER QUALITY ACT, PATH OF POLLUT W72-00941
WATER QUALITY, FLORIDA.: /UCTIVIT W72-01990
WATER QUALITY, ALGAE, AQUATIC LIF W72-01786
WATER QUALITY, CONIFEROUS FORESTS W70-03512
WATER QUALITY CONTROL, LAKES, WAT W70-03310
WATER QUALITY STANDARDS, GRAFTON(W70-03312
WATER QUALITY CHARACTERIZATION.: / W70-02764
WATER QUALITY CONTROL.: /TE-PRODU W70-02373
WATER QUALITY CONTROL, WINNISQUAM W70-02772
WATER QUALITY, NUTRIENTS, NORTHEA W70-02764
WATER QUALITY, *TROPHIC LEVELS, N W70-02772
WATER QUALITY, CATTLE.: /ODOR, TO W69-08279
WATER QUALITY CONTROL, WATER POLL W69-05705
WATER QUALITY: /G, GRANTS, FEDERA W69-06305
WATER QUALITY, WATER POLLUTION, W W69-05844
WATER QUALITY, WATER POLLUTION CO W69-06535
WATER QUALITY CONTROL, WATER POLL W69-05706
WATER QUALITY, WISCONSIN, LAKES,   W69-05868
WATER QUALITY, LAKE MORPHOMETRY,   W68-00470
WATER QUALITY, CHLORIDES, PHOSPHO W69-03948
WATER QUALITY, SEWAGE BACTERIA, E W69-04276
WATER QUALITY, *LIMNOLOGY, MINNES W69-00632
WATER QUALITY: /IES, PHYSICAL PRO W69-01273
WATER QUALITY, WASTE ASSIMILATIVE W68-00912
WATER QUALITY, ALGAL CONTROL, ALG W69-01977
WATER QUALITY PROPERTIES.:   W70-04475
WATER QUALITY NETWORK, ALGAE COUN W70-04507
WATER QUALITY CRITERIA, CYANIDE,   W70-04510
WATER QUALITY, SEWAGE EFFLUENTS,   W70-04507
WATER QUALITY.: /, MINNESOTA, WAT W70-04506
WATER QUALITY, WATER POLLUTION, N W70-04484
WATER QUALITY.: /NICIPAL WASTES,   W70-04266
WATER QUALITY, VIRUSES, ALGICIDES W70-03973
WATER QUALITY INDICATORS.:   PERI W70-05958
WATER QUALITY, AESTHETICS, RECREA W70-05645
WATER QUALITY, PHOSPHORUS, NITROG W70-04810
WATER QUALITY, ALGAE, SEWAGE EFFL W70-05084
WATER RECLAMATION, WASTE WATER TR W70-06053
WATER REQUIREMENTS, WATER CONVEYA W71-09577
WATER RESERVOIRS, *TURBID RESERVO W72-04761
WATER RESOURCES COMMISSION, PEST   W69-10155
WATER RESOURCES.: / CONSERVATION, W69-05844
WATER RESOURCES DEVELOPMENT, WATE W72-11906
WATER REUSE, FISH FARMING.: / SUL W69-05891
WATER SAMPLING, *WATER ANALYSIS,   W69-00659
WATER SAMPLING, ONTARIO WATER RES W69-10155
WATER SAMPLING, ALGAL GROWTH, OXY W70-00264
WATER SHORTAGE FILTERS, SNAILS, O W72-00027
WATER SPORTS.: /LAKE ONTARIO, WAT W71-09880
WATER SPORTS, ENVIRONMENT, AQUATI W69-04276
WATER STORAGE, EVAPORATION, WATER W69-09723
WATER SUPPLY, RESERVOIRS, WATER Q W69-10157
WATER SUPPLY, WATER QUALITY, WATE W69-10155
WATER SUPPLY, EFFLUENTS, STREAMS, W71-09880
WATER SUPPLY, WATER MANAGEMENT.: / W72-01867
WATER SUPPLY.: /, CHLORINATION, C W72-03562
WATER SUPPLY.: /N ALGAE, PHYTOPLA W69-05697
WATER SUPPLY, MUNICIPAL WATER, SE W69-05894
WATER SUPPLY.: /WATER POLLUTION E W69-07832
WATER SUPPLY, MUD, BENTHOS, HYDRO W69-07838
WATER SUPPLY, ARID LANDS, LIMNOLO W70-05645
WATER SUPPLY.: /HETICS, AEROBIC C W70-04430
WATER TECHNIQUES, PHYCOLOGY, MICR W70-04381
WATER TEMPERATURE, IRON.: /ER, AI W69-07096
WATER TEMPERATURE, OUTLETS, HYDRO W69-05023
WATER TEMPERATURE, DIATOMS, CHLOR W68-00860
WATER TEMPERATURE, ALKALINITY, *P W68-00248
WATER TEMPERATURE, HYDROGEN ION C W68-00891
WATER TEMPERATURE, HEAT BUDGET, T W72-01773
WATER TEMPERATURE, SEWAGE TREATME W71-08221
WATER TEMPERATURE, CHEMICAL ANALY W69-09676
WATER TEMPERATURE, DISSOLVED OXYG W70-00264
WATER TEMPERATURE, PHOSPHATES, TU W70-00667
WATER TEMPERATURE, SLUDGE WORMS,   W70-00475
WATER TEMPERATURE, *ALGAE BLOOMS, W70-09093
WATER TEMPERATURE, *EUTROPHICATIO W73-01446
WATER TREATMENT.: /NT, *NUTRIENT   W72-08133
WATER TREATMENT, ALGAE, PHOTOSYNT W72-10573
WATER TREATMENT, *ALGAE, *COPPER   W72-07442
WATER TREATMENT.: /ACTERIA, BACTE W72-12987
WATER TREATMENT.: /, DESIGN CRITE W72-06854
WATER TREATMENT, *POULTRY, INDUST W70-09320
WATER TREATMENT.: /Y, PHOTOSYNTHE W70-08416
WATER TREATMENT, *WATER POLLUTION W70-09186
WATER TREATMENT, CULTURES, CHLORE W70-09438
WATER TREATMENT.: / CHLOROPHYLL,   W70-08392
WATER TREATMENT, *BIOLOGICAL TREA W70-10437

, ACTIVATED SLUDGE, ALGAE, WASTE	WATER TREATMENT.: / DEMAND, COLOR	W70-10433
GAE, BIOLOGICAL TREATMENT, WASTE	WATER TREATMENT.: /PHOSPHORUS, AL	W70-07469
ATMENT, TRICKLING FILTERS, WASTE	WATER TREATMENT, PHOSPHATES, NITR	W70-06792
MIXING, SEWAGE TREATMENT, WASTE	WATER TREATMENT.: /OXYGEN DEMAND,	W70-06899
TOM SEDIMENTS, NUTRIENTS, *WASTE	WATER TREATMENT, *OXIDATION LAGOO	W70-07031
OXYGEN DEMAND, NUTRIENTS, WASTE	WATER TREATMENT, *FIBER(PLANT).: /	W70-06925
ISTRY, TASTE, ODOR, DEEP WELLS,	WATER TREATMENT, MECHANICAL EQUIP	W70-08096
S, ALGAE, ORGANIC LOADING, WASTE	WATER TREATMENT, WATER UTILIZATIO	W70-07508
TEMPERATURE, METABOLISM, *WASTE	WATER TREATMENT, LIGHT, MICROORGA	W70-06604
TENTIAL, SULFUR BACTERIA, *WASTE	WATER TREATMENT, *OXIDATION LAGOO	W70-07034
LORELLA, DAPHNIA, DETERIORATION,	WATER TREATMENT, DEGRADATION(DECO	W71-05719
LORELLA, DAPHNIA, DETERIORATION,	WATER TREATMENT, DEGRADATION(DECO	W71-03056
CONTROL, *DENIT/ *ALGAE, *WASTE	WATER TREATMENT, *WATER POLLUTION	W71-04557
TION, NUTRIENTS, NITROGEN, WASTE	WATER TREATMENT, PHOSPHOROUS, RUN	W71-04072
URRENTS, *WASTE TREATMENT, WASTE	WATER TREATMENT.: /MS, ELECTRIC C	W71-03775
CONTROL, *DENIT/ *ALGAE, *WASTE	WATER TREATMENT, *WATER POLLUTION	W71-04554
LIFORNIA, *PHOTOSYNTHESIS, WASTE	WATER TREATMENT.: /OORGANISMS, CA	W71-05267
CONTROL, *DENIT/ *ALGAE, *WASTE	WATER TREATMENT, *WATER POLLUTION	W71-04555
OXYGEN DEMAND, DELAWARE, WASTE	WATER TREATMENT.: /A, BIOCHEMICAL	W71-05013
, WATER POLLUTION EFFECTS, WASTE	WATER TREATMENT, PENNSYLVANIA.: /	W71-05026
CONTROL, *DENIT/ *ALGAE, *WASTE	WATER TREATMENT, *WATER POLLUTION	W71-04556
ZATION, LAGOONS, TOXICITY, WASTE	WATER TREATMENT, WASTE WATER DISP	W69-09454
LAGOONS, *DOMESTIC WASTES, WASTE	WATER TREATMENT, INDUSTRIAL WASTE	W70-01519
C POWER, SEWAGE DISPOSAL, *WASTE	WATER TREATMENT, *OXIDATION LAGOO	W71-09548
LAGOONS, *DOMESTIC WASTES, WASTE	WATER TREATMENT, INDUSTRIAL WASTE	W71-06737
E, *BIOLOGICAL TREATMENT, *WASTE	WATER TREATMENT.: /, AQUATIC ALGA	W71-09546
HETIC OXYGEN, FARM WASTES, WASTE	WATER TREATMENT.: /DGE, PHOTOSYNT	W71-08221
INISTRATION, MAINTENANCE, *WASTE	WATER TREATMENT.: /NS, ALGAE, ADM	W71-08970
ICATION,/ WASTE WATER TREATMENT,	WATER TREATMENT, *ALGAE, *EUTROPH	W72-01867
POSAL, *SEWAGE TREATMENT, *WASTE	WATER TREATMENT, *WASTE WATER DIS·	W72-01786
*ALGAE, *EUTROPHICATION,/ *WASTE	WATER TREATMENT, WATER TREATMENT,	W72-01867
ES, OXIDAT/ *WATER REUSE, *WASTE	WATER TREATMENT, *INDUSTRIAL WAST	W72-00027
TE IDENTIFICATION, ALGAE, *WASTE	WATER TREATMENT.: /VELOPMENT, WAS	W72-03299
ES, AMMONIA, *WATER REUSE, WASTE	WATER TREATMENT.: /RATURE, NITRAT	W72-06017
RPTION, *WATER REUSE, *WASTE	WATER TREATMENT, ALGAE, FILTRATIO	W72-03641
NS, *ALGAE, LIGHT INTENS/ *WASTE	WATER TREATMENT, *OXIDATION LAGOO	W72-06022
BIOCHEMICAL OXYGEN DEMAND, WASTE	WATER TREATMENT, OVERFLOW, SEWERS	W71-10654
, CHEMICAL OXYGEN DEMAND, *WASTE	WATER TREATMENT.: /L, DEGRADATION	W71-13335
ARBON DIOXIDE, DIFFUSION, *WASTE	WATER TREATMENT.: /NG, METHANE, C	W71-11377
S, ANALYTICAL TECHNIQUES, *WASTE	WATER TREATMENT, TREATMENT FACILI	W71-13313
X/ *FARM WASTES, *POULTRY, WASTE	WATER TREATMENT, SEDIMENTATION, O	W71-11375
RATORY TESTS, TEMPERATURE, WASTE	WATER TREATMENT.: /D OXYGEN, LABO	W71-12183
MONIA, WATER PURIFICATION, WASTE	WATER TREATMENT.: /FILTRATION, AM	W71-11393
GING./ *METAL CORROSION, COOLING	WATER TREATMENT, ON-STREAM DESLUD	W70-02294
ITROGEN, NITROGEN COMPOU/ *WASTE	WATER TREATMENT, *WATER REUSE, *N	W69-08053
CAL OXYGEN DEMAND, *COLOR, *WASTE	WATER TREATMENT, TREATMENT FACILI	W70-04060
E, BACTERIA, CULTURES, C/ *WASTE	WATER TREATMENT, *NUTRIENTS, ALGA	W70-05655
SES, PLANTS, TEMPERATURE, *WASTE	WATER TREATMENT.: /N, PHOTOSYNTHE	W70-06600
N, ODOR, ORGANIC LOADING, *WASTE	WATER TREATMENT.: /ISSOLVED OXYGE	W70-06601
← WASTE WATER RECLAMATION, *WASTE	WATER TREATMENT, *NUTRIENTS.: /TS	W70-06053
ESTUARIES, SANITARY ENGINEERING,	WATER TREATMENT, WASTE TREATMENT.	W70-05094
E/ *ALGAE, *FLOCCULATION, *WASTE	WATER TREATMENT, ELECTROLYTES, CH	W70-05267
NDICATORS, ODOR, TASTE, FILTERS,	WATER TREATMENT, DISSOLVED OXYGEN	W70-05264
S, WATER SPORTS./ *LAKE ONTARIO,	WATER TURBIDITY, INDUSTRIAL PLANT	W71-09880
L PLANTS, STRUCTURES, AFTERBAYS,	WATER TYPES, INTAKES.: /INDUSTRIA	W69-05023
*DOMESTIC	WATER USE.:	W68-00483
UALITY CONTROL, *BIODEGRADATION,	WATER UTILIZATION, FISH, ALGAE, M	W71-11881
LOADING, WASTE WATER TREATMENT,	WATER UTILIZATION, INDUSTRIAL WAS	W70-07508
URCES, *WATER POLLUTION EFFECTS,	WATER UTILIZATION, ANALYTICAL TEC	W71-03048
INDUSTRY, ODOR, TAST/ *AMERICAN	WATER WORKS ASSOCIATION, CHEMICAL	W71-05943
MANAGEMENT, HARVESTING OF ALGAE,	WATER ZONING.: /MENTS, WATERSHED	W68-00859
LEMENTS, ALGAE, SEA WATER, FRESH	WATER.: / *COPPER, *IRON, TRACE E	W71-11687
ETHODOLOGY, INSTRUMENTATION, SEA	WATER.: /ES, THERMAL POLLUTION, M	W70-10437
, *CANADA, NUTRIENT LEVELS, SOFT	WATER.: /E LAKE(CANADA), *ONTARIO	W70-02973
EMPERATURE EXPERIMENTS, RECYCLED	WATER.: /GAE, BLUE-GREEN ALGAE, T	W70-07257
ERSHEDS, SEWERAGE ALGAE, POTABLE	WATER.: /NATION, COAGULATION, WAT	W71-05943
, PHYSIOLOGICAL ECOLOGY, THERMAL	WATER.: /RBON DIOXIDE, CYANOPHYTA	W69-10160
APHICAL REGIONS, BAYS, BODIES OF	WATER.: /T REGION, REGIONS, GEOGR	W69-03683
AGULATION, ELECTROLYTES, POTABLE	WATER.: /TICIDES, TASTE, ODOR, CO	W72-13601
ALGAE, ALGAL ASSOCIATIONS, SHEAR	WATER(ENGLAND), AMPHORA OVALIS, C	W71-12870
*PHOSPHORUS SOURCES, *NU/ *WASTE	WATER(MUNICIPAL AND INDUSTRIAL),	W69-09340
ES, BEACHES, SANDS, BIOTA, WASTE	WATER(POLLUTION), CURRENTS(WATER)	W68-00010
TIES, PHYSICAL PROPERTIES, WASTE	WATER(POLLUTION), WATER QUALITY: /	W69-01273
KTON, *SELF-PURIFICATION, *WASTE	WATER(POLLUTION), *WATER PURIFICA	W70-05092
, PIGMENTS, PHYTOPLANKTON, WASTE	WATER(POLLUTION), PHOTOSYNTHESIS,	W70-04371
ALGAE, NUTRIENTS, SEWAGE, WASTE	WATER(POLLUTION), WATER POLLUTION	W71-05879
ECOMPOSING ORGANIC MATTER, WASTE	WATER(POLLUTION).: /, BACTERIA, D	W70-00664
LTS, PHOSPHATES, NITROGEN, WASTE	WATER(POLLUTION), SCENEDESMUS, CY	W70-02255
WASTE WATER(POLLUTION), CURRENTS(	WATER).: /BEACHES, SANDS, BIOTA,	W68-00010
, PHOSPHORUS, SULFATES, HARDNESS(	WATER).: /ATER QUALITY, CHLORIDES	W69-03948
RATES, PLANTS, AMMONIA, HARDNESS(	WATER), ACIDITY, OXYGEN, ALGAE, S	W69-07838
CENTRATION, ALKALINITY, HARDNESS(	WATER), CHLORIDES, ORGANIC MATTER	W71-04206
OGY, CULTURES, WASHINGTON, WAVES(	WATER), CONNECTICUT, RHODE ISLAND	W69-10161
CS, LAKES, INTERNAL WATER, WAVES(	WATER), DIURNAL, EPILIMNION, WIND	W70-09889
(BASINS), INSECTICIDES, HARDNESS(	WATER), DIATOMS, FOULING, PLANKTO	W72-01813
SMUS, BENTHOS, CONVECTION, WAVES(	WATER), FISH, SEWAGE, EUGLENOPHYT	W70-03969
ERMOCLINE, TEMPERATURE, CURRENTS(	WATER), HYPOLIMNION, ALGAE, NITRO	W70-09889
ENVIRONMENTAL EFFECTS, CURRENTS(	WATER), INDUSTRIAL WASTE, CORES,	W68-00470
, ALGAE, AQUATIC ALGAE, CURRENTS(	WATER), LIMNOLOGY, LAKES, *PILOT	W68-00479
, FLORIDA, THERMOCLINE, CURRENTS(	WATER), PHYTOPLANKTON, ZOOPLANKTO	W70-04368

OROMETRY, DYE RELEASES, CURRENTS(WATER), REMOTE SENSING, SEA WATER W70-05377
REAMS, LIGHT INTENSITY, CURRENTS(WATER), VELOCITY, DIATOMS, SEASON W70-02780
IMENTS, ORGANIC MATTER, CURRENTS(WATER), WATER POLLUTION SOURCES, W70-03501
LATION, DIURNAL, WINDS, CURRENTS(WATER), WATER POLLUTION, DRAINAGE W70-05404
ERIUM, RESPIROMETER, PROTOPLASM, WATER-BLOOMS, LAKE GEORGE(UGANDA) W70-04369
ONMENT, PONDS, BREEDINGS, SALINE WATER-FRESHWATER INTERFACES, COLL W73-00748
GONFLIES, ALGAE, SEDIMENTS, SOIL- WATER-PLANT RELATIONSHIPS, NUCLEA W71-09005
*ADDITIVES, *FLOW RATE, *WASTE WATER, *ALGAE, *SLUDGE, STORMS, T W70-05819
*ALGAE, *AQUATIC BAC/ *ACID MINE WATER, *AQUATIC MICROORGANISMS, W69-00096
UNGI, BACTERIA, BACTER/ *COOLING WATER, *BIOMASS, *SLIME, ALGAE, F W72-12987
, AEST/ *WATER REUSE, *RECLAIMED WATER, *CALIFORNIA, WATER QUALITY W70-05645
TION/ *WATER TREATMENT, *COOLING WATER, *CORROSION, *CHEMICAL REAC W70-02294
, TEMPERATURE, SALINITY/ *MARINE WATER, *COPPER, *ALGAE, SEA WATER W70-08111
ATER REUSE, COOLING TO/ *COOLING WATER, *ELECTRIC POWER PLANTS, *W W71-11393
EFFECTS, *LI/ *ALGAE, *ACID MINE WATER, *HABITATS, *ENVIRONMENTAL W70-02770
*MARINE ALGAE, *SEAWATER, *FRESH WATER, *HUMIC ACIDS, *COLOR, RIVE W70-01074
IA, *NEVADA, DEPTH, ALGAE/ *DEEP WATER, *LAKES, *PLANTS, *CALIFORN W70-00711
S, WATER POLLUTION EFFECT/ *OILY WATER, *MARINE ALGAE, *EMULSIFIER W70-01780
ROGEN CYCLE, *FARM WASTE, GROUND WATER, *NITRATES, EUTROPHICATION, W69-05323
OLLUTANTS, INHIBITIO/ *ACID MINE WATER, *NUTRIENT REQUIREMENTS, *P W70-02792
ION EFFECTS, *SALT MARSHES, OILY WATER, *OIL, *MARINE ALGAE, WATER W70-09976
ONS, PHOSPHATES,/ *ALGAE, *FRESH WATER, *PEPTIDES, COPPER, IRON, I W69-10180
SMUS, LIGH/ *ENVIRONMENT, *FRESH WATER, *PLANKTON, *ALGAE, SCENEDE W70-05409
PLANTS, *THERMAL STRESS, THERMAL WATER, *PROTOZOA, ALGAE, DIATOMS, W70-07847
*PHYSIOLOGICAL ECOLOGY, *SALINE WATER, *THERMAL WATER, ALGAE, AQU W69-03516
F, CULTURES, CH/ *BIOASSAY, *SEA WATER, *VITAMINS, VITAMIN B, ALGA W70-05652
LGAE, COPPER SULFATE, / *POTABLE WATER, *WATER QUALITY CONTROL, *A W71-00110
*HOT SPRINGS, HABITATS, ALKALINE WATER, ACIDIC WATER, TRAVERTINE, W69-10160
OGY, ACIDITY, ACID SOILS, ACIDIC WATER, ACIDS, PHOTOSYNTHESIS, PHO W70-02792
CONCENTRATION, WATE/ *ACID MINE WATER, ACIDIC WATER, HYDROGEN ION W70-05424
TE, *ODOR, *AMINO ACIDS, POTABLE WATER, ACTINOMYCETES, PLANT GROWT W72-01812
R BACTERIA, STABILIZATION, WASTE WATER, ALGAE, DIFFUSION, SULFUR.: W70-01971
RAPHIES, HERBICIDES, FISH, FRESH WATER, ALGAE, PONDS, WATER HYACIN W70-00269
ECOLOGY, *SALINE WATER, *THERMAL WATER, ALGAE, AQUATIC ALGAE, AQUA W69-03516
S, *NITROGEN UTILIZATION, *FRESH WATER, ALGAE, NITRATE AMMONIUM, N W69-04521
, ALGAE, PHYTOPLANKTON, ALKALINE WATER, ALKALINITY, FISH FARMING, W71-07731
IENTS, BACTERIA, ALGAE, BRACKISH WATER, ALKALINE WATER, SODIUM.: / W71-03031
POISONING, *MARSH PLANTS, *OILY WATER, AQUATIC ALGAE, MARINE ALGA W70-01231
PHORUS RADIOISOTOPES, *TRANSFER, WATER, BACTERIA, SEDIMENTS, MINNO W69-07861
MS, PONDS, RUNNING WATERS, FRESH WATER, BRACKISH WA: /S, CARP, CLA W70-09905
TION, *BIODEGRADATION, *DRAINAGE WATER, CALIFORNIA, DRAINAGE PROGR W71-04556
TION, *BIODEGRADATION, *DRAINAGE WATER, CALIFORNIA, DRAINAGE PROGR W71-04557
TION, *BIODEGRADATION, *DRAINAGE WATER, CALIFORNIA, DRAINAGE PROGR W71-04555
TION, *BIODEGRADATION, *DRAINAGE WATER, CALIFORNIA, DRAINAGE PROGR W71-04554
OLLUTION, *TEMPERATURE, *COOLING WATER, CHEMICAL PROPERTIES, ECOLO W69-05023
ENTS(WATER), REMOTE SENSING, SEA WATER, CHLOROPHYLL, OIL, ALGAE, P W70-05377
, CHRYSOPHYTA, BIOASSAYS, SALINE WATER, CHLORELLA, CHLAMYDOMONAS, W70-03974
CATION, DOMESTIC WASTES, SHALLOW WATER, CHLOROPHYTA, CHRYSOPHYTA, W70-00711
A, COL/ *ALGAL CONTROL, *POTABLE WATER, CYANOPHYTA, SULFUR BACTERI W70-08107
ON, RESERVOIRS, ALGAE, BODIES OF WATER, DISSOLVED OXYGEN, NORTH CA W70-00683
*SAMPLING, ALGAE TOXINS, SALINE WATER, EQUILIBRIUM, BIOLOGICAL CO W69-00360
IFORNIA, UNITED STATES, BRACKISH WATER, ESTUARIES, BACTERIA, MOLLU W70-05372
DE, IRON, FUNGI, CLAYS, BRACKISH WATER, FISHERIES, LIMNOLOGY.: /FI W70-05761
ON, DIATOMS, MARINE ALGAE, FRESH WATER, FLOW RATES, PUMPS, ELECTRO W70-09904
N, *CYCLING, *RADIOISOTOPES, SEA WATER, FOOD CHAINS, PHOSPHORUS RA W70-00708
IRON, TRACE ELEMENTS, ALGAE, SEA WATER, FRESH WATER.: / *COPPER, * W71-11687
SEDIMENTS, SAMPLING, DEPTH, SEA WATER, FRES WATER, LITTORAL, DISS W71-12870
WATER STRUCTURE, ATMOSPHERE, SEA WATER, FRESH WATER, SORPTION, SOD W70-04385
EMENT, TIDAL WATERS, MIXING, SEA WATER, FRESH WATER, SHORES, WATER W69-06203
VERTEBRATES, PLANKTON, FISH, SEA WATER, GAMMA RAYS, RADIOCHEMICAL W69-08269
LVOCALES, INDUSTRIAL USES, CLEAN WATER, GENUS, SPECIES, TOUCH SENS W70-05389
OLLUTION, SNOW, RUNOFF, DRAINAGE WATER, GRASSLANDS, SOIL EROSION,: W70-04193
GAE, CUTTING MANAGEMENT, POTABLE WATER, HERBICIDES, CHARA, ALKALIN W70-05263
, WATE/ *ACID MINE WATER, ACIDIC WATER, HYDROGEN ION CONCENTRATION W70-05424
CARBONATES, TEMPERATURE, SHALLOW WATER, HYDROGEN SULFIDE, COPPER S W70-02803
AGE EFFLUENTS, *ESTUARIES, FRESH WATER, LAKES, SILTS, PHOSPHATES, W70-02255
H PLANTS, *MARINE ANIMALS, *OILY WATER, LICHENS, TIDAL EFFECTS, DI W70-01777
SAMPLING, DEPTH, SEA WATER, FRES WATER, LITTORAL, DISSOLVED OXYGEN W71-12870
UTION, CLEANING, DISASTERS, OILY WATER, MARINE ALGAE, SHIPS.: /OLL W70-01470
NTS, PLANT POPULATIONS, WEATHER, WATER, MOVEMENT, CULTURES, BAYS, W70-06229
DATION, *WATER REUSE, *RECLAIMED WATER, NITROGEN COMPOUNDS, NITROG W69-08054
OUNDS, RESERVOIRS, ALGAE, GROUND WATER, NITRATES, PHOSPHATES, LAS W72-02817
ANGANESE, CULTURES, RECIRCULATED WATER, NITRATES, PHOSPHATES, SODI W70-07257
HLOROPHYTA, CYANOPHYTA, BRACKISH WATER, OLIGOTROPHY, TEMPERATURE, W70-05752
L WAST/ *WATER REUSE, *RECLAIMED WATER, ORGANIC COMPOUNDS, CHEMICA W70-10433
ECTS, SALINITY, TEMPERATURE, SEA WATER, OYSTERS, CLAMS, CRABS, NOR W69-08267
Y PRODUCTIVITY, *ALGAE, *THERMAL WATER, PERIPHYTON, BIOMASS, PHYSI W70-03309
DARY PRODUCTIVITY, SULFIDES, SEA WATER, PHAWOPHYTA, CRUSTACEANS, H W70-01068
PES, *MARINE ALGAE, *RESINS, SEA WATER, PHYSICOCHEMICAL PROPERTIES W69-08275
RE, PHYTOPLANKTON, ALGAE, HEATED WATER, POPULATION, WATER POLLUTIO W71-09767
OPES, LAKES, ORGANIC MATTER, SEA WATER, PRIMARY PRODUCTIVITY, LIGH W70-02504
H, STEAM, ELECTRIC POWER, HEATED WATER, PROTEINS, AQUATIC MICROORG W70-09905
N CONTROL, FISH, TOXICITY, FRESH WATER, PULP WASTES, PESTICIDE RES W70-10437
OIL POROSITY, IRRIGATION, GROUND WATER, RUNOFF, FROZEN GROUND, SOI W69-05323
IBUTION, BACTERIA, SAMPLING, SEA WATER, SEASONAL, PSEUDOMONAS, TEM W70-03952
ECHARGE, WATER SUPPLY, MUNICIPAL WATER, SEDIMENT LOAD, CONDUCTIVIT W69-05894
WATERS, MIXING, SEA WATER, FRESH WATER, SHORES, WATER CIRCULATION, W69-06203
BIOASSAY, AQUATIC LIFE, COOLING WATER, SITES.: /E, MODEL STUDIES, W71-11881
ALGAE, BRACKISH WATER, ALKALINE WATER, SODIUM.: /IENTS, BACTERIA, W71-03031
, *FUNGI, *CALCAREOUS SOIL, SOIL WATER, SOIL WATER MOVEMENT, LEACH W71-04067

561

), FLUSHING WATERWAYS, MILWAUKEE(	WIS), CHICAGO(ILL), ARTIFICIAL CI	W69-10178
KE NAGAWICKA(WIS), LAKE PEWAUKEE(	WIS), CHIRONOMUS, PROCLADIUS, COR	W70-06217
NDIPES PLUMOSUS, BEAVER DAM LAKE(	WIS), DELAVAN LAKE(WIS).: /SE, TE	W70-06217
Y, MICRONUTRIENTS, BADFISH RIVER(	WIS), DETROIT LAKES(MINN), NUTRIE	W70-04506
R CREEK(WIS), NINE SPRINGS CREEK(	WIS), DOOR CREEK(WIS), DUTCH MILL	W70-05113
E SPRINGS CREEK(WIS), DOOR CREEK(	WIS), DUTCH MILL DRAINAGE DITCH(W	W70-05113
SURE, PHENANTHROQUINONE, MADISON(	WIS), ELODEA, REDOX COMP: /L EXPO	W70-02982
VICULA PELLICULOSA, LAKE MENDOTA(	WIS), ENGLAND, LONDON(ENGLAND), E	W69-10180
DEPARTMENT OF INTERIOR, MADISON(	WIS), EUROPE, WATER POLLUTION ASS	W70-03975
DEPARTMENT OF INTERIOR, MADISON(	WIS), EUROPE, WATER POLLUTION ASS	W70-03975
EE(WIS), CHICAG/ MANURE, MADISON(	WIS), FLUSHING WATERWAYS, MILWAUK	W69-10178
ULDING'S POND(WIS), LAKE WAUBESA(	WIS), GREEN LAKE(WIS).: /VAL, SPA	W69-05868
ANESVILLE(WIS), SPAULDING'S POND(	WIS), HYDRODICTYON, NAJAS, ANACHA	W70-03310
TILIZATION SURVEY, MADISON LAKES(	WIS), LAKE MENDOTA(WIS), LAKE WAU	W70-02787
VERTICAL SAMPLING, LAKE MENDOTA(	WIS), LAKE MONONA(WIS), LAKE WAUB	W70-02969
, LAKE MENDOTA(WIS), LAKE MONONA(	WIS), LAKE WAUBESA(WIS), LAKE KEG	W70-02969
, LAKE MONONA(WIS), LAKE WAUBESA(	WIS), LAKE KEGONSA(WIS), WEED BED	W70-02969
MADISON LAKES(WIS), LAKE MENDOTA(	WIS), LAKE WAUBESA(WIS), LAKE KEG	W70-02787
LAKE MENDOTA(WIS), LAKE WAUBESA(	WIS), LAKE KEGONSA(WIS).: /(WIS),	W70-02787
SCIOTO RIVER(OHIO), LAKE MENDOTA(	WIS), LAKE MONONA(WIS), LAKE WAUB	W72-01813
, LAKE MONONA(WIS), LAKE WAUBESA(	WIS), LAKE KEGONSA(WIS).: /A(WIS)	W72-01813
, LAKE MENDOTA(WIS), LAKE MONONA(	WIS), LAKE WAUBESA(WIS), LAKE KEG	W72-01813
E WINNISQUAM(N H), MADISON LAKES(	WIS), LAKE WASHINGTON(WASH), NUIS	W70-07261
, ALGAL PHYSIOLOGY, LAKE MENDOTA(	WIS), LAKE MONONA(WIS), LAKE WAUB	W69-05867
TRIENT REMOVAL, SPAULDING'S POND(	WIS), LAKE WAUBESA(WIS), GREEN LA	W69-05868
URAL EUTROPHICATION, LAKE MONONA(	WIS), LAKE TAHOE(CALIF), LAKE WAS	W69-06858
NSA(WIS), LAKE WAU/ LAKE MENDOTA(	WIS), LAKE MONONA(WIS), LAKE KEGO	W69-06276
AKE WAUBESA(WIS), SPAULDING POND(	WIS), LAKE DELAVAN(WIS), PEWAUKEE	W69-05867
, PEWAUKEE LAKE(WIS), NORTH LAKE(	WIS), LAKE NAGAWICKA(WIS).: /WIS)	W69-05867
LAKE MENDOTA(WIS), LAKE MONONA(	WIS), LAKE KEGONSA(WIS), LAKE WAU	W69-06276
, LAKE MONONA(WIS), LAKE KEGONSA(	WIS), LAKE WAUBESA(WIS), MADISON(	W69-06276
, LAKE MENDOTA(WIS), LAKE MONONA(	WIS), LAKE WAUBESA(WIS), SPAULDIN	W69-05867
NOSTOC VERRUCOSUM, LAKE MENDOTA(	WIS), LAKE MONONA(WIS), LAKE WAUB	W69-08283
, LAKE MENDOTA(WIS), LAKE MONONA(	WIS), LAKE WAUBESA(WIS), ANABAENA	W69-08283
KE(ORE),/ *URBAN WASTES, MADISON(	WIS), LAKE ZOAR(CONN), KLAMATH LA	W70-04506
, LAKE MONONA(WIS), LAKE WAUBESA(	WIS), LAKE KEGONSA(WIS).: /H(WIS)	W70-05113
NSA(WIS), BLOOMING, LAKE MENDOTA(	WIS), LAKE WINGRA(WIS), ANABAENA	W70-05112
DRAINAGE DITCH(WIS), LAKE MONONA(	WIS), LAKE WAUBESA(WIS), LAKE KEG	W70-05113
(WIS), DUTCH MILL DRAINAGE DITCH(	WIS), LAKE MONONA(WIS), LAKE WAUB	W70-05113
ONSA/ *MADISON(WIS), LAKE MONONA(	WIS), LAKE WAUBESA(WIS), LAKE KEG	W70-05112
ESA(WIS), LAKE KEGONSA/ *MADISON(	WIS), LAKE MONONA(WIS), LAKE WAUB	W70-05112
, LAKE MONONA(WIS), LAKE WAUBESA(	WIS), LAKE KEGONSA(WIS), BLOOMING	W70-05112
LAKE MENDOTA(WIS), LAKE MONONA(	WIS), LAKE NAGAWICKA(WIS), LAKE P	W70-06217
WICKA(WIS), LAKE P/ LAKE MENDOTA(	WIS), LAKE MONONA(WIS), LAKE NAGA	W70-06217
LAKE MONONA(WIS), LAKE NAGAWICKA(	WIS), LAKE PEWAUKEE(WIS), CHIRONO	W70-06217
, LAKE MENDOTA(WIS), LAKE MONONA(	WIS), LAKE WINGRA(WIS), LAKE KEGO	W69-03366
), LAKE MONONA(WIS), LAKE WINGRA(	WIS), LAKE KEGONSA(WIS), LAKE WAU	W69-03366
, LAKE WINGRA(WIS), LAKE KEGONSA(	WIS), LAKE WAUBESA(WIS), PRECIPIT	W69-03366
S), NUISANCE ODORS, LAKE MENDOTA(	WIS), LAKE MONONA(WIS), LAKE WING	W69-03366
, NAJAS, ANACHARIS, LAKE WAUBESA(	WIS), MACROPHYTES, ALGAL GROWTH.:	W70-03310
LAKE KEGONSA(WIS), LAKE WAUBESA(	WIS), MADISON(WIS), UNIVERSITY OF	W69-06276
LAKE WINNEBAGO(	WIS), MENASHA(WIS).:	W72-01814
LAKE ERKEN(SWEDEN), LAKE MENDOTA(	WIS), NET PLANKTON, CHLOROPHYLL C	W70-03959
A RIVER(WIS), STORKWEATHER CREEK(	WIS), NINE SPRINGS CREEK(WIS), DO	W70-05113
LAKE DELAVAN(WIS), PEWAUKEE LAKE(	WIS), NORTH LAKE(WIS), LAKE NAGAW	W69-05867
A(WIS), LAKE MONONA(WI/ *MADISON(	WIS), NUISANCE ODORS, LAKE MENDOT	W69-03366
RSITY OF WISCONSIN, YAHARA RIVER(	WIS), NUTRIENT REMOVAL.: /, UNIVE	W69-06276
BLOOMS, GELATINOU/ *LAKE MENDOTA(	WIS), PELAGIC AREAS, HORIZONTAL,	W69-06229
PAULDING POND(WIS), LAKE DELAVAN(	WIS), PEWAUKEE LAKE(WIS), NORTH L	W69-05867
KASUKA POND(JAPAN), LAKE MENDOTA(	WIS), PHYTOPLANKTON CROPS, TROPHI	W69-05142
LAKE KEGONSA(WIS), LAKE WAUBESA(	WIS), PRECIPITATION.: /NGRA(WIS),	W69-03366
OPHELES QUADRIMACULATUS, MADISON(	WIS), PSOROPHORA CILIATA, ARTIFIC	W70-06225
, LAKE MONONA(WIS), LAKE WAUBESA(	WIS), SPAULDING POND(WIS), LAKE D	W69-05867
ABAENA, GLOEOTRICHIA, JANESVILLE(	WIS), SPAULDING'S POND(WIS), HYDR	W70-03310
TS, *MADISON LAKES, YAHARA RIVER(	WIS), STORKWEATHER CREEK(WIS), NI	W70-05113
WIS), LAKE WAUBESA(WIS), MADISON(	WIS), UNIVERSITY OF WISCONSIN, YA	W69-06276
LAKE WAUBESA(WIS), LAKE KEGONSA(	WIS), WEED BEDS, PHOSPHORUS, NITR	W70-02969
UVIATILIS, *LAKE BUTTE DES MORTS(	WISC), GROWTH STIMULATION, SURFAC	W72-00845
ANKTON, WATER POLLUTION EFFECTS,	WISCONSIN ZOOPLANKTON, TRIPTON, N	W70-03959
YOCOCCUS, PLANKTON DISTRIBUTION,	WISCONSIN LAKE SURVEY, NITROGEN C	W70-03959
FFECTS, WATER POLLUTION SOURCES,	WISCONSIN.: /S, WATER POLLUTION E	W69-06276
LLA, CHLAMYDOMONAS, SCENEDESMUS,	WISCONSIN.: /H SUBSTANCES, CHLORE	W72-00845
XICITY, DISSOLVED OXYGEN, ALGAE,	WISCONSIN.: /G ORGANIC MATTER, TO	W72-01797
ENTS, *LAKES, *ALGAE, *NITROGEN,	WISCONSIN.: /GEN-FIXATION, *NUTRI	W70-03429
ITIES, CLEANING, *ALGAL CONTROL,	WISCONSIN.: /STS, TREATMENT FACIL	W72-08859
S, *CYANOPHYTA, *EUTROPHICATION,	WISCONSIN, WATER QUALITY CONTROL,	W70-03310
IRS, CULTURES, SEWAGE EFFLUENTS,	WISCONSIN, ALGICIDES, HYDROLYSIS,	W70-02982
ON, LAKES, STREAMS, ALGAE, FISH,	WISCONSIN, SEWAGE, TROUT, NITROGE	W69-10178
ACIDS, CHLAMYDOMONAS, CHLORELLA,	WISCONSIN, LAKE MICHIGAN, PHYTOPL	W69-10180
USCORUN, NUTRIENT CONCENTRATION,	WISCONSIN, SULFUR, PHOSPHORUS.: /	W70-00719
NUTRIENT REMOVAL, LAKE RECOVERY,	WISCONSIN, LAKE MONONA, LAKE MEND	W69-09349
IS), MADISON(WIS), UNIVERSITY OF	WISCONSIN, YAHARA RIVER(WIS), NUT	W69-06276
OLLUTION EFFECTS, WATER QUALITY,	WISCONSIN, LAKES, EUTROPHICATION.	W69-05868
E ALGAE, PHOSPHORUS, WASHINGTON,	WISCONSIN, CALIFORNIA.: / NUISANC	W69-06858
CYANOPHYTA, ODOR, SCUM, FISHING,	WISCONSIN, BIBLIOGRAPHIES, LAKES,	W69-08283
OF SCIENCES(USA), UNIVERSITY OF	WISCONSIN, ASIA, NATIONAL RESEARC	W70-03975
NNING, RIVERS, LAKES, LIMNOLOGY,	WISCONSIN, URBANIZATION, CONNECTI	W70-04506
SWAMPS, FOREST SOILS, MINNESOTA,	WISCONSIN, MICHIGAN, PHOSPHORUS,	W70-04193
OF SCIENCES(USA), UNIVERSITY OF	WISCONSIN, ASIA, NATIONAL RESEARC	W70-03975
*HYDROLOGY, CHEMICAL PROPERTIES,	WISCONSIN, SEWAGE EFFLUENTS, NITR	W70-05113

BOTTOM SEDIMENTS, MUD, BENTHOS, WISCONSIN, SAMPLING, CONTOURS, DE W70-06217
ANCE ALGAE, SNAILS, LEGISLATION, WISCONSIN, CHEMCONTROL, PONDWEEDS W70-05568
MICROBIOLOGY, BALANCE OF NATURE, WISCONSIN, BIOASSAY, BIOCHEMICAL W69-03369
LGICIDES, TRACE ELEMENTS, CORES, WISCONSIN, BOTTOM SEDIMENT, BENTH W69-03366
YTA, CYANOPHYTA, NUISANCE ALGAE, WISCONSIN, NUTRIENTS, INDICATOR S W69-01977
NCE AL/ *EUTRIPHICATION, *LAKES, WISCONSIN, COPPER SULPHATE, NUISA W68-00468
, SPRA/ *EUTROPHICATION, *LAKES, WISCONSIN, ALGAE, *NUISANCE ALGAE W68-00488
POTAMOGETON CRISPUS, MICROSPORA WITTROCKII, SPIROGYRA, XANTHOPHYC W70-00711
BORSKII, RAPHIDIOPSIS, ANABAENA, WOLLEA, OSCILLATORIA, SPIRULINA, W70-05404
ASHINGTON, HOWARD A. HANSON RES/ WOOD, ALGAL COUNT, GREEN RIVER, W W72-11906
OCHOIDEUM, VINEYARD SOUND(MASS), WOODS HOLE(MASS). RADIOCARBON UPT W70-05272
TURNOVER TIME, TRIDACNA CROCEA, WOODS HOLE OCEANOGRAPHIC INSTITUT W69-08526
'BREAKING' OF / *MADISON(WIS), *' WORKING' OF LAKES, *WATER-BLOOM, W69-08283
TRY, ODOR, TAST/ *AMERICAN WATER WORKS ASSOCIATION, CHEMICAL INDUS W71-05943
*ADMINISTRATIVE AGENCIES, *WATER WORKS, *WASTE DISPOSAL, DOMESTIC W70-03344
*OHIO, WATER TEMPERATURE, SLUDGE WORMS, BLOODWORMS, AQUATIC BACTER W70-00475
PLANKTON, BENTHIC FAUNA, SLUDGE WORMS, BLOODWORMS, WATER POLLUTIO W68-01244
AE, DIATOMS, PROTOZOA, PLANKTON, WORMS, NEMATODES, ROTIFERS, LARVA W72-01819
ALGAE, INVERTEBRATES, TURBIDITY, WORMS, SUNFISHES.: /S, PLANKTON, W69-07861
*LAKE KLOPEINER(AUSTRIA), LAKE WORTHER(AUSTRIA), VERTICAL, CHALK W70-05405
ER LINE DISCRIMINATOR, RHODAMINE WT DYE, AIRBORNE TESTS, OIL SPILL W70-05377
MIDS, LAKE CONSTANCE, WHITEFISH, WURTTEMBERG(GERMANY), PLON(GERMAN W70-05761
LES, ANTIBIOTICS, AGARBACTERIUM, XANT: /ARY COMPOUNDS, MACROMOLECU W70-04280
ICROSPORA WITTROCKII, SPIROGYRA, XANTHOPHYCEAE, MYXOPHYCEAE, SALVE W70-00711
UCTS, ANABAENA CYLINDRICA, ZINC, XANTHOPHYCEAE, BACILLARIOPHYCEAE, W69-10180
TERRANEUS, NAVICULA PELLICULOSA, XANTHOPHYCEAE, BACILLARIOPHYCEAE, W70-02965
TER SHOCK WAVE, ANABAENA, PHYGON XL (DICHLONE), NEW HAMPSHIRE, LAKE W69-08674
N(WIS), UNIVERSITY OF WISCONSIN, YAHARA RIVER(WIS), NUTRIENT REMOV W69-06276
ISON) EFFLUENTS, *MADISON LAKES, YAHARA RIVER(WIS), STORKWEATHER C W70-05113
ECIES, DIVERSITY I/ DESICCATION, YAQUINA BAY(ORE), CAROTENOIDS, SP W70-03325
SCENEDESMUS, EUGLENA, BACTERIA, YEASTS, ACTINOMYCETES, PONDING, W W70-08097
TREATMENT, *AQUICULTURE,.ALGAE, YEASTS, BACTERIA, FISH, PROTEINS, W69-05891
DS, AMMONIA, NITRITES, NITRATES, YEASTS, BACTERIA, AQUATIC BACTERI W69-06970
TIVITY, WATER POLLUTION EFFECTS, YEASTS, CRUSTACEA, ALGAE.: /RODUC W69-10152
INTERTIDAL AREAS, CARBOHYDRATES, YEASTS, ENZYMES, CULTURES, HYDROG W70-03952
QUATIC BACTERIA, *AQUATIC FUNGI, YEASTS, HYDROGEN ION CONCENTRATIO W69-00096
DROGEN ION CONCENTRATION, FUNGI, YEASTS, PYRITE, ALGAE, CYANOPHYTA W68-00891
EPIPHYTIC BACTERIA, SUBLITTORAL, YELLOW MATERIAL, DU: /A HARVEYI, W70-01073
URVES, CALORIMETRY, SOLARIMETRY, YELLOWSTONE NATIONAL PARK, MT LAS W70-03309
DETERGENTS, FISH MANAGERS, FISH YIELDS, UNDERWATER SHOCK WAVE, AN W69-08674
ALGAE, BIOMASS, CHLOROPHYLL, NEW YORK.: /*PONDS, *EUTROPHICATION, W69-06540
TIVITY, *ZOOPLANKT/ *LAKES, *NEW YORK, *MEROMIXIS, *PRIMARY PRODUC W70-05760
NS, WATER POLLUTION EFFECTS, NEW YORK, CHLOROPHYTA, NUTRIENTS, ECO W70-03973
LICATES, CITRATES, BIOASSAY, NEW YORK, LAKES, INHIBITION, ALGAE, D W70-01579
S, *STREAMS, BIOINDICATORS, *NEW YORK, NUCLEAR WASTES, WATER POLLU W69-10080
OINDICATORS, NUISANCE ALGAE, NEW YORK, SEWAGE EFFLUENTS, SEWAGE TR W70-03974
OIDS, FLOW RATES, *FILTERS, *NEW YORK, TREATMENT FACILITIES.: /OLL W72-04145
, FILTRATION, NEW HAMPSHIRE, NEW YORK, VIRGINIA, ORGANIC COMPOUNDS W72-03641
CRITERIA, *WATER TREATMENT, *NEW YORK, WATER PURIFICATION.: /SIGN W72-08686
IATA, NAVICULA, FLORIDA CURRENT, YTTRIUM-90, KATODINIUM ROTUNDATA. W70-00707
AKE, NANNOCHLORIS, THORACOMONAS, YTTRIUM, 'BIOLOGICAL DILUTION', P W70-00708
COBBLES, EQUISETUM, THORIUM-232, YUKON RIVER, NEBESNA RIVER, MOOSE W69-08272
CYANOCOBALAMIN, EUGLENA GRACILIS Z, MICROBIOLOGICAL ASSAY, SPECIES W69-06273
GAE.: *NEW ZEALAND, SEWAGE TREATMENT POND AL W70-02304
IA), LAKE SCHLIER(BAVARIA), LAKE ZELL(AUSTRIA), LAKE HALL WILER(SW W70-05668
SEWAGE EFFLUENTS, ELECTROLYTES, ZETA POTENTIAL, BACTERIA, PROTOZO W70-01027
UBRA, CANDIDA DIDDENSII, CANDIDA ZEYLANOIDES, CANDIDA KRUSEI, CAND W70-04368
RON, TETRASTRUM, TISZA(HUNGARY), ZIMONA(HUNGARY), ALGAL TAXONOMY, W70-03969
PLANTS, NYLON-6, *VISCOSE RAYON, ZINC COMPOUNDS, CARBON DISULFIDE. W70-06925
ROPERTIES, ADSORPTION, COLLOIDS, ZINC RADIOISOTOPES, CALCIUM, POTA W70-00708
LLUSKS, STRONTIUM RADIOISOTOPES, ZINC RADIOISOTOPES, IODINE RADIOI W71-09850
TA, LIGHT INTENSITY, SUCCESSION, ZINC RADIOISOTOPES, CESIUM, COBAL W70-04371
US RADIOISOTOPES, TROPHIC LEVEL, ZINC RADIOISOTOPES, IRON, SEDIMEN W69-07853
CESIUM, STRONTIUM RADIOISOTOPES, ZINC RADIOISOTOPES, IRON, MANGANE W69-07845
S, POTASSIUM, IRON, BICARBONATE, ZINC, COBALT, CALCIUM, MAGNESIUM, W70-03978
PHOTOPHOSPHORYLATION, VANADIUM, ZINC, EUGLENA GRACILIS, ANABAENA W70-02964
, INHIBITION, SILICON, VANADIUM, ZINC, GALLIUM.: /INS, METABOLITES W69-04801
ONEMA OGLIMUN, UDOTEA FLABELLUM, ZINC, NICKEL, PUERTO RICO, LA PAR W69-08525
R PRODUCTS, ANABAENA CYLINDRICA, ZINC, XANTHOPHYCEAE, BACILLARIOPH W69-10180
TESTING, NUCLEAR SHIP SAVANNAH, ZIRCONIUM-95, RUBIDIUM-106, NIOBI W70-00707
14, ACTINIUM-228, SWIFT RIFFLES, ZIRCONIUM-95, NIOBIUM-95, RUTHENI W69-09742
ANS, *SARGASSUM NATANS, SEAWEED, ZIRCONIUM-95, NIOBIUM-95, PRE-BOMB W69-08524
ERTEBRATES, PLUTONIUM-239, RATS, ZIRCONIUM-95-NIOBIUM, EUROPIUM-15 W69-08269
URBAN WASTES, MADISON(WIS), LAKE ZOAR(CONN), KLAMATH LAKE(ORE), LA W70-04506
ZERLAND), KILCHBERG(SWITZERLAND), ZOLLIKON(SWITZERLAND), LAKE TEGER W70-05668
L LIMNOLOGY, CLADOCERAN, MONTANE ZONE, ALPINE ZONE, BOULDER(COLO), W69-10154
CLADOCERAN, MONTANE ZONE, ALPINE ZONE, BOULDER(COLO), TEA LAKE(COL W69-10154
K RESERVOIR(TENN), NAJAS, PHOTIC ZONE, DOUGLAS RESERVOIR(TENN), NO W71-07698
*EXTRACELLULAR PRODUCTS, PHOTIC ZONE, GLYCOLIC ACID, CHEMOTROPHY W70-02504
CYCLOTELLA MENEGHINIANA, PHOTIC ZONE, HELSINGOR(DENMARK), SALINO- W70-04809
CY, LIGHT MEASUREMENTS, EUPHOTIC ZONE, PLANT CAROTENOIDS, CARBON F W70-01933
TEMIA, CARTERIA, NORTH TEMPERATE ZONE, PLUTONIUM, HALF-LIFE(RADION W70-00708
TRANSACTIONS, BLACK SEA, CONTACT ZONES, ARTEMIA, BALANUS, DESORPTI W69-07440
IS, DEPTH, THERMOCLINE, CLIMATIC ZONES, ORGANIC LOADING, DISSOLVED W72-10573
MENT, HARVESTING OF ALGAE, WATER ZONING.: /MENTS, WATERSHED MANAGE W68-00859
MICAL PRECIPITATION, HARVESTING, ZONING, DIVERSION, DREDGING, IMPO W69-10178
, WEED CONTROL, *INVESTIGATIONS, ZOO-PLANKTON, SPRAYING, BIOCONTRO W68-00468
N, MEXICO, PREDICTIVE EQUATIONS, ZOOGEOGRAPHY, THERMOBIOLOGY.: /IO W70-03957
IGOTROPHY, DEPTH, PHYTOPLANKTON, ZOOPLANKTON, PHYSICOCHEMICAL PROP W70-03964
ICAL REGIONS, CYCLING NUTRIENTS, ZOOPLANKTON, ALGAE, AQUATIC PLANT W70-03975

565

```
LLICULOSA, GONYAULAX POLYEDRA, 3,    5-DIAMINOBENZOIC ACID DIHYDROCHLO    W69-08278
ALGAE, 2-4-D, AMINOTRIAZOLE, 2-4-    5-T, SCENEDESMUS, CHLAMYDOMONAS,     W70-03519
S, *MUSCULAR EXERTION, *2-METHYL-    5-VINYLPYRIDINE, *GLUE GREEN ALGA    W70-01788
S, *MAN-MADE FIBER PLANTS, NYLON-    6, *VISCOSE RAYON, ZINC COMPOUNDS    W70-06925
ION, DOW C-31, DOW A-21, NALCO N-    670.: /ARGE, ELECTROSTATIC REPULS    W70-05267
                          SIMAZINE   80 W, DACTHAL.:                      W70-03519
VICULA, FLORIDA CURRENT, YTTRIUM-    90, KATODINIUM ROTUNDATA.: /A, NA    W70-00707
, PLUTONIUM-239, RATS, ZIRCONIUM-    95-NIOBIUM, EUROPIUM-155, SILVER-    W69-08269
ASSUM NATANS, SEAWEED, ZIRCONIUM-    95-NIOBIUM-95, PRE-BOMB MATERIAL,    W69-08524
CONIUM-95, RUBIDIUM-106, NIOBIUM-    95, GYMNODINIUM SIMPLEX, FLAGELLA    W70-00707
UM-228, SWIFT RIFFLES, ZIRCONIUM-    95, NIOBIUM-95, RUTHENIUM-106, RH    W69-09742
S, SEAWEED, ZIRCONIUM-95-NIOBIUM-    95, PRE-BOMB MATERIAL, RUTHENIUM-    W69-08524
NUCLEAR SHIP SAVANNAH, ZIRCONIUM-    95, RUBIDIUM-106, NIOBIUM-95, GYM    W70-00707
T RIFFLES, ZIRCONIUM-95, NIOBIUM-    95, RUTHENIUM-106, RHODIUM-106, C    W69-09742
```

AUTHOR INDEX

In order not to complicate the computer program needlessly, only initials and the word "and" have been excluded from the alphabetic listing. Both first and last names of authors appear in the index. The numbers at the right are the accession numbers appearing at the end of the abstracts.

```
        E. B. COWELL, AND J. M.  BAKER.:                                /  W70-01231
        J. J. CONNORS, AND R. B.  BAKER.:                                /  W72-04495
                      K. J. M.  BAKER, AND E. B. COWELL.:                /  W70-09976
                      ROBERT C.  BALL.:                                     W68-00487
                      ROBERT C.  BALL, AND FRANK F. HOOPER.:             /  W69-09334
                          W. D.  BARLOW.:                                /  W70-01519
                          W. D.  BARLOW.:                                /  W71-06737
  WILLIAM CHAMBERLAIN, AND JUDITH  BARRETT.:            JOSEPH SHAPIRO,/  W70-08637
        C. DAVID MCINTIRE, AND  BARRY L. WULFF.:                        /  W70-03325
  ETH M. MACKENTHUN, AND ALFRED F.  BARTSCH.:                    KENN/  W72-01797
  THOMAS E. MALONEY, AND ALFRED F.  BARTSCH.:                         /  W70-07393
                          A. F.  BARTSCH.:                                 W68-01244
                          A. F.  BARTSCH.:                                 W68-00478
                          A. F.  BARTSCH.:                                 W68-00461
                          A. F.  BARTSCH, AND M. O. ALLUM.:             /  W72-01818
                        JOHN W.  BAUMEISTER.:                           /  W70-03068
  OVERFIELD, H. R. CRAWFORD, J. K.  BAXTER, L. J. HARRINGTON, AND I. /  W70-05819
  WILLIAM J. OSWALD, AND LOUIS A.  BECK.: /GOLDMAN, JAMES F. ARTHUR,/  W71-04557
                          A. M.  BEETON.:                                  W68-00253
       CORNELIUS I. WEBER, AND  BEN H. MCFARLAND.:                     /  W71-04206
                          H. D.  BENNETT.:                              /  W70-02770
  ORO, H. J. SCHNEIDER, AND E. O.  BENNETT.:            E. GELPI, J./  W69-08284
                     RICHARD J.  BENOIT.:                               /  W70-04385
                          R. J.  BENOIT, AND J. J. CURRY.:                 W68-00479
  I. SAX, PAUL C. LEMON, ALLEN H.  BENTON, AND JACK J. GABAY.:    N./  W69-10080
     ALBERTO M. WACHS, AND ANDRE  BEREND.:                             /  W70-04790
                    MAESTRINI.:      B. R.  BERLAND, M. G. BIANCHI, AND S. Y./  W70-04280
  C. SERRUYA, S. SERRUYA, AND T.  BERMAN.:                             /  W70-09889
  DEPOE.:           ERNEST C. NEAL,  BERNARD C. PATTEN, AND CHARLES E./  W70-04371
                                 BERNARD P. DOMOGALLA.:                /  W70-04494
  . MORAVCOVA, L. MASINOVA, AND V.  BERNATOVA.:                     A/  W72-05143
                                 BERNHARD DOMOGALLA.:                     W68-00488
                                 BERNHARD DOMOGALLA.:                     W68-00468
                             M.  BERNHARD, AND A. ZATTERA.:            /  W69-08275
                          A. E.  BERRY.:                                /  W72-03824
                      EDWARD C.  BERRY.:                                /  W71-02009
  RNST M. DAVIS.:               BERT A. FLOYD, E. GUS FRUH, AND E/  W69-06004
  GOODNIGHT.:                   BERT K. WHITTEN, AND CLARENCE J. /  W69-07861
                         NACHUM  BETZER, AND YEHUDA KOTT.:             /  W70-02364
                    SURINDER K.  BHAGAT, AND DONALD E. PROCTOR.:       /  W71-00936
          B. R. BERLAND, M. G.  BIANCHI, AND S. Y. MAESTRINI.:         /  W70-04280
                        HARTMUT  BICK.:                                 /  W71-03031
            VERA A. (DUGDALE)  BILLAUD.:                              /  W69-03515
                        JOHN W.  BISHOP.:                               /  W71-10098
  WOODS.:                      G.  BLUNDE, S. B. CHALLEN, AND D. L. /  W69-07826
                          R. H.  BOGAN.:                                   W68-00248
  NNART ENEBO, JITKA VENDLOVA, AND  BOHUMIR PROKES.: /D HEDENSKOG, LE/  W70-04184
                      BRENDA J.  BOLEYN.:                               /  W70-04469
                             P.  BOLTON.:                               /  W73-00758
                          A. D.  BONEY.:                                /  W70-01779
          H. S. AZAD, AND J. A.  BORCHARDT.:                           /  W70-03923
                          J. A.  BORCHARDT.:                            /  W69-07084
          E. DAVIS, AND J. A.  BORCHARDT.:                           /  W72-07334
      GERALD W. FOESS, AND JACK A.  BORCHARDT.:                        /  W71-03035
  PER, W. J. WEBER, JR., AND J. A.  BORCHARDT.:          G. R. PY/  W72-00917
                          J. A.  BORCHARDT, AND C. R. O'MELIA.:         /  W72-03703
                        JACK A.  BORCHARDT, AND HARDAM S. AZAD.:        /  W70-06604
                                 BOSTWICK H. KETCHUM.:                  /  W70-02804
    PIERRE GUEGUENIAT, PIERRE  BOVARD, AND JACQUES ANCELLIN.:        /  W71-09863
                 RUTH PATRICK,  BOWMAN CRUM, AND JOHN COLES.:          /  W70-07257
      T. D. ARMSTRONG, AND B. I.  BOYKO.:                              /  W71-08970
                      MORRIS L.  BREHMER.:                              /  W69-03695
                                 BRENDA J. BOLEYN.:                     /  W70-04469
                     PATRICK L.  BREZONIK.:                             /  W71-03048
                     PATRICK L.  BREZONIK, AND HUGH D. PUTNAM.:         /  W72-01990
  NNON, AND H. D. PUTNAM.:   P. L.  BREZONIK, W. H. MORGAN, E. E. SHA/  W69-09723
                                 BRIAN MOSS.:                           /  W70-02646
                                 BRIAN MOSS.:                           /  W72-03183
                      ROBERT M.  BRICE, AND CHARLES F. POWERS.:         /  W70-03512
                      ROBERT C.  BRIGHT.:                               /  W69-00659
                             G.  BRINGMANN, AND R. KUHN.:                  W68-00472
                             G.  BRINGMANN, AND R. KUHN.:                  W68-00475
                          F. J.  BRINLEY.:                                 W68-00483
                       FLOYD J.  BRINLEY.:                              /  W72-01788
      A. L. S. MUNRO, AND T. D.  BROCK.:                               /  W70-04811
    THOMAS D. BROCK, AND M. LOUISE  BROCK.:                            /  W69-09676
                     THOMAS D.  BROCK.:                                /  W69-10161
    M. LOUISE BROCK, AND THOMAS D.  BROCK.:                            /  W69-10163
     T. D. BROCK, AND M. LOUISE  BROCK.:                              /  W69-10160
                     THOMAS D.  BROCK.:                                /  W69-09755
    THOMAS D. BROCK, AND M. LOUISE  BROCK.:                            /  W69-10151
                          T. D.  BROCK, AND M. LOUISE BROCK.:           /  W69-10160
                     THOMAS D.  BROCK, AND M. LOUISE BROCK.:           /  W69-10151
                     THOMAS D.  BROCK, AND M. LOUISE BROCK.:           /  W69-09676
                      M. LOUISE  BROCK, AND THOMAS D. BROCK.:          /  W69-10163
                          A. J.  BROOK.:                                   W68-00255
                          D. D.  BROUSE.:                               /  W69-10155
                          J. G.  BROWN.:                                /  W72-08688
                        RANDALL  BROWN, AND JAMES ARTHUR.:             /  W70-03334
  JO/  JAMES F. ARTHUR, RANDALL L.  BROWN, BRUCE A. BUTTERFIELD, AND /  W71-04554
```

MILTON G. TUNZI.:	RANDALL L. BROWN, RICHARD C. BAIN, JR., AND /	W71-04555
ERSON.:	D. W. DAVIS, T. S. BROWN, S. SYROTYNSKI, AND E. HEND/	W72-08686
. JONES.:	BRUCE A. BUTTERFIELD, AND JAMES R/	W71-04556
MES F. ARTHUR, RANDALL L. BROWN,	BRUCE A. BUTTERFIELD, AND JOEL C./	W71-04554
PATRICK L. HUDSON, AND	BRUCE C. COWELL.: /	W70-05375
	BRUCE C. PARKER.: /	W69-06273
D. A. CULVER, AND G. J.	BRUNSKILL.: /	W70-05760
T. G.	BRYDGES.: /	W71-05879
W. E.	BULLARD, JR.: /	W71-00940
ROBERT L.	BUNCH, AND M. B. ETTINGER.: /	W71-04072
ROBERT L.	BUNCH, AND M. B. ETTINGER.: /	W70-04373
N. P.	BURMAN.: /	W72-06017
CAROLYN W.	BURNS.: /	W70-03957
ROBERT H.	BURRIS.: /	W70-03429
PHILIP A.	BUSCEMI.: /	W70-03501
IVONNE J. LE COSQUINO DE	BUSSY.: /	W70-10161
BRUCE A.	BUTTERFIELD, AND JAMES R. JONES.:/	W71-04556
THUR, RANDALL L. BROWN, BRUCE A.	BUTTERFIELD, AND JOEL C. GOLDMAN./	W71-04554
J. C.	BUZZELL, AND C. N. SAWYER.:	W68-00012
R.	CABRIDENC, AND H. LEPAILLEUR.: /	W69-07389
JOHN	CAIRNS, JR.: /	W71-03183
JOHN	CAIRNS, JR.: /	W70-07847
JOHN	CAIRNS, JR.: /	W70-07313
ERTOLDI MARLETTA.:	C. CALZOLARI, L. GABRIELLI, AND G. P/	W70-05178
R. HAYES, J. A. MCCARTER, M. L.	CAMERON, AND D. A. LIVINGSTONE.: /	W68-00858
D. M. ORCUTT, H. A. SCHWERTNER,	CARA L. MARTINEZ, AND HAZEL E. WI/	W70-07957
DALE A.	CARLSON, AND H. KIRK WILLARD.: /	W70-06792
A. F.	CARLUCCI, AND S. B. SILBERNAGEL.:/	W70-05652
	CAROLYN W. BURNS.: /	W70-03957
EDWARD J.	CARPENTER.: /	W70-09904
ROANN E. OGAWA, AND JOHN F.	CARR.: /	W70-05091
CHARLES R. GOLDMAN, AND RALF C.	CARTER.: /	W70-01933
HUBERT	CASPERS.: /	W70-00268
RICHARD W.	CASTENHOLZ.: /	W70-05270
RICHARD W.	CASTENHOLZ.: /	W69-03516
G. BLUNDE, S. B.	CHALLEN, AND D. L. WOODS.: /	W69-07826
WILLIAM	CHAMBERLAIN, AND JOSEPH SHAPIRO.:/	W70-05269
JOSEPH SHAPIRO, WILLIAM	CHAMBERLAIN, AND JUDITH BARRETT.: /	W70-08637
E. C. S.	CHAN, AND ELIZABETH A. MCMANUS.: /	W70-03952
E. C. S.	CHAN, AND ELIZABETH A. MCMANUS.: /	W70-04365
A.:	JU- CHANG HUANG, AND EARNEST F. GLOYN/	W70-08416
A.:	JU- CHANG HUANG, AND EARNEST F. GLOYN/	W71-12183
	CHARLES A. DAMBACH.: /	W70-01943
C. NEAL, BERNARD C. PATTEN, AND	CHARLES E. DEPOE.: ERNEST/	W70-04371
GOLUEKE, WILLIAM J. OSWALD, AND	CHARLES E. RIXFORD.: /CLARENCE G./	W71-11375
GARRISON, HARRY K. PHINNEY, AND	CHARLES E. WARREN.: /E, ROBERT L./	W70-00265
ROBERT M. BRICE, AND	CHARLES F. POWERS.: /	W70-03512
	CHARLES F. WURSTER, JR.: /	W70-05272
	CHARLES M. WEISS.: /	W69-09135
ARTER.:	CHARLES R. GOLDMAN, AND RALF C. C/	W70-01933
ZI, AND RICHARD ARMSTRONG.:	CHARLES R. GOLDMAN, MILTON G. TUN/	W70-03983
M. AUBERT, R.	CHARRA, AND G. MALARA.: /	W70-01466
A. K.	CHERRY.: /	W72-04155
A. E.	CHRISTIE.: /	W70-08097
A. E.	CHRISTIE.: /	W70-02248
A. E.	CHRISTIE.: /	W70-02973
A. E.	CHRISTIE.: /	W70-02795
A. E.	CHRISTIE.: /	W70-02968
A. E.	CHRISTIE.: /	W70-02198
N, M. F. P. MICHALSKI, AND A. E.	CHRISTIE.: M. G. JOHNSO/	W70-02792
N, M. F. P. MICHALSKI, AND A. E.	CHRISTIE.: M. G. JOHNSO/	W70-05424
S. P.	CHU.:	W68-00856
HENRYK	CHUDYBA.: /	W70-02784
	CLAIR N. SAWYER.: /	W70-02787
	CLAIR N. SAWYER.: /	W70-05565
HLER.:	CLARENCE E. TAFT, AND W. JACK KIS/	W70-04468
SWALD, AND CH/ GORDON L. DUGAN,	CLARENCE G. GOLUEKE, WILLIAM J. O/	W71-11375
BERT K. WHITTEN, AND	CLARENCE J. GOODNIGHT.: /	W69-07861
	CLARENCE M. TARZWELL.: /	W71-10786
D.: G. FRED LEE, NICHOLAS L.	CLESCERI, AND GEORGE P. FITZGERAL/	W69-03358
H. H. YEH, PIERRE S. WARREN, AND	CLIFFORD W. RANDALL.: / H. KING, /	W69-09884
	CLYDE EYSTER.: /	W70-02964
F. R. HAYES, AND C. C.	COFFIN.:	W68-00857
ROBERT A.	COLER, AND HAIM B. GUNNER.: /	W71-03775
H PATRICK, BOWMAN CRUM, AND JOHN	COLES.: RUT/	W70-07257
J. L.	CONFER.: /	W73-01454
J. J.	CONNORS, AND R. B. BAKER.: /	W72-04495
THOMAS D. WAITE, AND	CONSTANTINE GREGORY.: /	W71-05500
S. F.	COOK, JR.: /	W70-05428
WILLIAM B.	COOKE.: /	W72-01819
ETH M. MACKENTHUN, AND HAROLD L.	COOLEY.: KENN/	W70-06217
WILLIAMS.: O. HOLM-HANSEN, J.	COOMBS, B. E. VOLCANI, AND P. M. /	W70-05651
E. F.	CORCORAN, AND J. F. KIMBALL, JR.:/	W70-00707
TED C. FRANTZ, AND ALMO J.	CORDONE.: /	W70-00711
LEE, F. H. SCHRAUFN/ RICHARD B.	COREY, ARTHUR D. HASLER, G. FRED /	W69-10178
FARLAND.:	CORNELIUS I. WEBER, AND BEN H. MC/	W71-04206
IVONNE J. LE	COSQUINO DE BUSSY.: /	W70-10161
G. WATSON, C. E. CUSHING, C. C.	COUTANT, AND W. L. TEMPLETON.: /./	W69-07853
K. J. M. BAKER, AND E. B.	COWELL.: /	W70-09976
PATRICK L. HUDSON, AND BRUCE C.	COWELL.: /	W70-05375

E. B.	COWELL, AND J. M. BAKER.:	/ W70-01231
ROSALIE M.	COX, AND P. FAY.:	/ W70-04249
TTOE, L. B. POLKOWSKI, AND K. T.	CRABTREE.: /EL, E. MCCOY, O. J. A/	W69-05323
W. L. MINCKLEY, J. E.	CRADDOCK, AND L. A. KRUMHOLZ.:	/ W69-09742
RINGTON/ J. L. OVERFIELD, H. R.	CRAWFORD, J. K. BAXTER, L. J. HAR/	W70-05819
RUTH PATRICK, BOWMAN	CRUM, AND JOHN COLES.:	/ W70-07257
D. A.	CULVER, AND G. J. BRUNSKILL.:	/ W70-05760
R. J. BENOIT, AND J. J.	CURRY.:	W68-00479
GARY W. ISAAC, AND	CURTIS P. LEISER.:	/ W69-03683
C. E.	CUSHING, AND N. S. PORTER.:	/ W69-07862
TEMPLETON/ D. G. WATSON, C. E.	CUSHING, C. C. COUTANT, AND W. L./	W69-07853
T.: G. E. FOGG,	CZESLAWA NALEWAJKO, AND W. D. WAT/	W70-02504
ARD.:	DALE A. CARLSON, AND H. KIRK WILL/	W70-06792
CHARLES A.	DAMBACH.:	/ W70-01943
MOTO, TOMOTOSHI OKAICHI, LE DUNG	DANG, AND TAMAO NOGUCHI.: / HASHI	W70-08372
MOTO, TOMOTOSHI OKAICHI, LE DUNG	DANG, AND TAMAO NOGUCHI.: / HASHI/	W71-07731
	DANIEL F. JACKSON.:	/ W70-03974
	DANIEL F. JACKSON.:	/ W69-05697
	DANIEL F. JACKSON.:	/ W69-02959
LIN.:	DANIEL F. JACKSON, AND SHUNN-DAR /	W72-05453
	DANIEL J. NELSON.:	/ W70-02786
DANIEL F. JACKSON, AND SHUNN-	DAR LIN.:	/ W72-05453
P. H. JONES, AND TOM	DAVEY.:	/ W70-05084
TOERTZ, WILLIAM R. HEMPHILL, AND	DAVID A. MARKLE.: GEORGE E. S/	W70-05377
	DAVID C. FLAHERTY.:	/ W70-00264
	DAVID C. FLAHERTY.:	/ W69-05844
R. T. OGLESBY, AND	DAVID JARMISON.:	W68-00010
AS.: LLOYD L. MEDSKER,	DAVID JENKINS, AND JEROME F. THOM/	W69-00387
B. DWAIN VANCE, AND	DAVID L. SMITH.:	/ W70-03519
C.	DAVID MCINTIRE.:	/ W70-02780
C.	DAVID MCINTIRE.:	/ W71-00668
F.: C.	DAVID MCINTIRE, AND BARRY L. WULF/	W70-03325
N, HARRY K. PHINNEY, AND CH/ C.	DAVID MCINTIRE, ROBERT L. GARRISO/	W70-00265
ICKA.:	DAVID R. ZENZ, AND JOSEPH R. PIVN/	W70-09186
EARNEST F. GLOYNA, AND	DAVIS L. FORD.:	/ W72-03299
FLOYD, E. GUS FRUH, AND ERNST M.	DAVIS.: BERT A.	W69-06004
ERNST M.	DAVIS, AND E. F. GLOYNA.:	/ W69-09454
ERNST M.	DAVIS, AND EARNEST F. GLOYNA.:	/ W70-03928
ERNST M.	DAVIS, AND EARNEST F. GLOYNA.:	/ W70-09438
E.	DAVIS, AND J. A. BORCHARDT.:	/ W72-07334
E. ANGINO/ JOHN E. SIMEK, J. A.	DAVIS, C. E. DAY III, AND ERNEST /	W69-08524
, AND E. HENDERSON.: D. W.	DAVIS, T. S. BROWN, S. SYROTYNSKI/	W72-08686
OHN E. SIMEK, J. A. DAVIS, C. E.	DAY III, AND ERNEST E. ANGINO.: //	W69-08524
C. E. STEINMETZ, AND W. J.	DAY.:	/ W70-10433
IVONNE J. LE COSQUINO	DE BUSSY.:	/ W70-10161
ERNESTO ESPINO	DE LA O, AND E. F. GLOYNA.:	/ W71-11377
ERNESTO ESPINO	DE LA O, AND EARNEST F. GLOYNA.:	/ W70-07034
ROBERT Y. TING, AND V. ROMAN	DE VEGA.:	/ W69-08525
ERNARD C. PATTEN, AND CHARLES E.	DEPOE.: ERNEST C. NEAL, B/	W70-04371
G. P. FITZGERALD, AND M. E.	DERVARTANIAN.:	/ W69-10171
MAN M. VARMA, AND FRANCIS	DIGIAND.:	/ W70-06600
LEON S.	DIRECTO, AND MERLIN E. LINDAHL.: /	W69-05894
BERNHARD	DOMOGALLA.:	W68-00488
BERNHARD	DOMOGALLA.:	W68-00468
BERNARD P.	DOMOGALLA.:	/ W70-04494
D R. SPEEDY, NEIL B. FISHER, AND	DONALD B. MCDONALD.: RONAL/	W70-08107
	DONALD B. PORCELLA.:	/ W70-02779
	DONALD E. LEONE.:	/ W70-06053
	DONALD E. PROCTOR.:	/ W70-07491
SURINDER K. BHAGAT, AND	DONALD E. PROCTOR.:	/ W71-00936
D. HASLER.:	DONALD E. WOHLSCHLAG, AND ARTHUR /	W70-06229
	DONALD F. LIVERMORE.:	/ W69-06276
. EDMONDSON, G. C. ANDERSON, AND	DONALD R. PETERSON.: W. T/	W69-10169
AALTO.: NORBERT A. JAWORSKI,	DONALD W. LEAR, JR., AND JOHAN A./	W71-05407
JOHN M.	DONOHUE.:	/ W70-02294
MENACHEM RAHAT, AND INKA	DOR.:	/ W69-00360
ARR NICHOLS, THERESA HENKEL, AND	DOROTHY MCNALL.: M. ST/	W69-03366
TROY C.	DORRIS.:	/ W70-04001
R. L. MORRIS, J. D.	DOUGHERTY, AND G. W. RONALD.:	/ W72-13605
B. D. VANCE, AND W.	DRUMMOND.:	/ W72-04506
B. DWAIN VANCE, AND WAYMON	DRUMMOND.:	/ W70-03520
B. DWAIN VANCE, AND WAYMON	DRUMMOND.:	/ W69-08565
GORDON L.	DUGAN, AND WILLIAM J. OSWALD.:	/ W72-06854
AM J. OSWALD, AND CH/ GORDON L.	DUGAN, CLARENCE G. GOLUEKE, WILLI/	W71-11375
TER.: P. R.	DUGAN, J. I. FREA, AND R. M. PFIS/	W71-04758
VERA A. (	DUGDALE) BILLAUD.:	/ W69-03515
FISCHLER.: T.	DUKE, J. WILLIS, T. PRICE, AND K./	W69-08267
JACOB O.	DUMELLE.:	/ W71-06821
HASHIMOTO, TOMOTOSHI OKAICHI, LE	DUNG DANG, AND TAMAO NOGUCHI.: //	W71-07731
HASHIMOTO, TOMOTOSHI OKAICHI, LE	DUNG DANG, AND TAMAO NOGUCHI.: //	W70-08372
C. S.	DUTTON, AND C. P. FISHER.:	/ W71-09546
B.	DWAIN VANCE, AND DAVID L. SMITH.:/	W70-03519
: B.	DWAIN VANCE, AND WAYMON DRUMMOND./	W70-03520
: B.	DWAIN VANCE, AND WAYMON DRUMMOND./	W69-08565
	EARNEST F. GLOYNA.:	/ W70-04786
ERNST M. DAVIS, AND	EARNEST F. GLOYNA.:	/ W70-03928
JU-CHANG HUANG, AND	EARNEST F. GLOYNA.:	/ W71-12183
JU-CHANG HUANG, AND	EARNEST F. GLOYNA.:	/ W70-08416
ERNESTO ESPINO DE LA O, AND	EARNEST F. GLOYNA.:	/ W70-07034
JORGE AGUIRRE, AND	EARNEST F. GLOYNA.:	/ W70-07031

572

573

G. E.	FOGG, AND D. F. WESTLAKE.:	/ W69-10180
D. WATT.: G. E.	FOGG, CZESLAWA NALEWAJKO, AND W.	/ W70-02504
GEORGE P. FITZGERALD, AND	FOLKE SKOOG.:	/ W70-03310
GERALD C. GERLOFF, AND	FOLKE SKOOG.:	/ W69-05868
GERALD C. GERLOFF, AND	FOLKE SKOOG.:	/ W69-05867
M-HANSEN, GERALD C. GERLOFF, AND	FOLKE SKOOG.: OSMUND HOL/	W69-06277
TZGERALD, GERALD C. GERLOFF, AND	FOLKE SKOOG.: GEORGE P. FI/	W70-02982
RLOFF, GEORGE P. FITZGERALD, AND	FOLKE SKOOG.: GERALD C. GE/	W70-00719
RLOFF, GEORGE P. FITZGERALD, AND	FOLKE SKOOG.: GERALD C. GE/	W70-03507
EARNEST F. GLOYNA, AND DAVIS L.	FORD.:	/ W72-03299
EDWARD G.	FOREE, AND PERRY L. MCCARTY.:	/ W71-13413
A. C.	FOX.:	W68-00172
M. REBHUN, M. A.	FOX, AND J. B. SLESS.:	/ W70-02373
RE A. OLSON.: JACKSON L.	FOX, THERON O. ODLAUG, AND THEODO/	W70-05958
MAN M. VARMA, AND	FRANCIS DIGIAND.:	/ W70-06600
WARREN.: PAUL H. KING,	FRANCIS R. MCNEICE, AND PIERRE S./	W70-05470
ROBERT C. BALL, AND	FRANK F. HOOPER.:	/ W69-09334
TED C.	FRANTZ, AND ALMO J. CORDONE.:	/ W70-00711
	FRANZ RUTTNER.:	/ W70-03959
P. R. DUGAN, J. I.	FREA, AND R. M. PFISTER.:	/ W71-04758
G.	FRED LEE.:	/ W70-04382
G.	FRED LEE, AND E. GUS FRUH.:	/ W69-06535
D B. COREY, ARTHUR D. HASLER, G.	FRED LEE, F. H. SCHRAUFNAGEL, AND/	W69-10178
ND GEORGE P. FITZGERALD.: G.	FRED LEE, NICHOLAS L. CLESCERI, A/	W69-03358
R. W.	FREI, AND C. A. STOCKTON.:	/ W71-11687
C. R.	FRINK.:	W68-00511
TERRENCE P.	FROST.:	/ W69-08282
TERRENCE P.	FROST.:	/ W69-07818
TERRENCE P.	FROST.:	/ W69-08674
G. FRED LEE, AND E. GUS	FRUH.:	/ W69-06535
E. GUS	FRUH.:	/ W70-02251
E. GUS	FRUH.:	/ W70-04727
E. GUS	FRUH.:	/ W70-04765
BERT A. FLOYD, E. GUS	FRUH, AND ERNST M. DAVIS.:	/ W69-06004
RINNOSUKE	FUKAI.:	/ W71-05531
WALLACE H.	FULLER, AND JAMES E. HARDCASTLE.:/	W69-03491
TA.: WALLACE H.	FULLER, AND MICHAEL F. L'ANNUNZIA/	W71-04067
KAZUHIRO	FUTAGAWA, AND ERNEST F. GLOYNA.: /	W71-11381
ON, ALLEN H. BENTON, AND JACK J.	GABAY.: N. I. SAX, PAUL C. LEM/	W69-10080
TA.: C. CALZOLARI, L.	GABRIELLI, AND G. PERTOLDI MARLET/	W70-05178
A. R.	GAHLER.:	/ W72-06051
ND J. B. SLESS.: YEHUDA KOTT,	GALILA HERSHKOVITZ, A. SHEMTOB, A/	W70-02370
M. AUBERT, AND J. P.	GAMBAROTTA.:	/ W70-01467
H/ C. DAVID MCINTIRE, ROBERT L.	GARRISON, HARRY K. PHINNEY, AND C/	W70-00265
HAMMER.:	GARY L. HERGENRADER, AND MARK J. /	W72-04761
ER.:	GARY W. ISAAC, AND CURTIS P. LEIS/	W69-03683
D RO/ GEORGE W. REID, ROBERT A.	GEARHEART, JAMES M. ROBERTSON, AN/	W70-04074
ND E. O. BENNETT.: E.	GELPI, J. ORO, H. J. SCHNEIDER, A/	W69-08284
PHILIP J. SAWYER, JOHN H.	GENTILE, AND JOHN J. SASNER, JR.:	/ W69-05306
PHILL, AND DAVID A. MARKLE.:	GEORGE E. STOERTZ, WILLIAM R. HEM/	W70-05377
	GEORGE L. HARLOW.:	/ W69-01445
	GEORGE P. FITZGERALD.:	/ W69-03188
D LEE, NICHOLAS L. CLESCERI, AND	GEORGE P. FITZGERALD.: G. FRE/	W69-03358
	GEORGE P. FITZGERALD.:	/ W69-03364
	GEORGE P. FITZGERALD.:	/ W69-01977
	GEORGE P. FITZGERALD.:	/ W69-03186
	GEORGE P. FITZGERALD.:	/ W69-03185
	GEORGE P. FITZGERALD.:	/ W69-05704
	GEORGE P. FITZGERALD.:	/ W69-03374
	GEORGE P. FITZGERALD, AND THOMAS /	W69-03373
C. NELSON.:	GEORGE P. FITZGERALD, AND GERARD /	W69-03371
A. ROHLICH.: MARY M. ALLEN,	GEORGE P. FITZGERALD, AND FOLKE S/	W70-03507
KOOG.: GERALD C. GERLOFF,	GEORGE P. FITZGERALD, AND FOLKE S/	W70-00719
KOOG.: GERALD C. GERLOFF,	GEORGE P. FITZGERALD, AND FOLKE S/	W70-03310
KOOG.:	GEORGE P. FITZGERALD, GERALD C. G/	W70-02982
ERLOFF, AND FOLKE SKOOG.:	GEORGE P. GITZGERALD.:	/ W69-03362
	GEORGE TCHOBANOGLOUS.:	/ W70-01981
ROLF ELIASSEN, AND	GEORGE TCHOBANOGLOUS.:	/ W70-07469
ROLF ELIASSEN, AND	GEORGE W. REID, ROBERT A. GEARHEA/	W70-04074
RT, JAMES M. ROBERTSON, AND RO/	GEORGE W. SAUNDERS.:	/ W70-02510
G.: GEORGE P. FITZGERALD,	GERALD C. GERLOFF, AND FOLKE SKOO/	W70-02982
G.:	GERALD C. GERLOFF, AND FOLKE SKOO/	W69-05867
G.:	GERALD C. GERLOFF, AND FOLKE SKOO/	W69-05868
G.: OSMUND HOLM-HANSEN,	GERALD C. GERLOFF, AND FOLKE SKOO/	W69-06277
GERALD, AND FOLKE SKOOG.:	GERALD C. GERLOFF, GEORGE P. FITZ/	W70-00719
GERALD, AND FOLKE SKOOG.:	GERALD C. GERLOFF, GEORGE P. FITZ/	W70-03507
HARDT.:	GERALD W. FOESS, AND JACK A. BORC/	W71-03035
ALLEN, GEORGE P. FITZGERALD, AND	GERARD A. ROHLICH.: MARY M. /	W69-03371
SARLES.:	GERARD A. ROHLICH, AND WILLIAM B./	W72-01812
LEA.:	GERARD A. ROHLICH, AND WILLIAM L./	W70-02969
GEORGE P. FITZGERALD, GERALD C.	GERLOFF, AND FOLKE SKOOG.:	/ W70-02982
OSMUND HOLM-HANSEN, GERALD C.	GERLOFF, AND FOLKE SKOOG.:	/ W69-06277
GERALD C.	GERLOFF, AND FOLKE SKOOG.:	/ W69-05868
GERALD C.	GERLOFF, AND FOLKE SKOOG.:	/ W69-05867
G. C.	GERLOFF, AND K. A. FISHBECK.:	/ W70-02245
D FOLKE SKOOG.: GERALD C.	GERLOFF, GEORGE P. FITZGERALD, AN/	W70-00719
D FOLKE SKOOG.: GERALD C.	GERLOFF, GEORGE P. FITZGERALD, AN/	W70-03507
HYMAN H.	GERSTEIN.:	/ W70-00263
B. R. ALLANSON, AND J. M. T. M.	GIESKES.:	/ W70-06975
GEORGE P.	GITZGERALD.:	/ W69-03362

JOHN H. NEIL, AND	GLADWIN HILL.: / W70-04430
ERNST M. DAVIS, AND E. F.	GLENN E. OWEN.: / W69-04798
O ESPINO DE LA O, AND EARNEST F.	GLOYNA.: / W69-09454
EARNEST F.	GLOYNA.: ERNEST/ W70-07034
JORGE AGUIRRE, AND EARNEST F.	GLOYNA.: / W70-04786
ERNST M. DAVIS, AND EARNEST F.	GLOYNA.: / W70-07031
JU-CHANG HUANG, AND EARNEST F.	GLOYNA.: / W70-03928
KAZUHIRO FUTAGAWA, AND ERNEST F.	GLOYNA.: / W71-12183
ERNST M. DAVIS, AND EARNEST F.	GLOYNA.: / W71-11381
JU-CHANG HUANG, AND EARNEST F.	GLOYNA.: / W70-09438
RNESTO ESPINO DE LA O, AND E. F.	GLOYNA.: / W70-08416
EARNEST F.	GLOYNA.: E/ W71-11377
EARNEST F.	GLOYNA, AND DAVIS L. FORD.: / W72-03299
RUCE A. BUTTERFIELD, AND JOEL C.	GLOYNA, AND ERNESTO ESPINO.: / W70-01971
M. I.	GOLDMAN.: /R, RANDALL L. BROWN, B/ W71-04554
CHARLES R.	GOLDMAN, AND R. SHOOP.: / W72-00462
J. OSWALD, AND LOUIS / JOEL C.	GOLDMAN, AND RALF C. CARTER.: / W70-01933
HARD ARMSTRONG.: CHARLES R.	GOLDMAN, JAMES F. ARTHUR, WILLIAM/ W71-04557
ITZGERALD, JOSEPH SHAPIRO, C. R.	GOLDMAN, MILTON G. TUNZI, AND RIC/ W70-03983
. JAKOBS-MOGELIN.: H. L.	GOLDMAN, O. R. ARMSTRONG, AND JAC/ W69-06864
H/ GORDON L. DUGAN, CLARENCE G.	GOLTERMAN, C. C. BAKELS, AND J. J/ W70-03955
EL W. HEDGPETH, AND JEFFERSON J.	GOLUEKE, WILLIAM J. OSWALD, AND C/ W71-11375
H. S. SWINGLE, B. C.	GONOR.: JO/ W71-02494
BERT K. WHITTEN, AND CLARENCE J.	GOOCH, AND H. R. RABANAL.: / W70-05417
SWALD.:	GOODNIGHT.: / W69-07861
EKE, WILLIAM J. OSWALD, AND CH/	GORDON L. DUGAN, AND WILLIAM J. O/ W72-06854
R, AND W. K. KIM.: P. R.	GORDON L. DUGAN, CLARENCE G. GOLU/ W71-11375
HAROLD B.	GORHAM, J. MCLACHLAN, U. T. HAMME/ W70-00273
L. K. PORTER, AND A. R.	GOTAAS.: / W70-04465
G. M. P. MORDVINTSEVA, V. V.	GRABLE.: / W72-08110
ROSS E. MCKINNEY, AND ANDREW	GRABOVSKAYA, AND V. K. MARINICH.:/ W70-06925
MARGARET	GRAM.: / W72-01817
MARGARET	GREENWALD.: / W69-05706
PHILLIP E.	GREENWALD.: / W69-05705
THOMAS D. WAITE, AND CONSTANTINE	GREESON.: / W70-04721
G. K.	GREGORY.: / W71-05500
G. K.	GRUENDLING, AND A. C. MATHIESON.:/ W70-02764
JITKA VENDLOVA, AND BOHUMIR P/	GRUENDLING, AND A. C. MATHIESON.:/ W70-02772
CQUES ANCELLIN.: PIERRE	GUDMUND HEDENSKOG, LENNART ENEBO,/ W70-04184
S. P. MEYERS, D. G. AHEARN, W.	GUEGUENIAT, PIERRE BOVARD, AND JA/ W71-09863
ROBERT A. COLER, AND HAIM B.	GUNKEL, AND F. J. ROTH, JR.: / W70-04368
E.	GUNNER.: / W71-03775
E.	GUS FRUH.: / W70-02251
E.	GUS FRUH.: / W70-04765
E.	GUS FRUH.: / W70-04727
G. FRED LEE, AND E.	GUS FRUH.: / W69-06535
BERT A. FLOYD, E.	GUS FRUH, AND ERNST M. DAVIS.: / W69-06004
ROBERT A. COLER, AND	HAIM B. GUNNER.: / W71-03775
J. C. LACAZE, C.	HALLOPEAU, AND M. VOIGT.: / W72-02203
D. H.	HAMILTON, JR.: / W70-01579
GARY L. HERGENRADER, AND MARK J.	HAMMER.: / W72-04761
. R. GORHAM, J. MCLACHLAN, U. T.	HAMMER, AND W. K. KIM.: P/ W70-00273
F. J. HUMENIK, AND G. P.	HANNA, JR.: / W70-05655
OSMUND HOLM-	HANSEN.: / W69-08278
LKE SKOOG.: OSMUND HOLM-	HANSEN, GERALD C. GERLOFF, AND FO/ W69-06277
AND P. M. WILLIAMS.: O. HOLM-	HANSEN, J. COOMBS, B. E. VOLCANI,/ W70-05651
OSMUND HOLM-	HANSON.: / W69-03518
LOWELL D.	HANSON, AND WILLIAM E. FENSTER.: / W70-04193
JACK A. BORCHARDT, AND	HARDAM S. AZAD.: / W70-06604
WALLACE H. FULLER, AND JAMES E.	HARDCASTLE.: / W69-03491
GEORGE L.	HARLOW.: / W69-01445
KENNETH M. MACKENTHUN, AND	HAROLD B. GOTAAS.: / W70-04465
	HAROLD L. COOLEY.: / W70-06217
	HAROLD LEROY ALLEN.: / W71-00832
R. CRAWFORD, J. K. BAXTER, L. J.	HARRINGTON, AND I. W. SANTRY, JR./ W70-05819
RALPH H. WEAVER, AND	HARRY D. NASH.: / W69-00096
R.:	HARRY D. NASH, AND RALPH H. WEAVE/ W68-00891
	HARRY J. EBY.: / W71-08221
SHERRY L. VOLK, AND	HARRY K. PHINNEY.: / W69-04802
	HARRY K. PHINNEY.: / W70-03978
ID MCINTIRE, ROBERT L. GARRISON,	HARRY K. PHINNEY, AND CHARLES E. / W70-00265
ROBERT D.	HARTER.: / W69-03075
	HARTMUT BICK.: / W71-03031
R. S.	HARVEY.: / W69-07845
R. M.	HARWOOD.: / W72-09693
DUNG DANG, AND TAMAO N/ YOSHIRO	HASHIMOTO, TOMOTOSHI OKAICHI, LE / W71-07731
DUNG DANG, AND TAMAO N/ YOSHIRO	HASHIMOTO, TOMOTOSHI OKAICHI, LE / W70-08372
ARTHUR D.	HASLER.: / W69-06858
ARTHUR D.	HASLER.: / W68-00859
A. D.	HASLER.: W68-00680
ALD E. WOHLSCHLAG, AND ARTHUR D.	HASLER.: DON/ W70-06229
ARTHUR D.	HASLER, AND ELIZABETH JONES.: / W70-02249
FN/ RICHARD B. COREY, ARTHUR D.	HASLER, G. FRED LEE, F. H. SCHRAU/ W69-10178
A.	HAUGHEY.: / W70-02304
H. A.	HAWKES.: / W71-02479
F. R.	HAYES.: / W70-05399
F. R.	HAYES, AND C. C. COFFIN.: W68-00857
RON, AND D. A. LIVINGSTO/ F. R.	HAYES, J. A. MCCARTER, M. L. CAME W68-00858
CHWERTNER, CARA L. MARTINEZ, AND	HAZEL E. WICKLINE.: /UTT, H. A. S/ W70-07957
ENDLOVA, AND BOHUMIR P/ GUDMUND	HEDENSKOG, LENNART ENEBO, JITKA V/ W70-04184

575

MARY

BROWN, BRUCE A. BUTTERFIELD, AND
WILLIAM J. OSWALD, AND LOUIS /
. GONOR.:
WORSKI, DONALD W. LEAR, JR., AND
MAN.:   L. R. POMEROY, R. E.

RUTH PATRICK, BOWMAN CRUM, AND

RICHARD T. WRIGHT, AND
IGHT.:
IGHT.:
DAY III, AND ERNEST E. ANGINO/
ROANN E. OGAWA, AND

ER, JR.:   PHILIP J. SAWYER,

J. SAWYER, JOHN H. GENTILE, AND
AND RUDOLPH PRINS.:

HUGH F. MULLIGAN, AND

N.:

PAUL A. ERICKSON, AND

LOUCKS.:   MARY A. TIFFANY,
G. E. LIKENS, AND P. L.
A. E. CHRISTIE.:   M. G.
A. E. CHRISTIE.:   M. G.
PETUR M.
P.
UCE A. BUTTERFIELD, AND JAMES R.
ARTHUR D. HASLER, AND ELIZABETH
P. H.
L.:   R. G. STROSS, J. C.
YNA.:
ERIK G.
E. STEEMANN NIELSEN, AND ERIK G.
, JR., RONALD G. SCHUESSLER, AND
DAVID R. ZENZ, AND

WILLIAM CHAMBERLAIN, AND
RO.:
R. ARMSTRON/ G. P. FITZGERALD,
IN, AND JUDITH BARRETT.:
OYNA.:
OYNA.:
HAPIRO, WILLIAM CHAMBERLAIN, AND
OTIEV, L. V. KOLOSOVA, AND M. A.
OTIEV, L. V. KOLOSOVA, AND M. A.
V. G. KHOBOT'EV, V. I.
J. C. WARD, AND S.
GLOYNA.:
ONI, H. NUGENT MYRICK, AND E. C.

M MARCUS INGRAM.:
L. COOLEY.:
F. BARTSCH.:
INGRAM, AND RALPH PORGES.:
BOSTWICK H.
K. M. MACKENTHUN, L. E.
. RUKHADZE.:   V. G.
A. KAD/ N. S. STROGANOV, V. G.
A. KAD/ N. S. STROGANOV, V. G.
LACHLAN, U. T. HAMMER, AND W. K.
E. F. CORCORAN, AND J. F.
RRE S. WARREN.:   PAUL H.
, AND CLIFFORD W. RAND/ PAUL H.
HUGH F. MULLIGAN, AND JOHN M.
DALE A. CARLSON, AND H.
CLARENCE E. TAFT, AND W. JACK
TROGANOV, V. G. KHOBOTIEV, L. V.
TROGANOV, V. G. KHOBOTIEV, L. V.
EDWARD J.
NACHUM BETZER, AND YEHUDA
YEHUDA
TOB, AND J. B. SLESS.:   YEHUDA
CHIE J. MCDONNELL, AND R. RUPERT
CKLEY, J. E. CRADDOCK, AND L. A.

JO WITT.:   / W71-00110
JOEL C. GOLDMAN.: /R, RANDALL L. / W71-04554
JOEL C. GOLDMAN, JAMES F. ARTHUR,/ W71-04557
JOEL W. HEDGPETH, AND JEFFERSON J/ W71-02494
JOHAN A. AALTO.:   NORBERT A. JA/ W71-05407
JOHANNES, E. P. ODUM, AND B. ROFF/ W69-08274
JOHN A. LAMBIE.:   / W70-05645
JOHN C. NEESS.:   / W69-04804
JOHN CAIRNS, JR.:   / W70-07313
JOHN CAIRNS, JR.:   / W70-07847
JOHN CAIRNS, JR.:   / W71-03183
JOHN COLES.:   / W70-07257
JOHN E. HOBBIE.:   / W70-05109
JOHN E. HOBBIE.:   / W70-04194
JOHN E. HOBBIE, AND RICHARD T. WR/ W70-04284
JOHN E. HOBBIE, AND RICHARD T. WR/ W70-04287
JOHN E. SIMEK, J. A. DAVIS, C. E./ W69-08524
JOHN F. CARR.:   / W70-05091
JOHN G. STOCKNER.:   / W70-03309
JOHN H. GENTILE, AND JOHN J. SASN/ W69-05306
JOHN H. NEIL, AND GLENN E. OWEN.:/ W69-04798
JOHN J. PETERKA.:   / W70-09093
JOHN J. SASNER, JR.:   PHILIP/ W69-05306
JOHN K. REED, LAMAR E. PRIESTER, / W69-02782
JOHN M. DONOHUE.:   / W70-02294
JOHN M. KINGSBURY.:   / W69-06540
JOHN MCN SIEBURTH.:   / W70-00713
JOHN MCN SIEBURTH.:   / W70-01073
JOHN MCN SIEBURTH, AND ARNE JENSE/ W70-01074
JOHN MCNEILL SIEBURTH.:   / W70-01068
JOHN T. REYNOLDS.:   / W70-05264
JOHN T. REYNOLDS.:   / W69-00976
JOHN W. BAUMEISTER.:   / W70-03068
JOHN W. BISHOP.:   / W71-10098
JOHN W. VENNES.:   / W70-03312
JOHN W. WINCHESTER, AND RONALD H./ W69-08562
JOHNSON.:   / W69-08272
JOHNSON, M. F. P. MICHALSKI, AND / W70-05424
JOHNSON, M. F. P. MICHALSKI, AND / W70-02792
JONASSON.:   / W71-00665
JONATHAN YOUNG.:   / W70-00475
JONES.:   BR/ W71-04556
JONES.:   / W70-02249
JONES, AND TOM DAVEY.:   / W70-05084
JONES, F. M. UNGER, AND J. M. VAI/ W72-02417
JORGE AGUIRRE, AND EARNEST F. GLO/ W70-07031
JORGENSEN.:   / W70-05381
JORGENSEN.:   / W70-04809
JOSEPH L. PAVONI.: /. ECHELBERGER/ W70-05267
JOSEPH R. PIVNICKA.:   / W70-09186
JOSEPH SHAPIRO.:   / W70-01031
JOSEPH SHAPIRO.:   / W70-05269
JOSEPH SHAPIRO, AND ROBERTO RIBEI/ W70-02255
JOSEPH SHAPIRO, C. R. GOLDMAN, O./ W69-06864
JOSEPH SHAPIRO, WILLIAM CHAMBERLA/ W70-08637
JU-CHANG HUANG, AND EARNEST F. GL/ W70-08416
JU-CHANG HUANG, AND EARNEST F. GL/ W71-12183
JUDITH BARRETT.:   JOSEPH S/ W70-08637
KADINA.: /. STROGANOV, V. G. KHOB/ W71-03056
KADINA.: /. STROGANOV, V. G. KHOB/ W71-05719
KAPKOV, AND E. G. RUKHADZE.:   / W71-11561
KARAKI.:   / W72-01773
KAZUHIRO FUTAGAWA, AND ERNEST F. / W71-11381
KELLER, JR.:   R. H. T. MATT/ W71-05267
KENNETH M. MACKENTHUN.:   / W70-05568
KENNETH M. MACKENTHUN.:   / W69-09340
KENNETH M. MACKENTHUN, AND WILLIA/ W69-04801
KENNETH M. MACKENTHUN, AND HAROLD/ W70-06217
KENNETH M. MACKENTHUN, AND ALFRED/ W72-01797
KENNETH M. MACKENTHUN, WILLIAM M./ W70-06225
KETCHUM.:   / W70-02804
KEUP, AND R. K. STEWART.:   W68-00481
KHOBOT'EV, V. I. KAPKOV, AND E. G/ W71-11561
KHOBOTIEV, L. V. KOLOSOVA, AND M./ W71-03056
KHOBOTIEV, L. V. KOLOSOVA, AND M./ W71-05719
KIM.:   P. R. GORHAM, J. MC/ W70-00273
KIMBALL, JR.:   / W70-00707
KING, FRANCIS R. MCNEICE, AND PIE/ W70-05470
KING, H. H. YEH, PIERRE S. WARREN/ W69-09884
KINGSBURY.:   / W69-06540
KIRK WILLARD.:   / W70-06792
KISHLER.:   / W70-04468
KOLOSOVA, AND M. A. KADINA.: /. S/ W71-03056
KOLOSOVA, AND M. A. KADINA.: /. S/ W71-05719
KORMONDY.:   / W71-09005
KOTT.:   / W70-02364
KOTT, AND J. EDLIS.:   / W70-02363
KOTT, GALILA HERSHKOVITZ, A. SHEM/ W70-02370
KOUNTZ.:   AR/ W69-10159
KRUMHOLZ.:   W. L. MIN/ W69-09742

ANDREW J.	MARX.: / W72-01814
ER, AND RONALD H. LOUCKS.:	MARY A. TIFFANY, JOHN W. WINCHEST/ W69-08562
	MARY JO WITT.: / W71-00110
LD, AND GERARD A. ROHLICH.:	MARY M. ALLEN, GEORGE P. FITZGERA/ W69-03371
A. MORAVCOVA, L.	MASINOVA, AND V. BERNATOVA.: / W72-05143
J. HEMENS, AND M. H.	MASON.: / W72-08133
G. K. GRUENDLING, AND A. C.	MATHIESON.: / W70-02772
G. K. GRUENDLING, AND A. C.	MATHIESON.: / W70-02764
C. KELLER, JR.: R. H. T.	MATTONI, H. NUGENT MYRICK, AND E./ W71-05267
D.	MATULOVA.: / W72-12240
. LIVINGSTO/ F. R. HAYES, J. A.	MCCARTER, M. L. CAMERON, AND D. A W68-00858
EDWARD G. FOREE, AND PERRY L.	MCCARTY.: / W71-13413
S. A. WITZEL, ELIZABETH	MCCOY, AND RICHARD LEHNER.: / W71-05742
KI, AND K. T. CR/ S. WITZEL, E.	MCCOY, O. J. ATTOE, L. B. POLKOWS/ W69-05323
RICHARD P. W.	MCCUTCHEON.: / W69-01165
Y, NEIL B. FISHER, AND DONALD B.	MCDONALD.: RONALD R. SPEED/ W70-08107
ARCHIE J.	MCDONNELL, AND R. RUPERT KOUNTZ.:/ W69-10159
CORNELIUS I. WEBER, AND BEN H.	MCFARLAND.: / W71-04206
C. DAVID	MCINTIRE.: / W71-00668
C. DAVID	MCINTIRE.: / W70-02780
C. DAVID	MCINTIRE, AND BARRY L. WULFF.: / W70-03325
RY K. PHINNEY, AND CH/ C. DAVID	MCINTIRE, ROBERT L. GARRISON, HAR/ W70-00265
ROSS E.	MCKINNEY.: / W70-06601
ROSS E.	MCKINNEY, AND ANDREW GRAM.: / W72-01817
. KIM.: P. R. GORHAM, J.	MCLACHLAN, U. T. HAMMER, AND W. K/ W70-00273
E. C. S. CHAN, AND ELIZABETH A.	MCMANUS.: / W70-03952
E. C. S. CHAN, AND ELIZABETH A.	MCMANUS.: / W70-04365
JOHN	MCN SIEBURTH.: / W70-01073
JOHN	MCN SIEBURTH.: / W70-00713
JOHN	MCN SIEBURTH, AND ARNE JENSEN.: / W70-01074
OLS, THERESA HENKEL, AND DOROTHY	MCNALL.: M. STARR NICH/ W69-03366
PAUL H. KING, FRANCIS R.	MCNEICE, AND PIERRE S. WARREN.: / W70-05470
JOHN	MCNEILL SIEBURTH.: / W70-01068
E F. THOMAS.: LLOYD L.	MEDSKER, DAVID JENKINS, AND JEROM/ W69-00387
ROBERT O.	MEGARD.: / W69-00446
ROBERT O.	MEGARD.: / W69-00632
ROBERT O.	MEGARD.: / W69-10167
ROBERT O.	MEGARD.: / W70-04268
ROBERT O.	MEGARD.: / W70-05387
J. A. BORCHARDT, AND C. R. O'	MELIA.: / W72-03703
	MENACHEM RAHAT, AND INKA DOR.: / W69-00360
A.	MENNIE, AND E. TEHLE.: / W72-12987
LEON S. DIRECTO, AND	MERLIN E. LINDAHL.: / W69-05894
C.	MERVIN PALMER.: / W70-05389
C.	MERVIN PALMER.: / W70-07027
C.	MERVIN PALMER.: / W70-10173
C.	MERVIN PALMER.: / W70-04507
GRAM.: C.	MERVIN, PALMER, AND WILLIAM M. IN/ W72-01798
AND F. J. ROTH, JR.: S. P.	MEYERS, D. G. AHEARN, W. GUNKEL, / W70-04368
WALLACE H. FULLER, AND	MICHAEL F. L'ANNUNZIATA.: / W71-04067
M. G. JOHNSON, M. F. P.	MICHALSKI, AND A. E. CHRISTIE.: / W70-02792
M. G. JOHNSON, M. F. P.	MICHALSKI, AND A. E. CHRISTIE.: / W70-05424
J. A.	MIHURSKY.: / W70-09905
YU S POTAENKO, AND T M	MIKHEEVA.: / W70-08654
W. E.	MILLER, AND J. C. TASH.: / W72-06052
KATANI.: D.	MILLER, J. V. TOKAR, AND R. E. NA/ W71-11881
BROWN, RICHARD C. BAIN, JR., AND	MILTON G. TUNZI.: RANDALL L. / W71-04555
TRONG.: CHARLES R. GOLDMAN,	MILTON G. TUNZI, AND RICHARD ARMS/ W70-03983
A. KRUMHOLZ.: W. L.	MINCKLEY, J. E. CRADDOCK, AND L. / W69-09742
, C. C. BAKELS, AND J. J. JAKOBS-	MOGELIN.: H. L. GOLTERMAN/ W70-03955
W. D.	MONIE.: / W69-10157
G. W.	MOORE.: / W72-04145
RNATOVA.: A.	MORAVCOVA, L. MASINOVA, AND V. BE/ W72-05143
AND V. K. MARINICH.: G. M. P.	MORDVINTSEVA, V. V. GRABOVSKAYA, / W70-06925
WERNER STUMM, AND JAMES J.	MORGAN.: / W69-03369
PUTNAM.: P. L. BREZONIK, W. H.	MORGAN, E. E. SHANNON, AND H. D. / W69-09723
	MORRIS L. BREHMER.: / W69-03695
I.	MORRIS, AND J. AHMED.: / W70-05261
. RONALD.: R. L.	MORRIS, J. D. DOUGHERTY, AND G. W/ W72-13605
	MOSHE SHILO.: / W70-05372
BRIAN	MOSS.: / W70-02646
BRIAN	MOSS.: / W72-03183
HUGH F.	MULLIGAN.: / W70-10175
HUGH F.	MULLIGAN, AND JOHN M. KINGSBURY.:/ W69-06540
A. L. S.	MUNRO, AND T. D. BROCK.: / W70-04811
J.	MURRAY SPEIRS.: / W70-05263
OOD, WILLIAM S. SMITH, AND J. K.	MURRAY.: RICHARD W/ W70-05821
JACK	MYERS.: / W69-07442
LDMAN, O. R. ARMSTRONG, AND JACK	MYERS.: /JOSEPH SHAPIRO, C. R. GO/ W69-06864
R. H. T. MATTONI, H. NUGENT	MYRICK, AND E. C. KELLER, JR.: / W71-05267
	NACHUM BETZER, AND YEHUDA KOTT.: / W70-02364
. MILLER, J. V. TOKAR, AND R. E.	NAKATANI.: D/ W71-11881
C.	NALEWAJKO, AND L. MARIN.: / W69-10158
G. E. FOGG, CZESLAWA	NALEWAJKO, AND W. D. WATT.: / W70-02504
RALPH H. WEAVER, AND HARRY D.	NASH.: / W69-00096
HARRY D.	NASH, AND RALPH H. WEAVER.: / W68-00891
LES E. DEPOE.: ERNEST C.	NEAL, BERNARD C. PATTEN, AND CHAR/ W70-04371
JOHN C.	NEESS.: / W69-04804
ONALD.: RONALD R. SPEEDY,	NEIL B. FISHER, AND DONALD B. MCD/ W70-08107
JOHN H.	NEIL, AND GLENN E. OWEN.: / W69-04798

579

	NELSON L. NEMEROW.: / W70-09320
	NELSON L. NEMEROW.: / W71-05013
O. F.	NELSON.: / W73-02426
DANIEL J.	NELSON.: / W70-02786
RGE P. FITZGERALD, AND THOMAS C.	NELSON.: GEO/ W69-03373
A.	NELSON-SMITH.: / W70-01780
NELSON L.	NEMEROW.: / W70-09320
NELSON L.	NEMEROW.: / W71-05013
P. FITZGERALD.: G. FRED LEE,	NICHOLAS L. CLESCERI, AND GEORGE / W69-03358
THY MCNALL.: M. STARR	NICHOLS, THERESA HENKEL, AND DORO/ W69-03366
E. STEEMANN	NIELSEN, AND ERIK G. JORGENSEN.: / W70-04809
R.	NIEMEYER, AND G. RICHTER.: / W71-03033
OKAICHI, LE DUNG DANG, AND TAMAO	NOGUCHI.: / HASHIMOTO, TOMOTOSHI / W71-07731
OKAICHI, LE DUNG DANG, AND TAMAO	NOGUCHI.: / HASHIMOTO, TOMOTOSHI / W70-08372
	NORBERT A. JAWORSKI.: / W71-05409
AR, JR., AND JOHAN A. AALTO.:	NORBERT A. JAWORSKI, DONALD W. LE/ W71-05407
JR.: R. H. T. MATTONI, H.	NUGENT MYRICK, AND E. C. KELLER, / W71-05267
JACKSON L. FOX, THERON O.	ODLAUG, AND THEODORE A. OLSON.: / W70-05958
. POMEROY, R. E. JOHANNES, E. P.	ODUM, AND B. ROFFMAN.: L. R/ W69-08274
ROANN E.	OGAWA, AND JOHN F. CARR.: / W70-05091
R. T.	OGLESBY, AND DAVID JARMISON.: W68-00010
WALDEMAR	OHLE.: / W70-05761
N/ YOSHIRO HASHIMOTO, TOMOTOSHI	OKAICHI, LE DUNG DANG, AND TAMAO / W70-08372
N/ YOSHIRO HASHIMOTO, TOMOTOSHI	OKAICHI, LE DUNG DANG, AND TAMAO / W71-07731
	OLAV M. SKULBERG.: / W69-07833
	OLAV SKULBERG.: / W70-05752
	OLE A. SEATHER.: / W71-12077
LAWRENCE W.	OLINGER.: / W69-03730
HERON O. ODLAUG, AND THEODORE A.	OLSON.: JACKSON L. FOX, T/ W70-05958
MARTINEZ/ B. RICHARDSON, D. M.	ORCUTT, H. A. SCHWERTNER, CARA L./ W70-07957
ENNETT.: E. GELPI, J.	ORO, H. J. SCHNEIDER, AND E. O. B/ W69-08284
JAMES R.	ORR.: / W70-04810
D. W.	OSBORN.: / W71-11393
	OSMUND HOLM-HANSEN.: / W69-08278
LOFF, AND FOLKE SKOOG.:	OSMUND HOLM-HANSEN, GERALD C. GER/ W69-06277
	OSMUND HOLM-HANSON.: / W69-03518
GORDON L. DUGAN, AND WILLIAM J.	OSWALD.: / W72-06854
W. J.	OSWALD.: / W70-04787
CLARENCE G. GOLUEKE, WILLIAM J.	OSWALD, AND CHARLES E. RIXFORD.: / W71-11375
MAN, JAMES F. ARTHUR, WILLIAM J.	OSWALD, AND LOUIS A. BECK.: /GOLD/ W71-04557
	OTTO WASMER, JR.: / W69-00994
RALPH E.	OULTON.: / W71-13335
BAXTER, L. J. HARRINGTON/ J. L.	OVERFIELD, H. R. CRAWFORD, J. K. / W70-05819
JOHN H. NEIL, AND GLENN E.	OWEN.: / W69-04798
C. M.	PALMER.: / W70-01233
C. MERVIN	PALMER.: / W70-07027
C. MERVIN	PALMER.: / W70-10173
C. MERVIN	PALMER.: / W70-04507
C. MERVIN	PALMER.: / W70-05389
C. M.	PALMER.: / W72-03562
C. M.	PALMER.: / W72-03746
C. MERVIN,	PALMER, AND WILLIAM M. INGRAM.: / W72-01798
R. H. LUEBBERS, AND D. N.	PARIKH.: / W72-06022
C. D.	PARKER.: / W71-09548
BRUCE C.	PARKER.: / W69-06273
RICHARD A.	PARKER.: / W69-06536
C. D.	PARKER, AND G. P. SKERRY.: / W70-06899
	PATRICK L. BREZONIK.: / W71-03048
PUTNAM.:	PATRICK L. BREZONIK, AND HUGH D. / W72-01990
OWELL.:	PATRICK L. HUDSON, AND BRUCE C. C/ W70-05375
RUTH	PATRICK.: / W70-04510
RUTH	PATRICK.: / W71-02496
RUTH	PATRICK.: / W71-05026
RUTH	PATRICK.: / W72-01801
R.	PATRICK.: / W73-00748
LES.: RUTH	PATRICK, BOWMAN CRUM, AND JOHN CO/ W70-07257
ERNEST C. NEAL, BERNARD C.	PATTEN, AND CHARLES E. DEPOE.: / W70-04371
NOLDS.:	PAUL A. ERICKSON, AND JOHN T. REY/ W70-05264
ND JACK J. GABAY.: N. I. SAX,	PAUL C. LEMON, ALLEN H. BENTON, A/ W69-10080
AND PIERRE S. WARREN.:	PAUL H. KING, FRANCIS R. MCNEICE,/ W70-05470
. WARREN, AND CLIFFORD W. RAND/	PAUL H. KING, H. H. YEH, PIERRE S/ W69-09884
ALD G. SCHUESSLER, AND JOSEPH L.	PAVONI.: /. ECHELBERGER, JR., RON/ W70-05267
ROBERT W.	PENNAK.: / W69-10154
WINIFRED	PENNINGTON.: / W69-04521
EDWARD G. FOREE, AND	PERRY L. MCCARTY.: / W71-13413
CALZOLARI, L. GABRIELLI, AND G.	PERTOLDI MARLETTA.: C./ W70-05178
JOHN J.	PETERKA.: / W70-09093
N, G. C. ANDERSON, AND DONALD R.	PETERSON.: W. T. EDMONDSO/ W69-10169
	PETUR M. JONASSON.: / W71-00665
R. DUGAN, J. I. FREA, AND R. M.	PFISTER.: P./ W71-04758
	PHILIP A. BUSCEMI.: / W70-03501
, AND JOHN J. SASNER, JR.:	PHILIP J. SAWYER, JOHN H. GENTILE/ W69-05306
	PHILLIP E. GREESON.: / W70-04721
HARRY K.	PHINNEY.: / W70-03978
SHERRY L. VOLK, AND HARRY K.	PHINNEY.: / W69-04802
RE, ROBERT L. GARRISON, HARRY K.	PHINNEY, AND CHARLES E. WARREN.: / W70-00265
IN.: PIERRE GUEGUENIAT,	PIERRE BOVARD, AND JACQUES ANCELL/ W71-09863
AND JACQUES ANCELLIN.:	PIERRE GUEGUENIAT, PIERRE BOVARD,/ W71-09863
H. KING, FRANCIS R. MCNEICE, AND	PIERRE S. WARREN.: PAUL / W70-05470
RAND/ PAUL H. KING, H. H. YEH,	PIERRE S. WARREN, AND CLIFFORD W./ W69-09884

I. V. MAKSIMOVA, AND M. N.	PIMENOVA.: / W70-04195
DAVID R. ZENZ, AND JOSEPH R.	PIVNICKA.: / W70-09186
EL, E. MCCOY, O. J. ATTOE, L. B.	POLKOWSKI, AND K. T. CRABTREE.: // W69-05323
J.	POLTORACKA.: / W71-09767
L. R.	POMEROY, AND E. J. KUENZLER.: / W69-08526
UM, AND B. ROFFMAN.: L. R.	POMEROY, R. E. JOHANNES, E. P. OD/ W69-08274
K. K. VOTINTSEV, AND G. I.	POPOVSKAYA.: / W70-04290
DONALD B.	PORCELLA.: / W70-02779
UN, WILLIAM M. INGRAM, AND RALPH	PORGES.: KENNETH M. MACKENTH/ W70-06225
C. E. CUSHING, AND N. S.	PORTER.: / W69-07862
L. K.	PORTER, AND A. R. GRABLE.: / W72-08110
H. W.	POSTON.: / W69-03948
YU S	POTAENKO, AND T M MIKHEEVA.: / W70-08654
J. LEON	POTTER.: / W69-05891
M. TASSIGNY, G. LAPORTE, AND R.	POURRIOT.: / W71-06002
ROBERT M. BRICE, AND CHARLES F.	POWERS.: / W70-03512
G. D. AGRAWAL, AND A. V. S.	PRABHAKARA RAO.: / W70-08628
C. D. AGRAWAL, AND A. V. S.	PRABHAKARA RAO.: / W70-01027
G. W.	PRESCOTT.: / W70-02803
T. DUKE, J. WILLIS, T.	PRICE, AND K. FISCHLER.: / W69-08267
JOHN K. REED, LAMAR E.	PRIESTER, AND RUDOLPH PRINS.: / W69-02782
, LAMAR E. PRIESTER, AND RUDOLPH	PRINS.: JOHN K. REED/ W69-02782
DONALD E.	PROCTOR.: / W70-07491
URINDER K. BHAGAT, AND DONALD E.	PROCTOR.: S/ W71-00936
EBO, JITKA VENDLOVA, AND BOHUMIR	PROKES.: /D HEDENSKOG, LENNART EN/ W70-04184
LUIGI	PROVASOLI.: / W70-04503
HUGH D.	PUTNAM.: / W68-00912
PATRICK L. BREZONIK, AND HUGH D.	PUTNAM.: / W72-01990
MORGAN, E. E. SHANNON, AND H. D.	PUTNAM.: P. L. BREZONIK, W. H. / W69-09723
. BORCHARDT.: G. R.	PYPER, W. J. WEBER, JR., AND J. A/ W72-00917
I. L.	PYRINA.: / W71-10788
SWINGLE, B. C. GOOCH, AND H. R.	RABANAL.: H. S./ W70-05417
MENACHEM	RAHAT, AND INKA DOR.: / W69-00360
CHARLES R. GOLDMAN, AND	RALF C. CARTER.: / W70-01933
	RALPH E. OULTON.: / W71-13335
HARRY D. NASH, AND	RALPH H. WEAVER.: / W68-00891
H.:	RALPH H. WEAVER, AND HARRY D. NAS/ W69-00096
CKENTHUN, WILLIAM M. INGRAM, AND	RALPH PORGES.: KENNETH M. MA/ W70-06225
	RANDALL BROWN, AND JAMES ARTHUR.:/ W70-03334
FIELD, AND JO/ JAMES F. ARTHUR,	RANDALL L. BROWN, BRUCE A. BUTTER/ W71-04554
, JR., AND MILTON G. TUNZI.:	RANDALL L. BROWN, RICHARD C. BAIN/ W71-04555
IERRE S. WARREN, AND CLIFFORD W.	RANDALL.: / H. KING, H. H. YEH, P/ W69-09884
D. S.	RANWELL.: / W70-01777
AGARWAL, AND A. V. S. PRABHAKARA	RAO.: C. D. / W70-01027
AGARWAL, AND A. V. S. PRABHAKARA	RAO.: G. D. / W70-08628
J. E. G.	RAYMONT.: / W70-00161
S.: M.	REBHUN, M. A. FOX, AND J. B. SLES/ W70-02373
LPH PRINS.: JOHN K.	REED, LAMAR E. PRIESTER, AND RUDO/ W69-02782
M. ROBERTSON, AND RO/ GEORGE W.	REID, ROBERT A. GEARHEART, JAMES / W70-04074
ARNOLD W.	REITZE, JR.: / W69-06305
C. E.	RENN.: / W72-00027
JOHN T.	REYNOLDS.: / W69-00976
PAUL A. ERICKSON, AND JOHN T.	REYNOLDS.: / W70-05264
JOSEPH SHAPIRO, AND ROBERTO	RIBEIRO.: / W70-02255
T. R.	RICE.: / W70-00708
	RICHARD A. PARKER.: / W69-06536
R. GOLDMAN, MILTON G. TUNZI, AND	RICHARD ARMSTRONG.: CHARLES / W70-03983
R, G. FRED LEE, F. H. SCHRAUFN/	RICHARD B. COREY, ARTHUR D. HASLE/ W69-10178
G. TUNZI.: RANDALL L. BROWN,	RICHARD C. BAIN, JR., AND MILTON / W71-04555
	RICHARD C. BAIN, JR.: / W69-03611
	RICHARD C. BAIN, JR.: / W69-07520
	RICHARD C. BAIN, JR.: / W70-02777
	RICHARD J. BENOIT.: / W70-04385
A. WITZEL, ELIZABETH MCCOY, AND	RICHARD LEHNER.: S./ W71-05742
	RICHARD P. HERBST.: / W70-00667
	RICHARD P. W. MCCUTCHEON.: / W69-01165
JOHN E. HOBBIE, AND	RICHARD T. WRIGHT.: / W70-04284
JOHN E. HOBBIE, AND	RICHARD T. WRIGHT.: / W70-04287
BBIE.:	RICHARD T. WRIGHT, AND JOHN E. HO/ W70-04194
	RICHARD W. CASTENHOLZ.: / W70-05270
ND J. K. MURRAY.:	RICHARD W. CASTENHOLZ.: / W69-03516
CHWERTNER, CARA L. MARTINEZ/ B.	RICHARD WOOD, WILLIAM S. SMITH, A/ W70-05821
R. NIEMEYER, AND G.	RICHARDSON, D. M. ORCUTT, H. A. S/ W70-07957
	RICHTER.: ∠ W71-03033
ILLIAM J. OSWALD, AND CHARLES E.	RINNOSUKE FUKAI.: / W71-05531
:	RIXFORD.: /CLARENCE G. GOLUEKE, W/ W71-11375
IRWIN, E. L. ROBINSON, AND G. G.	ROANN E. OGAWA, AND JOHN F. CARR./ W70-05091
ER.:	ROBECK.: J. M. SYMONS, W. H. / W70-04484
ERTSON, AND RO/ GEORGE W. REID,	ROBERT A. COLER, AND HAIM B. GUNN/ W71-03775
	ROBERT A. GEARHEART, JAMES M. ROB/ W70-04074
	ROBERT A. LAUDERDALE.: / W70-09768
	ROBERT A. SWEENEY.: / W70-08652
	ROBERT C. BALL.: W68-00487
ER.:	ROBERT C. BALL, AND FRANK F. HOOP/ W69-09334
	ROBERT C. BRIGHT.: / W69-00659
	ROBERT D. HARTER.: / W69-03075
	ROBERT G. WETZEL.: / W71-10079
	ROBERT H. BURRIS.: / W70-03429
	ROBERT J. HENLEY.: / W71-09880
	ROBERT J. SHERWOOD.: W68-00256

581

ER.:
ER.:
NEY, AND CH/ C. DAVID MCINTIRE,
OWERS.:
ARHEART, JAMES M. ROBERTSON, AND

EGA.:
                    JOSEPH SHAPIRO, AND
D, ROBERT A. GEARHEART, JAMES M.
J. M. SYMONS, W. H. IRWIN, E. L.
                            WILHELM
    E. JOHANNES, E. P. ODUM, AND B.
    G. P. FITZGERALD, AND G. A.
RGE P. FITZGERALD, AND GERARD A.
                          GERARD A.
                          GERARD A.
OGLOUS.:
OGLOUS.:
        ROBERT Y. TING, AND V.
NNEY, WAYNE F. ECHELBERGER, JR.,
TIFFANY, JOHN W. WINCHESTER, AND
AND DONALD B. MCDONALD.:
RRIS, J. D. DOUGHERTY, AND G. W.

.:
G. AHEARN, W. GUNKEL, AND F. J.

K. REED, LAMAR E. PRIESTER, AND
OBOT'EV, V. I. KAPKOV, AND E. G.
    ARCHIE J. MCDONNELL, AND R.

HN COLES.:
                             FRANZ
                         WILLIAM R.
TER, L. J. HARRINGTON, AND I. W.
ERARD A. ROHLICH, AND WILLIAM B.
ER, JOHN H. GENTILE, AND JOHN J.
                          GEORGE W.
                           CLAIR N.
    J. C. BUZZELL, AND C. N.
                              C. N.
                           CLAIR N.
ENZ.:   C. N.
ENZ.:   C. N.
J. SASNER, JR.:   PHILIP J.
ON, AND JACK J. GABAY.:   N. I.
    E. GELPI, J. ORO, H. J.
UR D. HASLER, G. FRED LEE, F. H.
E F. ECHELBERGER, JR., RONALD G.
RICHARDSON, D. M. ORCUTT, H. A.
R. O. SYLVESTER, AND R. W.
                            OLE A.
            C. SERRUYA, S.
N.:   C.
                            W. I.
L. BREZONIK, W. H. MORGAN, E. E.
                           JOSEPH
WILLIAM CHAMBERLAIN, AND JOSEPH
                           JOSEPH
STRON/ G. P. FITZGERALD, JOSEPH
JUDITH BARRETT.:   JOSEPH
.:   T. F.
                               G.
UDA KOTT, GALILA HERSHKOVITZ, A.
NEY.:
                         ROBERT J.
                            MOSHE
    Y. HIYAMA, AND M.
    M. I. GOLDMAN, AND R.
    DANIEL F. JACKSON, AND

                          JOHN MCN
                          JOHN MCN
                      JOHN MCNEILL
                          JOHN MCN
WILLIAM T. HELM, AND WILLIAM F.
    A. F. CARLUCCI, AND S. B.
, AND ERNEST E. ANGINO/ JOHN E.
                             R. N.
                             H. N.

ROBERT L. BUNCH, AND M. B. ETTING/ W70-04373
ROBERT L. BUNCH, AND M. B. ETTING/ W71-04072
ROBERT L. GARRISON, HARRY K. PHIN/ W70-00265
ROBERT M. BRICE, AND CHARLES F. P/ W70-03512
ROBERT M. SWEAZY.: / ROBERT A. GE/ W70-04074
ROBERT O. MEGARD.:   / W70-05387
ROBERT O. MEGARD.:   / W70-04268
ROBERT O. MEGARD.:   / W69-00632
ROBERT O. MEGARD.:   / W69-00446
ROBERT O. MEGARD.:   / W69-10167
ROBERT T. JASKE.:   / W70-08834
ROBERT W. PENNAK.:   / W69-10154
ROBERT Y. TING, AND V. ROMAN DE V/ W69-08525
ROBERTO RIBEIRO.:   / W70-02255
ROBERTSON, AND ROBERT M. SWEAZY.:/ W70-04074
ROBINSON, AND G. G. ROBECK.:   / W70-04484
RODHE.:   / W70-05409
ROFFMAN.:   L. R. POMEROY, R./ W69-08274
ROHLICH.:   W68-00855
ROHLICH.:   MARY M. ALLEN, GEO/ W69-03371
ROHLICH, AND WILLIAM B. SARLES.: / W72-01812
ROHLICH, AND WILLIAM L. LEA.:   / W70-02969
ROLF ELIASSEN, AND GEORGE TCHOBAN/ W70-01981
ROLF ELIASSEN, AND GEORGE TCHOBAN/ W70-07469
ROMAN DE VEGA.:   / W69-08525
RONALD G. SCHUESSLER, AND JOSEPH / W70-05267
RONALD H. LOUCKS.:   MARY A. / W69-08562
RONALD R. SPEEDY, NEIL B. FISHER,/ W70-08107
RONALD.:   R. L. MO/ W72-13605
ROSALIE M. COX, AND P. FAY.:   / W70-04249
ROSS E. MCKINNEY.:   ·/ W70-06601
ROSS E. MCKINNEY, AND ANDREW GRAM/ W72-01817
ROTH, JR.:   S. P. MEYERS, D./ W70-04368
RUDOLPH N. THUT.:   / W70-04253
RUDOLPH PRINS.:   JOHN/ W69-02782
RUKHADZE.:   V. G. KH/ W71-11561
RUPERT KOUNTZ.:   / W69-10159
RUTH PATRICK.:   / W72-01801
RUTH PATRICK.:   / W71-05026
RUTH PATRICK.:   / W71-02496
RUTH PATRICK.:   / W70-04510
RUTH PATRICK, BOWMAN CRUM, AND JO/ W70-07257
RUTTNER.:   / W70-03959
SAMPLES.:   / W69-08053
SANTRY, JR.: /CRAWFORD, J. K. BAX/ W70-05819
SARLES.:   G/ W72-01812
SASNER, JR.:   PHILIP J. SAWY/ W69-05306
SAUNDERS.:   / W70-02510
SAWYER.:   / W70-02787
SAWYER.:   W68-00012
SAWYER.:   / W73-01122
SAWYER.:   / W70-05565
SAWYER, J. B. LACKEY, AND A. T. L/ W70-05112
SAWYER, J. B. LACKEY, AND A. T. L/ W70-05113
SAWYER, JOHN H. GENTILE, AND JOHN/ W69-05306
SAX, PAUL C. LEMON, ALLEN H. BENT/ W69-10080
SCHNEIDER, AND E. O. BENNETT.:   / W69-08284
SCHRAUFNAGEL, AND THOMAS L. WIRTH/ W69-10178
SCHUESSLER, AND JOSEPH L. PAVONI./ W70-05267
SCHWERTNER, CARA L. MARTINEZ, AND/ W70-07957
SEABLOOM.:   / W72-11906
SEATHER.:   / W71-12077
SERRUYA, AND T. BERMAN.:   / W70-09889
SERRUYA, S. SERRUYA, AND T. BERMA/ W70-09889
SHADIN.:   / W70-04475
SHANNON, AND H. D. PUTNAM.:   P. / W69-09723
SHAPIRO.:   / W70-01031
SHAPIRO.:   / W70-05269
SHAPIRO, AND ROBERTO RIBEIRO.:   / W70-02255
SHAPIRO, C. R. GOLDMAN, O. R. ARM/ W69-06864
SHAPIRO, WILLIAM CHAMBERLAIN, AND/ W70-08637
SHCHAPOVA, AND V. B. VOZZHINSKAYA/ W72-10623
SHELEF.:   / W73-02218
SHEMTOB, AND J. B. SLESS.:   YEH/ W70-02370
SHERRY L. VOLK, AND HARRY K. PHIN/ W69-04802
SHERWOOD.:   W68-00256
SHILO.:   / W70-05372
SHIMIZU.:   / W71-09850
SHOOP.:   / W72-00462
SHUNN-DAR LIN.:   / W72-05453
SIDNEY S. ANTHONY.:   / W70-08096
SIEBURTH.:   / W70-01073
SIEBURTH.:   / W70-00713
SIEBURTH.:   / W70-01068
SIEBURTH, AND ARNE JENSEN.:   / W70-01074
SIGLER.:   SUSAN S. MARTIN,/ W69-07846
SILBERNAGEL.:   / W70-05652
SIMEK, J. A. DAVIS, C. E. DAY III/ W69-08524
SINGH.:   / W70-05404
SINGH, AND B. S. SRIVASTAVA.:   / W69-10177

P. H. JONES, AND	TOM DAVEY.: / W70-05084
AND TAMAO N/ YOSHIRO HASHIMOTO,	TOMOTOSHI OKAICHI, LE DUNG DANG, / W70-08372
AND TAMAO N/ YOSHIRO HASHIMOTO,	TOMOTOSHI OKAICHI, LE DUNG DANG, / W71-07731
IASSEN, B. M. WYCKOFF, AND C. D.	TONKIN.: R. EL/ W72-04318
D. A. HOFFMAN, P. R.	TRAMUTT, AND F. C. HELLER.: / W72-02817
WILLIAM	TRELEASE.: / W69-08283
	TROY C. DORRIS.: / W70-04001
U. S.	TSCHORTNER.: / W70-08392
HARD C. BAIN, JR., AND MILTON G.	TUNZI.: RANDALL L. BROWN, RIC/ W71-04555
CHARLES R. GOLDMAN, MILTON G.	TUNZI, AND RICHARD ARMSTRONG.: / W70-03983
R. G. STROSS, J. C. JONES, F. M.	UNGER, AND J. M. VAIL.: / W72-02417
C. JONES, F. M. UNGER, AND J. M.	VAIL.: R. G. STROSS, J. / W72-02417
B. DWAIN	VANCE, AND DAVID L. SMITH.: / W70-03519
B. D.	VANCE, AND W. DRUMMOND.: / W72-04506
B. DWAIN	VANCE, AND WAYMON DRUMMOND.: / W70-03520
B. DWAIN	VANCE, AND WAYMON DRUMMOND.: / W69-08565
MAN M.	VARMA, AND FRANCIS DIGIAND.: / W70-06600
J. O.	VEATCH, AND C. R. HUMPHRYS.: / W69-08279
ROBERT Y. TING, AND V. ROMAN DE	VEGA.: / W69-08525
HEDENSKOG, LENNART ENEBO, JITKA	VENDLOVA, AND BOHUMIR PROKES.: /D/ W70-04184
V.	VENKATESWARLU.: / W69-07096
JOHN W.	VENNES.: / W70-03312
	VERA A. (DUGDALE) BILLAUD.: / W69-03515
G. G.	VINBERG, AND T. N. SIVKE.: / W70-05092
WILLIAM C.	VINYARD.: / W69-03517
C. LACAZE, C. HALLOPEAU, AND M.	VOIGT.: J./ W72-02203
O. HOLM-HANSEN, J. COOMBS, B. E.	VOLCANI, AND P. M. WILLIAMS.: / W70-05651
SHERRY L.	VOLK, AND HARRY K. PHINNEY.: / W69-04802
	VON HEINZ AMBUHL.: / W69-08668
K. K.	VOTINTSEV, AND G. I. POPOVSKAYA.:/ W70-04290
T. F. SHCHAPOVA, AND V. B.	VOZZHINSKAYA.: / W72-10623
ALBERTO M.	WACHS, AND ANDRE BEREND.: / W70-04790
THOMAS D.	WAITE, AND CONSTANTINE GREGORY.: / W71-05500
	WALDEMAR OHLE.: / W70-05761
ARDCASTLE.:	WALLACE H. FULLER, AND JAMES E. H/ W69-03491
L'ANNUNZIATA.:	WALLACE H. FULLER, AND MICHAEL F./ W71-04067
	WALTER ABBOTT.: / W69-04800
J. C.	WARD, AND S. KARAKI.: / W72-01773
RANCIS R. MCNEICE, AND PIERRE S.	WARREN.: PAUL H. KING, F/ W70-05470
HARRY K. PHINNEY, AND CHARLES E.	WARREN.: /E, ROBERT L. GARRISON, / W70-00265
UL H. KING, H. H. YEH, PIERRE S.	WARREN, AND CLIFFORD W. RANDALL.:/ W69-09884
OTTO	WASMER, JR.: / W69-00994
T. A.	WASTLER.: / W69-06203
ANT, AND W. L. TEMPLETON/ D. G.	WATSON, C. E. CUSHING, C. C. COUT/ W69-07853
G, CZESLAWA NALEWAJKO, AND W. D.	WATT.: G. E. FOG/ W70-02504
B. DWAIN VANCE, AND	WAYMON DRUMMOND.: / W70-03520
B. DWAIN VANCE, AND	WAYMON DRUMMOND.: / W69-08565
G. SCHUESSLER/ MARK W. TENNEY,	WAYNE F. ECHELBERGER, JR., RONALD/ W70-05267
HARRY D. NASH, AND RALPH H.	WEAVER.: / W68-00891
RALPH H.	WEAVER, AND HARRY D. NASH.: / W69-00096
CORNELIUS I.	WEBER, AND BEN H. MCFARLAND.: / W71-04206
G. R. PYPER, W. J.	WEBER, JR., AND J. A. BORCHARDT.:/ W72-00917
EDWARD J. MARTIN, AND LEON W.	WEINBERGER.: / W69-01453
CHARLES M.	WEISS.: / W69-09135
C. M.	WEISS.: / W70-10437
A. D.	WELANDER.: / W69-08269
EUGENE B.	WELCH.: / W70-04283
EUGENE B.	WELCH.: / W71-07698
.:	WERNER STUMM, AND JAMES J. MORGAN/ W69-03369
:	WERNER STUMM, AND MARK W. TENNEY./ W70-05750
W.	WESLEY ECKENFELDER, JR.: / W69-08054
G. E. FOGG, AND D. F.	WESTLAKE.: / W69-10180
ROBERT G.	WETZEL.: / W71-10079
T.:	WHITTEN, AND CLARENCE J. GOODNIGH/ W69-07861
BERT K.	WICKLINE.: /UTT, H. A. SCHWERTNER/ W70-07957
, CARA L. MARTINEZ, AND HAZEL E.	WILHELM RODHE.: / W70-05409
DALE A. CARLSON, AND H. KIRK	WILLARD.: / W70-06792
GERARD A. ROHLICH, AND	WILLIAM B. COOKE.: / W72-01819
	WILLIAM B. SARLES.: / W72-01812
	WILLIAM C. VINYARD.: / W69-03517
ARRETT.: JOSEPH SHAPIRO,	WILLIAM CHAMBERLAIN, AND JUDITH B/ W70-08637
HAPIRO.:	WILLIAM CHAMBERLAIN, AND JOSEPH S/ W70-05269
LOWELL D. HANSON, AND	WILLIAM E. FENSTER.: / W70-04193
	WILLIAM E. SLOEY.: / W72-00845
S. MARTIN, WILLIAM T. HELM, AND	WILLIAM F. SIGLER.: SUSAN/ W69-07846
GORDON L. DUGAN, AND	WILLIAM J. OSWALD.: / W72-06854
N L. DUGAN, CLARENCE G. GOLUEKE,	WILLIAM J. OSWALD, AND CHARLES E./ W71-11375
OEL C. GOLDMAN, JAMES F. ARTHUR,	WILLIAM J. OSWALD, AND LOUIS A. B/ W71-04557
GERARD A. ROHLICH, AND	WILLIAM L. LEA.: / W70-02969
C. MERVIN, PALMER, AND	WILLIAM M. INGRAM.: / W72-01798
ES.: KENNETH M. MACKENTHUN,	WILLIAM M. INGRAM, AND RALPH PORG/ W70-06225
KENNETH M. MACKENTHUN, AND	WILLIAM MARCUS INGRAM.: / W69-04801
	WILLIAM R. EBERLY.: / W69-03514
MARKLE.: GEORGE E. STOERTZ,	WILLIAM R. HEMPHILL, AND DAVID A./ W70-05377
Y.: RICHARD WOOD,	WILLIAM R. SAMPLES.: / W69-08053
	WILLIAM S. SMITH, AND J. K. MURRA/ W70-05821
IGLER.: SUSAN S. MARTIN,	WILLIAM T. HELM, AND WILLIAM F. S/ W69-07846
	WILLIAM TRELEASE.: / W69-08283
COOMBS, B. E. VOLCANI, AND P. M.	WILLIAMS.: O. HOLM-HANSEN, J. / W70-05651
.: T. DUKE, J.	WILLIS, T. PRICE, AND K. FISCHLER/ W69-08267

```
:        MARY A. TIFFANY, JOHN W.    WINCHESTER, AND RONALD H. LOUCKS./  W69-08562
                                     WINIFRED PENNINGTON.:              /  W69-04521
. H. SCHRAUFNAGEL, AND THOMAS L.     WIRTH.: /. HASLER, G. FRED LEE, F/  W69-10178
                        MARY JO      WITT.:                             /  W71-00110
  B. POLKOWSKI, AND K. T. CR/  S.    WITZEL, E. MCCOY, O. J. ATTOE, L./  W69-05323
ARD LEHNER.:              S. A.      WITZEL, ELIZABETH MCCOY, AND RICH/  W71-05742
:                       DONALD E.    WOHLSCHLAG, AND ARTHUR D. HASLER./  W70-06229
                          ABEL       WOLMAN.:                           /  W71-05943
                       E. J. F.      WOOD.:                             /  W69-03752
  MURRAY.:              RICHARD      WOOD, WILLIAM S. SMITH, AND J. K./  W70-05821
BLUNDE, S. B. CHALLEN, AND D. L.     WOODS.:                       G. /  W69-07826
                          E. T.      WRAY.:                             /  W72-00941
   JOHN E. HOBBIE, AND RICHARD T.    WRIGHT.:                           /  W70-04284
   JOHN E. HOBBIE, AND RICHARD T.    WRIGHT.:                           /  W70-04287
                       RICHARD T.    WRIGHT, AND JOHN E. HOBBIE.:       /  W70-04194
 C. DAVID MCINTIRE, AND BARRY L.     WULFF.:                            /  W70-03325
                       CHARLES F.    WURSTER, JR.:                      /  W70-05272
                          A. G.      WURTZ.:                            /  W70-04381
             R. ELIASSEN, B. M.      WYCKOFF, AND C. D. TONKIN.:        /  W72-04318
RD W. RAND/  PAUL H. KING, H. H.     YEH, PIERRE S. WARREN, AND CLIFFO/  W69-09884
           NACHUM BETZER, AND        YEHUDA KOTT.:                      /  W70-02364
                                     YEHUDA KOTT, AND J. EDLIS.:        /  W70-02363
 A. SHEMTOB, AND J. B. SLESS.:       YEHUDA KOTT, GALILA HERSHKOVITZ, / W70-02370
                          G. A.      YENAKI.:                           /  W71-00221
                          S.         YOSHIMURA.:                        /  W69-05142
CHI, LE DUNG DANG, AND TAMAO N/      YOSHIRO HASHIMOTO, TOMOTOSHI OKAI/  W71-07731
CHI, LE DUNG DANG, AND TAMAO N/      YOSHIRO HASHIMOTO, TOMOTOSHI OKAI/  W70-08372
                       P. JONATHAN   YOUNG.:                            /  W70-00475
                                     YU S POTAENKO, AND T M MIKHEEVA.:/  W70-08654
             M. BERNHARD, AND A.     ZATTERA.:                          /  W69-08275
                       DAVID R.      ZENZ, AND JOSEPH R. PIVNICKA.:     /  W70-09186
                          H.         ZIMMERMAN.:                        /  W70-08740
```

585